ELSEVIER'S DICTIONARY OF SCIENCE AND TECHNOLOGY

Russian - English

ELSEVIER'S DICTIONARY OF SCIENCE AND TECHNOLOGY

Russian - English

compiled by

G. CHAKALOV
Sofia, Bulgaria

ELSEVIER
Amsterdam – London – New York – Tokyo 1993

ELSEVIER SCIENCE PUBLISHERS B.V.
Sara Burgerhartstraat 25
P.O. Box 211, 1000 AE Amsterdam
The Netherlands

ISBN: 0-444-88924-8

This book is printed on acid-free paper.

Printed in The Netherlands

To Dany Chakalova

For a lifetime of unfailing self-denial, affection, support and devotion

From her husband

PREFACE

Regular users of dictionaries often come to realize that a large number of **self-explanatory composite terms or derivatives are included but rarely looked up.** To take the example of the following list of composite terms:

локальная аномалия (lokalnaya anomaliya)	local anomaly
магнитная аномалия (magnitnaya anomaliya)	magnetic anomaly
радиоактивная аномалия (radioaktivnaya anomaliya)	radioactive anomaly
региональная аномалия (regionalnaya anomaliya)	regional anomaly
рефракционная аномалия (refraktsionnaya anomaliya)	refraction anomaly

The three possible variants under which the user will consult the dictionary, say in the instance of "local anomaly", are the following: (1) both words are unfamiliar; (2) the word "local" is unfamiliar; and (3) the word "anomaly" is unfamiliar.

In case (1) he will find "local" where it belongs alphabetically (not under "anomaly"), then he will find "anomaly" in its place, and will not bother to go down the long list of composite terms to find out the meaning of the **self-explanatory** "local anomaly". In case (2) he will look for "local" and will not go to "anomaly". In case (3) he will look for "anomaly" but not for "local". The composite terms which **must** be included here are of the type "аномалия Буге" (anomaliya Buge) -- "Bouger gravity anomaly", which is **not** self-explanatory.

All such words, together with numerous plain and simple derivatives and self-explanatory composite terms, have been responsibly left out of this dictionary, making room for the incorporation of 135,000 valuable entries. Had we adhered to the principle of what one might call "lexicographic completeness", these terms would have had to be spread over three volumes with no less than 320,000 entries, most of which, being self-explanatory, would not be actively looked for. In the views of the author, this was a simple choice of priorities.

There also Russian terms ("ренин" -- renin, "реннин" -- rennin) whose pronunciation tells the reader what their English counterpart is, the spelling of the latter being mostly a matter of predictable transliteration.

Particular attention has been paid to such sectors as economics, finance, law, computer technology, social sciences and communications, in view of today's expanding ties between English-speaking and Russian-speaking countries.

This dictionary is a translation into Russian from an unpublished Bulgarian-English version. It would have never seen the light of day had it not been for the competence and erudition of a computer specialist, Dr. Hristo Penev, who helped in wresting a decent copy from the fairly modest equipment locally available for the production of such a manuscript in camera-ready form. The structure and the concept underlying the dictionary owe very much to the valuable instructions and suggestions of Elsevier's Dictionaries Department.

Needless to say, the author is solely responsible for any shortcomings that may be found in the dictionary.

The dictionary includes all new terms which have emerged in connection with the changes in the structure of the former Soviet Union.

REFERENCES

Англо-русский медицинский словарь, коллектив, "Русский язык", Москва, 1988
Англо-русский сельскохозяйственный словарь, коллектив, "Русский язык", Москва, 1986
Англо-русский словарь математических терминов, коллектив, "Издательство иностранной литературы", Москва, 1962
Англо-русский экономический словарь, коллектив, "Русский язык", Москва, 1977
Англо-русский Юридический словарь, Андрианов и Никифоров, "Международные отношения", Москва, 1964
Ботанический словарь, Давыдов, ФИЗМАТГИЗА, Москва, 1960
Русско-английский авиационно-космический словарь, Мурашкевич, "Военное издательство", Москва, 1971
Русско-английский военно-морской словарь, коллектив, "Воениздат", Москва, 1976
Русско-английский медицинский словарь, коллектив, "Русский язык", Москва, 1975
Русско-английский политехнический словарь, коллектив, "Русский язык", Москва, 1980
Русско-английский ядерный словарь, Восковойник и Циммерман, ФИЗМАТГИЗА, Москва, 1960
Русско-англо-немецко-французский словарь по вычислительной технике, коллектив, "Русский язык", Москва, 1990
Словарь иностранных слов, коллектив, "Русский язык", Москва, 1980
Словарь русского языка, 4 т., коллектив, Академия Наук СССР, Москва, 1961
Cybernetics. English-German-French-Russian, collective authorship, VEB Verlag Technik, Berlin, 1976
Data Preparation. Programming. English-German-Russian, E. Buerger, VEB Verlag Technik, Berlin, 1977
Elsevier's Russian-English Dictionary. 4 volumes, P. Macura, Amsterdam, 1990
Hydraulics. Pneumatics. English-German-Russian, G. Neubert, VEB Verlag Technik, Berlin, 1978
The Concise Oxford Dictionary of Current English, Oxford University Press, 1989

Also glossaries prepared for assisting translation and synchronous interpretation at about 450 scientific conferences, symposiums, meetings and seminars held during the last thirty years.

ABBREVIATIONS

*abbr.*eviation
*acc.*ounting
*acous.*tics
*adj.*ective
*adv.*erb
*aero.*dynamics
*agr.*iculture
*Am.*erican
*anat.*omy
*anthr.*opology
*arch.*itecture
art
*ast.*ronomy
*attr.*ibutive
*aut.*omobile
*av.*iation
*bibl.*ical
*biochem.*istry
*biol.*ogy
*biophys.*ics
*bot.*any
*build.*ing
*cart.*ography
*chem.*istry
chess
*cin.*ema
*coll.*ective
*colloq.*uial
*com.*munications
*comm.*erce
*comp.*uter technology
*constr.*uction
*cryst.*allography
*cul.*inary
*cyb.*ernetics
*dim.*inutive
*dipl.*omacy
*eccl.*esiastical
*ecol.*ogy
*econ.*omics
*educ.*ation
*el.*ectricity
*electr.*onics
*Eng.*lish
*eng.*ineering
*ent.*omology
*farm.*ing
*fig.*urative
*fin.*ance
food industry
*Fr.*ench
foundry
*game th.*eory
*geod.*esy
*geog.*raphy
*geol.*ogy
*geom.*etry
*gram.*mar
*hist.*ory
*hort.*iculture
*hydr.*aulics
laser
law
*leath.*er industry
*ling.*uistics
*lit.*erature
logic
*mach.*inery
*math.*ematics
*mech.*anics
*med.*icine
*met.*allurgy, metalworking
*meteor.*ology
*mil.*itary
*min.*ing
*miner.*alogy
*mus.*ic
*myth.*ology
n. noun
*naut.*ical
*nav.*igation
*nucl.*ear physics
*obs.*olete
*opt.*ics
*paleont.*ology
paper industry
*pathol.*ogy
*petr.*ography
*philos.*ophy
*phon.*etics
*phot.*ography
*phys.*ics
*physiol.*ogy
*plast.*ics
*pl.*ural
*pol.*itics
*polygr.*aphy
*print.*ing
*psych.*ology
*rad.*io, radar
*rel.*igion
*rock.*et technology
*sculpt.*ure
space technology
sport
*stat.*istics
*surg.*ery
*surv.*eying
*tech.*nical, technology
*tel.*ecommunications
*teleph.*one
*telev.*ision
*text.*iles
theatre
*theol.*ogy
*timb.*er industry
*top.*ography
v. verb
*vet.*erinary medicine
*weld.*ing
*wood.*working
*zool.*ogy

The use of the oblique has been helpful in saving much space in cases such as:
теплоотдача -- heat emission/irradiation/transfer/output/exchange/removal.

А, а

абабок *bot.* rough-stemmed boletus *(Boletus scaber)*
абазия *med.* abasia
~, атактическая atactic/ataxic abasia
~, дрожательная trepidant abasia, abasia trepidans
абака *bot.* abaca
абандон *naut.* abandonment
аббат abbot
аббревиатура abbreviation; acronym
абграт-пресс burring press
абграт-штамп trimming die
абелев Abelian
абелевость commutativity
абелия *bot.* abelia *(Abelia)*
абельмош *bot.* okra *(Abelmoschus esculentus)*
абернатиит *miner.* abernathyite
аберрация aberration
~, вековая secular aberration
~, внеосевая off-axis aberration
~ глаза ocular/eye/ophthalmic aberration
~ звезд aberration of fixed stars
~ оптических систем optical aberration
~ света aberration of light
~, сумочная diurnal aberration
~, цветовая colour/chromatic aberration
~ электронных линз electron-optical aberration
абзац paragraph; indention
абиогенез abiogenesis
абиссаль abyssal zone
абиссинец, абиссинский Abyssinian
абиссинка Abyssinian (woman)
абитуриент school-leaver; university entrant
аблактировать graft by approach/inarching
аблегер peat conveyer
аблаут ablaut
абляционный ablative
абляция ablation
~, интегральная cumulative ablation
~ испарением vaporizing ablation
~, ледниковая glacial ablation
~ сублимацией subliming/sublimation ablation
аболиционизм abolitionism
аболиционист abolitionist
абонамент, межбиблиотечный interlibrary loan scheme/system
абонент subscriber, user; party
аборальный *biol.* aboral
абордаж *naut.* boarding
абориген aboriginal, aborigene, aboriginal/autochthonous/indigenous inhabitant
аборт abortion, abortus; miscarriage
~, медицинский искусственный therapeutic/justifiable abortion
~, самопроизвольный spontaneous abortion
абортивный, абортный abortive
абрадировать abrade
абразив abrasive (material), abradant
~, искусственный manufactured abrasive
~, несвязанный loose abrasive
абразивность abrasivity, abrasiveness
абразивный abrasive, abrading
абразия abrasion
~, ледниковая glacial abrasion
~, морская marine erosion, wave abrasion
~, речная fluvial abrasion
абракадабра abracadabra
абрахиоцефал abrachiocephalus
абриаканит *miner.* abriachanite
абрис 1. contour, outline, sketch; 2. *print.* key transfer; 3. *el.* E-angle
абсент absinthe
абсентеизм absenteeism
абсолютировать *chem.* dehydrate
абсорбат absorbate
абсорбент absorbent
абсорбер absorber
~, насадочный packed absorber
абсорбировать absorb, take up
абсорбирующий absorbent, absorptive
абсорбциометр absorption meter, absorptiometer
абсорбциометрический absorptiometric
абсорбциометрия absorptiometry
абсорбционный absorptive, absorption *(attr.)*
абсорбция absorption
абстрагировать abstract
абстрактность abstractness, abstraction
абстракционизм abstract art
абстракционист, ~ка abstract artist
абстракционистский abstractionist *(attr.)*
абстракция abstraction
абсурд absurdity
абсцесс abscess
~, атероматозный atheromatous abscess
~, газосодержащий tympanic abscess
~, сопутствующий satellite abscess
~, холодный cold/Pott's abscess
абсцисса 1. abscissa, x-coordinate, horizontal coordinate; 2. *cart.* x-value/coordinate, difference of latitude
абулия abulia, aboulia
абштрих *met.* arsenic-antimony slag, scum
авалист *law* surety on a bill; guarantor
аваль *law* surety for a bill
авангардизм avant-gardism *(Fr.)*
авангардистский avant-garde *(Fr.)*
авандек end platform
аванзал 1. ante-room; entrance hall; 2. *min.* forehole
аванкамера *hydr.* forebay; forechamber
аванпорт outer harbour
аванпост outpost
авансцена proscenium
аварец Avar

аварийно in case of fault/trouble; under abnormal/emergency conditions
аварийно-спасательный 1. search-and-rescue; 2. survival
аварийность accident/breakdown/fault rate
аварийный emergency; breakdown; repair *(attr.)*
~ная бригада/команда emergency crew/team
~ная машина service van/lorry/truck
~ная сигнализация alarm system
авария 1. breakdown, failure; accident; wreck; crash; 2. emergency; 3. *nav., law* average
~, малая petty average
~, морская accident at sea
~ на взлете takeoff accident
~ на посадке landing accident
~, пусковая start-up accident
~ реактора reactor accident/disaster
~, частная particular/simple/partial average
~ ядерной установки nuclear accident
аварка, аварский Avar
аваруит *miner.* awaruite
авганит *miner.* auganite
авгиевы конюшни Augean stables
авгит *miner.* augite
авгит-бронзит *miner.* augite-bronzite
авгур augur, soothsayer
аверс obverse *(of a coin)*
Авеста Avesta
авиабаза air base
авиабензин aviation gasoline/petrol
авиабомба aerial bomb
авиагоризонт gyro/artificial horizon, attitude/flight indicator, bank-and-pitch indicator
~, выбивающийся toppling/tumbling horizon
~, фигурный all-attitude gyro horizon
авиадвигатель aircraft engine
авиадиспетчер air-traffic controller
авиаинженер aeronautical engineer
авиакомпания airline, air carrier
авиакомпас aircraft/flight compass
авиаконструктор aircraft designer
авиалиния 1. air-line/-route; airway; 2. airline, air carrier
~, внутренняя domestic air-line/-route
~, грузовая cargo airline; air-freight service
~, местная local-service air-line/-route
~, связующая feeder air-line/-route
авиамагистраль major air-route; major/trunk airline
авиаметеослужба aeronautical meteorological/weather service
авиамеханик aircraft mechanic, aircraftsman
авиамоделизм model aeroplane/aircraft construction
авиамоделист aeromodeller, model aeroplane/aircraft builder/constructor
авиамодель model aircraft/aeroplane
авианосец aircraft carrier
~, атомный nuclear-powered aircraft carrier
авиаопыливание aerial (crop) dusting
авиапарк aircraft/air fleet
авиапневматик aircraft tyre
авиаподкормка aerial top-dressing *(of plants)*
авиапочта air mail
авиаразведка air reconnaissance
авиастроение aircraft manufacturing/production/industry
авиасъемка aerial mapping/survey
авиационно-космический aerospace
авиационный aviation, aircraft, aeronautical, aerial, air; airborne
~ная школа flying school
авиация aviation; aircraft; *(science)* aeronautics
~, бомбардировочная bomber force, bombardment aviation
~, истребительная fighter/pursuit aviation, fighters
~, истребительно-бомбардировочная fighter-bomber aviation, fighter-bombers
~, санитарная ambulance aviation
~, штурмовая attack/assault aircraft
авиетка light aeroplane, baby-plane, aviette
авизо 1. letter of advice, (advice) note; 2. *naut.* aviso, advice boat
авионика avionics
ависта *econ.* at sight
авитаминоз avitaminosis
авитаминозный avitaminotic
авифауна birds *(collectively)*
авлос *mus., hist.* aulos
авометр multimeter, ampere-volt-ohmmeter, avometer
авост emergency stop; abend *(abnormal end)*
аврал *naut.* emergency/rush job; joint effort
аврикула *bot.* (primula) auricula
Аврора Aurora
австралиец, ~ийка, ~ийский Australian
австриец, ~ийка, ~ийский Austrian
Австро-Венгрия Austro-Hungary
автаркия *econ.* autarky, self-sufficiency
автархия *pol.* autarchy, absolute power
автиморф *miner.* authimorph
автобаза motor depot; car/truck/lorry fleet
автобензоцистерна gasoline/petrol tanker/truck
автобетономешалка agitator truck
автоблокировка automatic lockout/block system; automatic locking/hatching; safety interlock
автобрекчия *miner.* autobreccia
автобус bus; coach
~, воздушный airbus
~, двухэтажный double-decker
автогамия autogamy
автогенез autogenesis, spontaneous generation
автогенератор self-excited/-sustained oscillator; feedback oscillator
автогенный autogenous
автогенщик oxy-acetylene welder
автогетеродин autoheterodyne, autodyne
автогрейдер self-propelled grader, motor grader
автогудронатор tar paver

автодин autodyne receiver
автодиспетчер *comp.* supervisory/control computer
автодрезина rail motor trolley
автожир autogyro, gyroplane
автокамера inner tube (of tyre)
автокар industrial truck; shop truck
автокатализ autocatalysis
автокефалия autocephaly; autocephalous church
автоклав autoclave
~ для вытопки жира rendering tank
~ для выщелачивания leaching autoclave, digester
~ для гидрогенизации жиров oil converter
~, красильный pressure-dyeing machine
~, медицинский sterilizer
~, полимеризационный polymerizer
автоклавировать autoclave; sterilize; digest; retort
автоклавный steam(-and-pressure) cured, autoclave cured
автоклаза autoclase, autoclastic feature
автоковариация *stat.* autocovariance
автокод autocode
автоколебания self-excited/-sustained oscillations, self-sustained vibrations, natural oscillations
автоколлиматор autocollimator, tubullar collimator
автоколлимация autocollimation
автокомпенсатор self-balancing potentiometer
автокран truck crane
автократ autocrat
автократия autocracy
автол motor oil, automotive engine oil
автолавка sales van, travelling/mobile shop
автолебедка winch truck
автолиз autolysis
автолит *miner.* autolith, endogenous/cognate inclusion
автомагистраль superhighway, motor/truck highway, divided highway/carriageway, motor road, motor-way
автомат 1. automatic machine/device/apparatus; automatic control unit, controller; *cyb.* automaton, machine; 2. circuit-breaker; 3. slot-machine, vending machine; 4. *mil.* sub-machine gun
~, барометрический *av.* baroswitch, barometric switch
~, билетный ticket machine
~, бухгалтерский accounting/invoicing machine
~, воздушный air circuit-breaker
~, вычислительный automatic calculator; computing automaton
~, дозирующий automatic measuring machine
~ защиты сети automatic circuit-breaker
~, конечный finite-state machine
~, копировальный automatic tracer-equipped machine
~, масляный oil circuit-breaker
~, навигационный *av.* ground-position indicator, automatic navigator
~, патронный automatic chucking machine
~ переключения скоростей automatic speed-selection device
~ подачи топлива fuel-flow proportioner
~ с конечной памятью finite-memory automaton
~, сборочный automatic assembly machine
~ синтаксического анализа parsing automaton
~, ткацкий automatic loom
~, читающий reading machine
автоматизация automation; automatic industrial control
автоматизированный automated, automatic, automatically controlled
автоматизировать automate, render/make automatic; introduce automation into
автоматизм automatism; automaticity; automatic action
~, ассоциативный/идеаторный ideational automatism
автоматика 1. automatics; automation; control engineering; 2. automatic equipment/system/controls/mechanism/devices
~, познавательная *cyb.* cognitive automation
автоматический, автоматичный 1. automatic; self-acting; 2. unattended
автоматичность automaticity; automation
автомат-экзаменатор test-grading machine
автомашина = автомобиль
автомеханический automotive
автомобилепрокидыватель, автомобилеразгрузчик truck tipper
автомобилестроение motor-car industry/building/construction, automobile industry
автомобилизм motoring
автомобилист motorist
автомобиль 1. motor vehicle/car, automobile, car; 2. truck, lorry
~, безрамный frameless vehicle/automobile, unit-construction car
~ высокой проходимости cross-country vehicle/car
~, грузовой truck, lorry
~, грузовой легкий light-duty truck/lorry
~, грузовой с откидными бортами drop-side truck
~, грузовой тяжелый heavy-duty truck/lorry
~, закрытый saloon car, limousine
~, изотермический refrigerated truck/lorry
~, легковой (motor) car, (passenger) car
~, малолитражный economy/compact car
~, микролитражный baby car, mini-car
~, общего назначения utility vehicle
~, пожарный fire engine, fire-fighting vehicle
~, полноприводной all-wheel-drive vehicle
~, почтовый mail van/wag(g)on, postal car
~ с несущим кузовом frameless vehicle, unit-construction car

~ **с откидным верхом** convertible (car), soft-top car
~ **с передними ведущими колесами** front-wheel drive car/vehicle
~ **с четырьмя ведущими колесами** four-wheel drive car/vehicle
~, **санитарный** ambulance, medical vehicle
~ **скорой помощи** ambulance
~, **электрический** electric-battery car/truck
автомобиль-амфибия amphibian car/vehicle
автомобиль-пескомет sanding vehicle/truck
автомобиль-цистерна road tanker, tank truck
автомоделирование model-car construction
автомодуляция self-modulation
автомолит *miner.* automolite, zinc gahnite
автоморфный automorphic, euhedral, idiomorphic
~, **частично** hypidiomorphic, hypautomorphic, subhedral
автомотриса railcar, railway motor carriage/car
автонавеска automatic hitch/draftgear
автонастройка automatic frequency selection, automatic tuning; autotuning system
автономность 1. *cyb., math.* non-interaction; 2. independence; 3. *econ.* self-sufficiency
автономный 1. *cyb., math.* non-interacting; 2. autonomous, independent; 3. *comp.* off-line; 4. *econ.* self-sufficient; 5. *tech.* self-contained
автоокисление autooxidation, autoxidation
автооператор transfer arm
автоопрос autopoll
автоответчик automatic answer-back device
автопавильон enclosed bus stop
автопансионат motorists' camp/quarters
автопарк car fleet; truck/lorry fleet
автоперевозка road haulage; truck haulage, truckage
автопилот autopilot, automatic pilot
автопогрузчик 1. truck loader; 2. lift truck
~ **с вилочным захватом** fork-lift truck
автоподача automatic feed
автопоезд road train; articulated vehicle; lorry and trailers
автопоилка automatic water bowl
автопокрышка tyre, tire
~, **бескамерная** tubeless tyre
~, **легковая** car tyre
автополимер autopolymer
автополярный (self-)polar
автоприводка *print.* automatic register
автоприправка *print.* mechanical overlay
автоприцеп trailer
автопрограммирование self-programming
автопрокладчик automatic position plotter/tracer
авторадиограмма autoradiogram, autoradiograph, radioautograph
авторадиографический autoradiographic, radio-autographic
авторадиография 1. autoradiography, radioauto-graphy; 2. autoradiograph, radioautograph
авторадиолиз autoradiolysis
авторадиометр car-borne radiometer
авторегулируемый self-controlled
авторезина car tyres and tubes
авторемонт car repair
автореферат author's summary/abstract
авторефрижератор refrigerated truck
~ **с сухоледным охлаждением** dry ice truck
авторитаризм authoritarianism
авторотация *av.* autorotation; windmilling
~, **естественная** unassisted windmilling
~ **крыла** wing autorotation
~ **с работающим двигателем** power-on auto-rotation
~, **скрытая** latent autorotation
авторотировать autorotate; windmill
авторотирующий autorotating; windmilling, free-wheeling
авторское право copyright
авторулевой automatic helmsman, automatic steer-ing device
автосамосвал dump truck, tip-lorry
автосани snowmobile, motor-sleigh, sledge-car
автосервис car-care centre
автосин autosyn, synchro
автоскрепер motorscraper
автоспора autospore
автостоп 1. automatic braking gear, automatic stop; 2. hitch-hiking permit
автостоянка parking lot/ground
автострада superhighway, motor way, motor road/highway, trunk road
автосцепка automatic coupling/coupler; automatic hitch
автотека autotheca
автотележка (self-propelled) industrial truck
автотелеуправление automatic remote control, automatic telecontrol
автотипия autotype, half-tone
~, **глубокая** intaglio half-tone
~, **корешковая** grain screen autotype
~, **черная** black-and-white autotype
автотормоз automatic/mechanical brake
автотракторный motor and tractor *(attr.)*
автотранспорт motor/road transport
автотранспортер self-propelled conveyer
автотрансформатор autotransformer
~, **измерительный** instrument autotransformer
~, **регулируемый** variable-ratio autotransformer
автотрофный autotrophic
автофазировка automatic phasing; phase stability
автофреза self-propelled cultivator
автофургон van; pantechnicon
автохозяйство car fleet; truck/lorry fleet
автохром autochrome
автохтония autochthony
автохтонный autochthonous
автохтоны autochthons, aboriginals
автоцистерна tank truck/lorry, road tanker
~, **ассенизационная** cesspool emptier, cesspoolage truck

~, **изотермическая** insulated tank truck
~, **пожарная** fire-extinguishing tanker/wagon
автошина automobile tyre/tire
автоштурман automatic navigator, autonavigator, ground-position indicator, GPI
автоэмаль nitrocellulose paint for car bodies
автоэмиссия autoelectronic emission
автур automatic level gauge
авуары holdings, bank deposits (abroad); assets
авункулат *hist.* avunculate
агава *bot.* agave
агалактия agalactia, agalactosis
агальматолит *miner.* agalmatolite, steatite, soapstone, soap rock, pagodite
агама *zool.* agama
агамные *zool.* agamidae
агамный agamous, agamic
агамогония agamogony; agamogenesis
агар agar
~, **желчный** bile salt agar
~, **мясопептонный**; ~, **простой** plain agar
~ **Эндо** Endo's fuchsin agar
агар-агар (agar-)agar
агаровый agarinic
агат *min., print.* agate
~, **деревянистый** agatized wood
~, **моховой** moss agate
~, **яшмовый** agate-jasper
агвиларит aguilarite
агглютин agglutinum
агглютинативный agglutinative
агглютинация agglutination
~, **иммунная** immunoagglutination, immune agglutination
~, **перекрестная** cross agglutination
агглютинин agglutinin
~, **неполный** partial/minor agglutinin
агглютинировать agglutinate
агглютинирующий agglutinative, agglutinating
агглютиноген agglutinogen, agglutigen
аггравант *med.* malingerer
аггрaдация *geol.* aggradation, shore deposition
агенезический agenetic; sterile
агенезия agenesia, agenesis; sterility
агент *phys., chem.* agent; factor; medium
~, **антипенный** antifoaming agent
~, **вентилирующий** aerating agent/medium
~, **восстанавливающий** reducing agent
~, **вспенивающий** expanding/foaming agent
~, **вуалирующий** *phot.* fogging agent
~, **вымывающий** stripping agent
~, **высаливающий** salting-out agent
~, **дезактивирующий** decontaminant, decontaminating agent
~, **денатурирующий** denaturant, denature
~, **загрязняющий** pollutant
~, **комплексообразующий** complexing agent
~, **корродирующий** corrodent, corrosive agent
~, **охлаждающий** coolant, cooling agent
~ **по обеспечению престижа** publicity agent
~, **поверхностно-активный** surfactant, surface-active agent
~ **придающий клейкость/липкость** tackifier
~, **разобщающий** uncoupling agent
~, **регулирующий** control agent
~, **связующий** binding agent, binder
~, **сушильный** siccative
~, **тормозящий** inhibitor, inhibiting agent
~, **удерживающий** hold-back agent
~, **флокулирующий** flocculating agent
~, **холодильный** cooling/refrigerant agent, coolant, refrigerant
~, **экстрагирующий** extraction agent
~, **эмульгирующий** emulsifier, emulsifying agent
агломерат 1. *met.* sinter (cake); 2. agglomerate
~, **жерловой** vent agglomerate
~, **измельченный** crushed agglomerate
~, **окомкованный** pellet sinter
~, **офлюсованный** fluxed sinter
~, **раковистый** shelly agglomerate
~, **самоплавкий** self-fluxing sinter
агломерация 1. *met.* sintering, agglomeration by sintering; 2. urban sprawl, agglomeration
~ **в кипящем слое** fluidized-bed sintering
~ **в чашах** pan sintering
~ **крупных городов** megalopolis
агломерировать agglomerate, sinter
аглофабрика sintering plant
агнат *hist., law* agnate
агнатия agnathia, agnathy
агнозия agnosia
~, **болевая** analgia
~, **зрительно-пространственная** visual-spatial agnosia, apractognosia
агностерин agnosterol
агностик agnostic
агностицизм agnosticism
агностогенный agnostogenic, aphanogenic
агнотозойский Agnotozoic
агона agonic line
агональный agonal
агонист agonist
агпаит *miner.* agpaite
агранулоцит agranulocyte
агранулоцитарный agranulocytic
агранулоцитоз agranulocytosis, agranulosis
аграризация ruralization
аграфия agraphia
~, **психическая** mental agraphia
агрегат 1. assembly, set, unit; 2. aggregate; *stat.* grouping
~, **автономный** self-contained unit
~, **бензоэлектрический** gasoline engine generating set, petrol-electric generating set
~, **буровой** drilling rig and accessories
~, **ветровой электрический** wind turbine generator
~, **газотурбинный** gas turbine unit
~, **генераторный** generating set

~, гроздевидный botryoidal aggregate
~, зарядный charging set/unit
~ кристаллов druse
~, лудильный tinning stack
~, натечный sinter aggregate
~, отопительный unit heater
~, скорлуповатый conchoidal aggregate
~, сферолитовый axialite
~, травильный pickler
~, уборочный harvesting unit
~, электросварочный electric welding set/unit/machine
~, энергетический power-generating unit
агрегатирование *mach.* building-block design/principle/approach
агрегатировать 1. gang, team, connect, link, gear *(to);* 2. use building-block units
агрегатный modular, building-block
~ное состояние state of aggregation
агрегация 1. *phys.* aggregation, clustering; 2. formation of aggregates, agglomeration
агрегирование 1. aggregation, clustering; 2. pellet formation
агреман agrément
агрессивный *chem.* corrosive; harsh
агрессины aggressines
агриколит *miner.* agricolite
агрипния *med.* agrypnia, insomnia
агрогидрология agrohydrology
агролесомелиорация agricultural afforestation, agrosilviculture
агромелиорация irrigation and drainage, land improvement/reclamation, soil conservation
агрометеорология agricultural meteorology
агроминимум minimum of agricultural knowledge
агроном agronomist
агрономический agronomical, agricultural
агрономия agronomy
агропромышленный agroindustrial, agricultural and industrial
агротехник agricultural technician
агротехника field management, farming/agricultural technology, agrotechnics
агрофизика agricultural physics, agrophysics
агрофон agricultural background; soil fertility
агрохимик agrochemist
агрохимия agricultural chemistry
АГ-цемент anhydride cement
адажио adagio
адактилия adactylia, adactylism, adactyly
адалин adalin, carbromal
адамантит *miner.* adamantite
Адамар Hadamard
адамеллит *miner.* adamellite
адамин, адамит *miner.* adamite
адамит *hist.* Adamite
адамсит adamsite
адаптация adaptation
~, вкусовая gustatory adaptation
~ глаза ocular adaptation
~, слуховая acoustic adaptation
~, темновая dark adaptation
~, ферментативная enzymatic adaptation
адаптер 1. (lens) adapter; film-/plate-holder; 2. pick-up
~ межканальной связи channel-to-channel adapter
~ межпроцессорной связи on-line adapter
~ периферийных устройств peripheral adapter
~ сбора данных front-end adapter
~ сопряжения interface adapter
адаптивность adaptability
адаптивный adaptable; *cyb.* adaptive, self-adapting
адаптиогенез adaptiogenesis
адатом adsorbed atom, adatom
адвекция *meteor.* advection
адвентист Adventist
адгезия adhesion, adherence
аддитивность additivity
~ средних additivity of means
аддитивный additive
аддукт adduct, inclusion complex
адекватность adequacy; fit; correspondence
аделит *miner.* adelite
аделогенный *miner.* adelogenic, aphanitic
аденилаткиназа adenylate kinase, myokinase
аденин adenine
аденит adenitis
адения adenia, pseudoleukemia, lymphomatosis
аденовирус adenovirus
аденовирусный adenoviral
аденозин adenosine, adenine riboside
аденоидит adenoiditis
аденоидный adenoid
аденокарцинома adenocarcinoma
~ груди mammary/breast adenocarcinoma
аденома adenoma
аденоматоз adenomatosis
аденопатия adenopathy
адергнейс *miner.* adergneiss, venite
аджарец, ~рка Adzhar, Adzharian
аджарский Adzharian
адиабата adiabatic curve/path, adiabat
~, ударная impact/percussive/shock adiabat
адиабатический adiabatic
адиагностический *miner.* adiagnostic, dubiocrystalline
адиантум *bot.* maidenhair
адинамия adynamia
~, железодефицитная asiderotic adynamia
адипиновый adipic, adipinic
адипозит adipositis
Администрация ООН по вопросам помощи и послевоенного восстановления, ЮНРРА United Nations Relief and Rehabilitation Administration, UNRRA
Администрация Экономического сотрудничества Economic Cooperation Admin-

istration
адмирал admiral
~ флота admiral of the fleet
адмиралтейство the Admiralty
адмиральский корабль flagship
аднексит adnexitis, annexitis
Адонис *myth., bot.* Adonis
адренализм adrenalism, suprarenalism
адреналин adrenalin(e), epinephrine
адреналинемия adrenalinemia, adrenemia
адреналон adrenalone
адренергетик adrenergic substance/drug
адреноблокатор adrenoblocker
адренокортикотропин adrenocorticotrop(h)in, adrenocorticotrop(h)ic hormone
адреномиметик adrenomimetic drug/substance
адренопатия adrenopathy, adrenalopathy
адренотропин adrenotrop(h)in, adrenocorticotrop(h)in
адренотропный adrenotrop(h)ic, adrenocorticotrop(h)ic
адрес 1. *comp.* address; location number; 2. destination
~, абсолютный absolute/true/specific address
~ адреса indirect address
~, активный, буфера active buffer address
~, базовый base address
~ барабана drum address
~ блока данных data block address
~, внешний address out
~, внутренний home address, address within the block
~ возврата return address
~ восприятия sense address, sense information address
~, восьмеричный octal address
~, вспомогательний auxiliary address
~ входа в фазу phase entry point, phase start address
~ входа, заданный desired start address
~ входа, начальный entry-point address, start address
~ выборки из памяти access store address
~ вызова call address
~ выключения drop address
~ выполнения done address
~, вычисляемый calculated address
~ данных data address
~ двоичного разряда bit address
~, действительный absolute/explicit/actual/effective address
~, желательный пусковой start address
~ за пределами адресного пространства out-of-range address
~ заголовка head address
~ загрузки load-point address
~ записи record address
~, зашифрованный/закодированный coded address
~ зоны field/area address
~, индексированный indexed/subscripted address
~ индексного регистра address to index register
~, исполнительный execution/executive address
~, истинный actual/effective/absolute address
~, исходный start/initial address
~, канальный channel address
~ ключевого слова key word address
~, ключевой leading/key address
~ команды command/instruction/order address
~ команды логического сдвига logic shift command/instruction address
~ команды перехода address of the jump instruction
~, конечный end/final address
~ контрольной точки test point address
~ массива данных block address
~ матричной памяти address in a ferrite core
~ местоположения location address
~, начальный start/initial/beginning address, original address
~, недействительный invalid address
~, непосредственный direct address
~ непрерывной выборки consecutive access address
~, непрямой indirect address
~, нечетный odd address
~, неявный implicit address
~ обращения access address; call address
~, однозначный unique address
~, основной base/basic address
~ основной памяти main/primary store/memory address
~ останова stop/halt address
~, относительный relative/differential/offset address
~ параметра уровня level parameter address
~, первый leading address
~ передачи управления jump address
~, перемещаемый relocatable address
~ перехода jump address
~, плавающий floating address
~ поиска seek address
~, последний last field address
~, последовательный subsequent/sequential/sequence address; address of the subsequent instruction
~, постоянный fixed/constant/invariable address
~, постоянный пусковой fixed starting address
~, правый right address
~ программы program(me) address
~, прямой direct address, first-level address
~ разветвления branch address
~ разгрузки unload address
~ результата result/target address
~ сверхоперативной памяти rapid memory address
~, свободный blank address
~, символический floating/symbolic address,

pseudo-address
~ следующей команды sequent/sequence instruction address, next-instruction address
~ слова word address
~ смешения relocation address
~ со звездочкой asterisk address
~ со смещением offset address
~ сопряжения chaining address
~, текущий present address
~, условный symbolic address
~ устройства unit/device address
~ фазы, начальный phase entry point, phase start address
~, фактический actual/effective address; de facto address
~, фиксированный fixed/constant/invariable address
~, фиктивный dummy address
~ хранения store address
~ цели target address
~ цилиндра cylinder address
~ числа number address
~ электронной обработки данных EDP address
~, эффективный actual/effective address
~, юридический legal address; domicile
~, явный explicit address
~ ячейки памяти storage/memory location, memory/location address
адресат addressee; destination
адресация addressing; address information
~, абсолютная absolute addressing, reference by actual address
~, байтовая byte addressing
~ блоков block addressing
~, внутренняя intrinsic addressing
~ внутренней памяти addressing in primary memory store
~, гибкая flexible addressing
~, групповая group addressing
~ данных на магнитном барабане magnetic drum store/memory addressing
~, индексная indexed addressing, indexing
~ кодом команды inherent addressing
~, лавинная flooding
~, мандатная capability-based addressing
~, многоуровневая косвенная multilevel addressing
~, непосредственная immediate addressing
~, неявная implicit/implied addressing
~, относительная relative addressing/address
~, относительная индексная based relative indexing
~ пакетов, множественная multihoming
~ памяти на линиях задержки addressing in delay store
~ по базовой странице base-page addressing
~ по счетчику команд program(me) counter addressing
~ по текущей странице current-page addressing
~, побитовая bit addressing
~, повторная repetitive addressing
~, прямая immediate/direct addressing
~, распределительная broadcast addressing
~ с индексацией и базированием index and base register addressing, double register addressing
~ с переменной длиной адреса extensible addressing
~ с префиксам addressing/address with prefixing
~, символическая symbolic addressing
~, смысловая ingenious addressing
~, составная chained addressing; build-up addressing
~ ЭЛТ CRT addressing
~, явная explicit addressing
адресность number of addresses; addressness
~ ЦВМ number of addresses per instruction
адресовать address
адресограф addressing machine
адресуемость addressability
адресуемый addressable
~ по словам word addressable
~мая память content-addressable memory/store
~мые часы, ~мый счетчик времени program(me)-adressable clock
адсорбат adsorbate
адсорбент adsorbent
адсорбер adsorber
адсорбировать adsorb
адсорбционный adsorptive
адсорбция adsorption
~, избирательная preferential/specific adsorption
~ на ионообменной смоле ion-exchange resin adsorption
~, обратимая reversible adsorption
~, поверхностная surface adsorption
~, полимолекулярная multimolecular adsorption
~, химическая chemisorption
~, хроматографическая chromatographic/differential/stratographic adsorption
адусация *met.* malleablizing
адъюнкт(а) *math.* cofactor
адъюстаж *met.* auxiliaries
адыгеец, ~ейка, ~ейский Adygei
адюльтер *law* adultery
ажио agio
ажиотаж stock-jobbing
ажитато *mus.* agitato
ажур up to date
азалия *bot.* azalea
азбука alphabet, code
азеотропия azeotropy
азеотропная смесь azeotropic mixture
азербайджанец, азербайджанский Azerbaijani
азербайджанка Azerbaijani (woman)
азид azide

азимут azimuth; bearing; *rad.* radial
~, астрономический astronomical/true azimuth
~, геодезический geodetic/surveying azimuth
~, измеренный observed azimuth
~, имитированный simulated azimuth
~, истинный = астрономический
~, исходный/начальный initial/reference azimuth
~, обратный reverse/back azimuth
~, опорный reference azimuth
~ при смыкании final azimuth
~ привязки angle of attachment
~ простирания strike/trend azimuth
~, прямой forward azimuth
~ пуска launch azimuth
~, равный isoazimuth
~ радиомаяка radial (of a radio range)
~ светила star azimuth
~, упрежденный predicted/future azimuth
~ хода azimuth of a traverse
~ цели target azimuth
азимутальный azimuthal
азимут-квадрант azimuth dial
азин azine
азобелок azoprotein
азобензол azobenzene
азойский Azoic
азокраситель azo dye
азот nitrogen
~, мочевины blood urea nitrogen
~, общий total nitrogen
~, остаточный residual [non-protein] nitrogen
~, связанный bound/fixed nitrogen
~, усвояемый available nitrogen
~, фиксированный bound/fixed nitrogen
азотемия azotemia, uremia
азотизация nitrogenization
азотирование nitriding
азотировать nitride
азотисто-синеродистый *chem.* cyanonitride
азотистый nitrous
~тое железо iron nitride
азотноватистый hyponitrous
азотноватый hyponitric
азотнокислый nitrate
~ цинк zinc nitrate
азотный nitric; nitrogenous; nitrogen *(attr.)*
азотобактер Azotobacter, Azobacter, nitrogen-fixing bacteria
азотобактерин nitrogen-fixing fertilizer
азотометр nitrometer, azotometer
азотосодержащий nitrogenous
азотфиксатор nitrogen fixer
азофоска azophoska
азулен azulene
азур Azure blue
азурит *miner.* azurite; chessylite
аир *bot.* sweet flag *(Acorus)*
айдирлит *miner.* aidyrlite
айкинит *miner.* aikinite, needle ore
айкит *miner.* ajkite, ajkaite
айсберг iceberg
айсбимс *naut.* ice beam
академия academy
~, военная military academy
~, военно-воздушная airforce academy
~, военно-медицинская military medical academy/college
~, военно-морская naval academy
~, духовная theological academy/college
~ наук academy of sciences
~ сельскохозяйственных наук academy of agricultural sciences
~ художеств academy of arts
акадский Acadian
акажу *bot.* cashew
акалеф *miner.* acaleph
акальцикоз acalcicosis, acalcerosis
акант *bot., arch.* acanthus
акантит *miner.* acanthite, argentite, vitreous silver
акантоцит acanth(r)ocyte
аканье *ling.* akanje
акариды *Acaridae*
акаталепсия acatalepsy, acatalepsia
акать *ling.* fail to differentiate between unstressed back vowel-sounds *(in Russian and Byelorussian)*
акафист *church* acathistus
аквакобаламин aquacobalamine, hydroxocobalamin, vitamin B12
акваланг aqualung, SCUBA (self-contained underwater breathing apparatus)
аквалангист skin-diver, scuba-diver
аквамарин *miner.* aquamarine
акванавт deep-sea driver, frogman, aquanaut
акваплан aquaplane
аквариды *astr.* aquarids
акватинта *print.* aquatint
акватипия aquatone printing
акватория water area
акведук aqueduct
акинез akinesia, akinesis
акклиматизатор acclimatization house
акклиматизировать acclimatize
акколада *mus.* accolade
аккомодация accommodation
~, избыточная excessive accommodation
~, недостаточная subnormal accommodation
аккордный:
~ная плата payment by the piece, piece payment
~ная работа piece-work
~ная система piecework scheme/plan/agreement
аккредитив letter of credit
~, авизованный specially adviced (straight) letter of credit
~, автоматически возобновляемый revolving credit
~, безотзывный/авизованный irrevocable/straight letter of credit
~, делимый divisible letter of credit
~, компенсационный back-to-back letter of

credit
~, неподтвержденный unconfirmed letter of credit
~, отзывать revoke a letter of credit
~, отзывной revocable letter of credit
~, открывать open a letter of credit
~, подтвержденный confirmed letter of credit
~ с внесенным покрытием paid letter of credit
~ с оплатой тратт на предъявителя sight letter of credit
~ с предъявлением срочной тратты time letter of credit
~, совместный back-to-back letter of credit
~, товарный commercial letter of credit
~, туристский travel(l)ers' letter of credit
~, чистый open/clean letter of credit
аккумулирование accumulation, storage
аккумулировать accumulate, store
аккумулятор 1. (storage) battery; secondary/storage cell/unit; 2. *mech.* accumulator; 3. accumulating register
~, бортовой vehicle-borne battery
~, гидравлический hydraulic accumulator
~, железо-никелевый nickel-iron battery/cell
~, инерционный inertial energy-storage unit/device
~, кадмиево-никелевый nickel-cadmium battery/cell
~, кислотно-свинцовый lead-acid battery/cell
~, никель-цинковый nickel-zinc battery/cell
~, паровой steam accumulator
~, плунжерный ram accumulator
~, пневматический compressed-air accumulator
~ с металлокерамическими электродами sintered plate battery/cell
~ с пастированными пластинами pasted plate battery/cell
~ с формованными пластинами Planté cell
~, свинцовый lead-acid battery/cell
~, серебряно-кадмиевый silver (oxide) cadmium battery/cell
~, серебряно-цинковый silver (oxide) zinc battery/cell
~, стартерный/стартовый starter battery
~, цинково-ртутный zinc-mercury battery/cell
~, щелочной alkaline battery/cell
аккумуляция accumulation; accretion
аклина acline, aclinic line, magnetic equator
аклинальный aclinal
аклиновый acline
акме acme
акмеизм acmeism
аком acoustic ohm
аконит *bot.* aconite
акридин acridine
акрил acryl
акрилат acrylate
акриловый acrylic
акрилонитрил acrylonitrile; vinyl cyanamide
акролеин acrolein; acrylaldehyde
акромегалия acromegalia, acromegaly
акрополь acropolis
акростих *lit.* acrostic
акроцефалия acrocephaly
акселерограф accelerograph, recording accelerometer
акселерометр accelerometer, g-meter
~ больших ускорений high-level accelerometer
~, емкостный capacitance accelerometer
~, импульсный pulsed accelerometer
~ инерциальной системы наведения inertial-guidance accelerometer
~, крешерный crush accelerometer
~, максимальный maximum reading accelerometer, peak(-value) accelerometer
~ малых ускорений low-level accelerometer
~, маятниковый pendulous accelerometer
~, однокомпонентный single-component/-axis accelerometer
~ отрицательных ускорений decelerometer
~ поперечных ускорений lateral/transverse accelerometer
~ продольных ускорений longitudinal (fore-and-aft) accelerometer
~, регистрирующий/самопишущий recording accelerometer, accelerograph
~ со счетчиком counting accelerometer
~, тангажный pitch-sensing accelerometer
~, тензометрический strain-gauge accelerometer
~ ударных ускорений/перегрузок impact accelerometer
~, электронный vacuum-tube accelerometer
аксельбанты aiguillettes, aiglets, aglets
аксиальный axial
аксиология axiology
аксиома axiom; postulate; assumption
~ Архимеда Archimedean axiom/assumption, axiom of Archimedes
~ бесконечности axiom of infinity
~ выбора axiom of choice, choice axiom
~ математической индукции axiom of perfect/complete induction
~ параллельности axiom of parallelism, parallel axiom
~ сводимости reducibility axiom
~ сохранения retention axiom
~ счетности denumberability axiom
аксиоматизировать axiomatize
аксиоматизируемость axiomatizability, axiomatization
аксиоматизируемый axiomatizable
аксиоматика axiomatics
~, содержательная informal axiomatics
аксограф axograph
аксоид axoid
аксон axon(e), neurone
аксонометрия axonometry, perspective geometry
акт 1. act; *law* deed, document; statement;

2. commencement ceremony, speech-day; 3. *nucl.* event
~, аварийный survey/average report
~ деления *nucl.* fission event
~ захвата *nucl.* capture event
~ ионизации ionizing event
~ испытания test record/certificate
~, крепостной title deed
~ об испытании certificate of proof, test report
~, обвинительный (bill of) indictment
~, подзаконный by-law; subordinate legislation
~ приемки, ~ приемо-сдаточный acceptance certificate
~ рассеяния *nucl.* scattering event
~, рекламационный unsatisfactory equipment report, UER
~ судебной власти judicial act
актив I (Party) activists, the most active members
актив II assets, credit
~, забронированный/зарезервированный earmarked asset
~, замороженный frozen asset
~, истощимый depletable asset
~, материальный tangible asset
~, неликвидный fixed/permanent asset
~, нематериальный/неосязаемый intangible/fictitious asset; *pl.* intangibles
~, текущий current/circulating/liquid/floating asset; working capital
активатор activator; promoter; sensitizer; accelerator
~ катализатора catalyst promoter
~, основной dominant/principal activator
~ роста growth promoter
активизация, активирование activation
~ до насыщения saturation activation
~ охладителя/теплоносителя coolant activation
~, повторная reactivation
активировать, активизировать 1. activate; boost, bolster, spur; speed up; promote, stimulate; 2. *tech.* dope
активист person active in public/social affairs/life; activist
активность 1. activity; 2. radioactivity, intensity of radiation/radioactivity
~, вторичная second-order activity
~, дочерняя daughter activity
~ излучения intensity of radiation, radioactivity
~, изотопическая isotope activity
~ индикатора tracer activity
~ ионов ion activity
~, коррозионная corrosivity, intensity/severity of corrosion
~ образца sample strength/activity
~, объемная volumetric activity
~ окружающей среды environmental activity
~, осколочная *phys.* fission-fragment activity
~, остаточная residual activity
~, поверхностная surface activity
~, посторонняя extraneous activity
~, равновесная equilibrium activity
~, удельная specific activity
~, фоноэквивалентная background equivalent activity
~ частичек particulate activity
активный 1. active; 2. favourable
актинизм actinism
актиний actinium, Ac
актиничность actinism
актиния actinia, sea anemone
актинолит actinolith
актинометрия actinometry
актиномикоз actinomycosis
актиномицеты Actinomyces
актиномицин actinomycin, actinomycetin
актиноморфный actinomorphic, actinomorphous
актинон actinon, radon-219, actinum emanation
актиноскоп actinoscope, solar radiation meter
актиноуран actinouranium, actinium-uranium, uranium-actinium
актинофотометр actinic photometer
актуализация updating
актуализировать update; bring up to date
актуарий actuary
акулы *zool.* sharks, selachians (*Selachii)*
акуметр acumeter, acoumeter, acoutometer
акупунктура acupuncture
акустик *av.* soundman
акустика acoustics
~, волновая wave/physical acoustics
~, прикладная/техническая appllied/engineering acoustics
~, строительная acoustics of buildings
акустиметр acoustimeter
акутомия acusection
акцелерин accelerin, accelerator globulin
акцепт *fin.* acceptance
~, банковский banker's acceptance
~, безусловный general/clean acceptance
~, бланковый acceptance in blank
~, дружеский accommodation acceptance
~, неполный partial acceptance
~ с оговорками qualified/special acceptance
~, условный conditional/qualified/special acceptance
акцептант, акцептор acceptor
акцессория accessory
акцессорный accessory; auxiliary
акциденция *print.* jobbing work
акциз excise, excise-duty
акционер shareholder, stockholder
акционерный joint-stock; incorporated
~ капитал joint stock; capital stock
акция share; stock
~ без номинала no-par-value stock
~ выпускаемая со скидкой с номинала below par stock
~ дающая право голоса voting stock
~, директорская management stock

~, именная registered/nominal share
~ имеющая номинал par-value stock
~ как объект инвестиции investment stock
~ на предъявителя ordinary share
~, обыкновенная/обычная ordinary/common/ equity stock, equity
~, предъявительская bearer stock
~, привилегированная preference/preferential/ preferred/priority share
~ разрешенная к продаже transferable share
~ с накопляющимся дивидендом cumulative stock
~ с отсроченным дивидендом deferred stock
~ с твердым процентом fixed interest stock
~, частично оплаченная part-paid stock
аланин alanine
алантин alantin, inulin
алантол alantol, inulol
албанец, ~нка, ~нский Albanian
алгебра algebra
~, аннигиляторная annihilator algebra
~, банахова Banach algebra
~, булева Boolean algebra
~, векторная vector algebra
~ высказываний propositional algebra
~, высшая higher algebra
~ классов algebra of classes
~ конечного ранга algebra of finite order
~ Ли, свободная free Lie algebra
~, линейная linear algebra
~ логики algebra of logic, logical algebra
~ матриц matrix algebra
~ мер measure algebra
~ множеств algebra of sets
~ над полем algebra over a field
~ ограничений constraint algebra
~ простой характеристики algebra with prime characteristics
~ релейных схем switching algebra
~, реляционная relational algebra
~ с делением division algebra
~ с мерой measure algebra
~ схем circuit algebra
алгебраизация algebraization
алгебраист algebraist
алгебраический algebraic
алгеброидный algebroidal
АЛГОЛ ALGOL
~ компилирующая/генерирующая программа ALGOL-compiler
~ описание declaration ALGOL
~ рабочая программа ALGOL object program
Алгол *astr.* Algol
алгоритм algorithm
~, аппаратно-реализованный hardware algorithm
~ арифметической операции algorithm for arithmetic operation
~ ветвления branching algorithm
~ выбора оптимального маршрута best-route algorithm, optimal path algorithm
~ выбора решения decision algorithm
~ выполнения execution/performance algorithm
~, вычислительный computational algorithm
~, двойственный симплексный dual simplex algorithm
~, декодирующий decoding algorithm
~ деления algorithm for division, division algorithm
~, дополнительный auxiliary algorithm
~ замещения страниц paging algorithm
~ инверсии algorithm for the inversion
~ Квайна Quine algorithm
~, комплексный симплексный composite simplex algorithm
~ логического ввода inferencing algorithm
~ Мак-Класки McCluskey algorithm
~, максиминный max-min algorithm
~, машинный computer-oriented algorithm
~, минимаксный min-max algorithm
~, обобщенный generalized algorithm
~ перевода translation/interpretation algorithm
~, последовательный sequential algorithm
~ Поста/постовский Post algorithm
~ приведения reduction algorithm
~ проектирования algorithm for design(ing)
~ разложения decomposition algorithm
~ размещения placement algorithm
~ распределения scheduling algorithm
~ решения двойственной задачи dual algorithm
~ решения прямой задачи primal algorithm
~ решения транспортной задачи transportation algorithm
~ с неполной памятью partial-memory algorithm
~, самоизменяющийся self-adaptive algorithm
~, симплексный simplex algorithm
~, систолический systolic algorithm
~ сложения addition algorithm
~ составления графика/расписания scheduling algorithm
~, табличный table algorithm
~ трансляции compilation/translation algorithm
~ Тьюринга Turing algorithm
~ управления control algorithm
~ управляющего устройства controller algorithm
~, челночный shuttle algorithm
алгоритмизация algorithmic presentation, algorithmization
алеаторика *mus.* aleatorics
алебарда halberd
алебастр 1. plaster of Paris; 2. alabaster
алебастрит *miner.* alabastrite, compact gypsum
алеврит *miner.* aleurite, silt
алевролит *miner.* aleurolite, siltstone
алейрометр aleurometer
алейрон aleurone
александрийский Alexandrine; Alexandrian

~ **лист** senna
~ **стих** Alexandrine (verse)
александрит *miner.* alexandrite
алексия alexia
алеут, ~ка, ~ский Aleutian
алеф aleph
алжирец, ~рка, ~рский Algerian
алиби alibi
алигнин artificial cotton
алидада alidade
~ **квадранта/секстанта** index arm/bar
~ **теодолита** vernier plate
ализарин alizarin
аликвантный aliquant
аликвотный aliquot
алиментарный alimentary, nutritional
алиментный *law* alimentary
алименты alimony, maintenance
алитировать aluminize, calorize
алифатический aliphatic
алициклический alicyclic
алкализировать alkalize
алкалиметр alkalimeter
алкалоид alkaloid
алкалоидный alkaloidal
алкан alkane
алкен alkene
алкидный alkyd
алкил alkyl
алкилировать alkylate
алкоголиз alcoholysis
алкоголят alcoholate; alkoxide
алкоксикислота ether acid
алкоксисоединение alcoxy compound
алкоран = коран
аллантоин allantoin
аллегретто allegretto
аллегро allegro
аллели alleles
~, **парные** pair alleles
~, **потенциальные** potency alleles
аллеломорфизм allelomorphism
аллеломорфный allelomorphic
аллелотропический, аллелотропный allelotropic
аллелотропия allelotropism
аллель allele
аллеманда *mus.* alemande
аллен allene, propadiene
аллергический allergic
аллергия allergy, allergia
~, **кожная** skin/dermal allergy
~, **пищевая** *food* allergy
аллерголог allergologist, allergist
аллергология allerg(i)ology
аллил allyl
аллиловый allylic
аллобар allobar
аллод *hist.* al(l)odium, al(l)odial estate
аллозавр allosaurus
алломерия allomerism
алломорфный allomorphic, allomorphous
аллонж *law* allonge
аллопат allopath, allopathist
аллопластика alloplasty
аллополиплоидия allopolyploidy
аллотриоморфность *cryst.* allotriomorphism
аллотропический, аллотропный allotropic
аллотропия allotropy, allotropism
аллохтон *biol.* allochthone
аллювиальный alluvial, alluvian
аллювий alluvium, alluvion
аллюодит *miner.* alluaudite
алмаз 1. diamond; 2. glazier's diamond
~ **для резки стекла** diamond glass cutter
~, **ограненный** cut diamond
~, **сырьевой** rough diamond
~, **технический** industrial diamond
~, **черный (технический)** black/carbon diamond
~, **ювелирный** gem diamond
алмазный diamond *(attr.)*; adamantine
ални Al-Ni alloy, Alni
алнико Al-Ni-Co alloy, Alnico
алниси Al-Ni-Si alloy, Alnisi
алонж prolong; adapter
алопеция alopecia, acomia, baldness
алтайский Altai, Altaic, Altaian
алоэ *bot.* aloe; *med.* aloes
алунд aluminium-oxide abrasive, alundum
алунит *miner.* alunite, kalinite
алуноген *miner.* alunogen, feather alum; hair salt
алфавит alphabet
~ **верхнего регистра** uppercase alphabet
~, **входной** input alphabet
~, **выходной** output alphabet
~, **двоичный** binary alphabet
~, **машинный** computer alphabet
~ **нижнего регистра** lowercase alphabet
алфавитно-цифровой alphanumeric, alphameric, alphabetic-numeric
алыча cherry-plum *(Prunus cerasifera)*
альб *geol.* Albian stage
альбатрос albatross
альбедо albedo
~, **видимое** apparent albedo
~, **истинное** plane albedo
~ **нейтронов** neutron albedo
альбертит *miner.* albertite, asphaltic coal
альбигойцы *hist.* Albigensians
альбинизм albin(o)ism
альбинос albino
Альбион *poet.* Albion, Britain
альбумин albumin
~, **сывороточный** serum albumin
~, **яичный** albumen, egg albumin, ovalbumin
альбуминный albuminous
альбуминурия albuminuria
~, **истинная** intrinsic/true albuminuria
~, **неврогенная** neurotic albuminuria

альвейер wobble pump
альвеола alveolus
альвеолит alveolitis
альвеолярный alveolar, alveolate
альгология algology
альграфия *print.* aluminography, algraphy
альдегид aldehyde
  ~, анисовый anisaldehyde, anisyl aldehyde
  ~, бензойный benzaldehyde
  ~, глицериновый glyceraldehyde
  ~, кетоглютаровый ketoglutaraldehyde
  ~, лауриновый lauraldehyde
  ~, масляный butyl aldehyde
  ~, муравьиный formic aldehyde, formaldehyde
  ~, уксусный acetic aldehyde, acetaldehyde
альдегидоксидаза aldehyde oxidase
альдоза aldose
альдостерон aldosterone
альдрей Al-Mg alloy, Aldrey-type alloy
алькализация alkalization
алькализировать alkalify
алькатрон alcatron, unipolar field-effect transistor
альклед alclad aluminium alloy
альков alcove
алькусин Al-Cu-Si alloy, alcusin
альмалек Al-Mg-Si alloy, almalec
альпага = альпака
альпака I *zool.* alpaca
альпака II German silver
альпари *comm.* at par
альпинарий rockery, rock garden
альпиниада mass ascent, Alpiniad
альсекко *art* al secco
альсифер Al-Si-Fe alloy, alsifer
альт 1. alto; 2. viola
альтазимут altazimuth
альтиметр altimeter
Альтинг *pol.* Althing
альтист viola player
альфа-активность alpha-particle activity
альфа-ветвь alpha-branch
альфа-излучатель alpha-ray emitter
альфа-излучение alpha radiation/rays
альфа-линия alpha-peak
альфа-лучи alpha-rays/radiation
альфа-превращение alpha-alteration
альфа-распад alpha-decay/disintegration
альфа-спектрометр alpha(-ray) spectrometer
альфа-счетчик alpha-counter
альфатип *print.* alphatype
альфатрон alphatron, radioactive-ionization vacuum gauge
альфа-частица alpha-particle
альфетамин alfetamine
альфоль aluminium foil
алюмел alumel
алюминат aluminate
алюминий aluminium, aluminum, Al
  ~, брусковый aluminium bars
  ~, зерненый *print.* grained aluminium
  ~, прутковый aluminium rods
  ~, хлористый aluminium chloride
  ~, чушковый aluminium ingots/pigs
алюминировать = алитировать
алюминография *print.* algraphy
алюминотермия, алюмотермия aluminothermy
алюмогел alumina gel, alumogel
алюмогранат *miner.* aluminium garnet
алюмолит *miner.* bauxite
алюмосиликат aluminosilicate
амавроз *med.* amaurosis
амага *phys.* Amagat unit
амазонка Amazon
амальгама amalgam
амарант amarant(h)
амариллис amaryllis
амбивалентность ambivalence
амбидекстрия ambidexterity, ambidextrism
амбилевия ambisinistrism
амблигонит *miner.* amblygonite
амблиопия amblyopia
амблиподы amblipods
амбра ambergris
амбразура embrasure
  ~, фурменная tuyere arch
амброзия ambrosia; *bot.* ragweed
амбротипия *print.* ambrotype
амбус anvil bed/stake
амбушюр mouthpiece (of telephone)
амвон ambo
амеба amoeba, ameba
амебоцит amoebocyte
амелиорация land improvement, amelioration
аменорея amenorrh(o)ea
аменция amentia
Американская ассоциация стандартов American Standards Association, ASA
Американская федерация труда и Конгресс производственных профсоюзов, АФТ-КПП American Federation of Labor/Congress of Industrial Organizations, AFL/CIO
Американский стандарт United States standard, USS
Американское общество инженеров-механиков American Society of Mechanical Engineers, ASME
америций americium, Am
аметист amethyst
аметропия ametropia, ametropy
амиант *miner.* amiant(h)us
амигдалин amygdalin
амигдалит amygdalitis
амигдалоза gentiobiose, amygdalose
амигдалярный amygdaline
амид amide
  ~ муравьиной кислоты formamide
  ~ уксусной кислоты acetamide
амидин amidine
амидинотрансфераза amidinotransferase,

transaminidase
амидирование amidation
амидировать convert into amide, amidate
амидопирин amidopyrine, aminopyrine, pyramidon, dimethylaminoantipyrine
амидопроизводные amide derivatives/products
амиелинизация amyelination
амиелиновый amyelic, amyelous, amyeloic, amyelonic
амил, амиловый amyl
амилаза amylase
амилолиз amylolysis
амин amine
~ жирного ряда fatty amine
аминирование amination
аминировать aminate
аминоазот amino nitrogen
аминогруппа amino group
аминокислота amino acid
аминосмолы amino resins
амиокардия amyocardia
амиотрофия amyotrophy, amyotrophia
амитал, амиталовый amytal
амитал-натрий sodium amytal
аммиак ammonia
~, газообразный ammonia gas, gaseous ammonia
~, едкий aqua ammonia, ammonium hydroxide
аммиакат ammoniate
аммиачный ammoniac(al), ammonium, ammonia
~ная вода ammonia water, ammonium hydroxide
аммонал ammonal
аммонизатор ammoniation tank, ammoniator
аммонизация ammonation, ammoniation
аммонизировать ammonate, ammoniate
аммоний ammonium
~, азотнокислый ammonium nitrate
~, бензоинокислый ammonium benzoate
~, бромистый ammonium bromide
~, двууглекислый ammonium bicarbonate
~, иодистый ammonium iodide
~, лимоннокислый ammonium citrate
~, надсернокислый ammonium persulphate
~, роданистый ammonium rhodanate, ammonium thiocyanate
~, салициловокислый ammonium salicylate
~, углекислый ammonium carbonate
~, уксуснокислый ammonium acetate
~, фосфорнокислый ammonium phosphate
~, фтористый ammonium fluoride
~, хлористый ammonium chloride
~, щавелевокислый ammonium oxalate
аммонийный ammoniacal
аммонит 1. ammonite; 2. *min.* ammonia dynamite, quarrypowder
аммонолиз ammonolysis
аммофос ammophos
аммофоска ammophoska
амнезический amnesic, amnestic
амнезия amnesia
амнион amnion
амниоты amniotes *(Amniota)*
амнистировать amnesty
амнистия amnesty; free pardon
~, общая general amnesty/pardon
амок amok, amuck
амортизатор shock-absorber; damper, vibration absorber
~, воздушный pneumatic/air shock-absorber
~, гидравлический hydraulic shock-absorber; oleo leg/strut
~ двустороннего действия double-acting [two-way] shock-absorber
~, жидкостный liquid-spring [oleospring] shock-absorber
~, масляно-воздушный oil/oleo pneumatic shock-absorber
~, масляный oil/oleo shock absorber
~, пластинчатый block shock-absorber
~, пластинчатый резиновый rubber block shock-absorber
~, пневматический pneumatic [air-cushion] shock-absorber
~ подвески suspension bumper
~, поршневой piston-/lever-type shock-absorber
~, резиновый rubber anti-vibration/-shock mounting; rubber shock-absorber
~, рессорный concussion spring, snubber
~, фрикционно-пружинный spring-friction shock-absorber
~ шасси undercarriage [landing-gear] shock-absorber
~, шнуровой cord shock-absorber, shock-absorber cord
амортизационный depreciation *(attr.)*; amortization *(attr.)*
~ фонд/капитал sinking fund
~ные отчисления depreciation charges
амортизация 1. *econ.* depreciation; amortization; wear and tear; 2. *mech.* shock absorption; insulation from/against vibration
амортизировать 1. depreciate; charge off; 2. *mech.* absorb, damp, cushion
амортизованный 1. depreciated; 2. *mech.* shock-proof, shock-mounted
аморфность amorphism; amorphous state
аморфный amorphous
ампелография *bot.* ampelography
ампельные растения plants grown in hanging pots
ампер ampere, A
ампераж amperage
ампер-весы current balance
ампер-витки ampere-turns
~, активные effective ampere-turns
ампервольтваттметр amvoltwattmeter
ампервольтметр voltammeter
амперметр ammeter
~, детекторный rectifier ammeter
~, магнитоэлектрический moving-coil am-

meter
~ максимального спроса maximum-demand ammeter
~, перегрузочный overload ammeter
~, пиковый maximum-demand ammeter
~, самопишущий recording ammeter
~ тепловой системы hot-wire ammeter, thermal (expansion) ammeter
~, термоэлектрический thermo-emf ammeter
~, ферродинамический ferrodynamic ammeter, iron-cored electrodynamic ammeter
~, щитовой switchboard ammeter
~, электромагнитный moving-iron ammeter
амперметр-клещи bus-bar ammeter, split-electromagnet ammeter
амперометрия amperometry, amperometric titration
ампер-секунда ampere-second
ампер-час ampere-hour
ампир Empire style
ампирный Empire-style
ампициллин ampicillin
амплидин amplidyne, cross-field rotary amplifier
амплитрон amplitron, cross-field microwave amplifier tube
амплитуда amplitude; peak/crest value; range
~, двойная double/total amplitude, peak-to-peak value
~, действующая effective amplitude, root-mean-square amplitude
~, пороговая cut-off pulse height
~ прилива tidal range/amplitude
~, результирующая net amplitude
~, удвоенная double amplitude, peak-to-peak amplitude
~, эффективная = действующая
амплитудо-модулированный amplitude-modulated, AM, am
ампула 1. ampoule; 2. *anat.* ampulla
~ для проб sample ampoule
~ с отбиваемым концом break-seal ampoule
~ уровня level gauge
Амур *myth.* Cupid
амфетамин amphetamine
амфибии *pl.* amphibians *(Amphibia)*
~, безногие apodes
~, бесхвостые anurans
~, змееподобные dolichosomes
~, хвостатые urodelans
амфибия amphibian
амфибрахий amphibrach
амфибол *miner.* amphibole
амфиболит *miner.* amphibolite
амфион amphion, dipolar ion
амфодонт periodontum, dental periosteum
амфодонтит amphodontitis, periodontitis
амфолит ampholyte, amphoteric electrolyte
амфора amphora
амфотерность amphoterism
амфотерный amphoteric
амфоцит amphocyte
анагенез anagenesis, progressive development
анаглиф anaglyph
анаглифия anaglyphy, anaglyptics
анаграмма *lit.* anagram
анаконда anaconda
анакреонтический Anacreontic
анакруза, анакруса anacrusis
аналекты analects, analecta
анализ analysis; assay; evaluation
~, активационный (radio)activation analysis
~, ацидиметрический acid-base analysis
~ безубыточности break-even analysis; cost-volume-profit analysis
~ бесконечно малых *math.* infinitesimal calculus
~ в процессе эксперимента on-line analysis
~, валовой bulk/total/gross analysis
~, вариационный analysis of variance
~, векторный vector analysis
~, вероятностный probabilistic analysis
~, весовой gravimetric/weight analysis
~, вещественный substantial/material analysis
~ временного ряда time-series analysis
~, временной analysis in the time domain; time analysis
~ входных и выходных данных input-output analysis
~, выборочный sampling analysis
~, гармонический harmonic/Fourier analysis
~, глубокой in-depth analysis
~, гравитационный gravity analysis
~ граничных условий limit analysis
~, гранулометрический particle-/grain-size analysis
~ данных data evaluation/analysis
~ динамического ряда time-series analysis
~, динамометрический dynamic force analysis
~, дискретный sampling analysis
~, дисперсионный *math.* analysis of variance
~, дробный fractional analysis
~ затрат и технических характеристик cost-performance analysis
~ затрат на обеспечение качества quality cost analysis
~ издержек вследствие брака scrap-cost analysis
~ излома fracture test
~, изотопный isotopic analysis/assay
~, импульсный pulse analysis
~, инфракрасный спектральный analysis by infrared spectroscopy
~, капельный drop analysis
~, классификационный ranging analysis
~, ковариационный analysis of covariance
~, комбинаторный combinatorial analysis
~, кондуктометрический conductimetric analysis
~ конечных результатов end-point analysis
~, контрольный check analysis

~, кристаллохимический chemical analysis of crystals
~ крови blood count
~, кулонометрический coulometric analysis
~, люминесцентный fluorimetric analysis
~, масс-спектральный mass spectrometric analysis
~, машинный computerized analysis
~, межотраслевой intersectoral/interindustry analysis; input-output analysis
~ методом меченых атомов tracer analysis
~ методом моделирования simulation analysis
~ методом наименьших квадратов least-squares analysis
~ методом оплавления fusion analysis
~ методом площадей area analysis
~ методом титрования analysis by titration, titrimetric analysis
~, многомерный multivariate analysis
~, многошаговой multistage analysis
~ мочи analysis of the urine, urinalysis, uranalysis
~, общий global analysis
~, объемный volumetric analysis
~ операции activity analysis
~, опережающий anticipatory analysis
~ основанный на использовании ожидаемых/предполагаемых величин ex ante analysis
~ основанный на использовании фактических величин ex post analysis
~, пламефотометрический flame photometric analysis
~, последовательный sequential/successive analysis
~, предельный marginal analysis
~ предельных состояний limit analysis
~ приростной доходности incremental revenue and cost analysis
~, причинный analysis of causes, cause-and-effect analysis
~, пробирный assay, assaying
~ радиоактивности radioanalysis
~ разбиения на группы *stat.* cluster analysis
~ размерностей demensional analysis
~, растровый scanning analysis
~, регрессионный regression analysis, analysis of regression
~, рентгенографический radiographic analysis
~, рентгеноспектральный X-ray spectrometry
~, рентгеноструктурный X-ray diffraction analysis
~, рентгенофазовый X-ray phase analysis
~ с помощью вычислительной машины computer analysis
~, седиментометрический sedimetric/sedimentometric analysis
~ сжиганием combustion analysis
~, синтаксический parsing
~, системный systems analysis
~, ситовый mesh/screen/sieve analysis
~ состава семей household analysis
~, составной component analysis
~ состояние запасов stock analysis
~, спектральный spectrum/spectral analysis
~ схемы circuit analysis
~ таблицы адресов directory evaluation
~ тенденций рынка market trend analysis
~, тепловой/термический thermoanalysis
~, термомагнитный magnetothermal analysis
~, технико-экономический engineering-economic analysis
~ технических характеристик performance analysis
~, титровальный analysis by titration
~, углубленный in-depth analysis
~, фазовый phase(-shift) analysis
~, факториальный factor analysis
~, флуоресцентный fluorescence analysis
~ хозяйственной деятельности business analysis
~, частотно-временной time-and-frequency analysis
~ экономической эффективности cost-effectiveness analysis
~ эксплуатационных показателей performance analysis
~ эксплуатационных рекламаций field complaint/return analysis
~, электрографический *cryst.* electron diffraction analysis
~, элементарный elementary/ultimate analysis
~ эффективности затрат cost-effectiveness analysis
анализатор analyzer, analyser, estimation device
~, амплитудный pulse-height analyzer
~, векторный vectorlyzer, complex plane analyzer
~, вестибулярный vestibular analyzer
~ времени задержки/запаздывания time-delay analyzer
~ гармоник /гармонический harmonic/frequency/Fourier analyzer
~, дифференциальный цифровой digital differential analyzer; incremental computer
~ импульсов pulse analyzer
~, магнитный magnetic analyzer, analyzing magnet
~, обонятельный olfactory analyzer
~ переходных процессов transient response analyzer
~, ситовый testing sifter/sieve
~, слуховой/звуковой acoustic analyzer
~ спектра spectrum/spectral/spectroscopic analyzer
~, схемный circuit analyzer
~ формы сигнала waveform/wave analyzer
~, цифровой digital analyzer
~, частотный frequency response analyzer
~ частоты frequency analyzer

анализировать analyze, analyse; examine; assay, test
аналитик analyst
аналитика analytics
аналитический analytical
аналитичность analyticity
аналог analog(ue)
~, квантовый quantum analogue
аналогический analogical
аналогичный analogous; similar
аналогия analogy; similarity
~, строгая rigorous analogy
аналоговый analog(ue) *(attr.)*
анальгезия analgesia
анальгетический analgesic, analgetic
анальгия analgia
анальный anal
анальцим *miner.* analcime, analcite
анамнез anamnesis, case/medical history
~, отдаленный catamnesis
анаморфоз anamorphosis, anamorphism
анаморфот anamorphic lens, anamorphote
анапест anapest
анаплазмоз anaplasmosis
анапластика anaplasty, plastic surgery
анархо-синдикализм anarcho-syndicalism
анастигмат anastigmatic lens, anastigmat
~, полусимметричный convertible unsymmetrical anastigmat
анастигматизм anastigmatism
анастомоз anastomosis
анастомозировать anastomose
анастомозный anastomotic
анатоксин anatoxin, toxoid
анатом anatomist
анатомировать anatomize; dissect
анатомический anatomic(al)
анатомия anatomy
анафаза anaphase
анафема anathema
анафилаксия anaphylaxis
анафилактин anaphylactin, anaphylaxin
анафора anaphora
анафорез anaphoresis
анафродит anaphrodite
анахорет anchorite, anchoret
анаэроб anaerobe, anaerobian, anaerobion
анаэробный anaerobic
ангар hangar, shed
~, аэродромный flight hangar
~ для дежурных самолетов alert hangar
~ для дирижабля airship/dirigible hangar/dock
~ для контрольно-проверочных работ checkout hangar
~ для осмотра, ремонта и обслуживания dock/maintenance hangar
~ для реактивных самолетов jet hangar
~, куполообразный igloo hangar
~, надувной inflatable hangar
~ палатка tent hangar, tent-shed
~, палубный flight-deck hangar
~, раздвижной telescopic hangar
~, сборно-разборный portable/folding hangar
~ склад для грузов freight hangar
~, складной collapsable hangar
~, телескопический убирающийся retractable telescopic hangar
~ хранилище storage hangar
ангария *hist.* corvee, statute labour, task work
ангармонизм, ангармоничность anharmonicity
ангармонический anharmonic
ангидрид anhydride
~ азотной кислоты nitric anhydride
~ борной кислоты boric anhydride
~, кремневый silica, silicic anhydride
~, мышьяковистый arsenic oxide/trioxide, white arsenic
~, мышьяковый arsenic pentoxide
ангидрит anhydrite, anhydrous gypsum, karstenite
ангидрон magnesium perchlorate
ангина angina; tonsillitis, quinsy
~, грибковая mycotic angina
~, язвенная ulcerative angina, angina ulcerosa
ангинозный anginal, anginous
ангиография angiography
ангиома angioma
ангиоматоз angiomatosis
ангиомиома angiomyoma
ангиотензин, ангиотонин angiotensin, angiotonin, hypertensin
ангиотензиназа angiotensinase, hypertensinase
ангиотензиноген angiotensinogen, hypertensinogen
англезит anglezite, lead spar, lead vitriol
Английский банк Bank of England, B. of E.
англиканский Anglican
~кая церковь the Church of England, the Anglican Church
англист Anglicist, specialist in English philology
англистика English philology
ангоб slip, engobe
ангобиране *cer.* engobing
ангорская кошка Angora/Persian cat
ангстрем Angstrom, Å
андаксин meprobamite, equinal, meprospan
анданте andante
андезит *miner.* andesite
андрогиния androgyny, androgynism, female pseudohermaphroditism
андромания andromania, nymphomania
Андромеда *myth., astr.* Andromeda
андростерон androsterone
андрофилия androphylia
андрофобия androphobia
аневризма aneurism, aneurysm
аневризматический aneurysmal, aneurysmatic
анезин anezin, chloretone
анемия anemia, anaemia
~, злокачественная malignant anemia

~, **медикаментозная** drug-induced anemia
~, **местная** ischemic anemia, ischemia
~ **новорожденных** neonatal anemia
анемограмма anemogram(me)
анемограф anemograph, wind-speed recorder
анемоклинограф anemoclinograph
анемоклинометр anemoclinometer
анемометр anemometer, wind gauge, draught indicator/gauge
~, **крыльчатый** vane anemometer
~, **манометрический** pressure-tube anemometer
~, **ручной** hand-held/-operated anemometer
~, **сигнальный** alarm anemometer
~ **со счетчиком** counter anemometer
~, **тепловой/термометрический** hot-wire anemometer
~, **флажковый** swinging-vane anemometer
~, **чашечный** (revolving-)cup anemometer
анемометрия anemometry
анемона anemone
~, **морская** sea anemone, actinia
анемоскоп anemoscope
анемостат anemostat
анергический anergic
анероид aneroid (barometer)
анестезиолог anesthesiologist
анестезиология anesthesiology
анестезировать anesthetize
анестезирующий anesthetic
анестезия anesthesia, anaesthesia
~, **корешковая** radicular anesthesia
~, **местная** local anesthesia, toponarcosis
~, **общая** general anesthesia
~ **охлаждением** refrigeration anesthesia
~, **поверхностная** surface anesthesia
~, **полная** total anesthesia
~, **половая** sexual anesthesia, anaphrodisia
~, **спинномозговая** spinal/lumbar anesthesia
анестетик anesthetic
анетол anethol
анеуплоидия aneuploidy
анид anide
анизогамия anisogamy
анизол anisole
анизометр anisometer
~, **магнитный** torque magnetometer
анизоплоидия anisoploidy
анизотропия, анизотропность anisotropy, anisotropism
~, **кристаллографическая** crystal/crystalline anisotropy
анизотропный anisotropic
анизоупругость anisoelasticity
анил anil, indigo
анилин aniline
анилиновый aniline (*attr.*)
анион anion
~, **междуузельный** interstitial anion
анионирование anion-exchange process
анионит anionite, anion-exchange resin/material
анионный anionic
анионообмен anion exchange
анионообменник anion exchanger
анионопроницаемость anion permeability
анионотропия anionotropy
анион-терапия anionic therapy
анис anise; anise apples
анисовый anise; *chem.* anisic
анкер 1. stay, tie rod; anchor, anchorage; 2. pallet, strut
~, **винтовой** screw-in anchor
~ **оттяжки** backstay/guy anchor
~ **парашютного фала** parachute lanyard anchor
~, **поперечный** cross tie
~, **стяжной** tie anchor
анкерит *miner.* ankerite, brown spar
анкерный anchoring, anchor *(attr.)*
анкерок water breaker
анкилоз ankylosos
~, **искусственный** arthrodesis
анкилопоэтический ankylopoietic
анкилостома ankylostoma, trismus, lockjaw
анкилостомоз ankylostomiasis, ancylostomiasis
анклав enclave
анкорит anchorite
аннелид *zool.* annelid
аннигилировать annihilate
аннигилятор *math.* annihilator
аннотация abstract; summary
аннуитет annuity
~, **безвозвратный** nonrefund annuity
~, **возвратный** refund annuity
~, **делимый** apportionable annuity
~, **неделимый** nonapportionable annuity
~ **с единовременной уплатой взноса** single premium annuity
~, **условный** contingent annuity
аннулирование cancellation, revocation, rescission; nullification
аннулировать cancel, revoke, quash; rescind; nullify, abrogate, abolish, quash
аннулятор *math.* annihilator
анод anode; plate; anode circuit/lead/plate/target
~, **возбуждающий/дежурный** keep-alive anode
~, **второй** second/accelerating anode
~, **жидкий** liquid anode, anode layer
~, **зажигающий** ignition/starting anode
~, **оконечный** final anode
~, **первый** first/focussing anode
~, **полый** hollow anode
~, **послеускорительный/третий** post-acceleration electrode, intensifier electrode
~, **прострельного типа** through-/transmission-type target
~, **ускоряющий** accelerating/second anode
~, **фокусирующий** focussing/first anode
анодизация, анодирование anodization, anodizing, anodic oxidation/treatment
анодировать anodize

анодный anode *(attr.);* anodic, anodal
анододержатель anode mount
аноксит *miner.* anauxite
аноксия anoxia
~, высотная altitude anoxia
~, коматозная fulminant anoxia
анолит anolite
аномалия anomaly; abnormality
~ Буге Bouguer gravity anomaly
~, врожденная congenital anomaly
~, гравитационная, ~ силы тяжести gravity/gravitational anomaly
~, материковая continental anomaly
~ развития developmental anomaly
~, средняя mean anomaly
~, термическая temperature/thermal anomaly
аномалоскоп anomaloscope
аномальный anomalous
аномер *chem.* anomer
анорак anorak
анорексия anorexia
анормальность abnormality
анормальный abnormal
анортит anorthite
аносмия anosmia
анотрон anotron
анофелес anopheles
ансамбль *math., stat.* ensemble
~, гиббсовский Gibbs ensemble
~ сообщений set/ensemble of messages
антаблемент entablature
Антанта *hist.* Entente
~, Малая the Little Entente
Антарктида the Antarctic Continent
Антарктика the Antarctic; the antarctic regions
антенна 1. aerial; antenna; 2. *zool.* antenna
~ автоматического радиокомпаса ADF aerial, automatic direction finder aerial
~, антифединговая antifading aerial
~ бегущей волны travelling-wave aerial
~, бортовая vehicle-borne aerial; *av.* airborne aerial
~, веерная fan/spiderweb aerial
~, вибраторная dipole aerial
~, витковая loop aerial
~, волноводная wave-guide aerial
~, волноводно-щелевая slotted-guide aerial, leaky-pipe aerial
~, вращающаяся rotable aerial
~, всеволновая all-band aerial
~, всенаправленная omnidirectional aerial
~, встроенная built-in aerial
~, выпускная *av.* trailing aerial
~, выскакивающая pop-out aerial
~, выступающая protruding aerial
~, глиссадная glide-path aerial
~, Г-образная inverted L aerial
~, гофрированная corrugated-surface aerial
~, грибообразная mushroom aerial
~ дальней космической связи deep-space communications aerial
~ ДВ [длинных волн] LF [low-frequency] aerial
~, двунаправленная bidirectional/bilateral aerial
~, двускатная inverted V aerial
~, двухзеркальная Cassegrainian aerial
~, двухконусная biconical aerial
~, диапазонная wide-band untuned aerial, multiple-tuned aerial
~, директорная Yagi aerial
~, дисконусная discone aerial
~ доплеровского измерителя Doppler aerial
~, елочная fishbone aerial
~, емкостная capacitive/capacity aerial
~, зеркальная reflector-type aerial
~, измерительная pick-up antenna
~ Кассегрена Cassegrainian aerial
~, качающаяся nodding [rocking-horse] aerial
~ КВ [коротких волн], коротковолновая HF [high-frequency] aerial
~ КЛА, бортовая space-borne aerial
~, коаксиально-дипольная sleeve-dipole aerial
~, коллективная community/master aerial
~, коллинеарная collinear array
~, контрольная monitoring aerial
~, крестообразная cross/crossed aerial
~, круговая circular/ring aerial
~ курсового приемника *av.* localizer (receiving) aerial
~, ленточная ribbon aerial
~, линейная uniform linear array
~, линзовая lens aerial
~, логарифмическая/логопериодическая LP [log-periodic] aerial
~, лучевая beam aerial
~ маркерного приемника marker aerial
~, мнимая image aerial
~, многовибраторная dipole aerial
~, многорядная mattress array
~, многоярусная stacked dipole aerial
~, моноимпульсная monopulse aerial
~, наземная ground/surface aerial
~, направленная directional/directive aerial
~, невыступающая suppressed aerial
~, ненаправленная omnidirectional [non-directional] aerial, omniaerial
~, неподвижная fixed [non-scanning] aerial
~ нижнего питания base-fed aerial
~, ножевая/ножевидная blade [fin-type] aerial
~ обнаружения search aerial
~ обратноосевого излучения backfire aerial
~, обтекаемая streamlined aerial
~, однонаправленная unidirectional/unilateral aerial
~, однополюсная monopole aerial
~, остронаправленная narrow-beam [high-directional, pencil-beam] aerial
~, параболическая parabolic-reflector aerial

~, параболоидная paraboloid-reflector aerial
~, пеленгаторная/пеленгационная DF [direction finding/finder] aerial
~, пневматическая air-inflated aerial
~, поворотная steerable aerial
~, подводная underwater/submarine aerial
~, подземная underground/subsurface/buried aerial
~, подфюзеляжная *av.* bottom aerial
~, поисковая search aerial, scanning aerial
~, полуволновая half-wave aerial
~ помехозащищенная anti-interference aerial
~, приводная homing aerial
~ приемоответчика transponder aerial
~, приемопередающая transceiving [send-receive] aerial
~ продольного излучения end-fire aerial
~, противомеховая anti-interference aerial
~, профильно-лучевая shaped-beam aerial
~, профильно-отражательная shaped-reflector aerial
~, радиолокационная radar aerial
~ радиомаяка beacon aerial
~, радиотелеметрическая radiotelemetering/radiotelemetry aerial
~, рамочная loop aerial, frame/coil aerial
~, резонансная resonant aerial
~, ромбическая diamond aerial
~, рупорная horn/funnel aerial
~, рупорно-зеркальная horn-reflector aerial
~, рупорно-линзовая horn-lens aerial
~ с веерным лучом fanned-beam aerial
~ с неподвижной рамкой fixed-loop aerial
~ с остронаправленным лучом pencil-beam aerial
~ с параболоидным отражателям parabolic reflector aerial
~ с прижатым излучением low-angle aerial
~ с регулируемым наклоном tiltable aerial
~ с серединным питанием centre-fed aerial
~ с электронным сканированием electronically scanned aerial
~, самолетная aircraft/airborne aerial
~ СВ [средных волн] MF [medium frequency] aerial
~, сверхнаправленная superdirective/supergain aerial
~ СВЧ-диапазона microwave aerial
~, связная communications aerial
~ СДВ [сверхдлинных волн] VLF [very-low-frequency] aerial
~, сетчатая grid aerial
~, симметричная balanced aerial
~, синфазная broadside array
~ системы дальней космической связи deep-space communication aerial
~ системы инструментальной посадки ILS [instrument landing system] aerial
~ системы самонаведения homing aerial
~, сканирующая радиолокационная radar scanner, radar scanning aerial
~, скрытая suppressed/buried aerial
~, следящая track aerial
~ слежения/сопровождения tracking aerial
~, спаренная twin aerial
~, спиральная helical/spiral aerial
~, тарелочная dish-shaped aerial
~, телеметрическая telemetry aerial
~, телескопическая telescopic aerial
~, турникетная turnstile aerial
~, убирающаяся retractable aerial
~, угломестная elevation aerial
~, уголковая corner/angled aerial
~ УКВ [ультракоротких волн] VHF/UHF aerial
~, фазированная phase-locked aerial
~, ферритовая ferrite-rod aerial
~, фидерная TL aerial
~, цилиндрическая cage aerial
~, четвертьволновая quarter-wave aerial
~, щелевая slot aerial
~ Эдкока Adcock aerial
антенна-башня tower/pylon aerial
антенна-вибратор dipole aerial
антенна-линза lens aerial
антенна-мачта mast aerial
антенный aerial
~ная решетка aerial array
антеридий antheridium
антерозоид antherozooid, antherozoid
антефикс antefix
антиамбоцептор antiamboceptor, antiimmune body
антиантитело antiantibody
антиапекс antapex, antiapex
антиастматический antiasthmatic
антибакхий antibacchius
антибаллоник *text.* antiballooning ring
антибарион antibaryon
антибатный antibate
антибиоз antibiosis
антибионт antibiont
антибиотик antibiotic (drug)
~, противогрибковый antifungal antibiotic
~, противоопухолевый antineoplastic antibiotic
~, противотуберкулезный antitubercular/antituberculotic antibiotic
антибиотин antibiotin, avidin
антивещество antimatter
антивибратор damper; snubber
антивирус antivirus
антивирусный antiviral
антигелий antihelion
антигемолический antihemolytic, antihaemolytic
антиген antigen
~ видовой специфичности species-specific antigen
~ вирулентности Vi antigen, ''virulence antigen''

~, гетерогенный heterogeneous/heterophile antigen
~, жгутиковый flagellar antigen
~, меченый tagged antigen
~, неполный incomplete antigen, hapten(e)
~, полный complete antigen
~, соматический somatic antigen, O antigen
антигенный antigenic
антигиперон antihyperon
антигистамин antihistamine
антигормон antihormone
антигравитация antigravity, antigravitation
антидетонатор anti-knock agent/additive/dope
антидетонационный anti-knock *(attr.)*
антидюна antidune
~, тупая slow antidune
антиизоморфизм anti-isomorphism
антик antique
антиканцероген anticancerogen
антикатализатор anticatalyst
антикатод anticathode, target cathode
антиква *print.* antique, antiqua
антиклиналь anticlinal, anticline
~, килевая carinate anticline
~, косая inclined/asymmetral anticline
~, лежачая recumbent anticline
~, неразвитая arrested anticline
~, погружающаяся plunging anticline
~, прямая upright anticline
~, региональная regional anticline, geanticline
антиклинорий anticlinorium
антикоагулянт, антикоагулятор anticoagulant, antocoagulin
антикомпенсатор antiservo/antibalance tab
антикомпенсация antibalance
антикомплемент anticomplement, antialexin
антикоррозионный anticorrosive, anti-rust, rust preventive
антилогарифм antilogarithm, antilog
антилопа antelope *(Antilope)*
~, вилорогая prong-horned antelope
~, винторогая tragelaph
антилуна antiselena
антимагнитность antimagnetism
антиматерия antimatter
антимезон antimeson
антимикробный antimicrobial, antimicrobic
антимонит *chem.* antimonite, stibnite
антиморфный antimorphic
антимуссон antimonsoon
антинакипин descaler, antiscaling compound
антинейтрино antineutrino
антинейтрон antineutron
антиномия antinomy, paradox
антиобледенитель de-icer, de-icing [anti-icing] device
антиокислитель, антиоксидант antioxidant
антипассат anti-trade wind
антипирен fire-retardant
антипиретик antipyretic (drug)
антипирин antipyrine
антипироген antipyrogen
антипод antipode
~ эпицентра anticentre
антиподный antipodal
антиполимеризатор antipolymerization agent
антиполюс antipole
антипроизводная antiderivative
антипротон antiproton, negative proton
антирад *chem.* antirad
антиракета antimissile
антисвертывающий anticoagulative
антисегнетоэлектрический antiferroelectric
антисейсмический antiseismic, earthquake-proof/-resistant
антиселена antiselene
антисемит anti-Semite
антисемитский anti-Semitic
антисептик 1. antiseptic; 2. wood preservative
антисептика 1. antisepsis; 2. antiseptics
антисептирование wood preservation, preservative treatment
антисептировать treat with antiseptic/preservative
антисимметрия antisymmetry
антисовпадение anticoincidence
антисолнечная точка antisolar point
антиспазматический antispasmodic
антиспаст antispast
антиспутник antisatellite
антистаритель ageing/deterioration retarder
антистатик antistatic agent
антистрофа antistrophe
антисыворотка antiserum
антитеза antithesis
антитела antibodies
~, агглютинирующие agglutinating antibodies
~, антипочечные antikidney antibodies
~, антиэритроцитарные antierythrocytic antibodies
~, гетерогенные heterogenic/heterophile antibodies
~, холодовые cold antibodies
антитело antibody
антитетический antithetic
антитоксин antitoxin
~, столбнячный tetanus antitoxin
антитоксический antitoxic
антитон antitone
антиузел antinode
антифагоцитарный antiphagocytic
антифазный in antiphase, antiphased, in phase opposition, oppositely phased
антифебрильный antifebrile, antipyretic
антифермент antienzyme
антифибрин acetanilide
антифламинг *av.* flame damper; back-fire screen
антифон antiphone
антифриз antifreezing agent/compound/solution, antifreeze
антифрикционный antifriction *(attr.)*

антихолинергический anticholinergic
антихрист antichrist
антицепь antichain, inverse chain
антициклон anticyclone; area of high pressure
антициклонический anticyclonic
античастица antiparticle
антиэлектрон anti-electron, positron
антиэмульгатор demulsifier, demulsifying agent
антиэрмитовый *math.* anti-Hermitian
антоним antonym
антонимичный antonymous
антономазия *lit.* antonomasia
антофиллит anthophyllite
антоциан anthocyan(in)
антракоз anthracosis
~ легких pulmonary anthracosis, coalminers' disease
антраколит anthracolite
антракс anthrax
антрахинон anthraquinone
антрацен anthracene
антрацит anthracite, anthracite coal
~, высокоуглеродистый carbon-rich anthracite, methanthracite
антрвольт *constr.* spandrel
антресоль 1. mezzanine, entresol; 2. attic storey; ceiling cabinet, gallery
антропогенез anthropogenesis
антропозой *geol.* Anthropozoic era
антропоид anthropoid
антрополог anthropologist
антропологический anthropological
антропология anthropology
антропометрия anthropometry
антропоморфизм anthropomorphism
антропоморфический anthropomorphic
антропоморфный anthropomorphous
антропософия anthroposophy
антропофаг anthropophagus, canibal
антропофагия anthropophagy, cannibalism
антропоцентризм anthropocentrism
антрум antrum
анты *arch.* antae, parastades
антье *math.* greatest integer (in)
анурический, ануреэный anuretic, anuric
анурия anuria
~ новорожденных neonatal anuria, anuria of newborn
~, рефлекторная reflexogenic anuria
анус anus
~, искусственный artificial/preternatural anus
анфилада *arch.* suite (of rooms), enfilade
анфиладный *mil.* enfilade
анчар *bot.* upas-tree, antiar *(Antiaris toxicaria)*
анчоус anchovy *(Engraulis)*
аншлиф polished section
Аньези Agnesi
аньюм ainhum, spontaneous dactylosis
анэуплоидия aneuploidy
аорист aorist
аорта aorta
~, брюшная abdominal aorta
~, грудная thoracic aorta
аортальный aortal, aortic
аортит aortitis
апанаж appanage, apanage
апареон *astr.* apareon
апартейд, апартхейд apartheid
апатит apatite
апатитолит apatitolite, phosphorite
апатичность, апатия apathy
апатичный apathetic
апатрид *law* stateless person
АПВ (автоматическое повторное включение) automatic reclosure/recloser
апекс *astr.* apex
апеллировать appeal
апеллянт apellant
апелляционный appeal *(attr.)*
~ суд Court of Appeal
апелляция appeal
апериодический aperiodic, non-periodic, acyclic
апериодичность aperiodicity
апертометр *opt.* apertometer
апертура aperture; *phot.* f-number
~, звездообразная asteroid aperture
~, кольцевая annular aperture
~, лучистая radiate aperture
~, полутоновая tone control aperture
~, развертывающая scanning aperture
~, ситовидная cribrate aperture
~, сложная multiple aperture
апертурный orate
апертуровидный oroid
апикальный apical
апирексия apyrexia
апирогенный apyretic
апитерапия apitherapy
аплазия aplasia
~ бронха bronchial aplasia
~ кожи dermal aplasia
~ молочной железы mammary aplasia
~ почки renal aplasia
~ хрусталика lenticular aplasia
апланат aplanat, aplanatic lens
апноэ apnea, apnoea
апоапсида apoapsis
апоастр, апоастрий *astr.* apastron
апогей apogee
аподиктический apodictic, apodeictic
апокалипсис Apocalypse; the Revelation
апокалипсический, апокалиптический apocalyptic(al)
апокопа apocope
апокриф apocrypha
апокрифический apocryphal
аполид *law* stateless person
Аполлон Apollo
апологет apologist
апология excessive apologia, immoderate eulogiz-

ing
аполярность apolarity, non-polarity
аполярный non-polar
апоплексический apoplectic
~ удар apoplectic stroke/seizure
апоплексия apoplexy
~, бульбарная bulbar/pontile apoplexy
~, гипостазная congestive apoplexy
~, кровяная sanguineous apoplexy
~, молниеносная fulminating apoplexy
~ новорожденных neonatal apoplexy, apoplexy of the newborn
~, спазматическая spasmodic apoplexy
~, спинальная spinal apoplexy, hematorrhachis
~, тепловая heat apoplexy/stroke
апория aporia
~ Зенона *math.* Zeno's paradox
апоселений apolune, apocynthion
апосиндетический aposyndetic
апостема apostem, apostema, aposteme
апостериори a posteriori
апостериорный a posteriori *(attr.)*
апостильб apostilb, asb
апостол 1. apostle; 2. Books of the Apostles, the Acts of the Apostles
апостроф apostrophe
апострофа *rhet.* apostrophe
апосфера aposphere
апофема apothem
апоф(т)егма apophthegm, apophegm
апофеоз apotheosis
апофиз *med.* apophysis
апофиза *geol.* apophysis, tongue
апофизарный apophysial, apophyseal, apophysary
апохромат apochromat, apochromatic lens
апохроматизм apochromatism
апоцентр apocenter, apocentre; apofocus
аппарат 1. apparatus, instrument, device; equipment; unit; item/piece/unit of equipment; 2. *physiol.* apparatus, system; organs; 3. vehicle; aircraft, spacecraft
~, аэрозольный nebulizer
~, бродильный fermenter
~, буквопечатающий printer
~ Варбурга Warburg vessel/apparatus
~, выпарной evaporator; concentrator
~ Гольджи Golgi apparatus
~, государственный state machinery/organs/apparatus
~ грамзаписи phonograph recorder
~, двухпрочесный double card
~ для гипотермии hypothermic apparatus
~ для облета fly-by vehicle
~ для облета Луны circumlunar vehicle
~ для облета Марса Mars orbiter
~ для полета к Луне moon vehicle
~ для полета к Марсу Martian vehicle
~ для посадки на Марс Mars-landing vehicle, Mars lander
~ для растворения dissolver
~, дождевальный sprinkler, overhead irrigation equipment
~, доильный milking unit, milker
~, дробеструйный shot-blaster
~, дыхательный 1. respiratory system; 2. breathing apparatus/equipment, respirator
~, жевательный masticatory apparatus
~, железистый glandular system
~, звукозаписывающий sound recorder
~ "искусственная почка" artificial kidney (apparatus)
~ искусственного дыхания resuscitation apparatus
~ искусственного кровообращения artificial circulation apparatus, heart-lung apparatus
~, кассовый cash register
~, кинокопировальный film printer
~, кинопроекционный cine [motion-picture] projector
~, киносъемочный cine [motion-picture] camera
~, кислородный oxygen set
~, клубочковый glomerular system
~, контактный catalyst chamber
~, космический летательный spacecraft, space vehicle
~, космический летательный, возвращаемый recoverable space vehicle
~, космический летательный для исследований space probe
~, космический летательный для полета на Луну Moon-mission spacecraft
~, космический летательный, орбитальный orbiting/orbital space vehicle/spacecraft, orbiter
~, космический летательный с экипажем на борту manned spacecraft
~, красильный dyeing machine
~, кроветворный hematopoietic system
~, летательный, винтокрылый rotorcraft
~, летательный, легче воздуха lighter-than air aircraft; aerostat
~, летательный многократного пользования reusable vehicle
~, летательный невозвращаемый nonrecoverable vehicle
~, лимфатический lymphatic system
~, лудильный tinning stack
~, лущильный hulling unit
~, математический mathematical tools/techniques/apparatus; body of mathematics
~, мезенхимный mesenchymatous system
~ многоразового применения, летательный recoverable aircraft/spacecraft/vehicle
~, мочеполовой urogenital/genitourinary system, urinary apparatus
~ на воздушной подушке hovercraft, aircushion vehicle, cushioncraft, ground-effect vehicle
~, неспасаемый (летательный) non-recov-

erable [expendable] vehicle
~, опорно-двигательный locomotor system
~, орбитальный orbital/orbiting vehicle/spacecraft, orbiter
~, осадительный precipitator
~, островковый insular apparatus, islands of Langerhans
~, отстойный settler
~, парашютно-планирующий paraglider vehicle
~, пескодувный/пескоструйный sand blower, sandblaster
~, питающий feed unit
~, плавательный natatorial apparatus
~, полимеризационный polymerizer
~, посадочный landing vehicle
~, производственный production/productive/industrial facilities
~, пропиточный impregnator
~, разбрызгивающий sprayer
~, распылительный atomizer; sprayer
~, реакционный reactor, reaction vessel
~, режущий *agr.* cutterbar
~, рентгеновский X-ray unit/apparatus
~, ретикулоэндотелиальный reticuloendothelial system
~, рефлекторный reflex system
~, рецепторный receptor system
~, самопишущий automatic recorder
~, сварочный welding machine/apparatus/unit/set; welder
~, светокопировальный blue-printing machine
~, сердечно-сосудистый cardiovascular system
~, "сердце-легкие" artificial circulation apparatus, heart-lung apparatus
~, слуховой hearing aid; otophone
~, спасаемый (летательный) recoverable vehicle
~, спасательный rescue/recovery vehicle
~, спускаемый descent [re-entry] vehicle/module/capsule
~, стартовый launch vehicle
~, телеграфный telegraph apparatus/set; telegraph
~, телеграфный Юза Hughes telegraph
~, телефонный telephone set/instrument; telephone
~, телефонный, автоматический automatic telephone, dial telephone
~, тромбоцитарный thrombocytic system
~ тяжелее воздуха, летательный heavier-than-air aircraft; aerodyne
~, фотографический camera
~, фототелеграфный faximile apparatus
~ Холдейна Haldane's apparatus
~, чесальный card
~ "электросон" electric sleep apparatus
~, эндокринный endocrine system
аппаратная 1. control equipment/apparatus room; 2. equipment truck/vehicle
аппаратно *comp.* by/in hardware, under hardware control
~ реализованный *comp.* hardware-implemented
~ совместимый hardware-compatible
аппаратный *comp.* hardware-controlled; hard
аппаратура apparatus, equipment, outfit, instruments, instrumentation; gear; facilities; *comp.* hardware
~, бортовая vehicle-borne [airborne] equipment
~, дальномерная range instrumentation, range-finding equipment
~, индикаторная display equipment
~, коммутационная switchgear
~, контрольно-испытательная checkout equipment
~, моделирующая simulation equipment
~ на твердых схемах solid-state equipment
~ на транзисторах transistorized equipment
~, опознавательная *av.* identification equipment
~ радиопротиводействия electronic countermeasures equipment
~, самолетная airborne/aircraft equipment
~ самонаведения homing equipment
~, спасаемая recoverable [non-expendable] equipment
~, считывающая readout equipment
~, термоизвещательная overheat detection equipment
~, фототелеграфная facsimile equipment
аппаратчик 1. operator; 2. *polit.* apparatchik, executive party or government official
аппарель ramp; slipway
аппендикс appendix; appendage
аппендицит appendicitis
аппендэктомия appendectomy
апперкот *sport* uppercut
апперцепция apperception
аппликата *math.* applicate, z-axis/coordinate
аппликатура *mus.* fingering
аппликация *text.* appliqué work, appliqué; *tech.* application, administration, overlaying
апподжиатура *mus.* appogiatura, grace note
аппрет 1. *text.* finishing/dressing/sizing agent; 2. finish, dressing, glaze
аппретирование *text.* finishing, dressing, sizing
аппретировать finish, dress, glaze
аппретура *text.* finish(ing), dressing
~, водная water-based finish/dressing
~, глянцевая lustre, glaze
аппретурщик finisher, dresser
аппроксимативный approximate, approximative
аппроксимация, аппроксимирование approximation
~, линейная *econ.* linear approximation
~, сглаживающая *econ.* smoothing approximation
аппроксимировать approximate

аппроксимируемость approximability
апраксия *med.* apraxia
~, идеаторная ideatory/ideational apraxia
апрессин apressine, apresoline, hydralazine
априори, априорно a priori
априорный a priori; aprioristic
~ная вероятность prior probability
апробация appraisal; approbation
~ посевов yield checking/appraisal
апсида *arch.* apse
апсиды *astr.* apsides
аптерикс *zool.* apteryx
аптиализм *med.* aptialism, aptialia
аптих *paleont.* aptychus
апьезон apiezon wax
арабеск, арабеска arabesque
арабиноза arabinose
арабинозодегидрогеназа arabinose dehydrogenase
арабинокиназа arabinokinase
арабис *bot.* rock cress *(Arabis)*
арабист Arabist, Arabic scholar
арабистика Arabic studies/philology
аравановые *zool. pl.* osteoglossids, bonytongues *(Osteoglossidae)*
арак(а) arrack
аралиевые *bot., pl.* ginseng family *(Araliaceae)*
аралия *bot.* aralia *(Aralia);* mountain angelica
арамейский Aramaic
арат Mongolian peasant; cattle-breeder
арахидиловый arachidil
арахидоновый arachidonic
арахин arachin
арахиновый arachic
арахис *bot. coll.* peanuts *(Arachis)*
арахниды *zool.* Arachnida
арахнит arachnitis
арахнодактилия arachnodactyly, spider fingers
арахнодит arachnoditis
арахнология arachnology
арахнофобия arachnophobia
арахноэнцефалит arachnoencephalitis
арбалет *hist.* arbalest, crossbow
арбитр arbitrator; umpire
арбитраж arbitration
~, валютный arbitration of exchange, currency arbitration
~, многосторонний compound arbitration
арбитражный arbitration *(attr.);* arbitrated
~ суд arbitration tribunal, court of arbitration
арборицид silvicide
аргали *zool.* argali *(Ovis ammon)*
аргасовый argas
аргентит 1. *miner.* argentite; 2. *med.* argentitis
аргентометрия argentometry, argentometric titration
аргиллит *petr.* argillite, claystone
аргинин arginine
аргининдезиминаза arginine deiminase
аргининосукцинатлиаза argininosuccinatelyase
аргининфосфат phosphoarginine
аргирия, аргироз *med.* argyrosis, argyria
аргон *chem.* argon, Ar
аргумент argument; *law* submission
~ функции function argument, independent variable
~ широты argument of latitude
ареал area, range; natural habitat
~, внутренний inland area
~, высотный altitudinal area
ареа-синус *math.* inverse-hyperbolic [area-hyperbolic] sine
ареа-функция *math.* inverse-hyperbolic [area-hyperbolic] function
арека *bot.* areca palm *(Areca)*
ареколин arecoline
аренда lease; rent
~, бессрочная general tenancy
~ земли, бессрочная/долгосрочная *hist.* emphyteusis
~ исполу (с уплатой части урожая) metayage
арендатор (land) tenant; lessee, leaseholder
~, бессрочный tenant at will
~ земли на долгий срок *hist.* emphyteuta
~ платящий деньгами cash tenant
~, пожизненный life tenant
арендатор-издольщик cropper
арендаторство lesseeship
арендный lease *(attr.),* rent *(attr.)*
~ договор lease/tenancy agreement
~ная плата rent
~ное земледелие sharecropping
арендовать lease, rent
арендодатель lessor, bailor
ареометр hydrometer, areometer; densitometer
~ для рассола salinometer
~, кислотный acidometer, acidimeter
~, масляный oleometer
~, нефтяной petroleum hydrometer, oleometer
~, самопишущий density recorder, recording hydrometer
ареометрия hydrometry, areometry
ареометр-пикнометр areopicnometer
ареопаг *hist.* Areopagus; tribunal
арес *myth.* Ares
арест 1. detention, arest; 2. *econ.* attachment, seizure, sequestration; distraint, distrainment
~, налагать *econ.* distrain, attach, seize
~, ордер на *law* warrant of arrest
Ариадна *myth.* Ariadne
ариетта *mus.* arietta
ариец, арийца Aryan
арил *chem.* aryl
арилалкогольоксидаза aryl alcoholoxydase
арилациламидаза aryl acylamidase
арилирование arylation
ариллус, ариллю**с** *bot.* aril, arillus
арилэстераза arylesterase
аристократия aristocracy; nobility

~, землевладельческая land-owning aristocracy
~, рабочая worker aristocracy
~, финансовая plutocracy
аристотелевский Aristotelian
аритмический, аритмичный arrhythmic; *fig.* irregular; erratic
аритмичность arrhythmy
аритмия arrhythmia
~, мерцательная ciliary arrhythmia, cardiac fibrillation
~, нефазовая nonphasic arrhythmia
~, сердечная cardiac arrhythmia
~, синусовая sinus arrhythmia
~, фазовая phasic arrhythmia
арифметика arithmetic; arithmetical operations
~, адресная address arithmetic
~, двоичная binary arithmetic
~, двоично-десятичная BCD arithmetic
~, десятичная decimal arithmetic
~, машинная machine arithmetic
~ повышенной точности multilength arithmetic
~ порядков exponent arithmetic
~ с плавающей запятой/точкой floating-point arithmetic
~ с фиксированной запятой/точкой fixed-point arithmetic
~, целочисленная integer arithmetic
арифметико-геометрический arithmetic-geometrical
арифметический arithmetic
арифмометр calculating machine, desk calculator, comptometer, arithmometer
арка arch
~ ахтерштевня *naut.* stern-frame arch
~, бесшарнирная fixed/hingeless arch
~, буксирная towing arch
~, глухая blind arch
~, обратная/опрокинутая inverted arch
~, ползучая rampant/rising arch
~, разгрузочная relieving arch
~, шарнирная hinged arch
аркада *arch.* arcade
аркбутан *arch.* flying buttress, arch(ed) buttress
аркебуза harquebus, arquebus
аркебузир harquebusier, arquebusier
арккосеканс arccosecant, inverse cosecant
арккосинус arccosine, inverse cosine
арккотангенс arccotangent, inverse cotangent
аркообразный arcuate, arciform; archy
арксеканс arcsecant, inverse secant
арксинус arcsine, inverse sine
арктангенс arctangent, inverse tangent
аркус-функция, аркфункция inverse trigonometric function, arc-trigonometric function
арматура 1. fittings; fixtures; accessories; 2. reinforcement; steel framework
~, водопроводная plumbing fixtures
~ котла boiler fittings/accessories
~, осветительная lighting fittings/fixtures
~ работающая на растяжение tensile reinforcement
~ работающая на сжатие compression reinforcement
~ работающая на срез shear reinforcement
~, растянутая tensile reinforcement
~, сжатая compressive reinforcement
~, смазочная lubrication fittings
~, стальная steel rods/bars
~, стержневая bar reinforcement
~, тавровая tee bars
~, трубопроводная pipe fittings; valves
~ цилиндра cylinder fittings
~, электроустановочная wiring accessories
арматурное железо reinforcement steel
арматурщик *constr.* reinforcement fitter, steelman, steel erector, spiderman
армирование, армировка 1. reinforcement, reinforcing; 2. armouring, armour
~ стекловолокном fibre-glass reinforcement
~ тканью fabric reinforcement
армированный 1. reinforced; 2. armoured, sheathed
армировать 1. reinforce; 2. armour, sheathe
армко-железо armco-iron
армопенобетон reinforced foam concrete
армополибетон reinforced polymer concrete
армоцемент wire-fabric reinforced concrete
армяно-грегорианский Gregorian-Armenian
арника *bot.* arnica *(Arnica)*
~, горная mountain arnica/tobacco, wolf's/leopard's bane
арницин arnicin
арпеджио, арпеджо *mus.* arpeggio
арренома *med.* arrhenoma
арретир checking/caging device, stop, clamp, arrester, cage, catch
~ компаса clamp for compass card
арретирование caging
арретировать cage, arrest, clamp, stop
арсеназо arsen-azo indicators
арсенал arsenal, munition/ordnance depot; armoury; *fig.* store(s)
~, морской yard
арсенат arsenate
арсенид arsenide
арсенит arsenite
арсеницизм *med.* arseniasis
арсеносоединение arseno-compound
арсин *chem.* arsine
Артемида *myth.* Artemis
артериализация arterialization; vascularization
артерии arteries
~ большого мозга cerebral arteries
~, надсклеральные episcleral arteries
~, прободающие perforating arteries
~, тощекишечные jejunal arteries
артерио-венозный arteriovenous
артериография arteriography

артериола arteriole
артериопластика arterioplasty
артериосимпатэктомия arteriosympathectomy
артериосклероз arteriosclerosis
~, узелковый nodular arteriosclerosis
артериосклеротический arteriosclerotic
артериостеноз arteriostenosis
артерит arteritis
артерия artery
~, барабанная tympanic artery
~, бедреннная femoral artery
~ большого мозга cerebral artery
~, глазничная ophthalmic artery
~, завитковая helicine artery
~, затылочная occipital artery
~, лицевая facial artery
~, лобная frontal artery
~, локтевая ulnar artery
~, лучевая radial artery
~, малоберцовая fibular artery
~, маточная uterine artery
~, надблоковая supratrochlear artery
~, надглазничная supraorbital artery
~, основная basilar artery
~, питающая nutrient artery
~, плечевая brachial artery
~, подбородочная submental/mental artery
~, подглазничная infraorbital artery
~, подключичная subclavian artery
~, подколенная popliteal artery
~, подкрыльцовая axillary artery
~, почечная renal artery
~, промежностная perineal artery
~, селезеночная splenic artery
~, сонная carotid artery
~, срединная median artery
~, стекловидная hyaloid artery
~, щечная buccal artery
~, язычная lingual artery
~, яичковая testicular artery
артефакт artefact, artifact
артиллерист gunner
артиллерия artillery
~, воздушная aerial artillery
~, зенитная anti-aircraft artillery
~, корабельная shipboard artillery; naval ordnance
артишок *bot.* artichoke (*Cynara scolymus*)
артральгия arthralgia
артрит arthritis
~, рожистый erysipelatous arthritis
~, фунгозный arthritis fungosa
артрогрипоз arthrogryposis
артродез arthrodesis
артроз arthrosis
артропатия arthropathy
артропластика arthroplasty
архар *zool.* argali *(Ovis ammon)*
~, памирский Marco Polo's sheep
архив archives; records; record-keeping office
~ магнитных лент magnetic tape archives/stacker
архидиакон, архидьякон archdeacon
архиепископ archbishop
архиепископство archiepiscopacy, archiepiscopate
архиерей prelate, bishop, metropolitan
архимандрит archimandrite
архимедов 1. *math.* Archimedean; 2. *(of spiral)* Archimedes'
архипелаг archipelago
архитектоника architectonics
архитектура architecture *(also comp.)*
~, базовая foundation architecture
~, древовидная tree-structured architecture
~, кольцевая эстафетная token-ring architecture
~, конвейерная pipelined architecture
~, контролепригодная easy-to-test architecture
~, корабельная applied naval architecture
~, ландшафтная landscape architecture/gardening
~, многопользовательская multiuser architecture
~, многошинная multibus architecture
~, потоковая data-flow architecture
~, регистровая register-oriented architecture
~ с общей шиной unified-bus architecture
~ с побитовым отображением bit-map architecture
~, садово-парковая landscape gardening
~, секционированная slice architecture
~, сетевая network architecture
~ со сплошной адресацией flat-address architecture
~, шинная bus architecture
архитрав *arch.* architrave
архонт archon
аршин arshin *(28 inches)*
асбест asbestos
~, волокнистый fibrous asbestos
~, голубой/синий blue asbestos, crodolite
~, листовой asbestos sheets
~, продольноволокнистый matted asbestos
~, путано-волокнистый matted asbestos
~, рулонный asbestos rolls
асбест-амфибол amphibole asbestos
асбест-антофиллит anthophyllite asbestos
асбестин asbestine, fibrous talc
асбест-крокидолит blue asbestos, crocidolite
асбестовойлок asbestos felt
асбестоз *med.* asbestosis
асбестоцемент asbestos cement
асбест-тремолит tremolite asbestos
асбест-хризотил chrysotile asbestos
асбобумага asbestos paper
асбогетинакс asbestos paper laminate
асбозурит diatomaceous-silica asbestos
асбокартон asbestos sheet millboard
асболента asbestos tape

асболит asbestos sheet millboard
асбопластики asbestos-filled plastics
асбослюда asbestos-backed mica insulation
асботекстолит asbestos-cloth laminate
асботкань asbestos cloth
асботрубы asbestos cement pipes
асбофанера asbestos board
асбоцемент asbestos cement
асбошифер corrugated asbestos sheets
асемия *med.* asemia, asemasia
асептика asepsis
асептический aseptic
асимметричность asymmetry
асимметрический, асимметричный asymmetric(al)
асимметрия asymmetry
  ~ деления fission asymmetry
  ~ расположения configurational asymmetry
асимптота *math.* asymptote
асимптотический asymptotic(al)
асимптотичность *math.* asymptotic property
асиндетон *lit.* asyndeton
асинергия asynergia, asynergy
асинклитизм asynclitism
асинхронный asynchronous
асистолия asystolia, asystole
аскаридоз *med.* ascariasis
аскариды *Ascarides*
аскарицид ascaricide
аскомицеты *Ascomycetes*
аскорбинатоксидаза ascorbate oxidase
аскорбиновый ascorbic
аскофен ascophen
аспарагин asparagin(e)
аспарагиновый asparaginic, aspartic, asparagic
аспарагинсинтетаза asparagine synthetase
аспартат-аммиаклиаза aspartate ammonialyase
аспартаткиназа aspartate kinase
аспергиллез *med.* aspergillosis
  ~ кожи cutaneous aspergillosis
  ~, ушной aural aspergillosis, otomycosis
аспергиллезный aspergillous
асперматизм, аспермия aspermatism, aspermia
аспирант *educ.* postgraduate student; research student
аспирантура *educ.* postgraduate course/education/training; research course
аспиратор aspirator
аспиратор-пылеотделитель dust aspirator
ассемблер *comp.* assembler
  ~, двухпроходной two-pass assembler
  ~, обратный disassembler
  ~, "один к одному" one-to-one assembler
  ~, однопроходной single-pass assembler
  ~, перемещаемый relocatable assembler
  ~, построчный one-line assembler
  ~, резидентный resident assembler
  ~ символического языка mnemonic assembler
ассемблер-эмулятор assemulator
ассемблирование assembling
  ~, обратное disassembling
ассемблировать assemble
ассенизация sanitation; sewage disposal
ассигнование allocation, appropriation; allotment; grant
  ~ния, бюджетные budgetary provisions
  ~ния, неиспользованные undrawn allocations
ассигновать allocate, allot, assign, appropriate; earmark; grant
ассигновка = ассигнование
ассириец Assyrian
ассиро-вавилонский Assyro-Babylonian
ассортимент variety; choice, selection
  ~ изделий/продукции product-mix
  ~ товаров range of goods
ассоциативность associativity
ассоциативный associative; concomitant; *comp.* content addressable
  ~ное значение associated meaning
  ~ное обучение contiguity learning
ассоциация association
  ~ британских торговых палат Association of British Chambers of Commerce
  ~ для выработки технических стандартов, британская British Engineering Standards Association
  ~, звездная stellar association
  ~ ионов ion association/combination
  ~, некорпоративная unincorporated association
  ~ обществ ипотечного кредита Building Society Association
  ~ по борьбе с голодом, всемирная World Association for the Struggle against Hunger
  ~ по операциям с закладными National Mortgage Association
  ~ свободной торговли, европейская European Free Trade Association, EFTA
  ~, сельскохозяйственная кооперативная Agricultural Cooperative Association
  ~ содействия прогрессу цветного населения National Association for the Advancement of Colored People
  ~ стандартов, американская American Standards Association
  ~ транспортных фирм carrier association
ассоциировать associate
астазирование astatization
астазировать astatize
астатин *chem.* astatine
астатичность astaticism
астеник *med.* asthenic
астения *med.* asthenia
астенопия *med.* asthenopia
астериск *print.* asterisk
астероид asteroid, planetoid
астигматизм astigmatism
  ~, обратный reversed astigmatism
  ~, роговичный corneal astigmatism
  ~, сложный compound astigmatism

~, хрусталиковый lenticular astigmatism
астигматометр astigmatometer, astigmometer
астигматоскоп astigmatoscope, astigmoscope
астма asthma
~, бронхиальная bronchial asthma
~, вилочковая thymic asthma
~, сердечная cardiac asthma
~, спазматическая spasmodic asthma
астматический asthmatic
астрионика astrionics, astronomical electronics
астрографики star altitude curves
астродатчик star tracker
астродинамика astrodynamics
астроида *math.* astroid
астрокамера astronomical camera
астрокомпас astrocompass, celestial compass
астрокоррекция stellar monitoring
астрокупол astrodome
астролатрия *hist.* astrolatry
астролюк *space* astrohatch
астролябия astrolabe, equiangulator
астрометрия astrometry
астрон astron, relativistic electron beam
астронаведение stellar/celestial guidance
астронавигация astronavigation, celestial/astronomical navigation, star-tracking
астронавт astronaut
астронавтика astronautics, cosmonautics
астроном astronomer
астрономический astronomical
астрономия astronomy
~, внеатмосферная extraterrestrial astronomy
~, галактическая stellar/galactical astronomy
~, геодезическая field astronomy
~, звездная stellar astronomy
~, космическая space astronomy
~, мореходная nautical astronomy
~, нейтринная neutrino astronomy
~, радиолокационная radar astronomy
~, эфемеридная ephemerical astronomy
астроориентатор stellar/star tracker, astrotracker; astronavigator
астроориентация star-tracking, astroorientation
астрополяриметрия astropolarimetry, celestial polarimetry
астроспектрограф astronomical spectrograph
астроспектроскопия astronomical/stellar spectroscopy
астроспектрофотометрия astronomical spectrophotometry
астротелескоп star-tracking telescope
астрофизика astrophysics, physical astronomy
астрофотография astrophotography, astronomical photography
астрофотометрия astrophotometry, astronomical photometry
астрохимия astrochemistry
астроцит *anat.* astrocyte, spider cell
астроцитома astrocytoma
~ мозжечка cerebellar astrocytoma
астроэкология astroecology
астроэлектроника astroelectronics, astrionics
асфальт asphalt; bitumen; mineral pitch
~ из кислого гудрона acid sludge asphalt
~, твердый acid coke
~, эластичный elastic bitumen, elaterite
асфальтировать coat/cover/pave with asphalt
асфальтит asphaltite
асфальтобетон asphalt/bituminous concrete
~, горячий hot-lay asphalt concrete
~, холодный cold-lay asphalt concrete
асфальтоукладчик asphalt spreader/placer
асферический aspherical
асферичность asphericity
асфиксия asphyxia
~, белая asphyxia pallida
~, внутриутробная intrauterine/foetal asphyxia
~, застойная stagnant asphyxia
~, новорожденных postnatal asphyxia
~, синяя asphyxia livida
~, тканевая tissue asphyxia
асфиктический asphyxial, asphyctic
асцит ascites, abdominal dropsy, hydroperitoneum
~, хилезный chylous/milky ascites
атавизм atavism
атака 1. attack, assault, charge; *phys.* impingement
~ вдогон pursuit attack
~, имитируемая simulated attack
~, ложная false/dummy attack
~ с воздуха aerial/air attack
~ с малой высоты low-flying attack
~, условная simulated attack
~, химическая gas attack
атакоустойчивость impingement resistance
атаксиаграмма *med.* ataxiagram
атаксический atactic, ataxic
атаксия *med.* ataxia
~, лабиринтная labyrinthine ataxia
~, мозжечковая cerebellar ataxia
атараксия ataraxia
ателектаз *med.* pulmonary collapse
~ легких atelectasis
~ легких, очаговый focal atelectasis
~ легких у новорожденных postnatal atelectasis, atelectasis neonatorum
атермический athermanous, adiathermal
атерома atheroma
атеросклероз atherosclerosis
атетоз *med.* athetosis, posthemiplegic chorea
атетозный athetotic, athetosic, athetoid
атлант *arch., anat.* atlas; *myth.* Atlas
Атлантида Atlantis
атлас *geog., anat.* atlas
~, звездный celestial/stellar atlas
~ приливов tidal atlas
атлетика athletics
~, легкая track and field (athletics)
~, тяжелая heavy athletics; weightlifting
атмограф atmograph
атмометр atmometer

атмосфера 1. atmosphere; air; 2. *(unit)* atmosphere, atm
~, верхняя upper/higher/outer atmosphere
~, внешняя outer atmosphere
~, восстановительная reducing atmosphere
~, земная terrestrial/Earth's atmosphere
~, ионная ionic atmosphere/medium
~, искусственная artificial/simulated atmosphere
~, моделируемая simulated atmosphere
~, нижняя lower atmosphere
~, окружающая ambient air
~, разреженная rarefied atmosphere
~, условная 1. reference atmosphere; 2. hypothetical atmosphere
атмосферики *pl.* atmospherics, atmospheric noise
атмосферостойкий weather-proof/-resistant/-durable
атолл atoll
атом atom
~, вершинный apical atom
~ внедрения interstitial atom
~, выбитый knocked-on atom
~, горячий high-energy recoil atom, hot atom
~, делящийся fissionable atom
~, дочерний product/daughter atom
~, меченый labelled/tagged atom, tracer
~ мишени target atom
~ основного кристалла host crystal atom
~ отдачи recoil atom
~, подвижный labile atom
~, посторонний foreign atom
~, расщепившийся split/fissioned atom
~, расщепленный split/fissioned/disintegrated atom
~, расщепляющийся fissionable atom
~, связанный bound atom
~, смещенный displaced atom
~ способный делиться fissionable atom
~ топливного сырья fertile atom
~, чужеродный foreign atom
атом-акцептор acceptor atom
атомарный monoatomic
атом-донор donor atom
атомизатор sprayer, spray gun
атомизм, атомистика atomic theory, atomism
атом-мишень target/struck atom
атомность atomicity; valency
атомный atomic
~ котел atomic pile, nuclear reactor
~ная электростанция atomic power plant, nuclear-fuelled/power plant
атомовоз atomic locomotive, nuclear-powered locomotive
атомоход nuclear-powered vessel/ship
атония *med.* atony, atonia
атрезия atresia
атретический atretic, atresic
атрибут *comp.* attribute
~ возвращаемого значения return attribute
~ данных data attribute
~ количества number attribute
~ масштаба scaling attribute
~, обязательный mandatory attribute
~ по умолчанию default attribute
~, поисковый search attribute
~ размерности dimension attribute
~ распределенной загрузки scatter attribute
~ секретности security attribute
атрий, атриум *arch.* atrium
атропин atropine
~, сернокислый atropine sulphate
атропиновый atropinic
атрофия atrophy, atrophia
~ ногтей onychoatrophia
~, серозная serous atrophy
атташе attaché
~, военно-воздушный air attaché
~, военно-морской naval attaché
~, военный military attaché
~, культурный cultural attaché
~ печати press attaché
~, торговый commercial attaché
аттенюатор attenuator
~, индуктивный mutual-induction attenuator
~, поглощающий power-absorbing attenuator
~, предельный cut-off attenuator
~, рассеивающий dissipative attenuator
~, регулируемый variable attenuator
~, ступенчатый/шаговый step attenuator
аттестат certificate; diploma
~ годности operational certificate
~ зрелости general education certificate, school-leaving certificate
~, квалификационный certificate of competence
аттестационная комиссия examination board, certifying commission
аттестация attestation; certification; effort rating; *comp.* validation
~ программы program(me) validation
аттрибут = атрибут
аудиовизуальный audiovisual, videosonic
аудиограмма audiogram
аудиометр audiometer
аудиометрия audiometry
аудитор independent accountant
ауксанограф auxanograph
ауксин auxin
аукцион auction, public sale
аукционер, аукционист auctioner
аура *med.* aura
~, зрительная visual aura
~, надчревная epigastric aura
~, обонятельная olfactory aura
ауреомицин aureomycin
аускультационный auscultatory
аускультация auscultation
аустенизация austenization
аустенит austenite

аутоаллерген autoallergen
аутовакцина autovaccine
аутогамный autogamous
аутогемотерапия autoh(a)emotherapy, autogenous blood treatment
аутогенез autogenesis
аутогенетический autogenetic, autogenic
аутогенный autogenous
аутогипноз autohypnosis
аутодупликация self-duplication, autoduplication
аутоиммунизация autoimmunization
аутоинтоксикация autointoxication
~, обменная metabolic autointoxication
аутокатализ autocatalysis
аутолиз autolysis
аутолизат autolysate
аутопластика autoplasty
аутоплоидия autoploidy
аутополиплоид autopolyploid, autoploid
аутопсия autopsy, post-mortem (examination)
ауторадиография autoradiography, radioautography
аутоскоп autoscope
аутосома autosome
аутотрансплантат autoplastic graft, autotransplant, autograft
аутотрансплантация autotransplantation, autografting
аутохтонный autochthonous
афагия *med.* aphagia
афазия aphasia
~, акустическая acoustic/auditory aphasia, word deafness
~, амнестическая amnesic aphasia, aphemia
~, атаксическая ataxic aphasia
~, психосенсорная psychosensory aphasia
афелий *astr.* aphelion
Афина *myth.* Athena
афония aphonia
~, параноидная aphonia paranoica
африканс, африкаанс Afrikaans
Афродита *myth.* Aphrodite
афты *pl.* aphthae
афферентный afferent; centripetal
аффидевит *law* affidavit
аффинаж refining
аффинация affination
аффинитет affinity
аффинно-разрешимый affine-resolvable
аффинность *math.* affinity, affineness
аффинный *math.* affine; adjoining, related
аффинор *math.* affinor
аффриката *phon.* affricate
ахалазия *med.* achalasia
ахейлия *med.* acheilia, achilia
ахейрия *med.* acheiria, achiria
ахилия *med.* achylia
~, стойкая stable/permanent achylia
Ахилл, Ахиллес Achilles
ахиллорафия achillorrhaphy
ахиллотомия achillotomy
ахорион achorion
ахромат achromatic lens, achromat
ахроматизм achromatism, colourlessness
~, полный perfect achromatism
~ размера изображения achromatism of picture size
~ увеличения achromatism of magnification
ахроматин achromatin
ахроматический achromatic
ахромацит achromatocyte, achromacyte
ахромицин achromycin
ахромия *med.* achromia
ахтерлюк *naut.* afterhatch(way)
ахтерпик *naut.* afterpeak
ахтерштевень *naut.* sternpost; stern frame
ацетазин acetylpromasine
ацетальдегид acetic aldehyde, acetaldehyde
ацетамид acetamide
ацетанилид acetanilide
ацетат acetate
~ целлюлозы cellulose acetate
ацетил acetyl
ацетилгликоколь acetylglycine
ацетилен acetylene
ацетилирование acetylation, acetylization, acetylating
ацетилировать acetylate, acetylize
ацетилмочевина acetyl urea
ацетилсульфанилгуанидин acetylsulphaguanidine
ацетилхолин acetylcholine
ацетилхолинестераза acetylcholinesterase
ацетилцеллюлоза cellulose acetate
~, хлопьевидная cellulose acetate flakes
ацетилцистеин acetylcysteine
ацетиметр acetimeter, acetometer
ацетобутират целлюлозы cellulose acetate butyrate
ацетолиз acetolysis
ацетометр = ацетиметр
ацетон acetone
ацетонемия *med.* acetonemia
ацетоновый acetonic, acetone *(attr.)*
ацетопропионат целлюлозы cellulose acetate propionate
ацетоуксусный acetoacetic
ацетофеназин acetophenazine
ацефал *med.* acephalus
ацефалия acephalia, acephaly, acephalism
ацефалопод acephalopodius
ацефалорахия acephalorrhachia
ацидоз acidosis, oxyosis
~, изолированный outlying acidosis
~, почечно-канальцевый renal tubular acidosis
ацидометр acidometer, acidimeter
ацидофилия acidophilia
ацидофильность acidophilism
ацидофильный acidophilic, acidophilous
ацидоцитоз acidocytosis, eosinophylic leukocytosis

ацидоцитопения acidocytopenia, acidopenia, eosinophylic leukopenia
ацикличный acyclic; non-cyclic
ацилирование acylation
ацилхолин acylcholine
ацинарный acinar
ацинозный acinous
аэлотропный aelotropic
аэрарий *med.* aerarium
аэратор aerator
аэрировать aerate
аэроб aerobe, aerobion
аэробиоз aerobiosis
аэробный aerobic, aerobian
аэробус airbus
аэровокзал airport/air terminal (building)
аэрогазодинамика aerogasdynamics
аэрогенный aerogenic, aerogenous
аэрогеосъемка aerial geological survey
аэрогидродинамика aerohydrodynamics
аэродин aerodyne
аэродинамика aerodynamics
  ~, внутренняя internal aerodynamics
  ~ воздушного винта propeller aerodynamics
  ~ гиперзвуковых скоростей hypersonic aerodynamics, hypersonics
  ~ дозвуковых скоростей subsonic aerodynamics
  ~, индуктивная induced aerodynamics
  ~, линейная linearized aerodynamics
  ~, магнитная magnetoaerodynamics
  ~ на взлетных режимах take-off aerodynamics
  ~ на посадочных режимах landing aerodynamics
  ~, нестационарная non-steady [transient] aerodynamics
  ~ околозвуковых скоростей transonic aerodynamics
  ~, прикладная applied/engineering aerodynamics
  ~ сверхзвуковых скоростей supersonic aerodynamics, supersonics
  ~, техническая engineering/applied aerodynamics
аэродром aerodrome; *Am.* airfield, airdrome
  ~ базирования home aerodrome
  ~, базовый base aerodrome
  ~, грунтовый unpaved/natural/grass/ground aerodrome/airfield
  ~, запасной alternate/emergency airfield
  ~, клубно-спортивный flying-club airfield
  ~, промежуточный staging (air)field
  ~ снабжения supply/logistics aerodrome
  ~, фронтовой advanced landing field, advanced airfield
  ~, школьный flying-school airfield
аэрожелоб air-slide, pneumatic tube
аэрозоль aerosol
аэрозольный aerosolic
аэроизоклинальный aeroisoclinic
аэроионизатор aeroionizer
аэрокартография aerocartography
аэроклуб flying club, aeroclub
аэролаг air mileage unit, air log
аэролог aerologist
аэрология aerology
аэролоток pneumatic tube, air slide
аэрометр aerometer; densimeter
аэромеханика aeromechanics, air-flow mechanics
аэрон aeron
аэронавигационный aeronautical, air-navigational
аэронавигация air navigation
аэронавт aeronaut
аэронавтика aeronautics
аэроневроз aeroneurosis
аэрономия aeronomy
аэрообследование aerial survey(ing)
аэроотит *med.* aero-otitis
аэропленка aerial film
аэропоника *bot.* aeroponics
аэропорт airport
  ~ влета airport of entry
  ~, внутренний domestic airport
  ~ вылета airport of departure
  ~, гражданский civilian airport
  ~, грузовой freight/cargo airport
  ~ для вертолетов heliport
  ~ для самолетов короткого взлета и посадки short take-off and landing airport
  ~, запасной alternate/emergency airport
  ~ назначения airport of destination
  ~, узловой hub airport
аэропроектор aerial-photograph projector
аэропрофилограф airborne profile recorder
аэропыл airborne duster/pulverizer, aeroduster
аэрорадиометрия aerial radiometry
аэросани snowplane
аэросев aerial sowing
аэросинуит *med.* aerosinusitis, barosinusitis
аэроснимок aerial photograph
  ~, горизонтальный levelled aerial photograph
  ~, перспективный oblique aerial photograph
  ~, планово-перспективный vertical-oblique aerial photograph
  ~, плановый vertical aerial photograph
  ~, трансформированный rectified aerial photograph
  ~, щелевой slotted aerial photograph
аэростат aerostat, balloon, lighter-than-air aircraft
  ~, беспилотный unmanned balloon
  ~, дрейфующий free aerostat/balloon
  ~ заграждения barrage balloon
  ~, змейковый kite balloon
  ~, привязной captive/tethered balloon
  ~, управляемый dirigible aerostat/balloon, air ship
аэростатика aerostatics
аэростатный balloon-borne
аэросъемка aerial photography, aerophotography; aerial survey

аэротенк, аеротанк aeration tank, aerotank
аэротерапия aerotherapy
аэротермодинамика aerothermodynamics
аэротермометр vapour pressure thermometer
аэротит *med.* aerotitis
аэротория terminal airspace; aerodrome limits
аэротранспортабельный air-freightable; air-portable/transportable
аэроупругость aeroelasticity; aeroelastics
аэрофизика atmospheric physics, aerophysics
аэрофильтр aeration filter
аэрофинишер aircraft arresting equipment, arresting gear
аэрофобия *med.* aerophobia
аэрофотоаппарат aerial camera
~, многоспектральный multiband spectral aerial camera
~, плановый vertical aerial camera
~, щелевой split-image aerial camera, moving-film aerial camera
аэрофотограмметрия aerophotogrammetry, aerial photogrammetry
аэрофотокамера aerial camera
аэрофоторазведка aerial photographic reconnaissance
аэрофотоснимок aerial photograph
аэрофотосъемка aerial photography; photographic survey
~, перспективная oblique aerial/air photography
~, плановая vertical aerial photography
~, площадная general coverage photography
~ с перекрытиями mosaic photography
аэроэмболия *med.* aeroembolism
аэроэмфизема *med.* aeroemphysema

Б, б

баба ram
~ молота hammer head
баббит babbit(-metal), white metal
бабезиеллез *med.* babesiasis, babesiosis
бабка *mach.* head, stock; mandrel; *constr.* post
~, делительная indexing head
~, задняя tailstock
~ изделия work head
~, передняя headstock
~, расточная boring head
~, стропильная king post
~, шпиндельная headstock, spindle box
бабуин *zool.* baboon
багажник *(of car) Eng.* boot, *Am.* trunk; *(on car top)* rack; *(of bicycle)* carrier; *(in plane, etc.)* luggage compartment/hold
багасса bagasse, sugar cane trash
багатель *mus.* bagatelle
багер excavator, dredge, dredging shovel
багет *arch.* baguette, moulding, fillet
багор lifting/boat hook, grapnel, grappling iron; gaff
бадья bucket, pail, tub
база 1. base, basis; foundation; 2. depot, centre/center, station; *comp.* base; 3. *mach.* reference/datum surface; 4. resources, reserve
~, авиационная air base
~ адреса address reference number
~, военная military base
~, военно-воздушная airforce base
~, военно-морская naval base/station
~ входных аналоговых величин analogue input basis
~ геометрических данных geometric database
~ гидросамолетов seaplane base
~ горючего fuel depot
~ данных database
~ дьнных для хранения фактов factbase
~ данных, многопользовательская multiuser database
~ данных, наполненная populated database
~ данных, распределенная distributed database .
~ данных, реляционная relational database
~ данных с кольцевой структурой ring database
~ данных с простой списковой структурой direct list database
~ данных с сетевой структурой network database
~ данных с цепной структурой chain database
~ данных системы принятия решений decision support database
~ данных со смешанной структурой hybrid database
~, допустимая admissible/feasible basis
~ заправки горючим refuelling base
~ знаний knowledge base, base of knowledge
~, километражная *av.* ground speed course/base
~, кормовая fodder supplies, forage reserve
~, материальная material resources
~ материально-технического обеспечения logistics base
~ налогового обложения tax basis/base
~, плавучая depot/parent/mother ship
~ правил rule base
~ проектных знаний design knowledgebase
~, производственная manufacturing structure
~, разностная *math.* difference base
~, ремонтная repair depot/workshop
~, сбытовая sales depot
~, своя home base
~, сырьевая raw materials base, raw-material base
~ текстовых данных text database
~ технического обслуживания maintenance base
~, топливная fuel resources
~, учебная training station/centre

~ фактов factbase
~ цифрового ввода digital input base
~, энергетическая (electric) power base
базальт basalt
базедовизм *med.* Basedow's disease
базилик, базилика *bot.* basil *(Ocimum)*
~, душистый sweet basil *(Ocimum basilicum)*
базилика *arch.* basilica
базирование basing
базис basis, base; *cryst.* basal plane; *geod.* initial/base line
~, безусловный *math.* absolute basis
~, геодезический geodetic basis/base/length/distance
~, дополнительный patchline base
~ зрения interocular distance
~ индукции base of induction
~ окрестностей *math.* neighbourhood base
~, ортонормированный orthonormal basis/base, normal/normalized orthogonal basis
~ съемки survey base
~, съемочный camera base/baseline
~ триангуляции triangulation base
базисный basic; elementary; basal
~ная ставка *econ.* prime rate
базофил *med.* basophilic leukocyte, basophil
базофилия *med.* basophilia
базука *mil.* bazooka
байонет *tech.* bayonet joint
байпас by-pass, bypass; bypass pipe/line
байт *comp.* byte
~, адресный address byte
~, адресованный/вызванный addressed byte
~, адресуемый addressable byte
~ аргумента argument byte
~ графических данных graphic data byte, data byte for graphic devices
~ данных data byte, byte of data
~, значащий significant byte
~, идентифицирующий identification byte
~, информационный data byte
~, канальный channel byte
~ команды command byte
~, контрольный cyclic-checking byte, code check byte
~, младший low-order byte, low byte
~, младший значащий least significant byte
~ начального состояния initial status byte
~, нулевой zero byte, byte of zero
~ ошибки при анализе заголовка header analysis error byte
~, ошибочный error byte
~ памяти memory byte
~, префиксный prefix byte
~, пятиразрядный five-bit byte
~, рабочий effective byte
~, самый младший lowest byte, lowest-order byte
~, самый старший highest byte, highest-order byte
~ состояния status byte
~ состояния устройства device status byte
~ способа адресации address mode byte
~, старший high byte, high-order byte
~, управляющий control byte
~ устройства, стандартный standard device byte
байтовый byte-oriented
байцевать steep in mordant; stain
бак 1. tank, container, vat, cistern; *mil.* messdish; 2. *naut.* forecastle
~, аварийный напорный emergency header
~ аккумуляторной батереи storage battery container/tank
~ без дренажа unvented tank
~, белильный bleaching tank, bleacher
~, бензиновый *eng.* petrol tank, *Am.* gas/gasoline tank
~, вакуумный vacuum/depression tank
~, внешний outboard/external/side tank
~, водомерный water-gauge/-measuring tank
~, всасывающий suction tank
~ выдержки hold-up vessel/tank, holding tank
~, герметический sealed/leakproof tank
~, дозиметрический monitoring tank
~, дренированный vented tank
~ замедлителя moderator tank
~, красильный dyeing vat
~, маслосборный oil-catch tank, oil sump
~, мерный gauging tank
~, навесной strap-on tank
~, напорный pressure/head tank, header
~, отстойный settling/sedimentation tank
~, переливной overflow/decantation tank
~, питательный/питающий supply/feed tank
~, приемный receiver tank
~, промежуточный transfer tank
~, пропиточный saturation tank
~, проявочный *phot.* developing/developer tank
~, расходный feeder tank; service/day tank
~, расширительный surge tank
~ реактора reactor tank/vessel/container/housing/pool
~, сливной dump tank
~, топливный fuel tank
~, уравнительный surge tank
бак-аккумулятор storage tank
бакалавр bachelor, baccalaureate
~ естественных наук bachelor of science
бакаут *bot.* guaiacum, lignum vitae, rockwood *(Guajacum)*
бакелит phenolic resin plastic, bakelite
бакен, бакан *naut.* buoy; river/floating beakon
бакенщик light keeper, buoy-keeper
бак-мерник gauging tank
бактериемия bacteri(a)emia
бактерии *pl.* bacteria
~, автотрофные autotrophic bacteria
~, азотфиксирующие nitrogen-fixing bacteria
~, анаэробные anaerobic bacteria

~, аэробные aerobic/aerobian bacteria
~, веретенообразные fusiform bacteria
~, газообразующие gas-producing bacteria
~, гетеротрофные heterotrophic bacteria
~, гнилостные putrefactive bacteria
~, гноеродные pus-producing bacteria
~, грамотрицательные/грамположительные Gram-negative/-positive bacteria
~, колиподобные coliform bacteria
~, маслянокислые butyric acid bacteria
~, молочнокислые lactic acid bacteria
~, нитрофицирующие nitrobacteria, nitrifying/nitrate/nitrite bacteria
~, нитчатые filamentous bacteria
~, палочковидные rod-like/-shaped bacteria
~, психрофильные psychrophilic bacteria
~, светящиеся photogenic bacteria
~, уксуснокислые acetic acid bacteria
~, факультативно-патогенные elective pathogenic bacteria
бактериоз bacteriosis
бактериолиз bacteriolysis
бактериолитический bacteriolytic
бактериолог bacteriologist
бактериология bacteriology
бактериостаз bacteriostasis
бактериофаг bacteriophage
~, стрептококковый streptococcic bacteriophage
~, холерный choleraic bacteriophage
~, чумной plague bacteriophage
бактериофагия bacteriophagy, bacteriophagia
бактериофаготерапия bacteriophage therapy
бактерицид bactericide
бактерицидность bactericidal action
бактерицидный bactericidal
бактерия bacterium
бактриан *zool.* Bactrian camel *(Camelus bactrianus)*
бак-хранилище storage tank
бакштаг *naut.* backstay, guy
бакштов *naut.* guess-warp/-rope, boat line
баланит *med.* balanitis
~, диабетически balanitis diabetica
~, катаральный catarrhal balanitis
~, облитерующий balanitis obliterans
~, трихомонадный mycotic balanitis
~, язвенный ulcerative balanitis
баланс balance; balance weight; *acc.* balance sheet
~, активный favourable balance
~, благоприятный favourable/active balance
~, валовой gross balance
~, емкостный capacitance balance
~, заключительный summary account
~, материальный material balance
~, межотраслевой intersectoral balance
~, натуральный material/physical balance
~, омический resistance balance
~ основного капитала fixed assets balance
~, пассивный unfavourable/adverse balance
~, платежный balance of payments
~, прямой direct/net balance
~, рабочий пробный adjusted trial balance
~, сводный master balance sheet; consolidated statement
~, тепловой thermal/heat balance
~, топливный fuel balance
~, торговый trade balance
~ труда budget of labour
~ трудовых ресурсов manpower balance sheet, labour-force balance sheet
~, углеводный *med.* carbohydrate balance
~, физиологический physiological balance/equilibrium
~ часов balance-wheel
~, частичный пробный trial balance
~, энергетический/энергии energy balance
балансир balance/equalizing/rocking beam; balancer; rocker arm; *constr.* sway-arm; balance/balancing weight
~, весовой mass-balance beam
~ конуса *met.* bell beam
~ кривошипа crank back-balance
балансировать balance; *av.* trim
~ неверно *av.* mistrim
балансировка balance, balancing; *av.* trim, trimming
~, динамическая dynamic/running balancing/trimming
~, неправильная mistrim
~ по массе *av.* mass/weight balance/trim
~ по тангажу pitch trim
~, поперечная lateral trim
балансы *pl.* pulpwood, paper wood
балантидиаз *med.* balantidiasis
балка beam, girder, boom; timber; davit
~, анкеровая tie/anchorage beam
~, грузовая cargo davit
~, двутавровая I-/flanged beam, *(broader)* H-beam
~, двухпролетная beam with a central prop
~ жесткости stiffening girder
~, заделанная одним концом cantilever beam
~, защемленная restricted beam
~, коническая tapered beam
~, консольная cantilever beam
~ коробчатого сечения box beam
~, крановая crane jib
~, многопролетная multispan beam
~, направляющая guide rail
~, обвязочная binder
~, одноконсольная beam with overhang
~, однотавровая T-beam/girder, T-shaped beam, T-/tee-girder
~, продольная longitudinal girder/beam; longeron
~, решетчатая lattice/trellis beam
~, связующая bond beam

~, тавровая = однотавровая
~, трубчатая tubular beam
~, шарнирная hinged girder
~, швеллерная channel beam, U-beam
~, шпренгельная trussed beam
~, штампованная formed/pressed beam
балка-распорка *constr.* roof beam, purlin
балл number; score, grade, mark; point
~, коррозионный corrosion index/value; amount of corrosion
~ облачности cloud amount
баллас ballas diamond
балласт ballast
~, неподвижный/несбрасываемый fixed ballast
~, подвижный shifting/portable ballast
балластер *rail.* ballasting machine, ballaster
балластина *naut.* ballast pig, drift lead, drag anchor
баллер *naut.* spindle; stock
~ руля rudder head shaft, rudder pin/stock
баллиста *hist.* ballista, balista; mangonel
баллистик ballistician
баллистика ballistics
~, внешняя/внутренняя exterior/interior ballistics
~ гиперзвуковых скоростей hyperballistics
баллистический ballistic
баллистокардиография *med.* ballistocardiography
баллон cylinder, vessel, bottle, container; flask; bulb; *auto.* balloon tyre, low-pressure tyre
~ огнетушителя/противопожарный fire-extinguisher bottle/container
бальзам balsam
~, канадский Canada balsam/turpentine
~, копайский copaiba (balsam)
~, перуанский balsam of Peru
бальзамировать embalm
бальзамический balsamic, balmy
бальнеогрязелечение balneomudtherapy
бальнеолог balneologist
бальнеотерапия, бальнеолечение balneotherapy
балюстрада *arch.* balustrade; railing
балясина baluster, rung, rail post
бампер bumper
бамья, бамия *bot.* okra, gumbo *(Hibiscus esculentus)*
бандаж 1. shroud(ing), band(ing), strap(ping); *med.* bandage; 2. *rail.* tyre
~, безребордный flangeless tyre
~, грыжевый hernial bandage
бандажированный shrouded; tyred
бандаж-набрюшник abdominal bandage/band
банк *comp.* bank; *game th.* pot; *fin.* bank
~, акцептный acceptance house
~, акционерный joint-stock bank
~, взаимосберегательный nonstock-moneyed corporation
~, вносить в place on deposit
~, головной parent bank; main bank
~ данных data bank
~ данных, анкетный dossier data bank
~, инвестиционный investment bank
~ инкассатор collecting bank
~, ипотечный mortgage bank
~, коммерческий commercial/business bank
~ корреспондент correspondent bank
~ международных расчетов bank for international settlements
~, общий *comp.* common data bank
~ памяти memory/store bank
~ получателя payee's bank
~, расчетный clearing bank
~ реконструкции и развития, международный International Bank for Reconstruction and Development
~, учетный discount bank
~, чартерный chartered bank
~, эмиссионный bank of issue/circulation
банк-эмитент issuing bank
банка I 1. jar, can, tin, pot; *med.* cupping glass, cup; 2. *naut.* thwart
~, аптечная gallipot
~, консервная *Eng.* tin, *Am.* can
~, кровососная *med.* wet cup
~, лабораторная phial
~, лейденская *el.* Leyden jar
~ с аэрозолем spray can
~, шлюпочная boat thwart
банка II bank, shoal
банкаброш bobbin/flyer/fly/speed/spindle frame
~, перегонный intermediate frame
~, тазовый/толстый slubbing frame
~, тонкий roving frame
банкет raised platform; embankment; bank; *(of road)* banquette
банкноты paper money
~ крупного достоинства large denominations
~ мелкого достоинства small denominations
банкрот *fin.* bankrupt
~, злостный fraudulent bankrupt
банкротство *fin.* bankruptcy; insolvency
~, злостное fraudulent bankruptcy
банник tube/bore brush, scraper; rammer; cleaner, swab
~, воздушный air lance
~, проволочный wire brush
баня bath
~, воздушная air bath
~, масляная oil bath
~, паровая vapour bath
~, песочная sand bath
бар 1. *(phys. unit)* bar; 2. *mech.* jib; bar
~, отбойный breaking-down jib/bar
барабан *tech.* drum, cylinder, barrel; reel; *med.* tympanum; *arch.* dome drum
~, канатный rope drum
~, кодовый code drum

~, **купажный** *food* drum mixer/blender
~ **лебедки** winch drum
~, **окорочный** *wood.* barking drum, barker
~, **печатающий** printer drum; type drum
~, **полировальный** *mech.* burnishing barrel
~ **револьвера** revolver cylinder
~, **сортировальный** grading cylinder
~, **тормозной** brake drum
~, **тросовый** cable drum/reel
~, **цифровой** *comp.* digital drum
~, **швартовный** *naut.* warping drum
баратрия *law* barratry
барашек ear(ed) nut, wing nut, thumb screw, fly nut
барбамил barbamil, sodium amital/amobarbital/amilobarbitone
барбарис *bot.* barberry *(Berberis)*
барбитал barbital, barbitone, veronal
барбитураза barbiturase
барбитурат barbiturate
барбифен barbiphen, phenobarbital
барботаж bubbling; *Eng.* splashing
барботер bubbler
барботировать bubble
бареттер *el.* current regulator tube, ballast tube, barretter
баржа barge; lighter; scow
~, **беспалубная** open barge
~, **водолазная** diving lighter
~, **грузовая** cargo barge/lighter
~, **десантная** *mil.* landing barge
~, **жилая** berthing barge
~, **мусорная** garbage scow/lighter, rubbish barge; ash boat
~, **наливная** tank barge/lighter
~, **палубная** covered barge
~, **портовая** habrour utility craft/barge
~, **самоходная** self-propelled barge
~, **санитарная** ambulance barge
~, **сухогрузная** dry-cargo barge
баржевоз barge carrier
барий *chem.* barium, Ba
~, **азотнокислый** barium nitrate
~, **бромистый** barium bromide
~, **иодистый** barium iodide
~, **мышьяковистый** barium arsenate
~, **сернистый** barium sulphide/sulfide
~, **сернокислый** barium sulphate/sulfate, baryta white
~, **углекислый** barium carbonate
~, **уксуснокислый** barium acetate
~, **хлористый** barium chloride
~, **хромовокислый** chromate of baryta, barium chromicum, barium/baryta yellow
барион baryon, barion
барицентр barycentre
барический barometric, baric
баркас *naut.* (ferry) launch, long boat
барн *phys.* barn
барограмма barogram(me)
~ **подъема** climb barogram(me)
барограф barograph, altitude recorder, recording altimeter
~, **анероидный** aneroid barograph
барокамера (low-pressure) altitude chamber
~, **вакуумная** high-vacuum test chamber
бароклина barocline
бароклинность baroclinic distribution; baroclinity
барометр barometer
~, **жидкостный** liquid-column barometer
~, **регистрирующий/самопишущий** barograph, recording barometer
~, **ртутный** mercury barometer
~, **чашечный** cup/cistern barometer
барометр-анероид aneroid barometer
барометр-высотомер altimetric barometer; aneroid altimeter
барометрический barometric(al)
барометрия barometry
бароотит *med.* barootitis, barotitis
бароперключатель, барореле baroswitch, barometric switch
барорецептор baroceptor, baroreceptor
баросинусит *med.* barosinuitis
бароскоп baroscope, open-end manometer
бароспидограф velocity-height recorder
баростат barostat, baroswitch
баротерапия barotherapy
баротермограф barothermograph
баротравма barotrauma
баротропия barotropy
баротропность barotropic distribution
баротропный barotropic
барофорез barophoresis
бархоут binding/rubbing strake
барьер barrier
~, **высотный** altitude barrier
~ **деления** fission barrier
~, **звуковой** *av.* sound/sonic barrier
~, **тепловой/температурный** temperature/heat/thermal barrier
бассейн reservoir; pool; basin; pond
~, **бессточный** inland basin
~, **бревенный** *timb.* log pond
~, **брызгальный** cooling/spray pond
~, **буксировочный опытовый** experimental towing tank
~, **внутренний** enclosed dock
~, **водосборный** catchment basin
~, **водосливный** spillway basin
~ **выдержки** *phys.* cooling pond
~, **выравнивающий** compensating basin
~ **гидроаэродрома** seaplane basin
~, **гидродинамический опытовый** hydrodynamic test/experimental tank
~ **графика** *math.* basin of a graph
~ **для протаски, опытовый** towing tank
~, **каменноугольный** coal field/basin
~, **напорный** head-pond
~, **опытовый** testing/experimental tank, experimental model basin, model testing tank/basin

~, опытовый, буксировочный (experimental) towing tank
~ опытовый с волнопродуктором wave tank
~, отстойный sedimentation/settling basin
~, плавательный swimming pool
~ реки river basin
~, угольный coal field/basin
~, усреднительный equalizing basin
батальон *mil.* battalion
батан *text.* loom slay, lathe
~, пружинный flyer lathe
~, револьверный circular box lathe
батарейный *el.* battery-operated/powered
батарея *el., mil.* battery
~, аккумуляторная storage battery
~, анодная anode/plate battery, B-battery
~, железоникелевая nickel-iron battery
~, зенитная *mil.* anti-aircraft battery
~, кадмиево-никелевая nickel-cadmium battery
~ накала filament battery, A-battery
~, перезаряжаемая rechargable battery
~, свинцовая lead-acid battery
~, серебряная silver battery
~, сеточная grid battery, C-battery
~ сеточного смещения grid-bias battery, C-battery
~ смещения *el.* biasing battery
~, солнечная solar battery
~, стартерная starter/starting battery
~, сухозаряженная dry-charged battery
~, термоэлектрическая thermoelectric pile/battery
~, тяговая traction battery
~, угольно-цинковая zinc-carbon battery
~, щелочная alkaline battery
батат *bot.* batata, sweet potato *(Ipomoea batatas)*
батенсы *pl. timb.* battens
батиграмма bathogram
батиметрия bathymetry
батискаф bathyscaphe
батисфера bathysphere
батитермограф bathythermograph
батокс *naut.* buttock, bow and buttock line
~, кормовой buttock line, aftline
~, носовой bow line, foreline
батолиты *miner.* batholiths
батометр bathometer, water sampling device/bottle, water sampler/bottle
батопорт *nav.* caisson/flood/water gate; pontoon dock-gate
батохромный bathochromic
батрак farm/field hand, farm labourer
бафтинг *av.* buffeting, buffet
~ крыла wing buffet
~, элеронный aileron buffet
бахрома *telev.* fringe, fringing
бахтарма *leath.* flesh side *(of hide)*
Бахус *myth.* Bacchus
бацилла bacillus
бациллоноситель bacilli-/bacillus carrier, human carrier, carrier of bacilli
бациллоносительство carriage of bacilli
бациллы *pl.* bacilli
~, кислотоустойчивые acid-fast bacilli
~, сибиреязвенные anthracic bacilli, bacilli of anthrax
бачок canister, small tank
~, расширительный expansion tank
башмак shoe, boot
~, полюсный *el.* pole shoe
~ стрелы (крана) boom seat
башня tower; turret
~, водонапорная water tower
~ командного пункта control tower
~, лоцманская pilot tower
~, охлаждающая cooling tower
~, петлевая *met.* looping tower
~, поглотительная *met.* scrubber
~ с насадкой packed tower
~, силосная tower silo
~, сторожевая lookout tower
~, тушильная quenching/watering tower
~, улавливающая entrainment tower
бегство из тюрьмы breach of prison
бегун 1. runner, roll; pulley; 2. crushing/ grinding mill; 3. *rail.* free axle; *paper* kollergang, kollermill
бегунок 1. traveller; *(of slide rule)* runner; indicator; cursor; 2. *rail.* front bogie
бегущий travelling; progressive
беднота, сельская the rural poor
бедренный femoral
бедро femur, thigh
безаварийный failure-proof, failsafe, accident-free, trouble-proof
безадресный addressless
безазотистый anazoic, anitrogenous
безбрачие celibacy; *biol.* agamy
безбустерный unpowered, manual-operated
безветрие still/calm air, (wind) calm
безвихревой vortex-free, eddy-free; irrotational
безводный 1. *chem.* anhydrous; non-aqueous; 2. moisture-free, free of water, dry; 3. arid, dry
безвоздушный airless, air-free
безвозмездный *law* voluntary; free
безвременник *bot.* meadow saffron, autumn crocus *(Colchicum)*
безглазый eyeless, anophthalmous
безголовый acephalous
безголосый aphonic
бездействующий inoperative, idle
безденежный *fin.* cashless
бездефектный flawless, zero-defects
бездымный smokeless
беззубый edentulous, anodontous, toothless
безизбыточность nonredundancy
безизбыточный nonredundant, irredundant
безинерционный inertialess
безламповый tubeless

безлюфтовый backlash-free; anti-backlash
безмен steelyard
безмикробный amicrobic, abacterial
безмоментный *phys.* motionless, without momentum
безналичный *fin.* cashless
безногий legless, footless, limbless
безопасность safety, security
~, государственная state security
~ движения traffic safety
~, общественная social security
~, противопожарная fire safety
~, радиационная radiation protection
~, ядерная nuclear safety áesotsbhoé
безотзывной beyond/past recall
безотказность reliability; failure-free performance/operation
безотказный failure-free, trouble-free, troubleproof, no-trouble
безошибочный error-free
безработица unemployment
~, скрытая latent unemployment
~, частичная part-time idleness/unemployment
безразмерный nondimensional, dimensionless
безредукторный *tech.* direct-drive
безрукий abrachial, armless, limbless
безубыточность break-even
безубыточный break-even
~ ное ведение дела breaking-even
безусловнорефлекторный unconditioned-reflex; innate
безэховый anechoic
бейфут *naut.* truss
бейнит *met.* bainite
беккерель Becquerel, Bq
беккрос backcross(ing)
бел *(unit)* Bel, B
белена *bot.* henbane *(Hyoscyamus)*
бели *pl. med.* leucorrhea, leucorrhagia
~, влагалищные vaginal leucorrhea
~, маточные uterine leucorrhea/milk
~, трубные tubal leucorrhea
~, шеечные cervical leucorrhea
белила white pigment/paint; white
~, свинцовые lead white
~, цинковые zinc white
белильня bleaching works/installation, bleachery
белковый albuminous
белладонна *bot.* belladonna, deadly night shade *(Atropa belladonna)*
белок protein
~, молочный lactoprotein
~, неполноценный incomplete protein
~, неферментный nonenzyme protein
~ плазмы plasma protein
~, полноценный complete protein
~, растительный vegetable protein, phytalbumin
~, сложный conjugated protein
~ сыворотки крови serum protein
~, усвояемый assimilable protein
~, чужеродный foreign protein
~, эндогенный internal protein
белокровие leuk(a)emia
~, ложное pseudoleuk(a)emia
белокрыльник *bot.* calla *(Calla)*
~, болотный wild calla *(Calla palustris)*
белочный albugineous
бельканто *mus.* bel canto
бельмо *med.* leukoma
~, плотное dense leukoma
~, приросшее adherent leukoma
~ роговицы corneal leukoma
бенедиктинец *eccl.* Benedictine
бенефициар, бенефициарий *econ.* beneficiary
бенефиций *hist., econ.* beneficium, benefice, fief, feudal estate
бензель *naut.* seizing
бензидин benzidine
бензил benzyl
бензимидазолил benzimidazolyl
бензилцианид benzyl cyanide
бензин *Am.* gasoline, *Eng.* petrol
~, авиационный aviation petrol/gasoline/spirit
~, автомобильный motor petrol/gasoline
~ без присадки blank petrol/gasoline
~, высокооктановый high-octane petrol/gasoline
~, зимний/летний winter-/summer-grade petrol/gasoline
~, низкооктановый low-octane petrol/gasoline
~, отвержденный jellied petrol/gasoline
~, пусковой priming fuel
~, этилированный ethyled/leaded petrol/gasoline
бензиновоз petrol/gasoline tanker
бензиномер petrol/gasoline gauge
бензиноуловитель petrol/gasoline recuperator
бензоат *chem.* benzoate
~ калия potassium benzoate
бензобак petrol/gasoline tank
бензоил *chem.* benzoyl
бензоилхолин benzoyl choline
бензоилхолинэстераза benzoylcholinesterase
бензойный benzoic
бензоколонка filling station; *Am.* gas station
бензол benzene
бензоманометр petrol/gasoline pressure gauge
бензонасос petrol/gasoline/fuel pump
бензонитрил benzonitrile
бензопровод petrol/gasoline line/pipe/tube
бензорез oxy-benz cutting torch
бензфетамин benzphetamine
бензхинамид benzquinamide
бентиамин bentiamine
бентонит bentonite
бентос *biol.* benthos
бепаск calcium benzamidosalicylate
бердо 1. *text.* reed; 2. *naut.* loom
~, раздвижное expanding reed
берег coast, shore *(river)* bank; *naut.* waterside

~, **подветренный** lee shore
береговой coastal; shore-based; land *(attr.)*
береза *bot.* birch *(Betula)*
беременность pregnancy, gestation, gravidity
~, **брыжеечная** mesenteric pregnancy
~, **брюшная** abdominal pregnancy
~, **внематочная** extrauterine pregnancy/gestation, ectopic pregnancy/fetation, cornual pregnancy/gestation
~, **внутрибрюшная** intraperitoneal pregnancy
~ **двойней** bigeminal/twin/gemellary pregnancy, double gestation
~, **доношенная** full-term pregnancy
~, **заоболочечная** membraneous pregnancy
~, **интрамуральная/интерстициальная** interstitial/intramural/parietal/mural pregnancy
~, **истмическая** isthmian pregnancy
~, **ложная** pseudopregnancy, false/phantom/af(o)etal pregnancy
~, **маточная** uterine pregnancy
~, **маточно-брюшная** uteroabdominal pregnancy
~, **маточно-яичниковая** uteroovarian pregnancy
~, **мнимая = ложная**
~, **многоплодная** plural pregnancy, multif(o)etation
~, **одноплодная** single pregnancy
~, **париетальная = интрамуральная**
~, **переношенная** prolonged pregnancy
~, **трубная** fallopian/tubal pregnancy
~, **трубно-брюшная** tuboabdominal pregnancy
~, **трубно-маточная** tubouterine/uterotubal pregnancy
~, **тубо-лигаментарная** tuboligamentary/mesenteric pregnancy
~, **тубоовариальная** tubo-ovarian pregnancy
~, **шеечная** cervical pregnancy
~, **яичниковая** ovarian pregnancy
бери-бери *med.* beri-beri, beriberi
берилл *miner.* beryl
бериллид beryllide
бериллизация beryllium (diffusion) hardening
бериллий beryllium, Be
~, **металлический** beryllium metal
~, **сернокислый** beryllium sulphate
~, **фтористый** beryllium fluoride
~, **хлористый** beryllium chloride
берклианство Berkleianism
берклий, беркелий berkelium, Bk
берма bench, banquette, berm
берсальеры *mil.* bersaglieri
бескамерный *auto* tubeless
бескапсульный acapsular
бескаркасный frameless
бескислородный oxygen-free
бескислотный acid-free
бесклапанный valveless
бесклеточный acellular; cell-free
бесконечно infinitely; endlessly
~ **большая величина** infinite
~ **малая величина** infinitesimal
~ **малый** infinitesimal
бесконечнократный infinite-to-one
бесконечнолистный infinite-sheeted
бесконечномерный infinite-dimensional, of infinite dimensions
бесконечносвязный infinite-connective, of infinite connectivity
бесконечность infinity
~**и, до** at infinity, ad infinitum
бесконечный infinite; unbounded, unlimited
~**ная дробь** recurring decimal
бесконтактный contactless, non-contact
бескровный bloodless, exsanguinate
беспалубный undecked, open
беспалый adactylous, adactyl
бесперебойный uninterrupted, continuous; trouble-proof, trouble-free
беспилотный unmanned, pilotless
бесплодие sterility, infertility
~ **у женщин** female infertility, aphoria
~ **у мужчин** male sterility/infertility
бесплодный sterile, infertile, acyetic
беспозвоночные *pl.* invertebrates
бесполый agamous, agamic; asexual
~**лое размножение** apogamy
бесполюсный poleless
беспосадочный non-stop
беспошлинный *econ.* duty-free
беспорядок disorder; inversion
~ **в перестановке** *math.* inversion of a permutation
бесприбыльный *econ.* non-profit
беспроцентный *econ.* interest-free
бессальниковый packless
бессвязность incoherence
бессемерование bessemerizing; Bessemer process
бессемеровать bessemerize
бессеточный gridless
бессимптомный asymptomatic
бесспиновый spin-free, zero-spin
бессрочный permanent, termless
бесструктурный anhistous, anhistic, structureless
бесступенчатый stepless; fully/infinitely variable
бестурбулентный turbulence-free
бесфоновый no-background
бесцветный achromatic, colourless
бесшарнирный hingeless
бесшовный seamless; *med.* sutureless
бесщеточный *el.* brushless
бета beta
бета-ветвь beta-branch
бета-глобулин beta globulin
бета-дефектоскопия beta-ray flaw detection
бета-излучатель beta-emitter/radiator
бета-излучение beta-radiation/emission
бетаин betain
бета-лучи beta-rays/radiation

бета-превращение beta-transformation
бета-распад beta-disintegration/-decay
бета-съемка beta-survey
бета-терапия beta-ray therapy
бетатолщиномер beta-ray thickness gauge
бетатрон betatron, induction accelerator
~, безжелезный air-cooled betatron
~, двухлучевой dual-beam betatron
~ с подмагничиванием field-biased betatron
бета-уран beta-phase of uranium, beta-uranium
бета-цепь beta chain
бетацианин betacyanin
бета-частица beta particle
бетель *bot.* betel pepper *(Piper betle)*
бетон concrete
~, армированный reinforced concrete
~, баритовый barytes concrete
~, быстротвердеющий fast-hardening concrete
~, гидротехнический hydraulic concrete
~, неармированный plain concrete
~, опилочный sawdust concrete
~, предварительно напряженный prestressed concrete
~, с волокнистым заполнителем fibrous concrete
~, строительный structural concrete
~, тяжелый heavy-aggregate concrete, loaded concrete
~, ячеистый cellular concrete
бетонирование concreting; placement/pouring of concrete
бетонировать concrete, encase in concrete; place/cast/pour concrete
бетоновоз concrete-delivery truck
бетономешалка, бетоносмеситель concrete mixer
бетононасос concrete pump
бешенство *med.* rabies, hydrophobia
биатлон *sport.* biathlon
библиотека *comp.* library
~ алгоритмов algorithm collection
~ задания на пуск программ run library
~ исходных модулей source library, sourcestatement library
~ макрокоманд macro library
~ настраиваемых/перемещаемых модулей/программ relocatable library
~, общая link library
~ пользователя user-defined library
~ программ/подпрограмм program(me)/routine library; subroutine library
~ программ ввода loader library, loading routine library
~ программ ввода-вывода input-output library
~ программ на машинном языке code program(me) library
~ процедур procedure library
~ с прямым доступом direct-access library
~ стандартных модулей building block library
~ стандартных подпрограмм subroutine library
~ стандартных фигур shape library
~ ядер управляющих программ nucleus library, control program(me) nucleus
бивалент bivalent
бивариантный bivariant
бивектор bivector, divector
биг *print.* crease, score
бигамия bigamy
бигеминия *med.* bigeminy, bigeminal pulse/rhythm
~, желудочковая ventricular bigeminy
~, предсердная atrial bigeminy
биговка *print.* creasing, scoring
биде bidet
бидистиллят redistillate
бидистилляция redistillation
биекция *math.* bijection
биение *med.* beat, throb
~ пульса pulsation
~ сердца heart beat
биения *pl. tech.* beating, beats; wobbling, runout, play; *(of belt)* slack
~, радиальные radial runout
~, торцовые end play
бизань-мачта *naut.* after mast, miz(z)en mast
бикарбонат bicarbonate
~ натрия/кальция sodium/calcium bicarbonate
биквадрат biquadrate, fourth power
биквадратный biquadratic
бикомпакт *math.* bicompact(um)
билеты, казначейские *econ.* legal tender notes
билинейный bilinear
билинза *opt.* bilens, splitlens
билирубинемия *med.* bilirubin(a)emia
биллион *Am.* billion; *Eng.* thousand million, milliard
биллон, биллонная монета billon, base-metal coin
бильгарциоз *med.* bilharziasis, bilharziosis, schistosomiasis
бимера bi-measure
биметалл bimetal, composite/clad metal; bimetallic strip/plate
биметаллизм *fin.* bimetallism; double standard
биметаллический bimetallic
бимолекулярный bimolecular, dimolecular
бимомент *phys.* bitorque
бимс *naut.* (deck) beam
~, концевой hatch-end beam
~, люковый hatch beam
~, поворотный cant beam
~, продольный fore-and-aft beam
~, рамный deep/web beam
~, ростерный skid beam
~, трюмный hold beam
~, холостой undecked beam
бинарный binary
бинауральный binaural
бинейтрон bineutron, dineutron

бинокль binoculars
~, полевой field glasses
~, призменный prismatic field glasses
~ с сеткой graticulated glasses
~, театральный opera glasses
бинокулярный binocular
бином *math.* binomial
~ Ньютона binomial theorem/formula
биномиальный binomial
бинормаль *math.* binormal
бинт bandage; roller
~, марлевый gause bandage
~, резиновый rubber bandage
~, эластический elastic/crepe bandage
бинтовать bandage, dress
биоакустика bioacoustics
биоаппаратура bio-instrumentation
биоастронавтика bioastronautics
биогельминтоз biohelminthosis
биогельминты *pl.* biohelminth(e)s
биогенез biogenesis
биогения biogeny
биоголография bioholography
биодатчик biosensor
биоднозначный *math.* bi-unique, one-one
биокапсула biocapsule, biopack
биокатализатор biocatalyst
биокибернетика biocybernetics
биокоррозия biocorrosion
биокриминология criminal biology
биолизис biolysis
биолит biolith
биолог biologist
биологизм biologism
биология biology
~ клетки cell biology
~, космическая *space* biology, astrobiology
биолюминесценция bioluminescence
биомаса biomass
биомедицина biomedicine
~, авиационно-космическая aerospace biomedicine
биометрия biometry
биомеханика biomechanics
биомицин *med.* chlortetracycline
~, солянокислый chlortetracycline hydrochloride
бионика bionics
бионт biont
биополимер biopolymer
биопотенциал bioelectric potential
биопрепарат biological preparation
биопроба bio-assay; biological test
биопсия *med.* biopsy
~, аспирационная aspiration/needle biopsy
~, пункционная puncture biopsy
биоразъем bioplug; biopatch
биортогональность *math.* biorthogonality
биосинтез biosynthesis
биосистема biological system, biosystem
биоскоп bioscope
биосфера biosphere
биота *bot.* arborvitae *(Thuja)*
биотерапия biotherapy
биотехника bioengineering, biotechnology
биотин biotin
биотип biotype
биоток biocurrent, biological current
биотоп biotope
биотрон biotron
биофизика biophysics
биофор biophore
биохимия biochemistry
биоценоз bioc(o)enosis
биплан *av.* biplane
биполярность bipolarity
биполярный bipolar
биржа (stock) exchange
~, Американская фондовая American Stock Exchange
~, зерновая grain/corn market
~, лесная wood/log yard
~, товарная commodity exchange/market
~ труда labo(u)r/employment exchange
~, фондовая stock market/exchange; securities market
~, хлебная grain exchange
~, хлопковая cotton exchange
~, черная currency black market
биржевик (stock) jobber, stockbroker
биржевой stock-exchange
~ дилер dealer in stocks, exchange dealer
~ маклер stockbroker
бирка tally, tag *(for identification)*
бисексуализм bisexualism
бисексуальность bisexuality
бисистолия *med.* bisystole
биссектриса *mat.* bisectrix, bisector
биссиноз *med.* byssinosis
бистабильный bistable
бит bit, binary digit
~ адресной метки address mark bit
~ гашения erasing bit
~ данных data bit
~ защиты security bit
~ зоны/зональный zone bit
~, информационный data bit
~, контрольный check bit
~ маски файла file mask bit
~, младший low-order bit
~ памяти storage/memory bit
~ поиска, контрольный seek check bit
~, полезный effective bit
~, пустой ignore bit
~, самый младший lowest-order bit, least significant bit
~, самый старший highest-order bit
~ сек bits per second, bps
~ слова word mark bit, wm
~, стартовый start bit

~ **стирания** erase/erasing bit
~, **стоповый** stop bit
~ **считывания** sense bit
~, **управляющий** control bit
битартрат bitartrate
битенг *naut.* bitt
битенг-болт bitt pin
битенг-краспица cross-bitt
битум bitumen; asphalt
~, **дорожный** paving/road bitumen/asphalt
~, **мягкий нефтяной** semi-asphaltic flux
~, **нефтяной** petroleum/oil asphalt
~, **окисленный** air-blown asphalt
~, **природный** rock asphalt
~, **содержащий** bituminiferous
~, **угольный** coal bitumen
~, **флюсированный** cut-back asphalt
битумный bituminous; bituminiferous
бить *mech.* wobble, play, be/run out of true
биурет biuret, allophanamide
биуретовый biuretic
бифакторный bifactorial, bifactor
бифилярный bifilar; double-wound
бифуркация bifurcation; branching
бихевиоризм behavio(u)rism; behavio(u)ral science
бихевиористический behavio(u)ral
бихромат bichromate
бицепс biceps
бициклический bicyclic
бич *agr.* bar
биэквивалентность biequivalence
благо, общественное *law* public benefit/good
Благовещение *eccl.* the Annunciation, Lady Day
бланк 1. form, sheet, blank; 2. *com.* blank
~, **бесконечный** continuous paper form, endless tab form, continuous stationery
~, **буквенный** letter blank
~ **для получения деталей** parts issue voucher
~ **для программирования** *comp.* coding form
~ **задания/заказа** order ticket; job order sheet
~, **запросный** request sheet
~ **описания файлов** *comp.* file description specifications sheet
~, **рабочий** reference sheet
~ **сдельной работы** job ticket
~ **счета** account sheet, accounting form
~ **счета, индивидуальный** individual standard form
~, **телеграфный** message form/blank
бланк-ваучер *econ.* voucher apron
бланманже *cul.* blancmange
бланфикс *miner.* blanc fixe, barium white
бланширование *food* blanching; *leath.* blanching, whitening
бланширователь *food* blancher
бластема *med.* blastema
бластид blastid
бластин blastin
бластоген blastogene
бластогенный blastogenic
бластодерма blastoderm(a)
бластодермальный blastodermal, blastodermic
бластокинез blastokinesis
бластома *med.* blastoma
бластоматоз blastomatosis
бластомер blastomere
бластомикоз blastomycosis
~, **рассеянный** disseminated blastomycosis
~, **системный** systemic blastomycosis
бластомицета *bot.* blastomycete; *med.* Blastomyces, *pl.* Blastomycetes
~, **бразильская** Blastomyces brasiliensis
~, **кожная** Blastomyces dermatitidis
бластопор(а) blastopore
бластотомия blastotomy
бластоциста blastocyst
бластоцит blastocyte
бластоцитома blastocytoma
бластула blastula, blastocyst
бленда 1. *opt.* diaphragm; stop; lens hood; 2. *astr.* blend; 3. lead apron; blind, mask
бленкер *tel.* blinker light; magnetic/drop indicator
бленнорея *med.* blennorrhea, gonorrhea, gonorrheal conjunctivitis/ophthalmia
блефарит blepharitis
~, **паразитарный** blepharitis parasitica/ phthiriatica
блефарорафия blepharorrhaphy
блефоаденома blephoadenoma
близкодействующий short-range
близнец *math.* twin
близнецы *pl.* twins; *math.* prime pair, twins
~, **двуяйцевые/дизиготные** diovular/dizygotic twins
~, **монозиготные/однояйцевые** monozygotic/monovular twins
~, **разнояйцевые** hetero-ovular twins
~, **соединенные** conjoined twins
близорукий myopic, near-sighted
близорукость myopia, shortsightedness
~, **профессиональная** occupational myopia
~, **рефракционная** refractive myopia
блик speck/patch of light; dazzle; glare
блиндаж *mil.* shelter, dugout
блок 1. block, unit, set, assembly; package 2. *tech.* block, pulley; sheave; 3. *anat.* trochlea; 4. *econ., pol.* bloc; 5. lump
~, **автономный** self-contained unit
~, **азимутальный** azimuth unit
~, **арифметический и логический** arithmetical and logical unit; arithmetic-logical unit
~, **буферный (памяти)** buffer store block
~, **быстросменный** plug/pluggable unit
~, **валютный** *econ.* currency bloc
~ **ввода-вывода** input-output unit
~ **ввода данных** data input unit
~, **вентильный** gate unit
~ **веретена** *text.* wharve, pulley
~ **ветвления** decision block

~ внутренней памяти internal store unit
~ воспроизведения playback unit
~ времени unit of time, clock unit
~ вспомогательных запросов minor request block, minor RB
~, вставной plug-in unit, insert
~, входной input block/unit
~ вызова загруженных программ loaded-program(me) request block
~ вызова супервизора supervisor request block
~, выходной output block/element
~ генерации кодов code generating/investigaton unit
~, глухой dummy block
~ граничных адресов/пар boundary box, boundary address block
~ данных data block
~ данных, алфавитно-цифровой alphanumeric block of information
~ данных списковой структуры list data block
~ датчиков sensor unit, data-transmitter
~ дистанционной передачи данных data telecommunication unit
~ для массивов данных data extent block
~ для программирования programming block
~ для управления задачами task control block
~ для управления наборами/массивами/файлами данных data-set control block, file control block
~, добавочный *comp.* hook
~, замедляющий moderating block
~ записей record block
~ записи на ленту tape device/equipment, tape handling unit
~ запроса request block, RB
~ запроса программ program(me) request block, PRB
~ запросов на прерывание interruption request block
~ защиты памяти memory guardian
~ зубчатых колес *mech.* gear cluster
~ из нескольких записей multirecord block
~ импульсного генератора pulser box
~ индексации indexing unit
~ индикации indicating unit
~ индикации на электроннолучевой трубке CRT display unit
~ инерциальной системы inertial unit
~ информации на магнитной ленте magnetic tape data block
~ информации о заданиях job information block
~ информационный information block/group
~, канатный *mech.* rope sheave
~, карточный card unit
~ команд instruction/command unit
~ коммутации switching unit
~, коммутирующий multiplexing unit
~, компоновочный assembly unit; building block
~ контактных щеток brush station
~, контрольный check(ing) unit
~ контроля monitor (unit)
~ контроля данных data control block
~, конъюгационный pairing block
~ ламповой индикации lamp/light display unit
~ лексического анализа scanner
~, литой slip-cast block
~ магнитной ленты magnetic tape unit/block/device
~ магнитных головок magnetic-head assembly
~ магнитных дисков magnetic-disk unit
~, малоберцовый *med.* peroneal pulley
~, матричный matrix block
~ машины machine block/unit
~, мембранный bellows assembly
~ микропрограммного управления microprogram(me) unit
~, модульный modular unit, module
~, накопительный *comp.* stack
~ настройки tuning unit, tuner; *auto.* adjusting/setting unit
~, натяжной *mech.* tension block
~, начальный begin block
~, неавтономный on-line unit/equipment
~ невосстанавливаемый throw-away unit
~ обработки/операторный processing block
~ обработки команд instruction evaluation unit
~, общий помеченный labelled common block
~ объединения assembly unit
~ объяснения explainer
~, одношкивный single block
~ опроса и запоминания данных data storage and interrogation unit
~ опроса контактов contact inquiry group
~, основной вычислительной системы central calculating/calculator unit
~ оценки результата test answer evaluator
~ памяти memory/store unit
~ памяти, дополнительный backing memory/store unit
~ памяти на сердечниках core memory block/stack
~ параметров parameters pack
~ переключателей gang switch
~ переключения change-over unit, switch unit
~, переходный adapter, coupler
~ питания power supply unit, power package; *(of material)* feeding block
~ плечевой кости *med.* trochlea/pulley of the humerus
~, подвижный running block
~ подкачки информации для печати print spooler
~ подстройки tuning unit
~ полиспаста tackle block
~ постоянных constant block

~ преобразования conversion unit
~, преобразующий converter unit
~, приводной *mech.* driving/operating/hauling block
~ приема receive group
~, приемно-усилительный receiver-amplifier unit
~ программного управления program(me) control unit
~, программный program(me) block; basic block
~ программы code block
~, пружинный *mech.* spring cartridge
~ развертки sweep/scan unit
~ раскраски изображения на экране screen painter
~, расширительный extension chassis
~, резервный standby/spare/duplicate/redundant unit/block
~ решающих элементов computing block
~ самоконтроля self-checking block
~ самоликвидации self-destruction unit
~, связанный (сигнальный) on-line (signal) block
~, сдвоенный *mech.* twin pulley block
~, селекторный strobe unit, selector unit
~ селекции импульсов pulse gating unit
~, сельсинный synchro unit
~ сервопривода *mech.* servo unit
~ синхронизации clock unit
~, скуловой *naut.* bilge block; side block
~ сменной ленты alternate tape drive/unit
~, сменный/съемный plug-in unit
~, согласующий conditioning unit
~ сообщений message block
~ сравнения comparison unit, comparator
~, стандартный basic block, module
~, суммирующий summing/adder block
~ счетов на магнитных картах magnetic account card unit
~, талевый tackle block
~ таранной кости *med.* pulley of the talus
~, трехшкивный treble block, trispast
~ умножения multiplication device
~ управления control block/unit
~ управления выборкой команд command control block
~ управления основной памятью main store control element, MSCE
~ управления очередями queue control block
~ управления памятью memory management unit
~ управления периферийной/внешней памятью peripheral store control element PSCE
~ управления событиями event control block
~ управления томом data carrier control block, volume control block
~ управления устройством unit control block
~ управления файлами заданий job file control block
~ ускорителей *rock.* boost cluster
~ фиксированной длины fixed-length block
~ формирования изображения imager
~, функциональный operational unit
~ хранения адресов микрокоманд microinstruction address holder
~, цепной *mech.* chain pulley
~ цилиндров *auto.* cylinder block
~ цифровой индикации digital display unit
~ шестерен *mech.* gear cluster
~, электронный electronic building block
~ электропитания electrical power unit
~, элементарный стандартный single building block

блокада blockade, block
~, вагосимпатическая (автономной нервной системы) autonomic/vagosympathetic block
~, вагусная vagus block
~, новокаиновая novocaine/procaine blockade
~, паранефральная paranephric block
~, поясничная lumbar block
~ сердца heart block
~, синоаурикулярная sino-auricular block

блок-аппарат block instrument
блокинг-генератор blocking oscillator
блок-диаграмма block-diagram; *top.* relief diagram
блокинг-эффект blockage effect
блокирование blocking; interlocking; *econ.* freeze, freezing
блокировать 1. block, blockade; 2. interlock; 3. *econ.* freeze
блокировка blocking, interlock(ing); block-system; locking; lockout, lock; lockup
~ движения вперед carriage interlock
~ движения каретки carriage detent
~ записи write protection
~, защитная protective interlocking
~ клавиатуры keyboard lockout
~ контактных щеток brush interlock
~ памяти file interlock
~ переноса *comp.* carry suppression
~ по времени time-out
~, предохранительная safety interlock
~ раздела программы program(me) skip interlock
~, частичная weak lockout

блок-контакт interlock contact; auxiliary contact
блок-контактор interlock contactor; auxiliary contactor
блок-механизм block(ing) mechanism; interlock
блокнот notebook, writing pad; scratchpad
~, настольный desk pad

блокпост *rail.* signal box; blockhouse
блокреле holding/blocking relay
блок-схема block-diagram/chart/scheme; block skeleton; *(of process)* flow diagram/chart; routing diagram
~ вычислительной машины computer block-

diagram/scheme
~, иерархическая multilevel flow chart
~ материальных потоков material flow diagram
~, обобщенная general flow-chart
~ потоков информации information flowdiagram
~ программы flow chart/diagram, operational chart, program(me) flow chart
~, функциональная/функционирования functional block-diagram
блок-счетчик *naut.* meter wheel, registering sheave
блокшив, блокшип *naut.* hulk, blockship
блок-эффект blocking effect; self-blocking
блочноориентированный block-oriented
блочок slug, lump
блочный block-structured, modular, unitized, block (-type), in unit form
блуждание *stat., math.* walk; *rad.* wandering
~, случайное *stat.* random walk
блюдце *bot., zool.* patella; *geol.* minor depression
~, летающее flying saucer
блюм *met.* bloom
блюминг *met.* blooming (mill), bloomer
~ дуо two-high blooming mill
~, реверсивный reversing blooming mill
бляшка patch, spot, plaque
~, атеросклеротическая atherosclerotic patch/plaque
~ курильщиков smoker's patch
~, мокнущая oozing plaque
бобина reel, spool; coil; bobbin; *aut.* ignition/spark coil
~ для магнитной ленты magnetic tape reel
~ для намотки wind-up spool
~ для разматывания winding-off spool
~, подающая *comp.* feed reel/spool
~, приемная take-up reel/spool
~ с лентой tape reel/spool
бобр *zool.* beaver *(Castor)*
бобслей bob-sleigh
Бог *rel.* God
Богоматерь, Богородица *rel.* the Virgin, Our Lady
Богоявление *eccl.* Epiphany
богхед boghead (coal)
бод *comp.* baud, bits per second, bps
бодмерея *law, nav.* bottomry; respondentia
боевая часть, боеголовка warhead
~ направленного действия non-isotropic warhead, directional/aimed warhead
~ ненаправленного действия isotropic warhead
~ с инфракрасной системой наведения infra-red guided warhead
~ с радиолокационной системой наведения radar-guided warhead
~ с теплопоглощающим покрытием heat-sink warhead
боевой *mil.* combat, operational
боезапас ammunition
боек 1. block; hammer head; 2. *mil.* firing/hammer pin, striker
~ ударника firing pin, striker, nipple
боеприпасы ammunition
боеспособность fighting capacity/efficiency, operational capability
бозон, Бозе-частица *phys.* boson, Bose particle
бой 1. *mil.* battle, combat, fighting, action, engagement; fight; 2. breakage, scrap
~, воздушный aerial/air battle; fighter combat
~, встречный head-on fight/engagement
~, морской naval action/encounter, fleet engagement
~, шлаковый crushed slag
бойлер hot-water heater
бойскаут boy scout
бок 1. profile/sheer plan; 2. side, flank; *min.* wall
бокал *anat.* calyx
~, глазной eyecup
боковой lateral, sidelong, side *(attr.),* flank *(attr.); fig.* secondary, marginal, accessory
~вая линия collateral relationship
бокс I box, cell; bay
бокс II *leath.* box calf
боксит *miner.* bauxite
болванка *met.* blank, billet; ingot, pig
болевой *med.* algesic, painful
болезнетворный pathogenic
болезнь disease, illness, sickness; *med.* morbus, malum
~, Альцгеймера Alzheimer's disease
~, базедова Basedow's disease, exophthalmic goiter
~, бронхоэктатическая bronchiectatic disease, bronchiectasis
~, Бюргера Buerger's disease, thrombangiitis obliterans
~, воздушная air sickness
~, высотная high-altitude sickness, hypobaropathy
~, гипертоническая hypertensive disease, hypertension
~, гипотоническая hypotension
~, горная mountain sickness, hypobaropathy
~ Дауна mongolism
~, декомпрессионная = кессонная
~, желчнокаменная cholelithiasis
~, изнурительная consumptive/wasting/debilitating disease
~, кессонная caisson disease, compressed air disease/illness, decompression illness/sickness, aeremia, the bends, dysbarism
~, кистозная cystic disease/fibrosis
~, кишечнокаменная enterolithiasis
~, кожная skin/cutaneous disease
~, коронарная coronary disease, ischemic heart disease
~ крови blood disease, h(a)emopathy
~, лучевая radiation sickness/morbidity

~, **морская** seasickness
~, **мочекаменная** urolithiasis
~, **ожоговая** burn disease
~, **Паркинсона** Parkinson's disease
~ **печени** liver disease
~, **почечнокаменная** nephrolithiasis
~, **сахарная** diabetes mellitus
~, **серповидно-клеточная** sickle cell disease
~, **синяя/синюшная** morbus caeruleus
~, **слюннокаменная** sialolithiasis
~, **сонная** sleeping sickness, trypanosomiasis
~, **сывороточная** serum sickness
~ **Ходжкина** Hodgkin's disease, lymphogranulomatosis
~, **цейлонская** Ceylon sickness, beri-beri
~, **язвенная** peptic ulcer
болеутоляющие *med. pl.* analgesics
болеутоляющий *adj.* analgetic, analgic
боливар bolivar
болид *astr.* fireball, bolide
болограф bolograph, recording bolometer
болометр bolometer, barreter
болт bolt, pin
~, **анкерный** foundation/anchor bolt, stay bolt
~ **без нарезки** blank bolt
~, **вертлюжный** swivel bolt
~, **вилочный** shackle bolt
~, **двухсрезный** double-shear bolt
~, **зажимный** clamp bolt
~, **конический** taper/conical bolt
~, **контрящий** safety bolt
~, **крейцкопфный** crosshead/gudgeon pin
~, **крепежный/крепления** fastening/attachment bolt
~, **крышечный** bonnet/cap bolt
~, **монтажный** assembly/erection bolt
~, **натяжной** adjuster bolt
~, **разрывной** explosive bolt
~, **распорный** distance bolt; stay bolt
~, **расширительный** expansion bolt
~, **регулировочный** adjuster/adjusting bolt
~ **с двойной нарезкой** double-screw bolt
~ **с квадратной головкой** square-head bolt
~ **с отверстием под шплинт** clevis bolt
~ **с потайной головкой** countersunk-head bolt
~ **с ушком** eye bolt
~ **с цилиндрической головкой** cheese-/fillister-head bolt
~ **с шаровой головкой** ball-headed bolt
~ **с шестигранной головкой** hexagon-head bolt
~, **сквозной** through bolt
~, **стяжной** through/coupling/tie bolt
~, **транспортный** shipping/transport bolt
~, **упорный** stop bolt
~, **установочный** adjusting bolt
~, **шарнирный** swing bolt
~, **шплинтируемый** clevis bolt
~, **шпоночный** key bolt
болтанка air turbulence, turbulent/rough/bumpy weather, bumpy air conditions
болт-барашек butterfly bolt
болтик stud
боль pain, ache
~, **головная** headache, cephalalgia
~, **жгучая** burning pain, causalgia
~, **зубная** toothache, dentalgia, odontalgia
~, **мышечная** muscle pain/ache, myalgia
~, **родовая** pain of delivery, labour pain
~, **схваткообразная** spasmodic/cramping/colicy/paroxysmal pain
больница hospital
~, **венерологическая** VD hospital
~, **общесоматическая** general hospital
~, **психиатрическая** mental hospital, lunatic asylum
больной patient; sick; *med.* case
~, **амбулаторный** out-patient
~, **носилочный** litter patient/case
~, **стационарный** in-patient, hospital/indoor patient
большинство голосов *law* majority vote
бомба bomb
~, **агитационная** leaflet bomb
~, **аэрозольная** aerosol bomb
~, **атомная** atomic bomb
~, **водородная** hydrogen bomb
~, **глубинная** depth charge/bomb
~, **дейтериевая** deuterium bomb
~, **дейтериево-тритиевая** deuterium-tritium bomb
~, **зажигательная** incendiary bomb
~ **замедленного действия** delayed-action/-time bomb; time bomb
~, **кассетная** cluster bomb
~, **кобальтовая** cobalt bomb
~, **крылатая** winged bomb
~, **напалмовая** napalm bomb
~, **нейтронная** neutron bomb
~, **неядерная/обычная** conventional bomb
~, **осколочная** fragmentation bomb
~, **осколочно-фугасная** general-purpose bomb
~, **пирогелиевая** pyrohelium bomb
~, **планирующая** gliding/glide bomb
~, **плутониевая** fission bomb
~, **пристрелочная** sighter bomb
~, **противолодочная** anti-submarine bomb
~ **с урановой оболочкой, термоядерная** fission-fusion-fission bomb
~, **самонаводящаяся** homing bomb, self-guided bomb
~, **светящая** flare bomb
~, **термитная (зажигательная)** thermite bomb
~, **термоядерная** thermonuclear (fusion) bomb
~, **управляемая** guided/controlled bomb
~, **урановая** uranium bomb
~, **фотографическая** photoflash bomb
~, **фугасная** high-explosive bomb, blast/demolition bomb

~, **ядерная** nuclear bomb
бомбаж bloat(ing), puffer, buckling
бомбарда *hist.* bombard
бомбардирование bombardment, bombarding
~ **альфачастицами** alpha bombardment
~ **ионная/нейтронами/протонами/электронами** ion/neutron/proton/electron bombardment
бомбардировать bombard; bomb
бомбардировка bombardment; bombing; *see also* **бомбардирование**
~, **круглосуточная** round-the-clock bombing
~, **массированная** mass bombing
~, **челночная** shuttle bombing
бомбардировочный bombing; bombardment
бомбардировщик bomber
~, **ближний** short-range bomber
~, **дальний** long-range bomber
~, **палубный** ship-based bomber, carrier bomber
~, **пикирующий** dive bomber
~, **реактивный** jet bomber
бомбардировщик-разведчик reconnaissance bomber
бомбардировщик-ракетоносец rocket bomber, missile-carrying bomber
бомбардировщик-торпедоносец torpedo bomber
бомбардировщик-штурмовик strike/attack bomber
бомбодержатель bomb rack/carrier
бомбометание bombing
бомбонепробиваемый bomb-proof
бомбоубежище air-raid shelter, bombproof shelter, bombshelter
бон *naut., timb.* boom
~ **заграждения** defence boom
бонитер *agr.* culler
бонитировка *agr.* appraisal, valuation, graduation
бонус *econ.* superdividend
бор I *naut.* tidal bore, mascaret
бор II *chem.* boron, B
~, **фтористый** boron fluoride
бор III drill
~, **зубной колесовидный** wheel-shaped dental drill
~, **зубной шаровидный** spherical dental drill
бор IV *bot.* millet grass (*Milium*)
бор V *bot.* pine/coniferous forest
бора bora *(wind)*
борат *chem.* borate
боргес *print.* bourgeois
бордеро 1. invoice, account; 2. list of securities forwarded
бордюр curb, borderstone; border
борид boride
~ **кальция** calcium boride
борирование *met.* borating, boronizing
борировать borate, boronize
бормашина (dentist's) drill, drilling machine
борн *el.* terminal
борнеол *chem.* borneol, bornyl alcohol
борнокислый *chem.* tetraborate
боров flue, chimney intake
~, **сборный** collecting flue
бороводород boron hydride, borane
борогидрид borohydride
бородавка *med.* wart, verruca
~, **дегтярная** pitch wart
~, **остроконечная** pointed wart
бородавчатый warty, verrucose, verrucous
борозда 1. *agr.* furrow; 2. *med.* sulcus (*pl.* sulci); groove
~, **запирательная** obturator groove
~, **запястая** carpal groove/canal
~, **крыло-небная** pterygopalatine groove
~ **мозолистого тела** callosal sulcus
~ **морского конька** hippocampal fissure
~, **обонятельная** olfactory sulcus
~, **подбугорная** sulcus of Monroe
~, **разъемная** open furrow
~, **свальная** double furrow
~, **слезная** lacrimal groove
бороздка 1. groove, furrow, incision; 2. *med.* sulcus (*pl.* sulci)
бороздодел *agr.* furrow opener, furrower
бороздомер *agr.* furrow depth indicator
борона *agr.* harrow
~, **болотная** bog harrow
~, **дисковая** disk harrow
~, **зубовая** spike-tooth harrow
~, **лапчатая** chisel-type harrow
~, **шарнирная** flexible harrow
борона-волокуша toothless drag harrow
бороновать harrow
бороскоп borescope
борт 1. *naut.* side, board; 2. border, edge, rim; collar, flange
~, **левый** *naut.* port side
~, **наветренный** weather side
~, **надводный** freeboard
~, **откидной** collapsible/drop side; flap
~, **откидной задний** *auto* tailgate
~, **подветренный** lee board/side
~ **покрышки** tyre bead
~, **правый** starboard
бортжурнал log book; *av.* flight log
~, **вести** keep a log
бортинженер flight engineer
бортмеханик flight mechanic
бортовой *av.* airborne, on-board; *space* spaceborne; *naut.* shipborne, board *(attr.)*; side *(attr.)*; *auto.* vehicle-borne
бортпроводник air steward, cabin attendant
бортпроводница air hostess/stewardess
бортрадист flight radio operator
бортсестра flight nurse
борттехник flight technician
борштанга boring bar
борьба struggle, fight, combat; *fig.* countermeasures, control; *sport* wrestling

~, **вольная** free-style wrestling
~ **за выживание** struggle for survival
~ **за повышение заработной платы, острая** wage war
~ **за режим экономии** economy drive
~, **классическая** *sport* Greco-Roman wrestling
~, **классовая** *polit.* class struggle
~ **противоположностей** conflict of opposites
~ **с вредителями** *agr.* pest control
~ **с коррозией** corrosion control
~ **с пожаром** fire fighting/control
~ **с сорняками** weed control
~ **с шумом** noise control/abatement
босанова *mus.* bossanova
бот *naut.* boat; skiff
~, **водолазный** diving boat/tender
~, **лоцманский** pilot boat
~, **промерный** sounding boat
ботаника botany
ботдек *naut.* boat deck
ботулизм *med.* botulism, allantiasis
боулинг *sport* bowling
боцман *naut.* boatswain
~, **главный** ship's boatswain
бочка 1. barrel, cask; keg; 2. *naut.* buoy; 3. *av.* roll
~, **групповая** *av.* formation roll
~ **на мачте** *naut.* crow's nest
~, **причальная/швартовная** mooring buoy
~, **рейдовая** anchorage buoy
~, **сухотарная** slack barrel
~, **штопорная** flick roll; *Am.* snap roll
~, **элеронная** *av.* aileron roll
боярин *hist.* boyar(d)
брага *naut.* span
брадидиастола *med.* bradydiastole
брадикардия *med.* bradycardia
~, **вагусная** vagal bradycardia
~, **желудочковая** ventricular bradycardia
~, **идиопатическая** essential/idiopathic bradycardia
~, **узловая** nodal bradycardia
брадикинезия *med.* bradykinesia, bradykinesis
брадикинин bradikinin
брадилалия bradylalia
брадипноэ bradypnea
брадителия bradytely
брайтсток bright stock
брак I rejects, spoilage, waste, defective products
брак II marriage, matrimony, wedlock
~, **гражданский** civil (common law) marriage; de facto union, free union
~, **завершенный** consumated marriage
~, **морганатический** left-handed marriage, morganatic marriage
~, **неравный** misalliance
~, **пробный** trial marriage
~, **расторгнутый** broken/dissolved/disrupted marriage
~, **смешанный** mixed marriage; intermarriage
бракета *naut.* bracket (plate)
браковать *econ.* condemn, reject
браковка condemnation, rejection
браковщик quality checker/inspector
бракоспособный marriageable
брам-гордень *naut.* topgallant mastrope
брамсель topgallant sail
брандвахта guard ship, examination vessel; fire watch
брандер *hist.* fire-ship/-boat
брандмауэр fire-wall, fireproof wall
брандспойт 1. fire engine/pump; 2. fire nozzle
~, **моечный** wash gun
брас *naut.* brace
брасик *naut.* tiller becket
брас-шкентель *naut.* brace pendant
брат 1. brother; 2. *eccl.* lay brother, Friar, Brother
~, **двоюродный** (first) cousin
~, **единокровный** blood brother
~, **единоутробный** brother uterine
~, **молочный** foster brother
брахидактилия *med.* brachydactyly, brachydactylia
брахистохрона *math.* brachistochrone
брахицефалия *med.* brachycephaly, brachycephalism
брахман Brahmin
брахманизм Brahmanism
брахманский Brahmanic, Brahminic
брачность *law* nuptuality; marriage/matrimony rate
брашпиль *naut.* (anchor) windlass
бревно log; *pl.* lumber, timber
бревнокатка log roller/elevator
бревномер log scale
бревноспуск log chute/slide, chute raft
бревнотаска log hauler/conveyer/conveyor/haulway
бред *med.* delirium; delusion
~, **алкогольный** delirium tremens/alcoholicum
~ **величия** megalomania
~ **отношения** delusion of reference
~ **отрицания** delusion of negation, nihilistic delusion
~, **подражательный** ideo-onomatic delirium
~ **сумасшедшего** ravings of a madman
бредить *med.* delirate
брезент canvas, tarpaulin
брейд-вымпел broad pendant, burgee
брекватер breakwater, mole
брекер breaker (strip)
брекчия *miner.* breccia
бремза *text.* swivel tension
бремсберг *min.* gravity road/plane, brake incline, inclined drift/plane
бремя burden, load
~ **доказывания** *law* burden of proof/proving/evidence, bonus of proof/proving
Бреттонвудсское Соглашение *hist., econ.* Bretton Woods

брештук *naut.* breasthook, breast plate/knee
брешь breach, break-through, gap
бри *cul.* Brie cheese
бриг *naut.* brig
бригада 1. team, group, crew, gang, brigade; personnel; 2. *mil.* brigade; *naut.* subdivision
~, аварийная emergency repair team/crew
~, комплексная integrated crew/team, multiple-skill crew
~, маневровая shunting/switching crew
~, подсменная swing shift
~, рабочая working team
~, ремонтная repair/maintenance crew/team
~, ударная *obs.* shock brigade
бригадир 1. team/crew leader; foreman; 2. *mil.* brigadier
бригантина *naut.* brigantine
бридель *naut.* (mooring) bridle
~, буксирный towing bridle
бридер *nucl.* breeder (reactor)
бридждек *naut.* bridge deck
бриз breeze
~, береговой/ночной land breeze
~, морской/дневной sea breeze
бризантность brisance; shattering effect
бризантный brisant, highly explosive; shattering
брикет 1. briquette; brick; 2. *agr.* wafer, pellet
брикетировать briquette
бриллиант 1. (cut) diamond; 2. *print.* brilliant
бриология muscology, bryology
бриония *bot.* bryony *(Bryonia)*
бриош *cul.* brioche
Британская ассоциация для выработки технических стандартов British Engineering Standards Association
Британский институт научной организации управления British Institute of Management
бродильный fermentative
бродить ferment
брожение fermentation
~, лимоннокислое citric-acid fermentation
~, маслянокислое butyric fermentation
~, молочнокислое lactic-acid fermentation
~, направляемое controlled fermentation
~, окислительное oxidative fermentation
~, спиртовое alcoholic fermentation
~, уксуснокислое acetic fermentation
брокер *econ.* broker
~ по инвестиционным ценным бумагам investment broker
~ фондовой биржи stock broker, stock (exchange) broker
брокер-комиссионер commission broker
брокерские операции *econ.* brokerage
брокколь *bot.* broccoli *(Brassica oleracea botrytis)*
бром *chem.* bromine, Br
бромат bromate
~ натрия sodium bromate
броматометрия bromatometry
бромбензилцианид brombenzylcyanide
бромбутилкаучук brominated butyl rubber
бромид *chem.* bromide
бромизовал = бромурал
бромирование bromination
бромит bromite
бромкамфора camphor bromide
бромный bromine
бромопрен bromoprene
бромурал bromuralum, bromisoval
бронебойность armo(u)r-piercing capacity/capability
бронебойный armo(u)r-piercing
броневой armo(u)red, armour *(attr.)*
бронезащита armo(u)r protection
бронекатер *naut.* armo(u)red boat
бронеколпак *mil.* armo(u)red turret
бронелента *tech.* armouring plate
броненосец 1. *naut.* ironclad; battleship; 2. *zool.* armadillo *(Dasypus)*
броненосный armo(u)red, armo(u)r-plated, ironclad
бронеплита armo(u)r plate
бронепоезд armo(u)red train
бронеспинка back armo(u)r plate, armo(u)r backrest
бронестекло armo(u)red/bulletproof glass
бронетанковый *mil.* armo(u)red
бронетранспортер *mil.* armo(u)red personnel carrier
~, плавающий amphibious armoured personnel carrier
бронза bronze
~, адмиралтейская Admiralty bronze
~, алюминиевая aluminium bronze
~, безоловянная tinless bronze
~, бериллиевая beryllium bronze
~, колокольная bell metal
~, кремненикелевая nickel-silicon bronze
~, кремнистая silicon bronze
~, оловянная/оловянистая tin bronze
~, оловянносвинцовистая tin-leaded bronze
~, оловяннофосфористая tin-phosphor bronze
~, пушечная gun metal
~, свинцовистая leaded bronze
~, сурьмянистая antimonous bronze
~, фосфористая phosphor bronze
~, хромистая chromium bronze
бронзирование bronzing; bronze plating
бронзовый bronze *(attr.)*; bronzed
~вая болезнь Addison's disease
~ век the Bronze Age
бронирование 1. armo(u)ring, armo(u)r-plating; 2. (seat) reservation, booking
бронированный 1. armo(u)red; ironclad, steelclad, armo(u)r-plated
бронировать 1. armo(u)r; sheathe; 2. book, reserve; assure
бронтограф *meteor.* brontograph
бронтозавр brontosaur *(Brontosaurus)*
бронтофобия brontophobia

бронх *anat.* bronchus
~, главный primary bronchus
~, долевой lobar bronchus
~, надартериальный epiarterial bronchus
~, подартериальный hyparterial bronchus
бронхиальный bronchial
бронхиола bronchiole
бронхиолит bronchiolitis
~, гриппозный influenzal/grippal bronchiolitis
бронхиома bronchioma
бронхит bronchitis
~, гнилостный fetid bronchitis
~, гнойный purulent bronchitis
~, застойный congestive bronchitis
~, пленочный membranous bronchitis
~, туберкулезный tuberculous bronchitis
бронхоаденит bronchoadenitis
бронхоальвеолит bronchoalveolitis; bronchopneumonia
бронхоальвеолярный bronchoalveolar
бронховезикулярный bronchovesicular
бронхография bronchography
бронхолегочный bronchopulmonary
бронхомикоз bronchomycosis
бронхопластика bronchoplasty
бронхопневмония bronchial pneumonia, bronchopneumonia
бронхорасширитель bronchodilator
бронхорея bronchorrhea
бронхоскоп bronchoscope
бронхоскопия bronchoscopy
бронхоспазм bronchospasm, bronchoconstriction
бронхостеноз bronchostenosis
бронхосуживатель bronchoconstrictor
бронхотом bronchotome
бронхотомия bronchotomy
бронхоудерживатель bronchial forceps
бронхофония bronchophony
~, шепотная whispered bronchophony
бронхоэзофагоскоп bronchoesophagoscope
бронхоэктаз bronchiectasis, bronchiectasia
~, мешотчатый saccular bronchiectasis
бронхоэктатический bronchiectatic
броня armo(u)r
~, бортовая side armo(u)r
~, защитная protective/defensive armo(u)r
~, лобовая front/frontal armo(u)r
~, прозрачная transparent armo(u)r, bullet-proof armo(u)r
~, противопульная bulletproof armo(u)r
~, толстая heavy armo(u)r
бросок surge, dash, jump; inrush
~ обратного напряжения *el.* kickback voltage
~ тока current inrush
броуновское движение Brownian movement
брошюрование, брошюровка stitching
брудер *agr.* brooder (house)
бруксомания *med.* bruxomania
брунизем brunizem (soil)
брус 1. (squared) beam, timber; bar; girder; rod; 2. whetstone
~, бордюрный curb girder
~, грабельный rake bar
~, килевой keel piece
~, коньковый *constr.* ridge pole/piece
~, кормовой *naut.* stern post
~, направляющий slide bar; guide rail
~, обносный sponson beam
~, отбойный fender skid; bumper bar/beam
~, переводный *constr.* switch tie
~, привальный side fender, guard rail, fender guard
~, скуловой *naut.* chine bar/beam
~, сцепной draw bar
~, тентовый awning boom/rafter
~, хвостовой stern post
~, цилиндровый roller beam
бруски для закрепления груза bracing
брусок bar, rod; slug
~, абразивный honing stick, hone; whetstone
~, деревянный stud; block
~, диаметральный *naut.* king plank
~, измерительный calibration bar
~, магнитный bar magnet, magnetic bar
~ мыла bar of soap
~ с надрезом notched bar
~, топливный fuel slug
~, точильный whetstone
~, шлифовальный honing stick, hone
брус-помост scaffold plank
бруствер *mil.* parapet; breastwork
брусчатка paving block/stone
брутто gross (weight)
~ вознаграждение gross/total remuneration
брутто-вес gross weight
брутто-прибыль gross profit
брутто-тонна gross (register) ton
брутто-тоннаж gross(register) tonnage
брутто-формула *chem.* empirical formula
бруцеллез *med.* brucellosis
брыжеечный *anat.* mesenterial, mesenteric
брыжейка *anat.* mesentery
~ желудка mesogastrium
~ матки mesometrium
~ маточной трубы mesosalpinx
~ поперечноободочной кишки transverse mesentery
~ прямой кишки mesorectum
~ слепой кишки mesocecum
~ сухожилия mesotendon
~ толстой кишки mesocolon
~ яичника mesovarium
брызги *pl.* splashes, spray, spatter
~ металла metal splashes/spray
брызговик mudguard; splash apron/board/guard
~, резиновый *auto.* mud flap
брызгозащитный antispray; splashproof
брызгозащищенный, брызгостойкий splashproof, sprayproof

брызгонепроницаемый spraytight, splashproof
брызгообразование spraying
брызгоотбойник, брызгоотражатель spray deflector/guard
брызгоуловитель spray trap/separator
брюмер *hist.* Brumaire
брюсселька *bot.* Brussels sprouts *(Brassica oleracea gemmifera)*
брюстерит *miner.* brewsterite
брюховина *zool.* pauch; rumen
брюшина *anat.* peritoneum
брюшинный peritoneal
брюшко *anat.* belly, abdomen, venter
~, **мышечное** myogaster
~, **стебельчатое** petiolated abdomen
~ **хелицеровых** opisthosoma
~ **членистоногих** urosome
брюшно-влагалищный abdomino-vaginal
брюшно-генитальный abdominogenital
брюшно-грудной abdominothoracic
брюшной abdominal, ventral, enteric
~ **тиф** typhoid fever
~ **шов** ventral suture
~**ная полость** abdominal cavity
~**ная чешуйка** *bot.* amphigastrium
брюшно-маточный abdominouteral
брюшно-мошоночный abdominoscrotal
брюшно-пузырный abdominocystic, abdominovesical
брюшно-тифозный typhoid
~ **больной** typhoid patient
бубон *med.* bubo
~, **безболезненный** indolent bubo
~, **паховый** inguinal bubo
~, **туляремный** tularemic bubo
~, **чумной** malignant bubo
бубонный bubonic
~**ная чума** bubonic plague
бугель 1. clip, stirrup, clamp; *naut.* (ferrule) hoop; bandloop; 2. *el.* bow collector
~, **мачтовый** mast/hound band, mast hoop
~ **на стреле** boom band
~, **подъемный** stirrup, shipping ring, parrel
~, **предохранительный** safety stirrup
~, **пусковой** *rock.* launching lug
~ **сваи** pile ferrule/band
~ **эксцентрика** eccentric clip
бугор 1. knoll, hillock, mound, hammock; bump, lump; 2. *anat.* tuber (*pl.* tubera), protuberance, tubercle, tuberosity
~ **верхней челюсти** maxillary tuberosity
~, **лобный** frontal tuber/eminence
~, **пяточный** calcanean tuber
~, **седалищный** tuber of the ischium
~, **серый** ashen/gray tuber
~, **скуловой** malar eminence
~, **теменной** parietal tuber/eminence
~ **червя** tuber of the vermis
бугорок *anat., bot.* tubercle; knob; protuberance; boil; wart
~ **атланта, задний/передний** posterior/anterior tubercle of the atlas
~, **глоточный** pharyngeal tubercle
~, **губной** labial tubercle
~, **запирательный** obturator tubercle
~, **конический** conoid tubercle
~, **краевой** marginal tubercle
~, **лобковый** pubic tubercle
~, **надгортанный** epiglottic tubercle
~, **подсуставной** infraglenoid tubercle
~, **рожковидный** corniculate tubercle
~, **сонный** carotid tubercle
~ **турецкого седла** tubercle of the saddle
~, **яремный** jugular tubercle
бугристость tuberosity; eminence
~, **дельтовидная** deltoid eminence
~, **крестовая** sacral tuberosity
~, **крыловидная** pterygoid tuberosity
~ **локтевой кости** tuberosity of the ulna
Будда Buddha
буддизм Buddhism
буддийский Buddhist(ic)
буддист Buddhist
будка box, booth, cabin, cab, cubicle, stall
~, **кабельная** cable hut
~, **караульная** sentry box
~ **машиниста (локомотива)** (driver's) cab
~ **привратника** gatehouse
~, **проекционная** projection booth
~, **проходная** control post/boot
~, **собачья** doghouse
~, **суфлерская** prompt box
~, **телефонная** public-call box; *Am.* telephone booth
~, **трансформаторная** transformer vault
будуар boudoir
буек *naut.* anchor/life buoy
буер ice yacht, iceboat
буж *med.* bougie; probe
~, **конический** tapered bougie
~, **пищеводный** (o)esophageal bougie
бужирование *med.* bougieurage
буж-перфоратор bougie-perforator
буж-щуп bougie probe
бузина *bot.* elder (*Sambucus);* elderberry
буй buoy; float
~, **аварийно-спасательный** message/messenger buoy
~, **бочкообразный** barrel/cask buoy
~, **волноизмерительный** wave-measuring buoy
~, **волчкообразный** peg-top-buoy
~, **гидроакустический** sonobuoy, sonic (underwater) buoy
~, **гудящий** whistle buoy
~, **дымовой/дымный змоке** buoy, smoke float
~, **звучащий** sound(ing) buoy
~, **комбинированный** combination buoy
~ **левой стороны** port hand buoy

~, **маркерный** watch/marker buoy
~ **на месте затонувшего судна** wreck buoy
~ **направленного действия (радиогидроакустический)** directional sonobuoy
~, **несветящий** unlighted/blind buoy
~ **ограждающий** beacon buoy
~ **опорной сети на воде** survey buoy
~, **осевой** mid-channel buoy, fairway buoy
~, **поворотный** pivot/turning buoy
~ **правой стороны** starboard hand buoy
~, **промерный** surveying buoy
~, **радиогидроакустический/радиоакустический** sonoradio buoy, radio sonobuoy
~ **с гонгом** gong buoy
~ **с колоколам** bell buoy
~ **с ревуном** horn buoy
~, **светящий** light buoy
~, **сигнальный** signal/indicator buoy
~ **со свистком** whistle buoy
~, **спасательный** emergency/life buoy
~, **столбовидный** beacon/pillar buoy
~, **тральный** sweeping/dan buoy
~, **туманный** fog/position buoy
~, **фарватерный** channel buoy
~ **якорной стоянки** anchorage buoy
буйвол *zool.* buffalo (*Bubalus)*
буй-волнограф wave-recording buoy
буй-отметчик mark buoy
буйреп buoy rope/line
бук *bot.* beech (*Fagus)*
буква letter, character
~ **кода** code letter
~, **прописная** capital [upper-case] letter
~, **строчная** small [lower-case] letter
буквенно-звуковой alphabetic-phonetic; phonographic
буквенно-цифровой alphanumeric(al), alphameric, alphabetic-numeric
буквопечатающее устройство type printer
букет сфер union of spheres
букетировать *agr.* block, single
букетировка *agr.* (cross-)blocking, checkrowing, singling
букетировщик *agr.* (rowcrop) thinner, blocker
буккер *agr.* drill plough
букмекер bookie, bookmaker
букса hub, bush, sleeve; axle/journal box
~, **осевая** nave box, axle bearing
буксир 1. tow(ing) cable, towline, towrope, towing line/rope; 2. tug, tugboat, towboat
~, **базовый** harbour tug
~, **десантный** landing tug
~ **для буксировки барж** barge-handling tug
~ **для толкания и тяги** push-and-tow tug
~, **короткий** stope
~, **ледокольный** icebreaking tug
~, **морской** seagoing/oceanic/fleet tug, deep-sea/-water tug
~, **океанский** ocean tug, ocean-going tug, deep-water tug
~, **пожарный** fire-fighting tug
~, **портовый** harbour/dock/depot/yard tug
~, **рейдовый** harbour/estuary tug
~, **спасательный** rescue/salvage tug
~, **эскадренный морской** fleet ocean tug
буксир-кантовщик berthing/docking tug
буксирный tow *(attr.)*
буксировать tow, have in tow, haul
~ **борт о борт** tow alongside
~ **за кормой** tow aft
буксировка, буксирование towage; towing, haulage, hauling, tugging, tow, haul
~ **тросом** cable/line towing
буксировочный towing
буксировщик towing vehicle; tug
буксир-толкач push boat, pusher tug
буксируемый towable
булат Damascus steel, damascene
булеан Boolean, class of all subsets, potency class
булев(ский) Boolean, logical
були *pl.* bulges, blisters
булимия *med.* bulimia, food craving
булинь *naut.* bowline
буллезный bullous
булыжник cobblestone
бульб bulb
бульбарный bulbar
бульбкил *naut.* bulb keel
бульбо-кавернозный bulbocavernous
бульбо-спинальный bulbospinal
бульбо-уретральный bulbourethral
бульб-уголок bulb angle
бульдозер bulldozer/driver
бульдозерист bulldozer operator
бульон broth
~, **мясо-пептонный** beef-extract broth; meat infusion broth
~, **пептонный** peptone broth
~ **с углекислым кальцием** calcium-carbonate broth
~, **яичный** egg infusion broth
бум *econ.* boom
~, **биржевой** stock-market boom
~ **с последующей депрессией** *Am.* boom-and-bust
бумага 1. paper; 2. document; *pl.* official papers; securities
~, **акварельная** water-colour paper, aquarel
~, **александрийская** royal paper
~, **бактерицидная** antiseptic/disinfectant paper
~, **бандерольная** band stock
~, **бархатная** velour paper
~, **белая переводная** transfer paper
~, **бескопирочная** carbonless/action paper
~ **в клетку** squared paper
~ **в линейку** lined paper
~ **в форме непрерывной ленты** continuous paper
~, **веленевая** vellum paper

~, вероятностная *math.* probability paper
~ восковка stencil paper
~, вощеная wax(ed) paper
~, высококлееная hard-sized paper
~, газетная newsprint
~, гербовая *econ.* stamped paper
~, гигиеническая toilet paper
~, грунтованная pre-coated paper
~, гуммированная gummed paper
~, дважды логарифмическая log-log paper
~, диаграммная chart paper
~, диазотипная diazo-type paper
~ для глубокой печати copper-printing paper
~ для множительных аппаратов duplication paper
~ для мульчирования mulch paper
~ для обоев wall-paper
~ для печатающих устройств stationery
~, кабельная cable-insulating paper
~, клееная sized paper
~, клетчатая plotting/scale paper
~, конденсаторная *el.* capacitor paper
~, копировальная carbon/copying/copy paper
~, крепированная crepe paper
~, лакмусовая litmus paper
~, логарифмическая log paper
~, манильская manil(l)a paper
~, машинописная copying paper
~, мелованная chalk-coated paper
~, мешочная bag/sack paper
~, миллиметровая plotting/scale paper
~, наждачная emery/abrasive paper
~, нормальная вероятностная *math.* normal probability paper, Gaussian paper
~, оберточная wrapping/brown paper
~, обложечная cover paper
~, полулогарифмическая semilog paper
~, полюсная pole-finding paper
~, почтовая letter/post paper
~ проводящая *el.* conducting paper
~, промасленная oil(ed) paper
~, промокательная/пропускная blotting paper
~, реактивная reagent/test paper
~, рисовальная drawing paper
~, ротаторная mimeo paper
~, рулонная continuous stationery, web paper
~ с ведущими отверстиями pin-feed paper
~, светокопировальная heliographic paper
~, светочувствительная photosensitive/sensitized paper, light-sensitive paper
~, свинцовая lead acetate paper
~, сфальцованная гармошкой accordion-/concertino-folded paper
~, тонкая оберточная shop paper
~, упаковочная packaging/casing paper
~, фильтровальная filter paper
~, фальцованная fan-fold paper
~, хлопчатая cotton
~, чертежная (прозрачная) tracing paper
~, шлифовальная abrasive paper
~, электростатическая high-resistance paper
~, электрочувствительная electrosensitive paper
бумага-основа base/body/backing paper
бумаги *pl.* official papers; securities
~, ценные *econ.* securities; stocks
~, процентные ценные *econ.* interest bearing securities
~, ценные государственные public securities
бумагодержатель 1. paper clip; 2. *fin.* holder of securities, bondholder
бумагопрядение cotton spinning
бумагопрядильный cotton-spinning
~ная фабрика cotton mill
бумажный 1. paper *(attr.)*; 2. cotton *(attr.)*, made of cotton
~ная фабрика paper mill
бумеранг boomerang
буна Buna *(synthetic rubber)*
бунамидин bunamidine
бунгало bungalow
бунд *hist.* Bund
бундесвер Bundeswehr
бундестаг Bundestag
бундовец *hist.* Bundist
бункер 1. hopper, bunker; bin; silo; 2. underground shelter, air-raid shelter
~, верхний overhead bin/silo
~, весовой weighing bunker/hopper
~, делительный batcher
~, дозирущий metering/batching bin/hopper
~, завалочный filling hopper
~, загрузочный loading/feed hopper, feed bin
~, опрокидывающийся tipping bin
~, перегрузочный transfer bin/hopper
~, передвижной travelling bin/hopper
~, промежуточный transfer/holding hopper; *met.* surge bin
~, разгрузочный discharge/unloading hopper
~, расходный service bunker/bin
бункероваться coal, take on coal/oil
бункеровка, бункерование bunkerage, bunkering; coaling
бункеровщик *naut.* fueler
бункер-смеситель blending hopper/bin
бур I drill, auger, bore; bit
~, алмазный diamond drill
~, дробовой shot drill
~, канатный cable drill
~, колонковый core drill
~, корончатый crown drill
~, спиральный worm auger, auger drill
~, ударный percussion/hammer drill
~, шнековый = спиральный
бур II boer
бура *chem.* borax, sodium tetraborate
бурав = бур
~ Пресслера increment borer
буран blizzard, snowstorm

бурат trommel; reel, bolter
буревестник *zool.* petrel, shearwater *(Puffinus)*
бурение drilling; boring
~, вращательно-ударное rotary-percussion drilling
~, дробовое shot drilling
~, керновое core drilling
~, мокрое water-flush drilling, wet drilling
~, термическое flame-jet drilling
~, турбинное turbodrilling
~, шарошечное roller-bit drilling
~, шнековое auger drilling
буржуа bourgeois
буржуазия bourgeoisie; *hist.* third estate
~, крупная big/upper bourgeoisie
~, мелкая petty bourgeoisie; lower middle class
~, средняя middle class/bourgeoisie
бурильщик driller, drillman
буримость drillability
бурить drill
бурмистр *hist.* bailiff
буровая derick
буровой boring, drilling
~вая вышка derrick
~вая скважина borehole, well
бурoдержатель drill chuck
бурсит *med.* bursitis
~, известковый calcareous bursitis
~ коленного сустава knee joint bursitis
~ лопатки scapular bursitis
~, лучеплечевой radiohumeral bursitis, tennis elbow
~, надключичный supraclavicular bursitis
~ шахтеров miner's elbow
буртик *(of shaft)* collar, fillet, bead, shoulder; *(of boat)* fender bar, belting, (sheer) moulding
~, маслосбрасывающий oil collar
буртование *agr.* storing in clamps, clamping
бурун breaker, surf; feather, broken water
~, волновой surging breaker
~, носовой feather
~ образованный гребным винтом churning
~, поднимать cut a feather
бурундук *naut.* after-guy
буря storm; gale; tempest
~ и натиск *hist.* Sturm und Drang
буссоль surveyor's/azimuth circle/compass; *mil.* aiming circle, director
~ для определения магнитного наклонения dip compass/circle; declinometer
~, зеркальная mirror compass
~, мензульная box compass; plane-table compass
~, подвесная suspension compass
~ с диоптрами sight compass
буссоль-теодолит compass-theodolite
бустер booster; actuator; compensator
бустерный power-assisted/-operated, fully powered
бут rubblestone, rubble, quarrystone
бутаверин butaverine
бутадиен *chem.* butadiene
бутамид tolbutamide
бутан *chem.* butane
бутандиолдегидрогеназа butandioldehydrogenase
бутанол *chem.* butanol
бутара *min.* washing drum; sizing trommel
бутафория *theatre* props, dummy, fake
бутен butene, butylene
бутил *chem.* butyl
бутилацетат butyl acetate
бутилен butylene, butene
бутилкаучук butyl rubber
бутиловый butyl *(attr.)*
бутирометр butyrometer, cream gauge
бутобетон rubble concrete
бутон *agr.* (flower) bud
бутонизация bud-formation; budding
бутсы, бутцы, буцы football boots
бутылка bottle
~, плавучая drift bottle
буфер 1. buffer, bumper, shock-absorber; stop; dashpot; 2. *chem.* poising agent, buffer (solution)
~ ввода *comp.* input buffer
~ ввода-вывода *comp.* input-output buffer
~ графопостроителей graphic-device buffer
~ данных data buffer
~, динамический dynamically obtained buffer
~ для хранения сообщений communication buffer
~ записи write buffer
~, кадровый frame buffer
~ команд instruction buffer
~, межпроцессорный interprocessor buffer
~ на многоразрядное слово full-word buffer
~ на одну строку row buffer
~, паровой steam cushion
~, пересылочный send buffer
~ периферных устройств peripheral buffer
~, пневматический pneumatic/compression buffer
~ предыстории (процесса) look-aside buffer
~ регенерации refresh buffer
~ режима реального времени real-time buffer
~ сжатия (подвески) bump stop
~, скрытый hidden buffer
~ состояния status buffer
~ считывания read buffer
буферизация buffering
~ в последовательных сегментах chained segment buffering
~ пишущей машинки printer keyboard buffering
~ при обмене (обменная) exchange buffering
~ при передаче сообщений in-transit buffering
~, упреждающая anticipator buffering
буферизованный buffered
буфер-компенсатор compensating buffer

буферно-полосный buffer-strip
буферность *(of soil)* buffering; buffer state/capacity/value/action; poising action; *food* buffer value
буферный buffer *(attr.)*; buffering; buffered
~ное государство buffer state
буферовка *met.* buffing; polishing
бухгалтер bookkeeper, accountant
~ ведущий общую (главную) книгу general-ledger bookkeeper
~, главный management/managerial accountant, accountant in charge, accountant general
~, заводской factory/works accountant
~ по налогообложению tax accountant
~, присяжный *Am.* certified public accountant; *Eng.* chartered accountant
~ промышленного предприятия industrial accountant
бухгалтер-калькулятор cost analyzer
бухгалтер-контролер, главный comptroller
~ по исполнению сметы budget accountant
бухгалтер-ревизор auditor
бухгалтер-эксперт (дипломированный) *Am.* certified public accountant; *Eng.* chartered accountant
бухгалтерия 1. bookkeeping, accounting; 2. bookkeeping/accounting department
~, американская tabular bookkeeping
~, двойная double-entry bookkeeping
~, главная general accounting office
~, итальянская double-entry bookkeeping
~ коммиссионных commission accounting
~, общая general accounting office
~, производственная cost accountancy
~, простая single-entry bookkeeping
бухгалтерский bookkeeping, accounting
~ учет accounting
~кая задолженность ordinary debts
~кая книга account book; ledger
~кая машина bookkeeping/accounting machine
бухинолат buquinolate
бухта I coil, reel; hank; fake; bundle
~ троса hank, tier
бухта II bay, sea inlet, bight
бухтодержатель reel stand
бучарда bushhammer; granulating hammer; tooth axe
бучение *constr.* filling with rubble
бушприт *naut.* bowsprit
бык pier, buttress
~ с ледорезом sharp-noise pier
~, свайный pile pier
бык-гордень *naut.* buntline
быстровращающийся fast-rotating/-running, rapid-turning
быстровяжущий quick-hardening/-setting/-binding
быстрогорящий fast-burning
быстродвижущийся fast-moving, quick; high-speed
быстродействие speed of operation/response; fast response/operation, high-speed response
~, высокое high operation speed
~ памяти *comp.* memory speed
~ цифровой вычислительной машины digital computer speed
быстродействующий quick-acting/-operating, fast (-response)
быстроизнашивающийся rapidly wearing, of short operational life
быстролетящий fast-flying
быстроотпускающий quick-release
быстроразъемный quickly disconnecting, quick-disconnect
быстрореализуемый *econ.* liquid; readily marketable
быстрорастущий *agr.* fast-growing
быстрореагирующий quick-response
быстросменный quick-change
быстросохнущий fast-/quick-drying
быстросхватывающий quick-hardening/-setting
быстросъемный quick-detachable; quick-release
быстрота speed, velocity; quickness, rapidity; rate
~ ввода в режим entry rate
~ вывода recovery rate
~ изменения rate of change
~ нагрева heating rate
~ реакции *physiol.* reaction time
~ цикла cycling rate
быстротвердеющий quick-hardening/-setting
быстротекущий swift-/fast-flowing; *med.* galloping
быстроток high-velocity channel/chute/flow
быстроходность (specific) speed
быстроходный fast-/quick-running, fast; high-speed
бытие *philos.* being, existence, esse; reality
бытийно-квалификативный existential-qualificative
бытийный existential
~ное предложение sentence of being, existential sentence
бытовой household *(attr.)*, domestic; everyday, for everyday use; social
бычина *leath.* steer/ox hide, bulls
бьеф pool, pond, level, reach
~, верхний upper water/reach/pool; head water/race
~, нижний downstream water, lower/tail water
Бэйсик *comp.* Basic
Бэконианец *philos.* Baconian
бэкхенд *sport* backhand
бюгель clasp
~, двухкюветный two-flask clasp
бюджет budget
~, вещественный *phys.* mass balance
~, военно-морской navy estimates
~, военный defence/military budget
~, государственный State budget
~, кассовый консолидированный cash (-consolidated) budget

~ нейтронов *phys.* neutron budget
~, сводный consolidated budget
~, семейный family budget
бюджетный budgetary, budget *(attr.)*
~ год fiscal year
~ная комиссия *law* ways and means committee
~ная статья budget item, item in the budget
бюкса weighing/sample bottle
бюллетень 1. bulletin; report; chart; 2. ballot-paper, voting paper/slip
~ погоды weather report; meteorological bulletin
~ фондовой биржи stock-market report
бюргер *hist.* citizen; burgher
бюретка buret(te), dropping tube/glass
~, бескрановая pinch clamp buret(te)
~, измерительная measuring buret(te)
~ с краном stopcock buret(te)
бюро 1. bureau, office, department, service; 2. writing desk, bureau
~, вычислительное computation office
~ калькуляции *econ.* cost centre
~, конструкторское design/drawing office
~ налогов и сборов revenue office
~ находок lost and found department
~, патентное patent office
~ приема заказов order desk
~ путешествий travel/tourist agency
~ рекламы и информации publicity bureau
~, справочное information bureau/desk; inquiry desk
~ труда, Международное International Labour Office, ILO
~, чертежное drawing office
бязь *text.* coarse calico

В, в

Ваал *myth.* Baal
вавассор *hist.* vavaso(u)r
вавилонский Babylonian
~кая башня the tower of Babel
~кое столпотворение babel; disorder, confusion
вавилонянин Babylonian
вага 1. balance; public weighing machine; 2. lever, crowbar; splinter bar
вагальный *anat.* vagal
вагина *anat., bot.* vagina; sheath
вагинальный vaginal
вагинизм *med.* vaginismus
~, психогенный mental vaginismus
вагинит *med.* vaginitis, colpitis
~, трихомонадный, трихомональный mycotic vaginitis
~, эмфизематозный vaginitis emphysematosa
вагино-абдоминальный vaginoabdominal
вагино-везикальный vaginovesical
вагино-вульварный vaginovulvar
вагинодиния vaginodynia, colpodynia
вагино-лабиальный vaginolabial
вагиномикоз vaginomycosis
вагино-перинеальный vaginoperineal
вагинопластика vaginoplasty
вагинотомия vaginotomy, colpotomy
вагинофиксация vaginofixation, colpopexy, vaginal fixation
вагит vagitis
вагоглоссофарингеальный vagoglossopharyngeal
вагон 1. van, *Am.* car, coach; 2. carload, wagonload, wagonful
~, аккумуляторный battery-powered motor car
~, багажный baggage car
~, грузовой freight/pool car; goods van
~, динамометрический dynamometer car
~, жесткий railroad car with hard seats
~, измерительный track-testing car
~, изотермический refrigerator car
~, крытый (товарный) boxcar
~, купейный compartment car
~, монтажный installation car
~, моторный (electric) motor coach
~, мягкий railroad car with soft seats
~, пассажирский coach, *Am.* passenger car
~, пассажирский купейный sleeping carriage/ *Am.* car
~, пассажирский некупейный passenger coach/ *Am.* car
~, пассажирский плацкартный passenger coach/ *Am.* car with reserved seats
~, почтовый mail/parcel/postal van, *Am.* mail car
~, прицепной trailer car
~ прямого сообщения through carriage
~ прямой сборный consolidated car
~ с откидным дном bottom dump car
~, саморазгружающийся tipping car, self-clearing car
~, санитарный hospital car
~, сборный pool car
~, служебный business car, railway service car
~, спальный sleeping car
~, товарно-багажный package car
~, товарный goods van, *Am.* freight car
~, франко free on rail/car/ *Am.* board
вагон-вертушка merry-go-round wagon
вагон-весы scale car
вагон-вышка tower car
вагонетка industrial car, buggy; truck; trolley; lorry; car; transfer car; factory truck
~ внутризаводского транспорта industrial car
~, завалочная/загрузочная charging car
~, зарядная power/charging car
~, литейная casting bogie/buggy
~, полочная rack car
~, рудничная mine car

~, рудничная пассажирская man-rider/riding car
~ с донной разгрузкой drop-bottom car, hinged bottom car
~, саморазгружающаяся self-discharge car
~, самоходная self-propelled trolley
вагонетка-самосвал dump car
вагон-кран crane car
вагон-лаборатория track-test car
вагон-ледник ice-cooled car, iced car, *Am.* refrigerator car/van
вагон-мастерская repair car
вагонный парк rolling stock
вагоно-километр car kilometer
вагонооборот wagon traffic/turnover
вагоноопрокидыватель car dump(er), truck tipper
~, боковой tippler
вагонопоток car traffic volume
вагоноремонтный car-repair/-maintenance
~ завод railway-car repair works
вагоностроение car/carriage building
вагоностроительный car-/carriage-building
вагоносутки wagon-day (of 24 hours)
вагон-перегружатель transfer car
вагон-платформа flat/platform car, flatcar
вагон-ресторан dining car
вагон-рефрижератор refrigerator car
вагон-цистерна tank car/wagon
вагосимпатический *anat.* vagosympathetic
ваготомия *surg.* vagotomy
ваготония *med.* vagotony, vagotonia
вагранка *tech.* (blast) cupola, cupola furnace
~, малая cupolette
~ с передним горном cupola with receiver
~ с подогревом дутья hot-blast cupola
вагусный vagal
вад *geol.* wad, black ochre, bog manganese
~, медистый cupreous manganese, lampadite
~, черный black wad
вадеит *geol.* wadeite
вадозный vadose
ваер *naut.* warp
вазелин white petrolatum, vaseline, petrolatum ointment, petroleum/mineral jelly, white soft paraffin
~, борный borated petrolatum
~, жидкий liquid petrolatum
вазелиновый vaseline
~вое масло liquid paraffin
вазоанастомоз vasoanastomosis
вазоген oxyvaseline
вазодилятация vasodilatation
вазоконстриктор vasoconstrictor
вазомоторный vasomotor(ial)
вазоневроз vasoneurosis, angioneurosis
вазопресин vasopressin
вазопунктура vasopuncture
вазорафия vasorrhaphy
вазоспазм vasospasm, angiospasm
вазоспастический vasospastic, angiospastic
вазотонин vasotonin
ваия *bot.* 1. frond; 2. palm branch
~, вильчатая forked frond
~, спороносная fertile frond
~ий, неделя *eccl.* Palm Sunday
вайда *bot.* woad, pastel *(Isatis)*
~, красильная dyer's woad *(Isatis tinctoria)*
вайделот *hist.* Lithanian/Prussian priest
вакансия 1. opening, vacancy; *econ.* vacant job; *eccl.* voidance; 2. *phys.* vacancy, void, hole
~, в электронной оболочке shell vacancy
~, кислородная oxygen vacancy
~, кристаллической решетки lattice vacancy
~, электронная electron vacancy, hole
вакат *print.* blank page, white/blank strip
вакка *geol.* wacke
~, дымчатая smoke wacke
~, серая greywacke, graywacke
~, шаровая spheroidal wacke
вакуолизация vacuolization, vacuolation
вакуоль, вакуоля vacuole, tonoplast
~, пищеварительная food vacuole
~, сократительная contractile vacuole
вакуольный, вакуолярный vacuolar
вакуум vacuum
~, абсолютный perfect/absolute vacuum
~, высокий/глубокий high/fine vacuum
~, неполный partial vacuum
~, низкий low vacuum
~, предварительный initial vacuum, forevacuum
~, предельный ultimate vacuum
вакуум-аппарат vacuum pan/apparatus
вакуум-бак vacuum tank
вакуум-бетон deaerated concrete, vacuum (-processed) concrete
вакуумирование 1. vacuum processing treatment; evacuation; vacuuming,vacuum operation; 2. degassing
вакуум-ковш vacuum ladle
вакуумметр vacuum gauge
~, грубый vacuum indicator
~, жидкостный hydrostatic vacuum gauge
~, ионизационный ionization vacuum gauge
~, компрессионный hydrostatic-pressure vacuum gauge
~, термоэлектрический thermal-conductivity vacuum gauge
вакуум-насос vacuum/suction pump
~, поршневой reciprocating vacuum/suction pump
вакуум-непроницаемый/-плотный vacuum-tight
вакуумный vacuum, void, vacuous, evacuated
вакуум-проводка vacuum line/manifold
вакуум-упаковка vacuum packing/packaging
вакуум-сушка vacuum drying
вакуум-фактор vacuum factor; *el.* gas ratio
вакуум-фильтр vacuum/suction filter
вакуум-эксикатор vacuum extraction still,

vacuum desiccator
вакуум-экстрактор 1. *food ind.* vacuum extraction still; 2. *med.* vacuum-extractor
Вакх *myth.* Bacchus
вакханалия 1. *hist.* Bacchanalia; 2. orgy, orgiastic revel
вакханка Bacchante, maenad
вакцина vaccine
~, антирабическая rabies/antirabic vaccine
~, БЦЖ = противотуберкулезная
~, живая live vaccine, booster
~, коклюшная pertussis vaccine, whooping-cough vaccine
~, оспенная smallpox vaccine
~, поливалентная polyvalent/multipartial vaccine
~, порошковидная/пылевидная dust vaccine
~, противогриппозная influenza virus vaccine
~, противотифозная typhoid/antityphoid vaccine, typhobacterin
~, противотуберкулезная BCG vaccine, bacillus Calmette-Guerin vaccine, tuberculous vaccine
~, против полиомиелита poliomyelitis/ polio/ poliovirus vaccine
~, противочумная plague vaccine
~, сибиреязвенная anthrax vaccine
~, убитая inactivated vaccine
~, фенолизированная formaldehyde-treated vaccine
~, формалинизированная formolated vaccine, formolvaccine
~, яичная chick-embryo vaccine
вакцинальный vaccinal
вакцинация vaccination, inoculation
вакцинировать vaccinate
вакцинный vaccinal, vaccinial
вакцинопрофилактика vaccinal prevention
вакцинотерапия vaccinotherapy
~, внутрикожная intercutaneous vaccinotherapy
~, подкожная subcutaneous vaccinotherapy
вал I *tech.* shaft; spindle, axle, roller; drum, cylinder
~, боевой *text.* pick shaft
~, ведомый driven shaft
~, ведущий drive/driving shaft; drive capstan; input shaft
~, входной input/intake shaft
~, выводной/выходной output shaft
~, головочный *text.* rim shaft
~, горизонтальный lay shaft
~, гребной propeller(-drive) shaft, screw/tail/ propulsion shaft
~, карданный cardan shaft
~, кареточный *text.* back shaft
~, качающийся rocking shaft
~, коленчатый/кривошипный crankshaft
~, конусный taper shaft
~, кулачковый camshaft, eccentric/tappet shaft
~ обратного хода reverse shaft
~ отбора мощности power take-off shaft
~ отбора мощности, реверсивный reverse power take-off shaft
~, отжимный squeeze roll
~, передаточный transmission shaft
~, печатный *text.* printing roller
~, полый/пустотелый hollow shaft
~, приводной drive/power/transmission shaft
~, промежуточный intermediate/line/tunnel shaft
~, разрезной divided shaft
~, распределительный distributing shaft, camshaft
~, реверсивный reverse shaft
~ редуктора gearbox shaft
~, рифленый corrugated/ribbed shaft
~, рулевой steering-wheel shaft, control shaft
~ рулевой передачи steering rod
~, сукновальный fulling roller/cylinder
~ сцепления clutch shaft
~, торсионный torsion shaft/bar
~, трансмиссионный transmission shaft
~, трубчатый tubular shaft, torque tube
~, турбинный turbine spindle
~, удлиненный/удлинительный extension shaft
~, упорный thrust shaft
~, фланцевый flanged shaft
~, холостой idle shaft
~, червячный worm shaft
~, шарнирный articulated shaft
~ шестерни pinion shaft
~, шлифовальный grinding shaft
~, шлихтовальный sizing roller, size roll
~, шлицевой spline(d) shaft; castellated shaft
~, шпоночный splined/fluted shaft
вал II bank, embankment, earthen wall; bar; dam, weir; rampart; *geol.* swell arch
~, береговой beach barrier; offshore bar; (natural) levee
~, ветровой wind bar
~ коробления warping, swell
~, лавовый lava billow
~, ледниковый ice rampart
~, моренный dumped moraine, moraine line
~, морской breaker, swell
~, песчаный sand bar
~, прибойный swash bar
~, приливный tidal bar
~, огневой *mil.* barrage fire, creeping barrage
~, термокарстовый thermokarst mound
вал III *anat.* torus
~ носа ridge of the nose
вал IV *econ.* gross output
Валаам *bibl.* Balaam
валаит walaite, valaite
валансьенские кружева Valenciennes lace
валах, валахский Wallachian
валашка Wallachian (woman)
валгалла *myth.* Valhalla
валежник brushwood, windfallen trees/branches

валек 1. battledore; 2. roll, roller, cylinder; 3. swingletree; 4. *agr.* threshing flail; 5. *zool.* pilot fish, round whitefish *(Prosopium cylindraceus)*
валенбергия *bot.* rockbell *(Wahlenbergia)*
валенки *pl.* valenki *(felt boots)*
валентинит valentinite, white antimony, antimony bloom
валентность valency, *Am.* valence; basicity, atomicity
~, **множественная** quantivalence, quantivalency
~, **направленная** directional valency
~, **нормальная** normal/ordinary valency
~, **нулевая** null valency
~, **скрытая** latent valency
валентный valency, *Am.* valence
валериана valerian, Valeriana L.
валерианат valerianate
валерианелла, валерианица *bot.* corn salad *(Valerianella)*
~, **бугорчатая** Italian corn salad *(V. eriocarpa)*
валериановый valerian
валерьяна *bot.* valerian *(Valeriana)*
~, **аптечная/лекарственная** common valerian *(Valeriana officinalis)*
валерьяновые *pl. bot.* valerian family *(Valerianaceae)*
валерьяновый valeric, valerianic
~**вые капли** valerian drops, tincture of valerian
валет *(cards)* knave, jack
валец *tech.* roll(er)
~, **мяльный** breaker roll
~, **отрывающий** *agr.* snapping roll
~**льцы, мельничные/рифленые** embossed rolls
валидация validation
валидность validity
валидол validol, methyl valerate
валик I small earthen wall, small bank
валик II 1. *tech.* roll(er); spindle, shaft; drum, cylinder; bead; *comp.* platen roller; 2. bolster, long cylindrical pillow
~, **вальцмассный** *print.* composition roller
~, **ворсовальный** *text.* raising roller
~, **ворсовый** nap/friezed roller
~ **для усиления** reinforcing bead
~, **игольчатый** spiked/porcupine roller
~, **контактный** *comp.* contact drum
~, **красочный** *print.* inking roller
~, **лентопротяжный** tape-feed roll(er)
~, **нажимной** pressure roller
~, **накаточный** knurling roller
~, **обезрепеивающий** *text.* burring roller
~, **опорный** bearing/supporting roller
~, **отжигающий** annealing weld
~, **переставляемый** adjustable bail roll, adjustable platen
~, **пишущий** platen roller
~, **плющильный** *text.* calender roll
~ **подачи** feed shaft
~, **полировальный** burnishing roll
~, **приемный** receiving roll(er), take-up roll(er)
~, **прижимной** pinch roller
~, **промазочный** smearing roller
~, **раскатной** *print.* distributing roller
~, **рифленый** ribbed/corrugated roll
~ **с чекой** cotter bolt
~, **стопорный** stop/locking/retaining roller
~, **съемный** *text.* discharging roller, doffing roll
~, **товарный** combing/batching roll, cloth beam
~ **усиления** reinforcing bead
~, **фальцевальный** folding roller
~, **четвертной** *arch.* quarter-round fillet
~ **шва** bead joint, beaded weld
~, **щеточный** brush roller
валик III *anat.* torus, torulus, vallum
~, **глазной** eye/palpebral line, ocular ridge
~, **небный** palatine torus/protuberance
~ **ногтя** wall of the nail, nail wall
~, **трубный** Eustachian cushion
валики *pl.* toruli
~, **осязательные** tactile elevations
валиковидный, валикообразный spindle-shaped; vallate
валил valyl
валин valine
валиндекарбоксилаза valinedecarboxylase
валка fulling
валкий *naut.* crank(y), tender
валкирия = валькирия
валкователь swath collector
валковать swath, windrow
валкооборачиватель swath turner
валкоспособность felting power/ability
валкость *naut.* crankiness, tendency to roll
валлиец Welshman
валлийка Welshwoman
валлийский Welsh
валлиснерия *bot.* wild celery *(Valisneria)*
валлон, валлонский Walloon
валлонка Walloon (woman)
валовой 1. gross, wholesale; 2. general, overal
~, **анализ** bulk analysis
~, **доход** gross revenue
~**вая продукция** gross output
валогенератор shaft generator
валок 1. *agr.* swath(e), windrow; 2. *tech.* roll(er), cylinder
~, **боковой** breast roll
~ **для гофрирования** corrugating roll
~ **для холодной прокатки** cold roll
~, **консольный** overhanging roll
~, **опорный** back(ing)-up roll
~, **отделочный** finishing roll
~, **профилированный** cambered roll
~, **рабочий** driving/working roll(er)
~, **сортовой** grooved roll
~, **толстолистовой** plate roll
~, **тонколистовой** sheet roll
~, **фасонный** grooved roll
~, **черновой** roughing roll, breaking-down roll

~, чистовой finishing roll
валом in (the) gross
~ валить throng; heap/pile up
валонея *bot.* European Turkey oak *(Quercus cerris)*
валопровод shafting, shaft line
Валтасар Belshazzar
валторна *mus.* French horn
валуй *bot.* fetid russula *(Russula foetens)*
валун boulder, round stone; cobblestone, rubble (stone)
~, глиняный clay ball
~, исштрихованный striated/scratched boulder
~, ледниковый glacial boulder, morainal block
валунник cobble round stone
вальвассор *hist.* rear vassal, vavaso(u)r
вальвулит *med.* valvulitis
вальвулопластика valvuloplasty, valvoplasty
вальгусный valgoid, valgus
вальдшнеп *zool.* woodcock *(Scolopax rusticola)*
вальдштейния *bot.* barren strawberry *(Waldsteinia)*
валькирия *myth.* Valkyrie
вальма *arch.* hip
вальпургиева ночь Walpurgis Night
вальпургит *miner.* walpurgite
вальхия Walchia
вальцевание, вальцование *tech.* (forge-)rolling; flaring; milling; stretching; laminating; calendering
~, горячее hot rolling
вальцевать forge-roll, expand, flare, roll; mill, pass through rolls; work in rolls
вальцезагибочный roll(er)-bending
вальцмасса *print.* roller composition
вальцовка 1. rolling, roll forging; 2. expander; stretching rolls; rolling press/stock
вальцовщик miller, millman, mill operator, roller
вальцовщик-листовальщик sheeting mill operator
вальцуемость millability
вальцы *pl.* rolls
~, выпускные final mill
~, гибочные bending rolls/machine, roll-bending machine
~, листовальные sheeting mill
~, листогибочные plate-bending rolls/machine
~, листоправильные plate-levelling rolls/ machine
~, правильные straightening/levelling rolls, roll-straightening machine, roll straightener
вальян *text.* doffer
валюта *fin.* (foreign) currency
~, единая common currency
~, иностранная foreign exchange
~, ключевая key/vehicle currency
~, конвертируемая convertible currency
~, неконвертируемая/необратимая inconvertible/ nonconvertible/ soft currency
~, "плавающая" freely floating currency
~ расчета transaction currency
~, свободно обратимая иностранная freely convertible currency, free/hard currency
~, твердая hard/free currency, freely convertible currency
валютный currency *(attr.)*
~ (В.) комитет Европейского сообщества Monetary Committee of the European Community, MCEC
~ курс rate of exchange
~ная операция currency transaction
валяльный *text.* felting, fulling
~ное производство felting, fulling
валяние fulling, milling, felting
валяный felt(ed), fulled
валять full
вампир vampire
~, большой *zool.* vampire bat *(Vampyrus spectrum)*
вампум *hist.* wampum
ванадат *chem.* vanadate
ванадиевый, ванадовый vanadic
ванадий vanadium, V
ванадинит vanadinite
ванда *bot.* vanda *(Vanda)*
вандал *hist.* Vandal
вандеец *hist.* Vendean
ванденбрандеит vandenbrandeite, uranolepidite
ванесса *zool.* thistle butterfly *(Vanessa)*
ванилин *chem.* vanillin(e)
ваниль 1. *bot.* vanilla *(Vanilla)*; 2. vanilla beans
~, душистая/плосколистная Mexican vanilla *(Vanilla fragrans)*
ванна 1. *tech.* tank, van, tub; 2. *med.* bath; 3. bathtub
~, гальваническая (electro) plating bath/cell/ tank/ vat; galvanic tank
~, гипосульфитная hypo bath
~, грязевая *med.* mud bath, balneum luteum, illutation, moor/bog bath
~ для свинцевания lead bath
~ для цинкования spelter bath
~, железистая chalybeate bath
~, закалочная quenching bath
~, индифферентная tepid bath
~, красильная dye bath
~, лекарственная medical bath
~, лудильная tin(ning) pot
~, макательная dipping tank/bath
~, мягкая bland bath
~, мягчильная *text.* partial boiling bath, bate-drench
~, никелевая nickel-plating bath
~, осадительная/осадочная setting bath/tank
~, паровая vapo(u)r/steam bath
~, переменная alternating bath
~, песочная sand bath, arenation
~, потогонная sweat bath
~, световая light bath, electric cradle
~, светотепловая light-heat bath

~, серебряная silver-plating bath
~, соленая brine-bath
~, термостатируемая constant-temperature bath
~, травильная pickling bath
~, углекислая carbon dioxide bath, effervescent bath
~, успокаивающая sedative bath
~, хвойная pine bath
~, цинковальная galvanizing bath
~, электролитическая plating bath/cell, plating tank
ваннер *min.* vanner, slime washer
ванночка tray, dish
~, глазная eye bath
ванта *naut.* shroud, stay (rope), guy (rope)
вантуз 1. air escape valve, air hole, vent, air release (vent); 2. force cup
вапити *zool.* American elk, wapiti *(Cervus canadensis)*
вапор (steam-engine) cylinder oil
вапоризатор vaporizer, evaporator
вар 1. pitch, (pine) tar; cobbler's wax; 2. VAr, volt-ampere reactive
~, светлый shoemaker's paste, cobbler's wax
~, черный black pitch
варактор varactor diode, charge storage diode
варакушка *zool.* blue-throated robin, bluethroat *(Luscinia svecica)*
варан *zool.* monitor lizard *(Varanus)*
варваризм barbarism
вариабельность, вариабильность variability
вариак autotransformer, variable transformer
вариант variant, modification; alternative, alternate; derivative, version; mark; *comp.* option
~, всепогодный all-weather version
~ выхода за пределы сети off-grid option
~ выхода за пределы экрана off-screen option
~, тропический tropicalized version
~, увеличенный enlarged/growth version, scaled-up version
~, уменьшенный scaled-down version
~, упрощенный stripped version
~, усовершенствованный improved/developed/ advanced version
варианта variant; sequence
~, возрастающая increasing sequence
~, ограниченная bounded sequence
вариантность variance, variation; variability
вариантный variant, varying, alternate, alternative
вариатор 1. variable-speed drive; (speed) regulator; 2. buncher, input resonator, variator; variable-ratio transformer
вариационный variational, variant
~ное исчисление calculus of variations
вариация variation; alternation, change
~ показаний reading variation
~ постоянных *comp.* variation of constants
~ силы тежести gravity variation
~, суточная diurnal/daily variation
~, широтная latitude variation
~ции, звездные sidereal variations
вариетет variety
варикап variable capacity, varicap
варикозный varicose
варикaонд varicond, variable condenser
варикофлебит varicophlebitis
варикоцельэктомия varicocelectomy
варикс *zool.* varix
варингтонит waringtonite, brochantite
вариограф recording variometer, variograph
вариола *med.* smallpox, variola
вариолоид varioloid, alastrim
вариоль variole
вариометр 1. variometer, variocoupler; tuner; 2. vertical-speed indicator, rate-of-climb indicator/ meter/vertical-velocity indicator
~, горизонтальный магнитный H-variometer
~, двухрычажный two/double/beam balance
~, кабинный *av.* cabin rate-of-climb indicator
~, однонитный unifilar variometer
~ с сельсинной передачей transducing vertical-speed indicator
~ склонения declination/ H-variometer, horizontal intensity variometer
варисский Variscian
варистор varistor, variable resistor
~, карбидокремниевый silicon-carbide varistor
~, меднокупоросный copper-sulphate varistor
~, объемный bulk varistor
~, полевой field varistor
~, селеновый selenium varistor
варитрон varitron
варить 1. boil; cook; 2. pulpify; digest; 3. weld; found *(metal)*
~ не доводя до кипения simmer
~ пиво brew beer
варицелла *med.* alastrim
варка 1. boiling; cooking; 2. pulpifying, pulping; digesting, digestion
~ железа iron founding
~ на пару steam boiling
~ пива brewing
варница, варня 1. saltworks; 2. salt pan/oven
варочный (for) cooking/boiling; brewing
~ котел curing pan
варрант *econ.* warehouse warrant; customhouse license
~, доковый dock warrant
вартский Warthian
варфоломеевский *hist.* Saint Bartholomew's, Bartholomean
варьете variety show
варьирование variation
~, внутри группы batch variation
варьировать vary, modify, diversify
варяг *hist.* Varangian
василек *bot.* cornflower, centaurea *(Centaurea)*

василиск *zool.* basilisk *(Basiliscus)*
василис(т)ник *bot.* meadow-rue *(Thalictrum)*
васкулит *med.* vasculitis, angiitis
~, риккетсиозный rickettsial vasculitis
~, сыпнотифозный typhous vasculitis
~, узловатый vasculitis nodularis
васкуляризация vascularity, vascularization
васкулярный vascular
вассал vassal, liegeman
вассалитет, вассальство *hist.* vassalage
вассальный vassal
~льная зависимость vassalage
вассерглас water glass
вата cotton (wool); glass wool; wadding
~, асбестовая mineral cotton
~, гигроскопическая absorbent purified cotton
~, металлическая metal wool
~, минеральная mineral/rock wool/cotton
~, стеклянная glass/cinder/mineral wool
~, шлаковая slag wool
ватер, ватермашина ring-spinning frame; water frame
ватер-бакштаг *naut.* water backstay
ватервейс *constr.* waterway
ватер-вулинг, ватервулинг *naut.* grammoning
ватер-зейль water sail
ватерклозет lavatory, water closet, toilet; bathroom
ватерлиния waterline, floating line
~, грузовая load (water) line, (fully) loaded waterline
~, конструктивная design(ed)/construction waterline
ватерпруф waterproof coat, raincoat
ватер-штаг bobstay
ватин batting, sheet wadding
ватман Whatman paper
ватник quilted jacket
ватный cotton; quilted, wadded
~ное одеало quilt
ватообразный cottony; velvetlike
ваточник *bot.* milkweed, silkweed *(Asclepias)*
ватрушка *cul.* cheese cake
ватт I *el.* watt, W
ватт II tidal marsh
ваттметр wattmeter
~, воздушный air-flow wattmeter
~, многофазный polyphase wattmeter
~, реактивный idle-current wattmeter
~, самопишущий recording wattmeter
~, щитовой switchboard wattmeter
ватт-секунда watt-second
ватт-час watt-hour
ваучер *econ.* voucher
вафельница waffle/wafer iron/tongs
вафля waffle, wafer
вахмистр cavalry sergeant major
вахня *zool.* Pacific navaga *(Eleginus gracilis)*
вахта 1. watch, watch crew; 2. battle station
~ в море sea watch
~, вторая larboard watch
~, заступающая on-coming watch
~, палубная watch on deck
~, первая, ~ правого борта starboard watch
~, ходовая underway/steaming/bridge watch
вахта трилистная *bot.* bog/buck bean *(Menyanthes trifoliata)*
вахтенные watch on duty
вахтенный I: **~ журнал** logbook
~ командир officer of the deck
вахтенный II watch(man); watch officer/keeper; watch stander
вахтер janitor, porter
вахтовые *pl. bot.* buck-bean family *(Menyanthaceae)*
вашгерд *min.* gold washer, cradle, buddle
вашингтония *bot.* Washington palm *(Washingtonia)*
ваятель sculptor
ваять sculpture; carve, chisel; cast; model
вбивание driving/hammering in
вбивать drive/hammer/knock/ram in
~ мяч score a goal
вбивка driving/knocking in
вбирание absorbing, absorption
вбирать absorb/soak up, suck/drink in; inhale
вбок sideways, to one side
вброд: переходить - wade; ford
~ не переходимый unfordable
вбрызгивание injecting, injection, spraying/sprinkling in; splashing in
введение 1. advent, introduction; exordium; prolegomenon; 2. injection, insertion, breaking in; putting in operation; 3. leading in(to)
~ во храм *eccl.* Feast of the Presentation of the Virgin
~ дырок hole injection
~ изотопа incorporation/administration of an isotope
~ меченых атомов tagged-atom administration
~ параграфов *comp.* paragraphing
~ черно-белой программы monochrome programming
ввертывать screw in, insert; put in
вверх upward(s)
~ по реке/течению upstream
ввивание weaving in
ввивать weave in, interweave
ввинчивание screwing in
ввод 1. entrance, entry; lead-in; 2. input, insertion inlet; introduction, injecting; *el.* bushing
~, автономный/независимый *comp.* indirect input, off-line input
~, аналоговый analogue input
~, антенный aerial lead-in, antenna lead(-in)
~ бесконечного бланка/формуляра continuons form guide (and shelves)
~, бесформатный formatless entry, form-free entry
~ в бой *mil.* commitment to action

~ в действие putting into operation
~ в память, поблочный block storage
~ в реальном масштабе времени real time input
~ в штопор *av.* entry into the spin, spin entry
~ в эксплуатацию putting into operation, introduction/entry/putting into service
~, внешний peripheral input
~, высоковольтный high-voltage bushing
~ графических данных graphic data/information input
~ данных data acceptance, acceptance of data, data entry/input/insertion, insertion of data
~ данных в режиме он-лайн on-line data input
~ данных голосом vocal/voice input
~ данных, прямой/непосредственный direct data input
~ данных с перфоленты punched tape data input
~ даты date input, input of date
~, двоичный binary input
~ для вычислительной машины computer input
~ заданий/дистанционный remote job entry
~ календарного числа date input, input of date
~ карт card input
~ карт, непрерывный continuous card input
~ карт счетов account card feed(ing)
~, клавишный keystroke entry, keyboard input
~, кнопочный push-button entry
~ конденсаторного типа capacitance bushing
~, маслонаполненный oil-filled entrance bushing
~ начальной программы initial program loading
~, независимый off-line input, indirect input
~ носителя данных data carrier input
~ основной программы main program input
~, отсроченный deferred entry
~, пакетный batch entry
~, параллельный simultaneous input
~ перфокарт punched card feed/input
~, пословный input of single words
~ программы program loading/supply
~, прямой direct input
~, ручной hand/manual input
~ с клавиатуры keyboard entry/input
~ с магнитной ленты magnetic tape input
~, сенсорный touch input
~, токовый lead-in wire
~ указателя набора/массива данных dataset pointer entry, file pointer entry
~, форматированный form entry

вводит 1. introduce; insert; inject; move in; bring in; lead in; enter; incorporate; feed/put into; 2. *math.* interpolate
~ в действие put into operation
~ в заблуждение confuse, mislead
~ в оболочку invaginate
~ в производство put into operation
~ в расход put to expense
~ в смесь mix in
~ в эксплуатацию put into operation/use/service
~ во владение *law* put into possession
~ подпрограмму assemble subprogram/subroutine
~ торжественно в должность inaugurate

вводный 1. leading in; introductory; exordial; 2. *gram.* parenthetic(al)
~ное предложение parenthetic clause
~ное слово parenthesis, parenthetic word

ввоз importation; import(s)
ввозить bring in, import; cart, deliver, transport
ввозный import *(attr.),* imported
вгиб bend inwards
вгибание bending in (wards); incurvation
вгибать bend/curve inwards
вгон, вгонка herding/driving in
вдаваться jut out/into, run into
вдавление impression, dent, indentation; foveation
~, сердечное cardiac impression
~, пищеводное esophageal groove
вдавливание 1. pressing/forcing in(to); impressing; 2. indentation, indenting
вдавливать indent; press/force in; imbed; impress
вдалбливать hollow out, cut a mortise in
вдвижной insertable, sliding in
вдвойне double; twice; twofold; doubly
вдевятеро nine times, ninefold
~ больше nine times as much/many
~ меньше one-ninth
вдержка 1. pulling in/through; threading; 2. bodkin
вдесятеро ten times, tenfold
~ больше ten times as much/many
~ меньше one-tenth
вдовство widowerhood, widowhood; discoverture
вдовствование viduage, viduity; widowhood
вдовствующая императрица the Dawager Empress
вдовушка *zool.* nunbird *(Vidua serena)*
вдовушки *bot.* scabious *(Scabiosa)*
вдох breath, inhalation
вдув injection
~, интенсивный massive/large injection/blowing
вдувание insufflation, blowing in(to), inflation; *med.* pneumothorax
вдуватель insufflator
вдыхание 1. inhalation; 2. breath
вдыхательный inhaled, inspiratory
веберметр fluxmeter
вебстерит websterite
~, обманковый hornblende websterite
вегетарианец vegetarian
вегетарианский vegetarian
~ стол vegetarian diet
вегетарианство vegetarianism
вегетативно-нервный vegetative-nervous
вегетативно-сосудистый vegetovascular
вегетативный vegetative

вегетационный vegetative, vegetational, vegetal, vegetation *(attr.)*
~ домик greenhouse
~ период vegetative/growing season; vegetation period
вегетация vegetation, growth
вегетоневроз vegetative neurosis
ведаизм Vedaism
веданта Vedanta
ведение conducting, conduct, guiding, leading
~ архивов/картотек filing
~ бухгалтерских книг bookkeeping
~ дела conduct of an affair; transaction
~ записей record-keeping
~ мяча *sport* dribbling
~ отчетности cost accounting
~ родов management of birth
~ хозяйства housekeeping
веджер-бот wager boat
веди Slavonic name of the letter B
ведийский, ведический Vedic; Vedantic
ведовство sorcery, witchraft, Wicca
ведомость list, register, statement, report, record; log; form; *econ.* payroll, journal; 2. *pl.* gazette
~, вспомогательная supporting record/schedule
~, дефектная repair request form
~, дефектовочная statement of rejects, inspection record
~, дополнительная supporting schedule
~, дорожная motor waybill
~, зачетная academic record
~ издержек cost sheet
~, инвентаризационная inventory sheet
~, инвентарная statement of equipment
~, калькуляционная cost sheet/record
~, комплектовочная list of standard equipment, equipment component list
~, контрольная check list, inspection sheet
~ координат index of coordinates
~ материалов bill/schedule of materials
~ наличия на складе stock status report
~ наличных товаров statement of goods
~ осмотров и ремонтов inspection and maintenance record
~, отгрузочная shipping list
~ оценок report card
~, платежная paysheet, payroll
~, продовольственная victualling book
~, рассылочная mailing list
~, расходная expense/charge sheet; list of issues
~, расчетная pay list, paysheet, payroll
~, сопроводительная bill of lading
~, упаковочная packing slip/list
~ цеховых издержек department cost sheet
ведомственный departmental, institutional
ведомство service, office, department; authorities, authority
~, железнодорожное railway administration
~, налоговое fiscal administration
~, патентное patent office
ведомый *av.* wingman; supporting aircraft; *comp.* slave
ведро 1. bucket, pail; pailful; 2. vedro, twenty-one pints
~, дождемерное rain gauge receiver
~, заправочное pouring pail
ведущий 1. leading, principal; driving, guiding; *comp.* master; 2. *n.* leader, lead aircraft, leading plane
~ программу anchor man
~щее зубчатое колесо pinion, driving gear
~щее колесо driving wheel, drive sprocket
веды Veda
веер 1. fan; spread; 2. *bot.* rhipidium
~ дельтовых отложений deltaic fan
веерный fan-shaped, flabellate
вееровидный, веерообразный flabellate, fanlike,fan-shaped, flabelliform; radial
~ свод *arch.* fan tracery
веерокрылки *pl. zool.* many-plume moth *(Orneodidae)*
вееролистный *bot.* fan-leaved, flabellifoliate
веeронервный fan-veined/-nerved; diadromous
вездеприсутствующий omnipresent, ubiquitous
вездесущие omnipresence, ubiquitousness
вездесущий 1. omnipresent, ubiquitous; 2. *n.* God
вездеход cross-country vehicle, all-purpose/-terrain vehicle; landrover
вездеходный cross-country *(of vehicle)*
везикальный *med.* vesical, vesicant
везикария *bot.* bladderpod *(Vesicaria, Lesquerella)*
везикосигмоидальный vesicosigmoid
везикоуретральный vesicourethral
везикула vesicle, vesicule, vesicula
везикулезный, везикулярный vesicular
везикулит vesiculitis
~, трихомонадный mycotic vesiculitis
везувиан vesuvian(ite); idocrase
~, марганцевый mangan-vesuvianite
~, массивный californite
везувианит vesuvianite
везувит vesuvite
вейка purifier
вейник *bot.* reed grass *(Calamagrostis)*
вейнрута *bot.* common rue *(Ruta graveolens)*
вейсманизм *biol.* Weismannism
вейсманистский, вейсмановский Weismannian
век 1. age, stage, epoch, time; 2. century
~, атомный atomic age
~, бронзовый Bronze Age
~, вюрмский Wurmian stage
~, гюнцский Gunzian stage
~, древний каменный Paleolith
~, железный Iron Age
~, камазийский Kamasian
~, каменный Stone Age
~, неолитический Neolithic period
~, новый каменный Neolithic period

~, плювиальный pluvial period
~, рисс-вюрмский межледниковый Riss-Wurm interglacial age
~, рисский ледниковый Riss glacial age
~ человека human period, Recent
веко *anat.* eyelid, palpebra
~, верхнее/нижнее upper/lower eyelid
вековой age-long, age/centuries-old; *math.* secular
векодержатель eye speculum
векорасширитель blepharostat
векселедатель *econ.* drawer *(of a bill)*
векселедержатель payee, holder
векселеполучатель endorsee, indorsee
вексель *econ.* (promissory) note, bill of exchange; paper
~, безденежный accomodation bill/note/paper
~ в местной валюте inland/domestic bill
~ в одном экземпляре single/sole bill
~, домицилированный addressed/domiciled bill
~, дружеский accomodation bill /paper/note, pro forma bill
~, дутый kite
~ к оплате bill for payment
~, коммерческий commercial bill
~, краткосрочный простой short-term note
~ на инкассо bill for collection
~ на предъявителя draft at sight, sight/ presentation draft
~, недокументированный clean bill of exchange
~, обеспеченный collateral note
~, обратный переводный redraft, return draft
~ оплачивать meet a bill
~, опротестованный noted bill
~, переводный bill of exchange
~, погашаемый золотом gold note
~, подтоварный businness/commodity paper
~, предъявительский promissory note made out to bearer, *Am.* demand note; demand bill, bill at sight, sight bill of exchange
~, простой/прямой promissory note, note of hand
~ с бланковым индоссаментом skeleton bill
~ с платежом после предъявления after sight bill
~ с тремя подписями three-name paper
~ со сроком, исчисленным с момента выдачи after date bill
~, торговый commercial bill
~, учесть discount a bill
~ля, казначейские *Eng.* treasuries
~ля, учитивемые bankable securities
вексельный: ~ кредит paper credit
~ курс rate of exchange
вектор vector
~ аномалии disturbance vector
~ базиса basic/basis vector
~, безвихревой irrotational vector
~ Бюргерса Burgers/slip vector
~ ветра wind arrow/vector
~ вихря vortex vector
~ возмущения disturbance vector
~, волновой propagation vector
~ вращения spin vector
~ встречи collision vector
~ выигрыша *game th.* payoff vector
~, двойной divector
~, единичный unit/direction vector
~ импульса momentum vector
~, инкрементный incremental vector
~ интенсивностей intensity vector
~ конечного спроса final demand vector
~, координатный position vector
~ кривизны buckling vector
~, ненулевой non-vanishing vector
~, n-мерный n-tuple vector, n-dimensional vector
~ нормали normal vector
~, нормированный normalized/standardized vector
~, нулевой null/zero vector
~ ограничений constraint vector
~, осуществимый admissible vector
~ перемещения displacement vector
~ подъемной силы lift vector
~ Пойнтинга Pointing's vector, energy-flux vector
~, предельный limit(ing) vector
~, пространственный space vector
~ путевой скорости track velocity, ground-speed vector, track vector
~, свободный slack vector
~, связанный bound vector
~ силы тяжести gravitational vector
~ скольжения Burgers/slip vector
~, скользящий nonlocalized/sliding vector
~ скорости velocity vector
~, случайный random vector
~, собственный eigenvector, proper/latent vector
~ сопротивления *nav.* drag vector
~, составляющий component vector
~ состояния state vector
~, тангенциальный tangent vector
~ тяги thrust vector
~ угловой скорости angular-velocity vector
~ Умова-Пойнтинга energy flux vector, Pointing's vector
~ ускорения acceleration vector
~, экстремальный extreme vector
~ы, взаимные reciprocal vectors
~ы, ко(м)планарные coplanar/complanar vectors
вектор-градиент gradient vector
векториальный, векторный vector(ial)
векторизация vectoring
вектор-операнд operand vector
вектор-решение solution vector
вектор-столбец column vector
вектор-строка row vector
вектор-электрокардиоскоп vector electro-

cardioscope
веламен *anat., bot.* velamen
веленевый vellum, wire-wove
велень vellum (paper)
Велиар *theol.* Belial
велигер *zool.* veliger
великий 1. great; grand; 2. big, large; too big/old
~ князь grand prince/duke
~ четверг *eccl.* Maundy Thursday
~кие державы the Great Powers
великовозрастие *educ.* overageness
великоморавский Great Moravian
великомученик *eccl.* great martyr
великопостный *eccl.* Lenten
великоросс(ийский), великорус Great Russian
великосветский societarian; fashionable
величество majesty
~, ваше Your Majesty
величие grandeur, majesty, greatness
величина 1. value; magnitude; 2. quantity; amount; 3. size, dimension; scale, degree, extent, intensity
~, абсолютная absolute value, modulus
~, безразмерная non-dimensional/dimensionless quantity
~, бесконечно большая *math.* infinite (quantity), infinitely large magnitude
~, бесконечно малая infinitesimal/vanishing magnitude/quantity
~, выходная output quantity/value, output
~ вычислительной машины, входная computer input quantity
~, граничная boundary value
~, действующая virtual value
~, дополнительная complement
~, допускаемая allowable/permissible value
~, заданная prescribed/set/given value, *comp.* nominal/target value
~ задержки delay factor
~ затухания attenuation
~, именованная denominate quantity
~, искомая the unknown (quantity)
~, исходная basic value
~, колебательная oscillating quantity
~, конечная finite/final quantity/value
~ мертвой зоны skip distance
~, мнимая imaginary quantity
~, наблюденная observed value
~, наивыгоднейшая optimum value
~ насыщения saturation value
~, натуральная actual/full size/scale
~, натурная full-scale value
~, неизвестная unknown quantity
~, непрерывная analogue quantity
~, непрерывная случайная continuous random variable
~, ничтожная negligible quantity
~, номинальная nominal/rated value
~, нормированная standardized value
~, обратная reciprocal/inverse quantity/value
~, ограниченная bounded quantity
~, опорная reference value
~, оптимальная optimum/optimal value/quantity
~ ошибки extent of error, error amount
~, переменная variable (quantity)
~, периодическая periodic quantity
~, планируемая target variable
~ погрешности magnitude of error
~ помех noise level
~, пороговая threshold
~, постоянная constant (quantity)
~, постоянная во времени time-independent quantity
~ потерь энергии rate of energy loss
~, предельная limiting value
~ притока inflow rate
~, прогнозируемая predicted value
~, производная derivative
~, размерная denominate quantity
~, расчетная design quantity/variable/parameter
~, регулируемая controlled quantity/variable
~, регулирующая regulated condition, manipulated variable
~, синусоидальная harmonic quantity
~, скалярная scalar (quantity)
~, случайная random/stochastic/chance variable/variate, stochastic value
~, собственная eigenvalue
~, средневзвешенная weighted average/mean
~, среднеквадратичная root-mean-square value, rms value
~, средняя average/mean value
~, стохастическая stochastic/random variable
~, суммарная total value
~ тяги thrust value
~, удельная specific quantity
~, уравненная adjusted quantity
~, установившаяся steady-state value
~ ухода гироскопа gyro drift rate
~, целевая target variable
~, элементарная *comp.* elementary item
~, эмпирическая empirical value
~, эталонная reference quantity
~, эффективная effective/virtual value, root-mean-square value, rms value
веллингтония *bot.* giant sequoia *(Sequoia gigantea)*
веловентиль bicycle tube valve
велогонка bicycle race
велодром velodrome, bicycle track
велокамера bicycle tube
велокросс bicycle cross-country race
велопробег bicycle race
велорикша bicycle rickshaw
велосипед bicycle, cycle; tricycle
велоспорт cycling; bicycle racing
велотрек (bi)cycle track
велошина bicycle tire
~, гоночная bicycle racing tire
велум *bot.* velum, veil, integument

вельбот whaleboat, whaler
вельвет(ин) velveteen; (cotton) velvet
вельвет-корд corded/ribbed velveteen
вельветовый 1. velveteen; 2. *pl. zool.* velvet fishes *(Caracanthidae)*
вельвет-плющ plush velveteen
вельвет-рубчик twilled velvet
вельвичия *bot.* welwitschia *(Welwitschia)*
вельд *geol.* Wealden stage
вельзевул Beelzebub
вельможа grandee, great noble/lord
вельможество aristocracy, nobility
вельпс whelp
вельтшмерц *lit.* Weltschmerz
велюм *zool.* velum
велюр 1. velvet; 2. velvet leather, buff; nappa
веляризация *phon.* velarization
вена *anat.* vein, *see also* **вены**
~, бедренная femoral vein
~, воротная portal vein
~, глубокая шейная deep cervical vein
~, головная cephalic vein
~, диафрагмальная phrenic vein
~, затылочная occipital vein
~, кожная cutaneous vein
~ лица, глубокая deep facial vein
~, лицевая facial vein
~, межпозвоночная intervertebral vein
~, надглазничная supraorbital vein
~, надлопаточная suprascapular vein
~, нижняя брыжеечная inferior mesenteric vein
~, нижняя надчревная inferior epigastric vein
~, носо-лобная nosofrontal vein
~, общая основная common basal vein
~, подключичная subclavian vein
~, подколенная popliteal vein
~, подкрыльцовая axillary vein
~, подподбородочная submental vein
~, подреберная subcostal vein
~, подъязычная sublingual vein
~, позвоночная vertebral vein
~ полосатого тела striate vein
~, полунепарная hemiazygos vein
~ преддверия vestibular vein
~, предпривратниковая prepyloric vein
~, пузырная cystic vein
~, пупочная umbillical vein
~, селезеночная splenic vein
~ сердца, большая great cardiac vein
~ сердца, малая small cardiac vein
~, шило-сосцевидная stylomastoid vein
~ языка, глубокая deep lingual vein
~, язычная lingual vein
~, яичниковая ovarian vein
~, яремная jugular vein
вена-спутница accompanying vein
венгерец, венгр, венгерский Hungarian
венгерка 1. Hungarian(woman); 2. Hungarian dance; 3. *bot.* garden plum *(Prunus domestica)*
венгерское *n.* Hungarian wine
венд Wend, Sorb
вендетта vendetta
вендиспансер *abbr.* **венерический диспансер** venereal disease dispensary
венепункция *med.* venipuncture
Венера *astr., myth.* Venus
венерин волосок *bot.* maidenhair *(Adiantum)*
венерический *med.* venereal
венеролог venerologist
венерология venerology
венесекция venesection, venisection, phlebotomy
венесуэлец, венесуэльский Venezuelan
венеты *pl. hist.* Vends
венец I Viennese, inhabitant of Vienna
венец II 1. *tech.* ring, rim; edge; 2. *med.* crown; 3. *constr.* timber set; 4. *geol.* curb; 5. *eccl.* halo, aureole, corona
~, водонепроницаемый основной water curb
~, зубчатый gear ring/rim, toothing; girth/ring gear
~, зубчатый с внутренним зацеплением internal gear ring
~, каменный *geol.* stone/rock wreath
~, контактный corona, aureole; contact group/ring
~, лопаточный blading, blade ring
~, небесный *eccl.* aureole
~, опорный *min.* foundation curb
~, ресничный *anat.* ciliary crown
~, царский *bot.* crown imperial, imperial fritillary *(Fritillaria imperialis)*
~ цветка *bot.* corona
венецианец Venetian
венецианский Venetian
~ ярь verdigris
венечник *bot.* antherium *(Antherium)*
венечно-апертурный stephanoaperturate
венечно-бороздной stephanocolpate
венечный coronal, coronary; corolline; coronate(d)
~ шов coronal suture
вензель monogram
венипункция *med.* venipuncture
вениса *miner.* andradite, garnet
~, алабандская almandite, almandine
венит venite, veined gneiss
веничек 1. *bot.* panicle; 2. whisk
веничный *bot.* paniculate
венный, венозный *med.* venous
вено *hist.* bride price
вено-артериальный venoarterial
веночек *bot.* sertulum
венский Viennese
~ стул bentwood chair
~кое питье *pharm.* senna tea, black draft
вентилирование ventilation, aeration
вентилировать ventilate, air, fan
вентиль 1. *tech.* valve, vent; *el.* rectifier; tube/valve 2. *comp.* gate; *rad.* isolator; 3. *mus.* mute

~, **впускной** inlet/intake valve
~, **входной** in-gate
~, **выпускной** escape/release/discharge valve
~, **вытяжной** exhaust valve
~, **выходной** out-gate; read-out gate
~, **двухходовой** two-way valve
~, **декодирующий** decoding gate
~, **диафрагмовый** diaphragm valve
~, **диодный** diode gate
~, **дренажный** drain valve
~, **дроссельный** throttle valve
~, **дуговой** mercury-arc rectifier
~, **запирающий** locking gate
~, **запорный** stop valve
~, **запорный проходной** globe valve
~, **запорный угловой** angle valve
~ **запрета** inhibitory gate
~ **И** AND gate
~, **игольчатый** needle valve
~ **ИЛИ** OR gate
~ **ИЛИ-НЕ** NOR-gate
~, **импульсный** pulse gate
~ **И-НЕТ** inhibiting circuit
~, **интегральный** integrated circuit gate
~, **ионный** gas-filled rectifier valve, gas-filled tube
~ **камеры** tube air valve
~ **камеры шины** inner tube/tyre valve
~, **купроксный** copper-oxide rectifier
~, **ламповый** rectifier valve, *Am.* valve tube
~, **логический** logic(al) gate
~, **мажоритарный** majority gate, linear-input gate
~, **меднозакисный** copper-oxide rectifier
~, **многоходовой** multiway valve
~, **муфтовый** coupled valve
~ **НЕ** NOT gate
~ **НЕ И** NAND gate
~, **незадействованный** spare gate
~ **несовпадения** anti-coincidence circuit
~, **обводной** by-pass valve
~ **обмена информации** swap gate
~ **обратного хода** complement gate
~, **оксидный** metal-oxide rectifier
~, **оптический** optical isolator
~, **оптический бегущей волны** travelling-wave optical isolator
~ **переноса** carry gate
~, **поглотительный** absorption isolator
~, **полупроводниковый** semiconductor rectifier
~, **предохранительный** safety valve
~, **продувочный** blow-down/-off/-out valve
~, **проходной** straightway/globe valve
~ **прямого хода** true gate
~, **прямоточный** straight-through valve
~, **регулируемый** adjustable valve
~, **регулирующий** control(ling) valve
~, **ртутный** mercury-arc rectifier/tube/valve, mercury-arc rectifier tube
~, **ртутный стеклянный** glass-bulb mercury-arc rectifier
~, **ручной** manually-controlled valve
~ **с двумя входами** double-input gate
~ **с N входами** N-input gate
~ **с тлеющим разрядом** glow-discharge rectifier
~ **с тремя состояниями** tri-state gate
~, **сливной** drain valve
~ **сложения** add gate
~, **сухой** dry-disk rectifier
~, **счетверенный** quad gate
~ **считывания** read-out gate
~, **твердый** dry-disk rectifier
~, **титановый** titanium-dioxide rectifier
~, **точечно-контактный** point contact rectifier
~, **трехходовой** three-way valve
~, **управляемый** controlled rectifier; controllable gate, C-gate
~, **управляющий** control valve
~ **установки на нуль** zero gate
~, **фланцевый** flanged valve
~, **четырехслойный** four-layer semiconductor rectifier
~, **четырехходовой** four-way valve
~, **шиберный** gate valve
~, **электровакуумный** rectifier valve, *Am.* valve tube
~, **электролитический** electrolytic rectifier
~, **электрохимический** chemotronic rectifier

вентильность valve ratio
вентильный *comp.* gated
вентиль-формирователь gate driver
вентилятор (air) blower, ventilator; fan
~, **вдувной** induction/supply fan
~, **винтовой** helical fan
~, **винтовой крыльчатый** propeller-type fan
~, **всасывающий** drawing fan
~ **вытеснения** positive-displacement fan
~, **вытяжной** (air) exhauster, exhaust/suction fan, air-ejector fan, exhaust/suction blower
~, **дутьевой** blow/blast (air) fan, forced-draft blower/fan
~, **качающийся** pivoting fan/ventilator, oscillating-base fan/ventilator
~, **малооборотный** low-speed fan
~, **многолопастный** multiblade fan
~ **на валу** shaft-mounted fan
~, **нагнетательный** forced-draught fan, supply blower/fan, pressure fan
~, **нагнетающий** blower fan
~, **низконапорный** low-pressure/-ratio fan
~, **низкооборотный** low-speed fan
~ **обдува** blower
~, **объемный** positive-displacement fan
~ **одностороннего всасывания** single-suction fan
~, **осевой** axial(-flow) fan/blower
~, **отсасывающий** suction/exhaust fan
~ **очистки** *agr.* separator fan
~ **пневмотранспорта** material handling/con-

veying fan
~, поперечно-проточный cross-flow fan
~, потолочный ceiling fan
~ принудительной тяги forced-draught fan
~ приточной вентиляции forced-ventilation fan
~, приточный forcing fan
~, пылевой dust exhaust fan
~, радиальный centrifugal fan
~, реверсивный reversible fan
~, ротационный rotary blower
~ с редукторным приводом geared fan
~, туннельный *av.* ducted fan
~, тяговый thrust/propeller fan, fan propeller
~, шахтный mine fan
вентилятор-измельчитель *agr.* chopper fan
вентиляционный ventilating, ventilation *(attr.)*
вентиляция 1. ventilation, venting; 2. ventilation system
~, бесквозняковая draughtless ventilation, no-draught ventilation
~, вытяжная exhaust ventilation
~ диффузией diffusion ventilation
~, искусственная artificial/mechanical/induced ventilation
~, легочная lung ventilation
~, принудительная/приточная forced ventilation
~, приточно-вытяжная plenum-exhaust ventilation, balanced ventilation
~, регулируемая controlled/adjustable ventilation
~ с увлажнением air conditioning, humidity ventilation
~, смешанная compound ventilation
вентрикулит *med.* ventriculitis
вентрикулография ventriculography
вентрикулопункция ventriculopuncture
вентрикулоскопия ventriculoscopy
вентрикулотомия ventriculotomy
вентрикулярный ventricular
вентрогистеропексия ventrohysteropexy
вентромедиальный ventromedian, ventrimesal, ventromedial
вентроптоз ventroptosis
вентроскопия ventroscopy
вентрофиксация ventrofixation
венус *zool.* quahog *(Venus)*
венценосец *rel.* crowned head, wreath monarch, sovereign
венценосный *bot.* coroniferous, stephanophorous; wreath-bearing
венцеобразный crown-shaped
венчание wedding ceremony
~ на царство coronation
венчик 1. *med.* crown *(of tooth);* 2. *tech.* collar, edge, rim, bead; 3. halo, nimbus; 4. *bot.* corolla; 5. *geol.* crown, corolla, corona
~, внутренний stylostegium
~, губовидный labiate corolla
~, имеющий corolliferous
~, придаточный corolla appendage
~ споры corona
~, трубчатый tubular corolla
~ ударника collar of striker
венчиковидный, венчикообразный corolliform, corollaceous
~ околоцветник *bot.* corolla
венчикоплодный *bot.* stephanocarpous
венчикоцветный corollifeorous
вены *pl. anat.* veins
~, барабанные tympanic veins
~ большого мозга cerebral veins
~ век palpebral veins
~, верхнечелюстные maxillary veins
~, водоворотные vorticose veins
~, воспаленные phlebitis
~, глоточные pharyngeal veins
~, легочные pulmonary veins
~, локтевые ulnar veins
~, лучевые radial veins
~, маточные uterine veins
~, надблоковые supratrochlear veins
~, надсклеральные episcleral veins
~, печеночные hepatic veins
~, пищеводные esophageal veins
~, плечевые brachial veins
~ поджелудочной железы pancreatic veins
~, почечные renal veins
~, поясничные lumbar veins
~, расширенные varicose veins
~, решетчатые ethmoidal veins
~ сердца cardiac veins
~ шеи, поперечные transverse veins of the neck
вены-выпускники emissary veins
венэкстрактор vein stripper
вепрь wild boar
вепс, вепсский Veps(e)
веранда veranda(h); porch, terrace
верба *bot.* daphne willow *(Salix daphnoides)*
вербальный verbal
~ная нота *dipl.* verbal note, note verbale
вербейник *bot.* loosestrife *(Lysimachia)*
~, дубравный wood pimpernel *(L. nemorum)*
вербена *bot.* vervain, verbena *(Verbena)*
~, аптечная/лекарственная European verbena *(V. officinalis)*
вербигерация *med.* verbigeration
верблюд *zool.* camel *(Camelus)*
~, двугорбый two-humped camel, Bactrian camel *(C. bactrianus)*
~, одногорбый one-humped camel, Arabian camel *(C. dromedarius)*
верблюдка *bot.* tickseed *(Corispermum)*
вербный:
~ базар Palm Sunday fair
~ное воскресенье *eccl.* Palm Sunday
вербование, вербовка recruitment, recruiting, enlistment
вербовать enlist, enroll, recruit

вербовщик recruiting agent, recruiter; labo(u)r contractor
вербовый willow; osier *(attr.)*
~вая корзинка wicker basket
верболоз *bot.* bay-leaf willow, laurel willow *(Salix pentandra)*
вервь *hist.* community, commune; territorial division, district, region
вергельд blood money, bloodwit(e)
вердикт judgment, verdict
вердогемоглобин *med.* verdohemoglobin, choleglobin
веревка rope, cord; string, line
~, вытяжная extraction cord
~ для белья chothesline
~, привязная captive line
~, разрывная rip cord
~, спасательная rescue line
~, швартовочная mooring line
веревочный rope *(attr.)*; funicular
~ многоугольник circumsribed polygon
~ная лестница rope ladder
веревчатый *bot.* stringlike, cordlike, funiform
вередник *bot.* field pennycress *(Thlaspi arvense)*
вереск *bot.* heather *(Calluna)*
~ восковой wax myrtle *(Myrica cerifera)*
вересковидный *bot.* ericoid
вересковый heather, heathy, ericetal
веретеница *zool.* blindworm, slowworm *(Anguis fragilis)*
~, панцирная glass lizard *(Ophisaurus)*
веретенник *zool.* godwit *(Limosa)*
веретено 1. spindle; shaft, shank; 2. axis, axle
~ весла oar shaft
~ двойного кручения double-twist spindle
~, подвесное flexibly mounted spindle
~, самоцентрирующее gravity spindle
~ якоря *nav.* anchor shaft/shank
веретеновидный, веретенообразный fusiform, spindle-shaped, spindlelike
веретеноклеточный fusocellular, spindle-cell
верея *naut.* wherry
верзор *math.* versor
веризм verism, verismo
верительный *dipl.* accrediting
~ные грамоты letters credential, credentials, letters of credence
верификатор *comp.* verifier
верификация verification
~ временных соотношений timing verification
веркблей lead bullion; work lead
верлибр *lit.* free verse, vers libre
вермикулит *geol.* vermiculite
~, вспученный expanded vermiculite
вермикулитобетон vermiculite concrete
вермишель *cul.* vermicelli
вермут 1. *bot.* absinthium, common wormwood *(Artemisia absinthium)*; 2. Vermouth wine
вернерит scapolite, wernerite
вернеритит werneritite
вернисаж *art.* 1. private viewing, preview; 2. opening/first day
верноня *bot.* ironweed *(Vernonia)*
верноподданность allegiance, loyalty
верноподданный loyal, faithful
верноподданство loyalty, allegiance
верность 1. loyalty, faithfulness; 2. correctness, accuracy, truth; trustworthiness
верный 1. true, loyal, faithful; 2. correct, right; 3. sure, reliable; certain
верньер vernier; vernier dial; nonius, slide gauge
~ буссоли declination vernier
~, обратный retrograde vernier
~ теодолита transit vernier
вероисповедание creed, denomination; faith; religion
~ния, свобода freedom of worship/ conscience/ religion
вероисповедный confessional, denominational; religious
веронал *pharm.* barbital, veronal
веронец Veronese
вероника *bot.* speedwell *(Veronica)*
вероотступник *rel.* apostate
вероотступнический apostatic(al), apostate
вероотступничество apostasy
веропроповедание preaching the Gospel
веротерпимость religious toleration/tolerance
веротерпимый tolerant, latitudinarian
вероучение *rel.* religious doctrine, dogma, creed
вероучитель religious teacher, apologist
вероятностный probabilistic, probability *(attr.)*, probable, likely; stochastic
вероятность probability, likelihood
~, апостериорная posterior /inverse/empirical probability, a prosteriori probability
~, априорная prior/mathematical probability, apriori probability
~ безубыточности breakeven probability
~, безусловная unconditional probability
~ выживания *math.* survival probability
~ выигрыша *game th.* winning probability
~ вылета escape probability
~, геометрическая geometrical probability, probabilistic geometry
~ дефицита stockout/shortage probability
~, доверительная confidence level/probability/ coefficient, level of confidence
~ дожития chance of survival
~ дожития до определенного возраста survival probability
~ долговечности survival probability
~, заданная specified probability
~, истинная true probability
~ исчерпания запасов runout probability
~ наступления события event probability
~ ничейного исхода *game th.* probability of tying
~, обратная inverse probability

~ **отказа** loss/failure probability
~ **отсутствия требований** idleness probability
~ **ошибки** error probability, probability of error
~, **перехода** transition probability
~, **полная** composite/total probability
~ **попадания** hit probability
~ **появления интервала занятости** busy-period probability
~ **приемки** chance/probability of acceptance
~, **сложная** compound probability
~ **смерти** mortality probability
~, **суммарная** total/cumulative probability
~, **эмпирическая** empiric(al) probability
вероятный probable, likely; credible; stochastic; verisimilar
верп(-анкер) *naut.* kedge (anchor), stream anchor
веррукария *bot.* wartwort *(Verrucaria)*
веррукозный verrucous, verruciform
версальский Versailles *(attr.)*
~ **договор** *polit., hist.* treaty of Versailles
версификатор versifier, versemaker
версия version
верста verst *(3500 feet)*
верстак (work) bench, file/work bench
~, **гнутарный** bending trestle/horse
~, **слесарный** fitter's bench
~, **столярный** carpenter's/joiner's bench
~, **строгальный** planing/shaping bench
~, **формовочный** moulder's bench
верстальщик *print.* make-up man, maker-up; imposer, clicker
верстатка composing stick, galley
верстать I *print.* make up, compose, impose
верстать II *hist.* 1. conscript (for), recruit (for); 2. give, assign
верстка *print.* 1. page proof; form, made-up matter; 2. imposing, imposition
~, **наборная** composing stick, galley
верстовой:
~ **столб** milestone
вертебрально-базилярный vertebral-basilar
вертекс *astr.* vertex
вертел 1. skewer, roaster; spit, turnspit; 2. *anat.* trochanter
вертелка *text.* warp loom
вертеловидный spit-shaped
вертельный trochanterian, trochanteric; drilling, boring
вертиго *med.* vertigo; giddiness, dizziness
вертикал (descending)vertical flue; vertical circle
~, **первый** prime vertical circle
вертикализация vertical adjustment
вертикаль (the) vertical, vertical line; file *(on chessboard)*
~ **аэроснимка** parallel to the principal of line
~ **аэроснимка, главная** principal line/meridian
~, **гравитационная** gravity/gravitational vertical
~, **промерная** sounding vertical
~, **центральная гироскопическая** master vertical gyro
вертикальность verticality, perpendicularity
вертикальный vertical, erect, plumb, upright, perpendicular
вертиметр vertimeter
вертишейка *zool.* wryneck *(Jynx torquilla)*
вертлуг, вертлюг 1. *anat.* trochanter, head of the femur; 2. *tech.* swivel, shackle; 3. *mil.* pivot bracket/yoke
~, **канатный** rope swivel
вертлюжно-крестцовый cotylosacral
вертлюжно-лонный cotylopubic
вертлюжный swivel; acetabular, cuplike
~**ная впадина** *anat.* acetabulum
вертляница *bot.* Indian pipe *(Monotropa)*
вертодром heliport, helipad
вертолет helicopter
~, **беспилотный, телеуправляемый** drone helicopter
~, **боевой** combat/gunship helicopter
~, **всепогодный** all-weather helicopter
~ **вспомогательного назначения** utility helicopter
~, **грузовой** freight/cargo helicopter
~, **комбинированный** compound helicopter
~, **корабельный** ship-based helicopter
~, **многовинтовой** multirotor helicopter
~, **нескоростной** low-speed helicopter
~, **одновинтовой** single-rotor helicopter
~, **палубный** ship-based/-borne helicopter, carrier-based helicopter
~, **пилотируемый** manned helicopter
~ **ПЛО** anti-submarine helicopter
~, **поплавковый** float-type helicopter
~, **противопожарный** water-bomber helicopter
~, **санитарный** ambulance/hospital/casualty helicopter
~ **связи** liaison helicopter
~ **соосной схемы** co-axial(-rotor) helicopter
~, **спасательный** rescue helicopter
~, **сухопутный** land-based helicopter
~, **транспортно-десантный** troop-carrier helicopter
~, **турбореактивный** turbojet helicopter
вертолет-амфибия amphibious helicopter
вертолет-буксировщик towing helicopter
вертолет-корректировщик (artillery) observation helicopter
вертолет-кран crane helicopter
вертолетоносец helicopter carrier/ship
вертолет-ракетоносец missile-carrying helicopter
вертолет-самолет convertiplane, convertible helicopter
вертолет-такси helitaxi
вертолет-тральщик mine-sweeping/-countermeasures helicopter
вертопорт heliport
Вертумн *myth.* Vertumnus
вертунья, листовая *zool.* lesser bud moth *(Recurvaria nanella)*

вертушка 1. impeller; current meter; airstream propeller; 2. screw-propeller, screw rotator, vane, safety screw
~, цветная colo(u)r wheel
~ Экмана Ekman current meter
вертячка 1. *vet.* gid, turn sickness, blind staggers; 2. *zool.* whirligig beetle *(Gyrinus)*
верующий *rel.* believer
верфь shipyard, dockyard, yard
~, достроечная outfitting yard
~ речного судостроения river shipyard
~, судосборочная ship-assembly yard
~, судостроительная shipbuilding yard
верх 1. top, summit; head; 2. This Side Up *(label on a packing)*
~ обуви upper
~, фреатический *geol.* phreatic high
верхнебродильный top-fermented/-fermenting
верхнеглоточный upper glottalic
верхнегортанный superlaryngeal
верхнедолевой superlobar, supralobar
верхнеклапанный overhead-valve
верхнелужицкий Upper/High Sorbian
верхнемеловой Upper Cretaceous
верхненемецкий High German
верхнепропускающий high-pass
верхнереберный supercostal
верхнеузловой upper-nodal
верхнефюзеляжный dorsal
верхнечелюстной maxillary
верхний upper, top; apical, superior, overhead; surface, outer
~ регистр *mus.* highest register
~няя палата *polit.* upper chamber/house
верхняк capping; roof timber
~ дверного оклада cross-beam
верховный supreme; sovereign
~ главнокомандующий supreme commander in chief
~ Совет СССР *obs.* Supreme Soviet of the USSR
~ Суд Supreme Court
~ное существо *rel.* Supreme Being, God
верховодка top/leakage/vadose/ temporary water
верховые люди *hist.* Czar's retinue/suite
верховье riverhead; upper reaches
~ долины upper valley, valley head
~ реки river head, upper reaches/course, head-stream
верхоплодный *bot.* acrocarpous
верхоростный acropetal
верхоцветник *bot.* cyme
верхушечноцветковый acranthal
верхушечноцветный terminiflorous
верхушечный apical, terminal, tip *(attr.)*
~ побег terminal shoot
верхушка apex, tip; vertex; acme; top, summit
~, вегетационная *bot.* vegetative/apical cone
~ легкого apex of the lung
~, правящая *polit.* ruling clique
~ ушной раковины tip of the auricle
верчение *naut.* whirl(ing); *tech.* spinning, rotating, turning, revolving
верша fish bow net, fish trap
вершина peak, summit; top, crest; pinnacle; culmen; fastigium; *gram.* head; *geom.* apex, vertex
~, висячая dangling vertex
~ водораздела watershed head, interstream spur
~ волны wave crest
~ гиперболы vertex of a hyperbola
~, горная peak, mountain top, summit
~ графа vertex/node/ point of a graph
~ долины valley head
~ залива head of a bay
~ зуба tooth tip
~, интонационная crest of intonation curve
~, конечная terminal vertex
~ конуса apex/vertex of a cone
~, корневая root vertex
~ кривой peak of a curve
~ лопатки blade head
~ оврага gully head
~, опрокинутая inverted apex
~, пирамидальная horn
~ пирамиды vertex of a pyramid
~ побега shoot apex
~ подъема *av.* top of climb
~ резца tool nose/point
~ резьбы crest of (a) thread
~ складки apex of fold
~ стека top of stack
~ угла vertex of (an) angle
вершинкование tipping, topping
виршинник top log
вершинный apical
вершинообразующий crest-/peak-forming
вершок 1. vershok *(1. 75 inches)*; 2. peak, top
вес 1. weight; 2. *phys.* gravity
~, атомный atomic weight
~ брутто gross/packed/total weight
~ в концу периода беконного откорма *agr.* bacon weight
~, взлетный take off weight
~ взрослого животного *agr.* adult weight
~ во время откорма на пастбище *agr.* field weight
~ голого корпуса *naut.* bare hull weight
~, живой *agr.* body/live weight
~, (живой) после голодной выдержки *agr.* empty live weight
~ зеленой массы *agr.* green weight
~, изотопический/изотопный isotopic weight
~ индекса *stat.* index weight
~ коммерческой нагрузки pay load weight
~, кондиционный *agr.* conditioned weight
~, конечный final weight
~ конструкции structure/structural weight
~, контрольный check weight
~, молекулярный molecular weight
~ мытой шерсти, чистый *agr.* clean equiva-

lent weight
~ мясной туши *agr.* carcass/dead/slaughter weight
~, насыпной bulk weight
~ немытой шерсти *agr.* greasy weight
~, неполный short weight
~ нетто net weight
~, общий gross/total weight
~, объемный volume/unit weight, density
~, отгруженный shipped weight
~, относительный relative weight, weight-to-payload ratio
~, отъемный *agr.* weaner weight
~, оценочный estimated weight
~ ошибки error rate/frequency
~, первоначальный starting weight
~, первый полусредний *sport* light welter-weight
~, перегрузочный overload weight
~ платной нагрузки payload weight
~ полезной нагрузки payload weight
~, полетный flying/flight/airborne weight
~, полный gross/total weight
~, посадочный landing weight
~, предубойный *agr.* preslaughter weight
~ при отъеме *agr.* weaner weight
~ при разгрузке landed weight
~ при рождении birth weight
~, пустой empty/basic weight
~, расчетный design/estimated/calculated weight
~ с нагрузкой loaded weight
~ с полной нагрузкой gross weight
~ с тарой gross weight
~, скользящий *stat.* moving weight
~, собственный dead/empty/own/tare weight
~, спусковой/стартовый launch(ing) weight
~, тарифный estimated weight
~ тары tare weight
~ тары по тара тарифу schedule/customs tare
~ тары, установленный обычаем customary/uso tare
~, тройский troy weight
~, убойный *agr.* carcass/dead/slaughter weight
~, удельный density; specific weight/gravity, weight-to-power ratio
~, удельный, выраженный в градусах Боме Baume gravity
~, чистый net weight
~, эквивалентный weight equivalent
~, эксплуатационный weight in running order; *av.* operating weight, aircraft-prepared-for-service weight, A. P. S. weight
веса входов *comp.* input weights
веселка I spatula, stirrer
веселка II *bot.* : **~ вонючая/обыкновенная/рядовая** stinkhorn *(Phallus impudicus)*
весельный oared; rowing *(attr.)*
веселящий газ laughing/dental gas
весенник *bot.* winter aconite *(Eranthis)*
весенний spring, vernal
~ сев *agr.* spring sowing
вески *pl.* small scales/balance
веский weighty; heavy
~ довод telling argument
~кое доказательство *law* weighty/important evidence
весло 1. oar; paddle; 2. paddle-shaped stirrer
~, баковое bow oar
~, безвальковое single-bank oar
~, вальковое square-loom oar, double-bank oar
~, загребное stroke/aftermost oar
~, кормовое scull(ing) oar
~, парное scull
~, распашное single-bank oar
~, рулевое steering oar
~, среднее midship oar
веслоногий *zool.* 1. copepod; steganopodous; 2. *n. pl.* totipalmate birds *(Pelicaniformes);* copepods *(Copepoda)*
веслонос *zool.* paddlefish *(Polypodon spathula)*
весновспашка *agr.* spring plowing
веснушка *med.* freckle, ephelis, lenticula, lentigo (*pl.* lentigines)
веснянка I *bot.* faverel *(Erophila)*
веснянка II *zool.* stone fly *(Plecoptera)*
весовщик weigher, weigh master, scaleman; check-weigher
~ таможенный customs weigher
весовые *pl.* postage
весограф weightograph
весок plumb bob, (nivelling) plummet
весомость weightiness; ponderability
веспер Vesper, evening star
вест *naut.* west wind; *pl.* westerlies
Веста *astr., myth.* Vesta
весталка vestal (virgin)
вестгот *hist.* Visigoth
вестготский *hist.* Visigothic
вестерн western *(film, story)*
вест-зюйд-вест *naut.* west-south-west
вестибуло-мозжечковый *anat.* vestibulocerebellar
вестибулопатия *med.* vestibulopathy
вестибулотомия *surg.* vestibulotomy
вестибулярный vestibular
вестибюль 1. entrance hall, vestibule, lobby, foyer; 2. *zool., anat.* vestibule
вест-индиец West Indian
вест-норд-вест *nav.* west-north-west
вестовой 1. signal, warning; 2. *n. mil.* orderly; runner, messenger; guard, sentry; *naut.* seaman steward
~ кают компании wardroom servant
~ на мостике bridge messenger
~вая ракета signal rocket
вестовый west(ern), westerly
вест-тень-зюйд *nav.* west by south
вест-тень-норд *nav.* west by north
вестфалец, вестфальский Westfalian
весы 1. scales, balance; weigher; 2. *astr.* the Scales,

Libra
~, автомобильные truck scales
~, аналитические analytical balance
~, аэродинамические aerodynamic balance, (wind) tunnel balance, wind-tunnel balance
~, багажные luggage weigher
~, вагонные truck scales
~, грузовые freight/cargo weigher
~, дозировочные/дозирующие batching scales, feeder-weigher
~ для взвешивания грудных детей baby scales
~, десятичные decimal balance
~, ковшовые pan scales
~, кольцевые ring balance
~, конвейерные belt-conveyer weigher
~, корольковые button/assay balance
~, крутильные torsion balance
~, медицинские medical spring balance
~, мостовые weighbridge
~, настольные bench scales, table balance
~, незадемпфированные undamped balance
~, образцовые calibrating balance
~, общего назначения general-purpose balance
~, платформенные platform balance, weighbridge
~, пробирные assay balance
~, пружинные spring/spiral balance
~, рецептарные prescription balance
~, ручные hand scales
~, рычажные beam balance/scales
~, рычажные одноплечие single-arm beam balance
~, рычажные равноплечие equal-arm beam balance
~ с магнитной подвеской magnetic balance
~, самолетные aircraft weighbridge
~, самопишущие recording balance
~, сотенные centesimal balance
~, счетные totalizing/integrating scales
~, тензометрические strain-gauge balance
~, технические counter balance
~, токовые current/elecric/electrodynamic/Kelvin balance
~, универсальные general-purpose balance
~, утопленные dormant scales
~, химико-аналитические analytical balance
~, циферблатные dial(-indicating) scales
~, шестикомпонентные six-component balance
весы-автомат automatic scales
весы-дозатор weigh-hopper, hopper scales
ветвеволосый *bot.* cladotrichous
ветвеносный *bot.* ramiferous
ветвеобразный ramiform, branch-shaped
ветвецветковый ramiflorous
ветви *pl. anat.* rami, branches
ветвистоусые *pl. zool.* cladocers, cladocerans *(Cladocera)*
ветвистый branching, ramified; dendritic, ramose, spreading, branchy
ветвление branch(ing), ramification, forking
~, виргативное *geol.* virgate venation
~ вперед *comp.* forward branching
~, дихотомическое dichotomous branching/forking, forked branching
~ метелки ramification of ear
~ назад *comp.* backward branching
ветврач veterinarian
ветвь 1. branch, leg; arm, path; bough, limb; *med.* ramus; 2. channel
~, боковая lateral branch
~ волновода waveguide arm
~, восходящая ascending/rising branch
~ второго/первого порядка secondary/primary branch
~, выполняемая *comp.* execution path
~, гусеницы track run
~, маленькая ramule
~, пальчатая digitate branch
~, поровая pore-zone
~, программы program branch
~ разлома fault branch
~ распада branch of decay
~, спиральная spiral arm
~, стержневая main path
~ термопары arm of a thermo couple
~ траектории, восходящая upward leg, ascending branch
~ траектории, нисходящая downward leg, descending/falling branch
~ ценного привода chain strand
ветер 1. wind, breeze *see also* **ветры**
~, акустический/звуковой acoustic streaming/wind
~, береговой, ~ с берега off-shore wind
~, благоприятный helping/fair wind
~, боковой beam/cross/side wind
~ в верхних слоях атмосферы wind aloft
~ в нижних слоях тропосферы low-troposphere wind
~, восточный east/easterly wind
~, восходящий anabatic wind
~, встречно-боковой quartering headwind
~, встречный adverse/head/contrary/dead/foul wind, twelve o'clock wind, headwind, wind ahead
~, горный mountain breeze
~, катабатический katabatic/gravity/fall wind
~, крепкий high wind, moderate gale
~, легкий light breeze
~, лобовой head-on wind
~, морской onshore/sea wind
~, неблагоприятный baffling/unfavourable wind
~, очень крепкий fresh gale
~, попутно-боковой quartering tailwind
~, попутный helping/tail/fair wind
~, порывистый gusty wind
~, приземный ground/surface wind
~, противный foul wind

~, резкий bleak wind
~ с моря inshore wind
~ с суши continental/land wind
~, свежий fresh breeze
~, слабый gentle breeze
~, солнечный solar wind
~, термический thermal wind
~, умеренный moderate breeze
~, ураганный hurricane (force) wind
~, штормовой storm wind, gale
~, электрический electric wind, aura
ветеран veteran
ветеринар veterinarian, veterinary *(surgeon)*
ветеринария veterinary science/medicine
ветеринарный veterinary
~ пункт veterinary station
ветиверия *bot.* vetiver, khus-khus *(Vetiveria)*
ветка 1. branch; spray; 2. branch line, siding track, spur
~, железнодорожная spur track, branch line, railway branch
~, припортовая port siding
веткоцветный *bot.* ramiflorous
ветла *bot.* white willow *(Salix alba)*
веткорез branch cutter
вето veto
~, наложить veto
~, отлагательное *law* suspensory veto
веточка twig, branchlet, sprig; shoot; spur
~ нерва *anat.* nervule
веточный *bot.* ramal, rameal
ветреница *bot.* anemone, windflower *(Anemone)*
ветрогон *tech.* fan
ветродвигатель windmill, wind motor; wind-driven generator, wind turbine
~, барабанный turbine-type windmill
~, карусельный/роторный rotor-type windmill
~, крыльчатый propeller-type windmill
ветрозащитный windproof; wind-shelter(ing)
ветроколесо windwheel
ветролом windbreak, windfallen trees
ветромер wind gauge/dial, anemometer
ветронепроницаемый windproof
ветроопыление *bot.* wind pollination, anemophily
ветроопыляемый *bot.* wind-pollinated, anemophilous
ветроотбойник wind deflector
ветростанция, ветроэлектростанция windelectric power station/plant
ветроуказатель wind indicator/cone/sock; windsock; tetrahedron
ветроулавливатель, ветроуловитель *av.* rudder air scoop
ветрочет derivometer, drift meter, wind/drift computer, wind speed indicator; course corrector
ветроэнергетика wind-power engineering
ветры *pl.* 1. winds; 2. *med.* flatulance, flatulency, wind
~, восточные easterlies
~, западные westerlies
~ средних широт, западные variables
ветряк windmill, wind-driven generator, wind turbine
ветрянка *med.* chickenpox, varicella
ветряной wind(-powered)
~ная мелница windmill
ветряночный *med.* varicellous
ветряный wind(-powered)
~ная оспа *med.* chickenpox, varicella
ветхий 1. old, ancient; 2. decrepit; infirm
~ завет *rel.* the Old Testament
ветхозаветный Old-Testament, antiquated
ветчина ham
~, провесная *Am.* jerked ham
ветчинорезка ham slicer
вех *bot.* water hemlock *(Cicuta)*
вех(а) 1. *surv.* survey stake/peg/rod, surveyor's stake, surveying rod/stake; 2. *nav.* beacon, (spar) buoy; 3. *constr.* way-mark, bench mark, guide-post
~, волномерная wave (meter) stake, wave (height) pole
~, девиационная deviation pole/post
~, крестовая cross stake, cross spar buoy
~, мерная surveyor's stake
~, нивелировочная range pole/rod
~, пограничная landmark, boundary mark
~, привязочная identification post, witness stake
~, разбивочная *surv.* range/ranging pole
~, туманная fog spar
~, фундаментальная fundamental spar buoy
вехостав dan layer, danner, dan (planting) ship
вече *hist.* veche
вечерник, серебристый silver-haired bat *(Lasionycteris noctivagans)*
вечерница, рыжая I great bat *(Nyctalis noctula)*
вечерница II *bot.* rocket *(Hesperis)*
вечерницецветковый *bot.* rocket-flowered
вечерня *eccl.* vespers
вечеря *rel.* **~ тайная** the Last Supper
вечнозеленый *bot.* evergreen, sempervirent; indeciduous
вечномерзлый everfrozen, permafrost
вечноплавающий holoplanktonic
вешала rack; clips
вешальщик weigher, scaleman
вешение staking/marking out
вешенка *bot:*
~, обыкновенная/рожковидная oyster mushroom *(Pleurotus ostreatus)*
вешка *surv.* surveyor's stake; sureying post; landmark
~, нивелирная levelling pole
вещание broadcasting, radio broadcasting
~, звукозрительное/телевизионное television broadcasting, telecasting
вещать broadcast
вещественнозначный real-valued
вещественность corporality, corporeality; materiality, substantiality

вещественный substantial, material; corporal; real; hylic
~ная неопределенность emptiness
~ное доказательство material evidence
вещество substance, material, matter, stuff; agent
~, абсорбируемое absorbate
~, абсорбирующее absorbent, absorber
~, агглютинирующее agglutinating matter/agent
~, адсорбируемое adsorbate
~, адсорбирующее adsorbent
~, активирующее activating agent
~ активной зоны core material
~, антидетонационное anti-detonant, anti-detonation/-knock substance
~, антикоррозийное anticorrosive agent
~, антипенное anti-foam agent/substance
~, аттритовое attrital matter
~, белковое proteic substance
~, боевое отравляющее war gas
~, буферное buffer
~, взвешенное suspended matter
~, взрывчатое explosive, blasting agent
~, взрывчатое для горных работ mining explosive
~, взрывчатое инициирующее priming explosive
~, взрывчатое пластичное gelatin explosive
~, взрывчатое порошкообразное blasting powder
~, вкусовое flavo(u)ring (agent)
~, водоумягчающее water softener
~, возбуждающее stimulant
~, вспенивающее foaming/frothing agent, frother
~, вяжущее binder, binding material, cement
~, вязкое ductile substance
~, горючее inflammable, combustible
~, дегидратирующее dehydrating agent, dehydrator, dehydratant
~, дезинфекционное desinfectant
~, делящееся fissible/fissionable material/substance
~, дочернее daughter (product), descendant
~, дробящее взрывчатое blasting explosive
~, дубящее tanning agent
~, загрязняющее contaminant
~, задерживающее брожение antienzyme, antiferment
~, замедляющее 1. restrainer; 2. *nucl.* moderating/stopping material, moderator
~, звездное stellar matter/material
~, истирающее abrasive, abradant
~, исходное parent substance, source/starting material
~, клейкое adhesive (substance)
~, клеящее sizing agent
~, консервирующее preservative
~, курареподобное curare-like agent
~, легирующее doping material
~, летучее volative matter
~ мазера maser material/medium
~, межклеточное intercellular substance
~, металлоорганическое organometallic substance
~, моющее detergent, cleansing agent
~, нейтрализующее neutralizer
~, неорганическое inorganic matter
~, обедненное depleted material
~, огнеопасное inflammable, flammable
~, окисляющее oxidizer, oxidant, oxidizing agent
~, окрашивающее pigment
~, органическое organic matter
~, осаждающее precipitant, precipitating/flocculating agent
~, осушающее drying agent, desiccant
~, отбеливающее bleaching agent, bleacher
~, отверждающее hardener
~, отравляющее toxic agent
~, охлаждающее cooling medium/agent, coolant
~, очищающее purifier
~, пассивирующее passivating agent, passivator
~, пенообразующее frother, frothing agent
~, первоначальное source/initial material
~, питательное nutrient, nutritive matter
~, пищевкусовое flavo(u)ring (agent)
~, поверхностно-активное surfactant, surface-active substance/compound/agent
~, постороннее foreign matter/material/substance, extraneous agent
~, пропитывающее impregnant
~, простое elementary/simple substance
~, противокоррозионное corrosion inhibitor
~, рабочее working medium
~, разбавляющее diluent
~, разжижающее thinning agent, diluent
~, разъедающее corrodant
~, распадающееся decaying substance/matter, fissionable material
~, рассеивающее scattering material/substance, scatterer
~, растворенное solute, dissolved matter
~, растворяющее solvent
~, расщепляемое/расщепляющееся fissionable material
~, реагирующее reactant, reacting agent
~, сверхпроводящее superconducting matter
~, связующее binder, binding material
~, сгущающее thickener, coagulant
~, сенсибилизирующее sensitizer
~, серое gray substance
~, сигнальное *med.* warning agent
~, склеивающее adhesive
~, смазочное lubricant, lubricating substance, grease
~, смачивающее wetting agent
~, согревающее лекарственное calorifacient
~, сорбирующее sorbent
~, сохраняющее preservative

~, стойкое к действию излучения radiation-resistant material
~, студенистое/студенистообразное jelly-like matter, gelatinous substance
~, твердое solid substance/matter, solid
~, ускоряющее accelerator
~, химическое chemical
~, цементирующее 1. cement, bond, cement substance; 2. *met.* carbonizer
~, чистое pure/individual substance
~, эталонное reference substance
~, ядерное nuclear matter
~, ядерное взрывчатое nuclear explosive
вещество-носитель carrier material
вещный material, physical, real
~ное право law law of estate
вещь 1. object, thing, entity; 2. *pl.* belongings
~ в себе transcendental object
~, недвижимая *law* real thing
~, родственная *biol.* congener
веялка *agr.* winnower, fanning mill, winnowing machine/fan
веяльный winnowing, fanning
веяльщик *agr.* winnower
веяние 1. *agr.* winnowing; 2. blowing
веять *agr.* winnow, fan
вживление implantation
вживленный implanted
вживляемый implantable
взаимно reciprocally, mutually
~ активный interactive
~ выводимый interdeducible
~ заменимый interchangeable
~ замкнутый interlocked; intermeshed
~ однозначный *math.* one-to-one
~ простой *math.* coprime,relatively prime
~ противоположный mutually antithetic(al)
~ соединять interlock
взаимно-включенный mutually inclusive
взаимно-дополнительный mutually complementary
взаимномодуляция intermodulation
взаимно-моторный reciprocal-motion
взаимно-непрерывный bicontinuous
взаимно-обменный interchangeable
взаимно-однозначный *math.* one-to-one, mutually identical
взаимно-полярный polar-reciprocal
взаимно-проникающий interpenetrating
взаимно-простой *math.* relatively prime, coprime
взаимно-связанный conjugate, interrelated
взаимность reciprocity, mutuality
~ перемещения reciprocity of displacement
взаимный mutual, reciprocal
~ная зависимость interdependence
~ная помощь mutual aid
взаимоблокировка *comp.* deadlock
взаимовидимость intervisibility, mutual visibility
взаимовлияние interaction, mutual influence, interinfluence, interference
взаимовозгорание spontaneous combustion
взаимовыгодный mutually beneficial/advantageous
взаимодействие interaction, interreaction, reciprocal action, cooperation, coordination, interplay
~, аксиально-векторное axial-vector interaction
~, вигнеровское Wigner interaction
~, внеядерное extranuclear interaction
~, диполь-дипольное dipole-dipole interaction
~ Дирака Dirac interaction
~, инерционное inertial cross-coupling, inertial intercoupling
~, конструктивное structural interaction
~, кулоновское Coulomb interaction
~, мезон-мезонное meson-meson interaction
~, нуклон-нуклонное nucleon-nucleon interaction
~, обычное Wigner interaction
~, одностороннее one-way interaction
~ открытых систем *comp.* open-system interconnection
~, перекрестное intercoupling, interaction, interference, cross-coupling
~, распадное decay interaction
~ с пользователем user interaction
~, сверхобменное superexchange interaction
~ сетей *comp.* internetting
~ систем system interfaces
~ спинов spin-spin interaction
~, спин-орбитальное spin-orbit interaction
~, спин-решеточное spin-lattice interaction
~, спин-спиновое spin-spin interaction
~, тесное *mil.* close cooperation
~ Ферми/фермиевское Fermi interaction
~ человека и машины man-machine interface
~, ядерное nuclear interaction
взаимодействовать interact, interreact; cooperate
взаимозависимо interdependently
взаимозависимость interdependence, interdependency; inter connectedness
взаимозамена mutual substitution
взаимозаменяемый interchangeable, interconvertible, replaceable
взаимозаменямость interchangeability, interchange
взаимозаместимость, взаимозамещаемость interchangeabiblity, mutual substitution
взаимозамкнутость inter lock
взаимозамыкание *rail.* interlocking
взаимоиндуктивность, взаимоиндукция mutual inductance/induction
взаимоисключающий incompatible, mutually exclusive
взаимоисключение mutual exclusion
~ минералов mineral antipathy
взаимонезаменяемый noninterchangeable
взаимообмен interchange, exchange, replacing
взаимобратимый mutually, reversible; convert-

ible; interconvertible
взаимообратный inverse, reciprocal
взаимообращение mutual addressing
взаимообусловленность *philos.* interconditionality
взаимоотношение interrelation; relation
взаимопомощь mutual aid
  ~щи, касса benefit society, mutual aid/benefit fund
взаимопонимание mutual understanding; terms of reference
взаимопревращаемый interconvertible
взаимопревращение 1. *phys.* transmutation 2. *chem.* interconversion
взаимопридаточный *gram.* reciprocally subordinate
взаимопроникающий interpenetrating, interpenetrative
взаимопроникновение interpenetration, mutual penetration
взаимопроницание interpenetration
взаиморасположение mutual arrangement/position
взаиморастворение mutual solution
взаимосвязанность interconnectedness, interconnection, coherence
взаимосвязанный interconnected (inter)related; interlocked; conjugated, correlated, interdependent
взаимосвязывать interconnect, interrelate
взаимосвязь interconnection, (inter) relation(ship), interdependence, intercommunication, cross-coupling
  ~ массы и энергии mass-energy (inter)relation
взаимосмешиваемость intermiscibility
взаимосогласованность (inter)consistency
взаимотормозящий interinhibitive
взбивалка whipper, beater
взбивать whip, whisk, beat up; fluff/shake/puff up
взбалтывание agitation, shaking (up), stirring
взбрасывание *geol.* upthrusting; tossing up
взброс *geol.* upthrust, upcast, upthrow fault, overfault
  ~ при складчатости compression fault
взбрызгивание spraying, sprinkling, splashing, spatting
взведение setting, (automatic) cocking
  ~триггера *comp.* triggering
взведенный cocked
взвесь suspension, suspended matter
  ~, водяная water suspension
взвешенно-квадратический weighted-square
взвешенно-полиномиальный *math.* weighted polynomial
взвешенный *stat.* weigh(t)ed, suspended, suspension
  ~ное состояние suspension
  ~ное, среднее weighted average
взвешивание weighing; evaluation
взвешивать weigh; balance
взвод I *mil.* platoon
  ~, парашютно-десантный airborne platoon
взвод II *mil.* (cocking) notch
  ~, ставить на боевой cock; arm; unlock
  ~, ставить на предохранительный set to safe
взводный *mil.* platoon leader
взгляд 1. sight, glance; look; 2. view, opinion, notion
  ~ вперед *geod.* foresight, minus sight; forward sight/vision, front shot
  ~ вперед, допольнительный extra-foresight
  ~ назад reverse bearing, retracement, back sight/vision
взгонять *hunting* rouse, start, put up
вздваивание doubling up
вздувание swelling, intumescence, inflating, distention, puffing
  ~ почвы *geol.* heave, bulge
  ~ цен increase/inflation of prices
вздутие swell(ing), bulging, bulge, inflation, distention, flatulence, intumescence
  ~, антиклинальное anticlinal bulge
  ~ желудка gastric flatulence
  ~ живота 1. *med.* abdominal distention; celiectasia 2. *vet.* hoove
  ~ земной коры crustal warping/swell
  ~, лавовое lava tumour/swell
  ~, клубневидное tuber-like swelling
  ~, экваториальное Земли (equatorial) bulge of the Earth
вздутоплодник *bot.* Alexandrian laurel *(Danae racemosa)*
вздутоплодный *bot.* physocarpous
вздутый inflated, swollen, distended, bulging, flatulent, tumid; *bot., zool.* ventricose, ventricous
  ~ газами tympanous
вздымание upthursting, raising
вздыматься billow; rise, heave
взлет *av.* take-off, take off; (upward)flight
  ~, безаэродромный no-run take-off, vertical take-off
  ~ без разбега (вертикальный) vertical take off, no-run take off
  ~, групповой formation take-off
  ~ по ветру downwind take-off
  ~ по приборам instrument take-off
  ~, прерванный aborted/rejected take-off
  ~ при отсутствии видимости blind take-off
  ~ против ветра into-wind take-off, upwind take-off
  ~ с авианосца carrier launch
  ~ с аэродрома field take-off
  ~ с боковым ветром cross-wind take-off
  ~ с воды water take-off
  ~ с движения rolling start
  ~ с катапульты catapult launching
  ~ с коротким разбегом short take-off
  ~ с Луны lunar lift-off
  ~ с места stationary take-off

~ с палубы deck start
~ с полной нагрузкой full-load take-off
~ с разбегом running(-type) take-off
~ с ракетным ускорителем rocket-assisted take-off
~ строем formation take-off
взлетать take off; become airborne, lift off; fly up, take wing
~ по ветру take off down the wind
~ по приборам take off by instruments
~ против ветра take off into the wind
взлетно-посадочная полоса (ВПП) *av.* runway
~, дерновая grass/sod runway
~ для инструментальной/слепой посадки instrument(ed) runway
взлетно-посадочный takeoff-and-landing, landing-and-takeoff
взлетный take-off
~ная дорожка runway
~ная полоса take off strip
взлом breaking in/open; housebreaking
~ом, кража со burglary, housebreaking
взломщик burglar, housebreaker
~, компьютерный computer trespasser
взмах flap(ping); stroke, sweep, swing
взмачивание moistening, soaking, drenching, wetting
взмет flinging/throwing up; *agr.* first plowing/ploughing
~ почвы breaking of the sod
взметывать throw/toss/fling up
взморник *bot.* eelgrass *(Zostera)*
взморниковые *pl. bot.* eelgrass family *(Zosteraceae)*
взморье 1. seacoast, seashore, beach, seaside; 2. coastal waters
~, устьевое delta front
взмывание *av.* soaring up, shooting upwards (before landing)
взмымавть rocket, shoot/soar upwards
взнос deposit, payment, dues, fee, subscription, contribution
~, возвращаемый страховой return premium
~, вступительный registration fee
~, долевой по общей аварии general average
~ наличными деньгами cash down payment
~, начисленный depreciation cost(s)
~, очередной instalment
~, первый при покупке down/initial payment
~ по общей аварии general average deposit
~, страховой insurance premium
~, членский membership dues
вносить deposit, pay in(to), make payments; contribute
взрез 1. cutting open, incision; dissecting; 2. cut, opening
взрезать dissect, incise; cut open
~ землю plow up
взрослый 1. grown-up, adult; mature, fullgrown; 2. *n.* adult
взрыв explosion, detonation; burst(ing), outburst; blast; fulmination; rupture
~, атомный atomic bomb explosion
~ в карбюраторе backfire in the carburettor
~, демографический *econ.* population explosion
~, направленный pin-point blast(ing), pinpoint explosion
~, обратный back shooting
~, опытный trial blast
~, подземный subsurface/underground/buried explosion
~, породный rock burst
~ рудничного газа firedamp explosion
~ сверхновой звезды supernova explosion/outburst
~, солнечный solar outburst
~, сотрясающий quaker
~, управляемый controlled blast
~ ядра nuclear burst/fragmentation
взрываемость explosibility, explosivity, explosiveness
взрывание I blasting, exploding, blowing up, detonation
~ без вруба no-cut-hole blasting
~, беспламенное non-flame blasting, flamefree blasting
~ в скважинах well shooting
~, контактное contact firing
~, огневое (cap-and-)fuse blasting
~ с подрывной станции, управляемое observation fire
~ с помощью лазера laser blasting
~ шпура hole firing
~, электрическое electric firing
взрывание II *agr.* ploughing/plowing/turning/digging up
взрыватель fuse, *Am.* fuze, firing mechanism/device; detonator, detonating fuse
~, автономный *min.* self-contained firing mechanism
~, барометрический baro(meter) fuse, pressure fuse
~, безрезьбовый slide detonator
~, боковой body fuse
~, всюдубойный all-ways fuze/fuse, action fuse
~, гальваноударный electric(al) and percussion fuse
~, головной nose/point fuse
~, дистанционный time fuse
~, дистанционный пороховой burning-type fuse
~, минный mine timing device
~ с детонатором fuse block
~ ударного действия direct impact fuze/fuse
~, ударный concussion/impact/percussion fuse
взрывать I explode, blow up, burst, fire; demolish; detonate, blast, dynamite
взрывать II *agr.* plough, plow/turn/dig up

взрывной, взрывный 1. explosive, exploding, blasting; 2. *phon.* plosive; 3. *n. phon.* (ex)plosive
~ная волна blast wave
~ные работы blasting (operations)
взрывобезопасность explosion proofness
взрывобезопасный explosion-/blast-proof, blast-resistant
взрывоопасность explosion hazard/risk
взрывоопасный explosive
взрывостойкий, взрывоупорный explosion-proof
взрывчатость explosibility, explosiveness
взрывчатый, взрывочный explosive, detonating
взрыхление loosening, breaking up
взъезд 1. ascent, going up; 2. rising grade, upward slope
взъерошивание roughing
взыскание 1. punishment, penalty, fine; 2. enforcement, recovery, exaction; prosecution, vindication
~, дисциплинарное nonjudicial punishment
~, налагать award punishment, penalize
взыскательность exactingness, strictness; severity; exigence, stringency
взыскивать *law*. 1. exact (from/of); enforce, recover, vindicate; demand, extract; 2. penalize, call to account
~ долг recover a debt
взятие taking; capture
~ буровых образцов well/surface sampling
~ в плен taking prisoner
~ выборок sampling
~ дополнения *math.* taking the complement
~ керна core sampling
~ контрольных проб check sampling
~ крепости capture/seizure of a fortress
~ проб, вторичное resampling
~ проб(ы) sampling
взятка 1. hush money; bribe, graft; 2. honey gathering
взяток 1. bribe; 2. honey flow/gathering
взяточник bribetaker, grafter
взяточничество bribery, bribetaking, graft; corrupt practice, corruption
взять под стражу *law*. take into custody
виадук viaduct
виандот Wyandotte
вибракула *zool.* vibraculum
вибрант *phon.* vibrant, trilled sound
вибрато *mus.* vibrato
вибратор vibrator; *el.* oscillator, dipole
~, асинхронный non-synchronous vibrator
~, возбуждаемый в центре centre-driven dipole
~, возбуждать drive/feed a dipole
~, волновой wave dipole
~, вращающийся rotating vibrator
~, высокочастотный high-speed vibrator
~ Герца Hertz(ian) dipole
~, глубинный internal vibrator
~, горизонтальный диапазоновый band horizontal dipole
~, диапазоновый wide-band dipole
~, дипольный oscillating dipole
~, дополнительный complementary dipole
~, замкнутый closed vibrator
~, звуковой noise maker
~, кварцевый crystal transducer
~, магнитострикционный magneto-striction vibrator
~, магнитоэлектрический moving-coil vibrator
~, пассивный parasitic aerial, passive dipole
~, петлеобразный folded antenna
~, питаемый волноводом waveguide-fed dipole
~, П-образный bent dipole
~, полуволновой half-wave vibrator, doublet vibrator
~, поршневой reciprocating vibrator
~, приемный pick-up dipole
~, прямолинейный linear oscillator
~ ручки stick shaker
~ с симметрирующим устройством skirt dipole
~ с шаблоном screed vibrator
~, станковый table vibrator
~, стержневой needle vibrator
~, турникетный turnstill dipole
~, ударный impact vibrator
~, штыковой rod/poker vibrator
вибратор-сигнализатор срыва stall-warning vibrator
вибратрон vibratron
вибрационный vibrational, vibratory, oscillating, shaking
вибрация vibration; oscillation; shaking; flutter(ing); *av.* buffeting
~ валопровода shafting vibration
~, волновая wave-excited vibration
~ (конструкции) корпуса *naut.* structural/hull vibration
~, ответная/резонансная sympathetic vibration
~, поперечная transverse vibration
~ щетки *teleph.* wiper chatter
вибрион *med.* vibrio
~ Мечникова Gamaleia's spirillum
~, холерный comma bacillus, cholera germ
вибрирование vibration; oscillation
~ бетона compaction of concrete by vibration
вибрировать vibrate; oscillate; swing
вибрирующий vibrating, vibratory; oscillating; shaking
виброанализатор vibration analyzer
вибробулава spud/needle vibrator
вибробункер vibrating feed bin
вибробур vibratory drill, vibrodrill
вибробурение vibrodrilling
вибровозбудитель vibration exciter
вибровыпрямитель vibrating(-reed) rectifier

виброгаситель vibration damper
виброгенератор vibration generator
виброгладилка vibrating trowel
виброграмма vibration record
виброграф vibrograph, vibration measurer
виброгрейфер vibrating grab
виброгрохот vibroshaker, vibrocribble, racking screen
вибродатчик vibration/oscillation pick-up, vibration/oscillation transducer/sensor, vibration detector
~, зондовый vibration probe
вибродвижущий vibromotive
вибродозатор vibrating batcher/proportioner
виброжелоб vibrating chute
виброзатирка vibrating float
виброигла needle vibrator
виброизмерительный vibration-measuring
виброизолятор (anti-)vibration insulator
виброизоляция (anti-)vibration insulation/isolator/isolation
виброинвертор vibrator power supply
виброкардиограмма *med.* vibrocardiogram
виброкаток vibroroller
вибролитье vibratory casting
вибролопата vibrospade, spade vibrator, vibrator spade
вибролоток vibrating chute
вибромельница vibrating mill
виброметр vibration meter, vibrometer
~, синусный sine vibrometer
вибромолот vibrating hammer, vibrohammer
вибронный *phys.* vibronic
виброочиститель вагонеток car shaker
вибропитатель vibrating feeder
виброплощадка table vibrator
виброплуг oscillating plough
вибропоглотитель vibration absorber
вибропоглощающий vibration-absorbing
вибропогружатель vibratory pile driver
вибропогружение vibroimmersion
вибропреобразователь *el.* vibrator power supply, vibration rectifier, vibrator
вибропрессование vibrocompression
вибропрокат vibrorolling; vibration-rolled concrete
вибропрофилирование vibroforming
вибропрочность, вибростойкость, виброустойчивость vibration strength resistance, resistance to vibration
вибропрочный, вибростойкий, виброустойчивый vibration-proof/-resistant, shake-/shock-proof, shakeproof, shockproof
вибросито vibrosieve
виброскоп vibroscope
вибростенд vibration-testing machine
вибростол vibration table
вибротерапия *med.* vibrotherapy
вибротравма vibrotrauma
вибротрамбовка vibrating tamper
вибротранспортер oscillating conveyer
виброустройство для разгрузки вагонов car shaker
виброцентрифуга vibrating centrifuge
виброшаблон vibrating screed
виброштанга vibrating boom
виброштык pin vibrator
виброшумы *pl.* vibronoise
виброщуп vibroprobe
виварий vivarium, animal house
виваче *mus.* vivace
виверра *zool.* civet *(Viverra)*
виверровые *pl. n. zool.* civets *(Viverridae)*
вивианит *miner.* vivianite, blue ochre
~, землистый Prussian blue
вивидиффузия *physiol.* vividialysis, vividiffusion
вивипария vivipary
вивисектор, вивисекционист vivisectionist, vivisector
вивисекция biotomy, vivisection
виг *hist.* Whig
вигильность *psych.* vigilance
вигильный *psych.* vigilant
вигна *bot.* cowpea *(Vigna)*
виговский *hist.* Whig(gish)
вигонь 1. *zool.* vicuna, vicugna *(Lama vicugna);* 2. vicuna wool/cloth
вид I 1. view; appearance, look; air; aspect; 2. form, shape, mode; make; condition; state; 3. sort, type; kind
~ адреса *comp.* kind of address, address kind
~ в плане plan/top view
~ в полете in-flight view
~ в разрезе (cross-)sectional view/drawing
~ в трех проекциях three-view arrangement
~, внешний general/outside view, appearance
~, внутренний inside view, interior, cut-away view
~ вооруженных сил (fighting) service
~, главный front view
~, готовый ready for use
~ данных *comp.* data kind
~ запоминания *comp.* kind of storage
~ импульсов pulse mode
~ информации information kind
~, канонический *math.* canonical form
~ карты *comp.* card image
~ колебаний (oscillation) mode
~ операций *comp.* kind of operation
~ передачи transmission mode
~ по всему горизонту all-round view
~ поиска search pattern
~ распада mode of decay
~ рассеяния волн scatter mode
~ резонанса mode of resonance
~ с птичьего полета bird's-eye view
~ сбоку lateral/side view; outboard profile, profile plan
~ сверху plan/top view, view from above
~ сзади rear/back view
~, сквозной phantom view

~ **слева** left-side view
~ **снизу** bottom view, view from below
~ **спереди** front view
~ **справа** right-side view
~ **текучести** flow behavio(u)r
~ **шрифта** *comp., print.* type face
вид II 1. *bot.* species; *geol.* facies; 2. *gram.* aspect
~, **агамный** agamospecies
~, **второстепенный** minor species
~, **замещающий** vicarian/representative species
~, **карликовый** dwarfish/intersized species
~, **несовершенный** *gram.* imperfective aspect
~, **новый** species nova
~ **облака** cloud genus
~, **отдельный** distinct species
~, **повсеместный** cosmopolitan species
~, **прикрепляющийся** encrusting species
~, **реликтовый/суперститовый** relict species
~, **родовой** tribe
~, **родственный** allied species
~, **руководящий** marker
~, **случайный** casual species
~, **совершенный** *gram.* perfective aspect
~, **типичный** typical species, genoholotype
~, **эволюционирующий** evolving species
вид на жительство residence permit; identity card
видение 1. vision, viewing; 2. apparition, phantom; image; dream
видеоаппаратура video equipment
видеограмма visible record
видеоданные video data
видеозапись video recording
видеоимпульс video pulse, videopulse
видеоинформация video information
видеокассета video cassette
видеолента video tape
видеомагнитофон video tape recorder
видеометр transmissometer
видеонесущая video carrier
видеопередатчик visual transmitter
видеопомехи *pl.* video cross-talk
видеосвязь video communication
видеосигнал video signal
видеотелефон picturephone, video(tele)phone, phonovision system
видеоусилитель video amplifier
~, **многорезонансный** stagger-peaked video amplifier
~ **с расстроенными контурами** stagger(ed)-tuned video amplifier
~, **смесительный** video-mixing amplifier
видеочастота video/picture frequency
видер *agr.* weeder
виджерит *geol.* wiedgerite
видикон vidicon
видимость visibility; visual range
~ **в полете** flight visibility
~ **днем** daylight visibility
~, **малая/низкая** low visibility
~, **наклонная** slant/oblique visibility
~ **ночью** night visibility
~, **плохая** poor/bad visibility
~, **прямая** direct visibility, line of sight
~, **радиолокационная** radar contact
~ **у земли** (near-)ground visibility
~, **хорошая** high/fair visibility
видимый 1. visible; 2. apparent, obvious, evident
~ **в совершенстве** panoptic
~ **невооруженным/простым глазом** visible to an unaided eye, seen by an unaided eye
видность luminosity; luminosity/visibility factor
видный 1. conspicuous, observable, visible; 2. distinguished, prominent, eminent; important
видовой I view, appearance *etc.* *see* **вид I**
~ **объектив** landscape lens
~**вая кинокартина,** ~ **фильм** travel film
видовой II 1. *biol.* specific; 2. *gram.* aspectual
видоизменение 1. modification, transformation; alteration, change; metamorphosis; 2. variant, modification
видоизмененный modified, altered; revised
видоизменитель modifier
видоизменяемость mutability, variability, changeability
видоизменяемый modifiable, changeable
видоизменить modify, alter, change; refashion; mutate
видоискатель *phot.* (view-)finder, (optical) camera sight
~, **боковой** angular viewfinder
~, **зеркальный** reflex finder
~, **оптический** optical viewfinder
~, **рамочный** framefinder
~ **с визирной маркой** frame viewfinder
~ **с переменным фокусным расстоянием** zoom finder
видообразование *biol.* species formation, speciation, *see* **вид II**
~, **итеративное/повторное** recurrent speciation
~, **префиксальное** *gram.* preverbation
видоспецифичный species-specific
виза 1. visa; 2. official stamp
~, **въездная,** ~ **на въезд** entry visa/permit
~, **выездная** exit permit
~, **транзитная** transit visa
византиец *hist.* Byzantine
византийский Byzantine; Byzantinesque
византин *hist.* bezant *(coin)*
византинист, византолог Byzantine scholar, Byzantinist
византинистика Byzantine studies
визионер visionary, mystic
визионерство visionariness, mystical propensities
визионизм voyeurism
визир 1. *phot.* viewfinder; 2. *mil.* sight, hairline, visor; 3. cursor; tracker
~, **(аэро)навигационный** drift sight/indicator
~, **бортовой** driftmeter, drift sight/indicator

~ грубой наводки slewing sight
~, дульный muzzle bore sight
~, маршрутный strip finder
~, навигационный drift sight/indicator, driftmeter, course-setting sight
~, стыковочный docking sight
~ угла места angle of elevation sighting device
~ центральной наводки director-scope
визирка vane, slider
визирование I visaing, viseing
визирование II, визировка sight(ing), bearing, sight aiming; tracking; shot, pointing
~ вперед forward sight(ing), front shot
~, двустороннее reciprocal sight(ing)
~, назад/обратное backward sight(ing), reverse bearing, back sight(ing)
~, прямое direct sighting
визировать I V, visa, visé
~ паспорт visa/stamp a passport
визировать II level, take bearing(s), collimate, sight on/to; track
визирь vizier
~, Великий grand vizier
визит visit, call
~ вежливости courtesy visit/call
~ доброй воли good will visit
~, наносить официальный pay an official call
~, неофициальный informal/unofficial visit/call
~, ответный return call/visit
~, официальный formal/official visit/call
~, светский social visit
~, служебный duty/professional call/visit
визитация 1. call, visit; *med.* (doctor's) round; 2. *naut.* inspection
визитная карточка visiting card
визуализация visualization, viewing
~, вихреобразная vortex visualization
~ потока дымовым методом smoke-flow visualization
~ скачков уплотнения shock-wave visualization
визуально-звуковой visual-aural
визуально-моторный visual-motor
вик *abbr.* **(волостной исполнительный комитет)** volost executive committee
вика *bot.* vetch *(Vicia)*
викариант *bot.* vicarious species, vicariant
~, высотный altitudinal vicariant
викарий *eccl.* vicar; suffragan
викарный 1. *eccl.* vicarial, suffragan; 2. *physiol.* vicarious
викель rolled-up stock; roll, winding
викинг *hist.* Viking
виконт viscount
виконтесса viscountess
виконтство viscountcy
викторианец, викторианский Victorian
викторина quiz (game)
виктория 1. *hist.* victoria (carriage); 2. *eccl.* victory
виктория-регия *bot.* royal water platter *(Victoria regia)*
вилайет *hist., polit.* vilajet
вилка 1. fork; prong; *el.* (two-pin) plug, connector; *tech.* jaw, yoke; branching; 2. *mil.* bracket; 3. *zool., anat.* furcula;
~ безопасности safety fork
~, двухконтактная/двухштырьковая two-pin plug
~, двухлучевая two-prong(ed) fork
~, захватывающая gripping fork
~ кардана universal-joint fork/yoke
~, ловящая *print.* cylinder catch
~, мерная *for.* tree caliber
~, микротелефонная receiver rest, cradle switch
~, нитеводная *text.* traverse guide fork
~, ответвительная *el.* socket-outlet, adaptor
~, отводная disengaging fork
~ переключения передач gear shift(ing) fork
~, переходная *el.* plug adaptor
~, подрессорная spring fork
~, трехконтактная/трехштырьковая three-pin plug
~, тяговая towing yoke
~, хвостовая *anat., zool.* furculum
~ шатуна connecting rod fork
~, штепсельная (two-pin) plug
вилконосный *bot.* dicranophorous
вилкообразный forked, fork-shaped, forcate, furciform, bifurcate, Y-shaped
вилла *hist.* villa, vill
виллан *hist.* villein, bondman, bondar, serf
виллемит *miner.* willemite
виллиомит *miner.* viliaumite
виллис *mil.* jeep
вилльямсония *paleont.* williamsonia
вилообразный forked, furcate
вилорог *zool.* pronghorn *(Antilocapra americana)*
вилорогий pronghorned
вилохвост *zool.* springtail *(Collembola)*
~, буковый lobster moth *(Stauropus fagi)*
вилочка *anat.* furcula; wishbone, merrythought
вилочковый thymic
вилочный forked; bifurcate
вилт *pathol.* wilt
вилуит *geol.* viluite, wiluite
вилы *pl.* fork, pitchfork
~, грейферные grapple fork
~, камнепогрузочные stone fork
~, навозные manure/dung fork
~, перекопочные digging/spading fork
~, решетчатые sieve fork
~, сенные hayfork, pitchfork
вильчато-разветвленный/-раздвоенный forked, bifuracate
вильчатость (bi)furcation, dichotomous branching
вильчатохвостый fork-tailed
вильчатый (bi)furcate, forcipate, forked, Y-shaped
виляние wagging; rocking/side motion; wobble

~ **колеса** (wheel) wobble
вилять wag; wobble, twist, wind; meander
вина fault, guilt; responsibility; blame
~, **встречная/потерпевшего** *law* contributory guilt/negligence
виндзейль *naut.* windsail
виндикация vindication; *law* eviction, recovery
виндроза tail-vane
виндроуэр *agr.* windrower
~, **навесной** mounted windrower
~, **самоходный** self-propelled windrower
виндицировать *law* evict
винегрет *cul.* Russian salad
винил *chem.* vinyl
~, **хлористый** vinyl chloride
винилен *chem.* vinylene
винилирование vinylation
винипласт *chem.* rigid-vinyl plastic, rigid PVC, viniplast
~, **листовой** rigid PVC sheet
винительный *gram.* accusative
~ **падеж** accusative case
винкель *tech.* set square
виннокаменнокислый, виннокаменный *chem.* tartaric
виннокислый *chem.* tartrate, vinoacetous
~ **натрий-калий** sodium potassium tartrate
винный 1. wine, winy; 2. *chem.* tartaric; oenanthic
~ **камень** tartar
~**ная кислота** tartaric acid
~**ная ягода** dried fig
вино wine
~, **марочное** vintage wine
~, **отжившее** worn wine
~, **пресное** dead wine
~, **резкое** tart wine
~, **розливозрелое** bottling-ready wine
~, **сортовое** unblended wine
~, **сырое** young wine
виноватость 1. guilt; *law* culpability; delinquency; 2. consciousness of guilt
виновник culprit; author, initiator
виновность *law* culpability; delinquency; guilt; criminality
виновный *law* 1. guilty (of); culpable; delinquent; 2. *n.* the accused
~**ная сторона** party in/at fault
виноград 1. *bot.* grape, vine *(Vitis);* 2. *coll.* grapes
~, **девичий/дикий** creeper *(Parthenocissus)*
виноградарский vinicultural, viticultural, winegrowing
виноградарство winegrowing; viniculture, viticulture
виноградарь winegrower, viticulturist
виноградник vineyard
виноградный 1. grape, vine; 2. *n. pl. bot.* grape family *(Vitaceae)*
~ **сахар** glucose
~ **сезон** vintage
виноградовник *bot.* ampelopsis, virginia creeper *(Ampelopsis)*
виноградовые *n. pl. bot.* grape family *(Vitaceae)*
виноделие wine-making
винодельня winery
винокур distiller
винокурение (alcohol) distillation/distilling
виноматериал wine material/stock
виномер wine-level gauge
виноторговец wine merchant, vintner
виночерпий *hist.* cupbearer
винт 1. screw; 2. *av.* propeller, airscrew; 3. *av.* (carrier) rotor; 4. spiral, helix
~, **авторотирующий воздушный** windmilling propeller; autorotating rotor
~, **авторотирующий тормозной** rotochute
~, **архимедов** Archimedes/Archimedian/ Archimedean screw, hydraulic screw
~, **барашковый** wing-head screw, thumbscrew, eared/winged/butterfly screw
~ **без головки** grub/headless/set screw
~, **безредукторный воздушный** direct-drive(n) propeller
~, **бесшарнирный несущий** hingeless rotor
~ **большого шага, воздушный** high-pitch propeller
~ **вертолета** rotor
~ **вертолета, несущий** main/lift rotor, helicopter rotor
~ **вертолета, хвостовой** tail/steering rotor, anti-torque rotor
~, **водоподъемный = архимедов**
~, **воздушный** (air/aerial) propeller, air screw
~, **воздушный изменяемого шага** controllable-/variable-pitch propeller
~, **воздушный левого/правого вращения** left/right-handed propeller
~, **воздушный многолопастный** multiblade propeller
~, **воздушный неизменяемого шага** fixed-pitch propeller
~, **воздушный переставного шага** adjustable-pitch propeller
~, **воздушный реверсивный** reversible propeller
~, **воздушный трехлопастный** three-blade propeller
~, **воздушный туннельный** ducted propeller
~, **воздушный тянущий** tractor propeller
~, **воздушный фиксированного шага** fixed pitch propeller
~, **воздушный флюгерный** feathering propeller
~, **гребной** (screw) propeller, screw
~, **гребной вентилируемый** vented/ventilated propeller
~, **гребной реверсивный** reversible propeller
~, **гребной регулируемого шага** controllable/-variable-pitch propeller

~, гребной с поворотными лопастями feathering propeller
~, двухзаходный double-threaded screw
~, двухшаговый воздушный controllable (-pitch) propeller, selective-pitch propeller
~, динамический *mech.* wrench
~ изменяемого шага, воздушный variable-pitch propeller
~ изменяемого шага, рулевой variable-pitch tail rotor
~, исправительный adjuisting screw
~, контровочный lockscrew
~, крепежный, мелкий machine screw
~, крылевой, несущий wing-mounted rotor
~ левого вращения, воздушный left-handed propeller
~ малого шага, воздушный fine-pitch propeller
~, малооборотный несущий slow rotor
~, микрометрический fine adjustment screw, micrometer screw
~, многозаходный multiple-thread screw
~, многолопастный воздушный multi-blade(d) propeller
~, многолопастный несущий multi-blade(d) rotor
~, нажимной pressure screw
~, натяжной tightening/tension(ing) screw, slack-adjusting screw
~, невыпадающий captive screw
~, несущий (main) rotor, lifting rotor
~, нетеряющийся captive screw
~, однозаходный single-threaded screw
~, однолопастный воздушный single-blade(d) propeller
~, однолопастный несущий single-blade(d) rotor
~, околозвуковой transonic propeller
~, освобождать loosen/slacken a screw
~, поворотный воздушный tilt propeller
~, поворотный несущий tilt rotor
~, подающий lead screw
~, подстроечный vernier screw
~, подъемный *geod.* foot/levelling screw
~, потайной countersunk screw
~, потайной стопорный headless screw
~, реактивный воздушный jet propeller
~, реактивный несущий jet-/tip-driven rotor
~, регулировочный adjusting/correcting screw
~, рулевой auxiliary/steering/tail rotor, anti-torque rotor
~ с барашком butterfly thumb screw
~ с буртиком collar-head screw
~ с внутренним шестигранником Allen-head screw
~ с коническом концом cone-point screw
~ с левой/правой резьбой left-/right-hand(ed) screw
~ с накатанной головкой knurled-head screw
~ с непереставными лопастями, воздушный fixed-pitch propeller, non-adjustable propeller
~ с полукруглой головкой, крупный button-head screw
~ с полукруглой головкой, мелкий round-head screw
~ с потайной головкой counter-sunk screw, *Am.* flat-head screw
~ с правым ходом right-handed screw
~ с цилиндрическим концом flat-dog point screw
~ с цилиндрической головкой round-/fillisterhead screw, cap screw
~ с шестигранной головкой hexagon(-head) screw
~, самоконтрящийся self-lapping screw
~, самонарезающий self-tapping screw
~, свободновращающийся воздушный freewheeling propeller
~, силовой power-transmission screw
~, складывающийся несущий folding/foldable rotor
~ со шлицевой головкой headless set screw
~, стопорный stop/check/locking screw, clamp(ing) screw
~, стяжной swivel/coupling screw, tightening-up screw
~, тангенциальный tangent screw
~, толкающий воздушный pusher propeller
~, тормозной воздушный reverse-pitch propeller, braking propeller
~, тормозной несущий lifting rotor decelerator
~, транспортирующий 1. transit screw; 2. helix (of a conveyer)
~, транспортирующий бесконечный endless screw
~, транспортирующий ленточный ribbon helix
~, транспортирующий сплошной spiral helix
~, транспортирующий фасонный cut-flight helix
~, трехлопастный воздушный threeblade(d) propeller
~, трехлопастный несущий three-blade(d) rotor
~, упорный fixing/stop screw
~, установочный set(ting)/adjusting/positioning screw
~, фиксационный *med.* fixation screw
~, хвостовой tail/steering/antitorque/auxiliary rotor
~, ходовой lead/guide screw
~, цельнометаллический воздушный all-metal propeller
~, червячный worm
~, шарнирный swing screw
~, юстировочный adjusting screw
винт-автомат, гидравлический воздушный hydromatic propeller

винт-автомат, электрогидравлический воздушный electro-hydromatic propeller
винт-глухарь lag screw
винт-движитель, воздушный propulsion/thrusting propeller
винтовальный screw-cutting
~ная доска screw plate, thread-cutting plate
винтовальня thread-cutting plate, screw plate
винтовать *tech.* 1. rifle, thread; 2. screw together, fasten with screws
винтоверт screwdriver
винтовка rifle
~, боевая service rifle
~, магазинная magazine rifle
~, полуавтоматическая/самозарядная semiautomatic rifle
~, пристрелочная spotting rifle
~, штурмовая assault rifle
винтовой screw, spiral, helical; propeller-driven
~ пароход screw steamer
~вая лестница spiral staircase
~вая нарезка (screw) thread
~вая передача *tech.* helical gear
винтовочный rifle *(attr.)*
винтокрыл rotorcraft, rotary-wing aircraft
винтокрылый rotary-wing
винтолистный *bot.* spiral-leaved, helicophyllous
винтообразный screw-shaped, helical, spiral; coiled
винторез screw tap/plate, threading tool, thread-cutting die/plate
~ для труб pipe tap
винторезный screw-cutting
~ станок, ~ная машина screw cutter, screw-cutting lathe/machine
винтранец wing/helm transom
винчестер 1. Winchester rifle; 2. *comp.* Winchester disk drive
виньета, виньетка vignette; *print.* tailpiece
виньетирование *print.* vignetting, masking
виола *mus.* viol(a)
виолаксантин *chem.* violaxanthin
виолончелист *mus.* (violon)cellist
виолончель (violon)cello
виольдагамба viola da gamba
виомицин viomycin
випера *zool.* asp, viper
вира *hist.* wergild, wergeld
вираж I 1. turn; *av.* banked turn, bank; (*racing track*) banking, curve, bend
~, форсированный accelerated/wind-up (banked) turn
вираж II *phot.* toning; intensifier, toner
вираж-фиксаж *phot.* tone-fixing bath
виргалия *geol.* virga
виргация *geol.* virgation
виргинец Virginian
виргиний *chem.* virginium
виргула *paleont.* axial rod, virgula
виреле *lit.* virelay
виреон *zool.* vireo
виридин *chem.* viridin(e), andalusite
вирилизм virilism
~, гипофизарный hypophysial virilism
~, тиреоидный thyrogenous virilism
~, эпифизарный epiphysial virilism
вирирование *phot.* toning
вирник *hist.* 1. collector of wergeld/wergild; 2. judge dealing with criminal cases
вирола, суринамская *bot.* baboen *(Virola surinamensis)*
Вирсавия *bibl.* Bathsheba
виртуальный virtual
виртуоз virtuoso
виртуозность virtuosity
вирулентность *med.* virulence
вирулицид virucide, viricide
вирулицидный virucide, viricide
вирус virus
~ бешенства rabies virus, lyssin
~ вакцины vaccine virus
~, гемадсорбирующий hemadsorbing virus
~, гриппозный influenza virus
~, ослабленный/аттенуированный attenuated virus
~, парагриппозный parainfluenza virus
~, фиксированный fixed virus
~, фильтрующийся filterable virus, filter passer, filter-passing virus
вирусемия *med.* virusemia
вирусный viral
вирусолог virologist
вирусология virology
вирусоноситель virus carrier
виросоносительство carriage of viruses
вирусурия viruria
висбрекинг viscosity breaking
висение *av.* hover(ing)
~ на месте spot hover(ing)
висеть *av.* hover, overhang
вискоза *text.* viscose, viscous rayon
~, ацетатная acetate rayon
~, вакуумированная deaerated xanthate, spinning solution
~, кордная typo-cord viscose rayon
~, прядильная xanthate spinning solution
~, прядомая spinning viscose
~, штапельная staple viscose rayon
вискозиметр viscometer, viscosimeter, fluid-meter
~, емкостный variable-capacitance viscometer
~ истечения efflux/outflow viscometer
~, крутильный torsion/torque viscometer
~, ленточный band viscometer
~, маятниковый pendulum viscometer
~ падающим шариком falling-sphere viscometer
~, ротационный rotary/rotational viscometer
~, чашечный cup-and-ball viscometer
вискозиметрический viscosimetric(al)
вискозиметрия visco(si)metry

вискозин *chem.* viscosine
вискоскоп viscoscope
вискозность viscosity
вислокрылка *zool.* dobson *(Corydalus);* sialid *(Sialis)*
вислолистный *bot.* cremophyllous
вислоплодник *bot.* cremocarp
вислоплодный *bot.* cremocarpous
висмит *miner.* bismite
висмут *chem.* bismuth, Bi
~, азотнокислый bismuth nitrate
~, бензойнокислый bismuth benzoate
~, лимоннокислый bismuth citrate
~, молочнокислый bismuth lactate
~, основной азотнокислый bismuth subnitrate, bismuthyl nitrate
~, природный сернистый bismuthine
~, самородный native bismuth
~, сернистый bismuth sulfide
~, теллуристый bismuth telluride, tetradymite, wehrlite
~, углекислый bismuthite
висмутин bismuth hydride, bismuthine, bismuthinite, bismuth glance
висмутит *miner.* bismuthite, bismite
висмутный bismuthic
висмутовый bismuthal, bismuthic
висок *anat.* temple
високос leap year
високосный bissextile, intercalary
~ год leap year
~ день bissextile/intercalary day
височно-челюстной temporomandibular
височный temporal
~ ная доля temporal lobe
вист whist *(card game)*
вистария *bot.* wistaria *(Wistaria)*
вистлеры *pl. rad.* athmospherics
висцеральный visceral
висцерогенный viscerogenic
висцероптоз visceroptosis
висцерорецептор visceroreceptor
висцеросенсорный viscerosensory
висячий perched, pendent, hanging, pendulous; dangling; suspended
~ замок padlock
~ мост *arch.* suspension bridge
висящий hanging; pendent, suspended
витализм *biol., philos.* vitalism
виталист *philos.* vitalist
виталистический, виталистский vitalistic
витальный *biol.* vital
витамин vitamin, accessory food factor,advitant, food hormone
~ А, vitamin A, axerophthol, antixerophthalinic vitamin
~ А, концентрированный biosterin
~ В1 vitamin B1 thiamine, aneurin, catorulin; antineuritic vitamin
~ В1 синтетический betalin s
~ В2 vitamin B2, riboflavin, lactoflavin
~ В3 vitamin B3, pantothenic acid
~ В6 vitamin B6, pyridoxine, antiacrodynia factor, factor I, factor Y
~ В12 vitamin B12 cyanocobalamine
~ С vitamin C, ascorbic acid, antiscorbutic vitamin
~ С, кристаллический cebione, cevitamic acid
~ D vitamin D, antirachitic vitamin
~ D2 vitamin D2, calciferol, ergocalciferol
~ D3 vitamin D3, dimethylhydrocalciferol
~ Е vitamin E, tocopherd, antisterility
~ F vitamin F
~ G vitamin G, vitamin B2
~ Н vitamin H, factor H biotin
~ К1 vitamin K1, phytonadion
~ К3 vitamin K3, menadione
~ К4 vitamin K4, menadiol
~ К5 vitamin K5, synkamin
~ Р vitamin P, citrin, hesperidin chalcone
~, бактериальный bacterial/microbial vitamin
~, жирорастворимый fat-sobule vitamin
~, противоксерофтальмический antixerophthalmic vitamin, axerophthol
~, противоцинготный antiscorbutic vitamin, antscorbutin, ascorbic acid
витаминизация vitaminization
витаминность, витаминозность vitamin content
витаминный, витаминозный vitaminic, vitaminous
витаминология vitaminology
витаминотерапия vitamin therapy
витаминосодержащий vitaminous
витамит *geol.* withamite
витекс *bot.* chaste tree *(Vitex)*
витенагемот *hist.* witenagemot, witan
витиеватый florid, flowery, ornate
витилиго *med.* vitiligo, acquired leukoderma, piebald skin, white leprosy
витлокит *geol.* whitlockite
витнеит *geol.* whitneyite
витой twisted, coiled; spiralled; volute
~тая лестница spiral staircase
виток turn, coil; *rad.* loop; revolution, circuit, flight; spire, winding, convolution
~, заходной first turn
~ змеевика lap of coil
~, настроенный resonance loop
~, настроечный tuning loop
~ обмотки loop/turn of winding
~ обратной связи feedback loop
~ раковины whorl, convolution
~ связи coupling loop
~, торцовый end coil
~ транспортирующего винта flight of a helix
~ штопора *av.* turn of the spin
витраж stained-glass window panel, stained glass
витрен *geol.* vitrain
~, бесструктурный structureless vitrain
~, скрытоструктурный cryptostructural vitrain

~, структурный structural vitrain, telain
~ ясной растительной структуры anthraxylon
витрено-фюзен vitrain-fusain, vitrifusain
витрина 1. shop/show-window, shopwindow, display window; 2. showcase
~ для выставки товаров display case
витрификация vitrification
витропорфировый, витрофировый vitrophyric
витрофир *geol.* vitrophyre, glass porphyry
витрофирит vitrophyrite
~ жильный авгитовый sordawalite
витрофюзен vitrofusain
виттихенит wittichenite
витье twisting, winding, spinning; torsion, twining; weaving
витютень *zool.* ringdove, wood pigeon *(Columba palumbus)*
витязь *hist.* knight; hero
Вифлеем (City of) Betlehem, City of David
вихляние wiggle, wiggling
~ колес wheel wobble
вихревой vortical, vortex, turbulent; eddying, whirling
~вое поле curl/rotational field
вихрение vorticity
вихреобразный vorticose; eddy-type
вихреобразование vortex/eddy generation/formation
вихрепроводность eddy conductivity
вихрь 1. vortex, eddy, whirl; 2. *math.* curl; 3. whirlwind
~, атмосферный atmospherical eddy/vortex
~, векторного поля curl of vector function
~, воздушный whirlwind; air eddy
~ за телом wake vortex
~, замкнутый closed vortex
~, изолированный single/discrete vortex
~, колеблющийся oscillating vortex
~, кольцевой circular vortex
~, концевой tip/trailing vortex
~ лопасти/лопатки, концевой blade-tip vortex
~, магнитный magnetic whirl
~, нестационарный transient vortex
~, несущий lifting vortex
~, околополюсный/околополярный circumpolar vortex/whirl
~, осевой axial/axis vortex
~, песчаный sand whirl
~, П-образный/подковообразный horseshoe vortex
~ пульсирующий напряженности fluctuating-strength vortex
~, пылевой/пыльный dust column/whirl, whirlwind; dust devil/whirl
~, сбегающий trailing vortex
~ термического ветра vorticity of thermal wind
~, точечный point vortex
~, турбулентный turbulence eddy
вице-адмирал vice-admiral
вице-канцлер vice-chancellor
вице-король viceroy
вице-президент vice-president
вициналь *cryst.* vicinal form
вицинальный vicinal
вишарник, вишенник I *bot.* ground cherry *(Prunus fruticosa)*
вишенник II *bot.* plum clitopilus *(Clitopilus prunulus)*
вишист *hist.* Vichyite
вишневидный cherrylike
вишнеслива *bot.* myrobalan plum *(Prunus cerasifera)*
вишня *bot.* cherry (tree) *(Cerasus)*
~, виргинская common chokecherry
~, шпанская bigarreau, hard cherry
~, японская Chinese bush cherry
вкатка, вкатывание, вкачивание I rolling into/onto; wheeling in/into
вкачивание II pumping in(to)
ВКБ-метод Wentzel-Kramers-Brillouin treatment
вклад *fin.* 1. deposit, investment; 2. endowment; 3. advance payment, ante
~, беспроцентный noninterest bearing deposit
~, бессрочный sight/demand deposit
~, вынуть withdraw a deposit
~ деления *nucl.* fission contribution/bonus
~, краткровременный short deposit
~, сберегательный savings deposit
~, сделать deposit money
~, срочный fixed/time deposit
вкладка 1. *print.* inset, insert(ion), supplementary sheet; wraparound; 2. putting/placing in
вкладной 1. inserted, supplementary; 2. *bot.* overlapping; equitant
~лист loose leaf/sheet
вкладчик *econ.* depositor; investor
вкладывание insertion, laying/putting in; enclosing
вкладывать 1. *fin.* invest; 2. put in, insert; embed, imbed; inlay; enclose; nest
~ деньги в банк put money in(to) a bank
вкладыш 1. *tech.* shell, insert; bush(ing); (bearing) brass; inset; 2. *print.* supplementary sheet
~, графитовый graphite/carbon insert/lining
~, заливать babbit the brass
~ из карбида кремния silicon carbide insert
~ подпятника collar step
~ подшипника bearing liner/bush/brass/shell
~, сегментный segmented insert
~, сменный replaceable brass
~, сопловой/сопла nozzle/throat insert
~ стопора stop bushing
вклеенный glued-in
вклеивание gluing/pasting in
вклеивать paste/glue/stick in
вклейка 1. inset, insert, insertion; glued-in piece;

timing tape; spacer; 2. pasting/gluing in
вклепанный riveted-in
вклепка 1. riveting; 2. rivet
вклепывать rivet in, fasten with rivet(s)
вклинивание wedging-in
~, взаимное interdigitation
~ септ insertion of septa
вклиниваться wedge oneself (in)
включатель switch, circuit breaker; engaging switch
~ валика *comp.* platen release lever
~ вытяжного действия pull-on switch
~ зажигания ignition switch
~, инерционный acceleration/inertia switch
~, часовой clock starter
включатель-ликвидатор *mil.* disarming switch
включать 1. include, insert; incorporate; 2. *tech.* switch/turn on; energize, start, activate; plug in; connect, join up; put in gear
~ в работу put into operation
~ параллельно connect in parallel
~ последовательно connect/place in series
~ передачу put in gear
~ скорость engage a gear
~ совместно interconnect
включение 1. *tech.* switch(ing)/turn(ing) on; throwing in; cutting in, turning on; activation, engaging, connection; energizing, closing; 2. engagement, activation, starting; ignition; 3. *geol.* inclusion, enclave, enclosure, embedding; xenolith
~, алломорфное allomorph
~, амигдалоидное amygdaloid infilling
~ в гнездо занятой линии *tel.* overplugging
~ в множестве *math.* inclusion in a set
~ в породе *geol.* nest
~ в угле impurities in coal
~, видимое visible impurity
~, воздушное (en)trapped air
~, газовое occluded/entrapped gas
~, галечное boulder-like inclusion
~, гнездообразное admixture in pocket
~, гомогенное cognate/homogene enclave
~, гомологичное homologue, homologous enclave
~, дендритообразное *met.* fern-shaped precipitate
~, дробевидное bird-shot concretion
~ железа iron inclusion/staining
~, жидкое/жидкости liquid/fluid inclusion
~ записи *comp.* record insertion
~ зажигания ignition switch-on
~, каскадное cascading, cascade connection
~, кристаллическое *miner.* phenocryst
~, ксеногенное xenocryst
~, ленточное *geol.* lenticular intercalation
~ линии *comp.* line-up
~, марганцевое manganese nodule
~, минеральное concretion, mineral spot
~, многократное multiple switching
~ мостом bridge connection
~ на срабатывание circuit-closing connection
~, неметаллическое non-metallic inclusion
~ нефти oil pocket
~, параллельное parallel connection; connection in parallel
~, пластовое intraformational bed
~, плезиоморфное plesiomorphic inclusion/enclave
~, поикилитовое poikilitic inclusion
~, полигенное polygenetic enclave, polygenous inclusion
~, породное rock intercalation
~, последовательное series connection, connection in series
~, постороннее в вулканической породе enallogene enclave
~, прямое *tel.* straight multiple
~ связи *comp.* line-up
~ серы sulphur band
~ сланца в угле horseback
~, случайное accidental inclusion, enallogene enclave, exogenous enclosure
~, собственное strict/proper inclusion
~ сцепления clutch engagement
~, твердое *met.* hard spot
~, тонкое intercalation
~ тормозов braking on, brakes-on
~, червеобразное wormy/vermicular inclusion
~, шлаковое entrapped slag, slag inclusion
~, штекерное plug-in connection
~, экзогенное exegenous/exogenic/disomatic inclusion, exogenous enclosure
~, эндогенное endogenous enclosure, cognate inclusion, autolith
включенный 1. connected; included, incorporated; enclosed, embedded; inserted; 2. switched/turned on
~ в пласт inbedded, embedded
~ между пластами interbedded, intercalated
~ параллельно placed in parallel
~ последовательно connected/placed in series, series connected
включено switched/turned on; energized
~в эфир on-the-air
включительный inclusive
включить *comp.* enable
вколачивание 1. *anat.* peg and socket articulation, gomphosis, impaction; 2. knocking/hammering in
~ зуба dental impaction
вколачивать drive
~ до отказа drive home
ВКП(б) *abbr.* **Всесоюзная Коммунистическая Партия (большевиков)** *hist.* All-Union Communist Party (bolsheviks)
вкрапление 1. embedment, interspersing, sprinkling; 2. *geol.* dissemination, impregnation; inclusion, endomorph
вкрапленник *geol.* phenocryst, inset, impregnation
~, крупный megaphenocryst
~, рудный shot ore
вкрапленность *miner.* dissemination, impregna-

tion, shot
вкрапливать impregnate, disseminate
вкус 1. taste; flavo(u)r; 2. manner, style
~, вяжущий astringent taste
~, соленый saline taste
~, щелочной alkaline taste
вкусовой gustatory; taste *(attr.)*
~вое ощущение gustatory sense
~вые вещества flavo(u)ring substances
влага 1. *biol.* humo(u)r; 2. moisture; humidity, damp(ness)
~, внешняя air-dried moisture
~, внутренняя inherent moisture
~, водянистая aqueous humo(u)r
~, захваченная entrapped moisture
~, капиллярная cellular moisture
~, кристаллизационная crystal moisture/water
~, несвязанная free moisture
~, пленочная pellicular/film moisture
~, подпочвенная underground moisture
~, почвенная soil water/moisture
~, связанная bound moisture
~, стекловидная vitreous humo(u)r
влагалище 1. *anat., bot.* vagina, sheath; 2. *bot.* axil, ocrea, boot
~, волосяное hair shealth
~ глазного яблока shealth of the eyeball
~, двойное bipartite vagina, vagina septa
~, защитное endodermis
~, крахмалоносное starch sheath
~, нервное/нерва nerve tunic/sheath
~, сонное carotid sheath
~ сухожилия,синовиальное synovial sheath of a tendon
~, сухожильное tendon sheath
~, цветочное spathe
влагалищевидный,влагалищеобразный ocreaceous
влагалищно-прямокишечный *anat.* rectovaginal, vaginorectal
влагалищный *anat., bot.* vaginal, vaginate; sheathed
влагальце small vagina, vaginula
влагать 1. put/place in, insert, enclose; 2. *fin.* invest
влагодержащий moisture-holding
влагоемкость water capacity; *agr.* humidity ratio, moisture(-holding) capacity; hygroscopicity
~, весовая gravimetric moisture capacity
~, наибольшая maximum water capacity
~, наименьшая minimum water capacity
~, нормальная полевая normal field capacity
~, объемная volumetric moisture capacity
~, пленочная pellicular moisture capacity
~, полевая field water capacity
~, полезная available moisture capacity
~, полная maximum water capacity; total moisture capacity
влагозарядка moisture/water supply
влагоизоляция waterproofing, moisture-proofing, moisture insulation/seal, vapour barrier
влагомер moisture meter/tester/gauge, hydrometer, hygrometer
~, емкостный capacitance moisture meter
~, кондуктометрический conductivity moisture meter
~, нейтронный neutron (scattering) moisture meter
влагонепроницаемость moisture resistance/ impermeability
влагонепроницаемый,влагоупорный moisture-/damp-proof, moisture-impermeable/-resisting, waterproof, moistureproof, dampproof
влагообмен moisture exchange
влагооборот moisture rotation/circulation, moisture/ hydrological cycle
влагообразование moisture formation
влагоотдача water yielding capacity, moisture yielding ability
влагоотделитель moisture/water separator; water trap
влагоотталкивающий moisture-repelling/-repellent, water-repellent
влагопоглотитель desiccant; desiccator; dryer
влагопоглощение moisture/water absorption
влагопроницаемость moisture/humidity permeability
влагопрочность wet strength
влагопрочный moisture-resistant; water-fast
влагорегулятор humidistat
влагосниматель *food* humectant
влагосодержание moisture/water content; *agr.* humidity ratio
~, равновесное balanced moisture content
влагостойкий,влагоупорный,влагоустойчивый moistureproof, dampproof, moisture-resistant/-impremeable/-repellent
влагостойкость, влагоупорность, влагоустойчивость humidity/moisture resistance, moistureproofness
влагоудаление dehumidification
влагоудалитель moisture extractor, dehumidifier
влагочувствительный moisture-/humidity-sensitive, hygrosensitive
владелец *econ., law* possessor, owner; proprietor
~ акций stockholder
~ бенефиции/феода *hist.* beneficiary
~ долгового обязательства *econ.* debenture holder
~ жиросчета giro account holder
~ именной акции registered shareholder
~ линии связи *comp.* common carrier *(a company)*
~ ломбарда pawnbrocker
~ на правах неограниченной собственности tenant in fee(-simple)
~ недвижимости,пожизненный tenant for life
~ обыкновенных акций ordinary shareholder
~ прибрежной полосы riparian

~ привилегированной акции preference shareholder
~ с молчаливого согласия собственника tenant at sufferance
~, самостоятельный tenant in severalty
~, совместный tenant in common
владение 1. ownership; 2. possession, property; *law* sesin, seizin, estate, land; 3. command, mastery
~, временное tenure
~ землей land tenure
~ недвижимостью tenure, tenancy; seisin, seizin
~ обеими руками, одинаковое ambidexterity
~, отзывное precarions possession
~, пожизненное estate for life
~ с молчаливого согласия собственника estate at sufferance
~ с незапамятных времен immemorial possession
~, феодальное fief
~ языком mastery/command of a language
~ния, лишать *law* evict
~ния, лишение *law* ouster
владетель owner, possessor; ruler, sovereign
владеть possess, own; hold in demesne; control
владимирка 1. *bot.* morello, Vladimir cherry *(Prunus cerassus austera);* 2. *hist.* the road to Siberia, the Vladimir road
владыка 1. *eccl.* bishop, archbishop, metropolitan; 2. lord, master, ruler, sovereign
владычество sovereignty, dominion; sway
~, морское thalassocracy
владычица 1. *eccl.* Our Lady; 2. sovereign, misterss
влажнонеустойчивый moist-labile
влажность humidity, dampness, moisture (content), moistness
~, абсолютная absolute humidity
~, весовая gravimetric humidity, moisture content by weight
~, весовая в породе gravimetric rock humidity
~ воздуха air humidity, atmospheric moisture
~, горная quarry water, water of imbibition
~ завядания wilting moisture content
~, общая integral moisture content
~, оптимальная optimum water content
~, относительная relative humidity
~, полезная available moisture content
~, почвенная soil moisture
~, приземная surface humidity
~, равновесная equilibrium moisture
~, удельная *agr.* specific humidity, humudity ratio
~, условная reference humidity
влажный humid, damp, moist; wet
власоглав *zool.* whipworm *(Trichuris trichiura)*
власоед, собачий dog-biting louse
властолюбие craving/aspiration for power, love of power
власть 1. power; 2. authority; domination; *pl.* anthorities, administration
~, административная disciplinary powers; *pl.* administration
~, законная lawful authority
~, законодательная legislative power, legislature
~ злодеев kakistocracy
~, исполнительная executive (power)
~ мужчин virocracy
~, придержащая the powers that be
~, судебные judicial power
~, управляющая administering authority
власяница,власяная одежда *hist.* cilice, hair shirt
влах Vlach, Walachian
влашский Vlach, Walachian
влет *av.* entry, flying in
влетание flying; rushing
влечение inclination, bent; attraction
~, половое sexual appetite, carnal lust
~ событий *math.* implication of events
влечь draw, drag, trail
вливание *med.* infusion, injection; pouring
~, внутривенное intravenous injection, phelebaclysis
~, делать внутривенное administer intravenously
~, капельное drop infusion
~, капельное внутривенное drop/slow phleboclysis, intravenous drip
вливать pour, run in, infuse
~ по капле put in drop by drop; *med.* infuse
вливаться flow/stream/run *(into)*
влитие 1. infusion, injection; 2. replenishment, addition
влияние I influence, effect, impact
~, боковое lateral effect
~, взаимное interference
~ внешних полей (external) magnetic field influence
~, внешних условий environmental effect
~, возмущающее perturbation
~, вредное adverse/harmful/deleterious/ unfavourable effect
~ земли ground effect
~, индуктивное induction effect
~, контактное contact action/influence
~, лунно-солнечное lunisolar influence
~ масштаба scale effect
~ мертвых витков *el.* dead-end effect
~ местных условий local effect/influence
~ на окружающую среду environmental effect(s)
~ окружающей среды ambient effect, effect of the ambient conditions
~, осаждающее precipitating action
~ охлаждения chilling effect
~ помех interference effect
~, постороннее irrelevant influence
~ приливов tidal effect

~, **результирующее** net effect
~ **рельефа** topographic/terrain effect
~ **силы тяжести** gravity/gravitational effect
~ **соседней радиостанции** *tel.* kindred effect
~ **спутной струи** wake effect
~ **стенок** *naut.* wall constraint/effect, bank effect
~ **турбулентности** eddying effect
~ **условий залегания** *geol.* formation factor
влияние II flowing
влиять affect, have/produce an effect; influence
ВЛКСМ *abbr.* **Всесоюзный Ленинский Комунистический Союз Молодежи** *hist.* All Union Leninist Young Communist League
вложение 1. imbedding, embedding; *comp.* nest(ing); inclusion; enclosure; 2. *fin.* investment, placement
~ **в целом** global embedding
~, **капитальное** capital investment
~ **операторов** *comp.* nesting of statements
вложенный *comp.* nested; *math.* embedded, imbedded
~ **один в другой** *comp.* nested
вложимость embeddability
вмазка cementing, puttying
вменяемость *law* sanity, liability, responsibility, accountability
вменяемый *law* sane, of sound mind; liable, responsible, accountable
вменять impute, regard (as)
~ **в вину** charge with, accouse of
~ **в заслугу** regard as a merit
вмерзание freezing-in
вмерзать become ice-bound, freeze in
вместилище *med.* cistern; receptacle, conceptacle; container, vessel, (storage) tank; reservoir
~ **секретов** secretory sac/space
вместимость (holding) capacity, volume capacity; space, content, hold, room; tonnage
~, **аварийная** *naut.* life-saving capacity
~, **валовая/полная** gross (register) tonnage
~, **регистровая** register(ed) tonnage
~ **резервуара** tankage
~ **складских помещений** warehousing capacity
~, **удельная** specific/volume capacity
~, **чистая** net tonnage
вместительность capaciousness; spaciousness, capacity, roominess; contents
вместительный capacious, spacious, roomy
вмешательство intervention, interference; meddling; *tel.* cross-talk
~, **вооруженное** armed intervention
~, **хирургическое** surgical intervention
~ **человека** human intervention
вмещать 1. enclose, contain, encompass; hold; 2. seat, accommodate
вмещение *math.* embedding, imbedding; insertion
вмонтировать build/set/fit/mount in
вмонтированный 1. incorporated, built-in, embedded; 2. fixed, stationary
вмывание inwash, illuviation; washing in
вмятина hollow, dent; nick; dimple, depression; tool/sink mark
внавалку (in) bulk
внакладку overlap(ping), lapping
внакрой lap(ped), overlapped
внасыпную in bulk, loose
внахлестку overlap(ping), overlapped, lap(ped)
вначале in the beginning, at first
вне 1. out/outside of; beyond; exept; external/exterior to; without 2. in addition to
~ **закона** without the law
~ **закона, объявить** outlaw
~ **конкурса** hors concours
~ **времени и пространства** regardless of time and space
~ **помещения** out of doors
внеатмосферный extra-atmospheric
внеаудиторный extracurricular, out-of-class
внебрачность bastardy
внебрачный out of wedlock *(of child);* extramarital
внебрюшинный *med.* extraperitoneal
внебрюшной extra-abdominal
вневлагалищный extravaginal
вневойсковая подготовка military training for civilians
вневписанный escribed
вневременный supertemporal, nontemporal, timeless
внегалактический extragalactic, anagalatic
внегородской suburban, out-of-town
внегортанный extralaryngeal
внегравитационный gravity-free
внеграмматический extragrammatical
внегрудной extrapectoral
внедрение 1. *geol.* intrusion, invasion, injection, emplacement, irruption; 2. introduction; adoption; incorporation, inculcation, instillation; indoctrination; intercalation
~, **вакуумное** vacuum injection
~ **йонов** ion implantation
~ **передовых методов** introduction of advanced methods
внедрять 1. introduce, adopt, incorporate; 2. implant, embed; instil; inculcate
внеевропейский non-european
внежелудочковый extraventricular
внезапка *abbr.* **внезапная проверка** sudden prisoner check
внезародышевый extraembryonic
внеземной extraterrestrial
внекапиллярный extracapillary
внекапсульный, внекапсулярный extracapsular
внекишечный abenteric, extraintestinal, extraenteric
внеклассный out-of-class/-school; extracurricular
~**ное чтение** home reading
~**ные занятия** out-of-school/-class activities
внеклассовый *polit.* nonclass

внеклеточный ectocytic, exocellular, extracellular
внеконкурентный noncompetitive, uncompetitive
внеконкурсный hors concours
внеконтинентальный extracontinental
внеконтрольный out of control
внекостномозговой extramedullary
внекостный ectosseal, extraosseous
внелабораторный nonlaboratory
внелегочный extrapulmonary
внеледниковой extraglacial
внелингвистический extralinguistic; paralinguistic
внелинейный off-line
внелитературный nonliterary; extraliterary
внелунный extralunar
внематочный extrauterine
внемеридиональный exmeridian, extrameridional
внеметрический nonmetrical
внеминдаликовый extratonsillar
вненациональный non-national
вненосовой extranasal
внеосевой off-axis, extra-axial
внеочередной 1. extraordinary; extra, special; top-priority; 2. out-of-order/-turn
~ная сессия extraordinary session
внеочередность (top) priority
внепазушный *bot.* extra-axillary
внепарный unpaired, nonpaired
внепартийный *polit.* nonparty
внеперикардиальный extrapericardial
внепеченочный extrahepatic
внепиковой off-peak
внепирамидный extrapyramidal
внепланетный extraplanetary
внеплановой *econ.* extraordinary, not provided for by the plan
~ расход unforeseen expenses/expenditures
внеплевральный extrapleural
внеплодник *bot.* epicarp, exocarp
внеполовой extragenital
внепочечный extrarenal
внепузырный extravesical
внерейсовый *av.* non-scheduled
внеродовой nongeneric
внесезонный out-of-season, off-season
внесение 1. bringing/carrying in; importation; 2. introduction, submission, moving; 3. entry, entering, insertion; 4. deposit, paying in; 5. *agr.* application
~ в протокол entering in the minutes
~ на рассмотрение submitting for examination/inspection
~, припосевное application at seedling
~ поправок recitification
~ удобрений fertilizer applicaton
внесердечный exocardial, exocardiac
внесистемный exogenous, extrasystemic
внеслужебный leisure-time, out-of-office
внесметный unbudgeted, extraordinary
внесолнечный extrasolar
внесосудистый extravascular
внестудийный *telev.* live, outside, outdoor
внестеночный exothecal; extramural
внеструктурный nonstructural
внесудебный extrajudicial
внесуставной abarticular, extra-articular
внетабличный extratabular
внетычиночный *bot.* extrastamineous, extrastaminal
внеуниверситетский extramural, extra-university
внеурочный done outside of class hours
внеустьичный *bot.* extrastomatal
внеутробный extrauterine
внефокальный extrafocal
внефокусный out-of-focus, defocused
внефолликулярный extrafollicular
внефонологическй extraphonological, infraphonemic, nonphonological
внецветковый *bot.* extrafloral
внецентренный eccentric(al), excentric
внецентровой eccentric(al)
внечерепной extracranial
внешкальный off-scale
~ нуль inferred zero
внешкольный out-of-school, extrascholastic
~ная работа out-of-school work, campus activity
внешнеграмматический nongrammatical, extragrammatical
внешнее *n.* exterior(ity), externality, extrinsicalness
внешнелингвистический nonlinguistic, extralinguistic
внешнеполитический foreign policy *(attr.)*
внешнесекреторный exacrinous, exocrine
внешнесетчатый extrareticular
внешнеторговый foreign trade *(attr.)*
внешний 1. external, exterior, outer; outward; outboard, outside; extrinsic(al), extraneous; 2. foreign
~ вид outward appearance
~ покров *bot.* exothecium, outer anther case
~ угол *math.* external angle
внешность appearance; outside; exterior; surface; mien, air, look
внештатный supernumerary; nonsalaried, freelance
внешторг *abbr.* **Министерство внешней торговли** Ministry of Foreign Trade
внеэкономический noneconomic
внеэтический nonethical; amoral
внеядерный extranuclear; exocyclic
внеязыковый extralinguistic
вничью *sport* in a draw, drawn
~, сыграть draw
внос entry; introduction; application
вноситель mover, proposer
вносить introduce, bring/carry in; insert, enter; move
~ в банк *fin.* place on deposit
~ в инвентарь inventory

~ в таблицу tabulate
~ в список put/enter on a list
~ законопроект introduce a bill, bring in a bill
~ записи в журнал enter (notes) in the log
~ на рассмотрение submit for cousideration
~ поправку correct; insert a correction; *law* amend
~ предложение table/introduce a motion
~ смятение cause/spread confusion
~удобрения apply fertilizer, fertilize, manure
вноска 1. introduction; 2. insertion; entry
вносный 1. deposited, inserted; 2. carried over, transferred
~ные знаки inverted commas, quotation marks
внук grandson, grandchild
внутренне inwardly, internally; intrinsically; at heart
~ присущий immanent, inherent
внутреннепокровный *zool.* infrategillar
внутреннесетчатый infrareticulate, intrareticulate
внутреннеслоистопокровный infratectal
внутреннеточечный infrapunctate
внутренний 1. inside, interior, inner; *av., nav.* inboard, internal, intrinsicial, inward; indoor; *comp.* in-line; 2. home, inland, domestic; inherent
~ заем domestic loan
~ мир private world, inner life
~ рынок domestic/home/internal market
~ смысл inner meaning
~нее состояние inwardness, internality
~нее чутье intuition
внутренности *pl. anat.* viscera
внутренностный visceral, spanchnic
внутренность 1. interior, inside; core; 2. *pl. anat.* internal organs
~ угла *math.* angular domain / region
внутреродность *math.* endomorphism
внутри inside, within
~, порождать ingenerate
~, строго proper interior
внутриаортальный intra-aortal
внутриартериальный endarterial, intra-arterial
внутриатомный intratomic, subatomic, intra-atomic
внутрибрюшинный intraperitoneal
внутрибрюшной intra-aldominal
внутриведомственный in-house, intradepartmental
внутривенный, внутривенозный intravenous, endovenous
внутривидовой intraspecific
внутривлагалищный *anat., bot.* intravaginal
внутригалактический intragalactic
внутригаплоидный intrahaploid
внутригенный intragenic
внутриглагольный intraverbal
внутриглазничный *anat.* intraorbital
внутриглазной *anat.* intraocular
внутригнездник *bot.* endothecium
внутригорный intramontane
внутригородской intra-urban
внутригортанный intralaryngeal, endolaryngeal
внутригрудинный intrasternal
внутригрудной intrathoracic
внутригрупповой intragroup
внутридолевой intralobar
внутридольковый intralobular
внутридомовой inside the house
внутрижелезистый intraglandular
внутрижелточный intravitelline
внутрижелудочковый intraventricular
внутрижелудочный intragastric
внутризаводский intrafactory, interplant
внутризатылочный intra-occipital
внутрикапиллярный intracapillary
внутрикапсульный, внутрикапсулярный intracapsular
внутриклассный within the classroom
внутриклассовый *polit.* intraclass
внутриклеточный intracellular, endocellular
внутрикожный intracutaneous, intradermal, intradermic, endermic
внутрикольцевой endocyclic
внутрикомплексный *chem.* chelate(d)
внутриконтинентальный intracontinental
внутрикорковый intracortical
внутрикоровый intracrustal
внутрикостный intraosteal, intraosseous, endoosteal
внутрикратонный *geol.* intracratonal
внутрикристаллитный, внутрикристаллический intracrystalline
внутрилабиринтный intralabyrinthine
внутрилегочный intrapulmonary
внутриледниковый englacial, intraglacial
внутрилежащий inner, internal, lying in
внутрилингвистический intralinguistic
внутрилоханочный intrapelvic
внутримагматический intramagmatic
внутриматериковый inland
внутриматочный intrauterine, endometrial
внутриметаллический intrametallic
внутриминдаликовый intratonsillar
внутримицеллярный intramycellar
внутримозговой intracerebral, intramedullary
внутримозжечковый intracerebellar
внутримолекулярный intramolecular
внутримышечный intramuscular
внутринервный endoneural
внутриносовой intranasal
внутринуклонный subnucleonic
внутриобластной intraregional
внутриорбитальный intraorbital
внутриохлажденный internally cooled
внутрипартийный intraparty, inner-party
~ная дискуссия discussion within the party
внутриперегородочный intraseptal
внутрипеченочный intrahepatic
внутрипланетный intraplanetary
внутрипластовой intrastratal, intraformational

внутриплацентарный intraplacental
внутриплевральный intrapleural
внутриплодник *bot.* endocarp
внутриплодниковидный *bot.* endocarpoid
внутриплодный *bot.* endocarpous, endocarpic
внутрипозвоночный intraspinal
внутрипокровный intrategillar
внутриполитический home-policy
~ное положение internal political situation
внутриполостной intralocular; intra-cavity, intracavitary
внутрипочвенный subsurface
внутрипочечный intrarenal
внутрипредсердный intra-atrial
внутрипротоплазматический intraprotoplasmatic
внутрипузырный intravesical, intracystic
внутрирайонный intraregional
внутрирешеточный intralattice
внутриродовой intrageneric
внутриротовой intraoral
внутрирядный in-row
внутрисвязочный intraligamentous
внутрисегментный intrasegmental
внутрисекреторный incretory, endocrine, endocrinous, cryptorrhetic
внутриселезеночный intrasplenic
внутрисеместровый midterm, midsemester
внутрисептальный intraseptal
внутрисердечный intracardiac, endocardiac
внутрисерийный intraserial
внутрисетчатый intrareticulate
внутрисистемный internal; endogenous, intrasystem(ic)
внутриславянский intra-Slavic
внутрислизистый intramucous
внутрислоговой *ling.* (intra)syllabic
~ приступ syllabic cut
внутрисолнечный intrasolar
внутрисосудистый intravascular
внутрисоюзный inner-union
внутристенный, внутристеночный intramural; intraparietal
внутристолбчатый intrabaculate
внутрисуставной intra-articular
внутрисхемный *comp.* in-circuit
внутритабличный intralist, intratabular
внутритазовый intrapelvic, endopelvic
внутритканевый interstitial
внутритрахеальный endotracheal, intratracheal
внутритрубный intratubal
внутриуниверситетский intra-university
внутриутробный intrauterine
внутриучрежденческий within an/the institution
внутриушной intra-aural, intraaural, entotic
внутрифлективный internal-flexion
внутриформационный intraformational
внутрихромосомный intrachromosomal
внутрихрящевой intracartilaginous
внутричастичный intraparticle *(attr.)*
внутричерепной intracranial, encranial, endocranial, entocranial
внутричерешковый *bot.* intrapetiolar
внутришельфовый intrashelf
внутриягодичный intragluteal
внутриядерный intranuclear
внутриязыковый intralinguistic
внутрияремный intrajugular
внутриячейковый intracell(ular)
внушаемость suggestibility
внушать suggest, impress; instil
внушение 1. suggestion; 2. advice, counsel; admonition
~, гипнотическое hypnotic suggestion
~, сделать reprimand
внятность intelligibility, audibility
вобла *zool.* vobla, Caspian roach *(Rutilus rutilus caspicus)*
вовек(и) веков *eccl.* for ever and ever
вовлечение 1. entrainment, entrapment; dragging/drawing in, inclusion; 2. involving, implication
вогнуто-выпуклый concavo-convex
вогнутолистный *bot.* hollow-leaved
вогнутостворчатый *bot.* reduplicate
вогнутость concavity
вогнутый 1. concave; 2. caved-in; dished
~ вверх concave up
~ вниз concave down
вогнуть *math.* deform isometrically
вогул, вогульский Vogul
вода water; *see also* **воды**
~, абразивная abrasive/gritty water
~, агрессивная agressive/corrosive water
~, азотноуглекислая nitric-carbonated water
~, активная decomposition water
~, аммиачная ammonia liquor/water, caustic ammonia
~, артезианская deep-well water, artesian water
~, атмосферная meteoric/atmospheric water
~ атмосферных осадков precipitated water
~, баритовая baryta water
~, белковая albumen/albumin water
~, бидистиллированная bidistilled water
~ биологической защиты реактора *nucl.* shield water
~, бромистая/бромная bromine water
~, верхняя headwater, topwater, upper water
~, верховая upstream water
~, вешняя spring flood
~, внутрипоровая interstitial/pore water
~, водопроводная tap water
~, вторично используемая reused water
~, высокая full water, (high) flood
~, газированная aerated/soda/sparkling water
~, газовая (coal) gas liquor
~, газонапорная gas-zone water
~, гидратная hydration water
~, грунтовая (under)ground/phreatic/subsoil water
~, деаэрированная deaerated water

~, дистиллированная distilled water
~ для городского потребления municipal/urban water
~ для камбуза galley water
~ для питья drinking water
~ для промышленного потребления industrial water
~ для хозяйственно-бытовых нужд domestic water
~, добавочная make-up water, complementary water
~, дождевая rain/storm water, rain wash
~, жавелевая Javel(le) water, eau de Javelle
~, железистая chalybeate water
~, желтая *med.* glaucoma
~, жесткая hard water
~, задержанная entrapped/detained water
~, заиленная muddy water
~, закрытая shut-off water
~, замочная steep(ing) water
~, запруженная stagnant/impounded water, backwater, slack water
~ затворения tempering water *(of cement)*
~, защитная *nucl.* shield water
~, избыточная overflow/excess water
~, известковая lime water
~, кавернозная interstitial/cavern water
~, камфорная camphor water
~, кислая acidic water
~, капиллярная capillary water
~, кислотная acid water
~, конденсационная condensation water
~, конституционная constitutional water, water of constitution
~, контурная edge water
~, котловая boiler water
~, кристаллизационная crystal/crystallization/composition water
~, лавровишневая bay rum
~, легкая light/ordinary water
~, ливневая storm water
~, литиевая lithia water
~ лунного прилива, вторая полная antilunar tide
~, малая low tide
~, медленно текущая lagging water
~, междупластовая middle water
~, межслоевая interlayer water
~, мелкая shoal/shallow water
~, метаболическая *med.* metabolic water
~, минерализованная mineralized water
~, минеральная углекислая acidulous water
~, морская sea/salt/brackish/ocean water
~, мытьевая bath/wash(ing) water
~, мягкая soft water
~, мясная meat infusion
~, насыщенная газом aerated water
~, несвязанная free/uncombined/unbound water
~, нефтяная water of productive oil strata
~, нижняя bottom water
~, низкая low water
~, обессоленная desalted water
~, облученная activated water
~, оборотная circulating/reused water
~, оборотная сетевая *therm.* return heating water
~, обыкновенная ordinary/light water
~, опресненная desalinated/distilled water
~, орошающая reflux water
~, осветленная clarified water
~, открытая 1. water-front; 2. open water
~, отработавшая waste water
~, отработанная tail water
~, отстойная settling-vat water
~, отходящая discharge/waste water
~, охлаждающая cooling water, water coolant
~, охлажденная chilled water
~, паводковая flood water
~ первого контура *nucl.* primary water
~, первозданная primitive/eternal water
~, переохлажденная supercooled water
~, питательная feed water (of a boiler)
~, питающая feed water, make-up water
~, питьевая potable/drinkable/drinking water
~, пластовая stratal water
~, пленочная pellicular/film water
~, поверхностная surface/clay/land water
~, подземная subterranean/subsurface/underground water
~, подледниковая subglacial water
~, подпертая backwater
~, подпиточная make-up water
~, подпочвенная subsoil water
~, полая high flood/water; freshet
~, полная high/water tide
~, пресная sweet/fresh water
~, производственная process water
~, промойная wash water
~ промывания rinsing/flushing/wash water, eluate
~, проточная flowing/running/circulating water
~, прочносвязанная adsorbed water
~, радиоактивная activated/radioactive water
~, рекуперированная reclaimed water
~, сбросная/сбросовая/сливная effluent; waste water
~, связанная bound/combined water
~, сероводородная hydrogen-sulphide water, sulphuretted water
~, сероводородная углекислая sulphurous-carbonated water
~, соленая sea/salt/salty/brackish water
~, стоячая stagnant/stagnating water
~, темная *med.* amaurosis
~, техническая/технологическая process/service water
~, тритиевая tritiated water
~, трюмная bilge water
~, туалетная lotion, scented water
~, тяжелая heavy water

~, **уходящая** discharge/escaping water
~, **фильтрационная** seepage water
~, **химически связанная** chemically combined water
~, **хлорированная** chlorinated water
~, **хлорная** chlorine water
~, **хозяйственная** domestic/service/drinking water
~, **циркуляционная** circulating water
~, **шахтная** mine water
~, **щелочная** alkaline water
~, **энергетическая** power water
воданит *miner.* wodanite
водворение settlement, establishment
водевиль vaudeville, musical comedy, comic sketch
водилка *mech.* thread rail, traverse guide
водило 1. *mech.* carrier; *av.* tow bar, towing and steering arm, drawbar; arbo(u)r, mandrel; 2. leash, lead
водитель driver
~ **автобуса** bus driver
~ **ритма** pacemaker
водитель-механик driver mechanic
водительский driver *(attr.)*
водить lead; drive, conduct
~ **корабль** navigate/steer a ship
~ **поезд** drive a train
~ **самолет** pilot a plane
водка vodka
~, **крепкая** *chem.* aqua fortis, nitric acid
~, **царская** *chem.* aqua regia, nitrohydrochloric acid
водно-жидкостный liquid-water
водно-земной terraqueous
водноледниковый fluvio-glacial, aqueoglacial, glacifluvial
водно-масляный oil-water
водно-нерастворимый water-insoluble
водно-растворимость water solubility
водно-растворимый water-soluble
водно-смоляной water-tar
водно-сосудистый water-vascular
водно-спиртовой aqueous-alcoholic, hydroalcoholic
водность water content
воднофизический hydrophysical
водный 1. *chem.* aqueous, hydrous; *geol.* neptunian; 2. aquatic, water, watery
~ **путь** waterway
~ **раствор** aqueous solution
~ **спорт** aquatic sport, water sports
~**ная станция** aquatic sports station
водоаккумулирующий water-storing/-accumulating
водобой 1. (water) apron, rear apron; 2. water jet
водобоина pothole
водобоязнь *med.* rabies, hydrophobia
водовместилище water reservoir
водовместимость moisture storage capacity
водовмещающий water-bearing
водовод water (pipe-)line, water pipe/conduit
~, **безнапорный** free-/gravity-/flow conduit
~, **магистральный** water main
~, **напорный** pressure tunnel/pipeline
~, **подземный** culvert
~, **турбинный** penstock, power conduit
водо-водяной 1. water-to-water; 2. *nucl.* watermoderateal/-cooled
водовоз water carrier
водоворот vortex, whirlpool, eddy
водоворотный vorticose
водовосприимчивый water-sensitive
водовыпуск overflow spillway, water outlet/discharge
водовыпускной water-drain/-discharge
водогрейка water heater, boiling tank
вододействующий hydraulic
вододелитель water divider
вододисперсионный water-dispersible
водоем pond, pool, basin, (water) reservoir
~, **внутренний** inland water
~, **грунтовой** ground water reservoir
~ **для драги** flotation pond
~ **для регулирования паводков** flood control reservoir
~, **естественный** naturally impounded body
~, **запрудный** impounded/dammed reservoir
~, **застойный** stagnant basin
~ **оседания** sag-pond
~, **перемежающий** intermittent impounded body
~, **подземный** underground reservoir
~, **постоянный** perennial impounded body
~, **углубленный** depressed basin
водоемкий containing/absorbing much water
водоемкость water-retaining capacity
водозабор water intake/inlet, diversion of water
~, **башенный** intake tower
~, **безнапорный** free-flow (water) intake
~, **напорный** pressure (water) intake
~, **плотинный** dam (water) intake
водозаборник water scoop/inlet
~ **конденсатора** condenser scoop
водозадерживающий water-impounding/-retaining
водозапорный water-sealing
водозаправщик water bowser/loader
водозатор water intake
~, **башенный** water intake tower
водозащищенный waterproof, watertight; splash-proof
водозор hydroscope
водоизмеритель water meter
водоизмещение displacement
~, **весовое** displacement tonnage
~ **в полном грузу** full-load displacement
~ **корпуса** *naut.* hull displacement
~, **объемное** volume displacement, displacement volume
~, **полное** total displacement, full(-load) displacement

~ порожнем unloaded displacement, light(-ship) displacement
~, проектное designed displacement
~, стандартное standard/treaty displacement
водоисточник water source
водокачка pumping station, water-pump(ing) station, water tower
водокрас *bot.* frogbit *(hydrocharis)*
водолаз I diver
~ аварийно-спасательной службы underwater mechanic
~, глубоководный deep diver
~, дежурный standby diver
~, легкий frogman, skin diver
водолаз II newfoundland dog
водолазный diver, diver's, diving
~ колокол diving bell
~ костюм diving suit
водолаз-аквалангист *sport* scuba (self contained underwater breathing apparatus) diver
водолаз-матрос seaman diver
водолаз-разведчик/-подрывник frogman
водолей 1. *naut.* water(-tank) boat, water carrier; 2. (water) bailer, water pumper; 3. *astr.* Aquarius, Water Bearer
водолечебница balneary
водолечебный balneologic
водолечение balneotherapeutics, balneotherapy, hydratics, hydropathy, hydrotherapeutics, hydrotherapy, water cure
водолистниковые *n. pl. bot.* waterleaf family *(Hydrophyllaceae)*
водолом water break
водолюб *bot.* pennywort *(Hydrocotyle);* waterleaf
водолюбивый hydrophilous, water-loving
водолюбы *pl. zool.* water-scavanger beetles *(Hydrophilidae)*
водомаслозаправщик water-oil bowser/loader, oil-water servicing truck
водомер water meter, hydrometer, flowmeter
~, барабанный drum water meter
~, безнапорный free-flow water meter
~, винтовой propeller water meter
~, диафрагмовый diaphragm water meter
~, дисковый nutating disc water meter, wobbling-plate water meter
~, дроссельный differential-pressure water meter, constriction water meter
~, крыльчатый vane water meter, rotary-disc water meter
~, ротационный rotary water meter
~ с трубой вентури venturi(-tube) water meter
~, скоростной velocity water meter
водомер(ка) *zool.* water strider *(Hydrometra)*
водомерный water-measuring
~ кран gauge cock
~ная труба gauge tube
~ное стекло gauge glass
водомет water-jet (propeller); *hydr.* monitor
водометный движитель water-jet propeller
водомоина gully, gulley, deep rut, water cut, ravine
водонагреватель water heater
водонапорный water-pressure
~ная башня water tower
водонасыщенный water-saturated/-logged
водонепромокаемость, водонепроницаемость watertightness, waterproofness, water impermeability/resistance
водонепроницаемый water-tight/proof/- repellent, watertight, waterproof, impermeable (in water); impervious (to water)
~мая переборка *naut.* watertight bulkhead
водоносный water-bearing/-carrying, aquiferous
водообеспечение water delivery
водообеспеченность water supply
водообмен water exchange
водоопреснитель water-desalination apparatus; water-distilling plant
водоопыление *bot.* hydrogamy
водоотведение water removal/disposal
водоотвод water-way, drainage system; water diversion, discharge outlet; overflow pipe
водоотводный water-draining, drainage
~ная трубка waste pipe
водоотдача yield of water, water yield
~, удельная specific water yield
водоотделитель water separator /trap/eliminator/ extractor; dehydrator
водоотлив draining, drainage; water pumping/ drainage
водоотливный насос drainage/dewatering pump; *naut.* bilge pump
водоотстойник water-collecting sump; water sedimentation tank/vat
водоотталкиваемость water repellency
водоотталкивающий water repelling/repellent, waterproof, hydrophobe
водоохладитель water cooler
водоохлаждаемый water-cooled
водоохранная зона water conservation district
водоочиститель water purifier, water treatment plant
водоочистительный water-purifying
~ная станция waste water reclamation plant
водоочистка wastewater treatment, water purification/treatment
водоочищение water purification
водопад waterfall, cataract; cascade
~, грунтовой воды ground-water cascade
~ карельского типа Karelian-type waterfall
~, отступающий regressing/retreating waterfall
~, угасший/исчезнувший extinguished/obliterated waterfall
водоплавающий natatorial
~щая машина amphibious vehicle
~щие птицы *zool.* waterfowl
водоплавный floatable; floated; natatorial
водопоглотитель water absorber
водопоглощаемость water sorption/absorbability
водопоглощающий water-absorbing, hydroscopic

водопоглощение water absorption/uptake
водоподготовка water treatment
водоподогреватель water heater
~, абонентский domestic/apartment hot water converter
~, емкостный storage water heater
~, солнечный solar water heater
водоподъемник water elevator/lift
~, газоструйный air lift
~, цепной chain pump
водоподъемный water-lifting/-raising
водопой 1. watering *(of animals);* 2. watering place/trough; pond
водопоказатель water gauge
водопользование water management/consumption/use
водопользователь, водопотребитель water consumer/user
водопотребление water consumption/use
~, производственное industrial water consumption
~, сельскохозяйственное agricultural water consumption
~, хозяйственное domestic-water consumption
водопотребность water requirements
водоприемник water intake, water-inlet header; catch-water
~, поверхностный surface water intake
водопровод 1. water pipeline/conduit/ piping/main; plumbing; aqueduct; 2. water supply (system); running water
~ большого мозга *anat.* aqueduct of the cerebrum
~, городской urban water supply system
~ преддверия *anat.* aqueduct of the vestibule
~, промышленный/технический industrial water pipeline
~, противопожарный fire main/line
водопроводно-канализационный water-and-sewage
водопроводность water conductivity
водопроводный water-supply/-conducting
~ная магистраль water main
~ная сеть water supply (system)
~ная станция waterworks
водопроводчик plumber, pipelayer
водопроводящий water-conducting
водопроницаемость water permeability/penetration, permeability to water
водопроницаемый (water-)permeable
водопроток gutter waterway
водопрочность resistance to water, water stability
водопрочный water-stable
водоразборный кран hydrant
водоразборная колонка stand pipe, pillar fountain
водораздатчик water dispenser
водораздел watershed, water shed/parting, divide, interstream area
~, второстепенный subdivide
~, долинный valley divide
~ напорных вод pressure water divide
~, непостоянный shifting divide
~, поверхностный surface watershed
водораздельный water-dividing/-parting
водораспределитель water distributor, distributing/spreading baffle
водораспылитель water sprinkler/sprayer; spray(ing) nozzle
водорастворимость water solubility
водорастворимый water-soluble
водорез I *naut.* cutwater; *wood.* stem cutting bar, bobstay/lace piece
водорез II *zool.* skimmer *(Rhynchops)*
водорез III *bot.* water soldier *(Stratiotes alodies)*
водород *chem.* hydrogen, H
~, активный/атомарный active/atomic/nascent hydrogen
~, бромистый hydrogen bromide
~, газообразный hydrogen gas, gaseous hydrogen
~, легкий light hydrogen, protium
~, межгалактический intergalactic hydrogen
~, молекулярный molecular hydrogen
~, мышьяковистый hydrogen arsenide, arsine
~, сверхтяжелый tritium
~, сернистый sulphuretted hydrogen, hydrogen sulphide
~ содержащий hydrogenous
~, сурьмянистый antimony hydride
~, тяжелый heavy hydrogen, deuterium
~, фосфористый phosphorus hydride
~, фтористый hydrogen fluoride
~, хлористый hydrogen chloride
~, цианистый prussic acid
~, элементарный elemental hydrogen
~ом, насыщать hydrogenate
водородистый hydride; hydrogen *(attr.)*
водородкатионирование hydrogen zeolite softening
водородный hydrogen(ous)
~ная бомба hydrogen bomb
водородсодержащий hydrogenous, hydrogen-containing
водорослевидный *bot.* algalike, fucoid(al)
водорослевый algal
водоросль *bot.* alga, water plant; *pl.* eel/sea grass, weeds
~, бентическая benthos alga
~, бурая brown alga, phaeophycean *(Phaeophycea)*
~, выделяющая известь lime-secreting alga
~ диатомовая diatom *(Diatomaceae)*
~ жгутиковая flagellate alga
~, зеленая green alga, chlorophycean *(Chlorophycea)*
~, золотистая yellow-green alga *(Xanthophyceae)*
~, известковая calcareous alga
~, морская seaweed, kelp, tangle; *pl.* marine

algae

~, обволакивающая coating alga

~, переносимая течениями drift weed

~, породообразующая rock-building alga

~, прибрежная beach grass

~, сверлящая boring/penetrating alga

~, сине-зеленая blue-green alga, cyanophycean *(Schizophyceae, Cyanophyceae)*

~, сифониковая diplopora

~, харовая chara(-shell), charophyta

водосбор I water catchment area; watershed; reservoir; drainage system

водосбор II *bot.* columbine *(Aquilegia)*

водосборник 1. water(-collecting) header, catchment basin; water tank/catcher; 2. *min.* water/drain sump

~, рукавный collecting breeching

водосборный water-collecting

~ участок catch land

водосброс spillway, waste gate, dam outlet, (decanting) weir

~, береговой shore/coast spillway

~, верхний overflow

~, глубинный/донный bottom discharge

~, сифонный syphon spillway

~, шахтный shaft spillway

водосбросный spillway, weir *(attr.)*

~ная плотина flood (control) dam

водосвинка *bot.* capybara *(Hydrochoerus capybara)*

водосвязанный, водосвязный water-bound

водосвятие *eccl.* consecration of water

водоскат waterfall; rapids

водослив 1. spillway; weir (bulkhead); waste gate; 2. overflow, overfall

~, затопленный submerged weir

~, мерный weir gauge

~, открытый free/open spillway

водосливный spillway, waste-gate

~ная плотина overflow dam/weir

водоснабжение water supply/delivery

~, горячее hot water supply

~ и канализация water supply and sewerage

~, коммунальное water utilities, public water supply

~, оборотное circulating water supply

~, питьевое drinking water supply

~, противопожарное fire prevention water supply

~, техническое process water supply

водосодержание moisture/water content

водосодержащий water-bearing, aquiferous

водоспуск bottom water outlet, drainage/dam outlet, floodgate, spillway, drain weir, outlet conduit

водостойкий water-proof/-resistant/-resisting, waterproof, hydrostable

водостойкость water resistance, stability in water, waterproofness

водосток pipe drainage system; gutter, run-off, drain

~ двери автомобиля rain rail, roof drip rail

~, канализационный sewage, sewerage

~, уличный sewer

водосточный drain, gutter *(attr.)*

~ желоб gutter

~ная канава drain(ing) ditch, gully

~ная труба drainpipe

водоструйный water-ejecting/-jet

водотеплолечение hydrosudotheraphy

водотечность leakage, leaking

водоток channel, water course

~, поверхностный surface-stream flow

~ подземный subterranean stream

~, стесненный unfree water

водотрубный котел water-tube boiler

водоудержание,водоудерживание water retention/holding

~, удельное specific water retention

водоудерживающий water-holding/-retaining; moisture-holding

водоуказатель water gauge; water-level indicator

водоуловитель water catch

водоумягчение water softening/demineralization

водоумягчитель demineralizer, water softener

водоупор water-confining stratum/layer, aquiclude, impervious bed

водоупорность, водоустойчивость water resistance

водоупорный, водоустойчивый water-resisting/proof/-resistant, waterproof

водоходный navigation(al); amphibious

водоходство navigation

водохозяйство water supply (system), water distribution system

водохранилище water tank/storage, cistern impounded body, reservoir, water collecting area

водочерпалка water engine

водочерпательный water-lifting/-scooping

водочистка water purification

водоэкстракционный water-extracting

воды *pl.* waters; spa

~, внутренние inland/internal waters

~, железистые chalybeate waters

~, закрытые sheltered/closed waters

~, международные international waters, high sea

~, околоплодные *anat., zool.* amniotic/amnionic fluid/waters

~, покрывающие (континентальный шельф) superjacent waters

~, сбросные liquid waste effluent

~, сточные water-borne wastes, discharged/waste waters, sewage

~, судоходные navigable waters

~, таможенные jurisdictional waters

~, территориальные territorial waters, maritime belt

водян(и)ка *bot.* crowberry *(Empetrum)*

водянистость wateriness, aquosity

водянистый watery, aqueous

водянка *med.* edema, hydrops, dropsy
~ **барабанной полости** hydrotympanum
~ **беременных** hydrops gravidarum
~ **больших половых губ** hydrocele feminae
~, **брюшная** abdominal dropsy, ascites, hydroperitoneum, hydroperitonia
~ **влагалища** hydrocolpos, hydrocolpocele
~ **глаза** hydrophthalmos, hydrophthalmia
~ **головного мозга** hydrocephalus, hydranencephaly, dropsy of the brain
~ **желчного пузыря** hydrops of the gall-bladder
~ **живота** ascites, hydroperitoneum, abdominal dropsy
~ **почки** hydronephrosis
~ **семенного канатика** funicular hydrocele
~ **сустава** hydroarthrosos
~ **яичка** hydrocele, orchiocele
водяной I 1. water, aqueous, aquatic; 2. water-operated/driven
~ **знак** watermark
~**ная крыса** water-rat/-vole *(Arvicola amphibius)*
~**ная мельница** water mill
~**ное отопление** central heating, hot-water heating
~**ные птицы** *pl.* water fowl
водяной II *n. myth.* water elf/sprite
водяночный *med.* dropsical, hydropic
воевода 1. *hist.* voivode; 2. governor of province *(in Poland)*
воеводство 1. *hist.* voivodeship, office of voivode; 2. province *(in Poland)*
военачальник military leader, commander
военврач military/army doctor
военгиз *abbr.* **Военное государственное издательство** State Military Publishing House
военизация militarization
военизированный militarized; paramilitary
военком *abbr.* **военный комиссар** military commissar
военкомат *abbr.* **военный комиссариат** military registration and enlistment office
военкор *abbr.* **военный корреспондент** war correspondent
военмор *abbr.* **военный моряк** navy man, naval serviceman
военно-воздушный air-force
~**ные силы** Air Force(s)
военно-врачебный military-medical
~**ная экспертиза** military-medical examination
военно-медицинский military-medical
военно-морской naval
~ **флот** the Navy
военнообязанный *n.* person liable for call-up; reservist
военнопленный *n.* prisioner of war
военно-полевой battlefield, field
~ **суд** court-martial
военно-политический army-political
военно-полицейское государство police state, state run by the military and the police
военно-промышленный military-industrial, war-industry
~ **комплекс** industrial-military system/complex
военно-революционный revolutionary-military
военнослужащий *n.* soldier, seviceman; *pl.* armed forces personnel, military personnel
военно-стратегический strategic
военно-учебный military-training
~**ное заведение** military school
военно-химический chemical-warfare
военно-экономический military-economic
~ **потенциал** strategic potential, military-economic potential
военный 1. war, army, military; 2. *n.* serviceman, soldier
~ **врач** army doctor, medical officer
~ **завод** munitions factory
~ **коммунизм** *hist.* War Communism
~ **суд** court-martial
~**ная промышленность** war industry
~**ная служба** military service
~**ное положение** martial law
~**ные расходы** defence/military spendinf
военрук *abbr.* **военный руководитель** military instructor
военспец *abbr.* **военный специалист** military specialist
вожатый *n.* guide, leader
вождение navigation; driving; piloting, flying; leading
~ **по пересеченной местности** cross-country driving
вождь chief, leader
вожжи *pl.* reins
~, **отпустить** to give a horse the reins
воз 1. wag(g)on, cart; 2. cartload, carload
возбудимость excitability; irritability
~, **повышенная** hypererethism
~, **пониженная** hypoexcitability
~, **рефлекторная** reflex excitability
~, **чрезмерная** hyperexcitability
возбудимый *med.* erethitic; excitable; irritable
возбудитель 1. *tech.* exciter, driver; actuator, activator, inducer; energizer; 2. *med.* agent; 3. stimulant, stimulus
~ **болезни** pathogenic organism
~, **волновой** wave exciter
~, **кварцевый** crystal driver
~ **колебаний** vibrator
~ **модулятора** modulator driver
~, **противокомпаундный** differential compound exciter
~, **статический** static exciter
~ **шины** *comp.* bus driver
возбудительный stimulating, exciting, excitative
возбуждаемость excitability
возбуждаемый:
~ **высоким уровнем сигнала** *comp.* active-high
~ **низким уровнем сигнала** active-low

возбуждать give rise to, excite; *el.* energize; *med.* stimulate; activate, actuate, (a)rouse, call forth; provoke, agitate
~ аппетит stimulate the appetite
~ дело *law* institute proceedings
~ иск institute an action; bring in a lawsuit
~ колебания excite oscillations
~ тяжбу bruig a suit
~ уголовное преследование institute prosecution
~ ходатайство make an application
возбуждающий stimulating, exciting; actuating, energizing, driving
~щее средство *med.* stimulant
возбуждение excitement, agitation, arousal; *med.* stimulation; *el.* excitation, activation; energizing; *comp.* intitiation (of a line)
~ атома atomic excitation
~ излучением radiant/radiation excitement
~, кулоновское Coulomb excitation
~ лазера laser excitation
~, мотивационное motivation excitement
~, независимое separate excitation
~, нервное jitters
~, одночастичное single-particle excitation
~, охранительное inhibitory stimulation
~, параллельное shunt excitation
~ плазмы plasma generation
~, повышать build up (the) excitation
~, половое sexual stimulation
~ поля field excitation
~, последовательное series excitation
~, психическое psychic excitement
~, радиационное radiative excitation
~, разряда discharge initiation
~, резонансное resonant excitation
~, световое *med.* phonic stimulation
~, сейсмических колебаний *geol.* generation of seismic waves
~, сериесное series excitation
~, смешанное compound excitation
~, стоксово Stokes-law excitation
~, ступенчатое step-by-step excitation
~ схваток *med.* induction of labo(u)r
~, трехтактное three-phase excitation
~, ударное impact/impulse/collision excitation, shock excitation
~ фотонами photoexcitation, radiative/radiation excitation
~, ядерное/ядра nuclear excitation
возбужденный excited, stimulated, activated; high-strung, feverish
~ вид excited look/appearance
возведение raising; erection, construction
~ во вторую/третью степень raising to the second/third power
~ в квадрат squaring
~ в степень raising to a power; exponentation, involution, powering
~ обвинения *law* bringing of an accusation/ indictment
возвещать proclaim, announce, make known
возводить 1. *math.* raise; *law* bring, level; 2. derive, trace (to); 3. raise, erect, put up; elevate
~ в квадрат square
~ в куб cub
~ в третью степень raise to the 3rd power
~ на престол enthrone
возврат 1. reset(ting), restoration, return(ing); re-entry; recovery; repayment, reimbursement; 2. recycling; reject
~, автоматический automatic reset
~ агломерата sinter return
~ болезни *med.* relapse
~ в атмосферу atmospheric re-entry
~ в исходное положение reset(ting)
~ в точку рестарта *comp.* backout
~ для использования reutilization
~ из подпрограммы *comp.* return from a subroutine
~ из файла file return jump
~ к предыдущему состоянию backtracking
~ каретки carriage return
~ каретки, ускоренный accelerated carriage return
~ металлов recovery of metals
~ на одну позицию *comp.* backspacing
~ нуля zero reset
~ пошлины return duty
~ солнца *astron.* solstice
~ строки *comp.* line starve
~ тепла heat recovery
~ уноса cinder return
~ уноса в топку ash reinjection
~, циклический *comp.* wraparound
возвратимый retrievable, recoverable
возвратно-поступательный reciprocating
возвратность recurrence, reflexivity
возвратный 1. *med.* recurrent, recurring; returning, retrogressive; 2. *gram.* reflexive; 3. *econ.* repayable
~ тиф *med.* recurrent/relapsing fever
~ная пошлина drawback (duty)
возвращать 1. reset, return, restore, give/send back; refund, reimburse, pay back; 2. bring/call back, revert; retrieve
~ в исходное положение reset
~ в повторный цикл recycle
~ в процесс recycle
~ к жизни restore to life
~ к норме return to a normal level
возвращающийся на родину homeward-bound, homebound
возвращенец return migrant
возвращение redemption, reset(ting), return(ing), restoration, restitution, re-entry
~ безопасное *space* recovery
~ в атмосферу atmospheric re-entry
~ в круговорот rejuvenation
~ в твердое состояние resolidification; rege-

lation
~ **на родину** homecoming
~ **ступеней ракеты** rocket-stage recovery
возвышение 1. elevation, eminence, mound, prominence; altitude, height; 2. rise, increase, rising; gradient; 3. dais, platform
~, **альвеолярное** *anat.* alveolar yoke
~, **гвоздевидное** *geol.* peg-like prominence
~ **дна** knoll, eminence
~, **дугообразное** arcuate eminence
~ **ладьи** *anat.* eminence of the scapha
~, **лобоковое** pubic mound
~**ушной раковины** eminence of the concha
возвышенность height, hill(s), high/rising ground, upland; elevation
~, **небольшая подводная** sea-knoll
~, **плоская** plateau, tableland
~, **подводная** submarine rise/elevation
~, **скрытая** covered/buried hill
~, **топографическая** topographic elevation/high
возгон sublimate; sublimation
возгонка *chem.* sublimation, volatilization, dry distillation
~, **повторная** resublimation
возгонный subliming, sublimatory
возгоняемость sublimability
возгонять(ся) sublime, sublimate
~ **повторно** resublimate
~ **шлак** fume the slack
возгораемость ignitability, inflammability, combustibility
возгорание, возгорение inflammation, ignition; combustion
~, **самопроизвольное** spontaneous ignition/combustion
возгораться ignite, catch fire, begin to burn
воздвигание raising, erection
воздвигать build, erect, set/put up, raise
Воздвиженье *eccl.* Holy-Cross Day
воздействие effect, action; exposure (to); attack; control; influence; operation
~, **астатическое** astating response
~ **атмосферы** weather effect
~, **возмущающее** *comp.* perturbation (action); disturbing influence, disturbance
~, **входное** input action
~, **выходное** output action
~, **длительное** prolonged exposure
~ **звукового удара** sonic(-)boom effect
~ **мороза** frost action
~ **на окружающего среду** environmental impact
~, **паразитное** noise
~ **порыва** gust effect
~ **по скорости** rate action/response
~, **регулирующее** control action
~ **светопогоды** weathering
~, **сложное** compound action
~, **случайное** random input/function/action
~, **ступенчатое** step excitation/function/input
~, **управляющее** control (action)
~, **физическое** coercion, physical action; force
~, **циклическое температурное** thermal cycling
воздействовать act (on), affect; influence
возделывание *agr.* plowing, ploughing, tilling, cultivating; cultivation, growing
возделывать till; cultivate; grow
~ **землю** till the land
~ **рис** grow rice
воздержавшийся *n.* abstainer
воздержание continence, abstinence, abstention; moderation, temperance
~ **от действия** *law* act of omission
воздержанность, воздержность abstemiousness; abstinence; temperance
воздух I air
~, **вдыхаемый** *med.* inspired/respiratory air
~, **влажный** damp/humid/moist air, moisture-laden air
~, **взмывать в** *av.* ballon
~, **вовлеченный** entrained air
~, **возмущенный** turbulent air
~, **вольный** open air
~, **вторичный** auxiliary/secondary air
~, **жидкий** liquid air
~, **загрязненный** contaminated/polluted air
~, **засосанный** inducted air
~, **застойный** stagnant/dead air
~, **защемленный/захваченный** entrapped air
~, **ионизированный** ionized/conducting air
~, **кондиционированный** conditioned air, artificial atmosphere
~, **мертвый** *min.* black damp, dead air
~, **морской** maritime air
~, **наружный** outdoor air
~, **невозмущенный** undisturbed/free air
~, **неподвижный** stagnant/stationary/still air, air at rest
~, **окружающий** ambient/environmental/surrounding air
~, **осушенный** dehumidified/desiccated air
~, **отработанный** foul/used air
~, **почвенный** soil air
~, **предварительно сжатый** precompressed air
~, **приточный** incoming/induced/plenum air
~, **продувочный** scavenging air
~, **рабочий** actuating air
~, **рециркулирующий** return/recycled air
~, **рудничный/шахтный** mine air
~ **с пониженным содержанием кислорода** *med.* vitiated air
~, **сбросной** exhaust air
~, **сжатый** compressed air
~, **спертый** dead/close/stuffy air
~, **транспортирующий** carrier/conveying air
~, **уплотняющий** seal(ing) air
~, **форсуночный** injection air
~, **чистый** pure/clean/uncontaminated air
воздух II *eccl.* pall, paten

воздухобойный antiaircraft
воздуховместилище air chamber/conceptacle
воздуховод air duct/conduct/line
воздуховоз *min.* compressed-air locomotive
воздуховытяжной air-removing/-exhausting
воздуходув blast-engine operator
воздуходувка (air) blower, blower (tunnel), blast engine; ventilator, blast/draught blower/fan
~, газовая gas-engine blower, gas-driven blowing engine
~ доменной печи stove blower
~, коробчатая cased-in blower
~, лопаточная vane blower
~, нагнетающая pressure blower, blast engine
~, осевая axial-flow blower
~, поршневая piston blast engine, piston/reciprocating blower
~, смесительная blower-agitator, agitating blower
~, струйная jet blower
воздуходувный air-blower
~ные мехи bellows
воздухозаборник air intake/inlet/duct, air intake vent
~, безнапорный nonramming (air) intake
~, боковой side intake
~ вентиляционной системы ventilation intake
~, ковшевой air scoop
~, подфюзеляжный ventral intake
~, регулируемый controllable intake, variablegeometry/-ramp intake
~, туннельный ducted air intake
~, хвостовой aft intake
воздухозаборный air-intake/-inlet
воздухозаправщик compressor truck
воздухоизмещение air volume displacement
воздухолюбивый aerophilous
воздухомер aerometer, airflow meter
воздухонагреватель 1. air (pre)heater; 2. *met.* (hotblast) stove
~, вентиляторный heat blower
~ системы Каупера Cowper stove
воздухонагревательный air-heating
воздухонасытитель air saturator
воздухонепроницаемость air impermeability, airtightness
воздухонепроницаемый air-impermeable, airfast, airproof, airtight, airlocked
воздухоносный aeriferous; *med.* pneumatic
~ сосуд *bot.* trachea
воздухообмен air (ex)change
воздухообразный aeriform; gaseous, gasiform
воздухоосушитель air drier
воздухоотвод air offtake, air draw-off, air outlet port
воздухоотделитель air separator, deaerator
воздухоотражатель air baffle/deflector
воздухоотсасывающий air-removing/-exhausting
воздухоотсос air ejector
воздухоохладитель air cooler, air-cooling system
~, сухой dry coil air-cooling unit
воздухоохлаждаемый air-cooled
воздухоочиститель air filter/cleaner/strainer/scubber/purifier
~, жидкостный air washer
~, масляный oil-bath air cleaner
воздухоочистительный air-cleaning/-purifying
воздухоплавание aerostation, aeronautics
воздухоплаватель aeronaut; balloonist
воздухоплавательный aeronautic; aerostatic
воздухоподводка air supply/line
воздухоподогреватель air (pre)heater
~ вторичного воздуха main air heater
~ газовый gas-heated air heater
~, дробепоточный pebble air heater
~, многоходовой multipass air heater
~, одноступенчатый single-stage air heater
~, одноходовой single-pass air heater
~, паровой steam-coil air heater
~, пластинчатый plate air heater
~, ребристый gilled air heater
~, ребристый плитчатый plate-fin air preheater
водухоприемник air receiver; air inlet/intake; air scoop
воздухопровод air duct/conduit/pipe/main; air (pipe)line; blast (supply) line
~ воздухозаборника air intake duct/trunk(ing)
~, всасывающий suction air duct
~, главный main blast line
~ карбюратора carburator air duct
~, сборный air manifold
~ холодного дутья cold blast main(s)/line
воздухопроводка air-supply system
воздухопроводящий air-conducting
воздухопромыватель air washer
воздухопроницаемость air permeability/porosity, permeability to air
воздухопроницаемый air-permeable
воздухоразделительный air-deflecting/-fractioning
воздухораспределитель air distributor/manifold
воздухосборник air collector/holder/receiver/tank
воздухосос air sucker
воздухостойкий airproof, airtight; fast to aeration
воздухостойкость airproofness; fastness to aeration
воздухоуловитель air trap
воздухоэквивалентный air-equivalent
воздушник 1. air vent, vent pipe; 2. aerial communication specialist
воздушно-водный aero-aquatic
воздушно-водяной air-to-water
воздушно-гашеный air-slaked
воздушно-десантный airborne (-landing)
~ные войска airborne troops, paratroopers
воздушно-космический aerospace *(attr.)*
воздушно-масляный air-oleo

воздушно-охлаждаемый air-cooled
воздушно-плотный airtight
воздушно-подземный overhead-underground
воздушно-пылевой air-dust
воздушно-реактивный air-breathing/-jet
воздушность airiness, lightness; ethereality
воздушно-сухой air-dry/-dried
воздушно-эквивалентный air-equivalent
воздушный 1. aerial, air, aeriform; atmospheric; *med.* pneumatic; 2. air-driven/-operated; 3. airy, ethereal, light
~ **винт** airscrew, airplane propeller
~ **кабель** overhead cable
~ **пирог** *cul.* meringue
~ная железная дорога elevated/overhead railway
~ная линия airline
~ная разведка aerial reconnaissance
~ная тревога air-raid warning
~ная яма air pocket
~ное нападение air raid/attack
~ное сообщение air service
воздымание bulging up; uplift
возжигатель kindler
воззвание appeal
воззрение opinion, view, idea; attitude
возить convey, drive, carry, transport; draw, pull
возлагать entrust (with), charge (with), lay/place on
~ **вину** put the blame (on)
~ **поручение** entrust with a task
~ответственность make responsible
возмездие retaliation; retribution, requital
возместительный compensative, compensatory
возмещать compensate, make up (for); refund
~ **издержки** refund expenses
~ **убытки** compensate for losses
возмещение redress, compensation, indemnity; *law* damages; refund, reimbursement
~ **в натуре** recovery in kind
~ **издержек/расходов** reimbursement of expenses, cost recovery
~ **потерь** loss replacement
~, страховое insurance money
~ **убытков** compensation for losses/damages
~ **убытков от огня** fire indemnity
~ **ущерба** damages
возможность 1. possibility; feasibility; potentiality; chance, opportunity; 2. *pl.* means, resources; capabilities; *comp.* facilities; options, features
~ти, выразительные *comp.* expressiveness
~ти, вычислительные computational capabilities
~ти, диагностические diagnostic features
~ти, доступные для пользователя user option
~ти задания типов type ahead features
~ти исправления ошибок error-correcting capabilities
~ти организации сети *comp.* networking capabilities
~ти передвижения trafficability
~ти повторения retry features
~ти повторного входа reentrance features
~ти, производственные production potentialities
~ти, развитые comperehensive/advanced facilities
~ти расширения expandability
~ти рекуперации recuperability
~ти самоконтроля self-checking features
~ти структурирования structural features
~ти, факультативные options
~ти, функциональные functional capabilities, functionality
возможный potential, possible, feasible; contingent; likely
возмущение 1. disturbance, perturbation; 2. indignation, resentment
~, асимптопически затухающее subsidence
~, воздушное aerodynamic disturbance
~, земное terrestrial disturbance
~, импульсное pulse disturbance
~ **ионосферы** ionospheric disturbance
~, короткопериодическое short-period disturbance
~, линейно-нарастающее ramp disturbance
~, местное localized disturbance
~ **планет** perturbation of planets
~, постоянное time-independent perturbation
~, скачкообразное step disturbance
~, сложное compound disturbance
~, случайное random disturbance
~, солнечное solar disturbance
возмущенный 1. disturbed; pertubed; 2. indignant
вознаграждать recompense, reward, compensate, make up (for); indemnify
вознаграждение recompense, reward, compensation; remuneration, indemnity; *naut.* bonus; *law* grafitication; pay(ment), fee
~, добровольное gratification
~ **за сверхсрочную работу** pay for overtime work
~ **за спасение** *naut.* benefit of salvage, salvage remuneration
~ **за спасение груза** salvage (charges), (civil) salvage
~, поощрительное incentive fee
~, страховое insurance indemnity
вознесение 1. ascent, rise; *eccl.* Ascension (Day)
возникать occur, orginate, set in; spring/crop up, emerge
возникающий nascent, incipient, arising, emerging
~ **естественным путем** nomogenous
~щее вновь *philos.* occurent
возникновение initiation, origin, beginning; rise; emergence
~ **колебаний** onset of oscillations
~ **потока** initation of the flow
~ **права** *law* accrual of right(s)

~ **электрической дуги** arcing
Возничий *astron.* Auriga, Wagoner
возносить *rel.* raise, lift up; *fig.* praise
~ **до небес** praise to the skies
~ **молитву** offer up a prayer
вознощение lifting, elevation; pride
~ **даров** *eccl.* elevation (of the host)
возобновление restoration, regeneration; renewal, resumption, recommencement; revival
~ **договора** renewal of an agreement
~ **леса** re(af)forestation
~ **протектора** recapping/retreading of a tyre
возобновлять restore, renew, recommence, resume
~ **брожение** referment
~ **дело** *law* reopen a case
~ **занятия** resume/recommence lessons
~ **запасы** restock
~ **протектор** recap a tyre
возражать object, take an objection (to); take exception (to); retort, answer back
возражение *law* objection; plea, answer
~, **выдвигать** set up a defence/plea
~ **о прекращении обязательства** plea in discharge
~ **об отсутствии долга** plea of never indebted
~ **отклоняется** objection overruled
~, **отлагательное** dilatory defence/plea
~ **по существу дела** plea to the merits
~ **по существу иска** plea in bar
~ **правного характера** objection in point of law
~ **принимается** objection sustained
~ **против юрисдикции суда** plea to the jurusdiction
возраст age; stage
~ **брачный** age of consent
~ **взрослый** adult/ephebic stage
~, **детородный** fertile/childbearing age
~ **для выхода в отставку** retiring age
~, **зрелый** mature age
~, **интеллектуальный** mental age
~, **климактерический** climacteric age
~, **младший школьный** primary/elementary school age
~, **молодой** juvenile/neanic age
~ **нейтрона/нейтронов** neutron/Fermi age
~, **непроизводительный** dependant age
~, **образовательный** achievement age
~, **отроческий** boyhood
~ **отъема/отбивки** weaning age
~, **пенсионный** retirement/pension/pensionable age
~ **по аргону** argon age
~ **по гелию** helium age
~ **по радиоактивному углероду, ~ по радиоуглероду** radiocarbon age
~, **подростковый** juvenile age
~, **половой зрелости** reproductive age
~ **почвы** soil age/chronology
~, **предельный** age limit; limiting age
~, **престарелый** extreme old age
~, **призывной** draft/military age
~ **рубки** cutting/feeling age
~ **складчатости** *geol.* age of folding
~, **случной** breeding age
~ **смерти, наиболее вероятный** modal age of death
~, **средний** middle age
~, **средний и старший школьный** secondary-school age
~, **старческий** senility, gerontic/senile age
~ **суточного неравенства** *nav.* age of diurual inequality
~ **суточного прилива** diurnal (tide) age, age of diurnal tide
~ **трудовой деятельности** working age
~, **школьный** school(-going) age
возрастание increase, rise, growth, increment; angmentation; gain
возрастать increase, rise, crow
~ **в размерах** increase/grow in size
~ **неограниченно** increase beyond all bounds, increase without bound
~ **по линейному/экспоненциальному закону** increase/rise linearly/exponentially
~**тающая прогрессия** *math.* increasing progression
возрастной ценз age qualification
возрожденец revivalist
возрождение 1. revival; rebirth, regeneration; palingenesis; 2. *hist.* Renaissance
воин warrior, solder; fighter
воинский military; martial, warlike
~ **дух** martial spirit
~ **поезд** troop train
~ **устав** army regulations
~**кая обязанность, всеобщая** universal military service
~**кая повинность** compulsory military service
воинственность belligerence, bellicosity
воинственный aggressive, belligerent, bellicose; militant; martial
воинство *coll.* army, troops
воинствующий militant; aggressive
~ **атеизм** militant atheism
~**щих безбожников, Союз** *hist.* League of Militant Atheists
войлок felt(ing); mat; 2. *bot.* tomentum
~, **микролитовый** microlitic felt
~, **шлифовальный** grinding felt
войлокообразный feltlike
войлочный 1. felt(y); 2. tomentose
война war, warfare
~ **без применения ядерного оружия** non-nuclear warfare
~, **воздушная** air war(fare)
~, **воздушно-космическая** aerospace war(fare)
~, **глобальная** global war
~, **космическая** space war(fare)
~ **на истощение** war of attrition

~ на море war at sea, sea/maritime war
~, подвижная fluid warfare
~, ракетно-ядерная missile and nuclear warfare
~, термоядерная thermonuclear warfare
~, торговая trade war
~, формально объявленная *law* solemn war
~, холодная cold war
~ цен price warfare
~, экономическая economic(al) warfare
войска *pl.* troops, forces
~, аэромобильные air cavalry
~, воздушнодесантные air(borne) troops/forces
~, зенитные ракетные antiaircraft missile troops
~, мотомеханизированные light mechanized troops
~, наемные mercenaries
~, парашютнодесантные parachute troops, paratroops
~ ПВО air defence force(s)
~, ракетные rocket forces
~, сухопутные army, land forces
~, экспедиционные expeditionary force
войско *hist.* host (of Cossacks); army
~, запорожское the Zaporozhian host
войсковой 1. army, military; 2. *hist.* of the host
~ круг Cossack assembly
~ старшина lieutenant colonel *(of Cossack troops)*
войт *hist.* prefect, bailiff
вокабула headword *(in a dictionary)*; vocable
вокабулярий vocabulary, list of words
вокализ(а) *mus.* excercise in vocalization
вокализация *mus., phonet.* vocalization, vowelization; vocality
вокализм *phonet.* vowel system, vocalism
вокалист *mus.* vocalist, teacher of singing
вокеленит *miner.* vauquelinite
вокзал (railway) station, terminal; station building
~, железнодорожный railroad station; depot
~, морской seaport buildings
~, речной riverside station, river-boat station
вокодер *comp.* vo(co)der, voice coder, speesh processing system
~ основанный на разложении Лагерра Laguerre expansion vocoder
~ работающий по принципу сравнения с эталонными спектрами pattern-matching vocoder
~, тактильный tactile vocoder
вокоид *phonet.* vocoid
вокруг (a)round; about
~ заднего прохода *anat.* circumanal
~ Луны circumlunar
вокругартериальный periarterial
вокругзаднепроходный *anat.* circumanal
вокругсосудистый perivascular
ВОКС *abbr.* **всесоюзное общество культурной связи с заграницей** *hist.* All-Union Society for Cultural Relations with Foreign Countries
воксит *geol.* vauxite
волан 1. *text.* fancy roller; flounce; 2. *sport* shuttlecock
~, игра в battledore and shuttlecock, badminton
волвянка *bot.* wooly lactarius *(Lactarius torminosus)*
волдырник *bot.* campion *(Cucubalus)*
волдырь blister, bleb, wheal, urtica, bump, swelling
~, кровавый blood blister
~, лавовый *goel.* lava blister
волевой volitional, volitionary
~ акт volition, will
волеизъявление *law* declaration of will; will
волейбол volleyball
волжанка *bot.* 1. woolly lactarius *(Lactarius torminosus)*; 2. goatsbeard *(Aruncus)*
волитив *grom.* voluntative, volitive
волк I *zool.* (timber) wolf *(Canis lupus)*
~, азиатский серый Indian wolf *(Canis lupus pallipes)*
~, сумчатый tiger/Tasmanian wolf *(Thylacinus cynocephalus)*
волк II, волк-машина (rag) thrasher, willow(er)
волкамерия *bot.* glory bower *(Clerodendron)*
волкобой *bot.* February daphne *(Daphne mezereum)*
волкодав wolfhound
волластонит *geol.* wollastonite
волна wave; surge; wavelenght; *naut.* sea; *see also* **волны**
~, базисная base surge
~, бегущая travelling/mobile/progressive wave
~, блуждающая stray wave
~, большая billow
~, бортовая sea abeam, athwart/beam sea
~, взрывная blast (wave), burst/blast/detonating/explosive wave
~, внутренняя internal/subsurface wave
~, воздушная ударная air shock wave
~ возмущений Mach (shock) wave, disturbance wave
~, встречная *nav.* head/meeting sea
~, вынужденная forced wave
~, выходящая outgoing wave
~, головная nose/bow wave
~, граничная boundary/limiting wave, cut-off wave
~ давления pressure wave
~, двойная dual wave
~, дифрагированная diffracted wave
~, длинная long rolling sea
~, затухающая damped/decaying wave
~, звуковая acoustic/sound/sonic wave
~ звуковой частоты audio-frequency wave
~, земная ground/earthquake wave
~, зондирующая transmitted wave
~ изгиба *mech.* flexural wave
~, интерференционная interferential wave

~, ионосферная ionospheric/sky wave
~, кормовая quarter(ing) sea
~, короткая short sea
~ Коши-Пуассона Cauchy-Poisson wave
~, критическая boundary/critical/limiting/cut-off wave
~, левополяризованная left-handed polarized wave, counter-clockwise polarized wave
~, линейно-поляризованная plane-/linearly polarized wave
~, магнонная magnon wave
~ Маха Mach shock wave
~, многократно отраженная multiple reflection wave
~, морская sea wave, swell
~, набегающая incident wave
~ напряжений *mech.* stress wave
~ напряжения, прямоугольная rectangular voltage wave
~, незатухающая continuous/undamped/sustained wave
~, неоднородная inhomogeneous wave
~, неотраженная non-reflected wave, direct wave
~, нестационарная transient wave
~, несущая carrying/carrier wave
~, носовая *nav.* bow/nose sea
~, обратная back(ward)/inverse wave
~, обыкновенная ordinary wave
~, объемная bodily/body wave
~, одиночная solitary wave
~, опорная reference wave
~, основная fundamental/principal wave
~, отраженная reflected/ionospheric/sky/echo/back wave
~ паводка flood wave
~, падающая incident wave
~, первичная primary wave
~ перенапрежения voltage surge
~, плазменная plasma wave
~, плоская plane wave
~, плоскополяризованная plane-/linearly polarized wave
~, поверхностная surface wave
~ поднесущей subcarrier wave
~, полусуточная semidiurnal wave
~, поперечная transverse/transversal wave
~, попутная following sea, sea from aft
~, правополяризованная right-handed polarized wave, clockwise-polarized wave
~, преломленная refracted wave
~, прибойная surf
~, приземная ground/terrestrial wave
~, приливная tidal/tide wave
~, приходящая incident wave
~, пространственная ionospheric/sky/space wave
~, прямая *rad.* space/forward wave
~, прямая ударная normal shock wave
~, прямоугольная square/rectangular wave
~, рабочая operating/transmitting wave
~ резрежения rarefaction/suction wave
~, рассеяная scattered/dissipated wave
~, расходящаяся diverging wave
~ с траверза beam sea, sea on the beam, sea abeam
~ сжатия compessional wave
~, синусоидальная sinusoidal/sine wave
~, сопровождающая associated wave
~, спряженная associated/partial wave
~, спадающая decaying/collapsing wave
~, стоячая standing/stationary wave
~, убегающая outgoing wave
~, ударная blast/shock wave
~, ударная головная bow shock wave
~, ультразвуковая supersonic/ultrasonic wave
~, фокусированная beam wave
~, фононная phonon wave
~, холостая idler wave
~, центрированная centered/focused wave
~, шаровая spherical wave

волнение 1. *naut.* sea(s), rough sea/water(s); 2. riot, uprising, unrest; disturbance
~, беспорядочное random sea
~, значительное moderate/rough sea
~, исключительное confused sea/swell

волнисто-зубчатый *bot.* repandodentate
волнисто- лепестный *bot.* wave-petaled
волнисто-листный *bot.* wave-leaved, undulatifolious
волнисто-покровный unditegillate
волнистость 1. waviness, undulation; 2. ripple markings; buckles; corrugation
~ бумаги curl
~, мелкая *geol.* crenulation
~ пласта rollback

волнистый undulating, undulate, wavy, rolling; corrugated
~тая местность rolling country
~тое железо corrugated iron

волновод *electr.* waveguide
~, атмосферный atmospheric duct(ing), atmospheric waveguide
~, воздушный air-filled waveguide
~, гребневой ridge waveguide
~, гусеничный interlocked-type waveguide
~, естественный terrestial waveguide
~, змейковый sinuous waveguide
~, линзовый lens waveguide
~, литой cast waveguide
~, оптический light pipe
~ монолитной конструкции solid waveguide
~, Н-образный H-shaped waveguide
~ переменного сечения tapered waveguide
~, печатный printed-circuit waveguide
~, плазменный plasma waveguide
~, плоский planar/flat waveguide
~, П-образный channel-bar waveguide
~, полый hollow(-type) waveguide
~ постоянного сечения uniform waveguide

~ с одним типом колебаний standard-size waveguide, single-mode waveguide
~, скошенный tapered waveguide
~ со многими типами колебаний overmoded waveguide
~, согласованный matched waveguide
~, щелевой leaky/slotted waveguide
волновой wave *(attr.),* undulating, undulatory
волногаситель (beach) wave suppressor, wave absorber
волнограмма wave(-gauge) record
волнограф wave(-height) recorder
~ в открытом море offshore wave recorder
~, корабельный shipborne wave recorder
~, струнный continuous wire-gauge wave recorder
волноискатель wave detector
волнолом mole, breakwater; sea (defence) wall, pier
~, плавучий floating breakwater
волномер frequency/wave meter, wavemeter, ondometer, cymoscope
~, гетеродинный heterodyne wavemeter
~, зуммерный buzzer wavemeter
~, поглощающий absorption wavemeter
~, проходной transmission-type frequency meter
~, резонансный absorption frequency meter, resonant-circuit wavemetr
~ с цилиндрическим резонатором cylindrical-cavity wavemeter
волномерная веха wave staff/stake
волнообразный wavy, wave-like/-shaped, wave-like, undulatory, undulating, ripply
~ная обмотка wave winding
волнообразование undulation, wave making/formation
волноотвод 1. *naut.* manger board; 2. breakwater, mole, pier
волнопоглощение wave absorption
волноприбойный wave-cut/-built
волнопродуктор wavemaker, wave generator
~, плунжерный plunger/pneumatic(-type) wavemaker
волнорез breakwater, cutwater
волноуказатель, волноуловитель wave detector
волны *pl.* waves; sea(s)
~ де Бройля associated/matter/particle waves, de Broglie waves
~, дециметровые decimetre waves
~, инфракрасные intra-red waves
~, километровые long/kilometric waves
~, короткие short/decimetric waves
~, метровые metric waves
~, миллиметровые millimetric/millimetre waves
~, посторонние extraneous waves
~ Рэлея Rayleigh waves
~, сантиметровые centrimetric waves
~, сверхдлинные myriametric/infralong/superlong waves, very long waves
~ скручивания torsional waves
~, средние medium/hectometric waves
~, субмиллиметровые submillimetric waves
~, сходящиеся converging/convergent waves
~, тепловые heat waves
~ тяготения gravitational waves
~, ультракороткие ultrashort waves, very short waves, microwaves
~, фазовые associated/matter/particle waves, de Broglie waves
волнянка *zool.* pine brown-tail moth *(Euproctis terminalis)*
волован *cul.* vol-au-vent
воловик *bot.* bugloss *(Anchusa)*
володушка *bot.* thoroughwax, thorowax *(Bupleurum)*
володушколистный *bot.* bupleurifolious
волок logway
~, магистральный main/pricipal logway
~, трелевочный logway, gutter road, skidding trail
волока *tech.* draw/die plate, drawing/reducing die, drawhole
~ составная collapsible die
~ твердосплавная sintered-carbide die
волокнисто-зернистый fibrogranular
волокнисто-лучистый fibroradiated, fibrous radiated
волокнистообразный fibroid
волокнисто-сетчатый reticulate-fibrous
волокнистость fibrous structure, fibration,
волокнисто-эластичный fibroelastic
волокнистый fibrous, filamentous; fibrillar; sringy; threadlike
волокно fibre, *Am.* fiber, filament; staple; grain *(of wood); seism.* ribbon
~, акриловое acrylic fibre
~, акрилонитрильное acrylonitrile fibre
~, альгинатное alginate rayon
~, ацетатное cellulose acetate rayon
~, афферентное нервное *anat.* afferent fibre
~, базальтовое basalt fibre/wool
~, безмякотное нервное *anat.* nonmedullated/unmyelinated fibre
~ бора boron fibre
~, борографитовое boron and graphite fibre
~, вискозное viscose fibre/rayon
~, высокопрочное high-strength fibre
~, гетероцепное heterochain fibre
~, гидратцеллюлозное cellulose-regenerated fibre
~, двигательное нервное *anat.* motor fibre
~, диацетатное cellulose acetate rayon
~ древесины woody fibre, (wood) grain
~, древесное wood/lignous fibre
~ из фторопласта polytetrafluorethylene fibre
~, искусственное rayon
~, искусственное белковое regenerated protein fibre

~, искусственное целлюлозное regenerated cellulose fibre
~, исходное raw fibre
~, карбоцепное carbochain fibre
~, клеточное cell filament
~, коллагеновое collagen(ous) fibre
~, листовое leaf fibre
~, ломкое brittle fibre
~, лубяное bast fibre
~, льняное flax fibre
~, медно-аммиачное cyprammonium/copper rayon
~, мышечное *anat.* muscular fibre, myofibril(la)
~, мякотное нервное *anat.* medullary nerve fibre
~, незрелое unripe/immature fibre
~, нервное nerve fibre
~, нитратцеллюлозное cellulose nitrate fibre, nitrate rayon
~, оптическое optical fibre
~, пирографитовое pyrolytic graphite fibre
~, плодовое fruit-hair fibre
~, подкрепляющее reinforcing fibre
~, полиакриловое polyacrylic fibre
~, полиакрилонитриловое polyacrylonitrile fibre
~, полиамидное polyamide fibre
~, поливинилакриловое polyvinyl-acrylic fibre
~, поливинилацетатное polyvinyl-acetate fibre
~, поливиниловое polyvinyl fibre
~, поливинилспиртовое polyvinyl alcohol fibre, vinol fibre
~, поливинилхлоридное polyvinyl-chloride fibre, PVC fibre
~, полимерное polymeric fibre
~, полипропиленовое polypropylene fibre
~, полистирольное polystyrene fibre
~, полиуретановое polyurethane fibre
~, полихлорвиниловое polyvinyl-chloride fibre, PVC fibre
~, полиэтиленовое polyethylene fibre
~, полиэфирное polyester fibre
~, полудревесинное metaxylem
~, поперечное *mech.* transverse fibre
~, предварительно ориентированное pre-oriented fibre
~, рубить на штапель chop/cut/break into staple fibres
~, сжатое *mech.* compession fibre
~, синтетическое synthesized/synthetic(-base) fibre
~, смолистое gummy fibre
~, стеклянное glass fibre, fibreglass
~, техническое industrial fibre
~, торфяное peat fibre
~, триацетатное cellulose triacetate fibre
~, триклинное triclinic fibre
~, химическое chemical fibre, man-made fibre
~, хрупкое brittle fibre
~, штапельное staple fibre
~, эфферентное нервное *anat.* efferent fibre
волокнодержатель lint retainer
волокнообразный fibrous, fibriform, fibrelike, filamentary
волокноотвод lint flue
волокноотделение ginning
волокноотделитель (cotton)gin
~, валично-ножевой knife-roller gin
~, пильный saw gin
~, пневматический air-blast gin
волокноочиститель lint cotton cleaner, postginning cleaner
волокноочистка post-ginning cleaning
волокно-сырец raw fibre
волоком by traction/pulling/dragging
волоконце *anat., geol.* fibril(la); filament
волокончатый fibrillate, fibrillous; filamentary
волокуша 1. *agr.* drag harrow, sweep, sledge, smoothing harrow, sweep rake; float leveller; 2. tuck net/seine, dragnet
~ навесная тракторная tractor-mounted buckrake, haysweep, haycock lifter
~, сенная buckrake, haysweep
~, тросовая rope sweep
волокуша-подборщик buck/sweep rake
волонтер *hist.* volunteer
Волопас *astron.* Herdsman
волос 1. *anat., bot.* hair; 2. *el.* filament
~ бороды hair of the beard
~, конский horsehair
~ ноздрей vibrissa
~ подмышки hair of the axilla
~ уха tragus (*pl.* tragi)
волосатик *miner.* venus hairstone, maidenhair
волосатость hairiness
~, избыточная *med.* hypertrichosis, hypertrichiasis
~ языка hairy tongue, glossotrichia
волосатый hairy, villous, woolly, hirsute; crinite; shaggy; pilose
волосик *bot.* fuzz
волосинка filament, fibril(la)
волосистокорешковый *bot.* hairy-rooted, trichorhizous
волосистолистный *bot.* hairy-leaved, trichophyllous
волосистоплодный *bot.* hairy-fruited, trichocarpous
волосистостебельчатый *bot.* hairy-stalked
волосистость hairiness, pubescence; pilosity; villosity
волосистый 1. hairy, pilose, villous, comous; 2. *miner.* fibrous
волосной capillary, filiform
волосность *phys.* capillarity
волосовидный hairlike; capillary, capillaceous; filiform; trichoid
волосовина flake/hair/hairline crack, hair seam
волосок 1. hair; 2. hair spring

~, венерин *miner.* Venus hairstone, fleches d'amour
~, жгучий stinging hair, cnida
~, корневой root hair, rhizoid
~, плавательный natatory seta
~, ресничный cilium
~, слуховой auditory hair
~, спинной dorsal seta
~, чувствительный sensitive hair, tentacle
волосяной 1. hairy, hirsute; 2. capillary
~ покров *anat.* scalp
волочение drawing; draft; drag(ging), hauling, traction
~ без оправки sink drawing
~, горячее hot drawing
~ на длинной оправке mandrel/bar drawing
~, однократное single-pass drawing
~, осесимметричное axisymmetric drawing
~, подземное *geol.* subterranean dragging
~, холодное cold drawing
~, черновое first drawing
~, чистовое finishing drawing
волочильный drawing, pulling
~ станок wire-drawing bench
~ная доска drawplate
~ная машина drawing machine
волочильня drawing frame. drawplate; wireworks
волочить 1. drag, draw, trail; pull; haul; tow; 2. *tech.* draw
~ золото wiredraw gold
волтерианец freethinker, follower of Voltaire
волторна, валторна *mus.* French horn
волхв magician, sorcerer, wizard
~а, три *eccl.* the Three Wise Men, the Magi
волчанка *med.* lupus
~, туберкулезная lupus turberculosis
волчаночный *med.* lupoid, lupiform
волчец 1. *bot.* thistle *(Cnicus)*; 2. tungsten steel; 3. *vet.* mange, scab
волчеягодник 1. *pharm., med.* mezereum, mezereon; 2. *bot.* daphne *(Daphne)*
волчий lupine, wolf, wolfish
~ билет *hist.* passport issued to politically unreliable pesson
~чья пасть *med.* uranoschisis, cleft palate, uranischochasma, uranoschism, palatoschisis
~чья яма *bot., mil.* trou-de-loup
волчник *bot.* daphne *(Daphne)*
волчниковые *n. pl. bot.* mezereon family *(Thymelaeaceae)*
волчок I 1. *tech.* gyroscope; meat hasher/mincer, meat mincing machine; 2. *text.* willowing machine; 3. *bot.* root sucker, epicormic branch
~ для промасливания beater oiling roller
волчок II *bot.* 1. ghost pipe *(Anoplanthus, Thalessia)*; 2. broomrape *(Orobanche)*
волынка 1. bagpipes; 2. *rad.* droning
вольва *bot.* cap, veil, volva
вольвокс *biol.* volvox
вольвообразный, с вольвой *bot.* volvate
вольера open-air-cage
волькамерия *bot.* glory bower *(Clerodendron)*
вольница *coll., hist.* freemen, outlaws, freebooters
вольнодумец libertine, freethinker
вольнодумие, вольнодумство freethinking
вольноотпущенник *hist.* freedman, emancipated serf
вольноотпущенный *hist.* emancipated, free
вольнопрактикующий privately/freely practicing
~ врач private practicioner
вольнослушатель auditor, noncredit student
вольность freedom, liberty; *hist.* right, privilege
~стей, Великая хартия (The) Great Charter
вольный 1. free, unrestricted; *sport* free-style; 2. free employee, civilian
~ город free city
~ рынок free market
~ стиль *sport* freestyle (swimming)
~ые стихи verse libre, free verse
вольт I *el.* volt
вольт II *sport* volte
вольта *text.* thin cotton fabric
вольтаж voltage
вольтаметр *el.* voltameter
вольт-ампер volt-ampere, VA
~, активный active volt-ampere, watt
~, реактивный reactive volt-ampere
вольтамперметр volt-amperemeter, voltammeter
вольтеров(ский) Voltairian
вольтерьянец *hist.* freethinker, Voltairian
вольтерьянизм, вольтерьянство *hist.* Voltairianism, freethinking
вольтижер *sport* equestrian acrobat, vaulter
вольтижирование, вольтижировка acrobatics on horseback, equestrian vaulting
вольтметр voltmeter
~, амплитудный peak/crest voltmeter
~, амплитудный двойной peak-to-peak voltmeter
~, высокоомный high-resistance voltmeter
~, детекторной системы rectifier voltmeter
~, индуктивный induction voltmeter
~, коронный corona voltmeter
~, ламповый electronic voltmeter, vacuum-tube voltmeter
~, магнитоэлектрический moving-coil voltmeter
~, многопредельный multirange voltmeter
~, пиковый peak/crest voltmeter
~, с цифровым отсчетом digital voltmeter
~, самопишущий recording voltmeter
~, тепловой hot-wire voltmeter
~, усредняющий averaging voltmeter
~, цифровой digital voltmeter
~, электродинамический dynamometer/electrodynamic voltmeter
~, электроннолучевой cathode-ray voltmeter
~, электронный electronic voltmeter, *Am.* vacuum-tube voltmeter
вольтов voltaic, Volta's

~а дуга voltaic arc
вольтолизация voltolization, voltolizing
вольтоль voltolized/voltol oil
вольтомметр volt-ohmmeter
вольтоммиллиамперметр volt-ohm-milliammeter, multimeter
вольт-секунда volt-second
вольтцит *miner.* voltzite
вольфахит *geol.* wolfachite
вольфия *bot.* wolffia *(Wolffia)*
вольфрам 1. *chem.* tungsten; 2. tungsten steel
~, бромистый tungsten dibromide
~, ковкий ductile tungsten
~, порошкообразный tungsten metal powder
вольфрамат tungstate
~ калия potassium tungstate
вольфрамистый *chem.* tungstic
вольфрамит *miner.* wolframite, wolfram
вольфрамовокислый *chem.* tungstate
вольфрамовый tungstic; tungsten *(attr.)*
~, вая лампочка tungsten lamp
~, вая руда wolfram
волюметр volumeter
волюминометр, волюмометр volumenometer
волюнтаризм *philos.* voluntarism, libertarianism
волюнтарист *philos.* voluntarist, libertarian
волюнтаристский, волюнтаристический voluntaristic
волюта *arch.* scroll, volute
волютида *zool.* volutoid, volute
волютин *biol.* volutin
воля 1. will, volition; wish(es); 2. freedom, liberty
~ твоя, да будет *rel.* thy will be done
вомбат *zool.* wombat *(Vombatus)*
вомитофобия *med.* vomitophobia
вона hwan, won *(monetary unit in North Korea)*
воньерит *miner.* vaugnerite
вонючий fetid, putrid, stinking
вонючка 1. *zool.* skunk *(Mephitis mephitis)*; 2. *bot.* asafetida giant fennel *(Ferula asafoetida)*; bean trefoil
воображение imagination; fancy
вооружать arm; outfit
вооруженец armourer, ordnanceman
вооружение 1. armament, ordnance; fighting equipment; 2. arming, fitting out, equipment
~, авиационное air(craft) armament
~, бортовое airborne armament
~, бушприта bowsprit gear
~, для перехвата interception armament
~, класса земля-воздух surface-to-air armament
~, кэч,парусное *nav.* ketch rig
~, латинское парусное lateen rig
~, оборонительное defensive armament
~, основное primary/basic armament
~, парусное *nav.* rig(ging)
~, поступать на go into service, come into operation/service
~, принимать на introduce into the inventory, put into service, adopt
~, противопожарное fire-fighting equipment
~, пушечное gun/cannon armament
~, ракетное rocket/missile armament
~, с малой дальностью действия short-range armament
~, штатное standard armament
~ шхуны, парусное *nav.* schooner rig
вооруженность 1. equipment, outfit; installation(s); 2. armament
~ труда, техническая degree of mechanization
вооруженный equipped, fitted out; armed; in possession *(of)*
~ные силы armed forced
~ным глазом assisted by an optical instrument
вопиющий flagrant, gross, blatant; scandalous, crying
~ щая несправедливость crying/gross injustice
~ щее противоречие flagrant/glaring contradiction
вопиять *rel.* cry out, clamo(u)r
~ к Богу cry upon God
воплощение incarnation, embodiment, personification; realization
~ низости meanness incarnate
~ Христа *rel.* the Incarnation
вопрос 1. question; problem, matter; 2. query, inquiry
~, женский the question of women's rights
~ жизни или смерти matter of life or death
~, жилищный housing problem
~, поставить raise a question
~, требующий ответа да-нет *comp.* yes-no question
вопросник questionnaire
вопросно-ответный question-answer
вопросный containing questions
~ лист question sheet, form
вор 1. thief; burglar; 2. *hist.* traitor; criminal
~, карманный pickpocket
~, магазинный shoplifter
~, тушинский *hist.* the imposter of Tushino
ворвань menhaden/blubber/train/whale oil
~ дельфинья dolphin/porpoise oil
~ тюленья seal oil
воробей *zool.* sparrow *(Passer)*
воробейник *bot.* gromwell *(Lithospermum)*
воробей-самец cocksparrow
воробьиный 1. passerine, sparrow *(attr.)*
воровской 1. thievish, of thieves; 2. *hist.* illegal
~, жаргон/язык thieves' latin/cant
~кие приемы thievish methods
~кие деньги counterfeit money
воровство thieving, stealing; theft; pilfering; filching, pinching
~, литературное plagiarism, plagiary
ворон *zool.* raven *(Corvus corax)*
ворона *zool.* crow *(Corvus)*
воронение *tech.* blu(e)ing, browning

~ **в воздушной атмосфере** air blueing
воронец *bot.* baneberry *(Actaea)*
вороный corvine
вороника *bot.* crowberry *(Empetrum)*
воронило *print.* planing block, planer; bluer, bluing agent; burnisher
воронить *tech.* black-finish; blue
воронка 1. *tech.* funnel; chute; orifice; 2. *anat.* choana; infundibulum; 3. *geol.* cone, conical depression, sink(hole); 4. eddy, vortex, swirl; 5. *mil.* crater, shell hole
~, **вводная** feed/inlet funnel
~, **верхняя приемная** overhead hopper
~, **взрывная** explosion funnel, blasting cone
~, **вихревая** vortex (funnel)
~, **водосливная** drain box
~, **волочильная** drawing block
~, **вращающаяся** *text.* revolving funnel
~, **вулканическая** volcanic funnel
~, **выводная** *met.* outlet hopper
~, **выпускная** emptying funnel
~, **газовая** gas vent
~, **двойная** nested sink
~, **делительная** separating funnel
~ **депрессии** cone of depression/influence
~ **для обсадных труб** casing adapter
~, **дорожная** road crater
~, **загрузочная** feed box, filling funnel
~, **земляная** sink
~, **кабеля** cable pothead
~, **капельная** drop(ping) funnel
~, **карстовая** *geol.* 1. bedding cave; 2. sink hole, funnel sink
~, **концевая** sealing end bell *(of cable)*
~, **кратера** crater funnel
~, **литниковая** sprue/pouring cup
~, **мозговая** infundibulum
~, **наливная** filling funnel
~, **направляющая** collector/conductor funnel
~, **насыпная** powder funnel
~, **опрокинутая** inverted funnel
~, **осушения** funnel of desiccation
~, **питающая** feed hopper
~, **поглощения** funnel of absorption
~, **подбугорная** hypothalamic infundibulum
~, **провальная** slump hole, funnel of collapse
~, **прядильная** spinning funnel
~, **с раструбом кверху/книзу** upright/downward funnel
~, **сифонная** septal neck
~, **сортировочная** separating funnel
~, **срыва** funnel of shear
~, **уплотнительная** *text.* condensing/sliver funnel
~, **ушная** aural speculum
~, **фальцевальная** *print.* former
~, **Форда** Ford cup
~, **шлаковая** ash hopper
~, **экстракционная** extraction funnel
воронковидный, воронкообразный, воронковый cone/trumpet/funneal shaped; funneled, infundibuliform; funnel-form
вороновые *n. pl. zool. Corvidae*
воронок *zool.* martin(et) *(Delichon urbica)*
ворончатый funnel-/cone-/trumpet-shaped
воронь blue colo(u)r
ворот capstan, windlass, winch; reel
~, **цепной** chain winch
ворота *pl.* 1. gate(way), gate(s); 2. *anat.* hilus; 3. *sport* goal (posts)
~, **ангарные** hangar door
~, **временные** time gate
~, **габаритные** *rail.* clearance gate
~ **дока** tide/dock gate
~ **док-бассейна** flood-tide gate
~, **естественные** rock/natural arch
~, **задвижные** sliding door
~, **запанные** boom gate
~ **инфекции** infection atrium
~, **ледниковые** glacial grotto, glaciofluvial outlet
~ **лимфатического узла** hilus of a lymph node
~, **откидные дока** *naut.* flap(-type) gate
~ **печени** *anat.* portal fissure
~, **триумфальные** triumphal arch
~, **створчатые** flap(-type) gate
~ **шлюза, низовые** aft gate
~, **шлюзные** sluice/lock gate
~, **шлюзные поворотные** hinged gate
~, **шлюзные подъемные** vertical lift gate
воротник collar; *zool.* frill
~, **испанский** *med.* paraphimosis
~, **костный** frill of bone
~ **матросской рубахи** jean (collar)
~, **отложной** turn-down collar
~ **ствола** *min.* shaft collar
~, **стоячий** high collar, stand-up collar
~, **шкваловый** *meteor.* arcus, roll cloud, arc of dark cloud
воротничок collar, collaret(te); shirt collar/front; dickey
~, **крахмальный** starched/stiff collar
~, **синий** blue-collar worker
воротный *anat.* portal; gate *(attr.)*
~ **столб** gatepost
вороток I tap wrench; die stock; twisting pliers
вороток II front, shoulder
~, **рантовый** welting shoulder
ворох (thrashed) heap
ворохочиститель cotton extractor, precleaner
ворочание moving, shifting; turning, rolling
ворошилка, ворошитель agitator, stirrer; *agr.* turner
ворс(а) 1. pile, nap; 2. fluff; villus
~, **лоснящийся** shiny nap
~, **мохнатый** shaggy nap
~, **мшистый** fuzzy pile
~, **неразрезной** frieze
~, **плотный** closed nap
~, **прошивной** tufted pile
~, **разрезной** cut pile

ворсильный friezing, teaseling, carding, gigging
~, ная машина teaser, friezing machine
~, ная шишка teasel, gig
ворсильня, ворсовальня 1. cardroom, carding house; 2. teaser, wool card
ворсина fibre, fiber; hair
ворсинка fibre, fiber; hair; lint; *anat., bot.* villus (*pl.* -lli)
~, клеевая colleter, mucilaginous hair
~ки, синовиальные synovial villi/tufts
ворсинковидный villiform
ворсинчатость villosity
ворсинчатый shaggy; villiferous, villous; fleecy; lanate
ворсистость fluffiness, hairiness; nappiness; villosity
ворсистый 1. pubescent, setose, nappy, fluffy, fleecy; 2. *bot.* villous, lanate
ворсовальный carding, teasing, gigging, friezing, nap-raising
~ барабан gig barrel
~ная машина friezing machine, napper frame
~ная шишка gig, teasel
ворсование, ворсовка teasing; napping, raising
~, в сухом виде dry raising
~, во влажном виде wet raising
~, шишками teaseling
ворсовать *text.* tease, raise the nap
ворсовая ткань pile fabric
ворсянка *bot.* teasel *(Dipsacus)*
вортекс vortex
ворчун *zool.* (bull) grunter *(Pomadysus multimaculatum)*
воск wax
~, горный/буроугольный montan mineral/ fossil wax, ozocerite
~, животный animal wax
~, жиропотный suit, wool fat
~, ископаемый/земляной/минеральный mineral wax
~, карнаубский carnauba wax
~, пчелиный beeswax
~, растительный vegetable wax
воскировка waxing
воскобитие wax melting/refining
воскобой 1. wax refiner; 2. extraction of wax; residual beeswax
воскобойный wax-refining
воскобойная, воскобойня wax refinery
восковидный, воскообразный waxy, waxlike, ceraceous, cereous
восковина *zool.* cere, ceroma
восковка tracing/wax(ed)/greaseproof paper; stencil
восковник *bot.* bayberry *(Myrica)*; common wax flower *(Hoya carnosa)*
восковниковые *n. pl. bot.* bayberry family *(Myricaceae)*
восковница *bot.* bayberry, wax myrtle *(Myrica)*
восковой waxen, waxy, wax; cereous
~вая зрелость/спелость *agr.* gold/wax ripeness
восконосный ceriferous
воскоплодный *bot.* waxy-fruited
воскосемянный *bot.* waxy-seeded
воскотопка wax extractor; wax refining
воскоцветник *bot.* honeywort *(Cerinthe)*
воскресение resurrection
воскресенье Sunday
~, прощеное *eccl.* Shrove Sunday
~, сыропустное *eccl.* Quinquagesima Sunday
~, фомино *eccl.* Low Sunday, Quasimodo
воскресный Sunday *(attr.)*
воспаление *med.* inflammation
~, альтеративное degenerative/alterative inflammation
~, гнилостное putrid inflammation
~, гнойное purulent/suppurative inflammation
~ кишок inflammation of the intestines
~ легких pneumonia
~, нагноительное suppurative inflammation
~, некротическое necrotic/necrotizing inflammation
~, неспецифическое aspecific/nonspecific inflammation
~, пролиферативное proliferative/hyperplastic inflammation
~, рожистое erysipelatous inflammation
~, слипчивое adhesive inflammation
~, фибринозно-гнойное fibrinopurulent inflammation
воспаленный inflamed, sore; bloodshet *(of eyes)*
воспалительный inflammatory, phlogistic
воспитание education, training, upbringing; (good) breeding
~, половое sexual/sex education
~, физическое physical training
воспитанник 1. pupil; student, graduate; alumnus (*pl.* -ni); 2. ward, adopted child
~, Московского университета graduate from Moscow University
воспитательный educational, educative
~, дом foundling hospital
воспитывать 1. educate, bring up, rear; train; cultivate; foster; nurture; 2. breed; grow
воспламенение ignition, combustion, inflammation; firing, priming
~, замедленное delayed ignition
~ калильной трубкой hot-tube ignition
~ от сжатия compression ignition
~ от удара/толчка impact/shock ignition
~ от электрозапала grow-plug ignition
~, позднее delayed/retarded ignition
~, полное complete inflammation
~, последовательное multistep/multistage ignition, train-type ignition
~, раннее advance(d)/early ignition
~ с задержкой delay ignition
~ топлива в форсажной камере *av.* afterburner ignition
~, электрическое electric firing

воспламенимый combustible, (in)flammable, ignitable
~, легко deflagrable
воспламенитель ignitor, igniter; primer, detonator, ignition cartridge/charge
~, дополнительный supplementary/auxiliary igniter
~ маршевого заряда sustainer igniter
~, пиротехнический pyrotechnic/squib igniter, firing squib
~, пороховой powder(-charge) igniter
~, пружинный spring cap
~, ракетного двигателя rocket igniter
~, терочной friction igniter/primer
~, электрокапсюльный electric fuze
воспламеняемость ignitability, combustibility, (in)flammability
воспламенять ignite, inflame, kindle, set on fire; prime
воспламеняться catch fire, ignite, inflame, blaze up
воспламеняющийся ignitable, (in)flammable
воспретительный prohibitory, prohibitive
воспрещение prohibition, interdiction
восприемник 1. *law* adopter, adoptant; receiver; inheritor, heir; 2. godfather
восприимчивость 1. susceptibility, susceptiveness; sensitiveness; 2. receptivity, perceptivity
~, обратимая reversible susceptibility
~, обратная inverse susceptibility
воспринимаемое *philos.* sensum, sensible
воспринимаемость perceptibility, perceptivity
~ зрением visual
~ умом sensibility
~ чувствами externalism
воспринимаемый perceptive, perceptible; discernible; perceivable
~ на слух audible
~ объект perceptum
~ чувствами sensible, discernible
воспринимание *see* **восприятие**
воспринимать 1. perceive, apprehend; cognize; grasp, take in; 2. accept, sense, interpret; receive
~ неправильно misinterpret; misapprehend
воспринимающее *n.* percipient
восприятие sensing, perception, apperception; receptivity; comprehension; sensation
~, звуковое sound perception
~, зрительное visual perception
~ направления perception of direction
~ пробивок *comp.* (punched) hole sensing
~, пространственное space impression
~ речи speech perception
~, слуховое aural perception
~, стереоскопическое stereoscopic/space perception/sensation
~, умственное prehension
~, цветовое colour/chromatic sensation
~, чувственное sense perception
~, экстрасенсорное extrasensorial perception
воспроизведение 1. reproduction, reproducing; regeneration; 2. playback, replay
~ в масштабе scale reproduction/representation
~, высококачественное high-fidelity (sound) reproduction
~ голограммы hologram playback
~ данных data reproduction/presentation
~ записей record playback, replay of records
~ звука sound reproduction
~ звукозаписи playback
~ звукозаписи, прямое instantaneous recording playback
~ изображений, искусственное imagineering
~ изображения reproduction/reconstruction of the image, imaging
~ информации information display
~ речи voice reproduction
~ сигнала signal reproduction/presentation
воспроизводимость reproducibility
~ наблюдений repeatibility of observations
воспроизводить reproduce; display; playback; breed *(of nuclear fuel)*
~ в памяти recall, call to mind
~ данные present/reproduce data
~ магнитную запись playback a tape record
~ процесс reproduce/duplicate a process
воспроизводство *econ.* reproduction, regeneration; propagation; *nucl.* conversion
~ на быстрых нейтронах, расширенное *nucl.* fast breeding
~ на тепловых нейтронах thermal conversion
~, простое simple reproduction
~, расширенное *econ.* expanded/extended reproduction; *nucl.* breeding
~, суммарное complete breeding
~, ядерного топлива nuclear fuel conversation
воспроизводящий reproducing, fertile
восславлять *eccl.* extol, glorify, praise, hymn, belaud
воссоединение recombination; reunion, reunification
восставить перпендикуляр *math.* erect/draw a perpendicular
восстанавливаемость 1. restorability, recoverability; reducibility; 2. restorability rate
восстанавливать 1. restore, recover, reclaim; reestablish, renew, renovate; recondition, rebuild, recreate; reactivate; reinstate; *comp.* regenerate; 2. *chem.* reduce, deoxidize
~ в должность reinstate
~ в памяти recall, recollect
~ в правах *law* restore (someone's) rights, rehabilitate
~ дипломатические отношения resume diplomatic relations
~ промышленность rebuild the industry
~ протектор шины retread/recap (a tyre)
~ цикл reset a cycle

восстание 1. revolt, insurrection, (up)rising; 2. *eccl.* resurrection
~, вооруженное armed revolt
~, народное popular uprising
~ из мертвых rising from the dead
~ пласта *min.* seam pitch, rise of seam
~нию, по *min.* up-dip; to the rise, up the dip
восстановитель 1. restorer, renovator, regenerator; restituent, restorative; 2. *chem.* reducing agent, reducer, reductant, deoxidant
~ консистенции consistency reducing agent
восстановительный 1. regenerating, regenerative; 2. *chem.* reducing, deoxidizing
~ период period of reconstruction
~ная сукцессия *bot.* progressive succession
~ные работы restoration work
восстановление 1. restoration, restoring, reconditioning, re-establishment, renewal; recovery, regeneration; rehabilitation; reactivation, resumption; *law* reinstatement (in); restitution; *geod.* relocation; *comp.* reclaiming; 2. *chem.* reduction, deoxidation
~ в должности reinstatement
~ в кипящем слое fluidized-bed reduction
~ в памяти recall
~ в правах restoration of rights, rehabilitation
~ в стационарном слое fixed bed(-type) reduction
~ веса recovery of weight
~ данных data retrieval
~ двигателя engine reconditioning
~ здоровья recovery of health
~ изношенных деталей salvaging work components
~ импульсов pulse regeneration
~ информации data regeneration, information restoring
~ исходного положения *comp.* re-setting
~ исходного положения памяти memory reset
~, катодное cathodic reduction
~, косвенное gaseous/indirect reduction
~ ламп tube/valve reclamation/rejuvenation
~ местонахождения *miner.* resurrection of deposit
~ металлического стандарта *econ.* remonetization
~ несущей carrier reinsertion
~ оборудования reconditioning
~ памяти recollection period/phase
~ плавучести *naut.* refloating
~ после ошибок error recovery
~ почвы resoiling of land
~, программно-управляемое *comp.* software-controlled recovery
~ просвета *med.* repatency
~ протектора шины retreading, recapping
~ работоспособности readaptation to work, rehabilitation
~ разрушенных областей rehabilitation of devastated areas
~ сигнала signal regeneration
~ синхронизации *comp.* retiming
~, совместное coreduction
~ ствола shalt recovery
~ файла file reset
~ фронта волны wave-front reconstruction
~ функции, неполное *comp.* degraded recovery
~ химическое reduction
~ цикла cycle reset
восстановление-окисление reduction-oxidation
восстающий *min.* raise
~, вентиляционный air raise
~, передовой pilot raise
~, рудоспускной ore-pass raise
восток east; *geogr.* the East; the Orient
~, Ближний the Near East
~, Дальний the Far East
~, идущий на east-bound
~, на eastward
~, Средний the Middle East
востоковед orientalist
восток-юго-восток east-southeast, *abbr.* E.S.E.
восточный east, eastern, oriental; easterly *(of wind)*
~, самый easternmost
~ная церковь *rel.* the Eastern Church
востребование claim(ing), demand
~ния, до general delivery, poste restante
вострец *bot.* wild rye *(Elymus, Aneurolepidium)*
восхваление *eccl.* laudation, praising; eulogy, encomium
восхвалитель *eccl.* encomiast
восхвалять extol, (be)laud; eulogize; praise
восход rise, rising
~ луны moonrise
~ солнца sunrise
восходить 1. rise, ascend; climb, go up; mount; 2. *agr.* sprout, shoot
восходящий ascending, ascendant; rising, upward; *meteor.* anabatic
~щей линии, по in the line of ascent
восхождение *astr.* ascension; uprise, ascent
~, прямое right ascension
восшествие ascent, mounting
~ на престол accession to the throne
восьмеричный eightfold; octuple, octal, octonary
восьмерка *naut.* figure-eight knot; *av.* eight (maneuvre), figure eight
~, вертикальная vertical (stand-up) eight
~, кубинская Cuban eight
~, пологая gentle eight
восьмивалентность octavalence, octovalence
восьмигранник *geom.* octahedron
восьмигранный octahedral, eight-faced
восьмизначный eight-digit/-figure/-unit
восьмикратный eightfold, octuple
восьминог *zool.* octopus
восьминогие *pl. n. zool.* octopodans (*Octopoda*)
восьмиполюсник octopole, eight-terminal network

восьмиричный octal, octonary
восьмирядный octostichous, octofarious
восьмисложный octosyllabic
восьмиугольник octagon
восьмиугольный octagonal, octangular
восьмичашелистиковый *bot.* octosepalous
восьмичленный octomerous, eight-membered
Вотан *myth.* Woden, Wodan
вотировка voting
воткание *text.* interweaving
вотум vote
~, **доверия** vote of confidence
~, **недоверия** vote of no confidence
вотчина *hist.* ancestral lands, inherited estate; patrimony
вотчинник *hist.* owner of an ancestral estate, great landowner
вотчинный *hist.* patrimonial
вотяк Votyak
вотяцкий Votyak
воцарение accession (to the throne)
вочеловечение *theol.* incarnation
вочеловеченный *theol.* incarnate, embodied
вошь louse
~, **головная** head louse
~, **книжная** deathwatch
~, **лобковая** crab/pubic louse
~, **платяная** body louse
~, **травяная** plant louse, aphis, greenfly
вощанка 1. wax paper, oilpaper; waxed cloth; 2. cobbler's wax; 3. empty honeycomb
вощеный waxed; paraffin coated
вощить 1. wax, polish with wax; 2. install honeycomb in beehives
вояжер commercial traveller
впадение 1. discharging (into), falling (into); inflow(ing); influx; 2. confluence, mouth
впадина hollow, cavity; *geol.* depression, basin, trough; recess; dent, indentation
~, **аккумулятивная** *geol.* depression of accumulation
~, **в подошве** *geol.* swag
~, **вертлюжная** *anat.* cotyloid cavity, acetabulum
~, **височная** *anat.* temporal fossa/vacuity/fenestra
~, **вулкано-тектоническая** tectovolcanic basin
~, **глазная** eyehole, eyesocket, eye socket, orbit
~, **зубчатого колеса** tooth space
~, **карстовая** carst cavity
~, **коническая** conical depression
~, **краевая** foredeep
~ **кратера** volcanic sink
~, **ледниковая** glacial depression
~ **лопатки, суставная** glenoid cavity
~ **между зубьями** tooth space
~ **на горизонтальной части импульса** pulse valley
~, **нивальная** nivation hollow
~ **обрушения** collapse basin
~, **океаническая** oceanic trough
~, **окрайная океаническая** foredeep
~ **оседания** depression of subsidence, sag
~, **передовая** *geol.* fore trough, foredeep
~ **плавного отпускания** *geol.* down-warping
~, **подвздошная** iliac fossa
~, **подкрыльцовая/подмышечная** *anat.* axillary space, armpit, maschale
~, **породная** rock cavity
~ **профиля резьбы** root/bottom of thread
~, **сбросовая** fault trough
~, **синклинальная** trough
~, **структурная** structural basin
~, **термокарстовая** frost-thaw basin, thermokarst pit
~, **тыловая** back trough
~, **удлиненная** trench
~, **эоловая** *geol.* eolian deppression
впаивание, впайка soldering/sealing-in
впаивать solder/seal/braze in
впай 1. soldering/sealing-in; seal(-in) lead; 2. soldered-in part
~, **зажимный** hook lead
~, **плавкий** fused lead
~, **простой** tip lead
~, **трубочный** tubed lead
впалость concavity, hollowness
впалый concave, depressed; hollow, sunken
впараллель in parallel
вперед ahead, forward(s); fast *(of clocks)*
~, **дать очки** *sport* give odds/points
~, **малый** *nav.* slow (speed) ahead
~, **полный** full (speed) ahead
~, **самый малый** dead slow ahead
~, **средний** half ahead
впередисмотрящий *n. nav.* lookout (man)
вперекрой, вперекрышку lap(ped), overlap(ping)
впечатление imprint, cast; impression
~ **стереоскопической глубины** impression of depth
впечатлительность impressionability, impressibility
впивание imbibition, absorption
вписанный *math.* inscribed; *fin.* entered
вписка entry, entering; insertion, inserting
вписывать 1. *math.* inscribe; 2. enter/register/write in, make entries in
впитываемость absorptive/blotting capacity; absorbency, absorbability; saturation capacity
~ **чернил** ink holdout/receptivity
впитывание absorption, absorbing, seepage; soaking (up), imbibition
впитывать absorb, imbibe, soak (up)
впитывающий absorbing, imbibing, spongy, spongeous
вплавление fusing/melting in
впластованный embedded, interbedded, interstratified, intercalated
впластовывать embed, imbed, interbed, interstratify, intercalate

вплетание, вплетение interweaving, interlacing, intertwining
вплетать interweave, intertwine, interlace (with), plait (into)
вполнеприменимость complete applicability
впотай flush, level, even; countersunk
вправка *med.* (re)setting
вправление *med.* reposition, diaplasis; reduction
~ грыжи *surg.* taxis
вправлять *med.* (re)set, reduce
~ вывих set a bone
впрессовывать press/set/stamp in; embed
впритык butt-jointed; end-to-end; very close, flush, abutting, against
впрыск injection, injecting; squirting, spurting; shot
~, безвоздушный/бескомпрессорный solid injection
~ горючего fuel injection
~, комбинированный composite injection
~ с завихрением/вихревой swirl injection
~ топлива, двухфазный into-stage fuel injection
~ топлива, дробный intermittent fuel injection
впрыскивание injection; spraying in; spurting, squirting
~, внутрикожное intracutaneous injection
~, внутримышечное intramuscular injection
~, внутрисердечное intracardial injection
~, подкожное subcutaneous injection
впрыскиватель injector
впрыскивать inject; spray in
впуск inlet, intake, admission; inflow; admittance
~ воздуха air admission
впускание inlet, intake, admittance, admission
впускать let/take in, admit
впускной inlet, intake, admitting
~ клапан inlet/intake/charging valve
впячивание invagination, ingrowth; embole, emboly
враг enemy; foe
вражда hostility, enmity, animosity
враждебность hostility, enmity, animosity
враждебный hostile, inimical
~ иностранец *law* alien enemy
вражий hostile
~жья сила *eccl.* Satan, the Devil
вразброс, вразброску randomly, haphazardly
~, сеять *agr.* sow broadcast
вразнос, торговать peddle, hawk
врастание growing-in, intergrowth
~ внутрь ingrowth
врастать intergrow, grow in(to)
вратарь *sport* goalkeeper
врач physician, doctor, medical man; *mil.* surgeon
~, авиационный flight surgeon
~, ветеринарный veterinary
~, военно-морской surgeon, medical officer
~, военный army surgeon; medical officer
~, зубной dentist
~, лечащий attending physician
врач-универсалист general practitioner
врачебный medical
~ обход rounds
~ осмотр medical examination
~ная помощь medical assistance
вращатель rotator, spinner; (aircraft) scanner
~, антенный aerial spinner
~ бурового станка swivel head
~ рамки loop rotator
вращательно-качающийся rotary-vibratory
вращательно-колебательный rotational-vibrational
вращательный rota(to)ry, rotatable, rotative, rotational, revolving; vorticular
вращать rotate, turn, revolve; wind
вращаться rotate, turn, pivot; spin, circulate, revolve; whirl; circle, gyrate, roll; windmill
~ без биения run true
~ по часовой стрелке rotate clockwise
~ против часовой стрелки rotate anti/counter-clockwise
вращающий
~, влево *opt.* l(a)evo-rotatory
~, вправо *opt.* dextro-rotatory
вращающийся rotatable, rotating, revolving, turning, spinning; whirling; gyrating; swivelling, vorticose
вращение rotatation, spin(ning); revolution, revolving, urning; gyration; windmilling; whirling, circumvolution, rolling
~ антенны, принудительное (power) slewing
~ векторной функции *comp.* curl of a vector function
~ ветра veering of wind
~ влево counter-clockwise rotation
~ вправо clockwise rotation; dextrogyration
~, встречное counterrotation
~, кеплеровское Keplerian rotation
~, левое levogyration, counter-clockwise rotation
~, несобственное *math.* improper rotation
~, обратное reverse rotation; *astr.* retrograde rotation
~ по часовой стрелке clockwise rotation, right-hand rotation, dextrorotation
~, правое *opt.* dextrorotation
~, пространственное spatial rotation, three-dimensional rotation
~, против часовой стрелки counter-clockwise rotation, left-hand rotation, l(a)evorotation
~, собственное proper rotation
~, суточное *astr.* diurnal rotation
~, трехмерное three-dimensional rotation, spatial rotation
~, ускорять spin up
~, установившееся stedy-state rotation
~, холостое free-running
~, чистое pure/simple rotation
~, электрона, собственное spin
вращполе rotating field
вред injury, harm, hurt; detriment, damage

~, **гражданский** civil injury
~, **непоправимый** irreparable damage
~, **причинить/принести** (do) harm
вредитель 1. *agr.* infestant, pest, vermin; 2. *polit.* saboteur, wrecker
вредительский sabotaging, harmful, wrecking
вредительство sabotage, wrecking
вредность harm(fulness); injuriousness; insalubrity; nocuity, peniciousness
вредный noxious, harmful, deleterious, detrimental, injurious; dangerous; unhealthy, pernicious
~, **биологически** biologically damaging/noxious
~, **газ** noxious gas
~**ное учение** pernicious doctrine
вредоносный nocuous, damnific, hurtful
врез cutting, incision
врезание, врезка cutting(-in), incision; notch, gash
~ **горного амфитеатра** *geol.* circus cutting
~ **долины** downcutting of valley
~ **меандра** entrenchment
врезанный *geol.* entrenched, engrafted
врезать incise, cut/set/fit in
врезаться incise, entrench, engraft
врезной cut-/set-in; mortise, mortice
~ **замок** mortise-lock, dead lock
врезывание cutting/setting/fitting in
~ **реки** fluvial downcutting
временник chronicle, annals
временно-импульсный pulse-time
временной *tech.* time *(attr.)*; *philos.* temporal; *gram.* tense *(attr.)*
временный temporary; provisional; transitory, transitional; tentavite; makeshift
~ **комитет** interim committee
~**ная работа** temporary job
~**ное владение** *law* tenure
~**ное правительство** provisional government
временщик 1. *hist.* favo(u)rite; 2. temporary ruler
время 1. time(s), period; 2. *gram.* tense
~, **абсолютное** absolute time
~, **атлантическое стандартное** Atlantic/Eastern Standard Time
~, **атомное** atomic time
~, **ашельское** *geol.* Acheulean
~ **бездействия** unoccupied/idle/down time
~ **безотказной работы** time between failures
~ **безотказной работы, среднее** mean time between failures
~ **ввода** *comp.* input time
~ **взлета** *av.* take-off-time
~, **видимое** apparent time
~ **включения** (turn) on time; make-time
~ **возврата** reset time, resetting time
~ **возвращения** *game th.* recurrence time
~ **воздействия** exposure time
~ **воспроизведения** playback time
~ **восстановления** recovery time
~ **восстановления гироскопа** gyro erection time
~, **восточноевропейское** Eastern-European time
~, **всемирное** universal time
~, **вспомогательное** anxiliary/handling time
~ **встречи с целью** impact time
~ **выборки** *comp.* access/fetch/select time
~ **выборки записи** writing access time
~ **выборки информации** information read-out time
~ **выборки обращения** read-out time, access time
~ **выборки при записи** access time of writing
~ **выборки с барабана** drum access time
~ **выборочной реакции** discrimination reaction time
~ **выдачи информации** information access time
~ **выдерживания** decay/''cooling''/holdup time
~ **выдержки** curing/holding time
~ **выживания** survival time
~ **выжидания** wait time
~ **выключения** turn-off-time
~ **выполнения задачи** task time
~ **выполнения команды** execution/instruction time
~ **выполнения цикла** cycle/cycling time
~ **выпуска изделия** time of delivery
~ **выхода двигателя на режим** build-up time
~ **вычисления** computing/calculating/calculation time
~ **гашения обратного хода** blanking period
~ **года** season
~ **горения** combustion/*rock.* burning time
~, **горное поясное** *Am.* mountain stadard time
~, **гражданское** civil time
~, **гринвичское, звездное** Greenwich sidereal time
~, **гринвичское истинное** Greenwich apparent time
~ **действия** action time
~, **декретное** legal time
~ **дискретное** discrete time
~ **для вспомагательных работ** *comp.* miscellaneous time
~ **дня** time of day
~ **догорания** tail-off time
~ **доступа** access time
~, **единичное** unit time
~, **единое** common time
~ **жизни** life(-time), lifetime, survival time
~ **жизни нейтрона** neutron lifetime
~ **жизни по альфа-распаду** alpha-decay life-time
~ **жизни теплового нейтрона** thermal life-time
~ **загрузки** load time
~, **заданное** preset/planned/scheduled/desired/prescribed time, pre-set time
~ **задержки** delay/lag/deceleration time
~ **замедления** slowing-down time, deceleration/

retardation time
~ занятия *com.* holding time
~, западоевропейское Western-European time
~ запаздывания delay/lag/deceleration/response time, time lag
~ записи recording/writing time
~ запоминания storage time
~ запуска starting time, start-up period
~ зарядки charging time
~ затрачиваемое на ручную работу manual (working) time
~ затухания dacay time, die-away time
~ затухания импульса fall time
~, звездное sidereal time
~, земное terrestrial time
~ изодрома integral action time
~ искания *teleph.* selection time
~ исполнения объектной программой object time
~ использования usage time
~ использования станка, суммарное *tech.* machine load
~ исправного состояния up time
~, истинное apparent/true/real time
~, истинное звездное apparent sidereal time, sidereal true time
~, каменноугольное carboniferous time
~ кульминации Луны time of moon's transit
~ кульминации светила time of transit
~, летнее summer time, daylight saving time
~, машинное computer/machine time
~ между приливом и отливом slack water period
~, местное local time
~, мировое universe time
~ на одну штуку job/piece time
~ нарастания time of rise, rise time, build-up time, building-up period
~ начала time of origin
~, неопределенное (глагола) *gram.* indefinite tense
~, нерабочее down/idle time, off(-duty) time
~, нормативное рабочее standard machine time
~ нулевого пояса Greenwich mean time
~, ньютоново *astr.* ephemeris/Newtonian time
~ облучения exposure/irradiation time
~ обработки treatment/processing time
~ обратного хода retrace/return time
~ обращения при считывании reading access time, access time for reading; *el.* time of circulation
~ обслуживания servicing time, *math.* holding time
~ ожидания holding/waiting time; latency; *comp.* latent period
~ ожидания ответа станции *teleph.* answering interval
~ операции с фиксированной запятой fixed-point operation time
~, операционное operation time
~ опознания recognition time
~, опорное reference time
~ опроса sampling time
~ остановки outage time, shut-down time
~ отпускания *el.* release time
~ отсечки intercept time
~ отсчета readout time
~ отсчитываемое перед пуском count down time
~ отсчитываемое с момента пуска plus time
~, отсчитывать keep time
~, отсчитывать в обратном порядке count down (time)
~, отсчитывать от нуля вверх count up (time)
~ отхода/отшествия time of departure
~ отъема *agr.* weaning time
~ переключения switch(ing)/transit/turnover time
~ перемещения travel/motion time
~ переноса carry-over time
~ перерегулирования *comp.* overshoot period
~ перехода transition time
~ переходного режима transient period
~, подготвительное make-ready time
~ подготовки к работе setup time, warm-up time
~ поиска search time
~, полезное usable time
~ половинной выживаемости medium lethal time
~ полувыведения biological half-life
~, послесангамоское *geol.* post-Sangamonian time
~ послесвечения decay time; after-glow time
~, поясное zone/standard time
~ пребывания в воздухе airborne time
~ пребывания программы program residence time
~ пребывания топлива в реакторе fuel residence/stay time
~, предельное critical time
~, предстартовое prelaunch time
~ предупреждения warning time
~, принятое time kept
~ приработки running-in time
~ присутствия attendance time
~ пробега landing roll/run time; transit/travelling time
~ прогона turn-around time, run(ning) time
~ прогрева warm-up time
~, производительное production time
~ пролета flight/transit time, time of flight; time over
~ промывки flush/washing time
~ простоя downtime, down/bad/dead/lie/lost/outage/standby/idle time
~ прохождения time of transit/passage
~ пуска starting time, start-up time

~, пусковое duration of starting, starting period/time; *rock.* launch(ing) time
~, рабочее operating/operation/working/up time, worktime, on-duty time, hours of work
~ развертки sampling/scanning time
~ развертывания installation time, set-up time
~ разгона acceleration period/time
~ разогрева warm-up time
~, разрешающее/разрешения resolution/resolving time
~ разряда discharge time
~ раскрутки spin-up time
~ раскрутки гироскопа gyro(scope) run-up time
~ распада decay time
~, расчетное estimated/normal time
~ реакции reaction time, time lag
~, реальное real time
~ руления *av.* taxi time
~ сборки assembly time
~ свертывания крови *med.* coagulation/clotting time
~ свертывания,цитратное *med.* citrate clotting time
~, сверхурочное *educ.* overtime
~, свободное leisure time
~ складского хранения shelf time
~ сложения add(ition) time
~ случки/спаривания rutting season
~, солнечное solar time
~ спадания decay/drooping time, fall(-off) time
~ сохранения информации data-retention time
~ срабатывания response/operating/actuation time, pick-up time
~, среднее mean time
~, среднеевропейское Central-European time
~, среднее летальное/смертельное medium lethal time
~, среднее местное local mean time
~, сталийное *naut.* lay days
~ стоянки на приколе *naut.* duration of lay-up
~, стояночное *naut.* lay days
~ считывания reading/sensing/sampling/scanning time
~ считывания с перфокарты *comp.* card cycle
~ такта clock time
~, технологическое process time
~, тихоокеанское поясное *Am.* Pacific standard time
~ умножения multiplication time
~ успокоения transient/equilibrium time
~ установления сигнала *comp.* settling time
~ фиксируемое на кинокадре frame time
~ холостого хода lost motion time, no-load time; idle time
~ хранения storage time/period
~, центральное поясное *Am.* central standard time
~ цикла записи writing cycle time
~ цикла запоминающего устройства на сердечниках core store cyclic time
~ цикличности repetition time
~ чувствительности sensitive time/period
~, штилевое *naut., av.* still-air time, zero-wind time
~, штучное time per piece
~ экплуатации месторождения *geol.* life of field
~, эксплуатационное operating/running time
~ эксплуационной готовности usable time
~ экспозиции/экспонирования exposure time
~, Эрмосское *geol.* Hermosa time
~, эфемеридное *astr.* ephemeris time
~, ядерное nuclear (traversal) time
времяисчисление time reckoning, calendar; chronology
времянка *el.* temporary/haywire connection
врид *abbr.* **временно исполняющий должность** acting
вровень level/square/flush with
врожденный congenital, inherent, connate; innate, inborn; original, intrinsic(al)
вронскиан *math.* Wronskian (determinant)
вросток *geol.* ingrowith
~, лучистый radial growth
~, экзогенный exogenous growth
вросший intergrown
вруб *min.* cut, kerf; channel, groove, notch; hewing
~, боковой side(-draw) cut
~, воронкообразный cone cut
~, клиновой plough cut, wedge(-shaped) cut
~, косой slanting cut
~, нижний bottom cut, undercut
~ ножницы X-cut
~, уступный bench(ing) cut
~, щелевой slot cut
врубание cutting/hewing in
врубать cut; groove, notch
врубка joint; cut, notch, slot, groove
~ в ласточкин хвост dovetail joint, dovetailing
~ в ус mitred joint
~ вполдерева cross-halved joint, straighthalved joint, lapped joint, dovetail halving
~ гребнем cogged joint
~ зубом step joint
~, односторонняя dapping, gaining
~, стропильная bridle joint
~, треугольная birdsmouthing joint
~ шипом в гнезде housed joint
врубовка *min.* cutting machine, cutter
врубово-погрузочный cutting-loading
врубовый *min.* (coal-)cutting
~вая машина coal cutter
вручение delivery, handing, serving; investiture, presentation; *law* service
вручную manualy, by hand
всадник 1. *hist.* knight; equites; 2. horseman

всас suction
всасываемость suction capacity, blotting, adsorbability
всасываемый absorbable
всасывание suction, sucking; intake, inflow; absorption, absorbing, resorption; inbibition
 ~ выхлопных газов re-ingestion
 ~, двустороннее two-side suction
 ~, чрескожное cutaneous absorption
всасывать suck/draw/take in; inbibe; absorb, resorb
всебоковой omnilateral
всеведение *rel.* omniscience
всеведущий 1. omniscient; 2. *n. rel.* God
всевидец *myth., rel.* all-seer
всевластие omnipotence, unlimited/absolute power
всевобуч *abbr.* **всеобщее военное обучение** universal military training
всеволновой *rad.* all-wave/-band
всевышний sumpreme, the Most High
вседержитель *rel.* the Almighty
вседозволенность permissiveness
всезнание pansophy, universal knowledge
всезнающий omniscient, all-knowing
всекитайский all-China
вселенная universe, macrocosm, world
 ~, материальная cosmosphere
 ~, наблюдаемая visible/observable universe
 ~, нестационарная evolutionary/evolving universe
 ~, равновесная/стационарная steady-state inverse
 ~, расширяющаяся expanding universe
вселенский universal; *eccl.* (o)ecumenical
 ~ собор (o)ecumenical council
всемирный universal, world(-wide), global, cosmopolitan
 ~ почтовой союз Universal Postal Union, UPU
 ~ продовольственный совет (ООН) International Emergency Food Council, I. E. F. C.
 ~ная ассоциация туристических агенств World Association of Travel Agencies, WATA
 ~ная ветеринарная ассоциация World Veterinary Association, WVA
 ~ная организация здравоохранения World Health Organization, WHO, W. H. O.
 ~ная федерация профсоюзов World Federation of Trade Union, WFTU
всемогущество almightiness, omnipotence
всемогущий omnipotent, all-powerful; *rel.* Almighty
всенаправленный omnidirectional, omnibearing, all-directional; omnirange
всенародный nation/country-wide, national; public, general
 ~ная перепись national census
всенощная *eccl.* night service, vespers and matins
всеобуч *abbr.* **всеобщее обучение** universal education
всеобщее *philos.* the universal
 ~ идеальное в искусстве ethos
всеобщий general, common, universal, catholic
 ~щая забастовка general strike
 ~щая перепись населения general census of the population
 ~щее употребление common/general use
всеобщность generality, commonality, universality; catholicity
всеобъемлющий comprehensive, universal, all-embracing; overall; encyclop(a)edic; cosmic(al)
всеохватывающий compehensive, all-embracing
всепоглощающий all-absorbing/-consuming
всепогодный *av.* all-weather
всеопрощающий all-forgiving
всеопрощение *rel.* general pardon, amnesty
всережимный variable-speed, fully-variable
всероссийский all-Russian
всесветский cosmopolitian
всесилие unlimited authority, omnipotence, almightiness
всеславянский pan-Slav(ic)
всесовершенный, всесовершеннейший perfect, absolute, complete; total
всесоюзный all-union; all-Union
всесторонний omnidirectional, comprehensive, thorough, all-round, omniradial
 ~нее образование all-round education
всеуслышание, во publicly, for all to hear
всеядный omnivorous, pantophagous, euryphagous
всеядность (двигателя) multi-fuel capability
вскармливание feeding; rearing
 ~, искусственное artificial/bottle feeding
вскапывание excavating, digging(-)up
вскипание boiling (-up), effervescence, ebullition, ebullience, ebulliency; bubbling up
 ~ почвы soil effervescence
вскрывание 1. *bot.* bursting, dehiscence; 2. *med.* dissection
 ~ дырочками dehiscence by pores
вскрывать open (up), unseal; *min.* strip, develop; *med.* lance; dissect
 ~ дорожное покрытие rip up the pavement
 ~ злоупотребления expose abuses
 ~ нарыв/абсцесс *med.* lance/open an abscess
вскрытие opening, unsealing; baring; *min.* stripping; *met.* breack-down; *med.* lancing, cutting; post-mortem (examination)
 ~, весеннее spring break up
 ~ водоносного горизонта tapping of underground reservoir
 ~ дорожного покрытия ripping (-up) a pavement
 ~ контейнеров container opening
 ~ кровли deroofing
 ~ месторождения для разработки development srtipping of a deposit, stripping of a deposit
 ~ наносов stripping of an overburden
 ~, патологоанатомическое *med.* autopsy, autopsia, post-mortem examination

~ плода *bot.* dehiscence
~ реки debacle, break up (of ice)
~, судебно-медицинское *med.* medicolegal autopsy
~ трупа autopsy, post mortem examination
~ угольного пласта coal seam uncovering, coal winning
вскрытый laid bare
вскрыша *min.* 1. stripping, removal, baring; 2. overburden
~, мощная thick overburden
~ наносов overburden removal
вскрышные работы *min.* stripping
вспахивать plough/plow (up), till, cultivate
вспашка *agr.* ploughing, plowing, tilling
~, двухъярусная subbase plowing
~, плантажная trench plowing, ditching
~, поперечная всвал и вразвал balk plowing
вспенивание foaming, frothing, froth building
~, бурное effervescence
~ шлаков frothing of slags
вспениватель 1. frothing/foaming agent, foamer, frother, froth promoter; 2. froth-building machine, beater
вспениваться froth, foam; lather
всплеск splash (point); burst, eruption; surge, flash-up; bump (of current)
~ на импульсе spike
~ радиоизлучения солнца solar radio (out)burst
~ рентгеновского излучения X-ray burst
~ шлака slag splashing
всплываемость buoyancy, flotation
всплывание (up-)floating, floating(-up), surfacing, emersion, emerging; rising to the surface; buoyancy
~ заполнителя floating of aggregate
~ элеронов *av.* aileron up-floating
всплывать float (up), come/rise/buoy to the surface, *naut.* emerge, (break) surface, broach; become waterborne; refloat
~, заставлять bring up from a dive
всплывающий supernatant, buoyant, floating
всплытие *naut.* emersion, surfacing, rise
~, аварийное emergency surfacing
~ на поверхность surface exposure
вспомогательный ancillary, auxiliary, accessory, stand-by; booster; spare, emergency; subsidiary, subordinate, co-adjuvant; *med.* lacertus
~ адрес auxiliary address
~ выключатель auxiliary switch
~ глагол *gram.* auxiliary verb
~ файл scratch file
~ная память на сердечниках *comp.* auxiliary core store
~ная программа assistance program
~ные средства *comp.* aids
вспомоществование aid, assistance, relief
вспомоществователь sponsor, contributor
вспомоществовать (give) help, assist, aid; contribute; sponsor
впрыск 1. *med.* injection; 2. injector, sprayer
вспрыскивание (be)sprinkling; *med.* injecting, injection
вспухший tumid
вспученный bulged, bellied, swollen, inflated, heaved
вспучиваемость distensibility
~ угля *min.* coal blistering/bloating
вспучивание swelling, heaving, intumescence, upwarp(ing), bulging; *med.* flatulence; blistering, inflation,
~ грунта ground distending
~ пласта swell(ing), bulging
~ пород в скважине heaving trouble
~ почвы heave of soil
~ складки surging of fold
вспучиватель *constr.* bloater
вспучиваться swell up, bulge, intumesce, upwarp, blister, bloat
вспучина heave, swelling, bulge; *geol.* volcanic blister
вспыхивать catch/take fire; flash, flicker, scintillate; deflagrate, flame/flare/blaze up
вспышка 1. burst, explosion; flash, spark, ignition, deflagration, fulmination; flare/blaze-up; 2. eruption, spurt, outburst
~ активности burst of activity
~ видообразования *geol.* burst of speciation
~, запальная ignition flash
~ излучения burst of radiation
~ космического излучения cosmic-ray burst
~, магниевая *phot.* flashlight
~, метеорная meteor burst
~ новой звезды nova outburst
~, обратная backfire, backfiring
~ пламени, мгновенная instantaneous flash
~, преждевременная/предыскровая advanced fire, premature ignition/flash
~ света/световая light flash, scintillation
~ света, точечная point scintillation
~, солнечная solar eruption/flare
~, сумеречная twilight flash
~, тепловая/температурная thermal spike
~, хромосферная chromospheric flare
~, эмоциональная tantrum
~ ядерного взрыва nuclear flash
вставать *naut.* be/come aweigh *(of anchor)*
вставка insert(ion), insert, inlay; inset; spacer; intercalation; *comp.* stuffing, *el.* plug; embedding; fixing, framing; *print.* casing-in; *phon.* epenthesis
~, асбестовая asbestos insertion
~, амортизирующая shock-absorbing insert
~ битов *comp.* bitstuffing
~ блока *print.* block fitting, casing-in
~ в подъеме *leath.* apron
~, винтовая трубная twisted strip
~, инвариантная invariant embedding
~, письменная written completion
~, плавкая *el.* fuse (link)
~, промежуточная distance piece, spacer

~, **радиоактивная** radioactive insert
~, **сопловая** nozzle lining
~ **стекол** glazing
~, **цилиндрическая** *naut.* parallel middlebody
~ **штампа** die insert
вставление insertion, inserting, installation
вставлять insert, set/fit/put/fix/into; inlay; introduce, include; work in, embed
~ **в оправу** mount, set
~ **втулку** bush
~ **инструмент** enter the instrument
~ **стекла** glaze
~ **шпоны** *print.* lead
вставной inserted; plug-in; insertable; intercalary; detachable
~**ные зубы** dentures, false teeth
~**ные рамы** double window frames
вставной-вытяжной push-pull
вставочный inserted; insertable; *med.* intercalative; *phon.* epenthetic
~ **неврон** intercalary neuron
встраивать build in
встреча 1. encounter, rendezvous, meeting; reception; welcome; 2. *sport* match
~ **на высшем уровне** summit (meeting), meeting at the summit
~ **Нового года** New Year's Eve party
встречаемость occurrence; frequency
~, **совместная** co-occurrence
встречаться occur, happen
встречно-параллельный antiparallel
встречный 1. oncoming, approaching; head-on; 2. counter, contrary, opposite
~ **бой** *mil.* encounter battle
~ **ветер** head wind
~ **иск** *law* plea in reconvention, counter claim
~ **поезд** oncoming train
~**ное исполнение** *law* counter-execution
~**ное обвинение** *law* countercharge
~**ное право** countervailing equity
~**ное течение** counterflow
встроенный built-/set-in, integral, built/embedded into; *comp.* onboard, in-line
~ **в кристалл** *comp.* on-chip
встряхивание shaking / stirring (up); jolting; agitation, agitating; jarring
встряхиватель shaker
~ **для сит** seive/screen shaker
~ **с поворотным столом** turntable shaker
встряхивать shake (up); jolt, jig; agitate
вступать enter (into), join (in); start
~ **в бой** join battle
~ **в брак** marry
~ **в действие** come into force/effect/operation; become valid
~ **в дело** *law* intervene
~ **в контакт** come into contact, make contact
~ **в реакцию** react *(with)*
~ **в силу** *law* take effect, go into effect
~ **в соединение с** combine with
~ **в строй/эксплуатацию** become operational, go into operation production, come into service, go on line
~ **во владение** assume/take possession *(of)*
~ **на престол** ascend the throne
вступительный introductory, exordial; inchoative; ingoing, incoming
~ **взнос** entrance fee
~ **экзамен** entrance examination
~**ное слово** opening address
вступление entry, introduction, opening, preamble, prelude, exordium; entrance, opening, inlet; *law* accession
~ **в бой** engaging in combat, go into action
~ **во владение** taking of possession
~, **четкое** distinct arrival
всходить *agr.* shoot, sprout, come up
всходы *agr., pl.* shoots, sprouts, seedlings
~, **дружные** good sprouts
всюду определенный *math.* completely/everywhere defined
вталкивание pushing/shoving in(to); forcing/thrusting in
вталкиватель pusher
вталкивать push/force/thrust/shove in(to)
втекание influx, inflow(ing)
втекать flow/run/discharge into
втекающий influent, flowing in
втирание embrocation; inuncton
~ **мази** *med.* embrocation, fomentation
втирать rub in(to); anoint
втискивать press/thrust/force/squeeze into
втора *mus.* second voice/violin
вторжение invasion, intrusion, penetration; irruption, ingression, injection; inrush, infringement, encroachment, breaking in
~ **воды** water encroachment
~, **воздушное** air invasion
~, **глубокое** deep penetration
~, **магматическое** *geol.* magmatic wedging/injection
~ **теплого/холодного воздуха** warm/cold air invasion
~, **полярное** polar outbreak
вторично again, for the second time
~ **перегонять** rerun, redistil
вторичноизверженный *geol.* secondary-eruptive
вторичнопокровный *bot.* metachlamydeous
вторичнопорфировый secondary porphyric
вторичноразветвляющийся subdichotomous
вторичноротые *n. pl. zool.* deuterostomatous organism *(Deutrostomata)*
вторичность secondariness
вторичный secondary, deuterogenous, deuteromorphic, epigenetic; derived, derivative; subsidiary; additional
~ **продукт** by-product
~**ная перегонка** redistillation
~**ная ткань** secondary tissue

второзаконие Deuteronomy
второй second
 ~ экземпляр тратты *econ.* second of exchange
второразрядный, второсортный secondary, inferior, second-rate/-grade/-quality
второстепенный 1. minor, secondary; 2. *n.* collateral
втрое three times
 ~ меньше one-third as much/many great/big/large
 ~, увеличивать triple
втуз *abbr.* technical college
втулка *tech.* 1. bush(ing), sleeve; hub; liner; insert; 2. plug
 ~, бакаутовая lignum vitae bush(ing)
 ~, ввертная screw bushing
 ~, вводная entrance/inlet bushing
 ~ веретена bolster
 ~ воздушного винта propeller hub
 ~, динамометрическая dynamometric hub, hub dynamometer
 ~, дистанционная spacing/distance sleeve
 ~, закрепительная clamping/adapter/fixing sleeve
 ~, замедляющая copper slug
 ~, защитная protection sleeve
 ~, изолирующая insulating sleeve/bush(ing), lead-in insulator
 ~ колеса boss; hub of a wheel
 ~ коллектора commutator sleeve/shell
 ~, кольцевая packing sleeve
 ~, кондукторная jig bush(ing)
 ~, литниковая sprue bush(ing), feed bush
 ~ маховика flywheel boss/hub
 ~ надульника muzzle plug
 ~, направляющая guide/pilot bush(ing)
 ~ несущего винта (main-)rotor hub/head
 ~, осевая axle box
 ~, палубная deck socket/plate
 ~, переходная adapter, reducing bush(ing)
 ~, плавающая floating bush(ing)
 ~ поршня piston-boss bush(ing)
 ~ рабочего колеса impeller bush(ing)
 ~, проходная изолирующая grommet, grummet, gromet
 ~, разрезная/разъемная split bush(ing)
 ~, распорная distance/spacing/strut sleeve
 ~, резьбовая thread(ed) bush(ing)/insert
 ~ рулевого колеса steering-wheel hub
 ~ с прорезью slotted sleeve
 ~, стяжная clamping sleeve
 ~, уплотнительная sealing bush(ing)
 ~, упорная thrust bush(ing)
 ~ цилиндра cylinder liner / sleeve
 ~, шлицевая splined bushing
втулочка plug
втягивание moving in, pulling/dragging/drawing (in/into/up); retraction; intussusception; involving in
втягивать draw/pull/drag (in/into/up); retract
втяжка *print.* 1. (motto) indention; 2. concavity
втяжной *tech.* suction; pull-in; retractile
 ~ сердечник plunger
 ~ные когти retractile claws
втянутость retraction
 ~ барабанной перепонки *anat.* retraction of tympanic membrane
вуайеризм voyeurism
вуалирование *phot.* veiling (effect); fogging, masking effect, blooming, hazing
вуаль veil; turbidity, fuzziness; *phot.* fog, haze
вудвардия *bot.* chain fern *(Woodwardia)*
вудсия *bot.* woodsia *(Woodsia)*
вуз *abbr.* **высшее учебное заведение** institute of higher education
вулкан 1. volcano; 2. *myth.* Vulcan; 3. vulcan *(pigment)*
 ~, бездействующий quiet/inactive/dormant/ quiescent volcano
 ~, вздымающийся swelling volcano
 ~, второстепенный/побочный subordinate volcano
 ~ гавайского типа shield volcano
 ~, грязевой mud volcano/cone
 ~, двойной cone-in-cone volcano, nested crater
 ~, действующий active volcano
 ~, зараждающийся maar, embryonic volcano
 ~ извергающий соленую грязь salinelle
 ~, куполообразный dome-shaped volcano, lava dome
 ~, многоактный polygenic/multiaction volcano
 ~, одноактный monogenic volcano, single-action volcano
 ~, подводный submarine volcano
 ~, потухший extinct/dead/extinguished volcano
 ~, рассеченный/расчлененный dissected volcano
 ~, сателитовый/сопровождающий subordinate/satellite volcano
 ~, сложный/смешанный compound/composite volcano
 ~, слоистый bedded/stratified volcano
 ~, спящий dormant volcano
 ~, типа Везувия conical volcano
 ~, трещинного типа fissure-type volcano
 ~, угасающий/умирающий decadent/moribund/expiring volcano
 ~, файфского типа fife-type volcano
вулканизат vulcanizate
 ~, наполненный filled vulcanizate
 ~, перекисный peroxide vulcanizate
 ~, саженаполненный carbon black vulcanizate
 ~, смоляной resin-cured vulcanizate
вулканизатор vulcanizer; vulcanizing apparatus/ agent, curing agent
 ~, бандажный tread-band vulcanizer
 ~, барабанный drum vulcanizer
 ~, индивидуальный unit vulcanizer
 ~, камерный inner-tube vulcanizer
 ~ карусельного типа rotary vulcanizer
 ~, кольцевой retreader

вулканизаторщик vulcanizer operator
вулканизационный curing, vulcanizing
вулканизация curing, vulcanization, vulcanizing
~, без формы open vulcanization
~, бессерная non-sulphur vulcanization
~ в автоклаве autoclave vulcanization
~ в кипящем слое fluid bed vulcanization
~ в прессе press vulcanization
~, горячая hot vulcanization
~, камерная chamber vulcanization
~, котловая steam pan vulcanization
~, перекисная peroxide vulcanization
~, повторная repeat vulcanization
~, предварительная prevulcanization, precuring
~, прессовая press vulcanization
~, серная sulphur vulcanization
~, совместная covulcanization
~, ступенчатая step-up vulcanization
~, формовая mould vulcanization
вулканизировать cure, vulcanize
~ в воздушной/паровой среде vulcanize in air/steam
вулканизм *geol.* volcanicity, volcanism, vulcanism
~, многовыходной polyorifice volcanicity
~, молодой young volcanism
~, современный recent volcanicity/volcanism
вулканит *geol., tech.* volcanite
вулканический volcanic, igneous, eruptive; *myth.* Vulcanic
вулканичность volcanicity
вулканоид mud volcano
вулканолит volcanic ejectamenta
вулканолог volcanist, volcanologist
вулканология volcanology
вульва *anat.* vulva
вульвизм *med.* vaginismus
вульвит *med.* vulvitis
~, трихомонадный mycotic vulvitis
вульво-анальный *anat.* vulvoanal
вульвовагинит *med.* vulvovaginitis
вход 1. entrance, entry; inlet, port; access, orifice; intake; *el.* input; *anat.* aditus, introitus; 2. admission
~, адресный address input
~ без регистра-защелки unlatched input
~ бесплатный admission free
~ в гортань *anat.* superior aperture of the larynx
~ в долину *geol.* embouchure
~ в круг полетов над аэродромом *av.* (traffic) pattern entry, joining the circuit
~ в систему log-in, logon
~ воздуха air entry
~ воспрещен/воспрещается no admittance
~, главный grand entrance
~ глазницы *anat.* opening of the orbit
~ группового контроля *comp.* group control entry
~, дискретизированный sampled input
~ единицы one input
~, задний rear/back entrance
~ звукоснимателя *rad.* "phono" jack
~, имитированный simulated input
~, инверсный/инвертирующий inverting input
~, информационный data entry/input, information input
~, клиновидный wedge inlet
~, лабиринтный maze entrance
~, микрофонный microphone/speech input
~ насоса pump inlet
~, неинвертирующий uncomplemented input
~, нерегулируемый fixed inlet
~, неуправляемый uncontrollable entry
~, неправильный *comp.* invalid entry
~ описания description entry
~, отсроченный deferred entry
~, передний front entrance
~ переноса carrier/carry input
~ пилотируемого КЛА в плотные слои атмосферы manned (re-)entry
~, регулируемый variable-geometry inlet
~ сброса reset input
~ свободный open to the public
~, симметричный floating input
~ синхронизации sync/lock/clock input
~ сопла nozzle entry/entrance
~, стробируемый *comp.* gated input
~, счетный complementing/inverting input
~, шахтный entrance shaft
~, элементарный величины elementary item entry
входец *bot.* omphalodium
входимость повторная *comp.* reentrance, reentrability
входить enter, appear, occur, go/come/get/walk in(to); penetrate; fit in
~ в быт/обиход become the custom
~ в действие/силу come into force
~ в зацепление *mach.* mesh
~ в синхронизм drop/pull in synchronism
~ в состав form, be a part of
~ в суть дела get to the core of the mather
~ в уравнение appear in the equation, enter into the equation
~ в число be reckoned (among/with)
входная *eccl.* Introit
входной input; entrance; inlet *(attr.)*
~ная плата entrance fee
входной-выходной input-output
входящая incoming mail
входящий 1. incoming, entering; *math.* re-entrant; 2. forming, being a part of; 3. *mach.* male
~ журнал book of entries
вхождение entering, ingress; entry, intromission; intussusception
~ в атмосферу re-entry
~ в связь *rad.* establishment of communication/contact
~ в синхронизм coming into step/sychronism

~, явное explicit occurrence
ВЦИК *abbr.* **Всероссийский Центральный Исполнительный Комитет** *hist.* All-Russian Central Executive Committee
ВЦСПС *abbr.* **Всесоюзный Центральный Совет Профессиональных Союзов** *hist.* All-Union Central Soviet of Trade Unions
вчерне *met.* (in the) rough; *fig.* in draft form
вчистую clean, final, in final form
ВЧК *abbr.* **Всероссийская чрезвычайная комиссия по борьбе с контрреволюцией, саботажем и спекуляцией** *hist.* All-Russian Special Commission for Combating Counter-Revolution, Sabotage, and Speculations
вчувствование *psych.* projection
вшивица *bot.* pedicularis, lousewort *(Pedicularis)*
вшивость lousiness, lice infestation, pediculosis
въезд entrance, entry; approach; ramp; driveway
~, аппарельный ramp (approach)
~, наклонный ramp
~, паромный raft ramp
въездная виза entry visa
въездные ворота gates
въезжий entrance *(attr.)*
~, двор/дом *hist.* coaching inn
выбег stopway, stopping distance; running out/down, run-out/-down; overswing, overshoat; coasting
~ машины run-out/-down
~ стержня overtravelling of a rod
~ стрелки overshoot/overswing of a pointer
~ судна stopway, stopping distance
~ температуры excursion of temperature
~ частоты frequency drift
выбивание 1. knocking out, dislodging; chasing; embossment; 2. *agr.* overgrazing
выбивать dislodge; hammer/drive/knock out; emboss; coin, stamp
~ клеймо stamp a mark
~ пробку force a plug out
выбивка knock-/shake-out, knock(ing) out, dislodging
~ вагранки emptying/fettling the cupola
~, гидравлическая hydraulic knock-out
~ отливок/отливки casting knock-out, knockout of casting
~ форм mould knock-out
выбираемость selectivity
выбираемый *comp.* optional
выбирание 1. selecting, selection, choosing; election; 2. taking out, removal
выбиратель selector (switch)
выбирать 1. choose, select, pick (out); elect; 2. extract, remove; heave; *naut.* haul/take/veer in; rowse
~ время time
~ канат haul in a rope
~ конец veer in a rope
~ сердцевину core out
~ слабину take up slack, absorb slack
~ слабину троса haul in the bight of a rope
~рай ходом! *naut.* haul away!
выбленка (shroud) ratline
~, усиленная catch ratline
выбоина dint, dent, hollow, indentation; pit(ting); pothole, rut
выбойка *text.* print, printed linen; *tech.* stamping, striking
выбор 1. choice, selection; sampling, drawing; *comp.* option; 2. *pl.* election(s)
~, беспристрастный unbiased sampling
~, групповой cluster sampling
~ заданий/задач task selection
~ знаков *comp.* character selection
~ из бесконечной совокупности sampling from an infinite set
~ из двух альтернатив *comp.* binary decision
~ из конечной совокупности finite sampling
~ из кучи bulk sampling
~ информации information selection
~ канала данных data channel selection
~ катушек многократного использования multiple-reel option
~ кода code selection
~ конфигурации configuring
~, косвенный indirect sample
~ маршрута routing
~ масштаба scaling, choice of scale
~, многовариантный multiple-choice alternative
~, многомерный multivariate sampling
~, многоступенчатый multistage sampling
~ момента времени timing, time-point selection
~ на основе меню menu selection/picking
~, начальный initial selection
~, необъективный biased sampling
~, непрерывный continuous sampling
~, неслучайный nonrandom sampling
~, объективный unbiased sampling
~, однократный unitary sampling
~ операторов choice of operators
~ оптимального варианта trade-off
~ оптимальной программы program optimization
~ по умолчанию default option
~, повторный repeated/duplicate sampling, rotation sampling
~, последовательный time/successive sampling
~ при разветвления branch decision
~ приемного кармана *comp.* stacker select
~ присваивания giving option
~, пристрастный biased sampling
~ программной ветви branching decision
~, произвольный optional sampling
~, псевдослучайный quasi-random sampling
~, равновероятный equiprobable sampling
~ размеров sizing
~, расслоенный stratified sampling
~ режима работы mode selection

~ режима работы реактивного двигателя thrust selection
~, решетчатый lattice sampling
~ с возвращением sampling with replacement
~ символов character selection
~, случайный random sampling
~ строк, автоматический automatic line selection
~ строки choice of line, line selection
~, территориальный area sampling
~, усеченный truncated sampling
~ файла многократного использования multiple-file option
~ цели target selection
~ цифр digit selection
~ы в один тур *law* single ballot
~ы, дополнительные by-election
~ы кандидатов *law* primaries, primary elections
~ы с несколькими баллотировками successive ballot
~ы с одной баллотировкой single ballot

выборка 1. *stat.* sample, sampling; *comp.* access; selection; 2. *tech.* cutting/hollowing out
~ адреса address selection
~ без ограничений unrestricted sample
~, бесповторная sampling without replacement, non-repeated sampling
~, беспристрастная unbiased sample
~, вероятностная probability sample
~, гнездовая cluster sample
~ данных data sample/access
~, двойная double sample
~, детерминированная artificial sample
~, дефектная defective sample
~, жесткая nonrandom access
~ из выборки subsample
~ из памяти memory access
~ из памяти на дисках disk access
~ из серийных изделий production sample
~ из списка list sample
~ информации information readout, storage access
~, искусственная artificial sample
~, квазислучайная quasi-random sample
~ команды instruction fetching, order/command/instruction selection
~ магнитных карт magnetic card selection
~ методом совпадения coincident current selection
~ наблюдений sample of observations
~, начальная initial selection
~ не вполне случайная judgement sample
~, негруппированная nongrouped sample
~, непосредственная direct access
~, неразрешенная inadvertent access
~, неслучайная nonrandom sample
~, несмещенная unbiased sample/sampling
~, неупорядоченная unordered sample
~, объективная unbiased sample
~, однократная single sample
~, однократно цензурированная singly censored sample
~ операнда operand fetching
~, параллельная simultaneous access
~, повторная replicate sample
~ по приоритету priority selection
~, поразрядная bit access
~, последовательная sequential/serial selection
~, предварительная prefetching
~, преднамеренная purposive sample
~, произвольная random access
~, пропорциональная quota sample
~, прямая direct/random access
~, равновероятная equiprobable sample
~, расслоенная stratified sample
~ с перекрытием overlapped access
~ с повторением sampling with replacement
~, случайная random sample/sampling, probability sample
~, смещенная biased sampling
~, территориальная area sample
~, упорядоченная ordered sample
~, усеченная truncated sample
~, уравновешенная balanced sample
~, цензурированная censored sample
~, шумовая noise sample

выборочность 1. optionality; electivity; 2. appointment by election

выборный elective; electoral; elected; *n.* delegate
~ бюллетень voting paper
~ная должность elective office

выборочный sampling, selective, sampled; random; *comp.* optional
~ная проверка spot check

выбраковка *agr.* culling; discard, rejection, rejecting

выбранный electee

выбрасывание ejection, expulsion; throwing away, discarding
~ за борт *naut.* jettison
~ замедлителя expulsion of the moderator
~ тормозного парашюта drogue-chute ejection
~ урана в отвал *nucl.* uranium rejection

выбрасыватель ejector; extractor; lifting-out device; pick-up fingers
~ гильз shell extractor
~, правый right-hand extractor

выбрасывать throw out/away, discard, reject, emit, belch, ejaculate; expel, exclude
~ за борт throw overboard
~ колос/кисть/султан *agr.* ear, tassel
~ на берег cast ashore
~ отростки *bot.* sucker
~ стрелку *bot.* spear
~ усик *agr.* throw out a runner
~ флаг hoist a flag

выбрасывающий *med.* ejaculatory

выброс 1. ejection, ejaculation; rejection; effluent, exhaust; *geol.* outburst; outbreak; ejectamenta; 2.

overswing, overshoot
~, аутигенный authigenous outburst
~ в атмосферу discharge into the atmosphere
~, внезапный sudden outrush, sudden/instantaneous outburst
~, вулканический 1. volcanic outburst; 2. volcanic ejectamenta, volcanolite
~ газа fume-off; gas release/blowout/outburst/rush
~ дальности range pip
~ импульса, обратный backswing, undershoot
~ импульса, прямой overshoot
~ карт, автоматический *comp.* automatic card ejection
~, кластический *geol.* fragmental product
~, кротовый molehill
~, крупный вулканический ejected block
~ металла slopping/spitting of metal
~ мощности power excursion/surge
~, отдельный fragmentary ejecta
~ пены escape of froth
~, перемежающийся intermittent ejection/outburst
~, песчаный sandblow
~ пласта outbreak of seam
~ пород rock outburst, pressure bump
~, пусковой *rad.* trigger spike
~ реактивности reactivity excursion
~, родственный cognate ejecta
~, сильный violent outburst
~, сильный (породы) violent bump
~, шлаковый scoria ejection
выброска 1. ejecting; rejecting; discarding, throwing out; 2. *av.* airdrop(ping); 3. refuse, rubbish; discards, rejects
~ воздушного десанта airborne drop(ping)
~ корабля на берег beaching
выбуривать drill/bore out; spurt (out), well up
выбухание protrusion
~ пупка *med.* exumbilication, exomphalos, umbilical hernia
выбывать quit, leave
~ из строя *mil.* quit the ranks; *tech.* put out of commission
выбывающий *n.* dropout
вывал *min.* inrush
~ глины/породы/угля clay/rock/coal inrush
вываривать extract by boiling; decoct; *text.* scour
~ желатину degelatinize
выварка boiling extraction; *text.* scouring
выварной decocted, extracted by boiling; degummed
~ котел saltpan
выведенец *hist.* peasant forcibly removed to different area
выведение 1. bringing/taking/leading/pulling/drawing out; removal; extermimation (of pests); 2. deduction, conclusion; 3. breeding, raising, hatching (out); growing; 4. putting up, erection
~ формулы derivation of a formula
выверенный calibrated, checked
выверка calibration, check(ing), verification; regulation; alignment, adjustment, gauging, trueing
~ дистанции range calibration
~ курса heading alignment
~ на вертикальность plumbing
~ на горизонтальность levelling
вывернутый inverted, reversed; ectropic
~ наизнанку inverted, turned inside out
вывертывать undo, unscrew, screw/pull out
выверять align, adjust, straighten; rectify; calibrate, check, verify, gauge; rate
~ по отвесу plumb
~ по уровню level
вывеска 1. signboard; 2. *naut.* reballasting, (stop) trim dive
~ и балансировка trim and stability
~, предостерегающая caution sign board
выветренный weathered, (a)eolian, eroded, windblown
выветривавшийся weathered, rotten, decayed
выветривание 1. *geol.* weathering, erosion, disintegration, degradation, ablation, aeolation, decay, fading; 2. airing, ventilation, aeration; 3. *chem.* efflorescence
~, биохимическое biochemical weathering
~, ветровое aeolation, deflation
~ зернистых пород fretwork
~ кристаллов *chem.* salt efflorescence
~, кумулятивное cumulative/cumulose weathering
~, латеритное laterization
~, молодое immature weathering
~, морозное congelifraction, cryoturbation, frost weathering
~ почвы aeolation of soil
~, снеговое nivation
~ угля coal slacking
~, химическое chemical weathering/erosion
~, червяковое vermicular disaggregation
выветривать(ся) *geol.* weather, erode, deflate; *chem.* effloresce; decay, fade
вывинчивать screw out, undo, unscrew
вывих *med.* dislocation, luxation, abarticulation; dislocated part
~, акромиально-ключечный knocked-down shoulder
~ височно-челюстного сустава dislocation of the lower jaw
~, врожденный congenital dislocation
~, неполный incomplete dislocation
~, осложненный complicated dislocation
~, открытый/сложный compound dislocation
~ плеча shoulder-slip
~ позвонков vertebral dislocation
~ полового члена phallocrypsis
~ полулунной кости Kienböck's dislocation, dislocation of semilunar bone
~ таза spondylolisthetic/Prague/Rokitansky's pelvis
~ тазобедренного сустава dislocation of the

hip
~ тибио-тарзальный Nélaton's dislocation
~ хрусталика dislocation of (the) lens
вывихивать dislocate, sprain, luxate
вывихнутый dislocated, abarticular
вывод 1. removal, withdrawal, leading/bringing/moving out; *mil.* recovery; injection, insertion; 2. deduction, inference, implication; *math.* development, derivation; 3. output, outlet; *el.* lead; terminal; tap(ping); bushing; 4. *comp.* pin; lead-out
~, автономный off-line output
~, анодный anode lead
~ антенны aerial lead-out
~, бесформатный formless output
~ в буквенно-цифровой форме alphanumeric data read-out
~, верхний top cap
~, визуальный visual display
~ вычислительной машины computer output
~, гибридный hybrid output
~ графических данных graphic data/information output
~ данных data output
~ данных в режиме он-лайн on-line data output
~ данных, графический graphic(al) output
~ данных на печать printout
~ данных на экран data display
~ данных о процессе process data output
~ данных, скоростной immediate data output
~ данных, электрический electric data output
~, дедуктивный deductive reasoning
~, делать conclude, infer
~ заземления/заземляющий earthing terminal
~, земляной ground terminal
~ из вычислительной машины на микрофильм computer output on/to microfilm, micromation
~ из крена *av.* roll recovery
~ из петли *av.* loop recovery
~ из пикирования dive recovery, pull-out (from the dive)
~ из состава действующего флота inactivation
~ из строя rupture, abruption
~ из эксплуатации removal from service
~, информационный data pin
~, катодный cathode lead
~, контролируемый checked output
~, контрольный test lead
~ копий изображений output of image copies
~, краевой edge-assigned terminal
~, линейный line terminal
~, логический inference
~, ложный fallacy
~ микросхемы pinout
~ на бумажную ленту paper tape output
~ на высокую/низкую орбиту *av.* high/low orbit injection
~ на магнитную ленту magnetic tape output
~ на носитель данных data carrier output
~ на орбиту КЛА orbit(al) injection/insertion, launch into orbit
~ на перфоленту pinched-tape output
~ на печать printer output
~ на печать и восстановление информации dumping and restoring
~ на порт harbour approach
~, нестрогий nonmonotonic conclusion
~, нулевой neutral terminal
~, оборванный open lead
~, общий generalization
~ основной программы main program output
~ от средней точки centre tap
~, пистонный stack-mounting terminal
~ питания power lead
~, плавающий floating terminal
~, прямой direct injection
~ пучка extraction of particles, beam extraction/rejection
~ (разговорной) речи language/speech output
~ результатов прогона программы post-run output
~, результирующий resulting deduction
~ речи audio response, speech/language output
~ самолета на аэродром approach navigation
~, связный on-line output
~, сеточный grid lead
~ стержня withdrawal of a rod
~ счетного входа триггера information input terminal
~ судов лоцманами *nav.* pilotage outwards
~, точный precise injection
~ уравнения derivation of an equation
~ фиксированных изображений hard-copy output
~, формальный formal inference
~ формулы deduction, derivation
~, фотографический (данных) photographic readout
~ частиц extraction of particles *(from an accelerator)*
~, экранированный shielded/screened lead
~ электрода electrode lead
~ энергии, волноводный waveguide output/circuit
выводимость 1. deducibility, deductibility, derivability; 2. *agr.* hatchability
выводимый deducible, deductible, derivable
~, взаимно interdeducible
~ путем заключения deductive
~ путем рассуждения deducible
~ путем умозаключения inferential
выводить 1. remove, withdraw, lead/bring/send/force/turn out; terminate; 2. deduct, deduce, conclude, infer; 3. exterminate; 4. grow; breed, raise; hatch (out); 5. *nucl.* extract, eliminate; *mil.* recover
~ из заблуждения undeceive
~ из зацепления disengage, throw out of gear,

unmesh
~ **из каталога** uncatalogue
~ **из очереди** *comp.* dequeue
~ **из равновесия** disturb the equilibrium of, disturb from equilibrium
~ **из состава президиума** remove from the presidium
~ **из строя** disable, put out of action
~ **из цикла** remove from the cycle
~ **из эксплуатации** remove from service, take out of service
~ **на орбиту** place on/in orbit, put into orbit
~ **отметки/баллы** define the average grade
~ **сырость** extricate moisture
выводка *agr.* 1. breeding, raising; growing; 2. brood, hatch; litter
выводной 1. leading out, discharge; removing; 2. derived, inferential; 3. *anat.* excretory; 4. *n.* guard accompanying convicts
выводок *agr.* brood; hatch; litter
вывоз 1. taking out, removal; 2. *econ.* export, exported commodities
~ **золота** withdrawal of gold
вывозка removal; taking/carting out
~ **леса** removal of logs
вывозной *av.* introductory; *econ.* export *(attr.)*
выворачивание turning inside out; overturning; shaking loose
~ **века** *med.* ectropion(ization), ectropium
выворот 1. fold, twist; 2. *med.* inversion, extroversion
~ **век** ectropion, ectropium, abnormal inversion of the eyelids, ectropionization
~ **век, паралитический** flaccid/paralytic ectropion
~ **губы** eclabium, eversion/inversion of a lip
~ **матки** inversion of the uterus
выворотка *print.* reversing
выгарки *pl.* recrement, dross, slag
выгиб (outward/upward) bend, camber, curve; curvature; curving, warping
~ **вверх** *geol.* upward buckling
~ **горных пород** stepfold, flexure
~, **сбросовый** *geol.* faulted upfold
выгибание, выгибка warping, bending, arching, curving; wriggle
выгибать camber, hog, bend, curve, arch; buckle; warp
выгибаться warp, bend, curve, arch
выглаживание burnishing; smoothing (down); ironing
выгнутый cambered, convex, arched, curved
~ **вверх** upwarping
~ **вниз** downwarping
выговаривать enunciate, articulate
выговор 1. reprimand, reproof, rebuke; scolding, dressing down; 2. accent, pronunciation
~, **делать** reprimand
~, **объявлять** award a reprimand
выгода benefit, advantage; gain, utility; profit, interest
~, **взаимная** reciprocity
~, **личная** self-interest
~, **побочная/попутная** *econ.* incidental benefit
выгодность advantage, profitableness; efficiency, utility
~ **кода** efficiency of a code
выгодный advantageous, gainful, beneficial, profitable; favo(u)rable; remunerative, paying
выгодоприобретатель *law* beneficiary
выгон *agr.* (common) pasture, grazing (ground), grassland
выгонка 1. distillation; 2. *agr.* forcing
~ **рассады** seedling production
~ **растений, тепличная** forcing of plants
выгоночный *agr.* forced, forcing
выгонять 1. distil; 2. accelerate growing/blooming of plants; 3. drive/turn out, expel
выгорание burnout, burning/dying out; burn-up; pitting *(of contacts)*; parching
~ **топлива** (propellant) burn-out, fluxing, fuel burn-up
~ **цвета** fade, fading
~ **экрана** screen burn-out
~ **ядерного топлива** depletion of nuclear fuel
выгорать burn up/out/away
выгоревший burnt-up/-out, degraded
выгородка enclosure, recess; compartment
~, **кингстонная** *naut.* sea/flooding valve recess
~ **люка** trunked hatchway
~ **трапа** *naut.* stairway enclosure
~ **элеватора** hoist recess
выгреб rake-out; cesspool
выгружать unload, discharge; dump
выгрузка unloading, discharge
~, **выборочная** selective unloading
~ **навалом** dumping
~ **топлива** fuel discharge/unloading
выгул *agr.* pasture
выдавать produce, deliver; give out, issue
~ **зарплату** pay out wages
~ **плавку** produce a melt
~ **патент** grant a patent
~ **перебежчика** hand over a deserter
~ **преступника** extradite a criminal
~ **сигнал** generate a signal
выдаваться protrude, project, jut out
выдавка *met.* extrusion
выдавливание 1. expression, squeezing out; extrusion, squirting; 2. *tech.* chasing; embossing; stamping, pressing/squeezing-out
~ **на токарно-давильном станке** spinning
~, **ударное** impact extrusion
~, **холодное** die sinking
выдавливать press/squeeze out; extrude; emboss; stamp
выдалбливать hollow/gouge/scoop out; excavate, dig out, cave; slot
выдача 1. issue; payment; issuing, delivery; distribution; 2. output, yield, production; 3. presentation,

display; 4. extradition
~ **данных** presentation of information
~ **зарплаты** payment of wages
~ **слитка** ingot discharge
~ **целеуказания** target data transmission
выдающийся projecting, salient, protruding
~ **вперед** protuberant
выдвиг *nav.* advance; *geol.* outthrust, protrusion
выдвигать 1. promote, propose, nominate; 2. move/pull out; extend
~ **вверх** push up/over
~ **вперед** push forward
~ **на должность** nominate to an office
~ **обвинение** bring an accusation
~ **предложение** bring up a proposal
~ **убедительный довод** offer a convincing argument
~ **чью-нибудь кандидатуру** propose someone as candidate
выдвижение 1. moving out/forward; advancing, advance; promotion, advancement; 2. nomination
выдвижной retractable, telescopic, extendable, extension-type, sliding; draw-/pull-out
~ **ящик** drawer
выделение 1. extraction, segregation, separation; isolation, selection; 2. emission, expulsion, evolution, liberation, release; *pl. med.* discharge; secretion, excretion, transpiration, exudation; escape, emanation; ejection; 3. formation, precipitation; setting; 4. yield; recovery
~, **абсорбционное** absorption extraction
~ **влаги** exudation/oozing of moisture
~, **внезапное** sudden release, outburst
~, **вредное** poisonous/deleterious emanation
~, **вулканическое** mofette
~ **газа** gassing, evolution/liberation of gas, gas emanation/liberation/emission, exhalation
~ **газов из материала** outgassing
~, **гнойное** purulent discharge
~, **гетеротипное** heterotypic secretion
~, **гнездовое** pockety segregation
~ **дочерней активности** milking of daughter activity
~ **дочернего изотопа** (isotope) milking
~ **желчи** *med.* biliation, bilification
~, **жилообразное** streaky segregation
~, **изотипическое** isotypic secretion
~ **изотопов** isotope isolation
~ **информации** extraction of information
~ **из расплава** exsolution
~, **кислое** acid extract
~ **кислорода** deoxidation
~, **количественное** quantitative isolation
~ **кремния** silicon separation
~ **кристаллов** crystallization
~ **лавы** outflow of lava
~ **летучих элементов** escape of volatile constituents
~ **ложных целей** decoy discrimination
~, **магматическое** magmatic segregation
~ **металла** metal winning
~, **металлоносное** metalliferous exhalation
~ **метана** methane/firedamp emission/liberation
~ **минеральных зерен** removal of mineral grains
~ **мокроты** expectoration
~, **мономерное** monomeric segregation
~, **основное** basic extract/segregation
~ **памяти** *comp.* memory allocating
~, **полимерное** polymeric secretion
~, **порфировое** phenocryst
~ **признаков** feature extraction
~ **примесей** precipitation of impurities
~ **пыльцы** pollen recovery
~ **ресурсов** resource allocation
~ **сигнала (на фоне помех)** signal extraction
~ **сигнала цели на общем фоне** target discrimination
~ **сигнала цели на фоне шума** target-noise discrimination
~ **синхронизирующих импульсов** sync(hronizing pulse) separation
~ **соединения железа** deironing
~ **спор** spore separation
~, **стремительное** outburst, outrush
~ **тепла** heat liberation/release
~ **флюида** release/expulsion of fluid
~ **хлора** evolution of chlorine
~ **частоты** frequency discrimination
~ **через карбонаты** separation by carbonate digestion
~ **через кишечник** intestinal elimination
~ **через почки** renal elimination
~, **электролитическое** plating-out
~, **электрохимическое** electrochemical isolation
~ **энергии** energy release
~**ния, влагалищные** vaginal discharge
~**ния из организма** excreta, egesta, ejecta
~**ния, маточные** uterine discharge
~**ния, послеродовые** *med.* lochia, lochial discharge
выделенный isolated, separated(-out); evolved; liberated, released
выделитель 1. *math.* discriminant; 2. separator, discriminator
~ **семян** seed extractor
~ **синхронизирующих импульсов** sync(hronizing pulse) separator
~ **тактовых импульсов** discriminator
выделительный 1. excretory, secretory; 2. isolating, separating, eliminative; distinguishing; 3. emphatic, intensifying, intensive
~ **ход** secretary channel
выделка 1. manufacture; make; quality, workmanship; finish; 2. dress(ing)
выделывать 1. manufacture, make, produce, prepare; 2. dress
~ **кожу под замшу** give a suede finish to leather
~ **лайку/сыромять** taw

выделять 1. extract, separate, isolate, segregate; educe; 2. give off, evolve, exude, liberate, release; secrete, excrete, exude, discharge, emit, escape; 3. deposit, precipitate; yield; 4. allocate, assign, earmark; allot; apportion; 5. distinguish, discriminate; set off; give prominence to
~ в свободном виде set free, liberate
~ вид *geol.* distinguish/institute a species
~ курсивом *print.* put in italics
~ минералы из породы disaggregate, isolate
~ мочу urinate
~ серебро desilver
~ эманацию emanate
~ ядовитые газы give off noxious fumes
выделяться 1. deposit, plate out; 2. escape, emanate, exude, ooze/separate out; 3. *law* receive one's share (of a legacy)
~ в виде пузырьков effervesce
выдергивать pull/draw/pluck out; wrest
выдерживание 1. withstanding, holding(-off); keeping; storage; 2. ag(e)ing, maturing; weathering; curing, seasoning
~ в подвале cellaring
~ в термостате incubation
~ высоты *av.* height-keeping, constant-altitude control
~ для распада *nucl.* storage for decay, ''cooling''
~ заданного курса *av.* preset heading control
~ заданного посадочного курса localizer tracking
~ заданного режима datum keeping
~ курса (самолета) flying a heading
~ места в строю *av.* station-keeping
~ отходов hold-up of wastes
~ под открытом небом weather aging
~ при высокой температуре high-temperature storage
~ продуктов деления fission-product hold-up
~ радиоактивных веществ ''cooling''
~ траектории path control
~ числа М constant Mach number control
выдерживатель holder, holding tub, ag(e)ing tank
выдерживать 1. endure, withstand, stand up to; bear, stand, sustain; 2. season; cure; hold, soak; *nucl.* cool; 3. age, mature, lay up, keep; store
~ в термостате incubate
~ допуски hold tolerance
~ испытание stand/pass the test
~ нагрузку take/endure/withstand a load
~ направление *polit.* adhere to principles
~ несколько изданий run into several editions
~ перегрузку withstand overload, accept an overload
~ под арестом keep in custody
~ температуру withstand temperature
выдержка 1. stamina, endurance, stand-up, firmness; 2. ag(e)ing, curing; maturing; season(ing); *nucl.* cool; *min.* weather; 4. *phot.* (time of) exposure; shutter speed
~ бетона concrete curing
~ во времени retention interval
~ времени time lag/delay/interval
~ времени при отбое *teleph.* splitting time
~ на солнце solarization
~ на холоду exposure to cold
~, номинальная nominal exposure time
выдолбленный hollowed out
выдох exhalation, expiration, breathing out
~, полный complete/full expiration
~, удлиненный prolonged expiration
выдра *zool.* otter *(Lutra lutra)*
~, морская sea otter *(Enhydra lutris)*
выдувание blowing(-)out/off; deflation; inflation
~ почвы wind erosion, soil blowing
~ стекла glass blowing
~ через щель slot-blowing
выдувать blow (out/down); deflate, exhaust; inflate
выдувка *met.* blowing-out/-down
выдувной blown *(of glass)*; blow-through; inflatable
выдыхание breathing out, expiration
выдыхательный expiratory
выдыхать breath out, expire, exhale
выдыхающий exhalent
выедать attack, corrode, eat away
выезд departure; exit, egress, road out
~ каретки *text.* run-/drawing-out of carriage
выездка breaking-in, training *(of horses)*
выездной training; travelling, visiting, riding
~ матч *sport* away match
~ спектакль guest performance
~ная виза exit visa
~ная сессия суда *law* visiting assizes
выемка 1. removal; excavation; *min.* extraction; stoping; 2. hollow, groove, cavity, depression; pit, recess; excision; *naut.* roach; *arch.* fluting; 3. *rail.* gutter, ditch
~, аксиллярная axillary excision
~, биссусная byssal notch
~ грунта dredging
~ для берда groove of lay for the reed
~ для маховика flywheel pit
~, краевая emargination
~, кумулятивная charge hollow
~ платины *text.* sinker throat
~, прямоугольная rectangular recess
~, прорезающая вершину холма *geol.* hogbacked cutting
~, цокольная *rad.* base dimple
выемчато-апертурный sinuaperturate
выемчатый emarginate; sinuate, sinuous; grooved, fluted; retuse; hollowed out
выжечь burn out
выживаемость survival (rate)
выживание survival, survivability
~ наиболее приспособленных *biol.* survival of the fittest
~ при аварии/ударе crash/impact survival

выживать survive, live/pull through
выжиг 1. burning out, roasting; 2. yield from one firing
~ пламенем flame scarfing
выжигание 1. burning out; roasting, calcinating; scorching; branding; searing; 2. *med.* cauterization
~ каутером igniextirpation, ignioperation
~ кокса carbon burning
~ по дереву poker work
выжигать 1. burn out/off/down; scorch; calcine, roast; kiln; 2. brand; 3. *med.* cauterize
~ клеймо brand
~ по дереву do poker work
~ уголь char
выжидательный expectative, expectant; waiting, temporizing
~ная политика temporizing policy, wait-and-see policy
~ную позицию, занимать mark time, temporize
выжим *sport* press-up
выжимание 1. pressing (out), squeezing (out); wringing; expression; 2. *sport* pressing exercises, weightlifting
~, тектоническое *geol.* heaving, pinch, squeezing out
выжимать 1. squeeze/press out, pinch (out), strain off, wring (out); express; 2. *sport* press, lift
выжимка 1. squeezing (out), pressing (out); wringing; 2. release, rind, spew; marc, pomace
~ки, виноградные rape
~ки, льняные linseed cake
~ки, томатные tomato pomace
выздоровление convalescence, recovery, recuperation
~, гладко протекающее uneventful recovery
~, неполное incomplete/defective recovery
~, полное full/complete recovery
выздоравливать recover, recuperate
выздоравливающий 1. recovering, recuperating; 2. *n.* convalescent
вызванный induced, caused, provoked, invoked
~, неявно *comp.* implicitly called
~, явно *comp.* explicitly called
~ная процедура invoked procedure
вызов 1. ringing, signalling, call(ing); summons; 2. challenge; *comp.* fetching
~ абонента, осуществлять signal/ring to a subscriber
~, аварийный emergency/distress call
~ авиации air request
~ в суд summous to appear in court; *law* subpoena
~, взаимный reverting call
~ "всем, всем!", общий *rad.* general call ''to all stations''
~ звонком bell ringing
~, зуммерный phonic signalling
~, избирательный selective ringing
~, индукторный magneto ringing
~ информации из запоминающего устройства *comp.* retrieval of stored information
~ команды *comp.* instruction fetch
~ команды канала channel command call
~, машинный generator ringing
~ на бис encore
~, несанкционированный *comp.* unauthorized call
~ операционной системы generation of operating system
~ по внутреннему телефону intercom call
~ подпрограммы subroutine call
~ подпрограммы, неявный implict/implied subroutine call
~ по значению call by value
~ по имени call by name, name call
~ по номеру dialing; dial-up
~ по прерыванию interput call
~ по радио radio call
~ по сигналу тревоги alarm call
~ по ссылке call by reference
~, полуавтоматический manually started machine ringing
~, полуизбирательный semiselective ringing
~, принимать answer a call
~, принять accept a challenge
~ при стадии бедствия *rad.* distress call
~, служебный service call
~, срочный urgent call
~, телефонный (telephone) call
~, тональный voice-frequency ringing
~ функции *comp.* function reference
~, экстренный emergency call
вызывание 1. elicitation; 2. *eccl.* invocation
~ родов *med.* induction of labo(u)r
вызывать 1. induce, cause, give rise to, trigger; call forth, provoke, initiate, evoke, excite; *comp.* fetch; 2. summon, call on; 3. challenge
~ в милицию be summoned by the police
~ в суд summon(s), *law* subpoena
~ вражду antagonize
~ всеобщее восхищение win general applause
~ износ give rise to wear, result in wear
~ кровотечение cause bleeding
~ на дуэль challenge to a duel
~ на социалистическое соревнование *obs.* challenge to socialist emulation
~ ответную реакцию evoke response
~ послабление *med.* purge
~ прекращение процесса result in a cessation
~ рвоту cause vomiting
~ тошноту nauseate
~ (чрезмерное) напряжение put (unnecessary) strain on
~ эффект lead to effect
вызывающий inducing, causing, provoking
выигранные деньги *econ.* winnings
выигрыш gain, advantage, win(nings); prize; *game th.* payoff; *econ.* saving
~ в весе weight saving

~, главный first prize
~а, полиноминальная функция polynominal payoff
выигрышный winning; advantageous; effective
~ заем premium bonds
выйный *anat.* nuchal
выкалывать chip/prick out; tattoo
выкапывать dig up/out, excavate, exhume
выкатывание 1. rolling/smoothing out; mangling; 2. *av.* overrun; roll out
выкачивание pumping out
выкачивать pump out, exhaust, evacuate; deflate
выкашивание *agr.* mowing
выкид one-throw sample
выкидка 1. di(a)eresis; 2. casting/throwing/crossing out; ejection,deletion
выкидывание throwing/crossing/striking out; ejection; deletion
~ икры spawning, discharge of spawn
выкидывать discard, reject, throw out
выкидываться 1. jump/leap out; 2. rise, extend upwards; *bot.* put forth shoots/leaves
~ на мель/берег *nav.* run aground/ashore
выкидыш *med.* miscarriage, absortion; miscarried f(o)etus; *vet.* slink
~, искусственный abactio, artificial abortion
~, ранний ovular abortion
выкидышный abortifacient
выкипание boil(ing) off, boilover, boiling away
~ ракетного топлива propellant boiling-off
выкладка lining, facing, covering; 2. *math.* computation, calculation; 3. *mil.* kit; 4. ornamentation, decoration
~ бетона spreading of concrete
~ костра crib building
выкладывание 1. *tech.* facing; 2. taking/laying out; 3. *math.* calculation, computation
выкладывать 1. line, face; 2. lay/take out; decorate/embelish
~ дерном turf
~ изразцами tile
~ камнем pave
~ кирпичом line with bricks
выклепывать 1. remove by hammering, knock out; 2. whet, sharpen; hammer out; 3. rivet
выклинивание *geol.* thinning/petering/tapering/pinching out
~, взаимное/встречное intertonguing
~ на поверхность cropping out to the surface
выключатель cut-out switch, circuit breaker; on-off switch
~, аварийный emergency/safety breakdown/crash switch
~, автогазовый self-generated circuit breaker
~, автоматический automatic circuit breaker
~ автопилота *av.* autopilot disconnect switch
~ арретира cage switch
~, барометрический barometric switch, baroswitch
~, быстродействующий quick-break switch, highspeed switch
~ валика *comp.* platen release lever
~, взрывозащищенный explosion-proof switch
~, вибростойкий vibration-proof switch
~, включать close the switch
~, включать повторно reclose the circuit breaker
~, воздушный air(-break) switch, air(-break) circuit breaker
~, временной time switch
~, выключать open the switch
~, герконовый reed switch
~, герметический hermetically sealed switch
~, главный master switch
~, групповой group/cluster switch
~, двухпозиционный on-off switch
~, двухполюсный double-pole/-contact switch
~ демпфера damper switch
~, дистанционный remote(-control) switch, teleswitch
~, дифманометрический differential-pressure switch
~ для открытой проводки surface switch
~ для скрытой проводки flush(-mounting) switch
~ заземления/заземляющий earthing/grounding switch
~, закрытый enclosed/safety switch
~, импульсный impulse switch
~, инерционный inertia switch
~, кнопочный/нажимной push-/press-button switch
~, концевой limit switch; *av.* propeller angle limit switch
~ коррекции гироскопа torque/erection switch
~, кулачковый cam(-operated) switch
~, линейный line circuit breaker
~, максимальный overload switch, overcurrent/overvoltage circuit breaker
~, малоемкностный anti-capacitance switch
~, маломасляный small-oil volume circuit breaker, low-oil content circuit breaker
~, масляный oil(-immersed) switch; oil circuit breaker
~ мгновенного действия snap (action) switch
~, минимальный underload switch
~, многоконтактный multi(ple)-contact switch
~, многопозиционный multi(ple)-position switch
~ на ответвлении branch switch
~ нагрузки load-breaking isolator
~ накала filament switch
~, ножевой knife(-blade) switch
~, ножной foot(-operated) switch
~, нормально замкнутый/разомкнутый normally closed/open switch, n.c./n.o. switch
~ обходов by-pass switch
~, однополюсный single-pole switch
~, пакетный packet switch

~, **перегрузочный** load limit switch
~, **перекидной** tumbler, toggle switch, flipflop switch
~ **перепуска** by-pass switch
~ **питания** power (supply) switch
~, **пневматический** pressure(-operated) switch
~, **поплавковый** float switch
~, **предельный** limit switch
~, **предохранительный** safety switch
~, **пусковой** starting switch
~, **путевой** limit switch
~, **релейный** relay switch
~, **ртутный** mercury switch
~ **с плавким предохранителем** fuse(d) switch
~ **с роговым искрогасителем** horngap switch
~ **с часовым механизмом** clock-controlled switch
~, **секционный** sectionalizing switch, sectionalizing circuit breaker, sectionalizer
~, **сетевой** mains switch
~, **силовой** power switch
~ **стабилизации числа М** Mach hold switch
~, **ступенчатый** step switch
~, **тепловой** thermal cut-out
~, **шиносоединительный** busbar coupler, bus-tie switch
~, **штепсельный** plug-in switch
~, **щелчковый** snap switch
~, **экстренный** emergency switch/release
~, **элегазовый** sulphur hexafluoride switch
~, **электромагнитный** solenoid-operated switch
~, **электромеханический** motor-operated sswitch
~ **электропитания** electrical power (supply) switch

выключать 1. switch/turn/cut/shut off, switch out; de-energize; disconnect, disengage, release; 2. exclude, remove, delete; 3. *print.* justify
~ **мотор** shut down/off the engine
~ **радио** turn off the radio

выключение 1. disconnection, turning/shut(ting) off, shut-down, switching-off, disconnecting, disengaging, disengagement, release, uncoupling; switch(ing)-off, cut(ting)-off/-out; declutching; 2. excluding, exclusion
~, **аварийное** emergency cutout; *nucl.* scram *(of the reactor);* disaster/safety shut-down, emergency shut-off
~, **автоматическое** automatic shut-down
~, **мгновенное** *nucl.* momentary trip/scram; instantaneous switching-off
~ **печати** *comp.* print suppress(ion)
~ **по периоду** period trip
~ **по уровню мощности** power-level trip, overpower scram
~ **подачи горючего** fuel cut-off
~, **предохранительное** safety shut-down
~ **реактора, быстрое** scram
~ **реактора, неполное** junior scram

выключенный disconnected, turned-/shut/switched-off; deenergized; disengaged, uncoupled, released

выключено "switched off", "off"; out; open

выключка *print.* (line) justification, justifying

выколачивание knocking/beating out; tapping; hammering up

выколачивать knock/beat out; tap

выколашивание *agr.* earing, ear formation

выколашиваться *agr.* form ears, ear

выколотка 1. knock out, (driving) punch, drift; driftbolt; 2. (shoemaker's) block

выкорчевывать grub up, uproot, stub

выкос *agr.* mowing

выкраска dyeing, colo(u)ring; painting

выкрашивание I flaking, spalling; chunking; crumbling; breaking off, chipping; pitting
~ **валков** *met.* spalling of rolls
~ **брони** armo(u)r scabbing
~, **усталостное** fatigue flaking/spalling

выкрашивание II dyeing, colo(u)ring, painting

выкрашиваться flake, spall; break off, chip; crumble away

выкристаллизовывание *geol.* efflorescence

выкристаллизовываться efflorescence

выкройка pattern; cutout

выкружка 1. *arch.* cavetto, cove, scotia; *mach.* bossing; 2. circuilar cutting

выкручивание evulsion, exeresis; wrenching; wringing out

выкрывание *print.* stopping-out

выкуп *law* redemption; ransom
~ **закладной** redemption of mortgage

выкупаемый *law* redeemable

выкупать redeem, buy out/back

выкупное *n.* ransom money

выкупной redemption *(attr.)*
~**ное право** right of redemption
~**ные платежи** *hist.* redemption money

выкуривание smoking (out); fumigation, fumigating

выкусыватель cutting forceps
~, **полуциркулярный** semicircular cutting forceps

вылавливать catch; recover

вылазка 1. *mil.* sally, sortie; 2. excursion, outing

выламывать breack out/open/off
~ **руку** dislocate one's arm

вылежалый 1. seasoned, mature, aged; 2. stale

вылеживание, вылежка 1. ag(e)ing, maturing, ripening weathering; 2. storing

вылет 1. departure; flight; flying out; sortie; *av.* take off; 2. *nucl.* escape, emission, emergence; 3. *mach.* boom; gap; 4. sweep
~ **балки** overhang of a beam
~, **боевой** (combat) sortie
~ **выброса** radius of ejection
~, **групповой** group sortie
~ **крана** radius of operation of crane
~ **на поиск** search mission
~ **на прикрытие** coverage sortie

~ на разведку reconnaissance sortie / mission
~ на спасение rescue mission
~ нейтронов escape of neutrons
~, ночной night mission
~, одиночный single / solo sortie
~, первый maiden / first / initial flight
~, плановый planned mission
~ по приборам, стандартный standard instrument departure, SID
~ с авианосца carrier-borne mission
~, самостоятельный solo flight
~ сварочной машины throat of a welder, gap depth of a welder
~, срочный urgent sortie
~ станка sweep of a lathe
~, учебный training mission
~ частицы leaking-out of a particle, escape of a particle
~ челнока flight of shuttle
вылетание flying out/off; *av.* take-off
вылетать 1. fly out/off; *av.* take off; 2. escape, leak out; 3. pop/dash/dart/rush out
вылечивать cure, heal up, recover
выливание 1. pouring out; 2. casting
выливать 1. pour out; discharge, empty; 2. *tech.* cast, found, mo(u)ld
выливка 1. tapping, pouring out; 2. casting
вылом 1. breaking out/open/off; 2. gap; opening; breach
выломка breaking (off); break; gap; breach
вылуживание tin-plating
вылуживать tin(-plate)
вылущение enucleation
вылущивание husking, shelling
вылущивать 1. husk, shell; pod; enucleate; 2. *surg.* remove
вымачивание 1. steeping, soaking; 2. maceration; retting
~ в рассоле pickle curing
~ льна retting
вымачивать macerate
вымбовка *naut.* capstan bar, handspike
выменивать exchange, interchange; swap, barter
вымерзание *agr.* frost injury; freezing
вымерший extinct
выметывание *agr.* ear emergence, tasseling, heading
~ метелок heading of panicles
выметывать throw/cast/put out; *agr.* head, tassel
~ икру spawn
~ колос ear
вымешивание mixing, blending; stirrng
вымирание extinction, dying out
вымирать die out, be extinct
вымогание extortion; blackmail
вымойна gully, wash-out, water gall, ravine, gulch
вымокать get wet, be soaked
вымолачивать thresh (out)
вымолот threshing (out); total yield of grain
вымораживание freezing (out); frost killing
~ волокна *text.* freezing of the fibre
~ зерен металла shotting
~ носителей зарядов carrier freeze-out
вымочка 1. steeping, soaking; retting; maceration; 2. wet spot
вымпел 1. pendant, *Am.* pennant; 2. *av.* droppable message bag; *mil.* streamer
~, боевой action pendant
~ военного корабля ship/commissioning pendant, masthead pendant
~, кодовый code pendant
~ Международного свода сигналов code flag
вымпелфал *naut.* pendant halyard/tackle
вымыв outwash
вымывание 1. washout, leaching, washing (out); sluicing, flushing; 2. *geol.* erosion, outwash; 3. bleeding *(of dye)*
~ дождем rainwash
~ почвы soil washing
вымывать wash out/off/away; elute; leach out
вымывка массы paper brown-stock flushing
вынашивание 1. bearing (a child); 2. taming and training
вынесение 1. removing, carrying out; 2. announcement
~ арбитражного/судебного решения adjudication
~ за скобки *math.* factoring out
~ приговора adjudg(e)ment, pronouncement of sentence
вынесенный outlying, remote
вынимать remove, take/pull out, extract; withdraw
вынос 1. removal, carrying out / away; 2. debris cone; 3. subtraction; loss; 4. efflux; evacuation
~, делювиальный talus train
~ крыла stagger
~, ледниковый glacier outwash
~ лопасти blade deflection
~ материала evacuation of material
~ пыли flue dust escape
~ реки river drift
~ руды ore loss
~ солей carry-over
~ устья colluvial deposit
~, химический chemical subtraction
~, эоловый eolian debris cone, eolian evacuation
выносить 1. remove, carry out/away/off, take away; *naut.* boom out; 2. bear, endure, withstand, stand up to, hold out, tolerate; 3. submit (for); pass; pronounce; hand down
~ благодарность express gratitude, thank
~ вердикт *law* return verdict
~ за знак take/place outside
~ за скобки *math.* factor out/from; put outside the brackets
~ множитель из-под знака корня remove a factor from the radicand
~ на берег wash ashore
~ на обсуждение submit for discussion

~ **на рынок** bring to market
~ **на поля** make a marginal note of
~ **под строчку** make a footnote
~ **приговор** pronounce/pass sentence
~ **решение** pass a resolution; *law* pronounce judg(e)ment
выноска I 1. (marginal) note, footnote; 2. removal, taking/carrying out
выноска II taming and training
выносливость durability, hardiness; strength, faltigue strength/life/resistance/limit
~ **авиационной конструкции** air-frame fatigue life
~, **акустическая** sonic endurance, acoustic fatigue resistance
~, **вибрационная** vibration endurance
~, **динамическая** dynamic fatigue strength
~ **материалов** fatigue behaviour of materials
~ **при изгибе** bending fatigue strength
~ **при кручении** twisting fatigue strength
~, **радиационная** radiation tolerance
~, **статическая** *av.* low-endurance fatigue
~, **ударная** impact endurance
~ **человека** human tolerance
выносящий efferent, deferent
~ **семяпроток** deferent duct
выношенный 1. mature, ripe; 2. threadbare, worn-out
~ **ребенок** child born at full term
вынуждать force, compel, oblige; constrain, coerce
вынужденный forced, constrained, compulsory; induced
~**ная посадка** forced/emergency landing
выпад 1. precipitation, fallout; 2. *sport* lunge, thrust
выпадание precipitation, fallout, falling/dropping out
выпадать 1. precipitate; fall/come/drop/slip out; separate, be rejected/separated; 2. occur, turn out/up; 3. *sport* lunge, thrust
~ **из синхронизма** fall out of synchronism/step
~ **хлопьями** flocculate
выпадение 1. fall-out, fallout, falling out; precipitation, separation; 2. *med.* abaissement, falling (down), descent, decensus, prolapse, prolapsus; procidentia; shedding
~ **бровей** eyebrow shedding
~ **влагалища** vaginal prolapse, colpoptosis, colpoptosia, falling of the vagina
~ **в осадок** settling out
~ **зубов** dedentition
~ **из памяти** loss of memory, blackout
~ **матки** prolapse/prolapsus of the uterus, uterine prolapse, hysteroptosis, metroptosis, falling of the womb
~ **матки и влагалища** frank prolapse
~ **межпозвоночного диска** disk prolapse, slipped/prolapsed disk
~ **мозга** prolapsus/protrusio cerebri
~ **мочевого пузыря** exocystis, vesical prolapse
~ **ногтей** onichoptosis
~ **осадка/осадков** precipitation
~ **пластов** *geol.* omission of beds/strata
~ **полукадра из кадра** *telev.* displacement out of frame
~ **прямой кишки** rectum/rectal prolapse
~ **пуповины** prolapse of the umbilical cord
~ **радиоактивной пыли** radioactive-dust fallout
~ **радиоактивных осадков** radioactive fallout
~ **разряда** bit loss
~ **ресниц** shedding of cilia, deplumation, ptilosis
~ **ручки плода** arm plolapse/procidentia
~ **сажи** soot fall
~ **сигнала** signal loss
~ **хлопьев/хлопьями** flocculation
выпаливание *agr.* weeding out
выпар (flash) steam, vapo(u)r
выпаривание 1. evaporation, vaporization,volatilization; concentration, boiling down; 3. steaming
~**нием, отделять/разделять** boil off
~**нием, сгущать** boil down, concentrate
выпариватель evaporator, vaporizer; concentrator
выпаривать 1. vaporize, evaporate, boil down /off; concentrate; 2. disinfect (by steaming)
~ **досуха** boil dry
выпаритель vaporizer
выпарка evaporation, vaporization; concentration; steaming
выпарки *pl.* evaporated product, concentrate
выпас *agr.* grazing; pasture
~, **общинный** common/community pasture
выпахивание 1. ploughing/plowing (up); 2. digging out
~, **ледниковое** *geol.* glacial ploughing/scouring/ excavation
выпечка baking; batch
~, **опытная** test baking
выпиливать saw (out/up); cut out; file out
выпинг *naut.* whipping
выпирание *geol.* protrusion, bulging
выпирать protrude, bulge/stick out; shove out
выписать 1. extract, excerpt, copy out; 2. order *(by mail);* subscribe to, send/write for; 3. discharge, strike off the list
~ **из больницы** discharge from the hospital
~ **новые каталоги** order new catalogues
~ **отрывки из книги** copy out passages from a book
~ **счет** bill
~ **счет-фактура** invoice
выписка 1. extract, excerpt; copying, writing out; 2. ordering, subscription; 3. discharge
~ **из госпиталя/больницы** discharge form hospital
~ **счета/счетов** billing; invoicing
выписывать write out
~ **вексель** draw a bill
выпись copy, extract
выплавка 1. making, manufacture, production; (s)melting, liquation; 2. smelt, smelted metal;

3. spalling *(defect)*
~ подшипника burning out of bearing
выплавление (s)melting, extraction; fusion
~, выборочное selective fusion
выплавлять melt out/down, extract by melting; smelt; make, produce, manufacture
выплавляться melt out
выплавок *met.* button, regulus
выпластывать *tech.* roll, laminate
выплата payment, payout; disbursement
~ авансом prepayment
~ в рассрочку payment by installments
~ взамен (другой формы возмещения) in-lieu
~ взятки peccadillo
~, промежуточная progress payment
~, прямая премиальная direct bonus payment
~ страховой суммы benefit payment
~ субсидии subsidy payout
выплатить pay (out/off)
~ в рассрочку pay by installments
~ по чеку negotiate a cheque
выплеск splashing
выплод breeding
выплыв(ок) burl, excrescence
выполаживание flattening
выполаживать(ся) smooth/flatten out
выполнение 1. implementation, carrying-out, execution, executing; performance; accomplishment, realization; completion; achievement, fulfil(l)ment; 2. *geol.* (in)filling
~ алгоритмов execution of algorithms
~ взлета taking-off
~, жильное vein filling
~ контракта impementation of a contract
~, корковое crusted filling
~ микропрограммы microprogram run
~, миндалевидное amygdaloid infilling
~ обязательств fulfil(l)ment of commitments
~, одновременное concurrent/parallel execution
~ операции *comp.* operation flow
~, повторное reexecution
~ посадки making/effecting a landing
~ программы *comp.* running of a program, flow of the program
~ пустот cavity filling
~ ремонта repair work
~ требований meeting the requirements
~ трещин *geol.* crevice accretion, fissure occupation, cavity filling
~, циклическое looping/cyclic execution
выполненный fulfilled, made
~ за одно целое с, (made) integral with,
~ по заказу custom-built/-made
выполнимость practicability, feasibility, executability; fulfil(l)ment; *math.* satisfiability
выполнимый feasible, practicable, executable, workable, accomplishable
выполнить! *comp.* do!
выполнять 1. complete; carry out, execute, accomplish, perform; fulfil(l); discharge; 2. fill(up)
~ вычисления perform computations
~ на заказ made to order
~ предписания comply with orders
~ приказ carry out an order
~ свое обещание keep one's promise
~ согласно техническим условиям make in accordance/conformity with specifications
~ тест take a test
~ указания follow the directions/instructions
~ условие satisfy a condition
~ функцию act/serve/function/operate as
выпор 1. *met.* air gate; flow-/pop-off vent; air hole/ cavity; riser, shrink(age) head; 2. bulge, swelling
выпороток *leath.* still-born calf-skin, deacon; fur of an unborn animal
выпот 1. *med.* exudate, exudation, sweat; 2. *tech.* leaking, sweat, ooze; 3. *geol.* sweating; fumarolic sublimate
~ олова из чушек sweat tin bars
~ соли incrustation of salt
выпотевание 1. sweating, exudation; 2. *tech.* oozing; blooming
выправка 1. correction; 2. rectification, smoothing, straightening, flattening; 3. bearing, deportment; training
~, военная military bearing
~ пути, местная spot surfacing of track
выправление straightening; correcting, correction
~ матки *med.* basculation
~ расфокусирования debunching correction
выправлять 1. straighten/smooth/flatten out; even, level; 2. bring out/forward, display; 3. correct, set right, true up, improve
выпрессовка 1. *plast.* release, spew; flow, fin, (flash) overflow; 2. pressing/squeezing out
~ на шине tire buffing
выпрессовывание pressing off, squeezing
выпрессовывать press off, squeeze
выпрямитель 1. *el.* rectifier; rectifier circuit/unit; 2. straightener
~ анодного питания anode-/plate-supply rectifier
~, вибрационный vibrating rectifier
~, газотронный gas-discharge rectifier, gas-filled tube rectifier, gas-filled thermonic-diode rectifier/unit, gas-diode rectifier/unit
~, германиевый germanium rectifier
~, двухполупериодный full-wave rectifier, biphase rectifier
~, диодный diode rectifier
~, зарядный battery charger, battery charging rectifier
~, зарядный буферный trickle charger
~, импульсный pulse rectifier
~, кенотронный thermionic rectifier, high vacuum rectifier
~, кремниевый silicon rectifier
~, купроксный copper-/cuprous-oxide rectifier, copper-sulphide rectifier

~, ламповый tube/electron/valve/thermionic rectifier, vacuum-tube rectifier
~, магнетронный rectifying magnetron
~, мостиковый bridge rectifier
~, мощный heavy-duty rectifier, power rectifier
~, однополупериодный single-phase rectifier, half-wave rectifier
~, однофазный single-phase rectifier
~, плоскостной surface-contact rectifer
~, полупроводниковый semiconductor rectifier
~ по мостовой схеме bridge rectifier
~, путевой *rail.* track rectifer
~, ртутный mercury(-arc) rectifer (unit)
~ с запирающим слоем barrier-layer rectifier
~ с обратной связью feedback(-regulated) rectifer
~, сварочный welding rectifier
~, селеновый selenium rectifier
~ смещения bias rectifier
~, твердый dry rectifier
~, тиратронный thyratron rectifer
~, тиристорный thyristor rectifier
~ трехфазного тока three-phase rectifier
~, усредняющий averaging rectifier
~, электроконтактный mechanical rectifier
выпрямительный rectifying; righting
~ элемент valve/rectifying cell
выпрямление 1. *el.* rectification, rectifying; 2. unbending, straightening(-out); 3. *physiol.* erection
~, двухполупериодное full-wave rectification
~, диодное diode detection/rectification
~, однополупериодное half-wave rectification
выпрямлять 1. *el.* rectify; 2. *mech.* straighten, set upright/erect
выпукло-вогнутый convexo-concave
выпукло-изогнутый upwarped
выпукло-компактный convexo-compact
выпукло-плоский convexo-plane
выпукло-створчатый *bot.* induplicate
выпуклость 1. convexity, camber; 2. curvature, protuberance, bellying, bulge, bulging(-out), prominence, salience, swell(ing); 3. *art* relief; 4. *met.* side strain
~ боковой кромки *met.* side camber
~ валка *met.* crown
выпуклый 1. convex; embossed, raised, gibbous, domed, arched; pulvinate; 2. protruding, protuberant, prominent, eminentail, sailent; 3. *art* in relief
~ вверх convex upwards
~ вниз convex down(wards)
выпуск 1. release, discharge, freeing; 2. escape, outlet, exhaust, issue; drain; emission; 3. yield, output, production; delivery; 4. number, edition; issue; 5. *met.* tapping; 6. *educ.* graduating class, leavers, graduates
~ арматуры free length of reinforcement bars
~ воздуха air discharge
~, массовый large output
~ неуспокоенной плавки *met.* open tapping
~ парашюта parachute deployment
~ плавки heat tapping, hot metal tapping
~ стены projection of wall
~, текущий flow output
~ тормозного парашюта drogue-chute ejection
~ шасси undercarriage extension/lowering
~ шлака tapping of slag
выпускание discharging, discharge; evacuation; emission
выпускать 1. empty, discharge, drain, exhaust, release, set free, emit, liberate; 2. make, manufacture, produce, put/turn out, issue; 3. *met.* tap; 4. omit, skip, cut (out)
~ в обращение put into circulation
~ из тюрьмы release from jail
~ кинокартину release a film
~ листья put forth leaves, *bot.* frondesce
~ на рынок put on the market
~ новый заем float a new loan
~ серийно produce in lots/quantity
выпускаться be made/come in modifications
выпускник 1. graduating student; holder of school leaving certificate; 2. emissary
выпускной 1. release, exhaust, etc. *see* **выпуск**; 2. *n.* graduating student
~ класс graduation class
~ клапан exhaust pipe
~ кран discharge cock, bleeder
~ экзамен final exam(ination)
~ная труба exhaust pipe
~ная цена *econ.* market price
выпучивание buckling, bulging(-out), bulge, puff-up, upwarping; swelling
~ земной коры crustal warping
~ стержня buckling of a column
выпучина bulge
выпь *zool.* bittern *(Botaurus)*
выпячивание 1. *anat.* diverticulum; haustrum; 2. bulging up, bulge, crust warping, evagination; protrusion, protruding, protuberance
~, гипсовое gypsum bulge
~ куполообразное слепое cecum cupulare
~ куполов domical protrusion
~, ложное false diverticulum
~ ободочной кишки haustrum coli
~ радужной оболочки iridocele, hernia of the iris
~ спинного мозга myelocele
~ уретры, мешковидное urethrocele
~ хрусталика lenticonus
вырабатывание making, manufacture, production; development
вырабатывать 1. make, manufacture, produce, turn out; generate; 2. work out *(a mine);* 3. draw up; develop; 4. exhaust, deplete; 5. earn, make
~ иммунитет develop immunity
~ информацию produce/generate data/information
~ план draw up a plan

выработка 1. manufacture, production, making; formation; generation; 2. exhaustion, depletion; wear; 3. output, yield; 4. working out, drawing up; elaboration; development, preparation; 5. *min.* roadway, entry, heading, working, opening, mining, excavation
- **~, аккумулирующая** main roadway
- **~, буровая** drilling room
- **~, вентиляционная** air heading, airway
- **~, вентиляционная входная** intake airway
- **~, вертикальная** vertical opening
- **~, восстающая** entry, raise
- **~, годовая** *econ.* yearly output
- **~, горизонтальная** level/horizontal working
- **~, горная** underground working
- **~ для разминовки** branch heading
- **~, капитальная** permanent working/opening
- **~, наклонная** incline working slope
- **~, обходная** by-way
- **~, откаточная** haulageway
- **~, открытая горная** openacast mining, strip mining
- **~, погашенная** filled stope, gob working
- **~, подготовительная** development working
- **~, полевая** lateral working
- **~ политики** *econ.* policy formulation
- **~ понятий** concept formation
- **~, разведочная** exploring/exploratory working/ opening; prospecting/proving hole
- **~ рефлексов уклонения** avoidance conditioning
- **~ топлива** fuel utilization/usage
- **~, сборная** main roadway
- **~, скреперная** scraper working

выравнивание levelling, planation; flattening; levelling-off; balancing, alignment, matching, compensation; *comp.* allocation, adjustment; justification
- **~, боковое** lateral planation
- **~, волновое** wave planation
- **~ давления** pressure equalization
- **~ данных** data smoothing
- **~ емкостей** capacitance balancing
- **~ знаков** *comp.* character alignment
- **~, изостатическое** *geod.* isostatic adjustment/ compensation
- **~ команд** instruction alignment
- **~ кривизны поля изображения** flattening of image field
- **~, ледниковое** glacial planation
- **~ листа** *comp.* alignment of the sheet
- **~ отметок** *rad.* blip matching
- **~ по первому/последнему знаку** left/right (-hand) justification
- **~ потока** flux flattening
- **~, поперечное** *agr.* cross(wise) levelling
- **~ по расчетной траектории** *av.* computed flare
- **~, преждевременное** *av.* premature flare
- **~, продольное** *agr.* fore-aft levelling

выравниватель 1. *el.* equalizer; 2. *agr.* land leveller; levelling agent; dye leveller
- **~ затухания** attenuation equalizer
- **~ фазы** phase compensator

выравнивать 1. smooth out/down, level, make even, flatten, straighten; 2. line up, align, adjust, balance, trim; match, equalize; 3. *av.* flare out, level off, redress; correct, match
- **~ валки** *met.* align rolls
- **~ вешки по высоте** *geod.* bring to grade
- **~ давление** equalize pressure
- **~ дорогу** level the road
- **~ порядки** *math.* match exponents
- **~ поток** straighten the flow

выравниваться straighten/flatten out; become level/even; *mil.* form up, align; dress; *sport* equalize

выражать express; convey; voice; show
- **~ в процентах** express as a percentage, express in per cent

выражение expression; *math.* term
- **~, алгебраическое** algebraic expression
- **~, алфавитно-цифровое** alphanumeric expression
- **~, аналитическое** analytic(al) expression
- **~, билинейное** bilinear expression
- **~ в замкнутой форме** closed-form expression
- **~ в конечном виде** finite formula, expression in closed form
- **~ в скобках** bracketed/parenthetical expression
- **~, дифференциальное** differential expression
- **~, именующее** designational expression
- **~, истинное** true expression
- **~, конечноразностное** finite-difference expression
- **~, логическое** logical expression
- **~ над массивами** *comp.* array expression
- **~, образное** *ling.* trope
- **~, обратное** reciprocal expression
- **~, определяемое** definiendum
- **~, определяющее** definens
- **~, основное** *comp.* base record
- **~ отношения/сравнения** relational expression
- **~ периода компиляции/трансляции** *comp.* compile-time expression
- **~, подкоренное** radicand, expression under the radical sign, subradical expression
- **~, подынтегральное** integrand, element of integration, expression under the integral sign
- **~, приближенное** approximate expression
- **~, простое логическое** logical element, logical elementary expression
- **~ с индексами** *comp.* subscript expression
- **~, содержательное** semantic expression
- **~, степенное** exponentiation, exponential expression
- **~, физическое** physical terms
- **~, хорновское** Horn clause
- **~, численное** numerical expression

выраженный expressed; pronounced; marked;

salient
~ **резко** strongly pronounced
~ **слабо** feebly marked
~ **через** expressed in terms of
~ **четко** clearly defined/marked
выразительность expressiveness; distinctiveness
выразительный expressive; distinctive
~**ное чтение** *educ.* style of reading, elocution
выращивание 1. growing, growth; 2. *agr.* cultivation, culture, raising, bringing up, rearing; breeding
~ **в термостате** incubation
~ **кристаллов** crystal growth
~ **микробов** bacterial cultivation
~ **микробов в висячей капле** (hanging-)drop culture
~, **эпитаксиальное** *cryst.* epitaxial growth
~ **подпокровных культур** cover cropping
выращивать grow, raise, cultivate; rear, bring up; breed; incubate
вырез cut-out, cutting out, opening, notch, incision; groove; indent; undercut; emargination
вырезанный *med.* excised
вырезать cut out, excise; exsect; eliminate; carve, engrave
~ **зубцы** *mach.* scallop
~ **образцы (металла) для испытаний** remove coupons for test
вырезка 1. cutting out; 2. *med.* incisure, notch, emargination; 3. clipping; carving, engraving; 4. *food* tenderloin; fillet
~, **блоковая** trochlear notch
~, **ключичная** clavicular notch
~, **межкозелковая** intertragic notch
~, **надглазничная** supraorbital notch
~, **носовая** nasal notch
~, **седалищная** sciatic notch
~, **сердечная** cardial notch
~, **слезная** lacrimal notch
~, **синусная** sinus slit/notch
~, **сосцевидная** mastoid notch, digastric groove
~, **теменная** parietal notch
~, **яремная** jugular notch
вырезывание cutting out/off, excision, exsection; elimination; carving; engraving, grooving
выровненный smoothed, evened, levelled
выродившийся degenerate(d)
вырождаться degenerate (into); die out
вырождение degeneracy, degeneration; degradation; deterioration; regression
~ **крупной породы** dwarfing
~, **случайное** accidental degeneracy
~, **трехкратное** threefold degeneracy
вырожденный 1. *phys.* degenerate(d); 2. *math.* confluent, singular
вырост 1. emergence, protuberance, excrescence; apophysis, (out)growth; 2. growing
~ **апертуры у головоногих** *zool.* auricle
~, **бахромчатый** frill-like projecton, marginal crest
~, **волосовидный** cilium
~, **корневой** root nodule/tubercle
~, **крыловидный** alar prjection/prolongation
~ **на талломе** *bot.* enation
~, **нитевидный анастомозирующий** pseudopodium
~, **палочковидный** pilum
~, **пальцевидный** finger-like projection
~ **пыльцы** *bot.* pollen projection
~, **ресничатый** cilium
~, **спиральный у плеченогих** brachium
~ **у фораминифер** protuberance
выросток 1. *agr.* year-old calf; heifer; 2. *leath.* kip, dressed calf leather
вырубание 1. *tech.* punching, stamping; 2. cutting down/out
вырубать cut/punch/hack out, stamp, chisel; *min.* cut, dig
~ **дефекты сварного шва** chip out
~ **лес** fell trees
~ **отверстия** punch holes
вырубка 1. cutting down/out, felling; hewing out; indentation, notching; chiselling; punching, stamping; 2. *print.* die-cutting; 3. *pl.* scrap, ends, trimmings
~ **деталей обуви** blocking of a shoe
~ **дефектов на металле** chipping out
~ **заготовок** blanking
~ **леса** deforestation
~, **пневматическая** chipping with an air-operated chisel/gouge
вырубной cutting/stamping/punching out
~ **штамп** blanking/cutting die
выруливание *av.* taxy(ing)-out
~ **для взлета** taxying-out for take-off
выруливать *av.* taxi (out)
выручатель *eccl.* rescuer
выручать 1. rescue, bail out, come to the help of; 2. (re)gain, make, net; recover, get back
~ **из беды** help out of difficulty
~ **свои деньги** recover one's money
выручка 1. rescue, assistance, aid, bail-out; 2. gain; receipts, proceeds; earnings
~, **валовая** *econ.* gross proceeds
~, **годовая** annual receipts
~ **коммунальных предприятий** utility revenues
~ **от реализации** sales proceeds
~ **от текущих продаж** current sales proceeds
~, **чистая** net yield
вырывание 1. extraction, pulling out; drawing; uprooting; evulson; 2. digging (up)
вырывать 1. excavate, dig out; 2. break away/loose from; tear/pull/wrest out, extract; uproot; grab
вырытие exhumation; digging up
высадка 1. *av.* landing, debarkation, disembarkation; detrainment; 2. *met.* heading; upset(ting); 3. *hort.* transplantation, transplanting, planting out; transplanted/bedding plant(s)
~ **в грунт** bedding out
~ **воздушного десанта** airborne landing/assualt

~, **выборочная** *av.* selective landing
~ **десанта** landing operation/attack
~ **десанта, ложная** diversionary landing
~ **морского десанта** seaborne assualt/attack/invasion/landing
~ **рассады** transplanting of seedlings
~ **холодная/горячая** cold/hot heading, cold/hot upset(ting)
высадок *hort.* rootstock; bedding/transplanted plant
высаживание 1. disembarkation; landing; unloading of passengers; 2. *hort.* transplanting; 3. *chem.* precipitation
высаживать 1. put off, detrain, debus; help/set down; 2. *hort.* bed/set out, transplant
~ **головку болта** head a bolt
~ **десант** *mil.* make a landing
~ **с места службы** oust from a post
высаживаться debark, disembark, get off
~ **на берег** beach
высаливание I gresing, lubricaition
высаливание II salting out
~ **мыла,** ~ **мыльного ядра** soap graining
высасывание drawing/sucking out
высасывать exhaust, evacuate; suck out/dry; draw out/off
высачивание seepage, escape, exudation; filtration; oozing, leaking
~ **воды** water seepage, outcrop of water
~ **нефти** oil show/ filtration
высверливание drilling
высверливать drill (out), bore (out)
высвечивание de-excitation; fluorescence, luminescence, scintillation; afterglow
~ **гамма-квантов** gamma-quanta release
высвечивать de-excite
высвобождать (set) free, liberate; release; disentangle, disengage; make available
высвобождение liberation; detachment; *chem.* release
~ **электронов** detachment of electrons
~ **энергии** liberation of energy
высев(ка) sowing; seeding; quanitity of seed sown
~, **верхний/нижний** *agr.* upper/under seeding
высевки *pl.* screenings, siftings, bran; chaff; dross; refuse, fines
~ **коксовые** coke breeze
~ **щебня** crushed-stone screenings
высекать hew/cut (out), carve, sculpture; punch, stamp
выселение law ejectment, ejection; eviction, expulsion
высечка 1. hewing, cutting-out/off, carving; 2. *min.* trench, cutting; 3. sclupture; 4. *tech.* cutting instrument; blank; *print.* die-cutting
высиживание hatching, incubation, brooding
высиживать hatch (out), incubate; brood
выскабливание scraping/cleaning (out); abrasion; attrition; *med.* curetage
~ **полости** excochleation
выскабливать 1. scrape out/off; abrade; erase; *med.* curette; 2. adz(e)
высказывание 1. assertion, utterance, statement, dictum; opinion, view; 2. *math.* expression; sentence; *comp.* proposition
~, **истинное** true proposition
~, **основное/исходное** base assertion
~, **переменное** sentential variable
~, **сложное** *comp.* compound proposition
~, **составлять** form a sentence
~, **частное** particular proposition
~, **экзистенциальное** existential sentence
высказывать express, state; speak out, voice
выскакивать pop out (of button)
выслеживание shadowing operation; tracking down
выслуга period of service
~**гу лет, выплата за** long-service bonus
~**гу лет, за** for meritorious service, for length of service
выслушание *law* hearing
выслушивание audition, hearing, listening; *med.* auscultation
~ **стетоскопом** mediate auscultation
выслушивать lesten (to), hear out; *med.* auscultate, auscult, sound
высматривать scrutinize, examine, look through
высокий high; tall elevated; sublime; *mus.* high (-pitched), sharp
~ **гласный** *phon.* high vowel
~ **гость** hono(u)red guest
~**кие договаривающиеся стороны** *dipl.* the High Contracting Parties
~ **стиль** elevated style
высокоактивный highly (radio)active; *nucl.* hot
высокоблагородие:
~, **ваше** (Your)Hono(u)r/Worship
высокобонитетный high-bonitat
высокобортный high-sided
высокобризантный highly explosive
высоковакуумный high-vacuum
высоковирулентный highly virulent
высоковольтный high-voltage/-tension
~**ной передачи, линия** high-voltage line
высоковязкий high-viscosity
высокоглиноземистый high-alumina
высокогорный Alpine, mountainous, (high-)mountain; high-altitude
высокогорье highland, high mountains
высокоградусный high-grade/proof
высокогумусовый high-humic
высокодисперсный high-dispersoid; highly dispersive; micro-atomized
высокодоходный highly remunerative/profitable
высокое *n.* the sublime
высокозольный high-ash, very ashy
высокоидейный high-principled
высокоинтенсивный high-intensity
высококалорийный high-caloric/-calorific
высококачественный high-class/-grade/-quality/-performance

высококвалифицированный highly qualifed/skilled
высококипящий high-boiling
высококонтрастный high-contrast
высококоэрцитивный high-coercitivity
высококремнеземистый high-silica/-silicon
высококремнистый highly siliceous
высоколегированный high-alloy, highly alloyed
высокоманевренный highly manoevrable
высокометаморфизованный highly metamorphic
высокомолекулярный high/highly molecular
высокомощный high-power/-capacity; heavy-duty
высоконапорный high-pressure
высоконапряженный highly stressed
высоконасыщенный highly saturated
высокообогащенный highly concentrated/encriched
высокообразованный highly educated
высокооборотный high-speed
высокообъемный high-volume; bulky
высокоогнеупорный high-refractory, highly refractory
высокооктановый high-octane
высокоомный high-resistance/-impedance
высокоосновный high-base
высокоочищенный highly refined/purified
высокопитательный highly nutritious/nourishing
высокоплавкий high-melting/-fusing
высокоплан *av.* high-wing monoplane
~, свободнонесущий cantilever high-wing monoplane
высокопластичный high-plasticity, highly plastic
высокополимер high(-molecular) polymer, superpolymer, eupolymer
высокополимерный high-polymeric
высокопоставленный high-ranking, of high standing
~ное лицо very important person
высокопотенциальный *el.* high-potential
высокопревосходительство Excellency
высокопреосвященство Eminence, Grace
высокопреподобие Reverence
высокопробный high-grade/-standard; sterling
высокопродуктивный highly productive/efficient
~ное животное *agr.* high/target yielder
высокопроизводительный highly efficient/productive; high-production/-performance, heavy duty
высокопроницаемый high-permeability
высокопроходной high-pass
высокопроцентный high-percentage/-grade
высокопрочный high-strength/-grade
высокорадиоактивный highly radioactive
высокоразвитый highly developed, advanced
высокорезистивный high-resistance
высокорентабельный lucrative, highly remunerative/profitable
высокородие Hono(u)r, Worship
высокосернистый high-sulfur/sulphur
высокоскоростной high-speed/-velocity
высокосмоляной highly resinous; high-tar
высокосортный high-grade, of superior quality
высокоствольник high forest
высокоствольный high-standing, tall-trunked
высокостебельный, высокостебельчатый tall-stalked
высокостойкий highly stable
высокотвердый very hard, highly rigid
высокотемпературный high-temperature
высокотехничный high-technology
высокотоварный producing high marketable surplus
высокотоксичный highly toxic
высокотравный *agr.* high-/tall-grass
высокотравье *agr.* high/tall grass, tall herbaceous vegetation
высокоуважаемый highly esteemed, hono(u)red, respected
высокоуглеродистый high-carbon
высокоурожайный high-yielding
~ные культуры *agr.* high/target yielders, high-yield crops
высокофосфористый high-phosphorous
высокочастотный *el.* high-frequency
высокочувствительный highly/extremely sensitive, supersensitive; high-sensitivity *(attr.)*
высокоэластический, высокоэластичный highly elastic
высокоэнергетический high-energy
высокоэффективный high-performance, highly effective
высолаживание sweeting off; leaching
высолаживать desugar, sweeten off; leach
высота 1. altitude, height; 2. elevation, eminence; 3. high level/quality; 4. *mus.* pitch; *see also* **высоты**
~, абсолютная true altitude
~ антенны height of an aerial
~ антенны, действующая aerial effective height
~ апогея altitude of apogee
~ апоселения apocynthion/apolune altitude
~ бара cutting height
~, барометрическая barometric/pressure altitude
~, близмеридиональная (светила) exmeridian/circummeridian altitude
~ берда *text.* depth of the reed
~, большая high altitude
~ борта судна depth, hull height
~ борта, расчетная теоретическая moulded depth
~ в свету inner/free interval
~, вертикальная *geol.* vertical/normal throw, amount of throw, vertical displacement
~ висения *av.* hover height
~ волны wave height
~ всасывания suction head/lift
~, выдерживаемая *av.* base altitude

~, вычисленная calculated altitude
~, габаритная overall/clearance height
~ головки зуба шестерни (tooth) addendum
~ единицы передачи height of a transfer unit
~ завалки *met.* charging level
~ звука pitch of sound/tone
~ зоны охлаждения dew-point depression
~ зуба, рабочая working/contact depth of a tooth
~ инструмента *top.* elevation of sight, height/horizon of instrument
~, истинная absolute altitude
~, исходная reference altitude
~, командная commanding ground
~ крейсерская *av.* cruising height/altitude
~, магнитометрическая/магнитная magnetomeric high
~, малая low altitude
~, меридиональная meridian altitude
~, наблюдаемая apparent altitude
~ нагнетания pumping head
~ над поверхностью земли, истинная terrain clearance altitude
~ над уровнем моря sea level elevation
~ надводного борта *naut.* free-board
~, начальная initial/starting altitude
~ начинка *text.* length of a cop bit/bottom
~ небесного светила celestial altitude
~ ножки зуба (шестерни) (tooth) dedendum
~, нормальная (барометра) par line
~, нулевая zero altitude
~, общая overall/total height
~ ожидания посадки *av.* holding altitude
~, околомеридианная circummeridian altitude
~ орбиты orbital altitude/height
~ от поверхности геоида geoidal rise
~ очка (литеры) *print.* depth of a letter/body
~ паводка, максимальная flood level
~ парения *av.* hovering height
~ перигея altitude at perigee
~ периселения pericynthion/perilune altitude
~ платформы грузового автомобиля truck-bed height
~ по давлению pressure altitude
~ по масштабу/шкале scale height
~ подачи насоса lift of a pump
~ подъема 1. lifting height; 2. *naut.* height off water, clearance height; 3. hoisting height, height of lift, elevation
~ подъема домкрата lift of a jack
~ подъема клапана degree/height of valve lift
~ подъема крана stroke of a crane
~ подэтажа *min.* sublevel interval
~ полета flight altitude
~ помещения head room
~ поршня piston depth
~, предельная extreme/limiting altitude
~, приборная indicated altitude
~, приведенная reduced height
~, пьезометрическая piezometric head/height/level
~ разливки *met.* pouring height
~, разрешенная *av.* cleared altitude
~ самовсасывания насоса self-suction lift
~ сброса 1. *av.* release height; 2. *geol.* fault throw
~ сброса, нормальная к напластованию stratigraphic throw/separation
~ сброса по простиранию *geol.* total heave
~ сегмента *geod.* middle ordinate
~ сечения cross-sectional height
~ сечения рельефа contour/vertical interval
~ сечения, эффективная effective depth of section
~ сизигийного прилива spring(s) rise
~ складки *geol.* height of fold
~ слоя атмосферных осадков height of precipitation
~, смежная adjoining/adjacent height
~ смещения в плоскости сброса *geol.* lateral separation
~ Солнца solar angle
~ тела в положении сидя/стоя sitting/standing height
~ тона pitch of tone
~ тона биения pitch of beat note
~ тона эха echo pitch
~ треугольника altitude of a triangle
~, угловая angle of elevation/altitude; angular altitude
~, уравненная *geod.* standard elevation/height
~ устья скважины elevation of well
~ фермы depth of truss
~ флора *nav.* depth of floor
~ цели target altitude
~ центров *mach.* swing/ *Am.* height of centres/centers
~ шва, расчетная *weld.* (theoretical) throat
~ штабелевки piling height
~, эквивалентная equivalent altitude
~ этажной укладки stacking height
~, эффективная effective height
высотность altitude performance/capability/range; altitude, height
~ двигателя critical height of an engine
~ сопла design altitude of a nozzle
высотный 1. high-altitude/-rise/-level/-flying/pitched; altitudinal; 2. tall, high
высотомер *av.* altimeter; *mil.* height finder
~, абсолютный ground-clearance indicator
~, авиационный aircraft/airborne altimeter
~, акустический acoustic/sonic altimeter
~, анероидный aneroid altimeter
~, барометрический barometric/pressure/aneroid altimeter
~, высокочастотный/высокочувствительный (high-)sensitive altimeter
~, двухстрелочный two-pointer altimeter
~, двухшкальный dual-scale altimeter
~, емкостный (electrical-)capacity altimeter
~, звуковой sonar altimeter

~, импульсный pulse(d) altimeter
~, контактный contacting altimeter, altitude switch
~ малых высот low-altitude/-level/-range altimeter
~ на твердых схемах/телах (all-)solid-state altimeter
~, посадочный landing altimeter
~, радиолокационный radar altimeter
~, радиотехнический radio altimeter
~, резервный stand-by altimeter
~ с барабанно-стрелочным отсчетом drum-pointer altimeter
~ с вертикальной шкалой vertical-scale/-reading altimeter
~ с кодированием данных encoding altimeter
~ с круглой шкалой dial altimeter
~ с сельсинной передачей servo altimeter
~ с частотной модуляцией frequency-modulated altimeter
~, самолетный aircraft/airborne altimeter
~, самопишущий altigraph, altitude recorder; recording altimeter
~ скафандра suit altimeter
~ со стрелкой и счетчиком counter-pointer altimeter, single-needle altimeter, digital altimeter
~, счетно-стрелочный counter-pointer altimeter, digital altimeter, single-needle altimeter
~, тарировочный/эталонный calibartion altimeter
~, точный precise altimeter
~, широкодиапазонный full-range altimeter
~, штатный standard altimeter
~, электростатический electrostatic altimeter
~, эталонный calibration altimeter
высоты *pl.* altitudes
~, горные mountain altitudes
~, ежечасные hourly heights
~ от поверхности геоида values of the geoid
~, соответствующие equal altitudes
высотописец altigraph, altitude recorder; recording altimeter
высочайший imperial, royal; highest
~шая воля superme will
высочество Highness
~, Ваше Your Highness
выставка 1. alignment, adjustment, setting; 2. exhibition, show; display; show window
~, авиационная air show/display
~ в нулевое положение setting to zero position
~, всемирная world fair
~, грубая coarse alignment
~ картин exhibition of paintings
~ нулей zero adjustment
~ по азимуту azimuth adjustment
~ ракет misile display
~, точная fine alignment
выставление 1. showing, displaying; 2. nomination
выставлять 1. adjust, set; 2. display, exhibit, show; bring out/forward; set/put out; 3. *mil.* post, mount; 4. propose, bring up; lay down, adduce
~ аргументы adduce arguments
~ возражение raise an objection
~ караул post the guard
~ оценку assess
~ счет bill
~ требования lay down one's demands
~ часового set a sentinel
выставной 1. removable; 2. exhibited
~ное окно show window
выставочный exhibition, show, display *(attr.)*
~ зал showroom
~ комитет exhibition/hanging committee
выстилать pave, cover, lay, floor; line
выстилка 1. lining, covering, paving; bed; 2. corduroy
~ пода шлаковика slag bottom paving
~ фашинная fascine corduroy
~, хворостяная brush corduroy
~ хитиновая chitinous lining
выстраивание ядер nuclear alignment
выстрел 1. shot, fire, discharge; report; 2. *naut.* lower/swinging boom, bum, outrigger
~ беспламенный flashless discharge
~, боевой war shot
~, кормовой quarter shot
~ настильный grazing shot
~, предупредительный warning shot
~, пристрелочный ranging shot
~, учебный exercise shot
~, холостой blank shot
~, шлюпочный boat/riding boom
выстрел-балка boom davit
выстрел-брас lower-boom fore guy
выстреливание discharge
выстреливать launch; shoot, fire
выстрел-топенант lower-boom topping lift
выстукивание *med.* percussion, tapping
выстукивать *med.* tap, percuss
выступ 1. bulge, projection, prominence, protrusion; protuberance; flange; overhang; ledge; ridge; 2. *tech.* boss; lug, cam, cog
~, антиклинальный anticlinical nose/bowing
~, базальный *bot.* mentum
~, береговой shore cusp
~, гортанный *anat.* laryngeal prominence, Adam's apple
~ горловины shoulder
~ дупликатуры, внутренний selvage
~ для ануса anal tube
~ дна bottom swell
~, зубоподобный dentation
~, изоклинальный isoclinal ridge, hogback
~ кораллого рифа knoll
~, костный bony prominence
~ кулачка lobe of a cam
~ молоточка *anat.* mallear prominence
~, моноклинальный monoclinial nose
~ несущий анус anal sac

~, обтекаемый bulge
~, опорный stop lug
~, пальцевый digitation
~, подбородочный *anat.* mental protuberance/prominence/process
~, полюсный *el.* pole horn
~, роговидный horn-like process
~, сбросовый heaved block
~, секторный door lug
~ скалы bench, ledge
~, структурный nose
~, суставной condyle
~ ударника, боевой firing-pin plug
~ фланца flange ridge
~ фронта *mil.* salient
~ холма brow of hill
~, черный black ledge
~, эрозионный erosion scarp
выступание protrusion, protruding; jutting/sticking/crooping out; emerging
выступать 1. project, protrude, bulge/jut/come/break out; come/step forward; emerge, appear; 2. overhang; 3. appear publicly, advocate, favo(u)r
~ в защиту stand up (for), speak in defence (of)
~ вперед step forward
~ защитником *law* appear for the defence
~ на собрание address a meeting
~ по радио speak over the radio; broadcast
~ с докладом give a lecture
~ с протестом protest
выступающий prominent; protruding, projecting, outjutting; *naut.* off-lying; salient; sticking out
выступление 1. address, speech, statement; 2. appearance; performance
~, первое first performance, debut
высушивание, высушка drying (out), exsiccation, desiccation, dehumidification, dehydration
~ возгонкой cold drying
~ осушающими веществами cold drying
высушивать dry (out/up); exsiccate, desiccate; (case)-harden; *min.* cure
высчитывать compute; calculate
высший 1. high(er); highest; advanced; supreme; 2. *n.* superior
~ пилотаж aerobatics
~шая инстанция the highest authorirty; *law* court of last resort
~шая математика calculus, higher mathematics
~шая мера наказания supreme/death penalty
~шая школа university, higher school
~шее качество highest quality
~шее образование higher/university education
~шее общество (high) society
~шее учебное заведение university, higher education esablishment/institution
высылать 1. send (out), dispatch; 2. *polit.* deport, exile, banish
высылка 1. sending, dispatching; 2. deportation, exile, banishment
высыпание 1. spilling, emptying, pouring out; 2. *med.* eruption, breaking out; efflorescence
высыпать empty, pour out
высыпка *hunting* descent *(of birds)*
высыпь *med.* eruption, rash, breaking out
~, кожная skin eruption
высыхание desiccation, exsiccation, drying up/out
~ тела mummification
высыхать dry out/up, run/get dry
выталкивание forcing/pushing out, extrusion, expulsion, ejecting, ejection
выталкиватель pusher, ejector, push rod, extruder
~, печной furnace pusher
~ семян *agr.* seed ejector blade
~ слитков ingot pusher
~ центрального литника sprue ejector
выталкивать push/force out, eject, extrude
выталкивающий expelling
вытапливание 1. warming/heating up; 2. melting down, liquation; *min.* sweating (of ore)
вытапливать 1. melt out/down; 2. *met.* liquate
~ жир render/melt fat
~ масло clarify butter
вытаптывание trampling down
вытаскивать drag/pull/draw/take out; extract, haul out, withdraw
вытачивать turn (on a lathe); bore out; cut
вытачка tuck; fold
вытек bleeding, diffusion, run(ning) out
вытекание efflux, issue, escape, effluence; outflow, running/flowing out; bleeding; leak(age)
~ красителя bleeding of dye
~ металла metal run-out
~ сильной струей spurting
вытекать 1. flow/run out, discharge, escape, leak/drip out; 2. *math.* imply, follow, be a consequence; 3. *bot.* shed seeds
~ сильной струей sprut, spout
вытекающий 1. effluent; 2. resultant
~ логически consequential
вытеснение displacement, replacement; expulsion, dislodgement; ousting, exclusion, supplanting, substitution
~ катиона cationic displacement
~, метасоматическое metasomatism, metasomatosis
~ оснований depletion of bases
~ цинка кадмием substitution of cadmium for zinc
~ флюида fluid expulsion
вытеснитель 1. *chem.* extractant, displacer; 2. *tech.* plunger, ejector/expulsor pump
вытеснять displace, dislodge, force/push/drive out; expel; oust
~ вверх force up
вытирать wipe (out)
~ дочиста wipe clean
~ насухо wipe dry
~ пыль dust
вытиск 1. impression, impresing, stamping; 2. im-

print, impress(ion); proof
вытискивать 1. press/squeze out; 2. imprint, impress
вытопка fat/fallow rendering; liquation; smelting
~ жира в автоклаве pressure (tank) fat rendering
выточка 1. recess, neck; groove; undercut; 2. turnery; 3. maggot hole, wormhole
~, круговая circular recess
~ на валу neck journal
~ под прокладку gasket seat
вытравка discharge (printing); etching
~, восстановительная reduction discharge
вытравление 1. etching; corrosion; erosion; 2. *chem.* removal of spots; 3. eradication, extermination, destruction; 4. discharge
~ плода/зародыша induced abortion
вытравленный corroded, eroded; etched
вытравлять, вытравливать 1. etch; corrode, erode, pickle; 2. take/get out, remove; 3. eradicate, exterminate, destroy; damage *(crops);* overgraze; 4. *naut.* pay out
~ зародыш/плод induce an abortion
вытравной corrosive, corrodent; etching, erosive
вытрезвитель sobering-up station
вытчик *hist.* owner of a piece of land
вытягивание 1. extraction, extracting, drawing/pulling out; exhaustion; withdrawal; 2. extension, extending, lengthening, stretching, elongation
~ ремня stretching of a belt
вытягивать 1. extend, lengthen, stretch (out), elongate; 2. extract, withdraw, pull/draw/get out, pull up; exhaust; remove
~ в нити spin out
вытяжение *med.* extension, traction, drawing, pulling; stretching, lengthening
~ конечности *med.* leg lengthening
~ нервного ствола nerve stretching, neurectasis, neurectasia, neurectasy, neurotension
~, скелетное skeletal traction/extension
вытяжка 1. stretching, expansion, extension, elongation; 2. *met.* drawing; *text., leath.* draft; 3. *chem., med.* extract
~ в несколько переходов *met.* drawing in several runs
~, водная aqueous extract
~, глубокая deep drawing
~, кислая acid extract
~ нефти oil extraction
~, окончательная finishing draft
~, почвенная soil extract
~, спиртовая alcoholic extract
~, частная intermediate extract
~, эфирная ether extract
вытяжной 1. stretchable, extensible; 2. for extracting, for drawing out; exhaust *(attr.)*
~ вентилятор exhaust fan
~ пластырь drawing plaster
~ шкаф fume/exhaust hood.
~ная труба vent/exhaust pipe
выхват flay mark
выхлоп exhaust, discharge, expulsion; vehicle emmision
~, бездымный smokeless exhaust
выхлопной exhaust *(attr.)*
~ клапан exhaust/discharge valve
~ная труба exhaust pipe
выход 1. exit, outlet, egress; outflow, escape, emergence; discharge; 2. departure, leaving; going/coming out; 3. alternative, way out; 4. output, yield, production; *min.* outcrop; 5. publication, issue, appearance
~, аварийный emergency escape/exit
~ альфа-частиц yield of alpha-particles
~, артезианский artesian discharge
~ без защелки unlatched output
~ в космосе extravehicular activity, spacewalk
~ в лабораторных условиях laboratory yield
~ в отставку retirement
~ в стрелку *bot.* bolting
~ в трубку *bot.* shooting, stooling stage
~, валовой gross/total yield
~, великий *eccl.* great entrance
~, весовой yield by weight
~ воздуха air outlet
~ вулкана vent
~ газа gas show; gas vent/seep/discharge/yield/outlet/output
~ гамма-излучения gamma-ray output
~ годного из слитков ingot-to-product yield
~, грунтовой ground water discharge, phreatic discharge
~ группового контроля *comp.* group control exit
~ данных, прямой/непосредственный direct data output
~, двоично-кодированный binary coded output
~, двоичный binary output
~, дробный fractional yield
~ естественный, нефти oil show
~ желудка *anat.* pyloric end
~ за пределы transcendency
~ за пределы чувств supersensualism
~, закартированный mapped outcrop
~ замуж marriage *(of woman)*
~, запасной emergency exit/door; *min.* safety escape
~ зерна с каждого гектара *agr.* output per hectare
~ из блока программы *comp.* program step exit
~ из боя *mil.* withdrawal from action
~ из облаков *av.* cloud breaking
~ из пикирования *av.* pull-out
~ из-под контроля/управления runaway, running out of control
~ из системы *comp.* logout, logoff
~ из строя *tech.* abruption, rupture, failure
~ из ущелья *geol.* debouchement

~ из цикла *comp.* loop exit
~ излучения radiation yield
~ изогнутый arcuate outcrop
~ ионизации ionization yield
~, ионный ionic yield
~ ионов ion yield
~ источника *geol.* spring opening, point of issue
~, кварцевый quartz outcrop
~ к морю outlet to the sea
~ кокса coke yield
~ колонки *comp.* column exit
~, контролируемый checked output
~, коренной bedrock
~ лавы на штрек end face, face-on end
~ массы mass yield
~ мезонов/нейтронов meson/neutron yield
~, мультиплексный multiplexed output
~ на ветер *nav.* weathering
~ на маяк beacon approach
~ на поверхность *geol.* outcrop(ping), exposure, baring
~ нефти oil show
~ обедненного продукта stripped output
~ обогащенного продукта enriched output
~ обратного кода *comp.* complementary output
~, общий overall/total yield
~ Оже-электронов Auger yield
~ осей emergence of axes
~ осколков fission yield
~, относительный relative yield
~, отсроченный *comp.* deferred exit
~ охладителя coolant outflow
~ пар ионов ion-pair yield
~ пара steam escape/outflow
~ переноса carry output
~ пласта на поверхность seam outcrop
~ по напряжению voltage efficiency
~ по току current efficiency
~ по энергии power efficiency
~, подземный buried/subterranean outcrop
~, пожарный fire escape
~ при делении *nucl.* fission yield
~ продукции животноводства *agr.* livestock output
~, промышленный industrial output
~ прямого кода *comp.* true output
~ пучка beam escape
~, радиационно-химический radiation-chemical yield
~, разложившийся decomposed outcrop
~ N-разрядный N-bit output
~ рассеяния scattering yield
~ реакции reaction yield
~, речевой *comp.* voice output
~ руды ore outcrop
~ с защелкой *comp.* latched output
~ с тремя состояниями *comp.* three/tri-state output
~ сброса fault outcrop
~, световой light yield
~ сейсмического луча emergence of seismic ray
~ скота *agr.* livestock yield
~ сложения add output
~, смещенный misplaced outcrop
~ соли salt outcrop
~ сумматора adder output, counter total exit
~ суммы add/sum output
~ счетчика counter exit
~ счетчика для распечатки counter list printing exit
~ счетчика для списков counter list entry
~ сцинтилляции/сцинтилляционный scintillation response
~, тазовый pelvic outlet/strait
~, убойный dressing/killing percentage
~ угольного пласта на поверхность *geol.* coal smut, blossom of coal
~ умножителя multiplier output
~ усилителя записи *comp.* output of write amplifier, write amplifier output
~ флуоресценции fluorenscence/fluorescent output
~ фотоядерной реакции photonuclear yield
~ фракции fractional yield
~ фреатической воды на поверхность *geol.* phreatic water discharge
~, функциональный functioral output
~ чистой шерсти *agr.* wool yield
~, электронный (electron) emissivity
~ ягнят *agr.* lambs dropped alive
выходец emigrant; immigrant from different social class
~ из крестьянской среды peasant by origin/birth
~ из России of Russian extraction
выходить 1. go/come/walk out; get off; withdraw; 2. leave; emanate; 3. issue, be published, appear; 4. run out, expire
~ в колос *bot.* come into ear
~ в лист foliate
~ в трубку *bot.* spire
~ в море put out to sea
~ в отставку retire
~ в офицеры be commissioned
~ в тираж *fin.* be drawn
~ в финал *sport* reach the final
~ за пределы transcend; overstep the limits
~ замуж marry *(of a woman)*
~ из возраста pass the age limit
~ из обихода/употребления go out of use, fall into disuse
~ из печати come off the press
~ из строя break down, get out of commision
~ на ветер *nav.* get the wind, get windward of
~ на вызовы *theatre* take a curtain call
~ на орбиту go/settle into orbit
~ на позицию *mil.* get into position
выходной 1. outgoing, outbound; discharge; exit *(attr.);* 2. *pl.* *n.* severance pay

~ **день** day off, free day
~ **лист** *print.* title page
~**ная роль** *theatre* bit/supernumerary part
~**ное пособие** *econ.* serverance pay
выходящий outgoing
выхолаживание heat loss, cooling, chilling
выцвет bloom, efflorescence
~, **солевой** salt efflorescence
выцветание fading, discolo(u)ration, bleaching; efflorescence, blooming
выцветать lose colo(u)r, fade, bleach, bloom out; finish blossoming
выцеливать (take) aim at
вычеркивание crossing/striking out, deletion; *math.* cancellation; omission; obliteration
~ **по указаниям** *comp.* deletions by check(ing)
вычеркивать delete, cross/strike/blot out; expunge, erase
~ **из цепочки** *chem.* delete from the chain
вычерпывание bailing/scooping/taking out, ladling; dredging, excavation
вычерпывать bail/scoop/take out; drain, exhaust
вычерчивание drawing; plotting, mapping; laying out; sketching; linework
~ **графика** graphing
~ **инком** ink drafting
~ **кривой по точкам** curve fitting
~ **кривых** curve tracing
~, **полевое** *cart.* field sketching
~, **расчлененное** (colo(u)r-)separation drafting, separation drawing
~, **чистовое** line/fair drawing
вычерчиватель tracer, plotter, plotting device
вычерчивать draw, plot, protract, map out; trace/lay out
~ **горизонтали** *cart.* contour, draw contours
~ **карту** draw a map
~ **кривую по точкам** draw/plot a curve by the point-by-point method
вычесывать comb out
вычет *math.* residue; *fin.* deduction account bill, invoice
~ **в фонды пенсионного обеспечения** superannuation payment
~, **единый** block deduction
~ **из (облагаемого) дохода** income deduction
~ **квадратический/логарифмический** quadratic/logarithmic residue
~, **наименьший** principal remainder, least (non-negative) remainder
~,**наименьший положительный** least positive residue
~, **норменный** norm residue
~, **прямой** flat deduction
~ **степенной** power residue
вычисление computation, calculation; evaluation, rating; figuring; reckoning; *econ.* account, bill, invoice
~, **адресное/адресов** address calculation/computation/arithmetic
~ **базового адреса** basic address calculation
~ **в аналоговой форме** analogue calculation
~ **в истинном/реальном масштабе времени** real-time computation
~ **возраста** *stat.* age reckoning
~, **графическое** graphical calculation
~, **контрольное** check computation
~ **координат** position computation, measurement of coordinates
~ **курса** track analysis
~ **масштаба** scale computation
~ **методом Монте-Карло** Monte-Carlo calculation
~ **методом неявных функций** implicit computation
~ **методом фаз** phase-shift calculation
~, **многогрупповое** multigroup calculation
~ **на вычислительной машине** computer calculation
~ **направлений** computation of bearings
~, **одногрупповое** one-group calculation
~ **основного адреса** basic address calculation
~, **пошаговое** step-by-step computation
~, **приближенное** approximating, approximate calculation, calculus of approximation
~ **пути доставки** computaiton of delivery routes
~ **разностей** differencing
~, **рекурсивное** recursive computation
~ **с двойным просчетом** double calculation
~ **с обычной точностью** single-precision computation
~ **с плавающей запятой** floating point computation
~ **с фиксированной запятой** fixed-point computation
~, **символьное** symbolic computation
~, **табличное** spreadsheet calculation
~ **траектории** path/trajectory computation
~ **факториала** factorization
~ **частотности** statistical calculation
~ **элементов орбиты** orbit computation/calculation
~ **эфемерид** *astr.* ephemeris computation
~**ния, аналого-цифровые** hybrid computations
~**ния в пакетном режиме** batch computing
~**ния в темпе поступления данных** online computing
~**ния, камеральные** office computations
~**ния на местности** field computation
~**ния, отказоустойчивые** fault-tolerant computing
~**ния, полигонометрические,** ~ **полигонометрических ходов** traverse computations
~**ния по способу наименьших квадратов** least square calculations
~**ния, уравнительные** compensation/compensating computation

вычислимый computable
вычислитель computer, calculator, calculating/computing machine; *see also* **вычислительная машина**
~, **аналого-цифровой** analog(ue)-digital computer, hybrid computer
~, **бортовой** airborne computer, on-board computer
~ **дальности** ranging computer
~, **двухкомпонентный** two-variable computer
~ **курса** *av.* heading(-line) computer
~ **мощности** power computer
~ **на твердых телах** (all-)solid computer
~ **одноразового действия** single-shot computer
~, **оптический** all-optical computer
~ **параметров полета** air data computer
~ **периода** period computing circuit
~ **путевой скорости** ground-speed computer
~, **самонастраивающийся** (self-)adaptive computer
~ **системы самонаведения** homing computer
~ **системы сопровождения/слежения** tracking computer
~ **скорости счета** comunting-rate computer
~ **углов** angle computer
~ **упреждения ПУТС** *nav.* deflection computer
~, **цифровой** digital computer
~ **числа М** Mach-number computer
вычислительная машина computer, calculator, calculating/computing machine, *see also* **вычислитель**
~, **аэронавигационная** flight computer
~ **без печати** non-listing computer
~, **клавишная** keyboard computer
~, **моделирующая** computer simulator, simulator computing system
~, **быстродействующая** high-speed computer
~, **настольная** desk(-top) computer, lap-top computer
~, **настольная электронная** electronic desk/table calcualtor
~ **общего назначения** general purpose computer
~ **параллельно-последовательного действия** parallel-serial computer
~ **с многократным доступом** multiple-access computer
~, **специализированная** special-purpose computer, MAC
~, **универсальная** general(-purpose) computer
~, **цифровая** digital computer
~, **электронная, ЭВМ** (electronic) computer
вычислительный computational, computing, calculating; computer-oriented
вычислять compute, calculate; reckon
~ **с точностью до десятичных знаков** calculate to decimals
вычистка cleaning, cleansing
вычистки *pl.* brushings; cleanings
вычитаемое *math.* subtrahend
вычитание *math.* subtraction; deduction
~ **до нуля** below zero subtraction
~, **произвести** subtract
~ **цветов** *print.* subtractive mixing of inks
вычитатель subtractor, subtracter
~, **последовательный** serial subtracter
вычитать *math.* subtract; deduct, keep back
вычитка *print.* (proof)reading; editorial preparation
вычищать clean (out/up), scrub (out); *polit.* expel (from), purge
вычленение *med.* exarticulation, dearticulation
вычленять *math.* divide into parts; *med.* exarticulate, disjoint
вышедший emergent
~ **из строя** disabled, inoperative
~ **из употребления** obsolete, antiquated, gone out of use
вышеисчисленный above-calculated
вышележащий superposed, superincumbent, superjacent, overlying
вышелушивать shell, husk, hull; peel
вышеназванный aforenamed, aforesaid
вышеозначенный abovementioned, aforesaid
вышеприведенный given/shown above, aforecited
вышесказанный before-mentioned, aforesaid
вышеуказанный above-mentioned; above, foregoing
вышеупомянутый aforementioned, aforesaid
вышибной blow-out
вышивание needlework, embroidery, embroidering
~ **двойным швом** backstitch embroidery
вышивать embroider
~ **шелком** embroider in silk
вышивка needlework, embroidery
~ **бисером** beading, beadwork
~ **гладью** satin stitch embroidery
~ **по парче** brocade embroidery
~, **теневая** shadow embroidery
вышка tower, garret; attic stor(e)y
~ **аварийного освещения** emergency lighting tower
~, **авиадиспетчерская** air-traffic-control tower, ATC tower
~, **анемометрическая** wind tower
~, **буровая** drilling rig, (oil) derrick
~ **высоковольтной передачи** high-voltage transmission tower
~, **диспетчерская** *av.* (traffic) control tower
~, **наблюдательная** observation/lookout tower
~, **нефтяная** oil derrick
~ **обслуживания** servicing tower
~, **парашютная** parachute tower
~, **прожекторная** (flood)lighting tower
~, **пусковая** *space* lauching tower/rig, tower launcher
~, **садовая** orchard tower
~, **самоходная/передвижная** tower vehicle, mobile tower

~, **сигнальная** flag tower
~, **сторожевая** watchtower
~, **судейская** *sport* referees' tower
~ **управления** control tower
~ **управления воздушным движением** air-traffic control tower, ATC tower
вышлифовывать 1. *tech.* polish, grind out; make smooth; 2. give a polish to, smarten up, perfect
вышний 1. high(er); subcelestial; sky-high; 2. *eccl.* heavenly, divine; 3. *n. eccl.* Supreme Being, God
выштамповка *tech.* stamping/punching out, blanking
выштамповывать *tech.* stamp/punch/blank out, press
выщелачивание leaching, lixiviation; lixiviating; vatting
~, **безводное** non-aqueous leaching
~ **в автоклаве** pressure leaching
~ **в пульпе** leaching by agitation
~, **гидротермальное** hydrothermal leaching
~, **гипергенное** supergene leaching
~ **перемешиванием** leaching by agitation
~ **при метаморфизме** subtraction by leaching
~ **просачиванием** leaching by percolation
выщелачиватель 1. leaching vat; 2. leach; lixiviant, leaching agent
выщелачивать 1. *chem.* leach (out); lixiviate; extract; 2. steep/soak in lye, buck
выщипывание *print.* picking
выщупывать *med.* feel, touch
выя *anat.* nucha, nape/back of the neck
выявление 1. emergence; detection; discovery, revelation; 2. exposure; development; delineation
~ **вооружений** disclosure of armaments
~ **неисправностей** trouble-shooting
~ **ошибок** detection of errors
выявлять bring out, disclose, reveal, educe
~ **факт** elicit a fact
выявляться emerge, reveal itself, be disclosed/indicated/revealed, be brought up
выяснять clear up, explain, ascertain, find out; investigate, bring to light, look into
вьетнамец, вьетнамский Vietnamese
вьетнамка Vietnamese (woman)
вьюк pack; load
~, **седельный** saddle bag
вьюн *zool.* (mud) loach *(Misgurnus)*
вьюнец *bot.* alpine clematis *(Clematis alpina)*
вьюнковый *bot.* convolvulaceous; *pl. n.* morning glory family *(Convolvulaceae)*
вьюновые *n. pl. zool.* Cobitidae
вьюнок *bot.* glorybind, convolvulvus *(Convolvulvus)*
вьюрок 1. *zool.* finch *(Fringilla);* brambling; 2. *text.* scroll tube
~, **канареечный** serin *(Serinus canaria)*
~, **красный** red-breasted rose finch *(Pyrrospiza punicea)*
вьюшка *naut.* cable reel/drum; damper
~ **диплотлиня** deep-sea reel
~ **для лаглиня** log reel
~ **минрепа** mooring cable drum
~, **тральная** mine-sweeping reel
~, **тросовая** cable reel
~, **швартовная** hawser reel
~ **штерта** depth reel
вьющийся coiling, winding, twining, twisting; vinaceous; climbing
~**щееся растение** *bot.* climber, climbing plant
вяжечка *bot.* rock cress *(Turritis, Arabis)*
вяжущее *n.* binder, binding/bonding agent, cement
~ **на базе извести** calcareous cement
вяжущий 1. astringent, styptic; tart; 2. cementing, binding
вяз *bot.* elm *(Ulmus)*
вязальный knitting
~ **крючок** crochet hook
~**ная спица** knitting needle
вязание *text.* 1. knitting; crocheting; 2. knitwear; knitted garment; 3. tying, binding
вязанка bundle, bunch; sheaf
~ **хвороста** faggot
вязаный knitted; crocheted
вязать 1. *text.* knit; 2. bind
вязель *bot.* crown vetch, coronilla *(Coronilla)*
вязка 1. binding, joining, tying; 2. knitting; crocheting
вязкий 1. *tech.* ductile, malleable; tensile, resilient; 2. *med.* tough, tenacious; 3. *chem.* viscous, viscid; sticky, gluey, glutinous; ropy, stringy, treacly; smeary; slimy
~ **в толстом слое** viscous in body
вязкость 1. *tech.* ductility; toughness, tenacity; malleability, resilience; 2. *chem.* viscosity, stickiness; stringiness; adhesiveness; sliminess
~, **аномальная** non-Newtonian viscosity
~ **в медноаммиачном растворе** cuprammonium viscosity
~, **вихревая** eddy/turbulent viscosity
~ **внутренняя** intrinsic/internal viscosity, limiting visosity index
~, **горизонтальная** lateral viscosity
~, **жесткая** firmo-viscosisty
~ **истинная** intrisic/internal viscosity
~, **кажущаяся** apparent viscosity
~, **критическая** ultimate viscosity
~, **лучистая** radiative viscosity
~, **нормальная/ньютоновская** Newtonian viscosity
~, **объемная** volume/bulk viscosity
~, **остаточная** residual viscosity
~, **относительная** relative viscosity
~ **по воронке** viscosity by cup
~ **по Редвуду** Redwood viscosity
~ **по Сейболту** Saybolt viscosity
~ **по Энглеру** Engler viscosity
~ **под давлением** pressure viscosity
~ **разрушения** fracture toughness, strain crack toughness
~, **сдвиговая** shear viscosity

~, турбулентная eddy/turbulent viscosity
~ угля coal rheology
~, ударная impact strength/resistance, dynamic/impact ductility, resilience, toughness
~, удельная specific viscosity
~, упругая elastc viscosity
~, характеристическая intrinsic/internal viscosity; viscosity index
вязкотекучесть viscosity
вязкотекучий plastic, viscous-flow
вязкотекущий viscous/laminar-flow
вязкоупругий viscoelastic
вязкоупругость viscoelasticity
вязкоэластический elastoviscous, viscoelastic
вязовик 1. *bot.* dryad's saddle *(Polyporus squamosus)*; 2. hop tree *(Ptelea)*
вязовый *bot.* elm *(attr.)*
вязь ligature
вяление drying, (dry-)curing, jerking
вяленый (air-)dried, sun-cured, jerked
вялить dry, (dry-)cure, sun-dry; jerk
вялость flaccidity, flabbiness, anergy; limpness, inactivity, apathy, sluggishness
~ в работе slackness in work
~ кишечника sluggishness of the bowels
~ кровообращения circulatory inefficiency
вялый 1. *med.* flaccid, flabby; inert; limp; sluggish, lax; slack; 2. wilted, withered; faded
~ кишечник sluggish bowels
~лая рука limp hand
вянуть wilt, wither
вятич 1. inhabitant of Vyatka
вяха *bot.* water hemlock *(Cicuta)*
вяхирь *zool.* ringdove, wood pigeon *(Columba palumbus)*

Г, г

гаагский the Hague *(attr.)*
габара *naut.* gabbart; lighter
габардин *text.* gabardine, gaberdine
габарит *mach.* size, overall/external dimensions; clearance (limit)
~, вертикальный vertical clearance
~, вписываться в fall/lie within clearance limits
~ груза loading gauge
~, не вписываться в fall/extend beyond/outside clearance limits
~ по высоте height clearance, clear height
~ по глубине depth clearance
~ по ширине width clearance, clear width
~ погрузки loading gauge
~ подвижного состава *rail.* vehicle gauge
~, подмостовой 1. (under)bridge clearance, clear headway of a bridge, clear headroom; 2. clearance diagram for a bridge
~ приближения строений *rail.* obstruction clearance, clearance of obstructions
~ провода electrical/conductor clearance to earth
~ станков overall machine dimension
габаритный clearance *(attr.)*
~ные ворота *rail.* clearance gauge
габаритомер clearance/clearway gauge
габбро *geol.* gabbro
~, динамометаморфизованное flaser-gabbro
~, мелкозернистое жильное beerbachite, gabbro-aplite
~, оливиновое olivine gabbro
~, роговообманковое hornblende gabbro
~, рудное magnetite gabbro
~, слоистое layered gabbro
~, уралитовое uralite/hornblendite gabbro
~, шаровое spheroidal gabbro
габбро-амфиболит *geol.* gabbro-amphibolite
габбровидный gabbroid
габбровый gabbro(ic), gabbroitic
габион *geol.* gabion
габитуальный habitual
габитус 1. *cryst.* (crystal) habit, crystal line habit, habit of a crystal; 2. *med.* habit(us)
~, длинностолбчатый long-columnar habit
~, игольчатый/линейный acicular habit, needle-like habit
~, изометрический normal habit
~, листоватый lamellar habit
~, нитевидный filamentous/filamented habit
~, плоский lath-like habit, blade habit
~, прямостоящий erect habit
~, распростертый prostrate habit
гаваец, гавайский Hawaiian
гаванна Havana (tobacco/cigar)
гаванский Havana *(attr.)*
гавань harbo(u)r, port
~ в устье реки estuary harbo(u)r
~, внешняя outer harbo(u)r, outside port
~, внутренняя inner harbo(u)r
~, военная naval/war harbo(u)r, naval port
~, вольная free port/zone
~, зимовочная wintering harbo(u)r
~, приливная tidal/tide harbo(u)r
~, ремонтная (ship-)repair harbo(u)r
~, рыбачья fishing harbo(u)r
~, свободная free harbo(u)r
~, торговая merchant/commercial harbo(u)r
гавиал *zool.* gavial *(Gavialis)*
гавортия *bot.* haworthia *(Haworthia)*
гавот *mus.* gavot(te)
гага *zool.* eider(duck) *(Somateria)*
гага-гребенушка *zool.* king eider *(Somateria spectabilis)*
гагара *zool.* loon, diver *(Gavia)*
гагарка *zool.* razorbill *(Alca torda)*
гагаровые *n. pl. zool.* Gaviidae
гагат *geol.* gagate, jet
гагатизация gagatization
гагатовый jet(ty)
гагауз Gagauz

гад reptile, amphibian
Гадес *myth.* Hades
гадолиний gadolinium, Gd
гадрозавр *paleont.* hadrosaur *(Hadrosaurus)*
гадром(а) *bot.* hadrome, hadromestome
гадроптерус *zool.* gilt-garter fish *(Hadropterus evides)*
гадюка *zool.* viper, asp *(Vipera)*
~, рогатая horned adder *(Cerastes cornutus)*
гадюковые *n. pl. zool.* Viperidae
гаечный nut, screw *(attr.)*
~ ключ (screw) wrench
газ I 1. gas; 2. *pl. med.* wind, gases; *see also* **газы**
~, агрессивный corrosive gas
~, аэростатный lifting gas
~, баллонный bottled/cylinder gas
~, бедный lean gas
~, безокислительный noncorrosive gas
~, биологический sewage/sludge gas
~, благородный noble/inert gas
~, богатый летучими rich gas
~, бозе-эйнштейновский Bose-Einstein gas
~, болотный marsh gas, methane
~, буферный неразориентирующий non disorienting buffer gas
~, бытовой domestic gas
~ в пустотах пород occluded gas
~, ваграночный cupola gas
~, веселящий laughing gas, nitrous oxide
~, вдуваемый/введенный injected gas
~, взрывоопасный explosive gas
~, включенный included gas
~, влажный wet gas
~, возбужденный pumped gas
~, воздушный air-blast gas
~, восстановительный reducing gas
~, вредный harmful/noxious gas
~ вторичной очистки secondary gas
~, выделять evolve/liberate gas, give up gas
~, выделяющийся evolved gas
~, вырожденный degenerate gas
~, высококалорийный rich/high calorific gas
~, вытесняющий pressurizing gas, pressurant
~, выхлопной exhaust/burned/combustion gas
~, вязкий relaxing gas
~, генераторный producer/ power/ generator/ pressurization gas
~, генераторный эндотермический RX gas, generator gas
~ глубинного происхождения *geol.* gas of deep-seated origin
~, головочный casing-head gas
~, горновой(доменной печи) hearth gas
~, горный mine gas
~, горючий fuel/combustible gas
~, гремучий fire-damp, detonating gas
~ грубой очистки primary cleaned gas
~, грязный fume-laden gas; crude gas
~, двухатомный diatomic gas
~, делящийся fissionable gas
~, детонирующий detonating gas
~ для дыхания breathable gas
~ для коммунально-бытовых нужд gas for public-utility use(s)
~ для счетчиков counter/counting gas
~, доменный top gas, blast furnace gas
~, дутьевой blast gas
~, дымовой flue/slack gas
~, естественный natural/rock gas
~, естественный богатый нефтью combination gas
~, жидкий wet gas
~, заключенный occluded gas
~, запыленный dust-/ash-laden gas
~, заторможенный stagnated gas
~, защитный shielding gas
~, идеальный ideal/perfect gas
~ из головки Брадена Braden-head gas
~, излучающий emitting/radiating gas
~, инертный/индифферентный inert/ noble/ neutral/ inactive gas
~, ионизированный ionized gas
~, ионный ionic gas
~, искусственный artificial/manufactured gas
~, каменноугольный coal gas
~, канализационный sewer gas
~, карбюрированный carburetted gas
~, кислородно-конвертерный oxygen-converter gas
~, коксовый coke-oven gas
~, коксовый прямой volatile matter (of the coking process)
~, колошниковый blast-furnace gas, top gas
~, коммунальный commercial gas
~, конвертерный converter/waste gas
~, котельный boiler flue gas
~ легче воздуха lifting gas, lighter-than-air gas
~, магнитоионный magnetoionic gas
~, малый idle, idling
~, маслородный olefiant gas
~, межзвездный/межпланетный interstellar/ interplanetary gas
~, наземный малый *av.* ground idle
~, наполняющий filling gas
~, напряженный tensioned gas
~, наркотический anesthetic gas
~, невязкий inviscid gas
~, невырожденный nondegenerate gas
~, неконденсирующийся permanent/fixed gas
~, неочищенный raw/crude gas
~ нефтепереработки (oil-) refinery gas
~, нефтяной oil/petroleum gas
~, нефтяной попутный casing-head gas
~, нейтральный inert/neutral/inactive gas
~, нейтронный neutron gas
~, низкокалорийный poor gas, low calorific gas
~, обжиговый roaster gas
~, оборотный recycle gas
~, огнегасящий fire extinguishing/suppressing

gas
~, одноатомный mon(o)atomic gas
~, однородный simple/uniform gas
~, озоленный ash-laden gas
~, окклюдированный occluded gas
~, опасный harmful/noxious/dangeronus gas
~, освобожденный resurgent gas
~, остаточный residual gas
~, отходящий waste/flue/exit/effluent gas, off gas
~, охлаждающий cooling/refrigeration gas
~, переводить на малый put to idle
~, перманентный permanent/fixed gas
~, печной kiln gas
~, плотный solid gas
~, поглощающий absorbing gas
~, поглощенный occluded/absorbed gas
~ под большим/малым давлением high-/low-pressure gas
~, поддерживать use enough throttle
~, полетный малый *av.* flight idle
~, полный full power/throttle
~, полуводяной semi-water gas, Dowson gas
~, полусухой semi-dry gas
~, постоянный permanent gas
~ при обратной тяге, малый reverse idle
~ при прямой тяге, малый forward idle
~, природный natural/rock gas
~, пробный probe gas
~, проводящий conducting gas
~, продувочный *rock.* purge/flushing gas
~, псевдоожижающий fluidizing gas
~, рабочий working fluid; plasma-forming gas; *rock.* working/propulsion/power/test gas
~, разреженный rarefied gas
~, растворенный (в нефти) (oil-)dissolved gas
~, реагирующий reactive gas
~, реальный imperfect/real/actual gas
~, редкий rare gas
~, релаксирующий relaxing gas
~, рудничный detonating/mine gas
~, рудничный после взрыва after-damp
~, светильный illuminating/lighting/coal gas
~, светящийся glowing/luminous gas
~, свободный free gas
~, сернистый sulphur (di)oxide gas, sulphurous gas
~, серный sulphur trioxide gas
~, сжатый compressed/pressurized gas
~, сжижаемый coercible/compressible gas
~, сжиженный liquefied/condensed gas
~, силовой power gas
~, слезоточивый tear(-exciting) gas
~, смешанный mixed gas
~, совершенный perfect/ideal gas
~ содержащий бор boron-containing gas
~, содержащий дым fume-laden gas
~, сорбированный retained gas
~, спонтанный spontaneous gas
~, сухой residue gas
~, сушильный drying gas
~, сцинтиллирующий scintillating gas
~, счетный counting gas
~, теплопроводный heat-conducting gas
~, технологический process gas
~, токсичный toxic gas
~, толкающий driver gas
~ тонкой очистки finally cleaned gas
~, топливный fuel gas
~, топочный flue/furnace gas
~, торфяной peat gas
~, тощий lean gas
~, транспортирующий carrier gas
~, трехатомный triatomic gas
~, тяжелый heavy gas
~, увлеченный entrained gas
~, угарный *min.* whitedamp; carbon monoxide, carbonic oxide, carbonyl
~, углеводородный hydrocarbon gas, carburetted hydrogen gas
~, углекислый (gaseous) carbon dioxide, carbonic(-acid) gas
~, угольный coal gas
~, удушливый choke-gas/-damp, chokedamp
~, уплотняющий blanketing gas
~, устойчивый persistent gas
~, фреатический phreatic gas
~, фумарольный fumarolic/fumarole gas
~, фурменный tuyere gas
~, хвостовой tail gas
~, хранимый в баллоне bottle-stored gas
~, циркулирующий в скафандре suit-circuit gas
~, электронный electron(ic) gas/fluid
~, электроотрицательный electronegative gas
~, энергетический power gas
~, эруптивный вулканический eruptive volcanic gas
~, ювенильный juvenile gas
~, ядовитый poison(ous)/toxic gas
газ II *text.* gauze
газават holy war, jihad
газация aeration, gasification; see also **газификация**
газгольдер 1. gasometer; 2. gasholder/tank
~, мокрый, переменного объема wet gasholder
~ постоянного объема, сухой dry gas-holder
~, спиральный screw-type gas-holder
газель I *zool.* gazelle *(Gazella)*
~ антидорка sprinkbok *(Antidorcus marsupialis)*
~, обыкновенная mountain gazelle *(Gazella gazella)*
газель II *lit.* ghazal *(Arabic verse form)*
газета newspaper
~, стенная bulletin board newspaper
газетно-журнальный/-публицистический journalese
газетный newspaper *(attr.)*

~ная бумага newsprint
~ стиль journalese
газирование 1. aeration, gassing, gasification; 2. carbonation
газированный aerated, gassed; carbonated, effervescent, sparkling
~ная вода soda water
газировать 1. (emit) gas, aerate; 2. carbonate; bubble up
газификатор 1. gasifier, gas generator; 2. gasification engineer, gas fuel specialist
~ азота liquid nitrogen converter
~ жидкого кислорода liquid oxigen gasifier/converter
~, криогенный cryogenic evaporator
~ низкого давления low-pressure gasifier
~ ракетного топлива propellant gasifier
~, теплый high-pressure gasifier
~, холодный low-pressure gasifier
газификация 1. gasification, gasifying; 2. provision of gas supply, installation of gas (service), supplying with gas
~ бытовая gas supply for domestic uses
~ в слое fixed-bed gasification
~, подземная underground gasification
~, промышленная installation/provision of gas supply for industrial applications
~ углей, поточная continuous/stream gasification of coal
газифицировать 1. gasify; generate/extract gas (from); 2. install/provide gas (in), supply with gas
~ на месте gasify coal in place
газлифт gas/air lift
~, бескомпрессорный natural-pressure gas lift
газ-носитель carrier gas, gas carrier
газный gassy
газоанализатор gas/atmosphere analyser/analyzer
~, альфа-ионизационный alpha-ionization gas analyser
~, волю(мо)метрический volumetric gas analyser
~ выдыхаемого воздуха expired air analyser
~, денси(то)метрический densimetric gas analyser
~ для определения окиси и двуокиси углерода carbon monoxide and dioxide gas analyser
~ для определения углеводородов и окиси углерода certain hydrocarbons and carbon monoxide gas analyser
~, интерференционный interferometric gas analyser
~, магнитотермокондуктометрический magnetothermal conductivity gas analyser
~ на азот/метан/кислород nitrogen/methane/oxygen analyser
~, объемно-манометрический volumetric gas analyser
~, оптико-акустический optical-acoustic gas analyser
~ Орса Orsat apparatus
~, радиоизотопный radiological gas analyser
~ углекислого газа, фотоэлектрический photoelectric carbon dioxide gas analyser
~ Холдейна Haldane gas analyser
газоаппаратура gas fittings
газобаллон gas bottle/cylinder
газобезопасность 1. prevention of gas poisoning; 2. prevention of gas explosions
газобезопасный gasproof, gastight
газобетон aerated concrete
газование gassing, ebullition, bubbling up
газовзвесь suspension of matter in gas
газовидный gaseous
газовод gas passage
газовоз 1. gas carrier; 2. gas locomotive
газовоздуходувка gas-engine blower
газовоздухопровод gas-air duct/conduct/line
газовоздушный gas-and-air, air-gas
газовщик gasman, gas fitter; gas plant attendant
газовыделение 1. gassing, ebullition; 2. gas emission/evolution/liberation; generation of gas
~, внезапное sudden gas emission, gas outburst
газовыделяющий gas-yielding
газовый I gaseous, gassy, fiery
~ завод gasworks
~вая колонка *mech.* geyser
~вая плита gas oven/store
газовый II *text.* gauze *(attr.)*
газовырубка oxy-acetylene scarfing
газовырубщик scarfer
газогенератор gas generator/producer, producer furnace
~ для газификации топлива в слое fixed-bed gasifier
~ свободнопоршневой free-piston gas generator, power gas generator
~ с кипящим слоем fluidized-bed gas producer
~ с плотным слоем fixed-bed gas producer
~ со взвешенным слоем suspension-bed gas producer
~ твердого топлива solid-reactant gas generator
газогенераторный, газогенерирующий gas-generating/-producing
газогипс gas gypsum/alabaster
газодинамика gas dynamics
~, гиперзвуковая hypersonic gas dynamics
~ околозвуковых скоростей transonic gas dynamics
~ тонких тел thin-body gas dynamics
газодувка gas blower
~ ацетиленовая acetylene booster/torch
газоемкость gas capacity
газозолобетон gas-ash concrete
газозолосиликат gas-ash silicate
газоизмеритель gas meter
газойль gas oil
~, легкий каталитический light catalytic gas oil

газокалильный incandescent
~ колпачок gas mantle
~ная лампа incandescent gas lamp
~ свет,-ное освещение incandescent gaslight
газокамера fumigating chamber
газокинетика gas kinetics
газолин gasolene, gasoline, petrol
газомер gas(flow) meter, flowmeter, gasometer
~, мокрый/сухой wet/dry gas(flow)meter
газомет *mil.* gas projector
газометр gasometer, gas holder
газометрический gasometric
газометрия gasometry
газомотовоз gas-engine locomotive
газомотор gas engine
газон grassplot, lawn
газонаполнение gas filling
газонаполненный gas-filled
газонепроницаемость gastightness; gas impermeability
газонепроницаемый gas-proof/-tight/-impermeable, gasproof, gastight
газонокосилка *hort.* lawn mower
~, фронтальная front-mounted lawn mower
газоноситель carrier gas, gas carrier
газоносность presence of gas
газоносный gas-bearing/-containing
~ пласт gas-bearing bed/stratum
газообильность volume of gas
газообмен gas(eous) exchange, gases interchange, interchange of gases, gaseous metabolism
~, дыхательный respiratory metabolism
~, легочный pulmonary gas exchange
газообменник gas exchanger
газообразность gaseousness, gaseity
газообразный gaseous, gasiform; aeriform; gassy
газообразование generation of gas, gas formation/generation, gasification, gassing; boiling, ebullition; vaporization; vollatilization
~, бурное gassing, boiling
газообразователь blowing/foaming agent
газообразующий gas-forming/-making, aerogenic
газоопределитель gas analyzer/detector
газоосвещение gaslight(ing)
газоотвод gas vent/bleeder/outlet/conduit; vent
~, вертикальный/восходящий *met.* vertical offtake/uptake pipe
~ конвертера Кал-До fume hood
~, нисходящий downtake/downcomer pipe
газоотводный gas-conducting
газоотделение liberation of gases, gassing
газоотделитель gas(-)separator/trap/catcher/tank
~ с осевым потоком газа axial-flow gas separator
газоотравленный poisoned by gas, gassed
газоотражатель *rock.* jet flame/launcher/blast/exhaust deflector
~, неохлаждаемый uncooled deflector
~, односкатный single-jet deflector
газоотсасывающий gas-exhausting
газоохладитель gas condenser/cooler
газоочиститель gas purifier/cleaner/scrubber/washer; scrubber
~, мокрый gas washer, wet gas cleaner, scrubber
~, мокрый динамический/статический rotary/static spray scrubber
~, мокрый тонкой очистки secondary wet cleaner
~ полутонкой очистки primary cleaner
~, сухой dry gas cleaner
газоочистка gas cleaning/scrubbing/ purification
~, мокрая scrubbing, wet gas cleaning
газоочистный gas-cleaning/-purifying
газопламенный gas-flame
газопаропроницаемость permeability to gas and vapour
газоплотность gasproofness, gastightness
газоплотный gasproof, gastight
газопоглотитель gas absorber/absorbent; *el.* getter
газопоглощающий gas-absorbing
газопоглощение 1. gas absorption/occlusion, absorption of gas; 2. getting effect, getter action
газоподводящий gas-feed/-intake
газоподобный gasiform, gaslike, gaseous
газоподогреватель gas (pre)heater
газополный gas-filled
газопривод gas feed
газоприемник gas collector/receiver
газоприемный gas-collecting/-receiving
газопровод gas conduit, gas (pipe-)line, gas pipe(line), gas main
~, внутренний indoor gas (pipe-)line
~, магистральный cross-country gas (pipe-)line, gas main
~, местный local gas (pipe-)line
~, сборный gas-collecting main
газопроводка gas-supply system, gas line
газопроводный gas-supplying
~ная труба gas pipe
газопроводчик gas fitter
газопромыватель scrubber, wet gas cleaner, gas washer, gas-washing bottle
~, динамический dynamic spray scrubber
~ полутонкой очистки primary wet cleaner, primary (gas) washer, primary scrubber
~ тонкой очистки secondary/final (gas) washer/scrubber
газопромывочный gas-washing
газопроницаемость gas permeability, permeability to gas
газопроницаемый gas-permeable/-penetrable, permeable /penetrable to gas
газопылевой gas-and-dust
газоразрядный gas-discharge
газораспределение valve timing; gas distribution, distribution of gas
~, клапанное valve gear
газораспределитель gas distributor
газораспылительный gas-atomizing

газорассекатель *rock.* deflector cone/caps
газорезка gas/oxyacetylene cutting
газорезчик gas cutter
газосборник gas collector
 ~ турбины turbine plenum
 ~ турбины, входной turbine entry duct
газосборный gas-collecting
газосварка gas/oxyacetylene welding
газосварочный gas-welding
газосварщик gas welder
газосветная лампа gas-discharge lamp
газосигнализатор gas detector/alarm/indicator
 ~, автоматический automatic gas alarm/detector
газосиликат gas silicate
газосмеситель gas mixer
газоснабжение gas supply
 ~, городское urban gas utilities
 ~, коммунальное gas utilities
газостойкий gas-proof, gastight
газоструйный gas-jet
газосушитель gas drier
газосчетчик gas meter
газотехника gas engineering
газотрон gas-filled rectifier diode
газотурбина gas(-driven) turbine
газотурбовоз gas(-)turbine locomotive
газотурбогенератор gas turbine (driven) generator
газотурбонагнетатель gas turboblower
газотурбоход gas-turbine ship
газоубежище antigas shelter, gas-proof refuge/shelter
газоудерживающий gas-retaining
газоулавливание gas trapping
газоуловитель gastrap/collector/catcher
газоупорный gas-tight/-proof/-resistant
газоход gas conduct/duct
 ~, байпасный (котла) by-pass
 ~, опускной downtake (gas duct)
 ~, отстойный baffle collector
 ~, свободный open pass
газохранилище gas storage/tank; gasholder; gasometer
газошлакобетон gas-slag concrete
газошлакозолобетон gas-slag-ash concrete
газу, на полном at full throttle
газы *pl.* gases; *see also* **газ**
 ~ы, кишечные *med.* flatus
 ~ы крови *med.* gases of the blood
 ~ы отработавшие exhaust/discharge/waste gases
 ~ы пороховые powder gases
газящий gassy; gassing
гаитянин Haitian
гаитянский Haitian
гаичка *zool.* titmouse *(Parus)*
гайал *zool.* gayal *(Bibos frontalis)*
гайдамак *hist.* haydamak *(Ukrainian cossack)*
гайдамаковщина, гайдаматчина *hist.* revolutionary movement against the Poles in the Ukraine *(in 17 and 18 centuries)*
гайдингерит *geol.* haidingerite
гайдроп guide/trail rope
гайдук *hist.* haiduck, heyduck *(rebel against Turkish domination in the Balkans)*
гайдучество *hist.* national movement against Turkish domination in the Balkans
гайка nut; female screw
 ~, авиационная aircraft nut
 ~, анкерная anchor/plate nut
 ~, барашковая wing/butterfly/fly/thumb nut
 ~, быстронатяжная fast-tension nut, fast nut clamp
 ~, высокая thick nut
 ~, глухая blind/box/cover/cap nut, nut (screw) cap
 ~, зажимная lock-nut, clamp(ing)/locking/grip nut
 ~, колпачковая cap nut
 ~, контровая check nut
 ~, корончатая castle/castellated/crown/horned nut, grooved-head nut
 ~, крепежная fastening nut
 ~ крепления колеса hub/wheel nut
 ~, круглая circular/round nut
 ~, крыльчатая wing(ed)/butterfly nut
 ~, маточная clasp/split nut
 ~, накатанная knurled nut
 ~, накидная coupling nut
 ~, невыпадающая captive nut
 ~, неполная half nut
 ~, низкая thin nut
 ~, опорная back nut
 ~, опорно-коническая tapered back nut
 ~, опорно-сферическая spherical back nut
 ~, пазовая splined/clamp nut
 ~, получерная semi-finished nut
 ~, получистая semifinished [semi-bright] nut
 ~, потайная countersunk nut
 ~, предохранительная safety nut
 ~, прорезная slotted nut
 ~, простая plain nut
 ~, пружинная spring nut
 ~, разрезная/разъемная split nut
 ~, регулировочная adjusting nut
 ~ с буртиком collar(ed)/flanged nut
 ~ с левой резьбой left-handed nut, L. H. nut
 ~ с накаткой knurled/milled nut
 ~ с правой резьбой right-handed nut, R. H. nut
 ~ с прорезью clamp/clasp nut
 ~ с рымом eye nut
 ~ с ушком eye/finger/wing/fly/thumb nut
 ~ с фланцем collar nut
 ~ с шайбой flange(d)/collar nut
 ~ сальника gland nut
 ~, сгонная sleeve nut
 ~, самоконтрящаяся self-locking nut
 ~, самоуплотняющаяся self-sealing nut
 ~, самоустанавливающаяся self-aligning nut

~, свертная screw cap, cap screw
~ сгонная cap/blind/box nut
~, соединительная connecting/union nut
~, стопорная locking/safety/check/retaining/clamp/ grip nut
~, стяжная coupling/turnbuckle nut
~, упорная back nut
~, установочная adjusting/clamping nut
~, хвостовая tail nut
~, ходовая sliding nut
~, черная rough nut
~, чистая finished/bright nut
~, шариковая ball nut
~, шестигранная hex(agonal)/hexagon nut
~, шлицевая slotted nut
~, шлицованная splined/clamp nut
гайка-барашек butterfly/fly/finger/thumb/wing nut, ear(ed) nut
гайковерт (power) nut-driver, nut/speeder wrench
гайлюсакия *bot.* huckleberry *(Gaylussacia)*
гайлярдия *bot.* gaillardia *(Gaillardia)*
гайморит *med.* maxillary sinusitis
~, острый acute maxillary sinusitis
~, пристеночно-гиперпластический parietal hyperplastic maxillary sinusitis
гайморова полость *anat.* antrum of Highmore
гайморография roentgenography of the maxillary sinus
гайморотомия *med.* maxillary sinusotomy
гак *nav.* hook
~, буксирный towing hook
~, вертлюжный swivel hook
~, грузовой cargo/load hook
~, двурогий devil's claw
~, карабинный drop lip/safety/snap hook
~, несущий messenger hook
~, откидной slip hook
~, патентованный snatch hook
~, подъемный lifting/recovery hook
~ с защелкой trip hook
~ с коушем hook thimble, thimble hook
~, самовыкладывающийся (automatic) releasing/ tumbling hook, self-releasing hook
~, складной clasp/clip hook, match/sister hooks, crampoon
~, спусковой easing-out hook
~, тормозной arrester/arresting hook
~, швартовный mooring hook
~ шлюпочных талей davit hook
гакаборт taffrail
гакабортный taffrail *(attr.)*
~ лаг taffrail log
~ огонь poop lantern
~ счетчик taffrail recorder
гал gal *(unit of acceleration)*
галаго *zool.* galago *(Galago)*
галакс *bot.* galax *(Galax)*
галаксолистный *bot.* galax-leaved, galacifolious
галактаза *biochem.* galactase
галактан *chem.* galactan
галактика *astr.* galaxy
~ переходного типа transition-type galaxy
галактитол дегидрогеназа galactitol dehydrogenase
галактоз *physiol.* galactosis, lactation
галактоза *chem.* galactose
галактоземия galactosemia
~, наследственная/приобретенная hereditary/acquired galactosemia
галактозилтрансфераза galactosyltransferase, transgalactosylase
галактолипид galactolipid(e)
галактометр galactometer, lactometer, lactodensimeter
галактопоэз *physiol.* galactopoiesis
галакторея galactorrhea
галактоскоп galactoscope
галактостаз *med.* galactoschesia, galactoschesis, galactostasia, galactostasis, galactischia
галактотерапия galactotherapy
галактуроновый galacturonic
галалит galalith, artificial horn
галантерейный haberdashery, fancy-goods
~ магазин haberdashery, fancy-goods store
~ отдел notions department
~ные товары fancy goods
галантерея haberdashery, fancy goods, *Am.* small ware(s)
галантир, галантин *cul.* galantine
галантус *bot.* snowdrop *(Galanthus)*
галация *phot.* halation
галган *bot.* spiked ginger lily *(Hedychium spicatum)*
галгант galanga root, galingale
галеас *hist.* galleass, galliass
галево *text.* heald, heddle
~, металлическое wire heald
~, нитяное cotton heald, heald of knitted cotton
~, ремизное heald, heddle
галега *bot.* goat's-rue *(Galega)*
галезия *bot.* silver bell, snowdrop tree *(Halesia)*
галенит *geol.* galena, galenite, lead glance
галения *bot.* spur gentian *(Halenia)*
галеновый galenic
галенобисмутит, галеновисмутит *geol.* galenobismuthite, bismutoplagionite
галеон *nav.* galleon
галера *hist.* galley
галерейный gallery, tunnel *(attr.)*
~ лес gallery/fringe forest
галерея gallery, tunnel
~, бункерная bunker bay
~, картинная picture gallery
~, лавинозащитная snow shelter
~, минная *min.* mine gallery/tunnel
~, оросительная irrigation gallery
~ подачи агломерата sinter conveyer gallery
~, подземная adit
~, посадочная *av.* finger(-type gallery)
~, сводчатая arcade

~, слуховая listening gallery
~, смотровая inspection gallery
~, транспортная *met.* conveyer gallery
галета 1. (sea/ship/pilot) biscuit, galette; 2. *tech.* wafer, deck; disk/pancake coil
галетный wafer-type, flat
~ная катушка disk/pancake coil
~ная обмотка pie winding
галечник shingle, bench/coarse gravel, pebble-bed
~, крупный cobble roundstone
галечниковый, галечный pebbled, pebbly, shingly; pebble, shingle *(attr.)*
галикт *zool.* sweat bee *(Halictus)*
галилейский Galilean
галимодендрон *bot.* salt tree *(Halimodendron)*
галинзога, галинсога *bot.* quickweed *(Galinsoga)*
галиот *naut.* gal(l)iot, small galley
галипланктон haliplankton
галистерез *med.* halisteresis
галит *med.* halite, common salt
галичанин Galician
галка *zool.* (jack)daw *(Corvus monedula)*
галл I Gaul
галл II (plant) gall; cecidium
~, волосатый bedeguar
~, корончатый crown gall
~, орехоподобный gallnut
галлей-кабинет *print.* galley cabinet
галлей-реал *print.* galley rack
галлерта jelly-like animal glue
галлий *chem.* gallium, Ga
галликанец *eccl.* Gallican
галлиза *zool.* gall midge *(Contarinia)*
~, сорговая sorghum midge *(Contarinia sorghicola)*
галловой, галловый *chem.* gallic
~вая кислота gallic acid
галлон gallon
галльский 1. Gallic; Gaulish; 2. *hist.* French
галлюцинаторный hallucinative, hallucinatory
галлюцинационный delusional
галлюцинация hallucination
~ции, вкусовые gustatory hallucinations
~ции, гаптические haptic hallucinations
~ции, зрительные visual hallucinations
~ции, обонятельные olfactory hallucinations
~ции, рефлекторные reflectory hallucinations
~ции, слуховые auditory hallucinations
~ции, тактильные tactile hallucinations
галлюциноген hallucinogen
галлюциноз hallucinosis
~, алкогольный alcoholic hallucinosis
~, зрительный visual hallucinosis
~, сифилитический syphilitic hallucinosis
~, слуховой acoustic hallucinosis
гальярд galliard, calliard
галмей *miner.* galmei, galmey, calamine, hemimorphite
галмейный calamine
гало halo
~, полное circumscribed halo
~, радужное coloured halo
~, солнечное solar halo, parhelic ring
галобентос halobenthos
галобиоз halobiosis
галоболит halobolite, pelagite
галовакс halowax, chloronaphthalene wax
галоген *chem.* halogen, haloid
галогенангидрид *chem.* halogen(-)anhydride
галогенез halogenesis
галогенид halogenid
~ серебра silver halide
галогенидгидролаза halidohydrolase
галогенирование halogenation
галогенный halogenous, haloid
галогеноводород halogen hydride, halogen-hydrogen
галоид *chem.* haloid, halide, halogen
галоидалкил *chem.* aliphatic halide
галоидбензол *chem.* phenyl halide
галоидирование halogenation
галоидный halogen, haloid
~ное соединение halide
~ные соли halogen salts
галоидопроизводные *pl.* halogen derivatives
галоидоуглерод *chem.* halocarbon
галолит halogen rock
галометр halometer
галоморфизм halomorphism
галоморфный halomorphic
галоп 1. gallop; 2. galop *(dance)*
~, легкий canter
галопирование galloping
галопланктон haloplankton, saltwater plankton
галопрогестерон *med.* haloprogesterone
галотан halothane
галотрихит *miner.* halotrichite, iron alum
галофильный halophilous, salt-loving
галофит *bot.* halophyte
~, облигатный obligate halophyte
галофитный *bot.* halophytic
~ная серия halosere
галохимия halochemistry
галоша 1. overshoe, galosh; 2. galosha *(rubber solvent)*
~, водолазная weighted shoe
~, клееная, ~ ручной сборки hand-made overshoe, bonded overshoe
~, формовая moulded overshoe
галс *nav.* tack, ratch, board, traverse
~ в бакштаг free reach
~ в подветренную сторону lee tack
~, делать make a board
~, контрольный промерный check line of sounding
~, левый port tack
~, нормальный промерный normal sounding line
~, переменить tack

~, **поперечный** upsweep leg
~, **правый** starboard tack
~, **промерный** sounding tack/line, line of sounding(s)
галстук necktie; neckerchief
галстук-бабочка bow tie
галстучник *zool.* ring(ed) plover *(Charadrius hiaticula)*
галтель 1. fillet, hollow chamfer; flute; groove; 2. moulding/cornice/ogee plane
~ **вала** shaft fillet
~ **коленчатого вала** bearing fillet
~ **щеки коленчатого вала** crankshaft web fillet
галтовать tumble
галтовка barrel finishing/polishing, barreling, tumbling
галун galoon, lace
галургия mineral-salt production, halurgy, salt working
галфвинд *nav.* wind abeam/abreast, beam wind
гальванизация electroplating, galvanizing, galvanization
гальванизировать (electro)plate, galvanize
гальванически galvanically
~ **луженый** tin-electroplated
~ **покрытый** electroplated
гальванический galvanic
~ **элемент** galvanic cell
гальвано *print.* electrotype
гальваноглифия *print.* galvanoglyphy
гальванодиатермия *med.* galvanodiathermy
гальванокаустика galvanocautery
гальванокаустический galvanocauteric
гальванокаутер *med.* galvanocauter
~, **глазной** ophthalmic galvanocauter
гальваноклише galvano
гальванометр galvanometer
~, **апериодический** dead-beat/aperiodic galvanometer
~, **бифилярный** bifilar galvanometer
~ **Дарсонваля** d'Arsonval galvanometer
~, **зеркальный (с вращающейся катушкой)** (moving-coil) mirror/ reflecting galvanometer
~, **интегрирующий** integrating galvanometer
~, **магнитноэлектрический** moving-coil galvanometer
~, **осциллографический** oscillograph galvanometer
~, **петлевой** single-turn moving-coil galvanometer
~, **пружинный** spring galvanometer
~, **рамочный** multi-turn moving-coil galvanometer
~, **резонансный** vibration galvanometer
~ **с успокоенной стрелкой** dead-beat galvanometer
~, **самопишущий** recording galvanometer
~ **со световым отсчетом** multiple-reflection galvanometer
~, **струнный** string galvanometer
~, **тепловой** hot-wire galvanometer, thermal galvanometer
~, **термоэлектрический** thermo-emf galvanometer, thermocouple galvanometer
~, **шлейфовый** mirror/reflecting galvanometer
~, **экранированный** screened/shielded galvanometer
~, **электромагнитный** moving-iron galvanometer
гальваноотложение electrotype shell
гальванопластика galvanoplasty, galvanoplastics, electroforming, electrotypy, electroplating
~ **в псевдоожиженном слое** electroforming in fluidized bed
гальванопластический, гальванопластичный galvanoplastic, electrotyping, electroforming
гальванопокрытие (electro)plating, electrochemical facing, galvanic deposit, electroplated coating
~, **контактное** contact plating
гальваноскоп galvanoscope, current detector
гальваностег electrodepositor, electroplater
гальваностегический galvanostegic, electroplating
гальваностегия electroplating, electrodeposition, galvanostegy
гальваностереотип electrotype
гальваностереотипия electrotyping
гальванотаксис galvanotaxis
гальванотерапия galvanotherapy, galvanotherapeutics
гальванотермия galvanothermia
гальванотехник electrodepositor, electroplater
гальванотехника electrodeposition, galvanotechnics, electroplating technology, electrolytic metallurgy
гальванотипия electrotyping
гальванохромия electrocolouring
галька pebble(s), shingle, gravel (stones), rubble
~, **береговая** beach shingle/pebble
~ **величиной с горошину/орех** pea/hazelnut-size pebble
~, **граненая** faceted pebble
~, **исштрихованная** striated pebble
~, **кварцевая** quartz gravel
~, **крупная** cobble (round-stone)
~, **ограненная ледником** faceted pebble
~, **одногранная** one-edge pebble, einkanter
~, **песчаная** sand pebble
~, **прибрежная** beach pebble
~, **промытая** wash gravel
~, **сглаженная** soled cobble
~, **трехгранная** dreikanter, three-edge pebble
~, **удлиненная** stretched pebble
~, **черепицеобразно расположенная** imbricate pebbles
~, **эоловая** windcarved pebble, ventifact
гальковидный calculiform, pebble-shaped
гальпинит johannite

гальтония *bot.* summer hyacinth *(Galtonia)*
гальюн *naut.* (the) head(s), lavatory
гама *bot.* eastern gama grass *(Tripsacum dactyloides)*
гамадриада *myth.* hamadryad
гамадрил *zool.* hamadryas baboon *(Papio hamadryas)*
гамак hammock
~ на полочках *med.* hammock on poles
гамак-подстопник *med.* foot hammock
гамамелис *bot.* witch hazel *(Hamamelis)*
гамартобластома *med.* hamartoblastome
гамартома *med.* hamartoma
~ легких pulmonary hamartoma
~, неорганоидная anorganoid hamartoma
гамаша legging, gaiter
гамбир *bot.* Bengal gambir plant *(Uncaria gambir)*
гамбит gambit
Гамблий Gamblian
гамета *biol.* gamete, flagellated zoospore, sexual cell
~, женская macrogamete, megagamete
~, мужская microgamete
гаметангиальный, гаметангиевый *bot.* gametangial
гаметангий *bot.* gametangium
гаметический, гаметный *biol.* gametic
гаметогоний *biol.* gametogonium (*pl.* -gia)
гаметоид *bot.* gametoid
гаметоноситель gamete carrier
гаметофит *bot.* gametophyte
гаметофор(а) *bot.* gametophor
гаметоцид gametocide
гаметоцит gametocyte
гамильтониан *math.* Hamiltonian('s) function/ operator
гамлинит *geol.* hamlinite, bowmannite
~, барийсодержащий barium-hamlinite
гамма I 1. *mus.* scale; gamut; 2. *tech.* gamma
~ до мажор C major scale
~ ощущений the (whole) gamut/range of emotions
~, пентатоновая/китайская pentatonic scale
~ телевизионной передачи gamma
~ цветов colo(u)r gamut
гамма II Gamma *(letter of Greek alphabet)*
гамма-активный gamma-active/-emitting
гамма-актинометр gamma-ray actinometer
гамма-аминокислоты *pl.* gamma-aminoacids
гамма-аппарат *med.* gamma (therapy) unit, gamma apparatus
гамма-астрономия gamma-ray astronomy
гаммавлагомер gamma-ray moisture meter
гамма-глобулин *biol.* gamma-globulin
гаммаграфирование gamma-ray examination/ inspection
гаммаграфия gammagraphy, gamma(-ray) radiography
гамма-грунтомер gamma soil-density meter
гаммада ham(m)ada, rocky desert
гамма-дефектоскоп gamma-ray flaw detector, gamma-ray defectoscope
гамма-дефектоскопия (industrial) gamma(-ray) radiography, gamma-ray inspection /testing/ defectoscopy, gamma(-ray) flaw detection
гамма-дозиметр gamma-dosimeter
~, поисковый gamma survey meter
гамма-излучатель gamma-emitter/-radiator, gamma-ray source
гамма-излучение gamma radiation/rays, gammas, gamma-ray emission
~ высвечивания de-excitation gamma-rays
~ деления fission gammas
~, захватное capture gamma-rays
~, каскадное gamma-cascade, cascade gammas, successive/cascade gamma-rays
~, мгновенное prompt gammas
~ от неупругого рассеяния inelastic-scattering gammas
~ продуктов деления fission-product gammas
~ разрядки de-excitation gammas
~, чистое pure gamma-radiation
гамма-измеритель gamma-gauge
гамма-каротаж gamma(-)logging
гамма-квант gamma(-ray) quantum/photon
~ деления fission gamma quantum
~, захватный capture gamma-quantum
~ы, каскадные successively emitted gamma-quanta, cascade gamma-quanta
гамма-линия gamma-peak
гамма-лучи *pl.* gamma-rays
~, захватные capture gamma-rays
~, мгновенные prompt gamma-rays
гамма-облучатель gamma-irradiator
гамма-облучение gamma-irradiation
гаммапатия *med.* gammapathy, gammopathy
~, диклонова diclonal gammapathy
гамма-переход gamma(-ray) transition, radiative transition
~, мультипольный multipole gamma-transition
гамма-поле gamma-field
гамма-поток gamma-flux
гамма-прибор Гесса/Кольгерстера Hess/ Kohlhorster gamma-apparatus
гаммарит *geol.* hammarite
гаммарус *zool.* gammarid
гамма-снимок gammagraph, radiograph
гамма-спектрометр gamma(-ray) spectrometer
гамма-счетчик gamma-counter
гамма-съемка gamma-survey, gamma-ray surveying
гамма-телетерапия gamma(-ray) teletherapy
гамма-толщиномер gamma(-ray) thickness gauge
гамма-уран gamma-uranium, gamma phase of uranium
гамма-установка gamma(-ray)unit/source
~, терапевтическая therapeutic gamma unit
гамма-фактор gamma
гамма-фон gamma-background

~у, компенсированный по gamma-compensated
гамма-фотон gamma(-ray) photon/quantum
гамма-функция gamma function
гамма-цепь gamma-chain
гамобиум *zool.* gamobium
гамогенез *boil.* gamogenesis
гамомания eroticism
гамон *bioch.* gamone
гамонт *zool.* gamont
гамонтоцид gamontocide
гамопетальный *bot.* gamopetalous, sympetalous
гамосепальный *bot.* gamosepalous
~ная чашечка gamosepalous calyx
гамостела *bot.* gamostele
гамофиллия *bot.* gamophylly
гангамоптерис *paleont.* gangamopteris
ганглерон *anat.* ganglerone
ганглий *anat.* ganglion, nerve knot; (*pl.* ganglions, ganglia)
~ии, внутрисердечные intracardiac ganglia
~ии, межпозвоночные intervertebral ganglia
~ии, нервные nervous ganglia
~ии, парасимпатические parasympathetic ganglia
~ии, подкоровые subcortical ganglia
~ии, симпатические sympathetic ganglia
ганглиит *med.* ganglionitis, gangliitis
ганглиобласт *med.* ganglioblast, esthesioblast
ганглиобластома ganglioblastoma
ганглиоблокирующий ganglioblocking
ганглиоглиома *med.* ganglioglioma
ганглиозный, ганглионарный *anat.* ganglionic, ganglionary
ганглионеврит *med.* ganglioneuritis
ганглионевробластома *med.* ganglioneuroblastoma
ганглионеврома *med.* ganglioneuroma
ганглионейроматоз *med.* ganglioneuromatosis
ганглионит *med.* ganglionitis, gangliitis
ганглиоплегический ganglioplegic
гангрена *med.* gangrene, phagedena
~ влажная moist gangrene
~ газовая gas/emphysematous gangrene, gas phlegmon
~ гениталий, острая Fournier's disease
~, госпитальная hospital/nosocomial/pulpy gangrene, sloughing phagedena, phagedena nosocomidalis
~, застойная static /venous gangrene
~, инфекционная hot gangrene
~, ишемическая/белая white gangrene
~ кожи cutaneous gangrene
~ конечностей, симметричная Raynaud's gangrene/disease, gangrene of extremities
~, молниеносная Pirogoff's/malignant edema, fulminating gangrene
~ от пролежия bedsore, pressure gangrene, decubitus ulcer
~, ползучая phagedena gangrenosa
~, предстарческая presenile gangrena
~, самопроизвольная spontaneous /neurotic gangrene
~, старческая senile/Pott's gangrene
~, сухая dry gangrene/necrosis, mummification
~, тромботическая thrombotic gangrene
~, трофическая trophic gangrene
~, холодная cold gangrene
гангренозный gangrenous
~ процесс *med.* mortification
гангстер gangster
гангстерство gansterism
гандбол *sport* handball
гандикап *sport* handicap
гандшпуг *naut.* handspike, lever
ганец Ghan(a)ian
ганза I *hist.* Hanse, Hanseatic League
ганза II yellow pigments
ганзеатический, ганзейский *hist.* Hanseatic
ганистер *geol.* ganister
ганит *miner.* gahnite, zinc spinel
~, цинково-марганцовисто-железный dysluite, zinc- manganese-iron gahnite
ганнеит *miner.* hannayite
ганноверский Hannoverian
ганодонт *paleont.* ganodont
ганоид *zool.* ganoid
~, костный holostean
~, хрящевой chondrostean
ганоидный ganoid
ганоин *zool.* ganoin(e)
ганомалит *miner.* ganomalite
ганофиллит *miner.* ganophyllite
гантелеобразный dumbbell-like
гантель *sport* dumb-bell, dumbbell
гануман *zool.* entellus langur *(Presbytis entellus)*
ганушит *geol.* ganusite
ганшпуг *naut.* handspike
гаолян *bot.* kaoliang *(Sorghum japonicum)*
ГАПС гибкая автоматизированная производственная система *abbr.* fexible manufacturing system, FMS
гапаксантный *bot.* hapaxanthic, hapaxanthous
гаплобактерия haplobacterium
гаплобионт haplobiont
гаплогонидия *bot.* haplogonidium
гаплозис *biol.* haplosis
гаплоид *biol.* haploid
гаплоидия, гаплоидность *biol.* haploidity
гаплоидный hadploid(ic)
гаплоконидия *bot.* haploconidium
гаплолалия haplolaly, haplology
гаплостела haplostele
гаплостемонный haplostemonous
гаплотип *biol.* haplotype
гаплофир haplophyre
гаплофит *bot.* haplophyte, gametophyte
гаптера *bot.* hapteron
гаптически *psych.* haptic
гаптофор haptophore

гаптофорный haptophore, haptophoric
гараж garage
  ~, ставить (машину) в garage (a vehicle)
гарант *law*. guarantor, warrantor
гарантийный guarantee(d), warranted
гарантирование guaranteeing, securing
гарантированный guaranteed, warranted; assured; dependable
гарантировать guarantee, vouch for/against, secure (against); stand surety
  ~ возмещение убытков guarantee against boss
гарантия guarantee, guaranty, warranty
  ~, аварийная average bond
  ~, банковская bank/banker's guarantee
  ~ безопасности *law* safe conduct
  ~ возврата платы за товар money-back guarantee
  ~ золотого эквивалента gold-value guarantee
  ~ на год a year's garantee
  ~ подразумеваемая implied warranty
  ~тию, под against security
гарантоспособность dependability
гаргоилизм *pathol.* gargoylism, gargolism
гаргрот *av.* (middle) fuselage fairing
  ~, верхний upper/back fairing
  ~, нижний lower/ventral fairing
гардель *naut.* halyard; gear
гардемарин *hist.* naval cadet
гардения *bot.* gardenia *(Gardenia)*
гардероб 1. cloakroom, checkroom; *av., nav.* coat-stowage compartment; 2. wardrobe (piece of furniture); 3. clothes
гардеробная cloakroom, checkroom; *av., naut.* coat-stowage compartment
гардина curtain
гардистонит *geol.* hardystonite
гарем harem
гарибальди *zool.* garibaldi *(Hypsypops rubicundus)*
гарибальдиец *hist.* Garibaldian
гармала *bot.* peganum *(Peganum)*
гармалин *chem.* harmaline
гарматан harmattan
гармин *chem.* harmin(e)
гармонизация *mus.* harmonization
гармоника 1. *phys.* harmonic (wave); *math.* harmonic(s), harmonic (curve); 2. *tech.* (gangway) bellows; 3. *mus.* accordion; concertina
  ~, высшая higher/upper harmonic
  ~, дробная fractional harmonic
  ~, нечетная odd harmonic
  ~, нормированная normalised harmonic
  ~, объемная solid harmonic
  ~, основная/первая fundamental oscllation, first/fundamental harmonic/wave
  ~, поверхностная space/spatial harmonic
  ~, сферическая solid/spherical harmonic
  ~, четная even harmonic
  ~ки, кратные трем triplen harmonics
  ~ки крутящего момента torque harmonics
гармонист 1. accordion/concertina player; 2. *mus.* harmonist, specialist in harmony
гармониум *mus.* harmonium, reed-organ
гармоническая *n. phys.* harmonic
гармонический 1. *tech.* rhythmic; 2. *mus., math.* harmonic; harmonious
гармоничность harmoniousness, harmony
гармоничный harmonious, concordant
гармония 1. *tech., mus.* harmony; 2. accordion; concertina
  ~ управления harmony of controls
гарна *zool.* Anchian antelope *(Antelope cervicapra)*
гарнец 1. garnetz, garnets *(Russian dry measure = 3. 28 liters);* 2. container of one garnetz
гарниерит, гарньерит *miner.* garnierite, noumeite
гарнизон *mil.* garrison
  ~ом, ставить garrison
гарнир *cul.* garnish, trimmings, decoration
гарнирование, гарнировка garniture, garnishing; trimming, decoration
гарнисаж *tech.* skull, slag lining
гарнитур I suite; (head and chest) set; trimmings, fittings, mountings
  ~, головной *tech.* webbing cap, head-dress
  ~, спальный bedroom set
  ~, телефонный telephone headset assembly
гарнитур II *text.* gros de Tours (silk fabric)
гарнитура *tech.* fittings, mountings; outfit; *print.* set, type; *text.* fillet, clofhing
  ~, авиационная airman's headset
  ~, газетная newspaper type
  ~, головная *teleph.* headset
  ~, игольчатая card fillet/clothing/covering, filetting card, clothing wire
  ~, каллиграфическая hand/script type
  ~ котла boiler mountings
  ~, латинская Roman type
  ~, машинописная typewriter type
  ~, микротелефонная handset
  ~, пильчатая saw-toothed fillet
  ~, плакатная poster type
  ~, рубленая san-serif type
  ~ съемного валика *text.* stripper card
  ~, табличная Elzevir type
  ~ телефонистки operator's (telephone) set
  ~, шлемофонная *rad.* communication helmet
гарновка *bot.* durum wheat *(Triticum durum)*
гарный combustible, burning;
  ~ное масло lamp oil
гарньерит *miner.* garnierite, noumeite
гарпинс *nav.* harpin(g)
гарпиус *chem.* colophony (rosin), white rosin/resin
гарпия 1. *myth.* Harpy; 2. *zool.* harpy eagle *(Harpia harpyja);* harpy bat, hairy-winged bat *(Harpiocephalus harpia)*
гарполит *geol.* harpolith
гарпун harpoon
  ~ом бить harpoon
гарпунер harpooner

гаррельсит *miner.* garrelsite
гаррисит *miner.* harrisite
гарсоньерка bachelor's room
гарт *print.* type metal; printer's pie
гартблей hard/antimonial lead, type metal
гартлинг hard head
гартоварка *print.* remelting division
гартоплавилка *print.* melting furnace
гарцбургит *geol.* harzburgite
гарциния *bot.* garcinia *(Garcinia)*
гаршпен *zool.* jacksnipe *(Lymno cryptes minima)*
гарь 1. slash, burned-out forest; 2. *met.* ash(es), cinder(s); 3. smoke, fume(s); burning; 4. scale, soot formation, carbon deposit, caking
гасиенда hacienda
гаситель extinguisher; slaker, slaking apparatus; quencher, quenching agent/unit; damper, suppressor; flow damper
~ автоколебаний носового колеса *av.* anti-shimmy device, shimmy damper
~ вибраций/колебаний vibration damper/suppressor
~ гидравлического удара water-hammer alleviator
~ извести lime slaker
~ колебаний рыскания *naut.* yaw damper
~ крутильных колебаний torsional vibration damper
~ люминисценции quencher
~ пламени flame arrester
~ поля *el.* field killer
~ подъемной силы lift damper
~ поперечных колебаний *naut.* roll damper
~ продольных колебаний *naut.* pitch damper
~ путевых колебаний *av.* directional-control damper, yaw damper
~ тяги *av.* thrust spoiler
гасительный damping; extinquishing; quenching
гасить 1. extinguish, put out, turn off; 2. quench, damp(en), suppress; kill, stifle; 3. cancel
~ долг liquidate a debt
~ известь slake lime
~ кокс quench coke
~ мяч *sport.* kill the ball
гасконец Gascon
гаснуть go/die out, extinquish
гаспис *timb.* hawsepiece, hawse timber
гассулит *geol.* ghassoulite
гастальдит *geol.* gastaldite
гастелит *geol.* hastellite
гастеромицет *bot.* gasteromycete
гастингсит *geol.* hastingsite
гастральгия *med.* gastralgia
гастрит *med.* gastritis
~, алиментарный alimentary gastritis
~, антральный antral gastritis
~, гематогенный hematogenous/hematogenic gastritis
~, гипацидный subacid gastritis
~, гиперацидный hyperacid gastritis
~, гистаминный hystamine-induced gastritis
~, дифтерийный diphtheric gastritis
~, интерстициальный interstitial gastritis, initis plastica
~, катаральный catarrhal gastritis
~, коррозийный corrosive gastritis
~, опухлевидный tumorous gastritis, tumor-like gastritis
~, опухольсимулирующий tumor-imitating gastritis
~, очаговый focal gastritis
~, подострый subacute gastritis
~, полипозный polypous gastritis, polyposis gastrica
~, фиброзно-склеротический gastritis fibroplastica
~, эндогенный endogenous gastritis
~, язвенный ulcerative gastritis
гастрический *med.* gastric, stomachal
гастроаденит *med.* gastroadenitis, gastradenitis
гастроанастомоз *med.* gastroanastomosis, gastrostomy
гастробиопсия *med.* gastrobiopsy
~, прицельная aiming/directed gastrobiopsy
гастродискоидоз *med.* gastrodiscoidosis
гастродуоденит *med.* gastroduodenitis
гастрозооид *zool.* gastrozooid
гастроколит *med.* gastrocolitis
гастроколоптоз *med.* gastrocoloptosis
гастроколотомия *med.* gastrocolotomy
гастрокрин *med.* gastrocrine, ventriculin
гастролит 1. *med.* gastrolith; 2. *miner.* crab's stone/eye, gastrolith, lizardstone
гастроль *theatre* tour; guest performance, temporary engagement
~лях, на on tour
гастромаляция *med.* gastromalacia
гастромукорея *med.* gastroblenorrhea
гастроном I delicatessen store
гастроном II gourmet, epicure, gastronome
гастрономический gastronomic(al)
~ магазин delicatessen store, grocery shop
~ие товары delicatessen
гастропатия *med.* gastropathy
гастропексия *med.* gastropexy
гастропластика *surg.* gastroplasty
гастропликация *surg.* gastroplication
гастропод *zool.* gast(e)ropod
гастропора *zool.* gastropore
гастроптоз *med.* gastroptosis
гастрорафиотом *med.* gastrorrhaphiotom
гастроскоп *med.* gastroscope
~, жесткий rigid/hard gastroscope
~ с управляемым изгибом [панорамным обзором] variable flexible panoramic gastroscope
гастроскопия *med.* gastroscopy
гастроспазм *med.* gastrospasm
гастросуккорея *med.* gastrosuccorrhea
~, длительная/слизистая continuous/mucous gastrosuccorrhea

гастротом *surg.* gastrotome
гастротомия *surg.* gastrotomy
гастрофилез gastrofilosis
гастроцель *zool.* gastrocoel(e)
гастроэзофагит *med.* gastroesophagitis
гастроэнтеральный *anat.* gastroenteric
гастроэнтерит *med.* gastroenteritis
~, брюшнотифозный typhoid gastroenteritis
~, паратифозный paratyphoid gastroenteritis
~, стафилококковый staphylococcic gastroenteritis
~, холерный choleraic gastroenteritis
~, эпидемический вирусный epidemic virus gastroenteritis
гастроэнтероколит *med.* gastroenterocolitis
гастроэнтеролог gastroenterologist
гастроэнтерология gastroenterology
гастроэнтеропатия gastroenteropathy
гастроэнтеропластика gastroenteroplasty
гастроэнтероптоз gastroenteroptosia
гастроэнтероррагия gastroenterorrhagia
гастроэнтеростомия gastroenterostomy
гастроэнтеротомия gastroenterotomy
гаструла *embryol.* gastrula (*pl.* -lae)
гаструляция gastrulation, emboly
гастрэктазия gastrectasia, gastrectasis
гастрэктомия *surg.* gastrectomy
гаттерия *zool.* tuatara *(Sphenodon punctatus)*
гатура *cart.* hachure
гать 1. corduroy road, log-path/-road; 2. dam, levee
~, настилать lay a log-road
~, хворостная brushwood road
гаубица *mil.* howitzer
гаультерия *bot.* wintergreen *(Gaultheria)*
гауптвахта *mil.* guardhouse, guardroom
~, военно-морская naval detention/discipline barracks
гаур *zool.* gaur *(Bos gaurus)*
гаура *bot.* gaura *(Gaura)*
гаусзистор gaussistor
гаусманнит *miner.* hausmannite
гаусс *el.* gauss
гауссиан Gaussian
гауссметр gaussmeter
гаусторий, гаустория *bot.* sucker, haustorium
гаухекорнит *miner.* hauchecornite
гауч-вал *paper industry* couch roll
гауч-пресс *paper industry* couch press
гафель *naut.* gaff
~ для боевого/кормового флага battle/ensign gaff
~, постоянный standing gaff
гафельный *naut.* gaff-headed
гафний *chem.* hafnium, Hf
гафф haff
гацания *bot.* gazania *(Gazania)*
гач paraffin(e) slack wax, wax cake, cheese of paraffin(e)
гашение extinction, extinguishing; quenching; *comp.* erasing; cleaning, cancellation; putting out; damping, dampening
~ волн cancellation of waves
~ дуги arc extinction
~ знаков *comp.* character blanking
~ извести, воздушное air slaking
~ изображения display suppression
~ интенсивности cancellation of intensities
~ колебаний/вибраций vibration control/suppression/damping
~ луча beam suppression/blanking
~ магнитного поля field killing
~ механических колебаний dampening, suppression
~ обратного хода луча/развертки flyback/retrace suppression
~ памяти memory clearing
~ пены foam suppression
~, полное absolute damping
~ пучка beam suppression/blackout
~ разряда quenching of discharge
~ реактора shut-down
~ счетчика *comp.* clearance
гашеный slaked, extinguished, quenched; *comp.* erased, cleared
~ная известь slaked lime
гашетка firing button, trigger (switch)
~ку, нажать на pull the trigger
гашиш hasheesh, hashish, cannabis, marijuana, marihuana, Indian hemp
гашишизм hasheeshism, hashishism, cannabism
гашпиль *leath.* paddle-wheel (tank), reel
гашюра hachure, dash
гаэл *hist.* Goidel; Gael
гая *zool.* hooded cobra *(Naja haje)*
гаял *zool.* gayal *(Bibos frontalis)*
гвайфиллин guaifylline
гвайюла *bot.* parthenium *(Parthenium)*
гвайюлолистный *bot.* parthenium-leaved
гвардеец *mil.* guardsman
гвардейский *mil.* Guard's
~кая дивизия division of Guards
гвардиан(ец) *mil.* guardsman
гвардия *mil.* Guards, guard(s)
~, гражданская home guards
~дии, майор major of the Guards
~, старая old guard
гватемалец, гватемальский Guatemalan
гваюла *bot.* guayule *(Parthenium argentatum)*
гваяк *bot.* common lignum vitae *(Guaiacum officinale)*
гваякол *chem.* guaiacol
гваяктамин guaiactamine
гваяпат guaiapate
гваяцил guaiacyl
гвельф *hist.* Guelph
гвельфский *hist.* Guelphic
гвереца *zool.* colobus *(Colobus)*
гвизоция *bot.* niger seed *(Guizotia)*
гвинеец, гвинейский Guinean
гвоздевание nailing

гвоздевидный nail-shaped, claviform
гвоздевка *agr.* spike/nail harrow
гвоздик small nail, tack
гвоздика *bot.* carnation, pink *(Dianthus)*
~, бородатая sweet William, French pink *(D. barbatus)*
~, голландская/садовая clove pink *(D. caryophylus)*
~, китайская rainbow pink *(D. chinensis)*
гвоздиковидный *bot.* caryophyllaceous; clove-like
гвоздикоцветковый *bot.* pink-flowered, dianthiflorous
гвоздильный nail, nail-making
~ завод nailery
гвоздильня 1. nail puller/extractor; nailhead machine, nail-making machine, nail header; 2. nailery
гвоздимый nailable; nailing
гвоздичник *bot.* thrift *(Statice)*
~, мадагаскарский Madagascar clove *(Ravensara aromatica)*
гвоздичный *bot.* caryophyllaceous; *n. pl.* pink family *(Caryophyllaceae)*
гвоздодер nail extractor/puller, tack claw
гвоздок *zool.* kinglet *(Regulus)*
гвоздь nail; tack; pin; peg
~ без шляпки brad, wire nail
~, бесканальный трехлопастный uncannulated three-flange nail
~, бондарный barrel nail
~, бородчатый barbed nail
~ вытаскивать extract/draw a nail
~, гонтовый shingle nail
~, деревянный peg, dowel
~, заершенный barbed nail
~, забивать drive (in) a nail
~, загибать clinch a nail
~, затяжной *leath.* lasting tack
~, кованый wrought nail
~, кровельный roofing nail
~, машинный cut nail
~, обойный upholsterer's tack
~, отделочный finishing nail
~, паркетный brad
~, подковный horse-shoe nail, horseshoe nail
~, проволочный wire nail
~, резаный/реечный cut/batten nail
~ с большой шляпкой stud
~, сапожный shoe nail
~, стержневой core nail
~, стропильный rafter nail
~, тарный box/casing nail
~, трехлопастный three-flange nail
~, трубчатый tubular nail
~, формовочный moulder's brad/sprig
~, шиферный slating nail
~, шлюпочный boat/deck/spike nail, boat spike
~, штукатурный plaster/clout nail
гвоздь-холодильник chill nail
Геба, Гебея *myth.* Hebe
гебефренический *psychiatry* hebephrenic
гебефрения hebephrenia
гебраист Hebraist
гебронит *miner.* hebronite
гебуритодацит *miner.* geburite-dacite
гевер *bot.* hevea *(Hevea)*
~, бразильская Para rubber tree *(H. brasiliensis)*
гегельянец *philos.* Hegelian
гегельянство Hegelianism
гегемон leader, hegemon
гегемонизм *polit.* hegemonism
гегемония hegemony, supremacy
геденбергит *miner.* hedenbergite
~, гранулярный coccolite
~, темнозеленый baikalite
геджра hegira, Mohammedan calendar
гедихиум *bot.* ginger lily *(Hedychium)*
гедонизм hedonism
геенна Gehenna, hell
гезенк *min.* winze, blind shaft
~, проходить sink a winze
гейдельбергский Heidelberg
гейзер geyser
~, грязевой mud pot/geyser
~, ритмический intermittent geyser
гейзерит *miner.* geyserite, siliceous sinter
гейландит *miner.* heulandite
гейлюссакия *bot.* huckleberry *(Gaylussacia)*
гейм *sport.* game
геймит *miner.* heumite
гейтоногамический *bot.* geitonogamous
гейтоногамия *bot.* geitonogamy
гейтонокарпия *bot.* geitonocarpy
гейхера *bot.* heuchera *(Heuchera)*
гейша geisha
Геката *myth.* Hecate
гекатомба hecatomb
гекзаметр *lit.* hexameter
геккон *zool.* gecko *(Gekkon)*
~, плоскохвостый leaf-tailed gecko *(Uroplatus)*
~, тонкопалый sand gecko *(Stenodactylus petriei)*
гексагидрит *miner.* hexahydrite
гексагонально-дипирамидальный hexagonal dipyramidal, tripyramidal
гексагонально-пирамидальный hexagonal pyramidal, pyramidal hemimorphic
гексагонально-трапецоэдрический hexagonal trapezohedral
гексагональный hexagonal, six-sided
гексаграмма *lit.* hexagram
гексада hexad
гексадактилия *med.* hexadactylism
гексакисоктаэдр *geom.* hexakisoctahedron
гексакобальтнитрит натрия cobalt sodium nitrite
гексакоралл hexacoral
гексамерон hexameron

гексаметилентетрамин hexamethylenetetramine, methenamine
гексамидин hexamidine, primidone
гексамин hexamine, methanamine
гексан *chem.* hexane
гексантин hexantin, triaxial spicule
гексаплоид *biol.* hexaploid
гексаплоидия hexaploidy
гексархный *bot.* hexarch
гексастр *zool.* hexaster
гексатетраэдр hexatetrahedron
гексатетраэдрический hexatetrahedral
гексафилльный *bot.* hexaphyllous
гексафторид урана uranium hexafluoride
гексахлоран hexachlorane
гексахлорофен hexachlorophene
гексахлорциклогексан hexachlorcyclohexane
гексахлорэтан hexachlorethane, carbon trichloride
гексаэдр hexahedron
гексаэдрический hexahedral, cubic
гексит *chem.* hexite, hexitol
гексобарбитал hexobarbital/evipal sodium
гексоза *chem.* hexose
гексозамин hexosamine
гексоксидаза hexoxidase
гексозилтрансфераза hexosyltransferase
гексозодифосфатаза hexosediphosphatase
гексокиназа hexokinase
гексоктаэдр hexoctahedron
гексоктаэдрический hexoctahedral
гексоний hexamethonium
гексопирроний hexopyrronium
гектар hectare
гектический hectic
гектоватт *el.* hectowatt
гектограмм hectogram
гектограф hectograph
~, размножать текст на hectograph
гектографирование hectography
гектографировать (copy with a) hectograph
гектографический, гектографский hectographic
гектокотиль *zool.* hectocotylus (*pl.* -li)
гектолитр hectoliter
гектометр hectometer
Гектор *myth.* Hector
Гекуба Hecuba
геленит *miner.* helenite *(of coal);* gehlenite
гелениум *bot.* sneezweed *(Helenium)*
гелеобразный jellike, jellous, jellylike, jelly
гелеобразование gelation, gel-formation, gelling (action)
гелеобразующий gel-forming
гелеопланктон heleoplankton
гелеподобный jellylike, jelly, jellous, jellike
гелий *chem.* helium, He
~ вытеснительной системы *av.* pressurizing helium
~, газообразный helium gas, gaseous helium
~, радиогенный radiogenic helium
~, сверхтекучий superfluid helium
геликоид *math.* helicoid
~, наклонный oblique helicoid
геликоидальный helical, helicoid(al)
геликон *mus.* helicon; *myth.* Helicon
геликонида *zool.* Julia butterfly *(Colaenis julia)*
геликония *bot.* heliconia *(Heliconia)*
геликоптер *av.* helicopter
~, боевой combat helicopter
~, санитарный casualty helicopter
~, спасательный rescue helicopter
геликс helical element, helix
гелиобатарея solar (cell) battery
гелиогидрофит *bot.* heliohydrophyte
гелиогравюра *phot.* heliogravure, photogravure
гелиоводонагреватель solar water-heater
гелиограф heliograph
гелиография heliography, solar science
гелиодор heliodor
гелиокогира helical axis of symmetry
гелиоконцентратор *tech.* sun-light collector, solar-heat collector
гелиолит heliolite
гелиометр heliometer
гелион helion, alpha-particle
гелиоприемник sun-light collector, solar-heat collector
гелиопсис *bot.* heliopsis *(Heliopsis)*
гелиоскоп helioscope
гелиостат heliostat
гелиотаксис *biol.* heliotaxis
гелиотеплица helioglasshouse
гелиотерапия heliotherapy
гелиотехника solar-energy technology/engineering, heliotechnics
гелиотипия *phot.* heliotype
гелиотрипидин heliotripidine
гелиотроп 1. *miner.* heliotrope, blood-stone; 2. *paleont.* plasma; 3. *bot.* heliotrope *(Heliotropium)*
гелиотропизм *biol.* heliotropism, heliotaxis, phototropism
гелиотропин *chem.* heliotropin(e), piperonal
гелиотропист *geod.* heliotroper
гелиотропный *bot.* heliotropic
гелиоустановка sun/solar power plant, solar power generator
гелиофизик solar physicist
гелиофизика solar physics, physics of the Sun
гелиофильный heliophilous, sun-loving
гелиофит *bot.* heliophyte, sun plant
гелиофоб heliophobe
гелиофобия *med.* heliophobia
гелиофобный heliophobic, heliophobous
гелиохимия sun chemistry, cosmochemistry
гелиоцентрический heliocentric
гелиоцинкография heliozincography
гелиоэнергетика solar/sun power engineering
гелит *miner.* hoelite
гелификация gelification
гелифицированный gelified

гелихризум *bot.* everlasting *(Helichrysum)*
гелладотерий *paleont.* helladotherium
геллебор(ус) *bot.* hellebore *(Helleborus)*
гелоза *chem.* gelose, galactan
гелониас *bot.* swamp pink *(Helonias)*
гелотизм *biol.* helotism
гелофильный helophilous, bog-lowing
гелофит *bot.* helophyte, marsh plant
гель gel
~, акриловый acrylic gel
~, белковый protein gel
~, гидратированный hydrated gel
~, коллоидный colloidal gel
~, необратимый/обратимый irreversible/reversible gel
~, плотный firm/dense gel
~, рыхлый loose gel
~, силикатный silica gel
гельвет, гельветский *hist.* Helvetian
гельминт helminth
гельминтограф helminthograph
гельминтоз *med.* helminthism, helminthiasis
~, смешанный кишечный mixed intestinal helminth infestation
гельминтозооноз helminthozoonosis
гельминтолог helminthologist
гельминтологический helminthologic(al)
гельминтология helminthology
гельминтома helminthoma
гельминтоовоскопия helminthoovoscopy
гельминтоспориоз helminthosporiose
гельминтоцид helminthocide
гельминтоцидный antihelminthic, helminthagogue, helminthic, vermifuge
гельмпорт *nav.* helmport, rudder hole/port
гельский Goidelic
гем *bioch.* haem, heme
гемабарометр h(a)emabarometer
гемагглютинация h(a)emagglutination
гемагглютинины *pl.* h(a)emagglutinins
~, вирусные virus h(a)emagglutinins
гемангиобластома *med.* h(a)emangioblastoma
гемангиома h(a)emangioma
~, ветвистая/рацемозная racemous h(a)emangioma
~, кавернозная/пещеристая cavernous h(a)emangioma
~, склерозирующая sclerosing h(a)emangioma
гемангиоперицитома h(a)emangiopericytoma
~, злокачественная malignant h(a)emangiopericytoma
гемангиосаркома *med.* h(a)emangiosarcoma
гемангиофиброма h(a)emangiofibroma
гемангиоэндотелиобластома h(a)emangioendothelioblastoma
гемангиоэндотелиома h(a)emangioendothelioma
~ доброкачественная/злокачественная benign/ malignant h(a)emangioendothelioma
гемангиэктазия h(a)emangiectasia, h(a)emangiectasis
~, гипертрофическая hypertrophic hemangiectasia
гемантус *bot.* blood lily *(Haemanthus)*
гемартроз h(a)emartrosis, h(a)emartron, h(a)emarthros
~, гемофилический h(a)emophilic/bleeder's joint
гемартрома h(a)emarthroma
гематахометр h(a)ematachometer
гематеин h(a)ematein
гематин h(a)ematin
гематемезис h(a)ematemesis
гематит *miner.* hematite
~, блестящий specularite, specular hematite
~, бурый limonite, brown hematite
~, глинистый argillaceous hematite
~, губчатый iron froth, sponge hematite
~, плотный столбчатый compact columnar hematite
~, слоистый flag ore
~, слюдистый micaceous hematite
гематоаэрометр h(a)ematoaerometer
гематогенный h(a)ematogenic, h(a)ematogenous
гематодермия h(a)ematodermatosis
гематозис *physiol.* h(a)ematosis
гематокольпос h(a)ematocolpos
гематоксилин h(a)ematoxylin
гематолиз *physiol.* h(a)ematolysis
гематолит *miner.* hematolite
гематолог h(a)ematologist
гематология h(a)ematology
гематома *med.* h(a)ematoma
~, внутримозговая intracerebral h(a)ematoma
~, забрюшинная retroperitoneal h(a)ematoma
~, нагноившаяся h(a)ematic abscess, h(a)ematopostema
~, наружного уха othematoma, haematoma auris, insane ear
~, оссифицирующая ossifying h(a)ematoma
~, перитубарная peritubal h(a)ematoma
~, предбрюшинная anteperitoneal h(a)ematoma
~, пульсирующая aneurysmal/pulsative h(a)ematoma, false aneurism
~ твердой мозговой оболочки durematoma
гематометр h(a)ematometer
гематомиелия h(a)ematomyelia
гематонефроз h(a)ematonephrosis
гематоофтальмический h(a)emato-ophthalmic
гематорея h(a)ematorrhea
гематоррахис h(a)ematorrhachis, spinal apoplexy
гематосэльпинкс h(a)ematosalpinx
гематоскоп h(a)ematoscope
гематоспектроскопия h(a)ematospectroscopy
гематосперматоцеле h(a)ematospermatocele
гематофобия h(a)ematophobia
гематохром h(a)ematochrome
гематоцеле h(a)ematocele
гематурия *pathol.* h(a)ematuria
~, интермиттирующая intermittent

h(a)ematuria
~, **ложная** pseudo h(a)ematuria, false h(a)ematuria
~, **пузырная** vesical h(a)ematuria
гемера *geol.* hemera (*pl.* -rae)
гемералопия *med.* hemeralopia
~, **врожденная** congenital hemeralopia
~, **эссенциальная** essential hemeralopia
гемерокаллис *bot.* day lily *(Hemerocallis)*
гемерофильный hemerophylous
гемерофит *bot.* hemerophyte
гемиагевзия hemiageusia, hemiageustia
гемиальгия *med.* hemialgia
гемиамблиопия hemiamblyopia
гемиамиостения hemiamyosthenia
гемианальгезия hemianalgesia
гемианестезия hemian(a)esthesia
~, **альтернирующая** alternate hemian(a)esthesia
~, **перекрестная** crossed hemian(a)esthesia
гемианопия, гемианопсия *med.* hemianopia, hemianopsia, hemiablepsia, half-vision
~, **билатеральная** bilateral/binocular hemianopsia
~, **гетеронимная** heteronymous/crossed hemianopsia
~, **гомонимная** homonymous/equilateral/lateral hemianopsia
~, **двусторонняя** bilateral/binocular hemianopsia
~, **квадрантная** quadrantic hemianopsia, tetartanopia
~, **корковая** cortical hemianopsia
~, **одноименная** congruous hemianopsia
~, **односторонняя** unilateral/uniocular hemianopsia
~, **разноименная** incongruous hemianopsia
гемианоптический hemianoptic
гемиантиклиналь *geol.* hemianticline
гемиасинергия hemiasynergia
гемиатрезия hemiatresia
гемиатрофия hemiatrophy
гемиахроматопсия hemiachromatopsia
гемибазидий *bot.* hemibasidium
гемигепатэктомия hemihepatectomy
гемигидроз hemihidrosis, hemidiaphoresis
гемигипертония hemihypertonia, hemitonia
гемигипертрофия hemihypertrophy
~ **лица, перекрестная** crossed facial hemihypertrophy
гемигипотония hemihypotonia
гемигипоэстезия hemihyp(o)esthesia
гемиглоссит hemiglossitis
гемигнатия hemignathia
гемизигота hemizygote
гемизиготность hemizygosity
гемизиготный hemizygous
гемикарион hemicaryon
гемиклония hemiclonism
гемиколэктомия hemicolectomy
~, **правосторонняя** right hemicolectomy
гемикрания hemicrania, hemicephalia
hemimorphic
гемикриптофит *bot.* hemicryptophyte
гемикристаллический *geol.* hemicrystalline
гемиксерофит *bot.* hemixerophyte
гемиморфизм hemimorphism
гемиморфит hemimorphite, calamine
гемиморфия hemimorphy
гемиморфный hemimorphic
гемин *bioch.* h(a)emin
гемината double consonant
геминация gemination
гемиопия *med.* hemiopia
гемипаразит hemiparasite
гемипарез hemiparesis
гемиплегия *med.* hemiplegia
~, **альтернирующая** alternate/crossed hemiplegia, Gibler's hemiplegia/paralysis
~, **восходящая** ascending hemiplegia
~ **лица** facial hemiplegia
~, **наследственная** hereditary hemiplegia
гемипризма hemiprism
гемисапрофит *bot.* hemisaprophyte
гемисепта hemiseptum (*pl.* -ta)
гемисинклиналь *geol.* hemisyncline
гемисинклинорий *geol.* hemisynclinorium
гемисистолия hemisystole
гемискотома hemiscotosis, hemianopsia
гемиспазм hemispasm
гемиспороз hemisporosis
гемисфера hemisphere
гемитерпен hemiterpene
гемитропизм hemitropism
гемитропный hemitrope, hemitropous
гемихорея hemichorea
гемицеллюлоза hemicellulose
гемиэдр hemihedron
гемиэдризм *cryst.* hemihedrism, hemihedry
гемиэдрический hemihedral
гемиэндофит *bot.* hemiendophyte
гемлок *bot.* hemlock *(Tsuga)*
гемма I cameo, intaglio
гемма II *bot., zool.* gemma
геммаподобный gemmiform
геммация *biol., bot.* gemmation
геммула gemmule
гемобластоз h(a)emoblastosis
гемогения h(a)emogenia, pseudohemophilia
гемогистиобласт h(a)emohistioblast
гемоглобин *bioch.* h(a)emoglobin
~, **аномальный** abnormal h(a)emoglobin
~, **быстрый** fast-moving h(a)emoglobin
~, **восстановленный** reduced h(a)emoglobin
~, **зеленый** verdoh(a)emoglobin, green h(a)emoglobin
~, **медленный** slow-moving h(a)emoglobin
~, **мышечный** myoglobin, muscle h(a)emoglobin
~, **окисленный** oxyh(a)emoglobin, oxygenated h(a)emoglobin
~, **серповидноклеточный** sickle cell h(a)emoglobin
~, **фетальный** fetal/F h(a)emoglobin

гемоглобинемия h(a)emoglobin(a)emia
гемоглобинный h(a)emoglobinous, h(a)emoglobinic
гемоглобиноз h(a)emoglobinosis
~ S S-h(a)emoglobinosis/-h(a)emoglobinopathy, h(a)emoglobin S disease, sickle cell disease
гемоглобинолиз h(a)emoglubinolysis, h(a)emoglobinopepsia
гемоглобинометр h(a)emoglobinometer, hematinometer, hemameter
гемоглобинопатия h(a)emoglobinopathy
гемоглобинофильный h(a)emoglobinophilic
гемоглобинурийный h(a)emoglobinuric
гемоглобинурия h(a)emoglobinuria
~, иммуногемолитическая immunoh(a)emolytic h(a)emoglobinuria
~, малярийная malarial h(a)emoglobinuria, blackwater /h(a)emoglobinuric fever, bilious remittent fever
~, ночная nocturnal h(a)emoglobinuria, Marchiafava - Micheli syndrome
~, ожоговая burn h(a)emoglobinuria
~, послеродовая puerperal h(a)emoglobinuria, post-parturient h(a)emoglobinuria
~, экзогенно-токсическая toxic h(a)emoglobinuria
~, эпидемическая epidemic h(a)emoglobinuria, Winckel's disease
гемограмма h(a)emogram
гемодиализ h(a)emodialysis
гемодиализатор h(a)emodialyzer
гемодинамика h(a)emodynamics
гемодинамометр h(a)emodynamometer, h(a)emadynamometer
гемодромограф h(a)emodromograph, h(a)emadromograph
гемодромометр h(a)emodromometer, h(a)emadromometer
гемоконии *pl.* h(a)emoconia
гемокультура blood culture
гемокупреин h(a)emocuprein
гемолиз h(a)emolysis
~, ожоговый burn h(a)emolysis
~, повышенный increased h(a)emolysis
~, постгепатитный post-hepatitis h(a)emolysis
гемолизат h(a)emolysate
гемолизины *pl.* h(a)emolysins
~ змеиного яда venom h(a)emolysins
~, холодовые (warm-)cold h(a)emolysins
гемолизировать h(a)emolyze
гемолимфа *zool.* h(a)emolymph
гемолимфатический *zool.* h(a)emolymphatic
гемоль *pharm.* h(a)emol
гемомедиастинум h(a)emomediastinum, h(a)ematomediastinum
гемометр h(a)emometer, h(a)emoglobinometer, h(a)ematinometer
~ Флейшля Fleischl's h(a)emometer
гемометрий h(a)ematometra, h(a)emometra
гемоперикард h(a)emopericardium, h(a)ematopericardium
гемоперитонеум h(a)emoperitoneum, h(a)ematoperitoneum
гемопиррол h(a)emopyrr(h)ol(e)
гемопневмоторакс h(a)emopneumothorax
гемопоэз h(a)emopoiesis, h(a)ematopoiesis
гемопоэтин h(a)emopoietin, h(a)ematopoietin
гемопротеид h(a)emoprotein
геморрагин *bioch.* h(a)emorrhagin
геморрагия *med.* h(a)emorrhage
геморроидальный *med.* h(a)emorrhoidal
~ная шишка h(a)emorrhoid, pile
геморрой *med.* h(a)emorrhoids, piles
~, кожный cutaneous h(a)emorrhoids
~, кровоточащий bleeding h(a)emorrhoids
гемосидерин *bioch.* h(a)emosiderin
гемосидероз *med.* h(a)emosiderosis
~, пищевой nutritional h(a)emosiderosis
гемоскоп h(a)emoscope, h(a)ematoscope
гемоспермия h(a)emospermia, h(a)ematospermia
гемоспоридии *pl. H(a)emosporidia*
гемоспоридиоз h(a)emosporidiosis
гемостаз h(a)emostasis
гемостатический h(a)emostatic
гемотахометр h(a)emotachometer
гемотерапия h(a)emato therapy, h(a)emotherapy, h(a)emotherapeutics
гемотоксин *immun.* h(a)emotoxin
гемоторакс h(a)emothorax, h(a)ematothorax
гемотрансфузия h(a)emotransfusion, blood transfusion
гемофагоцитоз h(a)emophagocytosis
гемофилик *med.* h(a)emophiliac
гемофилия *med.* h(a)emophilia
~, почечная renal h(a)emophilia/epistaxis
гемофлавопротеид h(a)emoflavoprotein
гемофобия h(a)emophobia
гемофтальм h(a)emophthalmia, h(a)emophthalmos
гемохроматоз *med.* h(a)emochromatosis
~, вторичный/первичный secondary/primary h(a)emochromatosis
гемоцеле h(a)emoc(o)ele
гемоцитобласт h(a)emocytoblast
гемоцитометр h(a)emocytometer, globulimeter
гемэндотелиома h(a)emendothelioma
ген *biol.* gene *see also* **гены**
~, аллельный allelic gene
~, криптомерный cryptomere
~, мутирующий mutant gene
генвудит *geol.* henwoodite
генеалог genealogist
генеалогический genealogic(al)
генеалогия genealogy, line (of descent), pedigree
генез(ис) genesis, origin, source
генекен *bot.* henequen agave *(Agave fourcroydes)*
генерал-бас *mus.* figured bass
генерал-губернатор governor general
генерализация generalization
~ карты overlay compilation
генерал-лейтенант lieutenant general

генерал-майор major general
генерал-марш *mil.* assembly
генерал-полковник colonel general
генеральный general; basic
~ **консул** consul general
~ **секретарь** secretary-general
~ **штаб** general staff
~**ная линия партии** *polit.* Party general line
~**ная репетиция** *theatre* dress rehearsal
~**ная совокупность** *econ.* universal set
~**ное сражение** decisive battle
генеральский general's
~ **чин** rank of general, generalship
генеративный *biol.* genetic, generative
~**ная клетка** generative cell
~**ная почка** fruit bud
генератор generator; *rad.* oscillator
~, **аварийный** emergency/stand-by/back-up generator
~ **азимутальной развертки** azimuth-sweep generator
~, **асинхронный** induction/asynchronous generator
~, **аэрозольный** aerosol generator, fogger
~, **балансный** balanced oscillator
~ **бегущей волны** travel(l)ing wave-tube oscillator, TWT-oscillator
~ **белого шума** white noise generator
~, **бесколлекторный/бесщеточный** brushless generator
~ **биений** beat-frequency generator/oscillator
~, **брызгозащищенный/брызгонепроницаемый** splash-proof generator
~ **буквенно-цифровых кодов** alpha(nu)meric/symbol generator
~ **в двухтактный режим, переводить** convert an oscillator to push-pull operation
~ **Ван-де-Граафа** Van de Graaf generator, (electrostatic) belt generator
~ **векторов** vector generator
~ **видеосигналов** video generator
~ **видеочастоты** video-frequency signal generator
~ **вихря** vortex generator
~ **возбуждается** the generator builds up
~, **возбуждать** drive an oscillator
~ **возбуждающих импульсов** drive-pulse generator, driver
~, **вольтодобавочный** booster (generator)
~, **вспомогательный** auxiliary generator
~ **входных сигналов** input-stimulus generator
~ **вызывного тока** ringing generator
~ **высокой частоты** radio-frequency oscillator/generator
~ **выходной программы** output routine generator
~ **газа** gasifier, gas generator/producer
~, **ганновский** Gunn(-effect) oscillator
~ **гармоник** harmonic oscillator/generator
~, **гетеродинный тональный** heterodyne tone generator, beat-frequency oscillator
~, **гетерополярный** heteropolar generator, cross-field generator
~, **главный** *naut.* main/propulsion generator
~ **главных синхронизирующих импульсов** master clock
~, **горизонтальный/вертикальный** horizontal/vertical-shaft generator
~ **двойного тока** double-current generator
~ **двойных импульсов** double-pulse generator
~, **двухконтурный** tuned-input, tuned-output oscillator
~, **двухтактный** push-pull oscillator
~, **джозефсоновский** Josephson source
~, **диапазонный** variable-frequency oscillator
~, **динатронный** dynatron oscillator
~, **дуговой** arc converter
~, **емкостно-резистивный** RC-oscillator
~, **задающий** master/driving oscillator; *comp.* clock
~ **задержки** delay generator
~ **заднего полустроба** late-date generator
~, **закрытый** enclosed generator
~, **запирающий** blanking-pulse generator
~, **запускать** activate/enable/turn on an oscillator
~, **зарядный** (battery-)charging generator
~, **звуковой** tone/sound generator, audio-signal generator; audio/tone oscillator
~ **звуковой частоты, вызывной** voice-frequency ringing generator, low-frequency signaling set
~, **зуммерный** buzzer oscillator
~, **измерительный** test oscillator; signal generator
~ **изображений** pattern/image generator
~, **изотопный** radioisotope battery
~, **импульсный** pulse oscillator
~, **импульсный хронированный** tuned pulse oscillator
~ **импульсных кодов** pulse-pattern generator
~ **импульсных помех** discontinuous jammer
~ **импульсов** (im)pulse generator, pulser
~ **импульсов заданной формы** pulse waveform generator
~ **импульсов, маломощный** low-level pulser
~ **импульсов, мощный** power pulser
~ **импульсов отметки** notch generator
~ **индукторного вызова** subharmonic generator, ringing converter
~, **индукторный** inductor generator
~, **интегральный** integrated-circuit oscillator
~, **искровой** spark-gap converter
~ **кадровой развертки** vertical-scanning generator/circuit; frame-scan/-sweep generator/circuit; vertical/sweep generator/circuit
~, **калибровочный** calibration oscillator
~, **камертонный** tuning-fork oscillator
~ **качающейся частоты, ГКЧ** sweep-/swept-frequency generator; swept-signal source; sweep generator, wobbulator

~, квантовый quantum-mechanical oscillator
~, квантовый ИК-диапазона infrared laser, iraser, IR-laser
~, квантовый оптический, ОКГ laser
~, квантовый оптический жидкостный/импульсный liquid/pulse(d) laser
~, квантовый, оптический комбинационный Raman laser
~, квантовый оптический моноимпульсный giant-pulse laser
~, квантовый оптический полупроводниковый semiconduction laser
~, квантовый оптический с модуляцией добротности Q-switched laser
~, квантовый оптический с электронной накачкой electron-beam-pumped laser
~, кварцевый crystal oscillator; quartz generator, crystal-controlled oscillator
~, когерентный coherent oscillator, Coho
~ кодов *comp.* code generating/investigation unit
~ Кокрофта-Уолтона Cockroft-Walton generator
~ колебаний oscillator
~ (колебаний) с внешним возбуждением radio-frequency power amplifier, r. f. power amplifier
~ коротких импульсов narrow-pulse generator
~, ламповый valve oscillator, *Am.* vacuum-tube oscillator
~ линейно-возрастающего напряжения (тока) saw-tooth generator, voltage/current generator
~ линейно-изменяющегося напряжения linear/saw-tooth generator
~ линейно-падающего напряжения (тока) phantastron
~, магнитогидродинамический magnetohydrodynamic / MHD generator, magneto-fluid-dynamic generator, MFD generator
~, магнитострикционный magnetostriction oscillator
~, магнитоэлектрический permanent-magnetic generator
~ макрокоманд macrocommand generator
~ масштабных меток дальности calibration mark(er) generator
~ меток notch generator
~ меток времени time-mark(ing) generator
~, многопостовой (электро)сварочный multioperator welding generator
~, многотоковой multiple-current generator
~, многофазный polyphase/multiphase generator
~, моделирующий диаграмму направленности beam-pattern generator
~, молекулярный moleculer-beam maser, molecular oscillator
~, надтональный interpolation oscillator
~ накачки pump (oscillator)
~ напряжения voltage generator
~ начинает генерировать the oscillator is kicked into oscillations
~ незатухающих колебаний continuous-wave oscillator, CW oscillator
~ нейтронов neutron generator
~ несущей частоты carrier-frequency oscillator; *teleph.* carrier oscillator; *el.* generator
~ несинусоидальных колебаний nonsinusoidal oscillator
~, неявнополюсный implicit-pole generator
~ низкого напряжения low-tension generator
~ низкой частоты audio(-frequency) oscillator; audio signal generator
~, опорный frequency standard (assembly)
~, оптический квантовый laser
~, отказавший failed generator
~ отсчетов report generator, reporter
~ пара steam generator
~ параллельного возбуждения shunt (-wound) generator
~, параметрический parametric oscillator
~, педальный foot-operated generator, pedal generator
~ переднего полустроба early-gate generator
~ переменного тока alternator, alternating-/alternate-current generator, a. c. generator
~ переходит в режим электродвигателя the generator goes motoring
~ пилообразного напряжения (тока) saw-tooth voltage (current) generator; relaxation oscillator
~ пилообразной функции *comp.* ramp generator
~ плавного диапазона variable freguency oscillator, VFO
~, плазменный/плазмы plasma generator
~ повышенной частоты rotary frequency changer, rotary changer converter
~ помех interference generator, jammer
~ последовательного возбуждения series (-wound) generator
~ постоянного тока direct-current generator, d. c. generator, dynamo
~ постоянно-переменного тока double-current generator
~ по схеме емкостной трехточки Colpitts oscillator
~ по схеме индуктивной трехточки hartleg oscilator
~ по схеме моста Вина Wien-bridge oscillator
~ по схеме Шембеля electron-coupled oscillator, ECO
~ прикладных программ application generator
~ прямоугольных импульсов/сигналов square-wave generator, rectangular pulse generator
~ псевдослучайной последовательности PR sequence generator
~ пусковых импульсов trigger(-pulse) generator
~ развертки time-base (generator), time-base generator circuit; scanning/sweep generator

~ развертки дальности range-sweep generator
~,разгонять allow a generator to come up to speed, bring up a generator to speed
~, реактивный reluctance generator
~, резервный back-up generator, stand-by generator, reserve generator
~, релаксационный relaxation generator/oscillator
~, релейный relay pulse generator
~ речи speech generator
~ с кварцевой стабилизацией crystal-controlled oscillator
~ с независимым возбуждением 1. *el.* separately excited generator; 2. *rad.* power amplifier
~ с обратной связью feedback generator
~ с параллельным возбуждением shunt (-wound) generator
~ с поперечным полем cross-field generator, heteropolar generator
~ с последовательным возбуждением series(-wound) generator
~ с посторонним возбуждением separately excited generator; *rad.* power amplifier
~ с продольным полем homopolar generator
~ с самовозбуждением self-excited oscillator, feedback oscillator
~ с тормозящим полем retarding-field oscillator, positive-grid oscillator
~ с фазовым сдвигом phase-shift oscillator
~ с частотной модуляцией frequency-modulated oscillator
~, самовозбуждающийся self-excited generator
~, самохронирующийся self-pulsed oscillator
~ сантиметрового диапазона SHF oscillator
~, сварочный (arc-)welding generator
~, СВЧ microware oscillator
~, сельсинный selsyn/synchro generator
~, сериесный series(-wound) generator
~ сжатого воздуха, газотурбинный air generator, air compressor unit, turbocompressor
~ символов symbol/symbology generator
~ синусоидальных колебаний harmonic sinusoidal generator/oscilator, sine-wave generator/oscillator
~ синхронизирующих импульсов clock pulse generator, *comp.* timing generator
~ синхроимпульсов *telev.* sychonizing (-signal) generator, sync generator
~, синхронный *el.* synchronous generator; *rad., telev.* locked oscillator
~ случайных событий random event generator, randomizer
~ случайных чисел *math.* random(-)number generator, randomizer
~, смесительный collating generator
~ со скоростной модуляцией velocity modulated oscillator
~ со смешанным возбуждением compound (-wound) generator
~, стабилизированный кварцем crystal-controlled oscillator
~ стабильной частоты fixed-frequency generator
~, стояночный *nav.* harbour generator
~ строб-импульсов date generator
~ строб импульсов дальности range-date generator
~ строчной развертки horizontal-scanning generator/circuit, line-scan/sweep generator/cicuit, horizontal-sweep generator/circuit
~, струйный fluid oscillator
~ ступенчатой функции staircase generator
~ тактовый clock
~ тактовых импульсов clock(-pulse) generator
~, тахометрический tachometer generator, tachogenerator
~, твердотельный solid-state oscillator
~, термоэлектронный thermionic generator/converter
~, тиратронный thyratron oscillator
~ тока generator, dynamo
~, тональный audio-frequency oscillator, tone generator
~, точечный *comp.* dot generator
~, транзитронный transitron oscillator
~, трехточечный impedance voltage divider oscillator
~, трехточечный с емкостной обратной связью Colpitis oscillator
~, трехточечный с индуктивной обратной связью Hartley oscillator
~ трехфазного тока three-phase generator
~, углекислотный carbon-gas generator
~ ударного возбуждения shock-excited oscillator
~ ударных волн *aw.* shock-wave generator
~, ультразвуковой ultrasonic/supersonic generator
~, униполярный homopolar/unipolar generator
~, управляемый voltage-controlled oscillator
~ фиксированных строб-импульсов fixed-date generator
~, фотоэлектрический photoelectric generator, photocell
~ функций function(al) generator
~ Хола Hall generator
~, хронирующий timing generator
~ ЧМ sweep generator, swept-signal generator
~, электродуговой arc generator
~, электросварочный (arc-)welding generator
~, электростатический electrostatic generator
~, электрофорный influence machine
~, эталонный reference/standard oscillator
~, явнополюсный salient-/explicit-pole generator

генераторная *n.* generating plant, generator house; generating station
генераторный generator, oscillator *(attr.)*
~ газ producer/power gas

генератриса *math.* generatrix
генерационный generative
генерация generation, production; oscillation
~, альтернирующая alternate generation
~, бесполая asexual/nonsexual generation
~ в квантовом генераторе субмиллиметрового диапазона smasing
~ в лазере lasing
~ в лазере, послеимпульсная post-lasing
~ в мазере masing
~ второй гармоники, ГВГ second-harmonic generation, SHG
~ гармоник harmonic generation
~ дуги окружности circular arc generation
~, многомодовая multimode oscillation(s)
~ на 4-х дисках *comp.* four-drive generation
~ носителей carrier generation
~, одномодовая single-mode oscillation(s)
~, паразитная howling, spurious oscillation
~, побочная filial generation
~, половая sexual generation
~ электронно-дырочных пар electron-hole pair generation/production
генерирование generating, generation, production
~ импульсов pulsing
~ рентгеновских лучей X-ray production
~ отчетов *comp.* report generation
генерировать generate, produce; oscillate
~ продольные/поперечные колебания operate in axial/transverse modes
генетик geneticist
генетика genetics
~, космическая aerospace genetics
~, медицинская human/medical genetics
~, популяционная population genetics
~ растений plant genetics
~, статистическая statistical genetics, statisticogenetics
~, фармакологическая pharmacological genetics, pharmacogenetics
генетический genetic
~ код genetic code
генетта *zool.* genet(ta) *(Genetta)*
ген-индикатор *biol.* depressor
генионемус *zool.* roncador *(Genyonemus)*
генипа *bot.* genip tree *(Genipa)*
генипап *bot.* genipap, jagna *(Genipa americana)*
гениталии *pl.* genitalia, genitals
генитальный genital
генито-ано-реактальный *med.* genitoanorectal
генито-феморальный *med.* genitofemoral, genitocrural
ген-модификатор *biol.* modifier, modifying factor
генный genic, genetic
~ная хирургия genetic surgery
геноголотип genoholotype
геном genom(e)
геновид genospecies
геноним genonymum
ген-оператор *biol.* operator gene
геносинтип genosyntype
генотип genotype
генотипический, генотипный genotypic(al)
генофонд gene pool
геноцид genocide
генплан *abbr.* **генеральный план** general plan
ген-регулятор regulator gene
генри *el.* henry, H
генриметр henrymeter, inductance meter
генсовет *abbr.* General Council
гентианин *chem.* gentianin
гинтианоза gentianose
гентизин *chem.* gentisin
гентибиоза gentibiose
генуэзец Genoese
генуэзский Genoese
генциана *bot.* gentian *(Gentiana)*
генцианвиолет gentian violet, methylrosanilinum chloride
генцианин gentianin
генцианоза gentianose
генштаб *abbr.* general staff
гены *pl.* genes
~, аморфные amorphic genes
~, аутархные autarchic genes
~, буферные buffering genes
~, вещества substance genes
~, гипархные hyparchic genes
~, голандрические holandric genes
~, гомологические homologous genes
~, диагинические diagynic genes
~, изоморфные isomorphic genes
~, контролируемые полом sex-controlled genes
~, криптомерные cryptomeric genes
~, малые minor genes
~, межвидовые interspecific genes
~, неоморфные neomorphic genes
~, ограниченные полом sex-limited genes
~, полимерные polymeric genes, multiple/ cumulative/ additive factors
~ профагов prophage genes
~, сигнальные markers
~, сложные compound genes
гены-мутаторы *pl.* mutator genes
гены-ограничители *pl.* restriction genes
гены-усилители *pl.* extension genes
гэнэкология genecology
геобаллистика geoballistics
геобиоз geobios, terrestrial life
геобионт geobiont
геоботаник geobotanist, phytogeographer
геоботаника geobotany, plant geography
геогенезис geogenesis
геогнозия geognosy
географический geographic(al)
~ викариант geographical vicariant
~кая среда geographical environment
~кое замещение vicariation

~кое название place name
география geography
~, ботаническая botanical/plant geography, phytogrophy
~, физическая physiography
~, экологическая ecological geography
геодезист geodesist, (land) surveyor
геодезист-вычислитель geodetic computer
геодезист-землеустроитель/полевик field/land surveyor
геодезист-программист geodetic computer programmer
геодезист-триангулятор triangulator
геодезический geodetic, *math.* geodesic
геодезия geodesy
~, гравиметрическая/физическая physical geodesy
~ на плоскости plane geodesy
~, низшая elementary geodesy
~, полевая surveying
~, пространственная three-dimensional geodesy
~, спутниковая satelite geodesy
геодепрессия geodepression
геодиметр geodimeter
геодинамика geodynamics
геоид geoid
геоизобата geoisobath
геоизотерма geoisotherm
геоизотермический geoisothermal
геокарпический *bot.* geocarpic
геокарпия *bot.* geocarpy
геокорона geocorona
геокосмический geocosmic
геокосмология geocosmology
геокриологический geocryological
геокриология geocryology, cryopedology
геокриптофит *bot.* geocryptophyte
геолог geologist
~, рудничный/шахтный mining geologist
~, старший senior geologist
геологический geologic(al)
~кая серия geosere
геология geology
~ изверженных пород igneous geology
~, инженерная geological engineering
~ коренных пород bedrock geology
~, ледниковая glacial geology
~ моря marine geology, geology of sea
~, нефтепромысловая oil-field geology
~ нефти oil/petroleum geology
~, общая/региональная areal/regional geology
~ поверхности surface geology
~, подводная submarine geology
~, подземная subsurface geology
~, поисковая exploration geology
~, полезных ископаемых mining geology
~, прикладная practical geology
~, рекогносцировочная reconnaissance geology
~, рудничная mining geology
~ рудных месторождений mining/ore geology
~, стратиграфическая statigraphy, statgraphic geology
геолог-нефтяник oil/petroleum geologist
геологоразведка geological exploration
геологоразведчик geological prospector
геолог-подводник submarine geologost
геолог-стратиграф stratigrapher
геолог-четвертичник glacial geologist
Геракл *myth.* Herakles, Heracles
гераклеум *bot.* cow parsnip *(Heracleum)*
гераклитов Heraclitean, Heraclitic
гераклитовский Heracle(i)an, Heraclitean, Heraclitic
гераклический Heraclitean
гераклов Heracle(i)an
геральдик specialist in heraldry, armorist, herald
геральдика heraldry, blazonry
геральдический heraldic, armorial
геранецветные *n. pl. bot.* geranium family *(Geraniales)*
гераниевый *bot.* geranic, geranial; *n. pl.* geranium family *(Geraniaceae)*
геранил *chem.* geranyl
гераниол *chem.* geraniol
гераниум, герань *bot.* geranium, crowfoot *(Geranium)*
герб (coat of) arms; head *(of coin)*
гербаризатор herbalist, herborist
гербаризация herbarization, herborization
гербарий herbarium; exsiccatae
гербарный herbarial
гербера *bot.* gerbera *(Gerbera)*
гербицид herbicide
~, довсходовый/послевсходовый pre-/post-emergence herbicide
~ избирательного/контактного действия selective/ contact herbicide
~, листовой foliar herbicide
~ сплошного действия non-selective herbicide
гербицидный herbicidal
гербоведение heraldry
гербовый heraldic, armorial; bearing a coat of arms
~ сбор stamp tax/duty
~вая бумага stamped paper
~вая книга book of armory
~вая марка duty stamp
герд *min.* concentration table
геренук *zool.* gerenuk *(Lithocranius walleri)*
гериархия geriarchy
гериатр geriatrist, geriatrician
гериатрический geriatric
гериатрия geriatrics
геркогамия hercogamy, herkogamy
геркон *tech.* (hermetically sealed) reed relay, scaled contact reed relay
геркулес 1. *myth., astr.* Hercules; 2. *zool.* Herkules beetle; *(Dynastes hercules);* 3. *cul.* porridge; rolled oats

геркулесов Herculean
~вы столбы/столпы Pillars of Hercules; limit, boundary
геркулесовский Herculean
~ труд Herculean task
герма herma
германец Teuton, ancient German; *pl.* the Germanic peoples
германиевый *chem.* germanic, germanous
германизация Germanization
германизировать, германизовать Germanize
германий *chem.* germanium, Ge
~, легированный doped germanium
~ с собственной проводимостью intrinsic germanium
германист specialist in Germanic studies, Germanist
германистика Germanic studies
германофил Germanophile
германофоб Germanophobe
германофобия, германофобство Germanophobia
германский Germanic; Teutonic
~кие языки Germanic languages
гермафродит hermaphrodite
гермафродитизм *biol.* hermaphroditism, androgynism, androgyny
~, истинный/ложный true/false hermaphroditism
~, односторонний unilateral hermaphroditism
гермафродитический hermaphroditic
гермафродитный hermaphrodite, hermaphroditic; *bot.* teleianthous
герменевтика hermeneutics
Гермес *myth.* Hermes
герметизация, герметизирование (hermetic) sealing, watertight work; containment; canning; pressurization; making air-tight; encapsulation; potting
~ люков *naut.* hatch sealing
~ от воздействий влаги sealing against moisture
~ пыжом *naut.* sealing by the wad
герметизированный sealed(-in), canned; pressurized; potted
герметизировать seal (off/in), pressurize; can
герметик sealant, sealing compound
герметически hermetically
~ закрытый/закупоренный/запаянный hermetically sealed
герметический, герметичный hermetic, sealed, (air-)tight, canned, hermetically sealed; pressurized, pressure-/-water/ gas tight; leak-/air-/weather-proof
герметичность (air)tightness, leak-proofness, hermeticity, hermetic state
~ испытанный на leak-tested
~, испытывать/проверять на test for leak-proofness, test for leaks
герминальный germinal
гермицид germicide
гермицидный germicide, germicidal
гермобачок pressurized canister
гермоввод pressure-seal feed-through
гермовывод sealed lead(-out)
гермокабина pressure/pressurized cabin/cockpit
гермокостюм *av.* skintight suit
гермооболочка pressure shell
гермоодежда pressurized clothing
гермоперегородка pressure bulk head; aerodynamic seal
гермопереходник sealed adapter connector
гермоперчатки pressure gloves
герморазъем pressure-seal/-sealed connector
гермостекло *av.* pressure pan
гермоуплотнение pressure sealing
гермошлем pressure/pressurized helmet
~ высотного компенсирующего костюма partial-pressure helmet
~ высотного скафандра full-pressure helmet
~ космонавта space helmet
гермошпангоут *av.* pressure bulk head
~, полусферический pressure dome
гернзейский Guernsey
герниология *med.* herniology
грениопластика hernioplasty
герниотом *surg.* herniotome
герниотомия herniotomy
геродермия geroderma
геродонтика gerodontics
героизм heroism
героика heroics; heroic spirit/style
героин heroin, diacetylmorphine
~, хлористоводородный diamorphine hydrochloride
героинизм heroinism, heroin addiction
героический heroic
~кие меры drastic measures
герой hero; *theatre* principal character
~ дня *polit.* man of the hour
геройский heroic
~ подвиг heroic exploit
геройство (act of) heroism
герольд *hist.* herald
герольдия heraldic office, heraldry
герольдмейстер *hist.* president of the heraldic office
геронтический gerontic
геронтолог gerontologist
геронтология gerontology
геронтотерапия gerontotherapy
геронтофилия *psych.* gerontophilia
геронтофобия gerontophobia
герострат vainglorious person, fame-thiristy person
~а, лавры sad/disgraceful glory
герпангина *med.* herpangina
герпес *med.* herpes
~, лихорадочный herpes fibrilis
~ на губах/лице herpes labialis/facialis
~ на половых органов herpes genitalis
~, простой herpes simplex

герпетизм herpetism
герпетический *med.* herpetic
герпетофауна herpetofauna
герпидзолит *geol.* herpidzolite, herpidzolith
герполодия herpolhode
герренгрундит *geol.* herrengrundite
герундив *gram.* gerundive
герундий *gram.* gerund
герц *phys.* hertz, Hz, cycles per second
герцинидый *pl.* Hercynides
герцинит *miner.* hercynite, green spinel
герцинский *geol.* Hercynic, Hercynian
герцметр *tech.* frequency indicator
герцог duke
герцогиня duchess
герцогство duchy, dukedom
геснериевые *n. pl., bot. Gesneriaceae*
геснерия *bot.* gesneria *(Gesneria)*
гесперидий *bot.* hesperidium
гесперидин *chem.* hesperidin
гессенский Hessian
~кая муха *zool.* Hessian fly *(Mayetiola destructor)*
гессиан *math.* Hessian
гессит *miner.* hessite, telluric silver
гессонит *miner.* hessonite, cinnamon stone
гестапо Gestapo
гестаповец Gestapo policeman
гетевский Goethian
гетера hetaera, hetaira
гетерандрический *bot.* heterandrous
гетерандрия I *bot.* heterandry
гетерандрия II *zool.* least mosquito fish *(Heterandria formosa)*
гетерантера *bot.* mud plantain *(Heteranthera)*
гетерантерия heteranthery
гетеризм *biol.* hetaerism
гетероагглютинация heteroagglutination
гетероагглютинины *pl.* heteroagglutinins
гетероаллергия heteroallergy
гетероальбумозурия heteroalbumosuria
гетероантитело heteroantibody
гетероатом heteroatom
гетероаутопластика heteroautoplasty
гетеробрахиальный heterobrachial
гетеровалентный heterovalent
гетерогамета *biol.* heterogamete
гетерогаметангий *bot.* heterogametangium
гетерогаметность *biol.* heterogamety
гетерогаметный *biol.* heterogametic, digametic
гетерогамия *biol.* heterogamy
гетерогамный heterogamous
гетероген heterogeneous word
гетерогенез(ис) *biol.* heterogenesis
гетерогенетический heterogenetic
гетерогенизация heterogenization
гетерогенность heterogeneity, inhomogeneity
гетерогенный heterogeneous, inhomogeneous; heterogenic, heterogenetic, dissimilar
гетерогеномный heterogenomatic
гетерогеофит *bot.* heterogeophyte
гетерогинный heterogynic, heterogynous
гетерогологамия heterohologamy
гетерогония heterogony
гетерография heterography
гетеродин *rad.* heterodyne; (local) oscillator; conversion/ injection frequency source
~, калибрационный calibrating oscillator
~, кварцевый quartz/crystal heterodyne
~, клистронный klystron/local oscillator
~, когерентный coherent/local oscillator, Coho
~, навигационный beacon/local oscillator
~ телеграфного приема beat-frequency (local) oscillator
гетеродинамический heterodynamic
гетеродинамный heterodynamous
гетеродинирование heterodyning
гетеродинный *rad.* heterodyne
гетеродиод heterodiode
гетеродисперсный heterodispersed
гетеродистилия *bot.* heterodistyly
гетеродонтизм heterodontism
гетеродонтный *zool.* heterodont
гетеродромный *bot.* heterodromous
гетерозигота heterozygote
гетерозиготия, гетерозиготность *biol.* heterozygosis, heterozygosity, heterozygousness
~, вынужденная enforced heterozygosis
~ по транслокации translocation heterozygosis
~, простая/структурная simple/structural heterozygosis
гетерозиготный heterozygous, heterozygotic
гетерозис *biol.* heterosis
~, комбинационный combinational heterosis
~, одногенный one gene heterosis
~, однолокусный single locus heterosis
гетерозисный heterotic
гетерокариоз heterokaryosis, heterocaryosis
гетерокарион *biol.* keterokaryon, heterocaryon
гетерокариотический heterokaryotic, heterocaryotic
гетерокарпический *bot.* heterocarpous
гетерокарпия *bot.* heterocarpy, heterocarpism
гетерокарпный heterocarpous
гетерокатегорический heterocategorical
гетерокинез heterokinesia, heterokinesis
гетероклизия heteroclisis
гетероклинный *bot.* heteroclinous
гетероклитика heteroclite
гетероклитический heteroclitic
гетерокринный heterocrine
гетеролизин heterolysin
гетеролипиды *pl.* heterolipids
гетеролит heterolith, hair ball
гетеромезический heteromesical
гетеромерикарпия heteromericarpy
гетеромиксис heteromixis
гетероморфизм heteromorphism
гетероморфия heteromorphy, heteromorphism
гетероморфный heteromorphic, heteromorphous

гетероморфоз heteromorphosis
гетеронексуальный heteronexual
гетеронимия heteronymy
гетеропереход heterojunction
~, **дырочно-дырочный** p-p heterojunction
~, **прямосмещенный** forward biased heterojunction
~, **электронно-электронный** n-n heterojunction
гетеропетальный *bot.* heteropetalous
гетеропикноз heteropyknosis
гетероплазия *biol., med.* heteroplasia
гетероплазма heteroplasm
гетеропластика *biol., surg.* heteroplasty
гетероплоид *biol.* heteroploid
гетероплоидия *biol.* heteroploidy
гетероподный heteropodal
гетерополикислота *chem.* heteropolyacid
гетерополимер heteropolymer
гетеропольный, гетерополярный heteropolar
гетерополярность heteropolarity
гетеросексуализм heterosexuality
гетероскедастический *stat.* heteroscedastic
гетероскедастичность *stat.* heteroscedasticity
гетероскоп *med.* heteroscope
гетероспермия heterospermy
гетероспория *bot.* heterospory
гетероспоровый *bot.* heterosporic, heterosporous
гетеростилия *bot.* heterostyly, heterostylism
гетеростильный *bot.* heterostylous, heterostyled
гетеросфера heterosphere
гетеросыворотка heterogenous serum
гетеротаксис *bot.* heterotaxis, heterotaxy
гетероталлизм heterothallism, heterothally
гетероталломный heterothallic
гетеротермный heterothermal
гетеротип heterotype
гетеротоксин heterotoxin
гетеротоксический heterotoxic
гетеротопия heterotopy
гетеротрансплантат *med.* heterograft, heterotransplant
гетеротрансплантация *med.* heterografting, heterotransplantation
гетеротристилия heterotristyly
гетеротрихальный heterotrichous
гетеротропный heterotropous
гетеротрофия, гетеротрофность *physiol.* heterotrophy, heterotrophia
гетерофазия *med.* heterophasia
гетерофемия heterophemy
гетерофертилизация heterofertilization
гетерофиллия heterophylly
гетерофиллный heterophyllous
гетерофильный heterophil(e), heterophillic
гетерофит *bot.* heterophyte
гетерофитный heterophytic
гетерохламидный *bot.* heterochlamidous
гетерохрозия *zool.* heterochrosis
гетерохроматизация heterochromatization
гетерохроматин heterochromatin
гетерохромия *med.* heterochromia
~ **радужки** heterochromia iridis
~, **симпатическая** sympathetic heterochromia
гетерохромный heterochrome, heterochromatic
гетерохромосомы *pl.* heterochromosomes, allosomes
гетерохронизм *biol., physiol.* heterochrony, heterochronia, heterochronism
гетерохронический heterochronic, heterochronous
гетерохрония heterochrony, heterochronism
гетерохронность heterochronism
гетерохронный heterochronous
гетероцеллюлярный *biol.* heterocelluar
гетероцельный *zool.* heterocoelous
гетероцепной heterochain
гетероцеркальный, гетероцеркный *zool.* heterocercal
гетероцикл heterocycle
гетероциклический heterocyclic; heteronuclear
гетероциста *bot.* heterocyst, boundary cell
гетероцитотоксин heterocytotoxin
гетероэция *biol.* heteroecism, heteroecy
гетинакс micarta, paper-based laminate, synthetic-resin bonded paper laminate, SRB-paper laminate
~, **фольгированный** foil-clad paper-based laminate
гетит *geol.* goethite
гетман *hist.* hetman
гетманство *hist.* hetmanship, hetmanate
гетманщина *hist.* reign of hetmans; hetmanate
гетры *pl.* gaiters; spats
геттер *tech.* getter
~, **желобчатый** channel getter
~, **испарять** flash/fire a getter
~, **ленточный** ribbon getter
~, **направленный** directive getter
~, **спиральный** coil getter
~, **таблеточный** pellet getter
геттерирование gettering
гетто ghetto
геухера *bot.* alumroot *(Heuchera)*
Гефест *myth.* Hephaestus
гефироцеркальный *zool.* gephyrocercal
гештальтизм gestaltism
гештальтпсихология Gestalt psychology
гештальт-теория Gestalt (economic) theory
гиалин *bioch.* hyalin(e)
гиалиновый hyalin(c), translucent, transparent
гиалит 1. *med.* hyalitis; 2. *miner.* hyalite, water opal; 3. *petrography* hyalith
гиалография hyalography
гиалоидит *med.* hyaloiditis
гиалоидный *anat.* hyaloid, glassy, transparent
гиалома *med.* hyaloma, colloid millium
гиаломелан *geol.* hyalomelane
гиаломер hyalomere
гиалокристаллический *geol.* hyalocrystalline
гиалопилитовый *geol.* hyalopilitic

гиалоплазма *biol.* hyaloplasm, cytoplasmic matrix
гиалоплазматический *geol.* hyaloplasmatic
гиалопсит *geol.* hyalopsite
гиалосерозит *med.* hyaloserositis
~, множественный прогрессирующий multiple progressive hyaloserositis; polyorrhomeningitis
гиалосома *biol.* hyalosome
гиалоспора *bot.* hyalospore
гиалотипия *print.* glass printing
гиалотрахит *geol.* trachitic glass
гиалофан *miner.* hyalophane
гиалофобия *med.* hyalophobia
гиалофонолит *miner.* hyalophonolith
гиалурат hyalurate
гиалуронат hyaluronate
гиалуронидаза *bioch.* hyaluronidase, hyalase
гиалуроновый hyaluronic
гиатус hiatus; discontinuity of area, gap
гиацинт 1. *miner.* hyacinth, jacinth; 2. *bot.* hyacinth *(Hyacinthus)*
~, водяной water hyacinth *(Eichhornia)*
гиацинтовый hyacinthine
гиацинтоцветковый *bot.* hyacinth-flowered
гиб bend(ing)
~, двойной tangent bend
~ с перегибом bend(ing)-unbend(ing) test, backward and forward bending test
гиббенит *geol.* hibbenite
гибберелла *bot.* Gibberella
гиббереллин gibberellin
гиббереллoвый *bioch.* gibberellic
гибберилин gibberillin
гиббон *zool.* gibbon *(Hylobates)*
гиббсит *geol.* gibbsite, hydrargillite
гибель wreckage, wreck; (down)fall; death; destruction, ruin, loss
~, абсолютная полная *stat.* absolute total loss
~ в огне holocaust
~, действительная полная *stat.* actual total loss
~, конструктивная полная constructive total loss
~ посевов crop failure
~ судна loss/wreck of a ship
гибельный fatal, disastrous, catastrophic; ruinous, pernicious; baneful, holocaustal
гибернакула *bot.* hibernaculum, winter bud
гибернация hibernation
~, естественная/искуственная natural/artificial hibernation
гибернома *med.* hibernoma
~, доброкачественная/злокачественная benign/malignant hibernoma
~ средостения mediastinal hibernoma
гибисколистный *bot.* hibiscofolious, hibiscus-leaved
гибиск(ус) *bot.* hibiscus *(Hibiscus)*
гибка *tech.* bending
~ листового металла sheet bending
~, фасонная channeling
~, холодная cold bending
гибкий 1. flexible, pliant, pliable; ductile, resilient; lithe, lissom; elastic, springy; *bot.* virgate; 2. tractable, yielding; adaptable; *econ.* changeable, versatile
гибкость 1. flexibility, pliability, pliancy; ductility, elasticity, elastance; plasticity; 2. resourcefulness, versatility; *econ.* changeability
~ вычисления computing flexibility
~ стержня slenderness ratio of a column
гиблый rotten, bad, good-for-nothing
~лое дело bad job, lost cause
~лое место godforsaken place
гибнуть perish, disappear; fall to pieces, be destroyed
гибочный bending
гибралтарский Gibraltar *(attr.)*
гибрид hybrid, crossbred, mongrel; *pl.* *geol.* hybrid/ contaminated rocks
~, вегетативный vegetative hybrid
~, ложный false hybrid
~, межвидовой interspecific hybrid
~, межлинейный hybrid between lines
~, межродовой (inter)generic hybrid
~, межсортовой intervarietal hybrid
~, полигеномный polygenomatic hybrid
~, прививочный graft hybrid
гибридизатор hybridist, hybridizer
гибридизация hybridization, crossbreeding; contamination
~ валентностей valence hybridization
~ клеток cell hybridization
~, направленная directed hybridization
~, односторонняя unilateral hybridization
~, симпатрическая sympatric hybridization
~ состояний state hybridization
гибридизировать hybridize
гибридизм hybridism
~, мультипоступный mutipost hybridism
~, обратный inverted hybridism
~, паулопоступный paulopost hybridism
гибридность hybridity, hybridism
гибридный hybrid; crossbred; mongrel; contaminated
гибридогамия hybridogamy
гибридотрансплантация hybridotransplantation
гиг *naut.* gig
гигабайт *comp.* gigabyte
гигагерц gigahertz
гигант giant
гигантизм *med.* giantism, gigantism
~, гипофизарный pituitary gigantism
~, евнухоидный eunuchoid gigantism
~, истинный genuine/true gigantism
~, первичный primordial gigantism
гигантобласт *med.* gigantoblast
гигантоклеточный gigantocytous, gigantocellular
гигантолит gigantolite
гигантома gigantoma

гигантомастия *med.* gigantomasty, gigantism of the breasts
гигантопитек gigantopithecus
гигантоптерис gigantopteris
гигантостраки *pl.* gigantostracans
гигантоцит *anat.* gigantocyte
гигантский giant, gigantic, cyclopean; huge; enormous; titanic; gargantuan
~кие шаги *sport* giant stride
гигафлопс *comp.* gigaflops
гигаэлектрон-вольт gigaelectron-volt, Gev
гигиена hygiene, hygienics; environmental physiology
~ жилища housing hygiene
~, коммунальная municipal/communal hygiene
~, космическая space hygiene
~, личная personal hygiene
~ труда hygiene of work/labo(u)r
гигиенист hygienist
гигиенический, гигиеничный hygienic, health(ful); sanitary
~ навык health habit
гигрин *chem.* hygrine
гигрограмма hygrogram
гигрограф (self-)recording hygrometer, hygrograph
~, волосяной hair hygrograph
гигрокинез hygrokinesis
гигрология hygrology
гигрома *med.* hygroma
~, кистозная cystic hygroma
гигромезофит *bot.* hygromesophyte
гигрометр hygrometer
~, абсорбционный absorption/chemical hygrometer
~, весовой balance hygrometer
~, волосной hair hygrometer
~, конденсационный dew-point hygrometer
~, самопишущий self-recording hygrometer, hygrograph
гигрометрический hygrometric(al)
гигрометрия hygrometry, psychrometry
~, абсорбционная spectroscopic hygrometry
гигроморфизм hygromorphism
гигроморфный hygromorphic
гигропарестезия hygroparesthesia
гигроскоп hygroscope
гигроскопический hygroscopic(al), moisture/-water-absorbing
гигроскопичность hygroscopicity, water-absorbing capacity
~, максимальная maximum hygroscopicity
~ топлива fuel hygroscopicity
гигроскопия hygroscopy
гигростат humidistat, constant-humidity cabinet, hygrostat
гигросциафит *bot.* hygrosciaphyte
гигротермограф hygrothermograph
гигрофанный hygrophanous
гигрофил hygrophile
гигрофильность *biol.* hygrophily
гигрофильный *biol.* hygrophilous
гигрофит *bot.* hygrophyte
гигрофор *bot.* hygrophorus *(Hygrophorus)*
гигрофоровые *n. pl. bot. Hygrophoraceae*
гигрохазия *bot.* hygrochasy
гигроэлектрометр hygroelectrometer
гид guiding device/telescope; guide
~, фотоэлектрический photoelectric guiding device
гидальго *hist.* hidalgo
гидантоин *chem.* hydantoin
гидантоиновый hydantoic
гидатида *med., zool.* hydatid
гидатогенный *geol.* hydatogenic, hydatogenous
гидатода *bot.* hydathode
гидатоморфизм *geol.* hydatomorphism
гидатопирогенный *geol.* aqueo-igneous, hydatopyrogenic
гидатофит *bot.* hydatophyte
гидатохория *bot.* hydatochory
гиджра *hist.* hegira, hejira, hejra, hijra; flight
гидденит *geol.* hiddenite
гидирование *tech.* guiding
гидра *zool.* hydra *(Hydra); myth.* hydra
гидравлика 1. hydraulics; 2. hydraulic system/ controls/ drive
~, авиационная aircraft hydraulics
~, инженерная/сооружений hydraulics of construction, engineering hydraulics
~, судовая marine hydraulics
гидравлический hydraulic
гидравличность hydraulicity
гидраденит *med.* hidradenitis
~, гнойный hidradenitis suppurativa
~, подмышечный hidradenitis axillaris
гидраденокарцинома *med.* hidradenocarcinoma
гидраденома *med.* hidradenoma
гидраденофиброма *med.* hidradenofibroma
гидразид *chem.* hydrazid(e)
гидразин *chem.* hydrazin(e), diamide
гидрамнион *med.* hydramnion, hydramnios, dropsy of the amnion
гидрангея *bot.* hydrangea *(Hydrangea)*
гидрант I *tech.* hydrant, fire plug
~, пожарный (fire) hydrant
гидрант II *zool.* hydranth
гидраргиллит *geol.* hydrargillite, gibbsite
гидрартроз *med.* hydrarthrosis
~, интермиттирующий intermittent hydrarthrosis
гидрастин *chem.* hydrastine
гидрастинин hydrastinine
гидрастис *bot.* hydrastis, golden-seal, yellow/ jaundice root, Indian turmeric *(Hydrastis)*
гидрат *chem.* hydrate
~ брома bromine hydrate
~ закиси (lower) hydroxide, hydroprotoxide
~ закиси железа ferrous hydroxide/hydrate
~ закиси меди copper hydroprotoxide
~ кремнекислого магния hydrous magnesium

silicate
~ окиси hydroxide
~ окиси бария barium hydroxide, caustic baryte
~ окиси бериллия beryllium hydroxide
~ окиси железа hydroxide of iron, ferric hydrate
~ окиси железа, вторичный *geol.* transported limonite/iron
~ окиси калия potassium hydroxide
~ окиси кальция calcium hydrate
~ окиси марганца manganese hydroxide
~ окиси меди copper hydroxide
~ окиси натрия sodium hydrate, hydrate of sodium
~ перекиси hydroperoxide
~ перекиси кумола cumene hydroperoxide
~ хлора chlorine hydrate
гидратаза hydratase
гидратационный hydration, hydratational
гидратация hydration
~, избирательная selective hydration
~ масла oil degumming
~, предварительная prehydration
гидратировать hydrate
гидратируемость hydratability
гидратированный, гидратный hydrated
гидратор hydrator
гидратцеллюлоза hydrated/regenerated cellulose
гидрационит evaporite
гидремия *med.* hydr(a)emia
гидрид *chem.* hydride
~ калия/кальция potassium/calcium hydride
~ лития lithium hydride
гидрирование hydrogenation
гидрированный hydrogenated, hydrogenized
гидрировать hydrogenate, hydrogenize
гидрирующий hydrogenant
гидрирующийся hydrogenable
гидрия hydria
гидроавиатранспорт seaplane carrier
гидроавиация sea aviation, hydroaviation
гидроагрегат 1. hydrostatic unit, hydroelectric generating set; 2. hydraulic unit
гидроаккумулирование water storage, pumped (water) storage
гидроаккумулятор hydraulic accumulator
гидроакустик asdic/sonar operator; sonarman; soundman
~, авиационный airborne sonarman
~, базовый/береговой harbour defence soundman/ sonarman
~ надводного корабля surface sonarman
гидроакустик underwater acoustics, hydroacoustics, sonar technology
гидроакустик-подводник submarine sonarman
гидроакустический asdic, *Am.* sonar
~кая станция asdic, *Am.* sonar
~кая станция для поиска подводных лодок sub-search sonar
~кая станция, низкочастотная subsonic sonar
~кая станция, пассивная direct-listening (sonar) system, passive (sonar) system, underwater listening device
гидроаналогия hydraulic analogy
гидроаппарат hydrojet, jet pump
гидроаэродром seadrome, seaplane station, marine/sea aerodrome
гидроаэроионизатор hydroaeroionizer
гидроаэроионизация hydroaeroionization
гидроаэроионотерапия hydroaeroionotherapy
гидроаэроионы *pl.* hydroaeroions
гидроаэромеханика fluid mechanics, hydroaeromechanics, aerohydrodynamics
гидроаэроплан hydroplane
гидроаэропорт water airport
гидробак hydraulic reservoir
гидробактериология hydrobacteriology
гидробаллистика hydroballistics
гидробачок hydraulic reservoir
гидробиолог hydrobiologist
гидробиология hydrobiology
гидробионика hydrobionics
гидробионт hydrobiont, hydrocole
гидробиос hydrobios(is)
гидроборацит *miner.* hydroboracite
гидроботаника hydrobotany
гидробур hydroborer, water-jet borer
гидровзрывание hydraulic blasting
гидровскрыша hydraulic stripping
гидрогазификация hydrogasification
гидрогазодинамика fluid dynamics
гидрогалит *geol.* hydrohalite
гидрогель hydrogel
гидрогематит *geol.* turgite, hydrohematite
гидрогематоцеле hydrohematocele
гидрогеназа *bioch.* hydrogenase
гидрогенатор hydrotreater
гидрогенез hydrogenesis
гидрогенератор hydraulic(-turbine) generator, hydro-generator
~ с вертикальной/горизонтальной осью вращения vertical/horizontal(-shaft) hydraulic turbine generator
гидрогенизатор hydrogenator
гидрогенизация hydrogenation, hydrogenization
~, деструктивная hydrocracking, hydrogenative splitting, destructive hydrogenation, hydrogenolysis
~, недеструктивная addition/saturation hydrogenation
гидрогенизированный hydrogenized, hydrogenated
гидрогенизировать hydrogenize, hydrogenate
гидрогенизирующий hydrogenant
гидрогенный hydrogenic, hydrogenous
гидрогенолиз hydrogenolysis
гидрогеолог hydrogeologist
гидрогеологический hydrogeologic(al)
гидрогеология hydrogeology, ground-water hydrology
~, горная mining hydrogeology

~, инженерная engineering hydrogeology
гидрогеофит *bot.* hydrogeophyte
гидрогеохимия hydrogeochemistry
гидрогетит *geol.* hydrogoethite
гидроградус hydrograde
гидрогранат *geol.* hydrogarnet
гидрограф 1. hydrograph; 2. hydrographer
~, единичный unit hydrograph
~ паводка flood hydrograph
~, приточный inflow hydrograph
~ талого стока snow-melt hydrograph
~, элементарный unit hydrograph
гидрографический hydrographic
гидрография hydrography
~ моря nautical hydrography
гидрогроссуляр *geol.* hydrogrossular(ite)
гидродамал *zool.* rhytina, Steller's sea cow *(Hydrodamalis gigas)*
гидродвигатель hydraulic prime mover, fluid-power prime mover, hydraulic motor/engine; *see also* **гидромотор**
~, объемный positive-displacement hydraulic prime mover, positive-displacement fluid-power prime mover, positive-displacement hydraulic engine
~, поворотный rotary hydraulic prime mover, rotary fluid-power engine
гидродинамика hydrodynamics, fluid dynamics
~, квантовая quantum hydrodynamics
~ корабля ship hydrodynamics
~, магнитная magnetohydrodynamics, MHD
~, техническая engineering hydrodynamics
гидродинамический hydrodynamic(al)
гидродобыча hydraulic mining, hydraulicking
гидродомкрат hydraulic jack
~ двойного действия double-action hydraulic jack
гидродром seadrome
гидродроссель throttle valve
гидрожидкость hydraulic fluid
гидрожиобертит *geol.* hydrogiobertite
гидрозакись *chem.* (lower) hydroxide
гидрозакладка *min.* hydraulic filling/stowing
гидрозамок pilot controlled check valve
гидрозатвор hydro/water seal
гидрозолоудаление hydraulic ash disposal/handling
гидрозоль *chem.* hydrosol(e)
гидрозонд bathysonde
гидроид *bot., zool.* hydroid
гидроидные *n. pl. zool.* hydrozoans *(Hydrozoa)*
гидроиды *pl.* hydroids *(Hydroidea)*
~, бентонные benthonic hydroids
гидроизобата hydroisobath
гидроизогипса water-table contour, contour of water table, ground-water contour, hydroisohypse
гидроизоляционный dampproof
гидроизоляция 1. water/damp proofing, dampproofing, moisture insulation, hydroinsulation; 2. wet/hydraulic seal
гидроизоплета hydroisopleth
гидроизопьеза hydroisopiestic line
гидроизотерма hydroisotherm
гидроинкубатор water-heated incubator
гидроинтегратор hydrointegrator
гидроион hydroion
гидроионизатор hydroionizer
гидроионизатор-распылитель hydroionizer-nebulizer
гидроканал water/flow channel, tow(ing) tank/basin
~, карусельный water wheel
~, циркуляционный circulating water channel
гидрокарбонат *chem.* hydrocarbonate
гидрокарпический, хидрокарпный hydrocarpic
гидрокарпия hydrocarpy
гидрокаучук hydrorubber, hydrogenated rubber
гидрокианит *geol.* hydrocyanite
гидрокинетика kinetics of liquids, hydrokinetics
гидрокистома *med.* hydrocystoma
гидроклапан hydraulic valve
гидрокластический *geol.* hydroclastic
гидроклеис *bot.* water poppy *(Hydrocleis)*
гидроклиматический hydroclimatic
гидроклиногумит *geol.* hydrous clinohumite
гидрокольпос *med.* hydrocolpocele, hydrocolpos
гидрокомбинезон, гидрокостюм diving suit
гидрокортизон *med.* hydrocortisone, cortef; cortilan, cortril, hycortole, hydrocortone, Kendall's compound F, Reichstein's substance M, 17-hydroxy-corticosterone
гидрокортизон-ацетат *med.* hydrocortisone/hydrocortone/cortef acetate, hydrocortisone-21-acetate
гидрокотиль *bot.* pennywort *(Hydrocotyle)*
гидрокран hydraulic cock
гидрокрекинг hydrocracking, hydrogenative splitting
гидрокриптофит *bot.* hydrocryptophyte
гидрокрыло hydrofoil
гидрокрюк hydraulic pick-up hitch
гидрокс 1. Hydrox coal blasting; 2. hydrox (cartridge)
гидроксид *chem.* hydroxide
гидроксил *chem.* hydroxyl
гидроксилаза *chem.* hydroxylase
гидроксиламин *chem.* hydroxylamine
гидроксиламинредуктаза *chem.* hydroxylamine reductase
гидроксилизин *chem.* hydroxylysine
гидроксилирование *chem.* hydroxylation
гидроксилировать *chem.* hydroxylate
гидроксильный *chem.* hydroxyl(ic)
гидрол hydrol syrup
гидролабильность hydrolability
~, приобретенная acquired hydrolability
~, экзогенная/эндогенная exogenous/endogenous hydrolability
гидролабильный hydrolabile
гидролаза *bioch.* hydrolase

гидролакколит *geol.* hydrolaccolite, frost mound
гидролатерит *geol.* hydrolaterite, flavite
гидролебедка hydraulic hoist
гидролиаза hydrolyase
гидролиз *chem.* hydrolysis
~, кислотный acidolysis, acid hydrolysis
~ основанием basic hydrolysis
~, ферментативный enzymic hydrolysis
~, щелочной alkaline hydrolysis
гидролизат *chem.* hydrolysate
~, белковый protein hydrolysate
~, плазматический plasma hydrolysate
гидролизер hydrolyser
гидролизин hydrolysine
гидролизный, гидролитический hydrolytic
гидролизовать(ся) hydrolyse
гидролимфа hydrolymph
гидролит hydrolith; hydrolyte
гидролитический hydrolytic
гидролог hydrologist
гидрология hydrology
гидролокатор asdic (set/equipment), *Am.* sonar (system); hydrolocator
~, буксируемый вертолетом, поисковый helicopter-towed search sonar
~, вертолетный helicopter(-mounted) sonar
~, глубоководный deep-water/-sea sonar
~, инфразвуковой subsonic sonar
~ кругового обзора scanning sonar
~, опускаемый dip(ping) sonar
~, пассивный passive/listening sonar
~, погружаемый dunking/dip(ping) sonar
~ шагового поиска searchlight sonar
гидролокация sound fixing and ranging, sonar, hydrolocation
гидролоток open-top circulating water/flow channel; flume
гидролыжи *av.* hydroskі, hydro-/water ski
гидромагистраль water main
гидромагнезит hydromagnesite
гидромагнетизм hydromagnetism
гидромагнитный hydromagnetic
гидромания *med.* hydromania
гидромассаж hydromassage
гидромашина hydraulic machine/unit
~, обратимая combined/convertible pump-turbine unit
гидромегатерм *bot.* hydromegatherm
гидромедуза *zool.* hydromedusa
гидромезофильный hydromesophilic
гидромезофит *bot.* hydromesophyte
гидромель hydromel
гидроменингит *med.* hydromeningitis
гидроменингоцеле *med.* hydromeningocele, hydroencephalocele
гидрометаллургический hydrometallurgical
гидрометаллургия hydrometallurgy
~, автоклавная pressure hydrometallurgy
гидрометаморфизм *geol.* hydrometamorphism
гидрометеор hydrometeor
гидрометеорология hydrometeorology
гидрометр hydrometer
гидрометрий *med.* hydrometra
гидрометрист hydrometrist
гидрометрический hydrometric(al)
гидрометрия hydrometry
гидромеханизация hydraulic excavation/mining, hydromechanization
гидромеханизм hydraulic mechanism/actuator
~, единоагрегатный hydraulic (power) pack
~ регулирования высоты среза *agr.* hydraulic height control
гидромеханика hydromechanics, mechanics of liquids, fluid mechanics
~ корабля ship hydromechanics
гидромиелия *med.* hydromyelia
гидромиелоцеле *med.* hydromyelocele
гидромодуль rated water consumption, duty (of) water curve, hydromodulus
гидромонитор (hydraulic) giant monitor, moisture monitor, hydromonitor
~ ближнего боя *mil.* short-range giant
~, врубовый cutting/caving giant
~, поворотный swinging hydromonitor
~, подвесный suspended giant
~ разрушает породу *geol.* the giant loosens/ cuts/ caves the rock
~, смывной sluicing giant
гидроморфизм hydromorphism
гидроморфность hydromorphic feature
гидроморфный hydromorphic; hydrogenic *(of soil)*
гидроморфогенез hydromorphogenesis
гидромотор hydraulic motor, hydromotor
~ аксиально-поршневой swash-plate hydraulic motor, axial-piston hydraulic motor
~, лопастный guided-vane hydraulic motor
~, поршневой reciprocating hydraulic motor; piston(-type) hydraulic motor
~, радиально-поршневой rotary plunger hydraulic motor
~, роторный rotary hydraulic motor
гидромусковит *geol.* hydromica
гидромуфта (hydraulic) coupling, hydraulic clutch/coupling
гидронавт hydronaut
гидронавтика hydronautics
гидронасос hydraulic/fluid pump
гидронастический *bot.* hydronastic
гидронастия *bot.* hydronasty
гидронастуран *geol.* hydrous pitch blende, hydrouraninite
гидронефроз *med.* hydronephrosis, nephrohydrosis; uronephrosis
~, врожденный congenital hydronephrosis
~, ложный false hydronephrosis, pseudohydronephrosis
гидронефрома *med.* hydronephroma
гидронефелометр sea-water nephelometer
гидрообогащение *min.* hydraulic ore dressing

гидроокись *chem.* hydroxide, hydrated oxide
~ железа ferric hydrate, hydrous ferric oxide
~ калия caustic potash
~ кальция calcium hydroxide
~ натрия sodium hydroxide
~ основная basic hydroxide
гидроотвал hydraulic mine dump
гидроочиститель fluid cleaner
гидроочистка *met.* hydraulic cleaning; *chem.* hydrofining
гидропат hydropathist
гидропатический hydropathic(al)
гидропатия hydropathy
гидропередача hydraulic transmission
~, гидростатическая/объемная (positive-) displacement hydraulic transmission
~, динамическая hydrokinetic transmission
~, объемная с дроссельным/объемным регулированием valve-/pump-controlled hydraulic transmission
гидроперекись *chem.* hydroperoxide
гидроперикард *med.* hydropericardium
гидроперикардит *med.* hydropericarditis
гидропионефроз *med.* hydropyonephrosis
гидропиль *bot.* hydropyle
гидроплан hydroplane, seaplane
гидропланер hydroglider
гидропласт PVC/polyvinylchloride paste
гидропневматический hydropneumatic, oleopneumatic
гидропневматоз *med.* hydropneumatosis
гидропневмоаккумулятор pneumatic pressure tank, pressure storage tank, pressurizing tank
гидропневмогония *med.* hydropneumogony
гидропневмоперикард *med.* hydropneumopericardium
гидропневмоторакс *med.* hydropneumothorax, seropneumothorax
гидропогрузчик hydraulic loader
гидроподъем hydraulic handling, slurry hoisting (equipment)
гидроподъемник hydraulically operated hoist/lift, hydraulic jack
~, самолетный hydraulic aircraft jack
гидрополип *zool.* hydropolyp
гидропомпа hydraulic pump
гидропоник hydroponic bed
гидропоника *agr.* tray/tank agriculture, hydroponics, soilless growth of plants, gravel culture, chemical gardening
гидропонный hydroponic
гидропора hydropore
гидропорт hydroplane port, seaplane station
гидропот hydropote
гидропочта hydraulic rabbit
гидропреобразователь hydrotransducer
~, вальный reciprocating hydrotransducer
гидропресс hydraulic press/ram
гидропривод hydraulic drive/actuation, fluid (-power) drive; oil gear; hydraulic power *(as a method)*
~ вращательного движения rotary hydraulic drive
~, гидростатический/объемный positive-displacement hydraulic drive, pump-controlled hydraulic drive
~, динамический fluid-coupling drive, torque-converter drive
~, следящий hydraulic servomechanism, hydraulic servo system
~ тормозов hydraulic brake actuation
гидропровод hydraulic line
гидропульпа pulp, slurry
гидропульт hydraulic sprayer, spray gun, water cannon; stirrup pump
гидропята hydraulic balancing device
гидроразбиватель hydrobeater, hydraulic beater/disintegrator, hydropulper
гидрораспределитель hydraulic control valve
гидрорегулятор hydraulic control
~ заглубления *agr.* hydraulic depth control
гидроресурсы *pl.* water resources
гидрорея *med.* hydrorrhea
гидросальпинкс *med.* hydrosalpinx, salpingian dropsy
гидросамолет seaplane, hydroplane
~ вертикального взлета и посадки vertical take-off and landing air/sea craft, VTOL
~, двухпоплавковый twin-float (sea)plane
~, лодочный boat(-type) (sea)plane, flying boat
~, однопоплавковый single-float (sea)plane
~, поплавковый float(-type) (sea)plane
~ с тянущими винтами tractor hydroplane
~ типа "летающая лодка" flying boat
гидросбив окалины hydraulic scalebreaking/descaling
гидросепаратор hydroseparator
гидросепаратор-сгуститель hydroseparator-thickener
гидросернистый sulpohydrate
гидросеть hydraulic piping, drainage system
гидросеялка *agr.* hydroseeder
гидросиликат hydrosilicate
гидросирингомиелия *med.* hydrosyringomyelia
гидросистема hydraulic system, piping
~, аварийная emergency hydraulic system
~ навесного устройства hydraulic lift system
~, независимая live hydraulic system
~, основная primary/main hydraulic system
~, раздельно-агрегатная remote-cylinder hydraulic system
~, резервная stand-by hydraulic system
~у, прокачивать bleed the hydraulic system (of air); circulate fluid through the hydraulic system
гидроскоп hydroscope
гидрослюда hydromica, hydrous mica
гидрослюдизация hydromicatization
гидросмесь 1. slurry; 2. hydraulic fluid/liquid
~, авиационная aircraft hydraulic fluid
гидросома *zool.* hydrosome

гидроспира *paleont.* hydrospire
гидроспирометр hydrospirometer
гидроспуск seaplane ramp/slipway
гидростабилизация stabilization by hydrogenation
гидростанция hydroelectric (power) station
гидростат hydrostat
гидростатика hydrostatics
гидростатический hydrostatic(al)
гидростойка hydraulic prop
гидросульфат *chem.* hydrosulphate
гидросульфид *chem.* hydrosulphide
гидросульфит *chem.* hydrosulphite
гидросфигмограф *med.* hydrosphygmograph
гидротаксис hydrotaxis
гидротека *geol.* hydrotheca
гидротерапия *med.* hydrotherapy, hydrotherapeutics
гидротележка hydraulic power trolley
гидротерапия *med.* hydrotherapy, hydrotherapeutics
гидротерма thermal spring, hydrotherma
гидротермальный, гидротермический hydrothermal
гидротерпен *chem.* hydroterpene
гидротехник hydraulic engineer
гидротехника hydraulic engineering
гидротехнический hydraulic, hydrotechnic(al)
гидротипия hydrotypy
гидротипный *phot.* hydrotype
гидроторакс *med.* hydrothorax, serothorax, dropsy of the chest
~, хилезный chylous hydrothorax
гидроторит *geol.* hydrothorite
гидроторф hydropeat
гидротранспорт hydrotransport, hydraulic transportation
~, напорный pressurized hydrotransport
гидротранспортер hydraulic conveyer
гидротрансформатор (hydrodynamic) torque converter
гидротропизм *biol.* hydrotropism
гидротропия *biol.* hydrotropy, hydrotropism
гидротубация *med.* hydrotubation
гидротурбина hydraulic turbine, hydroturbine
~, активная impulse turbine
~, диагональная adjustable-blade Francis turbine, radial-flow Francis turbine
~, ковшовая Pelton turbine
~, модельная model(-test) turbine, turbine test model
~, напорноструйная reaction turbine
~, осевая axial-flow turbine
~, поворотно-лопастная propeller/Kaplan turbine
~, радиально-осевая radial-flow turbine, Francis turbine
~, реактивная reaction turbine
~, реактивная осевая paralled-flow reaction turbine
гидротурбогенератор hydroelectric generator
гидроударник hydraulic hammer
гидроузел (integrated) hydroscheme, hydroengineering complex, hydraulic power system
~, водосборный water catchment system
~, водохранилищный reservoir hygrosystem
~, высоко-/низко-/средненапорный high-/low-/medium-head water (power) development
~, энергетический hydroelectric power system
гидроуправление hydraulic control
~ навесными орудиями hydraulic lift control
~ сцепкой hydraulic draw-bar control
гидроупругость hydroelasticity
гидроуретер *med.* hydroureter
гидроусилитель *auto.* hydraulic booster; *av.* hydraulic actuator, power(ed)/servo control unit
~, последовательно включенный *av.* tandem actuator, tandem power unit, series servo (unit)
~ руля control surface servo (unit)
~ руля высоты elevator hydraulic actuator, elevator power unit
~ руля направления rudder power unit, rudder jack, rudder hydraulic actuator
~ системы управления power control jack
~ стабилизатора *av.* tailplane servo (unit)
~, счетверенный quadruplex actuator
~ цельноповоротного fin power unit
~, шестикамерный six-cylinder hydraulic actuator
~ элерона aileron power unit, aileron actuator/jack
гидроустановка hydroelectric power plant
гидрофизика hydrophysics
гидрофилит *geol.* hydrophilite, chlorocarlite
гидрофилия *bot.* hydrophily
гидрофил(лум) *bot.* waterleaf *(Hydrophyllum)*
гидрофильность hydrophilic nature/property/character; hydrophily, hydrophility
гидрофильный *bot.* hydrophilic, hydrophilous, water-receptive/-retaining
гидрофит *bot.* hydrophyte, aquatic plant
гидрофоб hydrophobe
гидрофобизатор water-repelling agent, water repellent
гидрофобизация water-repellency treatment, rendering hydrophobic
гидрофобия *med.* hydrophobia
гидрофобность hydrophoby, water-repellent property
гидрофобный 1. *med.* hydrophobic, hydrophobous; 2. hydrofuge, water-/moisture-repellent, waterproof
гидрофон hydrophone
~, буксируемый towed hydrophone
~, магнитнострикционный magnetostrictive hydrophone
~ направленного действия direction hydrophone
~, поршневой piston hydrophone
~, установленный на буе buoy-supported

hydrophone, buoyed hydrophone
~, шаровой spherical hydrophone
гидрофор hydrophore
гидрофранклинит *geol.* chalcophanite, hydrofranklinite
гидрофтальм *med.* hydrophthalmia, hydrophthalmos, hydrophthalmus
~, врожденный congenital hydrophthalmos
гидрохимия chemical hydrology, hydrochemistry
гидрохинин hydroquinin(e)
гидрохинон hydroquinone
гидрохлорид *chem.* hydrochloride
гидрохлортиазид *chem.* hydrochlorthiazide
гидрохория *bot.* hydrochory, water dispersal
гидрохорный *bot.* hydrochoric, hydrochorous
гидроцеле hydrocele
гидроцеллюлоза *chem.* hydrocellulose
гидроцентраль hydroelectric plant
гидроцефалия *med.* hydrocephalus, hydrocephaly
~, врожденная congenital hydrocephalus
~, желудочковая ventricular hydrocephalus
~, наружная external hydrocephalus
~, окклюзионная obstructive hydrocephalus
~, приобретенная acquired hydrocephalus
~, травматическая posttraumatic hydrocephalus
~, тромботическая thrombotic hydrocephalus
гидроцефалоид *med.* hydrocephaloid
гидроцианит hydrocyanite
гидроциклон hydraulic cyclone, hydrocyclone
гидроцилиндр hydraulic cylinder, hydrocylinder
~ автомата тяги *av.* throttle hydraulic actuator
~ воздушного тормоза airbrake hydraulic cylinder/jack
~, выносной *tech.* remote(-operation) hydraulic cylinder
~ замка (шасси) hydraulic lock cylinder
~ интерцептора *av.* spoiler hydraulic actuator
~, моментный single-vane hydraulic actuator, rotary ram
~ с запирающим клапаном-ограничителем hydrostop cylinder
~, силовой hydraulic power cylinder
~ створок жалюзи louvre hydraulic jack
~ створок реверса тяги reverser eyelid hydraulic cylinder
~ створок сопла nozzle hydraulic cylinder/jack
гидроцинкит *geol.* hydrozincite
гидроцистома *med.* hydrocystoma
гидрошахта hydraulic mine/colliery
гидроэкстактор *text.* hydroextractor, whizzer
гидроэкструзия hydrostatic extrusion, fluid-pressure extrusion
гидроэлеватор hydraulic elevator, jet pump
~, пожарный suction fitting
гидроэлектростанция, ГЭС *abbr.* hydroelectric (power) station/plant, water-power station/plant
~, аккумулирующая pumped-storage hydroelectric station
~, высоко-/низко-/средненапорная high-/low-/medium-head hydroelectric station
~, приливная tidal power station
гидроэнергетика hydraulic power engineering
гидроэнергия water power
гидроэнергостроительство hydropower construction work
гидрояркомер underwater luminance meter
гидрурез *med.* hydruria
гиена *zool.* hyena *(Hyaena)*
гиеновидный *zool.* hyeniform, hyenoid
гизекит *geol.* gieseckite
гик *naut.* (sail)boom, bum
гика-перты *naut.* boom horse
гика-топенант *naut.* topping lift
гикори *bot.* hickory *(Carya)*
гилления *bot.* false ipecac *(Gillenia, Portaramthus)*
гилея tropical (rain) forest
гилия *bot.* gilia *(Gilia)*
гилозоизм hylozoism
гиломорфизм hylomorphism
гиломорфический hylomorphic(al)
гилотеизм hylotheism
гилотеистический hylotheistic(al)
гилофит *bot.* hylophyte
гилофитный *bot.* hylophytic
гилум *bot.* hilum
гильберт *phys.* gilbert
гильдия *hist.* guild
гильза I 1. sleeve; liner; socket; bush(ing); case, shell; 2. cartridge case; (cigarette) wrapper
~, графитовая graphite/carbon liner
~ гнезда sleeve of a jack
~ для термометра thermometer pocket/well
~ доильного стакана *agr.* teat cup shell
~, зажимная clamping sleeve
~, замедляющая copper slug *(of relais)*
~, защитная protection sleeve, liner
~, катодная cathodic sleeve
~, нарезная screw socket
~, негорючая inert liner
~, папиросная cigarette wrapper
~, патронная cartridge case
~ ружейного патрона cartridge case
~, сменная removable liner
~, стреляная ejected cartridge/shell case
~ трубной заготовки hollow billet
~ трубной заготовки, прошитая pierced shell
~ цилиндра cylinder liner/sleeve
~ шпинделя quill
~, экстракционная extraction thimble
гильза II *zool.* giant herring *(Hilsa)*
гильзование lining
гильзовать line down *(a cylinder)*
гильзоотвод case ejection chute
гильзосборник case collector (box)
гильзоулавливатель shell bag
гильотина guillotine (clippers)
гильотинирование guillotinage, guillotinement
гильотинка cigar clipper
гильоширование *arch.* guilloche work

гильпинит *geol.* gilpinite
гилюс *anat.* hilus
гиляби *pl. geol.* bentonite
гималайский Himalayan
гимен *anat.* hymen
гименальный *anat.* hymenal
Гимен(ей) *myth.* Hymen
гимениевидный *bot.* hymenoid
гимений *bot.* hymenium, sporiferous layer
гименокаллис *bot.* hymenocallis *(Hymenocallis)*
гименолепидоз *med.* hymenolepiasis, hymenolepsis infestation
гименомицет *bot.* hymenomycete
гименоносец hymenophore
гименоспермы *pl. bot.* hymenosperms
гименоскоп *med.* hymenoscope
гименоспора *bot.* naked spore
гименотом *med.* hymenotome
гименотомия *med.* hymenotomy
гименофилловые *n. pl. bot.* filmy ferns family *(Hymenophyllaceae)*
гименофор *bot.* hymenophore
гимн hymn; doxology; anthem
  ~, государственный national anthem
гимназист high-school student
гимназия high/secondary school
гимнасиарх *hist.* gymnasiarch
гимнаст gymnast
гимнастика gymnastics; callisthenics; physical exercise
  ~, дыхательная respiratory gymnastics
  ~, земная standing exercises
  ~, лечебная therapeutic physical training
  ~, художественная rhythmic gymnastics
гимнастический gymnastic
  ~ зал gymnasium, gym (hall)
  ~кие упражнения gymnastics
  ~кие снаряды gymnastic apparatus
гимнобластический *zool.* gymnoblastic
гимногенные *n. pl.* gymnogens
гимнокарпный *bot.* gymnocarpous, gymnocarpic
гимнокладус, гимноклядус *bot.* coffe tree *(Gymnocladus)*
гимнопласт *biol.* gymnoplast
гимнософист *hist.* gymnosophist
гимноспора *bot.* gymnospore
гимностомный *bot.* gymnostomous, gymnostomatous
гимноцидий *bot.* gymnocidium
гимноцит gymnocyte
гимнура *zool.* moonrat *(Gymnura)*
гиназа gynase
гинандрический *bot.* gynadrous
гинандрия gynandry, gynandrism
гинандроморф *biol.* gynandromorph
гинандроморфизм *biol.* gynandromorphism
гинандроморфный *biol.* gynandromorphic
гинандрофор *bot.* gynandrophore
гинатрезия *med.* gynatresia
гингивальный *phon.* gingival
гингивит *med.* gingivitis
  ~, афункциональный afunctional gingivitis
  ~ беременных pregnancy gingivitis
  ~, гиперплазический hyperplastic gingivitis
  ~, гнойный suppurative gingivitis
  ~, калькулезный calculous gingivitis
  ~, катаральный catarrhal gingivitis
  ~, лейкемический гиперпластический leukemic hyperplastic gingivitis
  ~, острый acute gingivitis
  ~, фузоспирилезный fusospirillary gingivitis
  ~, цинготный scorbutic gingivitis
  ~, язвенно-мембранозный membranoulcerous gingivitis
  ~, язвенно-некротический necrotizing ulcerative gingivitis
  ~, язвенный ulcerative gingivitis
гингивоглоссит *med.* gingivoglossitis, uloglossitis
гингивостоматит *med.* gingivostomatitis
  ~, герпетический herpetic gingivostomatitis
гингивэктомия *surg.* gingiv(o)ectomy
гинекей *hist.* gynaeceum
гинекократический gyn(a)ecocratic
гинекократия gyn(a)ecocracy
гинеколог gyn(a)ecologist
гинекологический gyn(a)ecologic(al)
гинекология *med.* gyn(a)ecology
  ~, судебная forensic/legal gyn(a)ecology
гинекомастия *med.* gyn(a)ecomastia, gyn(a)ecomasty, gyn(a)ecomazia, gynomastia
гинекопатия *med.* gyn(a)ecopathy
гинекофилия gyn(a)ecophilia
гинекофобия *psychiatry* gyn(a)ecophobia
гинекофорный *zool.* gyn(a)ecophoric
гинецей *bot.* gynaeceum, gynoecium
гинея guinea *(coin)*
гини I *bot.* gean
гини II *pl. naut.* gin tackle, heavy/main purchase
  ~, грузовые cargo gin
гинкго *bot.* ginkgo, maidenhair tree *(Ginkgo, Salisburia)*
гинобазис *bot.* gynobase
гинобазический *bot.* gynobasic
гиногаметофит *bot.* gynogametophyte
гиногамоны *pl.* gynogamones
гиногенез *biol.* gynogenesis
гинодиэцизм *bot.* gynodioecism
гинодиэцичный *bot.* gynodioecious
гиномоноэцизм *bot.* gynomonoecism
гиномоноэционный, гиномоноэцичный *bot.* gynomonoecious
гиномоноэция *bot.* gynomonoecism
гиноспора *bot.* gynospore, megaspore, macrospore
гиноспорангий *bot.* gynosporangium, megasporangium
гиноспорофилл *bot.* gynosporophyll
гиностемий *bot.* gynostemium
гинофор gynophore
гинофорный gynophoric
гинсдалит *geol.* hinsdalite

гинтцеит *geol.* hintzeite
гинь *naut.* winding tackle
гинь-блок *naut.* gin block, whip gin
гиоид *zool.* hyoid
гиоидный *zool.* hyoid
гиолит *geol.* hyolith
гиортдалит *geol.* hiortdahlite
гиосциамин *chem.* hyosciamine
гиосцин *chem.* hyoscine, scopolamine
~, бромистоводородный hyoscine, scopolamine hydrobromide
гиохолевый *med.* hyocholic
гипабиссальный *geol.* hypabyssal
гипабисситы *pl. geol.* hypabyssites
гипавтоморфный *geol.* hypautomorphic, hypidiomorphic
гипаллага *gram.* hypallage
гипаллеломорфы *biol.* hypallelomorphs
гипальгезия, гипальгия *med.* hypalgia, hypalgesia
гипантий *bot.* hypanthium
гипантодий *bot.* hypanthodium, torus enlargement
гипархный hyparchic
гипафорин *med.* hypaphorine
гипеаструм *bot.* knight's-star *(Hippeastrum)*
гипекоум *bot.* hypecoum *(Hypecoum)*
гиперадипоз *med.* hyperadiposis, hyperadiposity
гиперадреналинемия *med.* hyperadrenalemia, hyperadrenalism, hyperadrenia
гиперазотемия *med.* hyperazotemia
гиперазотурия *med.* hyperazoturia
гиперакузия *med.* hyperacusia, hyperacusis
гиперальгезия *med.* hyperalgesia
~, слуховая auditory hyperalgesia
гиперальдостеронемия *med.* hyperaldosteronemia, (hyper)aldosteronism
гиперальконолемия *med.* hyperalcoholemia
гиперамилаземия *med.* hyperamilasemia
гиперамилазурия *med.* hyperamilasuria
гиперареальный hyperareal
гиперацидность *med.* hyperacidity, hyperchlorhydria
гиперацидный hyperacid, superacid
гипераэрация hyperaeration
гипераэродинамика hypersonic acrodynamics
гипербазит *geol.* ultrabasite
гипербаллистика hyperballistics
гипербаризм *med.* hyperbarism
гипербарический hyperbaric
гипербат(он) hyperbaton
гипербластический hyperblastic
гипербилирубинемия *med.* hyperbilirubinemia
~, врожденная/наследственная congenital/hereditary hyperbilirubinemia
~ новорожденных hyperbilirubinemia neonatorum
~, печеночно-клеточная hepatocellular hyperbilirubinemia
гипербрахицефалия *med.* hyperbrachycephaly
гипербола 1. *math.* hyperbola, hyperbolic line/curve; 2. *ling.* hyperbole
~, вырожденная collapsed hyperbola
~, равнобочная, равносторонняя equiangular/equilateral/rectangular hyperbola
гиперболический 1. *math.* hyperbolic; 2. *ling.* hyperbolical
гиперболоид *geom.* hyperboloid
~ вращения hyperboloid of revolution
гиперболоидный hyperboloid
гипербореец Hyperborean
гиперборейский hyperborean; northern
гипербулия *med.* hyperbulia
гипервакцинация *med.* hypervaccination
гипервентиляция *med.* hyperventilation, overventilation
гипервещественный hyperreal
гипервизор *comp.* hypervisor
гипервитаминоз *med.* hypervitaminosis
гиперволемический *med.* hypervolemic, plethoric
гиперволемия *med.* hypervolemia, plethora
гипергаммаглобулинемия *med.* hypergammaglobulinemia
гипергаплоиды *pl. med.* hyperhaploids
гипергевзия *med.* hypergeusia
гипергемоглобинемия *med.* hyperh(a)emoglobinemia
гипергедонизм *med.* hyperhedonia
гипергенез(ия) *med.* hypergenesia, hypergenesis
гипергенитализм *med.* hypergenitalism
~, эпифизарный epiphysial hypergenitalism
гипергенный hypergene, supergene
гипергидратация hyperhydration
гипергидроз *med.* hyper(h)idrosis
~, околоушно-височный periauriculotemporal hyperhidrosis
гипергидропексия *med.* hyperhydropexy, hyperhydropexis
гипергия hypergia
гипергликемия *med.* hyperglyc(a)emia, hyperglucosemia
~, рефлекторная reflex hyperglyc(a)emia
гипергликорахия *med.* hyperglycorrhachia
гипергликозурия *med.* hyperglycosuria
гипергонадизм *med.* hypergonadism
~ яичниковый ovarianhypergonadism
гиперграф *math.* hypergraph, set-system
гипердактилизм, гипердактилия *med.* hyperdactylism, hyperdactylia, hyperdactyly
гипердиплоиды *pl. biol.* hyperdiploids
гиперемированный *med., physiol.* hyperemic
гиперемия *med.* hyperemia
~, артериальная active/arterial/fluxionary hyperemia
~, венозная venous/passive hyperemia
~, застойная congestive hyperemia
~, очаговая focal hyperemia
~, рефлекторная reflex hyperemia
~, сосудистая vascular hyperemia
гиперергический hyper(er)gic

гиперестезия *med.* hyperesthesia
~, вкусовая hypergeusia, gustatory hyperesthesia
~, обонятельная hyperosmia, olfactory hyperesthesia
~, тактильная hyperaphia, tactile hyperesthesia, hyperpselaphesia
гиперзаряд hypercharge
гиперзвук hypersound
гиперзвуковой hypersonic, hyperacoustic
гипериммунизация *med.* hyperimmunization, hypervaccination
гипериммунный hyperimmune
гипериноз *med.* hyperinosemia, hyperinosis
гиперинсулинизм *med.* hyperinsulinism
гиперинфляция *econ.* runaway/galloping inflation
гиперит *geol.* hyperite
гиперкалиемия *med.* hyperkalemia, hyperkaliemia, hyperpotassemia
гиперкалиурез *med.* hyperkaluresis
гиперкальциемия *med.* hypercalcemia
гиперкальцинурия *med.* hypercalci(n)uria, hypercalcuria
гиперкапния *med.* hypercapnia, hypercarbia
гиперквадрика *math.* hyperquadric
гиперквантование hyperquantization
гиперкератоз *med.* hyperkeratosis
~, внефолликулярный extrafollicular hyperkeratosis
~, гонорейный gonorrheal hyperkeratosis
~, подногтевой hyperkeratosis subungualis
гиперкинез *med.* hyperkinesia, hyperkinesis
~, атетозный athetotic hyperkinesia
~, детский hyperkinesia of infants
~, мозжечковый cerebellar hyperkinesia
~, тиковый tic hyperkinesia
~, хореиформный/хореический choreiform/choreic hyperkinesia
гиперкогомология hypercohomology
гиперкомпаундирование overcompounding
гиперконъюгация hyperconjugation
гиперкортицизм *med.* hypercorticoidism, hypercortisonism
гиперлипемия *med.* hyperlipemia, hyperlipoidemia
~, наследственная hereditary hyperlipemia
гиперлордоз *med.* hyperlordosis
гипермастия *med.* hypermastia, polymastia
гиперматрица hypermatrix
гиперменорея *med.* hypermenorrhea
гиперметропия *med.* hypermetropia, hyperopia
~, осевая axial hypermetropia/hyperopia
гипермимия *med.* hypermimia
гипермиотония *med.* hypermyotonia
гипермнезия *med.* hypermnesia
гипернефрома *med.* hypernephroma, carcinoma of the kidney, hypernephroid carcinoma
гиперовария *med.* hyperovaria, hyperovarinism
гипероксалурия *med.* hyperoxaluria
~, первичная primary hyperoxaluria
гипероксемия *med.* hyperoxemia
гипероксидоз hyperoxidosis
гипероксия *med.* hyperoxia
гиперон *phys.* hyperon, superproton
гиперонихия *med.* hyperonychia
гипероскоАок hyperfragment, hypernucleus
гиперосмия *med.* hyperosmia, hyperosphresia
гиперосмотический hyperosmotic
гиперостоз *med.* hyperostosis
~, корковый cortical hyperostosis
гиперпаразит *biol.* hyperparasite
гиперпаратиреоз *med.* hyperparathyroidism
гиперперистальтика hyperperistalsis
гиперпигментация *med.* hyperpigmentation, superpigmentation
гиперпиезия *med.* hyperpiesis, hyperpiesia
гиперпирексия *med.* hyperpyrexia
гиперплавкий hyperfusible, superfusible
гиперплазированный hyperplastic
гиперплазический hyperplasic, hyperplastic
гиперплазия *biol., med.* hyperplasia
~, аденокистозная adenocystic hyperplasia
~, десенная gingival hyperplasia
~, системная systemic hyperplasia
гиперплоидия *biol.* hyperploidy
гиперплоскость hyperplane, hypersurface
гиперпноэ *physiol.* hyperpnea
гиперповерхность hypersurface
гиперполиплоиды *pl.* hyperpolyploids
гиперпространство *math.* hyperspace
гиперпротеинемия *med.* hyperprotein(a)emia
гиперрефлексия hyperreflexia
гиперсекреция *med.* hypersecretion
~, желудочная gastric hypersecretion
~, кишечная intestinal hypersecretion
гиперсенсибилизация hypersensitization, hypersensitizing, hypersensibilization
гиперсиндез hypersyndesis
гиперсомния *med.* hypersomnia
гиперсорбция hypersorption
гиперстен hypersthene
гиперстенавгит *geol.* hypersthenеaugite, calc-clinohypersthene
гиперстеник *med.* hypersthenic
гиперстенический, гиперстеновый hypersthenic
гиперстения *med.* hypersthenia
гиперстрофный *zool.* hyperstrophic
гиперсфера hypersphere
гипертелия *zool.* hypertely
гипертекст *comp.* hypertext
гипертензивный hypertensive
гипертензин hypertensin, angiotensin
гипертензиназа *med.* hypertensinase, angiotonase
гипертензиноген hypertensinogen, hypertensin precursor
гипертензия *med.* hypertension
гипертепловой hyperthermal
гипертермия *med.* hyperthermia, hyperpyrexia
~, коротковолновая short-wave hyperpyrexia
~, лихорадочная febrile hyperthermia

~, **предменструальная** premenstrual hyperthermia
гипертимизм *med.* hyperthymism, hyperthymization
гипертиреоз, гипертиреоидизм, гипертиреоидоз *med.* hyperthyroidism, hyperthyrea, hyperthyrosis
~, **скрытый** latent hyperthyroidism
гипертоник hypertensive patient
гипертонический hypertonic, hyperpietic
гипертония *med.* hypertension, supertension, hyperpiesia, hyperpiesis, hypertonia, high (arterial) blood pressure
~, **бледная** pale hypertension
~, **выраженная** marked hypertension
~, **доброкачественная** benign hypertension
~, **застойная** congestive hypertension
~, **злокачественная** malignant hypertension
~, **климактерическая** climacteric/menopausal hypertension
~, **красная** red hypertension
~, **легочная** pulmonary hypertension
~, **мышечная** muscular hypertension
~, **надпочечниковая** adrenal hypertension
~, **неврогенная** neurogenic hypertension
~, **послеродовая** postpartum hypertension
~, **почечная** renal hypertension
~, **почечно-сосудистая** renovascular hypertension
~, **системная** systemic hypertension
~, **транзиторная** transient/transitory hypertension
~, **упорная** persistent/refractory hypertension
гипертрихоз *med.* hypertrichosis, hypertrichiasis, hirsutism
~, **общий** hypertrichosis universalis
~, **пушковый** hypertrichosis lanuginosa
гипертромбоцитоз *med.* hyperthrombocytosis
гипертрофированный *biol.* hypertrophied, hypertrophic
гипертрофический hypertrophic
гипертрофия *biol.* hypertrophia, hypertrophy
~, **викарная** vicarious hypertrophy
~, **гормональная** hormonal hypertrophy
~ **крайней плоти** redundant prepuce
~, **ложная** pseudohypertrophy, false hypertrophy
~ **молочной железы** macromastia, macromazia
~ **ногтей** onychophyma
~ **почки** hypernephrotrophy
~ **предстательной железы** prostatic hypertrophy
~, **простая** simple/numerical hypertrophy
~, **псевдомышечная** pseudomuscular hypertrophy
~, **рабочая** work hypertrophy
~ **сердца** hypertrophy of the heart, cardiac hypertrophy
гиперурез *med.* hyperuresis
гиперфалангия *med.* hyperphalangism
гиперферремия *med.* hyperferremia
гиперфолликулинемия *med.* hyperfolliculin(a)emia, hyperfolliculoidism
гиперфрагмент hyperfragment, hypernucleus
гиперфункция hyperfunction, hyperactivity
гиперхлоремия *med.* hyperchlor(a)emia
~, **идиопатическая** idiopathic hyperchlor(a)emia, renal tubular acidosis
гиперхлорурия *med.* hyperchloruria
гиперхолестеринемия *med.* hypercholesterolemia, hypercholesteremia, hypercholesterinemia
гиперхромазия *med.* hyperchromasia
гиперхромия *med.* hyperchromia, hyperchromatism
~, **макроцитарная** macrocytic hyperchromia
гипершар hypersphere
гиперэкстензия hyperextension, superextension
гиперядро hypernucleus, hyperfragment
гипестезия *med.* hypesthesia
гипидиоморфный *geol.* hipidiomorphic, subhedral
гипновый *bot.* hypnaceous
гипноз hypnosis
~ **световым раздражением** photohypnosis
гипнонаркоз *med.* hypnonarcosis
гипнопедия hypnop(a)edia, sleep-learning
гипноспора *bot.* hypnospore, statospore, resting spore
гипнотерапия *med.* hypnotherapy
гипнотизация hypnotization
гипнотизер hypnotist, mesmerist
гипнотизировать hypnotize; mesmerize
гипнотизм hypnotism; mesmerism
гипнотик hypnotic, hypnotized person
гипнотический hypnotic, hypnagogue, hypnagogic; mesmeric; somnifacient
гипноцист *zool.* hypnocyst; *bot.* hypnospore
гипнум *bot.* Hypnum
гипобазальный *bot.* hypobasal
гипобаризм *med.* hypobarism
гипобатолитовый *geol.* hypobatholitic
гипобиоз hypobiosis
гипобласт *bot., zool.* hypoblast
гиповальва *bot.* hypovalve
гиповария *med.* hypovaria, hypovarianism
гиповентиляция hypoventilation
гиповитаминоз *med.* hypovitaminosis, vitamin deficiency
гиповолемия *med.* hypovol(a)emia, oligemia
гиполактия *med.* hypolactia
гипогаммаглобулинемия *med.* hypogammaglobulin(a)emia
гипогаплоиды *pl.* hypohaploids
гипогастральный *anat.* hypogastric
гипогеальный, гипогейный hypogeal; hypogean
гипогевзия *med.* hypogeusia
гипогенез(ия) hypogenesis
гипогенитализм *med.* hypogenitalism
~, **женский/мужской** female/male

hypogenitalism
гипогенный hypogenous; hypogenic, hypogenetic
гипогиалиновый *geol.* hypohyaline
гипогиний *bot.* hypogynium
гипогидроз *med.* hypohydrosis
гипогликемический hypoglyc(a)emic
гипогликемия *med.* hypoglyc(a)emia
гипогонадизм *med.* hypogonadism
~, тестикулярный/яичниковый testicular/ovarian hypogonadism
гиподактилия *med.* hypodactylia, hypodactyly, hypodactylism
гиподерма *bot.* hypoderma; *zool.* hypodermis
гиподермальный hypodermal; hypodermic
гиподермит hypodermitis
гиподинамия hypodynamia
гиподиэция *bot.* hypodioecism
гипоизостенурия *med.* hypoisosthenuria
гипокалиемия *med.* hypokal(a)emia, hypopotassemia
гипокальцинемия *med.* hypocalc(a)emia
гипокальциноз *med.* hypocalcinosis
гипокарпий hypocarp(ium)
гипокинезия *med.* hypokinesia, hypokinesis
гипокон *zool.* hypocone
гипокортицизм *med.* hypocorticoidism
гипокотиль *bot.* hypocotyl, stemlet
гипокотильный hypocotylous
гипокристаллический *geol.* hypocrystalline, merocrystalline, semicrystalline, hemicrystalline
гипоксантин *bioch.* hypoxanthine
гипоксемия *pathol.* hypoxemia
~, артериально-легочная arteriopulmonary hypoxemia
~, гемическая h(a)emic hypoxemia
гипоксилия *bot.* hypoxilia
гипоксия *med.* hypoxia, anoxia
~, внезапная sudden hypoxia
~, высотная high altitude hypoxia
~, гемическая h(a)emic hypoxia
~, застойная stagnant hypoxia
~, мнимая imaginary hypoxia
~, молниеносная fulminant hypoxia
~ новорожденных hypoxia neonatorum
~, повторная repeated hypoxia
~, постепенно развиваЮщаяся gradual hypoxia
~, тканевая tissue/histotoxic hypoxia
~, циркуляторная circulatory hypoxia
гиполейкоцитоз *med.* hypoleukocytosis
гипоменорея *med.* hypomenorrhea
гипометаморфизм hypometamorphism
гипомнезия *med.* hypomnesia
гипоморфный hypomorphic
гипонастический *bot.* hyponastic
гипонатринемия *med.* hyponatr(a)emia
гипонихий *anat.* hyponichium, nail bed
гипопаратиреоз *med.* hypoparathyrosis, hypoparathyroidism
гипопаратиреоидизм *med.* hypoparathyroidism
гипопатия *med.* hypopathy
гипопион *med.* hypopyon, hypopyum
гипоплазия *med.* hypoplasia, hypoplastic development
~ костного мозга bone marrow hypoplasia
~ молочных желез hypomastia
~ почки hypoplastic kidney
гипоплоиды *pl. biol.* hypoploids
гипоплоидия *biol.* hypoploidy
гипопротеинемия *med.* hypoprotein(a)emia
гипопротромбинемия *med.* hypoprothrombin(a)emia
гипоптиализм *med.* hypoptyalism
гипореактивность hyporeactivity
гипосаливация *med.* hypoptyalism
гипосиндез hyposyndesis
гипосинхронный hyposynchronous
гипоспадия *med.* hypospadias, hypospadia
~, мошоночная scrotal hypospadias
~, промежностная hypospadia perinealis
~, члено-мошоночная penoscrotal hypospadias
гипоспермия *med.* hypospermia
гипоспорангий *bot.* hyposporangium
гипостаз(а), гипостазис hypostasis
~, прижизненный vital hypostasis
~, трупный cadaveric hypostasis
гипостазирование hypostatization, hypostatizing
гипостасис *philos.* hypostasis
гипостатический, гипостатичный hypostatic(al)
гипостенический hyposthenic, weak
гипостенурия *med.* hyposthenuria
~, канальцевая tubular hyposthenuria
гипоталамический *anat.* hypothalamic
гипоталамус *anat.* hypothalamus
гипоталлий, гипоталлом, гипоталлус *bot.* hypothallus
гипотеза hypothesis; theory
~, астенолитная asthenolith hypothesis
~ волочения земной коры crust-dragging hypothesis
~ газового плавления gas-fluxing hypothesis
~ горения при поглощении engulfment-combustion hypothesis
~, допустимая admissible hypothesis
~ дробления fracture hypothesis
~ заполненных газом пустот gas-pocket hypothesis
~ зарядовой симметрии charge-symmetry hypothesis
~, изостатическая isostatic theory
~, кренетическая crenetic hypothesis
~ мишени target/hit theroy
~ мобилизма theory of drift
~, небулярная nebular theory/hypothesis
~, ненулевая/неосновная nonnull hypothesis
~ нулевая/основная null/main hypothesis
~ о пуассоновском распределении Poisson hypothesis
~, обобщаЮщая unifying hypothesis

~, осмотическая osmotic hypothesis
~, осцилляционная oscillation theory
~ отражения rebou und hypothesis
~ перемещения материков hypothesis of continental drift
~, планетезимальная planetesimal hypothesis
~, пластовая source-bed concept
~, приливная tidal hypothesis
~, синтетическая ликвационная synthethic-liquation theory
~ скольжения земной коры crust-sliding hypothesis
~, сложная composite hypothesis
~ случайности hypothesis of randomness
~ смещения полюсов hypothesis of polar shifting
~, ундационная undation hypothesis
~, целевая goal hypothesis
~, эргодическая ergodic hypothesis
~у, выдвигать make a hypothesis
~у, строить frame a hypothesis, hypothesize
гипотека I *bot.* hypotheca
гипотека II *econ.* mortgage
гипотензивный hypotensive
гипотензия *med.* hypotonia, hypotension, hypotony, hypotonicity
гипотенуза *math.* hypotenuse
гипотермальный, гипотермический hypothermal
гипотермия *med.* hypothermia
~, длительная prolonged/continuous hypothermia
~, контролируемая controlled hypothermia
~, лекарственная drug-induced hypothermia
~, хирургическая surgical/artificial hypothermia
гипотетический, гипотетичный hypothetic(al), postulatory, suppositional
гипотециальный *bot.* hypothecial
гипотеций *bot.* hypothecium
гипотиазид *pharm.* hydrochlorothiazide
гипотимия *med.* hypothymia, depression, blues
гипотиреоз *med.* hypothyrosis
гипотиреоидизм *med.* hypothyroidism, hypothyrea, hypothyroidea
гипотония *med.* hypotension, hypotonia, hypotonicity, hypotony, hypotonus
~, внутричерепная intracranial hypotension
~, ликворная hypotension of liquor
~, мозжечковая cerebellar hypotension
~, мышечная muscular hypotension
~, острая acute hypotension
~, управляемая controlled hypotension
гипотрихоз hypotrichosis
гипотрохоида *math.* epitrochoid, hypotrochoid
гипотрофия *biol., bot.* hypotrophy
~, врожденная congenital hypotrophy
~ новорожденных wasting disease of newborn
гипотрохоида *geom.* hypotrochoid
гипофалангия hypophalangism
гипофарингоскопия hypopharingoscopy
гипофиз *anat.* pituitary body
гипофиза *bot.* hypophyse, hypophysis
гипофизарный hypophysial, hypophyseal
гипофизопривный hypophysioprivic
гипофизэктомия *surg.* hypophysectomy
гипофилл *bot.* hypophyll(ium)
гипофория *med.* hypophoria
гипофсфатурия *med.* hypophosphaturia, oligophosphaturia
гипофосфит *chem.* hypophosphite
~ натрия sodium hypophosphite
гипофункция hypoactivity, diminished function
гипохерис *bot.* cat's-ear *(Hypochaeris)*
гипохилий *bot.* hypochil(ium)
гипохлоремия *med.* hypochloremia
гипохолестеринемия *med.* hypocholesteremia, hypocholesterolemia, hypochlolesterinemia
гипохондрический hypochondriac(al), atrabiliary
гипохондрия *med.* hypochondriasis, hypochondria
гипохромазия *med.* hypochromatism, hypochromasia
гипохроматичность hypochromacity
гипохромемия *med.* hypochromemia
гипохромия *med.* hypochromia
гипохромный hypochromatic, hypochromic
гипоцентр hypocentre, hypocenter, centre/center of origin
гипоцеркальный *geol.* hypocercal, reversed heterocercal
гипоэргия hypoergy, hypoergia
гиппарион *geol.* hipparion
гиппеаструм *bot.* knight's-star *(Hippeastrum)*
гиппогриф *myth.* hippogriff, hippogryph
гиппокамп *myth.* hippocampus; *anat.* hippocamp(us), Ammon's corn
гиппокампальный *anat.* hippocampal
гиппокентавр *myth.* hippocentaur
Гиппократ Hippocrates
Гиппокрин *myth.* Hippocrene
гиппология hippology
гиппопотам *zool.* hippopotamus *(Hippopotamus amphibius)*
~, карликовый pygmy hippopotamus *(Choeropsis liberiensis)*
гиппурит *geol.* hippurite
гипс 1. *miner.* gypsum; plaster (of Paris); 2. plaster (cast)
~, безводный anhydrous gypsum
~, волокнистый fibrous gypsum
~, вонючий fetid gypsum
~, глинистый clayey gypsum
~, жженый dried gypsum, plaster of Paris
~, землистый earthy gypsum
~, кристаллический crystallized gypsum, selenite
~, обжигать calcine/burn gypsum
~, плотный massive gypsum
~, пресноводный limnetic gypsum, fresh-water gypsum
~, строительный gypsum/cement plaster, hardwall plaster; stucco

~, **сырой** crude gypsum
~, **тонковолокнистый** satin spar
~, **тонкозернистый** alabaster
гипсаритмия *med.* hypsarrhythmia
гипсистафилия *med.* hypsistaphylia
гипсицефалический hypsicephalic
гипсобетон gypsum concrete
гипсование plastering; gypsum treatment; *chem.* gypsuming
гипсовать plaster; (treat with) gypsum
гипсовый plaster; gyps(e)ous, gypsiferous
~ **слепок** plaster cast
~**вая повязка** plaster bandage
гипсограф hypsograph
гипсография hypsography
гипсодонт *zool.* hypsodont
гипсолит gypsum
гипсолюбивый gypsophilous
гипсолюбка *bot.* gypsophila *(Gypsophila)*
гипсометр hypsometer
гипсометрия hypsometry
~, **цветовая** colo(u)r layering
гипсоносный gypsiferous, gypsum-bearing
гипсорасширитель plaster spreading forceps
гипсотермометр hypso(thermo)meter
гипсофил I *bot.* hypsophyll
гипсофил II gypsophilous plant
гипсофила *bot.* gypsophila *(Gypsophila)*
гипсофилл *bot.* hypsophyll
гипсофит *bot.* gypsophyte
гиптокинез *med.* hyptokinesis
гипуралия *zool.* hypural (bone)
гипургия hypurgia
гипюр guipure
гипфельфлюр *geol.* peak plain, gipfelflur
гиракодонт *paleont.* hyracodont *(Hyracodontida)*
гиракотерий *paleont.* hyracothere, hyracotherium
гиракс *zool.* hyrax *(Procavia)*
гиратор *tech.* gyrator
~, **емкостно-трансформаторный** capacitor-transformer gyrator
~ **четырехполюсный** two-port gyrator
гирационный gyratory
гиревик *sport* weight lifter
гирелла *zool.* girella *(Girella)*
~, **калифорнийская** greenfish *(Girella nigricans)*
гирка *bot.* awnless wheat *(Triticum vulgare muticum)*
гирло narrow strait; estuary, mouth of river; delta arm
гирлянда chain, garland, wreath, festoon
~ **изоляторов** insulator string
~ **изоляторов, натяжная** strain/tension insulator string
гироагрегат gyro(scopic) unit
гироазимут directional gyro(scope)
гироазимут-горизонт directional gyro and artificial horizon set
гироакселерометр gyro(scopic) accelerometer
гироблок gyro(scopic) unit
гировертикаль vertical gyro(scope), gyro(scopic) vertical
~, **арретируемая** cageable vertical gyro(scope)
~, **сдвоенная** twin/duplicated vertical gyro(scope)
~ **с маятниковой коррекции** pendulum-positioned vertical gyro(scope)
~, **центральная** master vertical gyro(scope)
гировоз inertia-type locomotive
гирогоризонт gyro/artificial horizon, azimuth gyro
гиродатчик gyro(scopic) sensor/unit/pickup
~ **вертикали** vertical signalling gyroscopic unit
~ **крена** roll(ing) gyro(scope)
~ **курса** heading gyro(transmitter)
~ **рыскания** yaw(ing) gyro(scope)
~ **системы стабилизации/индикации** reference gyro(scope)
~ **скорости крена** roll-rate gyro
~ **скорости разворота** turn-rate gyro
~ **скорости рыскания** yaw-rate gyro
~ **тангажа** pitch(ing) gyro(scope)
~ **угла** position gyro(scope)
~ **угловой скорости** rate-gyro sensor
~, **центральный** master (reference) gyro(scope)
гиродинамика gyrodynamics
гиродолгот longitude gyro
гироид gyroid
гироида rotoinversion axis
гироиндикатор gyro(scopic) indicator
гироинтегратор integrating gyro(scope), gyro(scopic) integrator
гирокомпас gyrocompass, gyroscopic compass
~, **маркшейдерский** survey gyrocompass
~ **с коррекцией по магнитному компасу** magnetically slaved gyrocompass
~ **с тремя гироскопами** triple gyrocompass
гирома *bot.* gyroma
гиромагнетический, гиромагнитный gyromagnetic
гиромаятник gyroscopic pendulum, gyropendulum
гирометр gyrometer
гиромотор gyromotor
гироориентатор gyroscopic altitude control unit/device
гироплатформа gyroplatform
гирополукомпас direction(al) gyro(scope), gyro direction indicator
гиропост gyro room/station
гироприбор gyroinstrument
гирорезонанс gyroresonance
гирорулевой *n.* gyropilot, autohelmsman
гиросекстан gyrosextant
гиросервомотор gyro-servo motor
гироскоп gyro(scope)
~ **автомата тангажа** auto-pitch gyroscope
~,**арретировать** cage a gyro
~, **астатический** balanced gyro
~ **без коррекции в азимуте** unslaved gyroscope

~, бескарданный gimballess gyroscope
~ в герметичном корпусе hermetically-sealed gyroscope
~, вибрационный vibratory/vibrating gyroscope
~, двухрамочный two-gimbal gyroscope
~, двухстепенной one-axis gyro(scope),rate gyro, single-degree-of-freedom gyro
~, демпфирующий rate/damping gyro(scope)
~ для инерциальных систем отчета inertial-reference gyroscope
~ для ракеты missile gyroscope
~, заарретированный caged gyroscope
~ изменений по крену roll-rate gyro(scope)
~, интегрирующий gyro(scopic) integrator,integrating gyro(scope)
~ крена roll gyro(scope)
~, криогенный cryogenic gyro
~, курсовой azimuth/heading/directional gyroscope
~, лазерный laser gyroscope
~, магнитогидродинамический magnetohydrodynamic/ MHD gyroscope
~, маятниковый pendulum gyroscope
~ на воздушных/газовых подшипниках air-/gas-bearing gyroscope
~, невыбивающийся non-toppling/tumbling gyroscope
~, некорректируемый/неуравновешенный unbalanced gyroscope
~, несвободный powered gyroscope
~, нулевой null gyroscope
~, одноосевой single-axis gyroscope
~, оптический/световой light/optical gyroscope
~, поплавковый floated/flotation(-type) gyroscope
~, прецессионный rate gyro(scope)
~, работающий на выбеге run-down gyroscope
~, разарретированный uncaged gyroscope
~ рассчитанный на работу при высоких перегрузках high-g-performance gyroscope
~ рыскания yaw gyro(scope)
~ с (вращающимся) ротором rotary/spinning gyroscope
~ с газовым прибором/приводом gas-driven gyroscope
~ с гидростатической подвеской ротора hydrostatic (all-suspended) gyroscope
~ с гравитационной коррекцией earth gyroscope, gravity-controlled/-erecting gyroscope
~ с двумя измерительными осями dual-/two-axis gyroscope
~ с двумя степенями свободы restrained gyroscope
~ с коррекцией в азимуте slaved gyroscope
~ с коррекцией вертикали vertically monitored gyroscope
~ с коррекцией от магнитного компаса compass- controlled/-slaved/-positioned gyro
~ с магнитной коррекцией magnetic slave gyroscope
~ с малым уходом low-wander gyroscope
~ с маятниковой коррекцией pendulum/pendulous gyroscope
~ с незначительным уходом drift-free gyroscope
~ с одной измерительной осью single-/one-axis gyroscope
~ с однорамочным карданным подвесом one-gimbal gyroscope
~ с пиротехническим приводом ротора cordite-driven gyroscope
~ с пневмоприводом air-driven gyroscope
~ с подогревом heated gyroscope
~ с ртутной ванной, поплавковый mercury-supported gyroscope
~ с системой коррекции tied gyroscope
~ с тремя степенями свободы three-frame gyroscope
~, свободный free/master gyro, two-degree-of-freedom gyroscope
~, силовой powered gyroscope
~, скоростной rate gyro(scope)
~ стабилизации антенны sightline gyro(scope)
~ тангажа pitch gyroscope
~, тяжелый heavy gyro
~ упреждения lead computing gyro
~, уравновешенный balanced gyro(scope)
~, ядерный nuclear gyro(scope)
гироскопический gyroscopic
гироскопия gyroscopy
гиростабилизатор gyrostabilizer, gyro(scopic) stabilizer
гиростабилизация gyroscopic stabilization, gyrostabilization
гиростабилизированный, гиростабилизируемый, гиростабилизованный gyro(scope)-stabilized, gyrostabilized
гиростат gyrostat
гиростатика gyrostatics
гиростатический gyrostatic
гиротахометр rate gyro
гиротеодолит gyrotheodolite
гиротрон vibrating gyro
гироузел rotor assembly, gyro unit
~ авиагоризонта horizon gyro unit
гироуказатель крена и поворота bank-and-turn gyroscope
гироуспокоитель gyro(scopic) stabilizer
гирофора *bot.* rock tripe *(Gyrophora)*
гироцентраль master/reference gyro
гирочастота gyrofrequency
гироширот latitude gyro
гироэдр pentagonal icositetrahedron
гирсутизм *med.* hirsutism, hirsuties
~ гипофизарный hypophysial hirsutism
гирудин *pharm.* hyrudin
гирча *bot.* selinum *(Selinum)*

гирчевник, гирчовник *bot.* hemlock parsley *(Conioselinum)*
гиря balance (scale) weight, dumbbell
~, заводная driving weight
~, передвижная (scale-beam) rider, sliding weight, poise
гиря-противовес tare
гисекит *geol.* gieseckite, dysyntribite
гистамин *bioch.* histamine
гистаминаза *bioch.* histaminase, diamine oxidase
гистаминемия *med.* histamin(a)emia
гистаминовый histaminic
гистаминоподобный histamine-like
гистаминоустойчивый, гистаминорефрактерный histamine-resistant
гистаминурия *med.* histaminuria
гистерезиметр hysteresis/histeretic meter
гистерезис *phys.* hysteresis
~ вращения rotating hysteresis
~ высотомера altimeter hysteresis
~ вязкий/ползучий viscous hysteresis
~ клапана valve lag
~, магнитный magnetic hysteresis/retardation
~ магнитострикции magnetostriction hysteresis
~, сорбционный sorption hysteresis
~, упругий elastic hysteresis
гистерогенный hysterogenetic, hysterogenic
гистерография *med.* hysterography, uterography
гистероманометр hysteromanometer
гистерометр hysterometer
гистеромиомэктомия *surg.* hysteromyomectomia
гистерон-протерон hysteron-proteron
гистеропатия *med.* hysteropathy
гистеропексия *surg.* hysteropexy
гистеропротерон hysteron proteron
гистероптоз *med.* hysteroptosis
гистеросальпингэктомия *surg.* hysterosalpingectomy
гистероскоп hysteroscope
гистероскопия *med.* hysteroscopy
гистеротом *surg.* hysterotome
гистеротомия *surg.* hysterotomy
гистерофор *med.* hysterophore
гистерэктомия *surg.* hysterectomy
~, брюшно-влагалищная aldominovaginal hysterectomy
~, брюшно-стеночная abdominal hysterectomy
~, влагалищная vaginal hysterectomy
~, паравагинальная paravaginal hysterectomy, Schuchardt's operation
гистидаза *med.* histidase
гистидин *bioch.* histidin(e)
гистидиндезаминаза histidine deaminase, histidase
гистиогенез *zool.* histogenesis, histogeny
гистиоцит *biol.* hist(i)ocyte
~, мозговой cerebral histiocyte
гистиоцитарный histiocytic
гистиоцитоз *med.* hist(i)ocytosis
~, липидный lipid hystiocytosis, Niemann-Pick disease
~, нелипидный nonlipid hystiocytosis, Letterer-Siwe disease
гистиоцитарный histiocytic
гистиоцитома *med.* histiocytoma
гистоаутoрадиография historadiography, histoautoradiography
гистобласт histoblast
гистогематин histohematin
гистоген *bot.* histogen
гистогенез *biol.* histogeny, histogenesis
~, постфетальный postf(o)etal histogenesis
гистогенный histogenetic, histogenic
гистограмма 1. bar/column chart, bar graph, column diagram, histogram; 2. *geol.* differential granulometric composition diagram
~ для нескольких признаков multiple bar chart
гистозим *bioch.* histozyme, hippuricase
гистолиз *bioch.* histolysis
~, ферментативный enzymic histolysis
гистолизат *bioch.* histolysate
гистолог histologist
гистологический histologic(al)
гистология histology
~, описательная descriptive histology
гистомониоз histomoniasis
гистоморфология histomorphology
гистон *bioch.* histon(e)
гистопатология *med.* histopathology
гистоплазмоз *med.* histoplasmosis
гисторадиография *med.* historadiography
гистотомия histotomy
гистофизиологический histophysiologic
гистофизиология histophysiology
гистохимический histochemical
гистохимия histochemistry
гит *sport* heat
гитаген githagen
гитален *chem.* gitalin
~, аморфный amorphous gitalin, gitalgin
гитара 1. *mus.* guitar; 2. *tech.* swinging arm, swing frame, bracket; quadrant; gear reduction/change unit
~ дифференциальных колес differential change gear train
~ подачи feed change gear train
~ станка bracket, quadrant
гитаровидный, гитарообразный fiddle-shaped, pandurate, panduriform
гитаролистный *bot.* pandurate-/fiddle-leaved
гитерс *naut.* skeet
гитлеризм Hitlerism; Naz(i)ism
гитлеровец Hitlerite, Nazi
гитлеровский Hitlerite, Nazi
гитов *naut.* clew line/rope; garnet; brail
~, коренной throat brail
~ нижнего паруса clew garnet
~ы, брать на brail
гитов-блок *naut.* strap-bored block, brail block

гитов-тали garnet purchase
гиттия *geol.* gyttja, black muck
гиф hypha (*pl.* - as)
гифа *bot.* hypha, (mycelial) filament
  ~, внедряющаяся penetrating hypha
  ~, воспринимающая receptive hypha
гифальный *bot.* hyphal
гифема *med.* hyph(a)ema, hyph(o)emia
  ~, тропическая (inter)tropical hyphemia
гифен hyphen
гиферезис *physiol.* hyphaeresis
гифиллин hyphylline
гифолит *geol.* hypholite
гифомикоз hyphomycosis
гифомицет *bot.* hyphomycete
гифообразный hypha-shaped, hyphal, hyphoid
гифоподий *bot.* hyphopodium
гичка *naut.* gig; outrigger
гишпанский *hist.* Hispanic
глабеллярный glabellar
глабель glabella
глава I 1. head, chief; 2. cupola *(of a church)*
глава II chapter
главбух *abbr.* **главный бухгалтер** chief accountant
главврач *abbr.* **главный врач** head physician
главенство supremacy, domination
главк *abbr.* **главный комитет** central board/administration/commitee, head office
главковерх *abbr.* **верховный главнокомандующий** Supreme Commander-in-Chief
главком *abbr.* **главнокомандующий** commander-in-chief
главнокомандование (general) headquarters
  ~, Верховное General Headquarters, High Command
главнокомандующий *n.* commander-in-chief
  ~ ВВС Commander-in-Chief of the Air Force
  ~, верховный Supreme Commander-in-Chief
главнополитпросвет *abbr. obs.* **Главный политико-просветительный комитет** Main Political Education Commitee
главный main, principal, chief, cardinal, major; primary; master; leading, predominant; head; senior
  ~ бухгалтер chief accountant
  ~ врач head physician
  ~ корень main root
  ~ штаб general staff
  ~ная квартира *mil.* headquarters
  ~ная книга ledger
  ~ное предложение *gram.* main clause
  ~ное управление central (administrative) board
глагерит *geol.* halloysite
глагол verb
  ~, моторный verb of motion
  ~, несобственно-переходный not proper transitive verb
  ~, основнопроизводящий stem verb
  ~, посредственно-переходный indirectly transitive verb
  ~, разноспрягаемый heteroclitic verb
  ~, хамзованный hamzaed verb
глаголать, глаголит *see* **говорить**
глаголица the Glagolitic alphabet
глаголический Glagolitic
глагол-связка catenative verb
глаголь 1. Church Slavonic name of the letter **г**; 2. L-shaped object; gib; gallows, crook; crane
глаголь-гак *naut.* pelican/mousing/slip hook
глагольно-видовой verbal-aspect
глагольно-именной (de)verbal-nominal
глагольный verbal
гладиатор *hist.* gladiator
гладилка *met.* sleeker, slicker, planishing hammer, steel burnisher polisher; *med.* smoother; *build.* flatter, smoothing trowel, float; *print.* planing block; planer
  ~, двусторонняя double-end smoother
  ~, механическая power trowel
  ~, торцевая cup-shaped sleeker
  ~, фасонная smoother
гладило burnisher, polisher
гладильный ironing; smoothing
  ~ная доска ironing board
гладиолус *bot.* sword lily, corn flag, gladiolus *(Gladiolus)*
гладить iron, press; smooth; plane
гладиус *geol.* gladius
гладкий 1. smooth; flat, even, level, plane; glabrous; sleek; slippery; straight; 2. *text.* plain, unfigured; *geol.* psilate; fluent
  ~ почти subpsilate
гладкоколосный *bot.* smooth-spiked
гладколистный *bot.* smooth-leaved, leiophyllous
гладконог *zool.* tailed frog *(Ascaphus truei)*
гладконосные *n. pl. zool.* verpertilionids *(Verpertilionidae)*
гладкопалубный *naut.* flush-deck(ed)
гладкопленчатый glabrous-chaffed
гладкоплодный *bot.* smooth-fruited, lissocarpous
гладкопокровный *geol.* psilotegillate
гладкопросветный *geol.* psiloluminate
гладкосемянный *bot.* smooth-seeded, lissospermous
гладкоспоровый *bot.* leiosporous
гладкостебельный *bot.* smooth-stalked/-stemmed
гладкостенный 1. *mil.* smooth-bore; 2. leiodermatous
гладкость 1. smoothness; evenness, flatness; sleekness; 2. fluency, volubility
  ~ бумаги, машинная machine finish of paper
  ~ кожи smoothness of skin
  ~, оптическая optical flatness
гладкоцветковый *bot.* smooth-flowered, leianthous
гладкошерст(н)ый sleek-/smooth-haired
глаженье ironing; smoothing; satining
глаз eye; eyesight

~, агрегатный aggregate eye
~ бури/циклона eye of storm/hurricane, wind's eye
~, бычий *med.* buphthalmos, buphthalmus, buphthalmia
~, ведущий sighting eye
~, вооруженный aided eye
~, вороний/обыкновенный *bot.* herb Paris *(Paris quadrifolia)*
~, дневной павлиний *ent.* peacock butterfly *(Vanessaio)*
~, дурной evil eye
~, жабий toad's eye tin
~, невооруженный unaided/naked eye
~, кошачий *miner.* cat's eye, cymophane
~, пиниальный pineal eye
~, простой simple eye, ocellus
~, сложный compound eye
~, соколиный hawk's eye
~, стебельчатый stalked eye
~, теменной/третий parietal/pineal eye
~ трилобита facet, ocellus
~, фасеточный faceted eye
~, четырехлистный *bot.* herb Paris *(Paris quadrifolia)*
~, шизохроический schizochroal eye
~, эмметропический emmetropic eye
глазет *text.* (silk) brocade
глазирование *cul.* frosting; glazing; glaze; varnish
глазировать glaze *(of paper)*; vitrefy *(of ceramics)*; *cul.* ice, glaze; candy; enrobe
глазировка glazing *(of paper)*; vitrefying *(of ceramics)*; *tech.* calendering; *cul.* icing; candying; glazing, enrobing
~ камнем flint glazing
глазировщик glazer; burnisher, polisher
глазница *anat.* orbital cavity, orbit, eye socket
глазнично-ушной *anat.* oculoauricular
глазничный *anat.* orbital
глазной ocular, ophthalmic; palpebral
~ врач oculist
~ нерв optic/ophthalmic nerve
~ное яблоко eyeball
~ные зубы canine teeth
глазовидный ocellate(d)
глазовращающий *anat.* oculogyric, ophthalmogyric
глазодвигательный *anat.* oculomotor
глазожелезистый *anat.* oculoglandular
глазок 1. eye; 2. peephole, inspection hole, eyesight; 3. *(pl.) bot.* bud; eye *(of potato)*; 4. *tech.* eye, eyelet; ocellus; ear, lug
~ волочильной доски drawing pass, die hole
~, латеральный stemma
~ нитеводителя/нитепроводника *text.* quiding eye, thread quide eye
~, простой stemma
~ ремизки mail
~, спящий *bot.* resting bud
~ фурмы tuyere eyesight
~, цокольный base eyelet
~ челнока shuttle eye
~ки, анютины *pl. bot.* wild pansy *(Viola tricolor)*
глазо-лицевой *anat.* oculofacial
глазомер eye/visual estimation, measuring by eye
~, плохой/хороший faulty/good eye
глазомерный estimated by eye/sight, approximate
глазо-носовой *anat.* oculonasal
глазообразный eye-shaped, oculiform
глазо-сердечный *anat.* oculocardiac
глазуревидный glaze-like
глазурование glazing
глазурованный glazed; vitrified; enameled
глазуровка glazing; engobe coating
глазурь 1. glaze; 2. *cul.* icing, frosting
~, глухая opaque glaze
~, землистая earthy glaze
~, легкоплавкая low-melting glaze
~, матовая opaque glaze
~,обжигать burn/fire the glaze
~, окрашенная pigmented glaze
~, поливная slip (glaze), slip glazing
~, прозрачная transparent/translucent glaze
~, свинцовая lead glaze
~, свинцово-оловянная leaden-tin glaze
~, соляная salt glaze
~, тугоплавкая hard glaze, high-melting glaze
~, фриттовая fritted glaze
глазчатый spotted; dotted, speckled; ocellate(d)
гланда *anat.* tonsil
глас 1. voice; 2. *eccl.* tune
~ вопиющего в пустыне the voice of one crying in the wilderness
глace *text.* glossy silk fabric
гласис *geol.* glacis
гласность openness, publicity; *polit.* glasnost
гласный I open, public; conspicuous, notorious
~ суд public trial
гласный II *attr.* vowel; vocalic
~, безразличный neutral vowel
~, опорный glide
гласный III *n. hist.* council(l)or
глауберит *geol.* glauberite
глаукодот *miner.* glaucodot
глауколит *miner.* glaucolite
глаукома *med.* glaucoma
~, врожденная congenital glaucoma, buphthalmos
~, вторичная secondary glaucoma
~, открытоугольная open angle glaucoma
~, первичная primary glaucoma
~, пигментная pigmentary glaucoma
~, простая атрофическая simple atrophic glaucoma, Donder's glaucoma
~, угрожающая glaucoma imminens
~, факолитическая phacolytic glaucoma
глауконит *miner.* glauconite
~, алюминиевый skolite

глаукопирит *miner.* glaucopyrite
глаукс *bot.* sea milkwort *(Glaux)*
глауциум *bot.* horn poppy *(Glaucium)*
глашатай *hist.* public/town crier
глеба *bot.* gleba
гледичия *bot.* honey locust *(Gleditschia caspia)*
глезер *met.* glazer; *print.* calender
глей *geol.* gley, gumbo (soil)
глейхения *bot.* *Gleichenia*
глеоцистида *bot.* gleocystidium
глет *geol.* litharge, massicot
~, свинцовый yellow lead
глетчер (mountain/Alpine) glacier
~, висячий cliff glacier
~ из каменной соли rocksalt glacier
~, континентальный glacier sheet, continental glacier, icekap
~ пиринейского типа hanging glacier
~, предгорный piedmont glacier
~, соляной salt glacier
~, фирновый névé glacier
глиадин *bioch.* gliadin
глиальный *anat.* glial
глигексамид *med.* glyhexamide
гликан *chem.* glycane
гликемия *med.* glyc(a)emia
гликогевзия *med.* glycogeusia
гликоген *bioch.* glycogen, animal starch
гликогеназа *bioch.* glycogenase
гликогенез, гликогения *bioch.* glycogenesis
гликогенный, гликогеновый *bioch.* glycogenetic, glycogenic
гликогеноз *pathol.* glycogenosis
гликозид *chem.* glycoside
~, генуинный genuine glycoside
~, сердечный cardiac glycoside
~, цианогенный cyanogenic glycoside
гликозил *chem.* glycosyl
гликозурия *med.* glycosuria
~, алиментарная alimentary/digestive glycosuria
~ беременных glycosuria of pregnancy
~, доброкачественная benign glycosuria
~, почечная renal glycosuria
гликокол *chem.* glycocoll, glycine
гликокортикоид *chem.* glycocorticoid
гликолевый *chem.* glycol(l)ic
гликолиз *bioch.* glycolysis
~, анаэробный/аэробный anaerobic/aerobic glycolysis
гликолитический *bioch.* glycolitic
гликоль *chem.* (ethylene) glycol
гликопротеид, гликопротеин *bioch.* glycoproteid, glycoprotein
гликофит *bot.* glycophyte
гликохолевый glycocholic
глина clay
~, абиссальная abyssal clay
~, автохтонная autochthonous clay
~, адсорбционная bentonite, Denver mud
~, активированная activated clay
~, анциловая Ancylus clay
~, белая (фарфоровая) white/potter's/china clay, kaolin, argil, bolus alba
~, бентонитовая bentonite clay
~, бокситовая bauxitic clay
~, болотная fen clay
~, бумажная paper clay
~, валунная boulder/drift clay
~, валунная морская (ice)berg till
~, ваяльная fuller's/modelling clay, cimolite
~ в кровле пласта muckle
~, вспучивающаяся effervescing/expansive clay
~, выгорелая сланцеватая burned oil shale
~, вязкая sticky clay
~, вязкая сланцевая sticky shale
~, гидрослюдисто-каолинитовая hydromica kaolin clay
~, гипсоносная gypsiferous clay
~, глубоководная красная red ooze
~, годовая ленточная annual varve
~, голубая blue clay/ground
~, гончарная argil, potter's clay
~, горшечная pottery clay
~, грубокерамическая brick clay
~ для удобрения fertilizer clay
~, донная bottomset clay
~, железистая ferruginous clay
~, железистоизвестковистая bole
~, железистосланцевая paint rock
~, желтая yellow earth/clay
~, жильная gouge
~, жирная fat/rich/soapy clay, loam
~ затвердевает/отвердевает the clay indurates/clunches
~, известковая (clay-)marl
~, известковая сланцеватая limestone shale
~, карманная pocket clay
~, квасцовая alum-clay
~, кимолийская cimolite
~, кирпичная brick clay/earth
~, кирпичная прессованная pressed brick earth
~, комовая ball clay
~, красная red clay
~, кремнистая siliceous clay
~, кристаллическая crystalline clay
~, купоросная alum earth
~, ледниковая glacial clay, till, alphitite
~, ленточная varved clay
~, листоватая book clay
~, мергелистая clay-marl
~, метаморфизованная converted clay
~, мономинеральная monomineral clay
~, монтмориллонитовая montmorillonitic clay
~, моренная morainic clay; till
~, морская marine clay
~, мягкая сланцеватая soft shale
~, мягкая черная углистая blackjack

~ набухает the clay swells
~, набухающая expansive clay
~, наносная drift clay
~, непроницаемая impervious clay
~, нижняя under clay
~, обезвоздушенная deaired clay
~, обожженная baked clay
~, огнеупорная refractory/fire clay
~, озерная lake/lacustrine clay
~, окремненная silicified shale
~, осадочная sedimentary clay
~, осажденная precipitated clay
~, остаточная primary/residual clay
~, отбеливающая bleaching clay
~, отвердевшая indurated clay
~, отложенная айсбергами (ice)berg till
~, отмученная elutriated clay
~, первичная primary clay
~, переотложенная redeposited/secondary clay
~, пестрая speckled clay
~, песчаная sandy clay
~, песчанистая сланцеватая sandy shale
~, пиритовая pyritic clay
~, пиритоносная pyritiferous clay
~, пластическая/пластичная plastic clay
~, плотная compact/heavy clay
~, плотная сланцеватая tough shale
~, поддувающаяся effervescing/expansive clay
~, подстилающая underclay
~, поименная жирная floodplain loam, floodloam
~, полосатая banded clay
~, почвенная soil clay
~, пылеватая silt clay
~, речная fluviatile clay
~, сапропелевая sapropel clay
~, сапропелитовая sapropelic clay
~, септариевая septarian clay
~, серая grey clay
~, скульптурная cimolite
~, сланцеватая schistous/slate clay; shale
~ слеживается the clay clumps
~ смыва colluvial clay
~, соляная saline clay
~, спекающаяся baking clay
~, стержневая core clay
~, сукновальная fuller's/Fuller's earth, fulling clay
~, сухарная kaolinic clay
~, сырцовая adobe clay
~, тектоническая attrition clay
~, тонкая fine clay
~, тонкодисперсная finely dispersed clay
~, тощая meagre/lean clay
~ трения gouge, leader stone
~, трубочная pipe-clay
~, тяжелая heavy/rich clay
~, углистая coal clay
~, уплотненная/утрамбованная consolidated clay
~, фарфоровая porcelain/China clay, kaolin
~, флоридиновая floridin(e)
~, цементирующая bond clay
~, цементная cement clay
~, черная углистая black chalk
~, чистая пластическая potter's clay
~, щебневатая stony clay
~, элювиальная eluvial clay
~, эоловая eolian clay
глина-наполнитель filler-clay
глинизация *build.* clay grouting; argillization; mudding off
глинизировать mud up/off
глиний *chem.* alumin(i)um
глинисто-железистый argillo-ferruginous
глинисто-известковый argillo-calcareous
глинисто-магнезиальный argillo-magnesian
глинисто-песчаный argillo-arenaceous
глинистость clayiness
глинистый 1. argillaceous, earthen, clayey, clayish; 2. clay colo(u)red
~ сланец shale
~тая почва loam, clay soil
глинит-цемент brick cement
глиница clay mine/pit
глинище clay mine/pit; clay soil
глинобетон sand-gravel-clay building mixture
глинобитный clay adobe *(attr.)*
глиновал clay kneader/treader
глинование addition of clay, claying
глинозем *chem.* alumina, aluminium oxide
~, (сырой) сернокислый alum cake, papermaker's alum
глиноземистый aluminiferous, aluminous
глиномешалка, глиномялка clay mixer/kneader, pug/clay mill, pugmill
глинообразование argillosynthesis, clay synthesis
глинт clint
глиняный 1. clay, earthenware; 2. argillaceous, clayey
~ная посуда pottery, earthenware
глиоархитектоника *med.* glioarchitectonics
глиобластома *med.* glioblastoma
глиоз *med.* gliosis
~, гипертрофический узловатый hypertophic nodular gliosis, epiloia
~, спинномозговой spinal gliosis, syringomyelia
~, туберозный tuberous gliosis
глиоза *med.* gliosa
глиоксилат *chem.* glyoxylate
глиоксилевый *chem.* glyoxylic
глиома *med.* glioma
~, веретенообразноклеточная spindle-cell glioma
~, гигантоклеточная giant cell glioma, astrocytoma
~ головного мозга cerebral glioma
~, диффузная diffuse glioma, oligodendroglioma
~ зрительного нерва glioma of the optic nerve

~ **носа** nasal glioma
~ **сетчатки** glioma of the retina
~ **сетчатки, внутренняя** glioma endophytum
~ **спинного мозга** glioma of the spinal cord
глиоматоз *med.* gliomatosis
глиоматозный gliomatous
глипталь glyptal resins
глиптика glyptic, engraving
глиптический glyptic(al)
глиптогенезис origin of the earth's surface/sculpture
глиптография glyptography, glyptic
глиптодонт *paleont.* glyptodont
глиптолит *geol.* glyptolith
глиптоморфоз crystal mould
глиссада *av.* glide path/slope; landing beam
~ **планирования** glide path/slope
глиссандо *mus.* glissando
глиссер *naut.* skimmer, planing/skimming boat, speedboat, hydroplaner, hydroglider
глиссирование (hydro)planing, gliding, aquaplaning
~ **колес/пневматика** tyre hydroplaning
глиссировать plane/skim on the water surface, glide, aquaplane, hydroplane
глист helminth, (intestinal) worm
~, **круглый** ascarid, roundworm
~, **ленточный** tapeworm
~, **подкожный** dracunculus
~, **пузырчатый** cysticercus
глистный helminthic, vermicular
глистовидный, глистообразный helminthoid, wormlike
глистогон *bot.* lavender cotton *(Santolina)*
глистогонный vermifugal, vermifuge, helminthagogic
~**ное (средство)** vermifuge, helminthagogue
глитч *comp.* glitch
глиф *arch.* glyph
глифталь glyphtal
глицерид *chem.* glyceride
глицерин *chem.* glycerin(e), glycerol, glycerite, trihydroxypropane
~, **щелочной** soap lye glycerin(e)
глицериновый *chem.* glyceric
глицерофосфат *chem.* glycerophosphate
~ **железа/кальция** ferrous/calcium glycerophosphate
глицерофосфатаза *bioch.* glycerophosphatase
глицерофосфорилхолин *chem.* glycerophosphorylcholine
глицилаланин *chem.* glycyl-alanin
глицилглицин *chem.* glycyl-glycine
глициллейцин *chem.* glycyl-leucine
глицин *chem.* glycin(e)
глициния *bot.* wistaria *(Wistaria)*
глицирризин *chem.* glycyrrhizin, glycyrrhizic acid
глия *med.* glia, neuroglia
~, **астроцитарная** macroglia
глобалистика global studies, study of global issues
глобигерина *zool.* globigerina *(Globigerina)*
глобин *biol.* globin
глобоидальный globoid(al)
глобосферит *geol.* globosphaerite
глобула *astr.* globule
глобулин globuline
~, **меченый** tagged globulin
~, **сывороточный** serum globulin
~, **человеческий** human globulin
глобулит *miner.* globulite
~, **нанизанный/цепочкообразный** margarite
глобуляриевые *n. pl. bot.* globe daisies *(Globulariaceae)*
глобулярия *bot.* globe daisy *(Globularia)*
глобус globe; sphere; *av.* automatic ground position indicator
~, **звездный [небесной сферы]** star/celestial globe
~ **земли [земного шара]** earth's globe
глог *bot.* red/bloody dogwood *(Cornus sanguinea)*
глокеншпил *mus.* glockenspiel
глокерит *geol.* glockerite
глоксиния *bot.* gloxinia *(Gloxinia, Sinningia)*
гломеропорфиритовый, гломеропорфировый *geol.* glomeroporphyritic, glomerophyric, cumulophyric
гломерулит *med.* glomerulitis
гломерулонефрит *med.* glomerulonephritis
~, **мембранный** membranous glomerulonephritis
~, **очаговый** focal glomerulonephritis
~, **смешанный** mixed glomerulonephritis
гломус *anat.* glome, glomus
глориоза *bot.* glory lily *(Gloriosa)*
глория 1. *opt.* glory, gloria, nimbus; 2. *text.* gloria (fabric)
глосса note of explanation, gloss
глоссарий *comp.* glossary
глоссит *med.* glossitis
~, **гуммозный** gummatous glossitis
~, **склерозный** sclerosed glossitis
~, **хронический поверхностный болезненный** Moeller's glossitis
глоссодиния *med.* glossodynia, burning tonque
глоссоптеридные *n. paleont.* glossopterids *(Glossopterides)*
глоссэктомия *med. pl.* glossectomy, glossosteresis
глотание swallowing, deglutition; ingestion, gulping
~ **воздуха** *med.* aerophagia, air swallowing
~ **слюны** sialophagia
глотательный swallowing, deglutitive, deglutitory
глотка *anat.* throat, gullet, pharynx
~ **одноклеточных** cytopharynx
глоточно-гортанный pharyngolaryngeal
глоточно-надгортанный pharyngoepiglottic
глоточно-небный pharyngopalatine
глоточно-носовой pharyngonasal
глоточно-пищеводный pharyngoesophageal
глоточно-ротовой pharyngooral

глоточно-челюстный pharyngomaxilary
глоточно-язычный pharyngoglossal
глоточный *anat.* pharyngeal
глохнуть *med.* become/grow/get deaf; *auto.* stall *(of engine)*
глубина depth
~, абиссальная abyssal depth
~ брода fording depth
~ вспашки *agr.* ploughing depth
~ высева *agr.* placement depth of drilling/ planting, depth of seed
~ горизонта dip of (the) horizon
~ детализации granularity
~ дыхания respiration depth, depth of breathing
~ заданная predetermined depth
~ заделки семян *agr.* depth of seeding
~ закаленного слоя depth of the hardening zone
~ залегания depth of occurrence, cover thickness
~ замирания *rad.* depth of fading
~ захода *mach.* working depth of tooth; depth of throat
~, измеренная sounding, sounded bottom
~, лимитирующая available depth
~, логическая *comp.* logical depth
~, малая *nav.* shallow/scant depth
~, наименьшая *nav.* least water
~, опорная reference depth
~ от уровня моря subsea depth
~ очага/фокуса землетрясения focal/focus depth, depth of seismic focus
~ памяти *comp.* memory depth
~ пахотного слоя ploughing horizon/depth, depth of plowing/top soil
~ погружения diving/submergence/submerged depth, depth of submergence/immersion, sinkage
~ погружения, малая shallow dive sensitivity
~ погружения, относительная submergence ratio
~ под килем under-keel clearance
~ по лоту sounding depth
~ последовательностная *comp.* sequentia depth
~ провара *weld.* depth of penetration
~ прокаливаемости depth of the hardening zone
~ промачивания wetted depth
~ промерзания freezing depth, frost penetration
~ проникновения альфа/бета-частиц alpha-/beta-penetration
~ проникновения поля *el.* depth of field penetration, skin depth
~ резкости *phot.* depth of focus
~ ремизки *text.* harness depth
~ тестирования *comp.* test throughness
~ трюма *naut.* depth of hold
глубинный abyssal, abysmal; deep; *geol.* bathygenic, plutonic, deep-seated, deep-laid, hypogeal; deep-water/-sea
~ная бомба depth charge
глубиномер depth gauge/indicator, depthometer; hydrobarometer
~, гидростатический pressure depth sensor
~, регистрирующий *agr.* depth recorder
глубокий 1. deep; 2. profound, penetrating, thorough(going), in-depth; 3. advaneed
~ вираж *av.* steep turn
~кая оборона *mil.* defence in depth
~кая печать intaglio printing
~кая старость extreme old age
~кое невежество profound/utter ignorance
глубокоблюдчатый *bot.* basin-/bowl-shaped, pelviform
глубоководный deep-water/-sea; *geol.* bathypelagic, abyssal; bathophilous, bathybic
~ное растение *bot.* benthophyte
глубокообрабатывающий deep-cultivating
глубокоразрезной *bot.* cleft; slit, jagged
глубокорыхлитель *agr.* chisel plow
глубокосидящий *naut.* deep-drawing
глубокоукореняющийся deep-rooting/-rooted
глубомер depthometer
глупыш *zool.* fulmar *(Fulmarus glacialis)*
глутамат *chem.* glutamate
глутаминовый *chem.* glutamic
глутатион *bioch.* glutathione
глухарь 1. *zool.* wood grouse, capercaillie *(Tetrao Urogallus)* 2. *zool.* yellow-billed serpent *(Ophisaurus apus, Pseudopus Palasii);* 3. *tech.* dump bolt, short-drive bolt; coach/lag/wood screw; 4. *nav.* deadlight
глухой deaf; *phon.* voiceless; *n. pl.* the deaf
~ сезон dead season
~ согласный voiceless consonant
~ тетерев wood grouse, capercaille *(Tetrao urogallus)*
глухонемой *n.* deaf-mute, deaf and dumb, surdomute
глухонемота deaf-mutism/-dumbness, surdomutism
глухота deafness, surdity
~, абсолютная/полная *med.* anacusis
~, врожденная congental deafness
~, лабиринтная labyrinthine deafness
~, неполная hypoacusis, disacusis
~, парадоксальная paradoxical deafness, Willi's paracusis
~, профессиональная occupational deafness
~, словесная verbal-auditory agnosia
~, старческая presbyacusis, presbyac(o)usia
глушение silencing, muffling, damping; *rad.* jamming
глушитель *tech.* silencer, antihum device
~ выхлопа (exhaust) silencer/muffler
~ радиолокатора radar jammer
~ шума noise suppressor, silencer, muffler
~ шума выхлопных газов exhaust silencer
~ шума реактивного двигателя jet (engine) silencer
глушить deafen, silence, muffle; stun, deaden; plug
~ боль dull pain

~ критику stifle criticism
~ мотор stop the engine
~ радиопередачи jam radiobroadcasts
глыба block, lump, clump, clod
~, висячая/качающаяся perched block, cornice
~, земляная clod
~, каменная boulder
~, клинообразная сбросовая fault wedge
~, окраинная marginal block
~, шлаковая lump/blob of slag
глыбовидный, глыбообразный clodlike, blocklike; cloddy
глыбодроб *agr.* clod breaker
глыбодробитель clod breaker/crusher
глюкан *chem.* glucan, glucosan(e)
глюкангидролаза glucanhydrolase
глюказа *bioch.* glucase, maltase
глюкоза *bioch.* dextrose, glucose, grape sugar
~, пищевая alimentary glucose
глюкозаминфосфат *chem.* glucosamine phosphate
глюкозид *chem.* glucosid(e)
глюкозил glucosyl
глюкозилсахароза glucosyl sucrose
глюкозофосфат glucose phosphate
глюкокинин glucokinin
глюкопираноза glucopyranose
глюкопротеин *bioch.* glucoprotein, glycoprotein
глюкуронил glucuronyl
глютамин *chem.* glutamin(e)
глютаминовый *chem.* glutamic
глютатион *bioch.* glutathione
~ восстановленный reduced glutathione
глютиновый gelatinous
гляделка peep-/inspection hole, eye-sight, peephole
глянец polish, gloss, lustre, glaze
~, придавать give polish
~, роговой clinkstone
глянцевание polishing; glossing, lustring, glazing
глянцеватель glazer
глянцевать polish; gloss, lustre; glaze; satine, shine
глянцевитый glossy, shiny, lustrous, polished; *bot.* vernicose
глянцевый glossy, brilliant, glazed, polished
глянцлак gloss(y)/brilliant/glazing varnish
глянцовка butting, polishing
гляриметр *print.* glossmeter
гляциальный glacial
гляцигенный glacigenous
гляциодислокация dislocation by glacier
гляциоизостазия glacial isostasy
гляциолог glaciologist
гляциология glaciology
гляциофлювиальный glacio-fluvial
гнатонемус *zool.* elephant fish *(Gnathonemus numenius)*
гнатопластика gnathoplasty
гнатостат gnathostat
гнатоцефал *med.* gnathocephalus
гнафалиум *bot.* cudweed *(Gnaphalium)*
гнедой *n.* bay horse
гнездный *bot.* locular
гнездо 1. *med., mach.* socket, seat(ing); jack, nest; receptacle, cavity; housing; *comp.* slot, snug; 2. *bot., geol.* nest, pocket, seat, recess; *med.* nidus, bunch, cluster; 3. nest, aerie, eyrie
~, антенное aerial jack
~ балки wall pocket
~, вить build a nest
~ в объединительной плате *comp.* bachplane slot
~ в панели *comp.* board slot
~, вызывное *tel.* calling jack
~ выреза *geol.* eye socket
~, газовое *min.* gas nest
~ гарнитуры телефонистки operator's telephone set jack
~ генов *med.* gene nest
~, двухразрывное *tel.* double-break jack
~ для БИС *comp.* LSI socket
~ для взрывателя fuze pot/well
~ для детонатора detonator housing
~ для внешних соединений *comp.* peripheral slot
~ для интегральной схемы IC socket
~ для кристаллодержателей chip-carrier socket
~ для монтажа накруткой *comp.* wire-wrap socket
~ для платы *comp.* card slot
~ для подслушивания listening jack
~ для расширительных модулей expansion slot
~ для термометра thermometer well/pocket
~, испытательное test jack
~, каблучное *leath.* heel seat
~ клапана/клапанное valve seat
~ коммутатора switchboard jack
~, контактное pin jack
~, контрольное test/monitoring jack
~, ласточкино *cul.* edible bird's nest
~ местного поля *tel.* answering/line jack
~ многократного поля *tel.* multiple jack
~ навоя *text.* warp beam support
~, ниппельное threaded socket
~, опросное *tel.* answering/line jack
~, параллельное *tel.* bridging/branch(ing) jack
~, подпружиненное *comp.* spring jack
~ подшипника bearing housing
~ початка *text.* cop bottom/bit
~ прослушивания monitoring jack
~ пружины spring seat
~, пулеметное machine-gun nest
~ пыльника anther lobe
~, разъединительное cut-off jack
~, расширительное *comp.* expansion slot
~ розетки socket
~, рудное *min.* nest/pocket of ore, ore-bunch/

pocket
~, свечное *ant.* spark-plug seat
~ сопротивления *mil.* pocket of resistance
~ стойки stake pocket/socket
~ стержня rod socket
~, телефонное jack
~ транзитного соединения through-line jack
~,формировать nest
~ цапфы trunnion seat/bed
~ челночной коробки *text.* shuttle compartment
~, штекерное pin jack, jack outlet
~ штепсельного разъема female contact/connector
гнездование nesting (habit), nidification
гнездовать *ling.* group words of the same root
гнездовидный nestlike; *bot.* loculiform
гнездовка *bot.* bird's nest orchid *(Neottia)*
гнездовковые *n. pl. bot.* bird's nest fungi *(Nidulariaceae)*
гнездовой nesting, nidulant; *bot.* locular
~ посев *agr.* cluster sowing
гнездовье nesting place
~ цапель heronry
гнездчатый *bot.* loculate, loculose, partitioned
гнезиогамия *biol.* gnesiogamy
гнейс *geol.* gneiss
~, астохитовый astochitic gneiss
~, жилковатый/жильный veined gneiss, adergneiss
~, кокардовый cocarde gneiss
~, ленточный banded gneiss
~, листоватый laminated gneiss
~, очковый augen-gneiss
~, первичный/первозданный fundamental/primary gneiss
~, плитняковый flaggy gneiss
~, послойно инъецированный lit-par-lit injected gneiss, leaf gneiss
~ с желваками knotted gneiss
~, сложный migmatite gneiss
~, слоистый layered/laminated gneiss
~, слюдяной mica/normal gneiss
~, средний amphoteric gneiss
~, стебельчатый columnar gneiss
~, флюидальный fluxion gneiss
~, шестоватый dendritic gneiss
~, шлифовый schlieren gneiss
гнейсит *geol.* gneissoid
гнейсовидный gneissic, gneissose, gneissy, gneissous, gneissoid
гнейсовый gneissic, gneissose, gneissy
гнетовые *n. pl. bot.* gnetaceous, gnetacean
гнет press
гнеток plunger
гнетум *bot.* joint fir *(Gnetum)*
гнида nit
гнидник *bot.* stavesacre larkspur *(Delphinium staphisagria)*
гниение putrefaction, rot(ting), decomposition, decay
~, вызывающий pultrefactive
гнилец *pathol.* bee pest
гнилогноекровие *med.* septicopyemia
гнилой decayed, putrid, putrefied, carious, rotten
~ зуб decayed tooth
гнилокровие *med.* septic(a)emia, sepsis
гнилостность putridity, rottenness, putrefactiveness, putrescence
гнилостный putrefactive, saprogenic, saprogenous; ichorous; putrid
гнилостойкий rotproof
гнилость putridity; rotenness
гниль 1. decay, rot, putridity, rotenness; rotten stuff, putrilage; 2. mould, mildew
~, вершинная top rot, blossom-end rot
~, грибная pythium rot
~, заболонная sap-wood rot
~, комлевая butt rot
~, склероциальная южная southern blight, red rot
~, стволовая stem/trunk rot
гнить decay, rot, decompose, putrefy
гниющий putrescent
гноевидный *med.* puriform, puruloid
гноекровие *med.* py(a)emia
гноение suppuration, purulence; rotting
~ глины aging of clay
гноеотделение *med.* discharge/flow of pus
гноеподобный *med.* pyoid
гноеродный *med.* pyogenic, pus-forming, suppurative
гноетечение *med.* pyorrhea, purulent discharge, suppuration
гноеточивый suppurating, suppurative
гноиться suppurate
гной pus, matter
~, злокачественный malignant pus
~, ихорозный ichorous pus
~, сливкообразный creamy pus
~, творожистый curdy/cheesy pus
гнойник abscess
гнойничковый *med.* pustular, impetiginous
гнойничок pustule
гнойно-воспалительный *med.* pyo-inflammatory
гнойный purulent, suppurative
~ная рана festering/suppurative wound
гном elf; gnome
гнома gnome, aphorism, maxim
гномический gnomic(al)
гномон *astr.* gnomon, sundial
гносеология *philos.* gnosiology, theory of knowledge
гностицизм gnosticism
гнотобиология, гнотобиотика gnotobiotics
гну *zool.* gnu *(Connochaetes)*
гнусавость *med.* rhinolalia, nasalized speech, nasal intonation; twang
~, закрытая/открытая rhinolalia clausa/aperta

гнусавый, гнусливый nasal, snuffling
гнуть flex, bend, curve, bow; deflect
гнутье bending, flexure
 ~ древесины wood bending
гнюс *zool.* electric ray *(Torpedo)*
гоацин *zool.* hoatzin *(Opisthocomus hoazin)*
гобелен gobelin, tapestry
гобель *print.* plough
гоби *geogr.* gobi
гобоист *mus.* oboe player, oboist
говельщик *eccl.* faster, person preparing for Communion
говения *bot.* raisin tree *(Hovenia)*
говенье *eccl.* fasting, preparation for Communion
говеть *eccl.* fast for confession and Communion
говорушка I *bot.* clitocybe *(Clitocybe)*
говорушка II *zool.* kittiwake *(Rissa tridactyla)*
гоголь *zool.* goldeneye *(Bucephala)*
гоголь-моголь *cul.* egg flip, beaten egg and sugar
год year
 ~, базисный basal/base year
 ~, бюджетный fiscal year
 ~ вступления в брак *law* year of marriage
 ~, високосный bissextile/leap/abundant/year
 ~, гражданский/календарный civil/calendar/legal year
 ~, грегорианский Gregorian year
 ~, драконический eclipse/draconitic year
 ~ затмения eclipse year
 ~, звездный/сидерический sidereal year
 ~, календарный calendar/civil/legal year
 ~, летный *agr.* swarm year
 ~, лунный lunar year
 ~, неурожайный *agr.* lean year
 ~, операционный *econ.* trading year
 ~, отчетный year under review; fiscal year
 ~ представления отчета *fin.* report year
 ~, полярный polar year
 ~, производственный working year
 ~, равноденственный equinoctial year
 ~ рождения year of birth, YOB
 ~, световой light year
 ~, сельскохозяйственный crop year
 ~, семенной *for.* mast year
 ~, сидерический sidereal year
 ~, синодический synodic year
 ~, смерти year of death, YOD
 ~, тропический equinoctial/tropical year
 ~, урожайный bumper-crop year
 ~, финансовый financial/fiscal year
 ~, хозяйственный business/production/working year
 ~, Юлианский Julian year
 ~ы между переписями intercensal years
година *rhetoric* time, period
 ~ бедствий calamitous time
 ~, тяжелая hard time
годичный annual, yearly; etesian; lasting a year
 ~ная смена annual change/fluctuation
 ~ное кольцо, - слой *bot.* growth/annual ring
годность fitness, workability, suitability; validity, serviceability
 ~ к плаванию seaworthiness
 ~ к полету airworthiness, flyability; *(of man)* flight fitness
 ~, летная *tech.* airworthiness, flyability
 ~, медицинская medical fitness
годный fit, suitable; serviceable; useful, valid
 ~ к плаванию seaworthy
 ~ к полетам, ~ к эксплуатации в воздухе airworthy
годовой annual, yearly; annotinous
 ~ доход annual income/revenue
годовщина anniversary
годограф hodograph, locus, time curve
 ~, амплитудный amplitude locus
 ~, вертикальный time-depth curve
 ~, вспомогательный auxiliary travel-time curve
 ~ корневой root locus
 ~, нагоняющий overtaking travel-time curve
 ~ орбиты orbit hodograph
 ~ передаточной функции transfer (function) locus
 ~ постоянного времени constant-time locus
 ~, сводный resulting travel-time curve
 ~, частотный frequency locus
 ~, частотный двойной dual locus diagram
годоскоп (counter) hodoscope
гойя *bot.* wax plant *(Hoya)*
гокко *zool.* curassow *(Crax)*
гол I *sport* goal
 ~, забить score a goal
гол II golfhole, cavity
голавль *zool.* chub *(Lenciscus cephalus)*
голарктический *geol.* Holarctic
голгофа *rel.* Calvary, Golgotha
голенастый *zool.* 1. grallic, gralline; ciconiiform; 2. *n. pl.* waders *(Ciconiiformes)*
голендофит *bot.* holendophyte
голенище 1. top *(of a boot);* 2. *(met.)* single lap, receding metal
голено-пальцевой *anat.* digitocrural
голеностопный *anat.* talocrural
голень *anat.* crus (*pl.* crura); shank, shin, tarsus
 ~, саблевидная saber shin
голец I *zool.* 1. loach, beardie *(Nemachilus)*; 2. char, common brooktrout *(Salvelinus)*
голец II *geol.* bald mountain/peak
голиаф 1. *zool.* goliath beetle *(Goliathus);* 2. *bibl.* Goliath
голкипер *sport* goalkeeper
голландер, голлендер *tech.* hollander
голландец Dutchman
голландка Dutchwoman
голландский Dutch
 ~ сыр Dutch cheese
 ~кая печь tiled stove
 ~кая черепица pantile
голобентос holobenthos
голобрюхий naked-bellied; apodous

голова 1. head; cephalon; upper end; top; 2. head, chief
~ волнолома mole head
~ грозового шквала thunderhead
~ городской mayor
~ медузы *med.* Medusa head, caput medusae
~ мола/пирса pierhead, jetty head
~ пласта *geol.* upper edge, basset, face/head of bed
~ сваи pile crown
~, собачья *bot.* shrubby plantain *(Plantago cynops)*
~ стеклянная *geol.* globular/kidney ore
~ трилобита *paleont.* cephalon, head shield
~ форштевня *naut.* stem head
~ шлюза, верхняя/нижняя upstream/downstream end of a lock
~ шлюпбалки *naut.* davit head
~ шпиля drum head, drumhead
головастик tadpole
головач *zool.* 1. chub *(Lenciscus cephalus);* 2. umbrella bird *(Cephalopterus);* common scheltopusik *(Ophisaurus apus);* bighead goby *(Neogobius kessleri)*
~, круглый *bot.* engraved/mosaic puffball *(Calvatia caelata)*
головастик tadpole
головка 1. head; cap, knob; 2. *med.* capitulum; 3. *bot.* glome, glomerule, flower head; capitulum
~ автосцепки automatic coupler
~, агрегатная *mech.* unit-type head
~ астронаведения star(-tracking) head
~ блока цилиндров *auto.* head (of cylinder block)
~ болта screw/gib/bolt head
~, болторезная threading die
~ бура bolt head
~ ветродвигателя headpiece, turntable
~ винта screw head
~, винторезная screw-cutting head
~, водораспылительная waterfog head
~, волноводная waveguide mount
~, волочильная drawhead, drawing head
~, воспроизводящая/воспроизведения reproducing head; *comp.* playback head
~, вращающаяся rotating/rotary head
~, выдувная blow head
~, высокочастотная radio-frequency head, RF head
~, гипсометрическая hypsometer head
~, гравировальная *print.* engraving head
~, гранулирующая *chem.* pelletizing head
~ графопостроителя plotting head
~, делительная dividing/index(ing) head
~, детекторная detector head/mount
~, диодная diode mount
~ дистиллята top fraction
~ ЖРД liquid-rocket-engine injector
~, закалочная quench head
~ заклепки rivet head
~ заклепки, высадочная rivet point, snap/closing rivet head
~, записывающая/записи recording/writing/write head
~, заточная grinding head
~, захватывающая grip head
~, звуковая/звукоснимателя sound head, pick-up
~, звукозаписывающая recording head
~ звукоснимателя pick-up head
~ зуба tooth tip/point
~, зуборезная gear-shaping cutter head
~, зубострогальная резцовая face-mill type gear cutter
~ иглы *text.* needle hook
~, измерительная measuring head
~ из неодимового стекла neodymium glass head
~ изолятора insulator cap
~, калильная hot bulb
~ клеймителя marking head
~ ключа *tel.* key button
~, коллиматорная collimating head
~, коническая steeple/cone head
~ кривошипа crank end
~, лазерная сканирующая laser scanning head
~, листующая slabbing/sheeting head
~ литеры *print.* beard
~ лука onion bulb
~ льна *bot.* flax boll
~, литьевая injection head
~, магнитная magnetic head
~ манипулятора manipulator head
~ мартеновской печи port end
~ микроскопа nosepiece of a microscope
~, многодорожечная *comp.* multitrack head
~, многоручьевая/многосекционная mutiple head
~, многошпиндельная multispindle head
~ молоточка *anat.* head of the malleus
~, накладная *mfg.* attachment head
~, неподвижная fixed head
~ ножа режущего аппарата *agr.* sickle head
~ обратной перемотки rewind knob
~, оживальная ogive/ogival head
~, пазовая (болта) slotted head
~, панорамная pan-and-tilt head
~ паяльника bit of a soldering iron
~, перезаписывающая read-record head
~, печатающая print(ing) head
~, плавающая floating head
~, пластицирующая *chem.* smear head
~ плечевой кости *anat.* head of the humerus
~, плоская форсуночная plate injector
~, поворотная swivel head
~ полового члена *anat.* glans penis
~, полукруглая button/cup/round head
~, полупотайная rounded countersunk head
~, полусферическая форсуночная domeshaped injector (assembly)

~ поршневая 1. piston head; 2. small end
~ поршня piston crown/top
~, потайная countersunk head
~ придатка яичка *anat.* head of epididymis
~, прицельная *mil.* sighting head
~, птичья *zool.* avicularium
~ пуансона nose of punch
~, разбрызгивающая shower head
~, развертывающая *print.* scanning head
~, распылительная injection/injector/spray head
~, расточная boring head
~, револьверная turret head
~, револьверная для крепления заготовок/инструмента turret stock/tool head
~, револьверная токарного станка capstan of a lathe
~, резцовая 1. tool head; 2. inserted blade milling cutter
~, резьбонакатная thread-rolling head
~, резьбонарезная thread- cutting head
~ рогульки *text.* flyer top
~ с барашком winged head
~ с воздушной подвеской, магнитная air-floated magnetic head
~ с накаткой/насечкой milled/knurled head
~ с переменным усилием прижима *comp.* ramp-load head
~ с распыливающими наконечниками *agr.* spray head
~ с ушками винта lug head
~ самонаведения homing head/device, seeker
~, самораскрывающаяся self-opening stud socket
~ самонаведения, инфракрасная/тепловая infra-red homing head, IR seeker, heat-seeking head
~ самонаведения, оптическая optical/light homing head/device
~ самонаведения, радиолокационная radar seeker, radar homing head
~ самонаведения, электрооптическая electro-optical seeker
~, сварочная welding head/tip
~ сверла drill bit
~, сверлильно-расточная drilling-boring head
~, силовая unit-type spindle head
~, следящая фотоэлектрическая *print.* electric eye scanner
~ следящего устройства tracker/tracking head
~ снаряда shell/cartridge nose, nose of explosive shell
~, стирающая erase/erasing head
~, стригальная *agr.* shear head
~, суставная articular head
~ счетчика counting/counter head
~ считывания *comp.* playback/read(ing)/sensing head
~ считывания с ленты *comp.* (punched) type reading head
~, съемная detachable head
~ таблицы box heading
~, телевизионная television/TV head
~, телеметрическая telemetering head
~ телефонного штепселя plug tip
~, универсальная *comp.* combined head, read-write/-record head
~, установочная setting head
~, учебная *mil.* dummy head
~, форсуночная burner plate, *Am.* injector (assembly)
~, фрезерная 1. milling head; 2. inserted blade milling cutter
~ фундамента footing cap
~ фурмы top of a tuyere/lance
~ хвостатого ядра *med.* head of the caudate nucleus
~ цветка *bot.* anthodium
~ цилиндра cylinder head
~ часов, заводная clock winder, watch-winding stem
~, четырехдорожечная four-track head
~, шарнирная swivel head
~, шарошечная cutter block
~ шатуна connecting-rod end
~ шатуна, нижняя/большая/кривошипная connecting-rod big end
~ шатуна поршневая/малая/верхняя connecting-rod small end
~ швейного аппарата stitching head
~, шипорезная tenoning head
~, шлифовальная 1. wheel head; 2. mounted grinding wheel/point
~, штативная tripod head
~, штативная шаровая ball-and-socket head
~ штепселя plug tip
~ щелевая смесительная ring-slot/-gap injector, concentric-ring injector
~, экструзионная die/extrusion/discharge head

головковидный, головкообразный head-shaped, capitate, capituliform, cephaloid
головной 1. *anat.* cephalic, cranial; 2. leading; 3. *n.* leader, head, chief
~ голос *mus.* head voice, falsetto
~ дозор *mil.* advance (guard) patrol
~ мозг *anat.* brain, cerebrum
~ отряд *mil.* vanguard
~ убор headdress, headgear
~ное гидротехническое сооружение headwork

головня I *pathol.* black rust, brand, smut
~, мокрая bunt
~ пшеницы, мокрая stinking smut of wheat
~, пыльная dust brand

головная II *zool.* common scoter *(Melanitta nigra)*
головогрудь cephalothorax
головодержатель head supporter
головокружение *med.* dizziness, vertigo, giddiness
~, высотное height vertigo

~, **рефлекторное** reflex vertigo
головоногий *zool.* cephalopod(an); cephalopodous; *n. pl.* cephalopods *(Cephalopoda)*
~**гие десятиногие** squids *(Decapoda)*
головообразный head-shaped; cephaloid
головохордовые *n. pl. zool.* cephalochords, cephalochordates *(Cephalochorda)*
головчатка *bot.* cephalaria *(Cephalaria)*
головчатоцветковый *bot.* head-flowered, cephalanthous
головчатый 1. bulbous, bulbaceous; capitate; cephalate, head-shaped; 2. many-headed
головщик *eccl.* choir leader, precentor
гологамета *biol.* hologamete
гологамия *biol.* hologamy, hologamia, macrogamy
гологенез hologenesis
гологиалиновый *geol.* holohyaline
гологинный hologynic
гологлаз *zool.* lidless skink *(Ablepharus)*
голограмма *opt.* hologram
~, **аберрационная** abberated hologram
~, **акустическая** acoustical hologram, acoustically formed hologram
~, **амплитудная** absorption/amplitude hologram
~, **амплитудно-фазовая** absorption-phase hologram, mixed hologram
~, **безлинзовая** lensless hologram
~, **бихроматическая** two-colour hologram
~ **в видимом свете** visible-light hologram
~ **в гамма-лучах** gamma-ray hologram
~, **внеосевая** off-axis hologram, offset/spatial hologram
~ **высокого разрешения** high-resolution hologram
~, **движущаяся** animated hologram
~, **двухчастотная** two-frequency hologram
~, **записанная** recorded hologram
~, **избыточная** redundant hologram
~, **машинная** computer/computed hologram
~, **монохроматическая** single-colour hologram, monochrome hologram
~, **мультиплицируемая** master hologram
~ **на несущей частоте** carrier frquency hologram
~, **неподвижная** still hologram
~, **объемная** volume/deep hologram, full-view hologram
~, **однослойная** single-layer hologram
~, **осевая** uniaxial hologram, in-line hologram
~, **отбеленная** bleached hologram
~, **отражательная** reflectance/reflection hologram
~, **полутоновая** gray hologram, half-tone hologram
~ **преобразования Фурье** Fourier-transform hologram
~, **пропускающая** transmission hologram
~, **проявленная** developed hologram
~, **рентгеновская** X-ray hologram
~ **с высокой избыточностью** highly redundant hologram
~ **с кодированным опорным пучком** coding reference beam hologram
~ **с оптическим стиранием** optically erasable hologram
~ **с фокусированного изображения** focused-image hologram, in-focus hologram
~, **синтезированная** synthesized/synthetic/synthetisizing hologram
~, **синтезированная ЭВМ** computer-synthesized/-generated hologram
~, **скалярная** scalar hologram
~, **телевизионная** television/TV hologram
~, **точечная** point hologram
~, **уплотненная** mutiplex(ed)/multiplexing hologram
~, **фазоконтрастная** phase-contrast hologram
~, **фраунгоферова** Fraunhofer (diffraction) hologram
~, **френелева** Fresnel (diffraction) hologram
~, **Фурье** Fourier-transform hologram
~, **чистофазовая** pure/only phase hologram, phase-only hologram
~, **экспонированная** exposed/exposure hologram
~, **эталонная** master hologram
голографирование holography, holographing
~ **теневым методом** holography by shadow casting
голографический holographic
голография holography
~, **амплитудно-контрастная** amplitude-contrast holography
~, **внеосевая** two-beam holography, side-band holography
~, **двухмерная** two-dimensional holography
~, **двухэкспозиционная** double-exposed/-exposure holography
~, **звуковая** sonic holography
~, **изобразительная** pictorial/picture holography
~, **ИК** infra-red holography
~, **импульсная** pulse holography
~, **импульсно-лазерная** pulsed laser holography
~, **многолучевая** multi-beam holography
~, **многоцветная** multicolour holography
~, **мультипликативная** product holography
~, **неоптическая** non-optical holography
~, **объемная** volume holography
~, **однополосная** single-sideband holography
~, **осевая** single-beam holography, in-line holography
~, **отражательная** reflecting(-type)/reflection holography
~, **панорамная** panoramic holography
~, **поляризационная** polarization/polarizing holography
~, **промышленная** commercial holography
~, **сверхвысокочастотная / СВЧ** microwave

holography
~ сфокусированных изображений focused(-image) holography
~, трехмерная three-dimensional holography
~, ультразвуковая ultrasonic holography
~, черно-белая black-and-white holography
~, широкоугольная wide-angle holography
голод hunger, starvation; famine; *fig.* shortage
~, сенсорный *med.* sensory deprivation
голодание starvation, deficiency, fasting, deprivation; lacking; malnutrition
~, белковое protein deficiency/deprivation
~, водное water deprivation
~, вынужденное enforced starvation/fasting
~, кислородное oxygen deficiency/lack
~, лечебное medicinal starvation, therapeutic fasting
~, минеральное mineral deficiency
~, патологическое pathologic starvation/fasting
~, полное complete starvation/fasting
~, солевое salt deprivation/deficiency
~, частичное partial starvation
голодающий starving, hungry, famished
голодный 1. hungry; esurient; caused by hunger/ famine/ starvation; 2. scanty, poor, meagre; 3. *n.* hungry person
~ год lean year
~ край barren region
~ обед scanty meal
~ поход hunger march
голодовка starvation; hunger strike
~ку, объявивший hunger striker
голожаберный *zool.* nudibranch(iate); *n. pl.* nudibranchiates *(Nudibranchiata)*
голозерный *bot.* nudigranular, hull-less
голозимаза *bioch.* holozymase
голокарпный *bot.* holocarpic
голокластический *geol.* holoclastic
голоколосый *bot.* bald-glume; naked-eared
голокорешковый *bot.* exorrhizal, naked-rooted
голокринный, голокриновый *anat.* holocrine
голокрылый *zool.* gymnopterous, naked-winged
гололед(ица) glazed frost, clear/glazed ice; *Am.* glaze, glare/glaze ice; freezing rain
гололейкократовый *geol.* hololeucocratic
гололинза hololens
гололист(н)ый *bot.* naked-leaved
голометаморфоза holometamorphosis
голометр holometer
голоморф *math.* holomorph
голоморфный holomorphic
голомянка *zool.* oilfish *(Comephorus)*
голомянка-лира *zool.* dracunculus, dragonet *(Callionymus)*
голоногий *zool.* bare-foot/-legged; nudiped; *n. pl.* Nudipeda
голономный holonomic
голопаразит *biol.* holoparasite
голоперый featherless, unfledged
голопестиковый, голопестичный *bot.* gymnogynous
голопланктон *biol.* holoplankton
голоплодный *bot.* gymnocarpic, naked-fruited
голопузка *zool.* dracunculus, dragonet *(Callionymus)*
голоростковый *bot.* gymnoblastic, gymnoblastous
голоротые *n. pl. zool.* gymnostomous forms *(Gymnostomata)*
голос 1. voice, vox; *mus.* part, voice; 2. *polit.* vote
~, решающий *law* . casting/deciding vote
~, синтетический voice operation demonstrator, voder
~а, право *polit.* the vote, suffrage
~ов поровну, разделение *law* tie
голосапрофит *bot.* holosaprophyte
голосеменной, голосемянный *bot.* gymnospermous, naked-seeded; *n. pl.* gymnosperms *(Gymnospermae)*
голословный proofless, unsubitantiated, unfounded/unsupported by evidence
~ное утверждение mere allegation
голосование voting; poll
~, всеобщее universal suffrage
~ вставанием rising vote, vote by sitting and standing
~, открытое voting by show of hands
~, поименное roll-call vote, call of the roll, vote by roll call
~, поразрядное *comp.* bit-by-bit voting
~, тайное ballot, secret ballot/vote
~, устное vote viva voce
~нием отвергать *law* . vote down
~ния, кабина для polling booth
голосоведение *mus.* harmonization of themes
голосовой vocal; phonic, phonal
~вая щель *anat.* glottis
~вые связки *anat.* vocal chords
голостебельный, голостебельчатый *bot.* nudicaulous, bare-stalked
голостомный *zool.* holostomatous, holostome
голосумчатый *bot.* exoascaceous
голосующий *n.* voter
голотелый gymnostomatous
голотип holotype, type specimen
голотурис, голотурия *zool.* sea cucumber, holothuria(n) *(Holothuria)*
голофермент holoenzyme
голофит *bot.* holophyte
голофитный *bot.* holophytic
голохроический *zool.* holochroal
голоцветковый, голоцветный *bot.* gymnanthous, nudiflorous, naked-flowered
голоцен *geol.* Holocene, Holocenic/Recent system, Recent Quaternary group
~, древний Paleoholocene
~, поздний Neoholocene
~, ранний Toholocene
~, средний Mesoholocene
голоценовый *geol.* Holocene
голоэдр *cryst.* holohedron

голоэдрический holohedral
голоэдрия *cryst.* holohedrism
голубеобразные *n. pl. zool.* *Columbiformes*
голубец I 1. *geol.* azurite, mountain blue; 2. *bot.* paxillus *(Paxillus)*
голубец II *cul.* stuffed cabbage
голубика *bot.* blueberry *(Vaccinium);* log bilberry *(V. uliginosum)*
голубой azury; light-/pale-/sky-blue
~, метиленовый methylene blue
~, горный mineral blue, blue verditer
голубок *bot.* European columbine *(Aquilegia vulgaris)*
голубушка *bot.* *Oxytropis*
голубь dove, pigeon *(Columba)*
~, декоративный fancy pigeon
~, полудомашний dovecot(e) pigeon
~, почтовый/связи carrier pigeon
~, сизый blue rock pigeon, rock dove *(Columba livia)*
~, трубастый/веерохвостый fantail pigeon
голубянка *zool.* 1. apefly *(Spalgis epius);* 2. brown argus butterfly *(Lycaena astrache);* 3. moth butterfly *(Liphyra brassolis)*
голый naked, nude, bare; glabrous, achlamydeous
~ провод naked/uninsulated wire
~ спирт pure alcohol
~лая истина the naked truth
~лое администрирование purely bureancratic methods
голыш *geol.* peble, gravel, scree; round flat stone; flint; shingle
голье (raw)hide, unhaired hide; skin; pelt
~, мягченное bated hide, pelt
~, набухшее swollen pelt
~, пикелеванное pickled skin
~, растеленное split hide
~, фасонированное scudded pelt
~, юфтовое unhaired Russian leather
гольмий *chem.* holmium, Ho
Гольфстрийм *geogr.* Gulf Stream
гольштинский *geol.* Holstein *(attr.)*
голян *zool.* minnow *(Phoxinus)*
голяшка *cul.* kunkle
гомбо *bot.* okra *(Hibuscus esculentus)*
гомеозис *biol.* homeosis
гомеоидный homeoidal; *geol.* seriate
гомеомерный *bot.* homeomerous, homoiomerous
гомеоморфа *cryst.* homeomorph
гомеоморфизм homeomorphism, topological isomorphism, diffeomorphism
гомеоморфный homeomorphous, homeomorphic
гомеопат homeopathist, homeopath
гомеопатия homeopathy
гомеополярность *chem.* homeopolarity, covalence
гомеополярный hom(e)opolar, covalent
гомеорезис *med.* homeorhesis
гомеостаз(ис) *biol.* homeostasis
гомеостат homeostat
гомеостатичность ultrastabilty
гомерический, гомеров, гомеровский Homeric
гоми *bot.* foxtail/Italian millet *(Setaria italica);* 2. green bristle grass *(S. viridis)*
гомилетика *eccl.* homiletics
гоминдан *polit.* Kuomintang
гоминдановец member/suporter of Kuomintang
гоминид *zool.* hominid
гомицид homicide
гомицидный homicidal
гомицидомания *med.* homicidomania
гоммоз *bot.* gummosis
гомоатомный homoatomic
гомогаметный *biol.* homogametic, monogametic
гомогамия homogamy
гомогенат homogenate
гомогенетический, гомогенетичный *biol.* homogenetic(al)
гомогенизатор homogenizer
~, двухступенчатый two-stage homogenizer
~, ультразвуковой ultrasonic homogenizer
гомогенизация 1. homogenization, homogenizing; 2. *met.* diffusion annealing
гомогенизировать homogenize
гомогенитальность homogenitality
гомогенность homogeneity, homogeneousness, uniformity
гомогенный homogeneous; homogenous; uniform
гомогеномный homogenomatic
гомогония *bot.* homogony
гомография *math.* homography
гомогруппа homogroup
гомодинамический, гомодинамичный *biol.* homodynamic
гомодромный *bot.* homodromous, homodromal
гомозигота *biol.* homozygote
гомозиготность *biol.* homozygosis, homozygosity
гомойогидрический homoiohydric
гомойотермный *zool.* homoiothermal, warm-blooded
гомоклиналь *geol.* homocline
~, плоская/ровная plane homocline
гомококкит *geol.* homokokkite, monomineral rock
гомолизин homolysin
гомолог homolog(ue)
гомологический, гомологичный homologous, homologic(al)
~, внутренне intrinsically homologous, cobordant
гомология homology
~, верхняя cohomology
~, внутренняя intrinsic homology, cobordism
~, инволютивная involutory homology
~, остаточная residual homology
гомолографический homolographic
гомомерный homomeral, homomerous
гомоморфизм homomorphism; homomorphy
~, естественный canonical homomorphism
~, инверсный antihomomorphism
~, скрещенный crossed homomorphism

гомоморфический, гомоморфный homomorphic, homomorphous
гомоним homonym
гомонимический homonymic, homonymous
гомонимия homonymy
~ **родов** generic homonymy
гомопластика homoplasty
гомополимер homopolymer
гомосексуализм homosexuality; uranism
гомосексуал(ист) homosexual; invert; p(a)ederast
гомосексуальный homosexual
гомосиллабический *phon.* tautosyllabic
гомоскедастический *stat.* homoscedastic
гомоскедастичность homoscedasticity
гомоспория *bot.* homospory, isospory
гомоспоровый *bot.* homosporous, isosporous
гомостилия homostyly
гомостильный homostylic, homostylous
гомосфера homosphere
гомотаксальность, гомотаксис *geol.* homotaxis
гомотаксальный homotaxial
гомотаксический, гомотаксичный homotaxic
гомотермия homothermal condition
гомотетичный, гомотетический *math.* homothetic, radially related, similar
гомотетия *math.* homothetic transportation, transportation of similitude; homethety, similarity, similitude; dilation, dilatation
гомотипия homotypy
гомотопический, гомотопичный *biol.* homotopic
гомотрансплантат *med.* homotransplant, homograft
гомотрансплантация homoplasty, homotransplantation
гомотропный *biol.* homotropal, homotropous
гомофил homophile
гомофильный homophyllous
гомофит *bot.* homophyte
гомофитный homophytic
гомохламидный *bot.* homochlamydeous
гомохромный homochromous
гомоцепной homochain
гомоцеркальный *zool.* homocercal
гомоцикл homoatomic ring
гомоцистин homocystine
гомоэлементный homoelemental
гомруль *law* Home Rule
гомстед homestead
гомункул(ус) homunculus, homuncle
гомфокарпус *bot.* *Gomphocarpus*
гомфрена *bot.* globe amaranth *(Gomphrena)*
гон 1. *tech.* grade, **g** (*hundreth part of the right angle);* 2. chase, hunt, pursuit; (hunting) area; 3. *zool.* heat; rut, rutling time
гонада *anat., zool.* gonad
гонадный gonadal
гонадотропин *bioch.* gonadotrop(h)in
гонадотропний gonadotrop(h)ic
гонг gong
~, **туманный** *nav.* fog gong
Гондвана Gondwana land
гондола 1. gondola; *rail.* gondola car; 2. *av.* pod, car, gondola, nacelle
~ **аэростата** aerostat/baloon car/nacelle
~, **внешняя** outbord/outer pod/nacelle
~, **внутренняя** inboard/inner pod/nacelle
~ **двигателя (поворотная)** (tiltable/tilting) engine nacelle
~, **двойная** paired nacelle
~ **дирижабля** airship car/gondola
~, **консольная крылевая** wing-tip pod
~, **консольно закрепленная** cantilevered nacelle
~, **крылевая** wing(-mounted) pod, wing nacelle
~, **моторная** engine pod/nacelle
~, **несъемная** nacelle
~, **пассажирская** passenger car
~, **подвесная** underslung nacelle
~, **подкрылевая/подкрыльная** underwing pod
~, **подфюзеляжная** ventral pod
~, **пушечная** gun pod
~, **сбрасываемая** drop-pod
~, **съемная** detachable/disposable/removable pod
~ **управления** control car
~, **хвостовая** aft-fuselage-mounted nacelle
~ **шасси** undercarriage housing
гондурасец, гондурасский Honduran(ean), Honduranian
гондураска Honduran(ean) (woman)
гонемия *med.* gonococcemia, gonohemia
гонение persecution
гонец *hist.* courier, messenger
гонидиальный *bot.* gonidial, gonidic
гонидий, гонидия *bot.* gonidium
гониокраниометрия *med.* goniocraniometry
гониолимон *bot.* sea lavender *(Limonium, Goniolimon)*
гониометр goniometer, direction finder
~, **азимутальный** azimuth indicating goniometer
~, **двукристальный** double-crystal goniometer
~, **двукружный** two-circle goniometer
~, **дифрактометрический** diffractometer goniometer
~, **отражательный** reflecting goniometer
~, **прикладной** contact goniometer
~, **рентгеновский** X-ray goniometer
гониометрический goniometric(al)
гониометрия *math.* goniometry
гонит *med.* gonitis
~, **гнойный** purulent gonitis
~, **гонорейный** gonorrheal gonitis
~, **ревматический** rheumatic gonitis
гонка 1. dashing, rushing; 2. distillation; 3. *sport* race, racing; 4. *hunting* chase, pursuit; 5. floated timber; raft(ing), float(ing)
~ **вооружений** arms race

~ **двигателя** engine run-up
~**ки, гребные** boat race, rowing regatta
~**ки, парусные** sailing regatta
~**ки фронтов сигналов** *comp.* (logic) race
гонконгский Hong Kong
гонкость *for.* rapidity of growth
гонобласт *biol.* gonoblast
гонобленнорея *med.* gonoblennorrhea
гонобобель *bot.* bog bilberry *(Vaccinium uliginosum)*
гонок *tech.* guide, slat; *text.* picker
гонококк *bact.* gonococcus
гонококкемия *med.* gonococcemia, gonohemia
гонококковый gonococcal, gonococcic
гонорант *econ.* payer for hono(u)r
гонорар fee, honorarium
~, **авторский** royalty *(often pl.)*
~ **за внешнюю ревизию** *econ.* public accounting fee(s)
гонор(р)ея *med.* gonorrh(o)ea
~, **осложненная/острая** complicated/acute gonorrh(o)ea
гоночный racing
гончар potter
гончар-глазуровщик glazer
гончарный potter's
~ **круг** potter's wheel
~**ное искусство** ceramic art
~**ное производство** pottery
~**ные изделия** pottery, earthenware
гончарня, гончарство pottery
гонщик 1. racer; 2. drover; *timb.* rafter
гоньба 1. dash, rush; racing; 2. *hist.* delivery by post horses; 3. chase, pursuit
гоньерит *geol.* gonyerite
гонять drive (away); chase
~ **двигатель** run up an engine
~ **лес** raft/float timber
~ **почту** *hist.* deliver mail
~ **скот** drive cattle
гопак hopak *(folk dance)*
гоплит *hist.* hoplite
Гор *myth.* Horus
гора mountain; hill
~, **аккумулятивная** composite mountain
~, **возрожденная** rejuvenated mountain
~, **глыбовая** block mountain
~, **дизъюнктивная** faulted mountain
~ **из тектонических покровов** overthrust mountain
~, **лысая** bald mountain
~ **надвиговая** overthrust mountain
~, **освобожденная** laccolite, laccolith
~, **остаточная** residual mountain; monadrock
~, **островная** island mountain, inselberg
~, **остроконечная** peaked mountain/hill; tor
~, **отдельная** isolated mountain; butte
~, **пликативная** folded mountain
~, **подводная** seamount, sea peak
~ **разгона** ski jump/hill
~ **с плоской вершиной** table mountain
~, **сбросовая** faulted mounain
~, **средневысокая** upland, highland
~, **старческая** old/senile mountain
~, **столовая** 1. table mountain, flat-topped mountain, mesa; 2. *astr.* Mensa, the Table
~, **ступенчатая** step(ped) mountain
~, **чешуйчатая,~, чешуйчатого строения** imbricated mountain
~, **юная** youthful mountain
Гора *hist.* Montagnard
горал *zool.* serow, goat antelope *(Nemorhaedus goral)*
гора-свидетель *geol.* island mountain, inselberg
горб 1. hump; *med.* humpback, hunchback; *rail.* hump (crest); 2. bulge, protuberance, gibbosity
~ **кривой** *math.* hump
~, **рахитический** rachitic humpback
~, **реберный** costal humpback
~, **сердечный** cardiac humpback
горбатик *timb.* spout plane
горбатки *pl. zool.* humpbacked flies *(Phoridael)*
горбатость *med.* gibbosity; humpiness
горбатый humpbacked, hunchbacked; gibbous
~ **мост** humpback bridge
~ **нос** aquiline/Roman nose
горбач 1. circular/compass plane; 2. *zool.* humpback (whale) *(Megaptera novaeangliae)*
горбоносый hook-/hawk-/Roman-nosed
горбыль I *timb.* slab, half log
горбыль II *zool.* jacknife fish *(Equetus lanceolatus)*
Горгона *myth.* Gorgon
гордеин *bioch.* hordein
гордень *naut.* fall-and-tackle, gantline, girt line; brail
~ **для подъема зарядов** cordite whip
~, **подъемный** whip (hoist), block and fall, lifting fall, hoisting/hauling line
Гордиев узел Gordian knot
гордовина *bot.* wayfaring tree *(Viburnum lantana)*
гордония *bot.* gordonia *(Gordonia)*
горелка burner; *weld.* (welding) torch
~, **атомно-водородная** atomic-hydrogen arc-welding torch
~, **ацетиленовая** acetylene burner/torch
~, **ацетилено-кислородная** oxyacetylene torch
~, **бензынжекторная** (balanced-)pressure torch
~, **беспламенная** radiant burner, surface combustion burner
~ **Бунзена/бунзеновская** Bunsen(-type) burner/torch
~, **водородно-кислородная** oxydric torch
~, **газовая** gas burner, blowpipe
~, **газовая Аргана** Argand burner
~, **газовая Мекера** Meker burner
~, **газовая Теклю** Teclu burner
~, **газомазутная** oil-gas burner
~, **двухпламенная** two-tip torch

~, двухулиточная two-scroll burner
~, двухходовая fishtail burner
~, дежурная pilot(-light) burner, flash pipe
~, дифузионная diffusion-flame burner
~ для воздушно-дуговой резки arc-air torch
~ для газопламенной очистки flame cleaning burner
~ для зачистки дефектов deseaming burner
~ для пламенной очистки descaling torch, flame cleaning/conditioning torch
~ для сварки и резки welding and cutting torch/blowpipe
~, дуговая обратной полярности transferred arch torch
~, дуговая прямой полярности non-transferred arc torch, plasma-jet torch
~, завихряющая turbulent burner, cyclone-type burner
~, инжекторная injector torch/blowpipe, low-pressure torch
~ испарительного типа vaporising/atomizing fuel nozzle/injector
~, кислородотопливная oxyfuel burner
~, кольцевая ring torch
~, комбинированная multifuel burner
~, круглая circular burner
~, мазутная/нефтяная oil burner
~, многопламенная/многофакельная multiflame/multijet torch/blowpipe
~, многощелевая multitip burner
~, мотылькообразная wing top burner
~ огневой зачистки scarfing torch
~, однопламенная/однофакельная single-flame torch/blowpipe
~, одноструйная rat-tail burner
~, паяльная glass blower's burner
~, плазменная plasma torch
~, поворотная tiltable burner
~, потолочная elevation/vertical burner
~, пылевая/пылеугольная pulverized-coal burner
~, пылегазовая pulverized-coal and gas burner
~, растопочная lighting-up burner, lighter
~, резательная cutting torch
~, сварочная welding torch
~, сварочная газовая gas welding torch
~ с вихревым пламенем turbulent burner, cyclone-type burner
~ с внутренним смесеобразованием premix burner
~, струйная jet burner
~, турбулентная turbulent burner, cyclone-type burner
~, угловая corner-fired burner
~ щелевая fantail burner
~, экономическая pilot burner
горелость *met.* hot tear
горелый burnt, scorched; *chem.* empyreumatic(al)
горельеф *art* high relief, alto-relievo
горение combustion, burning, blazing
~, бездымное smokeless combustion
~, беспламенное surface combustion
~, быстрое high-velocity combustion
~ в дозвуковом потоке *av.* subsonic combustion
~ в замкнутом объеме confined combustion
~ в сверхзвуковом потоке *av.* supersonic combustion
~ в установившемся потоке steady-flow combustion
~, взрывное detonation (combustion)
~ гибридного топлива hybrid burning
~, детонационное detonation combustion
~, диффузионное diffusion combustion
~ дуги arcing
~, замедленное delayed combustion
~, интенсивное high-intensity combustion
~, квазинейтральное/квазистационарное quasi-neutral burning
~, кратерное crater-like combustion
~, нестабильное unstable burning/combustion
~, неустоичивое irregular burning, unstable combustion
~, низкочастотное пульсационное chugging
~, погрузное immersion combustion
~, полное complete combustion
~, предкамерное precombustion
~, пульсирующее intermittent burning
~, равномерное uniform burning
~ с двумя ступенями тяги two-stage burning
~ с избытком горючего/окислителя fuel-/oxidizer-rich combustion
~ с торца,~, торцевое end/cigarette burning
~, стехиометрическое stoichiometric combustion
~ твердо-жидкого топлива hybrid burning
~ топлива в форсажной камере afterburning, tailpipe burning
~, установившееся steady-state combustion, sustained combustion
~, устойчивое stable combustion
горец *bot.* knotweed *(Polygonum)*
~, змеиный snakeweed *(P. bistorta)*
~, перечный water pepper *(P. hydropiper)*
~, птичий knotgrass, knotweed *(P. aviculare)*
горечавка *bot.* gentian *(Gentiana)*
горечавковые *n. pl. bot.* gentian family *(Gentianaceae)*
горздравотдел *abbr.* **городской отдел здравохранения** city health department
горжа gorge, defile
горизонт 1. skyline; *geol.* horizon, level; 2. *min.* gallery, adit, level, floor
~, абсолютный datum plane
~, артезианский artesian aquifer/horizon
~, базальный basal level
~, вентиляционный *min.* airway horizon/level
~ верхнего бьефа upstream water line, upper pond level
~, верхний higher/upper level

~, видимый skyline, apparent/visible/terrestrial/local/natural horizon
~ визирования *geod.* elevation head
~, водоносный aquifer, water-bearing strata/horizon/level
~, водоносный висячий perched water table
~ воды water table, elevation of water
~ воды, рабочей service water level
~, временной *comp.* time horizon
~ выпуска *min.* draw level
~ высоких вод high-water level
~, вышележащий overlaying formation
~, гироскопический artificial horizon, gyrohorizon
~, главный main level
~, глеевый gley
~, глубокий deep level
~ грунтовых вод ground-water level/table
~ грунтовых вод, верхний phreatic high
~ гумбо gumbotial horizon
~, гумосовый humus horizon
~ дробления *min.* grizzly level
~, дымчатый haze horizon
~, естественный natural horizon
~, закладочный *min.* waste floor
~ засыпи *met.* charging level
~, земной earth/terrestrial horizon
~, иллювиальный illuvial horizon, horizon C
~ инструмента *geod.* height of instrument, elevation head
~, искусственный *nav.* artificial/mercury horizon, gyrohorizon
~, истинный celestial/true/observer's horizon
~, истинный математический celestial/true horizon
~, исходный *geod., constr.* datum lane
~, маркирующий *geol.* reference/key horizon; *min.* marker
~, математический observer's/true/celestial horizon
~, маятниковый искусственный artificial pendulum horizon
~, мглистый hazy horizon
~, мнимый false horizon
~ наблюдателя, видимый *nav.* actual sea horizon
~ наблюдателя, истинный celestial/sensible horizon
~, нефтеносный oil horizon
~, нефтьсодержащий oil-bearing horizon
~ нижнего бьефа downstream water line, down pond level
~, нижний lower level
~, нижний грунтовых вод phreatic low
~ низкой воды low-water level
~, нормальный datum level
~, обнаженный outcropping horizon
~ огней приближения, световой *mil.* cross bar
~, ограничивающий confining bed
~ ограничительных огней, световой *mil.* wing-bar
~, окварцованный quartz reef
~ околоствольного двора *min.* shaft bottom level
~, опорный reference/key horizon
~, оптический optical horizon
~, ортштейновый iron pan
~, основной *min.* main level/floor/horizon; basal level, sole
~, основной водоносный basal water, basal aquifer horizon
~, отбойки *min.* mining floor
~, откаточный *min.* haulage horizon/level
~ относимости reference horizon
~, отражающий reflecting horizon
~, пахотный *agr.* ploughing horizon
~, перегнойно-аккумулятивный *geol.* horizon A1
~, перегнойно-элювиальный *geol.* horizon A
~, песчаный sand level
~, повторяющийся recurrent level
~, подвешенный водоносный perched water table
~ подпочвенной воды ground-water level/table
~ подрубки *min.* cutting horizon
~ подстилающих пород *geol.* basement floor
~, подэтажный *min.* sublevel
~, пониженный depressed horizon
~, почвенный soil horizon
~ почвы, погребенный buried soil
~ почвы, элювиальный *geol.* horizon B
~, преломляющий refracting horizon
~ приведения reference horizon
~, продуктивный producing horizon
~, промежуточный *min.* intermediate level, transition horizon
~ профиля profile horizon
~, рабочий *min.* working/producing level, mining floor
~, радиолокационный radar horizon
~, ртутный *astr.* mercury basin
~, рудный ore horizon
~, руководящий reference/key horizon
~, световой *av.* cross-bar
~, свободный от напряжений level of no strain
~ сернистых руд sulphide zone
~ скреперования *min.* scraping drift/level
~, смещенный disrupted horizon
~, сплошной continuous horizon
~, стандартный standard depth
~, стратиграфический stratum
~ съемки рудного тела surveyor's level
~, теоретический theoretical horizon
~, условный *geod.* reference level, datum line; *av., geol.* phantom horizon
~, физический *av.* wing bar
~, фонтанный gushing horizon

~, чистый clear horizon
~, эксплуатируемый production/mining level
~, элювиальный eluvial horizon
~ этажного штрека *min.* sill floor
горизонталь 1. (the) horizontal (line); level; *geogr.* contour line; *geol.* structure contour; horizon; 2. horizontal flue
~ аэроснимка a parallel to the axis of tilt
~ аэроснимка, главная principal/isometric parallel, centre line of aerial photograph
~, гипотетическая *geol.* approximate contour; *cart.* formline
~, гравитационная gravity contour
~,замыкать close a contour (line)
~ места local horizontal
~, нулевая initial level
~, подводная marine contour line
~, приближенная approximate contour
~, проводить run out a contour (line)
~, строительная longitudinal datum line
~, структурная structure contour
~, схематическая sketched contour
~ фюзеляжа, строительная fuselage datum
горизонтально-залегающий flat-lying
горизонтально-напластованный flat-bedded
горизонтальность horizontal position, horizontality
горизонтальный horizontal, aclinic, aclinal, level, flat-dipping/-lying, flat
горизонтирование levelling
горизонтировать level
горизонт-репер marker bed
горизонт-фантом phantom horizon
горилла *zool.* gorilla *(Gorilla)*
горисполком *abbr.* **городской исполнительный комитет** city executive committee
гористый upland, rangy, montiform, mountainous, hilly
горихвостка *zool.* redstart *(Phoenicurus; Diplootocus)*
горихвостка-чернушка *zool.* black redstart *(Phoenicurus ochruros)*
горицвет *bot.* adonis, pheasant's eye *(Adonis)*
горичник *bot.* hog's-fennel *(Peucedanum)*
горка *av.* 1. zoom(ing), steep climb; 2. *geol.* hill, hummock; 3. hump; *rail.* gravity yard; 4. *agr.* gleaner, grain-sorting machine
~, боевая combat hump
~, семяочистительная *agr.* friction/gravity seed separator/cleaner
~, семяочистительная, карусельная *agr.* horizontal-disk seed separator
~, сортировочная *rail.* gravity hump
горком *abbr.* **городской комитет** city/town committee
горлач *zool.* howling monkey *(Alovatta)*
горлец *bot.* European bistort *(Polygonum bistorta)*
горлица I *zool.* turtledove *(Streptopelia)*
горлица II gorlitsa *(Ukrainian dance)*
горло 1. *anat.* neck, throat; 2. *geol.* (sand) bar; wave-cut niche; bottle neck; 3. *math.* gorge circle
~ вулкана volcanic vent
~ входного диффузора inlet diffuser throat
~, дыхательное *anat.* trachea, windpipe
~ меандра neck of meander lobe
~ траловой сети trawl head
горловина 1. neck, nozzle, mouth, throat; orifice, crater, vent 2. filler; access hole; manhole, handhole, scuttle, hatch
~ вентиляции ventilating handhole
~, водонепроницаемая watertight manhole
~ воздушного баллона neck of the air-flask
~ вулкана vent, volcanic neck
~ гидробачка, заправочная hydraulic reservoir filler
~ для выхода отработанных газов escape hole
~ для заправки топливом самотеком gravity (re)fuelling point, gravity filler
~ для осмотра sighting hole
~, заливная/заливочная filler, filling hole
~ запального устройства throat of the primer device
~, заправочная filler neck, filler
~ кабельного колодца manhole frame
~ камеры сгорания combustion chamber throat
~ канала duct canal entrance
~ канистры fuel can nozzle
~ колбы bulb neck
~ колодца manhole chimney
~ конвертера *met.* (vessel) mouth
~ конвертера, симметричная concentric mouth
~, маслозаправочная/маслоналивная oil filler
~, наливная filler (neck)
~, сливная drain pipe
~, смотровая inspection handhole, access opening
~ сопла, входная nozzle entry/throat
~ станции *rail.* yard neck, rail bottleneck
~, топливозаправочная fuel filler
горловой *anat.* jugular, throaty; guttural; *math.* striction
~ голос guttural voice
~вая чахотка *med.* tuberculosis of the throat
горлюха *bot.* oxtongue *(Picris)*
горлянка *bot.* bottle/calabash gourd *(Lagenaria vulgaris /siceraria)*
гормогоний *bot.* hormogonium, hormogone
гормон *bioch.* hormone; *see also* **гормоны**
~, адренокортикотропный adrenocorticotrop(h)ic hormone, ACTH
~, антидиуретический antidiuretic hormone, vasopressin
~, гонадотропный gonadotrop(h)ic hormone
~ желтого тела progesteron
~, женский половой female hormone, estradiol, estrone, estriol

~ **задней доли гипофиза** posterior pituitary hormone
~ **коры надпочечников** adrenocortical hormone
~, **кринотропный** crinotrop(h)ic hormone
~, **лактогенный** lactogenic hormone, prolactin
~, **лютеинизирующий** luteinizing hormone, lutein-stimulating hormone
~, **меланофорный** melanophore-stimulating hormone
~, **мужской половой** male hormone, testosterone
~ **околощитовидной железы** parathyroid hormone
~ **передней доли гипофиза** anterior pituitary hormone
~, **плацентарный** placental hormone
~ **поджелудочной железы** pancreatic (hyperglycemic) hormone
~, **половой** sex hormone
~, **прогестативный** sexogen
~, **раневой** wound hormone, traumatic acid
~ **роста** growth hormone
~, **соматотропный** somatotrop(h)ic hormone
~, **стероидный** steroid hormone
~, **фолликулостимулирующий** follicle-stimulating hormone
~, **фолликулярный** follicular hormone
~, **хориальный гонадотропный** chorionic gonadotrop(h)ic hormone
~ **хромаффинных клеток надпочечника** chromaffin hormone, epinephrine
~ **щитовидной железы** thyroid hormone, thyroxine, thyroglobulin
~ **яичника** ovarian hormone, estrin
гормональный hormonal, hormonic
гормонный hormonic
гормоногенез hormonogenesis
гормонообразующий hormonogenic, hormonopoietic
гормонотерапия hormonotherapy
~, **заместительная** substitution hormonotherapy
~, **тормозящая** inhibitory hormonotherapy
гормоны *pl.* hormones
~, **андрогенные** androgenic hormones, androgens, male sex hormones
~, **глюкокортикоидные** gluco-corticoid hormones, gluco-corticoids
~, **кортикостероидные** corticosteroid hormones, corticosteroids
~, **минералокортикоидные** mineralocorticoid hormones, mineralocorticoids
~, **эстрогенные** estrogenic hormones, estrogens, female sex hormones
гормопротеин hormoprotein
гормоцист(а) *bot.* hormocyst
горн I hearth, kiln, furnace; forge; crucible
~ **вагранки** cupola crucible/well
~, **выносной** external forge
~ **для зейгерования** liquation hearth
~ **для обжига фаянсовых/фарфоровых изделий** China kiln
~, **зажигательный** ignition hearth
~, **калильный** annealing hearth
~, **кричный** refinery hearth, bloomery
~, **кузнечный** smith's/forge hearth, smith forge, forging furnace
~, **набивать** ram the hearth
~, **нижний** bottom hearth
~, **обжигательный** roasting hearth
~, **передний** forehearth, settler, receiver
~, **пробирный** assay hearth
~, **сварочный** bloomery hearth
горн II *mus.* bugle; horn
~, **пневматический с вибрирующей мембраной** reed horn
~, **туманный (язычкового типа)** (reed) fog horn
горнблендит *geol.* hornblendite
горнист *mil.* bugler
горнитос *geol.* hornito
горница *hist.* room, chamber
~, **большая** cenacle
горничная *n.* (house) maid; chambermaid; lady's made; *naut.* stewardess
горновой, горновщик furnace attendant, furnaceman
горнодобывающий mining
горнозаводский mining and metallurgical
горноколосник *bot.* stonecrop *(Orostachys)*
горно-луговой alpine-/mountain-meadow
горнолыжник mountain skier; downhill racer
горно-подзолистый mountain-podzolic
горнопромышленность mining industry
горнорабочий *n.* miner
горнорудный (ore) mining
горноспасательный mine-rescue
горностаевый ermine; ermine-colo(u)red
~ **мех** ermine *(fur)*
горностай 1. *zool.* ermine *(Mustela);* stoat *(Mustela ermina);* 2. ermine *(fur)*
горнотехнический mining-engineering
горношахтный mining
горный 1. mountain, upland, rangy; montane, mountainous; 2. mining; 3. mineral
~ **институт** College of Mines
~ **комбайн** cutting loading machine
~ **лен** asbestos
~ **пояс** montane belt
~ **хрусталь** rock crystal
~**ная растительность** mountain vegetation
~**ная порода** rock
~**ное дело** mining
~**ное солнце** artificial sunlight, sun lamp
горняк miner; mining engineer/student
город 1. town, city; 2. *hist.* walled/fortified town; burg; fortress, citadel; 3. *sport* base; home
~, **столичный** capital city
город-герой hero town/city
город-спутник satellite town

городничество *hist.* provostship, governorship
городничий *n. hist.* governor of a town
городовой 1. town, city *(attr.)*; 2. *n. hist.* policeman
~вое положение municipal statutes
городок small town
~, военный cantonment, military station
~, студенческий students' residence halls
~, университетский campus
городской 1. city, municipal; urban; 2. *n.* city/town dweller
~ голова *hist.* mayor
~ сад municipal gardens
~ транспорт municipal transport
~кая дума *hist.* municipal duma, town council
городчатый crenulate(d), crenate; notched
гороно *abbr.* **городской отдел народного образования** City Board of Education
горообразование orogenesis, orogeny, mountain making/folding/building
~, антильское *geol.* Antillean revolution
горообразовательный, горообразующий orogenic
гороскоп horoscope
~, составить cast a horoscope
горох 1. *bot.* pea *(Pisum)*; *coll.* peas; 2. *pl. text.* spots
~, крупный marrowfat peas
~, лущеный split peas
~, огородный field/garden pea *(P. arvense)*
гороховидный pea-shaped/-sized, pisiform, pisolitic
гороховник *bot.* 1. desert thorn, squaw-bush *(Lycium)*; 2. Siberian pea tree *(Caragana arborescens)*
горохолущитель *agr.* pea huller
горохомойка pea washer, pea washing machine
горохомолотилка *agr.* (pea) viner
горохоплодный *bot.* pea-fruited, pisocarpous
горошек 1. *bot.* vetch *(Vicia)*; 2. *text.* spotted design; 3. *geol.* pea coal
~, душистый sweet pea *(Lathyrus odoratus)*
горошина *text.* spotted design
горошинка *zool. Pisidium*
горсовет *abbr.,obs.* **городской совет** town/city soviet
горст *geol.* horst
~, наклонный tilted horst
горст-антиклиналь *geol.* horst-anticline
гортанно-глоточный laryngopharyngeal
гортанно-трахеальный laryngotracheal
гортанный 1. *anat.* laryngeal; 2. *phon.* guttural; glottal
~ взрывной glottal plosive
гортаноглотка *anat.* laryngopharynx
гортань *anat.* larynx
гортензия *bot.* hydrangea, hortensia *(Hydrangea)*
гортит *geol.* hortite
горчак I *bot.* 1. oxtongue *(Picris)*; 2. Russian knapweed *(Centaurea picris)*
горчак II *zool.* bitterling *(Phodeus)*
горчица *bot.* mustard *(Sinapsis, Brassica)*
горчичник *med.* sinapism, mustard poultice/plaster
горчичный mustard *(attr.)*
~ газ mustard gas
горшечник 1. *zool.* ovenbird *(Furnarius rufus)*; 2. potter
горшечный pottery; potting
~ промысел pottery
~ товар earthenware, pottery
~ное растение pot plant
горшок pot, pan; vessel; jug; vase
~, агломерационный sintering pan
~, выгоночный *agr.* planting pot
~, выхлопной muffler
~, конденсационный moisture/steam trap, steam separator, condensing pot
~, конденсационный с закрытым поплавком float(-type) steam trap
~, конденсационный с открытым поплавком inverted bucket(-type) steam trap
~, ночной chamber pot
~, отжигательный annealing pot
~, рассадный *aggr.* planting pot
~, сборный *met.* receiver
~, стекловаренный/стеклоплавильный glass pot/crucible
~, цветочный flowerpot
~, торфоперегнойный peat plant/pot, peat compost block
горшочкоделатель soil block/pot maker
горький bitter
~кая, английская/русская *n.* English/Russian bitters
горькозем bitter earth, magnesia
горькоплодные *bot.* bitter-fruited, picrocarpous
горькуха *bot. Lactarius rufus*
горючее *n.* fuel; combustible; gasoline
~, анилиновое aniline-base fuel
~, безуглеродное noncarbonaceous fuel
~, борсодержащее boron(-based) fuel
~ в бочках drummed fuel
~, взрывоопасное explosive fuel
~, жидкое liquid fuel
~, жидкометаллическое liquid-metal fuel
~, ископаемое mineral fuel
~, необычное exotic fuel
~, обычное conventional fuel
~, самовоспламеняющееся hypergolic fuel, self-ignition fuel
~, суспензированное slurry fuel
~, твердое solid fuel
~, термоядерное thermonuclear fuel
~, углесодержащее carbonaceous fuel
~, этилированное ethylized fuel
~, ядерное nuclear fuel
горючесть combustibility, inflammability
горючий combustible, inflammable
~ материал combustibles; fuel

горянка *bot.* barrenwort, epimedium *(Epimedium)*
горячекатаный hot-rolled
горячеломкий *tech.* hot-short
горячеломкость hot-short state, hot-shortness
горячетянутый *tech.* hot-drawn
горячка 1. *med.* fever; 2. rush, feverish haste/activity
~, белая delirium tremens
~, биржевая rush on the stock market
~, гнилая typhus
~, послеродовая puerperal fever
госаппарат *abbr.* **государственный аппарат** state machinery/apparatus
Госбанк *abbr.* **Государственный банк** State Bank
госбезопасность *abbr.* **Государственная безопасность** State Security
госбюджет *abbr.* **государственный бюджет** State Budget
Госдепартамент *abbr.* **Государственный департамент** Department of State
Госиздат *abbr.* **Государственное издательство** State Publishing House
госкапитализм *abbr.* **Государственный капитализм** state capitalism
госконтроль *abbr.* **Государственный контроль** State Control (Commission)
госкредит *abbr.* **государственный кредит** State credit
госларит *miner.* goslarite, zinc vitriol
го-сотерн *hist.* Haut Sauterne
госпитализация admission to a hospital, hospitalization
госпитализировать admit to a hospital, hospitalize
госпиталь (military) hospital
~, многопрофильный multipurpose hospital unit
~, плавучий hospital ship
госпитальер *hist.* Hospitaler
Госплан *abbr., obs.* **Государственный плановый комитет** State Planning Committee of the USSR
господарь *hist.* hospodar
господень *eccl.* the Lord's
~дня молитва The Lord's Prayer
господин 1. gentleman, sir; 2. master
~ президент Mr. President
~да 1. gentlemen; 2. ladies and gentlemen, Messrs
господний *eccl.* Lord's, dominical
господский 1. master's; 2. manorial, seigniorial
~ дом manor house
господство supremacy, superiority, hegemony, dominance
~ в воздухе air superiority/supremacy, control of the air
~ в воздухе и Космосе aerospace superiority
~, мировое world supremacy
~ на море naval/sea/maritime supremacy, command/control of the sea
господствующий 1. (pre)dominant, dominating, prevailing, prevalent; 2. ruling, hegemonic(al)
~ класс ruling class
~щее насаждение main/principal stand
госпожа 1. mistress, landlady; 2. Mrs.; Miss
госпредприятие *abbr.* **Государственное предприятие** state establishment/enterprise
госпромышленность *abbr.* **Государственная промышленность** state industry
госсипол *chem.* gossypol
госснабжение *abbr.* **Государственное снабжение** state supply
госстрах *abbr.* **Государственное страхование** state ensurance
ГОСТ *abbr.* **Государственный общесоюзный стандарт** All-Union standard; USSR state standard specification
госта *bot.* plantain lily *(Hosta)*
гостиница hotel, inn
~ для автотуристов motel
~, монастырская hospice
гость 1. guest, visitor; 2. *geol.* guest mineral, metasome
государственность state system, statehood
государственный state, public, government; government-owned, state-operated/owned
~ деятель statesman
~ долг public debt
~ переворот coup d'état
~ служащий civil servant
~ ум statesmanship
~ язык official language
~ная измена high treason
~ная служба public service
~ная собственность government property
~ное право public/constitutional law
~ные экзамены state/final examinations *(in higher education institutions)*
государство state; country, nation
~ всеобщего благополучия welfare state
~, морское maritime/seafaring nation, sea-going nation
~, невоюющее nonbelligerent (nation)
~ не имеющее выхода к морю noncoastal state
~, суверенное sovereign state
государствоведение political science
государыня empress, sovereign; Your/Her Majesty
~, милостивая *(as form of address)* madam
государь sovereign; *(as form of address)* Your Majesty, Sire
~, милостивый *(as form of address)* Sir
гот *hist.* Goth
готигляциал *geol.* Gotiglacial subepoch
готигляциальный *geol.* Gotiglacial
готика *arch.* Gothic style
готицизм Gothicism
готический Gothic
~ шрифт Gothic script
готовальня case/set of drawing instruments

готовность 1. availability; readiness, preparedness; *mil.* alert; stand-by; 2. willingness
~, боевая combat/operational readiness, alert
~ к полету flight readiness
~ к рейсу, немедленная *mil.* full operational status
~ номер 1 *mil.* readiness for action, condition 1
~,приводить в боевую *mil.* (place on the) alert
готовый ready, prepared; of-the-shelf, stock-produced; ready-made
~ для эксплуатации ready for use/service
~ к действию ready for action
~ к непосредственному использованию turnkey *(attr.)*
~ к погрузке load ready, ready for shipment
~вая продукция final/finished goods
~вые изделия finished manufacture/product
готский Gothic
~ язык the Gothic language
готтентот, готтентотский Hottentot
готтентотка Hottentot (woman)
гоустония *bot.* bluets *(Houstonia)*
гоутонит *geol.* houghtonite
гофер *zool.* 1. gopher *(Geomys);* 2. gopher tortoise *(Testudo polyphemus)*
гофмаклер *hist.* court broker
гофмаршал *hist.* Marshal of the (Emperical) Court
гофмейстер *hist.* steward of the household
гофр 1. crimp, goffer; 2. *met.* corrugation
~ы, поперечные *met.* buckles
гофрировальный goffering, fluting; crimping; corrugating
гофрирование *met.* corrugating; *text.* crimping, goffering; fluting; embossing
гофрированный corrugated, crimped; channelled; goffered, fluted; undulate(d); embossed
~ воротник goffered collar
~ное железо corrugated iron
гофрировать 1. *text.* crimp, goffer; plait, pleat; 2. *met.* corrugate; emboss; wave
Гоэлро *abbr., hist.* **Государственная комиссия по электрификации России** State Commission for Electrification of Russia
ГПС *abbr.* **гибкая производственная система** flexible manufacturing system, FMS
ГПУ *abbr., hist.* **Государственное политическое управление** State Political Administration, GPU *(the secret police of the USSR)*
грааль Grail
граб *bot.* hornbeam *(Carpinus)*
грабельник *pl.* **грабельки** *bot.* heron's-bill *(Erodium)*
грабен *geol.* graben, fault trough
~, взбросовый ramp valley
~, долина fault-block valley
~, котлообразный cauldron, fault-pit
~, круглый ring cataclase
~, сбросовой rift valley
~, срединный median graben
~, флексурный trough bend
грабен-синклиналь graben-syncline
грабинник *bot.* Oriental hornbeam *(Carpinus orientalis)*
грабли *pl.* rake
~, барабанные rotary rake
~, боковые (комбинированные) (combined) side(-delivery) rake
~, дистанционные *mil.* range rake
~, колесно-пальцевые finger-wheel rake
~, косоугольные oblique rake
~, мотовильные reel rake
~, навесные mounted rake
~, поперечные self-dump rake
~, прицепные trailed rake
~, сенные hayrake
грабли-волокуши *pl.* hay sweep, sweep rake
грабштихель engraving/engraver's tool, graver
гравелит *geol.* gritstone, grit, gravelite
гравер engraver; etcher
~ по дереву wood engraver, woodcutter
граверный engraving; engraver's
гравет *geol.* Gravettian
гравиевый gravelly
гравиемойка *tech.* gravel washer
гравий gravel, grit
~, бурый iron-stained gravel
~, грубый coarse gravel
~, дробевидный buckshot gravel
~, дробленый crushed gravel
~, карьерный pit gravel
~, крупный coarse gravel, shingle
~, мелкий pea-gravel, fine gravel, grit
~, наносный run gravel
~, насыпной loose gravel
~, окатанный gravel pebbles
~, остаточный residual/lag gravel
~, перемытый outwash gravel
~, промышленный pay gravel
~, речной bank/river gravel
~, террасовый bench-gravel
~, угловато-зернистый grit
гравийный gravelly
гравилат *bot.* avens, bennet, geum *(Geum)*
гравилатолистный *bot.* avens-leaved
гравиметр gravimeter, gravity meter
~, астазированный/астатический astatic gravimeter
~ Исинга Ising gravimeter
~, крутильный torsion gravimeter
~, маятниковый gravity pendulum
~, наземный surface gravimeter
~, скважинный down-/drill-/bore-hole gravimeter
~, статический static/stabilized gravimeter
~ с электрической измерительной схемой electric-gauge-type gravimeter
~ Тиссена/Хартли/Хойта Thyssen/Hartley/ Hoyt gravimeter

гравиметрический gravimetric
гравиметрия gravimetry
гравиразведка gravitation prospecting
гравировальный engraving
~ инструмент graver, burin
~ная доска steel/copper plate
~ная игла etching needle
гравирование engraving, cutting; etching
~, гальваническое galvanocautery
~ на стекле glass engraving, glass-etch process
~, фотографическое photoengraving
~, фотохимическое photochemigraphy
гравировать engrave; etch
гравис *phon.* grave accent, low pitch
грависфера gravisphere
гравитахеометр gravitacheometer
гравитационный gravitation(al), gravity *(attr.)*
гравитация gravitation
~, земная gravity, terrestrial gravitation
гравитирующий gravitating
гравитоплан *av.* gravity plane
гравюра 1. engraving, etching; plate; 2. (colour) print
~ на дереве wood block/cut, woodcut
~ на линолеуме linocut
~ на меди copperplate engraving
~ на стали steel engraving
~, плоская lithograph
~, цветная colo(u)red print/engraving
~ штампа die impression
град I hail; *fig.* volley, torrent, shower
~ камней volley of stones
~, рыхлый soft hail
~ пуль hail of bullets
град II *math.* grad(e)
градация gradation, grading; range, scale; graduation, shading
~ изображения image gradation
~ квантования quantizing step/level
~, нисходящая anticlimax
~, промежуточная intervening gradation
~ серого *phot.* shade/shading of grey
~ тонов range of tones, tonal range, shading
градиент gradient; degree
~, адиабатический adiabatic lapse rate
~ атмосферного давления baric gradient
~, барический/барометрический baric/barometric/pressure gradient
~, большой heavy/steep gradient
~, вертикальный lapse (rate), vertical gradient
~ ветра wind shear
~, влажноадиабатический saturated adiabatic lapse rate
~ влажности humidity/moisture gradient
~ водного зеркала gradient of water table
~ всплывания flotation gradient
~, геотермический geothermal gradient/depth/degree
~, гидравлический hydraulic (pressure) gradient
~ давления (меридиональный) (north-south) pressure/baric gradient
~ давления, положительный adverse pressure gradient
~, желудочковый *med.* ventricular gradient
~, знакопеременный alternating gradient
~ магнитного поля magnetic gradient
~, малый flat/low gradient
~ набора (высоты) climb gradient
~, напорный pressure gradient
~, напряжения tension gradient
~, нормальный широтный latitude gradient
~, нулевой zero gradient
~, обратный reciprocal gradient
~, первоначальный средний initial mean gradient
~ плотности density gradient
~ подъемной силы lift(-force) gradient
~, поперечный transverse gradient
~ порыва gust gradient
~ потенциала potential gradient
~, предсердно-желудочковый *med.* atrio-ventricular gradient
~ признаков *biol.* cline
~, продольный longitudinal gradient
~ рассеяния dispersion gradient
~ силы тяжести gravity gradient
~ скорости velocity gradient, lapse (rate)
~ скорости по размаху *av.* spanwise velocity gradient
~ снижения descent gradient
~, соединенный joint gradient
~, сухоадиабатический dry adiabatic lapse rate
~ температуры/температурный temperature/thermal gradient
~ температуры, вертикальный temperature lapse rate
~ температуры, меридиональный thermal slope
~ температуры, отрицательный вертикальный negative temperature lapse rate, temperature inversion
~, тепловой/термический thermal gradient, temperature depth gradient
~ усилий force gradient
~ усилия по перегрузке stick force per **g**
~, чередующийся alternating gradient
~ электрического поля electric-field gradient
градиент-зонд gradient-sonde
градиентный gradient, degree *(attr.)*
градиентометр, градиометр gradiometer
~, магнитный magnetic gradiometer, flux-gate magnetometer
градир(ня) 1. cooling tower, tower-cooler; 2. salt pan
~, атмосферная/открытая open-type cooling tower
~, башенная chimney-type cooling tower
~, брызгальная spray-type cooling tower

~, вентиляторная mechanical-draft cooling tower
~, капельная drip-type cooling tower, atmospheric cooling tower
~, пленочная film-type cooling tower
~, поперечно-точная crossflow cooling tower
~, противоточная counterflow cooling tower
~, секционная cell-type cooling tower
градирование, градировка evaporation of salt water
градобитие, градобой hail damage
градоначальник *hist.* town governor
градоначальство *hist.* 1. town, borough; 2. town governor's office
градоносный *meteor.* hail-bearing/-laden
градостроитель city/town/urban planner
градостроительный city-/town-/planning
градостроительство city/town planning, urban/town development
градуирование, градуировка graduation, graduating, calibration; rating, division
градуированный graduated, calibrated
градуировать graduate, gradate; calibrate; divide
~ шкалу calibrate a scale/dial
градуировка calibration
градус degree, grade
~ Боме Baume degree
~ долготы degree of longitude
~ дуги degree of arc
~ жесткости degree of (water) hardness
~ Кельвина Kelvin degree, **K**
~ по стоградусной шкале, ~ стоградусной шкалы centigrade/Celsius degree, degree centigrade
~ Реомюра Reaumur degree
~ угловой degree of circle
~ Фаренгейта Farenheit degree
~ Цельсия centigrade/Celsius degree, **C**
~ широты degree of latitude
~ Энглера Engler degree
градусный grade, degree *(attr.)*
~ная сетка *geogr.* grid
градусо-день *meteor.* degree-day, day degree
гражданин citizen, national
~ воюющей/невоюющей belligerent/nonbelligerent national
гражданка *hist.* Russian Civil War (1917-1921)
гражданский 1. civil; civilian; 2. civic
~ брак civil marriage; *law* common/free union, de facto union
~ долг civic duty
~ иск civil suit
~ кодекс civil code
~кая азбука/печать *hist.* Russian type
~кая война civil war
~кая панихида civil funeral rite
~кая смерть deprivation of civil rights
~кое платье civilian clothes
~кое право civil law
гражданственность 1. civil society; 2. civic spirit, civism
гражданство citizenship, nationality; *coll.* citizenry
~, двойное dual nationality
грамзапись disk/phonograph record(ing), sound(-on-disk) recording
граминология *bot.* graminology, agrostology
грамицидин gramicidin
грамматик, грамматист 1. grammarian; 2. *hist.* seminarian
грамматика grammar; grammar book
~, бесконтекстная context-free grammar
~, доминационная dominance grammar
~, исходная source grammar
~, контекстная context-sensitive grammar
~, контекстно-свободная context-free grammar
~, неоднозначная ambiguous grammar
~ непосредственных составляющих phrase-structure grammar
~ общего вида unrestricted grammar
~ ограниченного контекста bounded context grammar
~, однозначная unambiguous grammar
~, основная skeletal grammar
~, пополненная augmented grammar
~, порождающая generative grammar
~ составляющих phrase-structure grammar
~, трансформационная transformational grammar
грамм-атом *chem.* gram(me)-atom
граммема grammeme
грамм-ион gram-ion
грамм-калория gram calorie, small calorie
грамм-масса gram weight
грамм-молекула grammolecule, gram(me)-molecule, mole, mol
грамм-моль gram molecule, grammol
граммометр gram gauge
граммофон record player, gramophone, phonograph
граммофонный gramophone, record player *(attr.)*
~ная пластинка (phonograph) record
грамм-рад gram-rad
грамм-рентген gram-roentgen
грамм-сантиметр *phys.* gram-centimeter
грамм-сила gram-force
грамм-эквивалент *chem.* gram-equivalent
грамота 1. official document; missive letter; muniment; deed; 2. reading and writing
~, жалованная letters patent; charter
~, отзывная *dipl.* letter of recall
~, охранная charter of immunity
~, похвальная (school) certificate of good work and conduct
~, почетная diploma
~ы, верительные *dipl.* letters credential, diplomatic credentials
~ы, вручать свои верительные present one's credentials

~ы, ратификационные instruments of ratification
грамотность 1. literacy; 2. grammatical correctness; 3. competence
~, компьютерная computer literacy
грамотный 1. literate; 2. grammatically correct; 3. competent
~, политически politically aware/literate
грамотрицательный *med.* Gram-negative
грампластинка (phonograph) record
~, долгоиграющая long-play record, LP record; extended-play record, EP record
грамположительный *med.* Gram-positive
гран grain *(unit of weight)*
гранат I *bot.* pomegranate (tree) *(Punica granatum)*
гранат II *miner.* garnet
~, железоглиноземный iron-aluminous garnet
~, известковый carbuncle, red garnet
~, хромовый/черный chrome/black garnet
граната *mil.* grenade, shell
~, бронебойная armour-piercing grenade
~, осколочная fragmentation grenade
~, ручная hand grenade
гранатина *bot.* balausta
гранатник *bot.* pomegrenate (tree) *(Punica granatum)*
гранатный 1. pomegranate *(attr.)*; 2. garnet *(attr.)*; 3. grenade, shell *(attr.)*
гранатово-красный carmine, scarlet; punicaceous
гранатовый I pomegranate
гранатовый II garnet, garnetiferous
гранатомет *mil.* grenade launcher
гранатометчик *mil.* grenadier, grenade thrower
гранатсодержащий garnetiferous
гранатоэдр rhombododecahedron
гранд grandee *(Spanish nobleman)*
гранение cutting
~ алмазов diamond cutting
граненый cut, faceted; cut-glass
~ графин cut-glass decanter
~ стакан thick glass tumbler
~ное стекло cut glass
гранецентрованный, гранецентрический face-centered
гранильник (mezzotint) graver, graving instrument/tool
гранильный diamond-cutting, lapidary
гранильня lapidary workshop
~ алмазов diamond-cutting shop
гранильщик diamond cutter, lapidary
гранит granite
~, альпийский Alpine granite
~, аляскитовый alaskitic/alaskite granite
~, брекчиевидный brecciated granite
~, буквенный graphic granite, pegmatite, runite
~, выветрившийся disintegrated granite
~, гиперстеновый hypersthene granite, charnockite
~, двуслюдяной binary granite
~, еврейский jewstone
~, известковистый calcigranite
~, известково-щелочной calcalkali granite
~, натровый soda granite
~, пегматитовый pegmatophyre
~, первозданный primordial/primeval granite
~, письменный graphic granite, pegmatite, runite
~, плагиоклазовый plagioclase granite
~, полосатый banded granite
~, роговообманковый hornblende granite
~, сланцеватый schistose granite
~, узловатый knotted granite
~, шаровой spheroidal/orbicular granite
~, шлировой schlieren granite
гранитизация granitization
~, региональная regional granitization, anatexis
гранитит granitite
гранитный granitic, granite
гранитовидный granitoid
гранитогнейс *geol.* gneissose/gneissoid/gneissic granite
гранитоид granitoid
гранитоидный granitoid(al)
granитоль *text., leath.* nitrocellulose-treated cloth, granitol
гранит-порфир granite-porphyry
гранит-рапакиви rapakivi-granite
граница 1. border, frontier; boundary, end; borderline; *math.* bound, limit; 2. limitation, margin
~ байта byte boundary
~, верхняя снеговая upper snow line
~, вечного/постоянного снега perennial snow line
~ видов delimitation of species
~ вихревой зоны wake boundary
~, воздушная *av.* air boundary
~ возмущений Mach line
~, воображаемая imaginary/fictitious boundary
~ высотности *av.* critical altitude
~ газа и воды *geol.* gas-water surface
~ газа и нефти *geol.* gas-oil surface
~, государственная frontier
~ группы class interval
~ двойникования twin boundary
~ двойного слова *comp.* double word boundary
~ диффузии diffusion barrier
~ доверительная/достоверности *stat.* confidence limit/level
~ домена (magnetic) domain wall, domain boundary
~ допуска tolerance limit
~ древесной растительности tree limit, timberline
~ дымки haze boundary, limit of haze
~ жидкости fluid boundary
~ зерен/зерна grain boundary
~ знака *comp.* character boundary
~ зоны нечувствительности dead-zone boundary

~ зоны прибрежного плавания *nav.* inshore zone limit
~ зрительного ощущения limit of vision
~ зуммирования singing point
~ измерения массива *comp.* bound of an array
~ илов mud line
~ индекса *comp.* subscript bound
~ индекса, переменная adjustable dimension
~, интегральная alignment boundary
~ интервала *stat.* class boundary
~ интервала изменений переменной bound on a variable
~ кипения фракций cut point
~, клеточная cell boundary
~ колебательной устойчивости oscillation boundary
~ комптоновского поглощения Compton edge
~ конденсации на ионах ion limit
~ концевой полосы безопасности *av.* limit of stopway
~ K-полосы поглощения K-absorption edge
~ кристалла crystal boundary
~ лесов tree limit, timber line
~, литориновая Litorina limit
~ луча beam boundary
~ материковой отмели continental edge
~, межфазная phase boundary
~ наружной поверхности шва *met.* toe
~ невозмущенного потока free-stream boundary
~ нейтральной устойчивости neutral stability boundary
~, нижняя lower limit/boundary
~ облаков, верхняя cloud top/deck, top (edge) of the clouds
~ облаков, нижняя cloud base, ceiling of clouds
~ области *math.* domain boundary
~ облачности, фактическая actual cloud area edge
~ обледенения/оледенения glacial boundary, icing/glaciation limit
~, обоюдная mutual boundary
~, орографическая снеговая orographic snow line
~ отказоопасная *comp.* failure bound
~, отражающая reflecting boundary
~ ошибок error bound
~ пассата *nav.* trade-wind line
~ поглощения absorption edge
~ поглощения в кадмии cadmium cut-off
~ погрешности error bound
~ поиска *comp.* search limit
~ постоянного снега perennial snow line
~ постоянства характеристики isoperformance boundary
~ прилива tide line
~ прозрачности transmission cut-off
~ промерзания frost line
~ пропускания transmission cut-off
~, прямолинейная straight-line boundary
~ пучка beam boundary
~ радиоактивного заражения radioactive-contamination limit
~ раздела surface of separation/discontinuity interface, line of demarcation
~ раздела фаз phase boundary, interface
~ разрушения *chem.* interfacial failure
~ распространения льда ice limit
~ растворимости *met.* miscibility gap
~, резкая sharp cut-off
~, рэлеевская Rayleigh limit
~ сектора обстрела *mil.* limit of fire
~ сердечной тупости, ~ сердца *med.* border of the heart dullness, heart border
~, снеговая snow line
~ спутной струи *av.* wake boundary
~ срыва stall limit
~ струи jet boundary
~ судоходства line of navigation
~, теоретическая theoretical boundary
~ территориальных вод limit of the territorial waters
~ трехмильной полосы three-mile limit
~ турбулентности turbulence boundary
~ устойчивости stability boundary/limit/threshold
~, фазовая phase border
~ Ферми Fermi level/limit
~, фирновая névé line
~ фотопроводимости photoconduction limit
~ фотоэффекта, красная photoelectric threshold
~ эксплуатационных режимов operating limit
~ экстраполированная extrapolated boundary
~, электронно-дырочная p-n boundary
~ ядра и оболочки земли *geol.* core-mantle boundary
~ яркости brightness contrast border
~ы зоны регулирования control boundary
~ы массива, регулируемые *comp.* adjustable dimension bonds
~ы центровок centre-of-gravity limits
граничить border (on/upon), be contiguous (to/with), abut, adjoin
граничный boundary; limiting; marginal; interfacial
~ оператор *math.* face operator
гранка I *print.* 1. galley; 2. galley proof
гранка II faceting, cutting
гранкореал *print.* board rack
гранобластический *geol.* granoblastic, granulose
гранодиорит *geol.* granodiorite
граносферит *geol.* granosphaerite
гранула granule, (small) grain; pellet; particle, corpuscle
гранулезоклеточный granulocellular
гранулема *med.* granuloma, granulation growth
~, амебная amebic granuloma

~ Боровского Borovski's granuloma, granuloma endemicum, oriental boil
~, бруцеллезная brucellous granuloma
~, гигантоклеточная giant cell granuloma
~, грибовидная granuloma/mycosis fungoides
~, жировая lipoid granuloma
~, злокачественная malignant granuloma, lymphogranulomatosis, Hodgkin's disease
~, зубная dental granuloma
~, кольцевидная granuloma annulare
~, ревматическая rheumatic granuloma, Aschoff-Talalayev's granuloma
~, сыпнотифозная rickettsial/typhus granuloma
~, тифозная typhoid granuloma
гранулематоз *med.* granulomatosis
~, ксантоматозный lip(o)id granulomatosis
гранулирование granulation, graining, pelletizing
гранулированный papillate, tuberculate, granulated, granulous, granular, grained, grainy; pelletized, nodulized; pearly
гранулировать granulate, *met.* pelletize
гранулит *geol.* granulite
~, очковой ocellar granulite, augen-granulite
~, пятнистый spotted granulite
гранулометр granulometer
гранулометрический textural, granulometric, grain *(attr.)*
гранулометрия grading, grain sizing, gradation
~ заполнителей aggregate grading
~ по размеру size grading
гранулообразование pellet formation
гранулопоэз *med.* granulopoiesis
гранулоцит *med.* granulocyte
~ незрелый immature granulocyte
~, эозинофильный acidophilic granulocyte, eosinophil
гранулоцитопения granulocytopenia
гранулярный granulated, granular
гранулятор granulator, pelletizer
гранулятор-аммонизатор *agr.* granulator-ammoniator
грануляционный granular, granulating, granulation *(attr.)*
грануляция granulation, granulating; *met.* pelletizing, pelletization
~, мокрая/сухая (шлака) water/air-stream granulation
~ поверхностной скульптуры *geol.* surface granulation, raised boss
грань 1. face, side; facet, plane, edge; 2. border, wedge, brink; 3. *math.* period
~, верхняя точная *math.* least upper bound
~, доматическая dome plane
~ драгоценного камня bezel
~, единичная unit face
~ кирпича, нижняя bed of a brick
~ кристалла face of a crystal, crystal face
~ кристалла, скашивать bevel a face of a crystal
~ куба cube face
~ многоугольника face of a polyhedron
~, плоская plane face
~ плотины, напорная upstream face
~ плотины, низовая downstream face
~ подпорной стены, вертикальная vertical interior face
~, предельная termination
~ призмы, преобладающая prominent prism face
~ реактора reactor edge
~, треугольника triangular facet
граптолит *paleont.* graptolite *(Graptolithida)*
~, ветвистый dendroid graptolite
граптолитовый graptolitic
грат *tech.* flash, fin; burr, barb
~, снимать trim the fin
гратосниматель flash trimmer; burr-grinder
граус *zool.* grouse *(Lagopus scoticus)*
граф I count, earl
граф II *math.* graph
~ альгоритма flow chart
~ без циклов circuit-free graph
~, взвешенный weighted graph
~, вырожденный singular graph
~, двудольный bipartite graph
~, древовидный tree graph
~, информационный потоковый data-flow graph
~, конечный finite graph
~, направленный directed graph
~, неорентированный non-oriented/-directed graph
~ операций activity graph
~ описаний attributed graph
~, ориентированный oriented/directed graph
~ переходов transition graph
~, планарный/плоский flat/planar graph
~, помеченный labelled graph
~, реберный line graph
~ с петлями graph with loops
~, связный connected graph
~ сигнала signal-flow graph
~, синтаксический syntactic graph
~ состояний state graph
~, циклический cyclic graph
~ циклов loop graph
графа *ling.* graph
графа 1. column; 2. paragraph
графекон graphechon
графема *ling.* grapheme
график I 1. plot, chart, graph, graphic chart, diagram, table; 2. schedule, timetable
~, балансовый balance chart
~ баллистический ballistic chart
~ бухгалтерских операций bookkeeping schedule
~ в дважды логарифмическом масштабе log-log plot, double log (arithmic) chart
~ в логарифмических координатах по обеим осям double log (arithmic) chart, log-log

chart
~ в логарифмическом масштабе log plot, logarithmic graph
~ в полулогарифмическом масштабе semilog plot
~, винтовой *av.* propeller characteristic curve
~ времени timetable
~ временного ряда historigram
~ выходов на работу attendance chart
~ Ганта *econ.* Gantt/progress chart
~, градуировочный calibrated chart
~ движения запасов inventory schedule
~ движения (поездов) train schedule/diagram
~ движения поездов, диспечерский *rail.* dispatcher's train sheet
~ движения самолетов flight schedule
~, двусменный two-shift schedule
~, девиационный compass correction card
~ дежурства duty chart/table
~ длительности нагрузки *el.* load-duration curve
~, жесткий tight schedule
~ зависимости прибыли от объема производства *econ.* profit-volume chart
~ запуска программы *comp.* run schedule
~ затрат costing record chart, cost map
~ изодоз isodose chart
~, измерительный/расчетный measurement chart
~ исполнения работ work schedule, progress chart
~ испытаний test schedule
~, календарный schedule (chart), calendar progress chart
~, комплексный executive/master schedule
~ кривых равной прибыли *econ.* isoprof chart
~ крутостей *cart.* scale of slope
~, линейный line graph
~ набора высоты *av.* climb chart
~ нагрузки load curve
~, оптимальный optimal/optimum schedule
~, основной master schedule
~ остановок down-time schedule
~ отпусков leave schedule
~ ошибок diagram/graph of errors
~, периодический cycle chart
~, планово-контрольный progress chart
~, плотный tight schedule
~ погрузки schedule of loading
~ полета flight diagram/schedule
~ потребления consumption schedule/map
~ прибыльности profit graph
~ производства/производственный manufacturing/production schedule
~ пусков launch schedule
~ работ schedule of work, work schedule
~ работ, общий overall time scale
~ работы operating schedule, force-displacement diagram
~ разброса scatter chart
~ разгрузки unloading schedule, schedule of unloading
~, расчетный nomogram, conversion/alignment/measurement chart
~ ремонтных работ schedule of repairs
~ с логарифмическими координатами log(arithmic) chart
~, сетевой network (graph), control-flow chart
~ скороподъемности *av.* rate-of-climb chart/diagram
~ сменности shift timetable/schedule
~ состояния запасов stock chart
~ спроса demand schedule
~ стока, суммарный mass diagram of run-off
~ суммарной потребления электроэнергии integrated energy curve
~, тарировочный calibrated diagram/graph
~ текущего ремонта, ~ технического обслуживания maintenance schedule
~ температур(ы) temperature diagram
~ технологического цикла schedule of operating times
~, уплотненный tight schedule
~, учетно-плановый *econ.* Gantt/progress chart
~ Ферми-Кюри Fermi-Curie plot
~ функции plotted function
~ хода буровых работ drilling progress chart
~ хода выполнения работ progress chart
~, центровочный centre-of-gravity diagram
~ электропотребления *el.* load curve
~, энергетический energy diagram
график II graphic artist, black-and-white artist
графика 1. graphics, graphic art, drawing; 2. script, writing
~, буквенно-геометрическая *comp.* alphageometric graphics
~, буквенно-мозаичная *comp.* alphamosaic graphics
~, машинная computer graphics
~ монолитных тел solid graphics
~, растровая raster/scan graphics
~ с точечной адресации dot-addressed graphics
~, экранная on-screen graphics
график-квадрат chart-square
графин pitcher, decanter; carafe
графиня countess
графит 1. *miner.* graphite, black lead, plumbago; 2. pencil lead
~, волокнистый fibrous graphite
~, выделять reject/liberate graphite
~ высокой ориентации highly oriented graphite
~, глобулярный globular/spheroidal graphite
~, землистый black chalk
~, карандашный pencil lead
~, кристаллический crystalline/vein graphite
~, кусковой (artificial) bulk graphite
~, литой moulded graphite

~, пластинчатый/хлопьевидный flaked/flaky graphite
~, пропитанный impregnated graphite
~, распыленный powdered graphite
~, рассеянный disseminated graphite
~, реакторный reactor-grade graphite
~, связанный fixed graphite
~, скрытокристаллический cryptocrystalline graphite
~, сфероидальный/шаровидный globular/spheroidal graphite
~, точечный spotted graphite
~, чешуйчатый flaky graphite
графитизация graphitization, graphitizing
графитирование *print.* graphite coating, coating with graphite, graphitizing
графитировать graphitize, (coat with) graphite
графитный, графитовый graphitic
графитообразный graphitic, graphitoid
графитопласт graphite-reinforced plastics
графить draw lines, rule
~ в клетку check, checker
графический I graphic(al), diagrammatic, schematic
графический II *art* graphic
~кое искусство graphic art, graphics
графление ruling, drawing lines
графленый divided into columns, (vertically) ruled
графоаналитический graphical-analytic, analytic-tabular
графолог graphologist
графологический graphologic(al)
графология graphology
графоман graphomaniac
графопостроитель graph/curve/data plotter, plotting device
~, автоматический autoplotter
~, двухкоординатный X-Y (coordinate) plotter
~, многоперьевой multiple-pen plotter
~, перьевой pen-type plotter
~, печатающий printer-plotter
~, планшетный flatbed plotter
~, рулонный belt-bed plotter
~ с полистовой подачи (бумаги) sheet-fed plotter
~, цифровой-координатный digital point plotter
графорея *med.* graphorrhea
графостатика graphical statics, graphostatics
графостатистика graphical statistics
графство 1. county; 2. title of count
~ва, центральные the Midlands
графт-сополимер graft copolymer
Грация *myth.* Grace
грач *zool.* rook *(Corvus frugilegus)*
гребенка *av.* 1. rake, comb; 2. *tech.* rack; manifold, pipe header; *text.* comb; 3. hackle, hatchel
~ волокноотделителя *agr.* gin rib
~ волокноотделителя, семенная *agr.* seed grid
~, винторезная screw die
~, зуборезная rack-type gear
~ измерительных насадков *av.* rake system
~, колосовая *agr.* tail rake
~, лентонаправляющая *text.* sliver guide
~, раздаточная pipe-manifold valves
~, разравнивающая stripping comb
~, резьбовая thread chaser
~ хлопкоуборочной машины, съемная *agr.* doffer comb
гребенник *bot.* dog's tail (grass) *(Cynosurus)*
гребенной comb, rack; hackling; worsted
гребенчатоплодный *bot.* lophocarpous
гребенчатожаберные *n. pl. zool.* ctenobranchs *(Ctenobranchia)*
гребенчатый 1. comb-shaped, comblike; 2. *bot., zool.* cristate; pectiniform, pectinate
гребенщик *bot.* tamarisk *(Tamarix)*
гребенщиковые *n. pl. bot.* tamarisk family *(Tamaricaceae)*
гребень 1. *meteor.* ridge; 2. *text.* comb; card, hackle, hatchel; 3. *anat., bot., zool.* crest, crista, comb; 4. *tech.* flange; 5. (mountain) ridge, crest; 6. *arch.* rooftree, ridgepiece; 7. *agr.* ridge, restbalk
~, антиклинальный anticlinal ridge, crest line
~, аэродинамический aerodynamic vane, (stall/wing) fence
~, барический wedge, pressure ridge
~ водослива weir/spillway crest
~ волны crest/ridge of wave, wave crest
~ жилы *geol.* apex
~, задний слезный *anat.* posterior lacrimal crest
~, запирательный *anat.* obturator crest
~ затвора плотины crest of a weir shutter
~, зубовидный denticle
~, крестцовый *anat.* sacral crest
~, лобковый *anat.* pubic crest
~, лобный *anat.* frontal ridge/crest
~, льноочесывающий flax comb
~ моренного происхождения *geol.* coteau
~, мостовой crown of a pavement
~, наджелудочковый *anat.* supraventricular crest
~, небный *anat.* palatine crest
~, петуший 1. *bot.* common cock's comb *(Celosia cristata);* 2. yellow rattle *(Rhinantus crista-galli);* 3. *anat.* crista galli, cock's comb
~, петушинный cockscomb
~, подвздошный *anat.* iliac crest
~, предкилевой *av.* dorsal fin
~, раковинный *anat.* conchial crest
~, решетчатый *anat.* ethmoidal crest; agger nasi
~, сбросовый *geol.* fault/trail ridge
~ синклиналь trough line
~ складки apex of fold
~ уплотнения *therm.* seal strip
~ фланца flange ridge
~ холма hill crest
~, частый fine-tooth comb

~, шпунтового соединения tongue, feather piece
гребец oarsman, rower, *Am.* thwartman
~, баковый bow oarsman, bow (man)
~, загребной stroke/aftermost oarsman
гребешковый crested, cristate, pectineal
гребешок 1. comb; ridge, crest; 2. *zool.* scallop *(Pecten)*
~, ампульный *anat.* ampullary crest
~, зародышевый germ ridge
~ импульсов pulse comb
~, кожный *anat.* ridge of the skin
~ коллектора commutator riser/lug/neck
~ лобковой кости *anat.* pectineal line of the pubic
~, морской scallop *(Pecten grandis)*
~, носовой *anat.* nasal crest
~ окна улитки *anat.* crest of the cochlear window
~ петуший *bot.* common cockscomb *(Celosia cristata)*
~ подвздушной кости *anat.* crest of the ilium
гребля rowing, pulling
~, академическая boat racing
~, байдарочная canoe paddling
гребляк *zool.* boat bug, water boatman *(Corixa)*
гребневание *agr.* ridging
гребневидный, гребнеобразный cristiform; pectinate; comb-shaped; ridgelike
гребневик *bot.* dog's-tail (grass) *(Cynosurus)*
гребневики *pl. zool.* comb bearers, ctenophorans *(Ctenophora)*
гребневой 1. pectinal, pectinate; 2. *agr.* ridged, hilled
гребнедержатель *text.* comb holder
гребнежаберные *pl. zool. Ctenobranchia*
гребнекрылый *bot.* lophopterous
гребненосный comb-bearing, cristiferous, pectiniferous
гребнеобразователь *agr.* ridge former, ridger
гребнеперегородчатый *bot.* carinimurate
гребнерезательный comb-cutting
гребнеротные *n. pl. zool. Ctenostomata*
гребнеукладчик *agr.* ridge former/filler
гребнечесальная машина comber, hackling machine
гребнечесание *text.* combing, hackling
гребни *pl.* combs, cards, *etc. see* **гребен**
~ винограда grape stalks
~ параллельные плоскости завивания accessory skeleton, parachomata, chomata
~ радиального расположения *zool.* costules, costulae
~ спирального расположения *zool.* collumelar folds
гребник *bot.* dog's tail (grass) *(Cynosurus)*
гребнистый (high-)crested
гребной rowing *(attr.)*
~ спорт rowing
~ная лодка rowboat
~ное колесо paddle wheel
гребной винт (screw) propeller, propeller/propelling/water screw
~, бортовой side propeller
~ изменяемого шага adjustable (blade) propeller
~, кормовой stern propeller
~, левого вращения left-hand(ed) propeller
~, некавитирующий subcavitating propeller
~, носовой bow/forward propeller
~ постоянного шага constant-pitch propeller
~ правого вращения right-hand(ed) propeller
~ с нечетным числом лопастей odd bladed propeller
~ с радиально-переменным шагом varying pitch propeller
~ с четным числом лопастей even bladed propeller
~ фиксированного шага fixed-blade/-pitch propeller
~, частично погруженный partly submerged propeller, vane wheel
гребок 1. stroke, paddle; 2. rake (arm), rabble (arm), raker; 3. blade
гревиллея *bot.* grevillea *(Grevillea)*
гривия *bot.* grewia *(Grewia)*
грегамит *geol.* grahamite
грегарина *zool.* gregarine *(Gregarina)*
грейдер road grader/scraper
~, прицепной pull-type grader
~, ремонтный patrol grader
~ с откосником slope grader
~ с профилирующим ножом blade grader
~, самоходный motor/power grader
~ скребковый scraping grader
грейдер-бульдозер bullgrader
грейдер-элеватор elevating grader
грейдерист road grader operator
грейдерный grader *(attr.)*
~ная дорога dirt road
грейзен *geol.* greisen
грейпфрут *bot.* grapefruit (tree) *(Citrus paradisi)*
грейфер grab (bucket), clamshell bucket, grapple; claw
~ для захвата лома scrap grapple
~, канатный rope grab bucket
~, многочелюстный multijaw/multiprong/cactus grab
~, одноканатный single-line clamshell bucket
~, самозахватывающий automatic grab
грек Greek
грекофил Grekophil, philhellenist
грелка hot water bottle/bag; heater
~, паровая *naut.* steam heating pipe
~, садовая *agr.* orchard heater
~, электрическая electric/heating pad
гремучекислый fulminic, fulminate
гремучий rattling; thundering
~ газ firedamp, detonating gas

~ **студень** blasting gelatine, nitrogelatine
~**чая змея** *zool.* rattlesnake *(Crotalus)*
~**чая ртуть** fulminate of mercury
гремучник *zool.* rattlesnake *(Crotalus)*
гренадер *mil.* grenadier
гренадерка *zool.* crested tit *(Parus cristatus)*
гренадерский grenadierial
~ **полк** Grenadier regiment; Grenadiers
гренадин I *text.* grenadine
гренадин II *bot.* grenadine carnation, clove pink *(Dianthus cariophyllus)*
гренаж *agr.* silkworm breeding
гренарня *agr.* silkworm nursery
гренирование *print.* backing
гренландец Greenlander
гренландка Greenlander (woman)
гренландский Greenlandic
гренок *cul.* toast, sippet; crouton
~ **с сыром** Welsh rabbit
греп *naut.* gripe, cutwater
грех sin; offence
~, **первородный** *rel.* the fall of man
~, **смертный** *rel.* mortal sin
греховность *eccl.* sinfulness; peccability, peccancy; wickedness
греховный sinful; peccable, peccant
грехопадение *bibl.* the Fall
гречиха *bot.* buckwheat *(Fagopyrum)*
гречихоцветные *n. pl. bot. Polygonales*
гречишник *bot.* knotweed *(Polygonum)*
гречишный buckwheat; *n. pl. bot.* buckwheat family *(Polygonaceae)*
гриб fungus, mushroom
~, **базидиальный** basidiomycele
~, **воздушно-водный** aeroquatic fungus
~, **высший** eumycete
~ **головного мозга** *med.* fungus of the brain, fungus cerebri
~, **морской** marine fungus
~, **несъедобный** toadstool
~, **плесневой** saprophytic fungus
~, **почвенный** geophilous fungus
~, **сумчатый** ascomycete
~, **съедобный** edible fungus/mushroom
гриб-дождевик *bot.* puffball *(Lycoperdon)*
гриб-зонтик *bot.* lepiota, parasol mushroom *(Lepiota)*
грибковидный fungilliform; fungoid
грибковый fungous, mycotic, fungoid(al), fungal, fungic
грибница *bot.* mushroom spawn, mycelium
грибной fungous, fungal, fungaceous; mushroom *(attr.)*
~ **мицелий** *bot.* floccus
~**ная нить** hypha
~**ная флора** microflora
грибовидный, грибообразный mushroom (-shaped), fungiform, fungoid, mycetoid
грибоводня *bot.* mycotheca
грибоеды *pl. zool.* hairy fungus beetles *(Mycetophagidae)*
грибок 1. *biol.* fungus (*pl.* -gi), microorganism, mo(u)ld; 2. *tech.* cone, head
~, **дрожжевой** yeast (fungus) *(Saccharomyces)*
~, **вентиляционный** mushroom ventilator
~ **клапана** valve head
~, **нитевидный** filamentous fungus
~, **плесневой** mould fungus
~ **распыливания** atomizing/atomizer cone
грибокорень *bot.* mycorrhiza
грибоподобный mushroomlike, funguslike, fungaceous
грибостойкий fungus-proof
грибостойкость fungus resistance, resistance to fungi
грибоустойчивый fungus-resistant
грива 1. mane; 2. (wooded) ridge, crest; 3. *bot.* juba; coma
гривка *mil.* battle sight
~ **с прорезью** battle sight with notch
гривна *hist.* ten kopecks
гривоголовчатый *bot.* chaetocephalous
григорианский Gregorian
~ **календарь** Gregorian Calendar
гридистор gridistor
гридлик *el.* grid leak
гридница, гридня *hist.* quarters of bodyguard; reception hall
гридь *hist.* prince's bodyguard
гризеофульвин *pharm.* griseofulvin, fulvicin, grifulvin
гризли *zool.* grizzly bear *(Ursus horribilis)*
гризон *zool.* grison *(Grison)*
гризутин granulated dynamite
грикваит *geol.* griquaite
гриквалендит *geol.* griqualandite
~, **окремненный** tiger's eye
грильяж *cul.* candied roasted nuts/almonds
грим *theatre* make-up
гример *theatre* make-up man
гримерша *theatre* make-up woman
гримирование, гримировка making-up, application of make-up
griналит
Гринвич Greenwich
гринда *zool.* blackfish, pilot whale *(Globicephala melaena)*
гринделия *bot.* gumweed *(Grindelia)*
гринокит *geol.* greenockite, cadmium blende
гриот *bot.* All Saints' cherry *(Prunus cerasus semperflorens)*
грипп influenza, flu, grippe
~, **азиатский** Asian flu/influenza
~, **кишечный** intestinal flu/influenza
гриппозный influenzal
~**ное воспаление легких** influenzal pneumonia
~**ное заболевание** case of flu/influenza
гриф I 1. *myth.* griffin, griffon; 2. *arch.* griffon; 3. *zool.* vulture
~, **ушастый** lappet-faced vulture *(Torgos*

tracheliotus)
~, черный black vulture *(Aegypius monachus)*
гриф II 1. *mus.* finger board *(of stringed instrument);* 2. handle, haft
гриф III (signature) stamp, facsimile signature; seal, stamp
~ секретности security classification
~а, без nonclassiffied
гриф IV *sport* grip
грифелевидный styloid
грифель pencil lead; slate pencil
грифельный slate
~ сланец *geol.* grapholite
грифовый vulturine; *n. pl. zool.* vulture family
грифон 1. *arch., myth.* gryphon, griffin; 2. *zool.* griffon *(dog)*
~, грязевой *geol.* mud gryphon
гробница burial vault, sepulchre, tomb, grave; *hist.* cist, cistavaen, kist
гробокопатель gravedigger
гроза thunderstorm
~, внутримассовя air-mass thunderstorm
~, горная orographic thunderstorm
~, затухающая decaying thunderstorm
~, предфронтальная prefrontal thunderstorm, squall-line thunderstorm
гроздевидный, гроздеобразный grapelike, clustered, cluster-shaped, bunchlike, racemose
~ное тело *bot.* cystolith
гроздевой bunchy, clustery, clustered
~вая листовертка *zool.* European grape moth *(Polychrosis botrana)*
гроздовник *bot.* grape fern *(Botrychium)*
гроздь cluster, bunch; fascicle; fascicule, raceme
~ винограда bunch of grapes
~, виноградная *bot.* uva
грозовой, грозовый thunderous
~ разряд lightning discharge
~ удар lightning stroke
грозозащита lightning(-discharge) protection/ protector
грозозащитный lightning-protective
грозозащищенность lightning proofness
грозоразрядник lightning arrester/protector
грозорегистратор lightning recorder, sferics receiver
грозостойкий, грозоупорный lightningproof, surgeproof
грозоупорность lightning-surge proofness
гром thunder
громкоговоритель loudspeaker
~, выносной extension loudspeaker
~, динамический moving-coil loudspeaker, (electro)dynamic loudspeaker
~ для верхних частот tweeter
~ для низких частот woofer
~, конденсаторный capacitor/electrostatic/ condenser loudspeaker
~, контрольный monitor(ing) loudspeaker
~, пьезоэлектрический crystal loudspeaker
~, рупорный horn loudspeaker
~, электродинамический moving-coil loudspeaker, electrodynamic loudspeaker
~, электромагнитный moving-conductor loudspeaker
~, электростатический capacitor/electrostatic/ condenser loudspeaker
громкость loudness; volume (of sound); vociferance
громовержец *myth.* the thunderer; thunderbearer
громовик *bot.* onosma *(Onosma)*
громовой thunder *(attr.)*
~ отвод lightning conductor/arrester
громоздкий bulky
громоздкость *math.* awkwardness; bulkiness
громоотвод lightning arrester/conductor/rod
~, гребенчатый comb lightning arrester
громбух, гроссбух legder, account-book
гросмейстер, гроссмейстер 1. grand master *(at chess);* 2. *hist.* Grand Master
гросс gross, twelve dozen
грот I *naut.* mainsail
грот II grotto, cave
грота-брас main brace
грота-галс main tack
грота-гафель main gaff
грота-гик main-boom
грота-рей main yard
грота-стаксель mainstay yard
грота-шкот mainsheet
грота-штаг mainstay
гротеск 1. *print.* block type, nonserif; 2. *art* grotesque
грот-люк main hatch
грот-марс maintop
грот-марсель main topsail
грот-мачта mainmast
~, вторая main mizzenmast
грот-рей main yard
грохот (sizing) screen, grizzly, sifter, sieve
~, барабанный cylinder/revolving/rotary screen, trommel (screen)
~, валковый roller grizzly
~ грубого грохочения coarse screen
~, двухситный double-deck screen
~, дисковый revolving/disk grizzly
~, инерционный unbalanced-throw screen
~ забывать/замазывать clog a screen
~ картофелеуборочного комбайна potato separating grid
~, колосниковый fixed-bar grizzly, bar screen
~ кукурузной молотилки cob rack
~, ленточный belt screen
~, обдирочный scalping screen
~, обезвоживающий washing screen
~, однорядный single-desk screen
~, ополаскивающий rinsing screen
~, питающий feeder screen
~ поворотного грохочения rescreener
~, рудоразборный ore-sorting grizzly

~, ударно-кулачковый impact-cam grizzly
~, шламовый slurry screen
~, щелевой wedge wire screen
~, эксцентриковый shaker, shaking screen
~, электромагнитный electromagnetically vibrated screen
грохотать screen, sieve, sift, bolt, riddle
~ в барабанном грохоте trommel
грохочение screening, (screen) sizing; sieving, sifting, riddling, bolting
~ в барабанном грохоте trommeling
~, грубое/крупное coarse screening
~, мелкое fine screening
~, обезвоживающее dewatering screening
~, повторное rescreening
~, последовательное step sizing
~, предварительное scalping
грош 1. *hist.* two-kopeck piece (1675-1838); half-kopeck piece (1838-1917); 2. grosz *(Polish unit of currency)*
груббер grubber, grub hoe
груббер-культиватор grubber
грубоволокнистый coarse-fibred
грубоволосистый *bot.* rough-haired, shaggy
грубогумусовый coarse-humus, raw-organic
грубозернистый coarse(ly)-grained, coarse
грубоигольчатый macroacicular
грубомозаичный *geol.* granoblastic
грубообломочный *geol.* rudaceous, psephitic, coarsely fragmentated
грубообработанный roughworked
грубопористый macroporous, coarse-pored
груборебристый coarse-ribbed
грубость rudeness, roughness; callousness
~ти, допустить *sport* play rough
грубый 1. coarse, rough; crude, rude; 2. gross, flagrant; 3. preliminary
~бая ошибка gross/bad mistake
~бая работа crude workmanship
~бая сила brute force
~бая ткань coarse fabric
~бое нарушение договора *law* flagrant violation of a treaty
~бому подсчету, по as a rough/preliminary estimate
груда pile, cluster, heap, mass; stack
грудина 1. *anat.* breastbone, sternum; 2. breast
грудинка I brisket, breast *(of veal, etc.)*
грудинка II *bot.* sida *(Sida)*
грудино-ключично-сосковый *anat.* sternocleidomastoid
грудино-ключичный *anat.* sternoclavicular, sternocleidal
грудино-позвоночный *anat.* sternovertebral, vertebrosternal
грудино-реберный *anat.* sternocostal
грудино-трахеальный *anat.* sternotracheal
грудная клетка *anat.* chest, thorax; breast
~, воронкообразная funnel chest/breast, foveated chest/breast
~, килевидная keeled chest, pigeon/chicken breast
~, кифотическая kyphotic/cyphotic chest
~, паралитическая phthinoid/alar/pterygoid chest
~, узкая narrow chest, stenothorax
грудника *bot.* flowering maple, abutilon *(Abutilon)*
грудница I *text.* breast beam
грудница II *med.* mastitis, mammitis
грудница III *bot.* goldilocks *(Linosyris)*
грудной 1. thoracic, thoracal; pectoral, chest; breast *(attr.)* 2. *n.* baby, infant in arms
~ позвонок thoracic vertebra
~ ребенок baby, infant in arms
~ная жаба *med.* angina pectoris
~ная железа *anat.* mammary gland
~ная клетка *anat.* thorax, chest; breast
~ные мышцы *anat.* pectoral muscles
грудобрюшный *anat.* abdominothoracic, thoracoabdominal
~ная преграда diaphragm
грудь *anat.* 1. breast, chest; 2. mammary gland; bosom, bust; 3. shirt front
~, впалая hollow/sunken breast
~ насекомых *zool.* sternum
~ членистоногих *zool.* thorax
груз 1. *naut.* cargo; *rail., av.* freight; 2. weight, load; charge, burden; 3. *econ.* goods; commodity
~, авиационный air freight/cargo
~, акушерский *med.* obstetric weight
~, балансировочный (mass-)balance weight
~ большой скорости express/fast goods/cargo/freight, fast goods traffic
~, брать take in cargo/freight
~ в трюме, распределять trim the hold
~, выброшенный за борт *naut.* jettison
~, габаритный cargo/freight within loading gauge
~, громоздкий bulk freight/goods
~ для обратных рейсов backhaul cargo/freight
~, засланный astray freight
~, импортный inward cargo
~, инерционный bob-weight
~, каротажный *geol.* coring load
~, киповый bale cargo
~, комплектовать make up cargo
~, контейнерный container(ized) cargo/freight
~, крепить lash/bind/brace/fasten cargo
~, крупногабаритный bulk freight
~ малой скорости slow cargo/freight
~, маятниковый pendulous weight
~ навалом/навалочный/насыпью/насыпной bulk freight/cargo, load in bulk
~, наливной bulk cargo
~, негабаритный off-gauge load
~, неучтенный overfreight
~, обратный homeward cargo
~, объемный measurement cargo/goods
~, опасный hazardous freight, dangerous cargo
~, оплачиваемый pay load

~, охлажденный refrigerated/frozen cargo
~, пакетизированный palletized cargo/freight
~, палубный deck cargo/load
~ первой очереди essential cargo/freight
~, перевозимый авиатранспортом air freight
~, перевозимый автотранспортом motor freight
~, перевозимый в контейнерах containerized freight
~, платный payload
~, повагонный carload
~, подвесной store
~, полезный payload, useful/effective load
~, скоропортящийся perishable cargo/freight/goods/load
~, сосредоточенный concentrated/point/fixed/single load/weight
~, срочный priority freight, high priority shipment, time cargo/freight
~, стойкий к порче nonperishable cargo/freight
~, штучный package freight
~, экспортный outward freight
~ов, движение freight traffic
груздь *bot.* milk fungus/mushroom *(Lactarius)*
груз-заполнитель *min.* filler-cargo
грузик weight; plummet
~, балансировочный correction weight
~ весов, уравновешивающий rider
~ отвеса (plumb) bob, plumb
~ распределителя distributor weight
грузило plummet, sinker
грузин Georgian
грузинка Georgian (woman)
грузинский Georgian
грузить freight, load; *naut.* get aboard, embark; ship
~ навалом load in bulk
грузовик lorry, *Am.* truck
~ для доставки покупок на дом delivery truck
грузовик-самосвал dump truck
грузовместимость (cubic) cargo-carrying capacity; tonnage, freightage, freight capacity
~ автомобиля truckload
~ в обмерных тоннах freight tonnage
~, зерновая grain capacity
~, киповая bale capacity
~ насыпью dry bulk capacity
~, чистая payload capacity
грузовой cargo *(attr.)*, goods *(attr.)*, freight *(attr.)*
~ пароход, ~вое судно freighter, cargo boat
~вая марка loadline/freeboard mark, loadline (disk)
~вая марка для лесных грузов lumber/timber loadline
~вая машина truck
~вое движение goods traffic; freight trains; shipping
грузозахват load/weight handling device/fixture; hauling device
~, канатный (lifting) sling(s)
~, клещевой lifting tongs
грузозахватный load-lifting
грузонапряженность (density of) freight traffic, density of goods
~ перевозок freight density
грузооборот freight turnover, turnover of goods, cargo performance, (freight) ton-miles
~ железнодорожного транспорта rail freight turnover
~ морского транспорта seaborne freight turnover
грузоотправитель consigner/consignor of goods, shipper, sender of freight
грузоотсек freight compartment/hold
грузопассажирский passenger-and-freight
~кое судно *naut.* combination vessel
грузоподъемник load lifter, freight elevator, goods lift
грузоподъемность load-carrying/-lifting capacity, weight-lifting ability, lifting power, load-/weight-/cargo-carrying capacity, *naut.* (weight-)carrying capacity, tonnage; lifting/hoisting capacity
~, валовая deadweight (tonnage), gross deadweight, deadweight(-carrying) capacity
~ парома ferry capacity
~, паспортная marked capacity
~, полезная/чистая payload capacity
~, полная carrying capacity; (gross) deadweight
~, приведенная adjusted load class *(of a bridge)*
~, удельная deadweight efficiency, deadweight (-displacement) ratio
~, указанная marked capacity
~, чистая cargo deadweight
~ шины capacity of tire/tyre
грузоподъемный load-/weight-lifting
грузополучатель consignee
грузопоток freight traffic (volume), goods/freight/cargo/commodity traffic
~ов, мощность volume of goods traffic
грузоспособность cargo capacity
грузохранилище warehouse, stores
грузошина truck tire/tyre
грузчик loader, stevedore, *Am.* longshoreman; freight handler; docker
~, грейферный *min.* grab loader
~, портовый longshoreman
грундбукса main/neck bush; journal/axle box
грунт 1. ground, earth, soil, terrain; undersoil; bottom; subfondation; 2. prime coat/colo(u)r, primer; priming, undercoat, first/ground coat
~, акриловый acrylic primer
~, антикоррозионный anti-corrosive/-rust primer
~, болот(н)истый marshy ground
~, влажный низменный *geol.* cripple
~, водоносный water-bearing cripple
~, вывозить вынутый dispose of the spoil

~, вязкий slimy ground, sticky bottom
~, глинисто-песчаный argillaceous sand ground
~, глинистый clay(ey) bottom/ground, argillaceous ground, clay soil
~, жидкий илистый oozy bottom
~, известняковый chalky/limy soil
~, илистый slimy/muddy/mud ground, muck/muddy bottom
~, каменистый rocky bottom
~, кремнийорганический silicone primer
~, лакокрасочный priming paint
~, лессовый loessial soil
~, ложиться на settle on bottom
~, мергельный loamy ground
~, многолетнемерзлый permafrost
~, на *naut.* aground, ashore
~, наносный shifting bottom, alluvial soil
~, насыпной backfilling, fill-up soil/earth
~, незащищенный open ground
~, нескальный soil, earth
~, нетронутый intact ground, virgin soil
~, перекидывать вынутый rehandle the spoil
~, подстилающий bedrock
~, полускальный half-rock
~, проницаемый pervious soil
~, размокший sodden/soggy soil
~, разрушенный broken ground
~, разрыхленный mellow ground
~, рыхлый soil, earth, loose ground
~ с грузозернистыми включениями gritty soil
~ с наносом гравия waste-gravel soil
~, садиться на *naut.* take the ground
~, свинцово-суричный red-lead primer
~, скальный/скалистый rocky ground, rock
~, скелетный gritty soil/ground
~, сыпучий loose/uncemented ground
~, твердый hard soil/ground/bottom
~, топкий marshy/swampy ground
~ хорошо держащий (якорь) *naut.* good (holding) ground, holding bottom

грунт-гравий gravel(ly) bottom/ground
грунт-коралл coral bottom/ground
грунткраска primer paint
грунтлак priming varnish, filler
грунтобетон sand-clay concrete
грунтовальный priming, coating
грунтование *art* precoating, priming; *tech.* application of priming coat; padding
грунтовать prime, apply priming coat *(of paint); art* ground, prime, precoat
грунтовед soil scientist
грунтоведение soil science, pedology
грунтовка 1. primer; prime/first/ground coat; 2. *art* priming, precoating, prime coating
~ книжного переплета *print.* top-sizing
~, травящая wash primer
грунтовой phreatic, ground, underground
~вая дорога dirt/earth road
~вые воды subsoil waters
грунтовочный *art* priming
~ная краска priming paint; primary colo(u)r
грунтозацеп grouser, lug, cleat
грунтонос sample taker, sampler, formation tester; core lifter/barrel, coring tool
~, заряжать load a sample taker
~, опускать run a sample taker
~, стреляющий gun-type sample taker, gun sampler, sampling gun
грунтопровод hydraulic fill pipeline
~ земснаряда floating pipeline
грунтосмеситель soil/ground mixer
грунтосмесительный ground-batching
грунтоцемент soil-cement
грунтроп *naut.* ground-rope
группа group; series; *stat.* cluster; pool; *geol.* set, congestion, complex
~, абелева *math.* Abelian/commutative group
~, абонентов *comp.* subscriber group
~, авианосная carrier group
~, авиационная air group
~, автоморфизмов *math.* automorphism group
~, арильная aryl residue
~, архейская *geol.* Archean group
~, атомная atomic group
~ атомов cluster of atoms
~ без кручения *math.* torsion, free group, group without torsion
~, бесконечная *math.* infinite group, group of infinite order
~ блоков памяти memory bank
~ быстрых нейтронов fast-neutron group
~, видовая species group
~, винтомоторная aircraft power plant, propulsion unit
~, внутрикомплексная chelate group
~, возрастная age group/bracket
~ волн wave train/group
~ вольнослушателей voluntary captive audience
~, вторичная *com.* supergroup
~, выдыхательная *phon.* breath group
~, вырабатывающая политику policy group
~, высадочная *mil.* beach party
~ Галуа *math.* Galois group
~ горных пород, родственная *geol.* kindred rocks
~ даек *geol.* swarm of dikes
~, десантная *mil.* landing party/team
~, диверсионная raiding force/party
~, дозиметрическая radiation hazard evaluation group
~ дорожек *comp.* track group
~, дробно-линейная *math.* linear-fractional group
~, замещающая *chem.* substituting group
~ знаков character stack
~, знакопеременная alternating group
~ информационных обектов *comp.* data

group
~ ионов ionic cluster, cluster of ions
~ каналов *comp.* channel bank
~ каналов связи, магистральная trunk group
~, канальная, вторичная *tel.* (60-)channel supergroup
~, канальная, первичная (12-)channel bank
~, канальная, третичная (300-)channel master group
~, карбонильная *chem.* carbonyl group
~ карт *comp.* batch of cards, card group
~, киносъемочная film production staff
~ клетей *met.* train
~ колодцев *geol.* battery/gang of wells
~ колонок (перфо)карты card field
~ команд *comp.* block of instructions, instruction block
~ командных слов command word chaining
~, коммутативная *math.* commutative/Abelian group
~, коммутативная локальная *math.* locally compact group
~, конечная *math.* finite group
~, контактная contact set
~, контактная на замыкание/размыкание make/break contact springs
~, корневая stem group
~, котельная *mil., naut.* stoker division
~ крейсеров cruiser force
~, крейсерская raider group
~ крови blood group
~ мин nest of mines
~, морская береговая naval shore party
~, мультиоператорная *math.* group with multiple operators
~, мультипликативная *math.* multiplicative group
~, накрывающая *math.* covering group
~, научная research/scientific staff/team
~, некоммутативная non-commutative group
~, непрерывная *math.* continuous group
~, нильпотентная *math.* nilpotent group
~, нуклонная nuclear sub-unit, nucleon(ic) group
~ обеспечения бездефектности *econ.* zero defect team
~, обследуемая survey sample
~, объединенная общими интересами *econ.* interest group
~, однопараметрическая *math.* one-parameter group
~, оперативная task group/unit
~ осадочных пластов *geol.* sedimentary complex
~, отрядно-классная *geol.* ordinal-class group
~, переходная transition group
~, периодическая *math.* torsion group
~ пластов *geol.* series/group of strata
~ подстановок *math.* substitution/permutation group
~ поколений данных group of data generations
~, пожарная fire-fighting crew
~, поисковая hunting group, search team
~ пользователей user group
~ по размерам дохода *econ.* income bracket(s)
~ прикрытия cover group
~, примарная *math.* primary group
~ продовольственного снабжения commissary crew
~, проектная design team
~, пространственная space group
~, профессиональная *econ.* occupational category
~, работающая над проектом project team
~, рабочая working party
~, разведывательная *mil.* scouting force
~, разрешимая *math.* solvable/integrable group
~ регламентных работ schedule/preventive maintenance group
~, редкоземельная rare-earth group
~, ремонтная repair(ing) group
~, родительская *comp.* parent group *(of data)*
~, родовая generic group
~, рулевая *naut.* ship control crew/party
~ с кручением *math.* torsion group
~ сбросов *geol.* group of faults, fault bundle
~ свай clump of piles
~ связи liaison team
~, сетчатая reticulated group
~, силовая *av.* power plant (group), propulsion unit
~ слившихся столбиков *geol.* bacularium
~ слов *comp.* block of words
~ слоев переходная *geol.* transition series
~ сопротивлений bank of resistors
~, социальная *stat.* social stratum, social status category
~, спасательная rescue group/team; salvage group/party
~ стержней rod bank
~ ступенчатого включения *tel.* grading group
~ сцепления linkage group
~ счетчиков bank of counters
~, таксономическая *biol.* phylum
~, тематическая lexical set
~ тепловыделяющих частиц *nucl.* cluster of fuel elements
~ технического обслуживания maintenance group
~, типовая typical/typal group
~ точек cluster of points
~, точечная point group
~ угля *geol.* coal rank
~ ударная strike group, striking/attack force
~, унитарная *math.* unitary group
~, уреидная *med.* ureido group
~, фантомная *tel.* phantom group
~, целевая task group
~, цериевая *chem.* cerium metals

~, циклическая cyclic(al) group
~ цифр digit group
~ частиц particle group; cluster of particles
~ шрифта *print.* family of type
~, щелочная *chem.* alkaline suite
~ электронов bunch of electrons
~, энергетическая energy group
~, эскортная escort group
~пы, товарные (в ассортименте) merchandise lines
группетто *mus.* gruppetto, turn
группирование grouping; classification; batching; concentrating; *nucl.* clustering; bunching
~, идеальное ideal bunching
~, недостаточное *nucl.* underbunching
~ пучка *nucl.* beam stacking
~, сложное supermultiplication
~ четверок кабеля squaring
~ электронов electron bunching
~ электронов, чрезмерное overbunching
группирователь *nucl.* buncher, bunching system
~, каскадный cascade buncher
группировать group; classify
группировка group(ing), classification; alignment
~ данных data aggregate
~ классов *math.* pooling of classes
~, множественная *stat.* manifold classification
~(населения) по полу *stat.* sex distribution
~(населения) по полу и возрасту sex-age distribution
~, скобочная *math.* parenthetic grouping
~ строк *print.* typographic design
~ строк, блочная *print.* boxing of type lines
~, фазовая phase-focusing action
~, ударная *mil.* striking forces
~ членов *math.* grouping of terms
~ки, военные military groupings/alignments
~ки, политические political groups
группком *abbr.* **групповой комитет** group committee
групповой group *(attr.)*
~ полет formation flying
~ снимок group photograph
~вые занятия group study
~вые интересы cliquish interest
групповщина clique formation, cliquishness, clannishness
группоид *math.* groupoid
группорг *abbr.* **групповой организатор** group organizer
груша 1. *bot.* pear *(Pyrus);* 2. pyridion; 3. pear-shaped object
~, аллигаторова *bot.* American avocado *(Persea americana)*
~, земляная *bot.* Jerusalem artichoke *(Helianthus tuberosus)*
~, конвертерная (converter) vessel
грушанка *bot.* pyrola, shinleaf *(Pyrola)*
грушанковые *n. pl. bot.* pyrola/shinleaf family *(Pyrolaceae)*
грушевидный, грушеобразный pear-shaped, pyriform, piriform
грушевина pear tree
грушевый pear *(attr.)*
~ компот *cul.* stewed pears
грушина pear tree
грыжа *med.* hernia; rupture
~, бедренная femoral/crural hernia, femorocele, merocele
~ белой линии живота hernia of the linea alba, incisional hernia
~, брыжеечная mesenteric hernia
~ Вельпо Velpean's hernia
~, влагалищная vaginal hernia, colpocele
~, врожденная congenital hernia
~ головного мозга cerebral hernia, encephalocele
~ желудка gastrocele
~ живота abdominal hernia
~, жировая fatty hernia, hernia adiposa, adipocele
~ легкого pleurocele
~ матки metrocolpocele
~ межпозвоночного диска Schmorl's body/module, discal hernia
~ меккелева дивертикула diverticular/Littre's hernia
~ мозга и мозговых оболочек meningoencephalocele
~, мозговая cerebral hernia, encephalocele
~, мозговая водяночная hydroencephalocele
~ мочеточника ureterocele
~, мошоночная scrotal hernia, oscheocele
~, наружная external hernia
~, невправимая irreducible hernia
~, овариальная ovarian hernia, ovariocele
~, паховая inguinal hernia
~, пахово-бедренная inguinofemoral/inguinocrural hernia
~, плевральная pleurocele
~, позадислепокишечная retrocecal/Rieux's hernia
~, полная complete hernia
~ половой губы labial hernia, cremnocele
~, предбрюшинная паховая preperitoneal inguinal hernia, hernia en bissac
~, пристеночная кишечная parietal/Richter's hernia
~ прямой кишки rectal hernia, rectocele
~, пупочная umbilical hernia, omphalocele, exomphalos
~ сальника omental hernia, epiplocele
~, седалищная ischiatic/sciatic hernia, ischiocele
~ семенного канатика funicular hernia
~, сердечная cardiocele
~, синовиальная synovial/Birkett's hernia
~, скользящая sliding/slipped/parasaccular/exctrasaccular hernia
~ спинного мозга myelocele
~, спинномозговая meningocele, myelomeningocele

~ **Трейца** Treitz's/retroperitoneal hernia
~, **ущемленная** incarcerated/strangulated hernia
~ **хрусталика** phacocele
~, **эпигастральная** epigastric hernia
грыжевой, грыжевый hernial, herniary
~ **бандаж** hernial bandage, truss
грыжесечение *med.* herniotomy, celotomy, kelotomy
~, **наружное** external/Petit's herniotomy
грыжник *bot.* rupturewort, burstwort *(Herniaria)*
~, **гладкий/голый** common burstwort *(H. glabra)*
грызун *zool.* rodent; *pl.* rodent family, rodents *(Rodentia)*
грэемит *geol.* grahamite
грэс *abbr.* regional hydroelectric power plant
грюндер company promoter
грюнерит *geol.* grunerite
~, **асбестовидный** amosite
гряда 1. *agr.* (seed)bed, row; 2. *geol.* ridge, range; 3. *meteor.* bank (of clouds); series
~, **барханная** row of sand dunes, row of barkhans
~, **бэровская** Baer ridge
~, **изоклинальная** hog back, isoclinal ridge
~ **камней** ledge, stone ridge
~, **песчаная** sand ridge/wave
~, **паровая** hotbed
~ **торосов** pressure ridge
~, **холмистая** range of hills
грядиль *agr.* plough/plow beam
~, **крючковый** curved plough/plow beam
~ **с предохранителем** safety plough/plow beam
~, **трубчатый** cylinder plough/plow beam
грядка 1. *agr.* seedbed, row; 2. pole
~, **парниковая/рассадная** hotbed
~ **стремянки** ladder step
грядковый seedbed *(attr.)*, row *(attr.)*
~**вая культура** cultivation in beds
грядовой *hort.* growing/grown in beds
грядкоподъемник *agr.* ridge digger
грядоделатель *agr.* bed former, bedder, ridger
грязевик mud setter/collector, sediment trap, sludge pan, sump, mud drum/box
грязевой mud(dy)
~ **насос** mud/slush pump
~**вая ванна** mud bath
грязелечебница mud baths
грязелечение mud cure/therapy/treatment, fangotherapy, pelotherapy
грязеотделитель slime/sludge separator
грязеотстойник (mud/sludge) sump, mud settler/box, mud/sludge pan, sediment/settling tank
грязеочиститель mud scraper
грязесборник sludge pan
грязетерапия *med.* mud therapy, fangotherapy
грязеуловитель dirt/mud/sludge trap/collector; dirt excluder/pocket
грязи, лечебные mud, fango
грязно-бурый fulinginous, soot-colo(u)red
грязновато-белый canescent
грязно-желтый icterious, lurid
грязный 1. muddy, mud-stained, miry; 2. mud-gray; 3. garbage, refuse *(attr.)*
~**ное ведро** garbage/refuse pail
грязовик *zool.* broad-billed sandpiper *(Limicola falcinellus)*
грязь 1. mud; sludge, slurry; mire; 2. dirt, filth; 3. *pl. med.* fango, mud; mud baths
~, **буровая** bore mud, drill-hole mud, sludge
~, **вулканическая** volcanic/volcano mud
~, **въевшаяся** grime
~, **снятая** scummings
~, **тонкая** silt, slurry
ГТО *abbr.* **Готов к труду и обороне** *obs.* Ready for labo(u)r and defence
гуава, гуайова, гуайява *bot.* guava *(Psidium guajava)*
гуажара *bot.* icaco plum *(Chrysobalanus icaco)*
гуаназа *bioch.* guanase
гуанако *zool.* guanaco *(Lama guanachus)*
гуанетидин-сульфат guanethidine sulfate, ismelin
гуанидилмочевина guanyl urea
гуанидин *chem.* guanidine, carbotriamine
гуанизохин guanisoquine
гуано guano
~, **пещерное** bat guano
гуанозиндифосфат *chem.* guanosine diphosphate
гуанозинмонофосфат *chem.* guanosine monophosphate
гуара *zool.* maned wolf *(Canis jubatus)*
гуарана *bot.* guarana paullinia *(Paullinia cupana)*
гуахаро, гуачаро *zool.* oilbird, guacharo *(Steatornis caripensis)*
гуашь water-soluble colo(u)r; *art* gouache (painting)
гваюла *bot.* guayule *(Parthenium argentatum)*
гуаяк *bot.* lignum vitae *(Guaiacum officinale)*
губа I 1. *anat.* lip, labium (*pl.* -bia), labrum (*pl.* -ra); 2. *bot.* labellum, lip; mesopetalum; gripping jaw; 3. *pl.* pincers
~, **большая/малая срамная** large/small pudendal lip
~ **венчика** *bot.* lip of crown
~, **вертлюжная** acetabular lip
~, **верхняя** upper lip; labrum
~, **заячья** *med.* harelip
~, **нижняя** lower lip; labium
~, **срамная** pudendal lip
~ **стельки** *leath.* insole lip/channel
~ **цветка** *bot.* dorsal sepal
губа II bay inlet, firth, gulf
~, **карманообразная** pocket beach
~, **прибрежная** bay-head beach
губа III tree fungus
губа IV *hist.* guba
губан *zool.* 1. (cigar) wrasse *(Cheilio inermis)*; 2. guban goby *(Neogobius platyrostris)*
губановые *n. pl. zool.* rainbow fishes *(Labridae)*
губастик *bot.* monkey flower *(Mimulus)*

губастый *bot.* labiate, labiose
губач *zool.* sloth bear *(Melurus ursinus)*
губернатор governor
губерния *hist.* province
губительный destructive, ruinous, disastrous; pernicious; cankerous
губка I *tech.* gripping jaw, bit
~, вставная false/insert jaw
~, зажимная gripping jaw
~ тисков vise/vice jaw
~ тисков, верхняя/нижняя top/bottom nipper
~ гаечного ключа wrench jaws
губка II 1. sponge; porous piece; cellular rubber; 2. *zool.* sponge; 3. *bot.* tree fungus
~, березовая *bot.* birch polyporous *(Polyporus betulinus)*
~, бесскелетная myxospongian
~, ветвистороговая dendroceratine sponge
~, гемостатическая *med.* hemostatic sponge
~, горная agaric mineral
~, диктиороговая dictyoceratine sponge
~, древесная fungus
~, желтая *zool.* yellow sponge *(Euspongia agaricina)*
~, известковая calcareous sponge, calcisponge
~, известково-роговая calcareocorneous sponge
~, каменная/литистидная lithistid
~, кремневая siliceous sponge, silicisponge
~, кремнероговая corneosiliceous sponge
~, латексная latex foam
~, лиственничная *bot.* purging/white agaric *(Polyporus officinalis)*
~, металлическая metallic sponge, sponge metal
~, растительная rag gourd *(Luffa cylindrica)*
~, резиновая rubber sponge
~, роговая horny sponge
~ с диктиональным скелетом dictionine
~ с известковой спикулой calcarea
~, сверлящая *zool.* boring sponge *(Cliona)*
~, стеклянная glass sponge
~ титана titanium sponge, sponge titanium
~, туалетная bath sponge *(Euspongia officinalis)*
губком *abbr.* **губернский комитет** *hist.* provincial committe
губкообразный spongiform
губно-зубной *phon.* labiodental
губной 1. lip *(attr.)*; 2. *phon.* labial
~ согласный labial (consonant)
губно-небный *phon.* labiopalatal; *anat.* labiopalatine
губно-носовой *phon.* labionasal
губовидный, губообразный lip-shaped, labelloid; labiate
губоногие *n. pl. zool.* chilopods *(Chilopoda)*
губо-подбородочный *anat.* labiomental
губоротый *zool.* cheilostomatous, *n. pl.* cheilostomes *(Cheilostomata)*
губоцветковый *bot.* lip-flowered
губоцветный *bot.* labiate, lip-flowered, ringent; *n. pl.* mint family *(Labiatae)*
~ное растение labiate
губчатость sponginess
губчатый 1. spongiform, spongioid, spongeous, spongy, cellular, porous; 2. fungoid, fungus
~ каучук, ~тая резина foam rubber
~тая паренхима spongy/lacunose parenchyma
гувернантка governess
гувернер tutor
гугенот *hist.* Huguenot
гугенотский *hist.* Huguenot
гудайера *bot.* rattlesnake plantain *(Goodyera)*
гудение buzz(ing); hum(ming); hooting; honk
~ мотора drone of the engine
~ трала singing of the sweep
гудерманиан *math.* Gudermannian (function)
гудок I *auto.* horn; whistle; hooter, siren, hoot(ing); honk; toot
~, автомобильный automobile horn
~, сигнальный warning howler/hooter
~, электрический electric horn/hooter
гудок II *mus.* rebec(k)
гудрон flux, tar(mac), petroleum/residual asphalt, goudron, oil/petroleum/rod tar
~, асфальтированный asphalted oil tar
~, кислый acid tar/sludge
~, масляный/остаточный residual tar
гудронатор tar paver, road oiler, oil-tar spreader
гудронирование black topping, tarring, asphalting, bitumen spraying
гудронировать tar, asphalt
гужевой vehiculatory, cart, animal/horse drawn
~ транспорт animal-drawn transport, cartage
~вая дорога cartway
~вое движение horse-drawn traffic
гужон screw/stud/tap rivet, blunt bolt
гужрабсила *abbr.* **гужевая рабочая сила** tractive force
гуза *bot.* Levant cotton *(Gossypium herbaceum)*
гузка rear end, bottom part
~ стереотипа tang
гузнек *tech.* gooseneck
гузокорчеватель *agr.* cotton-plant puller
гуйаквилит *geol.* guyaquillite
гул drone, droning; boom(ing); hum(ming)
гуллит *geol.* hullite
гульден 1. *hist.* gulden *(coin)* 2. guilder
гулявник *bot.* sisymbrium *(Sisymbrium)*
~, аптечный/лекарственный hedge mustard *(S. officinale)*
гуляй(-город) *hist.* galapago, testudo
гуляш *cul.* goulash; stew
гумай *bot.* Aleppo/Johnson grass *(Sorghum halepense)*
гуманизм *hist.* the revival of learning; humanism
гуманист humanitarian, humanist
гуманистический humanitarian, humanistic(al)
гуманитарий humanist
гуманитарный 1. pertaining to the humanities; 2. humane

~**ные науки** humanities, liberal arts
гуманность humaneness, humanity
гумат *chem.* humate
гумбо *geol.* gumbo (clay)
гуменник 1. stackyard; threshing floor; 2. *zool.* bean goose *(Anser tabalis)*
гуменце *eccl.* tonsure
гумидность humidity
гумидостат humidifier, humidity controller
гумин humin; humic/humus coal
гуминизация huminification
гуминит *miner.* huminite
гуминификация huminification
гуминовый *chem.* humic
гумит *miner.* humite; humic coal
гумификация humification, ulmification
гумма *med.* gumma (*pl.* -as, -ata), gummy node
~**, кальцинозная** calcific gumma
~**, кожная туберкулезная** scrofuloderma gummosum
гумми gum elemi/mastic, resin, gum
гумми арабик gum arabic
гуммигут gamboge, gum ghatti(e)
гуммигутовый gambogian, gambogic
гуммилак shellac; gum-lac, lac wax
гуммиластик india rubber, eraser; gum elastic, rubber
гуммировальный gumming
гуммирование gumming, rubberizing, coating/lining with rubber
гуммировать gum, coat/line/cover with rubber, rubberize
гуммировка coating/lining with rubber, rubber coating
гумми-смола gummed tar
гуммоз *pathol.* gummosis, gumming
гуммозный *med.* gummatous; gummy
~ **пластырь** diachylon/litharge plaster
гумно *agr.* threshing floor; barn
гумолит liptobiolitic/humus coal
гуморализм *med.* humor(al)ism
гуморальный *biol.* humoral
гумулон *chem.* humulon(e)
гумус humus, soil ulmin
~**, лесной/морской** forest/marine humus
гумусность humusness, humus content
гумусный, гумусовый humic, humous
гумусообразование humus formation, ulmification
гунду *med.* goundou, henpuye
гунн *hist.* Hun
гуннера *bot.* gunnera *(Gunnera)*
гуньба *bot.* blue-white trigonella *(Trigonella coerulea)*
гура *bot.* sandbox tree *(Hura crepitans)*
гуран *zool.* roe *(Capreolus)*
гурд gourde *(monetary unit of Haiti)*
гурда *hist.* Gurda, old Caucasian sword
гурия houri
гурман gourmet, epicure
гурманство gourmandise
гурон Huron
гурт herd, flock; drove
~**, нагульный** fattening/fattened herd
гуртовщик herdsman, (cattle) drover
гуртом wholesale, in bulk
гуру gourunut, kola nut
гуссар *mil.* hussar; 2. *zool.* hussar monkey *(Erythocebius)*
гусек 1. *arch.* (moulding) ogee; cyma; cymatium; recta; 2. *tech.* gooseneck
~**, вентиляционный** goose/swan-neck ventilator
~ **у низа плотины** bucket of a dam
гусельник psaltery/gusla player; psaltery/gusla maker
гусеница 1. *zool.* caterpillar *(Palustra);* 2. *tech.* (caterpillar) track, crawler belt/track
~**, транспортная** crawler belt/track
гусеничный *zool., tech.* caterpillar
~ **трактор** caterpillar tractor
~**ная лента/цепь** caterpillar chain/track
гусиный *zool.* anserine, goose; *n. pl. Anserinae; Anseres*
~**ное перо** goose guil
~**ные потроха** giblets
гусит, гуситский *hist.* Hussite
гуситка *hist.* Hussite (woman)
гусли *pl. mus.* psaltery, gusla, gusli
гусляр plastery/gusla player
густера *zool.* silver/white bream *(Blicca bjoerkna)*
густеть thicken, get thicker/denser
густоветвистый *bot.* pycnocladous, densely branched
густоголовчатый *bot.* pycnocephalous
густозернистый close-/densely grained
густозубчатый *zool.* crebridentate
густоколосый *bot.* pycnostachous
густокрылый *bot.* pycnopterous, closely winged
густолиственный, густолистый *bot.* dasyphyllous, densely/tick-leaved, leafy
густоплодный *bot.* dasycarpous, pycnocarpous, closely fruited
густота 1. thickness, density; consistency; closeness; 2. depth, deepness, richness
~ **древостоя/лесонасаждения** *agr.* thickness of stand
~ **заселения** *econ.* residential density
~ **пассажиропотока** density of passenger traffic
~ **посева** *agr.* thickness of sowing, seed spacing
~ **сети опорных точек** control density
~ **травостоя** thickness of stand
~ **штриховки** *tech.* density of ruling
гусь *zool.* goose *(Anser)*
гусь-сухонос *zool.* Chinese/guinea goose *(Cygnopsis cygnoides)*
гусятница *bot.* prostrate knotweed *(Polygonum aviculare)*
гута *hist.* glass foundry/factory
гутта gutta
гуттаперча gutta-percha

гуттаперчевый gutta-percha *(attr.),* flexible, resilient; *n. pl. bot.* sapodilla family *(Sapotaceae)*
гуттация *bot.* guttation
гуттуральный *phon.* guttural
гуцмания *bot.* guzmannia *(Guzmannia)*
гуцул Huzul, Guzul
гуща 1. dregs, grounds, sediment, lees; 2. thicket
~, кофейная coffee grounds
~, пивная spent brewer's grains
гуэмал *zool.* guemal *(Hippocamelus)*
гэз *geol.* gaize, malm rock
ГЭС *abbr.* **Гидроэлектрическая станция** hydroelectric power plant
гюйс *naut.* (stem) jack, jack(-flag)
гюйсшток *naut.* jack staff
гюрза *zool.* blunt-nosed viper, kufi *(Vipera lebetina)*
гяур infidel, giaour

Д, д

давилка fulling board
~, виноградная wine-press
~ для сахарного тростника cane-crushing mill
давильник *met.* spinning tool
давильный пресс wine press
давильщик 1. presser; 2. *met.* spinner
давить press, crush; squeeze
давление pressure; thrust; compression; *fig.* stress
~ адиабатического торможения *aero.* adiabatic impact pressure
~, аксиальное axial/end thrust
~, атмосферное/барометрическое atmospheric/barometric pressure
~, баксовое *naut.* end-poppet pressure, pivoting pressure
~, боковое lateral/side/thrust pressure
~, брюшное abdominal pressure
~ в баке горючего fuel-tank pressure
~ в баллоне storage/bottle pressure
~ в воздушной системе pneumatic pressure
~ в гидросистеме hydraulic pressure
~ в пневматике/шине tyre/inflation pressure
~ в равновесной системе equilibrium pressure
~ в системе обратной связи feedback pressure
~, вакуумметрическое vacuum-gauge pressure
~, весовое equilibrium pressure
~ взрывной волны blast pressure
~, внешнее external/ambient pressure
~, внутреннее internal pressure
~, внутриглазное *med.* intraocular pressure
~, внутрипочечное *med.* renal pressure
~, внутрисердечное *med.* intracardiac pressure
~ воспламенения ignition pressure
~ впрыска топлива injection pressure
~ впуска inlet/intake pressure
~ всавывания suction pressure; inlet/intake pressure
~, высокое high/heavy pressure
~ выхлопа exhaust/outlet pressure
~ выходящих газов exhaust-gas pressure
~ выше атмосферного positive pressure
~, гидростатическое hydrostatic pressure/head; directionless pressure
~, горное rock pressure
~ грунта soil pressure
~, действующее/действительное virtual/effective pressure
~, диастолическое *med.* diastolic pressure
~, донное base pressure
~ жидкости fluid/liquid pressure
~, закритическое supercritical pressure
~, звуковое sound/acoustic pressure
~, избыточное 1. excessive/surplus pressure; overpressure; 2. differential pressure; pressure differential
~ излучения radiation pressure
~, индикаторное indicated pressure, pressure on diagram
~, инерционное inertia/mass pressure
~, инфляционное *econ.* inflationary pressure
~, испытательное test pressure
~ кавитации *naut.* cavitation pressure
~, капиллярное *med.* capillary pressure
~, касательное peripheral/circumferential pressure
~, когезионное cohesion/cohesive pressure
~, контактное contact pressure
~ кровли *min.* top/roof pressure
~, кровяное *med.* blood pressure
~, лобовое ram pressure
~, максимальное maximum/top/peak pressure
~, манометрическое gauge/manometer pressure
~, мгновенное instantaneous/dynamic pressure
~ на входе inlet/entrance/intake pressure
~ на выхлопе exhaust/outlet pressure
~ на выходе outlet/exit/discharge pressure
~ на выходе насоса pump-output pressure
~ на поверхности surface pressure
~ на уровне моря sea-level pressure
~ нагнетания discharge pressure
~ наддува boost/supercharge pressure
~, напорное delivery pressure
~ насыщения saturation pressure
~, недостаточное deficient pressure
~ ниже атмосферного subatmospheric/negative pressure
~, обратное back/return pressure
~, обрушающее *min., geol.* collapsing pressure
~, онкотическое *med.* oncotic pressure
~, опорное bearing/support pressure
~, ориентированное directed pressure
~, осевое axial/end thrust
~, остаточное residual pressure

~ отбора selection pressure; *therm.* extraction pressure
~, отрицательное negative pressure
~ пара steam/vapour pressure
~, парциальное partial pressure
~, пластовое *geol.* formational/rock pressure
~, поверхностное surface pressure
~, повышенное *med.* hypertension, increased pressure
~ подачи delivery/supply/feed pressure
~, пониженное *med.* hypotension, diminished pressure
~, поясное *constr.* circumferential pressure
~ прессования compacting pressure
~, пробное test pressure
~, пусковое starting pressure
~, рабочее operating/operational/working/effective pressure
~, равновесное equilibrium pressure
~, разрушающее collapsing/bursting pressure
~ разрушения collapse pressure
~ разрыва/разрывное bursting pressure
~, расчетное design pressure
~, реактивное reaction pressure
~ ртутного столба mercury pressure
~, сверхвысокое ultrahigh pressure
~, сверхкритическое supercritical pressure
~ сжатия compressive/compression pressure; compressive stress
~, систолическое *med.* systolic pressure
~, сминающее crumpling pressure
~ со стороны потребителей consumer pressure
~, сплющивающее collapsing pressure
~, стационарное steady-state pressure
~ сцепления 1. *phys.* cohesive pressure; 2. *auto* clutch pressure
~, тангенциальное circumferential pressure
~, упорное thrust
~, установившееся steady-state pressure
~ фильтрации percolation pressure
~, чрезмерно высокое over-pressure
~, эталонное reference pressure
давность *law* prescription
~, исковая limitation
~, погасительная extinctive prescription
~, приобретательная acquisitive prescription, usucapion
~и, право prescriptive right
дагерротипия *print.* daguerrotype
дайка *geol.* dike, dyke
~, аркообразная arcuate dike
~, кластогенная/обломочная clastic dike
~, кольцевая ring dike
~, питающая feeder dike
~, побочная satellite/associated dike; compensatory dike
~, подводящая feeding dike
~, секущая transverse dike
~, сложная composite/multiple/mixed dike
дакриоаденит *med.* dacryoadenitis, dacryadenitis
дакриолит *med.* dacryolith, dacriolite
дакриоцистальгия dacryocystalgia
дакриоцистит dacryocystitis
дакриоцистотомия *med.* dacryocystotomy
дактилальгия *med.* dactylalgia
дактилартрит *med.* dactylarthritis
дактилит *med.* dactylitis
~, сифилитический syphilitic dactylitis
~, туберкулезный tuberculous dactylitis
дактилограмма dactylogram, fingerprint
дактилодиния dactylodynia
дактилология dactylology, finger speech
дактиломегалия *med.* dactylomegaly
дактилосимфиз *med.* dactylosymphysis
дактилоскопия dactyloscopy, identification by means of fingerprints
дактиль *lit.* dactyl
далай-лама Dalai Lama
далматинец, далматинка, далматинский, далматский Dalmatian
дальнобойность range
дальнобойный long-range
дальнодействие long range, long-range action; remote control
дальнозоркий farsighted, longsighted, hypermetropic, presbyopic
дальнозоркость farsightedness, longsightedness, hypermetropia, hyperopia, presbytia, presbytism
~, старческая presbyopia
дальномер range-/distance-finder, distance gauge, distance-measuring equipment, diastimeter
~, акустический acoustic/sound range-finder, sound ranger
~, бинокулярный stereoscopic/binocular range-finder
~, вертикально-базисный depression range-finder
~, лазерный laser range-finder, ranging laser, laser ranger
~, монокулярный coincidence range-finder
~, облачный cloud telemeter
~, радиолокационный ranging radar; range-only radar
~, электронно-оптический, электрооптический geodimeter
дальнометрия distance measurement
~, лазерная laser ranging
~, радиолокационная radar ranging
дальность distance; range; radius
~ автоматического сопровождения *av.* autotracking range
~ активного самонаведения *av.* active homing guidance range
~ беспосадочного перелета non-stop range
~, боковая lateral range
~, большая long range
~ видения range of vision
~ видимого горизонта distance of the horizon, horizon distance

~ видимости visual/visibility/viewing range distance
~ видимости, ночная night-time visibility range
~, визуальная visual range
~ визуального обнаружения visual detection range
~ действия operational range; range of action; *rad.* coverage
~ действия маяка beacon range
~ действия радиомаяка radio range
~ действия радиосвязи radio coverage
~ действия системы обнаружения range of detection system
~ действия, эффективная useful range
~ крейсерского полета cruising range
~, малая short range
~ обнаружения acquisition/detection range
~ определенная РЛС radar range
~ перевозок length/distance of haul
~ плавания cruising range
~ плавания, надводная *naut.* surface range/radius
~ плавания, подводная submerged range/radius/endurance
~ поиска search range
~ полета flight/flying range
~ полета без заправки unrefuelled range
~ полета, перегоночная ferry range
~ полета с максимальной нагрузкой full-load range
~ поражения destruction range/radius
~, практическая operational/service range
~, предельная maximum/extreme/critical/ultimate range
~ привода homing range
~ приема сигналов signal range
~, прямая радиооптическая radio-optical range
~ прямой видимости line-of-sight range, visibility/visual range, range of visibility
~ радиолокационного обнаружения radar acquisition/detection range
~ радиолокационного сопровождения radar tracking range
~ слежения/сопровождения tracking range
~ слышимости audibility range
~ слышимости голоса hailing range
~ стрельбы *mil.* range of fire, firing range
~, текущая present range
дальноуправляемый remote-controlled
дальтон *phys.* dalton
дальтонизм *med.* colour blindness, daltonism
дальтоник *med.* achromate, coulour-blind
дамаск damascene, Damascus steel, damask
дамба embankment, dam, levee, barrage, dike, sea defence wall; jetty
~, естественная береговая natural levee
~, намывная alluvial dam
~, оградительная dike
дамп *comp.* dump
~, аварийный disaster dump
~, защитный rescue dump
~ измененных данных changed data dump
~, контрольный monitor control dump
~, мгновенный snapshot dump
~ оперативной памяти core/short memory dump; indicative dump
~, рабочий executive dump
~ содержимого основной памяти main memory dump, main store printout
~ содержимого экрана screen dump
данник *hist.* tributary
данничество *hist.* tributariness
данное datum
данные data; information; facts; (tabulated) readings
~, агрегированные aggregated data
~ азимутальные azimuth data
~, аналоговые analog(ue) data
~ без учета сезонных колебаний *econ.* deseasonalized data
~ большого объема mass data
~, буквенно-цифровые alphanumeric/alphameric data
~, буквенные alphabetic data
~ в двоичном коде (в двоичной форме) binary data
~ в реальном масштабе времени real-time data
~ в упакованном формате packed data
~ в форме кортежей tuple-structured data
~ в форме с плавающей/фиксированной точкой floating-/fixed-point data
~ в форме списка list-structured data
~ в форме файла filed data
~, введенные entry
~ ввода-вывода input-output data, i/o data
~, вводимые input data
~, возрастные age pattern
~, временные tentative data
~ вскрытия *med.* autopsy findings
~, вспомогательные auxiliary data; contributory evidence
~, входные input data; *econ.* input information
~, выборочные sampled data; *econ.* sample data
~, выдаваемые output data, read-out
~, выходные output data; output
~, двоичные binary data
~ динамических рядов *econ.* time-series data
~, дискретные discrete/sampled data
~ для служебного пользования *law* restricted data
~, достоверные valid data
~, доступные available data
~ задачи problem data; task data
~, запоминаемые stored data
~ измерений metric data
~ испытаний test/experimental data; test find-

ings
~, исходные initial/primary/basic data
~, клинические clinical findings/data/evidence
~, командные control data
~, лабораторные laboratory findings
~, машиночитаемые machine-readable data
~, начальные = исходные
~, недостоверные invalid data
~, незашифрованные clear data
~, ненужные garbage, gibberish
~, необработанные raw data
~, новейшие *econ.* up-to-date information
~ о внешних условиях environmental data
~ о местоположении locational data
~ о поставщике *econ.* source data
~ о состоянии *econ.* status data
~, обновляемые updatable data
~, обработанные processed data
~, общедоступные public data
~ ограниченного доступа classified data
~, оперативные on-line data
~ операции *med.* operative findings
~, опытные test/experimental/empirical data
~, основные basic/master/main data/specifications/information; main/principal particulars
~, оценочные estimated data; rating data
~, паспортные *tech.* nameplate data; rated values
~, первичные primary data
~, полевые field data
~ последовательной записи serial write data
~, посмертные *med., law* postmortem findings/evidence
~ поступающие в реальном времени real-time data
~ поступающие дистанционно remote data
~, предварительные tentative/preliminary data
~, прогнозированные *econ.* predicted/prediction data
~, проектные design data
~, противоречивые *comp.* incompatible data
~, протокольные *tech.* performance data
~, рабочие operational/working data
~, радиолокационные radar data
~, распакованные unpacked data
~, расчетные *econ.* estimated/calculated/calculation/computation/design/rated data
~, рентгенологические *med.* radiologic evidence
~, речевые speech data
~, сглаженные *econ.* smoothed data
~, секретные classified/restricted/confidential information/data
~, секционные *med.* autopsy findings
~, скалярные scalar features
~, справочные reference data
~, статистические statistics, statistical data
~, суммарные integrated/cumulative data; summarized data
~, табличные tabulated/tabular data
~, телеметрические telemetry/telemetered data
~, технические technical/engineering data, technical characteristics/specifications
~, упорядоченные *econ.* ranked data
~ файла file data
~, фактические actual data
~, хранимые stored data
~, цифровые digital/numerical data; figures
~, численные/числовые numerical data
~, эксплуатационные operational/service data
~ элемента изображения pixel data

дано *math.* given
дантист dentist
даосизм *rel.* Taoism
даосистический *rel.* Taoistic
дарвинизм Darwinism
дарвинист Darwinian, Darwinist
дарвинистский, дарвиновский Darwinian
дарение, прижизненное *law* absolute gift
дароносица, дарохранительница *eccl.* ciborium, pyx
дарси *miner.* darcy
дарственный акт *law* donation, grant
дата date
~ вступления в силу *law* effective date
~ выкупа *econ.* redemption date
~ выпуска *fin.* date of issue
~, заключительная closing date
~ закрытия книги *acc.* closing date
~ истечения срока expiration date
~, исходная *econ.* reference date
~ отправки date despatched/dispatched
~ погашения *econ.* redemption date; date of maturity
~ получения date received
~, фактическая *econ.* effective date

датирование, датировка dating; *miner.* age determination, dating
~ методом изотопных индикаторов isotopic dating
~ оплатой наложным платежом *econ.* c.o.d. dating
~ по радиоактивному углероду radiocarbon dating

датировать date; fix the date of
~ более поздним числом postdate
~ задним числом *law* antedate

датируемый datable
датский Danish
датчанин Dane
датчанка Danish woman/girl
датчик 1. sensor, detector; transducer, pickup; sensing device/element; 2. (data) transmitter
~ абсолютного давления absolute-pressure unit
~, автоматический automatic data unit; automatic transmitter
~ активного сопротивления potentiometric transducer
~, акустический acoustic pickup

~ астроориентации/астронавигационный star sensor, star-tracking sensor/pickup
~ барометрической высоты pressure-altitude transducer
~ бокового скольжения side-slip sensor
~, вибрационный vibration pickup
~ времени timer, clock
~ времени, контрольный watchdog timer
~ временных интервалов interval timer
~, время-импульсный cycle-repeat timer
~ высоты altitude data unit
~, газоразрядный gas-discharge transducer
~, генераторный self-generating transducer
~ гидролокатора sonar transducer
~ гироприбора gyro pick-off
~, гироскопический gyroscopic sensor/pickup
~ горизонта horizon sensor/scanner
~ давления pressure transducer/pickup/indicator/gauge
~ давления почвы soil-pressure cell/probe
~ дальности range sensor/keeper
~ деформации strain gauge
~, динамометрический force transducer
~, дистанционный remote sensor/pickup
~ дыхания *med.* respiration sensor
~, емкостный capacitance-type sensor/transducer, capacitive sensor/transducer
~ земной вертикали *av.* earth sensor
~, импульсный pulse transducer/transmitter
~ импульсов pulser, impulser
~, индукционный flux-gate sensor, variable-reluctance transducer/sensor/pickup
~, инерциальный inertial sensor
~, инерционный inertia sensor
~ команд orders transmitter
~, контактный sensing/sensor switch; impact sensor
~, контрольный monitor sensor
~ крена *naut.* roll sensor
~ кровяного давления blood-pressure sensor
~ крутящего момента torque sensor, torquemeter
~ курса heading sensor; course transmitter
~, магнитоэлектрический moving-coil transducer
~, малоинерционный fast-/high-response sensor
~, манометрический manometric transducer, filled-system transducer
~, матричный array sensor
~, мембранный diaphragm pickup/transducer
~ момента/моментный torque motor/generator
~, нейтронный neutron-sensitive element
~, омический resistive transducer
~ перемещения displacement transducer/pickup, movement pickup
~ перепада differential transducer
~ перепада давлений differential-pressure transducer, pressure-ratio transducer
~ пожарной сигнализации fire detector
~, позиционный position/attitude sensor
~ последовательности импульсов sequence timer
~ потока flow-sensing element
~, проточный flow-type transducer/pickup
~ прямого действия direct-acting transducer
~, пульсовый pulse sensor
~, путевой path-control transducer
~, пьезоэлектрический piezoelectric/crystal pickup/transducer
~, радиоактивный radiotracer
~, радиолокационно-акустический radar-audio sensor
~, сельсинный synchro/selsyn pickup
~ скольжения side-slip sensor
~, солнечный sun/solar sensor, sun-seeker
~, струнный vibrating wire sensor
~, тактильный touch sensor, tactile-type sensor
~ тангажа pitch sensor
~, телеметрический telemetering pickup
~, температурный temperature-sensitive/-sensing element
~, тензометрический strain-gauge transducer, wire-strain gauge
~ технической информации engineering sensor
~ угла рыскания *naut.* yaw-angle pick-off
~ уровня level detector/gauge
~ усилия force transducer/cell/gauge
~, цифровой digital transducer, digitizer
~ числа М. Mach transducer
~, электронно-лучевой electron-beam probe
Дафна *myth.* Daphne
дафнит *miner.* daphnite
дацит *miner.* dacite
дацито-липарит *miner.* dacite-liparite, dellenite
дача intake, input, feeding
~, суточная daily rate
двадцатигранник icosahedron
двадцатигранный icosahedral
двадцатиугольник icosagon
двадцатичетырехгранник icosatetrahedron
двенадцатигранник dodecahedron
двенадцатигранный dodecahedral
двенадцатиперстный *anat.* duodenal
двенадцатиричный duodecimal
двенадцатиугольник dodecagon
двенадцатиугольный dodecagonal
дверка, дверца (small) door; hatch
~ клапана drop shutter
~ люка access door
~, откидная flap/hinged door
~, раздвижная sliding door
~, смотровая inspection door
~, топочная furnace door
дверь door; gate
~, вращающаяся revolving door
~, выдвижная sliding door
~ гармошка, раздвижная concertina door

~, герметическая pressure-tight door, pressurized door
~, гладкая flush door
~, глухая blank door
~, двухпольная/двустворчатая double-wing [two-leaf, two-leafed, clamshell, double] door
~ для аварийного покидания escape door
~, жалюзийная louvred door
~, задняя *auto.* tail gate
~, запасная emergency door/exit
~, защитная safety door
~, клинкетная sliding door
~, навесная hinged door
~, однопольная single-wing/-leaf door
~ открывающаяся в обе стороны double-acting door
~ открывающаяся влево/вправо left-/righthand door
~, отпускная trap door
~, парадная street door; main entrance
~, пожарная fire exit/door
~, противопожарная fireproof door
~, раздвижная/сдвижная sliding door
~, складывающаяся accordion/folding door
~, шторная roller-blind door
дверь-навес canopy door
двигателестроение engine manufacturing
двигатель 1. *el.* motor; 2. *(internal combustion)* engine
~, авиационный aircraft engine, aeroengine
~, автомобильный car/automobile engine
~, асинхронный *el.* asynchronous motor, induction motor
~, атомный atomic engine
~, безредукторный direct-drive engine
~, бензиновый petrol engine, *Am.* gasoline engine
~, бескарбюраторный injection engine
~, бесклапанный valveless engine, sleeve-valve engine
~, бескомпрессорный 1. *auto.* airless injection engine, diesel engine; 2. *av.* compressorless/jet engine
~, бескрейцкопфный trunk-piston engine
~, бескривошипный axial engine
~, быстроходный high-speed engine
~ Ванкеля Wankel engine
~, вихрекамерный swirl-chamber engine
~ внутреннего сгорания internal-combustion engine
~ водяного охлаждения water-cooled engine
~ воздушного охлаждения air-cooled engine
~, воздушно-камерный air-chamber/-cell engine
~, воздушно-реактивный air-jet/-breathing chamber
~, воздушный air motor
~, всеядный multi-fuel engine
~, высокооборотный high-speed engine
~, высотный *av.* altitude engine
~, газогенераторный gas-generator engine, producer-gas engine
~, газотурбинный gas-turbine engine; *(in helicopter)* shaft turbine engine
~, герметический sealed motor
~, гидравлический hydraulic motor
~, гидрореактивный hydro jet engine
~, глубокопазный *el.* deep bar motor
~, горизонтальный *(of helicopter)* horizontal/propulsion/cruise engine
~ двойного действия double-acting engine
~, двухвальный two-shaft/-spool engine
~, двухклеточный *el.* double squirrel-cage motor
~, двухрядный double-/twin-/two-bank engine; two-/double-row engine
~, двухтактный two-stroke/-cycle engine
~, двухъякорный *el.* double-armature motor
~, дизельный Diesel engine; *Am.* diesel (engine)
~, длинноходный (поршневой) long-stroke engine
~, дозвуковой subsonic engine
~ жидкостного охлаждения liquid-cooled engine
~, забортный outboard engine
~, звездообразный radial engine
~, изотопный radioisotope rocket engine
~, индивидуальный ракетный astronaut manoeuvring unit, rocket belt
~, ионный ракетный ion engine
~, исполнительный actuating motor, servomotor; slave motor
~, калоризаторный hot-bulb engine
~, карбюраторный carburetted/carburettor engine
~, квантовый quantum/photon engine
~, комбинированный composite/compound/hybrid engine
~, компенсированный асинхронный *el.* compensated induction motor
~, компрессорный air-injection engine
~, конденсаторный *el.* capacitor motor
~, короткозамкнутый squirrel-cage induction motor
~, короткоходный (поршневой) short-stroke engine
~, крейцкопфный crosshead-type engine
~, кривошипный crank engine
~, крупногабаритный jumbo engine
~ левого борта port engine
~ левого вращения left-hand(ed) engine
~, лодочный, подвесной outboard engine
~, малолитражный small-displacement engine
~, малооборотный low-speed engine
~, малоресурсный short-life engine
~, многоблочный multibank/multirow engine
~, многокривошипный multicrank engine
~, многоскоростной *el.* multiple-/multi-speed motor

~, многотопливный multifuel engine
~, многофазный *el.* polyphase motor
~ на химическом топливе chemical engine
~, насосно-карбюрационный direct-injection engine
~, нефтяной oil engine
~ низкого сжатия low-compression engine
~, одновальный single-shaft engine
~, одноконтурный straight-through jet engine, one-flow engine
~, однофазный *el.* single-phase motor
~, опытный prototype/development/test engine
~, оребренный (поршневой) finned engine
~, охлаждаемый компонентами топлива propellant-cooled engine
~, первичный prime mover
~, пневматический pneumatic engine/motor, compressed-air engine/motor
~ повышенной тяги uprated engine
~, подвесный outboard motor
~, подъемный lift(ing) engine
~ пониженной тяги down-rated engine
~, пороховой ракетный powder rocket engine
~, поршневой piston/reciprocating engine
~ правого борта starboard engine
~ правого вращения right-hand(ed) engine
~, противоштопорный (ракетный) antispin (rocket) engine
~, прямоточно-ракетный ramjet rocket engine
~, пульсирующий реактивный pulse jet engine, pulsojet
~, пусковой starting engine
~, ракетно-прямоточный rocket-ramjet engine
~, ракетно-турбинный (air) turborocket engine
~, ракетный rocket engine
~, ракетный рулевой steering/control rocket engine
~ ракетный, тормозной retroengine
~ ракетный, ускорительный boost rocket engine
~ ракеты-носителя launch-vehicle engine, booster engine
~, реактивный 1. jet engine; reaction-propulsion unit; 2. *el.* reluctance/reaction motor
~, реактивный газотурбинный turbojet engine
~, реактивный жидкостный liquid-propellant rocket engine
~, реактивный ионный ion rocket engine
~, реактивный магнитоплазменный electromagnetic rocket engine
~, реактивный плазменный plasma-jet motor
~, реактивный фотонный photon rocket engine
~, реактивный электродуговой arc-heating rocket engine; plasma-jet engine
~, реактивный электротермический thermal-electric rocket engine
~, реверсивный reversible/reversing engine
~, редукторный geared engine
~, резервный standby/reserve/backup engine
~, репульсионный *el.* repulsion motor
~, ротативный rotary engine
~, рулевой control engine
~ с большой тягой high-thrust engine
~ с водным/воздушным охлаждением water-/air-cooled engine
~ с глубоком пазом *el.* deep bar motor
~ с малой тягой low-thrust engine
~ с малым расходом топлива low-consumption engine
~ с малым ресурсом short-life engine
~ с нагнеталем/наддувом supercharged engine, forced induction engine
~ с наддувом supercharged engine, pressure-charged engine
~ с непосредственным впрыском direct-injection engine
~ с низкой степенью сжатия low-compression engine
~ с перевернутыми цилиндрами inverted engine
~ с пересжатием supercompression engine
~ с переменным ходом variable-stroke engine
~ с постоянной нагрузкой constant-duty engine/motor
~ с постоянным числом оборотов constant-speed engine/motor
~ с принудительным воздушным охлаждением blower-cooled engine/motor
~ с расщепленной фазой *el.* split-phase motor
~ с регулируемой тягой controllable-/adjustable-thrust engine
~ с редуктором geared engine
~ с самовоспламенением self-ignition engine
~ с турбонаддувом turbocharged engine
~ с V-образным расположением цилиндров V-engine, vee-engine
~ с X-образным расположением цилиндров X-engine
~, сдвоенный twin-engine
~, серийный regular engine/motor
~, синхронизированный индукционный *el.* synchronous induction motor
~, солнечный solar engine
~, спаренный twin-engine
~ стартера *el.* starter motor
~, стартерный starting engine
~, стартовый *rock.* booster (engine)
~, судовой marine engine
~, тепловой/термический heat/thermal engine
~, тронковый trunk-piston engine
~, турбовальный shaft-turbine engine
~, турбовентиляторный turbofan engine
~, турбовинтовой turboprop/turboshaft engine
~, турбовинтовой реактивный turboprop engine
~, турбокомпрессорный реактивный tur-

bojet (engine)
~, турбопрямоточный реактивный turbo-ramjet/turboram engine
~, турбореактивный turbojet engine
~ ускорителя booster engine
~, фотонный photon engine
~, четырехтактный four-stroke/-cycle engine
~, электромагнитный ракетный plasma engine, electromagnetic jet engine
~, ядерный nuclear engine
двигатель-генератор motor-generator
двигательно-кинестетический moto-kinesthetic
двигательный *med.* motor; locomotory, locomotive
двигаться move; *tech.* travel
~ возвратно-поступательно reciprocate
~ поступательно travel translationally
движение 1. movement, motion; travel; flow; running; functioning; propulsion; 2. traffic
~, абсолютное absolute motion, no-relative motion
~, безвихревое steady/irrotational/streamline flow/motion, vortex-free motion
~, беспорядочное random/chaotic motion
~, боковое lateral movement
~ бортом *naut.* broadsiding
~, броуновское Brownian movement/motion
~ в бок sideward movement
~ в пространстве three-dimensional motion; space motion
~ вверх *tech.* upstroke, upward movement
~, винтовое helical/helicoidal/screw motion
~, вихревое vortex/swirl/rotational/circulatory/eddy motion; turbulence, eddying
~ влево *med.* sinistroduction
~ вниз *tech.* downstroke, downward movement
~, возвратно-поступательное reciprocating/alternating motion, back-and-forth motion
~ воздуха air movement/motion
~, воздушное air traffic
~, возмущенное disturbed/excited/perturbed motion
~, волновое/волнообразное undulatory/wave motion
~, восходящее ascending/upward motion
~ вперед forward movement/motion; advance
~ вправо dextroduction
~, вращательное rotary/rotational/angular motion; rotation, spin, gyration
~, встречное opposite traffic
~, вынужденное forced movement
~ глаз *anat.* ocular movement
~, грузовое goods/freight traffic
~, двустороннее/двухпутное two-way traffic
~, двухмерное two-dimensional motion, plane motion
~, двухрядное two-lane traffic
~ денежных средств *econ.* flow of funds
~, дыхательное *anat.* respiratory movement
~ задним ходом backward run/movement; *naut.* astern running, backing run
~ звезд motion of stars
~, знакопеременное reciprocating motion
~, интенсивное heavy traffic
~, истинное proper/true motion
~, капиллярное capillary flow
~, кеплеровское Keplerian motion
~ кзади dorsiduction
~ книзу subduction
~, колебательное oscillatory/vibratory/oscillating motion
~, коловратное gyration; spin; rotation
~, конвекционное convective motion
~, коническое coning motion
~ крена *naut.* rolling/banking motion
~, круговое/кругообразное circumduction, circular motion,circulation
~, круговращательное gyration; spin; rotation
~, ламинарное laminar flow
~, левостороннее *transp.* left driving; left-hand traffic
~, магистральное trunk-/main-line traffic
~ материков continental drift/migration
~, маховое flapping (motion)
~, маятниковое pendular/pendulum motion
~ назад backward motion/movement/running
~, напорное thrust motion
~, непрерывное continuous motion
~, неравномерное irregular/variable motion; non-uniform movement
~, несвободное forced/restricted motion
~, нисходящее descending/downward motion
~, обратное inverse/reverse/retrograde/counter motion
~, одномерное one-dimensional motion; unidimensional flow
~, однопутное/одностороннее one-way traffic
~, однорядное single-lane traffic
~, осесимметрическое axially symmetric motion/flow
~ передним ходом ahead running/travel
~, переносное translational motion; transient motion
~, периодическое periodic/cyclic motion
~ планет planetary motion/movement
~, плоское plane/planar motion, two-dimensional motion
~, плоскопараллельное plane-parallel flow/motion
~ по инерции coasting
~ по спирали spiral/helical motion
~ по часовой стрелке clockwise motion
~ полюсов Земли polar wandering/motion
~, попятное *astr.* retrograde motion
~, поступательное translational/progressive motion; advance motion
~, правостороннее *transp.* right-hand traffic
~, пригородное commuter traffic

~, пространственное three-dimensional motion
~ против часовой стрелки counter-clockwise motion
~, профсоюзное unionism, trade unionism
~, псевдопериодическое quasi-periodic movement
~, рабочее *econ., polit.* labour movement
~, равномерно замедленное/ускоренное *phys.* uniformly decelerated/accelerated motion
~, реактивное jet propulsion; reaction motion
~, рефлекторное reflex motion
~ рыскания *naut.* yawing (motion)
~, сложное compound motion
~, собственное proper motion
~, содружественное synkinesis, associated motion
~ Солнца Solar motion
~, среднее mean motion
~, суточное *astr.* diurnal motion
~ тангажа pitching motion
~, трамвайное streetcar/tram service
~, транзитное traffic flow; transit/through traffic
~, угловое angular motion
~ цен, повышательное *econ.* upward drift in prices
движимое имущество movable/personal property; movables; goods and chattels
движимость *law* movables, movable property
движитель propeller, propulser, propulsion/propulsive/propelling device
~, водометный/гидрореактивный water-jet(propeller)
~, конусный internal-cone propeller
~, крыльчатый vane propeller, rotating-blade propeller, vertical-axis propeller
~, прямоточный гидрореактивный water ramjet
~, судовой marine/ship propeller
движок 1. slide, runner, cursor, thumbpiece; 2. small engine, portable motor/engine
движущий motive, driving; propelling
двоеборец *sport* biathlonist
двоеборье *sport* double event, biathlon
двоебрачие bigamy
двоебрачный bigamic, bigamous
двоевластие diarchy
двоевластный diarchic, diarchial
двоеженец bigamist *(man)*
двоеженство bigamy *(of man)*
двоемужие bigamy *(of woman)*
двоемужница bigamist *(woman)*
двоение splitting
~ изображения *telev.* ghosting
двоецарствие diarchy; synarchy
двоить 1. split; 2. *telev.* ghost
двоично-десятичный binary-decimal
~ код обмена binary-coded decimal interchange code
двоично-кодированный binary-coded
двоично-пятеричный biquinary
двоично-рациональный binary-/dyadic-rational
двоичный binary; geminous; twin
~ разряд binary digit; bit
~ разряд, корректирующий correction bit
~ная цифра binary digit, bit
двойневость *agr.* twinning
двойник twin (crystal)
двойникование twinning
двойной double, dual; binary, duplex, binate; two-fold
~ная бухгалтерия double-entry bookkeeping
двойня twins
~, сросшаяся conjoint/conjoined twins
двойственность duality; dualism
двойственный dual; dualistic; adjoint; conjugate
~ себе *math.* self-dual
~ное число *gram.* dual number
двойчатый binary, binate; didymous; geminate
двор yard; court, courtyard
~, лесной wood yard
~, монетный mint
~, проходной communicating courtyard
~, птичий poultry yard
~, рудничный *min.* pit bottom
~, скотный/скотопригонный stockyard
дворянский nobiliary; of the nobility/gentry
~кие поместья manorial estates
двояковогнутый biconcave, concavo-/double-concave
двояковыпуклый biconvex, convexo-/double-convex
двоякодышащие *zool.* dipnoans, lungfishes, *(Dipnoi)*
двоякодышащий *zool.* dipnoan, dipnoid
двоякоизогнутый birecurvate
двоякоперистый *bot.* bipinnate
двоякопреломляющийся birefringent
двубокий bilateral
двубороздчатый disulcate, bisulcate, bisulcous
двубрачие bigamy
двубрачный bigamous
двубрюшный *anat.* digastric
двувалентность *chem.* bivalence
двувалентный bivalent, divalent
двувидный dimorphous, dimorphic
двухвидовой biaspectual
двуглавый two-headed, bicipital, bicipitate, bicephalous
~ орел double-headed eagle
двугласный *ling.* diphthong
двугорбый two-humped, double-humped
~ верблюд Bactrian camel *(Camelus bactrianus)*
двугранный dihedral
двугребенчатый bicruistate, bipectinate
двугубый *bot.* bilabiate
двудольный *bot.* dicotyledonous
~ное растение dicotyledon
двудомный *bot.* diclinous, dioecious
двудужный *zool.* diapsidan
двужаберный *zool.* dibranchiate

двужгутиковый biflagellate
двужилковый, двужилочный binervate
двужильный *tech.* twin-core; two-wire
двузамещенный *chem.* disubstituted
двузаостренный *bot.* biacuminate
двузиготный dizygotic
двузначность two-valuedness; ambiguity
 ~ импульса pulse ambiguity
двузначный 1. two-digit; 2. double-valued; 3. ambiguous
 ~ная буква digraph
двузубец two-pronged tool
двузубный bidentate
двузубый bidental, bidentate; two-pronged
двукамерный 1. two-chambered, bilocular, biloculate; 2. *polit.* bicameral
двукапсульный *bot.* bicapsular
двукарпельный *bot.* bicarpellate
двукислотный *chem.* diacid
двуклеточный bicellular
двуколосковый, двуколосный *bot.* two-spiked
двуконусный biconic(al)
двукопытный *zool.* bisulcate, cloven-footed
двукоробчатый, двукоробочный *bot.* bicapsular
двукрылый 1. *bot., zool.* dipterous, bialate; 2. *n.* dipteran; 3. *tech.* two-blade
двулепестковый *bot.* bipetalous, dipetalous
двулинейный bilinear
двулистный bifoliate, diphyllous
двулопастный two-bladed/-vaned; bilobate, bilobular
двулучевой biradiate
двумерность two-dimensionality
двумерный bidimensional; bivariate
 ~ная совокупность *stat.* bivariate population
двумолекулярный bimolecular
двунаправленный bidirectional
двунитный, двуниточный bifilar
двунога bipod
двуногий biped(al), two-legged
двуобмоточный *el.* double-winding
двуокись *chem.* dioxide
 ~ кремния silicon dioxide, silica
 ~ тория thorium dioxide, thoria
 ~ урана uranium dioxide, brown oxide
 ~ циркония zirconium dioxide, zirconia
двуосновный 1. *chem.* dibasic, diatomic; 2. *gram.* heteroclitic
двуосность biaxiality
двуосный, двуосевой biaxial
двупалость *zool.* didactylism
двупартийный *polit.* bipartisan
двуперистый bipinnate; *zool.* bipenniform
двуперстие *eccl.* the practice of making the sign of the cross with two fingers
двупестичный *bot.* digynian, digynous
двупластинчатый bilaminate, bilalamellate
двуплечий double-arm/-lever
двуплоскостной biplanar
двуполостный double-cavity
двуполость bisexuality, hermaphroditism
двуполый bisexual, hermaphroditic
 ~лое размножение amphigenesis, amphigony
двуполье *agr.* two-field crop rotation
двуполюсный, двуполярный bipolar
двуправорукость ambidexterity
двупреломление double refraction, birefringence
двураздельный bipartite; biseptate
двурогий bicorn, bicornuate
 ~гая луна crescent moon
двуротый distomatous
двусвязанный *math.* double-connected
двусеменодольный, двусемядольный dicotyledonous; *n. pl.* dicotyledons
двусемянный *bot.* dispermous
двусернистый bisulphide
двускатная крыша *arch.* gable roof
двусопряженный biconjugate
двустволка double-barrelled gun
двустворчатая дверь folding door, two-leaved door
двустолбчатый duplibaculate
двутавровый H-shaped, I-shaped
 ~вая балка I-beam
двуутлекислый *chem.* bicarbonate
двуугольник lune, gore
двуузловой binodal
двуустка *zool.* fluke
двуустый *zool.* distomatous
двуутробка *zool.* opossum *(Didelphis)*
двуутробный *anat.* amphidelphic, didelphic
двуушный binotic, binaural; biauriculate
двухадресный two-address
двуханодный *el.* double-anode
двухатомный diatomic, biatomic
двухбайтовый *comp.* double-byte
двухбалочный *av.* twin-boom
двухвалентный bivalent, divalent
двухвальный twin-shaft
двухвариантный bivariant
двухвинтовой twin-screw, twin-propeller, double-screw
двухволновой *el.* dual-frequency
двухвходовый two-input
двухгнездный bilocular, biloculate
двухгранный dihedral
двухдвигательный twin-engine, dual-motor
двухдольный *bot.* dicotyledonous
двухдорожечный *el.* dual-track
двухжильный *el.* two-core, twin-core(d)
двухзначный bivalent, binary; two-figure
двухканальный two-/twin-channel; double-track, twin-track; dual-link; two-groove
двухкаскадный two-stage
двухкилевой *naut.* twin-finned/-tail/-rudder
двухколейный *rail.* double-line/-track
двухкомпонентный bicomponent; binary
двухконечный double-ended; double-pointed
двухконтактный *el.* double-prong, two-pin;

double-contact
двухконтурный *el.* two-circuit/-stage
двухконусный biconical; double-cone
двухкоординатный:
~ное печатающее устройство data plotter, X-Y printer
двухкопытный cloven-footed
двухкорпусный *naut.* double-hulled
двухлетний *bot.* biennial
двухлинейный bilinear
двухлопастный two-blade, double-vane
двухлористый, двухлорный *chem.* bichloride, dichloride
двухлучевой two-beam, double-beam, biradial, biradiate; *(of star)* two-prong(ed)
двухмачтовый two-masted
двухмерный two-dimensional
двухместка two-seater
двухместный 1. binary, two-place; 2. two-seater, two-berth
~ функтор *math.* bifunctor
двухминеральный bimineral
двухобмоточный double-wound/-coiled; two-winding
двухосновный *chem.* dibasic
двухпалубный *naut.* two-decked; *av.* double-deck, twin-deck
двухпериодический biperiodic
двухпериодный *chem.* bicyclic
двухпозиционный *el.* on-off, push-pull *(of switch)*
двухполюсный bipolar, two-pole
двухполярный bipolar
двухпортовый dual-port
двухпоточный double-flow
двухпроводной two-wire/-line, double-wire
двухпроходной two-way
двухпутный two-way/-lane; double-track
двухразмерный two-dimensional
двухреберный bicostate
двухрежимный double-mode
двухромовокислый *chem.* bichromate
двухрядный double-row, two-row(ed); biserial
двухсегментный *zool.* dimeran
двухсекторный two-section/-sector
двухсерийный biserial; *(film)* two-part
двухсеточный twin-grid, double-grid
двухсложный *ling.* disyllabic
двухслойный double-layer, two-layer; two-ply; bistratal
двухстворчатый bivalve(d)
двухстепенный two-stage
~ые выбори indirect elections
двухстрочный double-row/-track; twin-track
двухступенчатый two-stage/-step, double-stage; two-level
двухтактный two-stroke; two-cycle
двухтарифный dual-tarrif, two-rate
двухтелый *biol.* disomatous
двухтрубный two-funneled
двухфазный, двухфазовый two-phase, diphase, diphasic, biphase, bi-phase
двухфокусный *opt.* bifocal
двухфюзеляжный *av.* twin-fuselage
двухходовой two-way; two-pass; double-thread
~ винт double-threaded screw
двухцветный dichromatic, bicolour(ed)
двухцелевой dual-purpose
двухцепной *el.* double-circuit; *mech.* double-chain
двухцилиндровый two-cylinder
двухчастный *mus.* duple
~ такт duple time; binary measure
двухшажный two-step
двухшарнирный double-hinged
двухщелевой double-slotted, two-slit
двухъядерный binuclear; *bot.* binucleate
двучлен, двучленный *math.* binomial
двуяйцовый dizygotic
деазотирование denitration
деазотировать denitride
деактивирование памяти memory inactivation
деалкилирование *chem.* dealkylation
деацетилирование *chem.* deacetylation
деаэратор deaerator
~, атмосферный open(-type) deaerator
~ барботажного типа direct-contact deaerator
~, брызгальный spray-type deaerator
~, пленочный film-type deaerator
~, струйный jet-type deaerator
~, тарельчатый tray-type deaerator
~, термический thermal deaerator, deaerating heater
деаэрация deaeration
~, вакуумная vacuum/mechanical deaeration
~, химическая chemical deaeration, scavenging
деаэрировать deaerate
дебай *phys.* Debye (unit)
дебайеграмма X-ray powder diagram/pattern
дебаланс *mech.* unbalance, unbalanced mass
дебаркадер landing stage; platform jetty
дебаркация debarkation, disembarkation
дебентура *econ.* debenture
дебет *econ.* debit
дебетовать, дебетировать *econ.* debit
дебил moron
дебильность debility, moronity
дебит discharge, output, yield; flow; capacity
~ скважины discharge of a well, flow rate of a well
дебитомер flowmeter
деблокировка unlocking; release
дебустер debooster
Дева 1. *rel.* the Virgin; 2. *astr.* Virgo
девалоризация devalorization, devaluation
девалоризовать devalorize, devaluate
девальвация *fin.* depreciation, devaluation
~ денег currency devaluation
девальвировать devaluate
девейлит *miner.* deweylite

девиация deviation
~, восточная easterly deviation
~, восьмерная octant(al) deviation
~, западная westerly deviation
~, истинная actual deviation
~, креновая heeling deviation
~ магнитного компаса magnetic deviation
~, поворотная turning error
~, полукруговая semicircular/polar deviation
~, четвертная quadrantal/intercardinal deviation/error
~, шестерная sextant(al) deviation
девиза *econ.* bill of exchange (payable in foreign currency)
девизы *econ.* foreign exchange/currency
~ в долларах dollar exchange
девиндит *miner.* dewindite
девон *geol.* Devonian (period)
~ верхний Upper Devonian (period)
~ новый Neodevonian (period)
девонский Devonian
девтоплазма *biol.* deutoplasm, deuteroplasm
девясил *bot.* elecampane, inula *(Inula)*
девятеричный nonary, novenary
девятисторонник, девятиугольник nonagon, enneagon
девятиугольный enneagonal
дегазатор degassifying agent, degassifier; decontaminator; *naut.* stripping vessel
дегазационный degassing; decontaminating
дегазация, дегазирование degassing, outgassing, degasification; decontamination
дегазировать degas, degassify; decontaminate
дегельминтизация *med.* dehelminthization
дегенеративность degeneracy, degeneration
дегенерация degeneration
~, амилоидная amyloid/hyaloid degeneration
~, водяночная hydropic degeneration
~ желтого пятна macular degeneration
~ жировая fatty/adipose degeneration
дегидраза *bioch.* dehidrase
дегидратаза *bioch.* dehydratase
дегидратация, дегидрация dehydration
дегидрирование dehydrogenation
~, ферментативное enzymatic dehydrogenation
дегидрировать dehydrogenate, dehydrogenize
дегидрогенизация dehydrogenation
~ нефтепродуктов hydroforming
~, окислительная oxidative dehydrogenation
дегидрокортизол dehydrocortisol
дегидрокортизон dehydrocortisone
дегидрокортикостерон dehydrocorticosterone, pregnenolotrone
дегидроситостерин dehydrosistosterol
дегидрохолевый dehydrocholic
дегидрохолестерин dehydrocholesterol
дегидроэметин dehydroemetine
дегоржарж, дегоржирование disgorge, disgorging
дегоржировать disgorge
деготь tar
~, каменноугольный coal tar
~, коксовый coke-oven tar
~, сланцевый shale tar
~, торфяной peat tar
дегра *leath.* tanning grease, degras
деградация 1. degradation; 2. longitudinal decalage
деградировать degrade, become degraded; degenerate; deteriorate
дегтбетон tar-macadam, tarmac
дегтеобразный tarlike, tarry
дегустатор taster; valuer
дегустация tasting
~ вин wine tasting
дегустировать taste, engage in the tasting of
дедвейт *naut.* dead weight
дедуктивно-описательный object-descriptive
дедуктивно равные interdeducible
деепричастие *gram.* adverbial participle; gerund
дееспособность *law* legal capacity; competence, capability, active capacity
дееспособный *law* competent, capable
дежурный 1. *adj.* on duty; duty *(attr.)*; 2. *n.* duty attendant/operator
~ врач doctor on duty
~ офицер, ~ по части officer of the day
~ по кораблю officer of the deck
~ная аптека drugstore open after working hours
дежурство duty, tour of duty; being on duty; attendance; *mil.* alert, watch; guard
~ в воздухе air alert
~ на аэродроме ground alert
дезагрегирование disaggregation
дезактиватор decontaminating apparatus/agent
дезактивация 1. *chem.* deactivation; 2. (radioactive) decontamination
~, ударная collisional deactivation
дезактивировать 1. *chem.* deactivate; 2. *radiol.* decontaminate
дезалкилирование *chem.* dealkylation
дезамидирование *chem.* deamidation, deamidization
дезаминаза *bioch.* desaminase
дезаминирование *chem.* deamination, deaminization
дезартикуляция *anat.* desarticulation, abarticulation
дезацетилаза deacetylase
дезацилаза deacylase
дезиминаза deiminase
дезинвагинация disinvagination
дезинсектировать disinfest
дезинсектор disinsector, insect exterminator
дезинсекционные средства insecticides
дезинсекция disinfestation, disinsectization, insect extermination; delousing
~ окуриванием fumigation
дезинтегратор disintegrator; shredder
~, центробежный centrifugal mill

дезинтегрировать disintegrate
дезинтоксикация disintoxication
дезинфектант disinfectant
дезинфектор disinfector; fumigator
дезинфекция disinfection; fumigation; decontamination
~, влажная wet disinfection
~, камерная plant disinfection
~, текущая concurrent disinfection
дезинфицировать disinfect; decontaminate
дезинфицирующий disinfectant, disinfecting
дезодорант deodorant
дезодоратор deodorizer; deodorant
дезодоризация deodorization
дезодоризировать deodorize
дезодоризирующий deodorant, deodorizing
дезоксиаденозин deoxyadenosine
дезоксигенация deoxygenation
дезоксигуанозин deoxyguanosine
дезоксикортизол desoxycortisol, hydroxydesoxycorticosterone
дезоксикортикостерон desoxycorticosterone, de(s)oxycortone, cortexone
дезоксирибоза desoxyribose
дезоксирибозилтрансфераза deoxyribosyltransferase
дезоксирибонуклеаза deoxyribonuclease
дезоксирибонуклеиновый deoxyribonucleic
дезоксирибонуклеотид deoxyribonucleotide
дезоксисахар deoxysugar
дезоксиуридиловый deoxyuridylic
дезоксихолевый deoxycholic
дезориентация disorientation
дезориентировать disorient
деионизатор deionizer
деионизировать deionize
дейдвуд *naut.* deadwood
действие 1. action, operation, functioning, running; performance; 2. effect, influence; 3. *law* act; 4. *math.* operation
~, алгебраическое algebraic operation
~, антилейкозное *med.* antileukemic effect
~, антиспастическое antispasmodic effect
~, арифметическое arithmetic operation
~, бесшумное silent operation
~, бризантное percussive effect/action
~, вентильное *comp.* gating
~, взаимное interaction
~ взрыва blast effect/action
~ внешней среды environmental influence
~, волевое volitional action
~, вредное deleterious/harmful/noxious effect; noxiousness
~, дистанционное remote action
~, длительное long-term action/effect
~, дробящее shattering effect/action
~, желчегонное *med.* cholagogic action
~, задерживающее inhibitory/retarding action
~, замедленное delay(ed) effect
~, замедляющее moderating effect
~, запорное blocking/barrier action
~, защитное protective/shielding/safety action/effect
~ змеиного яда *med.* envenomation, snake venom effect
~ излучения radiation effect
~, истирающее abrasive/grinding action
~, коррозионное corrosive action/effect
~, мешающее interference, disturbance
~, мутагенное/мутационное mutagenic activity/effect
~, логическое logical operation/action
~ на расстоянии remote operation
~, неправильное incorrect/irregular operation, maloperation, malfunction
~, непрограммируемое nonprogrammable action/operation
~, обратимое reversible action
~, обратное back action, reaction, retroaction
~, ответное response
~ по коду возврата action on return code
~ по ошибке error option, option of error action
~, побочное side effect, by-effect
~, поверхностное surface action/effect
~, полезное *tech.* efficiency
~ при наличии ошибок action in error condition
~, программируемое programmable action
~, процессуальное proceeding(s)
~ радиации radiation effect
~, разрушающее destructive effect; *chem.* attack
~, разрушительное destructive/damaging/disruptive effect
~, реактивное reaction
~, релейное on-off action/operation
~, рефлекторное reflex effect
~ рычага leverage
~, световое light action
~ сжатия pressure effect
~ силы тяжести gravity action
~ сифона siphonage
~, совместное joint/combined action
~, структурирующее *chem.* structure-forming effect/action
~, тепловое thermal/heating/temperature effect/action
~, токсическое toxic/poisonous action/effect
~ тормозов/тормозящее braking action
~, ударное impact, percussion; shock effect
~, эритемное erythema efficiency
~, эрозионное erosive action
~, юридическое legal act
действительно-значный *math.* real-valued
действия *pl.* operations, actions; functions; acts
~ авиации *mil.* air operations
~, боевые combat operations, warfare, fighting
~ истребительной авиации fighter operations
~, наступательные offensive action/operations
~, неправильные mishandling

~, оборонительные defensive operations
~, процессуальные *law* legal proceedings
~, тактические tactical operations/warfare
~, штурмовые *av.* strike sorties
действующий operational, operative; active, acting; *law* valid, effective, in force
~, постоянно standing
дейтерид deuteride
~ лития lithium deuteride
дейтеризация deuterating
дейтеризовать deuterate
дейтерий deuterium, heavy hydrogen
~, газообразный deuterium gas
дейтерирование deuteration
дейтероморфный deuteromorphic
дейтерон, дейтрон, дейтон deutron, deuton
дейтероокись deuterium oxide, heavy water
дейтеропатия *med.* deuteropathy
дейтероплазма, дейтоплазма *biol.* deuteroplasm, deutoplasm
дейтеросоединение deuterocompound
дейтеротокия *med.* deuterotoky
дейтон-сера deuton-sulphur
дека 1. *mus.* sounding board; 2. *agr.* grate, concave
~, защитная protection screen deck
декабрист *hist.* Decembrist
декагон *geom.* decagon
декада decade; ten-day interval/period
декалесценция decalescence
декалог *bibl.* Decaloge, the Ten Commandments
декалькирование transfer printing
декалькировать *art* transfer
декальцинация, декальцинирование, декальцификация decalcification
декальцинировать decalcify
декан I dean *(of university)*
декан II *chem.* decane
деканат dean's office
декантатор decanting vessel, decanter
декантировать decant, pour off, elutriate
деканюляция *surg.* decannulation, decannulization
декапир *met.* pickled sheet
декапировать *met.* pickle; scour, dip
декапитатор decapitator, decollator
декапитация decapitation, decollation
декаподы *pl. zool.* decapods *(Decapoda)*
декапсуляция decapsulation
~ почки nephrocapsectomy
декарбоксилаза decarboxylase
декарбоксилирование decarboxylation
~, окислительное oxidative decarboxylation
декарбюризация *met.* decarburization
декартелизация *econ.* decartelization
декартов, декартовский Cartesian
декатир *text.* decating/decatizing machine
декатировать *text.* decate, decatize
декатировка *text.* decating, decatizing
~, глянцевая/матовая lustre/dull decating
декатрон decade-counting tube
декаэдр *geom.* decahedron
декаэдрический *geom.* decahedral
декларация declaration; entry; *law* manifest
~, морская maritime declaration
~ о личных вещах экипажа *naut.* crew's effects declaration
~, таможенная *econ.* customs declaration
деклинатор, деклинометр declinometer, declination compass
дековилька *rail.* Decauville
декодер decoder
~, запоминающий *comp.* storage decoder
декодирование decoding
~ клавишного набора keyboard decoding
~, пороговое threshold decoding
~, последовательное sequential decoding
декодировать decode, decipher
декокт *pharm.* decoction
декомпенсация *med.* decompensation
~, сердечная cardiac decompensation
~ сердца, застойная congestive heart failure
декомпилятор *comp.* decompiler
декомпиляция *comp.* decompiling
декомпозиция decomposition
~ больших систем *comp.* large-scale decomposition
~, временная temporal decomposition
~, модульная modular decomposition
~, пространственная decomposition in space
декомпрессия decompression
~, взрывная explosive decompression
~, катастрофическая fatal decompression
~, фазовая phase decompression
декомпрессор decompressor, exhaust valve
декортикатор decorticator, stripping machine
декортикация decortication, stripping
декрейтер *food* decrater, uncrater, bottle decrating/uncrating machine
декремент decrement; decrease
~ затухания damping decrement/factor; decrement of attenuation
~ энергии energy decrement/loss
декрет *law* decree; edict
декреталия *eccl.* decretal epistle
декретный decretal, decretive, decretory
~ отпуск maternity leave
декрешендо, декрещендо *mus.* decrescendo
дексаметазон dexamethasone
дексель adz(e)
декстран *chem.* dextran(e)
декстрансахараза dextransucrase
декстрин *chem.* dextrin, starch gum
декстроверсия dextroversion
декстроза dextrose
декстрокардия *med.* dextrocardia
декстроторсия dextrotorsion
декурион *hist.* decurion
декурия *hist.* decury
делагил *pharm.* delagil, chloroquine
дележ *game th.* imputation
деление 1. division, dividing; splitting; *nucl.* fis-

sion; 2. *(on a scale)* graduation, point; degree; unit; reading
~ без восстановления остатка nonrestoring division
~ без остатка exact division
~, взрывное explosive division
~, двоичное *comp.* binary division
~, дифференциальное differential indexing
~ клеток *biol.* cell fission
~ круга cyclotomy
~ на две части bipartition; binary fission
~ на зоны zonation
~ на камеры concameration
~ на тепловых нейтронах thermal (neutron) fission
~ на три части tripartition, ternary fission
~ на участки sectionalization
~ нацело exact division
~, начальное original fission
~ под действием быстрых нейтронов fast (-neutron) fission
~ под действием гамма-квантов photofission
~ под действием дейтронов deutron-induced fission
~ под действием нейтронов/протонов neutron-/proton-induced fission
~ пополам halving; bisection; division by two
~ почкованием gemmation
~, преимущественное preferential fission
~, самопроизвольное spontaneous/natural fission
~, сотенное centesimal graduation
~, ускоренное/сокращенное short/abridged division
~ частоты scaling
~ шкалы scale division/graduation, graduation mark/line
~ шкалы в градусах sexagesimal division/graduation
~ шкалы, градовое centesimal circle graduation
~, эмиссионное emissive fission
~ ядер топлива fuel fission
~, ядерное/ядра nuclear fission
~ ядра нейтроном neutron-induced fission
~ ядра, тройное ternary fission
~ ядра урана uranium fission
~ ядра фотонами photofission
деликт *law* tort, delict; civil injury
~, имущественный property tort
деликтный *law* tortious
делимое *math.* dividend
делимость divisibiliy; partibility; *nucl.* fissionability, fissility
делинквент *law* wrongdoer, tortfeasor, offender, law-breaker, delinquent
делирий delirium
~, алкогольный delirium tremens/alcoholicum
~, лихорадочный febrile delirium
~, профессиональный occupational delirium
~, старческий senile delirium
делитель 1. *mech.* divider, separator; 2. *math.* divisor
~, многоремешковый *text.* multiple tape divider/condenser
~ на два 1. halver; 2. *comp.* scale-of-two circuit
~ на пять *comp.* scale-of-five circuit
~, наибольший общий *math.* greatest common divisor, GCD
~ напряжения *el.* voltage divider
~, нормальный *math.* normal/invariant subgroup, normal divisor
~ нуля zero divisor, divisor of zero
~ нуля, полный absolute zero divisor
~, общий common divisor/measure
~, общий наибольший greatest common divisor
~ пополам halver
~ потока *hydr.* flow dividing valve
~, простой prime divisor
~, ремешковый *text.* tape divider/condenser
~, строчный *telev.* line divider
~ частоты frequency divider/demultiplier; scaler
делительный dividing, separating; scaling
делить 1. divide; 2. classify, group, break down (into)
делиться 1. be divisible, divide; go into; 2. *nucl.* be fissionable, disintegrate, break down/into
дело 1. work, trade, practice, business, affairs; duty; occupation; concern; 2. file, records; 3. cause; deed; 4. matter, point; 5. *law* case; action; lawsuit, suit
~, банковское *econ.* banking
~, брокерское *fin.* broking
~, вести plead a case
~, военноинженерное military engineering
~, возбуждать *law* institute proceedings, bring an action *(against)*
~, возобновлять *law* reopen a case
~ выигранное защитой *law* case for the defence
~ выигранное обвинением *law* case for the prosecution
~, горное mining
~, гражданское civil case/suit
~, книгоиздательское publishing
~, морское seamanship
~ о признании брака недействительным *law* nullity suit
~, переплетное bookbinding
~, плотничье carpentry
~, слесарное bench work
~, столярное joinery
~, судебное case; trial; proceedings; litigation
~, транспортное *econ.* the carrying trade
~, уголовное *law* criminal case
~, штурманское navigation
деловитость efficiency, businesslike character
деловой business *(attr.),* businesslike
~ район города downtown business district

~вые круги business interests/circles
~ое время working hours/time
делопроизводитель filing clerk
делопроизводство office/clerical work; record keeping
делькредере *econ.* del credere
~, принять на себе stand surety
дельта *geog., math.* delta
~, бухтовая bay delta
~, веерообразная fan delta
~ волна delta wave
~, вторичная sub-delta
~, выдвинутая protruding/cuspate delta
~, древовидная dendritic delta
~ излучение delta radiation
~, лопастная lobate delta
~ образованная волнами wave delta
~ образованная приливами tidal delta
~ оператор *math.* Laplacian operator
~, отливная ebb delta
~, отступающая subsiding delta
~, полукруглая arcuate delta
~, приливная tidal delta
~ функция unit impulse function, (Dirac) delta function
дельтагеосинклиналь *geog.* deltageosyncline
дельтовидный *anat.* deltoid
дельтовый *geog.* deltaic, deltoid
дельтоид *math.* deltoid
дельтоид-додекаэдр, дельтоэдр *geom.* deltoid-dodecahedron
дельфин *zool.* dolphin *(Delphinus); sport* dolphin kick; *astr.* Delphinus
дельфинирование *naut.* porpoising
дельфинировать *naut.* jump clear of water, porpoise *(of vessel)*
делювий *geol.* talus (deposits), slide rocks, driftbeds, diluvium
~, горный mountain/hillside waste
~, движущийся/ползучий downhill creep, creeping waste, creep-wash
делянка plot, patch, parcel, lot *(of land)*
делячество empiricism; utilitarian approach
делящийся fissionable, fissile; *(process)* fissioning
дем *hist.* deme
демагнетизатор demagnetizer
демаркация demarcation; delimitation
демаскировать unmask; *fig.* disclose, expose
деменция *med.* dementia
демерредж *law, econ.* demurrage
демиелинизация demyelination, demyelinization
демилитаризация экономики civilian conversion
деминерализатор demineralizer
деминерализация demineralization; desalting
демиург demiurge; public official
демография demography
~ миграции *econ.* migration demography
~, формальная *econ.* demometrics
демодификатор demodifier
демодулировать demodulate
демодулятор demodulator
~, балансный ring/balanced demodulator
~, многоканальной системы связи multiplex demodulator
демодуляция demodulation
~ по времени time demodulation
~, частотная frequency demodulation
демократичность democratic character/nature; democracy
демократия, народная *hist.* people's democracy
Демокрит Democritus
демокритов, демокритовский Democritean
демометрика *econ.* demometrics, formal demography
демонтаж, демонтирование dismantling, disassembly, disassembling, stripping; tearing down
демонтировать disassemble, dismantle, strip, take apart; knock down
Демосфен Demosthenes
демосфеновский Demosthenic
демпинг *econ.* dumping
демпфер shock absorber, dash-pot; damper, deadener; silencer
~, инерционный inertia damper
~ крена *av.* roll/lateral damper
~, пневматический air dash-pot
~ рыскания *av.* yaw/directional damper
~ тангажа *av.* pitch/longitudinal damper
демпферирование damping
~ антенны (радиолокатора) anti-hunt
~, жидкостно-фрикционное fluid friction damping
~, избыточное overdamping
~ крутильных колебаний torsional damping
~, Кулоново Coulomb damping
~, недостаточное underdamping
~ по скорости rate damping
~ поперечных колебаний roll/lateral damping
~ посредством вихревых токов *el.* eddy damping
~, постоянное fixed-gain damping
~ продольных колебаний pitch/longitudinal damping
~ путевых колебаний yaw/directional damping
~ реактивной струей jet damping
~, самонастраивающееся/саморегулируемое self-adaptive damping
~, слабое poor/low damping
~, собственное natural damping
~ создаваемое крылом wing damping
~, чрезмерное overdamping
демпфировать damp; bring *(a system)* to rest
демультиплексирование demultiplexing
демультипликатор supplementary/auxiliary gear transmission/box; reducing gear; demultiplier
демурредж *econ.* demurrage
денатурант *chem.* denaturating agent, denaturant
денатурат *chem.* denatured alcohol
денатурация, денатурирование denaturing,

denaturation
денатурировать denaturate
денационализация 1. *econ.* denationalization, reprivatization; 2. *polit.* suppression/loss of national characteristics
денацификация *polit., hist.* denazification
дендрарий tree nursery
дендрит 1. *miner.* dendrite; 2. *anat.* dendrite, dendron
дендритовый dendritic; dendrite
дендрон *math.* dendron
денежный monetary; financial; money *(attr.);* pecuniary
~ знак bank note
~ перевод money order
~ рынок money market
~ штраф fine
~ ящик strongbox
~ное обращение currency of money
денитрирование, денитрация *chem.* denitration
денитрировать denitrate; denitrify
денитрификация, денитрогенация denitrification
денитрифицировать denitrify
деннаж *naut.* dunnage
денонсация *dipl.* denouncement, denunciation
денсиметр densimeter
денситометр densitometer, densimeter, opacity meter
денсограф recording densitometer
дентикула *arch.* dentil
дентин *anat.* dentin(e)
~, околопульпарный parapulpar dentin
денудация denudation
~, дождевая pluvial denudation/washout
~, механическая wind carving
день day
~, високосный leap/intercalary day
~ всех усопших All Souls' Day
~ выплаты зарплаты pay-day
~, выходной day off
~ духов/Троицын White Sunday
~, Иванов St. John the Baptist's Day
~ избиения младенцев *rel.* Holy Innocents' Day
~, короткий рабочий short hours
~, неполный рабочий *econ.* part time; short hours; undertime
~, неприсутственный *econ.* legal/public holiday, *law* dies non juridicus
~, нерабочий 1. day off; 2. legal/public holiday
~, рабочий *econ.* workday; working day
~, расчетный pay-day
~ рождения birthday
~ свободный от работы idle day
~, судный, ~ страшного суда the Day of Last Judgement, doomsday
деньги money; funds; currency
~, металлические coinage; specie
~, наличные *econ.* cash; effective/ready money
~, продовольственные victualling allowance
~, путевые travel allowance
~ с высокой покупательной силой hard/ tight/dear money
~, столовые subsistence allowance
~, счетные money of account, accounting money
~, тезаврированные money hoards; idle money
денье(-титр) *text.* denier
деонтология deontology
~, врачебная medical deontology/ethics
депарафинизировать *chem.* dewax
депектинизатор depectinization tank
депо 1. depot; shed; station; 2. *econ.* deposit
~ акций stock deposit
~, вагонное railway car shed/shop
~, пожарное fire station
~, трамвайное streetcar/tram shed
депозит deposit
~ до востребования *econ.* call/demand/sight deposit
~, почтовосберегательной postal savings deposit
~, срочный time/fixed deposit, deposit with fixed period
~ы, свободно переводимие transferrable deposits
депозитарий *econ.* depositary; *law* bailee
деполяризатор depolarizer
деполяризационный depolarizing; depolarization *(attr.)*
деполяризация depolarization, depolarizing
депонент *fin.* depositor; *law* bailor, bailer
депонентный *gram.* deponent
депонирование deposition, depositing
депонировать deposit; place on deposit
депорт *econ.* backwardation
депрессант *med.* depressant, depressing agent
депрессивный depressive; depressing
депрессиометр depression meter
депрессия 1. *econ.* depression; recession; slump; 2. *med.* depression; melancholy; dejection; 3. *meteor.* depression; low-pressure area, area of low pressure; *geog.* basin
~, барическая barometric depression
~, вогнутая impressed zone
~, вторичная secondary/satellite depression
~, вытянутая furrow
~ горизонта dip of the horizon
~, ипохондрическая hypochondriacal depression
~ подземных вод drop/decline of underwater ground level
~, психическая mental depression
депрессорный depressive; depressor *(attr.)*
дератизация deratization, rodent control
дерби *sport* Derby
дервиш *hist.* dervish
дерево 1. tree; 2. wood, lumber, timber; 2. *math.*

tree
~ включений *math.* inclusive tree
~ вывода derivation tree
~ грамматического разбора syntax tree
~, двоичное binary tree
~ игры *game th.* game tree
~, инвертированное upside down tree
~ исключений exclusive tree
~ исходов outcome tree
~, красильное dyewood
~, красное mahogany
~ лиственной породы broad-leaved tree
~ на корню standing timber
~ неисправностей fault tree
~, низкорослое bushy tree
~, обращенное inverted tree
~, окремненное silicified wood
~ относительной важности relevance tree
~ поиска search tree
~, пробковое corkwood
~ разветвленное на выходе fan-out tree
~, рангоутное *naut.* bum, spar
~ решений decision tree
~, родословное family/genealogical tree
~ совмещений overlay tree
~ хвойной породы coniferous tree, conifer
~, черное ebony
дерево-камень xylolite, xylolith
деревообрабатывающий woodworking
деревообработка woodworking (industry)
деревянный wooden, ligneous; wood *(attr.)*
держава *polit.* power; nation
~, мировая world power
~, морская naval/sea power, maritime nation
державка holder, support, carrier, stand; mounting; rack
~ для резца tool bit holder
~ карт card holder
~ ножа knife holder/carrier
~ предплужника colter clip
~, резцовая toolholder
держание дороги roadability, road stability
держатель 1. holder; clamp, grip; socket; support mount; carrier; adapter; bracket; 2. *anat.* retractor
~ инструмента tool holder
~, пружинный spring clip
деривация 1. diversion; diversion system/works/installation; channelling, canalization; drawing off; 2. drift; 3. *gram., math.* derivation
~, напорная pressure intake system/installation
~, отводящая tailrace system/installation
~, подводящая intake system/installation
дерма *anat.* derma, dermis, corium
дермамиотом dermamyotome
дерматин leatherette, imitation-leather (cloth)
дерматит *med.* dermatitis
~, контактный contact dermatitis
~, лучевой (рентгеновский) X-ray dermatitis, dermatitis skiagraphica
~, медикаментозный medicamental dermatitis
~, онхоцеркозный onchocercal dermatitis
~, профессиональный occupational dermatitis
~, ртутный mercurial dermatitis
~, себорейный seborrheic dermatitis
~, солнечный solar dermatitis
~, экзематозный eczematous dermatitis
~, эритематозный dermatitis erythematosa
дерматоаутопластика *med.* dermato-autoplasty
дерматогетеропластика dermatoheteroplasty
дерматоглифика *med.* dermatoglyphics
дерматограмма *med.* dermatogram(me)
дерматоз *med.* dermatosis
~, акаридозный acarine dermatosis
~, лихеноидный lichenoid dermatosis
~, профессиональный occupational dermatosis
~, световой photodermatosis
~, филяриозный filarial dermatosis
~, язвенный ulcerative dermatosis
дерматолог dermatologist
дерматом *med.* dermatome
дерматома *med.* dermatoma
дерматомер *med.* dermatomere
дерматомикоз *med.* dermatomycosis
дерматомиозит *med.* dermatomyositis
дерматомиома *med.* dermatomyoma
дерматоневроз *med.* dermatoneurosis
дерматопатия *med.* dermatopathy
дерматопластика *med.* dermatoplasty
дерматотропный *med.* dermatotropic, dermotropic
дерматофит dermatophyte
дерматофитоз dermatophytosis, dermophytosis
дермографизм skin writing, dermographism, dermography, dermatography
дермоид, дермоидный dermoid
дерн sod, block/piece of turf
дернина turf, vegetable layer; grassy turf
дерново-глеевый sod-gleyed
дерново-луговой sod-meadow
дерновый soddy, turfy, turfen
дерноснимм, дерноснимaтель sod knife, skimco(u)lter, skimmer
деррик(-кран) derrick(-crane)
~, вантовый guy(ed) derrick
~, мачтовый pole derrick
~, передвижной crab derrick
десант *mil.* landing force/party/troops; landing operation
~, вертолетный helicopter-borne/-landed party, helicopter assault force
~, воздушный airborne force
~, морской landing party
~, парашютный parachute landing
десантирование landing
десантник paratrooper; commando
десантный *mil.* landing; *naut.* amphibious; *av.* airborne
десемантизация *ling.* loss/shift of meaning
десенный *anat.* gingival
десенсибилизатор desensitizer

десенсибилизация desensitization, desensitizing
десенсибилизировать desensitize
десик(к)ант desiccant
десиликация desilication
десилицировать desiliconize
десквамация *geol.* desquamation; exfoliation; *med.* pealing
десмоид *med.* desmoid
десмолаза *bioch.* desmolase
десмолиз desmolysis
десмосома desmosome
десмостилида desmostylus
десмотропия *chem.* tautomerism, desmotropism
десмоцит desmocyte; fibrocyte
десмургия *med.* desmurgy; bandaging
десна *anat.* gingiva, gum
десорбция desorption
деспотичность, деспотия despotism
дестабилизатор destabilizer
дестиобиотин desthiobiotin
деструкция *phys.* destruction; decomposition; *chem.* breakdown; degradation
 ~, радиационная radiation destruction
 ~, термоокислительная thermal-oxidative breakdown
десульфитация *chem.* desulphitation
десульфуратор desulphurizer
десульфурация desulphurization, sulphur removal
десульфурировать desulphurize
десятерик measure/object of ten units
десятерица decuple
десятеричный denary, tenfold
десятигранник *geom.* decahedron
десятигранный *geom.* decahedral
десятизначный ten-character
десятизубый decemdentate
десятиклассник tenth-grader
десятиклассный tenth-grade
десятилетие decade, period of ten years; tenth anniversary
десятилетний ten-year, decennial, decadal
десятиместный ten-seater
десятимесячный decimestrial
десятина 1. *hist.* tithe, decima; 2. dessiatina, dessiatine *(2.7 acres)*
десятиногий decapodal, decapodan; *n. pl.* decapods
десятисловие *bibl.* Decalogue
десятисложный *ling.* decasyllabic
десятиугольник *geom.* decagon
десятиугольный *geom.* decagonal
десятично-двоичный decimal-binary
десятичный decimal
десятник 1. *hist.* commander of ten; junior commanding officer; 2. foreman
десяток ten; *pl.* dozens, scores, tens
детализация detailed elaboration/specification
 ~ знаний knowledge refinement
детализировать detail, present details of, elaborate, specify, work out in detail
деталировка detailed drawing
деталь 1. component, part; piece; element; member; 2. detail; feature
 ~, арматурная reinforcing member/piece
 ~, бракованная rejected part
 ~, готовая off-the shelf component/part
 ~, дистанционная spacer
 ~, запасная spare part; *pl.* spares
 ~ изображения detail, detailed feature
 ~, комплектующая component; accessory
 ~, конструктивная, ~ конструкции structural member
 ~, крепежная fastening part/piece, fastener
 ~, нормализованная standard component/part
 ~, ответственная essential/vital/critical part
 ~, охватываемая male part
 ~, охватывающая female part
 ~, переходная reducer
 ~, распорная spacer; distance piece
 ~, сварная welded part; weldment
 ~, сменная replacement part
 ~, соединительная fastener; fastening/connecting part; adapter
 ~, строительная structural element
 ~, съемная removable part/element
 ~, фасонная shaped part/element
детальный detailed; minute, fine
детандер pressure-reducing valve; expansion/reducer valve; expander
детектирование *rad.* detection, rectification
 ~, анодное anode-bend detection; *Am.* plate detection
 ~, диодное diode detection/rectification
 ~, квадратичное square-law detection
 ~, сеточное grid detection; *Am.* grid-leak detection
детектор detector; *el.* rectifier
 ~, активационный *nucl.* activation detector
 ~ альфа-частиц alpha-particle detector, alpha-radiation detector, alpha-detector
 ~ бета-излучения beta(-radiation) detector
 ~, быстродействующий *nucl.* fast detector
 ~ быстрых нейтронов fast-neutron detector
 ~ видеосигналов video-detector
 ~ вспышек flash detector
 ~, галеновый galena detector
 ~, галоидный halogen detector
 ~ гамма-излучения gamma(-radiation) detector
 ~, двухтактный push-pull detector
 ~ делений *nucl.* fission detector
 ~, диодный diode detector
 ~, дифференциальный difference detector
 ~ излучения radiation detector/probe
 ~ изменения адреса address-transition detector
 ~, изотопный isotopic detector
 ~, импульсный pulse-type detector
 ~ инфракрасного излучения infra-red detector
 ~, квадратичный 1. *phys.* square-law detector;

2. *stat.* quadratic detector
~, **кремниевый** silicon detector
~, **кристаллический** crystal/natural detector
~, **ламповый** valve [electrone tube] detector
~ **лжи** lie detector
~, **минный** *mil.* mine detector
~, **мощный** power detector, high-amplitude detector
~, **нейтронный/нейтронов** neutron detector
~, **огибающий** envelope detector
~, **пороговый** threshold detector
~ **с автопрерыванием** self-quenching detector
~, **сеточный** grid detector
~, **сигнальный** alarm/signal detector
~ **совпадений** coincidence detector
~, **твердотельный** solid-state detector
~ **теплового излучения** heat detector
~ **утечки** leak detector
~, **фазовый** phase discriminator/detector
~, **фазочувствительный** phase-sensitive detector; synchronous rectifier
~, **фольговый** foil detector, detecting foil
~, **фотоэмиссионный** photoemissive detector
~, **фотоэмульсионный** photographic detector
~, **частотный** frequency discriminator/detector; frequency-sensitive detector
~ **ядер отдачи** recoil detector

детекторный приемник *rad.* crystal receiver
детергент detergent
~, **амфотерный** amphoteric detergent
~, **анионоактивный** anionic detergent
~, **катионоактивный** cationic detergent
детерминантный determinantal
детерминатив *ling.* root determinative
детерминированность determinancy
детерминированный deterministic; determinate
детеубийство; детеубийца infanticide; filicide
детонатор detonator, primer, detonating charge; percussion cap
~ **замедленного действия** delay detonator
~, **промежуточный** booster, relay detonator
детонация 1. detonation; 2. *mech.* combustion knock, knocking, pinking
детоненавистничество misopedia
детонировать I 1. detonate, explode, burst; 2. *mech.* knock, pink
детонировать II *mus.* sing flat, sing out of tune
детонометр detonation meter
деторождение childbearing, reproduction, procreation; parity
~, **нерегулируемое** natural reproduction
~, **регулируемое** controlled reproduction
детрит detritus; detrital matter
детритовый detrital; deuterogenous
детритус *geol.* detritus
детумесценция detumescence
деутероплазма deuteroplasm, deutoplasm
дефазировка dephasing
дефекация *biol.* defecation, egestion; *tech.* purification, clarification, defecation
~, **ступенчатая** fractional liming/clarification
дефекосатурация defecosaturation
дефект defect, imperfection, flaw, fault; bug, breakdown, disturbance
~, **краевой** marginal defect
~, **конструктивный** design imperfection/defect
~ **конструкции** structural defect/fault
~ **кристалла** crystal/lattice defect/imperfection
~ **литья/отливки** casting defect
~ **лица** *leath.* grain defect
~, **непроверяемый** untestable fault
~ **по Френелю/Шотки** Fresnel/Shottky defect
~, **радиационный** radiation-induced defect
~, **скрытый** hidden fault/flaw; *law* latent defect
~, **угловой** *math.* angular defect
~ **укладки** *phys.* stacking fault
~, **хронический** *comp.* glitch
~ **ядра** *math.* nullity
дефектация flaw/fault detection, survey/supervision of defects/flaws
дефективность presence/existence of defects
дефективный 1. defective, faulty; 2. *med.* mentally deficient, defective, handicapped
дефектность defectiveness; *law* diminution
дефектный defective, faulty; imperfect
дефектовать inspect for defects; check
дефектолог *med.* specialist on mental and physical defects/handicaps
дефектоскоп flaw detector; non-destructive testing/inspection instrument/equipment/unit
~, **магнитно-порошковый** magnetic-particle flaw detector
~, **магнитный** magnetic-field flaw detector
~, **прокатный** scanner
~, **рентгеновский** X-ray flaw detector; X-ray inspection/test unit/instrument
~, **ультразвуковой** ultrasonic flaw detector
~, **феррозондовый** probe-type magnetic-field flaw detector
~, **фотоэлектрический** photo-electric scanner
~, **электроиндуктивный** magnetic field flaw detector
дефектоскопия non-destructive test(ing); flaw detection
~, **акустическая** sonic testing
~, **индукционная** eddy-current testing
~, **инфракрасная** infra-red inspection/testing
~, **капиллярная** liquid-penetrant test(ing)
~, **лучевая** radioexamination
~, **люминесцентная** fluoroscopic/luminescence flaw detection
~, **магнитная** magnetic(-field) testing, magnetic flaw detection
~, **магнитно-порошковая** magnetic-particle test(ing)
~, **рентгеновская** X-ray flaw detection; X-ray inspection/testing
~, **ультразвуковая** ultrasonic flaw detection; ultrasonic monitoring/inspection/testing
~, **электроиндуктивная** eddy-current testing

деферентит *med.* deferentitis
дефибрер fibre separator, defibrator; (pulp) grinder
дефибриллятор *med.* defibrillator
~, наружный external defibrillator
дефибрилляция *med.* defibrillation
дефибриновать *med.* defibrinate
дефибрировать defiber
дефицит *fin.* deficit; *econ.* shortage, scarcity; deficiency; stockout
~, бюджетный budget deficit/gap
~ строительных материал scarcity/shortage of building materials
~ влажности moisture/saturation deficit
~, внешноторговый trade deficit
~, водный water deficiency/shortage
~, кассовый cash deficit
~ мощности power shortage
~, непокрытый outstanding deficit
~ платежного баланса payments deficit, balance of payment deficit
~ пульса *med.* pulse deficiency
~ торгового баланса trade deficit
~ фермента enzyme deficiency
~ энергии energy gap/shortage
дефицитность shortage, deficiency, unavailability, shortage, short supply
дефицитный scarce, in short supply, critical, not readily available, difficult to obtain; *fin.* unremunerative
дефлегматор *chem.* reflux condenser, dephlegmator; fractioning column; partial condenser
дефлегмация, дефлегмирование *chem.* refluxing; partial condensation; fractionation, dephlegmation
дефлегмировать *chem.* reflux, dephlegmate, fractionate
дефлектор deflector, baffle; *el.* deflecting electrode
дефлокулировать deflocculate
дефлорация defloration
дефляционный *econ.* deflationary, disinflationary
дефляция 1. *fin.* deflation, disinflation; 2. *geol.* wind erosion, soil drifting, deflation
дефокусирование, дефокусировка defocusing
~ при отклонении deflection defocusing
дефолиант defoliant
дефолиация defoliation
дефонологизация *phon.* dephonologization, dephonemicization
деформация deformation; deformity; distortion; deflection; *phys.* strain
~ изгиба flexural/bending strain
~ конструкции structural deformation/distorsion
~, критическая ultimate/critical strain
~ кручения torsional strain
~, механическая strain, mechanical deformation
~, обратимая reversible deformation; recoverable strain
~, объемная volumetric/dilatational strain
~ основания foundation/bedding deformation
~, остаточноя residual deformation; permanent set
~ отливки casting strain
~, плоская plane deformation/strain
~ ползучести creep flow/strain, creep
~, поперечная lateral deformation/strain
~, продольная longitudinal strain; stretch
~, разрушающая breaking strain
~, разрывная *geol.* failure by rupture
~ растяжения tensile deformation/strain
~ сдвига shear deformation; shear(ing) strain
~ сжатия compressive/compression(al) strain
~ скручивания torsional/twisting strain
~ смятия crushing deformation
~, температурная temperature strain
~, тепловая/термическая thermal deformation/strain
~, термоупругая thermoelastic deformation/strain
~, удельная unit strain
~, упругая elastic/Hooke deformation
~, упруго-пластическая elastoplastic/plastoelastic strain/deformation
~, усадочная shrinkage strain/deformation; shrinkage
~, усталостная fatigue deformation
~ фотопленки film distortion
~, электрострикционная electrostrictive strain
деформировать deform, distort; strain
деформироваться become deformed; buckle; warp; change shape
деформируемость deformability
дефосфорация dephosphorization, phosphor removal
дефосфорилирование dephosphorylation, dephosphorylization
дефростер defroster, defrosting/thawing unit
дехлорирование, дехлорация *chem.* dechlorination, chlorine removal; dechloridation
дехлорировать *chem.* dechlorinate
дехолин *chem.* sodium dehydrocholate
децибел *acous.* decibel, dB
децил *chem.* decyl, decatyl
дециллион *math.* decillion
дециль *math.* decile
децима *mus.* decima
децимоль *mus.* decimole
децинормальный *chem.* tenth-normal, decinormal
дешифратор decoder, decipherer; decoding circuit; decoder matrix
~, диодный diode decoder/matrix
~ команд command/instruction decoder
~, разрядный bit decoder
~ с памятью storage decoder
~, стробированный gated decoder
~ строк row decoder
~, триггерный trigger decoder
дешифрация decoding; *el.* translation; transfer
дешифрирование, дешифрование decoding,

deciphering; interpretation; identification; cryptanalysis
~ **аэроснимков** aerial photo(graph) interpretation/identification/reading
дешифровать, дешифрировать decode, decipher, interpret
дешифровка deciphering, decoding
дешифровщик code breaker/cracker, cryptanalyst; decipherer
~ **аэроснимков** aerial photo interpreter
дешламация *geol.* desliming
деэлектризация de-electrization; deionization
деэмульгатор demulsifier
деэмульгировать demulsify
деэмульсация, деэмульгирование demulsification, demulsifying
де-юре *law* de jure
деяние act, deed
Деяния апостолов *rel.* the Acts of the Apostles
деятель figure; worker
~, **государственный** statesman
~ **искусства, заслуженный** Hono(u)red Artist
~ **науки** scientist
~ **науки, заслуженный** Hono(u)red Scientist
~, **общественный** public figure, person active in public life
~, **политический** politician
деятельность activities, activity; occupation, profession
~, **вулканическая** volcanic activity
~, **высшая нервная** higher nervous activity
~, **общественная** public activities/work, social work
~, **побочная** side work/occupation
~, **повышенная** hyperactivity, overactivity, superactivity
~, **подрывная** subversive activities
~, **пониженная** *med.* hypoactivity
~, **престижная** public relations
~, **психическая** psychic activity
~, **рекламно-пропагандистская** promotion
~ **сердца** operation of the heart
джайнизм *rel.* Jainism
джерсе *text.* jersey; jersey jacket
джигит Dzhigit *(Caucasian horseman)*
джигитовка *sport* trick/fancy riding
джин *agr.* gin
джинирование *agr.* ginning
джинн genie, jinn
джинсы jeans
джоуль *phys.* joule, J
джоульметр joulemeter
джут *bot.* jute
джутоткачество *text.* jute weaving
дзерен *zool.* dzeren *(Procarpa)*
~, **монгольский** Mongolian gaselle (*Procarpa gutturosa)*
~, **тибетский** Tibetan gazelle (*Procarpa picticaudata)*
дзета-потенциал *el.* zeta potential
дзета-функция *math.* zeta function
дзэн *rel.* Zen
дзю-до *sport* judo
дзюдоист *sport* judoist
диабаз *geol.* diabase
~, **игольчатый** acicular diabase
~, **пятнистый** spotted diabase
~, **шаровой** spheroidal diabase
диабет *med.* diabetes
~, **аллоксановый** alloxan diabetes
~, **бронзовый** bronzed diabetes
~, **гипофизарный** hypophysial diabetes
~, **голодный** starvation diabetes
~, **несахарный** diabetes insipidus
~, **островковый** insular diabetes
~, **перемежающийся** diabetes intermittens
~, **почечный** renal diabetes
~, **сахарный** diabetes mellitus
~, **скрытый** masked diabetes, diabetes decipiens
~, **старческий** senile diabetes
~, **фосфатный** phosphatic diabetes
~, **юношеский** juvenile diabetes
диагноз diagnosis
~, **окончательный** final/firm diagnosis
~, **ошибочный** misdiagnosis
~, **патологоанатомический** pathologic diagnosis; postmortem diagnosis
~ **при поступлении больного** admission diagnosis
~, **ставить** diagnose
диагност, диагностик diagnostician
диагностика diagnosis; diagnostics
~, **автономная** off-line diagnosis
~ **в процессе компиляции** compile-time diagnosis
~ **в рабочем режиме** on-line diagnosis
~, **дистанционная** remote diagnosis
~, **оперативная** on-line diagnostics
~ **отказов** failure diagnostics
~ **ошибок** error diagnosis
~, **серологическая** serum diagnosis
~, **ультразвуковая** ultrasound diagnosis
диагностикум diagnosticum
~, **вирусный** viral diagnosticum
~, **гриппозный** influenza diagnosticum
~, **паротитный** parotitis diagnosticum
диагностирование diagnostication, diagnosis
~ **сбоев** malfunction diagnosis
диагностировать diagnose, diagnosticate
диагональ diagonal (line)
~, **главная** principal/main/leading diagonal
~, **побочная** secondary diagonal
диаграмма diagram, chart, plot, graph; pattern; *el.* circuit
~ **в декартовых координатах** Cartesian diagram
~ **в полярных координатах** polar diagram
~, **векторная** vector diagram
~, **временная** time chart/graph/diagram
~ **вход-процесс-выход** input-process-output

diagram
~ дальности действия *rad.* coverage diagram
~ излучения radiation pattern/envelope
~ изодоз isodose chart
~ истечения effluogram
~ истинности *cyb.* truth diagram
~, коммутационная plugging chart
~, круговая circle/clock/locus diagram; pie chart; circular graph/chart
~, ленточная band/strip chart
~, линейная line diagram
~ нагрузки *com.* traffic diagram
~ Найквиста Nyquist diagram
~ направленности beam pattern, directional characteristic diagram/pattern
~ напряжении stress-strain diagram
~ неопределенности ambiguity diagram
~, номографическая nomographic/alignment chart
~, обзорная synoptic diagram
~ объект-отношение entity-relation diagram
~ остойчивости stability curve
~ палочковая stick diagram
~ переходов state transition diagram
~ потоков данных data flow chart/diagram
~, пространственная block-diagram, graphic model
~, рабочая operational chart; flow chart/diagram
~ распределения нагрузки loading chart
~ расхода flow diagram
~ ресурсов resource map
~, сводная cumulative/collective diagram
~, секторная circular/pie chart
~ сил stress diagram
~, столбиковая/столбчатая column/bar graph/chart/diagram
~, точечная dot chart; point-by-point diagram/plot
~ трафика traffic record/curve
~, узловая nodal diagram
~, функциональная action/operation chart
~, цикловая cyclogram
~, энтропийная enthropy-temperature diagram
диаграммный diagrammatic(al)
диада dyad
диадик, диадический *math.* dyadic
диадохокимограф diadochokymograph
диадохокинез diadochokinesis, diadochokinesia
диазоаминол diazoaminol dye
диазосоединение diazo compound
диазотипия diazotype (process)
диазотирование *chem.* diazotization
диазотировать diazotize
диак *el.* diac, bidirectional triode thyristor
диакарб *chem.* acetazolamide
диакаустика *phys.* diacaustic
диакаустический *phys.* diacaustic
диакинез *biol.* diakinesis
диаклаз *geol.* diaclase
диакритический *phon.* diacritic(al)
диакустика *phys.* diacoustics
диалектик *philos.* dialectician
диалектика *philos.* dialectics
диалектический, диалектичный *philos.* dialectic(al)
диалектолог *ling.* dialectologist
диалектология *ling.* dialectology
диализ dialysis
~, мембранный dialysis through a membrane/septum
~, перитонеальный peritoneal dialysis, vividialysis
диализат dialyzate
диализатор dialyzer, dialyzator
диализировать dialyze
диаллаг *miner.* diallage
диалог dialog(ue)
~ в запросно-ответном режиме question/answer dialogue
~, вложенный nested dialogue
~, вспомогательный subdialogue
~ с подсказками prompted dialogue
~ типа выбора меню menu dialogue
диалоговый *comp.* interactive, conversational
диальдегид *chem.* dialdehyde
диамагнетизм diamagnetism
диамагнетик diamagnetic material/substance
диамагнитный diamagnetic
диамант *print.* diamond
диамат *abbr.* dialectical materialism
диаметр diameter
~ без коры [в коре] *timb.* diameter inside [under] bark
~ в свету bore/inner/clear diameter, bore
~, внешний external/major/outer diameter
~, внутренний inside/internal/minor diameter
~ входа inlet diameter
~, габаритный full diameter
~ навивания провода coiling diameter of a wire
~, наружный outer/outside/full/major diameter
~ обточки turning/machining diameter
~ расточки bore diameter, bore
~ сверления drilled diameter
~ цилиндра cylinder bore
диаметрально противоположный diametrically opposite, antipodal
диаметральный 1. diametric(al); 2. diametral, of diameter
диамид *chem.* diamide
диамин *chem.* diamine
диаминопимелиновый diaminopimelic
диаминурия diaminuria
диаммофос *chem.* diammophos, diammonium phosphate
диапазон range, scope; band; spectrum, gamut; margin
~ адресов address range
~ амплитуд amplitude range
~ весов weight brackets

~, **вещательный** broadcast band
~ **возможностей** range of options
~ **волн** wave/frequency band
~ **высот** altitude range
~ **гармоник** frequency band of harmonics/ formants
~ **громкости** volume range
~ **допустимых значений** tolerance range
~, **закритический** supercritical range
~ **записи** recording band
~ **значений ошибки** error span, range of error
~ **излучения** radiation band
~, **инфракрасный** infra-red band
~ **коротких волн** short-wave band
~, **любительский** *rad.* amateur band
~ **метровых волн** metric-wave band, VHF band
~ **миллиметровых волн** millimetric-wave band, EHF band
~ **мириаметровых волн** miriametric wave band, very-low frequency band, VLF band
~ **мощностей** power range
~ **настройки** tuning range; *mech.* range of adjustment
~, **обзорный** search band
~ **ошибок** *comp.* error range
~ **порядков (чисел)** *comp.* order range
~, **рабочий** operating/operative/ effective/ working range
~, **радиовещательный** broadcast band
~ **сантиметровых волн** centimetric-wave band; superhigh/ultrahigh frequency band, SHF band
~ **сверхдлинных волн** superlong-wave band
~ **СВЧ** microwave range/band
~ **слышимости** range of audibility/hearing
~ **средних волн** medium-wave band
~ **температур** temperature span/interval/range
~, **узкий** narrow band; close range
~ **цен** price range/bracket
~ **частот** frequency range/band
~ **чисел** number range
~ **чувствительности** sensibility/sensitivity range, range of response
диапауза *zool.* diapause
диапедез *physiol.* diapedesis
диапозитив slide; transparency
диапроектор slide projector, diascope
диапсидный *paleont.* diapsid, diapsidian
диарея *med.* diarrh(o)ea
~, **вирусная** virus diarrh(o)ea
диартроз *anat.* diarthrosis
диархия diarchy
диас *geol.* Dyas
диаскоп diascope, slide projector
диаспор 1. *miner.* diaspore; 2. *bot.* diaspore, disseminule, dispersal unit
диастаз *med.* diastasis
диастаза *bioch.* diastase
диастальтика *med.* diastalsis
диастальтический diastaltic
диастема *biol.* diastem(a)
диастематомиелия *med.* diastematomyelia
диастола *biol.* diastole
~, **замедленная** bradydiastole, bradydiastolia
диастолический diastolic
диастрофизм *geol.* diastrophism
диасхиз *physiol.* diaschisis
диасхистический, диасхистовый diaschistic
диатез *med.* diathesis
~, **геморрагический** h(a)emorrhagic diathesis
~, **железистый** strumous diathesis
~, **контрактурный** contractural diathesis
~, **мочекислый** uric acid diathesis, uratic/lithic diathesis
~, **оксалемический** oxalic acid diathesis
~, **подагрический** gouty diathesis, goutiness
~, **спазматический** spasmodic diathesis
~, **туберкулезный** tuberculous diathesis
диатермический, диатермичный diathermic, diathermous
диатермичность diathermy; diathermance, diathermancy
диатермия *med.* diathermy
~, **щадящая** medical diathermy
диатермогрязелечение diathermomudtherapy
диатермоионтофорез diathermoiontophoresis
диатермокоагулятор diathermocoagulator
диатермотомия diathermotomy
диатомит *miner.* diatomite, diatomaceous earth
диатомовый diatomic; diatomaceous
диатрема *geol.* diatreme, volcanic pipe/neck/vent/ funnel
диафания diaphaneia
диафаноскоп diaphanoscope
~, **глазной** ophthalmic diaphanoscope
диафенилсульфон diaphenylsulphone
диафиз *anat.* diaphysis
диафизарный diaphysial, diaphyseal
диафизит diaphysitis
диафильм filmstrip; slide strip
диафон 1. diaphone; 2. audio-visual recorder
диафонема *phon.* diaphoneme
диафония 1. *mus.* diaphony; 2. *com.* crosstalk
диафрагма 1. diaphragm, membrane; midriff; 2. *phot.* stop; 3. *met.* orifice plate
~, **емкостная** *el.* capacitative diaphragm/window
~, **измерительная** orifice plate
~, **индуктивная** *el.* inductive diaphragm/window
~, **коллимирующая** collimating aperture
~, **мочеполовая** urogenital diaphragm/trigone
~ **объектива** diaphragm; lens stop; aperture
~ **полости рта** oral diaphragm
~, **развертывающая** scanning diaphragm
~, **разрывная** blow-out diaphragm/disk
~, **револьверная** turret/rotary/revolving diaphragm
~, **регулирующая** regulating orifice/valve
~, **световая** highlight stop
~ **сцепления** clutch diaphragm
~, **теневая** shadow stop
~, **щелевая** slit diaphragm

диафрагмальгия *med.* diaphragmalgia, diaphragmodynia
диафрагмально-желудочный phrenogastric
диафрагмально-печеночный phrenohepatic
диафрагмально-толстокишечный phrenocolic
диафрагмальный diaphragmatic, diaphragmic; *anat.* phrenic
диафрагматит diaphragmatitis, diaphragmitis
диафрагмировать 1. stop; diaphragm; 2. provide/adjust a diaphragm
диафрагмотомия *med.* diaphragmotomy
диацетил *chem.* diacetyl
диацетилморфин diacetylmorphine, heroine
диацетилхолин diacetylcholine
диашистовый geol. diaschistic
дибит *comp.* dibit
диборан *chem.* diborane
дивариантный bivariant, divariant
дивергент *phon.* divergent, allophone
дивергентный divergent, diverging
дивергенция divergence, divergency
дивергировать diverge
диверсант saboteur; wrecker
диверсионный sabotage *(attr.)*
диверсия 1. diversion; 2. sabotage; subversion
дивертикул diverticulum
~, аллантоисный allantoic diverticulum
~, глоточно-пищеводный pharyngo-esophageal diverticulum
~, пульсионный pulsion diverticulum
~ тощей кишки jejunal diverticulum
дивертикулез diverticulosis
дивертикулема diverticuloma
дивертикулит diverticulitis
дивертикулэктомия diverticulectomy
дивиденд dividend
~ выплаченный акциями stock dividend
~ выплачиваемый пропорционально покупкам dividend on purchases
~, денежный cash dividend
~, дополнительний superdividend
~ на акции share dividend
~ определенный уставом statutary dividend
~ по привилегированным акциям preferred dividend
~, предварительный dividend on account
~, финансовый fiscal dividend
~ы, накопленные accrued/accumulated dividends
~ы, отсроченные deferred dividends
дивизион *mil.* battalion; *naut.* squadron
дивизия *mil.* division
дивизор *math.* divisor
дивинил *chem.* divinyl
дигаметический *biol.* digametic; heterogametic
дигаплоид dihaploid
дигедральный dihedral
дигексагональный *geom., cryst.* dihexagonal
дигексаэдр *geom., cryst.* dihexahedron
дигексаэдрический dihexahedral
дигерирование *chem.* digestion
дигетерозигота *biol.* diheterozygote
дигетерозиготный diheterozygous
дигибридный dihybrid
дигидрат *chem.* dihydrate
дигидрит *chem., miner.* dihydrite
дигидробензол dihydrobenzene
дигидрокортизон dihydrocortisone
дигидроксикодеинон *pharm.* dihydroxycodeinone, dicodid
дигидропорфирин dihydroporphyrin
дигидрострептомицин dihydrostreptomycin
дигидроурацил dihydrouracil
дигидрофолевый dihydrofolic
дигиталин *chem.* digitalin
дигиталис 1. *pharm.* digitalis; 2. *bot.* foxglove *(Digitalis)*
дигитоксоза *med.* digitoxose
диглиф *arch.* diglyph
диглицерид *chem.* diglyceride
диграф, диграмма digraph
дидактик didactician
дидактика didactics
дидактичность didacticism
дидактичный didactic
дидодекаэдр *cryst.* didodecahedron
дидодекаэдрический didodecahedral
Дидона *myth.* Dido
диез *mus.* sharp
диены *chem.* dienes, diolefin(e)s
диета diet
~, ацидотическая acid-ash diet
~, безбелковая protein-free diet
~, белковая protein diet
~, бессолевая salt-free diet
~, высококалорийная high-calorie diet
~, голодная low-calorie diet
~, разгрузочная fasting diet
~, творожная cottage-cheese diet
~, углеводистая carbohydrate diet
~, щадящая sparing diet
~, щелочная alkaline diet
диетврач dietitian
диететика dietetics
~, заместительная replacement dietetics
диететический dietetic, dietary
~ магазин health food store
диетик dieter
диетический dietetic, dietary
диетолог dietitian; nutritionist
диетотерапия dietetics; nutritional/dietetic treatment
~, заместительная replacement dietetics
~, щадящая sparing dietetics
дизайн design
дизайнер designer
дизартрия *med.* dysarthria
~, мозжечковая cerebellar dysarthria
~, спастичная spasmodic dysarthria
дизеловоз *rail.* diesel locomotive

дизелестроение building of diesel engines
дизель Diesel engine; *Am.* diesel (engine); oil [compression-ignition] engine
~, безнадувный normally aspirated diesel engine
~, бескомпрессорный airless-injection diesel engine
~, быстроходный high-speed diesel engine
~, вихрекамерный swirl-chamber diesel engine
~, воздушнокамерный air-cell diesel engine
~, высокооборотный high-speed diesel engine
~ двойного действия double-acting diesel engine
~, двухтактный two-stroke diesel engine
~, двухтопливный dual-fuel diesel engine
~, компрессорный air-injection diesel engine
~, крейцкопфный crosshead-type diesel engine
~, многотопливный multifuel diesel engine
~, морской marine diesel engine
~, низкооборотный slow-speed diesel engine
~, рядный row [in-line] diesel engine
~ с газотурбинным наддувом turbo-charged diesel engine
~ с наддувом supercharged diesel engine
~ с неразделенной камерой open-chamber diesel engine
~, среднеоборотный medium-speed diesel
~, судовой marine diesel engine
~, тихоходный slow-speed diesel engine
~, тронковый trunk-piston diesel engine
~, четырехтактный four-cycle-/stroke diesel engine
дизель-генератор diesel driven electric set/generator
дизель-электровоз diesel electric locomotive
дизель-электроход diesel-electric ship
дизентерийный *med.* dysenteric
дизентерия *med.* dysentery
~, амебная am(o)ebic/walking dysentery
~, балантидийная balantidic dysentery
~, бильгарциозная bilharzial dysentery
~, спириллезная spirillar dysentery
дизестезия *med.* dysesthesia
дизодонтоз *med.* dysodontiasis
дизонтогенез dysontogenesis
дизонтогенетический dysontogenetic
дизопия dysopia, dysopsia
дизосмия dysosmia
дизостоз *med.* dysostosis, disosteogenesis
~, ключично-черепной cleidocranial dysostosis
~, рото-пальце-лицевой orodigitofacial dysostosis, oral-facial-digital syndrome
~, ушно-челюстный otomandibular dysostosis
~, челюстно-лицевой mandibulofacial dysostosis
~, черепно-лицевой craniofacial disostosis
дизурия *med.* dysuria, dysury
дизъюнктивный disjunctive
дизъюнктность disjunction, disjointness
дизъюнктный disjunct, disjoint; disjunctive; discontinuous
дизъюнкция disjunction
~, неразделительная inclusive disjunction
~, разделительная exclusive disjunction
~, строгая exjunction
~, условная conditioned disjunction
дизэндокриния *med.* dysendocrinism, dysendocrinia, dysendocriniasis
дииодид *chem.* diiodide
дииодохин diiodohydroxyquin, diiodoquin
дииодтирозин diiodotyrosine
дикаин tetracaine hydrochloride
дикарион *bot.* dikaryon
дикариофит *bot.* dikaryophyte
диккенсовский Dickensian
дикобраз *zool.* porcupine *(Hystrix)*
дикротия *med.* dicrotism
диктат *polit.* dictate
диктиокинез dictyokinesis
диктиома *med.* dictyoma
диктиосома dictyosoma (*pl.* dictyosomata)
диктиоспора *biol.* dictyospore
диктиостела *bot.* dictyostele
диктофон dictating machine, dictaphone
дикумарин dicoumarol, bishydroxycoumarin
дикция enunciation, articulation
дилататор *med., anat.* dilator
дилатограф dilatograph
дилатометрия dilatometry
дилер *econ.* dealer
~, биржевой dealer in stocks, exchange dealer
дилижанс stagecoach
~, почтовый mail coach
дилогия dilogy, ambilogy
дилювиальный *geol.* diluvial
дилювий *geol.* drift, diluvium
дилятор *anat.* dilator
димеризация *chem.* dimerization
димерность *bot.* dimery
димерный dimeric, dimerous
диметиламин *chem.* dimethylamine
диметилгидразин dimethylhydrazine
диметилкремный dimethylsilicone
диметилтриптамин dimethyltryptamine
диметилфталат dimethylphthalate
диморфный dimorphous, dimorphic
дина *phys.* dyne
динамик loudspeaker
динамика dynamics
~, газовая gas dynamics
~ жидкости fluid dynamics
~, звездная stellar dynamics
~ издержек *econ.* cost behaviour
~, квантовая quantum dynamics
~ конструкции structural dynamics
~ корабля ship dynamics; dynamics of naval architecture
~ летательных аппаратов aircraft/airframe dynamics
~ материальной точки particle dynamics
~ населения population/demographic develop-

ment
~ **нелинейных систем** non-linear dynamics
~ **полета** flight dynamics
~ **ракет** missile/rocket dynamics
~ **рождаемости** fertility dynamics
~ **системы материальных точек** dynamic of mass points
~ **системы управления** control system dynamics
~ **твердого тела** rigid-body dynamics
~ **точки** particle dynamics
~ **управляемых ракет** guided-missile dynamics
~ **цен** price development/dynamics
~ **экономической системы** behaviour of economic system
динамит dynamite
~, **желатинированный** gelatine dynamite
~, **морозостойкий** uncongealable dynamite
динамический dynamic(al)
~**кие денежные процессы** *econ.* monetary dynamics
динамограф recording dynamometer, dynamograph
динамокардиограф dynamocardiograph
динамо-машина dynamo
динамометаморфизм dynamic metamorphism
динамометр dynamometer; load gauge
~, **интегральный** integrating dynamometer
~, **крутильный** torque/torsion dynamometer
~, **мышечный** muscular dynamometer
~, **реверсивный** reversible dynamometer
~, **ручной** hand dynamometer
~, **становой** *med.* torso dynamometer
~, **стрелочный** dial dynamometer
~, **тензометрический** strain-gauge dynamometer
~, **тормозной** absorption/brake dynamometer
динамометрия dynamometry
динарий *hist.* denarius
динатрон dynatron
динги *naut.* dinghy
динго *zool.* dingo *(Canis dingo)*
динейтрон dineutron, bineutron
динистор dynistor, dinistor, diode thyristor
динитробензол *chem.* dinitrobenzene
динитротолуол *chem.* dinitrotoluene
динод dynode
динозавр dinosaur
~, **рогатый** ceratopsian
~, **утконосый** trochodont
~, **ящеротазовый** saurichian dinosaur
Диоген Diogenes
диод diode (tube)
~, **антизвонный** clamp(ing) diode
~, **бескорпусный** chip diode
~, **вакуумный** thermionic diode
~, **варакторный** charge storage diode, varactor diode
~, **волноводный** waveguide diode
~, **вплавной** alloyed diode
~, **выпрямительный** rectifier diode
~, **германиевый** germanium diode
~, **двойной** double/twin/dual diode
~, **двухбазовый** double-base diode
~, **диффузионный** diffused diode
~, **зенеровский** Zener/avalanche diode
~ **интегральной схемы** integrated (circuit) diode, IC diode
~, **интегральный** integrated diode
~, **ионный** gas-filled diode
~, **ключевой** switching diode
~, **кремниевый** silicon diode
~, **кристаллический** crystal/semiconductor diode, rectified diode
~, **лавинно-пролетный** avalanche transit time diode
~, **лазерный** laser diode
~ **маячкового типа** lighthouse diode
~, **обратносмещенный** reverse-biased diode
~, **обращенный** backward diode
~, **опорный** voltage-reference diode
~, **параметрический** varactor, parametric diode
~, **плазменный** plasma diode
~, **планарный** planar diode
~, **пленочный** thin-film diode
~, **плоскостный** junction diode
~, **полупроводниковый** semiconductor [solid-state, crystal] diode
~, **прямосмещенный** forward-biased diode
~, **разделительный** isolation diode
~ **с обратным смещением** back-biased diode
~, **сварной** bonded diode
~, **смесительный** mixer diode
~, **сотовый** honeycomb diode
~, **сплавной** alloyed diode
~, **таблеточный** pill diode
~, **тонкопленочный** thin-film diode
~, **точечный** point-contact diode
~, **туннельный** tunnel diode
~, **управляемый** controlled diode
~, **шумовой** noise diode
~, **электровакуумный** diode tube
диоксан dioxane
диоксид dioxide
диоктаэдрический *cryst.* dioctahedral
дионин ethylmorphine hydrochloride
Дионис Dionysos
диоптр *opt.* sight; diopter
диоптрика dioptrics
диоптрия *opt.* diopter; dioptry
~, **призменная** prism diopter
диорама diorama, composite picture
диорит *petr.* diorite
диотический diotic, binaural
диофантовый *math.* diophantine
дипептид *chem.* dipeptide
дипирамида dipyramid, bipyramid
диплакузис *med.* diplacusis
диплегия *med.* diplegia
~, **жевательная** masticatory diplegia

~ **лицевых нервов** facial diplegia
~, **мозжечковая** cerebellar diplegia
диплекс diplex working/operation; twinplex operation
диплогаплонт *biol.* diplohaplont
диплодиоз *bot.* diplodia dry rot
диплоид diploid
~, **стойкий** persistent diploid
~, **функциональный** functional diploid
диплоидия, диплоидность diploidy
диплоидный diploid
диплокарион diplokaryon
диплокаулический, диплокаульный *bot.* diplocaulescent
диплококк diplococcus
диплоксилический diploxylic
диплом diploma, certificate; degree
дипломант student working on a degree thesis; graduating student
дипломат diplomat; diplomatist
дипломатический корпус diplomatic corps
дипломированный certified, certificated, licensed; registered
дипломная работа degree thesis/work; graduate thesis/work
диплопия *med.* diplopia
~, **одноименная** homonymous diplopia
~, **перекрестная** heteronymous/crossed diplopia
~, **перекрестная двусторонняя** bilateral crossed diplopia
~, **физиологическая** physiologic/introspective diplopia
диплопора diplopore, double pore
диплоспория *bot.* diplospory
диплостемония *bot.* diplostemony
диплострептококк diplostreptococcus
диплот *naut.* sounding lead, deep-sea lead
диплотена diplotene
диплофаза *biol.* diplophase, diploid phase
диплофилл *bot.* diplophyll
диплофония diplophony
диплохламидный *bot.* diplochlamydeous
диплохория diplochory
диплохромосомы *biol.* diplochromosomes
диплоэ *anat.* diploe
диплоэдр diplohedron
диполь dipole, doublet; *mech.* double source, doublet; *el.* dipole aerial
~, **акустический** acoustic dipole; simple dipole source
~, **волновой** wave dipole
~, **направляющий** director dipole
~, **полуволновой** half-wave dipole
~, **широкополосный** wide-band dipole
~, **элементарный** elementary/infinitesimal dipole
дипольный dipole *(attr.)*
диполярный dipolar
дипразин *pharm.* promethazine hydrochloride
дипрофен diprofene
дипрофилин diprophylline
дипсомания *med.* dipsomania
дипсофобия *med.* dipsophobia
диптанк *naut.* deep-tank
диптих diptych
директива directive, instruction; *comp.* order
директор 1. manager; director; 2. *el.* passive aerial, director
~, **коммерческий** sales manager
~ **школы** principal
директор-администратор inside director
директорат board of directors/managers
директория *hist.* Directory, Directoire
директор-распорядитель managing/executive director, operations manager
директриса *math.* directrix (*pl.* directrices)
дирижаблестроение airship construction/building
дирижабль airship, dirigible
~, **жесткий** rigid airship
~, **мягкий** non-rigid airship
~, **полужесткий** semi-rigid airship
дирижизм *econ.* dirigisme *(Fr.)*
дисахарид *chem.* disaccharide
~, **редуцирующий/нередуцирующий** reducing/nonreducing disaccharide
дисбазия *med.* dysbasia
дисбактериоз *med.* dysbacteriosis
дисбаланс unbalance, disbalance; out-of-balance
дисбаризм *med.* dysbarism
дисбулия *med.* dysbulia
дисгевзия *med.* dysgeusia
дисгенезия *biol.* dysgenesis
дисгенитализм *med.* eunuchoidism
дисгенический, дисгеничный *biol.* dysgenic
дисгидроз *med.* dyshidrosis, dysidrosis
дисгипофизия *med.* dyshypophysia, dishypophysism
дисгликемия dysglycemia
дисграфия *med.* dysgraphia
дисдиадохокинез *med.* dysdiadochocinesia, dysdiadokokinesia
диссимметрия asymmetry, dissymmetry
диск disk, disc; *sport* discus; *mech.* wheel
~, **библиотечный** *comp.* library disk
~, **высевающий** seed disk/plate
~, **гибкий** floppy disk
~, **делительный** indexing disk/plate
~ **единиц** unit dial
~, **жесткий** hard disk
~ **зрительного нерва** optic disk/papilla
~, **кодирующий** code wheel
~ **коллективного пользования** shared disk
~, **компактный** compact disk, CD
~, **контрольный** monitor disk
~, **кривошипный** crank plate
~, **кулачковый** cam plate
~, **линзовый** *telev.* lens scanning disk
~, **магнитный** magnetic disk/plate
~, **магнитооптический** magneto-optic disk
~, **наборный** *teleph.* dial

~, несущий *av.* lifting disk, hub plate
~, ножевой cutting wheel
~ однократной записи write-once disk
~, односторонний single-sided disk
~, плужный plough disk
~, программно-секционированный soft-sectored disk
~, прядильный godet wheel
~, развертывающий scanning disk
~ Рэлея Rayleigh disk
~ с отверстиями apertured disk
~ с неподвижными головками fixed-head disk, multihead disk
~ с подвижными головками moving-arm/-head disk
~ с удвоенной плотностью записи double-density disk
~, системный *comp.* system residence disk
~, стираемый erasable disk
~, суставной *anat.* articular disk
~ сцепления clutch plate
~ сцепления, ведомый/ведущий clutch driven/drive plate
~, тормозной brake disk
~, упорный thrust/stop plate; bearing disk
~, форматированный formatted disk
~, фрикционный friction disk, clutch plate
~, шкальный dial
дискант *mus.* treble, descant
дискератоз *med.* dyskeratosis
~, врожденный congenital dyskeratosis
дискет, дискета diskette
дискинезия *med.* dyskinesia
дискование *agr.* disk harrowing, disking
дисковидный, дискообразный disk-shaped, disciform, discoid
дискоз *med.* discosis
дискоидальный discoid(al); *paleont.* patelloid
дисколит *geol.* discolith, discoidal coccolith
дискомицеты *Discomycetes*
дисконт *fin.* discount
~, банковский bank discounting
дисконтер discounter
дисконтинуум *math.* discontinuum
~, двоичный dyadic discontinuum
~ канторов Cantor set/discontinuum
дисконтировать *fin.* discount
~ вексель negotiate a bill of exchange
дископодобный disk-like, discoid
дискос *eccl.* paten, patin(e)
дископатия *pathol.* discopathy
дископлацента *anat.* discoplacenta
дискордантность discordance
дискория *med.* discoria, dyscoria, diplocoria
дискотека record library; discotheque, disco
дискоцветковый *bot.* discifloral, discoflorous
дискразия *med.* dyscrasia
дискрепантность discrepancy
дискретизация discretization, quantification; sampling; digitization
~ по времени (time) sampling
~ по уровню quantization, quantizing
дискретно *tech.* discretely, in steps/increments
дискретный *tech.* discrete, discontinuous; digital; quantified
дискреционный *polit.* discretionary, discretional
дискриминант *math.* discriminant
дискриминантный *math.* discriminantal
дискриминатор *electr.* discriminator; detector
~, амплитудный интегральный integrating pulseheight/amplitude discriminator
~, временной time discriminator
~, двоякопреломляющий birefringent discriminator
~, двухтактный push-pull discriminator
~, дробный ratio detector
~ импульсов pulse discriminator
~ импульсов по ширине pulse-width discriminator
~, однотактный single-ended discriminator
~, равноплечий equal-arm discriminator
~, стробированный strobed discriminator
~, фазовый phase discriminator
~, частотный frequency discriminator
~, широкополосный broadband discriminator
дискриния *med.* dyscrinism
дислалия *med.* dyslalia
дислексия *med.* dyslexia
дислогия *med.* dyslogia
дислокация dislocation; *geol.* displacement, shifting; *mil.* stationing
~, базисная basal dislocation
~, винтовая screw dislocation
~, дизъюнктивная disjunctive dislocation, faulting
~, краевая edge dislocation
~, ледниковая dislocation by glacier
~, оползневая landslide dislocation
~, сбросовая fault displacement
~, складчатая plicative dislocation
~, соляная salt tectonics
дислоцировать 1. *geol.* dislocate; 2. *mil.* station *(troops)*
дислюит *geol.* disluite
дисмегалопсия *med.* dysmegalopsia
дисменорея *med.* dysmenorrh(o)ea
~, оболочная membranous dysmenorrh(o)ea
~, первичная primary dysmenorrh(o)ea
~, спастическая spasmodic dysmenorrh(o)ea
~, трубная tubal dysmenorrh(o)ea
~, уретральная uretric dysmenorrh(o)ea
~, яичниковая ovarian dysmenorrh(o)ea
дисмеризм *biol.* dysmerism
дисметрия *med.* dysmetria
дисмимия *med.* dysmimia
дисмиотония dismyotonia
дисмнезия dysmnesia
дисоматический disomatic
дисомия disomy
диспанкреатизм *med.* dyspancreatism

диспансер *med.* health centre; outpatient department, dispensary
диспансеризация prophylactic medical examination/treatment
диспарейния *med.* dyspareunia
диспаша *law* average statement/adjustment; *fin.* claims adjustment
~, аварийная average statement
~шу, составлять make up the average, adjust the average
диспашер *law* average adjuster
диспепсия *med.* dyspepsia
~, гнилостная putrid dyspepsia
~, печеночная hepatic dyspepsia
~, рефлекторная reflex dyspepsia
диспергент, диспергатор dispersing medium/agent, dispersant
диспергировать disperse, scatter; diffuse
дисперсивный, дисперсионный dispersive, dispersing
дисперсия dispersion; scattering; *mat., stat.* variance, variability; standard deviation
~, водная aqueous dispersion
~ волн wave dispersion
~ воспроизводимости *math.* error mean square
~ выборки sampling variance
~, выборочная sample variance
~ генеральной совокупности population/universe variance
~ звука acoustic/sound dispersion
~ межгрупповая external variance
~, ожидаемая *stat.* desired/predicted variance
~ ошибок error variance/scatter
~, регенеративная reclaim dispersion
~, случайная accidental/random variance
дисперсность dispersity; dispersiveness, degree/extent of dispersion; degree/range of fineness
дисперсный dispersed, disperse, dispersive
дисперсоидний dispersoid
диспетчер 1. controller; (traffic) supervisor; 2. *comp.* dispatcher; executive programme, monitor (programme)
~, авиационный flight operations officer, air-traffic control officer
~ заданий *comp.* task dispatcher
~, наземный *av.* ground controller
~, поездной train dispatcher
~ посадки *av.* final controller
~ рабочих комплектов *comp.* working set dispatcher
~ службы движения *av.* air traffic controller, ATC officer
диспетчеризация introduction of monitoring/control/dispatcher system; production control; dispatching; traffic control
диспетчерский of controller/supervisor/monitor
~кая вышка *av.* control tower/room
дисплазия *med.* dysplasia
~, глиальная glial dysplasia
~, диафизарная diaphysial dysplasia
~ желчного пузыря gall-bladder dysplasia, cholecystosis
~ метафизарная metaphysial dysplasia
~, хрящевая chondral dysplasia
дисплей *electr.* display
~ без блока памяти nonstorage display
~, буквенно-цифровой alphanumeric/alphameric/character display
~, двухкоординатный two-dimensional display
~ зеленного свечения green-phosphor display
~, информационный data display
~, контрольный monitor display
~, матричный matrix display
~, монохроматический monochrome display
~, немерцающий flicker-free display
~, полиэкранный multiple window display
~, полноадресуемый all-point-addressable display
~, полностраничный full-page display
~, полустраничный half-page display
~, растровый raster(-scan) display
~ с блоком памяти storage display
~ с изображением вытянутым по вертикали portrait display
~ с изображением вытянутым по горизонтали landscape display
~ с клавиатурой keyboard display
~ с поэлементным отображением bit-map(ped) display
~ с программно-управляемым лучом calligraphic display
~ с регенерацией изображения refresh display
~ с речевым вводом voice-entry display
~, сенсорный touch display
~, узкоформатный thin-window display
~, цифровой numeric display, (all-)digital display
~, черно-белый black-and-white display
~, широкоформатный great display
~, экранный CRT display
дисплоидия dysploidy
диспноэ *med.* dyspn(o)ea, difficult breathing
диспозитивный *law* optional, non-mandatory
диспозиция disposition
диспондей *lit.* dispondee
диспраксия *med.* dyspraxia
диспрозий *chem.* dysprosium, Dy
диспропорциональность disproportionality, disproportion
диспропорционирование disproportionation
диспропорция disproportion
диспротеинемия *med.* dysprotein(a)emia
дисрафия *med.* dysraphia
диссектор *telev.* dissector tube, image dissector
диссеминула *bot.* disseminule, diaspore
диссертация dissertation, thesis
~, кандидатская master's thesis
диссидент dissident, dissenter; nonconformist

диссимилятивный dissimilative, dissimilatory
диссимметрия dissymmetry
диссимуляция *med.* dissimulation, dismalingering
диссинергия dyssynergia
диссипация dissipation; dispersion; scattering
диссонанс dissonance, disharmony; discord
~ сознания cognitive dissonance
диссоциация dissociation
~, клеточная cytogenic dissociation
~, клеточно-белковая albumino-cytologic dissociation
~, микробная bacterial/microbic dissociation
~ молекул [на молекулы] molecular dissociation
~ самопроизвольная spontaneous dissociation
~ сознания distractibility
~, фотохимическая photodissociation, photochemical dissociation
диссоциировать dissociate, split up, break up, break down
диссоциируемый, диссоциирующийся dissociable
дистальный distal
дистанционировать space, pitch
дистанционно *adv.* remotely
дистанционный remote, long-distance, distance-type, remote-controlled, remote-action, distant
~ метод доступа telecommunication access method
~ опрос telequiry
~ная передача данных data telecommunication, data remote transfer
~ная система учета и отчетности teleaccounting system
~ное измерение telemetering
~ное управление remote control
~но-управляемый remotely controlled
дистально-бороздной sulcate, anacolpate
дистально-поровый ulcerate, anaporate
дистанция distance; interval; clearance; range
~, взлетная take-off distance
~ выстрела range of a shot
~, короткая *mil.* close range
~, межзрачковая papillary distance
~, начальная datum range
~, посадочная landing distance
~ пробега *av.* landing/alighting run
~ разбега *av.* take-off run
~ разгона acceleration distance
~ самонаведения homing range
~ торможения braking/stopping distance
дистен *miner.* disthen(e), cyanite, kyanite
дистеррит *miner.* disterrite, brandisite
дистиллирование distillation, distilling
~, разрушительное destructive/dry distillation
~ с водяным паром steam distillation
дистиллировать distil
дистиллят distillate
дистиллятор still, distilling apparatus; distiller
дистилляция = дистиллирование

дистимия *med.* dysthymia
дистиреоз *med.* dysthyroidea, disthyreosis, dysthyroidism
дистоматоз *med.* distomatosis, distomiasis
дистония dystonia
дистопия *med.* dystopia
дисторсия distortion
~, бочкообразная/отрицательная barrel-shaped distortion
~, подушкообразная/положительная pin-cushion distortion
~, радиальная curvilinear distortion
дистрибутивность *math.* distributivity
дистрибутивный *math.* distributive, distributional
дистрофия *med.* dystrophy, dystrophia; degeneration
~, амилоидная amyloid degeneration, amyloidosis
~, белковая albuminoid/albuminous degeneration
~, гиалиновая hyaline/glassy degeneration
~, жировая adipose/fatty degeneration
~, зернистая granular degeneration
~, мышечная muscular dystrophy
дисульфат disulphate
дисульфид disulphide
дисульфокислота disulphonic acid
дисульфонат disulphonate
дисфагия *med.* dysphagia, dysphagy
~, амиотаксическая/сидеропеническая sideropenic/amyotaxic dysphagia
~, болевая painful dysphagia
дисфония *med.* dysphonia
дисфория *med.* dysphoria
дисфункция *med.* dysfunction, malfunction
дисхондроплазия *med.* dyschondroplasia
дисхромия *med.* dyschromia
дисциплина I discipline
~ без прерывания обслуживания noninterruption/nonpreemptive discipline
~, бюджетная *econ.* budgetary discipline
~ очереди *comp.* queue discipline
~ "первым пришел - первым обслужен" *comp.* first-in-first-out discipline, FIFO
~ "последним пришел - первым обслужен" *comp.* last-in-first-out discipline, LIFO
~ с абсолютным приоритетом *comp.* preemptive discipline
~ с динамическими приоритетами *comp.* dynamic priority discipline
~, сознательная self-discipline
~, трудовая labo(u)r discipline
~, учебная classroom discipline
дисциплина II subject; course (of study)
~, ведущая/основная major
~, зачетная course for credit
~, учебная subject
~, факультативная elective/optional course, elective
дисциплинарный disciplinary

~ **батальон** *mil.* penal battalion
дисциссия, дисцизия *med.* discission
дитетраэдр *cryst.* ditetrahedron
дитиазанин *chem.* dithiazanine iodide
дитилин diacetylcholine
дитриглиф *arch.* ditriglyph
диурез *med.* diuresis, urine excretion
~, **облигатный** obligatory/excretion urine
~, **солевой** salt diuresis
~, **суточный** diurnal urine excretion
диуретик *med.* diuretic (drug)
~, **ртутный** mercurial diuretic
дифенил *chem.* diphenyl
дифенин *pharm.* diphenylhydantoin sodium
дифиодонтизм *med.* diphyodontism
дифирамб dithyramb; eulogy
дифманометр differential manometer, differential pressure gauge
дифосген *chem.* diphosgene
дифосфопиридиннуклеотид *pharm.* diphosphopyridine-nucleotide, DPN
дифрагировать diffract
дифрактометр diffractometer
~, **рентгеновский** X-ray diffractometer
дифракция diffraction
~ **Брэгга** Bragg diffraction/scattering
~ **на краю** edge diffraction
~ **на кристалле** crystalline diffraction
~ **на круглом экране** diffraction by a disk
~ **на остром краю** knife-edge diffraction
~ **на отверстии** diffraction by an aperture
~ **на трехмерной решетке** diffraction by a crystal lattice
~ **частиц** particle diffraction
дифсистема *com.* hybrid set
дифтериеподобный *med.* diphtheroid
дифтерийный *med.* diphtherial, diphtheritic
дифтерия, дифтерит *med.* diphtheria
~, **злокачественная** malignant diphtheria
~, **молниеносная** fulminant diphtheria
дифтероид *med.* diphtheroid, pseudodiphtheria, false diphtheria
дифтеротоксин *med.* diphtherotoxin
дифтонг *phon.* diphthong
диффеоморфизм *math.* diffeomorphism, diffeomorphic mapping
дифферент *naut.* trim
~, **конструктивный** designed trim
~ **на корму** stern trim, trim by the stern
~ **на нос** trim by the bow/head/stem, forward trim
~, **проектный** designed trim
~, **ходовой** running trim
дифферентный trim, trimming
~**ная цистерна** trim(ming) tank
дифферентовка *naut.* longitudinal [fore and aft] trimming/compensation; ballast trim
~, **статическая** statical ballast trimming
дифферентометр *naut.* trim indicator, draught difference indicator
дифференциал 1. *mech.* differential gear; 2. *math.* differential
~ **второго/третьего порядка** second/third differential
~, **гармонический** harmonic differential
~, **двойной** *mech.* double differential gear
~, **конический** bevel gear
~, **межосевой** centre differential (gear)
~, **неполный** *math.* inexact differential
~, **полный** *math.* total/complete/exact/perfect differential
~, **счетный** *math.* computing differential
~, **частный** partial differential
~, **червячный** wormgear differential
дифференциальный differential; distinguishing, discriminative
~ **тариф** differential/discriminative duties
~**ное исчисление** differential calculus
~**ные ставки зарплаты** wage differentials
дифференциатор differentiator
~, **фрикционный (дисковый)** friction-plate differentiator
дифференциация differentiation
~ **доходов** *econ.* income differentiation
~ **заработной платы** wage differentiation
~, **клеточная** cytodifferentiation
~, **осадочная** sedimentary differentiation
~, **фракционная** fractionation
дифференцирование *math.* differentiation, derivation
~, **двукратное** double differentiation
~, **многократное** repeated differentiation
~ **неявной функции** implicit differentiation
~ **по области** derivation/differentiation over a region
~ **по параметру** parametric differentiation/derivation
~, **последовательное** successive differentiation
~ **при помощи логарифмирования** logarithmic differentiation
~ **сложной функции** indirect differentiation
~, **численное** numerical differentiation/derivation
дифференцировать differentiate
дифференцируемость differentiability
дифференцируемый *math.* differentiable
~, **непрерывно** continuously differentiable
диффлюэнция ice flow
диффузионный diffusion *(attr.),* diffusive
диффузия diffusion
~ **в твердой фазе** solid diffusion
~, **взаимная** interdiffusion
~, **вихревая** eddy diffusion
~ **внутри** inward diffusion
~ **газов** gas diffusion
~, **двуполярная** ambipolar diffusion
~ **к концам** end diffusion
~, **локальная** localized diffusion
~ **наружу** outward diffusion
~ **нейтронов** neutron diffusion
~, **нестационарная** unsteady-state diffusion
~, **обратная** reverse/back diffusion

~ по вакансиям diffusion by the vacancy mechanism
~ под давлением pressure diffusion
~, последовательная sequential diffusion
~, стационарная steady-state diffusion
~, турбулентная eddy/turbulent diffusion
~, управляемая controlled diffusion
~ через перегородку barrier diffusion
диффузность diffusivity, diffuseness
диффузный diffuse; diffusive; scattered
диффузор 1. diffuser; 2. *el.* loudspeaker cone, diaphragm
~,безлопаточный vaneless/open diffuser
~, входной inlet diffuser
~, выхлопной exhaust diffuser
~, глухой dummy diffuser
~, двухскачковый double-shock diffuser
~, кольцевой annular diffuser
~, лопаточный vaned diffuser
~, многоскачковый multishock diffuser
~, противоточный counter-current diffuser
~, расходящийся/расширяющийся divergent diffuser
~, регулируемый variable-area diffuser
~ с конической иглой conical-spike diffuser
~, свободно подвешенный *el.* floating cone
~, сужающийся convergent diffuser
~, щелевой open diffuser
диффундировать diffuse; scatter
~ взаимно interdiffuse
дихазий *bot.* dichasium, dichasial cyme
дихейлия *med.* dicheilia, dichilia
дихирия *med.* dicheiria, dichiria
дихлорамин *chem.* dichloramine; *pharm.* dichloramin(e) T
дихлорбензол dichlor(o)benzene
дихлорид dichloride, bichloride
дихлорофен dichlorophene
дихогамный dichogamic, dichogamous
дихотомический dichotomic, dichotomous
~кая совокупность *stat.* dichotomous population
~кое ветвление forked branching
дихотомия dichotomy
дихроизм *cryst.* dichroism, pleochroism
дихроит *cryst.* dichroite, iolite
дихромазия dichromasia, dichromasy, dichromatism
дихромат *chem.* dichromate, bichromate
диэдр *math.* dihedron
диэдральный, диэдрический dihedral
диэлектрик *phys.* dielectric, non-conductor; insulating medium/material
~, высокочастотный radio-frequency [r. f.] dielectric
~, кремнийорганический silicone dielectric
~, многослойный laminated dielectric
~, поглощающий absorptive/lossy dielectric
~ с потерями lossy dielectric
диэлектрический dielectric, non-conducting
диэтиламин *chem.* diethylamine
дланевидно-лопастный palmatilobate
дланевидно-нервный palmativeined
дланевидно-раздельный palmatipartite
дланевидно-рассеченный palmatisect
дланевидный palmate, palmated, palmiform
длина length
~ адреса *comp.* address size
~ буксира *naut.* towing span
~ вектора absolute value of a vector
~ витка *el.* length of a coil turn
~ волны wavelength
~ волны, граничная/критическая cut-off [critical] wavelength
~ волны, собственная natural wavelength
~ выбега coasting distance
~, габаритная overall length
~ двоичного слова *comp.* binary code word length
~ дуги *geod.* arc distance
~ захвата *agr.* cutting length
~ кодовой комбинации *comp.* word length
~, несущая bearing length
~ нулевых данных *comp.* zero data length
~ орбиты orbit circumference
~ очереди *math.* number in the queue, number of units waiting
~ поля *comp.* field length/size
~ пробега landing/alighting run
~ программы *comp.* program(me) size, code length
~, рабочая effective length
~ разбега *av.* take-off run
~ разгона acceleration distance
~, разрывная breaking length
~ распространения propagation distance
~ слова *comp.* word length/size/format
~ строки string length
~ траектории path length
~ хода stroke *(of piston)*
длинноголовость dolichocephalism, dolichocephaly
длинноголовый dolichocephalic, dolichocephalous
длиннопробежный long-range
длинноствольный 1. long-stemmed; 2. *mil.* long-barreled
длиннофокусный long-focus
длинномер surveyor's tape measure
длительность duration, time, length
~ включения on-time
~ жизни life span; lifetime; longevity
~ импульса pulse duration/length/persistence/width
~ искания *teleph.* selection time
~ переходного режима/процесса transient time
~ простоя down time
~ пускового периода start-up time
~ развертки scanning time
~ строба gate/strobe duration

~ торможения breaking/retardation time
дневник journal; diary
~ ошибок *comp.* bug diary
дни days
~, грационные/льготные *econ.* days of grace
~ отсрочки *econ.* days of respite
~, сталийные *econ.* lay days
днище bottom; floor; end-plate; face; end
~ котла boiler endplate
~ кузова floor, floorpan
~, откидное drop/loose/removable bottom
дно bottom; floor; *anat.* fundus
~ бочки barrel/cask head
~, глазное fundus of the eye, eyegrounds
~ кратера crater floor
~, морское sea bed/bottom
~ литорали littoral bottom
~ океана ocean floor/bed
~ пещеры cave floor
~ резьбы base of a thread
~ реки river bed
~ шлюза floor/bottom of a lock
~ экономического цикла business cycle trough
дноуглубитель, дночерпатель dredger
добавка addition; additive, admixture; dope
~, антидетонационная detonant
~, вспучивающая expansion/bloating agent
~, гасящая quenching admixture, quencher
~, гидрофобная waterproofing admixture
~, диспергирующая dispersing agent
~, пенообразующая foaming agent
~, пластифицирующая plasticizer
~, поверхностно-активная surfactant (admixture)
~, противоморозная antifreeze (admixture)
добавка-замедлитель retarder
добавка-ускоритель accelerator
добавленная стоимость *econ.* value added
добавочный additional, supplementary; accessory; extra
~ налог surtax
~ное время extra time
добалтийский Pre-Baltic
дображивание after-/secondary fermentation
добрачный antenuptial, premarital
доброкачественный 1. of good/high quality; 2. *med.* benign, nonmalignant
добросовестно conscientiously; *law* in good faith
добросовестность conscientiousness; *law* bona fides; good faith
добросовестный conscientious; scrupulous; *law* bona fide
добрососедский friendly; neighbourly
~кие отношения good-neighbour relations
добрососедство friendliness, neighbourliness
добротность quality (factor), Q(-factor)
добротный of high quality, high-quality
добывать mine, obtain; *(stone)* quarry; *(oil, gas)* produce
добывающий mining; extracting
~щая промышленность mining industry
~щая промышленность и сельское хозяйство *econ.* primary industries/sector
добыча mining; extraction; *(stone)* quarrying; *(oil, gas)* production; *econ.* output, yield
~, валовая gross output
~, гидравлическая hydromining, hydraulicking
~, карьерная/открытая open-pit mining/operation; strip mining, open-cast mining
~, морская off-shore operation
~ (нефти) вторичными методами secondary recovery operation/process
~ природного сырья resource output/production
доведение reduction
~ до наименьшего значения *math.* minimization
доверенное лицо trustee
доверенность *law* power/letter/warrant of attorney; procuration
~, общая general power of attorney
~ с правом передоверия power of attorney and substitution
доверитель *law* principal; mandator
доверительность *stat., math.* confidence
доверительный confidential; fiducial
~ интервал confidence/fiducial/reliable interval/region
~ уровень *stat.* confidence coefficient
~ная вероятность confidence probability
довод argument, reason; *law* submission
доводить *tech.* finish; refine; hone
доводка *met.* finishing; refining; honing, lapping; dressing; final adjustment
~, эксплуатационная operational development
довольствие *mil.* allowance
~, вещевое clothing
довулканизация aftercure, after-vulcanization
довыборы *polit.* by-election(s)
дог mastiff
~, далматский Dalmatian dog
~, датский great Dane
договор agreement, contract; *pol.* treaty, pact
~, арендный lease
~ аренды фермы farm tenancy contract
~, бодмерейный bottomry bond
~ высшего доверия *law* contract uberrimae fidei
~, двусторонний bilateral/bipartite treaty
~ купли-продажи *econ.* purchase-sale contract
~ личного найма labo(u)r contract
~, многосторонний multilateral/multipartite treaty
~ на реальный товар *econ.* spot contract
~ о буксировке *naut.* towage agreement
~ о взаимной помощи treaty of mutual assistance
~ о гарантии от убытков *econ.* contract of indemnity

~ о залоге *econ.* contract of pledge
~ о морской перевозке contract of affreightment
~ о ненападении *polit.* nonaggression pact
~ о поручении *econ.* contract of agency
~ о поручительстве contract of guarantee
~ о спасании *naut.* salvage agreement
~ о страховании contract of insurance
~, параллельный *econ.* associate contract
~ по нераспространению *polit.* nonproliferation treaty
~ с субподрядчиком subcontract
~, союзный treaty of alliance
~ страхования subrange policy
~, трудовой labo(u)r contract
~, трехсторонний tripartite treaty
~, фидуциарный, ~ требующий наивысшей добросовестности uberrima fides, contract uberrimae fidei
договоренность agreement, understanding; deal
договорный contractual; agreed; stipulated
~ная теория заработной платы bargaining theory of wages, bargaining-power theory
догорание after-combustion, delayed burning; *rock.* residual burning, torching, tail off
догрузка additional/supplementary charge
додекагон dodecagon
додекафония *mus.* dodecaphony, dodecaphonism
додекаэдр dodecahedron
додекаэдрический dodecahedral, dodecahedric
додекаэдро-пентагональный *cryst.* pyritohedral
доение milking
~ в молокопровод pipeline milking
~, машинное machine/mechanical milking
дождевалка sprinkler
дождевальный sprinkling
дождевание overhead irrigation
дождевой rain *(attr.);* pluvial
~ червь earthworm
~вое платье oilskins
дожделюбивый *bot.* rain-loving, ombrophilous
дождемер rain-gauge, pluviometer, pluviograph, ombrometer
~, самопишущий hyetograph
~, суммарный totalizer rain gauge
дождеприемник drain for rain water; rain gully; street inlet
дождь rain; shower
~, ледяной sleet
~, ливневый rainstorm, shower
~, метеорный meteor shower, meteoric rain
~, пепельный ash shower
~ радиоактивный radioactive fallout
~, силный/обильный heavy rain
~, слабый slight/gentle rain
доживаемость *stat.* survivorship
дожигание after-burning; reburning; reheating
дожитие time left to live
~, страхование на life insurance
доза dose; rate; dosage
~ бета-излучения, допустимая beta-tolerance dose
~ внешнего/внутреннего облучения external/internal radiation dose
~ для приема внутрь *med.* oral dose
~, допустимая permissible/tolerance dose
~, дробная divided dose
~ излучения radiation dose
~, индикаторная tracer dose
~, интегральная integral/cumulative dose
~, ионизационная ionization/ionizing dose
~ ионизирующего излучения ionizing radiation dose
~ ионизирующего излучения, экспозиционная exposure radiation dose
~, летальная/смертельная lethal/fatal dose; LD
~, кумулятивная cumulative/accumulated/integral dose
~, лечебная therapeutic/curative dose
~ малой мощности low dose rate
~, минимальная смертельная minimum lethal dose. M. L. D., m. l. d.
~, несмертельная nonlethal dose
~ облучения (ir)radiation dose/exposure
~ облучения всего тела whole-/total-body irradiation dose
~ облучения половинной выживаемости median lethal radiation dose
~ облучения, профессиональная occupational radiation dose
~ облучения, средняя смертельная median lethal radiation dose
~ облучения, суммарная total radiation dose
~ обратно рассеянного излучения backscattered dose
~, объемная volume dose
~, одноразовая/однократная single dose
~, подпороговая subliminal dose
~ половинной выживаемости median lethal dose
~, полусмертельная half-lethal dose
~, пороговая threshold dose
~, предельно допустимая maximum permissible/tolerance dose
~ профессионального облучения occupational radiation dose
~ радиоактивного облучения radiation/radiological dose
~, раздробленная fractionated dose
~, разовая допустимая single permissible dose
~, расчетная rated dose
~ рентгеновских лучей X-ray dose
~, сверхлетальная superlethal dose
~, смертельная lethal dose, LD
~, средняя интегральная mean integral dose
~, средняя смертельная median lethal dose
~, сублетальная sublethal dose
~, суммарная cumulative/total dose

~, суточная daily dose
~, тканевая tissue dose
~, толерантная tolerance dose
~, удельная specific dose
~, удельная поглощенная specific absorbed dose
~, чрезмерная overdose
~, эритемная erythema(l) dose
~, эффективная effective dose, ED
дозаправка refuelling; recharging; replenishment
~ топливом в воздухе/полете midair/aerial/inflight refuelling, air-to-air refuelling
дозаправлять refuel, refill; fill up; replenish
дозатор batcher, proportioner, metering device/apparatus, measuring hopper/pocket
~, весовой weigher, weighing machine/scale
~, весовой ленточный belt-conveyer weigher
~ для жидкостей metering pump/tank
~ для консервных банок can portioner
~ изотопов isotope dispenser
~ кормов feed meter
~, порционный, ~ периодического действия batch-type scale
~ соотношения proportioner
дозатор-наполнитель measuring and filling machine
дозатор-смеситель meter-mixer
дозиметр radiation monitor/meter/recorder, dosimeter, dose-meter
~ альфа-излучения alpha-dosimeter
~ гамма-излучения gamma-ray dosimeter
~, групповой stationary dosimeter
~, дверной doorway monitor
~ для жидкостей, проточный continuous-flow liquid monitor
~, загрязнений contamination monitor
~ излучений высокой интенсивности high-level dosimeter
~ излучения radiation dose meter
~, индивидуальный personal dosimeter/monitor
~, ионизационный ionization-type dosimeter
~ карандашного типа pen-type dosimeter
~, карманный pocket dosimeter
~, карманный пленочный pocket film dosimeter
~, кольцевой пленочный ring film meter
~, контрольный monitor
~, кристаллический crystal dosimeter
~, люминесцентный fluorescent dosimeter
~ местности area monitor
~ мощносты дозы dose-rate dosimeter
~, наперстковый thimble dosimeter
~, плоский пленочный film badge
~, поисковый survey dosimeter
~, полупроводниковый transistorized dosimeter
~ прямого отсчета direct-reading dosimeter
~, рентгеновский пленочный X-ray film pack
~ с ионизационной камерой ion-chamber dosimeter
~ с кварцевой нитью quartzfibre dosimeter
~, сцинтилляционный scintillation counter dosimeter
~, термолюминесцентный thermoluminescence dosimeter
~, узкодиапазонный low-range dosimeter
~, широкодиапазонный high-range dosimeter
дозиметрист radiation supervisor, radiological safety officer, health physicist
дозиметрический dosimetric
дозиметрия radiation control/monitoring, dosimetry; health monitoring
~, дистанционная remote monitoring
~, индивидуальная personal monitoring
~ личного состава *mil.* personnel monitoring
~ местности radiation area survey, area monitoring
~, термолюминесцентная thermoluminescence dosimetry
~, цериевая ceric-cerous dosimetry
дозирование proportioning, batching; measuring, metering out, weighing
дозировать proportion, batch; measure out, dose out, meter out; measure, weigh
дозировка 1. proportioning, batching; 2. dosage, amount
дозировочный batching, proportioning, dosing out; measuring
дознание *law* preliminary investigation, inquest; inquiry
~, производить hold an inquest
дозор *mil.* partol (duty), picket (duty)
~, воздушный aerial picket
~, океанский ocean sentry patrol
~, прибрежный offshore/inshore patrol
~, противовоздушный (anti-)aircraft picket
~, противолодочный anti-submarine patrol/picket
~, радиолокационный radar picket/control
~, свободный roving patrol
дозорный picket *(attr.)*, patrol *(attr.)*
дозревание after-ripening
дозреватель maturing tank
доильная машина/установка milking machine
дойная корова milch cow
док dock
~, балансирный balance dock
~, гидравлический подъемный hydraulic dock
~, гидравлический сухоподъемный hydraulic lift dock
~ для гидросамолетов, сухой seaplane dry dock
~, достроечный fitting-out dock
~, наливной flooding dock
~, плавучий floating dock
~, судостроительный building/construction dock
~, сухой dry dock
доказательный evidential, evidentiary; demon-

strative, conclusive, evincive
доказательство proof, evidence; demonstration
~, вещественное material/demonstrative evidence
~, внешнее extraneous evidence, evidence aliunde
~, косвенное indirect/inferential/circumstantial/presumptive evidence/proof
~, несводимое irreducible proof
~ основанное на слухе hearsay evidence
~ от противного proof by contradiction, *reductio ad absurdum* proof
~ по индукции inductive proof, proof by/in induction
~, строгое rigorous proof
доказать demonstrate, prove
~, что и требовалось *math.* quod erat demonstrandum, Q. E. D.
доказывание demonstration, substantiation, proof
~ния, средства *law* evidence
докапиталистический precapitalistic
докармливание supplementary feeding
док-бассейн wet basin/dock; basin dock
док-камера 1. docking basin; 2. testing/experimental tank; 3. launching dock
доклассический preclassical
доклассовый preclass *(attr.),* before the emergence of classes
доковать dock, drydock
доковый dock, dryoock *(attr.)*
~ые сборы dockage
докрасна to red heat, to redness
~, раскаленный red-hot
доктрина doctrine; teaching; principles
~, денежная *econ.* monetary doctrine
~, передаваемости *econ.* shiftability doctrine
доктринер doctrinaire, doctrinarian
доктринерство doctrinairism; dogmatism
документ document; paper; *law* deed, instrument
~, денежный оправдательный *econ.* voucher
~ для оптического считывания *comp.* optically-sensed document
~, контрольный checking/control document
~, машинный computer-based document
~, машиночитаемый *comp.* machine-readable document
~ на поставку продукции procurement document
~ на предъявителя bearer document
~, оборотный *fin., law* negotiable instrument/document
~, оправдательный/первичный *econ.* source document
~, передаваемый денежный negotiable document
~ с магнитном шрифтом magnetic writing document, document with magnetic characters
~ с магнитными знаками magnetic document
~, секретный security document
~, таможенный customs document
~, трансфертный instrument of transfer
~ты о ратификации instruments of ratification
~ты по отправке груза shipper's papers
~ты, погрузочные shipping documents
~ты, судовые ship's papers
~ты удостоверяющие личность identity papers
~ты, учетные records
документалистика documentation work/activities
документальный documentary; documental
~ фильм documentary (film)
документация documentation; documents, papers; file of documents
~, налоговая tax file
~, проектная design blueprints and specifications
~, техническая instruction manual; service record; production/manufacturing forms
док-эллинг *naut.* railway dry dock
долбить hollow, mortise, chisel, groove, slot
долбяк gearwheel/mortising cutter, gear cutter
~, втулочный hub-type gear cutter
~, косозубый helical gear cutter
долг debt; indebtedness
~ банку bank debt
~, государственный public/national debt
~, консолидированный consolidated/unified/funded debt
~, краткосрочный short-term debt
~, неконсолидированный floating debt
~, непогашенный outstanding/undischarged debt
~, нефундированный floating debt
~ обеспеченный залогом debt on pawn
~, облигационный bonded debt
~, отсроченный deferred debt
~, первоочередной preferential/preferred debt
~ подлежащий оплате due debt, debt due
~, фундированный consolidated/funded debt
долговечность durability; longevity; lifetime; survival time; *tech.* period of service, operation life; *med.* life span
~, гарантируемая specified life/longevity
~, заданная specified life
~ изделия product life
~, моральная obsolescence period
~ при хранении shelf life
~, расчетная design longevity; rated life
~, усталостная fatigue life
~, эксплуатационная operation life
долговечный durable, timeproof; long-lived, long-lasting
долговременный of long duration, durable, timeproof; long-term; *mil.* permanent
~ное укрепление permanent work/fortification
долгота 1. longitude; 2. length; duration
~, астрономическая celestial/astronomical longitude
~ в равноугольной проекции conformal longitude

~ в угловой мере, географическая longitude in arc
~ в часовой мере longitude in time
~, восточная east longitude
~ восходящего узла longitude of the ascending node
~, географическая terrestrial/geographical longitude
~, геодезическая geodetic/spheroidal longitude
~, западная west longitude
~, небесная celestial/eliptic(al) longitude
~ от гринвичского меридиана Greenwich value
~ периастра longitude of periastron
~ перигелия perihelion longitude
~ перигея perigee longitude
долготье *timb.* long log
долевой *med.* lobar
доледниковый preglacial
должник debtor
~, первоочередной primary debtor
~ по закладной mortgager
~ по ипотечному залогу mortgager
должностной official
~ное лицо official
~ное преступление malfeasance
должность post, job, office, duties; duty, obligation
~, свободная vacant job
~, штатная permanent appointment
~ти, по *law* ex officio
доли *anat.* lobes
долина valley
~, брошенная *geol.* dead valley
~ в откосе escarpment valley
~, вложенная valley in valley, trough-in-trough valley
~, врезанная incised/entrenched valley
~ грабен fault-block valley
~, закрытая blind valley
~, корытообразная trough valley
~, котлообразная cauldron valley
~, крутобокая steep-sided valley
~, ладиеобразная flat-floored valley, canoe valley
~, ледниковая glacial valley
~, мешкообразная blind valley
~, наложенная superimposed valley
~, подводная submarine valley
~ провала rift valley
~, разветвленная braided valley
~ размыва wash-out plain
~, сбросовая fault/rift valley
~, сбросовая без выхода cauldron valley
~, тектоническая structural valley
~, узкая coomb, comb(e)
~, циркообразная cirque
~, U-образная U-shaped valley
~, V-образная V-shaped valley
долихокефал, долихоцефал dolichocephal
долихокефалия, долихоцефалия dolichocephalism, dolichocephaly
долихоколия dolichocolon
долихосигма dolichosigmoid
долихостеномелия arachnodactyly, spider fingers
долихоцефалический dolichocephalic, dolichocranial
доллар dollar
~ с неизменной покупательной силой stabilized/compensated dollar
~ с постояннй покупательной силой (в отношении товаров) *econ.* commodity dollar
доломит *miner.* dolomite
~, железистый brown spar
~, пещеристый smoke wacke
~, пламенный fire/fiery dolomite
~ с железным шпатом brown spar
доломитизированный dolomitized
доломитовый dolomitic
доломитолит dolomite, dolostone
долото chisel; gouge
~, алмазное буровое diamond drill bit
~, буровое drill bit
~, буровое шестиугольное six-point drill bit
~, глазное *surg.* ophthalmic ghisel
~, долбежное ripping chisel
~, желобчатое fluting gouge
~, ложечное spoon gouge
~, полукруглое gouge
долька *anat.* lobule; *bot.* coccus
~ нижняя полулунная inferior lunate lobule
~, обонятельная olfactory lobule
~ плода mericarp
дольковый, дольчатый lobate, lobular
дольмен dolmen, table stone
доля fraction, portion, part; share; quota; *med., bot.* lobe; *econ.* interest
~, весовая weight fraction
~ выборки sampling fraction
~ мигрирующего населения migration ratio
~ мольная/молярная mole fraction
~, объемная volume fraction
~ печени, квадратная quadrate lobe
~ облигационного выпуска *econ.* tranche *(Fr.)*
~ полушария, лобная frontal lobe
~ почки renal lobe
~ участия contribution
~ участия компонента quota of component
дом home; house; dwelling; residence
~ для престарелых old people's home
~, исправительный *law* house of correction
~, многоквартирный block of flats
~, односемейный single-family house; detached house
~ отдыха holiday home; rest home
~, панельный house *or* block of flats
~, сборно-панельный prefabricated bearing-wall house
~, сборный prefabricated house
домен *phys., math.* domain

~, движущийся travelling domain
~ игольчатой формы spike-shaped domain
~, клиновидный tapered domain
~, полосовой strip domain
~, сегнетоэлектрический ferroelectric domain
~ сильного поля high-field domain
~, цилиндрический, магнитный magnetic bubble domain
домена *cryst.* domain
~ двойникования twinning domain
доместик *text.* domestic, cotton fabric
домик, вегетационный hothouse; greenhouse, glasshouse
доминант, доминанта dominant
доминантность, доминирование dominance
~, обусловленная conditioned dominance
доминиканец, доминиканский Dominican
домицилий *law* domicile
~ по происхождению natural domicile
домкрат (lifting) jack
~, автомобильный car jack
~, винтовой screw jack
~, выравнивающий levelling jack
~, педальный foot-operated jack
~, путевой track jack
~, реечный rack jack
~, ручной hand-operated jack
~, стояночный parking jack
~, телескопический telescopic/telescope jack
~, шарнирный articulated jack
домна *met.* blast furnace
домовая книга register of tenants/occupants
домол re-crushing; re-grinding; re-milling
домонополистический premonopolistic
донатор donor, donator
Донбасс Donets Basin/coalfields, Donbas
донесение report; dispatch, message
донка *naut.* donkey
донкихот Quixote
донкихотский quixotic
донкихотство quixotism, quixotry
донный benthonic, benthic; bottom *(attr.);* bottom-dwelling
~ лед ground ice
донор donor; blood donor
донорманский *hist.* pre-Norman
донорский пункт blood donor centre
доношенность плода *med.* maturity of the f(o)etus
доношенный *anat.* carried to the full term
доокисление further/complete oxidation
дооперационный *med.* preoperative
допамин *pharm.* dopamine, hydroxytyramine
допахинон *pharm.* dopaquinone
доплата additional payment/charge; afterpayment
~, гарантированная make-up pay
~ за сверхурочную работу overtime premium
доплачивание = доплата
дополнение supplement, complement; addition; appendix; *gram.* object
~, алгебраическое cofactor
~ географической долготы colongitude
~ географической широты colatitude
~ до двух two's complement
~ до единицы one's complement
~ до полного квадрата completing the square
~, звездное sidereal hour angle
~ множества complement of a set
~ нулями zero padding
~ пробелами padding
~ со знаком signed complement
~, точное radix/true complement
дополнитель complementer
дополнительность complementarity
дополнительный supplementary, additional, complementary; accessory; extra; ancillary; *law* appendant
~ налог/сбор surcharge
~ная статья, -ный пункт *acc.* rider
~ные цвета complementary colo(u)rs
дополнять complement, supplement, add to; complete
~ до полного квадрата *math.* complete the square
допотопный *geol.* antediluvial
допплерит *miner.* dopplerite, peaty pitch coal
допрашивать *law* examine; interrogate
допрос examination, interrogation, questioning
~, перекрестный cross-examination
допуск allowance, tolerance; limit
~, жесткий stringent/close tolerance
~, зазора fit/freedom tolerance
~, минусовый negative tolerance
~ на изготовление manufacturing tolerance
~ на износ wear tolerance/allowance
~ на отклонение фазы phase tolerance
~ на течь leak tolerance
~ на усадку shrinkage tolerance
~, плюсовый positive tolerance
~, производственный manufacturing/manufacturer's tolerance
допустимый permissible, admissible; acceptable; tolerable; feasible; *law* warrantable
дорабатывание, доработка finishing; correcting, modifying
доремонтный pre-repair
дорийский *hist.* Doric, Dorian
дорн mandrel; core; broach; drift-pin
~, жесткий rigid mandrel
~, калибрующий/полировальный burnishing broach
~, круглый annular core
~, прошивной broaching tool
~ центрирующий contering core
дорнование mandrelling, burnishing/polishing by mandrel
дорнодержатель mandrel/core/broach holder
дорога road; way; path
~, автомобильная motor/automobile road

~, **большая** highway
~, **грунтовая** dirt/country road
~, **гужевая** waggon/cart road, cartway
~, **железная** *Eng.* railway, *Am.* railroad
~, **железная двухпутная** double-track line
~, **железная пригородная** suburban railway
~, **железная узкоколейная** narrow-gauge railway
~, **железная ширококолейная** broad-gauge [fullgauge] railway
~, **зубчатая железная** rack railway
~, **кольцевая/окружная** belt line; circuit road
~, **лесовозная** forest road
~, **магистральная** arterial/backbone highway
~, **многорядная** multilane road
~, **монорельсовая** monorail
~, **подвесная монорельсовая** overhead monorail
~, **подъездная** access road
~, **проселочная** country/cart/dirt road
~, **скоростная** expressway; highway
~, **шоссейная** macadam road; highway
дороговизна high cost of living; scarcity
дорогостоящий high-priced; costly; expensive
дородовой *anat.* antenatal, prenatal, antepartum
дорожка track; channel, path, way, lane; strip
~ **беговая, подшипника** race
~ **беговая, протектора** tyre tread
~ **ведущей перфорации** sprocket channel, feed-hole track
~, **вихревая** vortex trail/street
~, **дефектная** bad track
~ **для выборки адресов** address selection track
~ **для записи данных** data track
~ **для посадки** landing runway
~, **закодированная** code track
~, **запасная** alternate track
~ **записи** recording/write track
~ **записи изображения** video track
~, **звуковая** sound track
~, **информационная** data track
~, **контактная** brush track
~, **направляющая** guide way; guide slot
~, **промежуточная** insertion track
~, **рулежная** *aero.* taxiway, taxitrack
~, **синхронизирующая** timing track
~, **тактовая** clock/timimg track, clock marker track
~, **цифровая** digital track
дорсальный dorsal
ДОСААФ Voluntary Society for Assistance to the Army, Air Force and Navy
досинхронный hyposynchronous
доска 1. panel, board; plate; slab; 2. *wood.* board, plank
~, **боковая** *wood.* outer plank
~, **винтовальная** screw plate
~, **волочильная** die/drawing plate; die
~ **для объявлений** bulletin board
~, **инструментальная** instrument panel/board
~, **калиберная** wire gauge (plate)
~, **классная** blackboard
~, **ключевая** *teleph.* keyboard
~, **коммутационная** switchboard; plugboard, patchboard
~, **комплексная приборная** integrated instrument panel/display
~, **мраморная** marble plaque/slab
~, **набивная** engraving plate, print stamp
~, **наборная** *print.* galley, letter board
~, **нестроганая** rough board
~, **обрезная** edged board
~, **полуобрезная** *timb.* outer plank
~, **посевная** seeding board
~, **приборная** instrument panel/board
~, **профильная** patterned timber/lumber
~ **с матрицами** dieplate
~, **стрясная** shuttle board, shaker pan
~, **счетная** calculating board; abac(us)
~, **транцевая** *naut.* transom board
~, **чертежная** drawing board
~, **шпунтованная** matched board
доследование *law* further inquiry/examination
досмотр supervision; inspection; examination
~ **и обыск** *naut.* visit and search
~, **таможенный** customs examination
досмотрщик inspector; customs official
досрочный ahead of schedule
~ **платеж** *law* anticipatory payment
доставка delivery; supply; conveyance
~, **конвейерная** conveying, conveyance
~ **на дом** home-delivery service
~ **по частям** delivery by instal(l)ments
~, **скреперная** scraper transportation/delivery
~, **срочная** express delivery
достаточность adequacy, sufficiency; ample supply; *math.* sufficiency
~ **дохода** adequacy of return
достижение achievement; breakthrough
достоверность authenticity; reliability; veracity; validity; confidence level
~**ти, критерий** *stat.* test of validity
достояние property, possessions
~, **общенародное** public ownership
доступ access; admission
~, **библиотечный** library access
~, **быстрый** fast access
~ **в пакетнои режиме** batch access
~ **в реальном времени** real-time access
~ **в режиме меню** menu-driven access
~ **в сеансе связи** session access
~, **взаимоисключающий** mutually exclusive access
~, **групповой** clustered access
~, **дистанционный** remote access
~, **индексный** indexed access
~, **индексопоследовательный** indexed-sequential access
~ **к базе данных** database access
~ **к данным** data access

~ к памяти memory access
~, коллективный shared access
~, множественный multiple access, multiaccess
~, монопольный exclusive access
~, непосредственный immediate access
~, непреднамеренный unintentional access
~, несанкционированный unauthorized access
~, ограниченный restricted access
~ по ключу keyed access, access by key
~ по линиям связи line access
~, последовательный sequential access
~, произвольный random access
~ с перекрытием overlapped access
~ с помощью клавиатуры fingertip access
~, санкционированный authorized acccess
~, свободный direct/random acccess
доступность accessibility, ease of access, free access; intelligibility; simplicity
~ данных data accessibility
~ маршрута route availability
~, телефонная telephone availability
досуха to dryness
досягаемость accessibility, reach; *mil.* range
~ по высоте altitude range
~ ракеты operational range of missile
дотация subsidy, grant-in-aid, approbation-in-aid; subvention
доход income, return, revenue, receipts, earnings
~ в форме заработной платы wage income
~, валовой gross income/earnings/return
~, вмененный imputed earnings
~, государственный state revenue
~, денежный money/cash income
~ до уплаты налогов before-tax income
~, национальный national income
~ некоммерческого характера nontrading income
~, нераспределенный retained income
~, нетрудовой unearned income
~ после уплаты налогов after tax [disposable] income/earnings
~, предельный marginal return/revenue
~, процентный interest yield
~ рантье rentier income
~, реальный effective/real income
~ облагаемый налогом income liable to tax
~ от имущества property/ownership income
~ от капитала capital income
~ от невидимых статей экспорта и импорта invisible income/revenue
~ от обыкновенных акций equity earnings
~ от собственности property/ownership income
~, рентный unearned income; investment income
~, скорректированный валовой adjusted gross income
~, трудовой earnings of labour; labour/earned income
~, условно-начисленный imputed earnings
доходы:
~, бюджетные budget receipts/revenues
~ за счет предоставления услуг service income
~, налоговые tax receipts
~ от прироста капитала capital gains
доходность profitableness, remunerativeness, earning power
~, сметная budgeted return
доходный remunerative, gainful, paying; lucrative, profitable
дохристианский pre-Christian
доцент reader, senior lecturer, associate/assistant professor
доцентура post of senior lecturer; assistant/associate professorship
дочерний daughter *(attr.),* subsidiary, branch *(attr.)*
дошкольный preschool
дощечка plate
~, калиберная templet
~, паспортная name plate
доэвтектический hypoeutectic
драга dredge; drag
~, гидровсасывающая hydrosuction dredge
~, грейферная clamshell/grapple dredge
~, землесосная с гидравлическим разрыхлителем hydraulic suction dredge
~, многочерпаковая chain-bucket dredge
~, одночерпаковая dipper dredge
~, пневмовсасывающая pneumatic-suction dredge
~, цепная chain drag
~, эжекторная jet dredge
~, эрлифтная airlift dredge
драгировать dredge, drag dut, scoop
драглайн dragline
~, вскрышной stripping dragline
~, шагающий walking dragline
драгун *hist.* dragoon
драже dragee; sugar-coated drop
дражирование *agr.* seed coating/pelleting
драйвер *el.* driver
~, графический graphics driver
~, программный software driver
~ режима разделения времени time-sharing driver
~, тестовый test driver
~ шины bus driver
дракон dragon; *astr.* Draco
драконовский Draconian, Draconic(al)
дракункулез *med.* dracontiasis, dracunculiasis, dracunculosis
драматический dramatic; *theatre (attr.)*
~ кружок theatrical circle
драматург playwright; dramatist
драматургический dramaturgical
драмкружок dramatic/theatrical circle
дрань lath, slat
~, штукатурная plasterer's slat/lath
драхма drachma
драч *tech.* jack plane

древесина 1. timber, lumber; wood; 2. wood pulp, xylem
~, балансовая pulp/paper wood, pulpwood
~, гнилая spunk
~, деловая commercial/industrial timber
~, дровяная firewood
~, заболонная sapwood
~, колкая cleavable wood
~, косослойная cross-fibred wood
~ лиственных пород hardwood
~, матерная/мелкослойная close-grained wood
~ на корню standing wood/forest
~, поделочная carpentry timber
~, пропареная steamed wood/timber
~, прямослойная straight-grown timber/wood
~ с наростами knaggy wood
~ с узорчатой текстурой veined wood
~, свежесрубленная green wood
~, свилеватая crooked-fibre wood/timber, wavegrown timber/wood
~, слоистая laminated timber; plywood
~, смолистая resinous wood
~, сплавная float wood/timber
~, сухостойная deadwood
~, сучковатая lopwood
~, товарная industrial/merchantable wood
~, хвойная softwood
~, широкослойная wide-ringed wood
~, экзархная exarch xylem
~, ядровая heartwood, duramen
древесиноведение wood science/technology
древесноволокнистый cellulose-fibre *(attr.)*
древесноплодный *bot.* xylocarpous
древесный arboreal, arboreous, arborescent; xyloid; ligneous
~ сахар xylose
~ уголь charcoal
~ная шерсть wood fibre
древнеанглийский Old English
древнеболгарский Old Bulgarian
древневерхненемецкий Old High German
древневулканический *geol.* paleovolcanic
древнегреческий ancient Greek *(attr.);* Grecian
древнееврейский Hebrew, Hebraic
древнекиевский Old Kievian
древнекрылые *paleont.* paleopters *(Paleoptera)*
древнеледниковье Old Glacial Period, old glaciation
древненемецкий Old German
древнерусский Old Russian
древнеславянский Old Slavic/Slavonic
древнецерковнославянский Old Church Slavonic
древнечетвертичный *geol.* Older Quaternary
древовал-кусторез land-clearing machine
древовидный dendroid, dendriform, arborescent
древостой stand of trees/timber, forest stand, standing timber
~, редкий sparse/open woodland
~, смешанный mixed stand of trees
дредноут *naut., text.* dreadnought
дрезина railway section car/trolley
дрейф drift; leeway
~, ветровой leeway, wind drift
~ льда ice drift
~ нуля zero drift, drift of zero
~ с отданным якорем *naut.* clubbing
дрейфовать drift, go adrift, drift away; lose ground/way
~ при отданном якоре club, drag the anchor
дрейфовый drift *(attr.)*
дрейфометр driftmeter
дрейфующий drifting
~щая научная станция drift-ice research station
дрек *naut.* grapnel, grappling/grapnel anchor, drag hook, grapple
дрель drill
~, грудная breast drill
~, лучковая bow drill
дрена drain
~, гончарная tile drain
~, гравийная rock-/stone-filled drain
~, закрытая subdrain, culvert
~, магистральная main/arterial drain
~, фашинная faggot/bush drain
дренаж 1. drain, drainage, draining; 2. vent, venting; 3. *med.* drainage tube, drain
~, клапанный valvular drain
~, кротовый mole drainage
~ маслобака oil-tank vent
~, открытый open-ditch drainage
~, самоточный gravity drainage/drain
~, шпалерный trellis drainage
дренажный drain *(attr.)*, draining
дренирование, дренажирование drainage, draining
~, межреберное *med.* intercostal drainage
дренировать drain, draw off
дренируемость drainability
дреноукладочный drain-installing/laying; drainpipe laying
дреноукладчик drainpipe laying machine; draintube layer
дрепаноцит *pathol.* drepanocyte, sickle/crescent cell
дрепаноцитоз drepanocytosis
дресва gruss, landwaste
дрессировка *tech.* skin/temper rolling; trimming, dressing off
дриада *myth., bot.* dryad
дрильбор burr
~, машинный mechanical burr
~, ручной hand-operated burr
дрифтер *naut.* drift boat, drifter
дробемет shot-blasting machine, shot blaster
дробеочистка shot-blast cleaning
дробилка crusher, crushing mill, breaker; grinder
~, агломерационная sinter breaker

~, **валковая** roll crusher, rolls
~, **валковая зубчатая** toothed-roll crusher
~, **валково-молотковая** roll-hammer crusher
~, **дисковая** disk crusher/breaker
~, **зерновая** grain grinder
~, **коксовая** coke mill
~, **конусная** gyratory-cone crusher
~, **кормовая** fodder grinder
~, **молотковая** hammer crusher/breaker
~, **мятниковая** blade crusher
~, **отбойная** baffle crusher
~, **ударная** impact crusher
~, **щековая** jaw crusher
дробина, дробинка small shot/pellet
дробление crushing; grinding; milling; granulation; fragmentation; splitting; cleavage; division; fractionation
~ **дозы** dose fractionation
~, **крупное** coarse/primary crushing
~, **мелкое** fine crushing
~ **частоты** frequency splitting
дробленка crushed/fractured particles
дробно-атомный subatomic
дробно-квадратичный/квадратный *math.* quadratic-fractional
дробнолинейный *math.* linear-fractional
дробность divisibility; fragmentation
дробный *math.* fractional
дробь I *math.* 1. fraction; 2. slant line, oblique stroke, oblique
~, **аликвотная** unit fraction
~, **бесконечная** infinite/unending fraction
~, **десятичная** decimal fraction
~, **конечная** terminating fraction
~, **непериодическая** non-repeating fraction
~, **неправильная** improper fraction
~, **непрерывная** continued fraction
~, **несократимая** irreducible fraction
~, **обратная** reciprocal fraction
~, **периодическая десятичная** repeating/recurring decimal, periodic decimal fraction
~, **подходящая** convergent of a continued fraction, approximant
~, **правильная** proper fraction
~, **простая** common/simple/vulgar fraction
~, **простейшая** partial fraction
~, **смешанная периодическая** mixed periodic fraction; mixed number/quantity
~, **сократимая** reducible fraction
~, **цепная** continued fraction
~, **чистая периодическая** purely periodic fraction
~, **элементарная** partial fraction
дробь II shot; grit
~, **буровая** drilling shot
~ **для очистки отливок** descaling grit
~, **мелкая** fine shot
~, **свинцовая** lead shot
~, **чугунная** iron shot
дрожание jitter, flicker, flutter; trembling, vibration; *pathol.* tremor, tremitus, thrill
~ **гидатид** hydatid tremitus/thrill
~, **голосовое** vocal tremitus
~ **изображения** *astr.* image tremor; *telev.* picture bounce, jitters
~ **импульса** pulse jitter
~ **радужной оболочки** iridodonesis
~ **стрелки** quiver of a pointer
~ **строк** *telev.* interline flicker
~ **частоты** frequency jitter
дрожательный trembling, tremulous
дрожжевание yeasting, fermenting
дрожжи yeast; leaven
~, **винные** wine yeast
~, **кормовые** nutrient yeast
~, **пекарские** bakers'/bakery yeast
~, **пивные** brewers' yeast, common beer yeast
~, **прессованные** yeast cake, compressed yeast
дрожь shivering, trembling
дрозд *zool.* thrush, ouzel *(Turdus)*
дромадер, дромедар *zool.* dromedary *(Camelus dromedarius)*
дромгед *naut.* drumhead
дрософила fruit fly *(Drosophila)*
дросселировать throttle; choke;
дроссель throttle (valve); *el.* choke (coil) *mech.* orifice plate
~, **высокочастотный** radio-frequency choke
~, **гидравлический** throttling valve
~ **звуковых частот** audio-frequency choke
~, **пневматический** air-flow constrictor/restrictor
~ **сверхвысокой частоты** *rad.* microwave choke
друза *miner.* druse, vug; *bot.* druse
друзовый drusy
дряхлость marasmus
~, **старческая** geromarasmus
дуайен *dipl.* dean
дуализм dualism
дуальность duality
дуант *phys.* duant, dee
дубильный tannic; tanning
~ **орешек** gallnut
~**ная кора** tan
дубильня tannery, tanyard
дубильщик tanner
дубитель tanning agent/material; tannin
~, **синтетический** syntan
~, **формальдегидный** aldehyde tanning agent
~, **хромовый** chrome tanning agent
дубление *leath.* tanning, tannage; *print.* hardening
~, **валонейное** valonia tannage
~, **жировое** oil tannage
~, **квасцовое** alum tanning
~, **несветовое/темновое** non-exposed hardening
~, **хромовое** chrome tannage
дублер stand-by *(operator, machine);* doubler; *tech.* doubling machine, doubler; *theatre* under-

study; stand-in
дублет doublet
~, линзовый *opt.* objective doublet
дублетность *ling.* dimorphism
дубликат duplicate, copy
дублирование duplication; doubling; *film* dubbing; *met.* doubling; *tech.* stand-by system
~, раздельное component duplication
дублировать duplicate; *film* dub; *met.* double
~ роль *theatre* to understudy a part
дублирующий duplicate; back-up; stand-by
дубль *film* take
дубль-бемоль/-диез *mus.* double flat/sharp
дубль-негатив/позитив duplicate negative/positive
дубльтон *print.* duotone ink
дубляж *cin.* dubbing
дубовина oak tanbark
дуга arc; arch; bow
~ аорты aortic arch, arch of the aorta
~, буксирная towing rail
~, вольтова voltaic/electric arc
~ графа *math.* edge of a graph
~, жаберная gill arch; branchial arch
~, защитная crash-bar, roll-/safety-bar
~, звучащая singing arc
~, короткозамыкающая short-circuiting arc
~ кривой *math.* arc of a curve
~, надбровная superciliary arch
~, небно-глоточная palatopharyngeal arch
~, небно-язычная palatoglossal arch
~, нижняя зубная mandibular dentition
~ окружности arc/segment of a circle
~ пантографа bow of a pantograph
~, подошвенная plantar arch
~ протектора tread arc
~, реберная costal arch
~, сопрягающая *math.* rounding-off curve
~, траловая (trawl) gallow
~, угломерная *naut.* bearing arc
~, электрическая electric arc
~, электрическая обратная reverse electric arc, arc-back
дуговидный, дугообразный arched, arcuate, archiform; bow-shaped
дугогаситель arc arrester/extinguisher
дугогашение *el.* arc blow-out, arc control
дугообразный arcuate, arched, bow-shaped
дугостойкий arc-resistant, non-arcing
дужка U-link, small arch, shackle; *av.* aerofoil, *Am.* airfoil
~ крыла aerofoil section
дукат *hist.* ducat
дуктилометрия ductilometry
дульцит dulcitol, dulcite
дум-дум пули dumdum bullet
думец *hist.* member of the Duma
думпкар dump car
дунайский Danubian; Danube *(attr.)*
дуоденальный *anat.* duodenal
дуоденит *med.* duodenitis
дуоденоеюностомия *med.* duodenojejunostomy
дуоденопластика *surg.* duodenoplasty
дуоденорафия duodenorrhaphy
дуоденоскопия duodenoscopy
дуоденостаз *med.* duodenal statis
дуоденостомия duodenostomy
дуоденохолецистотомия *surg.* duodenocholecystotomy, duodenocystotomy
дуодецима duodecimo tone
дуополия duopoly
дуопсония duopsony
дуостан two-high mill
дуплекс duplex; duplex system/operation, duplexing
~ дифференциальный differential duplex system, incremental duplex
~, мостовой bridge duplex system/station
~, одночастотный single-frequency duplex
дуплекс-автотипия duplex-autotype
дупликация duplication
дуралумин *met.* duralumin
дуст insecticide/disinfective powder; pesticide dust
дурман *bot.* thorn/mad apple, datura *(Datura)*, apple of Peru
дутье blowing, blast, forced draught/draft
~, вентиляторное forced draught/draft
~, верхнее top blast
~, острое overfire air
~, паровое steam-jet draught/draft
~, пескоструйное sand blasting
душ 1. shower bath, shower; 2. *med.* douche
~, боковой lateral douche
~, веерный fan douche
~, игольчатый needle-shaped douche
~, нитевидный threadlike douche
~, носовой nasal/Webers douche
~, паровой vapour douche
~, переменный alternating douche
душевая *naut.* crew's shower room
душевнобольной insane, deranged; mental patient, insane person
душегубка gas chamber
душеспасение *eccl.* salvation
душник air vent
дуэль *game th.* duel, game of timing
дым smoke; fume
дымарь, дымогенератор smoke generator, smoke unit/tender
дымка haze, mist
~, арктическая arctic haze
~, ледяная ice-crystal haze
дымление smoking, fuming
дымномер, дымомер smoke density indicator, visual smoke indicator
дымогарная труба fire/heating tube; flue
дымонепроницаемый smokeproof, smoketight
дымообнаружитель, дымоощутитель smoke indicator
дымообразующий smoke generating
дымоотвод offtake; flue, chimney, smokestack

дымоотводный smoke-exhausting
дымосжигание smoke combustion
дымосос smoke exhauster
дымостойкий, дымоупорный smoke-proof
дымоход smoke flue/stack; smoke/funnel uptake; chimney
~, вытяжный uptake flue
~, отводящий discharge flue
дыра, дырка hole; vacancy
~, электронная electron hole/vacancy
дыркоискатель hole-sensing pin
дырокол perforator; puncher
дыромер hole gauge
дыропробивной perforating, piercing, punching
~ станок punch press, perforator
дырчатый perforated, full of holes; foraminous, foraminated
дыхание respiration, breathing; breath
~, бронхиальное bronchial/tubular respiration
~, брюшное abdominal respiration
~, внешнее/внутреннее external/internal respiration
~, внутриутробное fetal respiration
~, грудное thoracic respiration/breathing
~, диафрагмальное diaphragmatic respiration
~, искусственное artificial/assisted/controlled respiration
~, кавернозное (vesiculo)cavernous respiration
~, кожное dermal/cutaneous respiration, transpiration
~, носовое nasal breathing
~ плода fetal respiration
~ по циркуляционной системе circle breathing
~, рефлекторное reflex respiration
~ рот в нос mouth-to-nose respiration
~ рот в рот mouth-to-mouth respiration
~, ротовое mouth breathing
~, стерторозное stertor/sonorous breathing
~, судорожное convulsive breathing
~, усиленное hyperpnea
~, хрипящее stertorous respiration
дыхательный respiratory; .breathing
дышло shaft, pole, rod, draft bar
~, сцепное coupling rod
дьюар Dewar flask/vessel, dewar
дюбель dowel; pin; key; expansion anchor/bolt
дюза nozzle
дюйм inch
дюймовик wooden rule *(marked in inches)*
дюкер inverted/return siphon/tunnel
дюна dune, sand drift
~, береговая coastal/strand/litteral dune
~, блуждающая travelling/wandering dune
~, валообразная ridge-like dune
~, движущаяся travelling dune
~, живая active dune
~, материковая inland dune
~, передовая foredune
~, песчаная sand dune/drift/hill
~, поперечная transversal dune
~, развеянная blown-out dune
дюраль, дюралюмин, дюралюминий duralumin
дюрен *miner.* durain, durite
дюрометр durometer
дятлина *bot.* red clover *(Trifolium pratense)*
~, белая white clover *(Trifolium repens)*

Е, е

Ева Eve
евангелие gospel; (**Е**) the Gospels
евангелист 1. evangelical; 2. Evangelist
евангелический evangelic(al)
евангельский gospel *(attr.)*
евгеника eugenics
евгенист eugenicist, eugenist
евкалипт = эвкалипт
евклидов Euclidean
евнух eunuch
евнухоид eunochoid
евнухоидизм eunochoidism, eunochism
евразиатсткий Euroasiatic
евразийский Eurasian
Евразия Eurasia
еврамерийский Euramerian
Евратом Euratom, European Atomic Energy Community
евроакция *econ.* Euroequity, Eurobond
евробанк Eurobank
евровалюта Eurocurrency
евроденьги *econ.* Euromoney
евродоллар *econ.* Eurodollar
еврооблигации *econ.* Eurobonds
Европейская ассоциация свободной торговли European Free Trade Association, EFTA
Европейская библиотека программ European Program(me) Library
Европейская валютная единица European Currency Unit, *see* **экю**
Европейская зона свободной торговли European Free Trade Area, EFTA
Европейская комиссия по атомной энергии,
Евратом European Atomic Energy Community, EURATOM
Европейский платежный союз European Payments Union, EPU
Европейский фонд развития European Development Fund, EOF
Европейское валютное соглашение European Monetary Agreement, EMA
Европейское объединение угля и стали European Coal and Steel Community, ECSC
Европейское сообщество European Community, EC

Европейское экономическое сообщество European Economic Community, EEC
европий *chem.* europium, Eu
еврофермеры Eurofarmers
евстахиев *anat.* Eustachian
евстахиит *med.* eustachitis
Евтерпа *myth.* Eutherpe
евфеника euthenics
Евфросина *myth.* Euphrosine
евхаристический *eccl.* eucharistic(al)
евхаристия *eccl.* Eucharist
егерь 1. *mil.* chasseur; 2. huntsman, gamekeeper, forester
египетско-арабский Egypto-Arabic
единица 1. one; *math.* unity; 2. unit; item
~, абсолютная absolute unit, abunit
~, автономная autonomous unit; self-sufficient unit
~ активности activity unit
~ активности дигиталиса *med.* digitalis unit
~ амортизации *econ.* depreciation unit
~ атомного веса atomic weight unit, awu
~ атомного веса, физическая/химическая physical/chemical atomic weight unit
~, боевая *mil.* combat unit
~, бонитировочная evaluation unit
~, британская тепловая British thermal unit, BTU
~, внесистемная off-system unit; arbitrary unit
~ времени unit time; time unit
~, двоичная binary unit; binary one; bit unit
~ длины unit length; length unit
~ дозы облучения rad unit
~, европейская валютная European currency unit, ecu
~ емкости памяти *comp.* memory size unit
~ изглашения phonational unit
~ измерения unit (of measurement)
~ измерения, акустическая acoustical unit
~ измерения громкости volume unit
~ информации information unit, unit of information, datum
~ информации, двоичная bit, binary digit
~ калькулируемой продукции *econ.* cost unit
~ калькуляции издержек *econ.* cost unit
~ количества информации *comp.* information content unit
~, кормовая fodder/feed unit
~, логическая logical ''1''
~ массы mass unit, unit of mass
~ массы, атомная atomic weight/mass unit, amu
~ масштаба computing unit
~, матричная unit matrix
~, международная international unit, I.U.
~ меры unit of measure
~, метрическая metric unit
~, мнимая imaginary unit
~, монетная monetary unit
~ мощности power unit
~, натуральная *econ.* physical unit
~ начисления износа *econ.* depreciation unit
~, нормированная normalized/reduced unit
~ облучения rad unit
~ объема volume unit
~ переноса transfer unit
~ площади unit of area, area unit
~ погрешности unit of error
~, потребительская *econ.* spending unit
~, приведенная reduced/normalized unit
~ продукция unit of output; production unit
~, производная derived unit
~, производственная *econ.* production/producing unit
~, разрушенная disturbed one
~, сборочная assembly unit
~, световая photometric unit
~ силы unit of force, force unit
~ стоимости *econ.* unit of value
~ счета count, tally
~, таксономическая taxonomic unit, taxon
~, тарифная charge unit
~ телесного угла *geom.* steradian
~, тепловая thermal/heat/caloricity unit
~, тяговая traction unit
~, условная conventional/standard/arbitrary unit
~ учета unit of account
~ ценности unit of value
~, эксплуатационная operational unit
~, электромагнитная electromagnetic unit, emu
~, электростатическая electrostatic unit, esu
~, энергетическая/энергии energy unit
~, ядерная nuclear unit
~, языковая unit of language
~цы, кратные *math.* multiple units
единобожие *rel.* monotheism
единобрачие monogamy
единобрачный monogamous; *bot.* monogamian
единовластие autocracy; absolute rule
единоженство monogamy
единокровный consanguineous; agnatic
~ брат half brother, brother-consanguinean
единоличник individual farmer
единоличный individual; personal; private
~ное крестьянское хозяйство individual/personal/family farm
единонаследие *law* primogeniture
единоначалие one-man management; undivided authority; combined administration/command
единообразие uniformity
единообразный uniform; *math.* einartig
единорог 1. *astr.* the Unicorn, Monoceros; 2. *myth.* unicorn; 3. *zool.* unicorn fish, narwhal(e)
единосущность *theol.* consubstantiality
единосущный *theol.* consubstantial; coessential
единоутробный uterine
~ брат uterine/half brother
единственность uniqueness; unicity
едкий caustic; corrosive

еж *zool.* hedgehog *(Erinaceus)*
~, морский sea urchin *(Echinida)*
ежевидный echinoid
ежевика *bot.* blackberry, bramble *(Rubus)*
ежегодник yearbook, annual; almanac
~, статистический statistical yearbook
ежовый echinoid
екклезиаст Ecclesiast
ектенья *eccl.* liturgical prayer; ektenia
екю *see* **экю**
елей anointing oil; uncion; chrism
елизаветинец, елизаветинский *lit.* Elizabethan
елисейский Elysian
елка fir, spruce
~, рождественская Christmas tree
ель *bot.* spruce, fir *(Picea)*
емкий capacious, roomy; large-capacity *(attr.)*
емкостно-резистивный *el.* capacitance-resistance
емкостно-связанный *el.* capacitance-coupled/connected
емкостный capacitive; capacitance *(attr.)*
емкость 1. capacity; volume; 2. *el.* capacitance; capacitor; 3. container, vessel, tank, reservoir
~ анода *el.* anode/plate capacitance
~ бака tank capacity
~, буферная buffer vessel
~ в байтах/битах *comp.* byte/bit capacity
~ в словах word capacity
~ внутренних баков *naut.* internal tankage
~ вспомогательной памяти auxiliary/secondary store capacity
~, входная input capacitance
~, выходная output capacitance
~, действующая effective capacitance
~, дифференциальная differential/incremental capacitance
~, дыхательная *anat.* inspiratory capacity
~, загрузочная/зарядная charge capacity
~ запоминающего устройства memory/storage capacity
~ ионного/катионного обмена ion-/cation-exchange capacity
~ канала связи channel capacity
~, конечная *teleph.* ultimate capacity
~, междувитковая/междуобмоточная *el.* interturn [inter-winding] capacitance
~, междупарная *teleph.* pair-to-pair capacitance
~ на "землю" *el.* capacitance to earth/chassis
~, номинальная rated capacity, rating
~ обратной связи feedback capacitance
~ оперативной памяти working store capacity;
~ памяти memory capacity/size; capacity of memory/store equipment
~ памяти в битах bit capacity
~, паразитная stray/spurious capacity
~, параллельная *el.* shunt capacitance/capacitor
~ перехода junction capacitance
~ перфокарты card capacity
~, подземная underground (storage) tank
~, полезная payload/available capacity
~, проходная transfer capacitance
~ регистра register length/capacity
~, резервная spare/idle capacity
~ сетки *el.* grid capacitance
~ сумматора accumulator volume/contents
~ счетчика counter capacity
~, топливозаправочная refuelling capacity
~ управляющего электрода gate capacitance
~, эксплуатационная service capacity
~, эффективная выходная effective output capacitance
енот *zool.* raccoon *(Procyon)*
епархиальный eparchial; diocesan
епархия *eccl.* diocese; bishopric; see
епископ bishop, diocesan
епископальный episcopal; episcopalian
епископский episcopal, diocesan
епископство 1. episcopate, bishopric; 2. *coll.* episcopacy
епитимья *eccl.* penance
епитрахиль *eccl.* stole
ересиарх *eccl.* heresiarch
ересь heresy; idolatry
еретик heretic
еретический heretical
еретичество heresy
ерш 1. *zool.* ruff *(Acerina cernua);* 2. *tech.* barbed nail/bolt; wire brush
естественник science teacher/student; naturalist
естественноисторический natural-historical
естественнонаучный natural-science *(attr.)*
естествовед natural scientist
естествоведение natural science/history
естествознание natural/physical science
естествоиспытатель natural scientist
ефес *mil.* hilt
ефесский, ефессянин *see* **эфесский**
ефрейтор *mil.* private first class, pfc; lance corporal
ехидна *zool.* echidna, porcupine anteater *(Tachyglossus)*
еюнит *med.* jejunitis
еюноеюностомия jejunojejunostomy
еюноилестомия jejunoilestomy
еюноколостомия jejunocolostomy
еюнолеит jejunoleitis
еюноэктомия jejunectomy

Ж, ж

жаба I *zool.* toad *(Bufo)*
~, вонючая/камышовая natterjack *(Bufo calamata)*
~, носатая burrowing toad *(Rhinophrynus dorsalis)*

жаба II *med.* quinsy; angina
~, брюшная angina abdominis
~, грудная angina pectoris; breast pang
жабернодышащий *zool.* gill-breathing
жаберный branchial; *n. pl.* branchiates *(Branchiata)*
жабо jabot
жаброногие *zool.* branchiopods *(Branchiopoda)*
жабры gills, branchiae; *av.* sponsons
~, листоватые book gills
~, кожные dermal gills
~, ложные pseudobranchs
~, перистые gill plume
~ прикрепляющиеся к основанию ног podobranchia, footgills
жаворонок *zool.* lark
~, полевой skylark *(Alauda arvensis)*
~, хохлатый crested lark *(Galerida cristata)*
жажда thirst
~, болезненная dipsosis
~, повышенная polydipsia
~, пониженная hypodipsia, twilight thirst
жаждоутоляющий *med.* adipsous
жаккард *text.* Jacquard loom
жако *zool.* gray parrot *(Psittacus erithacus)*
жалоба complaint, grievance
~, апелляционная *law* appeal
~бу, подавать lodge/make a complaint; *law* file an appeal
жалобщик *law* plaintiff; complainant
жалованье pay, wages, salary, remuneration
~ за проработанное время back pay
жалюзи Venetian blind, jalousie, louver(s)
~ радиатора radiator shutters/gills
жандарм gendarme
жандармерия gendarmerie, gendarmery
жанр *math.* genus, genre; *art* genre painting
жар heat; glow
~, белокалильный white heat; incandescence
~, лихорадочный *med.* fever heat
~, плавильный fusion heat/temperature
жарновец *bot.* broom *(Cytisus, Sarothamnus)*
~, метельчатый Scotch broom
жаровня fryer, brazier, roaster
жаровыносливый heat-resistant
жаропонижающий *med.* antipyretic; febrifugal
~щое средство alexipyretic, febrifuge
жаропрочность, жаростойкость, жароупорность high-temperature strength, heat fastness, thermal stability; resistance to heat, refractoriness, heat-proofness
жаропрочный, жаростойкий, жароупорный heat- resistant/-resisting, heatproof; refractory
жароустойчивость heat tolerance/resistance
жароустойчивый heat-tolerant/-resistant
жасмин *bot.* jessamine, jasmine *(Jasminum)*
жатка reaper, reaping machine, harvester
~, валковая windrower
~, грабельная self-rake reaper
~, лафетная windrow harvester
~, навесная mounted reaper
~, платформенная cutting platform/table
~, прицепная trailed/trailer reaper
~, прямоточная straight-through windrower
~, рядковая windrower
~ с боковым сбросом side-delivery reaper
~, фронтальная front-mounted reaper
~, широкозахватная wide-cut reaper/header
жатка-лобогрейка/самосброска self-rake reaper
жатка-молотилка harvester combine
жатка-сенокосилка reaper-mower
жатка-сноповязалка reaper-binder, corn binder
жвачный *zool.* ruminant, ruminating
жвачные *pl. zool.* ruminants *(Ruminantia)*
жгут braid; plait; gasket, packing, oakum; *el.* bundle, core; *med.* torniquet
~, вихревой vortex core
~ кудели tow
~, набивочный packing material; gasket
~, пеньковый hemp twist/cord/rope
~ электропроводов wire bundle, cable assembly
жгутик 1. small braid/plait; 2. cilium, flagellum
жгутиковый flagellate, fragellar
жгутовый rope-/braid-/plait-shaped
жгутораспрaвитель *text.* scutcher, guide rail
жевание chewing, mastication
жевательный *anat.* masticatory
жезл bar; staff, baton, token
желатина, желатин gelatin(e)
~, глицеринозированный glycerinated gelatin, glycerogelatin
~, питательный nutrient/edible/food gelatin
~, цинковый zinc gelatin, Unna's paste
желатинизация, желатинация, желатинирование gelatinization, gelation; gellification, gelling
желатинироваться gel, set to a gel
желатинный, желатиновый gelatinous
желвак concretion, nodule; bulge, lump, swelling; *med.* tomour; *bot.* struma
~ в пласте niggerhead
~, кремнистый flint nodule
~, миндалевидный amygdaloid/subglobular nodule
~ окиси марганца manganese nodule
~ пирита, блестящий cat face
~, роговиковый niggerhead
~, рудный ore nodule
~, сплющенный flattened globule
желвачный nodular, nodulous
желе jelly
железа *anat.* gland; *see also* **железы**
~, верхняя околощитовидная superior parathyroid gland
~, вилочковая thymus gland
~ внутренней секреции endocrine gland
~, добавочная надпочечная accessory suprarenal gland

~, малая преддверная lesser vestibular gland
~, молочная mammary gland
~, надпочечная suprarenal gland
~, нектароносная *bot.* nectar gland
~, нижняя околощитовидная inferior parathyroid gland
~, околоушная parotid gland
~, поджелудочная pancreas
~, поднижнечелюстная submandibular/submaxillary gland
~, подъязычная sublingual gland
~, предстательная prostate
~, слезная lacrimal gland
~, шишковидная pineal gland
~, щитовидная thyroid gland
железисто-плевральный pleuroglandular
железистый *med.* glandular
железистый *chem.* ferric *(trivalent);* ferrous *(bivalent);* ferruginous, ferriferous
железка *anat., bot.* glandule, gland
железковидный, железкообразный glanduliform, glandlike
железконосный glanduliferous, adenophorous
железная дорога railway; *Am.* railroad
железнение iron plating
железнодорожник railwayman, railroadman
железнодорожный railway, railroad, rail *(attr.)*
~ подвижной состав rolling stock
~ путь railway track
~ узел railway junction
~ная ветка branch line
~ное полотно railroad tracks
железнокислый *chem.* ferrate
железный 1. *met.* iron *(attr.);* 2. *chem.* ferrous
~ блеск *miner.* h(a)ematite
~ занавес *pol., hist.* Iron Curtain
~ купорос copperas, green vitriol
~ лом scrap iron
~ товар hardware
~ное дерево *bot.* lignum vitae *(Guaiacum officinale)*
железняк 1. iron ore, ironstone, blue brick, ironclay, pan; 2. *bot.* ironwood, parrotia
~, бобовый bean/pisolitic ore
~, болотный bog iron ore, iron pan
~, бурый bog/brown ore, limonite
~, глинистый clay/clayband ironstone, pennystone
~, гравийный buckshot gravel
~, желтый xanthosiderite
~, красный h(a)ematite, red iron ore
~, луговой бурый meadow brown ore, meadow limonite
~, магнитный magnetite, magnetic iron ore, native/natural magnet
~, натечный бурый sinter brown ironstone
~, озерный бурый lacustrine iron ore
~, полосатый banded ironstone
~, сфероидальный шпатовый spherosiderite
~, хромистый chromic iron, chromite
~, черный manganese limonite
~, шпатовый siderite, spathic/sparry iron ore, chalybite
железо iron; ferrum
~, бромистое ferrous bromide
~, вольфрамистое ferro-tungsten
~, глицерофосфорное ferric glycerophosphate
~, губчатое porous/spongy iron
~, двусернистое iron bisulphide
~, двутавровое I-iron, double T-steel
~, двухвалентное/закисное ferrous iron
~, закисное хлористое protochloride of iron
~, карбидное carburetted iron
~, кованое wrought/forge iron
~, коробчатое/корытное channel iron
~, кровельное roofing sheets/iron
~, метеорное meteoric/cosmic iron
~, молочнокислое закисное ferrous lactate
~, мышьяковое iron diarsenide, lollingite
~, мышьяковокислое ferric/iron arsenate
~, мягкое *(magnetically)* soft iron
~, некондиционное off-grade iron
~, оцинкованное galvanized iron
~, порошковое powdered iron
~, самородное native iron
~, сварочное welding iron
~, сернистое ferrous/ferric sulphide
~, сернокислое ferric sulphate, iron persulphate/sesquisulphate
~, сернокислое закисное ferrous/iron sulphate, green/iron vitriol, copperas
~, синеродистое ferrocyanogen
~, твердое *(magnetically)* hard iron
~, текстурованное oriented-grain iron
~, трехвалентное ferric iron
~, треххлористое trichloride of iron
~, углекислое ferric carbonate
~, углеродистое carbon/carburetted iron
~, фосфорноватистокислое ferric hypophosphite
~, хлористое ferrous-chloride, protochloride of iron
~, хлорное ferric chloride, iron trichloride/perchloride
~, янтарнокислое ferric succinate, basic iron succinate
железобетон reinforced concrete, ferroconcrete
~, легкий lightweight reinforced concrete
~, сборный precast/prefabricated reinforced concrete
~, тяжелый heavy reinforced concrete
~, ячеистый wire-mesh reinforced concrete
железографит ferrographite, iron graphite
железокирпич reinforced brick
железококс ferrocoke, ferrous coke
железолюбивый siderophilous, iron-requiring
железомагнитный ferromagnetic
железомарганцевый ferromanganese *(attr.)*
железообрабатывающий iron-working
железоплавильный iron-smelting

~ **завод** iron foundry
железоподобный ironlike, ferruginous
железопорфирин ferroporphyrin
железопрокатный завод/стан rolling mill
железосинеродистый ferrocyanide
железосодержащий ferruginous, ferriferous, iron-bearing
железостекло ferroglass
железотерапия *med.* ferrotherapy
железы *anat.* glands; *see also* **железа**
~, **альвеолярные** alveolar/acinous glands
~ **внешней секреции** exocrine glands
~ **внутренней секреции** endocrine/endocrinous/ductless glands
~ **внутренней и внешней секреции** endo-exocrine/exo-endocrine glands
~, **восковые** ceromінous glands
~, **глоточные** pharyngeal glands
~, **голокринные** holocrine glands
~, **гортанные** laryngeal glands
~, **двенадцатиперстнокишечные** duodenal glands
~, **желудочные** gastric glands
~, **кишечные** intestinal glands, follicles/crypts of Lieberkuen
~, **клубочковидные** glomiform glands
~, **конъюнктивальные** conjunctival/Krause's glands
~, **коренные** molar glands
~, **луковично-мочеиспускательные** bulbourethral/Cowper's glands
~ **матки, шеечные** cervical glands of the uterus
~, **мерокринные** merocrine glands
~, **передние язычные** anterior lingual glands
~, **пищеводные** esophageal glands
~, **половые** sexual glands
~, **потовые** sudoriferous/sudoriparous/sweat glands
~, **привратниковые** pyloric glands
~, **реоничные** ciliary glands
~, **ротовые** oral glands
~, **сальные** sebaceous glands
~, **синовиальные** synovial/mucilaginous/Havers' glands
~, **слизистые** mucous/muciparous glands
~, **слюнные** salivary glands
~ **хряща века** tarsal/tarsoconjunctival/Meibomian glands
~, **щечные** buccal glands
~, **экзокринные** exocrine glands
~, **эккринные потовые** eccrine sweat glands
~, **эндокринные** endocrine/endocrinous/ductless glands
~, **эруптокринные** eruptocrine glands
~, **язычные** lingual glands
желеобразный, желеподобный jelly-like, gelatiniform, gelatinous; pectous
желирование gelatinization, jellification
желоб chute, trough; flume; channel; groove, furrow; trench; slot; runner; *anat.* alveus; *bot.* vallecula
~, **водосточный** gutter; eaves trough
~ **выдувания** deflation groove
~, **выпускной** *met.* tapping spout
~ **для окалины** scale chute
~ **для сыпучих материалов** spill chute
~, **загрузочный** charging chute/trough; intake chute/apron; feed spout
~, **золотопромышленный** *min.* riffle
~, **кровельный** eaves trough
~, **подводящий** head race
~, **приемный** collecting launder
~, **разгрузочный** discharging chute/trough
~, **разливочный** pouring spout
~, **спускной** discharging chute/trough; drainage chute
~, **транспортирующий** conveying trough; chute conveyer
~, **шлаковый** *met.* slag runner
желобить flute, groove, slot
желобок groove, channel; furrow; riffle
~, **подносовый** *anat.* infranasal depression
желобчатолистный *bot.* canalifolious
желобчатый fluted, grooved, slotted, furrowed
желонка sludge/sand pump, bailer, sludger
~, **грейферная** clam-shell bailer
желток yolk
~, **питательный** food yolk, tropholecithus
~, **яичный** egg yolk
желтокожий yellow-skinned
желтокорень golden-seal *(Hydrastis)*
желтуха *med.* jaundice, hepatitis, icterus; *bot.* yellows
~, **вирусная** infectious/virus/catarrhal hepatitis
~, **деструкционная** destructive jaundice
~, **злокачественная** malignant jaundice, icterus gravis
~, **калькулезная/обтурационная** obstructive jaundice
~ **новорожденных** icterus of the newborn, inogenous jaundice, icterus neonatorum, pedicterus, Bull's/Winckel's disease
~, **печеночная** hepatogenous jaundice
~, **подострая** subacute jaundice
~, **семейная ахолурическая** familial acholuric jaundice
~, **скрытая** latent/occult jaundice
~, **сывороточная** homologous serum jaundice, serum hepatitis
~, **ядерная** kernicterus, nuclear jaundice
желтушник *bot.* erysimium; treacle mustard *(Erysimium)*
желтушный icteric, icteroid, icteritious; yellowish, bile-stained/-tinged
желудок stomach
~ **в виде песочных часов** hourglass/bilocular stomach
~, **грудный** thoracic stomach
~, **железистый** glandular stomach
~, **первый/второй** paunch/reticulum

~, третий (жвачных) omasum
~, четвертый (жвачных) abomasum
желудочек *anat.* ventricle; stomach
~ головного мозга ventricle of the brain
~ гортани laryngeal ventricle
~, павловский Pavlov pouch/stomach, miniature stomach
~ сердца ventricle of the heart
желудочковый ventricular
желудочно-диафрагмальный gastrophrenic
желудочно-кишечный gastrointestinal, gastroenteric
желудочно-легочный gastropneumonic, pneumogastric, gastropulmonary
желудочно-селезеночный gastrosplenic
желудочный gastric, stomachal; stomach *(attr.)*
желчевыносящий biliferous
желчегонный cholagogue, cholagogic, bile-expelling
желчеобразование biligenesis, cholopoiesis, bile production/formation
желчеобразовательный biligenic
желчеобразующий cholepoietic
желчеотделение bile secretion
желчнокаменный cholelithic
желчнопузырный cholecystic
желчный biliary, bilious
~ камень gallstone
~ пузырь gall bladder
желчь bile, gall, fel
~, бычья oxgall, fel bovis/tauri
~, дуоденальная A-bile
~, "печеночная" C-bile
~, "пузырная" B-bile
жемчуг, жемчужина pearl
жених 1. fiancé; 2. bridegroom; 3. suitor
женовластие gynarchy
женоложство Lesbianism
женоненавистник misogynist
женоненавистничество misogyny
женотдел women's section/department
женоубийство, женоубийца uxoricide
женщина woman; lady
~, бездетная nulliparous/zeroparity woman
~, незамужняя single woman
~, разведенная separated/divorced woman
~ родившая нескольких детей *med.* multiparous woman, multipara
женьшень ginseng
жеребейка chaplet; stud
~, двусторонняя double-headed chaplet/stud
жеребец stallion
~, племенной/производитель studhorse
жеребьевка *game th.* toss-up
жерминаль *hist.* Germinal, Revolutionary calendar
жерло orifice, mouth; crater; muzzle
жернов millstone
~ бегунов edge stone
~, верхний runner, upper millstone
~, нижний bed/nether millstone, groundsel
жерушник *bot.* nasturtium *(Roripa, Nasturtium)*
~, болотный marsh cress
~, земноводный brown cress
~, сидячецветковый sessile-flowered cress
жестепечатание tin printing
жесткокрылый hard-winged; coleopterous
жестколиственный *bot.* hard-leaved; sclerophyllous
жесткосемянный hard-seeded, sclerospermous
жесткость rigidity, stiffness; *(of water)* hardness
~, диэлектрическая elastance
~ крыла/лопасти wing/blade stiffness
~ пластины (цилиндрическая) flexural rigidity of a plate
~ подвески, угловая *auto*. roll stiffness of suspension
~ при изгибе flexural rigidity, rigidity in bending
~ при кручении torsional rigidity
~ при растяжении rigidity in tension
~ при сжатии rigidity in compression, compression stiffness
~ при срезе shear stiffness
~ рессоры spring rate
~, электрическая elastance
жесть sheet iron/metal, plate; tin-plate
~, белая tin-plate
~, горячелуженая hot-dipped tin-plate
~, лакированая lacquered tin-plate
~, луженая tin-plate
~ с двусторонним покрытием dual-coat tin-plate
~, черная black plate
жестяник tinsmith
живица oleoresin, soft resin, turpentine gum; barras
живопись painting
~, батальная battle painting
~ по стеклу glass painting
~, стенная murals
~, фресковая fresco
живородка *zool. Viviparus*
живородящий *zool.* viviparous; *pl.* viviparous animals
живорождение *zool.* vivipary, viviparity
живосечение vivisection
живот abdomen, venter, belly
~, впалый retracted belly
~, ладьеобразный navicular/scaphoid abdomen
~, отвислый pendulous abdomen
животворение vivifying, vivification
животворец vivifier
животновод livestock/animal/stock breeder
животноводство livestock/animal/stock raising, animal husbandry, livestock farming
~, молочное dairy farming
~, мясное beef cattle production
~, пастбищное grassland farming
~, племенное livestock breeding
животноводческий stock-raising
~кая ферма livestock farm
животное animal

~, **беспозвоночное** invertebrate animal
~, **всеядное** omnivore
~, **головноногое** squid
~, **двуногое** biped
~, **зарывающееся** burrowing animal
~, **ископаемое** fossil animal, zoolith
~, **млекопитающее** mammal
~, **многоклеточное** metazoan
~, **мягкотелое** malacazoon
~, **одноклеточное** unicellular animal
~, **окаменелое** zoolith
~, **отбракованное** cull
~, **пелагическое** surface swimmer
~, **плотоядное** carnivore
~, **подопытное** experimental/test animal
~, **позвоночное** vertebrate animal, vertebrate
~, **пойкилотермное** poikilothermic animal
~, **ползящее** crawler, sprawler
~, **простейшее** protozoan
~, **растениеядное/травоядное** herbivore
~ **содержащееся зимой на открытом воздухе** outlier
~, **сумчатое** marsupial
~, **толстокожее** pachyderm
~, **четвероногое** tetrapod, quadruped
~, **четверорукое** quadrumane
животнорастение zoophyte
животрепещущий topical, exciting, burning, stirring, of vital importance/significance
живоцерковник supporter/follower of Living Church
живучесть (probability of) survival, survivability; vitality, durability
~, **боевая** *mil.* invulnerability
живущий living; existent
~ **в крови** *biol.* hematobious
~ **в темноте** photophobe
~ **в ткани** histozoic
~ **на глубине** deep-dwelling
~ **на грибах** agaricolous
~ **на деревьях** arboreal
~ **на морском дне** benthonic
жидкий liquid, fluid
жидководородный liquid-hydrogen *(attr.)*
жидкометаллический liquid-metal *(attr.)*
жидкоплавкий fluid, liquid
жидкоплавкость fluidity, liquidity
жидкоподвижность flowability, fluidity
жидкостекловатый liquevitreous
жидкостно-воздушный liquid-pneumatic, liquid-air *(attr.)*
жидкостный fluid, liquid
жидкость liquid, fluid
~, **аллантоисная** allantoic fluid
~, **амортизаторная** shock-absorber fluid
~, **асцитическая** *pathol.* ascitic fluid
~, **белильная** bleach liquor
~, **буровая** *geol.* mud fluid
~, **внутриглазная** intraocular fluid
~, **внутриклеточная** intracellular fluid
~, **водяночная** *med.* hydropic fluid
~, **выщелачивающая** lixivant, leaching-out liquid
~, **вязкая** viscous fluid
~, **вязкоупругая** viscous-elastic liquid
~, **гидравлическая** pressure/hydraulic fluid
~, **дегазационная** chemical decontaminating fluid
~, **дезактивационная** decontamination fluid
~, **дезинфицирующая** sanitary fluid; disinfectant
~, **дезодорирующая** deodorant liquid, deodorant
~ **для гидросистем** hydraulic fluid
~ **для манометров** manometer liquid/fluid
~ **для охлаждения** coolant, cooling liquid
~ **для полоскания рта** mouth wash
~ **для хонингования** honing fluid
~, **закалочная** quench, quenching bath/medium
~, **зольная** lime liquor
~, **красящая** liquid dye
~, **культуральная** culture fluid
~, **легковоспламеняющаяся** inflammable fluid
~, **легкоподвижная** thin fluid
~, **легкопроникающая** penetrating/penetrant fluid/liquid
~, **маслянистая** oleaginous fluid
~, **мягчильная** bating liquor
~, **надосадочная** supernatant (fluid)
~, **невязкая** inviscid/non-viscous/frictionless liquid
~, **несливаемая** undrainable fluid
~, **несмачивающая** non-wetting liquid
~, **низкозамерзающая** low-freezing fluid
~, **низкокипящая** low-boiling fluid
~, **ньютоновская** normal/simple/Newtonian fluid
~, **огнегасящая** extinguishant (fluid)
~, **однокомпонентная** single fluid
~, **околоплодная** *anat.* amniotic fluid
~, **окрашивающая** liquid dye
~, **отечная** *med.* transudate
~, **отработавшая** spent/discharge liquid/liquor/water
~, **охлаждающая** cooling liquid, coolant
~, **паяльная** soldering liquid
~, **покоящаяся** *hydr.* liquid at rest
~, **покрывающая** supernatant (liquid)
~, **промывная/промывочная** wash/flushing liquid
~, **противообледенительная** de-icing/antiicing/anti-icer fluid
~, **противопожарная** fire-suppression fluid
~, **рабочая** working fluid
~, **резьбоуплотняющая** thread-sealing liquid
~, **семенная** *anat.* seminal fluid
~, **серозно-кровянистая** serosanguineous fluid
~, **синовиальная** synovial fluid
~, **смазочно-охлаждающая** lubricant-coolant

~, **смачивающая** wetting liquid
~, **спинномозговая** cerebrospinal fluid
~, **технологическая** process fluid
~, **тормозная** braking fluid
~, **уплотняющая** sealing/packing/compacting fluid
~, **упруговязкая** viscoelastic fluid
~, **этиловая** ethyl fluid, tetraethyl lead
жидкость-теплоноситель heat-transfer liquid
жидкотекучесть fluidity
жижеразбрасыватель *agr.* liquid-manure spreader/distributor
жижесборник *farm.* liquid-manure pit/tank
жизнедеятельность vital activity
жизнеобеспечение *av., space* life support/survival system
жизнепонимание world outlook/view, Weltanschauung
жизнеспособность viability
жизнеспособный viable
жиклёр jet, nozzle; fuel-injection/-metering jet/nozzle/orifice
~ **малого газа** idle/idling jet; low-speed jet/nozzle
~ **полной мощности** full-power jet/nozzle
~, **ускорительный** accelerating jet/nozzle
~ **холостого хода** idling/idle jet
жила 1. tendon; sinew; 2. *geol.* vein, ledge, lode; 3. *el.* core, strand
~, **богатая** pay streak
~, **боковая** dropper
~, **главная** mother/champion lode; master/main vein
~ **кабеля** cable wire/conductor
~, **камерная** chambered vein
~, **ленточная** banded/stringer vein/lode
~, **малоценная (рудная)** gangue
~, **мелкозалегающая** vein of shallow depth
~, **металлоносная** lode
~, **мощная** large lode
~, **непромышленная** dead/unkindly lode
~, **основная** mother lode
~, **пластовая** sheet/bed/bedded/blanket vein; interformational/intrusive sheet
~, **полиминеральная** compound vein
~, **постоянная** persistent lode
~, **простая** single vein
~, **рубцовая** gash vein
~, **сбросовая** slip/fault vein
~, **седловинная** saddle reef
~, **секущая** cross/rake vein, counter-lode
~, **скрученная** stranded conductor
~, **сложная** composite/compound/multiple vein
~, **трубчатая** linear/pipe vein
~, **четковидная** beaded vein
~, **чечевицеобразная** lenticular vein
~, **ячеистая** cavernous vein
жила-проводник lead vein, leader
жилет jacket, vest; waistcoat
~, **спасательный** life jacket
жилец lodger, tenant
~, **незаконный** squatter
жилистий fibrous, threadlike
жилище dwelling, living quarters, lodging; abode; habitation
~, **арендуемое** rented/tenure dwelling
~, **многосемейное** multifamily dwelling
~, **односемейное** single dwelling
жилищно-бытовой housing; domestic
~**вые условия** living conditions
жилищно-коммунальный housing-and-municipal
жилка veinlet, vein, string, nerve, fibre, nervure; *fig.* flair, streak
~ **листа, главная** *bot.* median nerve/vein/rib
~, **центральная** mid-rib
жилкование venation, nervation, neuration, veining
~, **дланевидное пальчатое** palmate venation
~, **дуговидное** camptodromous nervation
~, **краебежное** craspedodromous nervation
~, **обильное** flushed venation
~, **петлевидное** brochidodromous nervation
жилковатый veined, veiny; stringy; nervate; ribbed, costate
жилплощадь living space; floor area/space; accommodation, housing
жилстроительство house building
жилфонд housing fund, available housing
жир fat; grease; lard; tallow; oil
~, **животный** animal fat
~, **костный** bone fat/tallow
~, **пищевой** edible fat
~, **подкожный** blubber
~, **полутвердый** lard
~, **растительный** vegetable fat
~, **рыбий** fish oil; cod-liver oil
~, **твердый** hard fat, tallow
~, **технический** industrial fat
~, **топленый** rendered fat
~, **шерстный** dégras
~, **шерстяной** lanoline
жирант *econ., law* endorser, indorser
жират endorsee, indorsee
жираторный gyrating, gyratory
жираф *zool.* giraffe *(Giraffa camelopardalis)*
жирировать *econ., law* endorse, indorse
жирность fat content
~ **бетона** fatness of concrete
~ **почвы** fertility/richness of soil
жиро *econ., law* endorsement, indorsement
жировать grease, oil, lubricate
жировой *anat.* adipose, adipic, fatty; *physiol.* sebaceous, sebacic
~**вое перерождение** *med.* adipose/fatty degeneration
жировоск adipocere, lipocere
жирокомпас gyrocompass
жироловка grease trap
жиромасличный oleiferous
жиромер butyrometer

жиронда *hist.* the Gironde
жирондист *hist.* Girondist
жиронепроницаемость greaseproofness
жирооборот *fin.* paper transaction
жирообразующий lipogenetic, lipogenous
жироотстойник slush box; grease trap
жироприказ *fin.* banking order, endorsement
жирорастворимый fat-/oil-soluble, liposoluble
жирорасчет *fin.* noncash settlement
жирорасщепление fat cleavage/splitting, lipolysis
жироскоп gyroscope
жиросодержание fat content
жиростатика gyrostatics
жиротопление fat rendering/melting
жироуловитель grease trap; slush box
жиро-чек girocheque, girocheck
житель inhabitant, dweller, resident
 ~, временный nonresident
 ~, коренной aboriginal, autochthon
жмых oil cake, cake meal
 ~, молотый oil meal
 ~, сурепный rape-seed cake
жмыходробилка oil-cake grinder/mill
жнейка harvester, reaper
жом beat pulp, bagasse
жостер *bot.* buckthorn *(Rhamnus)*
жук beetle
 ~, картофельный/колорадский Colorado potato beetle
 ~, майский May bug, cockchafer
жук-долгоносик snout beetle
жук-навозник dung beetle, dorbeetle *(Geotrupes stercocarius)*
жук-олень, жук-рогач stag beetle *(Lucanus cervus)*
жупан *hist.* district chief
журавельник *bot.* 1. geranium *(Geranium);* 2. heron's bill *(Erodium)*
журавль 1. *zool.* crane; 2. *tech.* lever, arm; sweep
 ~, микрофонный microphone boom
журнал 1. magazine, journal; periodical; 2. register, log, diary, journal; account book; log-book
 ~, аппаратный *teleph.* message log/registry
 ~ базы данных database journal
 ~, башенный turret log
 ~, бортовый flight log
 ~, буровой drill log
 ~, вахтенный ship's/sea log-book; deck/duty/station log
 ~, гироскопический gyro-compass log
 ~ для записи пеленгов bearing book
 ~ заданий job log
 ~ издержек cost ledger
 ~, кассовый cash journal
 ~ кассовых операций cash book
 ~, компасный compass/deviation journal
 ~, котельный fireroom log
 ~, машинный *naut.* engine-room log/register
 ~, навигационный navigation data log
 ~, нивелирный field levelling book
 ~, оперативный log(-book)
 ~ повреждений trouble record book
 ~ программных испытаний program(me) test book
 ~, промерный sounding book/log/record
 ~ радиосвязи radio/wireless log
 ~, реферативный abstract journal
 ~, системный system log
 ~ строительных работ builder's diary/log
 ~, судовой ship's log(-book)
 ~ учета заданий job account book
 ~ учета работы operation log
 ~, хронометрический chronometer journal
 ~, штурманский navigator log-book

З, з

заальпийский transalpine
заанкеровать anchor
заарретированный caged
заатлантический transatlantic
заатмосферный extra-atmospheric
забастовка strike; stoppage; industrial action
 ~, итальянская sit-down strike
 ~, неофициальная unofficial/wildcat/outlaw strike
 ~, сидячая sit-down strike, strike on the job
 ~ солидарности sympathy/secondary strike
забвение neglect, disregard; contempt
 ~ долга dereliction of duty
забег 1. overshoot, overswing; 2. *sport* heat, round
заберег submerged off-shore ice; ice-free area (along the shore)
забивание hammering/ramming/knocking in; clogging, choking; *rad.* jamming
 ~ ошибок error erasure
 ~ фильтра filter clogging
 ~ формы *print.* filling-in of a form
забивать 1. ram/knock/hammer/knock in; clog/choke/stop up; pack, fill in, block; *rad.* jam; 2. slaughter
 ~ до отказа drive home
 ~ досками board up
забиваться clog (up), choke up, stop up
забивка stopping up; overprinting
заблуждение delusion; error; falacy
 ~, несущественное *law* error in accidentia
 ~ относительно основания *law* error in causa
 ~, существенное *law* error in essentia/substantia
забой I *min.* coal/pit face; heading; breast
 ~, глухой dead face
забой II 1. erasure; 2. slaughtering; driving in
забойник rammer, beetle, tamping stick/bar
забойщик coal-face worker, coal hewer/cutter/miner
заболачивание *geol.* bog formation; swamping
заболевание illness, disease, sickness, morbus
 ~, вирусное virus disease

~ волос trichopathy
~ вызванное токсином toxinosis
~, глистное helminthiasis, helminthic disease
~, грибковое mycosis, fungus disease/infection
~, декомпрессионное decompression/caisson disease/sickness
~, доброкачественное benign disease
~, заразное infective/infectious disease
~, злокачественное malignant disease
~, идиопатическое idiopathy, self-existing disease
~, лихорадочное febrile disease/illness
~, лучевое radiation sickness
~, местное local/localized disease
~, незаразное non-infective disease
~, нераспознанное undiagnosed illness, missed case
~ ногтей onychopathy
~, общее general/systemic disease
~, острое acute disease/illness
~, повторное recurrent disease
~, профессиональное occupational disease
~, психическое mental disease
~, рецидивирующее recurrent disease
~, семейное familial disease
~, системное systemic disease
~, соматическое bodily disease
~, хромосомное chromosome disease
заболонник *zool.* engraver *(Scolytus)*
заболонь *bot.* sapwood, splintwood, alburnum
заболоченный swamped
забор 1. fence, enclosure; hedge; 2. *tech.* intake; scooping, scoop
заборник intake, scoop; inlet
забортный *naut.* outboard *(attr.);* overboard; *tech.* extra-vehicular
заботливость *law* diligence
забраковывать condemn, reject
забрало *hist.* visor; *hydr.* boom
забрасыватель spreader
~ топлива, пневматический air-jet spreader
забрасывать cast, throw; toss
заброс throwing, casting; *aero.* overshoot
~ давления short/brief overpressure
~ импульса overshoot
~ оборотов overspeeding
~ регулируемого параметра hardover, runaway, overshoot, overcontrol
~ стрелки прибора off-scale sweep/swing
~ температуры short/brief overtemperature
~ элеронов aileron hardover
забрюшинный *anat.* retroperitoneal
забугорье *anat.* metathalamus
забутка *build.* backfill(ing)
завал 1. abatis, blockage; obstruction; 2. *min.* fall of roof; 3. *naut.* falling/tumble home; 4. *tech.* reverse camber; 5. steep slope
~ борта tumblehome, tumbling-in
~ гироскопа gyroscope tilting
~, поперечный transverse/lateral tilt(ing)
~, продольный fore-and-aft tilt(ing), longitudinal tilt(ing)
~ частотной характеристики roll-off of a frequency response curve
завалка *met.* charging; filling/heaping up
~, твердая cold charge/charging
заваривать *met.* weld up; *plast.* seal
заварка *met.* weld(ing); *plast.* seal(ing)
завгуж *abbr.* transportation manager
заведение institution, establishment
~, питейное public house *(Eng.)*
~, учебное educational institution
заведующий manager, administrator, director, head, in charge
~ отделом найма employment manager
~ отделом рекламы advertizing manager
~ отделом сбыта sales manager/director
~ отделом экспорта export manager
~ складом storekeeper, stores keeper
~ снабжением purchasing director; supply manager
~ учебной частью director of studies
заверка certification, authentication
завернутый convolute; reflexed
~ внутрь involute
~ назад retroflexed, revolute
завершение completion; end(ing); consummation
~ брака *law* consummation of marriage
завершенность finish; completeness
завеса curtain; screen
~, артиллерийская artillery barrage; curtain of fire
~, водяная water curtain
~, дымовая smoke screen; *(vertical)* smoke curtain/apron
~, горизонтальная дымовая smoke blanket
~, струйная jet curtain
завет behest; ordinance; covenant
~, Вехтий *rel.* Old Testament
~, Новый *rel.* New Testament
завещание *law* will, testament
~, собственноручно составленное olographic testament
~, устное nuncupative will
~ия, отсутствие intestacy
~ия, умерший без intestate
завещатель *law* testator, legator, testate, bequeather
~ недвижимости devisor
завещательница testatrix
завещательный testamentary
завещать bequeath; devise
завивка waving; curling; (hair) wave
~, холодная cold/finger wave
~, шестимесячная permanent wave
завинчивать screw up/down/in
~ до отказа screw tight home
зависание 1. *aero.* hovering; 2. congestion, sticking, bridging, hanging; *met.* scaffolding
~ клапана valve sticking

~ программы hang-up
~ элеронов aileron droop
зависимость dependence, dependency; relation, relationship; association
~, вассальная *hist.* vassalage
~, взаимная interdependence
~, временная time dependence
~, высотная height dependence
~, квадратическая/квадратичная quadratic dependence/relation
~, крепостная *hist.* serfdom, villeinage
~, линейная linear dependence
~ между прочностью и весом strength-weight ratio
~, монотонная uniform dependence
~ напряжений от деформаций stress-strain relation
~, нелинейная non-linear relation/dependence
~, непрерывная continuous dependence
~, обратная inverse relation/relationship
~, однозначная unique dependence
~ от времени/скорости time/velocity dependence
~ плотности от давления pressure-density relation
~, полукрепостная semi-serfdom
~, причинная *law* causal dependence
~, прямая direct relation
~, случайная random dependence
~, степенная *math.* exponential function
завиток 1. loop, ringlet; scroll; 2. *anat.* helix, whorl; 3. *math.* volute; 4. *(in wood)* knot
~, костный scroll of bone
~ раковины whorl, volution, coil
~ ушной раковины helix
завихрение swirling, whirling, eddy(ing); vortex, swirl; *(of gas)* looping
~ потока flow swirling
завихренность vorticity
завихритель swirl vane, swirler, vortex generator; turbulator
~, лопаточный vane swirler
~ форсунки nozzle swirler
завком *abbr.* factory committee
завод 1. plant, factory, works, mill(s); 2. winding mechanism, starter; 3. period of operation/running; 4. farm; breeding, raising
~, авиаремонтный aircraft repair plant
~, авиационный aircraft plant/factory
~, автомобильный automobile plant; *Engl.* motor-works
~, атомный atomic plant
~, аффинажный (metal) refinery
~, бетонный concrete-mixing plant
~, бумажно-целлюлозный cellulose and paper mill
~, военный munition factory
~, газовый gas works
~, двигателестроительный engine manufacturing plant
~, действующий *econ.* going plant
~, кабельный cable works
~, кирпичный brick yard/works
~, кожевенный tannery; skinnery
~, коксовый coking plant
~, коксохимический by-product coke plant
~, комбикормовый animal formula-feed plant, combined-fodder factory
~, конный stud farm
~, консервный cannery, canning factory
~, коптильный smoke-house
~, лесопильный saw-mill
~, ликеро-водочный distillery
~, литейный foundry
~, локомотивостроительный locomotive works
~, льнообрабатывающий flax-mill
~, маслобойный oilery, oil mill
~, маслодельный butter factory
~, машиностроительный engineering/machinebuilding plant/works
~, металлообрабатывающий metalworking plant
~, металлургический metallurgical works
~, метизный hardware plant
~, молочный dairy factory/plant
~, моторостроительный engine works
~, мыловаренный soap factory/plant
~, нефтеперегонный oil refinery
~, нефтеперерабатывающий petrochemical works, oil-processing plant
~, оловоплавильный tin smeltery
~, опытный pilot/experimental/development plant
~, пивоваренный brewery
~ по производству ядерного топлива nuclear fuel plant
~, плавучий рыбоконсервный fish-canning factory ship
~, рафинадный sugar refinery
~, рафинировочный *met.* refinery
~, рыбоводный fish farm/hatchery
~, рыбоконсервный fish cannery
~, самолетостроительный aeroplane/aircraft plant/works/factory
~, сахарорафинадный sugar refinery
~, сборочный assembly plant
~, смолоперегонный coal-tar processing plant
~, солеваренный saltworks
~, солеочистительный salt refinery
~, солодовенный malt-house
~, сталелитейный steel foundry
~, сталеплавильный steel works/plant
~, станкостроительный machine-tool plant
~, стекольный glassworks
~, судоремонтный ship repair works/plant/facility
~, судостроительный shipbuilding yard(s), shipyard, dockyard
~, утилизационный *(for fat)* rendering plant
~, химический chemical(s) plant/works

~, хлопкоочистительный ginnery
~, цинкоплавильный zinc smeltery
~, чугунолитейный/чугуноплавильный ironworks
завод-автомат automatic plant
завод-изготовитель manufacturing plant, manufacturer, factory of origin
заводнять inject water into, flood
завод-поставщик supplier/contractor plant, supplying factory
завод-смежник sub-contractor (plant)
заводь backwater, dead channel; creek
завоз delivery
~ якоря *naut.* laying out of an anchor
заволожский from/on the left bank of the Volga
~ие старцы *hist.* trans-Volga monks
заворот *med.* malrotation
~ век entropion, entropium, blepharelosis
~ желудка gastric torsion
~ кишок volvulus, twisting of the intestines
~, рубцовый cicatrical entropion
~, спазматический spastic entropion
завуч *abbr.* director of studies
завхоз *abbr.* assistant manager
завышать set/raise too high
~ стоимость overrate; overestimate
завышение стоимости cost overrun
завязь *bot.* ovary
~, зрелая pericarp
~, многогнездная compound ovary
~ плодов fruit inception
~ семян seed formation
загиб bend, fold; flange, lap
загибать bend, fold; flange, lap
заглазничный *anat.* postorbital
заглинизировать mud off/up
заглоточный retropharyngeal, postpharyngeal
заглушка stop/blank plug, stopper; dead end; blank/blind flange; end-piece, cap nut; screw cap; silencer, deadener
~, вышибная blow-out plug
~, нарезная/резьбовая screw cap/plug
~, сопловая nozzle dust cover
~ трубопровода pipe seal
загнутый bent, folded, turned down/up; recurvate
~ вниз deflexed
~ внутрь introflexed
~ вперед/вверх antrorse, incurved, inflexed, incurvate(d)
~ кзади recurved
~ крючком curved/bent backwards
~ назад retrocurved
заговенье *eccl.* last day before fast
заголовник head-rest
заголовок heading, head(er); preamble
загон *farm.* enclosure, pen; shelter; yard
загорание ignition; fire
~ электрической дуги arc initiation
~ электропроводки electrical-wiring fire
заготовитель government agent in charge of purchases/procurements
заготовительная *food* preparation room
заготовка 1. billet; blank; bloom; feed; slab; 2. (state) procurement/purchases; stockpiling; 3. stock, half-finished product
~ для проволоки wire rod
~, досковая board blank
~, кованая forged blank, forging
~, литая casting
~, накатанная rolled-up stock
~, обувная shoe upper
~, плоская sheet slab/bar
~, сортовая billet
~, трубная round billet
~, штампованная stamped blank, stamping
заготовщик person in charge of *(or* making) half-finished products
заготпункт storage place/depot
заградитель 1. *naut.* mine-layer/-planter; minecraft; net-layer/-tender; 2. wavetrap, rejector; stopper; *comp.* trap
~, линейний line trap
заграждение barrage, obstruction, blocking
~, аэростатное balloon barrage
~, боновое harbour boom, boom defence
~, минное mine field
~, полезащитное shelter belt, field windbreak
~, проволочное wire entanglement
~, снегозащитное snow fence
загрудинный *anat.* retrosternal, substernal
загружать load, charge; fill, feed, prime
загрузка load(ing), charge, charging, fill(ing), feed(ing), priming; batch; *econ.* utilization
~ без остановки реактора hot reactor loading
~ вразброс scatter loading
~ памяти memory loading
~, первичная *comp.* bootstrap
~, поблочная block loading
~, повторная начальная *comp.* reboot
~ программы program(me) loading
~, рабочая operating charge
~ реактора reactor loading/feed/charge
~ с последующим исполнением *comp.* load-and-go
~ системы *econ.* system utilization
~ станка machine load(ing)
~ страницы *comp.* page-in
~, эксплуатационная operating load
загрузчик *comp.* loader, loading routine/program(me), bootstrap; *mech.* loader
~, перемещающий relocating loader
~, связывающий linking loader
~, системный system loader
загрунтировать prime
загрязнение contamination, pollution, soiling, fouling; impurity, contaminant, pollutant
~ атмосферы atmospheric contamination
~ гамма-активным веществом gamma-contamination
~ клапана valve fouling

~, **минеральное** mineral impurities
~ **окружающей среды** environmental pollution/contamination
~ **продуктами деления** fission-product contamination
~, **радиоактивное** radioactive/radiological contamination, radiocontamination; radioactive contaminants
~, **сильное** severe contamination
~, **слабое** mild contamination
~ **фильтра** filter clogging
загрязненность impurity
загрязнять pollute, contaminate, soil, make dirty, foul
загс *abbr.* registry office
загустевание thickening, congealing, concentration, solidification, stiffening; gelation
загустевать thicken, condense; solidify, congeal
загуститель thickener, thickening agent
загущенность crowdedness, overcrowding
задавать *math.* give, assign, set; assume
задание task, assignment, job; mission; target; *math.* specification
~, **боевое** *mil.* combat assignment
~ **выполняемое в отсутствие пользователя** absentee-user job
~, **высокоприоритетное** foreground/priority job
~, **количественное** quantitative target
~ **начальных условий** *comp.* initialization
~, **пакетное** *comp.* batch(ed) job
~, **плановое** *econ.* planned task/target
~ **по реализации продукции** sales target
~, **производственное** production target
~ **размеров** dimensioning
~, **разовое** *comp.* one-shot job
~, **фоновое** background job
~ **формата** formatting
~ **функции** definition of a function
задаток *econ.* earnest/caution money, down payment, deposit
~, **давать** pay down
задатчик setting device, setter, selector, controller, reference-input element
~ **высоты** altitude selector
~ **давления** pressure selector
~ **курса** heading selector
~ **мощности** power controller
~ **последовательности операций** sequence unit
~ **температуры** temperature setter
задача problem; task, job, objective; target; mission; aim
~ **анализа операции** *econ.* activity analysis problem
~, **боевая** *mil.* combat/operational mission
~ **в реальном времени** real-time problem
~, **вариационная** variational problem
~, **вычислительная** computational problem
~, **граничная/краевая** boundary (value) problem
~, **двойственная** dual problem
~ **двух/трех тел** two-/three-body problem
~ **Дирихле** *math.* Dirichlet's problem
~ **для многоканальной системы** *econ.* multifacility problem
~ **для одноканальной системы** *econ.* simple (one-server) problem
~, **инженерная** engineering problem
~, **исходная** original/primal problem
~ **Коши** *math.* Cauchy problem
~, **краевая** boundary value problem
~, **логическая** logic problem
~ **массового обслуживания** *comp.* queueing problem, waiting-line problem
~ **нахождения седловой точки** *game th.* saddle-point problem
~ **Неймана** Neuman's (boundary value) problem
~, **неоднородная краевая** inhomogeneous/non-homogeneous boundary value problem
~ **о времени пребывания** sojourn time problem
~ **о разорении** *game th.* ruin problem
~ **о собственных значениях** eigenvalue problem
~ **о четвертой точке** *geod.* three-point problem/intersection; trilinear surveying
~, **обратная** inverse problem
~, **однородная краевая** homogeneous boundary-value problem
~, **оперативная** *mil.* operational mission
~ **оптимизации** optimization/optimum problem
~, **основная** primal problem
~ **оценивания** estimation problem
~ **оценки характеристик** benchmark problem
~ **планирования** planning/scheduling problem
~, **плоская** two-dimensional problem, plane problem
~ **поиска** search problem
~ **поиска информации** information retrieval problem
~, **пользовательская** user task
~, **прикладная** application/applied problem
~ **принятия решений** decision problem
~ **присваивания** assignment problem
~ **прогнозирования** prediction/forecasting problem
~, **прямая** primal problem
~ **размещения** location problem
~ **распределения** allocation problem
~ **с начальными условиями** initial-value problem
~ **с решением, учебная** worked problem/example
~, **системная** system task
~, **учебная** *mil.* training mission
~, **частная** particular problem
задвижка 1. slide/gate valve; 2. door bolt, latch, catch, locking bar, fastener
~, **газовая** gas shutter

~, **клиновая** wedge gate valve
~, **кольцевая** cylinder gate
~, **коническая** cone/plug valve
~, **оконная** window bolt/fastener/bar
~, **предохранительная** safety catch
~, **пружинная** spring bolt
~, **фланцевая** flanged gate valve
~, **фурменная** *met.* tuyere valve
~, **шандорная** sluice gate
~, **шиберная** gate

задел process stock/reserve; buffer/protective stock

заделка 1. closing, sealing, stuffing; 2. termination; fixation; restraint
~ **балки** *constr.* rigid fixing of beam
~, **жесткая** fully-fixed restraint
~ **кабеля, концевая** sealing end of cable, cable termination
~ **канала** duct sealing
~, **концевая** edge attachment/restraint
~ **концов** *constr.* end restraint
~ **кромок** edge restraint
~ **проводов** termination of wires
~ **семян** *agr.* covering of seeds
~ **удобрения** placement of fertilizers

заделывать close, seal, stuff, stop up, embed, bury; *agr.* plough/plow under
~ **в дуркупан** embed in durcupan
~ **в капсулу** encapsulate
~ **конец кабеля** seal off a cable
~ **под стерню** *agr.* stubble/plough in
~ **пробоину** seal/stop a hole
~ **семена** drill/embed seeds
~ **течь** stop a leak
~ **швы** *constr.* seal/stop/close joints

задержание retention, holding back; delay, retardation; *law* detention, arrest; *mus.* suspension
~ **мочи** *med.* retention/suppression of urine

задерживание реакции inhibition of reaction

задерживать retain, hold back; delay, retard; restrain; trap; withhold; suppress; inhibit; *law* detain, arrest; intercept; *(of property)* impound; arrest
~ **дыхание** hold breathing
~ **зарплату** withhold wages
~ **самое лицо** *law* seize the body of a person
~ **шаги** slow one's steps

задержка delay, lag; retardation, holdup; inhibition; retention, stoppage; entrapment; *law* detaining, detention, arrest; *phys.* latency
~ **азота** *med.* nitrogen retention
~ **в соединительных цепях** wiring delay
~ **в цепи передачи** transmission delay
~ **в цепи переноса** *comp.* ripple delay
~, **вентильная** gate delay
~ **во времени** time lag
~ **воспламенения** ignition delay
~ **впрыска** injection delay
~, **временная** time delay
~ **газов** *med.* gas retention, flatulence
~ **дыхания** breath holding
~, **единичная** unit delay
~ **желчи** *med.* ischolia
~ **импульса** pulse delay
~, **инструментальная** instrument lag
~, **компенсирующая** balancing delay
~ **менструации** ischomenia
~ **на один разряд** digit delay
~ **на передачу данных** communication delay
~ **на передачу маркера** buck delay
~ **мочи** retention/suppression of urine, ischuria
~, **нулевая** zero delay
~, **паразитная** stray delay
~ **переноса** *comp.* carry delay
~ **появления ошибки** error latency
~ **при передачи пакета** packet delay
~ **развития** arrest of development
~ **раскрытия парашута** parachute delay
~ **реакции системы** system response delay
~ **роста** growth inhibition
~, **транзитная** delay in transit
~, **фазовая** phase delay
~, **фиксированная временная** timed delay

задержник retainer; tripping device

задернение sod formation, turfing, grassing

задир scratch, score; scuffing, scoring
~ **зеркала цилиндра** scuffing of the bore

задирание scoring, scratching; *(of bearing)* seizing

задирать score; tear up; scratch

заднебоковой posterolateral

заднебрюшной abdominoposterior

задневерхний posterosuperior

задневисочный *anat.* posttemporal

задневнутренний posterointernal

заднегрудный postthoracic, metathoracic

заднегрудь postthorax, metathorax

заднежаберный *zool.* opisthobranchiate

заднелатеральный posterolateral, posteroexternal

заднелобный postfrontal

заднемедиальный posteromedial, posterointernal

задненавесной *tech.* rear-mounted

задненаружный posteroexternal

задненебный *phon.* postpalatal, guttural, velar

задненижний posteroinferior

заднепередний posteroanterior

заднепроходный *anat.* anal

заднеспинка *zool.* metatergum

заднесрединный posteromedian

заднетеменный postparietal

заднеязычность *phon.* gutturalization, velarization

заднеязычный *phon.* guttural, velar; back

задолженность indebtedness, liabilities; debts; arrears
~, **дебиторская** *econ.* receivables
~, **краткосрочная** floating debt
~ **по налоговым платежам** tax liabilities
~ **по работе** *econ.* arrears of work
~, **текущая** floating debt

задраивать *naut.* batten/dog down; clip up
задрайка *naut.* bolt bar, dog catch; snacket
~ **лаза** manhole dog
~ **люка** hatch dog, hatch-batten
задросселировать throttle down
задубливать *leath.* pretan, set the grain
задувка доменной печи blowing-in of a blast furnace
задымление *mil.* smoke screening
задымленность smoke content
задымлять screen with smoke
заедание jamming, seizing; sticking
заедать jam, seize; stick; grip
заезд *sport* lap, round, heat
заем *econ.* loan; *comp.* borrow
~**, внутренний** domestic loan
~ **до востребования** demand loan
~**, ипотечный** real estate loan; mortgage
~**, круговой** *comp.* end-around borrow
~ **на льготных условиях** low-interest loan
~**, ненадежный/сомнительный** problem loan
~**, облигационный** funded/bonded debt
~**, срочный** time loan
заемщик по ипотеке mortgage borrower
заершивать jag, rag
зажигание ignition; firing
~ **в мертвой точке** dead-centre ignition
~ **в форсажной камере** afterburner ignition
~**, внешнее/внутреннее** external/internal pilot-frame ignition
~ **двойное/двухискровое** two-spark/dual-spark ignition
~ **дуги** arc ignition
~**, искровое** spark-plug ignition
~**, калильное** hot-bulb ignition
~ **от запальной свечи** glow-plug ignition
~**, позднее** late/retarded ignition
~**, последовательное** train-type ignition
~**, предкамерное/форкамерное** precombustion chamber ignition
~**, принудительное** forced ignition
~ **пусковым топливом** starting fuel/fluid ignition
~ **пусковым факелом** pilot-flame ignition
~**, раннее** early/advanced ignition
~ **с емкостным разрядом** capacitor-discharge ignition
~ **с опережением** advanced ignition
~ **самовоспламеняющимися компонентами** self-energized (hypergolic) ignition
~ **тиратрона** firing of a thyratron
~**, факельное** pilot-flame ignition
~**, химическое** self-energized (hypergolic) ignition
~ **электрозапалом** glow-plug (electric-squib) ignition
зажигатель ignitor; starter; firing/starting electrode
зажигать ignite; light, kindle; set fire to
зажим clamp, fastener, grip, clip, cramp, clasp; *el.* terminal; *med.* forceps
~**, анкерный** dead-end clamp
~**, вертлюжный** swivel clamp
~**, винтовой** screw clamp/terminal
~ **для бюреток, пружинный** spring pinchcock
~ **для шланга** hose clamp
~**, желудочный** *med.* stomach forceps
~ **заземления** earth/ground terminal
~**, кабельный** cable clip
~**, кровоостановливающий** *med.* h(a)emostatic forceps; artery clamp
~ **"крокодил"** *el.* aligator clip
~**, лабораторный** test-tube clamp
~**, неподвижный** dead lock
~**, пневматический** air-operated clamp
~**, поворотный** swivel clamp
~**, подвесный/промежуточный/поддерживающий** suspension clamp
~**, сосудистый** artery forceps
~**, тормозной** brake clip
зажимать grip, cramp, squeeze; chuck, clench, cramp
заземление earthing; earth; *Am.* grounding; ground
~**, защитное** safety earthing/grounding
~ **переносное/ремонтное** temporary earth/ground
~ **самолета** aircraft grounding
заземлитель earth/ground electrode/rod/stake; earthing/grounding connection; earth, ground
заземлять earth, ground, connect to earth
зазор 1. clearance; gap; 2. *mech.* backlash, play
~**, боковой** side clearance/play; backlash
~ **в подшипнике** bearing clearance/space
~ **в свече зажигания** spark-plug gap
~**, допустимый** tolerable/safety clearance
~**, кольцевой** circular/annular clearance/gap
~**, компенсационный** expansion clearance
~ **развертки** reamer clearance
~**, регулируемый** adjustable gap
~**, сборочный** fit-up gap
~**, температурный** expansion clearance
~**, торцовый** end clearance
зазубренный serrate(d), toothed, notched, dentate(d)
зазубривать serrate, notch, jag, dent
зазубрина indentation; notch, jag; serration
заиливание silting, mud accumulation; sand-filling
заимодавец creditor, lender, debtee
заинтересованность interest; involvement
~**, материальная** material incentive
заирец, заирский Zairean
заказ order
~**, изготовленный на** custom-built/made
~**, индивидуальный** custom design
~ **мест** seat/space reservation
~ **на поставку** delivery order
~**, общий** blanket order
~**, повторный** repeat order
~**, последующий** follow-on order

~, **разовый** nonrepeat order
~ **с доставкой по почте** mail order
~, **сделанный на** made to order
~, **срочный** rush/pressing order
~**ы в данный момент** *econ.* work in hand
~**ы, невыполненные** backlog of business/orders
~**ы фирмы, внутренние** *econ.* company work orders
заказчик customer; client
~, **гражданский** civil/commercial customer
заказывать order, place an order for; book, reserve
~ **билет** book a seat/place
~ **телефонный разговор** place a call
закаливаемость hardenability, hardening capacity
закаливание, закалка hardening; quenching
~ **в воде/воду** water hardening/quenching
~ **в воздухе** air hardening
~ **в жидкой среде** quench-hardening
~ **в масле** oil hardening/quenching
~ **в штампе** die quenching
~, **высокочастотная/индукционная** induction hardening
~, **изотермическая** isothermal hardening, austempering
~ **магмы** magma chilling/quenching
~, **местная** point/local hardening
~ **на бейнит** bainite quenching
~ **на мартенсит** martempering
~ **с последующим отпуском** temper hardening
~ **с прокатного нагрева** quenching with rolling heat
~, **ступенчатая** graded hardening
закаливать harden; *(by abrupt cooling)* temper
закамуфлировать camouflage
закапотировать cowl *(an engine),* install a cowl on
закарстование *geol.* karsting
закарстованность cavernous porosity
закаспийский trans-Caspian
закат *met.* lap
закатка *mach.* beading; seaming
закатывать roll up; wrap
~ **металлические банки** seam cans
закваска starter; leaven; ferment, barm; leavening, fermenting
заквасочник starter vat
заквашивать sour, make sour; *(dough)* leaven; *(wine, milk)* start
закись monoxide, protoxide, lower oxide
~ **азота** nitrous oxide
~ **железа** iron monoxide/protoxide, lower ferrous/iron oxide
заклад имущества pawn
закладка laying; loading; *min.* fill(ing), gobbing
~ **выработанного пространства** gob flushing
~ **здания** stone-laying ceremony
~ **колоса/початка** initiation of ear
~ **породой** stone gobbing
~ **с уплотнением** stowing and packing
~, **самотечная** gravity stowing
закладная *law* mortgage (bond); *econ.* bill of sale
~, **именная** registered mortgage
~ **на движимое имущество** chattel mortgage
~ **под недвижимость** real estate mortgage
~, **совокупная** consolidated/aggregate/blanket mortgage
~, **таможенная** custom bond
закладывать put, place, lay; insert; embed; install; *law* mortgage, hypothecate; pawn
~ **выработанное пространство** fill with waste rock
~ **корабль** lay the keel of a ship
заклеивать glue/seal/stick up
заклепка rivet; riveting
~, **бондарная** cooper's rivet
~, **взрывная** explosive rivet
~ **впотай (с потайной головкой)** countersunk/flush rivet
~, **временная** tack/temporary rivet
~, **герметичная** sealing rivet
~, **дистанционная** spacer rivet
~, **закладная/односторонняя** blind rivet
~, **полая** tubular rivet
~, **полупустотелая/полутрубчатая** semitubular rivet
~, **потайная** countersunk rivet
~, **ременная** belt rivet
~, **стержневая** soft-collar rivet
~, **трубчатая** tubular shank rivet
~, **цепная** chain rivet
заклепывать rivet
заклинивание wedging; locking; seizing
~ **двигателя** engine seizing/seizure
заклинивать wedge, seize, jam; key
заключение 1. conclusion, end; winding/finishing up; 2. inference, corollary, conclusion; deduction; finding(s); 3. inclusion; 4. *law* confinement, imprisonment, detention; 5. conclusion, signature
~ **в капсулу** encapsulation
~ **в оболочку** jacketing; encasing
~ **в скобки** bracketing
~ **договора** conclusion of a treaty/contract
~, **пожизненное** *law* life imprisonment
~, **предварительное** *law* imprisonment pending trial
~ **публичного/внешнего ревизора** *econ.* audit certificate
~, **строгое тюремное** rigorous imprisonment
~ **эксперта** expert's report
заключенный prisoner; convict
~, **надежно охраняемый** securely guarded prisoner
заключительный closing, concluding, terminating, final
~ **аккорд** *mus.* finale
~**ное слово** concluding speech/address/remarks
заключка *print.* quoin
~, **роликовая/самовыравнивающаяся/**

шурушечная self-aligning quoin
закодировать code, encode
закоксование clogging with coke; coking; *(of engine)* gumming up; *(of catalyst)* carburization
закоксоваться coke, clog with coke; *(of engine)* be gummed, gum up
закольцовывание loopback
закон law; principle, rule; *law* statute, act
~ Авогадро Avogadro's hypothesis/law
~ Архимеда Archimedes' principle, law of Archimedes
~ ассоциативности associative law
~ аэродинамического подобия aerodynamic similitude law
~ Бернулли *math.* Bernoulli's theorem
~, биогенетический biogenetic/recapitulation law, law of ancestral repetition
~ Божий Holy Scripture(s); *educ.* divinity, scripture
~ Бойля-Мариотта Boyle's law, Boyle Mariotte's law
~ больших чисел law of large numbers
~ взаимозаместимости reciprocity law
~, военно-уголовный army act
~ всемирного тяготения law of universal gravitation
~ Гей-Люссака Gay-Lussac's law, law of volumes
~ Генри Henry's law
~, гражданский civil law
~ Грэма *chem.* Graham's law
~ Гука Hooke's law
~ Дальтона Dalton's law, law of partial pressures
~ двойного отрицания law of double negation
~, двучленный binomial law
~ действия и противодействия law of action and reaction
~ действующих масс mass action law
~ Джоуля-Ленца Joule's law
~ динамического подобия law of dynamic similitude
~ дистрибутивности distributive law
~, земельный agrarian/land law
~ идеального газа ideal gas law
~ излучения Планка Planck radiation/distribution formula
~ изменения по времени time law
~ изменения тяги thrust law
~ инерции квадратичных форм law of inertia for quadratic forms
~ исключенного третьего law of the excluded middle, *tertio non datur* principle
~, квадратичный square law
~, квантовый quantum law
~ Клапейрона ideal gas law
~ количества движения momentum conservation theorem
~ коммутативности commutative law
~ конвекции law of convection
~ косинуса cosine law, law of cosines
~ кратных отношений law of multiple proportions
~ Кулона Coulomb's law
~ Ленца Lenz law
~ малых чисел law of small numbers
~, материально-правовой *law* substantive law
~ Менделеева, периодический Mendeleev's law, periodic law
~ минимума допущений *econ.* Occam's/Ockham's razor
~ наименьшего действия principle of least action
~, неквантовый unquantized law
~ Ньютона Newton's law
~ о минимальной зарплате minimum-wage law
~ о минимальной продолжительности рабочего дня minimum-hour law
~ о мореплавании navigation act
~ о ростовщичестве usury law
~ о судоходстве shipping act
~ о торговом мореплавании merchant shipping act
~ обратимости *opt.* principle of reversibility
~ обратных квадратов inverse-square law
~, ограничительный restraining statute
~ Ома Ohm's law
~, органический *law* organic act; fundamental law
~, основной organic/fundamental law/act
~ Паркинсона Parkinson's law
~ площадей law of areas
~ подобия similarity/similitude/scaling law
~ постоянства весовых отношений law of constant/definite proportions
~ постоянства сумм тепла law of constant heat summation
~ постоянства углов law of constant angles
~ простых объемных отношений combining volumes principle, Charle's [Gay-Lussac's] law
~ равномерного распределения equipartition law
~ распределения distribution/partition law
~, распределительный distributive law
~ регулирования control law
~ рефлективности reflexive law
~ с обратной силой *law* retrospective law/statute
~ синусов/синусоидальный sine law
~ сложения addition law
~ сложных процентов law of compound interest
~ смещения displacement law
~ соответствия correspondence rule
~ сохранения количества движения law of conservation of momentum
~ сохранения массы/материи/энергии law of conservation of mass/matter/energy
~, сочетательный associative law

~ **спадания** decay law
~ **спроса и предложения** *econ.* law of supply and demand
~, **степенной** exponential/power law
~ **Стокса** Stokes' law
~, **сухой** *hist.* prohibition
~ **термодинамики** law of thermodynamics
~ **тождества** law of identity
~, **торговый** law merchant, commercial law
~ **транзитивности** transitive law
~ **трех вторых** three-halves power law
~ **тяготения** law of gravitation
~, **уголовный** *law* crown law, penal/criminal law
~ **управления** control law
~ **флага** *law, naut.* law of the flag
~ **химических эквивалентов** law of multiple proportions
~, **частный** private statute
законность legitimacy, legality; validity
законоведение jurisprudence, law
законодатель legislator; lawmaker, lawgiver
законодательство legislation; laws
~, **действующее** *law* Statute Books
~, **демографическое** population legislation
~, **морское** maritime/sea legislation/laws
~, **ограничительное** restrictive legislation
~, **охотничье** game-laws
законодательствовать legislate
закономерно normally, naturally, in conformity with relevant principles
закономерность 1. *law* conformity with the law(s); 2. *law* inherent principle/rule; regularity
~, **статистическая** statistical regularity
~, **тектоническая** tectonic relationship
закономерный 1. law -governed, in accordance with the law(s); 2. natural, regular
законопачивать caulk up, plug, pack
законоположение *law* regulation; enactment
законопроект *law* bill
~, **личный** private member's bill
~, **принять** pass a bill
~, **финансовый** appropriation bill
законосообразность 1. conformity with the law(s); legality, legitimacy; 2. regularity, normality
законосообразный 1. in conformity with the law; 2. expedient; necessary
законсервировать 1. put in prolonged storage, remove from service/operation; 2. apply rust preventive; 3. *food* preserve
законтривать lock; secure
законцовка tip
закорачивать *el.* short-circuit
закоротка *el.* short circuit
закраина flange, rim, ridge; bead
закрепитель fixing agent, fixative; conserver
закрепка fastener
закрепление fastening, fixation, securing, attachment, attaching, fixing; clamping
~ **анкерными болтами** anchoring
~ **грунтов** soil stabilization/grouting/consolidation
~ **грунтов, химическое** chemical grouting
~ **грунтов цементацией** soil stabilization by cement injection
~ **краски** *print.* ink fixing/fixation
~ **почвы** soil fastening/stabilization
~ **участка памяти** *comp.* memory attaching
~ **якорем** anchoring
закреплять fasten, fix, secure, attach; anchor; *fig.* consolidate
~ **болтом** bolt
~ **неподвижно** attach fixedly
закрепощение serfdom, serfhood; enslaving, slavery
закритический supercritical
закром cornbin; silo
закругление 1. curvature, curve, bend; chamfer; 2. rounding off, chamfering
закруглять round off, make round; chamfer
закрутка turning, twisting; torsion; twist; curling, twirling; *(gas, fluid)* whirl, swirl, vortex
закрученный convoluted, curled, coiled
~ **вниз** revolute
~ **внутрь** involute, intorted
~ **назад** revolute, retorted; twisted back
закручивание twisting, warping; curling; *(flow)* swirling
~ **крыла** wing-warping
закручивать twist, curl, twirl; *(flow)* swirl, whirl, spin
закрылок *aero., naut.* flap; tab
~, **выдвижной** extension flap
~, **разрезной** split flap
~, **струйный** jet flap
закрытие 1. closing, shutting, closure; shutdown, shutoff, turning down; 2. occlusion; 2. *mil.* cover
~ **клапана** valve closing/return
~ **кредита** withdrawal of credit
~, **люковое** hatch cover
закупка purchase; buying
~ **в период понижения курсов** *econ.* scale buying
~ **в целях поддержания цен** support buying
~ **по заказу** *econ.* order buying
~**ки, государственные** state/public purchases
~**ки, самопогашающиеся** *econ.* self-liquidating purchases
закупоривание plugging, stopping up; corking
закупоривать plug, stop; seal; encapsulate; cork, cap
закупорка clogging, blocking, stoppage; sealing, packing; *med.* occlusion, obstruction; embolism, thrombosis
~ **артерии** arterial occlusion/thrombosis
~ **бронха** bronchial obstruction
~ **вены** venous congestion/thrombosis
~ **кишок** intestinal obstruction
~ **мочеточника** ureteral obstruction
~ **труб** tubal obstruction
~ **эмболом** embolic occlusion
закупщик purchasing agent

зал hall, room
~, аппаратный instrument/apparatus room
~, выставочный/демонстрационный show-room, exhibition room/hall
~ гимнастический gymnasium, gym hall
~, доильный milking parlour
~ заседаний суда courtroom
~ лекционный lecture-hall/-room
~, машинный engine/machine room; *comp.* computer room
~ ожидания waiting room
~, проекционный screening room
~, тренировочный practice hall, training room
~, читальный reading hall/room
залегание *geol.* occurrence; bedding; stratification; seam; bed
~, запрокинутое tilted position
~, изоклинальное/моноклинальное isoclinal/nonoclinal folding
~, нарушенное broken bedding
~, облекающее overlap
~, опрокинутое overturned bedding
~, перемежающееся interbedding
~, регрессивное off-lap
залегать lie; occur
~ выше superpose
~ круто dip at high/steep angle
~ между пластами be interbedded, interlay
~ над пластом overlay
~ ниже underlay, underlie
залежалые товары *econ.* unsalable stock
залежь 1. *geol.* deposit, bed; ore-body; (oil) pool; 2. *agr.* fallow/idle land, layland
залет *aero.* landing, flying *(into); rare* photographic mission/sortie, photoflight
залив bay, gulf; cove
~, морской bay, gulf
~ озера arm of lake
заливать 1. fill in; prime; encapsulate; 2. flood; 3. extinguish
заливина *met.* flash, fin
заливка filling-in; flooding
~ асфальтом laying/spreading asphalt
~ двигателя engine priming
~ компаундом potting
~ насоса priming of a pump
~ подшипников babbitting, babbit(t) lining
~ раствора grouting
~ форм casting
~ форм под давлением pressure die casting
~ форм, свободная gravity casting
зализ *aero.* fillet, fairing
~, верхний dorsal fillet/fairing
~, нижний belly fillet/fairing
~ стабилизатора tailplane fillet
заличка *leath.* pretannage, preliminary tannage
залог I untilled land, virgin soil, layland
залог II *gram.* voice
~, действительный active voice
~, страдательный passive voice
залог III pledge, guarantee, security; deposit; pawning; *law* bail
~ в виде документов documentary pledge
~, изъятый/конфискованный forfeited pledge
~, ипотечный mortgage
~, невыкупленный unredeemed pledge
~, отдавать в pawn
залогодатель *law* pledger, pawnor, depositor
залогодержатель pledgee, pawnee, incumbrancer
залогополучатель *law* bailee
заложение location; laying, embedding
заложенное имущество pledged property
залп salvo, volley
~, бортовой broadside
залужение grassing, laying of grass/herb cover
замазка paste, putty, sealing compound, cement; mastic
~, влагоустойчивая waterproof cement
~, водопроводная plumbers' black
~, железная/железосуриковая iron putty
~, изоляционная insulating cement/putty
~, керамическая lute
~, масляная/меловая whiting putty
~, оконная glaziers' putty
~, суриковая/свинцово-суриковая red-lead putty/cement
~, тиоколовая thiocol mastic
~, уплотняющая sealing cement, sealant
~, фарфоровая porcelain cement
~, цоколевочная *el.* cap cement
замбиец, замбийский Zambian
замедление deceleration, slowing down; moderation; inhibition; retardation, delay; lag
~ взрывателя fuse delay
~ нейтронов neutron moderation/thermalization
замедлитель 1. moderator; inhibitor; decelerator; 2. *(device)* retarder; delay mechanism
~, водородосодержащий hydrogenous moderator
~ горения delayer
~ коррозии corrosion inhibitor
~ нейтронов neutron moderator
~ охладитель moderator-coolant
~, пороховый delay pellet
~ реактора moderator
~ реакции *chem.* inhibitor
~ схватывания setting retarder
замедлять decelerate, retard, slow down; *chem.* inhibit, moderate, retard, delay
замедляться decelerate, slow down
замедляющий decelerative, moderative, slowing down
замена substitution, replacement, change; interchange
заменимость interchangeability
заменимый interchangeable; replaceable
заменитель substitute; surrogate; spare; *tech.* eliminator
~ антенны *el.* aerial eliminator

~ **кожи** imitation leather, leather substitute
заменяемость interchangeability, substitutability, exchangeability
заменяемый interchangeable, replaceable
заменять replace; substitute; supplant; interchange; change
замер 1. measurement, measuring; determination; 2. reading
замерзание freezing, congelation; glaciation; solidification
~**ия, точка** freezing point
замерзать freeze, congeal; solidify
замерять measure; gauge; meter
замес 1. batch(ing), mix; 2. kneading
заместитель 1. deputy; proxy; substitute; 2. substituent, surrogate
~ **директора по административным вопросам** deputy director of administration
~ **министра** deputy minister
~, **первый** chief/first deputy
~ **председателя/президента** vice-chairman/president
замет casting
заметка mark, note; notice
замешивать mix; knead
замещать replace, substitute; exchange, change; take the place of, deputize for
замещение substitution; replacement
замирание dying down/away/out, fading
~, **интерференционное** interference fading
замирать die away/down; *rad.* fade
замкнутость isolation; closure
~, **экономическая** self-sufficiency
замкнутый closed; self-contained; secluded; select, exclusive
~, **ковариантно** coclosed
~, **накоротко** *el.* short-circuited
замок 1. lock, catch; latch; 2. joint; fastener; 3. *arch.* keystone
~ **автосцепки** automatic coupling lock
~, **американский (дверной)** Yale lock
~, **байонетный** bayonet lock
~, **быстродействующий** snap fastener
~, **быстрооткрывающийся** quick-release fastener
~, **висячий** padlock
~, **врезной** mortise lock
~, **гаечный** nut lock/retainer
~ **защиты** protection/safety lock
~, **игольный** *text.* needle cam
~ **капота** *av.* cowl fastener
~, **кольцевой** circlip
~ **лопатки** blade footing, blade locking piece
~ **на уровне блока данных** *comp.* block lock
~ **на уровне записи** *comp.* record block
~ **поршневого кольца (пружинный)** piston circlip
~, **предохранительный** safety block
~ **свода** *arch.* vault keystone/head
~ **секретности** privacy lock
~, **секретный** combination lock
~, **стрелочный** *rail.* point lock
~ **стыковочного устройства** *space* docking latch
~, **стяжной** turn buckle, swivel
~, **трубчатый** *text.* circular/tubular cam
~, **штифтовый** pivot [gudgeon-pin] attachment
замок-вертушка turnlock
замок-кнопка *leath.* snap-lock
замок-скоба clamp-fixing lock
замораживание freezing; congelation, glaciation; chilling
~ **заработной платы** *econ.* wage freeze
~ **орошением** spray freezing
~ **сухим льдом** dry-ice (carbon-dioxide) freezing
~ **углекислотой** carbon freezing
замораживать freeze; congeal; ice
замочка soaking, wetting; *text.* retting *(flax, etc.)*
замша suede, chamois/shammy leather
~, **искусственная** imitation chamois leather
~, **тканая** suedine, cloth suede
замшевание chamois/suede finish(ing)
замывание washing away/off/out; blurring
замыкание closure, closing; (inter)locking, fastening
~ **кольца** *chem.* ring closure, ciclizing, cyclization
~ **контактов** contact closing, contacting
~, **короткое** short circuit
~ **на землю** *el.* earth; *Am.* ground
~ **на корпус** body contact; earth/ground fault
~ **слоев** *geol.* closure of beds
замыкатель contactor, switch; lock, locking mechanism
~, **стрелочный** switch lock
замыкать lock; close; *math.* complete
~ **накоротко** short-circuit
замычка catch, lock; bolt
~ **на линейке** *rail.* locking dog
занесение 1. accumulation; piling; 2. recording, entering
~ **в файл** *comp.* filing
занзибарец, занзибарский Zanzibari
занижать underrate; underestimate
занос skidding; drift(ing); *med.* mole
~, **мясистый** carneous/fleshy mole
заносить skid; drift
зануление neutral earthing, earthing of the neutral wire
занятие 1. occupation, work, employment; profession; 2. *el.* holding, occupying, busying; 3. *pl.* lessons, studies
~, **дополнительное** secondary/subsidiary occupation
~, **доходное** *econ.* gainful employment
~, **обзорное** review lesson
~, **основное** main/major/principal/primary occupation
занято *el.* "busy", "engaged"; *(signal)* busy reject

занятость *el.* busy/engaged condition; *econ.* employment; work
~ в течение полного рабочего дня full-time employment
~, детская *sociol.* employment of children
~, неполная underemployment; part-time work, subemployment
~, первичная primary employment
~, полная full employment; full-time job/work
~, частичная partial employment; part-time job/work
заокеанский transoceanic; transatlantic
заострение tapering, taper; tip, point; sharpening
заочник *educ.* student taking correspondent course; extramural student
заочно 1. *educ.* by correspondence course; as extramural student; 2. *law* in absentia, in one's absence; by default
запад west; the West; the Occident
~, заходящий на westering
~, идущий на westbound
западение *med.* retraction
~ нижней челюсти mandibular retraction
~ языка tongue retraction
запаздывание delay, lag(ging), retardation, time lag
~ в действиях оператора operator response lag
~ воспламенения/зажигания ignition delay/lag/retard
~ восстановления restitution lag
~ клапана valve lag
~ по скорости velocity/rate lag
~ по фазе, ~ фазы phase lag
~ показаний прибора instrument lag
~ реагирования response lag
~ регулирования control lag, dead time of control (system)
~ управляющего воздействия control lag
запаздывать lag (behind), delay; be overdue
запаивать solder; seal up/off
запайка soldering; sealing
запал fuse, primer, igniter; *(action)* ignition, firing
~, гальванический electric fuse
~ замедленного действия delay primer
~, пороховой powder cartridge
~, преждевременный preignition
~, электрический electric primer
запальник горелки ignition torch
запань *timb.* holding/receiving boom/ground
~, ледозащитная ice boom
~, поперечная catch boom
~, продольная parallel holding ground
запараллеливать connect/link in parallel
запаривание, запарка steaming, stewing
запаривать steam
~ под давлением autoclave
запарник steaming plant/installation
запас 1. supply, stock, stockpile, supplies; reserve; store; 2. margin, allowance
~, буферный *econ.* buffer bank
~ водоизмещения margin of displacement
~, государственный government reserve
~ денег *econ.* monetary/money stock
~ древесины forest yield
~, золотой *econ.* gold stock/holding
~ и прирост (капитала) stock and flow
~, излишний *econ.* overstock
~, конечный *econ.* closing stock
~, межоперационный *econ.* buffer bank
~ мощности power reserve/margin
~ на рейс *naut.* sea stock
~ надежности safety margin
~, наличный stock in hand
~, неприкосновенный (аварийный) survival pack/kit
~, неснижаемый minimum level of supply
~, оборотный *econ.* turnover stock
~ остойчивости reserve of stability
~, переходный *econ.* transit stock
~, переходящий *econ.* carry-over
~ плавучести reserve (of) buoyancy, floatage
~ по ёмкости marginal capacity
~ по скорости speed margin
~ по ускорению acceleration margin
~ по усилению gain margin
~ по фазе phase margin
~ по характеристикам performance margin
~ прочности margin of strength; safety factor
~, располагаемый available stock
~, расходный retail stock
~ реактивности reactivity margin/excess
~, реальный *econ.* physical stock
~, резервный reserve stock
~, складской depot/store stock
~ слов vocabulary
~ статической устойчивости static margin
~, текущий everyday stores
~ топлива fuel reserve; fuel capacity/load
~ тяги thrust reserve/margin
~ усиления *rad.* gain margin
~ устойчивости margin of stability
~, эксплуатационный service stock
~ энергии energy reserve(s); store of energy
~ы в процессе транспортировки intransit stock
~ы, коммерческие товарные *econ.* business inventories
~ы, материально-производственные stores and supplies, stores
~ы, материальные *econ.* inventories; stocks
~ы полезных ископаемых, промышленные workable/commercial/payable reserves
~ы сырья raw-material stock
запасание stock-piling, storing, storage, accumulation
запасать store up, stockpile, accumulate
запатентовывать take out a patent for
запах odo(u)r, scent, smell
~, дурной/противный offensive smell

запахивать *agr.* plough in, plough/turn under (*Am.* plow)
запашник shallow plough/plow
запеленговывать take a bearing of/on
запесочивать oversand
запил notch, saw-cut slit; incision
запирание blocking, barring; obturation; shutting, closing, locking; barring; blanking, suppression, blackout; choking
~ приемника receiver suppression
~ пучка beam suppression/blanking
~ радиолокатора blanking the radar
запиратель obturator; choke
запирательный obturative; choking; locking
запирать bar, block, blank; lock (in), close (up), shut; cut off; choke; blockage
записка note; *pl.* proceedings, transactions
~, докладная report
~, люковая *naut.* hatch list
~, памятная *law* aide-memoire *(Fr.)*
~, путевая clearance card; *rail.* train ticket
~, служебная service record
~, увольнительная leave ticket; liberty card
записная книжка notebook
записывать write/put down, note; record; *acc.* enter, register
~ в протокол enter in the minutes
~ в расход enter as expenditure
записывающее устройство graphic instrument, recorder
запись record, notation, registration, recording, writing down; *comp.* write; *acc.* entry; *law* deed
~ акта о браке marriage statement
~ актов гражданского состояния civil registration; *stat.* demographic/vital record/registration
~ без возвращения к нулю non-return-to-zero recording/representation
~ без пояснения *econ.* blind entry
~, бесконтактная noncontact recording
~, бесскобочная Polish notation, parenthesis-free notation
~, блочная blocked record
~ в адресной книге directory recording
~ в двоичном коде binary recording
~ в дебет счета *econ.* debit entry
~ в вахтенном журнале log entry; logging
~ в кредит счета *econ.* credit entry
~ в операторной форме operator notation
~ в упакованном формате packed-format record
~ вектора vector plotting/notation
~ видеосигналов video recording
~, визуализируемая visual record
~, выборочная selective writing/record
~, главная master record
~ данных data record/recording; data acceptance, acceptance of data
~, голографическая holographic recording, hologram
~, двойная dual recording
~ двойной плотности double density recording
~, дебетовая *econ.* debit recording
~, дистанционная remote recording
~, дочерняя *comp.* child
~ единицы writing of a one
~, заключительная *econ.* closing entry
~, контактная contact recording
~, контрольная *comp.* reference recording; supervisory entry
~, корректирующая *econ.* adjusting entry
~, красная *acc.* red-ink entry
~, кредитовая credit record
~, логическая logical record
~ магнетографа magnetogram
~, многодорожечная multitrack record
~, многоканальная multichannel record(ing)
~ на борту *av., naut.* on-board recording
~ на видеоленту video-tape recording
~ на дорожке track/channel recording
~ на кинопленку filming, film recording
~ на ленту tape recording
~ на перфокарты card recording
~ на проволоку wire recording
~, неформатированная unformatted record
~ об отношении relationship record
~, обратная *acc.* reversing entry
~, однодорожечная single-track recording
~, операторная *math.* differential operator form
~ описания файла *comp.* file description entry
~ по времени recording against a time-scale, time history
~ подчиненного уровня member record
~ показаний гирокомпаса gyro log
~ показаний лага log record
~, полетная *av.* flight record
~, префиксная prefix notation
~, противоположная *acc.* reversing entry
~ прямого воспроизведения instantaneous recording
~, равновесная equilibrium writing
~, растеровая raster-scan writing
~ с возвратом к нулю return-to zero recording/representation, RZ recording
~ с высокой плотностью high-density recording
~ с наложением *cin.* dubbing
~ с насыщением saturation recording
~ с переменной плотностью variable density recording
~ с постоянной длиной fixed-length record
~ сальдо *acc.* balance forward
~, сложная *acc.* combined/compound entry
~, сокращенная abbreviated notation
~, ступенчатая incremental recording
~ текущего времени floating-time recording
~, телеметрическая telemetry/telemeter recording
~, точная accurate/instrumental recording
~ указателя набора данных file pointer entry,

data-set pointer entry
~ **фиксированной длины** fixed-length record
~, **цифровая** digital recording
~, **штриховая** stroke writing
~ **электронным лучом** electron-beam recording
~, **элементарная** unit record
~, **эталонная** master record
заплавлять seal/close by melting/fusion
заплата patch; *comp.* amendment, patch
заплетание, заплетка splicing
~ **троса** cable splicing
заплечики *mach.* shoulder, bead, fillet, swell; bosh; collet
заподлицо flush-mounted, flush *(with)*
заполнение filling (up); *(of screw)* solidity
~ **бланка** form filling
~, **местное** *av.* local solidity
~ **нулями** zero fill
~ **обмотки** closeness of winding
~, **повторное** refilling
~ **пустот** packing of voids
~ **файла** file population
заполнитель filling, filler, inert aggregate, filling/bolus material
~, **активный** reactive aggregate
~, **инертный** inert [non-reactive] aggregate
~, **искусственный** manufactured aggregate
~, **крупный** coarse aggregate
~, **мелкий** fine aggregate
~, **многофракционный** multiple-size aggregate
~, **нефракционированный** ungraded aggregate
~, **однофракционный** single-sized aggregate
~, **рядовой** run-of-the-pit aggregate
~, **фракционированный** graded aggregate
~, **шлаковый** cinder aggregate/filler
~ **швов** joint filler
~, **щебеночный** crushed-stone aggregate
запоминание memorizing, memorization; memory; *comp.* storage, storing
~ **алфавитной информации** alphabetic storage
~ **в дополнительном блоке памяти** additional storage
~ **в цифровом виде** digital storage
~ **данных** data storage
~ **изображений** image storage
~, **многоступенчатое** multi-level storage
~ **на магнитных дисках** disk storage
~ **переноса** carry-over storage
~ **сигнала разговорной речи** speech signal storage
запоминать memorize; *comp.* store
запоминающее устройство *comp.* store, storage, memory, memory bank *(see also* **память***)*
~ **большой емкости** high-capacity store, large-capacity memory, bulk memory/store
~, **кварцевое** quartz memory/store
~, **магнитодинамическое** magnetomotoric memory/store
~ **на магнитных стержнях** magnetic-rod store
~ **на микрофильмах** microfilm memory
~ **на сердечниках** core memory/store
~ **на стеклянных дисках** glass disk memory/store
~ **на электронно-лучевой трубке** CRT store/storage/memory
~, **основное** main/general/primary store/memory
~ **с подразрядной выборкой** bit access memory/store/storage
~ **(электростатическое) с запирающей сеткой** barrier-grid electrostatic store
запор latch, bolt, catch, stop, lock; fastener, bar; *med.* constipation
~, **привычный** habitual constipation
~, **условнорефлекторный** reflexogenic/reglexogenous constipation
запорожец *hist.* (Zaporozhian) Cossack
запорожский *hist.* Zaporozhian
заправка filling up, refuelling; priming, charging, loading; charge, batch; *met.* fettling
~ **горючим** fuelling, refuelling
~ **ленты** tape loading
~ **топливом в воздухе** in-flight refuelling, mid-air refuelling
запрессовывать press in/on; embed
запрет prohibition, ban; exclusion; *comp.* inhibit(ion)
~, **предварительный** *law* prior restraint
~ **прерываний** *comp.* interrupt inhibit
~, **судебный** injunction; restraining order
запретительный prohibitive, prohibitory
запретный prohibited, forbidden, restricted
запрещение ban, prohibition, interdiction; distraint; embargo; *comp.* inhibition
~, **судебное** injunction
запрос inquiry, query, request; demand, need; requirement, aspiration, interest
~, **ассоциативный** *comp.* associative query
~ **данных** data request
~, **дизъюнктивный** disjunctive query
~, **дистанционный** remote request
~, **информационный** request for information
~ **к базе данных** database query
~, **конъюнктивный** conjunctive query
~, **межбазовый** interdatabase query
~, **многоабонентский** multisite query
~ **на блокировку** lock request
~ **на ввод-вывод** I/O request, input-output request
~ **на обслуживание** service request
~ **на примерах** query by example
~ **на тестирование** test request
~, **непрограммируемый** ad hoc query
~, **пакетный** batch query
~ **пароля** password request
~, **письменный** *econ.* letter of inquiry
~ **повторения передачи, автоматический** automatic request for repetition, ARQ
~, **поисковый** retrieval request

~ **прерывания** interrupt request
~, **распределенный** distributed query
~ **с клавиатуры** keyboard request
~ **с подсказкой** prompting query
~ **с фиксированным порядком обработки** fixed-logic query
~, **случайный** spontaneous request
~ **соединения** call request
~, **устный** verbal request
~ **шины** bus request
~, **широковещательный** broadcast query
запросчик interrogator, inquiring station; challenger
~, **радиолокационный** interrogating radar
запруда dam, weir; barrage
запруживать impound, dam; block up
запуск start(ing), start-up; actuation; triggering; launch(ing)
~, **автоматический** unattended start-up
~, **автономный** self-contained starting, internal starting
~ **в воздухе** air starting/launching
~, **воздушный** pneumatic starting
~, **импульсный** pulse triggering
~, **ложный** false triggering/start
~ **модели в производство** model launch
~, **неавтономный** external starting
~, **пиропатронный/пиротехнический** cartridge starting
~ **по данным** data triggering
~ **по состоянию** status triggering
~, **повторный** restart
~ **программы** programme start
~, **пушечный** *rock.* full-flow start
~ **ракеты** rocket firing, missile launch(ing)
~, **ручной** hand starting
~, **тепловой** temperature triggering
запускать 1. start; trigger; actuate; 2. *rock.* launch
запчасти spare parts, spares
запыленность dust content/loading
запыленный dust-laden; dusty
запястно-лучевой *anat.* ulnocarpal
запястно-пястный *anat.* carpometacarpal
запястный *anat.* carpal
запястье wrist, *anat.* carpus
запятая comma; *math., comp.* point, decimal point
~ **в двоичном числе** binary point
~ **в позиционном представлении числа** arithmetic point
~, **двоичная** binary point
~, **десятичная** decimal point
~, **перевернутая** inverted comma
~, **плавающая** floating point
~, **регулируемая** adjustable point
~, **фиксированная** fixed/stated point
заработная плата wage(s), pay, labour remuneration; salary
~, **нищенская** starvation wages
~, **номинальная** *econ.* money/cash wages
~, **повременная** *econ.* time wages
~, **понедельная** *econ.* weekly wages
~, **поштучная** *econ.* piece wages
~ **рабочих и служащих** wages and salaries
~, **реальная** *econ.* real/actual wages
заражаемость susceptibility to infection
заражать infect; contaminate; *(with parasites)* infest
заражение infection; contamination; infestation
~, **внутрибольничное** intrahospital infection
~ **крови** blood poisoning
~ **местности** area/ground contamination
~, **повторное** repeated infection/contamination, reinfection
~, **профессиональное** occupational infection/contamination
~, **радиоактивное** radioactive contamination/poisoning
~, **химическое** chemical/gas contamination
~ **через кожу** transdermal infection
~ **через слизистые оболочки** transmucosal infection
зараженность contamination, infection
зарисовка sketch(ing)
зарисовывать sketch, draw a sketch of
зарница summer/sheet lightning
зародыш embryo; germ; *bot.* bud; *cryst.* nucleus; *biol.* f(o)etus
~ **кристаллизации** nucleation/nucleating centre
зародышевый germinal, embryonic, embryonal
зародышеотделитель *food* germ separator
зарождение generation, genesis, formation; origin, onset; conception, conceiving
~ **кристалла** nucleation
~, **самопроизвольное** spontaneous generation, abiogenesis
заросль brushwood, thicket, undergrowth; shrubs
~, **папоротниковая** fernery
~, **тростниковая** reed thicket
зарплата = заработная плата
зарубина, зарубка notch, incision, cut
зарубцованный *med.* cicatrized
зарубцовываться *med.* cicatrize, heal over
заруливать *av.* taxi-in
зарыблять stock with fish
зарываться bury oneself, burrow; *naut.* plunge, pitch (forward)
заря *mil.* reveille; retreat
~, **бить/играть** sound the reveille
заряд charge; load
~ **взрывчатого вещества** explosive payload
~, **воспламенительный** ignition/primer charge
~, **вышибной** burster/bursting charge
~, **инициирующий** initiating/detonating charge
~, **ионный** ionic charge
~, **метательный** propellant charge
~, **наведенный** *el.* induced charge
~, **осветительный** flare package/charge; *phot.* photoflash cartridge
~, **подрывный** demolition/blasting charge

~, разноименный *el.* unlike charge
~, разрывной bursting/explosive charge, burster
~, результирующий *el.* net charge
~ самоликвидатора destruct charge
~, скважинный deep-hole charge
~, сосредоточенный concentrated charge
~, составной segmented charge
~, точечный *el.* point charge
~, удельный specific charge
~, уравнительный *el.* equalizing charge
~ электрона electron/electronic charge
~ ядра nuclear/nucleus charge
зарядка 1. charging, loading; 2. setting-up exercises, morning exercises/drill
заряжение charging, loading
~ пусковой установки launcher loading
~ шпура blast-hole charging
заряжать charge, load
~ повторно recharge, reload
засаливать salt; pickle; corn
засасывать suck in, ingest
засахаренный candied, sugared, glacé
засахаривание saccharification
засахаривать candy, sugar; saccharify
засветка *rad.* gating, strobing; *phot.* illumination, exposure; flare spot
~ от земли ground clutter/return
~, фоновая background noise
~ экрана flare spot
засвидетельствовать authenticate; attest
~ нотариально notarize
заседание conference; session; meeting
~ за круглым столом round-table conference
~, закрытое closed session
~ совета *econ.* board meeting; council session
заседатель assessor
~, присяжный juriman, juror
засека abatis, felled-tree barrier/entanglement
засекать 1. notch, nick; 2. locate/determine by intersection; plot
~ время note down time
засекреченный classified, confidential, secret
засекречивать classify as secret, place on the secret list; restrict
засекречиватель речи speech scrambler
заселенность population
засечка 1. notch, cut; 2. intersection, cross bearing; 3. *print.* hair stroke, serif
~, боковая lateral intersection, side shot
~ времени clocking, timing
~, звуковая sound ranging
~, компасная compass cross bearing
~, обратная *geod.* resection
~, прямая intersection; direct sighting
заскакивать engage, catch, snap
заскок catching, snapping, blocking
заслонка shutter, valve, choke, damper, flap; *anat.* valve, valvule
~ венечного синуса coronary valve, valve of coronary sinus, Thebesian valve
~, воздушная air throttle/choke
~, дроссельная throttle valve/butterfly; baffle plate
~, задвижная sliding damper
~, карбюратора carburettor air door
~, лимфатическая lymphatic valvule
~, откидная hinged flap
~, отражательная deflector, baffle
~, поворотная butterfly valve
~, пружинная spring-loaded flap
~, подвздошно-слепокишечная ileoc(a)ecal/ileocolic valve
~, свободно падающая gravity-operated shutter
~ фурменного глазка tuyere cap
заслуженный distinguished, celebrated; hono(u)red
~ артист Hono(u)red Artist
~ профессор professor emeritus
засмаливать tar, pitch, gum; caulk
засов locking bolt/bar; catch; bolt
засоление *agr.* salinization, salting, salt accumulation
засоленность *agr.* salinity, salt content
засолка salting; pickling
засорение clogging, plugging, choking, blockage, stoppage; obstruction
засорять clog, block, obstruct, choke, stop up, plug
засос sucking in, inflow, intake
застава 1. *hist.* town gate; *mil.* outpost; picket
~, дорожная *hist., econ.* turnpike
~, пограничная frontier post
~, сторожевая outpost support
застаиваться stagnate; become stale
застежка fastener, clasp, buckle; hasp
~, механическая zipper
~, ременная belt fastener
застежка-молния zipper, zip fastener
~, герметическая *av.* pressure-sealing zipper
застеклять glaze
застой stagnation; standstill; *med.* congestion; *econ.* depression, stagnancy, slackness
~ в деловой активности stagnation of business
~ в спросе *econ.* stagnant demand
~ крови congestion, h(a)emostasia
~ крови, активный active congestion, hyper(a)emia
~ крови, мозговой brain congestion
~ крови, пассивный passive congestion, stagnation of blood
застойный stagnant; *med.* congestive, hypostatic
застопоривать lock, stop (up); check
застропить attach slings to, secure by slings
застудневание *chem., food* gelatinization, gelation; jellying, jellification
застудневать gel, gelate; *food* make/turn into jelly
застывание congelation, solidification, setting, gelation
застывать congeal, solidify, set

засуха 1. drought; 2. extinct lake
засуховыносливый drought-enduring
засухолюбивый xerophilous, xerophile
засухоустойчивость drought resistance, xerophytism
засухоустойчивый drought-resistant
засыпка backfilling, filling up; fill, backfill; *met.* batch, charge
затачивание sharpening, grinding
  ~, электроискровое electric-spark [electric-erosion] grinding/sharpening
затачивать sharpen, grind
  ~ сверло re-point a drill
затвердевание setting, solidification, induration, hardening, stiffening
затвердевать harden, solidify; set; congeal, cake, consolidate, indurate
затвердение induration; hardening; *med.* callus, callosity, induration
затвердитель hardener
затвор lock, stopper, bolt, gate, catch; plug; *phot.* shutter; *hydr.* seal; sluice
  ~, аварийный emergency gate
  ~, байонетный bayonet catch
  ~, барабанный roller gate
  ~, безопасный safety lock
  ~, бункерный bin gate
  ~, водяной water/hydraulic seal, waterlock
  ~, вращающийся *phot.* rotating shutter
  ~, газовый gas seal
  ~, глубинный deep [low-level, sluice] gate
  ~, дозирующий proportioning/metering gate
  ~ донного водоспуска bottom outlet gate, bottom dumpgate
  ~, дроссельный butterfly gate/valve
  ~, жидкостный liquid seal
  ~, клапанный swing valve, tilting gate
  ~, колошниковый top gate, throat stopper
  ~, конусный plug valve/cock
  ~, ламельный/лепестковый lamella/louvre shutter
  ~, лопастный rotary gate
  ~ лотка chute gate
  ~, опускаемый drop gate
  ~, откидной flap gate/lock
  ~, плавучий flood/water/caisson gate
  ~, плоский plain/lift gate
  ~, рабочий service gate
  ~, ремонтный duplicate gate
  ~, ртутный mercury cut-off; mercury seal
  ~, сегментный segment/radial gate/shutter
  ~, секторный sector/drum gate
  ~, скользящий slide/bulkhead gate
  ~, фотографический shutter
  ~, шандорный bulkhead gate
  ~, шарнирный hinged gate/breechblock
  ~, шаровой ball cock
  ~, шиберный slide gate
  ~, шпиндельный stem gate
  ~, шторный curtain shutter
  ~, щелевой slotted shutter
  ~, электронно-оптический electronic shutter
затворять 1. shut, close; 2. add water to, mix with water; *(lime)* slake
затек *geol.* pocket; intrusion; streak, flow
затекать flow, pour (into); leak
затемнение obscuring, shading, adumbration; darkening; *mil.* blackout, dim-out; *med.* loss of consciousness/memory, blackout; dark patch/spot
  ~, полное full blackout
  ~ стекол, регулируемое adjustable window darkening
затемнитель dimmer
  ~, сценический theatre/theater dimmer
затемнять obscure, dim, darken, shade; *mil.* black out
затенение shading, shadowing, blanketing; adumbration; blur
затененность opacity; dimness
затененный shaded
затенять shade, shadow; blanket; becloud, bedim
затесывать 1. rough-hew; sharpen; 2. jag, notch
затирать rub out/down/off/over; grind; trowel
затирка trowelling, trowel finish
затихать abate, calm/die down; fade
затишье calm; lull
  ~, деловое *econ.* slack business
затмение eclipse; *fig.* blackout, blank
  ~, кольцеобразное annular eclipse
  ~, лунное lunar eclipse
  ~, полное total eclipse
  ~, солнечное solar eclipse
  ~, частичное partial eclipse
затон backwater, cove, pool, weir, dam
затонуть sink, go down
затопить I inundate, flood
затопить II begin to heat, make/light a fire
затопление flooding, inundation; flood; submersion, submergence; *naut.* scuttling
  ~ плотины overdamming
затопляемый inundable, floodable; submersible
затоплять flood, inundate; overflow; submerge; *(ship)* scuttle
затор blockage, obstruction, congestion, jam(ming), blocking
  ~ движения traffic jam
  ~, ледяной, ~ льда ice jam(ming), blocking of ice
  ~ перфокарт muddle of cards
затормаживать 1. brake; retard, slow down; 2. *chem., physiol.* inhibit
заторможенность *psych.* inhibition; block
заточка 1. sharpening, grinding; 2. recess; notch; groove
затравка 1. *cryst.* seeding, inoculation; *chem.* seeding polymer; 2. primer, priming device/tube; detonator; 3. priming
затраты expenditure(s), outlay(s), expense(s), cost(s); overheads; charges
  ~, внеплановые unscheduled cost(s)

~ вычислительных ресурсов computational cost(s)
~, годовые annual charges/expenditures/costs
~, капитальные capital expenditure/cost(s)
~, конторские clerical/office cost(s)
~ на гарантийный ремонт warranty expenses
~ на единицу продукции, текущие *econ.* unit operating cost(s)
~ на корма food cost(s)
~ на материально-техническое снабжение *econ.* procurement cost(s)
~ на обработку данных processing cost(s)
~ на общественные нужды social expenditure
~ на сверхурочную работу overtime cost(s)
~ на складское хранение warehousing cost(s)
~ на содержание резервной мощности *econ.* standby charges
~ на транспортировку shipping/forwarding cost(s)
~, неосязаемые *econ.* intangible cost(s)
~, непроизводительные overheads
~, общие производственные burden
~, основные *econ.* prime costs
~, плановые *econ.* planned cost(s)
~ по гарантийному обязательству warranty cost(s)
~ по стимулированию сбыта promotional expenditure(s)
~, постоянные *econ.* fixed cost(s)
~, почасовые hourly cost
~ при полной загрузке (судна) waterline operating expenses
~ труда labour expenditure
~, фактические actual expenses
затраты-эффективность cost-effectiveness
затуманивать dim, cloud, fog; obscure
затупление bluntness; blunting
затуплять blunt, dull
затухание attenuation, dying down, fading, decay, sibsidence; damping; extinction; weakening
~ в атмосфере atmospheric attenuation
~ гармоник decay of the harmonics
~ звука sound/acoustic attenuation
~ излучения radiation damping/attenuation
~, километрическое attenuation per unit length
~ нейтронного потока neutron attenuation
~ несогласования mismatch attenuation
~ отражения reflection/return loss
~ отраженного сигнала echo attenuation
~, переходное, ~ по соседнему каналу crosstalk attenuation
~ потока частиц particle attenuation
~, пространственное free-space attenuation
~, собственное non-reflective attenuation; natural attenuation
~, ударное collision damping
~, циклотронное cyclotron damping
~, чрезмерное overdamping; excessive damping
затухать attenuate, damp, fade, die down, subside, be attenuated/damped
затыкать plug, stop
затыловать back off, relieve
затылок back of the head, *anat.* occipit; *rare* neck
затылочно-височный occipitotemporal
затылочно-лицевой occipitofacial
затылочно-лобный occipitofrontal
затылочно-подвздошный occipitoiliac
затылочно-теменной occipitoparietal
затылочный *anat.* occipital
затыльник backplate
затычка plug, stop, stopper
затягивание 1. tightening; 2. tailing, hysteresis effect, hangover; 3. *rad.* pulling
затягивать tighten; fasten; tie/lace up
затяжка brace, tie-bar/-rod/-beam; tightening
~, подвесная hanging brace
~, стропильная *constr.* stay piece/collar, footing beam
заусенец burr, fin, seam, flash, barb, spew
~ нцы, снимать file off/down burrs, deburr
заусеница agnail, hangnail
заутреня *eccl.* matins
~, пасхальная Easter Vigil
заушница *med.* (epidemic) parotiditis, mumps
заушный behind the ears; *med.* parotid
заформовывать mould
зафрахтовывать charter *(a vessel)*
зафронтальный post-frontal
захват 1. grip, clamp, catch, grab, trap, gripper, holding device, holder, hold; lock; 2. entrainment, entrapment, seizure, capture; 3. covering; *min.* web
~ быстрого нейтрона *phys.* fast-neutron capture
~, вредный non-productive capture
~, гарпунный harpoon fork; hay harpoon
~ дорожки track hold
~ замедлителем moderator capture
~, замковый locking grip
~ ионов ion pick-up
~, клещевой tongs
~ нейтронов neutron capture
~ нефти oil trapping
~, патронный socket tool
~ плазмы plasma trapping
~ радиолокационной станции radar lock-on
~ ракеты missile capture
~, рычажный lever tongs
~, самодержащий self-closing holder/catch
~ цели target lock-on
~ цикла cycle stealing
~, челюстной clamp
~ электромагнитный lifting magnet
~ ядром nuclear capture
захватка grip, catch, dog
захватывать 1. grip, catch, seize, grab, hold; clench; lock; clinch; 2. entrain, pick up, entrap; trap
~ цель lock on a target
захлебывание choke, choking; flood(ing)
~ ректификационной колонны flooding of a

column
захлестывание, захлествание lashing, whipping
захлопка flap/clack valve, flapper, flap
захлопывание 1. collapse; closure; 2. *phon.* implosion
заход stopping, call (at); *av.* approach; run-in; *astr.* setting
~, вынужденный emergency/forced call
~ Луны moonset
~ на бомбометание *mil.* bombing run
~ на посадку landing/final approach
~ на посадку, неудачный baulked approach
~ на посадку по ветру *av.* down-wind [tail-wind] approach
~ на посадку по приборам instrument approach
~ на посадку при ручном управлении manual approach
~ на посадку против ветра up-wind [into-the-wind] approach
~ на посадку с помощью радиолокатора radar-assisted approach
~ на торпедометание *mil.* torpedo run
~ на фотографирование photographic run
~ на цель target run, run-in
~ на штурмовку *mil.* strafing run
~ Солнца sunset
заходка *min.* stope
зацентровка centr(e)ing, centering
зацентровывать centre-drill, centre, center; drill centres
зацепка catch, hook, peg
зацепление engagement, mesh(ing), gearing; arrestment, catch(ing)
~, внешнее/внутреннее зубчатое external/internal gearing
~, зубчатое gearing, gears; engagement, mesh
~, коническое зубчатое bevel gearing
~, повышающее/понижающее зубчатое step-up/step-down gearing
~, постоянное constant mesh
~, прямозубое straight-tooth spur gearing
~, реечное зубчатое rack gearing
~, червячное зубчатое worm gearing
~, шевронное зубчатое herringbone gearing
зацеплять(ся) engage, mesh; catch; link
зацикливание cycling, recycling; infinite looping
зачаливать fasten, fix, moor, tie up/down
зачатие conception
~, непорочное *eccl.* Immaculate Conception
зачатковый germinal
зачаток germ; embryo; rudiment; primordium
~ листа leaf primordium
~ ости awnlet
зачаточный embryonal; primordial; incipient; rudimentary; germinative, germinal
зачеканивать caulk in
зачет (academic) credit; test
зачисление entry; enrol(l)ment; *mil.* enlisting
зачислять enter, put on the list; enlist, enrol(l)
~ в вуз enrol(l) at an university
~ в кредит счета *econ.* place to the credit
~ в щат take on the staff
~ на должность секретаря take on as secretary
зачистка cleaning, stripping, grinding, surface conditioning, scraping/filing bright; *met.* fettling; *leath.* paring
зачищать *met.* fettle, grind; strip; scrape/file bright; *leath.* pare
зашифровывать code, encode, put into code
зашкаливание off-scale reading
зашкаливать read/sweep/go off-scale
зашлаковывание slagging, clogging/blocking with slag/dross
зашпатлевывать, зашпаклевывать putty, fill with putty
зашплинтовывать splint
зашпунтовывать cut a tongue on, tongue
заштамповывать stamp, press, drop-forge
заштатный supernumerary; extra, not on the permanent staff
заштилеть *naut.* fall calm, be becalmed
заштриховать cross-hatch, shade
защебенивать fill up with rubble
защелка catch, stop, lock, latch, click, snap, arrester; interlock
~, безопасная/предохранительная safety latch
~, пусковая starting trigger
~, трещоточная/храповая ratchet lock
~ уровня сигнала level latch
защелкивать latch, snap, click, lock
защемление 1. restraint; 2. jamming, sticking, pinching; binding; jam, pinch
~, боковое lateral/sidewise restraint
защемлять fix, restrain, bind, stick; pinch, jam; squeeze
защита protection; safeguard; shielding, screening; defense/defence; screen, shield
~, аварийная emergency protection system, safety system
~, антикоррозионная corrosion inhibition/protection
~, бетонная concrete shielding
~, биологическая biological shielding; *mil.* biological warfare defence/defense
~ вод water protection/shielding
~ грунта soil conservation
~ данных data protection
~ данных по записи write protection
~, земляная earth/ground fault/leakage protection
~ магнитных полей magnetic screening
~, нейтронная neutron shielding
~ от взрыва blast protection; explosion suppression
~ от вибраций vibration protectton
~ от гамма-излучения gamma-shielding
~ от дурака foolproof guardian

~ от заморозков *agr.* frost protection
~ от запотевания (стекол) demisting, defogging
~ от излучения radiation protection/shielding
~ от коронного разряда *el.* corona prevention
~ от коротких замыканий short-circuit protection
~ от коррозии corrosion/rust protection; rustproofing
~ от коррозии, анодная/катодная anodic/cathooic protection
~ от крыс rat-proofing
~ от метеорных частиц micro-meteoroid protection
~ от неблагоприятных климатических воздействий climatic protection
~ от неблагоприятных метеорологических условий weather protection
~ от несанкционированного доступа illegal access protection
~ от несихронной работы loss-of-synchronism protection
~ от обледенения ice/icing protection, anti-icing
~ от облучения radiation protection/shielding
~ от ошибок error protection
~ от перегрева overheat protection
~ от перегрузки overload protection
~ от перенапряжений overvoltage protection
~ от помех interference protection
~ от радиоактивности radioactivity protection
~ от радиологического оружия radiological protection
~ от сверхтоков overcurrent protection
~ от ударов молнии lightning protection
~ от ядерных излучений nuclear shielding
~ памяти *comp.* memory/storage protection
~, первичная primary shielding
~ по напряжению, максимальная/минимальная over-voltage [under-voltage] protection
~ по питанию power protection
~, подводная *naut.* underwater shielding
~ при возвращении в атмосферу, тепловая reentry heat protection
~, противоатомная atomic defence/defense
~, противокоррозионная corrosion protection
~, противометеорная meteoroid protection
~, противообледенительная ice protection; anti-icing; de-icing
~, противоосколочная fragment protection
~, противопожарная fire protection
~, противоторпедная torpedo defence/shielding
~, противоударная collision protection
~, радиационная radiation protection/protection
~ реактора reactor shielding
~ рулей rudders guards
~ с блокировкой locking protection system
~ с помощью паролей *comp.* password protection
~ секретности *comp.* privacy protection
~, секционная sectionalizing protection
~, селективная discriminating protection
~ типа замка *comp.* lock-and-key protection
~ токовая, максимальная/минимальная overcurrent/undercurrent protection
~, шинная bus protection
защитник 1. *law* defence/defense counsel/attorney, counsel for the defence/defense; 2. *sport* full back; 3. protector; defender
защитный protective, protecting, shielding; guarding; defensive
~ цвет *mil.* khaki
защищать 1. protect, guard; shield, screen; 2. side with, champion; *law* defend; speak in support of
~ дело *law* plead a case
защищенность данных *comp.* data security
заявитель *law* deponent; declarent; applicant
заявка request, demand, claim; application; requisition
~, авторская author's/inventor's application
~ на материал requisition for supplies
~ на перевозку shipping request
~, патентная patent application
~, транзитная shipper's certificate
заявление statement, declaration, announcement; application
~, исковое *law* statement of claim
~ под присягой *law* statement on oath
~, подать put in [file] an application
заявлять announce, declare, state
~ о своей невинности *law* plead not guilty
~ претензию lodge a complaint
заявщик applicant
заякорить *naut.* to anchor
звание title; rank
~, ученое academic status/title
звезда 1. star; 2. *mach.* spider
~, двойная binary star
~, двухлучевая two-pronged star
~, морская *zool.* starfish
~, незаходящая circumpolar star
~ образованная протоном proton-produced star
~ первой величины first magnitude star
~, Полярная Pole Star, Polaris
~ расщепления disintegration star
~, слабая faint star
звезда-гигант giant star
звездолет spaceship, spacecraft; interstellar vehicle
звездолетчик, звездоплаватель astronaut
звездочка 1. stellule; 2. *mech.* sprocket, chain gear/wheel; 3. *print.* asterisk; *met.* pick
~, ведущая drive sprocket
~, натяжная tension sprocket
~, отжимная deflecting sprocket
~ цепной передачи chain sprocket
звездчатка sprocket, star wheel
звездчатолистный *bot.* star-leaved

звездчатоцветковый *bot.* star-flowered
звездчатый stellate, asteroid
звено 1. unit, element, component; member; 2. *mech.* link; 3. group, team; detachment; *av.* flight *(of three planes);* 4. *geol.* stratum, layer, bed, subformation
~, вертлюжное swivel link
~ гусеницы track link/block/shoe
~, дифференцирующее differentiating unit, differentiator
~, запаздывающее time-lag/-delay unit
~, зубчатое gear unit
~, интегрирующее integrating unit, integrator
~, исполнительное operating/executive element
~, множительное multiplier
~ приведения reduction link
~ с задержкой time-lag/-delay element/unit
~, связующее coupling element
~, соединительное connecting link, connector
~ сопряжения с абонаментом custom adapting element
~, стабилизирующее equalizer
~, суммирующее addition cascade, adder stage
~ цепи chain link
~, шарнирное swivel link
звенорг *abbr.* team/group/detachment organizer
звеньевой, звеньевод team/group leader
зверобой St. John's wort
звероводство fur farming
зверосовхоз state farm for fur-bearing animals
звероферма fur farm
звон ringing, ring, peal
~ в голове *med.* head noises, tinnitus cerebri
звонок bell; *el.* ringer
~, аварийный emergency/alarm bell
~, пожарный fire bell
звук sound; *med.* resonance
~, амфорический amphoric/cavernous resonance
~, высокий high-pitched sound
~, легочный pulmonary sound
~, металлический metallic ringing/tinkling
~, нетональный unpitched sound
~, низкий low-pitched sound
~, перкуторный *med.* percussion sound
~ распространяющийся в воде/воздухе waterborne/airborne sound
~, свистящий whistling sound
~, стационарный steady-state sound
~, стереофонический stereo sound
~, тупой/глухой dull sound
~, шипящий hissing sound
~, эталонный token sound
звукоанализатор sound analyzer
звукоблок sound reproducer
звукобоязнь *med.* acoustophobia
звуковидение sonic holography, acousto-optical imaging
звуковой sonic, acoustic, sound *(attr.),* audio; phonic
~вая дорожка sound track
звуковоспроизведение sound reproduction
звукогенератор audio-frequency generator
звукоглушитель sound damper; silencer, muffler
звукозаписыватель sound recorder
звукозаписывающий sound-recording
~ аппарат sound recorder
звукозапись sound recording
~ на пленку magnetic tape recording
~, стереофоническая stereo recording
звукозонд sonoprobe
звукозондаж echo sounding
звукозрительный audio-visual
звукоизлучатель sound projector, acoustic radiator/generator
звукоизолирование, звукоизоляция soundproofing, acoustic/sound insulation
звукоизоляционный soundproof
звуколокатор asdic; *Am.* sonar
звуколокация sound location/ranging, sound fixing and ranging
звуколюминесценция sonoluminescence
звукомаскировка sound camouflage
звукомаяк aural radio beakon
звукомер sound level meter; phonometer
звукомерный sound-measuring
звукометрический sound-ranging
звукометрия sound ranging; phonometry
звуконепроницаемость sound-proofness; sound/acoustic opacity
звуконепроницаемый soundproof, soundtight
звуконоситель sound recording material/medium; sound carrier/tape/wire
звукообразование phonation, sound formation
звукообразующий sound-generating/-producing
звукооператор sound engineer/technician/monitor, sound man
звукоотражение sound reflection
звукоощущение sound sensation/perception
звукопеленгатор acoustic ranger/goniometer; sound locator
звукопеленгация sound finding and ranging
звукопередатчик sound transmitter
звукопередача sound/speech transmission
звукописец sound/voice recorder
звукопись *mus.* sound painting
звукопоглотитель 1. sound-absorbing material/medium; 2. sound absorber/damper
звукопоглощательный, звукопоглощающий soundabsorbing
звукопоглощение sound absorption, muffling
звукоподражание onomatopoeia, sound symbolism
звукоподражательно onomatopoeically
звукоподражательный onomatopoeic(al)
звукопреломление acoustical refraction
звукоприемник sonic detector, sound locator/receiver
~, гидроакустический hydrophone
звукопровод acoustic line; sound duct

звукопроводимость, звукопроводность sound transmission/conduction
~, костная bone conduction
звукопроводный, звукопроводящий sound-conducting/transmitting
звукопроекция sound projection
звукопрожектор sound projector
звукопроизводный, звукопроизводящий soundproducing
звукопроизносительный aural-oral
звукопроницаемость acoustic permeability
звукопроницаемый transaudient, acoustically permeable
звукопропускание acoustical transmittivity
звукорассеивающий sound-scattering/-dispersing
звукорежиссер sound producer
звукорепродукция sound reproduction
звукоряд *mus.* scale
звукосигнал sound signal
звукосмеситель sound/audio mixer
звукосниматель sound pickup/head
~, емкостный capacitor pickup
~, пьезоэлектрический crystal pickup
~, угольный carbon-contact pickup
~, электрооптический light-beam pickup
звукосочетание sound combination
звукостирание sound erasure
звукосъемник sound pickup
звукосъемочный sound-pickup
звукотехника acoustic/sound engineering; phonics
звукоулавливатель, звукоуловитель sound detector/locator
звукоусиление sound amplification
звукоусилитель sound amplifier
звукоусилительный sound-amplifying
звукофикация (installation of) public loudspeaker system/network
звукочастотный audio-frequency *(attr.)*
звукочувствительный sound-sensitive
звучать sound, resound, be heard; ring
звучание sounding; phonation
здание building, edifice; structure
~, административное office building
~ аэровокзала air terminal building
~ без лифта walk-up building
~ для транзитных пассажиров transit building
~, жилое residential building, house; block of flats, apartment house
~, заводское factory building
~, каркасное skeleton-type structure/building
~, машинное power house
~ на сборных элементах prefabricated building/structure
~, нежилое nonresidential building
~, панельное bearing-wall building
~, сейсмостойкое earthquake-proof building
здравоохранение health protection; health service; public health
здравотдел health department
здравпункт health centre/center
зеаксантин zeaxanthin
зебра *zool.* zebra
зев 1. *tech.* mouth, jaw; opening; throat; 2. *anat.* fauces, orifice; pharynx; gorge; 3. *text.* shed, lease
~ валков roll gap
~ гаечного ключа mouth of a spanner, span
~ клещей tong jaw
зевгит *hist.* zygite
зевгма *rhet., gram.* zeugma
зевгогеосинклиналь *geol.* zeugogeosyncline
Зеведей *bibl.* Zebedee
зевообразование *text.* shedding
зевота yawning
Зевс *myth.* Zeus
зеин *chem.* zein
зейгерование liquation, seggregation
зейгеровать *met.* liquate, seggregate
зеландец, зеландка Zealander
Зеландия, зеландский Zealand
зеленка, зеленушка *bot.* 1. equestrian tricholoma *(Tricholoma equestre);* 2. yellow wort/centaury *(Blackstonia, Chlora);* 3. running pine *(Lycopodium clavatum)*
зеленоглазка *zool.* green-eyed fly *(Chlorops taeniopus)*
зеленоспинка *zool.* blue herring *(Pomolobus chrysochloris)*
зеленоцвет *bot.* chloranthus *(Chloranthus)*
зеленушка *zool.* 1. green linnet *(Chloris);* 2. greenfinch *(Chloris chloris);* 3. green wrasse *(Crenilabrus tinea);* 4. see **зеленка**
зеленчук *bot.* dead nettle *(Lamium)*
~, желтый yellow archangel, weaselsnout *(Laminum galeobdolon)*
зелень green colo(u)r; greenery
~, медная chrysocolla, aerugo
~, парижская Paris green
зелигманит *miner.* seligmanite
зелот *hist.* Zealot
земиндар *hist.* zemindar, zamindar
землеведение earth science; physical geography
землевладелец landowner; landlord; *pl.* landed interests/classes
~, крупный big landowner, landlord
~, мелкий smallholder
~, свободный *hist.* freeholder
землевладельческий landowning *(attr.)*
землевладение land holding/tenure/ownership; landed estate
~, общинное communal land ownership
земледелие agriculture, farming; cropping
~, богарное bogharic agriculture
~, буферно-полосное buffer-strip cropping
~, интенсивное intensive agriculture/cropping
~, контурно-полосное contour strip cropping
~, крупное large-scale agriculture
~, механизированное mechanized/machine agriculture
~, многоотраслевое diversified agriculture

~, **натуральное** subsistence farming
~, **неорошаемое** non-irrigated agriculture
~, **орошаемое** irrigated agriculture
~, **парцеллярное** parcelled-up farming; *Eng.* crofting agriculture
~, **плановое** planned agriculture
~, **полосное** strip cropping
~, **почвозащитное** conservation farming
~, **пропашное** hoe farming
~ **с ветроломными полосами** wind strip cropping
~ **с травосеянием** lea farming
~, **товарное** *econ.* commodity farming; *Am.* truck/commercial agriculture
~, **экстенсивное** extensive agriculture
землекоп excavator, digger; navvy
землемер land surveyer
землерой burrowing animal
землеройка *zool.* shrew *(Sorex)*
землепользование land use/utilization; land tenure
землесос suction/hydraulic dredge(r), dredge pump
землетрясение earthquake, earth tremo(u)r
~, **подводное** seaquake
землеустройство land management; system of land tenure
землечерпалка bucket dredger, excavator; dredging shovel
~, **многоковшовая** multibucket dredger
земли, редкие rare earths
Земля the Earth
земля earth, soil; ground; land; *el.* earth, *Am.* ground
~, **арендованная** leased land
~, **голубая** *miner.* blue earth/ground, kimberlite
~, **диатомовая** diatomaceous earth; tellurine
~, **заболоченная** waterlogged land
~, **залежная** layland
~, **засоленная** saline land
~, **инфузорная** infusorial/diatomaceous earth, kieselgur
~, **квасцовая** alum earth, alumina
~, **мелиорированная** improved land
~, **наносная** alluvion; derelict land
~ **не занятая под культурой** noncropped land
~, **незасеваемая** land out of crop
~, **немелиорированная** unimproved land
~, **необработанная** untilled/uncultuvated land
~, **необрабатываемая** noncultivated land
~, **неосвоенная** undeveloped land
~, **обрабатываемая** cultivated land
~, **общинная** common land
~, **орошаемая** irrigated land
~, **осушенная** reclaimed land
~, **парующая** fallow land
~, **пахотная** plowing/ploughing land
~, **пенистая** aphrite
~ **под культурой** land in/under crop
~ **под паром** fallow land
~, **прибрежная** riparian land
~ **пригодная для обработки** cultivable land
~, **растительная** overburden; sod kimberlite
~, **старопахотная** old arable land
~, **торфяная** peat soil, peaty earth
~, **улучшенная** improved land
~, **утомленная** sick land
~, **целинная** virgin land, old idle land
~, **щелочная** alkaline/alkali earth
земляника *bot.* strawberry *(Fragaria)*
~, **виргинская** scarlet strawberry *(Fragaria virginiana)*
земляничник *bot.* madrone *(Arbutus)*
~, **крупноплодный** strawberry madrone/tree *(Arbutus unedo)*
земляные работы excavations
земноводное *zool.* amphibian; *pl.* amphibia
земноводный amphibious; sea-land
земснаряд suction-tube dredge(r)
зенд Zend
зензивер *bot.* high mallow *(Malva sylvestris)*
зензубель rabbet/tooth/molding plane
зенит zenith; vertex
~ **наблюдателя** apparent zenith
зенитный 1. *mil.* antiaircraft *(attr.)*; 2. *astr.* zenithal
зенитчик *mil.* antiaircraft gunner
зенкер multiflute/vertical drill; hole-enlarging drill; reamer; counterbore, countersink
зенкерование counterboring, countersinking; hole-enlarging/-trueing
зенкеровать countersink, counterbore
зенкование countersinking, counterboring
зенковка (countersink) reamer, counterbore, countersink drill
зеркало mirror; reflector, retractor; *med.* speculum
~, **антенны** aerial/antenna reflector
~, **влагалищное** vaginal speculum
~, **вогнутое** concave mirror
~, **водное** water table/surface
~, **выпуклое** convex mirror
~, **глазное** ophthalmoscope
~, **гортанное** larynx speculum, laryngeal mirror
~ **для сердца** heart reflector
~ **заднего вида** rear-view mirror
~ **золотниковой коробки** slide-valve face
~, **зубное** dental mirror
~ **испарения** evaporation surface
~, **кривое** distorting mirror
~ **микроскопа** illuminating mirror
~, **нейтронное** neutron-reflecting mirror
~, **полупрозрачное** semitransparent/semireflecting mirror; half-silvered mirror; partially reflective mirror
~, **светоделительное** beam-splitting mirror
~, **светящееся** *med.* luminous speculum
~ **скольжения** slickenside(s), friction/gliding plane
~, **смотровое** surveillance/viewing mirror
~, **цветоизбирающее** colour-selective mirror
~ **цилиндра** *auto.* cylinder face

~, ушное *med.* ear speculum
зеркало-прожектор mirror searchlight
зеркальнолинзовый mirror-lens *(attr.)*
зеркально-отраженный specularly reflected
зеркальный mirror-like, specular; reflecting; glassy
зернение graining, granulation
зернистость granularity; grain-size; seediness; *med.* stippling; *food* grainy texture/structure
зерно 1. grain; seed; kernel; *fig.* core; *agr.* cereals, grain crops, corn, grain; *tech.* granule, nodule; *met.* crystallite
~, аустенитное austenite grain
~, вмещающее host grain/granule
~, жемчужное pearl
~, кормовое fodder grain
~, кристаллическое crystal grain
~, крупное coarse grain
~, ледниковое ice grain
~, мелкое fine grain
~, некондиционное non-grade grain
~, товарное commodity/marketable grain
~, тонкое fine grain
зерноанализатор grain tester
зерноаспиратор air-operated grain cleaner
зернобобовые *pl. bot.* puls crops
зернобобовый *agr.* leguminous
зерновидный granular, graniform
зерновка, фасолевая *zool.* bean weevil
зерновоз *(ship)* bulk grain carrier; *auto.* grain wag(g)on
зерновой grain, corn; cereal; frumentaceous
~вые культуры/злаки/хлеба grain crops, cereal crops
зерновые *pl.* grain crops, cereals
зернодробилка grain crusher
зернопогрузчик mechanical grain loader, grain conveyer
~, шнековый screw/auger grain conveyer/loader
зернопоставка grain delivery
зернопульт grain-throwing installation/plant
зерноуборочный grain-harvesting *(attr.)*
~ комбайн combine harvester
зернохранилище granary, grain storage/silo
зерноядный *zool.* granivorous
Зефир 1. *myth.* Zephyrus; 2. *poet.* Zephyr; 3. *text.* **(з)** zephyr cloth
зибелин *text.* zibelline cloth
зигаденус *bot.* fly poison, death camas *(Zygadenus)*
зигапофиза *anat.* zygaphophysis
зигзаг-машина zigzag sewing machine
зигмашина beading/creasing machine
зиговка creasing, bead-forming
зигогамия zygogamy
зигодактилия zygodactily
зигозооспора *bot.* zygozoospore
зигоматикомастоидит *med.* zygomaticomastoiditis
зигоматит zygomatitis
зигомицет zygomycete
зигоморфизм zygomorphism
зигоморфия, зигоморфность zygomorphy, zygomorphism
зигоморфный zygomorphic, zygomorphous
зигоспора *bot.* zygospore
зигоспорангий *bot.* zygosporangium
зигоспорофор *bot.* zygosporophore
зигота zygote
зигофит *bot.* zygophite
зигофор *bot.* zygophore
зигоциста zygocyst
зилот *hist.* Zealot
зимаза *chem.* zymase
зимняк 1. *zool.* rough-legged buzzard *(Buteo lagopus);* 2. *bot.* durmast oak *(Quercus sessiflora)*
зимование hibernation, wintering
зимовник 1. winter hive, apiary house; 2. *bot.* autumn crocus, meadow saffron *(Colchicum)*
зимогексаза zymohexase
зимоген zymogen
зимограмма zymogram
зимозан zymosan
зимозиметр, зимометр zymometer
зимолюбка *bot.* ground holly *(Chimaphila)*
зимонематоз zymonematosis
зимородок *zool.* halcyon *(Acedo)*
зимостерин zymosterol
зимостойкость winter-hardiness, resistance to cold
зимофор zymophore
зитатунга *zool.* sitatunga *(Tragelaphus spekii)*
злаки *pl.* cereals, gramineous plants
~, кормовые fodder/feed cereals
~, хлебные cereals
злаковидный graminiform, gramineous, graminifolious
злаково-бобовый, злако-бобовый grass-and-legume *(attr.)*
злаковые *pl. bot.* 1. grasses, the grass family; 2. cereals
злаковый graminaceous, gramineous
златка *zool.* charcoal beetle *(Melanophila consputa)*
~, зеленая beech argilus *(Argilus viridis)*
~, узкотелая argilus *(Argilus)*
златоглазка *zool.* golden-eyed fly, lacewing *(Chrysopa)*
златогузка *zool.* brown-tail moth *(Euproctis chrysorrhaea)*
златоискр *miner.* flamboyant quartz, goldstone, aventurine
златок *bot.* common Jacob's-rod *(Asphodeline lutea)*
златокорень *bot.* Spanish oyster plant *(Scolymnus hispanicus)*
златокрот *zool.* golden mole *(Chrysochloris)*
златолист *bot.* golden leaf *(Chrysophyllum)*
Златоуст *rel.* John Chrysostom; Goldenmouth
златоцвет *bot.* golden feather, pyrethrum *(Pyrethrum)*

зловоние feto(u)r, fetidity; mephitis
~, печеночное *pathol.* feto(u)r hepaticus
злокачественный 1. *pathol.* malignant; pernicious
~ная опухоль malignant tumour
~ное малокровие pernicious an(a)emia
злонамеренность evil intention/design, malevolence; *law* malice; dolus
злонамеренный ill-intentioned, ill-meaning; *law* malicious, dolose
злотый *n.* zloty
злоупотребление abuse, misuse
~ властью *law* misfeasance; abuse of administrative authority
~ доверием breach of confidence
~ налоговым обложением abuse of taxation, tax abuse
змеевидный serpentine; coiled; sinous
змеевик 1. *tech.* coil, coiled/spiral pipe; 2. *miner.* serpentine; 3. *bot.* snake-weed, European bistort *(Polygonum bistorta)*
~, благородный precious/noble serpentine
~, волокнистый chrysolite, picrolite
~, гладкотрубный plain-tube coil
~, листоватый antigorite
~, нагревательный heating coil
~, плотный massive serpentine
~, полосатый ricolite
~, ребристый finned/gilled coil
~, смолистый resinous serpentine, retinalite
змееголов *zool.* snakehead *(Ophicephalus argus)*
~, азиатский murrel
змееголовка *zool.* snake-eyed lizard *(Ophisops occidentalis)*
змееголовник *bot.* dragonhead *(Dracocephalum)*
змеерыбка *zool.* straight-nosed pipefish *(Nerophis ophidion)*
змеехвостка *zool.* ophiuroid, ophiuroidean, ophiurian; brittle star *(Ophiuroidea)*
змеешейка *zool.* snakebird, darter *(Anhinga)*
змееяд *zool.* short-toed eagle *(Circaetus gallicus)*
змееящерица *zool.* sand skink *(Scincus)*
змей kite
~, коробчатый box kite
змейка 1. *zool.* snake; 2. *zool.* greater weaver *(Trachinus draco);* 3. *aeron.* S-turn
~, зеленая grass snake
змейка-сортировка *agr.* helical/spiral/screw separator
змейковый serpentiform; sinuous
змея *zool.* snake; serpent
~, гремучая rattlesnake
~, зеленая grass snake
~, мокассиновая copperhead *(Ancistrodon contortrix)*
~, очковая cobra *(Naja naja)*
змий *obs.* the Serpent; satan
знак 1. sign; symbol; mark; signal; token; 2. *comp.* character; 3. *math.* digit, place
~, астрономический astronomical symbol
~, береговой landmark, coastmark
~, буквенно-цифровой *comp.* alphanumeric/alphameric character
~, буквенный alphabetic character
~ верхнего регистра uppercase character
~, водяной *print.* water-mark
~ возврата каретки carriage return character
~ возврата на одну позицию backspace character
~, волноприбойный ripple mark, wave-cut chasm
~, вопросительный question mark
~ вычитания minus sign
~, геодезический geodetic beacon
~, геометрический geometric mark
~, государственный опознавательный nationality mark/sign
~, двоично-кодированный binary-coded character
~, двоичный binary digit/character, bit
~ действия symbol of operation
~ деления division mark/sign
~, денежный banknote
~, дорожный traffic sign
~ замены replacement character
~ заполнения fill character
~, запрещенный *comp.* forbidden character
~ зодиака zodiac symbol
~ интеграла *math.* intrgral sign
~ квантора всеобщности *math.* universal quantifier, universal/all symbol
~ квантора существования *math.* existential quantifier/symbol
~ кода code character
~, контрольный check character
~, корректурный proofreader's mark
~ контрольного останова breakpoint symbol
~ корня *math.* radical sign
~, личный *mil.* identification tag
~, маркировочный mark, label
~ маркшейдерский survey mark; vane, sight
~, машиночитаемый *comp.* machine-readable character
~ молнии danger arrow
~, надстроечный superior character
~ нарастания, концентрический *bot.* annulation
~, несветящий *naut.* day beacon, day-mark
~, номерный registration/licence/number plate
~ объединения *math.* sign/symbol of aggregation/grouping
~, ограждающий clearing mark
~ окончания передачи *com.* end-of-transmission sign
~, опознавательный identification sign
~ отличия decoration, medal, distinction
~ отмены annul sign
~, отрицательный minus/negative sign
~ передачи цифр *com.* numerical sign
~, печатный printed character

~, **плавучий** *naut.* floating beacon/marker
~, **подбуквенный** sedilla
~, **подстрочный** inferior character
~, **положительный** plus/positive sign
~ **порядка** *math.* exponent sign
~ **порядка числа** *math.* exponent sign
~, **посадочный** landing mark/sign
~, **предупредительный** warning sign
~ **препинания** punctuation mark
~ **пробела** blank character
~ **произведения** product sign/symbol
~ **пунктуации** punctuation symbol
~ **равенства** equals/equality sign
~ **радикала** *math.* radical (sign)
~, **разделительный** separator, separating character/symbol
~ **редактирования** editing character
~, **рельефный** embossed character
~ **самого старшего разряда** *comp.* most significant character
~, **самый левый/правый** leftmost/rightmost character
~, **световой** light sign
~, **светящий** *naut.* nightmark
~ **сложения** plus/summation sign
~ **смены регистра** *comp.* escape character
~ **суммирования** summation symbol, sum sign
~ **Т, посадочный** *av.* landing T/tee
~, **товарный** trade mark
~, **торговый** trade sign
~, **триангуляционный** triangulation station mark
~ **удаления строки** line delete character
~, **умножения** multiplication sign
~ **управления форматом** format-control character
~, **управляющий** *comp.* control/functional character
~, **условный** (conventional) symbol
~, **фабричный** trade mark
~, **химический** chemical symbol
~, **цифровой** numeric/digital character
~ **числа** number sign
~, **штриховой** stroked character
знак-заполнитель fill/filler character
знаковыбиратель sign selector
знакоинвертор *comp.* sign inversor/changer/reverser
знакопеременный *math.* sign-variable, with alternating signs, alternating
знаменатель *math.* denominator
~ **геометрической прогрессии** ratio/quotient of a geometric progression
~ **масштаба** scale number/denominator
~, **общий наименьший** least common denominator
значащий significant
значение 1. value, magnitude; significance; 2. meaning, sense
~, **амплитудное** peak/crest value
~, **аномальное** *stat.* wild value
~ **атрибута** attribute value
~, **базисное** base-line value
~ **базового адреса сегмента** segment value
~, **взвешенное** weighted value
~, **выборочное** *econ.* sample value
~, **вырожденное** degenerated value
~, **двоякое** bifurcational value
~, **действующее** effective value
~, **допустимое** allowed/admissible/legitimate value
~ **единственное** unique value
~, **заданное** set/preset/given/predetermined/prescribed value
~, **запрещенное** forbidden/unpermitted value
~ **игры, верхнее/нижнее** upper/lower value of a game
~, **инверсное логическое** opposite logic value
~ **"истина/ложь"** truth/false value
~, **истинное** *stat.* ideal/true value; *log.* truth value
~, **классификационное** taxonomic value
~, **конечное** finite value
~, **контрольное** check value
~, **корреляционное** correlative value
~, **максимальное** maximum/peak value
~, **мгновенное** instantaneous value
~, **начальное** initial value
~, **ненулевое** non-zero value
~, **номинальное** rated/nominal value
~, **нулевое** zero value
~, **осредненное** averaged value
~ **поля** field data
~, **пороговое** threshold value
~, **предварительное** tentative value
~, **предвычисленное** predicted value
~, **предельное** limiting/critical value
~, **присваиваемое** assigned value
~, **равновесное** equilibrium value
~, **разрешенное** allowed/permitted value
~, **разрядное** place value
~, **расчетное** design/estimated/calculated value
~, **резко отклоняющееся** *stat.* wild value, wild shot
~, **сглаженное** smoothed value
~, **седловое** *game th.* saddle value
~, **собственное** eigen value; proper/characteristic value
~ **средневзвешенное** weighted mean value
~, **среднее** mean/average value, mean
~, **среднее арифметическое** arithmetical mean/average
~, **среднее геометрическое** geometrical mean
~, **среднеквадратичное** root-mean-square value, effective value
~, **среднесуточное** diurnal/daily mean
~, **стационарное/установившееся** steady-state value, stationary value
~, **стратиграфическое** stratigraphic significance/implication

~, **табличное** tabulated/tabular value
~, **текущее** current/running value
~, **требуемое** desired value
~, **установленное** pre-set value, datum
~, **уточненное** improved value; specified value
~, **характерное** representative/characteristic value
~, **целочисленное** integral value
~, **частное** particular value
~, **численное** numerical value
значимость (statistical) significance; validity
~ **адреса** address validity
значимый (statistically) significant; valid
значность *chem.* atomicity
~ **кода** word length of a code
значок check/distinctive mark
~, **дифференциационный** diacritical sign/mark
зоантарий *zool.* zoantharian *(Zoantharia)*
зоб *pathol.* goiter, struma
~, **базедов** exophthalmic/basedowian goiter
~, **базедовифицированный** basedoid goiter
~, **висячий** diver/wandering goiter
~, **внутригрудной** intrathoracic goiter
~, **загрудинный** substernal/retrosternal goiter
~, **злокачественный** struma maligna
~, **кистозный** cystic goiter
~, **лимфоидный** struma lymphomatosa
~, **первично-тиреотоксический** primary toxic goiter
~, **подвижный** wandering goiter
~, **позадипищеводный** retroesophageal goiter
~, **позадитрахеальный** retrotracheal goiter
~, **удушающий** suffocative goiter
~, **узловатый** nodular goiter, struma nodosa
~, **эутиреоидный** euthyroid goiter
~, **язычный** lingual goiter
зобный *med.* strumous
зобогенный *med.* goitrogenic
зодиак zodiac
Зоил *hist.* Zoilus; *fig.* (z) carping critic, Zoilist
зола ash(es), cinder(s)
~, **древесная** wood ash
~, **древесно-угольная** charcoal ash
~, **легкоплавкая** low-fusion-point ash
~, **летучая** fly/light ash
~, **тугоплавкая** high-fusing ash; refractory ash
золение *leath.* liming
золи sols
золить *leath.* lime, soak/drench in lime
золодообдуватель soot/ash blower
золоотвал ash dump
золоотстойник ash sump; ash-/dust-settling tank; slag settler
золотарник *bot.* goldenrod *(Solidago)*
золотистоплодный *bot.* chrysocarpous
золотистоцветковый *bot.* chrysanthous
золотник slide/gate/control valve; *auto.* inner valve, valve core
~, **коробчатый** plain/simple slide valve
~, **крановый** valve cock disk; rotary piston valve; cork plug
~, **отсечный** cut-off valve
~, **поршневой/цилиндрический** piston slide valve; spool valve
золото gold
~, **аллювиальное самородное** nugget
~ **в самородках** gold in nuggets
~ **в слитках** bar/ingot gold
~, **горчичное** mustard gold
~, **гремучее** fulminating gold
~, **иридистое** iridium gold
~, **крупнозернистое** coarse gold
~, **лигатурное** base bullion
~, **листовое** leaf gold; gold foil/leaf
~, **медистое** copper gold
~, **монетное** crown gold
~, **накладное** gold plate
~, **платинистое** platinum gold
~, **проволочное** wire gold
~, **пылеобразное** gold flour
~, **россыпное** placer/alluvial/stream gold
~, **самородное** native/free gold
~, **сусальное** mosaic gold; (artificial) gold leaf; tinsel
~, **теллуристое** gold telluride, calaverite
~, **упорное** rebellious gold
~, **хлорное** auric chloride
~, **чешуйчатое** flake gold
~, **чистое** fine bullion/gold
золотобородник *bot.* scented grass *(Chrysopogon)*
золотоглазка *zool.* goldeye *(Amphiodon alosoides)*
золотопромышленность gold mining/industry
золотосодержащий auriferous, goldbearing
золототысячник *bot.* centaury *(Centaurium)*
золотохворост *bot.* furze, gorse *(Ulex)*
золотоцвет *bot.* pyrethrum *(Pyrethrum)*
золотушник *bot.* toaoflax *(Linaria vulgaris)*
золотушный *pathol.* scrofulous
золоудаление ash disposal/removal/handling
золоулавливание ash collection/trapping; dust collection
золоуловитель ash/dust collector
~, **циклонный** cyclonic ash collector
золочение 1. gilding; 2. gilt
~, **гальваническое** gold plating
~ **обреза** *print.* edge gilding
золь sol, colloidal solution
~, **глинистый** clay sol
золь-каучук sol-rubber
зольник I ash bin/pit/dump/collector; *leath.* lime liquor; lime pit
~, **обжорный** white lime
~, **старый** stale lime
зольник II *bot.* common Russian thistle *(Salsola kali)*
зольность ash content
зольный ashy; cindery; high-ash *(attr.)*

зона zone, region, area; range; belt, band; section, sector; *space;* field
~, активная fuel/nuclear/fissile/reacting core *(of nuclear reactor);* core; active section
~, батиалная *geol.* bathyal zone
~, безопасная safe range
~ безопасности safety zone/area
~, береговая coastal/littoral zone
~, блестящая pellucid/clear zone, *zona pellucida*
~ боевых действий operational zone
~, болевая *med.* dolorogenic zone
~, буферная buffer zone/area
~, валентная valence band
~, валютная monetary area
~ ветров wind belt
~ видимости visibility zone/range, field of vision
~, вихревая eddy zone
~, водоохранная water conservation district/area
~ воспроизводства *(of nuclear reactor)* breeding blanket, blanket region
~ воспроизводства, внешняя/внутренняя/радиальная outer/inner/radial blanket
~ времени time zone
~ глиссады glide area
~ горения combustion zone
~, граничная boundary zone
~, двигательная motor zone
~ дезактивации decontamination area
~ действия *mil.* zone of operations
~, долларовая dollar area
~ досягаемости *mil.* fire area
~, запретная prohibited zone
~, зараженная contamination zone
~ затишья, экваториальная *meteor.* doldrums
~, климатическая climatic zone/region
~ конвергенции zone of convergence
~, круговая *anat.* orbicular zone
~, ликвационная *met.* segregation range
~ молчания, мертвая *rad.* zone of silence, dead zone; skip zone/distance; blind area
~ на диске disk extent, extent of disk
~ на магнитной ленте tape zone
~, надъядрышковая overnucleated region
~ неоднозначности *comp.* zone of ambiguity
~ нечувствительности dead zone/band, inert zone
~ обзора vision zone/area; scanned area
~ облучения (ir)radiation zone
~ обнаружения detection zone
~ обстрела field of fire
~, опасная danger zone
~ опознавания identification zone
~ освещенности illuminated zone
~ охранения protective area
~ памяти storage zone
~ памяти на сердечниках *comp.* core memory area
~ пассатов trade-wind zone, easterly belt
~ пеленгования bearing zone
~ переноса *(in text editing)* hot zone
~ перехвата interception area
~ перехода transition zone
~ по азимуту azimuth coverage
~, подпятовая *met.* gutter area
~ подхода *av.* approach area/zone
~, пожароопасная fire hazard zone
~ поиска field of search
~ полярных сияний auroral zone
~ помех interference zone
~, прибрежная coastal zone, littoral area
~ призабойная *min.* face zone
~, прикорневая *bot.* rhisosphere
~, приливная intertidal zone
~ приливов tidal zone
~, приротовая adoral zone
~ прямой видимости line-of-sight range, optical range
~, прямоточная активная straight-through core, one-pass core
~ пуска ракет firing bracket
~, пусковая launching area
~, рабочая operating zone, service area
~, равносигнальная equisignal zone
~, равнофазная equiphase zone
~ радиолокационного обзора radar surveillance zone
~ радиолокационной видимости radar coverage
~ рыболовства fishing limits
~, рыбоохранная fishery conservation zone
~ свободной торговли *econ.* free trade area
~ связи спутника satellite area of coverage
~, северная полярная north frigid zone
~, сейсмическая seismic area, belt of seismic activity
~, секретная closed security zone
~, складчатая *geol.* folded belt, zone of folding
~ слышимости audibility area
~ сплошного поражения lethal radius
~, стерлинговая *fin.* sterling area
~, сумеречная twilight zone
~ сходимости convergent zone
~, таможенная customs area
~, тарифная rate territory
~ тектонического разлома zone of fraction
~ тени shadow zone
~ турбулентности turbulent/turbulence area
~ умеренного климата temperate belt/zone
~, фаунистическая faunal/faunistic zone
~, флористичесная floristic/floral zone
~ франка *econ.* 1. free zone; 2. franc area
~, штилевая calm zone; still-air zone
~ экспресс-перевозок, тарифная express zone
~ эпицентра epicentral region
~, эротогенная *anat.* erotogenic zone
~ ячеек для данных *comp.* data cell array

зонально-апертурный zonaperturate

зонально-бороздчатый zonosulcate
зонально-поровый zoniporate
зональность zonality, zonation; zoning
~, **вертикальная** altitudinal/vertical zonality
~, **гипогенная** hypogene zoning
~ **кристаллов** zoning of crystals
~, **обратная** reverse zoning
~, **повторная** oscillatory zoning
зональный zonal, zoned; regional; zone *(attr.)*
~ **бит** *comp.* zone bit
зонд probe; sound; *geol.* sounding/earth borer; *aero.* sounding/weather balloon
~, **акустический** sound probe
~, **глазной** *med.* eye probe; lacrimal sound
~ **для внутреннего осмотра** intro-scope probe; boroscope
~ **для слезного канальца** lacrimal probe/sound
~, **желебоватый** grooved probe
~, **желудочный** gastric tube, stomach pump/sound
~, **зубной** dental explorer
~, **измерительный** measurement/measuring probe
~, **ионосферный** ionosphere sonde/sounder
~, **каротажный стандартный** normal sonde
~, **космический** space probe
~, **лунный** lunar probe
~, **маточный** uterine probe
~, **метеорологический** meteorological sounder/probe/sonde
~, **носовой** *med.* nasal probe
~, **обратный** inverse probe
~, **плазменный** plasma probe
~, **приемный** pick-up probe
~, **пуговчатый** bulbous-end probe
~, **радиометрический** radiometer probe
~ **с оливой** olive-end probe
~ **связи** *el.* coupling probe
~, **солнечный** solar probe
~, **соударяющийся** impact probe
~, **телескопический** telescope probe
~, **термоизмерительный** thermometer/pyrometer probe
~, **термоэлектрический** thermocouple probe
~, **электронно-лучевой** electron-beam probe
зондаж = зондирование
зонд-вилка fork-probe
зонд-игла probe needle
зондирование probing, sounding; exploration
~ **аорты** aortic catherizarion
~ **атмосферы** atmospheric/air sounding
~, **аэростатное** balloon sounding
~, **ветровое** winds sounding
~ **желудка** gastric intubation
~ **ионосферы** ionospheric sounding
~, **комплексное** composite sounding
~ **космического пространства** space probing/sounding
~ **мочевого пузыря** vesical catherization
~ **пищевода** (o)esophageal intubation
~, **проводить** take sounoing
~, **радиолокационное** radar sounoing
~ **сердца** cardiac catherization
~, **шаропилотное** balloon sounding
зондировать probe, sound; explore
зондировка probing, sounding
зонд-масс-спектрометр mass-spectrometer probe
зонирование zoning, zonation
~, **ограничительное/природоохранное** restrictive zoning
~, **сельскохозяйственное** rural zoning
зонированный zoned
зонт awning; hood; canopy; umbrella; *bot.* umbel
~, **вторичный** *bot.* umbellet
~ **над дверью** canopy
зонтиковидный umbelliform, umbellite; umbrella-shaped
зонтикоцветный *bot.* umbelliflorous
зонтичный *bot.* umbellate, umbelliferous
зонтовидный, зонтообразный umbraculiform; parasol-shaped
зонтоносный *bot.* umbraculiferous
зообентос *zool.* zoobenthos
зоогамета zoogamete
зоогенный zoogenic, zoogene
зоогеография zoogeography, faunology
зооглея zoogl(o)ea
зоогонидия *bot.* zoogonidium
зооеция *zool.* zooecium
зооид zooid
зоокарпий *biol.* zoocarp
зооксантелла *biol.* zooxanthella
зоолатрия *rel.* zoolatry
зоолит zoolith, zoolite
зоолог zoologist
зоологический zoological
~ **инстинкт** animal instinct
~ **парк/сад** zoological garden(s)
зоология zoology
зоолог-систематик taxonomist
зоомагазин pet shop
зоомариловый zoomaryl
зооморфизм zoomorphism
зооморфический, зооморфный zoomorphic
зооморфоз zoomorphosis
зоон *biol.* zoon
зооноз zoonosis, zoosis, zoontic disease
зоонозный zoonotic
зоопаразит zooparasite
зоопланктон zooplankton
зоопсихолог zoopsychologist
зоопсихология zoopsychology
зооспора *bot.* zoospore, planospore, swarm spore
~, **жгутиковая** flagellated zoospore
зооспорангиофор *bot.* zoosporangiophore
зоостерин *chem.* zoosterol
зоотеизм zootheism
зоотехник livestock specialist
зоотехника, зоотехния zootechny, zootechnics;

livestock breeding
зоотехнический zootechnic
зоотомия zootomy
зооферма stock-raising/-breeding farm
зоофилия zoophily, zoophilism
зоофильный zoophilous
зоофит zoophyte
зоофобный zoophobous
зоохимия zoochemistry; animal chemistry
зоохлорелла zoochlorella
зоохор *bot.* zoochore
зоохория *bot.* zoochory, animal dispersal
зоохорный *bot.* zoochoric
зооценоз zooc(o)enosis
зооцецидий zoocecidium
зооцид zoocide
зопник *bot.* Jerusalem sage *(Phlomis)*
Зороастр Zoroaster
зороастризм Zoroastrianism
зорька I *bot.* Maltese cross campion *(Lychnis chalcedonica)*
зорька II *zool.* scops owl *(Otus scops)*
зоря *bot.* garden lovage *(Levisticum officinale)*
зостера *bot.* eelgrass *(Zostera)*
зрачковый *anat.* pupillary
~ рефлекс iris contraction reflex
зрачок *anat., opt.* pupil
~, входной/выходной entrance/exit pupil
~, вяло реагирующий sluggish pupil
~, неподвижный fixed pupil
~, расширенный mydriatic pupil
~, суженный miotic pupil
зрелость ripeness; maturity
~, половая sexual maturity; puberty
~, преждевременная половая precocity, precocious puberty
зрение vision; eyesignt, sight
~, бинокулярное binocular vision
~, глубинное remote vision
~, дневное daylight vision
~, машинное computer/machine vision
~, монокулярное monocular vision
~, нормальное normal vision, emmetropia
~, пониженное depraved vision
~, стереоскопическое stereoscopic/solid vision, stereovision
~, сумеречное twilight vision
~, цветное colo(u)r vision
зрительно-двигательный visual-motor
зрительно-звуковой videosonic, visual-aural
зрительно-слуховой visual-oral, audio-oral
зрительный visual; optic; specular
~ нерв optic nerve
~ образ visual image; visualization
~ная труба telescope
зуав *hist.* Zouave
зуб tooth; cog; dent; *zool.* fang; *see also* **зубы**
~ берда *text.* reed blade, split/dent of reed
~, большой коренной molar (tooth)
~ бороны harrow tooth/point
~, грабельный rake finger/tine
~, живой vital tooth
~, замковый, замочный hinge tooth
~, искусственный artificial/false/dummy tooth
~, кариозный carious tooth
~, малый коренной premolar
~, мертвый nonvital tooth
~, молочный deciduous/milk/first tooth
~ мудрости wisdom/late tooth
~ на штифте pivot crown
~, непрорезывающийся unerupted tooth
~, пломбированный filled tooth
~, постоянный permanent/second tooth
~, полольный *agr.* weeding tooth
~, резцовой incisor tooth
~ с коронкой crowned tooth, jacket crown tooth
~, стопорный locking cog
~, храповой ratchet tooth
~, шевронный *tech.* herringbone tooth
~ шестерни gear cog/tooth
зубатка 1. bush hammer; 2. tooth ax(e); 3. *zool.* catfish
зубец tooth, cog, dent; cam; catch; tine; *el.* wave *(on ECG)*
~ вилки prong
~, двугорбый two-humped wave
~, патологический pathologic/abnormal wave
~, раздвоенный split wave
~, суженный narrow wave
~, углубленный notched/deep wave
~, уплощенный flat wave
зубило chisel; caulking iron; punch
~, канавочное grooving chisel; gouge
~, кузнечное anvil/forge chisel
~, ручное bench chisel
~, слесарное cold chisel
зубной dental; odontic
~ врач dentist
~ согласный dental consonant
зубовидный, зубообразный dentoid, dentate, ooth-shaped, dentiform, odontoid
зубовик *zool.* dentalium *(Dentalium)*
зубоврачебный кабинет dental surgery
зубодолбление gear shaping/cutting
зубозакругление gear-tooth chamfering
зуболечебница dental clinic
зуболечение dentistry; dental treatment
зубомер tooth/gear gauge
зубонарезание gear cutting
зубообкатка gear burnishing
зубоотделка gear finishing
зубопритирка gear lapping
зубопротезирование prosthetic dentistry
зуборезный gear-/tooth-cutting
зубострогальный gear-shaping/-planing
~ станок gear shaper, gear-shaping machine
зубострогание gear-shaping/-planing
зубофрезерный gear-milling *(attr.)*
зубохонингование gear honing
зубошлифование gear grinding

зубошевингование gear shaving
зубр *zool.* European bison *(Bison bonasus)*
зубрение notching, serrating
зубрить notch, serrate; indent
зубровка *bot.* sweet grass *(Hierochloe)*
~, душистая holy-grass, vanilla grass *(Hierochloe odorata)*
зубчатка I 1. rack wheel; pinion; gear; 2. indented chisel; serrated pick
зубчатка II *bot.* 1. eyebright *(Euphrasia);* 2. bartsia *(Bartsia, Odontites)*
зубчатовидный, зубчатообразный serratiform
зубчатоклювый *zool.* tooth-billed
зубчатость dentation; dentition; indentation, erration
зубчатый toothed, cogged; serrate(d), indented, otched, jagged; dentate, denticulate(d)
зубчик small tooth/cog/dent; *arch.* denticle; *bot.* crenation, leaf dent
зубы teeth; *anat.* dentition
~ вне дуги buck teeth
~ вне ряда snaggle-teeth
~, вставные artificial/false teeth
~, задние posterior teeth
~, мягкие malacotic teeth
~, передние anterior/front teeth
~, резцовые incisor teeth
~, фтористые fluoridated teeth
зубянка *bot.* toothwort *(Dentaria)*
зуд *med.* pruritis, itch
~, старческий senile pruritus
зудень *zool.* acarus
~, сырный cheese mite
~, чесоточный itch mite
зудящий *med.* pruritic, itching
зуек *zool.* plover *(Charadrius)*
зулус, зулусский Zulu
зуммер buzzer, hunmer; signal, tone; howler
~ занятости busy tone
~, микрофонный microphone hummer
~ ответа станции dial tone
зумпф *min.* sump; draining pit
зыбкость, зыбление vacillation, fluctuation; unsteadiness
зыбучий unstable, unsteady; quaking
~ие пески quicksands
зыбь rippling, ripple; *naut.* swell, aftertossing, surge, long rolling sea
~, беспорядочная confused swell
~, большая rather rough swell
~, крупная rough swell
~, крутая hollow swell
~, мертвая ground swell, aftertossing
~, неправильная abnormal swell
~, очень большая very heavy swell, rough swell
~, сильная heavy swell
зюзник *bot.* buglewood *(Lycopus)*
зюйд south; *fig.* southerly wind
зюйд-вест southwest, southwestward
зюйд-ост southeast; *fig.* southeaster wind
зюйд-тень-вест *naut.* South by West
зюйд-тень-ост *naut.* South by East
зяблик *zool.* chaufinch *(Fringilla coelebs)*
зябра *bot.* bee nettle *(Galeopsis speciosa)*
зябь autumn ploughing/plowing; land ploughed/plowed in autumn

И, и

иберииец, иберийский Iberian
иберийка *bot.* candytuft *(Iberis)*
ива *bot.* willow *(Salix)*
~, корзиночная osier *(Salix viminalis)*
~, плакучая weeping willow *(Salix babylonica)*
Иван-Купала St. John Baptist's day
иван-да-марья *bot.* 1. blue cowwheat *(Melampyrum nemorosum);* 2. heart's ease *(Viola tricolor)*
Иванов день Midsummer Day
Иванов хлеб *bot.* carob tree *(Seratonia siliqua)*
иван-чай *bot.* willow herb, bay willow; rosebay *(Chamaenerium angustifolium)*
иваси *zool.* iwashi, ivasi *(Sardinops)*
ива-шелюга *bot.* pussy willow *(Salix acutifolia)*
ивина *bot.* willow tree
ивишень *bot.* plum mushroom *(Clitopilus prunulus)*
ивовые *pl., bot.* willow family *(Salicaceae)*
иволга *zool.* oriole *(Oriolus)*
иврит Ivrit, Modern Hebrew
игла needle; stylus; pin; spike
~, воспроизводящая playback/reproducing stylus/needle
~, вязальная knitting/fillet needle/pin
~, гравировальная etching/engraving needle
~, двухголовочная/двухизнаночная double-headed needle
~, дисцизионная *med.* discission needle
~, длиннорылая *zool.* broad-nosed pipefish *(Syngnathus typhle)*
~ для взятия крови blood drawing needle
~ для переливания крови blood transfusion needle
~ для подкожных вливаний hypodermic needle
~ для спинномозговой пункции spinal needle
~, жаккардная вязальная Jacquard needle
~, звукозаписывающая recording stylus
~ звукоснимателя phono stylus, stylus/needle of a phonograph pick-up
~, изогнутая хирургическая surgeon's circle needle
~, инъекционная syringe needle
~, кеттельная вязальная looping needle
~, копировальная pricker
~, лигатурная *med.* ligature needle

~, морская *zool.* pipefish *(Syngnathus)*
~, носовая *av., naut.* nose spike
~, обкалывающая хирургическая abscission needle
~, ощупывающая sensing needle
~, полуизогнутая хирургическая surgeon's half-curved needle
~ поплавкового клапана float spindle/needle
~, препарировальная preparation needle
~, пробирная test needle
~, пункционная aspirating needle
~ распылителя injector nozzle needle
~ с ниткой для непрерывного шва *surg.* swagged needle
~ с предохранительной бусинкой bulb-guard needle
~, сбавочная вязальная narrowing needle
~, ушковая вязальная guide needle
~, хвойная acerose leaf
~, хирургическая suture/surgeon's needle
игла-вилка fork needle
игла-копье arrow-shaped blade
игла-рыба *zool.* pipefish *(Syngnathus)*
игла-скарификатор *surg.* blood lancet
иглица *bot.* butcher's broom *(Ruscus)*
иглобрюх *zool.* toadfish *(Tetraodon)*
игловидный acicular, acerate, acerous; spiniform; needle-shaped
иглодержатель *med.* needle-holder/carrier/forceps
иглокожие *pl., zool.* echinoderms *(Echinodermata)*
иглокожий *zool.* erinaceous
иглолистный *bot.* needle-leaved; coniferous
иглотерапия, иглоукалывание acupuncture, stylostixis
иглофильтр well-point
иглошерсть *zool.* porcupine *(Erethizon)*
иглу igloo
игнация *bot.* St.-Ignatius poison nut *(Strychnos ignati)*
игнипунктура *surg.* ignipuncture
игнитрон ignitron
иголка needle; spicule; spike
игольчатый acicular, aciculate, acerose, aciform; spiny
~ клапан needle valve
~ кристалл acicular crystal
~тая руда needle ore, aikinite
игра game; play, playing; *tech.* play, looseness, slack; backlash; clearance, freedom; *math.* game
~, азартная game of chance/luck; gambling
~, антагонистическая two-person zero-sum game
~ без ограничений free/unconstrained game
~, бескоалиционная non-cooperative game
~, бесконечная infinite game
~, боковая *tech.* side play
~ в нормальной форме game in normalized form
~ в кегли/шары bowling
~ валков backlash of rolls
~, военная war game
~, деловая business game
~, дифференциальная differential game
~, конечная finite game
~, кооперативная cooperative/coalition game
~, матричная matrix/rectangular game
~, минорантная minorant game
~ на бирже stock-market game
~ на истощение attrition game
~ на повышение *econ.* dealing(s) for a rise; bull speculation
~ на понижение *econ.* dealing(s) for a fall; bear speculation
~ не имеющая решения no-solution game
~, нормированная normalized game
~, осевая end/axial play
~, повторная iterated game
~, позиционная position(al) game
~ поршня *tech.* piston play
~, продольная axial clearance
~, прямоугольная matrix/rectangular game
~, рекурсивная game with feedback
~ с выбором момента времени game of timing
~ с ненулевой суммой [с ненулевым исходом] non-zero sun game
~ с нулевой суммой [с нулевым исходом] zero-sum game
~ с полной информацией perfect information game
~, сетевая network game
~, усеченная truncated game
~, честная fair play
~, экспериментальная game test
играть play
~ на бирже speculate on the Stock Exchange
~ на повышение deal for a rise; sell a bull
~ на понижение deal for a fall; sell a bear
игрек *(the letter)* 'y'
игрунка *zool.* marmoset *(Hapale, Callithrix)*
игрунок *zool.* tamarin *(Leontocebus)*
игуана *zool.* iguana *(Iguana)*
игуанодон *paleont.* iguanodon *(Iguandon)*
игумен *eccl.* Father Superior, hegumen
игуменья *eccl.* Mother Superior, hegumeness
ид *biol.* id
идант *biol.* idant
идеал *math.* ideal; nonpareil
~, главный *math.* principal ideal
~, двусторонний bilateral [two-sided] ideal
~, левый/правый left/right ideal
идеалистический, идеалистичный idealist, idealistic(al)
идейно-политический ideological and political
идейность 1. ideological content; ideology; 2. integrity, adherence to principle, moral substance/principle
идейный 1. ideological; 2. idealistic; of high moral/ideological standard, high-principled

идемпотент, идемпотентный *math.* idempotent
идемпотентность *math.* idempotency
идемфактор *math.* idemfactor
идентизм *philos.* identism
идентификатор identifier; descriptor
 ~ данных *comp.* data descriptor
 ~ записи данных data record name
 ~ класса class identification number
 ~ набора данных data-set identifier
 ~ описания файла file definition name
 ~ пользователя userid
идентификационный identification *(attr.)*
идентификация identification
идентифицировать identify; determine
идентифицируемость *stat.* identifiability
идентифицируемый identifiable
идентичность identity
идентичный identical
идеограмма ideogram
идеография ideographic writing, ideography
идеолог ideologist
идеомоторный ideomotor *(attr.)*
идеорея *med.* ideorrh(o)ea
идеофрения *med.* ideophrenia
идея idea; notion; concept
 ~, бредовая delusion
 ~, навязчивая obsession, fixed idea, monomania
идиллия *lit.* idyl(l)
идиоагглютинин idioagglutinin
идиогамия idiogamy, autogamy, self-pollination
идиогенетика idiogenetics
идиогенный *geol.* idiogenous
идиогеосинклиналь idiogeosyncline
идиогетеролизин idioheterolysin
идиоглоссия idioglossia
идиограмма idiogram, karyogram
идиодиахрония diachrony
идиоизоагглютинин idioisoagglutinin
идиоизолизин idioisolysin
идиом, идиома, идиоматизм idiom
идиоматика *ling.* idiomatics; *coll.* idioms, idiomatic expressions
идиоматический, идиоматичный idiomatic(al)
идиоморфизм *cryst.* idiomorphism
идиоморфно-гранулярный idiomorphic/automorphic-granular
идиопатический *med.* idiopathic, self-existing; essential
идиоплазма idioplasm
идиосинкразия idiosyncrasy, idiocrasy; allergy
 ~, врожденная congenital idiosyncrasy
 ~ к лекарственным веществам drug idiosyncrasy
 ~ к пищевым веществам food idiosyncrasy
идиосинкратический idiosyncratic, idiocratic
идиосинхрония synchrony
идиосома *biol.* idiosome
идиот *pathol.* idiot
 ~, монголоидный mongolian idiot
 ~, обезьяноподобный pithecoid idiot
идиотизм *pathol.* idiocy
идиотип idiotype
идиотический *pathol.* idiotic
идиотия *pathol.* idiocy
 ~, амавротическая семейная familial amaurotic idiocy
 ~, врожденная congenital idiocy
 ~, гемиплегическая hemiplegic idiocy
 ~, гидроцефалическая hydrocephalic idiocy
 ~, кретиноидная cretinoid idiocy, cretinism
 ~, монголоидная mongolian/mongoloid idiocy, mongolism
 ~, параплегическая paraplegic idiocy
 ~, сенсорная sensorial idiocy
 ~, эклamптическая eclamptic idiocy
 ~, эпилептическая epileptic idiocy
идиофанный *cryst.* idiophanous
идиохроматизм idiochromatism
идиохромосома idiochromosome
идиохрония idiochrony
идиш *ling.* Yddish
идоза idose
идолопоклонник idolator, image worshipper
идолопоклонничать idolize; idolatrize
идолопоклоннический idolatrous
идолопоклонничество, идолопоклонство idolatry
идумей *bibl., hist.* Edomite
идуроновый iduronic
Иегова *rel.* Jehovah
иеговисты *pl.* Jehovah's Witnesses
иезуит *eccl.* Jesuit
иезуитизм Jesuitism
иезуитский *eccl.* Jesuit, Jesuist
иезуитство *eccl.* Jesuitism, Jesuitry
иена *fin.* yen
иенит *miner.* yenite, ilvaite
иерарх hierarch
иерархический hierarchic(al)
иерархия hierarchy
иератический hieratic
иерей priest
иерейский priestly
иерейство priesthood
иеремиада jeremiad, lamentation
иерихонская труба trumpet of Jericho
иероглиф hieroglyph
иероглифика ideographic writing
иероглифический hieroglyphic(al)
иеродиакон hierodeacon
иеромонах hieromonach; celibate priest
иерусалимский Jerusalem *(attr.)*
иждивенец, иждивенка dependent
иждивение maintenance; support; means, funds
иждивенство support, keeping; dependence
иждивенчество dependence
изаллобара isallobar
изаллогипс isallohypse
изаллотерма isallotherm

изадрин *chem.* neoepinephrine
изанемона isanemone
изаномала isanomal, isanomalous line
изатин *chem.* isatin
изафет *ling.* postpositional attributive group
изба izba, peasant's log hut
избиение 1. *law* assault and battery; 2. slaughter, massacre; beating up, drubbing
~, младенцев *bibl.* Massacre of the Innocents
избежание avoidance; escape
избирание selection, selecting
избиратель 1. *tech.* selector; 2. *polit.* voter, elector
избирательно selectively
избирательность selectance, selectivity; discrimination
~, амплитудная amplitude selectivity
~ катализатора selective action of a catalyst
~ по направлению directional selectivity
~ по частоте frequency selectivity
~ радиоприемника selectivity of a receiver
~ радиоприемника по частоте frequency selectivity
~ фильтра discrimination of a filter
избирательный 1. *tech.* selective; 2. *polit.* electoral; elective
~ бюллетень voting/ballot paper
~ округ constituency, electoral district
~ список register of voters
~ участок polling district/station
~ ценз voting qualifications
~ное право suffrage; franchise
изборождение striation, furrow
~, ледниковое glacial striae/scratches
избранник elect, favoured/chosen one
~, народный chosen representative of the people
избыток excess, surplus; redundancy; spillover
~ спроса *econ.* demand surplus
~ плавучести reserve buoyancy
избыточно-возвратный redundant-reflexive
избыточность excessiveness, excess; redundancy, redundance
~, внутренняя *comp.* built-in redundancy
~, временная time duplication
~ кода code redundancy
~ схемы кристалла on-chip redundancy
избыточный redundant, excessive, surplus; superfluous
извержение eruption
известелюбивый *bot.* gypsophilous, lime-loving
известеразбрасыватель *agr.* lime-spreader
известие news; information; *pl.* proceedings, transactions
известитель *tech.* indicator; signalling device/apparatus; signal
~, маршрутный route indicator
~, пожарный fire-alarm box; break-glass station; street fire-alarm box
известка (slaked) lime
известкование lime (pre)treatment; liming, calcification; lime application; chalking
известковать lime, chalk
известково-гипсовый limestone-gypseous
известково-глинистый calceroargillaceous
известково-калийный lime-potash
известково-натриевый calcosodic
известково-роговой corneocalcareous
известково-силикатный calcium-silicate
известковость calcareousness; lime content
известково-удобрительный *agr.* lime-and-fertilizer
известково-щелочный calc-alkali
известковый calcigerous, calcic, limy; calcareous; calciferous
~ раствор slaked lime
известколюб *bot.* calciphile, calciphyte
известконосный calciferous
известное *math.* the known
известняк limestone, chalkstone
~, белый итальянский freshwater limestone, banded marble, travertine
~, битуминозный/вонючий bituminous limestone, stink stone, swinestone
~, глинистый cement stone, clayey limestone
~, горькоземистый magnesian limestone
~, доломитизированный dolomitic limestone
~, железистый ferruginous limestone
~, каменноугольный carboniferous limestone
~, коралловый coralline/coral limestone
~, кремнистый siliceous/cherty limestone, cornstone
~, литографский lithographic limestone
~, луговой limnetic/meadow limestone
~, магнезиальный magnesian limestüne
~, метаморфизованный crystalline limestone
~, окремненный metachert
~, оруденелый mineralized limestone
~, пенистый foamy/pumaceous limestone
~, песчанистый arenaceous limestone
~, почковатый amygdaloidal limestone
~, ракушечный shell limestone, muschelkalk
~, сланцевый shaly/sparry limestone
~, туфовидный kunkur, kunkar
~, черный кремнистый black-and-gold marble
известняковый, известчатый calcareous
известняк-ракушечник shell limestone
известь (quick) lime; calx
~, белильная bleaching powder, chlorinated lime, hypochloride/chloride of lime
~, гашеная slaked lime
~, гидравлическая hydraulic lime
~, гидратная hydrated lime
~, едкая caustic lime
~, жирная fat/rich lime
~, натровая/натронная soda lime
~, негашеная quicklime
~, побелочная whitewash
~, связанная combined lime
~, тощая lean/meagre lime
~, хлорная = ~, белильная

известь-кипелка (lump) quicklime
известь-пушонка powdered lime
извещатель signalling device; alarm installation/device; detector
~, **пожарный** fire alarm/detector
~, **тепловой** heat detector
извещение notice, notification, message, communication; announcement; *comm.* advice
~, **заблаговременное** advance notification
~, **инкассовое** *fin.* advice of collection
~ **о выкупе** *econ.* redemption notice
~ **о наличии** availability notice
~ **о платеже** advice of payment
~ **об отгрузке** shipping advice
~ **об отправке груза** shipping notice/advice
~, **письменное** notification in writing
~, **сигнальное** *tech.* signal indication
извилина sinuosity, curvature, winding; twist; *anat.* gyrus, convolution; *bot.* gyroma, helicoid cyme
~, **глазничная** orbital gyrus
~, **затылочновисочная** occipitotemporal gyrus
~, **зубчатая** dentate gyrus
~ **мозга** convolution of the brain
~, **надкраевая** supramarginal gyrus/convolution
~, **околоконечная** paraterminal gyrus
~ **пояса** convolution of the cingulum, callosal convolution
~, **прямая** gyrus rectus
~, **связочковая** fasciolar gyrus
~, **язычная** gyrus lingualis
извитость tortuosity, twist(ing), sinuation; *(fibre)* crimp, crimpiness
извлекать 1. extract; withdraw, draw off; remove; take out; 2. recover; isolate; derive; retrieve
~ **корень из** *math.* take/find the root of
извлечение 1. extraction, retrieval, recovery; 2. withdrawal, taking out; 3. excerpt, extract, abstract; summary
~ **влаги** dehydration
~ **данных** data extraction
~ **кислотой** acid recovery
~ **корня** *math.* extraction of a root, taking/finding the root of, evolution
~ **корня, квадратного** finding the square root of; square-rooting
~ **корня, кубического** cube-root extraction
~, **поглотительное** absorption extraction
~ **урана** uranium extraction
извращать 1. distort, misrepresent; 2. pervert
извращение 1. distortion; 2. perversion
~ **половое** sexual perversion
изгарь *met.* slag, dross, scoria
изгиб bend(ing), twist(ing), flexing, flexure, curvature, winding, crenulation, kink
~ **в упругой области** elastic bending
~ **внутрь** inflection
~ **земной коры** earth curvature
~, **знакопеременный** alternating bending
~ **крыла** wing bending/flexure
~, **мостовой** pontine flexure
~ **назад** dorsiflexion
~, **обратный** contraflexure, counter-camber
~ **пласта** seam contortion; bending of stratum
~, **поясничный** *anat.* lumbar flexion
~, **продольный** buckling
~, **синклинальный** synclinal flexure
~, **сложный** compound/complex flexure
~, **спинной** *anat.* dorsal flexure
~, **упругий** elastic flexure
~, **флексурный** monoclinal fold, stepfold
~, **шейный** *anat.* cervical flexure
изгибание bending; curving; deflection; deformation; inflection
изгибать bend; curve; deflect; flex
изгибный flexural
изглашение phonation
изгнание expulsion; exile
изгородь fence; enclosure; hedge
~, **живая** green hedge, hedgerow, hedge
~, **снегозащитная** snow-barrier
изготавливать manufacture, produce, make; fabricate
изготовитель manufacturer, producer, maker
~ **комплектного оборудования** original equipment manifacturer, OEM
изготовление manufacture, production, making, preparing; fabrication
~, **групповое** batch fabrication/manufacture
изготовленный manufactured, made, produced
~ **вручную** hand-made
~, **заранее** prefabricated; ready-made
~ **по заказу** custom-made/-built
издание 1. publication; *law* promulgation; 2. edition, issue
изделие item, article; product
~ **высшего качества** superior product
~, **готовое** finished/final/manufactured product; ready-made product
~, **забракованное** rejected/condemned product
~ **изготовленное по заказу** made-to-order product, custom-made/-build product
~, **кондиционное** *econ.* conforming/satisfactory product
~, **кузнечное** forging, forged piece/product
~, **кустарное** handicraft article/product
~, **литейное** foundry product
~ **многократного применения** multiple-use product
~, **недоработанное** incompletely engineered product
~, **некондиционное/нестандартное** nonconforming [off-standard, substandard] product
~, **неремонтируемое/неремонтнопригодное** unmaintainable item, pluck-out item
~, **обрабатываемое** *tech.* workpiece
~ **однократного применения** single-use product
~, **патентованное** proprietary product, restric-

tive proprietary item
~, полуобработанное semi-manufactured product, semi-finished piece
~, программное software product
~, прокатное rolled product
~, промежуточное semi-finished product
~, ремонтируемое maintainable item
~, серийное production-line item; item of serial manufacture; stock-produced item
~, унифицированное standard(ized) product
~, формованное moulding
~, чулочное hosiery
~ия, гончарные pottery, earthenware
~ия, железоскобяные hardware
~ия, канатные cordage
~ия, комплектующие component parts; accessories
~ия, металлические hardware
~ия, пресованные presswork
~ия, промышленные manufactured/industrial goods
~ия, скобяные locksmith's work; ironmongery
издержки cost(s), expense(s), expenditure(s); overheads; outlay(s)
~, валовые total costs
~, внерыночные extramarket costs
~ воспроизводства expenses of reproduction
~ вследствие простоя downtime/idletime costs; costs of idleness
~ вследствие рекламаций complaint cost(s)/expenditure(s)
~ вследствие ухудшения характеристик deterioration cost(s)
~, дополнительные incremental cost(s)
~ замещения выбытия replacement costs
~ истца *law* sue charges
~ конкуренции waste/cost of competition
~, конторские office expenses
~ материального характера tangible costs
~ на амортизацию depreciation expenses
~ на брак spoilage cost(s)
~ на единицу продукции *econ.* unit costs
~ на исправление брака rework/reoperation costs
~ на конструкторские работы design cost(s)
~ на материалы *econ.* material cost(s)
~ на моральный износ obsolescence cost(s)
~ на разработку development cost(s)
~ на техническое обслуживание и текущий ремонт upkeep/maintenance cost(s)
~ на топливо fuel cost(s)
~ наличными деньгами out-of-pocket cost(s)
~, невозвратные *econ.* sunk cost(s)
~, непериодические *econ.* nonrecurring cost(s)
~, непроизводственные nonmanufacturing cost(s)
~, нормативные standard cost(s)
~ обращения marketing/distribution cost(s); expenses of circulation
~, общезаводские overall/joint manufacturing cost(s)
~, периодические recurring cost(s)
~, плановые *econ.* planned cost(s)
~ по общей аварии average disbursement
~ подготовки производства preproduction cost(s), starting-load cost(s)
~, постоянные fixed cost(s), standing expenses
~, предельные marginal costs
~, представительские expenses of representation
~, приростные incremental cost(s)
~ производства expenses/cost(s) of production, operating/operation/operational/production/working cost(s)
~ производства в натуральном исчислении *econ.* real costs
~ производства, действительные actual cost price
~, производственные process cost(s)
~ производства в промышленности industrial costs
~, путевые travel(l)ing expenses
~, резко возрастающие *econ.* spiralling cost(s)
~, рекламные и торговые promotion and sales expenses
~ сбыта marketing cost(s)
~ связанные с пуском предприятия start-up cost(s)
~, совокупные aggregate cost(s)
~, сократимые *econ.* escapable cost(s)
~, сметные *econ.* budgeted cost(s)
~, среднеотраслевые *econ.* industry average costs
~, судебные legal costs; law expenses
~, суммарные aggregate cost(s)
~, торговые marketing/selling expenses, marketing cost(s)
~, транспортные shipping cost(s), cost(s) of transportation
~, удельные unit costs
~, удельные профакторные unit factor cost(s)
~, устранимые *econ.* escapable cost(s)
~ хранения cost(s) of storage
издольное хозяйство share farming
издольщик sharecropper, sharecrop tenant; metayer *(Fr.)*
издольщина sharecropping
изеикония *med.* iseiconia
иземический izemic
изентропический, изентропный isentropic
изерин *miner.* iserine
изерит *miner.* iserite
изжога *med.* heartburn, pyrosis, (water) brash
Изида *myth.* Isis
изидиевидный, изидиеобразный *bot.* isidioid, isidiose
изидиеносный *bot.* isidiiferous
изидиозный *bot.* isidiose
излечение cure, medical treatment; healing pro-

cess, recovery, recuperation
~, на under medical treatment
излечимость curability
излечимый curable, healable; benign
изливаться flow/pour out *(of),* issue, effuse, erupt *(from)*
излишек excess; surplus; remainder
~ веса overweight
~ имущества excess stock
~шки, товарные marketable surpluses
~ цемента excess of cement; excessive cement
излияние outflow, outpouring, discharge, issue
~ поверхностное *geol.* superfluent/superficial lava flow
~, трещинное fissure eruption
изложница (casting/ingot) mould
~, бутылочная bottle-top mould
~, глуходонная closed-bottom mould
~ заполняемая сверху/снизу top-/bottom-poured ingot mould
~, листовая slab mould
~, сквозная open-bottom mould
излом 1. fracture, break; 2. bend; *bot., anat.* geniculum
~, душевный mental unbalance
~ дымохода flue bend
~, зернистый granular fracture
~, нафталинистый silky fracture
~, оскольчатый chipping fracture
~, полураковистый subconchoidal fracture, tension fracture
~ при растяжении fracture in tension
~ при сжатии fraction in compression, compression fracture
~ при срезе fracture in shear, shear fracture
~, раковистый conchoidal fracture
~, сланцеватый slaty fracture
~, слоистый foliated fracture
~, усталостный fatigue fracture
~, шиферный fish-scale fracture
~, хрупкий brittle fracture
излучаемость emissive/radiating power, emissivity
излучатель radiator, emitter, radiating element/source
~, абсолютный perfect/complete/total radiator
~, акустический acoustic/sound source
~ альфа-частиц alpha-particle radiator; alpha emitter
~, волноводный waveguide radiator
~, звуковой acoustic radiator
~ инфракрасный infra-red source
~, магнитострикционный magnetostrictive source
~, многоэлементный multioutlet source
~, направленный directional source
~, ненаправленный nondirectional source; omnidirectional radiator
~ позитронов positron radiator
~, полуволновой half-wave radiator
~, пьезоэлектрический piezoelectric source
~ радиоволн, рамочный loop radiator
~ радиоволн, рупорный horn radiator
~ радиоволн, щелевой slot radiator
~ резонансный resonant source
~, точечный point radiator
~ фотонов photon emitter
~, черный black radiator
~, ультразвуковой ultrasonic vibrator; ultrasonic transducer
~, элементарный radiating element; elementary radiator
излучательный radiating, emitting; emissive
~ная способность radiation capacity/power
излучать radiate, emit, emanate, ray
~ вновь/обратно/повторно re-radiate
излучаться radiate
излучающий radiative, radiating, emitting; radiant
излучение radiation, emission, emanation; rays
~ абсолютно черного тела black-body radiation
~, альфа/бета/гамма alpha-/beta-/gamma-radiation
~, актиничное actinic radiation/rays
~, аннигиляционное annihilation radiation/gammas
~, безопасное nonhazardous emission/radiation
~, бетатронное betatron emission/radiation
~, ближнее инфракрасное near infrared radiation
~, видимое visible radiation/light
~, внеземное extraterrestrial radiation
~, внешнее/внутреннее тормозное outer/inner bremsstrahlung
~, восходящее upwelling radiation
~, всенаправленное omnidirectional radiation
~, вторичное secondary radiation
~, высокоинтенсивное high-intensity radiation
~ высокой энергии high-energy radiation
~, гетерогенное heterogeneous radiation
~ гидролокатора sonar transmission
~, гомогенное homogeneous radiation
~, далекое инфракрасное far infrared radiation
~, дипольное dipole radiation
~, длинноволновое long-wave radiation
~, жесткое hard [high-energy] radiation
~, запаздывающее delayed radiation/emission
~ заряженных частиц charged-particle radiation
~, захватное capture radiation
~, звездное stellar radiation
~, земное terrestrial/earth radiation
~, импульсное pulse(d) radiation
~, интегральное total radiation
~, интенсивное strong/intense radiation
~, инфракрасное ближнее/далекое near/far infra-red radiation
~, ионизирующее ionizing radiation
~, испускать emit radiation
~, исходящее emergent radiation

~, квадрупольное quadrupole radiation
~ кванта quantum emission
~, когерентное coherent radiation/emission
~, коротковолновое short-wave radiation
~, корпускулярное corpuscular/particle/particulate radiation/emission
~, космическое cosmic rays/radiation
~, краевое *rad.* fringe radiation
~ лазера laser radiation/emission/light
~, магнитно-тормозное cyclotron radiation
~ мазера maser radiation/emission
~ малой интенсивности low-level/-intensity radiation
~ малой энергии low-energy radiation
~, мгновенное prompt radiation/emission
~, митогенетическое mitogenetic radiation
~, монохроматическое/моноэнергетическое monochromatic/monoenergetic radiation
~, мультипольное multipole radiation
~, мягкое soft radiation/rays; low-energy radiation
~ накачки *(of laser)* pumping radiation/light
~, направленное beamed/directional radiation
~, нейтронное neutron radiation
~, некогерентное noncoherent/incoherent radiation
~, немонохроматическое heterogeneous/polyenergetic/polychromatic radiation
~, ненаправленное nondirectional/omnidirectional radiation
~, непрерывное continuous radiation/emission
~, непрерывное тормозное bremsstrahlung continuum
~, обратное backscatter radiation; reradiation
~, одномодовое unimode radiation
~, одночастотное monofrequency [single-frequency] radiation
~ остановленного реактора residual radiation
~, отраженное reflected radiation; back-scattered radiation
~, падающее incident/incoming radiation
~, паразитное spurious/stray radiation
~ поверхности surface emittance
~, позитронное positron emission/radiation
~, проникающее penetrating radiation
~, проходящее transmitted radiation
~, прямое direct radiation
~, прямонаправленное head-on radiation
~, радиоактивное radioactive radiation
~, радиочастотное radio-frequency radiation
~, рассеянное scattered radiation
~ реактора pile radiation
~, резонансное resonance radiation
~, рентгеновское roentgen radiation, X-radiation, X-rays
~ с непрерывным спектром continuous radiation
~, световое luminous radiation
~ серого тела gray-body radiation
~, синхротронное synchrotron radiation
~, слабопроникающее low-penetrating radiation
~, собственное self-radiation
~, солнечное solar radiation
~, слабое low-level radiation, weak radiation
~, стационарное steady-state radiation
~, стратосферное stratosphere radiation
~, суммарное total/integrated radiaton
~, температурное/тепловое thermal/heat/caloric radiation; caloradiance
~, тормозное braking radiation, bremsstrahlung, slowing-down radiation
~, тормозное непрерывное bremsstrahlung continuum
~, тропосферное tropospheric radiation
~, смешанное complex/mixed radiation
~, ударное impact/collision radiation
~, ультрафиолетовое ближнее/далекое near/far ultra-violet radiation/rays
~ урана uranium rays
~, уходящее outgoing radiation
~, фоновое background radiation
~ черного тела black-body radiation
~ электронов electron emission/radiation

излучина bend, curve, bight
~, речная meander

измаэлит *bibl.* Ismaelite

измельчаемость grindability, comminutability

измельчание grinding, breaking up, crushing; growing/becoming small; dwarfing, nanism; *fig.* degeneration

измельчать grind, crush, mill, crumble up, cut very small; comminute
~ в порошок pulverize, powder
~ в пыль pulverize
~ истиранием grind by attrition

измельчение grinding, milling, crushing, breaking up; comminution
~, тонкое fine grinding
~, тончайшее atomization

измельчатель shredder, mincer, chopper, crusher, grinder, pulverizer, atomizer; macerator

измена treason, treachery; betrayal; apostasy
~, государственная high treason
~, супружеская conjugal infidelity

изменение change, variation; alteration, modification; mutation; transformation
~ адреса address change
~, бесконечно малое *math.* infinitesimal change
~ в соответствии с новыми данными updating
~ в характере спроса *econ.* shift in demand
~, вековое secular variation
~ во времени variation with/in time, time variation/change, temporal variation
~, закономерное *econ.* regular variation
~ записи *comp.* (data) record change
~ комплекта оборудования configuration

change
~, конструктивное structural modification/change
~ маршрута route diversion
~ масштаба rescaling; scale change
~, объемное volumetric change
~ отношений reproportioning
~, периодическое cyclic variation
~ по ключу key change
~, прерывистое/прерывное discontinuous variation
~ пропорции *math.* shift in proportions; *math.* discontinuous change
~, ступенчатое step-like change; stepped change
~ расположения relocation
~, резкое abrupt/sudden change
~, скачкообразное step/sudden change;
~, случайное random change
~ теплосодержания enthalpy change/variation
~ тяги thrust variation
~ формата reformatting
~ шага *mech.* pitch change
изменчивость variability, changeability; mutability; variation
~, потенциальная potential/cryptic/concealed variability
изменяемость changeability, alternability, variability; mutability
~ видов transmutation of species
изменяемый changeable, alterable, variable, mutable; *gram.* inflected; inflectional
~мая величина *math.* variable
изменять change, alter; vary; modify
~ в соответствии с новыми данными bring up to date, update
изменяться vary, change; fluctuate
измерение 1. measurement, measuring, metering; sizing; gauging; *(of temperature)* taking; 2. dimension
~ в натуре, ~, натурное actual measurement; full-scale measurement
~ влажности psychrometry
~ времени chronometry; time-keeping
~ высот altimetry, height/altitude measurement/finding; hypsometry
~ высоты Солнца Sun sight
~ глубин depth measurement, sounding; *naut.* bathymetry
~ глубины эхолотом echo/sonic sounding
~, дистанционное remote measurement, telemetering; telemetry
~ дальности range/distance measurement, ranging
~ дальности, акустическое sound ranging
~ дальности, гидролокатором echo ranging
~, дозиметрическое health measurement
~ мутности turbidimetry
~ на борту on-board measurement
~ плотности density measurement; densitometry
~ по мостовой схеме bridge measurement
~ по точкам point-by-point measurement
~ поглощающей способности absorptiometry
~ осредненное по времени time average measurement
~ с помощью инерциальной системы inertial measurement
~ содержания соли halometry
~, термоанемометрические hot-wire measurement
~ толщины thickness gauging
~, точное precision measurement
~ угла места elevation measurement
~ частоты frequency measurement
~ щелочности alkalimetry
~ия, антропометрические body measurements
~ия, равноточные equal measurements
~ия, сцинтилляционные scintillation counting
измеримость measurability
измеримый measarable; subject to measurement
~, количественно quantifiable
измеритель meter, gauge, measuring instrument/device; measurer; *econ.* index
~ активности *radiol.* radiation meter
~ вибрации vibration meter, vibrometer
~ видимости visibility meter, videometer
~ высоты облаков ceilometer, cloud-base measuring instrument
~ выхода output meter
~ вязкости viscosimeter, viscometer
~ деформации strain gauge
~ добротности quality-factor meter, Q-meter
~, доплеровский Doppler sensor
~ емкости capacitance meter, capacitometer
~ жесткости rigidimeter
~ жесткости лучей radiation energy meter
~ импеданса impedometer
~ индуктивности inductance meter
~ интенсивности гаммаизлучения gamma-ray meter
~ интенсивности излучения radiation rate meter
~ интенсивности космического излучения cosmic-ray meter
~ испарения evaporimeter
~ кровотока *med.* blood flowmeter
~ крутящего момента torquemeter
~ коэффициента мощности power-factor meter
~ линейной деформации expansion tester/meter
~ мощности дозы радиации dose-rate meter
~ освещенности luxmeter
~ отрицательных ускорений decelerometer
~ перегрузки overload meter
~ плотности density meter, densitometer
~ поглощающей способности absorptiometer
~ поля зрения *med.* campimeter
~ прилива tide-gauge
~ расхода *tech.* flow meter

~ **силы звука** acoustimeter
~ **скорости** speed/velocity meter, speedometer
~ **скорости течения** current meter
~ **скорости потока** flow meter, rate-of-flow meter
~ **угла сноса** drift meter/indicator
~ **угла атаки** *phys.* incidence meter
~ **уклона** gradient indicator
~ **уровня** level gauge; depth gauge
~ **уровня масла** oil-depth gauge
~ **ускорения** accelerometer
измерительный measuring, metering, gauging
измерять measure, meter, gauge, rate, size; *geod.* survey
~ **глубину** take soundings, sound, plumb, fathom
измолот yield of threshed grain
изнанка inverse, inside; the wrong side
изначальный primordial, elemental, original
изнашиваемость wearing strength/capacity, wearability; depreciation
изнашивание wear, wear and tear; depreciation; erosion; deterioration
~, **усталостное** fatigue wear
~, **эрозионное** erosive wear
изнашиваться wear (out)
износ wear, wear and tear; attrition; abrasion
~ **денег** wearing-out of money
~ **истиранием** attrition wear
~, **моральный** *tech.* obsolescence, functional/technological depreciation
~, **начисленный** depreciation expenses
~, **планируемый** *econ.* planned depreciation
~, **полный** wear-out
~, **эксплуатационный** service wear
износостойкий, износоупорный, износоустойчивый wear-resistant; wearproof
износостойкость, износоупорность, износоустойчивость wear resistance
изнурение emaciation; exhaustion; wasting
изоагглютинация isoagglutination
изоагглютинин isoagglutinin
изоагглютиноген isoagglutinogen
изоазимут(а) isoazimuth
изоаллели *pl.* isoalleles
изоамиловый *chem.* isoamyl
изоанабаза isoanabase
изоантиген isoantigen
изоантитело isoantibody
изоатмический isoatmic
изобаза isobase
изобар *phys.* isobar
изобара *meteor.* isobar
~ **нормального/среднего давления** mesobar
изобата isobath, isobathic line/curve, depth/bathymetric contour, hydroisohypse
изобатитерма isobathytherm
изображать picture, depict, portray; represent; express; show
изображение 1. image, picture; figure; representation, portreal; display; pattern; 2. *math.* transform
~, **аксонометрическое** axonometric drawing
~, **внеосевое** off-axis image
~, **графическое** graphic representation, graph, diagram
~, **динамическое** *comp.* animated image
~, **зеркальное** mirror image
~ **знака** *comp.* character image
~, **искусственно создаваемое** synthetic image
~ **кода** code pattern
~, **контрастное** hard/contrast image
~, **лапласово** *math.* Laplace transform
~ **мнимое** false/virtual image
~ **на сетчатке** *anat.* retinal image
~, **неконтрастное** soft image
~, **неподвижное** static image; stationary pattern
~, **нерезкое** blurred/unsharp image
~, **обратное** inverted/reversed image; *anat.* real image
~, **объемное** stereoscopic picture, three-dimensional presentation/image
~, **остаточное** afterimage; image retention
~, **оцифрованное** digitized image
~, **плоское** two-dimensional image
~, **полутоновое** half-tone image, gray-scale picture
~, **пространственное** three-dimensional image, stereoscopic picture/presentation
~, **прямое** direct image; *phot.* erect image
~, **радиолокационное** radar display/image/picture
~, **раздвоенное** *telev.* split picture
~, **размытое** diffuse picture/image
~, **резкое** sharp/crisp image
~, **рельефное** relief image
~ **рельефа** relief representation
~, **скрытое** latent image
~, **стереоскопическое = пространственное**
~, **схематическое** diagrammatic representation; diagram; sketch
~, **теневое** skiagraph, (direct) shadow image, shadowgraph
~, **точечное** point image
~, **трехмерное = пространственное**
~, **цветное** colo(u)r image/display
~, **черно-белое** black-and-white picture
~, **четкое** sharp image
~, **штриховое** line image
изобразительный graphic; pictorial; descriptive; visual; decorative
~**ные искусства** visual arts; fine arts
изобретатель inventor
изобретать invent; contrive, devise
изобретение invention
изобронта *meteor.* isobront
изобутан isobutane
изовалериановый isovaleric

изовола isovol
изогала isogal
изогалина isohaline, isohalsine
изогамета isogamete
изогаметангий isogametangium
изогамия isogamy
изагамма isogam(me)
изогамный isogamic, isogamous
изогамонтия isogamonty
изогела *meteor.* isohel
изогемагглютинация isoh(a)emagglutination
изогемагглютинин isoh(a)emagglutinin
изогемолиз isoh(a)emolysis
изогемолизин isoh(a)emolysin
изогемотерапия isoh(a)emotherapy
изогемотрансфузия isoh(a)emotransfusion
изогенный isogenic
изогеномный isogenomatic
изогеотерма *geol.* isogeotherm
изогеотермический isogeothermal, geoisothermal, isogeothermic
изогидрический, изогидричный isohydric
изогидрометрический isohydrometric
изогиета *meteor.* isohyetal line, isohyet
изогипостенурия *med.* isohyposthenuria
изогипса isohypse, hypsographic curve, contour line, points of equal elevation
изогира isogyre, isogyra
изоглосса isogloss, isoglotic line
изоглютатион isoglutathione
изогнуто-ветвистый *bot.* curvirameous
изогнуто-перегородчатый curvimurate
изогнуто-пластинчатый curved-plated
изогнуто-стебельчатый *bot.* curvicaulis
изогнутость curvature, sinuosity; bend; camber; flexion
изогнуто-цветковый *bot.* curviflorous
изогнутый curved, bent; crooked; *bot.* campylotropous
~ кзади retroflexed
изогологамия isohologamy
изогона isogon, isogonic line, isogonal
изогональный isogonal, equiangular
изограда isograde, isogradic line
изоградный isograde, isofacial
изограмма isogram(me)
изограммиды *pl.* isogrammides
изограф 1. isograph; 2. icon painter
изография 1. isography; 2. icon painting
изогуанин isoguanine
изодактилия *zool.* isodactylism
изодиафер isodiaphere
изодиморфизм isodimorphism, double isomorphism
изодиморфный isodimorphous
изодина isodyne
изодинама isodynamic curve/line, isodynam
изодинамический isodynamic, isodynamous
изодицентрический isodicentric
изодоза isodose, isocount contour
~ нейтронного потока isoflux
изодозный isodose
изозиготность isozygoty
изоиммунизация isoimmunization
изоиммунный isoimmune
изокарбоксазид isocarboxazid, marplan
изокарпия *bot.* isocarpy
изокатабаза isocatabase
изокванта isoquant
изоклазит *geol.* isoclasite
изоклина isoclinic/isoclinical line
~, нулевая aclinic line
изоклиналь *geol.* isocline
изоклинальный, изоклинический isoclinal, isoclinic
изократия isocrasy
изокрима isocryme
изоланид *med.* isolanid, celanid
изолейцин *chem.* isoleucine
изолента insulating tape
изолепс *bot.* bulrush *(Isolepis, Scripus)*
изолецитальный *zool.* isolecithal
изолизин *chem.* isolysin
изолиния isoline, isometric line, isogram
~, светораспределения isocandela curve
~, чувствительности isosensitivity curve
изолированность isolation
изолировать 1. isolate, separate; seggregate; 2. *el., therm.* insulate, apply insulation on; seal off
изолировка insulation
изолитический isolytic
изолюкс isolux line, curve of equal illumination, isophote
изолятор 1. *tech.* insulator; insulating material; 2. *med.* isolation/probationary ward
~ антенного ввода aerial lead-in insulator
~, вводный lead-in insulator; wall insulator
~, гирляндный string insulator unit
~, горшковый pot insulator
~, двухъюбочный double-shed insulator
~, дистанционный stand-off insulator
~ для внутренней/наружной установки indoor/outdoor insulator
~, колпачковый cap insulator
~, крюковой swan-neck insulator
~, маслобарьерный oil-filled insulator
~, одноюбочный single-shed insulator
~, опорный pedestal/base insulator
~, орешковый egg insulator
~, подвесной suspension insulator
~, проходной bushing insulator
~, роликовый spool insulator
~, слаботочный electrocommunication-type insulator, weak-current insulator
~, стержневой stick/shackle insulator
~, тарельчатый disk insulator
~, шинный bus insulator
~, штыревой pin-type insulator
~, юбочной bell insulator
изоляты *pl.* isolates

изоляционизм *polit.* isolationism
изоляционист *polit.* isolationist
изоляционный 1. isolation *(attr.)*; 2. *tech.* insulation *(attr.)*
~ная лента insulating tape
изоляция 1. isolation; *med.* quarantine; 2. *tech.* insulation; insulator
~, абляционная ablation/ablative insulation
~, акустическая sound/acoustic insulation
~, антикоррозионная corrosion-resistant insulation
~, бумажная с пропиткой impregnated paper insulation
~, бумажно-масляная paper-oil insulation
~ вибраций vibration insulation
~, дугостойкая *el.* arc-resistant insulation
~, звуковая sound-proofing; sound insulation
~, комбинированная composite insulation
~, конденсаторная *el.* capacitor insulation
~, кремнийорганическая silicone insulation
~, криогенная cryogenic insulation
~, лакотканевая varnished-cotton insulation
~, маслобумажная oil-paper insulation
~, междуфазная *el.* phase-to-phase insulation
~, межвитковая *el.* turn insulation
~, огнеупорная/огнестойкая refractory [fire-resistant] insulation
~, пенокерамическая foamed-ceramic insulation
~, пенопластовая plastic-foam insulation
~ плазмы plasma containment/confinement
~ по напряжению voltage insulation
~, поглощающая absorbent insulation
~ пожара fire confinement
~, пробковая cork insulation
~, противопожарная fireproof insulation
~ тепловая thermal/heat insulation; *(of boilers, etc.)* lagging
~, теплостойкая high-temperature [heat-proof] insulation
~, стеклопластиковая fibre-glass insulation
~, хлопчатобумажная, провощенная waxed cotton insulation
~, экранно-вакуумная shield-vacuum insulation
изомагнитный isomagnetic
изомальтоза *chem.* isomaltose
изомальтотриоза isomaltotriose
изомасляный *chem.* isobutyric
изомер 1. *chem.* isomer; 2. *zool.* isomere
~ гамма-излучитель gamma isomer
~ с большим спином high-spin isomer
изомераза isomerase
изомеризация *chem.* isomerization
изомеризовать(ся) *chem.* isomerize
изомерия *chem.* isomerism
~ атомных ядер nuclear isomerism
~, валентная valence isomerism
~ положения position isomerism
~, поворотная rotational isomerism
~, пространственная stereoisomerism, space isomerism
~ сердцевины ядра core isomerism
~, скелетная skeleton isomerism
~, электронная electron isomerism
~ ядра nuclear isomerism
изомерный isomeric, isomerous
изометрический isometric(al), monometric(al), equidimensional, equant; *cryst.* tesseral
изометрия, изометризм isometry
изометропия isometropy
изоморфизм isomorphism
~, инверсный *math.* anti-isomorphism
~, топологический topological isomorphism/transformation
изоморфный isomorphous, isomorphic
изомочевина isourea
изонефа *meteor.* isoneph, isonephic line
изониазид *chem.* isoniazid
изонитрил *chem.* isonitril(e)
изонитросоединение isonitrocompound
изооктан iso-octane
изоосмотический iso-osmotic
изопага isopag
изопахита isopach, isopachous line
изопериметрия *geom.* isoperimetry
изопетальный *bot.* isopetalous
изопиеста isopiestic (line)
изопикна isopycnic (line), isopycnal
изопикнический isopycnic
изопикноэ isopycnosis
изоплета isopleth, index line
изоплотность isodensity
изопноз *med.* isopn(o)ea
изоповерхность isosurface
изополиплоидный isopolyploid
изопольный, изополярный isopolar
изопора isopor(e)
изопрен isoprene
изопространство isospace
изопьеза, изопьеста isopiestic (line)
изортоклаз *miner.* isorthoclase, isorthose
изосейсма, изосейста isoseismal (line)
изосейсмический isoseismic
изосепальный *bot.* isosepalous
изосеротерапия isoserotherapy
изосиндетический isosyndetic
изосинтагма isosyntagm
изосияние isoaurora
изосклера isosclere
изосоединение *chem.* isocompound
изоспин *phys.* isospin
изоспондильный isospondylous
изоспора isospore
изоспория isospory, homospory
изоспоровый isosporous, isosporic, homosporous; *pl.* isosporous plants
изостазия *geol.* isostasy; equilibrium, equal balance
изостата isostatic (curve), isostath

изостемонный *bot.* isostemonous
изостера isostere
изостудия fine arts studio
изотаха *meteor.* isotach, isovel, curve of equal velocity
изотера isothere
изотерма isotherm, isothermal line
изотермический isothermal, isothermic, homeothermal
изотермичность isothermality
изотермобата isothermobath
изотип *biol.* isotype
изотипность isotypism
изотома *bot.* shrub harebell *(Isotoma)*
изотипия isotypy
изотифен isotiphen
изотон *phys.* isotone
изотона isotone, isotonic line
изотонический, изотоничный, изотонный isotonic
изотония isotonicity
изотоп *chem.* isotope
~, активирующий(ся) activating isotope
~, альфа-/бета/-гамма-излучающий alpha-/beta-/gamma-emitting isotope
~, бета-стабильный beta-stable isotope
~ в основном состоянии ground-state isotope
~, гепатотропный hepatotropic isotope
~, делящийся fissile/fissionable isotope
~, долгоживущий long-lived isotope
~, дочерний daughter isotope
~, земной terrestrial isotope
~, избыточно-нейтронный neutron-rich isotope
~, искусственный artifical/synthetic isotope, man-made isotope
~, исходный original/parent isotope
~, короткоживущий short-lived isotope
~, лимфотропный lymphotropic isotope
~, материнский parent isotope
~, неактивирующий(ся) non-activating isotope
~, нейтронодефицитный neutron-deficient isotope
~, нейтроноизбыточный neutron-excess isotope
~, нейтронообедненный neutron-impoverished isotope
~, нейтронообогащенный neutron-rich isotope
~, нечетный odd isotope
~, обогащенный enriched isotope
~ образующийся в реакторе pile-produced isotope
~, осколочный fission-product isotope
~, остеотропный osteotropic isotope
~, остеофильный bone-seeking isotope
~, природный naturally occurring isotope, natural isotope
~, радиоактивный radioactive isotope, radioisotope, radionuclide
~ с избытком нейтронов neutron-rich isotope
~ с недостатком нейтронов neutron-deficient isotope
~, среднетяжелый intermediate isotope
~, транскюриевый transcurial isotope
~, трансплутониевый transplutonium isotope
~, тяжелый heavy isotope
~, четный even isotope
изотопический isotopic
изотоп-индикатор tracer isotope
изотоп-мишень target isotope
изотопия isotopy
изотрансплантат isotransplant, isograft
изотрисомия isotrisomy
изотропия, изотропность isotropy, isotropism
изотропный isotropic, isotropous
изоупругий isoelastic
изофаза isophase
изофан-инсулин isophane-insulin
изофациальный isofacial
изофема isofeme, isofemic line
изофен(а) *biol.* isophene, isophane
изофенный isophenic, isophenous
изофеногамия isophenogamy
изофермент isoenzyme
изофиллия *bot.* isophylly
изофона isophone, isophonic line
изофория isophoria
изофота isophote
изофотометр isophotometer
изохазма isochasm
изохинолин *chem.* isoquinoline
изохиона isohion, isonival
изохлора isochlor
изохора isochor(e) (line)
изохорический, изохорный isochoric
изохрома isochrome
изохроматический, изохроматичный isochromatic
изохромный isochromatic
изохромосома isochromosome
изохромата isochromatic curve/line
изохрона isochrone
изохронизм, изохронность isochronism
изохронический, изохронный isochronous
изоценоз isocoenosis
изоцентр isocentre, *Am.* isocenter; conjugate focal point, centre of homology
изоцианат *chem.* isocyanate
изоциановый *chem.* isocyanic
изоцитолизин isocytolysin
изоцитотоксин isocytotoxin
изоэндоморфизм isoendomorphism
изоэнтропический isoentropic
изоэйкония isoiconia
изоэлектрический isoelectric
изоэлектронный isoelectronic
изоэнергетический isoenergetic
изоэнтропический isoentropic
изоэнтальпа isoenthalpy

изоэтес *bot.* quillwort *(Isoetes)*
изоэтовые *pl. bot.* quillwort family *(Isoetaceae)*
изоэтолистный *bot.* quillwort-leaved
изразец ceramic tile
Израиль Israel
израильский Israeli; Israelite; *hist.* Israelitish, Israelitic
израильтянин Israeli; *hist.* Israelite
изрезанность indentation
~ береговой черты coastal indentation
изрезанный indented; jagged
изрезывать *geog.* indent
изумруд emerald
~, медный dioptase, copper emerald
~, никелевый zaratite, nickel emerald
изумрудный emerald *(attr.),* smaragdine
изучение study; investigation, exploration
~ возможностей применения feasibility study
~, глубокое in-depth study
~ конкретного случая case study
~, поверхностное *med.* anabrosis
~ повторяемости элементов frequency study
~ потребительского спроса *econ.* consumer study
~ рынка *econ.* market exploration/studies
изъязвление *tech.* pitting; *med.* ulceration, exulceration
изъязвленный ulcerous
изъян flaw, defect; imperfection
изъятие withdrawal; removal
~ из обращения withdrawal from circulation; redemption
изымать withdraw; remove
~ из обращения recall/withdraw from circulation
~ в пользу государства confiscate, seize
изыскание investigation, exploration, research, study; search; survey; prospecting
~, морское offshore prospecting
~, трассовое route surveying
изыскатель researcher; prospector
изэнтропия isentropy
изюбр *zool.* Manchurian wapiti *(Cervus elaphus xanthopygus)*
Иисус Jesus; **Иисус Христос** Jesus Christ
Икар *myth.* Icarus
икариец *hist.* Icarian
иконоборец *hist.* iconoclast
иконоборство *hist.* iconoclasm
инонография 1. iconography; 2. *art* portraits
иконописание, иконопись icon painting
иконописец icon painter
иконопоклонение, иконопочитание iconolatry, icon worship
иноноскоп iconoscope
иконостас iconostasis
иконотека *art* picture collection
икосаэдр *geom.* icosahedron
икосаэдральный, икосаэдрический icosahedral
икоситетраэдр *geom.* icositetrahedron
икота *med.* hiccup, hiccough; singultation, singultus
икотип *biol.* icotype
икотник *bot.* false alyssum *(Berteroa)*
ИК-приемник IR receiver/detector
икра 1. roe; spawn; 2. caviar(e)
икра ноги *anat.* calf
икромет, икрометание spawning
икс-единица *radiol.* X-unit, XU
иксиецветковый *bot.* ixia-flowered
икт *lit.* ictus
иктеричность bile-tinged skin
иктерогемоглобинурия icteroh(a)emoglobinuria
иктерогеморрагический icteroh(a)emorrhagic
ИК-система наведения IR heat-seeking system
ИК-спектр IR spectrum
иктидозавр *paleont.* ictidosaur
ил silt, slime; sludge; mud; ooze
~, активный activated sludge
~, аккумуляторный battery mud
~, гнилой decay ooze, putrid slime
~, жидкий slurry
~, зоогенный ooze
~, морской sea ooze
~, сырой raw/fresh sludge
иланг-иланг *bot.* ylang-ylang *(Cananga odorata)*
илеит *med.* ileitis
~, регионарный regional/terminal/distal ileitis; regional/chronic enteritis
илеоеюнит *med.* ileojejunitis
илеоилеостомия *med.* ileoileostomy
илеоколит *ned.* ileocolitis
илеоколостомия *med.* ileocolostomy
илеоректостомия *med.* ileorectostomy, ileoproctostomy
илеостомия *med.* ileostomy
илеотифлит ileotyphlitis, ileocecitis
илеотомия *med.* ileotomy
илеоцекальный *med.* ileocecal
илеоцекостомия *med.* ileocecostomy
илеус *med.* ileus
~ беременных ileus subparta
~, глистный verminous ileus
ИЛИ *comp.* OR
ИЛИ-ИЛИ OR-ELSE, EITHER-OR, EXCLUSIVE-OR
Илиада Iliad
илиосакральный *anat.* iliosacral
илиоспинальный *anat.* iliospinal
илисто-гумусовый clay-humus
илистость sliminess
илистый slimy, silty, oozy, uliginous, muddy
иллатив *gram.* illative
иллативно-соединительный illative-conjunctive
иллимеризация illimerization, lessivation
иллиний *chem.* illinium
иллириец, иллирийский Illyrian
иллит *miner.* illite

иллициум *bot.* anise tree *(Illicium)*
иллювиально-гумусовый humic-illuvial
иллювий illuvium
иллюзия illusion
~, **окулогиральная** oculogyral illusion
~ **противовращения** illusion of reversal motion
иллюминатор I. *naut.* porthole, portlight, light port; scuttle; *aero.* window
~, **бортовой** sidelight, portlight, side scuttle/port
~, **глухой** dead/fixed light
~ **крышки люка** hatch window
~, **палубный** deck light/scuttle
иллюминатор II. illuminator *(of books)*
иловатый uliginous, silty, oozy, slimy; muddy
иловка loam, siltstone
илонакопитель sludge tank
илоотделение desilting
илоотстойник silt-settling tank/basin
илоочиститель desilter
илосос ooze sucker
илот *hist.* helot
~, **освобожденный** neodamode
илоуловитель silt entrainer; silt pit
илька *zool.* fisher, pekan, black cat *(Martes pennanti)*
ильм, ильма, илем, илим *bot.* elm *(Ulmus)*
ильмовник 1. elm grove; 2. *bot.* Siberian elm *(Ulmus pumila)*
ильмовые *pl. bot.* elm family *(Ulmaceae)*
ильница-пчеловидка *zool.* drone fly *(Eristalis tenax)*
имам *eccl.* imam
имбецил *med.* imbecile
имбецильность imbecility
~, **моральная** moral imbecility/insanity
имбецильный imbecilic, imbecile *(attr.)*
имбирные *pl. bot.* ginger family *(Zingiberaceae)*
имбирь *bot.* ginger *(Zingiber)*
именной nominal
именослов *eccl.* litany
имид *chem.* imide
имидазол *chem.* imidazole
имин *chem.* imine
иминогидролаза *chem.* iminohydrolase
иминогруппа *chem.* imino group
иминодипептидаза *bioch.* iminodipeptidase
имитатор *tech.* simulator
~ **кабины космического корабля** spacecraft-cabin simulator
~ **курсовой системы** directional-system simulator
~ **невесомости** weightlessness simulator, null-G simulator
~ **полета** flight simulator
~ **ракеты** missile simulator
~ **ударных нагрузок** shock simulator
~ **условий окружающей среды** environmental simulator
имитация *tech.* simulation; mock-up; imitation; substitute
~ **атомного взрыва** simulated atomic explosion
~ **больших высот** high-altitude simulation
~ **кожи** leather substitute
~ **невесомости** weightlessness simulation
~ **полета** simulated/ground flight, flight simulation
имитировать 1. *tech.* simulate; 2. imitate
имманентность *philos.* immanence
имманентный *philos.* immanent; inherent
иммельман *av.* Immelmann turn, roll-off-the-top
~, **обратный/перевернутый** inverted Immelmann
иммобильность immobility
иммортель *bot.* strawflower *(Helichrysum bracteatum)*
иммунагглютинация immuneagglutination
иммунантиген immune antigen
иммунизация immunization
~, **плановая** planned immunization
иммунизировать 1. *med.* immunize; vaccinate; 2. *law* grant immunity to
иммунитет *med., law* immunity
~, **антивирусный** antiviral immunity
~, **антимикробный** antibacterial/antimicrobic immunity
~, **видовой** species immunity
~, **врожденный** congenital immunity
~, **гуморальный** humoral immunity
~, **дипломатический** diplomatic immunity
~, **естественный** natural/innate/inherent immunity
~, **инфекционный** infection immunity
~, **клеточный** cellular immunity
~, **приобретенный** acquired immunity
~, **судебный** legal/jurisdictional immunity
иммунность immunity
иммунный immune; resistant *(to disease)*
иммунобиологический immunobiological
иммунобиология immunobiology
иммунобласть immunoblast
иммуногематология immunoh(a)ematology
иммуноген immunogen
иммуногенез immunogenesis
иммуногенетика immunogenetics
иммуноглобулин immunoglobulin, immune serum grobulin
иммунодепрессант immunodepressant
иммунодиагностика immunodiagnosis
иммунодиффузия imnunodiffusion
иммунокомпетентный immunocompetent
иммунолог immunologist
иммунология immunology
иммуноморфология immunomorphology
иммунопатология immunopathology
иммунопротеин immunoprotein
иммуноселекция immunoselection
иммунотерапия immunotherapy
иммунотоксин immunotoxin
иммунотолерантность immunotolerance
иммунофлюоресценция immunofluorescence

иммунохимия immunochemistry
иммунохирургия immunosurgery
иммуноцит *biol.* immunocyte
иммуноцитохимия immunocytochemistry
иммуноэлектрофорез immunoelectrophoresis
импала *zool.* impala *(Aepyceros melampus)*
импеданс impedance
~ **внешний/внутренний** external/internal impedance
~ **контура/цепи** circuit impedance
~, **тепловой** thermal impedance
импеллер impeller, blade
императа *bot.* satin tail *(Imperata)*
империя empire
~, **небесная** *hist.* the Celestial Empire
имперский imperial
импетигинизация *med.* impetiginization
импетигинозный *med.* impetiginous
импетиго *med.* impetigo
импичмент *law* impeachment
имплантат implant
~, **эндометриальный** endometrial implant
имплантация implantation, ingrafting
импликанта implicant
имплозия implosion
импотенция *med.* impotence, impotency
~, **нейрорецепторная** neuro-receptory impotence
~, **психическая** psychic impotency
~, **эндокринная** endocrinal impotency
импрегнировать impregrate
импрессионизм impressionism
импрессионистический, импрессионистичный, импрессионистский impressionist
импульс 1. *el.* pulse; impulse; 2. *mech.* momentum; impetus, surge, burst
~, **антидромный** *med.* antidromic impulse
~, **афферентный** afferent impulse
~, **бланкирующий** blanking pulse
~, **ведущий** driving/master pulse
~, **вибрационный** vibratory impulse
~, **включающий** turn-on pulse
~, **волевой** volitional impulse
~, **вращательный** angular impulse
~, **временной** time pulse
~, **входной** input/incoming impulse
~, **выключающий** turn-off pulse
~, **выходной** output pulse
~ **вычитания** *comp.* subtract pulse
~ **гашения** quench/quenching pulse; blanking pulse; termination pulse
~ **давления** pressure surge
~ **дальности** range pulse
~ **данных** data pulse
~, **двухполярный** bidirectional pulse
~ **деления** fission pulse
~, **единичный** *rock.* specific impulse
~, **задающий** master pulse
~ **задержки** delay pulse
~, **запирающий** disable/disabling pulse; gate-closing pulse; blanking/blackout pulse
~ **записи** *comp.* write pulse
~ **заполнения** fill-in pulse
~ **запрета** inhibit/inhibitory pulse
~ **запроса** interrogation pulse
~, **запускающий** firing/triggering pulse; initiating/start pulse
~ **запускающий развертку** sweep-initiating pulse
~, **звуковой** acoustic/sound pulse
~, **зондирующий** transmitted/outgoing pulse
~, **излучаемый** transmitted pulse
~ **излучения** emission impulse
~, **измеримый** 1. measurable pulse; 2. *mech.* detectable momentum
~, **ионный** ionpulse
~, **интероцептивный** interoceptive impulse
~, **исполнительный** execute pulse
~, **испытательный** test pulse
~, **кадрирующий** framing pulse
~, **кадровый** frame/vertical pulse
~, **кадровый радиотелеметрический** complete pulse
~, **кадровый синхронизирующий** frame-synchronizing pulse
~, **калибровочный** calibration pulse
~, **кодированный** coded impulse
~ **кодового знака** digit pulse
~, **кодовый** code pulse
~, **колоколообразный** bell-shaped pulse, Gaussian pulse
~, **коммутирующий** switching pulse
~, **корковый** *anat.* cortical impulse
~, **короткий** brief/narrow pulse
~, **корректирующий** corrective/correcting pulse
~, **крутой** steep pulse
~, **кумулятивный** cumulative [pile-up] pulse
~, **линейный** linear momentum
~, **ложный** spurious pulse
~, **манипуляции** *com.* keying pulse
~, **маркерный** marking pulse; reference-time mark; marker/pip pulse
~, **механический** mechanical momentum
~, **мощный** high power [strong, giant] pulse
~ **набора номера** *teleph.* dialing pulse
~ **набора цифр** *teleph.* digit pulse
~ **на входе пересчетного устройства** unscaled pulse
~ **на выходе пересчетного устройства** scaled pulse
~ **напряжения** potential/voltage pulse
~, **нейтронный** neutron burst
~, **нервный** *anat.* nervous impulse
~ **несовпадения** anticoincidence pulse
~ **несущей частоты** carrier-frequency pulse
~, **обобщенный** *mech.* generalized momentum
~ **обратного хода луча** retrace/flyback pulse
~, **однополярный** single-polarity [unidirectional] pulse

~ опознавания цвета *telev.* colour-identification pulse
~ обратной полярности inverted pulse
~, опорный reference pulse; pedestal impulse
~ опроса *comp.* interrogation/interrogating/excitation pulse
~, останавливающий stop pulse
~, остаточный afterpulse
~, острый peaky [sharp-pointed] pulse
~ отдачи *mech.* recoil momentum
~ отметки marker pulse
~, относительный *mech.* relative momentum
~, отпирающий enable/enabling/antiparalyze pulse; gate-opening pulse
~, отраженный reflected/echoed pulse
~, отсчитываемый read-out pulse
~, парный paired pulse
~, перевернутый inverted pulse
~ переноса *comp.* carry pulse
~ переполнения overflow pulse
~, периодически повторяющийся repetitive pulse
~, пилообразный saw-tooth pulse
~, плоский square-topped pulse
~, П-образный square(d) pulse
~ подавления suppression/control pulse
~ полной выборки full-drive pulse
~, полный total/aggregate momentum/impulse
~, половой sexual impulse
~ полувыборки half-drive pulse
~ помехи interfering/disturb pulse
~, поперечный *mech.* transverse momentum
~, пороговой *mech.* momentum cut-off
~ последействия residual impulse
~, проверочный test pulse
~, программный program(ming) pulse
~, пропускающий gating/gate pulse
~, прямоугольный rectanguar/square pulse
~, пусковой initiating/starting/start/trigger/release/action/firing pulse
~, равноотстоящий equispaced pulse
~, радиочастотный radio-frequency pulse
~ развертки sweep pulse
~, разрешающий enable pulse
~, разрядный 1. *comp.* digit pulse; 2. discharge pulse
~ ракетного ускорителя boosting impulse
~, релятивистский *mech.* relativistic momentum
~ с врезками *telev.* serrated pulse
~ сброса *comp.* reset pulse
~, световой light pulse
~ СВЧ microwave pulse
~, сдвоенный double pulse
~ селекторный gate pulse, gate
~ силы impulse (of force), momentum
~, синхронизированный timed pulse
~, синхронизирующий clock/timing/synchronizing/sync pulse
~ сложения *comp.* add pulse
~, случайный random pulse
~ смещения displacement/shift/bias pulse
~ совпадения coincidence pulse
~, сопряженный *mech.* conjugate momentum
~, срезанный chopped pulse
~, стирающий erase pulse
~, стробирующий strobe/gate/sample/gating pulse
~, счетный count pulse
~ считывания *comp.* read pulse
~, тактовый clock/timing/strobe pulse
~, ударный collision/impact momentum; *el.* shock pulse
~, удельный specific impulse
~, управляющий control/drive/driving/steering pulse
~ управляющий пропусканием gating pulse
~ установки в состоянии единица/ноль set/reset pulse
~, уширенный stretched pulse
~, хронирующий timing/clock pulse
~ частицы *phys.* particle momentum
~, цифровой digit pulse
~ электрона *phys.* momentum of electron
~ эхолота *naut.* sounding pulse
импульс-врезка *telev.* serrated pulse, serration
импульсметр pulse meter
импульсник pulser
импульсно-модулированный pulse-modulated
импульсный impulsive; pulse *(attr.)*
импульсовидный pulse-like
импульсно-возбуждаемый pulse-actuated
импульсно-временный pulse-time
импульсно-кодовый pulse-code
импульсно-модулированный pulse-modulated
импульсно-регенеративный pulse-regenerating
имущественный property *(attr.),* possession *(attr.),* stock *(attr.)*
~ иск *law* real action
~ ценз property qualification
~ное положение property status
имущество property, possessions; assets; stock; belongings; *mil.* stores
~, движимое movable property, personal estate; personalty; (personal) chattels, things personal, goods and chattels
~ жены *law* paraphernal property
~, заложенное *econ.* pledged property
~, заповедное entailed property, entail, fee-tail
~, изнашиваемое *econ.* depreciable assets
~, малоценное и быстроизнашивающееся *econ.* supplies
~, материальное *law* tangible assets/property
~ наследуемое только по женской линии *law* estate tail female, tail female general
~ наследуемое только по мужской линии *law* estate tail male, tail mail general, tail male
~, недвижимое *law* real estate/assets, realty, immovable property/estate, immovables, *econ.* fixed assets; things real

~, **независимое** *law* estate in severalty
~, **незаложенное** property free from encumbrance
~, **нематериальное** intangible property
~ **несостоятельного должника** bankrupt's estate
~, **принадлежащее** *law* vested estate/interest
~, **самостоятельное** *law* estate in severalty
~ **управляемое по доверенности** trust estate
имя name; *fig.* reputation; *gram.* nomen, noun
~, **альтернативное** *comp.* alias
~ **входа** *comp.* entry name
~, **групповое** group name
~ **для ссылок** *comp.* reference name
~, **зарегистрированное** *comp.* login name
~, **индексированное** *comp.* subscripted name
~, **контекстное** *comp.* context name
~, **прилагательное** *gram.* adjective
~, **родовое** generic name
~, **системное** system name
~, **существительное** *gram.* substantive, noun
~, **уточненное** qualified name
~, **функциональное** *comp.* role name
~, **числительное** *gram.* numeral
~, **числительное количественное** *gram.* cardinal numeral
инакомыслие dissent; heterodoxy
инакомыслящий dissentient, differently minded; heterodox; *(noun)* dissident
инактивация, инактивирование inactivation; deactivation
инактивировать inactivate
инбирные = имбирные
инбредный *biol.* inbred
инбридинг *biol.* inbreeding
инвагинация *med.* intussusception, invagination
~, **еюногастральная/илеоцекальная** jejunogastric/ileocecal intussusception
~, **кишечная** intussusception
~, **ретроградная кишечная** retrograde intussusception
инвазивность invasiveness
инвазия *biol.* invasion; infestation
~, **амебная** am(o)ebic invasion, am(o)ebism
~, **внутриутробная** intrauterine invasion
~, **глистная** helminthic invasion
~, **протозойная** protozoal invasion
~, **эктопаразитная** ectoparasitic invasion
инвалид disabled person, invalid
~ **войны** disabled soldier; war invalid
~ **труда** disabled worker
инвалидность invalidism; disablement
~**ти, пенсия/пособие по** disability pension
инвалидный дом home for invalids
инвалюта *abbr.* foreign currency/exchange
инвар *met.* Invar (alloy)
инвариант, инвариантный invariant; nonvariant
~ **в себе** self-invariant
инвариантность invariance
~, **градиентная** gauge invariation
инвектива invective
инвентаризация stocktaking, inventory making, inventorying
~ **в натуре** physical stocktaking
инвентаризировать take stock; make inventory of
инвентарь stock; equipment; items; inventory
~, **живой** *agr.* livestock
~, **мертвый** *econ.* dead stock; equipment
~, **сельскохозяйственный** agricultural/farm equipment/implements
~, **составить** make [draw up] an inventory
~, **торговый** shop equipment; stock in trade
инвенция *mus.* invention
инверсивный, инверсионный inversive
инверсия inversion; *comp.* inverse
~, **временная** time inversion
~ **входных сигналов** input inversion
~ **двоичных цифр** binary inversion
~ **над горами** mountain inversion
~ **населенности** population inversion
~, **ночная** nocturnal/evening inversion
~ **переменных** inverse of variables, variables inversion
~, **приземная** ground/surface inversion
~, **сложная** compound inversion
~ **суммы** *comp.* not-sum
~, **температурная** temperature inversion
~, **частная/местная** local inversion
инверсор 1. *math.* inversor; 2. automatic focus(s)ing device
инвертаза *bioch.* invertase
инвертирование inversion, inverting; conversion
инвертировать invert
инвертор inverter, invertor; negator; *comp.* NOT circuit/gate/element, inhibitory gate/circuit; AND NOT gate, EXCEPT gate
~, **автономный** *el.* self-excited inverter
~, **зависимый, ~ ведомый сетью** *el.* grid-controlled inverter
~, **импульсный** *comp.* pulse NOT-gate, pulse inverter
~, **пневматический** fluidic NOT-gate, fluidic inverter
~, **ртутный** *el.* mercury-arc inverter
~, **самоуправляемый** *el.* free-running inverter
~ **со схемной совпадения** *comp.* gated inverter
инвеститор, инвестор investor
инвестиционные товары *econ.* capital goods; investment goods
инвестиция investment
~ **в материальные активы** tangible investment
~ **в жилищное строительство** residential investment
инволюта *math.* involute, evolvent
инволютный coiled, involute
инволюция involution

ингалятор *med.* inhaler
ингаляция inhalation
~, масляная oil inhalation
~, паровая steam inhalation
ингерентный inherent
ингибирование inhibition
ингибировать inhibit
ингибитор inhibitor, inhibitory/inhibiting factor/agent
~ детонации antiknock agent
~, конкурентный competitive inhibitor
~ коррозии corrosion inhibitor
~, летучий vapour-phase inhibitor
~, неконкурентный noncompetitive inhibitor
~ окислительных реакций antioxidant
~ ферментов enzymatic inhibitor
ингр Karelian
ингредиент ingredient, component, part
ингуш, ингушский Ingush
индантрен indanthrene (blue)
индау *bot.* rocket salad *(Eruca sativa)*
индеец (American) Indian
индейка 1. *zool.* turkey hen; 2. (flesh of) turkey
индейководство turkey farming
индейский петух turkey cock
индекс *math.* index; factor; classification symbol; index number
~ базы данных database index
~, верхний superscript
~, взвешенный *econ.* weighted index (number)
~ выживаемости survival index
~ выцветания/обесцвечивания fading index
~ вязкости viscosity index
~ Герфиндаля *econ.* Herfindahl index
~, главный master index
~, групповой *econ.* group/general index
~ Доу-Джонса *econ.* Dow-Jones index
~ дохода income term/index
~ единицы стоимости экспорта export unit value index
~ заработной платы wage index
~ карта master/guide/register card
~ команды command/instruction/order index
~ компенсации цен *econ.* price-compensation index
~ курсов акций *econ.* stock price average
~ курсов ценных бумаг index of securities
~, митотический mitotic index
~, младший *comp.* fine index
~, надстрочный superscript
~, немой *math.* dummy index
~, нижний subscript
~, нижний левый presub(script)
~, нижний правый post-sub(script)
~ Никкей Nikkei index
~, общий general/group index
~ очистки decontamination index
~, ограниченный bounded index
~, опсонофагоцитарный *med.* opsonocytophagic index
~ пересечения intersection coefficient/number
~, подстрочный subscript
~ покупательной способности *econ.* purchasing power index
~, поперечный cross index
~ потребности в рабочей силе *econ.* help-wanted advertizing index
~ прожиточного минимума index of cost of living
~ производственной мощности capacity index
~ с двойным взвешиванием *econ.* cross-weighted index
~ с неизменной базой fixed-base index
~ с переменной базой changing-base index
~ с поправкой на сезонность seasonally corrected index
~ с постоянной базой fixed-base index
~, сглаженный *math.* rectified index
~, сердечный *med.* cardiac index
~, составной aggregative index
~, средневзвешенный average weighted index
~, старший *comp.* master/main index
~ стоимости жизни index of cost of living, cost of living index
~ суммирования summation index
~, торакальный *med.* thoracic index
~, тотальный general/group index
~ умственных способностей aptitude score
~ цен price index
~ цен на потребительские товары consumer price index
~, цепной *stat.* chain index
индексация, индексирование indexing, indexation; notation
~, автоматическая autoindexing
~, буквенная/цифровая letter/figure notation
индекс-карта guide card
индексно-последовательный index(ed)-sequential
индентор indenter, penetrator
~, сферический sphere penetrator; hardness-tester ball
индетерминизм *philos.* indeterminism
индетерминистический, индетерминистский indeterministic
индианка 1. American Indian woman; squaw; 2. Indian woman, Hindu woman
индивид individual; person
индивидуализация individualization, individuation
~ продукции *econ.* product differentiation
индивидуализирование individuation
индивидуализировать individualize, individuate
индивидуалистический, индивидуалистичный individualistic
индиго 1. *bot.* indigo *(Indigofera);* 2. indigo (dye)
~, медное copper indigo, covellite, covelline
индигокармин carmine of indigo, indigo red
индигоноска, индигофера *bot.* indigo

(Indigofera)
~, красильная true indigo *(Indigofera tinctoria)*
индигорубин indigo red
индиготин *chem.* indigotin, indigo blue
индигоурия *med.* indigouria, indiguria
индиевый *chem.* indic
индиец, индийский Indian, Hindu
индий *chem.* indium, In
индийско-британский Indo-British
индикан *chem.* indican
индиканемия *med.* indican(a)emia
индиканурия *med.* indicanuria
индикатив *gram.* indicative
индикатор indicator, display, indicating device/instrument; tracer, marker; guide; index
~, акустический aural indicator
~ азимута и высоты azimuth-and-height display
~, бленкерный flag indicator
~ бокового уклонения *naut.* course deviation indicator, deviometer
~, бортовой *av.* airborne indicator
~, буквенно-цифровой alphanumeric indicator
~, визуально-звуковой audio-visual display
~ влажности hygrometer, moisture/humidity indicator
~, волокнооптический fibre-optic indicator
~, выносной/вынесенный remote display
~ газа gas detector/indicator
~, газоразрядный gas-discharge indicator
~ дальности range indicator
~ движущихся целей moving-target indicator
~ дозы излучения radiation dosage indicator
~, звуковой audio indicator
~, знаковый character display
~, жидкокристаллический liquid-crystal display, LCD
~ излучения radiation indicator
~, изотопный isotopic tracer/indicator; tracer element
~, инфракрасный infra-red indicator
~, ионизационный ionization indicator
~, кислотно-основный acid-base indicator
~ колебаний cymoscope, oscillation indicator
~, контрольный monitoring indicator
~, лазерный laser display/indicator
~ лжи lie detector
~, маршрутный route indicator
~ метана firedamp detector, methane indicator
~, многокомпонентный multiple tracer
~ момента окончания terminal-time indicator
~ мощности power(-level) indicator
~, многорежимный multi-mode display
~ множества *math.* characteristic function of a set
~, навигационный navigation(al) display, ground-position indicator
~ направления direction/bearing/azimuth indicator
~ настройки tuning indicator
~ неисправностей fault indicator
~, нулевой, ~ нуля null indicator
~ нулевого уровня zero level indicator
~ обледенения ice/icing indicator
~, обобщенный integrated display
~, окислительно-восстановительный oxidation-reduction indicator, redox indicator
~ останова программы *comp.* program(me) stop light
~ остающейся дальности range-to-go display
~ перегрева overheat indicator
~ пеленга и дальности range-and-azimuth indicator
~ перегрузки overload indicator
~ переключения диапазонов band-selector indicator
~, плоский flat/panel display
~, поисковый search display
~ положения position indicator
~ попаданий hit indicator, scorer
~ превышения температуры over-temperature indicator
~ превышения числа оборотов over-speed indicator
~, проекционный projected display
~ пуска start-up indicator
~ равновесных схем *el.* null indicator
~ радиоактивности radioactivity indicator/detector
~, радиоактивный radioisotopic tracer, radioactive indicator/tracer, radiotracer
~, радиолокационный radar indicator/display/scope
~ расхода flow indicator
~ режима работы mode/performance indicator
~ с движущейся картой moving/dynamic map display
~ с картой map/chart display
~ с перекрещивающимися стрелками cross-pointer display
~ с секторной разверткой sector-scan display
~, световой light indicator, indicator lamp
~, смешанный compound indicator
~, стрелочный pointer-type indicator
~, твердотельный solid-state indicator
~ текущих координат present-position indicator
~, точечный *comp.* dot-matrix display
~, трехмерный three-dimensional display, 3-D display
~ угла места elevation display
~, универсальный all-purpose indicator
~ условия *comp.* flag indicator
~ утечки leakage indicator
~, фотонный photon detector
~, химический chemical tracer
~, циферблатный dial indicator
~, цифровой digital/decimal/numerical display/indicator
~ широты и долготы display of latitude and

longitude
~, **электроннолучевой** electron-ray indicator
~, **яркостный** bright display
индикатриса *math.* indicatrix; indicator function; characteristic curve
~ **рассеяния** scattering indicatrix
индикация indication; presentation; display
~ **данных** data display; data read-out
~, **дистанционная** remote indication
~, **звуковая** aural/audible indication/presentation
~, **знаковая** alphanumeric indication
~, **комбинированная** combination/combining display
~ **курса** heading indication
~ **места** position indication
~, **наглядная** pictorial/direct/straightforward indication
~, **немерцающая** flicker-free display
~, **однозначная** unambiguous indication
~ **пеленга** bearing indication
~ **промаха** miss-distance indication
~, **радиолокационная** radar display
~ **разбалансировки** out-of-balance indication, out-of-trim indication
~ **рассогласования** error indication
~, **световая** indicating light/lamp; light indication
~, **стрелочная** pointer indication
~ **тяги** thrust indication
~, **цифровая** numerical/digital presentation; digital read-out; decimal display
индикт *hist.* indiction
индистор indistor
индифферентность indifference
~, **дозовая** *med.* dosage indifference
индитрон inditron tube
индоарийский Indo-Arian
индо-африканский Indo-African
индогерманский Indo-Germanic
индоевропеец, индоевропейский Indo-European
индоевропеистика Indo-European linguistics/studies
индоиранский Indo-Iranian
индокитайский Indo-Chinese
индоксил *chem.* indoxyl
индол *chem.* indole
индолил *chem.* indolyl
индолилацетилглутамин indolylacetylglutamine
индолилуксусный indolylacetic
индолог Indologist, Indologian
индологенный *bioch.* indologenous
индология Indology
индолурия *med.* indoluria
индо-малайский Indo-Malayan
индометацин indometacin
индонезиец, индонезийский Indonesian
индоссамент endorsement, indorsement; indorsation
~, **безоборотный** endorsement without recourse
~, **бланковый** blank endorsement
~, **дружеский** accommodation endorsement
~, **именной** special endorsement
~ **не содержащий определенного условия** unconditional endorsement
~, **полный** endorsement in full
~ **содержащий определенное условие** conditional endorsement
индоссант endorser, indorser; transferor
индоссат(ор) endorsee, indorsee; transferee
индоссирование *law* backing
индоссировать endorse, indorse
индофенол *chem.* indophenol
индри *zool.* indri *(Indris)*
индузий *bot.* indusium
индуизм Hinduism
индуктанс *phys.* inductance
индуктивно inductively, by induction
индуктивно-дедуктивный inductive-deductive
индуктивно-емкостный inductance-capacitance
индуктивность inductance; inductance coil
~ **вторичной обмотки** secondary inductance
~ **вывода** lead inductance
~, **кажущаяся** apparent inductance
~ **первичной обмотки** primary inductance
~, **последовательная** series inductance
~, **собственная** internal inductance
индуктивный inductive
индуктирование induction, inducing
индуктировать induce
индуктометр *el.* induction meter, inductometer
индуктопирексия *med.* inductopyrexia
индуктор inductor; induction/work coil; field structure/magnet
~, **земной** earth inductor
~, **магнитный** dip inductor
~, **нагревательный** induction heater
~, **путевой** *rail.* track inductor
~ **телефонный ручной** magneto/hand ringing generator
индуктосин inductosyn
индуктотермия inductothermy
индукционный inductive; induction *(attr.)*
~**ная катушка** induction coil
индукция induction
~, **взаимная** reciprocal/mutual induction
~, **зиготическая** zygotic induction
~, **магнитная** magnetic induction; magnetic flux density
~, **математическая** mathematical/complete induction
~ **насыщения** saturation induction; saturation flux density
~, **обратная** back induction
~, **остаточная магнитная** remanent/residual induction; remanence
~, **полная** complete/mathematical induction
~, **трансфинитная** transfinite induction
~, **униполярная** homopolar/unipolar induction
~, **электрическая** electric induction; electric

flux density
индульгенция *eccl.* indult; indulgence
индурация *med.* induration
~, фиброзная fibroid induration
индус Hindu; *obs.* Indian
~, ученый pundit
индусский Hindu; *obs.* Indian
индустриализация industrialization
индустрия industry
~, легкая light industry/industries
~ развлечений show/entertainment business
~ рекламы advertizing industry
~ туризма travel/tourist industry
индуцировать 1. *el.* induce; 2. stimulate, induce
индюк turkey cock
иней 1. *bot.* hoarfrost, frost, rime; 2. *bot.* pruina
инертность inertness; inertia; passivity
инертный inert
инерциальный inertial
инерционность lag; time lag/delay
~ зрительного восприятия persistence of vision
~ электроннолучевых трубок after-image
инерция inertia; momentum; (time) lag
~ вращения rotary inertia
~ зрительного восприятия/ощущения persistence of vision
~ матки *med.* uterine inertia
~, механическая mass inertia
~, психическая psychic inertia
~, световая afterglow
~, тепловая thermal lag/inertia
инжектировать inject
инжектор injector; jet pump
~, струйный jet injector
~, плазменный plasma injector
~ электронов electron injector
инжекция injection
~ дырок *el.* hole injection
~ носителей заряда *el.* charge/carrier injection
~ частиц particle injection
инженер engineer
~, авиационный aeronautical engineer
~, бортовой flight engineer
~, ведущий senior project engineer
~, главный chief/head engineer
~, дежурный shift engineer
~, дипломированный licensed engineer
~, горный mining engineer
~ ОТК quality-control engineer
~ по вычислительным машинам computer engineer
~ по двигателям и силовым установкам propulsion engineer
~ по наземной эксплуатации ground engineer
~ по приборно-измерительному оборудованию instrumentation engineer
~ по производительности труда efficiency engineer
~ по радиооборудованию radio engineer
~ по технике безопасности safety engineer
~ по технической эксплуатации maintenance engineer
~ проекта project engineer
инженер-атомник nuclear engineer
инженер-геолог geological engineer
инженер-гидротехник hydraulic engineer
инженер-звукотехник sound engineer
инженер-инструментальщик tool engineer
инженер-испытатель test(ing) engineer
инженер-исследователь research engineer
инженер-конструктор design engineer
инженер-консультант consulting/advisory engineer
инженер-механизатор agricultural machinery engineer
инженер-механик mechanical engineer
инженер-монтажник installation/assembly engineer
инженер-методист methods engineer
инженер-нефтяник oil/refinery engineer
инженер-нормировщик time-study engineer
инженерно-технический engineering and technical
инженерно-технические работники engineering manpower, technical and engineering employees
инженерный engineering
инженер-озеленитель landscape designer/architect
инженер-оценщик valuation/evaluation/appraisal engineer
инженер-прочнист stress engineer
инженер-сантехник plumbing engineer
инженерство engineering (profession)
инженер-строитель civil engineer; erecting/construction engineer
инженер-технолог process engineer, industrial/plant engineer; production design engineer
инженер-хронометражист time-study engineer
инженер-химик chemical engineer
инженер-эксплуатационник maintenance engineer
инженер-электрик electrical engineer
инжир *bot.* common fig *(Ficus carica)*
~, ослиный sycomore *(Ficus sycomorus)*
инзухт inbreeding
инициализация *comp.* initialization
инициатива initiative
~, законодательная *law* initiation of bills
инициатор-терминатор initiator-terminator
инициирование initiation
инициировать initiate; start; trigger off
инкапсулирование encapsulation, incapsulation; encystation
инкапсулировать encapsulate, incapsulate; encyst
инкарвиллея *bot.* incarvillea *(Incarvillea)*
инкарцерация *med.* incarceration
инкассатор *fin.* collector; cash messenger
инкассация, инкассирование collection, cash-

ing, encashment
инкассировать *fin.* collect, encash, cash, receive
~ деньги по векселю collect a bill
инкассо collection; encashment
~, агент по collection agent
~ тратты draft collection
ин-кварто *polygr.* in quarto, quarto
инквизиция *hist.* the Inquisition
инквилин *zool.* inquiline
инки the Incas
инклинатор (magnetic) inclinometer, inclination/dip compass
~, индукционный induction inclinometer, earth induction compass, flux-gate inclinometer
~, стрелочный dip-needle/-circle, dipping compass
инклинометр inclinometer
~, одноточечный single-shot inclinometer
инклинометрия inclinometer/directional surveying, determination of drill-hole inclination/dip and direction
инкогерентный incoherent
инкогнито incognito
инконгруэнтный incongruent, incongruous
инкор *abbr.* foreign correspondent
Инкотермы Incoterms
инкремент increment; growth
инкрементальный, инкрементный incremental
инкрет *biol.* incretion, internal secretion
инкреторный *biol.* incretory
~ная железа endocrine gland
инкретотерапия incretotherapy
инкреция incretion
инкриминация, инкриминирование incrimination, charging, accusation
инкриминировать charge *(with),* accuse *(of);* incriminate
инкрустация incrustation, inlay; inlaid work
инкрустирование incrusting, incrustation
инкрустировать encrust; inlay
инкуб incubus
инкубатор incubator; hatcher
~, секционно-проходной sectional incubator
инкубаторий hatchery, incubator/hatching house
инкубационный incubative, incubatory
инкубация incubation; *med.* incubation period
инкубировать incubate, hatch (eggs) by artificial heat
инкунабула *lit.* incunabulum
инкурабильный, инкурабельный incurable, not curable, immedicable
иннервация innervation; venation, nervation
~, вагусная vagal innervation
~, вазомоторная vasomotor innervation
~, дланевидная *bot.* palmate venation
~, миомоторная myomotor innervation
~, миостатическая myostatic innervation
~, сетчатая *bot.* netted venation
~, сосудодвигательная vasomotor innervation
инобытие *philos.* Anderssein
иноверец member of a different creed/faith
иноверие adherence to a different faith/creed; heterodoxy
иноверный belonging to a different faith/creed; heterodox
иноверческий heterodox
иноземный foreign, alien; ecdemic
инозин inosine
инозиназа *bioch.* inosinase
инозиндифосфат inosine diphosphate
инозит *bioch.* inositol, inosite
инозитурия inosituria, inosuria
инок *obs.* monk; cenobite
инокиня *obs.* nun
ин-октаво *polygr.* in-octavo, octavo
инокулировать inoculate
инокулят inoculum
инокуляция inoculation
иномыслие difference of opinion/views; dissent
инопексия *med.* inopexia
инoперабельный *surg.* inoperable
иноперкулятный *bot., zool.* inoperculate
инопланетный from another planet
инопланетянин, инопланетанец, инопланетник inhabitant/being/creature from another planet
иноплеменник member of a different tribe/nation
иноплеменный of a different tribe/nation; foreign
инородец *hist.* non-Russian
инородность heterogeneity
инородный 1. foreign, alien; allophylian; heterogeneous; 2. *petr.* ectogenic; *cryst.* epigene
инородческий *hist.* non-Russian; foreign
иносказание allegory
иносказательный alegorical, metaphorical; figurative
иносклероз *med.* inosclerosis
инославный *rel.* non-Orthodox
иностранец foreigner, alien
~, натурализованный *law* denizen
~ по рождению *law* alien born
иносущный *eccl.* Heteroöusian
инотдел *abbr.* foreign department/section
инотропный *physiol.* inotropic
иноформенный heteromorphous
иноходец ambler *(of horse)*
иноходь amble
инохондрит *med.* inochondritis
инохондрома *med.* inochondroma
иноцерам *paleont.* inoceram *(Inoceramus)*
иноческий *obs.* monastic
иночество *obs.* monasticism; monastic life
иноязычный speaking [belonging to] another language
~ное слово loan/foreign word
ин-плано *polygr.* in-plano
инсаливация *med.* insalivation
инсектицид, инсектисид insecticide
~, дыхательный fumigant insecticide

~, **кишечный** intestinal insecticide
~, **системный** systemic insecticide
инсектицидно-удобрительный *agr.* insecticide-fertilizer
инсектицидный insecticide, insecticidal
инсектофунгицид insectofungicide
инсоляция *phys., med.* insolation
инспектирование inspection
инспектор inspector; supervisor
~ **здравоохранения** health inspector/officer
~, **налоговый** tax officer
~ **по технике безопасности** safety inspector
~ **по качеству** quality inspector
~, **таможенный** customs inspector/officer
~ **счетчиков и установок** meterman
инспекторат inspectorate
инспекторский inspectorial
инспекторство inspectorate; inspectorship
инспекционный inspectorial; inspective
инспекция 1. inspection; 2. inspectorate
инспираторный inspiratory; *(of sound)* inhaled
инспирация 1. *physiol.* inspiration, inhalation; 2. inspiration, incitation
инсталляция installation
инстанция *law* instance; *mil.* echeleon (of command)
~, **командная** *mil.* chain/echelon of command
~, **первая судебная** court of first instance
~, **последняя судебная** court of last resort
инстилляция instillation, dropwise administration
инстинкт instinct
институт 1. *law* institution; 2. institute, institution, college, higher educational institution; 3. *hist.* girls' private boarding school
~, **горный** school/college of mines
~, **научно-исследовательский** research institute
~, **научно-исследовательский и опытно-конструкторский** research and development institute
~, **педагогический** teachers' college
институтка *hist.* boarding schoolgirl
институции *hist.* institutes
~ **Юстиниана** institutes of Justinian
инструктаж instruction; briefing
~, **предполетный** pre-flight briefing
инструктировать instruct, brief; direct
инструктор instructor
~ **морского дела** naval instructor
~ **физической подготовки** *mil.* athletic instructor
инструктор-методист organizer; adviser
инструкция instruction(s); manual, handbook; instruction sheet; directive; specification; *comp.* command
~ **вызова** *comp.* call command/instruction
~, **заводская** shop manual
~ **летчику** flight manual/handbook, pilot's note
~ **по ремонту** overhaul/repair manual
~ **по техническому обслуживанию** maintenance manual/regulation; service manual
~ **по установке** installation directions
~ **по эксплуатации** service manual, operating instructions; maintenance manual
~, **фирменная** shop manual
~ **сравнения** *comp.* compare instruction
инструмент tool; instrument, implement; *coll.* tools, tooling
~, **абразивный гибкий** coated abrasive (tool)
~, **алмазный** diamond tool
~, **безопасный** safety tool
~, **возимый** mobile tool(s)
~, **геодезичесний** geodesic instrument
~, **гравировальный** engraver; etching device
~, **делительный** indexing head
~, **деревообрабатывающий** woodworking tool(s)
~, **дистанционный** remote-operated/-handled instrument
~ **для разборки** dismantling tool
~, **дыропробивной** punch
~, **зажимный** clamping/gripping tool
~, **контрольный** inspection tool/instrument
~, **измерительный** measuring instrument
~, **концевой** point tool
~, **кузнечный** forging tools
~, **мерительный** measuring tool(s)
~ **металлокерамический** cermet-tipped tool
~, **металлорежущий** metal-cutting tool; machine tool for metals
~, **механизированный** machine tool, power tool
~, **монтажный** erection/erecting/installation/fitter's tool(s)
~, **музыкальный** musical instrument
~, **обрабатывающий** machining tool
~, **отделочный** finishing tool
~, **пассажный** passage/meridian/transit instrument; (astronomic) transit
~, **повторительный** repeating instrument
~, **прецизионный** precision instrument
~, **режущий** cutting tool(s)
~, **резьбонакатный** thread-rolling tool
~, **резьбонарезный** thread-cutting tool
~, **ручной** hand tool(s)
~ **с длинной ручкой** long-handed tool
~, **солнечный** solar instrument
~, **съемочный** *geod.* surveying instrument
~, **сборочно-монтажный** assembly tools
~, **твердосплавный** cemented-carbide tool(s)
~, **точный** precision tool(s)
~, **угломерный** angle gauge, anglemeter; angular/azimuth instrument; subtense instrument
~, **ударный** percussive/impact tool; *mus.* pulsatile/percussion instrument
~, **универсальный** universal tool; multipurpose tool
~, **чертежный** draftsman's/draughtsman's tools
~, **эксплуатационный** maintenance tools
~, **электрифицированный** electric (hand)

tools
инструментализм *philos.* instrumentalism, instrumental theory
инструментальная toolroom
инструментальщик toolmaker
инструментарий *coll.* tooling, tools; instruments; toolkit
инструментовка *mus.* instrumentation; orchestration
инструментодержатель toolholder
инсулин insulin
инсулиназа insulinase
инсулиновый insulinic
инсулинотерапия insulinization
инсулин-протамин protamine insulin
инсулома insuloma
инсульт *med.* insult, insultus; (apoplectic) stroke
~, геморрагический h(a)emorrhagic insult
~, ишемический isch(a)emic insult
~, мозговой cerebral insult
~, тромботический thrombotic insult
~, эмболический embolic insult
инсуфлятор *med.* insufflator
инсуфляция *med.* insufflation
инсценировка adaptation for stage/screen; stage/screen version
инталия intaglio
интарсия intarsia
интеграл *math.* integral
~, абелев Abelian integral
~ активации activation integral
~, биномиальный binomial integral
~ вероятности probability integral
~ вероятности ошибки error function integral, erf, erfi
~, внутренний inner/inside integral
~ во времени time integral
~, двойной double integral
~, двукратный twofold iterated integral, double(-iterated) integral
~ действия action integral
~ дифференциального уравнения integral/solution of a differential equation, primitive integral
~ дробного порядка fractional integral
~, контурный contour/circulatory integral
~ Коши Cauchy integral
~, кратный multiple integral
~, n-кратный n-fold multiple integral
~, криволинейный line/contour/curvilinear integral
~, круговой circuital integral
~ Лебега Lebesgue integral
~ Лебега-Стилтьеса Lebesgue-Stieltjes integral
~, линейный line integral
~, неопределенный indefinite integral, antiderivative, primitive
~, несобственный improper integral
~, нормировочный normalization integral
~, общий general integral; general solution
~, объемный volume integral
~, определенный definite/Riemann integral
~, особый singular integral
~ от показательной функции exponential integral
~ от тригонометрической функции trigonometric integral
~, относительный резонансный relative resonance integral
~ ошибок error function integral, erf, erfi
~ первого/второго рода, криволинейный line/curvilinear integral of the first/second kind
~ перекрывания overlap(ping) integral
~ по замкнутому контуру integral over a closed curve
~ по объему volume/space integral, integral over space
~ по области integral over a domain/region
~ по поверхности surface integral, integral through a surface
~, поверхностный surface integral
~, повторный iterated/repeated integral
~ поглощения, резонансный absorption resonance integral
~, полный complete integral
~ рассеяния scattering integral
~, расходящийся divergent integral
~ Римана Rieman integral
~ резонансного поглощения resonance-absorption integral
~ с бесконечным пределом infinite integral, integral with infinite limits
~ свертки/свертывания convolution integral
~ скалярный (криво)линейный scalar line integral
~ Стильтеса Stieltjes integral
~, сходящийся convergent integral
~, табличный table/fundamental integral
~, трехкратный threefold iterated integral
~, тройной triple integral
~, фазовый phase integral
~ Френеля Fresnel integral
~ Фурье Fourier integral
~, усредненный averaged integral
~, частный particular integral; particular solution
~, эллиптический elliptic integral
интегральный integral; integrated
~ное исчисление integral calculus
интегратор integrator; integrating circuit
~, идеальный perfect integrator
~, импульсный pulse integrator
~ ионного тока ion-current integrator
~ с обратной связью feedback integrator
~ светового потока light integrator
~, суммирующий summing integrator
~, цифровой digital integrator
~ частоты пульса pulse rate integrator
~, электронный electronic integrator
~, электрохимический solion integrator
интеграф integraph, analog(ue) integrator
интеграция integration

~, **вертикальная** upstream/vertical integration
~, **горизонтальная** horizontal/lateral integration
~ **на уровне приборов** devices integration
~ **на уровне систем** systems integration
интегрирование *math.* integration, integrating
~ **в комплексной плоскости** complex integration
~ **в конечном виде** integration in closed form
~ **в масштабе пластины** wafer-scale integration
~, **дискретное** digital integration
~, **дробное** fractional integration
~ **импульсов** pulse integration
~ **методом прямоугольников** rectangular integration technique
~, **многократное** multiple integration
~, **некогерентное** incoherent integration
~ **по точкам** point-by-point integration
~ **по частям** integration by parts
~ **подстановкой** integration by substitution
~, **почленное** term-by-term integration, termwise integration
~, **приближенное** approximative integration
~, **численное** numerical integration
интегрировать 1. integrate, bring/pool together; unite; 2. *math.* integrate
интегрируемость integrability
интегрируемый integrable
интегросумматор summing integrator
интегумент *zool., bot.* integument; tunicle
интегументальный integumental, integumentary
интеллект 1. intellect, mental power; 2. *cyb.* intelligence
интеллектуальность intellectuality
интеллектуально-коммуникативный intellective-communicative
интеллектуальный intellectual; intellective; *comp.* intelligent
интеллигент intellectual
интеллигентный educated, cultured; intellectual
интеллигенция intelligentsia; the intellectuals/educated
интендант *mil.* supply officer
интендантство *mil.* supply service/office/department
интенсивность intensity; strength; rate; scale, magnitude; force; intensiveness
~ **альфа/бета/гамма излучения** alpha-/beta/gamma intensity
~ **в кюри** *phys.* curiage
~ **вибраций** vibration severity/intensity
~ **вихря** vortex intensity
~ **входного потока** entrance/arrival rate
~ **выходящего потока** departure rate
~ **воздушного движения** air-traffic intensity
~ **горения** combustion rate
~ **группового поступления** *econ.* pooled arrival intensity
~ **движения** traffic intensity/density
~ **делений** fission rate
~ **дозы** dose/dosage rate
~ **дыхания** respiration intensity
~ **захвата** capture rate, rate of capture
~ **излучения** radiation/radiated/emitted intensity, radiation/emission power
~ **изнашивания** wear rate/intensity
~ **импульса** pulse intensity
~ **испускания** emission rate
~ **источника** strength of source, source strength
~ **нагрузки** density/intensity of loading
~ **отказов** failure rate; hazard rate
~ **нейтронного потока** neutron-flux level
~ **питания** feed rate
~ **подачи** input rate
~ **потока** flux level
~ **потока излучения** radiation flux
~ **потока информации** information density
~ **перевозок по данному маршруту** route density
~ **полетов** flying/flight intenfity
~ **помех** noise/jamming intensity
~ **пуассоновского потока** Poisson intensity
~ **радиации** radiation flux
~ **расходования запасов** stock depletion rate
~ **реверберации** reverberation strength
~ **свечения** luminous intensity
~ **сигнала** signal strength
~ **спонтанного деления** spontaneous fission rate
~ **страничного обмена** paging rate
~ **сушки** drying rate
~ **счета** counting rate
~ **теплопередачи** heat-transfer rate
~ **тормозного излучения** bremsstrahlung intensity
~ **утечки** leakage rate
~, **фоновая** background intensity
~ **шума** noise level
~ **эксплуатации** *mech.* operation rate
интенсивный intensive
интенсиметр ratemeter
~, **логарифмический** log-counting ratemeter
~, **поисковый** *min.* prospecting ratemeter
интенсионал, интенсия intension, intensional meaning
интенсификатор intensifier
интенсифицировать intensify; boost; step up
интерамбулякр *zool.* interambulacrum
интервал interval, space, diastem; range; gap, interspace, spacing; period
~ **безразличия** *stat.* indifference interval
~ **включения** on-interval
~, **временной** time span/interval
~ **выключения** off-interval, cut-off period
~ **группирования** *stat.* group interval
~ **гашения** blanking interval
~, **двусторонний доверительный** *stat.* two-sided confidence interval
~, **доверительный** *stat.* confidence interval; fiducial interval; reliable interval; confidence/

fiducial belt/band
~ **доступа** access interval
~, **замкнутый** closed interval
~, **защитный** guard interval/band/space
~, **защитный частотный** frequency guard space
~ **значимости** *stat.* significance band
~, **импульсный** pulse spacing/separation
~ **импульсов** 1. pulse time; 2. *mech.* momentum range
~ **кипения** *phys.* boiling range
~ **между знаками** character spacing
~ **между изолиниями** contour interval
~ **между колонками** column pitch/spacing
~ **между посадками, временной** landing time interval
~ **между строками** line/vertical spacing
~ **незанятости** idle/inaction period
~ **насыщения** saturation range
~ **незанятости** idle/inaction period
~ **плавления** melting/fusion range
~ **пластичности** plastic range
~, **полуоткрытый** *math.* half-open interval
~ **опроса** *comp.* polling interval
~ **по частоте** frequency interval/space
~ **повторения** repetition/recurrence interval
~ **превращения** *cryst.* transformation range
~, **рабочий** operating range
~ **регулирования** control interval
~ **сходимости** interval of convergence
~, **счетный** counting period
~ **температур/температурный** temperature range
~. **температур кристаллизации** crystallization/solidification range
~ **экспозиции** *phot.* exposure latitude
~, **энергетический**, ~ **энергии** energy range
интервaллюм intervallum, interval
интервент, интервенционист, интервенционистский interventionist
интервенция intervention; interference
интервидение intervision
интергляциал interglacial period
интергранулярный *petr.* intersertal
интердициплинарный interdisciplinary
интерес interest
~, **вещноправовой** *law* estate
интеркалярный intercalary; intercalate
интеркинез *biol.* interkinesis
интеркреция *geol.* intercretion, septarium
интеркурентный *med.* intercurrent
интерлиньяж space/distance between two lines
интерлюдия *mus.* interlude
интермедия *theatre, mus.* interlude
интерлобит *med.* interlobitis
интермедиарный intermediary
интермедин *bioch.* intermedin, chromatophorotrophic hormone
интерметаллид intermetallic compound
интермеццо *mus.* intermezzo
интермитоз intermitosis
интермицеллярный intermicellar
интернационал international; **И.** the Internationale
интернирование internment
интернированный 1. *adj.* interned; 2. *n.* internee
интернировать intern
интернатура *med., educ.* internship
интерорецептивный interoceptive
интерорецептор interoceptor
интерорецепция interoception
интерпариетальный *anat.* interparietal
интерпеллировать *polit.* interpellate
интерпеллянт *polit.* interpellator
интерпелляция *polit.* interpellation; question in Parliament
интерпол Interpol
интерполирование *math.* interpolation, interpolating
~ **вперед** forward interpolation
~, **двумерное** bivariate interpolation
~, **квадратичное** quadratic interpolation
~, **линейное** linear interpolation
~ **назад** backward interpolation
~ **по одному аргументу** single interpolation
~ **по трем аргументам** tripple interpolation
~ **с учетом вторых разностей** double interpolation
~, **тройное** tripple interpolation
интерполировать interpolate
иитерполятор interpolator
~ **дискретного действия** digital interpolator
~ **непрерывного действия** analog(ue) interpolator
~, **цифровой** digital interpolator
интерполяция = интерполирование
интерпретатор *comp.* interpreter; interpretive program(me)/routine
интерпретация interpretation
~ **данных** data interpretation
~ **команд** instruction interpretation
интерпретировать interpret
~ **неправильно** misinterpret
интерпретируемость interpretability
интерпункция punctuation
интеррадиум *zool.* interradium
интерсексуальность intersexuality
интерсексуальный intersexual
интерсексы *pl. biol.* intersexes
интерсенсорный intersensory
интерсертальный *geol.* intersertal
интерстадиал *geol.* interstadial
интерстициальный interstitial
интерстиций interstice
интерстициома *med.* interstitioma
интерстрофа interstrophe
интертексический intertexic
интертип *print.* intertype
интерфаза interphase
~, **постсиндетическая** postsyndetic interphase

интерфейс *comp.* interface
~ ввода-вывода input-output interface
~, дружественный user-friendly interface
~ канала прямого доступа DMA interface
~, канальный channel interface
~ командного типа command-driven interface
~ на основе меню menu-driven/-based interface
~, последовательный serial interface
~, программируемый programmable interface
~, прозрачный transparent interface
~ с пользователем user interface
~ с расширенным набором команд command-rich interface
~, связной communications interface
~, стандартный standard/standardized interface
~ типа главный-подчиненный master-slave interface
~, человеко-машинный human-computer interface
~ шины bus interface
интерференционный interferential, interference *(attr.)*
интерференция interference, interfering
~, вирусная viral interference
~ волн wave interference
~ гармонических колебаний harmonic interference
~, гасящая destructive interference
~ давления pressure interference
~ двух пучков/лучей two-beam interference
~ державки support interference
~ звуковых волн sound interference
~ когерентных волн coherent interference
~, межатомная interatomic interference
~, межсимвольная intersymbol interference
~, многолучевая multibeam/multipath interference, multiple-beam interference
~ незатухающих колебаний continuous-wave interference
~, ослабляющая destructive interference
~ от двух щелей two-slit interference
~, полезная favourable interference
~ при дозвуковых/сверхзвуковых скоростях subsonic/supersonic interference
~ радиоволн radio interference
~ рентгеновских лучей X-ray interference
~ света interference of light (waves); optical interference
~, усиливающая constructive interference
~, устойчивая coherent interference
интраабдоминальный intra-abdominal
интраартериальный intra-arterial
интраваскулярный intravascular
интравенозный intravenous
интравитальный *biol.* intravital
интравентрикулярный intraventricular
интрагеосинклиналь intrageosyncline
интрагранулярный intragranular
интрадермальный intradermal
интразвук intrasound
интразональный intrazonal
интракардиакальный intracardiac, intracardial
интраклинальный intraclinal
интракортикальный intracortical
интракраниальный intracranial
интраксилярный intraxylary
интракутанный intracutaneous, endermic
интралигаментарный intraligamentous
интралингвистический intralinguistic
интрамагматический intramagmatic
интрамеркуриальный *astron.* intramercurial
интрамолекулярный intramolecular
интрамуральный *anat.* intramural
интраназальный *anat.* intranasal
интранатальный intranatal
интраневральный intraneural
интранзитивный intransitive
интраорбитальный intraorbital
интраперитонеальный *anat.* intraperitoneal
интрапетиолярный *bot.* intrapetiolar
интраплевральный intrapleural
интраплювиальный intrapluvial
интрасептальный intraseptal
интраспинальный intraspinal
интрастаминальный *bot.* intrastamineal
интратекальный *anat.* intrathecal
интратрахеальный intratracheal
интрафасцикулярный intrafascicular
интрацеллюлярный intracellular
интроверсия introversion
интровертик *psych.* introvert
интронизация *eccl.* enthronization, enthronement
интрорзный *bot.* introrse, introrsal
интроскоп introscope
интроскопия introscopy
~, ультразвуковая ultrasonic visualization/imaging, sonoradiography
интроспективный introspective
интроспекция introspection
интрудировать *geol.* intrude, force
интрузив *geol.* intrusive
~, воронкообразный ethmolith
~, клинообразный sphenolith
~, цилиндрический neck
~ имеющий дно/почву floored intrusion
интрузивный intrusive, intercalary, plutonic, irruptive
интрузия *geol.* intrusion; intrusive body
~, воронкообразная funnel-shaped intrusion
~, глубинная deep-seated [plutonic] intrusion
~, дисконформная unconformable intrusion
~, дискордантная discordant/transgreve intrusion/injection
~, кольцевая ring intrusion
~, конформная conformable intrusion
~ направленная вверх uptrusion
~, пластовая intrusive sheet
~, побочная satellitic intrusion/injection
~, пронизывающая subjacent/riveting intrusion
~, сквозная subjacent intrusion

~, **сложная** composite intrusion
~, **слоистая** layered intrusion
интубатор *med.* intubator
интубация *med.* intubation
~, **носотрахеальная** nasotracheal intubation
~, **ротоtрахеальная** orotracheal intubation
интуитивизм *philos.* intuitionism
интуитивист *philos.* intuitivist
интуиционизм *philos.* intuitionism
интуиционистический, интуиционистский intuitionist(ic)
интуиция intuition
интурист *abbr.* foreign tourist
интуссусцепционный intussusceptive
интуссусцепция intussusception
инулаза *bioch.* inulase
инулин *chem.* inulin
инулосахараза inulosucrase
инфант infante
инфанта infanta
инфантилизм, инфантильность infantilism
~, **врожденный** congenital infantilism
~, **гипофизарный** pituitary infantilism
~, **дистиреоидальный** disthyroidal infantilism
~, **кахектический** cachectic infantilism
~, **микседематозный** myxedematous infantilism
~, **общий** universal infantilism
~, **первичный** primary infantilism
~, **плюригландулярный** pluriglandular infantilism
~ **половых органов** genital infantilism, infantilism of the genital organs
~, **психический** psychic infantilism
~, **спленогенный** splenogenic infantilism
~, **тиреогенный** thyrogenic infantilism
~ **токсемический** toxemic infantilism
инфантильный infantile
инфаркт infarct, infarction, heart attack
~, **анемический** an(a)emic/pale/white infarct
~, **геморрагический** h(a)emorrhagic infarct
~, **гипофизарный** pituitary infarction
~, **жировой** fatty infarct
~, **застойный** congestive infarct
~, **известковый** calcereous infarct
~, **ишемический** isch(a)emic infarct
~ **легкого** pulmonary infarction
~ **миокарда** myocardial infarction
~ **миокарда, боковой/задний/немой** lateral/posterior/silent myocardial infarction
~ **миокарда, диафрагмальный** diaphragmatic myocardial infarction
~ **миокарда, нижнебоковой** inferolateral myocardial infarction
~ **миокарда, нижний/очаговый** inferior/focal myocardial infarction
~ **миокарда, обширный** gross/vast myocardial infarction
~ **миокарда, переднебоковой/передненижний** anterolateral/anteroinferior myocardial infarction
~ **миокарда, переднеперегородочный** anteroseptal myocardial infarction
~ **миокарда, передний** anterior myocardial infarction
~ **миокарда, трансмуральный** transmural [through-and-through] myocardial infarction
~, **мочекислый** uric acid infarct
~ **сосочковых мышц сердца** papillary infarction
~, **эмболический** embolic infarction
инфект *gram.* infectum
инфектартрит *med.* rheumatoid arthritis
инфекциозность infectiosity, infectiousness
инфекционист *med.* specialist in communicable diseases
инфекционность infectivity
инфекционный infectious; contagious; infective
инфекция infection; contagion
~, **агональная** agonal/terminal infection
~, **аденовирусная** adenovirus infection
~, **анаэробная** anaerobic infection
~, **арбовирусная** arbovirus infection
~, **аутогенная** self-infection
~, **бактериальная** bacterial infection
~, **бессимптомная** inapparent/subclinical/silent infection
~, **вирусная** viral infection
~, **внутрибольничная** intrahospital infection
~, **внутриутробная** intrauterine infection; f(o)etal infection
~, **водная** water-borne infection
~, **воздушно-капельная** droplet infection
~, **вторичная** secondary/consecutive infection
~, **газовая** gas infection
~, **гематогенная** h(a)ematogenous infection
~, **гнилостная** putrid infection
~, **гноеродная** pyogenic infection
~, **гнойная** purulent infection
~, **гонорейная** neisserosis
~, **дремлющая** latent infection/microbism
~, **зоонозная** zoogenetic infection
~, **капельная** droplet infection
~, **кишечная** intestinal infection
~, **коли-бациллярная** colibacillosis, colibacteriosis
~, **лимфогенная** lymphogenic infection
~, **местная** local/localized infection
~, **множественная** polyinfection
~, **немая** silent infection
~, **общая** generalized infection
~, **одонтогенная** odontogenic infection
~, **очаговая** focal infection
~, **парагриппозная** parainfluenza infection
~, **первичная** primary infection
~, **послеоперативная** postoperative infection
~, **послеродовая** puerperal infection
~, **протозойная** protozoiasis, protozoosis
~, **профессиональная** occupational infection
~, **пупочная** umbilical infection

~, пылевая dust-borne infection
~, раневая wound infection
~, скрытая latent infection
~, стафилококковая staphylococcosis
~, стрептококковая streptococcosis
~, экзогенная/эндогенная exogenous/endogenous infection
инфигированный infixed
инфиксация infixing
инфильтрат infiltrate; infiltration
~, воспалительный inflammatory infiltration
~, гуммозный gummatous infiltrate
~, лейкемический leuk(a)emic infiltrate
~, эзинофильный eosinophilic infiltrate
инфильтрация infiltration; seepage
~, гликогеновая glycogenic infiltration
~, жировая fatty/adipose infiltration, steatosis
~, клеточная cellular infiltration
~, лейкоцитарная leukocytic infiltration
~, мочевая urinous infiltration
~, очаговая focal infiltration
~, пигментная pigmentary infiltration
~, капиллярная capillary seepage
инфильтрометр infiltrometer; seepage gauge
инфильтруемость infiltrability
инфинитив *gram.* infinitive
инфинитивно-безличный infinitival-impersonal
инфинитивно-глагольный infinitival-verbal
инфинитивно-именной infinitive-nominal
инфинитивно-назывной infinitival-nominal
инфинитивно-переходный infinitival-transitive
инфинитивный *gram.* infinitive, infinitival
инфицирование infection, infecting
инфицированный infected
инфицировать infect, contaminate
инфлексия *gram.* inflection
инфлектор inflector
инфлекционный *gram.* inflectional
инфлюация inflow
инфлюэнца, инфлуэнца *med.* influenza
инфляционный inflationary
инфляция *fin., econ.* inflation
~, безудержная/необузданная runaway/galloping inflation
~ вызванная ростом налогом tax-push inflation
~ обусловленная ростом издержек cost-push inflation; cost inflation
~, ползучая creeping inflation
~ спроса (excess-)demand inflation
ин-фолио *polygr.* folio, in-folio
информант *ling.* informant
информатика information science, theory of information; documentation science; informatics
информатор informant
информационный information *(attr.)*
информация information; data; intelligence; report
~, буквенная alphabetic information
~, буквенно-цифровая alphanumeric information, alphabetical and digital information
~, вводимая input/infed information
~, двоичная binary information
~ в графической форме graphical information
~ в машинном представлении computerized information
~, десятичная decimal information
~, дискретная discrete/digital information
~ для выполнения коррекции updating information
~ для контроля monitoring information
~, запрошенная *comp.* interrogated information
~,засекреченная classified information
~, знаковая alphanumeric information
~, избыточная excess/redundant/superfluous information
~, изобразительная pictorial information
~, исходная source information
~, итоговая summarized information
~, кодированная coded/symbolic information
~, коммерческая business data/information
~, контрольная checking information
~, наглядная pictorial information
~, научная scientific documentation/information
~, ненужная *comp.* garbage, gibberish
~, необработанная raw information
~, необязательная optional information
~, нечисловая non-numerical information
~, обновляемая updated information
~, новейшая up-to-date information
~ о курсе heading/directional information
~ о рейсе *av.* flight information
~ от доплеровского измерителя Doppler information
~ от потребителей *econ.* fleld feedback
~, обработанная processed information
~, отладочная *comp.* debugging information
~, первичная source information
~, пилотажная flight information
~ получаемая в истинном масштабе времени real-time information
~ поступающая от вычислителя computer output information
~, посторонняя extraneous information
~, считываемая read-out information
~, текущая current information
~, телеметрическая telemetry information
~, техническая technical/engineering/technological information/data
~, хранимая stored information
~, цифровая digital information
~, числовая numerical information
информбюро *abbr.* information office
инфразвук infrasound, infra-audible sound; infrasonics
инфразвуковой infrasonic
инфракрасный infrared
инфралюминесценция infraluminescence
инфрамаргинальный inframarginal, submarginal
инфрамногочлен *math.* infrapolynomial

инфранейстон *zool.* infraneuston
инфраокклюзия infraclusion, infraocclusion
инфрачерный infrablack
инфузия infusion
инфузория *zool.* infusorian
инфузорный infusorial
инфундибулэктомия infundibulectomy
инфундировать infuse
инхоатив *gram.* inchoative aspect
инхоативный *gram.* inchoative
инцептивный *gram.* inceptive
инцест incest
инцидент incident; occurrence; event
инцидентность *math.* incidence
инцизура incisure
инцистирование encystment, encystation
инцухт *bot., zool.* inbreeding
инъективный *math.* injective
инъекция injection; *geol.* intrusion
~, внутриартериальная intra-arterial injection
~, внутрибрюшинная intraperitoneal injection
~, внутривенная intravenous injection
~, внутрикожная intracutaneous injection
~, внутрикостная intraosseous injection
~, внутримышечная intramuscular injection
~, внутрисердечная intracardiac injection
~, многократная *geol.* multiple intrusion
~, конъюнктивальная conjunctival injection
~, несогласная *geol.* discordant injection
~, подкожная hypodermic/subcutaneous injection
~, послойная *geol.* leaf-by-leaf injection
~, скачкообразная *geol.* staccato injection
инъецировать inject
инъюнктивный *gram.* injunctive
иоакинф *arch.* hyacinth stone
иог, иога, иогурт *see* **йог, йога, йогурт**
иод, иодид, иодоформ, etc. *see* **йод, йодид,** etc.
иол *naut.* yawl
иольдия *geol.* Yoldia (clay)
иомен *hist.* yeoman, estatesman
иоменри *hist.* yeomanry
ион *phys.* ion
~, амфотерный amphoteric/dual ion, zwitterion
~, атомарный atomic/ionized ion
~, ближайший соседний nearest neighbour ion
~, блуждающий roaming/vagabond ion
~, быстрый high-speed ion
~ внедрения interstitial ion
~, возбужденный excited ion
~, вторичный secondary ion
~ высокой энергии high-energy ion, energetic ion
~, газовый gaseous ion
~, генерирующий lasing/laserable ion
~, гидратированный aquated ion
~, гидроксильный hydroxyl ion
~, двухзарядный doubly charged ion
~, захваченный trapped ion
~ индикатора tracer ion
~, исходный parent ion
~, кислотный acid ion, anion
~, магнитоактивный magnetic [magnetically active] ion
~, межрешеточный interlattice ion
~, многозарядный multiply charged ion
~, молекулярный molecular ion, ionized molecule
~ носителя carrier ion
~, однозарядный singly charged ion
~, осаждающий precipitating ion
~ отдачи recoil ion
~, отрицательный negatively charged [negative] ion
~, паразитный spurious ion
~, первичный primary ion
~, положительный positively charged [positive] ion
~, посторонний foreign ion
~ продукт деления fission-product ion
~, составной complex ion
~, тепловой thermal ion
~, трехзарядный triply/trebly charged ion
иониец *hist.* Ionian
ионизатор ionizer
ионизационный ionization *(attr.)*
ионизация ionization
~ атмосферы atmospheric ionization
~, атомарная/атома atomic ionization
~ газа gas ionizarion
~ гамма-излучением gamma-induced ionization
~, двукратная second-order ionization
~, дробная fractional ionization
~, естественная spontaneous ionization
~, излучением radiation ionization
~, колонная columnar ionization
~ космическими лучами cosmic-rays ionization
~, лавинная avalanche/cumulative ionization
~, лазерная laser-beam induced ionization
~, неравновесная non-equilibrium ionization
~ объемная volume ionization
~, однократная first-order [single] ionization
~, остаточная residual ionization
~ от полярного сияния auroral ionization
~ под действием излучения radiative ionization
~ под действием столкновений collisional/impact ionization
~, предварительная pre-ionization
~ светом photoionization
~ столкновением ionization by collision, collisional/impact ionization
~, ступенчатая step ionization
~, ударная collision ionization
~ ударного слоя shock-layer ionization
~, удельная specific ionization
~, фоновая background ionization
~ четвертого порядка fourth-order

[quarternary] ionization
~ **электронами** electronic ionization
ионизированный ionized
~ **дважды/однократно** doubly/singly ionized
~, **многократно** multiple-ionized
~, **предварительно** pre-ionized
~, **сильно** heavily ionized
~ **ударной волной** shock-ionized
ионизировать ionize
~ **ударами электронов** ionize by electron collision
ионизироваться be/become ionized; dissociate into ions
ионизируемый ionizable; subject to ionization
ионизирующий ionizing
~, **сильно** heavily-ionizing
ионизирующийся ionizable
ионизовать ionize
ионий *chem.* ionium, Io
ионийский Ionian, Ionic
ионик *print., lit.* Ionic
ионит ionite, ion-exchange resin, ion exchanger
ионический Ionian, Ionic
ионно-оптический ion-optical
ионность ionicity
ионный ionic
ионоген ionogen
ионография ionography
ионограмма ionogram(me)
ионозонд ionospheric sounder, ionosonde
ионоизбирательный ion-selective
ионоизлучающий ion-emitting
ионолюминесценция ionoluminescence
иономер *chem.* ionomer
ионометр ionometer
ионообмен ion exchange
ионообменник ion-exchanger, ion-exchange resin, ionite
ионообменный ion-exchange
ионообразование ion formation
ионопауза ionopause
ионопсидиум *bot.* diamond flower *(Ionopsidium)*
ионосфера ionosphere
~, **верхняя** upper ionosphere
~ **Земли** Earth's ionosphere
~, **невозмущенная** normal ionosphere
~, **нижняя** lower ionosphere
~, **слоистая** stratified ionosphere
ионосферный ionospheric
ионотерапия, ионофорез, ионтофорез iontophoresis, iontotherapy
ионофон ionic loudspeaker, ionophone
ионотрон ionotron
ионотропия ionotropy
иорданец Jordanian
иорданит *miner.* jordanite
иорданка Jordanian (woman)
иорданский Jordanian
иоркшир = йоркшир
иот, иота, etc. *see* **йот, йота,** etc.
иофобия iophobia, fear of poisons
иохимбин yohimbine
И-оператор AND-operator
ипекакуана *bot.* ipecac, ipecacuanha *(Uragoga)*
иподьякон *eccl.* subdeacon
ипомея *bot.* morning-glory *(Ipomoea)*
ипостась *theol.* hypostasis
ипотека mortgage
ипотечный mortgage *(attr.);* hypothecary
~ **залог** hypothecation
ипохондрик hypochondriac
ипохондрический hypochondriac(al)
ипохондрия hypochondria
ипподром racecourse, hippodrome
иппология hippology
иприт mustard gas, yperite, dichloroethyl sulphide
ипроклозид *chem.* iproclozide
ипрониазид-фосфат iproniazid/marsilid phosphate
ипсация *med.* ipsation, ipsism
ипсилон 1. upsilon; 2. *zool.* black cutworm *(Agrotis ypsilon)*
иразер iraser, infra-red [IR] laser
иракец, иракский Iraqi
иранец, иранский Iranian
иранка Iranian (woman)
ирачка Iraqi (woman)
ирбис *zool.* ounce *(Felis uncia);* snow leopard *(Pardus uncia)*
ирга *bot.* Juneberry *(Amelanchier)*
иргапирин irgapyrin
ирезина *bot.* bloodleaf *(Iresine)*
иренарх *hist.* irenarch
Ирида *myth.* Iris
иридемия iridemia
ириденклейзис iridencleisis
иридесценция iridescence
иридиевый *chem.* iridic
иридий *chem.* iridium, Ir
иридин *chem.* iridin
иридистый *chem.* iridous
иридодиализ iridodialysis
иридодиастаз iridodiastasis
иридодонез iridodonesis
иридопаралич *med.* iridoparalysis
иридопарез *med.* iridoparesis
иридоплегия *med.* iridoplegia
иридоптоз *med.* iridoptosis
иридотомия *surg.* iridotomy
иридохороидит *med.* iridochoroiditis
иридоциклит *med.* iridocyclitis
иридоциклохороидит *med.* iridocyclochoroiditis
иридоцит *zool.* iridocyte
иридэктомия *surg.* iridectomy
~, **оптическая** optical iridectomy
~, **периферическая** peripheral/stenopeic iridectomy
иризация iridescence, irisation
иринит irinite
ирис 1. *anat.* iris; 2. *bot.* iris, (sword)flag *(Iris);*

3. toffee
ирит *med.* iritis
~, гононокковый gonococcal iritis
~, папулезный iritis papulosa
~, подагрический gouty iritis
~, симпатический sympathetic iritis
ирландец Irishman
ирландизм Irishism, Hibernicism
ирландка Irishwoman
ирландский Irish
ирландско-кельтский Hiberno-Celtic
Ирод Herod (the Great)
ирокез Iroquois
ирокезка Iroquois (woman)
ирокезский Iroquoian
иррадиация irradiation
иррадиировать irradiate
иррациональность 1. irrationality; 2. *math.* irrational, surd
~, квадратичная/кубическая quadratic/cubic surd
~, присоединенная accessory irrational
ирредента irredenta
иррефлексивность irreflexivity, irreflexiveness
ирригатор 1. *agr.* irrigation specialist/technician; 2. *med.* irrigator; rectal tube; fountain syringe
Исида *myth.* Isis
иск *law* claim; action, suit; legal proceedings, lawsuit; plea
~, возбуждать institute an action, take legal steps, institute legal proceedings
~, вещный real action, action in rem
~, встречный counter-claim; cross-action, claim in return
~, гражданский civil action
~ за клевету libel action
~ за оскорбление действием action for assault and battery
~ из гражданского правонарушения action in tort; action ex delicto
~ из договора action in contract
~ из неосновательного обогащения action for money had and received
~ из присвоения движимой вещи action of trover
~, имущественный claim of ownership
~ о взыскании убытков action for damages
~ о возмещении вреда action in tort
~ о признании права action for declaration of right
~ об аннулировании plea in abatement
~, обязательственный action in personam
~ основанный на законе statutory action
~ основанный на праве справедливости equitable action
~, оспаривать meet a claim
~, перекрестный cross-claim
~, петиторный petitory action
~, поссесорный possessory action
~, право на title
~, предъявлять make a claim, claim, sue, commence an action
~, предъявлять встречный lay a counter-claim, counter-claim
~, признавать admit a claim; plead no defence
~, регрессный action over
~, уголовный criminal action
~, удовлетворять allow a claim
~, фиктивный fictitious action
~, судебный legal action, action at law
искажаемость distortability
искажать distort; deform; perturb
искажение distortion; deformation; disfigurement
~, амплитудно-частотное amplitude-frequency distortion
~, бочкообразное *telev.* barrel distortion
~, гармоническое harmonic distortion
~ за рельеф *geod.* deformation due to relief
~ импульсов набора *teleph.* dialing distortion
~ кадра *telev.* frame distortion
~ направлений *cart.* direction distortion//deformation/alteration
~, перекрестное *rad.* intermodulation distortion; *telev.* cross-colour
~ площадей *cart.* area deformation/distortion/deformation
~ по долготе/широте *cart.* error in longitude/latitude, longitude/latitude error
~, подушкообразное *telev.* pin-cushion distortion
~ при расширении expansion distortion
~ при передаче transmitter distortion, sending-end distortion
~ решетки lattice distortion
~, строчное *telev.* line/horizontal distortion
~, трапецеидальное *telev.* keystone distortion
~ углов angular distortion/alteration
~ формы deformation
~ формы импульсов pulse-shape distortion
искаженный distorted, deformed, disfigured
искание selection; dialing; switch action; hunting, finding
~, дальнее long-distance dialing
~ декады hunting for a level
~, прямое straight-through selection
~, шаговое step-by-step selection
искатель *teleph.* finder, selector
~ вызова call/line finder; finder switch
~, линейный final selector
~ машинной системы power-drive selector
~ обратного действия reverse motion selector
~, подъемно-вращательный two-motion switch
~ прямого действия direct-action selector/finder
~, сотенный hundreds selector
~, сцинтилляционный scintillation scanner
~, транзитный tandem selector
~, тысячный thousands selector
~, шаговый step-by-step selector switch

исключать exclude; expel; delete; eliminate
исключение 1. *math.* elimination; removal, deletion; exclusion; 2. exception
~ **неизвестных** elimination of unknowns
~ **параллакса** elimination of parallax
~ **символов** *comp.* character deletion
исковое заявление *law* statement of claim
искомое *math.* unknown/desired [sought-for] quantity
ископаемое fossil; mineral; mined material
~, **крупное** macrofossil
~, **мелкое** microfossil
~, **нерудное** non-metallic mineral
~, **окаменелое** petrified/mineralized fossil
~, **полезное** commercial/useful mineral; *pl.* mineral resources/wealth
~, **рудное** ore mineral, ore
искра spark
~ **зажигания/запальная** ignition spark
~, **лазерная** laser-induced spark
~ **при размыкании** break spark, spark at breaking
искрение sparking; arcing
~ **коллектора** commutator sparking
~ **щеток** brush sparking
искривление bend, bending; curving, curvature
~ **изображения** *telev.* curvature of the image field
~ **кпереди** procurvation
~ **назад** recurvation
~ **скручиванием** torsional warping
искрить spark, flash
искробезопасный sparkproof
искрогаситель sparkarrester/extinguisher/killer
искрогашение spark quenching/extinguishing/ suppression; arc suppression/quenching
искроловитель spark arrester/trap/catcher
искромер spark meter, scintillometer
искрообразование spark formation, sparking
искростойкий sparkproof
искростойкость arc/spark resistance
искротушитель spark quencher/extinguisher
искроудержатель spark arrester
искроуказыватель spark detector/indicator
искроуловитель spark trap/arrester/catcher
искусственный artificial, man-made
~ **каучук** synthetic rubber
~ **шелк** artificial silk
~**ное питание** bottle-feeding
~**ные зубы** false/artificial teeth
искусство 1. art; 2. skill, skil(l)fulness; craftsmanship
~ **верховой езды** horsemanship
~, **военное** art of war
~ **высшего пилотажа** aerobation
~ **гребли** oarsmanship
~ **кораблевождения** sea-craft
~, **летное** airmanship
~, **прикладное** applied art
~**ва, изобразительные** visual/pictorial arts
~**ва, изящные** fine arts
искусствовед art critic
искусствоведение study of art; art criticism
искусствоведческий art-study *(attr.)*
~ **факультет** department of fine art
исламизм Islamism
исламистский Islamic
исламовед Islamist, specialist in Islamic studies
исландец, исландка Icelander
исландский Icelandic
Исмаил *bibl.* Ishmael
исмаилит Ishmaelite, Ismaelite
исмаилитский Ismaelian, Ismalian
испанец Spaniard
испанизм Hispanism
испанист, испанистка Hispanist
испанистика Spanish studies, Hispanics
испанка Spanish woman
испано-американский Spanish-American
испанский Spanish, Hispanic
испарение evaporation, vaporization, volatilization; boil-off; fumes, vapo(u)r, effluvium; exhalation
~, **естественное** spontaneous evaporation
~ **из почвы** soil evaporation
~ **кислорода** oxygen boil-off
~, **мгновенное** flash vaporization; instantaneous evaporation
~, **повторное** re-evaporation
~ **с поверхности** surface evaporation
~ **твердых веществ** sublimation
~ **топлива** fuel vaporization
~, **устьичное** stomatal transpiration
~, **ядовитое** poisonous/noxious/deadly fumes
испаритель evaporator; vaporizer
~, **вакуумный распылительный** vacuum spray evaporator
~ **высокого давления** high-pressure evaporator
~, **газовый** gas-heated evaporator
~ **кислорода** liquid-oxygen converter
~, **кожухотрубный** shell-tube evaporator
~, **многоступенчатый** multiple-effect evaporator
~ **низкого давления** low-pressure evaporator
~, **пластинчатый** plate-type evaporator
~ **пленочного типа** film-type evaporator/boiler
~, **противоточный** counter-flow evaporator
~, **работающий под давлением** pressure evaporator
~ **с погруженными трубками** submerged-tube evaporator
~ **с принудительной циркуляцией** forced-circulation evaporator
~ **с ребристыми трубками** finned evaporator
~, **тонкопленочный** thin-film evaporator
~ **топлива** fuel vaporizer
~, **трубчатый** tubular evaporator
~, **утилизационный** heat-recovery evaporator
~ **холодильника** refrigerator evaporator
испарительный evaporating; evaporative
испарометр evaporimeter

испаряемость evaporativity, vaporability; volatility; evaporating capacity
испаряемый evaporable; volatile
испарять, испаряться evaporate, vaporize, volatilize; transpire
~ **мгновенно** flash out
испаряющийся evaporable, vaporable; volatile
испепелять incinerate, burn to ashes
испещреннолепестковый *bot.* stictopetalous
исповедание confession, creed
исполком *abbr.* executive committee
исполнение *1.* fulfilment, implementation, execution; realization; 2. design, modification, version, construction, embodiment, make; *law* feasance; *(of duty)* discharge
~ **герметическое** watertight construction/modification/design
~ **запроса** *comp.* query execution/solving
~ **тропическое** tropicalized construction/embodiment/modification/version
~, **пошаговое** *comp.* single-step execution
исполнимость feasibility, practicability
исполнимый feasible, practicable; realizable; possible
~, **вполне** perfectly feasible
исполнитель performer, executant; executor
~, **судебный** bailiff; law-enforcement officer
исполнительный executive
~ **лист** writ of execution, court order, writ
~ **комитет** executive committee
~ **приказ о вводе во владение** *law* writ of assistance
~**ная власть** executive (power)
исполняющий обязанности deputy, acting *(attr.)*
исполовник = испольщик
исполу half and half
~, **аренда** métayage, sharecropping
использование 1. utilization, use, employment; consumption; application; 2. recovery
~, **аварийное** emergency use
~ **в полевых условиях** field use
~ **контейнеров** containerization
~, **непотребительское** *econ.* nonconsumptive use
~, **потребительское** *econ.* consumptive use
~ **ресурсов, совместное** resourse sharing
~ **тепла** heat utilization/recovery
испольничать *hist.* farm land on the métayage system, sharecrop
испольничество *hist.* métayage,sharecropping
испольщик *hist.* métayer, sharecropper
испольщина *hist.* métayage, sharecropping
исправдом *abbr.* reformatory
исправительно-трудовой corrective-labour
~**вая колония** reformatory
исправительный corrective; remedial
~ **дом** reformatory
исправление correction, correcting; improvement; mending, repairing
~ **двойных ошибок** double-error correction
~ **дефектов** removal/rectification of defects
~ **единочных ошибок** single-error correction
~ **многократных ошибок** multierror correction
~ **ошибок** *comp.* error removing, debugging
~ **ошибок на перфокартах** (punched) card correction
~ **пакета ошибок** burst-error correction
исправник *hist.* district police superintendant
исправность good/working/operable/serviceable condition; serviceability
~**ти, в полной** in perfect order
исправный in good repair, in good working order serviceable, operable
испражнение defecation, evacuation
испражнения *pl.* excrements, faeces, stools
испускание emission, emitting, emanation; release; *(rays)* radiate
~ **термоэлектронов** thermionic emission, Edison effect
испускать emit, emanate, give off, let out, release
~ **лучи** radiate
~ **энергию** emit energy
испытание test, trial, testing; experiment, run; tryout; assay
~, **аттестационное** certification test
~ **без нагрузки** no-load test
~ **без разрушения образца** non-destructive testing
~, **буксировочное** towing test
~ **в арктических условиях** arctic test
~ **в аэродинамической трубе** wind-tunnel test
~ **в барокамере** altitude-chamber test
~ **в боевых условиях** combat testing; battle service test
~ **в гидробассейне** water-tank test
~ **в гидроканале** towing-basin test
~ **в космическом полете** space flight testing
~ **в натурных условиях** full-scale testing
~ **в непревывном режиме** continuous test
~ **в полевых условиях** field test
~ **в полете** in-flight testing, flight trial/test
~ **в потоке** flow test
~ **в прерывистом режиме** intermittent test
~ **в процессе доводки** development test
~ **в рабочих условиях** operation test
~ **в реверберационной камере** reverberant testing
~ **в свободном падении** free-fall testing
~ **в термобарокамере** thermal vacuum chamber test
~ **в тяжелых условиях работы** heavy-duty test
~ **в условиях невесомости** weightlessness test, zero-G test
~ **в условиях низких температур** winterization trial
~ **в штопорной аэродинамической трубе** spin-tunnel test

~ в эксплуатационных условиях field service test, test under operational conditions
~, внутриреакторное in-pile test
~, всестороннее comprehensive test
~, выборочное lot-by-lot testing, random/sample test
~, гарантийное guarantee test
~, гидравлическое hydraulic-pressure test
~ до разрушения breakdown test
~, доводочное developmental test
~ долговечности life service test
~, дорожное on-the-road test, road test
~, заводское factory/shop test, manufacturer's test
~ имитирующее аварию controlled crash test
~ имитирующее неисправности malfunctioning test
~, инерционное *naut.* stopway/stopping test
~, климатическое climate/environmental test
~, комплексное comprehensive test
~, контрольное routine check/test
~, лабораторное laboratory test; bench test
~, макетное mock-up test
~, маневренное *naut.* manoeuvrability trial
~ материалов material testing
~ материалов, неразрушающее non-destructive material testing
~ методом меченых атомов tracer test
~ методом моделирования simulation test
~, мореходное seakeeping/seaworthiness trial
~ на абразивный износ abrasion test
~ на адекватность *stat.* test for goodness of fit
~ на больших скоростях high-speed test
~ на борту космического аппарата space-borne test
~ на вибропрочность vibration-survival test
~ на вибростенде shake-table test
~ на виброустойчивость vibration-resistance test
~ на водоотталкиваемость *text.* water repulsion test
~ на воздействие внешних условий environmental test
~ на воздействие ускорений acceleration test
~ на вскипание effervescent test
~ на выдавливание dishing/cupping test
~ на выживание survival test
~ на выносливость endurance/fatigue test
~ на вытяжку cupping test
~ на вязкость 1. *(of fluid)* viscosity test; 2. *(of solids)* toughness test
~ на герметичность leakage/tightness/seal test; air-pressure test
~ на горячее кручение hot twist test
~ на долговечность service-life test, longevity/durability test, length-of-life test
~ на жидкотекучесть fluidity test
~ на загиб bending/angular test; bend-over test
~ на замораживание и оттаивание freeze-thaw test
~ на знакопеременный изгиб alternating bending test, bending-and-unbending test
~ на изгиб bending test
~ на излом fracture test
~ на износ wear/wearing/attrition test
~ на истирание abrasion test
~ на истирание при смятии *text.* crease-abrasion test
~ на кип boiling test
~ на коррозию corrosion test
~ на кручение torsional/twisting test
~ на крупномасштабной модели large-scale model test
~ на ломкость friability test
~ на макете mock-up test
~ на манекене dummy testing
~ на месте insitu testing
~ на месте установки/эксплуатации site testing
~ на модели simulation/analogue test
~ на моделях model/dummy test, mock-up test
~ на морозостойкость subzero/freezing test
~ на надежность reliability test; failure test
~ на надрыв tear test
~ на натурном объекте full-scale test
~ на непроницаемость tightness test
~ на обрабатываемость machinability test
~ на огнестойкость fire resistance test, fire-proofing test
~ на окисляемость oxidation test
~ на отказ fault testing
~ на перегиб bend-over test; backward-and-forward bending test
~ на перегрузку overload test
~ на плавучесть flotation test
~ на плавкость melting/fusion test
~ на плотность leak testing
~ на ползучесть creep test
~ на пробой breakdown test
~ на прогиб flexure test
~ на продолжительную работу endurance test
~ на продольный изгиб buckling test
~ на прочность strength test
~ на раздачу *(of pipe)* flare/flaring test
~ на разрыв breaking test; stress-to-rupture test
~ на раскалывание splitting test
~ на расплющивание flattening test
~ на растворимость solubility test
~ на растрескивание cracking test
~ на растяжение tensile test, tension testing
~ на сброс/сбрасывание drop test
~ на свариваемость weldability test
~ на сверхзвуковых скоростях supersonic test
~ на свойлачиваемость *text.* milling test
~ на сгораемость combustibility test
~ на сдвиг shear(ing) test
~ на сжатие compression test
~ на скороподъемность *av.* climbing test

~ на скручивание twisting/torsional test
~ на совместимость compatibility testing/trial
~ на сохраняемость storage test
~ на спекаемость sintering test
~ на сплющивание flattening test
~ на срез shear(ing) test
~ на срок службы life/service test
~ на срок хранения shelf-life test
~ на старение ageing test
~ на срыв stall(ing) test
~ на стенде bench test
~ на стойкость к атмосферным воздействиям weathering test
~ на твердость hardness test
~ на термический удар thermal-shock test
~ на трещинообразование cracking test
~ на тропические условия tropical-exposure test
~ на удар impact/shock test; mechanical-shock test(ing)
~ на ударную нагрузку dropimpact test
~ на упругость elasticity test
~ на усталость fatigue test
~ на устойчивость stability test
~ на утечку leakage test
~ на флаттер flutter test
~ на холостом ходу no-load test
~ на хрупкость friability test
~ на центрифуге centrifuge testing
~ на эксплуатационные показатели performance testing
~ на этапе разработки development trial(s)
~ навигационной системы navigational system trial
~, наземное ground test
~, натурное full-scale test
~, неразрушающее non-destructive testing
~, низкотемпературное low-temperature [cold-weather, subzero, winterization] test
~ облучением irradiation test
~, оценочное evaluation test
~ партии изделий batch test
~, повторное duplicate/repeated test
~ погружением immersion test
~ под давлениеи pressure test
~ под нагрузкой load(ing) test
~ под напряжением *el.* voltage test
~ подводное underwater test/ing)
~, полевое field test
~ полигонное proving-ground test
~ послеполетное post-flight test
~ предварительное preliminary test
~ предполетное pre-flight test
~ предпусковое pre-operational test; trial before commissioning
~ предстартовое pre-launch test
~ при высоких давлениях high-pressure test
~ при низкой температуре low-temperature test, subzero/cold test
~ при полной скорости full-speed trial
~ при повышенных температурах elevated-temperature test
~ при постоянной нагрузке steady/constant load test
~ приемо-сдаточное approval test
~ приемочное acceptability test; official acceptance test
~ проверочное test trial
~ промышленное factory/commercial/production test/trial
~ психотехническое ability test; personal performance test
~ пробное run-in test
~ прочности strength testing
~ простукиванием hammering test
~ разгонное overspeed test
~, рентгеновское X-ray test
~ с разрушением образца destructive test
~, с частичным разрушением образца semidestructive test
~ сварного соединения/шва weld test
~, сдаточное delivery trial
~ сжатым воздухом air-pressure test
~, склерометрическое scratch-hardness test
~, скоростное speed trial
~, случайное random trial/test
~, стендовое bench test
~, стопроцентное total-lot test, 100% test
~ судом божьим *hist.* trial by ordeal
~, тарировочное calibration test
~, тензометрическое strain-gauge test
~, техническое engineering/technical trial
~, типовое standard/routine/type test
~ травлением pickle test
~ трением friction test
~, тяговое pull test
~ ультразвуковым методом ultrasonic test
~ умственных способностей intelligence/mental test
~, усталостное fatigue test
~, форсированное accelerated test
~, ходовое road test; *naut.* sea trial
~ холодной штамповкой cold-pressing test
~ царапанием scratch test
~, швартовное dockside trial
~, эксплуатационное operational testing, service test/trial; performance test

испытанный tested, well-tried; proven
~ в полевых условиях field-tested/-proven
~ временем time-tested
~ на модели model-tested

испытатель 1. testing apparatus, tester, checker, test set; 2. tester, investigator; analyst; checker
~ изоляции insulation tester
~ проводки wiring tester
~ радиоламп tube tester/checker

испытательный test, trial, experimental; probationary
~ полигон proving/testing ground
~ срок trial period

~ **стаж** probation period
испытуемый, испытываемый under test, tested; experimental
испытывать test, subject to a test/trial, try out; assay
иссекать dissect; cut up
иссечение dissection; excision
исследование investigation, research, study; exploration; analysis; examination; survey
~ **атмосферы** atmospheric research
~, **ведущее** pilot research/study
~ **верхней атмосферы** upper-atmosphere [high-altitude] research
~, **влагалищное** *med.* vaginal examination
~ **возможностей применения** feasibility study
~ **выборки** *stat.* sample test
~, **глубоководное морское** deep-sea exploration
~ **изделия** *econ.* product analysis
~ **ионосферы** ionospheric research
~ **космического пространства** space exploration/research
~ **кривой** *math.* curve tracing
~ **крови** blood analysis
~, **макроскопическое** macroexamination
~, **медицинское** medical examination, medical/physical check-up
~ **методом изотопных индикаторов** labelled-atom investigation, tracer study/investigation
~, **микроскопическое** microscopic examination, microexamination
~ **мирового пространства** space research
~, **модельное** model research/study
~ **на моделирующей установке** simulator study/investigation
~, **научное** scientific research
~ **околоземного пространства** near-terrestrial [terrestrial-space] research
~ **окружающей среды** environmental study
~ **операций** operations research; operational/operations analysis/evaluation
~ **поведения потребителей** habit survey
~, **прикладное** applied research
~, **радиоавтографическое** autoradiographic study
~, **патологоанатомическое** postmortem examination/study
~, **перспективное** advanced research
~, **поисковое** exploratory research/study
~, **ректоскопическое** *med.* proctoscopic/rectoscopic examination
~, **рентгенологическое** X-ray examination
~, **рентгеноструктурное** X-ray diffraction analysis/study
~ **трупа** *law* post-mortem examination
~ **уноса массы** ablation study
~, **шаропилотное** ballooning
~, **эзофагоскопическое** *med.* (o)esophagoscopic examination
~, **электрокардиографическое** electrocardiographic examination
~, **электронно графическое** electron diffraction study
~ **языка** *med.* glossoscopy
исследования *pl.* studies, etc. *see* **исследование**
~, **газодинамические** gas-dynamics research
~ **и опытные разработки** research and development, R and D
~, **космические** space research
~, **межпланетные** interplanatary research
~ **фундаментальные** basic/fundamental research
исследователь researcher; investigator; examiner; explorer
исследовательский exploratory; investigatory; research *(attr.)*
исследовать investigate, study, examine; research into, inquire into; explore; prospect
исследуемый under study/investigation
иссоп *bot.* hyssop *(Hyssop)*
иссушение exsiccation; aridization
истерика hysterics; hysterical fit
истерический, истеричный hysteric(al)
истекать *law* fall due, expire; elapse
истерия *med.* hysteria
~, **военная** *fig.* war hysteria
истерокаталепсия *med.* hysterocatalepsy
истероневрастения *med.* hysteroneurasthenia
истероэпилепсия hysteroepilepsy
истец *law* plaintiff; demandant; suitor; complainant
~ **по апелляции** appellant; plaintiff in error
истечение 1. efflux, outflow, discharge; 2. expiration, expiry; lapse
~ **крови** *med.* haemorrhage
~ **срока аренды** expiration of lease
~ **срока страхования** expiration of insurance
~ **струи** jet efflux
истина, истинность *math.* truth; *comp.* true
истираемость wear, wearability
истирание abrasion, attrition, detrition, wearing away; *met.* galling, fretting
~ **ледниковое** ice abrasion
~ **пневматика** tyre scuffing/wearing
истирать(ся) wear (out), abrade; *met.* gall
истление disintegration, decay; deterioration
истмат *abbr.* historical materialism
истод *bot.* milkwort, polygala *(Polygala)*
истодовые *pl.* *bot.* milkwort family *(Polygalaceae)*
истмоплегия *med.* isthmoparalysis, isthmoplegia
истмэктомия isthmectomy
исток headwater; source; riverhead; efflux, issue, outflow
историзм historical method/approach; historism
историко-географический historico-geographical
историко-критический historico-critical

историко-сравнительный comparative-historical
историко-филологический historico-philological
историограф historiographer
историография historiography
историология historiology
исторический historical; *fig.* epoch-making
историчность historicity
история history
~ **болезни** *med.* case record
~ **древнего мира** ancient history
~ **искусства** history of art
~, **новая** modern history
источник source; supply; *pl.* resources
~ **аварийный** emergency source
~ **альфа-/бета-излучения** alpha/beta source
~, **внешний** external/extraneous source
~, **внутренний** internal source
~ **данных** *comp.* data source
~, **дуговой** arc source
~ **звука** acoustic/sound source; sound generator
~ **излучения** radiation source
~, **импульсный** pulsed source
~ **импульсов** impulse/pulse source
~ **инфекции** source of infection; focal point of infection
~ **инфракрасного излучения** infra-red source
~ **ионный** ionic source
~ **напряжения** *el.* voltage source
~, **нестационарный** transient source
~, **непрерывный** continuous source
~ **облучения** irradiation source
~, **открытый** bare source
~ **ошибки** error source, source of error
~, **пленочный жидкий** liquid film source
~, **постоянный** steady source
~ **поля** field source
~ **питания** *el.* power supply
~ **питания, аварийный** emergency power source
~ **радиации** radiation source
~, **радиевый** radium source, pack of radium
~ **радиоактивного излучения** radioactive source
~ **света** light source
~, **слоеный** sandwich(ed) source
~ **тока** current source; power supply
~, **точечный** point source
~ **электронов** electron(-emitting) source
~ **энергии** power source/generator
~, **эталонный** standard/reference source
истощать deplete, drain, exhaust; sap; emaciate
~ **почву** impoverish/exhaust/deplete the soil
истощение exhaustion, depletion; draining; petering out; attrition; emaciation
~, **война на** war of attrition
~ **вследствие голодания** inanition
~, **крайнее** *med.* skeletization
~ **месторождения** *min.* depletion of deposit
~ **нервной системы** nervous breakdown
~, **общее** *med.* cachexia
~ **почвы** soil depletion/exhaustion
~ **скважины** decline of well
истпарт *abbr.* history of the party
истребитель fighter (aircraft)
~ **беспилотный** unmanned/pilotless fighter
~ **вертикального взлета и посадки** vertical take-off and landing fighter, VTOL fighter
~, **высотный** high-altitude fighter
~, **легкий** lightweight fighter
~, **многоцелевой** multi-purpose/-mission fighter
~, **ночной** night fighter
~, **палубный** shipboard/carrierborne fighter
~, **реактивный** jet fighter
~, **сверхзвуковой** supersonic fighter
~ **сопровождения** escort fighter
истребитель-бомбардировщик fighter-bomber, strike fighter
истребитель-перехватчик interceptor (fighter)
истребитель-разведчик reconnaisance fighter
истребитель-ракетоносец missile-carrying fighter
исход outcome; way out; result
~ **беременности** pregnancy termination
~, **достоверный** *game th.* sure outcome
~, **ненаблюдаемый** *game th.* unobserved outcome
~, **смертельный** fatal outcome/termination
исходный initial, starting; original; parent
~ **падеж** *gram.* ablative case
~ **рубеж** starting point/line
~**ная конфигурация** source configuration
~**ные данни** master data
исхомения *med.* ischomenia
исхудание, исхудалость emaciation, haggardness
исцеление healing, cure; recovery
исцелимый curable
исчезновение disappearance, vanishing; extinction
~ **вида/рода** extinction of species/genus
исчерпывать exhaust, deplete; drain
исчерченность striation, lineolation
исчисление calculation, computation, estimation; calculus
~ **бесконечно малых** infinitesimal calculus
~, **вариационное** calculus of variations
~, **векторное** vector calculus/analysis
~ **вероятностей** calculus of probability/probabilities
~ **высказываний** propositional/sentential calculus
~, **дифференциальное** differential calculus
~ **задач** problem calculus
~ **издержек производства** *econ.* costing
~, **интегральное** integral calculus
~ **конечных разностей** calculus of finite differences
~, **матричное** matrix calculus/arithmetic/calculation

~, операционное operational/operator calculus
~ предикатов predicate/predicative calculus
~, предположительное *stat.* conjecture
~, пропозициональное propositional calculus
~, расширенное extended calculus
~, реляционное relational calculus
~, ситуационное situation calculus
~, спинорное spinor calculus
~, тензорное tensor calculus
исчислимый calculable, countable, numerable
исчислять calculate, compute; estimate; evaluate
исщербить dent, jag, serrate
италийский *hist.* Italic
италики *hist.* Italici
итало-византийский Italo-Byzantine
итало-греческий Italo-Grecian
итальянец Italian
итальянка Italian (woman)
итальянский Italian
~кая бухгалтерия double-entry bookkeeping
~кая земля burnt sienna
~кая забастовка sit-down strike
~кое окно Venetian/Palladian window
ительмен, ительменский Kamchadal, Itelmes
ительменка Kamchadal (woman)
итеология iteology
итератив *gram.* iterative aspect
итерационный iterational, iterative, repetitive
итерация, итерирование iterating, iteration process
итерировать iterate
итея *bot.* sweet spire *(Itea)*
итог 1. total, sum, balance; 2. summary; result
~, главный major total
~, общий *econ.* grand total, sum total
~, подводить/подвести totalize, total up, sum up
~, промежуточный subtotal, intermediate total
~, суммарный grand total, sum/summary total
~, частичный minor total
итоговый 1. total; grand; 2. final
иттербий *chem.* ytterbium, Yb
иттербовый *chem.* ytterbic
иттриалит yttrialite
иттриевый *chem.* yttric
иттрий *chem.* yttrium, Y
иттрогранат yttrium garnet
иттрогумит *miner.* yttrogumite
иттроколумбит *miner.* yttrocolumbite
иттрокразит *miner.* yttrocrasite
иттрофлюорит *miner.* yttrofluorite
иттроцерит *miner.* yttrocerite
Иуда Judas
иудей Jew, Israelite
иудейский Judaic
иудейство Judaism
иудино дерево *bot.* Judas tree *(Cercis siliquastrum)*
ихневмон *zool.* ichneumon *(Herpestes)*
ихнограмма ichnogram
ихнография ichnography
ихнология *geol.* ichnology
ихнофит ichnophyte
ихор ichor
ихорозно-гнойный *med.* ichoropurulent
ихорозный *med.* ichorous
ихороторакс ichorothorax
ихтиоакантотоксин ichthyoacanthotoxin
ихтиизм *med.* ichthyism; botulism
ихтиография ichthyography
ихтиоз *med.* ichthyosis, fishskin-disease, alligator-skin disease, xeroderma
~, ежовый ichthyosis hystrix
~, фолликулярный keratosis follicularis/vegetans
ихтиозавр *paleont.* ichthyosaurus
ихтиокол ichthyocol(la), isinglass
ихтиол *pharm.* ichthammol, ichthyol
ихтиолит *paleont.* ichthyolite
ихтиолог ichthyologist
ихтиология ichthyology
ихтиорнис *paleont.* ichthyornis
ихтиофауна ichthyofauna, fish fauna
ихтиофтальм *miner.* ichthyophthalmite, apophyllite
ихтиохория ichthyochory
ихтитерий ichthythere
ишемический *med.* isch(a)emic
ишак *zool.* donkey *(Equus asinus)*
ишемия *med.* isch(a)emia
~, компрессионная compression isch(a)emia
~ миокарда myocardial isch(a)emia
~ мозга cerebral isch(a)emia
~, обтурационная obstruction isch(a)emia
~, постуральная postural isch(a)emia
~ сетчатки isch(a)emia retinae
ишиальгия *med.* ischialgia
ишиас *med.* sciatica, ischias, ischialgia, ischiatica, ischioneuralgia
ишиатик sciatic patient
ишиатический *med.* sciatic
ишикаваит ishikawaite
ишиопаг ischiopagus
ишиопубиотомия ischiopubiotomy
ишурический ischuric, ischuretic
ишурия *med.* ischuria

Й, й

йеллоустонский Yellowstone
йельский Yale
Йемен Yemen
йеменец, йеменский Yemeni
йог yogi
йога yoga
йогурт yoghurt
йод *chem.* iodine, I

йодамин *chem.* iodamine
йодат *chem.* iodate
  ~ кальция calcinol
йодбензол *chem.* iodobenzene
йодид *chem.* iodide
  ~ калия potassium iodide
йодизм *med.* iodism, iodine poisoning
йодиназа iodinase
йодирит *miner.* iodyrite, iodite, iodargyrite
йодирование iodization, iodination
йодировать iodate, iodinate; iodize
йодистоводородный *chem.* hydroiodic
йодистокислый *chem.* iodous
йодистый iodide of
  ~ калий potassium iodide, iodide of potassium
йодит iodite
йодкрахмал iodized starch
йодль yodel
йодметан *chem.* iodomethane
йодноватокислый iodate of
йодный iodic
  ~ная кислота periodic acid
  ~ная настойка tincture of iodine
  ~ное число iodine number
йодобромит *miner.* iodobromite
йодоводород *chem.* hydrogen iodide
йодогност *pharm.* iodophthalein sodium
йододерма iododerma
йодолипол iodolipol, iodipin, iodized oil, lipiodol
йодометрия iodometry
йодопсин iodopsin
йодотерапия iodotherapy
йодотироглобин iodothyroglobin
йодотирин iodothyrin
йодотирозин iodothyrosine
йодотиронамин iodothyronamine
йодотиронин iodothyronine
йодофен iodophen
йодофильный iodophilic
йодоформ iodoform
йодэмболит iodembolite
йоркшир, йоркширский Yorkshire
йот J *(letter);* yod *(sound)*
йота iota
йотация, йотирование *phon.* iotacism; iotization, vowel softenig
йотизация *phon.* iotization, palatalization
йотированный, йотованный, йотовый palatized
йотнийский *geol.* Jotnian

К, к

кааба Kaaba
каама *zool.* Cape hartebest *(Bubalis caama)*
кабала *hist.* kabala (slavery); *see also* **каббала**
кабальный kabala *(attr.); fig.* enslaving
  ~ холоп kabala serf
кабан I *zool.* 1. wild boar *(Sus scrofa);* 2. boar, pig *(male)*
кабан II 1. *aero.* cabane; 2. block
  ~ несущего винта main rotor-pylon *(of helicopter)*
кабарагойя *zool.* Asiatic/water monitor *(Varanus salvator) zool.* musk deer *(Moschus)*
кабардинец, кабардинский Kabardian; Kabardian horse
кабардинка 1. Kabardian (wovan); 2. Kabardian dance; 3. Circassian goat
каббала cab(b)ala
каббалистика cab(b)alism
каббалистический cab(b)alistic, cab(b)alic
кабелеискатель cable locator/detector
кабелепровод cable duct/conduit
кабелепроводка cable network, cabling
кабелеукладчик cable layer, cable-laying machine
кабель cable
  ~, абонентный/абонента service cable
  ~, антенный aerial cable
  ~, бронированный armoured cable
  ~ в оболочке sheathed cable
  ~, ведущий leader/piloting cable
  ~, взрывной blasting cable
  ~, воздушный overhead/overground cable
  ~ высокого напряжения high-voltage cable
  ~, высокочастотный high-frequency cable
  ~, гибкий детонирующий flexible explosive cable
  ~ гидролокатора sonar cable
  ~, глубоководный deep-sea cable, subaqueous cable
  ~, городской телефонный exchange cable
  ~, двухжильный double-core cable
  ~, двухпарный double-D cable
  ~, детонирующий detonating/explosive cable
  ~ дистанционного управления remote control cable
  ~ для наружной установки outdoor/external cable
  ~ зажигания ignition cable
  ~ звездной скрутки spiral-quad cable, starquad cable
  ~, каротажный borehole/logging cable
  ~, коаксиальный coaxial/concentric cable
  ~, контролный control cable; pilot cable
  ~, коронкостойкий corona-proof cable
  ~, крарупизированный continuously loaded cable, Krarup cable
  ~, ленточный ribbon cable
  ~, лифтовой lift cable; *Am.* elevator cable
  ~, магистральный main/trunk cable
  ~, маслонаполненный oil-filled cable
  ~, междугородный long-distance cable
  ~, многожильный multiple-conductor [multicore] cable; stranded wire cable
  ~ наведения guiding cable
  ~, одножильный/однопроводный single-

core [single-conductor] cable
~, осветительный lighting cable
~, освинцованный lead-covered cable
~ парной скрутки paired cable
~, питающий feeder cable
~, плавучий buoyant cable
~, подвесной overhead cable
~, подводный submarine/underwater cable
~, подземный underground/buried/subterranean cable
~, проводной wiring cable
~, пупинизированный coil-loaded cable
~, пусковой firing cable *(of rocket)*
~, радиочастотный radio-frequency cable
~, размагничивающий degaussing cable
~ с бумажной изоляцией paper-insulated cable
~ с воздушно-бумажной изоляцией dry-core cable, air-space cable
~ с лаковой изоляцией varnished cable
~ с малыми потерями low-loss cable
~ с металлической оплеткой screened cable
~ с пластмассовой изоляцией plastic insulated cable
~ с поясной изоляцией belted cable
~ с резиновой изоляцией rubber-insulated cable
~ с хлопчатобумажной изоляцией cotton-covered/insulated cable
~ с центральным каналом annular cable
~ с эмалевой изоляцией enamelled-wire cable
~, сварочный welding cable
~ сверхвысокого напряжения extra-high-voltage cable
~, сверхпроводящий superconducting cable
~, световой light-conducting/transmitting cable
~ связи communication cable
~, силовой power cable
~, симметричный balanced cable
~ системы наведения guiding cable
~, слаботочный communication cable, weak-current cable
~, соединительный interconnect cable
~, станционный inside/office cable
~, ступенчатый graded cable
~, четверочный quadded cable
~, шахтный mine/shaft cable
~, шланговый flexible trailing cable
~, экранированный shielded cable
~, электрический conductor/wiring cable
кабель-кран cableway
кабель-план cable layout
кабельтов cable's length, cable (185, 2 m)
кабельщик cable technician/splicer/jointer
каберне Cabernet wine/grape
кабестан capstan
кабил, кабильский Kabyle
кабина booth, cabin, cab; *av.* cockpit
~ водителя *auto.* driver's cab
~, высотная high-altitide cabin
~, герметическая pressurized cabin
~, грузовая freight/cargo cabin/hold
~, двухпалубная double-decker compartment
~, дикторская announcer booth
~, душевая shower cubicle/room/compartment
~, задняя *naut.* rear/aft cabin
~, звукоизолированная sound-proof(ed) cabin
~, катапультируемая ejectable cabin
~, кондиционированная conditioned cabin
~, космическая space cabin/capsule
~ космического корабля spacecraft cabin
~ летчика cockpit
~ лифта lift cabin; *Am.* elevator car
~, отделяемая *av.* ejection/jettisonable cabin/capsule; escape module, detachable cabin
~, пассажирская passenger compartment/cabin
~, переговорная telephone booth
~ подъемника hoist box
~ радиста radio operator's cabin, wireless/radio room/cabin
~ стрелка *mil.* gunner's compartment
~ тренажера simulator cabin
~ управления двигателем engine control room
~ управления control cabin
~, хвостовая aft/rear/tail cabin
~ штурмана navigator's compartment
~ экипажа crew compartment/cabin; common cockpit
~ экскаватора excavator cab
~, экранированная screened room
кабинет study; consulting room; office *(private);* laboratory
~, зубоврачебный dental surgery
~, косметический beauty shop
~, рентгеновский X-ray room
~, физический physics laboratory
каблирование cable-laying, cabling
каблограмма cable message, cablegram
каблук heel
каболка *naut.* rope/cable yarn, hemp yarn
~, смоленая fox
кабомба *bot.* fanwort *(Cabomba)*
каботаж coasting/coastal navigation, cabotage
~, большой intercoastal navigation
~, малый inshore navigation
каботажный coastal; coastwise; coasting
~ное плавание coastal navigation/traffic
~ное судно coasting ship/vessel, coaster
кабриолет cabriolet
кабрирование *aero.* pitching-up, pitchup, nose-up
~ без крена symmetrical pull-up
~ с вращением вокруг продольной оси rolling pitchup
кабрировать *aero.* pitch up
кавалер *zool.* swallowtail *(Papilio)*
кавалергард horse guardsman
кавалерийский cavalry *(attr.)*
~кая лошадь cavalry horse, trooper

кавалерист cavalryman
кавалерия cavalry
 ~, легкая light horse
кавалла *zool.* king mackerel *(Scomberomorus cavilla)*
кавальер spoil bank; earth deposits
каватина *mus.* cavatina
каверна *med., geol.* cavern; cavity; pocket; *met.* flaw
 ~, бронхоэктатическая bronchiectasic cavern
 ~, верхушечная apical cavern
 ~, газовая gas pocket
 ~, костная osteal cavern
 ~, легочная lung cavern
 ~ почки renal cavern
 ~, туберкулезная tuberculous cavern
кавернозность cavernosity
кавернозный cavernous; porous; vesicular
кавернит *med.* cavernitis, cavernositis
каверноma *med.* cavernoma, cavernous angioma
каверномер (borehole) caliper, downhole gauge
кавернометрия borehole gauging
кавернoстомия *med.* cavernostomy, spleostomy
кавернотомия *med.* cavernotomy
 ~ почки renal cavernotomy
кавитация cavitation
 ~ в насосе pump cavitation
 ~, газовая gaseous cavitation
 ~, искусственная forced cavitation
 ~, начальная incipient cavitation
 ~, паровая vaporous cavitation
 ~, пленочная sheet cavitation
 ~, пузырчатая bubble cavitation
 ~, ультразвуковая ultrasonically-induced cavitation
 ~, установившаяся stationary cavitation
кавказец, кавказский Caucasian
кавказка Caucasian (woman)
кавказовед specialist in Caucasian studies
кавказоведение Caucasian studies
кавычки quotation marks, quotes
кагал *hist.* kahal
каган *hist.* chagan; khan
каганат *hist.* chaganate; khanate
кагор Cahors wine
кагу *zool.* kagu *(Rhinochetus jubatus)*
кагуан *zool.* common cabego *(Cynocephalus variegatus)*
кагуляр *polit.* Cagoulard
кадаверин *chem.* cadaverine
каданс cadence
кадастр cadastre
 ~, водный/земельный water/land cadastre
 ~, лесной forest cadastre
 ~, поземельный land cadastre
 ~, солнечный solar cadastre
кадастровый cadastral
каддиш Kaddish
каденция, каденца *mus.* cadenza; *lit.* and *mus.* cadence
кадет I *mil.* cadet
кадет II *abbr., hist.* Constitutional democrat, Cadet
кадетский корпус *hist.* military school
кади cadi, kadi, kadee
кадило I thurible, incensory, censer
кадило II *bot.* melittis, mellitis *(Melittis, Mellitis)*
Кадм *myth.* Cadmus
кадмеец, кадмейский *hist.* Cadmean
кадмиевый *chem.* cadmic
кадмий *chem.* cadmium, Cd
 ~ сернистый cadmium sulphide
 ~, сернокислый cadmium sulphate
кадмирование cadmium plating
кадмировать cadmium-plate
кадр I frame, cinema still; shot, exposure; scene
 ~ диафильма filmstrip frame
кадр II *mil.* cadre; *pl.* regulars;
кадрирование framing
кадры *pl.* 1. personnel, staff; manpower; skilled personnel/employees/workers; specialists; 2. *colloq.* personnel department
 ~, научные scientific manpower, brainpower, scientists
 ~, технические technical manpower, technicians, technical specialists
 ~ торгового флота maritime manpower
кадсура *bot.* kadsura *(Kadsura)*
кадуцей caduceus
кадык Adam's apple
кадь *obs.* vat, large tub; unit of measure equivalent to 104, 96 litres
каетинец, каетинский Cahetian
кажимость *philos.* appearance
казак Cossack
казарка *zool.* 1. brant *(Branta);* 2. snout beetle, leaf-roller beetle *(Rhynchites bacchus)*
казачок I 1. *hist.* boy servant; 2. kazachok (Ukrainian dance)
казачок II *zool.* mustard bug *(Euridema festival)*
казашка Kazakh (woman)
казеин *chem.* casein(e)
 ~, иодированный casein iodine, iodinated caseine, caseoiodine
казеинат *chem.* caseinate
казеиновый caseinic, caseic
казеиноген *biochem.* caseinogen
казеификация caseification
каземат *mil.* casemate
 ~, наблюдательный observation casemate
казенка *naut.* master's cabin; log deck
казеоза *biochem.* caseose
казеозный caseous
казна Treasury; public purse/coffers; *fig.* the State
 ~, государственная State Treasury
казначей treasurer; *mil.* paymaster; *naut.* purser
казначейский treasury *(attr.);* paymaster's
казначейство Treasury; *Engl.* Exchequer
казнить put to death, execute
казнь execution, death penalty, capital punishment
казолит kasolite

казуар *zool.* cassowary *(Casuarius)*
казуарина *bot.* beefwood, Australian pine *(Casuarina)*
казуариновые *pl., bot.* beefwood family *(Casuarinaceae)*
казуист casuist
казуистика casuistics, casuistry
казуистический casuistic(al)
казус *law* particular/special case; *med.* isolated/concrete case
~ белли casus belli
каик *naut.* caique
Каин *bibl.* Cain
каинит *miner.* kainite
кайенский cayenne
~ перец cayenne pepper
кайзер Caesar; kaiser
кайла *min.* pick, pickax(e), hack
кайма edge, border, margin, edging, hem, selvage, *med.* limbus
~ овальный ямки margin of the fossa ovalis
~, боковая lateral border
кайман *zool.* cayman *(Caiman)*
~, очковый spectacled cayman
кайнозит *miner.* cenosite
кайнозой *geol.* Cenozoic era
кайнозойский *geol.* Cenozoic
кайра *zool.* guillemot, murre *(Uria)*
кайянус *bot.* Congo pea, pigeon pea *(Cajanus indicus)*
Как *myth.* Cacus
какавелла ground cacao husks
какаду *zool.* cockato *(Kokatoe)*
какао 1. cocoa; 2. *bot.* cacao (tree) *(Theobroma cacao)*
какао-масло cocoa butter/fat
какилия *bot.* tassel flower *(Emilia flammea)*
какогевзия *med.* cacogeusia
какодилат *chem.* cacodylate
какодиловый *chem.* cacodylic
какология cacology
какосмия cacosmia, kakosmia
какофемистический cacophemous
какофонический cacophonous, cacophonic
какофония cacophony
кактиномицин cactinomycin
кактус *bot.* cactus *(Cactus)*
~, звездчатый star cactus *(Astrophytum)*
кактусовидный cactiform
кактусовый cactus *(attr.),* cactaceous
кактусовые *pl. bot.* cactus family *(Cactaceae)*
Какус *myth.* Cacus
кал f(a)eces, stool(s), excrement
~, ахолический acholic stool
~, кровавый f(a)eces cruentae, melena, bloody stools
~, первородный meconium
~ при брюшном тифе typhoid stool
~ при дизентерии dysenteric stool
~ при холере rice-water stool
~, твердый scybalum
кала-азар *pathol.* kala azar, tropical splenomegaly, Dundum fever
~, детский infantile kala azar
каладиум *bot.* caladium *(Caladium)*
каламин *miner.* calamine, hemimorphite
каламит *paleont.* calamite
каламондин *bot.* calamondin *(Citrus mitis)*
каламус *bot.* rattan palm *(Calamus)*
калан *zool.* sea otter, kalan *(Enhydra lutris)*
каландр *tech.* calender
~, гофрировочный goffer calender
~, креповый *text.* creping calender/machine
~, листовальный sheeting calender
~, матирующий *text.* mat calender
~, обкладочный coating/skimming calender
~, отделочный *leath.* finishing calender
~, паровой steam calender
~, подошвенный *leath.* sole calender
~, промазочный spreader calender
~, прослоечный interlayer calender
~, тиснильный embossing calender
~и, спаренные tandem calenders
каландриния *bot.* rock purslane *(Calandrinia)*
каландрирование calendering
каландрировать calender, glaze
каландровщик calender operator/attendant
каланча watchtower
~, пожарная fire observation tower
калатея *bot.* calathea *(Calathea)*
калган *bot.* tormentil(la) *(Potentilla erecta)*
калебаса *bot.* calabash gourd *(Lagenaria vulgaris)*
калевать channel, mould, bead
калевка 1. moulding, beading; fillet mould; 2. *(tool)* moulding plane
калевочный станок moulding machine
каледонский Caledonian
калейдоскоп kaleidoscope
калейдофон kaleidophon(e)
календарный calendar *(attr.);* calendarial, calendric
~ное имя Christian name
~ное планирание scheduling, calendar planning
~ план timing schedule
календарь calendar
~, астрономический astronomical calendar
~, грегорианский Gregorian calendar
~, морской nautical book
~, Юлианский Julian calendar
календула *bot.* calendula *(Calendula)*
~ лекарственная pot marigold *(Calendula officinalis)*
календулин *bioch.* calendulin
календы *pl. hist.* calends
каление incandescence, heat
~, белое white heat
~, красное red heat
каленый 1. hardened, tempered; 2. red-hot
кали *obs.* potash
~, едкое caustic potash, potassium hydrate

калибр gauge; caliber, calibre, bore
~, вспомогательный minor calibre
~, входящий male gauge
~, главный major calibre
~, гладкий plain gauge
~, двусторонний double-ended gauge
~, кольцевой ring gauge
~, контрольный master/check/reference gauge
~, конусный taper gauge
~, непроходной (контрольный) no-go gauge
~, обхватывающий female gauge
~, приемный inspection gauge
~, профильный contour gauge
~, проходной (контрольный) go-gauge
~, рабочий working gauge
~, резьбовой thread gauge
~, раздвижной slide gauge
калибратор calibrator, calibrating device
~ дальности range calibrator
~, кварцевый crystal calibrator
калибр-втулка ring-gauge
калибр-кольцо ring gauge
калибр-нутромер internal cauge
калибровать calibrate, graduate, size, gauge
калибровка, калибрование calibration; sizing; graduation; verification
~ акустических приемников *naut.* hydrophone calibration
~, взаимная intercalibration, reciprocity calibration
~ нуля дальности range-zero calibration
~ РЛС radar calibration
~ плодов sizing of fruit
~ по времени time calibration
~ по давлению pressure calibration
~ по дальности range calibration
~ по направлению/пеленгу bearing calibration
~ по теплу thermal calibration
~ по энергии energy calibration
~, повторная recalibration
~ прокатных валков roll groove/grooving/design pass
калибровочный calibrating, gauging
калибровщик calibrator
калибр-пробка plug/internal/male gauge
калибр-скоба external/snap gauge
калидонский Calydonian
калиево-медный *chem.* potassiocupric
калиево-натриевый potassiosodic
калиево-ртутный potassiomercuric
калиевый potassium *(attr.),* potassic
калий *chem.* potassium, K
~, азотистокислый potassium nitrite
~, азотнокислый potassium nitrate
~, бромистый potassium bromide
~, бромноватокислый potassium bromate
~, виннокислый potassium tartarate
~, двууглекислый potassium bicarbonate
~, двухромовокислый potassium bichromate
~, едкий caustic potash
~, железосинеродистый potassium ferrocyanide
~, иодистый potassium iodide
~, кислый виннокислый potassium acid tartarate
~, кислый сернокислый potassium acid sulphate, potassium bisulphate
~, кремнекислый potassium silicate, silicate of potash
~, ксантогеновокислый potassium xanthate
~, марганцовокислый potassium permanganate
~, обменный *agr.* exchange potassium
~, природный хлористый chloride of potassium, sylvite
~, роданистый potassium rhodanate
~, сернистокислый potassium sulphite
~, сернистый potassium sulphide
~, сернокислый potassium sulphate
~, углекислый potassium carbonate
~, ускуснокислый potassium acetate; *pharm.* sal diureticum
~, хлористый potassium chloride
~, хромовокислый potassium chromate
~, цианистый potassium cyanide
~, щавелевокислый potassium oxalate
~, фиксированный *agr.* fixed potassium
~, фтористый potassium fluoride
калийный *chem.* potassium *(attr.),* potassic
~ное удобрение *agr.* potash fertilizer
калийфиксирующий *agr.* potassium-fixing
калика *hist.* pilgrim; minstrel
каликант *bot.* strawberry shrub; sweet shrub *(Calycanthus)*
каликантемия *bot.* calycanthemy
каликс *bot.* calyx
калильный incandescent; glowing; *met.* tempering
калильщик *met.* temperer, annealer
калимерис *bot.* martinoe *(Calimeris)*
калина *bot.* viburnum *(Viburnum)*
~ красная/обыкновенная European cranberry bush, guelder-rose *(Viburnum opulus)*
калипсо *bot.* calypso *(Calypso); mus.* calypso
калиптра *bot.* calyptra, moss cap
калиптроген *bot.* calyptrogen
калистегия *bot.* glorybind *(Calystegia, Convolvulus)*
калиф caliph
калифат caliphate
калифорный *chem.* californium, Cf
калкан *zool.* brill *(Rhombus)*
калкановые *pl. zool. Bothidae*
калла *bot.* common calla lily *(Zantedeschia aethiopica)*
~, болотная wild calla *(Calla palustris)*
каллеза *biochem., bot.* callose
каллезный callous
каллиграф 1. calligraph; 2. calligrapher
каллиграфия calligraphy
калликарпа *bot.* French mulberry *(Callicarpa)*

каллима *zool.* leaf butterfly *(Kallima)*
каллиопа *myth.* Calliope; *mus.* calliope
каллистефус *bot.* China aster *(Callistephus)*
каллитрис *bot.* cypress pine *(Callitris)*
каллоза *biochem., bot.* callose
каллозный *bot.* callose, callosal
каллус callus, callosity
калмык, калмыцкий Kalmuck, Kalmyk
калмычка Kalmuck, Kalmyk (woman)
каловый f(a)ecal, stercorous, stercoral, stercoraceous
каломел calomel
калонг *zool.* kalong *(Pteropus vampyrus)*
калоризатор hot bulb; heater; ignition chamber
калоризация calorization, calorizing
калорийность calorific power/content/value; caloric content/value
калорийный high-calorie *(attr.);* caloric
калориметр calorimeter
~, **бомбовый** bomb calorimeter
~, **дроссельный** throttling calorimeter
~, **кислородный** oxicalorimeter
~, **непроточный** static calorimeter
~ **с постоянной температурой** constant-temperature calorimeter
~, **проточный** continuous-flow calorimeter
~ **с изотермической оболочкой** isothermal calorimeter
калометрирование calometric measurement/determination
калориметрия calorimetry, calorimetric measurement
~, **проточная** flow calorimetry
калорифер (air) heater
калоричесний caloric, calorific
калория calorie, calory
~, **большая/техническая** large calorie, kilocalorie, kilogram(me) calorie, Calorie
~, **малая** small/gram calorie
калотропис *bot.* calotrope *(Calotropis)*
калофиллум *bot.* beauty-leaf *(Callophylum)*
калохортус *bot.* Mariposa lily *(Calochortus)*
калуга *zool.* Amur sturgeon, great Siberian sturgeon *(Huso duricus)*
калужница *bot.* marsh marigold *(Caltha)*
~, **болотная** common marsh marigold, king-cup *(Caltha palustris)*
калуфер *bot.* costmary chrysanthemum *(Chrysanthemum balsamita)*
калышка kink, twist (in a rope), squirm
кальвиль calville (apples)
кальвинизм Calvinism
кальвинист, кальвинистка Calvinist
кальвинистический Calvinistic(al)
кальвинисткий Calvinist, Calvinistic(al)
кальдера *geol.* caldera, cauldron
~, **взрывная/эксплозивная** explosion caldera
~, **двойная** nested caldera
~, **погрузившаяся** sunken caldera
~, **эрозионная** erosion caldera, crater of erosion
~ **обрушения** collapse caldera
~ **опускания** subsidence/sunken caldera
калька 1. tracing paper/cloth; *Am.* vellum; 2. tracing, tracing-paper copy; 3. *ling.* calque, translation loan word
кальканеодиния *med.* calcaneodynia, painful heel
калькариурия, калькурия *med.* calcuria
калькгур rock milk, agaric mineral
калькирование tracinig
калькировать trace
калькулезный calculous
калькулировать calculate, estimate
калькулятор 1. tech. calculator; 2. cost accountant
~, **карманный** pocket calculator, hand-held calculator
~, **настольный** desk/desktop calculator
калькуляционная ведомость cost sheet/record
калькуляционная система учета cost accounting plan
калькуляция calculation, estimation; estimate; pricing; accounting; schedule of expenses, costing
~ **издержек, попартионная** batch costing
~ **издержек производства** costing
~ **издержек производства по процессам** process costing
~ **издержек поизводства продукта** product costing
~, **итоговая** final calculation
~ **расходов** costs statement
~ **себестоимости, нормативная** standard cost calculation
~ **себестоимости, позаказная** job costing, job-order costing, specific-order costing
~ **себестоимости продукта** product costing
~, **сметная** statement; precalculation
~ **стоимости** cost estimation
калькутский Calcutta *(attr.)*
кальмар *zool.* squid *(Loligo)*
кальмиецветковый *bot.* kalmia-flowered
кальмия *bot.* kalmia *(Kalmia)*
кальцеоловый *bot.* calceolate, slipper-shaped
кальцеолярия *bot.* calceolaria *(Calceolaria)*
кальцефил *bot.* calciphilous plant, calciphile
кальцефильный, кальциефильный calciphilous, calcicolous, calcicole
кальцефит *bot.* calciphyte, calcicolous plant
кальцефоб, кальцифоб *bot.* calciphobous plant, calciphobe, calcifuge
кальцефобный *bot.* calciphobous, calcifugous
кальциевый *chem.* calcium *(attr.),* calcic
кальций *chem.* calcium, Ca
~, **азотнокислый** calcium nitrate, nitrite of lime, calcerous nitre
~, **водный лимоннокислый** earlandite
~, **йодистый** calcium iodate, calcimol
~, **молочнокислый** calcium lactate
~, **пантотеновокислый** calcium pantothenate
~, **сернокислый** calcium sulphate; land plaster
~, **углекислый** calcium carbonate, carbonate of lime

~, **фосфорнокислый** calcium phosphate
~, **фтористый** calcium fluoride
~, **хлористый** calcium chloride, chloride of lime
~, **щавелевокислый** calcium oxalate
кальцикоз *med.* calcicosis
кальциметр calcimeter, limemeter
кальцинатор calcining funrace, calcinator
кальцинация, кальцинирование calcination, calcining, roasting; calcifying
кальцинировать calcine, roast, calcinate; calcify
кальциноз *med.* calcinosis
~, **дистрофический** dystrophic calcinosis
~ **кожи** calcinosis cutis
~, **обратимый** reversible calcinosis
~, **распространенный** calcinosis universalis
кальциоларсенит calcium larsenite
кальциоортосиликат calcioorthosilicate
кальциоплетный calciopletic
кальциоторит calciothorite
кальцит *miner.* calcite, calc-spar, calcareous spar
~, **альфа** elatolite
~, **волокнистый** satin spar
~, **глинистый** argillo-calcite
~, **магнезиальный** magnesian calcite
~, **мучнистый** rock meal; rock milk
~, **озерный** limnocalcite
~, **острокристаллический** dog-tooth spar
~, **пресноводный** limnetic calcite
~, **сталактитовый** stalactitic calcite, dropstone
кальциферол *biochem.* calciferol
кальцификация calcification
~, **дистрофическая** dystrophic calcification
~, **метастатическая** metastatic calcification
~ **пульпы зуба** pulp calcification
калюмет *hist.* calumet, pipe of peace
калютрон calutron, electromagnetic separator
кам *geol.* kame
камазий *geol.* Kamasian
камала *bot.* kamala *(Mallotus philippinensis)*
камарилья camarille, clique
камассия *bot.* camas *(Camassia)*
камбала *zool.* flatfish; flounder
~, **длинная** pole flounder *(Glyptocephalus cynogeossus)*
~, **морская** plaice *(Pleuronectes platessa)*
~, **песчаная** Pacific sand dab *(Citharichthys sordidus)*
камбала-ерш *zool.* rough dab *(Hippoglossides platessoides)*
камбалообразные *pl. zool. Pleuronectiformes*
камберлендский Cumbrian
камбий *bot.* cambium
~, **пробковый** cork cambium
камбио *comm., fin.* bill of exchange
камбист *econ.* cambist
камбоджиец, камбоджийкий Cambodian
камбоджийка Cambodian (woman)
камбуз galley, cook-house/-room/-galley; kitchen, caboose
~ **команды** crew's galley
камвольный *text.* worsted
камеденосный *bot.* gummiferous
камедеобразный gummy; gum-like
камедетечение *bot.* exudation of gum, gummosis
камедистый resinous; gummy
камедь resin, rosin, gum
~, **арбийская** gum arabic
~, **древесная** wood gum
~, **смолистая** gum resin
~, **трагакантовая** gum tragacanth
камелия *bot.* camellia *(Camellia)*
~, **маслоносная** Himalayan camellia *(Camellia oleifera)*
каменистый stony, rocky, petrean, stone *(attr.)*
каменка *zool.* wheatear *(Oenanthe)*
каменник *bot.* alyssum *(Alyssum)*
каменноугольный coal *(attr.);* coal-bearing, carboniferous; carbonaceous
~ **бассейн** coal-field; coal deposit
~**ная промышленость** coal-mining industry
каменный stone *(attr.);* stony; lithic; *med.* calculous
~ **век** Stone Age
~ **уголь** coal
~**ная кладка** stonework, masonry
~**ная соль** rock salt
~**ная болезнь** *med.* lithiasis
каменоломня stone quarry/pit, quarry
каменотес stone dresser/mason; stonecutter
каменушка *zool.* harlequin duck *(Ristrionicus histrionicus)*
каменщик 1. stomemason; bricklayer; 2. Mason
~, **вольный/свободный** Freemason
камень stone; rock; *(watchmaking)* jewel (bearing); *med.* calculus, concretion, stone
~, **адский** silver nitrate
~, **артериальный** arteriolith
~, **асфальтовый** asphalt stone/rock
~, **бордюрный/бортовой** curbstone, kerb stone; border stone
~, **бронхиальный** broncholith
~, **булыжный** cobblestone
~, **бутовый** quarrystone, rubble stone, rubble
~, **венный** phlebolith, phlebolite, vein stone
~, **винный** potassium acid tartarate, cream of tartar
~, **вонючий** stinkstone
~, **гороховый** 1. coarse pisolitic limestone, pea-grit; 2. aragonite oolite
~, **горшечный** pot stone
~ **для мостовой** paving stone
~, **драгоценный** precious stone, gem
~, **дробленый** crushed stone
~, **едкий** silver nitrate, caustic silver
~ **желудка** gastrolith, gastric/alvine calculus, concretion in the stomach
~, **желчный** gallstone, biliary calculus, cholelith
~, **жерновой** burrstone, buhrstone
~, **замочный** keystone
~, **звенящий** klinkstone, phonolite

~, **зубной** dental calculus, odontolith
~, **каловый** coprolith, fecalith, stercolith
~, **карьерный** run-of-the-quarry stone
~, **кишечный** enterolith, intestinal calculus
~, **кожи** cutaneous calculus, milium
~, **колотый** sledged stone
~, **коралловый** coral/dentritic calculus
~, **корешковый/корнованный** *polygr.* grained lithographic stone
~, **корнуольский** Cornwall/Cornish stone
~, **кровяной** h(a)emolith, h(a)ematolith, h(a)emic/blood calculus
~, **кулисный** *mech.* slide block
~, **легочный** pulmonary calculus
~, **листоватый** bibliolith, bibliolite
~, **литографский** lithographic limestone
~, **лунный** moonstone
~, **лучистый** actinolite
~ **матки** uterine calculus, hysterolith
~, **меловой** chalk rock
~ **мочевого пузыря** bladder stone, cystolith
~, **мочевой** urolith, urinary calculus, urolite
~ **мочеточника** uretorolith, ureteral calculus
~, **наждачный** emery stone
~, **необтесаный** untooled stone
~, **облицовочный** ornamental/facing stone
~, **оксалатный** oxalate calculus
~, **оловянный** tinstone, cassiterite; tin spar
~, **пенный/пенящийся** aphrolith, foam stone
~, **плитняковый** flagstone
~, **поделочный** semi-precious stone, industrial stone
~ **почки** renal calculus, nephrolith
~, **пробирный** touchstone, lyddite, lydite
~, **пробковый** rock cork
~, **рваный** quarry stone
~, **рвотный** tartar emetic, potassium antimony tartarate
~, **роговидный** stag-horn calculus
~, **синтетический** synthetic/reconstructed stone
~, **слоистый** ragstone
~, **слюнный** salivary calculus, sialolith
~, **солнечный** sunstone, aventurine feldspar
~ **сустава** arthrolith, articular/arthritic calculus, tophus, chalkstone
~, **тальковый** soapstone, talcum
~, **талькохлоритовый** potstone
~, **тесанный** "surface-tooled stone
~, **точильный** grindstone, oilstone
~, **ушной** aural calculus, otolith, otolite
~, **фасонный** figurate stone; shaped stone
~, **фиалковый** *bot.* rock violet (*Trentepohlia iolithus)*
~, **чернильный** inkstone, copperas stone
~, **шлаковый** slag brick
~, **штучный** dimension stone; ashlar
~ **яичника** ovarian calculus
камень-плитняк flagstone
камера 1. chamber; cell; compartment; 2. *phot.* camera; 3. inner tube; bladder *(of football)*
~, **автомобильная** inner tube
~, **акустическая** acoustic camera
~, **бетатрона** doughnut, donut
~, **быстродействующая** fast chamber
~, **вакуумная** vacuum chamber
~ **Вильсона** cloud chamber; expansion chamber
~ **вихревая** vortex/swirling chamber
~ **водоприемника** water intake chamber
~ **впускная** admission chamber
~, **выводковая** *zool.* ovicell
~, **высотная испытательная** altitude test chamber
~, **выхлопная** exhaust chamber
~, **газовая** gas chamber
~, **герметическая** pressurized/sealed chamber
~ **глазного яблока** chamber of the eye
~ **глазного яблока, задняя/передняя** posterior/anterior chamber of the eye
~ **глушителя** baffle chamber
~ **горения** combustion chamber
~ **давления** pressure chamber/vessel
~, **дезактивационная** *radiol.* decontamination chamber
~, **дезинфекционная** disinfecting chamber; fumatorium
~, **декомпрессионная** decompression chamber
~ **деления/делительная** *nucl.* fission chamber
~ **деления, импульсная** pulse-operated fission chamber
~ **деления, многоэлектродная** multiplate fission chamber
~ **деления, торцевая** end fission chamber
~ **для дневного и ночного фотографирования** day-night camera
~ **для моделирования космических условий** space chamber
~ **для ночного фотографирования** night camera
~ **для облучения** exposure cell/chamber, irradiation chamber
~ **для окуривания** fumigating chamber
~ **догорания/дожигательная** secondary furnace, secondary combustion compartment
~ **дожигания** *aero.* reheat/afterburning chamber
~, **дробеструйная** shot-blasting chamber
~, **дуговая** arc chamber
~, **дугогасительная** arc chute
~, **дутьевая** wind box
~ **жидкостного ракетного двигателя** liquid propellant thrust chamber
~, **жилая** *zool.* body chamber/whorl; living chamber
~ **завихрения** vortex chamber
~, **заглушенная** anechoic [echo-free] chamber
~, **закалочная** hardening/quenching chamber
~ **закручивания** vortex chamber
~, **звукоизолированная** acoustically-lined chamber
~, **звукопоглощающая** anechoic [dead-end] chamber

~, измельчительная *farm.* chopper chamber
~, импульсная pulse chamber
~, ионизационная дифференциальная differential ionization chamber
~ искусственного климата environmental/climatic chamber
~, испытательная test cell
~, киносъемочная motion-picture camera, cine camera
~, киносъемочная мультипликационная animation-cartoon camera
~, киносъемочная скоростная high-speed camera
~, климатическая climate chamber
~, контрольная monitor chamber
~, коптильная smoking/smoke chamber/cabinet
~, лазерная laser camera
~, лучевая *el.* beam chamber
~, мембранная bellows/sylphon chamber
~, минная mine/blasting/bursting chamber
~, морозильная freezing chamber/room
~, молотильная beating chamber
~ нагнетания/напорная pressure chamber
~, низкотемпературная cold chamber, low-temperature chamber
~ облучения exposure cell/chamber, irradiation chamber
~, огневая *(of boiler)* fire box
~, одиночная solitary confinement cell
~ окуривания fumigating chamber, fumigator room
~, осадительная/осадочная settling chamber/vessel
~ отрывающего аппарата *agr.* stripping chamber
~, отстойная settling chamber/vessel; antechamber
~ очистки decontamination cell
~, пекарная baking chamber
~, поворотная pivoted chamber
~, подвижная trailer-mounted chamber
~ переменных температур high-low temperature chamber
~ подавателя feeder chamber
~ подогрева preheating chamber
~ предварительного смешения premix chamber
~ предварительного прессования precompression chamber
~, поплавковая float chamber
~, предохранительная safety chamber
~, прессовальная *farm.* bale/baling chamber
~ прессования pressure chamber
~, пропарочная steam-curing chamber
~, проточная flow-type chamber
~, прядильная spinning cell
~, пузырьковая *nucl.* bubble chamber
~, радиальная radial-motion chamber
~ разложения decomposition chamber
~, разрядная discharge chamber
~, распарочная *plast.* steam-softening chamber
~, распылительная atomizing chamber; spray/jet chamber
~ рассеяния scattering chamber
~, расширительная expansion chamber
~, реакторная reaction chamber
~, рентгеновская X-ray camera
~ с искусственным климатом environmental/climatic chamber; climatizer
~ сгорания combustion chamber
~ сгорания, вихревая swirl-type [vortex, high-turbulence] combustion chamber
~ сгорания, двухконтурная double-flow combustion chamber
~ сгорания, двухступенчатая two-stage combustion chamber
~ сгорания, кольцевая annular combustion chamber
~ сгорания, одноконтурная single-flow combustion chamber
~ сгорания, поворотная swivelling combustion chamber
~ сгорания, противоточная reverse-flow combustion chamber
~ сгорания, прямоточная straight-flow [direct-flow] combustion chamber
~, сеточная mesh chamber
~ сжатия compression chamber
~, сливная overflow chamber
~, спаренная twin chamber
~, спиральная spiral/coiled chamber/case
~, сухая dry box
~, сушильная drying chamber/kiln
~, счетная counting/counter chamber
~, телевизионная репортажная field TV camera
~, технологическая process(ing) chamber
~, топочная furnace/combustion chamber; furnace
~, тороидальная toroidal chamber, doughnut, donut
~, туманная fog chamber
~, тюремная prison cell
~ увлажнения/увлажнительная humidifying chamber
~ уноса entrainment chamber
~, уравнительная surge tank
~ ускорения accelerating chamber
~ ускорителя accelerator chamber
~, успокоительная surge chamber; *aero.* plenum chamber
~, форсажная reheat unit; *Am.* afterburner
~ форсунки discharge-nozzle chamber
~, фотографическая camera
~, фоторегистрирующая recording camera
~, фоторепродукционная reproducing camera
~, холодильная refrigerating chamber/room
~ хранения багажа cloakroom; luggage room
~ циклонная vortex chamber
~, черно-белая monochrome camera

~, чувствительная detecting/detection chamber
~, швицевальная *leath.* sweat pit
~ шины tyre tube
~ шины, самозаклеивающаяся self-sealing (inner) tube
~, шлюзовая sluice/lock chamber
~, экологическая environmental cabinet
~, экранированная shielded chamber
~, эмульсионная *nucl.* emulsion chamber, pellicle stack
камера-сборник plenum chamber
камера-убежище *min.* refuge chamber/room
камергер chamberlain
камердинер valet
камерколлегия *hist.* board of finance/revenues
камерный *mus.* chamber *(attr.)*
камер-обскура camera obscura
камеростома *bot.* camerostome
камертон *mus.* tuning fork
~, настраиваемый adjustable tuning fork
камерунец, камерунский Cameroon
камерунка Cameroon (woman)
камея cameo
камзол camisole
камизар *hist.* Camisard
камилавка eccl. kamelaukion
камка I *bot.* common eelgrass *(Zostera marina)*
камка II *text.* damask
камин 1. hood; offtake; 2. fireplace
камнедробилка stone crusher
~, валковая roll stone crusher
~, щековая jaw crusher
камнедробитель *med., surg.* lithotrite, lithoclast, lithotriptor
камнедробление *med., surg.* lithotripsy, lithoclasty, litholapaxy
камнеискатель *med.* stone searcher, bladder sound
камнеломка *bot.* saxifrage *(Saxifraga)*
~ болотная yellow marsh saxifrage *(Saxifraga hirculus)*
камнеломковые *bot., pl.* saxifrage family *(Saxifragaceae)*
камнелюбивый *bot.* lithophilous
камнеоблицовка stone facing/lining
камнеобразование *pathol.* lithogenesis, lithogeny
камнеобразующий lithogenetic
камнеотделитель stone separator/extractor
камнепечатный lithographic
камнеплодник *bot.* tan oak *(Lithocarpus)*
камнерастение *bot.* lithophyte
камнерезка stone saw; stone cutter
камнесемянный *bot.* lithospermous
камнесечение *surg.* lithotomy
~, боковое lateral lithotomy
~, влагалищное vaginal lithotomy
~, высокое high lithotomy
~, надлобковое suprapubic lithotomy
~, почечное nephrolithotomy
~, промежностное perineal lithotomy
~, срединное median lithotomy
~, чрезбрюшинное transperitoneal lithotomy
камнеточец *zool.* 1. rock/stone borer *(Pholas dactylus)*; 2. wing shell *(Pholas campechiensis)*
камнешарка *zool.* turnstone *(Arenaria)*
камни:
~ аппендикса appendicolithiasis
~ желчного протока choledocholithiasis
~ желчного пузыря cholelithiasis
~ печени hepatolithiasis
камора *print.* superscript
кампанила *arch.* campanile
кампания campaign; drive; operating period, run
~, избирательная election campaign
~, обжиговая roasting period
~, посевная sowing season/campaign/drive; seedtime
~, рабочая operation period
~ реактора reactor lifetime
~, уборочная harvesting time/drive, harvest campaign
кампанула *bot.* bellflower *(Campanula)*
кампешевое дерево *bot.* logwood, campeachy wood *(Hematoxylon caupechianum)*
кампилотропный *bot.* campylotropous
кампиметр *med.* campimeter
кампиметрия campimetry
камподеовидный *zool.* campodeiform
камполон *med.* campolon
кампсис *bot.* trumpet creeper *(Campsis)*
кампучиец, кампучийский Kampuchean
кампучийка Kampuchean (woman)
камуфлет *mil.* camouflet
камуфлировать *mil.* camouflage
камуфляж *mil.* camouflage
камфан *chem.* camphane
камфен *chem.* camphene
камфоний trimethidinium methosulphate
камфора, камфара camphor
~, бромистая bromated camphor
камфорный camphoric; camphorated
~ное масло camphorated oil
камфороносный camphoraceous
камфоросма *bot.* camphor-fume *(Camphorosma)*
камчадал, камчадальский Kamchadal
камчадалка Kamchadal (woman)
Камчатка Kamchatka
камчатский Kamchatkan
камыш *bot.* bulrush; cane *(Scirpus)*
~, американский shore rush *(Scirpus americanus)*
~, болотный prairie bulrush *(Scirpus paludosus)*
~, дернистый deerhair *(Scirpus caespitosus)*
камышевка *zool.* warbler *(Acrocephalus)*
камышевидный *bot.* arundinaceous
камышекосилка reed cutter/mower
камышина reed stalk
камышит reed pressboard, reedfibre board
камышница *zool.* gallinule *(Gallinula)*; moor/water hen

камышовка *bot.* cane; *zool.* = **камышевка**
камышовый bulrush *(attr.),* cane *(attr.)*
канава ditch, gutter, trench, channel, trough
~, **водосточная** gutter
~, **дренажная** drainage ditch
~, **литейная** casting pit
~, **моечная** washing pit
~, **навозная** *farm.* dung channel
~, **оросительная** irrigation ditch
~, **осушительная** drainage ditch
~, **отводная** diversion ditch
~, **подводящая** tributary ditch
~, **разведочная** *min.* exploration trench
~, **разливочная** *met.* teeming pit
~, **смазочная** greasing/lubricating pit
~, **смотровая** inspection/service pit
~, **сточная** gutter
канавалия *bot.* jack bean *(Canavalia)*
канавка 1. *dim. of* **канава**; 2. groove, flute, mortise, chase, furrow, slot; riffle
~ **для смазки** oil groove
~, **кольцевая** circular/annular/ring groove
~, **круговая** circumferential/circular groove
~ **поршневого кольца** piston-ring groove
~, **смазочная** oil groove
~, **чугунная** iron runner
~ **шкива** sheave groove
~, **шлаковая** slag runner
~, **шпоночная** keyway, key seat
канавокопатель ditch excavator, trench digger, ditcher, trenching machine
~, **дренажный** drain digger
~, **многоковшовый** ladder-type trenching machine
~, **плужный** ditching plough/plow
~, **роторный** rotary excavator
~, **прицепной** trailer ditcher
~, **фрезерный** wheel-type trench digger
~, **черпаковый** bucket-type trenching machine
канавочиститель ditch cleaner
канавчатый grooved; fluted; slotted
канадец, канадский Canadian
канадка Canadian (woman)
канак Kanaka
канал canal; duct, passage, conduit; channel; *anat.* duct, canal; *(of gun barrel)* bore
~, **аварийный** emergency/safety channel
~ **без помех** noiseless channel
~, **безнапорный** gravity-flow channel/conduit
~ **безопасности** safety channel
~, **бедренный** *anat.* femoral canal
~, **боковой** side channel; lateral canal
~, **боковой полукружный** *anat.* lateral semicircular canal
~, **большой небный** greater palatine canal
~ **ввода данных** data-input channel
~ **ввода-вывода** input-output channel/connector
~ **ввода перфокарт** card input track
~, **виртуальный** virtual circuit
~ **вентиляционный** ventilation/air/venting channel/duct/passage
~ **веретена, спиральный** *anat.* spiral canal of the nodiolus
~, **вертикальный** *met.* downtake, uptake
~, **водоотливный** sluice way; drainage canal
~ **воздухозаборника** air inlet duct
~, **воздушный** air duct/passage/channel
~, **водосливный** overflow canal
~, **волновой** wave duct
~ **воспроизведения** playback channel/track
~, **впускной** intake/inlet/admission/entry duct
~, **временной** time channel
~, **входной** entrance/input channel
~, **вулканический** volcanic vent/chimney
~, **выводящий** channelway; effluent channel
~, **выделительный** secretory canal
~, **выносящий** excurrent canal/channel
~ **выключения** shut-down channel
~, **выпускной** exhaust/outlet port
~, **вытяжной** exhaust duct
~, **выходной** exit channel
~, **газовый** gas channel/passage/duct
~, **газоотводный** gas flue, gas-excape channel, exhaust-gas duct, gas-extraction/-deflection duct
~ **гидропочты** hydraulic shuttle hole
~, **глухой** thimble
~, **голеноподколенный** *anat.* crurompopliteal canal
~, **"горячий"** hot channel
~, **двоичный** binary channel
~, **двусторонней связи** two-way communication channel
~, **действующий** active/operative channel
~, **деривационный** diversion channel
~, **детекторный** detector channel/duct
~, **дипломатический** diplomatic channel
~, **диффузорный** diffuser duct
~ **для прохода морских судов** ship canal
~ **для вывода пучка** beam port/hole/tube
~ **для облучения** exposure hole, irradiation tunnel/port
~ **для производства изотопов** isotope-production tunnel
~, **дренажный** drain channel
~, **дублированный** duplicate/dual channel/duct
~, **дымовой** flue
~ **желудка** *anat.* ventricular canal
~ **желудка, выходной** pyloric canal
~, **заглушенный** blind duct
~, **заднепроходный** *anat.* anal canal
~, **задний полукружный** posterior semicircular canal
~, **запасный** *rad.* guard/spare channel
~, **запирательный** *anat.* obturator canal
~ **записи данных** data-recording channel
~ **запроса** *nav.* interrogation link/channel
~ **запястья** carpal canal
~, **заранее выбранный** pre-set channel
~ **звукового сопровождения** sound channel/track/path

~, звуковой audio/sound channel
~, зарядный charging tube
~, зеркальный image channel
~, зрительный *anat.* optic canal
~ измерений/измерительный measurement channel
~ изображения *telev.* video channel
~, изогнутый curved channel
~ индикации display channel
~, информационный information/data channel
~, кабельный cable duct; cable trench/trough
~, кольцевой annular channel; annulus
~ корня *(of tooth)* *anat.* root canal
~ коррекции correction channel
~, краевой marginal canal
~ крена roll channel; *aero.* roll axis
~ курса yaw channel
~, крестцовый *anat.* sacral canal
~, крыловидный *anat.* pterygoid canal
~, лесосплавный log chute
~, лимбербордовый *naut.* limber
~, литниковый sprue channel; gate
~, лицевой *anat.* facial canal
~, локтевой *anat.* ulnar canal
~, лопаточный blade passage
~, лучевой radial canal
~, маслопроводный oil duct
~, мелиоративный soil-reclamation canal
~, маслоотводящий oil-scavanging passage
~, маслоподводящий oil-supply passage
~, масляный lubrication duct; *bot.* vitta
~, межлопаточный vane channel; blade passage
~, межсептальный interseptal canal
~, морской ship/maritime/seaway canal
~, мочеиспускательный *anat.* urethra
~, мультиплексный байтовый/блоковый *comp.* byte/block multiplexor channel
~ мундштука *plast.* die channel
~, мышечно-трубный musculotubal canal
~, мыщелковый condylar/condyloid canal
~, небно-влагалищный *anat.* palatinovaginal canal
~ несущей частоты *el.* carrier channel
~ переменного сечения variable-area channel-tapered duct
~, перепускной by-bass channel/passage/duct
~, питательный *anat.* nutrient canal
~, пищеварительный *anat.* alimentary canal
~, плечемышечный humeromuscular canal
~ пневмопочты pneumatic shuttle hole
~ подачи топлива fuel-feed canal
~, подводящий intake conduit/duct
~, подглазничный *anat.* infraorbital canal
~ поддона *met.* runner
~, подъязычный *anat.* hypoglossal canal
~, позвоночный vertebral canal
~ поперечного управления lateral-control channel
~ постоянного сечения constant-area channel
~, приводящий *anat.* adductor canal
~ приема данных data acquisition channel
~ приема сигналов типа "да-нет" go-no-go channel
~ приема цифровых данных digital data acquisition channel
~, приносящий incurrent canal
~, приточный intake/influent duct/channel
~ продольного управления longitudinal-control channel
~, профилированный shaped/contoured passage
~ путевого управления yaw-control channel
~ прямого доступа к памяти *comp.* direct-memory access channel, DMA channel
~, пылеосадительный dust-precipitating duct
~ радиосвязи (вещательный) broadcast channel
~, радичастотный radio-frequency channel
~, резцовый incisive/incisor canal
~ релейного типа on-off channel
~, речевой voice/speech channel
~ рыскания yaw channel; *auto.* yaw axis
~ с буферной памятью *comp.* buffered channel
~ с высоким/большим усилением high-gain channel
~ с малым усилением low-gain channel
~ с помехами noisy channel
~ с обратной связью feedback (two-way) channel
~, самотечный *hydr.* gravity-flow conduit/channel
~ связи communication channel; link
~ связи, авиационный aeronautical service channel
~ связи, односторонний one-way communication channel
~ связи, симплексный simplex (one-way) channel
~ связи с частотным разделением frequency-division multiplexed channel/line
~ связи, уплотненный multiplexed channel
~, селекторный *comp.* selector channel
~ синхронизации *comp.* clock/timing channel
~ сжатого воздуха air-pressure duct
~, секреторный secretory canal
~ слежения tracking/following channel
~, слезно-носовой nasolacrimal canal
~, слезный *anat.* lacrimal duct
~, слепой пищеварительный closed alimentary channel
~, служебный service channel
~, слуховой acoustic duct
~, сонный *anat.* carotid channel
~, сопловой nozlle passage
~ сопровождения tracking/following channel
~, сошниково влагалищный *anat.* vomerovaginal canal
~ спинного мозга (спинномозговой) cere-

brospinal canal
~, срамной *anat.* pudendal canal
~ ствола винтовки *mil.* rifle bore
~ ствола, гладкий smooth bore
~, стекловидный hyaloid canal
~, сточный house drain, escape canal
~, ступенчатый stepped channel
~, судоходный navigation/navigable canal
~, сужающийся convergent channel/duct
~ тангажа pitch channel; *aero.* pitch axis
~, телевизионный television (TV) channel
~, тройной triplex channel
~ транспортировки перфокарт *comp.* card bed
~, транспортировочный transfer channel/hole
~ улитки, спиральный *anat.* spiral canal of the cochlea
~, уплотненный shared channel
~ управления control channel
~ управления релейного типа on-off channel
~ усиления amplifier channel
~, физический *comp.* physical link
~, хронирующий timing channel
~, цифровой digital channel
~, частотный frequency channel
~ шейки матки *anat.* cervical canal of the uterus
~, шляковый sluiceway
~ шлюза sluice-way
~, шлюзованный lock canal
~ ядерного реактора для образцов sample hole
~ ядерного реактора, рабочий reactor fuel tube/channel
~, эвакуаторный *anat.* egestorial canal
~, экспериментальный experimental port/hole/duct; test hole
каналец small canal/channel; canalicule
~, барабанный *anat.* tympanic canal
~, масляный *bot.* vitta
~, слезный *anat.* lacrimal duct
~, сонно-барабанный *anat.* caroticotympanic canal
~ сосцевидного отростка *anat.* mastoid canal; auricular canal
~ улитки cochlear canal
канализация 1. sewage [waste-water] disposal, sewerage (system); drainage system; 2. *el.* cable/conduit/duct system; 3. channelling
~, внутренняя waste piping
~, наружная sewerage
~, сточная waste-water drainage
каналокопатель trenching machine
каналикулит *med.* canaliculitis
канальный channel *(attr.),* canal *(attr.),* canaliculate, tubulate
канальцевый *anat.* canalicular, tubular
каналцы *anat.* small canals, tubules, ductules
~, зубные dental canals
~, извитые/прямые почечные convoluted/straight tubules of the kidney
~, семенные seminiferous tubules
~ слезной железы, выводные excretory ductules of the lacrimal gland
~ яичка, выносящие small seminal ducts
канальчатый canaliculate, canalicular, tubular; channelled, grooved
канамицин kanamycin
кананга *bot.* ylang-ylang *(Cananga odorata)*
канареечник *bot.* canary grass *(Phalaris)*
канареечный canary *(attr.)*
канарейка *zool.* canary *(Serinus canaria)*
канариум *bot.* canary tree *(Canarium)*
канаста canasta
канат rope; cable; hawser
~, буксирный towing/hauling rope, tow-rope; hawser
~, буровой drilling cable
~, витой twisted rope
~,выпускать pay out the cable
~, грузоподъемный hoisting rope
~ драглайна drag cable
~, крученый twisted cable
~ левой свивки left lay rope
~, многопрядный stranded rope
~ носовой буксирный bow hawser
~, подъемный hoist rope
~ полиспаста runner
~ правой свивки right lay rope
~, проволочный/стальной wire rope
~, чалочный pull/pulling rope
~, якорный anchor chain/cable
канатик 1. stranded wire/conductor; 2. *anat.* cord; funiculus
~, антенный antenna wire
~, пупочный umbilical cord
~, семенной spermatic cord
канатиковидный funicular
канатник *bot.* abutilon *(Abutilon)*
канатный cable *(attr.),* rope *(attr.),* funicular
~ная дорога cableway, ropeway
~ная железная дорога funicular railway
канатовидный ropelike
канатоемкость *(of drum)* rope/reeling capacity, coiling length
канаус coarse silken fabric
канва canvas
кандалы *pl.* shackles, fetters
~, ручные handcuffs
кандела *opt.* candela
канделябр candelabrum; sconce
кандидамикоз, кандидоз *med.* candidiasis
кандидат 1. candidate; applicant; *law* contender; 2. *educ.* Candidate; 3. nominee, candidate
~ в члены партии candidate-member of the party
~ в члены ЦК *obs.* alternate member of the Central Committee
~ медицинских наук Candidate of Medical Cience
кандидатский candidate *(attr.)*

~ минимум *educ.* minimum requirements at the qualifying examinations for the Candidate degree
~ стаж *obs.* probation term for Communist Party membership
~кая диссертация thesis/dissertation for obtaining the Candidate degree
кандидатура candidacy; nomination
кандидоз *med.* candidiasis
кандолюминесценция candoluminescence
кандык *bot.* fawn lily *(Erythronium)*
канегра *bot.* canaigre *(Rumex hymenosepalus)*
канель *bot.* sugar/custard apple, sweetsop *(Annona squamosa)*
канефора *arch.* canephora, caryatid
каникулы holidays, vacation, recess
канистра safety fuel tank, canister, jerry can
~, сменная replacement canister
канифас-блок *naut.* snatch/notch block
канифоль colophony, rosin, colophonium
~, гидрированная hydrogenated rosin
~, живичная gum rosin, gallipot colophony
~, сосновая pine rosin
~, талловая tall colophony, tall hard rosin
~, экстракционная wood rosin
канкан cancan
канкроид *med.* cancroid
канна I *bot.* canna *(Canna)*
канна II *zool.* eland *(Taurotragus oryx)*
каннабизм cannabism
каннелюра *arch.* flute, cannelure
каннибал cannibal
каннибализм cannibalism
канновый *bot.* cannaceous
каннановые *pl., bot.* canna family *(Cannaceae)*
канон canon
канонада cannonade
канонерка, канонерская лодка gunboat
канонизация eccl. canonization
канонизировать *eccl.* canonize
каноник *eccl.* canon
канонир *hist.* cannoneer, gunner
канонисса *eccl.* canoness
канонический, каноничный *eccl.* canonical
~ное право *eccl.* canon law
каноничность *eccl.* canonicity
канонник 1. prayer book; 2. reader of canons
канопа canopic vase
канопус *astr.* Canopus
Каносса Canossa
каноэ canoe
~, парное tandem canoe
каноэ-двойка canoe-pair
каноэ-одиночка canoe-single
кант I edge, edging; *print.* extended cover; *arch.* arris
кант II *eccl.* canticle
кантабиле *mus.* cantabile
канталупа *bot.* cantaloup(e) *(Cucumis melo cantalupensis)*
кантаридин *chem.* cantharidin(um)
кантата *mus.* cantata
кантеле *mus.* kantele
кантианец, кантианский Kantian
кантианство Kantianism
кантилена *mus.* cantilena
кантовальный *tech.* canting, tilting, tipping
кантование *tech.* canting, tilting; bevel(l)ing
кантователь tipper, tilter, turn-over device, tilting jig/table; positioner, manipulator
~, прокатный blooming-mill manipulator
~ рельсов rail turner
~, сварочный welding positioner
~ слитков ingot manipulator
кантовать I edge, border
кантовать II 1. turn over, turn on edge, tilt, cant; 2. *tech.* manipulate, position
кантовка I edging
кантовка II tilting, canting
кантовский Kantian
кантон canton
кантонец, кантонский Cantonese
кантонист *hist.* soldier's son
кантопластика canthoplasty
кантор cantor
канун 1. *eccl.* vigil; 2. eve
~ Нового года New Year's eve
~ рождества Christmas Eve
канцер *med.* cancer
канцероген *med.* carcinogen
канцерогенез *med.* carcinogenesis
~, радиационный radiation carcinogenesis
~, эндокринный endocrinous carcinogenesis
канцерогенный *med.* carcinogenic, cancerogenic
канцлер chancellor
~ казначейства *Engl.* Chanvellor of the Exchequer
канцлерство chancellery, chancellorship
канцона *mus.* canzone
канцонетта *mus.* canzonet
канчиль *zool.* moose deer *(Tragulus)*
каньон canyon
канюк *zool.* buzzard *(Buteo)*
канюля *med.* cannula
~ для промывания аттика attic cannula
~, интубационная intubation cannula
каолин porcelain/China clay, kaolin(e)
~, белый осадочный white sedimentary kaolin
~, мягкий lithomarge
~, отмученный washed kaolin
~, сырой crude kaolin
каолинизация kaolinization, loamification
каолинит kaolinite
каон *phys.* K-meson, kaon
кап 1. *naut.* cape, sliding hatch cover; 2. *bot.* burl, knag, burr, excrescence; wart
капелла 1. chapel; 2. choir
~ Богоматери Lady chapel
Капелла *astr.* Capella
капеллан chaplain
капеллирование cupellation

капеллировать cupel
капельмейстер *mus.* conductor; bandmaster
капельник 1. *geol.* drop-stone, stalagmite, stalactite; 2. dropping bottle
капельница dropper, dropping bottle/tube; pipette; drop oiler
капельно-жидкий liquid
капельный drop *(attr.)* dripping
капер *naut.* privateer, corsair, caper
каперс *bot.* caper *(Capparis); pl., cul.* capers
каперство *naut.* privateering
каперствовать *naut.* privateer
капибара *zool.* capybara *(Hydrochoerus capybara)*
капиллиций *bot.* capillitium, hairnet
капилляр *anat.* capillary; *phys.* capillary tube
~, кровеносный blood capillary
~, лимфатический lymph capillary
капиллярно-активный capillary-active
капиллярит *med.* capillaritis
капиллярность capillarity
капиллярный capillary; capillaceous, filamentous
капитал *fin.* capital; funds; assets
~, акционерный share/stock capital; joint stock; capital stock; shareholder/stockholder equity
~, банковский banking capital
~ в ликвидной форме liquid capital
~ в форме материальных активов physical capital
~ в форме титулов собственности property capital
~, валовой оборотный gross working capital
~, внешний outside venture capital
~, высвободившийся released capital
~, гибкий malleable capital
~, денежно-торговый financial capital
~, денежный money capital
~, долгосрочный long-term funds
~, заемный borrowed/debenture capital
~, закрепленный nonmalleable/unmalleable/ nonshiftable capital
~, замороженный frozen capital
~ и проценты principal and interest
~, избыточный surplus/redundant capital
~, израсходованный expended capital
~, инвестированный vested/invested capital
~, краткосрочный short-term funds
~, ликвидный available/account capital
~, мобилизовать raise funds
~, монополистический monopoly capital
~, невостребованный uncalled capital
~, негибкий unmalleable/nonmalleable/ nonshiftable capital
~, неиспользуемый idle/dead capital
~, нематериальный intangible/immaterial capital
~, непроизводительный nonproductive assets
~, оборотный circulating/working/current/fluid/ floating capital
~ обращения capital of circulation, circulating capital
~, объявленный stated capital
~, основной fixed assets; fixed capital; capital/ investment goods; fixed capital stock
~, переменный variable capital
~, постоянный оборотный permanent working capital
~, потребляемый consumed capital
~ предприятий business capital
~, привлеченный debt capital
~ приносящий проценты interest-bearing capital
~ приносящий ренту rent-paying capital
~, производительный productive assets
~, разводненный watered capital
~ рантье rentier's capital
~, реальный основной fixed/physical assets
~, ростовщический moneylender, moneylender's capital
~, свободный disposable/spare capital; idle money
~, скрытый денежный latent money capital
~, скрытый основной hidden assets
~, собственный ownership capital
~, совокупный aggregate capital
~, спекулятивный venture/risk capital
~, ссудный loan capital, loanable funds; interest-bearing capital
~ сформированный за счет акций low-geared capital
~ сформированный за счет облигаций highly geared capital
~, товарный commodity capital
~, торговый trading/merchant's/mercantile capital
~, уставный authorized capital stock
~ утекающий за границу refugee capital
~, фиктивный fictitious capital
капитализация capitalization, capitalizing
~ прибавочной стоимости capitalizing of surplus value
капитализировать capitalize
капитализм capitalism
~ всеобщего благосостояния welfare capitalism
~, государственно-монополистический state monopoly capitalism
~, домонополистический premonopoly capitalism
~, корпоративный collective capitalism
~, монополистический monopoly capitalism
~, промышленно развитый industrial capitalism
капиталист capitalist
капиталистический capitalist(ic)
капиталовложение capital investment
~, чрезмерное overinvestment
~ия, неокупаемые sunk cost(s)
~ия, плановые planned investment(s)
капиталовооруженность fixed capital per worker

капиталодержатель fund/assets holder
капиталоемкий capital-intensive
капиталоемкость capital intensity, capital-output ratio; capital coefficient
капиталоотдача output-capital ratio, capital productivity ratio
капиталь headband *(of book)*
капитальность soundness, solidity, stability
капитальный capital; substantial, fundamental, major, chief, principal, basic
~ **ремонт** major/comprehensive overhaul
~**ное оборудование** capital/basic equipment
капитан captain; master; *(rank)* captain
~ **дальнего плавания** sea captain
~ **торгового судна** ship master
капитан-исправник *hist.* district chief of police
капитан-лейтенант lieutenant-commander
капитанский мостик captain's bridge
капитанство captainship, captaincy
капитель 1. *arch.* capital; head; 2. *print.* small capitals
~ **колонны** column capital/head/cap
~ **пилона** pierhead
капительная буква small capital/cap
Капитолий Capitol
капитолийский Capitoline
капитул *eccl.* chapter
капитулировать capitulate, surrender
капитулянт capitulator; defeatist
капитулянтство capitulationism; defeatism
капитулярий *hist.* capitulary
капитуляция capitulation; surrender
капище heathen temple/shrine
каплевидный, каплеобразный droplike, drop-shaped
каплезащищенный, каплестойкий dripproof
капленепроницаемый driptight
каплепадение dripping
каплесобиратель drip pan
каплеуказатель drop sight glass
каплеуловитель, каплеулавливатель drop catcher; drip pan
каплеупорный dripproof
каплица chapel
капли *pl.* drops
капля drop, droplet
капнометрия capnometry
капноскоп capnoscope, smoke gauge
капок *bot.* silk-cotton tree, kapok tree *(Eriodendron anfractuosum)*; *(fibre)* kapok
капонир *mil.* 1. caponier(e); 2. revetment; aircraft shelter
капор bonnet; hood
капот 1. hood, bonnet, cowl, cowling; 2. *av.* nose over, overturn, turn over
~ **вентилятора** fan cowling
~ **из стеклопластика** glass-fibre cowling
~, **кольцевой** ring cowling
~ **радиатора** radiator hood
~ **с регулируемыми створками** gilled cowling
капотаж, капотирование *av.* nose-over, overturning
капотировать 1. cowl, hood, cover; 2. *av.* nose over, overturn
капрал *mil.* corporal
капремонт *abbr.* major overhaul
каприфоль, каприфолий, каприфолия *bot.* sweet honeysuckle *(Lonicera caprifolium)*
каприччио *mus.* capriccio
капролактам *chem.* caprolactam
капрон *text.* kapron, polycaprolactam fibre/fiber, nylon-6, perlon
капсаицин *chem.* capsaicine
капсантин *chem.* capsanthin
капсель sagger
капсикум *bot.* red pepper *(Capsicum)*
капсицин *chem.* capsicin
капсула capsule; casing, case, shell; theca
~, **аварийная спасательная** emergency escape capsule
~, **беспилотная** unmanned/pilotless capsule
~, **возвращаемая** recoverable/recovery/entry capsule
~, **герметизированная** pressurized capsule
~, **герметическая** sealed capsule
~ **гидропочты** rabbit
~, **двуместная** two-man capsule
~ **для облучения** irradiation capsule
~, **катапультируемая** ejection capsule
~, **космическая** space capsule
~, **лунная** lunar landing capsule
~, **отделяемая** jettisonable capsule
~ **печени, фиброзная** *med.* hepatic fibrous capsule
~ **пневмопочты** rabbit
~ **почки, жировая** *med.* perirenal fat
~ **почки, фиброзная** fibrous capsule of the kidney
~ **с приборами** instrument capsule
~ **с экипажем** manned/crew capsule
~ **спасяемая** recoverable/recovery capsule
~, **спасительная** rescue capsule
~, **суставная** *anat.* joint capsule
~ **щитовидной железы** sheath of the thyroid gland
капсулит *med.* capsulitis
капсулопластика capsuloplasty
капсулотом capsulotome
капсулотомия capsulotomia, capsotomia
капсульный capsular
капсюль, капсуль 1. percussion cap, (cartridge) primer, cap; 2. *el.* inset; 3. bottle cap/top, cork seal
~ **гильзы:** primer of cartridge
~, **запальной** ignition pellet
~, **микрофонный** microphone capsule
~, **пружинный** spring cap
~, **ударный** percussion primer
капсюль-воспламенитель percussion cap, primer; igniter

капсюль-детонатор blasting cap, detonating fuse/cap, fulminate detonator
капсюля = **капсула**
каптаж *hydr.* catchment, water catching; tapping; *min.* capping
~ скважины capping of well
каптакула *zool.* captaculum
каптал headband *(of book)*
каптенармус *obs., mil.* quartermaster sergeant
каптировать capture, catch, pipe
капустница *zool.* cabbage white butterfly *(Pieris brassicae)*
капуцин 1. Capuchin friar; 2. *bot.* nasturtium *(Tropaeolum);* 3. *zool.* capuchin monkey *(Cebus)*
капуциновые *pl. bot.* nasturtium family *(Tropaceolaceae)*
капюшон hood
капюшонница *zool.* hooded owlet moth *(Cucullia)*
кар *geol.* corrie, circus, cirque
карабах Karabagh riding horse
карабин 1. snap/spring hook, clasp; 2. *mil.* carbine
~, автоматический machine carbine
~, самозарядный self-loading carbine
карабинер 1. Italian policeman, carabiniere; 2. *mil.* carbineer, carabineer
каравайка *zool.* glory ibis *(Plegadis falcinellus)*
караван caravan; *naut.* convoy
караван-сарай caravansary, caravanserai
каравелла *naut.* caravel(le)
карагана *bot.* pea tree/shrub *(Caragana)*
карагач *bot.* elm *(Ulmus)*
караген *bot.* Irish moss *(Cladonia islandica)*
караиб Carib
караибка Carib (woman)
караибский Caribbean, Cariban, Caribal
караим, караимский Karaite
каракал *zool.* caracal *(Felis caracal)*
каракалпак, каракалпакский Kara-Kalpak
каракара *zool.* caracara *(Polyborus)*
каракатица *zool.* cuttlefish, devilfish *(Sepia)*
караколла *bot.* caracol *(Phaseolus caracolla)*
карактрон *comp.* alphanumeric/character display tube, Charactron
каракуль astrakhan (fur), karakul (fur)
каракульская овца karakul/astrakhan sheep
карамбола *bot.* carambola *(Averrhoa carambola)*
карамболь carom, cannon
карамелизировать caramelize; burn
карамель caramels *(coll.)*
карандаш pencil; crayon
~, гексахлорановый *med.* hexachlorocyclohexane pencil
~, глазной aluminated copper, lapis divinus
~, графитный lead pencil
~, квасцовый aluminous pencil
~, кровоостанавливащий styptic pencil, h(a)emostatic pencil
~, лекарственный medical pencil
~, литографский (lithographic) crayon
~, ляписный lapis pencil, silver nitrate stick
~, световой optical wand
~, химический indelible pencil
~, цветной crayon
~, чертежный drawing pencil
карандашный pencil *(attr.)*
~ное дерево cedar red juniper *(Juniperus virginiana)*
карандич *zool.* yellow gurnard *(Trigla lucerna)*
каранкс *zool.* caranx, horse mackerel *(Caranx)*
карантин quarantine
~, ветеринарный animal quarantine
~, полный absolute quarantine
~, фитосанитарный plant quarantine
~ экипажа космического корабля spacecraft-crew quarantine
~у, подвергать to quarantine
карантинное свидетельство (патент) bill of health
карапа *bot.* crab wood *(Carapa)*
карапакс *zool.* carapace
карапузики *pl. zool.* mimic/hister beetles *(Histeridae)*
карась *zool.* crucian carp
карат carat
караул sentry, guard; watch
~, почетный guard of honour
караульная будка sentry box
карбазид *chem.* carbazide
карбазоцин *chem.* carbazocine
карбамаза carbamase
карбамазепин carbamazepine
карбамат *chem.* carbamate
карбаматкиназа carbamate kinase
карбамид carbamide, urea
карбамилмочевина *chem.* biuret, carbamylurea
карбахолин *chem.* carbachol(um), carbiminoylcholine chloride
карбгемоглобин carboh(a)emoglobin, carbaminoh(a)emoglobin
карбензид *chem.* carbenzide
карбид *chem.* carbide
~ бария barium carbide
~ кальция calcium carbide
карболит *plast.* karbolit, Carbolite
карбоангиокардиография carboangiocardiography
карбогемоглобин *chem.* carboh(a)emoglobin
карбогидраза *biochem.* carbohydrase
карбогумин vegetable jelly
карбоксигемоглобин *chem.* carboxyh(a)emoglobin
карбоксигемоглобинемия *med.* carboxyh(a)emoglobinemia
карбоксил *chem.* carboxyl
карбоксилаза *biochem.* carboxylase
карбоксилиаза *biochem.* carboxylyase
карбоксилирование *chem.* carboxylation
карбоксилтрансфераза *biochem.* carboxyltransferase

карбоксильный *chem.* carboxylic
карбоксилэстераза *biochem.* carboxylesterase
карбоксиметилгидантоиназа *biochem.* carboxymethylhydantoinase
карбоксимиоглобин *biochem.* carboxymyoglobin
карбоксипептидаза *biochem.* carboxypeptidase
карбоксиполипептидаза *biochem.* carboxypolypeptidase
карболен carbolen(e), medicinal carbon
карболинеум carbolineum
карболит carbolite
карболовый carbolic
карболфуксин *chem.* carbolfuchsin
карбомицин *chem.* carbomycin
карбон Carboniferous period
карбонадо carbonado, black diamond
карбонарий, карбонар *hist.* Carbonarist
карбонат *chem.* carbonate
~ **бария и кальция** barytocalcite
~, **безводный** anhydrous carbonate
~, **бурый** brown carbonate
~ **железа** carbonate of iron
~ **извести** carbonate of lime
~ **кальция** calcium carbonate
~, **кислый** acid carbonate
~ **магния, основный** magnesium subcarbonate
~ **натрия** sodium carbonate
~ **никеля, водный** *miner.* nickel emerald, zaratite
~, **основный** basic carbonate
~ **свинца** lead carbonate
~, **средний** neutral carbonate
~ **щелочи** alkaline carbonate
карбонатность calcareousness
карбонатный carbonaceous, carbon-bearing
карбонизация carbonization, carbonation; coalification
карбонизировать carbonize, carbonify
карбонил *chem.* carbonyl
карбонильный carbonylic
карбонит carbonite, native coke
карбоновый 1. carboniferous, carbonaceous; 2. *geol.* Carboniferous
карборунд carborundum, silicon carbide (abrasive)
карбохолин *chem.* carbocholine
карбромал carbromal(um), adalin
карбункул 1. red precious stone; 2. *med.* carbuncle, carbunculus
~ **почек** renal carbuncle
карбюратор carburettor; *Am.* carburetor
~, **беспоплавковый** pressure carburettor
~, **впрыскивающий** injection carburettor
~, **двухдиффузорный** double-barrel carburettor
~, **диафрагменный** diaphragm carburettor
~ **для перевернутого полета** *aero.* inverted-flight carburettor
~, **добавочной** booster carburettor
~, **дополнительный** auxiliary carburettor
~, **испарительный** vaporizing carburettor
~, **основный** primary/main carburettor
~, **поплавковый** float-type carburettor
~, **пусковой** starter carburettor
~, **распылительный** atomizing carburettor
~ **с двумя смесительными камерами** double-barrel carburettor
~ **с термостатической регулировкой** thermostatic carburettor
~ **с экономайзером** economical carburettor
~, **сдвоенный** twin/double carburettor
карбюрация carburation
карбюризатор carburizing agent/compound, carburizer
~, **газовый** carburizing gas
карго *naut.* cargo
каргоплан *naut.* cargo-plan, stowage list
КАРД *comp.* compact automatic retrieval device, CARD
карда carding machine; card
кардамон *bot.* cardamom, cardamon *(Elettaria cardamomum)*
кардан universal/Cardan joint; gimbal
карданный Cardanic
~ **вал** Cardan shaft, shaft of universal joint
~**ная передача** gimble drive/gear
кардиазол *med.* cardiazol
кардиальгия *med.* cardialgia
кардиальный cardial
кардинал 1. *eccl.* cardinal; 2. *bot.* cardinal flower *(Lobelia cardinalis)*; 3. *zool.* ladybird, ladyclock *(Novius cardinalis)*; 4. *zool.* cardinal bird *(Richmondena cardinalis)*; 5. *zool.* cardinal fish *(Apogon)*
кардинализм *econ.* theory of cardinal utility
кардиналистская теория полезности *econ.* theory of cardinal utility
кардиналовые *pl. zool. Apogonidae*
кардиоангиология cardioangiology
кардиоаортальный cardioaortal
кардиоартериальный cardioarterial
кардиогемодинамография cardioh(a)emodynamography, dynamocardiography
кардиогенный cardiogenic
кардиограмма cardiogram
кардиограф cardiograph
кардиография cardiography
~, **ультразвуковая** ultrasonic cardiography
кардиодилататор cardiodilator
кардиокаирограф cardiocairograph
кардиолиз cardiolysis
кардиолог cardiologist
кардиоида *math.* cardioid
кардиологический cardiological
кардиология cardiology
кардиомаляция cardiomalacia
кардиомегалия cardiomegaly
~, **гликогенная** glycogen cardiomegaly
кардиомиопатия cardiomyopathy, myocardiopathy
кардионевроз cardioneurosis

кардионекроз cardionecrosis
кардиоментопексия cardio-omentopexy
кардиопатия cardiopathy
кардиоперикардит cardiopericarditis
кардиопластика cardioplasty
кардиоплегия cardioplegia
кардиопневмограмма cardiopneumogram
кардиопневмограф cardiopneumograph
кардиоптоз cardioptosia, cardioptosis
кардиопульмональный cardiopulmonary
кардиорафия cardiorrhaphy
кардиосептопексия cardioseptopexy
кардиосклероз cardiosclerosis
кардиоскоп cardioscope
кардиоспазм cardiospasm
кардиоспермиум *bot.* heartseed *(Cardiospermum)*
кардиостеноз cardiostenosis
кардиостимулятор cardiostimulant; cardioaccelerator; pacemaker
кардиосфигмограмма cardiosphygmogram
кардиотахограф cardiotachograph
~, мгновенный momentary cardiotachograph
кардиотахография cardiotachography
кардиотахометр cardiotachometer
кардиотахометрия cardiotachometry
кардиотоксический cardiotoxic
кардиофобия cardiophobia
кардиофон cardiophone
кардиоциклография cardiocyclography
кардит carditis
кардиохирургия heart surgery
кардиохирург heart surgeon
кардный carding, card *(attr.)*
кардобенедикт *bot.* blessed thistle *(Cnicus benedictus)*
кардолента *text.* card cloth, card
кардон I = картон
кардон II *bot.* cardoon *(Cynara cardunculus)*
кардочесание, кардочесальный carding
кардочесанный carded
кардуль *bot.* Spanish oyster plant *(Scolymus hispanicus)*
каре square
карел, карельский Karelian
карелка Karelian (woman)
карен *chem.* carene
карета coach, carriage
~, почтовая stagecoach
каретка carriage; slide; carrier; frame
~, буровая track-mounted drill
~ бухгалтерской машины accounting/booking carriage
~ зонда probe carriage
~ параллаксов *geod.* parallax slide
~ пера самописца pen carriage
~ ремизоподъемная dobby
~ суппорта main carriage
~ токарного станка lathe carriage
~, челночная shuttle carriage
~, эджерная *met.* edger carriage
каретка-держатель pen carriage
каретта *zool.* hawk's-bill turtle *(Eretmochelys imbricata)*
кариама, бразильская *zool.* seriema *(Cariama cristata)*
кариатида *arch.* caryatid
кариб Carib
карибка Carib (woman)
карибский Cariba, Cariban, Caribbean
Карибская ассоциация свободной торговли Caribbean Free Trade Association
карибу *zool.* caribou *(Rangifera tarandus)*
кариес *med.* caries
~, зубной dental caries
карийон *mus.* carillon
кариковые *pl. bot.* papaw family *(Caricaceae)*
карильон *mus.* carillon
карина *bot., zool.* carina
кариогамия karyogamy
кариоген karyogene
кариограмма karyogram
кариоз *med.* caries; decay
кариозность *med.* cariosity
кариозный *med.* carious
кариоид karyoid
кариокинез karyokinesis
кариокинетический karyokinetic
кариолемма karyolemma
кариолизис karyolysis
кариолимфа karyolymph
кариологический karyological
кариология karyology
кариомер karyomere
кариомерит karyomerite; karyomere
кариометр karyometer
кариометрия karyometry
кариомикросома karyomicrosome
кариомитоз karyomitosis
кариомитом karyomitome
карион karyon
кариопикноз karyopyknosis
кариопилит *miner.* caryopilite
кариоплазма karyoplasm
кариопласт karyoplast
кариопс *bot.* caryopsis
кариоптерис *bot.* bluebeard *(Caryopteris)*
кариорексис karyorrhexis
кариосома karyosome
кариосфера karyosphere
кариота *bot.* fishtail palm *(Caryota)*
кариотин karyotin
кариотип karyotype
кариотипирование karyotyping
кариотипический karyotypic
кариотолистный *bot.* caryotaefolious
кариофаг karyophage
кариофилидный caryophylid
кариоцерит caryocerite
каритив, каритивный *gram.* caritive

кария *bot.* hickory *(Carya)*
каркас I *tech.* frame, skeleton, framework
~, арматурный reinforcing cage
~ без заполнения *constr.* naked frame
~ без обшивки open construction
~, бетонный concrete frame
~, жесткий *constr.* rigid framework
~ катушки coil form
~ киля *av.* fin framework
~, клепанный riveted framework
~ крыла wing truss
~, надувной inflatable framework
~, несиловой secondary structure
~, несущий supporting/primary structure, load-bearing framework
~ покрышки, металлокордный [резиновый] steel-cord [rubber-cord] casing
~ самолета, силовой aeroplane primary structure
~, силовой load-bearing (primary) structure
~, стальной steelwork
~ трубчатый tubular framework
~, ферменный truss framework
~, шишечный *met.* core mandrel/rod
каркас II *bot.* hackberry *(Celtis)*
карлик dwarf, pigmy, pygmy, nanus
~, асексуальный/неполовой asexual dwarf
~, атиреоидный athyreoid dwarf
~, инфантильный infantile dwarf
~, микромелический micromelic dwarf
~, половой primordial/sexual dwarf
~, пропорциональный истинный proportional true dwarf
~, рахитический rachitic dwarf
~, фокомелический phocomelic dwarf
~, хондродистрофический chondrodystrophic dwarf
карликовость dwarfism, dwarfishness, dwarfness, pygmism, nanism, nanosoma, nanosomia, microsomia
карликовый dwarfish, pygmean, nanous
~ые племена Pygmies
карлинг *geol.* horn, horn peak
карлингс, карленгс *naut.* carling
карлист, карлистский *hist.* Carlist
карма karma
кармазин crimson
карман pocket; nest
~ брака reject bin
~, вакуумный pocket of reduced pressure
~, газовый gas nest/pocket
~ для золы *met.* cinder pot
~ для перфокарт, входной/выходной *comp.* card hopper/stacker
~ подачи карт *comp.* card feed hopper
~, приемный receiving hopper, reception pocket/stacker; *comp.* card stacker
~, рудный ore pocket, chamber/bag of ore
карманообразный pockety, nesty; pocket-shaped
карманьола *mus.* carmagnole
карматрон *el.* carmatron
кармелит Carmelite friar
кармелитка Carmelite nun
кармелитский Carmelite
кармеолит *miner.* carmeolite
кармин carmine
карминит *miner.* carminite
карминный, карминовый carmine
карминофильный carminophile, carminophilous
карнавал carnival
карналлитолит *miner.* carnallite rock
карнауба *bot.* Carnauba palm, Brazilian wax palm *(Copernica cerifera)*
карнаубский carnaubic
карнегиит, карнеджиит *miner.* carnegieite
карниз cornice; bench, ledge
~, снеговой snow-drift site
~, соляной overhang
карнитин carnitine
карнификация carnification
карнозавр carnosaurus
карнозин carnosine
карнозиназа carnosinase
карнотит carnotite, yellow ore
~, кальциевый calciocarnotite
каровый ледник cirque/corrie glacier
каролинг, каролингский *hist.* Carolingian
каронада *mil.* carronade
каротаж *geol.* logging, coring, log survey
~, газовый gas logging
~, гамма gamma-ray logging, gamma logging
~, интервальный long-interval logging
~, колонковый coring; sampling
~ по рассеянному излучению scattered-ray well logging
~, радиоактивный radioactivity logging, nuclear radiation logging
~ скважины well logging
~ скорости velocity logging
~ сопротивления resistivity/resistance logging
~, электрический resistivity/resistance logging
~, электрический боковой lateral resistivity logging, shielded-electrode logging
каротидный *anat.* carotid
каротин *chem.* carotene, carotin
каротиназа carotenase, carotinase
каротинемия carotenemia, carotinemia
каротинодермия *med.* carotinodermia
каротиноз *med.* carotinosis, aurantiasis cutis
каротиноид *biochem.* carotenoid, carotinoid
каротиноидемия *med.* carotinoidemia
карп, *obs.* **карпий, карпия** *zool.* carp *(Cyprinus carpio)*
карпаделий *bot.* carpadelium
карпальный *anat.* carpal
карпелярный *bot.* carpellary
карповый *zool.* carp *(attr.);* cyprinoid
карповые *pl. zool. Cyprinidae*
карпогон *bot.* carpogonium, carpogone
карпогонидия *bot.* carpogonidium

карпоед carp louse *(Argulus)*
карпозигота *bot.* carpozygote, zygospore
карпокарпальный *anat.* carpocarpal
карполит carpolith, carpolite
карпометакарпальный *anat.* carpometacarpal
карпообразные *pl. zool. Cypriniformes*
карпопедальный carpopedal
карпосперм(а) carposperm
карпоспора carpospore
карпоспорангий *bot.* carposporangium
карпоспорический carposporic, carposporous
карпоспорофит *bot.* carposporophyte
карпостом, карпостомий *bot.* carpostome
карпотропный carpotropic
карпофилл *bot.* carpophyll, carpel
карпофит carpophyte
карпофор carpophore
карр *geol.* swallow/sink hole, sink, corrie
~, снеговой pinnacle
карраген *bot.* carrag(h)een *(Chondrus crispus);* carrag(h)een moss *(Cladonia islandica)*
каррарский мрамор Carrara marble
карст *geol.* karst
~, (пустынно) глинистый clay karst
~, голый denuded karst
~, погребенный buried karst
~, покрытый mantled karst
карстовый karst *(attr.),* karstic
карта 1. map; chart; 2. card
~, анаглифическая anaglyphic map, anaglyph
~ анализа издержек вследствие брака *econ.* scrap cost analysis chart
~ анализа технологической операции operation analysis chart
~, артиллерийская fire control map
~, аэронавигационная aeronautical [air navigation] chart
~ барической топографии constant-pressure chart
~, батиметрическая bathymetric chart
~ без изображения рельефа planimetric map
~, бланковая blank/outline map
~ в азимутальной проекции azimuthal chart
~ в гномонической проекции gnomonic chart
~ в конической проекции conic chart
~ в конформной проекции conformal map, equal angle map
~ в косой проекции oblique chart
~ в меркаторской проекции Mercator chart
~ в перспективной центральной проекции great-circle chart
~ в полярной проекции polar chart
~ в поперечной меркаторской проекции inverse Mercator chart
~ в равновеликой проекции equal-area chart
~ в равноугольной проекции orthomorphic chart
~ в стереографической проекции stereographic chart
~ ветров wind chart
~ влажности воздуха moisture chart
~ воздушных масс air-mass chart
~, галсовая sounding lines chart
~, геоизотермическая geoisothermic map
~, гипсометрическая hypsometric/altimetric/layered chart
~, глазомерная *geod.* pace-and-compass traverse map
~, гравитационная gravity map
~ грунтовых вод depth-to-water map
~, депрессионная ventilation map
~ детали detail card; part card
~ для графической работы plotting chart
~ для ночных полетов night-flying chart
~ для радиопеленгации radio-direction finding chart
~, дорожная highway/road/route map
~ звездного неба star/celestial chart
~ земного магнетизма terrestrial magnetism chart
~ земного шара global map
~, идентифицирующая identification/personality card
~ изаллобар pressure-change chart
~, издаваемая chart on issue
~ издания Британского Адмиралтейства Admiralty chart
~ изобар isobaric chart
~ изобарической поверхности constant-level chart
~ изогипс isohypsic (constant-level) chart
~ изогон isogonic map
~ изодинам isodynamic map
~ изодоз isodose map
~ изоклин isocline/isoclinic chart
~ изопор isoporic map
~ изотерм isothermal chart
~ изохрон isochronic map
~, инструкционная instruction sheet
~, историческая period map
~, кадастровая cadastral map/plan
~ Карно *math.* Karnaugh map
~, климатическая climatic map
~, климатологическая climatographic chart
~, контурная blank/outline/contour chart
~ крупного масштаба large-scale map
~, лоцманская pilot chart
~, лунная lunar map
~ магнитных склонений isogonic chart
~, маломасштабная/мелкомасштабная small-scale map
~, маршрутная *tech., econ.* flow/process/route/travel chart/sheet
~, маршрутная полетная route flight chart
~, метеорологическая weather/meterorological chart
~, миллионная millionth map
~, многократно используемая fractional/multiple/tumble card
~, многоцветная full-colour chart

~, монтажная *el.* wiring card/chart
~, морская nautical chart
~, морская навигационная marine (nautical) chart
~, навигационная navigation/nautical chart
~ накопления данных data acquisition/collecting card
~ наличия attendance/presence card
~ неба sky chart
~, немая outline map
~ о наличии товаров *econ.* balance/stock/bin card
~ облачности cloud chart
~, обзорная general/review map; sketch
~, объемная рельефная relief model
~ одновременно выполняемых операций multiple acrivity process chart
~, одноцветная black-and-white map
~, оперативная/операционная flowsheet, operation/job/process sheet/chart/card
~, операционно-технологическая operation process chart/sheet
~, ортодромическая great circle chart
~ осадков precipitation/rainfall chart
~, основная master/guide/register card
~, палеографическая paleographical/paleoareal map
~, перфорационная (punch) card
~, перфорированная punched card
~, пилотажная pilotage chart
~, планшетная plane-table map
~, побережная coast chart
~ погоды weather/synoptic chart
~ подходов approach chart
~ поразрядного отображения информации bit-map
~, приходная receipt card
~, рабочая working map
~, радиолокационная radar chart/map
~, радионавигационная radiogoniometric [direction-finding] chart
~ распределения памяти *comp.* memory map
~ расхода материалов *econ.* material flow chart
~, рельефная relief map
~ с инструкциями *comp.* command/order/instruction card
~ с котидальными линиями cotidal chart
~ с сеткой квадратов squared chart
~ с сеткой координат grid chart
~ со штриховым рельефом hachure map
~ сборки *tech.* assembly chart
~, сборная composite chart
~, синоптическая synoptic/meteorological chart
~ смазки lubrication chart
~, структурная *geol.* structural/subsurface map
~, схематичная sketch/skeleton/schematic map
~ технического обслуживания service chart
~, технологическая flowsheet, operation/process/job/route chart; maintenance requirement card
~ технологического процесса operations/process sheet/chart/layout
~, топографическая topographic/ground/land map, surface contour map
~, точечная spot map
~, управляющая control/pilot card; master/guide/register card
~, учебная training/exercise chart
~, учетная *econ.* accounting card
~, фациальная facies map
~, хронометражная time card/chart
~ целей *mil.* target chart
~ часовых поясов time zone chart
~, экономико-географическая economic map
~, экранная radar chart/map
карта-врезка insert
карта-сетка lattice chart
карта-схема skeleton/outline chart
карт-бланш carte blanche
картвел, картвельский Georgian
картвелка Georgian (woman)
картезианец 1. *philos.* Cartesian; 2. *rel.* Carthusian monk
картезианка Carthesian nun
картезианский 1. *philos.* Cartesian; 2. *rel.* Carthusian
картезианство *philos.* Cartesianism
картелизм *econ.* cartelism
картелирование cartelizing, cartelization
картелировать form (combine into) a cartel, cartelize
картель *fin.* cartel
~, объединять в cartelize, combine into a cartel
~, открытый open cartel
~, принудительный compulsory/state cartel
~, ценовой price cartel
картер crankcase; casing; housing
~ ведущего моста axle housing
~ двигателя crankcase, engine case
~ двигателя, задний rear casing
~ заднего моста rear-axle housing
~ коленчатого вала, неразъемный barrel-type crankcase
~ коленчатого вала, разъемный two-piece (split-type) crankcase
~ компрессора compressor casing/housing
~ распределительного вала camshaft casing
~ редуктора gearbox casing
~ фрикциона clutch case
картечь, картечина *mil.* case/canister shot, grapeshot; buckshot
картина picture; pattern; *art* painting
~ быта scene from life
~ вихрей vortex pattern
~, волновая wave pattern
~, гидродинамическая flow pattern
~ движения motion pattern
~, дифракционная обратная back-reflection pattern

~, дифракционная прямая transmission pattern
~, дифракционная рентгеновская X-ray diffraction pattern
~, жанровая genre painting
~, интерференционная interference/fringe pattern
~, лучевая ray pattern/picture/diagram
~ масляными красками oil painting
~, непрерывная continuous picture
~ обтекания/течения flow pattern
~ распределения distribution pattern
~ установившегося течения steady-flow pattern
картинг go-cart race/racing; carting
картирование charting, mapping, plotting
~, аэрологическое aerial surveying/mapping
~, детальное close/detailed mapping
~, площадное areal mapping
~, почвенное soil mapping
картировать map
картограмма cartogram
картограф cartographer
картографирование charting, mapping, chart-/mapmaking
картографический cartographic, mapping
картография cartography, mapping
картодиаграмма diagrammatic chart/map; collation map
картометр cartometer
картометрия cartometry
картон pasteboard, cardboard, pressboard; card; paper
~, асбестовый asbestos board
~, билетный ticket stock/paper/board
~, битумный roofing board/felt
~, коробочный boxboard
~, кровельный roofing paper
~, макулатурный chipboard
~, матричный *print.* matrix paper
~, мелованный coated board
~, переплетный binders' board
~, плотный pressboard
~, прокладочный gasket/sealing board
~, строительный building paper
~, текстильный card
~, фильтровальный filter board/paper
~, электротехнический electrical pressboard
картонаж cardboard articles/products; paperboard containers/boxes
картонажная фабрика cardboard/paperboard box factory
картонка carton, cardboard/paperboard box
картосхема map-diagram/-scheme, diagrammatic/sketch map
картотека card file/index, file; filing cabinet
~, библиотечная card catalogue
~ кредитной информации *econ.* credit file
~ с визуальным поиском *comp.* batten/cardonnier system, peek-a-boo system
~ текущего учета active/live file
~ технического ремонта maintenance file
~, электронная electronic file cabinet
~, электрооптическая electro-optic file
~, эталонная reference file
картофелекопатель, картофелекопалка *agr.* potato digger
картофелерезка potato chopper
картофелесажатель, картофелесажалка potato planter
картофелесортировка potato grader/sizer
картофелесортировочный potato-grading/sorting
картофелетерка potato grater, potato grating mill
картофелеуборочный potato-picking *(attr.)*
~ комбайн potato harvester
картофелехранилище potato storage/cellar
картофелечистка potato peeler, potato-peeling machine
картофель *coll.* potatoes
~, молодой new potatoes
~, сладкий sweet potato *(Ipomoea batatas)*
~, хрустящий potato chips
карточка card; sheet, list; index card
~, абонаментная library card
~, бонитировочная *agr.* score card
~ бухгалтерского учета ledger card
~ вин wine card
~, итоговая summary card
~, каталожная index card
~, кредитная credit card
~, личная медицинская medical history sheet
~ образцов pattern/sample/show card
~, отчетная report card
~ проверок check list
~, продовольственная/промтоварная ration card
~, рабочая worksheet
~, регистрационная register card
~ сальдо balance/stock/bin card
~, семейная family record
~ учета запасов stock/record/stocktaking card
~ учета состояния запасов stock status card
~ учета яйценоскости egg record card
~, учетная record/account(ing) card
~, хронометражная time card
карточный card *(attr.)*
~ каталог card index/catalog(ue)
~ная система rationing (system)
картуш *arch.* cartouche
картушка compass card
~ с несколькими стрелками multineedle compass card
~ с одной стрелкой single-needle compass card
~, сферическая ball float
~, штормовая storm card
карункул *anat., bot., zool.* caruncle
~ мочеиспускательного канала urethral caruncle
карункула *bot.* caruncle

карусель *tech.* vertical lathe; turntable
карусельный станок vertical lathe
каруца Moldavian cart/waggon
карфагенский, карфагенянин Carthaginian
карфагенянка Carthaginian (woman)
карфология *pathol.* carphologia, carphology, floccillation, floccilegium
карцер punishment cell; lockup
карцерула *bot.* carcerule
карциноген *med.* carcinogen
карциногенез *med.* carcinogenesis
карциногенный = канцерогенный
карциноид *med.* carcinoid, argentaffin tumo(u)r, argentaffinoma
карциноидный carcinoid
карцинолизин carcinolysin
карцинология carcinology
карцинома *med.* carcinoma
  ~, аденоидно-кистозная adenoid cystic carcinoma, cylindromatous adenoma
  ~, базоцеллюлярная basal cell carcinoma
  ~, гигантоклеточная giant cell carcinoma
  ~, криброзная cribriform carcinoma
  ~, мелкоклеточная small cell carcinoma
  ~, мукоэпидермоидная mucoepidermoid carcinoma
  ~, онкоцитарная oncocytic carcinoma
  ~, слизистая mucinous carcinoma
  ~, хромофобная chromophobic carcinoma
карциноматоз *med.* carcinomatosis
карциносаркома *med.* carcinosarcoma
  ~, эмбриональная embryonal carcinosarcoma/nephroma, embryoma of the kidney
карьер I quarry; open pit
  ~, каменный/породный stone quarry
  ~, песочный sand pit
карьер II career, full gallop
карьера career
касание contact, touching; *math.* tangency; *geom.* osculation
  ~ земли *aero.* ground contact, touchdown
  ~ зубьев *tech.* gear contact
  ~ кривых *geom.* osculation of curves
  ~ по линии line contact
  ~, плавное *aero.* gentle touchdown
касательная *math.* tangent *(to a curve)*
  ~ в окружности tangential
  ~ в точке перегиба inflectional tangent
касательность connection, relation; linkage
касательный tangential, tangent
касатик *bot.* iris *(Iris)*
  ~ гребешковый crested dwarf iris *(Iris cristata)*
касатка *zool.* 1. swallow *(Hirundo rustica);* 2. falcated teal *(Anas falcata)*
касаться *math.* be tangential, touch
касик cacique
каска helmet, headpiece
  ~, защитная/предохранительная safety/crash helmet
каскад 1. cascade; chain, series; 2. *el.* stage
  ~ без смешивания non-mixing cascade
  ~, входной input stage
  ~, выходной output stage
  ~ вихревых колец vortex-ring cascade
  ~, задающий driver stage
  ~ защиты *rad.* protection cascade
  ~, квадратурный squared-off cascade
  ~, несмешивающий non-mixing cascade
  ~, оконечный output stage
  ~, предоконечный penultimate stage
  ~, пропускающий gate/gating stage
  ~, противоточный counter-current cascade
  ~, прямоугольный square cascade
  ~, регенеративный stripping cascade/section
  ~, смесительный mixer/adder cascade
  ~, стробирующий gating/gate stage
  ~, суммирующий addition/adder cascade
  ~, управляющий driving stage
  ~ усиления amplifier/amplifying stage
  ~, электронно-фотонный electron-photon cascade
каскадер *cin.* stuntman
каскадирование cascading; cascade operation
каскадировать cascade
Каскадия *geol.* Cascadia
каскадно in series
каскадский *geol.* Cascadian
каскариль *bot.* cascarilla, seaside balsam; *(Croton eleuteria);* leaf croton *(Codiaeum)*
каско *econ.* hull insurance; boxcar/tankcar insurance
каспиец sailor of the Caspian Fleet
каспийский Caspian
касса 1. cash register; cashbox; 2. *econ.* cash; 3. ticket office/window; cash/cashier's office; box office; booking office; 4. *print.* case
  ~ без кондуктора coin box
  ~ взаимопомощи mutual aid fund; benefit society
  ~, контрольная cash register
  ~, наборная type case
  ~ прописных литер upper case
  ~, сберегательная savings bank/fund
  ~ строчных литер lower case
касса-автомат slot machine
кассава *bot.* cassava *(Manihot)*
кассандра *myth.* Cassandra
кассатор *law* appellant, appealer
кассационный *law* cassation *(attr.),* of cassation
  ~ суд Court of Cassation
  ~ная жалоба appeal
кассация *law* cassation
  ~ии, подать appeal to a court of cassation
кассета cassette; magazine; cartridge; carrier; reel, spool; *phot.* plateholder, charger, loader
  ~, бомбовая *mil.* cluster bomb unit, bomb cluster, bomb cell/container
  ~ дисков disk cartridge
  ~ заряжаемая на свету daylight-loading magazine/cassette

~, **ленточная** tape cartridge
~ **магнитной ленты** tape cassette
~, **магнитофонная** tape spool/reel
~, **пленочная** film carrier/magazine
~, **подающая** feed/supply reel, pay-out reel
~, **приемная/принимающая** take-up spool/reel
~ **с данными** *comp.* data cartridge
~ **самописца** recorder cartridge/cassette
~ **светонепроницаемая** *phot.* light-tight drum container
~ **сигнальных ракет** signal-flare cassette
~ **тепловыделяющих элементов** *nucl.* fuel assembly
~, **фотографическая** cassette; plate holder; film cassette; *cin.* film magazine
кассиопа, кассиопея *bot.* cassiope *(Cassiope)*
кассиопей *chem.* crssiopeium
кассир cashier; teller; box-office clerk
~ **приходной кассы** receiving cashier/teller
~ **расходной кассы** paying cashier/teller
кассировать *law* reverse, quash, annul
касситерит *miner.* cassiterite
кассия *bot.* senna *(Cassia)*
кассовый cash *(attr.)*
~ **аппарат** cash register
~ **счет** cash account
~**вая книга** cashbook
~**вая наличность** cash balance
каста caste
Касталия *myth.* Castalia
кастанопсис *bot.* evergreen chinkapin *(Castanopsis)*
кастаньета castanet
кастелян *hist.* castellan
кастилец Castilian
кастилла *bot.* gum tree *(Castilla)*
~ **каучуконосная** Panama gum tree *(Castilla elastica)*
кастильский Castilian
кастовый caste *(attr.)*
кастор I *text.* castor
кастор II *miner.* castorite
Кастор *myth., astr.* Castor
касторовый 1. castor *(attr.)*; beaver *(attr.)*
~**вая шляпа** beaver hat
~**вое масло** castor oil
кастрат castrate, eunuch
кастрация castration, sterilization; emasculation, testectomy; gelding
~, **гормональная** functional castration
~, **паразитарная** parasitic castration
кастрировать castrate, emasculate; geld; caponize
кат *naut.* cat
~, **брать на** to cathead
ката *bot.* Arabian tea, khat, kat *(Catha edulis)*
катаболизм catabolism
катаболический catabolic
катависса *bot.* Welsh onion *(Allium fistulosum)*
катавотр *geol.* sink/swallow hole
катагенез *biol.* catagenesis
катагимений *bot.* catahymenium
катадиоптрика catadioptrics
катадромный *bot.* catadromous
катазона katazone
катаклаз *geol.* cataclasm
катаклазированный broken down
катаклазит *geol.* cataclasite
катакластический *geol.* cataclastic
катаклизм cataclysm; convulsion
~ **природы** convulsion of nature
катакомба catacomb
катакрота *physiol.* catacrotism
катакротический catacrotic
каталаза *chem.* catalase
каталепсия *med.* catalepsy
каталептический, каталепсический cataleptic
каталептик, каталептичка *med.* cataleptic
катализ *chem.* catalysis
~, **гетеролитический** acid-base catalysis
~, **гомолитический** oxidation-reduction catalysis
~, **кислотно-основный** acid-base catalysis
~, **окислительно-восстановительный** redox catalysis
~ **основаниями** base catalysis
катализатор catalyst, catalytic agent, catalyzer
~, **зерненый** beaded catalyst
~, **крекирующий** cracking catalyst
~, **отработанный** spent/dead catalyst
~, **отрицательный** negative catalyst, inhibitor
катализаторный, каталитический catalytic
катализация catalyzation
катализировать catalyze
каталитически catalytically
каталка wheelchair
каталог catalog(ue); directory
~, **алфавитный** author and title (alphabetical) catalogue
~, **алфавитно-систематический** classified catalogue
~, **генеральный** main/general/full catalogue
~, **звездный** star catalogue
~, **карточный** card catalogue
~, **корневой** *comp.* root directory
~ **массивов** *comp.* data-set catalogue, file catalogue
~, **предметный** alphabetic subject catalogue
~ **программного обеспечения** *comp.* software catalogue
~, **рабочий** working directory
~, **сводный** joint/repertory catalogue
~, **систематический** systematic/classified catalogue
~, **системный** system directory
~, **текущий** current directory
~, **тематический** topic(al) catalogue
~ **файлов** *comp.* file directory
~, **читательский** public catalogue
каталогизатор cataloguer, catalogist

каталогизация cataloguing
каталогизировать catalogue
каталожный catalogue *(attr.)*
 ~ ящик card drawer
каталонец, каталонский Catalan
каталонка Catalan (woman)
катальный rolling; felting, fulling; mangling
катальня fullery
катальпа *bot.* catalpa *(Catalpa)*
катальполистный *bot.* catalpifolious
катальщик felter, fuller; roller, calenderer
катамаран *naut.* catamaran
катаморизм *geol.* katamorphism
катамнез *med.* catamnesis
катаморфный catamorphic
катананхе *bot.* Cupid's dart *(Catananche)*
катание *met.* rolling; *text.* fulling, felting, mangling
 ~ верхом riding
 ~ в экипаже driving
 ~ на коньках skating
 ~ на лодке sailing, boating
 ~, фигурное figure skating
катанка *met.* wire rod, rolled wire
катаный rolled; *text.* mangled
катапетальный *bot.* catapetalous
катаплазия *med.* cataplasia, cataplasis
катаплазматический, катаплазматичный *bot.* cataplastic
катаплазма *med.* cataplasm
катаплексия *physiol.* cataplexy
катаплектический cataplectic
катапульта catapult
 ~, газовая gas-powered catapult
 ~, корабельная ship's catapult
 ~, наземная ground catapult
 ~, палубная deck catapult
 ~, паровая steam-powered catapult
 ~, пороховая powder-type catapult
 ~, пружинная spring catapult
 ~, реактивная jet catapult
 ~, самолетная aircraft catapult
 ~, стартовая catapult launcher
катапультирование catapulting; ejection, ejecting
 ~ аварийного самописца crash recorder ejection
 ~, аварийное emergency ejection
 ~ в пикировании ejection from a dive
 ~ вверх/вниз upward/downward ejection
 ~ на сверхзвуковой скорости supersonic ejection
 ~ обычным стреляющим миханизмом ballistic ejection
 ~, пиротехническое cartridge-fired ejection
 ~, подводное underwater ejection
 ~, последовательное sequenced ejection
 ~ при отрицательных углах тангажа nose-down ejection
катапультировать eject; *(of airplane)* catapult
катапультироваться eject oneself
катапультируемый ejectable
катар I *med.* catarrh
 ~, атрофический atrophic catarrh/rhinitis
 ~, весенний vernal catarrh/conjunctivitis
 ~ желудка catarrhal gastritis
 ~, острый acute catarrh
катар II *eccl., hist.* Catharist
катаракт 1. *tech.* dashpot, cataract, damper; 2. cataract, cascade; waterfall
 ~, масляный oil cataract/brake
катаракта *med.* cataract(a)
 ~, врожденная congenital/embryonal cataract
 ~, вторичная secondary cataract, after-cataract
 ~, зонулярная zonular/lamellar cataract
 ~, зрелая mature/ripe cataract
 ~, лучевая stellate cataract
 ~, мягкая soft cataract
 ~, набухающая swelling cataract
 ~, начальная primary cataract
 ~, незрелая immature/intumescent cataract
 ~, осложненная complicated cataract
 ~, перезрелая hypermature/overripe cataract
 ~, пирамидальная pyramidal cataract, coneshaped polar cataract
 ~, полная total cataract
 ~, профессиональная occupational cataract
 ~, рентгеновская radiation cataract
 ~, старческая senile cataract
 ~, сумочная capsular cataract
 ~, тетаническая tetany cataract
 ~, хрусталиковая lenticular cataract
 ~, центральная central/nuclear cataract
 ~, Юношеская juvenile/infantile cataract
катарактальный, катарактный *med.* cataractous
катаральный *med.* catarrhal
катарзис = катарсис
катаробный katharobic
катарометр *phys., chem.* katharometer
катарсис catharsis
катасетум *bot.* monkflower *(Catasetum)*
катастазис catastasis
катастат *bot.* katastate
катастрофа disaster, calamity, catastrophe; accident
 ~, авиационная air crash/disaster
 ~, автомобильная automobile/car accident
 ~, экономическая economic disaster
катастрофически catastrophically, disastrously
катострофический, катастрофичный catastrophic, disastrous
кататермометр catathermometer, katathermometer
катателий catathecium
кататонический *med.* catatonic, catatoniac, stuporous
кататония *med.* catatonia, catatony
катать 1. *tech.* roll; 2. *text.* mangle; felt, ull; 3. take for a drive
катафаза kataphase, cataphase

катафалк hearse; catafalque
катафилл *bot.* cataphyllary leaf, cataphyll
катафилльный *bot.* cataphyllary
катафиллярия *bot.* cataphyllum, cataphyllary leaf
катафорез cataphoresis
катафорезный, катафоретический cataphoretic
катафорит *miner.* cataphorite, amphibole
катафот *opt.* cat's eye
катафронт katafront
катахреза catachresis
катахромазия *biol.* katachromasia
катахтонный katachthonous
кат-балка *naut.* cat davit; cathead
кат-блок *naut.* cat block
кат-гак *naut.* cat hook
катгемоглобин cath(a)emoglobin
категорематический *logic* categorematic
категориальный categorial, categoric
категорически categorically; flatly
категорический, категоричный categorical; flat
категоричность categoricalness
категория category; class; rank; grade; rating
~, надродовая supergeneric category
~ срочности priority rank
~, ценовая price bracket
катедер-социализм academic socialism, academic socialist doctrine
катедер-социалист academic socialist
катексический cathectic
катенарный *math.* catenary
катеноид *geom.* catenoid
катенулин catenulin
катепсины *pl.* catepsins
катер *naut.* launch, boat; craft; barge; cutter
~ аварийно-спасательной службы salvage boat
~ авиации, спасательный aircraft rescue boat
~, адмиральский admiral's barge
~, артиллерийский gunboat, G-boat
~ береговой спасательной службы coastal escue boat
~, буксирный towing launch
~, быстроходный speed-boat, speed launch
~, водометный waterjet-propelled boat, hydrojet propelled boat
~, глиссирующий planing/skimming boat
~, грузовой cargo launch
~, десантный landing craft/boat/cutter, assault boat
~, гребной cutter
~, дозорный picket boat
~, заводский yard launch
~, лоцманский pilot launch/boat
~, лоцмейстерский buoy/dan boat
~, мореходный off-shore launch, seagoing craft
~, морской ocean-going launch
~, моторный motor/power boat
~ на воздушной подушке air-cushion boat/craft
~ на подводных крыльях hydrofoil boat/craft
~ обслуживания service craft
~, опытовый test craft
~, парусно-гребной cutter
~, перевозочный ferry launch
~, пожарный fire(-fighting) boat/launch
~, портовый power utility boat
~, посыльный dispatch launch
~, прогулочный pleasure boat/launch
~, промерный survey/sounding boat
~, противолодочный submarine chaser
~, рабочий utility boat/launch, motor work boat; ferry launch
~, разъездной crew's launch, crew boat
~, рейдовый estuary/inshore boat
~ с каютой cabin launch
~, санитарный ambulance/hospital boat
~, спасательный rescue motor launch, motor lifeboat
~, сторожевой seaward patrol craft; patrol boat/craft/vessel/gunboat
~, таможенный customs launch
~, торпедный motor torpedo boat
~, швартовный mooring launch
катер-толкач push boat
катер-торпеда explosive boat
катер-торпедолов torpedo recovery boat
катер-тральщик mine-sweeping boat
катет 1. *math.* leg, side *(of a right triangle);* 2. *weld.* leg
~, прилежащий adjacent/adjoining side/leg
~, противолежащий opposite side/leg
катетер *med.* catheter
~ для спринцеваний и орошений douche and irrigation catheter
~, микрофонный microphonic catheter
~, мочеточниковый urethral catheter
~, постоянный indwelling catheter
~ с двойным током two-way catheter
~, сердечный (intra)cardiac catheter
~, ушной ear/Eustachian catheter
катетеризация catheterization; catherism
~ сердца cardiac catheterization
катетеризировать *med.* catheterize
катетометр cathetometer
катехизис catechism
катехизический catechetic(al)
катехин *chem.* catechol, catechin
катехистический catechistic(al)
катехический catechetic(al)
катехоламин *biochem.* catecholamine
катеху catechu
катехумен *eccl.* catechumen
катехолоксидаза *biochem.* catechol oxidase
катион *chem., phys.* cation, basic ion, positively charged ion
~, заменяемый exchangeable cation
катионит cation-exchange resin, cationite
катионообменник cation exchanger

катоба *bot.* catawba *(Vitis labrusca)*
катогенный *geol.* catogene
катод *phys.* cathode
~, вторично-емитирующий/емиссионный secondary-emission cathode
~, горячий hot/thermionic cathode
~, действующий virtual cathode
~, дуговой [дугового разряда] arc cathode
~, жидкий/жидкометаллический pool cathode
~, карбидированный tungsten-carbide cathode
~ косвенного накала indirectly heated cathode, heater cathode
~, лучевой beam cathode
~, оксидный oxide-coated cathode
~, подогревный heater cathode
~, простой pure cathode
~ прямого накала directly heated cathode, filamentary cathode
~, пусковой starter cathode
~, ртутный mercury-pool cathode
~ с ионным нагревом ionic-heated cathode
~, сетчатый reticular cathode
~, термоэлектронный thermionic/hot cathode
~ тлеющего разряда glow-discharce cathode
~, торированный thoriated-tungsten cathode
~, точечный point cathode
~, трубчатый/туннельный sleeve/tunnel cathode
~, фотоэлектрический photoelectric cathode
~, фотоэлектронный photoemissive cathode, photocathode
~, холодный cold cathode, field-emission cathode
катодальный *bot.* kathodic, cathodic
катодно-лучевой cathode-ray
катодный *phys.* cathodic
катодолюминесцентный cathodoluminescent
катодолюминесценция cathodoluminescence
катодофосфоресценция cathodophosphorescence
каток roller; soil packer; roll, cylinder
~, болотный bog roller
~, вибрационный vibration roller
~, водоналивной water-filled roller
~, гребневой/грядной ridge roller
~, двухвальцовый two-wheel tandem roller
~, дорожный road roller, compactor
~, зубчатый toothed roller
~, кольчато-зубчатый sprocket packer
~, кольчатый disk/ring roller
~, кулачковый sheep's foot roller
~, многосекционный segmented roller
~, навесной mounted roller
~, пневмоколесный pneumatic-tyred roller
~, полевой гладкий plain roller
~, прицепной trailed (tractor-hauled) roller
~, ребристый sectional roller
~, самоходный self-propelled roller
~, статический static-weight roller
~, шпоровый crowfoot roller/packer
каток-глыбодроб *agr.* clod-crusher
каток-снегоуплотнитель snow roller/compactor
каток-тандем tandem roller
католик (Roman) Catholic
католикос *eccl.* Catholicos
католит *phys., chem.* catholyte
католицизм 1. (Roman) Catholicism; 2. Catholicity
католический (Roman) Catholic
католичество (Roman) Catholicism
католичка (Roman) Catholic
катонов, катоновский *hist.* Catonian
катоптрика catoptrics
каторга penal servitude/colony
каточек roller; press wheel
катран I *bot.* colewort *(Crambe)*
катран II *zool.* dogfish, dog shark *(Squalus)*
катрен *lit.* quatrain
катта *zool.* ring-tailed lemur *(Lemur catta)*
кат-тали cat tackle/purchase
каттлея *bot.* cattleya *(Cattleya)*
катушка coil, reel; bobbin; spool
~, биспиральная coiled coil
~, бифилярная bifilar coil
~, вогнутая concave bobbin
~ возбуждения field/exciting coil; launching/drive coil
~, воспринимающая pick-up coil
~ для магнитной ленты magnetic tape coil
~ для намотки wind-up spool
~ для разматывания winding-off spool
~ для сновки warping spool
~ для троса cable-drum
~, дроссельная choke coil
~, дугогасящая arc-suppression/quenching coil
~, дутьевья blow-out coil
~ зажигания ignition coil
~ индуктивности inductance coil
~, индукционная induction coil
~, искательная search coil
~ кабеля cable drum
~, кассетная *phot.* cassette spool
~, корзиночная basket (Lorenz, stagger-wound) coil
~, основная *text.* warp bobbin
~, отклоняющая deflecting coil
~ переменной индуктивности variable inductance coil, variable inductor
~, приемная 1. take-off roller; 2. pickup coil
~, пупиновская Pupin coil
~, пусковая *auto.* booster coil
~, расцепляющая trip coil
~, ровничная *text.* flyer bobbin
~, рукавная hose reel
~ с воздушным зазором air-gap coil
~ с лентой tape reel/spool
~ с отводами tapped coil
~ самоиндукции self-induction coil
~ связи coupling coil
~, сновальная *text.* warper's bobbin

~, сотовая honeycomb/lattice coil
~, тороидальная doughnut/toroidal coil
~, установочная current-limiting induction oil
~, фланцевая flanged bobbin
~ якоря armature coil
катэлектротон(ус) *physiol.* catelectrotonus
катюша *mil., colloq.* Katyusha, truck-mounted multirail rocket launcher
каудальный caudal; caudate
каузальгический *med.* causalgic
каузальгия *med.* causalgia, burning pain
каузальность *philos.* causality
каузальный *philos.* causal; causative
каузатив, каузативный *gram.* causative
каулифлория *bot.* cauliflory
каулифлорный *bot.* cauliflorous
каулом *bot.* caulome, stem axis
каулоцистида *bot.* caulocystidium
каупер hot blast stove, Cowper stove; blast furnace
каури *bot.* dammar/kauri pine *(Agathis)*
каустизация *chem.* causticization
каустический *bot.* caustic
каустика *opt.* caustic
каустичность causticity
каустобиолит *miner.* caustobiolith
~ животного происхождения caustozoolith
~ растительного происхождения caustophytolith
каутер *med.* cauter
каутеризация cauterization, cautery
~, химическая chemocautery, chemical cautery
~, холодовая cryocautery, cold cautery
~, электрическая electrocautery, galvanocautery, electric cautery
каутскианец follower of Kautsky
каучук rubber; caoutchouc
~, бутадиен-нитрильный butadiene-acrylonitrile rubber
~, бутадиен-стироловый butadiene-styrene rubber
~, дикий wild rubber
~, латексный latex rubber
~, кремнийорганический silicone rubber
~, маслонаполненный oil-extended rubber
~, минеральный mineral caoutchouc, elaterite
~, неопреновый neoprene rubber
~, несортный off-grade rubber
~, пенный scum rubber
~, плантационный estate/plantation rubber
~, полиэфирный ester rubber
~, сажевый carbon-black rubber, black-filled rubber
~, сажемасляный oil-black rubber
~, силиконовый silicone rubber
~, синтетический synthetic rubber
~, сополимерный copolymer rubber
~, сырой crude/raw rubber
~, товарный commercial rubber
~, хлоропреновый chloroprene rubber
~, цельный whole latex rubber
~, роговой hard rubber, ebonite
каучуковый rubber *(attr.)*
каучуконос *bot.* rubber plant
каучуконосный rubber-bearing
каучукообразный, каучукоподобный gum-like, rubberlike; gummy, rubbery
кафедра 1. *educ.* chair; department; 2. cathedra, episcopal see
кафедральный 1. departmental; 2. cathedral
~ собор cathedral
кафель ceramic/stone/Dutch tile
кафизма cathisma, kathisma
кафир 1. Kafir; 2. *bot.* kaffir corn *(Sorghum affrorum)*
кафолический *eccl.* (o)ecumenical, universal
кафр Kaffir
кафтан caftan
кахексия *med.* cachexia
~, гипофизарная hypophysial/pituitary cachexia
~, раковая cancerous cachexia
~, свинцовая saturnine cachexia
~, сердечная cardiac cachexia
~, церебрально-гипофизарная cerebrohypophysial cachexia
кахектический *med.* cachectic
кахрис *bot.* cachrys *(Cachrys)*
кацик cacique
качалка 1. rocking bar, rocker, swing/balance/reversing lever; 2. bell crank; 3. swing, seesaw; 4. two-wheeled racing cart
~, поддерживающая supporting lever
~, секторная quadrant
~ управления control crank
качание 1. rocking, swinging, swaying; oscillaton; 2. pumping
качателный rocking, swinging, swaying; shaking; scillating
качать 1. *rock.* swing; sway; dandle; oscillate; shake; 2. pump
качаться *rock.* swing, sway, shake; oscillate
качение rolling
~, чистое true roll
качества *pl.* qualities; characteristics
~, ездовые *auto.* rideability
~, инерционные *naut.* stopping ability
~, летные flight performance, flying qualities
~, мореходные seagoing/seakeeping/nautical gualities, seaworthiness, navigability
~, пилотажные *aero.* handling qualities
~, эксплуатационные performance qualities, seaworthiness, navigability
качественно-определительный qualitative-determinative
качественно-оценочный qualitative-evaluative
качественность high quality
качественный 1. qualitative; 2. high-quality
качество quality, grade; characteristic, nature, character, property; merit
~, аэродинамическое lift-to-drag ratio, aerodynamic efficiency

~ во время отгрузки shipped quality
~, высшее premium/prime/top/superior quality
~ изготовления workmanship
~, общее overall quality
~ оговоренное контрактом contract quality
~, однородное consistent quality
~, положительное merit; advantage
~ работы workmanship
~ топлива fuel grade/quality
~, физическое physicality
~ функционирования performance
~, хлебопекарное baking quality
~, худшее bottom quality
~, эксплуатационное functional quality
качим *bot.* chalk plant *(Gypsophila)*
качка rocking, swinging; *naut.* (oscillating) motions
~, бортовая/боковая rolling motion/oscillation, rolling, roll
~, вертикальная heaving motion/oscillation, heaving
~, килевая pitching motion/oscillation, pitching
~, нерегулярная random motions
~, плавная easy motion
~, поперечная lateral motion
~, поперечно-горизонтальная swaying (motion)
~, продольная longitudinal motion
~, продольно-горизонтальная surging (motion)
~, свободная бортовая free/natural rolling
~, сильная heavy pitching/rolling
~ судна ship's motions
качкар *zool.* Pamir argali *(Ovis ammon)*
качурка *zool.* petrel
~, малая stormy petrel *(Hydrobates pelagicus)*
~, морская frigate petrel *(Pelagodroma marine)*
качча *mus.* catch
каша, снеговая slush
кашалот *zool.* sperm whale, cachalot *(Physter catodon)*
каше *phot.* matte, mask, vignette
кашель *med.* cough, tussis
~, влажный wet/productive cough
~, лающий barking cough
~, нервный nervous cough
~, рефлекторный reflex cough
~, сухой dry/nonproductive cough
~, юношеский hebetic cough
кашемир *text.* cashmere
каширование 1. *phot.* masking; vignetting; 2. *print.* backing; 3. pasting; laminating
кашица paste, pulp; slurry
кашка *bot.* white clover *(Trifolium repens)*
кашкара *bot.* Siberian/snow rose *(Rhododendron chrysanthum)*
кашлять cough
кашмирец Kashmiri(an)
кашмирка Kashmiri(an) (woman)
кашмирский Kashmirian
кашпо cachepot
каштан *bot.* chestnut; chestnut tree *(Castanea)*
~, конский horse chestnut, buckeye *(Aesculus hippocastanum)*
каштановый chestnut *(attr.)*
каштаноспермум *bot.* bean tree *(Castanospermum ustrale)*
кашу catechu
кашубский Kashubian
каюк caique
каюта state cabin/room; cabin
~, двухместная double-berthed cabin
~, жилая harbour cabin
~ командира captain's state room/cabin
~, одноместная single-berth cabin
~, походная sea cabin
~, шифровальная coding office
кают-компания messroom; messing compartment; wardroom; chief cabin, passengers' lounge
каяк *naut.* kayak
каянус *bot.* pigeon/Congo pea *(Cajanus)*
квагга *zool.* bontequagga *(Equus quagga)*
квадр *constr.* cut stone
квадра *arch.* quadra
квадрант 1. *math.* quadrant; 2. *hydr.* single-vane hydraulic motor/actuator; 3. *mil.* (gunner's) clinometer, quadrant
квадрат *geom.* square; *math.* the square (of), number squared; *print.* quad
~, возводить в raise to the second power, (raise to the) square
~, греклатинский Greek-Latin square
~, наименьший least square
~, полный perfect square
квадратический square; quadratic
квадратично-косекансный cosecant-squared
квадратичность *rad.* square-law characteristic
квадратичный quadratic; quadric; square-law
квадратно-гнездовой *agr.* square-cluster *(attr.)*, heckrow *(attr.)*
квадратноскуловой *zool.* quadratojugal
квадратность squareness
квадратный square; quadratic; quadrate
квадратор *el.* square-law function generator, squarer
квадратриса *math.* quadratrix
квадратура quadrature, squaring
~ во времени time quadrature
~ в пространстве space quadrature
~ круга squaring the circle
~ луны quadrature of the moon
~, численная numerical quadrature
квадратурно-модулированный *el.* quadrature-modulated
квадратурный quadrature *(attr.)*
квадривий, квадривиум quadrivium
квадрига quadriga
квадригеминия *med.* quadrigeminia
квадрика quadric
квадрильон *math.* quadrillion *(US);* thousand

million *(UK)*
квадрипарез quadriparesis
квадриплегия quadriplegia
квадрирема quadrireme
квадрирование squaring
квадрируемость squareability
квадрируемый squareable
квадрицепс *anat.* quadriceps
квадруплекс *tel.* quadruplex
квадруплет quadruple
квадруполь quadrupole
квадрупольный quadrupolar
квазар *astr.* quasar
квазивосток *mil.* grid east
квазивременной quasi-time
квазивязаный quasi-bound
квазигеоид quasigeoid
квазигравитация quasi-gravity
квазидвуустойчивый quasi-bistable
квазизапад *mil.* grid west
квазизвезда *astr.* quasar
квазизвездный *astr.* quasar *(attr.)*
квазиимпульс quasi momentum
квазиклимакс quasi climax
квазиконформный quasi-conformal
квазикурс *mil.* grid course
квазиледяной quasi-ice
квазилинеарный, квазилинейный quasi-linear
квазимарксист quasi Marxist
квазимеридиан *mil.* grid meridian
квазимодо Quasimodo
квазинаправление *mil.* grid direction
квазинаучный quasi-scientific
квазиобратный quasi-inverse
квазиоднородный quasi-homogeneous
квазипарадигма quasi paradigm
квазипараллель *mil.* grid parallel
квазипериодичность quasi periodicity
квазиполе *math.* quasi field
квазипредложение quasi sentence
квазипродольный quasi-longitudinal
квазипроизвольный quasi-random
квазиравновесный quasi-equilibrium *(attr.)*
квазиразмах *math.* semirange
кзазирешетка quasi array
квазисевер *mil.* grid north
квазисоциалист quasi socialist
квазистатически quasi-static
квазистационарный quasi-stationarv; quasi-steady
квазитвердый quasi-solid
квазиупорядоченность quasi arrangement
квазиупругий quasi-elastic
квазиустановившийся quasi-steady
квазиустойчивость quasi persistence/stability
квазициклический quasi-cyclic
квазичастица quasi particle
квазиэкватор *mil.* grid equator
квазиюг *mil.* grid south
кваква *zool.* night heron *(Nycticorax)*
квакер Quaker
квакерка Quakeress
квакерский Quaker *(attr.);* Quakerish
квакерство Quakerism
квакша *zool.* hyla *(Hyla)*
квалиметр qualimeter, penetrometer
квалитативный qualitative
квалификационный qualification *(attr.);* classiying, evaluation, assessing
квалификация 1. qualification, expertise, proficency; 2. qualifications, skill(s); 3. grading
  ~ летчика pilot proficiency
  ~, производственная professional/manufacturing skill(s)
  ~, профессиональная job/vocational skill(s)
  ~ товаров grading of goods
квалифицированный 1. qualified; certificated; 2. skilled, trained, proficient
  ~, недостаточно semiskilled
  ~ рабочий skilled worker
  ~ труд skilled labour
квалифицировать 1. qualify; certify; 2. grade; 3. train, provide/offer further professional qualifications to
квалифицироваться acquire (further) professional skills
квамоклит *bot.* cypress vine, star glory *(Quamoclit)*
квант *phys.* quantum
  ~ времени time slot/slice
  ~ высокой энергии high-energy quantum
  ~ действия quantum of action
  ~ излучения quantum of radiation
  ~, индуцированный induced/stimulated quantum
  ~ малой энергии low-energy quantum
  ~, падающий incident quantum
  ~ плоской волны plane-wave quantum
  ~ поля field quantum
  ~, рассеянный scattered quantum
  ~, рентгеновский X-ray quantum/photon
  ~ света light quantum, photon
  ~, световой light quantum
  ~, стимулированный induced/stimulated quantum
  ~ тепловых колебаний quantum of thermal vibrations
  ~, электромагнитный electromagnetic quantum
  ~ энергии energy quantum
квантизатор sampler, quantizer
квантика *math.* quantic
квантиль *stat., math.* quantile, fractile
квантитативный quantitative
квантификатор quantifier
квантификация quantification
квантифицировать *math.* quantify
квантование quantization; sampling
  ~ времени *comp.* time slicing/slotting
  ~ по времени time sampling/quantization

~ сигнала, полное quantizing and sampling, amplitude and time quantization, digitization
~ сигнала по уровню amplitude quantization, quantizing
квантователь quantizer; digitizer
квантовать quantize
~ сигнал (по времени) sample; **(по уровню)** quantize
квантовый quantum *(attr.)*
~вая механика quantum mechanics
квантометр quantometer
квантор quantifier
~ общности universal/generality quantifier
~ существования existential quantifier
кварк quark
кварта *mus.* fourth
квартал 1. city block/quarter, district, ward; 2. quarter *(of year)*
~, жилой residential quarter
квартальный 1. of city quarter; 2. quarterly
квартана *med.* quartan
квартер *geol.* Quaternary system/period
квартердек *naut.* quarterdeck
~, повышенный/пониженный raised/dropped quartrdeck
квартерон quadroon
квартет quartet(te); *biol.* tetrad
квартиль *math.* quartile
квартира 1. apartment, lodging, accommodations, lat; 2. *pl. mil.* billets, quarters
~, главная *mil.* general headquarters
~ и стол room and board
~, коммунальная shared apartment
~, мебелированная furnished apartment/flat
квартирант-нахлебник boarder
квартирмейстер *mil., obs.* quartermaster
квартирный apartment *(attr.)*; housing *(attr.)*; *mil.* billeting; *pl.* lodging/hotel allowances
квартиронаематель tenant; lodger
квартирохозяин landlord
квартирьер *mil.* billeting officer
кварто *print.* quarto
квартование quartering, coning
квартплата *abbr.* rent
кварц *miner.* quartz; crystal; *el.* quartz crystal
~, аморфный fusible/fused quartz
~, белый стекловатый bastard quartz
~ в шапке capped quartz
~, волосистый hairstone
~, гребенчатый comb quartz
~, давленный stressed quartz
~, звездчатый asteriated/star quartz
~, зеленоватый prase
~, левый left-handed quartz
~, лучистый cross-course spar
~, молочный milky quartz
~, перистый feathered quartz
~, плавленый fused quartz
~, пористый floatstone
~, правый right-handed quartz
~, промежуточный interstitial quartz
~, розовый rose quartz, Bohemian ruby
~ с фантомом ghost quartz
~, сплавленный melted quartz; glassed silica
~, темный дымчатый morion
~, халцедоновый flint
~, червеобразный vermicular quartz
кварцевополевошпатовый quartzofelspatic
кварцевый quartz *(attr.)*, crystal *(attr.)*, quartzose, quartzous, quartzy
~вая лампа quartz lamp
кварцедержатель crystal holder
кварцин *miner.* quartzine
кварцит *miner.* quartzite
~, железистый ferruginous quartzite
~, известковый calcerous quartzite
~, мельничный millstone quartzite
~, пресноводный limnoquartzite, limnetic quartzite
~, пятнистый spotted quartzite
~, халцедонный chalcedony quartzite
кварцитовый *miner.* quartzitic
кварцоид *cryst.* quartzoid
кварцованный *el.* crystal-controlled
кварцсодержащий quartziferous, quartzose
квас kvass *(drink)*; fizz
~, хлебный fermented-bread kvass
квассия *bot.* quassia *(Quassia)*
квасцевание aluming
квасцеванный tanned/treated with alum, alumed
квасцевать, квасцовать alum, tan *(leather)*, soak/steep/treat in alum
квасцовый alum *(attr.)*; aluminous, aluminiferous
квасцы *pl. chem.* alum, alumen, potassium aluminium sulphate
~, аммиачные ammonia alum, tschermigite
~, горные rock alum
~, железные iron alum, halotrichite
~, железоаммиачные ferric ammonium alum
~, жженые alum flower
~, калиевые potassium/potash alum, kalinite
~, кристаллические alum glass
~, магниевые magnesia alum, pickeringite
~, марганцевые manganese alum
~, натровые soda alum
~, перистые feather alum, alumogen
~, хромовые chrome alum
~, хромокалиевые chromium-potassium sulphate, chrome alum
кватернион *math.* quaternion
кватроченто quatrocento
квашиоркор *med.* kwashiorkor, infantile pelagra, malignant malnutrition
квебрахо, квебрачо *bot.* red quebracho *(Quebrachia, Schinopsis)*
~, белое common white quebracho *(Aspidoperma quebrachoblanco)*
кверцетин *chem.* quercetin
кверцит *chem.* quercitrol, quercite
кверцитрин *chem.* quercitrol, quercitrin

квершлаг *min.* crosscut, crossheading
~, откаточный haulage crosscut
~, участковый district/auxiliary crosscut
квеста cuesta
квестор *hist.* qu(a)estor
квестура qu(a)estorship
квецал *zool.* quetzal *(Pharomacrus mocinno)*
квиетизм quietism
квиноя, квиноа *bot.* quinoa *(Chenopodium uinoa)*
квинта *mus.* fifth
~, чистая perfect fifth
квинтал quintal *(100 kg)*
квинтет *mus.* quintet(te)
квинтиллион, квинтильон quintillion *(US),* trillion *(UK)*
квинтиль *stat., math.* quintil(e)
квинтоль *mus.* quintole
квинтэссенция quintessence, fifth essence
квипрокво quid pro quo
квиринал Quirinal
квирит *hist.* Quirite
квисквалис *bot.* quisqualis *(Quisqualis)*
квислинг quisling
квистист *eccl., hist.* Hesychast
квит *bot.* common flowering quince *(Chaenoneles agenaria)*
квитанция receipt; *com.* acqnowledge(ment) signal
~, багажная luggage/baggage receipt/check
~ о подписке subscription receipt
~ об уплате страхового взноса premium receipt
~, отрицательная *com.* negative acqnowledgement
квитирование *com.* acqnowledgement (of receipt)
~ установления связи handshaking
кворум quorum
~, составлять *law* make house
квота quota; contingent; share
~ в международном валютном фонде IMF quota
~, иммиграционная immigration quota
~, общая global quota
~, рыночная marketing quota
квотация *econ.* quotation
кебаб *cul.* kebab, cabob
кегельбан bowling alley
кегельмашина batch roller
кегельная *print.* em
кегли *pl.* ninepins, skittles; tenpins
кегль *print.* point; size
~ шрифта type size
кедр *bot.* cedar *(Cedrus)*
~, гималайский deodar cedar *(Cedrus deodora)*
~, ливанский cedar of Lebanon *(Cedrus ibani)*
~ сибирский Siberian stone pine *(Pinus ibirica)*
кедровка *zool.* nutcracker *(Nucifraga)*
кедровый cedar *(attr.);* cedarn
~ странник Japanese stone pine *(Pinus pumila)*
кейнсианство *econ.* Keynesian theory; Keynesian economics
кек cake
~, фильтрационный filter/filtered cake
~, шламовый slurry cake
кеклик *zool.* partridge *(Alectoris)*
Кекроп(с) *myth.* Cecrops
кекс *cul.* fruitcake, cake
келарня cellar; dining hall and kitchen *(in a monastery)*
келарь *eccl.* cellarer; procurator
келейник *eccl.* lay brother; anchorite, hermite
келейница *eccl.* lay sister; anchoress
келерия *bot.* koeleria *(Koeleria)*
~ гребенчатая prairie June grass *(Koeleria cristata)*
келиоскопия *med.* celioscopy
келифитовый *geol.* kelyphitic
келлин khellin
келловей *geol.* Callovian
келоид *med.* keloid
кельвин *phys.* Kelvin, K
кельма, кельна *constr.* trowel
~, выпуклая circle trowel
кельрейтерия *bot.* gold rain tree *(Koelreuteria)*
кельт I Celt, Kelt
кельт II *archeol.* celt
кельтий *chem.* celtium, Ct
кельтицизм *ling.* Celticism
кельтолог Celtologist
кельтский Celtic, Keltic
келья *eccl.* cell
кемализм Kemalism
Кембридж Cambridge
кембриджский Cambridge *(attr.),* Cantabrigian
кембрий *geol.* Cambrian, Cambrian period/ystem
кембрик *text.* cambric (tape/fabric)
Кембрия Cambria, Wales
кемквот *bot.* round/Marumi kumquat *(Citrus japonica, Fortunella japonica)*
кемпедек *bot.* jack tree/fruit *(Arthocarpus heterophyllus)*
кемпинг camping site/ground; motorists' camp
кенаф *bot.* gambo/Deccan hemp, kenaf *(Kibiscus cannabinus)*
кенгуру *zool.* kangaroo
кендык *bot.* dog's-tooth violet, dogtooth fawnlily *(Erythronium dens-canis)*
кендырь *bot.* 1. dogbane, flytrap *(Apocynum);* 2. Indian hemp *(Trachomitum)*
~, коноплевидный/конопляный hemp dogbane *(Apocynum cannabinum)*
кенематика cenematics
кениец, кенийский Kenyan
кинийка Kenyan (woman)
кенкет Quinquet oil lamp
кеннель cannel (coal)
кеннинг *lit.* kenning
кенозойский *geol.* Cenozoic
кенозооид cenozooid
кенотаф, кенотафий cenotaph

кенотоксин *biochem.* kenotoxin
кенотрон *el.* kenotron, high-vacuum rectifier tube, hot-cathode rectifier tube
кенофобия kenophobia
кентавр *myth.* centaurus
Кентавр *astr.* Centaurus
кентер canter, easy callop
кентерберийский Canterbury *(attr.)*
кентия *bot.* kentia, curly palm *(Kentia)*
кентрантус *bot.* centranth *(Centranthus)*
кеньга *naut.* kink
керазин kerasin
керамзит expanded clay aggregate, keramzit
керамзитобетон lightweight concrete with expanded lay aggregate
керамик ceramist
керамика ceramics; *coll.* earthenware, stoneware
~, абляционная ablative ceramics
~, армированная reinforced ceramics
~, бытовая pottery
~, вспененная expanded ceramics
~, глазурованная enamelled stoneware/earthenare
~, жаростойкая heat-resistant (high-temperaure) ceramics
~, огнеупорная refractory ceramics
~, пьезоэлектрическая piezoelectric ceramics
~, радиотехническая radio ceramics
~, сегнетоэлектрическая ferroelectric ceramics
~, строительная structural ceramic goods, structural clay products
~, тальковая steatitic ceramics
~, теплопроводная heat-conductive ceramics
~, техническая engineering ceramics
~, электротехническая electric-grade ceramics
керавноид keravnoid, spherocrystal
керамицит *miner.* ceramicite
кераргирит *miner.* cerargyrite
кератенхима *bot.* ceratenchyma
кератин *biochem.* keratin
кератинизация keratinization
кератиновый *zool.* keratinous, horny
кератит *med.* keratitis
~, герпетический herpetic keratitis
~, грибковый mycotic keratitis
~, древовидный dendriform keratitis
~, лагофтальмический lagophthalmic keratitis
~, нейропаралитический neuroparalytic keratitis
~, поверхностный superficial/punctate keratitis
~, подагрический gouty keratitis
~, фликтенулезный phlyctenular keratitis
~, язвенный ulcerative keratitis
кератоакантома keratoacanthoma
кератогиалин keratohyalin, eleidin
кератоглобус keratoglobus
кератодерматит keratodermatitis
кератодермия keratodermia, ceratodermia
кератоз keratosis
~, подногтевой subungual/hyponychial keratosis
~, эссенциальный essential/idiopathic keratosis
кератоидный *biol.* keratoid, ceratoid
кератоиридоскоп keratoiridoscope
кератоирит *med.* keratoiritis
кератоконус *pathol.* keratoconus
кератоконъюнктивит *pathol.* keratoconjunctivitis
~, аденовирусный adenoviral keratoconjunctivitis
~, вирусный virus keratoconjunctivitis
~, фликтенулезный phlyctenous keratoconjunctivitis
кератолейкома keratoleukoma
кератома *med.* keratoma
кератомаляция keratomalacia
кератомикоз *med.* keratomycosis
кератоноз *med.* keratonosis
кератопластика keratoplasty
кератосклерит keratoscleritis
кератоскопия keratoscopy
кератотом keratotome
кератотомия keratotomy
кератэктомия keratectomy
Кербер *myth.* Cerberus
кервель *bot.* salad chervil *(Anthriscus erefolium)*
керенщина *polit., hist.* Kerensky's rule
керл *hist.* bondman
карлинг *sport .* curling
кермезит *miner.* kermesite
кермек *bot.* sea lavender *(Limonium, Statice)*
кермес *zool.* kermes *(Kermes)*
кермет metal ceramic, cermet, ceramet
керн drill/test core, drill sample, core; 2. centre-prick/-pop
керн(ер) centre/center punch, marking pin, rick punch
кернит centre-punch, centre-pop
корнодержатель core holder
керноизвлекатель core extractor
керноловка core-picker
кернорватель core gripper/catcher/lifter
кернохранилище core storage/shack
керн-пункт *top.* epipole
керн-функция *math.* kernel function
керосин kerosene, kerosine
~, авиационный aviation kerosene
~, осветительный burning kerosene
~, тракторный tractor-fuel kerosene
керосиновый kerosene *(attr.)*
керосинозаправщик kerosene fueller/loader, erosene loading vehicle
керосинорез kerosene cutter/torch, oxy-kerosene cutter
керсантит *miner.* kersantite
~, шаровой spheroidal kersantite
кесарев, кесарский Caesarian
кесарево/кесарское сечение *med.* Caesarian section
кесар *obs., rel.* Caesar; lord

кессон *naut.* caisson, cofferdam; *av.* torsion box, cell; *met.* water jacket
~, аварийний cofferdam
~, двухлонжеронный two-spar structural box
~, двухотсечный two-cell box
~ крыла wing torsion box
~ мартеновской печи dog-house
~ медеплавильной печи water jacket
~, подъемный float case
~, спасательный submarine rescue chamber
~, стреловидный *av.* swept-back box beam
~ центроплана *av.* wing centre-section box
кессонировать coffer
кессонный caisson *(attr.)*
~ная болезнь caisson disease, aeroembolism
кета *zool.* dog/calico salmon, keta *(Oncorhynchus keta)*
кетгут catgut
~, хирургический surgical gut
кетелеерия *bot.* keteleeria *(Keteleeria)*
кетен *chem.* keten(e)
кетогексоза *chem.* ketohexose, fructose
кетогенез *med.* ketogenesis
кетогенный ketogenic
кетоглюконокиназа ketogluconokinase
кетодезоксиглюконокиназа ketodeoxyglucono-kinase
кетоз *med.* ketosis
кетоза *chem.* ketose
кетокаин *chem.* ketocaine
кетокислота ketoacid, cetone acid
кетон *chem.* ketone, cetone
кетонемия *med.* ketonemia
кетонный, кетоновый ketonic, cetonic
кетонурия *med.* ketonuria
кетопантоальдолаза ketopantoaldolase
кетостероиди ketosteroids
кетоформа ketonic form
кетсал quetzal *(monetary unit of Guatemala)*
кетчуп ketchup
кефалевидные *pl. zool.* Mugiloidea
кефалевые *pl. zool.* gray mullets *(Mugilidae)*
кефалгематома cephalh(a)ematoma
кефализация *biol.* cephalization
кефалин *biochem.* cephalin
кефалотриб *med.* cephalotribe
кефаль *zool.* mullet *(Mugil)*
кефаэлин *chem.* cephaeline
Кефей *myth.* Cepheus
кефир kefir
кеч *naut.* ketch
кианг *zool.* kiang *(Equus kiang)*
кианит *miner.* cyanite, kyanite
кибермашина robot; computer
кибернетизация cybernation
кибернетик cybernetist, cybernetician
кибернетика cybernetics
~, техническая engineering cybernetics
~, экономическая economic cybernetics
~ электроэнергетических систем power-system cybernetics
кибернетический cybernetic
киборг *abbr.* cybernetic organism, cyborg
киборгизация introduction of cybernetic organisms; robotization
киборгизованный robotized; computerized
кивер *mil., hist.* shako
киви *zool.* kiwi *(Apteryx)*
~, обыкновенный brown kiwi *(Apteryx ustralis)*
киевлянин, киевлянка inhabitant of Kiev
киевский Kievan, Kiev *(attr.)*
киематогенез kyematogenesis
киематопатия kyematopathia, kyematopathy
кижуч *zool.* silver salmon, coho, silversides *(Oncorrhynchus kisutch)*
кизелгур kieselgu(h)r, infusorial earth, diatomite, mountain meal
кизерит *miner.* kieserite
кизил *bot.* dogwood, cornel *(Cornus)*
~, шведский dwarf cornel *(Cornus suedica)*
кизиловые *pl., bot.* dogwood family *(Cornaceae)*
кизильник *bot.* 1. cotoneaster *(Cotoneaster);* 2. *coll.* cornel thicket
кизляк *bot.* Naumburgia
киксия *bot.* fluellin *(Kickxia)*
кикуйю Kikuyu
кил kill, fuller's earth
килевание *naut.* careening, careenage
килеватость *naut.* deadrise
килевать *naut.* careen, heave down
килевидноперегородчатый *bot.* carinimurate
килевидный *bot.* keel-shaped, keeled, carinate, cariniform
килевой keel *(attr.); bot.* carinal
~вая качка pitching, pitching motions/oscillations
килегрудный cristate, cristiform, carinate
киленбалка *naut.* careenage
кило kilogram(me)
килоампер *el.* kiloampere, kA
килобайт kilobyte
килобар kilobar
килобарн kilobarn
килобит kilobit, K
килобод kilobaud
киловар kilovar, kiloVAr, kVAr
киловатт kilowatt, kW
киловаттметр kilowattmeter
киловатт-час kilowatt-hour, kWh
киловольт kilovolt, kV
киловольт-ампер kilovolt-ampere, kVA
~, реактивный kiloVAr, kVAr
киловольт-амперметр kilovoltammeter
киловольтметр kilovoltmeter
килогаусс kilogauss
килогерц kilohertz, kHz
килограмм kilogram(me)
килограмм-калория kilogram(me) calorie, large calorie, kcal
килограмм-молекула kilogram(me) molecule
килограммометр *mech.* kilogram(me)-meter,

kgm
килограмм-сила kilogram(me)-force, kg(f)
килоджоуль *phys.* kilojoule, large loule, kJ
килодина kilodyne
килокалория kilogram(me) calorie, large calorie, cal
килокюри kilocurie, kC
килолитр kilolitre
килолюмен kilolumen
киломегагерц kilomegahertz
киломегацикл kilomegacycle
киломегом *el.* kilomegohm
километр kilometre, kilometer, km
километраж mileage; distance in kilometers; kilometers travelled/flown
километровый kilometer *(attr.)*
киломоль kilomole, molecule kilogram(me), kmol
килоом *el.* kilohm, kilo-ohm
килопарсек *astr.* kiloparsec
килорентген kiloroentgen
килорезерфорд kilorutherford
килослово *comp.* kiloword
килотонна kiloton
килоцикл kilocycle
килоэлектрон-вольт kiloelectron-volt
киль 1. *naut.* keel; 2. *av. fin.* vertical stabilzer; 3. *anat.* carina
~, балластный ballast keel
~, боковой bilge keel
~, верхний dorsal fin
~ влагалища carina vaginae
~, выдвижной drop/sliding keel, centre-board
~, горизонтальный flat (plate) keel
~ дирижабля airship keel
~, доковый docking keel
~, имеющий carinate
~, коробчатый box/duct keel
~ летающей лодки flying-boat keel
~, надфюзеляжный dorsal fin
~, нижний/подфюзеляжный ventral fin, underfin
~, отпускной drop/sliding keel, centre-board
~, отрывной detachable keel
~, ровный even keel
~ синклинали syncline trough
~, скуловой bilge keel
~, стреловидный swept tail-fin
~, туннельный duct keel
~, центральный centre-line keel
кильблок *naut.* keel-block
~, доковый docking block
~, песочный sand block
~, подъемный hydraulic block
~, разборный/разбирающийся collapsible keelblock
~, шлюпочный boat block/chock
кильватер *naut.* wake
кильватерный wake *(attr.)*
~ная колона line ahead
кильпоручень keel rail
килька *zool.* sprat *(Sprattus)*
кильсон *naut.* keelson
~, боковой/бортовой side keelson
~, диаметральный/средний centre-line keelson, middle-line keelson
~, скуловой bilge keelson
КИМ *abbr., obs.* Communist Youth International
кимберлит kimberlite, blue earth/ground
кимвал *mus.* cymbal
кимвры *pl.* Cimbri
киммерийский Cimmerian
кимограмма kymogram(me)
кимограф kymograph
кимография kymography
кимоно kimono
кимоскоп kymoscope
кимоциклография kymocyclography
кимоцистография kymocystography
кимр *hist.* Cambrian
кимрский Cambrian; Cymric
киназа *biochem.* kinase
киназоген kinasogen
кингстон kingston/sea/flooding/outboard valve
~ продувания blow-off/-out valve
~ цистерны ballast flood valve
кинезальгия *med.* kinesalgia, kinesialgia, kinesalgia
кинезиметр kinesimeter, kinesiometer, cinesimeter
кинезиметрия kinesimetry
кинезиология kinesiology, cinesiology
кинезитерапия kinesitherapy, cinesitherapy, kinesiatrics, cinesiatrics
кинезия kinesia
кинезофобия kinesophobia
кинема kineme
кинематика kinematics
~, звездная stellar kinematics
~ материальной точки particle kinematics
~ механизмов engineering kinematics
~ тел(а) body kinematics
кинематический kinematic(al)
кинематограф 1. cinematography, motion pictures; 2. *obs.* motion-picture camera, cinematograph
кинематографический cinematographic; cinematic
кинематография cinematography, motion pictures
кинемика kinemics
кинемометр kinemometer
кинеморфема kinemorpheme
кинесика kinesics
кинескоп kinescope, television/picture tube
~ с электростатической фокусировкой electrostatically focused picture tube
~, трехпрожекторный tri-gun picture tube
~, трехцветный tricolo(u)r kinescope
~, черно-белый black-and-white kinescope
кинестезиометр *med.* kinesthesiometer, cinesthesometer
кинестезия *med.* kinesthesia, cinesthesia
кинестетический kinesthetic, cinesthetic

кинетика kinetics
~ реактора reactor kinetics
~ туманности nebular kinetics
~ химических реакций chemical kinetics
кинетин *chem.* kinetin
кинетический kinetic
кинетогенез *biol.* kinetogenesis
кинетография kinetography
кинетомер kinetomere
кинетометаморфизм *geol.* kinetometamorphism
кинеторадиограмма kinetoradiogram
кинетоскоп kinetoscope
кинетосома *biol.* kinetosome
кинетостатика kinetostatics
кинетофон kinetophone
кинетохор *biol.* kinetochore, centromere
кинжаловидный dagger-shaped, pugioniform
кинжалолистный *bot.* dagger-leaved
кинетофон kinetophone
кинизм *philos.* Cynicism
киник *philos.* Cynic
кинин kinin, phytokinin, cytokinin
кинкажу *zool.* kinkajou *(Potos flavus)*
кинкан *bot.* kinkan, kumquat *(Fortunella argarita)*
киннамон cinnamon
кино *bot.* kino *(Pterocarpus);* kino gum
кино II cinematography; cinema; film; motion picture
~, широкоэкранное cinemascopic film
киноактер film/movie actor
киноактриса film/movie actress
киноангиография cineangiography
киноангиокардиография cineangiocardiography
киноаппарат movie/cine camera, motion-picture camera
~, проекционный motion-picture projector
~, скоростной high-speed camera
киноаппаратная projector room
киноаппаратура cinematographic/cine equipment, motion-picture equipment
киноартист(ка) = киноактер, киноактриса
киноархив film archives/library
киноателье motion-picture studio
киноаудитория film audience
кинобласт kinoblast
кинобудка projection room
киноварно-красный vermilion-red, cinnabarine
киноварный cinnabaric, cinnabarine
киноварь cinnabar, vermillion
киноведение cinematology
киноведческий cinematological
кинодокумент documentary film
кинодраматург screen writer
киножурнал newsreel
киноискусство cinema(tographic) art
кинокадр movie/action still, cinema frame
кинокамера movie/cine camera, motion-picture camera
кинокефал *myth.* cynocephalus
киноклассика film classics
кинокольцовка filmstrip
кинокопировальный film-printing/-copying
кинолента film reel
кинология *zool.* cynology
киномеханик projectionist, movie technician, cinema operator
киномонтаж montage
кинооборудование cinema/cine equipment
кинооператор cameraman
кинопанорама cinerama
кинопередвижка travelling film projector
киноплазма *biol.* kinoplasm, kinetoplasm, kinetoplasm(a)
кинопленка motion-picture film, cine film
~, ацетилцеллюлозная cellulose acetate film
~, безопасная safety film
~, мелкозернистая fine-grain film
~, негорючая non-(in)flammable film
~, неэкспонированная raw film
~, нитроцеллюлозная cellulose nitrate film
~, обратимая reversal film
~, узкая narrow-gauge (cine) film
~, широкоформатная wide-gauge (cine) film
кинопозитив motion-picture positive
кинопроектор film projector
кинопроекционная projection room
кинопроекция film [motion picture] projection
кинопроизводство film production
кинопрокат film lending service
кинопрокатчик film distributor
кинопромышленность film [motion-picture] industry
кинопулемет cine-/camera-gun; gun/combat camera
киноработник cinema worker
кинорадиография cine-radiography
кинорежиссер film director
кинорежиссура film direction
кинорентгенография cinoroentgenography, cineradiography, cinematoradiography
кинорепортаж news film
кинорепортер television news reporter
киносеанс movie/film show
киносеть cinema network
киносенситометр scene tester
киностудия film [motion picture] studio
киносценарий film script; scenario
~, режиссерский screenplay
киносценарист scriptwriter
киносъемка filming, shooting; motion-picture photography
~, воздушная aerial photography/shooting
~, замедленная time-lapse filming
~, мультипликационная animation
~, павильонная stage shooting/filming
~, цейтраферная time-lapse filming
кинотеатр cinema/movie theatre, cinema
кинотеодолит cine-theodolite
кинотерапия cinematotherapy
кинотехника film/cine technique; cinematographic

engineering; cinematographic equipment
кинотоксин kinotoxin, fatigue toxin, ponogen
киноурография cineurography
киноустановка film/movie projector; camera unit
кинофестиваль film festival
кинофикация 1. extension of the network of cinema theatres; 2. adaptation for the screen
кинофильм film, motion picture, movie
~, документальный documentary
~, звуковой sound film
~, короткометражный short film
~, мультипликационный animation film, animated-cartoon-film
~, неигровой nonfiction film
~, панорамный panoramic film
~, полнометражный full-length film
~, рисованный cartoon/drawn film
~, узкопленочный small-frame [substandard, narrow-gauge] film
~, учебный training/educational film
~, хроникальный newsreel
~, художественный feature film
~, цветной colo(u)r film
~, черно-белый black-and-white film
~, широкоэкранный cinemascopic film, widescreen film
кинофицировать set up cinema theatres in, bring cinema to
кинофлюорограмма cinefluorogram(me)
кинофлюорография cinefluorography, cinematofluorography
кинофлюороскопия cinefluoroscopy
кинофобия cynophobia
кинофотоклей film cement
кинофотопулемет = кинопулемет
кинофрагмент filmstrip
кинохроника newsreel
кинохроникер newsreel cameraman
киноэкран cinema screen, cine projection screen
~, алюминированный aluminized screen
~, бело-матовый matt white screen
~, бисерный beaded screen
~, панорамный panoramic screen
~, просветный translucent screen
~, широкий wide screen
киноэпопея film epic
кинуренин *chem.* kynurenin(e)
кинурениновый kynurenic
кионотом *surg.* kionotome
киоск stall, booth, kiosk; stand
~, газетный newspaper stand/kiosk
~, книжный bookstall, bookstand
киот icon case
кип плавки *met.* boil
кип шкива pulley rim
кипа 1. bale; pack; bundle; 2. pile, stack
кипарис *bot.* cypress *(Cypressus)*
кипарисовник *bot.* false cypress *(Chamaecyparis)*
кипарисовый *bot.* cypress *(attr.)*
~вая шишка galbulus
кипарисовые *pl., bot.* cypress family *(Cypressaceae)*
кипение boiling, boil; ebullition, effervescence, bubbling; gassing; *met.* rimming
~ аккумулятора gassing; bubbling
~ в большом/свободном объеме pool boiling
~ на дне ванны bottom boil
~, объемное bulk boiling
~ плавки boil
~ плавки, безрудное *met.* oreless boil
~ плавки, известковое *met.* lime boil
~ плавки, рудное ore boil
~ плавки, углеродистое *met.* carbon boil
~ плавки, чистое *met.* lime boil
~ пленочное film /sheet boiling
~, пузырчатое nucleate boiling
~ с вынужденной циркуляцией forced-circulation boiling
~ электролита gassing, bubbling
кипоразборщик bale opener
кипп-реле multivibrator
Кипр Cyprus
кипрегель telescopic/sighting alidade, planeable alidade
кипрей *bot.* willow-herb/-weed *(Chamaenerium angustifolium)*
киприот Cyprian, Cypriot(e)
киприотка Cyprian (woman)
кипроминойский Cypriote-Minoan
кипрский Cypriot(e), Cyprian
кипсей *print.* ink trough/block
кипучесть ebullience
кипсы *leath.* kips
кипучий boiling, effervescent; seething
кипятильник hot-water boiler
~, дезинфекционный boiling-water sterilizer
~, дезинфекционный паровой steam-heated disinfection boiler
~, погружной immersion heater
кираса *hist., mil.* cuirass; coat of mail
кирасир *hist.* cuirassier
киргиз 1. Kirghiz; 2. Kirghiz horse
кирза *leath.* kersey, kirsa
кириллица Cyrillic alphabet
кирилловский Cyrillic
кирилло-мефодический Cyril-and-Methodius
Кирка *myth.* Circe
кирка pickaxe, pick
кирказон *bot.* birthwort *(Aristolochia)*
киркомотыга pick-mattock
кирпич 1. brick; 2. *math.* Hilbert's parallelotope/cube
~ воздушной сушки air-dried brick, adobe
~, гвоздимый nailable brick
~, графитовый graphite brick
~, динасовый silica brick
~, дорожный paving brick
~, дырчатый perforated/tubular brick
~, клиновой feather-edge brick

~, лекальный curved brick
~, лицевой facing/face/lining brick
~, намертво обожженный very hard-fired brick
~, недоенный pale brick
~, необожженный unburnt/green brick, adobe
~, облицовочный facing/lining/face brick
~, обожженный burnt brick
~, огнеупорный refractory brick
~, печной kiln brick
~, саманный adobe, green/unburnt brick
~ поставленный стоймя brick-on-end
~, силикатный lime-and-sand brick
~, сланцевый shale brick
~, футеровочный lining/facing brick
~, шамотный chamotte (fire-clay) brick
~, ячестый cellular brick
кирпичный brick *(attr.)*, built/made of bricks
~ завод brickyard, brickworks
кирпич-сырец adobe
киршвассер kirschwasser
киса bag
киселевидный jellylike, gelatinous
кислевание *leath.* drenching
кислевать *leath.* drench
кислица *bot.* oxalis *(Oxalis)*
~, обыкновенная shamrock *(Oxalis acetosella)*
кисличник *bot.* mountain sorrel *(Oxyria)*
кисличные *bot.* sorrel family *(Oxalidaceae)*
кислование acidification
кисловатость sourness, acidulousness
кисловатый sourish, acidulous
кисловка souring; acid treatment
кисломолочный fermented-milk, yoghurt *(attr.)*
кислород oxygen, O
~, атомарный atomic/elemental oxygen
воздуха atmospheric oxygen
~, газообразный gaseous oxygen, oxygen gas
~ для дыхания breathing oxygen
~, жидкий liquid oxygen
~, загрязненный contaminated oxygen
~, растворенный dissolved oxygen
~, медицинский oxygen for medical purposes
~, природный native oxygen
~, связанный *chem.* bound oxygen
~, сжатый compressed oxygen
~, технический industrial-grade oxygen
~, технологический process oxygen
~, тяжевый heavy oxygen
кислородно-ацетиленовый oxyacetylene
кислородный oxygen *(attr.)*; oxygenous
~ная подушка oxygen/breathing bag
кислородозаправщик oxygen dispenser
кислородосодержащий oxygen-containing, oxygenated
кислота 1. *chem.* acid; 2. sourness; acidity
~, агарициновая agaric/agaricic acid
~, адениловая adenylic acid
~, аденилпирофосфорная adenosine triphosphoric cid
~, аденозиндифосфорная adenosine diphosphoric acid
~, адипиновая adipinic acid
~, азотистая nitrous acid
~, азотистоводородная hydrazoic acid
~, азотная nitric acid
~, азотноватистая hyponitrous acid
~, аккумуляторная battery/sulphuric/electrolytic acid
~, аконитовая citridic acid
~, акриловая acrylic acid
~, альгиновая alginic acid, norgine
~, аминобензоилмуравьиная aminobenzoylformic acid
~, аминобензоилуксусная aminobenzoylacetic acid
~, аминобензойна aminobenzoic acid
~, аминовая amino acid
~, аминоглутаровая aminoglutaric acid
~, аминокапроновая aminocaproic acid
~, аминолевулиновая aminolevulinic acid
~, аминомасляная aminobutyric acid
~, аминоянтарная aminosuccinic acid
~, анисовая anisic acid
~, антраниловая anthranilic acid
~, арахидоновая arachidonic acid
~, аскорбиновая ascorbic/hexuronic/sevitamic acid, vitamin C
~, аспарагиновая aspartic/asparaginic acid, alpha-aminosuccinic acid
~, атроповая atropic acid
~, ауринтрикарбоновая aurintricarboxylic acid
~, ацетиламиноуксусная acetylaminoacetic acid
~, ацетилсалициловая acetylsalicylic acid, aspirin
~, ацетоускусная acetoacetic acid
~, барбитуровая barbituric acid
~, бензоилмуравьиная benzoylformic acid
~, бензоилуксусная benzoylacetic acid
~, бензойная benzoic acid
~, бензолдикарбоновая benzenedicarboxylic acid
~, бескислородная hydracid
~, борная boric acid
~, бромистоводородная hydrobromic acid
~, валериановая valeric/valerianic acid
~, варочная cooking/digester acid
~, винная tartaric acid
~, высшая жирная higher fatty acid
~, газообразная хлористоводородная hydrochloric acid gas
~, галактоновая galactonic acid
~, галактуроновая galacturonic/pectic acid
~, галловия gallic acid
~, гексозодифосфорная hexosodiphosphoric acid
~, гексозомонофосфорная hexosomonophosphoric cid

~, гексуроновая = аскорбиновая
~, гептановая heptanoic acid
~, гиппуровая hippuric acid
~, гликолевая glycolic acid
~, гликоневая glyconic acid
~, гликохолевая glycocholic acid
~, глиоксиловая glyoxylic acid
~, глутаминовая glutamic acid
~, глутаровая glutaric acid
~, глюконовая gluconic acid
~, глюкороновая glucoronic acid
~, гомогентизиновая homogentisic acid, alkapton
~, гремучая fulminic acid
~, гуанидинуксусная guanidinoacetic acid
~, гуанидин-рибозо-фосфорная = гуаниловая
~, гуаниловая guanilic acid
~, гуминовая humic acid
~, гумусовая humus acid
~, двухосновная dihydric acid
~, дегидрохолевая dehydrocholic acid
~, дезоксирибонуклеиновая desoxyribonucleic acid, DNA
~, дезоксихолевая desoxycholic acid
~, диаминовалерьяновая diaminovaleric acid
~, диаминопимелиновая diaminopimelic acid
~, дикарбоновая dicarboxylic acid
~, диоксивинная dihydroxytartaric acid
~, дисульфоновая disulphonic acid
~, дитионовая dithionic/hyposulphuric acid
~, древесно-уксусная wood vinegar
~, дубильная tannic/gallotannic acid, tannin
~, дымящая fuming acid
~, жасминовая jasmonic acid
~, желчная bile/taurocholic acid
~, жидкая карболовая liquefied carbolic acid, liquefied phenol
~, жирная fatty acid
~, изовалериановая isovaleric/isovalerianic acid
~, изолимонная isocitric acid
~, индолилуксусная indoleacetic acid
~, йодистоводородная hydroiodic acid
~, камфорная camphoric acid
~, каприловая caprylic acid
~, капроновая caproic/capronic acid
~, карбаминовая carbamic acid
~, карболовая carbolic acid, phenol
~, катехудубильная catechutannic acid
~, кетоглутаровая ketoglutaric acid
~, кетоизовалерьяновая ketoisovaleric acid
~, кислородная oxyacid
~, клофенамовая clofenamic acid
~, коевая kojic acid
~, коричная cinnamic acid
~, коричномуравьиная cinnamoyl formic acid
~, креатинфосфорная creatinephosphoric acid
~, кремневая silicic acid
~, кренитовая crenic acid
~, крепкая strong acid
~, кротоновая crotonic acid
~, ксантуреновая xanthurenic acid
~, кумаровая coumaric acid
~, левулиновая levulinic acid
~, ледяная уксусная glacial acetic acid
~, лигносульфоновая spruce wood tannin
~, лимонная citric acid
~, линолевая linolic/linoleic acid
~, линоленовая linolenic acid
~, липоевая lipoic acid
~, ложная pseudo-acid
~, малоновая malonic acid
~, масляная butyric acid
~, мевалоновая mevalonic acid
~, меркаптуровая mercapturic acid
~, метакремневая metasilicic acid
~, метафосфорная metaphosphoric acid
~, метилгуанидинуксусная methylguanidinoacetic acid, creatine
~, метилмалоновая methylmalonic/isosuccinic acid
~, мефенамовая mefenamic acid
~, миндальная mandelic/phenylglycolic/hydroxytoluic acid
~, минеральная inorganic acid
~, миристиновая myristic acid
~, молочная lactic acid
~, многоосновная polybasic acid
~, монобромуксусная monobrom(o)acetic acid
~, монохлоруксусная monochlor(o)acetic acid
~, мочевая uric/lithic acid
~, муконовая muconic acid
~, муравьиная formic acid
~, надсерная persulphuric acid
~, надхромовая perchromic acid
~, налидиксовая nalidixic acid
~, насыщенная saturated acid
~, нейраминовая neuraminic acid
~, ненасыщенная/непредельная unsaturated acid
~, неорганическая inorganic acid
~, несвязанная unbound acid
~, низшая жирная lower fatty acid
~, никотиновая nicotinic acid, niacin
~, нитрирующая mixed (nitric and sulphuric) acid
~, нонановая nonanoic acid
~, нуклеиновая nucleic acid
~, одноосновная monobasic acid
~, оксимасляная hydroxybutyric acid
~, оксиянтарная hydroxysuccinic acid
~, оксолиновая oxolinic acid
~, октановая octanoic acid
~, олеиновая oleic acid
~, опиановая opianic acid
~, органическая organic acid
~, орнитуровая ornithuric acid
~, оротовая orotic acid
~, ортокремневая orthosilicic acid

~, ортофосфорная orthophosphoric acid
~, отработанная waste acid
~, пальмитиновая palmitic acid
~, пантоевая pantoic acid
~, пантотеновая pantothenic acid
~, паяльная killed spirit
~, пеларгоновая pelargonic acid
~, пентозонуклеиновая pentosonucleic acid
~, перигнойная mould acid
~, пиколиновая picolinic acid
~, пикриновая picric acid
~, пимелиновая pimelic acid
~, пировиноградная pyruvic acid
~, пиросерная pyrosulphuric acid
~, пироурановая pyrouranic acid
~, пирофосфорная pyrophosphoric acid
~, плавиковая hydrofluoric acid
~, поликремневая polysilicic acid
~, предельная saturated acid
~, продукционная product acid
~, пропионовая propionic acid
~, птероиновая pteroic acid
~, рибонуклеиновая ribonucleic acid, RNA
~, рицинолевая ricinoleic acid
~, роданистоводородная rhodanic/sulphocyanic acid
~, розоловая rosolic acid, aurin
~, рубеановая rubeanic acid
~, салициловая salicylic acid
~, сантониновая santoninic acid
~, сахарная saccharic acid
~, связанная combined acid
~, себационовая sebacic acid
~, серная sulphuric acid
~, серная башенная Glower tower acid
~, серная контактная contact-process sulphuric acid
~, серная камерная chamber sulphuric acid
~, серная сухая dry sulphuric acid
~, сернистая sulphurous acid
~, серноватая hyposulphuric acid
~, серноватистая hyposulphurous acid
~, сероводородная hydrosulphuric acid
~, сиаловая sialic acid
~, синеродистая cyanic acid
~, синильная hydrocyanic/prussic acid
~, соляная hydrochloric/muriatic acid
~, стеариновая stearic/octodecanoic acid
~, сульфаниловая sulphanilic acid
~, сульфосалициловая sulphosalicylic acid
~, сульфоцианистая sulphocyanic/rhodanic acid
~, сырая raw/crude/tower acid
~, технологическая process acid
~, тимонуклеиновая thymonucleic acid
~, титрованная titrating/standard acid
~, триоксихолановая trihydroxycholanic acid
~, трихлоруксусная trichloracetic acid
~, троповая tropic acid
~, угольная carbonic acid
~, ульминовая ulmic/mould acid
~, ундекановая undecoic/undecylic acid
~, ундекеновая/ундециленовая undecylenic acid
~, ундециловая = ундекановая
~, урановая uranic acid
~, уридиловая uridylic acid
~, фенилакриловая phenylacrylic acid
~, фенилмасляная phenylbutyric acid
~, фенилмолочная phenyllactic acid
~, фенилпировиноградная phenylpyruvic acid
~, фенилпропионовая phenylpropionic acid
~, фенилсерная phenylsulphuric acid
~, фенилуксусная phenylacetic acid
~, формовочная forming acid
~, фосфатидиловая phosphatidic acid
~, фосфоглицериновая phosphoglyceric acid
~, фосфопировиноградная phosphopyruvic acid
~, фосфористая phosphorous acid
~, фосфорная phosphoric acid
~, фосфорноватистая hypophosphorous acid
~, фосфорная экстракционная wet-process phosphoric acid
~, фосфорная электротермическая dry-process hosphoric acid
~, фталевая phthalic acid
~, фтористоводородная hydrofluoric acid
~, фумаровая fumaric acid
~, хлористоводородная hydrochloric acid
~, хинная chinic acid
~, хинодубильная cinchotannic acid
~, холевая cholic/cholalic acid
~, хлорная perchloric acid
~, хлорноватая chloric acid
~, хлорноватистая hypochlorous acid
~, хлороуксусная chloracetic acid
~, холеиновая choleic acid
~, хондроитиновая chondroitic acid
~, хромовая chromic acid
~, цереброновая cerebronic/phrenosinic acid
~, цианистоводородная hydrocyanic/prussic acid
~, циановая cyanic acid
~, цианистая cyanide acid
~, цистеиновая cysteic acid
~, цитидиловая cytidylic acid
~, щавелевая oxalic acid
~, четырехосновная tetrabasic acid, tetracid
~, щавелевоуксусная oxaloacetic acid
~, эллаговая ellagic acid
~, этилсерная ethylsulphuric acid
~, яблочная hydroxysuccinic/malic acid
~, янтарная succinic acid
~ с прямой цепью straight-chain acid
кислоторастворимый acid-soluble
кислотность acidity
~, обменная exchange acidity
~, повышенная hyperacidity, superacidity

~, **пониженная** hypacidity, subacidity, hypoacidity
~ **топлива** acid content in fuel
кислотно-щелочный acid-alkali, alkali-acid
кислотный *chem.* acidic, acid *(attr.)*
кислотовыносливый acid-tolerant
кислотолюбивый acidophilous
кислотоизмерение acidimetry
кислотомер acidimeter, acidometer
~, **сифонный** hydrometer syringe
кислотонерастворимый acid-insoluble
кислотонеустойчивый acid-intolerant
кислотоносный acidiferous
кислотообработанный acid-treated
кислотообразование acid formation
кислоторастворимость solubilitv in acids
кислоторастворимый acid-soluble
кислотосодержащий acidiferous
кислотостойкий, кислотоупорный, кислотоустойчивый acid-resisting, acid-proof, acid-fast
кислотоупорность, кислотоустойчивость acid-resistance, acid-fastness
кислый acid *(attr.)*, acidic; acidiferous
киста *med.* cyst
~, **альвеоло-дентальная** alveolodental cyst, odontocele
~ **влагалища, ретенционная** hydrocolpos, hydrocolpocele
~, **вторичная** secondary/daughter cyst
~, **гидатидная** hydatid cyst
~, **дермоидная** dermoid cyst
~, **зубная** dental/odontogenic cyst
~, **костная** osseous cyst
~, **кровяная** h(a)ematocele
~, **ложная** cystoid
~, **молочная** milk/lacteal cyst
~ **молочной железы, ретенционная** galactocele, galactoma
~ **мочеточника** ureterocele
~, **паразитарная** parasitic cyst
~, **параовариальная** paraovarian cyst
~ **прямой кишки** encysted rectum
~, **сальная** sebaceous/atheromatous cyst, teatoma
~ **семенного канатика** spermatocele, gonocele
~, **слизистая** mucous cyst, mucocele
~, **сложная** compound cyst
~ **шеи** cervical hydrocele, hydrocele colli, hydrocelle of the neck, Maunoir's hydrocele
~, **эхинококковая** echinococcus/hydatid cyst
~ **языка** glossocele
~ **яичника** ovarian cyst; oophoritic cyst
кистевидный *bot.* racemose, clustery, botroydal
кистевик *bot.* penicillinum *(Penicillinum)*
кистеносный *bot.* racemiferous; racemose
кистеобразный penicillate, racemiform, racemose
кистер *eccl.* junior deacon
кистехвост *zool.* vaporer moth *(Orgyia antiqua)*
кистецветник starch grape hyacinth *(Muscari racemosum)*
кистецветный *bot.* thyrsiflorous, thyrse-flowered
кистистый *bot.* racemose; branchy
кистоаденома cyst(o)adenoma
кистограмма cystogram(me)
кистозный cystic
кистома *med.* cystoma
кистообразный *med.* gangliac, ganglial, gangliar, angliform
кисточка 1. *bot., zool.* tassel, tuft, penicillus; 2. brush
кисточковидный penicillate, penicilliform, brushike; aspergilliform
кисть 1. brush; 2. hand; 3. bunch, cluster; tassel
~, **белильная** lime brush
~, **малярная** paint brush
~ **винограда** bunch of grapes
~, **формовочная** water brush, swab
~, **штукатурная** whitewashing brush
~, **расщепленная/вилообразная** cleft/split hand
~, **когтеобразная** claw hand
~, **обезьянья** monkey/ape hand
кит whale; cete; *astr.* Cetus
~, **горбатый** humpback whale
китаевед Sinologist, Sinologue
китаеведение Sinology
китаец Chinese
китайка 1. *bot.* chinese apple *(Malus pruniolia)*; 2. *text.* nankeen
китайский Chinese
~ **идол** Chinese house idol, joss
~**кая тушь** India ink
китара *mus.* kithara, cithara
китаянка Chinese woman
китель *mil.* jacket
китобаза whale factory ship
китобоец whaling ship, whaler
китобой whaleman, whaler
китобойный whaling
китовидный cetaceous
китовый whale *(attr.)*, cetacean, cetic, cetaceous
~ **жир** blubber
~ **промысел** whaling
~ **ус** baleen, whalebone
китоглав *zool.* whale-billed stork, shoe-billed stork *(Balaeniceps rex)*
китолов whaleman, whaler
китоловный, китоловство whaling
китообразный *zool.* cetacean, cetaceous; *pl.* cetaceans *(Cetacea)*
кит-полосатик *zool.* finback *(Balaenoptera)*
кит-самец bull whale
кит-самка cow whale
киур stonemason's hammer
кифара = **китара**
кифоз *med.* kyphosis, cyphosis, hump/hunch back
~, **юношеский** juvenile kyphosis, vertebral piphysitis
кифосколиоз *med.* kyphoscoliosis
кифотический *med.* kyphotic, cyphotic
кичуа Quechua

кишечник *anat.* intestine(s); bowels
~, **толстый** large intestine
~, **тонкий** small intestine
кишечнодышащие *pl. zool.* enteropneustans, enteropneusts *(Enteropneusta)*
кишечно-маточный intestino-uterine
кишечнополостные *pl. zool.* coelenterates *(Coelenterata)*
кишечнопочечный enterorenal
кишечно-пузырный intestinovesical
кишечный intestinal, enteric, enteral
кишка intestine; gut
~, **восходящая ободочная** ascending colon
~, **двенадцатиперстная** duodenum
~, **нисходящая ободочная** descending colon
~, **ободочная** colon
~, **подвздошная** ileum, twisted intestine
~, **поперечная ободочная** transverse colon
~, **прямая** rectum, straight intestine
~, **сигмовидная** sigmoid colon
~, **слепая** c(a)ecum, blind gut, typhon
~, **толстая** large intestine
~, **тонкая** small intestine
~, **тощая** jejunum, empty intestine
кишнец *bot.* coriander *(Coriandrum)*
киянка mallet, (wooden) hammer
клавесин *mus.* harpsichord
клавиатура keyboard
~ **Бодо** Baudot keyboard
~, **буквенно-цифровая** alphanumeric/alphameric keyboard
~ **выбора** selection keyboard
~, **выносная** side-mounted keyboard
~, **двоичная** binary keyboard
~, **десятичная** decimal keyboard
~ **для ручного ввода информации** manual-entry keyboard
~, **итоговая** amount keyboard
~, **малая** keypad
~, **незаблокированная** live keyboard
~, **отделяемая** detachable keyboard
~, **откидная** fold-down keyboard
~ **печатающего устройства** printer keyboard
~ **пишущей машинки** typewriter keyboard
~, **плоская** low-profile keyboard
~, **рабочая** operation keyboard
~ **с блокировкой** lockable keyboard
~, **сенсорная** touch-control keyboard
~ **суммы** amount keyboard
~, **цифровая** digit/figure/numeric keyboard
клавикорды *pl. mus.* clavichord
клавир piano
клавиша key
~ **верхнего регистра** top key
~ **возврата в исходное положение** home/return key
~ **возврата на одну позицию** backspace key
~ **выключения** cut-out key
~ **выхода** escape key
~ **гашения** clear key
~ **коррекции** correction key
~ **обратного перемещения** backspace key, backspacer
~ **окончательного итога** key for totals, sum/final/total key
~ **освобождения устройства** release key
~ **останова** halt key
~ **переключения регистров** shift key
~ **перескока** skip key, column skipping key
~ **пробега** space key
~, **программируемая** *comp.* soft key
~ **промежуточного итога** partial (sub-total) key, key for partial total
~, **пусковая** start key
~ **с заданием функций через меню** menu-driven key
~, **светящаяся** luminous key
~ **смены режима** change mode key
~ **стирания** erase key
~ **суммы** sum key, key for totals, sum/final/otal key
~ **управления курсором** cursor control key
~, **управляющая** control key
~, **функциональная** functional/function key
клавишноуправляемый key-driven
клавишный keyboard *(attr.),* key-actuated, key-driven
кладантус *bot.* cladanthus *(Cladanthus)*
кладень ground beam/timber, sleeper
кладка 1. masonry; bricklaying, brickwork; laying, setting; 2. bond; 3. batch, clutch *(of eggs)*
~, **каменная** stonework, masonry
~, **кирпичная** brickwork
~ **кровли** laying of a roof
кладовая storeroom, warehouse, store; larder, pantry; stowage space; locker; storage
~ **запасных частей** spare parts stowage
~, **парусная** sail room
~, **раздаточная** issue room
~, **такелажная** *naut.* line stowage
~, **тентовая** *naut.* awning store
~, **тросовая** *naut.* cable/rope hold, cordage/line stowage
~, **шкиперская** *naut.* boatswain's store
кладовщик storeman, storeroom/warehouse man
кладогенез *biol.* cladogenesis
кладогенный cladogenous
кладогонидия *bot.* cladogonidium
кладодий *bot.* cladode, cladophyle
кладоискание treasure hunting
кладония *bot.* cladonia *(Cladonia)*
~, **альпийская** reindeer moss *(Cladonia angiferina)*
кладосифонный *bot.* cladosiphonic
кладоспориоз *astr.* cladosporiosis, black mould
кладофора *bot.* cladophora
кладрастис *bot.* yellowwood *(Cladrastis)*
кладь load; cargo, freight
~, **ручная** hand luggage
клайтония *bot.* spring beauty *(Claytonia)*

клаксон horn
клан clan
клапан 1. valve; valvule; flap; drop; 2. valve plate/disk/plug; 3. *mus.* vent, finger-hole
~, аварийный emergency valve
~, анероидный aneroid valve
~, антипомпажный blow-off valve
~, аортальный *anat.* aortic valve
~, байпасный by-pass valve
~, быстродействующий quick-acting/-action valve
~ вдоха inhalation/inlet valve
~, венозный valve of a vein
~, верхний overhead valve
~ вентиляции vent valve
~, воздушный air valve
~, впускной inlet/intake/admission/inflow valve
~ выхода expiratory/outlet/exhalation valve
~, выпускной release/discharge/exhaust valve
~, выхлопной exhaust valve
~, газовый gas valve
~, главный master/main valve
~ горючего fuel valve
~, грибовидный mushroom valve
~ двойного действия double-action valve
~, двухопорный double-beat valve
~, двухседельный double-seated valve
~, двухходовой two-way valve
~, двустворчатый bicuspid valve
~, дозирующий metering valve
~ дистанционного управления remote-operated valve
~ для жидкости fluid valve
~ добавочного питания make-up feed valve
~, дренажный draining/bleeding/bleeder/drain valve; vent valve
~, дроссельный throttle valve
~, дымовой chimney valve
~, дыхательный breathing valve
~, забортный *naut.* outboard/flooding/sea/kingston valve; sea connection
~, запирающий/запорный stop valve, full check valve, shut-off valve
~ запуска start valve
~, зарядный charging valve
~ затопления *naut.* flooding valve
~, захлопочный drop valve
~, золотниковый sleeve/slide/shuttle/spool alve
~, игольчатый needle valve
~, кавитирующий cavitating valve
~, качающийся swinging valve
~, контрольный control/chek/pilot valve
~, левый предсердно-желудочковый left atrioventricular/auriculoventricular valve
~, лимфатический lymphatic valvule
~, маневровый manoeuvring valve
~, мембранный diaphragm valve
~, митральный mitral valve
~, многоходовой multiwave valve
~ наддува pressurizing valve
~, наполнительный topping-up valve
~, напорный delivery/pressure valve
~, невозвратный nonreturn/check/clapper valve
~, обратный = невозвратный
~ одностороннего действия one-way valve
~, односедельный single-seated valve
~ отбора воздуха bleed air delivery valve
~, отливной discharge valve
~, отсечной/отсечный shut-off [cut-out, cut-off] valve
~, паровостопорный steam stop valve
~, парораспределительный steam-distributing valve
~, перегрузочный overload valve
~, переключающий change-/shift-over valve, switching valve; selector valve
~, перепускной overflow/relief valve, bypass valve; transfer valve
~, питательный feed valve
~, пластинчатый flap-type valve
~, подпружиненный spring-loaded valve
~, полулунный *anat.* semilunar valvule
~, поплавковый float valve
~, поршневой piston valve
~, правый предсердножелудочковый *anat.* right atrioventricular/auriculoventricular valve
~, предварительный pre-valve
~, предохранительный safety/relief valve
~, продувочный purging/scavanging valve; blow-down/-off/out valve
~ промывочной системы flush valve
~ противопожарной системы fire valve
~, проходной straight-way valve
~, пружинно-шаровой ball-and-spring valve
~, пружинный spring-loaded valve
~ прямого действия direct-acting valve; single-stage valve
~, пусковой starting valve
~, разгрузочный relief/unloading/discharge valve
~, разобщительный disconnecting (cut-out, cut-off) valve
~, разрывной explosion vent
~, распределительный distribution/distributing valve
~, расширительный expansion valve
~, регулируемый adjustable valve
~, регулирующий regulating/governor/pilot/control valve
~, редукционный pressure-reducing valve
~ с пневматическим приводом pneumatically operated valve
~ сброса давления pressure-relief valve
~, сбросной dump valve
~, сигнальный alarm valve
~, скальчатый plunger valve
~, сливной drain/dump/bleeding valve
~, соленоидный solendid-operated valve
~, тарельчатый poppet/disk valve
~ терморегулирования thermal control valve

~, топливодозирующий fuel-metering valve
~, трехстворчатый *anat.* tricuspid valve
~, трехходовой three-way valve
~, угловой angle valve
~, уравнительный compensation/equalizing valve
~, форсуночный nozzle valve
~, четырехходовой four-way valve
~, шариковый/шаровой ball valve
~, шариковый пружинный spring-and-ball valve
~, шиберный slide valve
~, электрический electric-actuated valve
~, электромагнитный solenoid-operated valve
клапан-мигалка flapper/swing valve, clapper
клапанный valve *(attr.);* valvate; valvular, valval, valvar
клапанообразный valvate, valve-shaped
кларен *geol.* clarain
кларено-дюреновый clarain-durenous
кларет claret wine
кларк *phys.* percentage of abundance, clark
~ концентрации abundance ratio
кларкия *bot.* clarkia *(Clarkia)*
кларнет clarinet
кларнетист clarinetist
класс 1. class; grade; sort; type; category; order *fig.* tier; 2. *math.* class, set; 3. classroom
~ бонитета *for.* quality class
~, второй *comm.* fine grade/quality
~ вычетов *math.* residue class
~, кристаллический/кристаллографический crystal symmetry class
~, младший elementary/primary grade
~ наемных рабочих wageworking class
~ нивелирных работ elevation class
~ предприятия по числу работающих employment size class
~, рабочий working class
~, пустой *comp.* empty class
~, смежный *math.* coset, residue class
~ событий *stat.* class of events
~ точности accuracy class/rating
~, третий *comm.* plain quality
~ триангуляции triangulation system; order of triangulation
~, туристский tourist class
~, четвертый *comm.* rough grade/quality
~ чистоты обработки surface finish class
~, экономический economy class
~сы, имущие propertied classes
классик classic; *(scientist)* classicist
классика the classics
классификатор classifier; *tech.* grader, sizer
~, винтовой spiral classifier
~, воздушный двухконусный double-cone classifier
~, гравитационный gravity classifier
~, двухреечный duplex classifier
~, дражный dredging classifier
~, конусный cone classifier
~, корытный launder classifier
~, пневматический air classifier
~ сифонного типа siphon classifier
~, чановый tank classifier
классификационный classification *(attr.);* classificatory, classificational
классификация classification, classing, classifying; grouping; grading, sizing, sorting; assortment; arrangement
~, агропроизводственная rural land classification
~ болезней nosology, nosotaxy
~ в кипящем слое fluidized-bed classification
~ грохочением screen sizing
~ команд *comp.* order/command/instruction classification
~, многосторонняя multiple/multiway classifiation
~, первичная prescreening, primary classifying
~, пневматическая air classifying
~ рассевом screen sizing
~, трехразрядная *econ.* ABC classification
~ по двум признакам two-way classification
~, прямая one-way classification
~, узкая close sizing
классифицирование classification, classifying, classing; grading; sorting
классифицировать classify, class; group; grade, size, sort; categorize
классицизм classicism
клеильный gluing, sizing
клей glue; gum, paste, adhesive; size; cement, bond
~, альбуминовый albumin glue
~, высокопрочный high-strength adhesive
~ горячего отверждения hot-setting adhesive
~, глютиновый hide-and-bone glue
~, гуммиарабиковый gum arabic size
~, декстриновый dextrin glue
~, желатиновый gelatin glue
~, казеиновый casein glue
~, канифольный rosin size
~, карбамидный urea-formaldehyde adhesive, carbamide adhesive
~ карбинольный carbinol adhesive
~, кожный leather/skin/hide glue
~, коллодиевый collodion (glue)
~, конструкционный structural adhesive
~, костный/костяной bone glue
~, крахмальный starch glue
~, креймнийорганический silicon adhesive
~, латексный latex adhesive/size
~, мездровый hide glue/adhesive
~, метилцеллюлозный methylcellulose size
~, морской marine glue
~, мочевино-формальдегидный urea-formaldehyde adhesive, carbamide adhesive
~ нитроцеллюлозный nitrocellulose size
~, оптический lens cement
~, пектиновый pectic size

~, пчелиный bee glue, propolis
~, почковый *bot.* blastocolla
~, растительный vegetable adhesive; *bot.* mucilage
~, резиновый rubber cement/adhesive/solution; elastomeric adhesive
~, резорциновый resorcinol resin adhesive
~, рыбий fish glue, isinglass
~, самовулканизирующийся/самоотверждаощийся self-vulcanizing/-curing adhesive
~, синтетический synthetic resin adhesive
~, теплостойкий heat-resistant adhesive
~ термореактивный thermosetting adhesive
~, универсальный all-purpose (multipurpose) adhesive
~, фенольно-каучуковый phenolic-elastomer adhesive
~ холодного отверждения cold-setting adhesive
~, эпоксидный epoxy adhesive
клейка gluing; bonding
клейкий viscous, gluey, gummy, adhesive; clammy, ticky; mucilaginous
клейдотомия *med.* cleidotomy
клейит *miner.* clayite
клейкость stickiness, adhesiveness; adhesion; glutinosity, gummosity; viscosity
клеймение branding, marking; sealing; stamping
~ животных branding (animals)
~, рельефное raised branding/marking
клеймить brand, mark; stamp; *fig.* stigmatize
клеймо brand, seal, stamp, mark; hallmark
~ монетного двора mint mark
~ на бревне bark mark
~, пробирное mark of assay, hallmark
~, фабричное trademark
клейстер sizing, dextrin glue
~, крахмальный starch glue
клейстогамия *bot.* cleistogamy
клейстогамный *bot.* cleistogamous
клейстокарпий *bot.* cleistocarp
клейстокарпия *bot.* cleistocarpy
клейстокарпный *bot.* cleistocarpous
клейстопеталия *bot.* cleistopetaly
клейстотеций *bot.* cleistothecium, cleistocarp
клекачка *bot.* bladdernut *(Staphylea)*
клекачковые *pl. bot.* bladdernut family *(Staphyleacea)*
клематис *bot.* clematis *(Clematis)*
клемма 1. (connecting) terminal; 2. clamp, clip
~, винтовая bolt-and-nut terminal
~ для кровеносных сосудов clamp
~, заземляющая earthing terminal
~ Павловича Pavlovicz's tong
клемник terminal block, block of terminals
клемный 1. terminal; 2. clamping, clipping
клен *bot.* maple *(Acer)*
~, белый/ложноплатановый sycamore maple, plane-tree maple
~, пенсильванский northern maple *(Acer pennsylvanicum)*
~, японский full-moon maple *(Acer japonicum)*
кленовник maple forest
кленовые *pl. bot.* maple family *(Aceraceae)*
кленовый maple *(attr.)*
кленолистный *bot.* maple-leaved
клеоме, клеома *bot.* spiderflower *(Cleome)*
Клеопатра *hist.* Cleopatra
клепало 1. riveting hammer; 2. *eccl.* wooden clapper *(used in monasteries instead of bell)*
клепальный riveting
~ молоток riveting hammer, riveter
клепать rivet, hammer; lap-rivet
~ косу hammer a scythe sharp
клепаный riveted
клепка 1. riveting; 2. *timb.* stave
~ бочки, донная head piece
~ взрывом explosion riveting
~ внакрой/внахлестку lap riveting
~ впотай flush/countersunk riveting
~ встык butt riveting
~ давлением compression riveting
~, двухрядная double-chain riveting
~, механизированная machine/power riveting
~, однорядная single-chain riveting
~, паркетная parquet stave
~, потайная flush/countersunk riveting
~, прессовая/ударная impact riveting
~, рядная chain riveting
~ цепным швом chain riveting
~ шахматным швом staggered/zigzag riveting
клепсидра clepsydra
клептобиоз *zool.* cleptobiobis
клептоман, клептоманка kleptomaniac
клептомания kleptomania
клептофобия kleptophobia
клерет clairette (wine)
клерикал 1. cleric, clergyman; 2. *polit.* clerical
клерикализм *polit.* clericalism
клерикальный clerical
клерк 1. office clerk; 2. clergyman
клеровальный, клеровочный *tech.* clarifying; defining
клерование *tech.* clarification; refining
клеровать *tech.* clarify; refine
клеродендрон *bot.* glory bower *(Clerodendron)*
клером not in code, not coded, in clear
клест *zool.* crossbill *(Loxia)*
клест-еловик *zool.* common crossbill *(Loxia curvirostra)*
клест-сосновик *zool.* parrot crossbill *(Loxia pytyopsittacus)*
клетен *naut.* keckling
клетка 1. *biol.* cell; 2. *anat.* chest, thorax; 3. cage, box, cell;. 4. *math., cyb.* mesh, square, cell; 5. *naut.* cribbing
~, адвентициальная adventitial cell
~, амебоидная am(o)eboid cell
~, анапластическая anaplastic cell

~, ацидофильная acidophil(e) cell
~, ацинозная acinar/acinous cell
~, базальная basal/basilar cell
~, базисная *phys.* elementary/unit cell
~, базофильная basophile/beta cell
~, беличья *el.* squirrel-cage
~, блуждающая wandering/am(o)eboid cell
~, бокаловидная goblet/beaker/chalice cell
~, веретенообразная spindle/fusiform cell
~, вкусовая gustatory/taste cell
~, волосковая hair cell, Corti's cell
~, выстилающая lining/littoral cell
~, ганглиозная/ганглионарная nerve/ganglion cell
~, гигантская giant cell
~, главная principal/chief cell
~, глиальная neuroglia cell
~ гноя pus cell/corpuscle, pyocyte
~, грамотрицательная Gram-negative cell
~, грамположительная Gram-positive cell
~, двигательная motor cell
~, дентинная dentinal cell
~ для облучения exposure cell
~ для подопытных животных animal cage
~, дочерняя daughter cell
~, дремлющая slumbering cell
~, железистая glandular cell
~ желудочной железы, главная chief cell of the stomach; pepsin-producing cell
~ желудочной железы, обкладочная parietal cell; oxyntic cell
~, жировая fat/adipose cell
~, замыкающая guard cell
~, зародышевая germ/germinal/formative cell
~, звездчатая stellate/star cell
~, зернистая/гранулярная granular cell
~ зернистой мембраны granulosa cell
~, зимогенная zymogenic/peptic cell
~, злокачественная malignant cell
~, зрительная visual cell
~, интерстициальная interstitial cell
~, камелоидная cameloid cell, elliptocyte
~, капсульная capsule cell, amphicyte
~, капсульно-лютеиновая theca-lutein cell
~, кариохромная karyochrome cell
~ кастрационная castration cell, signet ring cell
~, килевая *naut.* keel-track cribbing
~, клейкая adhesive cell
~, коллоидная colloid cell
~, комиссуральная comissural/heteromeric cell
~, конечная end cell
~, корзинчатая basket cell
~ кортиева органа hair cell, Corti's cell
~ коры cortical cell
~, костеобразовательная osteogenetic cell
~, костная bone/osseous cell, osteocyte
~ костного мозга bone marrow cell
~ крови blood cell
~ крови, белая white blood cell, leukocyte
~ крови, белая несегментированная stab/band/staff cell
~ крови, красная red blood cell, erythrocyte
~, кубовидная cuboid cell
~, купферовская Kupfer cell
~, ламеллярная lamellar cell
~, лепрозная lepra cell
~, лестничная *constr.* staircase, stairway
~, лимфатическая lymph cell
~, лимфоидная lymphoid cell
~ лифта elevator cage
~, люпоидная L. E. cell
~, лютеальная/лютеиновая luteal/lutein cell
~, макроглиальная macroglia cell
~, мантийная mantle cell
~, материнская mother/brood/parent cell
~, мезенхимная mesenchymal cell
~, мезоглиальная mesoglia cell
~, мезотелиальная mesothelial cell
~, мелкая midget cell
~, миелоидная myeloid cell
~, микроглиальная microglia cell
~, миоидная myoid cell
~, миоэпителиальная myoepithelial cell
~, митральная mitral cell
~, мишеневидная target cell
~ мозговой коры cell of the cerebral cortex
~, моторная motor cell; excitatory cell
~, мукозная mucous cell; goblet cell
~, мукосерозная mucoserous/mucoalbuminous cell
~, мышечная myoblast
~, невроглиальная neuroglia cell
~, недифференцированная indifferent/undifferntiated cell
~, мерцательная flickering/scintillating cell
~, нейромышечная meuromuscular cell
~, нейроэпителиальная neuroepithelial cell
~, нервная nerve/ganglion cell
~, нейросекреторная neurosecretory cell
~, неспециализированная nonspecialized cell
~, обонятельная olfactory cell
~, образовательная formative cell
~, овсяновидная oat-shaped cell
~, оксифильная oxyphyl cell
~, олигодендроглиальная oligodendroglia cell
~, опорная foot/pillar cell
~, остеогенная osteogenic cell, osteoblast
~, остеохондрогенная osteochondrogenic cell
~, островковая islet cell
~, осязательная tactile cell
~, палисадная palisade cell
~ паращитовидной железы wasserhelle cell, water-clear cell of the parathyroid gland
~, паренхимная parenchyma cell
~, паукообразная spider cell
~, первичная primitive cell
~, первичная кровяная h(a)emocytoblast, stem cell
~, первичная мужская половая spermatogone, spermatogonium

~, **первичная ретикулярная** primitive reticular cell
~, **первичная эмбриональная** primary embryonic cell, elementary cell
~, **переходная** transitional cell
~, **перителиальная** perithelial cell
~, **пессариевидная** pessary cell
~, **печеночная** hepatic cell
~, **пирамидная** pyramidal cell
~, **пироловая** pyrr(h)ol cell
~, **плазматическая** plasma cell, plasmacyte
~, **пограничная** border cell
~, **поддерживающая** supporting/sustentacular/floor cell
~, **покоящаяся** resting/durable/quiescent cell
~, **покровная** cover cell
~, **полихроматическая** polychromatic/polychromatiphil cell
~, **половая** sex cell, gamete, gametocyte
~, **полярная** pole cell, polar body
~, **предлюпоидная** pre-L.E. cell
~, **псевдогигантская** pseudogiant cell
~, **псевдоксантомная** pseudoxanthoma cell
~ **пульпы селезенки** pulpar/splenic cell, acrophage
~, **раздутая** balloon cell
~, **раковая** cancer cell
~, **растительная** plant cell
~, **ретикульно-эндотелиальная** reticuloendotheial cell
~, **роговая** horn cell
~, **саркогенная** sarcogenic cell
~, **светлая** bright/clear cell
~, **сегментированная** segmented cell
~, **сенсорная** sensory cell
~ **сердечного порока** heart failure cell
~, **серозная** serous cell
~, **серповидная** sickle/crescent cell
~, **эндогенная** endogenous cell
~, **эндотелиальная** endothelial cell
~, **энтерохромаффинная** enterochromaffin cell; argentaffin cell
~ **растра** resolution cell
~, **скуловая** *naut.* bilge cribbing
~ **Фарадея** Faraday cage
~, **эпидермальная** epidermic cell
~, **эпителиальная** epithelial cell
~, **эпителиоидная** epithelioid cell
клетка-спора *bot.* conidium
клетка-спутник *bot.* companion cell
клетка-спутница *bot.* satellite cell
клеткообразный cellular, cell-like, celliform; cytoid
клетневание *naut.* rope serving, parcelling
клетневать *naut.* serve, parcel (a rope)
клеточка *biol.* cellule; small cell/cage/room
клеточкообразный cellular, cellulated
клеточный cell (attr.), cellular, celled
клетра *bot.* clethra, white alder, summer sweet *(Clethra)*
клетчатка cellulose; *anat.* cellular tissue
~, **жировая** fat, fatty tissue
~, **параметральная** parametric fat
~, **парауретеральная** paraurеtic fat
~, **подкожная жировая** subcutaneous fat
~, **растительная** cellulose
~, **тазовая** pelvic fat
~, **водорослевая** algulose
клетчатый 1. tesselated; cellular; 2. checked; squared
клеть 1. *min.* cage; 2. *met.* stand
~, **трехвалковая** three-high stand
~, **черновая** roughing stand
~, **чистовая** finishing stand
~, **шахтная грузовая** cargo cage
~, **шахтная, людская** man-riding cage
клефт *hist.* klepht
клешневидный, клешнеобразный *zool.* chelate, claw-shaped, cheliform
клешня *zool.* claw, chela, nipping claw, nipper
клещ *zool.* tick; mite *(Trombicula irritans)*
~, **аргасовый** ear tick
~, **чесоточный** itch mite
клещевидный I *zool.* acaroid
клещевидный II forcipate, chelate
клещевина castor bean; castor-oil plant *(Ricinus)*
клещевинолистный *bot.* ricinifolious
клещевой acarian
клещи *pl.* tongs, pliers
~, **горновые** forge tongs
~, **кузнечные** blacksmith's tongs
~, **рычажные** lever tongs
~, **тигельные** crucible tongs
~, **токоизмерительные** tong-test [clip-on] instrument
~, **фальцовочные** folding tongs
клещик tick; mite
клиантус *bot.* parrotbill, parrotbeak *(Clianthus)*
кливаж cleavage, cleat
~ **волнистый/изогнутый** bent cleavage
~, **вынужденный** induced cleavage
~, **лицевой** face cleat
~, **параллельный** line/linear cleavage
~ **по напластованию** bedding cleavage
~ **разлома** fracture/fault cleavage
~ **размятия** crush cleavage
~, **секущий** transverse cleavage
~ **скольжения** strain-slip cleavage
~ **трещиноватости** jointing cleavage
кливер *naut.* jib
~, **штормовой** spitfire(-jib)
кливер-галс, кливер-леер jib-tack
кливер-фал jib-halyard
кливер-шкот jib-sheet
кливер-штаг jib-stay
кливия *bot.* Kaffir lily *(Clivia)*
клизма *med.* enema, clyster, clysma, clysis, rectal injection
~, **вяжущая** astringent enema
~, **капельная** proctoclysis, rectoclysis

~, **мыльная** soapsuds enema
~, **очистительная** cleansing/evacuant enema
~, **питательная** nutrient enema
~, **смягчающая** lubricating enema
клика clique, faction
кликун *zool.* whooper swan *(Cygnus cygnus)*
климаграмма climatic diagram, climagram
климакс 1. *physiol.* climacteric, climacterium, menopause; 2. climax
~ **леса** hylium
~, **мужской** climacterium virile
~, **ранний** premature climacteric
климактерий = климакс 1.
климактерический climacteric, menopausal
климат climate
~, **арктический** arctic climate
~, **влажный** humid climate
~, **искусственный** artificial/synthetic climate
~, **морской** maritime/marine/oceanic climate
~, **мягкий** mild climate
~, **сухой** arid climate
~, **тропический** tropical climate
~, **умеренный** temperate climate
климатизатор, климатизер room-air conditioner
климатип *bot.* climatype
климатический climatic
климатогенный climatogenic
климатография climatography
климатоизофит *bot.* climatoisophyte
климатолог climatologist
климатология climatology
климатопатология *med.* climatopathology
климатотерапия climatotherapy, climatic treatment
климатофизиология climatophysiology
климограмма climogram(me)
климограф *med.* climograph
клин wedge; key, cotter; *arch.* quoin
~, **буксовый** *rail.* axlebox wedge
~, **вогнутый** concave wedge
~, **выверочный** *opt.* correction wedge; *mech.* levelling wedge
~, **двойной** fox wedge
~, **зажимной** battening wedge
~, **кулирный** *text.* slur
~, **пазовый** slot wedge
~ **ползуна** crosshead cotter
~, **регулировочный** adjusting wedge/cotter
~, **ступенчатый** step wedge
~, **тангенциальный** tangent wedge
~, **тонкий** slender wedge
~ **тормозной колодки** *rail.* brake-shoe wedge
клинандрий *bot.* clinandrium
клинидий *bot.* clinidium
клиника 1. clinic; 2. clinical picture
клинико-патологический clinicopathologic
клиницист clinician, clinical physician/surgeon
клинический clinical
клинкер clinker
~, **дорожный** paving clinker
клинкет wedge-action valve/gate; gate/slide valve
~ **переборочный** bulkhead valve
клиновидный, клинообразный wedge-shaped, cuneate, cuneiform; V-shaped; sphenoidal, cineal
клиновой wedge *(attr.),* cotter *(attr.)*
~**вое соединение** cotter joint
клинолистный *bot.* wedge-leaved, sphenophyllous
клинометр clinometer, inclinometer
клиноось clino-axis
клинопирамида clinopyramid
клинописный cuneiform
клинопись cuneiform writing
клиноромбический clinorhombic; monoclinic
клиноромбоэдрический clinorhomboidal, triclinic
клиноскоп *med.* clinoscope
клиноэдрический *cryst.* clinohedral
клинтония *bot.* bead lily *(Clintonia)*
клинтух *zool.* stock dove *(Columba oenas)*
клинч *sport .* clinch, clinching
клипер *naut.* clipper
клиноклазит *geol.* clinoclase, clinoclasite
клинтонит *geol.* clintonite, seyberite
клипп *geol.* remnant, klippe
клир *coll.* the clergy
клиренс, клиранс clearance; road clearance
~, **креатининовый** creatinine clearance
~, **почечный** renal clearance
клирик *eccl.* clergyman, ecclesiastic
клиринг *fin.* clearing
~, **банковый** bank clearing
~, **односторонний/принудительный** unilateral clearing
клиринговое учреждение clearing agency
клирос choir *(in a church)*
клирошанин *eccl.* 1. chorist, chorister; 2. altar boy
клирфактор *el.* nonlinear distortion factor
клистир = клизма
клистрон klystron tube, Klystron
~, **бессеточный** gridless klystron
~, **генераторный** oscillator klystron
~, **двухконтурный, двухрезонаторный** double-cavity (double-resonator) klystron
~, **задающий** driver klystron
~ **с сетками** gridded klystron
~, **твердотелый** solid-state klystron
~, **усилительный** amplifier klystron
клитор *anat.* clitoris
клиторизм *med.* clitorism; clitoridauxe
клиторит clitoritis, clitobiditis
клиторомегалия clitoromegaly
криторотомия clitorotomy, clitoridotomy
клитория *bot.* butterfly pea, pigeonwings *(Clitoria)*
клифф (sea) cliff
~, **миниатюрный** nip
~, **моренный** glacial drift cliff
клица *el.* cleat
клише block, plate, cliche
~, **штриховое** line-process block

клоака *zool.* cloaca
клоакальный cloacal
клоачный cloaca *(attr.);* cloacal
клоачные *pl. zool.* monotremes *(Monotremata)*
клиширование block-/plate-making
клобук *eccl.* klobuk, headgear of Orthodox monk
клобучковый cuculiform, hood-shaped
клобучок cucullus; hood
клокичка *bot.* bladdernut *(Staphylea)*
клоктун *zool.* Baikal teal *(Anas formosa)*
клон *biol.* clon(e)
клональный clonal, clonic
клонировать *biol.* clone
клонический *med.* clonic
клоновый *biol., bot.* clonal
клонограф *med.* clonograph
клонорхоз clonorchiosis, clonorchiasis
клоноспазм *med.* clonospasm
клонус *med.* clonus
~, коленный knee (patellar) clonus
~ стопы foot clonus
клоп bug; bedbug
клоповник *bot.* peppergrass *(Lepidum)*
клопогон(ник) *bot.* bugbane *(Cimicifuga)*
клопфер (Morse telegraph) sounder; sounding relay
клоп-щитник *zool.* dusky stink bug *(Euschis ristigmus)*
клот(ик) *naut.* acorn button
клотоида *math.* clothoid, Cornu/Euler spiral
клостридии *pl.* Clostridium
клостридиопептидаза clostridiopeptidase
клоуновые *pl. zool.* frogfishes *(Antennariidae)*
клофен *chem.* clophen, diphenylchloride
клочень *zool.* bee moth *(Galleria mellonella)*
клочкование flocculation
клубеносный *bot.* tubiferous, bulbiferous
клубень *bot.* tuber; bulb
~, корневой root tuber
~, мелкий (клубенек) tubercle, tubercule, nodule
клубеньковидный tuberculiform, tubercular
клубеобразование *bot.* tuberization
клубковидный, клубкообразный clustered, glomerulate; club-shaped
клубковый glomal
клубневидный *bot.* tuberous, tuberoid, tuberiform
клубневой tuberous
клубнелуковица *bot.* corm, cormus, bulbotuber
~, воздушная cormlet
клубнелуковичный *bot.* bulbotuberiferous
клубнеплод *bot.* root crop; tuber-bearing plant
клубнеплодный *bot.* tuberous
клубника *bot.* strawberry *(Fragaria moschata)*
~ ананасная/ананасовая pine strawberry *(Fragaria ananassa)*
клубниковый tuberculate; nodular
~вые бактерии rhizobium
клубок 1. *text.* ball; 2. *anat.* glome; 3. *bot.* glome, glomerule
~, копчиковый *anat.* coccygeal gland
~, сонный carotid body
~, сосудистый choroid skein
клубочек *anat.* glomerule, glomerulus
~, мальпигиев Malpighian glomerulus
~, обонятельный olfactory glomerulus
~, почечный renal glomerulus
клубочковый glomerular, glomerulate; clustered
клупп 1. diestock, tap wrench; threading die holder; screw stock; 2. *text.* clip
клуша *zool.* lesser black-backed gull *(Larus uscus)*
клушица *zool.* clough *(Pyrrhocorax pyrhocorax)*
клык canine tooth; fang; tusk
~, молочный deciduous canine
клюв beak, rostrum
~ мозолистого тела beak of the corpus callosum
~ основной кости sphenoid beak
~ узловязателя knotter hook, knotting bill
клювик small beak, rostellum
клювиковидный rostelliform
клювовидно-ключичный coracoclavicular
клювовидно-плечевой coracohumeral
клювоголовые *pl. zool.* rhynchocephalians *(Rhynchocephalia)*
клювообразный, клювовидный coracoid; rostriform, ostrate; beak-shaped
клювонос *zool.* alligator gar, gar pike *(Lepidosteus)*
клювоносный rhynchophorous
клюворыл *zool.* beaked whale *(Ziphius)*
клюз *naut.* hawse(hole), hawse pipe
~, бортовой hawse pipe
~, буксирный towing port/pipe
~, открытый open chock
~, палубный deck pipe, chain-locker pipe
~, тральный stern chock
~, цепной chain locker pipe
~, швартовный mooring/hawser port/pipe
~, якорный hawsehole, hawse pipe
клюза *geol.* defile
клюз-сак *naut.* jackass
клюзия *bot.* clusia *(Clusia)*
клюзная пробка hawse plug
клюква *bot.* cranberry *(Oxycoccus)*
ключ 1. *el.* key; switch; gate; 2. *tech.* spanner, *Am.* wrench; 3. *mus.* clef; 4. spring; 5. legend;
~ альтовый *mus.* alto clef
~, басовый *mus.* bass clef
~, возрастающий ascending key
~, вызывной signalling/ringing key
~, гаечный spanner, wrench
~, гаечный двусторонний double-ended spanner
~, гаечный динамометрический torque-indicating spanner/wrench, dynamometric/torque spanner
~, гаечный с трещоткой ratchet wrench
~, гаечный тарированный torque wrench/spanner
~, гаечный торцовый socket wrench, box span-

ner
~, гаечный универсальный general utility [universal] wrench/spanner
~, гаечный коленчатый bent spanner/wrench
~, гаечный односторонний single-ended spanner
~, гаечный разводной adjustable spanner/wrench
~ дампа/дампинга *comp.* memory dump key
~ данных *comp.* data key
~ для доступа access key
~ зажигания ignition key
~ занятости busy key
~ защиты protection key
~, исходный source key
~ кода code key
~, коммутаторный (switchboard) key
~, маршрутный *rail.* route switch
~, общий general key, passkey
~, опросно-вызывной speak-ring key; listening key
~, опросный listening key
~, отбойный release/clearing key
~, патронный socket/chuck wrench
~, первичный primary key
~, поисковый search key
~, потенциальный candidate key
~, разъединительный cut-off [interruption] key
~ с арретиром locking key
~ свода *arch.* vault head
~, свечной *auto.* spark-plug spanner/wrench
~, системный system key
~, скрипичный *mus.* treble clef
~ сортировки sortkey
~, строительство под turnkey construction
~, телеграфный telegraph/Morse/manipulating/sender/operating key
~, тенорный *mus.* tenor clef
~, трубный pipe wrench/spanner
~, трубный аллигаторный alligator wrench
~, трубный цепной chain pipe wrench
~, убывающий descending key
~ управления control key
~, установочный setting tool
~, цепной chain wrench
~ шифра cipher key, key phrase
ключевой key *(attr.);* fountain, spring *(attr.);* fontanial
~ камень *arch.* keystone
~ отрасль key industry/branch
~вая вода spring water
~вая позиция key position
ключ-жезл *rail.* key token
ключики *pl. bot.* cowslip primrose *(Primula eris, Primula officinalis)*
ключица *anat.* clavicle, collar bone
ключично-акромиальный *anat.* acromioclavicular
ключично-грудный *anat.* cleidosternal
ключично-лопаточный *anat.* cleidoscapular
ключично-сосковый *anat.* cleidomastoid
ключичный *anat.* clavicular
ключ-трава moonwort *(Botrychium lunaria)*
клям(м)ера clamp iron, clamp, cleat
клямс *wood.* riser, sheer clamp; shelfpiece
кляссер stamp album
клятва oath, vow; *obs.* imprecation, curse
~ Гиппократа Hippocratic oath
~, ложная perjury
~, церковная excommunication
клятвенный on oath; oath *(attr.)*
клятвопреступление perjury
клятвопреступник perjurer
клятвопреступница perjuress
клятвопреступный perjured, oath-breaking
кляуза *law* barratry, vexatious litigation
кляузничество *law* barratry; *colloq.* pettifogging, cavilling
К-мезон K-meson, kaon
кмин *bot.* cumin *(Cuminum)*
~ тминный/тминовый cumin *(Cuminum cyminum)*
кнель *coll., cul.* meat/fish balls, quenelles
кнеорум *bot.* spurge olive *(Cneorum)*
кнессет Knesset(h)
кнехт *naut.* bollard, bitt; *obs.* farm labo(u)rer
~, буксирный towing bitt/bollard
~, выдвижной retractable bollard
~, крестовый cruciform bollard
~, мачтовый mast bitt
~, швартовный mooring/warping bollard/bitt
книга book; register, ledger, log
~, адресная address register
~, букинистическая secondhand book
~, бухгалтерская account book
~, бухгалтерская главная ledger
~ в обложке paperback
~ в переплете bound book
~, вспомогательная *econ.* subsidiary ledger
~ дебиторов debtors' ledger
~, домовая house/home register
~ заказов order book
~ записей актов гражданского состояния minicipal register
~ записи ссуд loan register
~, жалобная complaint book
~ инкассо *econ.* collection ledger
~, кассовая cashbook
~ кредиторов creditors' ledger
~ купонов coupon ledger
~ обязательств по акцептам acceptance liability ledger
~ параллельного учета *econ.* Boston/progressive ledger; collection ledger
~, поваренная cook-book
~, поземельная *law* land records
~ счетов подлежащих оплате accounts payable edger
~, телефонная telephone directory
~, торговая *econ.* business book

~ **учета обязательств** liability ledger
~ **учета земель** land register
~ **учета подписки на акции** subscription ledger
~ **учтенных векселей** discount ledger
~, **ходкая/сенсационная** best seller
~ **ценных бумаг** *econ.* securities ledger
книговед bibliologist; bibliographer
книговедение bibliology; bibliography; *econ.* bookkeeping
книгоед *zool.* bookworm, book louse *(Troctes divinatorius)*
книгоиздательский publishing
книгоиздательство publishing house
книголюб bibliophile
книгопечатание book printing
книгопечатный printing *(attr.);* typographic
~ **станок** printing press
книготорговец, книгопродавец bookseller
книготорговля book trade; bookshop
книготорговый book-selling *(attr.)*
книгохранилище book depository/stacks; library
книдобласт *zool.* cnidoblast
книдофор *zool.* cnidophore
книдоциль *zool.* cnidocil
книжечка booklet; notebook
книжечка-игольник needle book
книжка I booklet; pocket book; notebook; book; *(certificate)* card; record, log
~, **банковская расчетная** passbook
~, **депозитная** deposit pass book
~, **заборная** ration book
~, **записная** notebook; pocket book
~, **медицинская** health record
~, **продовольственная/промтоварная** *econ.* ration book
~, **профсоюзная** trade-union membership book/card
~, **расчетная** wage and payment book; employment book
~, **сберегательная** saving account passbook, savings-bank book
~, **трудовая** work record card
~, **чековая** checkbook
~, **членская** membership card
книжка II *zool.* omasum; psalterium
книжник 1. *hist.* scribe; 2. bibliophile
книжность bookishness; book learning
книжный 1. book *(attr.);* 2. bookish; abstract
~ **знак** bookplate
~ **шкаф** bookcase
~**ная вошь** *zool.* book louse *(Troctes divinaorius)*
кникус *bot. Cnicus*
~, **аптечный/благословенный** blessed thistle *(Cnicus benedictus)*
книфофия *bot.* torch lily, poker plant *(Kniphophia)*
кница *naut.* bracket, gusset, knee
~, **бимсовая** beam bracket
~, **горизонтальная** gusset plate
~, **кормовая** sternsom
~, **скуловая** bilge/wing bracket, margin-plate bracket
~, **транцевая** transom knee
кноп *naut.* knot
~, **буйрепный** buoy-rope knot
~, **простой вантовый** French shroud knot
~, **стопорный** stopper knot
кнопка (push) button, knob; finger contact; key
~ **аварийной защиты** scram/emergency button
~ **аварийного отключения** emergency cut-out button
~ **автоматичного поиска** automatic search button
~, **арретирующая** caging knob
~ **без арретира** non-locking key
~, **блокировочная** holding key
~, **висячая** pendant button
~ **вызова** call button
~ **выключения** cut-out key, disconnect/disengage button
~, **вытяжная** pull-knob
~ **загрузки** load button
~ **запроса** *comp.* request key
~ **запуска** start(ing) button
~, **нажимная** push/press button, press switch
~ **остановки/останова** stop button
~, **отжимная** pull button
~ **отключения** cut-out (disconnect, disengage) button
~ **отметка явлений** event button
~ **повторного пуска** restart button
~ **приведения к горизонту** level-flight button
~ **принудительного сброса** jettison button
~ **проверки** test button
~ **пуска/пусковая** start button; *rock.* firing/launching button; *comp.* trigger button
~, **разъединяющая** disconnect/release button
~ **с арретиром** lock button
~ **с принудительным возвратом** push-to-release button
~ **с фиксацией** push-to-lock button
~ **сбрасывания** cancel button
~ **сброса** release/jettison button
~, **световая** light button
~ **управления** control/operation knob/button
~ **фиксированной настройки** *rad.* preset-tuning utton
~, **цифровая** figure/numerical key
~, **чертежная** thumb tack
~, **шаровая** ball button
кнопочный button *(attr.),* knob *(attr.),* key *(attr.)*
~ **выключатель** push-button switch
кнопка-лампочка light button, illuminated push button
кнут whip; *hist.* knout
княгиня princess
~, **великая** Grand Duchess
княжата *pl.* prince's sons
княжение *hist.* reign

княженика *bot.* arctic bramble *(Rubus arcticus)*
княжески(й) princely
княжество principality
княжик *bot.* clematis *(Clematis)*
княжич prince *(unmarried)*
княжна princess *(unmarried)*
князек 1. princeling; 2. *zool.* azure tit *(Parus cyanis)*
князенок prince's young son
князь prince
~, великий grand prince; grand duke
кнюппель *met.* rod-shaped steel
коагель coagulated gel, coagel
коагглютинин coagglutinin
коагрегация coaggregation
коагулаза *biochem.* coagulase
коагулирование coagulation, coagulating
коагулировать coagulate; coalesce
коагулируемость coagulability
коагулируемый coagulable
коагулирующий coagulant
коагулограмма coagulogram
коагулометр coagulometer
коагулянт coagulant, coagulator, coagulating agent
коагулят coagulum, coagulate
коагулятор coagulant, coagulator
~, лазерный/мазерный laser/maser coagulant
коагуляционный coagulatory, coagulative
коагуляция coagulation
~, акустическая sonic coagulation
~, гравитационная gravity coagulation
~, лазерная/мазерная laser/maser coagulation
~, неограниченная unrestricted coagulation
~, самопроизвольная spontaneous coagulation
~, скрытая latent/dormant coagulation
~, тепловая thermal coagulation
~, явная evident coagulation
коадъютор coadjutor
коаксиальный coaxial
коала *zool.* koala *(Phascolarctos cinercus)*
коалесценция coalescence
коалиционер coalitionist
коалиция coalition; *math.* combination of players
коарктация coarctation
~ аорты coarctation of the aorta
~, ложная pseudocoarctation
коацерват *chem.* coacervate
коацервация *chem.* coacervation
кобальт cobalt, Co
~, азотнокислый cobalt nitrate
~, бромистый cobalt bromide
~, землистый earthy cobalt, asbolite, asbolan
~, мышьяковистий white cobalt
~, сернокислый cobalt sulphate
~, уксуснокислый cobalt acetate
~, хлористый cobalt chloride
~, шпейсовый grey/white cobalt
кобальтамин cobaltamine
кобальтит, кобальтин *miner.* cobaltite, cobalt glance
кобальтирование cobalt plating
кобальтистый cobaltous
кобальтникельпирит *miner.* cobaltnickelpyrite, hengleinite
кобальтово-синий smalt-blue
кобальтолигошпат *miner.* cobalt-oligospar
кобальтовый cobaltic, cobaltous
~вая краска cobalt blue
~вое стекло smalt
кобальтосодержащий cobaltiferous
кобея *bot.* cobea *(Cobea)*
кобза *mus.* kobza *(in the Ukraine)*
кобзарь *mus.* kobza player
КОБОЛ COBOL (Common Business-Oriented Language)
К-оболочка K-shell
кобура holster; leather case/bag
кобчик *zool.* red-footed falcon *(Falco espertinus)*
кобыла 1. mare; 2. vaulting horse; 3. *hist.* punishment rack
~, племенная studmare, pedigree mare
кобылка I *mus.* bridge *(of stringed instrument)*
кобылка II *zool.* locust, grasshopper
~, зеленополосая green-striped grasshopper *(Chortophaga viridifasciata)*
ковалентность covalence
ковалентный covalent
ковар *met.* Kovar
ковариант *math.* covariant
ковarianта covariance
ковариантность covariance
ковариантный covariant
ковариационный covariation *(attr.)*
ковариация covariation; covariance
~, выборочная *stat.* sample covariance
~, множественная *stat.* multiple covariance
~, нормированная *stat.* normalized covariance
~ оценок *stat.* covariance of estimates
~, предельная *stat.* limiting covariance
~ совокупности *stat.* population covariance
~ тренда *stat.* trend covariance
~, частная *stat.* partial covariance
ковать forge, hammer; press
ковбой cowboy
ковер carpet, rug; *tech.* blanket; *bot.* tapetum, tapetal layer, mat
~, белуджистанский Baluchistan carpet
~, иранский/персидский Persian carpet/rug
~, текинский Turkoman carpet
ковер-самолет magic carpet
коверсинус *geom.* coversed sine
ковка forging, hammering; smithing; *(of horses)* hoeing
ковкий ductile, malleable, forgeable
ковкость ductility, malleability, forgeability
ковроделие carpetmaking, rugmaking
ковроткачество carpet/rug weaving
ковчег 1. ark; 2. *eccl.* reliquary; shrine
~, Ноев Noah's Ark
~ завета ark of the covenant

ковш bucket, scoop; dipper, ladle; pan; *naut.* basin
~, бетонораздаточный placing bucket
~, двухчелюстной clamshell bucket
~ для остановки судов boat basin
~ драглайна dragline bucket
~, загрузочный charging ladle
~, золотопромывочный gold pan
~, литейный foundry ladle
~ механической лопаты shovel dipper
~, опрокидной tipping bucket
~ с коромыслом/подвеской crane bill ladle
~, сталеразливочный casting/pouring ladle
~ транспортера converter bucket
~, чугуновозный hot-metal transfer ladle
~, шлаковый slag/cinder ladle
ковшеобразный scoop-shaped; ladle-shaped
ковшовой scoop *(attr.),* bucket *(attr.)*
~ погрузчик loading shovel, scoop loader
ковыль *bot.* feather grass; needle grass *(Stipa)*
когезия cohesion
когерентность, когеренция coherence, coherency
~, временная time coherence
~, пространственная spatial coherence
~ света coherence of light
~, фазовая phase coherence
когерентный coherent
когерер *phys.* coherer
~ Бранли Branly/filings coherer
когнат cognate
когомология cohomology
когомотопия *math.* cohomotopy
когорта 1. *hist.* cohort; 2. *bot.* order
~, гипотетическая *stat.* synthetic cohort
когортатив *gram.* cohortative, hortatory mood
коготный clawed, claw *(attr.),* unguiculate
коготные *pl. zool.* unguiculates, clawed mammals *(Unguiculata)*
коготь claw; talon
коградиентность cogradiency
кограница *math.* coboundary
когтеобразный, когтевидный claw-shaped, unguiform
когтеходящий *zool.* unguligrade
когти *pl.* pole/lineman'g climbers, climbing irons
код code; code combination; key
~, авиасигнальный air-signals code
~, автоответчика answer-back code
~ адреса address code, address
~ адреса числа number address (code)
~, амплитудно-импульсный pulse-amplitude code
~ ассемблера assembly program(me) code, assembler code
~, блочный block code
~, буквенно-цифровой alphanumeric/alphameric code
~, буквенный alphabetic/letter/literal code
~, взвешенный weighted code
~, внутренний in-line code
~ восьмидорожечной перфоленты eight-channel code
~ восстановления данных data regeneration code
~, временный timing code
~ время-импульсный pulse-position code
~ Грея Gray code, cyclical/modified binary code; reflected binary code
~, двоично-десятичный binary-coded decimal code
~, двоично-пятеричный biquinary code
~, двоичный binary code; bit code; on-off code
~, двоичный натуральный/нормальный straight/normal/true binary code
~ двоичный рефлексный reflected binary code
~, двухпозиционный two-condition [double-current] code
~, двухадресный two-address code
~, двухрегистровый two-case code
~, десятичный decimal code
~, десятичный, число-импульсный decimal unit-counting code
~ длины инструкции instruction length code
~ для проверки check code
~, допольнителный/дополнения complement (code)
~, единично-десятичный decimal unit-counting code
~ записи данных data recording code, data record key
~ запроса interrogation code
~ знака sign code
~ знака суммы amount sign cooe
~, импульсный pulse code
~, интервально-импульсный pulse-interval code
~ интерпретирующей программы interpreter code
~ исключительного условия exceptional condition code, exception code
~ команд(ы) instruction/order/command code
~, контрольный check code
~, корректирющий error-correcting/detecting code
~, линейный linear code
~ ловушки trap code
~, машинный machine code
~, метеорологический meteorological notation
~ микрокоманд microinstruction code, microcode
~ Морзе Morse code
~ неисправности error status code
~, неравномерный variable-length code
~, несистематический non-systematic code
~, обратный reverse code; complement code
~ обнаруживающий ошибки error-detecting code
~, одноадресный single-address code
~ операции operation code, op code, opcode
~ опознавания/опознавательный identifi-

cation/identity/authentication code
~, переговорный language code, key book
~ ошибки error code
~, переменный alternate code
~, перестановочный permutation code
~ перфокарт (punched) card code
~ перфолент punched tape code
~, позиционный position(al) code
~, полный complete code
~, помехоустойчивый noise-immune code
~, последовательный serial code
~ преобразования conversion code
~, пятизначный five/digit/unit code
~, пятиразрядный (binary*)* five-bit code; five-digit/-unit code
~, равномерный equal-length code
~, рефлексный reflected code
~ с автоматическим исправлением ошибки self-correcting code
~ с дополнением до двух two's complement code
~ с избытком три excess-three code, three-excess code
~ с исправлением ошибок error-correcting code
~ с исправлением пакетов ошибок burst error correcting code
~ с контролем по четности parity-checking code
~ с минимальной избыточностью minimum redundancy code
~ с обнаружением ошибок error-detecting code
~ с самоконтролем self-checking code
~ с самопроверкой error-detecting code, self-checking code
~ с расстоянием единица unit-distance code, continuous progressive code
~, самодополняющийся self-complementing code
~, самопроверяющийся self-checking code
~, сверточный convolution code
~, сигнальный signalling code
~ семисегментного индикатора seven-segment code
~ символа/символьный character code
~, синоптический weather/synoptic code
~, сокращенный short/brevity code
~ соответствия correspondence code
~ состояния ожидания wait state code
~ строки line code
~ счета counting code
~ телетайпа teleprinter code
~ типа сообщения message type code
~, троичный ternary code
~, упрощенный brevity code
~ условия condition code
~, фазово-импульсный pulse-position code
~ Файра Fire code
~ Хемминга Hamming code
~, циклический recurrent/cyclic code
~, цифровой/числовой digital/numerical code; figure code
~, частотно-импульсный pulse-frequency code
~, частотный frequency code
~ числа coded number, number code
~, число-импульсный unit-counting code
~, широтно-импульсный pulse-width code, pulse-duration code
~, шифровальный cipher code; key book
~, штриховой bar/strip code
кода *mus.* coda
кодегидраза codehydrase
кодегидрогеназа codehydrogenase
кодеин *chem.* codeine
кодеиновый codeine *(attr.)*
кодекарбоксилаза codecarboxylase
кодекс law code; labour code
~, воздушный air code
~, военный Articles of War
~, гражданский civil code
~ законов о труде labour code
~, процессуальный code of practice
~, уголовный criminal code
~ чести law of honour
~и честной конкуренции *econ.* codes of fair competition
кодер coder, encoder
кодер-декодер codec
кодетта *mus.* codetta
кодиевый codiacean
кодирование coding, encoding
~ блоков в двоичном коде binary block (en)coding
~, буквенно-цифровое alphanumeric coding
~ данных data (en)coding
~ знаков character coding
~ обозначений notation coding
~ обращения access coding
~ при помощи функциональных двоичных разрядов functional bit coding
~, цифровое digital/numeric coding
кодировать code, encode; encypher
кодировка coding
кодировщик coder, code clerk, encoder; cryptographer
кодирующий coding, encoding
~щая программа code generating routine
~щее устройство coder, encoder, code equipment
кодификатор *law* codifier
кодификация codification, coding
кодифицировать *law* codify
кодицилл *law* codicil
кодово-импульсный pulse-code *(attr.)*
кодовой code *(attr.)*
кодограмма code message; pattern
кодопреобразователь code translator/converter
кодон codon

кодонопсис *bot.* Asian bell *(Codonopsis)*
кодосовместимость code-for-code compatibility
коелококкус *bot.* sago palm, ivory-nut palm *(Coelococcus)*
коечный berth *(attr.)*; cot *(attr.)*
кожа 1. skin; hide; *anat.* cutis; 2. leather; 2. *bot.* epidermis; peel, rind
~ аллигатора *med.* alligator skin, sauriasis
~ алюминиевого дубления alumed leather, alum-tanned leather
~, бланшированная buff leather
~, борушистая wrinkled leather
~, валичная roller leather
~, галантерийная fancy leather
~, гусиная gooseflesh; goose pimples; cutis anserina
~, депилированная dehaired skin
~ для верха обуви upper leather
~ для низа обуви bottom leather; sole leather
~, дряблая loose skin
~, искусственная artificial leather
~ квасцового дубления alum-tanned leather
~, лаковая patent/enamelled/japanned leather
~, липкая clammy skin
~, лицевая full-grain leather
~, лоснящаяся glossy skin
~, мраморная *anat.* cutis marmorate
~, недубленная untanned leather
~, неотделанная undressed leather
~, обветренная weather skin
~, одежно-галантерейная garment-and-fancy leather
~, опойковая calf leather
~, осклизлая pitted/pricked leather
~, переплетная book-binding leather
~, перчаточная gloving leather
~, подкладочная lining/facing leather
~, подошвенная sole leather
~, прокладочная gasket leather
~, ремневая band leather
~ с искусственным лицом artificial-grain leather
~ с нарезкой printed leather
~, свиная pigskin
~, свищеватая warbled leather
~, синтетическая synthetic leather
~ собаки dogskin
~, собственно *anat.* true skin, dermis
~, сыромятная rawhide
~, техническая industrial leather
~, тощая empty leather
~ хромового дубления chrome-tanned leather
~, чертова moleskin
~, шагреневая shagreen leather
кожан *zool.* large bat
кожанка *bot.* hop tree *(Ptelea)*
кожгалантерия haberdashery leather goods
кожевенный leather-dressing, leather *(attr.)*
~ завод tannery
кожевник I tanner, currier, leather dresser
кожевник II *bot.* sumch, sumc *(Rhus)*
кожеед *zool.* leather/dermestid beetle *(Dermestes)*
кожезаменитель, кожзаменитель artificial leather
кожелистный *bot.* leathery-leaved, scytophyllous
кожеподобный leatherlike; leathern
кожимит artificial/imitation leather, leatherette
кожистокрылые *pl. zool.* earwigs *(Dermaptera)*
кожистый leathery, leatherlike; coriaceous
кожица 1. *anat.* cuticle; 2. film, thin skin, pellicle, peel, rind
кожно-вегетативный skin-vegetative
кожно-венерический dermatovenereal
кожно-венерологический dermatovenerologic, skin-and-venerial *(attr.)*
кожно-гальванический cutaneogalvanic
кожно-мускульный dermomuscular
кожно-мышечный musculocutaneous
кожно-нарывной dermatovesical
кожно-электрический electrodermal
кожный skin *(attr.)*; cutaneous, dermal, dermitic
кожсырье hides; untanned/raw leather
~, тяжелое bovine hides and skins
кожура peel, rind, skin; *bot.* hull; husk; shell; aril; pod
~, внутренняя *bot.* endopleura, inner seed coat
~, семенная spermoderm, seed coat
кожух 1. housing, casing; jacket, case, mantle, envelope, enclosure; 2. sheepskin coat
~ авиационного двигателя aircraft engine boot
~ воздушного винта, кольцевой propeller shroud
~ гребного вала *naut.* propeller-shaft sleeve
~ горна *met.* hearth jacket
~ двигателя engine casing/cowling/cowl
~ для перевозки/транспортировки shipping cover
~ доменной печи blast-furnace jacket
~ дымовой трубы funnel casing
~ заплечиков *met.* bosh jacket/plates
~, защитный protective housing/casing/cover
~, колесный *naut.* paddle-wheel box
~, котельный boiler casing/jacket
~ маховика flywheel casing/housing/jacket
~, разъемный split casing
~ соплового аппарата турбины turbine-nozzle casing
~ печи furnace shell
коза goat
~, снежная mountain goat *(Oreamnos americanus)*
козел 1. he-goat; 2. *met.* plug, salamander, ear, saw; 3. *aero.* *(on landing)* bounce
~ безоаровый/бородатый bezoar goat *(Capra egagrus)*
~ в ковше ladle chill
~, гимнастический vaulting buck
~ горный альпийский Alpine ibex *(Capra ibex)*

козелец *bot.* salsify *(Scorzonera)*
козелок 1. trestle, support, horse, prop; 2. *anat.* ragus
козерог 1. *zool.* ibex *(Cabra ibex);* 2. (K.) *astr.* Capricorn
козий goat *(attr.),* caprine
козимаза *biochem.* cozymase, diphosphopyridine nucleotide, DPN, coenzyme I
козлиный goat *(attr.),* goatish, goaty, caprine
козлить bounce, jump, leap
козлобородник bot. goatsbeard *(Tragopogon)*
~, пореелистный vegetable-oyster plant *(Tragopogon porrifolius)*
козлоногий goat-hoofed/-footed
козлы *pl.* trestle; frame; gantry; sawbuck, saw-horse; jack; *sport* vaulting horse
~ (подъемного) крана crane gantry
~, рамные gallows frame
козляк *bot.* shallow-pored boletus, bovinus mush-room *(Boletus bovinus)*
козлятник *bot.* goat's-rue *(Galega)*
~ лекарственный/аптечный common goat's-rue *(Galega officinalis)*
козлятниколистный *bot.* galegifolious
козодоеобразные *pl. zool.* Caprimulgiformes
козодой *zool.* goatsucker, nightjar
козон(ок) (sheep's) knucklebone
козочка, морская *zool.* ghost shrimp *(Caprella)*
козуля *zool.* roebuck, roe *(Capreolus)*
козыльник *bot.* wormseed goosefoot *(Chenopodium ambrosioides)*
козырек screen, baffle (plate), deflector, windscreen, windshield; apron plate; deflector; cap peak, visor; camera hood
~ ветрового стекла sun shade; windshield visor
~ над кабиной водителя cab guard/shield canopy
~, солнечный sunvisor
~ фуражки cap peak
козырь trump; trump card
коикс *bot.* Job's-tears *(Coix lacryma-jobi)*
коипу *zool.* coypu *(Myopotamus coypus)*
коитофобия *med.* coitophobia
коитус *med.* coitus; copulation, coition, sexual union
~, затягиваемый coitus prolongatus
~, паранимфальный coitus paranymphalis
~, прерываемый coitus interruptus
койка *naut.* berth, bunk, cot; hammock; bed
~, подвесная hammock
~, раскладная camp cot/bed
койот *zool.* coyote *(Canislatrans)*
кой-тас *geog.* sheepback rocks
кок I *naut.* ship's cook
кок II *aero.* spinner; cone
~ воздушного винта propeller spinner
~, носовой nose cone
~, хвостовой tail cone
кока *bot.* coca, Huanocu cocaine tree *(Erythroxylon coca)*
кокаин cocaine
~, солянокислый cocaine hydrochloride
кокаинидин cocainidine
кокаинизация cocainization
кокаинизм cocainism, cocaine addiction
кокаинист(ка) cocaine addict
кокаиновый cocaine *(attr.);* cocainic
кокаиновые *pl. bot.* cocaine family *(Erythroxylaceae)*
кокаиномания *pathol.* cocainomania
коканцероген cocarcinogen
коканцерогенез cocarcinogenesis
коканцерогенный cocarcinogenous
кокарбоксилаза *biochem.* cocarboxylase, coen-zyme B
кокарда cap badge, cockade, cocarde
кокиль *met.* chill/ingot/metal mould; external chill
~, неразъемный one-piece metal mould
~, футерованный luted chill mould
кокильный *met.* chill *(attr.)*
~ное литье chill (permanent-mould) casting
кокк coccus *(pl. cocci)*
коккер-спаниель cocker spaniel
коккобацилла coccobacillus
кокковый coccal
коккогон *bot.* coccogone
кокколит *geol.* coccolith
~, дискообразный discolith
~, чашеобразный cyatholith
кокколитофор coccolithophore
кокколитофорида coccolithophorid
кокколоба *bot.* sea grape *(Coccolobis)*
кокконит *geol.* coccoconite
коккоподобный coccoid
коккосфера *geol.* coccosphere
коккулус, коккулюс *bot.* snail-seed *(Cocculus)*
коклюш *med.* whooping cough, pertussis
коклюшный, коклюшечный pertussal
кокон cocoon
коконировать(ся) cocoon
кокономотание reeling/unwinding of cocoons
коконопряд *zool.* velleda lappet caterpillar *(Tolype velleda)*
~, лиственничный larch lappet moth
коконосушилка cocoon roaster
кокорыш *bot.* aethusa, fool's-parsley *(Aethusa)*
кокос *bot.* 1. coconut palm *(Cocos nucifera);* 2. coconut
кокосовый coconut *(attr.)*
~ орех coconut
~вое молоко coconut milk
кокпит cockpit
кокс coke
~, доменный blast-furnace coke
~, задувочный bed-charge coke
~, естественный native coke, carbonite
~, литейный foundry/metallurgical coke
~, нефтяной petroleum coke
~, торфяной peat coke

~, **угольный** coal coke
кокса *anat., zool.* coxa
кок-сагыз *bot.* kok-saghyz, Russian dandelion *(Taraxacum kok-saghyz)*
коксальгический *med.* coxalgic
коксальгия *med.* coxalgia, coxodynia
коксальный *med.* coxal
коксартрит *med.* coxarthritis
коксиеллез Q fever
коксит *med.* coxitis, hip joint disease; *zool.* coxite
коксобрикет coke briquette
коксовальный, коксовый coke *(attr.),* coking
~**ная печь** coke oven
~**ное число** *phys.* coking value
коксование coking process, carbonization, coking, charring
~ **торфа** turf carbonization
коксовать(ся) coke, carbonize; become carbonized
коксовыталкиватель *met.* coke pusher
коксоцерит coxocerite
коксуемость coking capacitv
коксующийся coking *(attr.)*
коксообразование 1. coking, coke formation; *(of oil)* oil carbonization; 2. gumming
коксосортировка coke screen/screening
коксотомия coxotomy
коксохимия by-product coke industry
коктоантиген *biochem.* coctoantigen
кокушник *bot.* rein orchis *(Gymnadenia)*
кокцигодиния coccygodynia, coccy(o)dynia
кокцидий *bot.* coccidium, cystocarp
кокцидиоз coccidiosis
~, **общий** generalized coccidiosis
кокцидиоидный coccidioidal
кокцидиоидоменингит coccidioidomeningitis
кокцидиоидомикоз coccidioidomycosis
~, **бессимптомный** latent/asymptotic coccidioidomycosis
~, **вторичный** secondary coccidioidomycosis, coccidioidal granuloma
~, **первичный** primary coccidioidomycosis, valley/desert fever
кокцидия coccidium
кол post, picket, pole; stake; peg
~, **анкерный** anchor post
~, **нивелирный** levelling peg
~, **сажальный** planting stick/peg, dibble
кола *bot.* cola (nut) *(Cola)*
коламин *chem.* colamine, ethanolamine
колатитуда *geod.* colatitude
колба 1. flask; retort; 2. *el.* bulb; envelope; 2. *bot.* long-rooted onion *(Allium victorialis)*
~, **вакуум-перегонная** distilling vacuum flask
~, **длинногорлая** long-necked flask
~, **коническая** Erlenmeyer/conical flask
~, **круглодонная** round-botton flask
~, **лабораторная мерная неградуированная** blank volumetric flask
~, **лабораторная плоскодонная** flat-bottom(ed) flask
~, **лабораторная, промывная** washing/rinsing flask
~, **лабораторная трехгорлая** three-necked flask
~ **лампы, бесцветная** clear (glass) bulb
~ **лампы, кварцевая** quartz envelope
~ **лампы, матированная/матовая** frosted bulb
~ **лампы, металлизированная** silvered bowl
~ **лампы, полупрозрачная** translucent bulb
~ **лампы, стеклянная молочная** white bowl
~, **мерная** measuring/volumetric flask
~, **перегонная** distilling flask/retort
~, **широкогорлая** broad-mouth flask
колбовидный bulb-shaped; phialine
колбочка 1. small flask; 2. *anat.* cone
~ **сетчатки глаза** retinal cone
~ **сетчатки глаза, двойная** twin cone
колдоговор *abbr.* collective agreement
колдун sorcerer, wizard, magician, exorcist
колеантус *bot.* mud grass *(Coleanthus)*
колебание *phys.* oscillation, vibration; fluctuation; variation; *fig.* hesitation, vaccilation, wavering; *see* **колебания** *(pl.)*
~, **вековое** secular variation
~ **водного зеркала** water table fluctuation
~ **курса** *fin.* rate of exchange [stock exchange] fluctuations
колебания oscillations, vibrations; fluctuations; variations; vaccillations
~, **амплитудно-модулированные** *el.* amplitude-modulated oscillations
~, **аэроупругие** aeroelastic vibrations
~, **безвихревые** irrotational vibrations
~, **беспорядочные** random oscillations
~, **вынужденные** forced/constrained oscillations
~, **высокочастотные** high-frequency oscillations
~, **гармонические** harmonic oscillation(s)
~, **гидроупругие** hydroelastic oscillations
~, **годовые** annual variations
~, **деформационные** deformation vibrations
~, **длиннопериодические** long-period oscillations
~, **задемпфированные** damped oscillations
~, **затухающие** damped/decaying/convergent oscillations
~ **звуковой частоты** acoustic(al) oscillations
~, **звуковые** sound/acoustic(al) oscillations/vibrations
~, **знакопеременные** reversal vibrations
~, **изгибно-крутильные** flexural-and-torsional vibrations
~, **изгибные** flexural/bending vibrations
~, **изохронные** isochronal vibrations
~, **квазистационарные** quasi-static oscillations
~ **климата/климатические** fluctuations of climate
~, **когерентные** coherent oscillations

~, коллективные collective/cooperative oscillations
~, конъюнктурные *econ.* market fluctuations
~, короткопериодические short-period/-term oscillations
~ крена *naut.* rolling oscillation
~ крена, резонансные roll resonance oscillation
~, криптосейсмические cryptoseism
~ кристаллической решетки lattice modes
~, крутильные torsion(al) vibrations
~ курсов ценных бумаг на бирже stock market fluctuations
~, макросейсмические macroseism
~, маятниковые pendulum/pendular oscillations
~, микросейсмические microseism
~ модулированные по амплитуде *el.* amplitude-modulated oscillations
~ модулированные по фазе phase-modulated oscillations
~ модулированные по частоте frequency-modulated oscillations
~ нагрузки load variation
~, нарастающие divergent/diverging oscillations
~, незатухающие continuous/undamped/sustained oscillations
~, некогерентные noncoherent/incoherent oscillations
~, неплоские out-of-plane vibrations
~, неустановившиеся unsteady/transient oscillations
~, неустойчивые unstable oscillations
~, низкочастотные low-frequency oscillations
~, нормальные normal oscillatory modes
~, нулевые zero-point oscillations
~ обусловленные акустической нагрузкой acoustically-induced vibrations
~ объемного типа bulk-mode waves
~ объемные volume oscillations
~, односторонние unidirectional vibrations
~, остаточные residual oscillations
~, паразитные parasitic/spurious oscillations
~, пилообразные saw-tooth(ed) oscillations
~ плазмы, электронные electron plasma modes
~, поперечные transverse/lateral oscillations
~ по крену rolling oscillations
~ по срезу shear oscillations
~ по тангажу pitching oscillations
~ по толщине thickness modes
~ при входе в атмосферу re-entry oscillation
~, продольно-изгибные longitudinal-and-flexural vibrations
~, продольные longitudinal vibrations/oscillations
~ противодавления back-pressure vibrations
~, расходящиеся divergent/diverging oscillations
~ рыскания yaw oscillations/vibrations
~ с малой амплитудой small-amplitude oscillations
~ с переменной амплитудой/частотой variable amplitude/frequency oscillations
~, самовозбуждающиеся self-excited/induced vibrations
~ свободные free/natural vibrations/oscillations
~, связанные coupled oscillations
~ сдвига shear nodes
~, сдвиговые shear vibrations
~, сейсмические seism, seismic/earthquake vibration
~, синусоидальные sine-wave oscillations
~, синхротронное sinchrotron oscillations
~ случайного характера *econ.* accidental/chance fluctuations
~, случайные random vibrations
~, скрытые latent oscillations
~, собственные natural/free/proper oscillations
~, суточные diurnal/daily variations
~, тахисейсмические tachyseism
~, тепловые temperature/thermal vibrations/fluctuations
~, упругие elastic vibrations
~, установившиеся steady-state oscillations
~, устойчивые stable oscillations; *econ.* sustained fluctuations
~ цен на рынке *econ.* market price fluctuations
~, частотно-модулированные frequency - modulated oscillations
~, фазовые phase oscillations
~, флаттерные flutter oscillations
~, флюгерные weathercock oscillations
~, фреатические phreatic fluctuations
~, фугоидные phugoid(al) oscillations
~, шумовые noise waves
~ экономической активности tides in business activity
~, электромагнитные electromagnetic oscillations/modes

колебатель vibrator
колебательно-вращательный vibration-rotational
колебательно неустойчивый oscillatorily unstable
колебательный oscillatory, vibratory; oscillating, fluctuating, wobbling
колебаться 1. vibrate, oscillate; swing, sway; fluctuate, range, vary; wobble; 2. hesitate, vaccilate, falter, waver
колеблемость *math.* dispersion
колеблющийся oscillating, oscillatory, vibratory, vibrating; swinging, swaying; fluctuating
~ свободно *econ.* floating
коленкор calico
~, переплетный bookbinder's cloth
коленно-локтевой *anat.* genucubital, knee-elbow
коленно-грудной *anat.* genupectoral
коленно-пяточный *anat.* genucalcaneal, kneecalcaneal
коленный 1. *anat.* genicular, genual, knee *(attr.);*

2. *tech.* elbow *(attr.),* bend *(attr.)*
~ная чашка *anat.* kneecap, patella
колено 1. *anat.* knee; 2. *tech.* elbow, leg, bend, knee; 3. *bot.* geniculum, node, stemnode
~ вала crank
~ волновода waveguide elbow/bend
~ выгнутое кзади *anat.* genu recurvatum
~ кривой knee of a curve
~ мозолистого тела *anat.* knee of the corpus callosum
~, окулярное eye-piece slide/tube
~, отведенное knock-knee, in-knee, genu valgum/introrsum
~, подсемядольное *bot.* hypocotyl, caulicle
~, приведенное bowleg, out-knee, bandy leg, genu varum/extrorsum
~, прямое elbow bend
~ рычага bend of a lever
~ с фланцами flanged pipe bend
~, соединительное elbow piece
~ трубопровода pipe elbow
~ трубы knee/elbow pipe
коленовидный, коленообразный *anat.* geniculate
коленопреклонение genuflection, kneeling
коленопреклонный genuflectory, kneeling, genuant
коленце *anat.* geniculum
коленчатка *bot.* loment, lomentum
коленчатый 1. articulate(d), jointed; *anat.* geni-. culate(d); 2. elbow-shaped, cranked, knee *(attr.)*
~ вал crankshaft
колеоптер *aero.* coleopter
колеоптиль *bot.* coleoptile
колеориза *bot.* coleorhiza
колер I *art* colo(u)r, shade, tint
колер II *vet.* staggers
колесница chariot
~, погребальная hearse
колесный wheel *(attr.),* wheeled, on wheels
~ное судно paddle-wheel vessel
колесо wheel
~, ведомое driven wheel
~ ведомое зубчатое driven gear
~, ведущее driving/drive wheel; driving sprocket
~, ведущее зубчатое drive/driving gear
~, винтовое spiral/helical wheel
~, водяное water-wheel
~, геликоидальное зубчатое helical gear
~, гребное paddle wheel
~ двускатное *auto.* twin-wheel, double-tyre wheel, dual wheel
~, железнодорожное сцепное (railway) driver wheel
~, зубчатое gear; cogged wheel, cogwheel
~, зубчатое ведущее pinion
~, зубчатое, внутреннего зацепления internal gear
~, зубчатое гипоидное hypoid gear
~, зубчатое коническое bevel gear
~, зубчатое коронное crown gear
~, зубчатое лобовое crown bevel gear; crown gear
~, зубчатое планетарное planetary pinion
~, зубчатое реечной передачи rack pinion
~, зубчатое солнечное sun gear
~, зубчатое цилиндрическое spur gear
~, зубчатое шевронное herringbone (double-helical) gear
~, зубчатое эллиптическое elliptical gear
~ компрессора, рабочее compression impeller
~, кормовое гребное *naut.* sternwheel
~, косозубое angular gear
~, косозубое коническое angular bevel gear
~, лентопротяжное *comp.* feed wheel
~, лопастное blade wheel
~, лопаточное bladed wheel
~, маховое flywheel
~, многовенечное multirow wheel
~ нагнетателя, работное blower wheel
~, насосное (pump) impeller
~, одновенечное single-row wheel
~, печатающее *comp.* print wheel
~, планетарное planet wheel
~ плуга, бороздное furrow wheel
~ плуга, заднее rear furrow-wheel, (plough) trail wheel
~ плуга, копирующее gauge-wheel
~ плуга, полевое land-wheel
~, подрессоренное spring wheel
~, прямозубое цилиндрическое spur gear
~, рабочее *(of pump, etc.)* impeller; *(in turbine)* rotor
~, разъемное split wheel
~, рулевое steering wheel
~ с внешними зубьями external gear
~ с внутренними зубьями annular/internal gear
~ с круговой линией зуба, коническое skew bevel gear
~ с ребордой flanged wheel
~, самоориентирующееся/самоустанавливающееся castror/swivelling wheel
~, самоцентрирующееся носовое self-centering nose wheel
~, свободно вращающееся loose wheel
~ со спицами spoke wheel
~, солнечное main gear wheel; annular ring
~, счетное *comp.* counting wheel
~, тормозное brake wheel
~ турбинное turbine wheel; runner
~ турбины, активное impulse/impact wheel
~ турбины, рабочее turbine wheel
~ турбовоздуходувки turbo-blower impeller
~, убирающееся retractable whåel
~, управляемое steerable/steering wheel
~, фрикционное friction wheel
~, фрикционное коническое bevel friction-wheel

~, **хвостовое** tail wheel
~, **ходовое** running wheel
~, **храповое** ratchet wheel
~, **цепное** chain-wheel, sprocket
~, **цилиндрическое прямозубое** spur gear
~, **цифровое** *comp.* number wheel
~, **червячное** worm gear
~, **швыряльное** *text.* spinning/spinner wheel
~ **шасси** *av.* landing wheel
~, **шевронное зубчатое** herringbone gear
~, **шрифтовое** *comp.* type wheel
~, **штурвальное** control wheel
~, **эвольвентное зубчатое** involute gear
колесовидный, колесообразный wheel-shaped; wheel-like; trochoid
колесопрокатный стан wheel-rolling mill
колеус *bot.* coleus, flame nettle *(Coleus)*
колечко small wheel, ringlet; *bot.* annulus
колея 1. rut; run; 2. *rail.* gauge; track, wheelspan
~, **нормальная** standard-gauge line
~, **тормозная** braking track
~, **узкая** narrow-gauge line
~, **широкая** wide-gauge line
колибактерин colibacterin
колибактерия colibacillus, collon bacillus, *Escherichia coli*
колибациллез colibacillosis
колибациллемия colibacillemia
колибациллурия colibacilluria
колибациллярный colibacillary
колибри *zool.* hummingbird
коли-группа coli-group
Колизей Colosseum
колика *med.* colic, colica, abdominal pain
~, **аппендикулярная** appendicular/vermicular colic
~, **газовая** flatulent colic
~, **желудочная** gastric colic, stomachache
~, **желчная** biliary colic
~, **маточная** uterine colic
~, **медная** copper colic
~, **нефритическая** nephritic colic
~, **печеночная** hepatic/biliary colic, gallstone colic
~, **почечная** renal/nephritic colic
~, **прямокишечная** rectal colic
~, **свинцовая** lead/saturnine colic
коликообразный colic
коликоплегия *med.* colicoplegia
колипиелит *med.* colipyelitis
колировка *hort.* grafting
колисепсис *med.* colisepsis
колит *med.* colitis
~, **амебный** amebic colitis
~, **балантидийный** balantidial colitis
~, **перепончатый** membranous colitis
~, **слизистый** mucous/pseudomembranous colitis
~, **спастический** spastic colitis/colon
~, **язвенный** ulcerative colitis
колитерапия colitherapy
коли-титр coli-index
колитоксемия colitoxemia
колитоксикоз colitoxicosis
колиурия coliuria
колиформный *bact.* coliform
колицин colicine
колицистит colicystitis
колицистопиелит *med.* colicystopyelitis
количественно quantitatively, in quantitative terms
количественно-временной quantitative-temporal
количественно-именной quantitative-nominal
количественно-неизмеримый non-quantifiable
количественно-предметный quantitative-object *(attr.)*
количественно-пространственный quantitative-spatial
количественный quantitative
~ **анализ** quantitative analysis
~**ная теория денег** quantity theory of money
~**ное числительное** *gram.* cardinal number
количество quantity, amount; number
~ **влаги** moisture content; amount of moisture; humidity
~ **движения** momentum; quantity of motion
~ **дутья** blast capacity
~, **индикаторное** tracer amount
~ **материалов, наличное** material inventory
~ **мест** seating capacity
~, **ничтожное** *chem.* trace amount, traces
~, **равновесное** equilibrium amount
~, **ультрамалое** trace quantity
~, **фоновое** background
~ **цепей** *el.* circuit capacity, number of circuits
колиэкзотоксин coliexotoxin
колиэнтерит colienteritis
коллаборационизм *polit.* collaboration
коллаборационист *polit.* collaborationist, collaborator
коллаген *chem.* collagen, collogen
коллагеназа *med.* collagenase
коллагеновый collagenic, collagenous
коллагеноз collagenosis
коллагенолитический collagenolytic
коллаж collage
коллапс collapse
~ **легкого** pulmonary collapse
~, **ортостатический** orthostatic collapse
коллапсотерапия *med.* collapsotherapy, collapse therapy
колларгол colloid silver
коллатераль, коллатеральный collateral
коллегиальность collective spirit/nature
коллегиальный collegiate; collective, joint
коллегия, коллегиум board, collegium; college
~ **адвокатов** (the) bar
~, **избирательная** electoral college
~, **судейская** *law* judicial assembly
колледж college
~, **неполный/младший** junior college

~ **полного курса** senior college
коллежский *hist.* collegiate
~ **асессор** collegiate assessor
коллектив 1. team, body, group, association; 2. *math.* collective; 3. *opt.* collecting lens
~ **завода** workers and office personnel of a plant
~, **научный** team/body/group of scientists; research team
~, **студенческий** student body; (the) students
коллективизация collectivization; *el.* sharing
коллективизировать collectivize, reorganize on a collective basis [into collective establishments]
коллективизм collectivism, collectivity, collective spirit
коллективность collectivity
коллективный collective; joint; common; united; conmunal
~ **договор** agreement based on collective bargaining, collective agreement
~**ное владение** joint ownership
~**ное хозяйство** collective farm, kolkhoz
коллектор 1. collector; header; reservoir; 2. *el.* commutator; 3. *tech.* manifold; 4. collecting main; sewer; drainage ditch
~, **водяной** water drum/header
~, **воздушный** air manifold
~, **впускной/всасывающий** intake/suction manifold
~, **выпускной/выхлопной** exhaust manifold
~, **главный** trunk sewer
~ **горячего воздуха** *met.* hot air header
~, **канализационный** sewage header/collector
~, **кольцевой** collector ring; radial collecting main; ring header
~ **котла** boiler header
~, **ливневой** storm sewer
~ **нефти** oil trap
~, **паровой** steam drum/header/collector
~, **питающий** supply header
~, **топливный** fuel manifold
~, **угловой** angle header
~, **электромашинный** *(d.c.)* commutator; *(a.c.)* slip-ring assembly, collector
коллекционер collector
коллекционирование collection, collecting
коллекция collection
~ **образцов** collection/range of patterns
колленхима *bot.* collenchyma
~, **рыхлая** lacunar collenchyma
колленхиматический, колленхимный collenchymatic
колленхиматозный collenchymatous
коллетериум *zool.* colleterium
коллеция *bot.* colletia *(Colletia)*
колли collie, sheep dog
коллибия *bot.* collybia *(Collybia)*
коллидин *chem.* collidine
колизионный collision *(attr.)*
~**ное право** conflict of laws
коллизия collision, clash; conflict, strife
колликулит *med.* colliculitis
колликулэктомия colliculectomy
коллиматор collimator
коллиматорный collimating
коллимационный collimation *(attr.)*, collimating
коллимация collimation
коллимировать collimate
коллинеарность collinearity
коллинеарный collinear
коллинеация collineation
коллинзия, коллинсия *bot.* collinsia *(Collinsia)*
коллинсония *bot.* horseweed, horse balm *(Collinsonia)*
коллодий *chem.* collodion, collodium
~, **кантаридиновый** cantharidal collodion
~, **кровоостанавливающий** styptic/hemostatic collodion
~, **молочно-салициловый** milk salicylated collodion
~, **нарывной** blistering collodion, collodion vesicans
коллоид colloid
~, **гидрофильный** hydrophile/hydrophilic colloid
~, **гидрофобный** hydrophobic colloid
~, **жидкий** colloidal solution, sol
~, **лиофильный** lyophilic colloid
~, **лиофобный** lyophobic colloid
~, **радиоактивный** radiocolloid
коллоидно-дисперсный colloido-dispersion *(attr.)*
коллоидный, коллоидальный colloid, colloidal
коллоидоклазия colloidoclasia, colloidoclasis
коллоквиум 1. (academic) conference, colloquium; 2. oral examination
коллоксилин colloxylin, pyroxylin
колломия *bot.* collomia *(Collomia)*
коллотипия *print.* collotype
коллофан *geol.* collophane, collophanite
коллювиальный colluvial
коллювий colluvial deposits/soil, colluvium
колляция collation
колобома *med.* coloboma
~ **радужки** coloboma of the iris
~ **сетчатки** coloboma retinae
коловорот brace, bitstock, breast drill
~, **грудной** breast brace
~, **ручной** hand brace
коловратка *zool.* rotifer *(Rotifera)*
коловратный circular, rotary, revolving
кологарифм *math.* cologarithm
колода 1. block, chunk, log; 1. pack, deck; batch
~ **перфокарт** card deck/batch
~ **перфокарт с заданиями** job deck
~ **перфокарт с исходной программой** source deck
~ **с тестовой программой** test deck
колодезник well borer/sinker
колодец well; pit; shaft, manhole
~, **абиссинский** Abyssinian/tubular well

~, артезианский artesian well
~, водозаборный water intake well
~, водопонижающий *min.* water lowering sump
~, водоприемный water sump
~, водосточный catch pit
~, гравитационный gravity well
~, дозиметрический monitoring well
~, кабельный cable manhole
~, кабельный угловой L-type manhole
~, минный mine hole; demolition shaft
~, поглощающий seepage pit
~, приемный suction well
~, пусковой *rock.* launch(ing) silo
~, самоистекающий flowing well
~, сборный/водосборный bilge/drain/drainage well
~, смотровой inspection/access manhole/shaft/chamber
~ реактора reactor pit
~, субартезианский subartesian well
~, швертовый centre-board trunk/well
колодка shoe, pad, block, check; boot/shoe tree, chock; *el.* connector
~, зажимная gripping jaw
~, клемная/контактная terminal block
~, колесная wheel chock/block
~, переходная intermediate socket
~, переходная ламповая valve/tube adapter
~, переходная патронная plug adapter
~, тормозная brake shoe/block
~ упорного подшипника thrust pad
~, штепсельная receptacle
колоказия *bot.* elephant's ear (colocasia)
колокол 1. bell; 2. *aero.* tail slide, whip stall
~, водолазный diving bell
~, плавательный *zool.* swimming bell, nectocalyx
~, подводный submarine bell
~, сигнальный alarm/warning bell
~, туманный fog bell
колоколовидный, колоколообразный bell-shaped, campaniliform, campanulate
колокол-ревун bell-hawler
коло-колостомия *med.* colocolostomy
колокольный bell *(attr.)*
~ звон peal, chime
колокольня belfry, church/bell tower
колокольчик 1. handbell; 2. *bot.* bluebell *(Campanula)*
колокольчиковые *pl. bot.* bellflower family, *(Campanulaceae)*
коломбина 1. *bot.* European columbine *(Aquilegia vulgaris);* 2. *theatre* columbine
коломбо *bot.* snail-seed *(Cocculus)*
колометрия colometry
колон I *lit.* colon
колон II *fin.* colon
колон III *hist.* colonus
колонат *hist.* colonate
колониальный colonial
~ магазин grocer's shop
колонизатор colonizer; settler, colonist
колонист colonist, settler
колония colony; settlement
~, кустистая branching colony
~ мшанок *zool.* zoarium
~, слизистая mucoid colony
~, сложная *paleont.* synrhabdosome
~, стержневидная rhabdosome
~, шероховатая rough colony
колонка column; *min.* core sample; *bot.* columella
~, бензозаправочная/бензиновая gasoline/petrol pump
~, вибрирующая тарельчатая oscillating plate column
~, водомерная water-gauge column
~, водоразборная hydrant; standpipe
~, глубоководная deep-sea core
~ грунта core
~ донных отложений sediment core
~, заправочная filling station
~ изоляторов *el.* insulator stack
~, ионообменная ion exchange (resin) column
~, итоговая balance/total/summary column
~ капель band of drops
~, катионообменная cation-exchange column
~, колпачковая bubble plate/cap column
~, насадочная *chem.* packed column
~, обменная exchange column
~, очистительная decontamination column
~, перегонная distillation column
~, пожарная fire hydrant/standpipe
~, поперечная cross-column
~, промывная scrub column
~, противоточная counter-current column
~, пульсирующая pulse(d) column
~, распылительная spray column
~, ректификационная rectifying column
~, рулевая steering column
~ с механической пульсацией mechanically pulsed column
~ с набивкой/насадкой packed column
~ с сетчатыми тарелками sieve-plate column
~, следящая continue column
~, смесительная column mixer
~, составная compound column
~ суммы balance/total/summary column
~, тарельчатая plate column
~, термодиффузионная thermal-diffusion column
~ фракционная fractionating column
~, хроматографическая chromatograph(ic) column
~, штурвальная *av.* control-wheel column
~, экстракционная extraction column
~, экстракционно-промывная extraction-scrub column
колонна column; *chem.* tower; *arch.* pillar; *see also* **колонка**

~, абсорбционная absorption column/tower; absorber
~, аммиачная *chem.* ammonia still
~, анионообменная anion-exchange column
~, барботажная bubble column
~, бражная beer column
~, бурильная drill pipe string
~, графитовая graphite column
~, дегазационная stripping tower
~, дистилляционная distilling tower
~ дробной перегонки fractionating column/tower
~, концентрационная evaporation column
~, насадочная *chem.* packed column/tower
~, обсадная *min.* casing string
~, отпарная stripping tower
~, перегонная distillation tower/column
~, перколяционная percolation column
~ по три *mil.* column of three
~, походная marching column
~, разделительная separation column
~, ректификационная rectifying/fractionating tower; cracking fractionator
~ с колпачковыми тарелками bubble-cap column/tower
~ с ситчатыми тарелками perforated-plate/-tray tower/column
~, тепловая thermal/thermalizing column, sigma pile
~, ходовая *min.* mobile rig
колоннада colonnade
колон(н)ообразный column-shaped; columnar
колонный column *(attr.)*; columned; columnar
колонограмма *med.* colonogram
колонок 1. *zool.* Siberian weasel *(Mustela sibirica)*; 2. kolinsky fur
колонопатия colonopathy
колонорагия colonorrhagia
колонорея colonorrhea, colorrhea
колоноскоп *med.* colonoscope
колоноскопия *med.* colonoscopy
колонтитул *print.* running title/headline
колонцифра *print.* page number
колончатый columnar
колопексия *med.* colopexia, colopexy
колопроктит *med.* coloproctitis, proctocolitis
колопроктэктомия *med.* coloproctectomy
колоптоз *med.* coloptosia, coloptosis
колорагия *med.* colorrhagia
колорадский жук Colorado potato beetle *(Leptinotarsa decemlineata)*
колоратура *mus.* coloratura
колоректостомия, колопроктостомия *med.* colorectostomy, coloproctostomy
колоризация colo(u)ration, colo(u)ring
колориметр colorimeter, chromatometer
~, дифференциальный colo(u)r-difference meter
~ погружения immersion colorimeter
~, трехцветный trichromatic colorimeter
колориметрия colorimetry; colorimetric analysis
колориметр-нефелометр colorimeter-nephelometer
колорист *art* colo(u)rist, colo(u)r stylist; *print.* colo(u)r matcher
колорит colo(u)r, colo(u)ring, tint
колоритный colo(u)rful; picturesque; graphic
колос *bot.* ear, spike, spica
~, остистый awned spike
~, пахучий sweet vernal grass *(Anthoxanthum odoratum)*
колосеница *bot.* flaxseed plaintain, psyllium seed *(Plantago psyllium)*
колосистый *bot.* eared, spicate
колоситься *bot.* form ears, ear
колосковидный *bot.* spikelet-like
колосковый *bot.* spicate, spikelike
~ стержень rachilla
~вая чешуя glume; husk
колоскопия *med.* coloscopy
колосник fire/grate/furnace bar; rack, framework
колосняк *bot.* wild rye, line grass *(Elymus)*
колосовидный spike-shaped, spiciform, spicate
колосовые (культуры) *pl. bot.* cereals, grain crops, ear-forming crops
колосок *bot.* spikelet; spicule
~, душистый vernal grass *(Anthoxanthum)*
колосоносный *bot.* spiciferous
колосоподборщик *aggr.* ear retriever
колосоподъемник tailing conveyer/elevator; crop lifter
колосс colossus; Colossus
колостаз *med.* colostasis
колостомия colostomy
колотомия colotomy
колофон *print., hist.* colophon
колоцинт 1. *bot.* colocynth, bitter apple *(Citrullus colocynthis)*; 2. *pharm.* colocynth(is), colocynthidis pulpa
колоша *met.* charge, batch, burden, blast furnace charge
~, коксовая coke burden/charge
~, рудная ore charge
~, топливная fuel bed
~, чугунная iron charge
колошение *bot.* ear formation, earing, heading stage
колошник *met.* furnace top/throat
колпак cap; hood, cowl, bonnet; dome
~ взрывателя (предохранительный) fuse cap
~, вытяжной exhaust/vent/fume hood
~, дымовой funnel umbrella, chimney cap
~ клапана valve bonnet
~, паровой steam dome
~, предохранительный protective/safety/shielding cap/cover
~, пылезащитный dust cap
~, стеклянный bell glass/jar
колпачковидный, колпачкообразный

calyptriform; mitre-shaped
колпачок cap; hood; *bot.* calyptra
~, анодный anode/plate cap
~ взрывателя fuse cap
~, крончатый crown cap
~, предохранительный safety guard/cap
~, противозачаточный contraceptive cap
колпица *zool.* spoonbill *(Platalea leocordia)*
колтун *med.* plica polonica
колтунник *bot.* running pine, club moss
колумба *bot.* columba root *(Jatrorrhiza palmata)*
колумбиец, колумбийский Colombian
колумбий *chem.* columbium, Cb, Nb
колумбийка Colombian (woman)
колумбит *miner.* columbite
Колумбия Colombia; Columbia
колумелла columella
колун wood chopper, splitting ax(e), wood cleave
Колхида Colchis
колхамин *chem.* demecoline, colchamine
колхицеин colchicein
колхицин colchicine
колхоз *abbr.* collective farm, kolkhoz
колхозник, колхозница collective farm member
колхозный collective-farm *(attr.),* kolkhoz *(attr.)*
~ строй collective-farm system
~ное крестьянство collective-farm peasantry; collective farmers
~ное строительство organization of collective farms; setting up of the collective-farm system
колча *bot.* ephedra, joint fir *(Ephedra)*
колчак *bot.* hydnum *(Nydnum)*
колчаковец *hist.* follower of Kolchak
колчаковщина *hist.* counterrevolutionary movement led by Kolchak
колчан guiver
колчедан *miner.* pyrite, iron pyrites, kies firestone
~, белый никелевый rammelsbergite
~, волосистый fibrous/capillose/capillary pyrite, millerite
~, гребенчатый cockscomb pyrite, marcasite
~, железный iron pyrites, pyrite
~, железо-никелевый pentlandite, iron-nikel pyrite
~, желтый никелевый fibrous pyrite, millerite
~, кобальтовый/кобальто-никелевый linnaeite
~, копьевидный spear pyrite
~, красный никелевый copper nickel, arsenide of nickel
~, лучистый radiated/white iron pyrite, marcasite
~, медистый cupreous pyrite
~, медный chalcopyrite, copper pyrite
~, мышьяковистый sulpharsenide of iron
~ мышьяково-никелевый chloanthite
~, мышьяковый arsenopyrites, arsenical pyrites, arsenic iron
~, никелевый nickelglance, arsenical nickel
~, оловянный tin pyrites, stannite
~, разложившийся серный brownstone
~, серебряный sternbergite
~, углистый coal pyrites
колчеданистый pyritaceous
колчеданный pyritic, pyrite *(attr.)*
~ная руда sulphide ore, kies
колышек peg; picket; pin
~ для оттяжки guy pin
~, нивелирный levelling peg
~, разбивочный staking [stake-out, stake-off] peg/pin
колышка kink
кольдкрем cold cream
колье necklace
кольза *bot.* colza rape *(Brassica napus oleifera)*
кольматаж, кольматация *geol., eng.* colmatage, colmation, mud grouting/injection
кольматант colmatent, colmatage agent
кольпатрезия *med.* colpatresia
кольпит *med.* colpitis
~, старческий senile colpitis
~, трихомонадный colpomycosis, colpitis mycotica
кольпогиперплазия *med.* colpohyperplasia
кольпогистеропексия *med.* colpohysteropexy
кольпогистерорафия colpohysterorrhaphy
кольпогистерэктомия colpohysterectomy
кольпоклейзис *med.* colpocleisis
кольпомикроскоп colpomicroscope
кольпомикроскопия colpomicroscopy
кольпомиотомия colpomyotomy
кольпопексия colpopexy
кольпоперинеопластика colpoperineoplasty
кольпопоез colpopoiesis
кольпоптоз colpoptosis, colpoptosia
кольпорафия colporrhaphy
кольпоректопексия colporectopexy
кольпоррагия colporrhagia
кольпоскоп colposcope
кольпоскопия colposcopy
кольпоспазм colpospasm
кольпостат *med.* colpostat
кольпостеноз colpostenosis
кольпотомия colpotomy
кольпоуретеротомия colpo-ureterotomy
кольпоцеле colpocele
кольпоцелиотомия colpoceliotomy
кольпоцистит colpocystitis
кольпоцистоуретеромия colpocystoureteromy, colpoureterocystomy
кольпэктазия colpectasia
кольпэктомия colpegtomia
кольраби *bot.* kohlrabi *(Brassica oleraceae gognylodes)*
кольт Colt pistol; Colt machine gun
кольцевание girdling, ringing; notching; *(of birds)* banding
кольцевидный ring-shaped, annulate, annular
кольцевой ring *(attr.);* annular; annuloid
кольцо ring; loop, ferrule, collar; reel; eye; ring-shaped [circular] structure

~, абдоминальное *anat.* abdominal ring
~, азимутальное *geod.* azimuth dial, bearing circle
~, амниотическое amnion ring
~, ароматическое *chem.* aromatic ring
~, бандажное tread ring
~, барабанное *anat.* tympanic ring
~, белковое *chem.* albumen ring
~, бензольное *chen.* benzene ring
~, буксировочное towing eye/ring
~, ведьмино *bot.* fairy-ring mushroom *(Marasmius oreades)*
~, вихревое vortex ring/loop
~, воздухораспределительное *met.* blast/bustle pipe
~, гетероциклическое heterocyclic ring
~, гидравлическое liquid ring seal
~, глоточное *anat.* throat ring
~, глубокое паховое *anat.* deep inguinal ring
~, годичное *(of tree)* annual/growth ring
~, гомоциклическое homocyclic ring
~ дальности range ring
~, детское *med.* teething ring
~, дистанционное spacer ring
~ жесткости stiffening/reinforcing ring
~, замковое lock ring
~, защитное guard ring
~ защиты *comp.* protection ring
~, изоциклическое isocyclic ring
~ карданного подвеса gimbal ring
~ коллектора *el.* commutator ring
~, коллекторное *el.* collector ring
~, компрессионное поршневое compression piston ring, piston packing ring
~, контактное *el.* slip ring
~ конъюнктивы conjunctival ring
~, коронирующее corona ring
~, короткозамыкающее *el.* end ring *(of motor)*
~ коуша eye ring
~, круглое шкотовое ring clew
~, лабиринтное labyrinth packing ring
~, маслоотражательное oil deflector ring
~, маслосъемное/маслосбрасывающее (поршневое) oilscraper piston ring
~, мышечное muscular ring
~, набивочное packing ring
~ наводки *phot.* focussing ring/collar
~ обода колеса rim ring
~, обручальное wedding ring
~, окулярное eyepiece ring
~, отражателное thrower ring
~, охранное guard ring
~, паховое inguinal/abdominal ring
~, переходное adapter ring
~, пиррольное pyrrole ring
~, плевральное pleural ring
~, плеохроичное/плеохроическое pleochroic halo
~, подкрепляющее reinforcement ring
~, причальное mooring ring
~, порфириновое porphyrin ring
~, поршневое piston ring
~, поршневое маслоудерживающее oil-retaining ring
~, проставочное ring spacer, distance/spacer ring
~, протекторное *(of tyre)* tread ring
~, пупочное umbilical ring
~ радужки *anat.* ring of the iris
~ радужки, большое/малое greater/lesser ring of the iris
~, разрезное split ring
~, распорное spacer ring
~, распылительное spray ring
~, расширительное expansion ring
~ сальника gland/packing ring
~ сердца, волокнистое cardial fibrous ring
~, сжимающее constriction ring
~, смазочное oiler/lubricating ring
~, сопловое nozzle ring
~, сосудистое *anat.* vascular ring
~ спорангия sporangium annulus
~, стопорное lock/retaining ring, circlip
~, стыковочное mating/docking ring
~, стяжное clamping ring
~, сухожильное tendonous ring
~, счетное counter ring
~, токосъемное collector ring, slip-ring
~, тонзиллярное *anat.* tonsillar ring
~, трамвайное terminus, terminal ring
~, трахеальное tracheal ring/cartilage
~, уплотнительное/уплотняющее gasket/seal/packing/sealing ring
~, упорное thrust ring; stop ring
~, установочное setting/locating/set/stopper ring; set collar
~, форсуночное injector/burner ring
~, фурменное *met.* tuyere ring
~, цапфенное *met.* trunnion belt
~, цветочное *bot.* verticil, whorl
~, швартовочное *av.* tie-down ring
~, шейное лимфатическое Waldeyer's (throat) ring
кольцовка tape loop
кольцоид ringoid
кольчатый annulate(d), annuloid
~тые черви *zool.* annelids *(Annelida)*
кольчец *zool.* annelid
кольчуга chain armo(u)r, shirt of mail
кольчугалюминий Duralumin
колюр *astr.* colure
~ равноденствия equinoctial colure
~ солнцестояния solstitial colure
колюрия *bot.* coluria, clove-root plant *(Coluria)*
колютик *bot.* thistle *(Chamaepeuce, Cirsium)*
колючеветвистый *bot.* spiny-/thorny-branched
колючеволосый *bot.* hispid
колючеголовчатый acanthocephalous; spiny-/thorny-headed

колючезонтичник *bot.* prickly/sea parsnip, *(Echinophora)*
колючеколосый *bot.* prickly-eared/-spiked
колючелистник *bot.* 1. spine pink *(Acanthophyllum);* 2. carline thistle *(Carlina)*
колючелист(н)ый *bot.* acanthophyllous, spinifolious, prickly-leaved
колючеперые *pl. zool.* acanthopterigians (Acanthopterygii)
колючеперый *zool.* spiny-finned
колючеплодный *bot.* acanthocarpous, prickly-fruited
колючесть spinosity; prickliness
колючехвостые *pl., zool.* surgeon fishes *(Acanthuridae)*
колючехвостый *zool.* spiny-tailed
колючещетинистый *bot., zool.* strigose
колючий acanthous, spiny, prickly
~чая проволока barbed wire
колючка spike, spicule, prickle, spine, needle, acantha; thorn; barb
~, верблюжья *bot.* camel's thorn *(Alhagi camelorum)*
колючковидный spiniform, spinuliform
колюшковые *pl. zool.* sticklebacks *(Gasterosteidae)*
колющий cleaving, splitting; piercing, stabbing
~щая боль shooting pain
коляда Christmas carol, koliada
колядовать go Christmas carolling
коляска carriage; baby-carriage, perambulator; *(of motorcycle)* sidecar
~, больничная wheelchair
ком lump; chunk; clod; ball
~, каловый *med.* scybalum
~, пищевой globus
~, снежный snowball
кома I *astr., opt.* coma
кома II *med.* coma
~, алкогольная alcoholic coma
~, апоплектическая apoplectic coma
~, гипергликемическая hyperglycemic coma
~ ложноуремическая pseudouremic coma
~, надпочечниковая adrenal coma
~, печеночная hepatic coma
~, постгеноррагическая posthemorrhagic coma
~, тиреотоксическая thyrotoxic coma
~, хлоропеническая chloropenic coma
~ холерная choleraic coma
~, эклamптическая eclamptic coma
команда 1. order, command, instruction; 2. crew, party, detachment; team; gang; ship's company
~, аварийная *av.* crash team
~ ассемблера *comp.* assembly program(me)/instruction, assembler instruction
~ без адреса zero-address [addressless] command/instruction/order
~ безусловного перехода unconditional transfer/jump instruction
~ блокирования blocking order
~ блокировки disable/ignore instruction
~ вводимая с клавиатуры keyboard instruction
~ ветвления branch(ing) instruction
~ возврата return/reset instruction
~ вызова call instruction/command
~ выполнения математической операции math instruction
~, горноспасательная mine rescue crew/party
~, групповая block instruction
~ дублирования copy instruction
~, жесткая ad-hoc instruction
~ записи в память memory write command, store write instruction
~, запрещенная illegal instruction
~ избирания select command
~ "изменить" vary command
~ исполнения operate command/instruction
~ корабля hands, crew
~, курсовая directional command
~, ложная false/spurious command
~ маскирования файла file mask command
~, машинная 1. computer instruction; 2. *tech.* engineering gang
~ межрегистровой пересылки register-to-register instruction
~, минимальная *naut.* skeleton crew
~, многоадресная multiaddress instruction
~, морская высадочная naval beach party
~ на выходном языке object instruction
~ на машинном языке absolute/machine instruction
~ наведения guidance command
~, наземная (наземного обслуживания) ground crew
~, начальная initial/key instruction/order
~, неверная/неразрешенная illegal instruction
~, неиспользуемая unused instruction
~ обращения reference instruction
~ обращения к памяти memory reference instruction
~ обращения к регистру register reference instruction
~ обслуживания servicing crew
~, одноадресная single-address instruction
~, операторная operator command
~ останова halt/breakpoint/exit instruction
~ от вычислителя computer command
~ от ручки *av.* stick command
~ отбраковки reject command
~ ответа reply command
~ передачи данных data transmission instruction
~ передачи управления control transfer instruction
~ переполнения overflow instruction
~ пересылки transfer instruction
~ пересылки данных data movement instruc-

tion
~ перехода transfer/jump/alternate instruction
~ перехода при невыполнении условия jump if/or NOT instruction
~ перехода при переполнении jump on overflow instruction
~ печати printing command/instruction/order
~ по крену roll command
~ по рысканию yaw command
~ по тангажу pitch command
~ подводной лодки submarine crew
~, пожарная fire-fighting crew
~, подрывная demolition squad
~ привлечения внимания оператора attention command
~ пропуска skip/blank instruction
~ просмотра look-up instruction
~ работы со стеком stack instruction
~ разблокировки release command
~ разгрузки unload command
~ разминирования *naut.* clearing party
~, ремонтная repair team/gang
~ рестарта restart instruction
~, речевая voice command
~ с косвенной адресацией indirect instruction
~ с магнитной ленты magnetic tape command/intsruction/order
~ с одним операндом monadic instruction
~ с плавающей запятой floating-point instruction
~ с цифровым кодированием numerical coded instruction
~ сдвига shift instruction
~, символическая 1. mnemonic instruction; 2. pseudo-instruction/command
~ сложения add instruction
~ сортировки sort instruction
~, спасательная rescue team/crew
~ срввнения compare instruction
~, стандартная stereotyped/standard command
~, стартовая launch-control team
~ супервизора executive instruction
~ технического обслуживания maintenance crew
~ управления операцией ввода-вывода input-output instruction
~ управления/управляющая control instruction
~ условного перехода conditional jump/transfer/branch instruction
~, уставная standard command
~, фиктивная/холостая dummy instruction, no-op instruction, null
~ форматирования format instruction
~, футбольная *sport.* football team
~, циклическая loop/repetitive instruction
~ циклического сдвига rotate instruction
командарм *abbr.* army commander
командир commander, commanding officer, officer in command; *naut.* captain
~ воздушного корабля aircraft/plane commander, pilot-in-command, captain
~ космического корабля spacecraft commander
~ подводной лодки submarine captain/commander
~ соединения *mil.* force commander
~ соединения кораблей naval force commander
~ флагманского корабля flag captain
~ шлюпки boat officer
~ экипажа crew commander
командированный on business, on an official/business trip
командировать send/dispatch on an official/business trip, send on an assignment
командировка official/business trip, assignment, mission
~, научная study/research trip
командировочное удостоверение authority/warrant for official trip
командировочные деньги travel(l)ing allowance
командитное товарищество silent partnership
командный command *(attr.);* commanding
~ состав *mil.* the officers in command
командование 1. commanding, command; 2. headquarters, *coll.* command
~, авиационное air command
~, военно-морское naval command
~, главное General Headquarters
~, оперативно-тактическое tactical force command
~, учебное exercise command
командующий commander, commanding general/admiral; commander-in-chief
~ флотилией flotilla commander
~ флотом fleet commander
~ эскадрой squadron commander
комар mosquito, gnat
~, малярийный malaria mosquito *(Anopheles)*
комар-дергун *zool.* midge *(Chironomidae)*
комарник mosquito net
комаролов *zool.* gnat catcher *(Polioptila)*
коматозный *med.* comatose
~ное состояние coma, comatose condition
коматулида *astr.* comatulid
комбайн combine; multi-purpose machine, compound operation machine; *agr.* combine harvester
~ горный mining combine, cutter-loader
~, горный проходческий tunnelling/heading machine, tunnel/road borer
~, зерноуборочный/зерновой grain combine
~, картофелеуборочный potato combine
~, кукурузоуборочный maize/corn combine
~, кухонный kitchen gombine
~, навесной tractor-mounted combine
~, полунавесной semimounted combine
~, прицепной tractor-drawn combine (harvester)
~, самоходный self-propelled combine

~, силосоуборочный silage/forage combine harvester, field forage chopper, harvester-shredder
~, хлебоуборочный harvester combine
комбайнер, комбайнерка combine operator/driver
комбат *abbr.* 1. battalion commander; 2. battery commander
комбед *abbr., hist.* committee of poor peasants
комбижир *abbr.* combined/mixed fat
комбикорм *abbr.* compound/mixed/combined fodder; compound animal feedstuff
комбинат group of enterprises, integrated industrial complex; combine
~ бытового обслуживания integrated service establishment
~, домостроительный integrated house-building works
~, металлургический integrated metallurgical works
~, промышленный integrated industrial plant
~, целлюлозно-бумажный integrated pulp-and-paper mill
комбинаторика *math.* combination theory/analysis: theory of combinations
комбинаторный *math.* combinatory, combinatorial; combinative
комбинация combination
~, бытовая *comp.* bit combination
~, выпуклая convex combination
~, кодовая code combination
~, кодовая запрещенная forbidden/illegal/invalid code combination
~ приборов instrument cluster
~ с наименьшими издержками *econ.* least-cost [least-outlay] combination
комбинезон overalls
~, детский jumpers
~, летний flying suit
~, рабочий work clothes, dungarees
комбинировать combine, put together; arrange, rearrange
комбретовые *pl. bot.* myrobalan family *(Combretaceae)*
комбриг *abbr.* brigade commander
комдив *abbr.* divisional commander
комедиограф writer of comedies, comedist
комедия comedy
~ буфф comedie bouffe
комедокарцинома comedocarcinoma
комедон *med.* comedo; *pl.* comedones, blackhead
комель butt (end); shank, root
~ лопасти blade shank/root
~ сваи pile butt
комендант 1. *mil.* commandant; 2. warden; manager
~ береговой охраны coast guard commandant
~ города city major
~ погрузки naval embarkation officer
~ станции railway transport officer
комендантский час curfew
комендатура commandant's office
комендор *naut.* gunnery rating, seaman gunner
комета comet
кометарный, кометный cometary
коми Komi, Zyrian
комиксы *pl.* comics; comic books
комингс *naut.* coaming; sill
~ люка hatch coaming
кометоискатель comet seeker/finder
Коминтерн *abbr., obs.* Comintern, Communist International
Коминформ *hist., obs.* Cominform, Communist Information Bureau
комиссар 1. commissar; 2. commissioner
~, аварийный *econ.* average commissioner
~, верховный high commissioner
~, военный enlistment officer
~, народный *hist.* people's commissar
комиссариат commissariat
~, военный registration and enlistment office
комиссионер (commission) agent, broker, jobber
комиссионерство brokerage
комиссия 1. commission; (ad hoc) committee; board 2. *econ.* commission; charge
~, банковская *econ.* bank charges
~, бюджетная *law* ways and means committee
~, врачебно-контрольная medical control comission
~, врачебно-трудовая экспертная medical-and-labour examination commission
~ за акцептование *fin.* acceptance charge
~ за оформление протеста векселя protest charges
~ за трансферт *fin.* transfer fee
~, мандатная credentials committee
~, медицинская medical board
~, морская арбитражная maritime arbitration commission
~ ООН по атомной энергии United Nations Atomic Energy Commission
~ по атомной энергии Atomic Energy Commission
~ по защите от радиоактивного излучения, Международная International Commission on Radiological Protection
~ по качеству quality committee
~ по проверке технических характеристик engineering release committee
~ по разоружению disarmament commission
~, постоянная standing committee
~, ревизионная *econ.* auditing committee
~, следственная *law* committee of investigation; court of inquiry
комиссура *anat., bot.* commissure
комиссуротом *med.* conmissurotome
комитант *geom.* comitant
~, совместный concomitant
комитент *econ.* consignor, committent
комитет committee; board
~ действия committee of action, action commit-

tee
~, исполнительный executive committee
~, организационный organizing committee
~, отборочный select committee
~ по безопасности на море Maritime Safety Committee
~ по координации экспорта стратегических товаров между Востоком и Западом Coordinating Committee of East-West Trade Policy, COCOM
~ по проверке полномочий *law* credentials committee
~, подготовительный steering committee
~, постоянный standing committee
~, редакционный drafting committee
~, руководящий steering committee
~, специальный ad hoc committee
~, стачный strike committee
~, цеховой workshop committee
комифора *bot.* myrrh tree *(Commiphora)*
комма *med.* comma
коммандитное товарищество *econ.* silent/sleeping partnership
коммелина *bot.* dayflower *(Commelina)*
коммелиновые *pl. bot.* spiderwort family *(Commelinaceae)*
комменсал *biol.* commensal
комменсализм commensalism
комментарий comment; commentary; annotation
комментатор commentator
коммерсант businessman; merchant
коммерческий commercial; mercantile; business *(attr.)*
~ район города downtown business district
~ флот merchant marine
коммивояжер travel(l)ing salesman, commercial travel(l)er
коммиссура *anat., bot.* commissure
коммодор commodore
коммоция concussion, commotio
коммуна commune
коммунально-бытовой household *(attr.)*
коммунальный municipal, communal; household *(attr.)*; public-service *(attr.)*
~ная квартира shared apartment/flat
~ные отрасли промышленности и услуг public utility services
~ные предприятия municipal services/enterprises
~ные сооружения public utilities
~ные услуги public services/utilities
коммунар 1. *hist* Communard; 2. member of commune
коммунизм communism
коммуникационный communication *(attr.)*
~ная линия line of communication
коммуникации *pl.* communications; supply/serviclines; lines of communication; *naut.* lanes
~, жизненно важные life lines
~, морские sea/shipping lanes
~, пневматические pneumatic pressure lines
коммунист, коммунистка communist
коммунистический communist *(attr.)*
коммутант *math.* commutant
коммутативность *math.* commutativity, commutability
коммутатор *elec.* 1. commutator; 2. switchboard; multiplexer, (multiplexing) switch
~, кнопочный key switchboard
~ операции *comp.* operations commutator
~, телефонный бесшнуровой cordless switchboard
~, телефонный междугородный long-distance switchboard
~, цифровой digital multiplexer
коммутационный commutation *(attr.)*
коммутация commutation; switching, multiplexing
~, бездуговая/безыскровая arcless commutation
коммутировать commutate, switch
коммюнике communique; press release
комовой in blocks/lumps
~вая сера block/lump sulphur
комок clot; lump
комочек small lump/clot; pellet; ball
компакт *math.* compactum; compact space/set
компакт-диск compact disk
компактный compact; dense; concise
~ в себе *math.* absolutely compact
компания company; association
~, авиотранспортная airline(s), air carrier
~ автодорожных перевозок highway carrier
~, акционерная (joint) stock company
~ внутренних перевозок inland carrier
~, государственная public company
~ грузового автотранспорта motor freight company
~ грузовых перевозок cargo carrier
~, дочерняя subsidiary/affiliated company; subsidiary
~, материнская parent company
~ местных перевозок short-haul carrier; local service carrier
~, многоотраслевая multiinfustry company
~ с неограниченной ответственностью unlimited company
~ с ограниченной ответственностью limited liability company, limited company, Ltd
~, страховая insurance company; insurance carrier
~, строительная building company
~, судоходная navigation company
~, транспортная shipping company
~, фиктивная bogus company
~ чартерных перевозок charter operator
компановать put together; arrange in a set
компаратор comparator, comparer; comparison/standard base
~, амплитудный pulse-height comparer
компарирование *geod.* comparison; standardiza-

tion, calibration
компартия *abbr.* Communist party
компас compass
~, авиационный aircraft/flight compass
~, азимутальный azimuth compass
~, астрономический astrocompass, celestial compass
~, гироиндукционный gyro fluxgate compass
~, гироскопический gyroscopic compass, gyrocompass
~, главный master/standard compass
~, горный surveyors compass/dial
~, дистанционный remote-reading/-indicating compass; transmitting compass
~ для плавания под водой diving compags
~, индукционный earth inductor compass, fluxgate/-valve compass
~ магнитного склонения declination compass
~ наклонения inclination compass
~, наручный wrist compass
~, путевой steering compass
~ с жидкостным демпфированием liquid-damped compass
~ с неуничтоженной девиацией uncompensated compass
~ с уничтоженной девиацией compensated compass
~ с успокоителем колебаний aperiodic compass
~, солнечный sun/solar compass
~, судовой mariner's compass
~ штурмана navigator's compass
компаунд *met.* compound; filler
~, герметизирующий sealing compound
~, заливочный potting compound
~, пропиточный impregnating compound
компаундирование 1. compound-impregnation; compound-filling/-potting; 2. *el.* compound excitation
компендиум compendium; digest
компенсатор compensator, equalizer, balancer, balance gear, balance, compensating unit; expansion piece
~, весовой counterbalance
~ затухания attenuation equalizer
~ объема volume equalizer/compensator, pressurizer
~ сдвига фаз phase corrector/compensator
~, трубный температурный expansion joint/loop/bend; expansion U-bend; bellows expansion joint
~, электромагнитный compensation coils
~, электронный показывающий indicating potentiometer
компенсационный compensation *(attr.)*; compensatory, compensating; balancing; equalizing
~ тариф compensatory tariff
~ная сделка *econ.* 1. offsetting transaction/deal; barter deal; 2. compensation/compensatory transaction
компенсация 1. *econ.* compensation; indemnity; offset; reimbursement; 2. *tech.* balance, balancing, compensation, equalizing; correction; adjustment
~ астигматизма astigmatism correction/control
~, аэродинамическая aerodynamic balance
~, весовая mass balance
~ за убыток/ущерб allowance/compensation for damage; indemnity for damage
~ за увечье compensation/indemnity for injury
~ компасных ошибок conpass error compensation
~ крутящего момента torque compensation
~ угловой ошибки adjustment of angular error
компенсировать 1. *econ.* compensate, offset, make amends (for), idemnify; 2. *tech.* balance, equalize, equilibrate, make up (for), compensate (for), neutralize
компетентность competence
компетенция terms of reference, competence; *law* jurisdiction; cognizance
компилировать compile
компилятивный compilatory; compilation *(attr.)*
компилятор *comp.* compiler, compiling program(me), compiling routine
~, однопроходной single-pass compiler
компланарность coplanarity
компланарный coplanar
комплекс complex; system; set; structure
~, безразмерный *math.* dimensionless group
~, биотический biotic association
~, водохозяйственный water utilization scheme/system
~, военно-промышленный military industrial system/complex
~, военный military/defence establishment
~, вычислительный computer complex
~ государственных мероприятий public package
~, двухфазный diphasic complex
~, желудочковый ventricular complex
~, измерительный instrumentation complex
~, интрузивный *geol.* igneous complex
~, испытательный test complex
~, космический space complex
~, культурный complex culture establishment
~, навигационный integrated navigation system
~ неполноценности inferiority complex, feeling of inferiority
~, оборонный defence establishment
~ обработки данных data-processing complex
~, предсердный *anat.* atrial/auricular complex
~, промышленный industrial system
~ работ work package
~, стартовый launching complex
~, телеметрический telemeter complex
~ технических средств *comp.* hardware system
~ условий set of conditions
~, учебный training grounds/complex
~, эдипов Oedipus complex

комплексирование integration
~ информации information/data integration
~ систем system integration
комплекснозначный *math.* complex-valued
комплексный complex, integrated, composite, combined; overall *(attr.);* integral
~ная бригада multiple-skill crew
~ ная экспедиция combined expedition
комплексон *chem.* chelate, complexon, versene
комплексонометрия *chem.* chelatometry
комплексообразование *chem.* complexing
комплексообразователь complexing agent/ compound
комплексообразующий complexing
комплект complete/full set; assembly; kit, unit
~, аварийно-спасательный *av., naut.* survival kit/pack
~ дозиметрических приборов dosimeter kit
~ запасных частей set of spare parts, spare kit
~ инструментов tool kit/set
~ карт map outfit
~ кварцев set of crystals
~, микропроцессорный microprocessor set
~, монтажный installation/assembly kit
~ парусов set of sails
~ плат board/card set
~ программного обеспечения software package
~, ремонтный repair kit
~ хромосом genom(e)
~ шрифта type font
комплектация list of equipment; complete plant
комплектность completeness; complete unit
комплектный complete; unitized, in a set
комплектование 1. making up a set; 2. acquisition; 2. staffing; recruiting, regruitment
~ личным составом recruitment; staffing
комплектовать prepare/furnish/supply a complete set; replenish; *(of personnel)* staff
комплемент complement
комплементарность complementarity, complementariness
комплот plot, conspiracy
композит *gram.* composite/compound word
композиция composition; compound; arrangement
~, прессовочная moulding material/compound
комполка *abbr.* regimental commander
компонент component; constituent; ingredient
~, модифицирующий modifier
~, сопряженный associated constituent
~ электронной системы electronic component
компонента component
~, делящаяся fissile component
~ дисперсии *stat.* variance component
~, жесткая hard [high-energy] component
~, исходная primary/precursor component
~, мягкая soft [low-eneroy] component
~, проникающая penetrating/hard component
~, скачкообразная transition component
компоновать assemble, put together, set up; arrange, lay out
компоновка assemblage, assembling; arrangement, layout; configuration
~ аппаратуры preparing a complete set of equipment/apparatuses; packaging of equipment
~, базовая base-line configuration
~ карты map mounting/montage
~ приборов instrument configuration
~, стартовая launch configuration
компоновщик assembler; arranger
компост *agr.* compost, composted fertilizer
компостер ticket-stamping machine; ticket punch
компресс *med.* compress
~, горячий/согревающий hot compress
~, холодный cold compress/pack
компрессия compression
~ в картере *tech.* crankcase compression
~ головного мозга cerebral compression
~ легкого pulmonary compression
компрессометр compressometer, compression gauge
компрессор compressor
~ внутреннего контура *mech.* core compressor
~, водоструйный water-jet compressor
~, воздушный air compressor
~, встроенный integral [iternal, built-in] compressor
~ высокого давления (высоконапорный) high-pressure compressor
~ вытеснения positive-displacement compressor
~, газоструйный gas-jet compressor
~, газотурбинный gas-turbine compressor
~, гидравлический hydraulic air compressor
~ двойного действия double-acting piston compressor
~, двусторонний double-sided [double-entry] compressor
~, комбинированный mixed-flow [axial and centrifugal] compressor
~, лопастный/лопаточный rotary/vane compressor
~, многоступенчатый multistage compressor
~, наддувочный supercharging compressor, supercharger
~ низкого давления [низконапорный] low-pressure compressor
~ объемного типа positive-displacement compressor
~ одинарного действия single-action piston compressor
~, односторонний single-sided/-entry compressor
~, одноступенчатый single-stage compressor
~, осевой axial-flow compressor
~, пароструйный steam-jet compressor/ejector
~, поршневой piston/reciprocating compressor
~ с двусторонним входом double-entry/-sided compressor

~ **с механическим сжатием** positive-displacement compressor
~, **струйный** jet-type compressor
~, **трехступенчатый** three-stage compressor
~, **тронковый** trunk-piston compressor
~, **холодильный** refrigeration/refrigerant compressor
~, **центробежно-осевой** axial-centrifugal [mixed-flow] compressor
~, **центробежный** centrifugal [radial-flow] compressor
компрессорий compressorium
компрессорная compressor house/room/compartment
компримироват compress
компромисс compromise (solution, decision), tradeoff
комптония *bot.* sweet fern *(Comptonia)*
компургация *hist.* wager of law, compurgation
компьютер computer
~, **персональный** personal computer, PC
компьютеризация computerization, introduction of computing equipment
компьютеризованный computerized; computer-based
комсомол *obs.* Komsomol, Young Communist League
комсомолец, комсомолка Komsomol member
комсорг *abbr. obs.* Komsomol organizer
комсостав *abbr.* commanding personnel, commanders
комфрей *bot.* prickly comfrey *(Symphytum asperum)*
комчванство *abbr.* communist arrogance
комячейка *abbr., polit.* communist cell/group
конармеец cavalryman
конармия *abbr.* cavalry army
конатив *gram.* conative
конативность, конация *psych.* conation
конваллатоксин *chem.* convallatoxin
конвейер conveyer, *Am.* conveyor
~, **багажный** luggage conveyer
~, **вибрационный** vibrating conveyer
~, **винтовой** screw/spiral conveyer
~, **грузовой** cargo conveyer
~, **дисковый** disk conveyer
~, **забойный** *min.* face conveyer
~, **инспекционный** inspection belt/conveyer
~, **канатный** cable/rope conveyer
~, **ковшовый** bucket conveyer
~, **ленточно-цепной** chain-belt conveyer, belt/band conveyer
~, **магистральный** trunk conveyer
~, **напольный** floor-type conveyer
~, **перегрузочный** cross conveyer
~ **периодического действия** intermittent action conveyer
~, **пластинчатый** apron conveyer
~, **подвесной** overhead conveyer
~, **подъемный** elevating conveyer
~, **реверсивный** shuttle conveyer
~, **роликовый** roller conveyer
~, **роликовый гравитационный** gravity roller conveyer
~, **роликовый приводной** powered roller conveyer
~, **сборочный/сборки** assembly conveyer/line
~, **тележечный** car conveyer
~, **фартучный** apron conveyer
~, **цепной** chain conveyer
~, **штрековый** gate conveyer
конвейеризировать introduce conveyers into, equip with conveyers
конвейерщик conveyer operator
конвективный convective; convectional
конвектор convector
конвекция *phys.* convection
~, **вынужденная/принудительная/искусственная** forced convection
~, **естественная/свободная** natural/free convection
~, **замедленная/постепенная** tandem [step-by-step] convection
~, **мощная** deep convection
~, **прямая** single-step convection
~, **тепловая** thermal/heat convection
конвент *polit.* convention
конвенционализм conventionalism
конвенционный тариф agreed tariff
конвенция convention; agreement; accord
~, **консульская** consular convention
~ **о территориальном море и прилежащей зоне** Convention of the Territorial Sea and the Contiguous Zone
~ **об открытом море** Convention of the High Seas
~ **по денежным вопросам** monetary convention
~ **по охране человеческой жизни на море** Convention for the Safety of Life at Sea
конвергентность, конвергенция convergence, convergency
~ **глаз** *med.* ocular convergence
конвергировать(ся) converge
конверсия conversion
~, **взаимная** interconversion
конвертер *tech.* converter
~, **бессемеровский** *met.* acid Bessemer converter
~,**бессмеровский малый** side-blown Bessemer converter
~, **медеплавильный** copper converter
~, **парокислородный** steam oxygen converter
~ **с верхним дутьем** top-blown converter
~ **с нижним дутьем** bottom-blown converter
конвертер-реактор *nucl.* converter reactor
конвертин *biochem.* convertin
конвертирование converting, conversion; *met.* bessemerizing
конвертировать convert

конвертоплан convertiplane
конвоир escort; guard; *naut.* escorting ship
конвоировать escort, convoy
конвой escort, convoy
конволютный convolute
конвульсия convulsion
конгломерат 1. *geol.* conglomerate 2. conglomeration
~, золотоносный auriferous conglomerate
~, известняковый calcereus conglomerate, calcirudite
~, катакластический crushed conglomerate
~, ледниковый glacial conglomerate, tillite
~, озерный lacustrine conglomerate
~ ракушечный coquina
~, речной fluvial/fluviatile conglomerate
конгломератовый, конгломератный conglomeratic
конгломерация conglomeration
конглютинин *bioch.* conglutinin
конголезец, конголезский Congolese
конголезка Congolese (woman)
конгорот Congored
конгрегационализм Congregationalism
конгрегационалист congregationalist
конгрегация congregation
конгресс congress
~ производственных профсоюзов Congress of Industrial Organizations, CIO
конгрессист 1. *polit.* member of the Indian Congress Party; 2. congress/conference participant
конгруэнтность, конгруэнция congruence
конгруэнтный congruent, congruous
кондак *rel.* short hymn
конденсат condensate; condensation/condensed water
~, технологический process condensate
конденсатор 1. condenser; 2. *el.* capacitor
~, аммиачный ammonia condenser
~, блокировочный blocking [by-pass] capacitor
~, бумажный paper capacitor
~, воздушный air capacitor
~, двухходовой two-pass condenser
~, зарядный charging capacitor
~, змеевиковый coil condenser
~, импульсный pulse capacitor
~, ионный gas-filled capacitor
~, кожухотрубный shell-and-tube condenser
~, кремниевый silicon capacitor, varactor
~, масляный oil-filled capacitor
~, металлобумажный metalized paper capacitor
~, многоходовой multipass condenser
~, настроечный tuning capacitor
~ обратной связи feedback capacitor
~, оросительный reflux condenser
~, печатный printed capacitor
~, пленочный film-type capacitor
~, помехоподавляющий interference suppression capacitor
~, противоточный counter-current [contra-flow] condenser
~, проходной feed-through capacitor
~, пусковой starting capacitor
~, самопроточный scoop condenser
~, сегнетоэлектрический ferroelectric capacitor
~, струйный jet/ejector condenser
~, электролитический фольговый foil-type electrolytic capacitor
конденсация condensation
~, капельная dropwise condensation
~ пара steam condensation; vapour condensation
конденсировать(ся) condense
конденсор (optical) condenser; converging/burning lens
~, кардиоидный cardioid condenser
~, темнопольный dark-field condenser
кондилома *med.* condyloma
~, остроконечная pointed condyloma, fig wart
кондиломатоз *med.* condilomatosis
кондиломатозный *med.* condylomatous
кондиционер conditioner; air conditioner
~, автономный self-contained air conditioner
~, воздушный air conditioner
кондиционирование conditioning
~ воздуха air conditioning
кондиция quality, standard; condition
кондом condom
кондоминиум *law* condominium
кондор *zool.* condor *(Vultur gryphus)*
кондотьер *hist.* condottiere; soldier of fortune
кондуктивность *phys.* conductivity
кондуктометр *phys.* conductometer
кондуктометрия *phys.* conductometry
кондуктор 1. *tech.* jig; 2. *el.* conductor; railway guard, conductor
~, многоместный multiple jig
~, накладной lay-on jig
~, тормозной *rail.* brakesman
коневод horse breeder
конезавод stud farm
конек I little horse
~, морской *zool.* sea horse, hippocampus *(Hippocampus)*
~ крыши ridge
конек II *zool.* pipit *(Anthus)*
конесовхоз *abbr.* State stud farm
коneферма horse-breeding farm
конец 1. *naut.* line, rope; 2. end, terminal, termination, finish; tip
~, бросательный *naut.* heaving/hauling line
~ вектора extremity [end point, tip, extreme end] of a vector
~ долота nose of a chisel
~, задний posterior/rear end
~, капельный dripping point
~ картотеки *comp.* end of file, EOF
~ киля *av.* fin tip
~ крыла wing tip

~ **ленты** *comp.* end of tape EOT
~ **лопасти/лопатки** blade tip
~ **носителя данных** *comp.* end of volume, data carrier end
~ **передачи** *rad.* end of transmission, EOT
~, **передний** anterior/fore end
~, **подветренный** downwind end
~ **сеанса** log-off, log-out
~, **спасательный** lifeline
~ **текста** *comp.* end of text, ETX
~ **файла** *comp.* end of file, EOF
~ **шпинделя, передний** nose of a spindle
конечнокомпактный *geom.* finitely compact
конечнопорожденный *math.* finitely generated
конечнопокрываемый *math.* finitely covered
конечносвязный *math.* finintely connected
конечность 1. finiteness; 2. extremity, limb
~, **верхняя** upper extremity/limb
~, **листовая** leaf tip
~, **нижняя** lower extremity, inferior limb
~, **передняя** pectoral limb
~, **тазовая** pelvic limb
~, **фантомная** phantom limb
конечный final, terminal, last, ultimate, end *(attr.)*; *math.* finite
~**ная станция** last station, terminal
конидиеносец *bot.* conidiophore
конидиеносный *bot.* conidiiferous
конидий, конидия *bot.* conidium
кониин *chem.* coni(i)ne
коник *zool.* grasshopper; locust *(Stenobothrus)*
коникоид *math.* conicoid
конина horse hide/leather; horseflesh
кониоз *med.* coniosis
кониокортекс coniocortex
кониолимфостаз coniolymphostasis
кониометр coniometer, koniometer
кониотуберкулез anthracotic tuberculosis
кониофиброз *med.* coniofibrosis
конический conic(al); tapered
коничность conicity
конкистадор, конквистадор conquistador
конклав conclave
конкомитант concomitant
конкордат concordat
конкремент *med.* cencretion, concrement, calculus
конкретизация concrete definition/determination; concretization, concrete rendering; specification
конкретизировать render concrete
конкретность concreteness; substance
конкретный concrete; specific
конкреционный concretionary; nodulous
конкреция concretion, nodule
~, **глинисто-известковая** turtle stone, septarium
~, **известковая** calcereous concretion, lime nodule
~, **марганцевая** manganese nodule, halobolite
~, **сингенетичная** syngenetic/contemporaneous concretion
~, **чешуйчатая** scaly concretion
~, **шаровая** globular concretion
конкубинат concubinage
конкурент, конкурентка competitor
конкурентный competitive
конкурентоспособность competitiveness, competitive capacity/power/strength
конкуренция competition
~, **внутривидовая** *biol.* intraspecific competition
~, **внутриотраслевая** intraindustry competition
~, **добросовестная** *law* fair competition
~, **межвидовая** *biol.* interspecific competition
~, **межотраслевая** interindustry competition
~, **недобросовестная/нечестная** unfair competition, unfair competitive practices
~, **недозволенная** illicit competition
~, **острая** keen competition
~, **скрытая** latent competition
~, **ценовая** price warfare
~, **честная** fair [bona fide] competition
~, **явная ценовая** overt price competition
конкурс competition, contest
~ **проектов** design competition
конкурсный competition *(attr.)*; competitive, based on competition
~ **экзамен** competitive examination
конник cavalryman
конница cavalry
конногвардеец *mil., hist.* life guard; horseguardsman
коннозаводство horse breeding
конный horse *(attr.)*, equestrian; mounted; horse-drawn
~ **завод** stud farm
~ **спорт** equestrian sport
~**ная полиция** mounted police
~**ная статуя** equestrian statue
коновязь tethering post/rail
коноид 1. *geom.* conoid; 2. *text.* cone drum; three-dimensional cam
коноидальный conoid(al)
конопатить caulk; stop up, pack
конопатня caulking iron/material
конопачение caulking
коноплеводство hemp growing
коноплевые *bot.* hemp family *(Cannabinaceae)*
коноплевый hemp *(attr.)*, hempen *(attr.)*
коноплежатка hemp cutter
коноплемолотилка hemp thresher
коноплемялка hemp-crushing mill, hemp breaker/scutcher
конопля *bot.* hemp *(Cannabis)*
~, **китайская** ramie *(Boehmeria nivea)*
коноплянка *zool.* linnet *(Carduelis cannabina)*
конопляный hemp *(attr.)*, hempen; cannabic
~**ное масло** hempseed oil
конорыл *zool.* beaked salmon *(Gonorhynchus gonorhynchus)*
коносамент *econ.* bill of lading

~, бортовой on-board bill of lading, shipped bill of lading
~, именной straight bill of lading
~ на груз alongside bill of lading
~, нечистый claused/qualified/foul bill of lading
~, ордерный order bill of lading
~, парцельный parcel bill of lading
~ с оговорками claused/qualified bill of lading
~, сквозной through/transshipment bill of lading
~, чистый clean bill of lading
конофолис *bot.* cancerroot *(Conopholis)*
конрингия *bot.* hare's-ear (Conringia)
консервант preservative, conserving agent
консерватор 1. *polit.* conservative; 2. conserver, preserver, conservator
консерватория conservatory/academy of music
консервация 1. conservation, preservation, rust-preventive treatment, anticorrosion proofing; slushing; 2. laying-up, putting in prolonged storage
консервирование conservation, storage; preservation; canning
~ копчением preservatiom by smoke-curing
~ крови blood conservation/storage
~ маринованием preservation by vinegar pickling
~ обезвоживанием/сушкой preservation by dehydration
~ сыворотки крови serum conservation
~ шкур hide curing/cure
консервировать 1. conserve, preserve; *tech.* slush, provide with anticorrosion proofing; 2. lay up, put in prolonged storage
консервный can *(attr.),* tin *(attr.),* canning
~ нож can opener
~ная банка can, tin
~кая фабрика cannery
консервы *pl.* canned/preserved food
~, мясные canned meat
~, овощные canned vegetables
~, рыбные canned fish
~, фруктовые canned fruit
консигнат *comm.* consignor, consigner
консигнатор *comm.* consignee
консигнация consignment, consignation
консилиум *med.* consultation
консистенция consistency
консистометр consistometer; mobilometer
~ погружения dripping consistometer
консистория *eccl.* consistory
конский horse *(attr.);* equine
консоли *pl., fin.* consols, consolidated stocks
консоль cantilever; angle bracket
~ крыла *av.* outer/outboard wing panel
~, опорная bearing bracket
консольный cantilever *(attr.),* overhanging
консоме *cul.* consommé
консорций, консорциум, консорция consortium
консоциула *ecol.* consociule
консоция *ecol.* consocies
конспект synopsis, summary; abstract; outline
конспективный concise, synoptic(al)
константа constant *see also* **постоянная**
~, буквенно-цифровая *comp.* character constant
~ высоты elevation constant
~ диссоциации dissociation constant
~, ингибиторная inhibitor constant
~ ионизации ionization/dissociation constant
~ капиллярности capillarity constant
~, литеральная *comp.* literal
~ основности *chem.* basicity constant
~ проницаемости permeability constant
~ равновесия equilibrium constant
~ распада decay/disintegration coefficient
~ роста growth constant
~, символьная figurative constant
~ Фарадея Faraday's constant
~ химического равновесия chemical equilibrium constant
константан *met.* constantan
констатация finding; ascertaining, establishment
констелляция *astr.* constellation
конституанта Constituent Assembly
конституция 1. *polit.* constitution; organic law; 2. *med.* constitution, constitutio corporis, habitus status
~, астеническая asthenic constitution, habitus asthenicus
~, мускульная muscular constitution
~, циклотимическая cyclothymi(a)c constitution
~, чахоточная phthisic(al) constitution
конституэнт *math.* constituent
конструирование 1. building, constructing, erecting; 2. designing
~ изделия product engineering
конструировать 1. build, construct; 2. design
конструктивный 1. constructive; 2. structural, architectonic, constructional
конструктор designer
~, авиационный aircraft designer
~, генеральный/главный chief designer
~ судов naval architect
конструкторское бюро design office
конструкционизм constructionism, constructivism
конструкционный structural; constructional
~ная сталь structural steel
конструкция 1. structure; construction; 2. design, make, model, build
~, абстрактная abstraction
~, авиационная aircraft structure, airframe
~, агрегатная unitary (unit-type) design
~, безрамная *auto.* frameless structure
~, блочная en-block construction/structure; module/block design; panelized construction
~, верхняя superstructure
~, висячая suspension construction/structure
~, жесткая rigid/braced/trussed structure
~ защищенная от неправильного обращения foolroof design

~, кессонная cell/box structure
~, комплексная composite structure
~, крупнопанельная large-panel construction
~, металлическая metalwork
~, модульная modular design/structure/construction
~, моноблочная one-piece [integral] structure
~, монолитная *constr.* in-situ [cast-in-place] structure/construction
~, неразъемная one-piece structure
~, несущая supporting/carrying/bearing structure; framework
~, оболочечная/оболочковая shell structure
~, обшивно-стрингерная skin-stringer structure
~, однолонжеронная single-spar structure
~, панельная bearing-wall construction
~, плоская planar construction
~, подкосная strut-braced structure
~, подкрепленная reinforced structure
~, полужесткая semi-rigid structure
~, предварительно напряженная prestressed structure/construction
~ работающая на растяжение tension structure
~ равного натяжения isotensoid structure
~, расчалочная wire-braced structure
~, решетчатая lattice/grid structure
~ с заполнителем sandwich design/structure
~, сборная prefabricated/precast structure
~, сварная welded construction/structure
~, свободнонесущая cantilever structure
~, сейсмостойкая earthquake resistant structure
~, силовая load-bearing/carrying structure
~, складчатая folded-plate structure
~, статически неопределимая redundant [statically-indeterminate] structure
~, статически определимая non-redundant [statically-determinate] structure
~, типовая standard design
~, ферменная truss structure
~, цельная one-piece structure
~, цельносварная all-welded structure
консул consul
~, генеральный consul general
консульство 1. consulate; 2. consulship
~, генеральное consulate general
консультант consultant, adviser
~ по вопросам налогового обложения tax adviser
~ по рекламе publicity adviser
~ при судье assessor
консультативный consultative, advisory
консультационный consultation *(attr.),* advisory
консультация 1. consultation; expert opinion; 2. consultation service/bureau/clinic/room/office
~, детская child health/welfare centre, children's consultation dispensary/clinic
~ для беременных consultation centre/room for expectant mothers, prenatal consultation service/office/clinic
~, женская women's consultation/antenatal clinic/centre; women's dispensary
~, медицинская medical consultation/advice; medical consultation room/centre
~, юридическая legal advice bureau/office
контакт contact; (commutator) switch
~, акустический echo contact
~, блокировочный interlocking contact
~, вспомогательный auxiliary contact
~, гнездовой female contact
~ датчика sensor switch
~, замкнутый closed/live contact
~, замыкающий make contact, normally open (N. O.) contact
~ зубчатого зацепления gearing contact
~, катящийся rolling contact
~, кнопочный push button
~ на массу *el.* ground/frame connection
~, наклонный *geol.* plunging contact
~ ножевой knife-type contact; contact blade
~, нормально замкнутый break contact, normally closed (N. C.) contact
~, нормально открытый/разомкнутый make contact, normally open (N. O.) contact
~, переключающий changeover (transfer) contact
~, пусковой release contact; firing contact; starting switch
~, радиолокационный radar contact
~, размыкающий break contact, normally closed (N. C.) contact
~, разомкнутый open/dead contact
~, разрывной break/interrupting contact
~, розеточный cluster contact
~, скользящий sliding/variable switch
~, слоистый laminated contact
~, считывающий *comp.* readout contact
~, торцовый butt contact
~, точечный point contact
~ цифрового ввода digital input contact
~, штыковой bayonet contact
~, штыковый male contact
~, щеточный brush contact
контактировать make contact, come in contact, contact
контактный contact *(attr.)*
~ провод trolley wire
~ная сварка pressure contact welding
контактор *elec.* contactor; switch
~, кулачковый cam contactor
~, масляный oil-immersed contactor
~, переключающий changeover contactor
~, пусковой starting switch/contactor
~, реверсивный reversing contactor
~, тиристорный thyristor contactor
~, шиноприсоединительный *el.* bus tie contactor
контаминация contamination; impurity
контейнер container

~, герметизированный pressurized container
~, грузовой freight/cargo/shipping container
~ для криогенных жидкостей cryogenic storage container
~ для морских перевозок грузов shipping container
~ для насыпных грузов bulk container
~ для перевозки сыпучих грузов bulk carriage container
~ для упакованных грузов packaged-freight container
~, изотермический thermally insulated container
~, катапультируемый ejection capsule/pod
~ многократного использования re-usable container
~ одноразового использования expendable [non-reusable] container
~, отделяющийся ejection capsule
~, охлаждаемый refrigerated container
~, подкрыльный underwing pack
~, разборный collapsable container
~, сбрасываемый drop container
~, складной folding container
~, спасаемый recoverable/retrievable/recovery capsule/container
~, стандартный грузовой container standardized freight
~, транспортировочный shipping container
контейнеризация containerisation; introduction of containers; packing in containers
контейнеровоз *naut.* container ship/carrier; *auto.* container truck
контейнер-цистерна tank-container
контекст context
контингент contingent; *econ.* quota; *stat.* cohort; population
~, детский child population
~, мужской *stat.* male cohort
~ разведенных divorce population
~ рождающей *stat.* childbearing population
~ студентов student body/population
~ школьный school-age population
~ эмигрантов *stat.* migration cohort
контингентировать set the quota of/for; regulate quantitatively
континент continent; mainland
~, движущийся drifting continent
континентальность continentality
континуатив *gram.* continuative
континуум *mat.* continuum
конто *fin.* account
контокоррент account current
контора office, bureau
~, акцептная *fin.* acceptance house/bank
~, главная head/main office
~ завода factory office
~, ревизионная audit office
~, страховая insurance office/firm
~, транспортная shipping office
конторская книга account book, ledger
контрабанда 1. smuggling; contraband; 2. smuggled/contraband goods
контрабандист, контрабандистка smuggler, contrabandist
контрабандный contraband; smuggled
контрабараж *naut.* counterbarage
контрабас *mus.* contrabas, double bass
контрабасист *mus.* contrabasist
контравариант *math.* contravariant
контравариантность *math.* contravariance
контрагент partner; party to a contract; covenantor
контраградиентный *math.* contragradient
контр-адмирал rear admiral
контражур *phot.* back-lit exposure
контракт contract; agreement
~, аннулировать declare a contract null and void; cancel a contract
~ без срока действия open-end contract
~, брачный (об имуществе) marriage settlement
~, долгосрочный long-term contract
~, заключать conclude/sign a contract
~, засекреченный classified contract
~, исключающий exclusive contract
~ на перевозку contract of carriage/shipment
~ на разработку developmental contract
~, недействительный void contract
~, ограничительный exclusive contract
~ "под ключ" *constr.* turnkey contract
~, пожизненный life contract
~, последующий follow-up contract
~ с возмещением издержек cost-reimbursement (CR) contract
~ с твердой ценой fixed-price (FP) contract
~ с твердой ценой плюс поощрительное вознаграждение fixed-price incentive-fee [FPIF] contract
~ срочный terminal contract; *pl.* futures
контрактант contractor, contractant
контрактационный contract-stipulated
контрактация contracting, contract negotiation; contract, contractual agreement
контрактный contract *(attr.)*
контрактура *med.* contracture
~, артрогенная arthrogenic contracture
~, врожденная congenital contracture
~, ожоговая burn contracture
~, отводящая abduction contracture
~, приводящая adduction contracture
~, разгибательная extension contracture
~, рефлекторная reflex contracture
~, рубцовая cicatrical contracture
~, сгибательная flexion contracture
~ челюсти mandibular contracture
~, экстензорная extensor contracture
контральто *mus.* contralto; contralto singer
контрамарка *theatre* complementary ticket; free pass
контрапертура *med.* counter-opening/-incision,

contra-aperture/incision
контрапост *art* . contraposto
контрапункт *mus.* counterpoint
контрапунктический, контрапунктный *mus.* contrapuntal
контрассигнация, контрассигнировать countersign
контраст contrast
~, инфракрасный infra-red contrast/background
~, теневой shadow contrast
~, фазовый phase contrast
~, яркостный brightness contrast
контрастировать contrast, stand in contrast; compare
контрастность contrast; contrast range; detail resolution
контратака *mil.* counterattack
контратаковать counterattack
контратип duplicate negative, internegative
контратипирование *phot.* duplication
контрафагот *mus.* double bassoon
контрафакция *law* infringement
контрафорс = контрфорс
контрацептив, контрацептивный *med.* contraceptive
контрацептин contraceptin
контрбалансир counterweght, mass-balance weight
контрвинт counterpropeller, contra-propeller
контргайка lock/check/jam nut
контргалс *naut.* opposite tack
контргамбит *chess* countergambit
контрградиентный *math.* countergradient
контргруз counterweight, counterbalance, counterpoise
контрдоказательство *law* rebutting evidence
контрейлер contrailer
контржила *geol.* counterlode
контрзаговор counterplot
контрзатопление counterflooding
контрибуция (war) indemnity; contribution
контрить *tech.* lock; secure
контркалибр control/master/reference gauge
контрквадрант counterquadrant
коктрклин counterwedge
контркурс *naut.* opposite course
контрманевр *mil.* counter-manoeuvre, countermovement
контрмарш *mil.* countermarch
контрмера countermeasure
контрминирование *mil.* countermining
контрминоносец *nav.* destroyer
контрминировать, контрамина countermine
контрнаступление counteroffensive
контрнож *agr.* counterknife
контробвинение counteraccusation
контровка *tech.* 1. locking, securing; fixing; 2. screw-locking device; lockwire
~ гаек nut locking
~, проволочная wire locking
контроктава *mus.* contraoctave
контролеприспособленность monitoring convenience
контролепригодность testability
контролер inspector, supervisor, checker; *fin.* auditor; *tech.* controller, monitor
~, билетный ticket collector/inspector
~, технический quality inspector
контролировать inspect, monitor, check, verity; supervise, control
контролируемость controllability
контроллер *el.* controller; monitor; control unit
~, барабанный cylindrical/drum controller
~ ввода-вывода input-output (IO) controller
~ вертикального наведения elevation controller
~, главный master controller
~ горизонтального наведения train controller
~, групповой cluster control; multipath control unit
~, индивидуальный single-path controller
~, программируемый логический *comp.* program-logical controller, PLC
~ с программным управлением programmable controller
~, сварочный welding controller
~, связной comnunications controller
~, сетевой network controller
контроль supervision, inspection, monitoring; verification, test, check(ing), control
~, автоматический/автоматизированный automatic test/inspection/gauging
~, алфавитный групповой *comp.* alphabetic card-to-card comparing
~, активный on-line [in-process] inspection/gauging/testing
~, аппаратный *comp.* hardware check
~, аппаратурный built-in check
~, бортовой *av.* in-flight monitoring
~, бухгалтерский audit trial; accounting control
~ в процессе производства in-process inspection
~, валютный currency/exchange/monetary control
~, врачебный medical examination/monitoring
~ времени time check
~, встроенный built-in check
~, входной incoming inspection
~, выборочный random inspection, sampling inspection/check, percentage selection/sampling
~, выборочный многократный multiple sampling
~ готовых изделий end-item inspection
~, диспетчерский supervisory control
~, дистанционный remote monitoring/control
~, дозиметрический radiation monitoring
~, жесткий tight control
~ за загрязнением окружающей среды pollution control

~ за полетом flight monitoring
~ за состоянием окружающей среды environmental control
~ за страховыми операциями insurance control
~ за ходом работ progress control
~ качества guality control; product inspection
~ материалов material checking/testing
~ материалов вихревыми токами eddy-current inspection/testing
~ материалов гамма-просвечиванием gamma-ray inspection/radiography
~ материалов методом красок dye-penetrant testing
~, медицинский medical inspection/supervision/monitoring
~ методом обратной передачи loop check
~ на местах on-site control
~ на четность even-odd check, parity check
~ над судоходством shipping control
~ над ценами price control
~ над ценами и заработной платой wage-price control
~ наземный ground monitoring
~ наличие запрещенных команд *comp.* illegal command check
~, народный popular control
~, непрерывный радиолокационный continuous radar monitoring
~ нечетности odd/imparity control
~ облучения irradiation monitoring
~, обратный *com.* revertive monitoring
~, оперативный operational control
~ ошибок error check/control
~ переполнения *comp.* overflow check
~, периодический periodic(al)/cyclic inspection
~ по вызову *comp.* call control
~ по двоичным знакам *comp.* bit check
~ по избыточности redundance check
~ по флаговым разрядам flag check
~ по четности parity check
~, пооперационный step-by-step checking
~, приемочный acceptance inspection/testing
~, программный programmed check
~, произвоводственный manufacturing/plant supervision
~, профилактический preventive checking
~ рабочий workers' control/supervision
~, радиационный radiation control/monitoring
~ радиоактивности radioactivity monitoring
~, радиолокационный radar monitoring
~ размеров dimension inspection, gauging
~ рентгенопросвечиванием X-ray testing/inspection
~ рождаемости birth control; family planning
~ рублем financial control
~, санитарный sanitary control/supervision
~ системы system check
~ состояния продукции product status control
~ сроков исполнения follow-up control/action
~ судебный judical review
~, текущий monitoring, current/running control
~ технологического процесса process monitoring
~, технический engineering/technical supervision
~ типов данных *comp.* type-checking
~ уровня шума noise monitoring
~ хранения storage monitoring
~, цветной dye-penetrant testing
~ частоты frequency monitoring
~, эксплуатационный field inspection
контрольник verifier
~ карт *comp.* card verifier
контрольно-проверочный *tech.* gauge-checking
контрольно-распределительный check-and-distribution *(attr.)*
контрольно-ревизионный control-and-auditing
контрольный control *(attr.);* inspection *(attr.);* test *(attr.);* monitoring; supervisory
~ пуск check/test run
~ная сумма control total, check(ing) sum
~ное испытание performance test
~ые цифры estimated/provisional/scheduled figures
контрответ *sport* counterriposte
контроттяжка stub guy
контрпар countersteam, counterpressure/back steam
контрпоршень counterpiston
контрпретензия *law* counterclaim
контрпривод *mech.* countershaft
контрпружина counterspring, antagonistic/opposing/balance spring
контрпуансон counterpunch
контрразведка counterintelligence; counterreconnaissance
контрразведчик counterintelligence agent/officer
контрреволюционер, контрреволюционный counterrevolutionary
контррельс guard/safety/check rail
контррефлектор *(of aerial)* subdish
контрруль contrarudder, counter-rudder
контрсталия *law* demurrage
контртимберс countertimber
контртитул duplicate title
контругол contra-angle
контрудар counterblow, counterstroke
контрулитка *text.* counterscroll
контрупор matching dog
контрфланец counterflange
контрфорс buttress, abutment, counterfort
контршкив countersheave
контрштамп counterdie, female die
контршток *mech.* tail rod
контрэскарп *mil.* counterscarp
контузить bruise, contuse
контузия contusion, bruise; concussion
контур 1. outline, contour; profile; 2. circuit; loop
~ автопилота autopilot loop
~ активной зоны core circuit/loop

~, антенный aerial tuned circuit
~, балластный idler circuit
~, бортовой airborne loop
~, водный input tuned circuit
~ возбуждения exciting loop
~, второй/вторичный secondary circuit/loop
~ выключения shut-down circuit
~, выходный output tuned circuit
~, дешифрирующий decoding circuit
~ заземления grounding grid/mat
~, кинжаловидный dagger profile
~, колебательный oscillatory/oscillating/oscillation/resonant/tuned circuit
~ наведения guidance loop
~, незамкнутый open loop
~, нерадиоактивный cold loop/circuit
~ обратной связи feedback loop
~, отсасывающий absorption circuit
~ охладителя coolant circuit
~, охлаждающий cooling system/loop
~, первый/первичный primary circuit/loop
~ путевого управления yaw control loop
~, радиоактивный hot loop
~, разомкнутый open loop
~ регулирования control loop
~, резонансный resonant circuit
~ рыскания yaw loop
~ с обратной связью feedback circuit/cycle
~, сглаживающий smoothing circuit
~ сервоуправления servo loop
~, силовой power circuit
~ сопровождения tracking loop
~, спусковой trigger circuit
~ тангажа pitch loop
~ теплоносителя heat-transfer circuit/loop, coolant circuit/loop
~ теплообменника heat-exchanger loop
~, топливный fuel circuit
~ управления control loop
~ управления креном roll/bank control loop
~, усилительный amplifier circuit
~, утилизационный waste-heat loop
~, энергетический power circuit
контурный outline *(attr.)*; contour *(attr.)*
~ная карта contour/outline map
конунг *hist.* sea king
конурбация conurbation
конус cone; *tech.* bevel, taper; *av.* wind sock
~, аккумуляционный *geol.* conical/volcanic pile
~, артериальный *anat.* arterial cone, pulmonary conus
~, буксируемый trailing cone
~, ветровой *av.* wind cone/sleeve
~, вихревой vortex cone
~ воздухозаборника intake cone
~ возмущений/волновой Mach/shock cone
~ впадин *mech.* dedundum cone
~, вулканический volcanic pile/dome
~ выступов *mech.* addendum cone
~, затупленный blunt(ed) cone
~, инструментальный machine tool taper
~, косой scalene/oblique cone
~, круговой circular cone
~, мертвый blind cone
~, мозговой *anat.* medullary/terminal cone
~ мотки *text.* winding gone
~ нарастания *bot.* growing point, vegetative cone
~, ограничивающий collimating cone
~, острый pointed/sharp cone
~ прошивки *met.* tapered piercing mandrel
~, прямой right cone
~ раскатки *met.* reeling mandrel
~ распыла spray cone
~, распыляющий atomizer cone
~ рассеяния diffusion cone
~, световой light cone; *med.* cone of light
~ со сферическим затуплением spherically blunted cone
~, ступенчатый *mech.* gear cone
~, стыковочный *space* docking cone; drogue
~, теневой shadow cone
~, тонкий slender cone
~ трения friction cone/taper
~, усеченный frustum of a cone, truncated cone
~ фотоаппарата, внутренний inner lens cone
~, центрирующий centering/centring cone
конусность conicity, taper, angle of taper, conical shape
конусный conical; tapered, bevel
конусовидный, конусообразный cone-shaped, coniform conoid(al); tapered, conical
конфабуляция *path.* confabulation
конфедерация confederation; confederacy
конфекцион 1. production of ready-made clothing; 2. store selling ready-made clothes
конфекция piecing/putting together, building-up
конфервоидный *bot.* confervoid, confervaceous
конференция conference
~ в верхах *polit.* summit conference
~ ООН по торговле и развитию United Nations Conference on Treade and Development, UNCTAD
~ по стандартам standards conference
конфета candy, sweet
конфетчик confectioner
конфигурационный configurational; protective
конфигурация configuration; pattern, shape, outline; conformation
~, базовая basic [base-line] configuration
~, битовая *comp.* bit configuration
~ выводов pin configuration
~, двухнуклонная two-nucleon configuration
~, заказная custom configuration
~ занятых уровней configuration of filled levels
~, исходная source configuration
~, максимальная full configuration
~, минимальная entry configuratuon

~, настольная desktop configuration
~, однопроцессорная *comp.* uniprocessor configuration
~ поля field pattern
~, плоская planar configuration
~ программных средств software configuration
~ с прямым крылом *aero.* unswept configuration
~, седлообразная saddle-point configuration
~, средняя mean configuration
~ со стреловидным крылом *aero.* swept-wing configuration
~ технических средств *comp.* hardware configuration
конфиденциальность privacy
конфинальный *math.* cofinal
конфирмант *eccl.* confirmand
конфирмационный confirmation *(attr.);* confirmmatory
конфискат confiscated property/goods
конфискация seizure; forfeiture, appropriation, confiscation
~ во время войны war seizure
конфисковать seize; confiscate
конфликт conflict, collision, clash; contention; *game th.* conflict situation
~ на шине bus conflict/contention
~ по данным data collision/contention
~ по совпадению обращений для записи *comp.* write-write conflict
~ по совпадению обращений для чтения и записи read-write conflict
~ при обращении к памяти memory collision/contention
~ при совмещении операций concurrency conflict
конфликтный conflict *(attr.);* contention *(attr.)*
~ная комиссия grievance committee; arbitration tribunal/board
конфлюентность *math.* confluence
конфорка (gas) burner, ring
конформность conformity
конформный conformal; orthophormic
конфузор contraction cone; constrictor; reducer
конфуцианец Confucianist
конфуцианский Confucian
конфуцианство Confucianism
конх conch, shell
конха *arch.* concha
конхит *miner.* conchite
конхиферовые *pl. zool.* conchifers *(Conchifera)*
конхоида *math.* conchoid
~ Никомеда Nicomedes' conchoid
конхопорфирин conchoporphyrin
конхоскоп *med.* conchoscope
конхотом *surg.* conchotome
конхотомия *surg.* conchotomy
конхэктомия *surg.* conchectomy
концевой terminal, final, end *(attr.)*
концентр concentric circle
~, промежуточный intermediate level of education
концентрат concentrated product; concentrate; *(food)* concentrated food/meal, food concentrate
~, высокопротеиновый high-protein [protein-rich] concentrate
~, гранулированный pelleted concentrate
~, грубый rough/coarse concentrate
~ для молочного скота *agr.* dairy concentrate
~, мелкий fine concentrate
~, низкопротеиновый low-protein concentrate
~, пенный froth concentrate
~, рудный ore concentrate
~, суммарный ultimate concentrate
концентратомер *food* concentration meter; consistency meter
концентратор данных *comp.* data concentrator
концентрационный concentration *(attr.);* concentrating
~ лагерь concentration camp
концентрация concentration; *(of solution)* strength; density
~, безопасная safe concentration
~, весовая weight concentration
~, весовая долевая weight-fraction concentration
~ весовая процентная percent-by-weight concentration
~, долевая объемная volume-part concentration
~, долевая по массе weight-part concentration
~, допустимая permissible concentration
~, исходная initial/input concentration
~ загрязнения waste/impurity load
~ ионов ion concentration
~ ксенона, равновесная xenon equilibrium
~, максимально допустимая maximum allowable concentration
~, мольная/молярная molar concentration
~, мольная долевая mole-fraction concentration
~, мольная процентная mole-percent concentration
~, мольно-объемная volumetric molar concentration
~ напряжений stress concentration
~ нейтронов neutron concentration
~, ничтожная [ничтожно малая] trace concentration
~, неравновесная non equilibrium concentration
~, объемная volume concentration
~, опасная dangerous/intolerable concentration
~ по массе mass concentration
~, пороговая threshold concentration
~, предельно допустимая maximum permissible concentration
~ примесей impurity concentration
~, процентная объемная volume percent

concentration
~, **процентная по массе** weignt percent concentration
~, **равновесная** equilibrium concentration
~ **раствора** strength of solution
~, **смертельная** lethal concentration
~, **удельная** specific concentration
концентрирование concentrating, concentration; massing
~ **методом ионного обмена** ion-exchange concentration
концентрировать concentrate; mass; *fig.* focus
концентрироваться become (more) concentrated; concentrate
концентрический, концентричный concentric
концентричность concentricity
концепция concept; idea
концеравнитель trimming saw
концерн *econ.* concern
концерт *mus.* 1. concert; *(individual)* recital; 2. concerto
~, **камерный** chamber concert
концертант performer *(musical)*
концертировать give concerts
концертмейстер *mus.* 1. leader; 2. solo performer
концертный concert *(attr.)*
~ **рояль** grand piano
концессионер concessionaire
концессионный concessionary, concessional; concessive; concession *(attr.)*
концессия *econ.* concession
~, **железнодорожная** railway concession
~ **на добычу минерального сырья** mineral concession
~ **на эксплуатацию шахт** mining concession
концкорм *abbr.* concentrated fodder
концлагерь *abbr.* concentration camp
концовка *print.* tailpiece
кончик tip, apicle, spicule; point
~, **заостренный** *bot.* acumen
~ **носа/языка** tip of the nose/tongue
концировать *food* conche
коншмашина, конш conching machine, conch
конъектура conjecture, surmise, assumption
конъектурный conjectural
конъюгант *biol.* conjugant
конъюгата *anat.* conjugate, anteroposterior diameter of the pelvic inlet
~, **акушерская** obstetric conjugate
~, **истинная/ложная** true/false conjugate
~, **наружная** external conjugate
конъюгаты *pl. bot.* conjugates *(Conjugatae)*
конъюгация *biol.* conjugation
~ **бактерий** bacterial conjugation
~, **глициновая** glycine conjugation
~ **хромосом** chromosomal conjugation
конъюнктив *gram.* conjunctive
конъюнктива *anat.* conjunctiva
~ **век** palpebral conjungtiva
~ **глазного яблока** bulbar conjunctiva
конъюнктивальный *anat.* conjunctival
конъюнктивит *med.* conjunctivitis
~, **аденовирусный** adenovirus conjunctivitis
~, **актиномикозный** actinomycotic conjunctivitis
~, **банный** bath conjunctivitis
~, **весенний** spring/vernal conjunctivitis
~, **гнойный** purulent conjunctivitis
~, **гонорейный** gonorrheal conjunctivitis
~, **гранулярный/гранулезный** granular conjunctivitis, granular lids, trachoma
~, **катаральный** catarrhal conjunctivitis
~, **крупозный** croupous conjunctivitis
~, **личиночный** larval conjunctivits
~ **новорожденных** infantile conjunctivitis
~, **обызвествляющийся** calcerous conjunctivitis
~, **пленчатый** membranous conjunctivitis
~, **пустулезный** pustular conjunctivitis
~, **туляремический** squirrel plague conjunctivitis
~, **фликтенулезный** phlyctenular conjunctivitis
конъюнктор *comp.* AND gate/circuit
конъюнктура situation, conditions, state of affairs; juncture
~, **будущая экономическая** business outlooks
~, **деловая** business conditions
~, **кредитная** credit conditions
~, **политическая** political situation
~, **хозяйственная** business/economic conditions
конъюнктурный determined by the particular situation; time-serving; *econ.* market-oriented
~**ные цены** free market prices
конъюнкция 1. *logic* conjunction; 2. *comp.* AND function/operation
конь horse; *(chess)* knight
~, **гимнастический** vaulting horse
коньки *pl.* skates
~ **на роликах** roller skates
конюга *zool.* auklet *(Aethia)*
конюшня stable
~**ни, Авгиевы** *myth.* Augean stables
кообраз co-image
кооператив cooperative society
~, **жилищно-строительный** building society; housing cooperative
~ **по содержанию скота на пастбище** grazing cooperative
~, **потребительский** consumer cooperative
~, **производственный** producers' cooperative
~, **сбытовой** trading/seller cooperative
~, **сельскохозяйственный кредитный** agricultural credit cooperative
~, **сельскохозяйственный сбытовой** farmer marketing cooperative
~, **снабженческий** purchasing cooperative
кооперативный cooperative
кооператор member of a cooperative society
кооперация cooperative societies/system; *see also* **кооператив**

кооперирование introduction (setting up) of cooperative societies
кооперировать organize in a cooperative
кооперироваться form a cooperative organization
кооптация, кооптирование cooptation, coopting
кооптировать co-opt
координата coordinate; *see also* **координаты**
~, входная input coordinate
~, выходная output coordinate
~ точки снимка image/plate coordinate
~, угловая angular coordinate
координатник traverse/traversing gear/probe
координатограф coordinatograph, plotter
~, автоматический plotting table
координатомер coordinatometer; coordinate protractor/scale
координатор coordinator; position indicator
~, звездный celestial coordinator
~, навигационный ground-position [air-position] indicator
~, солнечный sun-angle [sun-vector] sensor
~ цели target seeker, homing head
~ цели, инфракрасный infra-red target seeker
~ цели, радиолокационный radar target seeker
~ цели, тепловой infra-red target seeker
координаты *pl.* coordinates
~, географические geographic/terrestrial coordinates
~, декартовы Cartesian/orthogonal coordinates
~, земные terrestrial/earth coordinates
~, косоугольные oblique coordinates
~, криволинейные curvilinear coordinates
~, лагранжевы Lagrange/material coordinates
~ места корабля *naut.* elements of fix
~ места корабля, уравненные *naut.* adjusted position
~, начальные reference coordinates; coordinates of the origin
~, небесные celestial coordinates
~, ограниченные фазовые bounded phase coordinates
~, плоские plane/planimetric coordinates
~ по карте map reference
~ приземления landing coordinates
~, пространственные space coordinates
~, прямоугольные rectangular coordinates
~, сферические spherical coordinates
~, текущие current/running coordinates, current position
~ точки position/fix of a point
~, угловые angular coordinates
~, фазовые phase coordinates
~ цветности chromaticity coordinates
~ цели target position/coordinates
координация coordination
~, безусловно-рефлекторная *biol.* instinctive coordination
~ движений coordination of movements
~, недостаточная *med.* hyposynergy
координационный coordination *(attr.)*, coordinating
координирование coordination, coordinating
~ промера survey coordination, coordination of soundings
~ съемки survey coordination
координированность coordination
координировать coordinate
копайба copaiba (balsam)
копайский copaibic, copaivic
~ бальзам copaiba balsam
копайфера *bot.* copal tree, copaifera *(Copaifera)*
копал copal; copal varnish
копаловый лак copal varnish
копание digging, excavating
копатель digger, excavator
копать dig, excavate
копач digger, lifting plough
~, вильчатый digging prong
~, дисковый lifter wheel
~, лемешный share lifter
~, шнековый screw lifter
копеечник *bot.* sweet vetch *(Hedysarum)*
копейка kopeck
копейщик *hist.* lancer, lanceknight
копеподы *pl. zool.* copepods *(Copepoda)*
копер 1. pile driver; 2. impact testing machine; 3. *min.* headwork, headframe, pit head
~, бойный drop hammer; scrapbreaker
~, надшахтный pit/mine headframe, mine headgear
~, паровой steam pile driver
~, плавучий floating pile driver
~ с бабой drop-hammer pile driver
~, ударный impact pile driver
~, шахтный проходческий sinking headframe
коперниция *bot.* copernicia *(Copernicia)*
~, восконосная carnauba palm *(Copernicia cerifera)*
коперщик piledriver operator
копигольд *hist.* copyhold
копилка coin-box
копильник *met.* forehearth; receiver
копир template, former, master cam/form; tracer, contour follower
~, резьбовой threading block
копировальный copying, duplicating
~ная бумага carbon paper
~ная машина duplicator, copying machine
копирование 1. copying, diplicating, duplication; 2. tracing, profiling; 3. *mach.* template machining; 4. *phot.* printing; 5. imitation
~ ленты *comp.* tape duplication/replication
~ на кальку tracing
~, обводное *mech.* circumferential copying
~ по координатам *mech.* coordinate copying
копировать 1. copy, duplicate; 2. trace, profile; 3. *phot.* print; 4. imitate
копия copy, duplicate; replica
~, заверенная certified copy

~, **машинописная** carbon copy
~ **чертежа** tracing; blueprint
копка *agr.* digging; lifting
копланарность coplanarity, complanarity
копланарный coplanar, complanar
копна сена haycock
копнитель *agr.* harvester-stacker, shocker; straw collector/piler
копнить heap up, stack
копновоз *agr.* shock hauler, cocklifter
копр *zool.* dung beetle, tumble bug *(Copris)*
копра copra
копровой pile-driver *(attr.)*
~**вая баба** drop hammer; rammer
копролагния *med.* coprolagnia
копролалия meg. coprolalia
копрогенный coprogene. coprogenous, coprogenic
копролит *gegl.* coprolite, fossilized excrement
копропорфирин coproporphyrin
копростаз *med.* coprostasia, constipation
копростан coprostane
копростерин coprosterol, coprostanol, coprosterin
копрофаг *zool.* coprophage
копрофагия coprophagy
копрофилия coprophilia
копрофильный coprophilic
копрофит *bot.* coprophyte
копт copt
коптилный smoking *(attr.),* smoke-curing, fumatory
~ **завод** smoke-curing factory
коптильня smokehouse, fumatory
коптить cure in smoke, smoke-cure, smoke
коптский Coptic
копула 1. *gram.* copula; 2. *mus.* coupler
копулировать 1. copulate; 2. *hort.* graft by branching, whipgraft
копулятивный copulative
копуляция 1. copulation; 2. *mus.* coupler
копчение smoke-curing, curing in smoke, smoking
~, **холодное** cold smoking
копченый smoke cured/-dried, smoked
копчик *anat.* coccygeal bone, coccyx
копчиковый *anat.* coccygeal
копыл *naut.* (launching) poppet
~, **кормовой** after poppet
~, **носовой** forepoppet
копылок joggled wedge
копытень *bot.* wild ginger *(Asarum)*
копытные *pl. zool.* ungulates *(Ungulata)*
копытный 1. hoofed, ungulate; 2. hoof *(attr.)*
копытоходящий *zool.* unguligrade
копытчатый 1. hoofed, ungulate; 2. hoof-shaped
копь pit, mine
копье lance, spear; *sport* javelin
копьевидный lanceolate, spear-/lancet-shaped
копьеметатель *sport* javelin thrower
копьеносец *hist.* lance bearer, lancer
кора 1. crust; 2. *bot.* bark; rind; 3. *anat.* cortex
~, **дубильная** tanbark
~, **земная** earth's crust
~ **мозга** cerebral cortex
~ **мозжечка** cerebellar cortex
~ **мыльного дерева** soapbark, quillaia bark
~, **мыльная** *bot.* soapbark tree *(Quillaia saponaria)*
~ **надпочечника** adrenal cortex
~ **почки** renal cortex
~ **хрусталика** cortex of lens
корабельный shipborne, shipbased; ship *(attr.)*
~ **инженер** naval architect
~**ная авиация** shipborne aircraft
кораблевождение (marine) navigation
~ **подводных лодок** submarine navigation
кораблекрушение shipwreck
~, **терпеть** be shipwrecked
кораблестроение shipbuilding, ship construction
~ **военное** naval construction
кораблестроитель shipbuilder, ship constructor
кораблик 1. small/toy boat; 2. *zool.* nautilus *(Nautilus)*
корабль ship, vessel; *see also* **судно**
~, **атомный** nuclear-powered ship
~, **буксируемый** tugee, towed ship
~, **военный** warship, war/naval vessel
~, **воздушный** airliner; airship
~, **гладкопалубный** flush deck ship, flushdecker
~, **десантный** landing ship/vessel/craft
~, **дозорный** lookout ship
~, **концевой** rear ship
~, **космический** spaceship, spacecraft, space vehicle
~ **ледовой разведки** ice patrol ship
~, **линейный** battleship
~, **многоцелевой** multimission ship
~, **мореходный** seaworthy/seagoing ship, heavy-weather vessel
~ **на воздушной подушке** hovership, air-cushion ship, ground-/surface-effect ship
~ **на подводных крыльях** hydrofoil ship
~, **надводный** surface ship
~ **общего назначения** general purpose ship
~, **однотипный** twin/sister ship
~, **орбитальный** orbital spaceship
~ **охранения** escort/screening vessel; convoying vessel
~, **патрульно-эскортный** patrol escort (ship)
~ **первой линии** first-line ship
~, **подводный** underwater/undersurface/undersea ship/vessel
~, **поисково-спасательный** search-and-rescue ship
~, **поисковый** scout ship/vessel
~ **пополнения запасов** replenishment ship
~, **разведывательный** reconnaissance ship
~, **ракетный** missile ship
~, **речной** riverine craft
~ **с большой дальностью плавания** long-range vessel
~ **с малой дальностью плавания** short-range

ship
~ **связи** communication ship
~, **связной** linking ship
~, **сторожевой** destroyer escort
~, **торпедный** torpedo craft/vessel
~, **транспортный космический** transport/delivery spacecraft
~, **ударный** attack ship
~ **управления** command ship
~, **учебный** training/school ship
~, **флагманский** flagship, flag officer's command ship
~ **флота, флагманский** fleet flagship
~, **штабной** headquarters/staff/command ship
~, **шхерный** sound ship
корабль-вертолетоносец helicopter ship
коразмерность codimension
коразол *pharm.* corasole
коракан *bot.* ragi millet *(Eleusine coracana)*
коракоид, коракоидный coracoid
коралл coral
~, **рифообразующий** reef-building coral
кораллин *chem.* corallin, rosolic acid
~, **красный** red corallin, peony red, peonin
кораллина nullipore, coralline alga
коралловидный coralliform, coralloid, corallike
коралловый coral *(attr.)*, coralline; coral-red
Коран the Koran
корвет *naut.* corvette
корд cord; *text.* ribbed woollen fabric
~, **бедфордский** Bedford cord
~, **вискозный** viscose/rayon cord
~, **стальной** steel-wire cord
~, **шинный** tyre cord
коргликон corglycon
кордаитовый *paleont.* cordaitean
кордаиты *pl. paleont.* cordaites, cordaiteans *(Cordaitales)*
кордебалет corps de ballet
кордельер *hist.* Cordelier; Franciscan
кордиамин cordiamine
кордилина *bot.* dracena *(Cordyline)*
кордилит *miner.* cordylite
кордит *mil.* cordite
кордия *bot.* cordia *(Cordia)*
кордон cordon
корд-шнур cord; rope
коревидный *med.* morbilliform
коревой *med.* measly, morbillous
кореец, корейский Korean
коренизация *polit.* indigenization
коренной 1. radical, basic, fundamental; 2. indigenous, aboriginal, native
~ **житель** native, indigene, aboriginal/indigenous inhabitant
~ **зуб** molar
корень root; *math.* root; radical; radix
~, **арифметический** positive root
~, **воздушный** aerial root
~, **двойной** double root
~, **зародышевый** seed root
~ **из единицы, первообразный** primitive root of unity
~, **извлекать** *math.* extract/take the root
~, **ирный** *bot.* sedge root, sweet flag *(Acorus calamus)*
~, **квадратный/квадратичный** *math.* square root
~, **кратный** *math.* multiple/repeated root
~, **кубический** cube root, third root
~, **лишний** *math.* extraneous root
~ **многочлена** root/zero of a polynomial
~, **находить** *math.* find/evaluate the root
~, **неразветвленный** *bot.* simple root
~, **первичный** primordial root
~, **первообразный** primitive root
~, **придаточный** adventitious root
~, **репчатый** napiform root
~, **сопряженный** *math.* conjugate root
~ **третьей степени** cubic/cube root
~, **тройный** *math.* triple root
~ **уравнения** root of an equation
кореометр coreometer, corometer, pupillometer
кореопсис *bot.* coreopsis *(Coreopsis)*
кореподобный *med.* morbilliform
корешок 1. radicle, rootlet, root, radix; 2. counterfoil, stub; *(of book)* back
~, **брюшной** *anat.* ventral root
~, **грудной** thoracic root
~, **задний** posterior root
~ **зрительного тракта** root of the optic tract
~, **крестцовый** sacral root
~, **нервный** nervous root
~, **нижний** inferior root
~, **передний** anterior root
~, **пояснично-крестцовый** lumbosacral root
~, **поясничный** lumbar root
~, **спинной** dorsal root
~, **чувствительный** sensory root
~, **шейный** cervical root
кореянка Korean (woman)
корзина basket
~ **аэростата** balloon basket
~ **с откидным днищем** dump-bottom basket
~, **спасательная** rescue basket
корзинка 1. small basket; 2. *bot.* calathide, calathium
корзиночка *zool.* corbicula
кориандр *bot.* coriander *(Coriandrum)*
кориария *bot.* coriaria *(Coriaria)*
корибант *myth.* Corybant
коридалин *chem.* corydalin
коридор corridor, passage, alley; *naut.* alleyway
~, **воздушный** air corridor
~ **входа в атмосферу** *space* re-entry corridor
~ **выведения** injection corridor
~ **гребного вала** *naut.* shaft alley
~ **набора высоты** climb corridor
~, **пусковой** launch corridor
~ **снижения** *aero.* entry/re-entry corridor

корилопсис *bot.* winter hazel *(Corylopsis)*
коринка currants, corinths
коринфийский, коринфянин Corinthian
кориумкарцинома coriumcarcinoma
корифа *bot.* corypha (Corypha)
корифей coryphaeus
корифена *zool.* dorado *(Coryphaena hippurus)*
корица cinnamon
коричневорубашник *pol., hist.* Brown Shirt, Nazi
коричнезем cinnamonic soil, korichnezem
коричник *bot.* cinnamon (Cinnamomum)
коричный cinnamon *(attr.),* cinnamonic, *chem.* cinnamic
~ное масло cinnamon oil
корка crust, scale; *met.* scum; skin
~, апельсинная orange peel
~ выветривания weathering rind
~, защитная protective/desert varnish
~, мерзлая frozen crust
~, спутанноволокнистая matted fibrous crust
~ хлеба bread crust
корковидный barklike, crustate, corticate; crustaceous
корково-корковый corticocortical
корково-подкоровый corticosubcortical
корково-рубральный corticorubral
корково-таламический corticothalamic
корковый crustal, crustose, crusted; crustaceous *anat.* cortical
~ слой bast/cortical layer
корконосный *bot.* corticiferous
коркообразование crustification, crust-formation
корм fodder; feed; forage
~, гранулированный/брикетированный pelletized feed
~, грубый coarse fodder, roughage
~, жидкий feeding slop
~, комбинированный combined fodder/feed, formula feed
~, подножный grazing fodder, pasturing grass
~, силосный ensilage fodder
~, сочный succulent feed/fodder/food
корма *naut.* stern
~, бульбообразная bulb(ous) stern
~, крейсерская cruiser stern
~, круглая round/apple stern
~, ложкообразная spoon stern
~, транцевая transom stern
~, туннельная tunnel (arch-type) stern
кормление feeding, alimentation
~ вволю feeding to appetite
~, исскуственное artificial feeding
~, мнимое sham/fictitious feeding
~, обильное liberal feeding
~, парентеральное parenteral feeding
~, принудительное forced feeding
~, смешанное compound feeding
~ через желудочный зонд gavage, suralimentation, superalimentation
кормовой I 1. stern *(attr.);* aft, after *(attr.);* 2. *n.* stern rope
~, задержать check the stern rope
~, крайний sternmost
~, отдать let go aft
~ флаг colo(u)r, ensign
~вое весло scull
кормовой II fodder *(attr.);* forage *(attr.)*
~вая свекла mangel-wurzel, stock/fodder beet
~вые культуры fodder/forage crops
кормодробилка fodder crusher/grinder
кормозагрузчик feed loader
кормозапарник feed/fodder steamer, feed steaming plant/installation
кормоизмельчитель fodder/feed chopper, feed mincer
кормокухня feed-preparation kitchen/house
кормомешалка fodder/feed mixer
кормораздатчик fodder/feed distributor; mechanized feeder
~, тросовый cable-type feeder
~, шнековый auger/screw feeder
кормрезка fodder/feed chopper chaffcutter
кормофилогения *bot.* cormophylogeny
кормофит *bot.* cormophyte
кормоцех food preparation shop/section
кормчая (книга) nomocanon
корнеальный corneal
корневидный rootlike, radiciform; rhizomorphous
корневище *bot.* rhizome; rootstock
~ мужского папоротника aspidium, filix mas, male fern
корневищный *bot.* rhizomatous
~ отпрыск stem sucker
корневой root *(attr.)*
~ое влагалище coleorhiza
корнеголовые *pl. zool.* rhizocephalans *(Rhizocephala)*
корнеед *zool.* root borer
корнеизмельчитель root cutter/shredder
корнеискатель *math., comp.* root solver/finder
корнеклубнеплод tuberous root
корнелистный *bot.* root-leaved, rhizophyllous
корненожки *pl. zool.* rhizopods *(Rhizopoda)*
корнеобразующий root-forming, rhizogenous
корнеплодный root *(attr.);* rhisocarpous
~ные культуры root crops
корнерасположение root arrangement, rhizotaxis
корнеродный *bot.* rhizogenic, rhizogenous
корнесеменный root-seeded, rhizospermous
корнесистема root sysem
корнет *mus.* cornet
корнецветковый *bot.* radiciflorous
корниловец *hist.* Kornilovets
корниловщина *hist.* Kornilov's counterrevolutionary movement
корнишон gherkin
корнуоллский Cornish
корнцанг dressing forceps, packer
~, изогнутый curved packer

короб 1. duct; 2. box, chest, basket; pot; bucket
~ **грохота** screen box
~, **загрузочный** *met.* charging box
~, **напорный** delivery duct
~, **отсосный** suction duct
коробиться buckle; warp; pucker
коробка box, case; chest; casing, housing
~, **анероидная** evacuated capsule, aneroid/diaphragm capsule
~ **выводов** *el.* terminal box
~, **дверная** door frame/casing
~ **для кинопленки** film can
~, **кабельная** cable/splice/junction box
~, **кингстонная** sea chest
~, **клапанная** 1. valve box/chest; 2. steam chest; nozzle box/chamber
~, **клемная** *el.* terminal box
~, **консервная** can, tin
~, **манометрическая** manometric/barometric capsule
~, **мембранная** bellows assembly, sylphon
~, **набивочная** packing/stuffing box
~, **огневая** fire-box
~, **оконная** window casing/frame
~, **паровая** steam chest
~, **парораспределительная** steam box
~ **передач** gearbox
~, **распределительная** distribution/junction box
~, **сильфонная** bellows element/box
~ **скоростей** (speed) gearbox
~, **соединительная** junction/connection box
~, **сопловая** nozzle chamber/box
~ **спичек** box of matches
~, **топочная** firebox
~, **челночная** *text.* shuttle box
~, **штепсельная** plug box
коробление buckling; warping, warpage, cupping; deformation, distortion
~, **поперечное** (transverse) warping; kink
~, **продольное** twisting, bowing; buckle
коробоватость warpage; camber
коробоватый warped, buckled
коробок 1. small box; 2. small basket
~ **спичек** box of matches
коробочка 1. small box/chest/case/chamber; 2. *bot.* pod, capsile, boll
~, **растрескивающаяся** elater
~, **семенная** seed capsule/vessel
коробочковидный capsular
коробочконосный *bot.* capsuliferous
коробчатый boxlike, box-shaped; capsular
корова cow
~, **дойная** milch cow, milker
~, **недойная** dry cow; cow not in milk
~, **стельная** cow with calf, calver
коровий cow *(attr.)*; bovine
~**вье масло** butter
коровка 1. small cow; 2. *zool.* ladybird *(Coccinella)*; 3. *bot.* edible boletus, cepe *(Boletus edulis)*
коровник cow shed/barn, cowhouse
коровой cortical, crustal
коровяк *bot.* 1. mullein *(Verbascum)*; 2. bovinus mushroom *(Boletus bovinus)*
короед *zool.* bark beetle
короед-древесинник *zool.* timber beetle *(Xyleborus)*
королева queen
~, **вдовствующая** queen dowager
королевич king's son
королевна king's daughter
королевский royal, kingly, regal, king's, queen's; basilic(al)
~ **слон** *chess* king's bishop
~**кая кобра** *zool.* king cobra *(Ophoiphagus)*
королевство 1. kingdom; realm; 2. royal authority
королек 1. *zool.* kinglet *(Regulus)*; 2. blood orange; 3. petty king, kinglet; 4. assay button, bullion bead; 5. *met.* shot
король king
~, **сельдяной** *zool.* oarfish *(Regalecus glesne)*
~ **шлюза** pointing sill
коромысло 1. rocker/rocking arm, balance beam, rocker, rocking shaft; 2. *zool.* blue darner *(Aeshna)*
~ **весов** weigh beam
~ **клапана** valve rocker
корона crown; *el.* corona discharge, corona
~, **дворянская** coronet
~ **звезды** stellar corona
~, **земная** terrestrial corona, geocorona
~, **солнечная** solar corona
коронарит *med.* coronaritis
~, **ревматический** rheumatic coronaritis
коронарный *med.* coronary
коронарография *med.* coronarography
коронаросклероз *med.* coronary sclerosis, coronarosclerosis
~, **стенозирующий** stenosal coronarosclerosis
коронаспазм coronary spasm
коронатромбоз coronary thrombosis
коронация coronation, crowning
короний coronium
коронирование *el.* corona discharge
коронировать *el.* show a corona discharge
коронка 1. (small) crown; 2. *anat.* crown; 3. drilling/crown bit; 4. *bot.* coronet, coronule
~, **алмазная** diamond drill bit
~, **бескерновая** full-hole drill bit
~ **для выбуривания керна** coring drill bit
~, **звездчатая** star/rose drill bit
~ **зуба** crown of a tooth
~, **зубильная** chisel drill bit
~, **крестовая** cross drill bit
~, **перфораторная** hammer drill bit
~, **твердосплавная** hard-alloy drill bit
коронковидный, короновидный crown-shaped, coroniform; coronal
коронограф *astr.* coronograph
короностойкий *el.* corona-proof

короностойкость *el.* corona resistance
коронтин prenylamine
корообдирка 1. debarker, barker; 2. debarking, bark-stripping
короставник *bot.* scabious *(Scabiosa)*
коростелек *zool.* marsh crake *(Porzana pusilla)*
коростель *zool.* crake *(Crex)*
короткий short; brief
коротокобороздный *bot, zool.* brevicolpate
коротковолновый short-wave *(attr.)*
коротковолосистый *bot., zool.* stupulose
коротковолосый short-haired
короткоголовость *anat.* brachycephaly, brachycephalism
короткогребневой short-crested
короткогривистый brachychaetous
короткодействующий short-range *(attr.)*
короткоживущий short-lived
короткозамкнутый short-circuited
короткозамыкатель short-circuiting switch/plug/ device
коротноклювый brevirostrate
короткоколосистый *bot.* short-eared
короткокрылый brachypterous, brevialate, shortwinged
коротконогий short-egged/-footed, breviped
коротконожка *bot.* false brome *(Brachypodium)*
короткоостистый *bot.* short-awned, breviaristate
короткопалость *med.* brachydactilia
короткопалый brachydactilic
короткофокусный short-focus *(attr.)*
короткохвост *zool.* 1. uacari *(Cacajao)*; 2. stump-tailed lizard *(Trachysaurus rugosus)*
короткоходный *mech.* short-stroke *(attr.)*
корофилия *med.* corophilia
корпорант member of a student fraternity *(in Germany)*
корпоративность corporationism; corporational nature/spirit
корпоративный corporative; corporational
корпорационный collegiate
корпорация corporation
~, акционерная stock corporation
~, государственная public corporation, state-owned/run corporation
~, дочерняя subsidiary corporation
~, международная/многонациональная multinational corporation; multinational
~, многопрофильная conglomerate
~, муниципальная public corporation
~, некоммерческая non-profit corporation
~, отечественная domestic corporation
~, подконтрольная subsidiary corporation
~, торговая trading corporation
~, транснациональная multinational corporation
~, частная private [privately-owned] corporation
корпус 1. body; torso; corpus; 2. *tech.* body, carcass, frame, hull, housing, case, casing; shell; 3. *el.* chassis ground/earth; 4. *print.* long primer; 5. building, block; 6. *mil., dipl.* corps
~, авиационный air corps
~, армейский army corps
~ вентиля valve body
~ вентилятора fan case/casing
~, внешний external case/casing/housing
~, внутренний centre case/casing; *naut.* inner hull
~, герметизированный pressurized housing/ body
~ гидросамолета seeplane hull
~ глушителя *auto.* muffler/silencer body
~, голый *naut.* bare/naked hull
~ двигателя engine case/housing
~, дипломатический diplomatic corps/body
~, кадетский military school
~ камеры сгорания combustion-chamber casing
~ клапана valve body
~ компаса compass bowl
~ компрессора compressor casing
~ котла boiler shell/casing
~, морской naval college
~, наружный outer casing
~ насоса pump casing/case
~ отопительного радиатора radiator box
~ печи furnace body/housing
~ подшипника bearing housing/case
~ поршня piston barrel
~ прибора instrument case
~, прочный pressure vessel; *naut.* pressure/ strength hull
~ ракеты rocket body; rocket airframe
~ реактора reactor vessel; reactor shell
~ редуктора reduction-gear housing/casing
~ судна ship hull
~ турбины turbine cylinder/casing
~ фотокамеры camera body
~ челнока *text.* shuttle body
~ штепселя plug sleeve
корпускула corpuscle
корпускулярный corpuscular
корпусной corps *(attr.)*
корпусный frame *(attr.)*, body *(attr.)*, hull *(attr.)*, housing *(attr.)*
корразия *geol.* corrasion
~, ветровая wind corrasion/carving
~, ледниковая glacial corrasion
~, речная river/fluvial corrasion
коррегирование correction, correcting
коррегировать correct; read proofs
корректив correction; amendment
коррективный remedial
корректирование, корректировка correction, adjustment; readjustment; rectification;
~ цен price adjustment
~ дальности adjustment/correction in range
корректировать correct, rectify; adjust; *(of text)* proofread; *tech.* update
корректор 1. proofreader; 2. *tech.* corrector, adjuster

~ дальности и высоты range and altitude corrector
~ нуля zero adjuster
~, фазовый phase compensator
~, частотный frequency compensator
корректура 1. *print.* proof sheet, proof; 2. proofreading; 3. *mil.* correction, adjustment
~ в гранках galley proof
~ в листах page proofs, proof in sheets
~, подписная foundry proof
~, цветоделительная colo(u)r correction
корректурный оттиск proof sheet, proof
коррекция correction; remedy; adjustment; compensation; updating; slave; erection
~, амплитудная *com.* amplitude equalization
~, высотная altitude correction; *naut.* altitude compensation
~ гирокомпаса, инерциальная *naut.* inertial slaving
~ гироскопа gyro erection/monitoring/slaving
~ гравитационная gravity-controlled erection
~ движения motion correction
~ на маршевом участке mid-course correction
~ орбиты orbit(al) correction
~ трапецеидального искажения *telev.* keystone correction
~, хроматическая colo(u)r correction
коррелирование correlation
коррелированность correlation
коррелировать correlate
коррелятивность correlativity; correlation
коррелятор, коррелометр correlator
корреляционный correlational; correlative
корреляция correlation
~, взаимная cross correlation
~, внутригрупповая intraclass correlation
~, временная time correlation
~, выборочная sample correlation
~, геномная genomic correlation
~, ложная spurious correlation
~, межгрупповая interclass correlation
~, множественная multiple correlation
~, обратная invert correlation
~, порядковая/ранговая rank correlation
~, совместная joint correlation
~, упорядоченная ordered correlation
~, частная partial correlation
корреспондент 1. correspondent; 2. *rad.* distant station
корреспонденция 1. correspondence; mail; 2. (correspondent's) report
~, заказная registered mail
~, коммерческая business correspondence
корригиола *bot.* corrigiole, knot-grass *(Corrigiola)*
корригировать = корректировать
корригум *zool.* korrigum *(Damaliscus corrigum)*
коррида corrida; bullfight
корродирование corroding; corrosion
корродировать corrode
коррозиеустойчивость corrosion resistance
коррозиестойкий, коррозиеустойчивый, коррозионностойкий, коррозийностойкий corrosion-resisting; noncorroding, corrosion-proof; rustproof
коррозионный corrosion *(attr.);* corrosive
коррозия corrosion; rust; attack
~, аэробная aerobic corrosion
~, анодная anodic/anode corrosion
~, внутрикристаллинная intragranular corrosion
~ водородная hydrogen corrosion/attack
~, газовая gas/gaseous corrosion
~, защитная sacrificial corrosion
~, избирательная selective corrosion
~, интеркристаллитная intercrystalline/intergranular corrosion
~, катодная cathodic corrosion
~, кислородная oxygen corrosion
~, кислотная acid corrosion/attack
~, контактная contact/galvanic (two-metal) corrosion
~, кристаллитная crystalline corrosion
~, линейная line corrosion
~, межкристаллитная intercrystalline/intergranular corrosion
~, местная local/localized/isolated corrosion
~, морская seawater corrosion
~, неравномерная 1. non-uniform corrosion; 2. nonequal-rate corrosion
~, нитевидная thread-line corrosion
~, оспенная tubercular corrosion
~ под напряжением stress corrosion
~, пористая pinhole corrosion
~, послойная layer corrosion
~ пятнами spot corrosion
~, радиационная radiation-initiated corrosion
~, сернистая sulphur attack
~, сквозная penetration/through corrosion
~ солевая salt attack
~, точечная point corrosion
~, транскристаллитная transcrystalline corrosion
~, ударная impingement attack
~, усталостная fatigue corrosion
~, щелевая crevice corrosion
~, электрохимическая electrochemical/galvanic corrosion
~, язвенная pitting
корсаж bodice
корсак *zool.* corsac fox *(Vulpes corsac)*
корсар corsair, privateer; pirate
корсарство piracy; privateering
корсет corset, stays; *med.* jacket, surgical bandage
~, гипсовый plaster-of-Paris jacket
корсиканец, корсиканский Corsican
корсиканка Corsican (woman)
кортесы *pl. polit.* Cortes
кортизол hydrocortisone, corfet, cortril, cortifan, hydrocortone, hycortole

кортизон cortisone
кортик dirk
кортикопетальный corticopetal
кортикостероид corticosteroid
кортикостерон corticosterone
кортико-таламический cortico-thalamic
кортикотропин corticotropin, corticotrophin, adenocorticotrophic hormone, ACTH
кортикотропный corticotrophic
кортикофугальный corticofugal
кортин *biochem.* cortin
кортол *biochem.* cortol
кортолон *biochem.* cortolone
кортуза *bot.* bear's-ear sanicle *(Cortusa)*
корунд *miner.* corundum
~, бесцветный white sapphire
~, желтый oriental topaz
~, зеленый oriental emerald
~, искусственный manufactured corrundum
~, красный oriental/true ruby
~, пурпуровый oriental amethyst
~, синий sapphire
корчевалка stump puller
корчевальный uprooting, root-extracting
корчевание rooting out, stubbing, grubbing
корчеватель grubber, uprooter; stumper, stump-puller
корчеватель-бульдозер tree-doser
корчевать uproot, root/stump out, stub/grub up
коршун *Zool.* kite *(Milvus, Elanus)*
корыто trough; tray, pan; *geog.* inverted saddle
~ для стружки chip pan
корытообразный trough-shaped, U-shaped
корь *med.* measles
~, геморрагическая h(a)emorrhagic/black measles
корье *coll.* bark (stripped from trees)
корьедробилка bark mill/shredder/crusher
корьеломка bark breaker
корьерезка bark-cutter/-chopper
корюшка *zool.* smelt *(Osmerus)*
корюшковые *pl. zool.* smelts *(Osmeridae)*
коса I scythe
коса II *geog.* (shallow) spit, (sand) bar, tongue, bay-bar, shoal head
~, береговая spit of land
~ в заливе bay bar
~, крылатая winged headland, hammerhead spit
~, кулисообразная overlapping/echelon spit
~, морская off-shore bar, outer bar
~, намывная spit
~, островная connecting bar
~, отчлененная flying bar
~, прибрежная bay-head beach
~, серповидная cuspate spit
косатка *zool.* killer whale, orca *(Orcinus orca)*
косатковый *pl. zool. Bagridae*
косач *zool.* blackcock *(Lyrurus tetrix)*
косвенно-возвратный indirect-reflexive
косвенно-дополнительный indirect-complimentary
косвенно-объектный indirect-object
косвенность obliqueness, obliquity, slant
косвенный indirect, oblique
~ падеж *gram.* oblique case
~ные улики *law* circumstantial evidence
косеканс *math.* cosecant, cosec, csc
~, гиперболический hyperbolic cosecant
косекансоида cosecant curve
косеть *top.* conet
косилка mower, mowing machine
~, навесная mounted mower
~, прицепная trailed mower
~, самоходная self-propelled mower
косилка-силорезка fodder/forage harvester
косинус *math.* cosine, cos
~ амплитуды amplitude cosine
~, интегральный cosine integral, integral cosine
~ направления *geod.* direction cosine, cosine of direction
~, обращенный versed cosine
~ преобразование Фурье Fourier cosine transform
косинус-версус versed cosine
косинус фи *el.* power factor
косицы флага fly of a flag, swallow tail
косметик cosmetician
косметика 1. *coll.* cosmetics; 2. cosmetology; beauty treatment
косметический cosmetic
~ кабинет beauty shop
косметичка cosmetiste, beautician
косметолог 1. cosmetologist; 2. cosmetician, beautician
косметология cosmetology
космический space *(attr.),* cosmic; outer-space
~ аппарат (летательный) spacecraft
~ корабль spaceship, space vehicle, spacecraft
~кое пространство (outer) space
космичность cosmicality
космовидение space television
космогенез cosmogenesis
космогеология cosmic geology
космогония cosmogony
~, звездная stellar cosmogony
космодром space-launch site
космолетчик space pilot
космолог cosmologist
космология cosmology
космонавигация astronautics
космонавт astronaut, cosmonaut; spaceman
космонавтика astronautics, cosmonautics, interplanetary navigation
космонавтка woman astronaut
космоплавание space navigation/flight/travel
космополит cosmopolite, cosmopolitan
космополитизм cosmopolitanism, cosmopolitism
космополитический, космополитичный cosmopolitan, cosmopolitic
космос I (outer) space, cosmos

~, **ближний** adjacent/near space
~, **дальний** far-out space, far/deep space
космос II *bot.* cosmos *(Cosmos)*
космотеизм cosmotheism
космотетический *philos.* cosmothetic
космофизик space physicist
космохимия space chemistry
косноязычие *med.* ankyloglossia, tongue-tie
косоворотка type of Russian shirt
косоглазие *med.* strabismus, heterotropia, squint
~, **альтернирующее** alternating squint
~, **мнимое** sham strabismus
~, **монолатеральное** unilateral squint
~, **моноокулярное** uniocular strabismus
~, **неаккомодационное** nonaccomodative squint
~, **расходящееся** divergent/external squint
~, **скрытое** latent squint, heterophoria
~, **содружественное** concomitant squint
~, **сходящееся** convergent/internal squint
косоглазый squint-/cross-eyed, cockeyed
косоголовость plagiocephaly, plagiocephalism
косогорник *bot.* rattlesnake root *(Prenanthes)*
косой oblique, skew, slanting, slant, inclined, sloping
~ **парус** *naut.* fore-and-aft sail
~ **почерк** sloping handwriting
~ **угол** *math.* oblique angle
косолапость *med.* talipes, clubfoot, cyllosis, pes contortus
~, **внутренняя** pes varus
~, **наружная** pes valgus
~, **полиомиелитная** talipes poliomyeliticus
косолапый taliped, club-footed
косолистный *bot.* plagiophyllous
косоплодный *bot.* loxocarpous
косорукость *med.* talipomanus, club hand
косослоистый *geol.* obliquely laminated; crossbedded
косослой 1. *(wood)* cross/diagonal grain; 2. oblique bed
косоугольник rhomb(us)
косоугольный oblique-angled, bevel
косоур *constr.* bridgeboard, string(er), stair carriage
косоустьицевый *bot.* plagiostomatous
костедержатель *med.* bone-holding forceps
~, **двухдорожковый** two-pronged bone forceps
костедробилка bone crusher/grinder/mill
костенеть ossify; grow stiff
костенец *bot.* mouse-ear *(Holosteum)*
костеобразование osteogenesis
~, **несовершенное** disosteogenesis, achondroplasia
~, **энхондральное** enchondral/endochondral osteogenesis
костеобразующий osteogenic, osteogenetic, osteoplastic; ossific, ossifying, bone-forming
костеподобный osteoid; bone-like
костер I woodfire; pyre; *geol.* cog, chock
костер II *bot.* bromegrass, cheat *(Bromus)*
костеязычные *pl. zool.* osteoglossids *(Osteoglossidae)*
кости *pl. see* **кость**
костистые *pl. zool.* teleosts *(Teleostei)*
костно-кожно-мышечный osteodermomuscular, osseodermomuscular
костномозговой medullar(y)
костно-мышечный osteomuscular, osseomuscular
костнопластический osteoplastic
костнопузырные *pl. zool.* ostariophysans *(Ostariophysi)*
костно-суставной osteoarticular
костнотуберкулезный osteotuberculous
костно-фиброзный osteofibrous
костно-хрящевой osteocartilaginous, osseocartilaginous, osteochondrous
костнощекие *pl. zool.* *Cottoidei*
костнощитковые *pl. zool.* osteostracans *(Osteostraci)*
костный osseous; osteal, bone *(attr.)*, bony
~ **клей** bone glue
~ **мозг** (bone) marrow
костоеда *med.* caries
~, **вторичная** secondary/backward caries
~, **зубная** dental caries
~, **межзубная** interdental caries
~, **первичная** primary caries
~ **позвоночника** spinal caries, tuberculosis of vertebra
костоедный *med.* carious
костолом *bot.* asphodel *(Narthecium)*
костоломный *med.* osteoalgic
костостернопластика *med.* costosternoplasty
костотомия *med.* costotomy
косточка ossicle; kernel, stone; baleen, whalebone
~, **слуховая** auditory ossicle
косточковый *bot.* drupaceous
~ **плод** stone fruit
костра shive, shave, boon, chaff
костус *bot.* spiral flag *(Costus)*
костыль 1. spike/dog nail; *aero.* skid; 2. crutch
~ **лопасти** *naut.* float lever
~, **рельсовый (винтовой)** sleeper bolt
~, **хвостовой** tail skid
кость bone
~, **барабанная** tympanic bone
~, **бедренная** femur, thighbone
~, **безымянная** innominate bone
~, **боковая клиновидная** lateral cuneiform bone
~, **большеберцовая** tibia, shinbone
~, **венечная** coronary bone
~, **верхнечелюстная** maxilla, upper jawbone, superior/upper maxilla
~, **височная** temporal bone
~, **воздухоносная** pneumatic bone
~, **головчатая** capitate bone
~, **грудинная** sternum, breastbone
~, **заднеглазничная** postorbital bone
~, **затылочная** occipital bone
~, **игральная** die *(pl.* dice)

~, **клиновидная** sphenoid bone; cuneiform bone
~, **копчиковая** coccygeal bone
~, **крестцовая** sacrum, sacred bone
~, **крыловидная** pterygoid bone
~, **крючковатая** hooked bone
~, **кубовидная** cuboid bone
~, **лобковая** pubic bone
~, **лобная** frontal bone
~, **локтевая** ulna, elbow bone
~, **лучевая** radius, spoke bone
~, **малоберцовая** fibula, calf-bone, peroneal bone
~, **мамонтова** *geol.* fossil ivory
~, **межтеменная** interparietal bone
~, **надгрудная** episternal bone
~, **надзатылочная** supraoccipital bone
~, **небная** palatine bone
~, **непарная** unpaired bone
~, **нижнечелюстная** mandible, inferior/lower maxilla
~, **носовая** nasal bone
~, **парная** twin bone
~, **переднеушная** prootic bone
~, **пластинчатая** splenial bone
~, **плечевая** humerus
~ **плюсны** metatarsal bone
~, **подвздошная** iliac bone, flank-bone
~, **подъязычная** hyoid bone
~, **полулунная** semilunar/lunate bone
~ **предплюсны** tarsal bone
~ **предплюсны, ладьевидная** navicular bone
~, **пястная** metacarpal bone
~, **пятнистая** spotted bone
~, **пяточная** calcaneus, calcaneum, heel bone
~, **резцовая** incisive bone
~, **решетчатая** ethmoid bone
~, **рыбья** fishbone
~, **седалищная** ischial bone
~, **сесамовидная** sesamoid bone
~, **скуловая** zygomatic bone
~, **слезная** lacrimal bone
~, **слоновая** 1. ivory; 2. elephant bone; 2. *bot.* cohune palm *(Orbignya cohune)*
~, **тазовая** hipbone, innominate bone
~, **таранная** talus, astragalus, anklebone
~, **теменная** parietal bone
~, **трапециевидная** trapezoid bone, small multiangular bone
~, **треугольная** triangular bone
~, **трехгранная** triquetral bone, three-cornered bone
~, **черепная** cranial bone
~, **чешуйчатая** squamosal bone

кость-трапеция large multiangular bone

костэктомия *med.* costectomy

костюм suit; dress, clothes; overalls
~, **автономный защитный** survival suit
~, **водолазный** diving suit
~, **водонепроницаемый** immersion suit
~, **высотный** pressure suit
~, **высотный компенсирующий/компенсационный** high-altitude [partial pressure] suit, pressurized suit
~, **герметический** pressure suit
~, **защитный** protective/safety suit
~, **космический/космонавта** space suit, astronaut's suit
~, **летный** flight garment
~, **морской спасательный** sea rescue suit
~, **надувной** inflatable suit
~, **непромокаемый** rainsuit
~, **огнестойкий/противопожарный** fire-fighting suit
~ **с капюшоном** suit-and-hood unit
~, **штормовой** storm suit

костюмерная *theatre* wardrobe

костяника *bot.* stone bramble, roebuck berry *(Rubus saxatilis)*

костянка I *bot.* stone fruit

костянка II *zool.* common British centipede *(Lithobius forficatus)*

костянкоплодник *bot.* bony-berry *(Osteomeles)*

костяной bone *(attr.)*; osseous
~ **клей** bone glue
~ **ножик** ivory knife
~**ная мука** bone meal

косуля *zool.* roe deer *(Capreolus capreolus)*

косынка gusset/junction/corner/knee plate

косьба mowing

косяк jamb; post, pillar; cant; slope
~, **дверной** doorjamb
~, **оконный** window post

кот tomcat
~, **барханный** sand cat *(Felis margarita)*
~, **камышовый** jungle cat *(Felis chaus)*
~, **морской** 1. fur seal *(Callorhinus ursinus)*; 2. sting ray *(Dasyatis)*; 3. spotted dogfish *(Scyliorhinus caniculus)*

котангенс cotangent, *abbr.* cot, ctnh
~, **гиперболический** hyperbolic cotangent, *abbr.* coth, ctnh

котангенсоида *math.* cotangent curve

котел 1. boiler; 2. kettle, cauldron, copper; 2. *geol.* pothole, kettle
~, **агрегатированный** packaged boiler
~, **атомный** atomic pile/reactor
~, **бланшировочный** *food* blanching boiler, blancher
~, **водогрейный** hot-water boiler
~, **водотрубный** water-tube boiler
~, **газовый** gas-fired boiler
~, **двухколлекторный** double-headed boiler
~, **двухконтурный** double-pressure boiler
~ **для вытопки жира** rendering tank
~, **дымогарный** fire-tube boiler
~, **жаротрубный** fire-tube boiler, flue boiler
~, **комбинированный** composite boiler
~, **ледниковый** *geol.* glacial pothole
~, **мазутный** oil-fired boiler
~, **мыловаренный** soap pan/copper/kettle

~, огнетрубный fire-tube (smoke-tube) boiler
~, паровой steam boiler/generator
~, пиковый peak-load boiler
~, пропарочный autoclave
~, пропиточный impregnating vessel
~, прямоточный straight-through (once-through) boiler
~, пылеугольный pulverized-coal fired boiler
~ с принудительной циркуляцией forced-circulation boiler
~, секционный sectional-header boiler
~, судовой marine boiler
~, упарочный evaporator, concentrating tank
~, урановый uranium reactor/pile
~, утилизационный (паровой) exhaust-/waste-heat boiler, heat-recovery boiler
~, цилиндрический barrel boiler
~, шатровый *heat.* A-type boiler
~, экранный *met.* baffle boiler
~, энергетический power-generating boiler
~, ядерный nuclear reactor/pile
котелок small pot/kettle; bowl
~ компаса compass bowl
котел-утилизатор heat-recovery boiler
котельная boiler room/house
котельщик 1. boilermaker; 2. boiler operator/attendant
котерия coterie
котик 1. small tomcat; 2. *zool.* fur seal (Callorhinus ursinus)
котики *pl. bot.* hare's-foot clover *(Trifolium arvense)*
котиледон *bot.* cotyledon
котилозавр *paleont.* cotylosaur
котип cotype, syntype
котировать *fin.* quote
котироваться be quoted; *fig.* be rated
котировка *fin.* quotation
~, биржевая exchange/market quotation
~ во внебиржевом обороте over-the-counter quotation
~, заключительная closing quotation
котировочный quotation *(attr.)*
котирующийся quotable
котлета *cul.* cutlet
~, отбивная chop
~ты, картофельные potato cakes
~ты, рыбные fish cakes
котлетный *cul.* cutlet *(attr.)*
~ное мясо minced meat
котлоагрегат boiler unit
котлован *constr.* foundation ditch, excavation; fosse, trench
котловина *geog.* basin, kettle, trough, hollow
~, вершинная headwater basin
~ выдувания deflation basin
~, депрессионная kettle depression
~ опускания cauldron subsidence
~, сброшенная fault basin
котлообразный 1. cauldron-/kettle-shaped; 2. *geol.* basin-shaped
котлостроение boiler manufacture
котлостроительный boiler-making/-manufacturing
котлотурбинный boiler-and-turbine *(attr.)*
котовник *bot.* nepeta *(Nepeta)*
кот(т)онизация cottonization, cottonizing
кот(т)онин cottonized fibre/fiber, cottonin
котула *bot.* brass buttons *(Cotula)*
котурн *hist.* cothurnus, buskin
котятник *bot.* common ground ivy *(Glechoma hederacea)*
коуш *naut.* thimble, deadeye, eye, eyelet, glut
~ для троса/перлиня hawser thimble
~, треугольный heart-shaped thimble
~ шкотового угла clew ring
кофактор cofactor
кофе 1. coffee; 2. *bot.* coffee tree *(Coffea)*
~ в зернах coffee beans
кофеварка coffee maker, coffee-making machine; percolator
кофеин *chem.* caffeine, trimethylxanthine
кофеин-бензоат натрия *chem.* caffeine and sodium benzoate
кофеинизм *med.* caffeinism; caffeine poisoning
кофеиновый caffeinic
кофель-бугель *naut.* spider band/hoop
кофель-нагель belaying/tack/jack pin, rack-pin
кофель-планка belaying pin rack; range cleat
кофемолка coffee grinder/mill
кофермент coenzyme
кофункция *math.* cofunction
коффердам cofferdam
кохинхинка Cochin-China fowl, cochin
кохия *bot.* summer cypress, kochia *(Kochia)*
кохлеарный *anat.* cochlear
кохлеит *Pathol.* cochleitis, cochlitis
кохлео-вестибулярный cochleovestibular
кохлео-пальпебральный colcheopalpebral
кохлеопатия cochleopathy
коцепь *math.* cochain
кочевник nomad; migrant, migrator
кочегар stoker, fireman
кочегарка *naut.* stokehold, stokehole, fire/boiler room
кочедыжник *bot.* spleenwort *(Athyrium)*
кочерга poker, rake, fire iron
~, пудлинговая *met.* rabble
кочка mound, hillock, hummock
кошачий cat's *(attr.),* catlike; feline
~ глаз *miner.* cat's eye
~ лемур *zool.* ring-tailed lemur *(Lemur catta)*
кошачьи *pl. zool.* felids *(Felidae)*
кошельки *pl. bot.* calceolaria *(Calceolaria)*
кошениль 1. *zool.* cochineal *(Coccus cacti);* 2. cochineal (dye)
кошенильный cochineal; carmine; crimson
кошка 1. *zool.* cat *(Felis);* 2. *tech.* crab, trolley; crane carriage; pulley block; 3. *naut.* grapple, grapnel, drag; 4. mountaineering boot

~, барханная *zool.* sand cat *(Felis margarita)*
~, виверровая/рыбья fishing cat *(Felis viverrina)*
кошка-девятихвостка cat-o'-nine-tails
кощунство blasphemy, sacrilege; heresy
коэнергия co-energy
коэнзим coenzyme
коэнзометр coenzometer, coenzymometer
коэрцитивность coercivity, coercitivity
коэрцитивный coercive, coercitive
коэрцитометр coercimeter, coercive force meter
коэффициент coefficient. factor; ratio; modulus; rate
~ абразивности abrasion factor
~ абсорбции absorptance, absorptivity, absorption factor/rate
~ аварийности accident rate
~ амплитуды amplitude/peak/crest factor
~ аналогии analogy factor
~ анаморфозы *opt.* anamorphic ratio, anamorphizing factor
~, баллистический ballistic coefficient/factor
~ безопасности safety factor, margin of safety
~, безразмерный dimensionless [non-dimensional] coefficient
~ безубыточности *econ.* break-even load factor
~ бесплодности index of sterility
~, биномиальный binomial coefficient
~ блокировки blocking factor
~ брачности nuptiality rate
~, буквенный *comp.* literal coefficient
~ вариации/вариационный coefficient of variation, variation coefficient
~, вероятностный probabilistic factor
~ вертикальной полноты *naut.* vertical prismatic coefficient
~, весовой weight(ing) coefficient/factor
~ ветвления branching ratio
~ видимости visibility factor
~ влажности water-content coefficient
~ водопользования water-use efficiency
~, возрастной age-adjusted/-specific rate
~ волнового сопротивления wave-resistance [wavedrag] coefficient
~ воспроизводимости repeatability factor
~ воспроизводства *nucl.* conversion/reproduction/regeneration factor, breeding/conversion ratio; *econ.* reproduction/replacement rate
~ воспроизводства населения replacement index, index of reproduction
~ восстановления *mech.* coefficient of restitution
~ выигриша gain/advantage factor
~ выживаемости survival index
~ выносливости endurance ratio
~ выхода yield factor
~ вязкости viscosity factor, coefficient of viscosity
~ готовности availability
~ девиации deviation coefficient
~ демпфирования damping coefficient/factor/level
~ дефлегмации reflux ratio
~ диафрагмирования orificing factor
~ диффузии diffusion coefficient/constant, diffusivity
~ диффузного отражения diffuse-reflection factor
~ добротности factor/figure of merit
~ доверия *stat.* confidence coefficient/factor
~ доходности margin of profit
~ дублирования duplication factor
~, дыхательный respiratory coefficient/quotient
~ емкости, температурный temperature coefficient of capacitance
~ естественной освещенности daylight factor
~ жесткости *mech.* rigidity/stiffness coefficient
~ живучести survivability factor
~ загрузки load(ing) factor
~ загрязнения contamination/fouling factor
~ задержки/запаздывания lag factor; hindrance factor
~ занятия/занятости occupance rate; operative efficiency; *teleph.* call fill
~ запаса safety factor
~ запаса прочности reserve strength factor
~ заполнения space factor, block coefficient; permeability factor/coefficient; surface permeability
~ заполнения поверхности surface permeability factor
~ затрат *stat., econ.* cost/input coefficient
~ затухания damping/attenuation factor, attenuation constant/coefficient, decay/extinction coefficient
~ звукоизоляции acoustical reduction factor
~ звукоотражения acoustical/sound reflection factor/coefficient
~ звукопоглощения acoustical absorptivity, sound absorption coefficient
~ излучения radiation factor/constant; emissivity
~ изменения coefficient of variation
~ изменяющийся во времени *stat.* time-trending/varying coefficient
~ инверсии *math.* power of inversion
~ искажений distortion factor
~ использования utilization/use/usage factor, utilization rate/parameter, coefficient/measure of utilization; efficiency
~ использования производственных мощностей capacity factor; capacity utilization rate
~ использования системы system utilization rate
~ использования топлива fuel-utilization factor
~ кислотности acid number
~ корреляции correlation coefficient, coefficient of correlation
~ контрастности *phys.* gamma

~ кратности группы group repeat count
~ лага log factor
~ летальности lethal coefficient/range
~ ликвидности liquidity ratio
~ линейного расширения coefficient of linear expansion
~ лучеиспускания radiation factor/constant
~ кривизны buckling factor
~ массообмена/массопередачи mass-transfer coefficient
~, масштабный scale factor
~ материалоемкости input-output coefficient
~ мертворождаемости stillbirth rate
~ множественной корреляции multiple correlation coefficient
~ модуляции modulation factor/index
~ момента torque coefficient
~ мощности 1. capacity rate; 2. *el.* power factor
~ нагрузки load coefficient/factor
~ надежности reliability index
~ накопления build-up factor
~ напора head coefficient
~ направленности *rad.* directivity factor
~ нарастания grouth factor
~ насыщения saturation ratio
~ неодновременности diversity factor
~, нормирующий normalizing factor
~, обобщенный generalized coefficient
~ обогащения enrichment factor
~ обратного потока reflux ratio
~ обратного рассеяния back-scattering factor
~ обратной связи feedback factor
~ общей полноты *naut.* block coefficient
~ объединения по входу fan-in
~ объемного расширения coefficient of volumetric expansion, cubic/volumetric expansion coefficient
~ окупаемости капиталовложения *econ.* rate of return, return ratio
~ ослабления attenuation factor; *(of light)* extinction coefficient
~ отдачи output factor; yield efficiency
~ отклонения departure factor
~ ослепления dazzling coefficient
~ отражения reflection index/factor/coefficient; reflectance, reflectivity
~ очистки/очищения decontamination factor; clearing coefficient
~, переводной conversion factor
~ перегрузки load factor
~ передачи gear/transmission/gain/linkage/propagation coefficient; gear ratio
~ передачи преобразователя transducer gain
~ переноса массы mass-transfer coefficient
~ перекрытия (gear) engagement factor
~ пересчета scaling/scale/conversion factor
~ перехода/переходный conversion factor
~ плодовитости fertility index
~ плотности density coefficient
~ по току *el.* current gain
~ повторения repetition factor
~ поглощения absorption coefficient; absorptivity, absorptance
~ поджатия contraction coefficient
~ подобия ratio of similitude, homothetic ratio
~ подъемной силы lift coefficient/factor, coefficient of lift
~ полезного действия (кпд) efficiency coefficient/factor, efficiency; degree of efficiency; performance factor; input-output ratio
~ полезного действия, общий net/overall/total efficiency
~ полезного действия, результирующий net efficiency
~ полезного действия, средний mean/average efficiency
~ полезного действия, тепловой heat/thermal efficiency
~ полезного действия, тяговой propulsion efficiency
~ полезного действия, чистый net efficiency
~, полиномиальный *math.* coefficient of polynomial form, polynomial coefficient
~ полноты fullness coefficient
~ полноты водоизмещения *naut.* block coefficient
~ полноты сгорания combustion efficiency
~ поля *el.* field-form factor
~, поправочный correction factor/index
~ пористости voids ratio
~, пороговой threshold coefficient
~ порывистости gust factor
~ потерь loss factor coefficient, coefficient of losses
~ преломления refraction/refractive index
~ преобразования conversion/transformation factor/coefficient
~ прибыльности margin of profit
~ проводимости conductivity factor
~ производительности *econ.* productivity factor
~ проницаемости permeability coefficient
~ пропорциональности proportionality factor/constant/coefficient, factor of proportionality
~ пропускания transmission coefficient
~ простоя downtime ratio/rate
~ прочности strength factor
~ Пуассона Poisson's ratio
~ пустотности void ratio
~ разбавления dilution ratio
~ разброса *stat.* scatter coefficient
~ разветвления по выходу fan-out
~ разделения separation factor
~ размагничивания demagnetization factor
~ размножения multiplication factor
~ ранговой корреляции *stat.* rank correlation coefficient
~ распада decay/disintegration coefficient
~ распознавания coefficient of resolution/discrimination; recognition differential/coefficient

~ распределения distribution/partition coefficient
~ распространения propagation factor/constant
~ рассеяния scattering coefficient; dissipation factor
~ расхода flow/discharge coefficient
~, расчетный design factor
~ расширения coefficient of expansion, expansion factor/ratio
~ регрессии *stat.* regression coefficient
~ регулирования control factor
~ резонанса tuning factor
~ рентабельности *econ.* net profit ratio
~ рефракции refraction coefficient, refractive index
~ рождаемости birth-rate coefficient/index; *stat.* birth/childbearing rate
~ самоиндукции coefficient of self-induction, self-inductance
~, световой light/luminous coefficient
~ связи coupling coefficient
~ сглаживания smoothing factor
~ сжатия contraction/constriction coefficient
~ сжимаемости compressibility coefficient
~ скольжения coefficient of sliding/kinetic friction
~, скорректированный *econ.* adjusted rate
~ слышимости audibility factor
~ смертности mortality rate coefficient; *stat.* death rate; *(from accidents)* fatality rate
~ смертности, суммарный all-age death rate
~ сокращения coefficient of contraction
~ сопротивления resistance/drag coefficient; specific resistance
~ стабилизации stabilization factor
~, старший *math.* leading [highest-degree] coefficient
~, стехиометрический stoichiometric mixture ratio
~ стока runoff coefficient
~ стоячей волны standing-wave ratio
~, суммарный *stat.* cumulative rate
~, температурный temperature coefficient
~ теплового использования thermal utilization factor
~ теплового расширения coefficient of thermal expansion
~ теплоотдачи/теплопередачи heat-transfer coefficient
~ теплопроводности heat conduction factor, coefficient of thermal conductivity
~ торможения braking coefficient
~ трансформации *el.* transformation/voltage ratio
~ трения coefficient of friction, friction factor/coefficient
~ трения движения coefficient of sliding/kinetic friction
~ трения покоя coefficient of static friction
~, трехцветный trichromatic coefficient
~ тяготения gravity constant
~ тяги/упора thrust coefficient
~, угловой *math.* slope, angular coefficient
~ удержания retention coefficient
~ удлинения index of elongation
~ упругости coefficient of elasticity
~ усадки shrinkage factor/ratio
~ усиления amplification/intensification factor/coefficient; gain factor/coefficient, gain
~ усиления контура *el.* loop gain
~ усталости fatigue ratio
~ устойчивости stability coefficient
~ утечки leakage factor/coefficient
~ фазы phase constant/factor
~, феноловый phenol coefficient, hygienic laboratory coefficient
~ формы form factor; *(wood)* diameter quotient
~ фондоемкости *econ.* capital coefficient
~ Фурье *math.* Fourier coefficient
~, холодильный refrigerating factor
~,числовой numerical coefficient
~ чистой скорости index of net velocity
~ шероховатости roughness factor/coefficient
~ шума noise factor/figure
~ экономии трудовых затрат labour-saving ratio
~ экранирования screening constant/number, shielding factor
~ экспозиции exposure factor
~ экстинкции extinction coefficient
~ электровооруженности труда electric power (installed, available) per worker
~ эффекта масштаба scale factor
~ эффективности effectiveness coefficient; efficiency coefficient, efficiency; power efficiency; efficiency/performance factor
~ эффективности затрат cost-effectiveness ratio
~ эффективности поиска *comp.* hit ratio
~ яркости luminance factor/coefficient, coefficient of brightness

коядро *math.* cokernel
КПСС *abbr.* Communist Party of the Soviet Union
краб *zool.* crab
краг *geol.* crag
~, белый/коралловый coral crag
крагуй *zool.* sparrow hawk *(Accipiter nisus)*
краеведение study of local history, culture, etc,
краекучник *bot.* lip fern *(Cheilanthus)*
кража theft; larceny
~, квалифицированная *law* aggravated theft
~, мелкая petty theft/larceny; pilferage
~ со взломом burglary
край 1. edge, border, rim, fringe; margin; boundary, verge; brim; brink; 2. country, land, region, area; 3. krai, territory *(in former USSR)*
~ века margin of the eyelid
~, верный *(of Sun or Moon)* upper limb;
~ губы vermilion border
~, десенный gingival margin

~, зубчатый (листа) toothed/dentate margin
~, извилистый sinuate margin
~ крыши verge of a roof
~ лопасти blade tip
~, межкостный interosseous margin
~, неровный ragged edge/rim
~, нижний *(of Sun or Moon)* lower limb
~ перфокарты, ведущий *comp.* leading edge
~ перфокарты, нижний *comp.* nine edge
~, пильчатый serrate margin
~, режущий cutting/incisal/shearing edge
~, ресничный ciliary margin
крайисполком *abbr.* executive committee of a krai/territory
крайком *abbr.* krai/territory committee
крайт *zool.* krait *(Bungarus)*
кракен *myth* kraken
краковиан *math.* cracovian
кракс(а) *zool.* curassow *(*Crax; *Mittu)*
крамбе *bot.* colewort *(Crambe)*
крамбол *naut.* anchor davit
кран I cock, valve; tap, faucet
~ аварийный emergency cock
~, водопробный water-gauge cock
~, водопроводный (water) tap, faucet, cock
~, впускной inlet cock
~, выпускной outlet cock
~ герметизации pressurization cock
~, двухходовой two-way cock; angle cock
~, декомпрессионный pressure-release cock
~, дренажный vent cock
~, дроссельный throttle valve/cock
~, запорный shut-off cock, stop cock
~ кольцевания *av.* cross-feed cock/valve
~, конусный lifting-plug cock
~, многоходовой multiposition cock
~, муфтовый coupler-joined cock
~ перекрестного питания cross-feed cock
~, перекрывной shut-off cock, stop valve
~, пожарный 1. fire cock; 2. *av.* fuel emergency shut-off cock
~, продувочный blow-off cock
~, распределительный selector/distributing cock
~, ручной manual-operated valve
~, сальниковый gland cock
~, сливной drain cock
~, тормозной brake valve
~, трехходовой three-way valve
~ управления control cock
~, фланцевый flanged cock
~, цапковый threaded-end cock
~, шаровой ball cock
кран II crane
~, аварийный crash/wreck crane
~, автомобильный crane truck, truck-mounted crane
~, башенный tower/pillar crane
~, береговой *naut.* quay crane
~, грейферный clamshell/grab crane
~, грузовой/грузоподъемный cargo crane
~, достроечный fitting-out crane
~, железнодорожный rail crane
~, козловой ga(u)ntry crane, travelling bridge crane
~, консольно-поворотный bracket-swing crane
~, консольный jib/cantilever crane
~, консольный на колонне column jib crane
~, корабельный ship('s) crane
~, мачтовой mast/derick crane
~, минный mine derick
~, монтажный erecting crane
~, мостовой overhead travelling crane, bridge crane
~, мостовой двубалочный double-girder travelling crane, double I-beam travelling crane
~ на колесному ходу wheeled crane
~, палубный deck(-mounted) crane
~, передвижной travelling crane
~, плавучий pontoon/floating crane; crane ship
~, поворотный rotary crane
~, подвесной, однобалочный monorail hoist
~, подъемный hoisting crane
~, полноповоротный full-swing crane
~, полукозловой semi-gantry crane
~, полупортальный half-portal crane
~, портальный gantry/unloader/portal crane, overhead travelling crane
~, рельсовый rail/locomotive crane
~, самоходный плавучий self-propelled floating crane
~, универсальный multipurpose crane
кран-балка overhead-track hoist/crane; *naut.* anchor davit; cat-crane/-davit
кран-брызгалка spray cock
кранец *naut.* fender, bumper
~, задний/кормовой aft/stern fender
~, круглый rolling fender
~, носовой bow/stem fender
~, плетеный mat fender
~, шлюпочный boat pad
краниальный *anat.* cranial
краниит *med.* craniitis
краниограф craniograph
краниография craniography
краниоклазия *med.* cranioclasia, cranioclasis
краниология craniology
краниометрия craniometry
краниопластика cranioplasty
краниорахисхиз craniorrhachischisis
краниосклероз *med.* craniosclerosis
краниоспинальный craniospinal, craniorrhachidian
краниостеноз craniostenosis
краниотабес craniotabes
краниотомия *surg.* craniotomy, trephining
краниотрактор *med.* cranioclast
краниофарингиома craniopharyngioma, craniopharyngeal adamantionoma

краниэктомия craniectomy
крановщик crane operator/driver
кран-штабелеукладчик stacking crane
кран-штатив camera crane
крапива *bot.* nettle *(Urtica)*
~, глухая white dead nettle *(Laminum album)*
крапивна *bot.* cockspur flower *(Plectranthus fruticosus)*
крапивник *zool.* wren *(Troglodytes)*
крапивница *med.* urticaria, nettle rash, hives
~, гигантская/отечная giant urticaria, angioneurotic (o)edema
~ холодовая cold/congelation urticaria
крапивные *pl. bot.* nettle family *(Urthicaceae)*
крапп *bot.* common madder *(Rubia tinctorum)*
крапчатость spottiness; mottling; *agr.* mosaic/mottle disease
красавица, дневная *bot.* dwarf gloribind *(Convolvulus tricolor)*
красавица, ночная common four-o'clock *(Mirabilis jalapa)*
красавка 1. *bot.* deadly nightshade, belladonna *(Atropa belladonna);* 2. *zool.* demoiselle crane *(Anthropoides virgo)*
красильный tinctorial, dyeing; colo(u)rific
~ завод dye works
~ная лента typewriter ribbon
краситель dye, dyestuff, pigment, colour, stain
~, активный reactive dye
~, витальный vital stain
~, кубовый vat dye/colour
~, легколиняющий unfast dye
~ натуральный natural dyestuff
~, несмываемый washfast dye
~, пищевой food dye/colour
~, протравной mordant dye
~, растительный vegetable dyestuff/colo(u)r
~, светопрочный light-fast dye
~, спирторастворимый spirit-soluble dye
~ холодного/ледяного крашения azoic dye
~ хромовый chrome dye
красить dye, stain, colour; paint
краска paint; dye, colo(u)r, pigment
~, авиационная aircraft paint
~, алюминиевая aluminium varnish
~, акварельная watercolo(u)r
~, антикоррозийная anticorrosive/anticorrosion/antirusting paint
~, водоотталкивающая water-repellant paint
~, грунтовочная priming paint
~, гуашевая gouache
~ для морских судов marine paint
~ для подводной части корпуса bottom paint
~, жароупорная high-temperature paint
~, защитная protective paint/varnish
~, звукоизолирующая sound-deadening paint
~, кислотостойкая acid-proof paint
~, клеевая size paint, colo(u)r wash, distemper, calcimine
~, латексная latex paint
~, маскировочная camouflage paint
~, масляная oil paint
~, масляная густотертая oil-paint paste
~, необрастающая antifouling paint
~, несмываемая indelible paint
~, огнестойкая fire-resistant paint
~, огнеупорная refractory paint
~, печатная printing ink
~ по дереву wood paint
~, погодоустойчивая weatherproof paint
~, покровная *leath.* pigment finish
~ полиграфическая printing ink
~, противокоррозийная = антикоррозийная
~, противорадиолокационная antiradar paint
~, светящаяся luminous paint
~, теплостойкая heat-resisting/-resistant paint
~, теплоупорная high-temperature paint
~, типографская printing ink
~, флюоресцентная fluorescent paint
~, формовочная foundry mould wash; parting mixture
краскодувка, краскомет, краскопульт paint sprayer/blower, spraying gun, pneumatic paint brush
красная *n. zool.* sockey red salmon *(Oncorhynchus nerka)*
красноармеец Red Army soldier/man
краснобородка *zool.* common/jumping mullet *(Mugil cephalus)*
красноволоска *bot.* water starwort *(Callitriche)*
красногвардеец *hist.* Red Guard
краснозем krasnozem, red soil, terra rossa
краснознаменец holder of the Red Banner medal
краснозобик *zool.* curlew sandpiper *(Calidris testacea)*
красноклоп *zool.* stainer *(Dysdercus)*
красноножка *zool.* redshank *(Tringa totanus)*
красконос *zool.* red-crested pochard *(Netta rufina)*
красноперка *zool.* redeye *(Scardinius)*
красноплодный *bot.* red-fruited, erythrocarpous
краснопузырник *bot.* bittersweet *(Celastrus)*
краснопузырниковые *pl. bot.* staff-tree family *(Celastraceae)*
красноречие eloquence
краснота redness; *med.* erythema, rubor
краснотал *bot.* sharp-leaved willow *(Salix acutifolia)*
краснофлотец Red Navy sailor
краснохвост *zool.* pale tussock moth *(Dasychira pubidunda)*
краснохвостка *zool.* common redstart *(Phoenicurus phoenicurus)*
краснуха *med.* rubella, German measles, epidemic roseola
~, коревая rubeola
красный red; ruddy, rubicund
~ная доска *fig.* board of hono(u)r
~ная икра red caviar
красовласка *bot.* water starwort *(Callitriche)*

красоднев *bot.* day lily *(Hemerocallis)*
красоля *bot.* common masturtium *(Tropaeolum majus)*
красотел *zool.* European ground beetle *(Colosoma)*
красотка *zool.* black-winged damsel fly *(Calopteryx)*
краспица *naut.* crosspiece
кратер crater; vent
~, бездействующий dormant vent
~, замкнутый closed crater
~, концентричный cone-in-cone crater
кратное *math.* multiple
~, наименьшее общее least common multiple
~, общее common multiple
~, целое integral multiple
кратность multiplicity (factor); ratio, rate, frequency
~ дробления size reduction ratio
~ резервирования *comp.* redundancy rate
~ уменьшения reduction ratio
кратный *math.* multiple *(of),* divisible *(by);* aliquot
~, бесконечно infinite-to-one
крафт-бумага Kraft paper
крахмал starch
~, животный animal starch, glycogen
~, кукурузный corn starch
крахмалистый starchy, amylaceous
крахмалить starch
крацевание scratch-brushing
крачка *zool.* tern, tarrock *(Sterna)*
крашение dyeing, colouring
~, кубовое vat-dyeing
~, ледяное/холодное azoic dyeing
~, протравное mordant dyeing
креатин *biochem.* creatine
креатинин *biochem.* creatinine
креатинкиназа creatine kinase
креатиновый creatinic
креатинурия creatinuria
креатинфосфат creatine phosphate
креатинфосфокиназа creatine phosphokinase
креветка *zool.* prawn *(Palaemon serratus)*
кредит credit
~, автоматически возобновляемый revolving credit
~, акцептно-рамбурсный reimbursement credit
~, акцептный acceptance credit
~, банковский bank credit
~ без обеспечения [бланковый] blank credit
~, в on credit
~ в рассрочку installment credit
~ в розничной торговле retail credit
~ в форме открытого лимита open-end loan
~, вексельный paper credit
~, государственный public credit
~, дешевый low-interest credit
~, дисконтный discount credit
~, долгосрочный long-term credit
~ из низкого процента low-interest credit
~, ипотечный credit on mortgage (on property, on real estate)
~ используемый при необходимости standby credit
~, исчерпанный exhausted credit
~, краткосрочный short-term credit
~, ломбардный collateral loan
~ на льготных условиях credit on easy terms
~, неограниченный unlimited/open credit
~, открывать/предоставлять allot/supply credit
~ по контокоррентному счету current account credit
~ по открытому счету charge account
~ под недвижимость credit on real estate [on property/mortgage]
~, подтоварный credit against/on goods
~, потребительский consumer credit
~, пролонгированный extended credit
~, резервный standby credit
~ с единовременным погашением noninstallment credit
~ с погашением в рассрочку instalment credit
~, среднесрочный intermediate-term credit
~, стесненный *econ.* credit stringency
~, удорожить to tighten credit
кредитив *fin.* letter of credit
кредитивная грамота credentials, letter of credence
кредитирование, специальное special credit facilities
кредитно-денежный credit-and-monetary *(attr.)*
кредитный credit *(attr.)*
кредитовать give credit(s) to, finance, credit
кредитор creditor
~ по векселю bill creditor
~ по закладной [по ипотечному залогу] mortgagee
~ по ипотеке mortgage creditor
~ с предпочтительным правом требования preferential/preferred creditor
~, совокупный joint creditor
кредитоспособность credit worthiness/capacity/standing; borrowing power; solvency
кредитоспособный creditworthy; solvent
кредо creed, credo
крез Croesus
крезол *chem.* cresol, tolyl hydroxide
крейсер *naut.* cruiser
~, атомный nuclear-powered cruiser
~ ПВО anti-aircraft cruiser
~, подводный undersea cruiser
~, ракетный guided-missile cruiser
~ с артиллерийским вооружением gun cruiser
~, учебный training cruiser
крейсер-авианосец *naut.* cruiser carrier
кейсер-вертолетоносец helicopter cruiser

крейсер-разведчик scout cruiser
крейсерский cruising, cruise *(attr.)*
~кая скорость cruising speed
крейсерство cruiser warfare; cruise, cruising
крейсирование *naut.* cruising
крейсировать *naut.* cruise
крейтер *food* crater, crating machine
крейцкопф crosshead; slide block
крейцмейсель grooving/groove/cape chisel
крекер cracker
крекинг cracking *(of oil);* petroleum cracking process; cracking plant, *colloq.* cracker
~, глубокий heavy cracking
~ на неподвижном слое катализатора fixed-bed catalytic cracking
~ на псевдоожиженном слое катализатора fluidized-bed catalytic cracking
~, однократный single-phase cracking
~ под высоким давлением high-pressure cracking
крекинг-бензин cracked gasoline/petrol
крекинг-битум cracked asphalt
крекинг-дистиллят cracked distillate
крекинг-лигроин cracked/pressure naphtha
крекинг-мазут cracked fuel oil
крекинг-остаток cracking residue
крекинг-печь cracking still
крекинг-процесс, крекирование cracking process
крекинг-смола cracking tar
крекинг-установка cracking plant
крекировать crack
крекируемость crackability
крем cream
~, сапожный shoe polish
~, сладкий/яичный *cul.* custard
кремальера rack-and-pinion (gear); automatic lock
~ курсозадатчика set-course knob
кремастер cremaster (muscle)
кремастер-рефлекс cremasteric reflex
крематорий crematorium, crematory
кремационная печь incinerator
кремация cremation
кремень flint
~, известковистый calcareous silex/flint
кремировать cremate; incinerate
кремль 1. citadel; 2. the Kremlin
кремневка kremnyovka, kaolinic refractory clay
кремневодород *chem.* hydrosilicon, silane
кнемнезем silica, earth silicon, silicon earth
~, мучнистый rock meal
~, стекловатый glassy silica
~, студенистый silica jelly
кремнеземнистый, кремнеземный siliceous, silicic
кремнеизвестковый silicicalcerous, siliceocalcerous
кремнекислота *chem.* silicic acid
кремнекислый *chem.* silicic
~ калий potassium silicate
кремненосный siliciferous
кремнероговой corneosiliceous
кремнефторид *chem.* silicofluoride
кремнефтористоводородный hydrofluorosilic
кремниевый silicic
кремний silicon, Si
~, азотистый silicon nitride
~, водородистый silicomethane
~ пассивированный нитридом/окислом nitride-/oxide passivated silicon
кремнийборид *chem.* silicoboride
кремнийсодержащий siliceous
кремнистый siliceous; flinty
кремокарпий *bot.* cremocarp
кремометр *agr.* creamometer
крен *naut.* heel, list; roll; careen(ing); *av.* bank, roll
~ без груза, имеющий lopsided
~, большой heavy list
~, внезапный snap roll
~, глубокий steep bank
~, мелкий flat/gentle bank
~ на левый/правый борт list to port/starboard
~, небольшой light list
кренгельс *naut.* cringle
кренгование *naut.* careening, careenage
кренговать *naut.* careen, heave down
кренение banking, rolling
кренить(ся) *naut.* heel, careen; *av.* bank, roll
кренование *naut.* 1. heeling, inclining; 2. heeling/inclining/stability test
креномер, кренометр *naut.* inclinometer
креозол *chem.* creosol
креозот *chem.* creosote
креол Creole
креолин Creolin
креолка Creole (woman)
креольский Creole; creolized
креп 1. crepe (rubber); 2. crepe
~, траурный (mourning) crepe
крепдешин crepe de Chine
крепеж 1. fasteners; brackets; *min.* timber, pit posts; 2. bracing, bracketing; supporting, propping, timbering
крепежный fastening; reinforcing; propping
~ лес *min.* pit posts; timber
крепильщик *min.* timberer
крепитация *med.* crepitation, crepitus, crepitatio, crackling
крепитель binder, bond, binding agent
крепительный 1. strengthening, reinforcing, fortifying; 2. *med.* astringent
крепитирующий crepitant
крепить secure, fasten, fix; brace; shore up, prop; *min.* timber, case
крепкоплодник *bot.* euclidium *(Euclidium)*
крепление 1. fastening, securing, bracing, propping, fixing; attaching; 2. *min.* timbering, casing, roofing

~, **амортизирующее** anti-vibration mounting
~, **жесткое** rigid mounting
~ **крыла** wing attachment
~, **резьбовое** thread attachment
~ **стяжным хомутом** collar fastening
~ **троса** cable attachment
~, **фланцевое** flange fastening/attachment
~, **шомпольное** piano-hinge attachment
крепостная serf *(woman)*
крепостничество serfdom
крепостной serf *(attr.)*, serfdom *(attr.)*; 2. *n.* serf; villein
~ **акт** *econ.* title deed
~ **крестьянин** serf
~**ная зависимость** serfdom
~**ное право** serfoom
крепость I fortress; stronghold
крепость II strength; hardness, toughness; tenacity; stability
крепость, купчая *hist.* deed of purchase
крепсатин *text.* satin-backed crepe
креп-фай *text.* silk faille crepe
крепь support; *min.* timbering, props
~ **анкерная** *min.* roof bolting
~, **арочная** *min.* arched support
~, **забивная** spills, spilling, forepoling
~, **забойная** *min.* face support
~ **передвижная** self-advancing support
~, **шахтная** *min.* shaft mining
~, **штанговая** *min.* roof bolting
крескограф *med.* crescograph
кресло armchair, easy chair; seat
~, **авиационное** aircraft seat
~ **Барани, вращающееся** *med.* Barany's chair
~, **вращающееся** swivel/revolving/pivoting chair
~, **гинекологическое** gyn(a)ecologic chair
~, **катапультное/катапультируемое** ejection seat
~, **стоматологическое** dental chair
кресло-каталка wheelchair
кресло-коляска wheeled chair
кресс *bot.* English peppergrass *(Lepidium campestre)*
кресс-салат *bot.* garden cress *(Lepidium sativum)*
крест 1. cross; 2. four-way junction piece
крестец *anat.* sacrum, sacred bone
креститель baptizer, baptist; **Иоан Креститель** *rel.* John the Baptist
крестник godchild, godson
крестниuа goddaughter, godchild
крестный 1. cross *(attr.)*; 2. baptismal
~ **отец** godfather
~ **путь** road to Calvary
~ **сын** godson, godchild
~ **ход** religious procession
~ **ная дочь** goddaughter, godchild
~**ная мать** godmother
~**ное целование** oath-taking
~**ные дети** godchildren
крестовидный cruciform, cross-shaped, cruciate
крестовик 1. *zool.* garden spider, common cross spider *(Aranea diadema)*; 2. *geol.* macle, chiastolite
крестовина crosspiece, crosshead, cross; four-way piece; *el.* spider; *rail.* frog
~ **ротора** field spider
~, **поворотная** *rail.* reversible frog
~, **трубная** four-way pipe union
~ **якоря** *el.* armature spider
крестовка *zool.* pelodytid *(Pelodytes)*
крестовник *bot.* groundsel, ragwort *(Senecio)*
крестовница *bot.* crosswort *(Crucianella)*
крестовый cross *(attr.)*
~ **поход** *hist.* crusade
крестоносец 1. *hist.* crusader; 2. *zool.* wheat chafer *(Anisoplia austriaca)*
крестоносный 1. cruciferous; 2. crusader's, crusader *(attr.)*
крестообразно crosswise
крестообразный cruciform, cross-shaped, crucial; *bot., zool.* cruciate
крестопоклонная неделя third Sunday in Lent
крестоцветные *bot.* cabbage family *(Cruciferae)*
крестоцветный *bot.* cruciferous
крестцово-копчиковый *anat.* sacrococcygeal
крестцово-маточный sacrouterine
крестцово-подвздошный sacroiliac
крестцово-поясничный sacrolumbar
крестцово-седалищный sacrosciatic
крестцовый *anat.* sacral
крестьянин peasant
крестьянка peasant woman/girl
крестьянский peasant; rural; village *(attr.)*
крестьянство peasantry; the peasants
кресчендо = крешендо
кресченция *bot.* calabash tree *(Crescentia)*
кретин *med.* cretin, idiotic dwarf
кретинизм cretinism
~, **гипофизарный** pituitary cretinism
кретиноидный, кретиноподобный cretinoid
кретинический cretinistic, cretinous
кретон *text.* cretonne
кречет *zool.* Arctic falcon, gerfalcon *(Falco gyrfalco)*
кречетка *zool.* sociable plover *(Charadrius gregaria)*
крешендо, крещендо *mus.* crescendo
крещение 1. baptism, christening; 2. Epiphany, Twelfth-day
кривая curve; datum line
~, **адиабатическая** adiabatic curve
~, **аклиническая** acline
~, **алгебрическая** algebraic curve
~, **балансировочная** trim curve
~, **батиграфическая** bathygraphic curve
~ **большого радиуса** wide/easy curve
~ **вероятностей** probability curve
~, **волнообразная** wave curve
~ **второго класса** *math.* line conic
~ **второго порядка** *math.* quadratic curve,

conic
~ выживания survival/survivorship curve
~, высотная altitude curve
~ высшего порядка *math.* higher plane curve
~ выхода yield curve
~ Гаусса Gaussian curve
~, гипсографическая hypsographic curve
~, гладкая smooth curve
~, градуировочная calibration curve
~, двугорбая double-humped curve
~ деформации stress-strain curve
~, дискриминантная discriminant curve, p-discriminant locus
~ заболеваемости sick rate curve
~, замкнутая closed curve
~ извлечения elution curve
~, изоквантная isoquant
~, изоклинальная isocline, isoclinic line
~, интегральная integral/accumulation curve
~, исправленная fitted curve
~, калибровочная standard/calibration curve
~, колоколообразная bell-shaped [Gaussian] curve
~ крена *naut.* roll trace
~ критерия мощности power curve
~, крутая steep/abrupt curve
~, локсодромическая loxodromic curve, rhumb line
~, многосвязная multiply connected curve
~ мышечного сокращения *med.* myogram, muscle curve
~ набора высоты climb curve
~ намагничивания magnetization curve
~ нарастания growth curve, build-up curve
~ насыщения saturation curve
~, незамкнутая open curve
~, непрерывная continuous curve
~, несимметричная skew curve
~, несущая carrier curve
~, обратная reverse/return curve
~, огибающая envelope curve
~ определения determinative curve
~ ошибок error curve
~, остроконечная peaked curve
~, пилообразная sawtooth curve
~, плоская plane curve, curve in a plane
~, пограничная boundary curve
~, показательная index curve; exponential curve
~ поправок correction/calibration curve
~, пространственная space curve, three-dimensional curve
~ пуассоновского распределения Poisson chart
~, пунктирная dotted-line curve
~, равновесная/равновесия equilibrium curve
~ равных вероятностей equiprobability curve
~ распада decay curve
~ распределения distribution/percentage curve
~, расчетная estimated/calculated/predicted curve
~, резонансная resonance curve/characteristic
~, результирующая resultant curve
~, сглаживающая fitting curve
~, сигмоидальная sigmoid curve
~, синусоидальная sine curve, sinusoid
~, S-образная sigmoid curve
~, сплошная full curve
~ срока службы life curve
~, степенная exponential curve
~, тарировочная calibration curve
~ термического цикла time-temperature curve
~ третьего порядка *math.* cubic curve
~ тяги thrust curve
~, усеченная truncated curve
~, фугоидная phugoid curve
~, характеристическая characteristic/performance curve
~, центровочная *naut., aero.* centre-of-gravity curve
~, штриховая dashed curve
~, экспоненциальная exponential curve

кривизна curvature; flexure, bending; tortuosity
~ желудка, большая/малая greater/lesser curvature of the stomach
~ земной поверхности curvature of the Earth
~ с переменой знака reversed curvature
~, средняя mean curvature
~ траектории trajectory/path curvature

криволесье elfin wood
криволинейность curvilinearity
криволинейный curvilinear, curved
криволистный *bot.* curvifoliate, cyrtophyllous
криволопастный *bot.* cyrtolobous
криворядный curviserial
кривоцвет *bot.* wild bugloss *(Lycopsis)*
кривошея *med.* torticollis, wry-neck, stiff-neck
~, мышечная spastic torticollis
~, ревматическая rheumatic/symptomatic torticollis
~, рубцовая cicatrical torticollis
~, спастическая intermittent/spastic torticollis
~, стойкая fixed torticollis

кривошип *mech.* crank
~, ведущий driving crank
~, внутренний/концевой inner/outer crank
~, спаренный twin crank

криз *med.* crisis
~, вазодилаторный vasodilator crisis
~, гипертонический hypertensive crisis
~, гипотонический hypotensive crisis
~, желудочный gastric crisis
~ кишечника intestinal crisis
~, надпочечниковый suprarenal crisis
~, ретикулоцитарный reticulocytic crisis
~, солярный acute inflammation of solar plexus, acute solaritis

кризанол *chem* chrisanolum, oleochrysine
кризис crisis; *econ.* depression; slump

~, **валютный** monetary crisis; foreign-exchange crisis
~, **денежный** monetary crisis
~**ы, периодические экономические** recurrent economic crises
крикет *sport* cricket
криминалист *law* criminalist
крикотиреотомия *med.* cricothyreotomy
криминалистика *law* criminalistics
криминальность criminality; crime rate
криминолог criminologist
криминология criminology
криноидея *zool.* crinoid
кринолин 1. crinoline, hoop skirt; *2. naut.* propeller/screw guard, crinoline
кринум *bot.* crinum *(Crinum)*
криоген cryogen, refrigerant
криогеника cryogenics
криогенин cryogenine
криогенный cryogenic
криогидрат cryohydrate
криоконит *miner.* cryoconite
криолит *miner.* cryolite
криоаэротерапия cryoaerotherapy
криогипофизэктомия cryohypophysectomy
криоглобулин cryoglobulin
криоглобулинемия cryoglobulinemia
криокаутер cryocauter; cryocautery
криология cryology
криопатия cryopathy
криопрофибрин cryoprofibrin
криоскопия cryoscopy
криостатирование cryostatting
криоталамэктомия *surg.* cryothalamectomy
криотехника cryogenic engineering
криотрон cryotron
криотурбация *geol.* cryoturbation, congeliturbation
криофибриногенемия cryofibrinogenemia
криофилактический cryophylactic, crymophylactic
криофиллит *miner.* cryophyllite
криофильный cryophylic, crymophylic
криофит *bot.* cryophyte
криохимия cryochemistry
криохирургия cryosurgery
криоэлектроника cryoelectronics
крип creep
крипоустойчивость creep resistance/strength
крипта *hist., anat.* crypt
криптит *med.* cryptitis
криптобиоз cryptobiosis
криптогамия *bot.* cryptogamy
криптогамный *bot.* cryptogamic, cryptogamous
криптоген *bot.* cryptogene
криптогенный cryptogenetic, cryptogenic
криптограмма 1. cryptogram; 2. *bot.* rock brake *(Cryptogramma)*
криптография cryptography
криптозойский *zool.* cryptozoic
криптоклаз cryptoclase
криптококкоз *med.* cryptococcosis
~, **кожный** dermal cryptococcosis
~, **костный** osteal/bony cryptococcosis
~, **легочный** pulmonary cryptococcosis
~, **мозговой** cerebral cryptococcosis
криптококкома cryptococcoma
криптоксантин cryptoxanthine
криптолит cryptolith
криптоменорея *med.* cryptomenorrh(o)ea
криптомитоз *biol.* cryptomitosis
криптомнезия cryptanamnesia, cryptomnesia
криптон *chem.* krypton, Kr
криптопертит *miner.* cryptoperthite
криптоподия cryptopodia
крипторхид *med.* cryptorchid
крипторхизм cryptorchidism, cryptorchism
криптостемонный *bdt*. cryptostemonous
криптостом(а) *bot.* cryptostoma, cryptostome
криптофит *bot.* cryptophyte
криптофталм(ия) cryptophthalmia, cryptophthalmus
криптоциста cryptocyst
криптэктомия cryptectomy
криптэстезия cryptesthesia
кристалл crystal; *comp.* chip
~, **ажурный** skeleton crystal
~, **базовый** *comp.* master chip
~, **бескорпусный** *comp.* bare/unpacked chip
~, **включенный** embayed crystal
~, **гомеополярный** homeopolar/atomic crystal
~, **гранецентрированный кубический** face-centered cubic crystal
~, **двойниковый** twin crystal
~, **двоякопреломляющий** birefringent crystal
~, **дендритный/древовидный** arborescent/dendritic crystal
~, **дугообразно изогнутый** arcuate crystal
~, **жидкий** liquid crystal
~, **зарождающийся** incipient crystal
~, **затравочный** seed/inoculating crystal
~, **игольчатый** acicular [arrow-head] crystal
~, **идеальный** perfect crystal
~, **искаженный** distorted crystal
~, **искривленный** curved crystal
~ **конвейерного процессора** pipeline chip
~, **кремниевый** *comp.* silicon chip
~, **левовращающийся/левый** left-handed crystal
~, **материнский** mother crystal
~ **микропроцессора** microprocessor chip
~, **монохроматизирующий** monochromatizing crystal
~, **недоразвившийся** immature crystal
~, **недоразвитый** skeleton crystal
~, **неидеальный/несовершенный** imperfect crystal
~, **неявный** cryptocrystal
~, **нитевидный** filamentary crystal
~, **объемноцентрированный кубический**

body-centered cubic crystal
~, одноосный uniaxial crystal
~, пластинчатый tabular/platy crystal
~, полупроводниковый/полупроводящий semiconductor/semiconducting crystal
~ правовращающийся/правый right-handed crystal
~, проросший intergrown crystal
~, пьезоэлектрический piezoelectric crystal
~, равноосный equiaxed crystal
~, реальный real/imperfect crystal
~, решетчатый/скелетный skeletal/skeleton crystal
~ с водородной связью hydrogen-bonded crystal
~ с изменяемой структурой restructable chip
~, сдвойникованный twinned crystal
~, сегнетоэлектрический ferroelectric crystal
~ систолической матрицы *comp.* systolic chip
~, столбчатый columnar crystal
~, сцинтиллирующий crystal scintillator, scintillating crystal
~, таблитчатый tabular crystal
~, униполярный homopolar crystal
~, центральносимметричный centrosymmetrical crystal
~, эвгедральный euhedral crystal
кристалл-вкрапленник phenocryst
кристалл-гость chadacryst
кристаллизатор crystallizer, crystallization vat/tank/disk/mould, crystallizing evaporator
~, сборный/составной *met.* composite mould
~ периодического действия batch crystallizer
кристаллизационный crystallization *(attr.),* crystallizing
кристаллизация crystallization, crystallizing
~ в движении crystallization in motion
~ в равновесных условиях equilibrium crystallization
~, второстепенная minor crystallization
~, направленная oriented crystallization
~, последовательная successive crystallization
~, принудительная forced/coercive crystallization
~, раствор-расплавная solution-melt crystallization
кристаллизировать(ся) crystallize
кристаллизуемость crystallizability, crystallizing capacity/tendency
кристаллит *miner.* crystallite
кристаллический crystalline
клисталличность crystallinity
кристаллогенез crystallogenesis
кристаллогидрат crystalline hydrate
кристаллография crystallography
~, рентгеновская X-ray crystallography
~, структурная structural/structure crystallography
кристаллодержатель crystal mount/holder; *comp.* chip carrier
кристаллоид crystalloid
кристаллоносный crystalliferous
кристаллооптика crystal optics
кристаллофизика crystallophysics, crystal physics
кристаллохимия crystallochemistry, crystal chemistry
кристалл-фантом phantom crystal
кристалл-хозяин host crystal
кристальный crystalline
кристобалит *miner.* cristobalite
критерий criterion; standard; *stat.* test
~ безопасности safety criterion
~, двусторонний double-tailed test, two-sided/-tailed test
~, двухвыборочный two-sample test
~ достаточности *stat.* criterion of sufficiency
~ достоверности *stat.* test of validity
~ значимости significance test
~ качества работы performance measure/criterion
~ Лапласа criterion of rationality
~ максимакса *math.* maximax criterion
~ максимального правдоподобия maximum likelihood criteriîn
~ максимина *math.* maximim criterion
~ минимакса *math.* minimax criterion
~ минимаксного риска minimax risk criterion
~ минимума среднего квадрата ошибки integral square error criterion
~, многовыборочный multisample test
~ надежности reliability index
~, односторонний single-tailed test
~ оптимальности optimality criterion/test
~ отношения правдоподобия *stat.* likelihood ratio criterion
~ оценивания *stat.* criterion of estimation
~ ошибок, квадратичный error-squared criterion
~ подбора *math.* fitting criterion
~ подобия similarity/similitude criterion
~ полезности measure of utility
~, последовательный sequential test
~ равновесия *chem.* equilibrium number
~, расчетный design criterion
~ Рейнольдса Reynolds number
~ сглаживания *math.* fitting criterion
~ случайности *stat.* test of randomness
~, смещенный *stat.* biased test
~ согласия fitting criterion
~, состоятельный consistency test
~, статистический statistical test
~ Стьюдента Student-test
~ сходимости convergence criterion
~ точности fidelity criterion
~ устойчивости stability criterion
~ хи-квадрат chi-square criterion
~ эффективности measure of effectiveness
критериум *sport* championship, contest

критика 1. criticism; 2. critique
~ критики metacriticism
критический critical; marginal
критичность criticality; criticalness
крито-микенский Creto-Mycenean
критский Cretan
критмасса *abbr.* critical mass
критрадиус *abbr.* critical radius
критразмер *abbr.* critical size/dimension
крица *met.* bloom; bar; (iron) ball
~, пудлинговая puddle bar
кровавик *miner.* blood stone; h(a)ematite
кровать-весы bed balance
кровезаменитель blood substitute
кровельщик roofer
кровеносный sanguiferous; circulatory
~ сосуд blood vessel
~ная система circulatory system
кроветворение h(a)emopoiesis, h(a)ematopoiesis, h(a)ematogenesis, h(a)ematoplasty, h(a)emoplasty
~, внекостно-мозговое extramedullary h(a)emopoiesis
~, желточное yolk h(a)emopoiesis
~, костно-мозговое medullary h(a)emopoiesis
~, эмбриональное embryonic/f(o)etal h(a)emopoiesis
кроветворный h(a)emopoietic, h(a)ematopoietic, h(a)ematogenetic, h(a)ematogenic, h(a)emogenic, h(a)matogenous, h(a)emoplastic, h(a)ematoplastic
кровля *constr., min.* roof, roofing; top
~, висящая roof pendant; hanging layer
~ водоносного слоя *geol.* confining bed
~, глинистая agrillaceous roof/top
~, гонтовая wood shingle roofing
~ жилы back roof of vein, wall of vein
~, обрушающаяся deteriorating roof; collapsible top
~ пласта top of bed, superface of stratum
~ свежеобнаженная fresh roof; green rocks
~, слабая soft/loose/weak roof, bad top
~, устойчивая firm/fast/solid roof
~, черепичная tile roofing
~, шиферная slate roofing
~, этернитовая asbestos-shingle roofing
кровный 1. blood *(attr.); fig.* vital, deep; deadly 2. thoroughbred
кровоизлияние *med.* h(a)emorrhage, bleeding; h(a)ematoma; apoplexy
~ в глаз h(a)ematopsia, h(a)morrhage into the eye
~ в мозг cerebral h(a)morrhage, h(a)ematencephalon
~ в надпочечник adrenal apoplexy
~ в полость плевры h(a)ematothorax, h(a)emothorax
~ в почку renal apoplexy
~ в сетчатку retinal apoplexy
~ в спинной мозг h(a)matomyelia
~ в стекловидное тело vitreous h(a)emorrhage
~ в сустав h(a)emarthrosis, h(a)emarthron
~, внутреннее internal h(a)morrhage/bleeding
~, внутриполостное h(a)matocele, h(a)matocelia
~, внутричерепное intracranial h(a)ematoma
~, забрюшинное retroperitoneal h(a)emorrhage
~, точечное punctate/petechial h(a)emorrhage
кровообращение blood circulation
~, аллантоидно-пупочное umbilicoallantoic circulation
~ в системе портальной вены portal circulation
~, искусственное artificial/assisted/extracorporeal circulation
~, кожное cutaneous circulation
~, коронарное coronary circulation
~ малого круга lesser/pulmonary circulation
~, мозговое cerebral circulation
~, перекрестное cross circulation
~, периферическое peripheric/peripheral circulation
~, печеночное hepatic circulation
~, плацентарное placental circulation
~ плода f(o)etal circulation
~, шунтовое derivative circulation
кровоостанавливающий h(a)emostatic, styptic; astringent
кровоочистительный h(a)emocatharsic
кровоподтек ecchymosis; bruise
~, трупный cadaveric bruise
кровопереливание blood transfusion
кровопотеря loss of blood, h(a)emorrhage
кровопускание bloodletting
~ из артерии arteriotomy
~ из вены phlebotomy, venesection
кровосмесительный incestuous
кровосмесительство, кровосмешение incest
кровоснабжение blood supply
кровосос *zool.* 1. vampire bat *(Desmodus);* 2. bloodsucker lizard *(Calotes ophiomachus)*
кровотечение bleeding, h(a)emorrhage, h(a)emorhea, h(a)ematorrhea
~, бронхиальное h(a)emoptysis, bronchial h(a)emorrhage
~ в грудную полость h(a)emothorax, h(a)ematothorax
~, венозное venous h(a)emorrhage, phleborrhagia
~, внутрибрюшинное intraperitoneal h(a)emorrhage
~, желудочное gastrorrhagia
~ из глаза opthalmorrhagia
~ из глотки pharyngorrhagia
~ из мочеиспускательного канала urethrorrhagia
~ из почки renal h(a)emorrhage, nephrorrhagia
~ из прямой кишки proctorrhagia
~ из прямой кишки, викариирующее proctomenia
~ из спинного мозга, викариирующее myelomenia
~, кишечное intestinal h(a)emorrhage, enterorrhagia

~, **легочное** pulmonary h(a)emorrhage, pneumorrhagia, h(a)emoptysis
~, **маточное** uterine bleeding, metrorrhagia
~, **маточное функциональное** menstruation
~, **межменструальное** intermenstrual bleeding
~, **носовое** nasal bleeding, epitaxis
~, **пищеводное** (o)esophageal h(a)emorrhage
~, **плацентарное** placental apoplexy
~, **полостное** cavitary h(a)emorrhage
~, **послеродовое** postpartum h(a)emorrhage
~, **почечное** renal h(a)emorrhage, h(a)ematuria
~, **предродовое** antepartum h(a)emorrhage
кровоток blood flow
~ **в малом круге** lesser blood flow
~, **мозговой** verebral blood flow
~, **обратный** retrograde blood flow
~, **почечный** renal blood flow
~, **тканевый** tissual blood flow
кровоточащий bleeding
кровоточивость h(a)emorrhagic diathesis, h(a)emophilia, bleeding sickness
~ **десен** stomatorrhagia
кровоточивый h(a)emorrhagic
кровоточить bleed
кровохарканье h(a)emoptysis, bloody expectoration
~, **туберкулезное** tuberculosis h(a)emoptysis
кровохлебка *bot.* burnet *(Sanguisorba)*
~, **лекарственная** great(er) burnet *(Sanguisorba officinalis)*
кровь blood
~, **артериализованная** arterialized blood
~ **в моче** h(a)ematuria
~, **гепаринизированная** heparinized blood
~, **густая** thick blood
~, **дефибринированная** defibrinated blood
~, **запекшаяся** gore, cruor, clotted blood
~, **катионитная** cationitic blood
~, **консервированная** ampuled/banked/stored/conserved blood
~, **лаковая** laky blood
~, **периферическая** peripheric blood
~, **плацентарная** placental blood
~, **цельная** whole blood
~, **цитратная** citrated blood
кровянисто-гнойный sanguineopurulent
кровянисто-пенистый sanguineofoamy
кровянистый sanguinolent; bloody
кровяной sanguineous; blood *(attr.)*
~ **шарик** blood corpuscle
крокет croquet
кроки rough sketch, area/field sketch
крокирование sketching; outlining
крокус *bot.* crocus *(Crocus)*
~, **осенний** common autumn crocus *(Colchinum autumnale)*
кролиководство rabbit breeding/husbandry
кролист *sport* crawlist
кроль *sport* crawl (stroke)
~ **на груди** front crawl
~ **на спине** back crawl
кроманьонец, кроманьонский Cro-Magnon
кромвелевский Cromwellian
кромка edge, rim, border, brim; bead
~, **входящая** entering/leading edge
~, **выходящая** trailing/following edge
~ **ковша, рабочая** attacking edge of a bucket
~ **лопасти** blade edge/tip
~ **облаков** cloud border/edge; cloud level
~, **отбортованная** flanged/beaded edge
~, **передняя** leading edge; lip
~, **скошенная** bevelled/scarfed edge
~ **ткани** *text.* selvage, selvedge, hem
~, **фланжированная** flanged edge
кромкозагибающий edging, flanging
кромкообрубочный edge-trimming
кромлех cromlech, dolmen
крон I chrome yellow
крон II crown glass
крона 1. crown, top; 2. *fin.* crown, krone
кронблок crown block
кронверк *mil.* crownwork
кронглас crown glass
кроненпробка crown cap
кронпринц crown prince
кронпринцесса crown princess
кронциркуль caliper(s)
~, **внутренний** inside calipers
~, **двусторонний** outside-inside calipers
~, **наружный** outside calipers
кроншпен *zool.* curlew *(Numenius)*
кронштейн bracket, arm, cantilever; strut
~ **крепления** mounting bracket
~, **однолапый** single-arm strut
кропило *eccl.* aspergillum
кропильница *eccl.* 1. aspersorium; 2. aspergillum
кросс I *telev.* distributing/distribution frame; terminal room
кросс II *sport* cross-country race/ride
~, **лыжный** cross-country ski race
кроссинговер *biol.* crossing-over
~, **межаллельный** interallelic crossing-over
~ **неправильный** illegitimate crossing-over
~ **неравный** unequal crossing-over
~, **сложный** compound crossing-over
кроссирование *fin.* crossing *(of cheque)*
кроссировка cross connection
кросс-коунтри cross-country race
кросс-курс *fin.* cross rate
кроссовер crossover
кроссоверный crossover *(attr.)*
крот *zool.* mole *(Talpa europaea)*
кротализм *vet.* crotalism
кроталин *biochem.* crotalin
кроталярия *bot.* crotalaria, rattlebox *(Crotalaria)*
кротование *agr.* mole ploughing/draining
кротователь *agr.* mole plough
кротон *bot.* croton *(Croton)*
кротоназа *chem.* crotonase, crotonyl hydrase
крохаля *zool.* fish duck *(Mergus)*

кроше crochet
крошение breaking up; crumbling
крошить(ся) break up, crumble
крошка crumb; grit
~, **мраморная** marble aggregate
круг circle; wheel; ring; disk; cycle; orbit
~ **абразивный** abrasive disk
~, **гончарный** potter's wheel, jigger
~, **делительный** indexing dial
~, **заточный** tool-grinding wheel
~ **кривизны** circle of curvature
~, **наждачный** emery wheel
~, **поворотный** swivel plate; *rail.* turntable
~, **полярный** polar circle
~, **северный полярный** arctic circle
~, **сегментный** segmental wheel
~ **склонения** circle of declination
~, **спасательный** life-buoy ring, circular buoy
~ **сходимости** *math.* circle of convergence
~, **точильный** grinding wheel
~, **часовой** *geod.* hour circle, circle of declination
~, **шлифовальный** *met.* grinding wheel
~, **Южный полярный** antarctic circle
круги *fig.* circles, quarters; community
~, **деловые** business quarters/community/interests
~, **монополистические** monopoly interests
~, **правящие** the Establishment
~, **сельскохозяйственные** agricultural/farming community
~, **финансовые** financial circles; moneyed interests
кругискатель *geod.* finder-circle
кругловращательный circulatory
круглоголовый trochocephalous
круглогубцы round-nose pliers
круглозубчатый *bot.* crenate
круглозубчиковый *bot.* crenulate
клуглокрылый cyclopterous, round-winged
кругломер out-of-round gauge
круглоротые *pl. zool.* cyclostomes *(Cyclostomata)*
круглоротый *zool.* cyclostomatous
круглосуточный round-the-clock, twenty-four-hour, day-and-night
круглоязычные *pl. zool.* bell toads *(Discoglossidae)*
круглый round, circular; globular, spherical
круговой circular, annular, orbicular
~**вая дорога** roundabout route
~**вая порука** mutual guarantee/responsibility
круговорот rotation, turnover, circulation, cycle; circular motion
~, **биологический** biocycle
круговращательный, куговоротный circulatory, rotary, gyrating; circular
круговращение circulation, rotation, gyration, circular motion
кругом round, around, about; *mil.* about face
~ **марш!** to the rear, march!
~ **через правое плечо!** right about face!
кругообзорный omnidirectional
кругооборот circulation, cycle; rotation; circle
~ **капитала** circulation of capital, capital cycle
кругообразно in a circle, circle-wise, circularly
~ **свернутый** *bot.* circinate
кругообразный circular, round; orbicular
кругообращение rotation, circulation; cycle; circular motion
кругосветный round-the-world; *naut.* circumnavigatory
кружало *constr.* bow member; curve piece
кружевница *text.* lace-maker
кружево lace
кружение gyration, whirling, spinning; eddy; dizziness
кружок 1. small circle, disk; 2. study/hobby group
~, **литературный** literary circle
~, **околососковый** mammary areola
~, **ресничный** ciliary disk
крумциркуль double-bowed caliper
крунода *math.* crunode
круп I *med.* croup
~, **вирусный** virus croup
~, **гортанный** laryngeal/catarrhal croup
~, **дифтерийный** diphtheric croup
~, **ложный** pseudocroup, false croup
~, **скарлатинозный** scarlatinal croup
круп II *(of horse)* croup, crupper
крупа 1. groats; cereal(s); 2. granular snow, sleet, graupel
~, **ледяная** small hail
~, **снежная** snow pellets, granular/tapioca snow, soft hail
крупинка granule, small grain
крупитчатый granular, grainy
~**тая мука** fine wheet flour
крупка I grist, middlings; semolina; whole meal
крупка II *bot.* draba, whitlowwort *(Draba)*
крупнобугристый macrotuberous
крупноветвистый *bot.* macrocladous
крупноволокнистый coarse-fibred
крупногабаритный large-sized; bulky
крупноголовчатый, крупноголовый macrocephalous
крупнозем coarse earth
крупнозернистый coarse-/large-grained, macrogranular; coarse fibred
крупнозубчатый large-toothed
крупнокалиберный large-caliber *(attr.)*
крепноклеточный macrocellular
крупноколос(н)ый *bot.* large-eared/spiked, macrostachyous
крупнокорневой *bot.* macrorhizous
крупнокристаллический macrocrystalline, coarse-crystalline
крупнокусковой lumpy
крупнолист(н)ый *bot.* large-leaved
крупномасштабный large-scale
крупномолотый coarsely ground
крупноочаговый macrofocal

крупнопанельный large-panel
крупноперый *bot.* with long spikes and awns
крупнопестичный *bot.* large-styled
крупнопластинчатый macrolaminar
крупнопористый macroporous, coarse-pored
крупнопузырчатый, крупнопузырный macrovesiculous, coarse-vesiculous
крупноразмерный large-sized
крупнорогий large-horned
крупносеменной, крупносемянный large-seeded
крупносерийное производство large-scale/lot/isssue manufacture/production
крупносланцеватый *geol.* platy
крупностебельный *bot.* macropodal, macropodous
крупность size; lump/grain size; coarseness
крупнотоннажное судно large-capacity vessel, super-ship
крупнотравье tall (herbaceous) vegetation
крупнотычинковый *bot.* macrandrous
крупноузелковый macronodular
крупноцветковый large-flowered
крупночашечколистиковый *bot.* macrosepalous
крупноягодный *bot.* large-berried
крупноячейный large-/wide-meshed
круповейка *agr.* middlings purifier
круподерка *agr.* hulling machine/mill
крупозный *med.* croupous
крупон *leath.* butt, crop, corps
крупонировать *leath.* crop
крупопропариватель *agr.* groats parboiling unit
крупорезка groat(s) grinder
крупорушка *agr.* sheller, grits cutter; hulling/peeling mill
крупосортировка groats grader
крупье croupier
крура *zool., anat.* crus (*pl.* crura)
крустификационный, крустифицированный incrustate
крустификация *geol.* crustification
крутизна 1. steepness; slope; tangent; 2. transconductance
  ~ кривой slope of a curve
  ~ нарастания нейтронного потока neutron transconductance
  ~ преобразования conversion transconductance
  ~ склона slope angle
  ~ фильтра filter discrimination
крутило knob; hand wheel
крутильный torsional; twisting; *text.* doubling
  ~ная машина twisting machine, twiner
крутильщик *text.* twister
крутить twist; turn, wind; roll up
крутка twist(ing)
  ~ крыла wing twist
  ~ крыла, отрицательная wash-out
  ~ крыла, положительная wash-in
  ~ лопасти blade twist
  ~, остаточная residual/permanent twist
  ~, правая/левая right/left twist
  ~, слабая loose/soft twist
круткомер twist tester/counter
крутой steep; sharp, abrupt, sudden; *fig.* stern, severe, drastic
крутость gradient, slope
крутящий twisting, turning, winding
  ~ момент torque
кручение 1. torsion, twist; rotation; 2. *text.* twisting, spinning, doubling
крученый twisted; tortile; *sport* spinning, with a spin on
крушение wreck, wreckage, ruin; collapse; accident
  ~ судна shipwreck
крушилка grinder, crusher, grinding/crushing mill
крушина *bot.* buckthorn *(Rhamnus)*
  ~ ломкая/ольховидная alder buckthorn
крушинница *zool.* brimstone butterfly *(Gonepteryx rhamni)*
крушиновые *pl. bot.* buckthorn family *(Rhamnaceae)*
крыж hilt, cross guard; handle; crosspiece
  ~ на талях *naut.* racking stopper
  ~ у якорных цепей *naut.* elbow in the hawse
крыжовник *bot.* gooseberry *(Grossularia)*
крыланы *pl. zool.* fruit bats *(Megachiroptera)*
крылатка 1. *tech.* wing nut; vane; 2. *bot.* winged fruit; samara; 3. *zool.* ribbon seal *(Histriophoca fasciata)*
крылатковидный *bot.* samariform, samaroid
крылатоокаймленный *bot.* peripterate
крылатоцветный *bot.* wing-flowered
крылатый winged; alate, alar, alary
  ~ болт butterfly bolt
  ~тая гайка wing nit
  ~тая ракета cruise missile
крылач fan, blower, ventilator; impeller
крыло wing; aerofoil; blade, vane, arm; leaf; *geol.* limb, leg, side, wall
  ~, авиационное air wing
  ~ автомобиля wing; *Am.* fender
  ~, верхнее top wing
  ~, выдвижное extensible wing
  ~, высокорасположенное *aero.* high-mounted/set wing, shoulder wing
  ~, заднее rear wing/fender
  ~ здания aisle, wing
  ~ изменяемой геометрии variable(-geometry) wing
  ~ изменяемой стреловидности variable-sweep wing
  ~ клиновидной кости, большое/малое *anat.* great/lesser wing of the sphenoid bone
  ~, консольное cantilever wing
  ~, короткоширокое stub wing
  ~, крестообразное crossed/cruciform wing, X-wing
  ~ кривой wing of the curve

~, **крышеобразное треугольное** caret wing
~ **летательного аппарата** aerofoil, wing
~ **малого сопротивления** low-drag wing
~, **машущее** flapping wing
~, **многолонжеронное** multi-spar wing
~, **многостеночное** multi-web wing
~ **мостика** *naut.* bridge extension/wing
~, **мягкое парашутирующее** parawing
~, **натурное** full-scale wing
~ **несимметричного профиля** cambered wing
~, **нижнее** lower wing
~, **низкорасположенное** low-set/mounted wing
~ **носа** *anat.* wing of nostril
~ **обратной/отрицательной стреловидности** forward-swept [swept-forward] wing
~, **однолонжеронное** monospar wing
~ **отъемное** detachable wing
~ **парапланера** paraglider wing, parawing
~ **переменного размаха** variable-span wing
~, **переставное** adjustable wing
~, **плоское** planar/plane/uncambered wing
~, **поворотное** tiltable/tilting wing
~, **подводное** hydrofoil
~, **прямое** straight/upswept wing
~ **прямой стреловидности** back-swept [swept-back] wing
~, **раздвижное** telescopic wing
~ **ромбовидного профиля** double-wedge wing
~, **ромбовидное** diamond-shaped wing
~ **с большой стреловидностью** highly-swept wing
~ **с изменяемым углом установки** variable-incidence wing
~ **с круткой** warped wing
~ **с отрицательной круткой** wash-out wing
~ **с отрицательной стреловидностью** forward-swept [swept-forward] wing
~ **с положительной круткой** wash-in wing
~ **с положительной стреловидностью** back-swept [swept-back] wing
~, **свободнонесущее** cantilever wing
~, **сводовое** *geol.* arch limb
~ **семафора** semaphore arm/blade
~, **складывающееся** folding wing
~ **смыкающее/соединительное** *geol.* common limb
~, **стреловидное** swept wing
~, **треугольное** triangular/triangle/delta wing
~, **трехлонжеронное** three-spar wing
~, **цельное** one-piece wing
~ **цветочное** *bot.* spathe
~ **чечевицеобразного профиля** biconvex wing
крыло-верхнечелюстной *anat.* pterygomaxillary
крыловидный, крылообразный pterygoid, aliform, winglike, wing-shaped
крыло-небный *anat.* pterygopalatine
крыло-нижнечелюстной *anat.* pterygomandibular
крылоногие *pl. zool.* pteropods *(Pteropoda)*
крылоногий *zool.* aliped, pteropodous
крылоносный aliferous
крылоорешина, крылоорешник *bot.* wing nut *(Pterocarpa)*
крылопанцерные *pl. paleont. Pterichthys, Pterichthyodes*
крылоплодный *bot.* pterocarpous
крылосемянный *bot.* pterospermous
крылотычинник *bot.* stone cress *(Aethionema)*
крылоухие *pl. zool.* pterygote insects *(Pterygota, Pterygogenea)*
крылоцветковый *bot.* wing-flowered
крылоягодный *bot.* pterococcous
крылышко *bot. zool.* alula
крыльный alar
крыльчатка impeller, blade wheel
~, **двусторонняя** double-sided/entry impeller
~ **нагнетателя** supercharger impeller
~, **односторонняя** single sided/entry impeller
крыльчатый winged
~**тая гайка** fly/wing nut
крым-сагыз *bot.* krym-saghyz *(Taraxacum hybernum)*
крымский Crimean
крымотатарский Crimean Tatar
крыночка *bot.* pyxidium
крыса *zool.* rat *(Rattus)*
крыша 1. roof; 2. *anat.* tectum, tegmen, cover, tectorium, roof
~ **автомобиля** automobile top
~ **барабанной полости** *anat.* roof of the tympanum
~ **башни** turret roof/top
~, **двускатная** gable roof
~, **застекленная** skylight
~, **мансардная** mansard/curb roof; gambrel roof
~, **односкатная/шедовая** shed roof
~ **черепа** *anat.* skull cap
~ **четвертого желудочка** roof of the fourth ventricle
~, **шатровая/вальмовая** hip roof
~, **щипцовая** gable roof
крышевидный, крышеобразный roof-/chevron-shaped; tectiform
крышка 1. cover, cap, lid, top; 2. *anat.* tegmen
~ **аварийного люка** emergency hatch cover
~, **броневая** armoured cover
~, **герметическая** sealed cover
~ **горловины** scuttle cover
~ **заливной/заправочной горловины** filling hole [filler] cover
~ **иллюминатора** sidelight shutter
~ **клапана** valve bonnet
~ **колодца** manhole cover
~ **лаза** manhole cover
~ **люка** hatch door
~, **откидная** flap/hinged cover
~, **отрывная** tear-off cap

~ **подшипника** bearing cap
~, **притертая стеклянная** ground stopper
~ **сальника** gland cap
~, **сдвижная** sliding cover
~ **тигля** crucible lid
~ **цилиндра** cylinder cover
~, **черепная** *anat.* calvaria, calvarium
крюойсов *hist.* kevel, clamp
крюйс-пеленг *naut.* cross-bearing
крюк hook
~, **буксирный/буксировочный** towing hook
~, **вертлюжный** swivel/shackle hook
~, **двурогий** sister hook
~, **задерживающий** arrester/arresting hook
~, **запирающий** locking hook
~, **однорогий** single hook
~, **посадочный** *av.* landing/arrester hook
~, **рулевой** rudder pintle/hanger
~, **тормозной** arrester nook
~, **хвостовой (тормозной)** tail hook
~, **цепной** chain hook
крюкоблок block hook, hook block
крюковатый hooked, hooklike
крюкоголовый *zool.* rhyncocephalian
крючковатый hamose, hamate, hook-shaped; hamular
крючковидный, крючкообразный hooklike, hook-shaped; hamiform; unciform
крючкорог *zool.* Arctic sculpin *(Artediellus uncinatus)*
крючок hook; catch, clasp; *surg.* tenaculum, retractor; *bot., zool.* uncus, hamulus
~, **акушерский** blunt hook
~, **вязальный** crochet hook
~, **глазной** ophthalmic tenaculum
~, **зубчатый хирургический** pronged surgical retractor
~, **крыловидный** pterygoid hamulus
~, **пластинчатый хирургический** plate/laminar surgical retractor
~, **спусковой** trigger
~, **трахеотомический** tracheotomy hook
~, **хирургический** surgical retractor
крючок-переключатель *el.* hook switch
крючок-подъемник hook-elevator
крюшон cold fruit punch; cup
кряж 1. ridge; mountain range; 2. block, log, stump; *bot.* jugum
кряква *zool.* mallard, wild duck *(Anas platyrhynchos)*
ксанталин *chem.* xanthaline
ксантат *chem.* xanthate
ксантеин *chem.* xanthein, anthochlorin
ксантелазма *med.* xanthelasma, xanthoma palpebrarum
~ **век/кожная** dermal/skin xanthelasma
ксантелазмоидия xanthelasmoidea, dermal xanthomatosis
ксантемия xanthemia, carotinemia
ксантин *bioch.* xanthine
ксантиндегидраза, ксантиноксидаза xanthine oxidase/dehydrogenase
ксантинурия xanthi(n)uria, xanthuria
Ксантиппа Xanthippe
ксантогенератор xanthator
ксантогенирование xanthation
ксантогранулема *med.* xanthogranuloma
ксантогранулематоз xanthogranulomatosis
ксантодермия xanthodermia, xanthodermy, xanthoplasty
ксантоз xanthosis
ксантозин xanthosine
ксантозома = ксантосома
ксантокреатин xanthocreatine, xanthocreatinine
ксантоксилум *bot.* prickly ash *(Xanthoxylum, Zanthoxylum)*
ксантома *med.* xanthoma, Rayer's disease
~ **век** xanthelasma, xanthoma palpebrarum
~, **ложная** pseudoxanthoma elasticum
~, **множественная** xanthomatosis, xanthoma multiplex
~, **плоская** xanthoma planum
~, **резорбтивная** resorption xanthoma
~, **узелковая** xanthoma tuberosum
ксантоматоз *med.* xanthomatosis
~ **глаз** xanthomatosis bulbi
~ **кожи** dermal xanthomatosis, urticaria pigmentosa, xanthelasmoidea
~, **костный** bony xanthomatosis
~ **легких** pulmonary xanthomatosis
~ **суставов** articular xanthomatosis
ксантоматозный, ксантомный xanthomatous
ксантомицин xanthomycin
ксантон *chem.* xanthone, xanthenone, benzophenone oxide
ксантопатия xanthopathy
ксантопия xanthopia
ксантопикрин *chem.* xanthopicrin(e)
ксантопротеин *chem.* xanthoprotein
ксантопротеинемия xanthoproteinemia
ксантопротеиновый xanthoproteic
ксантопсин xanthopsin, visual yellow
ксантопсия xanthopsia, yellow vision
ксантопуцин xanthopuccine, canadine
ксанторамнин *chem.* xanthorhamnin
ксанториза bot. yellowroot *(Xanthorrhiza, Zanthorrhiza)*
ксанторрея *bot.* grass tree *(Xanthorrhoea)*
ксантосаркома *med.* xanthosarcoma
ксантосома *bot.* malanga *(Xanthosoma)*
ксантофан xanthophane
ксантофиброма xanthofibroma
ксантофил xanthophyll
ксантофор *zool.* xanthophore
ксантохромия xanthochromia
ксантохромный xanthochromatic, xanthochromic, xanchromatic
ксантоцианопия xanthocyanopia, red-green blindness
ксантуреновый xanthurenic

ксантурия xanthuria, xanthinuria, xanthiuria
ксенартральный *zool.* xenarthral
ксеногамия *bot.* xenogamy, cross-fertilization
ксеногамный *bot.* xenogamous
ксеногенный *biol.* xenogenic
ксеногибридизм xenohybridization
ксенодиагностика xenodiagnosis
ксенолалия xenolalia
ксенокристалл xenocryst, exogenic/foreign inclusion
ксенолит xenolith
ксеномения *med.* xenomenia, vicarious menstruation
ксеноморфный xenomorphic, allotrimorphic
ксенон *chem.* xenon, Xe
ксеноновый xenon *(attr.)*
ксенопаразит xenoparasite
ксенопатичесний xenopathic
ксенотрансплантат xenograft, heterograft
ксенофоб xenophobe
ксенофобия xenophobia
ксенофтальмия xenophthalmia
ксеричесний xeric
Ксеркс *hist.* Xerxes
ксерогалофит *bot.* xerohalophyte
ксерогеофит *bot.* xerogeophyte
ксерограмма xerogram
ксерография, ксерографирование xerography
ксерографический xerographic
ксеродерма *med.* xeroderma, xerodermia
ксероз *med.* xerosis
~ кожи xeroderma, xerodermia
~ конъюнктивы xerophthalmia, xerophthalmus
ксерозный *med.* xerotic
ксероклейстогамия *bot.* xerocleistogamy
керокопия xerox copy
ксерома *med.* xeroma
ксероморфизм, ксероморфность xeromoriphism, xeromorphy
ксероморфный xeromorphic, xeromorphous
ксероморфоз *bot.* xeromorphosis
ксерорадиограф xeroradiograph
ксерорадиография xeroradiography
ксеростомия *med.* xerostomia
ксеротермический xerothermic
ксерофагия xerophagia, xerophagy
ксерофил *bot.* xerophil(e)
ксерофилия *bot.* xerophily
ксерофильный xerophile, xerophilous
ксерофит *bot.* xerophyte
ксерофобный xerophobous
ксероформ xeroformium
ксерофтальмия *med.* xerophthalmia
ксерохазия xerochasy
ксерохазный, ксерохастический xerochastic
ксикаин *chem.* lidocaine hydrochloride
ксилан *chem.* xylan
ксиланаза xylanase
ксилема *bot.* xylem
ксилен *chem.* xylene
ксилоген *bot.* xylogen
ксилогенный xylogenous
ксилогидролаза xylohydrolase
ксилогравюра xylograph, woodcut
ксилограф xylographer
ксилография 1. xylography, block printing; 2. xylograph, woodcut
~, многоцветная chromoxylography
ксилоза *chem.* xylose
ксилоизомераза xylose isomerase
ксилокарп xylocarp
ксилокарпный xylocarpous
ксилол *chem.* xylene, xylol
ксилолит xylolith, xylolite
ксилометр *for.* xylometer
ксилоподий xylopodium
ксилофильный xylophilous
ксилофит *bot.* xylophyte
ксилофон *mus.* xylophone
ксилулоза xylulose, pentulose
ксилулокиназа xylulokinase
ксироспазм *med.* xyrospasm, shaving cramp
ктенидий *zool.* ctenidium, gill comb
ктенофор *zool.* ctenophore
ктитор churchwarden
ктырь *zool.* robber fly *(Asilus)*
куб I 1. *math.* cube; 2. boiler; still; vat
~ непрерывной перегонки continuous still
~, элементарный unit cube
куб II indigo plant/tree
кубанец Kuban Cossack
кубанка *bot.* hard wheat *(Triticum durum)*
кубатура cubic content, volume, cubature
кубатурник *for.* volumetric table
кубеба *bot.* cubeb pepper *(Piper cubeba)*
кубик (small) cube; block
кубинец, кубинский Cuban
кубинка Cuban (woman)
кубический cubic; cube-shaped
~ корень *math.* cube root
кубовидный cubiform, cuboid, cube-shaped
кубовый indigo *(attr.)*
кубоид *math.* cuboid
кубок bowl, beaker, goblet; *sport* cup
~, глазной optic cup
~ Нептуна *zool.* Neptune's cup/goblet
~, переходящий challenge cup
кубометр cubic meter
кубрик *naut.* crew's accommodation/quarters/berthing/space, bunk room
кубышка *bot.* cow lily *(Nuphar)*
кувалда sledge hammer
кувейтец, кувейтски Kuwaiti
кувейтка Kuwaiti (woman)
кувшинка *bot.* 1. water lily *(Nymphaea);* 2. cow lily *(Nuphar)*
кувшинковые *pl. bot.* water-lily family *(Nymphaeaceae)*
куга *bot.* bulrush *(Scripus lacustris)*

кугуар *zool.* cougar, puma *(Felis concolor)*
кудель tow
~, льняная flax tow
~, прочесанная carded tow
кудри curls
~, царские *bot.* martagon lily, Turk's cap lily *(Lilium martagon)*
кудрявец *bot.* Jerusalem oak, feather geranium *(Chenopodium botrys)*
кудрявоцветковый *bot.* crispifloral, curly-flowered
куду *zool.* koodoo, kudu *(Tragelaphus)*
Кузбасс *abbr.* the Kuznetsk Basin [coal fields], the Kuzbas
кузбасс-лан coal-tar varnish
кузнец blacksmith; farrier
~, ковочный horseshoer
кузнечик *zool.* 1. grasshopper; 2. Mormon cricket *(Anabrus);* 3. coulee cricket *(Peranabrus scabricollis)*
кузнечно-прессовый forge-and-press
кузница blacksmith's shop, forge, smithy
кузов *auto.* body
~, безрамный/несущий frameless/unitized body
~ с откидными бортами drop-sided body
кузовок 1. *bot.* cyathium; pyxidium; cap; 2. *zool.* rock shellfish, chapin *(Lactophrys tricornis)*
кузов-холодильник refrigerator body
кузу *zool.* true phalanger *(Trichosurus)*
кузыка *zool.* grain beetle *(Anisoplia)*
куизм Coueism
кукабурра *zool.* kookaburra *(Dacelo gigas)*
куколица *bot.* silene *(Silene)*
куколеотборник *agr.* cockle separator, cockler
куколка *zool.* chrysalis, pupa
~, ложная coarctate larva
куколь I corn cockle *(Agrostemma)*
куколь II *eccl.* cowl
кукольван *bot.* cocculus, snail-seed *(Cocculus);* Malay fishberry *(Anamirta cocculus)*
кукуруза maize *(Zea);* (Indian) corn *(Zea mays)*
~, одичавшая wild corn *(Zea canina)*
~, воздушная/лопающаяся popcorn
кукурузный corn *(attr.),* maize *(attr.)*
~ початок corncob
кукурузодробилка corncrusher, maize crusher
кукурузолущилка corn sheller
кукурузоуборочный corn-picking
кукушка *zool.* cuckoo *(Cuculus canorus)*
кукушкин cuckoo's, of a cuckoo
~ лен haircarp (moss) *(Polytrichum commune)*
~ы слезки spotted orchis *(Orchis maculata)*
кукушкообразные *pl., zool.* *Cuculiformes*
кукушник *bot.* ragged robin *(Lychnis floscuculi)*
куку-яман *zool.* blue sheep, bharal *(Pseudois nayaur)*
кукша *zool.* Siberian jay *(Cractes/Perisoreus infaustus)*
кулак I cam; knuckle; fist
~, поворотный *auto.* steering knuckle
кулак II kulak, rich peasant
кулакоплодный *bot.* condylocarpous
кулан *zool.* koulan, Asiatic wild ass *(Equuus hemionus)*
кулачество the kulaks, the rich peasants
кулачок cam, cog; catch, detent, pawl
~, запирающий locking cam
~, кардиоидный heart cam
~, спусковой releasing cam
кулежа *zool.* salmon trout *(Salmo trutta)*
кулик *zool.* sandpiper; stint *(Calidris)*
кулинар culinary specialist; expert cook
кулинария the culinary art, cookery
кулинарный culinary
кулиса rocker (arm), connecting/reversing link
~, поступательно-движущаяся reciprocating link
кулисы *theatre* side scenes, slips, wings
кулиус *zool.* brown gudgeon *(Culius fuscus)*
кулич Easter cake
куличок *zool.* sandpiper *(Calidris)*
Ку-лихорадка *med.* Q-fever
кулон *phys.* coulomb, ampere-second, C
кулонметр *el.* coulomb meter, coulometer
кулуары lobby
кулоновский Coulombian, Coulombic; coulomb
кульбаба *bot.* hawkbit *(Leontodon)*
кульверт culvert
кульдоскоп *med.* culdoscope
кульдоскопия *med.* culdoscopy
кульминация culmination
~, верхняя upper/superior culmination, upper medidian passage
~, нижняя lower culmination/transit, lower meridian passage
культ cult; worship; religion, denomination
~ личности personality cult
культивар *agr.* strain, cultivar, variety
культиватор cultivator; tiller; scarifier
~, дисковый disk cultivator
~ для междурядной обработки inter-row cultivator
~, лапчатый duckfoot cultivator
~, многорядный gang cultivator
~, пропашной inter-row cultivator/hoe
~, штанговый rod weeder
~, фрезерный gyrotiller
культиватор-букетировщик thinner-weeder
культиватор-плоскорез subsurface cultivator
культиватор-рыхлитель scarifier
культивация cultivation, tillage, tilling; growing, raising
культивировать cultivate; grow, raise
культивируемый tillable
культкомиссия *abbr.* commission for cultural activities/affairs
культмассовая работа cultural work among the masses; education of the masses
культовый religious; cult *(attr.)*

культпроп *abbr.* cultural/educational propaganda
культработа *abbr.* cultural and educational work
культработник *abbr.* person engaged in cultural and educational activities
культура 1. culture; 2. *agr.* crop; 3. *biol.* culture; 4. *agr.* cultivation, growing, raising
~, **агаровая** agar culture
~, **бактериальная** (bacterial) culture
~, **беспочвенная** *agr.* gravel culture
~, **бессеменная** *agr.* continuous cropping
~, **бобовая** leguminous crop
~, **богарная** dryland culture
~, **бульонная** broth culture
~, **вирусная** virus culture
~, **выгоночная** forcing crop
~, **высокоурожайная** *agr.* high yielder
~, **глубинная** submersed culture
~, **гравийная** *agr.* gravel culture
~, **засухоустойчивая** drought-resistant [short-rain] crop
~, **зерновая** cereal crop
~ **земледелия** efficiency in agriculture
~, **злаковая** grain crop
~, **интенсивная** intensive [high-value] crop
~, **исходная** *biol.* stock culture
~, **кормовая** *agr.* feed/fodder crop
~, **кровяная** blood culture
~ **лизогенная** lysogenic/lytic culture
~, **масличная** oil-bearing crop
~, **микробная** germ culture
~, **многолетняя** *agr.* perennial crop
~ **на зеленый корм** *agr.* haylage
~ **на корню** standing crop
~, **неорошаемая** non-irrigated crop
~, **низкоурожайная** *agr.* low yielder
~, **озимая** winter crop
~, **орошаемая** irrigated crop
~, **пастбищная** grass/grazing/sod crop
~, **пластообразующая** sod-forming crop
~, **плодовая** orchard crop
~, **поверхностная** surface culture
~, **погруженная** submerged culture
~, **подвижная** *biol.* motile culture
~, **пожнивная** stubble crop
~ **покоящихся клеток** resting-cell culture
~, **поливная** irrigated crop
~, **почвоистощающая** soil-depleting crop
~, **почвоочищающая** cleaning crop
~, **почвоулучшающая** soil amendment/improving crop
~, **продовольственная** food crop
~, **пропашная** tilled/cultivated crop
~, **противоэрозионная** erosion-resisting crop
~, **рядовая** drill culture
~, **самосевная** self-seed crop
~, **сахароносная** sugar crop
~, **севооборотная** rotating crop
~, **селективная** selective culture
~, **сельскохозяйственная** (cultivated) crop
~, **силосная** ensilage/silage crop
~, **скороспелая** short-duration/season crop
~ **со сменой среды** replacement culture
~ **сплошного посева** close-growing crop
~, **техническая** industrial crop
~, **теплолюбивая** warm weather crop
~ **ткани** *biol.* tissue culture
~ **ткани, однослойная** monolayer tissue culture
~, **уплотненная** compaction/crowded crop
~, **физическая** physical education/training
~, **чистая** pure culture
~, **яровая** spring/summer crop
культуризм body building
культурист body builder
культурно-бытовой cultural and social
культурно-воспитательный cultural and educational
культурно-исторический cultural-historical
культурно-массовый for spreading culture among the masses
культурность culture, cultural standads, level of culture
культурный 1. cultural; 2. cultivated, cultured; educated; 3. *agr.* cultivated, for cultivation cultivar, cultivated variety
культуры *agr.* crops
~, **ведущие/основные** staple crops
~, **тепличные/парниковые** crops under glass
культя *med.* stump
куманика *bot.* European blackberry *(Rubus fruticosus)*
кумафос coumafos
куметр quality-factor meter, Q-meter
кумжа *zool.* brown trout *(Salmo trutta)*
кумкват *bot.* kumquat *(Fortunella margarita)*
кумыс koumiss, kumiss
кумысолечение k(o)umiss therapy/treatment
кунджа *zool.* Fast Siberian char *(Salvelinus leucomaensis)*
кунжут *bot.* sesame *(Sesamum)*
куница *zool.* marten *(Martes)*
куннингамия *bot.* China fir *(Cunninghamia)*
купаж blend, blending; mixture
Купала, Купало *rel.* St. John the Baptist
купальница *bot.* globeflower *(Trollius)*
купе *rail.* compartment
купелирование, купеляция cupellation
купель *eccl.* font
купена *bot.* Solomon's-seal *(Polygonatum)*
куперит *med.* inflammation of Cowperian glands
купирование *med.* arresting, cupping, stopping
купировать *med.* arrest, cut short, cup off, stop
купол dome; cupola; *met.* crown
~, **астрономический** astrodome
~, **лавовый** lava dome, tumulus
~, **ледниковый** glacial dome, ice cap
~ **неба** the canopy of heaven
~ **парашюта** parachute canopy
куполовидный, куполообразный dome-/cupola-shaped
купон coupon; ticket

~ **за получение дивиденда** dividend coupon
~ **на оплату процентов** *econ.* interest coupon
~, **облигационный** bond coupon
~, **премиальный** premium coupon
купорос *chem.* vitriol, copperas
~, **железный** iron vitriol
~, **медный** blue/copper vitriol
~, **цинковый** zinc vitriol
купоросный vitriolic; vitriol *(attr.)*
~**ное масло** vitriol oil
куприт *miner.* cuprite
~, **волосной** capillary cuprite, chalcotrichite
~, **землистый** earthy cuprite
купроотунит cuproautonite
купроуранит cuprouranite
купула *bot.* cupule
купчая *econ.* bill of sale/purchase
купырник *bot.* hedge parsley *(Torilis)*
купюра 1. *fin.* denomination; 2. abridgement, cut
курай *bot.* Russian thistle *(Salsola)*
куракосборщик *agr.* cotton boll harvester
кураре curare, curari
курареподобный curarelike, curariform
курариазция curarization
курарин curarine
куратор 1. curator; 2. medical student in charge of a patient
курбет *sport* curvet
курвиметр curvimeter, curvometer; hodometer
курган burial mound; hill, tumulus
курганник *zool.* long-legged buzzard *(Buteo rufinis)*
курд Kurd
курдский Kurdish
курдянка Kurdish woman
курзал Kursaal
курильница incense burner, censer
курица hen; chicken
~, **шлемоносная** *zool.* helmeted curassow *(Pauxi pauxi)*
куркума *bot.* turmeric *(Curcuma)*
куроводство poultry/chicken farming, poultry breeding
курок trigger, firing hammer
~, **спусковой** *naut.* launching trigger/pawl
куропатка *zool.* partridge *(Perdix)*
куропаточник *bot.* Lapland willow *(Salix lapponum)*
курорт health resort
~, **бальнеогрязный** mud-bath health resort
~, **бальнеологический** spa (resort), balneologic health resort
~, **грязевой** mud-cure resort
курортотерапия health-resort treatment
курослеп *bot.* tall buttercup *(Ranunculus acris)*
курочка chicken, pullet
~, **водная** moorhen, gallinule *(Gallinula chloropus)*
курс 1. course, heading; 2. *fin.* exchange rate; 3. *educ.* course
~ **акций/биржевой/биржи** share/stock/market price; exchange/market quotation
~, **валютный** rate of exchange, exchange rate
~, **вексельный** exchange rate
~ **ведущий к столкновению** collision course
~, **взлетный** *av.* takeoff headway
~, **встречный** head-on course
~ **дня** quotation of the day
~, **заданный** set/predetermined course
~, **заочный** *educ.* correspondence course
~, **истинный** true/datum course/heading
~ **на бирже** Exchange price
~ **на закрытие биржи** closing quotation/price
~ **на открытие биржи** opening quotation/price
~ **нового выпуска ценных бумаг** issue price
~, **новый** *hist.* New Deal
~, **обменный** exchange rate
~, **обратный** reciprocal course/heading; opposite course; reverse road
~ **обучения** curriculum
~ **отплытия/отхода** course at departure
~, **плавающий** *fin.* floating/fluctuating/free rate of exchange
~ **по кассовым сделкам** spot rate
~ **по срочной сделке** forward rate
~ **по стеллажным сделкам** put-and-call price
~ **подхода/прибытия** approach course
~ **покупателей** bid quotation; buying rate
~ **полета** flight course
~ **популярных лекций** extension program(me)
~ **продавцов** asked quotation
~, **принудительный** *econ.* forced exchange
~ **самолета** (aircraft) heading
~, **устойчивый** *naut.* steady course
~ **ценных бумаг** security price
~ **черного рынка** curb rate
~ **ы усовершенствования** postgraduate courses
курсант cadet; student, member of a course
~ **авиационной школы** aviation cadet
~ **военно-морского училища** naval cadet
курсив italics, italic type
курсировать cruise, run, ply
курсовой course *(attr.)*
~ **угол** *naut.* course angle, relative azimuth/bearing/direction; angle on the bow
курсограмма course recording
курсограф course recorder
курсозадатчик course selector/setter
курсопрокладчик (automatic) ground-position indicator; course plotter
курсоуказатель *av.* course indicator/detector
куртаж *comm.* brokerage
куртизанка courtesan
куртка jacket
~, **ветронепроницаемая** windbreaker
~, **летная** flight/flying jacket
~, **надувная спасательная** inflatable life vest/jacket
куртье broker

курум stone stream/river
курухтан *zool.* ruff *(Philomachus pugnax)*
курфюрст *hist.* elector
курфюршество *hist.* electorate
курцгалоп canter
курчавка *bot.* goat's-wheat *(Antraphaxis)*
курчатовый kurchatovium, Ku
курчявость curliness; curl
курьер courier, messenger
кусачки nippers, cutting pliers/forceps
~, боковые oblique/side cutting pliers
~, комбинированные combination pliers
~, костные bone forceps
~, стоматологические dental forceps
кусачки-плоскогубцы flatnose cutting pliers
кусковой lumpy, in lumps, lump *(attr.)*
кускус I couscous
кускус II *zool.* phalanger *(Phalanger)*
куст 1. *bot.* bush, shrub; tuft; bunch; 2. *miner.* pocket, bunch; 3. group, cluster
~ точек cluster of points
кустарник bushes, shrubs, shrubbery
~, древовидный arbuscle
кустарничество handicraft
кустарничковый *bot.* fruticulose
кустарный handicraft *(attr.);* homemade
кустарь artisan; handicraftsman
кустование *el.* interconnection
кусторез brush cutter, bush/shrub saw
кутас tassel
кутикула *anat., bot.* cuticle
~ зуба dental/enamel cuticle, skin of the teeth
кутровые *pl. bot.* dogbane family *(Apocynaceae)*
куттер *naut.* cutter
кутум *zool.* kutum *(Ratilus fristi kutum)*
куфея *bot.* cuphea *(Cuphea)*
куфический Kufic, Cufic
кучево-дождевой *meteor.* cumulo-nimbus
кучевой *meteor.* cumulus
кучерявка *bot.* babies'-breath *(Gypsophila paniculata)*
кучность compactness; concentration
кучный compact; concentrated; *geol.* glomeroplasmatic, glomeroplastic
кушитский Cushite
кхмер, кхмерский Khmer
кхмерка Khmer (woman)
кьянти Chianti wine
кэпрок *geol.* cap rock
кэрри *cul.* curry
кэт catboat
кэч *naut.* ketch
кюбель bucket
кювет roadside ditch
кювета container, vessel, cell, cuvette; flask; *phot.* bath
~ для проявления developing dish
~, зуботехническая dental flask
~, качающаяся tilting tray
~, фотографическая processing dish
кюветокопатель roadside ditch excavator
кюльдоскопия *med.* culdoscopy
кюльдотомия *med.* culdotomy
кюльдоцентез *med.* culdocentesis
кюммель kummel
кюрасо curacao
кюретаж *surg.* curettage, curettement
кэш *comp.* cache
~ с упреждающей выборкой forward-looking cache
кюретка curette
~, гортанная laryngeal curette
кюри curie, Curie unit, C
кюриграф curiegraph
кюрий *chem.* curium, Cm
кюрит curite
кюритерапия curietherapy, radium therapy
~, внутритканевая interstitial curietherapy
кюри-эквивалент curie-equivalent

Л, л

ла *mus.* A, la *see also* **ля**
лабаз grain merchant's shop/warehouse
лабазник I grain dealer/merchant, flour dealer
лабазник II *bot.* meadowsweet *(Filipendula)*
лабардан *cul.* salted/dried cod filet
лабданум *bot.* gum rockrose *(Cistus ladaniferus)*
лабеллум *bot.* labellum
лабиализация *phon.* labialization
лабиализм *med.* labialism
лабиальный labial
лабидометр *med.* labidometer, labimeter
лабильность lability, instability
~ психики psychic(al) lability/instability
лабильный labile, unstable
~ к низкой температуре *med.* frigolabile
лабиодентальный *phon.* labiodental
лабиомансия lip reading, labiomancy, visual hearing
лабиоментальный labiomental
лабиопластика *med.* labioplasty, chiloplasty
лабиохорея *med.* labi(o)chorea, labiochoreic stuttering
лабиоцервикальный *anat.* labiocervical
лабиринт labyrinth; maze
~, кортикальный *med.* cortical/Ludwig's labyrinth
~, костный osseous/bony labyrinth
~, перепончатый membranous labyrinth
~, прямоточный straight-through labyrinth
~, решетчатый ethmoidal labyrinth
~, ступенчатый stepped labyrinth
~ Шеннона Shannon's maze
лабиринтит *med.* labyrinthitis, otitis interna/labyrinthica
~, гнойный purulent/suppurative labyrinthitis

~, менингогенный meningogenic/meningogenous labyrinthitis
~, ограниченный limited labyrinthitis
лабиринтный labyrinthine, labyrinthiform
лабиринтодонт *paleont.* labyrinthodont
~, брахиопидный brachyopid labyrinthodont
~, рахиотомный rachiotomous labyrinthodont
лабитом *surg.* labitome, cutting forceps
лаборант laboratory assistant/technician; laboratorian
~, ведущий опробование assayer
лаборатория laboratory
~, авиационная aeronautical laboratory
~ авиационного оборудования airborne/aircraft equipment laboratory
~, аэродинамическая wind-tunnel laboratory, aerodynamic laboratory
~, больничная hospital laboratory
~ высокогорная mountain-top laboratory
~ высоко(радио)активных материалов *nucl.* hot laboratory
~ высотная *av.* high-altitude laboratory
~, гидроакустическая underwater sound laboratory
~, гистологическая histologic(al) laboratory
~, горячая *nucl.* hot laboratory
~, дозиметрическая radiation detection laboratory
~, заводская works/factory laboratory
~, зуботехническая dental laboratory
~ излучений radiation laboratory
~ изотопных индикаторов tracer laboratory
~ испытаний и контроля качества testing and quality control laboratory
~, испытательная testing laboratory
~, космическая space laboratory
~ механических испытаний materials testing laboratory
~, наземная ground based laboratory
~, научно-исследовательская (scientific) research laboratory
~ невесомости weightlessness laboratory, zero-gravity laboratory
~, обитаемая орбитальная manned orbital laboratory
~, отраслевая applied research laboratory
~, плавучая floating laboratory, technical research ship
~, подвижная бактериологическая bacteriological laboratory lorry
~, подвижная стоматологическая dental laboratory truck
~, полевая field laboratory
~, полевая химическая chemical agent analyser kit
~, проблемная fundamental/basic research laboratory
~, производственно-контрольная industrial laboratory; materials-testing laboratory; quality-control laboratory
~ радиоактивных материалов hot laboratory, radioactive material laboratory
~, санитарно-бактериологическая sanitary and bacteriological laboratory
~, санитарно-гигиеническая sanitary-hygienic laboratory, public-health laboratory
~ слабоактивных материалов *nucl.* "warm" laboratory
~ среднеактивных материалов *nucl.* "semi hot" laboratory
~, судебно-медицинская forensic medical laboratory
~, учебная school/college/ university laboratory
~, фотографическая dark-room
~, фотоиспытательная film testing laboratory
~, химическая chemical laboratory
~, центральная заводская main laboratory
лабораторный laboratory, laboratorial
лабрадор *miner.* labradorite
лаброцит *biol.* labrocyte
лабрум *zool.* labrum
лабфермент *bioch.* chymosin, rennet, rennin
лава I *geol.* lava
~, автобрекчированная autobrecciated lava
~, анортитовая anorthite lava
~, базальтовая basaltic lava
~, волнистая ropy/dermolithic lava
~, вязкая viscous/viscid lava
~, глыбовая block-lava, aphrolithic lava
~, грубая пузырчатая rough cellular lava, asperite
~, грязевая mud lava, moya
~, губчатая slaggy/scoriaceous lava
~, донная bench lava
~, жидкая fluid lava
~, занозистая splintery lava
~, застывшая/затвердевшая solidified lava
~, канатоподобная ropy/corded lava
~, кишкообразная pahoehoe lava
~, криптолейцитовая crypto-leucitic lava
~, мельничная mill(stone) lava
~, миндалекаменная amygdaloid lava
~, обломочная block-lava, aphrolitic lava
~ пахоэхоэ pahoehoe lava
~, подушечная pillow lava
~, пористая porous/vesiculated lava
~, потроховидная visceral lava
~, пузырчатая blister/cellular/vesicular lava
~, сверхподвижная hypermobile lava
~, светящаяся luminous lava
~, столбчатой отдельностью columnar lava
~, текущая flowing/effluent lava
~, трубчатая tubular lava
~, туфовая tuf-lava
~, тягучая viscous/viscid lava
~, шаровая spherulitic/globular lava
~, шехлунитовая sheklunite lava
~, шлаковая slag lava
~, эллипсоидальная ellipsoidal/spheroidal lava
~, языковидная tongue-like lava

~, ячеистая blister/cellular/vesicular lava
лава II *min.* wall, (long) face, longwall (face)
~, безлюдная manless face
~, комбайновая cutter-loading face
~, машинная cutter face, cutter longwall (face)
~, молотковая pick hammer face
~, спаренная double face
~, струговая plough face
лава III *mil.* (Cossack) cavalry charge
лава IV footbridge, planked footway
лава-этаж *min.* horizon-face
лаванда *bot.* lavender *(Lavandula)*
лавандовый lavender-blue
лаватера *bot.* tree mallow *(Lavatera)*
лаваш unleavened wheat cake, unleavened flat bread
лавина avalanche; *geol.* shower
~, весенняя/грунтовая ground avalanche
~ вулканического песка sand avalanche/flow
~, зимняя powdery avalanche
~, ледниковая glacial avalanche, stream of ice
~, снежная snow slide/slip/avalanche
~, сухая dust/powdery avalanche
~ Таунсенда/таунсендовская Townsend avalanche
лавинообразный avalanche-like/-type
лавинообразование avalanching, avalanche formation
лавирование *nav.* tacking
лавировать *nav.* go about, beat (up), tack
~ против ветра beat against the wind, beat to windward
лавировка *nav.* beating (against the wind)
~ против ветра windward work
лавис *print.* copperplate engraving/printing
лавка I bench
лавка II store, shop
~, корабельная canteen
~ со смешанным ассортиментом mixed/general shop
лавопад lava-fall, lava cascade
лавоподобный lavic, lavalike, lavatic
лавочник shopkeeper, shopster
лавр 1. *bot.* laurel *(Laurus)*; 2. *pl. fig.* laurels
~, благородный Grecian laurel *(L. nobilis)*
лавра monastery
лаврак *zool.* bass *(Morone labrax)*
лавровишня *bot.* cherry laurel, laurocerasus *(Laurocerasus)*
лавровый laurel *(attr.),* laureate
~ венок laurel wreath
~ лист *cul.* bay leaf
лавсан *chem.* lavsan; therylene, daeron, tergal
лавсония *bot.* henna *(Lawsonia inermis)*
лаг *naut.* log
~, бросать heave the log
~, вертушечный taffrail/propeller log
~, воздушный air log, (air-)distance recorder
~, гидравлический pitometer log, pressure type log
~, гидродинамический dynamic pressure log, Pitot/pitometer log
~, голландский Dutchman's log
~, доплеровский Doppler log
~, механический буксируемый patent log
~, ручной chip/hand log
~, самопишущий recording log
лагенария *bot.* gourd *(Lagenaria)*
лагер *geod.* bearing
лагерный camp *(attr.)*
лагерстремия *bot.* crape myrtle *(Lagerstroemia)*
лагерь camp
~, концентрационный concentration camp
~, раскинуть pitch a camp
лаглинь *naut.* log line
~, травить pay/run out the log line
лагом *nav.* beam-on, broadwise (on/to), sidewise
~ к ветру broadside to the wind
~ к волне beam-on to the sea/swell
~, поворачиваться swing broadside
~, становиться (к волне) fall athwart
лагосома *zool.* common lizard *(Lagosoma)*
лагофтальм *med.* lagophthalmia, lagophthalmos, hare's eye
лагофтальмический lagophthalmic
лагохилин lagochilin
лагохилус *bot.* lagochilus *(Lagochilus)*
лагранжиан *math.* Lagrangian
Лагтинг Lagting
лагун *naut.* scuttlebutt, scuttle cask
лагуна *geol.* lagoon; crateral lake
~, атолловая atoll lake
~, замкнутая landlocked embayment
~, широкая sound
лагунный lagoonal, lagoon *(attr.)*
лад 1. *mus.* mode, key; harmony, concord; 2. *mus.* fret; stop; 3. manner, way
~, мажорный major key
~, петь в sing in tune
~, петь не в sing out of tune
ладан 1. incense, frankincense, olibanum; 2. *bot.* Bible frankincense *(Boswellia carteri)*
~, конский *bot.* European wild ginger *(Asarum europaeum)*
~, росный *pharm.* benzoin
ладанка 1. incense burner; 2. amulet
ладанник *bot.* rockrose *(Cistus)*
ладанниковые *n. pl. bot.* rockrose family *(Cistaceae)*
ладаноносный thuriferous, incense-bearing
ладонный *anat.* palmar, volar
ладонь *anat.* palm
ладьевидный, ладьеобразный navicular, boat-shaped, scaphoid, cymbiform, cymbate; crinate
ладьеногие *n. pl. zool.* scaphopods *(Scaphopoda)*
ладья I 1. *anat.* scapha; 2. barge, rowboat, sailboat
ладья II *chess* castle, rook
ладьян *bot.* coralroot *(Corallorrhiza)*
лаж *econ.* agio, premium
лаз I 1. access hole, manhole, porthole; scuttle; *av.* crowlway; 2. *hunting* track

~ в крышке люка hatch scuttle
~, водонепроницаемый watertight manhole
~, смотровой inspection hole
~, туннельный tunnel hatch
лаз II Laz(e), Lazi *(member of ethnic group)*
лазанье, лазенье (mountain) climbing
лазарет *mil.* (field) hospital; sick quarters, dispensary; *naut.* sick bay
~, полевой casualty clearing station, infirmary, field ambulance station, field hospital
лазер laser, optical maser
~ бегущей волны travelling wave laser
~ ближнего действия close-range laser
~ взрывного типа explosion laser
~, возбуждаемый постоянным током dc-excited laser
~, гелий-неоновый helium-neon (gas) laser, He-Ne-laser
~, двухмодовый double-/two-mode laser
~, двухпримесный double-doped laser
~, двухрезонаторный dual-cavity laser
~, двухфотонный two-photon laser
~ для слежения за космическими объектами space track(ing) laser
~, задающий initiating laser
~ ИК диапазона infra-red laser, iraser, IR laser
~, импульсный pulsed/pulsing laser
~, инжекционный injection/diode laser
~, комбинационный Raman laser
~, моноимпульсный giant-pulse laser
~, многомодовый multimode/multimodal laser
~, мощный high-power/-light laser
~ на двуокиси углерода carbon-dioxide laser
~ на неодимовом стекле neodymium-doped glass laser
~ на хелатах chelate laser
~ непрерывного излучения continuous wave laser
~, несинхронизованный free-running laser
~, одномодовый unimode laser, single-mode laser
~, перестраиваемый tunable laser
~, резонаторный cavity/resonant laser
~ с внешней модуляцией externally modulated laser
~ с возбуждением электронным пучком electron-beam pumped laser
~ с ВЧ накачкой RF excited laser
~ с высокочастотной генерацией high-PRF laser
~ с модуляцией добротности Q-switching/-switched/-spoiled laser
~ с непрерывной генерацией CW laser
~ с оптической накачкой optically pumped laser
~ с передачей накачки cross-pumped laser
~ с резонатором Фабри-Перо Fabry-Perot laser
~, самосинхронизирующийся self-locked laser
~, сверхизлучающий superradiant laser
~ со связанными модами mode-coupled/-locked laser
~, твердотельный solid-state laser
~, термостатированный temperature-controled laser, termostatted laser
~, трехуровневый three-level laser
~, узкополосный narrow-band laser
~ ультрафиолетового диапазона uraser, ultraviolet/UV laser
~, частотно-модулированный frequency-modulated laser, FM laser
~, эпитаксиальный epitaxial(-grown) laser
лазерь-дальномер laser range finder
лазерпициум *bot.* laserwort, woundwort *(Laserpitium)*
лазка Laz(e), Lazi (woman)
лазоревка *zool.* blue tit *(Parus caerulus)*
лазоревый azure, sky-blue
~ камень lapis lazuli
лазорник *bot.* dayflower *(Cammelina)*
лазский Laz(e), Lazi
лазулит *miner.* lazulite
лазурит *miner.* lazurite
лазурник *bot.* laserwort, woundwort *(Laserpitium)*
лазурный azure, sky-blue
лазурь azure, blue, blue pigment
~, берлинская/ ~, прусская синяя Berlin/ Prussian blue, iron earth, ferric ferrocyanide
~, железная Prussian/milori blue
~, медная azurite
лайба sailing boat
лайка I husky, Eskimo/wolf dog
лайка II kid(leather), kidskin, taw
лайм *bot.* (native) lime *(Citrus aurantifolia)*
~, китайский myrtle lime, limeberry *(Triphasia trifolia)*
лайма *bot.* Lima bean *(Phaseolus limensis)*
лайнер *naut., av.* liner
~, быстроходный express liner
~, воздушный airliner
~, грузовой cargo liner
~, пассажирский passenger liner
~, реактивный воздушный jetliner, jet airliner
~, турбовинтовой воздушный propjet airliner
лайнсмен *sport* linesman
лак I lacquer; varnish
~, авиационный aircraft dope
~, автомобильный automobile/automotive lacquer
~, асфальтовый/битумный bituminous varnish
~, ацетатный acetate dope
~, ацетилцеллюлозный cellulose-acetate lacquer
~, бакелитовый bakelite varnish, phenol-formaldehyde lacquer
~, бесцветный clear varnish/lacquer

~, битумный bituminous/alsphalt varnish
~, влагостойкий moisture-proof varnish
~ воздушной сушки air-drying varnish
~, глифталевый glyceryl-phtalate varnish
~ горячей сушки stoving varnish
~, грунтовочный priming/undercoating varnish
~, даммаровый dammar varnish
~ для наружных работ exterior/outdoor varnish
~ для ногтей nail polish
~, защитный protective varnish; stop-off lacquer
~, звукоизоляционный soundproofing varnish
~, каменноугольный coal-tar varnish
~, канифольный colophonic/rosin varnish
~, кислотостойкий/кислотоупорный acid-resistant/-proof varnish
~, кремнийорганический silicone lacquer
~, кумароновый cumarone resin varnish
~, летучий volative solvent varnish
~, люминесцентный luminescent lacquer
~, маслостойкий oil-resistant varnish
~, масляный oleoresinous/oil varnish
~, матовый flat/mat varnish
~, мебельный furniture lacquer
~, мочевиноалкидный urea-alkyd lacquer
~ на масляной/спиртовой основе varnish
~ на целлюлозной/полимерной основе lacquer
~, нитроцеллюлозный nitrocellulose lacquer, nitrate dope
~, отделочный finishing varnish
~, пековый coal-tar varnish
~, перхлорвиниловый perchlorovinyl lacquer
~, печной stoving varnish
~, покровный coating/overprint varnish
~, полиакриловый polyacrylic lacquer
~, поливинилхлоридный polyvinyl chloride lacquer, PVC lacquer
~, полировочный polishing varnish
~, полистирольный polystyrene lacquer
~, полиэфирный polyester lacquer
~, пропиточный impregnation/impregnating varnish
~, пустынный *geol.* desert varnish
~, светочувствительный light-sensitive varnish
~, синтетический synthetic(-resin) lacquer
~, смоляной syntethic varnish
~, срывной *print.* parting/stripping varnish
~, фенольно-масляный spar varnish
~ холодной сушки air-drying varnish
~, цапоновый nitrocellulose lacquer
~, цветной pigmented lacquer
~, шеллачный (shel)lac varnish
~, шлифовальный flatting varnish
~, шпаклевочный putty/filling varnish
~, щелочестойкий alkali-resistant varnish
~, электроизоляционный insulating varnish
~, эмалевый enamel lacquer
~, эмульсионный emuslion varnish
~, эпоксидный epoxy/epoxide lacquer
~, эфироцеллюлозный cellulose lacquer
~, янтарный amber varnish
~, японский black japan
лак II Lak *(member of ethnic group)*
лакедра *zool.* rudder fish *(Seriola)*
лакировальный for varnishing/lacquering
лакировальщик varnisher, lacquerer
лакирование varnishing, lacquering, japanning, glazing
лакированный varnished, lacquered
~ная кожа patent leather
~ные туфли patent-leather shoes
лакировать lacquer, varnish, japan; apply varnish/lacquer
лакировка 1. varnishing, lacquering, japanning; 2. varnish/lacquer finish
лакировочный varnishing, lacquering
лакказа *bioch.* laccase
лакколит *geol.* laccolith
~, межформационный interformational laccolith
~, разорванный upfaulted laccolith
~, сложный complex/multiple/composite laccolith
~, смешанный compound laccolith
~ типа кедрового дерева cedar-tree laccolith
~, эруптивный eruptive/extrusive laccolith
лакколитит *geol.* laccolithite, laccolith-forming rock
лакколитообразный laccolithlike
лакмус *chem.* litmus; litmus paper
~, красильный *bot.* dyer's croton, turnsole *(Croton tinctorium)*
лакмусовый litmus *(attr.)*
~вая бумага litmus (paper)
лаколента varnished tape
лаконический, лаконичный short-spoken, laconic; terce
лаконичность laconicalness, laconi(ci)sm
лаконос *bot.* pokeberry *(Phytolacca)*
лаконоска *bot.* common pokeberry *(Phytolacca americana)*
лаконосные *n. pl. bot.* pokeweed family *(Phytolaccaceae)*
лак-опоек patent calf leather
лакостеклоткань varnished glass cloth
лакоткань varnished cloth/fabric/cambric
~, электроизоляционная varnished cambric/cloth
лакримальный *anat., zool.* lacrimal, lacrymal, lachrymal
лакриматор *chem.* lacrimator, lachrymator, lacrimatory/tear gas
лакриматорный lacrymatory, lacrimatory, lachrymatory
лакримация *med.* lacrimation, lachrymation
лакримотом *med.* lacrimotome
лакрица, лакричник *bot.* licorice, liquorice *(Glycyrrhiza)*
лакросс *sport* lacrosse

лактаза *bioch.* lactase
лактальбумин lactalbumin, milk albumin
лактальдегидредуктаза *bioch.* lactaldehyde reductase
лактарин lactarene, lactarine
лактат *chem.* lactate
~ железа/кальция ferrous/calcium lactate
лактацидемия *med.* lactacidemia, lacticacidemia
лактационный lactational, lactagogue, galactagogue, lactogenous, galactopo(i)etic, lactiferous, galactophorous, lactic
лактация *med.* lactation, lactogenesis, galactopoiesis
~, длительная prolonged lactation
~, избыточная (ga)lactorrhea
~, недостаточная oligogalactia
~, обильная galactosis
~, патологическая galactacrasia
~, подавленная galactischia, galactoschesis, galactostasia
лактемия *med.* lactemia
лактенин *bioch.* lactenin
лактобациллин 1. lactobacillin; 2. yoghurt
лактобутирометр lactobutyrometer
лактогенный lactogenous, galactopo(i)etic
лактоденсиметр lactometer, lactodensimeter, milk-gauge
лактоза *chem.* lactose, lactobiose, lactosum, milk sugar
лактозурия *med.* lactosuria
лактокультура lactic culture
лактометр lactometer, galactometer, milk-gauge
лактон *chem.* lactone
лактоникрил lactonitrile
лакторея *med.* lactorrhea, galactorrhea
лактоскоп lactoscope, galactoscope
лактотерапия *med.* lactotherapy, galactotherapy
лактотропин *bioch.* lactotrophin, prolactin, mammotrop(h)in
лактотропный lactotrop(h)ic
лактофлавин *bioch.* riboflavin, lactoflavin
лак-шевро patent kid
лакуна lacune, lacuna, gap, hiatus
~, мелкая lacunula
~ мочеиспускательного канала *anat.* urethral lacuna
лакунарный lacunal, lacunar(y)
лакфиоль *bot.* wallflower *(Cheiranthus)*
лаллеманция *bot.* lallemantia *(Lallemantia)*
лалоневроз laloneurosis
лалорея *med.* lalorrhea, lagorrhea
лалофобия lalophobia
лама I *zool.* llama *(Lama)*
лама II *rel.* lama
ламаи(с)т lamaist
ламаистский, ламаитский lamaistic
ламантин *zool.* manatee *(Trichechus)*
ламаркизм Lamarckism, Lamarckian theory
ламаркистский Lamarckian
ламаркия *bot.* goldentop *(Lamarckia)*
ламбада lambada
ламбдацизм lambdacism(us)
ламбдовидный *anat.* lambdoid(al)
ламберт *phys.* lambert
ламбрекен lambrequin
ламе *text.* lamé
ламелла lamella
ламеллибранхиаты *n. pl. zool.* lamellibranchs *(Lamellibranchiata)*
ламеллярный lamellar, lamellate
ламель 1. lamel(la), lamina; *el.* bar; 2. *text.* dropper
~ анодного блока anode segment
~ коллектора commutator bar/segment
~ переключателя commutator bar/strip
~ потенциометра potentiometer section
ламельный lamellar, lamellate
ламентация complaint, lamentation
ламинальный laminar, laminal
ламинаран *bioch.* laminaran
ламинариевые *n. pl. bot.* kelps *(Laminariaceae)*
ламинария *bot.* blade kelp, laminaria *(Laminaria)*
~, пальчатая/пальчато-рассеченная sea staft/tangle/wand *(Laminaria digitata)*
ламинаризация laminar control, laminarization
ламинарный laminal, laminar
ламинация lamination
~ покровов sheet lamination
ламинирование lamination
ламинировать press into sheets, laminate
ламинит *med.* laminitis
ламинограмма laminagram, laminogram
ламинография laminagraphy, laminography
ламинэктомия *surg.* laminectomy
ламия *myth.* lamia
лампа 1. light, *el.* lamp bulb; 2. *electr., rad.* valve; *Am.* tube
~, аварийно-сигнальная, ~ аварийной сигнализации alarm light/lamp, trouble lamp, emergency/warning light
~ "бак пуст", сигнальная tank empty caution light
~, бактерицидная bactericidal/dermicidal/uviol lamp
~ бегущей волны travelling-wave /valve/tube
~, биспиральная coiled-coil lamp
~, вакуумная vacuum tube/valve/ bulb
~, вентильная rectifier valve/tube, gate/valve tube
~, взрывобезопасная (flame) safety lamp
~, вибростойкая shock-resistant lamp; rough-/vibration-service lamp
~, вихревая vortex valve/tube
~, вольфрамовая tungsten lamp
~, входная input tube
~, вызывная *tel.* calling/line lamp
~, выпрямительная rectifier tube/valve, rectifyng tube
~, высоковакуумная high-vacuum valve/tube, hard valve/tube

~, высоковольтная high-voltage lamp
~, высокочастотная high-frequency valve, h-f-valve, radio-frequency tube, r-f tube
~, выходная output tube
~, газополная/газонаполненная gas-filled lamp/tube
~, газоразрядная 1. gaseous (discharge) lamp, (gas/gaseous) discharge lamp; 2. gas-discharge tube
~, газосветная glow-discharge lamp
~, генераторная transmitting valve/tube
~, главная master caution light
~, глазная щелевая *med.* ophthalmic slit lamp
~, головная *min.* head/cap lamp
~ "дверь не заперта", сигнальная door unlocked warning light
~, двуханодная double-anode valve, double-plate tube
~, двухлучевая double-beam valve/tube
~, двухнитевая twin-filament lamp
~, двухсветная *anat.* double-/two-filament bulb, bifilar bulb
~, двухцокольная double-base/-ended tube/valve
~, декадная decade counting valve/tube
~, демпферная damper valve/tube
~, детекторная detector valve/tube
~ для инфракрасного облучения infrared irradiation lamp, IR lamp
~ дневного света daylight/flourescent lamp
~, дуговая arc lamp
~ Дэви *min.* Davy lamp, Davy's safety lamp
~, жесткая hard valve/tube, high-vacuum valve/tube
~, задающая driver valve/tube
~ занятости *tel.* busy/engaged lamp
~, занятости всех регистров *tel.* all-senders-busy lamp
~, запасная spare lamp
~, запускающая *rad.* trigger tube
~, защитная guard lamp
~, зеленая сигнальная green indicator light
~, импульсная flash lamp/tube/bulb
~, инверторная inverter valve/tube
~, индикаторная indicator tube, (indicating) light, display/indicating/signal lamp
~, индикаторная цифровая numerical read-out tube, digital indicator tube
~ индикации включения автопилота auto-pilot engage light
~ индикации захвата цели *mil., av.* radar lock-on indication light
~ индикации режима работы mode indicator light
~ инфракрасного излучения, ~инфракрасная infra-red lamp, IR lamp
~, ионная gas-filled valve/tube
~, кварцевая quartz/sun lamp
~, керосиновая kerosene lamp
~, керосинокалильная incandescent oil vapour lamp
~, комбинированная multi-unit valve/tube, multiple/multisection valve/tube
~, коммутаторная switchboard lamp
~, контрольная включения дальнего света фар high-/main beam headlight, main-beam indicator, headlight indicator, blue control lamp, beam indicator lamp
~ контроля времени *teleph.* time-check lamp
~, копировальная printer lamp
~, коронирующая corona tube
~, лобная *med.* head lamp
~, люминесцентная luminescence/ luminescent/ fluorescent lamp
~, манипуляторная keying/keyer valve/tube
~ маркерного приемника, сигнальная marker beacon light
~, матированная frosted lamp
~, маячковая/маячная lighthouse tube/lamp, disk-seal tube
~ Минина *med.* Minin light
~, многосеточная multigrid valve/tube
~, мягкая soft/gassy valve/tube, low-vacuum valve/tube
~ накаливания incandescent/filament lamp
~ накаливания, авиационная aircraft filament bulb
~ накачки (лазера) pump(ing) lamp, pump tube
~ накачки (лазера), импульсная pump(ing) flashtube
~ направленного освещения flood, flood lamp, light
~, натриевая sodium(-vapour) lamp
~, неоновая neon-filled/-glow lamp, neon tube/lamp
~, неразборная permanently sealed tube
~ номерного знака *auto.* licence plate lamp
~ обратной волны backward-wave valve/tube, BWT
~, общая сигнальная master caution light
~, общевызывная pilot lamp
~, однонитевая single-/one-filament bulb
~, операционная *med.* operative lamp, operative illumination device
~ оптической накачки (лазера) optical pump(ing) lamp/tube
~, осветительная illuminating/lighting lamp, bulb
~ освещения подножки *auto.* courtesy lamp
~ освещения отсека compartment lamp
~ освещения приборной доски instrument panel light
~ освещения пульта управления *av.* console light
~, отбойная *teleph.* clearing/supervisory lamp
~, пальчиковая small-button glass tube, bantam tube
~, паросветная metal vapour lamp
~, паяльная blow/soldering/brazing torch, soldering lamp

~, перекальная photoflood
~ переменной крутизны variable-mu valve/tube
~, переносная utility light, extension/ hand/ inspection/ portable lamp
~, пневмоэлектрическая compressed air electric lamp
~, подсвечивающая bright-up lamp, exciter lamp
~ подсчета шкалы dial lamp
~ последовательного включения series lamp
~, предохранительная *el.* safety lamp
~, преобразовательная converter tube, heterodyne conversion transducer
~ приборного щитка panel lamp/light, dash (-board) light, facial light, instrument (cluster) lamp
~, приемно-передающая широкополосная broadband TR valve/tube
~, приемно-усилительная receiving valve/tube
~, промежуточная intertube
~, пропускающая gate valve/tube
~ прямого накала directly heated valve/tube, battery valve/tube, filamentary-cathode valve/tube
~, пятиэлектродная five-electrode tube, pentode
~, радиально-лучевая radial-beam valve/tube
~, радиоусилительная amplifying valve/tube
~, разборная demountable valve/tube, knock-down valve/tube, assembled tube
~, разделительная pulse separator valve/tube
~, разрядная discharge lamp
~, реактивная reactance valve/tube
~ регулируемого усиления fading tube/hexode
~, регулирующая control valve/tube
~ резнатронного типа resnatron
~, резонансная resonance (fluorescence) lamp
~, ртутная mercury(-vapo(u)r) lamp
~, ртутно-кварцевая qurtz-mercury lamp, mercury vapo(u)r lamp, Kromayer's lamp, U-shaped quartz lamp
~, рудничная miner's lamp
~ рулежной фары taxying lamp
~ с автоэлектронной эмиссией cold-cathode valve/tube
~ с верхним выводом Anode-cap valve, plate-cap tube
~ с водяным охлаждением water-cooled valve/tube
~ с вторичной эмиссией secondary-emission valve/tube
~ с накаленным катодом hot-cathode valve/tube, thermionic valve/tube
~ с плоскими электродами planar-electrode valve/tube
~ с подогревным катодом cathode-heater valve/tube
~ с поперечным управлением babitron valve/tube
~ с простой спиральной нитью single-coil lamp
~ с сеточным управлением grid-control valve/tube
~ с тормозящим полем retarding field valve/tube
~ с тусклым накалом dull-emitter valve/tube
~ с угольной нитью carbon (filament) lamp
~ с удлиненной характеристикой remote cut-off valve/tube
~ с холодным катодом cold-cathode valve/tube
~, светодиодная light-emitting-diode lamp, LED lamp
~, светомаркировочная black-out lamp
~ СВЧ microwave valve/tube
~ сигнализации о приближении к срыву stall warning light
~ сигнализации остатка топлива fuel quantity-low warning light
~ сигнализации отключения автопилота autopilot-off light, disengage light
~ сигнализации перегрузки ovreload light
~, сигнальная свободной линии idle indicating lamp, free line signal
~ сигнализации состояния (системы) (system) status warning light
~, сигнальная warning/pilot light
~ системы управления пограничным слоем, сигнальная boundary layer control light, BLC light
~, склеральная *med.* scleral lamp
~ смесительная frequency changer, mixing tube
~ смещения bias valve/tube
~ "Соллюкс" *med.* Sollux lamp
~, софитная double-ended lamp
~, спектральная spectral/spectroscopic lamp
~, стробирующая gate valve/tube
~, суммирующая adder valve/tube
~, счетная counter/counting tube
~, термокатодная hot-cathode valve/tube, thermionic valve/tube
~, термоэлектродная thermionic valve/tube
~ тлеющего разряда glow-discharge tube
~ тревожной сигнализации alert/warning light
~, угольная carbon (filament) lamp
~, ударопрочная rough-service lamp
~ ультрафиолетового облучения/излучения ultra-violet lamp, UV lamp
~, управляющая control valve/tube
~, усилительная amplifyer, amplifying valve/tube
~, фиксирующая clamping valve/tube
~, флюоресцентная fluorescent lamp
~ "цель захвачена", индикаторная *mil.* lock-on indication light
~, цельнометаллическая all-metal tube

~ цистоскопа *med.* cystoscopic/mignon lamp
~, четырехэлектродная four-electrode valve/tube, tetrode
~, читающая exciter lamp
~, шаговая step valve/tube
~, шкальная dial light
~, шнуровая *teleph.* cord lamp
~, шумовая *teleph.* noisy valve/tube
~, щелевая (Gullstrand's) slit-lamp
~, электрическая electric lamp/bulb
~, электрометрическая electrometer valve/tube
~, электронная vacuum valve/tube; electron(ic) tube
~, электроннолучевая electron-ray tube, "magic eye"
~, эритемная sunlamp, erythematous lamp
~, эталонная comparison/standard lamp
лампа-вспышка flash bulb, flashtube
лампа-желудь acorn valve/tube
лампад(к)а icon lamp
лампара lampara (net)
лампас *mil.* stripe
лампион chinese lantern
ламповая *min.* lamproom
ламповый lamp, tube *(attr.)*
~ выпрямитель tube rectifier
~ генератор vacuum tube generator
~ приемник tube receiver
~вое стекло chimney *(of kerosene lamp)*
ламповыниматель *teleph.* lamp extractor/puller
ламподержатель tube socket/holder, valve/tube retainer
лампоизмеритель, лампоиспытатель tube tester/analyzer
лампочка light bulb
~ накаливания incandescent lamp
~, сигнальная signal/pilot lamp
лампридовые *n. pl. zool. Lampridae*
лампрофир *geol.* lamprophyre
~, дайковый hamrongite
лампрофировый lamprophyric
ламут, ламутский Lamut
ламутка Lamut (woman)
лан 1. lan *(Chinese or Korean measure of weight = ounce);* 2. lan *(old Chinese coin = 25 kopecks)*
лангобард *hist.* Langobard, Lombard
лангобардский Langobardic, Lombard(ic)
лангур *zool.* langur *(Presbytis)*
лангуст *zool.* rock/spiny lobster, sea crayfish *(Palinurus vulgaris)*
ландвер Landwehr
ландграф *hist.* landgrave
ландграфиня *hist.* landgravess, landgravine
ландграфство *hist.* landgraviate; landgraveship
ландо *hist.* landau
ландольфия *bot.* gumvine, landolphia *(Landolphia)*
ландрат Landrat(h)
ландскнехт *hist.* landsquenet, landsknecht, mercenary
ландсмол Landsma(a)l
ландстинг Landst(h)ing
ландтаг *hist.* dietine; Landtag
ландшафт 1. *geogr.* landscape, landform, topography, country; 2. *math.* surface, relief; 3. *art*. landscape
~, аналитический modular surface/relief
~, ванновый *geol.* pond landscape
~, вулканический volcanic topography
~, горно-долинный ridge-and-valley topography
~, карстовый karst landscape/topography, cockpit country
~, неразвитый juvenile landscape
~, пересеченный/рассеченный dissected/rugged landscape
~, слаженный tame/subdued landscape
~, ювенильный/юный juvenile landscape
ландштурм *hist.* Landsturm
ландштурмист member of Landsturm, homequard, militiaman
ланды *pl.* landes; dune coast
~, вересковые calluma heath
ландыш *bot.* lily-of-the-valley *(Convallaria)*
ландышник *bot.* lily turf *(Ophiopogon, Mondo)*
ланка *zool.* hind
ланкастерский Lancasterian
ланкорд, ланкорт *text.* lankord
ланолин lanolin
~, безводный anhydrous lanolin
~, водный hydrous wool fat
ланостерин lanosterol
лансада *sport* gambade, gambado
лансиум *bot.* langsat *(Lansium)*
лансмол Landsma(a)l, Nynorsk
лантан *chem.* lanthanum, La
лантан(о)иды *pl.* lanthanides
лантионин lanthionine
ланцет *med.* lance(t); *tech.* spatula
~ для кровопускания blood lancet, fleam
ланцетник *zool.* lancelet *(Amphioxus)*
ланцетно-ланцетовидно-линейный lance-linear
ланцетовидный, ланцетообразный, ланцетолистный *bot.* lanceolate; lancet-shaped
лань *zool.* 1. fallow dear *(Cervus dama, Dama dama);* 2. doe
лаосец, лаотянец, лаотянин Laotian
лаоска Laotian(woman)
лаосский Laotian
лапа 1. *zool.* paw; pad; foot; 2. *tech.* grip, claw; lug, leg, boss, foot; tenon, dovetail; 3. *agr.* blade, shovel, foot, sweep, tine, point; 4. *leath.* shank; 5. *naut.* fluke
~, ветвящаяся forking arm
~ бурового долота drilling bit leg
~ для разделки гребней sweep blade
~ для ячеистой пахоты pit blade
~, загребающая gathering arm

~, загрузочная loading arm
~ клети *min.* cage shoe
~ крепления *mach.* mounting foot
~ культиватора/культиваторная cultivator blade/share/tooth, hoe (blade), tine, shovel
~ маркера *agr.* marker point
~, медвежья *bot.* spiny acanthus *(Acanthus spinosus)*
~, односторонняя side shovel
~, односторонняя плоскорежущая L-hoe, scuffle knife
~, окучивающая earthing blade
~, опорная *mach.* pick-up; *agr.* stabilizer blade
~, откидная tiltable claw
~, перепоночная *zool.* webbed foot
~, полольная *agr.* weed shovel, weeder tine
~ почвоуглубителя subsoiling blade, subsoil shank/tine
~, прополочная weed-cutting blade
~ прореживателя thinner/gapping blade
~, пружинная resilient tine
~, рыхлительная cultivator/chisel point, cultivating tine, ripping tooth
~, рыхлительная копьевидная spear-head/-point shovel
~, рыхлительная оборотная double-pointed shovel
~ с ногтями *zool.* claw
~ с твердосплавной наплавкой hard-faced sweep
~, стрельчатая A(-hoe) blade, duckfoot shovel/tine, V-shaped sweep
~, чешуйчатая волчья *bot.* swine's-cress *(Coronopus squamatus)*
~ якоря anchor blade/fluke/palm/flue
лапагерия *bot.* Chile bells *(Lapageria)*
лапарогастростомия *med.* laparogastrostomy, celiogastrostomy
лапарогепатотомия *med.* laparohepatotomy
лапарогистеросальпинго-оофорэктомия *med.* laparohysterosalpingo-oophorectomy, celiohysterosalpingo-oothecectomy
лапарогистерэктомия *med.* laparohysterectomy, celiohysterectomy
лапароколотомия *surg.* (laparo)colotomy
лапарокольпотомия *med.* laparocolpotomy, celioelytrotomy
лапароколэктомия *surg.* (laparo)colectomy
лапаромиомотомия *med.* laparomyomotomy, celiomyomotomy
лапарорафия *med.* laparorrhaphy, celiorrhaphy
лапаросальпинготомия *med.* laparosalpingotomy, celiosalpingotomy
лапаросальпингэктомия *med.* laparosalpingectomy, celiosalpingectomy
лапароскоп *med.* laparoscope, celioscope
лапароскопия *med.* laparoscopy, cel(i)oscopy, peritoneoscopy
лапаротифлотомия *med.* laparotyphlotomy
лапаротомафилия *med.* laparotomaphilia, Muenchhausen's syndrome
лапаротомия *surg.* laparotomy, celiotomy, abdominal section
~, влагалищная vaginal celiotomy
~, диагностическая/пробная exploratory laparotomy
лапароцеле *med.* laparocele, abdominal/ventral hernia
лапароцистотомия *med.* laparocystotomy, laparocystidotomy
лапароэлитротомия *med.* laparoelytrotomy, celioelytrotomy
лапароэктомия *surg.* laparectomy
лаперузия *bot.* false freesia *(Lapeirousia)*
лапидарность conciseness, terseness
лапилли *pl. geol.* lapilli, ejectamenta
~, волосистые/нитеобразные filiform lapilli, Pele's hair
~, пористые porous cinders
лапина *bot.* wing nut *(Pterocarya)*
лапка 1. *zool.* tarsus (*pl.* -si); 2. *tech.* claw, clutch; lug; pawl; tab; toe; tonque
~, боковая *naut.* side lug
~, воронья *bot.* waler cress *(Coronopus)*
~, гусиная *bot.* silverweed cinquefoil *(Potentilla anserina)*
~ для (при)крепления fastening lug
~, захватывающая grip pawl
~, заячья *bot.* 1. hare's-foot sedge *(Carex leporina)*; 2. *pl.* rabbit-foot clover *(Trifolium arvense)*
~, кабельная *el.* cable grip
~, кошачья *bot.* pussytoes *(Antennaria)*
~, нажимная *agr.* clip
~, С-образная cramp iron
~ рогульки *text.* leg/arm of flyer
~ сверла drill tang
~ки для натяжки проводов *tech.* draw-vice/-tongs
лапки *pl. print.* double quotes, inverted commas
лапландец Lapp(lander)
лапландка Lapp (woman)
лапландский Lappish, Lapponian
лапласиан *math.* Laplacian, Laplace operator, buckling
~, геометрический geometrical buckling
лаповидный, лапообразный paw-shaped
лапортея *bot.* wood nettle *(Laportea)*
лапорты *pl. met.* coke-oven doors
лаппарантит *geol.* lapparentite
лаппингование *tech.* lapping (process)
лаппинговать lap
лапсердак lapserdak, gaberdine
лапчатка *bot.* cinquefoil *(Potentilla)*
лапчатоногий *zool.* 1. web-/fin-footed, palmiped; 2. *n. pl.* finfoots *(Heliornithidae)*
лапчатость *bot.* palmateness
лапчатый 1. *zool.* paw-shaped, web-footed, webby, webbed; 2. *bot.* palmate(d), palmiform, pedatiform
лапша 1. *cul.* noodles; noodle soup; 2. *bot.* cla-

varia *(Clavaria)*
лапшерезка noodle cutter
ларва *rel.* larva
ларвальный larvaceous, larval, larvate
ларвацид, ларвицид larvicide
ларвицидный larvicide, cercaricidal
ларгетто *mus.* larghetto
ларго *mus.* largo
ларингал *phonet.* laryngeal
ларингальный, ларингеальный laryngeal
ларингит *med.* laryngitis, angina laringea
~, абсцедирующий flegmonous laryngitis
~, атрофический/сухой atrophic/dry laryngitis, laryngitis sicca, Turck's trachoma
~, отечный edematous laryngitis
~, подострый subacute laryngitis
~, подсвязочный subglottic laryngitis
~, псевдомембранозный membranous laryngitis
~, разлитой diffuse laryngitis
~, рецидивирующий язвенный relapsing ulcerative laryngitis
~, спазматический laryngitis stridulosa, crowing convulsion,glottic spasm, laryngospasm, false/spasmodic croup
~, туберкулезный tuberculous laryngitis laryngophthisis
~, удушающий suffocative laryngitis
ларинголог laryngologist
ларингологический laryngologic(al)
ларингоплегия *med.* laryngoplegia, laryngoparalysis, laryngeal paralysis
ларингоптоз *med.* laryngophthisis
ларингоскоп *med.* laryngoscope
~, хирургический operative/surgical laryngoscope
ларингоскопия *med.* laryngoscopy
~, непрямая/подвесная indirect/suspension laryngoscopy
ларингостеноз *med.* laryngostenosis
ларинготомия *med.* laryngotomy
~, верхняя subhyoid/thyrohyoid/superior laryngotomy
~, нижняя inferior laryngotomy
~, поперечная transverse laryngotomy
~, срединная median laryngotomy, laryngofissure
~, частичная laryngocentesis
ларинготрахеальный laryngotracheal
ларинготрахеотомия *surg.* laryngotracheotomy
ларингофарингит *med.* laryngopharyngitis
ларингофиссура *med.* complete pharyngotomy, laryngofissure
ларингофон laryngophone, laryngaphone, throat microphone
ларингоцеле *med.* laryngocele
ларингэктомия *surg.* laryngectomy
лары *pl. myth.* lares
ласка I *mach.* scarf
ласка II *zool.* least weasel *(Mustela nivalis)*
ласкирь *zool.* sargus, sea bream *(Sargus annularis)*
ласковец *bot.* clasping pepperwort *(Lepidium perfoliatum)*
ласковцы *pl. bot.* round-leaved thoroughwax *(Bupleurum rotundifolium)*
ласт I fin, flipper
ласт II last *(two tons)*
ластик 1. *text.* lasting *(material);* elastic *(fabric);* 2. eraser, India rubber
~, чернильный ink eraser
ластовень swallowwort *(Cynanchum)*
ластовица I *bot.* greater celadine *(Chelidonium majus)*
ластовица II = **ласточка** 1.
ластовневые *n. pl. bot.* milkweed family *(Asclepiadaceae)*
ластовое судно cargo/transport ship
ластоногий *zool.* pinniped, fissiped, cloven-footed
ласточка 1. *zool.* swallow *(Hirundo rustica);* 2. *sport* arabesque
~, береговая bank swallow, sand martin *(Riparia riparia)*
~, городская martin(et) *(Delichon urbica)*
~, деревенская common swallow *(Hirundo rustica)*
ласточник *bot.* common milkweed *(Asclepias cornuti/syriaca)*
лата *naut.* latten
латвиец, латвийский Latvian
латвийка Latvian (woman)
латекс latex; milky juice
~, карбоксилатный carboxilated butadiene latex
латентность 1. latency; dormancy; abeyance; 2. quiescence, quiescency
латентный 1. latent, dormant; abeyant, in abeyance, concealed, delitescent, hidden; dumb; 2. quiescent
латеральность laterality
~, перекрестная crossed laterality
латеризация, латеритизация *geol.* laterization
латерит *geol.* laterite
латероид fish paper
латерофлексия *med.* lateroflexion, lateroflection
латимерия *paleont.* coelacant, latimeria *(Latimeria)*
латинец *hist.* inhabitant of Latium
латинизация Latinization
латинизировать Latinize, Romanize
латинизм 1. Latin construction, Latin loan word, Latinism; 2. Latin/Roman influence
латинист Latin scholar, Latinist
латиница Roman alphabet/letters/type
латиноамериканский Latin-American
~кая ассоциация свободной торговли Latin-American Free Trade Anociation, LAFTA
латинский I Latin(ic); Roman
~ шрифт Roman type/character
латинский II *naut.* lateen
латинянин *hist.* inhabitant of Latium; Roman
латиризм *med.* lathyrism, githagism, lupinosis

латифундист large landowner
латифундия *hist.* latifundium
латник *hist.* armo(u)r-clad warrior; cuirassier
латный *hist.* cuirass, armo(u)r *(attr.)*
~ные доспехи (suit of) armo(u)r
латук *bot.* lettuce *(Lactuca)*
латук-салат cultivated lettuce
~, кочанный head lettuce
латунирование brassing, brass plating
латунь brass, orichalc(h)
~, деформируемая/ковкая wrought brass
~, литейная casting brass
~, марганцевистая manganese bronze
~, низкоцинковая/высокоцинковая low/high brass
~, рольная brass for rolled flat products
~, свинцовая leaded brass
~, специальная/часовая alloy/clock brass
~, судостроительная/морская naval/admiralty brass
латы *pl. hist.* cuirass, armo(u)r
латыш Lett
латышка Lett (woman)
латышский Lettish, Latvian
лауданин *chem.* laudanin(e)
лаун-теннис *sport* lawn tennis
лауреат laureate; prize winner
лаурин *chem.* laurin
лауриновый lauric
лаурон *chem.* aurothioglycanide, lauron(e)
лауэграмма Laue pattern/photograph, Laue X-ray diffraction pattern
лафет *mil.* gun carriage/mount
~, неоткатывающийся stay carriage
~ с раздвижными станциями split-trail carriage
~, тумбовый column mount
лафит Lafit(t)e (wine)
лахар *geol.* lahar
лахтак *zool.* 1. bearded seal *(Erignatus barbatus);* 2. sea hare *(Aplysia)*
лацпорт *naut.* port (hole)
~, бортовой side port
~, входной entrance/gangway port
~, грузовой cargo port, side loading port
лебеда *bot.* 1. orach(e), saltbush *(Atriplex);* 2. goosefoot *(Chenopodium)*
лебединый swan *(attr.)*
~ная шея swan neck; *tech.* S-bend pipe
лебедка I female swan, pan
лебедка II *tech.* winch, windlass, hoist
~, автоматическая automatic/self-tensioning winch
~, антенная antenna reel, aerial winch
~, грузоподъемная lifting/cargo winch
~ для спуска и подъема шлюпок, ручная lowering drum
~, кабельная cable winch/hoist
~ лифта lift /*Am.* elevator drive
~ лифта, барабанная rope/winding drum
~, маневровая auxiliary winch
~, откаточная *min.* haulage winch
~, передвижная/переносная portable hoist, crab (winch)
~, подъемная hoist(ing) winch
~, реверсивная reversible winch
~ с грейфером bucket hoist
~, скреперная scraper winch/hoist
~, стенная bracket winch
~, трелевочная skidder
~, цепная chain winch
~, швартовная docking/mooring/warping winch
~, шпилевая capstan winch
лебедолистный *bot.* halimifolious, saltbush-leaved
лебедчик winch operator, wincher
лебедь 1. *zool.* swan *(Cygnus);* cob (swan); cygnet; 2. *astr.* Cygnus
лебедь-кликун *zool.* whooping swan *(Cygnus cygnus)*
лебедь-трубач *zool.* trumpeter swan *(Cygnus buccinator)*
лебедь-шипун *zool.* mute swan *(Cygnus olor)*
леборист=лейборист
лев I *zool.* lion *(Panthera leo)*
~, Малый *astr.* Leo Minor
~, морской sea lion *(Zalophus)*
~, муравьиный ant lion *(Myrmeleon; Dendroleon)*
лев II lev *(Bulgarian monetary unit)*
левада 1. riparian woodlands; 2. meadow; garden/orchard
левак *polit.* left-winger, leftist
леван *chem.* lavan
леваллуа Levallois
левант Levant
левантин *text.* levantine (fabric)
левантинец Levantine
левацкий *polit.* leftist; ultraleft
~ загиб leftist deviation
левачество *polit.* leftism
леветь *polit.* shift to the left
левиафан I leviathan; *philos.* Leviathan
левиафан II wool-washing machine
левизна *polit.* leftism; radicalism
левират levirate
левит *hist.* Levite; *rel.* Leviticus, the third book of the Old Testament
левкой *bot.* stock, gillyflower *(Mathiola)*
лево *nav.* to port
~!, больше Better/More port!
~ на борт All/Hard aport! *Am.* Hard left rudder!
~ не ходить! Nothing (to) port! *Am.* Nothing to the left!
~ помалу! Steady aport! *Am.* Left handsomely!
левовращающийся left-rotating, levorotatory, levogyrate; *bot.* sinistrorse
левозавернутый laeotropic, sinistral
левоидеал *math.* left ideal
левоинвариантный left-invariant

левомицетин *pharm.* chloramphenicol, chlormycetin
левонаправленный laeotropic
левооднородный left-homogenous
леворазрешимый left-solvable
леворукий left-handed, sinistromanual, sinistral
леворукость *med.* mancinism, left-handedness, sinistrality
левосторонний sinistral, left-side/-handed
левоуничтожающий left-annihilating
левоупорядоченный left-ordered
левофланговый *mil.* left-flank/-wing
левретка Italian grayhound
левулеза *chem.* levulose, fructose, fruit sugar, levoglucose
левулеземия *med.* levulosemia
левша left-handed person, left-hander, sinistral
левый 1. left; left-hand; *nav.* port; 2. *polit.* left (wing); 3. *n.* leftist, left-winger
~ **уклон** left deviation
~**вая сторона** left(-hand) side; wrong side *(of material)*
легализация legalization; *nav.* certification
легализм lawfulness, legalism
легализовать legalize
легальность legality
легальный legal, lawful; legitimate
легат specific legacy; legate
легатарий *law* legatee
легато *mus.* legato, slur
легатский legatine
леггорн *agr.* Leghorn (chicken)
легенда legend, descriptive data/statement
~ **(карты)** legend, (explanatory) notes
легион legion (in various senses)
~**, иностранный** the Foreign Legion
легионер legionary
легирование *met.* alloying, alloyage; doping
~ **методом ионного внедрения** ion-implanted doping
легировать *met.* alloy; compound; dope
легислатива *hist.* Legislative Assembly
легислатура legislature, term of office
легист *hist.* legist
легитимация legitimization, legitimation
легитимизм *hist.* legitimism
легитимизовать legitimize, legitimate
легитимировать legitimize
легитимист, легитимистка *hist.* legitimist
легитимный legal, lawful, legitimate
легкие *pl.* lungs
~**, железные** iron lungs, Drinker respirator
легкий light
~ **случай (заболевания)** *med.* mild case
~**кая атлетика** track and field athletics
~**кая индустрия** light industry
~**кие нравы** lax morals
легкоатлет (track-and-field) athlete
легкобронированный lightly armo(u)red
легковес *sport* lightweight
легковесный lightweight *(attr.)*; light
легководолаз skin-diver, shallow water diver
легковой passenger *(attr.)*
~ **автомобиль** car, automobile
~ **извозчик** cab(man)
легковооруженный light-armed
легковоспламеняющийся (highly) inflammable, deflagrable
легкодоступный easily accessible
легкое *n. anat.* lung
~**, водное** water lung
~**, сотовое** honeycomb lung
легкокипящий low-boiling
легкомоторный low-powered
легкоплавкий low-melting, easily melted, fusible, eutectic
легкопораженный slightly injured
легкоразличимый easily distinguishable
легкораненый slightly wounded
~ **ходящий** walking case/wounded
легкорастворимый readily soluble
легкосжижающийся easily liquefied
легкость lightness; ease
~ **пуска** launching flexibility, ease of launching
~ **управления** driving ease, ease of control
легкосъемный readily detachable
легкоходовой free-/smooth-running, smoothly operating
легость rope; cable, hawser
легочница *bot.* lungwort *(Pulmonaria)*
легочно-аортальный *anat.* pulmoaortic
легочно-брюшинный *anat.* pulmo(no)peritoneal
легочножаберные *n. pl. zool.* branchiopulmonates *(Branchiopulmonata)*
легочно-сердечный *anat.* cardiopulmonary, cardiopneumatic
легочный pulmonary, pulmonic; pulmoniferous, air-breathing; lunged; *n. pl. zool.* pulmonates, dipnoi *(Pulmonata)*
легумин *bioch.* vegetable casein, legumin
лед ice
~**, береговой** ice foot/belt/wall
~**, битый** broken/slack ice
~**, глетчерный/ледниковый** glacier ice
~**, донный** anchor/ground/bottom/depth ice
~**, древний полярный** paleocrystic ice
~**, дрейфовый/плавающий/плавучий** drift/floe ice, ice floe
~**, зернистый** granular ice, firn, névé
~**, ископаемый/каменный** fossil ice, ice-formed rock
~**, искусственный** manufactured ice
~**, материковый** land/inland/continental ice
~**, мелкобитый** ice brush, brash
~**, многолетний** perennial/old ice
~**, молодой** newly-formed/-frozen ice, fresh/new ice
~**, однолетний** winter ice
~**, оторванный/свободный** loose ice
~**, отступающий** waning ice

~, пищевой food service ice
~, поровый interstitial ice
~, припайный fast ice
~, снеговой snow/slush ice
~, сплошной ice field, consolidated ice
~, стоячий ground ice
~, сухой solid carbon dioxide, dry ice
~, торосистый hummocky ice
~, фирновый firn/granular ice, névé
~, шельфовый ice barrier/shelf, shelf ice
~, шхерный fjord ice
ледгилит *geol.* ledhillite
ледебурит *met.* ledeburite
леденец *cul.* fruit drop, hard/sugar candy, lollipop; sugarplum
~, мятный peppermint drop
ледерин leather-cloth, leatherette, artificial leather
~, бумажный leather-paper
лѐдник icehouse; icebox
~, домашний icebox
ледни̇к glacier
~, альпийский/горный Alpine/mountain glacier
~, возрожденный recemented/reconstructed glacier
~, второго порядка glacieret
~, главный trunk glacier
~, диффлюэнтный diffluent/through glacier
~, достигающий моря tidewater glacier
~, зарождающийся nascent glacier
~, звездообразный stellate/asteriated glacier
~, ископаемый/окаменелый fossil glacier
~, континентальный ice cap, continental glacier
~, маляспинский Malaspina glacier
~, материковый sheet glacier, inland ice
~, мертвый/неподвижный dead/stagnant glacier
~, новорожденный nivation-type glacier
~, остаточный relic/remnant glacier
~ подножья piedmont glacier
~ с расширенным концом bulb glacier, expanded-foot glacier
~, сливающийся confluent/coalescing glacier
~, смежный juxtaposed glacier
~, стационарный equilibrium glacier
~, фирновый névé/firn glacier
ледниково-морской glacial-marine
ледниково-подпрудный glacier-dammed, glacier-ice-blocked
ледниковый glacial, englacial
~ период ice age, glacial period
ледниковье *geol.* glaciation, glacial period
ледобрекчия ice breccia
ледоведение glaciology
ледовитый icy; ice-covered
~ океан, Северный the Actic Ocean
ледовый ice *(attr.)*
~вая разведка ice patrol
~вое побоище *hist.* the Battle on the Ice *(April 15,1242)*
ледокол icebreaker (ship), iceboat
~, атомный nuclear-powered icebreaker, atomic icebreaker
ледокол-буксир ice-breaking tug
ледолом ice apron, icebreaker
ледораздел ice/glacial divide
ледорез 1. *naut.* ice cutter; 2. ice-apron/-breaker
ледоруб ice-ax(e)
ледосброс, ледоскат, ледоспуск ice chute
ледостав ice formation, freezing, freeze-up, freezing over
~, полный complete ice coverage
ледостойкий sleetproof
ледоформа ice mould/form/tray
ледоход ice drift/floe, drifting/floating of ice
лед-сало ice slush, grease ice
ледянка ice-boat/canoe
ледянник *bot.* ice plant *(Mesembryanthemum crystallinum)*
ледяной (of) ice; glacial, icy; icecold, chilling
~ная сосулка icicle
леер *naut.* handrail, (grab)rail, grab/hand/taut rope; manrope; rope,line
~, бельевой cloth(es) line
~, бортовой life rail/line
~, верхний (леерного устройства) life line
~, нижний footline
~ с мусингами knotted life line
~, спасательный life rail/line
~, средний (леерного устройства) housing line
леерсия *bot.* cut-grass *(Leersia)*
лежак *naut.* deck chair, chaise longue
лежание lying, resting
~ в дрейфе lying-to
лежать lie, be (situated)
~ в дрейфе *nav.* lie to
~ в основе underlie
лежачий 1. recumbent, lying; 2. unused
~ больной bed patient
~ ниже underlaying
лежбище 1. breeding ground; rookery; 2. bed, lair
~ тюленей seal rookery
лежень ground brace, groundsel; (mud)sill, sole; stay block; ledger; foundation beam; sleeper
~, береговой bank/abutment sill
~, опорный saddle beam, sleeper
~, продольный longitudinal sill
~, якорный anchor log
лежкоспособность shelf/storage life
лежняк bottom millstone; bed stone
лезвие blade, edge
~ головки паяльника tip of a soldering bit
~ долота nose of a chisel
~ режущего инструмента cutting-tool edge
~ резца steel edge
~ струга plane bit
~, ударное impact-knife edge
~ фрезера mill blade
лезвиеобразный blade-shaped

лезгин Lezghian, Lezghin
лезгинка I Lezghian (woman)
лезгинка II lezghinka *(Caucasian dance)*
лезгинский Lezghian, Lezghin
лей leu, ley *(monetary unit of Romania)*
лейас *geol.* Lias
лейасовый *geol.* Liassic
лейб-гвардеец life-guardsman
лейб-гвардия *hist.* Life Guards
лейб-медик *hist.* physician in ordinary
лейбницевский Leibnitzian
лейборизм *polit.* Labo(u)rism
лейборист *polit.* Labo(u)rite
лейбористский *polit.* Labo(u)r *(attr.)*
лейденский Leyden
~кая банка Leyden jar
лейка I 1. *hort.* watering can/pot; 2. *naut.* (boat) bailer
лейка II Leica (camera)
лейканемия *med.* leukanemia, proleuk(a)emia
лейкемический leuk(a)emic
лейкемия *med.* leuk(a)emia, leukosis
~, алейкемическая aleukemic/aleucocythemic leuk(a)emia, aleukemic myelosis
~, базофильная basophilic/basophilocytic leuk(a)emia, mast cell leuk(a)emia
~, гиперцитарная high cell count leuk(a)emia, hypercytic leuk(a)emia
~, гистиоцитарная histiocytic leuk(a)emia, (acute) monocytic leuk(a)emia
~, кожная leuk(a)emia cutis, leuk(a)emid
~, лимфатическая lymphatic leuk(a)emia
~, малопроцентная low-percentage leuk(a)emia
~, миелогенная myelogenic/myelogenetic/myelogenous leuk(a)emia
~, миелоидная myeloid/myelocytic leuk(a)emia
~, нейтрофильная nature cell leuk(a)emia, neutrophilic leuk(a)emia
~, острая acute leuk(a)emia/leukosis
~, плазмоклеточная plasma cell leuk(a)emia/leukosis
~, полиморфноклеточная polymorphocytic leuk(a)emia
~, смешанноклеточная mixed (cell) leuk(a)emia
~, спленогенная splenogenous/splenetic leuk(a)emia
~, спленомиелогенная splenomyelogenous/splenomedullary leuk(a)emia
~, эмбриональная stem cell leuk(a)emia, embryonal leuk(a)emia
~, эозинофильная eosinophilic/eosinophilous/eosinophilocytic leuk(a)emia
лейкин *med.* leukin, endolysin
лейкобилин *med.* leukobilin, white bile
лейкобласт *anat.* leukoblast, proleucocyte
~, гранулярный granular leukoblast, premyelocyte, promyelocyte
лейкобластоз *med.* leukoblastosis
лейкодерма *med.* leukoderma, leukodermia, leukopathia, achromia, achromoderma
~, врожденная congenital leukoderma, albinism
~, вторичная/ложная pseudoleukoderma
~, приобретенная acquired leukoderma, vitiligo
~, сифилитическая syphilitic leukoderma, leukodermia colli, vitiligoid fox, pigmentary syphiloderma
лейкодистрофия *med.* leukodystrophy, leukoencephalopathy, poliencephalopathy
лейкоз *med.* leukosis, leuk(a)emia, leucosis
~, врожденный congenital leukosis/leuk(a)emia
~, миелоидный myeloid/myelocytic leuk(a)emia
~, острый acute leukosis/leuk(a)emia
~, эозинофильный eosinophilic/eosinophilous/eosinophilocytic leuk(a)emia
лейкокератоз *med.* leukokeratosis
~, врожденный congenital/oral leukokeratosis, familial white folded dysplasia
лейкоксен *miner.* leucoxene
лейколиз *med.* leukolysis, leukocytosis
лейколизин *med.* leukolysin, leukocytolysin
лейколитический *med.* leukolytic, leucocytolytic
лейкома *med.* (adherent) leukoma, leucoma
лейконекроз *med.* leukonecrosis, white gangrene
лейконихия *med.* leukonychia, white/gift spots
лейконоидный *zool.* leukon-type
лейкооснование leuco base
лейкопения *med.* leukopenia, leukocytopenia, hypoleucocytosis
~, доброкачественная benign leukopenia
~, злокачественная malignant leukopenia, agranulocytosis
~, лимфоцитарная lymphocytic leukopenia, lymphopenia, lymphocytopenia
~, лучевая radiation leukopenia
~, моноцитарная monocytic leukopenia, monocytopenia
~, нейтрофильная neutrophilic leukopenia, neutrophilopenia
лейкоплакия *med.* leukoplakia, leukoplasia
~, бородавчатая leukoplakia verrucosa
~ губы leukoplakia labiorum, keratosis labialis, smoker's patch
~ языка ichthyosis/tylosis linguae
лейкопласт *bot.* leucoplast
лейкопластырь *med.* adhesive plaster/tape/bandage
~, молескиновый moleskin adhesive plaster
лейкоррея *med.* leucorrh(o)ea
лейкосапфир *geol.* leucosapphire
лейкосаркома *med.* leuco(lympho)sarcoma
лейкосоединение *tech.* leuco compound
лейкотаксис *med.* leuco(cyto)taxis, leuko(cy)taxia
лейкотоксин *med.* leuco(cy)toxin
лейкотоксичность *med.* leukotoxicity
лейкотомия *surg.* leukotomy, lobotomy
лейкотрихия *med.* leucotrichia, whiteness of the hair
~, кольцевая leukotrichia annularis, ringed hair
лейкофан *miner.* leucophane

лейкоцит *anat.* leukocyte, white blood cell, white globule/corpuscle
~, агранулярный agranular/nongranular leukocyte
~, базофильный basophilic leukocyte, basophil, basocyte, basophilocyte, mast leukocyte
~, гиалиновый hyaline/transitional leukocyte, monocyte
~, зрелый mature(d) leukocyte
~, нейтрофильный neutrophilic leukocyte, neutrophil, neutrophilic granulocyte
~, неподвижный nonmotile leukocyte
~, пигментированный melanocyte
~, полиморфоядерный poly(morpho)nuclear/multinuclear leukocyte
~, эндотелиальный endothelial leukocyte/phagocyte, endotheliocyte
~, эозинофильный eosinophilic leukocyte, eosinophil, eosinocyte, acidocyte
лейкоцитарный, лейкоцитный leukocytic
лейкоцитоз *med.* leukocytosis
~, агональный agonal/terminal leukocytosis
~, базофильный basophilic leukocytosis, basophilia
~, лимфоцитарный lymphocytic leukocytosis, lymphocytosis
~, медикаментозный medicament/drug leukocytosis
~, моноцитарный monocytic leukocytosis, monocytosis
~, нейтрофильный neutrophilic leukocytosis, neutrophilia
~, пищеварительный digestive leukocytosis
~, токсогенный toxogenous/toxogenic leukocytosis
~, эозинофильный eosinophilic leukocytosis, eosinophilia
лейкоэнцефалит *med.* leukoencephalitis
лейнер (gun)liner, lining tube
~, запасной replacement tube
~, полый cavity liner
~, свободный removable/loose liner
лейодермия *med.* leiodermia, smooth/glossy skin
лейомиома *med.* leiomyoma
~ кожи leiomyoma cutis
~, ценкеровская Zenker's leiomyoma, leiomyosarcoma
лейомиосаркома Zenker's leiomyoma, leiomyosarcoma
лейста *min.* lath, forepole
лейстовидный lath-like
лейтенант lieutenant
~, младший junior/second lieutenant
~, старший senior/first lieutenant
лейтмотив *mus.* leitmotif; *fig.* leading idea
лейцит *miner.* leucite
лейцитит *geol.* leucitite
~, мелилитовый melilite-leucitite
лейцитоид *cryst.* leucitoid
лейцитолит *geol.* leucitolith
лейчестер *agr.* Leicester (sheep)
лейшманиоз *med.* leishmaniasis, leishmaniosis
~, висцеральный visceral leishmaniasis, kala-azar, Assam fever
~, висцеральный средиземноморский canine leishmaniasis, canine/Mediterranean kala-azar
~, детский infantile leishmaniasis, infantile kala-azar
~, кожно-слизистый mucocutaneous leishmaniasis
~, кожный cutaneous/dermal leishmaniasis, oriental/Bagdad/ Biskra/Delhi boil
лейшманоид *med.* leishmanoid
~, кожный dermal leishmanoid
лек lek *(monetary unit of Albania)*
лекаж *com.* leakage
лекало (curve) template, templet; gauge, mould, pattern;
~, резьбовое profile gauge
~, фасонное curve gauge
~, чертежное drawing curve
лекальщик gauge/pattern maker, gauger
леканора *bot.* lecanora *(Lecanora)*
лекарственноустойчивый drug-resistant
лекарственный medicinal, officinal(is), medicamentous
~ное растение medicinal plant
лекарство medicine, drug, medicament, pharmacy, physic
лекисты *pl. lit.* Lake Poets/School
лекиф *bot.* lecythus, lekythos
лекпом *abbr.* physician's/medical assistant
лексема lexeme
лексика vocabulary
лексикализация lexicalization
лексико-грамматический lexical-grammatical
лексикограф lexicographer
лексикографический lexicographic(al)
лексикография lexicography
лексиколог lexicologist
лексикология lexicology
лексико-морфологический lexical-morphological
лексикон vocabulary, lexicon, dictionary
лексико-семантический lexico-semantic
лексико-синтактический lexical-syntactical
лексикостатистика *ling.* lexicostatistics
лексико-стилистический lexical-stylistic
лексический lexical
лектика sedan; litter
лектор lecturer
лекторий 1. speakers bureau; centre organizinig public lectures; 2. lecture hall
лектория reading room
лекторский lecturer's
лектотип *biol.* lectotype
лекционно-семинарский lecture-discussion
лекция lecture
лемех *agr.* plowshare, (plough) share
~, безремонтный throw-away share

~, **выкопочный** undercutter
~, **двусторонний** reversible share
~ **для каменистых почв** cobblestone share
~, **зубчатый** notch-edged share, toothed share
~, **корытообразный** channel share
~, **отвальный** *min.* loading blade
~, **подкапывающий** lifting/digging (plow)share, digger shovel
~ **почвоуглубителя** subsoiling share, furrowing plowshare
~ **предплужника** fore/skim share
лемма *math.* lemma
лемминг *zool.* lemming *(Lemmus)*
леммобласт *biol.* lemmoblast, lemnoblast
леммобластома *med.* lemmoblastoma, neurinoma, neurilemmoma
лемнесит *geol.* lemnaesite
лемниската *math.* lemniscate
~ **Бернули** lemniscate of Bernoulli, Bernoulli's lemniscate
~ **Бута** Booth's lemniscate
~ **Героно** lemniscate of Gerono
лемонграс *bot.* lemon grass *(Cymbopogon citratus)*
леморан *pharm.* levorphanol
лемур *zool.* lemur *(Lemur)*
~**ы, летающие** dermopterans *(Dermoptera)*, flying lemurs
лемурия *hist.* Lemuria *(Roman festival)*
лемуры *pl. rel.* lemures
лен *hist.* fief, fee
~, **отдавать в** enfeoff, give in fee
~ **подвассала** vavasory
лен *bot.* flax *(Linum)*
~, **горный** *miner.* asbestos
~, **замачивать** ret the flax
~, **слабительный** fairy flax *(L. catharticum)*
~, **теребить** pull the flax
~, **трепать** scutch/dress the flax
~, **чесать** hackle the flux
ленд-лиз lend-lease; Lend-Lease Act
лендлорд landbord, landowner
лендлордство *coll.* landlords
лен-долгунец/-посевной *bot.* fiber/common flax *(Linum usitatissimum)*
ленец *bot.* bastard toadflax *(Thesium)*
ленивец 1. *tech.* idling sprocket, idler; 2. *zool.* sloth *(Bradypus)*
~, **гигантский/наземный** ground sloth *(Megalonyx)*
ленивый 1. lazy, idle; slothful; 2. *cul.* prepared quickly
~**вые щи** fresh cabbage soup
леникс *tech.* tension/pinch roller; idle wheel, idler, belt tightener
ленинградец *abs.* inhabitant of Leningrad
ленинец Leninist
ленинизм Leninism
ленинский Leninist, Lenin *(attr.)*
лен-кудряш *bot.* crown/oil flax
лен-моченец retted flax
ленник *hist.* feudatory, liege, vassal; feofee
ленный *hist.* feudatory; feudal
леновые *n. pl. bot.* flax family *(Linaceae)*
ленок *bot.* 1. European goldilocks *(Linosyris vulgaris)*; 2. coreopsis *(Coreopsis)*
лен-прыгунец *bot.* dehiscent common flax *(Linum usitatissimum crepitans)*
ленский расстрел *hist.* Lena shooting
лен-стланец dew-retted flax
лен-сырец raw-flax
лента 1. *tech.* tape, ribbon, band, strap; belt; strip; *text.* sliver; 2. *med.* t(a)enia
~ **аварийного самописца** *av.* crash recorder tape, accident data recorder tape
~, **агломерационная** sinter belt
~, **аннинская** ribbon of order of St. Anne
~, **асбестовая** asbestos tape
~, **ацетатная магнитная** acetate base (magnetic) tape
~ **барографа** barograph chart
~, **бегущая** *comp.* streaming tape
~ **бортового самописца** *av.* flight recorder tape
~, **броневая кабеля** band iron
~, **бронирующая** restricting tape
~, **бумажная** paper tape
~, **буферная** dump tape
~ **в виде бесконечной петли** endless tape
~ **в кассовом аппарате** cash-register tape
~ **ввода программы** *comp.* program tape
~, **видеомагнитофонная** video tape
~ **восьмиканальная/-дорожечная** eight-track tape
~ **второго поколения** *comp.* father tape
~ **входных данных** input tape
~ **выходных данных** output tape
~, **герметизирующая** sealing tape
~ **гребенного прочеса** *text.* combed sliver, tops
~, **гусеничная** caterpillar track
~, **двойная** twin band/tape
~, **диаграммная** chart strip
~ **для звукозаписи** recording/audio tape
~ **для пишущих машин** typewriter ribbon
~ **зарядная** charging belt
~ **зрительного бугра** *anat.* t(a)enia of the thalamus, thalamic t(a)enia
~, **игольчатая** *text.* card clothing/fillet/covering, filleting card
~ **из фольги** foil tape
~, **изоляционная** insulating tape
~, **кардная** card clothing
~, **карманообразующая** taschengurt
~, **кембриковая** cambric tape
~, **кинематографическая** (reel of) film
~, **киперная** surgical tape
~, **клейкая** Scotch/adhesive tape
~ **конвейера/конвейерная** conveyor belt
~, **контрольная мерная** reference/standard tape

~ корректур *comp.* amendment tape
~, красящая *tel.* inking/inked/ink ribbon
~, крепежная attachment strap
~ курсографа course graph
~, липкая (self) adhesive/Scotch tape; masking tape
~, лотковая *min.* troughed belt
~, магнитная (magnetic/sound) tape
~, магнитная измерительная reference tape
~, магнитная оперативная scratch tape
~, магнитофонная audio tape
~, мерная measuring tape/band, (steel) tape line/measure
~, мерная геодезическая geodetic surveying tape
~ металлизации (electrical) bonding strip
~, многодорожечная mutichannel tape
~, многослойная multiply band/tape; sandwich tape
~, наждачношлифовальная emery belt
~, неразмеченная virgin tape
~, нивелирная/нивелировочная levelling tape
~ ободочной кишки *anat.* t(a)enia of Valsava, t(a)enia coli
~, опознавательная identification tape
~, орденская 1. ribbon; 2. *zool.* underwing moth *(Catocala)*
~, основная master tape
~, патронная ammunition/cartridge belt
~, паузная timing tape
~ первого поколения *comp.* grandfather tape
~ перепуска (воздуха) *av.* by-pass band; bleed valve
~, перфорированная punched/perforated tape
~, пильная saw blade
~, пильчатая *text.* saw-toothed card clothing
~ подпрограммы *comp.* subroutine tape
~, полулотковая *min.* semitroughed belt
~, породоотборная *geol.* rock collecting belt/band
~, пробная *tel.* specimen/trial slip
~, проволочная wire tape
~, программная *comp.* program/instruction tape
~, протекторная *auto.* tread strip
~, пустая blank/raw tape
~, пятидорожечная five-channel tape
~ распечатки *comp.* printing-out tape
~, регистрирующая strip/record chart
~ речевого самописца voice recorder tape
~, рудоразборная/рудосортировочная ore-picking belt
~ руды ribbon of ore
~ с быстрым остановом и разгоном *comp.* fast-stop-go tape
~ с данным data band/tape
~ с магнитым покрытием ferrous coated tape
~ с управяющей программой steering tape
~ с файлом file tape
~ самописца recorder strip/chart
~ свода *anat.* t(a)enia of the fornix
~, сменная alternate tape
~, сортировочная picking belt
~ сосудистой основы *anat.* band of the tela
~ стандартных подпрограмм subroutine tape
~, стыковочная butt strap
~ сульфида *geol.* ribbon of sulphide
~, телеграфная telegraph tape
~, тонкая *anat.* t(a)eniola
~, тормозная brake band/strap
~, транспортерная conveyer belt
~, упаковочная/уплотнительная sealing tape
~ управления графопостроителем plotter control tape
~, управляющая control/pilot tape
~, управляющая программная master instruction tape, MIT, master program file
~, усиливающая reinforcing ribbon
~, ферромагнитная magnetic tape
~, холоднокатаная cold rolled strip
~ хронографа registering ribbon
~, цветная ink ribbon
~, чесальная *text.* card sliver, cardend
~, чистая (неразмеченная) *comp.* virgin tape
~ ЧПУ numerical tape
~ шлифовальная abrasive belt
~, эталонная master tape
~ эхолота sounding graph
лентец *zool.* tapeworm *(T(a)enia)*
~, широкий fishworm, broad tapeworm *(Diphyllobotrium latum)*
лентигиноз *med.* lentiginosis, mutliple freckles
лентикулярный lenticular, phacoidal; lentil-/lens-shaped
ленто *mus.* lento
лентовидный, лентообразный ribbonlike, ribbon-shaped; *anat.* t(a)eniate, t(a)eniform, t(a)enioid, tape-form
лентонаправитель *text.* sliver guide
лентообмоточный taping
лентопротяжка tape drive system
лентопротяжный tape-transport/-drive/-moving/-winding; film-transporting
лентосклеивающий (tape-)splicing
лентоткачество ribbon/tape weaving
лентоукладчик *text.* coiler
ленточка *tech.* margin; land
ленточница *zool.* underwing moth *(Catocala)*
ленточно-спиральный tape-helix
ленточный band, tape, ribbon *(attr.)*; *anat.* t(a)enial
~ глист/червь tapeworm
~ конвейер belt conveyer
~ погрузчик belt loader
~ тормоз band brake
~ная пила band saw
леонинский *lit.* leonine
~ стих leonine/Latin verse, leonines
леонотис *bot.* lion's-ear *(Leonotis)*
леонтиаз *med.* leontiasis

леонтица *bot.* lion's-leaf *(Leontice)*
леонтоподиум *bot.* edelweiss *(Leontopodium)*
леопард *zool.* leopard *(Pardus pardus)*
~, морской sea leopard *(Hydrurga leptonyx)*
леотропный *zool.* laeotropic, leiotropic, sinistral
лепестковидный *bot.* petal-shaped, petaliform, petaloid, petaline
лепестковый petal(l)ed, petaline
~вое шпорце foliar spur
лепесток 1. *bot.* petal, leaflet; 2. *mach.* tab, lug; *rad.* lobe
~ антенны, основной antenna lobe
~, бахромчатый fimbriate petal
~, боковой/вторичный side/minor lobe
~, второстепенный minor lobe
~, главный major/main lobe
~ диаграммы направленности антенны antenna lobe
~ интеференции involute petal
~ ламповой панельки interference lobe valve/*Am.* tube socket lug
~, лучистый rayed petal
~стки, имеющий petal(l)ed, petalous, petaliferous
~стки, имеющий разные heteropetalous
~стки, образующий petaliferous
лепешка 1. lump, stag, biscuit; clot; 2. troche, lozenge, pastil, tablet; 3. *cul.* flat cake; cookie; scone; griddlecake
~, водорослевая algal biscuit
~ из угольных осколоков cake of coal fragments
~, сдобная muffin
~, сулемовая bichloride tablet
~, сухая shortbread
~, шлаковая slag pancake, lump/blob of slag
лепешкообразный tablet-like; oblate
лепидин *chem.* lepidine, methylquinoline
лепидодендрон *paleont.* lepidodendron
лепидома *med.* lepidoma
~ второго порядка secondary/transitional lepidoma, lepidoma of second order
лепидомелан *miner.* lepidomelane
лепидопус *zool.* frostfish *(Lepidopus caudatus)*
лепидофит *geol.* lepidophyte
лепка model(l)ing, sculpturing; mo(u)lding; stucco(work)
лепление model(l)ing, mo(u)lding, fashioning
лепнина stucco mo(u)lded
лепной modelled, mo(u)lded
~ное украшение stucco mo(u)lding
лепра *med.* leprosy
лепрозный leprose, leprous, lepric
лепрозорий leprosarium, leprosery, leper-house/home/lazaret/ colony
лепролог leprologist
лепта 1. mite, bif; contribution; 2. lepton *(bronse coin in ancient Greece)*
лептинит *geol.* leptynite
лептогонидия leptogonidium
лептодермальный *bot.* leptodermous, leptodermatous
лептоклаза *geol.* leptoclase
лептоксилема *bot.* leptoxylem
лептом(а) *bot.* leptome
лептомедузы *pl. zool.* leptomedusans *(Leptomedusae)*
лептоменингит *med.* leptomeningitis, piaarachnitis
~, наружный leptomeningitis externa, archnitis
лептомный leptomatic
лептон *phys.* lepton
лептонема *biol.* leptotene, leptonema
лептосперм(ум) *bot.* tea tree *(Leptospermum)*
лептоспира *bact.* leptospira
лептоспироз *med.* leptospirosis
~, безжелтушный bovine leptospirosis, leptospirosa grippotyphosa
~, желтушный leptospiral jaundice, ictero-hemorrahagic spirochetosis, leptospirosis ictero-hemorrhagica, Vasilyev-Weil disease
~ собак canine leptospirosis/fever, Stuttgart dog plaque
лептоспирозный leptospiral, leptispirochetal
лептоспироносительство carriage of Leptospira
лептоспорангиевые *n. pl. bot.* leptosporangiates *(Leptosprangiatae)*
лептосфериоз *pathol.* leaf spot, leptosphaeria
лептотена *biol.* leptotene
лептофилльный *bot.* leptophyllous
лептофлоэма *bot.* leptophloem
лептохерида leptochoerid
лепчица *bot.* catchweed bedstraw *(Galium aparine)*
лепщик modeller, sculptor
лерка die/thread chaser, (threading) die
лерцит *geol.* lherzite
лерцолит *geol.* lherzolite
лес 1. wood, forest, silva; 2. *only sg. coll.* timber, lumber
~ в плотах timber float
~, выпасаемый grazing forest
~, высокоствольный high forest
~, дубово-грабовый hornbeam oar-forest
~, заповедный/защитный forest reserve, protection forest
~, ископаемый fossil forest
~, корабельный ship timber
~, летнезеленый deciduous summer forest
~, лиственный/черный leaf wood, green wood, deciduous forest
~, листопадный deciduous forest
~, мангровый mangrove/mangle forest
~, мелкий underwood, underbrush, coppice, copse
~ на корню standing timber
~, низкорослый stunted forest
~, окаменелый petrified wood
~, перестойный overmature/overripe forest
~, поделочный carpentry timber
~, порослевый copse, coppice (forest)
~, пьяный forest growing on landslides

~, **разреженный** sparse wood, thinned-out wood
~, **смешанный** mixed/mingled wood
~, **сортовой** commercial timber
~, **спелый** mature forest
~, **сплавной** raff wood
~, **строевой** timber/lumber wood
~, **темнохвойный** taiga forest
~, **трелевать (заготовленный)** transport (felled) trees/logs/timber
~, **тропический влажный** tropical humid forest
~, **тугайный** flood-plain forest
~, **хвойный/красный** coniferous forest, wood of coniferous trees
~, **хвощовый** horsetail forest
~, **широколиственный** broad-leaved forest
леса I *pl.* scaffold(ing), falsework, framework, staging
~, **выпускные** outrigger scaffolding, cradle scaffold
~, **инвентарные** knock-down scaffolding
~, **коренные** pole scaffold
~, **механизированные** mechanized staging/scaffolding
~, **передвижные** mobile staging, travelling scaffolding
~ **рамной конструкции** saddle scaffold
~, **сомоподъемные** climbing scaffolding
~, **трубчатые** tubular scaffolding
леса II *naut.* fishing line
лесбианизм, лесбийство lesbianism, lesbian love; sapphism
лесбийка, лесбиянка lesbian, female homesexual; sapphist
лесбийский, лесбосский Lesbos; lesbian
~кая любовь lesbianism, lesbian love
лесник forest ranger, forester
леснина *bot.* common/wild apple *(Malus sylvestris/pumila)*
лесничество 1. forestry departmant, forest ranger station; 2. forestry, forest area
лесничий *n.* forester; forestry officer, forest ranger/warden
лесной silvan, forestal, forest
~ **войлок,** ~ **ная подстилка** forest floor/litter
~ **двор/склад** lumber yard
~ **заповедник** forest reserve, protection forest
~ **институт** school of forestry
~ **материал** lumber
~ **питомник** mursery forest
~ная промышленность timber industry
~ные насаждения afforestation
лесоавтопоезд timber/lumber/log truck
лесоведение silvics, dendrology
лесовод silviculturist, forestry specialist
лесоводство forest management, silviculture, forestry; forest-growing
лесоводческий silvicultural
лесовоз *naut.* timber carrier/ship, lumber ship
лесовозобновление, лесовосстановление reforestation, reafforestation
лесогон *min.* rafter, raftsman
лесозавод lumber mill
лесозаготовитель woodsman, lumber-jack
лесозаготовка timber cutting; *pl.* logging, lumbering
лесозащита forest conservation/protection
лесоинвентаризация forest valuation
лесокультура forestation, planting of trees
лесоматериал timber, lumber, wood
~, **выдержанный** seasoned timber
~, **колотый** cleaving/cleft timber
~, **крепежный** mine timber, props
~, **необработанный** unsawn timber, logs
~, **обработанный** sawn timber, lumber
~, **пиленый/пильный** saw(ed) timber
~, **строганный** dressed timber
~, **строительный** structural/construction timber
~, **тесаный** hewn timber
лесомелиорация forest amelioration/reclamation/improvement
лесонасаждение 1. forest stand/plantation; 2. afforestation, forestation
~, **высокополнотное** fully stocked forest stand
~, **высокоствольное** long-boled forest stand
~, **низкополнотное** incomplete forest stand
~, **нормальное** all-aged forest stand
~, **перестойное** overmature forest stand
~, **полезащитное** shelterbelt forest
лесообразующий forest-forming
лесоохранение forest protection
лесоочистка deforestation
лесопарк forest park
лесоперевалка transshipment of lumber
лесопильный wood-sawing
~ **завод** sawmill
~ная рама log frame
лесопитомник tree nursery, nursery forest, arboretum
лесоповал tree felling
лесоповальщик tree feller; lumberjack
лесопогрузчик logger, machine for loading logs
лесополоса forest/woodland belt; windbreak
лесопосадка afforestation, forestation
лесопровод logway
лесопромышленность timber industry
лесопункт logging unit, lumber camp
лесоразведение forest-growing, planting of forests
лесоразработка 1. forest exploitation (area), logging area; 2. forest working, timber logging
лесорама frame saw, saw frame
~, **двухпоставная** double log frame, double blade frame saw
~, **однопоставная** single deal frame
лесоруб lumberjack, feller, woodcutter, lumberman, logger
лесосека cutting/felling area; clearing
лесосплав timber rafting
~, **молевой/россыпью** loose floating
лесопуск timber slide/chute/raft
лесостепь forest-steppe

лесосушилка timber dryer; lumber kiln
лесотаска log chain/conveyer/hauler; timber grab
лесотехнический forestry *(attr.)*
~ институт school of forestry
лесоторговец lumber dealer
лесотранспортер log conveyer/haul/chain
лесотундра forest-tundra
лесоукладчик timber-stacking machine
лесоустройство forest managemont
лесоучасток forest plot/area
лесохимия wood/forest chemistry, silvichemistry, dendrochemistry
лесохозяйство forestry
лесохранилище log storage ground
лесоэкспорт timber export
леспедеза, леспедеца *bot.* lespedeza *(Lespedeza)*
леспромхоз *abbr.* integrated logging-lumbering enterprise, (State) timber industry enterprise
лесс *geol.* loess, (a)eolian soil
~, озерный lacustrine loess
~, поздний ergeron
лессиваж lessivation, lessivage, pseudo-podzolization
лессиве *geol.* leached soil
лессировка *art* scumbling; glazing
лессовый loessial, loessic
лессообразующий loess-forming/-yielding
лестница 1. stair; staircase, stairway; ladder; 2. *anat.* scala
~, веревочная rope ladder
~, винтовая winding stairs
~, встроенная self-contained ladder
~, движущаяся/самодвижная escalator
~, запасная emergency ladder
~, иерархическая hierarchy, the scale of rank
~, монтажная erecting ladder
~, парадная main/front staircase
~, пожарная fire escape
~, предгорная piedmont steps
~, приставная (access) ladder
~, складная stepladder
~, служебная/черная backstairs, service stairs
лестница-стремянка access ladder, stepladder
лестницеобразный scalariform
лестничный ladder-type, scalariform; *anat.* scalene
~ная клетка staircase, well *(of stairs)*
лестовка *rel.* rosary
лесхоз *abbr.* district/regional forestry (administration)
Лета *myth.* Lethe
леталь *biol.* lethal gene
летальность lethality
~, клеточная cell lethality
летальный lethal, fatal
летаргический lethargic
летаргия lethargy; hebetude; *nucl.* mean logarithmic energy decrement/loss
~, африканская African lethargy, sleeping sickness, lethargus
~, гипнотическая induced/hypnotic lethargy, hypnosis
~, неполная hemilethargy
летательный flying
~ аппарат aircraft
~ная перепонка *zool.* wing membrane
летать fly
~ в простых метеоусловиях fly VFR (visual flight rules), fly under VFR conditions
~ в сложных метеоусловиях fly IFR (instrument flight rules), fly under IFR condition
~ ведомым fly wing
~ горизонтально fly level
~ на приводную станцию fly inbound
~ по приборам fly on the instruments
~ с убранным шасси fly gear up
летероид fish paper insulation
летка *met.* tap-hole, tapping hole; (foundry) gate
~ для штейна charging thimble
~, козловая salamander-drain hole
~, чугунная iron tap-hole, iron notch
~, шлаковая slag/dross hole/notch
летне-осенний (a)estivo-autumnal
летний (a)estival, summer *(attr.)*
~ период покоя summer dormancy
летник *bot.* annual
летний flyable, flying, flight *(attr.)*
~ состав flight personnel
~ная погода flying weather
~ное дело flying
~ное поле airfield
лето summer
~, бабье Indian summer, Old Wives' Summer
летование *zool.* (a)estivation
летовать *zool.* (a)estivate
летоисчисление chronology
леток *agr.* entrance *(in beehive)*
летописание chronicle, annals; writing of annals
летописец annalist, chronicler
летопись annals, chronicle
~, геологическая geological records
~, палеонтологическая fossil record
летосчисление chronology, time-table
леттсомит *geol.* lettsomite, cyanotrichite
летучее *n.* volatile matter
летучесть *chem.* volatility; fugacity
~, терять devolatilize
летучие *n. pl.* volatiles
летучий 1. flying; *fig.* ephemeral, passing; 2. *chem.* volatile; fugacious, fugaceous; 3. *med.* shifting
~ листок leaflet
~ митинг briefing; short meeting
~ петух *zool.* flying gurnard *(Dactylopterus)*
~чая мышь *zool.* bat *(Verspertilio murinus)*
летучка 1. briefing; short meeting; 2. *mil.* mobile detachment; 3. *bot.* egret, pappus; winged fruit; 4. *zool.* flying fish *(Exocoetus)*
~, авторемонтная automotive repair truck
летчик pilot, aircrew member, flier, airman

~ **аэростата** balloon pilot
~ **ВВС** air force pilot
~, **ведомый** wingman
~, **военный** military/service pilot
~, **второй** copilot, second pilot
~ **гражданской авиации** commercial/airline pilot
~, **дежурный** alert/duty pilot
~, **запасной** reserve pilot, stand-by pilot
~, **классный** rated pilot
~, **морской** naval pilot
~, **обучаемый** trainee/pupil pilot
летчик-инструктор instructor-pilot, flight/flying instructor
летчик-испытатель test pilot
летчик-истребитель fighter pilot
летчик-командир корабля pilot-in-command
летчик-космонавт (aero)spce pilot, pilot-cosmonaut, astronaut
летчик-курсант student/cadet/pupil pilot, pilot cadet
летчик-любитель private pilot
летчик-наблюдатель aircraft observer
летчик-новичок neophyte/novice pilot
летчик-планерист glider pilot
летчик-приемщик (самолетов) acceptance pilot
летчик-стажер practice pilot
летчик-штурман navigator pilot
летчик-штурмовик strike pilot
летчица woman/female pilot, aviatress
летяга *zool.* flying squirrel *(Pteromys)*
~, **большая сумчатая** taguan *(Schoinobates volans)*
леукофиллум *bot.* silverleaf *(Leucophyllum)*
лехитский Lekhitic
лецидеевый, лецидеиновый *bot.* lecideine
лецитальбумин *bioch.* lecithalbumin
лецитис *bot.* monkeypot tree *(Lecythis)*
лечащий врач physician in charge
лечебница clinic
лечебно-охранительный treatment-and-protective
лечебно-профилактический treatment-and-prophylactic
лечебно-трудовой treatment-and-labo(u)r
лечебный medical, medicinal, medicative, therapeutic, curative, sanative, iatric
~**ная физкультура** physiotherapy exercises, recuperative gymnastics
лечение (medical) treatment, therapy, therapeutics, therapia, cure
~, **амбулаторное** outpatient treatment, ambulant therapy
~, **больничное** hospital treatment/cure
~ **вибрационным массажем** seismotherapy
~ **внушениями** suggestion/mental therapy, pithiatism
~ **голодом** fasting/hunger/starvation cure, peinotherapy, eimotherapy, nestotherapy
~, **грязевое** fangotherapy
~, **диетическое** diet therapy, dieting
~, **диспансерное** follow-up care
~, **консервативное** conservative/expectant treatment
~ **массажем** massotherapy
~, **медикаментозное** pharmacotherapy, drug therapy
~, **местное** topical/local treatment
~ **минеральными водами** crenotherapy, crounotherapy
~ **облучением** (ir)radiation treatment
~, **последующее** aftertreatment
~, **предупредительное/профилактическое** preventive treatment
~, **принудительное** compulsory/forced therapy
~, **симптоматическое** symptomatic/expectant therapy
~ **сном** sleep/narcosis therapy, narcotherapy
~ **стариков** gerontotherapy
~, **стационарное** hospital/institutional treatment, hospital cure
~ **сухоядением** dipsotherapy, thirst cure
~ **теплом** thermotherapy
~, **терапевтическое** internal treatment
~ **терренкуром** terrain cure
~ **точечными прижиганиями** pyropuncture
~ **фарадическим током** faradotherapy, faradization
~, **хирургическое** surgical treatment/management
~ **холодом** cryotherapy, frigotherapy, psychrotherapy
~ **шоком** shock therapy/treatment
~ **эндокринными препаратами** endocrine/glandular/hormonal therapy
~, **этиологическое** (a)etiological treatment
лечить treat, heal, cure
лешательерит *geol.* lechatelierite
леший *myth.* wood goblin
лещ *zool.* bream *(Abramis)*
лещадь *met.* (blast-furnace) bottom; hearth bottom/block
~ **вагранки** cupola hearth/bottom
~ **из углеродистой набойки** carbon paste bottom
лещина *bot.* hazel *(Corylus)*
лещинолистный *bot.* corylifolious
лжеакация *bot.* black locust, false acacia *(Robinia pseudoacacia)*
лжеапельсин *bot.* Osage orange *(Maclura aurantiaca)*
лжегрибница *bot.* pseudomycelium
лжедемократия pseudo democracy
лжедурнишник *bot.* sump weed *(Iva)*
лжеискусство pseudo art
лжекипарис *bot.* false cypress *(Chamaecyparis)*
лжеклассик pseudoclassic
лжелиственница *bot.* golden larch *(Pseudolarix)*
лженарцисс *bot.* common daffodil *(Narcissus*

pseudo-narcissus)
лженаука pseudo science
лженаучный pseudoscientific
лжеприсяга *law* perjury
лжепропаганда false propaganda
лжепророк false prophet
лжесвидетель *law* perjurer, false witness
лжесвидетельство false evidence/testimony, perjury
лжесвидетельствовать give false evidence
лжесоциалистический pseudosocialistic
лжетеория false theory
лжетсуга *bot.* Douglas fir *(Pseudotsuga)*
лжеучение false doctrine
лжеучитель false teacher
лжец liar
лживость falsity, inveracity, mendacity
~, патологическая pathologic(al) lying, pseudologia phantastica
ЛЗП *abbr.* **линия заданного пути** *av.* track required; *Am.* course (line)
ли li *(Chinese unit =576 metres)*
лиаза *bioch.* lyase
лиана *bot.* liana, liane
лиановидный, лианоидный lianoid
лиатрис *bot.* button snakeroot, gayfeather *(Liatris)*
либерал liberal; freethinker
либидо *psych.* libido, sexual impulse
либоцедрус *bot.* incense cedar *(Libocedrus)*
либрация *astr.* libration
~, суточная diurnal/daily libration
либреттист librettist
либретто libretto
либриум librium, chlordiazepoxide
либриформ *bot.* libriform
ливанец, ливанский Lebanese
ливанка Lebanese (woman)
ливень (heavy) shower, downpour; cloudburst
~, боковой lateral/inclined shower
~, внутримассовый air-mass shower
~, вызванный электронами electron-induced shower
~, грозовой thundershower
~, ионный avalanche of ions
~, космический, ~ космического излучения (cosmic-ray) shower
~ Оже Auger shower
~, проникающий penetrating/hard shower
~, широкий атмосферный extended air shower, Auger shower
~, шквалистый cloudburst
~, электронно-фотонный photon-electronic shower
~, электронно-ядерный electron-nuclear shower
~, электронный electronic shower
ливер I *cul.* pluck
ливер II *tech.* pipette, siphon
ливерная колбаса liverwurst
ливиец, ливийский Libyan
ливийка Libyan (woman)
ливистона *bot.* fan palm *(Livistona)*
ливневый showery; torrential; cumulo-nimbus *(clouds)*
ливнеобразование *phys.* shower formation
ливнесброс rain-water drainage
ливнесток storm water inlet
ливный water-filled/-laden
~ дождь pouring rain
ливонский *hist.* Livonian
ливр 1. *hist.* livre; 2. pound, livre
ливрея livery
лига I league
~ наций *hist.* Leaque of Natons
лига II *mus.* ligature, bind, tie
лига III league *(measure of distance)*
лигамент ligament
~, внешний *zool.* alivincular ligament
лигаментит *med.* tendovaginitis, tenosynovitis
~ кольцевидной связки пальца trigger/spring/stuck/snap finger, jerk-finger
~, стенозирующий stenotic tendovaginitis
лигатура 1. *met.* alloying composition, hardener; *chem.* (master) alloy, foundry alloy, base metal; 2. *print.* ligature; 3. *mus.* ligature, tie, slur
~ в золоте base (metal)
~, медная copper hardener
лигирование ligation
лигнин lignin
~, замещенный substituted lignin
~, пиролизный pyrolized lignin
лигниннообразование lignification
лигнино-протеиновый lignin-protein
лигнино-целлюлозный ligno-cellulosic
лигнит *miner.* lignite, wood/bovey coal
~, волокнистый древесный xylite
~, древесный woody lignite
~, землистый earthy lignite
~, молодой peat coal
~, низкосортный brown coal
~, превращать в lignify
~, рыхлый moor coal
~, черный black lignite
лигнитный, лигнитовый lignitiferous, lignite-bearing, lignitic
лигногумат lignohumate
лигноза woody vegetation, lignose, lignosa
лигносульфонат lignosulfonate
лигнофоль laminated lignin plastic, wood impregnated laminate, lignefol
лигноцеллюлоза *bot.* lignocellulose
лигноцериновый lignoceric
лигодий, лигодиум *bot.* climbing fern *(Lygodium)*
лигроин *chem.* ligroin(e), naphta (fraction)
лигулярный *bot.* ligular, ligulate
лидаза *med.* lydasa, hyaluronidase
лидар lidar
~, бортовой airborne lidar
лиддит *chem.* lyddite

лидер 1. *polit.* leader; 2. *sport* front runner; 3. *naut.* flotilla leader
~ эскадренных миноносцев destroyer leader
лидерство 1. *polit.* leadership; 2. *sport* lead, first place
~ в ценах price leadership
лидийский Lydian
лидирование *sport* leading
лидит *miner.* Lydian stone, lydite, touchstone
лиеврит *geol.* lievrite, ilvaite
лиенальный *anat.* lienal, splenic
лиенография *med.* lienography, splenography
лиеномедуллярный *anat.* lienomedullary, splenomyelogenous
лиенопанкреатический *anat.* lienopancreatic, splenopancreatic
лиенопатия *med.* lienopathy, splenopathy
лиенотоксин lienotoxin, splenotoxin
лизгольдер *hist.* leaseholder
лизергиновый *bioch.* lysergic
лизигенный lysigenous, lysigenic, lysigene
лизидин *chem.* lysidine, methylglyoxalidine
лизикарпный *bot.* lysicarpous
лизиметр lysimeter, evapotranspirometer
лизин lysine *(amino acid);* lysin *(antibody)*
лизинг *econ.* leasing
лизиндекарбоксилаза *boich.* lysine decarboxylase
лизинрацемаза lysine racemase
лизис *bioch., med.* lysis
~ микробов bacteriolysis
лизогенез lysogenesis
лизозим, лизоцим *bioch.* lysozime
лизоклин lysocline
лизол *chem.* lysol
лизунец 1. *agr.* salt lick, dairy salt; 2. *pharm.* lincture
лизхолдер *hist.* leaseholder
лик 1. face image; 2. *eccl.* assembly, host
ликантропия *med.* lycanthropy
ликбез *abbr.* 1. elimination of illiteracy; 2. school for illiterates
ликват segregated material
ликвация *met.* segregation, liquation
~ дендритная dendritic segregation
~, капельная droplet segregation
~, обратная inverse/negative segregation
~ по удельному весу gravity segregation
ликвидамбар *bot.* sweet palm *(Liquidambar)*
ликвидатор 1. *mil.* destructor, destruction mechanism/device; sterilizer; 2. liquidator
~ имущества *econ.* receiver
ликвидаторство *polit.* liquidationism
ликвидация *econ.* liquidation; *polit.* wiping out; elimination, abolition; *mil.* destruction
~ долгов settlement of debts
~, принудительная *law* winding-up by the court, compulsory liquidation
~ции, приступить к go into liquidation
ликвидирование liquidation; elimination, eliminating
ликвидировать liquidate, remove, eliminate; abolish, wipe out; *econ.* wind up
~ безграмотность do away with illiteracy
~ дефицит *econ.* pull(oneself) out of the red
~ опозицию liquidate opposition
~ предприятие wind up a business
~ убыточное предприятие mop up
ликвидность *fin.* liquidity
~ баланса solvency
ликвидный *fin.* liquid
~ные средства liquid assets, ready money
ликвидус *phys. chem.* liquidus (curve)
ликвор *med.* liquor
ликворея *med.* liquorrhea
~, желудочковая ventricular liquorrhea
ликование *rel.* exultation, triumph
ликовать *rel.* rejoice, exult, triumph
ликоподиевые *n. pl. bot.* club-moss family *(Lycopodiaceae)*
ликоподий 1. *bot.* lycopodium, vegetable sulphur, club moss *(Lycopodium);* 2. *pharm.* lycopodium powder
ликтор *hist.* lictor
ликтрос *naut.* boltrope
~ задней шкаторины after-leech rope
~ передней шкаторины luff rope
ликуала *bot.* licuala palm *(Licuala)*
ликующий *rel.* exultant, jubilant; triumphant
лилейный *bot.* liliaceous; *n. pl.* lily family *(Liliaceae)*
лилиецветковый, лилиецветный *bot.* lily-flowered
лилипут Lilliputian; homunculus; dwarf
лилия 1. *bot.* lily; *(Lilium);* 2. *zool.* crinoid
~, белая/бледная Madonna lily *(L. candidum)*
~, водяная water lily
~, двупоясковая/дициклическая dicyclic crinoid
~, личинкообразная larviform crinoid
~, мелкая морская microcrinoid
~, морская crinoid, sea lily *(Crinoidea)*
лима *bot.* 1. Lima bean *(Phaseolus limensis);* 2. mauritius, cabuyao *(Citrus hystrix)*
лиман estuary, firth, frith; liman, (brackish) lagoon; coastal salt lake
лиманда *zool.* dab *(Limanada limanada)*
лимацид limacide
лимб 1. *tech.* limb; dial, graduated circle; 2. *paleont.* border; 3. *zool., bot.* limbus, edge, border
~, азимутальный azimuth dial
~, буссоли declination arc
~, головной cephalic limb
~ дальности range dial
~, микрометрический micrometer dial
~, нониусный, ~ с верньером vernier dial
~ отсчета курсов (по компасу) grid-ring
~ установки скорости подачи feed dial
лимберсы *pl. naut.* limbers
лимбо limbo *(dance)*

лимбусный limbic
лиметта *bot.* lime *(Citrus aurantifolia/limetta)*
лимит *econ.* limit; quota; *pl.* allocations
лимитатив *logic* limitative
лимитирование limiting, limitation
лимитный limital; limitative
лимитроф limitrophe, border state
лимитрофный adjacent, on the frontier, limitrophe
лимнантемум *bot.* floating heart *(Nymphoides, Limnanthemum)*
лимнантес *bot.* meadow foam *(Limnanthes)*
лимниграмма limnogram
лимниграф limnigraph, limnimeter, limnometer
лимнобентос limnobenthos
лимнобиоз *biol.* limnobiosis
лимнолог limnologist
лимнология limnology
лимнонектон limnonekton
лимнопланктон *biol.* limnoplankton, fresh-water plankton
лимнофильный *zool.* limnophilous
лимнофит *bot.* pond plant, limnophyte
лимодорум *bot.* grass-pink orchid *(Limodorum, Calopagon)*
лимон *bot.* 1. lemon tree; 2. lemon (fruit)
~, персидский wood apple *(Feronia)*
лимонад lemonade; lemon soda
лимонарий lemon tree mursery
лимонен *chem.* limonene
лимонит *miner.* brown hematite, limonite
~, бурый землистый brown umber
~ в чечевицеобразных агрегатах bean iron ore, pea ore
~, губчатый bog iron ore, morass ore
~, мягкий землистый yellow ochre
~, оолитовый oolitic iron ore
лимонник *bot.* magnolia vine *(Schizandra)*
лимонница *zool.* brimstone futterfly *(Gonepterix rhamni)*
~, гладкая *bot.* Asian Molucca balm *(Moluc(c)ella laevis)*
лимоннокислый *chem.* citrate (of), citric acid *(attr.)*
лимоннолистный *bot.* citrifolious
лимо(н)нообразный, лимоновидный lemon-shaped, limoniform
лимонноцветковый *bot.* citriflorous
лимонный citric; lemony
~ сок lemon juice
~ная кислота citric acid
~ного цвета citreous
лимский Lima *(attr.)*
лимузин limousine
лимфа *physiol.* lymph
~, апластическая aplastic/corpuscular lymph
~, бычья bovine lymph
~, вакцинная vaccine/vaccinia lymph
~, вакцинная человеческая humanized lymph
~, воспалительная inflammatory/plastic lymph
~ Коха Koch's lymph, tuberculin
~, кровяная blood lymph
~, прививочная/оспенная inoculation lymph
~, тканевая tissue/true lymph
~, фибринозная fibrinous/euplastic/croupous lymph
лимфаденит *med.* lymphadenitis, adenolymphitis
~, гнойный suppurative lymphadenitis
~, доброкачественный вирусный benign viral lymphadenitis
~, злокачественный malignant lymphadenitis/lymphoma
~, небактериальный регионарный nonbacterial regional lymphadenitis, cat scratch disease
~, паратуберкулезный paratuberculous/paracaseous lymphadenitis
~, паховый inquinal lymphadenitis
~, прикорневой hilar adenopathy/lymphadenitis
лимфаденоз *med.* lymphadenosis
~, доброкачественный benign lymphadenosis, infectious mononucleosis
~, злокачественный malignant lymphadenosis/lymphoma
~, острый acute lymphadenosis
лимфаденома *med.* lymphadenoma
лимфаденоматоз *med.* lymphadenomatosis
~, общий костный general lymphadenomatosis of bones, Kahler's disease
лимфаденопатия *med.* lymphadenopathy
лимфангиома *med.* lymphangioma, angio lymphoma, angioma lymphaticum
~, истинная true lymphangioma
~, кавернозная lymphangioma cavernosum, cavernous lymphangiectasia, cavernoma lymphaticum
~, кистовидная/кистозная cystic lymphangiectasia/lymphangioma
лимфангит, лимфангоит *med.* lymphangitis, lymphangiitis, angiolymphitis, lymphangeitis
~, острый acute lymphangitis
~, рожистый erysipelar/erysipelatous lymphangitis
~, сетчатый reticular lymphangitis
~, стволовой truncular lymphangitis
~, эпизоотический epizootic lymphangitis
лимфангитический lymphangitic
лимфатизм *med.* lymphatism
лимфатический, лимфатичный *physiol.* lymphatic
лимфобласт *anat.* lympho(cyto)blast
лимфобластома *med.* lymphoblastoma
~, макрофолликулярная macrofollicular lymphoblastoma, giant follicular lymphoma
лимфогранулема lymphogranuloma
~, плазмоклеточная plasma cell lymphogranuloma
лимфогранулематоз *med.* lymphogranulomatosis
~, злокачественный Hodgkin's disease, malignant lymphoma
~, паховый inguinal lymphogranulomatosis,

lymphogranuloma venereum/inguinale, climatic/nonvenereal/tropical bubo
лимфография lymphography
лимфоидоцит *anat.* lymphoidocyte, h(a)emocytoblast, h(a)ematocytoblast, h(a)ematogonia
лимфоидэктомия *surg.* lymphoidectomy, adenoidectomy, tonsillectomy
лимфолейкоз *med.* lymphoid leukosis/leuk(a)emia
лимфома *med.* lymphoma
~, атипическая atypical lymphoma, lymphosarcoma
~, африканская/Буркитта African/Burkitt's lymphoma
~, злокачественная malignant lymphoma, Hodgkin's disease
~, множественная multiple lymphoma
лимфообразование lymphopoiesis
лимфообразующий lymphopoietic
лимфообращение flow of lymph
лимфоотток lymph drainage
лимфопения *pathol.* lympho(cyto)penia
лимфоретикулоз *med.* lymphoreticulosis
~, доброкачественный benign lymphoreticulosis, cat-scratch fever/disease
лимфорея *med.* lymphorrhea, lymphorrhagia
лимфосаркома *med.* lymphosarcoma
~, забрюшинная retroperitoneal lymphosarcoma
лимфосаркоматоз *med.* lymphosarcomatosis
~ кожи lymphosarcomatosis cutis
лимфоток lymph flow
лимфоузел lymphonodus
лимфоцеле *physiol.* lymphocele, lymphocyst
лимфоцит *physiol.* lymphocyte, lymph cell/corpuscle
~, крупный macrolymphocyte, large lymphocyte
~, малый microlymphocyte, small lymphocyte
лимфоцитарный lymphocytic
лимфоцитоз, лимфоцитемия *pathol.* lymphocytosis, lymphocyt(a)emia, achroacytosis
лимфоэндотелиома *med.* lymph(angio)entothelioma
лимфоэпителиома *med.* lymph(o)epithelioma
линамарин *chem.* linamarin
линария *bot.* toadflax *(Linaria)*
лингафон linguaphone
лингвист linguist
лингвистика linguistics
~, вычислительная computational linguistics
~, пространственная spatial linguistics
лингулойд linguloid
линдерния *bot.* false pimpernel *(Lindernia)*
линдсейит *geol.* lindsayite
линеаризация linearization
~, вибрационная linearization by oscillation
~, принудительная forced linearization
линеаризовать linearize
линеарность linearity
линеарный linear
линейка I 1. rule, ruler, straight edge; 2. line; 3. *print.* rule
~, акцидентная *print.* border (rule)
~, биговальная *print.* score rule
~, вечерняя evening parade
~, визирная *geod.* aiming/sight rule, azimuthal lever
~ высот *geod.* height arm/slide
~, гартовая *print.* lead rule
~, гибкая чертежная spline
~, глазная скиаскопическая *med.* ophthalmic skiascope
~, дешифраторная *com.* code bar
~ для вычисления боковых поправок azimuth adjustment rule
~, кегельная *print.* em rule
~, комбинаторная *tel.* selector bar
~, лагерная camp line(s)
~, лекальная toolmaker's straight-edge
~, логарифмическая/счетная slide rule
~, масломерная oil dipstick
~, масштабная scale rule
~, масштабная измерительная dimension scale
~, мерная dipstick; *for.* log scale, dividing/measuring rule
~, наборная *print.* setting rule
~, навигационная (счетная) computer slide-rule
~ направлений *geod.* direction arm
~, направляющая 1. *for.* guide fence; 2. *met.* guard
~ параллаксов *geod.* parallax arm/measurer, parallactic bar
~, поверочная/рихтовальная straight-edge
~, разметочная *mach.* marking rule
~, синусная *mach.* sine bar
~ слова *comp.* word location
~, типографская line gauge, printer's rule
~, тонкая *print.* fine-face rule
~, тупая *print.* bold-face rule
~, формовочная *met.* straight-edge
~, центровочная centre-of-gravity computer slide rule
~, чертежная draftman's rule
~ штурманская navigation(al) (slide-rule) computer
~ки, нотные *mus.* staves
линейка II wagonette
линейка-пеленгатор sighting-vane bar
линейкорубилка *print.* rule cutter
линейно linearly; arcwise
~ связанный arcwise connected
линейно-компактный *math.* linearly compact
линейно-ланцетный/ланцетовидный *bot.* lance-linear; linear-lanceolate
линейно-мечевидный *bot.* linear-ensate
линейно-продолговатый linear-oblong
линейность lineation, linearity
~ по кадрам *telev.* vertical linearity
~ развертки *telev.* sweep/scan(ning) linearity
~ характеристики linearity of response

линейный 1. *math.* linear; lineal; 2. *mil.* of the line; 3. boundary, frontier *(attr.)*
~ корабль battleship
~ная скорость linear speed
~ные меры long measures
~ные сооружения on-line installations
линейчато-геометрический line-geometric
линейчатость lineation
линейчатый lineate, lineolate; striate; linear; lined, ruled
линек *nav.* rope's end, colt
линеолированный lineolate
линза 1. lens; 2. *geol.* lentil, lenticle
~, акустическая acoustic/sound lens
~ Бертрана Bertrand lens
~ видоискателя finder magnifier
~, вогнутая concave/minus/reducing lens
~, вогнуто-выпуклая concave/concavo-convex lens
~, входящая в оптическую систему lens element
~, выпуклая convex/plus/magnifying lens
~, вытянутая elongated lens
~, глазная контактная ophthalmic contact lens
~, глинистая *geol.* clay gall
~, двояковогнутая biconcave lens, double-concave lens, concavo-concave lens
~, двояковыпуклая biconvex lens, double-convex lens, convexo-convex lens
~, длиннофокусная long-focus lens
~, замедляющая decelerating lens
~ замещения replacement lens
~, квадрупольная quadrupole lens
~, коллективная collecting lens
~, коллиматорная collimating lens
~ колчедана *geol.* sulphur kidney
~, контактная contact lens
~, короткофокусная short-focus lens
~ Люнеберга Luneberg lens
~, магнитная magnetostatic/magnetic lens
~, менисковая meniscus lens
~, насадочная adapter/attachment/accessory lens anamorphic adapter/attachment, anamorphoser
~, насадочная диффузионная diffusion attachment, diffuser
~, обратная inversion lens
~ объектива objective lens
~ окуляра ocular/eye(piece) lens
~, оптическая optical/light lens
~ пирита *geol.* pyrite lens, lenticular pyrite
~, плоско-вогнутая flat-/plano-concave lens
~, плоско-выпуклая flat-/plano-convex lens
~, плоскоцилиндрическая plano-cylinder lens
~, положительная collecting lens
~, поясная drum lens
~, проволочная wire lens
~, просветленная coated/antireflection lens
~, рассеивающая diverging/dispersing/defocusing/negative lens
~, решетчатая lattice lens
~ с седлообразным полем saddle-field lens
~, светосильная high-dispersing/-aperture lens
~, собирательная/собирающая converging/collecting lens
~, составная composite/compound lens
~, ступенчатая Fresnel lens
~, увеличительная magnifying lens
~, узколучевая narrow-angle lens
~, ультразвуковая ultrasonic lens
~, фокусирующая focusing lens
~ Френеля Fresnel lens
~, цилиндрическая cylindrical (drum) lens
~ шкалы dial lens
~, щелевая slit lens
линзенштихель engraver's tool
линзовидный, линзообразный phacoid(al), lenticular, lentiform, lens-shaped
линзовыниматель cap extractor
линиаризация linearization
линиатура растра screen line number
линиевыбиратель *teleph.* selecting switch
линимент liniment(um), embrocation
линин *biol.* achromatin, linin
линит *med.* linitis
~, пластический Brinton's disease, chronic sclerosing gastritis
линия 1. line; 2. *biol.* strain, variety
~, абонентская subscriber('s)/individual/ exchange line, subscriber's loop
~, абонентская индивидуальная individual line, direct exchange line, one-party telephone
~ абриса *cart.* planimetric line
~ АВ energizing/current/power line
~, автоматическая *mach.* (automatic) transfer line
~, автоматическая комплексная *mach.* integrated complex line, integrated manufacturing system
~, автоматическая с гибкой связью *mach.* non-synchronous transfer line
~, автоматическая с жесткой связью *mach.* synchronous transfer line
~, автоматическая со спутниками *mach.* pallet type transfer line
~, автоматическая с управлением от ЭВМ *mach.* computer-controlled transfer line
~, агониатитовая лопастная *geol.* agoniatite lobe line
~, агоническая *geod.* agonic/zero line
~, адресная *comp.* address line
~ азимута azimuth line
~, аммонитовая перегородочная *geol.* ammonite suture, ammonoid line
~, антистоксова anti-Stokes line
~ апсид *astr.* apse/apsidal line, line of apsides
~, атмосферная *heat* air evacuation line
~, базисная *math.* reference line; *av.* base line; *opt.* air base line
~, базовая base/datum/centre line
~ бедра, шероховатая *anat.* rough femoral

line
~ бедренной кости, межмыщелковая *anat.* intercondylar femoral line
~, береговая shore/strand/coast line, water front, coastline
~, береговая опускания shore line of depression
~, береговая погружения shore line of submergence
~, береговая поднятия shore line of emergence
~, береговая смешанного характера compound shore line
~, бесконечная *math.* line of infinity
~, бесседельная лопастная *geol.* asellate lobe line
~, боковая lateral line
~ бросания *mil.* line of departure
~, брэгговская Bragg peak/maximum
~ в состоянии готовности hot line
~ вакуумная vacuum line
~ вала/валов line of shafting, shaft line
~ вектора скорости speed line
~ видимого горизонта horizon line, skyline
~, визирная (indicator) hair-line
~ визирования *geod.* line/axis of sight, sight line
~, винтовая helix, helicoid, spiral, helical line
~, вихревая vortex line; *math.* rotation curve/line
~ возмущений Mach/sonic line
~ впадин/выступов (зубчатого колеса) dedendum line
~, вращательная rotational line
~, входная *comp.* input line
~ входящей связи incoming/inward line
~ выборки *comp.* select(ion)/access line
~, выпускная exhaust line
~ выруливания *av.* lead-off line
~ выходов *min.* outcrop line; course of outcrops
~, геодезическая geodetic/geodesic line/curve, great-circle line
~, гибридная hybrid strain
~ горизонта skyline, horizon line; *nav.* sealine
~, горловая *math.* striction line
~, граничная boundary line
~, гребенчатая *el.* comb (transmission) line
~, гребешковая *anat.* pectineal line
~, грудинная *anat.* sternal line
~, грузонапреженная heavy-trafic line
~ дальности range line
~ движения (частиц) trajectory
~ двоякой кривизны twisted/skew curve, line of double curvature
~, двусторонняя соединительная both-way trunk
~ действительного горизонта true-horizon line
~ действия силы force line
~ действия силы тяжести gravitational vertical (line)
~ действия тяги thrust line, axis of thrust
~, действующая hot/live line
~, демаркационная demarcation line
~, диаметральная (longitudinal) centreline, middle /*naut.* midship line
~, дренажная *av.* vent line
~, дугообразная *anat.* arcuate/semicircular line, line of Douglas
~, железнодорожная railway line
~, железнодорожная, двухпутная double-track railway line
~, железнодорожная однопутная single-track railway line
~ живота, белая *anat.* white line
~, жирная heavy line
~ забоев faces line
~ загоризонтной связи over-the-horizon link/line
~, заданная datum line
~ заданного пути, ЛЭП *av.* (desired) course, course line; required/intended track
~ задержки dalay/lag line
~ задержки, акустическая acoustic/sonic delay line
~ задержки, компенсированная equalized delay line
~ задержки, кварцевая quartz delay line
~ задержки, многократная multiple delay line
~ задержки, проволочная wire delay line
~ задержки, регулируемая variable delay line
~ задержки, ртутная mercury delay line
~ задержки, твердотельная solid-state delay line
~ задержки, ультразвуковая ultrasonic delay line, ultrasonic storage cell
~, задняя срединная *anat.* posterior median line, posterior midline
~, задняя ягодичная *anat.* posterior gluteal line
~, замкнутая магнитная closed magnetic line
~ записи trace
~, заправочная filling line
~ заруливания *av.* lead-in line
~ зарядки charging line
~ зацепления contact/pressure line
~ зацепления головок head-line of contact, top line of action
~ зацепления ножек зубьев dedendum line of contact
~, звуковая sonic/Mach line
~ зрения line of sight, sight line
~, зубчатая береговая crenulate shore line
~ изгиба *rail.* curvature line
~, измерительная *electr.* slotted/measuring line, standing-wave meter
~, изобатическая isobath, isobathic line
~, изогиетная isohyetal line
~, изогональная isogonal/isogonic line, isogonic

curve
~, изоклинальная isoclinal/isoclinic line
~, изопьестическая isopestic line
~, информационная data-path/bus
~ искаженных масштабов zero line
~ искуственного горизонта horizon bar
~ исполнительного старта *av.* departure line
~, исходная starting/datum line
~ исходящей связи *tel.* outgoing/outward line
~ камбаловидной мышцы *anat.* soleal/popliteal line, line for the soleus muscle
~ касания line of contact
~ качения line of rolling contact
~ кильватерных колонн *nav.* parallel lines ahead
~, колебательная vibrational line
~, коллимационная collimating/collimation/sight(ing) pointing line, line of collimation
~ кольцевания *av.* cross-feed line
~, контрольная check(ing)/control/test line
~, контурная outline, contour line
~, координатная coordinate line
~ координатной сетки grid line
~, косеисмическая coseismic line
~, косая *anat.* oblique line
~ космической связи space communication link
~, котидальная *naut.* co-tidal line
~ крестца, поперечная *anat.* transvese sacral line
~, кривая curve, curved line
~ кривизны line of curvature
~ критических точек stagnation/attachment line
~ курса/курсовая *av.* heading (line), course (line)
~, локсодромическая rhumb line
~, ломаная broken line, open polygon
~ Людерса *met.* Luder('s)/slip line
~, магистральная trunk/main line, trunk
~, магнитная силовая magnetic line of force
~ магнитной индукции magnetic flux line, line of magnetic flux
~ Маки Mackie line
~ малой воды low-water (shore) line
~, мантийная *geol.* pallial line
~ масштабной сетки grid line
~, межвертельная *anat.* intertrochanteric/spiral line
~, междугородная автобусная intercity bus service
~, мерная measuring line, *nav.* trial course
~, многоотводная *comp.* multidrop line
~, мутантная mutant strain
~ набора высоты *av.* climbing line
~ нагнетания discharge/delivery line
~ наибольшего ската *math.* line of maximum inclination, line of greatest declivity
~, наивысшая выйная *anat.* highest nuchae line
~ наименьшего сопротивления line of least resistance
~ напластования *geol.* stratification line
~, напорная pressure line
~, направляющая directrix
~, наступающая береговая invading shore line
~ начала отсчета reference/zero/datum line, line of origin
~, начальная generating line
~ невидимого контура invisible/hidden (out)line
~, неискаженная line of no distortion
~, нейтральная neutral line
~, неоднородная *tel.* non-uniform line, heterogeneous line
~, несобственная *math.* ideal line
~, нулевая zero/null line
~ нулевого склонения, ~ нулевых значений *geod.* zero/agonic line
~, обводная by-pass line
~ обметания *naut.* sweep line
~ обрушения *min.* line of caving, *geol.* line of break/collapse
~ обрыва break line
~, обсеквентная *geol.* obsequent line
~ обтекания stream-/flow line
~ одинаковой прибыли *econ.* isoprof
~ одинаковой стоимости *econ.* isoval
~ одновременных приливных течений *nav.* co-current line
~, однопутная single-track railway
~, однородная *com.* uniform line
~ Оже Auger-line
~, околопозвоночная *anat.* paravertebral line
~ окончательной сборки final assembly line
~, ортодромическая great-circle line
~, осевая axis/axial/centre line
~, осевая складки *geol.* fold axis
~ основания картины *top.* axis of homology/perspective, perspective axis, ground line
~, основная *nav.* base-line
~ отвеса *geod.* plumb line
~, отвесная the vertical (line); plumb line
~, отводящая outlet line
~ отдачи recoil line
~ откоса shoulder/slope line
~ отсчета reference/datum line
~ отсчета данных, контрольная reference trace
~ падения *min.* line of dip/incidence
~ палубы deck/sheer line
~ пеленга bearing line
~, перегородочная suture
~ передачи *rad.* transmission line
~ передачи данных *comp.* data link/line, data communication line
~ передачи команд command link
~ передачи, печатная *electr.* printed line
~ передачи, пневматическая air-pressure line
~ передачи, полосковая strip (transmission)

line
~ передачки transfer line
~ перелива overflow line
~ перелома fracture line
~ пересечения line of intersection
~ перспективы *top.* perspective line/ray
~ питания (power) supply line, feeder/supplying line
~ питания дуантов dee line
~, питающая feeder, supply/feed/fill line
~, плазовая loft line
~ поверхности земли terrain line
~ поглощения absorption line
~ поглощения, вращательная rotational absorption line
~ погружения trim line
~ погружения, предельная *naut.* margin line
~, подводящая *hydr.* feeding line
~ подгонки зубов *anat.* alignement
~, подкрыльцовая *anat.* axillary line
~, подымающаяся береговая *geol.* shore line of elevation
~ полета flight path, line of flight
~ полной воды landwash
~ положения, ЛП *nav.* line of position, LOP, position line, LP
~ положения, астрономическая celestial line of position, altitude line
~ положения, высотная Sumner (position) line
~ положения самолета, ЛПС *av.* aircraft position line, APL
~, полуденная *geod.* magnetic north/meridian line, noon line
~, полудуплексная *comp.* half-duplex line
~, полулунная *anat.* semilunar line, line of Spigelius
~ поля line of force/field, field line
~ постоянной интенсивности вихрей isocurlus
~ постоянной скорости isovel
~ потока line of flow
~, поточная flow/product/production/belt line
~, правая окологрудинная *anat.* right parasternal line
~ предварительного старта *av.* prestart line
~, предельная boundary line
~, приемная receiving line
~, причальная *naut.* mooring line, berthage, frontage
~, проверочная reference/datum line
~ прогиба bending/deflection line
~, производственная product/production line
~ прокатки rolling train
~ прокладки кабеля cable run
~ промерзания *build.* frost line
~ простирания *min.* strike line; level course, line of strike/ bearing
~ профиля, средняя median/centre line, camber line
~, прямая straight line
~ прямой видимости line of sight
~, пунктирная dotted line
~ пути track/course line, track
~ пути цели *mil.* target track(line)
~ равного интервала isochore, isochore line, equal-space line
~ равного потенциала equipotential line
~ равного содержания воды isohume
~, равноденственная/равноденствия equinoctial line
~ равной солености isosalinity line
~ равносигнальная *rad.* equisignal line
~ равных высот *geod.* line of equal elevation
~ равных значений isoline
~ равных пеленгов самолета, ЛРПС line of bearings
~ равных потенциалов equipotential line
~ равных приливов co-tidal line
~ равных разностей расстояний hyperbola, loran line of position
~ равных расстояний circle of position
~ равных склонений isogonic line
~ равных скоростей isotach
~, радиопроводная combined radio and wire link
~, радиорелейная radiorelay link/line
~ радиосвязи radio circuit/link
~, радиотелеметрическая radio-telemetry link/line
~ радиотелефонной связи radiotelephone circuit
~ развертки sweep/scanning/beam trace
~ разворота (на стоянке) *av.* turning line
~, разграничительная dividing line
~ раздела boundary (line), frontier, frontal line
~, размерная dimension line
~ размыва *geol.* denudation line
~, разрешенная 1. resolved peak; 2. permissible/ allowed line
~ разрыва line of rent
~, разрядная *comp.* bit/digit line
~ разъема split/break line
~ разъема формы *foundry* parting/joint line of a mould
~, раскаленная line of incandescence
~ раскалывания *geol.* line of split/rift
~, распадающаяся *math.* decomposed line
~ регрессии regression line
~, резонансная resonance peak/line
~ рентгеновского спектра X-ray line
~, реперная reference/datum line
~, риасовая береговая *geol.* ria shore line
~ родства, боковая collateral relationship
~ рулежной дорожки, осевая *av.* taxiway centre line
~ самоопыленная self-polinated line, inbred line
~ сборки (на конвейре) assembly line
~ сбрасывания drop/release line

~ сброса *geol.* fault line/trace
~ связи communication line/link/circuit
~ связи, арендуемая *comp.* leased line
~ связи, воздушная aerial line
~ связи, входящая incoming line
~ связи Земля-Космос uplink (service)
~ связи, исходящая outgoing line
~ связи, комбинированная composite communication link
~ связи, коммутируемая *comp.* dial-up line
~ связи Космос-Земля down-link (service)
~ связи, природная *tel.* suburban line
~ связи, пупинизированная coil-loaded line
~ связи, ретрансляционная relay link
~ связи, спутниковая satellite communication link
~ связи, столбовая pole line
~ связи, уплотненная multiplexed line
~, секущая secant
~, силовая line of force, field line
~, силовая магнитная magnetic line of force
~ скольжения *geol.* glide line; *met.* slip/Lüder(s) line
~ слабости земной коры *geol.* line of crustal weakness
~, сливная drain/dump line
~, слоевая *cryst.* layer line
~ смены дат *geogr.* date line
~ смещения displacement line
~ смыкания створок *geol.* commissural line
~ снижения descending line
~, соединительная junction route/circuit; *Am.* trunk (line)
~ соединительная, транзитная *tel.* through-traffic junction (route)
~ Сомнера curve of altitude, Sumner (position) line
~, сосковая *anat.* mammillary line
~, спайная *anat.* commissural line
~, спектра spectrum/spectral line
~ спектра испускания emission-spectrum line
~ спектра поглощения absorption line
~ спектра, разрешенная resolved spectrum line
~, спектральная spectrum/spectral line
~, спектральная, колебательная vibrational spectral line
~, спектральная, ложная ghost spectral line
~, спектральная, размытая diffuse spectral line
~, сплошная full/solid line
~, справочная *tel.* information/inquiry circuit
~, срединная median (line)
~ средней толщины medium line
~, средняя midline, centre line
~ старта *av.* start line, take-off line
~, стоксова (спектра) Stokes line
~ стоянки parking/rest line
~, судоходная *nav.* shipping line
~, судоходная трамповая tramp line
~, сутурная suture, sutural line
~ сходимости convergence line
~ таза, пограничная *anat.* (pelvic) terminal line, isopectineal line
~, тектоническая tectonic/trend line
~ текучести flow line
~, телеметрическая telemetry line/link
~ телеуправления telecontrol line, remote-control line
~, телефонная telephone line/connection
~, технологическая production line
~ течения flow line, stream, course
~ тока flow/stream line, line of flow/current, streamline
~, толстая heavy line
~, тонкая light line
~, транспортная регулярная scheduled service
~ труб run of pipes
~ тяги draft/thrust line
~ тяги, смещенная offset thrust line
~ уборки (шасси) *av.* retraction line
~ удара impact line, line of impact
~, узлов(ая) nodal line
~, уравнительная *therm.* equalizing line
~ упругого рассеяния elastic-scattering peak
~ уровня level/contour line/curve
~, утолщенная accentuated line
~ фактического пути *av.* (flight) track
~, фидерная *el.* feeder line
~, филогенетическая lineage, phylogenetic sequence
~ фокусов aerodynamic centre line/axis
~, форвакуумная roughing-down line
~ форм рельефа *geod.* landform line
~, характеристическая characteristic line/curve
~, ходовая *geod., top.* computation course/line, route
~, холостая *el.* unloaded line
~ хорды *av.* chord line
~ цели target line
~, центральная central line
~ центров line of centres, centre line
~, цепная *math.* catenary line/curve, chain/circuit line, catenary
~, челюстно-подъязычная *anat.* mylohyoid line, internal oblique line
~, цилиндрическая винтовая circular helix, (cylindrical) helix
~, шаговая pitch line
~, широкодельная лопастная *geol.* latisellate lobe line
~ широты *nav.* line of latitude
~, шкваловая squall line
~, шовная suture line
~, штриховая пунктирная dash-and-dot line
~, штриховочная cross-hatching line
~, штрихпунктирная dash-dot line
~, эвтектическая eutectic boundary line

~ **электропередачи, ЛЭП** (electric) power line
~ **электропередачи, воздушная** aerial power line
~ **электропередачи, подземная** underground/ buried power line
~ **электропитания** power line
~, **эталонная** standard line
линкольн *agr.* Lincoln (sheep)
линкор *abbr.* **линейный корабль** battleship
линкруст flexible paper-base laminate
линнеевский Linn(a)ean
линнеит *miner.* linn(a)eite
~, **никельсодержащий** siegenite
линнеум cobalt pyrite
линнея *bot.* twinflower *(Linnaea)*
линобатист *text.* cambric, lawn
линование ruling, marking with lines
линовка ruling, lining
линогравюра linocut, linoleum engraving
линолеат linoleate
линолевый *chem.* linoleic
линоленовый *chem.* linolenic
линолеум linoleum; floor covering
~, **безосновный** unsupported floor covering
~, **глифталевый** alkyd floor covering
~, **огнестойкий** fire-retardant linoleum
~, **резиновый** rubber linoleum
линон *text.* lawn *(fabric)*
линотип linotype composing machine, line casting machine
линотипист linotypist, linotyper
линофировый linophyric
линт *bot.* lint
линтер linter
линтерование, линтеровка lintering
линтоотделитель delinter
линч lynch
линчевание lynching, lynch law
линь I *zool.* tench *(Tinca tinca)*
линь II *naut.* line, marline; ratline
~, **белый** cod line
~ **для бензелей** seizing line/twine
~ **морской пеньковый** sea hemp ratline
~, **несмоленый** white line
~, **тонкий** jackline
линька mo(u)lt, mo(u)lting
лионский Lyonese
лиогель lyogel
лиозоль lyosol
лиотропный *phys. chem.* lyotropic
лиофилизация lyophylization, freeze drying, lyophilic drying
лиофильный *chem.* lyophyl(e), lyiphylic
лиофобный *chem.* lyophobe, lyophobic
липа *bot.* lime(-tree), linden(-tree) *(Tilia)*
липаза *bioch.* lipase
~ **желудочного сока** gastric lipase
~ **панкреатического сока** pancreatic lipase, steapsin
липарис I *zool.* sea snail *(Liparis)*
липарис II *bot.* twayblade *(Liparis)*
липарит *geol.* liparite, rhyolite
~, **щелочной** comendite
липемия, липидемия *med.* lip(a)emia, lipidemia, lipohemia, piarrhemia
~, **алиментарная** alimentary/postprandial lip(a)emia
липид, липин *bioch.* lipid(e), lipin
липкий sticky, gluey, glutinous, (self-)adhesive
~**кая железка** *anat.* retinaculum, viscid gland
липкометр *print.* tack(iness) meter, ink meter
липкопластырный adhesive-plaster
липкость stickiness tackiness, adhesiveness; glueyness; viscidity
липнуть adhere, stick, clam, cleave
липоатрофия *med.* lipoatrophia
~, **ограниченная** lipoatrophia circumscripta
липобласт lipoblast
липовый linden *(attr.); n. pl.* linden family *(Tiliaceae)*
~ **мед** white honey
~ **цвет** linden blossom
липогенный *physiol.* lipogenous, adipogenous, lipogenic, producing fat
липогранулема *med.* lipogranuloma
липогранулематоз *med.* lipogranulomatosis, lipid/lipoid granulomatosis
липодермоид *med.* lipodermatoid, lipodermoid
липодистрофия *med.* lipodystrophia, lipodystrophy
~, **прогрессирующая кишечная** progressive intestinal lipodystrophy, Whipple's disease
липодолит *geol.* clinkstone, phonolite
липоевый lipoic
липозитол lipositol, inositol
липоид *bioch.* lipoid, lipid(e), lipin
липоидоз *med.* lipoidosis
~ **кожи и слизистых оболочек** lipid proteinosis
~, **мозговой** cerebral lipoidosis
~ **роговицы** lipoidosis corneae, arcus senilis
~, **фосфатидный** lipid histiocytosis, sphingomyelinosis
липокаин *bioch.* lipocaine
липокальциногранулематоз *med.* lipocalcinogranulomatosis, chalk gout, interstital calcinosis
липокортикотропный *med.* lipocorticotrop(h)ic
липоксантин lipoxanthine, yellow lipochrome
липоксения *bot.* lipoxeny
липолиз *med.* lipolysis, lipoclasis, lipodieresis
липолитический lipolytic, lipoclastic
липома *med.* lipoma
~, **обызвествляющаяся** calcified/calcifying lipoma
~, **саркоматозная** liposarcoma
~, **слизистая** myxolipoma
~, **твердая** hard lipoma, lipoma durum
~, **узловая** nodular lipoma
~, **фетальная/фетальноклеточная** fetal-fat

cell lipoma
~, фиброзная fibrolipoma
липоматоз *med.* lipomatosis, liposis, adiposis
липомиома *med.* lipomyoma, leukomyoma
липопексия *med.* lipopexia, adipopexia
липопротеид, липопротеин *med.* lipoprotein
липосаркома *med.* liposarcoma
липофор lipophore, xanthophore
липофрения *med.* lipophrenia
липофусцин lipofuscin
липохондродистрофия *med.* lipochondrodystrophy, Hurtler's disease, gargo(y)lism
липоцит *med.* lipocyte, fat cell
~, многопузырчатый multivesicular lipocyte
~, пузырчатый vesicular lipocyte
липучка 1. *bot.* stickseed *(Lappula);* 2. flypaper, adhesive paper, sticker
липушник *bot.* cleavers, goose grass, catchweed *(Galium aparine)*
лира I 1. *mus.* lyre; 2. *zool.* lyrebird *(Menura);* 3. *astr.* the Lyra
лира II lira *(monetary unit)*
лирелла *bot.* lirella
лиризм lyricism
лирик lyric poet, lyrist
лирика lyric poetry; lyrics
лирико-драматический lyrico-dramatic
лирико-эпический lyrico-epic
лириодендрон *bot.* tulip tree *(Liriodendron)*
лирический lyric; lyrical
лировидно-раздельный lyrately partite/pinnate
лировидный lyrate, lyriform, lyre-shaped
лирообразный, лироподобный lyrate, lyriform
лирохвост *zool.* lyrebird *(Menura)*
лиса fox(fur)
лисель *naut.* stud(ding) sail, stunsail
~, наставной drab(b)er
лисель-индигент *naut.* bobstay/lace piece
лисий foxy, vulpine
~ мех fox fur
~ хвост 1. fox brush; 2. *bot.* foxtail *(Alopecurus)*
лисица *zool.* fox; vixen *(Vulpes)*
~, кафрская bat-eared fox *(Otocyon megalotis)*
лисичка I *astr.* Little Fox, Vulpecula
~, морская *zool.* sea poacher *(Agonus)*
лисичка II *bot.* cantharellus, chanterelle *(Cantharellus)*
лисичковые *n. pl. bot.* cantharellus family *(Cantharellaceae)*
лисохвост(ник) *bot.* foxtail *(Alopecurus)*
лиссабонский Lisbon (wine)
лист 1. *bot.* leaf; blade; *see also* **листья**; 2. sheet, leaf *(of paper);* plate; *geol.* lamina; sheet quadrandle
~, авторский author's sheet, signature *(40000 printed characters),* unit of 40000 ens
~, анкерный tie plate
~, александрийский *bot.* (Alexandria) senna *(Cassia acutifolia)*
~, биметаллический composite/clad sheet
~, броневой armour plate
~, бумажный standard sheet
~, винтовой verticillate leaf
~, вкладной *print.* loose leaf
~, влагалищный sheathing leaf
~ водоотделителя baffle plate
~, волнистый *bot.* undulate leaf; *met.* corrugated sheet
~, волокнистый fibrous leaf
~, вторичный leaflet
~, глянцеватый sleek leaf
~, гофрированный corrugated sheet/plate
~, групповой переписной institutional schedule
~, дважды перистосложный bipinnate leaf
~ двоякой кривизны double/doubly curved plate
~, двупильчатый double-serrate leaf
~, декапированный pickled sheet
~, декартов *math.* Cartesian/Descartes' folium, folium/leaf of Descartes
~, дланевидно рассеченный palmately parted leaf
~ днища *auto.* floor-pan, bottom plate
~, дольчатый cleft leaf
~, дорожный waybill
~, дырчатый perforated sheet/plate
~ жесткости stiffening plate
~, закладной *econ.* mortgage bond
~, зародышевый blastophyllum
~, защитный quard plate
~, игольчатый needle-like/-shaped leaf, acerous leaf
~, интеркостальный intercostal plate
~, исполнительный *law* (writ of) execution
~ карт, сборный chart index, index chart
~ карт map sheet
~, клинообразный cuneate leaf, wedge-shaped leaf
~ комингса coaming plate
~, контрольный check list
~, копьевидный hastate leaf
~, ланцетовидный lanceolate leaf
~, ловчий vasculum
~, лопатообразный palaceous leaf, spade-shaped leaf
~, макулатурный waste/spoiled sheet, spoil
~ Мебиуса *math.* Moebius band/strip/loop
~, мелкозубчатый denticulate leaf
~, мечевидный ensiform leaf, sword-shaped leaf
~, многократно-рассеченный repeatedly branched leaf, multibranched leaf
~, многолетний persistent leaf
~, монтажный *print.* lay-out sheet, draw sheet
~, надрезной incised leaf
~, накладной *naut.* cover/doubler/doubling plate
~, направляющий guide vane/baffle
~ наружной обвивки shell/skin plate
~ настила *auto.* floor plate
~, непрерывный continuous leaf

~, нижний basal leaf
~ обшивки skin panel
~ обшивки доменной печи mantle sheet
~, обшивочный *heat* lagging sheet, skin
~, объемлющий clasping leaf
~, однолопастный unifoliate leaf
~, округлый orbicular/orbiculate leaf
~, опорный base plate
~, опросный census schedule, questionnaire
~, опушенный hirsute leaf
~, остроконечный cuspidate leaf
~, отбойный dash/swash plate
~, отражательный *heat* deflector, baffle
~, оцинкованный galvanized sheet
~ пальмы palm branch
~, пальчатый digitate leaf
~ папоротника/папоротникообразных frond
~, параллельнонервный parallel-veined leaf
~, парноперистосложный abruptly pinnate leaf
~, первичный initial leaf
~, переписной *stat.* schedule, census-paper/schedule
~, перистонервный feather-veined leaf
~, перистораздельный pinnately divided/lobed leaf, pinnatifid leaf
~, печатный printer's sheet, quire
~, пильчатый serrate leaf
~, плакированный clad sheet
~ планшетов, сводный index map
~, подписной subscription list
~, почковидный reniform leaf, kidney-shaped leaf
~, поясной flange plate
~, продолговато-ланцетовидный oblong-lanceolate leaf
~, прокладной slip sheet
~, прокладочный flitch sheet
~, путевой *auto.* driver's trip ticket; *rail.* train list
~, разделительный diaphragm plate
~, рассеченный divided leaf
~, ремнеобразный strap-like leaf
~, рессорный *auto.* spring leaf/blade/plate
~ римановой поверхности *phys.* Rieman surface sheet
~, рифленый fluted sheet, checkered plate
~, ромбический rhomboidal leaf
~ с матовой поверхностью tarnished leaf
~, сводный планшетов index map
~, сердцевидный cordate leaf, heart-shaped leaf
~, сетчатонервный anastomosing leaf
~, сетчатый net-veined leaf
~, сидячий sessile leaf
~, скорбный *med.* case sheet, hospital history sheet
~, скуловой *naut.* bilge plate
~, соединительный *constr.* joint plate
~, списочной формы, переписной *econ.* collective/enumerator's schedule
~, стеблевый cauline leaf
~, стреловидный arrow-form/-shaped leaf
~, съемный *naut.* detachable/removable plate
~, титульный title page
~, толстый plate
~, тонкий sheet
~, транцевый *naut.* transom plate
~, тройчатосложный trifoliate leaf
~, тройчатый ternate leaf
~, упаковочный packing list/slip
~, учетно-издательский publisher's signature
~ фальшборта *naut.* bulwark plate
~, фанерный plywood sheet
~, фасонный *met.* sketch plate
~, хронометражный timing sheet
~, цветочный floral leaf
~, цельный complete leaf
~, цилиндрический terete leaf
~, черешковый petiolate/petiolar/petioled leaf
~ чертежной бумаги drawing sheet
~, чистый *print.* blank leaf, blind print, clean sheet
~ шпунтового пояса *naut.* garboard plate
листаж number of sheets *(in a book)*
листва *coll.* foliage, leafage
~, спороносная fertile foliage
лиственница *bot.* larch *(Larix)*
лиственный *bot.* deciduous; (broad-)leafed, leafy, foliar, foliaceous
~ное дерево deciduous tree
листвяк *bot.* thallus
листель *arch.* listel
листер *agr.* lister-plough
~, дисковый disk lister
листереллез, листериоз *vet., pathol.* listeriosis
~ плода f(o)etal listerosis
~ половых органов genital listerosis
листерование *agr.* listering
листинг *comp.* listing
листоблошка *zool.* sucker, psylla *(Psylla)*
листование sheeting, forming into sheets
листоватость *geol.* lamination, foliation, fissility
~, волокнистая fibrous foliation
~, ложная pseudo-lamination
~ напластования bedding fissility
~, тонкая fine lamination, fissile foliation
листоватый foliaceous, foliate(d), lamelliform; phylliform; micaceous
листовертка *zool.* leaf-rolling moth; tortricid, tortrix *(Tortrix)*
~, виноградная grape-berry moth *(Polychrosis viteana)*
листоветвь *bot.* cladophyll, phylloclade
листовидка *zool.* leaf butterfly *(Kallima)*
листовидный 1. foliaceous, foliate, leaflike, leaf-shaped, foliiform; phylloid; frondescent; 2. lamellar, sheetlike; tabular
~ стебель phylloclade, cladophyll
листовка 1. *print.* leaflet; fly sheet; 2. *bot.* fol-

licle, follicular fruit; parmelia *(Parmelia)*
листовковидный *bot.* follicular
листовой 1. foliar, foliate(d), foliose; leaf *(attr);* 2. lamellar, sheeted, in sheet form
~ бугорок leaf primordium
~ зачаток primordial leaf
~ рубец *bot.* cicatrix
~вая паренхима mesophyll
~вая рессора laminated spring
~вая сталь steel plate, sheet steel
~вое почкосложение vernation
листоголовчатый *bot.* phyllocephalous
листогрыз *zool.* leaf miner *(Tineoidea)*
листоед *zool.* leaf beetle *(Chrysomela)*
листок 1. *bot.* leaf, foliole, leaflet; 2. *anat.* folium; 3. *print.* form, leaflet
~ бумаги slip of paper
~, зародышевый germinal/germ layer, blastophyllum
~, зародышевый, внутренний entoderm, entoblast
~, зародышевый, наружный ectoderm, ectoblast
~, зародышевый, средний mesoderm, mesoblast
~ нетрудоспособности medical disability certificate, medical certificate of incapacity for work
~ пленки film wafer
~, рекламный handbill
~, температурный temperature chart
~ технических данных technical data sheet
листоколосник *bot.* phyllostachys *(Phyllostachys)*
листоногие *n. pl. zool.* phyllopods *(Phyllopoda)*
листонос *zool.* phyllostome *(Phyllostomidae)*
листообразный leaflike, lamellate, lamellar, tabular
листообразующий *bot.* phyllogenetic, phyllophorous
листообъемлющий *bot.* amplexifoliate
листопад (autumn) leaf fall
~, искусственный defoliation
листоплодный *bot.* phyllocarpous
листоправильный plate-/sheet-straightening
листопрокатка rolling of sheet metal
листопрокатный sheet-rolling; laminating
~ стан sheet mill
листоразположение *bot.* phyllotaxy, phyllotaxis, leaf arrangement
~, накрест-супротивное decussate leaf arrangement
листорез(ка) *print.* sheet cutter, sheet cutting machine
~, ротационная rotary cross-cutting machine
листорезный sheet-/plate-cutting
листосложение *bot.* vernation, foliation, convolution
~ в почках prefoliation
листостебельный *bot.* cormophytic, frondiferous
~ное растение cormophyte
лист-оттиск *print.* sheet-copy
~, печатный printer's sheet-copy
~, учетно-издательский publisher's sheet-copy
листоукладчик sheet piler
листоцветковый *bot.* foliiferous
листоцветок *bot.* phylloclade
листочек 1. leaflet, foliole, lamel(la); pinna; 2. *geol.* flake
~, вторичный pinnule
~ околоцветника tepal
~, слюдяной mica scale
~ сложного листа leaflet, pinna
~, спайный cleavage flake
листочковый laminar, laminal
листоядный browsing, leaf-eating
листья *pl.* leaves
~, имеющие круглые rotundifolious
~, имеющие узкие stenophyllous, angustifoliate
~, мутовчатые whorled leaves
~, раскидистые spreading leaves
~, сбрашивающий deciduous
~, черепитчато расположенные overlapping leaves
литавра *mus.* ketedrum, timbal; atabal
литания *eccl.* litany
литарг *chem.* litharge
литвин Lithuanian
литвинка Lithuanian (woman)
литейная foundry, smelting house
литейный casting, founding
~ двор casting bed, mould yard
~ завод/цех foundry
литейщик founder, foundryman, smelter, caster
литемия *med.* lithemia, uricacidemia
литер travel warrant
литера *print.* type(body), character, letter
~, беспольная large type character
~, конусная bottle-necked type
~, прописная upper-case letter, capital
~, строчная lower-case letter
литерал *comp.* literal
~ постоянных constant literal
литератор literary man, man of letters
литература literature
~, изящная/художественная belles-lettres
литературно-книжный literary-bookish
литературность literary character, literariness
литературный literary
~ кружок literary circle
~ная собственность copyright
литературовед literary critic, specialist in the study of literature
литературоведение study of literature, literary criticism
литеровыталкиватель *print.* type pusher
литиаз *med.* lithiasis
литиевый *chem.* lithic
литий *chem.* lithium, Li
~, бензойнокислый lithium benzoate
~, бромистый lithium bromide

~, **йодистый** lithium iodide/iodate
~, **салициевокислый** lithium salicylate
литификация *geol.* lithification
литический lytic; *med.* acritical
лития *rel.* litany; prayer said on the parvis
литка bar, ingot
литкружок *abbr.* literary circle
литмонтаж collage, montage of excerpts
литник *tech.* gate, gate runner, flow/pouring gate; pouring channel; sprue
~, **вертикальный** downgate, downsprue
~, **вилкообразный** finger gate
~, **впускной** ingate; feed gate/orifice
~, **клиновидный** wedge sprue, finger gate
~, **пальчиковый** finger gate
~, **подводящий/разводящий** runner
~, **разветвленный** spray gate, branched runner
~, **рожковый** horn gate
~, **центральный** central gate/runner/pipe
~, **центровой** *foundry* fountain, trumpet
~, **щелевой** lap/knife gate
литовец, литовский Lithuanian
литовка Lithuanian (woman)
литогенез(ис) *geol.* lithogenesis, lithogeny
литогенетический, литогенный lithogen(et)ic, lithogenous; lithisolic
литоглифика lithoglyph
литограф lithographer
литография lithograph; lithography
~, **цветная** chromolithograph(y)
литографский lithographic
~ **оттиск** lithoprint
литодесма *zool.* lithodesma
литоз *med.* lithosis, pneumonoconiosis
литоидальный, литоидный lithoid(al), stony
литой cast; *agr.* filled out, plump *(of grain)*
~**тая сталь** cast/ingot steel
литокатена lithological sequence
литоклаз *geol.* lithoclase
литология lithology
литомарж *miner.* lithomarge, compact china-clay
литопедион *med.* lithopedion, lithopedium, osteopedion, ostembrion
литопон *chem.* lithopone
литоралит *geol.* littoralite, littoral rock
литораль littoral, intertidal zone
литоральный intertidal; coastal, littoral
литорея old criptographic writing
литосидерит lithosiderite
литосфера *geol.* earth('s) crust, geosphere, lithosphere
литота, литотес *lit.* litotes
литотамния lithothamnium
литотекс lithotex
литотомия *surg.* lithotomy
~, **высокая** high/suprapubic lithotomy, epicystotomy
~, **надлобковая** suprapubic/high lithotomy
~, **чрезбрюшинная** perineal lithotomy
литотриптор *med.* lithotrite, lithoclast, lithotriptor
~, **окончатый** fenestrated lithotrite
литофация lithofacies
литофиза *geol.* lithophysa, stone bubble
литофильный lithophylic, lithophylous
~**ное растение** lithophyte
литофит *bot.* lithophyte
литофитный *bot.* lithophytic, lithophytous
литохолевый *bioch.* lithocholic
литоцист lithocyst
литоцистотомия *surg.* lithocystotomy, vesical lithotomy
литр litre, liter
литраж capacity in litres, cylinder/litre capacity; *eng.* piston displacement
~ **двигателя** engine (cubic) capacity/displacement, total piston displacement
литредактор *abbr.* literary editor
литургийный, литургический liturgical
литургия liturgy
литуус *math.* lituus
литфак *abbr.* department of literature
литцендрат Litzendraht, Litz wire
лить 1. cast, found, mo(u)ld; 2. pour; shed, spill
литье 1. *met.* casting, mo(u)lding, founding; 2. *coll.* casts, catings, mo(u)ldings
~ **в глиняные формы** loam casting
~ **в землю** sand casting
~ **в кокиль, кокильное** chill/die casting
~ **в открытые формы** open casting
~ **в переворачиваемую форму** slush casting
~ **в песчаные формы** sand(-mould) casting
~ **в постоянные формы** permanent-mould casting
~ **в сырую форму, ~ всырую** green-sand casting
~ **в ямах** pit casting
~ **вмокрую** green-sand casting
~ **восходящей струей** uphill casting, rising pouring
~ **всасыванием** suction casting
~ **всухую** dry sand casting
~ **затвора** seal casting
~, **красное** red brass
~, **латунное** cast brass
~, **машинное** machine casting/moulding
~ **пластмасс** plastic moulding
~ **под давлением** die casting, *plast.* injection moulding
~ **по-сухому** dry sand casting
~ **по-сырому** green-sand casting
~, **стержневое** core-mould casting
~, **фасонное** cast/moulded shapes
~, **художественное** art/ornamental castings
~, **цветное** non-ferrous castings
~, **шаблонное** template casting
~, **шликерное** slip casting
лиф bodice
лифт lift, *Am.* elevator
~ **без проводника** unattended lift/elevator
~, **грузовой с проводником** attended goods

lift, attended freight/cargo elevator
~, камбузный *naut.* dumb waiter
~, пассажирский passenger lift/elevator
~ с верхним приводом top-drive lift/elevator
~ с канатоведущим шкивом Vee(-sheave) drive lift, traction(-sheave) elevator
~ с кнопочным управлением automatic lift/ elevator
~ с нижним приводом bottom-drive lift-/elevator
~, скоростной express lift
лифтер *agr.* grain (saving) guard
лифтер lift/elevator operator
лифт-экспресс express lift/elevator
лихва interest
лихен *med.* lichen *see also* **лишай**
лихенин *chem.* lichenin(e), lichen starch
лихенификация, лихенизация *med.* lichenification, lichenization
лихено(в)идный *med.* lichenous
лихенология *bot.* lichenology
лихнис *bot.* lychnis *(Lychnis)*
лихниск six-rayed spicule
лихорадка *med.* fever, pyrexia, febris, ague; fever/ cold sore; *fig.* rush
~, алиментарная food fever
~, ассамская Assam fever, kala-azar, visceral leishmaniasis
~, Биамба, комариная Bwamba fever
~, белковая protein fever
~, болотная swamp fever
~, возвратная relapsing/recurrent fever
~, волынская five-day fever, Volhynia fever, Werner-His disease
~, гемоглубинурийная (острая) h(a)emoglubinuric/blackwater/h(a)ematuric fever
~, гнойно-резорбтивная purulent resorptive fever
~ дальневосточного типа, геморрагическая Far Eeast h(a)emorrhagic fever
~ денге dengue, dandy/date/breakbone fever
~ долины Рифт Rift Valley fever
~, ежедневная quotidian fever
~, железистая glandular fever, infectious mononucleosis, benign lymphadenosis
~, желтая yellow fever, febris flava
~, золотая *econ.* gold run/rush
~, искусственная therapeutic fever, pyretotherapy
~ кишечного происхождения intestinal fever, enteroidea
~ Конго, красная Congo(lian) red fever, murine typhus
~, крапивная urticarial fever, nettle rash, Asiatic schistosomiasis
~ Ку Q/Queensland fever, ''nine mile'' fever
~ литейщиков zinc-fume fever
~, мальтийская Malta/Mediterranean fever, brucellosis
~, малярийная malarial fever, acute malaria
~, марсельская Marseilles/eruptive fever, tick typhus
~, молочная milk fever
~, москитная sandfly/pappataci/phlebotomus fever, three-day fever
~ нервного происхождения neuropyra
~, однодневная ephemeral fever, one-day fever
~ Оройя Oroya fever, Carrion's disease, bartonellosis
~ паппатачи sandfly/pappataci/phlebotomus fever
~, перемежающаяся/ремиттирующая remittent/intermittent fever
~, ползучая slow fever
~, послеоперационная surgical fever
~, послеродовая puerperal fever/pyrexia
~, претибиальная pretibial fever, Fort Bragg fever
~, пятидневная=волынская
~, семидневная seven-day fever, autumn fever
~ Семлики, комариная Semliki forest fever
~, сенная hay fever, autumnal catarrh
~, септическая septic fever, septicemia
~ Скалистых гор, пятнистая Rocky Mountain spotted fever
~, солевая salt/thirst/inanition fever
~, тепловая heat/thermic fever, heat pyrexia
~, техасская Texas fever
~, травматическая traumatic/symptomatic fever
~, трехдневная малярийная tertian aque/ fever, tertian malaria
~, узловатая nodal fever, erythema nodosum
~, уретральная urethral/catheter/urinary fever
~, шестидневная sextan fever
~, японская речная Japanese river fever, mite/ tropical typhus, tsutsugamushi
лихорадочник *bot.* drug hedge hyssop *(Gratiola officinalis)*
лихорадочный feverish, fevered, febrile, pyre(c)tic, pyrexial
лихтер *naut.* lighter
~, беспалубный open lighter
~, грузовой cargo lighter
~, десантный landing/amphibious lighter
~, несамоходный dumb lighter
~, нефтеналивный oil/fuel lighter
~, палубный covered lighter
~, портовый harbour lighter/barge
лихтеровка *naut.* lighterage
лица *text.* heald, heddle
лицевать face, cover (with)
лицевой 1. *anat.* facial; 2. exterior; *print.* illuminated, illustrated
~ счет personal account
~вая сторона facade, front; right side; obverse *(of coin)*
лицеист Lyceum student
лицемер hypocrite, dissembler
лиценэиар licensor
лиценэиат licensee

лицензия licence, license, permit; concession
~ импортная import license
~ на перегрузку товара tran(s)shipment license
лиценциат licentiate
лиций *bot.* wolfberry, desest thorn, boxthorn *(Lycium)*
лицо 1. face; *anat.* facies; 2. exterior; 3. right side; *leath.* grain; 4. person
~ без подданства stateless person
~, "бычье" *med.* cow face, facies bovine
~, высокопоставленное high-ranking person
~, государственное должностное public official
~, действующее *theatre, lit.* character, personage
~, доверенное trustee
~, должностное official, officer, office holder
~ кожи, гладкое smooth grain
~ интеллигентного труда professional worker
~ кожи, искусственное simulated/artificial grain
~ кожи, тусклое dull grain
~ комадного состава (на корабле) officer
~, лунообразное moon face
~, львиное *med.* leontiasis
~, маскообразное *med.* mask-like face, Parkinson's facies
~ медицинского персонала, сопровождающее medicial attendant
~, назначенное assignee
~, находящееся на борту occupant
~ не имеющее гражданства stateless person
~, обслуживающее operator, serviceman, attendant
~, одутловатое tallow face
~, официальное official
~, перемещенное displaced person
~, подотчетное reporter; accountable person
~, подставное figurehead
~, потерпевшее кораблекрушение shipwrecked person
~, принимающее receiving person/party
~, производящее осмотр inspecting officer, inspector
~, производящее расследование (accident) investigator
~, светское layman; *pl.* the laity
~ свободной профессии professional (worker)
~, сдающее transferring party
~, терпящее бедствие person in distress/emergency
~, третье third man /*gram.* person /*law* party
~, физическое *law* natural person
~, юридическое legal/juridical/juridict(al)/person, body corporate
личина I mask; *fig.* guise
личина II *tech.* escutcheon, metal shield (around a keyhole)
личинка *zool.* larva (*pl.* -vae); grub, maggot
~, безногая larvina
~ комара mosquito larva
~, молодая larvule
~ насекомого maggot larvule
~ ручейника caddis case
~ с ресничным покровом ciliated larva
~, филяриевидная filariform larva
личинкообразный *zool.* larviform
личинник *bot.* scorpion's tail *(Scorpiurus)*
личиночный larval, larvaceous, larvate(d)
личной *anat.* facial; face *(attr.); leath.* right-side up
личность 1. personality, person, individual, figure; 2. *pl.* personal remarks, personalities
~, антисоциальная antisocial personality
~, двойственная dual personality
~, историческая historic figure
~, неприспособленная inadequate personality
~, собственная the self
~, шизоидная schizoid personality
~ти, удостоверение identity card
личный personal, individual, private; hypostatic, personalistic
~ секретарь private secrtary
~ состав personnel, staff; officers and men; crew, complement; force
~ состав запаса body of reserve
~ состав машинного отделения engine-room complement
~ состав , обслуживающий operating personnel
~ состав службы наблюдения и связи signal force
~ная охрана bodyguard
~ная собственность personal property
~ное дело personal records; dossier
лишаевидный *bot.* lichenoid; *med.* herpetic
лишай *bot.* lichen; *med.* herpes, pityriasis herpes
~, блестящий lichen nitidus, Pinkus disease
~, волосяной lichen pilari
~, золотушный lichen scrofulosorum
~, мокнущий salt rheum, eczema
~, опоясывающий (herpes) zoster, zone, shingles, girdle
~, отрубевидный pityriasis versicolor
~, стригущий tinea trichophytina, trichophitosis, ringworm
~, тропический miliaria
~, чешуйчатый psoriasis, psora
лишайник *bot.* lichen *(Lichen),* Iceland moss *(Cetraria islandica)*
~, бородатый bearded usnea *(Usnea barbata)*
лишайниковидный lichenoid, licheniform, lichenous
лишайниковый lichenaceous
~ покров lichen cover
~вая пустошь lichen heath
лишать deprive (of)
~ владения *law* evict, dispossess
~ гражданских/избирательных прав de-

prive of civil rights, *law* disfranchise
~ **крова** make homeless; evict
~ **материальности** unsubstantiate
~ **места** *law* unseat
~ **наследства** disinherit
~ **права** disqualify
~ **простоты** sophisticate
~ **себя жизни** take one's life
~ **слова** deny (someone) the right to speak
~ **собственности** dispossess
~ **ядовитых свойств** detoxicate
лишенец *hist.* nonvoter, disfranchised person *(in USSR up to 1936)*
лишение deprivation, loss
~ **владения** *law* eviction, dispossession, ouster
~ **гражданских/избирательных прав** disfranchisement
~ **гражданства** denaturalization
~ **наследства** *law* disherison, disinherison, disinheritance
~ **права** disqualification, deprivation of rights
~ **права возражения** *law* estoppel
~ **права собственности** dispossession
~ **свободы** confinement, imprisonment
~ **судоводительских прав, временное** *law* suspension of the master's certificate
~ **юридической силы** invalidation
лишенный deprived, devoid (of), lacking (in); destitute
~ **киля** ecarinate
~ **чешуек** esquamate
~ **яичек** *med.* anorchus
лишний superfluous, excessive; extra, spare, odd; *anat.* epactal
Ллойд Lloyd
ллойдия *bot.* mountain spiderwort *(Lloydia)*
лоаза *bot.* Chile nettle *(Loasa)*
лоаоз *med.* lo(i)asis; Calabar/fugitive swellings
лоб *anat.* forehead, brow
~, **атака в** *mil.* frontal attack
~ **надвига** *geol.* front of thrust
~, **покатый** receding forehead
~ **покрова/складки** *geol.* crown
~, **стрелять в** *mil.* fire point-blank
лобан *zool.* common mullet *(Mugil cephalus)*
лобарный *anat.* lobar
лобби *polit.* lobby
лоббизм *polit.* lobbyism
лоббистский lobbist
лобелиевые *n. pl. bot.* Lobelia family *(Lobeliaceae)*
лобелин *pharm.* lobeline
~, **сернокислый** lobeline sulphate
~, **хлористоводородный** lobeline hydrochlorate, alpha-lobeline
лобелия *bot.* cardinal-flower, lobelia *(Lobelia)*
лобзик *tech.* fretsaw, jig/scroll/coping/compass saw
лобия *bot.* hyacinth dolichos, lablab *(Dolichos lablab)*
лобковый *anat.* pubic
~**вая кость** pubis, mons veneris
лобно-височный *anat.* frontotemporal
лобно-задний *anat.* frontoposterior
лобно-затылочный *anat.* frontooccipital
лобно-носовой *anat.* frontonasal
лобно-передний *anat.* frontoanterior
лобно-теменный *anat.* frontoparietal
лобный *anat.* frontal
~**ная пазуха** frontal sinus
~**ное место** *hist.* place of execution/death
лобовой frontal, front; head-on
~ **фонарь** headlight
~**вая атака** *mil.* frontal attack
~**вое сопротивление** *av.* drag, head resistance
лобогрейка *agr.* dropper, reaper, harvester
лобок *anat.* pubic region, pubis, mons veneris
лобоподий, лобоподия lobopodium, thick lobose pseudopodium
лоботомия, лобэктомия *surg.* lobotomy
~, **префронтальная** prefrontal lobotomy, leukotomy
лов fishing
~, **дрифтерный** drift fishing
~, **кошельковый** seine-net fishing, purse seining
~ **крючковыми орудиями** pole and line fishing, long-line fishing
~ **сетью** netling
~, **траловый** trawling, trawl(er) fishing
ловердек *naut.* lower deck
ловец fisher(man); hunter
ловитель 1. *mach.* catcher; pilot, guide pin; 2. *min.* extractor
~ **вагонеток** car/wag(g)on arrester, wag(g)on retarder
ловить catch; fish
ловкость adroitness, dexterity, deftness, agility
~ **рук** sleight of hand; operating skill
ловля 1. catching, hunting; 2. fishing ground
~ **западней** trapping
~ **птиц** birdcatching, fowling
~, **рыбная** hishing
ловушка trap, catcher, decoy
~, **адресная** address catcher
~, **всасывающая** suction trap
~ **для раствора** solution catcher
~, **дрейфующая** drifting trap
~, **дырочная** hole-trap
~, **ионная** ion-trap
~, **литологическая** *geol.* depositional trap
~, **незамкнутая** misclosure
~, **нефтяная,** ~ **для нефти** oil trap
~ **прерываний** interrupt trap
~, **пустая/свободная** empty/free trap
~, **стеклянная сферическая** glass-ball nonreturn valve
~, **тепловая** thermal trap
~ **увлекаемых частиц** entrainment trap
~, **электронная/электронов** electron trap
ловчий *n. hist.* huntsman, master of hounds
лог *geol.* ravine

логарифм *math.* logarithm
~, гиперболический hyperbolic/natural/Napierian logarithm
~, двоичный binary logarithm
~, двойной log-log
~, десятичный decimal/common/Brigg's logarithm, logarithm to the base 10, log, lg
~, натуральный natural/hyperbolic/Napierian logarithm, ln
~, Неперов Napierian/natural/hyperbolic logarithm
~, обращенный inverse logarithm, antilogarithm
~ отношения температур, средний logarithmic mean temperature difference
~, повторный iterated logarithm, log-log
~, пятизначный five-place logarithm
логарифмика logarithmic curve
логарифмирование taking the logarithm of, taking logs/logarithms
логарифмировать take the logarithm of, take logs
логарифмически logarithmically
~ нормальный lognormal, logarithmic-normal
~ экспоненциальный log-expontial
логарифмический logarithmic
~, дважды/повторно log-log
~кая линейка slide rule
логаэд logaoedic, logaoedic verse
логгер *naut.* lugger
логизм logism
логик logician
логика logic
~, арбитражная arbitration logic
~, бинарная binary logic
~, большинства majority logic
~, булева Boolean logic
~, вероятностная probalilistic logic
~ высказываний propositional logic
~, двоичная/двухзначная binary logic, two-valued logic
~, диалектическая dialectical logic
~, двухканальная dual-rail logic
~, запрещенная illogical logic
~, зашитая *comp.* hardwired logic
~, интуиционистская intuitionistic(al)/intuitional logic
~, командная instruction logic
~, мажоритарная majority logic
~, математическая mathematical logic
~, многозначная multivalued logic, many-valued logic
~ на основе ячеек cellular logic
~ на трансфлюксорах multi-aperture device logic, MAD-logic
~ платы/встроенная on-board logic
~ получения выводов inferential logic
~, пороговая treshold logic
~ предикатов predicate logic
~, прерывная interrupt logic
~, программируемая programmable logic
~, символическая formal/mathematical logic
~, схемная circuit/mathematical logic
~, теоретическая mathematical logic
~, троичная ternary logic
~, управляющая control logic
логик-метафизик logicalist
логико-информационный logical-informational
логико-математический logical-mathematical
логистика *math.* logistic(s), symbolic logic
логит logit
логицизм logicism
логически logically
~ вытекающий consequential
~ последовательный consequent
логический logic(al), Boolean
~кая последовательность logical order
логичность logic; logicality
логичный logical; sound
логнормальный lognormal
логограмма logogram
логогриф logogriph
логоклония *med.* logoclonism
логометр ratiometer, logometer, quotient meter
логопатия logopathy, speech defect
логопед logop(a)edist, speech therapist
логопедия logop(a)edics, speech therapy
логорея *med.* logorrhea, garrulousness
логоспазм *med.* logospasm, stuttering, explosive speech
логотип *print.* logotype
лоджия *arch.* loggia
лодикула *bot.* lodicule, lodicula
лодка boat
~, атомная подводная atomic submarine, nuclear-powered submarine
~, гидросамолета seaplane hull
~, гоночная gig
~, гребная rowing boat
~, двухвесельная pair-oar
~, канонерская gunboat
~, моторная motorboat, launch
~, надувная inflatable boat
~, надувная спасательная inflatable lifeboat
~ на полозьях ice canoe
~, подводная submarine
~, спасательная lifeboat
~, штурмовая assault boat
лодкообразный cymbiform, boat-shaped
лодочка 1. *bot.* carina, keel; 2. *chem.* (combustion) boat
лодочный boat *(attr.)*
~ спорт boating
~ная станция boating station; boat-house
лодыга trunnion plate
лодыжка *anat.* malleolus, anklebone
~, латеральная lateral/external/outer malleolus
~, медиальная medial/internal/inner malleolus
~, откидная folding platform
лодыжковый *anat.* malleolar
ложа 1. *theatr.* box; 2. *(masonic)* lodge; 3. gunstock

ложбина 1. *geol.* hollow, rill, narrow/shallow gully; 2. *meteor.* trough
~, барическая, ~ низкого давления trough of low pressure
~, барическая теплая trowel(l)
~, безрусловая стока run-off dell
~, лесная wooded hollow
~, моноклинальная monoclinal ravine
~ на склоне вулкана, радиальная barranco
~, орографическая orographical trough
~ послеледникого стока rill of postglacial discharge
~ размыва wash-out rill
ложбинка stria, rill, hollow, furrow, groove, vallecula
~, перегородочная fossula
ложбище *hunt.* resting place of wild animals
ложе 1. bed, channel; 2. *bot.* receptacle, bed, thalus; 3. gunstock
~, брачное nuptial bed
~ долины valley floor
~, игольное *text.* needle bed
~, капиллярное capillary bed
~, лавовое lava-bed
~ ногтя *anat.* nail matrix, nailbed, onychostroma, hyponychium
~ океана ocean bed, sea/ocean floor
~, плодовое torus
~ потока stream bed
~, прокрустово *myth.* bed of Procrustes
~ реки river bed/channel, runway
~ сброса foot of fault
~, сосудистое vascular bed
~ соцветия phoranthium, anthoclinium
ложеобразователь *agr.* scraper bar
ложецветковый *bot.* thalamiflorous
ложечка spoon; *foundry* scoop
~ высаживающего аппарата planting cup, picking hand
ложечница *bot.* scurvy weed/grass *(Cochlearia)*
ложечный spoon-shaped
~ная трава scurvy weed/grass *(Cochlearia)*
ложиться lie down; *av., nav.* set
~ на новый курс set a new course
~ в дрейф heave/lie to
ложка spoon; *met.* laddle, skimmer; spoonful
~, буровая *min.* earth auger
~, глазная *med.* ophthalmic curette
~ для экстракции катаракты cataract spoon/curette
~, литейная casting ladle
~, острая acute spoon, scoop, scraper, pointed curette
~, оттискная impression tray
~, разливательная soup ladle
~, формовочная sleeker
ложковидный, ложкообразный spoon-shaped, cochleariform; concave
ложница *hist.* bedroom (suite)
ложнодвухрядный pseudobiserial
ложнодождевик *bot.* hard-skinned puff-ball *(Scleroderma)*
ложнодырчатый quasi-perforate
ложножабра *zool.* pseudobranch
ложноклассицизм pseudoclassicism
ложнокольчатый pseudogyrate
ложнокороеды *pl. zool.* powder-post beetles *(Bostrichidae)*
ложнокристаллический pseudocrystalline
ложнокузнечик *zool.* Ierusalem cricket *(Stenopelmatus)*
ложноледниковый pseudoglacial
ложнолиственница *bot.* golden larch *(Pseudolarix)*
ложнолуковица *bot.* pseudobulb
ложноногие *n. pl. zool.* boas and pythons *(Boidae)*
ложноножка pseudopod(ium)
~, широкая массивная lobopodium
ложнооднодольный *bot.* pseudomonocotyledonous
ложнооднолистный *bot.* pseudomonofolious
ложнооднoплодный *bot.* pseudomonocarpous
ложноопенок *bot.* hypholoma *(Hypholoma)*
ложноперегородчатый pseudoseptate; septulate
ложноперепончатый pseudomembranous
ложнопесчанистый sandlike
ложнопокровица *bot.* cloak fern *(Notholaena)*
ложноравнолопастный *zool.* homocercal
ложносеменной *bot.* pseudospermous, pseudospermic
ложносетчатокрылые *n. pl. zool.* pseudoneuropterans *(Pseudoneuroptera)*
ложнослоники *pl. zool. Platystomidae, Anthribidae*
ложный false, spurious, fallacious, dummy, sham; phoney; fictitious
~ ариллус *bot.* arillode
~ная луковица pseudobulb
ложь lie, falsehood; *law* false pretence; *comp.* false
~, заведомая falsism
лоза 1. rod; witte; 2. vine, runner
~ виноградная grapevine
лозина osier, willow; withe, osier twig
~, миндальная almond-leaved willow *(Salix triandra)*
лозоискатель water diviner, dowser
лозоносный *bot.* sarmentiferous, sarmentose
лозообразный *anat.* pampiniform
лозоподобный grapelike, vinelike
лозунг catchword, slogan; bill, poster
лойник *bot.* selfheal *(Prunella)*
локализатор localizer, detector; finder
~ облучения X-ray therapy applicator
~ повреждений fault indicator/detector
локализация localization; location, site; *zool.* gregarization
~, избирательная selective localization
~, излюбленная site of predilection/election
~ инородных тел localization of foreign bodies

~ **источника звука** sound localization
~ **пучка рентгеновских лучей** collimation
локализировать localize
локально-бикомпактный locally bicompact
локально-выпуклый locally convex
локально-гомеоморфный locally homeomorphic
локально-эвклидовый locally Euclidean
локальность locality
локальный local; autochthonous, edaphic
локатор locator, position finder, radar (set)
~, **звуковой** sonar, sound locator
~, **инфракрасный** infrared/IR radar
~, **лазерный** lidar, laser radar
~ **муфт, магнитный** magnetic collar locator
~, **ультразвуковой** ultrasonic sensing probe *(for the blind)*
локаут lockout, shutout
локаутированный dismissed
локаутировать lock out
локация detection and ranging, location, locating, finding
~, **звуковая** sonic/sound detection and ranging
~, **инфракрасная** infrared/IR ranging and detection
~, **лазерная** laser range and detection
локва *bot.* loquat *(Eriobotrya japonica)*
локер locker, refrigerated compartment
локер-морозильник frozen food locker
локер-холодильник refrigerated locker
локомобиль locomobile, traction/portable engine
локомотив locomotive, engine
~, **атомный** atomic/A locomotive
~, **ведущий** hauling locomotive
~, **грузовой** freight/goods locomotive
~, **магистральный** main-line locomotive
~, **маневровый** switching/shunting/yard locomotive
~, **сборочный** *min.* gathering locomotive
~, **узкоколейный** narrow-gauge locomotive
локомотивостроение locomotive engineering
локомоция locomotion
локон curl, lock
~ **Аньези** *geom.* witch of Agnesi, versiera
~, **фальшивый** toupee
локотник elbow rest; arm *(of a chair)*
локоть 1. *anat.* elbow, forearm, ulna; 2. *hist.* cubit, ell *(measure)*
локсодрома *math.* loxodrome (curve), loxodromic curve/line/spiral
локсодромия 1. *math. see* **локсодрома**; 2. *nav.* Mercator chart/track, rhumb line
локсоклаза *geol.* loxoclase
локтевой *anat.* cubital, ulnar
~**вая кость** funny bone, ulna
локте-лучевой *anat.* ulnoradial
локула *bot.* locule, loculus
локус 1. *biol.* (gene) locus; 2. *math.* locus
лолинь *naut.* lead-line
лоллард *hist.* Lollard

лом 1. crowbar, digging/pinch bar; 2. *coll.* scrap, junk
~, **амортизационный** obsolete/obsolescent/old/dormant scrap
~, **бытовой** household scrap
~, **ваграночный** foundry/cupola scrap
~, **валковый** roll scrap
~, **габаритный** standard-size scrap, sized scrap
~ **для вскрытия летки** *met.* tapping iron/bar
~ **для выпуска шлака** skimmer bar
~ **для переплавки** (re)melting scrap
~, **железный** scrap/waste iron
~, **измельчат** reduce/shred/macerate the scrap
~, **кантовальный** pinch bar
~, **крупногабаритный** bulky scrap
~, **крупный** capital scrap
~, **кусковый** lump scrap
~, **мартеновский** open-hearth scrap
~, **металлический** scrap metal
~ **металлообработки** industrial scrap
~, **негабаритный** oversize/irregular scrap, off-size scrap
~, **оборотный** plant/return scrap
~, **пакетировать** bale/bundle/faggot the scrap
~, **разделывать** upgrade/shred/macerate the scrap
~, **сложный** mixed scrap
~, **товарный** saleable scrap
~, **шуровой** *met.* poking bar, poker
ломаная *math.* broken/polygonal line, open polygon
~ **Эйлера** Euler's polygonal line
ломано-линейный dog-leg(ged)
ломать 1. break, fracture; 2. destroy, wreck
ломбард 1. pawnshop, pawn office; 2. *hist.* loan bank
~, **заложить в** pawn
ломбардец Lombard
ломбардный pawnshop *(attr.)*
~**ная квитанция** pawn ticket
ломбардский Lombard(ic)
ломбер omber, ombre *(card game)*
ломбрицинкиназа *physiol.* lombricine kinase
ломик short crowbar; hook, pinchbar, jimmy; forcer
~ **для открывания люков** manhole hook
ломка breaking; demolition, destruction
~ **чушек** *met.* crushing of pigs
ломкий brittle, fragile, crumbly; friable; short
~ **в холодном состоянии** cold-short
~ **плод** *bot.* loment(um)
ломкость fragility, brittleness; frangibility, friability
~ **в холодном состоянии** cold shortness
~ **капилляров** capillary fragility
~ **лица** *leath.* crackiness, cracking of the grain
~ **костей** osteopsathyrosis, fragility of the bones
~ **костей, врожденная** constitutional fragility of the bones
~ **ногтей** brittleness of the nails
ломонос *bot.* clematis *(Clematis)*
ломонтит *geol.* laumontite, caporcianite

ломтерезка *food* slicing machine, slicer
лонгобард *see* **лангобард**
лонгуолл *min.* longwall
лонгхорн *agr.* longhorn
лонгшез deck chair
лондонец Londoner
лонжа lunge, longe
лонжерон *av.* wing spar; longeron, beam; *auto.* side member/rail, frame girder
~, балочный beam spar
~, вспомогательный auxiliary/secondary spar, shear-web spar, shear beam
~, главный mean beam/longeron/spar
~ гондолы nacelle longeron
~ двутаврового сечения I-section spar
~, изогнутый cambered spar
~ киля fin spar/post
~, коробчатый/полый box spar/beam
~ крыла wing spar, mean beam
~ крыла, вспомогательный wing subsidiary spar
~, литой cast spar
~, ложный false spar, shear beam
~, мощный massive spar
~, неразъемный one-piece spar
~, прессованный extruded spar
~ руля высоты elevator spar
~, сварной welded spar
~, сплошной solid spar
~ стабилизатора tailplane spar
~, ферменный trussed spar
~ фюзеляжа fuselage longeron
~, цельный one-piece spar
~ центроплана centre-wing spar
~ швеллерного сечения channel-section spar
~ элерона aileron spar
лонжерон-стенка shear beam
лонно-бедренный *anat.* pubofemoral
лонно-копчиковый *anat.* pubococcygeal
лонно-предстательный *anat.* puboprostatic
лонно-пузырный *anat.* pubovesical
лонный *anat.* pubic
лоно *rel.* bosom, lap
лоносечение *med.* symphysiotomy, symphiseotomy
лонхокарпус *bot.* lancepod *(Lonchocarpus)*
ло̀парь *naut.* fall
~ грузовых талей cargo fall
~ подъемного горденя whip
~, раздвоенный split fall
~ фиш-талей painter's fall
~, цепной lifting chain
~ шлюпочных талей boat's fall
лопа̀рь Lapp, Laplander
лопастеобразный fan-/paddle-/beavertail-shaped
лопастник *bot.* turbantop, helvella *(Helvella)*
лопастнолистный *bot.* lobophyllous
лопастность lobation
лопастный lobed, lobate, lobulate; *bot.* laciniate
лопасть 1. *tech.* blade; vane, fan; (wheel) paddle; 2. *bot.* lamina; 3. *geol.* lobe
~ батана *text.* slay sword
~, боковая side/lateral lobe
~ весла oar blade/palm
~, воздушного/гребного винта propeller blade
~, всебоковая omnilateral lobe
~, глазная palpebral lobe
~ гребного колеса float, paddle
~, дополнительная adventive lobe
~, закручивать twist the blade
~ крыльчатки impeller blade
~ ледника, предгорная piedmont lobe
~, месильная *food* kneading blade
~ мешалки agitator blade, mixing paddle
~ несущего винта rotor blade
~ разбрасывателя *agr.* spreader blade/arm
~ рулевого винта tail-rotor blade
~, саблевидная scimitar-/sweptback blade, sabre-shaped blade
~, съемная detachable blade
~, цельнометаллическая all-metal blade
~, шарнирно подвешенная articulated/hinged blade
~ швырялки *agr.* impeller blade
лопата 1. spade, shovel; scoop; 2. blade; vane, fan; paddle
~, балластная *rail.* square shovel
~ для удаления воды с палубы squeegee
~, дренажная drain spade
~, метательная *met.* throwing shovel
~, механическая power/mechanical shovel, excavator
~, механическая, обратная back digger, backhoe
~, механическая, поворотная rotating power shovel
~, механическая, прямая face/dipper power shovel, shovel dipper
~ одноковшового экскаватора power shovel
~, паровая steam shovel
~ с треугольным лотком trangular spade
~, сажальная *agr.* planting spade
~, тракторная tractor shovel, tractor-mounted bucket
~, формовочная *foundry* moulding spade, moulder's shovel
~, шахтерская mining shovel
~ экскаватора dipper assembly
лопатить *agr.* shovel, *naut.* squeegee
лопатка 1. spade; *mech.* blade, vane; shovel; scoop, skimmer; trowel; 2. *med.* depressor, spatula; 3. *anat.* scapula, shoulder blade; 4. *bot.* unripe/flat/long pod; 5. *arch.* pilaster
~, активная impulse blade
~, бандажированная shrouded blade
~ вентилятора fan blade
~, вставная inserted blade
~, входная inlet/guide vane
~, двухопорная (рабочая) tip-shrouded blade

~ для шин (монтажная) tyre iron
~, завихривающая/завихрительная swirl vane
~, закрученная twisted blade
~ компрессора compressor blade
~ компрессора, направляющая compressor guide vane
~, крыловидная *anat.* winged scapula, scapula alata
~, ложкообразная spoon-shaped blade
~, направляющая guide vane, stator vane/blade
~, одноопорная/консольная unshrouded blade
~, поворотная variable(-incidence) blade
~ позвоночного scapula, shoulder blade
~, полая/пустотелая hollow blade, cored-out blade
~, рабочая moving/rotating/working blade
~ рабочего колеса impeller vane
~ с испарительным охлаждением transpiration-cooled blade
~, скафоидная scaphoid scapula
~, спрямляющая exhaust stator blade, straightener vane
~, фрезерованная machined blade
~, хирургическая (surgical) spatula, (surgical) depressor
~, цельнофрезерованная integral blade
~ штукатура plasterer's trowel
~, щелевая slotted blade
лопатки *pl.* buckets, blading
~ высоконапорной ступени компрессора high-pressure blading
~, рабочие rotor blading
~ решетки cascade blades
~ с малым шагом low-staggered blading
лопат(к)ообразный, лопатовидный spade-/shovel-shaped; spatulate
лопатоногий *zool.* scaphopod(ous); *n. pl.* scaphopods *(Scaphopoda)*
лопатонос, американский *zool.* shovel-nosed sturgeon *(Scaphirhynchus)*
лопаточно-задний *anat.* scapuloposterior
лопаточно-ключичный *anat.* scapuloclavicular
лопаточно-передний *anat.* scapuloanterior
лопаточно-плечевой *anat.* scapulohumeral
лопаточный scapular; skatulate
лопатчатый 1. skatulate spade-/shovel-shaped; 2. paddled, with blades/vanes
лопинит *geol.* lopinit *(fossil coal)*
лополит *miner.* lopolith
лопух, лопушник *bot.* burdock *(Arctium, Lappa)*
лоп-штаг *naut.* bowsprit man-rope
лорантовые *n. pl. bot.* mistletoe family *(Loranthaceae)*
лорд lord
~ов, Палата House of Lords
лорд-канцлер Lord Chancellor
лорд-мэр Lord Mayor
лордоз *med.* lordosis
~, поясничный lumbar lordosis
~, шейный cervical lordosis
лордосколиоз *med.* lordoscoliosis
лордство lordship; *coll.* lords
лорд-хранитель печати Lord Privy Seal
Лорелея Lorelei
лори *zool.* lori(s) *(Loris);* lory *(Lorius, Domicella)*
лорнет lorgnette
лор-футокс *naut.* second buttock
лосина elk skin, chamois leather
лоск gloss, shine, lustre, polish; *text.* sheen
~ бумаги paper gloss
лоскут 1. *surg.* flap, graft; 2. scrap, rag, shred
~, гетерологический heterograft
~, десенный gingival flap
~, закрепленный на обоих концах, стебельчатый rope graft
~, изогенный isogenic graft, isograft
~ из фасции fascia graft
~, кожевенный *leath.* waste/scrap leather
~ кожи без подкожной клетчатки Wolfe's graft
~, кожный cutaneous/dermal/skin graft, integumentary flap
~ надкостницы periosteal graft
~ на ножке, кожный pedicle graft
~, несвободный flap
~, покровный *geol.* outlier, detached mass of overthrust sheet
~, скользящий accordion graft
~ слизистой оболочки mucous graft
~ собственно кожи cutis graft
~, стебельчатый pedicle graft
~, странствующий jump flap
~, сухожильный tendon graft
~ Тирша Thiersch's graft
~, трубчатый sleeve/tube graft
~, филатовский Filatov's graft
лоскутный scrappy; made of scraps
~ная ампутация flap amputation
~ное одеяло patchwork quilt
лосняк *bot.* twayblade *(Liparis)*
лососевый salmoneous; salmon-colo(u)red; salmon *(attr.); n. pl. zool.* salmon family *(Salmonidae)*
лососина *cul.* salmon (flesh)
лososка samlet, salmonet
лосось *zool.* salmon *(Salmo salar)*
лосось-таймен *zool.* brown/salmon trout *(Salmo trutta)*
лось *zool.* elk *(Alces alus)*
~, американский moose *(Alces americanus)*
лосьон lotion
лосятина *cul.* elk (flesh)
лосятник 1. elk hunter; 2. elk reserve
лот I 1. *naut.* sounding lead/line, sea gauge; lead-and-line; plumb bob, depth finder; 2. plummet
~, бросать swing the lead
~, гидроакустический sonic fathometer, sonic echo-sounder

~, глубоководный deep sea (sounding) lead
~, механический sounding machine
~, ручной sounding line/lead, lead-and-line
лот II lot *(half an ounce)*
лотаппарат optical plum(b)-bob
Лотарингия Lorraine
лотарингский Lotharingian
лотерея lottery, raffle
~, составная compound lottery
лотлинь *naut.* lead/sounding line, leadline; sounding wire
~, обносить pass the leadline
~ ручного лота hand line
лото lotto
лотовой, лотовый *naut.* leadsman
лоток 1. (hawker's) tray/stand, trough, pan; 2. chute, channel, shoot, gutter
~ Вентури, измерительный Venturi flume
~, вибрационный vibrating chute
~, водосточный spouting chute
~, гидрометрический measuring flume
~ для деталей *mach.* tote box; component tray
~ для засыпки в мешки bagging chute/spout
~ для спуска грузов cargo chute
~, жалюзийный (сушилки) louvre bed
~, загрузочный feed chute
~, золотопромывочный gold pan
~, инструментальный tool tray
~ канала canal cavity
~, качающийся reciprocating tray
~, лесосплавной log-floating channel, log chute
~, мельничный mill-race/-course, millrace
~, открытый flume, open channel
~, питающий/подводящий feed tray
~, приемный charging trough
~, сборочный assembly tray
~, сотрясательный shaking tray
~, спускной chute
~, травильный etching trough
лотос *bot.* lotus *(Nelumbium)*
~, египетский white Egyptian lotus *(Nymphea lotus)*
лоточник hawker, huckster, pedlar, *Am.* peddler
лоуренсий *chem.* lawrencium, Lr
лофант *bot.* giant hyssop *(Lophanthus)*
лофиодонт *paleont.* lophiodont
лофолатилус *zool.* tillfish *(Lopholatilus)*
лофофор *zool.* lophophore
лох I *bot.* oleaster *(Elaeagnus)*
лох II egg-laying salmon, salmon during spawning/blushing period
лоханка 1. *anat.* pelvis; 2. basin; (wash)tub
~, почечная renal pelvis
лохии *pl. med.* lochia, discharge after childbirth
~, белые lochia alba
~, кровянистые lochia cruenta/rubra
лохиометрит *med.* puerperal metritis, lochiometritis
лохиорея *med.* lochiorrhea
лоховые *n. pl. bot.* oleaster family *(Elaeagnaceae)*
лохолистный *bot.* elaeagnifolious
лоциация *ecol.* local variant of association, lociation
лоция I *ecol.* locies, lociation
лоция II *nav.* pilot, (pilot's) sailing direction
лоцман 1. *nav.* pilot; 2. *zool.* pilot fish *(Naucrates ductor)*
~, ледовый ice master/pilot
~, прибрежный coast pilot
~, старший chief pilot
лоцманство *nav.* pilotage, piloting
лоцмейстер *nav.* chief pilot
лошадиный equine; horsy; *med.* jumentous; *n. pl. zool.* equids *(Equidae)*
~ная сила horsepower
лошадность *agr.* availability of horses
лошадь *zool.* horse *(Equus caballus)*
~, беговая/скаковая race horse
~, верховая saddle/riding horse
~, выносная fore-horse
~, заводская studhorse
~, коренная shaft/thiller horse
~, ломовая dray horse
~ Пржевальского Prshevalski's horse *(Equus przewalskii)*
~, пристяжная outrunner
~, упряжная draught horse
лощение glossing; *leath.* glazing
лощеный glossy, polished; glazed
~ная пряжа glazed yarn
лощилка *met.* slicker
лощило *met.* burnisher
лощина ravine, gulch; hollow, depression
~, образованная водой gully
лоялист loyalist
лояльный loyal, fair, honest
ЛСД LSD, lysergic acid diethylamide
Л-система laboratory system
луб 1. *bot.* (lime) bast, phloem; 2. *coll.* hemp/flax/nettle fiber
~, лубодревесный phloem ray
лубовой, лубовый *see* **лубяной**
лубоед *zool.* bark beetle *(Epidae)*
лубок 1. strip/article of bast; 2. *med.* splint; 3. woodcut, (primitive) wood block
лубоотделитель defibrer
лубочный basten, made of bast
~ная картинка cheap popular print
~ная литература popular/cheap literature
лубрикатор automatic lubricator, oil cup/feeder, oiler
лубянка article made of bast
лубяной 1. basten, made of bast; 2. fibrous
~ная паренхима bast parenchyma
~ные культуры fibre/fiber crops
лувийский Luwian, Luian, Luish
луг *agr.* meadow
~, временный lea
~, выпасаемый pasture meadow

~, двухокосный meadow of two cuts
~, заливной water meadow, flood plain
~, искусственный cultivated meadow
~, сенокосный hay meadow
луговик, луговник *bot.* hair grass *(Deschampsia)*
луговка *zool.* lapwing, peewit *(Vanellus vanellus)*
луговодство *agr.* grassland husbandry, grassland/meadow management
луговой meadow *(attr.);* pratal
~ берег lower bank
~вая собачка *zool.* prairie dog *(Cynomys)*
лугово-пустынный meadow-desert
лугово-сероземный meadow-sierozem
лугово-степной meadow-steppe
лугово-черноземный meadow-chernozem
лугомелиорация *agr.* grassland improvement, meadow cultivation
лугопастбищный pasture meadow *(attr.)*
~ные сорные растения meadow/pasture weeds
луда I tinning alloy
луда II *geol.* luda
луддит *hist.* Luddite
лудильный tinning
лудильщик tinsmith, tinman, tinner
лудить tin-plate, tin
лужа pool, puddle
лужайка grassplot, lawn; sward; (forest) glade, swale
лужайник, лужница *bot.* mudwort *(Limosella)*
лужанка *zool.* river/pond snail
лужение tinning, tin-plating; blanching
~, горячее hot-dip tinning
~, электролитическое tin-plating, electrolytic tinning
луженый tinned, tinplated
лужицкий Lusatian, Sorbian
лужичанин Lusatian, Sorbian
лужичанка Lusatian, Sorbian (woman)
лузговейка *agr.* hull aspirator
луидор louis d'or
лук I *bot.* onion *(Allium)*
~, зеленый spring onion, scallion
~, многоярусный Egyptian onion *(A. proliferum)*
~, репчатый bulb onion *(A. cepa)*
лук II bow
лука 1. bend, sinuosity; 2. pommel, arch
~, задняя cantle, rear arch
~, передняя front arch
~, речная meander, river bend
лукавый 1. crafty, sly; 2. playful; *n. rel.* the Evil One
лук-батун *bot.* Welsh onion *(Allium fistulosum)*
луковица 1. *anat.* bulb, bud, cap; 2. *bot.* bud; onion; 3. *arch.* cupola, onion(-shaped) dome
~ аорты aortic/arterial bulb
~, вкусовая taste bud/bulb
~, воздушная *bot.* bulbil, bulbet
~ волоса hair bulb
~ двенадцатиперстной кишки duodenal bulb/cap
~, дочерняя daughter bud
~, обонятельная olfactory bulb
~ полового члена bulb of the penis
~ чувствительного нерва end/sensory bulb
~ яремной вены jugular bulb, bulb of jugular vein
луковицевидный, луковицеобразный bulbiform, bulbar, bulbous, bulbaceous
луковиценосный bulbiferous, bulb-bearing
луковичка *bot.* clove, bulbil, bulbet
луковичкообразный bulbillate
луковичный bulbar, bulbous, bulbaceous
~ная чешуя bulb scale
луковый aniony; alliaceous, cepaceous
лукообразный I bulbiform, bulbilike; onion-shaped
лукообразный II arched; bow-shaped
лукоцветковый *bot.* onion-flowered
лук-порей *bot.* leek *(Allium porrum)*
лук-резанец/-скорода *bot.* chive *(Allium schoenoprasum)*
лук-рокамболь *bot.* giant garlic, rocambole *(Allium scorodoprasum)*
лук-татарка *bot.* Welsh onion *(Allium fistulosum)*
лук-чеснок *bot.* garlic *(Allium sativum)*
лукуллов(ский) Lucullean, Lucull(i)an
лук-шалот *bot.* shallot *(Allium ascalonicum)*
лумахель lumachel(la), lumachelle
лумбанг *bot.* candlenut tree *(Aleurites moluccana)*
лумп lump sugar
лумпен-пролетариат Lumpenproletariat
луна 1. moon; 2. *myth.* Luna; Cynthia
~ в первой/последней четверти half moon
~ в ущербе *astr.* gibbous phase
~ "времени" slow-motion picture
~, ложная mock moon
~, нарастающая waxing moon
~, полная full moon
~, убывающая/ущербная waning moon
луна-парк amusement grounds/park
лунарий lunarium
луна-рыба *zool.* sunfish, moonfish *(Mola mola)*
лунатизм somnambulism, sleepwalking
лунатик somnambulist, sleepwalker
лунатический noctambulant, somnambulistic
лунация *astr.* lunation
лунит lunar matter
лунка 1. *tech.* indentation; turbine bucket; 2. *geol.* alveole, alveolus, lunule; socket; small cavity; hole
~ износа (на резце) crater(ing)
~, опорная *min.* bearing hitch
~, посадочная *agr.* hollow
~, семенная *agr.* seed hole
~, серебристая *zool.* buff-tip moth *(Phalera bucephala)*
лункоделатель, лункокопатель *agr.* dibb(l)er, dibble, dibbing machine, hole digger
лункообразователь *agr.* basin harrow
лунник I lunar probe, moon rocket

лунник II *bot.* honesty *(Lunaria)*
лунниколистный *bot.* honesty-leaved
лунно-полусуточный lunar-semidiurnal
лунно-солнечный lunisolar
лунный lunar, moon *(attr.)*
~ камень moonstone
~ модуль lunar module
луновидный, лунообразный moon-shaped, moonlike; lunate
луноглазка *zool.* toothed herring *(Hiodon tergisus)*
лунодром lunar space-vehicle launching site
луносемянник *bot.* moonseed *(Menispermum)*
луносемянные *n. pl. bot.* moonseed family *(Menispermaceae)*
луноход moon rover/car, lunar vechicle, moon research vehicle
луноцвет *bot.* moonflower *(Ipomoea bonanox)*
луночка lunule, lunulet; alveole, alveolus; *anat.* lunula
луночный alveolar, alveolate
лунулитовидный *paleont.* lunulitiform
лунчатозубые *n. pl. zool.* selenodonts *(Selenodonta)*
лунь *zool.* harrier, ringtail *(Circus)*
лупа magnifying/reading glass, loupe, magnifier, magnifying lens
~, бинокулярная binocular magnifier
~ для чтения карт chart magnifier/reader, map reading lens
~, отсчетная *geod.* reading magnifier
~, очковая eyeglass loupe
~, ткацкая web glass, cloth prover
лупинин *chem.* lupinine
лупиться peel, flake
лупулин *bot.* lupulin
лур *abbr.* language laboratory
лут *zool.* leatherback (turtle) *(Dermochelys coriacea)*
луток *zool.* smew *(Mergus albellus)*
луфарь *zool.* bluefish, greenfish *(Pomatomus saltatrix)*
луциан *zool.* snapper *(Lutianus)*
луч 1. ray, beam; pencil (of light); 2. *bot., zool.* ray; *anat.* radius; *see also* **лучи**
~, азимутальный azimuth beam
~, бегущий scanning beam
~, веерный fan(-shaped) beam
~, визирный *geod.* collimating ray
~, глиссадный glide-path beam, glide slope
~, граничный limiting ray
~ жаберной перепонки branchiostegal ray
~, записывающий recording/writing beam
~ зрения line of sight
~ из точки надира nadir radial
~, инфракрасный infra-red beam, IR beam
~, исходящий emergent beam
~, карандашный pencil beam
~, катодный cathode ray/beam
~, керновый *top.* epipolar ray
~, курсовой (посадочный) localizer beam/course
~, лазерный laser beam/ray
~, ленточный *comp.* ribbon beam
~, митогенетический mitogenetic ray
~ наведения guide/guidance beam
~, направленный directed/directive beam/ray
~, ножевой knife-edge beam
~, остронаправленный narrow-directive beam, pencil beam
~, острый sharp/pencil beam
~, отклоненный diffracted/deflected beam
~, отраженный reflected beam/ray
~, падающий incident beam/ray
~, поляризованный polarized light
~, преломленный refracted ray/beam
~, приводной approach/homing beam
~, проецирующий *math.* projector, projecting ray
~, пространственный space ray
~, радиолокационный radar beam
~ радиомаяка radio(-range) beam
~, развертывающий scanning beam
~, расходящийся divergent beam
~, световой light beam/ray, beam/pencil of light
~, сердцевидный medullary/wood/xylem ray
~, сложный *bot.* aggregate ray
~, считывающий reading beam
~ угла *math.* arm of an angle
~, электронный cathode/electron beam, electron ray
~, эпиполярный *top.* epipolar ray
~, эталонный reference beam
лучебежный actinodrome, actinodromous
лучевиднолистный *bot.* actinophylous
лучевидный, лучеобразный beamlike, radial, raylike
лучевики *pl. zool.* radiolarians *(Radiolaria)*
лучевод optical waveguide, light/beam guide
~, волоконный fibre-optic waveguide
~, отражающий reflecting beam waveguide
лучевой radial, ray *(attr.)*
~вая болезнь *med.* radiation sickess
~вая кость *anat.* radius
лучезапястный *anat.* radiocarpal
лучезарность radiance, radiancy
лучезвуковой radiophonic
лучеиспускаемость emittance, emissivity, radiativity
лучеиспускание *phys.* radiation, emission
лучеиспускательный radiating, emitting
лучеиспускать radiate, beam, emit rays
лучелистный *bot.* actinophyllous
лучелоктевой *anat.* radioulnar
лучеобразнорасходящийся radiating, radial
лучеобразный raylike, beamlike, radial
лучеобразующий beam-forming
лучеотражаемость radiant reflectivity/reflectance
лучеперые, лучеплавниковые *n. pl. zool.* actinopterygians *(Actinopterygii)*
лучепреломление *phys.* refraction, refringence,

refringency
~, **двойное** birefringence, double refraction
~, **низкое** birefringence, weak refraction
лучепреломляющий refractive
лучепрозрачный *phys.* diathermic
лучи *pl.* rays
~, **актиничные** actinic/chemical rays
~ **Беккереля** Becquerel rays
~, **вторичные рентгеновские** secondary/fluorescent X-rays
~, **граничные рентгеновские** borderline/grenz rays
~ **Гурвича** Gurwitsch/mitogenic rays
~ **Дорно** Dorno rays
~, **жесткие** hard/penetrating rays
~ **Ленарда** Lenard rays
~ **космические** cosmic rays, ultrarays
~, **мягкие** soft rays
~, **невидимые** invisible/obscure rays
~, **рентгеновские** Roentgen/X-rays
~ **рентгеновские**, **сверхбыстрые** ultraspeed X-rays
~ **рентгеновские**, **средней жесткости** medium-hard X-rays
~, **сверхжесткие** ultra-hard rays
~, **солнечные** sunlight, sunbeams
~, **терапевтически активные** dynamic rays
лучинки *pl. print.* pins, fingers
~, **выводные** shooflies
лучисто-волокнистый divergently/radial-/radiating fibrous, radiating/radial-columnar
лучистогрибковый *bot.* actinomycetic
лучисто-симметричный actinomorphic, actinomorphous, radiosymmetrical
лучистость radiant intensity, triboluminescence; radiance, radiancy; radiation
~, **полярная** polar radiation
лучистый 1. radiate(d), radial, radiant, radiolate(d); 2. radiating
~ **грибок** Actinomyces
~**тая энергия** radiant energy
лучица *bot.* stonewort *(Chara)*
лучковый bow-shaped
~**вая пила** whipsaw, frame/bow/sash saw
лучок 1. *tech.* small bow; 2. *mus.* violin bow
~ **дрели** drill bow
~ **пилы** saw bow
лущевка *agr.* hoeing, loosening
лущение *agr.* 1. shallow plowing/ploughing; hoeing, loosening; scuffling, paring; 2. shelling, husking, hulling, peeling
лущеный 1. loosened *(of soil)*; 2. hulled, shelled
лущилка sheller, huller, hulling mill
лущильник 1. shallow/paring/stubble/skin/hoeing plough/plow; 2. sheller, huller, shelling machine
~, **дисковый** disk stubble plough/plow, disk tiller
~, **лемешный** share scuffler
~, **прицепной** trailed stubble plough/plow
~, **тракторный** tractor disk tiller
лущильный hulling, husking, shelling
луяврит *geol.* lujaurite, lujavrite
лыжа 1. ski; snowshoe; slide shoe; 2. *el.* collecting shoe
~ **на бюгеле** collector strip
~, **неубирающаяся** fixed ski
~, **носовая** nose ski
~ **пантографа** *rail.* pantograph slide shoe
~ **токоприемника** collecting slide shoe
~ **убирающаяся** retractable ski
~ **угольного комбайна** skid of a cutter-loader
лыжа-отражатель deflection ski
лыжник skier
лыжник-гонщик/скоростник racing skier
лыжный ski *(attr.)*, skiing
~ **спорт** skiing
~**ная база** skiing lodge/centre
лыжня ski track/run
лыко bast, libre, phloem
~, **волчье** *bot.* mezereon *(Daphne mezereum)*
лыковый basten, made of bast
лысина *med.* bald head/spot, calvities, alopecia, baldness
лысоплодный *bot.* naked-fruited
лысостебельный *bot.* naked-stalked
лысун *zool.* harp/Greenland seal *(Histriophoca groenlandica)*
лысуха *zool.* coot *(Fulica atra)*
лысый bald(-headed), calvous
лычный basten, made of bast
львиность *med.* leontiasis
~, **костная** leontiasis ossea, craniosclerosis
львиноязычковый *bot.* lion-tongned
львиный leonine, lion's
~ **зев**, ~ **ная пасть** *bot.* snapdragon *(Antirrhinum)*
~**ная доля** the lion's share
льгота *econ.* advantage, privilege
~, **налоговая** tax rebate/privilege/concession/credit
~ **на налог** tax allowance
льготный favo(u)rable, privileged; preferential; reduced
~ **билет** privilege/free ticket; ticket at reduced price
~ **срок** grace period
~**ные дни** *econ.* days of grace
~**ные пошлины** preferential duties
~**ные условия** preferential ferms
льдина ice-floe, block/flake of ice
~ **дрейфующая** drift ice
~, **плавучая** ice-floe
льдистый icy; ice-covered
льдогенератор ice(-making) machine
~, **дисковый** disk ice maker
~, **рассольный** brine ice maker, can-type ice maker
~ **чешуйчатого льда** flake ice maker
льдодробилка ice crusher
льдообразование ice accretion/formation
льдоскат ice chute

льдоудаление deicing
льдоудалитель *av.* deicer
льдоудалительный ice-removing, deicing
льдохранилище icehouse, ice storage/store
лье league *(unit of distance)*
льезон *phon.* liaison
льновидный linoid, flaxlike
льноводство flax growing/production/cultivation
льноволокно flax fibre
льновый flax; *n. pl. bot.* flax family *(Linaceae)*
льнозавод flax treating plant, rettery, (flax-) scutching plant
льнокомбайн flax harvester, flax-harvesting machine
льноконоплемолотилка flax-and-hemp thresher
льноконоплемялка flax-and-hemp breaker
льнолистный *bot.* linophyllous, flax-leaved
льномолотилка flax-seed thresher
~, вальцовая roll-type flax thresher
льномялка scutcher, swingle, flax-breaking machine
льномяльный flax-breaking
льнообрабатывающий flax-processing/-manufacturing
льноподборщик-очесыватель flax gathering-and-combing machine
льнопрядилка, льнопрядильня flax(-spinning) mill
льнопрядильный flax-spinning
~ная фабрика flax mill
льносемя flaxseed
льносеющий flax-sowing
льносеялка flax-seed drill, flax-sowing machine
льносолома deseeded flax
льносушилка flax (fibre) drier
льнотеребилка flax puller
льноткачество linen weaving
льнотрепалка flax scutcher
льнотрепальный flax-scutching
льнотреста flax straw
льноуборка flax harvesting/pulling
льноуборочный flax-harvesting/-pulling
~ная машина flax puller
льноутомление *agr.* flax-culture fatique
льноцветковый *bot.* liniflorous, flax-flowered
льночесалка flax-hackler
льнянка *bot.* toadflax *(Linaria)*
льнянколистный *bot.* toadflax-leaved
льняной 1. flaxy, of flax; flaxen; flax *(attr.)* 2. linen *(attr.)*
~ная промышленность linen industry
~ное масло linseed oil
~ное семя linseed, flax seed
льюзия *bot.* lewisia *(Lewisia)*
льюисит *miner.* lewisite
льяло 1. *naut.* bilge course, bilgeway, draining bilge, hold/bilge/drainage well; 2. *tech.* casting mo(u)ld
льячка *met.* ladle
лэндлордизм landlordism
любим-трава *bot.* common avens *(Geum urbanum)*
любисток *bot.* lovage *(Levisticum)*
любитель amateur; ham; fan; lover
~ музыки music lover
~ спорта sports fan
любитель-коротковолновик short-wave radio ham
любительский 1. amateur(ish); avocational; 2. choice *(attr.)*
любительство dilettantism, amateurishness, dabbling
любка *bot.* rain orchid *(Platanthera)*
любник *bot.* winter aconite *(Eranthis)*
любовник lover, love mate
~, первый *theatre* jeune premier
любовница mistress, concubine; hetaera
любовь *educ.* love (for/of); devotion, fondness
~ к ближнему love for one's neighbour
~ к Богу и человеку theophilanthropism
любознательность *educ.* inquisitiveness, intellectual curiosity
~, познавательная epistemic curiosity
любопытство *educ.* curiosity
люверс *naut.* glut, lacing eye; eyelet; cringle
люгер *naut.* lugger
люд *coll.* people, folk
~, мелкий petty townsfolk
~, рабочий working people
людвигит *geol.* ludwigite
~, корейский collbranite
людвигия *bot.* seedbox *(Ludwigia)*
люди I *pl.* 1. people; 2. *mil.* men
~, рядовые rank and file
люди II Slavonic name of the letter "**л**"
людоед cannibal; ogre
людоедский canibalistic, hominivorous, anthropophagous
людоедство cannibalism, anthropophagy
людской human
~ род human race, manking
~ состав *mil.* personnel
люес *med.* syphilis
люизит *chem.* lewisite
люк 1. manhole, access hole; 2. *naut.* hatch, hatchway; port; 3. *min.* chute; *av.* escape/emergency hatch
~, астрономический astrohatch
~ аэрофотоаппарата camera window/hatch
~, вентиляционный *auto.* ventipane, pivoted ventilation window
~, входный entrance port, access hatch
~, выпускной *min.* pull chute, chute door
~, выходной exit port, egress hatch
~, грузовой *av., nav.* cargo/freight hatch
~, двойной *min.* two-way chute
~, задраивать *naut.* batten down the hatch; *av.* seal a hatch
~, коллекторный *el.* commutator inspection hole
~, прочистной cleanout
~, светлый/световой *naut.* skylight

~, **смотровой** access/inspection hatch/hole
~, **спасательный** *naut.* escape hatch
~, **сходный/сходной** companion hatch(way), companionway
~ **трюма** hold hatch
~, **эксплуатационный** *av.* maintenance/servicing hatch
~ **экстренного покидания** quick-escape hatch
люковый *n. min.* chuteman
люкс I *phys.* meter-candle, lux/lx
люкс II first class, de luxe
Люксембургский Luxembourg
люкс-метр lux(o)meter, photometer, light meter
люкс-секунда lux second, lx s
люлька 1. cradle; 2. *constr.* cage; bucket; cable car/chair
~ **слитковоза** *met.* ingot chair
~ **хлебопекарной печи** swinging tray
люмбаго *med.* lumbago, backache, osphyalgia, osphyitis
люмбализация *med.* lumbarization, assimilation sacrum
люмбальный *anat.* lumbar
люмбоабдоминальный *anat.* lumboabdominal
люмбоколостомия *surg.* lumbocolostomy
люмбоишиалгия *med.* lumbar ischialgia
люмен *phys.* lumen, lm
люменметр lu(men)meter, integrating photometer
люмен-секунда lumen-second, lm s
люмен-час lumen-hour
люминал *pharm.* luminal, phenobarbital, phenobarbitone
люминал-натрий *pharm.* soluble phenobarbital
люминесцентный luminescent
~**ная лампа** fluorescent lamp
люминесценция *phys.* luminescence
~, **катодная** cathod luminescence
~ **кристаллов** crystalloluminescence
~, **сенсибилизированная** sensitized luminescence
~, **ударная** impact luminescence
~, **ультразвуковая** ultrasonic luminescence
люминесцировать luminesce
люминоген luminogen, phosphorogen
люминография lumiprinting
люминоскоп luminoscope
люминофор luminophore, phosphor
~, **жидкий** liquid phosphor
~, **рентгеновский** X-ray phosphor
~, **равноэнергетический** equal-energy phosphor
~ **с быстрым затуханием** short-persistance/-lag phosphor, rapid-decay phosphor
~ **с длительным послесвечением** long-persistence/-lag phosphor
~, **сложный** composite/multiple phosphor
~ **с малым временем послесвечения** short-persistance/-lag phosphor, rapid-decay phosphor
люминофорный luminiferous
люмистерин lumisterol
люмитайп lumitype
люмнит lumnite *(shielding material)*
люмпеновые *n. pl. zool.* Lumpenidae
люмпенпролетариат *polit.* Lumpenproletariat
люнебургит *geol.* luneburgite
люнель Lunel *(wine)*
люнет 1. *mach.* steadyrest; follow-rest; support, collar plate; 2. *mil.* lunette, demilune; 3. *arch.* lunette
~, **двухкулачковый** two-point steadyrest
~ **для токарных работ** turnining steadyrest
~, **неподвижный** steadyrest
~, **плашечный** jaw-type steadyrest
люнетодержатель steady/stay holder
люпин *bot.* lupine *(Lupinus)*
люпинин *chem.* lupinine
люпозорий *med.* lupus sanatorium
люпулин *bot.* lupulin
люпус *med.* lupus
люрик *zool.* little auk *(Alle alle, Plautus alle)*
люстр flash
~ **крахмала** starch brilliance
люстра chandelier
~, **проходческая** cluster floodlight
люстрин *text.* lustrine, lutestring
люстрирование *leath.* lustring
лютеин *bioch.* lutein
лютеинизация luteinization
лютеома *med.* luteoma
лютеостерон *bioch.* luteosterone
лютеостимулирующий *med.* luteinizing
лютеотропный luteotrop(h)ic
лютеранин, лютеранский *rel.* Lutheran
лютеранство *rel.* Lutheranism
лютеций *chem.* lutecium, Lu
лютик *bot.* buttercup, crowfoot *(Ranunculus)*
лютиковидный *bot.* ranunculoid
лютиковые *n. pl. bot.* buttercup family *(Ranunculaceae)*
лютит lutite, lutyte
лютнист lute player
лютня *mus.* lute
люфа, люффа *bot.* 1. dishcloth/rag gourd, luffa *(Luffa cylindrica);* 2. vegetable sponge, luffa(h)
люфт *tech.* backlash, (free) play, creep; gap, clearance
~, **боковый** side play
~, **выбирать** take up backlash/play
~, **осевой** end/axial play
люцерна *bot.* alfalfa, medic *(Medicago)*
люцерноутомление *agr.* lucerne sickness *(of soil)*
Люцифер Lucifer, Satan
люцифераза *bioch.* luciferase
лючина *naut.* hatch cover/board
люэс *med.* syphilis
люэтический syphilitic
ля *mus.* A; la
~ **бемоль** A flat
~ **диез** A sharp

лягерстремия *bot.* grape myrtle *(Lagerstroemia)*
лягодон *zool.* robin *(Lagodon rhomboides)*
лягушечник *bot.* frogbit, frog's bit *(Hydrocharis)*
лягушечный *zool.* anuran
лягушка 1. *zool.* frog *(Rana);* 2. *tech.* (draw) tongs; *el.* frog, cable grip
~, зажимная wire grip
лягушкосемянный *bot.* batrachospermous
ляда *min.* folding/trap door
лядвенец *bot.* deervetch *(Lotus)*
ляллеманция *bot.* lallemantia *(Lallemantia)*
лямбда lambda
лямбдиаграмма lambdagram
лямблия *biol.* lamblia, giardia
лямблиоз *med.* giardiasis, lambliasis
лямина *geol.* lamina
ляминария *bot.* blade kelp, sea tangle *(Laminaria)*
лямка (shoulder) strap
~, нагрудная chest strap
~, парашютная parachute strap
~, поясная waist strap
ляпис *chem.* argentic/silver nitrate, argentum nitricum, caustic silver, lunar caustic
ляпис-лазурь lapis-lazuli
ляпсус blunder, lapsus, slip
лярва larva, ghost, spectre, spirit
лятилямина *zool.* latilamina
лях *hist.* Pole
ляшка *hist.* Polish (woman)
лятский *hist.* Polish

М, м

маар *geol.* 1. maar, embryonic volcano, volcano embryo; 2. explosion/gas vent; 3. crater/explosion lake
мавзолей mausoleum
мавр Moor
мавританка 1. Moor (woman); 2. Moresque
мавританский 1. Moorish; *arch.* Moresque; 2. Mauritanian
маг 1. *hist.* Magus; Magian; 2. magician, wizard; theurgist
магазин 1. store, shop; 2. magazine *(in firearms);* 3. depot, store, warehouse; 4. *mach.* hopper; dispenser; box; stack *(of punch cards);* 5. periodical, magazine
~, антикварный antique store
~ бакалейных товаров grocery
~ бакалейных товаров с продажей мяса combination store, *Am.* grocery
~, барабанный ammununition drum
~ без продавцов self-service store
~, галантерейный variety store
~ гастрономических товаров, ~, гастрономический delicatessen store, provision shop, fine-food store
~ готового платья ready-to-wear store
~, декадный decade box
~, дисковый 1. disk magazine; 2. cartridge drum
~ для новобрачных bridal shop
~ емкостей capacitance box
~, затуханий attenuation box
~ индуктивностей inductance box
~, коробчатый box magazine
~ мелких товаров variety store, five-and-dime store, five-and-ten-cent store
~, мясной butcher shop; *Am.* meat market
~ на дому у торговца parlor shop
~, обувный shoe store
~ перфокарт card hopper, input magazine
~, писчебумажный stationery shop
~ подержанных вещей *Am.* secondhand store
~ проводимостей conductance box
~, продовольственный без продажи свежего мяса dry-groceries store
~ продовольственных товаров food shop/ *Am.* store
~ разной одежды family clothing store
~, рыбный fish market
~ рыбных и других морских продуктов sea food market
~ с прилавками counter-service shop
~ самообслуживания self-service store
~ самообслуживания для автовладельцев *Am.* drive-in store
~ со смешанным ассортиментом *Am.* general (merchandise group) store
~ сопротивлений resistance box
~, торгующий по сниженным ценам *econ.* discounter
~, торгующий по телефонным заказам *Am.* telephone supermarket
~, трубчатый tubular magazine
~, универсальный (department) store, general shop
~ челноков *tech.* shutter magazine
~ шунтов shunt box
магазинирование *geol.* stoping; shrinkage
~, наступающее advancing shrinkage
~, отступающее retreating shrinkage
~, полное block shrinkage
~, частичное skeleton shrinkage
магазинный magazine *(attr.)*
~ная винтовка repeating rifle
~ная коробка magazine *(of firearm)*
магалебка *bot.* mahaleb cherry *(Padus/Prunus mahaleb)*
магараджа maharaja(h)
магарани maharanee
маггемит *geol.* maghemite
магдебургский Magdeburg
~кие полушария Magdeburg hemispheres
магистерство 1. master's degree; 2. *hist.* magistracy
магистр 1. holder of master's degree; 2. master's degree; 3. *hist.* head of knightly/monastic order

~, **великий** grand master
~ **искусств** Master of Arts
магистраль 1. main/arterial road, highway; 2. *el. com.* trunk (line), main (line), pipe-line
~, **адресная** *comp.* address line
~, **балластная** ballast main
~, **внутригородская** urban highway
~, **водопроводная** water main
~, **водяная** water main
~, **воздушная** air(-pressure) manifold/line/main/duct/pipe/route
~, **впускная** induction main
~, **газопроводная** gas main/trunk
~ **данных** data highway
~ **движения** traffic artery
~, **дифферентная** trimming main
~, **дорожная** arterial road; main line
~, **дренажная** drain collecting main
~, **железнодорожная** main (railroad) line
~, **кабельная** cable trunk
~, **командная** *comp.* instruction/command line
~, **кольцевая** ring main
~, **масляная** oil line
~, **нагнетающая** pressure/feed line
~, **напорная** pressure/delivery line
~, **осветительная** lighting trunk
~, **откачивающая** scavenge line
~, **паровая** steam main, steam supply line
~, **питающая** supply main/line
~, **пожарная** fire main
~, **распределительная** distributing manifold
~, **сигнальная** *comp.* signal/control line
~, **силовая** *el.* power trunk
~, **сливная** drain line/pipe; return line
~, **смазочная** lubrication manifold
~, **топливная** fuel line, fuelling main
~, **транспортная** thoroughfare
~, **числовая** *comp.* number line
магистральный main, arterial
~**ная дорога** main/trunk road
~**ная линия** main line, trunk
магистрант candidate for the master's degree
магистрат 1. city/town council; 2. magistrate, magistracy
магистратский magisterial; municipal
магистратура magistrature; magistracy; *coll.* body of magistrates
магический magic(al); theurgic(al)
~ **круг** vicious circle
магия magic; theurgy
~, **черная** black magic; diabolism
магма *geol.* magma
~, **активная** aggressive magma
~, **бедная/богатая летучими** dry/wet magma
~, **вторичная** submagma, hybrid magma
~, **жидкая** liquid/molten magma
~, **застывшая** chilled magma
~, **кислая** acid/granitic magma
~ **кратера** crateral magma
~, **материнская** parent magma
~, **первичная** primary/primordial magma
~, **ультраосновная** ultra-basic magma
магматический *geol.* magmatic, plutonic, pyrogenic
магматогенный magma(to)gene
магнавольт two-stage rotary amplifier
магналий magnalium, aluminium-magnesium alloy
магнамицин *pharm.* magnamycin, carbomycin
магнат 1. *hist.* member of Upper House of Diet in Poland/Hungary; 2. magnate, captain of industry; *Am.* baron, tycoon
магнезит *miner.* magnesite, magnesium carbonate
~, **студенистый** gelatinous magnesite
магнезия magnesia, magnesium carbonate
~, **белая** magnesia alba
~, **водная** magnesia hydroxide
~, **жженая** calcin(at)ed/light magnesia, dead-burned magnesia, magnesium oxide, magnesia levis
~, **сернокислая** magnesium sulfate, Epsom salt
магнесин *el.* magnesyn, permanent-magnet rotor synchro
магнетизм 1. magnetism, magnetic effect; 2. *phys.* magnetics
~, **естественный** natural/spontaneous magnetism
~, **земной** terrestrial/Earth's magnetism
~, **индуцированный/наведенный** induced magnetism
~, **остаточный** remanent/residual/retentive/remnant magnetism, retentivity, magnetic remanence
~, **связанный/скоростной** condensed magnetism
~, **собственный** intrinsic magnetism, retentivity
~ **южного/северного магнитного полюса** red/blue magnetism
магнетит *geol.* magnetite, magnetic iron ore, lodestone, loadstone
~, **пылевидный/твердый** pulverized/solid magnetite
магнетитный, магнетитовый magnetitic
магнетический magnetic
магнето *tech.* magneto
~, **пусковое** starting/booster magneto
магнетограф *phys.* magnetograph
магнетометр *phys.* magnetometer
магнетон *phys.* magneton
~ **Бора/боровский** Bohr/electronic magneton
~, **ядерный** nuclear/proton magneton
магнетопроводимость magnetoconductivity
магнетохимия magnetochemistry
магнетоэлектрический magnetoelectric
магнетрон *electr.* magnetron, ionic centrifuge
~ **бегущий волны** travel(l)ing wave magnetron, TW magnetron
~, **блочный** packaged magnetron
~, **двухцокольный** double-ended magnetron
~, **импульсный** pulsed magnetron
~, **многокамерный/многорезонаторный** multicavity magnetron
~, **непрерывной генерации** continuous-wave magnetron, CW magnetron

~, обращенный inverted magnetron
~, одноцокольный single-end magnetron
~ с разрезным анодом split-anode magnetron
~ типа беличьего колеса squirrel-cage magnetron
~, щелевой slot magnetron
магний *chem.* magnesium, Mg
~, бензойнокислый magnesium benzoate
~, виннокислый magnesium tartrate
~, основный magnesium hydroxide
~, салициловокислый magnesium salicylate
~, сернокислый magnesium sulfate, epsomite, epsom salt
магнийкатионирование magnesium zeolite softening process
магнико magnico
магнистор hybrid-loop-core saturable reactor, magnistor
магнит magnet
~, анализирующий energy-analyzing magnet, magnetic analyser
~, безжелезный air-core magnet
~, возбуждающий actuating/field magnet
~ вращения rotary magnet
~, глазной *med.* ophthalmic/Haab's magnet
~, девиационный *nav.* corrector magnet
~, демпфирующий damping magnet
~, естественный natural/native magnet, lodestone, loadstone
~, направляющий control magnet
~ отбоя *rail.* release magnet
~, отклоняющий deflecting magnet
~, пластинчатый laminated magnet
~, подковообразный horseshoe-/C-magnet
~ подъема vertical magnet
~, подъемный lifting magnet
~, постоянный permanent magnet
~, природный natural/native magnet, loadstone, lodestone
~, С-образный C-magnet
~ со слоистым сердечником laminated magnet
~, стержневой bar magnet
~ стоп-стержня scram-rod magnet
~, тактовый cadence magnet, time tapper
~, удерживающий (rod-)holding magnet
~, ферритовый ceramic magnet
~, элементарный molecular magnet
~, эталонный reference/standard magnet
~, юстировочный positioning magnet
магнитик molecular magnet
магнитно-возмущенный magnetically disturbed
магнитно-гидродинамический magnetohydrodynamic
магнитно-дипольный magnetic-dipole
магнитно-жесткий magnetically hard
магнитно-мягкий magnetically soft
магнитность magnetization, magnetizability; magnetism
~, постоянная permanent magnetization
магнитный magnetic
~ железняк magnetite, loadstone, lodestone
магнитоакустический magnetoacoustic
магнитогазодинамика magnetogasdynamics
магнитогидродинамика magnetohydrodynamics
магнитограмма magnetogram
магнитограф magnetograph
магнитодержатель magnet carrier/cradle/holder/support
магнитодинамический magnetodynamic(al)
магнитожесткий magnetically hard
магнитоионный magneto-ionic
магнитокардиограмма magnetocardiogram
магнитола combined radio and tape recorder
магнитология magnetology
магнитометр magnetometer
~, вибрационный vibrating-coil magnetometer
~, генераторный generating magnetometer
~, квантовый quantum-mechanical magnetometer
~, магнитомеханический/магнитостатический moving-magnet magnetometer
~, магнитонасыщенный saturable-core magnetometer
~, морской marine magnetometer
~, оптический optical/absorption magnetometer
~, поднимаемый *av.* airborne magnetometer
~, скважинный borehole magnetometer
~, электронно-парамагнитно-резонансный electron paramagnetic resonance magnetometer, EPR magnetometer
~, ядерно-магнитно-резонансный nuclear magnetic resonance (NMR) magnetometer
~, ядерно-прецессионный spin-precession magnetometer
магнитометрия magnetometry
магнитооптика magneto-optics
магнитопауза magnetopause
магнитоплазма magnetoplasma
магнитопровод magnetic circuit; core; yoke
~, броневого типа shell-type magnetic circuit/core
~, двухстержневой two-leg/-limb core
~, неразветвленный single-path magnetic circuit
~, разветвленный multipath magnetic circuit
~, сложный composite magnetic circuit
~ стержневого типа core-type magnetic circuit, C-type magnetic circuit
~ стыковой конструкции butt-end core
~, шихтованный laminated/imbricated/interleaved core
магнитопроводимость magnetoconductivity
магниторазведка magnetic prospecting
магниторезистивность magnetoresistivity
магниториум device for the study of magnetic field of the Earth
магнитоскоп magnetoscope
магнитосопротивление magnetoresistance
магнитостабильный magnetically stable

магнитостриктор *geol.* magnetostrictive drill
магнитострикция magnetostriction
~, объемная bulk/volume magnetostriction
магнитосфера magnetosphere
магнитотепловой magnetothermal
магнитотерапия magnetotherapy
магнитотропизм magnetotropism
магнитотурбулентность magnetoturbulence
магнитоуловитель magnetic detector
магнитоупругий magnetoelastic
магнитофильм tape recording
магнитофон magnetic tape recorder (audio)tape recorder
~, двухдорожечный dual-/double-track tape recorder
~, кассетный cassette magnetic tape device, casette tape recorder
~, монофонический mono tape recorder
~ на дисках magnetic disk recorder
магнитофорез magnetophoresis
магнитохимия magnetochemistry
магнит-уничтожитель *naut.* magnetic compensator
~креновой девиации компаса heeling adjuster/corrector
магнитоэлектрический magnetoelectric
магнификат *rel.* Magnificat, Song of the Virgin Mary
магнолиевые *n. pl. bot.* magnolia family *(Magnoliaceae)*
магнолит magnesite flooring cement
магнолия *bot.* magnolia *(Magnolia)*
маго, магот *zool.* magot, Barbary ape *(Macaca sylvana)*
Магомет Mohammed
магометанин Mohammedan
магометанка Mohammedan woman
магометанский Mohammedan
магометанство Mohammedanism
магония *bot.* mahonia *(Mahonia)*
мадаполам, мадеполам *text.* madapollam *(cloth)*
мадемуазель, мадмуазель mademoiselle
мадера Madeira (*wine)*
мадистор madistor
мадия *bot.* madia, tarweed *(Madia)*
мадлен *geol.* Magdalenian
мадонна madonna
мадресе madrassa(h)
мадрепора madrepore
мадригал *mus.* madrigal
мадригальный madrigalian
мадронья *bot.* (Pacific) madrone *(Arbutus menziesii)*
мадьяр, мадьярский Magyar
мадьярка Magyar (woman)
маевка *hist.* illegal May-day meeting *(in prerevolutionary Russia)*
маестро *see* **маэстро**
мажара long rack waggon
мажор *mus.* major key
мажоранта *math.* majorant (series); majorant function; numerical majorant
мажорантный *math.* majorizing, majorant, dominant
мажордом major-domo
мажорирование majorization
мажорировать majorize, dominate
мажоритарный *math.* (based on) majority
мажорный *mus.* major
маз I additional stake *(at cards)*
маз II mace, billiard cue
мазание greasing, lubricating, oiling; anointing; daubing
мазанка mud hut, cottage of daub and wattle
маздеизм Mazdaism, Mazdeism
маздеистский Mazdean
мазевый (prepared) with ointment
мазеобразный unctuous; salvelike
мазер *phys.* maser, Microware Amplification by Stimulated Emission of Radiation
~ на аммиаке ammonia maser
~ на твердом теле solid-state maser
~, резонаторный cavity maser
мазница 1. tar vessel/container; 2. small brush
мазок *med.* smear; dab, touch, stroke of the paintbrush
~, вагинальный vaginal smear
мазонит masonite *(shielding material)*
мазохизм *med.* masochism
мазохист masochist
мазур Mazur(ian)
мазурий *chem.* masurium, Ma
мазурка I maz(o)urka *(dance and music)*
мазурка II Mazur(ian) woman
мазут mazut, black oil, petroleum residue; fuel/residual/boiler oil
~, высокосернистый high-sulphur residual oil
~, малосернистый low-sulphur residual oil
~, топочный furnace/fuel oil
~, топочный судовой marine residual/bunker oil
~, флотский bunker fuel
мазутохозяйство fuel oil handling system
мазутохранилище fuel oil storage tank
мазь 1. ointment, unguentum, salve, liniment(um), embrocation; 2. grease
~, автоловая lubricating oil/ointment
~ Вилькинсона Wilkinson's ointment
~, гидрокортизоновая hydrocortisone (acetate) ointment
~, глазная eye ointment, oculentum
~, жидкая liniment(um), embrocation
~, колесная axle grease
~, летучая ammoniated soap liniment
~ Листера Lister's ointment
~ от ожогов burn ointment
~, серая ртутная blue ointment, diluted mercurial ointment
~, феноловая carbolic acid ointment, phenol

ointment
~, цинковая zinc ointment
маис *bot.* maize, Indian corn *(Zea mais)*
маисовый maize *(attr.)*
~вая каша polenta, hominy
майдан burial mound
майка I T-/sports/football shirt; undershirt
майка II *zool.* oil beetle *(Meloe)*
майна I *zool.* myna(h) *(Acridotheres)*
майна II *naut.* heave ho!
майник *bot.* bead-ruby *(Maianthemum)*
майолика majolica
майонез *cul.* mayonnaise
майор major
майоран *bot.* marjoram *(Majorana)*
майорат *law* 1. right of primogeniture; 2. entailed estate, entail
майоратство *law* right of primogeniture
майордом major-domo
майорство majorate
майский May(-Day)
~ жук *zool.* May beetle *(Melolontha)*
майшфильтр *tech.* mash filter
майя I maya; Maya(n)
майя II *text.* fine cotton fabric
мак *bot.* poppy *(Papaver); coll.* poppy seed
макадам macadam
макадамия *bot.* Australian/Queensland nut *(Macadamia ternifolia)*
макак(а) *zool.* macaque *(Macaca, Macacus)*
макао *zool.* macaw *(Ara)*
макаронизм *lit.* macaronism
макассар macassar oil
маквис *bot.* macchia, macchie, maqui
македонец, македонский Macedonian
македонка Macedonian (woman)
македонянин *hist.* Macedonian
макелий *geol.* Macalian
макет 1. model, pattern, dummy, mock-up; 2. *el., comp.* breadbord construction/model ; 3. *print.* layout
~, архитектурный scale model
~ в натуральную величину full-scale/-size model/mock-up
~ вертолета helicopter mock-up
~ данных data layout
~ для испытаний на совместимость compatibility test unit
~ местности terrain model
~ перфокарты punch(ed) card design
~ печати printer layout
~, полупромышленный brassboard model
~, технологический engineering mock-up
~ топливного канала simulated fuel channel
~, эскизный *print.* visual, rough
макетирование breadboarding; prototyping; mocking-up
макетировать breadboard, mock up, make a model
макетчик modelist, maker of models/dummies
маки *zool.* macaco, macaque *(Macacus)*
маки 1. *coll., bot.* maquis, evergreen shrub thicket; 2. maquis, French underground forces during WW II
макиавеллевский, макиавеллистический Machiavellian
макиавеллизм Machiavellism; Machiavellianism
макиавеллист Machiavellian
макинтош mac(k)intosh, waterproof overcoat
макитра earthenware pot
маклейя *bot.* plume poppy, bocconia *(Macleya, Bocconia)*
маклер broker; stock broker
~, биржевой share/stock broker
~, вексельный bill broker
~, внебиржевой curb broker
~, денежный money dealer
~ на денежном рынке money broker
~, неофициальный coulissier *(Fr.)*
~ по операциям с недвижимостью real estate broker
~ по операциям с ценными бумагами securities broker
~ по приисканию грузов loading broker
~ по фрахтованию судов ship(ping) broker
~, страховой insurance broker
~, таможенный customhouse broker
~, фондовый share broker
маклерство brokerage
маклюра *bot.* Osage orange *(Maclura)*
маковидный papaveraceous, poppylike
маковка 1. poppyhead; 2. *cul.* sweetmeat of poppy seed and sugar
маковый *bot.* papaverous, papaveraceous; *n. pl.* poppy family *(Papaveraceae)*
макоцветковый, макоцветный *bot.* poppy-flowered
макрандриевый *bot.* macrandrous
макрандрий *bot.* macrander, macrandre
макрелещука *zool.* saury *(Scombresox saurus)*
макрель *zool.* mackerel *(Scomber)*
макроагрегат macroaggregate
макробиотика *biol.* macrobiotics
макробактерия macrobacterium, megabacterium
макробиоз macrobiosis, longevity
макробласт macroblast
макровозмущение broadscale perturbation
макровызов *comp.* macrocall
макровязкость macroviscosity
макрогамета *biol.* macrogamete, megagamete
макрогаметангий *bot.* macrogametangium
макрогаметофит *bot.* macrogametophyte
макрогамия *biol.* macrogamy
макрогастрия macrogastria, megalogastria, megastria
макрогенератор *comp.* macrocommand generator
макрогенитосомия *med.* macrogenitosomia precox, Pellizzi's syndrome
макроглобулинемия *med.* macroglobulin(a)emia
~ Вальденстрема Waldeström's macroglobulin(a)emia
макрогонидий, макрогонидия *bot.* macro-

gonidium
макрография macrography, megalography
макрогруппа macrogroup
макродактилия *med.* macrodactyly, macrodactylism, macrodactylia, megalodactylia, dactylomegaly
макродемография macrodemography
макродентизм *med.* macrodontia, macrodontism, macrodentism
макродинамика macrodynamics
~ **плазмы** continuous plasma dynamics
макрозооспора *bot., zool.* macrozoospore
макроиндикатор high-speed indicator
макроинструкция *comp.* macroinstruction
~, **описательная** descriptive macro, declarative macroinstruction
макроисследование macroexamination
~ **на растворимость** macrosolubility test
макрокардия *med.* macrocardia, megalocardia, cardiomegaly
макроклимат macroclimate
макрокоманда *comp.* macroinstruction, macrocommand, *pl.* macros
~ **времени дня** time of day macro
~ **генерации данных** data generation macroinstruction
~, **описательная** declarative macroinstruction
~, **системная** system macroinstruction
макроконидий, макроконидия *bot.* macroconidium
макрокосм macrocosm
макрокосмический macrocosmic(al)
макроконцентрация macroscopic concentration
макрокосмос macrocosm(os)
макрокристалл macrocrystal
макрокристаллический macrocrystalline
макромания *med.* macromania, megalomania
макромастия *med.* macromastia, macromazia
макромер(а) macromere
макромерный macromeric, macromeral
макромир macroworld, macrocosm
макромицеты *pl.* macromycetes
макромолекула *chem.* macromolecule
макронапряжение macrostress
макронеоднородность macroheterogeneity
макронихия *med.* macronychia, megalonychosis
макрообъявление *comp.* macrodeclaration
макрооператор *comp.* macroinstruction
~, **описательный** declarative macroinstruction, descriptive macro
макроопределение *comp.* macrodefinition
макроось *cryst.* macroaxis
макропалеонтология macropal(a)eontology, megalopal(a)eontology
макропаразит macroparasite
макроперекрытие *geol.* powerful thurst
макропеременная macrovariable
макроплазия *med.* macroplasia, giantism
макропланктон macroplankton
макропод *zool.* paradise fish *(Macropodus, Polynemus paradiseus)*
макрополость macrocavity, macrovoid
макропора macropore
макропористость macroporosity
макропрограммирование *comp.* macroprogramming
макропроталлий macroprothallium
макропроцесс large-scale process
макропсия *med.* macropsia, megalopia, megalopsia
макрорадикал macroradical
макрорадиография radiomacrography; radiomacrograph
макрорасширение macroexpansion
макрорельеф macrorelief
макрореология macrorheology
макросейсм macroseism
макросигма macrosigmoid
макросиноптический macrosynoptic
макросинтагматика macrosyntagmatics
макросистема macroscopic system, macrosystem
макросклера macrosclere, megasclere
макросклереида macrosclereid
макроскопический macroscopic, megascopic
макроскопия macroscopy
макроснимок macro(photo)graph
макросомия macrosomia, megasomia
макроспора macrospore, megaspore
макроспорогенез macrosporogenesis
макростатистика macrostatistics
макростилоспора macrostylospore
макроструктура macroscopic structure, macrostructure
макросхема macrocircuit
макросъемка ultra-close-up photography
макротагмема macrotagmeme
макротвердость macrohardness
макротомия *bot.* prophet flower *(Macrotomia)*
макротрещина macrofissure
макротурбулентность macroturbulence
макроуровень macrolevel
макрофаг macrophage, macrophagocyte
~, **оседлый** stable macrophage, histiocyte
~, **свободный** free macrophage, polyblast
~, **соединительнотканный** conective-tissue macrophage
макрофит macrophyte
макрофлора macrophytic plants, macroflora
макрофотография photomacrography; photomacrograph
макрофотоснимок photomacrograph
макрохейлит *med.* macrochilia
макрохейрия *med.* macrochiria, chiromegaly
макрохет *zool.* macrotrichium
макрохимия macrochemistry
макроцефал macrocephalus
макроцефалия *med.* macrocephalia, macrocephaly, macroencephalia, macrocephalism
макроцист(а) macrocyst
макроцит macrocyte, macroerythrocyte
~, **астроглиальный** macrogliocyte

макроцитаза *bioch.* macrocytase
макроцитоз *med.* macrocytosis, macrocythemia, megalocytosis, megalocythemia
макрочастица macroscopic particle, macroparticle
макрошлиф macrosection
макроэволюция macroevolution
макроэкономика macroeconimics
макроэлемент 1. *tech.* macrocell; *comp.* macro; 2. *biol.* macronutrient element, macroelement
макроязык *comp.* macrolanguage
макроячейка *comp.* macrocell
макрурус *zool.* marlinespike *(Macrourus bairdii)*
мак-самосейка *bot.* field/corn poppy *(Papaver rhoeas)*
максвелл *el.* maxwell
~ **виток** maxwell-turn
максвеллметр maxwellmeter, fluxmeter
максвелловский *el.* Maxwell's
максилла *anat.* maxilla, superior/upper maxilla, upper jaw
максима 1. maxim, general principle; adage; 2. *mus.* maxim
максимализм *hist.* maximalism
максималист *hist.* maximalist
максимальный maximal, maximum, largest, peak; highest possible, greatest
максимизация, максимизирование maximization
~, **векторная** vectorial maximization
~ **полезности** utility maximization
максимизировать maximize
максимин *game th.* maximin
максиминимизирующий maximinimizing
максиминный maximin
~**ная полезность** maximin utility
максимум maximum; peak
~, **абсолютный** alltime high, *math.* absolute/global maximum
~, **безусловный** unconstrained maximum
~, **брегговский** Bragg peak/maximum
~ **Брейта-Вигнера** Breit-Wigner peak
~, **второстепенный** submaximum
~, **вулканический** peak eruption
~, **глобальный** overall maximum
~, **горбообразний** hump
~, **гравитационный** gravity high/maximum
~, **локальный** relative/local maximum
~ **максиморум** peak-peak
~ **нагрузки** peak load
~, **обособленный** isolated maximum
~ **острый** sharp/acute maximum
~, **побочный** subordinate maximum
~ **поглощения** maximum absorption
~ **правой** curve peak
~, **размытый** diffuse maximum, broad peak
~, **резонансный** resonance peak
~, **температурный** thermal peak
~, **тепловой** thermal maximum
~, **условный** constrained maximum
макси-юбка maxiskirt
макула macule, macula
макулатура *print.* spoiled sheet(s); waste paper
макулоцеребральный maculocerebral
макулярный macular, maculate
макуха oil cake/meal
макушечный apical, umbonal
макушка apex, top; umbo, peak; vertex
~, **конечная/терминальная** terminal apex
~, **спирогирная** spirogyrate umbo
макьявеллизм Machiavellism
малага Malaga (wine)
малагасиец, малагасийский Malagasy
малагасийка Malagasy (woman)
малаец, малайский Malay(an)
малайка Malay (woman)
малакия onanism
малакогамия malacogamy
малакозоология, малакология malacology
малакон *miner.* malacon
малакоплакия *med.* malakoplakia, malacoplakia
малакофилия *bot.* malacophily
маланизм malanism
малатдегидрогеназа malate dehydrogenase, malic acid dehydrogenase
малахит *miner.* malachite, green copper, mountain green
~, **кремнистый** chrysocolla, copper pitch ore
~, **свинцовый** plumbomalachite
~, **синий** blue carbonate of copper, blue malachite, azurite
~ **содержащий свинец** lead-bearing malachite, bleimalachite
малахитовый malachite
~**вая зелень** malachite green
малахра *bot.* wild ochra/okra *(Malachra)*
малеиновый *chem.* maleic
малигнизация *med.* malignancy
малиец Mali
малийка Mali (woman)
малина *bot.* raspberry *(Rubus)*
малиновка I *zool.* robin, redbreast *(Erithacus rubecula)*
малиновка II raspberry flavo(u)red vodka
малица deerskin parka
малка *tech.* bevel (square/protractor), protractor
~, **острая** close/shut bevel
~ **под 45°** mitre square
~, **разводная/тупая** open/standing bevel
~, **сводная** close/shut bevel
~, **шпангоутная** frame bevel
малковка beveling
маллеин *vet.* mallein
маллеинизация mallein inoculation, malleinization
маллотус *bot.* kamala *(Mallotus philippinensis)*
малоазиатский, малоазийский Asia Minor *(attr.)*
малоактивный low-activity
малоалкогольный of low alcoholic content
малоамперный low-amperage/-current

малобелковый protein-deficient
малоберцово-пяточный *anat.* fibulocalcaneal
малоберцовый *anat.* fibular, peroneal
~вая кость fibula, peroneal bone
малоблагоприятный scarcely favo(u)rable/conducive, unfavo(u)rable
маловажность insignificance
маловаттный low-wattage
маловероятность unlikeliness
маловероятный unlikely, hardly probable; improbable
маловеский, маловесный light-weight; light, short-weight
маловетвистый, маловетвящийся *bot.* pauciramose, subramous
маловодие *med.* oligoamnios, oligohydramnios
маловодность lack/deficiency of water; shallowness
маловодный with little water, shallow; dry *(of land)*
маловодье 1. lack/shortage of water; 2. shallowness, low water level
маловыгодный of small profit, not remunerative, unprofitable
маловыразительный not very expressive
маловысотный low-level
маловязкий of low viscosity
малогабаритный small(-scale/-size), miniature, compact; midget
малоголовый microcephalous
малограмотность crudity, ignorance; semiliteracy
малограмотный crude, ignorant, unskilled; half-educated, semiliterate
малогумусный, малогумусовый low-humic, oligohumous
малодаровитый of meager gifts/talents, poorly endowed
малодейственный, малодействительный ineffective
малодоказательный not convincing, lacking strength
малодостоверный improbable, unlikely; not wellfounded
малодоступность limited availability
малодоступный not readily available
малодоходный not very profitable/remunerative
малодушный fainthearted, chicken-hearted, white-livered; weak-kneed/-spirited
малоемкий small-capacity/-sized, low-capacity
малоемкостный *el.* low-/small-capacitance
маложирный of low fat content
малозаселенный sparsely populated
малоземелье, малоземельность shortage of (arable) land
малозлокачественный *med.* semi-malignant
малозначительность insignificance, unimportance, little importance
малозольный of low ash content, low-ash
малозубчатый, малозубый oligodontous
малоизведанный little surveyed, hardly investigated
малоизогнутый slightly bent
малоизученный insufficiently studied/explored
малоимущий indigent, needy, poor
малоиндукционный low-induction
малоинерционный fast-/high-/quick-response, low-inertia
малоинтеллигентный not very refined/cultured
малоинтенсивный low-intensity, of low intensity
малоисследованный scantily explored/studied
малокалиберный small-caliber; small-bore
малокалорийный low-calorie, low in calories
малоквалифицированный semiskilled, barely qualified
малокомнатный having few rooms, consisting of few rooms
малокомпетентный not possessing full rights; having insufficient competence
малокорневой *bot.* oligorhizous
малокровие *med.* an(a)emia, olig(ha)emia, hypovol(a)emia
~, злокачественное malignant/pernicious an(a)emia
~, местное local an(a)emia, isch(a)emia, hypo(a)emia
~ с серповидными эритроцитами sickle-cell an(a)emia
малокровный an(a)emic, isch(a)emic, olig(a)emic
малокультурный lacking culture
малолетний *law* juvenile, minor; *n.* infant; juvenile, minor
малолетство *law* infancy; nonage; minority
малолистный *bot.* oligophyllous, paucifolious
малолитражный of small (cylinder) capacity
~ автомобиль midget car, car with small cylinder capacity
малолюдность, малолюдство poor attendance *(of meetings, etc.)* scarcity of people
маломасштабный small-scale
маломерка small size *(of shoes/dresses)*
маломерный of small size, undersized
малометражный not spacious/roomy
маломощный *tech.* low-power(ed)/-capacity/duty; *econ.* poor, lacking power; *geol.* shallow *(of soil)*
маланадежный undependable, not very reliable
малонаселенный sparsely/thinly populated, underpopulated
малонат *chem.* malonate
малоновый *chem.* malonic
малообеспеченный of scanty means, not sufficiently provided for
малообитаемый sparsely populated, having few inhabitants
малооблачный slightly overcast
малоблученный slightly irradiated
малообогащенный insufficiently/slightly enriched
малообоснованный ill-founded
малоопытный inexperienced, having little experience, semiskilled

малоосвоенный insufficiently assimilated/mastered
~ **район** underdeveloped area
малоостистый *bot.* few-awned
малопитательный not very nutritious
малоплодовый *agr.* bearing little fruit
малоплодородный low-productive
малоподвижный slow-moving
~ **образ жизни** sedentary life
малопоместный *hist.* yeomanly
малопонятный abstruse; enigmatic(al); obscure
малоприбыльный bringing little profit
малопригодный of little use
малопродуктивный, малопроизводительный not efficient/fruitful/productive; unproductive, of little efficiency; *agr.* low-yielding/output
малопроцентный low-grade *(of ore)*
малорадиоактивный weakly radioactive
малоразвитость backwardness, underdevelopment
малоразвитый undeveloped, underdeveloped, half-educated, backward, late in growth/development
малоразрядный low-discharge
малореберный rare-costate
малорентабельный not sufficiently remunerative/profitable
малорослость *med.* nanism, dwarfishness, dwarfism
малорослый dwarfish, undersized, stunted
малоросс, малороссиянин, малорус Little Russian, Ukrainian
малорусский, малороссийский Little Russian, Ukrainian
малосведущий ill-informed
малосемянный *bot.* oligospermous, few-seeded
малосильный *tech.* low-powered
малоскоростной low-speed/-velocity
малосольный slightly/inadequately/fresh-salted
малостойкий *chem.* not stable
малость smallness, littleness, paucity
малосущественный not substantial, immaterial
малотиражный of small circulation
малотоннажный of small tonnage; light
малотычинковый *bot.* oligandrous, oligostemonous
малоумный feeble-minded
малоурожайный *agr.* low-yield
малоуспевающий hardly succeeding; *n. pl. educ.* underachievers *(of students)*
малоустойчивый unstable, unsteady
малохлебный *agr.* yielding small amount of grain
малоцветковый *bot.* few-flowered, pauciflorous, oliganthous
малоценный of little value, of inferior quality
малочисленность paucity; small number
малочник *tech.* bevel gauge/tester
малочувствительный of little sensitivity
малошумный, малошумовый quiet, low-noise; noiseless
малоэффективный not very effective
малхит *geol.* malchite, dioritic lamprophyre
малый I small, little
~, **бесконечно** infinitesimal
~, **пренебрежимо** negligible
~ **ход** *naut.* low speed (ahead)
~**лая скорость** *rail.* slow freight, low speed
~**лые формы** *theatre* short plays, variety entertainment
малый II fellow, chap, guy, boy, lad
~, **бесшабашный** happy-go-lucky fellow
мальбек Malbeck *(grapes and wine)*
мальва *bot.* mallow *(Malva)*
мальвазия malmsey *(wine)*
мальваструм *bot.* false mallow *(Malvastrum)*
мальвовый *bot.* malvaceous; *n. pl.* mallow family *(Malvaceae)*
мальвообразный *bot.* malvaceous
мальвоцветковый *bot.* mallow-flowered
малькольмия *bot.* Malcolm stock *(Malcolmia)*
мальм *geol.* Malm; malm, marl
мальма *zool.* Dolly Varden trout *(Salvelinus malma)*
мальпигия *bot.* malpighia *(Malpighia)*
мальта maltha, viscid bitumen, malthite
мальтаза *biochem.* maltase
мальтиец, мальтийский Maltese
мальтодекстрин maltodextrin
мальтретировать maltreat, ill-treat
мальтузианец Malthusian
мальтузианский Malthusian
мальтузианство Malthusianism
мальчишник stag party *(on eve of wedding)*
маляр house painter
малярийный malarial, malarious, aguish
маляриолог malariologist
маляриология malariology
малярить paint, whitewash
малярия *med.* malaria, ague
~, **врожденная** congenital malaria
~, **злокачественная** pernicious malaria
~, **искусственно привитая** impf malaria, malaria inoculata
~, **летне-осенняя** estivoautumnal malaria
~, **острая** acute malaria
~, **трехдневная** tertian fever, benign tertian malaria
~, **тропическая** tropical malaria, malignant tertian malaria
~, **хроническая** chronic malaria, malarial cachexia
~, **четырехдневная** quartan fever/malaria
малярная paint store/locker
малярный *anat.* malar
маляция *med.* malacia, malacoma, malacosis
~, **миелопластическая** myeloplastic malacia, osteogenesis imperfecta
мамалыга hominy, polenta
мамба *zool.* mamba *(Dendroaspis angusticeps)*
мамелюк *hist.* Mameluke
маметр Machmeter, Mach indicator

мамиллярия *bot.* mamillaria *(Mamillaria)*
маммея *bot.* mamey *(Mammea)*
маммология mammology
маммотомия *med.* mammotomy, mastotomy
маммотропин *med.* mammotrop(h)in, prolactin
мамона Mammon
мамонт mammoth *(Elephas primigenius)*
мамонтовый mammoth *(attr.)*
~вое дерево *bot.* giant sequoia *(Sequoia gigantea)*
мамура *bot.* arctic bramble *(Rubus arcticus)*
манакин *zool.* manakin *(Pipridae)*
мангал(ка) mangal, brazier (pan)
манганат *chem.* manganate
манганит *miner.* manganite, brown manganese ore
мангифера, манго *bot.* mango *(Mangifera)*
мангольд *bot.* chard, leaf beet *(Beta vulgaris cicla)*
мангостан *bot.* mangosteen *(Garcinia mangostana)*
мангрова mangrove forest
мангуст(а) *zool.* mongoose *(Herpestes)*
мандант mandant, mandatory
мандарин I *hist.* mandarin, Chinese official
мандарин II *bot.* mandarin orange *(Citrus nobilis/reticulata)*
мандаринка *zool.* mandarin duck *(Aix galericulata)*
мандаринник mandarin(e) plantation
мандаринный, мандариновый mandarin-orange
мандаринский mandarin *(attr.)*
мандат 1. warrant; 2. *polit.* term of office, mandate; credentials
мандатарий *law* mandatary, mandatory, bailee
мандатный mandatory
~ная комиссия credentials committee
~ная территория mandated territory
манджур Manchurian
манджурка Manchurian (woman)
манделатрацемаза *med.* mandelate racemase
манджурозавр mandschurosaur *(Mandschurosaurus)*
мандибула *anat.* mandibula, mandible, mandibulum, inferior maxilla
мандибулофарингеальный *anat.* mandibulo-pharyngeal
мандибулэктомия *surg.* mandibulectomy
мандибулярный mandibular
мандола *mus.* mandola
мандолина *mus.* mandolin(e)
мандрагора *bot.* mandrake, mandragora *(Mandragora officinalis)*
мандрен *med.* mandrin
мандрил *zool.* mandrill *(Mandrillus sphinx)*
маневр manoeuvre
~, боевой combat manoeuvre
~, быстрый snap/fast manoeuvre
~ для уклонения от атаки avoiding movement, evasive action
~, заатмосферный exo-atmospheric manoeuvre
~, ложный feint/false manoeuvre
~, предпосадочный let-down procedure(s)
~, противоракетный antimissile (evasive) manoeuvre
~, путевой yaw manoeuvre
~, резкий violent/abrupt manoeuvre
~ сближения closing/rendezvous manoeuvre
~, синергетический synergetic turn
~ спуска descent manoeuvre
~ стыковки docking manoeuvre
~ торможения braking/slowdown/deceleration manoeuvre
~ уклонения avoiding/evasive action/ manoeuvre, escape manoeuvre
~ ухода escape/breakaway manoeuvre
маневренность manoeuvrability
~ в условиях невесомости zero-g manoeuvrability
~ по скорости speed changing
маневренный manoeuvrable
~ная война war of movement
маневрирование 1. manoeuvring, *Am.* maneuvering; 2. *rail.* switching, shunting
~, боковое cross-range manoeuvring
~, групповое formation manoeuvring
~ при сближении rendezvous manoeuvring
~ с целью уклонения evasive steering
~, учебное practice manoeuvres
маневрировать 1. manoeuvre, *Am.* maneuver; 2. *rail.* shunt, switch
маневровый *rail.* shunting, switching
манеж 1. riding school/house/hall; manege; 2. ring, arena *(in a circus);* 3. playpen, baby's pen
~, детский creeping pen
манекен (tailor's) dummy, mannequin; model
~ с приводкой wired dummy
манекенщик mannequin producer
манера I style, manner, way; *pl.* manners
манера II stencil; printing block
манерка *mil.* canteen, mess tin
манжет(а) 1. cuff; 2. *tech.* collar, sleeve, packing/sealing ring; seal; gland; cup (packing); flap
~ кабеля cable guard
~, кожаная leather cup
~ коллектора sleeve
~, огнеупорная *foundry* brick cup
~ поршня piston cup
~, прорезиненная/резинотканевая rubber-fabric cup
~, уплотняющая sealing collar/boot, packing seal
манжетка I *print.* book band
манжетка II *bot.* lady's mantle *(Alchemilla)*
манжур, манжурский Manchurian
манжурка Manchurian (woman)
маниакально-гебефренический manic-hebephrenic
маниакально-депрессивный manic-depressive
~ психоз affective insanity
маниакально-меланхолический manic-melan-

cholic
маниакальный manic, maniacal
манизм manism
манилла *bot.* abaca (banana) *(Musa textilis)*
маниловщина *lit.* Manilovism, futile daydreaming and smug inactivity
маниок(а), маниот 1. *bot.* cassava, manioc *(Manihot);* 2. cassava starch
манипул *hist.* maniple
манипулирование manipulation, handling; operation; *rad.* keying
~ графическими объектами graphics manipulation
~, дистанционное remote manipulation/handling
~ изображениями на экране screen manipulation
~, полудистанционное semiremote handling/manipulation
~, ручное handling, manipulation
манипулировать handle, manipulate; operate, key
манипулятор manipulator; operator; handler; *rad.* manipulating key; mechanical arms
~, безрельсовый off-track manipulator
~ ближнего действия handler
~ ввода-вывода input-output handler
~, гидравлический hydraulic manipulator, hydroman
~, координатный *nucl.* rectilinear manipulator
~, копирующий (master-)slave manipulator
~, рельсовый track-riding manipulator
~ с дистанционным управлением remote-control manipulator
~, сварочный welding positioner
~, телеграфный keyer
~, универсальный multipurpose/general-purpose manipulator
манипуляция 1. handling, manipulation; operation; *rad.* keying; 2. *fig., polit.* intrigue, machination
~, амплитудная on-off keying, CW telegraphy
~, анодная anode; *Am.* plate keying
~ в первичной цепи primary keying
~ на сетку grid keying
~ нулями zero manipulation
~ обеими руками *med.* conjoined/bimanual manipulation
~, фазовая phase-shift keying, PSK
~, частотная frequency-shift keying, FSK
манифест 1. manifesto; 2. *naut.* (ship's) manifest
~, грузовой shipping manifest
~, судовой summary of bills of lading, manifest
манифестант *polit.* demonstrator
манифестация *polit.* demonstration
манифестировать demonstrate, take part in a demonstration; show, manifest
маниха neap/dead tide
манихеизм Manicheism, Manicheanism
манихей Manich(a)ean, Manichee
манихейский Manich(a)ean
манихейство Manich(a)eism, Manich(a)eanism
манишка *naut.* (diver's) corselet
мания *med.* mania
~ величия megalomania
~ выражения hypermania
~, периодическая periodic/recurrent mania
~ преследования delirium of persecution
~, преходящая transitory mania
манко *fin.* cash shortage; leakage, shortage
манна 1. *bot.* manna lichen/lecanora *(Lecanora esculenta);* 2. *rel.* manna; 3. tree sap
маннан *chem.* mannan
манная *cul.* semolina
манник *bot.* manna grass (*Glyceria)*
маннит *chem.* mannitol
маннитоль mannitol, mannite
манноза *chem.* mannose
мановакуумметр (compound) pressure-and-vacuum gauge
манограф recording manometer, recording pressure gauge, manograph
манодетандер pressure reducing valve, pressure regulator
манок hunter's whistle, decoy, lure
манокриометр manocryometer
манометр manometer, pressure gauge
~, анероидный aneroid/dial manometer
~, бесшкальный transmitting pressure gauge, non-indicating pressure gauge
~, вакуумный vacuum gauge/manometer, vacuummeter
~, водяной water-pressure gauge, water manometer
~, воздушный air-pressure gauge
~, вязкостный viscosity pressure gauge
~, газоразрядный gas discharge pressure gauge
~, дефомарционный elastic element pressure gauge
~, дистанционный distant-/remote-reading manometer
~, дифференциальный differential manometer, differential pressure gauge
~, жидкостный liquid (-column/-filled) manometer
~, ионизационный ionization gauge/manometer
~, Кнудсена radiometric pressure gauge
~, кольцевой ring-balance manometer
~, крешерный crusher pressure gauge
~ Ленгмюра Langmuir/viscosity pressure gauge
~, магнитострикционный induction-type pressure gauge
~ Мак-Леода McLeod gauge
~, масляный oil pressure gauge
~, мембранный diaphragm pressure gauge
~, механический elastic element pressure gauge, mechanical manometer
~, наддува boost gauge
~ Пирани Pirani gauge, hot-wire gauge
~ пневматической системы compressed air gauge

~, поплавковый float-type manometer
~, пружинный spring-element pressure gauge
~ прямого действия direct-acting pressure gauge
~, пьезоэлектрический piezoelectric pressure gauge
~, радиометрический radiometer manometer/gauge, radiometric gauge
~, ртутный mercurial/mercury pressure gauge, mercurial/mercury manometer
~ с вялой мембраной slack-diaphragm pressure gauge
~ с жесткой мембраной stiff-diaphragm pressure gauge
~ с открытой трубкой open-tube manometer
~ с перевернутым колоколом inverted bell manometer
~ с силовой компенсацией force-balance pressure gauge
~, самопишущий recording pressure gauge, pressure recorder
~, сильфонный bellows pressure gauge
~ сопротивления resistance pressure gauge
~, стеклянный glass-tube manometer
~, стрелочный indicating pressure gauge
~, тепловой hot-wire gauge, Pirani gauge
~, термоэлектрический thermocouple gauge
~, технический industrial pressure gauge
~, топливный fuel-pressure gauge, fuel indicator
~, U-образный U-tube manometer
~ Филлипса Phillips ion gauge
~, чашечный cup/well-type manometer
~, шинный tyre-pressure gauge
~, щитовой panel-mounted pressure gauge
~, электроразрядный electric-discharge manometer
манометрический manometric(al)
манометрия manometry
~, ушная ear manometry
манометровый pressure-gauge, manometer *(attr.)*
манор manor (estate)
манориальный manorial
маностат pressure control instrument, manostat
мансарда attic, garret, mansard
мансардный, мансардовый mansard(ed), attic *(attr.)*
манси, мансиец Mansi
мансийка Mansi (woman)
мансонеллез *med.* mansonelliasis, Ozzard's filariasis
мантелет *mil.* mant(e)let
мантель light summer overcoat
мантийный *geol.* mantle; *zool.* pallial
мантикула manticula
мантилья mantilla
мантисса *math.* mantissa
~ числа fractional/significant part, mantissa
мантия 1. cloak, mantle; 2. *geol.* mantle; 3. *paleont., zool.* pallium
~ земли Earth's mantle
~, новая neopallium
манто (lady's) coat, mantle, manteau
мантуанец, мантуанский Mantuan
мануал *mus.* manual
манубриум manubrium
манул *zool.* manul, Pallas's cat *(Felis manul)*
манускрипт manuscript
мануфактура 1. manufactory; textile mill; 2. fabrics, textiles
мануфактурист 1. owner of a textile mill; 2. dealer in textiles
манхеттен Manhattan cocktail
манцинелла *bot.* manchineel *(Hippomane mancinella)*
манципация *law* mancipation
манчестерец *econ.* Manchesterist
манчестерский Manchestrian; Mancunian
манчестерство *econ.* Manchesterism
манчжур *see* **маньчжур**
маншинелла *bot.* manchineel *(Hippomane mancinella)*
маньеризм *art* mannerism
маньерист *art* mannerist
маньчжур, маньчжурский Manchurian
маньчжурка Manchurian (woman)
маньяк maniac
маньячеcкий maniac(al)
маори 1. Maori; 2. *zool.* rainbow fish *(Ophthalmolepis lineolatus)*
мара *zool.* mara *(Dolichotis patagonica)*
марабу *zool.* marabou(t) *(Leptoptilus)*
марабут Marabout, Mohammedan hermit
маразм *med.* marasmus
~, старческий senile marasmus, dotage
маразматический *med.* marasmic, marantic
марака(с) maraca, dance-band instrument
марал *zool.* red deer, maral, Siberian stag *(Cervus elaphus)*
мараловодство maral breeding
маральник enclosure/nursery for marals
маран marrano
марант *bot.* arrowroot *(Maranta)*
марантический *med.* marantic, marasmic, wasting away
марантовые *n. pl. bot.* arrowroot family *(Marantaceae)*
мараскин *cul.* maraschino (liqueur)
маратор deck/lintel ring
мараттиевые *n. pl. bot.* marattians *(Marattiaceae)*
маратхи, мараты *pl.* Maratha, Mahratta, Sanskritic language
марафон Marathon
марафонец *sport* marathon runner
марафонский Marathonian
~ бег *sport* marathon race
марашка mottle, slur, smear
марганец *chem.* manganese, manganum, Mn
~, виннокислый manganese/manganous tartrate
~, иодистый manganese iodide

~, молочнокислый manganese lactate
~, роданистый red manganese
~, сернистый manganous sulphide
~, сернокислый manganese/manganous sulfate
~, углекислый manganese carbonate
~, фосфорнокислый manganese phosphate
марганецсодержащий manganiferous
марганцевый, марганцовый manganic, manganesian
марганцовистокислый manganate (of)
марганцовистый manganous, manganic, manganiferous
марганцовка potassium permanganate, manganese crystals
марганцово-кислый permanganate (of)
~ калий potassium permanganate
марганцово-цинковый manganese-zinc
маргарин margarine
маргарит *miner.* margarite, brittle mica
~, калиевый potash margarite
~, натриевый soda margarite, ephesite
маргаритка *bot.* daisy *(Bellis)*
маргаритколистный *bot.* daisy-leaved
маргиналия marginal note; *pl.* sidenotes, marginals, marginalia
маргинальный marginal
марго margin *(of document/page)*
марево 1. (superior) mirage, looming; 2. (heat) haze
маревые *n. pl. bot.* goosefoot family *(Chenopodiaceae)*
маревый boggy, marshy, swampy
мареграф tide/depth gauge, marigraph
маремма maremma, seaside marshland
марена *bot.* madder *(Rubia)*
маренго *text.* pepper-and-salt fabric
мареновидный madderlike, rubioid
мареновые *n. pl. bot.* madder family *(Rubiaceae)*
мареограмма marigram, tidal curve
мареограф tide/depth gauge, marigraph
~, береговой/прибрежный off-shore depth gauge
~, поплавковый float depth gauge
маржа equity
марзан *print.* clump, furniture, bearer
~, ростовый type-high clump
марзанка *bot.* squinancy *(Asperula cynanchica)*
мари *pl.* mari, larch forests with peat moss litter
марикультура sea farming, mariculture
маримба *mus.* marimba, African xylophone
маримонда *zool.* marimonda *(Ateles belzebuth)*
марина *art* seascape
маринад *cul.* marinade, pickles
маринадный pickling
маринист I *art* painter of seascapes, marine painter
маринист II advocate of the domination of the seas
маринка *zool.* marinka (fish) *(Schizothorax)*
мариновать pickle, marinate
марионетка marionette, puppet
марионетный, марионеточный puppet *(attr.)*
марипоза *bot.* maripoza (lily) *(Calochortus)*
марихуана marijuana, marihuana, mariguana, maria-juana
марка I 1. (postage) stamp; 2. mark *(monetary unit in Germany);* markka *(monetary unit in Finland);* 3. mark, sign; brand; make; 4. chip, counter; 5. grade, sort
~, базисная base mark
~, визирная sighting mark/target; index/floating mark
~, высшая best brand
~, гербовая revenue stamp
~, грузовая load mark, free-board mark
~, заводская manufacture mark
~, известная name brand, well-known brand
~, измерительная floating/gliding/wandering/ *geod.* adjusting mark
~ магнитного пункта magnetic-station mark
~ масла oil specification
~ на лотлине fathom mark
~, нивелирная fixed point, bench mark
~ осадки *nav.* draft/displacement mark
~, почтовая stamp
~ производителя manufacturer's brand
~, производственная maker's label
~ стали steel quality
~ судна, грузовая *nav.* Plimsoll mark
~ топлива fuel specification, grade of fuel
~, торговая/фабричная trade/emblem/manufacture mark
~, триангуляционная triangulation mark
~ углубления *nav.* draft/displacement mark
~ угля rank of coal
~, установочная *opt.* focussing/ sighting mark
~, фирменная label
~ цемента cement brand
~, юбилейная commemorative stamp
~ки глубин на лотлине marks and deeps
марка II *hist.* mark *(medieval German territorial unit)*
марказит *miner.* marcasite, white iron pyrite, cellular/radiated/hepatic pyrite
~, гребенчатый cockscomb pyrite
~, лучистый radiated marcasite
маркграф *hist.* margrave
маркграфина *hist.* margravine
маркграфский margraval
маркер *ling.* marker
маркер I 1. *comp.* label, tag, mark; 2. marker; *nucl.* tracer
~, аэронавигационный air(-navigation) marker
~ блока block mark(er)
~, красящий dye marker
~ конца ending character, end-of-block mark
~ на ленте tape label/mark/tag
~ начала beginning mark/character
~, начальный на ленте beginning reel/tape label
~, неотобранный unselected marker

~, **радиолокационный** radar marker, ramark
~, **тележный** *agr.* cart-type marker
маркер II billiard marker/scorer
маркерный marking
~ **(радио)маяк** marker beacon
маркерство marking, scoring
маркетинг marketing
маркиз marquis, marquess
маркиза I marchioness; marchesa; marquise
маркиза II sun blind; awning; markee
маркизет *text.* marquisette
маркирование marking, stamping; stencil(l)ing; label(l)ing
маркировать label, stencil, identify, mark; indicate, stamp; lay/mark out
маркировка 1. marking, stamping, identification; label(l)ing, tagging; branding; 2. grading; brand, mark, (symbols of) grades; 3. time-mark
~ **бракованной дорожки** *comp.* defective track flagging
~ **линии** tracing
~, **монтажная** match marking
~ **о бережном обращении** precaution mark
~, **радиоактивная** radioactive labelling/tagging
маркировщик marker; marking machine
маркитант *hist.* camp follower/-hawker, sutler
марковость *math.* Markov behaviour
марксизм Marxism
марксизм-ленинизм Marxism-Leninism
марксист Marxist
марксистский Marxist, Marxian
марксистско-ленинский Marxist-Leninist
маркшейдер mining/mine surveyor
маркшейдерия mine surveying
марлерезка gauze cutter
марлин *zool.* marlin (*Makaira)*
марлинь *nav.* marline
марля gauze; cheesecloth
~, **кровоостанавливающая** h(a)emostatic gauze
~, **нестерилизованная** nonsterilized/plain/absorbent/unmedicated gauze
~, **синтомициновая** chloramphenicol/chloromycetin gauze
мармелад *cul.* candied fruit jelly
мармит steam(-heated) table, food warmer
мармитница warmer container
мармозетка *zool.* plumed marmoset *(Callithrix)*
марморация marbleization
марморировать *tech.* marble
мародер marauder, pillager, looter
мародерство marauding, pillage, looting
марокен, марокин *text.* marocain (fabric); *leath.* morocco (leather)
марокканец, марокканский Moroccan
марокканка Moroccan (woman)
мар(р)он *hist.* fugitive slave, maroon
Марс *astr., myth.* Mars
марс *nav.* (round-)top
~, **верхний** masthead top
марсала Marsala (wine)
марса-топенант *naut.* toplift
марселин *text.* marceline *(fabric)*
марселиновый (made of) marceline
марсель *naut.* square topsail
~, **прямой** topsail
марсельеза Marseillaise
марсельский Marseilles
марсианин, марсианский Martian
марсилиевые *n. pl. bot.* pepperwort family *(Marciliaceae)*
марсилия *bot.* pepperwort *(Marsilia, Marsilea)*
марсовед Martianologist
марсовой, марсовый *nav.* round-top; *n.* topman
марсоход Mars rover, Mars research vehicle
марсоцентрический *astr.* areocentric
марстохрон *econ.* marstochron
марсупиализация *surg.* marsupialization, pocketing, infolding
мартагон *bot.* martagon lily *(Lilium martagon)*
мартен *tech.* 1. open-hearth furnace; 2. open-hearth steel
мартеновский *tech.* open-hearth, Martin
~ **процесс** Martin process
~ **цех** open-hearth plant/shop/section
мартенсит *tech.* martensite
~, **игольчатый** needle-shaped martensite
~, **мелкоигольчатый** fine-needled martensite
мартенситный martensitic
мартингал *math.* martingale
мартин-гик *naut.* dolphin striker, martingale
мартинизм Martinism
мартинист Martinist; Mason
мартиния *bot.* devil's claws, martynia *(Martynia)*
мартин-штаг *naut.* martingale
мартиролог(ий) *eccl.* martyrology
мартирологический martyrological
мартышка I *zool.* marmoset *(Cercopithecus)*
мартышка II *zool.* tern
марчеван *constr.* forepole
марципан *cul.* marzipan; marchpane
марш I march
~, **похоронный** funeral march
~, **торжественный** march in review
марш II *mil.* forward! march!
~, **бегом** at the double
~, **шагом** forward/quick march
марш III marsh(land)
~, **осушенный** reclaimed marsh
марш IV *constr.* flight (of stairs)
маршал *mil.* marshal
маршаллизация *hist.* Marshall Plan, European Recovery Program
маршальство rank of marshal, marshalship, marshalcy
маршанция *bot.* marchantia *(Marchantia)*
маршевик *mil.* reservist called into active duty
маршевый march *(attr.)*
~ **двигатель** cruise engine
марширование, маршировка marching

маршрут route, itinerary, track; course, heading
~ аварийной посадки emergency landing pattern
~, авиационный air route
~, аэрофотосъемочный flight strip, air survey (photographic) strip, aerial traverse
~, внерейсовый non-scheduled route
~, внетрассовый off-airway route
~, внутренний domestic route
~, воздушный air route
~, двухточечный *comp.* point-to-point route
~ захода на посадку holding/approach pattern
~, маневровый *rail.* shunting route
~, отправительский *rail.* exit route
~ подхода inbound/approach route
~ поиска search pattern; *comp.* retrieval route
~ полета route of flight, flight course
~ приема *rail.* entrance/incoming route
~, разрешенный службой движения air-traffic control clearance track
~ с малой густотой перевозок thin route
~ с небольшой плотностью движения low-traffic density route
~, технологический flow route
~ челночных полетов shuttle route
маршрутизация, маршрутирование routing; running traffic in blocks
~ без задержки *comp.* hot-potato routing
маршрутный enroute
~ автобус express bus
~ная карта flow/process/route/travel chart, process sheet, from-to chart
~ное такси fixed-route taxi
марь I mari, larch peatmoss bog forest
марь II *bot.* goosefoot *(Chenopodium)*
маска mask; shield; screen, covering; face guard
~, быстронадевающаяся quick-donning mask
~, вертикальная vertical screen
~, дорожная road screen
~, дыхательная face-mask
~, защитная face guard, protective mask, shield
~, кадровая *telev.* framing mask
~, кислородная oxygen mask
~, марлевая для лица gauze face veil
~, наркозная an(a)esthesia mask
~, сварочная welder's/welding helmet
~, теневая *telev.* aperture/shadow mask
маска-противогаз smoke mask
маскарад masquerade, masked ball, fancy(-dress) ball
маскаре, маскарэ tidal surge, maskaret
маскирование masking; disguising, concealment; *mil.* camouflaging
маскировать(ся) *mil.* camouflage
маскировка 1. masking; screening; veneering; disguise; 2. *mil.* camouflage
~, акустическая acoustic blanket
~ войск field camouflage
~, радиолокационная radar deception
~, световая darkening, blackout
маскировочный masking, disguising; camouflaging
масковидный *bot.* masked, masklike
маскообразный masklike
маскосеть camouflage net
маскхалат *mil.* camouflage cloak
масленица *eccl.* Shrovetide; carnival
масленка 1. *tech.* lubricator, oil grease cup, oiler, greaser, oilcan; lubricating nipple; 2. pasteboard paper; 3. butterdish
~, автоматическая self-acting lubricator
~, буксовая axle lubricating box
~, игольчатая needle lubricator
~, капельная drop lubricator/oiler, drop-fed oil cup
~, колпачковая screw pressure lubricator, Stauffer's lubricator, grease cup
~ под шприц nipple
~, пружинная spring grease cup
~, ручная наливная (hand) oilcan
~, фитильная wick oiler
~ цепной передачи chain oiler
~ Штауфера Stauffer's lubricator, screw pressure lubricator
масленок *bot.* butter mushroom *(Boletus luteus)*
масленщик greaser, lubricator, oiler
масленый buttered; oily, oiled; greasy
~ная неделя *eccl.* Shrovetide
маслина *bot.* olive (tree) *(Olea)*
маслинные, маслиновые *n. pl. bot.* olive family; oil-bearing plants *(Oleaceae)*
маслить butter; oil; grease
масличность oil content/percentage
масличный oil-bearing/-yielding, oleiferous, oleaginous *n. pl.* oil-bearing plants *(Oleaceae)*
~ные растения oleaginous/oil plants
масличный olive
~ная ветва olive branch
масло 1. oil; 2. butter; 3. *art.* oil (paints); oil painting
~, авиационное aviation/aero oil
~, автотракторное motor oil
~, аирное calamus oil
~, анисовое oil of anise
~, арахисовое peanut/arachis oil
~, вазелиновое petrolatum, petroleum jelly, vaseline oil
~, веретенное spindle oil
~, всесезонное *auto.* multigrade oil, allweather oil
~, вязкое thick oil
~, газовое gas oil
~, гарное crude naphtha
~, гвоздичное oil of cloves
~, горное naphtha, petroleum
~, горчичное mustard oil
~, грецкого ореха walnut oil
~, густое thick/viscous oil
~, дегтярное tar oil
~, дизельное diesel oil

~, диспергирующее dispersion oil
~, дистиллятное distillate oil
~ для волочения *met.* drawing oil
~, животное animal oil
~, жидкое thin oil
~, закалочное quenching oil
~, земляного ореха peanut/arachis oil
~, имбирное ginger oil
~, каменноугольное coal-tar oil
~, камфорное camphorated oil, camphor liniment
~, канифольное resin oil
~, карболовое carbolic/carbolized oil
~, касторовое castor oil
~, кедровое cedar(-wood) oil, oleum cedrelae
~, кожевенное currier's oil
~, кокосовое coco(a)nut oil
~, компаундированное compounded/blended oil
~, конденсаторное condenser/capacitor oil
~, конопляное hempseed/cannabic oil
~, кукурузное corn oil
~, кунжутное oil of sesame, benne/teel/gingili oil
~, купоросное oil of vitriol
~, лавровое oil of bay/myrcia
~, легкое light/thin oil
~, летучее volatile oil
~, льняное linseed/flaxseed oil
~, льняное, лаковое varnish linseed oil
~, маковое poppy(-seed) oil
~, машинное machine/engine/lubricating oil
~, миндальное almond oil
~, минеральное coal/mineral/petroleum oil
~, можжевеловое oil of savin, oil of juniper berries
~, морозостойкое non-freezable oil
~, моторное motor oil
~, мятное oil of peppermint
~, непищевое inedible oil
~, неочищенное, нерафинированное crude oil
~, неразжиженное undiluted oil
~, нефтяное petroleum oil
~, оливковое olive/sweet oil
~, ореховое hazelnut oil
~, осевое axle oil
~, осветительное illuminating oil
~, остаточное residual oil
~, отработавшее used oil
~, отстойное sump oil
~, пальмовое palm oil
~, парафиновое paraffin oil
~, перилловое perilla oil
~, печатное lithographic oil
~, пищевое edible oil
~, поглотительное absorbent/absorption oil
~, подвижное thin oil
~, подсолнечное sunflower(-seed) oil
~, прованское olive/sweet oil
~, противокоррозийное slushing oil
~, разжиженное diluted oil
~, рапсовое oil of rapeseed, rapeseed oil
~, растительное vegetable oil
~, рафинированное refined oil
~, реактивное jet-engine oil
~, регенерированное refiltered oil
~, резиновое rubber oil
~, розовое oil/attar of rose, oleum rosae
~ с присадками additive oil
~, сандаловое santal/sandalwood oil
~, сафлоровое safflower oil
~, сезамовое oil of sesame, benne/teel/gingili oil
~, сивушное fusel oil, amyl alcohol
~, скипидарное turpentine essence
~, скрубберное wash oil
~, сланцевое shale oil
~, смазочное lubricant, lubricating/lube oil, luboil
~, смазочное фильтрованное filter stock
~, смолистое resinous oil, oleoresin
~, смоляное resin/tar oil
~, соевое soybean oil, soya(-bean) oil
~, соляровое diesel fuel oil; straw/solar oil
~, сосновое pine oil
~, специально приготовленное tailor-made oil
~, сырое crude oil
~, терпентинное oil of turpentine
~, технологическое process oil
~, тминное oil of caraway
~, тунговое tung oil
~, турбинное turbine oil
~, тыквенное pupmkin seed oil
~, тяжелое (из нефти) pyronaphtha
~, углеводородное hydrocarbon oil
~, укропное dill(seed) oil
~, уплотненное bodied oil
~, фенольное carbolic oil
~, фенхелевое oil of fennel
~, флотационное flotation oil
~, хлопковое cottonseed oil
~, цилиндровое cylinder oil
~, цитрусовое citrus oil
~, швейное sewing(-machine) oil
~, эвкалиптовое oil of eucalyptus
~, электроизоляционное electrical insulating oil
~, эмульсионное cutting oil
~, эфирное ethereal/essential oil
маслобак oil tank/reservoir
маслобензостойкость oil-and-petrol resistance
маслобитня 1. creamery; 2. oil mill
маслобой 1. buttermaker, butterman; oiler, oil manufacturer; 2. oil mill worker
маслобойка 1. churn; 2. oil press/mill
маслобойный завод 1. creamery; 2. oil mill
масловка oiling
масловлагоотделитель oil-and-moisture separator
масловлагоуловитель oil-and-moisture trap

маслогасимый oil-quenched
маслодел butter/oil manufacturer
маслоделие butter manufacture/making; oil manufacturing
маслодельный butter/oil manufacturing
~ **завод** creamery
маслодобывающий oil-extracting/-extraction
маслоемкость oil absorption
масложировой fat-and-oil
маслозавод 1. creamery, butter dairy; 2. (vegetable) oil-mill
маслозаправщик oil dispenser, oil(-servicing) truck
маслоизготовитель buttermaking machine, churn
масло-какао cacao butter
маслокран oil cock
масломер oil(-level) gauge, oil level indicator, oil meter; dipstick
масломанометр oil pressure indicator/gauge
масломерный oil measuring
маслонагреватель oil heater
маслонасос oil pump
маслонепроницаемый, маслостойкий oil-proof/-resistant, oiltight, oilproof, greaseproof
маслообразный oillike, oily, greaselike
маслообразующий olefiant, oil-forming
маслоотделитель oil/lubricant separator
маслоотражатель oil (splash) guard, oil deflector/baffle
маслоотстойник (oil) sump
маслоочиститель oil filter/strainer/cleaner/purifier/clarifier
~, **полнопоточный** full-flow oil cleaner
маслоочистка oil cleaning/refining
маслопоглощаемость oil absorbing capacity, oil absorption
маслоподогреватель oil heater
маслоприемник oil pick-up, oil header
маслопровод oil conduit/line/duct; oil pipe(line), oleoduct
~ **всасывающей линии** feed (oil) pipe
~ **откачивающей линии** scavenge (oil) pipe
~, **отсасывающий** oil-return line
маслопроводный oil-piping
маслопропитанный oil-impregnated
маслопрочность oil resistance
маслорадиатор oil cooler (assembly)
маслоразбрызгиватель oil splasher/thrower
маслораспределитель oil distributor/dispenser
маслораспылитель oil splasher
~, **капиллярный** capillary lubricator
маслорастворимый oil-soluble
маслородный olefiant
маслосборник oil catcher/collector/pan, (oil) drip pan
маслосистема lubrication/oil system
маслослив oil drain
маслостойкий oil-resistant/-proof, oilproof, oiltight
маслоуказатель oil gauge/indicator
~, **трубчатый** tubular oil indicator
маслоулавливатель, маслоуловитель oil catcher/trap/collector/interceptor
маслоуплотнительный oil-seal(ing)
маслофильтр oil filter
маслохранилище oil storage tank
маслюк *zool.* gunnel *(Centronotus)*
маслянистость oiliness, greasiness, unctuousness
маслянистый unctuous, oleaginous, emulsive; buttery; butyraceous
масляница *eccl.* Shrovetide
масляно-бальзамический butyrobalsamic
масляно-воздушный air-oil
маслянокислый butyric, butyrate (of)
масляно-пневматический oleo-pneumatic
масляный oily; buttery; oleic; butyric, butyrous
~**ная кислота** butyric acid
~**ная краска** oil paint
масон Freemason, Mason
масонский Masonic
масонство Freemasonry, Masonry
масса 1. mass; body *(in various senses) pl. polit.* the masses; 2. *el.* frame; earth, *Am.* chassis ground; *av.* aeroplane structure; 3. pulp, paste, compound, substance, matter; *geol.* aggregation
~, **автохтонная** autochthonous/edaphic mass
~, **активная** active mass/material
~ **античастицы** antiparticle mass
~ **атома** atomic mass
~, **бесконечная** infinite mass
~, **бифилетическая основная** biphyletic groundmass
~, **бумажная** paper-making stock
~, **вальцовая** *print.* composition
~, **водная** water body/mass
~, **воздушная** air mass
~, **возмущающая** disturbing mass
~, **войлокоподобная** felted mass
~, **волокнистая** *cell.* pulp
~, **вращающаяся** gyrating/rotating/spinning mass
~, **голая** bare mass
~ **двигателя, удельная** engine mass-to-thrust ratio
~, **денежная** *econ.* monetary/money stock, money supply
~ **для заполнения швов** jointing paste
~, **древесная** wood pulp
~, **древесная рафинированная** refined groundwood pulp
~, **единичная** unit mass
~, **желатино-глицериновая** *print.* composition
~, **жильная** *geol.* vein filling/matter/rock/stuff, lode rock
~, **завалки** *foundry* charge size
~, **заливочная** *el.* filling compound, filling-in paste
~, **затравочная** bare mass
~, **избыточная** excess/extra mass
~, **изоляционная** insulating paste/compound

~, инертная, инерциальная/инерционная intertial/inert mass; stationary/steady mass
~, интрузивная/инъецированная intrusive/injected mass
~, кабельная cable compound
~, караваеобразная bun-/loaf-shaped mass
~, карборундовая carborundum plastic
~, колеблющаяся oscillating/seismic/sloshing mass
~, комковатая основная lumpy groundmass
~, конечная final mass
~ конструкции structural mass
~, крахмальная starch composition
~, леточная *met.* tap-hole clay/loam
~, литьевая moulding material
~, материнская parent mass
~ мезона meson(ic) mass
~ молекулы molecular mass
~, молярная molar mass
~, набивная ramming/tamping mass
~, надвинутая overriding mass
~, надкритическая supercritial mass, above-critical mass
~, наследственная *law* mass of the succession
~, нечетная odd mass
~, нижележащая subjacent/underlying mass
~, обмазочная *met.* coating mixture
~, огнеупорная refractory body
~, основная groundmass, matrix, paste
~, относительная relative mass, mass ratio
~, относительная атомная atomic weight
~, относительная молекулярная molecular weight
~, отторженная detached mass
~, переменная variable mass
~ планеты planetary mass
~, пластическая plastic mass; plastic
~, плитовидная slab-like mass
~, подкритическая subcritical mass
~, поднимающаяся воздушная ascending/rising air mass
~, подрессоренная sprung mass/parts/weight
~, подстилающая subjacent/underlying mass
~ покоя (zero)rest mass, stationary mass, mass at rest
~ полезной нагрузки payload mass, useful load mass
~, полная total/gross mass
~ породы body of rock
~, постоянная stationary mass
~ по физической/химической шкале mass on physical/chemical scale
~, предельная critical mass
~, прессованная moulding compound/powder
~, приведенная reduced mass
~, присоединенная associated/additional/added/entrained mass
~, пропиточная impregnating compound
~ ракетного топлива propellant mass
~, расплавленная molten mass
~, расходуемая cousumable mass
~, рафинированная refined screenings
~, реакционная *chem.* stock
~, рудная lode rock/stuff, rock mass/bulk
~, сбросовая *geol.* block-faulting mass
~, сверхкритическая above-critical mass, supercritical mass
~, связущая binder/cementing mass
~, сосредоточенная localized mass
~, сплошная block mass, continuum
~, стартовая *space techn.* launching mass
~, твердая каловая *med.* scybalum, scybalous mass
~, текучая основная fluidal groundmass
~ тела body mass
~ топлива рабочая fuel as received
~, тяжелая gravitational mass
~ упаковки package mass
~, уплотняющая sealing material/mass
~, устойчивая steady/stationary/stable mass
~, фильтровальная filter stock
~, фосфорная friction compound
~, хромитовая chrome plastic
~, целлюлозная pulp
~ частицы particle mass
~, четная even mass
~, шлифовальная grinding paste/compound
~, электродная anode paste
~ электрона electron(ic) mass
~ ядра nuclear mass
~, ячеистая honeycomb mass, cellular mass
массаж massage
~, аппаратный mechanical massage
~ болезненых точек *med.* nerve-point massage
~ десен gingival massage
~ легкого pneumomassage
~ подводный underwater massage
~ сердца cardiac/chest massage
массажелечение massotherapy
массажист masseur
массажистка masseuse
массажный massage *(attr.)*
масс-анализатор mass-analyzer
массив 1. *geol.* massif; block, (solid) mass; expanse; 2. *comp.* array; 3. (data) file, data bank; body, group
~, автохтонный *geol.* terrane
~ адресов *comp.* address note-book, directory
~, бетонный mass concrete
~, буферный buffer pool
~, вулканический pile; volcanic cluster
~, горный massif, mountain mass
~ горных пород rock mass
~ данных *comp.* data array/file
~ данных, большой mass data
~ данных, обновлять update a file
~, двумерный two-dimensional array
~ документов document file/collection
~, древний old land (mass)
~, жилищный residential quarter/district, com-

plex/row of apartment buildings
~, **закладочный** *min.* filling mass
~, **интрузивный** typhon, boss, stock, plutonic plug
~ **информации** data file
~ **команд** *comp.* instruction array/file, command field
~ **компонентов** component file
~, **конический вулканический** conic(al) pile
~ **констант** constant pool
~, **континентальный** land mass
~, **кристаллический** crystalline rock mass
~, **куполообразный экструзивный** swelling dome
~, **лесной** tract of forest, large forest, forest expanse, big stand
~, **многомерный** multidimensional array
~, **нетронутый** unmined block
~, **неупорядоченный** *comp.* heap
~ **породы, ненарушенный** virgin rock mass
~, **срединный** median mass
~ **суши** land mass
~ **угля** coal massif/block
~ **указателей** *comp.* pointer array
~, **устойчивый** resistance mass
массивный massive; solid, compact; bulky, substantial
массикот *geol.* plumbic ocher, yellow lead, massicot
массирование I massaging, masseuring
массирование II *mil.* concentration, massing
массированный *mil.* concentrated, massed
массировать I massage, rub, knead
массировать II *mil.* concentrate, mass
массировка *mil.* massing, concentration
массовик *polit.* 1. political worker (among the masses); 2. organizer of popular cultural/recreational activities
массовка *polit.* 1. mass meeting; 2. propaganda leaflet/newspaper *(published in large numbers);* 3. *cin.* crowd scene
массовость mass character
массовый mass, bulk, wholesale; popular
~ **митинг** mass meeting
~**вая литература** popular literature
~**вые товары** staples, consumer goods
массомер *cell.* basis weight gauge
массомойка pulp washing machine
массообмен, массопередача mass transfer/exchange
масс-сепаратор mass-separator
масс-спектр mass spectrum
масс-спектрограф mass-spectrograph
масс-спектрометр mass-spectrometer/analyzer
~, **ионный** ion mass-spectrometer
~, **трохоидальный** trochotron, trochoidal mass-spectrometer
~, **цуговый** tandem mass-spectrometer
масс-спектроскопия mass-spectroscopy, mass spectroscopic work
масс-эквивалент mass equivalent
мастер 1. foreman; master (craftsman), skilled worker; overseer; 2. expert, master
~ **в забое** *min.* face deputy
~, **дамский** hair stylist
~ **золотых дел** goldsmith
~, **котельный** boiler maker
~, **монетный** minter
~, **оружейный** gunmaker, gunsmith; ordnance artificer
~ **по гальваноклише** electrotyper, electrotypist
~ **своего дела** good workman
~, **сапожный** shoemaker
~, **сменный** shift foreman
~, **сменный горный** shift attendant/foreman
~, **старший** head foreman
~, **тележный** wain wright
~, **тросовый** rope maker
~, **часовой** watchmaker
~, **шлюпочный** boatwright
мастерица 1. milliner; seamstress; 2. skilled worker; a good hand (at)
~, **кружевая** lace maker
мастер-копир master cam
мастер-модель master die model
мастеровать be a foreman, be master craftsman
мастеровой *n.* artisan, tradesman; workman, factory hand
мастерок trowel, float, darby
мастер-пуансон master punch
мастер-ремесленник master
мастерская *n.* workshop; workroom; studio; shop *(in factory)*
~, **авиаремонтная** aircraft repair shop
~, **авторемонтная** automobile/motor/car repair shop, motor workshop
~, **вулканизационная** tyre repair shop
~, **деревообделочная** woodworking shop
~, **инструментально-штамповая** tool-and-die shop
~, **компасная** compass loft
~, **кустарная** handicraft/artisan workshop
~, **механическая** machine shop
~, **парусная** sail loft
~, **переплетная** bindery
~, **плавучая** floating workshop, repair ship
~, **плотницкая** carpenter shop; shipwright's workshop
~, **пошивочная** tailor shop
~, **разметочная** marking shop
~, **ремонтная** repair shop
~, **сапожная починочная** shoe-repair shop
~, **скорняжная** skinnery
~, **слесарная** locksmith's workshop; *naut.* plumber's shop
~, **столярная** joiner's shop
~, **судоремонтная** ship repair facility
~, **такелажная** rigging loft/shop
~ **художника** artist's studio
~, **шиноремонтная** tyre repair shop

~, **шлюпочная** boat shop
~, **электротехническая** electrician's shop
мастерской masterly, skilled, skilful
~**кое произведение** masterpiece
мастер-станок master tool
мастерство (handi)craft, craftsmanship, workmanship; skill, mastery, masterhood
~, **высокое летное** expert airmanship
~ **высшего пилотажа** precision flying skill
мастер-штамп master-die
мастика 1. mastic, resin; paste, putty, cement; 2. floor polish
~, **бакелитовая** bakelite cement
~, **битумная** bitumen mastic, bituminous cement
~, **битумно-резиновая** bituminous-rubber cement
~, **герметизирующая** sealing compound
~ **для заполнения швов** joining paste
~ **для натирки полов** floor polish
~, **цементная** mastic cement
~, **цоколевочная** base cement
мастикатор *tech.* masticator
мастикация mastication
мастиковый mastic, resinous
~**вое дерево** *bot.* mastic tree *(Pistacia lentiscus)*
мастит 1. *med.* mastitis, mammitis, breast inflammation; 2. *vet.* garget
~, **гиперпластический** gargantuan mastitis
~, **гнойный** suppurative mastitis
~ **новорожденных** mastitis in the newborn
~, **острый** acute mastitis
~, **паренхиматозный** parenchymatous/glandular mastitis
~ **плазмацитоклеточный** plasmacell mastitis
~, **разлитой** diffuse mastitis
~, **раковый** carcinous mastitis
~, **субмаммарный** submammary/retromammary mastitis, paramastitis
~, **узелковый** nodular disease of the mamma, Tillaux's/Phocas/Schimmelbausch's disease
~, **флегмонозный** phlegmonous mastitis, broken/gathered breast
~, **юношеский** puerperal mastitis
маститость old age; venerableness
мастихин spatula
мастичный lutaceous, mastic, resinous
мастодиния *med.* mastodynia, mastalgia, mastadenitis, mazodynia, mammary neuralgia
мастодонзавр *paleont.* mastodonsaurian, mastodonsaurus
мастодонт mastodon
мастодонтовый mastodont(ic)
мастодонтообразный mastodontoid
мастоидит *med.* mastoiditis
~, **огнестрельный** bullet mastoiditis
~, **первичный** primary/Bezold's mastoiditis
~, **рецидивирующий** relapsing mastoiditis
мастоидотомия *med.* mastoidotomy, Wilde's incision
мастопатия *med.* mastopathy, mazopathy
~, **кистозная** mastopathia cystica, Copper's disease
мастопексия *med.* mastopexy, mazopexy
мастурбатор masturbator, onanist
мастурбация masturbation, self-abuse, onanism
мастурбировать masturbate
мастэктомия *med.* mastectomy, mammectomy
масть colo(u)r
масштаб scale
~ **аэроснимка** (aerial) photograph scale
~ **Бонжана** *nav.* Bonjean diagram/curves
~ **в данной точке** point scale
~ **в метрах** meter scale
~ **вероятности** probability scale
~ **времени** time scale
~, **главный** base/principal/nominal/general scale
~, **долготный/длины** scale of longitude
~, **заводской** plant scale
~ **заложений** scale of slope
~ **записи** record(ing)/recorder scale
~ **изображения** *telev.* image scale, scale of image
~ **искажений** exaggeration scale
~ **карты** chart/map scale
~, **крупный** large scale
~, **линейный** scale line, bar/graphical scale, linear scale
~, **мелкий** small scale
~, **натуральный** full scale
~, **основной** base/principal/general/nominal scale
~, **плавающий** floating scale
~ **по меридиану** scale in meridian, meridional scale
~ **по параллели** scale in parallel, parellel scale
~ **по экватору** equatorial scale
~, **поперечный** diagonal/transversal scale
~, **постоянный** fixed scale
~ **построения** plotting scale
~, **приведенный** corrected scale
~, **производственный** industrial scale
~, **произвольный** arbitrary/undefined scale
~, **промышленный** commercial scale
~ **съемки** scale of survey(ing)
~, **условный** arbitrary scale; representative fraction
~ **уменьшения** reduction scale
~, **частный** particular/ralative/local scale
~, **численный** numerical/natural/fractional scale; representation fraction; equivalent scale *(of map)*
~, **широтный** scale of latitude
~**е, в маленьком** on a small scale; in a small way
~**е, в увеличенном** scaled up
~**е, в уменьшенном** scaled down
масштабирование scaling
~ **по времени** time scaling
~ **химических реакторов** reactor scale-up
масштабность (large) scale, range
масштабный scale *(attr.)*; large-scale, vast, extensive

~ная линейка (measuring) rule, scale
мат I (floor/door) mat; matting
~, диэлектрический insulating mat
~, защитный protective blanket
~, облицовочный surface mat
~, плетеный wrought mat
~ под швартовы *naut.* mooring mat
~, тканый sword matting
~, шпигованный thrum(med)/quilted mat
~, электронагревательный electothermal blanket
мат II mat (finish); frosted *(glass)*
мат III *chess* checkmate, mate
матадор matador
мате *bot.* mate, Paraguay tea *(Ilex paraguayensis)*
мателот *nav.* adjacent ship
~, задний ship next astern
~, левый ship next aport
~, передний ship next ahead, leader
~, правый ship next astarboard
математизация mathemati(ci)zation
математик mathematician
математика mathematics
~, актуарная actuarial mathematics
~, высшая higher/advanced/college mathematics
~, вычислительная computing/numerical mathematics
~, прикладная applied mathematics
~, чистая pure/abstract mathematics
математический mathematical
~кое обеспечение software
~кое ожидание theoretical mean value, expectation value, mathematical expectation
матереубийственный matricidal
матереубийство matricide *(act)*
матереубийца matricide *(agent)*
материал 1. material; product stock; stuff; 2. cloth, fabric; *see also* **материалы**
~, аблятивный ablator, ablative/ablating material
~, абразивный abrasive (material); abradant
~, авиационный конструкционный airframe material
~, автоклавный steam-cured material
~, агломерированный sintered material
~ активной зоны *nucl.* core material
~, амортизирующий damping/cushioning material, shock-absorbing material
~, антикоррозионный rust inhibitor
~, армированный reinforced material
~, армирующий reinforcing/reinforcement material
~, асбестовый рулонный asbestos blanket
~, асбестоцементный cementitious material; asbestos-cement material
~, битумный asphaltic material/product
~, бронирующий inhibitor/inhibiting material
~, бутовочный filler
~, взвешенный реки river load/burden
~, вибропоглощающий vibration-absorbing/absorptive material
~ внеземного происхождения extraterrestrial material
~, волокнистый fibrous material
~, воспроизводящий *nucl.* breeder material
~, восстановительный (зубной) restorative material
~, впитывающий absorbent material
~, вспучивающий bloater, bloating agent
~, высокой чистоты high-pure material
~, вяжущий binder, binding/cohesive/cementing material
~, вязко-упругий viscoelastic material
~, герметизирующий sealing material; potting compound; encapsulant
~, горючий combustible material
~, грубый обломочный coarse waste/detritus
~, губчатый spongeous/spongy material
~, демпфирующий damping material
~, дефицитный material of short supply, scarce material
~, диэлектрический dielectric
~ для защиты от радиации radiation-shielding material
~ для КЛА, конструкционный spacecraft construction material
~ для летательных аппаратов aerospace material
~ для оболочки *nucl.* sheathing/canning material
~ для покрытия cladding material
~, жаростойкий heat-resisting/resistant material
~, жесткий inflexible/rigid/stiff material
~, жильный *geol.* vein material/rock/stone/matter
~, жировальный tanner grease, tanning oil
~, закладочный *min.* stowage material
~, защитный protective/shielding material
~, защищающий от радиации antiradiation material
~, звукоизолирующий sound-proofing/-resistant material
~, звукоизоляционный sound insulator, sound-insulating material
~, звукопоглощающий sound suppressant (material), noise-attenuating material, sound-absorbent/-absorbing material, acoustical absorbent
~, золотосодержащий для промывки wash(ing) stuff, wash gravel
~, извергаемый вулканический ejectamenta, ejecta
~, изоляционный insulating/insulation material; *el.* non-conducting material; insulator, insulation
~, инертный inert material
~, инородный foreign material/matter
~, искусственный man-made material
~, исходный source/parent material
~, картографический cartographic(al) document
~, керамический ceramic/sintered material; *constr.* (structural) clay product
~, керамический уносимый ceramic ablator

~, кислотостойкий acid-resisting/-resistant material
~, кислотоупорный acid-proof material
~, кластический *geol.* clastic/fragmental material
~, кластический изверженный clastic ejectamenta
~, композиционный composite (material)
~, конструкционный structural/construction material
~, коррозионностойкий non-corrodible material, corrosion-resistant material
~, кремнийорганический silicone material
~, кровельный roofing material
~, кроющий covering material, coating, cover
~, кусковой lump material
~, легковоспламеняющийся inflammable material
~, лесной timber
~, листовой sheet/plate material
~, люминесцирующий luminescent material
~, магнитно-жесткий-твердый hard magnetic material
~, магнитно-мягкий soft magnetic material
~, магнитнострикционный magnetostrictive material
~ малой плотности, абляционный low-density ablator
~, металлокерамический cermet (material), sintered powder metal
~, металлокерамический пористый porous cermet
~, металлокерамический фрикционный friction cermet
~, моренный morainic material
~ на основе каучука rubbery material
~, набивочный packing, stuffing; padding
~, наборный type matter
~ навалом bulk material
~ накладки facing/lining material
~, нанесенный drift(ed) material
~, неактивный inert material
~, негорючий incombustible material, noncombustible material
~, необугливающийся non-charring material
~, необычный exotic material
~, неоднородный heterogeneous material
~, непроводящий non-conducting material
~, непрозрачный opaque material
~, нерасщепленный unfissioned material
~, несжимаемый incompressible material
~, нескользящий non-slip material
~, несцементированный incoherent/uncemented material
~, оберточный wrapping material, wrap(per)
~, обработанный finished material/stock
~, обтирочный cleaning/wiping material, wiping rags
~, обугливающийся charring material, charforming material
~, обшивки skin material
~, огнегасящий fire-suppressant material
~, огнестойкий fire-resistant material
~, огнеупорный fireproof material, refractory
~, однородный homogeneous material
~, основной вулканический essential ejectamenta
~, отвалочный *geol.* debris, talus
~, отделочный leather finishing agent/material
~, отфильтрованный filtered matter
~, пенокерамический foam-ceramic material
~, первичный raw material
~, перевязочный first-aid material
~, перенесенный drift load
~, перспективный advanced material
~, печатный printed matter
~, пламестойкий flameproof material
~, пластический plastic (material)
~, пластично-вязкий yielding material
~, пластичный ductile material
~, повторно используемый recycled material
~, поглощающий absorbent/absorbing material
~, подкисляющий acidifier
~, подклеечныий adhesive backer
~, подложки backing material
~, подошвенный *leath.* soling
~, полировальный polish, polishing compound
~, полупроводниковый semiconducting/semiconductor material
~, порошковый powdered material
~, посадочный planting stock
~, посевной seed grain, seeds; *med.* inoculum
~, применяемый в космической технике space(craft) material
~, применяемый в технике engineering material
~, пробельный *print.* space, spacing material
~, пробельный и обкладочный furniture
~, пробельный междустрочный leads
~, пробельный полосный leads and slugs
~, пробельный строчной quads and spaces
~, пробный sampling material
~, проводящий conducting material, conductor
~, прозрачный transparent (material)
~, прокатный rolled stock
~, прокладочный sealing/packing/gasket material, leak-proofing material
~, промывочный flushing material
~, профильный sections, shapes, section material
~, противоэрозионный erosion-protection/-resistant material
~, рекламный advertising matter/material, publicity material
~, родственный cognate material
~ с избирательным поглощением frequency-selective damping material
~, сверхпроводящий superconductor
~, светочувствительный light-sensitive material
~, связывающий jointing/binding material,

binder
~, селекционный breeding material
~, семенной seed grain, seeds
~, склеиваемый adherent
~, слоистый stratified material, laminated/sandwich/multilayer material, laminate
~, смазочный lubricant
~, сотовый honeycomb material
~, стойкий к окислению oxidation-resistant material
~, сублимирующий subliming material, sublimator
~, сыпучий loose material
~, сырой raw material
~, теплозащитный thermal-protective material, heat-shield material, thermal protector
~, теплоизоляционный heat-insulation material, thermal-insulating material, lagging material
~, теплостойкий heat-resistant material
~, термореактивный thermosetting material
~, тонколистовой sheet material
~, тонкопленочный thin-film material
~, трудновоспламеняемый flash-resistant material
~, тугоплавкий refractory material
~, ударопрочный impact material
~, упаковочный packaging material, fillers-in
~, уплотняющий/уплотнительный sealing/packing material, leak-proofing material, sealant
~, упрочненный strengthened/reinforced material
~, упругий elastic material
~, урансодержащий uranium-bearing material
~, флюсующий fluxing agent
~, формовочный moulding material
~, футеровочный lining/facing material
~, хрупкий brittle material
~, шлифующий abradant
~, шумопоглощающий noise-absorbing/-damping material
~, экранирующий shielding material
~, ячеистый cellular material
материал-заменитель substitute material
материализация materialization
~ общественного богатства *econ.* embodiment of social wealth
материализм materialism; hylism, hylicism
~, диалектический dialectical materialism
материализовать materialize; pragmatize; hypostatize
материалист materialist; hylicist; *philos.* physicist
материалистический *philos.* materialist, materialistic
материалистский materialist *(attr.)*
материал-носитель carrier-material
материаловедение materials science, materiology, study of commercial materials
материалоемкий raw-material intensive
материалоемкость *econ.* materials-output ratio; specific consumption of materials, materials consumption
материалы *pl. coll.* material(s)
~ активной зоны *nucl.* core material
~, атомные сырьевые atomic source/raw material
~ в метрических размерностях metric material(s)
~ в обработке in-process stock
~, воспроизведенные bred material
~, воспроизводящие breeder material
~, вспомогательные accessory/indirect material
~, вспомогательные производственные factory/manufacturing supplies
~, горюче-смазочные fuels and lubricants/oils
~, делящиеся fissionable material
~, дефицитные critical materials
~ для ядерного реактора reactor/pile/nuclear material(s)
~, искусственные радиоактивные man-made radioactive materials
~ научные scientific material(s)
~, отравляющие poisonous material
~, расщепляющиеся fissionable material
~, рекламные promotional materials
~, слабопоглощающие low-capture material
~, ядерные сырьевые nuclear raw/source material
материально-ответственный *fin.* accountable
материально-правовой *law* substantive
материальность materiality; corporeality, substantivity
материально-технический material-and-technical
~ное обеспечение *mil.* logistic/material support, logistics
материальный 1. material; corporeal, substantial; *law* substantive; hylic; 2. financial, pecuniary
~ мир material world, world of matter
~ склад warehouse
~ная заинтересованность material incentive(s)
~ная ответственность financial responsibility
~ная часть equipment, hardware; material
~ные затруднения financial difficulties
материк *geogr.* mainland, continent; 2. *geol.* country rock, bedrock, subsoil, native soil; 3. *hydr.* horizon D
~, древний old land/continent
материковый inland, continental
материнский 1. mother(ly), maternal; 2. *geol.* parent(al), native, source *(attr.)*
~кая клетка mother cell
~кая порода *geol.* parent/source rock
~кое насаждение *for.* shelterwood
материнство maternity, motherhood; parenthood
материя 1. matter; *philos.* substance; 2. *med.* pus, matter; 3. cloth, fabric, material
~, вырожденная *astr.* degenerate matter
~, живая living matter
~, звездная stellar matter

~ **на рубашки** shirtling
~, **солнечная** solar matter
~, **темная** *astr.* dark matter
~, **шерстяная** woolen fabric
матерка *bot.* pistillate hemp
матерь *rel.* Mother of God
матерья *see* **материя**
матиола *bot.* Grecian stock, night-scented stock *(Matthiola bicornis)*
матирование mat finishing; frosting *(of glass)*
матировать dull, give a mat surface, (render) mat; frost *(glass)*
матировка frosting; matting; delustring, flatting
матица 1. *tech.* joist, tie beam; girder; 2. *sport* poke net *(fishing)*
матка 1. *anat.* uterus, womb, matrix; 2. female *(of animals);* queen (bee); 3. *econ.* parent stock; physical source; matrix; 4. master batch; 5. *naut.* (submarine) tender; aircraft carrier
~ **гирокомпаса** master gyrocompass
~, **двойная** double/bipartite uterus
~, **двурогая** bifid uterus
~, **зародышевая** gravid/pregnant uterus
~, **однорогая** one-horned uterus
~, **племенная** broodmare
~, **пчелиная** queen bee
матлот 1. matelote *(dance and music);* 2. *cul.* matelote
матобеспечение software
~, **прикладное** application software
~, **резидентное** resident software
матовость dullness, frosting; dimness
~, **лица** *leath.* dulling of grain
матовый mat(t), mat finished, lusterless; dull, dim; tarnished
~ **блеск** dull/mat finish
~**вая лампочка** frosted bulb
~**вое золото** dead gold
маточка *anat.* urticle
~, **предстательная** prostatic urticle
маточник *chem.* mother liquor; master batch; 2. *zool.* queen bee's cell; 3. *bot.* ovary, style; parent (plant); 4. *tech.* lead screw, master tap; 5. *agr.* enclosure for female animals; 6. *bot.* Moldavica dragonhead *(Dracocephalum moldavicum)*
~, **болотный** *bot.* bog angelica *(Angelica palustris)*
маточно-брюшной *anat.* uteroventral, uteroabdominal
маточно-брюшностеночный *anat.* uteroparietal
маточно-влагалищный *anat.* uterovaginal
маточно-крестцовый *anat.* uterosacral
маточно-пузырный *anat.* uterovesical
маточно-тазовый *anat.* uteropelvic
маточно-трубный *anat.* uterotubal
маточно-шеечный *anat.* uterocervical
маточно-яичниковий *anat.* utero-ovarian
маточный 1. *anat.* uterine; 2. female
~ **лист** starting sheet
~ **метчик** master tap
~ **раствор** mother solution/liquor
~**ная порода** *geol.* mother/parent rock
~**ные дрожжи** barm
матрас mattress
матрац 1. mattress; 2. *geol.* pillow lava
~, **надувной** inflatable air-bed
~, **теплоизоляционный** heat-insulating blanket
матриарх matriarch
матриархат matriarchy
матрикс matrix
~ **хромосомы** chromosomal matrix
матрикул(а) *educ.* student's record card
матрица 1. *math.* matrix; 2. *tech.* (female) die, mould, matrix; *comp.* array
~ **Адамара/адамаровая** Hadamard matrix
~ **без повторения** non-recurrent matrix
~, **булева** Boolean matrix
~, **вентильная** gate array
~, **весовая** weighting matrix
~, **вещественная** real matrix, matrix of real elements
~, **волновая процессорная** wavefront array
~ **выигрышей** payoff matrix
~, **вырожденная** singular matrix
~, **гипсовая** plaster mould
~, **двухрядная** 2-by-2 matrix, two-row matrix
~, **двухсрочная** two-row matrix
~, **дешифраторная** decoder matrix
~, **диагонализирующая** polar matrix
~ **Дирака** Dirac matrix
~ **для двоичного преобразования** binary conversion matrix
~, **единичная** identity/unit matrix
~, **железобетонная** reinforced concrete jig
~, **запоминающая,** ~ **запоминающего устройства** memory array/matrix
~ **запоминающего устройства на сердечниках** core plane/matrix
~ **игры** payoff matrix
~ **иммитансов** immittance matrix
~, **импримитивная** imprimitive matrix
~ **инцидентности/инциденции** incidence matrix
~, **квадратная** square matrix
~, **квазиопределенная** quasidefinite matrix
~, **клеточная** array, tableau; hypermatrix
~, **ковариационная** covariance matrix
~, **кодирующая** encoder matrix
~, **коммутационная** interconnect matrix
~, **конденсаторная** capacitor array
~, **корреляционная** correlation matrix
~, **кососимметричная** skew(-symmetric) matrix
~, **логическая кремниевая** silicon-diode logic matrix
~ **магнитного запоминающего устройства** magnetic memory matrix/plane
~, **марковская** Markov matrix
~, **масочно-программируемая** mask pro-

grammable logic array, MPLA
~ межотраслевых потоков transaction matrix
~ Метцлера Metzler matrix
~ Минковского-Леонтьева Minkovski-Leontief matrix
~, модальная polar matrix
~ назначений assignment matrix
~ на тороидальных сердечниках toroidal core matrix
~, невырожденная nonsingular/regular matrix
~, неокаймленная unbordered matrix
~, неособенная nonsingular matrix
~, неприводимая irreducible matrix
~, несобственная improper matrix
~, нулевая zero/null/vacuous matrix
~, обратимая invertible/nonsingular matrix
~, обратная inverse/reciprocal matrix
~, обратная Леонтьева Leontief inverse matrix
~ ограничений constraint matrix
~, окаймленная bordered matrix
~, опознающая recognition matrix
~, особенная singular matrix
~ памяти memory array/matrix
~, перестановочная permutation matrix
~ печатающего устройства printing matrix
~ планирования planning matrix
~, платежная/платежей *game th.* payoff matrix/array
~, полная entire matrix
~, положительно-определенная positive-definite matrix
~, полярная polar matrix
~ порядка matrix of order
~ потерь *game th.* regret matrix
~, приведенная reduced matrix
~, приводимая reducible matrix
~, присоединенная adjugate matrix, classical adjoint (matrix)
~, проверочная parity-check matrix
~, программируемая логическая programmable logic array, PLA
~, производная derivative matrix
~, производящая generator matrix
~, процессорная волновая wavefront array
~ пуансонов block/matrix of punch pins, punching matrix
~, распознающая recognition matrix
~ рассеяния scattering matrix
~, расширенная augmented matrix
~ результатов outcome matrix
~ рудной жилы *geol.* veinstone, vein-stuff, gangue
~ с регулярной структурой cellular array
~ с целочисленными элементами integer matrix, matrix with integer entries
~, светодиодная light-emitting-diode matrix, LED-matrix
~, сердечников core mat(rix)
~, симметрическая symmetric(al) matrix
~, систолическая systolic array
~ смежности adjacency matrix
~, сопровождающая companion matrix
~, сопряженная adjoint of a matrix
~, составная *mach.* sectional die; *math.* composite matrix
~, сотовая mesh-connected array
~, твердосплавная hard-alloy matrix
~ точек dot matrix
~ транспонированная transpose of a matrix, transposed matrix
~, унитарная unitary matrix
~, унитарно-эквивалентная unitarily equivalent matrix
~, ферритовая ferrite/magnetic/core matrix/plane
~, целлулоидная celluloid mould
~, целочисленная integral matrix
~ четного ранга matrix of even order
~, четырехредная квадратная 4-by-4 matrix
~, шифраторная encoder matrix
~, эрмитова Hermitian matrix
~, эталонная master die
~, N-рядная квадратная N-by-N matrix

матрица-вектор matrix-vector
матрица-строка row matrix
матрица-функция matrix-function
матрицедержатель *mach.* die-holder
матрицирование 1. *math.* matrixing; 2. *print.* moulding; matrix making/punching
~ ручное brush moulding
матрицировать make matrix moulds
матричнозначный matrix-valued
матричный *math.* matrix *(attr.)*
матроклиния *bot.* matrocliny
матроклинный matroclinous, matriclinous
матрона matron
матроним matronymic
матронство matronage
матрос (sea)man, sailor
~, баковый forecastle man
~, принимающий швартовы line runner
~, старший senior seaman
~, торгового флота merchant seaman
матрос-артиллерист quarter rating
матрос-торпедист seaman torpedoman
маттиола *bot.* stock *(Matthiola)*
матуар *eng.* mattoir
матурограф maturograph
матурометр maturemeter
матч *sport* match
~, выездной away match
~ на первенство мира (по) match for the world title (in)
~, ответный return match
матчасть *abbr.* **материальная часть** materiel, equipment
мать 1. mother; 2. *zool.* female
~, крестная godmother
~, приемная foster mother
мать-и-мачеха, мать-мачеха *bot.* coltsfoot

(Tussilago farfara)
мать-казначея cellaress, procuratrix
мать-одиночка unmarried mother
матэ *bot.* Paraguay tea, mate *(Ilex paraguariensis)*
маузер *mil.* Mauser *(pistol/rifle)*
маун *bot.* valerian *(Valeriana)*
мауновые *n. pl. bot.* valerian family *(Valerianaceae)*
мафист Mafia member, mafioso
мафический *geol.* mafic
мафия, маффия maf(f)ia, Cosa Nostra
Мафусаил *bibl.* Methuselah
мах Mach (number)
махабхарата Mahabharata
махагони *bot.* West Indies mahogany *(Swietenia mahagoni)*
махайрод *paleont.* saber-toothed tiger *(Machairodus)*
махалка *bot.* panicle
махаон *zool.* swallowtail *(Papilio machaon)*
махараджа maharaja(h)
махатма mahatma, wise and holy leader
махе Mache unit
махизм *philos.* Machism
махинатор wheeler-dealer, machinator
махинация intrigue, machination; artifice;
махист *philos.* Machist
махметр *av.* Machmeter
махновец *hist.* Makhnoite
маховик flywheel; handwheel; knob
~, зубчатый, ~ с зубчатым венцом cogged/toothed flywheel
~ поворотного механизма traversing handwheel
маховик-вентилятор fan/spoked flywheel
маховик-регулятор flywheel governor
маховичок handwheel, knob, (pilot) wheel
~ реверса reversing handwheel, manoeuvring wheel
~ с ручками star wheel
~, установочный adjusting handwheel, setting knob
маховой wagging, waving
~ момент moment of gyration, flywheel moment
~вое колесо flywheel
~вые перья *zool.* flight/wing feathers
махорка 1. *bot.* Aztec tobacco *(Nicotiana rustica)*; 2. shag, makhorka *(inferior kind of tobacco)*
махровость *bot.* doubling, doubleness
махровый 1. *bot.* double(-petalled); 2. *text.* terry
~вая ткань terry cloth
махры *pl.* fringes, tassels, frayed edges
мацерал *geol.* maceral
~, водорослевый algite
~ дюренового угля D-coal
мацерация, мацерирование maceration, macerating
мацони *cul.* matzoon, sour milk
мачек *bot.* horn poppy *(Glaucium)*
мачеха, мачиха stepmother
мачта mast; pole, post; tower
~, ажурная lattice/trellis/cage mast
~, антенная aerial/antenna mast
~ ветроуказателя wind-cone mast
~, временная jury mast
~ грузовой стрелы cargo/derrick mast
~, заваливающаяся *naut.* hinged/collapsible/lowering mast
~, двуногая bipod/sheer mast
~, кормовая after mast
~, монтажная erecting mast
~, причальная mooring mast
~ с брам-стеньгой bend mast
~, треногая tripod mast
~, четырехногая quadruped mast
мачта-антенна tower-type antenna
мачта-однодеревка pole mast, single-spare mast, one-piece mast
мачта-труба mack, stack mast
мачты *pl.* masts
~, парные goal-post masts, twin/pair masts
мачтовой *n. nav.* mastman
мачтовый mast *(attr.)*
~ лес mast timber
~вое дерево spar
мачуэла *zool.* thread herring *(Opisthonema)*
маш *bot.* mung bean *(Phaseolus aureus)*
машбюро *abbr.* **машинописное бюро** typing pool
машина 1. machine, mechanism, apparatus; engine; 2. car; truck; vehicle; 3. *pl. coll.* machinery
~, агломерационная sintering machine
~, агрегатная integrated machine
~ базы данных database machine
~ базы знаний knowledgebase machine
~ без программного обеспечения bare machine
~, бланшировочная blancher, blanching machine
~, бланшировочно-очистительная peeling scalder, scalder-peeler
~, боевая combat vehicle
~, буквонаборная/буквопечатающая typesetter
~, буквофотографирующая photocomposing machine, photosetter
~, бурильная rock-drill, drilling machine
~, бурильная ручная hand-held drill, jackhammer drill
~, бухгалтерская billing/invoicing/accounting/bookkeeping machine
~, вакуум-закаточная vacuum seaming machine
~, вакуум-укупорочная vacuum closing/capping machine
~, вакуум-упаковочная vacuum packer, vacuum packing machine
~, веечная winnower, winnowing machine
~, вездеходная all-terrain vehicle, crosscountry car

~, взбивальная *food.* beater, beating/whipping/whisk(ing) machine
~, военная *mil.* defence establishment
~, ворсовальная carding/raising/nepping machine
~, врубовая coal cutter
~, врубовая дисковая disk coal cutter
~, врубовая многобаровая multijib coal cutter
~, врубо(во)-навалочная cutting-loading machine
~, встроенная *el.* built-in machine
~, высадкопосадочная *agr.* planter
~, вычислительная, ВМ calculating/computing machine, computer, calculator:
ВМ, автокорреляционная autocorrelation computer
ВМ, аналоговая, АВМ analog computer
ВМ, аналоговая пневматическая pneumatic analog computer
ВМ, аналого-цифровая analog-digital computer, hybrid computer
ВМ, безадресная no-/zero-address computer
ВМ, бионическая bionic computer
ВМ второго поколения second-generation computer
ВМ, гибридная analog-digital computer, hybrid computer
ВМ, главная host (computer)
ВМ, двоичная binary computer, radix two computer
ВМ, двухадресная two-address computer
ВМ для машинного перевода translation machine
ВМ для коммерческих расчетов business computer
ВМ для обработки нечисловой информации non-numeric machine
ВМ для обработки результатов переписи census computer
ВМ для управления производственными процессами process control computer
ВМ, игровая game(-playing) machine
ВМ, информационно-логическая information-logical machine
ВМ, клавишная keyboard computer; key-driven calculator
ВМ, координатная point-to-point computer
ВМ космического корабля spaceborne computer
ВМ, ламповая electron(ic) tube computer
ВМ на микроминиатюрных блоках micro-circuit computer
ВМ на модулях modular(ized) computer
ВМ на перфокартах punch(ed)-card computer
ВМ, настольная desk(-top) computer, lap-top computer; desktop calculator
ВМ на струйных элементах fluid/fluidics computer
ВМ на твердых элементах solid-state computer
ВМ на транзисторах transistor(ized) computer
ВМ, обучающая teaching machine
ВМ, одноадресная single-address computer
ВМ параллельного действия parallel computer
ВМ, периферическая peripheral computer
ВМ, перфорационная punched-card computer
ВМ, полупроводниковая transistor computer
ВМ последовательного действия serial computer
ВМ работающая в истинном масштабе времени real-time computer
ВМ с жесткой программой fixed-program computer
ВМ с наборной программой wired program computer
ВМ с плавающей запятой floating-point computer
ВМ с программным управлением program-controlled computer
ВМ с программным управлением от перфокарт card-programmed computer
ВМ с совмещением операции simultaneous computer
ВМ с фиксированной запятой fixed-point computer
ВМ, самовоспроизводящаяся self-replication machine
ВМ, самоорганизующаяся self-organizing computer
ВМ, самопрограммирующая self-programming computer
ВМ, сверхмалая minicomputer
ВМ, статистическая statistical machine
ВМ, универсальная general-/all-purpose computer, universal computer
ВМ, цифровая, ЦВМ digital computer
ВМ, цифровая интегрирующая incremental computer
ВМ электронная electronic computer
~, вязальная 1. *agr.* bale tie machine 2. *text.* knitting machine
~, вязальная овальная oval knitting machine
~, вязальная штучная body-length machine
~, гозоопаливающая *text.* gas-singeing machine
~, газорезательная gas-/flame-cutting machine
~, газорезательная ацетилено-кислородная oxyacetylene-cutting machine
~, газорезательная прямоугольно-координатная cross carriage cutting machine
~, газосварочная gas welder, gas welding machine
~, герметически закрытая *el.* airtight machine
~, гибочная bending/forming machine
~, гидравлическая одноколесная single-impeller machine
~, гидравлическая осевая axial-flow machine

~, гидравлическая рулевая hydraulic servo motor
~, гидроочистная hydroblasting machine
~, гладильная ironer
~, глянцевальная glazing machine
~, горная mining machine
~, государственная machine(ry) of the state
~, гофрировочная *text.* crimping machine; *met.* corrugator
~, гравировальная engraver, engraving machine
~, гребнечесальная comb(er), combing machine, nip comb(er)
~, гребнечесальная двупольная double-action comber
~, гребнечесальная с круглыми гребнями circular comber
~, грузовая lorry, *Am.* truck
~, грузоподъемная hoisting/lifting machine(ry)
~, грунтовальная *text.* coating machine
~, грунтосмесительная soil/ground mixing machine
~, гусеничная tracked vehicle
~, двоильная *leath.* splitter, splitting machine
~, двусторонняя точечная duplex spot welder
~, двухадресная two-address machine
~, декортикационная decorticator, stripping machine
~, делительная *met.* dividing/ *opt.* ruling machine
~, десантная *mil.* landing vehicle
~, детерминированная *cyb.* determinate machine
~, диагонально-закройная bias-cutting machine
~ для ацетилено-кислородной резки oxy-acetylene-cutting machine
~ для внесения аммиакатов *agr.* ammonia solution applicator
~ для внесения гербицидов *agr.* herbicide applicator
~ для внесения удобрений *agr.* fertilizer placer
~ для газовой резки труб (gas) pipe-cutting machine
~ для газопламенной резки по копиру flameprofiling machine
~ для гидропескоструйной очистки sand-water cleaner
~ для дуговой сварки arc-welding machine
~ для зачистки *met.* deseamer
~ для земляных работ earth-moving machine
~ для испытаний на выносливость endurance-testing machine; fatigue (test) apparatus/machine
~ для испытаний на износ wear-testing machine
~ для испытаний на разрыв tensile-testing machine
~ для испытаний на ударную вязкость droptesting machine
~ для испытаний на удар impact-testing machine
~ для испытаний на усталость fatigue-testing machine
~ для испытаний прочности strength-testing machine
~ для кислородной резки oxygen-cutting machine, gas-/flame-cutting machine
~ для литья под давлением injection moulding machine
~ для набивки перфокарт punch-card printer
~ для обработки данных, электронная electronic data processing machine
~ для обработки перфокарт, гидравлическая fluid-operated(punched) card processor/machine
~ для объемной штамповки die-forging machine
~ для подметания улиц street-sweeper
~ для подпочвенного внесения удобрений *agr.* subsurface applicator
~ для подрезки живой изгороди hedgecutter
~ для профильной резки shape cutting machine
~ для разлива в бутылки bottle filler
~ для разливки металла, карусельная casting wheel
~ для раскатки труб expanding/reeling machine
~ для резки арматурных стержней bar cutter/cropper
~ для рельефной сварки projection welding machine
~ для сварки под флюсом submerged arc welding machine
~ для составления счетов invoice typewriter, billing/invoicing machine
~ для статических испытаний load testing machine
~ для точечной сварки spot-welding machine
~ для удаления сердцевины плодов corer, coring machine
~ для упаковки в ящики case packer
~ для экономических задач business machine
~, дождевальная sprinkling machine, (overhead) sprinkler
~, доильная milker, milking machine
~, дорожно-строительная road-building machine
~, дробеструйная shot blasting machine
~, дробеструйная для очистки от окалины blast descaler
~, жаккардовая jacquard loom/machine
~, завалочная charger, charging machine
~, заварочная sealing-in machine
~, заверточная wrapper, wrapping machine
~, загрузочная loading/charging machine, charger
~, закаточная 1. seamer, edge-curling machine;

can-sealing machine; 2. *met.* beading machine
~, закладочная packing/stowing machine, stower
~, запаечная *food* end soldering machine
~, землеройно-фрезерная excavating milling machine
~, зерноочистительная grain cleaner, graincleaning machine
~, зерносортировочная grain grader
~, золотопромывочная gold washer
~, измельчительная *food ind.* mincer, shredder, mincing machine
~, инерционно-ударная impact power tool
~, информационная data-processing machine
~, испытательная tester, testing machine
~, испытательная для материалов materials testing machine
~, кабелепрокладочная *naut.* cable engine
~, калибровочная 1. size grader, size-grading machine; 2. *foundry* gauging machine
~, кардочесальная card, carding machine, carder
~, картонорезальная cardboard cutter
~, кетлевочная/кеттельная binding-off machine, looping/linking machine
~, клавишная key-driven machine
~, клеемазальная/клеенамазывающая gluer, glu(e)ing/gumming machine
~, клеймильная/клеймовочная stamping/marking machine
~, клепальная rivet(t)ing machine, rivet(t)er
~, кокильная chill casting machine
~, колесно-гусеничная wheeled-caterpillar vehicle
~, коллекторная *el.* commutator machine
~, колотильная *leath.* beetle
~, кормораздаточная food distributor
~, корновальная grainer, graining machine
~, корчевальная root extractor, stamper
~, косточковыбивная pitter, pitting/stoning machine
~, кремовзбивальная creamer, cream beating machine
~, кромкозагибочная edging/beading machine
~, кругловязальная circular knitter, circular knitting machine
~, круглочулочная circular footwear machine, circular hose knitter
~, крупосортировочная groats grader
~, крупошлифовальная pearl mill, pearler
~, крутильная *text.* twisting frame
~, крутильная реверсивная *mech.* reversed torsion machine
~, куделеприготовительная flax tow preparing machine, tow cleaner/scutcher
~, лентосоединительная *text.* sliver lap machine
~, ленточная drawing frame
~, лесопогрузочная timber loader
~, лесопосадочная tree planter
~, листогибочная sheet-/plate-bending machine, sheet bender
~, листоподборочная *print.* collator, collating machine
~, листоправильная sheet straightener/leveller
~, литейная для литья под давлением die casting machine
~, литейная для центробежного литья centrifugal casting machine
~, литографская lithographic press
~ логического вывода inference machine
~, ломтерезальная slicer, slicing machine
~, лущильная huller, hulling machine
~, льнотрепальная flax scutcher, flax scutching machine
~, месильная kneader, (dough-)kneading machine
~, множительная duplicator, duplicating machine
~, множительная офсетная offset duplicator
~, моделирующая вычислительная computer simulator, simulator computing system
~, моечная washer, washing machine
~, моечная вибрационная shaker washer
~, моечная роторно-элеваторная squirrel cage washer
~, моечная шприцевая spray washer
~, моечная щеточная brush washer
~, мотальная *text.* winder, winding machine
~, мотальная горизонтальная drum winding machine, winding frame with horizontal spindles
~, мягчильная *text.* cloth mellowing machine
~, мяльно-трепальная breaker and scutcher, scutching-and-breaking machine
~, набивная *food* stuffer, stuffing machine
~, наборная typesetter, typesetting/composing machine
~, наборно-фотографирующая phototypesetting machine
~, накатная *text.* rolling-up machine; *met.* knurling machine
~, наполнительная filler, filling machine
~, обдирочная stripping/shucking machine
~, оберточная wrapper, overwrapping machine
~, обжарочная roaster, roasting machine
~, обоечная hulling/scouring machine, huller, scourer
~, обучающая teaching/learning machine
~ общего применения *el.* general-purpose machine
~, овсюгоотборочная oat separator
~ огневой зачистки *met.* (flame) scarfer, scarfing machine
~, ополаскивающая rinser, rinsing machine
~, опрыскивающая sprayer, spraying machine
~ оптической печати optical printer, optical printing machine
~, основовязальная *text.* warp loom, warp knitting machine
~, основовязальная трикотажная raschel machine

~, отбраковочная rejecting machine, quality-control machine
~, отделочная finisher, finishing machine
~, отсадочная *met.* jig(ger), jigging machine
~, отсадочная лопастная paddle jig
~, отсадочная поршневая piston/plunger jig
~, очистительная cleaner, cleaning machine
~, паркетоотделочная floor-polisher machine
~, паркетострогальная floor dresser
~, паровая (reciprocating) steam engine
~, паровая компаунд compound steam engine
~, паровая прямоточная uniflow steam engine
~ переменного тока *el.* alternating-current machine
~, пескоструйная sandblast machine
~, петлеобметочная buttonhole machine
~, печатная printing machine/press
~, печатная валичная roller printing machine
~, печатная высокой печати letter (printing) press, relief printing press
~, печатная глубокой печати intaglio (printing) press
~, печатная для двусторонней печати perfecting machine/press, perfector
~, печатная листовая sheet-fed press
~, печатная офсетная offset press
~, печатная ротационная rotary press
~, печатная рулонная web(-fed) press
~, печатная тигельная platen press
~, печатная универсальная multipurpose press
~, печатная цилиндрическая *text.* cylinder printing machine
~, плавающая amphibious vehicle
~, подъемная hoist engine
~, подъемная на копре tower winder/hoist
~, подъемная шахтная mine hoist
~, пожарная fire tender/engine, fire(fighter) truck
~, поперечно-стригальная cross-cutting/-shearing machine
~, породопогрузочная rock-loader, mucker
~, породоуборочная mucker
~, посадочная planting machine, mechanical planter
~ последовательного действия sequential/serial machine
~, почвообрабатывающая tilling/cultivating machine
~, почвообрабатывающая фрезерная rotary cultivator
~, правильная straightening/flattening machine, leveller, straightener
~, правильная, многовалковая multiroll straightener
~, пробкоукупорочная corker, corking machine
~, проволокошвейная *print.* stapler, wire stitcher, stapling machine
~, промывная scourer, scouring machine
~, просеивающая screening/sifting/bolting machine, bolting mill, bolter
~, протирочная *agr.* triturator; *food* pulper, pulping machine
~, прошивная *leath.* Blake (sewing) machine
~, проявочная processing/developing machine
~, прядильная spinning frame
~, пуговичная button machine
~, разведывательная *mil.* reconnaissance/scout vehicle
~, разливочная *met.* casting machine; *food* (liquid) filling machine, filler
~, раскладочно-подборочная *comp.* collator
~, распылительная pulverizer, pulverizing machine
~, рассадопосадочная seedling planter, planting machine
~, расшифровочная punch card printer
~, резальная cutter, cutting machine
~, резервная backup machine
~, рельсогибочная rail-cambering machine
~, рельсоправильная rail straightener
~, рисообдирочная rice-hulling machine, rice mill/scourer
~, рихтовальная flattener
~, ровничная *text.* roving frame
~, ровничная перегонная *text.* intermediate roving frame
~, ровничная тазовая slubbing frame
~, росткоотбойная germ separating machine, germ separator, degerminator
~, рубильная chipping machine, wood chipper
~, рулевая actuator, servo unit/motor; steering engine
~ с автоматическим вводом печатного текста scanning machine
~ с конечной памятью finite memory machine
~ с элементами искусственного интеллекта artificial intelligence machine
~, садочная charger, charging machine
~, сажальная planter, plant setter
~, самообучающаяся learning machine
~, сваебойная pile driver
~, сварочная welder, welding machine
~, сварочная для стыковой сварки butt welder
~, сварочная точечная spot-welding machine
~, свеклорезальная beet slicer, beet slicing machine
~, светокопировальная blueprint(ing) machine
~, семеочистительная seed dresser/separator/cleaner/sorter, seed-cleaning machine
~, сеноуборочная grass harvester, hay-making machine
~, сливковзбивальная creamer, cream beating machine, creaming machine
~, смешивающая mixing/blending machine, mixer, blender
~, снеговальная snow making machine

~, сновальная beaming/warping machine/mill
~, сокоотжимная juicing machine
~, сортировальная *comp.* sorter; *agr.* sizer, grader, grading/sorting machine
~, сортировальная ременная *agr.* belt sorter
~, сортировальная решетная *agr.* screen grader/sizer
~, сортировально-подборочная *comp.* collator
~, сортировочная sorter, sorting/grading machine, grader; classifier
~, сортоправильная *met.* section straightener
~, стиральная washing machine, washer
~, строгальная chipper, shaver, chipping/shaving machine
~, счетная calculating machine
~, счетная автоматическая automatic calculator
~, счетная портативная/ручная manual calculator
~, счетная электронная electronic calculator
~, счетно-аналитическая tabulating machine, punched-card installation/machine
~, (счетно-)бухгалтерская accounting bookkeeping machine
~, счетно-перфорационная punched-card installation/machine
~, счетно-решающая computer, computing/calculating machine
~, счетно-сортировочная counting-and-sorting machine
~, считывающая reading machine
~, сшивная stitcher
~, таблетировочная powder tableting machine
~, теленаборная teletypesetter
~, тепловая heat engine
~, терочная grater, grating machine
~, тестовальцовочная dough-rolling/-sheeting machine, dough roller/sheeter
~, тестомесильная dough-making/-kneading machine, dough mixer/kneader
~, травильная *print.* etcher, etching machine; *met.* pickling machine
~, трепальная scutcher, spreading machine
~, тянульная *leath.* staker, staking machine
~, уборочная *agr.* harvester, picker
~, уборочная бункерная tanker harvester
~, уборочная навесная linkage-mounted harvester
~, уборочная пневматическая suction harvester
~, угароочищающая waste cleaner, waste cleaning machine
~, униполярная *el.* unipolar/homopolar/acyclic machine
~, углепогрузочная coal loader
~, узловязальная knotting machine
~, укупорочная capping/corking machine, capper
~, упаковочная packer, packing machine
~, уплотняющая вибрационная vibrotampering machine
~, учебная training vehicle
~, фактурная accounting/invoicing machine
~, фальцевальная folder, folding machine
~, фасовочная weighing-and-filling machine
~, фасовочно-упаковочная weighing and packing machine
~, фланжировочная flanger
~, фольгозаверточная foiling machine
~, формовочная moulder, moulding/forming machine
~, фотонаборная photosetter, photocomposing machine
~, франкировальная franker, franking machine
~, хлопкоуборочная cotton picker/harvester
~, хмелеуборочная hop picker
~, холодильная refrigerating machine
~, циклевочная floor grinder
~, чаеподрезочная tea pruner
~, чаеполировальная tea polishing machine
~, чертежная drafting/drawing machine
~, чесальная hackle, *text.* card(er), carding machine
~, чесальная ватная wadding card
~, чесальная игольчатая wire card
~, чесальная камвольная worsted card
~, чесальная тонкая finisher card
~, числовая calculating machine
~, чулочновязальная hosiery/stocking knitter
~, швейная sewing/stitch(ing) machine; stitcher; *leath.* seamer
~, шелушильная hulling/scouring machine, huller, scourer, sheller
~, шерстомойная wool-scouring/-washing machine
~, шинковальная chopping machine
~, шифровальная cipher device/machine; converter, cryptogadget
~, шкуросъемная skinning machine, hide pulling machine, skinner, dehider
~, шлеечная *text.* thread extractor/picker
~, шлихтовальная *text.* sizing machine
~, шпунтовочная *wood.* grooving machine
~, шрифтолитейная typecasting machine
~, штамповочно-ковачная forging-and-stamping machine
~, штанцевальная punching machine
~, электронно-вычислительная, ЭВМ electronic computer
~, электросварочная electric welder, electric welding-machine
~, этикетировочная labeller, labelling machine
~, явнополюсная *el.* salient-pole machine
машина-двигатель engine
машина-дублер doubling machine
машинальный mechanical
машина-орудие machine tool
машина-шлюз *comp.* gateway
машинерия machinery
машинизация mechanization

машинизированный mechanized, equipped with machines
машинист 1. *rail.* engine-driver; *min.* operator; machinist, mechanic, engineman; 2. *theatre* sceneshifter
~, котельный stoker
~, палубный deck artificer
~, скиповой ямы skipman
~, старший котельный leading stoker
машинист-старшина *naut.* artificer
машинка 1. operating gear, machine; 2. typewriter 3. sewing machine; 4. (hair) clipper(s)
~, взрывная blasting machine, shot exploder
~, пишущая typewriter
~, подрывная blasting apparatus/machine; blaster, blasting exploder
~, рулевая space control-surface actuator; *av.* control-surface servo
~, стригальная (farm)clipper, shearer, shearing machine
~, якорная boatswain's finger
машинно-тракторный machine-and-tractor
машинно-электротехнический mechanical and electrical engineering
машинный mechanical; automatic; computer-assisted
~ язык computer language
~ная обработка machining
~ное оборудование mechanical equipment
машинобоязнь computer phobia
машиноведение theoretical engineering, science of machines/mechanics
машинообразный machinelike
машиноориентированный machine-oriented
машинописный typewritten, typed, typing
~ное бюро typing office
машинопись typewriting, typing, typescript
машинораспознаваемый machine-recognizable
машиностроение machine building/manufacturing, mechanical engineering, engineering industry
~, сельскохозяйственное agricultural machinery industry
~, энергетическое power engineering industry
машиностроитель machinery manufacturer, mechanical engineer, machinebuilder
машиностроительный machinebuilding
машиносчетная станция data processing centre
машино-час machine-hour
машиночитаемый machine-readable/treatable
машкарад, машкерад *see* **маскарад**
маш-салат *bot.* European corn salad *(Valerianella locusta/olitoria)*
маштаб *see* **масштаб**
маэстозо *mus.* maestoso
маэстро *art* maestro; master
маюскул majuscule
маяк beacon, lighthouse
~, акустический/звуковой sound/aural(-type) beacon
~, бортовой anticollision light
~, ведомый slave beacon
~, ведущий master beacon
~, заградительный (световой) hazard beacon
~, затмевающий occulting beacon
~, курсовый course light
~, линейный airway beacon
~, маркерный marker beacon
~, маркерный входной approach marker beacon
~, маркерный пограничный boundary marker beacon
~, наземный ground beacon
~, ориентирный landmark beacon
~, плавучий lightship, light-vessel, flotation beacon
~, посадочный landing beacon
~, приводной homing beacon
~, проблесковый flashing beacon
~, радиогидроакустический radiosonic ranging system
~, радиолокационный radar beacon, racon
~ с вращающейся рамочной антенной rotating loop beacon
~, свайный pile lighthouse
~, световой заградительный hazard (light) beacon
маяк-ответчик responder/transponder beacon
маятник pendulum
~, астазированный astatic pendulum
~, гравиметрический gravity pendulum
~, изохронный isochrone pendulum
~, крутильный torsion pendulum
~, математический simple pendulum
~, оборотный reversible/Kater's pendulum
~, секундный length of seconds pendulum
~, торпедный torpedo balancer
~, уравнительный compensator, compensation pendulum
~, фрикционный/Фруда Froude pendulum
~ Фуко Foucault pendulum
маятникообразный pendulumlike
маячник lighthouseman
МГБ *hist.* **Министерство государственной безопасности** USSR Ministry of State Security (1946-1953)
МГД-генератор *abbr.* **магнитогидродинамический генератор** MHD-generator
МГД-гироскоп MHD-gyroscope
мгла dry fog, haze, mist
~, белая whiteout
~, дымная smog, smoke haze
~, ледяная snow mist
мглистый misty, hazy; foggy; brumous
мгновение instant, moment
мгновенный instantaneous, prompt
меандр 1. meander; 2. *arch., art* ornamental pattern of lines
~, блуждающий wandering meander
~, врезанный incized/entrenched/cast/enforced

meander
~, **изолированный, отрезанный** cut-off meander
~ **поймы** *geol.* flood-plain lobe/scroll
меандрический meandrine, meandriform, meandrous
мебелевоз furniture-moving van
мебель furniture
~, **корпусная** cabinet furniture
меблирование furnishing
меблировка furnishing; furniture
мевалонаткиназа *bioch.* mevalonate kinase
мегабар megabar
мегабит megabit
мегабрекчия megabreccia
мегаварметр megavarmeter
мегаватт megawatt
мегаватт-час megawatt-hour
мегавольт megavolt
мегагалактика galaxy of galaxies
мегагамета *biol.* megagamete
мегагаметофит *bot.* megagametophyte
мегагаусс megagauss
мегагерц megacycle per second, megahertz
мегаджоуль *phys.* megajoule
мегадина megadyne
мегадуоденум *med.* mega(dolicho)duodenum
мегазооспора *bot.* megazoospore
мегакариобласт *anat.* megakaryoblast, megacaryoblast
мегакариоцит *anat.* megakaryocyte, megacaryocyte
мегаколон *anat.* megacolon, pelvirectal achalasia
~, **врожденный** megacolon congenitum, Hirschsprung's/Mya's disease
~, **приобретенный** acquired megacolon
мегакулон megacoulomb
мегакюри megacurie
мегалиния megaline
мегалит *archeol.* megalith
мегалобаза megalobase
мегалобласт megaloblast
~, **ортохромный** orthochromatic megaloblast
мегалогонидия *bot.* megalogonidium
мегалозавр *paleont.* megalosaur *(Megalosaurus)*
мегаломания megalomania
мегалопенис *med.* megalopenis, macrophallus
мегалоспления *med.* megalosplenia, splenomegaly
мегалоцефалия *med.* megalocephalia, megalocephaly, megacephaly
мегалоцит *med.* megalocyte
~, **полихромный** polychromatic megalocyte
мегапорфировый megaporphyritic
мегарад megarad
мегарезерфорд megarutherford
мегарентген megaroentgen
мегасейсм megaseism
мегасинклиналь megasyncline
мегаслово *comp.* megaword
мегаскоп megascope
мегаспора *bot.* megaspore, macrospore
мегаспорангий *bot.* megasporangium
мегаспорогенез *bot.* megasporogenesis
мегаспорофилл megasporophyll
мегаспороцит *biol.* megasporocyte
мегастробил *bot.* megastrobile, megastrobilus
мегасферический megaspheric, megalospheric
мегатектоника megageotectonics
мегатерий, мегатериум *paleont.* megathere *(Megatherium)*
мегатерм megatherm
мегатонна megaton
мегатроф megatrophic plant
мегафанерофит *bot.* megaphanerophyte
мегафарада *el.* megafarad
мегафит megaphyte
мегафлопс *comp.* megaflop
мегафон megaphone, speaking trumpet
мегаэволюция megaevolution
мегаэзофагус *med.* megaesophagus, megaloesophagus
мегаэлектрон-вольт megaelectron-volt, Mev
мегаэрг *phys.* megaerg
меггер *el.* megohmmeter, megger
мегера *myth.* Megaera
мегом *el.* megohm
мегом(м)етр *el.* megohmmeter, megger
мед 1. honey; 2. mead
~, **дикий** *bot.* golden-shower senna *(Cassia fistula)*
~, **медицинский** mel, clarified honey
медалист, медальер medal winner, medal(l)ist
медаль medal
~, **памятная** commemorative medal
медальон medallion, locket
медбрат male nurse
медведеобразный bearlike
медведина bearskin, bearhide
медведица 1. she-bear; 2. *zool.* milkweed moth *(Euchaetias egla); pl.* tiger/tussock moths *(Arctiidae)*
~, **Большая** *astr.* the Great Bear, Ursa Major
~, **Малая** *astr.* the Little Bear, Ursa Minor
медведка 1. *tech.* bear punch, punch press; 2. bearskin, bearhide; 3. *zool.* mole cricket *(Gryllotalpa);* 4. handcart; *mil.* truck carriage
медведь 1. *zool.* bear *(Ursus);* 2. bearskin
~, **белый** polar bear *(Euarctos maritimus)*
~, **серый** grizzly bear *(Ursus arctos)*
~, **цепкохвостый** kinkajou *(Potos flavus)*
медведь-губач *zool.* sloth bear *(Melursus ursinus)*
медвежий ursine, bear *(attr.)*
~**жье ушко** *bot.* common mullein *(Verbascum thapsus)*
медвежина bearskin, bearhide
медвуз *abbr.* **медицинское высшее учебное заведение** medical college
медвытрезвитель medical facility for alcoholics

медвяный melliferous
~**ная роса** honeydew
меделянка, меделянская собака mastiff
меденосный cupriferous, cupr(e)ous
медеплавильный copper-smelting
медепрокатный copper-rolling
~ **завод** copper mill
Медея *myth.* Medea
медж(и)лис Majlis
медиальный medial, median, central
медиана *math.* median; median value
~, **выборочная** *stat.* sample median
медиановый median
медианта *mus.* mediant
~, **нижняя** submediant
медиастинит *med.* mediastinitis
~, **гнойный** suppurative mediastinitis
медиастиноперикардит *med.* mediastino-pericarditis
~, **рубцовый/слипчивый** cicartizant/adhesive mediastinopericarditis
медиатизация *hist.* mediatization
медиатор I *immun.* mediator, amboceptor
~, **тормозной** inhibition mediator
медиатор II *mus.* plectrum
медиатриса *math.* midperpendicular
медиация mediation; arbitration
медиеваль medieval (type)
медиевист medievalist
медиевистика medieval history
медийский *hist.* Median
медик medical student
медикамент drug, medicine, medicament, remedy
медикаментозный medicamental, medicamentous
медико-биологический medicobiologic
медико-санитарный medicosanitary
медико-судебный medicolegal
медико-хирургический medicochirurgical, medicosurgical
медико-юридический medicolegal
мединал *pharm.* barbital sodium
мединститут *abbr.* **медицинский институт** school of medicine
медиовентральный medioventral
медистый cupriferous, cupr(e)ous
~**тое золото** copper gold
медитация meditation; thought
медиум I medium
медиум II *mus.* middle register
медицейский *hist.* Medicean
~**кая Венера** Venus de Medici
медицина medicine
~, **арабская** Arabic medicine
~, **военная** military/war medicine
~, **военно-морская** naval medicine
~, **восстановительная** rehabilitation medicine
~, **городская** urban medicine
~, **государственная** public/state medicine
~, **домашняя** domestic medicine
~, **космическая** space medicine
~, **народная** folk medicine
~, **профилактическая/предупредительная** preventive medicine
~, **радиационная** radiation medicine
~, **сельская** rural medicine
~, **судебная** forensic/legal medicine, medical jurisprudence
~, **схоластическая** eclectic medicine
медицинбол *sport* medicine ball
медицинский medical; medicinal
~**кая сестра** (hospital) nurse
~**кое свидетельство** certificate of health
медиэвист *see* **медиевист**
медкомиссия *abbr.* **медицинская комиссия** medical board
медленновращающийся slowly rotating
медленнодействующий slow-operating/-acting
медленный 1. slow; slack; tardive, tardy; 2. gently sloping
медлитель temporizer, protractor
медлительный slow, tardy; sluggish; anergic
меднение copper plating
медненый copper-plated
медник coppersmith, whitesmith; tinker
медницкая *n.* copper smithy
меднозакисный cuprous
меднозеленый aeruginous
меднокислый cuprate
меднолитейный copper-founding/-smelting
~**ная** *n.* copper foundry
медноплавильный copper-smelting
~ **завод** copper works
медноплакированный copper-plated
меднопрокатный copper-rolling
~ **завод** copper-rolling mill
меднорудный copper-ore
медный 1. copper(y); brazen; 2. *chem.* cupric, cupreous; 3. copper-colo(u)red
~ **век** the Age of Copper
~ **колчедан** chalcopyrite, copper pyrite
~ **купорос** blue vitriol
~**ные деньги** copper coins/money
медобслуживание *abbr.* **медицинское обслуживание** medical care
медоварение mead brewing
медоварня, медоваренный завод mead brewery
медовик *bot.* 1. nectary, honeycup; 2. bristlestem hemp nettle *(Galeopsis tetrahit)*
медовка 1. *bot.* balm *(Melissa);* 2. *zool.* apple psylla/sucker *(Psylla mali)*
медовый honeyed; melleous, nectar(e)ous
~ **камень** *miner.* mellite
~ **месяц** honeymoon
~**вая вода** hydromel
~**вая ямка** nectarostygma
медогонка honey extractor/centrifuge
медоед *zool.* honey badger *(Melivora capensis)*
медок I *bot.* nectar
медок II Medoc (wine)

медонос *bot.* honey/nectariferous plant
медоносный nectariferous, melliferous; mellifluous
  ~ цветок melliferous flower
  ~ная пора honey pit/pore
  ~ная чешуя nectariferous scale
медосбор honey gathering (by bees); honey harvest; yield of honey, honey yield
медосмотр *abbr.* **медицинский осмотр** medical examination
медосос *zool.* honeysucker, honey eater *(Meliphagidae)*
медососущий mellisugent, honey-sucking
медоточивый mellifluous, mellifluent
медохранилище nectarotheca
медоцветковый melianthous
медперсонал *abbr.* **медицинский персонал** medical staff
медпомощь *abbr.* **медицинская помощь** medical aid/service
медпункт *abbr.* **медицинский пункт** first aid station; medical centre/station, aidpost; *pl.* health facilities
медресе madrasa(h), madrassah, madrasseh; Mohammedan college, mosque school
медсанбат *abbr.* **медико-санитарный батальон** 1. medical sanitary battalion; 2. field hospital; battalion's first-aid station
медсестра *abbr.* **медицинская сестра** (hospital) nurse, sister
медуза 1. *zool.* medusa (*pl.* -sae), jellyfish; 2. *myth.* Medusa
медузовидный, медузообразный medusoid, medusiform
медузовый medusan, medusal
медуллизация medullization
медуллоза medulla
медуница *bot.* lungwort *(Pulmonaria)*
медфак *abbr.* **медицинский факультет** department of medicine; university medical school
медь 1. copper, cuprum, Cu; 2. *coll.* copper coins, coppers; copper utensils
  ~, азотнокислая copper nitrate
  ~, белая German silver
  ~, бескислородная oxygen-free copper
  ~, водная кремнекислая dioptase, copper emerald
  ~, деревянистая wood copper, fibrous olivenite
  ~, железистая ferric copper
  ~, желтая orichalc, brass
  ~, зейгеровать liquate copper
  ~, йодистая cuprous iodide
  ~, катодная брикетированная coalesced copper
  ~, конвертерная blister copper
  ~, кремнекислая copper silicate
  ~, лимоннокислая cupric/copper citrate
  ~, листовая sheet copper, copper sheets
  ~, литейная casting copper
  ~, мышьяковистая copper arsenide
  ~, нагартованная cold-worked copper
  ~, прутковая bar copper
  ~, самородная native copper
  ~, сернистая cupric sulfide/sulphide
  ~, сернокислая copper/cupric sulphate/sulfate
  ~, технически чистая tough pitch copper
  ~, ускуснокислая copper acetate
  ~, фосфорнокислая copper phosphate
  ~, хлористая cuprous chloride
  ~, хлорная cupric chloride
  ~, черновая blister/black copper
  ~, чушковая pig copper
  ~, штыковая cake copper
  ~, электролитическая electrolytic copper
медьсодержащий cupriferous, copper-bearing
медянистый coppery, cuprous; copper-colo(u)red
медяница *zool.* blindworm, slowworm *(Anguis fragilis)*
  ~, грушевая pear psylla *(Psylla pyricola)*
  ~, яблочная apple sucker *(Psylla mali)*
медянка I *chem.* olympic green, verdigris
медянка II *zool.* grass snake *(Tropidonatus natrix)*
межа boundary, balk
межагрегатный interaggregate, intergranular
межальвеолярный interalveolar
межамбулякр *zool.* interambulacrum
межамериканский inter-American
  ~ банк развития Inter-American Development Bank, IADB
межатомный interatomic
межбедренный interfemoral
межбороздный intercolpate, intercolpar
межведомственный interdepartmental, interagency
межвертельный *anat.* intertrochanteric
межвидовой interspecific
  ~ гибрид interspecific/genetic hybrid
  ~вая конкуренция interspecific competition
межворосинковый, межворосинчатый intervillous
межвузовский interuniversity; intercollegiate
межгалактический intergalactic
межганглионарный interanglionic
межглазничный *anat.* interorbital
межгорный intermontane, intermountain
межгорье intermontane region
межгосударственный intergovernmental
междиалектный interdialect(al)
междисциплинарный interdisciplinary
междолевой interlobar
междольковый, междольчатый interlobular
междометие *gram.* interjection
междоузельный interstitial, internodal
междоузлие interstitial site; *bot.* internode; joint, neck
междоусобие, междоусобица civil/intestine strife; civil war
междоусобный intestine, internecine
междуатомный interatomic
междублочный interblock, interunit

междуведомственный interdepartmental, interagency
междувидовой interspecific
междугородный interurban, intercity, intertown
междугранулярный intergranular
междужилковый interveinal
междузвездный interstellar
междукаскадный interstage
междуквартильный *math.* interquartile
междуклапанный intervalvular
междуклеточный intercellular
междуколонный *arch.* intercolumnar
междукоммутаторный interswitchboard
междукристаллический intercrystalline
междуледниковый interglacial
междулежащий intermediate
междумолекулярный intermolecular
междумускульный intermuscular
международник specialist on international law/affairs
международный international
~ный банк реконструкции и развитии, ИБРР International Bank for Reconstruction and Development, IBRD
~ный валютный фонд, МВФ International Monetary Fund, IMF
~ный институт общественно-политических наук International Institute of Social and Political Sciences, IISPS
~ный союз защиты детей International Union for Children Welfare, IUCW
~ный союз судовладельцев International Shipping Federation, ISF
~ный союз телекоммуникации International Telecommunication Union, ITU
~ная ассоциация развития, МАР International Development Association IDA
~ная комиссия по механизации сельского хозяйства International Commission of Agricultural Engineering, ICAE
~ная конвенция по рыболовству International Fisheries Convention, IFC
~ная организация гражданской авиации International Civil Aviation Organization, ICAO
~ная организация по стандартизации International Standardization Organization, ISO
~ная организация предпринимателей International Organization of Employers, IOE
~ная организация труда, МОТ International Labo(u)r Organization, ILO
~ная палата по судоходству International Chamber of Shipping, ICS
~ная стандартная отраслевая классификация International Standard Industrial Classification, ISIC
~ная торговая палата International Chamber of Commerce, ICC
~ная финансовая корпорация International Finance Corporation, IFC
~ное общество почвоведения International Society of Soil Science, ISSS
~ное общество почвоведов International Association of Soil Science, IASS
~ное право international *law*
~ное профсоюзное объединение Trade Unions International, TUI
междуначалие interim period
междуобмоточный interwinding
междуосный interaxial
междупарье *agr.* interfallow time
междупланетный interplanetary
междуплеменной intertribal
междуполюсный interpolar
междупроводной wire-to-wire
междупутье *rail.* track spacing
междуреберный *anat.* intercostal
междурельсовый intertrack, between rails
междуречный interfluvial
междуречье interfluve, interstream area
междурядный *agr.* interrow, between rows
междурядье *agr.* row-spacing/-width, planting width, space between rows
междуслойный interlaminar, interlayer *(attr.)*
междусоюзнический interallied, inter-Allied
междустрочие *print.* interlinear space
междустрочный interlinear
междуузелковый interganglionic
междуцарствие interregnum
междуширотный interlatitudinal
междуэлектродный interelectrode *(attr.)*
междуэтажный interfloor *(attr.)*
междуядерный internuclear
межевальный measuring, surveying
межевание, межевка *geod.* (land) surveying, chorometry
~, генеральное ordnance survey
межевик land surveyor
межевой boundary *(attr.); n.* land surveyor
~ знак boundary mark, landmark
меженный low-water
~ расход dry-weather discharge
~ сток low-water flow
межень low-water, normal water level/season
межжаберный interbranchial
межжелудочковый interventricular
межзаводский, межзаводской interfactory, between factories
межзвездный interstellar
межзвеневой interlinkage
межзональный interzonal
межзубный interdental
межионный interionic
межкамерный interlocular, intercameral
межкаскадный *el.* interstage
межклапаннный *anat.* intervalvular
межклассовый interclass; intergroup
межклетник, межклеточник intercellular duct/space
межклеточный *biol.* intercellular
~ ход intercellular duct/channel

межклоновый interclonal
межключица *anat.* interclavicle
межколхозный interkolkhoz
межкостный interosseous, interosseal
межкристаллический intercrystalline
межледниковый interglacial
  ~, афтонское Aftonian Interglacial
  ~, последнее Ultimate/Last Interglacial
  ~, предпоследнее Penultimate Interglacial
межлежащий intermediate
межлиночный intermolt
межличностный interindividual
межлобковый, межлонный interpubic
межлонжеронный interspar
межлопаточный *anat.* interscapular; *tech.* blade-to-blade, between blades
межмерзлотный intrapermafrost
межмышечный intermuscular
межмыщелковый intercondylar, intercondyloid
межнациональный international
межневральный interneural
межневронный, межневрональный interneuronal
межножечковый, межножечный interpeduncular
межножковый intercrural
  ~ узел intercrural ganglion
межнуклонный internucleon(ic)
межобластной interregional
межорбитальный *anat.* interorbital
межосевой interax(I)al
межостистый interspinal, interspinous
межостровной interisland
межпальцевой interdigital
межпарламентский interparliamentary
межпаровой *agr.* interfallow
межперегородочный interseptal
межпланетный interplanetary
межпластиночный, межпластинчатый interlamellar
межпластовой interstratal, interleaved
межплеменной intertribal
межплодник *bot.* mesocarp
межплоскостной interplanar
межплюсневой *anat.* intermetatarsal
межповерхностный interfacial
межпозвонковый, межпозвоночный *anat.* intervertebral
межполивной between irrigation
  ~ период irrigation interval
межполовой intersexual
межпородное скрещивание *biol.* crossbreeding, crossing
межпотоковый interstream
межправительственный intergovernmental
  ~ная консультативная организация по вопросам мореплавания Intergovernmental Maritime Consultative Organization, IMCO
межприливный intertidal
межпучковый *bot.* interfascicular
межрайонный interdistrict
межрасовый interracial; transracial
межреберье *anat.* intercostal space
межремонтный between repairs
межрешеточный interlattice
межродовой intergeneric
межручной interarm
межсвязочный interligamentous
межсегментарный, межсегментный intersegmental
межседалищный intersciatic
межсезонный between seasons, off-season
межсезонье period between seasons
межсемядольный *bot.* intercotyledonary
межсессионный interim, intersession
межслойный interlaminar; between layers
межсовхозный serving several state farms
межсоединение interconnection
межсортовой intervarietal
межсосковый intermammillary
межсосочковый interpapillary
межсосудистый intervascular
межсоюзный interunion, between unions
межстрочный interlinear
межсуставный *anat.* interarticular
межтеменной *anat.* interparietal
межточечный interdot *(attr.)*
межтрапповый *geol.* intertrappean
межтычиночный interstaminal
межуниверситетский interuniversity; intercollegiate
межуточный *physiol.* interstitial; intermediate
межфабричный interfactory
межфазный, межфазовый interphase *(attr.)*
межфаланговый *anat.* interphalangeal
межхрящевой *anat.* interchondral, intercartilaginous
межцеховой intershop, interdepartmental
межчастичный interparticle
межчелюстной *anat.* intermandibular, intermaxillary
межчерешковый *bot.* interpetiolar
межчерпаловидный *anat.* interarytenoid
межштатный interstate
межъягодичный *anat.* intergluteal
межъядерный internuclear
межъязыковой interlingual, interlanguage
межэлементный interunit, interelement
мезальянс mismarriage, mesalliance
мезадиод mesa diode
меза-область mesa
мазархный *bot.* mesarch
мазаструктура mesa structure
  ~, диффузионная diffused mesa structure
мезатранзистор mesatransistor
  ~, сплавной alloy mesa transistor, AM transistor
мезга 1. *bot.* alburnum; sapwood; 2. vegetable pulp
мездра scrapings, parings *(of hide)*, hide shreds, inner side of hide
мездрить scour, scrape, skive

мездряк fleshing/paring/scraping knife
мезембриантемум *bot.* ice plant, mesembryanthemum *(Mesembryanthemum)*
мезентер(и)альный mesenterial, mesenteric
мезентерий *anat.* mesentery
мезенхима mesenchyme, mesenchyma
~, внезародышевая extraembryonic mesenchyme
~, эстрогенпродуцирующая (o)estrogen-producing mesenchyme
мезенхимальный, мезенхимный mesenchymal
мезит *geol.* intermediate rock, semi-basic rock
мезитил *chem.* mesityl
мезоатом *phys.* mesic/mesonic atom, meson
мезобласть mesoblast, mesoderm
мезогамия *bot.* mesogamy
мезогеосинклиналь mesogeosyncline
мезогидрофит mesohydrophyte
мезогидрофитный mesohydrophytic
мезоглея *zool.* mesogloea
мезоглиома *med.* mesoglioma, oligodendroglioma
мезогнейс *geol.* mesogneiss
мезоглия *med.* mesoglia, oligodendroglia
мезодерма *biol.* mesoderm
~, внезародышевая extraembryonic mesoderm
мезодермальный, мезодермический mesodermal, mesodermic
мезозавр *paleont.* mesosaur *(Mesosaurus)*
мезозой *geol.* Mesozoic (era)
мезозойский Mesozoic
мезозона mesozone
мезокарп(ий) *bot.* mesocarp
мезоколопликация *med.* mesocoloplication, mesocolopexy
мезокольпиум mesocolpium, intercolpium
мезокотиль *bot.* mesocotyle
мезократовый *geol.* mesocratic
мезоксалмочевина *med.* mesoxalylurea, alloxan
мезоксерофильный *bot.* mesoxerophilous
мезоксерофит *bot.* mesoxerophyte
мезол *geol.* mesole, tromnosite
мезолимфоцит *med.* mesolymphocyte
мезолит *geol.* mesolite, mesole; Mesolithic (period)
~, волокнистый antrimolite
мезолитический, мезолитовый mesolithic
мезология mesology, ecology
мезомер mesomer
мезомерия mesomerism
мезомитоз mesomitosis
мезомолекула mesomolecule, mesonic molecule
мезоморфный mesomorphous, mesomorphic
мезон *phys.* meson
~, фоторожденный photoproduced meson
мезоневрит *med.* mesoneuritis
мезонефрос *med.* mesonephros, Wolffian body
мезоний mesonium
мезонин 1. attic; 2. mezzanine; 3. penthouse
мезонный *phys.* mesonic
мезонэкзина mesonexine
мезопауза mesopause
мезопик mesopeak, stratopause
мезоплейстоцен Mesopleistocene
мезоподий, мезоподиум *bot., zool.* mesopodium
мезопора mesopore
мезопориум mesoporium, interporium
мезопорода *geol.* mesorock
мезорельеф mesorelief
мезосигмопликация *med.* mesosigmoidopexy, mesosigmoidoplication
мезоскаф mesoscaph(e)
мезосома mesosome, mesosoma
мезоспора, мезоспорий *bot.* mesospore, mesosporium
мезостык mesojuncture
мезосфера mesosphere
мезотека mesotheca
мезотелий *anat.* mezothelium
мезотендиний *anat.* mesotendon
мезотермальный, мезотермический mesothermal
мезотермный mesothermic, mesothermal
мезотеций *bot.* mesothecium
мезотип mesotype
мезотрон *phys.* meson, mesotron
мезотроф *bot.* mesotrophic plant
мезофанерофит *bot.* mesophanerophyte
мезофизин *med.* intermedin
мезофилл *bot.* mesophyll
мезофилльный mesophyllous
мезофильный mesophile, mesophilic, mesophilous
мезофит mesophyte; Mesophytic era
мезофитный, мезофитовый mesophytic
мезофлоэма *bot.* mesophloem
мезофрагма *anat.* mesophragma, Hensen's line
мезофреатофит mesophreatophyte
мезохилий *bot.* mesochilium
мезоцефалический *anat.* mesocephalic, mesocephalous
мезоцефалия *anat.* mesencephalon, mesocephalon, mesocephaly
мезоцикл mesocycle
мезоцит mesocyte
мейергофферит *miner.* meyerhofferite
мейоз *biol.* m(e)iosis; meiotic division
~, двухстадийный two-step miosis
~, исходный initial miosis
мейоспора meiospore
мейоспорангий *bot.* meiosporange, meiosporangium
мейотаксис, мейотаксия meiotaxy
мейофиллия *bot.* meiophylly
мейстерзингер Meistersinger, mastersinger
мекзина mexine
Мекка Mecca
меконий *biol.* meconium
меконопсис *bot.* blue Tibetan poppy *(Meconopsis)*
мексиканец, мексиканский Mexican
мексиканка Mexican (woman)

мел 1. chalk; whitewash, whiting; 2. Cretaceous period/system; 3. *acous.* mel
~, озерный lacustrine chalk
~, чистый ground/true chalk
мелаконит *miner.* melaconite, black copper
мелалеука *bot.* melaleuca, cajeput *(Melaleuca)*
мелальгия *med.* melalgia, meralgia, sciatica
меланезиец Melanesian
меланезийка Melanesian (woman)
меланезийский Melanesian
меланж melange
меланжер 1. melangeur; 2. blood-count pipette
меланжирование colour blending
меланин *bioch.* melanin
меланит *miner.* melanite
~, серовато-черный pyreneite
меланобласт *biol.* melanoblast
меланогенез melanogenesis
меланоглоссия *med.* black tongue, melanoglossia
меланодермия *med.* melanoderma, melasma
~, паразитарная parasitic melanoderma, vagabondia
~, старческая melanoderma senile
меланоз 1. *med.* melanosis, melanism; 2. *plant pathol.* Septoria black leaf blight
~, врожденный congenital melanosis
~, диффузный disseminated melanosis
~ кишечника intestinal melanosis
~ кожи dermal melanosis
~, профессиональный occupational melanosis
маланокарцинома *med.* melanocarcinoma, black cancer
меланома *med.* melanoma, melanocarcinoma
~, злокачественная malignant melanoma, melanoma de novo, melanotic malignant tumo(u)r
~, юношеская juvenile melanoma
меланоплакия melanoplacia, melanoplakia
меланоптоховый melanoptochous, hololeucocratic
меланостомиевые *n. pl. zool.* scaleless dragon fishes *(Melanostomiatidae)*
меланофор *biol.* melanophore
меланхолик melancholiac, melancholic person
меланхолический, меланхоличный melancholic, atrabilious, atrabilarious
меланхолия *med.* melancholia, melancholy, athymy, barythymia
~, возвратная recurrent melancholia
~, гипохондрическая hypochondriacal melancholia
~, климактерическая climacteric melancholia
~, простая melancholia simplex
меласса molasses, blackstrap
мелассообразователь molasses-maker/-former
мелатонин *bioch.* melatonin
мелена *med.* melena
~, истинная melena vera, true melena
~, ложная melena spuria
~, новорожденных melena of the newborn
меление chalking
мелецитоза *chem.* melezitose
мелиант *bot.* honeyflower, honey bush *(Melianthus)*
мелиациевые, мелиевые *n. pl. bot.* mahogany family *(Meliaceae)*
мелибиаза raffinase, melibiase
мелибиоза *chem.* melibiose, isolactose
мелизм(а) *mus.* grace note
мелинит *chem.* melinite, perlite
мелиоидоз *med.* melioidosis, malleoidosis, Whitemore's disease
~, подострый subacute melioidosis
мелиоративный, мелиорационный land-improvement/-reclamation, (a)meliorative
мелиоратор specialist in land reclamation/improvement
мелиорация *agr.* land reclamation/improvement/cleaning, drainage and irrigation, (a)melioration
~, агротехническая cultivating reclamation
~ земель improvement of land
~ земли land reclamation
~, обводнительная irrigation amelioration
~, осушительная drainage amelioration
~ почвы soil development
~, техническая engineering reclamation
мелиорирование land reclamation
мелиорировать reclaim, ameliorate, improve (soil)
мелис granulated/sand sugar
мелисса *bot.* (sweet) balm, melissa, lemon lobelia, sweet Mary *(Melissa)*
~, аптечная, лекарственная common balm *(M. officinalis)*
мелитемия *med.* melitemia, glycemia, glycosemia
мелитурия *med.* mel(l)ituria, glycosuria
мелить chalk; polish with chalk
мелия *bot.* chinaberry *(Melia)*
мелкий 1. small, petty; light, fine; 2. shallow; shoaly, fordable; 3. fine, minute
~ дождь light drizzle
~ песок fine sand
~ рогатный скот goats and sheep
~ чиновник minor official
~ шаг *tech.* fine pitch
~кая буржуазия petty bourgeoisie
~кая жилка veinlet
~кая кисть raceme, racemule
~кая кража *law* petty larceny
~кие разходы petty expenses, pocket money
~кое хозяйство small/petty farming
мелко fine, minutely
мелкобахромчатый fimbrillate
мелкобородавчатый verruculose
мелкобородчатый barbellate, barbellulate, barbulate
мелкобрикетированный pellety
мелкобугристый pit-and-mound *(of microrelief)*
мелкобуржуазный *polit.* petty-bourgeois
мелковееробразный *bot.* flabelliform
мелковетвистый *bot.* ramulose
мелковильчатий furcellate

мелковкрапленный *geol.* disseminated
мелководный shoaly, shallow(-water)
мелководье shallow/shoal water, shoaliness
мелковолокнистый fine-fibrous, close-grained, fibrillate
мелковолосистый microtrichous
мелковыемчатый finely sinate, sinuolate
мелкоголовчатый *bot.* capitellate
мелкогорный low-/small-mountain
мелкогородчатый crenellate, crenulate
мелкогорье low-mountain topography
мелкогребешковый small-combed/-crested, finely cristate
мелкоделяночный *agr.* small-/micro-plot
мелкодисперсный finely divided
мелкодольчатый lacinulate
мелкодонный shallow-bottomed *(boat/river)*
мелкодробленный finely ground/fragmented
мелкожилковатый venulose, nervulose
мелкозазубренный crenulate, serrulate, crenellate
мелкозвездчатый stellular, stellulate
мелкозем silt, aleurite, fine earth
мелкозернистый fine-/close-/small-grained, fine (-granular), parvigranulate, microgranular
мелкозубчатый crenulate, denticulate, serrulate; fine-toothed *(of saw)*
мелкозубый fine-/close-toothed, microdontous
мелкоиглистый small-acanthous/-spinous, finely acanthous
мелкоигольчатый *bot., zool.* echinulate
мелкокистевидный *bot.* racemulose
мелкоклеточковый parvicellular
мелкоклеточный microcellular
мелкоклетчатый close-meshed
мелкоклочковатый flocculose
мелкоклювистый rostellate
мелкоколосый *bot.* small-spiked
мелкоколючковый spinulose, spinulate
мелкокомковатый *agr.* crumbly
мелкокорневой shallow-rooted
мелкокристаллический fine(ly) crystalline, microcrystalline, microcrystallitic
мелкокрупитчатый *bot.* grumose, knobby
мелкокрючковатый hamulate
мелкокустарниковый *bot.* fruticulose
мелколепестник *bot.* fleabane *(Erigeron)*
мелколепестный *bot.* small-petalled, micropetalous
мелколесье brush, low/brush forest, scrubs, thicket of half-grown trees
 ~, ковровое tapestry forest
мелколиственный small-leaved, parvifoliate
мелколистность *plant pathol.* frenching
мелколистный small-leaved, microphyllous
мелколопастный lobulate
мелкомасштабный small-scale
мелкоморщинистый *bot., zool.* rugulose
мелкоморье shallow (sea)
 ~, неподвижное stabile shelf
 ~, подвижное labile shelf
мелкомшистый *bot.* cobwebbed
мелконадрезной lacinulate
мелкоперегородчатый septulate
мелкоперистый pinnulate
мелкопестичный *bot.* small-pistilled
мелкопетлистый close-/fine-meshed
мелкопыльчатый serrulate, serratulate
мелкоплодник *bot.* mountain ash *(Micromeles, Sorbus)*
мелкоплодный *bot.* microcarpous, small-fruited
мелкопокровный *bot.* microchlamydeous
мелкополосчатый finely striate, striolate
мелкопоместный small *(of landowners)*
 ~ное дворянство small gentry
мелкопористый fine-pored/porous, finely porous, fine-perforate
мелкопродырявленный acusporide
мелкопятнистый finely spotted
мелкораздробленный finely divided, comminuted
мелкоребристый fine-ribbed
мелкореснитчатый ciliolate
мелкорешетчатый cribrate, cribral, cribellate
мелкорослый undersized, small, little, tiny
мелкосемянный small-seeded, microspermous; *n. pl. bot. Microspermae*
мелкосерийный small-scale/-lot/-series; short-run
 ~ное производство small-scale/-lot manufacture, short-run/-scale production, limited manufacture/production
мелкосетчатый close-/fine-meshed
мелкосидящий *nav.* shallow-draught
мелкоскладчатый *bot., zool.* rugulose
мелкослоистый finely layered, thinly laminar
мелкослойный fine-grained *(of wood)*
мелкособственнический of small/petty owners
 ~кая психология small-proprietor mentality
мелкосопочник hillocky area
мелкостебельчатый *bot.* small-stalked
мелкоструктурный fine-structure
мелкость 1. shallowness; 2. fineness, smallness; 3. *math.* mesh
мелкотемие narrow specialization
мелкотоварный *econ.* small-scale
мелкоузловатый micronodular, minutely nodular
мелкоцветковый, мелкоцветный *bot.* micranthous, parviflorous, small-flowered
мелкочашечковый, мелкочашечный microcalycular
мелкочешуйчатый scurfy, finely squamosed, squamulose, squamulate
мелкошерстистый lanuginose, lanuginous
мелкошерст(н)ый short-wooled
мелкошиповатый micracanthous, subechinate, echinulate, aculeolate
мелкоштрихованный striolate, finely striate
мелкощелистый rimulose
мелкощетинистый setulose, strigillose
мелкощитовидный clypeolate
мелкоязычковый finely ligulate, small-tongued

мелкоямчатый foveolate; punctulate(d)
мелкоячеистый close-/fine-meshed; foveolate
меллит *miner.* honeystone, mellite
мелованный chalked
меловой 1. *geol.* Cretaceous; 2. chalky
~вая бумага art paper
мелодраматичный, мелодраматический melodramatic
мелокактус *bot.* melon cactus *(Melocactus)*
меломан lover of music
мелотипия music-printing
мелочь *min.* fines, chips
~, буровая borings, (drilling) cuttings
~, брикетная briquet fines
~, каменная stone chips
~, карьерная pit-run fines
~, комковая braize, breeze
~, рудная ore fines, fine ore
~, слюдяная scrap mica
~, угольная dust coal, coal fines
мель shoal, (sea) shallow; bank; sand bar
~, морская sea shallow/shoal
~, песчаная sand bar/bank
~ посредине фарватера *nav.* middle ground
~, речная каменистая cripple
мельзавод flour mill
мелькание flicker(ing), flashing, sparkling
~ строк line flicker
~ цвета colo(u)r break-up
~ цветов colo(u)r flicker
мелькать flicker; flash
мелькомбинат milling/grinding plant
мельница mill; grinder
~, аэробильная paddle-type mill
~, ветряная wind mill
~, вибрационная vibratory mill
~, водяная water mill
~, глетчерная *geol.* glacier-mill, glacial pot-hole, moulin
~, грубого помола coarse-grinding mill
~, дисковая disk mill
~ для помола пробы sampling mill
~, жерновая burr(-stone) mill, millstone
~, кофейная coffee grinder
~, ледниковая glacier-mill, glacial pot-hole, moulin
~ мелкого помола pulverizer, pulverizing mill
~, молотковая hammer mill
~, мукомольная flour(ing) mill
~, паровая steam-operated mill
~ тонкого помола pulverizer, pulverizing mill
~, ударная impact mill
~, шаровая ball crusher/mill
~, шахтная hammer mill
мельница-вентилятор pulverizing fan
Мельпомена *myth.* Melpomene
мельхиор German silver, cupronickel
мельчайший submicroscopic(al)
мембрана diaphragm, membrane; film
~, базальная basilar membrane
~, бороздковая colpar membrane
~, волнистая undulate membrane
~, гармониковая sylphon, bellows
~, гофрированная corrugated diaphragm
~, ионоизбирательная ion-selective membrane
~, клеточная cell membrane
~, коническая cone diaphragm
~, легочная pulmonary membrane
~, лицевая *leath.* grain membrane
~ микрофона microphone diaphragm
~, плавникообразная membranaceous ala
~, плазматическая plasma membrane
~, полупроницаемая (для анионов) (anion-)semipermeable membrane
~, разделительная partitive membrane
~, разрывная blowout/bursting diaphragm
мембранипоровидный membraniporine
мембранный membranous, membranaceous, membranate
мемброновидный membranoid, membraniform, membranelike
мемноскоп memnoscope
мемориал 1. monument, memorial; 2. *econ.* day-book
мемориальный memorial *(attr.)*
~ная доска memorial plaque
~ная книга daybook
мемотрон memotron storage tube
мемфисский Memphian
мена *econ.* barter, exchange
менада *myth.* maenad; bacchante
менажер *sport* manager
менархе *physiol.* menarche, first menstruation
менгир *archeol.* menhir, monumental (standing) stone
менделевий *chem.* Mendelevium, Md
менделевский, менделистский *biol.* Mendelian
менделизм *med.* Mendelism, Mendelian inheritance
менеджер manager; impressario
менек *zool.* cusk *(Brosme brosme)*
менестрель *hist.* minstrel
мензула plane/surveyor's table, surveying panel, topographic drawing board
мензульный plane-table
~ метод plane-table method
мензурка laboratory/graduated beaker, measuring glass, graduated cylinder
менингеальный *anat.* meningeal, meningic
менингизм meningism
менингиома *med.* meningioma
~, злокачественная malignant meningioma
менингит *med.* meningitis
~, асимптомный/бессимптомный asymptomatic meningitis
~, гнойный purulent meningitis
~, листереллезный listeria meningitis
~, ограниченный local meningitis, arachnoiditis
~ основания мозга basilar meningitis

~, отогенный otitic meningitis
~, разлитой disseminated meningitis
~, серозный serous/alcoholic meningitis, wet brain, serous apoplexy
~, синегнойный pyocyanic meningitis
~, торулезный cryptococcal meningitis
~, эпидемический церебральный epidemic cerebrospinal meningitis, typhoid meningitis
менингококк *med.* meningococcus
менингококцемия *med.* meningococc(a)emia
менингоцеле *med.* meningocele
~, травматическое spurious/traumatic meningocele
менингоэнцефалит *med.* meningoencephalitis, meningocephalitis, encephalomeningitis
~, двухволновой diphasic meningoencephalitis
мениск *anat., math., opt.* meniscus; meniscus lens
~, вогнутый concave meniscus
~, выпуклый convex meniscus
~, осязательный tactile meniscus, (Merkel's) tactile disk
~, отрицательный negative/diverging meniscus
~, положительный positive/converging meniscus
~, суставной articular meniscus
менисковидный meniscoid, menisciform
менисковый meniscal
менискотерий *paleont.* meniscothere *(Meniscotherium)*
менискоцит *med.* meniscocyte, sickle/crescent cell
менискоцитоз *med.* meniscocytosis, sickle cell an(a)emia
мениспермин *chem.* menispermine
мениспермовые *n. pl. bot.* moonseed family *(Menispermaceae)*
мениспермум *bot.* monseed *(Menispermum)*
меннонит, менонит Mennonite
меновой *econ.* exchange
~вая стоимость exchange value
~вая торговля barter
менонитство Mennonitism
менопауза *med.* menopause, change of life
меноррагия *med.* menorrhagia, (hyper)menorrhea
меностазия *med.* suppressed menstruation, menostasia, amenorrhea
менструальный *physiol.* menstrual, menstrous, catamenial
менструации *physiol. pl.* menses, menstruations, menorrhea
~, викариирующие vicarious menstruations, menoplania, menoxemia, menometastasis
~, климактерические emmeniopathy, paramenia
~, неправильные menoxenia
~, редкие opsomenorrhea
~, слабые hypomenorrhea
менструация *physiol.* menstruation, menses, catamenia
~, болезненная painful menstruation, dysmenorrhea, menorrhalgia
~, ранняя precocious/premature menstruation, menstruatio precox
менструировать menstruate
ментализм mentalism, mentalistic theory
менталистический, менталистский mentalistic
ментол *chem.* menthol, pepermint camphor
менуэт *mus.* minuet
менхаден *zool.* menhaden *(Brevoortia)*
менцизия *bot.* menziesia, skunkbush *(Menziesia)*
меньшевизм *polit.* Menshevism
меньшевистский *polit.* Menshevist
меньшинство minority
меню *comp.* menu
~, главное root/main menu
~, древовидное tree-coded menu
~, консультационное help menu
~ с вытеснением нижней строки pull-down menu
~, экранное on-screen menu
менять change, alternate; barter
~, деньги (ex)change money
мепробамат *pharm.* meprobamate, equinal, meprospan, meprotabs, miltown
мера 1. measure, size, dimension; rate; 2. degree, extent; standard; gauge; 3. undertaking; *see also* **меры**
~ Бореля Borel measure
~ вероятности/вероятностная probability measure
~ веса measure of weight
~ вместимости measure of capacity
~, градусная grade/degree/angular/sexagesimal measure
~ длины linear measure
~, дуговая arc measure
~ Жордана Jordan measure
~ Лебега Lebesgue measure
~ Лебега-Стильтеса Lebesgue-Stieltjes measure
~, линейная linear measure
~, метрическая metric measure
~ надежности measure of reliability
~ наказания, высшая *law* capital punishment
~, образцовая reference standard
~ объема volume measure
~ площади square measure, measure of area
~ потребления measure of consumption
~ предосторожности precaution, precautionary measure, safeguard
~, предупредительная precaution, safeguard, preventive measure
~, рабочая working standard
~, радианная radian/circular measure
~ рассеяния measure of dispersion/scatter
~ скорости velocity/speed measure
~, стимулирующая expansionary action
~ стоимости *econ.* measure/standard of value
~, счетноаддитивная absolutely/countably additive measure

~ **сыпучих тел** dry measure
~ **точности** modulus/measure of precision
~, **угловая** 1. *geom.* angular measure; 2. angle standard
~ **эксцесса** *stat.* measure of kurtosis
~ **электрической величины** precision resistor/capacitor
мербромин *chem.* merbromin, mercurochrome
мерва waste product of beewax processing
мергелевание *agr.* marling, marl application
мергелевать *agr.* (apply) marl
мергелистый marly, marlaceous
мергель *geol.* marl; chalky clay
~, **глинистый** clay marl
~, **известковый** calcareous marl
~, **луговой** meadow marl
~, **пресноводный** limnetic marl, fresh-water marl
~, **пятнистый** variegated marl
~, **ракушечный** shell marl, crag
~, **суглинистый** loamy marl
мергельный marlaceous, marly
мердочит *geol.* murdochite
мережа trammel net
меренга *cul.* meringue
меренхима *bot.* merenchyma
меренхимный *bot.* merenchymatous
мереология mereology, Lesniewski system
мерея *leath.* grain (surface/pattern)
~, **накатанная** boarded grain
~, **нарезанная** embossed grain
мерзлота congelation, congealment, frost, frozen ground/state
~, **вечная/многолетняя** permafrost, everfrost, perpetually frozen ground, pergelisol
~, **реликтовая** relict frozen ground
~, **сухая** dry frost
мерзлотно-перегнойный frozen-humic
мерзлотно-подзолистый frozen-podzolic
мерзлотный cryogenic
~ **пояс** frost belt
мерзлотовед specialist in geocryology
мерзлотоведение geocryology, permafrostology
мерзлотомер cryopedometer
мерзлый frozen; *agr.* damaged by frost
меридиан meridian (line)
~, **географический** terrestrial/geographical meridian
~, **главный** principal/reference meridian
~, **гринвичский** Greenwich/initial/principal/prime/first meridian
~, **истинный** true meridian
~, **исходный** meridian of origin, base meridian
~, **начальный** principal/initial/Greenwich/prime/first/base meridian, meridian of origin
~, **небесный** celestial meridian, principal vertical circle, circle of declination
~, **нулевой** prime meridian
~, **осевой** central/mean/prime meridian, initial line
~, **основной** principal/reference meridian
меридианный, меридиональный meridional, meridian
меридионально-слитнобороздный syncolpate
меридионально-слитнополубороздный syndemicolpate
мерикарп(ий) *bot.* mericarp
мериклинальный mericlinal
~**ная химера** mericlinal chim(a)era
мерило criterion, standard
меринос merino (sheep/wool)
меристела *bot.* meristele
меристема *bot.* embryonic tissue, meristem
~, **вставочная** intercalary meristem
~**мы, развивающиеся из** meristogenetic
меристематический *bot.* meristematic
~**ная ткань** embryonic/meristematic tissue
меристемный *bot.* meristematic
меристический *biol.* meristic; segmented
меристогенный meristogenous
мериталлий, мериталлус *bot.* internode, merithallus
меритель measuring instrument
меритерий *paleont.* moerithere *(Moeritherium)*
меритократия meritocracy, meritocratic system
мерить measure, gauge
~ **глубину** sound the depth
мерка 1. measure; measurements; 2. gauge, yardstick, measuring rod
меркантилизм *econ.* commercialism, mercantilism
меркантилистический, меркантилистский mercantilistic
меркантильность commercialism, mercenariness
меркантильный mercantile, mercantilistic
меркаптан *chem.* mercaptan
меркаторский *cart.* Mercator's, Mercatorial
~**кая проекция** Mercator's projection
меркузал *pharm.* mersalyl, mercusal, salyrgan
меркуриализация mercurialization
меркуриальный mercurial, mercurous; containing mercury
меркурий *chem.* mercury; *myth., astr.* Mercury
меркурирование *chem.* mercuration, introduction of mercury (into)
меркурометрия mercurometry
меркурохлор (mild) mercurous chloride, calomel
меркурохром merbromin, mercurochrome
мерлан(г) *zool.* whiting *(Odontogadus merlangus)*
мерлуза *zool.* hake *(Merluccius)*
мерлушка morling, dressed lambskin
~, **крымская** crimmer, krimmer
мерник gauging/measuring/calibrated tank, metering vessel, batcher, batch box/meter, gauge box
мерный 1. measuring, gauging; mensural, dimensional; 2. uniform, rythmic(al)
~**ная речь** metric speech
мерогенез merogenesis
мерогенетический merogen(et)ic
мерогония *embryol.* merogony

мерозигота merozygote
мерозоит merozoite
мерокринный, мерокриновый *anat.* merocrine
~ные железы merocrine glands
мерокристаллический merocrystalline, hemicrystalline, semicrystalline
мероксен *miner.* meroxene
меромикросомия *med.* local dwarfism, meromicrosomia
меромиксис meromixis
мероморфизм *math.* meromorphic mapping, meromorphism
мероморфный *math.* meromorphic, fractional
мероопределение metrization; metrics; mensuration; definition of measure
меропланктон *biol.* meroplankton
мероприятие arrangement, procedure; provision, practice; legislative enactment; *pl.* action, measures, operations
~, противопожарное antifire measure
~, профилактическое preventive action
~тия, аварийно-профилактические accident preventive measures
~тия, организационно-технические administrative and technical measures
меросинклиналь merosyncline
меросистолический merosystolic, partially systolic
мероспорангий *bot.* merosporangium
меростоматы *pl., zool.* merostomes *(Merostomata)*
мероэдрический merohedric
мерсеризация mercerization, mercerizing
мерсеризировать mercerize
мерсийский Mercian
мертвец corpse, cadaver
мертвецкая *n.* mortuary, morgue
мертвоеды *pl., zool.* carrion/burying beetles *(Silphidae)*
мертворождаемость *med.* mortinatality, natimortality, stillbirth rate
~, антенатальная antenatal mortinatality
мертворожденный mortinatus, stillborn, deadborn; *n.* stillborn child
мертвый 1. dead, lifeless, cadaver(ic); 2. *n.* corpse, dead body
~ инвентарь *agr.* dead stock
~ фрахт *econ.* dead freight, DF
~ час quiet time, siesta
~ штиль *naut.* dead calm
~ якорь *naut.* permanent mooring anchor
~вая зыбь *naut.* swell
~вая петля *av.* loop
~вая точка *tech.* dead point
~вая хватка death grip
~вое пространство *mil.* dead ground
мертель lime mortar
~, огнеупорный refractory mortar
~, шамотный fireclay mortar
мертензия *bot.* smooth lungwort, bluebells *(Mertensia)*
мерцание 1. *tech.* flicker(ing), twinkling, (flickery) scintillation; glitter, shimmering; 2. *med.* fibrillation
~ звезд stellar scintillation, twinkling of stars
~ желудочков ventricular fibrillation
~ катода flicker effect
~ предсердий atrial/auricular fibrillation, atrial/auricular flutter
мерцательный *biol.* fibrillar(y), vibrating, vibratile; scintillating, twinkling
мерцать scintillate, twinkle, flicker; glimmer, glitter, flash
мерщик measurer
меры *pl.* measures, arrangements, provisions, steps
~ безопасности safety measures
~, ветеринарные animal quarantine measures
~ взыскания disciplinary measures
~, дюймовые *econ.* imperial dimensions
~, крутые crackdown
~ по предотвращению летных происшествий (air) accident prevention effort(s)
~ предосторожности security measures, safety precautions
~, санитарные sanitary/health measures
~ техники безопасности safety measures
~, фитосанитарные agricultural quarantine measures
месилка mixer, kneading machine; malaxator
мескалин *chem.* mescaline, peyote
мескито *bot.* mesquite *(Prosopis)*
мескитовый, мескитский *bot.* mesquite
месмеризм biomagnetism, mesmerism, animal magnetism
месмерист mesmerist, magnetist
месса *rel.* mass
Мессалина Messaline; profligate
мессдоза pressure capsule, force cell
мессерштихель knife tool
мессианизм Messianism
мессианский Messianic
мессидор *hist.* Messidor
мессия Messiah
местком *abbr.* **местный комитет** local (trade union) committee
месткомовец member of local (trade union) committee
местничать *hist.* observe order of precedence; contend for a position on hereditary grounds
местничество 1. *hist.* order of precedence/seniority; 2. *polit.* departmentalism, sectionalism; localistic tendencies
местнодействующий *med.* topical
местность locality, district, region; terrain, ground
~, асейсмичная non-seismic/aseismic region
~, волнистая undulating/rolling country
~, горизонтальная level ground
~, гористая highland, upland, mountainous region/area
~, зараженная contaminated area/ground

~, засушливая arid land
~, населенная (densely) populated area/locality
~, низменная low country/-land
~, низменная болотистая swamp, march
~, открытая open country
~, пересеченная rugged/broken ground, rugged terrain, broken country
~, пустынная desert, arid region
~ с уклоном sloping site
~, сеисмичная seismic/earthquake region, seismic country
~, холмистая hilly country/ground, undulating ground

местный 1. local, native, indigenous, edaphic, aboriginal, autochthonous; domestic; *med.* topical; 2. *gram.* locative
~ говор local dialect
~ житель local resident
~ная флора indigenous flora
~ное самоуправление municipal government

место 1. location; station, site, place, position, spot; 2. room, space; 4. job, post; 4. *theatre* seat; *rail.* berth; 5. *pl.* the country, the provinces
~ аварии crash scene
~, асимптотическое *math.* asymptotic spot
~ в боевом порядке *mil.* place in formation
~ в оболочке, вакантное/свободное shell vacancy
~ в решетке *cryst.* lattice site
~ в строю, держать свое *mil.* keep station
~, вакантное lattice/shell vacancy, vacant position
~ взлета take-off point
~ водителя driver seat
~ вставки *med.* insertion region
~ выброса seat of bump; ejection opening
~ высачивания seep(age)
~ выхода источника spring orifice
~, геометрическое вершины apex locus
~, геометрическое точек *math.* (geometric) locus
~, геометрическое точек прикосновения tac-locus
~, голубое *med.* blue place
~ дежурства *mil.* duty station
~ действия/происшествия scene of action
~, детское *med.* placenta
~ для гонки двигателей runup point
~ для хранения багажа luggage stowage (place)
~, доходное well-paid job, lucrative appointment
~ жительства place of residence, dwelling place
~ загрузки feed end
~ заключения jail
~ залегания (place of) occurrence
~ захоронения repository, cemetery, place of burial
~ захоронения радиоактивных отходов burial/dumping ground, graveyard
~ звезды, географическое subastral/substellar point
~ изгиба *geol.* bending point; *met.* bent section
~ излома breaking point
~ изображения, светлое *telev.* highlight
~ изображения, темное *telev.* low light
~, испытания test site
~, истинное *astr.* true place, *nav.* true position
~, исходное reference position
~ комплектования грузов consolidating point
~ корабля (в ордере) ship's station
~ Луны, географическое sublunar point/spot
~ наблюдений view point
~ назначения (point of) destination, place of destination
~, населенное populated place
~ ожидания stand-by station
~ ожидания при рулении *av.* taxi-holding position
~, определенное гидролокатором sonar fix
~, определенное графическим счислением graphical position
~, определенное по крюйс-пеленгу running fix
~, определенное радиолокационным способом radar fix
~, определенное способом Сомнера fix by Sumner's method
~ останова программы *comp.* break-point
~ отправления point of departure
~, отхожее latrine
~, пассажирское passenger seat/place
~ повреждения damaged area
~ погружения *naut.* point of submergence
~ погрузки loading berth
~ погрузки угля coaling place
~ получения багажа luggage claim area
~ посадки landing area
~ преступления scene of the crime
~ причала moorage
~ прогнозированное forecast position
~ происхождения place of origin
~ пуска *space tech.* launching site
~, пустое blank (space)
~ работы place of work, workplace
~, рабочее (work) station
~, рабочее учебное practice/learner's position
~ распространения habitat, place of occurrence
~, расчетное calculated/estimated position
~ ремонта и проверки place of repairs and check
~ рождения place of birth
~, рыбное fishing spot
~ самолета, МС *av.* aircraft/ground position, fix (point)
~ сбора данных data acquisition location, data collecting position
~ светила, видимое apparent place of star
~ светила, географическое stellar point
~, священное sacrosanct place
~, сидячее seat, place

~, **слабое** weak point
~ **совершения действия** *law* venue
~ **Солнца, географическое** sub-solar point/spot
~, **сообщать свое** *nav., av.* lend bearings
~, **ставить мину на** *naut.* draw home
~ **стоянки** parking lot/area; *nav.* berthing, berth; moorings; *av.* tarmac
~ **стоянки судна** moorage
~ **схождения путей** bottleneck
~, **счислимое** *nav.* dead reckoning/reckoned position, DR-position
~, **счислимо-обсервованное** running fix
~ **течи** leak spot, leaky area
~ **точек, геометрическое** locus
~ **углубленное якорное** dredged berth
~, **узкое** bottleneck; *econ.* gooseneck
~ **у причала** berth
~ **у стенки** quay berth
~, **фактическое** actual/true position
~ **хранения** stowage
~ **члена экипажа, рабочее** crew (member) station
~, **якорное** anchorage, anchor ground, anchoring (place)
~**та, лишать** *law* unseat
~**та, послать инструкции на** *polit.* mail out instructions to the provinces
местоблюститель *eccl.* locum tenens
местожительство dwelling/abiding place, place of residence
~, **постоянное** domicile
местоимение *gram.* pronoun
~, **личное** personal pronoun
местоимение-подлежащее pronominal subject
местоимение-числительное pronominal numeral, indefinite numeral
местоименно-вопросительный pronominal-interrogative
местоименно-соотносительный pronominal-correlative
местоименный *gram.* pronominal
местом(а) *bot.* mestome
местоназначение destination
местонахождение occurrence; whereabouts; locality, location, position, site, spot, seat
~ **программы** program residence
местообитание habitat, locality, site
местоопределение position finding, fixing
местоположение location, position, seat; situation; locality, site, spot; emplacement
~, **определять** locate
~ **скважины** well location
местопребывание abode, abiding place, residence, whereabouts
местопроизрастание *bot.* site, habitat
местопроисхождение provenance, origin
месторасположение 1. location, locality, site, situation; locus; 2. *geol.* occurrence, deposit, field
~, **газовое** gas field
~, **ликвационное** liquation deposit
месторождение *geol.* deposit, occurrence, field, pool, layer
~ **в гребне антиклинали** saddle reef
~, **вкрапленное** disseminated deposit
~, **вскрытое** developed field/deposit
~, **горизонтально залегающее** flat/sheet ground
~, **жильное** lode (deposit), vein deposite
~ **замещения** substitution/replacement deposit
~ **золота** gold field
~, **ископаемых** mines-and-carriers
~, **истощенное** depleted deposit
~, **коренное** primary deposit
~, **метасоматическое** replacement/metasomatic deposit
~ **минералов, ~ полезных ископаемых** mineral deposit
~, **напластованное** bedded/stratified deposit
~, **нарушенное** dislocated/disturbed/displaced deposit
~, **неравномерное** patchy/sporadic deposit
~, **нетронутое** virgin field, virgin/maiden deposit
~, **нефтяное/нефти** oil deposit/pool/field
~, **обводненное** water-bearing/-logged deposit, watered field/deposit
~, **пластовое** bedded/stratified/sheet deposit
~, **пластообразное** blanket deposit/formation/vein
~, **плоское** flat sheet/mass, bedded vein
~, **поверхностное рудное** hypotaxic/surface deposit
~ **полезного ископаемого** mineral deposit, mine field
~, **пригодное для открытой разработки** strippable deposit
~, **промышленное** commercial/payable deposit
~, **разведанное** proven territory, proved field
~, **рассеянное** scattered/disseminated deposit
~, **россыпное** placer /gravel deposit
~, **рудное** ore deposit/body/bed, metalliferous deposit
~, **седиментогенное** sedimentogene deposit
~, **слепое** concealed deposit, blind vein
~, **торфяное** peat moor/bog
~ **угля/угольное** coal deposit/field
~, **эрратическое** erratic ore body
месть revenge, vengeance
~, **кровная** blood feud
месяц 1. month; 2. moon
~, **високосный** leap month
~, **драконический** draconitic/nodical month
~, **звездный/сидерический** sidereal month
~, **молодой** new moon
месячина *hist.* monthly payment in kind
месячник 1. month (of); 2. *hist.* serf receiving monthly payment in kind
месячный 1. monthly; *med.* mensal; 2. lunar(y)
мета mark, sign
метабелевый *math.* metabelian

метабиоз *biol.* metabiosis
метабитум metabitumen
метаболизм metabolism
метаболический, метаболичный metabolic(al)
метавоксит metavauxite
метагалактика metagalaxy
метагенез metagenesis
метадромный *bot.* metadromous
метазид metazid, metazamide
метазнания *pl.* metaknowledge
метакинез *biol.* metakinesis
метакод meta code
метаколлоидный metacolloid(al)
метакоманда metainstruction
метакремниевый metasilicic
метакрилат *chem.* metacrylate
метакристалл metacrystal
метакристаллический metacrystalline
метакритика metacriticism
металингвистика metalinguistics
металл metal
~, белый babbit, white metal
~, благородный noble/precious metal
~, ваграночный cupola metal
~ высокой чистоты high-purity metal
~, высокосортный high-test metal
~, жаропрочный high-temperature metal
~, износостойкий wear-resistant metal
~, кислотоупорный acid-proof metal
~, кислотоустойчивый acid-resistant metal
~, конструкционный structural metal
~, коррозионностойкий non-corrosive metal
~, легкоплавкий low-melting metal
~, листовой волнистый corrugated sheet metal
~, листовой оцинкованный galvanized sheet metal
~, многосернистый polysulphide metal
~, неблагородный base metal, non-precious metal
~ обрабатывать (давлением в холодном состоянии) cold-work a metal, work a metal cold
~, одновалентный univalent/monad metal
~, основной base/parent/basic metal
~, пенистый foamed metal
~, плакированный clad/coated metal
~, полосовой flats
~, пористый foamed metal
~, порошковый powdered metal
~, присадочный added/filter metal
~, пробирный test metal
~, редкоземельный rare-earth metal
~ с низким коэффициентом расширения low-expansion metal
~, самородный native metal
~ сварочного шва weld metal
~, сверхредкий trace metal
~, сверхчистый ultra-pure metal
~, тугоплавкий refractory metal
~ усиления шва weld-reinforcement metal
~, химически активный reactive metal
~, цветной non-ferrous metal
~, черновой crude metal
~, черный ferrous metal
~, чушковый pig metal
~, щелочноземельный alkali-earth metal
~, щелочный alkali metal
металл-заменитель emergency metal
металлид metallide
металлизатор metal spraying gun/torch/pistol, metallizing torch, metal sprayer
~, газопламенный gas-stabilized plasma spraying pistol
~, порошковый powder pistol
металлизация, металлизирование deposition of metal coating; bonding, plating, metallization, metallizing
~ бумаги metallization of paper
~, горячая metal spraying
~, диффузионная surface impregnation/absorption, metallic cementation
~, плазменная plasma spraying process
~ распылением spray coating, metal spraying
~ самолета aircraft bonding
~ электролитическим способом electroplating
металлизировать apply a metal coating on/to; bond
металлизм *econ.* metal(l)ism
металлист metalworker, metal(l)ist
металлический metallic, metal *(attr.),* metalline
металличность metallicity
металл-микроскоп metallographic microscope
металлобумажный metallized paper, metallic-paper
металловед metallographist, metal scientist/physicist
металловедение physical metallurgy, metal studies, metallurgical science, science of metals
металловидний metalline, metalliform, metallic
металловолокно metal wool
металлогенетический *geol.* metallogenetic
металлогенический *geol.* metallogenic, metalliferous
металлогения *geol.* metallogeny
металлограф metallographer
металлография metallography
металлоемкий metal-intensive/-consuming
металлоемкость metal-intensity
~, удельная specific metal content
металлоид metalloid, non-metal
металлоидальный, металлоидный metalloid(al); nonmetal
металлоизделие hardware
металлоискатель locator/detector
металлокерамика cermet, ceramal, ceramet, powder metallurgy, metal ceramics
металлокерамический ceramic(-to)-metal, cerametallic, metal-ceramic, cermet, ceramet
металлоконструкция metalwork, steelwork
металлолом scrap metal

металломер metallometer
металлометр metal tester, metallometer
металлометрия metallometric survey, metallometry
металлоносность metal content; presence of metal
металлоносный metal-bearing, metalliferous
металлообрабатывающий metal-working
металлообработка metalwork
металлообразный metalline, metal-like
металлооптика metal optics
металлоплавильный smelting, metal-melting
~ **цех** foundry
металлопласт metal-base laminate
металлопластинчатый metal-plate
металлопокрытие metal coating
металлополупроводниковый metal-semiconductor
металлопродукция metal products
металлопрокатный metal-rolling
металлопромышленность metal industry
металлопротеид metalloprotein
металлорганический organometallic, metal-organic, metalloorganic
металлорежущий metal-cutting/-machining
металлосодержащий metalliferous, metal-containing, metallic
металлостеклянный metal-glass
металлотерапия metallotherapy
металлотермия metallothermy, metallothermic process
~, **внепечная** out-of-furnace metallothermy
металлоотражатель speculum metal
металлофермент metalloenzyme
металлоуловитель metal remover/catcher/trap
металлофизика metallophysics, physics of metals
металлохимия chemistry of metals, metal chemistry
металл-связка binder metal, metal binder
металлосодержащий metalliferous, metal-containing
металлург metallurgist, metal-maker/-worker
металлургия metallurgy, metallurgic engineering
~ **материалов реактора** reactor metallurgy
~, **порошковая** powder metallurgy
~, **цветная, -цветных металлов** nonferrous metallurgy
~, **черная** ferrous metallurgy, iron and steel industry
металогика metalogic
металогия metalogy
метальдегид *chem.* metaldehyde
метаматематика metamathematics
метамер *zool.* metamere; *chem.* metamer(ide)
метамеризм *zool.* metamerism, metameric segmentation
метамерия metamerism, metamery
~, **гомономная** homodynamy, homonomic metamerism
метамерность metamerism, merism
метамерный *chem., zool.* metameral, metameric
метамиктный, метамиктовый metamict
метаморфизация *geol.* metamorphization
метаморфизм *geol.* metamorphism
~, **внешний** exomorphism
~, **внутренний** endomorphism
~, **возвратный** retrograde/retrogressive metamorphism
~, **глубинный** plutonic metamorphism, deep-seated metamorphism
~, **инъекционный** injection/penetration metamorphism
~, **сложный** composite metamorphism
метаморфизовать, метаморфизировать metamorphose, metamorphize
метаморфоз(а) metamorphosis
~, **полный** complete metamorphosis
метан methane, marsh gas
~, **сорбированный** occluded methane
метанал methanal, formalin
метанализ metanalysis
метандрия metandry
метанефрогенный metanephrogenic, metanephrogenous
метанефрос *embryol.* hind kidney, metanephros, metanephron
метание 1. casting, toss(ing), throwing
~ **икры** spawning
~ **копья** *sport* javelin throwing
метанизация methanization
~ **нефти** oil methanization
метанол *chem.* methanol, methyl alcohol
метапепсис metapepsis, paroptesis
метапирогеновый metopyrogenic, metapyrogenous
метаплазический *physiol.* metaplastic
метаплазия *physiol.* metaplasia, metaplasy
~, **кишечная** intestinal metaplasia
~, **прямая** direct metaplasia
метаплазм(а) metaplasm
метапластический metaplastic
метаполный metacomplete
метаположение metaposition
метаправило metarule
метасемиотика metasemiotics
метасепта major/cardinal septum
метасикула *paleont.* metasicule, metasicula
метасиндезис metasyndesis
метасистема metasystem
метасома *zool.* metasoma; *geol.* metasome, guest mineral
метасоматизм *geol.* replacement, metasomatism
метасоматоз *geol.* metasomatosis, metasomatism
метастабильность metastability
метастабильный metastable
метастаз 1. *med., zool.* metastasis; 2. *phon.* off-glide, final glide
~, **известковый** calcareous metastasis
~, **пульсирующий** pulsating metastasis
~, **смешанный** mixed/compound metastasis

~, узловатый nodosal metastasis
метастазирование *med.* metastatic spreading
метастазировать *med.* metastasize
метастатический *med.* metastatic(al)
метастома *zool.* metastome, metastoma
метаструктура metastructure, secondary structure
метасурьмяный *chem.* metaantimonic
метатаксис *geol.* metataxis
метатарзалгия *med.* metatarsalgia, Morton's disease
метатарзальный *anat.* metatarsal
метатеза, метатезис metathesis
метатель *sport* hurler, thrower
~ копья javelin thrower
метательный missile, throwing, hurling, flinging
~ снаряд projectile
метатеория metatheory
метатерий *zool.* metatherian
метатетический metathetic(al)
метатрахеальный *bot.* metatracheal
метатрофный *bot.* metatrophic
метать throw, cast, fling
~ бисер перед свиньями *bibl.* throw pearls before swine
~ икру spawn
~ пар(ы) *agr.* fallow
~ сено *agr.* stack hay
~ стог *agr.* stack
метаустойчивость metastability
метаустойчивый metastable
метафаза *biol.* metaphase
метафизик metaphysician, metaphysicist
метафизика metaphysics
метафизичность metaphysic, metaphysical character/nature
метафит *bot.* metaphite
метафлоэма *bot.* metaphloem
метафония metaphony
метафора *lit.* metaphor
метафорический, метафоричный metaphoric(al), figurative
метахимия metachemistry
метахламидный *bot.* metachlamydeous
метахромазия *physiol.* metachromasy, metachromasia
метахронизм metachronism
метахронность metachronism, parachronism
метацентр *phys.* metacenter, metacentre
метацеркария *biol.* metacercaria
метациклический metacyclic
метаязык meta-language
метгемальбумин *med.* meth(a)emalbumin, pseudometh(a)emoglobin
метгемоглобин *med.* meth(a)emoglobin, ferrih(a)emoglobin
метгемоглобинемия *med.* meth(a)emoglobin(a)emia
~, вторичная secondary/enterogenous meth(a)emoglobin(a)emia
~, первичная primary/congenital meth(a)emoglobin(a)emia
метектика methectics, group dynamics
метелемер blizzard/snowdrift gauge
метелка *bot.* panicle, tassel, juba
~, полупоникшая semidrooping panicle
~, скученная crowded panicle
~, щитковидная umbellike panicle
метелковидный, метелкообразный *bot.* paniculiform, panicle-shaped; coremiform
метель snowstorm, drifting/blowing snow, blizzard
~, низовая low drifting snow
метельник *bot.* weaver's-broom *(Spartium)*
метельчатость *plant pathol.* hexenbesen, witches'-broom
метельчатый *bot.* panicled, paniculate
~тое соцветие panicled fluorescence, anthela
метемпсихоз *psych.* transmigration of souls, metempsychosis
метеограмма *meteor.* meteorogram
метеоданные *pl.* weather data
метеодонесение weather report
~, полетное in-flight weather report
метеозонд meteorological sounding balloon
метеоинструктаж meteorological briefing
метеоинформация meteorological information/data
метеоминимум *av.* weather/meteorological minimum
метеонаблюдение meteorological/weather observation
~, самолетное aircraft ascent
~, шаропилотное balloon ascent
метеоневроз *med.* meteoroneurosis
метеообстановка meteorological conditions, weather conditions/situation
метеоприбор meteorological instrument
метеор meteor
метеорадиолокатор weather radar
метеоразведка weather reconnaissance
метеоризм *med.* meteorism, tympanism, tympanites, drum belly
~, механический obturation meteorism
~, трупный cadaveric meteorism
метеорит meteorite
~, железосетчатый lithosiderite
~, железный iron meteorite, siderite
~, каменный stony meteorite, aerolite, falling/meteoric stone
~ промежуточного состава siderolite
метеоритика *astr.* meteoritics
метеоритный, метеоритовый meteoritic
метеоритолог *astr.* meteoriticist
метеорный 1. meteoric; 2. atmospheric
метеорограмма meteorogram
метеорограф meteorograph
~, аэростатный captive-balloon meteorograph
~, змейковый kite-lifted meteorograph
~, зондовый balloon-borne meteorograph, sounding meteorograph
~, самолетный airborne meteorograph

метеороидный meteoroidal
метеоролит meteorolite, aerolite
метеоролог meteorologist
метеорологический meteorologic(al)
метеорология meteorology
~, авиационно-космическая aerospace meteorology
~ больших пространств macrometeorology
~, физическая meteorological physics
метеоролог-наблюдатель meteorological observer
метереолог-прогнозист, авиационный aeronautical meteorological forecaster
метеосводка weather report, meteorological message
~, свежая/последняя latest weather report
метеосигнал meteorological signal
метеослужба weather/meteorological service
метеосообщение weather report
метеоспутник weather satellite
метеостанция meteorological/weather station, weather bureau
метеотопография meteotopography
метеоусловия *pl.* weather/meteorological conditions
~ для визуальных полетов visual meteorological conditions, VMC
~ для полетов по приборам instrument meteorological conditions, LMC
~, простые VFR conditions
~, сложные IFR conditions, difficult weather conditions
метеоцентр meteorological office
метеошар meteorological balloon
метизация *biol.* crossbreeding
метизный hardware, metalware *(attr.)*
метизы *pl. abbr.* **металлические изделия** hardware, metal products, metalware
метил *chem.* methyl
~, бромистый methyl bromide
метилаза methylase
метилаль methylal
метиламин *chem.* methylamine, aminoethane
метилен *chem.* methylene
~, йодистый methylene iodide
метиленовый methylene
~ синий methylene blue, methylthionine chloride
метилирование *chem.* methylation
метилметакрилат *chem.* methyl metacrylate
метиловый methylated, methylic
~ красный methyl red
~ спирт methanol, methyl alcohol
метилоз(ис) *geol.* methylosis
метилоранж *chem.* methyl orange
метилфиолетовый methyl-violet
метилцеллюлоза methylcellulose, celothyl, methocell
метилэтилкетон *chem.* methyl ethyl ketone
метионин *bioch.* methionine
метис 1. *biol.* halfbreed, crossbreed, mongrel; half-blood/-breed; 2. *anthr.* metis, mestizo
метисация crossing
метиска mestiza
метить brand, tag; *radiol.* label, tag
метка *comp.* mark(er), label, tag
~ блока block marker
~ в виде точки *comp.* dot mark
~, временная time mark
~, головная header label
~ дальности range marker
~ для оптического считывания photo-sensing mark
~ для ссылок reference tag
~ дорожки track label
~ заголовка header (label)
~ записи record label
~, контрольная check mark
~ контрольной точки check point label
~ конца end mark/character, end-of-block mark
~ конца ленты end-of-tape label
~ конца массива данных end-of-data-set record
~ конца файла ending file label, end-of-file label, data-set trailer, file end (trailer) label
~, концевая trailer label
~, координатная collimating mark/point
~, курсовая радиолокатора heading marker
~, монтажная matchmark
~ на ленте tape label/tag/mark
~ начала заправочного конца ленты load mark
~ начала (ленты) beginning tape label, beginning mark/character
~, начальная header (mark/label)
~, необязательная optional label
~ оператора *comp.* statement label
~ передачи управления target label
~, пользовательская user label
~, прицельная *mil.* aiming reticle
~, сборочная matchmark, location mark
~, символьная symbolic label
~, служебная control mark
~ тома volume label; data carrier label
~, установочная *cart.* adjusting/guide mark
~, факультативная optional label
~, юстировочная adjustment mark
меткость marksmanship; accuracy, precision
~ глаза keenness of vision
метлица *bot.* bent grass *(Agrostis)*
метловидный *bot.* coremiform, scoparious
метод method; procedure, technique(s); manner, mode; way; system
~ Адамса Adams-Moulton method; Adams interpolation/extrapolation (method)
~, алгебраический algebraic process
~ амплитудного анализа kick-sorting technique
~ анализа, количественно-минералогический Rosiwal's micrometric method
~ анализа рождаемости, когортный *econ.*

cohort fertility method
~ аннуитета при начислении износа *econ.* annuity method of depreciation, compound interest depreciation method
~, апостериорный a posteriori method
~, априорный a priori method
~ ассигнования appropriation method
~ Бекке Becke method
~ биений beat method
~ быстрейшего спуска *stat.* steepest descent method
~ вариации постоянных *math.* method of variation of parameters/constants
~, вариационный variation(al) method
~ ввода данных data input method
~, веерообразный fan shooting
~ Вентцеля-Крамерса-Бриллюэна Wentzel-Kramers-Brillouin treatment
~ Вернейля *rad.* Verneuil method
~, весовой gravimetric/gravity/gravitational method
~ ветвей и границ branch-and-bound method
~ взаимных пеленгов method by reciprocal bearings
~, вихревой vortex method
~ возврата backtracking method
~ воздушной проекции aero-projection method
~ возможных направлений *econ.* method of feasible directions
~, волнометрический ondometric method
~ Волтона-Эшби Walton-Ashby method
~ вращающихся колб roll-bottle method
~ вращающихся пробирок roll-tube method
~ вращений *math.* Jacobi's rotation
~ вращения method of revolution
~ врезания plunge-cut method
~ выбега *el.* retardation method
~, выборочный *comp.* sampling (method); sampling technique; selective method
~, выборочный статистический statistical sampling method
~ выведения на орбиту orbital injection method
~ вывода recovery method
~ выживания survival technique
~ выметания *math.* sweep(ing)-out method
~ высокочастотного резонанса radioresonance method
~ Гарднера fiducial-time method
~ Гаусса Gaussian algorithm
~, голографический holographic method
~ градусных измерений meridian-arc method
~, графико-аналитический graphic analysis method
~, графический graph method
~ графического трансформирования *top.* grid method
~, графоаналитический semigraphical method
~ графов *math.* graph method
~ Дарбишайра Derbishire method
~ датирования, радиоуглеродный radiocarbon method of dating
~ двойной бухгалтерской записи double entry method
~ двух возрастных групп two-ages method
~ двухгодичного максимума biennial maxima method
~ двух реек *geod.* two-staff method
~ двух узлов *el.* nodal-pair mathod
~ дезактивации deactivation/depletion method, negative activation method
~ Дельфи Delphi technique
~ деформаций deflection method, method of slopes
~ динамического моделирования dynamic simulation technique
~ диполей doublet distribution method
~ дисков disk sensitivity method
~ дистанционного доступа к данным (basic) telecommunication access method
~ догона stern-chase method
~ доступа access method
~ доступа, библиотечный partitioned access method
~ доступа, графический graphical access method
~ доступа, индексно-последовательный indexed-sequential access method
~ доступа к данным по разделам partitioned data access method
~ доступа к массиву по разделам positioned file access method
~ доступа, последовательный sequential access method
~ доступа, резидентный resident access method, RAM
~ доступа с очередями queued access method
~ дробных шагов method of fractional steps, method of splitting
~ естественного поля self-potential field method
~ естественного тока natural current method, self-potential method
~ засечек resection method
~ измененных частот displaced frequency method
~ измерения по точкам point-by-point method
~ измерения углов по аэроснимкам photogoniometric method
~ изображения рельефа штрихами hachure method
~ изотопных индикаторов tracer method
~, индексно-последовательный, доступа с очередями *comp.* queued indexed sequential access method
~ индексов index method
~, индикаторный tracer method
~, индуктивный электроразведки inductive method of electrical prospecting, electromagnetic

method
~ интегрирования Якоби *math.* Jacobi('s) method (of integration)
~ исключения method of elimination/exclusion
~ исключения переменных *math.* elimination approach
~ испытаний в аэродинамической трубе wind-tunnel technique
~ испытаний, кислотный acid test
~ испытаний с разрушением образца destructive testing technique
~ истечения efflux method
~ итерации/итерационный iteration/iterative method/technique, iteration procedure, iterative process
~ календарного планирования scheduling method
~ калькуляции costing method
~ калькуляции по процессам process/method of cost accounting
~ касательных *math.* Newton's method
~, капельно-чашечный drop-plate method
~ качаний swing method
~, качественный qualitative method
~ Кендалла Kendall's method
~, кессонный caisson method
~ классификации игр game classification method
~ когорт *econ.* cohort analysis
~ кодирования encoding method
~ колеблющегося диска oscillating-disk method
~, количественный quantitative method
~ кольца и шара ball-and-ring method
~ комбинекс Combinex method
~ компонентов, продольный cohort-component method
~, кондуктометрический conductance-measuring method
~ конечных вихрей finite-vortex method
~ конечных разностей finite-difference method
~ конечных элементов finite-element method
~ консервирования curing method
~ контроля, неразрушающий non-destructive testing method
~ контроля, дифференцированный differential control method
~ контурных токов mesh-current method, loop method
~ конуса cone method
~ корневого годографа *cyb.* root-locus method
~ красок (в дефектоскопии) dye-penetrant method
~ кратчайшего пути *math.* shortest route approach
~ критического пути critical path method
~ культивирования, чашечный plate culture method
~, кумулятивный accrual method
~ Лауэ Laue method
~ линии положения intercept method
~, люминесцентный fluorimetric/luminescent method
~ Ляпунова Lyapunov's method
~ магнитного порошка magnetic particle/ powder method
~, магнитно-люминесцентный fluorescent magnetic particle method
~ максимального правдоподобия maximum likelihood method
~ малого параметра method of small parameter
~ малых возмущений perturbation method, method of small perturbation
~, масс-спектрометрический mass-spectrometric method
~, матричный matrix method
~, маятниковый pendulum method
~, мензульный plane-table method
~ ментора mentor method
~ меченых атомов tracer method
~ микроразливов microplating method
~ минимакса min-max method
~, многогрупповой multigroup method
~ множественно-равномерного начисления износа multiple straight-line method of depreciation
~ множителей Лагранжа Lagrangian multiple method, Lagrange's method of undetermined multipliers
~, мокрый wet process
~ моментных площадей *mech.* area moment method
~ моментов moment(um)/moments method, method of moments
~ монтажа *comp.* wire-wrap method
~ Монте-Карло *math.* Monte Carlo technique/ method
~ мушиных точек fly-speck method
~ мягкой посадки soft-landing technique
~ наименьшего отклонения minimum deviation/variance/ procedure/method
~ наименьших квадратов least-squares technique/method, least/minimum squares, method of least squares
~ наименьших квадратов Гейгера Geiger's least square method
~ наискорейшего спуска steepest descent method
~ накачки(лазера) pumping/excitation method
~ накопления *nucl.* ''backing-space'' method
~ наложения superposition method
~ насыщения *game th.* saturation method
~ начисления износа по каждому отдельному предмета *econ.* unit/item method of depreciation
~ (неопределенных) множителей Лагранжа Lagrange/Lagrangian multiplier

method
~, непоследовательный non-sequential method
~ непрерывной инвентаризации perpetual inventory method
~, неразрушающий non-destructive method/testing
~, низкочастотный audiofrequency method
~, нулевой null/zero-deflection method
~ нулевых биений zero-beat method
~ Ньютона Newton's method
~ обезвоживания dewatering method
~ обработки (данных) (data) processing/reduction method
~ обратных матриц inverse matrix method
~ обращения access method
~ обследования, корреспондентский mail-out-mail-back procedure
~ обслуживания в обратном порядке last-in-first-out method
~ обслуживания в порядке поступления first-in-first-out method
~, объемный volumetric method
~, однополюсный electrode probe method
~ окраски бактерии по Граму Gram's method
~, окислительно-восстановительный redox method
~ опорного времени fiducial-time method
~ опроса polling procedure/method
~ опроса, корреспондентский mail questionnaire method
~ организации поиска, словарный *comp.* dictionary technique
~ ореолов рассеяния haloes of dispersion method
~ осреднения averaging/smoothing method
~ остаточных чисел residual method
~ отбора проб sampling method/technique
~ отдачи ядра nuclear-recoil method
~ отклонения deflection method
~ отопления *met.* fuel practice
~ отраженных волн reflection method/survey
~ оценки costing method
~ оценки запаса на основе нормативов затрат *econ.* standard cost method of inventory
~ оценки и пересмотра планов PERT-method
~ оценки на принципу "первая (партия) на приход - первая в расход" *econ.* first-in-first-out costing method, FIFO costing method
~ оценки по принципу "последняя (пар тия) на приход - первая в расход" *econ.* last-in-first-out costing method, LIFO costing method
~ оценки по фактическим затратам actual cost method
~ очистный выемки *min.* stoping method
~ парного сравнения paired comparison method
~ первого приближения first approximation method
~ перевала *math.* saddle-point method
~ переменного шага variable-cell method
~ пересекающихся лучей crossed beam method
~ перехвата interception technique/method
~ перспективных сеток *top.* grid method
~ ПЕРТ PERT-method
~ петли loop method
~, петротектонический petrofabric method
~ пламени flame method
~ пламенной спектрофотометрии flame spectrophotometric analysis
~ пластинчатых разводок plate method
~ повторного опроса reinterview method
~ поглощения absorption method
~ погони stern/tail-chase method (of guidance), pure pursuit method
~ подбора try-and-error method, cut-and-try method
~ подгонки, кривой curve-fitting method
~ подобия similitude method
~ подориентирования *top.* setting on points of control
~ поиска search(ing) method/procedure
~ полного перебора exhaustive method
~ полной деформации total-strain method
~ полос, оптический schlieren technique/method
~, полунепрерывный semicontinuous method
~ полярных координат polar method
~, порошковый powder method, Debye-Scherer-Hull method
~ последовательного доступа по индексам basic-indexed sequential access method, BISAM
~ последовательного приближения method of successive approximation
~, последовательный trail method, *stat.* sequential procedure
~ последовательных подстановок substitution process
~ последовательных приближений step-by-step method, method of successive approximations, approximation/iteration method
~ последовательных элиминаций method of exhaustion
~, поточно-конвейерный flow-line conveyor method
~, приближенно-двойственный almost-dual method
~ приведения к настоящему моменту present value method
~ проб (и ошибок) trial/try-and-error method/technique, cut-and-try method
~ пробегов range method
~ прогонки sweep method
~ продольного анализа longitudinal method
~ проекции градиента gradient projection method

~ производства, поточный production on the line
~ просечек cross-cutting method
~ прямого доступа (basic) direct-access method
~ пузырков Сорби Sorby bubble method
~ равномерного начисления износа straight-line method of depreciation
~, радиоактивный для определения возраста radioactive dating
~, радиоактивный разведки radioactive prospecting
~ радиоактивных индикаторов radioactive-tracer technique
~ радиопротиводействия jamming method
~ разделения переменных method of separation of variables
~ раздельного исследования факторов one-factor-at-a-time method
~ разложенного времени resolved-time method
~ разностей difference method
~, разностный finite-difference method
~ разработки программного обеспечения software development
~ ранжирования работ ranking method
~ распределения пропорционально прямым затратам *econ.* prime cost method
~ растра *telev.* grid method
~ расчета на прочность structural analysis technique
~ регулирования рождаемости contraceptive measure
~ резерва на амортизацию retirement reserve method of depreciation
~ релаксации relaxation method (of Southwell)
~ рентгеновских порошкограмм X-ray powder method
~ решения транспортной задачи MODI-method
~ решета *math.* sieve method
~ Рога Roga method
~, родоидный rhodoid method, dry-peel method
~ ротации rotation method
~ Рунге-Кутта Runge-Kutta method
~ самонаведения method of homing
~ сближения *av.* rendezvous technique
~ сборных карт *meteor.* composite map method
~ свилей schlieren technique/method
~ сглаживания (данных) smoothing technique, graduation method
~ сейсмический определения падения dip shooting
~ секущих *math.* method of false position, secant method
~ секущих плоскостей cutting-plane method
~ Сент-Илера Saint Hilaire method
~ серого клина *phys.* gray-wedge method
~, сетевой network method
~ сеток *math.* net(-point) method, finite-difference method
~ скользящих средних moving average method
~ совмещения line-on-sight method (of guidance), matching curve method
~ совпадений coincidence method
~ создания (радио)помех, силовой brute-force jamming
~, сокращенный short-cut method
~ сопротивления resistivity method
~ сопутствующих изменений method of concomitant variations
~, спектральный spectrographic determination method
~ спрямленных лучей straight-path method
~ сравнения коэффициентов coefficient comparison method
~ средних (значений) method of average
~ средних квадратов midsquare method
~, статистический statistical technique/procedure
~ статистических испытаний Monte-Carlo method/technique
~ стохастической аппроксимации stochastic approximation technique/method
~ структурных коэффициентов composition ratio method
~, ступенчатый step-by-step method
~ суммарного счета total-count method
~ сцепленных Х-хромосом *med.* attached X-method
~ Сциларда-Чалмерса Szillard-Chalmers method
~, счетно-импульсный pulse-counting method
~, табличный diagram method
~ Тамма-Данкова Tamm-Dancoff method
~ тарировки calibration technique
~, теневой (direct-)shadow method
~, термоанемометрический hot-wire method
~ титрования фага путем подсчета стерильных пятен plaque counting method
~, толстослойных эмульсий nuclear emulsion/photo plate method
~ торможения braking/arresting/deceleration method
~, точечный point method
~ точечных оценок point-estimation method
~, углеродный carbon dating
~ угловой деформации slope-deflection method
~ угловых несогласий interval change method, angular divergence method
~ узлов *constr.* method of joints
~ узловых потенциалов node-voltage method
~ упорядочения *stat.* ranking method
~ уравнивания по направлениям *geod.* method of directions, direction method
~ уравнивание по углам *geod.* method of angles, angle method
~ усреднения smoothing/averaging method

~ **учета естественного движения населения** vital statistics technique
~, **федоровский** Fedorov method
~, **ферментный** enzymatic method
~ **фиктивной партии** *game th.* method of fictitious play
~ **фотоэмульсий** emulsion technique
~ **характеристик** method of characteristics, method of characteristic curves
~ **хлористых солей** chlorine method
~, **хроматографический** chromatographic-column method
~ **хронометрирования по отдельным отсчетам** *econ.* snapback method
~ **частотных характеристик** frequency-response method
~, **численный** numerical method
~ **Чохральского** *cryst.* Czochralski method, vertical pulling technique
~ **Шора** Shore hardness
~ **штафельного учета** *fin.* balance method
~ **штрафных функций** penalty function method
~ **шунтировки** shunt method
~ **щуповой** stylus method
~, **эвристический** heurism
~ **экспертных оценок** Delphi technique
~, **экстракционный** extractive method
~, **электрометрический** electrical resistivity method
~ **электроразведки Билера-Ватсона** Bieler-Watson method
~ **электрофореза** electrophoretic method
~, **эмпирический** try/trial-and-error method
~, **эргатический** *comp.* interactive/conversational technique
~ **ядерных эмульсий** nuclear-emulsion/photoplate method
~ **Якоби** Jacobi('s) method
~**ы, неколичественные** *econ.* nonquantitative methods
~**ы производства, капиталоемкие** capital-intensive methods

методизм I *rel.* Methodism
методизм II methodical procedure, methodism
методика technique, procedure, method(s), system; principles; methodology
~ **взятия образцов** sampling technique
~ **вычислений** computational technique
~ **измерений** measurement procedure
~ **летных испытаний** flight-testing technique, test-flying technique
~ **проверки** test procedure
~ **прогнозирования погоды** forecasting technique
~ **расчета** design procedure
~, **стандартная** standard procedure
~ **эксперимента** experimental technique

методист I methodologist, specialist in teaching methods; guidance councel(l)or
методист II *rel.* Methodist
методистский *rel.* Methodist
методический methodical, systematic(al); method(s) *(attr.)*
~**кое совещание** conference on (teaching *etc.*) methods

методкабинет *abbr.* **методический кабинет** department of methodology
методолог methodologist
методология methodology; method(s)
метол *chem.* metol
метонимический, метонимичный *lit.* metonymic(al)
метонимия *lit.* metonymy
метоп *arch.* metope
метопланктон *biol.* metoplankton
метр 1. metre, *Am.* meter; 2. meter ruler
~, **квадратный** square metre
~, **погонный** running metre
~, **складной** folding rule
~, **усадочный** shrinkage gauge; measure of shrinkage

метраж metric area; length in metres; footage *(of films)*
метралгия *med.* metralgia, hysteralgia
метр-ампер *rad.* metre-ampere
ментранпаж *print.* make-up man, maker-up
метрдотель headwaiter
метрейриз *med.* metreurysis
метрейринтер *med.* metreurynter, hystereurynter, obstetrical balloon
метризационный metrization *(attr.)*
~**ное кольцо** valuation ring

метризация, метризование *math.* metrization; metrication, metrification
метризуемость metrizability
метризуемый metrizable
метрика I 1. *math.* metric(s), distance function; valuation; 2. *lit., mus.* metrics
~ **программного обеспечения** software metrics
~ **пространства-времени** *phys.* space-time metric
~, **эквиареальная** equiareal metric

метрика II birth certificate
метрит *med.* metritis, hysteritis
~, **отслаивающий** exfoliative metritis
~, **послеродовой** postpartum metritis

метрически-двойственный metrically-ambiguous
метрический I metric
~**кая система** metric system

метрический II *lit.* metrical
~**кое ударение** metrical stress, ictus

метрический III:
~**кая выпись,** ~**кое свидетельство** birth certificate
~**кая книга** register of births, marriages and deaths

метро *abbr.* **метрополитен** subway, metro
метролог metrologist

метрология metrology
~, законодательная legal metrology
метроклист *med.* metroclyst, uterine irrigator
метромаляция *med.* metromalacia, metromalacoma, metromalacosis
метроном metronome, rhythmometer
метрополитен underground/metropolitan railway, *Am.* subway
метрополия 1. mother/parent/home country, parent state; 2. capital city, metropolis
метроптоз *med.* metroptosis, metroptosia
метрорея *med.* metrorrh(o)ea
метрорагия *med.* metrorrhagia
метросальпингография *med.* metrosalpingography, hysterosalpingography, uterotubography
метросидерос *bot.* iron tree *(Metrosiderosis)*
метроэндометрит *med.* (metro)endometritis
~, послеабортный postabortion endometritis
метр-свеча metre-candela, candle-meter, lux
метршток sounding rod/stick
метрэктомия *surg.* metrectomy, hysterectomy
метчик (screw) tap, tap (borer), threading tool; cutting gimlet, twist drill
~, гаечный nut tap
~ для конической резьбы, ~ для нарезки труб pipe tap
~, маточный master tap
~, плашечный die tap
~ с направляющей цапфой pilot tap
метчик 1. marker; 2. marking tool, stamp, punch
меум *bot.* baldmoney, spicknel *(Meum)*
мефенезин-карбамат *pharm.* mephenesin carbamate, tolseram
мефентермин *pharm.* mephentermine, wyamine
мефистофелевский, мефистофельский Mephistophelian, Mephistophelean
Мефистофель Mephistopheles
мефобарбитал *pharm.* mephobarbital, mebaral
мефолин *pharm.* phenmetrazine
мех 1. fur; *coll.* furriery, furs; 2. bellows
~, кузнечный smith's bellows
механизатор 1. *agr.* mechanic, machine operator/servicer; technician; 2. mechanization expert
механизация mechanization; machines
~, комплексная integrated/comprehensive/overall mechanization, large-scale mechanization
~ крыла *av.* high-lift device(s)
~ погрузочно-разгрузочных работ materials-handling mechanization/machines
~ сельского хозяйства farm mechanization; introduction of farming machinery
механизировать mechanize
механизируемость mechanizability
механизм 1. mechanism; device, gear; 2. *pl., coll.* machinery
~, автономный self-reacting device
~, азимутальный azimuth gear/mechanism
~, арретирующий caging mechanism
~ блокировки interlock/locking/lockout mechanism
~, блокировочный latching mechanism
~, блокирующий lock gear, blocking mechanism
~ быстрого хода rapid-traverse mechanism
~, быстродействующий quick-acting mechanism, high-speed mechanism
~, ведущий drive/driving mechanism
~, взрывной detonating mechanism
~, винтовой screw-type mechanism
~, винтовой подъемный screw-jack mechanism
~ включения clutch/engaging mechanism
~ возврата тележки carriage return mechanism
~ возвратно-поступательного движения reciprocating mechanism
~ вращения антенны scanner assembly
~, временной timer, timing/time mechanism, time-delay mechanism, timing equipment/device
~, встроенный программный built-in programmer
~ выборки *comp.* access mechanism
~ выборки из памяти на дисках disc access mechanism
~, выключающий shut-off mechanism
~ газораспределения valve gear
~, государственный machinery of government
~, грузоподъемный hoisting device
~, делительный divider
~, дифферинциальный differential (gear)
~, дозирующий batching device
~ загрузки топлива fuel charger
~, загрузочный charger, charging device
~ задержки (time-)delay mechanism
~, захватывающий gripping mechanism
~, защитный defence/defense mechanism
~, зубчатый gear train
~ извержения *geol.* mechanism of eruption
~ изменения шага pitch/control mechanism
~, исполнительный actuator (mechanism), actuating mechanism, slave *(of a manipulator)*
~, исполнительный дискретный digital actuator
~, исполнительный пневматический air actuator
~, исполнительный поршневой piston actuator
~, кассирующий coin collector
~, красящий inker
~ красящей ленты ribbon device
~, кривошипно-коромысловый crank-and-rocker mechanism
~, кривошипно-кулисный oscillating crank gear
~, кривошипно-ползунный slider-crank mechanism
~, кривошипно-шатунный, ~, кривошипный crank mechanism
~, кулачковый cam mechanism/gear
~, кулачковый распределительный tappet gear
~, кулисный link gear, rocker mechanism

~, лентопротяжный 1. *comp.* tape drive/transport; 2. strip feeding apparatus, tape transmitter, tape-moving/-transport mechanism
~, лентопротяжный кинокамеры film-pulling mechanism
~, мембранный diaphragm-operating mechanism
~, намоточный winding machine
~, направляющий карты *comp.* card guide/bed
~ обратной связи feedback mechanism, follow-up mechanism
~ отказа failure mechanism
~, передаточный transmission mechanism, transfer device
~ переключения скоростей gear shift(ing) mechanism, speed control mechanism
~, печатающий printer
~ питания supply/feed mechanism
~, питающий feeding mechanism, feeder
~, планетарный planetary gear/train
~, плоский plain mechanism
~, поворотный *rail.* traversing mechanism, slewing gear
~ подачи/подающий feed (mechanism)
~ подачи бланков *comp.* chart drive
~ подачи ленты *comp.* tape feed device/mechanism, tape/strip feeding apparatus
~ подъемника jack mechanism
~, подъемный hoist, lifting mechanism
~, предохранительный safety device/gear/mechanism
~ преобразования структур данных data structure machine
~, приводной operating/driving/drive/actuating mechanism, actuator, driving gear
~ программирования полета *av.* flight programmer
~ протяжки бумаги paper-advance mechanism
~, пусковой starting/trigger mechanism
~, рабочий operating/working mechanism
~ разрешения конфликтов contention mechanism
~ распознавания recognition machine
~ распространения propagation mechanism
~, расцепляющий tripping device, release gear
~ реагирования response mechanism
~, реверсивный reversing mechanism/gear
~, регулирующий control mechanism, adjusting gear
~, редукционный reduction/reducing gear
~, режущий *agr.* cutter bar
~, рулевой steering gear
~, рулевой с усилителем power-assisted steering gear
~, рычажный leverage, linkage, lever motion
~ самовзрывания self-destroying mechanism
~ самонаведения homing mechanism
~ свободного хода валика platen release mechanism
~, следящий follow-up mechanism
~, сотрясательный *agr.* shaker mechanism
~, стержневой link mechanism
~, стопорный arrester, arresting/locking device/gear, lock(ing) mechanism
~, стреляющий firing gear
~ стыковки docking mechanism
~, тангенсный cross-slide mechanism
~, тормозной brake/braking gear
~ указателя воздушной скорости *av.* airspeed indicaror mechanism
~ управления control (mechanism)
~, храповой ratchet-and-pawl mechanism/gear
~ цен *econ.* price mechanism
~, часовой clockwork (timer/drive), clock mechanism
~, шаговый step-by-step mechanism
~ы, главные main/propelling/propulsion/propulsive machinery
~ы, палубные *naut.* deck machinery

механик 1. mechanic, mechanical engineer; 2. student of mechanics; specialist in mechanics
~, бортовой *av.* flight mechanic
~, второй *naut.* second engineer
~, младший *naut.* assistant engineer
~ паровых машин steam engineer
~, старший *naut.* chief engineer

механика mechanics
~ абсолютного твердого тела mechanics of rigid bodies
~ грунтов soil mechanics
~ жидкостей и газов mechanics of fluids, fluid mechanics
~, квантовая quantum mechanics
~ материальной точки particle mechanics
~, небесная celestial mechanics
~ неизменяемых систем mechanics of rigid bodies
~, ньютоновая Newtonian mechanics
~, общая theoretical mechanics
~ полета flight mechanics
~ развития *physiol.* developmental mechanics
~ развития народонаселения *econ.* population mechanics
~, релятивистская relativistic mechanics
~ сплошных сред mechanics of continua, continuum mechanics
~, строительная structural mechanics
~ твердого тела mechanics of rigid body
~, точная fine mechanics
~, физико-химическая physical-chemical mechanics, physiochemical mechanics

механико-акустический mechanical-acoustic
механико-электрический mechanical-electrical
механист *philos.* reductionist, mechanist
механистический, механистичный, механистский *philos.* mechanistic
механицизм *philos.* reductionism, mechanism
механический 1. power-driven, mechanical; 2. *philos.* mechanistic
~кая мастерская machine shop

~**кая обработка** machining
~**кое оборудование** machinery, mechanical equipment
механогастрограмма *med.* gastromechanogram
механокрекинг mechanical cracking, mechanocracking
механоморфизм *philos.* mechanomorphism
механопрочность crushing strength
механо(ре)цептор *anat.* mechanoreceptor, mechanicoreceptor
механосборочный machine-assembling
механосинтез mechanical synthesis, mechanosynthesis
механострикция mechanostriction
механотерапия *med.* manipulative treatment, mechanotherapy
механотрон movable-electrode tube
~**, ионный** movable-electrode gas-filled tube
~**, электронный** movable-electrode vacuum tube
механохимия mechanical chemistry, mechanochemistry
мехоед *zool.* hide beetle *(Attragenus)*
мехообразный furry; fleecy
меценат Maecenas, patron of art/literature
меценатство patronage of arts/literature
меццо-сопрано *mus.* mezzo-soprano (voice/singer)
меццо-тинто *print.* mezzotint
меч sword
~**, Дамоклев** sword of Damocles
мечевиднолистный *bot.* ensifoliate, xiphophyllous, sword-leaved
мечевидный *bot.* ensate, ensiform, sword-shaped; swordlike, gladiate; *anat.* xiphoid; *zool.* ensate
мечелепестный *bot.* xiphopetalous
мечелистный *bot.* xiphophyllous, sword-leaved
мечение label(l)ing, tagging; marking; *zool.* branding
~ **изотопным индикатором** tracer-labelling
~ **отдачей** recoil tagging
меченосец *hist.* knight of the sword; sword-bearer
меченосница *bot.* hatchet vetch *(Securigera coronilla)*
меченый label(l)ed, tagged, marked; stigmatic
~**ные атомы** radioactive tracers, tagged atoms
мечеобразный ensate, ensiform; sword-shaped; *anat.* xiphoid
мечехвост *zool.* horse crab, xiphosuran *(Xiphosurus)*
мечехвостые *n. pl. zool.* xiphosurans *(Xiphosura)*
мечник *hist.* sword-bearer
меч-рыба *zool.* sword fish *(Xiphias)*
меч-трава *bot.* twig rush, saw grass *(Cladium, Mariscus)*
меш mesh
мешалка mixer, agitator, stirrer, stirring rod; blender; kneading machine
~ **для красок** paint/colo(u)r mixer
~ **дозатора** agitator
~**, лопастная** impeller mixer/agitator
~**, мотыльковая** butterfly agitator
~**, скребковая** scraper-type agitator
~**, струйная** jet/injection mixer
~**, ступенчатая** gravity mixer
~**, якорная** anchor agitator, anchor-impeller mixer
мешалка-охладитель mixing-and-cooling tank
мешальный mixing, stirring, agitating
мешальщик mixer, stirrer
мешание mixing, stirring, agitation
мешковидный, мешкообразный sacculate(d), sacciform, saccate, bursiform; baggy; pouchlike; baglike; cystlike; *anat.* glenoid
мешковина bagging, sacking, sackcloth, burlap, hopsacking
мешковый *bot.* saccate
мешкозагрузочный bag-filling
мешкозагрузчик sack loader, sack handling machine
мешконаполнитель bag filler
мешконасыпатель bagger, bag filler
мешкопогрузчик bag loader
мешкоподъемник sack hoist/lifter
мешкоротые *n. pl. zool.* gulpers *(Saccopharyngidae)*
мешкоспуск sack chute/shoot
мешкотаска sack conveyer
мешок 1. bag, sack; packet; 2. bursa; vesicle; 3. bagful, sackful; 4. pocket, cul-de sac
~ **амниона** sac of amnion
~**, амортизирующий** impact/attenuator bag
~**, балластный** sandbag
~**, бумажный** paper bag
~**, ветровой** windsack
~**, вещевой** knapsack, haversack
~**, водяной** water pocket
~**, воздушный** (air)bag; *anat.* air sac, sacculus alveolaris, air pocket
~**, воздушный (птиц)** pneumatocyst
~**, вытеснительный** expulsion/expellant/pressurization bag
~**, грыжевый** *med.* hernial sac
~**, двураздельный** bilobed sac
~**, дезинсекционный** delousing bag
~**, джутовый** gunny
~**, дыхательный** breathing bag
~**, жаберный** gill pouch/sac
~**, желточный** yolk/vitelline sac
~**, замкнутый** cyst, closed case
~**, зародышевый** embryo sac
~**, защечный** *zool.* cheek pouch
~**, зубной** dental/tooth sac
~**, легочный** lung book
~ **Листера** *med.* Lister bag
~**, матросский вещевой** kit bag
~**, надувной** inflatable bag
~**, обонятельный** nasal sac
~**, огневой** *mil.* fire pocket
~**, плодный** *anat.* f(o)etal/gestation sac
~**, походный** knapsack, haversack

~, пылеосадительный dust collecting bag
~, пыльниковый pollen/anther sac
~, слезный *anat.* lacrimal/tear sac
~, спальный sleeping bag
~, срединный median sac
~, чернильный ink bag
~ом, окруженный perisaccate
мешотчатый cystlike; follicular, folliculate; saccate, saccular, *bot.* utricular
мешочек *biol.* follicle, (small) sac, saccule; *bot.* utricle
~, альвеолярный *anat.* alveolar sac
~, волосяной hair follicle
~ гортани saccule of the larynx
~, пленчатый perisporium
~ преддверия *anat.* sacculus vestibuli/proprius
~ спорангия capsule
мешковидный follicular, utricular, utriculiform, sacciform, saccular, sacculiform
мешконосный sacculate(d), sacciferous
мещанин *hist.* petty bourgeois
мещанский *hist.* petty-bourgeois
мещанство *hist. coll.* petty bourgeoisie, lower middle class
мещина *hist.* monthly payment in kind
миаз(ис) *med.* myiasis
~, глазной ocular myiasis
~, кишечный intestinal myiasis
~, кожный subcutaneous myiasis
~, полостной cavitary myiasis
~, раневой wound myiasis
миазма miasma; mephitis
миазматический, миазменный miasmatic, miasmal
миальгический myalgic
миальгия *med.* myalgia
~, эпидемическая epidemic myalgia, Bornholm disease
миаролитовый *geol.* miarolitic
миарсенол sulpharsphetamine, sulpharsphenobenzene
миастения *med.* muscular weakness, myasthenia
~, злокачественная myasthenia gravis
миатония *med.* myatonia, myatony
миатрофия *med.* myatrophy, myoatrophy
мибора *bot.* mibora *(Mibora)*
миг I moment, instant
миг II *abbr.* **Микоян и Гуревич** Mig *(Soviet fighter aircraft)*
мигание nictitation, nictation, winking; blinking; flicker; flashing; flickering, twinkling
~ изображения image flicker
мигательный nictating, nictational; flickering, blinking; wink *(attr.)*
мигматит *geol.* injection gneiss, migmatite
мигрант *bot., zool.* migrant
миграционный migratory
миграция migration, travel
~ внутри города intraurban migration
~ долин *geol.* valley shifting
~ дырок hole migration
~, маятниковая *econ.* push-pull migration
~ междоузлий interstitial migration
~, обратная return migration
~ рабочей силы labor migration
~, сельско-городская rural-urban migration
мигреневый migrainous
мигрень *med.* migraine; sick headache, hemicrania, megrim, hemicephalalgia
~, глазная ophthalmic migraine
мигрировать migrate
МИД *abbr.* **Министерство иностранных дел** Ministry of Foreign Affairs, Foreign Office
Мидас *myth.* Midas
мидель midsection, maximum cross-section
мидель-шпангоут middle/midship frame
мидиеподобный *zool.* mytiliform
мидия *zool.* mussel *(Mytilis)*
мидриаз *med.* mydriasis
~, паралитический paralytic mydriasis, cycloplegia
~, спазматический smasmodic/spastic mydriasis
мидриатический mydriatic, cycloplegic
миелалгия *med.* myelalgia
миелапоплексия *med.* spinal apoplexy, myelapoplexy
миеластения *med.* spinal neurasthenia, myelasthenia
миелин *bioch., anat.* myelin(e)
миелинизированный *anat.* medullated
миелиновый myelinic, myelinate(d)
миелит *med.* myelitis
~, апоплектический apoplectiform/foudroyant myelitis
~, восходящий ascending myelitis
~, гнойный purulent myelitis
~, контузионный concussion myelitis
~, лучевой radiation myelitis
~, шейный cervical myelitis
миелобласт *biol.* eosinoblast, myeloblast
миелогенный myelogenous, myelogenetic, myelogenic
миелоз *med.* myelosis, myeloid leukemia
миелоидный myeloid, myeloic
миелолейкоз *med.* myeloleukemia
миелома *med.* myeloma
~ бедра femoral myeloma
~, множественная multiple(x)/myeloid myeloma, Kahler's disease, myelomatosis, plasmacytic sarcoma
~, эндотелиальная endothelial myeloma, Ewing's tumo(u)r
миеломный myelogenetic, myelogenous, myelogenic
миелопаралич *med.* spinal paralysis, myeloparalysis
миелофтиз *med.* myelophthisis, myelanalosis, tabes dorsalis
миелоцит myelocyte

мизандрия misandry
мизансцена *theatre* staging, stage setting
мизантропия misanthropy
мизерере *eccl.* Miserere
мизиды *pl. zool.* opposum shrimps, mysids *(Mysidacea)*
мизинец *anat.* little finger/toe
мизогамия misogamy
мизонеизм extreme conservatism, misoneism, misocainia
мизофобия *med.* dread of filth, mysophobia
микалекс Mycalex *(insulating material)*
микалента mica tape
миканит micanite
~, формовочный moulding micanite
микарта micarta
микенский Mycenaean
микобактерия *bact.* mycobacterium
~ паратуберкулеза Johne's bacillus
~ туберкулеза tubercle bacillus
микодерма mycoderm
микоз *med.* mycosis
~ глотки pharingomycosis
~ кожи dermatomycosis, mycosis cutis
~ ротовой полости thrush
микоза *chem.* mycose, trehalose
микозный mycotic
микология mycology, fungology
микориза *plant pathol.* myco(r)rhiza
~, ложная pseudomyco(r)rhiza
микостатический fungistatic, mycostatic
микостерин mycosterol
микотоксикоз mycotoxicosis
~, пищевой alimentary mycotoxicosis
микотроф *bot.* mycotrophic plant
микофлора mycoflora
микроавтобус minibus
микроагрегатный microaggregate
микроампер *el.* microampere
микроамперметр microammeter
микроанализ microanalysis
~, объемный microvolumetric analysis
микроанализатор microanalyzer
~, рентгеноспектральный X-ray microanalyzer
микроанаэростат anaerobic culture apparatus
микроатмосфера artificial atmosphere, microatmosphere
микроб microorganism, microbe, germ
микробиология microbiology
~, пищевая food microbiology
~, почвенная soil microbiology
микроблок microassembly
микробный microbic, microbial, microbian
микробоноситель microbe/germ carrier
микробостатический bacteriostatic
микробоцидный microbicidal, microbicide, bacteri(o)cidal
микробрекция, микробрекчия *geol.* microbreccia
микробюретка microburet(t)e
микроватт *el.* microwatt
микровертушка pygmy current meter
микровесы *pl. phys.* microbalance
~, кварцевые silica microbalance
~, крутильные/торзионные torsion microbalance
микровзрыв microexplosion
микровид microspecies
микроволны *pl.* microwaves
микроволосок microvillus
микровплавление microalloying
микровспышка *astr.* microflare
микровыключатель microswitch
микровязкость microviscosity
микрогамета *biol.* microgamete
микрогенри *el.* microhenry
микрогеометрия microgeometry, micro-irregularities, microgeometrical deviations
микроглазковый *geol.* microaugen
микроголограмма microhologram, holomicrogram
микрогорелка microburner
микрограмм microgram
микрографика, микрография (photo)micrography
микрогруппа microgroup
микрогруппировка microphytocoenosis, microcommunity
микрогубчатый microspongeous
микрогэс *abbr.* **микрогидроэлектрическая станция** microhydroelectric power plant
микродвигатель micromotor, miniature motor
микроделянка *agr.* small plot, microplot
микродеталь microelement
микродеформация microstrain, microdeformation
микродиск *comp.* microfloppy
микродобавки *pl.* microadditives, trace additives
микродозировка micrometering
микрозавр *paleont.* microsaur(ian)
микрозапись microrecording
микрозернистый microgranular
микрозоид *biol.* microzoid
микрозона zonule
микрозонд microprobe
микрозондирование micrologging, contact log
микроигла *med.* microneedle
микроизображение micro-image
микроинтерферометр interference microscope
микроизмельчитель microgrinder
микроинфаркт *med.* microinfarction
микроископаемое *n.* microfossil
микроисследование microexamination, microanalysis
~ на растворимость microsolubility test
микрокамера microchamber, Sievert chamber
~, сварочная welding glove box
микроканал pin-hole
микрокаротаж micrologging, microlog/micro-

resistivity survey
микрокаутер micro(thermo)cauter
микрокиносъемка cine(photo)micrography
микрокипение microboiling
микроклизма microclyster
микроклин *miner.* microcline
~, зеленый amazone stone, amazonite
микрокнига microfilmed book
микрококк micrococcus
микрококкит *geol.* granule-shaped microlith
микроколебание microoscillation
микроколичество tracer/trace amount, microquantity, microgram quantity
микрокоманда *comp.* microinstruction
микрокомпонент microcomponent
~, водорослевый algal microcomponent
~, смоляной resinite
микрокомпьютер microcomputer
микроконденсатор microcapacitor
микроконтакт microswitch
микроконцентрация trace concentration, microconcentration
микрокопирование microphotography, microcopying
микрокопия microphotograph, microcopy
микрокосм(ос) microcosm(os)
микрокосмический microcosmic, microcosmal
микрокристалл microcrystal
~, волокнистый hair-shaped microcrystal
микрокристаллический microcrystalline
микрокристаллоскопия crystalline microchemical analysis
микрокулон *el.* microcoulomb
микрокюри microcurie
микроландшафт *microbiol.* microsite
микролимфоцит *med.* micromyelolymphocyte, microleucoblast
микролит *geol.* microlith, microlite
микролитический, микролитовый microlit(h)ic
микролучевой *phys.* microray *(attr.)*
микром *el.* microhm
микроманометр micropressure gauge, micromanometer
микромастия *med.* micromastia, micromazia
микромасштаб microscale
микромер *tech.* (contact) comparator
микромер(а) *embryol.* micromere
микромерия *bot.* micromeria *(Micromeria)*
микроместообитание microhabitat
микрометод microprocedure, micromethod
микрометр micrometer (gauge)
~, бесконтактный non-contact thickness gauge
~, емкостный capacitance micrometer
~, зубомерный gear-tooth micrometer
~, интерференционный interferometer micrometer
~, кольцевой annular/ring micrometer
~, нитяный filar/crossbar micrometer
~, окулярный eyepiece/ocular micrometer
~, отсчетный reading micrometer
~, радиоактивный radioactive thickness gauge
~, рычажный lever-type micrometer
микромир microcosm, microworld
микромодуль microcircuit module, micromodule, microassembly
~, плоский two-dimensional microcircuit module
~, таблеточный pellet/microminiature/microcircuit module
микромозоль microcallus
микрон micron
микронапряжение microstress
микронаселение micropopulation
микронезийский Micronesian
микронеоднородность microinhomogeneity
микронивелир microlevel
микронутромер inside micrometer
микрооблучение microirradiation
микрообработка micromachining
~ лазерным лучом laser-beam microprocessing
микрообъем microscopic volume, microvolume
микрообъемный microvolumetric
микроозоление microincineration, microdigestion
микрокаменелость microfossil
микроом *el.* microhm
микроомметр microhmmeter
микрооползень *geol.* microslickensiding
микроопределение microdetermination
микроорганизм *biol.* microorganism
~, дрожжеподобный yeast-like microorganism
~, животный microzoon
~ почвы soil microb/microorganism
~, растительный microphyte
~, растительный анаэробный anaerophyte
микроостаток microfossil
микроосциллограф microoscilloscope; microoscillograph
микроочковый microaugen
микропенис *anat.* micropenis, microphallus
микропередатчик miniature transmitter
микропереключатель microswitch
микроперелом microfracture
микропеременная microvariable
микроперепись microcensus
микропечать microprint
микроплакит *geol.* microplakite, tabular microlith
микропластинка microlamella
микропленка microfilm
микроподвид microsubspecies
микроползучесть microcreep
микрополость microcavity
микропора *geol.* microscopic capillary; micropore
микропочвоведение micropedology
микропрепарат microslide, slide mount, microsection
микропрерывание *comp.* microbreak
микропрерыватель microchopper
микропривод miniature drive/motor
микропровод microwire

микропрограмма *comp.* microcode
микропрограммирование microcoding, microprogramming
микропрорастание microintergrowth
микропучок microbeam
микроразмельчитель microgrinder
микрорайон 1. small district, microdistrict; neighbourhood unit; 2. building estate
~, жилой housing estate
микрорезание microcutting
микрорезерфорд *phys.* microrutherford
микрорельеф underfeature, microrelief
микрорентген microroentgen
микрорентгенография microroentgenography, microradiography
микросветовод microguide
микросдвиг microshift, microdisplacement, minute movement
микросекунда microsecond
микросеточка microreticule
микросеть *med.* microrete
~, клеточная cellular microrete
микроскоп microscope
~, автоэлектронный field-emission microscope
~, горизонтально повернутый inverted microscope
~, измерительный measuring/comparator microscope
~, инструментальный tool-room microscope
~, инфракрасный infrared microscope
~, ионный field-ion microscope
~, люминесцентный fluorescence microscope
~, оптический light/optical microscope
~, отсчетный reading microscope
~, растровый (электронный) scanning (electron) microscope
~, рентгеновский X-ray microscope
~ с большим увеличением high-power microscope
~ с малым увеличением low-power microscope
~ с цифровым отсчетом digitized microscope
~, световой light/optical microscope
~, стереоскопический stereomicroscope, stereoscopic microscope
~, теневой shadow microscope
~, фазоконтрастный phase-contrast microscope
~, шкаловый reading microscope
~, электронный просвечивающий transmission electron microscope
~, электронный растровый scanning electron microscope
микроскопирование microscopy
микроскопический microscopic(al)
~ная флора microflora
~кое растение microphyte
микроскопия microscopy
~ биологических объектов biological microscopy, biomicroscopy
~, витальная vital microscopy, macromicroscopy
~ живого глаза corneal microscopy
~, люминесцентная fluorescence/luminescent microscopy
~, оптическая optical/light microscopy
микроскоп-микрометр micrometric microscope
микросланцеватость microfoliation
микрослоистость microlamination
микроснимок micrograph
микросоленометр microsalinometer
микросплавление microalloying
микроспора *bot., zool.* microspore
~, толстостенная/тонкостенная thick-/thin-walled microspore
микроспория *med.* Grudy's disease, microsporia
микроспороз *med.* ringworm of the skin, microsporosis
микроспороцит pollen mother cell, microsporocyte
микросреда microhabitat, microenvironment
микросрез microslide, microsection
микростереосъемка stereo-micrography
микростык microjuncture
микросхема microcircuit, *comp.* chip; integrated circuit, IC
~, аналоговая analog chip
~ в корпусе packaged IC
~, заказная custom microcircuit/chip
~, интегральная integrated microcircuit, IC
~, печатная тонкопленочная printed thin-film microcircuit
~, полупроводниковая semiconductor/monolithic IC, solid-state IC
~, цифровая digital chip/microcircuit/IC
микросхемотехника microcircuitry
микросъемка (cine)photomicrography
микротвердость microhardness
микротипия photomicrograph print
микроток microcurrent
микротом *biol.* section cutter, histotome, microtome
~, замораживающий freezing microtome
~, санный sliding microtome
микротомный microtome *(attr.)*
~ нож section knife
микротрещина microcrack
микроудобрение trace-element fertilizer, microfertilizer, micronutrient
микроузел microassembly, microunit
микроукол microinjection
микроуровнемер microlevel gauge
микрофаг *anat.* microphag(e), microphagocyte
микрофарада *el.* microfarad
микрофизика microphysics
микрофильм microfilm, microfilm record/copy
микрофильтрация microseepage
микрофительный, микрофитовый microphytic, microphytal
микрофитоценоз microphytocoenosis, microcommunity
микрофиша microfiche

микрофон microphone
~, бесмембранный no-diaphragm microphone
~, гарнитурный/нагрудный breastplate microphone
~, губной lip microphone
~, дипольный dipole transmitter
~, конденсаторный capacitor microphone
~, ненаправленный non-directional microphone
~, односторонне направленный cardioid microphone
~ переменного магнитного сопротивления variable-reluctance microphone
~, пищеводный *med.* (o)esophageal microphone
~, противошумовой/помехозащищенный antinoise microphone, noise-cancelling microphone
~ сердечных звуков *med.* cardiomicrophone
~, стереофонический stereo microphone, stereotuner
~, телефонный telephone transmitter
~, фазовый phase-shift microphone
микрофония microphony, microphonia, microphonism, microphonic effect
микрофотография 1. microphotography, photomicrography; 2. photomicrogram, microphotograph
микрофотокамера photomicrographic camera
микрофотокопия microphotograph, microfilm print
микрофотонасадка photomicrographic attachment
микрофотопленка (negative) microfilm
микрофотоснимок microphotograph
микрофотоустановка photomicrographic set-up
микрохимия microchemistry
микрохирургия microsurgery
микроцефалия *med.* microcephaly, microcephalism
микроцист(а) microcyst
микрочастица microparticle
микрошприц *med.* microsyringe
микроэкономика microeconomics
микроэлектроника microelectronics
~, полупроводниковая solid-state microelectronics
~, тонкопленочная thin-film microelectronics
микроэлемент 1. *tech.* microcell; microcomponent; 2. minor/trace/micronutrient element
~ питания растений minor plant nutrient
~, чувствительный microsensor
~, электронный microelectronic element
микроявление microphenomenon, microeffect
микругия *biol.* micrurgia
~, ультрамикроскопическая ultramicroscopic micromanipulation/micrurgia
миксаденит *med.* myxadenitis
~, лабиальный Baelz's disease
миксамеба *bot.*, *zool.* myxamoeba
микседема *med.* myx(o)edema, Gull's disease
~, врожденная congenital myxedema, cretinism
~, детская infantile myxedema
~, послеоперационная operative myxedema
миксер stirrer, mixer, agitator, blender
~, лопастный blade mixer
~, лопаточный shovel mixer
миксина *zool.* hagfish *(Myxine glutinosa)*
миксобактерии *pl.* slime bacteria, myxobacteria
миксовирус myxovirus
~, парагриппозный paragrippal myxovirus
миксома *pathol.* myxoma
~ предсердия *med.* atrial myxoma
миксоплоидия *biol.* mixoploidy
миксоподий, миксоподия *zool.* myxopodium (*pl.* -dia)
микст *tennis* mixed doubles
микстура mixture, potion, lotion, suspension
миктогаплонт *biol.* mictohaplont
микшер *rad.* 1. mixer, fader; 2. monitor man
~, ламповый electron-tube mixer, electron fader
~, немой dummy fader
мил mil
~, угловой angular mil
миланец, миланский Milanese
мильдю mildew, honey-dew
миледи milady
милетский *hist.* Milesian
милиарный *med.* miliary
~ туберкулез galloping consumption
милитаристский militarist(ic)
милицейский *n.* militiaman; policeman
милиционер 1. militiaman; 2. policeman *(in the former USSR)*
милиция 1. militia, police force; 2. territorial army
~ции, отделение police station
милленаризм millenarianism
миллефоль *bot.* common yarrow *(Achillea millefolium)*
миллиампер *el.* milliampere
миллиамперметр *el.* milliammeter
миллиампер-секунда milliampere-second
миллиард milliard, thousand million, *Am.* billion
~ битов billibit
~ электрон-вольт billion/Giga-electron-volt, Bev, Gev
миллиардер billionaire, multimillionaire
миллиарий *hist.* milliary, Roman milestone
миллибар millibar
милливольтамперметр millivolt-ammeter
миллигал milligal
миллигенри *el.* millihenry
миллиграмм milligram(me)
милликюри millicurie, m-curie, mC
миллиламберт millilambert
миллилитр milliliter
миллиметр millimetre, *Am.* millimeter
миллиметровка squared/graph paper
миллиметровый millimetric
~вая бумага squared paper
миллимол millimole
миллимикрон millimicron
миллион million

~ электрон-вольт million electron-volt, Mev
миллирезерфорд millirutherford
миллисекундомер millisecond timer
миллиэквивалент milliequivalent
милорд milord
милосердие *rel.* charity; clemency, mercy
~дия, сестра (medical) nurse
милосердный *rel.* charitable, merciful
~ная, сестра (medical) nurse
милостивый gracious, kind
~ государь *(form of address)* sir; *(in letters) Dear Sir*
~вая государыня madam; *(in letters)* Dear Madam
милостыня alms, charity
милость 1. mercy, charity; 2. favo(u)r, grace; good graces
~, ваша *(form of address)* your worship
~тью Божией *rel.* by the Grace of God
мильеранизм Millerandism
мильрейс milreis *(coin)*
мильтон *text.* Milton (cloth)
мильтония *bot.* pansy orchid *(Miltonia)*
миля mile
~, английская сухопутная (уставная) statute/land mile (1, 609 km)
~, географическая geographical mile (7, 4204 km)
~, морская sea/nautical mile (1, 852 km)
мим *theatre* mime (actor)
мимеограф duplicating machine, mimeograph
миметизм *biol.* mimetism, mimesis, mimicry
миметический mimetic, mimic, simulating, imitative
мимик, мимист mime, mimic
мимика mimicry, facial expression
мимикрия *zool.* mimicry, living camouflage
мимический mimic
мимоза *bot.* mimosa *(Mimosa)*
~, стыдливая sensitive plant *(M. pudica)*
мимозка *bot.* common mesquite *(Prosopis juliflora)*
мимозовидный mimosoid, mimosaceous
мимозовые *n. pl. bot.* mimosa family *(Mimosaceae)*
мимулус, мимулюс *bot.* monkey flower *(Mimulus)*
мимусопс *bot.* bulletwood *(Mimusops)*
мина I 1. *mil.* mine; torpedo; 2. bomb, mortar shell
~, авиационная aircraft-laid/-launched mine, aerial mine
~, акустическая acoustic/sound/sonic/listening mine
~, всплывшая free mine
~, гальваноударная electro-chemical/-contact mine, chemical/horn mine
~, донная ground/bottom mine
~, инженерная shore-controlled mine
~, лодочная submarine(-laid) mine
~, самодвижущаяся mobile mine, self-propelling mine
~, якорная anchored/moored/buoyant mine
~ами, заграждать mine in
~ами, усеянный mine-infested/-strewn
~ну, наскочить на strike a mine
~ну, обезвреживать neutralize/deactivate a mine
~ну, ставить lay/plant a mine
~ну, ставить на мертвый якорь moor a mine
мина II expression, countenance, mien
мина III *hist.* mina *(ancient coin of Greece);* 2. mina *(unit of weight)*
мина IV pencil, strip of black lead
минарет minaret
миндалевидный, миндалеобразный amygdaloid, amygdaliform, almondlike, almond-shaped
~ная железа *anat.* tonsil
миндалелистный *bot.* almond-leaved
миндалик, миндалина *anat.* tonsil
~, небная palatine tonsil
миндалина 1. almond; 2. *anat.* tonsil; 3. *geol.* amigdule
миндаль *bot.* almond *(Amygdalus, Prunus amygdalus)*
миндальный amygdalic, amygdaloid, amygdalate; *n. pl. bot.* Amygdalaceae
~ камень *geol.* almond rock, amygdaloid
миндель *geol.* Mindel
миндель-рисский Mindel-Rissian
минер 1. *mil.* miner, mine layer; 2. *zool.* miner
минераграфия mineragraphy, chalkography; ore microscope
минерал mineral
~ альпийских жил Alpine cleft mineral
~, взаимоисключающий antipathetic mineral
~ выщелачивания leach mineral
~, господствующий dominant mineral
~, жильный vein mineral
~, искусственный artificial/manufactured mineral
~, кластический clastic/detrital mineral
~, метасоматический metasomatic/guest mineral
~, неразличимый/скрытый occult mineral
~, нерудный non-metallic mineral
~, определяющий essential mineral
~, первичный/первоначальный primary/original mineral
~, положительный positive mineral, plus-mineral
~, природный naturally occurring mineral
~, режущий sectile mineral
~, рудный ore mineral
~, руководящий index/guide/diagnostic mineral
~, самородный native/natural mineral
~, синтетический synthetic/artificial mineral
~, слюдистый micaceous mineral
~, столбчатый prismatic mineral, rod-like mineral
~, устойчивый stable/resistant/persistent mineral

~, шлифующий abrasive
минерал-гост guest mineral, metasome
минерализатор mineralizing agent, mineralizer
минерализация *geol.* mineralization; *geochem.* salination
~ воды salt content, salinity of water
минерал-индикатор indicator mineral
минералит asbestos cement
минералог mineralogist
минералогия mineralogy
~ раковин shell mineralogy
минералоид colloid mineral
минералообразование mineral formation, minerogenesis
минерал-спутник accessory/associate mineral
минерал-указатель indicator/tracer mineral
минерал-хозяин host mineral
минеральный mineral
~ воск mineral wax, ozocerite
~ная шерсть mineral/rock wool
Минерва *myth.* Minerva
минетта *geol.* minette
~, шаровая spheroidal minette
минея euchology
~ четья menology; martyrology
минзаг *abbr.* **минный заградитель** *mil.* mine layer
минздрав *abbr.* **Министерство здравоохранения** Ministry of Public Health
миниатюра *art.* miniature (painting); *lit.* short story
миниатюрист miniaturist, miniature painter
мини-диск *comp.* minifloppy
мини-компьютер mini (computer)
мимимакс *math.* minimax, min-max, saddle point
минимал minim
минимально-обнаружимый minimum detectable
минимальность minimality, minimum
минимальный minimal, minimum; smallest, least
~ уровень жизни subsistence level
миниметр minimeter, mechanical/contact comparator
минимизация minimization
~, безусловная unconstrained minimization
~, многошаговая stepwise minimization
~, условная conditional/constrained minimization
мини-мода mini fashion
минимум minimum; low, depression
~, абсолютный absolute/global minimum
~, безусловный unconstrained minimum
~ временного ряда trough of time series
~, гравитационный gravity low/minimum
~ заработной платы minimum wage
~ зоны проводимости conduction band valley
~, исследовать *math.* test for a minimum
~, локальный local/relative minimum
~, магнитный magnetic/magnetometer low
~, прожиточный cost of living, subsistence level, living/minimum wages
~, технический essential technical qualifications
~, условный constrained minimum
~а к минимуму, от peak-to-peak
~ы, метеорологические weather minima
минирецессия *econ.* minirecession
минирование *mil.* mine laying, mining
минированный mine-infested/-strewn, mined
минировать *mil.* lay/plant mines
миниспад *econ.* minirecession
министерство ministry; department
~ внутренних дел Ministry of the Interior; **(Великобритании)** Home Office
~ здравоохранения Ministry of Public Health
~ иностранных дел Ministry of Foreign Affairs; **(Великобритании)** Foreign Office
~, морское Navy Department
~ обороны Ministry of Defence/Defense
~ путей сообщения Ministry of Railways
~ сельского хозяйства Ministry of Agriculture
~ социального обеспечения Ministry of Social Welfare
~ торговли Ministry of Trade; **(Великобритании)** Board of Trade
~ торгового флота Ministry of Shipping
~ финансов Ministry of Finance; Department of the Treasury
министр *polit.* minister, Secretary
~ без портфеля minister without portfolio
~ внутренних дел Minister/Secretary of the Interior
~ здравоохранения Minister of Health
~ иностранных дел Minister of Foreign Affairs, Foreign Minister
~ обороны Defence Minister, *Am.* Secretary of Defense
~, полномочный Minister Plenipotentiary
~а, товарищ deputy minister
министр-президент prime minister, premier
миниум minium, red lead
мини-юбка miniskirt
миннезанг *hist.* minnesong, Minnesang
миннезингер *hist.* minnesinger
минный *mil.* mine *(attr.)*, mine-laying
~ное заграждение/поле mine field
минога *zool.* lamprey *(Petromyzon)*
миноговые *n. pl. zool.* lampreys *(Petromyzonidae)*
миноискатель 1. mine detector/locator *(apparatus)*; 2. *mil.* sapper
минойский Minoan, Cretan
минологий menology
миномет *mil.* mortar, mine thrower; torpedo tube
~, гвардейский multirail rocket launcher
~, дульнозарядный muzzle-loaded mortar
минометание aerial mine-laying
минометный *mil.* mortar *(attr.)*
~ расчет mortar detachment
минометчик *mil.* mortar man/gunner
миноносец *naut.* torpedo boat
~, эскадренный destroyer

миноносный armed with torpedoes
~ катер torpedo boat
минор *math.* minor, minor determinant, sub determinant
~, главный principal minor
~, дополнительный complementary minor
минорант(а) *math.* 1. minorant (series); 2. minorant (function); numerical minorant
минорит Minorite, Franciscan friar
минорный *mus.* minor
Минос *myth.* Minos
миносбрасыватель mine rack, mine-release mechanism
минотавр *myth.* Minotaur
минофировый minophyric
минреп *nav.* mine mooring (cable), mooring rope
минтай *zool.* pollack, pollock, Alaska/walleye pollack *(Theragra chalcogramma)*
минус *math.* minus, negative/minus sign
~, общий *el.* common return/earth
минускул minuscule, Gothic writing
минусный minus
минусовый minus; *el.* negative
минута minute
~, дуговая/дуги minute of arc
~, метрическая centesimal angular minute
~, тарифная *com.* chargeable/paid minute
~, угловая angular minute
минутный minute *(attr.)*
~ная стрелка minute hand
минхер Mynheer
миньон *print.* minion *(7-point type)*
миобласть *zool.* myoblast
миобластомиома *med.* granular cell, myoblastoma
миоген *biol.* myogen
миогенный myogenic, myogenetic, myogenous
миогеосинклиналь *geol.* miogeosyncline
миоглобин *bioch.* myoglobin
миограмма myogram, myographic tracing
миодистрофия *med.* myodystrophy, myotonia atrophica
миоз *biol.* miosis, meiosis
миозит *med.* myositis, myitis, inflammation of a muscle
~, гнойный purulent myositis
~, инфекционный infectious myositis, polymyositis
~, прогрессирующий disseminated myositis
~, профессиональный occupational myositis
~, тропический tropical myositis/pyomyositis, muma
~, эпидемический Bornholm disease, epidemic myositis
миокард(ий) *anat.* myocardium, cardiac muscle
миокардит *med.* myocarditis
~, гигантоклеточный giant cell myocarditis
~, острый изолированный acute isolated myocarditis, Fiedler's myocarditis
~, очаговый focal/nidal myocarditis
~, смешанный mixed/compound myocarditis
миокардиоз myocardiosis
миоклония *med.* myoclonia
~, велопалатинная palatal myoclonia
~, инфекционная infectious myoclonia, chorea
миоклонус *med.* myoclonus
~, сложный polyclonia, paramyoclonus
миокристаллический *geol.* myocrystalline
миома *med.* myoma
~ из гладких мышечных волокон leiomyoma
~, саркоматозная myosarcoma
миомаляция *med.* myomalacia
~ миокарда cardiomalacia
миопатия *med.* myopathy
~, наследственная hereditary/familial myopathy
~, поздняя tardive myopathy
миопия *med.* myopia, short-/near-sightedness
~, злокачественная malignant/progressive myopia
миоплегия myoplegia
миоплиоценовый Mio-Pliocene
миотония *med.* myotonia
~, атрофическая Steinert's disease
~, врожденная Eulenburg's/Thomsen's disease
~, приобретенная Talma's disease
миофибрилла *anat.* myofibril(la)
~, поперечнополосатая cross-striated myo(protro)fibril
миоцен Miocene
миоценный, миоценовый Miocenic
миоцит myocyte
~ Аничкова caterpillar cell
миоэпителий myoepithelium
мир I world, universe; kingdom
~, деловой business constituency
~, животный fauna, animal kingdom
~, литературный world of letters
~, неорганический mineral kingdom
~, растительный flora, plant/vegetable kingdom
~, финансовый moneyed interests
мир II peace
~, почетный peace mith hono(u)r
мир III *hist.* mir, Russian village/community
мира *opt.* test object, mark, mire
мирабель *bot.* mirabelle *(Prunus insititia syriaca)*
мирабилис *bot.* four-o'clock, marvel-of-Peru (Mirabilis)
мирабилит *miner.* mirabilite, Glauber salt
мираж mirage, fata morgana, looming; optical illusion
миракль *lit.* miracle play
мирза Mirza, Prince
мириаватт *el.* myriawatt
мириаметр *el.* myriameter
мирика *bot.* bayberry *(Myrica)*
мирикария *bot.* false tamarix *(Myricaria)*
мирингодерматит *med.* myringitis bulbosa, myringodermatitis

мириофиллюм *bot.* parrot's-feather *(Myriophyllum)*
мирицил *chem.* myricil, melissyl
мирлитон *bot.* chayote *(Sechiumedule)*
мирмекофил *bot.* myrmecophilous plant, myrmecophil
мирмекофит *bot.* myrmecophyte
мирмекохор *bot.* myrmecochore
мирмидонец *myth.* Myrmidon
мирница *eccl.* chrismatory
мирный peace *(attr.)*, peaceful; irenic(al)
 ~ договор peace treaty
 ~ное сосуществование peaceful co-existence
миро *eccl.* Holy oil, chrism, oil of consecration
миробалан *bot.* myrobalan plum *(Prunus cerasifera)*
мировая *n.* peaceful settlement; amical agreement
мировидение global television
мировоззрение ideology, Weltanschauung, philosophy of life
мировой I mondial, universal; world, mundane
 ~ рынок world market
мировой II 1. conciliatory; mediatory; 2. *n.* Justice of the Peace
 ~ посредник *hist.* arbitrator, conciliator
 ~ судья Justice of the Peace
мироздание *rel.* the universe
мирозин *chem.* myrosin
мирок community, restricted circle
мироксилон *bot.* balm tree *(Myroxylon)*
миролюбец supporter of peace, pacifist
мирон *zool.* barbel *(Barbus)*
миронарушитель peacebreaker
мироносица *rel.* balsam-bearer
мироописание cosmography
мироощущение attitude, disposition
миропомазание *eccl.* chrismation, chrismatory, anointing, unction; consecration to office
миропомазанник *hist.* anointed sovereign
миропонимание, миросозерцание *philos.* Weltanschauung, ideology; world outlook
миротворец *rel.* reconciler, peacemaker
миротворный, миротворящий conciliatory, pacificatory
мирохранительница *eccl.* oilstock
мирр(а) myrrh
миррис *bot.* myrrh *(Myrrhis)*
мирт(а) *bot.* myrtle *(Myrtus)*
миртовидный *bot.* myrtiform
миртовые *n. pl. bot.* myrtle family *(Myrtaceae)*
мирянин *rel.* layman; *pl.* people
мисдиминор *law* misdemeano(u)r, bad conduct
мискантус *bot.* silver/sword grass *(Miscanthus)*
миссионер, миссионерский missionary
миссионерство missionary work
миссия mission, delegation; assignment
 ~, военная military mission
 ~, торговая trade delegation
мистагог mystagogue
мистерия *rel.* mystery; *lit.* mystery/miracle play
мистик mystic; visionist
мистика mysticism; cabalism
мистицизм mysticism
мистический mystic(al)
мистраль mistral *(wind)*
мисюрка *hist.* medieval helmet, military headpiece
мителла *bot.* miterwort, bishop's-cap *(Mitella)*
митинг (political) rally, (mass) meeting
миткалевый, миткальный made of calico
миткаль *text.* calico
митогенетический *biol.* mitogenetic, mitogenic
митоз *biol.* mitosis, karyokinesis, mitotic division
 ~, многополюсный multipolar mitosis
 ~, патологический morbid mitosis
митохондрия *biol.* mitochondrion, plastosome, chondriosome
митра I *eccl.* mitre, miter
Митра II Mithra(s); Mitra
митраизм mithraism
митральез *print.* roller mould
митральеза *mil.* mitrailleuse
митральный *anat.* mitral
митрон voltage-tunable magnetron
митрополит *eccl.* metropolitan
митрополия *eccl.* metropolitan see
митрофорный *eccl.* mitered, mitred, entitled to wear a miter
миттель *print.* English, 14-point type
миттельшмерц *med.* mittelschmerz
миттельшпиль *game th.* mid-game
митчелла *bot.* partridgeberry *(Mitchella)*
миф myth; *pl.* mythical literature
мифический mythic(al)
мифология mythology
мифотворческий mythogenic
мифотворчество mythogenesis, formation of myths
мицелиальный, мицелиевый *bot.* mycelial, mycelian
мицелиевидный *bot.* mycelioid, mycelium-shaped
мицелий *bot.* mycelium
 ~ грибов spawn, fungal mycelium
 ~, зачаточный pseudomycelium
 ~, клеточный/членистый septate mycelium
 ~, многолетний perennial mycelium
 ~, неклеточный nonseptate/coenocytic mycelium
мицелла *biol.* micella, micelle
мицелярно-карбонатный mycelial-calcareous
мицетогенный mycetogenic, mycetogenetic, mycetogenous
мицетология mycology, mycetology
мицетома *med.* mycetoma
 ~, ложная pseudomycetoma
мичман *naut.* warrant officer; midshipman
мишень target
 ~, бумажная target sheet
 ~ для холодной пристрелки bore-sight target
 ~, неспасаемая expendable target
 ~ рентгеновской трубки target anticathode

~, **сверхзвуковая** supersonic target
~, **спасаемая** recoverable target
~, **стрелковая** gunnery target
~, **теплоизлучающая** infrared target
мишень-конус sleeve target
мишень-оболочка casing target
мишен-отражатель reflection target
мишень-полотнище banner target
младенец baby, infant
~**нцев, избиение** *rel.* the Massacre of the Innocents
младенчество infancy, childhood, babyhood
младогегельянец *philos.* Neo-Hegelian
младогегельянство *philos.* Neo-Hegelianism
младограмматик neogrammarian, Junggrammatiker
младограмматический neogrammatical
младший 1. the youngest; 2. junior, younger
~ **командный состав** noncommissioned officers
~ **лейтенант** second lieutenant
~ **научный сотрудник** junior research worker
~ **партнер** junior partner
млековидный chyliform
млеконосный lactiferous
млекопитающие *n. pl.* mammals, mammalians
~, **высшие** higher mammals
~, **первобытные** prototherians
~, **плацентарные** placental mammals, eutherians
~, **травоядные** herbivora
~, **хищные** carnivora
млекопитающий *zool.* mammalian
млечник I 1. *bot.* sea milkwort *(Glaux);* 2. latex tube/vessel, lactiferous cell, lactifer
млечник II *bot.* lactarius *(Lactarius)*
млечность lactescence, milkiness
млечный milk(y), lacteous, lactic, lacteal; *med.* chylous; galactophorous
~ **сок** *bot.* milk, latex; *physiol.* chyle
~ **сосуд** latex vessel
~ **Путь** *astr.* Milky Way, the Galaxy
М-метр Machmeter, Mach indicator
мнема *psych.* mneme
мнемокод mnemonic code
мнемоник mnemonist
мнемоника mnemonics, mnemotechny, mnemotechnics
мнемонический mnemonic
мнемотехника mnemotechny, mnemonics
мнение opinion, assessment, judgement
~, **общественное** public opinion
~, **общепринятое** orthodox opinion
~, **особое** *law* dissenting/separate opinion
мнимоисторический pseudohistoric(al)
мнимость ostensibility
мнимый imaginary, sham; feigned; ostensible, alleged, putative
многоадресный multiple-address, multiaddress
многоактный multievent
многоамперный *el.* heavy-current
многоанодный multianode
многоаспектный *gram.* multiaspectual
многоатомный polyatomic
многобайтовый multibyte
многобожие polytheism
многоборец *sport* multiathlonist, all-round athlete
многобородник *bot.* polypogon *(Polypogon)*
многобороздный, многобороздчатый polysulcate, multisulcate
многоборье *sport* multiathlon
многобратный, многобратственный *bot.* polyadelphian, polyadelphous
многобрачие *law* polygamy
многобрачный polygamous; *bot.* androgynous
многобугорчатый multituberculate, multicuspid
многобуквенный multiletter
многовалентность multivalence, polyvalence
многовалентный multivalent, polyvalent, plurivalent
многовариантный multiple(-path), multivariant
~ **анализ** multiple-path analysis
многоваттный *el.* high-watt
многовековой centuries-old
многовершинный multipeak, multivertex; polyconic, multiple-cone
многоветвистый *bot.* polycladous, multiramose
многовидный multiform, multimode
многовитковый multiloop, multispiral; multiple-turn
многовластие *law* multiple authority
многоводие *med.* hydramnion, hydramnios
многоводье 1. (time of) high water level; 2. *med.* hydramnion, hydramnios
многоглавый polycephalous, multicipital; hydra-headed; many-domed
многогласие 1. *mus.* polyphony; 2. disagreement, different opinions
многогнездный 1. *tech.* multijack; multislide; 2. *biol.* multilocular; polycystic; many-celled; multicamerate
~**ная завязь** *bot.* compound/multilocular ovary
~**ная костянка** nuculanium, baccate fruit
многоголовчатый, многоголовый polycephalous, multicapitate, many-headed
многоголосие *mus.* polyphony
многогранник polyhedron, polytope
~, **эоловый** ventifact
многогрупповой multigroup
многодвигательный multiengine(d), multimotor, multiple-generator
многодетный having many children
многодиапазоновый multirange
многодисковый *tech.* multidisc, multiplate
~**вое сцепление** multiplate clutch
многодолевой, многодольный, многодольчатый multilobar, multilobate, multilobular
многодомность *bot.* polygamy
многодомный *bot.* polygamous, polyoicous, polyoecious
многодорожечный multi-track, multipath

многодырочный multiaperture, multihole
многодырчатый polyforate; perforated
многоемкость multicapacity
многоженство polygamy
многожилковатый, многожилковый multinerved
многожильный *tech.* multicore, multiple-core; polycore; multiwire, multiple-cord/-strand
~ **кабель** multicore cable
многозазорный multigap
многозародышевый polyembryonic, multigerm
многозарядный multicharge, multiple-charge, multiply charged
~**ная винтовка** repeating rifle
многозаходный *tech.* multiple-thread
~**ная резьба** multiple(-start) thread
многозвучный polyphonic, multisonous
многозначность 1. *math.* multivalence, polyvalence, multiple valuedness, multivaluedness; 2. polysemy, polysemanticism, ambiguity
многозначный 1. multivalent, polyvalent; multiple-valued, multivalued; 2. polysemantic, polysemous
~**ное число** multiplace number
многозонный multizone, multirange, multiregion
многозуб *zool.* paddlefish *(Polyodon spathula)*
многозубчатый multidentate; multipronged
многоиглистый multispinous
многоименный polyonymous
многоинерционный multilag
многокамерный multicellular, multiseptate, multichamber, many-chambered, multicamerate, polythalamous, multiple-cavity
многоканальный multichannel, multiway, multiduct, multitrack, multiplex
многокаскадный *rad.* multistage, multiple-stage
многокатодный multicathode, polycathode
многоквартирный having many apartments
~ **дом** tenement/apartment house
многокилевой *bot.* multicarinate
многокислотный polyacid
многоклеточный *biol.* multicellular, pluricellular, multilocular, plurilocular; *n. pl. zool.* metazoans *(Metazoa)*
многоковшовый multiscoop, multibucket
~ **экскаватор** chain-bucket excavator
многоколейный multiple-track/-line
многоколесный multityre, multi-wheel, many-wheeled
многоколосковый multispiculate
многоколосый *bot.* polystachyous
многокомпонентный complex, multicomponent
многоконтурный multiloop, multiple-loop, multicircuit, multistage
многоконцевой multiterminal
многокоренник *bot.* duck's-meat *(Spirodela)*
многокорешковый *bot.* polyrhizous
многокорневой multirooted
многокоробочный multicapsular
многокорпусный multiple(-unit)
~ **плуг** *agr.* gang/multiple plow/plough, many-bottom plough
многокосточковый *bot.* polypyrenous
многокрасочность polychromatism, diversity/variety of colo(u)rs; variegation
многокрасочный polychromatic, multi-/many-colo(u)red, variegated
многократно repeatedly, many times
~ **дифференциальный** multidifferential
~ **используемый** reusable
~ **объявленный** *comp.* multideclared
многократно-складчатый *geol.* (multi)plicate, having many plicae
многократносложный supradecompound
многократность recurrence, multiplicity; repetition; *rad.* multiplex(ing)
~ **видами импульсов** pulse(-mode) multiplex
многократный 1. multiple, multifold, manifold; reiterated, repeated, recurrent; 2. *gram.* iterative, frequentative
многокруговой polycyclic
многокрылый polypterous, multialate
многокультурный *agr.* multiple-crop
многоламповый multi-lamp, multitube, multiple-bulb
многолезвийный multiblade, multipoint
многолемешный *agr.* multishare
~ **плуг** multiple-furrow plow/plough
многолепестный *bot.* polypetalous, pluripetalous, apopetalous
многолетнемерзлый everfrozen, eternally frozen, permafrost
многолетний 1. *bot.* perennial, perennating; 2. long-term, persistent
~**нее растение** perennial plant
многолетник *bot.* perennial (plant)
многолинейный multicircuit; multiple-line
многолистный 1. *bot.* polyphyllous, multifoliate, plurifoliate; 2. *math.* multivalent
многолисточковый *bot.* multifoliolate
многолопастный *tech.* multiblade(d); *bot.* multilobed, multilobate, multilobar, multilobular, palmate
многолучевой mutlibeam, multipath, multiple-wave, multiwave; polyradiate, multiradiate
~ **верхоцветник** *bot.* pleiochasium, polychasium
многолучевость *rad.* multipath propagation/effect
многолучистый multiradiate, polyradiate
многолюдье throng
многомерность multidimensionality, pluridimensionality
многомерный multidimensional; multivariate, multivariable
~ **выбор** multivariate sampling
многоместный multiseater, multiplace; polyadic
многоморфный multiform
многомоторный multiengine(d), multiple-motor
многомужество, многомужие polyandry
многонадрезный polytomous, multifid
многонаправленный multidirectional, poly-

directional
многонациональный multinational
многоначалие polyarchy, multiple authority
многонитный, многониточный multiple-thread, multifilament
многоногий multiped, many-legged
многоножка I *bot.* polypody *(Polypodium)*
многоножка II *zool.* centipede, myriapod
многоножковые *n. pl. bot.* polypody family *(Polypodiaceae)*
многоножковый *zool.* myriapodan, myriapodous
многономенклатурность *econ.* diversity
многонуклеоновый many-nucleon
многообмоточный multiwinding, multiwound
многообразие, многообразность diversity, multiformity; *math.* variety, manifold, multiplicity
~, алгебраическое algebraic variety/manifold
~, получистое *math.* unirational/semipure variety
~ разветвления *math.* ramification manifold
многообразный manifold, multiform, varied, diversified; *bot.* polymorphous, polymorphic
многообъективный multilens, multiple-lens
многоопорный *tech.* multiple-seated
~ная балка *constr.* continuous beam
многоосевой multiaxial
многоосновность *chem.* polybasicity
многоосновный *chem.* polybasic
многоосный multiaxial, polyaxial, pluriaxial
многоотраслевой encompassing several branches/fields, diversified
~вое хозяйство diversified economy
многопалость *med.* polydactylism, hyperdactyly
многопальчатый multidigitate, polydactyl
многопарный multijugate, multijugous
многопартийность multiparty system
многоперегородчатый pluriseptate, multiseptate
многопестичный *bot.* polygynous
многопланный, многоплановый having many plans, multifarious; multiple-plan
многопластинчатый multilaminate, multiplate
многопластовый multilayer
многоплечий *tech.* multiple-arm
многоплодие 1. *med.* multiple pregnancy; 2. *bot.* polycarpy
многоплодник *bot.* manyseed *(Polycarpon)*
многоплодниковый *bot.* polycarpous, polycarpic
многоплодный 1. *med.* multiparous; 2. *bot.* polycarpous
~ная беременность multiple pregnancy
многопозиционный multiposition; multipoint
многополье, многопольная система *agr.* multicourse system, many-field system *(of crop rotation)*
многопольный *agr.* pertaining to the many-field system
~ севооборот multiple crop rotation
многополюсник multi-terminal circuit/network, multi-port
многополюсный multipolar, multipole, multiple-pole, multiterminal
многопоровый polyporate, poriferous
многопортовый *comp.* multiport
многопоясковый multizonate
многопредельный multirange *(of instrument)*
многопризнаковый multifeature
многопроводниковый multiple-conductor
многопроводной multiwire, multiconductor
многопрограммный multiprogram(me); multiprogram(m)ing
многопродуктовый *econ.* multicommodity
~вая система multiitem system
многопроцессорный multiprocessing
многопутный multitrack, multilane; multiple-line
многопучковый *bot.* multifasciculate
многораздельный multifid, multipartite; decompound
многоразмерный multidimensional
многоразовый non-expendable; repeated, frequent
многорезной multislot; multisegmental
многоразрядный multidigit; multiple-order
многоребристый multicostate, many-ribbed
многорежимность multiplicity/variety of operation conditions/states/levels
многорежимный multimode
многорезцовый *tech.* multiblade(d); multicutting; *n. pl. zool.* polyprotodonts *(Polyprotodontia)*
многореснитчатый, многоресничный multiciliate
многоречивость loquacity, garrulity
многородящий *zool.* multiparous
многорусловый braided *(of river)*
многорыльцевый *bot.* polystigmatic
многорядник *bot.* holly fern *(Polystichum)*
многорядный multiserial, multifarious, polyserial; multiple-row, many-ranked; *tech.* multibank; *bot.* multiseriate, pluriseriate, polystichous
многосвязный multiply connected
многосегментный multisgment; *zool.* polysomitic
многосекционный multisection(al), multipart
многосеменный *bot.* polyspermous
многосемядольный *bot.* polycotyledonous
многосемянка *bot.* polyachenium
многосемянный *bot.* polyspermous, plurispermous, many-seeded
многосерийный serial
~ телефильм TV serial
многосеточный multigrid
многоситовый multiscreen
многоскачковый multishock, multiple-hop
многоскоростной multiple-speed, multispeed, multivelocity *(attr.)*
многословие verbosity, prolixity, verbiage; *med.* polyphrasia
многословность prolixity
многосложность 1. polysyllabism; 2. complexity; comlicatednes
многосложный polysyllabic; complex, complicated, intricate; *bot.* decompound

многослойный multilayer(ed), laminated; sandwich-type, stratified
многосоставный complex; polysynthetic
~ **язык** polysynthetic language
многосоюзие polysyndeton
многоспектральный multispectral
многоспоровый polysporous, plurisporous
многосрезный multiple-shear
многостадийный multistage, multiphase, many-stage
многостандартный multistandard
многостаночник workman operating a number of machines simultaneously
многоствольный 1. *bot.* deliquescent; many-stemmed; 2. *tech.* multibarelled
многостворчатый multivalved, multivalvar
многостебельный *bot.* multicaulis
многостепенный multistage, multigrade
многостолбиковый polystylous
многостолбовой multiple-pole
многосторонний multilateral; *math.* polygonal
многосторонность versatility; *math.* polygonality
многострочный *comp.* multitrack
многострунный many-stringed, polychord
многоступенный multistage, polystage
многоступенчатый multiple-/many-stage, multistage, polystage; multilevel, multistep, multiple-phase
многосуставной, многосуставный *med.* multiarticular, polyarthric
многотактный *math.* multiple, multistage
многотерминальный *comp.* multiterminal
многотипный polytypic(al)
многотиражка factory newspaper
многотоковый *el.* multicurrent
многотомный multivolume; voluminous
многотональный polytonic
многотоннажный large-tonnage
многотонный I polytonic
многотонный II large-tonnage; very heavy
многоточечный multipunctate; multipoint(ed)
многоточие *math.* dots; marks of omission, omission points
многотрубный multitubular
многотрубчатый polysiphonous, polysiphonic
многотычинковый *bot.* polyandrous, multistaminate
многотычиночный *bot.* polystemonous, polyandrous
многоугольник polygon
~, **веревочный** funicular/string polygon
~, **вихревой** closed vortex polygon
~, **круговой** arc polygon
~, **правильный** regular polygon
~, **пространственный** skew/space polygon
~ **сил/силовой** *mech.* polygon of forces, force polygon
многоугольный polygonal; *bot.* polyangular, multiangular
многоузелковый, многоузловой multinodal
многоуровневый multilevel
многоустка *zool.* polystome *(Polystomum)*
многофазный, многофазовый *el.* polyphase, multiphase
многофакториальный multifactorial
многофакторный multiple-factor
многофидерный *tech.* multifeed
многофонемный polyphonemic
многоформность polymorphism
многофилиальный *econ.* multidivisional
многофункциональный multifunctional
многоходовой multipath, multiway, multipass; multiple-start; *(of screw)* multiworm, multithread
многоцветковый *bot.* polyanthous, multiflowered
многоцветность polychromy, pleochromatism
многоцветный *geol.* polychromatic, polychrome, multicolo(u)red; *bot.* multiflorous; *zool.* hetorochromous
многоцелевой multipurpose, multimission, multirôle, multifunction
многоцентрический polycentric
многоцентровой multicentric
многоцепной *el.* multicircuit
многоциклический polycyclic; pleiocyclic
многоцикловой polycyclic
многоцилиндровый multicylinder, multiple-cylinder, polycylindrical
многочастотный multifrequent, multifrequency *(attr.)*
многочелночный multishuttle
многочерпаковый multibucket
~ **экскаватор** chain-and-bucket excavator
многочешуйный, многочешуйчатый many-scaled, polylepidous
многочисленность multiplicity; plurality; numerical strength, numerosity
многочисленный numerous, multitudinal, multitudinous
многочлен *math.* polynomial, multinomial
~ **деления круга,** ~, **круговой** cyclotomic polynomial
~, **интерполяционный** interpolation/interpolating polynomial
~ **Лагерра** Laguerre polynomial
~ **Лежандра** Legendre polynomial
~, **неприводимый** irreducible/prime polynomial
~, **нормированный** monic polynomial
~ **от нескольких переменных** polynomial in several variables
~ **от x** polynomial in *x*
~, **приводимый** reducible polynomial
~, **элементарный симметрический** elementary symmetric polynomial/function
~ **Эрмита** Hermite('s) polynomial
~ **Якоби** Jacobi('s) polynomial
многочленистый multiarticulate, many-segmented
многочленный polymeric(al), polymerous; *math.* polynomial, multinomial; polyadic
~**ное предложение** *gram.* multinuclear sen-

tence
многошаговой multistep, multistage
многошкальный multiscale, multidial, multirange
многошпиндельный multiple-spindle, multispindle
многоштанговый multibar
многоштыревой polyrod, multiple-rod
многощелевой multislot, multislit
многощетинковый *zool.* polychaetous, multisetose
многоэлектродный multielectrode
многоэлементный multicomponent, multielement; multiple-unit
многоэтажный many-storied, multistory/multistorey, multistoried
многоэтапный multistage
многоядерность polynucleosis, multinucleosis
многоядерный polynuclear, multinuclear, multinucleate, plurinucleate; *chem.* polycyclic, multiring
~ная гамета coenogamete
многоядность *biol.* polyphagia, polyphagism
многоядный polyphagous
многоязычие polylingualism, plurilingualism
многоязычковый *bot.* multitongued
многоязычный 1. polylingual, multilingual; 2. polyglot
многоярусный multilayer(ed), pluristratal
многоячеистый, многоячейковый multicellular, many-celled, multi-chamber(ed)
многоячейный multiple-chamber, multichambered
множественность plurality, multiplicity; numerosity
~ значений multiple meaning
множественный plural
~ное число *gram.* plural (number)
множества *pl. math.* sets
~, непересекающиеся mutually exclusive sets, nonoverlapping/disjoint sets, nonintersecting sets
~, пересекающиеся overlapping sets
~, равномощные/эквивалентные equivalent sets, sets of the same cardinality
множество 1. *math.* set; 2. aggregate, assemblage, population, collection
~, ациклическое primitive set
~, бесконечное infinite set
~, бесконечное счетное countably infinite set
~, борелевское Bore/Borelian set
~ внутренних точек interior of a set
~, вполне упорядоченное well-/normally ordered set
~ всех подмножеств set of all subsets
~, выборочное *econ.* sampling population
~, выпуклое convex set
~, допустимое admissible set
~, замкнутое closed set
~ значений функции range of a function
~, измеримое measurable set
~, конечное finite set
~, линейно-упорядоченное linearly/simply/serially ordered set
~, многогранное polyhedral set
~, неизмеримое non-measurable set
~, непрерывное continuum
~, непустое nonempty/nonvoid set
~, несчетное uncountable set
~, нечеткое *comp.* fuzzy set
~, нульмерное, ~ меры нуль set of measure zero, null set, zero-dimensional set
~, общерекурсивное general recursive set
~, открытое open set
~, перечислимое enumerable set
~, полуупорядоченное quasi-ordered set
~, производное derived/derivative/cluster set
~, прямоугольное rectangle set
~, пустое empty set
~, связное connected set
~ стратегий *game th.* set of strategies
~, счетное countable/denumerable/enumerable set
~, точечное point set, set of points
~, упорядоченное ordered set, serially/simply ordered set
~ частиц cloud of particles
~, частично упорядоченное partially ordered set
~ элементарных событий aggregate of simple events
множимое *math.* multiplicand, first factor
множитель *math.* multiplier, factor, coefficient
~ асимметрии asymmetry factor
~, буквенный literal (factor)
~, весовой weight(ing) factor
~ времени, масштабный time-scale factor
~, дополнительный complementary factor
~, Ланде Lande (splitting) factor, g-factor
~, логический Boolean/logical factor
~, масштабный scale factor, scaling multiplier
~, нормировочный normalization factor/constant
~, нормирующий normalizing/normalization factor
~, общий common multiplier
~, одноквадратный one-quadrant multiplier
~, переводной conversion/reduction factor, conversion coefficient
~ по входу, нагрузочный fan-in
~ по выходу, нагрузочный fan-out
~, поправочный correction factor
~, постоянный multiplicative constant
~, простой prime factor
~, редуцирующий scale factor
~, статистический statistical weight factor
~, убывающий decreasing factor
~, экспоненциальный exponential factor
множительный multiplicative, multiplying; *print.* duplicating
~ аппарат duplicating machine
множить increase, augment; *math.* multiply
мо *el.* mho, siemens
моа *zool.* moa *(Dinornis novazealandiae)*
моаре *text.* moire *(fabric)*

мобилизатор *physiol.* mobilizer
мобилизация mobilization
~, поголовная *law* levy en masse
мобилизованность complete readiness for action
мобильность mobility
~ программного обеспечения software portability
могар *bot.* foxtail millet *(Setaria italica)*
могер *text.* mohair (fabric)
могиканин, могиканский Mohican
могиканка Mohican (woman)
могила grave
~, братская common grave
могильник 1. *archeol.* sepulcher, burial ground; 2. *zool.* imperial eagle *(Aquila heliaca);* 3. *bot.* African/Syrian rue
~ радиоактивных отходов burial
могильница *bot.* periwinkle (Vinca)
могильный sepulchral
~ холм (sepulchral) mound; tumulus
~ная плита gravestone, tombstone
могильщик *zool.* 1. burying beetle *(Necrophorus)*; 2. imperial eagle *(Aquila heliaca)*
могол Mogul
~, Великий *hist.* the Great Mogul
мода *tech., stat., math.* mode
~, акустическая acoustical mode
~, антистоксовая anti-Stokes mode
~, волноводная waveguide mode
~, вырожденная degenerate mode
~, генерируемая лазером lasing mode
~, генерируемая мазером masing mode
~, излучаемая radiating mode
~ колебаний mode of vibration, oscillation mode
~, осевая/продольная axial/longitudinal mode
~, паразитная spurious mode
~, собственная eigen/natural mode
~, соседняя adjacent mode
~, спиральная twisted mode
~, угловая angular mode, off-axial mode
модально-волевой modal-volitional
модальность *math.* modality
моделизм model(l)ing
~, авиационный model aeroplane flying
моделирование 1. model(l)ing; model test; 2. simulation, simulating; model making/building
~, аналоговое analog(ue) simulation
~, аналоговое в истинном масштабе времени real-time simulation
~, аналоговое электронное electronic simulation
~ в полете in-flight simulation
~ внешней среды environment simulation
~, дискретное digital simulation
~, игровое gaming simulation
~ команд command simulation
~ летных испытаний simulated flight testing
~, математическое mathematical modelling/simulation, math model(l)ing
~ методом черного ящика black-box modelling, input-output modelling
~ на наземной установке ground-based simulation
~ поражающего действия *med.* damage simulation
~ условий невесомости weightlessness simulation
~ условий полета ракеты missile environment simulation
~ хозяйственной деятельности business simulation
~, цифровое digital simulation
~, численное computational modelling
~, экономическое economic model building
моделировать model, shape, design; simulate
~ процесс на ЭВМ simulate a process on/with an electronic computer
моделировка *art* model(l)ing
моделист modelmaker, model(l)er, fashion designer
модель model, pattern; mock-up; make, type; *comp.* analog(ue)
~ анализа деловой активности activity analysis model
~ атома, боровская Bohr atom model
~ атома, томсоновская Thomson's atom model
~ без двигателя *nucl.* non-burning model, unpowered model
~ в масштабе scale model
~ в натуральную величину full-scale model
~ в разрезе sectioned model
~, вариационная variational model
~, вероятностная stochastic/probabilistic model
~ взаимодействия interaction model
~ возрастной диффузии age-diffusion model
~ вселенной model universe, world model
~ входящего потока, пуассоновская Poisson model
~, выставочная display/show model
~, геометрически подобная geometrically similar model
~, головная pilot model
~, графовая graph model
~, двухмерная two-dimensional model
~, двухступенчатая two-echelon model
~, дебаевская Debye model
~, действующая working model
~, детерминированная deterministic model
~,деформировать strain a model
~, динамически подобная dynamic/inertial model
~ для непрерывного времени continuous-time model
~, доводочная development model
~ дуэли, неадаптивная *game th.* nonadaptive duel model
~, закрытая/замкнутая closed model
~ затрат и доходов cost-benefit model
~ затраты-выпуск input-output model
~ изменчивости pattern of variability

~ из уносимого материала ablating model
~, имитационная simulation model
~, каркасная wire-frame model
~ Кейнса Keynesian model
~, кордовая control line model
~ корпуса hull model
~, корпускулярная corpuscular picture
~ линейного программирования linear programming model
~, литейная casting pattern
~, литейная пустотелая hollow pattern
~ марковской системы массового обслуживания Markovian queueing model
~ массового обслуживания waiting line model
~ минимизации затрат cost-minimizing
~, многоступенчатая multiechelon model
~, многошаговая multistage model
~, натуральная natural pattern
~, натурная full-scale model
~ независимых частиц independent-/individual-particle model
~ непрерывного замедления continuous slowing-down model, Fermi-age model
~, непуассоновская non-Poisson model
~ нуклонной группы sub-nuclear unit model
~, однопродуктовая one-commodity model
~, односкоростная one-velocity/-group model
~, одношаговая single-stage model
~ перфокарты punched card simulation
~, плоская planar model, two-dimensional model
~ поведения behavioral model
~, полупуассоновская semi-Poisson model
~ потока flow model, simulated flow
~, причинно-следственная causal model
~, прогнозирующая predictive model
~, программная software model
~, пространственная three-dimensional model
~ равновесия запасов inventory balance model
~, разомкнутая open model
~ распределения allocation model
~ реактора reactor model/simulator, dummy reactor, reactor mock-up
~, реляционная relational model
~ решетки mocked-up lattice
~ рынка marketing model
~ с вогнутой функцией затрат concave cost model
~ с возможностью перерывания обслуживания interruption model
~ с двигателем *nucl.* powered model
~ с задалживанием спроса backlogging model
~ с приоритетами, зависящими от состояния системы endogenous priority model
~ с резервным запасом buffer-stock model
~ с учетом противодействия *game th.* retaliation model
~ сбыта marketing model
~ семиуровневая seven-layer model *(of open systems)*
~, серийная production model
~, сетевая network model
~ скользящего среднего moving-average model
~, термоупругая thermoelastic model
~ технической осуществимости feasibility model
~ типа объект-отношение entity-relationship model, E/R model
~ топливного стержня simulated fuel slug
~, трубная (wind-)tunnel model
~ Уилсона Wilson model
~, фермиевская возрастная Fermi-age model
~, флаттерная flutter model
~, фреймовая frame-based model
~ экономического цикла trade-cycle model
~, электрическая/электронная electric/electronic analog(ue)
~, эталонная standard model
~ эффективности затрата cost-effectiveness model
~ ядра, статистическая continuum/statistical nuclear model
модель-болванка dummy model
модельер (hair) stylist, model(l)er; modelmaker; fashion designer
модельная *n.* pattern shop
модельщик patternmaker
модель-эталон master pattern
модем modem
модер twin mull, moder
модерато *mus.* moderato
модератор moderator; speed regulator
модерн *art* modernist style
модернизатор modernizer
модернизация modernization, updating
модернизировать bring up to date, update, modernize
модернист *art* modernist; neoteric
модернистический, модернистский modernist(ic); neoteric
модий *hist.* modius *(ancient measure of weight)*
модильон *arch.* modillion
модиола *zool.* horse mussle *(Modiola)*
модистка milliner, modiste
модификатор modifier, modifying factor/agent
~ масштаба scale modifier
~ порядка exponent factor/modifier/key
~ состояния status modifier
модификация 1. modification, change; alteration; 2. derivative, version
~ адреса команды command/instruction address change
~, длительная persistent modification
~ команд instruction/command/order modification
~ навыка habit modification
~, наследственная heritable modification

модифицирование modifying, modification
~ чугуна inoculation
модифицировать modify; change; *met.* inoculate
модифицируемость modificability
модный fashionable, stylish
~ные товары style-and-fashion goods
модулирование modulation, *see also* **модуляция**
модулировать *mus., tech.* modulate, control
модулируемость modulability, modulation capability
модулометр modulation meter/monitor
модуль 1. *math.* modulus, absolute value; 2. *tech.* module
~, аппаратный hardware module
~ вектора length/modulus/magnitude of a vector, absolute/numerical value of a vector
~, вставной/сменный plug-in module
~, выполнимый executable module
~ загрузки/загрузочный load module
~ затухания decay modulus
~ ЗУ *comp.* memory module
~ изгиба bending modulus
~, интегральный integrated circuit module, IC module
~, командный command module
~ комплексного числа complex number modulus, absolute value of a complex number
~ крупности *min.* gradation factor
~ кручения torsion(al) modulus, modulus of torsion
~, логический *logic* module
~, лунный *space* lunar module
~, многократно используемый reusable module
~ над алгеброй algebra module
~, независимый self-contained module
~, объемный, ~ объемного расширения bulk modulus (of elasticity)
~ основания *const.* foundation modulus
~, основной *space* orbital module
~, перемещаемый relocateable module
~ печатным монтажом printed circuit board/card
~ полного сопротивления *el.* scalar impedance
~ полной проводимости *el.* scalar admittance
~ поперечной упругости modulus of shear(ing)/rigidity, modulus of transverse elasticity
~ преломления refractive modulus
~, программный software modulus
~ продольной упругости Young's/tension modulus, modulus of elongation, modulus of longitudinal elasticity
~, пьезоэлектрический piezoelectric modulus
~ размерности modulus of dimensionality
~ разрыва modulus of rupture
~ расстояния *astr.* distance modulus
~ расширения add-in/-on module
~ сдвига shear(ing)/rigidity modulus, modulus of transverse elasticity
~, сменный plug-in module
~ сравнения modulus of congruence; equals module
~ стока *hydr.* run-off rate, modulus of flow
~, схемный circuit module
~, управляющий control module
~ упругости elastic modulus, modulus of elasticity
~ упругости второго рода shear(ing)/rigidity modulus
~ упругости, касательный tangent modulus of elasticity
~ упругости первого рода, ~ упругости при изгибе Young's/tension modulus, modulus of elongation, modulus of longitudinal elasticity
~, фиктивный dummy module
~ цифрового ввода-вывода digital input-output module
~ шестерни pitch of a gear
~ эллиптического интеграла modulus/module of an elliptic integral
~ этажерного типа cordwood module
~ Юнга Young's/tension modulus, modulus of elongation, modulus of longitudinal elasticity, modulus of elasticity
модульный modular, modularized
~ коэффициент modulus coefficient
~ принцип mechanical assembly technique
модулярность, модульность modularity
модулярный modular
модулятор modulator; control grid; driving device, keyer
~, балансный balanced modulator
~, время-импульсный pulse-position modulator
~, двухтактный push-pull modulator
~ добротности лазера Q-spoiler, Q-switch
~, импульсный pulse modulator
~, кодово-импульсный pulse-code modulator
~, кольцевой ring modulator
~ лазера, активный/пассивный active/passive shutter
~, мостовой bridge-circuit modulator
~ света, частотный frequency light modulator
~, сегнетоэлектрический ferroelectric modulator
~, фототелеграфный facsimile modulator
~, частотно-импульсный pulse-frequency modulator
~, широтно-импульсный pulse-width modulator
модуляция 1. *tech.* modulation, keying; 2. *mus.* transition, inflexion, tone control, modulation
~, амплитудная, АМ amplitude modulation, AM
~, временная, ~ время-импульсная, ВИМ pulse position modulation, PPM
~ голосом voice control
~, двухтактная push-pull modulation

~ **дефазированием** outphasing modulation
~ **добротности лазера** Q-switching/-modulation/-spoiling
~, **дробная** fractional modulation
~, **звуковая** audio modulation
~, **идеальная** perfect modulation
~ **излучения лазера** laser-beam modulation
~, **импульсная** pulse modulation
~, **квадратурная** quadrature modulation
~, **кодово-импульсная, КИМ** pulse-code modulation, PCM, quantized pulse modulation
~ **многократная** multiple/compound modulation
~ **на большой мощности** high-power/-level modulation
~ **на малой мощности** low-power/-level modulation
~, **недостаточная** undermodulation
~, **однополосная** single-sideband modulation
~, **однотактная** single-cycle modulation
~ **отклонением** deflection modulation
~, **паразитная** extraneous/spurious modulation
~, **перекрестная** cross modulation
~ **по амплитуде** amplitude modulation, AM
~ **по времени** time modulation
~ **по фазе** phase modulation, PM
~ **по частоте** frequency modulation, FM
~ **по крутизне импульса** pulse-slope modulation
~ **поглощением** absorption/loss modulation
~, **поляризационная**, ~ **по поляризации** polarization modulation, PM
~, **помеховая** jamming modulation
~ **пучка** beam modulation
~, **речевая** voice/speech modulation/control
~ **с качанием частоты** sweep-frequency modulation
~, **сеточная** grid modulation
~ **толщины базы** base thickness/width modulation
~, **тональная** tone/buzzer modulation
~, **фазово-импульсная, ФИМ** pulse-position modulation, PPM
~ **фоном** hum modulation
~, **частотная, ЧМ** frequency modulation, FM
~, **частотно-импульсная, ЧИМ** pulse-frequency modulation, PFM
~, **широтно-импульсная, ШИМ** pulse-duration/-width modulation, PDM, PWM
~ **шумом** noise modulation
~ **яркости** intensity modulation
~, **частотно-временная** variable duration-frequency modulation

модус mode (of precedure)
моевка *zool.* kittiwake *(Rissa)*
моечный washing
~**ная машина** washer
можжевеловка juniper-flavo(u)red vodka, gin
можжевеловые *n. pl. bot.* juniper family *(Juniperoideae)*
можжевелоягодник *bot.* dwarf/scaly mistletoe *(Arceuthobium)*
можжевельник *bot.* juniper *(Juniperus); coll.* juniper berries
мозазавр *paleont.* mosasaur *(Mosasaurus)*
мозаик *biol.* mosaic
~, **гоносомный** gonosomic mosaic
~, **хромосомный** chromosome mosaic
мозаика 1. mosaic; inlay; tesselation;. 2. *plant. pathol.* mosaic
~, **глыбовая** block mosaic
~, **деревянная** marquetry
~, **полосчатая** streak mosaic
~, **сбросовая** fault mosaic
~, **складчатая** striate mosaic
~, **смалитовая** smalto
~, **табака, белая** tobacco bleaching mosaic
~, **шахматная** tessel(l)ation
мозаиковый mosaic *(attr.);* inlaid
мозаист, мозаичист worker in mosaic; mosaicist
мозаичность variegation, mottling; tessellation; mosaic structure/nature; *plant pathol.* mosaic disease
мозаичный mosaic, inlaid, tessellate(d); granoblastic; honeycomb
мозг brain; *anat.* marrow; medulla
~, **большой** cerebrum
~, **головный** brain
~, **добавочный** myelencephalon, afterbrain, medullary brain
~, **желтый костный** yellow bone marrow
~, **задний** metencephalon
~, **зрительный** thalamic/tween brain, interbrain
~, **искусственный** artificial brain
~, **конечный** endbrain
~, **костный** medulla, (bone) marrow
~, **передний** forebrain
~, **продолговатый** medulla oblongata
~, **промежуточный** diencephalon
~, **ромбовидный** hindbrain
~, **спинной** spinal cord/marrow
~, **средний** midbrain
~, **электронный** electronic brain
мозговидный cerebriform
мозговик encephaloid cancer
мозговой cerebral, medullar(y), brain *(attr.)*
мозель(вейн) Moselle (wine)
мозжечково-вестибулярный *anat.* cerebellovestibular
мозжечково-мостовой *anat.* cerebellopontile
мозжечково-оливный cerebello-olivary
мозжечково-спинальный cerebellospinal
мозжечковый cerebellar
мозжечок *anat.* cerebellum
~, **древний** archicerebellum
мозжуха *bot.* ground cherry *(Physalis)*
мозолевидный callous
мозоленогие *n. pl. zool.* tylopods *(Tylopoda)*
мозолистолистный *bot.* tylophyllous, blisterleaved
мозолистость callosity

мозолистый callosal, corneous, callous(ed), horny, toil-hardened
мозоль callositas, callosity, callus, clavus, corn, horn
~, водяная soft corn
~, кожная cutaneous horn, cornu
~, костная callus, clavus
~, псориатическая psoriatic clavus
~, твердая hard corn
Моисеев Mosaic
Моисей Moses
мойва *zool.* capelin *(Mallotus villosus)*
мойка 1. washing; 2. washer, washing machine; sink
~, автомобильная car-washing installation
~, кухонная kitchen sink
~ погружением immersion washing
~, струйная spray washing
мокасиновый moccasin
~вая змея moccasin snake *(Agkistrodon mokasen)*
мокасины *pl.* Indian shoes, moccasins
мокет Brussels carpet
мокка, мокко mocha *(coffee)*
мокрец 1. *vet.* malanders; 2. *bot.* common chickweed *(Stellaria media)*; 3. *pl. zool.* biting midges *(Ceratopogonidae)*
мокрица 1. *zool.* sow bug, wood louse *(Oniscus)* 2. *bot.* common chickweed *(Stellaria media)*
мокрота *med.* sputum, expectoration
~, кровянистая sanguinolent sputum, bloody expectoration
~, слизистая sputum crudum
~, с (про)жилками крови blood-streaked sputum
мокрый wet, damp, soggy, sodden, moist
мол mole, (quay) pier, jetty, breakwater (pier)
~ без сплошной стенки open piling pier
моласса *geol.* Molasse, molassa
молдаванин Moldavian
молдаванка Moldavian (woman)
молдаванский, молдавский Moldavian
молебен, молебствие *eccl.* public prayer, church service
молебствовать *eccl.* say/celebrate mass, conduct a service
молектроника molecular electronics, molectronics
молекула molecule
~, двухатомная diatomic/binary molecule
~, дипольная dipole molecule
~, иониз(ир)ованная ionized molecule
~, меченая labelled/tagged molecule
~, многоатомная polyatomic molecule
~, одноатомная mon(o)atomic molecule
~ растворенного вещества solute molecule
~, связанная tie molecule
~, сложная supermolecule, complex molecule
~, хелатная chelate molecule
молекула-акцептор acceptor molecule
молекула-донор donor molecule
молекулы *pl.* molecules
~, неодинаковые unlike molecules
молекулярно-дисперсный molecular-disperse
молекулярность molecularity
молекулярный molecular
молелистовертка *zool.* mimosa webworm *(Homodaula albizziae)*
моление *eccl.* praying, prayer
молескин *text.* moleskin *(fabric)*
молибдат *chem.* molybdate
~, естественный *geol.* wulfenite, yellow lead ore
~ свинца lead molybdate
молибден *chem.* molybdenum, Mo
~, сернистый molibdenum sulphide
молибденирование molybdenizing
молибденит molybdenite
~, второстепенный minor molybdenite
молибденовый molybdenous, molybdic, molybdenic
~ блеск molybdenite
молибдит molybdite, molybdic ochre
молизация molecule formation (from ions)
молиния *bot.* moor grass, molinia *(Molinia)*
молитвенник *eccl.* prayer book
молитвенный prayful
~ дом chapel
молли(енезия) *zool.* mollie *(Mollienesia)*
моллуго, моллюго *bot.* carpetweed *(Mollugo)*
моллюск 1. *zool.* mollusc, mollusk; shellfish; 2. *med.* molluscum
~, боконервный amphineuran mollusc
~, брюхоногий gastropod
~, головоногий cephalopod
~, голый naked mollusc
~, двустворчатый bivalve mollusc
~, желебобрюхий solenogaster
~, легочный air-breathing mollusc, lung mollusc
~, лопатоногий scaphopod
~, панцирный armo(u)red mollusc, loricate
~, сальный molluscum sebaceum/pseudo-carcinomatosum, keratoacanthoma
~, червеобразный vermiform mollusc
моллюсковидный, моллюскообразный molluscoid(al)
моллюсковый molluscous, molluscan
моллюскоядный molluscivorous
молниезащита lightning(-discharge) protection; lightning protector
молниеносный fulminant, fulminating, lightning
~ удар lightning stroke
~ная война *mil.* blitzkrieg
молниеотвод, молниеуловитель lightning rod/arrester/conductor
~, линейный leakage conductor
~, угольный carbon protector
молния 1. lightning; 2. *tel.* express telegram, lightning priority; 3. news flash
~, ленточная ribbon lightning
~, линейная streak lightning
~, разветвленная/зигзагообразная forked

lightning
~, **сплошная** sheet lightning
~, **четочная** beaded/pearl lightning
~, **шаровая** ball/globe/globular lightning
~, **штриховая** streak lightning
молодежь *coll.* youth; young people
~, **учащаяся** students
молодило *bot.* houseleek *(Sempervivum)*
молодняк 1. young growth, saplings, undergrowth; 2. young animals; cubs
~, **ремонтный** *agr.* replacements
молозиво *physiol.* colostrum, foremilk, beestings
молокан *bot.* lettuce *(Lactuca)*
молокан(ин) *rel.* Molokan
молоко milk
~, **горное/лунное** *geol.* rock milk, agaric mineral
~, **известковое** cream/milk of lime, lime milk
~, **крахмальное** starch slurry
~, **сгущенное** condensed milk
~, **снятое/обезжиренное** skim(med) milk
~, **цементное** laitance
молоковоз milk tank truck
молокогонный (ga)lactagogue, galactic, galactopo(I)etic, lactiferous, lactigerous, lactigenous
молокомер milk gauge/meter
~, **записывающий** milkograph
молокоотделение lactation, lactogenesis, galactopoiesis, milk production
~, **избыточное** *med.* polygalactia, (ga)lactorrhea
~, **недостаточное** *med.* oligogalactia
молокоотсос breast glass/pump
молокоподобный lacteal, lactecent, milklike
молокопоставка milk delivery
молокопровод milk line
молокосборник milk tank
молокосодержащий lactiferous, milk-containing
молот hammer
~, **высокоскоростной** rapid-action hammer
~, **гидравлический** hydraulic hammer
~ **двойного действия** double-acting hammer
~, **двухбоевой** double-faced hammer
~, **дробильный** granulating/crushing hammer
~, **кузнечный** forging hammer; sledge (hammer)
~, **кулачковый** cam-operated hammer
~, **механический** power hammer
~, **падающий** drop hammer
~, **пневматический** compressed air hammer
~, **свайный** pile-driving hammer
~, **шаботный** anvil hammer
~, **штамповочный/штамповый** die-forging hammer, swage hammer
молотилка thresher, thrasher, threshing machine
~, **вальцовая** roller thresher
~, **двухбарабанная** two-cylinder thresher
~ **комбайна** combine separator
~, **кукурузная** corn sheller, husker-sheller
~, **кукурузная с измельчителем** husker-shredder
молотило *agr.* swingle; flail
молотильный threshing
молотильня *agr.* threshing floor/machine
молотить thresh
молоткообразный hammer-shaped
молотоглав *zool.* 1. hammerhead, hammer-headed bat *(Hypsignathus monstrosus);* 2. umbretta *(Scopus umbretta)*
молоток hammer; mallet; beater
~, **бондарный деревянный** mallet
~, **бурильный** jack hammer, hammer drill
~, **гвоздильный** spike driver
~, **геологический** prospecting/geological hammer, geologist's pick
~, **гладильный** planishing hammer
~, **дробильный** grinding hammer
~ **забойщика** miner's hammer
~, **камнетесный** boss hammer
~, **клепальный** riveting hammer, riveter
~, **кузнечный** smith's hammer
~, **маркировочный** *for.* log stamp, marking hammer
~, **отбойный пневматический** jack/air/pneumatic hammer
~, **пневматический** air-operated hammer
~, **правильный/рихтовальный** *met.* straightening hammer
~ **с круглым бойком** round set hammer
~, **слесарный** bench hammer
~ **со скошенным бойком** bevel set hammer
~, **столярный** joiner's hammer
~, **хирургический** surgical mallet
~, **чеканочный** caulking hammer
молотообразный hammer-shaped
молоточек 1. *anat.* hammer, malleus; 2. clapper
~ **звонка** bell clapper
~, **печатающий** striking/printing hammer
молот-рыба *zool.* hammerhead (shark) *(Sphyrna)*
молотьба threshing, harvest (season)
молох I *bibl.* Moloch
молох II *zool.* moloch
молочай *bot.* 1. devil's-milk, euphorbia, spurge *(Euphorbia);* 2. orange-brown lactarius *(Lactarius volemus)*
молочайные *n. pl. bot.* spurge family *(Euphorbiaceae)*
молочная 1. *agr.* creamery, dairy; 2. milk shop
молочница *med.* thrush, white mouth, mycotic stomatitis
молочно-восковой milk-wax
~**вая спелость** *agr.* milk-wax stage of ripeness
молочнокислый lactic, lactate (of)
молочность 1. *agr.* yield *(of cow);* milking capacity; 2. lactescence, milkiness; *pl.* milk products
молочноцветковый *bot.* lactiflorous
молочный milk(y); dairy; lactic; lactaceous, galactophorous; mammary
~ **альбумин** *med.* lactalbumin
~ **брат,** ~**ная сестра** foster brother/sister
~ **режим** milk diet
~**ная железа** *anat.* mammary gland

~ная кухня infant-feeding center
~ная ферма dairy farm
~ное животноводство/хозяйство dairy farming
~ное стекло frosted/opal glass
молчаливый tacit, unspoken
~вое согласие *law* tacit consent
молчальник *eccl.* silentiary, one who has taken a vow of silence
молчание silence; *med.* mutism
моль I *chem.* mole, gram molecule
моль II, платяная моль *zool.* (clothes) moth *(Tineola biseliella)*
мольбище pagan place of prayer/sacrifice
мольва *zool.* ling *(Molva molva)*
мольность *chem.* molarity, molality
мольный I *chem.* mole, molar, molal
мольный II *mus.* minor
моль-чехлоноска *zool.* casebearer *(Coleophora)*
молюск *see* **моллюск**
молюцелла *bot.* Molucca balm *(Moluccella)*
моляльность *chem.* molarity, molality
моляр *anat.* molar/multicuspid tooth
молярность molarity
молярный *chem., phys.* molar
Мом *myth.* Momus
момент 1. instant, moment; 2. *phys.* momentum; 3. feature, factor, element
~, балочный girder moment
~ включения time of engagement
~ воспламенения ignition time
~, возмущающий disturbing/exciting moment
~, восстанавливающий restoring/stability/righting moment; *av.* righting couple
~, вращающий/вращения torque, moment of rotation, turning/torsion(al)/rotational moment
~ времени, начальный zero time
~ выброса instant of ejection
~ выделения *chem.* nascent state, emission moment
~, дипольный dipole moment
~ диполя moment of double force
~, дополнительный excess torque
~ зажигания firing point/position
~, изгибающий bending moment, moment of flexure
~ импульса angular momentum, moment of impulse/momentum
~ инерции moment of inertia, inertia/second moment
~ инерции массы second moment of mass
~ инерции, смешанный/центробежный product of inertia
~ килевой качки *nav.* pitching moment
~, кинетический, ~ количества движения angular momentum, moment of momentum
~ количества движения, внутренний intrinsic angular momentum, spin
~ контроля состояния запасов *econ.* inventory control point
~ крена rolling moment
~, кренящий *nav.* heeling/inclining moment
~, крутящий/кручения torque, torsion(al)/twisting moment, moment of torsion
~ крыла wing moment
~ кульминации (светила) *astr.* time of transit
~ массы, статический moment of mass
~, маховой flywheel moment, moment of gyration
~ нагрузки load moment/torque
~, неспиновый магнитный extra-spin magnetic moment
~, неуравновешенный unbalanced/unstable moment
~, нулевой zero moment
~, опорный support/restraining moment
~, опрокидывающий *naut.* overturning/capsizing/tilting/tipping moment
~, орбитальный orbital/angular momentum
~ от постоянной нагрузки dead-load moment
~ пары сил (moment of) couple
~ пикирования *av.* nose-down moment, pitching moment
~, полуцелый half-integer momentum
~ поступления entry point
~ приземления *av.* touchdown
~, противодействующий counter/restoring torque
~ против штопора, ~ противоштопорный *av.* antispin(ning) moment
~, пусковой *el.* moment of starting, starting torque
~, путевой yawing moment
~, разрушающий breaking moment, moment of rupture
~, расчетный factored moment
~, реактивный (крутящий) contra-rotating moment
~ рыскания *av., naut.* yawing moment
~, сваливающий *av.* stalling moment
~ силы force moment, moment of force
~ силы плавучести, статический *naut.* moment of buoyancy
~ силы тяги thrust moment
~, скручивающий twist(ing)/turning moment
~ случайной величины *math.* moment of a random variable
~, смешанный *math.* product/multivariant moment
~, собственный intrinsic moment
~ сопротивления resisting moment, moment of resistance
~ сопротивления вращению antitorque moment
~ сопротивления шпангоута *naut.* frame modulus
~ спина momentum of spin
~, спиновый spin moment
~ срыва break-away torque

~ старта initial time
~ тангажа pitching moment
~, тормозной braking/drag/retarding torque
~ трения friction(al) torque
~ тяги thrust moment
~ упругости moment of elasticity
~ успокоения damping torque
~ устойчивости формы form-stability moment
~ финиша terminal time
~ центробежной силы centrifugal moment
~, штопорный spin(ning) moment
~ элерона aileron moment
~, ядерный орбитальный nuclear angular moment
~ ядра nuclear spin; nuclear magnetic moment
~, ядровый core moment
моментальность momentariness, instantaneousness
моментальный instant, momentary, instantaneous
~ снимок snapshot
моментомер torque meter
моментометр *stat.* planimeter
момордика *bot.* balsam apple *(Momordica balsamina)*
Момус *myth.* Momus
монада *philos.* monad
монаднок *geol.* monadnock
монадология *philos.* monadism, monadology
монаксиальный uniaxial, monaxial
монакский Monacan, Monegasque
монандрия *bot.* monandry
монарда *bot.* bee palm, horsemint *(Monarda)*
монартикулярный *med.* uniarticular, monarticular, monarthric
монарх monarch; sovereign
монархизм monarchism
монархист monarchist
монархистский monarchist(ic)
монархический monarchic(al), monarchal
монархия monarchy
монархный *bot.* monarch
монастер *biol.* mother star, monaster
монастырь monastery; cloister
~, женский convent, nunnery
монах 1. *rel.* monk; friar; 2. *geol.* earth pillar, (rock-)pinnacle
монахиня *rel.* nun
монахиня-экономка *rel.* procuratrix
монацит *geol.* monazite
монашенка *zool.* nun moth *(Lymantria monacha)*
монашество monasticism; monkhood; *coll.* regular clergy, monks
монгол I Mongol(ian)
монгол II Mongolian horse
монголизм *med.* mongolianism, mongolism, Mongolian idiocy
монголист, монголовед specialist in Mongolian studies
монголистика, монголоведение Mongolian studies
монголоид *pathol.* mongolian, mongoloid
монгольский Mongolian
монгольфер montgolfier, fire-balloon
монель-металл Monel(-metal), monel
монета coin
~, звонкая effective coin; hard cash
~, разменная (small) change, loose change
~, фальшивая spurious coin
~, ходячая currency
~, юбилейная commemorative coin
монетаризм *econ.* monetarism
монетизация *econ.* monetization
монетка *bot.* collybia *(Collybia)*
монетник, монетница coin box, collection receptacle
монетный 1. monetary; 2. coin-operated
~ двор mint
монеторазменник coin changer
монетовидный coin-shaped, nummular
монжус *food* montejus, air lift
монизм *philos.* unitism, monism
монилиаз, монилиоз *med.* moniliasis, moniliosis
монист *philos.* unitarian, monist
монистический *philos.* unitarian, monistic(al)
монитор monitor
~, камерный *telev.* camera/preview monitor
~, нейтронный neutron monitor
~, отладочный *comp.* debug monitor
~ пакетного режима *comp.* batch monitor
~ реального времени *comp.* real-time monitor
~, сердечный cardiac monitor
~, фольговый monitor foil
моноазокраситель monoazo dye
моноалфавитный monoalphabetic
монобиотический monobiotic
моноблок monoblock (unit)
монобрахия *med.* monobrachism
моновакцина monovalent vaccine
моновокалический one-vowel
моноволокно monofilament, monofiber
моногамический, моногамный monogamic, monogamous
моногаплоиды *pl.* monohaploids
моногенез monogenesis
моногенетический monogenetic
моногенический monogenic
моногенный monogenous, monogene, monogenic; *geol.* monomineralic
моногеосинклиналь monogeosyncline
моногибрид *biol.* monohybrid
моногидрат *chem.* monohydrate
моногинецейный *bot.* monogynoecial, monothalamic
моногляциализм *geol.* monoglacialism, monoglaciation
монограмма monogram
монография monograph(y)
монодактилия *med.* monodactylism
монодоминантный monodominant
~ное (растительное) сообщество

monodominant community
монодромия *math.* monodromy
монозамещенный monosubstituted
монозиготный *biol.* monozygotic, uniovular
моноидальный *math.* monoidal
моноидеизм *med.* mono-ideism
моноимпульс single giant pulse *(of laser)*
моноимпульсный monopulse
монойодтирозин monoiodotyrosine, MIT
монокаин monocaine/butethamine hydrochloride
монокарион monocaryon, monokaryon
монокарпий, монокарпик *bot.* monocarp, monocarpic plant
монокарпический, монокарпный *bot.* monocarpous, monocarpic
монокаузализм mechanical determinism
моноклимакс *ecol.* monoclimax
моноклиналь *geol.* monocline, homocline
моноклинный monoclinal; monoclinic; *bot.* monoclinous
моноклональный monoclonal
монокок *av.* monocoque
монокольпатный monocolpate
монокомпаратор monocular comparator
монокристалл single crystal, monocrystal
~, затравочный seed crystal
монокристаллический monocrystalline, single-crystal
моноксенический *biol.* monoxenous
моноксильный monoxylic, monoxylous
монокультура *agr.* one-crop farming/system, monocultural farming, single-crop farming
монокуляр monocular
монокулярный one-eyed, monocular
монолингвизм monolingualism
монолит monolith
монолитность *polit.* monolithic character, firmness; solidity
монолитный monolithic
монолог *theatre* soliloquy, monologue
~ под занавес curtain speech
монологический monologic(al)
~кая речь monologue
моном *math.* monomial
мономер *chem.* monomer
~, исходный initial monomer
мономерный *bot., zool.* monomerous; *chem.* monomeric(al)
монометаллизм *econ.* single standard, monometallism
мономиальный *math.* monomial
мономиктный *geol.* monomict
мономинеральный monomineral(ic)
мономолекулярный monomolecular
мономорфизм, мономорфия *math.* monomorphism, isomorphy
мономорфный monomorphous, monomorphic
мононить monofilament
мононуклеоз *med.* mononucleosis
~, инфекционный glandular fever, infectuous mononucleosis
моноокись *chem.* monoxide
~ фтора oxygen difluoride
моноплакофора *zool.* monoplacophoran mollusk
моноплан *av.* monoplane
~, подкосный strut-braced monoplane
~ с высокорасположенным крылом shoulder-/high-wing monoplane
~, свободнонесущий cantilever monoplane
монополизация, монополизирование monopolization
монополизировать monopolize
монополистический, монополистский monopolist(ic)
монополия *econ.* monopoly
~ банкнотной эмиссии note-issue monopoly
~, государственная government/state monopoly
~, групповая consolidated/group/shared monopoly
~ покупателя buyer's monopoly
~лии, международные multinationals
монополь *math.* monopole; *econ.* sole power, monopoly
монопольный monopolistic, exclusive
~ное право exclusive rights, monopoly
монопсихоз *med.* monopsychosis; monomania
монопсония monopsony
монорельс monorail; overhead monorail conveyer/carriage
монорема monoreme
монорхизм *med.* monorchism, monorchidism
моносахарид *chem.* monosaccharide
моносемия monosemy, univocity
моносиллаб monosyllable
моносимметричный monosymmetric(al); *cryst.* monoclinic; *bot.* zygomorphic
моносимметрия monosymmetry; *cryst.* monoclinic symmetry
монослой monolayer
моносома *genetics* monosome
моноспермический monospermous, one-seeded; monospermic
моноспираль single-coil filament
моностабильный monostable
моностела *bot.* monostele
моностих monoverse, monostich
монотеизм monotheism
монотеистический monotheistic
монотип monotype
монотипист monotype operator, monotyper
монотипический, монотипичный monotypic(al)
монотонический monotonic, monotone
монотоничность monotony, monotonism
монотонность *math.* monotonicity, monotony
монотонный monotone, monotonous, monotonic
монотопливо monopropellant
монотычинковый *bot.* monandrous
монофаг monophagous parasite; monovalent phage

монофагия monophagy; *med.* monophagism, monophagia
монофакториальный monofactorial, unifactorial
монофизит *eccl.* Monophysite
монофилетизм monophyletic evolution, monophiletism
монофилетический, монофилетичный monophyletic(al)
монофилия monophyletic origin, monophyleticism
монофилльный *bot.* unifoliate, monophyllous
монофобия *med.* monophobia
монофония monophony
монофтонг *phon.* monophthong
монофтонгический monophthongal
монохазиальный *bot.* monochasial
монохазий *bot.* monochasium, simple cyme
монохлордифторметан *chem.* monochlorodifluoromethane
монохлортрифторэтилен *chem.* trifluorochloroethylene
монохром monochrome
монохроматичность monochromaticity, monochromatism
монохроматизировать monochromate
монохроматор *phys.* monochromator
~ с дифракционной решеткой grating monochromator
~ с прерывателем chopper monochromator
монохромность monochromaticism
монохромный, монохроматический monochrome, monochromic, monochroic
монохронный monochronous
моноценоз monocoenosis
моноцефальный monocephalous
моноцикл monocycle
моноцитарный monocytic
моноцитоз monocytosis
моноэдр monohedron
моноэдрический monohedral, monohedric
моноэнергетический monoenergetic; *phys.* monochromatic
моноэнергетичность energy homogeneity
монсеньор *eccl.* monsignor, monseigneur
монстера *bot.* monstera *(Monstera)*
монструозность monstrosity
монтаж 1. assembly, assemblage, fitting, mounting; installing, installation; *el.* wiring; *constr.* erection; 2. *mus., art* arrangement; *cin.* montage
~ аэроснимков aerial/photographic mosaic, adjustment of image
~, беспаечный solderless wiring
~ в стойке rack mounting
~ в ячейках cell mounting
~, выдвижной drawout/telescopic mounting
~ кабельной проводки cabling
~ на месте field assembly
~ на объединительной плате motherboard wiring
~ на салазках skid mounting
~, навесной point-to-point wiring
~, объемный space wired interconnections; three-dimensional wiring
~, открытый *el.* exposed wiring
~, панельный panel mounting
~, печатный printed wiring/circuitry
~, силовой power wiring
~, скрытый concealed wiring
~, стоечный rack mounting
~ схемы connecting-up, wiring
~ трубопровода pipe laying
~ шины tyre mounting/fitting
монтажер, монтажист montage specialist; *print.* stripper
монтажная *cin.* cutting/editing/clipping room; *mech.* assembly room/shop
монтажник fitter, mounter, installer, errector; rigger; assembly worker; millwright; *pl.* errection gang
монтажный assembling, fitting; interconnecting
~ная мастерская assembly room/shop; fitting shop
монтан-воск montan-wax
монтан-селитра montan-salpeter
монтаньяр *hist.* moutagnard
монтбреция *bot.* tritonia *(Montbretia; Tritonia)*
монте xerophytic scrub, shrubby tract, monte
монтер 1. fitter, adjuster; mounter, assembler; serviceman, repairman, mechanic; 2. electrician
~, кабельный cableman, cable splicer
~, линейный line(s)man
~ по уходу maintenance/service man
~ связи communication technician
монтер-линейщик line(s)man
монтер-эксплуатационник maintenance/service man
монтикула *geol.* monticulus, monticle, monticule
монтирование *tech.* fitting, mounting; assembling, errecting; *cin.* editing; arranging
монтировать *tech.* assemble, errect, install, mount, set, fit (up); *cin., telev.* edit
монтировка (telescope) mount(ing)
~, азимутальная altazimuth mount(ing)
~, американская fork-type mount(ing)
~, параллактическая/экваториальная equatorial/parallactic mount(ing)
монтировщик *cin.* 1. montage specialist, film editor; 2. *tech.* fitter, mounter, assembler, rigger
монументалист sculptor of monuments
монция *bot.* Indian lettuce *(Montia)*
мопед moped, motorized bicycle
МОП-накопитель *comp.* MOS-memory
мопс pug (dog)
~, китайский Pekinese, Pekingese (dog)
МОП-структура *comp.* MOS (-structure)
мораво-паннонский *hist.* Moravian-Pannonian
моралите *hist.* morality (play)
мораль (code of) morals, ethics; morality
морально-политический moral and political
моральный moral, ethical
~ износ, ~ное изнашивание машин obso-

lescence of machines
мораторий, мораториум *law, econ.* moratorium
морг I morgue, mortuary
морг II *hist.* morg *(0. 56 hectares)*
морганатический morganatic
морганизм *econ.* morganism, Thomas Morgan's theories
морганист follower of morganism
морген morgen *(land measure of about 2 acres)*
мордвин, мордовец Mordvinian
мордвинка, мордовка Mordvinian (woman)
мордент *mus.* mordent
~, верхний inverted mordent, pralltriller
мордовник *bot.* globe thistle *(Echinops)*
мордовский Mordvinian
море sea
~, береговое shelf sea
~, в открытом off-shore
~, внутреннее/закрытое enclosed/inland/intracontinental sea, land-locked sea
~, выходить в открытое get on offing, gain the offings
~, идущий в seabound
~, лунное lunar mare
~, мелководное shallow sea
~, наступающее invading/transgressing sea
~, неприливное tideless sea
~, окраинное fringing/border/marginal sea
~, открытое open sea, high sea(s)
морель *bot.* pie/sour cherry *(Prunus cerasus acida)*
морелюбивый thalassophilous, sea-loving
морена *geol.* moraine, glacial clay
~, береговая flank moraine, morainal/morainic terrace
~, валунная boulder wall
~, верхняя upper/superglacial moraine
~, внутренняя englacial/internal moraine
~, донная basal/subglacial/ground moraine
~ напора push/shoved moraine
~, перемытая washboard moraine
~, стадиальная recessional moraine
моренный morainal, morainic
мореный stained *(of wood)*
~ дуб fumed oak
~ная известь slaked lime
мореплавание navigation, seafaring; sailing
~, общее public navigation
мореплаватель seafarer, navigator
мореплавательный nautical, navigational
мореход(ец) seafarer, navigator
мореходность seaworthiness, seagoing/seakeeping ability, navigability
~ в водоизмещающем режиме displacement/hullborne seakeeping
мореходный seagoing, seaworthy, seafaring; nautical, navigational
мореходство navigation, seafaring
морея *bot.* moraea *(Moraea)*
морж 1. *zool.* walrus *(Odobenus rosmarus);* 2. *sport* winter swimmer *(in open-air pools)*
Морзе morse
~, азбука Morse code
морзист Morse code signaller
морилка wood preservative/stain
моринга *bot.* horse-radish tree *(Moringa oleifera)*
морион I *hist., mil.* morion, open helmet
морион II *geol.* black quartz, morion
мориск *hist.* Morisco, Moor
морковник 1. *bot.* silaus *(Silaus);* 2. *cul.* carrots
моркововодный, морковообразный carrot-shaped, carrotlike
морковь *bot.* carrot *(Daucus)*
мормон *rel.* Mormon
мормонизм, мормонство *rel.* Mormonism
мормонка Mormon (woman)
мормонский Mormon
моровой *hist.* pestilent
~вая язва plague, pestilence
морожение freezing; chilling
мороженица ice-cream freezer; freezing machine
мороженое *n., cul.* ice-(cream)
~, сливочное (vanilla) ice cream
мороженое-ассорти *cul.* Neapolitan ice cream
морозилка freezing machine, freezer, froster; refrigerator
морозильник deep-freezer
морозильный freezing
~ная камера freezing chamber
морозник *bot.* hellebore *(Helleborus)*
морозобоина *geol.* frost cleft/fissure/hollow/crack
морозобой *geol.* winterkilling
морозобойный frost-shattered, ice-break *(of soil)*
морозостойкий, морозоустойчивый frost-resistant/-hardy, nonfreezing, winterproof, frostproof
морозостойкость, морозоустойчивость, морозоупорность cold/frost resistance/hardiness
морось drizzle
~, переохлажденная freezing drizzle
морошка *bot.* cloudberry *(Rubus chamaemorus)*
морской 1. sea, marine, maritime, nautical; seagoing, seaborne; 2. naval, navy
~ ангел *zool.* monkfish *(Squatina)*
~ конек *anat.* Ammou's horn, hippocampus *(Hippocampus)*
~ офицер naval officer
~ планктон marine/saltwater plankton
~ флот navy, fleet
~кая болезнь seasickness
~кая водоросль seaweed
~кая растительность marine/maritime vegetation
мортира 1. *mil.* (artillery) mortar; 2. *mach.* boss(ing)
~, осадная siege mortar
~ судового винта boss(ing)
мортирка mortar
~, ружейная grenade launcher, rifle-grenade discharger
морула *embryol.* morula
морфа *biol.* morph

Морфей *myth.* Morpheus
морфема *ling.* morpheme
морфема-связка linking morpheme
морфематический, морфемный morphemic
морфемография morphemography
морфизм *math.* morphism
  ~, дифференцируемый diffeomorphism
морфий morphia, morphine
морфин *chem.* morphine
  ~, хлористоводородный morphine hydrochloride
морфинизм morphine addiction, morphinism
морфинист morphine addict
морфоархитектура morphological architecture
морфогенез *biol.* morphogenesis, morphogeny
морфоз *biol.* morphosis, morphogenesis
морфолексический morpholexical
морфолог morphologist
морфологическо/морфолого-синтаксический morphological-syntactic
морфология morphology; physiography
  ~ беспозвоночных invertebrate morphology
  ~ долин *geol.* fluvial morphology
  ~, патологическая pathomorphology
  ~ развития developmental morphology
  ~ растений structural botany, plant morphology
  ~, физиологическая physiologic/normal morphology
морфонема *ling.* morphoneme
морфонематический *ling.* morphonematic, morphonemic
морфонология *ling.* morphophonemics, morphonology
морфосинтаксис morphosyntax
морфотактика morphotactics
морфотип morphotype
морфотропизм *chem.* morphotropic effect, morphotropism
морфотропия *chem.* morphotropy
морфофизиология morphophysiology
морфофонематический *ling.* morphophonemic, morphophonematic
морфофонемический, морфофонемный *ling.* morphophonemic
морщина corrugation, wrinkle, crumple; crease, fold; *bot., zool.* ruga
  ~ на отливке casting lap
  ~ раковины shell unda
  ~ течения flow wrinkle
  ~ны бумаги paper cockling
  ~ны, влагалищные *anat.* ridges of the vagina
морщинистолистный *bot.* wrinkle-/rugose-leaved
морщинистолопастный wrinkle-/rugose-lobate
морщинистоплодный *bot.* wrinkle-/rugose-fruited
морщинистосемянный *bot.* wrinkle-seeded
морщинистость wrinkling, rugosity; crinkle
морщинистый corrugate, wrinkled, crumpled, rugose, rugate, rugous; lined, shrivelled; creasy, creased
морщино-апертурный rugorate
морщино-бороздный rupate
моряк seaman, sailor(man), mariner, seafarer
  ~, военный navy man
  ~ торгового флота merchant seaman
морянка *zool.* old squaw, long-tailed duck *(Clangula hyemalis)*
МОС *comp.* management operating system
москатель *coll.* commercial chemicals
москательня oilery, paint store
~ная лавка oilery, paint store
  ~ная торговля drysaltery
москвич 1. Muscovite, inhabitant of Moscow; 2. Moskvich (car)
москвичка Muscovite (woman), inhabitant of Moscow
москит mosquito
москитка *med.* pappataci/sandfly fever, threeday fever
москиты *pl., zool. Phlebotomus*
  ~ переносчики бартонеллеза *Phlebotomus noguchi*
  ~ переносчики москитной лихорадки *Phlebotomus pappatacii*
московка *zool.* coal tit *(Parus ater)*
московский (of) Moscow
  ~кая Русь *hist.* Muscovy
Моссовет *abbr., hist.* **Московский Совет народных депутатов** Moscow Soviet (of People's Deputies)
мост 1. bridge; 2. *tech.* axle; 3. *anat.* pons
  ~, автодорожный highway bridge
  ~, арочный arch bridge
  ~, балансовый balance cock *(of watch)*
  ~, балочно-консольный cantilever beam bridge
  ~, балочный beam bridge
  ~, ведущий *auto.* drive/driving axle
  ~, висячий кабельный cable suspension bridge
  ~, воздушный *mil.* air lift; *min.* air crossing
  ~ головного мозга *anat.* pons
  ~, двукрылый double-leaf bridge
  ~, двухпутный two-lane bridge
  ~, двухступенчатый *auto.* double-reduction axle
  ~, двухъярусный double-deck bridge
  ~, естественный *geol.* natural bridge, rock arch
  ~, железобетонный монолитный cast-in-site reinforced concrete bridge
  ~, железобетонный сборный prefabricated reinforced concrete bridge
  ~, задний *auto.* back/rear axle
  ~, измерительный *el.* electrical/measuring bridge
  ~, измерительный Вина Wien bridge
  ~, измерительный двойной double/Kelvin bridge
  ~, измерительный емкостный capacitance bridge

~, измерительный магазинный *el.* box (-type) bridge
~, измерительный многоплечий multiple-arm bridge
~, измерительный полного сопротивления impedance bridge
~, измерительный равноплечий equal-arm bridge
~, измерительный реохордный slidewire bridge
~, измерительный Томсона double bridge
~, измерительный Уитстона Wheatstone bridge
~, измерительный уравновешенный null-type bridge
~, измерительный Шеринга Schering bridge
~ индуктивностей, измерительный inductance bridge
~, колошниковый *met.* top trestle
~, консольный cantilever bridge
~, косой skew/oblique bridge
~, малопролетный short-span bridge
~, многопролетный multiple-span bridge
~, наводить bridge, lay a bridge
~, наплавной floating/boat bridge
~ на рамных опорах trestle bridge
~ на свайных опорах pile bridge
~, однокрылый single-leaf bridge
~, однопролетный single-span bridge
~, перевесный weighbridge
~, перегрузочный *met.* transfer trestle
~, передний *auto.* front axle
~, пешеходный pedestrian overpass
~, поворотный swing bridge
~, подъемный (vertical-)lift bridge, drawbridge
~, понтонный pontoon bridge
~ полного сопротивления *el.* admittance/impedance bridge
~, разводить draw a bridge
~, разводной drawbrigde, leaf/bascule bridge
~, раскрывающийся bascule bridge
~, свайный pile bridge
~, смыкать couple a bridge
~, совмещенный combined bridge
~ сопротивления *el.* resistance/Wheatstone bridge
~, тензометрический strain-gauge bridge
~, транспортно-отвальный *min.* transport and dumping bridge
~ Уитстона Wheatstone/resistance bridge
~, управляемый *auto.* steering/guiding axle
~, фирновый snow-drift site
~ через суходол dry bridge
~, шинный *el.* busbar bridge
~, штурмовой assault bridge
мостик 1. foot-path, pathway/catwalk; overpass; 2. *el.* bridge (circuit); 3. *anat.* ponticulus; 4. *chem.* cross-link; 5. *naut.* deck bridge
~, емкостный capacitance bridge
~, измерительный постоянного тока direct current bridge, d. c. bridge
~, капитанский/командирский captain's bridge
~, кормовой after bridge
~, межклеточный *anat.* intercellular ponticulus
~ перекидной gangway footbridge
~, переходной *naut.* catwalk/flying bridge
~, поперечный *anat.* transverse ponticulus
~, рукавный hose ramp
~, сложноэфирный *chem.* ester cross-link
~, флагманский admiral's/flag bridge
~, ходовой/штурманский navigating/conning bridge
~, швартовный docking/warping bridge
~, штурманский с рубкой charthouse flat
мостки, мостики *pl.* brow, gang plank, gangway, planked footway, boardwalk
~, висячие suspended catwalk
~, подвесные flying gangway
~, рабочие gangway
мостовая *n.* pavement, paving, causeway, roadway
~, асфальтовая asphalt pavement/road
~, брусчатая (stone-)block pavement
~, булыжная boulder/cobblestone pavement, cobblestone road
~ гигантов *geol.* giant's causeway
~ из плит flagstone pavement
~, изборожденная striated pavement
~, каменная bouldering, boulder flat
~, клинкерная brick pavement
~, торцовая block pavement
мостовидный bridgelike
~ эпителий tessellated epithelium
мостовик, мостостроитель bridge-builder, bridgebuilding specialist
мостовой bridge *(attr.)*
~вая ферма bridge girder
~вые весы weigh bridge
мосто-мозжечковый *anat.* cerebellopontile
мостооперкулиевый pontoperculate
мостостроение bridge building/construction
моталка *cin.* rewind(er); *met.* coiling machine, coiler, reel(er), winder
~ для проволоки coil-winding machine
~, ленточная ribbon reel
~, многопазовая multiple-slot coiler
~, надпольная upcoiler
~, натяжная pull/tension reel
~, подпольная downcoiler
~, разматывающая uncoiler
~, сматывающая delivery reel
~, сортовая системы Гаррета Garret/pouring reel
~, сортовая системы Эденборна Edenborn/laying reel
~, штрипсовая strip reel
мотальщик winder, reeler
мотание winding, reeling
мотель motel
мотет *mus.* motet

мотив I motive; ground, reason, cause
мотив II *mus.* tune; theme
мотивация motivation; drive
~, **высшая** higher motivation
~, **сложная** complex motivation
мотивирование, мотивировка motivation, reason(s), justification; motives
мотивированность motivation, justification, reason
мотивировать motivate, give reason (for), justify
мотобетонолом motorized concrete breaker
мотобол *sport* motorcycle football
мотобот motorboat
мотобур portable drilling rig
мотовелосипед motor bicycle, motor bike, moped
мотовило reel, reeling frame; coiler
~, **барабанное** bird-cage reel
~, **звездообразное** star reel
~, **катушечное** spool reel
~, **коконоомотальное** bobbin reel
~, **копирующее** ganging reel
~, **лопастное** bat reel
~, **эксцентриковое** cam-action reel
мотовоз rail car, (small) diesel shunting locomotive
мотогондола engine nacelle
мотогонки *pl.* motorcycle race
мотодивизия motorized/mechanized division
мотодрезина motor car/trolley *(on rails)*
мотодром motorcycle racing track, motordrome
мотоинструмент power tool
моток bundle, hank, skein; reel
~ **крестовой намотки** cross hank
~ **пряжи/ниток** skein, hank
~, **спутанный/склеенный** sticky skein
мотокамера motorcycle tyre/tire/tube
мотоколона motorized column
мотоколяска cycle-car, motorized carriage/wheel-chair
мотокомпрессор engine-compressor unit
мотокросс cross-country motorcycle race, motocross (race)
мотокультиватор motor cultivator/hoe
мотокультура *agr.* mechanized tilling of land
мотолодка motorboat
мотолопатка motor/power shovel
мотомеханизированный mechanized; vehicularized
мотонарты *pl.* snow scooter
мотоневрон, мотонейрон *anat.* motoneuron
мотопед moped
мотопехота motorized infantry
мотопехотинец motorized infantry soldier
мотопила motor/power saw
мотопланер powered glider
мотопомпа engine-driven pump
~, **пожарная** fire power pump
мотопробег motocross, cross-country motorcycle race
мотор motor; engine *see also* **двигатель**
~, **герметизованный** canned motor
~ **самописца** recorder motor
~ **следящей системы** follow-up motor
~ **триммера** trim motor
~ **угломестного привода** elevating motor, elevation-drive motor
мотоpaлли motorcycle rally
моторама engine mount(ing)
мотор-вагон motor coach
мотор-генератор motor generator set, dynamo, dynamotor
моторесурс service life/hours, overhaul period; motor potential/capacity
~, **межремонтный** overhaul period
моторизация motorization, mechanization
моторизованный motorized, vehicularized
моторика *physiol.* motility
моторист motor/engine mechanic; engineer; *naut.* machinist
~, **авиационный** air fitter/mechanic
моторно-вращательный machine-rotary
моторный I motor *attr.*
~ **вагон** *rail.* front car, motorcar
~**ная лодка** motor boat
~**ная установка** power plant/unit
моторный II *physiol., psych.* motive, motor; motion *(attr.)*
~ **глагол** *gram.* verb of motion
мотороллер (motor) scooter
моторостроение building of motors/engines
моторостроительный motor-/engine-building
~ **завод** motor [engine-building] works
мотор-редуктор geared motor
мотосани *pl.* snowmobile, motor sledge
мотоспорт motorcycle sport
мототерапия *med.* exercise therapy
мототурист tourist travel(l)ing by motorcycle
мотоцикл(ет) motorcycle
~, **лидерский** *sport* pacemaker motorcycle
~ **с коляской** motorcycle with sidecar
мотоциклист motorcyclist
моточасть *mil.* motorized unit
мотошлем crash helmet
мотто motto (*pl.* -toes)
мотушка bobbin, reel; hank, skein *(of thread)*
мотыга 1. *agr.* (grub)hoe; 2. pickax(e), pick, hack, mattock
~, **вращающаяся** rotary harrow, crumber
~, **конная** horse hoe
мотыжение hoeing, spudding, hacking
мотыжно-огневой *agr.* hoe-and-burn
мотылек *zool.* butterfly, moth
~, **кукурузный** European corn borer *(Ostrinia nubilalis)*
мотыль I *tech.* crank, handle
~, **коленчатый** angle/joint lever
мотыль II mosquito grub/larva
мотыльковый *bot.* papilionaceous; *n. pl.* papilionaceous plants *(Papilonaceae)*
мофетта *geol.* mofette, effervescent spring
мох *bot.* moss *(Muscus)*

~, **белый болотный** peat moss, sphagnum
мохер *text.* mohair
мохнатоплодный *bot.* eriocarpous, dasycarpous, shaggy-fruited
махнаторукий *zool.* woolly-handed
мохнатость shagginess, hairiness
мохнатоцветковый *bot.* woolly-flowered
мохнатый shaggy, hairy; long-napped, woolly; *bot.* villous, pilose, tomentose, tomentous
~**тое полотенце** *text.* Turkish towel
мохноногый *zool.* shaggy-/rough-/featherlegged, rough-footed
мохновидный *bot.* muscoid, bryophytic, mossy, mosslike
моховик 1. *bot.* mossiness mushroom *(Boletus variegatus/circinus)*; 2. *geol.* mocha stone, moss agate; 3. *zool.* wood grouse *(Tetrao urogallus)*
моховой mossy
~ **агат** moss agate, mocha stone
~ **торф** moss peat
мохово-лишайниковый *bot.* moss-and-lichen
мохолюбивый *bot.* bryophilous, moss-loving
мохообразный *bot.* bryophytic, muscoid, moss-like, mossy; *n. pl.* bryophytes *(Bryophyta)*
мохры *pl.* fringes, tassels
моцион physical exercise
моча *med.* urina
~, **алкаптоносодержащая** alcaptonuria
~, **амебосодержащая** ameburia
~, **ацетоносодержащая** acetonuria
~, **белоксодержащая** albuminuria
~, **бесцветная** achromaturia, albiduria, albinuria
~, **гноесодержащая** pyuria
~, **жирная** chyluria, galact(os)uria, adiposuria, lipuria
~, **кислая** aciduria, oxyuria
~, **кровавая** hematuria, hematuresis, bloody urine, urina cruenta
~, **млекоподобная** chyluria, chylous/milky urine
~, **мутная** cloudy/nebulous urine
~ **с высоким удельным весом** baruria
~ **с низким удельным весом** hydruria
~, **сахаросодержащая** glycosuria, inosituria; lactosuria
~ **содержащая мочевую кислоту** lithuria, uricaciduria
~, **уратосодержащая** uraturia, lithuria
~, **фосфатосодержащая** phosph(at)uria
~, **щелочная** alkalinuria
~, **эритроцитосодержащая** erythrocyturia
мочалка 1. bast; 2. wisp of bast fibres
~, **проволочная** wire/steel wool
мочало shredded bast; soaked bast fibres
мочевина *chem.* urea, carbamide
мочевой urinary, uric, urinous; uriniferous
~ **пузырь** *anat.* bladder
мочевыделение urination
~, **через прямую кишку** *med.* urochesia
мочегонный diuretic, emictory, uragogue, uretic
мочеизнурение, мочеистощение *med.* diabetes
мочеиспускание *med.* uresis, urination, micturition, emiction
~, **атипичное** planuria, uroplania
~, **замедленное** bradyuria
~, **затрудненное** paruria, dysuria
~, **непроизвольное** enuresis, nycturia
~, **ночное** nycturia, nocturnal enuresis; urinating at night, bed wetting
~ **по каплям** strangury
~, **частое** pollakiuria, thamuria
мочеиспускательный *anat.* urethral
~ **канал** urethra
мочекаменный urolithic
мочекислый urate, uratic
~**лая соль** *chem.* urate
мочекровие *med.* uremia
мочение steeping, soaking, maceration; retting
~ **льна** retting of flax
моченка *bot.* narrow-leaved cattail *(Typha angustifolia)*
моченый steeped, soaked, wetted, retted
мочеобразование uropoiesis, urine formation
~, **повышенное** polyuria, urorrhagia, diuresis
~, **пониженное** oliguria, oliguresis, hypourocrania
мочеобразующий ur(in)ogenous
мочеотделение urination, urinary excretion
мочеотделительный urinary
мочеполовой *anat.* genitourinary, ur(in)ogenital
мочеприемник urinal, urine glass
мочеточник *anat.* ureter
~, **первичный** primary ureter
~, **посткавальный** postcaval ureter
мочеточниково-влагалищный *anat.* uretero-vaginal
мочеточниково-кишечный *anat.* uretero-enteric
мочеточниково-пузырно-влагалищный *anat.* uretero-vesico-vaginal
мочеточниково-толстокишечный *anat.* ureterocolic
мочеточниково-шеечный *anat.* ureterocervical
мочеточниковый *anat.* ureteral, ureteric
мочило steeper, retting box/crate/pond
мочильный steeping, soaking, wetting, retting
мочить steep, soak, ret; macerate
мочиться urinate, micturate, pass urine
мочка I steeping; retting, macerating
мочка II 1. *anat.* lobule, small lobe; 2. *bot.* root hair, fibril; 3. bundle of fibres/fibers; tow
~ **уха** lobule of the auricle/ear
мочковатость fibrousness
~ **корней** bearding of roots
мочковатый fibrillose, fibrillar, filamentous
~ **корень** fibrous root
мошенник sharper, swindler, chiseller
~, **крупный** big-time operator
мошенничество fraud, fraudulence, swindle, cheating *(at games)*; sharp practice

мошка *zool.* midge, gnat *(Chironomida)*
мошонка *anat.* scrotum
мошонковидный, мошонкообразный scrotiform, pouch-shaped
мошоночный *anat.* scrotal, oscheal
~ная грыжа scrotal hernia
мощение paving, surfacing
мощеный paved, surfaced; cobbled
мощи *pl. rel.* relics
мощность 1. power; strength; force; 2. *tech.* capacity; output rating; horsepower; 3. *min.* thickness; depth; *geol.* magnitude; 4. *math.* cardinal number, cardinality, potency
~, асимптотическая *stat.* asymptotic power
~ без впрыска *av.* dry power
~ в ваттах watts
~ в лошадиных силах horsepower
~, взлетная (maximum) take-off power
~, видимая *geol.* apparent/observed thickness
~ вскрыши *min.* thickness of stripping, cover thickness
~, входная input power
~, выходная output power, power output
~, вычислительная computer/processing power
~ генератора generator capacity
~, генерируемая generated capacity
~, гидравлическая liquid horsepower, fluid power
~ двигателя power/rating of an engine, engine output/power
~ двигателя ракеты thrust of a rocket engine
~, действительная/истинная *geol.* true thickness
~, длительная continuous/permanent power/output
~ дозы (облучения) dose/dosage rate
~ дозы, допустимая tolerance/permissive dose rate
~, допустимая power-carrying capacity
~ звука sound/acoustic power, volume
~ и тяга, взлетные take-off power and thrust rating
~, избыточная/неиспользуемая производственная *econ.* excess/redundant capacity
~ излучения emissive/radiation/radiated power
~ индикаторная indicated (horse)power/output
~ источника source strength/power
~, кажущаяся apparent output/power
~ котла boiler capacity
~ короткого замыкания short-circuit power
~ критерия *stat.* strength/power of test
~, крюковая draught power
~ ледниковой шапки *geol.* depth of ice cap
~, литровая *aut.* power-to-volume ratio, power per litre of displacement
~, максимальная, эксплуатационная maximum continuous power
~, мгновенная instantaneous/momentary power
~ множества *math.* cardinality of a set, cardinal number of a set
~ на валу shaft (horse) power/outpt, S.H.P.
~ на единицу веса mass power
~ на единицу объема power per unit volume, power density
~ на режиме малого газа *av.* idle/idling power
~ нагрузки power loading
~ накачки pump(ing) power
~, номинальная rated power, (power) rating, nominal or rated output/horsepower/capacity; *el.* nominal rating
~ облучения exposure/irradiation rate
~, общая *geol., min.* aggregate/summary/total thickness
~, остаточная *nucl.* shut-down power *(of a reactor)*
~, отдаваемая power delivered, output power
~, паспортная nameplate/rated capacity, capacity rating
~, передаваемая power flow, transmitted power
~ пласта *min.* seam/bed thickness
~ по заводским данным nameplate rating
~ по теплу heat power, thermal capacity
~ подъемных двигателей lifting power
~, полезная useful/net/actual/effective power, useful horsepower; net capacity
~, полная full/gross/total power, *el.* apparent power
~, постоянная uniform thickness
~ потока rate of flow
~, потребляемая power consumption, intake/consumed power, energy/power/rated input
~, потребляемая в ваттах wattage, watt consumption
~, потребная power demand, required power
~ почвы depth/thickness of soil
~, проектная design output, designed/rated capacity
~, производственная production/productive/producing/manufacturing capacity
~, промышленная *min.* minable thickness
~, рабочая operating power
~, разрывная *el.* breaking/rupturing capacity
~, располагаемая available power/output/horsepower
~, распределяемая power flow
~, рассеяния/рассеянная/рассеиваемая dissipated power, power dissipation
~ рассеяния на аноде anode power dissipation
~, расчетная design(ed)/rated/power, rated capacity
~, реактивная reactive/wattless power
~, резервная spare/reserve capacity, reserve/standby power
~, средняя mean/average power
~, суммарная total power; aggregate/overall capacity; *geol.* aggregate/total thickness
~, тепловая heat output/power; thermal capacity/output, caloric/calorific power
~, тормозная brake (horse)power

~ **турбины** turbine capacity
~ **турбины, электрическая** generator output of a turbine
~, **тяговая** *aut.* tractive/propulsive power; *naut.* towrope/thrust (horse)power, T.H.P.
~ **угольного пласта** coal height
~, **удельная** specific power, power density, specific output/horsepower, power-to-weight ratio
~ **установки** plant capacity
~, **установленная** installed capacity
~, **форсированная** boost (horse)power
~ **холостого хода** shut-off capacity
~, **эксплуатационная** service (horse)power
~, **энергетическая** generating capacity
~, **эффективная** effective/brake (horse)power
~**ти, производственные** *econ.* production/productive/industrial facilities
мощный powerful, high-power(ed), high-/heavy-duty, heavy; *geol.* thick; *geophysics* competent
~ **пласт** thick seam/bed
~**ная жила** large lode
мощь power, force, strength
~, **авиационно-/воздушно-космическая** aerospace power
~, **морская** sea power
~, **огневая** firepower
~ **флота, ударная** striking power of the fleet
моющий detergent; washing
~**щие средства** detergenta
мракобес obscurantist
мракобесие obscurantism
мракобесный obscurantist, obscurant(ic)
мрамор marble
~, **крупнозернистый** coarse-grained marble
~, **мелкозернистый** fine-grained marble
~, **огненный** fire marble
~, **разноцветный** variegated marble
мраморизация, мраморирование marmorization, marbling, marbleization
мраморировать marble(ize)
мраморник *zool.* blue cod *(Scorpaenichthys marmoratus)*
мраморность marbleness, marbly nature
~ **зубной эмали** *med.* mottled enamel of teeth
мраморный marmoraceous, marmoreal; mottled
мраморовидный marbly, marmoric, marblelike, marmoreal
мраморщик marble cutter/polisher, marbler
мрачность gloom; *med.* barythmia
мрежа fishing net, dragnet, trammel
мстительность *law* revengefulness, vindictiveness
МТС *abbr.* **машинно-тракторная станция** machinery and tractor station/depot
муар *text.* moire; watered silk/effect
муарирование moireeing, watering
муаровый *text.* moire; watered
мувиола *cin.* moviola
мудрость wisdom
мудрствование *med.* sophistication
~, **болезненное** morbid sophistication, sophistical reasoning
муж husband; man
~, **государственный** statesman
~, **ученый** man of science
мужебоязнь *med.* androphobia
мужевластие virocracy
мужелож(е)ство sodomy, buggery, pederasty
мужеложец sodomite, bugger
мужелюбие androphilia
мужененавистничество misandry
мужеподобность *med.* virilism, masculinity
мужеподобный android, manlike; andromorphous; masculine
мужеубийство *law* mariticide; homicide
мужик muzhik, Russian peasant
мужской male; masculine; virile, mannish; *bot.* staminate
~ **пол** male sex
~ **род** *gram.* masculine gender
~**кая гамета** *bot.* pollinoid
мужчина man, male
муза Muse; muse
музеевед museologist
музееведение museology
музей museum; repository
музейный museum *(attr.)*
~**ная редкость** museum piece
музей-усадьба memorial estate, country-house museum
музыка music
музыкальность musicalness, musicality
музыкальный musical; melodious; *phon.* pitch
~ **слух** an ear for music
музыкант musician
музыковед musicologist
музыковедение, музыкознание musicology
мука suffering, torment; torture; *pl.* pangs, throes
~**кам, хождение по** *rel.* Purgatory
~**ки, вечные** *rel.* eternal damnation
~**ки Тантала/танталовые** *myth.* torments of Tantalus
мука meal, flour
~, **буровая** bore meal, borings, drillings
~, **горная** rock flour
~, **инфузорная** infusorial/diatomaceous earth
~, **кварцевая** ground quartz
~, **костная** bone fertilizer/meal
~, **ледниковая** glacier milk
~, **пробковая** cork dust
мукоед *zool.* meal beetle *(Laemophloeus testaceous, Silvanus surinamensis)*
мукозный mucous
мукомешатель flour blender/mixer
мукомол miller
мукомольный flour-grinding
~ **завод** flour mill
мукор *bot.* mucor *(Mucor); med.* mucus, mucor
мукоровый *bot.* mucoraceous
мукосмеситель flour blender/mixer
мукоцеллюлоза *bioch.* mucocellulose

мукронатовый *bot., zool.* mucronate
муксун *zool.* muksun, whitefish *(Coregonus muksun)*
мул, мулица mule
мулат mulatto
мулинетка fan brake/dynamometer; club propeller
мулла mullah, moolah, mollah, Moslem schoolmaster/teacher
муль *text.* mull *(fabric)*
мульга *bot.* mulga *(Acacia aneura)*
мульда 1. *geol.* trough, basin; syncline, synclinal fold; 2. *tech.* (ingot) mould, charging box/pan
~, второстепенная minor trough
~, ковшообразная cap-shaped hearth
~ оседания cauldron
~с пустой породой waste-filled basin
~, сбросовая fault(ed) trough
мультибайт multibyte
мультивалентный *chem.* multivalent
мультивариантный multivariant
мультивектор *math.* multivector
мультивибратор multivibrator
~, автоколебательный free-running multivibrator
~, ждущий/заторможенный single-shot multivibrator, flip-flop
~, задающий master multivibrator
~, задержанный/задержки delay multivibrator
~, однотактный one-/single-shot multivibrator, start-stop multivibrator, monostable multivibrator
~ с двумя устойчивыми положениями bistable multivibrator
~ с одним устойчивым положением monostable multivibrator
~ с положительным смещением positive-bias multivibrator
~ узких/широких стробов narrow-/wide-gate multivibrator
мультигруппа *math.* multigroup
мультидифференциальный multidifferential
мультидоступ *comp.* multiaccess
мультиклавиатура multi-keyboard
мультикогерентный multicoherent
мультиколлинеарность *stat.* multicollinearity
мультиномиальный multinomial
мультиоператорный *math.* multiple-operator
мультиобработка multiprocessing
мультиотношение multirelation
мултипара *med.* multipara
мультиплан *av.* multiplane
мультиплексор *comp.* multiplexer
~ передачи данных terminal multiplexer
мультиплет *phys.* multiplet
мультиплетность multiplicity
мультипликатив numeral adverb, multiplicative
мультипликатор 1. *tech.* speed-increasing gear, step-up gear; booster; intensifier; duplicating machine; 2. *math.* multiplier; 3. *cin.* animator, animated cartoon artist; 4. *opt.* multiple-lense camera, multiplying camera
мультипликационный multiplicating, multiplier
~ фильм (animated) cartoon film
мультипликация *math.* multiplication; *cin.* animated cartoon (film)
мультиплицировать *tech.* gear up
мультипольный, мультиполюсный, мультиполярный multipolar
мультипрограммирование multiple programming, multiprogramming
~ с переменным числом задач multiprogramming with a variable number of tasks, MVT
~ с фиксированным числом задач multiprogramming with a fixed number of tasks, MFT
~ с фоновой и приоритетной работами foreground/background multiprogramming
мультипрограммный multiprogramming
мультипроцессор *comp.* multiprocessor, multi-PU
мультиротация *med.* multirotation, mutarotation, birotation
мультисенсорный multisensory
мультисписок *comp.* multilist
мультиструг multiplough
мультиструктура *math.* multilattice
мультиустойчивость multistability
мультиустойчивый multistable
мультифакторный multifactorial
мульт(и)фильм (animated) cartoon film
мультициклон multistage cyclone, multiclone, multicyclone *(gas cleaning apparatus)*
мультстанок *cin.* animation stand
мультопостумный multoposthume
мульча *agr.* mulch
мульчирование mulching
~ навозом top manuring
~ стерней stubble mulching
мульчировать *agr.* mulch
мулюра *arch.* mo(u)lding
муляж 1. mo(u)lding, model(l)ing, casting; 2. mo(u)ld, (plaster) cast; moulage
муляжист mo(u)lder
муме *bot.* Japanese apricot *(Prunus mume)*
му-металл mumetal
мумификация mummification
мумифицировать mummify
мумифицирующий mummificating
мумия I mummy, embalmed corpse
~мию, превращать в mummify
мумия II *miner.* mummy, Prussian red, ruddle, reddle, red ocher, colcothar
мунго 1. *zool.* banded mongoose *(Mungos mungo)*; 2. *bot.* mungo bean, black grain *(Phaseolus mungo)*
мундир *mil.* (full-dress) uniform, dress coat
~, полковой regimentals
~а, честь the hono(u)r of the regiment
мундшенк *hist.* cupbearer; butler
мундштук 1. tip, nozzle; 2. *mus.* mouthpiece; 3. cigerette/cigar holder; 4. (curb) bit
~ резака torch tip, cutting nozzle

~ **семяпровода** *agr.* grain conductor tip
~, **фильтрующий** filter tip
муниципализация municipalization
муниципализировать municipalize
муниципалитет municipality, municipal administration; town council
муниципий, муниципиум *hist.* Roman municipium
мунсель *naut.* moonsail
мунтжак *zool.* barking deer *(Muntiacus)*
мунц-металл muntz metal
мур *geol.* stone run, rock river
мурава *tech.* glaze, glazing, enamel
муравей *zool.* ant *(Formicidae)*
~, **белый** termite *(Isoptera)*
муравей-древоточец *zool.* carpenter ant *(Camponotus)*
муравейник 1. anthill; 2. *zool.* ant bear *(Murmecophaga jubata)*
муравей-фуражир *zool.* foraging ant
муравить *tech.* glaze (pottery)
муравление *tech.* glazing
муравленый *tech.* glazed
муравьед *zool.* anteater, ant bear *(Murmecophaga)*
муравьежук *zool.* ant beetle *(Thanasimus formicarius)*
муравьиный 1. ant *(attr.)*; 2. *chem.* formic
~ **альдегид** formaldehyde
~**ная куча** anthill
мурашеед *zool.* numbat *(Myrmecobius)*
мурена *zool.* moray *(Muraena)*; white moray *(Echidna nebulosa)*; black eel *(Muraena pseudothyrsoidea)*
мурза *hist.* murza
мурит *geol.* mooreite
мурлыканье *med.* purr, purring thrill/fremitus
~, **кошачье** cardiac thrill
муррея *bot.* jasmine orange *(Murraya)*
мусинг *naut.* diamond knot; mouse; mousing
мускардина muscardine, calcino
мускари *bot.* starch grape hyacinth *(Muscari racemosa)*
мускарин *chem.* muscarine
мускат 1. *bot.* common nutmeg *(Myristica fragrans)*; nutmeg (seeds); 2. muscat, muscadine *(grapes)*; 3. muscatel *(wine)*
мускатель 1. muscat *(grapes)*; 2. muscatel *(wine)*
мускатник *bot.* nutmeg *(Myristica)*
мускатница *bot.* muskroot, moschatel *(Adoxa moschatellina)*
мускатный nutmeg *(attr.)*
~ **орех** nutmeg
~ **цвет** mace *(spice)*
мусковит *miner.* muscovite, white mica
~, **водный** damourite
мускул muscle
~, **жевательный** masseter, chewer
мускулатура musculation, musculature; *coll.* muscles; sinews; muscular system
~, **висцеральная** visceral [non-striated] muscles
~, **сифональная** retractor muscles
~, **соматическая** somatic muscles
мускулистый muscular, brawny, sinewy
мускульный muscular
мускус moschus, musk
мускусница *bot.* muskroot *(Adoxa moschatellina)*
мускусный musky, moschus, musk *(attr.)*; moschate
~ **бык** musk ox *(Ovibos moschatus)*
мускусовый musky; moschate
муслин *text.* muslin
~, **небеленый** raw muslin
~, **шерстяной** (muslin) delaine
муслин-де-лен *text.* (muslin) delaine
мусор 1. refuse, garbage, rubbish; sweepings, dust; 2. debris
~, **бытовой** household rubbish, domestic garbage
~, **городской** town refuse
мусорный refuse, garbage *(attr.)*; *bot.* ruderal
~ **ящик** garbage can, dustbin
мусоровоз garbage-removal truck, refuse/garbage collector, garbage/sanitation truck
мусороприемник garbage receiver
мусоропровод refuse/rubbish chute
мусоросборник *av.* garbage container
мусоросжигание garbage incineration
мусоросжигатель incinerator
мусоросжигательный (garbage) incinerator *(attr.)*
~**ная установка** garbage disposal plant
мусороуборочный refuse/garbage collection
мусорщик 1. garbage collector/man; 2. scavenger
мусс *cul.* mousse
муссон monsoon
~, **влажный/летний** summer/wet monsoon
~, **внетропический** extratropical monsoon
~ **зимний/сухой** winter/dry monsoon
муссонный monsoon(al)
~ **лес** deciduous rain forest
мустанг I *zool.* mustang
мустанг II *bot.* mustang grape *(Vitis candicans)*
мустероид Mouster(i)oid
мустье Mousterian
мустьерский Moust(i)erian
мусульманин Mohammedan, Moslem
мусульманка Mohammedan/Moslem woman
мусульманский Mohammedan, Moslem
мусульманство Mohammedanism, Islam
мутабильность mutability
мутаген *bioch.* mutagen
~, **физический/химический** physical/chemical mutagen
мутагенез mutagenesis
~, **гетерокариотический** heterocaryotic mutagenesis
мутагенность mutagenicity
мутагенный mutagenous, mutagenic
мутант *biol.* mutant
~, **безадениновый** adenineless mutant
~, **безгуанидиновый** guanidineless mutant

~, **безинозитоловый** inositolless mutant
~, **бесцистеиновый** cysteineless mutant
~ **фага с изменением стерильного пятна** plaque-type mutant
~, **фагоустойчивый** phage-resistant mutant
мутаротация *chem., phys.* mutarotation, multirotation, birotation
мутатор *el.* mutator
мутационный mutative, mutational
мутация 1. *biol.* mutation; 2. sound shift(ing)
~, **блуждающая** floating mutation
~, **взаимодействующая** interacting mutation
~, **внутригенная** intragenic mutation
~, **вредная** detrimental mutation
~, **генная/геномная** genic/genome mutation
~, **главная** major mutation
~, **индуцированная** induced mutation
~ **истощения** depletion mutation
~, **ключевая** key mutation
~ **крупных отличий** large scale mutation
~, **ложная** false mutation
~, **межгенная** intergenic mutation
~ **недостаточности** loss mutation
~, **обратная** back/reverse/return mutation
~, **почковая** bud mutation
~, **точечная** point mutation
мутизм *psychiatry* mutism; muteness
~, **акинетический/хореический** akinetic/choreic mutism
мутиляция mutilation
мутномер nephelometer; turbidimeter
мутность 1. *opt.* turbidity, feculence; dullness, opacity, cloudiness; 2. sediment concentration, silt content
мутный turbid, feculent; muddy, cloudy; *phon.* dark
мутовка *bot.* whorl, verticil
~, **ложная** verticillaster, false whorl
~, **цветковая** flower whorl
мутовчатка *bot.* carpetweed *(Mollugo)*
мутовчатый *bot.* verticillate, whorled
~**тое листорасположение** verticillate leaf arrangement, whorl leaf phyllotaxy
мутуализм *biol.* symbiosis, mutualism
мутулы *pl. arch.* mutules
муть turbidity, sludge, feculence, slime, dregs, lees; haze; murk
~, **белковая** albuminous cloudiness
~, **ледниковая** glacier milk
~, **моечная** washings
~, **толчейная** ore slime
мутьевой slimy, cloudy, feculent
~ **поток** turbidity current, mud flow
муфель *tech.* muffle, retort
~, **закалочный** quenching muffle
муфельный muffle, retort *(attr.)*
~**ная печь** muffle/retort furnace
муфлон *zool.* (Corsican) moufflon *(Ovis musimon)*
муфта 1. *tech.* coupling; clutch, connecting piece, joint; 2. *el.* (junction) box, head, sleeve; socket (joint), union
~, **вертлюжная/вращающаяся** swivel union
~, **виброизолирующая** vibration-absorbing coupling
~, **винтовая** threaded sleeve
~, **винтовая стяжная** turn-buckle
~, **втулочная** sleeve coupling, sleeve joint box
~, **гидравлическая** hydraulic/fluid coupling/clutch
~, **гидродинамическая** hydrodynamic/fluid coupling
~, **глухая** rigid/closed coupling
~, **дисковая** disk clutch
~, **жесткая** rigid coupling, sleeve-type coupling
~, **жидкостная** hydraulic coupling/clutch
~, **зубчатая** gear(-type) clutch
~, **кабельная** cable box/head/sleeve
~, **кабельная концевая** cable sealing box/end, cable shoe
~, **кабельная соединительная** cable connector/sleeve, cable joint box
~, **компенсирующая** flexible coupling
~, **коническая/конусная** cone clutch
~, **кулачковая** jaw/claw clutch/coupling
~, **ламельная** plate/disk coupling
~, **ленточная** band clutch
~, **магнитная** magnetic clutch
~, **многодисковая/многотарельчатая** multidisk clutch
~, **неразъемная** fixed coupling
~, **нереверсивная** non-reversible clutch
~, **обгонная** overrunning clutch, free-wheel clutch
~, **однодисковая** single-plate/-disk clutch
~ **отбора мощности** power take-off coupling
~, **ответвительная** tee/branch/joint connection
~, **пальцевая** bolt/pin coupling
~, **переходная** adapter
~, **пластинчатая** plate/disk coupling
~ **подшипника** bearing sleeve
~, **поперечно-свертная** flange coupling
~ **предельного момента** torque-limiting clutch
~, **предохранительная** safety clutch
~, **пружинная** spring coupling, coil clutch
~, **разъединительная/разобщающая** decoupler
~, **расцепная** disengaging clutch
~, **реверсивная** reversing clutch
~, **резьбовая** threaded sleeve
~, **самоуправляемая** automatic clutch
~, **самоустанавливающаяся** self-aligning coupling
~ **свободного хода** overrunning/overriding (freewheel) clutch
~ **скольжения** slip clutch/coupling; hydraulic coupling, eddy-current clutch
~, **соединительная** muff, sleeve coupling, union
~, **спирально-кулачковая** spiral jaw clutch
~, **стержневая** pin/bar coupling
~, **стяжная** turn-buckle
~ **сцепления** clutch

~ **сцепления, пневматическая** air clutch
~, **тарельчатая** plate/disk clutch
~ **трения** friction coupling/clutch
~, **тройниковая** T-joint
~, **управляемая** clutch
~, **упругая** elastic/flexible clutch/coupling
~, **фланцевая** flange coupling
~, **фрикционная** friction coupling/clutch
~, **храповая** pawl/ratchet coupling, jaw clutch
~, **центробежная** expanding clutch, centrifugal clutch/coupling
~, **шарнирная** articulated/joint coupling
~, **шаровая** ball socket
~, **шестеренная** gear coupling
~, **штифтовая** bolt/pin coupling
~, **штыковая** bayonet socket
~, **электроиндукционная** eddy-current clutch
муфти(й) *rel.* mufti, Muslim priest
муха *zool.* fly *(Musca)*
~ **цеце** tsetse (fly) *(Glossina palpalis)*
муха-жигалка *zool.* stable fly *(Stomoxys calcitrans)*
Мухаммед Mohammed
мухаммеданин, мухаммеданский Mohammedan, Muhammadan, Muslim
мухаррем Moharram, Muharram, Muharrem, the first month of the Mohammedan year
мухобойка fly swatter
мухолов *zool.* flycatcher *(Muscicapa)*
мухоловка 1. *bot.* drosophyllum *(Drosophyllum);* 2. *zool.* flycatcher *(Muscicapa);* 3. flytrap; flypaper
мухомор *bot.* fly agaric/amanita *(Amanita muscarina)*
мухоморовые *n. pl. bot.* amanita family *(Amanitaceae)*
мухосед *plant pathol.* flyspeck, sooty blotch
мухояр *text.* cotton fabric *(with wool/silk)*
мучение torture, torment
мученик martyr
мученица martyress
мученичество martyrdom
мучитель torturer, tormentor; tyrant
мучнеросные, мучнисторосные *n. pl. bot.* powdery mildews *(Erysiphaceae)*
мучнистый mealy, farinaceous, floury, farinose; *n. pl.* farinose plants
~**тая роса** *bot.* powdery mildew, oidium
мучное *n.* farinaceous/starchy food(s)
мучной farinaceous, floury, mealy
мушка I foresight, front sight
~, **крупная/мелкая/нормальная** full/fine/normal sight
мушка II gnat, midge
~ **шпанская** Spanish fly, cantharis *(Lytta vesicatoria); pharm.* cantharides
мушкарь caulking hammer, driving mallet, making/setting maul
мушкат *see* **мускат**
мушкель (caulking) mallet, maul, wooden hammer
мушкет *hist.* musket
мушкетер *hist.* musketeer
мушкетон *hist.* musketoon
мушмула *bot.* medlar *(Mespilus)*
муштабель maulstick
муштра drill; regimentation
муштрование *mil.* drilling
муштровать drill
муштровка drill(ing), discipline; regimentation
муэдзин, муэззин muezzin, Muslim religious crier
мха *bot.* moss *(Muscus)*
мхи *pl.* mosses
~, **зеленые** true mosses *(Bryales)*
~ **сфагмовые/торфяные** peat mosses *(Spagnaceae)*
мцыри novice, lay brother
мшанка I *zool.* moss animal, bryozoan *(Bryozoa)*
~, **губоротая** cheilostomatous bryozoan *(Cheilostomata)*
мшанка II *bot.* pearlwort (Sagina)
мшара mossy swamp forest, mshara
мшистость mossiness; *text.* fuzziness
мшистый moss-clad/-begrown, mossy; sphagnous; *text.* fuzzy, cottony
мыза farm(stead), country house
мызник owner/lessee of a farmstead
мыкание *agr.* hackling, rippling *(of flax)*
мыкать hackle, ripple
мылистый soapy, saponaceous; covered with lather *(of horse)*
мыло 1. soap; 2. foam, lather *(of horse)*
~, **бактерицидное** germicidal soap
~, **борное** borax soap
~, **горное** *geol.* mountain soap, saponite
~ **горячей варки** hot-process soap
~, **дегтярное** (coal-)tar soap
~ **для морской воды** marine soap, salt-water soap
~, **зеленое** (medical) soft/green soap, linseed-oil soap
~, **марсельское** Castile/Marseilles soap
~, **медицинское** medicated/medicinal soap
~, **мраморное** mottled soap
~, **татарское** *bot.* Maltese cross campion *(Lychnis chalcedonica)*
~, **техническое** industrial soap
~, **хозяйственное** laundry/household soap
~ **холодной варки** cold-process soap
~, **ядровое** grained soap
мыловар soap boiler, soapmaker
мыловарение soap manufacture/boiling, soapmaking
мыловаренный soap-manufacturing
~ **завод** soap factory
мыловарня soap factory/work(s)
мыловка *miner.* soapstone, steatite, French chalk
мылонафт naphthenate/naphthene soap
мылообразный soaplike
мыльница soap-box/-dish/-tray

мыльный soap(y), saponaceous
~ камень *miner.* soapstone, steatite, French chalk
~ корень *bot.* soaproot *(Gypsophila, Saponaria)*
мыльнянка *bot.* soapwort *(Saponaria officinalis)*
мыс 1. *geol.* headland, foreland, cape, promontory; 2. *anat.* promontory
~ барабанной полости promontory of the tympanum
~ Горн Cape Horn
~ делителя *agr.* divider point/shoe
~, остроконечный cuspate foreland
~, подмытый wave-cut cape
~, структурный plunging anticline
мыслете Slavonic name of the letter M
мыслитель thinker; notionalist
мысль thought; cogitation; notion, conception
~, навязчивая notion of compulsion
~лей, образ way of thinking, views, frame of mind
мысок toe
мыт I *veter.* glanders, strangles
мыт II, мыто *hist.* toll
мытарство trying experience, ordeal, hardship, affliction
мытарствовать suffer hardship/afflictions; wander
мытарь *hist.* publican, farmer of the taxes and public revenues; toll collector
мытник *bot.* lousewort, wood betony, pedicularis *(Pedicularis)*
мытный I *veter.* glanderous, glandered
мытный II *hist.* toll *(attr.)*
~ сбор toll collection, collection of transit dues
мышей *bot.* bristle grass *(Setaria)*
мышелов 1. *zool.* rough-legged buzzard *(Archibuteo lagopus);* 2. mouse catcher
мышеловка mousetrap
мышеобразный mouselike
мышехвостик *bot.* mousetail *(Myosurus)*
мышечно-кожный *anat.* musculocutaneous
мышечно-скелетный musculoskeletal
мышечно-сухожильный musculotendinous
мышечный *anat.* muscular, myogenetic, myogenic, myogenous; sarcous
мышиный mousy, murine
~ горошек *bot.* cow vetch *(Vicia cracca)*
мышка armpit, oxter
мышление thinking, thought
~, замедленное sluggish mentality, bradyphrenia
~, понятийное conceptual thinking
мышовка *zool.* birch mouse *(Sicista betulina)*
мышца *anat.* muscle; *see also* **мышцы**
~, веретенообразная fusiform muscle
~, выпрямляющая erector (muscle)
~ вырезки завитка Santorini's muscle, muscle of the notch of the helix
~, гладкая smooth/nonstriated/unstriated/involuntary muscle
~ глаза, круговая orbicular muscle of the eye
~, глазная ocular/eye muscle
~, глазничная orbital/Muller's muscle
~ голени, трехглавая triceps of the calf
~ головы, верхняя косая superior oblique muscle of the head
~, грудинная sternal muscle
~, грудино-ключично-сосковая sternocleidomastoid muscle
~, двубрюшная digastric muscle
~, двуглавая biceps (muscle)
~, двуперистая bipennate muscle
~, дельтовидная плечевая deltoid
~, жевательная chewer, masseteric muscle
~, задняя большеберцовая musculus tibialis posterior
~, закрывающая abductor
~, затылочно-лобная sculp muscle
~, кожная cutaneous/skin muscle
~ козелка muscle of the tragus, Valsalva's muscle
~, кольцевая circular muscle
~, круговая orbicular muscle
~, лестничная scalene muscle
~, лобково-копчиковая pubococcygeal muscle
~, лобково-прямокишечная puborectal muscle
~ локтя, суставная articular muscle of the elbow
~, одноперистая unipennate muscle
~, опускающая depressor (muscle)
~, отводящая abductor (muscle)
~, подлопаточная subscapular muscle
~, поперечнополосатая striated/stripped/voluntary muscle, cross-striated muscle
~, портняжная tailor's muscle
~, приводящая abductor (muscle)
~, продольная longitudinal muscle
~, произвольная voluntary muscle
~ разгибатель extensor (muscle)
~, ресничная Bowmann's muscle
~ рта, круговая orbicular muscle of the mouth
~ сгибатель flexor (muscle)
~, сердечная cardiac muscle
~ сжиматель sphincter (muscle)
~, скелетная skeletal muscle
~, сосочковая papillary muscle
~, трехглавая triceps
мышцы *pl.* muscles
~, вращающие rotators, rotator muscles
~ гортани muscles of the larynx
~ грудной клетки, вращающие thoracic rotators
~, длинные/короткие вращающие long/short rotators
~ копчика muscles of the coccyx
~, надподъязычные suprahyoid muscles
~ спины dorsal muscles
~ шеи, вращающие cervical rotators
мышь *zool.* mouse *(Mus); comp.* mouse
~, летучая bat *(Vespertilio murinus)*
~, многоклавишная *comp.* multibutton mouse
~, одноклавишная single-button mouse
~, песчаная shrew *(Sorex)*

~ **Шенона** *cyb.* Shannon's mouse
мышь-малютка *zool.* harvest mouse *(Micromys minutus)*
мышь-полевка vole, field mouse *(Microtus)*
мышьяк *chem., miner.* arsenic
~, **белый** arsenic oxide, poison flour, arsenite
~, **желтый** arsenic trisulphide, arsenic yellow
~, **красный/односернистый** ruby sulphur, realgar
~, **металлический** arsenic metal, metallic arsenic
~, **самородный** native arsenic
~, **селенистый** arsenic selenide
~, **сернистый** arsenic disulphide
~, **серый** arsenic metal, metallic arsenic
~, **сурьмянистый** arsenical antimony, allemontite
~, **трехйодистый** arsenic triiodide
мышьяковистый *chem.* arsen(I)ous, arseniferous
мышьяковый *chem.* arsenic(al), arsenious
~ **ангидрид** arsenic pentoxide
мыщелка *anat., zool.* condyle
мыщелковый condylar, condyloid, malleolar
мыщелок *anat., zool.* condyle
~, **затылочный** occipital condyle
мэнский Manx
мэр mayor; *hist.* demarch
мэрия mairie, mayoralty
мэтр teacher; master
мю mu
мюзет *mus.* small bagpipe; musette *(kind of oboe)*
мюзикл *mus.* musical
мюзик-холл music hall
мюль(-машина), мюльная машина mule (-jenny), spinning jenny
мю-мезон, мюон *phys.* mu-meson, muon
мюоный *phys.* muonium
мюрид murid; Sufi disciple
мюридизм muridism, Suf(i)ism
мюскадина *bot.* muscadine grape *(Vitis rotundifolia)*
мютюэлизм *philos.* mutualism
мютюэлист mutualist
мягкий soft, pulpy; mellow, pliant
~ **вагон** *rail.* first-class car, soft-seated car; sleeping car
~**кая мебель** upholstered furniture
~**кое кресло** easy chair
мягкокожий soft-skinned/-shelled, malacodermal
мягколистный *bot.* soft-leaved, malacophyllous
мягконебный *phon.* velar, soft-palatal
мягконогие *n. pl. zool.* malacopods, onychophorans *(Malacopoda, Onychophora)*
мягкоотожженный soft-annealed
мягкоперый *zool.* anacanthine
мягкоплодный *bot.* soft-fruited
мягкосемянный *bot.* malacospermous, soft-seeded
мягкотелый soft-bodied; *n. pl. zool.* molluscs, shellfish
мягкотянутый *tech.* soft-drawn
мягкоцветковый *bot.* soft-flowered
мягкошанкерный *med.* chancroidal
мягкошерст(н)ый soft-haired
мягчение softening, mollification; *(of silk)* soupling
~ **кожи** (a)bating
~ **кожи в киселе** drenching
мягчитель softener, softening agent; mollifer, mollifier
мягчить soften, mollify, emolliate
~ **кожу** bate
мязга pulp, septum
мякина chaff, husk, glume
мякиш *zool.* pad
мякотница *bot.* adder's-mouth orchid *(Malaxis)*
мякотный pulpous, pulpy; sarcoid
мякоть flesh; pulp
~ **зуба** dental/dentinal pulp
~ **листа** *bot.* mesophyll
~ **пальца** digital/finger pulp
~, **пороховая** meal(ed) black powder
~ **селезенки** splenic pulp
мялка *tech.* brake(r); pulper; crusher
мяльщик braker, scutcher
мясистоветвистый *bot.* flesh-twigged
мясистоголовчатый *bot.* sarcocephalous, flesh-headed
мясистолистный *bot.* sarcophyllous, flesh-leaved
мясистоплодный *bot.* sarcocarpous
мясистость pulpiness, fleshiness, meatiness, succulence
мясистоцветковый *bot.* flesh-flowered
мясистый carn(e)ous; pulpy, fleshy, meaty; succulent
~ **межплодник, мясистая часть** *bot.* sarcocarp
мясной meat(y), carneous, sarcous
~ **сахар** *chem.* inositol
мясо 1. flesh; meat; 2. (fruit) pulp
~ **в живом весе** meat on hoof
~, **заготовленное** cured meat
~, **пушечное** cannon fodder
~, **сладкое** (stomach) sweetbread
~, **фасованное** weighed and wrapped meat
мясоед *eccl.* time from Christmas to Shrovetide
мясозаготовка state purchase(s) of meat
мясокомбинат meat-preserving/-processing/-packing plant
мясоконсервный meat-packing/-preserving
мясопродукты *pl.* meat products
мясопункт slaughterhouse
~, **полевой** travelling slaughterhouse
мясопуст *eccl.* Shrovetide
мясорезка meat cutting machine, mincing machine, meat cutter/grinder
мясорубка meat hasher/mincer/grinder/chopper
мясосовхоз state farm raising beef cattle
мясохладобойня meat-packing house, combined slaughterhouse and refrigeration of meat

мясцо *anat., zool.* caruncula, caruncle
мята *bot.* mint *(Mentha)*
~, перечная peppermint *(Mentha piperita)*
мятеж mutiny, insurgence, revolt; sedition
мятежник mutineer, rebel, insurgent; seditionist
мятежный mutinous, rebellious; seditious
мятлик *bot.* bluegrass *(Poa)*
мятличник bluegrass meadow
мятный mint *(attr.)*
~ные капли mint drops
~ные леденцы peppermints
мятый crumpled, crushed; waste *(attr.)*
~ пар *tech.* exhaust/waste steam
мятье kneading, crushing, scutching; crumpling, creasing
мяч ball
~ в игру, ввести *sport* start the ball in play
~, теннисный tennis ball
~, футбольный football

Н, н

набаб nabob
набавка addition, increase; extra charge
набавлять add (to), increase
~ цену raise the price; bid up
набат warning signal, alarm; alarm bell, tocsin
набатный warning; alarming
~ колокол alarm bell, tocsin
набег raid, invasion, incursion, inroad
набегание moving to the high part/point, climbing, run-on
~ волны climb of wave
~ знаков *com.* tailing
набегать move to the high part/point, climb
набережная embankment, quay, wharf, pier; sea front
~, достроечная outfitting quay
набивание 1. padding, stuffing, filling; packing; 2. gasket, filler; 3. *text.* printing
набиватель wadboard packer
набивать 1. stuff, pack, fill (with); pad; ramin; 2. fix, fasten, attach *(with nails);* 3. *text.* print
~ цену bid up, raise the price
набивка 1. stuffing, packing; ramming; filler, gasket; gland; pad(ding), wad(ding), lining; packing material; 2. *text.* printing
~, асбестовая asbestos packing
~, батиковая wax printing
~, войлочная felt packing
~ жгутовая/шнуровая cord/rope packing, cordyarn packing
~ сальника gland packing
набивной 1. *text.* (fast-)printed; 2. packing, stuffing
~ ситец printed calico
набивочный padding, stuffing, ramming, packing
набирать 1. *print.* set up, compose; 2. contract (workers); collect, gather; *mil.* enlist, recruit; 3. take on, assemble
~ высоту *av.* climb
~ вразрядку *print.* space out
~ новых работников hire new workers
~ номер (телефона) dial a (telephone) number
~ скорость gain/gather speed
набла *math.* differential operator, nabla
набла-оператор *math.* del operator
наблюдаемость observability
наблюдаемый observed, observable, experimental; under observation
наблюдатель observer; look-out, spotter; inspector, supervisor, surveyor
~, береговой *mil.* coast watcher
~ в бою *mil.* battle look-out
~, воздушный air(borne) observer
наблюдатель-метеоролог weather observer
~, старший supervisory meteorologist
наблюдательность (power of) observation
наблюдательный observation(al); viewing, lookout; observant, inspective
~ пункт observation post
наблюдатель-радист radio observer
наблюдать observe, watch; scrutinize; supervise; superintend
наблюдение 1. observation, examination, surveillance, lookout, watch(ing); supervision; 2. *nav.* sight
~, астрономическое astronomical/celestial observation
~ в отраженном свете observation by incident light *(in microscopy)*
~, вести keep watch
~, взять под put under observation
~, визуальное direct/visual observation
~, врачебное *med.* medical supervision
~, гравитационное *geol.* gravity measurements
~, дистанционное remote viewing/observation
~ за эфиром *rad.* listening watch
~, натурное prototype observation
~, одностороннее *geod.* one-way observation
~, повторное duplicate observation
~, полевое land survey, field observation
~, радиолокационное radar surveillance
~ снаряда *mil.* tracking the shell
~, солнечное sun/solar observation
~, уровнемерное tide-gauge observation
~, шаропилотное pilot-balloon observation
набожность devotion, devoutness, devotionalism; piety, godliness, theopathy
набожный pious, devout; pietistic, theopath(et)ic
набойка 1. padding, packing, stuffing; 2. *text.* printed (pattern on) cloth; 3. lining, facing; 4. heeltap *(of footwear);* 5. mallet
набойник calking iron/tool; mallet
набойщик 1. *text.* calico/linen/cloth printer; 2. stuffer, filler, packer

набор 1. set, kit, collection, assembly; stack, batch, outfit; pool; 2. collecting, gathering; dialing; 3. enrol(l)ment; number of students enrolled; admission; levy; recruitment, conscription; engagement, selection *(of workers);* 4. typesetting, types and spacing (material); composed matter; 5. *nav.* framework, framing
~, акцидентный job composition
~ атомных электронов complement of atomic electrons
~ в аспирантуру degree-credit enrolment
~ высоты *av.* climb(ing), ascent
~ высоты, вертикальный vertical climb
~ высоты, ступенчатый step(ped) climb
~ высоты, установившийся steady climb
~ генов gene pattern
~, дальний long-distance dialing
~ данных sample/assembly of data, data set
~ для оказания первой помощи emergency kit
~, записываемый *comp.* write set
~ знаков character set, set of symbols
~ зубчатых колес set of gears
~ инструментов set of instruments, tool kit/set
~, интерфейсный interface kit
~ квантовых чисел quantum number set
~ кода/клавишный push-button dialing
~ команд set of instructions
~ констант constant pool
~ корпуса судна ship framing
~ лент tape desk
~ объективов range/set of lenses
~ операций set of operations
~, палубный deck framing
~ переменных set of variables
~ подтональной частоты low-frequency dial(l)ing, subaudio dial(l)ing
~, рекрутский recruiting, recruitment
~, ручной *print.* hand setting/composition
~, силовой primary structure, load-bearing structure
~ скорости acceleration
~, строкоотливной *print.* line composition
~, считываемый readset
~, тональный voice-frequency dial(l)ing, V.f. dial(l)ing
~ шестерен set/train of gears
~ шрифтов type set
наборка 1. assembly, gathering, collecting; 2. recruitment
наборная *print.* typesetting room/office
наборный 1. typesetting; 2. combined, made from many parts
~ цех composing room
~ная доска galley
~ная касса (type) case
~ная линейка typesetting rule
~ная сбруя harness decorated with copper/silver
набортный on-board
наборщик 1. recruiter; 2. compositor, typesetter
набрасывать 1. sketch, outline, draft, draw; 2. jot down; 3. pile/heap up; throw about
наброска 1. *constr.* fill(ing); 2. *geol.* talus
~, каменная rock fill(ing), enrockment
~, шлаковая slag fill
набросок sketch, rough drawing/copy, draft, outline; layout; adumbration
~, первоначальный first/rough draft
~, черновой rough plan
набрызг *constr.* sprayed plaster base; sputtering
набрызгивание sputtering, splashing
набрюшник (abdominal) binder; abdominal band/supporter
набухаемость expansibility, swelling (ability)
набухание swelling, intumescence; turgidity, turgescency; *leath.* plumping; torus; *agr.* upswelling *(of seeds)*
~ котловой воды, ~ уровня boiler water swell
~, мутное *med.* cloudy/turbid/albuminous swelling
~ почвы *geol.* swelling of soil
~ почек *bot.* gemmation
набухать swell, bulk up; *leath.* plump, turgesce
набыль thrust
навага *zool.* navaga *(Eleginus)*
наваждение delusion; nympholepsy; obsession
Навал *bibl.* Nabal
навал bulk; *leath.* dung
наваливать heap up; pile (on)
наваливаться *naut.* fall foul (of), foul
~ на весла lay out on the oars
навались! *nav.* pull away/hard!
навалка loading, heaping, lading, piling
навалом in bulk, piled up
навалоотбойка *min.* coal hewing
навалочник *naut.* bulk carrier, bulker
навалочный heaping, loading
~ная машина mechanical loader
навальщик loader, shoveller, car filler
навар 1. infusion, decoction; brew; 2. built-up metal, boss, shoulder
наваривание building up, welding on
наваривать build up, weld on
~ покрышку *auto.* recap/retread a tyre
наварка building up; welding on; weld bead
наваха clasp knife
навахи *pl.* Navaho Indians
наведение 1. *mil.* guidance, guiding, homing action; pointing, laying
~, автономное onboard/internal guidance
~, астроинерциальное inertial-celestial guidance
~, ближнее short-range guidance
~ в вертикальной плоскости guidance in elevation
~, вертикальное elevation
~, вручную manual guidance
~, инфракрасное infra-red homing/guidance
~ истребителей fighter control/direction/guidance

~ методом погони pursuit guidance
~ методом трех точек line-of-sight guidance
~ на конечном участке terminal guidance
~ на маршевом/среднем участке mid-course guidance
~ на стартовом участке launch guidance
~ на цель homing on a target
~ по азимуту azimuth guidance/orientation
~ по дальности range guidance
~ по лазерному лучу laser guidance
~ по лучу beam guidance
~ при заходе на посадку approach guidance
~ при стыковке docking guidance
~, программное pre-set guidance, programmed guidance
~ с земли earth-/ground-based guidance
~ с оптическим сопровождением optical-track guidance
навек(и) *rel.* for ever (and ever)
навербовывать enroll, enlist, recruit; hire
навертка gimlet
навертывание winding/twisting (round); screwing
навертывать wind/twist (round); screw on
наверх upward, upstairs
~ ! все *naut.* all hands on deck!
навес 1. shed, hangar; 2. overhang, shelter
~ выдувания aeolian rockshelter/recess
~ из парусины awning
~, ледяной hanging glacier
~, скалистый rock shelter
~, соляной salt overhang
навеска 1. hinge; mounting, lift linkage; 2. *chem.* sample, (dose by) weight, weighed amount
~ в трех точках three-point hitch
~, плавающая flexible hitch, free linkage
~, фронтальная front-end mounting
навесной hinged, suspended, hook-on, mounted
~ная дверь door on hinges
~ная петля hinge
навесный overhanging, shelter *(attr.);* hinged, suspended
~ огонь, ~ная стрельба *mil.* plunging/curved fire
наветренный upwind, windward(ly), exposed to the wind
~ная сторона weather side
навечно for ever/good; in perpetuity
навешивание 1. suspension, hanging on; 2. weighing out
~ бирок tagging
~ квантов quantification
Навзикая *myth.* Nausicaa
навивание winding on
~ влево sinistrality
~ вправо dextrality
навивать wind (on), spool, reel, *el.* coil
~ основу *text.* beam the warp
~ пружину coil/wind a spring
навивка coiling; winding(-on); reeling; wrapping
~ труб *therm.* tube arrangement
навигатор navigator; pilot
навигационный navigation(al); nautical
навигация 1. navigation; 2. shipping season
~, автономная autonomous/independent navigation
~, астрономическая celestial navigation, astronavigation
~, ближняя short-range navigation, close navigation
~ в космическом пространстве space navigation, astrogation
~ в районе аэродрома approach navigation
~, визуальная landmark/visual navigation; *av.* ground-contact navigation
~, воздушно-космическая aerospace navigation
~, всепогодная all-weather navigation
~, дальняя long-range navigation
~, космическая space navigation, astrogation
~, межзвездная interstellar navigation
~, межпланетная interplanetary navigation
~ методом счисления пути DR navigation, deadreckoning navigation
~, наземная ground navigation
~ по приборам blind navigation
~ по условной координатной сетки grid navigation
~ с помощью доплеровской РЛС Doppler navigation
~ с помощью РЛС radar navigation
навикулоидный naviculoid; naviform
навинчивать screw on
~ от руки screw on finger-tight
нависерт, навицерт navicert, navigation certificate
наводить 1. guide, home, vector, control, lead; aim, point (at); 2. *el.* induce; *cin.* focus; 3. apply, cover, coat
~ глянец/лоск polish, gloss, glaze
~ мост (build a) bridge
~ оружие point a gun
~ справки make inquiries
~ шлак build up slag
наводка 1. *mil.* aiming, laying, pointing, directing; 2. *el.* inducing; *cin.* focusing; 3. covering, coating; 4. construction, building
~, зеркальная foil
~ моста bridging
~ на резкость focus(s)ing
~ помех *el.* stray pick-up; noise inductance
~ помех от энерголиний power-line interference
~ попаромно construction by rafts
~ понтонно construction by successive pontoons
~ шлака slag formation
наводнение flood, inundation, freshet
наводной floating
наводчик 1. *mil.* sighter, spotter, pointer, gun layer,

gunner; 2. coater, spreader
~, **авиационный** air controller
наводящий aiming, directing, guiding; inducing
~ **вопрос** leading question
навоз *agr.* manure, dung
~, **перепревший** decomposed manure
навозник 1. *zool.* dung beetle *(Geotrupes);* 2. *bot.* ink mushroom, ink(y) cap *(Coprinus);* 3. manure carter; 4. box/container for dung
навозниковые *n. pl. bot. Coprinaceae*
навозно-земляной manure-soil
навозно-минеральный manure-mineral
навозный 1. *physiol.* stercoraceous; 2. *bot.* coprophilous, fimicolous
навозолюбивый coprophilous
навозопогрузчик manure/dung loader
навозохранилище *agr.* manure/dung pit, dungyard
навоз-сыпец *agr.* dry-fine manure
навой 1. *text.* warp/back beam, weaver's beam; 2. winding, reeling, spooling
наволакивание *met.* galling, fretting
наволок 1. *geol.* overthrust folding, nappe; 2. water meadow
Навсикая *myth.* Nausicaa
Навуходоносор *bibl.* Nebuchadnezzar
навык acquired habit, practice; skill, experience
~, **двигательный** motor skill
~, **речевой** speech habit
~**и устной речи** speaking proficiency
навь *Russ. myth.* ghost, spirit; phantom
навязчивость insolence; importunity
навязчивый importunate, obtrusive
~**вая идея** obsession, idée fixe
нагайка whip; lash
наган revolver
нагана *vet.* nagana
нагар (carbon) deposit, soot formation; caking
~, **пороховой** powder/fouling residue
нагарообразование carbonization, deposition/formation of carbon, caking
нагартовка work/strain/cold/wear hardening, cold working
~, **дробеструйная** shot peening
нагель dowel, pin, peg
нагельфлю *geol.* nagelfluh, gompholite
нагибание bending
наглазник (horse) blinder, blinker
наглазный for the eyes
~ **щиток** eyeshield, eyeshade
наглухо hermetically, tightly
~ **закрепленный** *el.* dead-ended
наглядно graphically, visually
наглядно-геометрический descriptive-geometric
наглядность 1. visualization, obviousness; 2. use of visual aids/methods
~, **зрительная** seeing imagery
~, **сохранять** keep in evidence
наглядный graphic, descriptive, visual; pictorial
~ **пример** graphic example
~ **урок** object lesson
~**ное доказательство** clear proof, visual evidence
~**ное обучение** visual teaching methods
~**ные пособия** teaching aids
нагнетание discharge/pumping under pressure, injection, infusion; focrcing, pressing in
~ **воды** *min.* water injection
нагнетатель blower; supercharger
~ **волновода** waveguide pump
~, **двухскоростной** two-speed supercharger
~, **лопаточный** vane-type blower
~, **механический** gear-driven blower, geared supercharger
~, **многоступенчатый** multi-stage supercharger
~, **объемный** positive-displacement blower
~, **продувочный** scavenge blower
~, **смазочный** lubricating pump
~ **теплого воздуха** heating blower
~, **центробежный** centrifugal supercharger
нагнетательный forcing, pressing, charging; supply *(attr.)*
~ **насос** force/delivery pump
~ **клапан** delivery/pressure valve
нагнетать discharge/pump/feed under pressure, force
нагноение *med.* suppuration, pylsis, pyosis, pyopoiesis, fester, purulency, apostemation
~ **кожи** cutaneous suppuration
~ **сустава** pyarthrosis
нагноительный suppurative, suppurating, suppurant
наголоватка *bot.* jurinea *(Jurinea)*
наголовник *constr.* head, cap; crown
~ **вышки/деррика** guy cap, derrick crown
~ **копыла** dagger plank
~ **сваи** pile/driving cap
~ **стойки** post cap
нагон pileup, surge, piling up; run-up; onset
~, **ветровой** wind onset
~, **штормовой** storm surge, wind-induced surge
нагонка 1. distillation; 2. *hunting* training a dog to chase an animal
нагорный 1. high(land), upland; hilly, mountain(ous)
~**ная проповедь** *bibl.* Sermon of the Mount
нагорье highland, upland, mountain land, plateau
~, **моноантиклинальное** arched mountain
награда 1. recompense, reward; 2. award; decoration; prize
~, **денежная** gratuity; money award
наградка foxtail/dovetail saw
наградные *n. pl.* gratuity, bonus
награждать 1. recompense/reward (with); 2. award (to), confer (on); decorate (with)
награждение 1. rewarding; decorating; 2. gratuity, reward, remuneration; award
нагрев heating(-up), warming(-up), heat build-up
~ **вихревыми токами** eddy-current heating

~ **гиперзвуковым потоком** hypersonic heating
~, **джоулев** Joule heating
~ **джоулевым теплом** resistance/ohmic/Joule heating
~ **излучением/лучистый** radiant/radiation heating
~, **повторный** reheating
~ **погружением** immersion heating
~, **последующий** subsequent heating, postheating
~, **предварительный** preheating
~ **при сжатии** compressional heating
~ **при трении** frictional heating
~, **садочный** *met.* batch (re)heating
~ **токами высокой частоты** high-frequency heating
~ **электрической дугой** arc heating
~, **электроннолучевой** electron-beam heating
нагревание calefaction; heating, warming
~, **предварительное** preheating
нагреватель radiator, (resistive) heater, heating apparatus
~, **лучистый** radiant heater
~, **погружной/погружаемый** immersion heater
~, **солнечный** solar(-heat) collector
~, **точечный** spot heater
нагревательный heating, warming
нагревать(ся) heat (up), warm (up)
~ **до красного/белого каления** bring to red/white
нагревный heating, warming
нагревостойкость heat resistance, thermal endurance
нагромождать pile/heap (up), accumulate
нагромождение pile, accumulation, heaping
~ **вулканических продуктов** volcanic pile
~ **льда** pack ice
нагрудник 1. breastplate; breast collar; 2. jacket, vest
~, **спасательный** life(-saving) jacket/vest
нагрудный chest-type, breast *(attr.)*
~ **значок** badge
нагружаемость load(-carrying) capacity
нагружать load (with), charge, freight
~ **внецентренно** load eccentrically
~ **не полностью** underload
~ **статически** load quiescently
нагружение load(ing), load application; charge, charging
~, **аэроупругое** aeroelastic loading
~, **бигармоническое** biharmonic loading
~, **знакопеременное** alternative loading, reversal of loading
~, **нестационарное** transient loading, variable-cycle loading
~ **подъемной силой** lift loading
~ **по размаху** spanwise loading
~, **рациональное** practical loading
~, **сложное** complex/combined loading
~, **стационарное** steady-state loading
~, **ступенчатое** step(ped) loading
~, **циклическое** repeated loading
нагрузка 1. loading, filling, lading; 2. load, freight, shipment; charge, burden; capacity; duty *(of machine);* 3. *tel.* traffic
~ **аварийного режима** emergency load
~, **активная** *el.* resistance/active load
~, **анодная** anode/plate load
~, **балластная** dummy/dry load
~, **боевая** *mil.* combat/weapon/ordnance load
~, **боковая** side/lateral/transverse load
~, **бытовая** *el.* appliance/residential load
~ **в контейнерах** containerized load
~, **весовая** gravity load
~, **внепиковая** off-peak load
~, **внецентренная** *mech.* eccentric load
~, **входная** input load
~, **выходная** output load; terminator
~, **гололедная** *el.* sleet/ice load
~, **действительная** actual load
~, **длительная** constant/continuous load
~, **длительно действующая** sustained load
~, **доковая** docking load
~, **допускаемая/допустимая** allowable/permissible/safe load
~, **емкостная** capacitive load
~, **знакопеременная** alternate/alternating load
~, **изгибающая** bending/buckling load
~, **изменяющаяся по времени** time-dependent load
~, **индуктивная** lagging/inductive load
~, **инерционная** inertia(l) load
~, **испытательная** test/trial load
~ **карты** map details, amount of (map) details
~, **коммерческая** payload, commercial load
~, **конечная** terminal/final load
~, **консольная** cantilever load
~, **косонаправленная** oblique load
~, **кратковременная** instantaneous/momentary load
~, **критическая** critical load
~, **критическая продольная** buckling load
~, **крутящая** torsion(al) load
~, **лобовая** drag load
~, **максимальная** maximum/peak load
~ **на висении** *av.* hover load
~ **на грунт, удельная** *av.* footprint loading
~ **на единицу площади** unit area loading
~ **на колесо** wheel load
~ **на конструкцию** structural load
~ **на лопасть** blade loading
~ **на шасси** landing-gear load
~, **наибольшая** maximum/peak load
~, **неполная** underload
~, **неравномерная** uneven load, non-uniform load
~, **несимметричная** unsymmetric(al)/unbalanced load

~, несъемная fixed load
~, неустановившаяся unsteady/transient load
~, номинальная rated load, power rating
~, ограничиваемая объемом capacity load
~, осветительная *el.* lighting demand/load
~, осевая thrust/axial load, axial loading
~, основная primary/basic/base load
~, осредненная average(d) load
~, переменная variable/varying load
~, пиковая peak load
~, платная commercial load, payload
~, повторяющаяся repeated load
~, погонная *constr.* linear load, load per unit (of) length
~, подвижная live/moving/travelling/rolling load
~, полезная useful/net/actual/service load, payload
~, полная total/full/gross load
~, посадочная *av.* landing load
~, постоянная static/constant/permanent/steady/dead/fixed load
~, предельная limit/ultimate/load
~, предельная разрывная ultimate tearing load
~, предстартовая pre-launch load
~, преподавательская academic/teaching load
~, пробная test/trial/proof load
~, продольная longitudinal loading
~, проектная design load, load rating
~, пульсирующая pulsating/intermittent load
~, рабочая work(ing)/operating/service load, workload
~, разрушающая breaking/crushing/cracking/collapse/collapsing load, fully-factored load
~, разрывающая tearing stress
~, растягивающая extension/tensile/tension/stretching load; tensile stress
~, распределенная distributed/spread load
~, расходуемая expendable/consumable load
~, расчетная design/calculated/specified/assumed load
~, реактивная *el.* reactive load
~, сжимающая compressive/compression/pressure load
~, скручивающая torsional load/stress
~, случайная random load
~, согласованная matched load
~, солевая *med.* salt load(ing)
~, сосредоточенная concentrated/point/fixed/single load
~, сплошная distributed load
~ спутника, полезная satellite payload
~, срезающая shear load
~, стартовая launch load
~, статическая мышечная static muscular load
~, телефонная telephone traffic
~, тепловая/термическая heat/thermal load
~, тормозная brake load
~, тяговая *transp.* traction load; *av.* thrust loading
~, ударная impact/shock load, impact stress
~, удельная specific/unit load
~ уплотнения consolidation/packing load
~, условная substitute load
~, усталостная fatigue load
~, учебная academic/teaching load
~, фактическая actual load
~, частичная fractional load
~ шины, оконечная *comp.* bus terminator
~, эксплуатационная operational/operating/service load
нагуливать acquire, develop
~ жир *agr.* fatten up, put on weight
нагур *zool.* bharal, nahoor *(Pseudois mayaur)*
надбавка allowance
~ за выслугу лет, ~ за стаж long-service bonus, seniority pay
~ за переутомление fatigue allowance
~ к цене за вес *econ.* price-for-weight differential
~ к цене за качество grade price differential
~ летному составу за полеты flying/flight pay
~ на дороговизну *econ.* cost-of-living allowance, COLA
~ на обрастание *nav.* fouling allowance
~ на цены increase in prices
~ на шероховатость roughness allowance
~ обеспечивающая только прожиточный минимум *econ.* subsistance allowance
~ки, семейные family subventions
надбавочный additional, supplemental, extra
надбарабанный *anat., zool.* supratympanic, epitympanic
надблоковый *anat.* supratrochlear
надбровница *anat.* superciliary arch
надбровный *anat.* supraciliary, superciliary
надбровье *anat.* supercilium, superciliary arch
надбрюшие *anat.* epigastrium
надбрюшный *anat.* epigastric
надбугорный *anat.* supratrochlear
надвертельный *anat.* supratrochanteric
надвиг *geol.* (over)thrust, overlap, nappe
~, антиклинальный break-thrust
~, крутой upthrust
~, наволоченный drag thrust
~, пластовой bedding thrust
~, пологий low-angle overthrust
~ разлома break/fracture thrust
~ срыва strip thrust
~, чешуйчатый scaly/imbricated thrust
надвигание *geol.* encroachment
~ краевой воды encroachment of edge water
надвигать(ся) thrust over, move upon, (over)thrust
надвигообразование *geol.* thrusting
надвлагалищный *anat.* supravaginal
надводный above-water, overwater, surface *(attr.)*; supraqueous

~ борт *nav.* freeboard
~ корабль surface ship
надворный outdoor
~ советник *hist.* court counsel(l)or
~ суд *hist.* court council
~ние постройки farm buildings
надглавок *arch.* decoration on a church cupola supporting a cross
надглазничный *anat.* supraorbital
надглазный *anat.* superciliary; supraocular
надглоточный *anat.* suprapharyngeal
надгортанник *anat.* epiglottis
надгортанниковый, надгортанный epiglottal
надграфик *math.* epigraph
надгробный graveside, funeral, placed on/over a grave
~ камень tombstone
~ная надпись epitaph
~ная речь, ~ное слово graveside oration
надгрудинный *anat.* supersternal, episternal
~ная вырезка suprasternal notch
надгруппа *math.* supergroup
наддалбливать chizel, gouge
наддача 1. outbidding; 2. increase, increased value
наддиагональный off-diagonal
наддиалектный supradialectal
наддув 1. supercharging, supercharge, pressure charging, boost; 2. pressurization
~, безагрегатный non-compressor supercharging
~ кабины *av.* cabin pressurization
~, скоростной ram effect
~, форсированный power boost
надежность 1. reliability, fail-safety, (margin of) safety, dependability; 2. credibility, trustworthiness
~ в работе/эксплуатации reliability of operation/service
~ измерения measurement accuracy
~ набора данных data-set security, file security
~, расчетная design reliability
~, резервируемая redundant reliability
~, собственная inherent reliability
~, функциональная dependability
~ шифрпереписки cryptographic security, cryptosecurity
~, эксплуатационная operational/operative reliability, serviceability
надежный 1. reliable, secure, dependable; (fail-) safe, unfailing; 2. trustworthy
~ное убежище safe hiding place
наделка 1. addition; extension; 2. *nav.* guide stud
~, верхняя top lug
~ для защиты от льда, носовая *nav.* bow-grace
~ для подвески suspension lug
~, противоминная blister, bulge
~, рулевая wood lock
~, Т-образная idle hook bracket, tee guide
~, транспортировочная transporting guide stud
надельный additional
~ная земля *hist.* allotment
наделять provide/invest with, allot, give
наджаберный suprabranchial
надзвуковой supersonic, superaudio
надземный aboveground; overhead, aerial, elevated; epigeal, epigeous, superterranean
~ клубень *bot.* tuberidium
~ное растение epiphyte
надзиратель inspector, overseer, superintendent, supervisor
~, полицейский *hist.* police inspector
~, тюремный warder, warden
надзирательница supervisor, inspector
~, тюремная wardress, prison matron
надзирательство inspectorate, inspectorship, superintendency
надзирать supervise, superintend, oversee
надзор 1. supervision, superintendence, surveillance; 2. *coll.* inspectorate
~, прокурорский Directorate of Public Prosecutions
~, санитарный sanitary inspection/inspectors
~, строительный building inspection
~, технический technical/engineering supervision
~, школьно-санитарный hygienic service in schools
надзорный supervising, supervisory
надзубный supradental
надир *astr.* nadir
надкислота *chem.* peracid
надкласс superclass
надклассовый *polit.* independent of class, transcending class, standing above classes
надклеивать stick/glue/paste on, lenghten *(by pasting)*
надклепывать rivet on, lenghten *(by riveting)*
надключичный *anat.* supraclavicular
надковывать add, join, lenghten *(by hammering)*
надкожица *anat., bot.* (outermost) skin, cuticle
надкол cleft; crack, split; chink
надколенник *anat.* kneepan, kneecap, patella
~, смещающийся slipping patella
надколенный *anat.* patellar
~ная чашка patella, kneecap, kneepan
надкольцо *math.* overring, extension ring
надкопытье *zool.* fetlock
надкостница *anat.* periosteum
~, альвеолярная peridontium
~, краниальная pericranium
надкостничный, надкостный *anat.* periosteal
надкраевой supramarginal
надкрестцовый *anat.* suprasacral
надкритический supracritical, above-critical
надкритичность *nucl.* supercriticality
надкрыловидный elytroid, elytriform
надкрылье *zool.* sheath, elytron, shard; wing case
надкрыльный *av.* overwing
надледниковый superglacial
надлежащий appropriate, suitable; proper, fitting

~ **срок, в** in due time
надлинейный superlinear
надлистный *bot.* epiphyllous, suprafoliaceous
надлобковый *anat.* suprapubic
надлобный *anat.* frontal
надлом 1. incipient fracture/break; 2. *med.* infraction
надлопаточный *anat.* suprascapular
надлунный translunary, superlunary
надминдаликовый *anat.* supratonsillar
надмножество *math.* superset, including/comprehending set
надмодель hypermodel
надмыщелковый *anat.* epicondylic, epicondylian
надмыщелок *anat.* epicondyle
наднациональный supranational
наднильпотентный *math.* hypernilpotent
надниточник *mach.* upper/faller wire
надобность exigency, need, necessity; requirement
надой *agr.* yield (of milk), milk output/yield
надолба pillar, stake, post
~**, бревенчатая** log post
~**, рельсовая** rail obstacle
надомник *econ.* homeworker, outworker
надостный *anat.* supraspinal, supraspinous
надотряд *biol.* superorder
надпазушный superaxillary
надпалубный *naut.* superstructure *(attr.)*
надпаразит *biol.* superparasite, hyperparasite
надпереносье *anat.* glabella
надпестичный *bot.* epigynous, syngynous
надпилка, надпиловка sawing, notching (slightly)
надпись inscription, supercript(ion), notice; lettering; address; legend; caption
~**, безоборотная** *econ.* endorsement without recourse
~**, маршрутная** *auto.* destination indicator
~**, освещаемая** illuminated sign
~**, передаточная** *econ.* endorsement
~**, предостерегаlющая** warning/caution/danger notice, caution(ary) note
~**, светящаяся, ~ с подсветом** illuminated sign
~**, экплуатационная** legend
надплодник *bot.* epicarp
надподъязычный *anat.* suprahyoid
надполе *math.* overfield, superfield, extension field
надпочвенный above the surface, above-soil
надпочечник *anat.* adrenal/suprarenal gland
надпочечниково-корковый *anat.* adrenocortical
надпочечниковый *anat.* adrenal
надпузырный *anat.* supravesical
надпяточный *anat.* above the heel
~**ная кость** ankle bone
надреберный *anat.* supracostal
надрез nick, notch; indentation; gash, groove; *med.* incision, cut
надрезание incision, incising
надрезанно-зубчатый incisodentate
надрезанно-листный cut-leaved
надрезать nick
~ **провод/изоляцию** score wire/insulation
надрезка incomplete cut; notching, slitting
надрезывание notching; indenting, cutting-in
надрезывать notch, incise; scarify
надруб notch, kerf; incision, cut
надрубка run-/lead-in cut, nick, notch, incision
надрудный supra-ore
надрыв surface tear/rupture; *foundry* contraction/shrinkage crack
~**, горячий** hot tear
надсекать incise slightly; scarify
надсемейство *zool.* superfamily, legion
надсемядольный *bot.* epicotyledonary
~**ное колено** epicotile
надсернокислый *chem.* persulphate
надсерный *chem.* persulphuric
надсеть *math.* supernet(work)
надсечка notch, cut, incision
надсинхронный supersynchronous, hypersynchronous
надсистема supersystem
надслуховой supersonic, ultraphonic
надсмотр supervision, surveillance
надсмотрщик overseer, supervisor; jailer
надставка 1. extension (piece); top, built-in part; adapter; 2. honey chamber; 3. *ling.* added-on particle
~ **рычага** extension arm
~ **тигля** crucible top
~ **штока** extension rod
надставлять extend, lenghten, add on
надставной lenghtened, patched/put on
надстроечный *philos.* superstructural
надстройка 1. superstructure, additional stor(e)y; *nav.* (deck) erection; 2. *philos.* superstructure
~**, баковая** forecastle (erection)
~**, кормовая** stern superstructure
~**, маячная** hut
~**, носовая** bow/forward superstructure
~**, средняя** bridge erection/superstructure
~**, ютовая** poop (erection)
~ **энергетической установки** superposed/topping plant
надстрочный superlinear
~**ные знаки** diacritical marks
надтепловой epithermal, above-thermal
надтональный supersonic, ultrasonic, superaudio; supertonic, hyperacoustic
надув inflation, inflating
надувательство *law* sharp practice, cheating, imposture
надувать inflate, blow up, distend, fill with air
надувка inflating, inflation
надувной inflatable; pneumatic
~**ная лодка** inflatable rubber dinghy
надульник *mil.* barrel mouthpiece, muzzle attachment, gas preserver; compensator

надутый 1. inflated, bloated; swollen; blown up, distended; 2. *lit.* bombastic, pompous; 3. *bot.* torous, torose
надфиль needle/broaching file, rat-tail file
надфункция *math.* majorant
надфюзеляжный dorsal
надхвостье upper fail coverts
надхромовый *chem.* perchromic
надхрящница *anat.* perichondrium
надчелюстной *anat.* supramaxillary
надчеревный, надчревный *anat.* epigastric
надчерепной *anat.* epicranial
~ная оболочка pericranium
наезд 1. *cin.* zoom(ing), dolly-in; 2. flying visit; *mil.* raid, incursion, inroad
наездник 1. *tech.* rider; 2. equestrian, horseman; *mil.* raider; 3. *zool.* ichneumon fly *(Ichneumonidae)*
наездничество *mil.* (cavalry) raid; partisan raid
наезжий *n.* newcomer
наем hire, hiring; renting
~, взять в rent
~ рабочей силы recruiting drive
~, сдать в let (out), let for rent
наемник hireling; *hist.* mercenary
наемничество 1. wage earning, hiring of manpower; 2. mercenaryness
наемный rented, hired; mercenary
~ дом rented house
~ труд wage labo(u)r/worker
~ убийца hired assassin
наемщик 1. lessee, tenanat, lodger; 2. hired man/worker
нажатие depression, pressure
~ на клавишу *tel.* keystroke
наждак emery
наждачный emery *(attr.)*
нажива gain, profit
~, легкая easy money
наживание acquisition; gaining
наживать gain, acquire
~ болезнь contract an illness
~ состояние make a fortune
наживной acquirable, gainable; lucrative, profitable
нажигать 1. heat red-hot, overheat; 2. brand; burn, scald; scorch
нажим 1. pressure, stress, push; 2. *tech.* clamp
~, закулисный *polit.* wirepulling
нажимать 1. press, depress; clamp, pinch; 2. stress, emphasize
нажимной, нажимный *tech.* clamping, pressing, pressure *(attr.)*
~ винт adjusting/thrust screw
~ ролик pressure roller
~ная кнопка push button
нажимно-отжимный push-and-pull
нажин *agr.* amount of reaped crop
нажор *leath.* plumping, swelling; plumpness
~, зольный lime plumpness
назад back(wards)
~, полный ход *nav.* full speed astern
назализация *phon.* nasalization
назальный nasal
назарей *rel.* Nazarene
назареянин *hist.* Nazarene
назатыльник neckguard, neck flap
название name, title; term, designation, appellation
~, видовое specific name
~, географическое geographical/place name
~, родовое generic/tribal name
~, торговое trade name
названый adopted, sworn
~ сын/брат adopted son/brother
наземный (sub-)aerial, terrestrial, superficial, land *(attr.),* ground *(attr.); bot.* terraneous, terrene
~ стебель *bot.* corm
~ное животное land animal
~ные войска ground troops
назидание exhortation, edification
назидать edify, instruct, teach
назначать 1. fix, set, appoint; specify; 2. nominate; institute; 3. allot, allocate, assign; 4. prescribe
~ высоту полета *av.* assign a flight-level
~ заседание set a meeting
~ оплату fix the rate of pay
~ цену charge a price
назначение 1. appointment; assignment; nomination; 2. duty, function, purpose, use; scope; mission, role; 3. prescription; fixing, setting
~ выборов calling an election
~ допусков specification of tolerances
~ карты function of chart/map
~, комплексное integral use
~ корабля mission of a ship
~ лекарства prescription/prescribing of a medicine
~ приоритетов priorization
~ устройств device allocation/assignment
~ эшелона *av.* flight-level assignment
~нию врача, по on doctor's orders
~ния, двойного dual-puprose, double-duty
~ния, место destination
назначенный assigned, designated, allotted; appointed, nominated
назо-антральный *anat.* nasoantral
назо-лабиальный *anat.* nasolabial
назо-лакримальный *anat.* nasolacrimal
назо-оральный naso-oral
Назорей *rel.* the Nazarene, Jesus Christ
назревать mature, ripen, become ripe; gather head
назревший urgent, pressing
~шие вопросы urgent questions
называние naming, designation
называть name, announce; call, designate
назывной nominative; naming
наиб substitute, deputy; proxy
наибольший (the) largest, greatest, maximal
~ делитель *math.* highest common factor
наивероятнейший the most probable

наивность naivety, naiveté
наивный naive, unsophisticated
наивысший the highest/utmost
наизусть by heart/rote
~, заучивание *educ.* rot drill
~, читать recite
наилование *geol.* silt deposition
наименование name, title; appellation, designation
~, торговое trade name
~ числа denomination of a number
наименовывать label, name, designate
наименьший the least/smallest, minimal
~шее кратное, общее *math.* lowest common multiple
наискось aslant, obliquely
наистрожайший the strictest/severest
наитие inspiration
найлон(овый) nylon *(attr.)*
наймодатель letter
найтов *nav.* (span)lashing, seizing, mousing
найтовить *naut.* frap, seize, lash (down)
наказ 1. instructions; order; direction; 2. mandate
наказание penalty; punishment; chastisement
~ господне holy terror
~ отбывать serve time
~, условное *law* suspended sentence
наказной, наказный punishment *(attr.)*
~ атаман appointed/interim hetman
наказуемость punishability
наказуемый punishable, disciplinable
наказывать I punish, chastise, castigate; discipline
наказывать II bid, instruct, order
накал 1. glow, incandescence; intense heat; 2. filament, heating
~, белый/красный white/red heat
~, выключать turn off filament voltage
накаленный incandescent, glowing; heated
~ докрасна red-hot
накаливание incandescing; incandescence, heating, glowing
накаливать glow, heat, incandesce
~ добела bring to white heat
~ докрасна make red-hot
накаливаться glow, get hot, incandesce
накалывать pin down/on; prick, pierce
накапливание agglomeration, storage, accumulation
~ процентов accumulation of interest
накапливать accumulate, amass, store
~ запас stockpile
~ заранее prestore
накат 1. *constr.* counter/dead/false floor, false ceiling; 2. layer *(of planks),* overhead logs; 3. scab, (back)fin; 4. *print.* application of ink, inking; 5. slope, run-up, uprush, landwash; 6. packed/smooth road; 7. reeler, reeling apparatus, winder; 8. *mil.* countercoil
~ волн wave run-up
накатанный rammed, packed, made smooth
накатать roll (in), move by rolling
накатина joist, trimmer; *constr.* sound board
накатка 1. knurl(ing); knurling tool; 2. packing, ramming, rolling smooth; 3. roller
~, косая сетчатая cross knurling
~ резьбы thread rolling
накатник 1. *mil.* countercoil mechanism, recuperator; 2. layer *(of beams),* overhead logs
~, воздушный pneumatic recuperator
накатывание 1. roll forming, rolling (on/onto); 2. milling; knurling
~, калибрующее size rolling
~ краски inking
~ резьбы thread rolling
~, формообразующее form rolling, (k)nurling
~ шариком ball burnishing
накачивание pumping; inflation
накачивать pump (up); inflate
накачка *el.* pump(ing), excitation
~, двухтактная push-pull pump(ing)
~, многочастотная multiple-frequency pump(ing)
~, оптическая optical pumping/injection
~, световая light pumping
~, синфазная in-phase pump(ing)
накернивать centrepunch, prick-punch
накипеобразование *tech.* scaling, scale formation
накипеобразователь *tech.* scaleforming constituent/substance
накипеочиститель *tech.* scaler
накипеудаление *tech.* scale/incrustation removal
накипь 1. (boiler) scale, incrustation, sinter, fur, coating, deposit; 2. scum; spume
~, жемчужная/перламутровая *miner.* pearl sinter, geyserite
~, известковая calcareous tufa, calc sinter
~, котельная boiler scale
~, кремневая quartz/siliceous sinter
~, медная tyrolite
~, снимать descale
~, фумароловая fumarole incrustation
накладка 1. (but) strap, (cover) plate, lap plate, lining, gusset, cover bar, facing, overlapping; *geol.* (over)lap, superposition; 2. hairpiece, false hair, switch; 3. trimming *(of a dress);* 4. *theatre* stage wait
~ бампера bumper guard
~, безраспорная kicking strap
~ грядиля *agr.* beam block
~ для сосков nipple shield
~, липкая adhesive patch
~, лицевая flush plate
~, мультипликационная overlay animated device
~, отсасывающая suction cup
~, продольная longitudinal butt strap
~, ромбовидная diamond plate
~, стыковая butt/cover/lap/finish plate
~, тормозная brake pad

~, угловая corner strap, gusset plate
~, усиливающая doubler (plate), reinforcing plate
~, фрикционная/тормозная friction/brake lining
~ щетки *el.* top insert
накладная *n.* bill, note, invoice, waybill, bill of parcels
~, железнодорожная waybill, *Am.* bill of lading
~ на груз, транспортная cargo manifest, consigument/delivery note, (motor) waybill, (railroad/railway) bill of lading
~, ордерная транспортная order bill of lading
накладной 1. superimposed, laid/put on, applied; 2. false; artificial
~ное серебро plated silver
~ные расходы *econ.* overhead cost(s)/charges, overheads
накладывание 1. super(im)position, superimposing, laying on; 2. plating, coating
накладывать 1. super(im)pose, put/lay on, lay over; 2. plate, coat, apply; 3. load, pack; pile up; fill, put, lay *(a quantity of)*
~ пластырь plaster
~ повязку apply a landage
~ швы put in stitches
наклевывание *bot.* sprouting
наклеивать paste/stick/glue on, cement
наклейка 1. label, sticker; patch; 2. gluing/pasting on
~ надписей lettering
~, отрывная pull tab
наклейной for pasting/sticking on; pasted/glued on
наклеп wear/strain/cold/hammer/mechanical hardening; cold working; riveting
наклепка 1. riveting, peening; 2. rivet; 3. cold-working
наклепывание riveting, peening; cold-working
наклепывать rivet, peen; hammer, cold-work
наклон inclination, slope, dip, pitch; cant, rake, incline, slant, grade, gradient; tilt(ing); deviation; *geol.* hade
~ аэроснимка tilt
~, боковой side rake
~ в градусах degree of inclination
~ вперед tilt
~ жилы hade of vein
~ кривой slope of a curve
~, куполообразный quaquaversal dip
~, небольшой gentle dip/slope
~ плоскости сброса fault-plane dip
~, поперечный lateral/X-tilt
~, продольный longitudinal/Y-tilt
~ скважины drift of a borehole
~ стрелы крана pitch of a boom
~ характеристики slope of a characteristic
~ шрифта slope of lettering
наклонение I inclination, tilting, obliquity, dip; nutation; intortion
~ видимого горизонта dip/depression of the horizon
~, магнитное magnetic inclination/dip, dip of the needle
наклонение II *gram.* mood
~, повелительно-предположительное imaginary imperative
~, сослагательное subjunctive mood
наклоненный inclined, tilted
~ вперед bent forward
наклонная *n. math.* inclined/oblique line
наклонность 1. inclination, leaning, bent, proneness, propensity, proclivity; tendency; 2. obliquity
наклонный inclined, sloping, oblique, slanted, slanting, tilting; declivate
наклономер tiltmeter, dipmeter, clinometer
наклонять(ся) tilt, incline
наковальня 1. anvil; 2. *anat.* incus (*pl.* incudes)
~, верстачная/настольная bench anvil
~, однорогая beak iron
~, правильная elevating anvil
~ пули bullet anvil
~, ручная (dressing) stake
~ среднего уха *anat.* anvil, incus
наковывать forge/hammer into shape
накожник *zool.* itch mite; external parasite *(Psoroptes)*
накожный *med.* epicutaneous, epidermic, epidermal
~ная болезнь skin disease
накол prick; impression
наколенник 1. *tech.* kneepiece; 2. *hist.* kneecap, kneepiece; 3. kneepad, knee guard
наколка pricking, piercing, puncturing
наконечник *tech.* tip, lug, tag; head, point, cap; clip; pole shoe/terminal; lug; ferrule
~, бронебойный armor-piercing cap
~, буровой drill bit; drilling end
~ зуба рыхлителя *agr.* cutter point
~, кабельный cable shoe/thimble
~, каменный стрелы *archeol.* fairy stone, stone arrowhead
~ клапана valve bonnet
~, костяной bone point
~ провода wire lug/tag
~ сварочной горелки torch tip
~ спицы spoke head
~ стрелы arrowhead
~, шланговый hose nozzle
~ шнура cord terminal
наконечный final, terminal
накопитель 1. *comp.* storage element/circuit, storing device, store; memory; 2. hoarder, money-maker
~, адресный address file
~, винчестерски (дисковый) Winchester
~ вычислительной машины computer storage
~, дисковый disk drive
~ заряда charge integrator
~ информации information storage

~ **на бегущей ленте** streaming-tape drive
~ **на гибких магнитных дисках** floppy drive
~ **на гибких микродискетах** microfloppy-diskette drive
~ **на дисках, касетный** cartridge disk drive
~ **на дискетах** diskette drive
~ **на магнитной ленте, касетный** cartridge tape drive
~**, односторонний** read-only memory, ROM
накопительный storage, reservoir; cumulative
накопление storage, accumulation, pile-up, stockpiling, piling; aggregation, conglomeration
~ **в цифровом виде** digital storage
~ **данных/информации** data/information storage/accumulation
~ **камней, береговое** *geol.* rampart
~**, коническое** conic pile
~ **на магнитной ленте с аналоговой записью** analogue magnetic tape storage
~ **осадков** deposition of sediment
~ **ошибок округления** roundoff accumulation
~**, первоначальное** original/previous/primary/primitive accumulation
~ **погрешности** accumulation of errors
~ **продуктов деления** fission-product buildup, fission-product accumulation
~ **тепла** heat localization/accumulation
накоплять(ся) accumulate, amass; heap/mass up
накос *agr.* mow, amount mowed
накостница *vet.* splint, bony excrescence
накрашиваемость dyeability
накренение wing dropping, banking, rolling
накренивать *naut.* cause to list
накренять(ся) bank, roll (in); list
накрест cross(wise), crossways
~ **лежащий** opposite; *bot.* decussate
~ **лежащие углы** alternate angles
~ **супротивный** *bot.* decussate
накрывать 1. cover (with), lay over; 2. *mil.* hit, destroy
~ **залпом** straddle
~ **на месте преступления** *law* catch redhanded
~ **сводом** vault over
накрывка plastering; finishing/setting coat (of plaster); finish/white/skin coat
накрытие 1. covering; 2. *mil.* straddle
нактоуз *naut.* binnacle
налагать impose; inflict; lay, put
~ **арест на имущество** *law* seize property
~ **епитемью** *rel.* discipline
~ **контрибуцию** impose an indemnity; require an indemnity
~ **резолюцию (на заявление)** endorse an application
наладка adjustment, adjusting, alignment; tuning-up, setting up; setup; debugging; presetting; repairing; arrangement
~ **ЭВМ** (computer) debugging
~ **ЭВМ, автономная** off-line debugging
~ **ЭВМ, комплексная** system debugging
~ **ЭВМ, первичная** preservice debugging
~ **ЭВМ, профилактическая** computer preventive maintenance
~**, эксплуатационная** field adjustment
наладчик *tech.* adjuster, setter-up, troubleshooter
налаженность order, harmony, co-ordination
налаженный adjusted, tuned up, set; fixed, repaired; well organized
налаживать 1. arrange, put in order, set straight/going, organize; *comp.* debug; fix, repair, adjust, regulate; 2. dispose, incline
~ **повторно** readjust
~ **дела** set things straight
налегание superposition, juxtaposition, covering
налегать pressure, bring pressure (on); urge
наледь ice body/mound, icing
~**, висячая** ice overhang
~**, речная** river/stream icing
налет I deposit, film; residue, (powdery) coating (on); bloom, pruinescence; tarnish; cloudy/milky appearance
~ **в горле** *med.* patch
~**, восковой** *bot.* wax coating, vegetable wax, waxy bloom
~**, зубной** dental deposit
~ **на коже** bloom, taint, spew, mud
~ **на коже, жировой** fish oil bloom/spew
~ **на коже, солевой** vegetation of salt
~ **на угле** incrustation
~ **на языке** *med.* fur
~**, пустынный** desert varnish
~ **ржавчины** rust film
~**, сизый** glaucous bloom, glaucousness
~ **солей** salt efflorescence, pellicular salt
~**, твердый** hard rime
~ **шлама** residue of slime
налет II *av.* flying/flight time; raid, attack, strike; foray, incursion
~ **авианосной авиации** carrier strike
~**, безаварийный** accident-free time
~**, воздушный** air attack raid
налетный 1. blown/drifted in; flying in; 2. migratory, passing *(of birds)*
налив 1. impletion; plumping, ripening, swelling *(of fruit);* filling, forming *(of grain);* 2. pouring in, filling; bulk; 3. *vet.* swelling
~ **в бочки** barrelling
~ **зерна** *agr.* grain filling
~ **через рукав** hosing
наливной 1. (for) pouring, (for) filling; for conveying liquids; 2. juicy, ripe
~ **док** wet dock
~**ное судно** tanker
наливом in bulk
налим *zool.* burbot, eelpout *(Lota lota)*
налипание sticking, adherence, adhering; caking
налипать stick, adhere, cling
налицо present, available, on hand
наличие availability, presence; existence, occur-

rence
~ завещания *law* testacy
наличник 1. casing, jambs and lintel of a door/window; platband; stile strip, finishing stile; 2. shoe, liner; pallet, (anvil) plate; escutcheon
~, буксовый *rail.* axle box liner; jaw liner
~, дверной/оконный door/window casing
~, подоконный window apron
наличность *econ.* amount on hand, cash in hand
~, банковская cash at bank
~ в кассе counter/till cash
~ в кассе, денежная cash in hand
~ в расходной кассе vault cash
~, денежная cash holdings
~, золотая gold reserve(s)
~, кассовая/кассы cash balance; cash in safe/hand, balance on hand
~, товарная, ~ товаров в магазине stock in trade
наличный available, on hand; *n. pl.* cash, ready money
~ расчет cash payment
~ расчет, за *econ.* for cash down
~ состав available personnel; *mil.* effectives
~ные деньги cash, ready money
~ные товары goods on hand
налобник 1. headrest; head band; 2. frontlet, headstall
налог *econ.* tax, duty
~, аккордный lump-sum tax
~, взимаемый по единой ставке proportional tax, flat-rate tax
~, добвочный/дополнительный surtax, surcharge, supertax
~ за пользование use tax
~, коммунальный/местный rate
~, косвеный indirect tax
~ на все виды собственности general property tax
~ на дарения gift tax
~ на добавленную стоимость value-added tax
~ на заработную плату wage tax
~ на имущество wealth tax
~ на капитал capital levy
~ на наследство inheritance tax/duty *Am.* legacy; death tax
~ на недвижимое имущество (real) estate tax
~ на нераспределенную прибыль *law* undistributed profit tax
~ на покупки purchase tax
~ на потребление consumption tax
~ на предметы роскоши sumptuary tax
~ на продажу предметов роскоши luxury tax
~ на сверхприбыль excess profits duty/levy
~ на стоимость ad valorem tax
~ на увеличение рыночной стоимости капитала capital-gains tax
~, натуральный tax in kind
~, подоходный income tax
~, поголовный/подушный head/poll/capitation tax; *hist.* capitation
~, поземельный land-tax
~, поимущественный property levy/tax
~, пропорциональный flat-rate tax, proportional tax
~ с дифференцированной ставкой graded/graduated tax
~ с доходов компании company income tax
~ с избирателей, подушный poll tax
~ с оборота turnover/receipts tax
~, уравнительный equalization tax
~и, коммунальные и государственные rates and taxes
~и на наследство estate/death duties
налогонеплательщик tax dogder
налогообложение taxation
~, уравнительное egalitarian taxation
налогоплательщик taxpayer, rate payer
налогоспособность taxable capacity
наложение 1. superposition, superimposing, imposition; apposition, juxtaposition; application, laying on, overlap(ping); *comp.* overlay(ing); *math.* covering; 2. infliction
~ ареста *law* seizure, attachment
~ бандажа bandaging
~ запрета *hist.* intercession
~ импульсов pulse pile-up
~ лигатур на сосуды *med.* ligation
~ маски *comp.* masking
~ новой беговой дорожки recapping of tyre
~ первичной повязки first-aid dressing
~ проб *geol.* application of samples
~ противоестественного ануса *med.* proctotoreusis
~ резонансов resonance overlap
~ течений superposition of flow patterns, flow superposition
~ швов *med.* suturation, suturing, stitching
~ шины splinting
~ щипцов *med.* application of forceps
наложенный superimposed, overlaying
~ платеж pay/cash on delivery, P./C.O.D.
наложимость applicability, superposition, covering
наложимый *math.* applicable, superposable
наложить superimpose, lay (on/upon)
~ эмбарго lay embargo
наложница concubine; hetaera
наложничество cohabitation, concubinage
налокотник elbow guard/cushion; *hist.* cubitiere
намагничение magnetization, magnetizing
намагниченность (intensity of) magnetization
~, результирующая net magnetization
~, собственная intrinsic magnetization
намагничиваемость magnetizability
намагничивание magnetizing, magnetization
~ до насыщения saturation magnetization
~, мгновенное slash magnetization

~, остаточное residual/remanent magnetization, remanence
~, плоское lamellar magnetization
~, поперечное transverse/perpenducular/ *naut.* athwartship magnetization
намагничивать magnetize
намаз *rel.* (Mohammedan) prayer
намазка 1. filler; coat; 2. smearing, spreading
~ клеем *leath.* solutioning
намазной coating, pasting, spreading
намазывание coating, covering (with); pasting, smearing
намазывать smear, coat, cover, paste, daub
намазь *leath.* depilatory paint
наматывание winding, reeling, coiling, spooling; rolling up
наматывать reel (up), coil
намачивание soaking, steeping, wetting, moistening
~, предварительное presoaking
намачивать soak, steep, moisten, wet
намащивать anoint, rub with balm/balsam
намеливать rub/whiten with chalk
намереваться intend, mean, contemplate
намерение intent(ion), design; purpose; contemplation
~, возыметь form the intention
~нием, с заранее обдуманным deliberately; *law* of malice prepense
намеренный intentional, deliberate, premeditated; purposive; *law* wilful
наместник 1. *hist.* governor general, vicegerent; 2. *eccl.* vicar; superior
наместнический 1. vicegeneral, vicegerent; 2. *eccl.* vicarial, vicarian
наместничество 1. *hist.* vicegerency; region ruled by governor general; 2. *eccl.* vicariate
намет *constr.* coat(ing) layer; plaster coat
~, второй brown plaster coat
~, накрывочный finish plaster coat
~, первый scratch plaster coat
~ штукатурки plaster coat
наметка I first/preliminary outline, rough draft
наметка II *text.* basting (thread), tacking (thread)
~, направляющая top basting
наметка III *naut.* sounding pole/rod, mast clamp/ hasp; wrapper
~, размечанная calibrated rod
наметка IV casting net
наметка V trunnion cap
наметывать I 1. outline, make a rough draft; 2. throw together, pile up
~ икры spawn
наметывать II *text.* tack, baste
намечать I 1. plan, project, have in view; contemplate; 2. select, nominate
~ план shape a plan
намечать II (make/set a) mark; mark/lay out
наминка gall, sore spot
намокаемость water absorption
намол *agr.* flour/milling yield, amount of grain ground
намолот *agr.* threshing yield, amount of threshed grain
~ зерна grain output/yield
намордник muzzle; noseband *(of a harness)*
~, надеть muzzle
намотанный wound, coiled
~ на ребро edgewise-wound
~ по шаблону form-wound
намотка winding, spooling, coiling
~ нити filament winding
намоточный winding, spooling; coiling; reeling
намушник *mil.* (fore)sight cover
намыв 1. *geol.* alluvium, aggradation, shore deposition, inwash; 2. panning
~ грунта hydraulic deposition of soil
~ плотины silting of a dam
~ русла aggradation of stream channel
намывание aggradation, deposition, alluviation, wash(ing)
~ почвы warping of land
намывать 1. alluviate, aggrade, wash in/ashore, deposit; 2. pan out
намывной 1. *geol.* alluvial, washed-in, wavebuilt, water-borne; leveed; 2. obtained by panning
нанаец, нанайский Nanaian
нанайка Nanaian (woman)
нанду *zool.* rhea, nandu *(Rhea)*
нанесение 1. deposition, application, applying; 2. heaping, drifting; 3. plotting, marking, drawing, charting; tracing; 4. strike, attack, infliction
~ газопламенным способом *met.* flame plating
~, гальваническое electroplating
~ горизонталей contouring
~ данных на карту plotting
~ делений graduation
~ краски на печатную форму inking
~ на график plotting
~ на карту/планшет plotting, charting, mapping
~ опорных пунктов plotting of controls
~ фольги foil cladding
нанизм *biol., bot.* nanism; dwarf(ish)ness, dwarfism
~, надпочечниковый *med.* adrenal nanism
нанизывание stringing, threading
нанизывать string, thread
наниматель 1. leaseholder, lessee, leaser, tenant; 2. hirer, employer
~ судна affreighter
нанимать 1. employ, hire, engage; 2. rent
~ рабочих take on workers
нанка *text.* nankeen *(cloth)*
нанкин Nanking china, Nankeen porcelain
нанноропс *bot.* mazari palm, *(Nannorrhops)*
наноампер nanoampere
нанопланктон *biol.* nanoplankton
нанорельеф nanorelief
нанос accretion, alluvion, accumulation; drift

~, снежный snowdrift
наносекунда nanosecond
наносить 1. pile/heap up, deposit; wash in, drift; 2. coat, apply; 3. dash (against); inflict; 4. mark (down); plot, draw, chart; superimpose
~ деления graduate, make divisions on a scale
~ намазь на кожу paint, paste
~ поражение inflict defeat
~ удар strike, deliver a blow
наносный *geol.* alluvial, aggradational
наносотранспортирующий silt-carrying, sediment-transporting
наносоулавливающий sediment-detention
наносоуловитель silt/sediment trap
наносы *geol., pl.* detrital deposits, detritus, overburden
~, аллювиальные alluvium, alluvial fan
~, ветровые wind-deposited matter
~, взвешенные (в потоке) turbid matter
~, галечные rubbish detritus, pebble beds
~, донные bed load, settled sediments
~, ледниковые diluvium, glacial drift
~, новые top-set beds
~, остаточные eluvium, residual sediments
~, речные stream-borne material, river drift/load/burden
нанофанерофит *bot.* nanophanerophyte
нанофарада nanofarad
наноцефал nanocephalus
наноцефалия *med.* nanocephalia, nanocephaly
нансук *text.* nainsook
наоборот inside out, the wrong side out; the wrong/other way round; per contra
наоткос aslant, slantwise, obliquely
наотрез point-blank, bluntly
нападать attack, assault; fall (upon); infest; assail
нападающий *n. sport* forward; *av.* attacker
~, левый/правый крайний outside left/right
нападение 1. attack, assault; aggression, invasion; 2. *sport* forwards, forward line; 3. infestation
~ с воздуха air attack/assault
~ния, центр centre forward
напаивать I solder (on/to); fuse (on); braze (to)
напаивать II water *(an animal),* give to drink
напайка soldering (on); welding on; brazing (to); building up
напалм napalm
наперсный *eccl.* pectoral
наперсток thimble
наперстянка *bot.* foxglove, digitalis *(Digitalis)*
напильник *tech.* file
~, бархатный superfine/supersmooth/barette file
~, вращающийся, дисковый circular file
~, грубый/драчевый bastard/coarse file
~, личной smooth(-cut) file
~, ножовочный knife/hack file
~, полукруглый half-round file
~, полуличной second-cut file
~, ромбический lozenge file
~ с рашпильной насечкой rasp-cut file
~, трехгранный three-square file
напильник-брусовка rough-cut file, coarse file
напиток beverage, drink
~, газированный effervescent beverage
~, крепкий hard liquor
~, прохладительный soft drink
наплав *sport* float, cork *(angling)*
наплавка 1. welding on, cladding, surfacing, build(ing)-up; hard-facing, overlaying; 2. beading, weld seam
~ бронзой bronze surfacing
~, газовая oxyacetylene surfacing, gas-torch surfacing
~ напылением hard-facing by spraying
~, плазменная plasma-jet hard-facing
~ электрической дугой arc deposition
наплавление building up
наплавлять weld/fuse on, overlay, surface, build up; hard-face
наплавной floating
~ мост floating/pontoon bridge
напластование *geol.* bedding, stratification, deposition, superposition, lamination
~ горных пород rock stratification
~, дюнное dune bedding
~, скрытое blind bedding
~, согласное concordant/conformable bedding
~, тонкое lamination, foliation
~, угловое несогласное clino-unconformity
напластованный *geol.* stratified, bedded
~ диагонально cross-/false-bedded
напластовываться superpose, stratify
наплечник 1. shoulder strap; epaulet(te); 2. *sport* shoulder guard, bracer; 3. *hist.* brassart, brassard
наплыв 1. influx, abundance; inundation; 2. *bot.* wart, excrescence, burr, burl, callus; 3. *telev., cin.* dissolve (view), fade over; 4. defect, sag; roll
~, сталактитообразный congelation
~ термитного шва collar
наплывать run/dash against; *telev.* dissolve, fade up and down
наполеодор napoleon *(gold coin)*
наполеонизм Napoleonism
наполеоновский Napoleonic
наполнение filling; charge, charging; inflation; packing
~ доверху topping up
~ дубителем *leath.* surtannage
~, максимальное full admission
~, смысловое meaningfulness
наполненность saturation, fullness, completeness
наполненный full, filled (out), replete; inflated
наполнитель *tech.* 1. filler, filling agent/material; additive; 2. feeder, charger, filling machine
~, активный/инертный active/inert filler
~, армирующий reinforcing filler
~, волокнистый fibre filler
~, инертный inert filler, excipient
~ удобрений fertilizer additive
наполнительный filling, feeding; charging

наполнять fill, stuff; inflate; charge; suffuse
напольный I outdoor; ground-type
напольный II (standing/situated on the) floor
напор (water) head; pressure; compression; (in)rush; thrust
~, геометрический elevation head
~, гидравлический hydraulic (pressure) head
~ на всасывании/отрицательный suction head
~, общий total/gross head
~, полезный/нетто effective/net head
~, скоростной velocity/kinetic head, dynamic pressure
напорный pressure *(attr.)*
направительный directing, directional, directive, guiding; referential
направление 1. direction; course, heading, route; sense; (compass) bearing; 2. trend, tendency; tenor, way; 3. warrant, order; 4. *mil.* sector
~, азимутальное azimuth bearing
~ вдоль хорды chordwise direction
~ вектора sense of a vector
~ волокна (металла) grain flow
~ движения traffic direction/route
~ делу, дать *law* take action on a matter
~ и величина сноса set and drift
~ истинного ветра true wind angle
~ кажущегося ветра relative wind angle
~ колебаний direction of vibration
~, кристаллографическое crystallographic line
~, литературное literary school
~ на работу, дать assign to a job
~ надвига *geol.* direction of thrust
~ наибольшего сокращения least strain axis
~, начальное *geod.* base/initial/reference/original direction
~, небрэгговское non-Bragg direction
~ обхода *math.* sense of rotation, direction of traversal; *el.* direction of summation
~ осей координат arrangement of axes
~ откуда дует ветер *naut.* wind's eye
~ падения *geol.* direction/line of hade; incident direction
~ падения, общее *geol.* general/average hade
~ пласта trend
~ плоскостей напластования bedding/stratification plane
~ по часовой стрелке clockwise direction
~ по штопору pro-spin direction
~ подлета flying-on course
~ посадки *av.* runway heading, landing direction
~ потока stream course
~ приливо-отливного течения set of tide
~ простирания strike (line), direction of strike, trend
~ против штопора antispin direction
~ развертки, левое/правое reverse/normal direction of scanning
~ раскола *geol.* cleat, cleave, reed
~ распространения line of propagation
~ связи route, routing
~ связи, запасное sidetrack
~ связи, обходное alternative trunking, sidetrack(ing), by-pass routing
~ скольжения glide direction, direction of slip
~ ума turn of mind
~ штриховки orientation of stria
направленность 1. directivity, directionality, directional effect; trend, tendency; 2. purposefulness
направленный directed, directional, directive; oriented, headed; orientational
~ в одну сторону unidirectional
~ внутрь inward
~ к верхушке acroscopic
~ наружу outward
~ обратно antagonistic
направлять 1. direct (to/at), guide, aim, level (at); head (toward); 2. send; route; refer; 3. sharpen, strop
~ бритву sharpen/strop a razor
~ на работу assign/send to work
~ ошибочно misroute
направляющая 1. *tech.* guide (patch), rail, runner, slide (way); track; 2. *math.* directrix
~, буксовая *rail.* axle(-box), guide
~ гусеницы track skid
~ закрылков *av.* flap rail/guide/track
~ крейцкопфа crosshead guide
~ ласточкин хвост wedge slide
~ предкрылка *av.* slat rail/guide/track
~, призматическая vee-slide
~ сидения *auto.* seat slide/rail
направляющий guiding, guide, directing, leading; aligning
~ валик/ролик guide roller
напрасный 1. unavailing, ineffective; vain, idle; 2. false, unjust, wrongful, unbased
напрокат for/on hire, for rent
~, брать hire
~, дать hire/rent out
напряжение 1. *tech.* stress, strain; *el.* tension; voltage, potential; 2. effort, exertion, intenseness
~, акустическое acoustic intensity
~, амплитудное peak voltage
~, включать energize
~ возбуждения driving/excitation/exciting voltage
~ волочения draw stress
~ вольтдобавки *telev.* boost voltage
~ впадины valley voltage
~, входное input voltage
~, высокочастотное radio-frequency voltage
~, выпрямленное rectified voltage
~, выходное output voltage
~ гашения blanking voltage
~ двойникования twinning stress
~, деформирующее distortional/deformational stress/strain
~, добавочное boost(ing) voltage

~, допустимое allowable/admissible stress
~, дробящее crushing stress
~ зажигания firing voltage/potential; critical grid voltage
~, закалочное hardening/quenching/cooling stress
~, зарядное charging voltage
~, знакопеременное alternate/alternating stress
~, запирающее cut-off voltage, blanking voltage, black-out voltage
~, изгибающее bending stress/force
~, индуцированное induced stress/voltage
~ искрения sparking/arcing voltage
~, касательное tangent(ial)/shear(ing) stress
~, кольцевое *naut.* hoop/circumferential stress
~, коммутирующее switching voltage
~, контактное contact stress/voltage
~ короткого замыкания short-circuit voltage
~, критическое critical/buckling/ultimate/breaking stress
~ кручения torsional stress
~, линейное *el.* line voltage; *mech.* linear stress
~, ломающее breaking stress
~, магнитное magnetic difference of potential, m.d.p.; magnetic stress
~, механическое pressure, stress, tension
~, модулирующее pulsing voltage
~ на обходе loop voltage
~ накала filament/heater voltage
~ накачки pumping voltage
~ насыщения saturation voltage
~, начальное initial stress/voltage
~, невращательное irrotational stress
~, неустановившееся transient voltage
~, номинальное voltage rating, rated/nominal voltage
~ обратного зажигания fire-back voltage
~, обратное reverse/inverse voltage
~, объемное volumetric stress
~, однородное homogeneous stress
~, окружное hoop/tangential stress
~, опорное standard/reference/comparison voltage
~, осесимметричное axisymmetrical stress
~, остаточное residual voltage/stress
~, отклоняющее deflection/deflecting voltage
~, отпирающее starting/trigger voltage
~ отражателя repeller voltage
~ отсечки cut-off voltage
~ переброса turnover voltage
~ перекрытия изоляции flashover voltage
~ переменного тока alternating/a.c. voltage
~, переменное alternating stress/voltage
~, периодически действующее intermittent stress
~ пика/пиковое peak point voltage
~, пилообразное sawtooth voltage
~, питающее supply voltage
~, поверхностное surface stress
~, повышенное over-voltage
~ погасания extxinction potential/voltage
~ под действием центробежных сил centrifugal stress
~ под нагрузкой load stress
~ ползучести creep stress
~, полное 1. *el.* total voltage; 2. *mech.* combined/compound/composite stress
~, поперечное transverse strain/stress
~, полное касательное shear(ing) stress
~, пониженное under-voltage
~, пороговое threshold voltage/potential
~ постоянного тока direct/d.c. voltage
~, постоянное constant voltage/stress; direct (-current) voltage
~, предварительное prestressing
~, предельно допустимое stressing limit
~, предельное limit/breaking stress; ultimate stress/strength
~ при изгибе bending/flexural stress
~ при кручении torsional/twisting stress
~ при перегибе *naut.* hogging stress
~ при прогибе *naut.* sagging stress
~ при разрыве rupture stress
~ при растяжении tensile stress
~ при сдвиге shear(ing) stress
~ при сжатии compressive/compressing/compression stress
~ при скручивании torsional/twisting stress
~ при слеминге slamming stress
~ при срезе shearing stress
~ при ударе impact stress
~, приливное tidal strain
~, пробивное puncture/breakdown/disruptive voltage
~ пробоя breakdown voltage
~ прокола punch-through voltage
~, псофометрическое psophometric voltage
~, пусковое starting/trigger voltage
~, рабочее operating/running/working voltage, working/operating stress
~ развертки sweep/scanning voltage, time-base voltage
~, раздробляющее crushing strain
~, разрушающее ultimate/breaking/failure/crushing stress/strain
~, разрывное rupture/tearing stress
~, растягивающее/растяжения tensile/tension stress
~, расчетное design/computed stress
~, реактивное reactance/reactive voltage
~ сбивки нуля anti-stickoff voltage
~, сверхвысокое supervoltage
~ сдвига shear(ing) stress
~, сдвинутое по фазе out-of-phase voltage
~ сети mains voltage, *Am.* (supply-)line voltage
~ сжатия compressing/compression/compressive stress
~ сигнала ошибки error voltage
~ силы тяжести intensity of gravity
~, синфазное common-mode voltage
~, скалывающее shear(ing)/cleavage stress

~, скручивающее torsional/twisting stress
~, сложное combined stress
~, смещающее dislocation/shifting stress; bias(s)ing voltage
~ смещения bias voltage
~ смыкания punch-through voltage
~, снимать deenergize
~, собственное inherent/initial stress, locked-in stress
~ среза/срезывающее shear(ing) stress, scission strain/stress
~ сцепления bond stress
~, тангенциальное tangential stress
~, температурное/тепловое heat/temperature/thermal stress
~, термоупругое thermoelastic stress
~, ударное shock/impact stress
~, удельное specific stress
~, управляющее control/driving voltage
~, упругое elastic stress/strain
~, усадочное shrink(age) stress
~, усталостное fatigue stress
~ устранения ложного нуля anti-stickoff voltage
~ холостого хода on-loan voltage, opencircuit voltage
~, хронирующее timing voltage
~, циклическое cyclic(al) stress
~, чрезмерное overstrain, overstress, excessive strain
~ шумов noise/hum voltage
~ электронного луча beam voltage
~, электрострикционное piezoelectric stress
~, эффективное effective/r. m. s. voltage
напряженность *el.* intensity, tension; *mech.* strain, stress (level); (in)tenseness
~ магнитного поля Земли intensity of terrestrial magnetism, geomagnetic field intensity
~ поля field intensity/stregth
~ трафика density/rate of traffic
~ электрического поля electric field strength, electric intensity
напряженный intense, intensive; tense, strained, stressed
~ гласный narrow vowel
~, предварительно prestressed
напуск 1. *tech.* lap(ping), overlap(ping); bleeding-/leak-in; *agr.* flooding, irrigation; 2. filling (with), letting in, admitting; 3. raid, foray; *hunting* letting loose, slipping *(from leash)*; 4. *arch.* projection, jutty; awning
напускать fill (with), let in, admit; let loose, set on
напутствие parting words, farewell speech
напыление (deposition by) spraying, sputtering, deposition
~, вакуумное vacuum deposition
~, газопламенное flame spraying/plating
~, геттерное getter sputtering
~, катодное cathode sputtering
~, лазерное laser spraying
~ металла metal spraying
~, плазменное plasma (jet) spraying
~, электродуговое (electric-)arc deposition/spraying
напыливать spray, dust; spray-coat, sputter on
нара *bot.* narras *(Acanthosicyos horrida)*
наработка (nonfailure) operating time; *av.* run(ning) time; *comp.* lifelength
~, гарантийная warranty life
~ двигателя operating/running time of engine
~ до капитального ремонта overhaul life
~ до первого отказа, средняя *econ.* mean-time-to-first-failure, MTFF
~ на отказ mean-time-between-failures, MTBF, failure interval, mean-cycles-between-failures, MCBF
~ на стенде bench running
~, суммарная accumulated operating time
наральник *agr.* (cultivator) point, furrow opener
~ сошника сеялки drill coulter point, drill furrow-opener point
нарамник *eccl.* omophorion; amice
нарастание build(ing)-up, growth, increase, rise, accumulation, accretion; increment, intensification
~ в толщину thickening
~, вторичное secondary growth
~ коралла gemmation; budding
~ кристалла intumescence of crystal
нарастать build up, grow, increase; rise, swell, accrete, accumulate, accrue
нарастающий growing, increasing, rising, mounting; positive-going; increscent; *physiol.* auxotonic
~ верхушкой *bot.* acrogenous
наращение 1. growth, increase, growing, accumulation, augmentation, accretion, building up; *math.* increment; 2. *gram.* augment
~, отрицательное negative increment
~ суши *geol.* accretion of land
наращивание increase, growing, building up; *tech.* joint; *med.* grafting
~ вооружений armaments build-up
~ хромированием electroplating with chromium
наращивать 1. splice *(of cable)*; 2. extend, advance, lengthen; *tech.* plate; graft (on); 4. grow, build up; increase; accumulate, augment; intensify
нарвал *zool.* narwhal(e), unicorn fish *(Monodon monoceros)*
наргиле narg(h)ile, nargile(h), hooka(h)
нард *bot.* citronella grass *(Cymbopogon nardus)*; 2. spikenard, matweed, nard *(Nardus stricta)*; 3. spikenard/nard ointment
нардовый nardine
нарез 1. incision, cut(ing); notch, groove; thread *(of screw)*; *pl.* rifling grooves; 2. *hist.* plot of land
нарезание cutting, threading; tapping, rifling, grooving; toothing
~ пазов grooving
~ резьбы фрезой thread-hobbing
~, черновое rough cutting

нарезать 1. cut (into pieces), slice; *tech.* thread; rifle; 2. *hist.* allot, parcel out *(land)*
~ борозды furrow
~ выработку *min.* cut an entry
~ зубья cog, tooth
~ кристалл dice a crystal
нарезка 1. *tech. (*screw) thread, worm; cut(ting), threading, rifling; slicing; 2. plot, lot *(of land)*
~ болта male thread
~, винтовая screw thread
~, внутренняя female thread
~, глубокая deep-cut rifling
~ карт на сборном листе *cart.* division into sheets
~, левая left-handed rifling
~ мерей *leath.* embossing, printing
~, наружная male screw
~ ниш *min.* formation of stables
нарезной threaded, rifled, chased, cut; notched
~ ствол rifled barrel
наречие I dialect
наречие II *gram.* adverb
~, необстоятельственное qualitative adverb
~ образа действия adverb of manner
~, присловное word-modifying adverb
~ следствия illative adverb
~, счетное numeral adverb
наречно-обстоятельственный adverbial-circumstantial
нарзан Narzan, kind of mineral water
нарицательный appellative; *econ.* nominal
~ная стоимость nominal cost
~ное имя common noun
Наркисс *myth.* Narcissus
наркоген *chem.* trethylene
наркогипния *med.* narcohypnia
наркоз *med.* an(a)esthesia, narcosis; 2. drug, an(a)esthetic
~, внутривенный intravenous narcosis, infusional an(a)esthesia, phlebonarcosis, phleboan(a)esthesia
~, внутрикишечный rectal an(a)esthesia
~, газовый gas/vapo(u)r an(a)esthesia
~, общий general/stupor/surgical an(a)esthesia
~, смешанный mixed/crossed an(a)esthesia
~, фторотановый fluoroxene an(a)esthesia
~, хлороформный chloroform/trichloromethane an(a)esthesia
~, эфирный ether an(a)esthesia
нарколепсия *med.* sleep epilepsy, narcolepsy
нарколептический *med.* narcoleptic
нарколог expert in narcology
наркология narcology, addictology
нарком *abbr.* **народный комиссар** *hist.* people's commissar
наркоман narcomaniac, (drug) addict
наркомания *med.* narcomania, narcotism, drug addiction
~, алкогольная alcoholism
~, гашишная hashishism
~, морфинная morphinism
Наркомат *abbr.* **Народный комиссариат** *hist.* people's commissariat
Наркомвнешторг *abbr.* **Народный комиссариат внешней торговли** People's Commissariat of Foreign Trade
Наркомвнудел *abbr.* **Народный комиссариат внутренних дел** People's Commissariat of Internal Affairs
Наркомздрав *abbr.* **Народный комиссариат здравоохранения** People's Commissariat of Public Health
Наркомзем *abbr.* **Народный комиссариат земледелия** People's Commissariat of Agriculture
Наркоминдел *abbr.* **Народный комиссариат иностранных дел** People's Commissariat of Foreign Affairs
Наркомпрос *abbr.* **Народный комиссариат просвещения** People's Commissariat of Education
Наркомюст *abbr.* **Народный комиссариат юстиции** People's Commissariat of Justice
наркотизатор an(a)esthesist
наркотизировать narcotize, an(a)esthetize
наркотик 1. narcotic, dope; an(a)esthetic; 2. drug addict
наркотикораздражающий narcotico-irritant
наркотин *chem.* narcotina, narcotine
наркотический narcotic
~кое средство narcotic
наробраз *abbr.* **отдел народного образования** department of education
народ nation; people, folk
~, черный *hist.* peasants and artisans
народ-богоносец *rel.* the Chosen People
народник *hist.* populist, narodnik
народничать *hist.* propagate/preach populism
народничество *hist.* populism, narodnik movement
народно-демократический people's democratic
народно-песенный folk-song
народно-разговорный popular-colloquial
народность 1. nationality; nation, people; 2. national/folk character
народнохозяйственный national-economic
народноэтимологический folk-etymological
народноязыковый popular-language
народный 1. national; 2. of the people, popular; public; demotic; 3. folk
~ заседатель juriman
~ комиссар *hist.* people's commissar
~ комиссариат *hist.* people's commissariat
~ обычай folk custom
~ следователь examining magistrate
~ суд People's Court
~ судья judge in People's Court
~ фронт Popular/National Front
~ная воля *hist.* Narodnaya Volya (The People's Will)

народовед ethnographer; ethnologist
народоведение ethnography, ethnology
народоведческий ethnographic(al); ethnologic(al)
народовластие sovereignty of the people, democracy
народоволец *hist.* member of the movement Narodnaya Volya
народоволь(че)ство *hist.* movement of Narodnaya Volya
народонаселение population
народоописание demography
народоправие sovereignty of the people, democracy
народоправство *hist.* democracy
нарождение nascence, nascency; birth
~ месяца (appearance of) new moon
нарост 1. *tech.* (erosion) scab, cut, wash; sand buckle; scale, incrustation; 2. excrescence, knot, wart; node, tubercle, tumo(u)r, (out)growth, burr, knar; *bot.* struma
~, костный exostosis
~ на резце built-up edge
~, свилеватый knag
~, смоляной gummy excrescence
~, твердый incrustation
~, шлаковый slag crust
наросток *bot.* epiblast
нарочный express/special messenger; courier
~, конный estafette
нарпит *abbr.* **народное питание** *hist.* state-supported eating facilities
нарсуд *abbr.* **народный суд** People's Court
нартекс *arch.* narthex
нартеций *bot.* bog asphodel (*Narthecium)*
нарты *pl.* reindeer/dogs sledge
нарубка 1. *tech.* cut, notch incision; 2. chopping, cutting
наружное *n., med.* (medicine) for external use
наружнораковинные *n. pl. zool.* ectocochlians (*Ectocochlia)*
наружность outward appearance/aspect; exterior
наружный exterior, external, outside, outer, outdoor, outward, distal; surface; extraneous, extrinsic
~ белок perisperm
~ паразит external parasite, ectoparasite
~ная оболочка outer cover, integument
нарукавник oversleeve; armlet; cuff
нарукавный (worn on the) sleeve
~ная повязка arm band
наручи *pl. hist.* vambrace, vantbrace
наручники *pl.* handcuffs, manacles
нарушать 1. break (down), disturb; upset, disrupt; *geol.* dislocate; 2. infringe, violate, transgress; infract; interfere
~ договор break the agreement
~ присягу violate/break one's oath
нарушение break (down), violation, fault, infringement, infraction, delinquency, dereliction, nonobservance; perturbation, disturbance, disruption, breach; irregularity; *geol.* dislocation, diastrophism
~ авторских прав на программное обеспечение software piracy
~ балансировки unbalance, unbalancing
~ герметичности leakage, loss of sealing
~ дисциплины misconduct
~ доверенным лицом *law* break of trust
~ договора breach of contract
~, дорудное *geol.* pre-mineral/-ore deformation/dislocation
~ земной коры crustal disturbance
~ контакта contact fault
~ логики paralogism
~ международных правил breach of international rules
~ морских сообщений disorganization of shipping, commerce raiding
~ нормального порядка inversion
~ общественного порядка *law* breach of peace
~ питания nutritional disturbance
~ пород *geol.* rock failure
~ последовательности *comp.* desequencing
~ пускового режима start-up accident
~ работоспособности malfunction
~ работы operating trouble
~ равновесия unbalance, unbalancing
~, радиационное radiation damage
~ радиосвязи blackout
~ регулировки upsetting
~, сбросовое *geol.* faulting, fault dislocation
~ слуха hearing disorder
~, случайное occasional disturbance; *geol.* chance dislocation
~, структурное structural disturbance/failure
~ структуры disarrangement of the structure
~ целости disintegration
~ эксплуатации breakdown of service, operating trouble
нарушитель transgressor, trespasser, infringer; disturber; infractor
~ общественного порядка *law* nuisancer
нарцеин *chem.* narcein(e)
нарцисс 1. *bot.* daffodil, narcissus (*Narcissus)*; 2. *myth.* Narcissus
нарциссизм *med.* narcism, narcissism, autophilia
нарциссин narcissine
нарциссовые *n. pl. bot.* amaryllis family (*Amaryllidaceae)*
нарциссоцветковый *bot.* narcissiflorous
нары *pl.* plank bed, bunk
~, двухъярусные double-deck bunk
нарыв *med.* abscess, aposteme; boil
нарывание *med.* blistering, apostemation
нарывник *zool.* blister beetle (*Meloidae)*
нарывной *med.* blistering, epispastic, vesicatory
~ пластырь blistering/drawing plaster
нарывный abscessing
наряд I dress; costume, attire; finery
наряд II 1. order, warrant; 2. *mil.* (duty) detail,

squad; duty
~, **караульный** guard detail
~ **милиции** squad of policemen
~ **на работу** work order; fatigue duty
~, **рабочий** work card
~, **суточный** day's duty
наряд III *hist.* artillery
наряд-заказ order
~, **заводской** factory/shop order
~ **на выполнение услуг** service order
нарядность smartness, elegance
нарядный 1. smart, elegant, dressy; 2. (of the) duty
насадитель *hist.* propagator; advocate; inculcator
насадка 1. extension (piece); 2. *tech.* head, nozzle, mouthpiece, cap(ping), adjustage; 3. *chem.* filling, packing, checker (work); 4. putting/setting/fitting on, building up, built-up part; attachment, adapter
~, **вставная** inserted nozzle
~, **дистилляционная** distilling tube
~ **из ионообменной смолы** resin bed
~ **из колец Рашига** pack with Raschig rings
~, **кольцевая** ring packing
~, **многоходовая** multi-pass checker
~, **направляющая/Корта** shroud(ing), accelerating/Kort nozzle, thrust-augmenting nozzle
~, **фотографическая** camera attachment
насадной, насадный put/fitted on, mounted
насадок nozzle; probe, head, orifice, spout
~ **Борда** Borda mouthpiece
~, **воздухоприемный** intake spout, air scoop
~ **выпускной трубы** *auto.* exhaust extension
~ **карбюратора** *auto.* trumpet, *Am.* stack
~ **Лангмюра** Langmuir probe
~, **малогабаритный** miniature probe
~, **расширяющийся** divergent nozzle
~, **реактивный** propelling nozzle
~, **самоориентирующийся** swivel(l)ing tube
~, **эжекторный** ejector shroud
насадочный put/fitted on
~**ная линза** supplementary lens
насаждение plantation, planting; (*forest*) stand; wood
~, **ветроломное** windbreak planting
~, **кустарниковое** shrubbery stand
~, **материнское** shelterwood
~, **перестойное** overmatured/declining stand
насаживание mounting, setting, fixing; hafting, putting on
насаживать fit/put/set on
~ **бандаж на колесо** set a tyre
насаливание salting
наседка *agr.* brood hen
насека notch, incision
насекать 1. make incisions (on/in); incise, notch; 2. damascene, damask; emboss
насекомое *n.* insect
~, **бескрылое** wingless insect (*Apterygota*)
~, **крылатое** winged insect (*Pterygota*)
~, **сетчатокрылое** neuropterous insect
~, **ядовитое** venomous insect
насекомообразный insectiform
насекомоопыление *bot.* insect pollination, entomophily, entomogamy
насекомоопыляемый *bot.* insect-pollinated, entomophilous
насекомоуловитель insect trap
насекомоядный *bot., zool.* insectivorous, entomophagous; *n. pl.* insectivores (*Insectivora*)
насекомые-переносчики *pl.* transmitting insects, insects vectors
население 1. population; inhabitants; 2. settling, peopling
~, **гражданское** civilians
~, **коренное** aboriginal/autochthonous/native/indigenous population
~, **промышленное** manufacturing population
~, **рыбное** ichthyofauna
~, **самодеятельное** active population
~ **старших возрастов** aged population
~ **трудоспособного возраста** working age population, able-bodied citizens
~, **фиктивное** ghost population
~, **экономически активное** gainfully occupied population
населенность (density of) population; occupancy
~, **избыточная** *el.* excess population
~, **накапливать инверсную** build up a population inversion
населенный (densely) populated, inhabited; populous
~ **пункт** settlement, populated/ *mil.* inhabited area
насельник, насельщик inhabitant
населять populate, settle
насечка 1. cutting (of file teeth); notching, denting; 2. cut, incision; file teeth; groove, slot; knurl; hatch
~, **бархатная** superfine/supersmooth cut
~, **брусовая** rough/coarse cut
~, **драчевая/крупная** bastard cut
~, **личная** smooth cut
~, **рашпильная** rasp cut
насилие 1. violence, force, coercion, constraint; 2. rape
насилование 1. coercion, constraint; 2. rape
насиловать 1. force, coerce, constrain; 2. rape
насильник 1. oppressor, user of force; tyrant; aggressor; 2. rapist; violator
насильнический violent, tyrannical, forcible
насильный, насильственный violent, forced, forcible, invasive
наскальный 1. *bot.* saxatile; rupicoline, rupicolous; 2. *hist.* pertroglyphic, rupestrian
~**ное растение** rupicolous plant
наслаивание stratification, stratifying; super(im)position; lamination; overlapping
наслаивать stratify, layer; laminate; overlap
наследие heritage; legacy
наследник *law* heir, legatee, inheritor; successor
~, **возможный/предполагаемый** heir pre-

sumptive
~, **кровный** heir of the blood
~, **непосредственный/прямой/первый** heir apparent
~ **по боковой линии** heir collateral
~ **по договору** heir conventional
~ **по завещанию** legatee, testamentary heir, heir by devise
~ **по закону** heir general
~ **по нисходящей линии** heir of the body
~ **по прямой нисходящей линии** natural heir
~, **подназначенный** expectant/substitute heir
~, **прямой** apparent/lineal heir
наследница *law* heiress, inheritress; successor
наследный hereditary
~ **принц** crown prince
наследование inheritance, succession; heredity; inhereting
~, **комплексное** complex inheritance
~ **по завещанию** *law* testamentary succession
~, **совместное** *law* estate in coparcenary
наследователь *law* testator, legator
наследовательница *law* testatrix
наследовать inherit; succeed (to)
наследственность *biol.* heredity
наследственный 1. hereditary; (in)heritable, inherited; 2. ancestral
наследство legacy, inheritance; *law* succession; *fig.* heritage
~, **культурное** cultural heritage
~, **неистребованное** vacant succession
наследуемость *biol.* heritability, hereditability
наследуемый heritable, hereditable, hereditary, inheritable
наслоение 1. *geol.* bedding, lamination, stratification; deposition; layer, stratum; 2. later development (*of culture, etc.*)
~ **в долине** valley filling
~, **волнообразное** wavy/undulating bedding
~, **диагональное** cross-bedding/-lamination
~, **косое** oblique bedding
~, **первичное** original/primary bedding/stratification
насморк *med.* rhinitis, nasal/running cold
~, **зловонный** ozena, coryza foetida
~, **острый** coryza, acute rhinitis
~, **профессиональный** occupational rhinitis
~, **сенной** hay fever
~, **сифилитический** syphilitic rhinitis
~, **спастический** spasmodic rhinorrhea
насолаживание *geol.* sweeting on
насос I pump
~ **аварийного слива (топлива)** jettison(ing) pump
~, **аварийный** emergency pump, stand-by pump
~, **аварийный водоотливный** *naut.* salvage pump
~, **аксиально-кулачковый** swash plate operated pump
~, **аксиально-поршневой ротационный** axial-plunger rotary pump
~, **багерный** dredging pump
~, **балансирный** beam pump
~, **бензиновый** fuel pump
~, **бессальниковый** glandless pump
~, **буровой** *geol.* slush/mud pump
~, **вакуумный ртутный** mercury vacuum pump
~, **вакуумный струйный** ejector
~, **водоотливной** *naut.* bilge/drain(age) pump
~, **водоструйный** water-jet pump
~, **возвратно-поступательный** reciprocating/piston pump
~ **впрыска/впрыскивающий** injection pump
~, **вращательный** rotary pump
~, **всасывающий** suction/aspiring pump
~, **вспомогательный** auxiliary/booster pump
~ **высокого давления** high-pressure pump, HP pump
~ **высокой производительности** high-capacity pump
~, **высоконапорный** high-lift pump
~, **герметизованный/герметический** canned/sealed/leakproof pump
~, **геттерно-ионный** getter-ion pump
~ **главного контура, циркуляционный** *nucl.* main-loop circulating pump
~ **горючего** fuel pump
~, **грунтовой** dredging pump, suction dredge
~, **грязевой** mud/sludge pump
~ **двустороннего/двукратного действия** double-acting pump
~, **двусторонний** duplex pump, double-entry pump
~, **двухступенчатый** rwo-stage pump
~, **двухцилиндровый** duplex pump, two-cylinder pump
~, **диагональный** mixed-flow pump
~, **диафрагменный** diaphragm pump
~, **динамический** rotodynamic pump
~, **дистанционно управляемый** remote-controlled pump
~, **дифферентный** *naut.* trim(ming) pump
~ **для впрыска топлива** fuel injection pump
~ **для холодильного агента** refrigerant pump
~, **дозировочный** proportioning pump
~, **жидкостный** fluid-flow pump
~, **заливной/заливочный** priming pump, primer
~, **зачистной (в танкере)** stripping pump
~, **ионно-сорбционный** getter-ion pump
~, **ионный** ion(ic) pump
~, **коловратный** lobe/lobar rotary pump
~ **Комовского** *med.* Komovsky's vacuum-and-pressure pump
~, **консольный** cantilever pump
~, **креновый** *naut.* heeling pump
~, **кривошипный** crank pump
~, **криогенный** cryogenic pump
~, **криосорбционный** cryogetter pump
~, **крыльчатый** semi-rotary pump, vane pump

~, лопастный impeller/wing pump
~, магнитогидродинамический MHD-pump
~, масляный lubricating/lubrication/oil pump
~, мембранный diaphragm/membrane pump
~, многоступенчатый multistage pump
~, нагнетательный/нагнетающий force/forcing/pressure/delivery pump
~ низкого давления low-pressure pump, LP pump
~ общего назначения general service pump
~, объемный positive-dispplacement pump
~ одинарного действия single-acting pump
~ однократного/одностороннего действия single-acting pump
~, одноступенчатый single-stage pump
~, осевой axial-flow pump
~, откачивающий scavenge/backing/exhaust pump
~, отливной discharge/dewatering pump
~, отсасывающий suction pump
~, паромасляный oil-vapour pump
~, перекачивающий/перекачки transfer pump
~, питающий feed/supply pump
~, плунжерный plunger(-type) pump, displacement/ram pump
~, погружной immersion/immersible/immersed/submersible pump
~, подающий *av.* delivery pump
~, подкачивающий *av.* booster pump
~, подпиточный *therm.* make-up pump
~, пожарный fire service pump
~, поршневой piston/reciprocating pump
~, приводной power pump
~ приемистости *av.* accelerator pump
~, продувочный scavenge/scavenging pump
~ противообледенительной системы de-icer pump
~, прямодействующий direct-acting pump, piston pump
~, рабочий normal pump
~, реверсивный reversible pump
~, регулируемый variable capacity pump
~, резервный emergency pump, stand-by pump
~, роторный/ротационный rotary pump
~, ручной hand(-operated) pump, manual pump
~ с герметизированным ротором canned-rotor pump
~ с дистанционным управлением remote-controlled pump
~ с масляным уплотнением oil-sealed pump
~, сомовсасывающий self-priming pump
~, свободновихревой torque flow pump
~, сдвоенный dual/duplex pump
~, секционный stage chamber pump
~, сильфонный bellows pump
~, смазочный lubricating pump
~, спасательный salvage pump
~ сточной системы sanitary pump
~, струйный jet(-type) pump, ejector
~, судовой marine pump
~, тепловой thermal pump, thermocompressor
~, топливный fuel(-oil) pump
~, топливоподкачивающий fuel-oil priming pump
~, тормозной *rail.* brake pump
~, трехцилиндровый triplex pump, three-cylinder pump
~, трюмный bilge pump
~, фановый *naut.* sewage(-disposal) pump
~, фекальный sewage/sanitary pump
~ флюгирования feathering pump
~, форвакуумный fore/backing pump, roughing-down pump
~ форсажного топлива *av.* afterburner fuel pump
~, футерованный lined pump
~, цементационный grouting pump
~, центробежный centrifugal pump
~, центробежный с лопаточным отводом turbine centrifugal pump
~, циркуляционный circulating pump, circulator
~, шахтный mine/shaft pump
~, шестеренный/шестеренчатый gear(ed) pump, gear-type pump
~, шиберный guided/sliding/vane pump
~, шламовый slurry/slush/mud pump
~, шланговый hose-type pump
насос II *vet.* lampas(se)
насос-дозатор metering pump, controlled volume pump
насос-лягушка diaphragm pump
насос-мотор pump-motor
насосная *n.* pump house, pumping station
насосный pump *(attr.)*
насос-смеситель pump-mixer
насос-триплекс triplex pump
насос-форсунка unit ejector, pump and ejector unit
насос-эжектор jet/ejector pump
наспинник 1. *hist.* backplate, backpiece *(of armour);* 2. backstrap *(of harness)*
наст frozen ice
наставка 1. extension (piece); 2. addition, lengthening
наставление 1. manual, handbook; 2. instruction, direction(s); precept, exhortation, admonition
~ по закупкам procurement regulation
~ по эксплуатации maintenance/service/operating manual
наставлять I extend, lengthen, put/add/piece on
наставлять II teach, edify, exhort, instruct; *rel.* catechize
наставник tutor, teacher, instructor; imponent authority; mentor
~, духовный *rel.* the nurse of souls
~, классный supervising teacher/instructor
наставница tutoress, schoolmistress
наставнический tutorial, instructive; edifying

наставничество tutorship, preceptorship
наставной set on, lengthened, extended, added
наставочный for lengthening/extending
настаивание I insistence, persistence
настаивание II steeping, soaking, infusion
насталивание steeling, acierage, steel facing
настали(ва)ть plate/edge/point with steel
настенный wall(-type), mural; parietal
~ная живопись mural (painting)
настил 1. boarding; plating, planking, floor(ing), deck(ing); 2. laying, spreading; 3. frame; trestles
~, временный false floor
~ второго дна *naut.* inner-bottom plating, tank-top plating
~, защитный protective matting
~ крыши roof deck
~ моста bridge deck
~ палубы deck planking/plating; decking
~ пола floor boarding, flooring
~, шпунтовый matched floor
настилать overlay, overlap, cover; floor, deck, plank, pave
настилка 1. flooring, planking, boarding; 2. laying, spreading
настилочный flooring, paving, planking, boarding
настильность *math.* horizontality, flatness
~ кривой flatness of a curve
настой infusion, extract
~ валерианы infusion of vandal root, infusion of garden heliotrope, valerian infusion
~ горицвета/адониса lychnis infusion
~ корня ипекакуаны ipecacuanha infusion
~ ромашки camomile infusion
~ с кипячением infusodecoction
~ шиповника sweetbrier infusion
настойка 1. tincture, extract, infusion; 2. nastoyka *(kind of liqueur)*
~ безвременника colchicum tincture, tincture of meadow saffron
~ белены henbane/hyoscyamus tincture
~ белладонны belladonna (leaf) tincture
~ бензойной смолы, сложная compound benzoin tincture, Friar's/Turlington's balsam
~ боярышника hawthorn/crataegus tincture
~ василистника thalictrum tincture
~, водно-спиртовая hydroalcoholic tincture
~ горечавки gentian(a) tincture
~ женьшеня ginseng/panax tincture
~ имбиря ginger tincture
~ ландыша convallaria radix/majolis tincture
~ мирриса myrrh(is) tincture
~, мятная mint/mentha tincture
~ наперстянки digitalis/foxglove tincture
~, опийно-бензойная camphorated opium tincture, paregoric
~ перечной мяты peppermint/piperita tincture
~ подснежника squill/galanthus tincture
~, разбавленная weak tincture
~ рвотного корня ipecacuanha tincture
~ сабура aloes tincture
~, сливовая sloe gin
~, сложная compound tincture
~, спиртовая alcoholature
~ стручкового перца capsicum tincture, tincture of red/cayenne pepper
~, хинная cinchona tincture
~ чемерицы veratrum tincture
~ чилибухи tincture of strychnos sead
~ шалфея salvia/sage tincture
~ шпанских мушек cantharides tincture
~, эфирная ethereal tincture
~ яблочнокислого железа ferric mallate tincture
настольный table-/desk-type; desktop
~ная лампа desk/table lamp
~ное руководство handbook, reference book, manual
настоятель *eccl.* 1. Father Superior; superior, prior; 2. archpriest, senior priest *(of a church)*
настоятельница *eccl.* Mother Superior; superioress, prioress
настоятелство dignity of a superior, rank of prior/prioress; priorate, priorship
настоящее *n.* the present (time)
настоящий 1. present, current; 2. real, genuine, veritable
~щее время *gram.* present tense
~щим *law* hereby
настраиваемый adjustable, tunable
настраивать I 1. *mus.* tune (up), attune; set tune in; 2. adjust, tool, true up, aling; 3. incite, instigate, set
настраивать II build on, construct
настраиваться tune
настриг, настригание, настрижка *agr.* clipping, shearing
настроение mood, disposition, frame of mind
~ умов state of public opinion, public mood
настроенность mood, humo(u)r
настроенный adjusted; tuned
настрой 1. *mus.* pitch; 2. mood, humo(u)r
настройка 1. *tech.* adjustment, setup, tuning, alignment, tooling, setting; 2. *mus., rad.* tuning; syntonization
~ автоматического регулятора controller setting
~ автопилота autopilot adaption
~, бесшумная quiet tuning
~ выдержки времени timing
~, грубая coarse/flat tuning
~ диапазона range adjustment
~, клавишная key tuning
~ на нуль zero adjustment
~ на частоту, острая sharp tuning
~ на частоту, тупая broad tuning
~ неправильная misadjustment
~, плохая maladjustment
~, полосовая band-pass tuning
~, предварительная initial tune-up
~ реле adjustment of relay

~ реле с преобладанием biased adjustment of a relay
~, тонкая fine tuning
~ точки регулирования control point adjustment/setting
~, точная fine/sharp/hairbreadth tuning
~, тупая broad/flat tuning
настройка II superstructure; building-on
настройщик tuner, adjuster
наступательно-оборонительный *mil.* offensive-defensive
наступательный *mil.* offensive; *fig.* aggresive
наступать I 1. step/tread (on); 2. *mil.* advance; be on the offensive
наступать II begin, approach, come; ensue, set in
наступающий 1. advancing, invading, attacking; 2. *n.* attacker; 3. approaching, beggining, coming
наступление I *mil.* offensive, attack
наступление II advance; encroachment; incursion, invasion, ingression, onrush; onset
~ ледника advance/spread of glacier
~ моря sea encroachment/ingression/invasion
~ моря на сушу marine transgression
~ ночи nightfall
~ события *stat.* occurence; onset of an event
настуран *miner.* pitchblende, nasturan, uranitite
настурциевые *n. pl. bot.* nasturtium family (*Tropaeolaceae)*
настурция *bot.* nasturtium (*Tropaeolum)*
настыль *met.* accretion, (slag) buildup, crust, skull, incrustation
~, коксовая coke tree
~ цинковая cadmia
насущность importance, urgency; vitality
насущный urgent, vital, needful; essential, fundamental
~ хлеб *rel.* daily bread
насыпать 1. pour(in/into), fill(with); 2. spread, scatter, sprinkle; 3. pile up
насыпка *constr.* fill
насыпной (in) bulk; piled(up); poured
~ вес bulk weight
~ная дамба pile dike
насыпь 1. embankment, bank, fill, dam, dike; mound; 2. loading/storing in bulk
~, железнодорожная railroad embankment
~, земляная earth bank/embankment
~, намывная hydraulic fill
насыпью in bulk
насыщаемость saturating capacity, saturability
насыщать 1. *chem.* saturate, impregnate, soak, imbue; 2. sate, satiate
~ минеральными солями mineralize
~ фосфором phosphorize, phosphorate
насыщаться 1. *chem.* become saturated; 2. be sated/full
насыщение 1. saturation, impregnation; 2. satiation, satiety
~ бетона водой concrete inundation
~ перед расширением pre-expansion saturation
~, предварительное presaturation
~ тока current saturation
насыщенность (degree of) saturation; impregnation; saturated state
~, видовая richness in species
~ отпечатка *print.* chroma saturation
~ шрифта weight of type
насыщенный saturated, impregnated
~ наносами silt-laden
~ недостаточно undersaturated
наталкиваться run/come across
натек 1. *geol.* deposit; dripstone, stalactite; sinter; 2. *met.* incrustation, icicle; 3. pool; 4. *med.* (o)edema, dropsy
~, глинистый clay flow
~, известковый calc sinter
~, кремнистый siliceous sinter, geyserite
~, перламутровый pearl sinter
~, фумароловый fumarole incrustation
натекание (in)leakage, leak-in
натекатель leak
натечка 1. inleakage, infiltration; 2. fine-calibrated orifice
натечник *med.* tuberculous/cold/congestive abscess
~, заглоточный retropharingeal tuberculous abscess
~, подзатылочный suboccipital tuberculous abscess
натечный *geol.* sinter *(attr.),* travertine; *med.* congestive
нативизм *philos.* doctrine of innate ideas, nativism
нативист *philos.* nativist
нативистический *philos.* nativistic
нативный native, natural, inherent, innate
натирание rubbing (in); polishing
натирка polishing; rubbing; grating
~ воском waxing
~ для пола floor polish(ing)
натиск 1. impact; onset, charge; inrush; 2. *print.* printing pressure, impress(ion)
~, легкий *print.* kiss impression
НАТО *abbr.* **Североатлантический союз** NATO
натовский of NATO
наточка sharpening; grinding
натр *chem.* soda, natron
~, едкий sodium hydrate/hydroxide, caustic soda
натравка etching
натравливать etch
натравочный etching
натравщик etcher
натрапник *naut.* ladder screen
натриево-солонцеватый sodium-solonetzic
натриевый sodium *(attr.);* sodic
натрий *chem.* sodium, natrium, Na
~, азотистокислый sodium nitrite
~, азотнокислый sodium nitrate
~, бромистый sodium bromide
~, водородистый sodium hydride

~, вольфрамокислый sodium tungstate
~, двууглекислый baking soda, sodium bicarbonate
~, едкий caustic soda, sodium hydroxide
~, йодистый sodium iodide
~, йодноватокислый sodium iodate
~, кремнефтористый sodium silicofluoride
~, мышьяковокислый sodium arsenate
~, надсернокислый sodium persulphate
~, радиоактивный rachomate, sodium-24
~, сернистокислый sodium sulphate
~, сернистый sodium sulphide
~, серноватистокислый sodium hyposulphite/thiosulphate
~, серноватокислый hyposulphate of soda
~, сернокислый sodium sulphate
~, углекислый sodium carbonate, sal/washing soda
~, урановокислый sodium uranate
~, фтористый sodium fluoride
~, хлористокислый sodium chlorite
~, хлористый sodium chloride, common/table salt
~, хлорноватокислый sodium chlorate
~, цианистый sodium cyanide
~, щавелевокислый sodium oxalate
натрий-катионирование sodium zeolite softening process
натрийкобальт *chem.* cobaltonitrite
натрийурез natriuresis
натристый sodic
натрит natrite
натрокальцит *miner.* natrocalcite, gaylussite
натролит *miner.* natrolite
~, тольчатый needle zeolite
натронный sodic
~ная селитра nitratine
натруска 1. spilling, scattering; sprinkling; 2. vessel for pouring/sprinkling; powder flask/horn
натура 1. nature; *agr.* natural weight *(of grain);* 2. *art* model, sitter; 3. *econ.* kind; 4. *cin.* outdoor scene; 5. character, disposition
натураванс advance in kind
натурализация naturalization
натурализм *philos.* naturalism; *art* realism
натурализованный naturalized
~ иностранец *law* denizen
натурализовать naturalize
натуралист (field) naturalist; natural historian
натуралистический, натуралистичный naturalist(ic); descendental
натуральный 1. natural, native; 2. real, genuine; true; crude, unrefined; 3. *econ.* in kind
~ мед pure honey
~ обмен barter
~ная оплата payment in kind
~ное крестьянское хозяйство self-subsistence peasant economy
~ный масштаб времени *comp.* real time
натурный 1. natural; *art* plein-air, from life; full-scale; *cin.* outdoor, on location; 2. *econ.* in kind
~ вес natural weight *(of grain)*
натуроплата *abbr.* **натуральная оплата** payment in kind
натурфилософ natural philosopher
натурфилософия physiophilosophy, natural philosophy
натурфилософский physiophilosophical, natural philosophic
натурщик, натурщица (artist's) model, sitter
натюрморт *art* still life
натюрмор(т)ист still life painter
натяг 1. interference, negative allowance, oversize; 2. tightness (of fit); tension; 3. hoop cramp
~ клина taper of wedge
~, предварительный preload
~ резьбового соединения threaded joint tightness
~ резьбы thread tightness
~, средний mean oversize
натягивание tightening; stretching; spanning, pulling, stringing
натягивать tighten, stretch, span; tension, pull in, draw tight, taut(en)
~ лук draw a bow
натяжение tension(ing), tightening; stress, strain; pull
~, вторичное *logic* second intension
~, капиллярное capillary tension, capillarity
~ на поверхности раздела interfacial tension
~, поверхностное surface tension
~, предварительное prestressing
~ ремня belt pull/tension
~, чрезмерное excessive tension, overstrain
натяжка 1. tensioning, stretching, tightening; 2. strained meaning/interpretation; forced explanation
~ кожи зажимами *leath.* toggling
натяжной tension(ing), tightening, stretching
~ замок stretching/tension screw
~ ролик idler pulley, tension roller
натянутый spanned; tightened; drawn, tense
наугад *math.* at random, randomly
науглероживание carbonization, carbonizing, carburizing, carburization
~, повторное recarbonization, recarburization, *met.* carbon pick-up
наугольник (try) square, back/bevel square; corner plate
~, складной level square
наугольный angular, corner *(attr.)*
наука 1. science; scholarship; learning, study; 2. knowledge
~ и техника science and technology
~ о драгоценных камнях gemmary
~ о камнях petrography, lithology
~ о травах agrostology
~, сельскохозяйственная agricultural science
~ управления management science
~ки, гуманитарные humanities
~ки, естественные natural sciences

~ки, общественные social studies
науковедение research-on-research, science of science
науковидный pseudoscientific
наукоемкий science-intensive, high-technology *(attr.)*
наукообразность pseudo science/scholarship, scientism
наукообразный, наукоподобный pseudo-scientific
Наум *bibl.* Nahum
науплиус *zool.* nauplius
наутиликон *paleont.* nautilicone
наутилоидея *paleont., zool.* nautiloid
наутилоидный nautiloid
наутофон *naut.* (fog) nautophone, electric-diaphragm emitter
научно-исследовательский scientific-research, research(-and-development), R&D
~ институт research institute
научно-консультативный scientific-advisory
научно-политический political-scientific
научно-популярный popular-scientific
научно-публицистический scientific-journalistic
научно-теоретический scientific-theoretical
научно-технический scientific-technical; research-engineering, scientific and technological/technical
научно-фантастический science-fiction
научный scientific, scholarly
~ работник scientist, scholar; researcher
~ная фантастика science fiction
наушники *pl.* 1. earphones, headphones; headset; 2. earmuff, earflap
~, контрольные monitoring earphones
~ типа клипсы earclip headphones
нафталин *chem.* naphthalene, naphthalenum, naphthalin(e); albo-carbon
нафталинный, нафталиновый naphthalenic, naphthalic
нафтен *chem.* naphthene
нафтойный *chem.* naphthoic, naphthalic
нафтология naphthology
нафтохинон *chem.* naphthoquinone
нахимовец nakhimovite, nakhimovets *(student of Nakhimov naval college)*
нахлестка lap (joint), overlap(ping)
нахлестывание (over)lapping
находить find; discover, detect
находиться exist, occur, be found/located
~ под судом stand trial, be brought up for trial
находка occurrence, find
нахождение occurrence; finding, locating
~ информации information retrieval
нахур *zool.* bharal, nahoor *(Pseudois nayaur)*
нацело *math.* without remainder
~, делиться divide evenly/integrally, divide without remainder
наценка *econ.* extra charge
~, транспортная transportation margin
наци Nazi
нацизм Nazism
национал-демократ national democrat
национализация nationalization
национализировать nationalize, naturalize
националист nationalist
националистический, националистичный, националистский nationalist(ic)
национал-социализм National Socialism
национал-освободительный national-liberation
национальность 1. nationality, ethnic group; 2. national character
национально-языковый national-linguistic
национальный national
~ная администрация по аэронавтике и исследованию космического пространства, НАСА National Aeronautics and Space Administration
~ное меньшинство national minority
нацист Nazi; Hitlerite
нацистский Nazi
нация nation, people
~, наиболее благоприятствуемая most favo(u)red nation
нацмен member of national minority
нацменьшинство national minority
начало 1. beginning, onset, outset, start, inception, commencement; 2. source, origin(ation); 3. principle, basis
~, брать originate (from, in)
~ военных действий outbreak of hostilities
~ воспламенения ignition initiation
~ вывода start of recovery, recovery initiation
~ детонации inception of knock
~ канала *nav.* canal head
~ координат point of origin, origin (of coordinates)
~ ленты *comp.* beginning of tape, BOT
~ обмотки start of winding
~ отсчета *math.* reference point, origin of scale; *stat.* (computing) origin, origin of reading
~ отсчета времени zero/reference time, origin of time
~ отсчета, произвольное arbitrary origin
~, положить begin, commence, initiate
~ прилива *nav.* young flood
~ программы *comp.* program start
~ процесса *tech.* start-up
~ разбега *av.* start of take-off run
~ сеанса *comp.* log-in
~ судоходной части реки navigation head
начальник head, chief, boss, superior, master; *mil.* commander, commandant
~ аэропорта airport manager
~ военного порта captain of naval port
~ военно-морского училища superintendent of naval school
~ дока dock master; *mil.* docking officer
~ инструментального цеха toolroom superintendent

~ **караула** officer of the guard
~ **лоцманской службы** pilot master
~ **механического цеха** machining superintendent
~ **милиции** chief of police
~ **отдела боевой подготовки** training officer
~ **отдела кадров, ~ отдела личного состава** personnel officer/manager
~ **ОТК** chief quality inspector
~ **поезда** trainmaster
~ **порта** port authority
~ **пристани** wharf superintendent
~ **причала** berthing master
~ **производства** director of (plant) operations, plant superintendent
~ **разведки** reconnaissance officer
~ **связи** chief signal officer
~ **смены** shift supervisor/superintendent
~ **станции** station master
~ **транспортного отдела** traffic and transportation supervisor
~ **тысячи** *hist.* chiliarch
~ **цеха** shop foreman
~ **штаба** chief of staff
~ **штаба ВВС/ВМФ** chief of the air/naval staff
начальный 1. first, initial; original; starting, incipient; primordial; 2. elementary; 3. *n.* chief, head, superior
~**ная школа** primary school
начальство 1. *coll.* management, anthoriries, superiors; 2. command, direction; 3. boss
начальствование command(ing)
начальствовать be in command/charge
начальствующий 1. commanding, leading, chief *(attr.);* 2. *n.* head, chief, superior; commander
начеканивать coin/mint (a quantity of)
начеканка coining, coinage, minting
начертание sketch, outline; rough draft
~ **логических схем, условное** logic symbol, logic(al) representation
начертательный graphic, descriptive
~**ная геометрия** descriptive geometry
начес *text.* 1. pile, raising; nap (of material); 2. carding, (back)combing
начет *fin.* unauthorized expenditure (to be recovered)
начетчик person well-read in Scriptures; *fig.* dogmatist, doctrinarian; pedant
начинание undertaking; beginning, inception
начинатель initiator, originator; author
начинательный initiating; *gram.* inceptive, ingressive, inchoative
~ **глагол** inchoative/inceptive verb
начинать initiate, undertake; begin, start
начинающий *n.* beginner; novice
начинка *cul.* filling, stuffing
начинщик filler, stuffer
начисление extra charge, additional sum
~ **фиктивной стоимости перевозок** *econ.* phantom freight charging
~ **износа** depreciation deduction
начислять add (on); *fin.* charge extra
начитанность erudition
начитанный well-/widely-red; erudite, learned
начитывать *fin.* charge extra
нашатырный ammoniac(al)
~ **спирт** liquid ammonia
нашатырь *chem.* sal ammoniac, ammonium chloride
нашейник *hist.* 1. gorget; 2. necklace, neck ornament
нашествие invasion, incursion, inroad; descent
нашивка *mil.* stripe, chevron (on sleeve); tab
~, **золотая** gold-lace stripe
~, **средняя/узкая** broad/narrow stripe
~, **широкая золотая** full width gold lace
нашильник chokestrap *(of harness)*
нашлемник crest *(of helmet)*
нашпиговывать *cul.* lard
нашумевший sensational, much-talked-of
наэлектризовывать electrify
наяда 1. *myth.* water nymph, naiad; 2. *bot.* naiad (Najas)
НГМД *abbr.* **накопитель на гибких магнитных дисках** *comp.* floppy drive
неабелевый *math.* non-Abelian
неабсорбирующий nonabsorbent, unabsorbent
неавтоматический nonautomatic; manual
неавтономный nonautonomous; co-operative
~ **выходной самописец** *comp.* on-line graphical output recorder
неавторитетный unauthoritative, inauthoritative
неагрессивный nonaggressive
неадаптированный unadapted; *biol.* unspecialized
неаддитивность non-additivity
неадекватность lack of fit
неадекватный, неадэкватный inadequate; incommensurate (with, to)
неадиабатический nonadiabatic, nonisentropic
неадресованный unaddressed
неакадемический, неакадемичный unacademic; nonacademic
неаккуратность 1. inaccuracy, carelessness; 2. unpunctuality; 3. untidiness
неаккуратный 1. inaccurate, careless; inexact, imprecise; 2. unpunctual; 3. untidy
неактивированный *nucl.* non-activated
неактивный inactive, unreactive, inert; passive, indolent; idle
неактуальный out-of-date; inoperative
неакцептование nonacceptance
неакцептованный unaccepted
неалгебраический nonalgebraic
неаминный nonamino
неаналитический nonanalytic, unanalytical
неаналогичный unanalogous
неанглийский non-/un-English
неандерталец *anthr.* Neanderthaler, Neanderthal man

неандертальный Neanderthaloid
неандертальский Neanderthal
неанический *zool.* neanic
неаннулированный uncancelled, unabrogated, unrepealed; unreversed
неаполитанец, неаполитанский Neapolitan
неаполитанка Neapolitan (woman)
неаппетитный unappetizing; uninviting
неаргументированный unreasoned, unsubstantiated
неарестованный unarrested, unapprehended
неаристотелевый non-Aristotelian
неарифметический nonarithmetical
Неарктида *geogr.* Nearctica
неароматический nonaromatic; aliphatic
неартикулированный inarticulate; unarticulated
неартроз *med.* nearthrosis, pseudoarthrosis
неархимедовский, неархимедовый non-Archimedean
неассигнованный unappropriated
небаланс inbalance, unbalance
небалансный out-of-balance
небеременеющая *n. med.* nulligravida
небеременная nonpregnant
небесный celestial, heavenly; divine, supernal;
~ свод firmament
~, Царь *rel.* Heavenly Father
~ная империя *hist.* China, the Celestial Empire
~ные светила heavenly bodies
небиблейский unbiblical; unscriptural
небинарный nonbinary
небиогенный non-biogenic
небитый nubeaten; unwhipped
~ скот *agr.* live cattle
неблаговидность unseamliness, unappropriateness
неблаговидный 1. improper, unseemly; 2. unsightly; unattractive
неблагозвучие disharmony, dissonance; discord(ance); cacophony
неблагонадежный unreliable; unfaithful; *polit.* suspect
неблагополучие ill-being; trouble, setback
неблагоприятность unfavo(u)rableness, disadvantageousness; adversity
неблагоприятный unfavo(u)rable, disadvantageous; adverse, inauspicious, unpropitious
небно-альвеолярный alveopalatal
небно-глоточный palatopharyngeal
небно-челюстной palatomaxillary
небно-язычный palatoglossal
небный *anat.* palatine; *phon.* palatal
~ная занавеска uvula
небо *anat.* palate; 2. sky; heaven
~, костное bony palate
~, ледовое/ледяное iceblink
~, мягкое/твердое soft/hard palate
небоевой *mil.* non-tactical; *nav.* non-combatant
неболезнетворный *med.* nonpathogenic, avirulent, nonvirulent
небосвод dome, firmanent, sky, the canopy, vault/roof of heaven
небосклон sky (immediately over horizon)
небоскреб skyscraper
небрежность *law* negligence, carelessness; inadvertency
~, виновная culpable negligence
~, встречная supervening negligence
~, грубая gross negligence
~ истца contributory negligence
~, преступная criminal/culpable negligence
небрежный negligent, careless; inadvertent, lax
небулиты *pl.* nebulites
небулярный *astr.* nebular
небьющийся shatterproof, unshatterable
~щееся стекло safety glass
неведение ignorance; nonacquaintance; nescience
неведомый unforeknowable, unknown; unfamiliar
неверие *rel.* lack of faith, scepticism; unbelief, irreligion
неверный 1. wrong, erroneous; incorrect, inexact; unveracious; 2. disloyal, unfaithful; deceptive, delusive, unreliable; *rel.* infidel; 3. *n. rel.* infidel; the Evil one, devil
~ Фома *rel.* doubting Thomas
неверующий 1. irreligious, nullifidian; 2. *n.* nonbeliever, unbeliever, nullifidian, atheist
невесомость *phys.* weightlessness; imponderability; zero-G
невесомый *phys.* weightless, agravic; imponderable
невеста fiancée; bride
~, Христова *rel.* nun
невещественность immateriality; immaterialness; incorporeity
невещественный immaterial, incorporeal; *law* nonsubstantial
невзаимный nonreciprocal, unilateral
невзаимодействующий noninteracting
невзвешенный unweighted
невзрачница *bot.* parsley-piert (*Aphanes arvensis)*
невзрываемость non-explosiveness
невзрывчатый, невзрывающийся inexplosive, unexplosive
невидимый invisible
невинность innocence, guiltlessness
~сти, заявлять о своей *law* plead not guilty
невиновный guiltless, innocent (of); *law* not guilty
~, фактически innocent in fact
невихревой nonturbulent; irrotational
невключение noninclusion
невменяемость *law* irresponsibility; lunacy
~ти, в состоянии not answerable for one's actions
невменяемый *law* irresponsible; unimputable; *n.* distracted person
невмешательство noninterference, nonintervention
~ства, политика policy of noninterference
невнимание lack of attention, inattention; negli-

gence, carelessness
невод seine, (sweep/casting) net, trammel
~, выбирать/поднимать *naut.* boat a seine
~, донный ground seine
~, закидной/закидный beach/shore seine
~, закольный stake net
~, заметный trammel net
~, кольцевой ring net
~, кошельковый purse/tuck seine
~, пелагический deep-sea seine
~, тяговый drag seine
неводник fishing boat for seining; seiner
неводный nonaqueous
неводостойкий hydrolabile
невоенный 1. nonservice, nonmilitary; 2. *n.* noncombatant, civilian
невозбужденный unexcited; uninflamed
невозвратимый irrevocable, unrecoverable, irretrievable, irreparable; irredeemable
невозвратность irrevocability, irretrievableness
невозвратный 1. irrevocable, irretrievable; 2. *gram.* nonreflexive
невозвращенец *polit.* defector
невозвращенство *polit.* refusal/failure to return
невоздержность intemperance; lack of selfcontrol/-restraint
невозмещенный uncompensated; unredressed
невозможность impossibility; inability
~ выполнения (договора) impossibility of performance
~ исполнения, фактическая *law* impossibility in fact
~ исполнения, юридическая impossibility in law
невозмущенный unperturbed, undisturbed
невоидный *med.* nevoid
невокарцинома *med.* nevocarcinoma
невоксантоэндотелиома *med.* nevoxanthoendothelioma, juvenile xanthogranuloma
неволокнистый nonfibrous
невооруженный unarmed
~ цепень hookless tapeworm
~ным глазом with the naked eye
невоспалительный *med.* noninflammatory
невоспламеняемость noninflammability, incombustibility
невоспламеняемый incombustible, noninflammable, uninflammable
невоспламеняющийся non(in)flammable, flameproof, nonignitable, incombustible
невосприимчивость *med.* insusceptibility, immunity
невосприимчивый 1. *med.* nonsusceptible, immune (to); 2. irreceptive; unperceptive, unresponsive, insensitive
невоспроизводимый irreproducible
невосстанавливаемый unrestorable, irretraceable, nonrecoverable
невостребованный not called for, unclaimed; dead *(of letter)*
невправимый *surg.* irreducible
невправленный *surg.* nonreduced, unset *(of broken bone)*
невралгический *med.* neuralgic, neuralgiform
невралгия *med.* neuralgia
~, возвратная reminiscent neuralgia
~, затылочная occipital neuralgia
~ культи stump neuralgia
~ лицевого нерва geniculate neuralgia
~ молочной железы mammary neuralgia, mastodynia
~, односторонняя hemialgy, hemialgia
~ плечевого сплетения brachialgia
~ плюсне-фалангового сочленения metatarsalgia, Morton's foot/toe/neuralgia
~ седалищного нерва sciatic neuralgia, sciatica
~ троичного нерва trifacial neuralgia
~, упорная obstinate/persistent neuralgia
~, эпилептиформная tic douloureux
неврастеник neurastheniac, neurotic
неврастенический, неврастеничный neurasthenic, neurotic
неврастения *med.* neurasthenia, nervous exhaustion, Beard's disease
~, ангиопатическая angiopathic/pulsating neurasthenia
~, тяжелая затяжная neurasthenia gravis
невращающийся nonspinning, nonrotating, irrotational
неврилемма *med.* neurilemma, neurolemma
невринома *med.* neurinoma, perineural fibroblastoma, schwannoma
~ глотки pharyngeal neurinoma
~ кожи cutaneous neurinoma
~ средостения mediastinal neurinoma
неврит *med.* neuritis
~, внутриглазной intraocular neuritis
~, восходящий ascending neuritis
~, диссеминированный polyneuritis
~, затылочный occipital neuritis
~ зрительного нерва optic neuritis
~, корешковый radicular neuritis
~, множественный multiple neuritis
~, нисходящий descending neuritis
~, простудный cold neuritis
~, сегментарный segmental/segmentary neuritis
~ седалищного нерва sciatic neuritis
~, эндемический endemic neuritis, beri-beri
невроастроцитома *med.* neuroastrocytoma
неврогипофиз *anat.* neurohypophysis
неврогипофизарный neurohypophysial
невродермит *med.* neurodermatitis, atopic dermatitis
~, ограниченный limited neurodermatitis
неврожденный acquired, noncongenital
невроз *med.* neurosis
~, военный war neurosis, shell-shock
~, истеротравматический conversion hysteria neurosis

~ **конечности** acroneurosis
~ **мочевого пузыря** cysterethism
~ **навязчивых состояний** impulsive neurosis
~ **полости рта** oral neurosis
~, **профессиональный** occupational/professional neurosis
~ **прямой кишки** rectal neurosis
~, **сердечно-сосудистый** cardiovascular neurosis, neurocirculatory asthenia
~, **сосудистый** angioneurosis
~, **трофический** trophoneurosis
~, **старческий** neurosis tarda
~ **утомления** neurasthenia, psychasthenia, fatigue neurosis
невролог neurologist
неврологический neurologic(al)
неврология neurology
неврома *med.* neuroma
~, **истинная** true neuroma, ganglioneuroma
~ **кожи** neurofibroma of the skin
~, **ложная** pseudoneuroma
~, **сетевидная** plexiform neuroma
невромер *anat.* neuromere
невромиозит *med.* neuromyositis
неврон *anat.* neuron(e)
~, **вставочный** internuncial/intercalary neuron
~, **периферический двигательный** lower motor neuron
~, **регулирующий** adjuster neuron
~, **тормозной** ckecking neuron
~, **центральный двигательный** upper/higher motor neuron
~, **центробежный** efferent neuron
невронофаг neuronophag(e)
невропатия *med.* neuropathy, neuronosus
~, **корешковая** radicular neuropathy
~, **наследственная** heriditary neuropathy
невропатолог neuropathologist
невропиль *anat.* neuropile
невропластика *surg.* nerve grafting
невроптероидные *n. pl. paleont.* neuropteroids (*Neuropteroidea)*
неврорафия neurorrhaphy
невротик neurotic (person)
неврофиброматоз *med.* neurofibromatosis
~, **очаговый** incomplete/abortive neurofibromatosis
невроэпителиома *med.* neuroepithelioma
~, **обонятельная** olfactory neuroepithelioma
невручение nondelivery, failure to deliver/ *law* serve
невсхожесть *agr.* inability to sprout; nonemergence
~ **семян** blind seeds disease
невсхожий 1. *agr.* nonemerging, not sprouting; 2. *cul.* not rising *(of dough)*
~**жие семена** unfruitful seeds
невус *med.* nevus
~, **бородавчатый** nevus verrucosus
~, **волосяной** hairy mole, nevus pilosus
~, **ихтиозиформный** nevus linearis/nervosus/papllaris/lichenoides
~, **липоматозный** nevus lipomatodes/lipomatosus, nevolipoma
~, **паукообразный** spider nevus/mole/angioma
~, **пигментный** pigmented nevus, mole
~, **плоский** flat mole, nevus spillus
~, **пограничный** junction nevus, melanoma
~, **сосудистый** nevus vascularis/vasculosus/angiectodes/flammeus/sanguineus
невыветренный unweathered, unaerated
невыгодность 1. disadvantageousness, unfavo(u)rableness; 2. unprofitableness
невыгодный 1. disadvantageous, unfavo(u)rable; 2. unprofitable, unremunerative, unlucrative
невыдача lack of issue/distribution/payment
невыделанный unmanufactured; *leath.* undressed *(of hides)*
невыдержанный 1. intemperate, short-tempered; 2. *food* unmatured, new
невыплата nonpayment; *econ.* arears
невыполнение nonperformance, inexecution, nonfulfilment; *law* defanult, failure to carry out
~ **договора** default in performance of the contract
~ **международных обязательств** default on international obligations
~ **обязанностей** default, delinquency, *law* neglect of duty, failure in duties
~ **приказания** neglect of duty
~ **требований закона** failure to comply
невыполненный 1. unfulfilled, undischarged, unexecuted, unperformed; *econ.* outstanding; 2. unfilled, hollow
невыполнивший *law* in default
невыполнимость unfeasibility, impossibility, impracticability
невыполнимый unrealizable, unachievable, unfeasible, inexecutable
невыправленный *el.* unrectfied
невыровненный nongraded, ungraded, unlevelled
невырожденный *math.* non-degenerate; regular
невытраливаемый *nav.* unsweepable
невычет *math.* nonresidue
невыщелачивающийся unleachable
невыявленный latent; unrevealed, unexposed
невязка *geod.* (error of) closure, misclosure; discrepancy, disparity
~, **абсолютная линейная** absolute linear misclosure
~ **баланса (энергосистемы)** imbalance
~, **высотная** elevation discrepancy
~, **конечная** discrepansy in closing
~, **координатная** position error/misclosure
~ **треугольника** closing/triangular error, closure of triangle
~, **угловая** angular difference/discrepancy
невязкий inviscid, nonviscous
негабарит oversize
негабаритность bulkiness
негабаритный oversize, excessive in seize; off/

gauge, out of gauge
негазовый non gaseous, gas-free
негазящий non-gassing
негатив *phot.* negative
~, вялый/мягкий/неконтрастный flat/soft negative
~, контрастный hard negative
~, первоначальный original negative
~, полутоновый/растровый half-tone negative
~, стеклянный plate negative
~, цветоделенный (color-)separation negative
~, штриховой *print.* line negative
негативист *philos.* negativist
негативистский negativist(ic)
негативносетчатый areolate
негативный negative; negatory
~ гелиотропизм *bot.* negative phototropism, apheliotropism
негатоскоп viewing screen, X-ray film viewer
негатрон negatron, electron
негашеный unslaked
~ная известь quicklime, unslaked lime
негерметизированный unsealed, unpressurized
негерметический non-tight, non-wateright, nonhermetic, leaking
негерметичность non-tightness, leakage, leaking, seepage, faulty sealing, seal failure
негибкий rigid, stiff, inflexible, unpliant
негигиенический, негигиеничный unhygienic, unsanitary
негигроскопичность poor moisture absorption
негидратированный unhydrated
негниючка *bot.* eastern arborvitae *(Thuja occidentalis)*
негниючник *bot.* marasmius, toadstool *(Marasmius rotula)*
негной-дерево *bot.* yew *(Taxus)*
негнойный apyetous
негодность unfitness; worthlessness
~ к военной службе unfitness for military service
~, привести в put out of commission
неголономность *math.* anholonomity, nonholonomicity
негомогенный inhomogeneous, nonhomogeneous
негоризонтальность off-levelling, dislevelment
негорючий incombustible, non-combustible/-burning/-flammable, flameproof, fire-resistant
негоциант merchant
негр Negro; *fig.* ghost writer
неградуированный ungraded
неграмотность illiteracy; ignorance
неграмотный 1. illiterate, ungrmmatical; 2. ignorant (of), not versed (in); 3. *n.* illiterate (person)
негритос Negrito
негритоска Negrito (woman)
негритянка Negress; Negro girl
негритянский Negro, Negritic
негроид, негроидный Negroid
негрофобия Negrophobia
негрский Negro, Negritic
негуманный inhumane
негус Negus
негэнтропия negative entropy, negentropy
недавний late, recent, neoteric
недальновидность lack of foresight, shortsightedness
недальнозоркость shortsightedness
недвижимость real property/estate, immovable property/estate; *law* realty, things real, real/fixed assets
~, брошенная vacant possession
недвижимый immovable, pr(a)edial; *law* real
~мая собственность real property/estate, immovable property/estate
~мое имущество real estate, things real, real/fixed assets
недвижущийся stationary
недгедс *timb.* knighthead, bollard timber
недееспособность inability, incapability, incapacity, *law* disability
недееспособный *law* incapable
недейственный inefficient, effectless
недействительность inefficacity, ineffectiveness; *law* nullity
~, юридическая invalidity
недействительный 1. ineffective, inefficacious; 2. unreal; imaginary; nonobjective; 3. *law* void, null, null and void; invalid; nugatory
~ным, делать *law* vitiate, invalidate
недействующий inoperative, non-operating, nonfunctioning, inofficious; inert, passive, idle
неделимость indivisibility, inseparability; *law* impartibility
неделимый indivisible, inseparable; *law* impartible
~мое, призматическое prismatic individual
неделя week
~, масленая/мясопустная *rel.* Shrovetide
~, рабочая workweak
~, страстная *rel.* Holy/Passion Week
~ Фомина *rel.* Low Week
неделящийся non-fissionable
недержание *med.* irretention, incontinence
~ кала incontinence of feces, rectal incontinence
~ мочи urinary incontinence, enuresis
~ мочи, ночное nocturnal enuresis, nycturia
недетерминированный nondeterminate; nondeterministic
недеформированный undeformed; unwarped
недеформируемый unyielding, nondeformable
недеятельность inactivity, passivity, inertness
недипломированный uncertificated, nonprofessional, without a diploma/degree
недистиллированный undistilled
недоброжелательный malevolent, hostile
недоброкачественность poor/bad quality
недоброкачественный of inferior quality, inferior, low-grade

недобросовестно unscrupulously, unconscientiously; *law* in/from bad faith
недобросовестность 1. bad faith, unscrupulousness, unconscientiousness; *law* mala fides; 2. negligence, carelessness
недобросовестный 1. unscrupulous, unfair; *law* mala fide; 2. careless
недоверие distrust, mistrust
~рия, вотум vote of no confidence
недовес underweight, short weight, shortage in weight
недовольный 1. displeased, discontented, dissatisfied; 2. *n.* malcontent
недоговоренность lack of agreement/understanding
недогрев underheating
~ охладителя *nucl.* subcooling
недогруженный underloaded
недогруз underloading
недоделка off-standard work, subquality/unfinished work, fault in workmanship
недодержка *phot.* underexposure; underdevelopment
недоедание malnutrition, undernourishment, denutrition
недожог incomplete burning/combustion; underfiring
недозволенный unallowed, unlawfull, impermissible, illicit
недозрелый underripe, unripe; immature; *biol.* abortive
недоимка arrears, amounts overdue
недоимщик defaulter
недоказанность failure to prove
недоказанный unproven, unsubstiantiated
недоказуемый unprovable; inadequate, indefensible
недокалка *tech.* insufficient tempering
недокорм(ка) underfeeding
недолговечный temporary, transient, short-lived
недолет *mil.* short shot, falling short
~, большой way short
~ при посадке undershoot
недомерок undersize (object)
недомогание *med.* indisposition, fatigue, malaise
недомол undermilling, undergrinding
недонесение *law* withholding information
~ о совершенном/готовящемся государственном преступлении misprision of treason
недоноситель *law* accessory after the fact; one guilty of negative misprision
недоносительство *law* misprision of felony, negative misprision
недоносок 1. prematurely born child; slink *(of animals);* 2. *fig.* crude/unfinished work
недоношенный premature, prematurely born
недоокисленный insufficiently oxidized
недоокись *chem.* suboxide, protoxide
недоохлаждение undercooling
недооценивать underestimate, underrate, undervalue
недооценка underestimation, underrating, undervaluation
недопекание *met.* undersintering
недоплата underpayment; unpaid sum
недопоставка incomplete/short delivery
недопотребление underconsumption
недопрессовка faulty moulding, moulding flaw, undermoulding
недопроизводство underproduction
недопроявленный *phot.* underdeveloped
недопустимый inadmissible, intolerable; unallowable; illegal
недопущение nonadmittance, nonadmission; barring, banninhg
недоразвитие *med.* underdevelopment, hypoplasia, infantilism, atresia
~ половой системы hypogenitalism
~ семяпочки *bot.* ovule abortion
~, умственное mental deficiency
недоразвитость 1. underdevelopment, juvenilism; 2. backwardness
недоразвитый 1. *med., bot.* underdeveloped, partially developed, stunted, vestigial, hypoplastic, infantile, rudimentary, immature, retarded; 2. backward
~ столбик abortive style
~ цветок abortive/mutilated flower
недоразумение 1. misunderstanding; 2. argument, disagreement
недорасход underconsumption
недорасширение underexpansion
недород *agr.* crop/harvest failure, poor harvest
недородный *agr.* lean, low-yield *(of harvest)*
недоросль *hist.* minor
недосев insufficient sowing; underfulfilment of sowing plan
недосека *text.* miss of weft
недосмотр oversight, inadvertence; negligence
недоснабжение insufficient supply, undersupply
недоставка nondelivery
недостаток 1. deficiency (in), deficit, shortage, insufficiency (of), lack (of); nonavailability; 2. disadvantage, drawback, demerit, shortcoming, limitation, flaw, defect, imperfection; *Am.* bug
~ вакансии *econ.* job shortage
~ денег money stringency
~ нейтронов neutron deficiency
~ питания nutritional deficiency
~, психический mental infirmity
~ рабочих сил job shortage, shortness of hands
~ речи speech abnormality
~, физический physical defect
~ формы defect of form
недостаточность insufficiency, deficiency, failure, inadequacy, shortage; *med.* incompetence, incompetency
~, аортальная aortic incompetence
~, белковая protein deficiency

~, гипофизарная pituitary insufficiency
~ двустворчатого клапана mitral incompetence
~ клапана аорты aortic incompetence
~ клапана сердца valvular insufficiency/incompetence
~ кровообращения circulatory insufficiency
~, легочная pulmonary insufficiency/incompetence
~ надпочечников adrenal insufficiency, Addison's disease
~, острая acute insufficiency/failure
~ паращитовидной железы hypoparathyroidism
~ печени hepatic insufficiency
~ питания malnutrition
~ почек renal insufficiency
~, психическая aphronia
~ роговой ткани aceratosis
~ сердечная/сердца cardiac insufficiency, heart failure
~ спроса *econ.* deficiency of demand
~ щитовидной железы hypothyroidism, hypothyroidea, thyroid insufficiency
недостаточный insufficient, inadequate, wanting, deficient; scanty, meager, lean
~ глагол *gram.* defective verb
недостижимость inaccessibility, unattainability
недостижимый inaccessible, unachievable, unattainable
недостоверность unauthenticity; uncertainty; invalidity
недостоверный unauthentic, unreliable, doubtful; unveracious; apocryphal
недоступный 1. unobtainable, unavailable, beyond reach; prohibitive; 2. inapproachable
недосылка nondelivery
недосягаемость inaccessibility; unattainability
недотрав *met.* underpickling
недотрога *bot.* snapweed, touch-me-not
недоукомплектованный understaffed, undermanned
недофокусировка underfocusing
недочет deficit, shortage; defect, shortcoming
недра *pl.* 1. mineral resources; 2. interior part of the earth, bowels
недуг disease, ailment; illness, malady
неевклидовость *math.* non-Euclideanism, non-Euclidicity
неевклидовый non-Euclidean
нееврей gentile, non-Hebrew
нееврейский gentile, non-Hebrew/-Jewish
неевропеец non-European
неживой non-living, dead, inanimate, abiotic
нежизнеспособность lack of vital capacity; nonvitality; helplessness
нежизнеспособный nonviable, lacking vitality; feeble, helpless
незабудка *bot.* forget-me-not *(Myosotis)*
незабудочник *bot.* eritrichium *(Eritrichium)*
незавершенность incompletion, unfinished state
незавершенный unfinished, incomplete, unconcluded, unaccomplished
~ное производство work in process/progress, goods in process
независимо-возбуждаемый separately excited
независимость independence; sovereignty
~, аппаратная hardware independence
независимый independent; self-contained; standalone; *biol.* autonomic
незадействованный idle, spare, vacant; dead
незадросселированный unthrottled
незаземленный *el.* unearthed, ungrounded
незаконнорожденный illegetimate, adulterine; love-begotten; out of wedlock
незаконность illegality, unlawfulness
незаконный illegal, illicit, unlawful, illegitimate, wrongful, unjudicial
незакономерность irregularity, exceptionality
незакономерный irregular, exceptional; unjustified, unfounded
незаконченность incompletion, unfinished state
незаложенный unhypothecated, unmortgaged, unpawned, unencumbered
незалупа *med.* phimosis
незамедленный unmoderated
незаменимость indispensability
незаменимый indispensable, irreplaceable
незаменный invariable, unchanging; unexchangeable
незаменяемый irreplaceable, unexchangeable
незамерзающий uncongealable, non-freezing; frost-proof, ice-free; *tech.* antifreeze
незамкнутый open, unclosed, unlocked; unenclosed
незамужняя 1. unmaried, unwed, single; *law* discovert; 2. *n.* unmarried woman, maiden lady
незанятость *tel.* availability, free condition; inoccupation
незанятый available, free, not occupied/busy/engaged; unfilled; vacant; idle; *econ.* unemployed
незапланированный unplanned, unscheduled; unarranged, undevised
незаполнение *met.* underfilling
незаполненный unoccupied, not filled; unstuffed; blank, empty, vacant
незаприходованный *fin.* unrecorded, unlisted, not entered
незараженный uninfected
незаразный noncontagious
незаращение *med.* nonclosure
~ канала Гартнера persistent Gartner's duct
~ мягкого неба fissured palate
незаращенный open, expanded; *med.* patent
незарегистрированный unregistered; unlicensed
незасеваемый unsown, unplanted, *agr.* out of crop
незаселенный unpopulated, uninhabited
незасеянный *agr.* unsown, uncropped

незасоленный nonsaline, sweet *(of soil)*
~ бекон uncured bacon
незасоренный clear, clean, unobstructed; *agr.* weedless
незастекленный unglazed
незатвердевающий incongealable; unconsolidated
незащищенный unprotected, unsheltered, unshielded, unguarded, exposed; unfortified
~ со стороны моря open to the sea
незванный uninvited, uncalled
нездоровье indisposition, ill health, illness, malaise
незлокачественный *med.* nonmalignant, benign
незнакомый 1. unfamiliar, unknown; unacquainted; 2. *n.* stranger
незнание ignorance, nescience
~, априорное prior ignorance
~, виновное *law* voluntary ignorance
незначительный insignificant, unimportant, negligible, inconsiderable, inappreciable
незрелость immaturity; infantility, impuberty
~, духовная spiritual nonage
незрелый unripe, immature; *fig.* verdant
незримый invisible, imperceptible; *n., rel.* the Supreme Being, God
незрячный/незрящий blind
незыблемый immovable, unshakeable; firm, stable, steady
~мые правила set rules
НЕ-И *comp.* NOT-AND, NAND
неизбежность imminence, inevitability
неизбежный imminent, inevitable, unavoidable, inescapable, unevadable
неизбыточный *comp.* irredundant
неизвестное *n.* the unknown (quantity)
неизвестность uncertainty, obscurity
неизвестный 1. unknown, uncertain; unfamed; obscure; 2. *n.* unknown person, stranger
неизлечимость incurability
неизлечимый incurable, irremediable, unhealable
неизлучающий non-radiative, nonradiating
неизменяемый invariant, permanent, constant, invariable, unalterable, unmodifiable
~, грамматически uninflected
~ по времени time-constant
неизмеримый immeasurable, unfathomable, fathomless; *bot.* anisomerous
неизолированный *el.* non-insulated, naked, bare
неизотропность anisotropy, anisotropism
неизученный unexplored, unstudied; unknown
неизъяснимый inexplicable, ineffable
НЕ-ИЛИ *comp.* NOT-OR, NOR
неимущий poor, indigent
неискаженный undistorted, undisturbed, undisfigured; unmutilated
неисполнение nonperformance, nonfulfilment, nonexecution; *law* failure to perform
неисполненный unfulfilled; *law* unsatisfied
неисправность defect, fault, trouble, malfunction, breakdown; *Am.* bug
~ аппаратуры наведения guidance malfunction
~ в работе breakdown of service, operating trouble
~ двигателя engine trouble/fault
~, находить и устранять trouble-shoot
~, неустойчивая transient fault
~, перемежающаяся intermittent fault
~, скрытая latent fault
~ типа "константная единица" *comp.* stuck-at-1-fault
~ типа костантного обрыва *comp.* stuck-open fault
~ типа неопределенного состояния *comp.* stuck-at-X fault
~, устойчивая permanent fault
~, устранять clear/rectify a fualt, remove/shoot a trouble
неисправный faulty, inoperable, defective, incorrect, out of order, damaged, in bad repair, in disrepair
неисследованный unexamined, unexplored; *nav.* unsounded, uncharted
неисчезающий nonvanishing
неисчерпаемый inexhaustible, undraina ble; unfailing
нейзильбер *met.* white copper, German silver
нейлон nylon
нейрин *chem.* neurine
нейристор *el.* neuristor
нейрит *anat.* neurit(e)
нейроатропатия *med.* Charcot's joint
нейровисцеральный *med.* neurovisceral, neurosplanchnic
нейроганглиома *med.* ganglioneuroma
нейрогенный neurogenetic, neurogenic, neurogenous
нейроглиальный neurogliac, neurogliar
нейрогормон neurohormone
нейродермит *med.* neurodermite
~, диссеминированный Besnier's prurigo
нейроз neurosis
нейроиндукция *med.* neuroinduction, suggestion
нейролептик neuroleptic, tranquilizer
нейролимфоматоз *med.* neurolymphomatosis
~ птиц fowl paralysis
нейромезенхима ectomesenchyma
нейромиастения *pathol.* neuromyasthenia
нейрон *anat.* neuron(e), nerve cell
~, моделировать simulate/model a neuron
~, чувствующий receptor neuron
нейронный neuronal, neuronic
нейрорецидив neurorelapse, neural recurrence
нейросекрет, нейросекреция neurosecretion
нейроспонгиома neurospongioma, medulloblastoma
нейротраст iophendylate
нейротропизм neurotropism, neurotropy
нейрофизиология neurophysiology
нейрохирург neurosurgeon
нейрохирургия neurosurgery

нейроэктазия neurectasia, neurectasis, neurectasy
нейроэндокринология neuroendocrinology
нейстон neuston, floating vegetation
нейтрализатор neutralizer, neutralizing agent, eliminator
нейтрализация neutralization, neutralizing; sweetening *(of soil)*
~ неисправности *comp.* fallback
~ топлива decontamination of fuel
нейтрализм neutralism; nonaligment
нейтрализировать, нейтрализовать neutralize; sweeten *(of soil)*
нейтралитет *polit.* neutrality
нейтраль *el.* neutral nonconductor/wire/point; zero line
нейтральность neutrality
нейтральный neutral; *polit.* nonaligned; *geol.* sweet *(of soil); rel.* adiaphorous
~ные растения к длине дня day-neutral plants
нейтральрот neutral red
нейтрино *phys.* neutrino
~, мю-мезонное,~ мюонное neutretto, myontype neutrino
~, электронное electron-type neutrino
нейтриса *math.* neutrice
нейтрон *phys.* neutron
~, блуждающий stray neutron
~, бомбардирующий bombarding/impacting/impinging neutron
~, быстрый fast neutron
~ в момент испускания nascent neutron
~, влетающий oncoming neutron
~, вылетающий escaping neutron
~ высокой энергии high-energy neutron
~, деградированный degraded neutron
~ деления fission neutron
~ деления, замедленный moderated fission neutron
~, дочерний daughter neutron
~, закадмиевый epi-cadmium neutron
~, замедленный до тепловой энергии thermalized neutron
~, запаздывающий delayed neutron
~, зарегистрированный recorded/detected neutron
~, избыточный surplus/excess neutron
~ из реактора reactor/pile neutron
~, испущенный expelled/ejected neutron, given-off neutron
~ малой энергии low-energy neutron
~, мгновенный prompt/instantaneus neutron
~, медленный slow neutron, low-velocity neutron
~, надрезонансный epiresonance neutron
~, надтепловой epithermal neutron
~, незарегистрированный undetected neutron
~, непарный unpaired neutron
~, нетепловой non-thermal neutron
~ низкой энергии low-energy neutron
~, обедненный stripped neutron
~, отраженный reflected neutron
~, падающий incident neutron
~, первичный virgin/primary neutron
~, подтепловой sub-thermal neutron
~, потерянный wasted neutron
~, преломленный refracted neutron
~, промежуточный intermediate neutron
~, рассеяный stray/scattered neutron
~ с пороговой энергией treshold neutron
~ с уменьшенной энергией degraded neutron
~, связанный latent neutron
~ со спином вверх/вниз up-/down-spin neutron
~, сталкивающийся impinging neutron
~ утечки leakage neutron
~, эпитепловой/эпитермический epithermal neutron
~, ядерный nuclear neutron
нейтронный neutron *(attr.)*
~ная физика neutronics
нейтронограмма neutron diffraction pattern
нейтронограф neutron-diffraction camera
нейтронография neutron-diffraction study/analysis, neutron radiography
~, структурная neutron diffraction structure study
нейтронодефицитный neutron-deficient
нейтроноизбыточный neutron-excess
нейтрононепроницаемый neutron-tight
нейтронотерапия neutron therapy
нейтропения *med.* neutropenia, neutrophilopenia, neutrophilic leukopenia
нейтрофил *anat.* neutrophil(e), polymorphonuclear leucocyte
~, палочкоядерный stab/band/immature neutrophil
~, сегментоядерный segmented/mature neutrophil
~, юный juvenile neutrophil, metamyelocyte
нейтрофилез *med.* neutrophilia
нейтрофильный neutrophilic, neutrophilous
неканонизированный *rel.* uncanonized, unsainted
неканонический noncanonical, apocryphal
некапиталистический noncapitalistic
некатор *zool.* American hookworm *(Necator americanus)*
некатороз *med.* hookworm disease, necatoriasis
некачественный low-quality, of poor quality
неквадратный nonquadratic, nonsquare
неквалифицированный unskilled, unqualified; incompetent
~ рабочий unskilled worker
неквантованный, неквантовый unquantized, nonquantized
неклассифицированный unclassified; ungraded, unsorted
неклен *bot.* box elder, negundo *(Acer negundo)*
неклеточный *biol.* noncellular
некоаксиальный misaligned, uncoaxial, noncoaxial

нековкий inductile, unmalleable
некогерентность incoherence, noncoherence
~, зависящая от спина spin-dependent incoherence
некогерентный incoherent, noncoherent; scrambled
неколебательный non-oscillatory/-oscillating
некоммерческий nonprofit
некоммунистический noncommunist(ic)
некомпенсированный uncompensated, unrecompensed, unredressed
некомпетентность incompetency, incompetence
некомпетентный incompetent, unqualified
неконвертируемый, неконвертабельный *econ.* inconvertible
некондиционный sub-standard, off-grade
неконкурентный noncompetitive
неконституционный unconstitutional
неконтактный 1. noncontact(ing); 2. *med.* avoidant
неконтрастный *phot.* flat, lacking contrast
неконтролируемый 1. uncontrolled; unguided; unruled; 2. uncontrollable
неконфессиональный *rel.* undenominational
неконфирмованный *rel.* unconfirmed
некоординированный incoordinate, uncoordinate(d)
некоррелированность *math.* lack of correlation
некорродируемый non-corrodible/corrosive, corrosion-resistant
некратный *math.* aliquant
некредитоспособность *econ.* insolvency
некремия *med.* necr(a)emia
некрещенный *rel.* unbaptized, unchristened
некристаллический non-crystalline
некритичность *nucl.* noncriticality
некробиоз *med.* necrobiosis
некрогормон necrohormone
некроз *med.* necrosis
~, влажный liquefactive/colliquative necrosis
~, восковидный Zenker's necrosis/degeneration
~ головки плеча Panner's disease
~, десенный gingival necrosis
~, коревой measles necrosis
~ кости osteonecrosis
~ культи stump necrosis
~ мышц myonecrosis
~ поджелудочной железы, острый Balser's necrosis, acute pancreonecrosis
~ стеблей *bot.* stem blight
~, сухой coagulation necrosis
~, творожистый caseous/caseation necrosis
некролог obituary, necrologue
некрологический necrologic(al), epitaphic(al)
некромантия necromancy
некропневмония *med.* necropneumonia
некрополь necropolis
некротизированный necrotized; sphacelous
некротический necrotic
некрофил *med.* necrophile
некрофилия *med.* necrophilia, necrophilism
некроценоз necrocoenosis
~, гравийный pebble necrocoenosos
нектар nectar
~, выделяющий nectariferous
нектарин *bot.* nectarine *(Prunus persica nectarina)*
нектарник *bot.* nectary, honeycup, nectar gland
~, внецветковый extrafloral nectary
нектарница *zool.* sunbird *(Nectarinia metallica)*
нектарный nectar(e)ous, nectarean; nectarial
нектаровместилище, нектарохранилище *bot.* nectarotheca
нектаронос nectar plant
нектароносный *bot.* nectariferous
~ная железа nectar gland, nectary
нектароподобный nectarlike, nectarous
нектофор *zool.* nectophore, nectocalyx
нектрия nectria, coral spot
некультурный 1. uncultured, uneducated; rough (-mannered); 2. *bot.* wild, uncultivated
некурящий nonsmoking; *n.* nonsmoker
нелегированный unalloyed
нелейкемический *med.* aleukemic
нелетучий nonvolatile, fixed, nonfugitive
неликвид *econ.* nonconvertible stock, unmarketable goods
неликвидированный *econ.* uncleared, unliquidated *(of debt)*
неликвидный illiquid, nonliquid, unmarketable
~ные активы long-term assets
нелинеарность, нелинейность nonlinearity
~, несущественная incidental nonlinearity
~, плавная continuous/smooth nonlinearity
~ с гистерезисом backlash nonlinearity
~, слабая small nonlinearity
~, существенная esential/intentional nonlinearity
~ характеристики nonlinearity of response
нелогичность illogicality, irrationality
нелогичный illogical, irrational
нелокальный illocal
нелояльность disloyalty
нелояльный disloyal
нельма *zool.* Siberian white salmon *(Stenodus leucichthys nelma)*
нельсон nelson *(wrestling)*
нелюдим loner; unsociable person
нелюдимый unsociable; offish; lonely
неманевренный *av.* non-aerobatic
немаркированный unmarked
немарксист non-Marxist
немарксистский non-Marxist, unmarxist
нематериальный nonmaterial; nonsubstantial, incorporeal; supersubstantial
нематеций *bot.* nemathecium
нематод(а) *zool.* nematode, nematoid
нематоды *n. pl. zool.* nematodes *(Nematoda)*
нематоз *med.* nematoda infestation, nematodiasis
нематофикус, нематофитон *paleont.* nematophyton

немахровый *bot.* single *(of flower)*
Немезида, Немесида *myth.* Nemesis
немезия *bot.* figwort, nemesia *(Nemesia)*
немейский *myth.* Nemean
неменяющийся unchanging
немертины *pl. zool.* nemertine(an)s *(Nemertinea)*
неметалл nonmetal, metalloid; mineral
неметаллический nonmetallic, non-metalliferous
~кие товары *econ.* soft goods/wares
неметрический unmetrical, nonmetrical
немец German
немеченный unlabelled, untagged
немка I German (woman)
немка II cow/velvet ant *(Mutilla)*
немнoголистный *bot.* paucifolious
немногоцветковый *bot.* pauciflorous
немой 1. dumb; *n.* dumb man, mute; 2. barren; blank, unfilled
~ согласный *phon.* mute consonant
~ фильм silent film
~мая азбука deaf-and-dumb alphabet
немореходный unseaworthy
немота mutism, dumbness, muteness
немощь infirmity, feebleness
~, старческая infirmities of age
немузыкальный unmusical, tone-deaf
ненаблюдаемый unobserved, unwatched; unobservable
ненависть hate, hatred
~ к богам *rel.* misotheism
ненавязывание *rel.* nonintrusion
ненагруженный unloaded, empty; uncharged
ненадежность unreliability, untrustworthiness; insecurity
~, эксплуатационная unserviceability
ненадежный unreliable, insecure, unsafe, undependable, untrustworthy
~ в эксплуатации unserviceable
ненаказуемый nonpunishable, unpunishable
неналоговой nontaxable
ненамеренный unintended, unintentional, involuntary
ненапластованный *geol.* unstratified, ataxic, massive
ненапряженный strainless, unstrained
ненарушимый inviolable, inviolate, infrangible
ненаселенный unpopulated, uninhabited, unpeopled
ненасильственный *law* nonforcible, nonviolent
ненастроенный untuned, unadjusted
ненасыщенный unsaturated, nonsaturated; unsated
~ основаниями base-deficient/-unsaturated
ненец, ненецкий Nenets
ниниль nonnil, nonzero
ненка Nenets (woman)
ненормальность abnormality, irregularity, anomaly
ненормальный abnormal, irregular, anomalous
ненулевой nonzero
~ вектор *math.* nonvanishing vector
неоартроз *med.* ne(o)arthrosis, pseudoarthrosis
необветренный unweathered
необедненный undepleted, unstripped
необеспеченный 1. *econ.* uncovered, unsecured; 2. needy, withous means, unprovided for
~ заем unsecured loan
необлагаемый not taxable, untaxed, taxfree
необлицованный *tech.* unlined, not faced
необложенный *med.* uncoated, unfurred
необлученный unirradiated, nonirradiated
необнародованный unpublished, unpromulgated
необнаруженный undetected, unfound, undiscovered
необогащенный unprepared, crude, unenriched; *min.* undressed *(of ore)*
необозначенный unmarked, undesignated
необорудованный unequipped; unfurnished
необоснованность groundlessness, baselessness
необоснованный groundless, unfounded, illfounded, baseless; unreasoned
необработанный 1. *agr.* untilled, uncultivated; 2. raw, crude, rough; *tech.* unmachined, unfinished; untreated, undressed
необразованность lack of education; ignorance
необразованный uneducated; unlettered, illiterate
необрастающий *nav.* nonfouling
необратимость irreversibility, inconvertibility
необратимый irreversible, nonreversible; *fin.* inconvertble
необращающийся *econ.* nonnegotiable
необрезанный *agr.* unclipped, untrimmed; *timb.* undressed; *rel.* uncircumcised
необсужденный undiscussed, undebated
необусловленный unstipulated, unconditioned
необученный untrained, unskilled; *mil.* undrilled
необходимость necessity, need, indispensability
~, крайняя emergency
~, острая exigency
необходимый necessary, needed, essential, indispensable
необъективность partiality, bias(s)ed nature
~, политическая political bias
необъяснимый unexplainable, unexplicable, unaccountable
необъятность immensity, boundlessness
необязанный unliable
необязательный optional, facultative, unrequired, not obligatory
неовулканический *geol.* neovolcanic
неогегельянец, неогегельянский Neo-Hegelian
неогегельянство Neo-Hegelianism
неоген, неогеновый Neogene, Neocene
неогея Neogaea
неогнатический *zool.* neognathous
неоголотип *biol.* neoholotype
неограмматик neo-grammarian
неограниченный unrestricted, unlimited, unbounded, unrestrained
~ная монархия absolute monarchy

~нно маневренный *av.* aerobatic
неодадаизм *art* neo-Dadaism
неодарвинизм Neo-Darwinism
неодевон *geol.* Neo-Devonic period, Upper Devonian
неодим(ий) *chem.* neodymium, Nd
неодинаковый unlike, dissimilar
неоднозначность ambiguity, lack of univalence
~, устранять resolve an ambiguity
неоднозначный nonunique, ambiguous, multivalued, multiple-valued
~ное решение ambiguous solution
неоднолистность *math.* multivalence
неоднолистный *math.* multivalent, nonschlicht
неоднородность heterogeneity, inhomogeneity, discontinuity, non-uniformity, dissimilarity
~ потока flow non-uniformity
неоднородный heterogeneous, inhomogeneous, non-homogeneous, non-uniform; dissimilar
неодушевленный inanimate, insensate, insentient
неозой *geol.* Neozoic/Tertiary group
неозойский *geol.* Neozoic
неоимпрессионизм Neo-Impressionism
неоимпрессионистический, неоимпрессионистский Neo-Impressionist
неокантианец, неокантианский Neo-Kantian
неокантианство Neo-Kantianism
неокейнсианский Neo-Keynesian
неокейнсианство Neo-Keynesian theory
неокисленный unoxidized
неокисляемый inoxidizable, unoxidizable
неокисляющийся anti-oxidant, nonoxidizing
неокладной untaxable, nontaxable
~ сбор unassessed tax
неоклассик neoclassicist
неоклассицизм neoclassicism
неоклассический neoclassic
неоколониалистский neocolonial(ist)
неоком *geol.* Neocomian (stage)
неокончательный not final, inconclusive; *law* interlocutory
~ вид глагола *gram.* imperfective aspect
неокопанный *mil.* unentrenched
неокортекс *med.* isocortex, neocortex
неоламаркизм Neo-Lamarckism
неоламаркистский Neo-Lamarckian
неолит 1. *archeol.* Neolithic stage, Neolithic Age, New Stone Age; 2. neolith
неолитический neolithic
неомальтузианец, неомальтузианский Neo-Malthusian, Neo-Malthusianist
неомальтузианство Neo-Malthusianism
неомарксист Neo-Marxist
неомицин *chem.* neomycin sulfate
неомический *el.* nonohmic
неоморфный *biol.* neomorph(ic)
неон 1. *chem.* neon, Ne; 2. neon lamp/tube
неонацизм Neo-Nazism
неонацист, неонацистский Neo-Nazi
неопалеозой, неопалеозойский Neo-Pal(a)eozoic
неопалимый that cannot be scorched/singed
~мая купина *bibl.* the burning bush
неопифагореец, неопифагорейский Neo-Pythagorean
неопифагорейство Neo-Pythagoreanism
неоплазия, неоплазма *med.* neoplasia, neoplasm
неопластицизм *art* Neo-Plasticism
неоплатный irredeemable, irremunerable; insolvent
~ долг irredeemable debt
~ должник insolvent debtor
неоплатоник Neoplatonist
неоплатонический Neoplatonic
неоплейстоцен *geol.* Neo-Pleistocene
неопозитивизм Neo-Positivism
неополе *math.* neofield
неоправданный unjustified, unwarranted
неопределенность 1. *math.* indeterminacy, indeterminancy, uncertainty, indeterminate expression/form; 2. vagueness, indefiniteness
неопределенный 1. indefinite, indeterminate, undetermined, not fixed; 2. vague, unsure; ambiguous
~ член *gram.* indefinite article
~ная форма глагола *gram.* infinitive
~ное наклонение *gram.* infinitive mood
неопределимость indefinability
неопределимый indefinable, indeterminable
неопровержимый irrefutable, irrebutable
~мая презумпция *law* conclusive presumption
неопрокидывающийся *nav.* uncapsizing, noncapsizable
неопущение *pathol.* lack of prolapse
неорганизованный 1. unorganized; uncoordinated; nonunion *(of workers);* 2. wild, primitive
~ рабочий nonunionist
неорганический inorganic, anorganic
~кая химия inorganic chemistry
неореализм Neo-Realism
неореалист, неореалистический Neo-Realist
неориентированный unoriented, non-directional/-aligned
неоромантизм Neo-Romanticism
неоромантик, неоромантический Neo-Romantic
неорошаемый nonirrigated, unirrigated, unwatered
неортодоксальный nonorthodox, unorthodox, heterodox; dissenting
неосальварсан *pharm.* neoarsphenamine
неосмотрительный incautious, imprudent
неосновательный unfounded; insubstantial; *law* unmaintainable
неосновной nonbasic
неосознанный inconscient; unperceived; unconscious
неососсюрианский neo-de-Saussurian
неоспариваемый indisputable, unquestionable
неоспоримый incontestable, irrefutable, undeniable; *law* cogent; unchallengeable

неосталинизм Neo-Stalinism
неосторожность carelessness, incautiousness, imprudence, indiscretion
неосхоластизм Neo-Scholasticism
неосхоластический Neo-Scholastic
неосциллирующий nonoscillating, nonoscillatory
неосязаемый intangible, intactile, impalpable; insensuous
неоталамус *anat.* neothalamus
неотвратимый inevitable, inescapable, unpreventable
неотделимый inseparable, inherent, intrinsic
неотело *math.* neoring
неотения neoteny, neote(i)nia
неотип *biol.* neotype, hypotype
неотожженный unannealed
неотомизм Neo-Thomism
неотпущенный 1. *met.* untempered; 2. *rel.* unremitted, unpardoned *(of sin)*
неотработанный unfinished; unspent, unexhausted
неотражающий nonreflective, nonreverberatory
неотрегулированность *tech.* maladjustment
неотрегулированный *tech.* maladjusted
неотрицательность *math.* non-negativity/-negativeness, non-negative character/nature
неотрывный continuous, uninterrupted
неотступный insistent, persistent, relentless, unremitting
~ное преследование relentless pursuit
неотчуждаемость *law* inalienability
неотчуждаемый *law* inalienable, unalienable
неотъемлимый integral, inherent; inalienable
~мая част integral part
~мое право inalienable right
неофашизм neo-fascism
неофашист, неофашистский neo-fascist
неофилолог neophilologist, neogrammarian
неофит *rel.* new convert, newly baptised, neophyte
неофитовый neophytic, neomorphic
неофициальный unofficial, informal; unceremonious
~ные переговоры informal talks
неохлажденный uncooled, unchilled
неоцен, неоценовый *geol.* Neocene
неочищенный 1. unpurified, unrefined, unclarified 2. *chem.* unrectified; undistilled; 3. *agr.* unpeeled; unhulled
неощутимый imperceptible, impalpable, insensible inappreciable
неощущаемый nonsensory
непалец Nepali, Nepalese
непалка Nepalese, Nepali(woman)
непальский Nepalese, Nepali
непараллельность *comp.* nonconcurrency, lack in parallelism, nonparallelism
непарламентский unparliamentary
непарник *zool.* gypsy moth *(Ocneria dispar)*
непарнокопытный, непарнопалый *zool.* odd-toed, perissodactyle
непарноперистый *bot.* imparipinnate, odd -pinnate
непарный unpaired, nonpaired, odd, azygous, single
непатентованный unpatented; unlicensed
непахотный nonarable
непахотопригодный *agr.* unarable, unsuitable for tilling
непедагогический unpedagogic(al), uneducational
непентес *bot.* pitcher-plant, nepenthes *(Nepenthes)*
непер *phys.* neper, napier
непереваренный undigested
непереводимый untranslatable, unrenderable
непередаваемый inexpressible, indescribable; *econ.* nonnegotiable
неперемежающийся nonalternating, unintermitting
непереносимость intolerance
непереносимый intolerable, unendurable, unbearable
непереносный stationary, nonportable
непересекаемость *math.* disjointness
непересекающийся noncrossing, nonintersecting; *math.* disjoint
непереставляемый, неперестановочный noninterchangeable
непереходной immovable *(of holidays)*
непереходный *gram.* intransitive
непериодический, непериодичный nonperiodic(al), aperiodic(al), acyclic, noncyclic
непериодичность aperiodicity
непериферичный nonperipheral, nonmarginal
неперметр neper meter
неперов *math.* Napierian
непетолистный *bot.* nepeta-leaved
непилотируемый unmanned, pilotless, unpiloted
неписанный unwritten
непитательный innutritious, unnourishing
непифагоровый non-Pythagorean
неплавкий nonfusible, infusible, nonmelting
непланарный nonplanar
непластичный inductile, nonplastic
неплатеж non-payment, default in payment
неплатежоспособность *econ.* insolvency, inability to pay
неплатежоспособный insolvent, bankrupt, unable to pay
неплательщик nonpayer, defaulter
~, злостный incorrigible defaulter
неплодоносный unfertile, unfruitful
неплодоносящий *bot.* nonfruiting, acarpous, fruitless
неплодородие, неплодородность infertility, sterility
неплодородный infertile, barren, sterile
неплодущий infertile; sterile
неплоскостность nonflatness
неплотность noncompactness; leakage, leakiness
~ отливки casting leakage
неплотный 1. incompact, loose; 2. leaky, leaking, nontight; 3. low-density

неповиновение insubordination, insubmission, disobedience; *law* contumacy
~, гражданское civil disobedience
~, открытое defiance
неповинующийся insubordinate, insubmissive
неповреждаемый undamageable
~ пламенем flameproof
~ при тряске shakeproof
~ пылью dustproof
неповрежденный unimpaired, undamaged, intact
неповторимый unique, unrepeatable
непогашенный uncancelled, unused *(of stamps);* unslaked *(of lime)*; *econ.* uncleared *(of debt)*
непогода bad/ugly/inclement weather
непогрешимость impeccability, infallibility, inerrancy
непогрешимый infallible, inerrant, impeccable
неподатливый inflexible, unyielding; unmanageable; rigid, immalleable; uncompliant
неподатной *hist.* exempt from capitation, tax-exempt
неподвижность 1. immobility, nonmotility, fixity, stiffness; 2. *med.* acampsia, akinesia; catatonia
неподвижный statonary, fixed, immobile, motionless; stagnant
~ взгляд fixed stare
~ная спора *bot.* resting spore, hypnospore
неподвластный independent
неподкупный incorruptible, unbribable
неподпруженный *tech.* unsprung
неподрессоренный *tech.* springless, unsprung
неподсудность *law* nonjurisdiction, incompetence
~ по антитрестовскому законодательству *econ.* antitrust immunity
неподсудный *law* incompetent, not under the juisdiction (of)
неподтвержденный unconfirmed, uncorroborated, unsubstantiated; unverified
неподчинение insubordination, insubmission; noncompliance
непознаваемость *philos.* unknowableness, unapprehensiveness
непознаваемый *philos.* unapprehensive, unknowable, metagnostic, nescient
непокорность, непокорство disobedience, insubmission, insubordination; recalcitrance
неполадка defect, breakdown, failure, trouble
~, эксплуатационная operating trouble
неполагаемость *agr.* standing power
неполезность insalubrity
неполезный unwholesome, insalubrious
неполногласный *phon.* nonpleophonic
неполнозубый *zool.* edentate
неполнокристаллический hemicrystalline, hypocrystalline
неполномерный undersized
неполноправный non possessing full rights
неполнота imperfection, incompleteness; *law* diminution
~ наблюдений undercoverage
неполноценность inferiority
~, психическая mental deficiency
неполноценный inferior, inadequate, defective
неполный incomplete, insufficient; partial; imperfect
~ вес short weight
~ рабочий день part time, undertime
~ная занятость underemployment; part time
~ное использование underutilization
неполовозрелый *biol.* impuberal, preadolescent, immature
неполовой *biol.* asexual
неполучение nondelivery; nonreceipt
неполяризуемый, неполяризующийся nonpolarizable
неполярный nonpolar, apolar
непомерный excessive, exorbitant, immoderate
непомеченный unlabelled, unmarked
непонимание incomprehension, lack of understanding
непопадание *mil.* miss, missing the target
непоправимый irreparable, irremediable, irretrievable
~мая ошибка fatal mistake
непораженный *med.* unaffected
непорочность chastity, virginity, immaculacy
непорочный immaculate, chaste, pure
~ное зачатие *rel.* Immaculate Conception
непортящийся nonperishable, nonfouling
непосильный excessive
~, труд back-breaking toil
непоследовательность inconsequence, inconsistency
непоследовательный inconsequent, inconsecutive, unsuccessive; incoherent
непосредственно directly; immediately
~ доказывающий deictic
непосредственный direct, immediate
непостный *eccl.* ferial
непотопляемость *nav.* resistance to flooding, insubmersibility
непотопляемый unsinkable, insubmersible
непоявление *math.* nonoccurrence
неправильность irregularity, abnormality, anomaly
неправильнотакситовый *geol.* ataxitic
неправильный 1. irregular, abnormal, anomalous; 2. incorrect, wrong, fallacious, erroneous
~ глагол irregular verb
~ цветок *bot.* irregular/zygomorphic flower
~ная дробь *math.* improper fraction
неправоверный unorthodox
неправомерный illegal, illegitimate; wrongful
неправомочность *law* incompetence
неправомочный *law* incompetent, unauthorized
неправоспособность *law* disability, incapacity
неправоспособный *law* disabled, disqualified, incapable
неправосудный *law* improper, unjust, wrong
неправый 1. unjust; 2. wrong, mistaken

~ **суд** unfair trial
непраздна *rel.* pregnant
непредельный *chem.* unsaturated
непреднамеренный unintentional, undeliberate, involuntary
непредотвратимый inevitable, unavertible, impreventable, unavoidable
непредсказываемый unpredictable
непредубежденный unprejudiced, unbias(s)ed
непредумышленный unpremeditated
~**ное убийство** manslaughter
непредъявление *law* nonpresentment
непреломляющий *geol.* aclastic
непреодолимый insurmontable; irresistible
~**мая сила** *law* act of God, force-majeure
непрерываемый uninterruptable; uninterrupted
непрерывность continuity; *phys.* continuum
~**, равностепенная** equicontinuity
непрерывный continuous, uninterrupted, incessant, persistent, unbroken
~**, (псевдо)равностепенно** (pseudo) equicontinuous
непривилегированный unprivileged, unfavo(u)red
неприводимость *math.* irreducibility, indecomposability
неприводимый *math.* irreducible, indecomposable; irredundant
непригодность unsuitability, unfitness, inaptitude
~ **к военной службе** *mil.* ineligibility
~ **к плаванию** *nav.* unseaworthiness
~**, физическая** physical unfitness
непригодный unfit, unsuitable, unserviceable; *mil.* ineligible
неприемливый unacceptable, unsuitable, objectionable
неприкосновенность immunity, inviolability
~ **жилища** inviolability/sanctity of the home
~ **личности** personal immunity
~ **парламентская/дипломатическая** parliamentary/diplomatic immunity
неприкосновенный untouchable, inviolable, immune
~ **запас** reserve(d) funds; *mil.* emergency rations, iron ration
неприкрепленный detached; drifting, floating
неприменимый inapplicable; unusable; irrelevant
неприсоединение *polit.* nonalignment
неприсоединившийся *polit.* nonaligned
неприспособленный maladjusted, unadjusted, unadapted; *biol.* unspecialized
неприспособляемость inadaptability
неприсутственный *law* nonjudicial
~ **день** public holiday; *law* nonjudicial day
неприсущий extrinsic
неприсягатель *hist.* nonjuror
неприцельный *mil.* unaimed
непричастный not privy (to), not implicated (in)
~**ное прилагательное** *gram.* adjectival participle
непричащающийся *rel.* noncommunicant
неприятель enemy
непровар *tech.* 1. faulty/incomplete/poor fusion; 2. lack of penetration, incomplete/poor penetration
непроводимость *phys.* nonconduction
непроводник *phys.* nonconductor, nonconducting material, dielectric
непроводящий nonconductive, nonconducting
непроданный unsold; *econ.* outstanding
непродуктивный unproductive; barren *(of soil)*
непрозрачный opaque, nontransparent; cloudy *(of liquids)*
непроизводительный unproductive, nonproductive; wasteful
непроизводный *gram.* primary, underived, nonderivative
непроизводственный unearned
~**ные услуги** software
непроизвольный involuntary, unintentional; contravolitional
непромокаемый waterproof, impermeable
~ **плащ** mackintosh, raincoat
непромышленный uneconomic, unpayable, *min.* unprofitable *(of ore)*
непроницаемость impermeability
~ **для излучения** *nucl.* radio-opacity, radiopacity
непроницаемый unpenetrable, impermeable, impervious; leakproof; tight
~ **для звука** soundproof
~ **для пули** bulletproof
непропечатка *print.* missing type
непропорциональный disproportionate
~**ное сложение** malconformation
непропускающий nontransparent, nontransmitting; impervious, tight
непрореагировавший *chem.* unreacted
непротивление nonresistance
непротиворечивость *math.* consistency, noncontradiction, compatibility
~ **данных** consistency of data
непроходимость obstruction, impaction; impassability, imperviousness; blockage, clogging
~ **кишечника/кишечная** *med.* ileus, enterocleisis, intestinal obstruction, obstipation, volvulus
непроходимый impassable, impervious, impenetrable
~ **вброд** *mil.* unfordable
непрочность instability; fragility; frailty
~ **к истиранию** *leath.* liability to rub-off
непрочный undurable, fragile, unsolid; not fast *(of colours)*
непрощенный *rel.* unabsolved, unpardoned; *econ.* unremitted *(of debts)*
непрощупываемость *med.* impalpability
~ **пульса** acrotism
непрямолинейность misalignment; nonlinearity
непрямолинейный nonlinear
непрямоходящий *zool.* pronograde

Нептун *astr., myth.* Neptune
нептуний *chem.* neptunium, Np
нептунический *geol.* neptunian, neptunic
нептуния *bot.* sensitive brier, neptunia *(Neptunia)*
неработоспособность disablement, disability
нерабочий nonworking; inactive; *tech.* inoperative
~ день free/nonworking day
неравенство 1. disparity, inequality, nonequivalence; 2. *math.* inequation
~, абсолютное absolute/unconditional inequality
~, жесткое nonslack inequality
~, линейное linear inequality
~, нежесткое slack inequality
~, нестрогое conditional inequality
неравновесие disbalance; disequilibrium, imbalance
~сия, состояние disequilibrium/nonequilibrium state
неравногранный scalenohendral
неравнозначность *philos.* inadequacy
неравнозубчатый inequidentate, unequally-toothed
неравнокрылый *bot.* anisopterous
неравнолистный *bot.* anisophyllous
неравнолопастный anisolobous, inequilobate
неравномерность irregularity, nonuniformity, variation, unevenness
~ вершины импульса overshoot on the pulse top
~ в паузе между импульсами undershoot between pulses
неравномерный irregular, uneven, nonuniform
~ное насаждение uneven stand
неравнополярный *bot.* anisopolar
неравноправие social inequality
неравноправный disparate, unequal
неравносемядольный *bot.* anisocotyledonous
неравносторонний scalene, inequilateral
неравный unequal, disparate
~ брак mesalliance, misalliance
~ные шансы long odds
неразвертывающийся *math.* nondevelopable
неразвитость (intellectual) backwardness, (mental) retardation, underdevelopment
неразвитый, неразвитой undeveloped, backward; immature, rudimentary
нераздельнокипящий *chem.* azeotropic
нераздельноплодный *bot.* monoclinic, monoclinous
нераздельнополый *bot.* monoclinous
нераздельный impartible, inseparable; unsegmented
~ное совладение *law* tenancy in common
неразличимый indiscernible, imperceptible
неразложимый indecomposable, irresoluble, indissoluble
неразлучник *zool.* lovebird *(Agapornis)*
неразрешенный unsolved; banned, prohibited, unallowed
~ная выборка *comp.* inadvertent access
неразрешимый insoluble, unsolvable, irresolvable
неразрывность continuity
неразъемный nondetachable, all-in-one, onepiece, single-part
нераковый *med.* nonneoplastic, noncancerous
нерасплывающийся nondeliquescent
нераспознаваемый undecipherable, adiagnostic
нераспространение nonploliferation
нерастворимый in(dis)soluble, indissolvable, indissolving
нерасторгнутый *law* undissolved *(of marriage)*
нерасторжимый inseverable; *law* indissolvable
нерастрескивающийся *bot.* undehiscent
нерасчетный off-design
нерасчлененный *geol.* unrugged, whole; nonarticulated, nonsegmented
нерасщепленный aschistic, diaschistic, undifferentiated; unsplit
нерафинированный impurified, unrefined; crude
нерв *anat.* nerve; *bot.* rib, vein
~, барабанный tympanic nerve
~ бедра/бедренный femoral nerve
~, блоковый trochlear/pathetic/fourth nerve
~, блуждающий vagus/pneumogastric nerve; tenth cranial nerve
~, главный *bot.* midrib
~, глоточный pharyngeal nerve
~, диафрагмальный phrenic nerve
~ листа leaf rib
~, лицевой facial nerve, seventh cranial nerve
~, седалищный (great) sciatic nerve
нерватура, нервация *bot., zool.* venation, nervation
~, дланевидная palmate venation/nervation
нервация nervation, venation
~, вилообразная furcate nervation/venation
~, дланевидная palmate nervation/venation
~, перистая pinnate nervation/venation
~, сетчатая reticulate nervation/venation
нервизм *med.* nervosism, nervosity
нервно-артритический neuroarthritic
нервнобольной nerve/nervous patient, neurotic
нервно-мышечный neuromuscular, nervimuscular
нервно-психический neuropsychic
нервность nervousness; irritability
нервный nerval, neural; nervous; nerve *(attr.)*
~ припадок fit of nerves
~ узел ganglion
~ центр nerve center
нервюра 1. *bot., zool.* nervure, nerve; 2. *av.* rib; 3. *arch.* nervure
~, балочная girder(-type) rib
~ крыла wing rib
~, ложная false/form(er) rib
~, рамная braced rib
~, силовая stiff/strong rib
~ со сплошной стенкой plate-web rib
~, ферменная truss rib
~, хвостовая trailing-end rib

нерегулируемый non-adjustable
нередактируемый not editable
нерезидент nonresident
нереида 1. *myth.* sea/water nymph; 2. *zool.* nereid *(Nereis)*
нерезкость *cin.* unsharpness, blur
Нерей *myth.* Nereus
нерентабельный unprofitable, unremunerative, uneconomic(al)
нерест *zool.* spawning
нерестилище spawning ground
нерестование *zool.* spawning
неречевой nonverbal, nonvocal, nonspeech
нержавейка stainless steel
нержавеющий stainless, rustless, rust-resistant/-proof, nonrusting, noncorroding
нерине *bot.* Guernsey lily/flower, nerine *(Nerine)*
нерка *zool.* blueback salmon, sockeye (salmon) *(Oncorhynchus nerka)*
неровность 1. unevenness, roughness, ruggedness, irregularity; 2. inequality
~, волнообразная corrugation, undulation
~ краев jagged edges, edge scalloping
~ рельефа roughness/ruggedness of relief
неродственный unallied; unrelated
нерожавшая *philos.* nonparous
Нерон *hist.* Nero
неронов, нероновский Neronian, Neronic
нерпа *zool.* phoca, seal *(Phoca vitulina)*
нерпуха *zool.* eared seal *(Eumetopias jubatus)*
нертера *bot.* bead plant *(Nertera granadensis)*
нерудный nonmetalliferous, nonmetallic
нерукотворный *rel.* not of human making
нерусский non-Russian
нерушимый indestructible, inviolable, indissoluble
несварение желудка *med.* indigestion
несвоеобразный *math.* nonsingular
несвойственный unusual (for/to); extrinsic; incidental, accidental
несвязанный disconnected, unbound; uncohesive, incoherent; *chem.* uncombined
несвязность incoherence, incohesion, disconnectedness
~ мышления rambling thought
~ речи rambling speech, incoherence, divagation
несвязный incoherent, disconnected; desultory
несгораемый fireproof, incombustible
~ шкаф safe
несезонный out-of-season, off-season
несейсмический aseismic, nonseismic
несекретный unclassified, nonclassified
несение *mil.* execution, performance
~ вахты *nav.* watch keeping/standing
~ службы, повседневное routine
несессер toilet/dressing case
несжимаемый incompressible, incondensable; *phys.* incoercive
несимметричность, несимметрия asymmetry, dissimmetry, unsymmetry
~ распределения случайностей skewness
несимметрический, несимметричный asymmetric(al), dissimmetric(al); irregular, anisomerous
несинхронизованный free-running, out-of-step; nonsynchronized; astable
несинхронный asynchronous, nonsynchronous
несистематизированный unstructured, unsystematized
несистематический unsystematic, unmethodical
несклоняемый *gram.* indeclinable; aptotic, uninflected
нескользящий skidproof, nonskid, antiskid, nonslip(ping)
неслежавшийся noncaking
несливной *med.* discrete *(of rush, eruption)*
неслучайность *stat.* nonrandomness
несмачиваемый nonwettable
несменяемость irremovability (from office)
несменяемый 1. unrelieved, irremovable; 2. unexchangeable, unchangeable, constant
несметный incalculable, innumerable, countless, numberless
несмешиваемость immiscibility
несмыслоразличительный *ling.* nondistinctive
несмычность *phon.* nonplosiveness, nonocclusiveness
несмягченный *phon.* hard, nonpalatized
несоблюдение inobservance, nonconformity
~ закона *law* infraction
несобственный *math.* improper; non-intrinsic, singular; ideal
несовершение *law* failure
несовершеннолетие *law* infancy, minority, nonage
несовершеннолетний *law* underage; *n.* infant, minor
несовершенный 1. imperfect, incomplete; inadequate; *law* uncommitted *(of crime); bot.* declinuous; 2. *gram.* imperfective
~ вид imperfective aspect
~ная ткань *bot.* merenchyma
несовместимость incompatibility; incongruity; irreconcilability
~ лекарственных средств drug incompatibility
~, обоюдная cross incompatibility
~, тканевая tissue incompatibility
несовместимый incompatible, incongruous; irreconcilable
~, попарно mutually exclusive
несовместный *math.* incompatible, inconsistent
несовмещение misfit, mismatch
~ цветов colour misregistration
несовпадение discrepancy, noncoincidence; lack of convergence; *comp.* mismatch(ing); incongruity
~ красок colour misregister
~ осей misalignment
~ фазы out of phase
несовременный out-of-date
несогласие 1. dissent; *math.* variance; disagree-

ment; discord(ance); disconformity, unconformity; 2. refusal, noncompliance
~, видимое apparent disconformity
~, тектоническое/дислокационное tectonic/structural unconformity
~, угловое angular unconformity/displacement, clino-unconformity
несогласование *gram.* nonagreement
несогласованность mismatching; lack of coordination, incoordination; inconsistency, discord(ance)
несогласованный mismatched, uncoordinated; atactic; not in agreement, discordant
несократимость *math.* irreducibility
несократимый, несокращаемый *math.* irreducible
несокращенный unabridged; unabbreviated
несокрушимый undestroyable, indestructible; invincible
несоосность malalignment, misalignment, nonaxiality
несоосный nonaxial, misaligned
несоответствие mismatch, discrepancy, disparity, incongruity
~ характеров incompatibility of temperaments
несоответствующий inappropriate, inadequate, unsuitable, incommensurate (with), incongruent
несопоставимость incommensurability; incompatibility
несоразмерность disproportion; inadequacy
несоразмерный disproportionate, incommensurate, inadequate
несостоятельность 1. *econ.* insolvency, bankruptcy; 2. *med.* incompetence, inability, incapacity; 3. groundlessness, unfoundedness
~ дыхательного центра failure of the respiratory centre
несостоятельный 1. *econ.* insolvent, bankrupt; 2. unprosperous, of modest means; 3. groundless, unfounded; *math.* inconsistent, untenable
несотканный *math.* webless
несохранение nonpreservation
~ четности nonconservation of parity
несочетаемость incongruity
несочлененный inarticulate(d), unjointed
неспастический *med.* aspastic
неспаянный unwelded, unsoldered; *fig.* nonadhering
неспекающийся noncaking, nonsintering
неспелый unrife; green
неспециальный general-purpose, universal
неспецифический unspecific, nonspecific; paraspecific
несплавление *tech.* incomplete/faulty/poor fusion
неспособность inability, incapability, incapacity
~ к детерождению incapacity of bearing
несправедливость injustice, inequity, unfairness
несправедливый inequitable, unjust, unfair
несрастание, несращение *med.* nonunion
несродный lacking affinity, uncongenial
Несс *myth.* Nessus
нестабильность instability, unstable state
нестабильный unstable, nonsteady, transient
нестандартный substandard, nonstandard
нестационарный nonstationary, transient
нести 1. carry, bear; support; 2. perform
~ дежурство be on duty
~ караул *mil.* be/stand guard
~ обязанности perform the duties
~ потери *mil.* suffer losses, incur casualties
~ убытки *fin.* incur losses
нестроевой *n., mil.* noncombatant
несудимость *law* absence of previous conviction
несуммируемость *math.* nonsummability
несуммируемый *math.* nonsummable
несуммирующий nonadditive
несущая *rad.* carrier
~ изображения picture/video carrier
~, контрольная pilot carrier
несущий carrying, bearing; supporting
~ венец/коронку *bot.* coronated
~ винт *av.* (main) rotor *(of helicopter)*
~щая поверхность carrying area, lifting/supporting surface; *av.* airfoil
несходимость divergence, nonconvergence
несходность dissimilarity, dissimilitude, unlikeness; disparity
несходство disparateness; discrepancy
несчетно-бесконечный uncountably infinite
несъедобный uneatable; inedible
~ гриб inedible mushroom
несяк floeberg
нетабельный *mil.* unauthorized; *tech.* expedient, improvised
нетабличный abnormal
нетанниды *pl.* nontanning matter/substances, nontannins
нетленность imperishability
нетождественный nonidentical; inadequate
нетождество nonidentity
нетопырь *zool.* bat *(Verspertilio)*
неточность inaccuracy, inexactness; laxity
нетрадиционный unconventional; nontraditional
не-тронь-меня *bot.* snapweed, touch-me-not
нетрудовой not derived from labo(u)r
~ доход *econ.* unearned income
нетрудоспособность disability, disablement, incapacity (to/for work), invalidity
~, временная/частичная temporary/partial invalidity
нетрудящийся *n.* nonworker
нетто *com.* net
~ импорт net-import surplus
~ процент pure/net interest
~ чартер *law* net charter
~ экспорт net-export surplus
неубранный *agr.* unharvested
неубывающий non-decreasing
неуважение disrespect, irreverence; *law* contempt
~ к флагу disrespect to colours
неуд *abbr.* **неудовлетворительно** *educ.* unsat-

isfactory (school) grade
неудача failure, reverse, rebuff; misfortune
~ при поиске *comp.* miss
неудачный unsuccessful, unfortunate; poor
неудовлетворящий unsatisfactory
~ стандартам substandard
неукомплектованный shorthanded, undermanned, understaffed
неуместный 1. inappropriate, improper, misplaced; 2. *law* irrelevant
неумышленный inadvertent, unintentional, unpremeditated
~ное убийство *law* manslaughter
неуплата *law* default of payment, nonpayment
~ задолженности по кредиту *econ.* credit default
~ процентов default of interest
неуплаченный unpaid, outstanding; delinquent
неуполномоченный unanthorized; unaccredited
неупомянутый unnamed, unmentioned; unstated
неупорядоченность disorder, disarray
неупорядоченный disordered, unsystematic; random(ized)
неупотребление disuse
неуправляемый out-of-control/-command, uncontrollable, nonsterable; *nav.* disabled, unnavigable; *av.* unpiloted; unguided *(of rocket)*
неупряжнение nonexercise
~ния, закон law of disuse
неуравновешенность unbalance, unbalancing, disbalance, non-equilibrium; instability
~, статическая static(al) unbalance
неуравновешенный unbalanced, out-of-balance, unstable
неурожай *agr.* crop/harvest failure, bad/poor crop
неуспеваемость *educ.* poor progress, underachievment
неуспевающий 1. lagging behind, underachieving; 2. *n.* underachievier, weak student, slow lerner
неустановившийся transitional, unsteady, transitory; *el.* transient
неустойка *law* forfeit *(for breach of contract);* smart money
неустойчивость instability, unsteadiness; inconsistency; fluctuation; *geol.* astaticism, impersistence
~, нервная/психическая nervous/mental instability
~ по крену, ~, поперечная lateral/roll(ing) instability
~, путевая *av.* directional/yaw instability
~ системы ориентации attitude control instability
~, спиральная corkscrew/helical instability
~ типа изгиба wriggle instability
~ типа перегиба kink instability
неустойчивый unstable, unsteady, astatic; fluactuating, labile; *med.* unbalanced; *geol.* mobile, yielding
неучет neglect, failure to take account of

неф *arch.* nave
нефанализ nephanalysis
нефелин *miner.* nepheline, nephelite
нефелиум *bot.* lychee *(Nephelium; Litchi)*
нефелогилея cloud/mossy forest
нефелометр nephelometer, turbidimeter
нефелометрия nephelometry, turbidimetry
неферментный *bioch.* nonenzymatic
нефиксированный unstable; floating
неформатированный *comp.* nonformatted
нефоскоп nephoscope
~, грабельный (Besson) comb nephoscope
~ прямого визирования grid nephoscope
~, решетчатый grating nephoscope
нефрит I *med.* nephritis (*pl.* -tides), Bright's disease
~, военный trench/war nephritis
~, восходящий ascending nephritis
~, гнойничковый suppurative nephritis
~, злокачественный malignant nephritis
~, канальцевый tubular nephritis
~, лучевой radiation nephritis
~, межуточный interstitial nephritis
~, острый acute (glomerulo)nephritis
~, очаговый focal nephritis
~ у беременных nephritis gravidarum
нефрит II *geol.* nephrite
~, белый tremolite
~ Боуэна Bowen nephrite, bowenite
нефритит nephritic stone
нефробластома *med.* Wilm's tumo(u)r, nephroblastoma
нефрогенный nephrogenic, nephrogenous
нефроз *med.* nephrosis
~, генуинный congential nephrosis
~, гидремический hydronephrosis
~, желтушный cholemic nephrosis
~, мышьяковый arsenous nephrosis
~, ртутный mercury nephrosis
~, сулемовый sublimate nephrosis
нефрокапсулэктомия nephrocapsectomy
нефролепис *bot.* sword fern *(Nephrolepis)*
нефролитиаз *med.* nephrolithiasis
нефрома nephroma
~, эмбриональная embryoma of the kidney
нефрон *anat., zool.* nephron
~, корковый cortical nephron
нефропатия *med.* nephropathy
~, волчаночная lupus nephropathy
~, лихорадочная febrile nephropathy
~, миеломная myelonephropathy
нефросклероз *med.* nephrosclerosis
~, доброкачественный/злокачественный benign/malignant nephrosclerosis
~, сосудистый vascular nephrosclerosis
нефростомия *med.* nephrostomy
~, внутрилоханочная intrapelvic nephrostomy
~, временная temporary nephrotomy
~, постоянная constant nephrotomy
нефротиф *med.* nephrotyphoid

нефроэктомия *surg.* nephrectomy
~, **внебрюшинная/латеральная** paraperitoneal nephrectomy
~, **поясничная** lumbar/posterior nephrectomy
~, **чрезбрюшинная** abdominal/anterior nephrectomy
нефтеаппаратура oil-production equipment
нефтебаза oil bulk plant, tank farm, oil tank yard/installation
~, **морская** off-shore reloading store
~, **перевалочная** oil handling terminal
~, **призаводская** crude-oil yard, oil-stock yard
~, **распределительная** oil distributing center, distribution tank farm
~, **тупиковая** terminal tank farm
нефтевоз oil carrier/tanker/truck, oiler, oil-carrying ship
нефтегазоводоотделитель three-phase separator
нефтегазовый oil-and-gas
нефтегазонакопление oil and gas accumulation
нефтегазоносность oil and gas content
нефтегазопровод oil and gas pipeline
нефтегиль *geol.* neft de gil, neft-gil
нефтегруз cargo oil
нефтеденсиметр oleometer
нефтедобывание oil extraction
нефтедобыча oil production/extraction
нефтеемкость oil capacity, capacity for oil
нефтезавод oil refinery
нефтезалежь oil pool
~, **козырьковая** peaked oil pool
нефтеловушка oil trap
нефтемер oil meter
нефтеналивной oil-loading; bulk-oil
~**ное судно** oil tanker
нефтенепроницаемый oiltight
нефтеносность presence of oil, oil-bearing capacity, oil content
нефтеносный petroliferous, oil-bearing
нефтеотдача oil recovery
~, **суммарная** cumulative/ultimate oil recovery
нефтеотстойник oil-settling tank
нефтеочистка, нефтепереработка oil/petroleum refining
нефтеперегонка oil distillation
нефтеперерабатываiощий oil-/petroleum-refining
~ **завод** refiner
нефтеподогреватель oil heater
нефтепорт oil harbo(u)r
~ **тупикового типа** terminal oil harbo(u)r
нефтепровод oil pipeline
нефтепродукт oil/petroleum product/derivative
нефтепроизводящий oil-producing/-yielding
нефтепромысел oil field
~, **морской** offshore oil field
нефтепромышленность oil industry
нефтепроницаемый oil-permeable
нефтепроявление oil show/seepage/token
нефтеразведка oil prospecting/exploration
нефтесборник crude (oil) collector, initial oil storage tank
нефтесборщик oil skimmer
нефтесодержание oil content
~ **объемное** volumetric oil content
нефтестойкий oil-resistant/-resisting
нефтетопливо fuel oil
нефтехимик petrochemist
нефтехимия petrochemistry, petroleum chemistry
нефтехранилище oil storage/tank/dump
~, **морское** offshore oil storage
~, **наземное** surface oil storage
~, **промысловое** field oil storage
нефтецемент diesel oil cement
нефть (mineral) oil, petroleum, crude (oil)
~, **асфальтовая** asphalt-base petroleum
~ **в пласте** oil in place/situ
~, **едкая** sour oil
~, **легкая** low-density oil
~, **мигрировавшая** migratory oil
~, **натуральная/неочищенная** crude oil/petroleum
~, **первичная** primary oil, protopetroleum
~, **поглощенная** absorbed/occluded oil
~, **сырая** crude oil/petroleum, (petruleum) crude
~, **топливная** fuel oil
~, **трещинная** crevice oil
~, **черная смолистая** maltha, viscid bitumen
~, **чистая** pure/clean oil
нефтяник 1. oil-industry worker, oilman; 2. oil-well expert
нефтянка hot-bulb engine
нефтяной oil *(attr.)*
~ **источник** oil well
~ **фонтан** oil gusher
~**ная вышка** derrick
нехватка lack, shortage, deficiency, scarcity
~ **жилья** housing demand
~ **ликвидности** strain on liquidity
~ **ликвидных ресурсов** liquidity bottleneck
~ **наличных денег** cash deficit
~ **продовольствия** food shortage
~ **производсвенных мощностей** capacity bottleneck
~ **рабочей силы** labo(u)r stringency, manpower tightness
нехворощ *bot.* sagewort wormwood *(Artemisia campestris)*
неходкий *econ.* unsaleable, unmarketable, not in demand
неходовой *tech.* 1. not running, not in working order; 2. unmarketable, not in demand
нехристь *hist.* unbeliever, pagan, infidel
нецветковый *bot.* cryptogamic, cryptogamous
нецелесообразный unsuitable, inexpedient, unpractical
нецелое *n. math.* nonintegral, noninteger
нецелочисленный, нецелый nonintegral, fractional

нецензурный unquotable, obscene
нецентр nonnucleus
нецентральный off-centre, noncentral, acentral
нецентрированность eccentricity, misalignment
нецентрированный eccentric
нецерковный unclerical
нециклический noncyclic, acyclic
нечеловеческий 1. inhuman; 2. preterhuman; superhuman
нечеткий illegible, indecipherable; indistinct
нечетно-кратный odd-multiple
нечетно-мерный odd-dimensional/-measured
нечетно-нечетный odd-odd
нечетность *math.* oddness, imparity
~ти, контроль imparity check, odd control
нечетно-четный odd-even
нечетный odd, uneven
нечисловой nonnumerical
нечистокровный half-bred, mongrel
нечистый 1. unclean, soiled, dirty; 2. impure, adulterated; 3. *rel.* unholy; 4. *n.* the Evil Spirit
~тая пища *rel.* unclean food
~тая порода impure breed
~тая сила the Evil Spirit/One
~ное произношение defective pronunciation
нечленораздельный inarticulate, unintelligible
~ная речь *med.* anarthria
нечувствительность insensitivity, insensibility, numbness
нечувствительный 1. insensitive, nonresponsive; *med.* indolent; 2. insensible
~ к цвету colo(u)rblind
нешоссированный *constr.* unmacadamized
нештампованный *tech.* unimpressed
нештатный supernumerary, not on the regular staff; free-lance
нешунтированный unshunted
неэвклидов(ый) non-Euclidean/Euclidian
неэквипотенциальность nonequipotentiality
неэкранированный unshielded, unscreened
неэластичность inelasticity, rigidity
неэндемичный *med.* ecdemic
неэрмитовый *math.* non-Hermitian
неэтилированный unleaded, unethylated
неэффективность inefficiency, ineffectiveness
неэффективный inefficient, ineffective, inefficacious; inoperative
неявка absence, nonapperance; *law* default of appearance, failure to appear
~ в суд default; contumacy
~ на работу absenteeism
неявный unobvious, unapparent, covert; *math.* implicit
неядовитый nonpoisonous, unvenomous; *chem.* nontoxic
неязыковый extralinguistic; nonlinguistic
неясновыраженный poorly defined, indistinct
неяснораковистый *zool.* subconchoidal
неясыть *zool.* barred owl *(Strix varia)*
ниацин *bioch.* niacin, nicotinic acid
нивальный nival
~ пояс perpetual snow belt, nival belt
нивация nivation, nival erosion
нивелир *tech.* (geodetic) level, levelling box, spirit level
~, глухой dumpy level
~ с перекладной трубкой Y-level
~ с элевационными винтами tilting level
~ со зрительной трубкой stadia level
нивелирование, нивелировка level(l)ing, grading, alignment, evening-out
~, гипсометрическое hypsometric levelling
~, магистральное primary levelling
~ теодолитом theodolite levelling
нивелировать level, bring into level position
нивелировщик level(l)er, levelman, surveyor,
нивелир-теодолит level-theodolite
нивяник *bot.* chrysanthemum, white weed *(Leucanthemum)*
нигериец, нигерийский Nigerian
нигерийка Nigerian (woman)
нидерландец Dutchman
нидерландка Dutchwoman
нидерландский Dutch, Netherlandic, Netherlandian
нижележащий underlying, subjacent
нижнеальпийский subalpine, alpestrine
нижнебоковой inferiolateral
нижнеглазничный *anat.* infraobital
нижнеглоточный lower-glottalic
нижнезадний inferoposterior
нижнелужицкий Low(er)-Sorbian
нижненемецкий Low-German
нижнереберный *anat.* costoinferior, infracostal
нижнесилурийский Lower-Silurian
нижнесерединный inferomedian
нижник *mach.* bottom part/section; dolly
нижний 1. lower; 2. *geol.* Lower, Early *(of periods)*
~ задний inferoposterior
~ чин *mil.* ranker
~ этаж ground floor
~нее белье underwear
~няя губа/челюсть lower lip/jaw
~няя палата Lower Chamber/House
~няя юбка petticoat
низ 1. lower part; bottom; ground floor; 2. *auto.* underbody
~ отливки drag side of a casting
~, фреатический *geol.* phreatic low
низведение debasement, bringing down
~ верхушки *med.* apicolysis *(of a lung)*
низвергать(ся) rush/crash down
низвержение overthrow, subversion
низина lowland, depression, low place
~ приливно-отливной полосы mud/tidal flat
~, топкая callow, everglade, low-lying meadow, marshy meadow
~, топографическая topographic low
низинный low-lying, lowland *(attr.)*

~ **луг** *agr.* lowland meadow
низкий 1. low; 2. base, abject, dishono(u)rable, mean
~ **голос** deep voice
низковольтный low-voltage
низковязкий low-viscosity
низкоглиноземистый low-alumina
низкогорье low-mountain relief
низкожелезистый low-ferruginous
низкозольный low-ash
низкокалорийный low-calorie
низкокачественный low-grade/-quality
низкокипящий low-boiling
низколегированный low-alloy(ed)
низколесье low/shrub forest
низколетающий low-flying
низконапорный low-pressure
низкооборотный low-speed
низкообъемный low-volume
низкоомный *el.* low-resistance
низкооплачиваемый *econ.* low-/poorly paid
низкоплавкий low-melting
низкоплан *av.* low-wing monoplane
низкопоклонство *polit.* cringing, obsequiousness, servility, toadyism
низкопробность inferior quality, low grade
низкопробный base, low-grade
низкопродуктивный low-yielding
~**ное животное** low-yielder
низкопроцентный low-percentage/-grade
низкорамный *tech.* low-built
низкорасположенный low-set/-sited/-lying
низкорослость dwarfism, dwarfishness; short stature
низкорослый low(-growing), nanous; stunted; undersized, dwarfish, shortish
низкосортный low-trunk, short-stemmed
~ **лес** coppice forest/wood
низкоствольник coppice forest/wood, shrubwood
низкотравный short-grass
низкотравье short grass
низкоурожайный *agr.* low-yielding
~**ная культура** low yielder
низкочастотный subsonic, audio-/low frequency
низменность lowland, low ground/country
~**, болотистая** callow, everglade
низменный low(-lying)
низмянка *bot.* chaffweed *(Centunculus)*
низовик *polit.* worker in basic organization
низовка *poilt.* basic organization
низовой I 1. *geogr.* downstream, lower, on the lower reaches; 2. bottom, basilar, basal; 3. deep, grave *(of sound)*
~ **ветер** wind blowing from downstream
~ **злак** low grass
~ **лист** bottom/basal leaf
низовой II 1. local; *polit.* grass-roots; 2. *n. min.* bottomer
~ **работник** local worker
~**вая организация** basic/local organization
~**вая печать** local press
низовье *pl. geogr.* the lower course/part of river
низший 1. inferior, lower; 2. most primitive; 3. *educ.* primary; 4. subaltern, subordinate; junior
~**шее образование** primary education
~**шее растение** lower plant
~**шие организмы** *biol.* the lowest organisms
низы *pl. geol.* lowermost strata
Ника, Нике *myth.* Nike
никандра *bot.* apple of Peru *(Nicandra)*
никеверин niceverine
никейский Nicene
никеленосный nickeliferous
никелин nickeline
никелирование nickel plating/facing
~**, химическое** electroless nickel plating
никелировать nickel(-plate), nickelize
никелировка nickel plating; coating of nickel
никелистый containing nickel
никель nickel, Ni
~**, зерненый** nickel shot
~**, сернокислый** nickel sulphate
~**, теллуристый** tellurnickel, melonite, nickel tellurite
~**, хлористый** nickel chloride
никодуст nicotine sulphate dust
николаевка *hist.* bank note *(during the reign of Nicholas II)*
николаит nicolayite
николь *cryst.* nicol(prism)
~**, скрещенный** crossed nicol
никонианец *hist., rel.* follower of Nikon *(patriarch of the Russian Orthodox Church)*
никонианство official Russian Orthodox Religion *(practiced by the Old Believers in czarist Russia)*
никотин nicotine
никотиновый nicotinic, nicotinian
никтагиновые *n. pl. bot.* four-o'clock family *(Nyctaginaceae)*
никталгия *med.* night pain, nyctalgia
никталопия *med.* night blindness, nyctalopia
никтантес *bot.* night jasmine, tree of sadness *(Nyctanthes)*
никтериния *bot.* night phlox *(Nycterinia)*
никтинастия *bot.* nyctinasty, sleep-movement
никтофилия *med.* nyctophilia, scotophilia
никтофобия *pathol.* morbid fear of darkness, nyctophobia
никтурия *med.* nocturnal enuresis, nocturia, nycturia
нилас nyl(l)as, black ice
ниль-группа *math.* nil-group
ниль-кольцо *math.* nil-ring
ниль-потентность *math.* nilpotency
ниль-радикал *math.* nil-radical
ниль-ряд *math.* nil-series
нильсборий *chem.* nielsbohrium, Ns
нильский Nilotic
нильстепенный, ниль-потентный *math.* nilpotent

ниль-элемент *math.* nil-element
нимб aureole, nimbus, halo
Нимрод *bibl.* Nimrod
нимфа 1. *myth.* nymph; 2. *zool.* pupa, immature insect, nymph; 3. *zool.* nympha, ligament support; 4. *pl., anat.* labia minora, nymphae
 ~ источника Arethusa
 ~, морская Nereid
 ~, связочная chondrophore, resilifier
нимфалида *zool.* 1. lilac nymph *(Crenis rosa);* 2. tree nymph *(Crenis);* 3. *pl.* brush-footed butterflies, numphalids *(Nymphalidae)*
 ~, виргинская painted beuaty *(Vanessa virginiensis)*
нимфальный *zool., anat.* nymphal
нимфейник *bot.* floating heart *(Nymphoides)*
нимфейные *n. pl. bot.* water-lily family *(Nymphaeaceae)*
нимфея *bot.* water-lily *(Nymphaea)*
нимфит *med.* nymphitis
нимфоманиакальный *pathol.* nymphomaniac(al)
нимфомания *pathol.* nymphomania, morbid sexual desire
нимфоманка *pathol.* nymphomaniac
нингидрин *chem.* ninhydrin
ниневиянин *hist.* Ninevite
Ниоба, Ниобея *myth.* Niobe
ниобат *chem.* niobate, columbate
ниобиевый *chem.* niobic, niobous
ниобий *chem.* niobium, Nb, columbium, Cb
ниппель *tech.* nipple, sleeve, adapter, union
 ~, вдвижной slip/push nipple
 ~, переходный reducing nipple
 ~, резьбовой screwed nipple
 ~, смазочный lubricating nipple
 ~, соединительный union nipple, connector
 ~, трубный pipe nipple
 ~ шланга hose fitting
нирал *nav.* downhaul
нирвана nirvana, blissful state
нирембергия *bot.* cupflower *(Nierembergia)*
ниспослание *rel.* sending down *(from heaven)*
ниспосланный *rel.* granted, sent down
 ~ богом godsent
ниспос(ы)лать *rel.* grant, send down *(from heaven)*
нисса *bot.* tupelo, gum tree *(Nyssa)*
нистагм *med.* nystagmus
 ~, возвратно-поступательный see-saw nystagmus
 ~, вращательный rotatory nystagmus
 ~ горнорабочих miner's nystagmus
 ~, калорический Barany's sign
 ~, качательный rocking nystagmus
 ~, крупноразмашистый large-swinging nystagmus
 ~, лабиринтный labyrinthine nystagmus
 ~, маятникообразный pendular nystagmus
 ~, наследственный inherited nystagmus
 ~, поездной railroad nystagmus
 ~, произвольный voluntary nystagmus
 ~, толчкообразный jerk nystagmus
 ~, установочный adaptive nystagmus
 ~ у шахтеров miner's nystagmus
 ~, физиологический end-position nystagmus
 ~ Чейн-Стокса Cheyne's nystagmus
нистагмоподобный nystagmoid, nystagmiform
нистатин *pharm.* nystatin, mycostatin
нисходяще-восходящий falling-rising
нисходящеперистый *bot.* decursively-pinnate
нисходящий descending; *bot.* decurrent, decursive; *phon.* falling
нит nit *(unit of luminous intensity; unit of information)*
нитазол *chem.* aminitrazole
нитевидночерешковый *bot.* filicauline
нитевидный, нитеобразный filiform, capillary, threadlike, filamentary, filamentous
нитевод thread guide
нитеводитель *text.* thread/yarn carrier
 ~, заправочный кольцевой lacing ring guide
 ~, проволочный wire thread carrier
нитедержатель thread holder
нителла *bot.* nitella *(Nitella)*
нителовка *text.* thread picker
нитенаправитель thread/yarn guide
нитеносный filiferous
нитеобразование filamentation
нитепроводник thread/yarn guide
нитеразделитель *text.* thread separator
нитестебельчатый *bot.* thread-stalked, filicauline
нити *see* **нить**
нитка 1. thread; yarn; fibre; 2. string; 3. *chem.* stream, train
 ~, вышивальная embroidery yarn/thread
 ~, штопальная darning thread
нитон *chem.* niton, radon, Rn
ниточный filamentary; filamentous
 ~ное производство spinning
нитрагин inoculant fertilizer
нитраллой *met.* nitralloy
нитранол nitretamine phosphate
нитрат *chem.* nitrate
 ~ ртути mercurious nitrate
 ~ свинца lead nitrate
нитратредуктаза nitrate reductase
нитрид *chem.* nitride
нитрирование nitriding
нитрировать *chem.* nitrate; nitride
нитрит *chem.* nitrite
 ~ натрия sodium nitrite
нитрификатор nitrifier
нитрифицировать nitrify
нитробактерия nitrifying bacterium, nitrobacterium
нитробензол *chem.* nitrobenzene
нитрование nitration, nitridation
нитроглицерин nitroglycerin, glonoin, glyceryl trinitrate, blasting oil
нитрогумус nitrohumus

нитрозамещение nitrosubstitution
нитрозокраситель nitrozo dye
нитроиприт *chem., mil.* nitrogen mustard (gas)
нитроклетчатка *chem.* nitrocellulose
нитрокраситель nitro dye
нитрокраска nitrocellulose paint/enamel
нитроксигемоглобин nitroxyh(a)emoglobin
нитролак nitrocellulose lacquer/varnish, nitrolacquer
нитролевулеза nitrolevulose
нитроманнитол *chem.* nitromannitol, mannitol hexanitrate
нитромерзол *chem.* nitromersol, metaphen
нитромочевина nitrourea
нитрон Nitron, Orlon, Acrylon
нитропленка nitrate film
нитросахар nitrosugar
нитросоединение nitrocompound
нитросорбид isosorbide dinitrate
нитроспирт nitrated alcohol
нитросульфатиазол nitrosulphathiazole, nisulfazole
нитротолуол nitrotoluene
нитрофил, нитрофит *bot.* nitrophyte, nitrophilous plant
нитрофильм nitrate film
нитрофоска *agr.* nitrophoska, combined fertilizer *(nitrogen, phosphoric acid and potassium)*
нитрофуразон nitrofurazone, furacin
нитрофурантоин nitrofurantoin, furadantin
нитрохлороформ nitrochloroform, chloropicrin
нитроцементация carbonitriding
нитрошелк nitrosilk, nitrorayon
нитроэмаль nitrocellulose enamel
нитроэтаноксидаза nitroethane oxidase
нитчатка 1. *bot.* conferva; crow-silk; 2. *zool.* filaria, tapeworm *(Filaria)*
нитчатолистный *bot.* thread-leaved, nematophyllous
нитчатостебельчатый *bot.* thread-stalked
нитчатый filamentous, filiform; *bot.* confervoid
нить filament; fibre, thread; yarn; wire
~, асбестовая asbestos thread
~ Ариадны *myth., fig.* Ariadne's clue
~, волластоновская Wollaston wire
~ диоптра hair of sight vane
~, искусственная man-made fibre
~, капроновая capron thread
~, клеточная cellular filament
~, коконная cocoon filament
~, кордная cord thread
~, крученая twisted fibre
~ лампы накалывания, бифилярная coiled-coil filament
~, льняная linen thread
~ микрометра, неподвижная fixed line/wire
~ накала/накаливания incandescent filament
~, натуральная native fibre
~, путеводная clue
~, семенная *anat.* spermatic/seminal filament
~ спинного мозга, концевая *anat.* terminal thread, nervus impar
~, текстильная fibre
~ теодолита stadia wire/hair
~, тычиночная *bot.* anther stalk, stamen filament
~, хвостовая caudal filament
~, хирургическая surgical thread
~ти в зрительной трубе, крестообразные haircross of surveying telescope
нифе *geol. nife*
нихром *met.* Ni-Cr alloy, Nichrome
ницелоз *bot.* creeping willow *(Salix repens)*
ницшеанец, ницшеанский *philos.* Nietzschean
ницшеанство *philos.* Nietzscheanism
ничейный *sport* drawn; neutral
~ная игра drawn game
~ная полоса *mil.* neutral zone
ничтожность negligibility; insignificance; *law* nullity
ничтожный 1. inconsiderable, insignificant; negligible; infinitesimal; 2. *law* (null and) void, null
ниша niche, recess; pocket; *arch.* bay; *av.* well; *min.* hold
~, береговая shore-line nip
~, волноприбойная wave-cut notch/recess, sea cave
~ в перилах моста bridge refuge
~ в фюзеляже fuselage well
~ колеса шасси wheel well
~, оконная window niche/recess
~ при язве желудка *med.* Haudek's niche
~, экологическая ecological niche
~, якорная anchor recess/pocket
нишеобразный niche-shaped
нищенский insufficient; poor, scanty
~кая заработная плата *econ.* starvation wages
нищий 1. destitute, indigent, poverty-ridden/stricken
~ духом *rel.* poor in spirit;
нобелевский Nobel *(attr.)*
нобелий *chem.* nobelium, No
нобилитет *hist., coll.* the nobility/aristocracy; patriciate
нобиль *hist.* nobleman; aristocrat, patrician
новарсенол *pharm.* neoarsphenamine, neosalvarsan
новатор innovator, innovationist, trail blazer
новаторский innovatory, innovative
~кая работа pioneering work
новаторство innovation, innovating
новация 1. *law* novation; 2. innovation
новейший latest, up-to-date, modern; neoteric
новелла 1. *lit.* short story, novella; 2. *Roman law* novel
новеллетта 1. *lit.* little novel, novelette; 2. *mus.* novelette
новеллист short-story-writer
новик *hist.* 1. young courtier; 2. novice, tyro, tiro
новина *agr.* 1. virgin soil; 2. newly harvested grain
новобиоцин *pharm.* novobiocin
~, кальциевый albamycin/cathomycin calcium
новобранец recruit
новобрачная *n.* bride

новобрачные *n. pl.* newlyweds
новобрачный *n.* bridegroom
нововведение innovation
~**ния, технологические** technological developments
новогреческий Modern-Greek
новозаветный *rel.* of the New Testament
новозеландец New Zealander
новозеландский New-Zealand
новоиндийский Modern-Indian
новокаин procaine hydrochloride, novocain
новокаин-пенициллиновый procaine-penicillic
новокрещенец convert, newly baptized person
новокрещенная newly baptized woman
новокрылые *n. pl. zool.* neopterans *(Neoptera)*
новоледниковый *geol.* neoglacial
новоледниковые *geol.* New Glacial period; new glaciation
новолуние change of the moon, new moon
новомежледниковые New Interglacial period
новонабранный 1. newly drafted/recruited; 2. *print.* newly composed
новообразование *med.* neoplasm, tumo(u)r, neoformation; *geol.* neogenesis, new formation/growth; *ling.* neologism
~, **генерализованное** generalized neoplastic disease
~, **доброкачественное/злокачественное** benign/malignant tumo(u)r
~, **истинное** blastoma
~ **минералов** neocrystallization
новообразованный neogenic, neogenetic, neoplastic; newly formed; hysterogenic
новообращенный newly converted; *n.* convert, catechumen, neophyte
новоподобный *astr.* nova-type
новопреставленный *rel.* recently deceased/departed
новоприбывший, новоприехавший *n.* newcomer
новоприобретение new acquisition
новорожденный *n.* newborn (child), infant, neonatus, neonate
~, **доношенный** mature infant
~, **недоношенный** premature infant
новосел 1. new settler/tenant; 2. *hist.* neodamode
новотел *agr.* first/new calving
новотельный *agr.* newly calved
новшество novetly, innovation
~**ва в машиностроении** new developments in engineering
новый 1. new; fresh; 2. modern; recent
~ **завет** *rel.* the New Testament
~**ые языки** modern languages
новь *agr.* 1. virgin soil; 2. newly harvested grain
нога 1. leg; foot; 2. *tech.* brace, stand; *see also* **ноги**
~, **задняя** hing leg, posterior foot
~, **передняя** forefoot, anterior foot
~, **плавательная ракообразных** pleopod
~, **стропильная** rafter, spar, principal truss
~ **трехножного станка** tripod leg
~, **ходильная** walking foot; ambulatory limb
~ **шасси** undercarriage leg/unit
ногавка 1. vervel, varvel; 2. ankle boot
ноги *pl.*
~, **молочные** *med.* milk legs, phlegmasia alba dolens
~, **неодинаковые по длине** badger legs, anisomalia
~, **непропорционально крупные** macroscelia, megalomelia, macromelia
~, **ножницеобразные** scissor/cross legs
~, **О-образные/саблевидные** bandy/bow legs
~, **слоновые** elephantiasis
~, **Х-образные** baker's legs, knock-/in-knee(s)
ногодержатель leg holder
ноголист(ник) *bot.* May apple *(Podophyllum)*
ногоперый *zool.* pediculate
ногоплодник *bot.* podocarp *(Podocarpus)*
ногоротые *n. pl. zool.* gigantostracans *(Gigantostraca)*
ноготковидный unguiform, unguiculate
ноготок, ноготки *pl. bot.* calendula, (pot-)marigold *(Calendula)*
ноготь 1. nail; 2. *med.* onyx
~, **вросший** onychocryptosis, ingrowing toenail, ingrown nail
ногохвостка *zool.* 1. springtail, collembolan *(Collembola);* 2. armo(u)red springtail *(Achorutes armatus)*
ногочелюсть *zool.* foot jaw, maxilliped
ногтевидный onychoid, unguiculate, unguiform
ногтевой ungual, unguinal
ногтеобразующий onychogenic
ногтоеда *med.* whitlow, paronychia, panaris, panoritium; hangnail
нодозный nodular, nodose
ноев *bibl.* Noachic(al)
~ **ковчег** Noah's arc
~ **потоп** Noachic deluge
Ноеминь *bibl.* Naomi
нож knife; blade
~ **аттенюатора** flap
~, **бритвенный** razor blade
~, **брюшистый** single-bladed knife
~, **двубрюшистый** double-bladed knife
~ **для грыжесечения** hernia knife, herniotome, slender-bladed knife
~ **для рассечения роговицы** Graefe's knife
~ **для съемки шкур** *leath.* cleaver
~ **ковша** bucket lip
~, **консервный** can opener
~, **корообдирочный** spudder
~ **косилки** mower knife
~ **лущильного станка** *wood.* veneer knife
~, **многолучевой** multiblade knife
~, **мозговой** cerebrotome, brain knife
~, **монтерский** lineman's knife
~, **насекательный** pruning hook

~, обоюдоострый double-edged knife
~, отбойный *text.* mote knife
~, перочинный picker knife
~ плуга skim cutter/coulter
~, подающий picker knife
~, подрезочный *agr.* pruning knife
~, промазочный spreading knife
~, разрезной paper knife
~, реберный *surg.* costotome
~, садовый garden/pruning knife
~, секционный postmortem knife
~, скрытый гортанный *surg.* concealed laryngeal knife
~, снимающий shaving/scraping knife
~, строгальный *wood.* plane knife, planing cutter
~ струга plough blade
~ Тирша *surg.* Thiersch's knife, broad hollowground knife
~, хирургический operating/surgery knife
нож-вальвотом valvotomy/Brock's knife
нож-долото osteotome
ножевидный knife-edged/-shaped, cultrate, cultriform
ножевой knife *(attr.)*
~ мастер cutler
~вые товары cutlery
ножедержатель knife holder
ножеклюв *zool.* skimmer *(Rhynchops)*
ножка 1. leg, foot; stem; 2. *anat.* limb, crus; peduncle; stipe; 3. *bot.* pedicle, stalk, stipe; 4. *paleont.* podium
~, амбулакральная ambulacrum, ambulacral/tube foot
~ большого мозга cerebral peduncle
~ брахи(о)поды pedicle
~ иглокожего podium
~ камертона tine of a tuning fork
~ катода cathode stem
~, пестичная gynopodium
~ свода *anat.* crus of the fornix
~ стремени, передняя *anat.* anterior limb of the stapes
~ целика *min.* pillar stump/residue
ножки *pl.* limbs, crura
ножковидный stipi(ti)form; stalklike
ножкодержатель mount support
ножколистный *bot.* podophyllous
ножкоплодный *bot.* podocarpous
нож-литотом lithotome
ножницевидний, ножницеобразный scissor-shaped; *zool.* forficulate
ножницы *pl.* 1. scissors; shears; clippers; shearing machine, cutter; 2. *econ.* discrepancy
~, аллигаторные alligator/crocodile shears
~, анатомические dissecting scissors
~ блюминга bloom shears
~, вырубные nibble shears, nibbler
~, гильотинные throat guillotine/gate shears
~, дисковые circular shears
~ для ногтей nail clippers
~ для перевязочного материала gauze shears
~ для пересечения пуповины umbilical cord scissors
~ для разрезания гипсовых повязок plaster shears
~ для стрижки волос hair clippers
~, изогнутые curved scissors
~, кишечные intestinal/enterotomic scissors
~, коленчатые angle shears
~, кромкообрезные side trimming shears
~ Листона *med.* Liston's shears
~, механические power shears; shearing machine
~, многопарные gang shears
~, обрезные cropping shears
~, проволочные wire cutting shears
~, прутковые bar shears
~, рычажные lever/alligator/crocodile shears
~ с верхним резом downcut shears
~ с нижним резом upcut shears
~, садовые garden/pruning shears
~, сосудистые bload vessel scissors
~, стригальные *agr.* sheep shears
~, стуловые bench shears
~, толстолистовые plate shears
~, тупоконечные blunt scissors
~, уточные wett-cutting scissors
~ цен *econ.* price scissors
~, шпалерные hedge trimmer/shears
ножной pedal, foot *(attr.); zool.* podal, peduncular; *tech.* foot-operated
~ная скамейка footstool, foot bench
ножнолистный *bot.* coleophyllous
ножны *pl.* scabbard, sheath
~ штыка bayonet sheath
ножовка hacksaw, hand-saw
~, приводная power hacksaw
~, садовая hacksaw, pruning saw
~, узкая keyhole saw
нож-распатор knife with raspatory
нож-скальпель scalpel
нож-шпатель knife-spatula
нож-электрокаутер cautery knife
ноздреватость poriness, porosity, cavernosity, sponginess
ноздреватый porous, pory, spongy, spongeous, honeycombed, cavernous
ноздревидный nariform
ноздря nostril
~, внутренняя internal opening
нозеан *miner.* nosean, noselite
нозобык *zool.* oxbird *(Tringa variabilis)*
нозометрия nosometry, pathometry
нозономия nosonomy; nosology
Ной *bibl.* Noah
нок *nav.* butt; guy/boom end
нокаут *sport* knockout
нокаутировать knock out
нокдаун *sport* knockdown

нокотница *zool.* piked/spiny dogfish *(Squalus acanthias)*
ноксирон *pharm.* glutethimide
ноктамбулизм somnambulism, noctambulism
ноктовизор snooperscope
ноктюрн 1. *mus.* nocturne; 2. *art* night scene/piece
ноль zero; naught; nil, null
ном I nome, nomos
ном II *mus.* nome
нома *med.* noma, gangrenous stomatitis
номад nomad
номада *zool.* cuckoo bee *(Nomada)*
номарх nomarch
номархия nomarchy
номен *hist.* nomen, nome
номенклатура 1. nomenclature; 2. catalog(ue), list; range; schedule; 3. *polit.* nomenklatura
~, анатомическая Nomina Anatomica
~, изделий listed products
~, двойная binary/binomial nomenclature, double-naming
~ карты map sheet designation
~ листа sheet number
~ продукции product mix
~ текущего запаса service stock list
~ товаров bill of goods, commodity classification
~, тройная ternary nomenclature
номенклатурный nomenclative
~ работник appointee
номер 1. number; *math.* place; 2. issue; 3. size, gauge; calibre, caliber; 4. item on the program; 5. tally, label, ticket; check; 6. room *(in a hotel);* 7. *mil.* member
~, атомный atomic number
~, бортовой hull/pendant/side number
~ ботинок size of shoes
~, вызываемый call number
~ готовности *mil.* degree of alert
~ детали part number
~ журнала, февральский February issue of a magazine
~, заводской works/serial/manufacturer's number
~ заказа job/order number
~ звезды *astr.* Flamsteed's number
~ иглы needle size
~, инвентарный inventory/accession number
~, исходящий/исходный (expedite) reference number; *rad.* station serial number
~ картотеки/массива *comp.* data-set (reference) number, file (reference) number
~, личный *mil.* order number
~ партии lot/job number
~ позиции item number
~, порядковый sequence/serial/ordinal number
~ проволоки wire gauge
~ пряжи (fibre/yarn) count
~ пряжи, метрический metric count
~ пряжи, номинальный nominal count
~ пять, трамвай the number five trolley(-car)
~, регистрационный *auto.* license plate number
~ рейса *av.* flight number
~ семейства гипербол hyperbolic number
~ сита size of a sieve
~, складский stock number
~, старый back issue/number
~ строки *math.* number of a row, line number
~ элемента, порядковый atomic number
~ эшалона *av.* number of flight level
номератор 1. numerator, numbering machine; 2. *el.* annunciator, indicator board; *see also* **нумератор**
номерация numeration, numbering
номерник annunciator, indicator
~, телефонный drop annunciator/indicator, number switch
номерной 1. numbered; 2. *n.* valet *(in a hotel)*
номерок check, tag, label, ticket
номеронабиратель telephone dial
~, дисковый rotary dial
~, кнопочный push-button dial
номероуказатель display panel, *Am.* call indicator
номинал rating, nominal/rated/face/denominational value; *econ.* par
~ акции par value
~, кратковременный short-time rating
~ облигации par value
~, подгонять adjust to value
~, прерывистый intermittent duty rating
~ резистора, подгонять trim a resistor
номинализм *philos.* nominalism, terminism
номиналистический *philos.* nominalist(ic)
номинальный nominal, titular; *tech.* rated, indicated
~ная стоимость denominational/nominal/face value
номинатив *gram.* nominative (case)
номинация nomination, naming
номограмма *math.* nomogram, nomograph, alignment/nomographic chart
~ из выравненных точек *math.* alignment nomogram
~, механическая calculating board
~, сетчатая net chart
номографирование *math.* nomographic representation, nomography
номографируемость *math.* nomographic representability, nomographability
номографический nomographic(al)
номография *math.* nomography
нона *mus.* ninth
нониллион quintillion, *Am.* nonillion
нониус nonius, vernier (scale), sliding caliper
нонконформист nonconformist
нонпарель *print.* 6-point type, nonpareil
ноны *pl.* nones *(in Roman calendar)*
норадреналин norepinephrine
норвежец, норвежский Norwegian
норвежка Norwegian (woman)
норд *nav.* north, north wind

норд-вест *nav.* northwest; northwester, northwesterly wind
норд-вестовый northwest(erly), northwestward
нордический Nordic
норд-норд-ост *nav.* north-northeast, northnortheast wind
нордовый northern, northerly
норд-ост *nav.* northeast; northeaster, northeasterly wind
норд-остовый *nav.* northeast(ern), northeasterly
норичник *bot.* figwort *(Scrophularia)*
нория noria, bucket chain/elevator, chain bucket elevator
норка 1. *zool.* mink *(Mustela);* 2. mink fur/pelt
норлейцин *bioch.* norleucine, amino caproic acid
норма 1. norm, standard; 2. rate; quota; allowance, target figure; normal; *see also* **нормы**
~ амортизации amortization quota, rate of depreciation
~ брака acceptance rate
~ внесения удобрений *agr.* fertilizer rate
~ времени time rate/value, standard time
~ времени, расчетно-аналитическая synthetic time standard
~, временная draft/temporary/presumptive standard
~ выработки production/piece rate, output/performance standard, rated/standard output
~ выработки машин machine output
~ выручки earning rate
~ высева sowing/planting rate
~ грузовых работ handling rate
~ довольствия, суточная day of supply
~, заводская factory standard
~, заниженная loose standard
~ издержек на станко-час machine-hour rate
~ инвестирования investment rate
~, минимальная minimum standard
~ накопления rate of accumulation
~ неприкосновенного запаса emergency level of supplies
~, оросительная irrigation rate
~ отказов failure rate
~ отпуска продовольствия food scale
~ отходов wastage rate
~ погрузки loading rate, rate of loading
~ потребления consumption/usage rate/factor
~ потребления воды water allowance
~ прибавочной стоимости rate of surplus value
~ прибылы rate of return, profit rate
~ прибылы, общая general rate of profit
~ продовольствия ration allowance
~ производительности output/performance/productivity standard
~ производственных отходов scrape rate
~ профилактики preventive maintenance rate
~ процента interest rate, rate of interest
~ разгрузки rate of discharge, unloading rate
~ расхода кормов *agr.* feed input standard
~ расхода материала material consumption rate
~ расхода семян seeding rate
~ сбережений saving rate
~ снабжения level of supply, issue allowance
~ содержания запасов level of supply/stock
~, суточная day's quota
~, техническая engineering/practical standard
нормалемер *mach.* gear-tooth micrometer
нормализатор *math.* normalizer
нормализация normalization, normalizing; standardization
нормаль 1. *math.* normal; 2. standard
~ Бриггса Briggs standard
~ к базисной линии centre line
~ к кривой normal to curve
~, цеховая shop practice standard
нормальный 1. normal; standard; natural; 2. *med.* sane
нормандец Norman
нормандский Norman
~кие острова Channel Islands
норманизм Normanism
норманн *hist.* Norseman, Northman
норманнско-французский *hist.* Norman-French
норматив norm, (established) standard; criterion; amount per unit, requirement
~ оборотных средств quota of current assets
~ы, гигиенические hygienic regulations
~ы затрат input normals
~ы цен price normals
нормативность standardization, normalization
нормативный 1. standard(ized); corresponding to norm; 2. prescriptive, normative
нормирование 1. normalization, normalizing, standardization; introduction of norms; rate setting/fixing; 2. rationing; 3. *math.* valuation
~ удобрений fertilizer rationing
~ продовольствия food rationing
~ работ job standardization
нормированный normalized, standardized; fixed; set
~ рабочий день fixed working hours
нормировать 1. normalize, standardize; 2. ration
нормировка *math.* normalization
нормировочный normalizing, standardizing
~ множитель *math.* normalization factor
нормировщик time-and-cost clerk, official in charge of work-quotas, time-study man, norm/rate setter, rate fixer
нормируемость normability
нормообразующий/нормоустановительный *law* rule proclaiming/establishing
нормо-час standard hour
нормы *pl.* regulations, standards, specifications
~, действующие current standards/specifications
~ допустимых загрязнений permissible contamination standards
~ летной годности *av.* airworthiness regulations/standards

~ **медицинского освидетельствования** regulations for physical examination
~ **по охране окружающей среды** environmental standards
~, **правовые** legal regulations
~, **процессуальные** (rules of) procedure, practice
~ **прочности** strength/stress criteria/standards/requirement
~, **стоимостные** standard cost(s)
~, **технические** technical regulations

нормоцит *anat.* normocyte, normoerythrocyte

Норна *myth.* Norn

норсульфазоль *pharm.* sulphathiazole
~, **растворимый** sulphathiazole sodium

нортумберлендский, нортумбрийский Northumbrian

нос 1. *anat.* nose; 2. beak; 3. *nav.* bow, head; prow; 4. *geol.* ness, promontory, headland, naze
~, **горбатый** aquiline/hooked nose
~, **заостренный** pinched nose
~, **ледокольный** icebreaker bow
~, **ложкообразный** shovel/spoon bow
~ **наковални** horn of an anvil
~, **отпускать/поднимать** *av.* nose down/up
~, **приплюснутый** flat nose
~, **седлообразный** saddle nose
~ **складки** nose of fold
~ **судна** bow of a ship

носатка *zool.* 1. lantern fly *(Fulgora);* 2. snout butterfly *(Hypatus bachmanni)*

носатый 1. (big-/long-)nosed; *zool.* nasute; 2. beaky

носач *zool.* proboscis monkey *(Nasalis)*

носик tip; spout; beak, bill; *bot.* rostellum, rostrum
~ **лампы, отпаечный** exhaust tip
~ **насадка** probe/tube nose
~ **сосуда** spout, beak
~**и, комаровы** *bot.* branching/forking larkspur *(Delphinium consolida)*

носилки *pl.* 1. litter(s), stretcher; sedan (chair); 2. *constr.* handbarrow
~, **вьючные** cacolet
~, **колесные** wheel litter
~, **спасательные** rescue stretcher

носилки-каталка wheel litter

носильщик porter, carrier

носитель 1. *tech.* carrier; bearer; medium; support; 2. *med.* carrier; 3. *chem.* (carrier) vehicle
~ **биологического катализатора** biocatalyst carrier
~ **входных данных** input medium
~ **гена серповидноклеточности** sickle cell trait
~ **данных** data carrier/medium; *comp.* EDP data carrier
~ **(для)записи (данных)** (data) recording medium, data-storage medium
~ **заряда** charge carrier
~, **захваченный** trapped carrier
~ **информации** data/information carrier
~ **катализатора** catalist support
~, **магнитный** magnetic medium
~, **машиночитаемый** automated data medium, machine-readable medium
~, **неразмеченный для записи** *comp.* virgin medium
~, **осажденный** carrier precipitate
~ **текста, широкоформатный** *comp.* wide text carrier
~ **тока** charge/current carrier
~, **удерживающий** hold-back carrier
~, **управляющих данных** control volume

носительство *med.* carriage

ноский *agr.* heavy-laying *(of fowl)*
~**кая курица** a good layer

носкость *agr.* egg production; egg-laying ability

носовой *phon.* nasal; *nav.* foremost, fore, bow *(attr.)*

носоглотка *anat.* nasopharynx, rhinopharynx

носоглоточный *anat.* nasopharyngeal, rhinopharyngeal

носо-губной nasolabial

носок front part; spout, lip, tip; nozzle; point; nose(piece); *av.* leading-edge; *naut.* bill, pick, pea
~ **ковша** pouring lip, ladle spout
~ **крыла** *av.* wing leading-edge, leading-edge subassembly
~ **лапы якоря** anchor bill/pick/pea
~ **лемеха** share point, plough/share tip
~, **переливной** overflow lip
~, **съемный** *av.* removable leading-edge

носолистный *bot.* beak-leaved

носо-лобный *anat.* nasofrontal

носо-небный *anat.* nasopalatal, nasopalatine

носорасширитель *med.* nasodilator

носорог *zool.* 1. rhinoceros *(Rhinoceros);* 2. unicorn beetle *(Oryctes nasicornis)*

носо-ротовой *anat.* nasooral

носо-слезный *anat.* nasolacrimal

ностальгия *med.* nostalgia; homesickness

ностоковые *n. pl. bot.* fresh-water algae *(Nostocaceae)*

носуха *zool.* (brown/red) coati *(Nosua narina/rufa)*

НОТ *abbr.* **научная организация труда** industrial engineering, scientific organization of labo(u)r

Нот *myth.* Notus

нота I 1. *mus.* note, tone; 2. *pl.* (sheet) music, score
~, **восьмая** quaver
~, **двойная целая** double whole note
~, **тридцать вторая** demisemiquaver
~, **шестнадцатая** semiquaver

нота II (diplomatic) note
~, **вербальная** verbal note

нотабена, нотабене nota bene, N. B.; mark well, take notice

нотабль *hist.* Notable

нотариально notarially
~ **засвидетельствовать** notarize

нотариальный notarial
~ная кантора notary's office
нотариат *law* notariate
нотариус notary
~, государственный notary public
нотация notation; *mus.* tablature
~, бесскобочная *comp.* operator notation
~, буквенная alphabetical notation
нотировать *mus.* note
нотис *nav.* notice of readiness
нотификация notification
нотифицировать notify
нотный note *(attr.),* music *(attr.)*
~ стан *mus.* staff, stave
~ная тетрадь music book
Нотогея *biogeogr.* Notogaea
нотозавр *paleont.* nothosaur *(Nothosaurus)*
нотокайнозой notocenozoic
нотоносец *mus.* staff, stave
нотопечатание music printing, printing of sheet music
ноторность notoriety
ноторный *law* notorious
нототерий *paleont.* notothere *(Nototherium)*
нотофагус *bot.* false beech *(Nothofagus)*
нотропис *zool.* redfin *(Notropis)*
ноумен *philos.* noumenon (*pl.* -mena), unknown/unknowable substance/thing
ноуменальность *philos.* noumenality
ноу-хау know-how
ночецвет(ник) *bot.* common four-o'clock, marvel-of-Peru *(Mirabilis jalapa)*
ночница I *zool.* 1. noctuid, owlet moth *(Agrotis segetum);* 2. bat *(Myotis)*
ночница II *bot.* night phlox *(Nycterinia)*
ночь night; *myth.* Nyx, Nox
~, белая white/polar night; midnight sun
~, Варфоломеевская *hist.* Massacre of St. Bartholomew
~, новогодняя New Year's Eve
ноябрь November
ноябрьский November *(attr.);* Novemberly, Novemberish
нрав temper; nature, disposition; 2. *pl.* customs, manners, ways
~, кроткий meek disposition
~, крутой stern temper
нравоучение moral admonition, moralizing, lecture; *lit.* moral
нравоучительный didactic, moralizing
нравственность morals; morality
нравственный ethical, moral
нубиец, нубийский Nubian
нубийка Nubian (woman)
нувистор *el.* nuvistor
нувориш nouveau riche
нуг *bot.* niger-seed *(Guizotia)*
нуга *cul.* nougat
нудизм nudism
нукер *hist.* bodyguard
нуклеарный nuclear
нуклеация nucleation
нуклеиновый nucleic
нуклеогиалоплазма *biol.* nuclear sap
нуклеогистон *bioch.* nucleohistone
нуклеозид *bioch.* nucleoside
нуклеозиддифосфатаза *bioch.* nucleoside diphosphatase
нуклеола nucleolus
нуклеоплазма *biol.* nucleoplasm; karyoplasm
нуклеоплазматический *biol.* nucleoplasm(at)ic; karyoplasmic
нуклеосома *biol.* nucleosoma, karyosome
нуклеотид *bioch.* nucleotid(e)
нуклеотидгидролаза *bioch.* nucleotid(e) hydrolase
нуклеотидилтрансфераза *bioch.* nucleotidyl transferase
нуклеофильный *chem.* nucleophilic
нуклеохилема *biol.* nuclear sap
нуклеус nucleus; queen cell
нуклид *phys.* nuclide
нуклон *phys.* nucleon
~, бомбардирующий incident/incoming/bombarding nucleon
~ отдачи recoil(ing) nucleon, nucleon recoil
~, упругорассеянный elastically scattered nucleon
нуллиплекс *biol.* nulliplex
нуллификация nullification
нуль zero; null; nil; naught, nought, cipher, cypher
~, абсолютный absolute zero
~, ведущий *comp.* leading zero
~, внешкальный inferred zero
~ высот *cart.* datum level, datum of heights
~ высот на карте chart datum
~ глубин *geod.* hydrographic/sounding/chart datum
~ глубин, установленный reference datum
~ государственной нивелировки ordnance datum
~ графика datum plane; actual zero point
~ дальности range zero
~ карты chart datum
~ Кронштадского футштока datum of Kronshtadt gauge
~, машинный computer zero
~, незначащий *comp.* nonsignificant zero
~ оси времени time zero
~, плавающий *comp.* floating zero
~ поста *nav.* datum
~, разрушенный disturbed zero
~ рейки zero of pole
~ слышимости aural null
~, структурный *comp.* placeholder
~, тождественный *math.* zero/null function
нуль-адресный zero-addess
нульарный null
нуль-валентность zero valence
нуль-вектор *math.* null/zero vector

нуль-гипотеза null hypothesis
нуль-индикатор zero-reader, left-right indicator, null-indicator
нуль-индикатор, стрелочный cross-pointer (indicator)
нуль-мерность *math.* zero-dimensionality
нуль-многообразие zero variety
нуль-множество *math.* null/empty set
нуль-орган null detector/element
нуль-полугруппа *math.* zero semigroup
нуль-последовательность *math.* null/zero sequence
нуль-пространство null-space
нуль-пункт *geod.* zero-mark
нуль-стабильный nilstable
нуль-тензор *math.* zero tensor
нуль-форма *math.* null-form
нуль-функция *math.* zero/null function
нумен *philos.* noumenon (*pl.* -mena), unknown/unknowable substance/thing
нумератор 1. numbering apparatus/machine/unit, numerator; 2. *el.* annunciator, indicator board
~, блинкерный drop annunciator
~, стрелочный needle annunciator
~ с хлопушкой clap-stick
нумерация numeration, numbering
~ групп пакетная *econ.* batch numbering
~ пряжи yarn numbering/counting
~ страниц pagination
нумеровальный numbering
нумерование numbering; indexing
нумерованный numbered, indexed
нумеровать number, index
нумизмат numismatist
нумизматика numismatics
нуммулит *zool.* nummulite
нунация *gram.* nunnation
нунций nuncio, legate
~, папский papal nuncio
нут *bot.* garbanzo, gram chick-pea *(Cicer arietinum)*
нутация nutation
~ наклона эклиптики nutation in obliquity of ecliptic
нутрия *zool.* coypu, nutria *(Myocastor coypus);* nutria fur
нутромер inside callipers/micrometer, internal gauge
нутч-фильтр *chem.* nutsch/suction/Nutsche filter
нуцеллус, нуцеллюс *bot.* nucellus
нырок I *boxing* duck(ing)
нырок II *zool.* diving duck *(Aythya)*
ныряло plunger
ныряльщик diver
ныряние diving
~ рудного тела *geol.* plunge/pitch of ore body
ньютон *phys.* newton, N
ньютонов Newtonian
ньюфаундленд Newfoundland (dog)
ньяла *zool.* nyala *(Tragelaphus)*
нэксина nexine
НЭП *abbr. hist.* **новая экономическая политика** NEP, New Economic Policy
нэпман nepman, profiteer *(during period of NEP)*
нюанс *art* nuance, shade; subtle difference
нюдизм nudism
нюх sense of smell, scent; *fig.* flair
няня *med.* nurse's aid, (hospital) nurse

О, о

оаз(ис) oasis
оазисный oasal, oasean, oasitic
обанкротиться turn/become/go bankrupt, go broke
обвал *geol.* collapse, downfall, failure, caving, cavein, crumbling, foundering
~, внезапный/выработки collapse
~, горный rock fall
~, снежный avalanche, snowslide
обваливаться collapse, cavein, fall
обвалование embankment, bounding; diking, dyking; banking-/damming-up
обвертка 1. *bot.* husk, spathe, involucre, envelope; 2. wrapping up
~, общая periclinium
~ соцветия/цветка spathe
обвертковидный *bot.* involucriform, hull-shaped
обверточка *bot.* involucel
обверточный 1. *bot.* involucral; 2. (for) wrapping
обвес *naut.* weather cloth, (wind) dodger
обветренный weather-beaten, wind-blown
обвехованный marked
обвивка wrapping; winding round
обвинение 1. charge, accusation; indictment; delation, inculpation; 2. verdict; 3. *law* the prosecution
~, возводить/выдвигать accuse, charge, indict
~, встречное *law* countercharge
~, вынести declare a verdict of guilty
~, предъявить indict
обвинитель accuser; denunciator; *law* prosecutor
обвинительный accusatory, accusatorial; denunciatory; inculpatory
~ акт indictment
~ приговор verdict of guilty
~ная речь speech for the prosecution
обвиняемый *n.* the accused/defendant; *law* prisoner at the bar
обвинять *law* charge, prosecute
обвод 1. contour, outline; profile; *pl.* shape; 2. bypass, roundabout way
~ контуров contour follower, contouring
~ профиля profiling
~ы корпуса корабля ship's/hull lines
~ы, обтекаемые streamlined form/shape
обводка border, edging, outline; inking
~, выводная *met.* escapement repeater
обводнение 1. *tech.* water-supply development,

augmentation of water supply; 2. flooding, inundation; water invasion/influx
~, **недостаточное** *leath.* poor rehydration
~, **пласта** *geol.* water encroachment
обводнительный irrigation *(attr.)*
обводный surrounding, encircling; by-passing
~ **канал** by-pass canal/channel
обвойник *bot.* silk vine *(Periploca)*
обволакивание *el.* encapsulation; covering; *geol.* coating
обволакивать coat, envelop, wrap, encapsulate (in)
обвязка 1. buiding, strapping, fastening; 2. framework, framing, brace; 3. fitting, connection, piping
~ **дверная вертикальная/горизонтальная** stiles/rails
~ **резервуара** tank connection
обвязывание binding, tying; *surg.* circumclusion
обвязывать bandage, tie (round); *arch.* truss
обгалдер *naut.* upholder
обгон overtaking, passing; outstripping
обгонный for passing
~ **пункт** *rail.* passing station
обгорание burning, scorching; *el.* spark-wearing
~ **изоляции** insulation burning
~ **клапана** valve scorching
обгорать 1. burn, scorch; 2. *el.* be pitted, spark-wear, wear by sparking
обделка lining, jacketing, casing
~ **грузового порта** cargo port brows
~, **монолитная** cast in-situ lining
~, **сборная** precast/prefabricated lining
обделывать mount, set
обдир 1. hulling, shelling, peeling, stripping; 2. *tech.* roughing, roughworking; 3. *timb.* barkless spot, scar *(on a tree trunk)*
обдирать *tech.* rough; shin, strip; *agr.* hull, shell
~ **абразивом** snag
обдирочный hulling, peeling; stripping; abrasive
~**ная мельница** scouring mill, scourer
обдув forced air cooling
обдувать *tech.* blast; blow over
обдувка *mach.* blasting; airflow; air blasting/cleaning/purging/cooling
~ **металлической дробью** shot blasting
~ **песком** sand blasting
~ **сжатым воздухом** air blasting
обдумывать consider, ponder, reflect, think over
обед dinner, meal; dinnertime, mealtime, lunchtime
~, **званный** dinner party
обеднение depletion; impoverishment, strip(ping)
~ **горючей смеси** leaning
~, **равновесное** equilibrium depletion
~ **фауны** depauperation of fauna
обедня *eccl.* mass
~**ню, служить** say mass
обеднять deplete, impoverish; strip
обезболивание *med.* an(a)esthesia, an(a)esthetization
~, **комбинированное** balanced anesthesia
~, **общее** general anesthesia, narcosis
~ **охлаждением** refrigeration anesthesia, cryoanesthesia
~ **родов** labo(u)r pain relief, painless labo(u)r
обезболивать an(a)esthetize, induce an(a)esthesia, render an(a)esthetic
обезболивающий an(a)esthetic, analgetic, analgesic
~**щее средство** an(a)esthetic
обезвоживание dehydration, dewatering, drainage, deaquation; exsiccation
~, **изотермическое** dehydroxylation
обезвоживатель dehydrator, desiccator, dewaterer
~, **скребковый** scraper dewaterer
обезвоженный dehydrated, dewatered
обезвоживать dehydrate; desiccate; dewater, drain
обезволашивание *leath.* unhairing, scouring, unwooling, dehairing; depilation
~, **ферментативное** enzyme unhairing
~, **швицеванием** unhairing by sweating
обезволашивать *leath.* unhair, unwool, depilate
обезвреженный decontaminated; safe, harmless
обезвреживание rendering safe/harmless; neutralization; decontamination
~ **взрывоопасных предметов** explosive disposal
~ **змеиного яда** evenomation
~ **неразорвавшихся бомб** bomb disposal
~ **ядов** detoxication
обезвреживать render safe, neutralize; decontaminate; detoxify
обезгаживание degassing, degasification, deaeration, outgassing
обезгаживать degas, outgas, deaerate
обезглавление beheading, decapitation
обезглавливание decapitation
~ **вулкана** *geol.* cone degradation/decapitation
~ **реки** stream capture
обездвиживать immobilize
обездубливать *leath.* detan
обезжелезивание iron removal
обезжиренный defatted, degreased, fatless, fat-free; skimmed
~**ное молоко** skim(med) milk
обезжиривание degreasing, defatting, deoiling, fat extraction
обезжириватель degreasing agent, degreaser
обезжиривать degrease, ungrease, deoil; remove fat from; skim
обеззараживание decontamination, disinfection, sterilization
обеззараживать decontaminate, disinfect
обезземливать dispossess of land
обеззоливание deashing, ash removal; decalcification
обезлесение, обезлесивание deforestation
обезлиствение defoliation, exfoliation
обезлиствитель defoliant
обезличение depersonalization

обезличенный impersonal; general-purpose; depersonalized
обезличивать deprive of individuality, depersonalize, render impersonal
обезлюдеть become depopulated/deserted
обезмедивание decopperization
обезображение, обезображивание disfiguration; deformation, defacing, mutilation
обезображенный disfigured; deformed, mutilated
обезопасить(ся), обезопашивать(ся) secure/guard (against)
обезоружение disarming
обезоруживание disarmament
обезуглероживание decarbonization, decarburization
обезыливание desilting, desliming
обезьяна monkey; ape, simian *(Simia)*
обезьяний apish, apelike; *zool.* simian
 ~ **мех** nutria (fur)
обезьяноподобный apelike, pithecoid
 ~ **человек** ape-man
обезьяночеловек ape-man *(Pithecanthropus erectus)*
обезьянский aplike, monkey *(attr.)*
обелиск *math.* obelisk
обельный *hist.* unassessed, untaxed
 ~ **холоп** bond slave
обелять 1. whitewash; 2. *hist.* exempt from paying taxes
обер-камергер *hist.* lord high chamberlain, archchamberlain
обер-кельнер headwaiter
обер-кондуктор *rail.* chief guard
обер-офицер subaltern officer
обер-полицмейстер *hist.* head police master
обер-прокурор *hist.* chief procurator; *law* attorney general
обер-секретарь chief secretary
обертка 1. wrap(per); envelope, sheath; dust jacket, cover; 2. *paleont.* involucre, integument
 ~, **метацентрическая** curve of metacentres
 ~, **общая** general involucre
обертон *mus.* harmonic, overtone
обертональный *mus.* harmonic
обертывание wrapping, envelopment, lapping
обескровливать drain of blood, bleed white, exsanguinate
обеспаривание vapour evacuation, devaporation
 ~, **влажное** wet pack, water compress
 ~ **со льдом** ice pack
обеспечение 1. guaranteeing, securing, ensuring; 2. providing/supplying (with); support; 3. guarantee; insurance; *law* collateral; *mil.* protection, security; 4. maintenance
 ~, **авиационное материально-техническое** air logistic support
 ~, **аппаратное** hardware
 ~, **аэродромно-техническое** airfield maintenance support
 ~ **безопасности** safety control, safeguarding
 ~ **валюты** *econ.* backing of curreney
 ~, **вещественное** property/real security
 ~ **высадки десанта** *mil.* amphibious support
 ~, **денежное** cash security
 ~ **живучести** *mil.* casualty control
 ~, **заказное** *comp.* custom-made software
 ~, **защищенное авторским правом** *comp.* copyrighted software
 ~ **золотом** gold guarantee
 ~, **информационное** dataware
 ~ **качества изделий** quality control
 ~, **коммерческое программное** business software
 ~, **математическое** software
 ~, **материально-техническое** logistics, logistic support
 ~, **под** *econ.* against security
 ~ **принятия управленческих решений** decision support
 ~, **программное** programming support; software
 ~ **работой** job security
 ~, **социальное** social security/service/welfare
 ~ **старости** old-age security
 ~, **судебное** legal tender
 ~, **техническое** hardware
 ~, **универсальное программное** heavy-duty software
обеспеченность 1. provision (of/with); 2. prosperity, comfortable/easy circunstances, material wellbeing/security
 ~ **водой** water supply
 ~ **почвенной съемкой** soil survey coverage
обеспыливание dust removal/control/elimination
обеспыливатель dedusting apparatus, deduster
обессахаривание desugarization
обессеребрение desilverization, extraction of silver
обессеривание desulphurization, sulphur removal
 ~, **внедоменное** ladle desulphurizing
обессмоливание pitch extraction/removal, resin extraction, deresination; detarring
обессоливание demineralization; desalination, desalting, salt elimination
 ~, **анионитовое/катионитовое** anion-/cation-exchange demineralization
обесточение, обесточивание *el.* de-energizing, killing
обесточивать *el.* de-energize, disconnect, cut off current
обесфосфоризация dephosphorization
обесфторивание deflourination
обесцвечивание decolo(u)ration, discolo(u)ration, fading, bleaching; depigmentation
 ~ **выветриванием** *geol.* weather stain, parachrosis *(of minerals)*
обесцвечивать decolo(u)rize, discolo(u)r; fade, bleach; *leath.* tone down
обесценение, обесценивание depreciation, price reduction, loss of value, devaluation
обесцененный depreciated

~ные ценные бумаги *econ.* wildcat securities
обесценивать depreciate, cheapen; devalue; reduce the price of
обесцинкование dezincing, dezincification
обесшламливание desludging, desliming, deslurring
обесшумливание silencing, noise control/reduction
обет vow, promise
~ безбрачия celibacy; vow of chastity
~, монашеский monastic vow(s)
обетованный *rel.* promised
~ край, ~ ная земля the Promised Land
обечайка drum, ferrule, cowling, shell; *arch.* course
~ диффузора diffuser cowl(ing)
обещание promise
обжа 1. plough beam/handle; 2. *hist.* obzha *(measure of arable land)*
обжаривание *cul.* grilling; browning
обжатие reduction; compression, pressing, cogging
~, абсолютное reduction in thickness
~ пневматика tyre deflection
~, полное *av.* bottoming
~ стойки шасси *av.* strut compression
обжиг 1. *tech.* roasting; burning, firing; 2. annealing, calcination, glazing
~ в кипящим слое fluidized-bed roasting
~ в кучах heap roasting, roasting in piles
~ до спекания, ~ намертво dead burning/roasting
~, хлорирующий chloridizing roasting
обжигала annealer; kiln worker
обжигательный *tech.* roasting; firing, baking; annealing, glazing; calcining
~ная печь roasting/calcining furnace; kiln
обжигать 1. *tech.* roast; 2. fire, bake, kiln; calcinate, anneal
~ кирпич kiln brick
обжим 1. reduction, pressing (out); compression; *met.* upsetting; squeezing; 2. crimper
~, высокий high rate of reduction
~ литейной формы crush in a mould
обжимать 1. *met.* swage; upset; cog; shingle; 2. break down, rough
обжимка 1. swage; crimper, crimping tool; hobby; riveting die; 2. *met.* upsetting; shingling; 3. pressing /squeezing/wringing out
~, заклепочная riveting set
~ клепального станка snap set
~, круглая dolly
~, патронная (cartridge) crimper
обжимный *tech.* blooming, pressing; cogging; shingling
~ пресс shingling press
~ стан blooming/cogging mill
обзол *wood.* wane, cant
обзор 1. observation, observing; 2. field of vision/view; 3. coverage, survey, synopsis
~, конический conical scanning
~ круговой all-round vision, panoramic view
~ неба sky survey
~, обобщающий summary review
~ печати press roundup/review
~ приборов *av.* instrument scanning
~, радиолокационный radar scanning
~ состояния рынка *econ.* market report
обзорность visibility, sweep of vision
обзорный survey *(attr.)*
~ пункт observation point
~ная статья review (article)
обивка upholstery, lining, trimming
~ сиденья seat upholstery
обилие abundance, copiousness, plenitude, profusion
~, видовое range of species
обильно-нервный many-veined
обитаемый inhabited, manned, habitable
обитатель inhabitant, dweller, resident
~ океана oceanophilus
~ почвы geobiont
обиход 1. custom, practice, use; way of life; 2. paraphernalia; 3. rules of church singing; devotional manual, ordinary
обкат(ка) *tech.* running/breaking in; rolling, running(-in)
~ двигателя engine running in
~ нового двигателя green run
обкатывать 1. roll (in); roll smooth; 2. *tech.* break/run in
обкладка 1. cover(ing), coat(ing), encasing; lining; 2. plate
~ подножки *auto.* running board/step/mat, step pad
обкладывание facing, covering, lining; coating
обкладывать 1. face, cover, line; coat; *med.* fur; 2. surround, edge; 3. (impose a) tax
обком *abbr.* **областной комитет** oblast/regional committe
обкомовец employee of oblast commitee
облава roundup; raid, swoop; *hunting* battue, beating up
~, полицейская police raid, manhunt
облагаемый taxable, leviable
~ таможенной пошлиной *econ.* customable
облагораживание *tech.* purification, refining, enriching
~ бумаги paper surface treatment
~ древесины wood modification
облагораживать upgrade; purify, refine
облака *pl.* clouds; *see also* **облако**
~, высококучевые alto-cumulus
~, высокослоистые alto-stratus
~, кучево-дождевые cumulonimbus
~, кучевообразные хлопьевидные flockiform cloudlets
~, кучевые cumulus, heap clouds
~, ложноперистые tonitro-cirrus, false cirrus
~, перисто-кучевые cirrocumulus
~, перисто-слоистые stratocirrus
~, перистые cirrus

~, **рассеянные** scattered clouds
~, **слоисто-дождевые** nimbostratus
~, **слоисто-кучевые** stratocumulus
~, **слоистые** stratified clouds, stratus
~, **фронтальные** frontal clouds
облако cloud
~ **верхнего яруса** high(-level) cloud
~, **вихревое** eddy cloud
~, **волнистое** billow/wave/undulated cloud
~, **воронкообразное** funnel(-shaped) cloud, pendant cloud
~ **восходящего скольжения** anabatic/upglide/upslide cloud
~, **вулканическое** volcanic blast, volcanic explosion cloud
~, **вымеобразное** mammatus (cloud)
~, **грибовидное** *nucl.* mushroom cloud
~, **дающее осадки** precipitating cloud
~, **изверженное** eruptive cloud
~, **курчавое** cauliflower cloud
~ **нижнего яруса** inferior/lower/low cloud
~, **перистое башенковидное** cirrus castellatus
~, **пылевое** dust/particulate cloud
~, **разорванно-дождевое** fractonimbus
~, **разорванно-кучевое** cumulus fractus
~, **разорванно-слоистое** fractostratus, stratus fractus
~, **серебристое** noctilucent cloud
~ **тумана** fog patch
~, **феновое** foehn cloud
~, **шкваловое** squall cloud
облакомер cloud-range meter/indicator
облакообразование cloud formation
обламывание breaking off
~ **ледника** calving
областком *abbr.* **областный комитет** *see* **обком**
областной provincial, regional; oblast *(attr.)*
~ **съезд** oblast convention
область 1. province, region, area, district; zone, oblast; 2. *anat.* region; 3. field, sphere, domain; space; range
~, **адресуемая** *comp.* addressable area
~, **асейсмическая** aseismic/nonseismic region
~, **безвалунная/зандровая** *geol.* driftless/outwash area, outwash plain
~, **бессточная** drainless area
~, **височная** *anat.* temporal region
~, **вихревая** vortex region
~, **вогнутая** downwarp
~ **вторжения воды** water-invaded region
~, **выпуклая** convex region/domain
~ **высоких энергий** high-energy region
~ **высот полета** height envelope
~, **гейгеровская** Geiger region
~ **действия адресной метки** address marker area
~ **диагностического экранирования** diagnostic scan-out area
~, **доверительная** *stat.* confidence interval/region
~, **докритическая** subcritical area/region
~ **допустимых значений** *comp.* tolerance range
~ **дрейфа** *el.* driftspace
~ **живота, хрящевая** *anat.* hypochondriac region
~ **закадмиевых энергий** *nucl.* epicadmium-energy region
~, **запальная** seed region
~, **закритическая** superstall/supercritical area/region
~, **заполненная** *comp.* fill area
~, **запрещенная** *phys.* forbidden region
~, **затопляемая** flood area/zone/region
~ **захвата** capture/trapping spot
~ **значений** *math.* range of values
~ **значимости** *stat.* significant range
~ **изменения давления** allobar
~ **изменения переменных** *math.* variables domain
~, **индексов, замкнутая** embedded index area
~, **карстовая** karst region; cockpit area
~, **комагматическая** comagmatic area, petrographic province/area/region
~, **критическая** *stat.* critical region
~ **критической точки** stagnation-point region
~, **лобковая** *anat.* pubic region
~, **лобная** *anat.* frontal region
~ **малых энергий** low-energy region
~, **материковая** land region
~, **многосвязная** *math.* multiply connected domain/region
~, **мочеполовая** *anat.* urogenital region
~ **надтепловых нейтронов** epithermal-neutron region
~, **надчревная** *anat.* epigastric region
~, **невозмущенная** undisturbed area
~ **неопределенности** *econ.* domain of uncertainty; *math.* area of ambiguity
~ **непринятия решений** *stat.* indeterminate zone
~ **нерасчетных режимов** off-design area
~ **неустойчивости** instability region, unstable area
~, **нефтеносная** oil-bearing region
~ **низкого давления** low, cyclone, depression, area of low pressure
~ **обильных дождей** hyetal region
~ **обитания** locality, habitat
~, **однолистная/однолистности** *math.* univalent/schlicht domain, one-sheeted domain, domain of univalence
~, **односвязная** *math.* simply-connected domain
~ **определения** *comp.* domain of definition
~ **оптимальности** *stat.* optimality range
~ **падения (барометрического) давления** katallobaric area
~ **памяти** *comp.* memory space, store area
~ **памяти, зарезервированная** *comp.* pot

~ памяти на сердечниках core memory area
~ пассатов trade-wind region
~, переполненная *comp.* overflow area
~ питания region of alimentation
~ питания ледника *geol.* glacier reservoir
~, пограничная *phys.* border zone, boundary region
~, подветренная leeward country, lee area
~ полетных режимов *av.* flight region/envelope
~, полярная polar region/cap
~ полярных сияний auroral region
~, поясничная/поясная *anat.* lumbar/abdominal region
~, предгорная piedmont
~, предметная *comp.* enterprise *(in a database);* knowledge domain
~ применения field of application; *econ.* range of use
~ принятия гипотезы *math.* (hypothesis-) acceptance region
~ принятия решения *stat.* determinate zone
~, промежностная *anat.* perineal region
~ промежуточного хранения *comp.* holding area
~ пропорциональности proportional region/band, proportionality band
~, пупочная *anat.* umbilical region
~, рабочая working/work/operating area/region
~ распространения range, distribution/ propagation area
~ рассуждения *logic* universe of discourse
~ режимов *av.* envelope
~ роста давления anallobaric area, pressure-rise area, area of rising pressure
~ с молодым рельефом region of youthful topography
~ сбросов *geol.* area of faulting
~, скуловая *anat.* zygomatic region
~ следа за телом *av.* wake region
~, слуховая *anat.* otic region
~ сноса source area
~ соударения impingement area
~ сохранения *comp.* save area
~ срыва потока stall(ing) region
~ старения *met.* ageing range
~ стока *geol.* run-off area, drainage basin
~ стока ледника flowing tongue of ice
~ страничного обмена *comp.* paging area
~ существования domain of existence
~ сходимости *math.* convergence set, domain of convergence
~, тектоническая неустойчивая mobile/unstable region
~, теменная *anat.* parietal region
~ тепловой энергии thermal-energy region/range
~ тумана fog patch, area of fog
~, тыловая backland, hinterland
~, углепромышленная coal province/field
~, ультрафиолетовая ultra-violet region, the ultra-violet
~ устойчивости stability region, stable area
~ хранения команд *comp.* instruction area
~ целостности *math.* integral domain, domain of integrity
~, челюстная *anat.* jaw region
~ чувствительности sensitivity area, sensitive region
~, щечная *anat.* buccal region
~, ягодичная *anat.* gluteal region

облатка 1. *pharm.* capsule; *cul.* wafer; 2. *eccl.* host, wafer
облачение array, clothing; *eccl.* robes, vestments
облачность cloud covrer(age), cloudiness, overcast
~, высокая high (cloud) ceiling
~, мощная/густая thick clouds, cloud cover
~, переменная variable sky
~ под самолетом undercast
~ с просветами broken clouds
~, сплошная continuous [ten-tenths] cloud(iness), totally overcast sky
облачный cloudy, nebulous
облегчать alleviate, relieve, ease; *law* commute
облегчение alleviation, mollifcation; relief; simplification, easing
~ посадки *av.* landing support
облегченный 1. alleviated; relieved; 2. lightly-built/constructed; light-weight
обледенение ice formation/accretion, icing (cover)
~, предупреждать provide anti-icing (facilities); anti-ice
облепиха *bot.* sea buckthorn *(Hippophae)*
облесение (af)forestation
облессование, облессовывание *geol.* loessivation
облет (test) flight; flyoff, fly-by/around, flyover
~ Луны circumlunar flight/pass
обливка 1. glazing; 2. glaze
обливщик glazer
облигатный *biol.* obligate
облигато *mus.* ob(b)ligato
облигационный bond *(attr.)*
облигационер *econ.* bondholder
облигация *fin.* bond; *pl.* stock
~ без специального обеспечения debenture/plain bond
~, беспроцентная passive bond
~, временная interim bond
~ военного займа, ~ военных займов war bond
~ выигрышного займа premium bond
~, заграничная external/foreign bond
~, именная registered (coupon) bond/debenture; *pl.* inscribed stock
~, казначейская Exchequer bond/bill; *pl.* stock
~ коммунальной компании public utility bond
~ на предъявителя bearer debenture, coupon bond, bond (payable) to bearer

~, неименная coupon bond, bond (payable) to bearer
~, неподлежащая погашению до наступления срока irredeemable/noncallable bond
~, обеспеченная закладной под недвижимость mortgage bond, real estate bond
~ по долгосрочному займу long-term bond
~ по которой отсрочен платеж deferred bond
~, погашаемая золотом gold bond
~, пролонгируемая на следующий период continued/extended bond
~ с правом досрочного погашения optional/callable/redeemable bond
~ с приоритетом prior/first/lien bond
~ со специальным обеспечением secured bond
~, рентная annuity bond
~, срочная к оплате payable bond
~ции, комунальные municipal stock(s)
~ции национализированных предприятий compensation/nationalization stock(s)
~ции, правительственные government bonds
облик 1. apperance, (out)look; habit; aspect; 2. cast of mind; character, temper
~, внешний *med.* habitus, appearance
~, волокнистый fibrous habit
~, доскообразный/дощатый platy habit
облиствление foliation, leafing; frondescence
облиственный leaf-bearing/-clad, leafy, leaved, foliate(d)
облицовка *constr.* facing, revetment; coating, lining, liner; jacketing; panelling
~ блоками block facing
~ гребнего вала shaft liner
~ порога *geol.* sill apron
~ радиатора radiator case/shell
облицовывать 1. *constr.* face/line/revet (with); 2. *met.* fettle
обложение 1. *fin.* faxation, rating; assessing, assessment; 2. impost, tax; 3. *mil.* besieging
~, налоговое tax incidence/load
обложка dust cover/jacket; case; folder
~, мягкая paperback
облом *arch.* 1. profile; mould(ing); 2. broken off piece; breaking off
обломки 1. *pl.* wreck(age); 2. debris, fragments, rubble
~ в породе *geol.* cognate enclosures, homogenous enclaves
~, вулканические ejecta, volcanic fragments
~ горных пород *geol.* detritus, rock debris/waste
~ судна shipwreck
~, шлаковые slags
обломовщина *lit.* "oblomovism"
обломок fragment, fraction
~, окатанный cobble
~, остроугольный sharp-edged fragment
~, случайный accidental/stray/chance fragment
обломочнозернистый granular-fragmental
обломочный detrital, clastic, fragmental
облоно *abbr.* Oblast/Regional Board of Education
облопачивание *therm.* blading *(of turbine)*
облучатель radiation source; irradiator
~, антенный aerial/antenna feed, excitation source
~, бактерицидный *med.* bactericidal lamp
~, осколочный *nucl.* fission-fragment irradiator
~, рентгеновский X-ray machine
облучать irradiate, expose to radiation
облучаться be exposed to radistion; *med.* be treated with radiation
облучение 1. (ir)radiation; exposure; 2. illumination; 3. *tech.* radiation processing/treatment
~ альфа-частицами alpha-irradiation
~ до насыщения saturation irradiation
~, допустимое allowable/permissible irradiation
~, корпускулярное particle bombardment
~, летальное lethal irradiation
~ малой дозой light exposure
~, непрофессиональное non-occupational exposure
~, повторное re-irradiation
~ полной дозой full-scale irradiation
~, постоянное maintained/persistent irradiation
~ потоком высокой плотности high-flux irradiation
~, профессиональное professional/occupational irradiation
~ радиолокатором/ радиолокационное radar illumination
~ рентгеновскими лучами *med.* roentgenization, X-ray irradiation, X-raying
~ солнечным светом insolation
~, суммарное cumulative/total/integral irradiation
~ топлива в активной зоне *nucl.* corefuel irradiation
~ ультразвуковыми волнами insonifying, insonification
~, ультрафиолетовое ultra-violet illumination
~, фоновое background irradiation
облученность irradiancy, irradiance; exposure rate
~, пространственная scalar irradiance
облысение *med.* alopecia, baldness, acomia, calvities
~, круглое alopecia areata
обмазка 1. covering, coating, plastering; 2. plaster, lute
~, огнезащитная fire-resisting coating
обмазывание coating, plastering, putying
обман 1. deception, fraud; bluff; 2. delusion, illusion
~, оптический optical illusion
~, страховой *econ.* insurance fraud
~ чувств hallucination, illusion
обманка *miner.* blende, glance
~, бархатная pribramite, sammet blende
~, висмутовая bismuth blende, eulytite
~, медная copper blende, sandbergite

~, **роговая** hornblende, amphibole
~, **смоляная урановая** pitchblende, nasturan, uranitite
~, **угольная** anthracite, hard coal
обманный fraudulent; deceitful, deceptive
обматывание winding/wrapping round, lapping
обмелевший shoaled
обмеление shoaling, shallowing
обмен 1. exchange, interchange; trade, barter; 2. traffic, communication; 3. *med.* metabolism
~, **азотный** nitrogenous metabolism
~, **бартерный** *econ.* swap(ping) metabolism
~, **белковый** protein
~ **веществ** metabolism
~ **веществ, повышенный** hypermetabolism
~ **воздуха** air change, perflation
~, **газовый** gaseous exchange
~ **данными** *comp.* swapping, data communication/ exchange
~, **двусторонний** two-way traffic
~, **жировой** fat metabolism
~, **изотопный** isotope exchange
~, **ионный** ion(ic) exchange
~, **исходящий** calling traffic
~ **натуральный** *econ.* swap(ping)
~ **по паритету** exchange at par
~ **по шине** bus exchange
~, **промежуточный** intermediate metabolism
~, **стериновый** sterol metabolism
~, **телеграфный** telegraph traffic
~, **углеводный** carbohydrate metabolism
~ **энергии** interchange of energy
обменивать exchange (for); trade, swap, barter
обменный 1. *med.* metabolic; metathetical; 2. exchangeable, interchangeable
обмер measure, (ad)measurement
~ **вместимости** tonnage (ad)measurement
обмерщик measurer
~ **груза** bulker
обмолвка lapsus linguae, slip of the tongue
обмолот *agr.* threshing yield
обморожение frostbite; chilblain
обморок *med.* syncope; swoon; fainting (spell)
~, **высотный** altitude syncope
~, **падать в** faint
~ **при мочеиспускании** micturition syncope
обморочный *med.* syncopic, syncopal
~ **приступ** fainting spell
обмотка 1. *el.* winding, coil; 2. sheating, taping, wrapping; 3. puttee
~, **безындукционная** noninductive winding
~ **возбуждения** *el.* field/excitation/exciting winding
~, **главная** main coil
~, **катушечная** coil winding
~, **левая** left-hand winding
~, **многофазная** polyphase winding
~, **накальная** *rad.* heater/filament winding
~, **одноходовая** simplex winding
~ **подмагничивания** bias winding
~, **правая** right-hand winding
~, **разрядная** *comp.* bit winding
~, **ускоряющая** accelerating coil
~, **успокоительная** damper winding
~, **шаблонная** preformed/diamond winding
~ **якоря** *el.* armature winding
обмундирование *mil.* uniform (dress), clothing, outfit
~, **повседневное** undress uniform
~ **подводников, специальное** submarine kit
~, **рабочее** undress blue
~, **тропическое** tropical rig
~, **шерстяное** woolens
обмуровка brickwork, setting, masonry
~, **огнеупорная** refractory setting
~, **панельная щитовая** bagging
обмуровывать brick-line
обнажаться crop out, denude, lie bare, be exposed, come out to the day
обнажение 1. *geol.* outcrop(ping), baring, exposure; 2. uncovering, stripping
~ **горных пород** rock exposure/outcrop
~ **кольцевое (замкнутое)** (closed) annular exposure/outcrop
~ **кровли** roof exposure
~, **поверхностное** surface exposure
обнаженный exposed, bare, denuded, denudate(d), stripped
~ **меч** unsheathed sword
обнародование publication, manifestation; *law* promulgation
обнародовать publish; divulge; *law* promulgate
обнаружение detection, acquisition, finding, discovering
~, **дальнее** early warning, long-range detection
~ **на слух** aural detection
~ **неисправности** fault finding, malfunction detection
~ **отказа** failure detection
~ **ошибок** error detection
~ **цели** target acquisition
обнаруживаемость detectability
обнаруживать 1. detect, discover, uncover, locate, trace, spot; 2. manifest, reveal. display
обнаружитель locator, detector, finder, feeler
~, **магнитный** magnetic anomaly detector
~ **трещин** crack detector
~ **турбулентности, бортовой** airborne turbulence sensor
обнищание *econ.* immiserization, impoverishment, depauperation
обновление renewal, rejuvenation, renovation; updating
~ **данных** (data) updating
~ **данных, групповое** *comp.* batch updating
~ **изображения на экране** screen redraw
~ **памяти** storage regeneration
~ **покрытия** *constr.* re-paving
~ **реки** river rejuvenation
обнубиляция *med.* beclouded mental state,

obnubilation
обобщать generalize; combine, unite, unify
обобщение *math.* generalization, extension
обобщенный *math.* generalized; self-contained
обобществление collectivization; socialization
~ средств производства *econ.* socialization of means of production
обобществленный socialized; collectivized
обобществлять socialize; collectivize
обогатитель ore dresser; enriching agent; concentrator
обогащать enrich; *min.* dress, concentrate
обогащение enrichment; *min.* dressing, concentration
~, гравитационное gravity, separation/concentration
~ кислородом oxygenation
~ минерала/минералов mineral dressing
~ руд ore dressing/concentration/benefication
~ углекислот carbonation
~ угля, мокрое coal washing
~ фауны fauna enrichment
~ флотационное flotation
обоготворение idolization, deification
обогрев heating, warming
~, огневой firing
~ отработавшим паром waste-steam heating
обогреватель heater
~ ветрового стекла windshield defroster/heater
~, керосиновый kerosene-fired heat exchanger
обод rim, ring, band; hoop; felly
~, зубчатый toothed rim
~, разъемный split rim
~ шкива pulley rim
ободок 1. ring, ferrule; 2. annulus; bandelet; fillet
ободочный *anat.* colonic
обоеполость bisexualism, hermaphroditism; androgynism; monoeciousness
обоеполый bisexual, hermaphroditic, androgynous; monoecious, monoclinous
~лое растение androgyne
обоесторонний two-way, bilateral; double-sided
~няя ткань double-sided fabric
обожание 1. adoration; hero worship; 2. deification; idolatry
обожествление deification, idolization, apotheosis; worshipping
~ человека anthropolatry
обоз string of carts/sledges, wagon train, transport; *mil.* train, transport
обозначать mark, designate; label; define, indicate, symbolize
обозначение 1. (de)notation, mark(ing), designation; 2. code; label sign, symbol
~, буквенное lettering, letter designation
~ государственной принадлежности nationality marking
~, двоичное binary notation
~, условное (conventional) sign, symbol, legend
~, цифровое numerical designation
~ния на карте plotting symbols
обозник, обозный *n. mil.* transport driver
обозревание surveying, inspection, examining, examination
обозреватель commentator, reviewer, columnist
~, спортивный sports columnist
обозрение review, commentary; survey; theatre revue
~, ежемесячное monthly review
обозримый visible, foreseeable
обои *pl.* wallpaper
~, наклеивать hang wallpaper
обойка 1. upholstery (material); flax/hemp/corn waste; 2. scouring machine, huller, scourer
обойма 1. *tech.* yoke, collar stirrup, casing, housing; cage; (iron) ring; holder, shackle; 2. *mil.* cartridge clip
~ блока pulley fork/casing
~ подшипника bearing race
~ пружины spring holder
~, роликовая roller cage
~ сваи pile casing
~, шариковая ball cage
~ щеткодержателя *el.* brush box
обол *hist.* obol(e), obolus
оболочка 1. coat(ing), sheath(ing), cover(ing); envelope, jacket, mantle, case, casing; *constr.* shell; 2. *anat., bot., zool.* tunic, membrane, skin, layer; integument; operculum; 3. appearance, outer covering
~, аллантоисная allantoid membrane, allantois
~ атома atom shell
~, бороздчатая striated sheath
~, вакуумная vacuum envelope
~, влагалищная vaginal tunic/sheath
~, внешняя outer shell
~, воздухонепроницаемая airtight envelope
~, волокнистая fibrosal tunic
~, ворсинчатая (allanto)chorion
~, выпуклая convex hull/closure
~, газовая gas blanket, gaseous envelope
~, герметизирующая containment (vessel)
~, герметическая hermetical/pressure shell
~ глаза, белочная sclera, sclerotica, sclerotic coat, white of the eye
~ глаза, внутренняя retina
~ желудка, слизистая mucous membrane
~, замкнутая closed/completed/filled shell
~, защитная integument; protective envelope/shell;
~, защитная, реактора contaiment/protective envelope
~ Земли Earth's shell/mantle
~, измеримая *math.* measurable hull
~, изолирующая insulation envelope
~, ионная ion sheath
~ кабеля cable sheath(ing)
~, капсульная, отпадающая deciduous capsular membrane
~, каркасная grid-work shell

~, клеточная/клетки cell membrane/wall, cytoderm
~, колбасная sausage casing
~ космического корабля spacecraft hull/shell
~, линейная *math.* linear span/closure
~, липоидная lipoid envelope
~, ложная accidental/false membrane, pseudomembrane
~ мины mine shell/case
~ мозга, мягкая pia mater
~ мозга, паутинная arachnoid, meninx serosa
~ мозга, твердая dura mater, meninx fibrosa, pachymeninx
~, мозговая medullary sheath; meninx
~, многослойная multilayer/laminated/multiwalled envelope
~, мышечная muscular layer, myometrium
~, наружная outer shell
~, незамкнутая open/unfilled shell
~, околоплодная amnion
~, осесимметричная axisymmetrical shell
~, отпадающая deciduous membrane
~, паутинная arachnoid (sheath)
~, плазменная plasma sheath
~, плеохроничная pleochroic halo
~, плодная f(o)etal membrane
~, подкрепленная *av.* stiffened shell
~, пологая shallow shell
~, проводящая conducting shell
~ пыльцевого зерна, наружная (pollen) ex(t)ine
~, радужная iris of the eye
~, роговая cornea
~ рта, слизистая mucous membrane of the oral cavity
~ световедущей жилы cladding, coating
~ сердца, внутренняя endocardium
~ сердца, мышечная myocardium
~ сердца, наружная epicardium
~, слизистая mucous membrane/coat, mucosa
~ сосуда/сосудистая tunic, vascular layer
~, спорангиевая sporangial envelope, operculum
~, споровая spore extine/vessel/casing
~ споры spore vessel/casing
~, стекловатая glassy skin
~, стекловидная hyaloid membrane
~ тепловыделяющего элемента fuel-element jacket/can
~ теплообменника heat-exchanger shell
~, урановая uranium blanket
~, циркониевая zirconium can
~, электронная electron(ic) shell; electronic cloud
~, ядерная/ядра nuclear membrane/shell
~ яйца, желточная ovular membrane
оболочники *pl. zool.* tunicates *(Tunicata)*
обоняние olfaction, sense of smell
обонятельный olfactory, olfactive
оборачиваемость *fin., econ.* turnover; *naut.* turnaround
~ вагонов wagon turnaround
~ оборотных средств working capital turnover
~ основного капитала plant turnover
~ танкера tanker turnaround
~ товарных запасов stock/inventory turnover
оборачивать 1. turn (round); revert; circumvolve; invert, turn over; 2. put into circulation; 3. convert, transform
оборона defence, *Am.* defense; defensive positions
~, активная offensive defence
~, гражданская civil defence
~, необходимая *law* private defence
~, противовоздушная antiaircraft/air defence(s)
~, противокосмическая anti-cosmic defence, space-air defence
оборонительный, оборонный defensive; *zool.* aposematic
обороноспособность defence/defensive capacity, defence potential, defensibility
обороноустойчивость defensive power
обороняющийся *n.* defender
оборот 1. revolution, rotation, turnaround; 2. *fin., econ.* turnover, cycle; 3. circulation; 4. back, reverse side; 5. *ling.* turn (of speech), expression, phrase; 6. *zool.* volute; *see also* **обороты**
~ варки *paper* cooking cycle
~, вводить в put into circulation
~, годовой annual turnover
~ капитала turnover of capital
~ между отделениями фирмы interdivisional transfer
~, полный complete revolution
~ раковины volution, whorl, turn of a shell
~ речи phrase, turn of speech
~, торговый amount of business, trade turnover
оборотень werewolf, lycanthrope; zombi(e)
оборотность *tech.* rotation, revolution
оборотный 1. reverse, reversible, back; 2. circulating, working
~ капитал working capital
~ное е *name of the letter* **э**
обороты *pl.* speed, revolutions (per minute), r.p.m.
~, большие/высокие high speed/r.p.m.
~ двигателя engine speed
~, максимальные peak r.p.m.
~ малого газа idle speed/r.p.m.
~, малые low speed/r.p.m.
оборудование 1. equipment, installation(s), outfit; facilities, instrumentation; machinery, fixtures; 2. equipping, outfitting
~, авиаремонтное aircraft repair facility/equipment
~, автономное *comp.* off-line equipment
~, аппаратное hardware
~, ассенизиационное sanitary equipment
~, бортовое airborne equipment
~, бытовое furnishing equipment
~, внешнее *comp.* peripherals

~, водоподготовительное water-treatment equipment
~, вскрышное *min.* stripping equipment
~, вспомогательное service/support/accessory/auxiliary/ancillary equipment
~, вытесняющее рабочую силу labo(u)r-displacing equipment
~, готовое off-the shelf equipment
~, дальномерное distance measuring equipment, DME
~ двухкратного применения dual shot facility
~, дезактивационное decontamination equipment
~, десантное landing gear
~ дистанционного управления remote-control equipment
~ для обнаружения радиоактивности activity-sensing equipment
~ для ядерных исследований nucleonic equipment
~, дозиметрическое radiation-monitoring/-measuring equipment
~, дополнительное additional/optional/extra equipment
~ жизнеобеспечения life-support/-saving equipment
~, забойное *min.* face equipment
~, машинное machinery
~, заводское manufacturing/production facility
~, запасное spares, spare equipment, stand-by equipment
~, зарезервированное back-up equipment, redundant equipment
~, защитное protective equipment
~, измерительное metering/measuring equipment
~, капитальное durable/capital(ized) equipment
~ контрольно-измерительными приборами instrumentation
~, мобильное mobile equipment
~, модернизированное updated equipment
~, модульное modular hardware
~, наземное ground facility/equipment, (above-) ground installation
~, нивелировочное level-and-rod equipment
~, океанографическое oceanographic instrumentation
~, опытное test equipment
~ оргсвязи, оконечное data terminal equipment
~ отечественного производства home equipment
~, передвижное/подвижное mobile equipment
~, погрузо-разгрузочное materials-handling equipment
~, подъемно-транспортное (material-) handling equipment
~, полностью совместимое *comp.* plug-compatible equipment
~, портовое погрузочно-разгрузочное stevedore equipment
~ причала *naut.* terminal/quay facilities
~, программное *comp.* software
~, промежуточное (в системе) interface equipment
~, промышленное industrial plant equipment
~, пусковое launching/firing equipment
~, раздаточное dispensing equipment
~, резервное standby/reserve equipment
~, ремонтное repair facilities
~ с коротким сроком службы short-lived equipment
~, сберегающее живой труд labo(u)r-saving machines
~, связное communication(s) equipment
~, серийное standard/serial equipment, off-the-shelf equipment
~, снегоуборочное snow-removal equipment
~, сновальное *text.* beaming machine
~ собственного производства home equipment
~, спасательное salvage/resque gear; lifesaving/survival equipment
~, съемное removable equipment
~, уточно-мотальное *text.* cop/weft winding machinery
~, учебное training device
~, швартовочное tie-down equipment
~, шлихтовальное *text.* sizing machinery
~, эксплуатационное production/maintenance equipment
~, энергетическое power-generating plant

оборудовать equip, fit out

обоснование evidence; basis, ground; proof; substantiation
~ вида foundation of species
~, высотное *geod.* height/vertical/elevation control
~, логическое *comp.* rationale
~, планово-высотное (field) compilation survey
~, строгое rigorous substantiation
~, технико-экономическое feasibility study

обоснованность validity; *law* sufficiency
~, юридическая sufficiency in law

обоснованный substantiated, well-founded/-grounded; valid, proved; *law* sufficient

обосновывать substantiate, base, ground, found

обособление separation, isolation; setting apart
~ ветвей branching
~ видов isolation of species
~ языков linquistic segregation

обособленный individual, isolated, solitary; enclosed; set off; self-contained

обособлять isolate, separate, single out

обострение intensification; sharpening; accentuation; enhancement; exacerbation; worsening
~ сигнала *rad.* signal peaking

обостренный 1. pointed, sharp; 2. aggravated, strained; acute
~ные отношения tense relations
обочина roadside, shoulder, wayside; edge, curb; berm
~ дороги road shoulder
обоюдность reciprocity, mutuality
обоюдный (com)mutual, reciprocal
обоюдовогнутый biconcave, concavo-concave
обоюдовыгодный mutualy benefical
обоюдовыпуклый biconvex, convexo-convex
обоюдоострый two-/double-edged; *bot.* ancipital, ancipitous
обрабатываемость workability; machinability
обрабатывать 1. work, process, treat; machine; 2. *agr.* cultivate, till; 3. perfect, polish, dress
~ грузы handle goods
~ предварительно pretreat
обрабатывающий processing, treating, working; machining
~щая промышленность manifacturing/processing industry
обработанный 1. processed, treated; machined; 2. *agr.* cultivated, tilled; 3. dressed, finished; trimmed
обработка 1. processing, treatment, working (-up); machining; turning; handling; 2. *agr.* cultivation, tilling; 3. finishing; adaptation
~, автономная *comp.* off-line processing
~ багажа *av., rail.* luggage handling
~ брюшной стенки *med.* abdominal cleansing preparation
~ в диалоговом режиме conversational processing
~ в патроне chucking work
~ в реальном (масштабе) времени real-time processing
~ в режиме трансляции translational processing
~ в центрах machining between centres
~ воды water treatment
~, волновая wavefront processing
~, горячая hot machining/working
~ грузов freight handling
~ давлением metal forming, pressure/noncutting shaping
~ данных data processing/handling
~ данных, автономная off-line data processing
~ данных без накопления in-line data processing
~ данных в темпе поступления real-time data processing
~ данных, комплексная integrated data processing
~ данных на месте on site data processing
~ данных, оперативная on-line data processing
~ данных, пакетная basic stacked job processing
~ данных по запросам demand (data) processing
~ данных по требованию (on-)demand (data) processing
~ данных, предварительная front-end processing, data reduction/transformation
~ данных, прямая on-line data processing
~ данных, электронная electronic data processing
~, дистанционная remote processing
~, довсходовая *agr.* pre-emergence treatment
~ документации processing of documents
~, дробеметная/дробеструйная shotblasting
~ заданий job processing
~ запросов query processing
~, идеологическая brain washing
~ изображения image/picture processing
~, интенсивная *agr.* intensive cultivation
~ информации information/data processing
~ и отбор данных data reduction
~, камеральная indoor/laboratory investigation, office studies/processing
~ кож leather working
~ кож, первичная taking-up of hides
~, конвейерная pipelining
~, машинная machine processing
~ металлов резанием machining
~ металлов, слесарная benchwork
~ металлов, химико-термическая surface impregnation
~, металлоструйная grit-blasting
~ метеоданных meteorological data processing
~, механическая machining (operation)
~, многократная multiprocessing
~, многопоточная multithread processing
~, мультипроцессорная multiprocessing
~, начальная head-end treatment
~, неавтономная on-line processing
~, окончательная tail-end treatment; *tech.* finishing operation
~ он-лайн on-line processing
~, оперативная in-line (data) processing
~, отсроченная deferred processing
~ отходов waste treatment
~ ошибок error handling
~, пакетная batch processing; stacked job operation
~, параллельная concurrent processing
~, первичная primary treatment; pre-processing
~, пескоструйная sandblasting
~, пирометаллургическая pyroprocessing
~, пожнивная *agr.* stubble tillage
~, последующая aftertreatment
~, предварительная pretreatment
~ прерываний interupt processing
~ радиоактивных отходов processing of radioactive waste
~ распределенная distributed processing
~ с перекрытием overlapped processing
~, санитарная cleansing
~ символьной информации symbolic processing

~ текстовой информации word processing
~ термическая *met.* heat/thermal treatment; *hydr.* hot processing
~, токарная turning
~, фоновая backgrounding
~, холодная cold working
~ чеков check processing
~, чередующаяся/сменная *agr.* land rotation cultivation
~, черновая roughing
~, чистовая finishing
~ экипажа, санитарная crew decontamination
~, экстенсивная *agr.* extensive cultivation
~, электроэрозионная/электроискровая spark(-erosion) machining
образ 1. image, representation, portrayal; appearance, shape, form; 2. fashion, style, mode, manner; kind; *comp.* pattern; *eccl.* icon
~ действия modus operandi; course of action
~ жизни way/mode of life, manner of living, life style, habit
~ жизни, планктонный planktonic habit
~ жизни, ползающий crawling habit
~ жизни, сидячий sedentary life
~, запрестольный *eccl.* altarpiece
~ множества *math.* image set
~, мысленный visuality
~, обучающий *comp.* training pattern
~ Пуркинье *med.* Purkinje's image
~ элемента *math.* image of an element
~, эталонный reference pattern (for recognition)
~ом, случайным at random, in a random way
образец 1. sample, model specimen, test piece; pattern; 2. type, version
~ аппаратуры, макетный breadboard (model)
~, боевой *mil.* operational version
~ без цены sample post; sample without value
~, бесплатный free sample
~, биологический biological specimen
~, выбранный picked sample/specimen
~, высокоактивный strong sample
~ для анализа assay sample
~ для испытания test piece/specimen
~ изделия item/product sample
~, испытуемый/испытываемый test sample/specimen/piece
~, контрольный reference sample/specimen, keysample
~, летный flight model
~ лунного грунта sample of lunar soil, lunar sample
~, опытный pilot sample, prototype (model)
~ поведения behavio(u)r pattern
~ подписи *fin.* signature card
~ породы *geol.* rock/formation sample, specimen rock
~, промышленный industrial design, production prototype
~, серийный production model/copy
~, смешанный composite sample
~ топлива test fuel; fuel sample
~, утвержденный standard piece
~, экспериментальный experimental prototype
~зцы, парные matched samples
образная *n. eccl.* icon room
образник, образница *eccl.* icon room/case
образование I formation, generation, initiation, creation; develipment; morphosis; evolution
~ айсбергов calving
~, ветвистое dendritic formation, treeing
~ видов speciation
~ воздуха в тканях aerosis
~ волдырей blistering
~ двойников *cryst.* twinning
~ дополнения *comp.* complementing, complementation
~ жилы veining
~ зародышей nucleation
~ каверн cavitation
~ камня lithogenesis, lithogeny
~ кожи cutification
~ колосьев *bot.* earing, ear formation
~ коры crusting, crustification
~ костной ткани osteogenesis, osteogeny
~ кратера cratering
~, ледниковое ice-laid deposit, diluvium, drift deposit
~ лейкоцитов leucopoiesis
~ льда ice formation
~ млечного сока chylosis, chylification
~ надрезов *met.* collaring
~, наносное alluvium, alluvion
~ настылей *met.* incrustation
~, неправильное malformation
~ окалины scaling
~ осадков sedimentation
~, остаточное residual formation, perluvium
~ очередей queueing
~ пены foaming
~ плода f(o)etation
~ поперечных связей crosslinking
~ пузырей bubbling
~ пыли dust generation
~ пятен maculation, spotting
~ разрывов fracturing
~ рельефа relief formation; landscape/scenery making
~ сахара glycogeny
~, сетевидное *anat.* reticular/Deiter's formation
~ складок folding, plication
~ узелков nodulation
~ трещин fracturing, fissuring, cracking, crack initiation/ formation
~ центров кристаллизации nucleation
~ цепочек catenation
~ циклонов cyclone formation, cyclogenesis
~ ядер конденсации nucleation
образование II education, training; schooling
~ в области управления management education

~ **в объеме колледжа** college education
~, **всеобщее** compulsory education
~, **всеобщее начальное** universal primary education
~, **высшее** higher/tertiary education, third-level education, tertiary level
~ **для взрослых** adult education
~, **коммерческое** business education
~, **начальное** primary/elementary education, grade school education, first level education
~, **неполное среднее** postprimary education
~, **обязательное** compulsory education
~, **продолженное среднее** postsecondary education
~, **профессиональное** trade/vocational education/training
~, **среднее** secondary education/level
~, **субсидируемое** subsidized education, below-cost education
~, **университетское** university education/training

образованность erudition; education; culture
образованный educated; cultured
образовательный morphotic, formative
~**ная ткань** formative/nascent tissue
образовывать form, build, generate, produce; make up, constitute
образующая *n. math.* generatrix, generator
~ **конуса** generator of a cone
~, **прямолинейная** rectilinear generatrix
образующий forming, generating, evolving
~ **камбиальный слой** *bot.* cambiogenetic
образцово-показательный model *(attr.)*
образцовый exemplary; model *(attr.)*
~ **порядок** perfect order
~**вое хозяйство** model farm
обральдрук *print.* duplicating process
обрамление 1. frame(work); 2. bezel; 3. framing, mounting, bordering
обрастание *naut.* fouling, sea-crust
обрастать *naut.* be encrusted with marine growth; be/become overgrown
обрат I skim(med) milk
обрат II *text.* reworkable waste
обратимость 1. reversibility, invertibility; 2. *econ.* convertibility
~, **поведенческая** *econ.* behavioural reversibility
обратимый 1. reversible, invertible; 2. *econ.* convertible
обратно inversely, conversely; back; counter-
обратнобулавовидный inversely clavate, obclavate
обратногетероцеркальный *zool.* hypocercal, reversed heterocercal
обратногородчатый obcrenate, inversely crenate
обратное *n. math.* inverse
~ **левое/правое** left/right inverse
обратнозависимый inversely related/dependant
обратноконический, обратноконусовидный obconic(al), inversely conical
обратнокубарчатый obturbinate
обратноланцетный, обратноланцетовидный oblanceolate
обратнопирамидальный obpyramidal
обратно пропорциональный inversely proportional
обратносердцевидный obcordate
обратносжатый obclavate
обратноязычковый obligulate
обратнояйцевидный obovate, obovoid
обратный 1. reverse; back; return; 2. opposite; 3. *math.* inverse, converse; reciprocal; 4. *bot.* anatropous
~ **адрес** return adress
~ **код** reverse code
~ **ход каретки** *comp.* carriage return
~**ная семяпочка** *bot.* anatropous ovule
~**ная сила закона** retroactive law
обращаемость *fin.* negotiability
обращающийся *fin.* negotiable
обращение 1. address(ing), appeal; allocution; *comp.* reference, call; 2. rotation, revolution, circulation, turn; 3. application; manipulation, handling, treatment; use, usage; 4. transformation, conversion, reversal; reversion
~, **банкнот** paper circulation
~, **безопасное** safe handling
~ **в христианство** conversion to Christianity
~ **внутрь** introversion
~ **времени** time reversion
~, **грубое** hard usage
~ **движения** motion reversing
~, **денежное/денег** currency of money, currency/money/monetary circulation
~ **для выборки операндов** *comp.* access for operand selection
~ **Земли** revolution of the Earth
~ **к массиву** access to the file
~ **к памяти** *comp.* storage/memory access, reference to storage
~ **к памяти посредством окон** windowed memory reference
~ **к подпрограмме** subroutine call
~ **матрицы** matrix inversion
~, **неверное** invalid call
~ **негатива** *cin.* negative reversal
~, **неправильное** maloperation, mishandling
~, **несанкционированное** unauthorized reference
~, **неумелое** fumbling
~ **по значению/имени** reference by value/name
~, **простое товарное** *econ.* simple circulation of commodities/goods
~ **ряда** *math.* inversion/reversion of a series
~ **с горючим** fuel handling
~, **товарное** commodity/goods circulation
~ **фазы** phase reversal/inversion
обращенный inverse, inverted
~ **вверх** *med.* supine

обрез 1. cut; edge; 2. sawn-off gun; 3. half-barrel
~ страницы, нижний lower/bottom edge
обре́зание 1. cut-off; 2. *med., rel.* circumcision
~ импульсов clipping
~, профилактическое preventive circumcision
обреза́ние cutting, clipping; trimming, paring, pruning; edging
обрезанный *n.* circumcised person
обрезать 1. clip, pare; prune, trim; cut(off); *rel.* circumcise; 2. *tech.* quit *(of engine)*
обрезинивание rubberizing, rubber coating
обрезка cutling(-off), clipping, paring; trimming, edging
~ выпрессовок trimming of flash, deflashing
~, зимняя *agr.* dormant pruning
~ строк *print.* slug cutting
~ сучьев *timb.* abnodation
обрезный cutting, trimming; edging
обрезок scrap, cut, length; *pl.* ends, trimmings, cuttings, clippings; odds and ends
обрезчик cutter, trimmer
обрезь *met.* discard, crop ends, cuttings
обременение burdening
~ вещное *law* real obligation
обременительный burdensome, onerous; excessive
обременять burden, load
~, чрезмерно overburden; overtax
обрешетка 1. lathwork; lathing of roof; 2. guard rail
обрешетник batten ceiling
обриеция *bot.* aubrietia *(Aubrietia)*
обрисовка depicting, depiction, portrayal; delineation
оброк *hist.* quitrent
~, денежный quitrent
~ натурой pr(a)edial tithe
~, натуральный métayage
оброчник, оброчный *hist.* peasant on quitrent
обрубка I chipping; chiselling(-off)
~, косая bevelling
~ отливок chipping of castings
обрубка II hemming
обрубная *n.* *(foundry)* cleaning shop
обрусить Russianize, Russify
обруч hoop; collar, ring; ampyx
~и на бочку набивать hoop a barrel/cask
обручальный betrothal *(attr.)*
~ обряд betrothal
~ное кольцо wedding ring
обручаться become engaged *(to)*
обручение betrothal
обрушение 1. *min.* caving, cave-in, collapse; *geol.* downfall, failure, avalanche; stoping; 2. demolition, pulling down
~ горной породы rockfall
~ кровли overhead stoping; roof caving
~, постепенное piecemeal stoping
~ руды ore caving
обрушивание I hulling, scouring
обрушивание II falling/pulling/bringing down, collapsing
обры *pl. hist.* the Avars
обрыв 1. (e)scarp; bluff, cliff; precipice; 2. break(age), disruption, breaking, discontinuity; 3. cutoff, disconnection
~, береговой (sea) cliff; bluff
~ в схеме/цепи *el.* open(-circuit)
~, горный mountain scarp
~ ряда rupture/discontinuity of a series
~ сброса *geol.* fault scarp/ledge, tectonic termination
обрываться *math.* terminate, stop
обрывистый precipitous, bluff(y)
обрывок scrap; shred; *pl.* odds and ends
обрызг *build.* scratch coat
обрызгивать spray, sprinkle
обряд rite, ritual, ceremony; *eccl.* ordinance
обрядность rites, ceremonial
обрядный ritual, ceremonial
обсадка 1. *met.* upsetting; casing; 2. planting round
~ скважины well casing
обсаженный planted round, lined
~ лесом afforested
обсеменение dissemination; *bot.* going to seed
обсеменяться *bot.* go to seed
обсерватория observatory
~, космическая space observatory
~, орбитальная orbiting/orbital observatory
обсечение, обсечка cutting off; trimming; dressing *(of stone)*
обсечки *pl.* ends, remnants; cuttings, clippings
обсидиан *geol.* obsidian
~, волосистый/гавайский Pele's hair
~, расстеклованный devitrified obsidian, apobsidian
обследование investigation, survey, exploration; inspection; observation; *med.* examination
~ в один этап *econ.* single-round inquiry
~, визуальное contact investigation, visual examination
~, выборочное sampling observation/inquiry, sample/sampling survey
~ грунта *naut.* examination of bottom
~, дозиметрическое protection survey, radiation exposure survey
~, контрольное test survey
~, медицинское medical examination
~ на добровольных началах voluntary inquiry
~ на месте field inquiry/survey
~, периодическое periodical survey
~, повторное reinterview study
~, предварительное previewing, look-up; exploratory/pilot survey
~, пробное pilot inquiry
~ ребенка p(a)ediatric examination
~, социологическое sociologic(al) survey
обследователь investigator; inspector

обследовать examine, investigate, inquire into
обслуживание service, servicing; *tech.* maintenance, technical servicing, attendance, care; upkeep; catering
~, агентское ship's agency service
~ без приоритета nonpreferential treatment
~, бесплатное медицинское free medical service
~, больничное hospital service
~ в обратном порядке last-in-first-out rule, first-in-last-out rule
~ в порядке поступления first-come-first-served rule, first-in-first-out discipline
~, внеплановое техническое unscheduled maintenance
~, групповое service in batches/bulk
~, дистанционное remote maintenance
~, ежедневное daily maintenance
~, культурно-бытовое welfare
~, материально-техническое material upkeep
~, медицинское medical service/care/attendance
~, плановое routine maintenance
~, портовое harbo(u)r/dock service
~, профилактическое preventive maintenance
~ с приоритетом preferential treatment
~, стендовое bench servicing
~, техническое maintenance, servicing
~, техническое по вызову on-call maintenance
обслуживать 1. service, maintain, take care of, attend to; 2. operate, handle
обслуживающий servicing; operating
~ персонал (operating) staff
~ сектор *econ.* tertiary sector
обстановка 1. situation, condition(s), circumstances; atmosphere, environment; setup; 2. furniture; theatre set
~, аварийная emergency (situation)
~, боевая *mil.* combat situation
~, метеорологическая meteorological/weather conditions/situation
~, окружающая suroundings environment
~, сложившаяся situation at hand
~ учебной работы learning environment
~ фарватера *naut.* buoyage (and beaconage)
обстановочный situational
~ная пьеса costume play
обстипация *med.* intestinal obstruction, intractible constipation, obstipation
обстоятельственный *gram.* adverbial
обстоятельство I *gram.* adverb
~ образа действия adverbial modifier of manner
обстоятельство II condition, situation; *pl.* circumstances
~, отягчающее *law* matter of aggravation
~, смягчающее *law* mitigating circumstance, extenuation
~ва, денежные pecuniary circumstances
~ва, непредвиденные contingencies
~ва, отягчающее aggravating circumstances
обстрел fire, firing, shelling
~, артиллерийский (artillery) bombardment
~, круговой all-round fire
обстругивать plane (off)
обструкционизм *polit.* obstructionism
обструкция *polit.* obstruction; filibuster
обсуждать discuss, analyse, scrutinize; evaluate
обсуждение discussion; consideration, deliberation
обтачивание I whipstitching, stitching round
обтачивание II 1. *tech.* turning, machining; 2. *geol.* wearing away, corrosion, erosion, grinding
обтачивать I whipstitch, stitch round
обтачивать II 1. *tech.* turn, machine; 2. *geol.* wear away, corrade, erode, grind
~ на конус taper-turn
~ по копиру turn to template
обтекаемость streamline form; streamlining
~, нарушить offend the wind/air shape
обтекаемый streamline(d); aerofoil-type
обтекание flow(-over/-around); streamlining; circumfluence
~, безотрывное attached flow
~, отрывное separated flow
~, плавное streamline(d) flow
обтекатель *naut.* fairwater; *av.* fairing; cowl; streamlined unit
~, антенный antenna cap, aerial fairing/cap
~ гидролокатора asdic dome, *Am.* sonar dome, sonodome
~, защитный protective cover
~, носовой *av.* nose cone/fairing
~, хвостовой *av.* fail cone
обтекать flow about/(a)round/over/past; by-pass
обтеска dressing, trimming; hewing; squaring
~, грубая rough dressing, coarse carving
~ камня stone hewing
обтирание rubdown; wearing down/out; rubbing
обточка *tech.* machining, turning, rounding off
~ в патроне chuck turning
~ в планшайбе turning on a faceplate
~ в центрах turning between centres
~, круглая cylindrical turning
~ на конус taper turning
~ на размер turning to size
~ по копиру turning to template
~ по шаблону turning from guide
~, фасонная profile turning
~ черновая/чистовая rough/finish turning
обтурация, обтюрация obturation, sealing, stopping up
обтюратор *tech.* obturator; *phot.* shutter
~, двухлопастный two-vane rotating shutter
~, плавающий floating obturator
~, цилиндрический barrel shutter
~, шторный curtain shutter
обтягивать 1. stretch, elongate; 2. *naut.* bouse down/taut
обтяжка jacketing, covering; *av.* skin; *met.* stretch forming

~ **заготовки** *leath.* hoisting
~, **защитная** protective footwear
~, **ортопедическая** orthopedic footwear
обуглероживание carbonization, carburization
обугливание charring, coalification
обуржуазивание embourgeoisement
обуржуазить make bourgeos
обусловленность causation, causality; conditionality, conditionalism
обусловливать make dependent (on), condition, stipulate
обутый shod; *bot.* booted, peronate
обух *naut.* pad, lug, eyebolt, butt, head
~ **для жвака-галса** cable clench
~ **для оттяжки** guy ring
~ **для подвески** suspension lug
~, **подъемный** lifting/hoisting pad/lug
обучаемый *n.* trainee
обучение training, instruction; education; learning, study
~ **без отрыва от производства** in-service education, on-the-job training
~, **заочное** home study; correspondence course/school
~, **наглядно-зрительное** figural learning
~, **очное** resident instruction
~ **по месту работы** on-the-job training, in-service training
~, **поэлементное** task-component training
~, **программированное** programmed instruction, computer-aided instruction
~, **производственное** on-the-job training, in-service training; industrial training
~, **профессиональное** occupational training
~, **совместное** co-education
обхват 1. strap; clasping, grasp, hold; 2. circumference; girth; compass; *math.* perimeter; 3. *mil.* outflanking
~, **ножной** leg strap
обхватывать 1. encompass, surround, envelop; include, cover; 2. seize; embrace; clasp
обход 1. by-pass, detour, circumvention; diversion, alternate route, roundabout way; 2. round, visit(ing); beat; *mil.* turning movement; patrol(ling); 3. *math.* traversing, traversal, traverse
~ **контура** *math.* path-tracing
~, **направлять в** *comp.* divert [re-route] the traffic
~ **препятствий** *av.* terrain avoidance
~, **путевой** *rail.* track patrolling
обходить 1. go/walk round; surround, encircle; 2. by-pass, avoid, leave out, get round; 3. *mil.* outflank, turn; 4. make a beat/round
~ **закон** get round a law
~ **фланг противника** turn the enemy's flank
обходной circuitous, indirect; *med.* collateral
обходный 1. round about, encircling, circuitous; 2. *mil.* turning, outflanking; 3. *n.* guard, patrol (lineman)
обходчик *rail.* track walker, patrol lineman, trackman
обшивка panelling
обшивать 1. edge, border; sew round/up; 2. face, line, cover; sheathe, jacket
~ **досками** plank, board up
~ **листами** plate
~ **панелями/панелью** panel; dado
обшивка 1. casing, covering, facing, lining; plating, boarding, planking, panelling; sheathing; jacketing; 2. skin, jacket, shell; sheath
~, **армированная** reinforced skin/shell
~, **бортовая** side plating/planking
~ **вгладь** flush plating; carvel planking
~, **защитная** protective covering
~, **звукопоглощающая** acoustic lining
~ **из стеклоткани** glass-cloth skin/covering
~, **ледовая** ice plating/sheathing/lining
~, **листовая** sheeting, sheet covering
~, **матерчатая** cloth covering
~, **медная** copper sheathing
~, **многослойная** honeycomb skin
~, **наружная** outer/upper/top skin
~, **несущая** bearing/structural skin
~, **панельная** panelling
~ **полотняная** fabric covering
~ **с заполнителем,** ~, **слоистая** sandwich/laminated skin/covering
~ **траншеи** shoring, poling
~ **фальшборта** bulwark plating
~, **фанерная** plywood covering/skin
обшивной bordered, edged; trimmed
обширный extensive, vast; comprehensive; voluminous
общеармейский *mil.* common to the whole army, all-army
общевозвратный *gram.* general-reflexive
общевосковой *mil.* common to all arms, all-arms, all-army
общевосточнославянский Common-East-Slavic
общегосударственный nationwide, national
~**ные интересы** interests of the country at large
общедоступность accessibility
общедоступный 1. accessible, public, open to general use; 2. *econ.* of moderate price; 3. popular, comprehensible
общеевропейский common-European
общежитейский everyday, customary, ordinary
общежитие 1. boarding house; residence hall; dormitory; 2. community, society
~, **студенческое** student dormitory
общезаводский, общезаводской all-factory; of/for the whole factory
общезначимый generally/universally valid
общеизвестный well-known, notorious, reputed; popular
общеиндоевропейский Common-Indo-European
общеклассовый *polit.* common to all classes
общекрестьянский of all peasants, all-peasant
общелингвистический general-linquistic
общеметодологический general-methodological

общемировой universal, world-wide
общенародный national, nationwide
общенациональный nationwide
общение relations, contacts, dealings; intercourse; *rel.* communion
общеобразовательный of general education; comprehensive
~ные предметы general subjects
общеобязательный obligatory, compulsory for all
общепартийный all-party
общеполезный universally beneficial; socially useful
общеполитический general-political
общепонятный generally/easily understood
общепризнанный universally recognized/admitted
общепринятый generally accepted; conventional; othodox
общероссийский, общерусский All-Russian
общеславянский Common-Slavic/Slavonic; all-Slavic/Slavonic
общесоюзный All-Union
общественник public figure; socially active person
общественно-исторический sociohistorical
общественно-опасный *law* injurious to the public
общественно-полезный socially useful
общественно-политический sociopolitical
общественно-публицистический sociojurnalistic
общественность 1. public circles; the public/community; 2. public opinion
~, научная scientific community/world
общественно-экономический socioeconomic
общественный 1. public, social, societal; communal; 2. voluntary
~ деятель public figure
~ обвинитель public prosecutor
~ная работа voluntary work
~ное питание public kitchen/catering
~ные здания и сооружения public facilities
~ные науки social sciences/studies
общество 1. society; community; association; circle; 2. *econ.* company, corporation; 3. *hist.* peasant commune
~, акционерное joint-stock company
~, бесклассовое classless society
~ взаимного кредита mutual loan society
~ изобилия affluent society
~, классовое class society
~, кредитно-строительное building-and-loan association
~, модное the fashionable set
~, научное scientific organization
~ независимых государств, ОНГ Commonwealth of Independent States
~ по атомной энергии, Европейское European Atomic Energy Community, EURATOM
~, потребительское consumer society
~, спортивное sports society
~, транспортное carrier
~, фешенебельное the smart set
обществовед social scientist, specialist in social sciences
обществоведение social science
общетеоретический general-theoretical
общепотребительный in general use; customary, current
общеустановленный generally-established
общечеловеческий, общечеловечный common to all manking
общеязыковедческий general-linquistic
общий general; common; mutual; total; aggregate
~, абсолютно transcendental
~ азот *agr.* total nitrogen (content)
~ итог grand/sum(mary) total
~ рынок European Economic Community, EEC, the Common Market
~щая собственность common property
~щее дело common cause
~щее образование general education
~щее соглашение о торговле и тарифах, ГАТТ General Agrecment on Tariffs and Trade, GATT
~щее содержание питательных веществ total of nutrients
община 1. community; commune; 2. commonwealth; brotherhood, 3. herd, flock; colony
общиник *hist.* member of a commune
общинный common, communal; collective
общительный communicable, sociable, companionable
общность 1. community; generality; 2. affinity, similarity, relationship
~ генотипа *biol.* homogeny
~ имущества упружеская *law* estate in entirety
~, языковая linquistic community
объединение 1. union; unification, merging, joining; consolidation; 2. association; society
~ в блок(и) *comp.* blocking
~ в сеть networking
~ вида *biol.* mergence of species
~ данных chain of data, data chain(ing)
~ крестьянских хозяйств consolidation of peasant holdings
~, многофазное polyphase merge
~, множеств *math.* union of sets
~, производственное production amalgamation
~ энергосистем power system interconnection
объединенный united, joint; amalgamated
~ комитет начальников штабов (США) Joint Chiefs of Staffs
~ная энергосистема electric power pool
объединительный uniting, unifying
объединять unite, unify; join, merge, integrate; pool
~ ресурсы pool resources
объездка *sport*. breaking in *(of horses)*
объездной 1. roundabout, circuitous; 2. patrolling;

3. *n.* mounted posse/patrol
~ная дорога roundabout way
объездчик 1. mounted posse/patrol; 2. horsebreaker
объект 1. object, unit, entity; 2. project, works, site; installation; 3. *mil.* objective, target; 4. *gram.* object
~, абстрактный abstaction
~, береговой shore installation
~ восприятия percept
~, геометрический geometric(al) object/entity
~ действия *math.* operand
~ изучения object of study
~, информационный data object
~ исследования subject of inquiry
~, ложный decoy installation
~, неопознанный летаlющий, НЛО unidentified flying object, UFO
~ поиска object of search
~, промышленный industrial object/target
~ ссылки referent
~, двухлинзовый doublet objective lens
~, длиннофокусный long-focus/-focal lens/objective
~, зеркальный reflecting objective
~, иммерсионный (oil) immersion objective/lens
~, короткофокусный short-focus lens
~, панкратический zoom lens
~ с большой светосилой high-power objective
~ с исправлением аберрации corrected lens
~ с переменным фокусным расстоянием zoom lens
~, светосильный high-speed lens, fast lens
~, сменный accessory/interchangeable lens
~, трехлинзовый triplet lens
~, четырехлинзовый four-element lens
объектив-апланат aplanat(ic) lens
объективация, объективизация objectivation, objectification; exteriorization
объективизм objectivism; objectivity
объективистический, объективистский objectivist(ic)
объективность objectivity, objectiveness; impartiality
объективный objective; impartial, unbias(s)ed
объект-код object code
объект-микрометр stage micrometer
объектно-пространственный objective-spatial
объектный *gram.* objective
~ падеж objective case
объем 1. volume, space; 2. amount, size, scope, extent, capacity
~ выборки *comp.* sample size
~ данных data amount/bulk/quantity/volume
~ двигателя, рабочий engine displacement
~, единичный unit volume
~ информации body of data/information
~, кажущийся apparent volume
~ места груза bulk
~, молярный gram-molecular volume, molar volume
~, остаточный *med.* tidal volume/air
~ памяти memory/store size
~ партии (production) lot size
~ пассажирских перевозок passenger/traffic capacity
~ полезной нагрузки payload volume
~, полезный payload/net volume
~, полный total volume/capacity
~ продукции (предприятия) factory load
~ производства (amount of) output
~ производства, годовой yearly output
~ производства, общий total production, overall output
~ промышленного производства industrial output
~, рабочий working volume
~ рабочий цилиндра piston displacement
~, связанный covolume
~ сердца, минутный cardiac output
~ текущего потребления current consumption
~ ткани, чрезмерный *med.* hypertrophic size/volume
~ тракторных работ *agr.* tractor work load
~, удельный specific volume
~, фактический real volume
~ цилиндра, рабочий displacement/swept volume
~, элементарный *math.* volume element; *stat.* elementary unit
~ ядра nuclear volume
объемистый bulky, voluminous
объемлющий comprehensive, inclusive, enveloping; *bot.* equitant, amplectant
объемно-центрированный body-centered
объемный 1. volumetric(al), by volume; *hydr.* positive-displacement; 2. siz(e)able, voluminous; 3. *math.* three-dimensional
объемометр volumeter, volumenometer
объявление 1. announcement; *comp.* declaration; 2. advertisement; notice; poster
~ войны declaration of war
~ подписки subscription offer
объявлять 1. annonce, proclaim, propagate; *comp.* declare; 2. advertise
~ вне закона proscribe
~ перемирие declare truce
объяснение explanation, elucidation; clarification
объяснительный explanatory, interpretative, elucidative, expository
объяснять explicate, explain, define, interpret; clarify
объясняlющее *n. philos.* explanans
обывательский *bibl.* Philistine
обывательщина Philistinism
обыгрывать 1. turn to advantage; *theatre* play up, use to the best effect; 2. *mus.* (make) mellow *(of musical instrument)*
обыкновение habit, custom; *law* usage
обыкновенная *n.* ordinary priority *(of telegram)*

обыкновенный usual, common(place); average, simple; plain
обыск search
обыскивание search(ing), perquisition
обычай custom; usage
~, торговый practice of the trade; *pl.* trade usage
обычайный usual, ordinary, routine, of common occurrence; *law* consuetudinary
~ное право customary law
обязанность duty, obligation; responsibility
обязательность obligatoriness; compulsoriness
обязательный obligatory; compulsory, binding; *law* sine qua non
~ное обучение compulsory education
обязательство 1. obligation; commitment, bond; 2. *law* liability
~, выполнить discharge a liability
~, гарантийное warranty policy
~, генеральное comprehensive bond
~, долговое debenture, promissory note
~, долгосрочное long-term obligation
~, ипотечное mortgage note
~, недепозитное nondeposit liability
~, первоклассное денежное *econ.* gilt-edged paper, prime paper
~, подразумеваемое implied obligation
~, потенциальное contingent liability
~, принимать на себя assume/incur liability
~, совокупное и раздельное joint and several liability
~, солидарное joint liability
~, условное contingent liability
~, чрезмерное overcommitment
~ва, денежные pecuminary obligations
~ва, договорные treaty obligations
~ва, долговые credit instruments
~ва, долгосрочные fixed liabilities
~ва, краткосрочные floating liabilities
~ва по счетам ностро nostro liabilities
овал oval
~, декартов *math.* Cartesian oval, oval of Descartes
~ Кассини Cassinian oval
овалоклеточный ovalocellular
овалоцит ovalocyte, elliptocyte
овалоцитоз *med.* ovalocytosis, elliptocytosis, elliptocytic an(a)emia
овальбумин *bioch.* ovalbumin, egg albumin
овальноланцетовидный lance-oval, ovate-lanceolate
овально-круглый ovate-orbicular
овально-продолговатый ovate-oblong
овальность ovality, out-of-roundness
овально-эллипсоидальный ovate-ellipsoidal
овальный oval(-shaped), oviform, ovate; elliptic(al)
овариальный *anat.* ovarian
овариоалгия *med.* ovar(i)algia, oothecalgia
овариолитический ovariolytic, oothecolytic
оварионевралгия *med.* ovarian neuralgia, ovariodysneuria
овариорексис *med.* ovariorrhexis, oothecorrhexis
овариотомия *surg.* ovariotomy
~, нормальная Battey's operation
овариоцеле *med.* ovariocele, oothecocele
овариоэктомия *surg.* ovariectomy, oothecectomy
овация applause; ovation
Овен *astr.* Aries, the Ram
оверарм *sport* overarm stroke
овердрафт *econ.* overdraft
~, непокрытый unsecured overdraft
оверлей *comp.* overlay
оверкиль turnover
оверштаг *naut.* tack
овершют *min.* overshot
~, многоступенчатый multiple-bowl overshot
овес *bot.* oat *(Avena); pl.* oat field
овечий ovine
~ загон sheepwalk
овеществление objectification; reification
~ труда embodiment of labo(u)r
овеществлять substantiate, give material form to; reify
Овидий Ovid
овладение 1. capture; acquirement; seizure; 2. mastering, mastery
овоапогамия oopogamy
овобластема ooblastema
ововакцина ovovaccine, egg vaccine
ововителлин *bioch.* (ovo)vitellin
овогамия *biol.* oogamy
овогенез *biol.* ovogenesis, ovigenesis, oogenesis
овоглобулин *bioch.* ovoglobulin, egg globulin
овод *zool.* botfly *(Oestrus)*
овоидальный ovoid(al)
овокинез *biol.* ookinesis
оволемма *zool.* oolemma
оволосение pilosis
овоскоп ovoscope, egg tester
овоскопировать candle
овосыворотка ovoserum
овотида *zool.* ootid
овоцит *biol.* ovocyte, oocyte
овощ vegetable
~, плетневой vine vegetable
овощевод vegetable grower
овощеводство vegetable farming/growing, olericulture
~, промышленное *agr.* market gardening
~, товарное olericulture, market gardening
овощеводческий vegetable-growing/-raising
овощеперерабатывающий vegetable-processing
овощерезка vegetable slicer/cutter
овощесушилка vegetable dehydrator
овощехранилище vegetable storehouse
овощник greengrocer
овощная *n.* greengrocery
овощной vegetable, vegetal; oleraceous
~ стол vegetarian cooking
овраг *geol.* gorge, defile, coomb; ravine, gorge,

gully
оврагозакрепительный gully-stabilization/-control
оврагообразование gullying
овражистый raviney, gullied
овсец oat (grass)
~, луговой *bot.* meadow oat grass *(Avenastrum pratensis)*
овсодробилка oat grinder
овсорушка oat clipper, oats milling machine
овсошелушитель oat huller
овсюг *bot.* wild oat *(Avena fatua)*
овсюжник *agr.* wild oat grader/cylinder
овсяница *bot.* fescue *(Festuca)*
овсянка *zool.* bunting *(Emberiza)*
овсяной oaten, avenaceous
овуляция *biol.* ovulation
~, преждевременная premature ovulation
овца *zool.* sheep; ewe *(Ovis)*
овцебык *zool.* musk ox *(Ovibos moschatus)*
овцевод sheep farmer/breeder
овцеводство sheep breeding/raising/farming/husbandry
овцематка ewe
овцеферма sheep farm
овчина sheepskin
овчина-голяк sheared sheepskin
огам og(h)am, ogum, ancient British alphabet
огамический og(h)amic
огеливание gel(atin)ation, gelling
огибание bending (around), rounding, skirting
огибающая *math.* (the) envelope (curve)
огибающий cicumflex, bending around
~ импульсов pulse envelope
огива *stat.* percentile/ogival curve, ogive (curve)
~ плоскости скалывания/скольжения shear plane ogive
огигиев *myth.* Ogygian
оглавление table of contents, directory
~ тома *comp.* data carrier contents, volume table of contents
оглаголивание verbalization
оглашать announce, proclaim; make public, publicize
оглашение publication, announcement, proclaiming
~ приговора pronouncement of a verdict
оглеение gleying, gleization, gleyification
оглинение argillization, loamification, clayization
оглинивание argillic alteration, preargillization
оглобля *naut.* dagger plank, poppet stringer/ribband
оглушающий deafening; (ob)stupefacient
оглушительный ear-splitling, deafening, stunning
огнебезопасный fireproof
огневик *med.* anthrax, carbuncle
огневка *zool.* 1. pyralid(id) *(Pyralis);* 2. red fox
огневой *geol.* igneous, pyrogenous
~ бой firing
~вая мощь fire power
~вая позиция gum/firing position
огнегасительный fire-extinguishing
огнезащита fire barrier/curtain/protection; fireproofing
огнезащитный fire-resistant/-retardant/-protective
огнемет *mil.* flame thrower
~, ранцевый knapsack flame thrower
огнеметный fire-/flame-throwing
огненножидкий *geol.* liquid-molten
огненнолистный *bot.* pyrophyllous
огненный igniform; flamy; lurid
~ бой *mil.* firing, exchange of fire
огнеопасность fire hazard/risk, inflammibility
огнеопасный fire-hazardous, inflammable
огнепоклонник fire worshiper, ignicolist, pyrolater
огнепоклонничество fire worship, pyrolatry
огнепостоянный fireproof, refractory, fire-resistant
огнеприпасы *pl.* ammunition
огнепровод cross-ignition duct; portfire
~, пороховой powder train
огнепроводный portfire *(attr.)*
~ шнур Bickford fuse
огнесмесь burning mixture
огнестойкий fire-resistant/-resisting, fireproof, nonignitable
огнестойкость refractoriness
огнестрельный *mil.* firing
~ная рана bullet wound
огнетушитель fire extinguisher
огнеупор refractory brick/material
~, глиноземистый aluminia brick
~, динасовый silica refractory brick
~, связанный bonded refractory
~, шамотный fireclay
огнеупорность refractoriness, resistance to fire, fireproofness
огнеупорный fireproof, flameproof, apyrous, refractory; recalcitrant; nonignitable
огни *pl.* lights; *see also* **огонь**
~, авиационные navigation(al)/positional lights
~, глиссадные *av.* angle-of-approach lights
~ подхода приближения lead-in lights
~ святого Эльма Saint Elmo's fire
огниво *min.* cap
оговорка 1. stipulation, reservation, provision; *law* proviso; qualification, clause; warranty; 2. slip of the tongue, lapsus linguae
~, арбитражная compromissary note
~, мультивалютная multiple currency clause
~ о наибольшем благоприятствовании most favo(u)red nation clause
~ об обмене акций conversion provision
оголение *geol.* baring, denudation
оголенный denuded, exprosed, naked, (laid) bare; glabrate, glabrous
оголовье headband; headgear
оголять 1. denude, lay/make bare; strip, uncover; 2. *mil.* expose
~ провод strip a wire
огон *naut.* eye (splice)
~ ванты shroud eye

~, голландский Flemish/selvage eye
~, разрубной cut splice
~ штага eye of a stay
огонек (small) light; (glimmer of) light, flicker
~, блуждающий ignis fatuus (*pl.* ignes fatui), will-o'-the-wisp
огонь 1. fire, flame; 2. light, lamp; *see also* **огни**
~, антонов *med.* St. Anthony's fire, gangrene
~, артиллерийский gunfire
~, главный сигнальный master caution light
~, заградительный interdiction/ptotective barrage
~, зенитный anti-aircraft fire
~, изофазный rhythmic light
~, кильватерный formation keeping light
~, кодовый code light
~ кругового освещения allround/globular light
~, мигающий blinker light
~, опознавательный identification/recognition light
~, постоянный fixed light
~, проблесковый flashing light
~, пушечный gunfire
~, топовый *naut.* masthead/steaming/top light
~, фарватерный fairway light
~, фарный floodlight
огоньки *pl. bot.* Siberian globe flower *(Trollius asiaticus)*
огораживание enclosing; enclosure, inclosure
огород kitchen/vegetable garden; truck farm
огородник truck farmer, market gardener
огородничество truck farming, market gardening
огородный vegetable *(attr.)*
ограбление burglary; robbery
ограблять plunder, pillage, rob; strip
ограда 1. fence, enclosure; 2. *arch.* balconet
оградитель guard, protector
оградительный protective, guarding, enclosing
ограждение barrier, enclosure, guard(rail), railing; fence, fencing; *naut.* fairwater; 2. screen
~ буями/вехами buoyage
~, леерное *naut.* guard railings, life rails
~, предохранительное safety barrier
~, решетчатое grill guard
ограничение restriction, limit(ation), restraint, constraint; confinement, abridg(e)ment
~, аппаратное hardware restriction
~ в виде неравенства *math.* inequality constraint
~ в виде равенства *math.* equality/equated constraint
~ в снабжении logistic limitation
~, верхнее upper-bound constraint
~, возрастное age limit
~ вооружений weapons freeze
~, вынужденное forced constraint
~ двойственной задачи *math.* dual constraint
~, двумерное *math.* two-dimensional constraint
~, допустимое feasible constraint
~, жесткое hard/active constraint
~, замещенное *math.* surrogate constraint
~, избыточное redundant constraint
~ импульсов pulse clipping
~, интегральное *math.* integral constraint
~, квадратическое *math.* quadratic constraint
~ кредита credit/money squeeze
~, линейное *math.* linear constraint
~, математическое/программное software restriction
~, нежесткое soft/inactive constraint
~, обоюдное mutual boundary
~ по памяти memory limitation
~ по перегрузке load-factor limitation
~ по числу М Mach limitation
~ по шуму noise restriction
~, проектное design constraint
~ прямой задачи *math.* primal constraint
~ рождаемости *econ.* birth/fertility/family/population control
~ сигнала *el.* (signal) limiting
~ сигнала по верху/максимуму positive limiting
~ сигнала по низу/минимуму negative limiting
~ сигнала по амплитуде signal clipping
~, совместное joint constraint
~, составное *math.* composite constraint
~, целочисленное *math.* integer-value constraint
~ния, жесткие stringent restrictions
~ния, эксплуатационные operational/operating limitations
ограниченно-линейный linear-limited, limited-linear
ограниченно-маневренный *av.* semi-aerobatic
ограниченно-слабый bounded-weak
ограниченность 1. boundedness, limitedness, limitation; scantiness; 2. *fig.* narrow-mindedness
ограниченный 1. limited, confined, restrained, restricted, bounded; qualified; 2. *fig.* shallow-brained
~ почти всюду *math.* essentially bounded
~ная монархия constitutional monarchy
ограничивать bound, restrict, confine, circumscribe, limit, cut down
~ сигнал по сетке *el.* clip the signal at the grid
ограничитель *tech.* limiter, stop; catch(er); clipper, delimiter
~ мощности power limiter/governor
~ предохранителя safety stop
~ слова word delimiter
~ хода restraining arm
ограничительно-временной restrictive-temporal
ограничительный restrictive, restraining, limiting, limitative; arresting
огранка 1. cutting, facetting; 2. *cryst.* facet, face; cut
~, ступенчатая trap cut

огрех *agr.* lapse, flaw *(in ploughing, sowing, etc.)*
огульный 1. wholesale, indiscriminate, blanket; 2. unfounded, groundless
~ная торговля wholesale trade
огурец *bot.* cucumber *(Cucumis sativus)*
~, морской *zool.* holothurian, sea cucumber *(Holothuria)*
~, соленый *cul.* dill pickle
огуречник 1. *bot.* borage *(Borago);* 2. *agr.* cucumber plot/field; 3. *zool.* pond smelt *(Hypomesus olidus)*
огурцевидный cucumiform
ода *lit.* ode
~, пиндарическая Pindaric ode
одалиска odalisk, odalisque
одевание clothing, dressing
одевать 1. dress, clothe; 2. cover
одежда 1. dress, clothing, clothes; garments; outfit, uniform; 2. *tech.* jacket, lining; *constr.* pavement, top dressing
~, асфальтовая asphalt/surface top
~, битуминозная black/bituminous top/surfacing
~, бревенчатая log revetment
~, верхняя overcoat, outer garment
~, ветрозащитная windproof outer suit
~, защитная protective/safety clothing
~, огнезащитная fire retardant/-protective clothing
~ пола floor covering
~, производственная overalls
~, рабочая fatigue dress
~, форменная uniform
одеревенелость numbness
одеревенение 1. lignification; 2. stiffness, hardness
одерживать 1. gain; 2. *naut.* ease, meet (the rudder), check the helm
одержимый 1. possessed, obsessed; maniac; 2. *n.* madman, maniac
одессит inhabitant of Odessa
одеяло blanket
~, байковое flannelette, cotton blanket
~, жаккардовое fancy blanket
~, стеганое quilt
одеяние apparel, attire; garb, garment
Один *myth.* Odin, Woden
одинаковолистный *bot.* homophyllous
одинаковоплодный *bot.* homocarpous
одинаковость identify, uniformity, sameness
одинарный single, unitary, unary
одиннадцатиплоскостной hendecahedral
одиннадцатиугольник hendecagon
одинокая *n.* single woman
одинокий *n.* bachelor, single man
одиночка 1. unmarried/single mother; 2. one-man cell, solitary confinement; 3. *sport* single sculls; 4. one-horse carriage/sleigh
~, академическая single scull
~, гоночная race skiff
одиночник *sport* 1. sculler; 2. solo figure scater
одиночноцветущий *bot.* single-flowering
одиночный individual, separate; one-man; single, isolated; sole, solitary
Одиссей *myth.* Odysseus, Ulysses
одиссея Odyssey
одичавший having grown/run wild, feral
однакоаксиально-двоякопреломляющий uniaxially birefringent
одноадресный single-/one-address
одноактный one-act
одноатомность monovalence
одноатомный mon(o)atomic
однобазовый single-base
однобаковый single-tank
однобатарейный single-battery
однобокий lopsided, one-sided; *bot.* unilateral, secund
однобороздной, однoборозчатый monocolpate, monosulcate
однобратственный *biol.* monadelphous
однобрачие monogamy
одновалентность *chem.* monovalence, univalence
одновалентный monovalent, univalent
одновариантный monovariant, univariant
одноверец co-religionist
одновершинный *math.* one-crest/-peak/-vertex, unimodal; unimucronate
одновесельный one-oared
одноветвистый uniramous
одновибратор univibrator, monovibrator, one-shot multivibrator, kipp-oscillator, flip-flop oscillator
одновидовой one-aspect
одновинтовой single-screw/-rotor
одновитковый single-coil/-turn
одноводный *chem.* monohydrate
одновозрастной even-aged, coeval
одноволновой single-wave
одновременность simultaneity; coincidence; synchronism, isochronism
одновременный simultaneous, concurrent; synchronous, isochronous, contemporaneous
одноглазие *med.* cyclopia, synophthalmia
одноглазка *zool.* cyclops
одноглазный monophthalmic, monocular, one-/single-eyed
одногнездный unicellular, monothalamous; monothecious, unilocular, one-celled
одноголовчатый *bot.* monocephalous
одноголосие *mus.* monotony
одногорбый one-humped
~ верблюд Arabian camel, dromedary
одногубый *bot.* unilabiate
однодворец *hist.* smallholder
однодеревка 1. dugout (boat); 2. pole mast
однодиапазонный *rad.* single-band/-range
однодифференцируемый *math.* once differentiable
однодневка *zool.* ephemerid
однодневный ephemeral, ephemerous; one-day
однодолевой unilobar

однодольный *bot.* monocotyledonous
однодомность *bot.* monoeciousness
однодомный *bot.* monoecious
однодорожечный one-/single-track
однодревка pole mast
однодужные *n. pl. zool.* synapsidans *(Synapsida)*
одножаберный *zool.* monobranchiate
одножгутиковый monotrichous, uniflagellate, monociliated
одножелезистый uniglandular
одноженец monogynist
одножилковатый, одножилковый uninervate, uninerved
одножильный *el.* single-core
однозамещенный *chem.* monosubstituted
однозародышевый *bot.* monogerm, monembryonic
однозарядный single-charged/-loading/-loader
~ное ружье single-loader
однозаходный single-thread *(of screw)*
однозвучный homophonous, monotonous
однозерный *bot.* monospermous
однозернянка *bot.* einkorn *(Triticum monococcum)*
однозначащий synomymous, identical; monosemantic
однозначность monosemy, univocacy, univocity; synonymity; uniqueness, unambiguity
однозначный 1. monosemantic, monosemous, synoymous; 2. *math.* single-/one-valued, univalent, simple, unique, unambiguous
~, взаимно one-(to-)one, biunique
~ное число digit, simple number
~ное решение univalent/unique solution
однозубчатый unidentate
однозубый monodont(al)
одноидемпотентный *math.* one-idempotent
одноизотопный monoisotopic
одноименно-полюсный homopolar
одноименность homonymy
одноименный cognominal, homonymous
одноимпульсный monopulse
одноиндексный single-index
однокадровый *phot.* single-frame
однокамерный unilocular, monothalamous; *tech.* single-chamber(ed)/-barrel/-compartment; *med.* unicameral, unicamerate
одноканальный single-channel/-duct
однокапсульный *bot.* unicapsular
однокаскадный one-/single-stage
однокачественный *math.* isomorphic
одноквантовый one-quantum
одноклапанный one-valve, single-valved
одноклеточный *biol.* unicellular, monocelled, single-cell(ed); *n. pl.* protozoa *(Protozoa)*
одноклубник *sport* clubmate
одноколейный single-track
одноколенчатый *tech.* single-jointed
одноколесный one-wheeled
~ велосипед monocycle, unicycle
одноколонковый unicolumellate
одноколонный single-column
одноколосый *bot.* single-spiked, monostachyous
однокомпонентный single-component, monocomponent
одноконтурный *el.* single-circuit, single-/one-loop
однокопытный *zool.* soliped, solidungulate, solid-/one-hoofed
однокоренной, однокорневой single-root
однокоробковый, однокоробочный *bot.* unicapsular
однокорпусный single-unit/-hull
однокрасочный monochrome, monochromatic
однократноцветущий *bot.* hapaxanthic, hapaxanthous
однократный single, once-through, unitary, onefold, univariate
однокристальный *comp.* one-/single chip
однокровный consanguineous
однокрылый monopterous, one-winged
одноламповый *rad.* single-tube/valve
однолемешный *agr.* single-share
однолепестковый, однолепестный *bot.* monopetalous
однолетний *bot.* annual, monocyclic
однолетник annual (plant)
однолистность *math.* univalence
однолистный 1. *bot.* monophyllous, unifoliate; 2. *math.* univalent, one-sheet
однолисточковый *bot.* unifoliolate
одноличный unipersonal, monopersonal; personal
однолопастный unilobate, unilobed, single-lobed/-blade
однолучевой uniradiate, single-beam
~ верхоцветник *bot.* monochasium, simple cyme
~вое событие *phys.* one-pronged event
одномачтовый one-/single-masted
одномерный one-dimensional, unidimensional; univariate
одноместный 1. single-seat(er), single seated; 2. one-position/-place, monadic
одноминеральный *geol.* monomineral, monomict, monogenic
одномодальный unimodular
одномолекулярный monomolecular, unimolecular
одномоторный single-motor/-engine, one-motor
одномускульный *zool.* monomyarian
однонаправленный monodirectional, unidirectional, one-way; unilateral, unicursal; homodromous
однонатриевый *chem.* monosodic
однонервный single-/one-nerved, uninerved
однонит(оч)ный single-thread; unifilar
одноногий monopode, one-legged/-footed
однообмоточный single-coil/-winding
однооборотный single-revolution/-turn/-cut/-thread
однообразный 1. uniform, homogeneous; 2. monotonous; *math.* monotonic

одно-однозначный one-(to-)one, single-identical
одноокись *chem.* monoxide
одноосевой uniaxial, monaxial; *bot.* haplocaulescent
~, заметно sensibly uniaxial
одноосновный *chem.* monobasic, single-base
одноосный single-axle/-pin; uniaxial, single-axis
одноотверстный one-entry/-hole; single-duct
однопалатный *polit.* single-chamber, unicameral
однопалость *med.* monodactylism
однопалубный single-/one-decked
однопарнорезцовые *n. pl. zool.* simplicidents *(Simplicidentata)*
однопарный uniparous, unijugous, unijugate
однопартийный *polit.* one-party
однопестичный *bot.* monogynous
одноплатный *comp.* single-board
одноплечий *tech.* one-armed/-lever
одноплодный *bot.* monocarpous, monocarpic
одноплодолистиковый *bot.* monocarpellary, unicarpellary
однопокровный *bot.* monochlamydeous
однополосный single-band
однополостной 1. unilocular; 2. single-/one-cavity
однополостный *geom.* of one sheet
однополость unisexuality
однополый *biol., bot.* unisexual, monosexual
однопольный 1. *agr.* without rotation of crops; single-crop; 2. single(-leaf)
однополюсник unipole
однополюсный, однополярный *phys.* unipolar, homopolar, monopolar; single-pole
однопотенциальный univoltage, unipotential
однопроводный single-wire/-line
однопроходный 1. one-pass; 2. *n. pl. zool.* monotremes *(Monotremata)*
однопутный single-line/-track, one-way
одноразмерный unidimensional
одноразрядный one-column/-digit; single-order
однорельсовый monorail, single-rail
однореснитчатый monociliated
однорогий unicornous, monocerous, one-horned
однороднопокровный homochlamydeous
однородность homogeneity, homogeneousness; uniformity; *biol.* homogeny
однородноячеистый homobrochate
однородный homogeneous, uniform; isotropic; *biol.* congeneric(al), congenerous
~ное качество *econ.* consistent quality
~ные величины *math.* like quantities
однорыльцевый *bot.* monostigmatic, monostigmatous
однорядка *hist.* single-breasted caftan
однорядный uniserial, unilinear; single-row, one-row/-tier; *math.* universal; *bot.* monostichous, unifarious
односвязность simple connectivity/connectedness
односеменодольный, односемянодольный *bot.* monocotyledonous; *n. pl.* monocotyledons *(Monocotyledones)*
односемянный one-/single-seeded, monospermous
односимвольный single-character
однослоговой monosyllabic
однослойный single-/one-layer(ed), monostratal, monolayer, unilaminate, unistrate; single-/one-ply
односоставный 1. homogenous, uniform; 2. *ling.* monomial, unitary
односпоровый monosporous, one-spored
одноствольный 1. *bot.* monocormic; 2. single-barrel(led)
одностворчатый *zool.* univalve(d)
одностебельный *bot.* monocaulous, unicaulous
одностержневой single-bar/-rod
одностолбиковый *bot.* monostylous
одностопный one-foot
~ стих *lit.* monometer
одностороннецветковый *bot.* secundiflorous
односторонний 1. one-/single-sided, unilateral; *fig.* biased; *bot., zool.* secund; 2. one-direction/-way, unidirectional
одностручковый *bot.* single-podded
однострунный single-/one-string
одноступенчатый single-step/-stage/-level, one-stage
односумчатый *bot.* monosaccate
односуставной uniarticulate, single-/one-jointed
однотактный *tech.* single-ended/-cycle/-phase, one-stroke
однотарифный one-rate, single-fee
однотипный uniform, of the same kind/type, homotypic; monotypic
~ корабль sister/twin ship
однотомник one-volume edition
однотонный I 1. single-tone, monophonic; 2. monotone, sole-colo(u)red
однотонный II one-ton
однотрубчатый *bot.* monosiphonous
однотычинковый, однотычиночный *bot.* monandrous
одноугольный one-angled/-cornered
одноутробный 1. born of the same mother; *med.* uterine 2. monodelphian
однофазный monophase, single-phase
однофокусный confocal, monofocal, single-focus(s)ing
однофонемный monophonem(at)ic
одноходовый single-pass/-cut/-thread; one-pass, straight-through
~ винт single-thread screw
однохозяйственность *bot.* autoecism
однохозяйственный *bot., biol.* ametoecious, autoecious
одноцветковый *bot.* uniflorous, monanthous, single-flowered
одноцветный monochromatic, monochrome, homochromous, unicolo(u)r, one-colo(u)r; *bot.* uniflorous
одноцепной single-circuit/-chain
одноцилиндровый single-cylinder, mono-cylindrical

одночастотный one-/single-frequency; monofrequent, unifrequent
одночелночный *text.* one-shuttle
одночлен *math.* monomial (term), term
одночленный 1. *math.* monomial; 2. *bot.* monomerous
одношарнирный single-joint/-link
одношкальный single-scale
однощелевой single-hole/-slot(ted), monolete
одноэлементный single-element/-unit
одноягодник *bot.* herb Paris *(Paris quadrifolia)*
одноядерный mononuclear, mononucleate, uninuclear, uninucleate
одноязычный monolingual, unilingual
однояйцовый unioval, monogerminal, monozygotic, uniovular
~вые близнецы identical twins
одноярусный single-tier/-stage; one-deck/-story/-layered
~ное растительное общество unistratal community, one-layered community
одноячейковый one-/single-celled; unicellular, monocellular
одограф odograph
одомашнение, одомашнивание domesti(fi)cation, taming
одомашнивать domesticate, tame
одометр odometer, cyclometer
одонтальгия *med.* odontalgia, toothache
одонтатрофия *med.* odontatrophia, odontatrophy
одонтобласт *anat.* dentinal cell, odontoblast
одонтогенез *biol.* odontogeny, odontosis
одонтолит 1. *miner.* odontolite, bone turquoise; 2. *med.* odontolith
одонтома *med.* odontoma, odontome
~, обызвествляющаяся calcifying odontoma
одонтономия dental nomenclature, odontonomy
одонтопараллакс *med.* odontoloxia, odontoloxy, odontoparallaxis
одонторрагия odontorrhagia
одревеснение lignification
одряхление ag(e)ing; senility, decrepitude, (con)senescence
одубина *leath.* spent bark/tan, tanbark waste
одуванчик *bot.* dandelion, blowball *(Taraxacum)*
одухотворение spiritualization, animation; vivification
одухотворенный inspired; spiritual
одухотворять inspire; animate, spiritualize
одушевленный living, animated
одушевлять inspirit, animate, inspire
одышка *med.* dyspn(o)ea, short wind, shortness of breath
~, инспираторная inspiratory/Traube's dyspnea
~, сердечно-легочная cardiopulmonary dyspnea
оевропеивать Europeanize
ожеледь glazed frost; ice-covered ground
ожелезнение 1. ferrugin(iz)ation; 2. iron-plating
ожелезненный ferruginous
~ная глина iron clay
оже-переход Auger transition
ожерелье necklace; gorget
~ Венеры *med.* Venus/Casal's necklace
~ из раковин *hist.* wampum
оже-электрон Auger electron
оже-эффект Auger effect
ожива *arch.* ogive
~ с протоком ducted ogive
оживало ogival head
оживление 1. resuscitation, reanimation, revivification; 2. regeneration, revival, recovery; *econ.* rally, buoyancy
~, экономическое buoyancy of the economy
оживленный animated, vivacious; lively
~ током *el.* electrically excited
~ная торговля brisk trade
оживляться liven up; *econ.* rally *(of demand)*
ожидание anticipation; *math.* expectation; *av.* hold(ing)
~, математическое (mathematical) expectation; *stat.* probabilistic/stochastic mean, probabilistic/assembly/stochastic average, expectation value
~ обслуживания *econ.* queueing
~ния, время latent period, waiting time
ожижение 1. liquefaction, liquefying; destructive hydrogenation *(of coal);* 2. *meteor.* liquation
ожижитель liquefier, liquidizer
ожика *bot.* wood rush *(Luzula)*
ожина *bot.* European dewberry *(Rubus caesius)*
ожирение *med.* obesity, adiposity, general adiposis/liposis, corpulence, polysarcia, fatness
~, атрофическое fatty atrophy
~, болезненное paratrophy, Dercum's disease
~ легкого fatty lung
~, пубертатное pubertal adiposity
~, эндокринное endocrinal obesity
ожог 1. *med.* burn; scald; 2. *plant pathol.* blight, heatrot, firing
~, бактериальный bacterial blight
~ верхушки top-end firing
~, лучевой radiation burn
~, монилиальный monilia
~ от трения brush/mat/rope burn
~ при ядерном взрыве flash burn
оз *geol.* esker, os(ar)
~, галечный rubblestone os
~, питающий/вспомогательный feeding esker
озвончение *phon.* vocality, vocalization, voicing, sonorization
озвучение sonication; *cin.* scoring; sound recording
озвученный *cin.* scored
~ фильм sound film
озвучивание *tech.* sonic treatment, insonification
озвучивать 1. *cin.* score, record sound; 2. wire for sound; insonify
оздоровительный health-improvement, sanitary
оздоровление sanitation

~ **леса** forest hygienics
озеленение landscape gardening
озеленитель landscaper
~, **инженер** landscape designer
озена *med.* oz(a)ena, fetid polypus
озерно-речной fluviolacustrine
озерный laky, lacustrine, limn(et)ic
озеро lake
~, **альпийское** tarn
~, **береговое** coastal/marginal/border lake
~, **бессточное** endorheic lake, lake without outflow/outlet
~ **в грабене** *geol.* rift-valley lake, graben lake
~, **заболоченное** bog lake
~, **каровое** corrie/cirque lake
~, **карстовое** solution/sink lake
~, **ледниковое** glacial/glacier lake
~, **провальное** plunge-basin lake
~, **проточное** running-water lake
~, **разливное** flood lake
~, **рапное** brine lake
~, **слезное** *med.* lacrimal lake
~, **смолянное** pitch lake
~, **соляное** salt/saline lake, salina
~, **старичное** bayou/oxbow lake
~ **сточное** exorheic lake, lake with outflow/outlet
~, **четководное** beaded lake
~, **щелочное** alkaline/alkalitrophic lake
озероведение linmology, lake hydrology
озимый *agr.* winter(-annual); fall-seeded; *n. pl.* winter crops
~ **посев** fall seeding
~**мые хлеба** winter crops/cultures
озимь *agr.* winter crop/culture
Озирис *myth.* Osiris
ознакомление familiarization, acquaintance
~, **предварительное** familiarity
ознаменование commemoration, marking
ознаменовывать commemorate, mark, celebrate
означать 1. denote, indicate, designate; 2. mean, stand for, signify
озноб *med.* rigor, chill, shiver(ing), fever, cold fit, algor, algidity
~, **лихорадочный** febrile chill
~, **потрясающий** shaking chill
ознобление *med.* chilblain, frostbite, pernio, kibe, chimetlon
озокерит *miner.* ozocerite, ozokerite, mineral/earth/lignite/ader/fossil wax, mineral fallow
~, **волокнистый** fibrous wax
озоление ashing, incineration, combustion
~, **каталитическое** catalytic ashing
~, **мокрое** wet combustion/oxidation/ashing
озолять (reduce to) ash, incinerate
озон *chem.* ozone
озонизация, озонирование ozonization, ozone treatment
озонинклин ozone incline
озоностойкость ozone resistance
озотипия ozotype process
озубление dentition
~, **постоянное** permanent dentition
озубленный denticulate
оидиум *plant pathol.* powdery/vine mildew, oidium
окаймление 1. margin, border, rim, fringe, edge; 2. bordering, rimming, edging, margination
~, **светлое** halo, halation (ring)
окаймленнолистный *bot.* lomatophyllous, marginate-leaved
окаймленный marginate, margined, bordered, edged, banded; *biol., bot.* limbate, halonate
~, **дважды** bimarginate
окаймлять fringe, border, flank, rim, edge, hem
окалина (hammer) scale, (anvil) dross; slag, scoria, scum
~, **отстающая** exfoliating scale
~, **прокатная** mill scale
окалиноломатель scalebreaker
окалинообразование scaling
окалиностойкий scale-resistant
окаменевать fossilize, pertify, lithify
окаменелость fossil, pertification
~, **окатанная** rolled/rounded fossil
~, **руководящая** index/reference/guide fossil
окаменелый petrified, fossil(ized)
окаменение petrification, fossilization, calcification, lithification
~ **осадков** lithification of sediments
окантовка 1. shaping strip, edge former; edging; 2. fringe effect; 3. edge piping
~ **клише** half-tone block boxing
~ **лаза** manhole frame
~ **низков** bottom piping
окапи *zool.* okapi, okapu *(Okapia)*
окапывать bank up (earth), dig round; *agr.* mould up
окарстовывание karstification
окатывать hose down
окатыш *met.* pellet
окаянный 1. cursed, damned; 2. *n. rel.* the devil, impious wretch
окаянство *eccl.* sinfulness, godlessness, ungodliness
окаянствовать *eccl.* live a life of sin
окварцевание silicification, silication
океан ocean
~, **Атлантический** Atlantic Ocean
~, **Мировой** world(-wide) ocean
~, **открытый** mid-ocean, open sea
~, **Северный Ледовитый** Artic Ocean
~, **Тихий** Pacific Ocean
океанавт oceanaut
океанарий, океанариум oceanarium, seaquarium
океанида *myth.* sea/water nymph, Oceanid
океанический oceanic, neptunian, pelagic, thalassic
Океания Oceania
океанограф oceanographer
океанография oceanography, thalassography
океанология oceanology, oceanics

океанский ocean(ic), ocean-going
окиление *bot., zool.* carina, keel
окиленный *bot., zool.* carinate(d), cristate, keeled, keel-shaped
окисель *chem.* oxide
~, промежуточный amphoteric oxide
окисление *chem.* oxidation, oxidating, oxidization; acidification, acetification
~ валков *met.* blueing
~, клеточное cell oxidation
~, медленное quiescent oxidation
~, межкристаллитное intergranular/intercrystalline oxidation
~, неполное/частичное partial oxidation
~, последующее subsequent oxidation
~, постепенное smooth oxidation
окисление-восстановление reduction-oxidation, oxido-reduction, redox
окисленный oxidized, oxygenated
окислитель *chem.* oxidizer, oxidizing/acidulating agent, oxidant, acidifier
окислительный *chem.* oxidizing, oxidative
окисляемость *chem.* oxid(iz)ability
~ воды water oxidation susceptibility
окислять oxidize, oxygenate, oxydate; acidify
~ электролитически anodize
окисносеребряный silver-oxide
окисный oxidic
окись *chem.* oxide
~ азота nitric/nitrogen oxide
~ алюминия aluminium oxide, alumina
~, безводная anhydride
~, водная hydroxide, hydrate
~ вольфрама tungsten oxide
~, двойная dioxide, double oxide
~ железа, безводная iron/ferric oxide; red hematite
~ железа, водная limonite
~ кальция quicklime, calcium oxide
~ магния periclase, magnesium oxide
~ меди copper oxide
~ металлоида semimetal oxide
~ натрия natron, sodium oxide
~, низшая protoxide, suboxide
~, полуторная sesquioxide
~ свинца, желтая lead monoxide, plumbic ocher, massicot
~ углерода carbon monoxide, carbonic oxide
~ цинка zinc oxide/white/bloom
~ цинка, белая zinc white
~ цинка, красная red zinc oxide/ore, zincite
окказионализм *philos.* occasionalism
окклюдатор *med.* articulator
окклюдирование occlusion
окклюзия *med., meteor.* occlusion
~, боковая lateral occlusion
~, выступающая protrusive occlusion
~, дистальная disto-occlusion, distocclusion
~, загнутая recurved occlusion
~, неправильная malocelusion
~, пропорциональная balanced occlusion
~, скользящая gliding occlusion
~, скошенная torsive/torso-occlusion
окклюзометр occlusometer, guatodynomometer
оккультизм occultism, theosophy
оккупант occupationist, occupier, invader
оккупация *mil.* occupation
оклад I 1. *min.* set; 2. *rel.* setting, framework *(of icon);* 3. contour of the face; 4. *hunting* encirclement, cornering
~, дверной cap-and-leg drift, tunnel set
~, половинчатый corner set
оклад II *econ.* (rate of) pay, salary; tax
~, временный provisional allowance
~, твердый fixed/stated salary
оклейка glueing; (material for) pasting; papering
~ корешка (case) lining
оклик call, hail
окно 1. window; window sill; 2. opening, hole, slot, port, gap, aperture
~ ахтерштевня *naut.* propeller aperture
~, брюстеровское *opt.* Brewster(-angle) window
~, венецианское *arch.* picture window
~, вентиляционное 1. *auto.* ventipane, pivoted ventilation window; 2. *min.* air/ventilation/check door
~, верхнее eyebrow window
~ в лесу glade, opening
~ в облаках break in the clouds, cloud gap
~, впускное inlet/intake/admission port
~, выпускное/выхлопное *auto.* exhaust window, exit/discharge/exhaust port
~, выходное *electr.* output window
~, глухое false window
~, завалочное *met.* charging door
~, защитное *nucl.* shielding/radiation window
~ золотника slide-valve port
~, кадровое *cin.* gate aperture, film gate
~, кассовое teller's window
~, клапанное *auto.* valve port
~, контактное *comp.* contact window/hole
~ на передачу *comp.* send window
~, отпускное drop window
~, паров(ы)пускное steam port
~, перепускное by-pass port
~, подъемное sash/sliding window
~ преддверия *anat.* verstibular window
~, продувочное scavenging port
~, раздвижное sliding window
~, слуховое dormer/gable window
~, смотровое peephole, observation/inspection/viewing/sighting port/window
~, створное cased window
~, створчатое folding/casement window
~, торцевое end window
~ улитки *anat.* cochlear window
~, шлаковыпускное slag hole, cinder tap/notch
~ щеткодержателя *el.* brush holder cavity
~, эркерное *arch.* jut/oriel window

окномойка window-washing machine
око *hist.* oko *(unit of weight)*
оковка binding *(with metal);* ironwork; fittings
оковы *pl.* fetters, chains; *polit.* bondage
околачивание hammering
околоаксилярный periaxillary, circumaxillary
околоанальный perianal, circumanal, periproctic
околовенозный paravenous, perivenous
околовлагалищный paravaginal
окологлазной periocular, circumocular
окологортанный perilaryngeal
окологрудинный adsternal
окологрыжевой perihernial, periherniary
околозвездный circumstellar
околозвуковой trans(s)onic
околоземный circumterraneous, circumterrestrial
околозубной peridental, periodontal
околоканальцевый pericanalicular
околоклеточный pericellular, pericytal
окололепестный *bot.* peripetalous
околоматочный periuterine, perimetric
околомозговой pericerebral
околоневральный perineural
околонервный epineural
околоосевой paraxial, periaxial
околопериодонтальный pericemental
околопестичник *bot.* perigynium
околопеченочный perihepatic
околопищеводный periesophageal
околопланетный circumplanetary
околоплодник *bot.* pericarp, seed-vessel
околоплодный *med.* amniotic
околопозвоночный paravertebral, perispondylic
околополюсный circumpolar, peripolar
околополярный circumpolar
околопочечный perirenal, circumrenal, perinephric
околопредсердный periatric, periauricular
околопузырный *anat.* paravesical, paracystic, perivesical, pericystic
околопупочный paraumbilical, paraomphalic, periomphalic
околоселезеночный perisplenic
околосердечный pericardial, pericardiac
~ная сумка pericardium
околослепокишечный pericecal, perityphlic
околососковый peripapillary
~ кружок mammary areola
околососудистый perivascular
околосуставной circumarticular, periarticular
околоток 1. *rail.* line section; 2. town district, precinct
околотолстокишечный pericolic
околоустье *bot.* peristome
околоушный *anat.* parotid; *zool.* periotic
околоцветник *bot.* floral envelope, perianth
околоцветный *bot.* perianthial
околочелюстной admaxillary
околоядерный *biol.* paranuclear, perinuclear
околояичниковый parovarial, parovarian
окольничий *n.* *hist.* okolnichy, member of a high rank of boyars
окомкование *min.* nodulizing, pelletizing
окомплексивание *math.* compexification
оконечность tip, end; termination; *anat.* extremity
~, кормовая aft(er) end
~ ледника termination/terminus/front of glacier
оконтуривание 1. delimitation, contouring, delineation; 2. *min.* mapping
окончание 1. finish; termination; end(ing); 2. *gram.* ending
~, аварийное *comp.* ABEND, abnormal ending/termination
~, нервное teleneuron, nerve ending
~, подударное stressed ending
~ния, чувствительные нервные sensory nerve endings
~ния, гроздевидные нервные grape/racemose nerve endings
окончательный final, definite, definitive, conclusive, ultimate
окончатый *anat.* fenestrate(d)
окоп *mil.* trench, entrenchment
~ для миномета mortar pit
~, одиночный/парный стрелковый one-/two-man foxhole
окопник *bot.* comfrey *(Symphytum)*
окопокопатель trench/foxhole digger
окостенение *med.* ossification
~ аневризмы osteoaneurysm, bone aneurysm
~ артерии arteriostenosis
~ перикарда armo(u)red heart, ossification of the pericardium
~ плода lithopedion, lithopedium, ostembryon, osteopedion
~, полое hollow ossification
~, соединительнотканное connective ossification
~ сухожилий tenostosis
окот *agr.* parturition; lambing time
окоченение rigidity, stiffening
~, посмертное/трупное *med.* rigor mortis, cadaveric rigidity
окошечко *biol.* foramen
окраина 1. margin, marginal/border land; 2. outskirts, outlying/border district, subrub; 3. *bot.* peristome
~ геосинклинали geosynclinal margin
~ материка, подводная continental margin
окраска 1. paint(ing), dyeing, colo(u)r(ing), stain(ing); 2. tint, hue, paintwork
~, антикоррозийная rust-preventative paint, rustproof coating, anti-corrosion paint(ing)
~, бросающаяся в глаза anti-crash paint
~, двойная double staining
~, дикая *zool.* agouti
~, дополнительная counterstaining
~, желтушная icteric discolo(u)ration
~, защитная protective colo(u)ring
~, искажающая dazzle painting

~, **камуфляжная** baffle/dazzle painting
~, **контрастная** contrastive staining
~, **маскировочная** baffle/camouflage paint, protective painting
~, **маскировочная, ночная** blackout doping
~ **обреза** *print.* edge staining/treatment
~ **по Граму** Gram's staining
~, **покровительственная** *biol., zool.* protective/assimilation colo(u)ration
~ **пород, нефтяная** oil staining of rocks
~ **распылением** spray painting
окрашенный colo(u)red, stained, tinted, dyed
~ **до сушки** *leath.* straight-dyed
~ **окислами железа** iron-stained
~, **сильно** hyperchromatic
окрашивание colo(u)ring, colo(u)r reaction, painting, dyeing
~, **неравномерное/пятнами** patchy colo(u)ration
~ **окунанием** dip painting
~ **распылением** paint spraying, spray painting
окрашивать colo(u)r, paint, tinge, tint, stain
окремление silicification, silicifying
окремненный silicified, siliceous
окрестность 1. vicinity, region; *math.* neighbourhood; 2. outskirts, ambit, surroundings
~ **критической точки** stagnation region
~ **резонанса** near-resonance region
окристаллизованность crystallinity
окропление *eccl.* perfusion
окрошка *cul.* okroshka, cold kvass soup
округ okrug, district, region
~, **военно-воздушный** air defence district
~, **жизненный** natural region
~ **штата** *Am.* county
округление *math.* rounding-off
~ **с избытком** rounding up
~ **с недостатком** rounding down
округленно *math.* in round figures
округленно-острый subactive
округленно-угловатый subangular
округленный rounded(-off)
~ **до секунды** *math.* rounded-off to the nearest second
округло-зубчатый round-/scalloped-toothed, crenate
округло-куполовидный round-arched
округло-листный *bot.* round-leaved, rotundifoliate
округло-лопастный round-lobate/-lobed
округло-многоугольный rounded-polygonal
округлый round(ed), spheroidal, spherical, orbicular
~ **спорофилл** clypeole, clypeola
округлять round (off), approximate
окружающий circumflex, encircling, peripheral
~ **землю** circumterraneous
~ **орбиту** circumorbital
~ **сосуд** circumvascular
~ **сустав** circumarticular
~ **ядро** circumnuclear
окружение surroundings, environment, background, milieu, circumambience
окружной 1. district *(attr.);* 2. peripheral, circumferential
окружность circle, circumference (of a circle) periphery
~, **вневписанная** excircle, escribed circle
~, **вписанная** incircle, inscribed circle
~, **вписывать** inscribe a circle
~ **выступов** *mach.* addendum/top circle
~ **ножек/впадин** *mach.* dedendum circle
~, **делительная** pitch circle
~ **кривизны/соприкасающаяся** osculating circle, circle of curvature
~, **описанная** circumscribed circle, circumcircle
~, **суставная** *med.* articular circumference
оксалат *chem.* oxalate
~ **уранила** uranium oxalate
оксалемия *med.* oxal(a)emia
оксалилмочевина *chem.* oxalylurea, parabanic acid
оксалис *bot.* oxalis, wood sorrel *(Oxalis)*
оксамид *chem.* ox(al)amide
оксамицин oxamycin, cycloserine
оксаммит *miner.* oxammite
оксиакойа *med.* oxyacoia, oxyakoia
оксигевзия *med.* oxygeusia
оксигемоглобин *med.* oxyhemoglobin
оксигемограф *med.* recording oximeter
оксигемометр *med.* oxi(hemo)meter
~, **ушной/комбинированный** ear oximeter
оксигемэритрин oxyhemerythrin
оксиген *pharm.* oxygen(ium)
оксигенатор oxygenator
~, **барабанный** drum oxygenator
~, **мембранный** diaphragmatic oxygenator
~, **пластинчатый** lamellar oxygenator
~, **пленочный** filmy oxygenator
оксиген(из)ация *chem.* oxygenation
оксигениум *pharm.* oxygenium
оксигидроген oxyhydrogen
оксидаза *bioch.* oxidase, oxydase
~, **завершающая** terminal oxidase
оксидиморфин oxymorphone hydrochloride, numorphan
оксидирование oxidizing, oxidation
~, **электрохимическое** electrolytic oxidation
оксидированный oxidized, oxide-coated
оксидировать *chem.* oxidize
оксидировка oxidizing, oxidation, oxide coating
оксидоредуктаз *bioch.* oxidoreductase
оксикамфора *med.* oxycamphor
оксикефалия *pathol.* turricephaly, oxycephaly
оксикислота *chem.* ox(y)acid, hydroxy acid
оксикринин oxykrinin, secretin
оксиликвит oxyliquit
оксилофильный humus-loving, oxylophyte
оксиметазолин oxymetazoline
оксимирование oximation
оксинафталин *chem.* oxynaphthalin, betanaphthol

оксинейрин *chem.* oxyneurin(e), betaine
оксиоксобутиратальдолаза hydroxyoxobutyrate aldolase
оксиопия oxyopia, oxyopy, oxyblepsia
оксиосфрезия *med.* oxyosphresia, extreme acuteness of the sense of smell
окситензиометрия oxitensiometry
~, амперметрическая ampermetric oxitensiometry
окситетрациклин *pharm.* oxytetracycline, terramycin
~ хлористоводородный oxytetracycline hydrochloride
окситиамин oxythiamine
окситон oxytone
окситонный oxytone *(attr.)*, oxytonical
окситоцин *bioch.* oxytocic hormone, oxytocin
оксиурия *med.* oxyuria(sis), oxyuriosis
оксифедрин oxyfedrine
оксифил *biol.* oxyphil(e), acidophil
оксифильный *biol.* oxyphilous, oxyphilic
оксифир *geol.* oxyphyre
оксифит *bot.* oxyphyte
оксихинолин *chem.* oxyquinoline, hydroxy quinoline
~, сернокислый oxyquinoline sulphate/sulfite
оксихолин *chem.* oxycholine, muscarine
оксихроматин *biol.* oxychromatin, oxyphile chromatin
оксицеллюлоза *chem.* oxycel(lulose)
оксицефалический oxycephalous, oxycephalic
оксицефалия *med.* oxycephaly, oxycephalia, hypsicephaly
оксиэтилтиазолкиназа *bioch.* hydroxyethylthyazole kinase
оксоизомераза oxoisomerase
оксология *biol.* auxology
оксосинтез oxo process
оксфордский Oxford *(attr.)*
оксюморон *lit.* oxymoron
октава 1. *mus., lit.* octave; 2. *mus.* low bass
~, итальянская *lit.* ottava rima
октада *math.* octad
октакорд *mus.* octachord
октан *chem.* octane
октан-корректор *auto.* octane selector
октант *math.* octant
октантный octantal
октастиль *arch.* octastyle
октаэдр *math.* octahedron
октаэдрит *miner.* octahedrite
октаэдрический octahedral
октет *mus.* octet(t), octette
октил *chem.* octyl
октильон, октиллион thousand quardillion, *Am.* octillion
октод *electr.* octode
октоих *eccl.* octoichos
октруа *econ.* octroi, octroy
октуполь, октупольный octopole
октэстрол *chem.* benzestral
октябрины *pl. obs.* communist ceremony of baptising
октябрист *hist.* Octobrist
октябрь October; *fig.* Russian revolution of October 1917
окукление *zool.* nymphosis
окукливание *zool.* pupation
окукливаться *zool.* pupate
окулирование *hort.* inoculation, (en)grafting
окулировать *hort.* inoculate, (en)graft
окулировка *hort.* inoculation, graftage, grafting, budding
~ щитком shield budding
окулировочный grafting
окулировщик *hort.* grafter
окулист eye specialist, ophthalmologist, oculist
окуломоторный *anat., zool.* oculogyric, oculomotor, ophthalmogyric
окультуривание *agr.* cultivation, amelioration, development, reclamation
окультуривать *agr.* cultivate, ameliorate, reclamate, domesticate
окулянт *hort.* grafted tree
окуляр eyepiece, eye glass, ocular
~ Гюйгенса Huygens eyepiece
~, дающий прямое изображение erecting eyepiece
~, демонстрационный demonstration eyepiece
~ Рамсдена Ramsden eyepiece
~, угловой offset eyepiece
окуляр-гониометр eyepiece goniometer
окунание dipping, immersing
окунеобразные *n. pl. zool.* perciforms *(Perciformes)*
окунь *zool.* perch *(Perca)*
окупаемость *econ.* cover (of expenditure), payback; return
~ затрат economic return
окупать repay, make up (for), compensate
окупаться pay for itself, pay off
окуривание fumigation
окускование *min.* agglomeration
~ брикетированием briquetting
окучивание *agr.* hilling/banking/earthing up, covering, ridging
окучивать *agr.* hill/earth up
окучник *agr.* furrower, hiller, shovel/ridging plough, ridger, banker
окципитализация *anat.* occipitalization
окципитальный *anat.* occipital, cervical
оладья *cul.* pancake, fritter, flapjack
олдермен *law* alderman
олеандомицин *chem.* oleandomycin phosphate
олеандр *bot.* oleander *(Nerium)*
олеандролистный *bot.* oleander-leaved, nereifolious
олеандроцветковый *bot.* oleander-flowered
олеария *bot.* daisybush *(Olearia)*
олеат *chem.* oleate

~ ртути mercury/mercuric oleate
оледенение 1. *geol.* glaciation, congelation, congealment; 2. numbness
~, айовское Iowa Glaciation
~, афтонское Aftonian Glaciation
~, бюльское Buehl Glaciation
~, вартинское Warthe Glaciation
~, великое восточное Great Eastern Glaciation
~, висконсинское Wiskonsin Glaciation
~, вислинское Weichsel/Vistula Glaciation
~, глючское Glutsch Glaciation
~, гуронское Huronian Glaciation
~ дагу Taku Glaciation
~, джерсейское Jerseyan Glaciation
~, допредпоследнее Antepenultimate Glaciation
~, древнее ancient glaciation
~, заальское Saale Glaciation
~, индуцированное sympathetic glaciation
~, канзанское Kansan glacial stage
~, лихвинское Likhvin Glaciation
~, максимальное maximum ice
~, малое уэльское Little Wales Glaciation
~, миндельское Mindelian Glaciation
~, мюльбергское Muehlbergian Glaciation
~, небрасское Nebraskan Glaciation
~, окское Oka Glaciation
~, последнее Last Glaciation
~ поянху Poyang Glaciation
~, предпоследнее Penultimate Glaciation
~, раннее Early Glaciation
~, рисское Riss Glaciation
~, скандинавское Inland/Sheet Glaciation
~, сканийское Scanian Glaciation
~, тимискаминское Timiskaming Glaciation
~ Хесле Hessle Glaciation
~, эльбское Elbe Glaciation
~, эльстерское Elster Glaciation
~, ярославское Yaroslavl Glaciation
олеиновый oleic
оленебык *zool.* eland *(Taurotragus)*
оленевод reindeer breeder
оленеводство reindeer/stag breeding, deer-raising
оленек *zool.* chevrotain *(Tragulus)*
оленеобразные *n. pl. zool.* cervoids *(Cervoidea)*
оленерогие *n. pl. zool.* cervicornians
олений cervine, elaphine
оленина venison
олень *zool.* (rein)deer *(Cervus)*
~, американский caribou *(Rangifer caribou)*
~, благородный red deer *(Cervus elaphus)*
~, индийский axis deer *(Cervus axis)*
~, канадский wapiti *(Cervus canadensis)*
~, пятнистый sika deer *(Cervus nippon)*
~, северный reindeer *(Rangifer tarandus)*
олеогранулема *med.* paraffinoma
~ легких pulmonary paraffinoma
~ молочной железы mammal paraffinoma
~ новорожденных paraffinoma neonatorum
олеография 1. chromolitography, oleography; 2. oleograph
олеотерапия oleotherapy, eleotherapy
олеоторакс *med.* oleothorax, eleothorax
олеофильность oil-receptivity; water-repellance
олефин *chem.* olefin(e)
олива 1. *bot.* olive (tree) *(Olea europea);* olive; 2. *anat.* olivary eminence/body; 3. *zool.* olivella *(Olivella)*
~ верхняя superior olive
оливенит *miner.* olivenite
~, волокнистый wood copper
оливин *miner.* olivin(e)
~, драгоценный peridot
~, желизистый fayalite
~, кальциевый calcio-olivine
оливковидный, оливкообразный olive-shaped, oliviform
оливник *bot.* spurge olive *(Cneorum)*
олигарх oligarch
олигархический oligarchic(al)
олигархия oligarchy
олигемический oligemic
олигемия *med.* oligemia, hypovolemia, olig(o)hemia
олигидрия *med.* olig(h)idria
олигоглия *med.* oligo(dendro)glia
олигодактилия olygodactylia, olygodactyly
олигозооспермия oligozoospermatism, oligizoospermia
олигокардия oligocardia, bradycardia
олигоменорея *med.* oligomenorrhea
олигопноэ *med.* oligopnea
олигополия *econ.* oligopoly
олигоспермия *med.* oligospermatism, oligospermia
олиготрихия *med.* oligotrichia, oligotrichosis
олиготроф *bot.* oligotrophic plant
олиготрофия *ecol.* oligotrophy
олигофрения *pathol.* oligophrenia, mental weakness, feeblemindedness
~, фенилпировиноградная phenylketonuria, phenylpyruvic oligophrenia
олигоцен *geol.* Oligocene
олигоцитемия *med.* globular an(a)emia, oligocythemia
олигурия *med.* oliguria, oliguresia, oliguresis
Олимп *myth.* Olympus
олимпиада 1. *hist.* olympiad; 2. Olympic games; 3. competition, contest
олимпиец *myth.* Olympian; *sport* Olympic athlete
олимпийский Olympic; Olympian
~кие игры Olympic games
олифа drying oil; varnish
олицетворение personification, typification, embodiment
олицетворенный personified, incarnate
олицетворять personify, typify, embody
олово *chem.* tin, Sn
~, аммиачнохлорное pink salt
~, двусернистое disulphide of tin, mosaic gold

~, двухлористое stannous chloride
~, деревянистое wood tin
~, металлическое tin metal
~, паяльное soldering tin
~, прутковое bar/rod tin
~, самородное native tin
~, сернистое tin/stannous sulphide
~, твердое pewter
~, черновое pig tin
~, четыреххлористое stannic chloride
~, чешуйчатое flake tin
оловоносный tin-bearing, stanniferous
оловосодержащий stanniferous
оловянирование tin plating
оловянистый stannous
оловяннокислый stannate (of)
оловянный 1. tin *(attr.),* stannic, stannous; 2. lead-colo(u)red
олограф *law* (h)olograph
олометр holometer, altimeter
Олоферн Holofernes
олуша *zool.* booby, gannet, solan goose *(Sula)*
ольбанит *geol.* albanite
ольдгаминидный oldhaminid
ольдгамит oldhamite
ольдермен *law* alderman
ольпа *hist.* olpe, Greek jug
ольха *bot.* alder *(Alnus)*
ольховый aldern
~вое криволесье subalpine alder elfinwood
оляпка *zool.* water ouzel *(Cinclus)*
ом *el.* ohm
~, тепловой thermal/heat ohm
омар *zool.* lobster *(Homarus)*
омартрит *med.* omarthritis
~, гнойный purulent omarthritis
омасливание oil treatment
омбилический *math.* umbilical
омброграф recording rain gauge, hyetograph, pluviograph, ombrograph
омброметр ombrometer, hyetometer, pluviometer
омброфил *bot.* ombrophil(e), ombrophilous plant
омброфильный *bot.* rain-loving, ombrophilous, ombrophilic
омброфит *bot.* ombrophyte
омброфоб *bot.* ombrophobous plant, ombrophobe
омег *bot.* hemlock *(Conium)*
омега omega
омеднение coppering, copper plating
омеднять copper
омежник *bot.* 1. water dropwort *(Oenanthe);* 2. hemlock *(Conium)*
омела *bot.* mistletoe *(Viscum)*
омеловые *n. pl. bot.* mistletoe family *(Loranthaceae)*
оментопексия *med.* omentopexy, epiplopexy, Talma's operation
омертвелость stiffness, numbness; *med.* necrosis, mortification
омертвелый *med.* necrotic, gangrenous
омертвение *med.* necrosis, gangrene, sphacelation
омертвлять deaden, necrotize; *econ.* immobilize, withdraw from circulation
омикрон omicron
омический *el.* ohmic
омлет *cul.* omelet(te)
омметр *el.* ohmmeter
омнопон *pharm.* omnopon, pantopon
омницид omnicide
омовение laving, ablution
~ рук *eccl.* lavabo
омография homography
омозолелость callosity, callositas
~, кожная cutaneous callosity
омозолелый callous
омозоление callosity
омолаживание, омоложение *agr., econ.* (re)juvenation, rejuvenescence
омоложенный rejuvenated
омоморф(а) homomorph
омоним *ling.* homonym
омонимический homonymous, homonymic
омофонный homophonous
омофор *eccl.* omophorion
омоформа homophorm
омоформный homophorm *(attr.)*
омуль *zool.* omul *(Salmo omul)*
омфалит *med.* omphalitis
омфалодес *bot.* navelseed *(Omphalodes)*
омфалорагия *med.* omphalorrhagia
омфалорея *med.* omphalorrhea
омфолотрипсия omphalotripsy
омфалофлебит omphalophlebitis
омфацит *miner.* omphacite
омываемый circumfluent, circumfluous
омывание circumfluence
омыватель washer
~ ветрового/лобового стекла *auto.* windshield squirter/washer
омыление *chem.* saponification
~, кислотное/щелочное acid/alkalic saponification
омыляемость saponifiability
омыляемый saponifiable
омылять saponify
он *hist.* archaic name of the letter **о**
онагр *zool.* onager *(Equs onager)*
онагра *bot.* evening primrose *(Oenothera)*
онанизм onanism, masturbation
онанировать masturbate
ОНГ *abbr.* **Общество независимых государств** Commomwealth of Independent States, CIS
онгстрем *see* **ангстрем**
ондатра 1. *zool.* muskrat, musquash *(Ondatra zibethica);* 2. muskrat (fur)
ондограф ondograph
ондулятор (telegraph/Morse) ink recorder, telegraphic register, undulator
онегит *miner.* onegite

онейродиния *med.* oneirogynia, nightmare
онейроид oneiroid
онейроидно-кататонический oneiro-catatonic
онейроидный oneiric, oniric
онемение numbness, dumbness, muteness
онемечение Germanization
оникс *miner.* onyx
~, мраморный onyx marble, Oriental alabaster
ониризм *med.* oneirism
онихалгия *med.* onychalgia
онихия *med.* onychia, onychitis
~, грибковая onychia parasitica, onychomycosis
~, дрожжевая yeast onychia
~, латеральная onychia lateralis/peringualis, paronychia
~, сухая brittle nails, onychia sicca
онихомаляция onychomalacia
онихоптоз *med.* onychoptosis
онихофагия *pathol.* onychophagy, nailbiting
онкилонит *miner.* onkilonite
онкоген oncogene
онкограф oncograph
~, электрический electrooncograph
онкология *med.* oncology; carcinology
онколь *fin.* call money/account, money at call
онкольный *fin.* call (*attr.*)
~ счет (an-)call account
~ная операция call transaction
~ная ссуда demand loan, DL, d/l
онкометр oncometer
онкометрия oncometry
онкосфера *zool.* oncisphere
онкоцит *med.* oncocyte
оноклея *bot.* ostrich fern (*Struthiopteris)*
онома(с)тика onoma(s)tics
ономатологический onomatologic(al), onomastic
ономатомания *med.* onomatomania
ономатопея onomatopoeia
ономатопоический onomatopoeic
ономатопоэз onomatopoeia, onomatopo(i)esis
ономатопоэтический onomatopoe(t)ic
онопордум *bot.* cotton thistle (*Onopordum)*
оносма *bot.* onosma (*Onosma)*
онтогенез, онтогенезис, онтогения *biol.* ontogenesis, ontogeny
онтологический ontologic(al), ontologistic
онтология *philos.* ontology
онхоцеркоз *med.* onc(h)ocerciasis, oncocercosis
онуча onucha (*cloth wrapped round feet)*
oобластемный ooblastic
оогамия *bot.* oogamy
оогоний, оогония *biol.* o(v)ogonium
оокинез, оокинезия *biol.* ookinesia, ookinesis
оолемма *zool.* oolemma
оолит *geol.* oolite
~, железистый ferruginous oolite
~, зачаточный embryonic oolite
~, тонкозернистый ammite, roe-stone
оолитовый oolitic, pisolitic
оомицеты *bot.* oomycetes (*Oomycetes)*
ООН *abbr.* **Организация Объединенных Наций** United Nations Organization, UNO
ооплазма o(v)oplasm
ооспермa *zool.* oosperm
ооспора *biol., bot.* oospore
оосфера *biol.* oosphere
~, оплодотворенная carposperm
оотека *bot., zool.* ootheca
оотит *zool.* ootype
оофит *bot.* oophyte
оофоридий *bot.* oophoridium
оофорит *med.* oophoritis, oothecitis
~, гонорейный gonorrheal oophoritis
~, острый acute oophoritis
оофоропатия *med.* oophoropathy, oothecopathy
оофороцитоз *med.* oophorocytosis, oothecocitosis
ооцентр oocentre
ооциста oocyst
ооцит *med.* oocyte
~ первого порядка primary oocyte
ооэция *zool.* ooecium
опад fallen leaves
~, лиственный leaf debris
~, мертвый наземный dead soil cover
опадание falling off; *leath.* pulling-/bringing-down
~ ветвей cladoptosis
~ листьев leaf fall, phylloptosis
~ плодов carpoptosis
~ цветков anthoptosis
опадать fall off/away; *leath.* pull/come down
опадающий deciduous; not permanent, transitory
~ лист deciduous leaf
~, рано caducous
опадение 1. diminishing, subsidence; 2. *bot.* abcission, cast, drop
~ листьев leaf fall, phylloptosis
опак-вещество opaque matter
опал *miner.* opal
~, бесцветный hyalite, water opal
~, благородный precious opal
~, восточный harlequin opal
~, деревянистый xylopal, wood opal
~, недрагоценный botch, worthless opal
~, огненный fire/sun opal
~, пористый floatstone
опалесценция opalescence
~, звездная stellate opalescence
опаливание, опалка singeing, burning
опаловый opal(ine), opalescent, opal-like
опалубка shutter(ing), shutters; form(work); timbering, casing; forms
~, крупнощитовая large-panel/-section forms
~, несъемная retained formwork, permanent/retained forms
~, переставная progressive/travelling shuttering/forms
~ подъемная lift shutters, climbing forms
~, разборная knock-down forms
~ ствола *min.* shaft sheathing
опара *cul.* leaven, sourdough; leavened dough

опасность danger, hazard, risk, peril; jeopardy
~, биологическая biological hazard
~ вдыхания breathing hazard
~ взрыва explosion risk/hazard
~, возможная hazard potential
~ возникновения пожара, ~ пожарная fire hazard/risk
~ загрязнения contamination hazard
~ лучевого поражения radiohazard
~, метеорная meteorid hazard, meteor danger
~, мнимая imaginary danger
~ появления солнечных вспышек solar-flare hazard
~, радиационная (ir)radiation hazard, radiohazard
опасный dangerous, hazardous, perilous; critical
~, общественно *law* injurious to the public
опах *zool.* opah *(Lampris)*
опахало *bot.* fan-shaped cyme, rhipidium; *zool.* vane, eb
опахивать plough/plow (a)round
опдалит *geol.* opdalite
опека 1. tutelage, custody, wardship, guardianship, trusteeship; *sport*. guarding; 2. *coll.* guardians; 3. care, surveillance
опекаемый under wardship; *n.* ward
опекун *law* tutor, guardian, trustee, custodian
~, естественный *law* guardian by nature
~, законный tutor-at-law
~, назначенный завещанием tutor nominate
~ по завещанию testamentary guardian
опекунский tutorial, tutelary
опекунство tutorship, guardianship
опенок *bot.* honey agaric *(Armillariella mellea)*
опера 1. opera; 2. opera house
операбельный *med.* operable, amenable to surgical treatment
опера-буфф *mus.* opera bouffa
операнд *math.* operand
~ из нескольких слов *comp.* multiword operand
~, неявный implied operand
оперативка *polit.* briefing, emergency/extraordinary meeting
оперативность efficiency, drive
оперативно-технический operating-technical
оперативный 1. *med.* operative, surgical; 2. *mil.* operational, operation(s); 3. executive, operative, operational; 4. efficient
~ отдел operations section, operational staff
~ная память *comp.* main memory/store, core memory/store, read/write memory
оператор 1. operator, operative; *comp.* statement; 2. *cin.* cameraman; 3. *med.* surgeon
~, арифметический calculation/compute statement, arithmetic instruction
~ бесформатного считывания unformatted read statement
~ Буля logical connective/connector, Boolean (logical) operator
~ ветвления branch statement
~ включения include statement
~, вложенный nested statement
~ возврата return (jump) statement
~ вполне непрерывный completely continuous operator/map(ping), compact operator
~ входа enter statement
~ выбора case statement
~ выбора зоны/массива field select statement
~ вызова call statement
~, вызывающий caller
~, выполняемый executable statement
~ Гамильтона *math.* Hamiltonian (operator), del/nabla operator
~ Даламбера Dalambertian, d'Alembertian (operator), d'Alembert('s) operator
~, двоичный bit handler
~, двуместный infix operator
~, дежурный *rad.* duty operator
~ деления divide statement
~, дифференциальный differential operator
~ если if-statement
~ заголовка heading/header statement
~ задания job statement
~, замкнутый *math.* closed operator/map(ping)
~ звукозаписи sound recordist
~ идентификации файлов file identification statement
~, компактный completely continuous map(ping), compact operator
~ конец задания end-of-job statement
~ конец файла end-of-(data-)file statement
~ конца файла endfile statement
~ Лапласа Laplace operator, Laplacian
~, логический Boolean (logical) operator
~ над знаком sign operator
~ над массивами operator on files
~, непрерывный continuous operator/map(ping)
~ объявления declarer
~, ограниченный bounded operator
~ описания файла file definition statement
~, описательный/описания declarative instruction, descriptive statement
~ определения функции function defining statement
~ отбора select statement
~, отладочный debugging statement
~ перемотки rewind statement
~ перестановки координат space-exchange operator
~ пересылки данных move statement
~ перехода *comp.* go-to-statement; *math.* transition operator
~ периода компиляции compile-time statement
~, порожденный generated statement
~, последний terminal controller
~ приема accept statement controller
~ примечаний comment controller
~ присваивания assignment controller

~ **проверки** examine statement
~ **продолжения** continuation statement
~, **пустой** dummy/null statement
~ **рассеяния** scattering operator
~ **редактирования** edit statement
~ **РЛС** radar operator/controller
~, **самосопряженный** self-adjoint/-conjugate operator
~ **систем радиопротиводействия** electronic counter measure operator
~ **следования** sequential operator
~, **сменный старший** *com.* supervisor of the shift
~, **составной** compound statement
~ **станции перехвата** intercept officer
~ **считывания** read statement
~, **трассировочный** trace statement
~ **уничтожения** annihilation/destruction operator
~ **управления огнем** fire-control operator
~, **управляющий** control statement
~, **условный** conditional (statement)
~ **шифровальной машины** crypto operator
оператор-выражение expression statement
оператор-гидроакустик asdic/*Am.* sonar operator
операторно-гомоморфный operator-homomorphic
операторная *n.* control room
операторный operational, operating, symbolic(al)
оператор-ограничитель delimiter statement
оператор-планшетист plotter, plotting-table operator
оператор-функция function statement
операционная *n.* operating room/theatre
операционно-перевязочный *med.* operative-dressing
операционно-совместимый interoperable
операционный operative, surgical, operating, operational
~ **стол** operating table
операции *pl.* operations; *see also* **операция**
~, **арифметические** arithmetics
~, **банковские** banking
~, **биржевые** stock exchange operations
~, **брокерские** brokerage
~, **валютные** exchange transactions
~, **выполняемые на борту** *naut.* on-board procedures
~, **клиринговые** clearing transactions
~, **крейсерские** cruiser warfare
~, **стандартные** routine work
~ **фондовой биржи** stock exchange dealings
операция operation; *econ.* transaction; *see also* **операции**
~, **алгебраическая** algebraic operation
~, **арифметическая** arithmetic(al) operation, arithmetics
~, **бинарная** binary/dyadic operation
~ **блокировки** block operation
~, **бухгалтерская** accounting transaction
~, **висячая** *math.* hanging activity
~ **выбора** *med.* operation of choice
~ **вывода** exit/output operation
~, **вычислительная** computation
~ **Г** *math.* Pierce stroke
~, **гинекологическая** genital operation
~, **двухмоментная** double-stage operation
~, **десантная** landing operation
~, **длинная** time-consuming operation
~ **запрета** inhibit operation
~ **И** AND-operation
~ **ИЛИ** OR-operation
~ **инициируемая нажатием клавиши** keystroke operation
~ **исключающее ИЛИ** exclusive OR operation
~, **логическая** logic(al) operation/function, Boolean operation
~ **логического вывода** inference operation
~, **мажоритарная** majority operation
~, **машинная** computer operation
~, **многоцелевая** multi-mission operation
~, **набеговая** (hit-and-run) raid
~ **над базой данных** database operation
~ **над внутренним адресом** home address operation
~ **над массивом** array operation
~ **над множествами** set operation
~ **над скалярными величинами** scalar operation
~ **над строками данных** string operation
~ **над числами двойной величины** double-length operation
~, **наладочная** setting-up, adjustment, tuning
~, **наступательная** offensive operation
~ **НЕ** NOT operation
~, **незавершенная** activity in progress
~ **НЕ-И** NAND-operation
~ **НЕ-ИЛИ** NEITHER-NOR operation
~, **необратимая** irreversible operation
~, **неэлементарная** nonprimitive operation
~, **обратная** inverse operation
~, **онкольная** *fin.* call transaction
~, **отделочная** finishing operation
~, **производственная** manufacturing/processing operation
~, **пустая** *comp.* void
~ **развертывания** *math.* expansion
~, **регистровая** register operation
~, **ручная** manual operation
~ **с одним операндом** monadic operation
~ **с плавающей запятой/точкой** floating-point operation
~ **сброса и запирания** clear-and-gate operation
~ **симметризации** operation of symmetrization
~, **служебная** bookkeeping/housekeeping operation
~, **совместная** joint operation
~, **спасательная** rescue (operation)
~, **срочная** *med.* emergency operation
~, **стековая** stack operation

~, стратегическая strategic(al) operation
~ стыковки docking operation
~ считывания read operation
~, технологическая production operation
~ удаления крайней плоти *med.* circumcision
~, унарная *math.* one-place operation, unary operation, operation of rank 1
~ управления control operation
~ фиксированной запятой/точкой fixed-point operation
~, фондовая stock exchange transaction
~, хирургическая surgical operation
~, целочисленная integer operation
~ цифровой передачи move digit operation, digit movement operation
~ эквивалентности identity operation
~, элементарная *math.* subjob
опережение 1. lead, advance; 2. forestalling, anticipation
~ впрыска injection advance
~ зажигания ignition/spark advance
~ зажигания, постоянное set spark
~ открытия клапана valve-opening lead
~ подачи горючего fuel lead
~ по фазе phase lead/advance
~, угловое angular advance
оперение 1. *zool.* feathering, plumage, indumentum; 2. *av.* empennage, tail unit/assembly/surfaces, fin assembly, vane
~, вертикальное (хвостовое) vertical tail/blade
~, изолированное isolated tail/unit
~, кормовое stern control surface
~, наклонное canted fins
~, носовое canard/nose surface
~, однокилевое single-fin tail (unit)
~ ракеты rocket/missible stabilizer, rocket fin unit
~ рыбы *zool.* fins and rudders
~ самолета empennage
~ самолета, V-образное butterfly tail/unit, V-tail
~ типа утка canard-type surfaces
~, трапециевидное tapered tail surfaces
~, треугольное delta/triangular tail surfaces
~, хвостовое tail vane (unit); tailfins
~ хвостовое Т-образное T-(shaped) tail
~, хвостовое V-образное Vee tail/unit, butterfly-type tail unit
оперенный feathered, finned
оперетта *mus.* musical comedy, operetta
оперирование 1. handling, manipulation, operation; 2. *med.* operating, operation
оперкулум, оперкулюм *bot., zool.* operculum
оперкулянтный operculate, opercular *(attr.)*
опермент arsenic trisulphide, orpiment *(pigment)*
опечатка corrigendum, erratum, misprint, writing/write error
опечение *med.* hepatization
~, желтое/серое yellow/gray hepatization
опианин *chem.* narcotin(e)
опиат opiate
опий opium
опийный thebaic
опиливание filing (round); sawing round
опилки 1. *pl.* filings, sawdust; 2. *bot.* ramenta, chaffy scales
опилковидный, опилкообразный scobiform
опиловка filing, lopping off
опиокурение opium smoking
опиофагия opium eating, opiophagism, opiophagy
опирать *mech.* support
~ шарнирно pivot
опираться *mech.* bear, bean *(against);* rest (on), be carried *(by)*
описание definition, description, depiction, characterization
~ готового изделия end-item report
~, камеральное office description
~, климатологическое climatological summary
~ массива array declaration
~ обращения к файлу *comp.* access file attribute
~ огней light list
~ патента patent specification
~, повторное re-description
~ трав agrostography
описанный described; *math.* circumscribed
описатель *comp.* descriptor, specificator, attribute
~ масштаба scaling attribute
~ назначения файла file function attribute
~, стандартный default attribute
~ элемента данных data element descriptor
описательный descriptive, depictive, narrative; *ling.* periphrastic
опистогирный opistigyral
опистодетный *zool.* opisthodetic
опистопарный *paleont.* opisthoparian
описторхис *med.* opisthorchis
~, кошачий cat liver fluke, opisthorchis felineus
описторхоз *med., vet.* opisthorchiasis
опистотонус *med.* opisthotonos
опистоцельный *zool.* opisthocoelic, opisthocoelous
описывание description, depiction; delineation; *lit.* portrayal
описывать 1. describe, depict; delineate; explain, present, set forth; 2. list, make an inventory; 3. *law* distrain, arrest, seize; 4. *math.* circumscribe
опись 1. list, inventory, schedule; 2. survey
~ груза loading list
~ имущества *law* distraint
~, инвентарная inventory, stocktaking
~, морская exploratory/running survey
~ недостач list of shortages
~ содержимого note of (the) contents
опиум opium
оплавление *met.* flash(ing), burn-off; glazing, fire-polishing, vitrification, partial melting; melting away, fusing
оплата pay(ment), remuneration, defrayal

~ в натуральной форме remuneration in kind
~, заработная salary, wages, labo(u)r remuneration
~ натурой pay in kind
~ недоимок back pay
~ по времени charging by time, time-work payment
~, поощрительная incentive pay
~, поэтапная progress payment
~ предварительная prepayment
~ при доставке collect on delivery
~, сверхурочная overtime pay
~ труда labo(u)r remuneration
~ труда, повременная pay by the day/week, time payment
~ труда, прогрессивная incentive wages
~ труда, сдельная piece/productivity wages, piecework, piece-rate pay
~ чека check/cheque payment
оплатить (re)pay, defray, hono(u)r *(of cheque)*
~ по предъявлении pay on presentation
оплаченный (pre)paid
оплачиваемый paid, payable
~, хорошо well-paid
оплачивать pay (for), repay, defray
~ убытки pay damages
оплетка *el.* braid(ing), sleeving; *naut.* keckling, keckle
~ двухслойная double-braid covering
~ кабеля cable braiding
~ конца троса back splice
~, металлическая wire braiding
~, нитяная textile braiding
~, пропитанная impregnated braid
~, противогнилостная imputrescible braid
~, расцвеченная colo(u)red braiding
~, татарская graft
~, трубчатая braid sleeve
оплодотворение fertilization, impregnation, ingravidation, insemination, fecundation; *bot.* fructification
~, близкородственное inbreeding
~, вегетативное vegetative fertilization
~, искусственное artificial fertilization/ impregnation/insemination
~, перекрестное cross-fertilization, xenogamy, allogamy
оплодотворенный impregnated, fertilized
~ное яйцо oosperm
оплодотворитель *biol.* fertilizer
оплодотворять impregnate, fecundate, fertilize; *bot.* fructify
~ перекрестно cross-fertilize
оплот 1. bulwark, stronghold; 2. *hist.* barrier, rampart
оплывина mud stream/flow/avalanche
оповеститель annunciator; signal device
оповестительный announcing, informing, signal(l)ing
оповещать give warning, let know, announce, notify
оповещение informing, announcing, notice; warning
~ звуковым сигналом audible warning
~ о воздушном налете air-raid warning
~ о патенте patent advertizing
~ штормовое storm warning/information
опогружение *math.* embedding
оподельдок *hist., med.* opodeldoc, soap-liniment
оподзоленность presence of podsol
оподзоленный podzolic; podsolized
~ чернозем podzolic chernozem
оподзоливание podsolization, development of podsol/ podzol
опоек *leath.* calf skin/leather
опоек-велюр suede calf (leather)
опоек-нубук nubuk calf leather
опоек-склизок slunk
опознаваемость *stat.* identifiability
опознавание identification, recognition
~ знаков character recognition
~ образов pattern recognition
~ путем непосредственного наблюдения contact identification
~ речи speech recognition
~ самолетов aircraft identification
~ свой-чужой *av.* identification friend-or-foe
~ слов word recognizing
~ тома (data) volume recognition, data carrier recognition
опознаватель recognizer, identifier
опознавательные *n. pl. naut.* challenge, identification signals
~, показывать challenge
опознавательный identifying, distinguishing; episematic
опознак *geod.* topobeacon, control survey point
опознание identification
опозорение *law* defamation
опока I *met.* flask; moulding/casting box, moulding frame
~, верхняя cope, top box
~, нижняя drag, nowel
~ разъемная snap flask, pop-off flask
~, ручная one-man flask
~, сборная/составная built-up moulding box
~, съемная slip flask
опока II *geol.* silica clay, gaize, opoka
оползание *geol.* slump(ing), slipping, sliding
оползать slump, slip, slide
оползень *geol.* landslide, landslip, landfall, slump, earth-/rock-slide, soil creep/flow
~, глинистый clay slide
~, древний ancient rock-slide
~, мелкий shallow landslide
~, мелкоземистый earth creep
~, многоярусный multicycle landslide
~, подводный subaqueous slump(ing)
~, скользящий sliding landslip
половник *zool.* bottle lit *(Aegithalos caudatus)*

ополченец home guard, civil guardsman; militiaman
ополчение home guard, militia
ополячение, ополячивание Polonization
опора support, bearer, bearing, prop, leg, foot, footing, foothold, shore, rest, mount; pier; *med.* sustentaculum; *bot., zool.* fulcrum
~, башенная tower pier
~ в сборе support assembly
~ вала shaft bearing/support
~, деревянная wooden pole
~ для ноги/руки footrest/handrest
~ инструмента *geod.* instrument base
~, катковая roller bearing
~ кузова body support/bracket
~, ножевая knife-edge support
~ подшипника bearing support/stand/block
~, посадочная landing leg
~, призматическая knife-edge support
~, пружинная spring/flexible support/cushion
~, ряжевая crib pier
~, свайная pile trestle/pier
~ таранной кости *anat.* support of the ankle bone
~, телескопическая spur leg
~ трубопровода pipeline saddle
~, упругая elastic support, resilient mounting
~, цапфовая trunnion bearing
~, шарнирная hinged support
~, шаровая spherical bearing
~ шасси leg, undercarriage (member)
опоражнивание discharging, emptying, draining, dumping
опоражнивать discharge, empty, dump, void, deplete
опорный 1. key, index, guide, reference; 2. bearing, supporting, pivotal; *bot.* fulcral, fulcraceous
~ пункт *mil.* strongpoint
~ная плита base/bearing plate
~ная плоскость plane of support
~ная точка reference point
опорожнение emptying, dump(ing); voiding
~ желудка gastric emptying/evacuation
~ кишечника def(a)ecation
~, ненормальное abevacuation
опорос *agr.* farrow
опорочивать 1. *law* vitiate, discredit; 2. defame, dishono(u)r
опосредованный, опосредствованный *philos.* mediated
опосредственность, опосредствование *philos.* mediacy
опоссум *zool.* opossum *(Didelphis)*
опоясанно-апертурный/-поровой zonaperturate, zonorate
опоясанно-бороздковый zonisulculate
опоясывающий circular
опоясывать gird(le), engird(le), surround
оппозиционер opositionist
оппонент opponent
оппортунизм opportunism
оппортунистическиий, оппортунистичный, оппортунистский opportunist(ic)
оправа holder, case, mounting, frame, rim
~ объектива lens mount/cell
~, очковая spectacle/eyeglass frame
~ пенсне pince-nez holder
оправдание 1. *law* acquittal, discharge; 2. justification, excuse
оправдательный *law* absolutory, justificative, apologetic(al)
~ документ *econ.* voucher
~ приговор verdict of "not guilty"
оправдывать 1. excuse, find excuses for; warrant, justify; 2. *law* exculpate, acquit
оправдываться 1. clear/vindicate oneself; 2. prove correct, come true
~ незнанием *law* plead ignorance
оправка 1. mandrel, holder; 2. mounting, setting
~ для волоки die holder/insert
~, калибрующая swage
~, разжимная expanding mandrel
~, установочная aligning bar
~, фасонная former
опрашивание interrogation
опрашивать interrogate; *law* (cross-)examine, guestion
~ триггеры *comp.* sample (the) flip-flops
определение 1. determination, definition; 2. finding, spotting, sensing; 3. computation, calculation, estimation, reckoning; 4. *law* decision; 5. *gram.* attribute
~ ареала geographical demarkation
~, астрономо-геодезическое astrofix(ation)
~, весовое gravimetric determination
~ возраста age determination, (radiocarbon/helium) dating
~ времени time determination
~ высот *geod.* heighting, relief determination
~ высоты height finding
~ гранулометрического состава grading test, grain-size analysis
~ грунта *geol.* terrain interpretation
~ дальности ranging, range measurement/determination
~ дальности, глазомерное range estimation
~ зольности ash test
~, индикаторное tracer determination
~, количественное quantitative estimation/determination
~ коэффициента полезного действия efficiency test
~ места location, fix, position fixing
~ места доплеровским методом Dopplex fixing
~ места повреждения fault loca(liza)tion
~ места по светилам celestial fixing
~ места по счислению reckoning of position
~ места по трем светилам three-star fix
~ места утечки leak location

~ **местонахождения** location
~ **местоположения** fix(ing), position finding/fixing
~ **местоположения расчетным путем** dead reckoning
~ **неисправности** fault location, trouble shooting
~, **непосредственно-примыкающее** *gram.* immediate adjunct
~, **объемное** volumetric determination
~ **окаменелостей** identification of fossils
~, **органолептическое** sensory detection
~ **оруднения** ore localization
~, **повторное** re-definition
~ **погрешности** error estimation
~, **полярографическое** polarography
~ **приоритета** *comp.* priority resolution
~ **пространственного положения** attitude determination/sensing
~ **размеров** sizing, dimensioning
~ **силы сцепления** cohesion test
~ **сорта** grade assessment
~ **твердости по Бринеллю** Brinell hardness test
~ **твердости по Виккерсу** Vickers/diamond (pyramid) hardness test
~ **твердости по Герберту** Herbert cloudbust (hardness) test
~ **твердости по Моосу** Mohs scratch-hardness test
~ **твердости по Роквеллу** Rockwell hardness test
~ **твердости по Шору** Shore (scleroscope) hardness test
~ **температуры вспышки** flash-point test
~ **угла места** elevation finding
~ **частотности ошибок** error-frequency count

определение-местоимение pronominal attribute
определение-прилагательное attribute adjective
определение-приложение appositional adjunct
определенно-количественный definite-quantitative
определенно-несовершенный determined-imperfective
определенно-отрицательный negative-definite
определенно-положительный positive-definite
определенность definit(iv)eness, definability, distinctness; exactitude, accuracy
~, **комбинаторная** *math.* combinatorial definability
определенный 1. definite, (well-)defined, determined, determinate, fixed; decided; 2. certain, spectific, distinct, concrete
определимый definable, determinable, determinate; assayable, identifiable
~, **статически** statically determinate
определитель 1. *math.* determinant; 2. detector, finder, locator; 3. key, guide, index
~ **Вандермонда** Vandermonde('s) determinant
~ **Вронского** Wronskian (determinant)
~ **Грама** Gram('s) determinant, Gramian
~, **раскрывать** expand a determinant
~ **Фредгольма** Fredholm('s) determinant
~, **функциональный** functional determinant, Iacobian
определительно-временной determinative-temporal
определительно-обстоятельственный attributive-adverbial
определительный attributive, determinative
~ **подвид** determinative subaspect
определить determine, define; establish, ascertain
определяемое *n.* determinatum, definiendum
определяемый defined, determined
определять 1. define, determine, establish, identify; 2. detect, locate, fix; appoint, specify, assay; 3. allot, assign, appoint
~ **на глаз** determine by sight
определяющее *n. math.* determinandum, definiens
определяющий determinative, determinant
опрелость *med.* intertrigo
~, **дрожжевая** yeast intertrigo
опрелый *med.* intertriginous
опреснение demineralization, desalination, desalting, freshening
опреснитель desalting kit/unit, desalter
опреснительный freshening, desalting, desalination *(attr.)*
опреснять freshen, desalt, desalinate
опрессованный pressurized, press-fitted, pressure-tested
опрессовать, опрессовывать press, shape, mould
опрессовка pressure moulding, hydraulic/hydrostatic test
~ **водой** water pressure test(ing)
~ **сжатым воздухом** air pressur test
опричник *hist.* member of the oprichnina, oprichnik
опричнина *hist.* oprichnina, special administrative elite
опробкование *bot.* suberization
опробковение *bot.* suberification, corky core, corking
~, **наружное** external corking
опробование 1. test(ing), try-out, checking; 2. sampling, assaying
~, **валовое** bulk sampling
~, **вторичное** resampling
~ **двигателя** *auto.* (engine) run-up, engine check
~ **двигателя в полете** *av.* in-flight engine check
~ **двигателя на земле** engine ground run-up
~, **задирковое** sampling by stripping
~, **контрольное/проверочное** check sampling
~, **радиометрическое** radiometric assay
~, **шпуровое** pit sampling
опробовать test; sample; check
~ **повторно** resample

опровержение denial, disproof, refutation; *law* rebuttal, rebutment
опровержимый refutable
опрокидывание 1. tipping/tilting (over), overturning; *naut.* capsizing, tumbling; 2. inversion
~ **изображения** reversal of the image
~ **фазы** phase reversal
опрокидыватель tipper, dumper; tripping gear
~ **вагонеток** car dumper
опрокидывать 1. turn/tip over, tilt; 2. invert, reverse
опрокинутый overtruned, tilted; inverted
опрос inquiry, interrogation, questioning, quiz; *law* (cross-)examination; *comp.* poll(ing), survey
~, **автоматический** autopoll
~, **дистанционный** telequiry
~, **поквартирный** door-to-door poll
~, **сплошной** house-to-house poll
~**а, корреспондентский способ** *stat.* mail survey
опросный interrogatory, polling
~ **лист** questionnaire
опросчик interrogator, questioner
опротестование act of protect
опрыскивание spraying, sprinkling
~, **авиационное** aerial spraying
~, **довсходовое** pre-emergence spraying
~, **мелкокапельное** atomized/mist spraying
~, **предпосевное/предуборочное** *agr.* pre-planting/harvest spraying
опрыскиватель (crop)sprayer, spray(ing) machine, sprinkler, atomizer
~, **аэрозольный** aerosol sprayer, fogger
~, **полевой** field(-crop) sprayer
~, **ранцевый** knapsack sprayer, back-pack sprayer
~, **самолетный** aircraft/air/aerial sprayer
опрыскиватель-культиватор sprayer-tiller
опрыскиватель-опыливатель sprayer-duster
опрыскивать (be)sprinkle, perfuse
опрятность tidiness, neatness, cleanliness
ОПС *abbr.* **однополосный сигнал** single-side-band signal, SSB signal
~, **формировать** derive/develop/generate SSB signal
опсанус *zool.* toadfish *(Opsanus)*
опсиноген, опсоген *immun.* ops(in)ogen
опсонизация *immun.* opsonization
опсонин *bact.* opsonin
~, **гонококковый** gono-/gonococeal opsonin
~, **нормальный** normal/common opsonin
опсонотерапия opsonotherapy
опсоно-фагоцитарный opsonocytophagic
опт *econ.* wholesale, goods sold in large quantity
опталический opthalic
оптант *law* optant, person having right of option *(of citizenship)*
оптатив *gram.* optative (mood)
оптация *law* option, optation *(of citizenship)*
оптик optitian, optometrist
оптика optics; *coll.* optical instruments
~, **волновая** wave/physical/optics
~, **волоконная** fibre optics
~, **геометрическая/лучевая** geometrical/ray optics
~, **длиннофокусная** long-focal-lenght optics
~, **зеркальная** reflecting optics
~, **используемая для сближения** *space techn.* rendezvous optics
~, **офтальмологическая** ophthalmologic optics
~, **сменная** changeable optics
~, **физическая** physical/wave optics
~ **Шмидта** reflective optics
~, **электронная** electron optics, optronics
оптикатор optical gauge
оптико-гемиплегический opticochemiplegic
оптико-кинетический opt(ic)okinetic
оптикомиелит *med.* opticomyelitis
оптико-тепловой optitherm
оптикоэлектроника optoelectronics
оптимальность optimality
оптимальный optimal, optimum *(attr.)*
оптиметр telescope/optical cal(l)iper
оптимизатор optimizer
оптимизация optimization, optimizing
~, **байесовская** Bayesian optimization
~, **безусловная** unconstrained optimization
~, **векторная** multiobjective optimization
~, **глобальная** overall optimization
~ **затрат на обеспечение качества** quality cost optimization
~ **путем расчета** theoretical optimization
~ **управления по минимаксным критериям** minimax control
~, **условная** constrained optimization
~, **целочисленная** integer optimization
~ **цепей, (машинная)** (computer-aided) network optimization
оптимизировать optimize
оптимизм optimism
~, **лечебный** therapeutic optimism
оптимум optimum (*pl.* -ma); the best, ideal
~, **безусловный** unconditional/unconstrained optimum
~, **граничный** boundary optimum
~ **Парето** Pareto optimum
~, **послеледниковый** post glacial optimum
~, **условный** conditional/constrained optimum
оптировать *law* opt(for), choose, make one's choice (for)
оптический optic(al), visual
~ **обман** optical illusion
оптовик *econ.* wholesale dealer, wholesaler
оптовый wholesale
оптом by/in (the) gross, wholesale
~ **и в розницу** wholesale and retail
оптометр *med.* optometer
оптотехника optical equipment/technology
оптотранзистор optoelectronic transistor
оптоэлектроника optoelectronics
~, **твердотельная** solid-state optoelectronics

оптрон photon-coupled pair, opto-isolator
оптроника optronics
опубликование publication, promulgation *(of law)*
опубликовывать publish, make public; *law* release, promulgate
опудривание dusting, powdering
опудривать dust, powder
опунция *bot.* prickly pear *(Opuntia)*
опус *mus.* opus, composition
опуск omission
опускание 1. lowering, descending, sinking; *geol.* subsidence, settling, dipping, descent, depression; 2. dropping, leaving out, omission
  ~, изостатическое isostatic settling
  ~ индекса *math.* lowering of an index
  ~ колоши *met.* drop/slip of the charge
  ~ континентов subsidence/sinking of continents
  ~, предельное ultimate subsidence
  ~, самопроизвольное spontaneous sinking
  ~ сидения seat lowering
  ~ суши land subsidence
  ~, тектоническое tectonic subsidence/settling
опускать 1. lower, sink, dip; let down; 2. leave out, omit
  ~ перпендикуляр drop a perpendicular
  ~ шаг *comp.* skip/bypass a job step
опускаться descend, subside, sink, go down, fall
опускающий *med.* depressor *(attr.)*
опускной lowerable; movable
  ~ная дверь trap door
опустение emptiness; *tech.* cavitation
опустошать 1. devastate, ravage, lay waste; 2. deplete
опустошение 1. devastation, ravage, havoc, ruin; 2. depletion
опустошительный devastating; disastrous, ravaging
опустыненный desertificated
опутывать entangle, enmesh
опухание *med.* swelling, tumefaction, (in)tumescence
опухлость swelling, intumescence; turgidity
опухолевидный tumour-like, carcinomatous
опухолевый tumoral, neoplastic
опухоль swelling; *med.* tumour, neoplasm, neoformation
  ~, амилоидная amyloid tumour
  ~ Бреннера/Варсина/Вильмса Brenner/Warthin's/Wilms tumour
  ~, вирусная virus/viral tumour
  ~, генерализованная злокачественная generalized malignancy
  ~, гигантоклеточная giant cell tumour
  ~ Гравица Grawitz's tumour
  ~, доброкачественная benign/innocent tumour
  ~, жировая adipose tumour, lipoma
  ~, злокачественная malignant tumour/neoplasm
  ~ из клеток Лейдига/Сертоли Leydig/Sertoli cell tumour
  ~ Крукенберга Krukenberg's tumour
  ~, меланическая melanotic tumour
  ~, невроэпителиальная neuroepythelioma
  ~, нервная neuroma, nerve tumour
  ~, профессиональная occupational tumour
  ~, смешанная mixed tumour
  ~, соединительнотканная histoid tumour, connective tessue tumour
  ~, сосудистая angioma, vascular tumour
  ~, фиброзная/фиброидная fibroid tumour
  ~, хрящевая chondroma, cartilaginous tumour
  ~, эхинококковая echinococcal cyst
опушение 1. pubescence, downiness; 2. *bot.* indumentum
  ~ початка кукурузы silking of maize
опушеннозубчатый *bot.* downy-toothed
опушенноплодный *bot.* downy-fruited
опушенностолбиковый *bot.* ptilostylous
опушенность *bot.* pubescence
опушенноцветковый *bot.* downy-flowered, pubiflorous
опушенный pubescent, tomentose, tomentous, nappy
  ~, коротко puberulent
  ~ мехом fur-trimmed
  ~ ресницами ciliated
опушка I trimming, edging
опушка II edge, skirt *(of forest)*
  ~ леса forest margin
опущение 1. *med.* ptosis, falling; prolapsus; 2. lowering, letting down; 3. omission
  ~ века, хорнеровское ptosis sympathetica, Horner's syndrome
  ~ верхнего века blepharoptosis
  ~ влагалища colpoptosis
  ~ внутренних органов visceral ptosis, splanchnoptosia
  ~ диафрагмы phrenoptosis
  ~ желудка gastroptosis
  ~ кишечника enteroptosis
  ~, ложное pseudoptosis, false ptosis
  ~ матки falling of the womb
  ~ толстых кишок coloptosis
  ~ яичка descent of the testis
опущенный 1. *med.* ptosed, descended, fallen; 2. missing, omitted
опцион, опция option
опыление 1. *bot.* pollination; 2. dusting; spraying
  ~ ветром anemophily
  ~, внутрисортовое/межсортовое intra-varietal pollination
  ~ насекомами entomophily, insect pollination
  ~, перекрестное cross-pollination
  ~ птицами ornithophily
  ~, родственное close pollination
опыливание *agr.* dusting
  ~, авиационное aerial dusting
  ~ посевов crop dusting
  ~, увлажненное wet dusting
опыливатель *agr.* (crop) duster, dusting machine/

apparatus, powder blower
~, **самолетный** airplane duster
опылитель 1. *bot.* pollinator, pollinizer; 2. *agr.* insecticide dust sprayer
опыляемый pollinated
~ **ветром** anemophilous, wind-pollinated
~ **животными** zoidiophilous, zoophilous
~ **насекомыми** entomophilous, entomophilic
~ **птицами** ornithophilous, bird-pollinated
~ **улитками** malacophilous
опылять I *bot.* pollinate
~, **перекрестно** cross-pollinate
опылять II *agr.* spray with insecticide dust
опыт 1. experiment; test, trial; assay; 2. experience, practice
~, **административный** administrative experience/background
~ **без разрушения образца** non-destructive experiment/test
~, **боевой** fighting/combat/battle experience
~ **в лабораторных условиях** laboratory experiment
~ **вне реактора** out-of-pile experiment
~, **дополнительный** *stat.* replicate experiment
~, **достоверный** significant experiment
~, **контрольный** check experiment
~ **кренования** *naut.* heeling/inclining experiment/test
~, **критический** critical test
~, **лабораторный** laboratory trial
~, **летный** in-flight experience
~, **натурный** full-scale experiment
~, **параллельный** replica test
~, **показательный** model experiment
~, **полевой** field experiment/test
~, **полномасштабный** full-scale experiment
~, **полупромышленный** pilot experiment/trial
~, **проводимый** experiment in progress
~ **работы** operating/operational/service experience, operational/service background
~ **эксплуатации** operating/operational experience, in-service experience
опытник experimenter
опытно-конструкторский developmental
опытно-показательный experimental-demonstrative
опытность experience; proficiency
опытный 1. expert, experienced, competent; 2. experimental; developmental; tentative; test *(attr.)*
~ **самолет** prototype
~**но-промышленная установка** pilot plant
опьянение intoxication, inebriation, drunkenness
~, **алкогольное** alcoholic intoxication, alcoholism
~, **кокаиновое** cocaine inebriation
~, **мескалиновое** mescaline inebriation, mescalism
~, **патологическое** pathologic intoxication
опьянять intoxicate, inebriate, make drunk
орабочение proletarianization
орабочивать proletarianize
оракул oracle
оракульский oracular, oraculous
орангутанг *zool.* orangutan, orangoutang *(Pongo pygmaeus, Simia satyrus)*
оранжада orangeade
оранжерея greenhouse, glasshouse
оранжист *polit.* Orangenean
оранжит *miner.* orangite
орарь *eccl.* maniple, orarion
оратор (public) speaker, orator; spokesman, mouthpiece
оратория *mus.* oratorio; *eccl.* oratory
ораторный *mus.* oratorial
ораторский oratorical, oratorial
орбикулярный orbicular, spheroidal
орбита 1. orbit, orbital path; 2. *anat.* eye socket, orbital cavity, orbita
~, **боровская** Bohr orbit
~, **валентная** valence orbit
~, **возмущенная** perturbed orbit
~ **встречи** rendezvous orbit
~, **заданная** scheduled/planned/desired orbit
~, **замкнутая** closed orbit
~ **захвата** capture orbit
~, **кеплеровская** Keplerian/inertial orbit
~, **межпланетная** interplanetary orbit
~, **незаполненная** uncompleted orbit
~, **одночастичная** single-particle orbit
~ **ожидания/парковочная** parking/staging/waiting orbit
~, **околоземная** near-earth orbit
~, **окололунная** circumlunar orbit, near-moon orbit
~, **околополярная** near-polar orbit
~, **переходная** transfer orbit
~, **промежуточная** intermediate/parking/staging orbit
~ **сближения** rendezvous orbit
~, **случайная** random orbit
~, **спиральная** spiral(ling) orbit
~ **стыковки** docking orbit
орбиталь orbital
~, **несвязывающая** nonbonding orbital
орбитально-затылочный *anat.* orbitoccipital
орбитально-носовой *anat.* orbitonasal
орбитальный orbital
орбитирование orbiting
оргазм *physiol.* orgasm
орган 1. *biol.* organ; 2. member, element; device, instrument; 3. agency, body; 4. publication, organ; *see also* **органы**
~ **вкуса** gustatory organ, organ of taste
~ **власти** authority
~ **государственной власти** government agency, public authority
~, **дроссельный** constrictor, restrictor
~ **дыхания** respiratory organ
~, **заделывающий** *agr.* covering device
~, **законодательный** legistature

~, **исполнительный** 1. *tech.* effector; actuator (device), servo element; 2. excutive organ/agency
~, **контролирующий** monitoring element
~, **кортиев** *anat.* Corti's/spiral organ
~, **мочевой** uropoietic organ
~, **наблюдательный** *law* supervisory authority
~, **надземный** aboveground organ
~ **обоняния** olfactory organ, organ of smell
~ **пищеварения** digestive organ
~, **постоянный** *law* standing body
~, **рабочий** *tech.* tool
~, **регулировочный** control
~, **рудиментарный** rudimentory/vestigial organ
~, **сошниково-носовой** Jacobson's organ
~, **спиральный** spiral/Corti's organ
~ **управления** (operating) control
~ **чувств(а)** sense organ
~, **эмалевый** *anat.* enamel organ
орган *mus.* pipe-/barrel-organ
органди *text.* organdy, organdie, fine muslin
органелла *biol.* organella, organelle, organoid
организация organization; society; agency
~ **американских государств** Organization of American States, OAS
~, **байтовая** byte-wide organization
~, **бесприбыльная** nonprofit organization
~, **благотворительная** charity organization
~, **Всемирная метеорологическая** World Meteorological Organization
~, **Всемирная океанографическая** World Oceanographic Organization
~ **европейского экономического сотрудничества** Organization of European Economic Cooperation, OEEC
~, **закупочная** procurement agency
~ **запоминающего устройства** word-organized memory
~ **здравоохранения, Всемирная** World Health Organization, WHO
~, **конвейерная** conveyer/pipeline organization
~ **лесопольного хозяйства** forest field management
~ **ловушек** *comp.* trapping
~ **материально-технического обеспечения** logistic support organization
~, **Межправительственная морская консультативная** Intergovernmental Maritime Consultative Organization, IMCO
~ **многозадачного режима** *comp.* cotasking
~, **неавтономная** *tech.* on-line organization
~, **некоммерческая** nonprofit organization
~, **неправительственная** nongovernmental organization
~, **неформальная** informal organization
~ **Объединенных Наций, ООН** United Nations Organization, UNO
~ **Объединенных Наций по вопросами просвещения, науки и культуры, ЮНЕСКО** United Nations' Educational, Scientific and Cultural Organization, UNESCO, Unesco
~ **Объединенных Наций по промышленному развитию, ЮНИДО** United Nations' Industrial Development Organization, UNIDO
~ **Объединенных Наций, продовольственная и сельскохозяйственная, ФАО** Food and Agricultural Organization, FAO
~ **окон** *comp.* windowing
~ **очередей** *comp.* quen(e)ing
~ **памяти, (по)разрядная** bit-organized memory
~ **памяти, последовательная** logical/control/sequential storage organization
~ **памяти, трехмерная** three-coordinate memory
~ **пастбищного хозяйства** *agr.* grassland/pasture management
~ **по стандартизации, Международная** International Standardization Organization, ISO
~, **подрядная** contracting organization
~ **полетов** flight operations organization
~ **полиэкранной работы дисплея** multiwindowing
~, **правительственная** government agency
~ **предпринимателей, Международная** International Organization of Employers
~ **производства** production/industrial engineering
~ **прохождения задач** *comp.* tasking
~ **рандеву** *comp.* rendezvousing
~ **регионального сотрудничества в целях развития** Organization of Regional Cooperation for Development, ORCD
~ **сети ЭВМ** computer networking
~ **стран-экспортеров нефти, ОПЕК** Organization of Petroleum Exporting Countries, OPEC
~ **труда** human engineering; job management
~ **труда, Международная** International Labo(u)r Organization, ILO
~ **труда, научная** industrial engineering
~ **тыла** logistics
~ **файла** filing
~ **фоновой обработки** *comp.* backgrounding
~, **хозрасчетная** self-sustained organization
~ **циклов** *comp.* looping
~ **экономического сотрудничества и развития ОЭСР** Organization for Economic Cooperation and Development, OECD
~, **эксплуатирующая** operator
~ **ядерных исследований, Европейская** European Organization for Nuclear Resarch
организация-исполнитель (договорной работы) contractor
организация-разработчик development agency
организм organism, body
~, **аэробный растительный** aerophyte
~, **бентонный** benthos organism
~, **болезнетворный** causative organism
~, **высший** higher organism
~, **донный** botton-dwelling organism

~, земной terrestrial organism
~, многоклеточный metaphyte; metazoon
~, низший lower organism
~, одноклеточный/одноячеистый unicellular organism, monad, unicell, protist
~, породообразующий rock-forming/-building organism
~, простейший растительный protophyte
~ хозяина host
~, эврибатный eurybathic organism
~, эвригалинный euryhaline organism
~, эйрифотический euryphotic organism
~ы, криофильные crymophilic/cryophilic organisms
~ы, стенотермные stenothermic organisms
~ы, термофильные thermophilic organisms
~ы, эвритермные eurythermic organisms
организованность orderliness; organization
организованный 1. organized, managed, arranged; 2. orderly
организовать organize; arrange
органик specialist in organic chemistry
органика 1. organic chemistry; 2. *agr.* organic fertilizer
органист organist; organ player
органический, органичный organic
органоген *chem.* energy element, organogen
органогенез, органогения *biol.* organogeny, organogenesis
органогенный organogen(et)ic, organogenous
органограф(ист) *biol.* organographist
органографически organographic(al)
органоид *zool.* organoid, organelle
органолит *geol.* organolith, organolite, biolith, organic rock
органо-минеральный organomineral
органон *philos.* organon
органопластика organoplasty
органотерапия organotherapy
~, ингибиторная inhibitory organotherapy
~, стимулирующая stimulation organotherapy
органсин organzine
органум *mus.* organum
органчик 1. *zool.* organ-pipe coral *(Tubipora);* 2. musical box; 3. small organ
органы *pl.* control devices; *see also* **орган**
~, аэродинамические aerodynamic controls, aerodynamic control surfaces
~ глаза, вспомогательные *anat.* accessory organs of the eye
~ дыхания respiratory system
~, исполнительные executing agencies
~ кроветворения h(a)emopoietic organs
~ кровообращения circulation organs
~ межправительственные intergovernmental bodies
~, мочевые uropoietic organs
~ пищеварения digestive system
~, половые genital organs
~ управления controls
~ управления летательным аппаратом flight controls
~ устойчивости stabilizing surfaces
оргбюро *abbr.* **организационное бюро** organization bureau
оргвывод *abbr.* **организационный вывод** practical conclusion
оргиастический orgiastic(al)
оргия orgy; debouch(ery)
оргкомитет *abbr.* **организационный комитет** organizing committee
оргработа *abbr.* **организационная работа** organizational work
оргстекло acrylic plastic
оргтехника clerical/managerial aids, office equipment
орда horde; crowd, band
~, Золотая the Golden Horde
ордалия *hist.* ordeal
орден I medal, decoration
~ славы первой степени Medal of Glory, first grade
орден II order, religious/knightly fraternity
~ францисканцев the Franciscan Order
орденоносец medal bearer
ордер warrant, order; *law* writ
~, боевой fighting formation/order
~, грузовой broker's order
~ на арест warrant of arrest
~ на ботинки shoe coupon
~ на задержание general warrant
~ на погрузку shipping note
~, погрузочный *naut.* loading/broker's order, shipping note
~, походный cruising order/disposition/formation
~, фрахтовый *naut.* chartering order
ординализм theory of ordinal utility
ординар zero-water level, normal level
ординарец *mil.* orderly
ордината *math.* ordinate, y-coordinate, y-axis
~, плазовая *naut.* offset
~, промежуточная half-spaced ordinate
ординатор *med.* staff/hospital physician, intern
ординаторская *n.* staff room in a hospital
ординатура 1. appointment as intern, internship; permanent appointment; 2. clinical studies
ординация ordination
ордовик *geol.* Ordovician system
ордовикский Ordovician
ордонанс *hist.* ordonnance
ордынец member of the Golden Horde
ореада *myth.* mountain nymph, oread
оребрение finning, fins, ribbing
оребренный ribbed, finned
орел *zool.* eagle *(Aquila)*
ореодонт *paleont.* oreodont *(Oreodontida)*
ореол 1. halo, halation, aureole, corona, nimbus; 2. *geol.* reaction rim, envelope
~, газовый gas aureole
~ рассеяния dispersion halo; dissemination

nimbus
~, **цветной** pleochroic halo
ореолообразование *cin.* halation
ореопитек *paleont. Oreopithecus*
орех 1. *bot.* nut tree, walnut *(Juglans);* 2. nut; 3. *timb.* walnut *(wood)*
~, **американский** Brazil nut *(Betholletia excelsa)*
~, **водяной** water chestnut *(Trapa natans)*
~, **грецкий** English/European walnut *(Juglans regia)*
~, **земляной** peanut, groundnut *(Arachis hypogaea)*
~, **кедровый** pine nut
~ **кола** cola nut
~, **обратнояйцевидный** obovoid nut
~, **разделать под** give a walnut finish
ореховидка *zool.* nuculid *(Nucula)*
ореховидный nuciform, nut-shaped
ореховка *zool.* nutcracker *(Nucifraga caryocatactes)*
ореховый hazel-colo(u)red, avellaneous; nut *(attr.); n. pl.* walnut family *(Juglandaceae)*
~ **лес** hazel coppice/grove
орехоплодный *bot.* nut-bearing, caryocarpous
орешек nutlet, nucule
~, **дубильный/чернильный** oak nut/gall, nutgall
~, **крылатый/парусный** wing nut
орешковидный nutlet-shaped, nucamentaceous
орешник 1. *bot.* hazel(nut) *(Corylus avellana);* 2. hazel grove/coppice, filbert orchard
орешниколистный *bot.* corylifolious
ориби *zool.* oribi *(Ourebia ourebi)*
оригинал original (copy); subject/master copy; type/authentic specimen
~ **дифракционной решетки** *cryst.* master grating
~ **карты, полевой** field/base sheet, map manuscript, surveyor's/basic plate
~ **матрицы** master
~, **тоновый** continuous-tone original/copy
~, **цветоделенный** colo(u)r-separated copy
~, **черно-белый** black-and-white copy
оригинал-макет *print.* make-up page
оригиналодержатель *print.* copy-holder
оригинатор breeder, originator
ориенталистика oriental studies
ориентатор *av.* orientation device, attitude control unit
~, **астрономический** stellar/star tracker, astrotracker
~, **гироинерциальный** gyroscopic attitude control unit/device
~, **оптический** optical orientation instrument
~, **солнечный** astrotracker, sunseeker
ориентация *math.* orientation; *av.* attitude control; *naut.* attitude manoeuvre
~, **астрономическая** astroorientation
~ **в пространстве** *med.* spatial orientation
~ **КЛА** spacecraft attitude control
~ **напластования** *geol.* attitude of bedding
~, **обратная** inverse/negative/reverse orientation
~ **по небесному своду** celestial orientation
~, **преимущественная** preferred orientation
~, **пространственная** spatial orientation
~, **профессиональная** occupation guidance
~, **прямая** direct/positive orientation
~ **радиолокационного изображения** radar display stabilization
~ **радиолокационного изображения по азимуту** azimuth stabilization
~, **ручная** manual attitude control
~ **спина** spin orientation/direction
~ **частей электрического соединения** connector polarization
ориентир 1. reference/orientation point; prominent featured landmark; 2. orientator
~, **видимый** visible mark
~, **визуальный** *av.* visual reference
~, **внешний** external fixture
~, **главный топографический** major datum point
~, **контрольный** check reference (point)
~, **наземный** *av.* (prominent) landmark, terrestrial reference
~, **ночной** nightmark
~, **основной** main reference point
~, **световой** illuminated reference (point), light marker
~, **точечный** pin/reference point
ориентирование orientation, orienting; guidance
~, **внешнее** exterior/outer/absolute orientation
~ **колеса шасси** *av.* wheel castoring
~ **на местности** terrain orientation
~ **планшета** orientation of plane table
~ **по географическому/истинному меридиану** true orientation
~ **по странам света** cardinal points reference
~ **сопоставлением** relational guidance
~, **стереоскопическое** spatial orientation
~ **теодолита** orientation of transit
ориентировать orient(ate); acquaint (with); direct
ориентироваться orient oneself
~ **на местности** get/take one's bearings in the locality
ориентировка arrangement, orientation
~, **линейная** lineation
~, **незакономерная/случайная** random orientation
ориентировочный 1. reference *(attr.);* position-finding; 2. approximate, tentative
ориктоценоз *miner.* oryctocoenosis
ориньяк *hist.* Aurignacian
Орион *astr.* Orion
орифламма *hist.* oriflamme, royal standard of France
орицикл *math.* horocycle, oricycle
оркан *meteor.* tropical cyclone

оркестр 1. orchestra, band; 2. orchestra pit
~, духовой brass band
~, струнный string orchestra
оркестрант member of an orchestra/band
оркестрование, оркестровка orchestration
оркестровый orchestral
оркнейский Orcadian, of Orkney *(Islands)*
орлан *zool.* sea eagle *(Haliaetus)*
орлан-белохвост *zool.* erne, white-tailed (sea) eagle *(Haliaetus albicilla)*
орлеан *bot.* achiote, anatto tree *(Bixa orellana)*
орлеанист *hist.* Orleanist
орлец 1. *miner.* manganese spar, rhodonite; 2. *eccl.* round hassock/carpet
орлиный aquiline
орлон *text.* orlon
орляк *bot.* bracken *(Pteridium)*
Ормузд *rel.* Ormuzd, Ormazd; the good principle
орнаментика *mus.* ornament, grace
орнаментирование, орнаментировка, орнаментовка ornamenting, ornamentation
орнид *med.* bretylium tosylate, darenthin
орнитин *bioch.* ornithine
орнитогалум *bot.* star-of-Bethlehem *(Ornithogalum)*
орнитоз *vet.* ornithosis
орнитология ornithology
орнитомимус *paleont.* ornithomimus
орнитоптер *av.* ornithopter, orthopter
орниторинхус *zool.* duckbill, ornithorhynchus *(Ornithorhynchus)*
орнитофауна ornithologic fauna, ornithofauna
орнитофил *bot.* ornithophilous plant, bird-pollinated plant, ornithophile
орнитофилия *bot.* bird pollination; ornithophily, ornithogamy
орнитофильный *bot.* ornithophilous
~ное опыление ornithophilous/bird pollination
~ное растение ornithophilous plant, ornithophile
орнитохория *bot.* ornithochory
орноит *geol.* ornoite
орогенез(ис), орогения orogenesis, orogeny
орогенический orogen(et)ic
орогеосинклиналь orogeosyncline
ороговевать keratinize, become horny
ороговевший *med.* keratosic, cornified, horny, horned
ороговение *med.* keratinization, cornification, hornification, keratosis
~, патологическое hyperkeratinization
~, чрезмерное hyperkeratosis
орография orography
ороклин orocline
оронтиум *bot.* golden club *(Orontium)*
оросинклиналь orosyncline
ороситель 1. sprinkler; irrigation ditch/channel; 2. irrigation specialist
~, линейный spray-line irrigator
оросительный 1. irrigating, irrigatory, irrigative; 2. bedewing
~ канал irrigation canal
орошаемый irrigable; irrigated
~мое земледелие irrigation agriculture
орошать 1. *agr.* irrigate; 2. sprinkle, spray
орошение irrigation; bedewing
~ дождеванием sprinkling, sprinkler irrigation
~, капельное drip irrigation
~, лиманное flood/inundative irrigation
~, подъемное overhead irrigation
~ сточными водами sewage/wastes irrigation
~, чековое check irrigation
орпингтон *agr.* Orpington
орс *abbr., obs.* **отдел рабочего снабжения** worker's supply department
орсель archil, orchel(la), orchil(la)
орсин *chem.* orcin(e), orcinol
орт 1. *math.* basis/unit vector; 2. *min.* breakthrough, crosscut, cross/drift/heading/entry
~, выпускной drawpoint crosscut
ортикон *telev.* (image) orthicon
ортит *miner.* orthite, allanite
ортоамитоз *med.* orthoamitosis
ортоводород *chem.* orthohydrogen
ортогелий *chem.* orthohelium
ортогенез(ис), ортогения *biol.* orthogenesis
ортогенный orthogenic, orthogene
ортогеосинклиналь *geol.* orthogeocyncline
ортогнатический *anat.* orthognathic, orthognathous
ортогнатия *anat.* orthognathia, orthognathy
ортогнейс *geol.* orthogneiss
ортогонализация *math.* orthogonalization
ортогонально-ассоциированный *math.* orthogonally associated
ортогональность *math.* orthogonality
ортогональный *math.* right-angled, orthogonal
ортодактилия *med.* orthodactylia
ортодиаграфия *med.* orthodiagraphy
ортодейтерий orthodeuterium
ортодихлорбензол *chem.* orthodichlorobenzene
ортодокс conformist, orthodox
ортодоксальность, ортодоксия orthodoxy
ортодоксальный orthodox(al)
ортодонтия orthodontia, orthodontics
ортодополнение *math.* orthocomplement
ортодромия *geod.* orthodrome, orthodromic line
ортокератоз *med.* orthokeratosis
ортокислота *chem.* orthoacid
ортоклаз *miner.* orthoclase
~, бариевый barium orthoclase, hyalophane
~, натриевый soda orthoclase
~, прозрачный adular(ia)
~, стекловатый glassy felspar
ортокремниевый orthosilicic
ортолит ortholith, ortholite
ортомиксовирус orthomyxovirus
ортонормализация, ортонормирование *math.* orthonormalization
ортонормированный orthonormal
ортоось orthoaxis, orthodiagonal

ортопед(ист) orthop(a)edist, orthop(a)edic surgeon
ортопедия *med.* orthop(a)edy, orthop(a)edics, orthomorphia
ортоперкуссия orthopercussion
ортоплазия *biol.* organic selection, orthoplasy
ортопластиночка ortholamella
ортоплоидия *med.* orthoploidy
ортопноэ *med.* orthopn(o)ea
ортоповерхность orthosurface
ортопозитроний orthopositronium
ортопорода orthorock
ортопраксия *med.* orthopraxia, orthopraxy; *rel.* orthopraxis, orthodox praxis in religion
ортрорадиоскопия orthoradioscopy
ортоpомбический (ortho)rhombic, prismatic
ортосимметрический, ортосимметричный orthosymmetric(al)
ортоскоп 1. *med.* orthoscope; 2. *mil.* aiming device, aim corrector
ортосланец *geol.* orthoschist, magnetic schist
ортосоединение orthocompound
ортостернальный orthosternal, orthosternous
ортостиль *arch.* orthostyle
ортостиха, ортостихия *bot.* orthostichy
ортостиховый orthostichous
ортотеломитический orthotelomitic
ортотропный *bot.* orthotropous, orthotropic, orthotropal
ортотрубка orthotubule
ортофир *geol.* orthophyre, orthoclase porphyry
ортофория *med.* orthophoria
~, астеническая asthemic orthophoria
ортофотический orthophotic
ортофотоплан orthophotomap
ортофотоснимок orthophoto(graph)
ортохроматический orthochromatic, isochromatic
ортоцентр *math.* orthocentre; *Am.* orthocenter
ортоцефалия orthocephaly, orthocephalism
ортоэлювий orthoeluvium
ортштейн *geol.* ortstein, hardpan
ОРУД *abbr.* **отдел регулирования уличного движения** street traffic control department
оруденелость *geol.* mineralization
оруденение *geol.* mineralization, metallization
~, рассеянное dispersion mineralization
орудие 1. implement, tool; instrument; equipment; 2. gun, cannon
~, артиллерийское artillery piece
~, башенное turret gun
~ в сборе gun assembly
~, вспомогательного калибра minor-calibre gun
~, дисковое *agr.* disker, disk implement
~, дренажное draining implement
~, зенитное anti-aircraft gun
~, каменное stone implement
~, кормовое after gun
~, многоцелевое all-purpose gun
~, неолитическое neolith
~, отвальное mouldboard implement
~, палеолитическое paleolith
~, почвообрабатывающее tillage/tilling implement
~, прицепное trailed/drawn implement
~ производства instrument/implement of production
~, рыболовное fishing gear/tackle
~, сельскохозяйственное farm/agricultural implement
~, сигнальное warning/signal gun
~, тяжелое heavy artillery piece, large gun
~, широкозахватное wide-cut implement, broad-gang implement
~дия древнего человека artifacts, artefacts
~дия, ремесленные handicraftman's implements
~дия труда tools
орудийный tool *(attr.)*; gun *(attr.)*
~ расчет gun crew
~ падеж *gram.* instrumental (case)
оруднение *geol.* mineralization
оружейник gunsmith, gunmaker; armo(u)rer
оруженосец *hist.* sword/armo(u)r-bearer
оружие weapon(s), arm(s), armament(s)
~, авиационное aircraft weapon; aerial warfare weapon
~, воздушно-космическое aerospace weapon
~ для уничтожения спутников antisatellite/countersatellite weapon
~ класса воздух-воздух air-to-air weapon
~ класса воздух-поверхность air-to-surface weapon
~, космическое cosmic/space weapon
~, личное/ручное hand arm(s)
~, магазинное repeating firearm
~ массового поражения (mass) destruction weapon
~, морское naval ordnance/weapons/armament
~, наступательное offensive weapon
~, оборонительное defensive weapon
~, обычное conventional weapon
~, огнестрельное firearm
~, ракетное missile weaponry, rocket weapon
~, ракетно-ядерное nuclear-missile weapon
~, стрелковое small arms
~, тактическое ядерное tactical nuclear weapon
~, термоядерное thermonuclear weapon
~, управляемое guided weapon
~, холодное cold steel
~, ядерное nuclear weapon
орфа *zool.* orf(e) *(Leuciscus idus)*
орфарион, орфеорион *mus.* orpharion, orpheorion
Орфей *myth.* Orpheus
орфизм *art.* Orphism
орфический Orphic
орфограмма orthogram
орфография spelling, orthography

орфотон orthotone, orthotonesis
орфоэпический orthoepic(al)
орфоэпия correct pronunciation, orthoepy
орхестра *hist.* orchestra
орхидалгия *med.* orchi(d)algia
орхидейный *bot.* orchidaceous, orchidean
орхидея *bot.* orchid *(Orchis)*
орхидный *bot.* orchidaceous
орхидопексия *med.* orchidopexy
орхидорафия *med.* orchi(d)orrhaphy
орхидэктомия *surg.* orchi(d)ectomy, orchectomy
орхит *med.* orchitis
~, гнойный purulent orchitis
~, острый acute orchitis
орхотомия *surg.* orch(i)otomy
орцин *chem.* orcin(e), orcinol
оршад *cul.* orgeat
оса *zool.* wasp *(Vespa)*
осада siege
осадитель *chem.* precipitator, precipitant, precipitating agent, settler
осадка 1. settling, settlement; set; sag(ging); sinking; subsidence; 2. *naut.* draught, draft; 3. stopping, halting; 4. *met.* upset(ting), swaging
~, большая deep draught
~, габаритная navigational draught
~, гидравлическая hydraulic pressurizing
~, грузовая loaded draught
~, зимняя winter draught
~ колоши *met.* charge slip
~ кормой aft(er)/stern draught
~, предельная limiting draught
~ при миделе midship draught, mid-length draught
~ при ходе на крыльях foilborne draught
~ рессоры spring camber
~ судна draught
~, теоретическая mo(u)ld(ed) draught
~ шихты burden slip
~, эксплуатационная service draught
осадки *pl. meteor.* precipitation, rainfall; sediments, deposits
~, водные water-borne sediments
~, глинистые argillaceous sediments
~, годовые annual precipitation
~, грубые coarse sediments/detritus
~, дождевые/жидкие rain precipitation, rainfall
~, древние old/ancient sediments
~, известковые calcareous sediments
~, изотопные isotopic deposits
~, комковатые lumpy sediments
~, ледниково-морские glacio-marine sediments
~, ледниковые glacial deposits/drift, ice-borne sediments
~, ливневые shower-type precipitation, showers
~, материнские source sediments
~, моросящие drizzle(-type precipitation)
~, навеянные winnowed residue
~, обложные widespread precipitation
~, радиоактивные (radioactive) fallout
~, слоистые bedded/stratified sediments
~, хлопьевидные flocculated sediments
~, эоловые (a)eolian sediments, wind-borne sediments
осадкомер 1. *naut.* draught gauge; 2. precipitation/rain/snow gauge, hyetometer
осадконакопление sedimentation, sludging
осадкообразование sedimentation, sludging; *geol.* lithogenesis
~, морское marine sedimentation
осадный obsidional
~ное положение state of siege
осадок sediment, precipitate, precipitation, settling; residue, sludge, deposit
~, аккумуляторный battery mud
~, глобальный радиоактивный global fallout
~, донный bottoms, bottom sediment
~ из уратов urate deposit
~, мочевой urinary sediment
~ на фильтре (filter) cake
~, отфильтрованный strained sediment
~ при титровании flocks
~, пригорелый burnt deposit
~, промытый washed precipitate
~, пушистый fluffy precipitate
~, радиоактивный radioactive precipitate/fallout/deposit; fallout
~, студенистый gelatinous/jellylike precipitate
осадочный 1. precipitating, sedimentary; *med.* hypostatic; 2. upsetting
осаждаемость sedimentation capacity; precipitability
осаждать I besiege, lay siege to; beset, beleaguer
осаждать II precipitate, settle, deposit
осаждаться deposit, settle, fall
осаждение deposit(ion), sediment(ation), settling, precipitation, lodging
~, вторичное reprecipitation
~, гравитационное sedimentation by gravity
~, дробное fractional precipitation
~ кристаллов crystal settling
~ отстаиванием settling, sedimentation
~, совместное/одновременное coprecipitation
~ центрифугированием spinning-down
осажденный I besieged, beleaguered
осажденный II deposited, precipitated, settled
осаживание *met.* 1. swaging; upsetting; 2. clenching, clinching
осаживать I precipitate, deposit, settle
осаживать II 1. force back; check, halt; 2. *met.* upset; swage out; clinch, clench; 3. move/pull down; 4. cause to settle/fall/subside
осаживаться subside, settle, form a sediment
осанка posture, attitude
осанна hosanna, praise to God
осахаривание saccharification
осахариватель hydrolyser

~ **крахмала** starch converter
осваивать 1. *agr.* bring under cultivation; *bot.* acclimatize; 2. master, cope with; 3. *tech.* bring in, introduce
~ **земли** reclaim land; bring new land under cultivation
~ **целину** develop virgin land
осведомитель informant, informer, intelligencer; stool/decoy pigeon, snitch
осведомление informing; information, notification
освежать refresh, freshen up
~ **знания** *educ.* brush up one's knowledge
освежение freshening (up)
~ **дернины** turf reconditioning
освежитель freshener
осветитель lighter, illuminator; illuminant
~ **лазера** (laser) pump
~, **лобный** frontal reflector
~, **фототелеграфный** exciter lamp of facimile-transmitter scanner
осветление clarification; clearing; bleaching
осветлитель clarifying agent, clarifier
осветлять 1. clarify *(liquids),* defecate; 2. illuminate, light; 3. *for.* thin out
освечивание *phot.* candela-second
освещать light (up), illuminate; expose to light; *fig.* elucidate, throw light on
освещение illumination, light(ing); *fig.* elucidation
~, **аварийное** emergency lighting
~, **безбликовое** glare-free illumination
~, **боковое** side/lateral/oblique illumination
~, **верхнее** overhead lighting
~, **внутреннее** indoor illumination, interior lighting
~ **взлетно-посадочной полосы** runway lighting
~, **встроенное** integral lighting
~, **дежурное** emergency lighting
~, **дневное** daylight illumination
~ **камеры Вильсона** cloud-chamber illumination
~, **контровое** *cin.* back lighting
~, **люминесцентное** fluorescent lighting
~, **маскировочное** camouflage lighting
~, **наружное** outdoor/exterior illumination/lighting
~, **неактиничное** *cin.* safe light
~, **недостаточное** underillumination
~, **полное дневное** broad daylight
~ **приборной доски** panel lighting
~, **прожекторное** floodlight
~, **прямое** direct/vertical illumination
~, **рекламное** display lighting
~, **сквозное** trough illumination, transillumination
~ **солнечными лучами** insolation, solarization
~, **центральное** central/axial illumination
~ **шкалы** dial lighting
освещенность illuminance, illumination (intensity); optical exposure
освидетельствование examination, inspection, survey
~, **первичное** original examination
~ **повреждений** damage survey
~, **повторное** reexamination
освидетельствовать examine, survey
освинцо(вы)вание lead coating/plating/lining
освинцовывать lead(-plate), cover/coat with lead
освободительный liberating, emancipating, emancipative
~**ное движение** liberation movement
освобождать 1. release, set free, liberate; *law* set at liberty; absolve; affranchise; emancipate; 2. dismiss, relieve, exonerate; exempt; 3. clear (out), empty; vacate
~ **на поруки** release on bail
~ **от военной службы** exempt from military service
~ **от дежурства** excuse from duty
~ **от рабства** release from slavery, manumit
~ **под честное слово** release on parole
освобождаться *math.* eliminate, clear, get rid of
~ **дробей** clear fractions
~ **от иррациональности** rationalize
освобождение 1. liberation, deliverance, emancipation; release; 2. disengagement, unblocking; 3. vacation, clearing; 4. dismissal; exemption
~ **крестьян** liberation of peasants
~ **от налога** immunity from taxes
~ **от наказания** absolution
~ **от оболочки** de-canning/-jacketing/-cladding
~ **от обязанности** release from obligation
~ **от обязательств** *law* waiver
~ **от рабства** manumission, liberation of slaves
~ **от таможенных пошлин** exemption from customs duty
~ **от уплаты налога** tax exemption
освобожденный released, freed, liberated; exempt
освод *abbr.* **Общество содействия развитию водного транспорта и охраны жизни людей на водных путях** Society for Development of Water Transport and Passenger Safety
освоение assimilation, familiarization, mastery; development
~ **земель** land development
~ **космического пространства** development of space research
освящение *rel.* consecration, blessing, sanctifying
освященный blessed, consecrated
~ **временем/давностью** time-hono(u)red
осеверение *agr.* northering
осевой axial; *bot.* axile
оседание 1. precipitation, sedimentation, deposition, settling; 2. *geol.* sinking, subsidence, sag(ging), settlement
~ **горных пород** subsidence of rocks
~ **грунта** land subsidence
~, **котлообразное** cauldron subsidence
~ **лейкоцитов/эритроцитов** leucocyte/eryth-

rocyte sedimentation
~ **радиоактивного пепла** radioactive ash fall-out
~, **стесненное** crowded precipitation
оседать 1. settle down; sediment; deposit, lodge (on); sink, sag, yield; 2. settle *(of people)*
оседлость settled way of life; *hist.* the Jewish Pale
оседлывать saddle; straddle
осел *zool.* donkey, ass *(Equus asinus)*
оселедец *hist.* topknot, long tuft of hair on shaven head
оселок whetstone, hone, oilstone; touchstone
осеменение (in)semination
~, **искусственное** artificial insemination
осенник *bot.* meadow saffron, common autumn crocus *(Colchicum autumnale)*
осенчук *bot.* bluestem, broom sedge *(Andropogon)*
осень fall, autumn
~, **золотая** Indian summer
осесимметрический, осесимметричный axiosymmetrical, axisymmetric(al)
осетин(ец) Osset(e)
осетинка Osset(e) (woman)
осетинский Ossetic
осетр *zool.* sturgeon *(Acipenser)*
осетровые *n. pl. zool.* acipenserids *(Acipenseridae)*
осецентробежный axial-centrifugal
осецилиндрический axiocylindrical
осечка misfire
оси *pl.* axes; *see also* **ось**
~ **вращения самолета** *av.* body axes, axes of aircraft
~ **деформации, главные** principal axes of strain
~ **инерции, центральные** central axes of inertia
~ **координат** axes of coordinates, reference axes
~, **полусвязанные** *av.* wind-body axes, stability axes
~, **поточные** wind axes
~, **связанные** body axes
оси, державы *hist.* Axis Powers
осина *bot.* (European) aspen, trembling poplar *(Populus tremula)*
осиновик *bot.* aspen mushroom, orange-cap boletus *(Boletus vesipellis)*
осиный vespine
Осирис *myth.* Osiris
осквернение *rel.* profanation, desecration; defilement
осквернитель *rel.* profaner, desecrator; defiler
осквернять desecrate, profane; defile
осколкообразование splitting
осколок fragment, splinter, chip
~ **деления** *nucl.* fission fragment
~ **отдачи** recoil fragment
оскольчатый splintered
~ **перелом** splintered/comminuted fracture
оскомина *med.* soreness/drawing of the mouth; nausea
оскопление castration, emasculation
оскоплять castrate, emasculate
оскорбительный offensive, offending, abusive, insulting; outrageous
оскорбление offence; *Am.* offense; insult; opprobium, outrage; insulting, offending
~ **величества** *law* lese majesty
~ **действием** *law* assault and battery
~, **словесное** *law* assault
~ **суда** criminal contempt, contempt of court
оскорблять insult, offend, injure; hurt, humiliate
оскудевать become impoverished; grow scarce; fall into decline
оскудение impoverishment
оскула, оскулюм *zool.* oscule, osculum
оскуляторный, оскуляционный osculatory
ослабевание weakening, lessening, slackening; abatement
ослабелый enfeebled; atonic
ослабитель 1. *el.* attenuator; attenuation network; 2. *phot.* optical wedge; clearing agent
~, **поглощающий** absorbing attenuator
~, **предельный** cut-off attenuator
~ **света** light attenuator
~, **согласованный** matched attenuator
~, **фотографический** reducer
ослабление 1. *el.* attenuation; 2. *print.* reduction, clearing; 3. *mech.* slackening, loosening; *geol.* release, relaxation; 4. *phon.* softening
~ **антициклона** high weakening
~ **вакуума** loss of vacuum
~ **ветра** abating/decrease of wind
~ **волоса** *leath.* hairslipping
~ **давления** release/drop of pressure
~ **излучения** attenuation of radiation
~ **контрастности** *telev.* contrast decay
~ **нагрузок от порывов** *av.* gust alleviation
~ **накачки** depumping
~ **напряжения** stress release; slackening of tension; *polit.* detente
~ **нейтронного потока** neutron(-flux) attenuation
~ **осадков** rainfall attenuation
~ **поля** *telev.* field reduction
~ **породы** rock slackening
~ **потока** 1. flux depression/attenuation; 2. flow decay
~ **пучка** beam attenuation
~ **сварного шва** fillet weld concavity
~ **сжатия** decompression
~ **шторма** decrease of gale
ослабленный 1. *el.* attenuated; 2. relaxed, reduced; loose, lax; 3. weakened; hyposthenic
ослаблять 1. *el.* attenuate; 2. *telev.* reduce; *phot.* clear; 3. *mech.* slacken
~ **дисциплину** relax discipline
~ **темпо** slacken the pace
ославяниваться become Slavicized
осланцевание rock/stone dusting

ослепление blinding, blindness, dazzling
~ вспышкой flash blindness
ослеплять dazzle, blind
ослизнение sliming, mucilagination
ослизный mucilaginous
ослинник *bot.* evening primrose *(Onagra, Oenothera)*
осложнение complication
ослюденение *geol.* micatization, micasization
осляк *zool.* 1. onager *(Equus hemionus onager);* 2. mule
осман Osmanli Turk, Ottoman
османский Osmanli, Ottoman
осмат *chem.* osmate
осматривание examination, inspection; inspecting
осмидроз *med.* osmidrosis, bromidrosis
осмий *chem.* osmium, Os
~, иридистый iridosmium, iridosmine
осминник *hist.* osminnik (0. 675 acres)
осмол *for.* resinous wood
осмоление gumming; gum formation; *chem.* resinification
осмоленный pitched, tarred
осмолка tar coating, tarring
осмоляемость tar value/amount
осморегуляция osmotic regulation, osmoregulation
осмос osmose, osmosis
~, внутриклеточный endosmosis
осмотерапия *med.* osmotherapy
осмотический osmotic
осмотр examination, inspection, check(ing); survey
~, ветеринарный veterinary inspection
~, врачебный medical examination
~, выборочный spot-check inspection, sampling inspection
~, контрольный check
~, медицинский physical/medical examination
~, наружный visual examination/inspection, exterior check
~, периодический routine/periodical inspection
~ повреждений damage survey
~, приемо-сдаточный acceptance inspection
~, профилактический preventive inspection/examination
~, регулярный routine/periodical inspection
~ самолета летчиком walk-around inspection
~, таможенный customs inspection
~, текущий routine inspection
~, технический check-up
~, эпизодический casual inspection
осмотрительность circumspection, circumspectness; discreetness
осмотрительный careful, cautious; discreet, wary
осмотрщик inspector
осмоцептор osmo(re)ceptor
осмунда *bot.* royal fern *(Osmunda regalis)*
осмысливание interpretation, comprehension, conceptualization; renedering meaningful
осмысливать interpret, comprehend, coneptualize
оснастка equipment; rigging, hardware; instrumentation
~ грузовой стрелы derrick rigging
~, инструментальная tooling
~ опор pole attachments
~ приборами instrumenting
~, производственная production tooling
~, такелажная rigging
~, технологическая machining attachments, industrial equipment
~ цеха shop auxiliaries
оснастчик rigger
оснащать equip; *naut.* rig
оснащение rigging, equipment; instrumentation; fitting out, outfit
основа 1. basis, base; foundation; 2. *pl.* principles, fundamentals; 3. *anat.* matrix, tela; 4. *gram.* stem; 5. *text.* warp
~, бобинная spool warp
~ вкладыша подшипника backing, supporting shell
~, выборочная *stat.* sample basis
~, геодезическая geodetic base/datum
~, ковровая carpet warp
~, общеотраслевая industrywide basis
~, повременная time-rate basis
~, полиефирная polyester base
~, рациональная rationale
~, сдельная piece-rate basis
~, хрящевая cartilage matrix
~вы грамоты basic literacy
~вы расчета design principles
основание basis; basement, floor, foot(ing), foundation; *chem.* base; *constr.* pedestal bed(ding), bottom
~ антенны antenna mounting
~ веретена *anat.* base of the modiolus
~, десятичное decimal base
~ для предъявления иска cause of action
~ зуба *mech.* root of a tooth
~, каменистое rock bedding
~ клетки cellular base
~ кода number of symbols; size of a code alphabet
~ колонны column foot
~ кузова underframe
~ листа leaf base
~ логарифма radix, logarithmic base
~, морское offshore unit/platform
~ надвига *geol.* fault sole
~ орудийной установки stand assembly
~ перпендикуляра foot of a perpendicular
~ печатной платы substrate
~, плакированное metal-clad base material
~ подины sub-hearth
~ покрытия *constr.* base/bed course
~ резьбы thread bottom
~, рельсовое rail support
~ рубчика семени *bot.* omphalodium
~, рыхлое soil base, supporting soil
~ свиты *geol.* base of a series

~ **семяпочки** *bot.* chalaza
~ **системы счисления** *math.* radix/base of a number system
~, **скальное** bedrock (base)
~ **степени** *math.* base of the/a power
~ **стремени** *anat.* base of the stapes
~ **фундамента** subfoundation
~ **шкалы** dial
основательность validity, soundness, solidity
основательный 1. solid, substantive, substantial; 2. well-grounded, justified, sound
основной 1. principal, main, fundamental, basic, radical, cardinal, primary, major, chief, essential; elemental; 2. basal, basilar; *geol.* subsilicic; *med.* basylous
~ **вопрос** basic question
~ **закон** organic law
~ **капитал** fixed capital, capital stock
~ **продукт** staple
~**ная причина** principal cause
~**ная ткань** parenchyma
~**ные данные** master data
основно-небный *anat.* sphenopalatine
основность *chem.* basicity
основный *chem.* basic
основовязальный *text.* warp-knitting
основоглавный dominant
основовязание *text.* warp/tricot knitting
основообразование *gram.* stem derivation
основообразующий *gram.* stem-forming
основополагающий fundamental, basic
основоположник initiator, founder
основоположный fundamental, principal, basic
основосложение *gram.* stem formation, composition, compounding
основщик *text.* warper
основывать 1. found, set up, base; establish; 2. constitute
осоавиахим *abbr.* **Общество содействия обороне, авиационному и химическому строительству** Society for the Promotion of Aviation and Chemical Defence
особенность feature, (characteristic) property, characteristic; distinction; *math.* singularity
~ **конструкции** structural/design feature(s)
~, **устранимая** apparent/removable singularity
особняк *arch.* detached house; private residence
~, **большой** mansion
особоуполномоченный *n.* special plenipotentiary
особый 1. special; specific; distinct; 2. particular; peculiar; *math.* singular
~**бое мнение** *law* dissenting opinion
особь *biol.* individual, specimen; zooid
~, **взрослая** imago; adult
~, **женская** female
~, **конъюгирующая** conjugant
~, **материнская** parent individual
~, **молодая** juvenile individual
~, **мужская** male
~ **с задержанным/ускоренным развитием** retarded/accelerated individual
осовременивание *rel.* aggiornamento
осовременить bring up to date, modernize
осоед *zoll.* pern, honey buzzard *(Pernis apivorus)*
осознавание perception, realization
осознавать perceive, realize
осознанность consciousness; deliberateness
осока *bot.* sedge *(Carex)*
осоковые *n. pl. bot.* sedge family *(Cyperaceae)*
осоколистный *bot.* sedge-/carex-leaved
осокоцветный *bot.* cyperaceous
осолаживание malting
осолаживатель *bioch.* malting substance
осолоделый solodic, solodized
~**лая почва** *geol.* alkaline podzol
осолодение solodization, solothization
осолонение salinization
~ **почвы** alkalization of soil
осолоненность salinity
осолонцевание alkalinization, solonetzification, solonetzization
осот *bot.* sow thistle *(Sonchus)*
оспа *med.* pox
~, **белая** alastrim, white pox
~, **вариолоидная** varioloid smallpox
~, **ветряная** chicken pox, water-pox, varicella
~, **коровья** cowpox, vaccinia
~, **ложная** pseudopox
~, **натуральная** smallpox, variola
~, **черная натуральная** black smallpox
оспенный variolous, variolar, variolic, variolitic; *n.* affected with smallpox
оспина pockmark
осповакцина variolovaccine
осповидный *med.* varioliform
оспоподобный *med.* varioloid, varioliform
оспопрививание vaccination
~ **по Дженнеру** Jenner vaccination, arm-to-arm vaccination
оспоримость *law* contestability, voidability
оспоримый *law* voidable, contestable, disputable, traversable
осреднение *math.* averaging, taking the mean, smoothing
~ **данных** *comp.* data smoothing
~ **по времени** time averaging (operation)
~ **по множеству** ensemble averaging
осредненный averaged, mean
~ **по времени/множеству** time/ensemble-averaged
осреднитель averager
оссеин 1. *bioch.* ossein(e); 2. *anat.* bone cartilage
оссификация *physiol.* ossification
ост *naut.* east; east wind
ОСТ *abbr.* **общесоюзный стандарт** All-Union Standard
останавливать stop; arrest; shut down/off
~, **быстро** scram *(a reactor)*
останавливаться stop, haul up, come to a stand-

still
останец outlier, remnant; *geol.* residual hill
~ кровли roof pendant
~, подземный subsurface outlier
~, столовый butte
~ тектонического покрова nappe outlier
останец-скала skerry
останов 1. stop(ping), halting; 2. safety/stop/arresting gear; shut-down, outage; catch, lock, detent; *comp.* breakpoint, halt
~, аварийный *comp.* abend; *tech.* emergency shut-down
~ без возможности повторного запуска dead halt
~ двигателя engine stopping/stoppage
~, жесткий uncushioned stop
~, контрольный *comp.* breakpoint halt
~, кулачковый clutch stop
~ курка hammer stop
~ по требованию request stop
остановка stop(age), stopping, halt; shut-down/-off, outage
~, аварийная emergency shut-down
~ автобуса bus stop
~ доменной печи blowing-out of a blast furnace
~ дыхания respiratory standstill
~ дыхания новорожденного *med.* apnea neonatorum
~ кровотечения *med.* arrest of bleeding, h(a)emostasia, h(a)emostasis
~, плановая scheduled outage
~ реактора, быстрая scram
~ реактора, медленная rundown, pile-down
~ сердца cardiac arrest
остатки *pl.* 1. remains, relics, remnants; 2. left-overs, odds and ends *see also* **остаток**
~, зоогенные zoogene remains
~, ископаемые fossils, fossil remnants
~, кубовые (still) bottoms
~, органические organic remains
~ продуктов деления residual fission products
~, современные recent relics
~, технологические process residue
~, фитогенные phytogenic/phytogene remains
остаток 1. remainder, residue; residual; *geol.* residuum; 2. *fin.* balance, rest; *see also* **остатки**
~ водный *chem.* hydroxyl
~ кассы cash balance
~, кислотный acid residual/residue
~ к получению balance on order
~, кубовый vat/still residue, (still) bottoms
~, нелетучий bottom ash, solid constituent
~ непроданных товаров stock in trade
~ окалины drawn-in scale
~ от деления *math.* remainder (of devision)
~, очаговый furnace/ash residue
~, переходящий *econ.* carry-over
~ по счету ностро nostro balance
~, порубочный shash(es), felling residue
~, причитающийся balance due
~, реальный actual balance
~ ряда *math.* remainder of a series
~, среднедневной average daily balance
~, сухой solid/dry residual, solids
~ счета balance of an account
~ топлива *av.* remaining fuel
~ топлива, неиспользуемый *av.* trapped unusable fuel
~, фактический actual balance
~ флюса residual flux
остаточно-карбонатный residual-carbonate
остаточный relic, residual, detrital, indigenous; vestigial, remanent
~ная стоимость основного капитала *econ.* sound/residual value, written-off value
остгот *hist.* Ostrogoth
остготский *hist.* Ostrogothic, East Gothic
остеит *see* **остит**
остекление, остеклование vitrif(ic)ation, transparent panelling, glazing; *av.* transparency
~ кабины летчика glass cockpit windows
~, силикатное glass panelling
~ фонаря canopy transparency
остекленный, остеклованный glass-panelled, glazed; vitreous, vitrified
остекляться become vitreous/glassy
остеоартрит *med.* osteoarthritis
~, гиперпластический hyperplastic osteoarthritis, pulmonary osteoarthropathy
~ позвоночника, туберкулезный tuberculous spinal disease
~, старческий senile osteoarthritis
остеоартроз *med.* osteoarthrosis
~, генуинный congenital osteoarthrosis
остеоартропатия *med.* osteoarthropathy
~, гипертрофическая легочная pneumogenic/pulmonary osteoarthropathy
остеобласт *anat.* bone-forming cell, osteoblast
остеогенез *physiol.* osteogenesis, osteogeny
~, несовершенный dysosteogenesis
остеогенный osteogen(et)ic, osteogenous
остеодисплазия *med.* dysplasia of bone, osteodysplasia
~, врожденная congenital osteodysplasia
~, диафизарная diaphyseal dysplasia
~, наследственная hereditary osteodysplasia
~, челюстно-лицевая mandubulofacial dysplasia/dysostosis
остеодистрофия *med.* osteodystrophy, osteodystrophia
~ голодающих starvation osteodystrophy
~, желчная bilious osteodystrophy
~, легочная pulmonary osteodystrophy, Marie's disease
~, эндокринная endocrinous/endocrinic osteodystrophy
остеоидный osteoid, ossiform,osseus, osteal, bony
остеоклазия *surg.* osteoclasia, osteoclasis
остеолиз *med.* osteolysis
остеолит *geol.* osteolith, osteolite

остеома *med.* osseous growth, osteoma
~, ветвистая branchy osteoma
~, гигантская остеоидная osteoblastoma
~, костномозговая osteoma medullare
~, саркоматозная osteogenic sarcoma, osteosarcoma
остеомалятический osteomalacic
остеомаляция *med.* softening of bones, osteomalacia
~ голодающих starvation osteomalacia
~, старческая senile osteomalacia
~, юношеская infantile/juvenile osteomalacia, rickets, rachitis
остеомиелит, остеомиэлит *med.* osteomyelitis
~, гнойный purulent osteomyelitis
~, гумозный gummatous osteomyelitis
~, злокачественный malignant osteomyelitis, myelomatosis
~, множественный multiple osteomyelitis
~, опухлевидный tumo(u)rous osteomyelitis
~, склерозирующий condensing osteomyelitis
остеон osteon(e)
остеонекроз *med.* osteonecrosis
~, лучевой radiation osteonecrosis, osteoradionecrosis
остеонефропатия *med.* osteogenic nephropathy, osteonephropathy
остеопатия osteopathy, osteopathia
~ голодающих starvation osteopathy
~, желудочная/желчная gastric/bilious osteopathy
~ новорожденных, геморрагическая infantile scurvy, Barlow's/Moeller's/Cheadle's disease, acute rickets
~, склерозирующая рассеянная disseminated condensing osteopathy, osteopetrosis, osteopoikilosis, marble bones
остеопериостит *med.* osteoperiostitis
остеопороз *med., physiol.* bone rarefication, osteoporosis
~ культы stump osteoporosis
~, равномерный homogenous/uniform osteoporosis
~, распространенный generalized osteoporosis
~, регионарный regional osteoporosis
остеопорозный osteoporotic
остеопсатироз *med.* osteopsathyrosis
остеорафия *surg.* osteorrhaphy, osteosuture
остеосиновит *med.* osteosynovitis
остеосинтез *med.* osteosynthesis
~, внекостный extraosseous/extramedullary osteosynthesis
~, внутрикостный intraosseous/intramedullary nailing osteosynthesis
остеосклероз *med.* osteosclerosis
~, генуинный congenital osteosclerosis, achondroplasia
~, распространенный disseminated osteosclerosis, disseminated condensing osteopathy
остеоспонгиома *med.* osteospongioma, spina ventosa
остеотделитель (de)awner, debearder
остеотомия *surg.* osteotomy
~, клиновидная cuneiform osteotomy
~, корригирующая correcting osteotomy
остеофил bone seeker, bone-seeking element/substance
остеофит *pathol.* osteophyte
~, сегментарный segmental osteophyte
остеофлебит *med.* osteophlebitis
остеофон osteophone, bone-conductor receiver
остеохондрит *med.* osteochondritis
~, гнойный purulent osteochondritis
остеохондродистрофия *med.* osteochondrodystrophy
~, деформирующая Morquio's disease, Brailsford-Morquio disease
остеохондроз *med.* osteochondrosis
~, межпозвонковый intervertebral osteochondrosis
~, юношеский Osgood's/Schlatter's disease
остеохондрозный osteochondrous, osseocartilaginous
остеохондропатия *med.* osteochondropathy
~, множественная multiple osteochondropathy
~, частичная клиновидная partial cuneiform osteochondropathy
остеоцистома *med.* bone cyst, osteocystoma
остерегание admonition, caution, warning
остзеец *hist.* Baltic German
остинато *mus.* (basso) ostinato, ground bass
ост-индийский East-Indian
ости *pl., anat.* spines
~, небные palatine spines
остиофолликулит *med.* ostial folliculitis
остисто-бугристый tuberositospinal
остистолистный *bot.* awn-leaved
остистый *anat.* spinous, spinal; *bot.* awned, bearded, barbate, aristate
остит *med.* ost(e)itis
~, казеозный tuberculous caries, caseous ost(e)itis
~, множественный кистозный туберкулезный Juengling's disease
~, ограниченный фиброзный Albright's syndrome
~, склерозирующий condensing/formative osteitis
~, фиброзный диссеминированный Albright's disease
~, центральный osteomyelitis, endosteitis
остов 1. *constr.* skeleton, frame(work); *tech.* body, carcass; hull, shell, casing; 2. *math.* spanning set; 3. *anat.* skeleton, stroma; 4. *naut.* wreck
~ двигателя engine framework
~ замка breechblock; lock plate
~, костный bony frame, skeleton
~ кузова body frame
~ симплекса *math.* skeleton of a simplex
~ ядра core of the nucleus

остовый east(ern), easterly
остоз *med.* ost(e)osis
~, паращитовидный parathyroid osteosis
остойчивость *naut.* (metacentric) stability
~, аварийная damaged/impaired stability
~ веса pendulum/weight stability
~ в море sea stability, seaworthiness
~, динамическая dynamic/metacentric stability
остойчивый *naut.* stable; seaworthy
осторожность caution, guardedness, prudence; circumspection
осторожный cautious, circumspect, prudent; wary
остракизм, острацизм *hist.* ostracism, expulsion from society
остракода *zool.* ostracod(e)
острец *bot.* sedge *(Carex)*
острие 1. point, tip, edge; spike; 2. *bot., zool.* cusp, mucro
~ бойка point of a striker
~ долота nose of a chisel
~, коронирующее corona point
~ кривой *math.* cusp
~, многогранное diamond point
острица I *bot.* madwort *(Asperugo)*
острица II *zool.* pinworm, seat worm *(Enterobius vermicularis)*
остров 1. island; isle; 2. grove (of trees)
~, барьерный barrier island
островершинность peakedness, pointedness
~ распределения *stat.* peakedness of distribution
островок islet; eyot, ait; *anat.* island
~ безопасности pedestrian/safety/traffic island
~, кровяной sanguineous island
~, прибрежный holm
~, речной ait, holm
~, шхерный rocky isle, skerry
острог 1. *hist.* stockade, palisade; 2. stockaded town
острога (fishing) spear, harpoon; gaff, grappling hook
острогангренозный *med.* acute gangrenous
острогнойный *med.* acute purulent/suppurative
острогубцы, острозубцы *pl.* nippers, cutting pliers
остродефицитный *econ.* 1. showing extreme loss/deficit; 2. of very short supply
острозубчатый *bot.* sharp-toothed
остроконечие 1. sharp point; *bot.* mucro, cusp, acumination; 2. *anat.* spine
~, межщечное intergenal spine
~, щечное genal spine
остроконечный peaked, sharp(-pointed), tapered, arrowy, cuspidate(d); turbinate; *bot. biol.* apiculate, acuminate, mucronate, spiniform
остролист *bot.* English holly *(Ilex aquifolium)*
остролиственный, остролист(н)ый *bot.* sharp-leaved, acutifoliate
остролодочник *bot.* crazyweed *(Oxytropis)*
остролучевой pencil-beam
остронаправленный narrow-beam, pointed; ultradirectional, high(ly)-directional
остронос(ик) *zool.* (little) mullet *(Mugil saliens)*
остроплодный *bot.* sharp-fruited, oxycarpous
острота sharpness; acuity; acuteness; acumen; acrimony, acridity; pungency
~ зрения visual acuity, acuteness/acuity of vision
~ зрения, повышенная hyperacuity
~ настройки sharpness of tuning/resonance
~ ощущения sensory acuity
~ слуха auditory/hearing acuity
остроугольник *math.* acute-angled figure
остроугольно-глыбистый subangular-blocky
остроугольный acute-/-sharp-angled
острофокусный fine-/sharp-focussed
остročелюстный *zool.* oxygnathous
острочешуйный *bot.* oxyglumose, oxyglumous
острощетинистый *bot.* exasperate
оструктурирование structure formation
острый 1. sharp, pointed, acute; 2. acerate, aciform; 3. acrid, pungent, strong; spicy
~ недостаток acute shortage
~ пар live steam
остряк *rail.* tongue, ramp, point
~, направляющий guiding tongue
~ стрелки rail point
остудневание jellification, gelatification, gelling, gelation
ость 1. *anat.* spine; 2. *bot.* arista, awn, barb, beard
~, блоковая trochlear spine
~, верхняя dorsal awn
~, зубчатая scabrous awn
~, надпроходная Henle's spine
~, перистая plumose awn
~, подбородочная mental spine, genial tubercle
~, седалищная sciatic spine, spine of the ischium
~, угловая spine of the sphenoid
осуждать 1. condemn, censure, blame; 2. *law* sentence, convict
~ на смерть sentence to death
осуждение 1. censure, condemnation, blame; 2. *law* conviction
осужденный condemned, convicted; *n. law* convicted person/offender, convict
осумкование encystenent, encystation, encapsulation, sacculation
осумкованный encysted, encapsulated, sacculated
осушать drain, dewater; dry; desiccate; dehumidify; reclaim
осушение drying; drainage, draining; dewatering, dehumidifying, dehumidification
~ земель drainage
~ самотеком gravity drainage
осушитель 1. drier, drying agent, desiccant, siccative; dehydrator, dehumidifier; 2. lateral/drainage ditch
~, химический drying reagent
осуществимость feasibility, practicability, realizability
~, техническая technical/engineering feasibility

осуществление realization, actua(liza)tion; accomplishment, fulfilment; implementation
~ **операции ИЛИ** *comp.* ORing
осуществлять effect, actualize, carry out, fulfil, accomplish, implement
~ **план** carry out a plan
осциллограмма oscillogram, oscillograph trace, oscilloscope record; CRO display
осциллограф oscillograph, trace recorder
~, **гальванометрический** galvanometer oscillograph
~, **двухлучевой** two-gun oscillograph, double oscillograph
~, **шлейфовый** loop/bifilar oscillograph, rotating-mirror oscillograph, mirror-galvanometer oscillograph
~, **электроннолучевой** cathode-ray oscillograph
осциллоскоп oscilloscope
~, **двухлучевой** dual-trace oscilloscope
~, **стробирующий** sampling oscilloscope
осциллятор oscillator, wave generator
~, **квантовый** quantum oscillator
осцилляторный, осцилляционный oscillatory
осцилляция oscillation
осыпь *geol.* talus, scree, hillside waste; debris
~, **береговая** coastal debris
~, **коническая** conoid talus, debris cone
~, **мелкоземистая** silt talus
~, **подводная** subaqueous talus
ось 1. axis (*pl.* axes); centre line; 2. axle; spindle, pin, shaft, pivot; *see also* **оси**
~ **абсцисс** X-/abscisse-/abscissa axis, axis/scale of abscissae/abscissas
~ **аппликат** Z-/applicate axis, axis of applicate
~ **бандажированного валка** *met.* roll mandrel
~, **большая** *geod.* longer/major axis
~ **валка** *met.* arbor, spindle
~, **ведущая** driving axle
~ **взлетно-посадочной полосы** runway centerline/centreline
~, **визирная** sight(ing)/collimation axis
~, **винтовая** spiral/helical axis
~ **вихря** eddy axis, axis of vortex
~ **ВПП** line of runway, runway centre line
~ **вразвилку** forked axle
~, **вращающаяся** live axle
~ **вращения** *math.* axis of rotation/revolution; rotation(al) axis; *av.* spin axis; hinge axis/line; transit axis
~ **вращения, вертикальная** veer axis
~ **времени** time axis/base, timing axis
~, **выделенная** preferred axis
~, **главная** major/principal axis; *geol.* critical axis
~ **гребня** *meteor.* axis of high pressure
~, **двойниковая** *cryst.* twinning axis
~ **двойной симметрии** axis of twofold symmetry
~, **действительная** *math.* real(-number) axis, axis of reals
~ **деформации** axis of strain
~, **диагональная** clinoaxis
~ **дифференциала** differential shaft
~, **длинная** long axis; macroaxis
~ **дороги** centre-line of a road
~ **жесткости** elastic axis
~ **Земли/земная** earth's/terrestrial axis
~ **зигзага** mean pattern course
~, **зрачковая** *anat.* pupillary axis
~, **зрительная** optic axis, line of vision
~, **измерительная** sensitive/input axis
~ **инерции** inertia axis, axis of inertia
~ **качания** axis of oscillation
~ **кисти** *bot.* shaft of raceme
~ **клетки** *biol.* cellular axis
~ **колоса** *bot.* ear axis, rachis
~ **координат** *math.* axis of coordinates
~, **короткая** *cryst.* short axis; *paleont.* brachyaxis
~ **крена** *av.* roll(ing) axis
~ **мира** *astr.* celestial axis
~, **мнимая** *math.* imaginary axis
~ **Монжа** *math.* Monge('s) axis
~ **мульды** *geol.* trough axis, bottomline
~ **наклона** tilt axis
~, **начальная** basis/reference/initial axis
~, **нормальная** Z-/normal axis
~, **общая** common axis, coaxis
~, **опорная** axial rod
~, **оптическая** optic(al) axis
~ **ординат** vertical scale, Y-axis, axis/scale of (the) ordinates
~ **отсчета** reference axis
~, **первичная** protaxis
~, **побочная** secondary/auxiliary axis
~, **подчиненная** sub-axis
~, **поперечная** Y-/lateral axis
~ **прицела** axis of sight
~, **продольная** longitudinal/X-axis; *av.* fore-and-aft axis
~ **пространственных координат** solid axis
~ **развертки** timing axis, time base
~ **регулятора** governor shaft
~ **рыскания** *av., naut.* yaw(ing) axis, axis of yaw
~ **симметрии** axis of symmetry
~ **синклинали** synclinal/trough/saddle axis
~, **спряженная** associated axis
~, **строительная** construction line
~ **тангажа** *av.* pitch(ing) axis
~, **тройная** triad
~ **управления** *av.* control axis, axis of no feathering
~ **фарватера** midchannel
~, **четверная** tetrad
~ **шарнира** hinge axis/pin, articulation, link pin, fulcrum
~, **шарнирная** pivotal axis
~, **шестерная** hexad

осьминог *zool.* octopus *(Octopodus)*
осьмогласник *eccl.* antiphonary
осязаемость palpability, tangibility
осязаемый tangible, tactile, palpable
осязание taction, sense of touch, tactile sense
~, врачебно-диагностическое tactus eruditis/expertus
осязательный tactile, tactual; touchable; haptic; tangible
~ные результаты tangible results
осязать touch, feel
отава *agr.* eatage, aftergrass, aftermath
отакелаживать *naut.* rig(out)
оталгия *med.* earache, otalgia, otodynia
~, геникулярная geniculate otalgia/neuralgia
~, рефлекторная reflex otalgia
отапливать heat, warm
отбел *met.* hard spots, chill(ing)
~, обратный internal/inverse chill
отбеливание bleaching
~ под крашение dyers bleaching
~ (серого) чугуна chilling of (gray) iron casting
отбеливатель bleacher, whitener, decolorant
отбеливать bleach, whiten; *tech.* refine; chill; blanch
отбелка bleaching, whitening; *tech.* refining; chilling; blanching
~ бельевого полотна market/white bleaching
~, перекисная peroxide bleaching
отбельный bleaching; *tech.* blanching; chilling, refining
~ раствор liquid bleach
~ная глина bleaching clay
отбензинивание *chem.* topping *(of petroleum);* stripping
отбивать 1. draw down, hammer (out); sharpen, whet; 2. repel, repulse, beat off; 3. chip/break/knock off
отбивка 1. whetting, sharpening; 2. delineation, marking out; 3. *print.* spacing
отбивная *n. cul.* chop
отбивной *cul.* pounded, tenderized
~ная котлета chop
отбирать 1. select, screen; pick out; 2. extract, tap, bleed
~ воздух от компрессора *av.* bleed air off the compressor
~ мощность take off power
отблеск gleam, reflection
отбой 1. retreat; all-clear (signal); *teleph.* ring-off, ringing off, clearing signal; 2. repulsion, repelling
~, автоматический automatic release
~, бить beat a retreat
~, давать *teleph.* ring off
отбойка *min.* breaking (ground); cutting (of coal)
~, валовая breaking out in bulk
~, гидровзрывная hydraulic blasting
~ по породе/руде rock-/ore-breaking
отбойник 1. impingement plate; baffle (plate); 2. chipper *(tool)*
отбойный breaking, cutting
~ молоток pick hammer
~ молоток, пневматический pneumatic hammer/pick
~ная свая fender pile
отбойщик *min.* cutter, breaker, pickman
отбор selection, picking out, sampling; cull(ing); screening; *chem.* extraction; *tech.* bleed-off
~, вероятностный probability sampling
~, взвешенный *math.* weighted sampling
~ воздуха air bleeding/offtake, air bleed-off
~ давления pressure tap
~ данных data sampling
~, естественный *biol.* natural selection
~, исчерпывающий exhaustive sampling
~ мощности power take-off
~, направленный purposive selection
~ пара steam bleeding/extraction
~ пара, регенеративный regenerative bleed-off
~ по линиям line sampling
~, половой sexual selection
~ проб sampling
~, профессиональный occupational selection
~ тепла heat removal
~ щелока waste liquor bleeding
отборка 1. picking (out), selection; 2. mo(u)lding/cornice plane
отборник 1. separator, sifter; sorter, sampler; 2. mo(u)lding/cornice plane
отборочный selection, elimination *(attr.)*
~ная комиссия selection board
отбортовка bead(ing), flanging; flange
отборщик sorter, grader; selector
отбраковка rejection, sorting, grading
отбраковщик scanner; quality checker
отбраковывать reject, throw out, scrap
отбрасывание rejection; deletion; dropping, omission, discarding, throwing away, casting out
~ незначащих нулей zero-suppression
отбрасывать 1. push away, jettison; *math.* reject, drop, omit, discard; 2. *mil.* repel, repulse
~ нули suppress zeros
~ перенос discard the carry
~ тень cast a shadow
отбросы *pl.* waste(s), refuse, garbape; residue, offal; tailings
~, боенские slaughterhouse waste
~, городские/промышленные town/industrial refuse
~ общества scum of society
отбытие departure
отвал I 1. disposal area, dump; spoil bank/heap; dirt pile, rubbish heap; 2. blade; mouldboard
~ бульдозера blade of bulldozer
~, носовой V-blade
~, карьерный spoil bank
~ плуга breast, plows hare, plough-share, mouldboard, earthboard

~ шлака slag dump
отвал II *naut.* casting/pushing/putting off
отваливать *nav.* cast/push/putt off
отвалообразование piling, refuse disposal
~, бестранспортное direct dumping
отвалообразователь *min.* stoker, stacker, (overburden) spreader
~ драги stacker
~, передвижной travel(l)ing stacker
отвальный *n. min.* stoker
отвар decoction, apozem; broth
~, мясной *cul.* beef tea
отваривание 1. *chem.* decoction, digesting; boiling off/out; 2. *tech.* unwelding
отваривать 1. decoct; boil (off); 2. *tech.* unweld
отварка decoction, decocting
отведение 1. *med.* abduction, lead; *tech.* removal, rejecting, drawing off; 2. allocating, assigning
~, биполярное bipolar lead
~, затылочное occipital lead, occipital area take-off
~, полупрямое semidirect lead
~, прекордиальное precordial/chest lead
~, теменное sincipital contact point
отвергать refuse, reject; override; *law* repudiate
~ голосованием vote down
отвердевание consolidation, hardening, solidification, congelation, setting; crystallization; freezing
~ пыльцевых зерен *bot.* stiffening of pollen grains
отвердевать solidify, harden, consolidate, indurate, congeal, set
отвердение hardening, solidification; induration
отвердитель *chem.* curing/hardening agent, hardener, solidifier
отверждать(ся) cure, harden; consolidate; set
отверждение solidification; hardening, setting; cure, curing
~, горячее hot hardening/curing
~, последующее after-bake/-hardening
~, предварительное pre-curing
отверстие 1. opening; hole; gap, break; perforation; port; vent; 2. *anat., zool.* aperture, pore; foramen (*pl.* -mina); orifice, ostium (*pl.* -tia); hiatus; stoma (*pl.* -mata); fenestra; ostiole
~, анальное *anat.* anus, anal orifice
~ аорты aortic ostium
~, большое (затылочное) great foramen
~ в защите shield opening
~ в перегородке septal defect
~ в пульпе зуба root canal orifice
~ в бумаге pin-hole *(a defect)*
~, ведущее guide/feed/sprocket hole
~, вентиляционное ventilation opening, vent hole
~, височное temporal fossa
~, вкусовое gustatory pore
~ влагалища vaginal opening, ostium vaginae
~, водоотливное water outlet; *naut.* boat's bale
~ волоки drawing die orifice
~, впускное admission/inlet/intake/induction opening, inlet/intake port
~, всасывающее suction hole
~, входное inlet (opening); access port
~, выпускное exhaust port, discharge opening
~, высевное *agr.* seed hole, feed opening
~, выхлопное *aut.* exhaust port, discarge opening/outlet
~, выходное outlet (hole), discharge hole, vent
~, глухое blind hole
~ губок, выходное osculum
~ диафрагмы, аортное aortic opening
~ для образца sample port
~ для пучка beam port/hole
~, дозирующее metering orifice
~, дренажное/водовыпускное drainage hole, drain opening/hole, vent (hole)
~, дульное firing port
~, дыхательное stigma, spiracle; gill slit
~, загрузочное charging/opening port
~, заднепроходное anus, anal orifice
~, заклепочное rivet hole
~, заливное/заправочное filling orifice
~ зева isthmus of the fauces
~, зрительное optic foramen
~ истечения discarge orifice
~, калиброванное calibrated/gauged orifice
~, клоачное vent, cloaca
~, кодовое code hole
~, коллимирующее collimating aperture
~, кольцевое annulus
~, контрольное inspection/locating hole
~ кратера vent
~, крепежное mounting hole
~, круглое *anat.* round foramen, circular opening
~ литника *met.* runner opening
~, литниковое sprue (opening)
~, маркерное sprocket/guide/feed hole
~, маслоналивное oil-filling hole
~ матки ostium of the uterus, vaginal opening of the uterus
~, матричное die hole
~, межжелудочковое interventricular foramen
~, мерное measuring orifice
~, метализированное plated hole
~, монтажное lead hole
~, надглазничное supraorbital foramen
~, наливное filling hole/orifice, filler (opening)
~, направляющее pilot hole
~, наружное слуховое external acoustic/auditory pore
~, небное palatine foramen
~ объектива lens aperture/opening
~, откачное exhaust hole
~, относительное relative aperture, aperture ratio, F-valve
~ охлаждения cooling hole
~, перепускное by-pass orifice
~, переходное via-hole
~, перфорационное punched hole

~, питательное nutrient foramen
~, поглощающее sink/swallow hole
~, подглазничное infraorbital foramen
~, позвоночное vertebral foramen
~, поровое foraminal aperture
~, потовое sweat pore
~, приемное feed opening
~, промывное flushing port
~, промывочное flushing orifice/hole
~, прочистное cleanout port
~, пупковое umbilical opening
~, пусковое starter opening
~, рваное lacerate foramen
~, ротовое mouth, stoma (*pl.* -mata)
~, разгрузочное discharge opening
~, разгрузочное насоса equalizing hole
~, резьбовое threaded/tapped/screw hole
~, свечное spark-plug port
~, свободное clear opening
~ сита mesh, screen opening
~ синхродорожки *comp.* feed hole
~, сквозное (reach-)through hole
~, слепое blind opening, thimble; *anat.* c(a)ecal foramen
~, слуховое ear opening
~, смазочное lubricating hole, oil port
~, смотровое inspection/peep hole, inspection opening, observation port, eye-sight
~, спускное drain hole/opening, drainage hole
~, установочное locating/*comp.* indexing hole
~, форсуночное spray(-type) orifice, injection port
~, фурменное tuyere/lance hole
~, штекерное jack outlet
~, шуровочное poking hole
~, щелевое slit
~, яремное *anat.* jugular foramen
отвертка screwdriver
~, антимагнитная nonmetallic screwdriver
~ с трещоткой ratchet screwdriver
отвертывать unscrew, screw/twist off
отвес plumb line, plumb (bob), bob, plummet
~, оптический vertical collimator
отвесность verticality, perpendicularity
отвесный vertical, perpendicular; precipitous, abrupt; plumb, upright
ответ answer, reply, response
~, автоматический *comp.* autoanswer
~, выборочный multiple-choice answer, selected response
ответвитель branch; *rad.* coupler
~, направленный directional coupler
ответвление 1. branch, branching (off), leg, arm; bypass channel; *el.* tap(ping), shunt(ing); 2. *bot.* offset, ramification, offshoot, side branch
~ жилы stringer, offshoot
~ линии tap
~ магистрального трубопровода leg of main
~ от трубопровода tee of a pipeline, branch pipe
~ трубопровода, сделать tap a (pipe)line
ответвлять branch/spur/set off; tap; *el.* shunt
ответвляться spur off, branch out, bifurcate
ответный answering, in answer/reply; responsive; reciprocal; sympathetic
~ огонь *mil.* counterfire, return fire
ответственность responsibility; liability, accontability; onus; *law* amenability
~, административная management responsibility
~, гражданская civil liability
~, денежная pecuniary liability
~ за порчу имущества impeachment of waste
~ за убытки liability for damages
~, материальная liability for breakage; financial liability
~, ограниченная limited liability
~, принимать на себя assume responsibility, incur liabilities
~, солидарная *law* joint and several liability
~, уголовная criminal responsibility
ответственный 1. responsible, liable, accountable; 2. critical, crucial
~ в первую очередь primarily liable
~ перед законом amenable
~ работник responsible worker; executive
~ редактор editor-in-chief; executive editor
ответчик 1. *electr.* transponder, responder; 2. *law* respondent, appellee, accountant, defendant, avowant, libellee
~, бортовой *av.* airborne ATC transponder, airborne responder
~ по апелляции appellee
отвечать answer, reply, respond; be responsible (for)
~ стандартам be in conformity/compliance with standards
отвешивать weigh out
отвлекать draw off, withdraw; divert *(of resources)*
отвлекающий distracting; diversionary
отвлечение 1. diversion, distraction; 2. withdrawing, removal; 3. *med.* derivation
отвод 1. drain, outlet; discharge; 2. diversion; withdrawal; extraction, removal; 3. allocation, allotment *(of land);* 4. *law* recusation; challenge; rejection; 5. *el.* tap(ping); shunting; 6. *tech.* (pipe)bend, elbow, branch; outlet, bleed
~ в землю shunting off to ground
~ воды water drainage
~ воды для орошения irrigation diversion
~ воздуха extraction of air
~ газа gas withdrawal/extraction
~ жидкости fluid extraction/removal
~ за борт overboard discharge
~, литой cast branch/outlet
~ остаточного тепла afterheat removal
~, процессуальный *law* demurrer
~, прямой elbow bend

~ **речного стока** run-of-stream diversion
~ **свидетелей** *law* demurrer of evidence
~ **статического электричества** static discharge
~ **тепла** heat removal/extraction/withdrawal/rejection/release, abstraction of heat
~, **трубный** lateral
~ **трубопровода** pipe branch, pipeline bend
~, **фланцевый** flange(d) tap
отводитель *naut.* otter
~, **решетчатый** multiplane otter
отводить 1. abstract; reject, remove; 2. allocate, assign; 3. divert, draw aside/off; deflect; 4. *law* disqualify, challenge
~ **руль** *naut.* ease the rudder
~ **свидетелей** disqualify witnesses
~ **тепло** abstract/reject/remove heat
отводка 1. shifting device; (belt) shifter; 2. branch pipe
~ **ремня** shifting fork
~ **сцепления** clutch throw-out yoke, clutch release fork
отводный drain(age), by-pass, branch *(attr.)*
~ **трубопровод** discharge piping; bypass pipeline
отводок *hort.* offset, offshoot; cutting, layer
~ **от пня** stool layer
отводящий 1. *med.* abducent, abducting; 2. deflecting, diverting
~ **трубопровод** bleeding pipe
отвоевывание reconquering, reconquest
~ **земли у моря** sea bottom reclamation
отвоз transportation, carting away
отволаживание *food* softening; *leath.* moistening
отволожка *leath.* rehydration
отворачивать turn aside/away
отвращение aversion, disgust, repugnance; loathing
~ **к браку** misogamy
~ **к пище, патологическое** *med.* cibophobia
отгематома, отематома *med.* othematoma
отгиб 1. crease, fold, flange, bend; 2. *bot.* limb; 3. *hort.* cutting; layer
отгибание 1. bending away/aside; unbending; 2. diffraction; deflection
~ **кромок** flanging
отгибать rib, bend off; unbend, turn back/down, fold; straighten (out)
отглагольный *gram.* (de)verbal, deverbative, postverbal
отгон I *chem.* distillate, distillation product; distillation
отгон II *agr.* pasturing
отгонка *chem.* distillation, rectification; driving-off
~ **легких фракций** skimming, topping, flashing off
~ **летучего вещества** sublimation of volatile substance
отгонять drive away/off; *av.* ferry out
отгонять(ся) distil off, top
отгружать dispatch, despatch, ship
отгрузка shipment, shipping, dispatching; unloading
отдавать *naut.* cast (off), let go *(of anchor); mil.* recoil, kick *(of firearms)*
отдание 1. giving back, returning; 2. *eccl.* observing *(of a festival)*
~ **чести** *mil.* saluting
отдача 1. *tech.* efficiency, performance; rebound; 2. *econ.* output, return, yield; 3. emission, evolution; 4. *naut.* casting off; letting go; 5. *mil.* recoil, kick
~, **ампер-часовая** ampere-hour efficiency
~, **весовая** load ratio
~ **внаем** letting, renting
~ **животноводства** return from animal husbandry
~ **отбойного молотка** back-blow of a jackhammer
~ **по емкости** capacity efficiency
~ **по сравнению с затратами** payoffs compared to costs
~, **полезная** net output
~ **протона** proton recoil
~ **с коротким ходом ствола** short recoil
~, **световая** luminous efficiency
~ **флюоресценции** fluorescence yield
~ **электрона** loss of an electron
~ **ядра** nuclear recoil
отдел 1. department, division, sector, branch; 2. part, section; *see also* **отделение;** 3. *biol.* series, formation
~ **здравоохранения** public health department
~ **кадров** personnel department
~, **коммерческий** sales department
~, **научно-исследовательский** research department
~, **оперативный** operational staff; (department of) operations
~ **оснастки** tooling department
~, **плановый** planning department
~, **приборно-измерительный** instrumentation section
~, **продовольственный** victualing department, food-supply section
~, **разведывательный** intelligence office
~ **растениеводства** plant industry division; crop-raising section
~ **ремонта и технического обслуживания** maintenance and repair department
~ **санитарного инспектора** health officer's office
~ **сбыта** sales department
~ **светской хроники** gossip column
~ **снабжения** supply division/office/department
~ **снабжения и перевозок** logistic supply group
~ **спорта** sport section
~ **технического контроля** quality control department

~, технологический engineering division
~ труда и зарплаты labo(u)r and wages department
~ тыла *mil.* logistic supply group
отделение 1. separation, segregation; isolating; dissociation, disengagement, detaching; 2. evolution, emission; excretion; 3. compartment, room; section; branch; division; bay; 4. department; 5. *mil.* squad
~, амбулаторное out-patient department/ward
~, багажное luggage/baggage section/room/hold
~, бельевое lingerie department
~ больницы hospital department/unit
~, вахтенное watch squad
~ вещества separation/evacuation of material
~ вихрей vortex detachment
~ вспомогательных механизмов *naut.* auxiliary machinery room
~ второй ступени *rock.* second-stage separation
~ генераторов generator room
~, грузовое freight/cargo compartment/hold
~, дозиметрическое radiation control room
~, инфекционное department of infectious diseases
~, кожное dermatologic department/ward
~, количественное quantitative isolation
~ корня *math.* root location, isolation of a root
~, котельное fireroom, boiler room/compartment/space
~, машинное engine/machine room; machinery space
~ плазмы plasma separation
~ под действием силы тяжести gravitational/gravity segregation
~ порта subport
~ порта, санитарное port sanitary authorify
~, рентгеновское X-ray department
~, роди(те)льное maternity/obstetric department/ward
~, румпельное *naut.* steering/tiller room/compartment
~, самолетное aeroplane division
~ санитаров-носильщиков litter squad
~ слоев exfoliation
~ слюны, чрезмерное *med.* salivation, excessive secretion of saliva
отделенный isolated, segregated, separated, detached; *n.* *mil.* squad leader
~ командир *mil.* squad leader
отделитель 1. separator; *el.* isolating switch; 2. sifter
~ жидкости liquid trap
отделительный separating, separation *(attr.)*
~ слой abscission/separation layer
отделка 1. finishing, aftertreatment; dressing, trimming; 2. finish, decoration; onlay
~ кожи, водостойкая waterproof leather dressing
~ под мрамор marble finish
~, противоусадочная *text.* shrink-proof finish
~, чистовая smoothing, final finishing
отделочный finishing, trimming
отдельность 1. *geol.* jointing, parting, cleavage; 2. separability; singularity; individuality
~, зияющая fissure/gaping/open jointing
~, концентрическая concentric cleavage/jointing
~, линейная linear parting
~, плитчатая platy/slab parting
~, почвенная soil particle
отдельный 1. individual, separate; 2. distinct, single; independent
~ вход private entrance
~ случай isolated instance, individual case
отделяемое *n.* discharge
~, гнойное *med.* suppurative/purulent discharge
~, серозное serous discharge
отделяемый separable, detachable
отделять 1. segregate, separate, isolate; detach; 2. divide, part; 3. secrete; 4. allot
~ корень *math.* isolate a root
~ перегородкой partition off
отдраивать *naut.* undog; polish, swab
отдулина blister; bulging, bulge
отдушина air vent/hole/funnel, venthole; air drain
отдушник air vent/hole
отдушка perfume (additive)
отдых relaxation, rest; holiday
~ вулкана repose period of volcano
~ металлов recovery of metals
отек 1. *med.* (o)edema, hypostasis, dropsy, hydrops, anasarca; 2. stain, stripe *(of running liquid)*
~, агональный agonal (o)edema
~, ангионевротический/Квинке/сосудистый angioneurotic edema, giant urticaria/hives, periodic/Quincke's edema, Bannister's disease
~, воспалительный inflammatory edema
~, голодный nutritional/starvation/famuine/hunger/war edema
~, застойный congestive edema
~, затяжной lingering edema
~, кахектический cachectic/marantic edema
~ кожи cutaneous dropsy
~ конъюнктивы chemosis
~ легких pulmonary edema
~ мозга brain edema/swelling
~ мошонки hydroscheocele, oscheohydrocele
~, общий general/anasarca edema
~, ограниченный circumscribed edema
~ подкожной клетчатки, общий subcutaneous dropsy, anasarca, general edema
~, почечный renal (o)edema/dropsy
~ при недостаточности питания nutritional/famine/hunger edema
~, сердечный cardiac dropsy
~, слизистый myxedema
~ стопы podedema
~, хилезный chylous dropsy
отекать 1. swell, become/get dropsical; 2. run/flow off

отел *geol.* calving
~ ледника glacier calving
отель 1. hotel; 2. manor house
отенит *miner.* autunite
отеплять winterize, make winterproof, protect against cold
отеска roughhewing, trimming; (rustic) dressing
отесывать square, roughhew; trim
отец father
~, крестный godfather
~, приемный foster father
~, святейший Most Holy Father
отечественный domestic, home(-produced); native; patriotic
отечество native land, fatherland, homeland, country
отечность *med.* puffy swelling
отечный *med.* (o)edematic, (o)edematose, (o)edematous, dropsical
отжиг 1. annealing; firing; 2. fritting
~, низкотемпературный stress-relief annealing
~, предварительный prefiring
~, предварительный в водороде prehydrogen firing
~ ртутного вентиля bake-out of a mercury-arc rectifier
~ стали, диффузионный steel homogenizing (annealing)
~ чугуна на ковкость malleablizing of cast iron
отжигательный annealing
отжигать anneal; malleablize; *el.* fire
отжим, отжимание expression, pressing, squeezing/wringing/pressing out; centrifuging
отжимать squeeze/press/force/wring out
отжимный squeezing/wringing/pressing out
отзвук echo; reverberation; *fig.* response; repercussion
отзовист recallist
óтзол *leath.* process of liming; tannery waste
óтзыв 1. opinion, judgement; reference, comment; review; 2. *mil.* counterparole; reply
отзы́в *dipl.* recall
отзывной recall *(attr.)*; *fin.* revocable
~ные грамоты letters of recall
отиатр otologist, aurist, ear specialist
отиатрия otiatries, otiatrics, otiatria
отит *med.* otitis, inflammation of the ear
~, внутренний otitis interna/intima/labyrinthica, labyrinthitis
~, гнойный purulent/suppurative otitis
~, наружный клещевый ear/otodectic mange, otacariasis, parasitic otitis
~, средний inflammation of tympanum
отит-антрит *med.* otoantritis
ОТК *abbr.* **отдел технического контроля** inspection department, quality control department
отказ 1. refusal, rejection; denial; disallowance; 2. *tech.* failure, fault; breakdown, nonoperation; 3. *law* relinquishment, renunciation; 4. *mus.* natural
~ аппаратуры equipment failure
~ взрыва misfire
~ в помощи denial of assistance
~ в предоставлении ссуды loan rejection
~ в процессе эксплуатации in-service failure, operational failure
~ в работе breakdown
~ вследствие износа wear-out failure
~, вторичный/зависимый dependent/secondary failure
~ двигателя engine failure
~ из-за занятости busy reject
~ из-за неправильного обращения misuse failure
~, категорический flat refusal
~, конструкционный design error failure
~ материальной части materiel failure
~, независимый/первичный independent/primary failure
~, обнаружимый detectable failure
~ от исполнения команды *comp.* command reject
~ от опциона/премии *fin.* abandonment of option
~ от платежа refusal of payment; dishono(u)r
~ от права *law* waiver, disclaimer
~ от протеста (векселя) *law* waiver of protest
~ от разговора *teleph.* cancellation/withdrawal of a call
~, первичный primary/independent failure
~, первый initial failure
~ периода переработки *tech.* infant failure
~ по вине обслуживающего персонала human failure
~ по (несовпадению) четности *comp.* parity failure
~, полный total/complete failure
~, постепенный creeping(-type) failure, degradation of performance
~, приработочный running-in failure, infancy failure
~ резервированной системы redundant failure
~, самоустраивающийся transient failure
~, скрытый latent failure
~, случайный random/unpredictable failure
~, усталостный endurance/fatique failure
~, эксплуатационный operational/field failure
~ электропитания power failure
отказоустойчивость fault-tolerance
отказывать 1. reject, refuse, deny, decline; 2. dismiss, discharge; give notice to; 3. *tech.* fail, break down
отказываться 1. decline, give up, turn down; 2. abnegate; abdicate, relinquish; *law* waive, disclaim
~ от веры *rel.* apostatize
~ от исполнения преступления *law* countermand the execution of the crime
откалывание chipping-/flaking-/breaking-/splitting off, spalling; *geol.* calving
откат 1. recoil; 2. rollback, rolling back/away; *naut.*

backrush
~ **волны** back draft/sweep of wave
~ **с повторением обработки** *comp.* rollback-and-retry
~ **ствола (орудия)** barrel/gun recoil
откатка *min.* haulage, hauling; trucking
~, **безрельсовая** trackless haulage
~, **ручная** hand tramming
~, **самокатная** gravity haulage
откаточный haulage *(attr.)*
~ **штрек** haulage drift
откатчик dragsman, haulage man, trammer, wheeler
откатывать roll aside/away; wheel; tram; haul
откачивание exhaustion, evacuation, pumping out; scavenging
~ **поршнем** swabbing
откачивать pump out, evacuate, exhaust
откачивающий scavenging, suction *(attr.)*
откачка pump(ing)-down/-out, evacuation, exhaustion, scavenging
~ **из оперативной памяти** *comp.* rollout
~, **криогенная** cryopumping
~, **повторная** re-evacuation
откашливание *med.* expectoration, coughing up
откашливающее *n. med.* expectorant
откидывать throw/cast away
~ **вниз** swing down
~ **на петлях** hinge out
откидыш *bot.* bulbil
откладка *zool.* oviposition
откладывать 1. deposit, lay down, lay off; *biol.* lay; 2. set aside; defer, postpone
~ **вынесение решения суда** *law* reserve judgement
~ **исполнение приговора** *law* sustain conviction/sentence
~ **по оси X** lay off as abscissa, plot on the abscissa
~ **яички** flyblow
~ **яйца** lay eggs, oviposit
откладываться 1. *geol.* deposit, settle, precipitate; 2. detach oneself; *polit.* secede
отклик response; *fig.* repercussion, echo
~, **ожидаемый** *comp.* predicted response
отклонение 1. deflection, deviation, departure, divergence; declination; 2. *stat.* digression, variance, aberration, discrepancy; 3. refusal, disallowance
~, **акустическое** acoustic deviation
~ **атомного пучка** atomic beam deflection
~, **баллистическое** ballistic deflection
~, **боковое** *av.* lateral deviation
~ **в процентах** percentage deviation
~ **вызванное действием силы тяжести** gravity deflection
~, **гасить** correct deviation
~, **допустимое** permissible/admissible divergence/deviation, acceptable variation, tolerance
~ **до упора ручки управления** full movement
~ **жилы** *geol.* vein departure
~, **квартильное** *stat.* semi-interquartile range
~ **магнитной стрелки** deviation of magnetic needle
~ **матки в сторону** *med.* lateroversion of uterus
~ **матки кзади** retroversion of uterus
~ **матки кпереди** anteroversion of uterus
~ **меандр** *geol.* swinging of meanders
~, **многократное** multiple deviation
~ **назад** retroversion
~, **накопленное** cumulated deviation
~, **нулевое** zero drift, drift of zero
~ **отвеса за рельеф** topographical effect
~ **от (заданного) маршрута** track excursion
~ **от (заданной) траектории** trajectory deviation/error
~ **от истины** prevarication
~ **от курса** course deviation, yaw(ing), sag
~ **от линейной зависимости** departure from linearity
~ **от прямой** cant
~ **от типа** *biol.* variance
~ **оценки, стандартное** standard error of estimate
~ **по крену** roll deviation
~ **по тангажу** pitch deviation
~ **под действием силы Кориолиса** Coriolis deflection
~, **путевое** yaw deviation
~ **самолета** aeroplane diversion
~ **скважины** hole deviation
~, **среднее** mean/average deviation
~, **среднее арифметическое** mean absolute error/deviation
~, **среднеквадратическое** (root) mean square deviation, standard deviation
~ **струи** jet deflection
~, **угловое** angular deviation/deflection
~ **цены от стоимости** divergence of price from value
~, **чрезмерное** hard-over
~ **электрической оси сердца** *med.* axis shift
~ **элеронов** aileron deflection
~, **явное** manifest deviation
отклоненный deflected, diverted, deviated
~ **вниз** declinate, bent downward
~ **назад** reclinate
отклонитель deflecting tool, deflector; whipstock
~, **извлекаемый** retrievable whipstock
отклоняться 1. diverge, deviate, digress; 2. *naut.* swerve, yaw, sheer, edge away; 3. slant, tilt; decline
~ **влево** *naut.* swerve out to port
~ **вправо** *naut.* swerve out to starboard
~ **от истины** prevaricate
~ **от отвесной линии** get out of plumb
отклоняюшийся aberrant
отключатель circuit breaker
отключать disconnect; turn/switch off; deenergize; disengage, detach
~ **питание** deenergize, switch off
отключение disconnection, disengagement, cut-out/-off, switch(ing)-off; deenergization, shutting down

~, **аварийное** emergency cut-out, emergency shutting down
~ **питания** power cut-off
откорм *agr.* fattening; cramming
~, **быстрый** short feeding
~ **птицы** fowl cramming
~, **станковый** stable fattening
откормка fattening (up)
откормочная *agr.* stable for feeding
откос 1. slope, bank, (e)scarp; aclivity, declivity, inclination; shelving; 2. slant; *constr.* strut
~, **береговой** bank slope
~, **естественный** slope of repose
~ **из каменной наброски** talus slope
~ **плотины, верховой** back of a dam
~, **подводный** sea scarp
~, **предконтинентальный** continental rise
откосник back sloper
откосный declivous, sloping; slanting
открепление detachment, removal
~ **ресурса** resource deallocation
откровение reveal, revelation, apocalypse
~ **Иоанна Богослова** *eccl.* the Revelation
откручивать untwist, untwine; unscrew, screw off
открыватель 1. discoverer; 2. opener
открывать 1. detect, discover, find, uncover; 2. reveal, disclose; 3. open
~ **жилу** *geol.* strike a lode
~ **счет** *econ.* open an account; *sport.* score first
~ **месторождение** discover/detect a deposit
открытие 1. inauguration, opening; 2. discovery, detection; disclosure
~ **замков** unclocking, lock release
~ **месторождения** strike
~, **последовательное** chained opening
открытка (picture) post card
открытоплодные *n. pl. bot.* discomycetes *(Discomycetes)*
откуп 1. ransom; 2. *hist.* (tax) farming
отлагательство procrastination; delay
отлагать *geol.* deposit, settle
отлагаться precipitate, deposit, lodge (on)
отладка adjusting, tuning up; check-out
~ **программы** *comp.* debugging
~ **систем** test of systems, check out of systems
отладчик *comp.* debugger
отлаживание debugging
отлаживать adjust, tune up; *comp.* debug; check out, rectify
отлив I 1. ebb (tide), falling tide; reflux; 2. pouring off; pumping/bailing out; 3. discharge; 4. casting, founding
~ **и прилив денег** flux and reflux of money
~, **квадратурный** dead/neap tide
отлив II hue, shade, tint; *miner.* play, chatoyancy
отливка 1. casting, founding; 2. *met.* cast, ingot, mo(u)lding; 3. pouring/bailing/pumping out; pouring off
~, **бесстержневая** coreless casting
~ **в землю** loam casting
~, **недолитая** short-run casting
~, **опочная** box casting
~, **отбеленная** chilled casting
отливной cast, mo(u)lded, founded
отливное течение ebb current/tide
отливочный casting, founding, mo(u)lding
отливщик caster, founder, mo(u)lder
отлип tackiness, tack
отличие difference, dissimilarity; contrast; differentiation; distinction
~ **за храбрость** decoration for bravery
~, **конструктивное** structural difference
~**м, диплом с** hono(u)rs degree
отличительный distinctive, distinguishing; differential, discriminative
отличник 1. *educ.* hono(u)rs pupil/student; high achiever; 2. standout; outstanding worker; first-ranker
отличный 1. excellent, perfect; 2. distinct; diverse; dissimilar
отлог inclination, slope
отлогий (gently) sloping
отложение 1. deposit(ion), sediment(ation), precipitation; 2. laying *(of eggs)*; 3. postponing; 4. *hist.* secession
~ **белков в организме** proteopexy
~ **жира** *med.* adiposis, adiposity, adipositas, adipopexia, lipopexia, obesity
~ **красящего вещества** pigmentation
~ **нагара** carbon build(ing)-up
~ **наносов** alluviation
~ **осадков** sedimentation, deposition of sediments
~, **повторное** recurrent deposition, redeposition
~, **подагрическое** gouty deposit
~, **радиоактивное** radioactive deposit
~, **удельное** specific deposition
отложения *pl.* sediments, deposits
~ **атмосферной пыли** dust deposits
~, **береговые** beach/shore/littoral deposits
~, **грубообломочные** coarse deposits
~, **доледниковые** preglacial drift
~, **донные** bottom deposits/sedimentation
~, **каменноугольные** carboniferous deposits, coal measures
~, **камерные** cameral/chambered deposits
~, **континентальные** continental detritus
~, **корковые** crusted deposits
~, **красноцветные** red beds
~, **кремнистые** siliceous deposits
~, **ледниковые** glacial deposits/drift
~, **лимнические** lacustrine deposits
~, **мелкозернистые донные** fine sediment(s)
~, **наносные** silting deposits, driftbeds
~, **напластованные** bedded deposits
~, **нефтеносные** oil deposits, petroliferous strata, oil-bearing beds
~ **образованные айсбергами** berg-till
~, **одновозрастные** coeval deposits
~, **озерные** lacustrine/lake deposits
~, **окаймляющие** skirting/flanking deposits

~, перемежающиеся alternating deposits
~, пещерные cave/cavern deposits
~, подстилающие underlying deposits
~, пойменные flood-plain deposits
~, покрывающие cover beds
~, почвенные soil deposits
~, прибрежные inshore/coastal deposits, near-shore deposits
~, речные fluvial/fluviatile deposits
~, связанные coherent deposits
~, слоистые stratified/bedded deposits
~, современные recent beds/deposits
~, эоловые (a)eolian deposits
отлуп *for.* circular shakes, internal annual shake
отлучать remove, separate, exclude
~ от церкви *eccl.* excommunicate
отлучение *eccl.* excommunication
отлучка absence
~, самовольная *mil.* absence without leave
отматывание *print.* setoff, offsetting
отмелый beachy, shelfy; shoaly; shallow
отмель (offshore) bank, shallow, sandbar, high bed, shoal border/stretch
~, береговая in-shore shoal
~, береговая вокруг группы островов archipelago apron
~ в устье реки river mouth bar
~, глинистая argillaceous bank
~, каменистая rocky shoal
~, континентальная continental fringe
~, материковая (continental) shelf
~, надводная emerged shoal
~, намывная overwash flat
~, песчаная sand bank
~, подводная submerged shoal
~, устричная oyster-bed
~, устьевая river bar
отмена cancellation, annulment; abrogation, repeal; abolition; *law* . revocation, rescission; circumduction, defeasance, avoidance
~ договора (купли-продажи) покупателем redhibition
~ золотого стандарта suspension of gold standard
~ рейса *av.* flight cancellation
~ решения revocation of a decision
отменение abolition, abolishment, abrogation
~ крепостного права *hist.* abolition of serfdom
отменять annul, cancel, abolish; *law* circumduct, rescind, quash, countermand, overrule; abrogate
~ комендантский час lift the curfew
~ приказ countermand an order
отметка 1. sign, mark(er), note, grade, check point; date; 2. marking, indexing
~ времени time/timing mark, timing (line), time break
~, высотная elevation mark, level
~ высоты над уровнем моря bench mark, elevation
~ горизонтали, цифровая contour number
~, исходная высотная datum level
~, калибровочная calibration mark(er)/ring
~, контрольная reference mark
~ крена, линейная roll bar
~ курса (корабля) *nav.* (ship's) heading marker
~, линейная index line
~, масштабная scale marker; calibration mark(er)
~ прибоя, верхняя swash mark
~, прицельная aiming mark, tracking index
~, проектная design elevation
~ режима event marker
~ репера/реперная bench mark (elevation), fixed datum
~, точечная dot
~ уровня, опорная ordnance datum
~ цели (target) blip; target spot; target marker circle
~ шкалы scale/graduation mark, scale graduation
~, юстировочная adjustment mark
отметчик 1. recorder; marker; indicator
~ времени time/timing marker; *electr.* time-mark generator
~ курса heading marker
отмечать mark; note, register; record
отмирание *med.* dying off, atrophy, disappearance; necrosis, fading/withering away
~ верхушек *bot.* dieback, perishing
~ государства *polit.* withering away of the state
отмирать die off/away/out, disappear, wither away, perish
отмока *leath.* soak(ing), softening, soaking process
отморожение *med.* frostbite
отмотка *text.* unwinding; backing-off
отмучивание decantation, elutriation, clarification
отмучивать elutriate, wash, clarify, levigate, decant
отмывка 1. panning, washing (for gold); 2. *cart.* tinting
~ рельефа relief shading, wash-drawing
отнесение reference; attribution
относительность relativity, relativeness
относить(ся) *math.* refer, relate, apply
отношение 1. attitude; treatment; 2. correlation, relation(ship); 3. *math.* ratio, quotient
~, ангармоническое anharmonic/cross ratio
~, безразмерное dimensionless ratio
~ биологического эффекта к физическому biological-physical ratio
~, бридерное breeding ratio
~ вероятностей альфа- и бета-распада alpha-to-beta-decay ratio
~ веса к емкости weight-to-capacity ratio
~ включения inclusion relation
~ гомотетии *math.* ray ratio
~ графит/уран graphite-uranium ratio
~, двойное double/cross/anharmonic ratio
~, двучленное binary/dyadic relation
~ действительных и случайных отсчетов real-to-random-count ratio

~ длительности импульса к периоду повторения pulse-duration/-time ratio, duty ratio
~ дохода к издержки benefit-cost ratio
~ жидкого к твердому liquid-to solid ratio
~ заряда к массе charge-to-mass ratio
~ ионизирующих способностей ionization ratio
~, количественное magnitude relation
~ максимума к минимуму peak-to-valley ratio
~ масс осколков mass ratio of fission/fragments
~, многозначное multivalued relation
~, модельное model ratio
~, молярное molar ratio
~ непосредственного предшествования immediate predecessor relation
~ непосредственного следования immediate successor relation
~, обратное reciprocal/inverse ratio
~, передаточное transmission/gear ratio
~ плотностей деления fission-density ratio
~ подчиненности owner-member relation
~ порядка order(ing) relation
~ правдоподобия *stat.* likelihood ratio
~ предпочтения preference relation
~, причинно-следственная causal relation
~, пробочное mirror ratio
~, разностное difference quotient/ratio
~ сигнал-шум signal-to-noise ratio
~ скоростей velocity ratio
~ сущность-связь entity/relationship relation
~ торий-топливо Th-to-fuel ratio
~ тяги к весу thrust-to-weight ratio
~ тяги к лобовому сопротивлению thrust-to-drag ratio
~ фон-сигнал hum/noise-to-signal ratio, background ratio
~, характеристическое aspect ratio
~, шаговое pitch ratio
~ эквивалентности equivalence relation
~ния, деловые dealings; business relations
~ния обратных связей feedback relations
~ния, общественные social relations
~ния, производственные production relations
~ния, родственные kinship relationship
~ния, рыночные товарные exchange relations
~ния, финансовые fiscal/financial relations
отнятие removal, elimination; deprivation; ablation
~ от груди weaning
отоантрит *med.* mastoidal antritis, otitis mastoidea
отображать represent; *math.* map
~ гомоморфно map homomorphally
отображение representation, reflection; display; imaging; *math.* map(ping)
~, антиконформное anti-isogonal conformal mapping
~ в себя self-mapping
~, вероятностное probabilistic display
~ данных *comp.* data display/(re)presentation
~, зеркальное mirror transformation
~, знаковое alpha-numeric display, character (data) display
~, классифицирующее classifying map
~, конформное conformal/isogonal transformation, equiangular transformation; *math.* conformal mapping
~ множеств mapping of sets
~ обстановки situation display
~, однолистное univalent mapping
~ первого рода isogonal conformal mapping
~ расслоений *math.* fibre map
~, топологическое topological map(ping), homeomorphism
отогнутый bent, recurved, deflected, declinate; canted
~ вниз *bot.* progeoesthetic, bent downward
~ назад *bot.* replicate, replicative; *med.* retroflexus
отогревать reheat; thaw out
отож(д)ествление identification
отож(д)ествлять identify; homologate
отозвание *polit.* recall
отоларинголог *med.* ear specialist, otolaryngologist
отолит *anat., zool.* otolith, otolite, ear dust/stone, otosteon, otoconium
отолог otologist
отоокципитальный oto-occipital
отопиорея *med.* otopyorrhea
отопитель heater
~, бензиновый/мазутный combustion heater
отопление heating
~, внутреннее interior heating
~, воздушное hot-/warm-air heating
~, калориферное hot-air heating
~, лучистое radiant/direct heating
~ от теплоцентрали district heating
~ отбросным теплом waste heating
~ отходящими газами waste gas heating
~, печное stove heating
~, радиационное direct/radiant heating
оторея *med.* otorrh(o)ea, discharge from the ear
оториноларинголог *med.* oto(rhino)laryngologist
оторочка 1. margin, fringe, edge; 2. *bot.* armilla; limb(us); 3. trimming
~, водяная water fringe
~, лопастная flaps of skin
~, пленчатая filmy margin
~, плотная dense margin
оторрагия otorrhagia
оторфление, оторфование *geol.* peat formation
оторцовка *for.* trimming, cross-cutting
отоэнцефалит *med.* otoencephalitis
отпадающий deciduous
отпаивать 1. unsolder; 2. *el.* (provide a) tap; 3. seal off
отперфорированный *comp.* punched, keyed
отпечаток (im)print, impress(ion); replica; copy; mould, cast
~ большого пальца thumbprint

~ валков *met.* roll mark
~, пробный proof
~, фотографический photographic print
отпечатывать 1. (im)print; type; 2. unseal, open (up)
отплата remuneration, retribution, repayment; revenge
отплывать *nav.* (set) sail; get underway; swim/float off
отплытие sailing, departure
отпор rebuff; repulse
~, дать drive back, repulse
отпотевание sweat(ing); fogging, misting
отправитель *com.* sender, transmitter; shipper, consignor, originator
отправка shipping, dispatch(ing), sending off; *com.* transmission
~ груза shipment/shipping of cargo
~ на консигнацию consignation
~ навалом shipping in bulk
~ партиями shipping in lots
отправление I 1. transmitting, transmission; sending, dispatching; shipping, consignment; 2. departure
~, почтовое (item of) mail
отправление II exercise, performance; function
~ должности exercise of one's duties
отправлять send, dispatch ship; *com.* transmit
~ бандеролью/посылкой send by parcel post
~ водным/железнодорожным транспортом dispatch/ship by water/rail
отправной initial, starting
~ная точка datum, starting point; point of reference
отправочный forwarding, consigning, dispatching, shipping
отпрыск 1. *bot.* offshoot, scion; 2. offspring, descendant; brat
~ жилы *min.* spear
~, корневой root sucker
отпрысковый *bot.* proliferous, pullulant
отпрысковый, отпрысконосный *bot.* sarmentose, sarmentiferous
отпугивание frightening away
~ птиц от аэродрома bird scaring
отпугивающий *zool.* aposematic, repugnatorial; deterrent
отпуск I 1. leave (of absence), holiday(s), vacation; 2. *eccl.* Nunc Dimittis
~, внеочередной emergency leave; *mil.* special liberty
~, оплачиваемый holidays with pay
~ по беременности и родам maternity leave
~ по болезни sick leave
~ по домашним обстоятельствам compassionate leave
отпуск II 1. issue, issuance, delivery, distribution; 2. *tech.* tempering, drawing(-back)
~ в масле oil tempering
~ (стали), высокий/низкий high/low (-temperature) tempering
отпускание release, freeing
~ реле drop-out of a relay
отпускать 1. release, set free, let go/off; 2. loosen, unfasten, slacken, relax, ease (off); 3. *tech.* anneal, temper, draw (back); 4. *eccl.* forgive, remit
~ грехи *eccl.* give absolution
~ на волю *law* disenthral(l), manumit
отпускная *n.* leave pass; *law* letter of affranchisement, manumission
отпускник, отпускной vacationer, vacationist, holiday-maker; *mil.* serviceman on leave
отпуст *eccl.* Nunc Dimittis
отпущение *rel.* remission, remitment
~ грехов absolution
~ на волю enfranchisement, release
отпущенник *hist.* freedman
отрабатывать 1. work off; 2. develop; optimize; perfect
~ норму/время/срок complete/serve one's term/time/quota, fulfil one's task
отработавший spent, used, waste
~ пар exhaust/spent steam
отработанный worked out; exhaust(ed), used, waste, spent
~ газ exhaust gas
~ное масло used/waste oil
отработка 1. paying by work, working off; 2. finishing, development (work), optimization, perfection; 3. *mil.* training, drill
~ аппаратуры final adjustment of equipment
~ совместных действий exercising in coordinated operations
отравление poisoning, intoxication; venenation
~ алкоголем alcohol(ic) poisoning, alcoholism
~, долговременное long-term poisoning
~, медикаментозное drug poisoning
~, общее systemic poisoning
~ осколками fragment poisoning
~, пищевое food poisoning, bromatotoxism
~ трупным ядом *med.* ptomaine poisoning
~, эндогенное endointoxication, autointoxication, self-poisoning
отравляющее вещество (war)gas, toxic agent
~, боевое (warfare) agent, war gas
~, жидкое toxic liquid
~ кожно-нарывного действия vesicant/blister gas
~, мышьяксодержащее arsenic war gas
~ нервно-паралитического действия nerve gas/agent
~ общеядовитого действия systemic poison
~ рвотного действия vomiting gas
~ слезоточивого действия tear gas, lacrimator
~ удушающего действия choking/suffocating gas
отравляющий poisoning, toxic, intoxicating
отражатель 1. reflector, mirror; 2. deflector, baffle, repeller; 3. ejector

~, антенный antenna/aerial reflector
~, вдвижной move-in reflector
~, дипольный *rad.* window, *Am.* chaff
~ звука sound baffle, sounding board
~, зеркальный specular/mirror reflector
~, клистронный klystron repeller
~, лазерный laser ranging retro-reflector
~, многозонный multiregion reflector
~, многослойный multilayer reflector, multireflector
~ молотильного аппарата cut-off plate/bar, cylinder cut-off
~ не воспроизводящий топливо *nucl.* non-fertile reflector
~ нейтронов neutron reflector/mirror
~, пружинный spring-actuated ejector
~, радиолокационный radar reflector
~ факела *rock.* flame/blast/exhaust/jet deflector
~ частиц particle repeller
отражательный reverberatory, reverberative *(of furnace);* reflecting, reflective, deflecting; catoptric
отражать 1. reflect; repell; mirror; 2. reverberate; rebound; 3. display, indicate, represent; 4. repel, repulse, rebuff; parry, avert
~ атаку/нападение repulse/rebuff an attack
~, повторно re-echo
отражение 1. reflection, mirror image; reflectance; 2. reverberation; echo; display, representation, indication; 3. rebuff, repulse; warding off, parry
~ атаки repulse of an attack
~, брэгговское Bragg/constructive reflection
~, двунаправленное bidirectional reflection
~, диффузное diffuse/scattered reflection
~ звука sound reflection
~, зеркальное mirror reflection/image, specular/regular reflection
~, ложное false echo
~, многократное multiple reflection
~, многократное радиоволн multi-hop reflection
~, нерегулярное sporadic reflection
~ от дна bottom reflection
~ от земли ground echo/reflection, ground/land return
~ от местности terrain echo
~ от морских волн sea clutter
~ от фона clutter
~, полное внешнее total external reflection, TER
~, полное внутреннее total internal reflection, TIR
~, пространственное reflection, inversion
~, рассеянное scattered/diffuse reflection
~ скачка уплотнения shock-wave reflection
~, скользящее sliding/glancing reflection
отраженный reflected, reverberated; echoed
отразитель *polit.* rejector of foreign influence/ideas
отрасль 1. branch; field; 2. sprout, shoot, sucker; offspring, scion
~, капиталоемкая capital-intensive industry/branch
~, необрабатывающая nonmanufacturing industry
~ промышленности trade, (branch of) industry
~ли, производящие орудия производства instrumental industries
~ли промышленности и услуг, коммунальные public utility industries
отреагирование *psych.* abreaction; catharsis
отрезать cut (off); sever; clip
~ по шаблону cut to template
отрезка cutting-off
отрезок 1. cut, length, leg, section, piece; fragment; *math.* (line) segment; *hist.* plot (of land); 2. distance, interval, stretch
~ воздушной дистанции над ВПП runway air distance
~ времени space of time
~ интегрирования interval/range of integration
~ корня *hort.* root cutting
~ кривой portion of a curve
~ маршрута leg/section/segment of a route, route segment/sector
~ маршрута между пунктами посадки *av.* stage
~ маршрута полета leg (of the flight)
~, отсекаемый на оси абсцисс X-intercept
~, прямой (straight) line segment
~ пути stretch of road
~ ряда Тейлора *math.* truncated Taylor series
~ с противопосадочным курсом *av.* down wind leg
~ следа track segment
~ цилиндра *math.* ungula of the cylinder
отрекаться renounce; disavow; repudiate; fore-swear
~ от престола abdicate
отречение abnegation; disclaimer; renunciation (of), repudiation, disavowal; abdication
~, клятвенное *law* abjuration
отреченный rejected; apocryphal; false
отрицание (ab)negation; denial; disavowal; *philos.* sublation
~ дизъюнкции *comp.* nondisjunction
~ импликации *math.* nonimplication
~ конъюнкции *comp.* nonconjunction
~ переноса *comp.* no/not-carry
отрицательный negative; unfavo(u)rable, deleterious, bad
~ герой *lit.* anti-hero
~ фототропизм aphototropism, negative phototropism
отрог *geol.* spur, offshoot; *meteor.* ridge
~, обработанный trimmed spur
~ озера arm of lake
отрок *hist.* servant; page
отростки *pl. med.* processes
~ клеток, пальцевидные digitate/fingerlike cellular processes

~, клеток, полиповидные polypoid cellular processes
отросток 1. *anat., zool.* process, projection, outgrowth, apophysis, appendix; 2. *tech.* branch, extension; 3. *bot.* shoot, sprout
~, боковой offshoot, offset
~, височный temporal process
~, имеющий appendiculate
~, имеющий шпоровидный calcarate
~, замочный hinge/cardinal process
~, зубовидный odontoid process
~, клювовидный coracoid process
~, крыловидный alar projection
~, крючковидный hamulus, hamus; uncinate/unciform process
~ линии spur-line, T-line
~, локтевой funny bone, tip of the elbow
~, мечевидный xiphoid process/cartilage, ensiform cartilage, metasternum
~, остистый spinous process
~, плечевой acromial process
~, поддерживающий sustentaculum
~ поковки *tech.* tag
~, серповидный falciform process/ligament
~ слепой кишки appendix
~, сосцевидный mastoid process
~, суставной articular process
~, хвостатый caudate process
~, шиловидный styloid process
~, югальный/яремный jugal/jugular process, jugum
отруб *for.* butt (of tree)
отрубать chop/cut off, clip
~ зубилом chisel (off)
отрубевидный *med.* branny, branlike; scurfy, scaly, furfuraceous, furfurous
отрыв 1. breaking off/away/loose, separation; detachment, removal; breakdown, break-away; 2. *med.* abruption, avulsion; 3. *fig.* alienation, loss of contact
~ двигателя *av.* engine break-away
~ нейтрона от ядра removal of a neutron from the nucleus
~ носового колеса nose-wheel lift-off
~ от земли *av.* takeoff, liftoff
~ плазмы detachment of plasma
~ плаценты *med.* abruptio/ablatio placentae
~ потока flow separation
~ при/на взлете unstick, *Am.* lift-off
~ электронов detachment of electrons
отрывать I break off/away/loose; detach, separate
отрывать II dig (up/out), excavate, unearth
отрываться break away, detach, separate; *av.* unstick; lift off
~ от земли *av.* get/become air-borne
~ от причала *naut.* break adrift/loose, fetch away
~ от противника gain distance, dart off
отрыжка *med.* eructation, belch(ing)
~ газами gaseous eructation
~, кислая sour eructation
отряд 1. squadron, force, group, detachment, unit; party; 2. *biol.* order
~, аварийно-спасательный rescue party
~ водолазов-подрывников underwater demolition unit
~, военно-морской naval party
~ десантно-высадочных средств landing-craft squadron
~, десантный landing party
~, минный mine force
~, оперативный task unit
~, опытовый development unit
~, передовой advance party; vanguard
~, пожарно-спасательный fire and rescue party
~, поисковый scouting/sweeping force
~, противолодочный anti-submarine unit
~, спасательный salvage/rescue party/unit
~ судов снабжения logistic support unit
~, учебный training detachment
~ эскортных кораблей escort force
отсадка 1. *min.* jigging, ore dressing; 2. *hort.* transplanting; 3. separation, isolation
~, контрольная check jigging
~, крупная coarse jigging
~, мокрая wet jigging, hotching
отсадок transplanted plant
отсаживать 1. *min.* jig; 2. *hort.* plant out, transplant; 3. set apart
отсасывание suction, sucking/drawing off, drawoff; exhaustion; aspiration
~, длительное/непрерывное prolonged/continuous suction
отсасыватель *med.* suction unit/device, aspirator
отсасывать suck/draw off, aspirate, draw away
отсвет reflection, sheen
отсвечивание glare; reflection, sheen
отсвечивать reflect, shine, gleam
отсев 1. riddling, screening, sifting; elimination; dropout; 2. chaff; riddlings, screenings, siftings
отсевки *pl. agr.* siftings, screenings
отсеивание sifting; *fig.* elimination, screening
~ элементов массива *comp.* sifting
отсек compartment, room, space, cubicle, bay, cell; *space techn.* module
~, аккумуляторный accumulator/battery room/space
~, багажный baggage compartment; *rail.* luggage cupboard; *av.* luggage section/hold
~, баковый tank section
~, боевой warhead
~, бомбовый *av.* bomb bay/cell
~, водяной water compartment/vessel
~ вспомогательного оборудования accessory bay
~, газовый gas cell
~, герметический/герметизованный pressurized compartment/bay/section
~, грузовой freight/cargo compartment/hold
~ двигателя *av.* engine bay/compartment

~, движущийся по орбите *space techn.* orbiting module
~ для работы и отдыха work/rest module
~, жилой living compartment/module
~ измерительных приборов instrumentation bay/compartment
~ картера crankcase compartment/section
~, кондиционированный conditioned bay
~ космонавта astronaut's compartment
~ космонавтов command module
~, лунный lunar module
~ маслобака, циркуляционный hopper, hotwell
~, машинный machinery compartment
~, наблюдательный viewer/viewing box
~ оборудования equipment bay/compartment
~, огневой *rock.* test cell, firing bay
~, орбитальный orbital module/section
~, передний *rock.* forebody
~, переходный adapter module
~ плавучего дока, сухой buoyancy compartment
~, пушечный gun bay
~, расстыкованный undocked module
~, реакторный reactor compartment/cubicle
~, сбрасываемый *rock.* detachable section
~ систем service module
~ с полезной нагрузкой payload capsule
~, стыковочный docking module
~, суживающийся tapered section
~, топливный *rock.* propellant/fuel compartment/section
~ управления control module
~, успокоительный anti-surge compartment
~, центральный *av.* centre-body section
~ шасси undercarriage bay, landing-gear bay
~, шлюзовой airlock compartment
~ экипажа crew compartment/module
~ электронной аппаратуры electronics compartment/bay
~ электрооборудования electrics bay
отсек-убежище escape/refuge compartment
отсек-хранилище storage bay
отсечение severance, cutting off, amputation
~ части графического изображения *comp.* scissoring
отсечка 1. cut(ting)-off, clipping; shutoff, shutdown; 2. interception; *pl. print.* serifs
~ двигателя engine cut-off
~ (подачи) топлива fuel cut-off
отскакивание rebound, ricochet, bouncing back
~ контактов *el.* contact bounce
отскакивать recoil, rebound, bounce
отскок recoil, rebound
отслаивание exfoliation, desquamation, delamination; *tech.* peeling (off), flaking, spalling
~ краски flaking
~ многослойного материала ply separation
~, чешуйчатое flaking, spalling, spallation
отслаивать 1. exfoliate, peel/strip off; 2. deposit
отслоение exfoliation; layer, strip
отслойка exfoliation; detachment, separation
~ плаценты *med.* placental detachment
~ сетчатки *med.* detached retina
отсоединение 1. *mech.* detachment, disengagement, separation; 2. *el.* disconnection, switching/turning off, isolation
отсоединять *mech.* detach, disengage, release; *el.* disconnect, switch/turn/cut off
отсортированный sorted, picked; graded
отсос suction; drain
~ под разрежением vacuum suction
~, щелевой slot suction
отсрочка postponement, delay; *econ.* (appropriation) deferment; grace (period); *law* respite, adjournment; prolongation, moratorium
~ налога tax deferment
отстав *print.* lining
отставание 1. lag(ging); *phys.* hysteresis; 2. delay, retardation; 3. detachment, peeling off, exfoliation
~ выхлопа exhaust lag
~ зажигания ignition lag
~ по времени time lag
~ раската от валка backward creep
отставать lag/fall/drop behind; detach itself, separate (from); *naut.* fall astern
отставка 1. dismissal, discharge; 2. resignation, retirement
отстаивание 1. settling, gravity sedimentation, deposition; 2. upholding, standing up for
отстаивать defend; advocate, dispute, assert; uphold, stand up for
отстаиваться settle, precipitate
отсталость backwardness; immaturity
~, умственная *med.* mental retardation/deficiency/debility, feeblemindedness, moronity, morosis
отсталый *n.* backward/retarded person; straggler
отстегивание unfastening, unbuttoning
~ привязных ремней *av.* unstrapping
отстой sediment, deposit, residue, bottom(s), dregs, sludge
~ вина lees
отстойник settling basin/tank, precipitation/septic tank, sedimentation basin/tank, storage pond, settler, sump, cesspool
~, грязевой mud settler/sump
~ коксовой батареи hot drain tank, tar well
~, масляный oil sump
~, шламовый slurry/slime pond, slime tank
отстойный settling, desilting
отстояние clearance, space, interval; *top.* distance, range
отстраиваться *rad.* tune out/away (from)
отстранение dismissal, dicharge
~, временное (от должности) suspension
отстранять dismiss, dicharge; suspend
отстрел 1. shooting, firing; 2. *rock.* blast-off, separation; 3. *av.* jettisoning
~ взрывателей fuse calibration

~ (крышки) люка hatch(-door) jettisoning
~ начальных скоростей velocity fire
~ орудия gun calibration fire
~, пробный proof firing
~ фонаря canopy jettisoning
отстреливать *hunting* kill, shoot (down)
отстройка *rad.* detuning, mistuning; tuning out
отступ 1. *print.* space, break off; 2. *phon.* final glide
~, абзацный *print.* (paragraph) indention
отступание recession, regression; waning
~ берега retreat of the shore
~ ледника deglaciation, ice waning
отступательный *mil.* retreat *(attr.),* withdrawal *(attr.)*
отступать 1. recede, regress; step/move back; *mil.* fall back, retreat; 2. give up; back down; 3. *print.* indent
~ от веры *rel.* belie one's faith
отступаться renounce, give up; apostatize
отступление 1. digression, deviation, shift; re(tro)gression, recession; 2. *mil.* retreat
~ береговой черты retrogression of beach
~ от договора recession from a contract
~ от правил deviation from the rules
отступник secessionist; apostate, recreant; turncoat
отступничество apostasy; abjuration, recreancy
отсутствие absense, nonexistence; deficiency, lack
~ видимости zero/nil visibility
~ доказательства failure of proof
~ завещания *law* intestacy
~ знаний inscience
~ избыточности *comp.* nonredundancy
~ листьев *bot.* aphylly
~ связанности incoherence
~ сигнала no-signal (condition/state)
~ хода standstill
отсутствующий absent, missing; wanting; *n.* absentee
~, безвестно *law* untraceable
отсчет 1. reading, readout, count; display, marking; 2. *comp.* sample
~, брать/производить/снимать take a reading, read
~ в реальном масштабе времени real-time readout
~, визуальный visual/eye reading, visual readout
~ времени time reading/reference; *rock.* countdown
~ времени, централизованный central timing
~, грубый coarse readout
~ дальности range/distance readout
~, дистанционный remote readout/reading
~, зеркальный mirror indication/reading, light-beam indication
~ курса (самолета) heading readout
~, ложный spurious count
~, непосредственный direct reading/readout/indication
~, неправильный/ошибочный misreading
~, нулевой zero reading
~ по барометру pressure reading
~ по компасу compass reading
~ по нониусу vernier reading
~ по прибору instrument reading
~ по рейке, дальномерный (stadia) rod intercept
~ по секстану, брать make a reading with sextant
~ по шкале scale readout
~ показаний приборов, разовый one-shot reading
~, производить take a reading
~ секстана sextant reading
~ угла места elevation readout
~, фоновый background count
~, цифровой digital readout
~ы, накопленные stored counts
отсчитывание reading/counting off
отсчитывать take a reading, read/count off/out
отсылать 1. refer; 2. forward, send away/off, dispatch
отсылка 1. dispatch(ing), sending off; 2. reference
~, перекрестная cross reference
отсыпка *min.* fill
~, каменная riprap
оттаивание thawing (out); defrosting
отталкивание repulsion, repelling
~, взаимное mutual repulsion, repulsive interaction
~, кулоновское Coulomb repulsion
~, обменное exchange repulsion
~ центромер centromere repulsion
~, электростатическое electrostatic/Coulomb repulsion
отталкивать push away/aside/back/off; *naut.* fend off; *nucl.* repulse, repel; *fig.* alienate, antagonize
оттачивать sharpen, whet
оттенение *cin.* shading; shadowing; adumbration
оттенок tinge, tint; shade; hue
~ кожи, желтушный *med.* yellowish discoloration of skin, xanthochromia
~, слабожелтушный subicteric tint
~ травления etching shade
оттепель thaw(period); thawing weather
оттертрал *naut.* otter trawl
отеска *timb.* trimming, squaring
оттеснять drive back/away, push/shove aside; press back
оттесывать *timb.* trim, square
оттирать rub off/down/out
оттирка rubbing off/out; attritioning
оттиск 1. impress(ion); imprint; mo(u)ld; 2. reprint, offprint; ectype
~, корректурный/пробный (page) proof
~, рельефный raised printing, imitation embossing
~, совмещенный registered print
~, черновой brush proof

отток flow-/run-off; outflow, discharge, drainage; withdrawal
~ прибыли profit outcome
оттопывать *hunting* scare, put up, start
оттопыренный bulging out; protruding, protuberant; sticking out; *bot., zool.* squarrose
отторженец *geol.* erratic/detached mass
отторжение tearing(away); *med.* rejection; *econ.* divestiture, divestment
отточие *print.* (row of) dots, leaders
отточить *print.* insert row of dots
отточка sharpening, whetting
оттрелит *miner.* ottrelite
оттрубить *mil.* beat retreat; beat the reveille
оттягивание 1. retraction; pulling away/back; 2. prolonging, delaying
оттягивать 1. retract; draw/pull back; pull away/aside; 2. extend, expand, stretch; *tech.* forge out
~ под молотом forge out
оттяжка 1. delay, procrastination; 2. *naut.* guy (rope); guy-wire; stay, strut, brace; haulback, inhauler, outhauler; 3. guying, anchoring
~, анкерная anchor guy/rope/stay, stub guy
~, ветровая guy wire
~, внутренняя inner guy rope
~ грузовой стрелы boom/derrick guy/vang
~ опоры pole guy
~, поперечная side guy
~, проволочная anchor wire
~, продольная head guy
~, страховая safety guy (line)
~ трала sweep upright
~ углубителя depressor span pendant
~ щелока liquor withdrawal
оттяжной delaying, bracing
~ная пружина back-moving spring
отунит *miner.* autunite
отупение *med.* stupor, stupefaction, torpor, torpidity; dullness; apathy, inactivity, sluggishness; insensibility, unconsciousness, lethargy
отучать break (of) wean (from)
~ от привычки disaccustom
отфактуровывать *econ.* invoice
отфильтровывать filter off/out, strain
отформовывать form, mo(u)ld, shape
отфрезеровывать *tech.* mill, cut
отхаркивание *med.* expectoration
~ гноя pyoptysis
отхаркивать expectorate, cough out
отхаркивающее *n.* expectorant
отход 1. backward movement, withdrawal, retreat; 2. departure, sailing; leaving; deviation; *see also* **отходы**
~ сосунков *agr.* pre-weaning mortality
~, стратегический strategic withdrawal
отходить 1. withdraw; retire, fall back, swing off, clear; 2. deviate, digress, diverge; 3. fork, branch (off); 4. fall/go/pass to *(by inheritance);* 5. pass (away), come to an end; veer off
отходная *n. eccl.* prayer for the dying
отходник *agr.* seasonal/itinerant worker
отходы *pl.* waste (material), wastes, scrap; by-products; *see also* **отход**
~, бытовые domestic wastes, (domestic/household) garbage
~, высокоактивные hot waste, high-level/-activity waste
~, газообразные gaseous waste, off-gas
~, городские municipal waste
~, горячие hot waste
~ грохочения screen reject
~, жидкие liquid waste, waste liquid, effluent
~ жизнедеятельности человека human waste
~ лесопиления raw waste lumber
~, малоактивные cold/cool waste, low-level/-activity waste
~, оборотные utilizable waste
~, отвальные nonutilizable waste
~ переработки топлива fuel process(ing) waste
~ перфорации (punched) card scrap
~ производства waste products
~, производственные industrial/process(ing) waste
~, радиоактивные radioactive/atomic/A-waste, radioactive residues
~, реализуемые salable refuse
~ сельскохозяйственного производства *agr.* farm waste
~, случайные non-recurrent waste
~ сортирования screening
~, топливные waste fuel
~, химические chemical waste
отхождение moving away, leaving
~ вод *med.* bursting of waters
~ газов *med.* passage of flatus
отхромировать plate with chromium
отцеживатель *food* straining tank
отцеживать strain off, filter
отцеп uncoupling, unhooking, release
~ вагонов *rail.* cut of cars
отцепка release, uncoupling, disengaging
~ парашюта parachute release
отцепление detachment, unhooking, uncoupling
отцеплять release, uncouple, disengage, disconnect; unhook; unhitch
отцепной detachable
отцеубийство patricide; parricide
отцеубийца patricide; parricide
отцовский paternal; fatherly; *bot.* staminate
~кое наследство *law* patrimony
отцовство paternity; fathership, fatherhood; parenthood
~, спорное disputed paternity
~, устанавливать *law* affiliate, filiate
отчаливать *naut.* unmoor, cast/push off
отчеканивать coin, mint
отчество patronymic
отчет account, report; statement

~, бухгалтерский accounting report
~, ежесуточный day-to-day report
~, кассовый cash statement
~ о командировочных расходах travel expence report
~ о кредитных операциях credit report
~ о проверке состояния качества quality survey report
~ о производственной деятельности manufacturing report
~ о простое idle-time report
~ о состоянии запасов stock status report
~ о состоянии счетов statement of accounts
~ о рабочей силе labo(u)r report
~ о трудозатратах timekeeping report
~ о ходе работ progress report
~ об ущербе damage report
~, промежуточный interstatement
~, ревизионный audit(or's) report
~, сводный consolidated/general report
~, технический engineering/technical report
отчетливость distinctness, itelligibility
отчетно-выборочный:
~ное собрание meeting held to hear a sheduled report and elect new officials
отчетность 1. accounting, bookkeeping; 2. paperwork; accounts
~, бухгалтерская accounting control
~, заполнять make out paperwork
отчетный accountable; current; financial
~ год financial/current year
~ доклад report
~ная карточка record/report card
отчисление 1. deduction; assignment; allocation; 2. dismissal; 3. *pl.* charge(s)
~, амортизационное depreciation charge/expense
~ в резерв allocation to reserve
отчислять 1. deduct; assign, allot; 2. discharge, dismiss; expel
отчленение *med.* abjunction; abstriction
отчуждаемость *law* alienability
отчуждать 1. alienate, estrange; 2. *law* confiscate, expropriate
отчуждение 1. alienation, estrangement; 2. *law* expropriation, dispossession
отчужденность alienation, estrangement
отшвартовывать *naut.* unmoor, cast off/loose
отшельник eremite, hermit, anchoret, anchorite; ascetic
отшельничество life of a hermit/recluse
отшествие *naut.* departure
~, восточное easting
~, западное westing
отшлифовка grind(ing); polish(ing)
отшнурование disjunction, separation
отшнуровывание constriction, pinch
отшпиливать unfasten, undo
отштамповывать stamp, impress, imprint
отщеп *min.* flake
отщепенец 1. renegade; 2. heretic, apostate
отщепление splitting/chipping off; detachment, detaching
отщеплять ship/split off; detach; *math.* segregate
отъезд departure, leaving; *cin.* pull-back
~, делать dolly out
отъявленный inveterate, arrant, thorough
отыскание detection, tracing, localization
~ максимума maximizing
~ неисправностей/повреждений fault loca(liza)tion, trouble shooting
~ повреждений в схеме circuit-disturbance test
~ решений finding of solutions
~ экстремумов extremization
отэктомия *surg.* otectomy, ossiculectomy
отягчать burden; aggravate, oppress
отягчающий aggravating
~щие (вину) обстоятельства *law* aggravating circumstances
оуэнизм Owenism
оферт(а) offer
офидизм *med.* ophidiasis, ophidism
офиопогон *bot.* snake's beard, lily-turf *(Ophiopogon, Mondo)*
офис *see* **оффис**
офиуры *pl. zool.* ophiuroids *(Ophiuroidea)*
офицер officer
~, вахтенный officer of the watch, watch officer/keeper
~, дежурный officer of the day, duty officer
~ запаса reserve officer
~ интендантской службы supply officer
~, кадровый regular officer, officer of the active list
~ медицинской службы medical officer, surgeon
~, младший junior officer
~ наведения aircraft controller
~ плавсостава officer serving afloat
~ по боевой подготовке training officer
~ разведки reconnaissance/intelligence officer
~ связи liaison officer
~ службы безопасности security officer
~, старший senior officer
~, строевой line officer, officer of the line
~ финансовой службы accountant officer
~ штаба staff officer, officer of the staff
офицер-корректировщик spotting officer
официальный official, formal
официант waiter
официантская waitersroom
официоз semiofficial organ
оф-лайн off-line
оформитель decorator; *print.* layout man
~ книги book designer
~ основ *ling.* morpheme
оформление 1. design; layout; 2. shaping, forming; 3. (official) registration, legalization
~ архитектурное styling

~, **грамматическое** grammatical expression
~ **программы** program scheduling
~ **процесса, аппаратурное** instrumentation/mechanization of a process
~ **результатов** presentation of the results
~ **служебной документации** execution of forms, records and reports
~, **художественное** styling
оформлять 1. shape; finalize; design; *print.* format; 2. legalize, register officially; formalize
~ **книгу** design a book
~ **служебную документацию** execute service records
офорт *print.* copper-plate engraving, etching
офортист aquafortist, etcher
офранцуживать Frenchify, Gallicize
офрис *bot.* twayblade, listera *(Ophrys, Listera)*
офсайд *sport* off side
офсет offset (process)
~ **высокий** letterpress/relief offset, indirect relief/letterpress
~, **глубокий** deep-etched offset
офтальмальгия *med.* ophthalmalgia
офтальматрофия *med.* ophthalmatrophia, ophthalmatrophy
офтальмия, офтальмит *med.* ophthalmia, ophthalmitis
~, **весенняя** spring ophthalmia, vernal conjunctivitis
~, **гнойная** purulent ophthalmia/conjunctivitis
~, **гранулярная** granular ophthalmia/lids
~, **египетская** Egyptian ophthalmia, trachoma
~, **мигрирующая** migratory ophthalmia
~, **нодозная** caterpillar-hair ophthalmia
~, **периодическая** recurrent iridocyclitis, moon blindness, periodic ophthalmia
~, **симпатическая** sympathetic/transferred ophthalmia
~, **слизистая** mucous ophthalmia
офтальмодиния *med.* ophthalmodynia, optic neuralgia
офтальмол ophthalmolum
офтальмолог ophthalmologist, oculist, eye specialist
офтальмологический ophthalmologic(al)
офтальмология ophthalmology, ophthalmiatrics
офтальмометрический ophthalmometric(al)
офтальмомиаз ophthalmomyasis, ocular myasis
офтальмомикоз ophthalmomycosis, oculomycosis
офтальмоплегия ophthalmoplegia
~, **врожденная** congenital ophthalmoplegia
~, **мигренеподобная** migraine-like ophthalmoplegia
~ **Парино** Parinaud's ophthalmoplegia
~, **приобретенная** acquired ophthalmoplegia
~, **пучковая** fascicular ophthalmoplegia
офтальмоскоп *med.* ophthalmoscope
~, **зеркальный** ghost/mirror ophthalmoscope
офтальмоскопия ophthalmoscopy
~, **безрефлексная** reflexless ophthalmoscopy
~ **в обратном виде** indirect ophthalmoscopy
офтальмостазия *surg.* ophthalmostasis, ophthalmostasia
офтальмохирургия ophthalmosurgery
офферент offerer
оффис office
охарактеризование characterization, description
охват 1. *stat.* coverage; inclusion; 2. envelopment, wrap; 3. scope, range, reach, sweep
~ **с обоих флангов** *mil.* double envelopment
~**а, неполнота** *stat.* undercoverage
охватывать 1. embrace, envelop, encompass; enclose; comprise, include, cover; 2. grip, seize; 3. comprehend, involve; 4. *mil.* surround
~ **с фланга** outflank
охвостье *agr.* chaff, husks
охладитель 1. cooler, refrigerator; 2. coolant, cooling agent
~, **башенный** cooling tower
~, **водный** water coolant, cooling water
~ **водяной/воздуха** water/air cooler
~, **вторичный** secondary coolant, recooler
~ **второго контура** secondary coolant
~ **выпара** vapour/damp condenser
~ **дымовых газов** stack-gas cooler
~, **жидкометаллический** liquid-metal coolant
~, **змеевиковый** coil-in-box cooler
~, **кожухозмеевиковый** shell-and-coil cooler
~, **кожухотрубный** shell-and-tube cooler
~, **многоярусный** multitier cooler
~, **натриевый** cooling sodium, sodium coolant
~, **обратный** reflux condenser
~, **оросительный** trickle cooler, drip-type cooler
~ **пара** attemperator, desuperheater
~, **паровой** steam coolant
~, **пластинчатый** plate cooler
~, **предвключенный** precooler
~ **приемника (излучения)** detector cooler
~ **продувочной воды** blow-down cooler
~, **промежуточный** intercycle cooler, intercooler
~, **противоточный** countercurrent cooler
~, **прямоточный** uniflow cooler
~ **реактора** reactor coolant
~, **ребристый** finned cooler
~ **с электрическим оттаиванием** electric defrost cooler
~, **трубчатый** multitube cooler
~, **форсуночный** spray cooler
~, **циркулирующий** recirculation cooler
охлаждаемый cooled
охлаждать cool (off/down); chill, refrigerate
охлаждающий cooling, refrigerating, algefacient
охлаждение 1. cooling, refrigeration, chilling; 2. quenching; 3. *fig.* alienation
~, **абляционное** ablation/ablative cooling
~, **аварийное** emergency cooling
~, **брызгальное** spray cooling
~ **в печи** furnace cooling
~, **вентиляторное** fan cooling

~, внезапное *med.* shock chilling
~ водой под давлением high-pressure water cooling
~ водяной рубашкой (water-)jacket cooling
~ вынужденной конвекцией forced-convection cooling
~ выпотеванием sweat cooling
~ высотного скафандра (full-)pressure-suit cooling
~ газом/газовое gas cooling
~ газовой пленкой gaseous-film cooling
~, гелиевое helium cooling
~ груза cargo refrigeration
~ двигательного отсека engine-bay cooling
~ двигателя engine cooling
~, динамическое (двигателей) regenerative cooling
~, дутьевое *el.* fan cooling
~, емкостное *rock.* capacitative cooling
~, естественное self-cooling, natural cooling
~ естественной конвекцией free-convection cooling
~, жидкостное/жидкостью fluid/liquid cooling
~, искусственное artificial cooling
~, испарительное transpiration/evaporative cooling
~, конвекционное convective/convection cooling
~ конструкции *av.* structural cooling
~, краевое marginal cooling
~ лучеиспусканием radiation cooling
~, льдосоляное ice-salt cooling
~, местное local cooling
~ набегающим потоком воздуха ram-air cooling
~, направленное directional cooling
~, натриевое sodium cooling
~ обдувом blast/blower cooling
~, общее general/total cooling
~ отсека bay cooling
~ по замкнутому циклу closed-cycle cooling
~, поверхностное surface cooling
~, повторное recooling
~ погружением immersion cooling
~, пористое *rock.* sweat/porous cooling
~, предварительное precooling
~, принудительное forced cooling
~, промежуточное intercooling
~ противотоком/противоточное counter-current/-flow cooling
~, проточное one-through cooling, direct-flow cooling
~, прямоточное direct-flow cooling, uniflow cooling
~ путем массобмена mass-transfer cooling
~ путем разбрызгивания лития lithium-spray cooling
~ путем уноса массы mass-transfer cooling, ablation cooling
~ разбрызгиванием spray cooling
~ расплавленным металлом molten-metal cooling
~, рассольное brine cooling
~ расширением cooling by expansion
~, регенеративное closed-cycle cooling
~, резервное stand-by cooling
~ с печью furnace cooling
~ туманом mist cooling
~, туннельное ducted cooling
~, чашевое pan cooling

охлопок 1. tuft; 2. *pl.* waste, lint, combings
охлопье waste, combings, lint
охота I 1. hunt(ing); chase; 2. *mil.* free-lance operation, lone-wolf operation
~ на медведей bear hunting
~ на птицу fowling
~, ружейная, ~ с ружьем shooting
~, соколиная falconry
охота II 1. wish, desire; 2. *physiol.* heat, rut-time
~, половая oestrum, oestrus *(in female animals);* rut *(in male animals)*
охотник I 1. hunter; sportsman; trapper; 2. *mil.* chaser
~ за крупной дичью big-game hunter
~ за подводными лодками submarine-hunting ship/unit
~, морской submarine chaser
~, соколиный falconer
охотник II 1. enthusiast; amateur; 2. volunteer
охотничий 1. hunting, sporting; shooting; venatic(al), venatorial; 2. *mil.* volunteer *(attr.)*
~ домик shooting box/lodge
~ заповедник warren
охра *miner.* ochre, *Am.* ocher
~, бурая brown ochre
~, висмутовая bismuth ochre, bismite
~, вольфрамовая tungstic ochre, tungsten trioxide, tungstite
~, желтая yellow ochre/earth
~, красная red ochre, raddle
~, купоросная vitriol ochre, glockerite
~, молибденовая molybdic ochre
~, свинцовая lead ochre
~, черная black ochre, wad
охрана 1. protection, guarding; security; preservation, conservation; 2. guard(s), escort, defence; 3. *hist.* Secret Police Department
~, береговая patrol forces
~ детей раннего возраста infant protection/welfare
~ детства child welfare
~ дичи preservation of game
~ здоровья health protection; medical care
~ качества quality protection
~ леса forest conservation
~, личная bodyguard
~ материнства maternity protection/service
~ младенчества child welfare
~, морская пограничная coast guard

~ окружающей среды environment(al) control/protection
~ природы wildlife management/preservation
~ рыболовство fishery protection
~ труда safety arrangements and precautions, labo(u)r protection, prevention of accidents
охранение (safe)guarding; *mil.* protection; *nav.* escort (screen), convoy escort; screen
~, активное offensive screen
~, ближнее/дальнее close/outer screen
~, дальнее воздушное distant air(craft) screen
~ из военных кораблей warship escort
~ из легких крейсеров light cruiser screen
~ конвоя convoy screen/escort
~, непосредственное close-in protection, close escort
~, передовое advanced screen
~ походного ордера cruising screen
~ при входе в порт/гавань entry screen
~, радиолокационное radar picket defence/screen
~, сторожевое outposts
охранитель 1. guardian, protector; 2. *naut.* paravane
охранительно-карантинный protective-and-quarantine
охранительно-стимулирующий protective-and-stimulant
охранка *abbr.* **охранное отделение** *hist.* Okhranka, Secret Police Department
охранник 1. (security) guard, watchman; 2. *hist.* member of Okhranka, secret police agent, okhrannik
охранный guarding, protective
~ лист, ~ная грамота safeconduct, pass
~ ная зона *mil.* restricted area
~ное отделение *hist.* Secret Police Department
охранять (safe)guard, protect; defend; screen; shelter
охренный ocherous, ochreous
охриплость *med.* hoarseness
охристый ocherous, ochreous; ochre-colo(u)red
охродермия *med.* ochrodermia, ochrodermatosis
охролит *geol.* ochrolite
охрома *bot.* West Indies balsa *(Ochroma lagopus/pyramidale)*
охроноз *med.* ochronosis
~, глазной ocular ochronosis
охрупчивание embrittlement
оцелот *zool.* ocelot *(Felis pardalis)*
оценивание evaluation, estimation, rating, appraisement; scoring
~ выборочным методом sample estimation
~, точечное point estimation
оценивать 1. evaluate, value, rate, estimate, appraise, grade; assess, judge; 2. appreciate
оценка estimation, (e)valuation, guess value, rating, appraisal; score; *stat.* estimate, estimator, bound
~, аналитическая analytical estimation
~, асимптотическая asymptotical estimate
~, асимптотически гауссова asymptotically Gaussian estimate
~, асимптотически несмещенная *math.* asymptotically unbiased estimate
~, байесова/байесовская *math.* Bayes(ian) estimation/assessment
~, балльная (numerical) score
~ в целом overall evaluation
~ возраста *geol.* age assessment
~, всесторонняя thorough assessment/evaluation
~ выборки/выборочная sample estimate
~ выхода yield estimation
~, граничная верхняя/нижняя upper/lower bound
~, грубая eye/ocular/rough/crude estimate
~ данных data evaluation
~ дебита скважины *geol.* rating of well
~, доверительная *math.* confidence estimate
~, завышенная overestimation
~, инвариантная invariant estimate
~, интегральная integral criterion
~, интервальная interval estimate/estimation
~, кадастровая cadastral valuation
~, квадратическая quadratic estimate
~ ковариации *stat.* estimated covariance
~, количественная quantitative estimate
~ кредитоспособности credit rating
~ максимального подобия *math.* maximum likelihood estimate
~ массы mass estimation
~ метеообстановки evaluation of weather conditious
~ местонахождения *geol.* valuation of deposits, deposit evaluation/estimation
~, непереводная failing grade
~, несмешанная unconfounded estimate
~, несмещенная unbiased estimate
~, нижняя *stat.* lower bound
~, нормированная normalized estimate
~ обстановки estimate of the situation
~ обученности performance score
~ оптимума estimated optimum
~, основанная на обработке динамических рядов *econ.* time-series estimate
~ осуществимости feasibility evaluation
~, параметрическая parametric evaluation
~ параметров сигнала estimation of signal parameters
~, перспективная projection
~ по максимальному правдоподобию maximum likelihood estimation
~ по максимуму апостериорной вероятности maximum a posteriori estimate
~ по методу максимального приближения maximum likelihood estimate
~ по методу наименьших квадратов least-squares estimate
~ по минимуму средней квадратической ошибки minimum mean-square error estimate

~ по порядку величины order-of-magnitude estimate
~ по состоянию на середину года *econ.* midyear estimate
~, приближенная rough/coarse estimate
~ программы, экспертная *comp.* peer code review
~ производительности performance rating
~, псевдоусловная *econ.* pseudoshadow price
~ работы job rating
~ равномерно взвешенная uniformly weighted estimate
~ руд ore evaluation
~ с минимальной дисперсией minimum-variance estimate
~ случайной величины *stat.* bound
~, смешанная confounded estimate
~, смещенная biased estimate
~, совместная joint estimation
~, совокупная collective estimate
~, состоятельная consistent estimate
~, средняя average grade
~, статистическая statistical estimation, statistic(al) estimate; estimating function, estimator
~ стоимости cost estimation/appraisal
~, страховая insurance valuation
~, субъективная value judgement
~, техническая engineering estimate
~ товарно-материальных ценностей *econ.* inventory costing
~ точечная point estimate
~ успеваемости *educ.* gain score
~ финансового положения *econ.* financial rating
~ характеристик performance evaluation
~, эффективная efficient estimate/evaluation/estimator, efficient estimating function
оценочный evaluative, evaluating, estimating
оценщик valuator, valuer, estimator, appraiser
~, официальный qualified appraiser
~ размера страхового убытка *law* insurance adjuster
оцепенелый numb, benumbed; torpid
оцепенение *med.* rigidity, numbness, stupefaction, stupor, torpor, catalepsy, cataplexy
оцепенять (be)numb; *fig.* stun
оцинкование galvanizing, galvanization, zinc-plating
~ парами цинка vapour galvanizing
оцинкованный zinc-plated/coated, galvanized
оцинковывать galvanize, (coat with) zinc, zinc-plate
оцифровка numbering, graduation; calibration; marking
~ дорожки (самописца) trace numbering
~ компаса compass marking
~ через число alternate numbering
~ шкалы scale marking
оцифровывать digitize
очаг 1. centre, seat; 2. hotbed, breeding ground; nidus; *med.* focus; 3. fireplace, range, hearth; 4. nucleation site
~, вулканический volcanic focus
~, гнойный suppurative focus
~, гоновский *med.* Ghon's primary lesion
~ движения focus of movement
~, домашний home
~ заражения infection focus
~ инфекции focal point of infection
~, коррозионный corrosion centre
~, магматический magmatic chamber/hearth
~ омертвения *med.* infarct(ion)
~ отслоения point of separation
~ пожара fire area
~ промежуточного типа intermittent vent
~ разрушения site of breakdown, seat of destruction
~ сопротивления centre/island of resistance
~и, сливающиеся *med.* confluent foci
очаговый focal
очанка *bot.* eyrbright *(Euphrasia)*
очевидец eyewitness
очевидность manifestation, evidence, obviousness, patency
очеловечение humanization, hominization
очеловечивать humanize, hominify, make human
очередной 1. next (in turn), successive; alternate; 2. routine, regular, recurrent
очереднолистный *bot.* alternate-leaved, alternifoliate
очередность order (of priority), sequence, (regular) succession; priority, line of seniority; *comp.* queue
~ замены личного состава succession of reliefs for personnel
~ операций sequence of operations
~ передачи радиограмм messages precedence
~ посадки landing sequence
~ председательствования rotation of presidency
~ претензий ranking of claims
~ требований по долгам priority of debts
очередь 1. turn; series, row; 2. (waiting) line, queue; lineup; 3. *mil.* salvo, burst
~, батарейная battery salvo
~, бесконечная unlimited/unrestricted/infinite queue
~ ввода/входная input (work) queue
~, дискетная sampled queue
~ заданий job queue
~ запросов к устройству device queue
~ к каналу channel waiting queue
~, коррелированная correlated queue
~ логического канала logical channel queue
~ магазинного типа push-down queue
~ на обработку process queue
~ обычных требований ordinary queue
~, основная major queue
~ передаваемых сообщений communication queue

~ **с приоритетам** priority queue
~ **строительства** phase of a project
~ **частично обработанных сообщений** work-in-process queue
очерет *bot.* ditch reed *(Phragmites communis)*
очеретник *bot.* beak sedge/rush *(Rhynchospora)*
очерк sketch, essay, study; outline
очеркивать outline; place a circle round
очеркист essayist; feature writer
очертание contour, profile, shape, outline, delineation; circumscription
очес *text.* waste, combing(s); flocks, hurds, tow
~, **гребенной** comber waste, noil
~, **торфяной** peat dust
очески *pl.* combings, flocks; hurds, tow
очесывание combing (out); hackling
очесыватель combing machine
очесывать comb (out), hackle, hatchel
очиститель cleaner, purifier, clarifier; refiner; (gas) scrubber
~, **ионитный** deionizer
~ **решет** screen cleaner
очистительный cleaning, cleansing, purifying; cathartic(al)
очистка cleaning, cleansing, purification, purifying; refinement; filtration, distillation, clarification; decontamination; clearance
~ **активным илом** activated-sludge decontamination
~, **аспирационная** suction cleaning
~ **в барабане** drum cleaning, tumbling
~, **водоструйная** hydroblasting
~ **воды** water purification/clarification
~ **воздуха** air purification
~, **воздушная** pneumatic/air cleaning
~ **газов** gas cleaning
~ **газов, мокрая** gas scrubbing
~ **газов, центробежная** cyclone gas cleaning
~, **газопламенная** flame cleaning
~, **гидравлическая** hydroblasting, hydrocleaning
~ **груза** clearance of cargo
~, **дробеструйная** shot-blast-cleaning, shot blasting, blast cleaning
~ **забоя** face cleaning
~, **изоморфная** isomorphous purification
~ **литья** dressing a casting
~ **масла** oil cleaning
~ **масла, щелочная** alkali-refining of oil
~ **металла перед сваркой** preweld metal cleaning
~, **механическая** mechanical cleaning, power-tool cleaning
~ **от кожуры (плодов/овощей)** peeling
~ **от кожуры в автоклаве** retort peeling
~ **от кожуры обжигом/термическая** flame/heat peeling
~ **от мин** mine clearing
~ **от пошлины** (customs) clearance
~ **от препятствий** obstruction clearing, removal of obstruction
~ **от (радиоактивных) загрязнений** decontamination
~ **от семян** seeding
~ **от смолы** detarring
~ **от снега** snow clearing
~ **от сучьев, естественная** natural pruning
~ **от таможенных формальностей** customs formalities clearance
~ **памяти** memory clearing
~ **паром** steam clearing
~ **пассажиров** clearance of passengers
~, **пескоструйная** sand-blast cleaning, sand-blasting
~, **пламенная** flame cleaning
~, **пневматическая** air-blasting
~ **поверхности** surface cleaning
~ **почвы** *agr.* weeding
~ **промышленных вод** sewage/waste treatment/disposal
~, **противоточная** counter-current scrubbing
~ **растворителем** solvent cleaning
~ **сварного шва (от шлака/брызг)** weld cleaning
~ **семян** seed cleaning
~ **сточных вод** waste (water) treatment, sewage treatment/disposal, purification of sewage
~, **сухая** dry cleaning
~, **термическая** flame cleaning
~, **химическая** chemical purification, chemical cleaning
~ **центрифугированием** centrifugal purification
~ **шлаковика** slag removal
~ **шпура** hole cleanout
~ **щелочами** alkaline cleaning
~, **электролитическая** electrolytic cleaning
очистки *pl.* peelings
очисток *bot.* stonescrop *(Sedum)*
очищать clear; cleanse, clean, scour; purify; peel; deterge; filtrate; distil, rectify
~ **от масла** unoil
~ **от накипи** remove scale
~ **от ржавчины** derust, clean from rust
очищение *see* **очистка**
~ **кишечника** *med.* purgation, evacuation of the bowels
~ **мочевины** urea clearance
очищенный cleaned, purified; clarified; decontaminated; scrubbed
очки *pl.* (eye)glasses, spectacles; *av.* Savoy knot, spectacles
~, **бифокальные** bifocal/pantoscopic spectacles
~, **дырчатые** stenop(a)eic spectacles
~, **защитные** (safety) goggles
~, **защитные химические** splash goggles
~, **зеркальные** mirror glasses
~, **контактные** contact spectacles/lenses
~, **корригирующие** corrective/corrigent spectacles

~, **поляроидные** polarizing spectacles
~, **предохранительные** (safety) goggles
~, **предохранительные/противопылевые** antidust goggles
~, **сварочные** welding goggles
~, **солнцезащитные** sun glasses
~, **стенопические** stenop(a)eic spectacles
~, **стереоскопические** 3-D viewer
~, **телескопические** telescopic spectacles
~, **царские** *bot.* yellow marsh saxifrage *(Saxifraga hirculus)*
~, **цветные** tinted eyeglasses/spectacles
очко 1. *sport* point; 2. hole, eye; mesh; 3. spot, pip; 4. *hort.* bud, eye; 5. *print.* face
~, **волочильное** drawhole
~, **сигнальное** *rail.* signal field
~, **смотровое** peephole
~, **фурменное** tuyere aperture
~, **шлаковыпускное** cinder notch
очкование *hort.* budding
очковый I spectacle(d)
~**вая змея** *zool.* cobra *(Naja)*
очковый II point *(attr.)*
~**вая система** point system of scoring
очная ставка *law* confrontation
очный *educ.* internal, intramural
~**ное обучение** *educ.* intramural studies, classroom instruction
очный цвет *bot.* pimpernel *(Anagallis)*
очувствление sensitization; activation
ошак *bot.* Bombay sumbul *(Dorema ammoniacum)*
ошвартовать, ошвартовить *naut.* moor, make fast
ошибаемость error frequency/rate
ошибаться be mistaken, be at fault, be (in the) wrong, err
ошибка error, mistake; fault; inaccuracy; fallacy; blunder; *Am.* bug
~, **абсолютная** absolute error
~, **азимутальная** azimuth(al) error, closure error of azimuths
~, **аэродинамическая** *av.* position/pressure error
~, **базисная** *geod.* base error
~, **боковая** lateral error
~ **в вычислении** computational error, miscalculation
~ **в данных** data error
~ **в исправлении** patch(ing) error
~ **в исходных данных** datum error
~ **в ответах, систематическая** *stat.* response bias
~ **в отсчете** datum error
~ **в праве** *law* error of law
~ **в предмете договора** *law* error in rem
~ **в прицеливании** aiming error
~ **в расчетах/расчете** computational error, miscalculation
~ **в результате округления** round-off error
~ **в системе программного обеспечения** software error
~ **в словах** *law* error in verbis
~ **в смещении нуля** bias error
~ **в согласии** *law* error in consensu
~ **в суждении** *law* error in judgement
~ **в существенных признаках** *law* error in essentia/substantia
~ **в сущности** *logic* material fallacy
~ **в фактах** *law* error of fact
~, **вероятная** probable error
~ **визирования** sighting error
~, **виражная** banked-turn error, turning error
~ **воспроизведения** replication error
~ **вследствие дрейфа нуля** zero-drift error
~ **вследствие запаздывания** drift/creep/lag/aftereffect error, time-effect error
~ **второго рода** *math.* error of the second kind
~ **выборки, выборочного обследования** *stat.* sampling error
~ **гистерезисная** hysteresis error
~ **Госсена** error of Gaussin
~ **градировки** graduation/calibration error
~, **грубая** gross error, blunder
~, **действительная** actual error
~, **диагностическая** *med.* diagnostic error/mistake
~, **дисквалифицирующая** *sport* disqualifying foul
~, **допустимая** permissible/admissible/allowable/tolerated error
~, **допущенная** introduced error
~, **достоверная** significant error
~ **записи** recording error
~ **за эксцентриситет** eccentricity error
~, **значительная/заметная** appreciable error
~ **игнорирования человеческого фактора** human factor error
~ **измерения** measurement/measuring/metering error
~, **индивидуальная** individual/human/personal error
~ **индикации** display error
~, **карданная** gimbal error
~, **квадрантная** quadrantal error
~, **квадратическая** square error
~ **квантования** quantization error
~ **квантования по времени** sampling error
~, **коллимационная** error of collimation; *geod.* index error
~, **курсовая** heading error
~, **люфтовая** backlash error
~, **локализованная** isolated/localized error
~, **максимальная** maximum error
~, **масштабная** scale error
~ **места** error in position, position error
~ **местоположения, круговая** circle of uncertainty
~ **метода** methodic error
~, **методическая** measurement-method error
~, **минимальная** minimum error

~, многократная multiple/repeated error
~ монтажа *comp.* wiring error
~ на параллакс paralactic error
~ на (перфо)карте (punched) card error
~ на символ error per digit
~ наблюдения observation(al) error, error of observation
~ наведения guidance/vectoring error
~, накопленная cumulative/accumulated error
~ настройки tuning error
~, начальная initial error
~, независимая indenpendent error
~ неопределенности ambiguity error
~, неопределимая indeterminate error
~, несистематическая accidental/random error
~, обнаружимая detectable error
~, одиночная single/isolated error
~ округления round-off error
~ оператора human mistake, bust
~, орфографическая misspelling
~, остаточная residual error
~ от воздействия ускорения acceleration error
~, относительная relative error
~ отсчета reading/readout error, error of reading
~, параллактическая parallax/parallactic error
~ пеленгования bearing error
~ первого рода *math.* error of the first kind
~, перемежающаяся intermittent error
~ по положению position error
~ по ускорению acceleration error
~, поворотная turn(ing) error
~, повторяющаяся repetitive error
~ поиска seak/search error, seek failing
~, полная aggregate/total error
~, полукруговая semi circular error
~, постоянная fixed/biased error
~, предвнесенная inherent/inherited error
~, предельная limiting/maximum error, limit of error
~ при исправлении программы *comp.* patching error
~ при объявлении переменной declare error
~ при поиске информации false drop
~ при работе с клавиатурой keying mistake
~ при сортировке missort
~ при считывании с карт card read error
~ приближения error of approximation
~, приборная device/instrument(al) error
~ прицеливания aiming error
~ прицеливания из-за бокового ветра sight crosswind error
~, процентная per cent error
~ разности error of difference
~ рассогласования misalignment error
~, результирующая resultant/consequent error
~, симметричная balanced error
~, систематическая systematic(al) bias; *math.* bias, systematic/fixed error, biased error
~ слежения/сопровождения tracking error
~, случайная random/accidental/occasional/unbiased error
~ состояния машины machine state error
~, среднеквадратичная mean(-root)-square error
~, средняя mean/average error
~, статистическая statistical error/uncertainty
~, статическая static error, steady-state error
~, субъективная individual/human/personal error
~, судебная *law* miscarriage of justice
~, суммарная cumulative/aggregate/total error
~ сцепления chaining error
~ счета counting error, miscount
~ считывания reading error
~ считывания кода ambiguity error
~, траекторная trajectory error
~, угловая angular error
~, угломестная elevation error
~, умышленная intentional error
~ управления faulty operator intervention, operating error
~ усечения truncation error
~ установки коэффициента coefficient setting error
~, установочная installation/alignment error; lubber-line error
~, фактическая *law* ignorance of fact
~ центровки centering error
~, четвертная quadrantal error
~ эксперимента experimental error
~ эталонирования error of standardization
~, явная appreciable error

ошибневые *n. pl. zool.* cusk eels *(Ophidiidae)*
ошибочность fallibility, fallaciousness, fallacy; inaccuracy, erroneousness
ошибочный mistaken, wrong, fallacious, erroneous, inaccurate, incorrect, faulty
ошиновать 1. *el.* bus, install bus bars; 2. tyre, *Am.* tire, provide with tyres
ошиновка 1. *el.* leads; buses, bus arrangement; 2. providing with tyres
ошкуривать 1. (rub with) emery; 2. sand, give a sanding
ошлакование slagging
ошлаковывать (form) slag
ошлифовывать polish, burnish/grind round
ошпаривание *food* scald(ing)
оштукатуривание plastering, stuccowork
оштукатуривать (cover with) plaster
ощелачивание alkalization, alkalizing
ощелачивать alkalize, alkalify
ощупывание palpation, touching, feeling, perceiving; groping
ощупывать probe, feel; grope; palpate
ощупь touch(ing), feeling
~, влажный/горячий/сухой на wet/hot/dry to (the) touch
ощутимость visualization, perceptibility; palpability

ощутимый perceptible, tangible, palpable
ощущать sense, feel, percept, perceive; sensate
ощущение *med.* sensation; esthesis, esthesia; perception
~ ампутированной конечности, ложное stump hallucination
~, анормальное спонтанное abnormal spontaneous sensation
~, ассоциированное synesthesia
~ глубины *opt.* sense of deptth, deep perception
~ громкости loudness sensation
~ давления, последующее afterpressure
~, ложное false sensation, pseud(o)esthemia, hallucination
~ неприятного прикосновения haptodysphoria
~ первых движений плода *med.* quickening
~ повышенной интенсивности hyperesthesia, oxyesthesia, acroesthesia
~ пониженной интенсивности hypesthesia, bradyesthesia, obtusion
~, последующее aftersensation
~, последующее вкусовое aftertaste
~, последующее слуховое afterhearing, aftersound
~ с неправильной локализацией allesthesia, alloesthesia, allachesthesia, allochiria, achiria, dyschiria
~ сердцебиения heart consciousness
~, слуховое auditory sensation, sensation of hearing
~, субъективное зрительное entoptic sensation
~, субъективное слуховое entacoustic sensation
~, тактильное tactile sensation, sense of touch
~ тепла caumesthesia
~ тяжести heavy feeling
~ укуса odaxesmus, biting sensation
~ холода cryesthesia
оэций *zool.* ooecium
оямалит *geol.* oyamalite
ояременный yoked

П, п

пава *zool.* peahen
павиан *zool.* baboon *(Papio)*
павильон pavilion; film studio
~, выставочный exhibition hall
павликианин, павликианка, павликианский *rel., hist.* Paulian
павлин *zool.* peacock, peafowl *(Pavo); astr.* Pavo
павлиний peacock *(attr.),* pavonine
павлиноглазка *zool.* saturniid *(Saturniidae);* sheep moth *(Pseudohazis eglanterina)*
павловния *bot.* paulownia *(Paulownia)*
павлония *bot.* paullinia *(Paullinia)*
паводок flood, high water; flash flood
~, расчетный design/projected/predicted flood
павой *bot.* greenbrier *(Smilax)*
павун *bot.* bog bean *(Menyanthes)*
пагинация pagination
пагода pagoda
падать fall, drop; decline, decrease; *phys.* be incident on; *geol.* dip
~ в обморок faint
падающий falling; decreasing; *phys.* incident
~щая звезда shooting star
па-де-де pas de deux
падеж *gram.* case
~, винительный accusative (case)
~, дательный dative (case)
~, именительный nominative (case)
~, количественно-выделительный/отделительный partitive genitive (case)
~, косвенно-орудийный mediative case
~, объектный objective case
~, общий common case
~, предложный prepositional case
~, притяжательный possessive (case)
~, родительный genitive (case)
~, становительный translative case
~, творительный instrumental (case)
падение fall(ing), drop(ping), subsidence, decrease, reduction; lowering; *(slope)* grade, gradient; *phys.* incidence; *geol.* dip, pitch
~ давления pressure drop
~, крутое *geol.* heavy pitch, steep/high dip
~ курсов акций fall of stocks
~, наклонное oblique incidence; *geol.* gentle/easy dip
~ на поверхность, косое oblique incidence
~ напряжения *el.* voltage drop; *polit., fig.* easing of tensions
~ пласта *geol.* seam pitch/inclinatiun
~ по касательной grazing incidence
~ по нормали normal incidence
~ под действием силы тяжести gravity drop/fall
~ потенциала potential drop
~, резкое (курса) slump
~ реки stream gradient
~ температуры temperature drop
~ тяги thrust decay
падеспань pas d'Espagne
па-де-труа pas de trois
падишах padishah
падуб *bot.* holly *(Ilex)*
падубовые *pl. bot.* holly family *(Ilicaceae)*
падчерица stepdaughter
падь creek valley, small river valley
паек ration; daily allowance
~, аварийный emergency ration
~, летный/бортовой *av.* flight ration
~, полноценный *med.* balanced ration
пажитник *bot.* trigonella *(Trigonella)*

паз groove, slot, mortise, recess, channel, slit, notch; (edge) seam
~ в виде ласточкина хвоста dovetailed groove, V-slot
~, кольцевой circular groove
~, скошенный skewed/bevelled slot
~, тавровый T-slot
~ шлицевого соединения spline groove
~, шпоночный key slot/way
~, якорный *el.* armature slot
пазания *bot.* tan oak *(Pasania, Lithocarpus)*
пазник *bot.* cat's-ear *(Hypochoeris)*
пазуха 1. *anat.* sinus; 2. *bot.* axil; 3. *tech.* gullet
~ аорты aortic sinus
~, барабанная tympanic sinus
~, венечная venous sinus
~, затылочная occipital sinus
~, лобная frontal sinus
~, миндальная tonsillar fossa
~, мочеполовая urogenital sinus
~, околоносовая accessory nasal sinus
~, основная sphenoidal sinus
~, пещеристая cavernous sinus
~, почечная renal sinus
~ преплюсны tarsal sinus
~ сердца, венечная coronary sinus
~, сонная carotid sinus
~, S-образная сигмовидная sigmoid sinus
пазушный 1. *anat.* sinusal; 2. *bot.* axillar(y), axile, alar
пай *econ.* share
пайена *bot.* payena *(Payena)*
пайза *bot.* barnyard grass, Japanese millet *(Echonichloa frumentacea)*
пайка soldering, brazing; solder; seal
~ в атмосфере водорода hydrogen brazing
~ в вакууме vacuum brazing
~, вакуумная vacuum seal
~ волной припоя flow/wave soldering
~ заливкой припоем flow soldering/brazing
~, индукционная induction brazing
~, мягкая [мягким припоем] soldering
~ погружением dip brazing
~ серебряным припоем silver brazing
~, твердая [твердым припоем] brazing
пайол *naut.* bottom/floor ceiling, bilge boards/ceiling, timber flooring
пайщик I пайщица shareholder
пайщик II solderer
пак 1. pack ice, icepack; 2. *el.* power pack
пака *zool.* paca *(Cuniculus paca)*
пакгауз warehouse, storehouse, repository, entrepot; freight/goods shed/house
пакер *min.* packer
пакет packet; package, parcel; pack, stack, pile; batch; *econ.* block
~, волновой wave packet
~ данных *comp.* packet
~ акций shareholding(s), block of shares
~ векселей stock of bills of exchange
~ дисков, съемный removable disk pack
~ из восковой оболочки wax-dip package
~ импульсов/импульсный pulse packet
~, индивидуальный first-aid kit
~ катализатора catalyst pack
~, контрольный *econ.* controlling interest
~ отладки *comp.* debugging packet
~ ошибок *comp.* error burst
~ программ *comp.* software package
~, трубный tube bank
пакетирование packaging; piling; *el.* stacking
пакетировать pack, package; pile; *el.* stack
пакетная обработка batch processing
пакет-пресс *met.* packing/faggoting press
пакистанец, пакистанский Pakistani
пакистанка Pakistani (woman)
паклен *bot.* hedge/common maple *(Acer campestre)*
пакля oakum; tow; fibre/fiber packing; packing yarn
паковка packing
паксилла *zool.* paxilla
пал I *naut.* pawl, pall; bollard, dolphin; mooring post
~ для крепления палубного груза lashing post
пал II (site of) fire
палаквиум *bot.* nato tree *(Palaquium)*
палата chamber; house; *med.* ward, room
~ депутатов Chamber of Deputies
~ мер и весов Board of Weights and Measures
~, Оружейная Armoury
~ по судоходству, Международная International Chamber of Shipping
~, послеоперационная postoperative/recovery ward
~, послеродовая postnatal ward
~, предродовая *med.* prenatal ward
~ представителей *Am.* House of Representatives
~, пробирная assay office
~, расчетная clearing/settlement office
~, родильная delivery ward/room
~, Торговая Chamber of Commerce
~, Торгово-промышленная Chamber of Commerce and Industry
палатка 1. tent; 2. *anat.* tentorium
~, кислородная oxygen tent
~, фумигационная fumigation/gas tent
палатограмма palatogram(me), palatograph
палатограф palatograph, palatomyograph
палатолалия palatolalia
палатопластика palatoplasty, uranoplasty
палатоплегия palatoplegia, uranoplegia
палаторафия palatorrhaphy, uranorrhaphy
палафит palafitte
палафоксия *bot.* palafoxia *(Palafoxia)*
палгед *naut.* pawl head
палгун *naut.* pawl rim/ring/bed
палеарктика, палеарктический Pal(a)earctic
палеоазият, палеоазиатский Paleo-Asiatic

палеоантрополог paleoanthropoligist
палеоантропология paleoanthropology
палеоботаник paleobotanist, paleophytologist
палеоботаника paleobotanics, paleophytology
палеоген , палеогеновый Paleogene
палеогенез *biol.* paleogenesis
палеогенетика paleogenetics
палеогнаты *pl.* paleognaths *(Palaeognathae)*
палеозавр paleosaur *(Palaeosaurus)*
палеозой, палеозойский *geol.* Paleozoic
палеоихтиология paleoichthyology
палеоконх paledconch
палекортекс paleocortex, rhinencephalon
палеолит Paleolith(ic) age
палеолитический Paleolithic
палеонтолог paleontologist
палеопитек Palaeopithecus, Sivapithecus
палеопорфир paleoporphyry
палеопочвоведение paleopedology
палеосибирский, палеосибириец Paleo-Siberian
палеостриатум paleostriatum, pale globe
палеотерий paleothere, paleotherium
палеофир paleophyre
палеофит paleophyte
палеофитология paleophytology, paleobotany
палеоцен, палеоценовый Paleocene
палеоценоз paleobiocoenosis
палеоэхиноидеи *pl.* paleoechinoids *(Palaeoechinoidea)*
Палестина Palestine
палестинец, палестинский Palestinian
палестинка Palestinian (woman)
палестра *hist.* pal(a)estra, wrestling school
палета pallet stone/jewel
палетка surveyor's table/plane; transparency; measuring grid
  ~, точечния dot chart/grid
палец finger; *tech.* cam, bolt, cog, pin, peg, stud, catch, tooth; dog; detent; *anat.* actyl, digit
  ~, безымянный ring finger
  ~, большой thumb; *(of foot)* big/great toe
  ~, выбирающий *teleph.* selecting finger
  ~ гусеницы track pin
  ~, кольцевой ring/fourth/medical finger
  ~, копирный tracer pin
  ~ кривошипа cranck pin
  ~ мотовила reel tine
  ~, наименьший little/fifth finger; little toe
  ~, направляющий guide pin
  ~ ноги/стопы toe
  ~, поршневой piston pin
  ~ прицепного шатуна wrist pin
  ~, средний middle/third finger
  ~, указательный index/second finger
  ~, шарнирный link/knuckle pin
палея *hist., lit.* Palaea
паликар *hist.* palikar
палилалия palilalia, paliphrasia
палилка *text.* cloth-singeing machine, singer
  ~, валичная revolving singer
  ~, плитная plate singer
палимбакхий palimbacchius
палимпсест palimpsest
палингенез(ис) palingenesis, palingeny
палисад palisade, paling; *mil.* stockade
палисадник 1. front garden; 2. palisade, paling
палисадный palisade *(attr.)*
  ~ная ткань *bot.* palisade tissue
  ~ная хлоренхима *bot.* palisade parenchyma
палисандр *bot.* palisander, black rosewood *(Dalbergia nigra)*
палитра palette
палить burn, singe, scorch
палиурус *bot.* paliurus *(Paliurus)*
  ~ держи-дерево Christ's-thorn paliurus *(Paliurus aculeatus)*
палия *zool.* lake char *(Salvelinus lepechini)*
палка stick, staff, rod, cane
Паллада *myth.* Pallas
палладий *chem.* palladium, Pd
  ~, губчатый palladium sponge
палладированный palladized, palladium-coated
палланестезия *med.* pallanesthesia, apallesthesia
паллиатив, паллиативный palliative
паллидо-рубральный pallidorubral
паллидо-стриальный pallidostrial
паллидотомия *surg.* pallidotomy
паллидум pallidum
паллидумальный pallidal
паллестезия *med.* pallesthesia
палоделатель *agr.* checker, bund former, landshaper
палоло *zool.* palolo worm *(Eunice viridis)*
паломник pilgrim
паломничество pilgrimage
палочка 1. small stick/rod; 2. *med.* bacillus
  ~, барабанная drumstick
  ~, болгарская Massol's bacillus *(Lactobacillus bulgaricus)*
  ~, брюшнотифозная typhoid bacillus
  ~ газовой гангрены gas bacillus, Welch's bacillus
  ~, дирижерская conductor's baton
  ~, кислотоупорная acid-fast bacillus
  ~, кишечная coli/collon bacillus, *Escherichia coli*
  ~, ляписная silver nitrate stick
  ~, молочнокислая *Lactobacillus*
  ~, паракишечная paracolon bacillus
  ~ проказы leprosy/Hansen's bacillus
  ~ сапа glanders bacillus
  ~, сенная grass/hay bacillus
  ~, смоляная resin stick
  ~, стеклянная глазная glass ophthalmic spatula
  ~, столбнячная drumstick/tetanus bacillus
  ~, туберкулезная tubercle/Koch's bacillus
  ~, эстафетная *sport* baton
палочковидный 1. rod/club-shaped, rodlike; 2. bacilliform, bacillary

палочковый *med.* bacillary
палочник I stick insect
палочник II *bot.* reed mace, cattail *(Typha)*
палтус *zool.* halibut, turbot *(Psetta maxima)*
палуба deck
~, ангарная hangar deck
~ бака forecastle deck
~ безопасности safety deck
~, верхняя upper deck
~, взлетная take-off deck
~, главная main/bulkhead deck
~, грузовая cargo deck
~, жилая accommodation deck
~, карапасная turtle-/whale-back deck
~, кормовая afterdeck
~, минная mine/mining deck
~, мостиковая bridge deck
~, навесная shelter deck
~ надстройки superstructure/erection deck
~, нижняя lower deck
~, открытая weather/summer deck
~, полетная flight deck
~ полубака forecastle deck
~ полуюта poop deck
~, посадочная landing [flying-on] deck
~, прогулочная promenade deck
~ с погибью cambered/crowned deck
~ с покрытием sheathed deck
~ с седловатостью sheered deck
~ средней надстройки bridge(-house) deck
~, средняя middle deck
~, съемная removable deck
~, тентовая awning/shade deck
~, тральная sweeping deck
~, угловая полетная angled/canted flight deck
~ юта poop deck
~, ящичная trunk deck
палубник deck planking
палубный deck *(attr.)*; carrier-based/-borne
пальма *bot.* palm *(Palmae)*
~, веерообразная flabelliform palm
~, кокосовая coconut palm *(Cocos nucifera)*
~, персиковая peach palm *(Guilielma speciosa)*
~, финиковая date palm *(Phoenix dactylifera)*
~, чилийская coquito palm *(Jubaea spectabilis)*
Пальмира Palmyra
пальмира *bot.* palmyra palm *(Borassus flabellifer)*
пальмирский Palmyrenian
пальмитин *chem.* palmitin
пальмитиновый *chem.* palmitic
~вая кислота palmitic acid
пальмовидный, пальмообразный palmiform, palmlike
пальмовый *bot.* palmaceous, palm *(attr.)*
~вая роща palm grove/forest
пальмовые *pl., bot.* palm family *(Palmaceae)*
пальмолистный *bot.* palmifolious, palm-leaved
пальпа palp, palpus
~, ротовая labial/oral palp
пальпация *med.* palpation; touch
~, баллотирующая balloting palpation
~ голосового дрожания vocal fremitus palpation
~ живота abdominal palpation/touch
~, перкуторная palpatopercussion
~, поверхностная surface palpation
~, скользящая sliding palpation
~, толчкообразная balloting palpation
пальпировать palpate
пальпируемый palpable
пальстаб *archeol.* palstave
пальцевидный, пальцеобразный digitate, digitiform, finger-shaped
пальцевой digital; digitate
пальцеходящий *zool.* digitigrade
пальцеходящие *pl., zool.* digitigrades *(Digitigrada)*
пальцы fingers *(of hand),* toes *(of foot)*
пальчатка *bot.* hairy crab grass *(Digitaria sanguinalis)*
пальчатожилковатый digitinervate, digitinerved
пальчатолопастный palmatilobate
пальчатонадрезный palmatisected
пальчатонервный digitinervate, digitinerved, finger-veined
пальчатоперистый digitipinnate, digitate-pinnate
пальчатораздельный digitipartite, palmatipartite
пальчатосложный palmately compound
пальчатость *bot.* digitation; palmateness
пальчатый digitate, palmate, dactylate, fingered
пальщик *min.* blaster; *text.* singer, burner
палюс *zool.* palus
пампас pampas
пампасовый, пампасный, пампасский pampean, pampas
пампельмус *bot.* pummelo, pompelmous *(Citrus maxima/decumana)*
памфлет lampoon, pamphlet
памфлетист, памфлетер lampoonist, pamphleteer
памятка 1. memento; check list; 2. instructions
память 1. *comp.* memory, storage, magnetic store; 2. memory, recollection; consciousness; mind;
~, автономная off-line storage
~ адресов address storage/store
~, адресуемая addressable memory
~, ассоциативная associative memory/store; content-addressable memory
~ большого объема [большой емкости] high-/large-capacity memory/store, bulk/mass memory
~, буферная buffer/intermediate/temporary memory
~, быстродействующая high-speed memory, fast-/rapid-/zero-/quick-access memory, rapid/fast memory
~ ввода, буферная input buffer store/memory
~ видеоданных video memory
~, внешняя external/secondary/file memory
~, внутренняя internal memory/storage

~, главная адресная addressable bulk storage
~, главная безадресная non-addressable bulk storage
~ графических данных graphics memory
~ данных data memory/store
~ данных о наличии inventory memory
~ данных, основная main data store
~, двухпортовая dual-ported memory
~, двусторонняя read-write memory
~, двухуровневая two-level memory
~ динамического типа dynamic memory
~, диодно-конденсаторная diode-capacitor store
~, дисковая disk store/file, disk-type store
~ дисплея, буферная off-screen memory
~ для данных data memory/store
~ для расширения системы add-on memory
~, долговременная long-term memory; read-only [fixed] storage
~, дополнительная backing [back-up] memory
~, зрительная visual/eye memory
~ из ячеек cell memory/store
~, имеющаяся available/availability store
~ клавиатуры, буферная keystroke memory
~, криоэлектронная cryoelectronic memory
~, магазинная stack/nesting memory
~, магнитная magnetic memory
~, магнетооптическая magneto-optical memory
~, матричная matrix memory
~, медленнодействующая slow-access memory/store
~ микропрограмм microcode memory
~ на больших интегральных схемах large-scale integration memory, LSI memory
~ на конденсаторах capacitor memory
~ на криотронах cryotron memory
~ на линиях задержки delay-line memory
~ на магнитной проволоке plated (magnetic) wire memory/store
~ на магнитном барабане (magnetic) drum store
~ на магнитных дисках magnetic disk memory
~ на магнитных пленках magnetic-film memory
~ на магнитных сердечниках magnetic-core memory/store
~ на магнитных стержнях magnetic-rod memory
~ на перфокартах card store/memory
~ на сердечниках, прямо адресуемая direct addressable core memory
~ на твисторах twistor memory
~ на туннельных диодах tunnel-diode memory
~ на ферромагнитных сердечниках (magnetic-) core memory/storage
~, нестирающаяся non-erasable storage
~ обратного магазинного типа FIFO memory
~, односторонняя read-only storage
~, оперативная on-line [temporary, working, short-time] memory/store
~, оптическая постоянная optical fixed store
~, основная main/primary memory
~, перезаписываемая writable memory
~, постоянная read-only memory, ROM, fixed/permanent storage
~ прямых и обратных пакетов (сообщений) piggyback memory
~ с алфавитной индикацией alphabetic indication store/memory
~ с бесконечной магнитной лентой continuous-loop recorder
~ с быстрой выборкой = ~, быстродействующая
~ с высоким разрешением high-resolution store
~ с дистрибутивной логики distributed-logic memory
~ с защитой от записи write-protected memory
~ с малым временем выборки short-/quick-/fast-/rapid-access memory
~ с мгновенной выборкой = ~, быстродействующая
~ с нулевым временем ожидания no-wait memory
~ с последовательным доступом sequential-access memory/store
~ с произвольной выборкой random-access memory
~ с прямым доступом direct-access memory
~ с файловой организацией file memory
~ со стиранием scratch-pad memory
~ со страничной организацией paged memory
~, сверхбыстродействующая ultrahigh-access memory
~, слуховая aural memory
~ спин-эха spin echo memory
~, старческая *med.* senile/anterograde memory
~, стирающаяся/стираемая erasable memory
~, ферритовая пленочная laminated ferrite memory
~, цифровая digital memory
~, чувственно-образная affect memory
Пан *myth.* Pan
панагия *eccl.* panagia
панагглютинация panagglutination
панагглютинины panagglutinins
паназиатский Pan-Asiatic
панакс *bot.* ginseng *(Panax)*
~ женьшень Asiatic ginseng *(Panax ginseng)*
паналлоплоиды pan-alloploids
панама 1. Panama hat; 2. *text.* panama; 3. Panama
панамериканизм Pan-Americanism
панамериканский Pan-American
панамский Panamanian
~ная шляпа Panama hat
панангиит *med.* panangiitis

панангликанский Pan-Anglican
панаортит *med.* panaortitis
панарабизм Pan-Arabism
панариций *med.* panaritium, panaris, felon, whitlow
~, костный osteal/bony panaritium
~, межпальцевой frog felon
~, ногтевой unguinal/ungual panaritium
~, околоногтевой paraungual panaritium, paronychia
~, перекрестный crossed panaritium
~, подногтевой subungual panaritium
~, суставной articular panaritium
~, сухожильный thecal whitlow
панартериит *med.* panarteritis
панартрит panarthritis
панатрофия *med.* panatrophy
панафриканизм Pan-Africanism
панафриканский Pan-African
панацея panacea, cure-all
панбархат *text.* panne velvet
панген *biol.* gemmule; pangene
пангенез pangenesis
пангерманизм Pan-Germanism
пангерманист, пангерманский Pan-German
панголин *zool.* pangolin *(Manis)*
панда *zool.* panda *(Ailurus)*
пандажметр dipmeter, pendage meter
пандан, панданус *bot.* screw pine, pandanus *(Pandanus)*
пандекты *pl. law, hist.* Pandects, the Digest
пандемический, пандемия pandemic
пандемичность *med.* pandemicity
панджабец, панджабский Punjabi
Пандора *myth.* Pandora
пандус fixed ramp
пан-Европа Pan-Europe
панегирик panegyric, eulogy
панегирист panegyrist, eulogist, laudator
панегирический panegyrical, eulogistic, laudatory
панель 1. panel; board; panelling, wainscot; 2. footpath, sidewalk
~ аварийного запуска emergency start panel
~, амортизированная shock-mounted panel
~, бликоподавляющая antiglare panel
~, гнездовая *com.* jack panel
~ двигателя, пусковая engine starting panel
~, двухпролетная two-bay panel
~, диспетчерская dispatcher's/controller's supervision board
~ дистанционного управления remote-control panel
~ измерительных приборов instrumentation panel
~, индикаторная display panel
~, испытательная test panel
~, коммутационная switchboard/switching panel, switchboard, plugboard
~, контрольная supervisory panel
~ крыла wing panel
~, клавишная keyboard, keypad
~, ламповая *rad.* valve panel/holder
~, многопролетная multibay panel
~, монолитная integral panel
~, облицовочная facing panel
~ обшивки skin panel/plate
~, объединенная integrated panel
~ отображения данных/информации data-display panel
~ переключений switchboard/switching panel
~, печатная printed-circuit board
~ пола *auto.* floorpan
~, потолочная ceiling/overhead/roof panel
~ приборов instrument panel/board
~, программная program(me) table
~, пусковая launch/firing panel
~ речевой записи voice deck
~ с визуальными приборами visual panel
~ с мнемонической схемой mimic diagram panel
~ световой сигнализации annunciator panel
~, сдвижная sliding panel
~, сенсорная touch panel
~ сигнализации warning/caution panel; indicator panel
~, слоистая sandwich panel
~, смотровая inspection/access panel
~ солнечных элементов solar-cell panel
~, съемная removable panel
~ управления control panel/board
~, усиливающая reinforcing panel
~, шарнирная hinged panel
~, штекерная plugboard
~ электропитания power panel
панентеизм panentheism
панер *naut.* apeak, up and down, .
панзоотический, панзоотия panzootic
паникадило *eccl.* (church) chandelier
панировать *cul.* fry in egg and breadcrumbs
панисламизм Pan-Islamism
панихида requiem, mass for the dead, memorial service; burial rites, funeral ceremony
~, гражданская civil requiem
панкардит *med.* pancarditis
панкреасфиброз *med.* pancreatofibrosis
панкреатин *chem.* pancreatin
панкреатит *med.* pancreatitis
~, абсцедирующий abscessing pancreatitis
~, гнойный purulent pancreatitis, propancreatitis
~, гуммозный gummatous pancreatitis
~, калькулезный calcereous pancreatitis
панкреатический *anat.* pancreatic
панкреатолиз pancreatolysis, pancreolysis
панкреатолитиаз pancreatolithiasis
панкреатолитотомия pancreatolithotomy
панкреато(то)мия pancreatotomy, pancreatomy
панкреатэктомия pancreatectomy, pancreectomy
панкреозимин pancreozymin
панкреонекроз pancreatonecrosis
панкреопривный pancreoprivic

панмиелоз panmyelosis
панмиолопарез panmyeloparesis
панмиелопатия panmyelopathia, panmyelopathy
панмиелофтиз panmyelophthisis
панмиксис, панмиксия *biol.* panmixia, panmixy
панмицин panmycin
панневрит panneuritis
панникулит panniculitis
~, рецидивирующий relapsing pannicukitis
панно 1. painting; picture; 2. panel
паннус *med.* pannus
~, сосудистый vascular pannus
~, толстый/тонкий thick/thin pannus
~, фликтенулезный flictenular pannus
паноза *med.* panose
панорамирование scanning, panoraming, panning
панорамировать scan, pan
панофтальмит *med.* panophthalmitis, panophthalmia
панплегия panplegia, pamplegia
пансексуализм pansexuality, pansexualism
пансерозит panserositis
пансину(с)ит pansinu(s)itis
панславизм Pan-Slavism
панславист Pan-Slavist
панславистский, панславянский Pan-Slavic
панспермия, панспермaтизм panspermia, panspermatism
пантанкилоблефарон pantankyloblepharon
пантатрофия pantatrophia, pantatrophy
пантафобия pantaphobia, absolute fearleness
пантахроматический pantachromatic
пантеизм pantheism
пантеист, пантеистка pantheist
пантеистический pantheistic
пантеон pantheon
пантера *zool.* panther *(Panthera pardus)*
пантетеин pantetheine
пантетеинкиназа pantetheine kinase
пантограф pantograph
пантодрель pantodrill
пантокарены cross curves of stability
пантократия pantocracy
пантомим, пантомима pantomime
пантомимика pantomimicry
пантомимический, пантомимный pantomimic(al)
пантоморфизм pantomorphism, pantomorphia
пантоморфный pantomorphic
пантонин pantonine
пантопон *pharm.* pantopon
пантоскоп pantoscope
пантотенат *chem.* pantothenate
~ натрия sodium pantothenate
пантотенаткиназа pantothenate kinase
пантотеновый pantothenic
пантотерии *pl. paleont.* pantotheres *(Pantotheria)*
пантофобия pantophobia, pan(o)phobia
панфлегмона panphlegmona
пантюркизм Pan-Turkism
панцирноголовые *pl. zool.* stegocephalians *(Stegocephalia)*
панцирные *pl. zool.* loricates *(Loricata);* testaceans *(Testacea)*
панцирный testaceous
панцирь 1. armour; coat of mail; 2. *zool.* testa, lorica, armo(u)r, shell; 3. *bot.* frustule
~, кожный *med.* dermal carapace
панцитопения pancytopenia
панчен-лама Panchen Lama
панщина *hist.* corvée, villein service
панъевропейский Pan-European
панэллинизм Panhellenism
панэллинский Panhellenic
папа Pope
~, римский (the) Pope, (the) Supreme Pontiff
папаверин papaverine
~, хлористоводородный papaverine hydrochloride
папаин, папайотин papain, papayotine
папайя *bot.* papaya, pa(w)paw *(Carica papaya)*
паперть parvis, church porch
папизм 1. papistry; 2. Roman Catholicism
папилла papilla
папиллит *med.* papillitis, papilledema
~, десенный gingival papillitis
~ почки renal papillitis
папиллозный papillose, papillous
папиллома *med.* papilloma
~, базально-клеточная basal-cell papilloma
~, внутрипузырная intracystic papilloma
~, ворсинчатая villous papilloma
~ молочных протоков duct papilloma
~, плоскоклеточная squamous cell papilloma
~ сплетения plexus papilloma
~ хороидального сплетения papilloma of choroid plexus
папилломатоз papillomatosis
папилломатозный papillomatous
папиллотония miosis, myosis, coreclisis
папилля papilla
папиллярный papillary
папироса cigarette
папирус 1. papyrus; 2. *bot.* papyrus *(Cyperus papyrus, Papyrus antiquorum)*
папоротник *bot.* fern
~, мужской male/shield fern
папоротниковидный *bot.* fernlike, ferny, filicoid, filiciform; pteroid
папоротниковые *pl. bot.* ferns *(Filicinae)*
папоротниковый fern *(attr.),* ferny, filicial
~вая заросль fernery
папоротникообразные *pl. paleont.* pteridospermaphytes *(Pterydospermaphyta)*
папоротникосемянный *bot.* pteridospermous
паппус 1. *bot.* pappus; 2. *zool.* coronule
паппусовидный *bot.* pappose, pappous
папский papal, pontifical
~ престол St. Peter's chair

папуас, папуасский Papuan
папуаска Papuan (woman)
папула *med.* papule, papula, pimple
~, мокнущая moist/mucous papule
~, плотная solid papule
~, фолликулярная follicular papule, folliculo-papule
папулезный papular
папулопустула papulopustule
папшер cardboard cutter
папье-маше papier-maché
пар I vapour; steam; fumes; exhalation
~, влажный wet steam
~, голый open steam
~, мятый exhaust steam
~, насыщенный saturated steam/vapour
~, острый live steam
~, отработанный waste/exhaust/spent steam
~, перегретый superheated steam
~, переохлажденный supercooled steam/vapour
~, технологический process steam
~, энергетический power-generating steam
пар II *agr.* fallow (land)
~, чистый/черный bare fallow
пара pair, dyad; couple
~ аэродинамических сил aerodynamic couple
~, витая twisted pair
~, гальваническая galvanic/voltaic couple
~, граничная *comp.* boundary
~ двоичных разрядов bit pair, pair of bits
~, золотниковая spool-and-sleeve
~, зубчатая pair of gears, gear set
~, ионная/ионов ion pair
~, кинематическая kinematic couple/pair
~, колесная wheel pair/set
~, мю-мезонная mu-pair
~ объективов paired objectives
~ окуляров paired eyepieces
~, плоская planar pair
~, поступательная rectilinear pair
~ самолетов two-plane formation
~, связанная coupled pair
~ сил couple (of forces)
~, сопряженная conjugate pair
~ стратегий *game th.* strategy pair
~, электронная electron pair
~, электронно-дырочная electron-hole pair
~, электронно-позитронная electron-positron pair
параксон *anat.* paraxon(e)
параактиномикоз para-actinomycosis, pseudo-actinomycosis
парааллокортекс para-allocortex
парамилоидоз para-amyloidosis
парааминобензойный para-aminobenzoic
парааминосалициловый para-aminosalicylic
параамфиболит para-amphibolite
параанальгезия para-analgesia, paranalgesia
парaанестезия para-anesthesia, paranesthesia
парааппендицит para-appendicitis, perityphlitis
парабеллум automatic pistol
парабиоз *biol.* parabiosis
парабола I *math.* parabola
~, кубическая cubic(al) parabola
~ невесомости *space* zero-lift parabolic trajectory, zero-G parabola
~, полукубическая semicubical/Neil's parabola
~ убегания/освобождения *space* parabola of escape
парабола II parable
параболизировать make paraboloidal, parabolize
параболический I *math.* parabolic
параболический II parabolical
параболоид paraboloid
параболоидальный, параболоидный paraboloidal
паравагинит *med.* paravaginitis
паравакцина paravaccinia
параван *naut.* paravane, kite otter
~, многокрылый multiplane kite otter
параван-балка paravane/dropping davit
параван-охранитель *naut.* protector paravane
параван-трал A-sweep
параван-шкентель kite pendant
паравезикальный paravesical, paracystic
паравезикулит paravesiculitis, paracystitis
параводород parahydrogen
паравыпуклый paraconvex
параганглий *anat.* paraganglion
~, нехромафинный nonchromaffin paraganglion
~, хромафинный chromaffin paraganglion
параганглиома *med.* paraganglioma
парагваец, парагвайский Paraguayan
парагвайка Paraguayan (woman)
парагевзический parageusic
парагевзия parageusia
парагелий parahelium
парагематин parahematin
парагемофилия parahemophilia
парагенез(ис) *geol.* paragenesis
парагенный paragenous, paragene
парагеосинклиналь, парагеосинклинальный parageosyncline
парагепатический parahepatic
парагидроз parahidrosis, paridrosis
парагинный paragynous
парагипноз parahypnosis
парагипофиз parahypophysis
параглиома paraglioma
параглобулин paraglobulin
параглобулинурия paraglobulinuria
парагнейс paragneiss
парагога *gram.* paragogue
парагонимоз paragonimiasis, distomiasis
парагонит *miner.* paragonite
парагонорейный paragonorrheal
парагормон parahormone
парагранулема paragranuloma
параграф paragraph; item

парагрип paragrippe, parainfluenza
парагриппозный paragrippal
парад parade; display; review
~, авиационный/воздушный air display/parade
~, военно-морской naval review
~, принимать review the troops
парадейтерий paradeuterium
параденит paradenitis
парадизентерия *med.* paradysentery
парадигма *gram.* paradigm
парадигматика paradigmatics
парадизея *bot.* St.-Bruno's lily *(Paradisea)*
парадизка *bot.* paradise apple *(Pyrus malus paradisiaca)*
парадокс paradox
~ близнецов twin paradox
~ часов clock paradox
парадоксалист paradoxer
парадоксальность paradoxicality, paradoxicalness
парадоксальный paradoxical
парадонтоз paradontosis, periodontosia
паразит parasite; *tech.* idle gear, idler
~, внутриклеточный intracellular parasite
~, внутренний endoparasite
~, гетероксенный heteroxenous parasite
~ кожи, животный dermatozoon
~ кожи, растительный dermatophyte, dermophyte
~, кровососущий bloodsucking parasite
~, разнохозяйный plurivorous/generalized parasite
~, растительный biophyte, entophyte
паразитарный parasitic; parasitogenic
паразитемия parasitemia
паразитизм parasitism
~, постоянный permanent parasitism
паразитировать parasitize
паразитицид parasiticide
паразитический, паразитный parasitic(al)
паразитолог parasitologist
паразитологический parasitologic
паразитология parasitology
паразитоноситель parasite carrier
паразитоносительство *med.* parasitosis
паразитотропизм parasitotropy
паразитотропный parasitotropic
параизомер para-isomer
параиммунитет paraimmunity
параинфекционный parainfectious
паракаллус paracallus
параканаликулит paracanaliculitis
паракардиальный paracardiac, paracardial
паракарпий *bot.* paracarpium
паракарпный paracarpous
паракаучук para rubber
паракератоз *med.* parakeratosis
паракинез parakinesia, parakinesis
паракислота para-acid
параклаза *geol.* paraclase
паракокцидиоидальный paracoccidioidal
паракокцидиоидоз paracoccidioidosis
паракокцидиоидомикоз paracoccidioidomycosis
параколит paracolitis
паракольпит paracolpitis
паракристалл paracrystal
паракристаллический paracrystalline
паралиагеосинклиналь paraliageosyncline
парализатор paralyzer, paralyzant
парализация, парализование paralyzation, paralyzing
парализировать, парализовать paralyze
парализованность paralysis; palsy
парализованный paralyzed; paralytic, palsied
парализующий paralyzing, paralyzant
паралингвистика paralinguistics
Паралипоменон Paralipomenon, the Chronicles
паралитик, паралитический paralytic
паралич *med.* paralysis, palsy
~ акушерский obstetric paralysis, birth palsy
~, атрофический atrophic/creeping paralysis, creeping palsy, progressive muscular atrophy
~ Богерта-Бертрана Bogaert-Betrand paralysis
~, вазомоторный vasomotor paralysis, vasoparesis, angiopalsy
~ века blepharoplegia, blepharoptosia
~, временный temporal/transient paralysis
~, врожденный congenital paralysis/palsy
~, вялый flaccid paralysis
~ глазодвигательного нерва oculomotor paralysis, ocular palsy
~ гортани laryngoparalysis, laryngoplegia
~, двигательный motor paralysis
~, декомпрессионный decompression/diver's paralysis, caisson disease
~, детский infantile paralysis, acute anterior poliomyelitis
~, детский инфекциозный/спинальный/эпидемический infantile spinal paralysis
~ диафрагмы phrenoplegia
~, дрожательный shaking/trembling palsy, Parkinson's disease, ballism
~ желудка gastroplegia
~, ишемический ischemic paralysis
~ кишечника enteroplegia, ileus
~, клещовый tick paralysis
~ лица, мимический mimetic paralysis
~ лицевого нерва facial paralysis/palsy, facioplegia, prosoplegia
~, мышечный myoparalysis, myoparesis
~, неполный paresis
~, ночной night palsy, waking numbness
~ нижних конечностей paraplegia
~, общий general paralysis, holoplexia
~ одной конечности monoplegia
~ половины тела hemiplegia
~ речи laloplegia
~, свинцовый lead paralysis/palsy, saturnine

palsy, wristdrop
~ **сердца** cardioplegia
~, **тактильный** stereoagnosis, astereognosis
~, **частичный** paresis
~ **языка** glossolysis, plossoplegia
паралический *geol.* paralic
параличный *med.* paralytic
~ **больной** paralytic
параллакс parallax; parallax effect
~, **вековой** secular parallax
~, **гелиоцентрический** heliocentric/annual parallax
~, **геоцентрический** geocentric/diurnal parallax
~, **годичный** annual/heliocentric parallax
~, **звездный** stellar parallax
~, **лунный** lunar parallax
~ **по высоте** paralax in altitude
~ **по склонению** parallax in declination
~, **перекрестный** heteronymous/crossed parallax
~ **прибора** instrumental parallax
~, **прямой** homonymous/direct parallax
~ **Солнца** solar parallax
~, **стереоскопический** stereoscopic/binocular parallax
~, **суточный** diurnal parallax
параллактический, параллаксный parallactic(al)
параллелепипед *math.* parallelepiped, cuboid
~, **наклонный** oblique parallelepiped
~ **повторяемости** recurrence/repetition parallelepiped
~, **прямой** right parallelepiped
~, **прямоугольный** rectangular parallelepiped/cuboid
параллелепипедальный parallelepipedal, parallelepipedous
параллелизация parallelization; correlation
параллелизм parallelism; concurrency
~, **линейный/плоскостной** linear/plane parallelism
параллелизуемый parallelizable
параллелограмм *math.* parallelogram
~ **сил** parallelogram of forces
~, **шарнирный** parallel link mechanism
параллель parallel; parallel line
~, **географическая** parallel of latitude, terrestrial parallel
~, **геодезическая** geodetic parallel
~, **земная** parallel of latitude
~ **карты, главная** standard parallel
~ **касания** parallel of tangency
~, **магнитная** magnetic parallel, isoclinic line
~ **на картографической сетке** grid parallel
~ **наблюдателя** parallel of observer
~, **небесная** parallel of altitude
~ **светила/суточная** diurnal circle
~ **склонения** parallel of declination
параллельно in parallel; parallel *(with, to)*; *fig.* simultaneously; concurrently
~ **по битам/байтам** parallel-by-bit/byte
параллельно-включенный connected in parallel
параллельно-волокнистый parallel-fibrous
параллельно-жилковатый/нервный parallel-veined/nerved, parallelinervate
параллельно-полосчатый parallel-banded
параллельно-последовательный parallel-serial/series, series-parallel
параллельно-стебельчатый parallel-acicular
параллельно-столбчатый/шестоватый parallelcolumnar
параллельность parallelism
параллельный parallel; collateral; concurrent
параллергический *med.* parallergic
параллергия parallergy
паралогизм *logic* paralogism
паралогика paralogia, paralogy; paralogism
~, **тематическая** thematic paralogia
паралогический paralogical; paralogistic
паральбумин paralbumin
паральдегид paraldehyde
паралямбдацизм paralambdacism
парамагматический paramagmatic
парамагнетизм, парамагнитность paramagnetism
парамагнетик *phys.* paramagnet
парамагнитный paramagnetic
парамастит *med.* paramastitis
парамедиальный paramedian, paramesial
параменингококк parameningococcus
параметр parameter; factor; variable; property
~, **безразмерный** dimensionless parameter
~, **видовой** generic parameter
~ **винтового движения** screw parameter
~, **волновой** *rad.* wavelength parameter
~ **включения** on-parameter
~, **входной** input parameter
~, **газодинамический** gasdynamic property
~ **генеральной совокупности** *math.* population parameter
~, **геометрический** size-shape factor
~ **группирования** bunching parameter
~ **движения** motion parameter
~, **зависящий от времени** time-dependent parameter
~, **зеркальный** image parameter
~, **изменяющийся во времени** time-varying parameter
~, **квантованный** quantized parameter
~, **ключевой** *comp.* keyword parameter
~, **конструктивный** design parameter/variable, structural parameter
~ **кривизны** buckling
~ **масштаба** scale parameter
~ **моделирования** simulation parameter
~, **необязательный** optional parameter
~, **несущественный** incidental parameter
~, **обобщенный** combined parameter
~, **обязательный** mandatory parameter
~, **определяющий** diagnostic variable
~, **оптимизируемый** objective variable

~ подобия similarity parameter
~ полета flight parameter
~ положения location parameter
~, прогнозирующий *comp.* predictor
~, расчетный design condition /parameter
~, регулируемый controlled parameter/value
~, режимный operating condition
~ решетки lattice parameter/constant
~ сложности complexity index
~, случайный random parameter
~ состояния state variable
~ соударения/столкновения/удара impact/collision parameter
~ течения flow parameter
~, фактический actual parameter
~, физический physical property
~, фокальный latus rectum
~ы, внешние outer [out-of-line] parameters
~ы, временные time parameters/transients
~ы, кинематические kinematic/linkage parameters
~ы пара steam conditions
~ы реактора reactor variables
~ы, сосредоточенные lumped parameters
параметриальный *anat.* parametric, parametrium *(attr.)*
параметрий *anat.* parametrium
параметрика parametrics
параметрит *med.* parametritis, parametrial inflammation, pelvic cellulitis
~, акушерский obstetric parametritis
~, задний posterior parametritis
~, послеродовой postnatal parametritis
параметритный *med.* parametritic, parametritis *(attr.)*
параметрический *math.* parametric(al)
параметрон parametron, parametric oscillator
~ на ферритовых сердечниках ferrite core parametron
~, тонкопленочный thin-film parametron
парамеция *zool.* paramecium, Slipper animalcule *(Paramecium)*
парамиелобласт paramyeloblast
парамиксовирус paramyxovirus
парамил paramylon, paramylum
парамилоид paramyloid
парамилоидоз paramyloidosis
парамиозиноген paramyosinogen
парамиоклонус paramyoclonus, myoclonus, polyolonia
парамион paramyon
парамиотония paramyotonia
~, атаксическая ataxic paramyotomia
парамицетома paramycetoma
парамнезия paramnesia, pseudomemory
парамолочный *chem.* paralactic
параморфный paramorphic
параморфоза *miner.* paramorph; paramorphism
паранг parang, barong
парангон flawless precious stone
паранджа paranja, yashmak
паранекроз paranecrosis
паранефрит *med.* paranephritis
~, абсцедирующий paranephric/perinephric abscess
~, кистозный paranephric cyst, pseudohydronephrosis
параноид paranoid, paranoiac
параноидный paranoid
параноик paranoiac
параноический paranoid; paranoiac
паранойя paranoia, paranoid schizophrenia
~, алкогольная alcoholic paranoia
~, инволюционная involutional paranoia
паранойяльный paranoiac
парономия *med.* paranomia
парантгелий, парантелий *astr.* paranthelion
парантез parenthesis
парануклеарный paranuclear, paranucleate
парануклеин paranuclein, pyrenin
парапарез *med.* paraparesis
~, пирамидный piramidal paraparesis
парапет parapet, balustrade; coping, cope
парапитек *paleont. Parapithecus*
параплазма paraplasm
параплан, парапланер paraplane, paraglider, parawing
~, грузовой parawing utility vehicle
парaплеврит parapleurisy, parapleutitis
~, огнестрельный gunshot parapleurisy
параплегический paraplegic
параплегия *med.* paraplegia
~, верхняя superior paraplegia
~, разгибательная paraplegia in extension
~, сгибательная paraplegia in flexion
~, судорожная tetanoid paraplegia
параплектенхима paraplectenchyma
парапневматогенный parapneumatogene
параподий, параподия *zool.* parapod(ium)
парапраксия parapraxia
парапроктит periproctitis, perirectitis, paraproctitis
~, внекишечный extrarectal periproctitis
~, гнилостно-некротический necroputrefactive periproctitis
~, гнилостный putrefactive periproctitis
~, ишиоректальный ischiorectal periproctitis/abscess
~, подковообразный shoe-shaped periproctitis
~, подслизистый submucosal periproctitis
~, позадипрямокишечный retrorectal periproctitis
~, седалищно-прямокишечный ischiorectal periproctitis
~ со свищами fistulous periproctitis
~, тазово-прямокишечный pelviorectal periproctitis
~, флегмонозный flegmonous periproctitis
парапростатит paraprostatitis
парапротеин paraprotein
парапротеинемия paraproteinemia

парапротеинурия paraproteinuria
парапротеиноз paraproteinosis
парапсихоз parapsychosis
парапсихолог parapsychologist
парапсихология parapsychology
парапсориаз parapsoriasis
парапузырек paravesicle
параритмия pararrhythmia
парасальпингит parasalpingitis
параселена *meteor.* paraselene
парасерозем paraserozem
парасигматизм parasigmatism, lisping
парасимболия parasymbolia
парасимпатикопаралитический parasympathoparalytic
парасимпатикотония parasympathonia, vagotonia
парасимпатикотропный parasympathotropic
парасимпатический parasympathetic
парасимпатолитический parasympatholitic
парасимпатомиметический parasympathomimetic
парасинапсис *biol.* parasynapsis
парасиндез *biol.* parasyndesis
парасиновит *med.* parasynovitis
парасинтез parasynthesis
парасинусный parasinoidal
парасистолический parasystolic
парасистолия parasystole
парасифилис parasyphilis, parasyphilosis
парасифилитический parasyphilitic
парасол *aero.* parasol
парасостояние parastate
параспазм paraspasm
  ~, лицевой facial paraspasm
парастерильность parasterility
парастернальный parasternal
парастиха *bot.* parastichy
парасцений parascenium
паратгормон parathyroid hormone, parathormone
паратенонит *med.* paratenonitis
паратеций parathecium
паратион parathion
паратип paratype
паратиреоидин parathyroidin
паратиреоидный parathyroid
паратиреоидэктомия parathyroidectomy
паратиреокрин parathyrocrin
паратиреотропный parathyrortop(h)ic
паратиреоцит parathyreocyte
паратиф paratyphoid (fever)
паратифлит paratyphlitis
паратифобациллез paratyphobacyllosis
паратифозный paratyphoid
паратонзиллярный paratonsillar
паратрахеальный paratracheal
паратрахома paratrachoma, granular conjuctivitis
паратуберкулез paratuberculosis
паратуберкулезный paratuberculous
паратрофический, паратрофный paratrophic
параунитарный para-unitary
парауретральный paraurethral
парауретрит paraurethritis
  ~, гонорейный gonorrheal paraurethritis
  ~, трихомонадный mycotic paraurethritis
параф initials
парафазия paraphasia
  ~, тематическая thematic paraphasia
парафазный paraphase
парафарингеальный parapharyngeal
парафарингит parapharyngitis
парафемия paraphemia
парафиз *zool.,* **парафиза** *bot.* paraphysis
  ~ водорослей paranema
парафилия paraphilia
парафиллий *bot.* paraphyllium
парафимоз paraphimosis
парафин paraffin (wax)
  ~, борированный borated paraffin
  ~, жидкий liquid paraffin
  ~, дейтеризованный D-paraffin
  ~, натуральный native paraffin
  ~, неочищенный crude paraffin
  ~, нефтяной oil paraffin, petroleum wax
  ~, твердый hard paraffin
  ~, технический commercial paraffin/wax
парафинированный waxed with paraffin, paraffined, paraffinized
парафинистость paraffin content, paraffinicity
парафинистый paraffinous, paraffinaceous
парафиновый paraffinic, paraffin *(attr.)*
парафинолечение paraffinotherapy
парафиноотделитель wax/paraffin separator
парафинома *med.* paraffinoma
парафинопластика paraffinoplasty
парафировать initial
параформальдегид *chem.* paraformaldehyde
парафраза, парафразировать paraphrase
парафрастический paraphrastic
парафрения paraphrenia
парафренный paraphrenic
парахимозин parachymosin
парахинон paraquinone
парахолера paracholera
парахолия paracholia
парахор *chem.* parachor
парахордальный parachordal
парахрома parachroma, parachroia
парахроматин parachromatin
парахроматоз parachromatosis
парахроматофор parachromatophore
парахронический parachronistic
парацеллюлоза paracellulose
парацентез *med.* paracentesis
парацетамол paracetamol, apamide
парацистит paracystitis
  ~, предпузырный precystic paracystitis
парашют parachute
  ~, аварийный emergency parachute
  ~ в сборе parachute assembly

~, введенный deployed parachute
~ военного образца service-type parachute
~, вспомогательный auxiliary parachute
~, вытяжной auxiliary/pilot parachute
~ вытянувшийся в трубку squid
~, грузовой freight/cargo parachute, supply-dropping parachute
~, двухкупольный double-canopy parachute
~, десантный assault parachute
~ для метеоприборов meteorological parachute
~ для снижения/спуска космических аппаратов space vehicle parachute
~ для торможения при входе в атмосферу планет planetary entry parachute
~, запасной emergency/reserve parachute
~, кольцевой annular parachute
~, нагрудный chest-pack parachute
~, наспинный back-pack/-type parachute
~, нераскрывшийся streamed parachute
~, нераскрытый unopened parachute
~, неуправляемый unguided parachute
~, планирующий gliding(-type) parachute
~, подкладной seat-pack parachute
~, посадочный landing parachute
~ принудительного раскрытия automatic parachute
~, противоштопорный anti-spin parachute
~, ранцевый pack-type parachute
~, раскрывшийся developed parachute
~ с быстрой отцепкой quick-release parachute
~ с куполом из трех полотнищ three-gore parachute
~ с независимым вводом free parachute
~ с принудительным вводом automatic parachute
~ с пружинным вводом, вытяжной spring-ejected pilot parachute
~ свободного раскрытия free parachute
~ системы спасения *rock.* recovery parachute
~, спасательный emergency [life-saving] parachute; recovery parachute
~, тормозной brake/drag/retarder parachute
~, управляемый guided/steerable/manoeuvrable/controllable parachute
~, хвостовой тормозной tail brake parachute
~, щелевой ring-slot/sail parachute
парашютизм parachute activity/diving/sport
парашютирование parachuting; *aero.* pancaking
парашютировать 1. parachute; drop by parachute; 2. *aero.* pancake
парашютист parachutist, parachuter, parachute jumper; *mil.* paratrooper
парашютнодесантный *mil.* paraborne, paradrop
парашютноспасательный pararescue *(attr.)*
парашютный parachute *(attr.)*
паращитовидный *anat.* parathyroid
паргелий parhelion
паргелический parhelic
парез *med.* paresis
~ желудка gastroparesis
~ кишечника enteroparesis
~ Лиссауэра Lissauer's paresis
~, мозжечковый cerebellar paresis
~, родовой labo(u)r paresis
парейазавр *paleont.* pareiasaurian *(Pareiasaurus)*
парейния pareunia
паремия, паремья paroemia
парение I steaming; letting out steam, emission of steam
парение II hovering, soaring
~ планера glider soaring
парентеза parenthesis
парентеральный parenteral
паренхима *bot., anat.* parenchyma
~, губчатая spongy parenchyma
~ листа mesophyll
~, лубяная phloem parenchyma
~ яичка parenchyma of the testis
паренхиматит *med.* parenchymatitis
паренхиматический parenchymatic
паренхиматозный parenchymatous, parenchymal
паренхимный parenchymal
паренхимула parenchymula
парестезический paresthetic
парестезия *med.* paresthesia
~, вкусовая gustatory paresthesia
париетальный parietal
париетовисцеральный parietovisceral
париетомография parietomography
парижанин Parisian
парижанка Parisian (woman), Parisienne
парижский Parisian
~кая зелень Paris green
~кая лазурь/синь Paris blue
паризит *miner.* parisite
парик wig
~, алонжевый full-bottomed wig
~, пудренный periwig
париковое дерево *bot.* common smoke tree *(Citinus coggygria)*
парило *bot.* common agrimony *(Agrimonia eupatoria)*
парильный for steaming/stewing
парильня steam shop/room; sweating room
парировать arrest, stop; counteract
Парис *myth.* Paris
паритет parity, par value, par
~, валютный par/parity of exchange
~, золотой gold parity
~, интервалютарный par/parity of exchange
~, монетный mint parity/ratio, mint par of exchange
~, обменный exchange parity
~ покупательной способности purchasing-power parity
~, по at par
~, скользящий sliding parity
паритетный parity *(attr.),* par *(attr.)*
парить I steam, scald; stew; emit [let out] steam

парить II soar, hover
пария pariah, outcast, castoff
парк 1. park; 2. depot, yard; *mil.* park; 3. stock, pool, fleet
~, авиационный air/aircraft fleet
~, автомобильный stock/fleet of cars
~, артиллерийский ordnance depot
~, дендрологический arboretum
~, локомотивный engine yard/depot
~ отдыха и культуры recreation park
~, отраслевой промышленный single-trade industrial unit/estate
~, понтонный pontoon equipment/train
~, разбивать layout a park
~, сортировочный marshalling yard/tracks
парка I steaming; *cul.* stewing
парка II parka
Парка *hist. rel.* Parca
паркеризация parkerization, parkerizing
паркет parquet floor/flooring, parquet; parquetry
~, наборный glued-laminated parquet(ry)
~, штучный block parquet(ry)
паркинсонизм parkinsonism, Parkinson's disease
~, свинцовый lead parkinsonism
паркинсоник *med.* parkinsonian
паркия *bot.* nitta tree *(Parkia)*
парламент parliament
парламентарий parliamentarian
парламентарный parliamentary, parliamentarian
парламентер *mil.* truce envoy
парламентерский флаг flag of truce
парламентский parliamentary
пармазан, пармезан Parmesan cheese
пармелия *bot.* parmelia *(Parmelia)*
Парнас(с) Parnassus
парнасец, парнасский Parnassian
парник hotbed, forcing bed/frame; hothouse
парниковый hotbed *(attr.)*, hothouse *(attr.)*
~ эффект hothouse effect
парнобороздный geminicolpate
парножаберный *zool.* dipleurobranchiate
парнокопытные *pl. zool.* artiodactyls *(Artyoactyla)*
парнокопытный cloven-footed/hoofed, artiodactyl
парнолистник *bot.* bean caper *(Zygophyllum)*
парнолистный *bot.* zygophyllous
парнолистные, парнолистниковые *pl. bot.* beancaper family *(Zygophyllaceae)*
парнопалый *zool.* even-toed
парноперистосложный, парноперистый *bot.* paripinnate
парнопоровый diploporous
парность paired relationship; *math.* twoness
парноцветковый *bot.* geminiflorous
парный *bot.* conjugate; geminate; *zool.* didymous
парование *agr.* fallowing
паровариальный *anat.* parovarian
пароварка double boiler
паровик (steam) boiler
пароводяной steam-and-water *(attr.)*
паровоз steam engine, locomotive
паровоздушный steam-air
паровозоремонтный locomotive-/engine-repair
паровой steam *(attr.)*; steam-driven/operated
паровпуск steam admission/inlet (pipe)
паровыпускной steam-exhaust *(attr.)*
парогазогенератор steam (-and-gas) generator
парогенератор steam generator
~, башенный tower-type steam generator
~, противоточный counter-flow steam generator
~, прямоточный once-through steam generator
парогидротермы steam hydrothermae
пародийно-героический mock-heroic; Hudibrastic
пародинамо steam-driven dynamo
пародонт *med.* parodontium
пародонтит *med.* parodontitis
пародонтоз parodontosis
пародонтопатия parodontopathia, parodontopathy
парозапорный steam-cutoff
пароизоляционный steamtight
паройкия paroecia
пароконный two-horse
пароксизм paroxysm; fit
пароксизмальный paroxysmal
пароле, пароли paroli
пароль watchword; countersign; parole; *comp.* password
паром ferryboat, ferry, ferrycraft, passage boat
~, автомобильно-пассажирский car-and-passenger ferry
~, большегрузный heavy ferry
~, железнодорожный railway/train ferry, ferry bridge, sea-train
~, морской рейсовый cruise-ferry vessel
~ на воздушной подушке hover-ferry
~ на подводных крыльях hydrofoil ferry
~, самоходный motor ferry
~, цепной chain ferry
паром-амфибия amphibious ferry
паромер steam flowmeter
паром-ледокол icebreaking ferry
паромщик ferryman
паронепроницаемый steam-tight, steam-proof; vapo(u)rtight
пароним *gram.* paronym
паронимический paronymic, paronymous
паронимия paronymy
паронит paronite, rubberized asbestos fabric
паронихия *med.* paronychia, paraungual panaritium
~, нагноительная suppurative paronychia
~, роговая corneous/horny paronychia
~, язвенная ulcerous paronychia
парономазия paronomasia, annomination
парообработка *agr.* fallow tillage
парообразный steamlike; vaporous
парообразование steam formation/generation; vaporization
пароосушитель steam drier/desaturator
пароотвод steam discharge pipe, steam bleeder

пароотводный steam-exhaust/discharge
пароотделитель steam separator/trap
пароотсекатель steam cutoff valve
пароохладитель steam cooler/attemperator, desuperheater
пароперегреватель steam superheater
~, вторичный reheater
~, двухходовой two-pass surerheater
~, огнетрубный smoketube superheater
~ с противотоком counter-current superheater
~ с прямотоком co-current superheater
параподвод steam supply
паропровод steam pipeline/line/pipe
паропроизводитель steam generator
паропроизводительность steam-generating capacity, steam output/rating; evaporating capacity
паропроницаемость permeability to steam
парораспределитель steam distributor/header
парораспылитель steam atomizer
паросборник steam collector/receiver/accumulator
паросский Parian
паростеит *med.* parosteitis, parostitis
паростеоз, паростоз *med.* parosteosis, parostosis
пароструйный steam-jet, steam-ejecting
паротеплоход ship with steam and Diesel engines
паротерма *geol.* steam jet
паротидный *med.* parotid
паротидоскирр parotidoscirrhus
паротит parotiditis, parotitis
~, гнойный purulent parotiditis
~, ложный pseudoparotiditis
~, рецидивируlющий relapsing parotiditis
~, серозный hydroparotiditis
паротурбина steam turbine
паротурбовоз steam-turbine locomotive
паротурбогенератор steam-turbine generator/alternator
паротушение steam smothering/extinguishing
пароход steamship, steamer
~, винтовой screw steamer
~ внутреннего плавания inland steamer
~, грузовой cargo steamer
~, колесный paddle(-wheel) steamer
~ с бортовыми гребными колесами side-wheel steamer
~, сухогрузный dry-cargo steamer
пароходство shipping company/agency, steamship line/company
пароэлектростанция, пароэлектроцентраль steam power plant
паррия *bot.* parrya *(Parrya)*
парс Parsi, Parsee
парсек *astr.* parsec
парсизм Parsism, Parseeism
партактив *abbr.* party activists
партбилет *abbr.* party membership card
партбюро *abbr.* party bureau
партвзнос *abbr.* party dues
партгрупорг *abbr.* party group organizer
партгруппа *abbr.* party group
партенапогамия parthenapogamy
партеногамета parthenogamete
партеногамия parthenogamy
партеногенез parthenogenesis
~, сезонный season parthenogenesis
партеногонидий parthenogonidium
партенокарпический parthenocarpic, parthenocarpous
партенокарпия *bot.* parthenocarpy
партеноспермий parthenosperm
партер parterre; *theatre* orchestra, parquet
партесное пение *eccl.* part singing
партиец Communist Party member
партийность party spirit/affiiation/principle
партийный party *(attr.)*; Communist Party *(attr.)*
партитура *mus.* score
партия 1. lot, consignment, batch, series, run; 2. group, team, gang, party, detachment; *pol.* party; 3. *sport, game th.* play; 4. *mus.* part
~, аварийно-спасательная rescue party
~ для эксплуатационных испытаний service test batch
~, контрольная *(of goods)* inspection lot
~, крупная large/long run
~, небольшая short run
~, несходящаяся nonconvergent play
~ с нулевой суммой zero-sum play
~ товаров/груза consignment
парткабинет *abbr.* (Communist) Party study centre
партколлектив (Communist) Party unit/group
партком *abbr.* (Communist) Party Committee
партмаксимум *abbr., obs.* maximum salary for Communist Party members
партминимум *abbr., obs.* minimum requirements for Communist Party membership
партнер partner; associate
партнерство partnership
парторг *abbr.* (Communist) Party organizer
парторганизация *abbr.* (Communist) Party organization
партстаж *abbr.* length of membership in the Communist Party
партсъезд *abbr.* Communist Party congress
партячейка *abbr.* Communist Party cell, local party group/committee
парус 1. *naut.* sail; canvas, cloth; 2. *anat.* velum; 3. *bot.* flag, vane, vexillum
~ без гика, косой free-footed sail
~, верхний мозговой superior medullary velum
~, гафельный guff sail
~, задний after sail
~, косой fore-and-aft sail
~, латинский lateen sail
~, нижний мозговой inferior medullary velum
~, разрезной divided/split sail
~, рейковый lug sail
~ с гиком boom sail
~ свода/купола *arch.* pendentive(s)
~, шпринтовый spritsail

~, штормовой storm sail, smiter
~а, везущие working canvas
~а, гротовые main sails
~а, косые gaff and stay sails
парусина sailcloth, canvas; tarpaulin; duck
парусинник canvas coat/jacket
парусник 1. sailboat, sailer, sailing vessel; 2. sailmaker, sail sewer; 3. *zool.* sailfish, spikefish *(Istiophorus);* 4. *zool.* veliger
парусный sailing, sail *(attr.)*
парусовый made of canvas/duck/tarpaulin
парусоносный *bot.* vexillary, vexillate
Парфенон *hist.* Parthenon
парфянин, парфянский *hist.* Parthian
парфянка *hist.* Partian (woman)
парцелла parcel (of land)
парциальный partial; fractional
парша *med. pathol.* favus; scab
пасека apiary
пасечник beekeeper, hiver
паскаль pascal, Pa
пасквиль lampoon, pasquil; squib
паслен *bot.* nightshade *(Solanum)*
пасленовые *pl. bot.* nightshade/potato family, *(Solanaceae)*
пасленовый nightshade *(attr.);* solanaceous
пасмо *text.* lea; skein
паспарту passe-partout
паспорт 1. certificate; record; 2. passport
~, заводской factory certificate
~, корабельный ship's warrant
~ массива *comp.* array ticket
~, нансеновский Nansen certificate
~ прибора certification of calibration
~, технический technical certificate
паспортизация 1. certification; rating; 2. introduction of passports [of identification cards]
пассаж passage
пассажир passenger
пассажиропоток passenger traffic
пассакалия, пассакалья passacaglia
пассат trade wind
пассатижи combination pliers
пассив 1. *econ.* liabilities; 2. *gram.* passive voice
пассиватор passivating agent, passivator
пассивация, пассивирование passivation, passivating treatment
пассивноплавающий *zool.* epiplanctonic, planctonic
пассивность passivity, passiveness
пассивный passive; inactive; inert
~ баланс unfavo(u)rable balance
~ное избирательное право *polit.* eligibility
пассирование *cul.* pan-frying, browning
пассифлора *bot.* passionflower, passiflora *(Passiflora)*
паста paste; compound
~, ихтиоловая ichthyol paste
~, мышьяковая arsenical paste
~, наждачная emery paste
~, необрастающая *naut.* antifouling paste
~, паяльная soldering paste
~, противозачаточная contraceptive paste
~, уплотняющая sealing paste, sealant; joint
пастбище pasture, grassland, range
~, общинное common /community pasture
пастбищный pasture *(attr.);* pascual; pasturable
паства *eccl.* congregation, flock
пастель crayon, pastel
пастельный pastel *(attr.),* crayon *(attr.);* drawn in pastels/crayons
пастереллез pasteurellosis
пастеризатор pasteurizer
~ мгновенного действия flash pasteurizer
пастеризация pasteurization
~, лучевая radiation pasteurization, radiopasteurization
пастеризовать pasteurize
пастернак *bot.* parsnip *(Pastinaka)*
~, посевной garden parsnip *(Pastinaca sativa)*
пастеровский Pasteur's
пастиччо pasticcio
пастораль 1. *lit.* pastoral; 2. *mus.* pastorale
пасторальный pastoral, bucolic
пастух shepherd; herdsman; cowherd
пастуший shepherd's; shepherdly
~шая сумка *bot.* shepherd's purse *(Capsella bursa- pastoris)*
пастушок *zool.* rail *(Rallus)*
пастырский shepherd's; *eccl.* pastoral, vicarial
пасть mouth, jaws
~, львиная *bot.* snapdragon *(Antirrhinum)*
Пасха 1. Passover; 2. Easter; 3. *cul.* paskha
~, сырная Easter cheese cake
пасхалия *eccl.* paschal cycle/tablets
пасхальный paschal; Easter *(attr.)*
пасынок 1. stepchild; stepson; 2. *bot.* side shoot, epicormic branch; 3. *tech.* stub
пасьянс solitaire
пасюк *zool.* Norway rat *(Rattus norvegicus)*
пат *chess* stalemate
патагонец, патагонский Patagonian
патагонка Patagonian (woman)
патан, патанский Pathan
патанка Pathan (woman)
пателла *hist., bot.* patella
пателлопластика patelloplasty
пателлэктомия *surg.* patellectomy
пателлярный *anat.* patellar
патент patent
~, выдавать grant a patent
~, действующий unexpired patent
~ на промышленний образец design patent
~, отменять cancel/revoke a patent
~, получать take out a patent (for)
~ с истекающим сроком действия expired/lapsed patent
~, судовой *naut., econ.* certificate of registry
~а, владелец patentee
патентованный patented; patent *(attr.);* propri-

etary
~ное изделие restrictive proprietary [patented] item/product
патентовать take out a patent for, patent
патентовладелец, патентообладатель, патентодержатель patentee
патентоспособность patentability
патентоспособный patentable
патер *eccl.* Father
патергия *med.* pathergia, pathergy
патерик *eccl.* Lives of the Fathers
патерство *eccl.* priesthood
патефон phonograph
патина patina
патинирование, патинировка patination
патинироваться patinate, patinize
патиссон *bot.* bush pumpkin *(Cucurbita pepo melopepo)*
патоархитектоника *med.* pathoarchitectonics
патобиология pathobiology
патогенез pathogenesis, pathogenesy, pathogeny
патогенность pathogenicity
патогенный pathogenic
патогномонический, патогномичний pathognomonic
патока treacle; molasses
патолог pathologist
патологический, патологичный pathologic; morbid
патология pathology, pathobiology
~, **гуморальная** humoral pathology
~ **животных** animal pathology
~ **конечностей** acropathology
~, **профессиональная** occupational pathology
~ **растений** phytopathology, plant/vegetable pathology
~, **частная** special pathology
патологоанатом *med.* pathoanatomist
патологоанатомический pathoanatomical
патологогистологический pathohistological
патометаморфоз pathometamorphism; pathometamorphosis
патомимезия pathomimesis
патоморфоз pathomorphism
патопсихология pathopsychology
патопсихофизиология pathopsychophysiology
патотропизм pathotropism
патофизиолог pathophysiologist
патофизиология pathophysiology
патофобия pathophobia
патриарх *eccl.* patriarch
патриархальность patriarchal character/nature
патриархальный patriarchal
патриархат patriarchy
патриархия, патриаршество *eccl.* patriarchate
патриаршеский, патриарший patriarchal
патрилокальность patrilocality
патримониум, патримоний patrimony
патристика patristic studies, patristics
патрициат *hist.* patriciate; patrician system
патриций *hist.* patrician
патрогенез *biol.* patrogenesis
патроклинность, патроклиния *biol.* patrocliny
патролог eccl. patrologist
патрология eccl. patrology
патрон 1. cartridge; round; 2. *mach.* chuck; 3. *el.* socket, lampholder; 4. patron; *eccl.* patron saint; 5. pattern; design
~, **боевой** live cartridge
~, **быстросменный** quick-change chuck
~, **винторезный/болторезный** screw-cutting chuck
~, **запальный** primer cartridge
~, **кулачковый** jaw chuck
~, **поводковый** driver/centre chuck
~, **разжимный** expanding chuck
~, **резьбонарезной** thread-cutting chuck
~, **самоцентрирующий** self-centering chuck
~, **токарный** lathe chuck
~, **трехкулачковый** three-jaw chuck
~, **цанговый** collet chuck
~, **штыковый** bayonet lampholder
~ **электромагнитный** magnetic chuck
патронаж home nursing
патронат patronage
патроним, патронимический patronymic
патронимия patronymy
патронка design, pattern
патронник cartridge/breech/gun chamber
патронодержатель cartridge clip
патрон-осушитель dehydrator plug/cartridge
патронташ ammunition belt, bandoleer, bandolier
патроциний, патроциниум patrocinium, patronage
патрубок branch pipe, stub-pipe; socket; sleeve; nipple; nozzle
~, **впускной** inlet pipe/connection
~, **всасывающий** suction/inlet pipe
~, **выпускной** outlet pipe/connection
~, **коленчатый** elbow pipe
~, **маслоналивной** oil filler
~, **напорный/нагнетательный** outlet/delivery/discharge pipe/connection
~, **сливной** drain pipe
патрулировать patrol
патруль patrol
~ **воздушный** air/airborne patrol
~ **зеленый** forest patrol
патуа patois
пауза pause, break, interruption, stop, interval; spacing; *mus.* interval
паук *zool.* spider
паукообразные *pl. zool.* arachnids *(Arachnida)*
паулопостумный *miner.* paulopost, deuteric
паутина spider's web; gossamer
паутинный *zool.* araneous, arachnoid(al)
паучник *bot.* spider plant, spiderflower *(Cleome)*
пауэллизм Powellism
пауэллист Powellite
пах *anat.* groin

пахание, пахарство plowing, ploughing, tilling
пахать plow, plough, till
пахиакрия *med.* pachyacria
пахиблефарон pachyblepharon
пахиглоссия *med.* pachyglossia, macroglossia
пахидактилия pachydactyly
пахидерматоцеле pachydermatocele
пахидермичесний pachydermatous, pachydermic
пахидермия *med.* pachyderma, pachydermia
пахидермы *pl. zool.* pachyderms *(Pachydermata)*
пахикарпин pachycarpine
пахименингит *med.* pachymeningitis
~, гиперпластический/гипертрофический hyperplastic pachymeningitis
~, гнойный purulent pachymengingitis
пахинема pachynema
пахиноз *bot.* pachynosis
пахиостоз pachyostosis
пахиперитонит pachyperitonitis
пахиплеврит pachypleuritis
пахитена *bot.* pachytene
пахлава *cul.* pakhlava
пахлук *bot.* musk yarrow *(Achillea moschata)*
пахово-бедренный *anat.* femoroinguinal, inguinofemoral
паховой, паховый *anat.* inguinal
пахово-мошоночный inguinoscrotal
пахота plowing, ploughing, tillage; tilth
пахотнопригодность *agr.* arability, arable capability, plough capacity
пахотный arable, cultivable
пахта buttermilk
пахтать churn
пахучеколосник *bot.* sweet vernal grass *(Anthoxanthum odoratum)*
пахучка *bot.* calamint *(Calamintha);* 2. savory *(Satureia)*
пачка packet, pack; bunch; pile, deck, stack; *mil.* (cartridge) clip
~ газет stack of newspapers
~ импульсов pulse burst
~ ошибок error burst
~ папирос pack of cigarettes
~ слойков packet of laminae
~ слюды book of mica
~ угля patch/band/batch of coal
~ фольг foil stack, stacked foil
~ эмульсий emulsion stack
пачули *pl. bot.* patchouli *(Pogostemon patchouli)*
пашечный 1. growing in cornfields; 2. arable
пашня arable/plowed/ploughed land
паяльник soldering iron/bit
паяльный soldering; brazing
~ная лампа blowlamp, blowtorch
паяние soldering; brazing
паять solder; braze
ПВО *abbr.* antiaircraft defences
ПВХО *abbr.* antiaircraft and antichemical defence
пеганка *zool.* shelduck *(Tadorna)*
Пегас *myth. astr.* Pegassus
пегматит *miner.* pegmatite
~, голлаитовый hollaitic pegmatite
~, письменный graphic pegmatite
~, слюдоносный micaceous pegmatite
пегматитовый pegmatic
пегматоид, пегматоидный pegmatoid
пегматофир *miner.* pegmatophyre
пегматофировый pegmatophyric, granophyric
пегматофитовый pegmatophytic
пегнитогенный pegnitogenous, pegnitogene
педагогика pedagogy, pedagogics
педагогический pedagogic(al); educational
педагогичный pedagogic; pedagogically correct
педализация *mus.* pedal(l)ing
педаль pedal; foot lever
~ акселератора accelerator/gas pedal
~, ножная foot/toe pedal
~, правая *mus.* forte pedal
~, средняя *mus.* sostenuto pedal
~, левая *mus.* piano pedal
~ сцепления *auto.* clutch pedal
~ тормозная (toe) brake pedal
педантичность, педантизм pedantry, pedantism
педвуз *abbr.* higher educational institution in pedagogy
педераст p(a)ederast, sodomite
педерастия pederasty, sodomy
педиатр pediatrician, pediatrist
педиатрия pediatrics, pediatry
педикулез pediculosis, lousiness
педилантус *bot.* slipperplant *(Pedilanthus)*
пединститут *abbr.* teachers' training college
педипальпа *zool.* pedipalp(us)
педиплен *geol.* pediplain
педогамия pedogamy
педогенез 1. *zool.* pedogenesis; 2. soil formation
педолог pedologist
педология I pedology, soil science
педология II p(a)edology
педофилия *med.* p(a)edophilia
педсовет *abbr.* pedagogical/teachers' council
педтехникум teachers' training college
педучилище training college for elementary-school teachers
педфак *abbr.* teacher-training department
пезета peseta
пейзажист landscape painter
пейслиизм Paisleyism
пек pitch
~, буроугольный lignite tar pitch
~, каменноугольный coal-tar pitch
~, нефтяной petroleum pitch
~, связующий pitch binder
пекан I *zool.* pekan, black cat *(Martes pennanti)*
пекан II *bot.* pecan *(Carya pecan)*
пекари *zool.* peccary *(Tayassu)*
пекинский Peking *(attr.),* Pekinese
пектатлиаза *bioch.* pectatelyase
пектенит *med.* pectenitis
пектиновый pectinic; pectic, pectinaceous

пекториловкия *med.* pectoriloquy, pectorophony
~ **шепотная** whispered pectoriloquy
пелагиаль pelagic zone
пелагит *geol.* pelagite, halobolite
пеламида *zool.* bonito *(Sarda)*
пеларгония *bot.* pelargonium, garden geranium
пеласг *hist.* Pelasgian
пеласгический Pelasgic
пелейский *geol.* Pelean
пелена veil; shroud; sheet
~, **вихревая** vortex/vorticity sheet
~, **красная** *av., med.* red-out
~, **облачная** cloud veil, pallium
~, **престольная** Communion/altar cloth
~, **серая** *av., med.* gray-out
~, **черная** *av., med.* black-out
пеленг bearing; *nav.* echelon
~, **акустический** acoustic/sonic/sound bearing
~, **бортовой** relative bearing
~, **гидролокационный** echo-bearing
~, **истинный** true bearing
~, **контрольный** check bearing
~, **левый/правый** left/right echelon
~, **локсодромический** Mercatorial/rhumb bearing
~, **начальный** datum bearing
~, **обратный** reciprocal/reverse bearing
~, **ограждающий** clearing bearing
~, **опасный** danger bearing
~, **определять** observe a bearing
~ **ориентира** bearing of the object
~, **ортодромический** great circle bearing
~, **прокладывать** plot a bearing
~, **прямой** direct bearing
~ **створа** transit bearing
~, **уменьшать** close the bearing
пеленгатор direction finder; compass locator, azimuth vane
~, **авиационный магнитный** aircraft bearing compass
~, **визуальный** bearing plate
~, **инфракрасный** infra-red seeker unit
~, **коротковолновый** HF direction-finder
~, **магнитный** bearing compass
~, **самолетный** airborne direction-finder
~, **средневолновый** MF direction-finder
~, **ультракоротковолновый** VHF [very-high-frequency] direction-finder
пеленгация, пеленгование direction finding
~ **по минимуму** aural null direction finding
пеленговать take the bearing(s) of; set a course
пелеципода *zool.* pelecypod *(Pelecypoda)*
пеликозавр pelycosaur *(Pelycosauria)*
пелиом *geol.* peliom, iolite, cordierite
пелит *petr.* pelite
~, **глинистый** argilaceous pelite
~, **диатомовый** diatomaceous pelite, diatomite
~, **кварцевый** quartz pelite
пелитовый pelitic, lutaceous
пелитолит lutite, lutaceous clay
пеллагра *med.* pellagra, maidism
~, **детская** infantile pellagra, kwashiorkor
пеллагрический, пеллагрозный pellagrous
пелликулярный pellicular, pelliculous
пелоидотерапия *med.* pelotherapy
пелопоннесский Peloponnesian
пелорический *bot.* peloric, pelorian; peloriate
пелорус *naut.* bearing plate, dumb compass, pelorus
пелтария *bot.* shieldwort *(Peltaria)*
пельвиоперитонит pelviperitonitis, pelvioperitonitis
~, **микробный** bacterial pelviperitonitis
пель-компас bearing/azimuth compass, azimuth finder
пельмени *pl. cul.* dumplings filled with minced meat, pelmeni
пельматозой *zool.* pelmatozoan
пельтатный *bot.* peltate, scutiform
пельтифиллум *bot.* saxifrage *(Peltiphyllum)*
пелюшка *bot.* field pea *(Pisum arvense)*
пемза pumice (stone)
пемзобетон pumice concrete
пемзовать polish with pumice
пемфигус *med.* pemphigus
пена foam, froth
~, **высокократная** high-expansion foam
~, **жидкая** aqueous foam
~, **латексная** latex froth
~, **огнегасящая** fire-fighting foam
~, **полиуретановая** polyurethane foam
Пенелопа *myth.* Penelope
пенеплен *geol.* peneplain
~, **вскрытый** stripped peneplain
~, **древний** paleopeneplain, paleopeleplane
~, **ископаемый** fossil peneplain
~, **реликтовый/бывший** past-peneplain
пенепленизация *geol.* peneplanation
пенис *anat.* penis
пенистость frothiness, spumescence, foaminess
пенистый foamy, frothy, spumescent; lathery
пенитенциарий, пенитенциарный *low* penitentiary
пенить(ся) foam, froth
пеницилл, пеницилий, пенициллиум penicillium
пенициллин penicillin
пенициллинамидаза penicillin amidase
пенка scum, skimmings; skin
~, **медная** copper froth/scum
~, **морская** sepiolite, meerschaum
пенни penny
пенобетон foam concrete
пеногаситель foam suppressor/breaker; antifoaming agent
пеногенератор foam generator
пеногипс foamed gypsum; foam plaster
пенографит cellular graphite
пенокерамика ceramic foam, foam ceramics
пенококс coked cellular material
пенообразование foaming, frothing

пенообразователь foaming/frothing agent, foamer, frother; foam generator
пеноотделитель froth/foam separator; skimmer
пенопласт foam/foamed plastic, expanded plastic
~, полиуретановый polyurethane foam
~, целлюлозный cellulose foam
пенопластмасса expanded plastic
пенополивинилхлорид polyvinylchloride foam
пенополистирол expanded polystyrene, polystyrene foam, styrofoam
пенополиуретан polyurethane foam
пенополихлоропрен polychloroprene foam
пенополиэпоксид polyepoxy foam, foamed epoxy resin
пенорезина cellular/foam rubber
пеносиликат cellular silicate
пеносниматель skimmer, froth/foam remover
пеностекло foam/foamed glass, cellular glass
пеностойкость froth resistance
пенотушение foam fire-fighting/extinguishing
пеночка *zool.* warbler *(Phyllospcopus)*
пенс penny
пенсионарий *hist.* pensionary
пенсия pension
~ без предварительных взносов noncontributory pension
~ за выслугу лет retirement pension; superannuation
~ по инвалидности disability pension
~ по нетрудоспособности incapacity pension
~ по старости old-age [superannuation] pension
~, пожизненная life pension
пенсне pince-nez
пентагон 1. *math.* pentagon; 2. the Pentagon
пентаграмма *math.* pentagram
пентада *med.* pentalogy
пентакринусовый *zool.* pentacrinoid
пентамерный pentamerous
пентаметр, пентаметрический pentameter
пентапирамида pentagonal pyramid
пентаплоид, пентаплоидный pentaploid
пентаполь *hist.* pentapolis
пентапризма pentaprism
пентастиль *arch.* pentastyle
пентатлон *sport* pentathlon
пентатоника *mus.* pentatonic scale
пентахорд *mus.* pentachord
пентаэдрический pentahedral
пентелийский Pentelic(an)
пентер-гак *naut.* shank-painter hook
пентод *el.* pentode
~, двойной duo-pentode, twin/double pentode
пентобарбитал pentobarbital, nembutal
пентоза *chem.* pentose, pantaglucose
пентозан *chem.* pentosan
пентозоальдолаза *bioch.* pentosealdolase
пентрит *chem.* penthrite
пентстемон *bot.* beardtongue *(Pentstemon)*
пенька hemp (fibre)
~, индийская sunn crotalaria
~, манильская Manila hemp, abaca
~, смольная tarred hemp
пеньковый hempy; hempen
пенькотрепальный hemp-hackling
пенькочесальный hemp-skutching/combing
пепелина *bot.* oidium
пепельник *bot.* cineraria *(Cineraria)*
пепельнолопастной tephrolobed, tephrolobate
пепельноплодный *bot.* tephrocarpous
пепин, пепинка pippin *(apple)*
пепсин pepsin
~ мочи uropepsin
пепсиновый pepsin *(attr.),* peptic
пептидсинтетаза peptide synthetase
пептизатор peptizer, peptizing agent
первитин pervitin, methamphetamine/methylamphetamine hydrochloride
первичнобескрылые *pl. zool.* primitively wingless *(Apterygota)*
первичноводные *pl. zool.* protoaquatic forms
первичнопокровный *bot.* archichlamydeous
первичнорадиальный perradial
первичноротые *pl. zool.* protostomes, archeostomatous organisms
первичнотрахейные *pl. zool.* onychophorans, protracheans
первичночерепной *zool.* procephalic
первичный primary, primigenial; primitive; primordial
~ корешок radicle
~ная клетка initial cell
первичнощечные *pl. zool.* protoparians
первоапостолы *eccl.* pillar/first apostles
первобытнообщинный primitive-communal
первозверь *zool.* prototherian, protothere
первокурсник first-year student, freshman
порвомученик protomartyr
первообразная *math.* antiderivative
первоочередность precedence
первопричина *philos.* initial/original cause
перворазрядник *sport* first-rank sportsman
первородный грех original sin
первородство *law* primogeniture
первородящая *med.* primapara
первоEnd
первосвященник chief/high priest; pontiff
первосвященство pontificate
первотелка first-calf heifer
первоцвет *bot.* primrose *(Primula)*
первоцветные *pl. bot.* primrose family *(Primulaceae)*
перга beebread, ambrosia
пергамен(т) parchment
пергаментирование parchmentizing
пергаментная бумага oilpaper
пергамин paper parchnent, pergamyn
пергиалиновый perhyaline
пергидрат *chem.* perhydrate
пергидрид *chem.* perhydride
пергидроль *chem.* perhydrol, hydrogen peroxide

переадресация readdressing; address modification/substitution; redirection
~, циклическая cyclic address change
переаминирование *bioch.* transamination
переаранжировать *mus.* rearrange
переассигновывать reassign, reallocate
перебазировать redeploy, redislocate; relocate
перебаллотировка second ballot
перебег *tech.* overtravel
перебежка *mil.* dash, rush
перебивание *rad.* breaking in; jamming
перебивать break in; interrupt; jam
перебивка interruption; *cin.* cut-in
перебой interruption, intermission; irregularity; trouble, outage, stoppage, failure
~ согласных consonant shift
перебор 1. looking through/over; 2. reassembly, reassembling; 3. surcharge
~, полный exhaustive search
переборка 1. *tech.* overhauling; reassembly; 2. *naut.* bulkhead; 3. *print.* resetting
~, ахтерпиковая afterpeak bulkhead
~, бортовая side/wing bulkhead
~, газонепроницаемая gas-tight bulkhead
~, диаметральная centreline bulkhead
~, конструктивная structural bulkhead
~, кормовая aft/after bulkhead
~, нефтенепроницаемая oil-tight bulkhead
~, отбойная swash/wash bulkhead
~, проницаемая nontight bulkhead
~, противопожарная/огнестойкая fireproof bulkhead
~, таранная collision bulkhead
~, форпиковая forepeak bulkhead
перебрасывание, переброс 1. overthrowing, throwing over, throw-over, flip-over; 2. transfer; shifting; 3. *geol.* faulted overfold, overthrust
переброска transfer; transportation, lift, ferry
~ на вертолетах helicopter airlift/transfer
~ по воздуху airlift, air ferry
~ по море sea lift
~ стока diversion of run-off
перебуривание redrilling
перевал passing, crossing; pass; *met.* bridgewall, sand dam
перевалка transfer; reloading; transshipment
переваривание digestion
переваривать 1. digest; 2. boil/cook anew
перевес 1. overweight; 2. preponderance
перевод 1. conversion, transformation; 2. transfer, shift, changeover; switch; 3. translation; interpretation
~, банковский bank transfer, bank money order
~ в почтовой системе жиросчетов giro postal order
~, взаимный interconversion
~ двоично-десятичных чисел BCD translation
~, денежный money order; cash/money remittance
~ долга conversion of debts
~, машинный computer/machine translation
~, почтовый money order
~ румбов *naut.* conversion of course
~ стрелки *rail.* shunting, switching
~ строки *comp.* line feed, LF
~, телеграфный cable/wire transfer
переводить 1. transfer, shift; move; 2. convert; 3. shunt, switch off; change over; 4. translate; 5. remit, transmit
перевозбуждение overexcitation, overexcitement
перевозить transport, ship, carry, haul
перевозка transportation, transport, carriage, haulage, lift, shipment, cartage; movement
~ автотранспортом trucking, carriage by road/truck
~, безрельсовая road haulage; cartage, carting
~, бестарная bulk carriage/haulage/shipment
~, ближняя/дальняя short/long haul
~ войск trooping; troop movement
~ грузов, транзитная transshipment
~, железнодорожная carriage by rail; railage
~, морская carriage/shipment by sea
~ навалом bulk carriage/haulage/shipment
~ "туда и обратно" out-and-back haul
~ холодильным транспортом refrigerated transport/shipment
~ки баржами barging
~ки внутрь страны inland transport
~ки, воздушные air transportation; airlift
~ки, встречные *econ.* cross freight; backhaul
~ки грузов/грузовые freight/cargo traffic
~ки грузов, внутригородские local trucking service
~ки, контейнерные container shipment
~ки легких грузов light freighting
~ки, морские overseas transport, seaborne/waterborne transportation; marine freight transportation
~ки, нерегулярные non-scheduled transport services; tramp operations
~ки по бездорожью off-road haulage
~ки по воздуху air transportation; *mil.* airlift
~ки, сухопутные overland transport
~ки, челночные воздушные shuttle air services
перевозчик 1. carrier; ferryman, boatman; 2. *zool.* common sandpiper *(Actitis hypoleuca)*
~, морской sea carrier
перевооружать 1. rearm; 2. reequip, refit
перевооружение 1. rearmament; 2. refitting
перевоплощение reincarnation
переворачивание inversion; overturning; capsizing
переворачиватель *agr.* tedder
~ покосов hay tedder
переворот overturn; radical change; coup; *geol.* cataclysm; *av.* half-roll
~, государственный coup d'ètat
~, двойной *av.* roll

~ **на горке** *av.* wingover
~ **на спину** *av.* upside-down turn
~ **через крыло** *av.* half-roll and split-S
перевскрытие *med.* repeated autopsy
перевыполнять overfulfil, exceed, overimplement
~ **норму** exceed the quota
перевязка 1. bond; binding, tie, band; bandage, dressing; ligation; 2. *anat.* vinculum
~ **в елку [колосовая]** herring-bone bond
~ **варикозных вен** *med.* cirsodesis
~, **крестовая** *constr.* cross bond
~, **ложковая** stretcher bond
~ **пуповины** umbilical ligation
~, **тычковая** header bond
перевязочная *med.* dressing room/ward
перегиб 1. bend(ing), twist(ing); fold; kink; 2. *naut.* hog(ging); 3. *math.* inflexion, inflection
~ **к переди** anteflexion
~, **коленчатый** geniculation
~ **кзади** retroflexion
~ **мочеточника** uretral kink
перегибать(ся) bend, fold; turn up, double
перегибный *math.* inflexional, inflectional
перегласовка *phon.* mutation
перегной humus; mull
переговоры negotiations, talks
~, **вести** conduct/hold negotiations
~ **за круглым столом** round-table talks
~ **на высшем уровне** summit talks, negotiations at the highest level
~, **неофициальные** informal talks
~ **о заключении коллективного договора** collective bargaining
~ **о размере (повышении) заработной платы** pay talks, wage bargaining
перегон span, section, leg
перегонка *chem.* distillation; rectification
~, **азеотропная** azeotropic distillation
~ **в кипящем слое** fluidized distillation
~, **дробная** fractional distillation
~ **нефти** oil/petroleum refining
~, **однократная** flash distillation
~, **периодическая** batch distillation
~ **под давлением** compressive distillation
~, **противоточная** counter-current distillation
~, **прямая** straight-run distillation
~ **с кренингом** cracking distillation
~, **селективная/экстракционная** extractive distillation
перегонщик *av.* ferry(ing) pilot
перегородка partition; dividing wall; bulkhead; baffleplate; *anat.* septum; diaphragm
~, **бедренная** femoral/crural septum
~, **герметическая** pressure bulkhead
~, **глазничная** orbital septum
~, **междолевая/междольковая** interlobular septum
~, **мышечная** myoseptum
~, **отражательная** baffle
~, **противопожарная** fireproof wall/partition
~, **тепловая** thermal baffle
перегородковидный, перегородкообразный septiform
перегородчатый septate; murate
перегрев overheating; superheat
~, **местный** hot spot, thermal spike
перегреватель superheater
перегревать overheat; superheat
перегружать 1. reload; 2. overload, overburden
перегрузка overload(ing); reloading
перегруппировывать regroup; rearrange; *mil.* redeploy
передатчик transmitter
~, **бортовой** airborne transmitter
~, **задающий** master transmitter
~, **импульсный** pulse/impact transmitter
~, **курсовой** localizer transmitter
~, **ламповый** valve/tube transmitter
~, **направленный** directional transmitter
~ **помех** jamming transmitter, jammer
~, **приводной** homing transmiter
~, **радиолокационный** radar transmitter
~, **ретрансляционный** relay transmitter
~, **самолетный** airborne transmitter
~ **сигналов бедствия** distress signal(s) transmitter
~, **управляющий** control transmitter
~, **фототелеграфный** fascimile transmitter
~, **цифровой** digital transmitter
передача 1. transmission; transference; communication; broadcast; 2. delivery, transfer, passing; *phys.* conduction; *sport* pass; 3. *mech.* drive, gearing, transmission; gear; 4. *law* assignment
~ **азбукой Морзе** keying
~ **алфавитной информации** alphabetic transmission
~, **бесступенчатая** *mech.* infinitely variable transmission
~, **главная** main drive
~ **данных** data transmission/communication
~ **данных в реальном времени** real-time communication
~ **данных, дистанционная** data telecommunication
~ **данных, параллельная** data parallel transfer
~, **жесткая** rigid transmission
~, **зубчатая** gear transmission, gearing, gear drive/train
~ **импульса** momentum transfer
~, **импульсная** pulse transmission
~, **канатная** rope transmission
~, **карданная** cardan transmission; gimbal drive/gear
~, **клиноременная** V-belt transmission/drive
~, **коническая зубчатая** bevel gear
~, **многократная** multiplex transmission
~, **межрегистровая** interregister transfer
~, **многоканальная/мультиплексная** multiplex transmission, multiplexing
~ **мощности** power transmission

~, **направленная** beam transmission
~, **нереверсивная** *mech.* unidirectional drive
~, **обратная** back transfer; retransmission
~, **одноступенчатая** single-reduction gear
~, **пакетная** batch communication
~, **планетарная** planetary gear/transmission
~, **плоскоременная** flat-belt/band transmission
~ **по адресу** address transfer/transport
~ **по наследству** congenital transmission
~ **по шине** bus trnsfer
~, **побитовая** *comp.* bit-by-bit transfer
~, **поблочная** *comp.* block-by-block transfer
~, **повышающая зубчатая** step-up gear(ing)
~, **познаковая** character-oriented communication
~, **понижающая зубчатая** step-down gear(ing)
~, **последовательная** serial transfer
~, **реверсивная** reversing gear
~, **реечная** rack-and-gear drive, rack gear
~, **ременная** belt transmission/drive
~, **рычажная** linkage
~ **с раздвоением мощности** divided train gear
~ **с частотной манипуляцией** frequency-shift-keying transmission, FSK transmission
~ **сообщений** messaging
~, **телеграфная** telegraph transmisbion; keying
~ **телеметрических данных** telemeter data transmission
~ **тепла** heat transfer
~ **технологии** technology transfer
~, **тороидная зубчатая** toroidal worm gear
~, **угловая** angle drive/transmission
~ **управления** *comp.* transfer of control
~, **фототелеграфная** facimile transmission
~, **фрикционная** *mech.* friction transmission/drive/gearing
~, **цепная** chain gear/drive; chain-and-sprocket transmission
~, **червячная** worm gearing
~, **шестеренчатая** gear train
~ **энергии** 1. *phys.* energy transfer; 2. power transmission
передвижение movement, shifting; locomotion; travel, transfer, transportation
переделка remaking, alteration; conversion
передержка *phot.* overexposure
передислокация redeployment, relocation
передислоцировать redeploy, relocate
переднебоковой anterolateral
переднебрюшной *anat.* abdominoanterior
передневерхний anterosuperior
передневнутренний anterointernal
переднезадний anteroposterior
переднезатылочный occipitoanterior
переднелатеральный anterolateral
переднемедиальный anteromedial
передненавесной front-mounted
передненаружный anteroexternal
передненебный *phon.* prepalatal
передненижний anteroinferior
переднесрединный anteromedian
переднешовные *paleont.* proparians *(Poparia)*
переднеязычный *phon.* apical, predental, prevelar
передозировка overdosage, overdose
переедание 1. overeating; 2. *chem.* corrosion
переезд 1. crossing; 2. passage, moving, transit
~ **на одном уровне** level/grade crossing
пережевывать chew, masticate
пережим *geol.* pinch, nip; contraction; squeeze
пережог overburning, overbaking, burn-out
перезапись recording; transcription; rewriting, dubbing
перезапуск restart(ing)
перезарядка 1. supercharging, overcharging; 2. recharging, reloading
переиздание re-edition; reprint, new edition
переизлучение reradiation, re-emission
переисследование re-examination
перекаливать overtemper, overharden; overheat
перекармливать overfeed
перекат *met.* re-rolled products
перекачка transfer; pumping over
переквалификация occupational/professional retraining, training for a new job; skill conversion
переквалифицировать train for a new occupation/job; retrain
перекидывать throw over
перекись peroxide
~ **водорода** hydrogen peroxide
~ **магния** magnesium peroxide
~ **урана** uranium peroxide/tetroxide
переклад *min.* cap/ceiling girder, roof bar
перекладина crosspiece, crossbar, crossbeam
перекладка resurfacing, relaying
переклепка reriveting
переключатель switch, commutator; selector
~, **барабанный** drum/barrel switch
~, **барометрический** baroswitch
~, **бдительности** *rail.* forestalling/acknowledgement switch
~, **бесконтактный** static switch
~ **ближнего/дальнего света** *auto.* antidazzle/beam switch
~, **блокировочный** interlock switch
~, **вращающийся** rotary switch
~, **вытяжной** pull switch
~, **галетный** wafer switch
~, **главный** main/master switch
~ **диапазонов** range/band switch; band selector
~, **дистанционный** remote switch
~ **зажигания** ignition switch
~ **запуска** starting switch
~, **клавишный** keyswitch, piano-key switch
~, **кнопочный** push-button switch
~, **координатный** crosb-bar switch
~, **кулачковый** cam switch
~ **на два направления** double-throw switch
~ **на N направлений** n-way selector
~ **на три положения** three-position switch
~, **нажимной** push-button switch

~ **ножевого типа** knife-blade switch
~, **ножевой пружинный** spring blade switch
~, **педальный** foot-operated switch/commutator
~, **поворотный** rotary switch
~ **полярности** polarity reverser
~, **поплавковый** float-operated switch
~ **режимов работы** mode selector/switch
~, **ртутный** mercury switch
~, **рычажный** toggle/lever switch
~ **с автоматического режима на ручное** auto/automatic-manual switch
~, **сдвоенный** double switch
~, **силовой** power/PWR switch
~, **скользящий** slide switch
~ **скоростей** gear lever
~, **твердотельный** solid-state switch
~, **тумблерный** toggle switch
~ **частот** frequency selector
~, **шаговый** stepping switch
переключать switch; commutate
переключение switching, commutation, change-over; selection; reversing, resetting
~ **передач/скоростей** *auto.* gear shift/change
~ **со звезды на треугольник** star-delta switching
перековка 1. reforging; 2. reshoeing
перекодирование, перекодировка recoding
~ **данных** translation of data, data conversion
перекомпенсация 1. *aero.* overbalancing, overneutralization; 2. *el.* overcompensation
перекомпоновка rearrangement
переконструировать redesign, remodel; reconstruct; rebuild
перекоп cross ditch
перекос skewness, skew, misalignment; curving, bending; twist, distortion
перекрест *anat.* chiasm; decussation
~ **зрительных нервов** optic chiasm
~ **пирамид** decussation of the pyramids
~, **частичный** semidecussation
перекрестноопыленный *bot.* cross-pollinated
перекрестноопыляющийся *bot.* allogamous
перекрестноочередной crosswise-alternate
перекрестнопарный crosswise-opposite
перекрестнослоистый cross-stratified/laminated
перекрестный crossed; cross *(attr.)*
перекрестье *opt.* cross-hairs, reticle
перекрещиват(ся) intersect, cross, decussate
перекристаллизация recrystallization
~, **зонная** floating-zone refining
перекручивать twist; intertwist
перекрывать overlap, overlay, lap over; cover; span; bridge over
перекрытие 1. lap, overlap(ping); bridging, spanning; 2. *el.* spark-/flash/arc-over; *constr.* floor; ceiling
~ **аэроснимков** aerial photographs overlap
~, **балочное** beam ceiling
~, **безбалочное** girderless floor
~ **дугой** flashover
~ **импульсов** pulse overlap
перекус *bot.* haircap moss *(Polytrichum commune)*
перелеска *bot.* hepatica *(Hepatica)*
перелет 1. transit/transfer flight; 2. overshoot
~, **беспосадочный** non-stop flight
~, **круговой** circular transit flight
~ **на посадке** landing overshoot
~, **соколий** *bot.* cross gentian *(Gentiana cruciata)*
перелетный *zool.* migratory
~**ная птица** bird of passage
перелив overflow; *fig.* iridescence
~, **белый/молочный** opalescence
~ **рабочей силы** exit mobility of labour
переливание 1. pouring over; overflowing, overrunning; 2. decantation; 3. *med.* transfusion
~, **внутрибрюшинное** peritoneal transfusion
~, **внутрикостное** intraosseous transfusion
~ **иммунной крови** immunotransfusion, reciprocal transfusion
~, **капельное** drip transfusion
~ **крови** blood transfusion, h(a)emotransfusion
~ **крови в мозговые пазухи** sinusal transfusion
~ **крови, полное обменное** exchange/total/exsanguination/substitution transfusion
~, **непрямое** mediate/indirect transfusion
~, **прямое** immediate/direct transfusion
переливать 1. transfer, decant, pour; 2. overfill; 3. *met.* recast; 4. *med.* transfuse
перелог fallow (land)
перелом 1. beak(ing), fracture; discontinuity, sudden change/turn; 2. *math.* point of inflexion
~, **вдавленный** *med.* depressed fracture
~, **вколоченный** impacted fracture
~, **метафизарный** metaphysial/metaphyseal fracture
~, **множественный** multiple/composite fracture
~, **огнестрельный** gunshot fracture
~ **от растяжения** strain/sprain fracture
~, **поднадкостничный** subperiosteal fracture
~, **сложный** compound fracture
~, **чрезмыщелковый** transcondylar fracture
~, **щелевой** capillary/fissured fracture
перемагничивание 1. magnetic polarity reversal; 2. remagnetization
перематывать rewind, rereel
перемежающийся intermittent; alternating
перемена change, alteration, transformation, reversal; variation
~ **знака** sign inversion
~ **полярности** reversal of polarity
~ **хода** *mech.* stroke reversal
переменная variable (quantity)
~, **автомодельная** similarity variable
~, **базисная** basic/dependent variable
~, **булева** Boolean variable
~, **вещественная** real vabiable
~, **внесистемная** exogenous variable
~, **внутрисистемная** endogenous variable

~ выводимая outgoing variable
~, входная/выходная input/output variable
~, двоичная binary variable
~, двойственная dual variable
~, зависимая dependent variable
~, заданная fixed variable
~, заранее определенная predetermined variable
~, знаковая character variable
~ интегрирования integration variable, variable of integration
~, конструктивная design variable
~, машинная machine variable
~, моделируемая problem variable
~, небазисная nonbasic variable
~, независимая independent variable, argument
~, ненаблюдаемая latent variable
~, необъявленная undeclared variable
~, неопределенная undefined variable
~, непрерывная continuous variable
~, неслучайная nonrandom variable
~, неуправляемая uncontrollable variable
~, ограниченная bounded variable
~, опорная based variable
~, опущенная missing variable
~ периода компиляции compile-time variable
~, пространственная spatial variable
~, разделяемая separable variable
~, разрывная discontinuous variable
~, регулируемая controlled variable
~ с индексами subscripted variable
~, свободная unrestricted/free/nonbasic/independant/slack variable
~, связанная bound variable
~, скрытая latent variable
~, случайная random variable
~, совместно используемая shared variable
~, сопутствующая concomitant variable
~ состояния status variable
~ типа метки label variable
~ типа переключатель switch variable
~ типа событие event variable
~ типа строка string variable
~ типа указатель pointer variable
~, управляемая controlled variable
~, управляющая manipulated variable
~, фиктивная dummy/slack variable
~, фазовая *game th.* state variable
~, целочисленная integer variable
переменная-указатель pointer variable
переменно-замедленный variable-decreasing
переменно-полярный heteropolar
переместительность commutativity
перемет snowdrift
перемешивание mixing, stirring, agitation; intermixing, intermingling; dispersion
~, взаимное intermixing
~, приливное tidal mixing
~ радиальное *chem.* radial dispersion
перемешивать agitate, stir; mix, blend
перемещать move, transpose, shift, displace, translocate; dislocate, displace; relocate
~ обратно backspace
перемещаться move, shift, travel; migrate
перемещение movement, transposition, shift(ing); translocation, displacement, dislocation, relocation; travel, migration; drift, move
~, винтовое screw translation
~, вязное viscous displacement
~, горизонтальное *sociol.* horizontal mobility
~ данных data movement
~ дюн dune migration/travel
~ земной коры crustal movement
~ каретки carriage advance
~ колеса wheel travel
~ континентов continental drift
~, конвекционное convective overturn
~, косое oblique displacement
~ назад rearwards translation
~ полюса polar wandering/shifting
~, поперечное *naut.* sway displacement
~, поступательное translation
~ перфоленты *comp.* punched tape advance
~, поступательное/прямолинейное *mech.* translation
~ программы program(me) relocation
~, продольное *naut.* surge displacement
~, угловое angular displacement
~ центра тяжести centre-of-gravity [c. g.] shift
~ файла moving of file/data set
~ хромосомы chromosome translocation
перемножать *math.* multiply together/with
перемножение *math.* multiplication
перемножитель *math.* multiplier
перемол *agr.* regrinding
перемонтаж reinstallation; *el.* rewiring
перемотка; перемоточный rewinding, rereeling
перемыкать connect; span, bridge; *el.* shunt
перемычка 1. cross connection, crosspiece, connecting link; *el.* jumper; connector; 2. cofferdam; bulkhead; dam; 3. *zool., bot.* jugum; septum
~, аккумуляторная cell connector
~, гибкая *el.* flexible jumper/strap
~, дверная door lintel
~, дугообразная U-link
~, заземляющая earth strip
~, закорачивающая/короткозамыкающая shortcircuiting jumper, short
~ металлизации *el.* bonding strip/jumper/braid
~, набросная rock-fill cofferdam
~, насыпная earth-fill cofferdam
~, одиночная single-wall cofferdam
~, оконная window lintel
~, рельсовая track jumper
~, ряжевая crib(-type) cofferdam
~, свайная pile stockade
~, шахтная противопожарная stopping
перенагревание overheating, superheating
переназначение reappointment; renomination
перенакопление overaccumulation

переналадка readjustment
перенапряжение 1. *el.* overvoltage, excess voltage; surge; 2. *mech.* overstrain, overtension
перенаселение overpopulation, surplus population
перенаселенность overcrowding; conjestion
перенастройка retuning; readjustment
перенасыщение, перенасыщенность oversaturation, supersaturation
перенацеливание *mil.* reaiming; reorientation
перенормировка renormalization
перенос 1. transfer, transport; carry(ing) over, transferring, transference, carry(ing) over, translocation; moving, movement; 2. *comp.* carry; hyphen; 3. *math.* translation, transposition
~ без соблюдения грамматических правил discretionary hyphen
~ в сумматоре adder carry
~ вещества mass transfer
~, групповой *comp.* block carry
~, двоичный binary carry
~, диффузионный diffusive transfer
~, задержанный *comp.* delayed carry
~, запоминаемый *comp.* stored carry
~ заряда charge transfer
~ излучения radiation transfer; radiative transport
~ из предыдущего разряда *comp.* previous carry
~ инфекции transmission of infection
~ ионов ion(ic) migration
~, каскадный cascaded [step-by-step] carry
~ количества движения momentum transfer
~, конечный end carry
~, лучистый radiative transport; radiation transfer
~ массы mass transfer
~, межфазный interphase transfer
~ нейтронов neutron transport/transfer
~ осей координат translation of coordinate axes
~, поразрядный cascade carry
~, последовательный successive carry
~ при сложении add carry
~ системы координат translation of coordinates
~, сквозной ripple-trough carry
~ слова, автоматический hyphenation
~ тепла heat transfer
~ тепла, радиационный heat transfer by radiation
~, ускоренный *comp.* high-speed carry, look-ahead carry
~, циклический *comp.* end-around carry, cyclic carry
~, частичный partial carry
~ частоты frequency translation
~ электронов electron transfer
~ энергии energy transfer
переносица *anat.* bridge (of the nose)
переносный portable, transportable
переносчик carrier; *biol.* vector; *bot.* retinaculum
~ инфекционных заболеваний infection carrier, vector
~ кислорода oxygen carrier
переоблучение overirradiation, overexposure
переоборудовать re-equip; refit, retool; convert
переобучение retraining; conversion training
переопыление transpollination; repollination
переориентация reorientation
переосмысление reinterpretation
переоснащение re-equipping; retooling
переотправка, переотправление reshipment; transshipment
переохлаждение supercooling, overcooling
перепад difference; drop; differential
перепелица, перепелка female quail
перепелятник *zool.* (sparrow) hawk *(Accipiter nisus)*
перепись census; listing, inventory
~, выборочная sample census
~ жилищного фонда census of housing
~, контрольная pilot census
~ населения, проводимая по почте mail-back [mailout] census
~, обязательная compulsory census
~, поголовная house-to-house survey
~, пробная pilot census
~, проводить take the census
переплет 1. binding; book cover; 2. casement, window sash, transom
~, бесшвейный loose-leef binding/binder
~, мягкий paper back
~, твердый hard cover
переплетать 1. interweave, interlace, intertwine; 2. bind *(books)*
переплетение 1. interweaving, interlacing, intertwining; *text.* weave
~, атласное satin /sateen weave
~, вафельное honeycomb/towel weave
~, вельветовое corduroy weave
~, ворсовое pile weave
~, жаккардовое Jacquard weave
~, камчатное damask weave
~, ластичное rib weave/stitch
~, махровое terry weave
~, миткалевое plain/tabby weave
~, основонастилочное warp weave
~, репсовое rep weave
~, саржевое twill weave
~, суконное woollen weave
~, трикотажное loop
~, шашечное check weave
переподъем overhoisting
переполнение 1. overflow(ing), overfilling; *comp.* overflow; thrashing
~ дорожки track overflow
~ при делении divide overflow
~ при сложении add overflow
~ (разрядов) порядка exponent/characteristic overflow
~ сумматора accumulator overflow
пиреполюсовка polarity reversal

переполяризация repolarization
перепонка membrane, diaphragm; septum; *zool.* web
~, барабанная tympanic/drum membrane, membrane of the tympanum, drumhead
~, запирательная obturator membrane
~, зрачковая pupillary membrane
~, межреберная intercostal membrane
~, покровная tectorial/Corti's, membrane
~, сетчатая reticular membrane
~, стекловидная hyaloid membrane
перепончатокрыл(ат)ый *zool.* hymenopterous
перепончатый membranaceous; membraniform
переправа crossing, ferrying; passage; ford
переприем retransmission, relay operation
перепряжка *hist.* changing of horses; stage, lap
перепуск crossover; bypass; overflow
переработка processing, treatment; reprocessing, reworking; refining; conversion; remaking, thorough revision
~ нефти crude oil [petroleum] refining
~ угля coal conversion
перерасход overexpenditure, overconsumption, excessive consumption; *fin.* overdraft
перерасходовать spend/use/consume too much; spend to excess; *fin.* overdraw
перерасширение overexpansion
перерезка cutting, section, dissection
~ пуповины omphalotomia
~ сухожилия tenotomy
перерождение 1. degeneration; 2. regeneration
~, восходящее ascending degeneration
~, гиалиновое hyaline/glassy degeneration
~, жировое adipose/fatty degeneration
~, известковое calcareous degeneration
~, казеозное caseous degeneration, caseation
~, нисходящее descending degeneration
~, подагрическое gouty degeneration
~ сетчатки retinopathy
~, стекловидное vitreous degeneration
перерожденный degenerative
перерыв break, recess, pause, intermission, interruption, discontinuity, hiatus, gap
пересадка transplantation, graft(ing); *bot.* replanting; transplanting
~ роговицы keratoplasty
~ сердца heart transplantation, cardiac transplant operation
пересаживать transplant, graft; *bot.* replant, transplant
переселенец migrant; settler; emigrant, immigrant
~ внутри города intracity migrant
пересечение intersection; crossing; traverse; *math.* crosscut
~ дорог road crossing/intersection
~ железнодорожных путей на одном уровне grade railway crossing
~ потока flux traverse
~, сейсмическое seismic traverse
перескок jump; skip; *phys.* migration
~ искры sparkover
~ ленты tape skip
переслаивание *geol.* interbedding, interstratification, alternation of strata
~, линзовидное interlensing
пересмотр inspection; revision; *law* review; retrial
переспрос repeated inquiry; request for repetition; re-examination
перестановка 1. transposition, rearrangement; reshuffle; readjustment; 2. *math.* permutation
перестраивать reconstruct, rebuild; redesign, refashion; reorganize; 3. retune; change the tuning of; *rad.* switch over
перестрахование, перестраховка reinsurance
~, квотное quota share reinsurance
~ с эксцедентом убытка stop loss reinsurance
~, эксцедентное surplus reinsurance, except-of-loss reinsurance
перестраховывать reinsure
перестройка rebuilding, reconstruction; rearrangement, realignment; conversion; *pol.* perestroyka; *mus. rad.* retuning
переступень *bot.* bryony *(Bryonia)*
пересчет 1. *math.* scaling; 2. recalculation, conversion; 3. recount(ing), counting again/anew
~, бинарный/двоичный binary scaling
~, десятичный decade scaling
~ на год *econ.* annualization
пересчитывать 1. recalculate; recount, count over/again; 2. scale; convert
пересылка sending, forwarding; remittance
~ данных data transfer
~ данных, конвейерная pipelining of data
~ по адресу address transfer/transport
~ принятого сообщения hop
~ страниц *comp.* page migration
~, числовая numeric move
пересыльный transit *(attr.);* for deportees
~ пункт transit camp
пересыщение supersaturation, oversaturation
переток crossflow
переточка regrinding, resharpening
переустройство reconstruction; modification
перетяжка constriction; strangulation
переучет inventory; stocktaking; *fin.* rediscount
перехват 1. interception; 2. *anat.* isthmus
~ кольцевой anular isthmus
~ телефонных разговоров wire tapping
переход transition, transit, passage; passing/changing/switching over; *comp.* junction; jump; escape
~, агрегатный *phys.* change of state
~, безусловный unconditional jump
~ в абсолютных адресах *comp.* absolute jump/skip
~ в полупроводниковом приборе transition region/layer; junction
~ в программе program(me) jump/skip
~, внешный *comp.* far jump
~, волноводный waveguide junction

~ на ветвь branching
~ на подпрограмму *comp.* subroutine jump
~ Оже Auger radiationless transition
~, оптический запрещенный optically forbidden transition
~, пешеходный pedestrian overpass/underpass; subway
~ по минусу negative jump
~ по переполнению *comp.* jump if overflow
~ по плюсу positive jump
~ по условию "больше" jump if above/greater
~ по условию "не больше" jump if not above/greater
~ по условию "не меньше" jump if not below/less
~ при множественном ветвлении multiway jump
~, прямой cross-over transition
~, радиационный radiative transition
~, резонансный resonant transition
~, сверхразрешенный superallowed transition
~, слабозапрещенный unfavoured transition
~, условный conditional jump
~, фазовый phase transition/transformation, change of phase
переходник adapter; reducer, reducing piece
~, стыковочный *space* docking adapter
переходный 1. transitional, transitory, transient, passing; intermediate; 2. *gram.* transitive
переходящий transient, transitory; passing
~ кубок *sport* challenge cup
перец pepper *(Piper)*
~ гвоздичный/душистый allspice, pimento *(Pimenta officinalis)*
~, однолетний/кайенский/стручковый Cayenne pepper, red pepper *(Capsicum annuum)*
~, черный black pepper *(Piper nigrum)*
перечеканивать recoin; coin
~ в монету monetize
перечень list; enumeration
~, контрольный check list
~ технологических операций route card
~ товаров inventory
перечник *bot.* candytuft *(Iberis)*
перечные *pl. bot.* the pepper family *(Piperaceae)*
перешеек 1. *geog.* neck of land, isthmus; 2. *anat.* isthmus
~ матки isthmus of the uterus
~, подводный submarine isthmus
~ слуховой трубы isthmus of the auditory/Eustachian tube
переярок *agr.* yearling
периантий, перианций *bot.* perianth(ium)
периаортальный *anat.* periortic
периастр, периастрий, периастрон *astr.* periastron
периблема *bot.* periblem, envelope tissue
перивагинит *med.* perivaginitis, pericolpitis
пуриваскулит *med.* perivasculitis, periangitis
перивисцеральный perivisceral, perisplanchnic
перигастрит *med.* perigastritis
~, слипчивый adhesive perigastritis
перигей *astr.* perigee
перигейный *astr.* perigeal
перигелий *astr.* perihelion
перигепатит *med.* perihepatitis
~, склерозирующий condensing/sclerosing perihepatitis
перигиний, перигиниум *bot.* perigynium
перигиния *bot.* perigyny
перигландулит *med.* periglandulitis
перигломерулит *med.* periglomerulitis
перигоний 1. *bot.* perigonium; 2. *zool. bot.* perigone
перидактилия peridactylia
перидерма 1. *zool.* periderm; 2. *bot.* periderm, outer bark
перидермий, перидермиум *bot.* peridermium
перидий *bot.* peridium
перидиола *bot.* peridiolum, peridiole
перидуоденит *med.* periduodenitis
~, надбрыжеечный supramesenteric periduodenitis
перизоний *bot.* perisonium
перикамбий *bot.* pericambium
перикард *anat.* pericardium
перикардиальный pericardial, pericardiac
перикардиорафия pericardiorrhaphy
парикардиостомия pericardiostomy
перикардиотомия pericardiotomy
перикардит pericarditis
~, гнилостный putrefactive pericarditis
~, гнойный purulent pericarditis, empyema of the pericardium
~, сдавливающий constrictive pericarditis
~, серозно-фиброзный serofibrinous pericarditis
~, серозный hydropericarditis
~, слипчивый adhesive pericarditis
~, фибринозный fibrinous pericarditis, hairy heart
~ хилезный chylous pericarditis
~, частичный limited pericarditis
перикарион perikaryon
перикарп, перикарпий *bot.* pericarp
~пия, без gymnospermous
периколит *med.* pericolitis, pericolonitis
перикорнеальный pericorneal, perikeratic
перила banisters, balustrade, parapet, (hand)rail
перилимфа perilymph(a)
перилимфатический perilymphangial, perilymphatic
перименингит perimeningitis, pachymeningitis
периметр *geom. opt.* perimeter
~, проекционный *opt.* projecting perimeter
~, смоченный/подводный *geol.* wetter perimeter
периметрий perimetrium
перимиелит *med.* perimyelitis
перимизий *med.* perimysium

~, внутренний endomysium
~, наружный epimysium
периневрий *anat.* perineurium
перинеопластина *med.* perineoplasty
перинеорафия perineorrhaphy
перинефрит *med.* perinephritis
~, фиброзно-липоматозный fibrolipomatous perinephritis
периний *bot.* perinium, perine
период period; time interval, time; phase, stage; spell; cycle
~ альфа-распада alpha-decay period
~, базисный *stat.* standard/reference period
~, вегетационный vegetative/vegetation period
~ выборки access cycle
~, гарантийный guarantee period
~, детеродный childbearing/reproductive period
~ дождей pluvial period
~ за который начисляется процент interest period
~ задержки/запаздывания delay period
~ записи write period
~ износа wear-out period
~ импульсов pulse interval/period
~, испытательный probation period; test period
~ кадра *telev.* frame period/time
~ компиляции *comp.* compile-time
~, льготный *econ.* grace period
~, ледниковый glacial age
~ малой водности low-flow period
~ межремонтный time between overhauls, TBO
~, нерабочий standby period;
~ обзора *rad.* scanning interval
~ обкатки/освоения running-in [breaking-in] period
~ обновления *tech.* replacement cycle; turnover period
~ обращения period of revolution, orbital period
~, отчетный accounting period; period under review
~ периодической дроби *math.* repetend
~, плановой planning period
~ половодья high-water season, overflow stage
~ полувыведения half-life period
~ полуобмена half-time of exchange
~ полураспада half-life (period)
~ прилива tidal period
~ приработки shake-down period
~ прогрева warm-up period
~ простоя down-time
~, пусковой start-up period
~ работоспособного состояния uptime
~ радиоактивного распада radioactive-decay period
~ развертки scan period
~ разгона boost/accelerating period
~ распада decay period
~, расчетный accounting period
~ решетки lattice spacing/pitch
~, скрытый latent period
~, сметный budget period
~ стажировки probation period
~ сохранности *comp.* retention period
~, суточный diurnal period
~, сухостойный *agr.* interlactation period
~ хранения файла file retention period
~ эксплуатации operating period; operational life
периодический peridoic(al), intermittent, cyclic
~кая система элементов periodic system
периодичность periodicity; cyclicity; frequency
~ загрузки feed rate
периодонт *med.* periodontium
периоофорит *med.* perioothecitis, perioophoritis
периорхит *med.* periorchitis
~, гнойный purulent periorchitis
периост *anat.* periosteum
периостальный periosteal, periosteous
периостит *med.* periost(e)itis, cortical osteitis
~, гнойный purulent/suppurative periostitis
~, гуммозный gummatous periostitis
~, серозный albuminous periostitis
~, траншейный trench periostitis
периостоз *med.* periost(e)osis
~, оссифицирующий ossification periosteosis
перипанкреатит *med.* peripancreatitis
перипатетизм Peripateticism
перипатетик, перипатетический Peripatetic
перипахименингит *med.* peripachimeningitis
перипиэстический peripiestic, anthropopiestic
перипнейстический *zool.* peripneuistic
перипорит *med.* periporitis
перипроктит *med.* periproctitis, perirectitis
периптер *arch.* peripteros
периптерический *arch.* peripteral
периректит *med.* perirectitis, periproctitis
перирenальный perirenal, perinephric
периселений *astr.* perilune, pericynthion
перисинусит *med.* perisinu(s)itis
перисинуситный *med.* perisinous
перисистола *med.* perisystole, presystole
перисистолический perisystolic, presystolic
перископ periscope, periscopic viewer
~, астронавигационный star tracker periscope
~, коленчатый corner periscope
~, панорамный scanning periscope
~, П-образный saddle periscope
~, стереоскопический binocular periscope
перисперм, периспермий *bot.* perisperm
периспланхнит *med.* perisplanchnitis
периспленит *med.* perisplenitis
периспондилит *med.* perispondylitis
периспорий, периспора *bot.* perispore
~, пленчатый filmy perispore
перистальтика *physiol.* peristalsis
~, замедленная bradyperistalsis
~ кишечника intestinal peristalsis
~, усиленная hyperperistalsis
~, форсированная mass peristalsis
перистиль *arch.* peristile

перистоветвистый *bot.* pinnately branched
перистоголовник *bot.* winghead *(Pterocephalus)*
перистодольчатый pinnatilobate, pinnatilobed
перистожаберные *pl. zool.* pterobranchs *(Pterobranchia)*
перистожилковатый pinnately veined, pinninervate
перистокрылые *pl. zool.* caddis flies *(Trichoptera)*
перистокрылый *zool.* trichopterous
635-кучевой *meteor.* cirrocumulus
перистолистный pinnatifolious, pinnate
перистолопастный pinnately lobed, pinnatilobate
перистом(а) peristoma, peristome
паристомальный, перистомный, перистоматический peristomal, peristomatous, perioral
перистонервный pinnately veined, pinninervate
перистораздельный pinnately parted, pinnipartite
перисто-слоистый *meteor.* cirro-strative, cirrostratous
перистрофа *bot.* peristrophe *(Peristrophe)*
перистый pinnate; plumose; feathery; cirrous
перитека peritheca
перитектика peritectic alloy
перитециевидный *bot.* perithecoid
перитеций *bot.* perithecium
перитонеоцентез *med.* peritoneocentesis, abdominocentesis
перитонизация peritonization, peritoneoplasty
перитонит peritonitis
~, газовый gas peritonitis
~, гнилостный putrid peritonitis
~, желчный bile/biliary peritonitis, choleperitonitis
~, послеродовый puerperal peritonitis
~, прободный perforative peritonitis
~, склерозирующий condensing peritonitis
~, слипчивый adhesive peritonitis
~ хилезный chyle peritonitis
~ язвенный ulcerative peritonitis
перитрихиальный peritrichous, peritrichic, peritrichiate
перитрофичесний, перитрофный peritrophic
перифиз(а) *bot.* periphysis
перифлоэма *bot.* pericambium, periphloem
перифрастический periphrastic
перихондрий *anat.* perichondrium
перицеллюлярный pericellular, pericytial
перицементит pericementitis, periodontitis
~, верхушечный apical pericementitis
~, гранулирующий granular pericementitis
перициклоида *math.* pericycloid
перицистит *med.* pericystitis; pericholecystitis
перицит pericyte
перицитарный pericytial, pericellular
перка flat drill/bit; drill point, cutter
перкаль percale, cambric muslin
перкислота *chem.* peracid
перкносома *bot.* percnosome, perknosome
перкуссия percussion
~ ключицы clavicular percussion
~, непрямая mediate percussion
перкутировать *med.* percuss
перламутр mother-of-pearl, nacre
перламутровка 1. *zool.* pearl butterfly *(Argynnis paphia)*; 2. *bot.* silver weed *(Argyreia)*
перлвейс pearl white
перлинь *naut.* cable warp; hawser; tow line
~, кокосовый grass hawser
~, спасательный salvage hawser
перлит *1. petr.* perlite; 2. *met.* pearlite
~, вспученный expanded perlite
перлитобетон perlite concrete
перловица *zool.* pearl oyster/shell *(Margaritana margaritifera)*
перловник *bot.* melic, melic/onion grass *(Melica)*
перлювий perluvium
перманганат permanganate
перль *print.* pearl, 5-point type
пермаллой *met.* permalloy
пермокарбон Permo-Carboniferous
пермский Permian
пермутация, пермутирование permutation
пермь *geol.* Permian
перо 1. stylus, pen; 2. feather, plume, pinna; 3. blade
~, линовальное ruling pen
~, лучевое beam pen
~, морское *zool.* sea pen *(Pennatula)*
~, рулевое *zool.* retrix
~ руля *naut.* rudder plade/plate
~ самописца stylus, recording pen; tracer
~ стабилизатора *aero.* stabilizer fin
~ стрелки *rail.* switch tongue
~, чертежное drawing pen
перовидный pinniform
пероксидаза *biochem.* peroxidase
пероксидазный peroxidatic
перонин *chem.* benzylmorphine hydrochloride
перпендинуляр perpendicular
~, восставить erect a perpendicular
~, кормовой *naut.* aft(er) perpendicular
~, косовой fore/forward perpendicular
~, миделевый/средний middle perpendicular
~, опускать drop a perpendicular
перпендикулярность perpendicularity
перпетуум-мобиле perpetuum mobile
перрон 1. *rail.* platform; 2. *av.* ramp; apron
перс Persian
персеиды *astr.* Perseids
Персей *myth. astr.* Perseus
персея *bot.* persea *(Persea)*
персидский Persian
персик peach tree *(Prunus persica)*; peach
персимон *bot.* persimmon, date-plum *(Diospyros)*
персоль *chem.* persalt
персона person
~ грата persona grata
~ нон грата persona non grata
персонал personnel, staff; employees

~, **вспомогательный** support/additional personnel
~, **конторский** office staff
~ **магазина** store employees
~, **обслуживающий/оперативный/технический** operating personnel/staff; attendants
~, **производственный** factory personnel
~, **эксплуатационный** operating personnel
персональный personal, individual
~ **пенсионер** person receiving merit pension
перспектива 1. *art* perspective; 2. prospect(s), outlook
перспективный 1. promising, with good prospects; 2. long-term, over a long period
перстневидный ring-shaped, annular; *anat.* cricoid
перт *naut.* footrope
пертубация *med.* tubal insufflation, Rubin test
перуанец Peruvian
перуанка Peruvian (woman)
перуанский Peruvian
~ **бальзам** Peru balsam
Перун *myth.* Perun
перфокарта punched/perforated card, (punch) card
~, **ведущая/главная** master card
~, **микрофильмовая** aperture/microfilm card, film aperture card, microcard
~, **многократная** multiple/tumble card
~ **оборота** accounting detail card, movement card
~, **парная/непарная** matched/unmatched card
~ **с двоичной перфорацией** binary punched card
~ **с краевой перфорацией** edge-perforated/punched card
~, **щелевая** slit punched card
перфолента punch/punched/perforated tape
~ **контрольная** check punched tape
~, **основная** master punched tape
~, **пятидорожечная** *comp.* five-channel tape
~, **управляющая** control tape
перфоратор punch, puncher, perforator, punch block; drill; punch-card machine
~ **вращающийся** rotary drill
~, **вычислительный** calculating punch
~, **дублирующий** reproducing punch, reproducer
~, **итоговый** summary punch
~, **карточный** card punch
~, **клавишный** keyboard perforator, key punch
~ **ленточный** paper tape punch
~, **множительный** multiplier punch
~, **переписывающий** transcript punch
~, **скважинный** *min.* borehole perforator
~, **суммирующий** summary/adding punch, add-punch machine
перфораторий *zool.* perforatorium, acrosome
перфорация 1. perforation, punching; drilling, boring; 2. punched/perforation hole
~, **ведущая** designation/control punching; feed holes
~, **итоговая** summary punching
~, **краевая** marginal perforation
~, **направляющая краевая** guide hole edge
~, **повторная/последующая** afterpunching
~ **текущего номера** consecutive number punching
перфорировать punch, perforate; drill, bore
перцеяд *zool.* toucan *(Rhamfastos)*
перципировать perceive
перчатка 1. glove; gauntlet; mitten; 2. *zool.* glove sponge *(Euspongia tubulifera)*
~, **диэлектрическая** insulating glove
~, **хирургическая** surgeon's glove
перышковидный pinnuliform
пес (male) dog
~, **Большой** *astr.* Canis Magor
~, **Малый** *astr.* Canis Minor
песета peseta
песец *zool.* arctic/polar fox *(Alopex lagopus)*
песиголовец *myth.* dog-headed person
пескарка *zool.* dragonfish *(Callionymus)*
пескарь *zool.* gudgeon *(Gobio)*
пескодувка sandblast unit, sand-jet blower
пескожил *zool.* lobworm, lugworm *(Arenicola)*
песколовка sand trap; grit catcher
песколюб *bot.* beach grass *(Ammophila)*
пескомойка sand washer
пескорой *zool.* digger/fossorial wasp *(Fossores)*
пескоройка *zool.* ammocoete larva
пескосортировка sand screen
пескоуловитель sand trap/collector
пескочерпалка sand drag
песо peso
песок sand; *med.* gravel
~ **для раствора** mortar sand
~, **золотоносный** gold dust, auriferous sand
~, **зыбучий** drift sand
~, **крупный/крупнозернистый** coarse sand; grit
~, **мелкий/мелкозернистый** fine sand
~, **мозговой** acervulus, brain sand
~, **мочевой** gravel
~, **почечный** renal sand
~, **сахарный** granulated sugar
~, **сыпучий** quicksand
~, **формовочный** moulding sand
песочник *zool.* sandpiper *(Calidris)*
пессарий *med.* pessary
пест, пестик I pestle; pounder; stamp
~, **падающий** gravity stamp
пестик II *bot.* pistil
пестиковидный *bot.* pistillate
пестиковый, пестичный pistillary, pistillate
пестрец *bot.* vanilla grass *(Hierochloe borealis)*
пестроовсяничник *bot.* variagated fescue *(Festuca varia)*
пеструшка *zool.* 1. lemming *(Lemmus);* 2. rainbow/speckled trout *(Salmo irideus)*
пестряк *zool.* checkered beetle *(Trichodes)*
пестрянка *zool.* forester *(Ino)*
песчаник I *bot.* dune wild rye *(Elymus arenarius)*

песчаник II sandstone
~, глинистый clayey/argillaceous sandstone
~, крупнозернистый/грубозернистый coarse-grained sandstone
~, мелкозернистый fine-grained sandstone
~, слоистый laminated sandstone
песчаник III *zool.* large-toothed [yellow] souslik *(Citellus fulvus)*
песчанистый sandy, arenaceous
песчанка *zool.* 1. gerbil(le) *(Gerbillus);* 2. sandpiper *(Calidris);* 3. sand eel *(Ammodytes);* 4. sanderling *(Crocethia alba);* 5. *bot.* sandwort *(Arenaria)*
песчано-глинистый sandy-argillaceous
песчано-лессовый sandy-loessic
песчано-литоральный psammolittoral
песчаный sandy, arenaceous, arenarius; psammitic
~ное растение psammophyte, arenaceous plant
петарда 1. *mil.* petard; 2. *rail.* detonating cartridge, torpedo
петехия *med.* petechia
петит *print.* brevier *(8 points)*
петлевидный, петлеобразный loop-shaped, looplike, ansate, ansiform; *bot.* brochidodromous
петлевание *met.* looping
петление *met.* kinking
петли *anat.* ansae
петля 1. loop; ansa; sling; lemniscus; 2. hinge
~, бактериологическая ansa, loop
~, бесконечная endless/continuous loop
~, внутриреакторная in-pile loop
~, ворсовая pile loop
~ гистерезиса hysteresis loop
~, групповая *av.* formation loop
~, нисходящая descending/diving loop
~, обратная *av.* outside loop
~, перевернутая *av.* inverted loop
~, пружинная spring hinge
~, рулевая *naut.* rudder gudgeon/hinge/brace
петрокаллис *bot.* rock beauty *(Petrocallis)*
петрофит *bot.* petrophyte, rock plant
петрохимия petrochemistry
петрушка *bot.* parsley *(Petroselinum)*
петуния *bot.* petunia *(Petunia)*
петушки *bot.* Hungarian iris *(Iris variegata)*
петушник *bot.* hemp nettle *(Galeopsis)*
петушок *el.* riser, neck *(of commutator)*
пехота *mil.* infantry
~, воздушнодесантная airborne infantry
~, морская marines, Marine Corps; *(in Britain)* Royal Marine
пехотинец infantryman
пехштейн *petr.* pitchstone
печатать 1. print; 2. *(on typewriter)* type
печатающее устройство printer
~, быстродействующее data high-speed [fast, rapid] printer
~ для магнитного шрифта magnetic ink imprinter
~, посимвольно character-at-a-time printer
~ с буферной памятью block printer
~ с литерными колесами type wheel printer
~ с литерными рычагами type bar printer
~ точечное dot-printer
печатник 1. printer; 2. *hist.* Keeper of the Seal
печатный 1. printing; 2. printed
~ знак printed character
~ лист printer's sheet, quire
~ станок printing press
печать 1. printing, print; 2. seal, stamp; 3. the press; 4. type, print
~, безударная nonimpact printing
~ вразрядку space printing
~, высокая relief/letterpress printing
~, глубокая intaglio printing; photogravure
~, глубокая ротационная rotogravure
~, двойная dual printing
~, должностная seal of office
~, знаковая character/symbol/sign printing
~, малая государственная *law* privy seal
~ машинописного качества letter-quality printing
~ на пишущей машинке type printing/print
~, периодическая periodicals
~, постраничная page printing
~, построчная line printing
~ строчных букв lower-case printing
~, струйная ink-jet printing
~, точечная dot printing
~, убористая close print
~, электрографическая electrosensitive printing
печенег, печенежский Petcheneg
печеночник *bot.* liverwort *(Hepatica)*
печеночница *bot.* 1. hepatica *(Hepatica);* 2. liver agaric, beefsteak fungus/mushroom *(Fistulina hepatica)*
печеночно-двенадцатиперстный hepatoduodenal
печеночно-желудочный hepatogastric
печеночно-кишечный hepatoenteric
печеночно-почечный hepatonephric, hepatorenal
печеночно-пузырный hepatocystic
печеночно-селезеночный hepatolienal
печеночный hepatic
печень liver
~, амилоидная amyloid/lardaceous/waxy liver
~, блуждающая wandering liver, hepatopsia, hepatoptosis
~, бугристая tuberous/nodular liver
~, дольчатая lobular river
~, жировая lardaceous/amyloid/waxy liver
~, застойная congestive liver
~, сахарная frosted/icing liver
~, цирротическая cirrhotic liver, billiary/Glissons's cirrhosis
печерица *bot.* agaric *(Agaricus)*
печь furnace; oven; kiln; stove
~, агломерационная sintering furnace
~, ванная tank/bath furnace

~, возгоночная subliming furnace
~, восстановительная reduction furnace
~, вращающаяся rotary kiln; rotating furnace
~, гипсообжигательная plaster kiln
~, горшковая pot furnace
~ для пайки brazing furnace
~, доменная blast furnace
~, дуговая electric arc furnace
~, закалочная hardening furnace
~, известеобжигательная lime kiln
~, индукционная induction furnace
~, камерная chamber kiln/furnace/oven
~, канальная hot-air oven
~ кипящего слоя fluidized-bed furnace
~, кирпичеобжигательная brick kiln
~, коксовая coke oven
~, колодезная pit furnace
~, колпаковая bell-type furnace
~, кольцевая ring/annular furnace
~, контактная catalyst furnace
~, коптильная smoking kiln
~, купеляционная cupeling furnace, cupel
~, мартеновская open-hearth furnace
~, мусоросжигательная incinerator
~, муфельная muffle furnace
~ непрерывного действия continuous firnace/kiln
~, обжарочная/обжигательная/обжиговая roasting furnace, roaster; firing kiln
~, опалочная singeing oven
~, отжигательная annealing furnace
~, отражательная reverberatory furnace/kiln
~ периодического действия batch furnace/kiln
~, плавильная melting furnace
~, пламенная combustion furnace
~, пробирная assay furnace
~, проходная continuous furnace
~, рудоплавильная smelting furnace
~, сварочная welding furnace
~ сопротивления resistance/resistor furnace
~, стекловаренная glassmaking furnace
~, сушильная drying oven/kiln
~, термическая heat-treatment furnace
~, тигельная crucible furnace
~, трубчатая tube furnace
~, углевыжигательная charcoal kiln
~, форсуночная burner-fired furnace
~, хлебопекарная baking oven
~, цементационная case-hardening furnace
~, цементообжигательная cement kiln
~, шахтная shaft furnace
~, электродуговая electric arc furnace
пещера cave; cavern; *anat.* antrum
~, прибойная sea cave
~, сосцевидная mastoid/tympanic antrum
пещеристый *anat.* cavernous
пещерник cave-dweller
пещерный cave *(attr.)*; spelean; cave-dwelling
~ человек cave man, troglodyte
пещероведение speleology
пи *math.* pi
пиассава *bot.* piassava *(Attalea funifera)*
пиастр piaster
пивовар brewer
пивоварение brewing, beer manufacture
пигалица *zool.* pewit, lapwing *(Vanellus)*
пигмей pygmy
пигмент pigment, colouing agent, colo(u)r
~, гемоглобиногенный hematogenous pigment
~, гепатогенный hepatogenous/bile pigment
~, дыхательный respiratory pigment
~ старения lipofuscin
пигментация pigmentation
~ кожи, желтая xanthochromia, xanthopathy
пигментный pigmentary, pigmental
пигментообразующий chromogenic
пигментофор pigmentophore, chromatophore
пиелит pyelitis
пиелография *med.* pyelography
~, внутривенная intravenous/excretion pielography, pyelography by elimination
~, восходящая/ретроградная retrograde/ascending pyelography
пиелонефрит pyelonephritis
~, гнойный purulent pyelonephritis
~, калькулезный calculous/calculary pyelonephritis
~, послеродовый puerperal pyelonephritis
пиелоцистит *med.* pyelocystitis
пиелоэктазия *med.* pyeloectasis
пиемия *med.* py(a)emia, pyosapremia
~, портальная portal/supparative pylephlebitis
пижма *bot.* tansy *(Tanecetum)*
пизанг *bot.* plantain banana *(Musa paradisiaca)*
пик peak, high, apex, crest, spike
~ импульса pulse peak
~, ложный ghost peak
~ нагрузки peak load, load peak, maximum demand
~ смещения displacement spike
~, температурный thermal spike
пика I 1. lance, pike; 2. poker, slice bar
пика II spade; **дама пик** queen of spades
пикап pickup truck
пике *text.* quilting; piqué
пикет I stake, peg; *rail.* 100-metre mark
пикет II picket; strike picket
пикет III piquet *(card game)*
пикетаж pegging/staking/marking out
пикетирование picketing
пикирование dive, diving
~ на цель diving on target
~ по спирали spiral dive
~, пологое flat/shallow dive
пикировать I dive, swoop
пикировать II *agr.* thin out *(seedlings)*
пикировка I diving, dive
пикировка II *agr.* thinning out, singling
пиккер *agr.* picker

пикколо piccolo
пиклоксидин picloxydine
пикнантемум *bot.* mountain mint *(Pycnanthemum)*
пикнида *bot.* pycnidum, pycnid
пикний *bot.* pycnium
пикник picnic
пикнит *miner.* pycnite
пикнический pyknic
пикногон *zool.* pycnogonid
пикногонидия *bot.* pycnogonidium
пикноз pycnosis, pyknosis
пикноконидия *bot.* pycnoconidium
пикноксилический, пикноксиличный, пикноксильный pycnoxylic
пикнолепсия *med.* pyknolepsy, pyknoepilepsy
пикнометр pycnometer
пикноспора *bot.* pycnospore
пикностиль *arch.* pycnostyle
пикнотеций pycnothecium
пикнотический pyknotic
пико picot
пиковый I peak *(attr.)*
пиковый II of spades, spade *(attr.)*
~вая дама queen of spades
пикофарада *el.* picofarad, pF
пикрасма *bot.* quassia wood *(Picrasma)*
пикрат chem. picrate
пикриновый *chem.* picric
пиксидий, пиксидиум *bot.* pyxidium
пиктография pictograph
пикульник *bot.* hemp nettle *(Galeopsis)*
пикша *zool.* haddock *(Melanogrammus aeglefinus)*
пила saw
~ ажурная fretsaw, scroll saw
~, балансирная swing saw
~, бочарная barrel saw
~, бугельная bow sow
~, выкружная лучковая scroll saw
~, гонторезная shingle saw
~, двуручная trump saw
~, дисковая circular saw
~, канатная rope saw
~, кантовальная bevel saw
~, кромкообрезная edging saw
~, круглая ring saw
~, ленточная band/belt saw
~, лобзиковая fretsaw
~, лучковая bow saw
~, медицинская bone saw
~, механическая power saw
~, многодисковая gang saw
~, ножовочная hacksaw
~, обрезная edging saw
~ по дереву woodworking saw
~ по металлу metalworking saw
~, подвесная swing saw
~, подрезная coping saw
~, поперечная cross-cut saw
~, проволочная wire/thread/chain saw
~, продольная cleaving/ripping saw
~, рамная frame/mill/gate saw
~ с пластинками из твердых сплавов carbide-tipped saw
~, салазковая sliding-frame saw
~, стоматологическая disk saw
~, торцовочная trimming saw
~, цепная chain saw, link-tooth saw
~, цилиидрическая barrel/cylinder saw
~, циркулярная disk/circular saw
~, шипорезная dovetail saw
~ электромоторная electric saw
пилав *cul.* pilaw, pilau, pilaff
пила-рыба *zool.* sawfish *(Pristis)*
Пилат Pilate
пиленый sawn; cut
~ лес timber, lumber
~ сахар cube sugar
пилетромбоз *med.* pylethrombosis
пилефлебит pylephlebitis
~, гнойный purulent pylephlebitis
~, слипчивый adhesive pylephlebitis
пилея *bot.* clearweed *(Pilea)*
пилигрим pilgrim
пилигримство pilgrimage
пилидий *zool.* pilidium
пилильщик *zool.* sawfly *(Tenthredinidae)*
пилить saw; file
пилка 1. (fret) saw; 2. small file; nail file; 3. sawing
пиллерс *naut.* pillar, deck stanchion
~, откидной hinged pillar/stanchion
~, подвесной hanging pillar
~, трубчатый hollow/tubular pillar
~, трюмный hold pillar/stancheon
пиловидный saw-shaped; serriform
пилозубный saw-toothed; serrate
пилокарпин *chem.* pilocarpine
~, азотнокислый pilocarpine nitrate
~, хлористоводородный pilocarpine hydrochloride
пилокарпус bot. jaborandi leaf *(Pilocarpus)*
пилолистник *bot.* lentinus *(Zentinus)*
пиломатериал sawn timber, lumber
~, бессортный assorted lumber
~, необрезной unedged [edge-unsurfaced] lumber
~, обрезной edge-surfaced lumber
~, поделочный factory lumber
~, профильный patterned lumber
~, строительный structural lumber
~, тарный cratewood, crate lumber
пиломоторный *physiol.* pilomotor *(attr.)*
пилон pylon; tower
~, внешний outboard pylon
~, внутренний inboard pylon
пилонос *zool.* saw shark *(Pristiophorus)*
пилообразный saw-shaped, sawlike; serrate
пилорит *med.* pyloritis
пилорический pyloric

пилоропластика pyloroplasty
пилороптозия pyloroptosia
пилоростеноз pylorostenosis
пилорыл *zool.* sawfish *(Pristis)*
пилот pilot
  ~, второй copilot
пилотаж piloting, flying, pilotage
  ~, высший acrobatic/aerobatic/stunt flying, higher/advanced aerobatics, aerial acrobatics
  ~, групповой formation flying
  ~, групповой высший formation aerobatics
  ~, сложный advanced flying
  ~, фигурный stunt/advanced flying
пилотировать fly, pilot
пилотируемый piloted, manned
пилоцистида pilocystidium
пильтдаунский Piltdown *(attr.)*
пильчато-зубчатый serrate-dentate
пильчатолистный *bot.* serrate-leaved, serratifolious
пильчатоплодный *bot.* serratocarpous
пильчато-реснитчатый serrate-ciliate
пильчатость serration
пильчатый saw-toothed, dentate; serrate; notched
пилюльница *bot.* pillwort *(Pilularia)*
пилюля pill; bolus
пилястр, пилястра *arch.* pilaster; alette
пи-мезоатом pi-mesic atom
пи-мезон pi-meson, pion
пимелея *bot.* rice flower *(Pimelea)*
пимелиновый *chem.* pimelic
пимент *cul.* pimento
пимпинель *bot.* small burnet *(Sanguisorba minor)*
пинагор *zool.* lumpfish, sea hen *(Cyclopterus lumpus)*
пинакоид *cryst.* pinacoid
  ~, основный basal pinacoid
пинакотека picture gallery, pinakotheke
пинанг(а) *bot.* pinanga palm *(Pinanga)*
пингвин *zool.* penguin
пиния *bot.* Italian stone pine *(Pinus pinea)*
пинка *naut.* pink
пиноцит *med.* pinocyte
пинта pint
пинцет forceps; pincers, tweezers
  ~, глазной ophthalmic forceps
  ~, зубный dental forceps/plier
  ~, зуболапчатый tenaculum forceps
  ~, кровоостанавливающий h(a)emostatic forceps
  ~, нейрохирургический brain forceps
  ~, окончатый fenestrated forceps
  ~, остроконечный pointed forceps
  ~, стоматологический dental forceps/plier
  ~, хирургический thumb forceps
  ~, эпиляционный epilating forceps
пиогемоторакс *med.* pyoh(a)emothorax
пиогенный piogenic, pyogenous, pyogenetic
пиодерматит *med.* pyodermatitis, pyodermitis
пиодерматоз *med.* pyodermatosis
пиодермия *med.* pyodermia, cutaneous suppuration
  ~, ротовая oral pyodermia
  ~, стрептостафилококковая streptostaphylococcal pyodermia
  ~, хирургическая operative pyodermia
  ~, язвенная ulcerative pyodermia
пиококк *med.* pyococcus
пиокольпос *med.* pyocolpos, pyocolpocele
пиоксантин pyoxanthin
пиометрит pyometritis
пиомиозит *med.* pyomyositis
  ~, тропический tropical pyomyositis, bungpagga
пион I *bot.* peony *(Paeonia)*
пион II *phys.* pi-meson, pion
пионер 1. pioneer; precursor; *bot.* pioneer plant; 2. young pioneer
пионервожатый *obs.* young pioneer leader
пионерлагерь pioneer camp
пионеротряд *abbr.* young pioneer detachment
пионефрит *med.* pyonephritis
пионефроз *med.* pyonephrosis
  ~, брюшнотифозный typhoid pyonephrosis
  ~, перемежающийся intermittent pyonephrosis
  ~, частичный limited pyonephrosis
пиопиелэктазия *med.* pyopyelectasis
пиопневмоперикард pyopneumopericardium
пиопневмоторакс *med.* pyopneumothorax, pneumopyothorax
  ~, гнилостный putrefactive pyopneumothorax
  ~, ограниченный localized/circumscribed pyopneumothorax
  ~, поддиафрагмальный subdiaphragmatic/subphrenic pyopneumothorax
пиопоэз *med.* pyopoiesis, pyogenesis, suppuration
пиоптиз *med.* pyoptisis, purulent expectoration
пиорея pyorrhea
  ~ альвеолярная periodontitis
пиосальпингит *med.* pyosalpingitis
пиосальпинкс pyosalpinx, pus tube
пиосептицемия *med., veter.* pyosepticemia
пиоторакс *med.* pyothorax, empyema
пиоурахус *med.* pyourachus
пиофизометра pyophysometra
пиофтальмит pyophthalmia, pyophthalmitis
пиоцеле pyocele
пиоцефалия pyocephalus, pyencephalus
  ~, ограниченная circumscribed pyocephalus
пиоцианин *bioch.* pyocyanin
пиперазин *chem.* piperazine
пиперидин *chem.* piperidine, hexahydropiridine
пипетка pipette; (medicine) dropper
  ~, измерительная measuring pipette
  ~, капельная drop pipette
пиразинамид pyrazinamide
пиразол pyrazole
пираканта *bot.* fire thorn *(Pyracantha)*
пирамида 1. pyramid; 2. *mil.* rifle stack
  ~, возрастно-половая population pyramid
  ~, правильная *math.* regular pyramid
  ~, трехгранная *geod.* tripod signal

~, усеченная *math.* truncated pyramid, frustrum of a pyramid
пирамидально-метельчатый *bot.* bunchlike, thyrsoid
пирамидка 1. small pyramid; 2. *bot.* thyrsus, thyrse
пирамидный pyramidal; conoid
пирамидон *pharm,* amidopyrine, pyramidon
пиранометр *phys.* pyranometer
пираргирит *miner.* pyrargirite
пират pirate
~, воздушный air pirate, plane hijacker
пиратство piracy
~, воздушное air piracy, aircraft hijacking
пиратствовать freeboot, engage in piracy
пиргелиометр *astr.* pyrheliometer
пиргеометр *phys.* pyrgeometer
пирекс pyrex
пирексия *med.* pyrexia, fever
пирен *chem.* pyrene
пиренемия *med.* pyrenemia
пиренейский Pyrenean
пирений pyrenium
пиренин pyrenin, paranuclein
пиреноид, пиреноидный *bot.* pyrenoid
пиренокарп *bot.* pyrenocarp
пиренокарпный *bot.* pyrenocarpic, pyrenocarpous
пиреномицет pyrenomycete
пиретический *med.* pyre(c)tic, pyrexial, feverish
пиретогенный pyretogenic, pyretogenous
пиретрум 1. *bot.* pyrethrum *(Pyrethrum);* 2. *pharm.* pyrethrum, Dalmatian insect powder
пиридин *chem.* pyridine
пиридиннуклеотид pyridine nucleotide
пиридинсульфонамид pyridine-sulphonamide
пиридоксалькиназа pyridoxal kinase
пиридоксаминфосфатоксидаза pyridoxamine-phosphate oxidase
пиридоксин pyridoxin(e), vitamin B6
пиримидин pyrimidine
пирит pyrite
~, волосистый capillary pyrite
~, оловянный tin pyrite, stannite
~, почковидный nodular pyrite, brass balls
пиритиамин pyrithiamine
пиритный, пиритовый pyritic
пиритогедральный pyritohedral
пиритоносный, пиритосодержащий pyritiferous
пиритоэдр *cryst.* pyritohedron
пирке *med.* tuberculine test
пиробитум pyrobitumen
пироболт explosive bolt
пировинный *chem.* pyrovinic
~ная кислота pyrovinic/pyrotartaric acid
пировиноградный pyruvic
пирог 1. *met.* cake; 2. *cul.* pie
~ агломерата sinter cake
пирога *naut.* pirogue
пирогалловый *chem.* pyrogallic
пирогаллол *chem.* pyrogallol, pyrogallic acid
пирогелиометр pyroheliometer
пирогенетический pyrogenetic
пирогенизация *chem.* pyrogenation
пирогенный 1. pyrogenic, pyretogenic; 2. igneous
пироглобулин pyroglobulin
пироголовка explosive cap
пиографит pyrolitic graphite
пирозамок (rocket) separation charge
пирозапал cartridge igniter, activation squib
пирозаряд explosive charge
пирозома *zool.* pyrosome
пирокатехин pyrocatechol, pyrocatechin
пирокатехинамин pyrocatechinamine
пирокатехинурия pyrocatechinuria, catecholuria
пирокислота *chem.* pyroacid
пироклапан explosive/pyrotechnic valve
пирокластолит volcanic tuff
пироксен *miner.* pyroxene
пироксилин pyroxylin, pyroxylon, guncotton, collodion wool, colloxylin
пиролатрия pyrolatry, fire worship
пиролиз *chem.* pyrolysis
пиромантия pyromancy
пирометр pyrometer
~, погруженный immersion pyrometer
~, рефлекторный mirror pyrometer
~, рефракторный lens pyrometer
~, самопишущий recording pyrometer
~, цветовой colo(u)r pyrometer
пиромеханизм pyrotechnic device/mechanism, gun
~ принудительного сброса *av.* ejection gun, explosive jettisoning device
пирон *chem.* pyrone
пиропатрон pyrotechnical cartridge, explosive charge, ignition cylinder
~ катапультного кресла ejection seat cartridge
пиропроводимость pyroconductivity
пиросвеча igniter squib
пиросерный *chem.* pyrosulphuric
пиростартер cartridge starter
пиротехника pyrotechnics
пирофосфат pyrophosphate
пирофосфатаза pyrophosphatase
пирофосфорилаза pyrophosphorylase
пирохимия pyrochemistry
пирохлор *miner.* pyrochlore
пироцилиндр explosive actuation cylinder
пирриха pyrrhic (dance)
пиррихий *lit.* Pyrrhic foot
пиррова победа Pyrrhic victory
пиррол *chem.* pyrrol
пиррокаин *chem.* pyrrocaine
пирролидин pyrrolidine, tetrahydropyrrole
пирролин pyrroline, dihydropyrrole
пирронизм Pyrrhonism
пирронистский Pyrrhonian
пирронический Pyrrhonic
пирросидерит *miner.* pyrrhosiderite

пирс *naut.* pier, jetty
~, достроечный fitting-out pier
~, свайный pile pier
пируват *chem.* pyruvate
пируэт pirouette
писание writing; screed
~, священное Holy Scripture
писарь clerk; *naut.* writer
пистик *bot.* common horsetail *(Equisetum arvense)*
пистиллодий *bot.* pistillode
пистия *bot.* water lettuce *(Pistia)*
пистолет pistol
~ бензозаправщика/заправочный fuelling pistol/nozzle
~, боевой service pistol
~, ракетный flare pistol
~, дюбельный dowel-driving gun
~, сварочный weldind gun
пистолет-краскораспылитель paint-spraying gun
пистолет-пулемет submachine gun
пистоль pistole *(gold coin)*
пистон percussion cap, primer
писчебумажный магазин stationery shop
писчебумажные принадлежности stationery
письмо letter; writing, script
~, гарантийное letter of guarantee; bond/letter of indemnity
~, заказное registered letter
~, залоговое letter of deposit/lien/trust; letter of hypothecation
~, кассовое cash letter
~, повторное/-напоминание follow-up letter
~, ценное insured letter
письмоводство office/clerical work
питание 1. feeding, nourishment, nutrition, alimentation; diet; 2. *tech.* (power) supply; delivery; feed, feeding
~, аварийное emergency (power) supply
~, автономное independent/autonomous power supply; self-contained power supply
~, безбелковое protein-free diet
~, витаминизированное high-vitamin diet
~, двойное dual supply/feed
~, искусственное artificial feeding/nutrition
~ котла boiler feeding
~, ленточное belt feed
~, лечебное dietotherapy, alimentotherapy
~, общественное public catering
~ переменным/постоянным током a.c./d.c. supply
~, резервное силовое stand-by power supply
~ топливом fuel supply
~, трехфазное three-phase power supply
~, электрическое electric power (supply)
питатель *tech.* feeder, feeding device/magazine
~, бункерный bin feeder
~, вибрационный vibrating feeder
~, винтовой screw/spiral feeder
~, гравитационный gravity feeder
~ грохота screen feeder
~, ленточный feed belt, belt-conveyer feeder
~, лопастный paddle/vane feeder
~, лотковый chute feeder
~, ременно-пальцевый tined belt feeder
~, пульсирующий pulsefeeder
~, скребковый scraper feeder
~, цепной chain feeder
~, черпаковый scoop feeder
~, шнековый screw feeder
питательность nutrient value, nutritiousness
питательный 1. nutritious, nourishing; alimenative; 2. *anat.* alimental, alimentary; 3. *tech.* feeding, feed *(attr.);* power-supply
~ пункт refreshment place
~ная среда *biol.* nutrient/culture medium
питать 1. feed, nourish; nurture; *fig.* cherish, entertain; 2. *tech.* feed; supply
питекантроп pithecanthrope *(Pithecanthropus)*
питеколобиум *bot.* ape's earring *(Pithecolobium)*
питириаз *med.* pityriasis
питон *zool.* python *(Python)*
питоцин *pharm.* pitocin
питтинг *met.* pitting; corrosion
питтоспоровые *pl. bot.* tobira family *(Pittosporaceae)*
питуитрин *pharm.* hypophysin, pituitrin
питуицит *med.* pituicyte
питч *tech.* pitch
пиурия *med.* piuria
пифагореец Pythagorean
пифагореизм Pythagoreanism
пифагорейский, пифагорийский Pythagorean, Pythagoric
пифагорейство Pythagoreanism
пифагоров Pythagorean
~а теорема Pythagorean proposition
пифиада Pythiad
пифийский Pythian
пифия *hist.* the Pythian, Pythoness
пифон *myth.* Python
пифос pithos
пи-фотомезон photo-pion
пихта *bot.* fir *(Abies)*
пихтарник fir forest
пицеин picein wax
пиццикато, піччикато *mus.* pizzicato
пишущий writing; recording
~щая машинка typewriter
~щее устройство для кредитных и чековых карт credit-card and cheque writer
пищаль *hist.* harquebus, arquebus; cannon, culverin
пищеварение digestion
~, внеклеточное extracellular digestion
~, внутриклеточное intracellular digestion
~, желудочное gastric/peptic digestion
~, межклеточное intercellular digestion
~, неправильное maldigestion
~, пристеночное parietal digestion
~, слюнное salivary digestion

пищеварительный digestive
~ **канал** digestive tract, alimentary canal
пищевкусовой food-processing/producing
пищевод (o)esophagus, gullet
пищеводный (o)esophageal
пищевой food *(attr.)*; alimentary; nutritional
~**вая промышленность** food industry
~**вое растение** food plant
пищик fife; reed
пищуха *zool.* 1. tree creeper *(Certhia)*; 2. pika, coney, piping hare *(Ochotona)*
пиявка *zool.* leech *(Hirudinea)*
плав 1. fusion cake; 2. float
~**у, на** afloat, floated
плавание 1. swimming; 2. navigation, sailing, cruise, cruising, voyage
~ **в надводном положении** surface sailing/run
~ **в океане** ocean voyage
~ **в ордере** cruising in formation
~ **в подводном положении** submerged sailing/run/travel, submarine cruising
~, **внутреннее** inland cruise
~, **гарантийное** guarantee cruise; *Am.* shakedown cruise
~, **каботажное** coasting/coastwise navigation, cabotage
~, **кругосветное** global/world cruise, circumnavigation
~ **на парусной яхте** yacht sailing
~, **первое** maiden voyage/trip
~ **по счислению** dead reckoning sailing/navigation
~ **по фарватеру** channel navigation
~, **прибрежное** coasting/coastwise navigation
~, **речное** river navigation
~, **пробное** trial trip/run
~, **учебное** training/practice/exercise cruise
плавать 1. float; drift; 2. sail, navigate; 3. swim
~ **под парусом** sail
плавбаза mother ship; floating fish-processing factory
плавень flux; fusing agent
плавиковий fluoric, fluor *(attr.)*
~**вая кислота** hydrofluoric acid
плавильник melting pot, crucible
плавильный melting, smelting
~ **тигель** crucible
плавильщик smelter, founder, furnace attendant
плавить smelt; melt
плавиться melt, fuse
плавка 1. melting, smelting; fusion, fusing; liquation; melt, cast; 2. heat
~, **бросовая** lost heat, off-heat
~ **в кипящем слое** fluid bed smelting
~, **ваграночная** cupola heat
~, **взвешенная** flash melting
~, **восстановительная** reduction smelting
~, **гарнисажная** autocrucible melting
~, **доменная** blast-furnace smelting
~, **дуговая** arc melting
~, **захоложенная** cold heat
~, **избирательная** selective fusion
~, **испорченная** lost heat
~ **на роштейн** raw matte smelting
~, **отражательная** reverberatory process/smelting
~, **разделительная** top-and-bottom smelting
~, **тигельная** crucible process
~, **ушедшая** off-heat
~, **шлакующая** slag smelting
~ **шлихов** concentrate smelting
~, **электроннолучевая** electron-beam melting
плавкий meltable, fusible
плавкость fusibility
плавление fusion, melting
плавленый melted, fused
~ **сыр** processed cheese
плавник fin
плавнорегулируемый continuously adjustable/variable
плавность smoothness, evenness; continuity
плавсостав seafaring personnel
плавун I *zool.* water opossum *(Chironectes minimus)*
плавун II *geol.* quick ground
плавунец *zool.* water tiger/beetle *(Dityscus)*
плавунчик *zool.* phalarope *(Phalaropus)*
плавучесть buoyancy, floatage, floatability
плавучий floating; buoyant
~ **док** floating dock
~ **дом** houseboat
~ **маяк** lightship
~**чая льдинка** ice floe
~**чая мастерская** repair ship
плавучая база parent/mother/base/depot ship; *Am.* tender
~ **вертолетов** helicopter support ship
~ **для личного состава** barrack ship
~ **для самолетов** floating base for aircraft
~ **подводных лодок** submarine mother/depot ship; submarine tender
~, **портовая** harbour tender
~, **ракетная** floating missile base
~ **спасательных судов** salvage tender
плагиат plagiarism
плагиатор plagiarist
плагиаторство plagiarism
плагиировать plagiarize
плагиоклаз *miner.* plagioclase
плаз loft, mould(ing) room/loft
плазма *phys., biol., miner.* plasma
~ **газового разряда** gas-discharge plasma
~, **гелиевая** helium plasma
~, **горячая** hot/thermal plasma
~ **горячих электронов** hot-electron plasma
~, **дейтериевая** deuterium plasma
~ **дугового разряда** arc plasma
~, **зародышевая** germ plasma
~, **звездная** stellar plasma
~, **излучающая** emitting plasma

~, ионная ion plasma
~ крови blood plasma
~, ламинарная quiescent plasma
~, мышечная muscle plasma
~, неравновесная nonequilibrium plasma
~, несветящаяся nonluminous plasma
~, нестационарная unstable/transient plasma
~, нетепловая nonthermal plasma
~, ограниченная confined plasma
~, однородная uniform plasma
~, полностью ионизированная stripped plasma
~, равновесная equilibrium plasma
~, разреженная rarefied [low-density] plasma
~, распадающаяся decaying plasma
~, самобразующаяся self-generated plasma
~, сжатая pinched plasma
~ тлеющего разряда glow-discharge plasma
~, ударная shock-generated plasma
~ ударного слоя shock-layer plasma
~, удержанная confined plasma
~, электронная electron plasma
~, электронно-дырочная electron-hole plasma
плазмаген plasmagene
плазмалемма plasmalemma
плазмалоген *bioch.* plasmalogen
плазмаль plasmal
плазматический plasmatic, plasmic
плазматрон plasma generator/facility, plasmatron
~, многодуговой multiple arc plasma facility
плазменноядерный cytonuclear
плазменный plasma *(attr.)*
плазмида *biol.* plasmid
плазмовый *miner.* plasma *(attr.)*
плазмогамия *biol.* plasmogamy
плазмодесма *biol.* plasmodesm
плазмодий plasmodium
плазмодинамика plasma dynamics
плазмозаменитель plasma substitute/expander
плазмолиз plasmolysis
плазмома plasmoma, plasmacytoma, plasmocytoma
плазмон plasmon
плазмоптиз *biol.* plasmoptysis
плазморексис *med.* plasmorrhexis, plasmatorrhexis
плазморея *med.* plasmorrhea
плазморрагия plasmorrhagia
плазмотрон = плазматрон
плазмоферез plasmapheresis
плазмохимия plasma chemistry
плазмохин *plarm.* Plasmochin, pamaquine
плазмоцид plasmocid, pamaquin
плазмоцит *anat.* plasmocyte, plasmacyte
плазмоцитоз plasmacytosis, plasmocytosis
плазмоцитома plasmacytoma, plasmocytoma, plasmoma
~, экстраоссальная extraosseous plasmacytoma
плазома *biol.* plasome
плакат poster, bill, placard
~, учебный educational/wall chart
плакировать plate, clad; veneer
плакировка, плакирование plating, cladding
~ наплавкой fusion cladding
плакода placode
~, обонятельная olfactory placode
~, ушная auditory placode
~, хрусталиковая lens placode
плакосинклиналь placosyncline
плакун-трава *bot.* purple loosestrife/lythrum *(Lythrum salicaria)*
плакучая ива *bot.* weeping willow *(Salix babylonica)*
пламегаситель flame arrester/damper/eliminator; flame-quenching additive
пламенникоцветковый *bot.* phlogiflorous, phlox-flowered
пламеотражатель flame deflector
пламестойкий flame-resistant
пламеуловитель flame trap/bowl
пламя flame
~, ацетилено-кислородное oxyacetilene flame
~, вихревое cyclonic flame
~, водородно-кислородное oxyhydrogen flame
~ газовой горелки torch flame
~, дульное muzzle blast/flash
~, запальное ignition/pilot flame
~, контрольное pilot flame
~, некоптящее sootless flame
~, несветящееся nonluminous flame
~, ячеистое cellular flame
пламяфотометр flame-photometer
план 1. plan; project, scheme, design; program(me); 2. layout, plan, map; 3. plane
~ в горизонталях *geod.* contour plan/map, plan in profiles
~, встречный *econ.* 1. response plan; 2. counter plan
~ выборки sampling scheme
~, генеральный general/master plan; general layout
~, государственный state plan
~, грузовой loading/stowing plan; vessel stowage list
~, задний background
~ занятости рабочей силы employment plan
~, календарный (timing) schedule
~ капиталовложений investment plan
~, комплексный package plan; comprehensive/overall plan
~ компоновки assembly schedule
~, краткосрочный short-term plan
~, крупный *phot.* close-up, close shot
~ Маршалла Marshall Plan, European Recovery Program(me), ERP
~ накладных расходов overhead budget
~, народнохозяйственный state economic plan, macroeconomic plan
~ научных исследований research plan

~, несмещенный *stat.* unbiassed plan
~, общий 1. master/general plan; 2. *phot.* long shot
~, опережать be ahead of schedule
~, опорный basic plan
~, осуществимый practicable/feasible plan
~ передний *phot.* foreground
~, перспективный long-term/-range plan
~ полета flight plan
~ порта port plan
~, примерный model pattern
~, производственный production plan/schedule
~, пятилетний five-year plan
~ работ work schedule
~, рабочий working/operative plan
~ развития народного хозяйства economic development plan
~ размещения/расположения layout
~ разработки development plan
~, реальный practicable plan
~, сводный master plan
~, ситуационный site plan
~, смещенный *stat.* bias(s)ed design
~, средний *phot.* medium shot
~ страхования от болезней health insurance plan
~ стройки building schedule
~, схематический diagrammatic plan/layout
~ счетов chart of accounts
~ съемки survey plan; *phot.* camera shot
~ телефонной сети exchange area layout
~ технического обеспечения technical-support plan
~ технического обслуживания maintenance plan
~, типовой model plan/pattern
~, укрупненный outline/overall plan
~, уточненный detail plan
~, учебный curriculum
~, факторный factorial design
~ хода работы operation plan/schedule/record
~ хранилища storage layout
~ эксперимента design of an experiment
~, эскизный sketch plan
планвектор *math.* plane vector
план-график schedule
планер glider, sailplane; airframe
~, бесхвостый tailless glider
~, буксируемый towed glider
~ крылатой ракеты winged-rocket airframe
~, моторный powered glider
~, надувной inflatable glider
~, пилотажный aerobatic-type glider
~ самолета airframe
~ со складывающимися крыльями folding-wing glider
~, спортивный performance-type glider
~, транспортно-десантный troop-carrier glider
~, учебный training glider
планер-вертолет rotary-wing glider
планер-мишень target glider
планер-паритель sailplane, soaring plane/glider, performance-type glider
планера *bot.* water elm *(Planera)*
планеризм gliding, glider flying
планерист glider pilot
планеродром glider airfield, gliderport
планет *agr.* weeder, weedhook; hand cultivator
планета planet
~, большая major planet
~, верхняя superior planet
~, внешняя outer planet
~, внутренняя interior/inner planet
~ земной группы terrestrial planet
~, малая minor planet; planetoid, asteroid
~, нижняя inferior planet
~ солнечной системы solar system planet
планетарий planetarium
планетарный, планетный planetary; global
планетоведение, планетология planetology
планетолог planetologist
планетоход planet research vehicle
планиметр planimeter
планиметрирование planimetration, planimetry
планиметрия planimetry; *(school subject)* plane geometry
~ Эвклида Euclidean plane geometry
планирование I *av.* gliding, glide
~ по спирали spiral glide
планирование II planning; scheduling
~, встречное response planning; counter planning
~ добычи production planning
~, долгосрочное long-term/range planning; forward planning
~, заблаговременное early planning
~, календарное scheduling
~, краткосрочное short-term/range planning
~ народного хозяйства macroeconomic planning
~ номенклатуры изделий product planning
~, оперативно-производственное day-to-day production planning
~, отраслевое sectoral planning
~, перспективное long-term planning, forward planning
~ прибыли profit planning
~ работы, календарное job planning
~ ресурсов resource planning
~ рождаемости family planning
~, сетевое network planning
~, системное system engineering
~, скользящее календарное rolling scheduling
~ трудовых ресурсов manpower planning
планировать 1. plan; design; 2. lay out; 3. *av.* glide; 4. grade, level
планировка 1. grading, levelling; flattening;

2. layout; laying out
~ **веерная** fan layout
~ **рабочего места** workplace [work-station] layout
планировщик 1. planner; scheduler; 2. *tech.* grader, planer, land level(l)er
~ **заданий** job scheduler
~ **основания** subgrade planer, subgrader
планиторакс *med.* planithorax
планка plate, plank; strip, bar, lath, strap
~, **дистанционирующая** spacer bar
~ **для измерения** measuring rod
~, **зажимная** clamping plate
~, **киповая** *naut.* mooring chock
~, **кольцевая** ring-rail
~ **крена** roll bar
~, **монтажная** *el.* wiring strip
~ **мотовила** reel bat/board/slat
~ **нитепроводников** *text.* thread-guide rail
~, **пазовая** *naut.* seam strip
~, **палубная** deck pad
~, **прижимная** hold-down strap
~ **с обухом** eye plate
~ **с рымом** ring plate
~ **скуловой обшивки** bilge plank
~, **стопорная** lock strip
~, **стыковая** *naut.* butt strip
~ **тангажа** pitch bar
~, **указательная** index/name/label plate
~, **штормовая** *naut.* storm/fiddle rack
планктон plankton
~, **водорослевый** algal plankton
~, **морской** haliplankton
~, **настоящий** euplankton
~, **озерный** limnoplankton
~, **прудовой** heleoplankton, pond plankton
~, **речной** potamoplankton
~, **сетной** mesoplankton, net plankton
~ **солоноватых вод** haliplankton
~ **стоячих вод** stagnoplankton
планктонный planktonic
планктофит planktophyte
плановик planner
планово-предупредительный routine-preventive
плановость planned character/nature; planning
плановый planned, according to plan; systematic; planning *(attr.)*
планогамета planogamete
планозигота planozygote
планомерно according to plan, systematically
планомерность regularity; systematic character; conformity to plan
планомерный according to plan, systematic
планофировый planophyric
плантаж deep/trench ploughing, trenching
планшайба *mach.* faceplate, chuck (plate)
планшет 1. plotting board, plotter, surveyer's table, chart board; plotting sheet; 2. topographic map/sheet; 3. map case
~, **авиационный** aircraft chart board
~, **автоматический** automatic plotter, autoplot board
~, **гидроакустический** sonar plot
~, **главный** master plotting board
~ **для карт** chart board
~, **наколенный** knee-pad/board
~, **огневой** *mil.* artillery board; fire-direction chart
~, **поисковый** search plotting board
~, **полевой** field/topography board
~ **преобразования координат** conversion board
~, **радиолокационный** radar plotting board
~ **с гониометрической сеткой** standard circle sheet
планшетист plotter; plot observer; board keeper
планшир(ь) *naut.* bulwark/wood rail, gunwale, gunnel; washboard, covering board
пласт layer, stratum; seam; *geol.* bed, formation
~, **вертикальный** vertical/upright seam
~, **водоносный** water-bearing stratum/bed, aquifier
~, **вышележащий** overlying bed
~, **газовый** gassy/fiery seam
~, **каменноугольный** coal bed/seam
~, **крутопадающий** steep bed, steeply pitching/dipping bed
~, **лежачий** recumbent bed
~, **маломощный** thin bed
~, **мощный** thick seam
~, **налегающий** superincumbent bed
~, **нерабочий** unworkable seam
~, **нефтеносный** oil-bearing bed
~, **нижележащий** subjacent/lower bed
~, **осадочный** sedimentation bed
~, **плывучий** running bed
~, **поднимающийся** acclivous seam
~, **подстилающий** subjacent/underlying bed
~, **покрывающий** overburden; capping
~, **рабочий** workable/commercial seam
~, **сброшенный** faulted seam
~, **сложный** multiple/composite bed
~, **сопряженный** conjugated/associated bed
~, **средний** median seam
~, **угольный** coal bed/seam
пластбетон plastoconcrete
пластида *biol.* plastid(ium)
пластик 1. plastic material; polymeric material
~, **армированный** reinforced plastic
~, **армированный стекловолокном** glass-fibre reinforced plastic
~, **вспененный** foamed/expanded plastic
~, **кремнийорганический** silicone plastic
~, **слоистый** laminated plastic, plastic laminate
~, **термопластический** thermoplastic (material)
~, **термореактивный** thermosetting plastic
~, **ударопрочный** shock-/impact-resistant plastic
~, **ячеистый** cellular plastic

пластика 1. plastic arts; 2. *surg.* -plasty; 3. polymeric material
~ аневризмы aneurysmoplasty
~ влагалища vaginoplasty, colpoplasty
~ клапанов сердца valvuloplasty
~, кожная epidermatoplasty
~, костная bone/osseous graft, osteoplasty
~ молочной железы mastoplasty
~, мышечная myoplasty
~ носа rhinoplasty
~ подбородка genioplasty
~ роговицы keratoplasty
~ рта stomatoplasty
~ сосуда angioplasty
~ сустава arthroplasty
~ челюсти gnathoplasty
~ черепа cranioplasty
пластикат plastic compound, flexible PVC
~, кабельный cable compound
пластикатор masticator, kneading machine, kneader, plasticator
~, червячный screw plasticator
пластикация plasticization, mastication
~ на вальцах mill plasticization
пластиковый plastic
пластилин plasticine, plasticene, modelling wax
пластина plate
~, аккумуляторная battery plate
~, базовая master plate
~, биметаллическая bimetallic plate
~, выпуклая dished plate
~, дистанционирующая spacer plate
~, завихрительная whirl/swirl plate
~, заземляющая earth/ground plate
~, консольная cantilever plate
~, монокристаллическая single-crystal wafer
~, отбойная bumper plate
~, отклоняющая deflecting/deflector plate
~, отражающая baffle plate
~, охранная guard plate
~ полупроводникового материала wafer; slice
~, подкрепленная stiffened plate
~ радиатора radiator fin/gill
~ ротационного насоса vane
~ с заполнителем sandwich plate
~, светоделительная *opt.* beam spiller
~ сердечника *el., mach.* core plate/lamination
пластинка 1. (small) plate, lamina, lamella; wafer; flake; 2. (phonograph) record, disk; 3. photographic plate; 4. gill, fin; blade
~, агаровая agar plate
~, ганглиозная ganglious lamella
~, грамофонная долгоиграющая long-playing record, LP record
~ гриба gill
~, двигательная motor plate
~, кожная cutis plate, dermatome
~ кристалла полупроводника wafer
~, кровяная blood plate, platelet
~ Луммера-Герке Lummer-Gercke plate
~, медуллярная medullary/neural plate
~, мышечная muscle plate
~ печатным монтажом printed-circuit board/card
~, полуволновая half-wave plate
~, пограничная terminal plate
~, решетчатая cribriform lamina
~, ротовая oral plate
~, сегментарная segmental plate
~, серповидная tympanic plate
~, сосудистая vascular layer
~, спиральная костная osseous spiral lamina
~, танталовая tantalum screen
~, хориальная chorionic plate
~ эпифизиальная epiphyseal plate
пластинковидный, пластинкообразный laminiform
пластинник gill fungus
пластинниковые *pl. bot.* gill fungi *(Agaricaceae)*
пластиножаберные *pl. zool.* elasmobranchs *(Elasmobranchii)*
пластинокожие *pl. zool.* placoderms *(Placodermi)*
пластинокожий placodermal, placodermoid
пластинообразный lamellar, platelike; tabular
пластиночка platelet; lamellule
пластиночник gill fungus
пластинчатовидный lamelliform
пластинчатожаберные *pl. zool.* lamellibranchs *(Lamellibranchia)*
пластинчатожаберный *zool.* lamellibranchiate
пластинчато-звездчатый lamellar-stellate
пластинчатость lamination
пластинчатоусые *pl. zool.* leaf-horned beetles, lamellicorn beetles *(Lamellicornia)*
пластинчатоусый *zool.* lamellicorn; dynastid
пластинчатые *pl. bot.* gill fungi *(Agaricaceae)*
пластинчатый laminal, laminar, laminate, lamellar, lamellate, laminated; platy, flaky; foliate(d); sheety; tabular; *zool.* placoid
~ гриб *bot.* gill fungus
пластификатор plasticizer, plasticizing/softening agent, softener, elasticizer
~, гидрофилизующий hydrophyle plasticizer
~, гидрофобизующий hydrophobe plasticizer
~, полиэфирный polyester plasticizer
~, химический peptizer, peptizing agent
~, эпоксидированный epoxy plasticizer
пластификация plasticization
пластифицировать plasticize; masticate
пластический plastic
пластичность plasticity; flexibility; pliability; *met.* ductility
~ в холодном состоянии cold ductility
~ при изгибе bend ductility
~ при кручении torsion ductility
~ при растяжении tensile ductility
~ при статической нагрузке static ductility

пластичный plastic; pliable, pliant; *met.* ductile
пластмасса plastic (material)
~, авиационная aircraft plastic
~, акриловая acrylic plastic
~, армированная волоком fibre-reinforced plastic
~ для крайне низких температур cryogenic plastic
~, газонаполненная gas-expanded plastic
~, композиционная compounded plastic
~, конструкционная structural plastic
~, слоистая laminated plastic, plastic laminate
~, термопластичная thermoplastic (material)
~, термореактивная thermosetting plastic
пластовой stratal; stratified
пластомер *biol.* plastomere
пластометр plastometer
пластохондрия *biol.* plastochondrion
пластун *hist.* Cossack infantryman
пластырь *med.* plaster; *naut.* patch, mat
~, аварийный collision mat
~, горчичный mustard plaster
~, жесткий/мягкий hard/soft patch
~, липкий adhesive plaster/tape
~, свинцовый lead/litharge/diachylon plaster
~ стручкового перца capsicum plaster
~, шпигованный quilted mat
плата I 1. pay, payment; remuneration; salary, wages; 2. charge, fee; toll; dues
~, арендная rental payment
~, входная admission fee
~ за квартиру rent
~ за обслуживание service charges
~ за пастьбу agistage
~ за перевозку груза freight/hauling charge, freightage
~ за переноску porterage
~ за подключение connection charge
~ за посадку *av.* landing fee
~ за проводку судна pilotage/pilot fees
~ за проезд в один конец single fare, one-way fare
~ за случку *agr.* serving cost
~ за стоянку (судна) quay dues; moorage
~ за обучение tuition fee
~, заработная wage(s); salary
~ на базе суточных ставок per diem charge
~ натурой payment in kind
~ по расстоянию *teleph.* distance rate/tariff
~, подписная subscription price
~, поразговорная *teleph.* message/measured rate
~, поштучная piece price
плата II plate, sheet, board, card
~ более низкого уровня daughter board
~, вставная plug-in card
~ для монтажа накруткой wire-wrap board
~, зажимная terminal box/plate
~ модульной конструкции *el.* circuit card
~, монтажная wiring board
~, многослойная multilayer card
~, односторонняя single-sided board
~, печатная printed circuit board, p-c board
~, расширительная expansion/extender board
~ с логической схемой logic card
~, системная system board
~, сменная plug-in card
~, соединительная interconnect board
платан *bot.* plane (tree), sycamore *(Platanus)*
плата-основание baseplate
платеж payment; settlement
~ в рассрочку payment by instalments
~, взыскать enforce payment
~, досрочный anticipatory/advance payment
~ звонкой монетой specie payment
~ наличными деньгами cash payment, down payment
~, наложенный cash/collect/pay on delivery
~ натурой payment in kind
~, отсроченный deferred/delayed payment
~, первоначальный down/initial payment
~, первоочередной preferential payment
~ по клирингу clearing payment
~ по предъявлении грузовых документов sight payment
~ по урегулированию расчетов settlement payment
~, побочный side payment
~, поощрительный incentive payment
~, просроченный late/overdue/back payment
~ процентов interest charge(s)
платежеспособность solvency, paying capacity, ability/capacity to pay; financial responsibility
платежеспособный solvent
платежный payment *(attr.)*
~ баланс balance of payments
~ день payday
~ная ведомость pay sheet, payroll
~ное средство, законное lawful/legal tender
плательщик payer
~ налогов taxpayer
платибазия *med.* platybasia
платимисциум *bot.* macawood *(Platymiscium)*
платина I platinum, Pt
~, губчатая platinum sponge
~, мышьяковистая platinum arsenic, sperrylite
платина II 1. *text.* sinker; 2. frame/mounting plate *(of clockwork)*
платинирование platinum plating, platinization
платинировать platinize
платинит 1. *miner.* platinite; 2. *met.* iron-nickel alloy
платинка *hist.* platinum coin
платинный, платиновый platinic
платиносодержащий platiniferous
платинотрон *el.* platinotron
платировка *leath.* slicking
платиспондилия *med.* platyspondylia
платить pay
~ в счет причитающейся суммы pay on

account
~ **вперед** pay in abvance, prepay
~ **наличными** pay in cash, pay ready money
~ **натурой** pay in kind
~ **при предъявлении (векселя)** pay on demand/presentation
платифиллин platyphyllin
платицериум *bot.* common staghorn fern *(Platycerium alcicorne)*
платицефалия *med.* platycephaly
плато plateau, elevated plain, upland
платонизм Platonism
платоник Platonist
платонический Platonic
платформа platform; ramp
~, **буровая** drilling platform
~, **весовая** platform scales/balance
~, **грузовая** cargo platform
~ **грузовика** truck bed
~, **железнодорожная** flat-car
~, **космическая** space platform
~ **обслуживания** servicing platform/trestle
~, **поворотная** turntable
~, **погрузочная** loading platform
~ **подъемника/подъемная** elevating platform
~, **пусковая** launching ramp/platform
плаун *bot.* club moss *(Lycopodium)*
плауновидные *pl. bot.* lycopsids *(Lycopsida)*
плауновые *pl. bot.* club-moss family *(Lysopodi aceae)*
плаунок *bot.* selaginella, resurrection plant *(Selaginella)*
плафон 1. *arch.* plafond; 2. dome lamp/light
плац *hist.* parade/drill ground
плацдарм *mil.* bridgehead; *naut.* beachhead; *av.* airhead
плацебо *med.* placebo
плацента placenta
~, **гемохориальная** hemochorial placenta
~, **двойная** placenta duplex
~, **добавочная** accessory/succenturiate placenta
~, **кольцевидная** annular/zonary placenta
~, **неотпадающая** nondeciduous placenta
~, **плодная** f(o)etal placenta
~ **семенных растений** sporophore
~, **ущемленная** incarcerated/trapped placenta
плацентарные *pl., zool.* placentals *(Placentalia)*
плацентарный placental
плацентит placuntitis, placentitis
~, **гнойный** purulent placuntitis
плацентома *med.* placentoma, deciduoma
плацкарта reserved seat ticket
плашка threading die, screw/cutting die
~, **винторезная** bolt die; thread-rolling die
~, **накатная** thread-rolling die
~, **нарезная** thread-cutting die
плашкодержатель die stock
плашкоут, плашкот *naut.* scow, hulk
~, **десантный** landing scow
плашкоутный мост pontoon bridge
плащаница *eccl.* holy shroud, shroud of Christ
плебей, плебейка, плебейский plebeian
плебейство, плебс *coll. hist.* plebs
плебисцит plebiscite
плева 1. *bot.* aril; seed coat; 2. *anat.* membrane,
~, **девственная** hymen
~, **крыловидная** *anat.* pterygium
плевел *bot.* 1. weed; 2. ryegrass *(Lolium)*
~, **многолетний** English ryegrass *(Lolium perenne)*
плевра *anat.* pleura
~, **легочная** pulmonary pleura
~, **околосердечная** pericardial pleura
~, **реберная** costal pleura
~, **средостенная** madiastinal pleura
плевральгия *med.* pleuralgia
плевральный pleural
плевренхима *bot.* pleurenchyma, woody tissue
плеврит *med.* pleurisy, pleuritis
~, **гнилостно-гнойный** putrefactive-and-purulent pleurisy
~, **двусторонний** double pleurisy
~, **осумкованный** encapsulated/blocked/encysted pleurisy
~, **пристеночный** parietal pleurisy
~, **сердечный** cardiac pleurisy
~, **слипчивый** adhesive pleurisy
~, **хилезный** chylous pleurisy
~, **экссудативный** exudative/wet pleurisy, pleural effusion
плевритический, плевритный pleuritic
плевродиния *med.* pleurodynia
плеврокарпический, плеврокарпный pleurocarpous
плевроклиз *med.* pleuroclysis
плевролегочный pleuropulmonary
плевротиф *med.* pleurotyphoid
плевроцистида *bot.* pleurocystidium
плезиозавр plesiosaur(us) *(Plesiosaurus)*
плезиозавровый plesiosaurian, plesiosauroid
плейас *geol.* playa
плейотропия pleiotropy, pleiotropism, pleiotropia
плейстоцен, плейстоценовый Pleistocene
плексиглас acrylic plastic, plexiglass
плексит *med.* plexitis
~, **плечевой** brachial plexitis
~, **родовой** labo(u)r plexitis
плектенхима plectenchyma
плектр *mus.* plectrum
племенной 1. thoroughbred, pedigree; 2. tribal
~ **скот** breeding cattle
племхоз *abbr.* pedigree-cattle breeding farm
плена 1. *met.* scab, blister, flaw; 2. film
~, **прокатная** rolling skin
пленка 1. film; pellicle, membrane; *bot.* glume, husk; *el.* (magnetic) tape
~, **ацетилцеллюлозная** acetate film
~, **высокочувствительная** high-speed film
~, **двусторонняя/двухэмульсионная** sandwich film, double-coated film

~, **диазотипная** *phot.* diazosensitized film
~, **диапозитивная** transparency film
~, **дозиметрическая** radiation monitoring film
~, **клейкая** adhesive film
~, **малочувствительная** slow film
~, **мелкозернистая** fine-grain film
~, **напыленная** sputtered film
~, **невоспламеняющаяся** *phot.* safety film
~, **обратимая** *phot.* reversible/reversal film
~, **односторонняя** single-coated/emulsion film
~, **паровая** steam blanket/film
~, **пластмассовая** plastic film
~, **покровная** *leath.* coating film
~, **полимерная** polymer film
~, **растрированная** screened film
~, **рентгеновская** X-ray film
~ **с подложкой** backed film
~, **светочувствительная** photosensitive film
~, **слоистая** laminar film
~ **фонограммы** sound film
~, **цветная** colo(u)r film
~, **черно-белая** black-and-white film
пленковидный filmlike, filmy; chafflike
пленконосный membraniferous; glumiferous
пленкообразование film formation
пленкообразователь film-formimg material/agent
пленум plenary session, plenum
пленчатовидный membranelike; hymenoid
пленчатокорешковый *bot.* hymenorhizous
пленчатолистный *bot.* hymenophyllous
пленчатоцветковый *bot.* glumiflorous
пленчаточашелистиковый *bot.* hymenosepalous
пленчатый membranaceous, filmy; pellicular, pelliculate
~**тая луковица** *bot.* tunicated bulb
пленэр *art*. plein-air painting
пленэрист plein-air artist/painter
плеопода *zool.* pleopod, swimmeret
плеохроический, плеохроичный pleochroic, pleochromatic
плером(а) *bot.* plerome
плесень mo(u)ld; mildew
плесневеть grow mo(u)ldy/musty
плессиметр *med.* plessimeter, pleximeter
плетельный braiding
плетизмография *med.* plethysmography
~, **импедансная** rheocardiography
плетора *med.* plethora
плечевой *anat.* humeral, brachial
плечеголовной brachiocephalic
плечелоктевой humero-ulnar
плечелопаточный humeroscapular
плечелучевой brachioradial, humeroradial
плеченогие *pl. zool.* brachiopods *(Brachiopoda)*
плечо 1. shoulder; *anat.* (upper) arm, humerus; 2. *tech.* lever, arm, leg
~ **восстанавливающего момента** righting arm/lever
~, **задающее** master arm
~, **исполнительное** slave arm
~ **кренящего момента** heeling arm/lever
~ **кривошипа** crank arm/web
~ **остойчивости** arm/lever of stability
~, **рабочее** actuating arm
~ **рычага** lever arm
~ **силы** arm of force
~ **статической остойчивости** metacentric arm
~ **хвостового оперения** *aero.* tail arm/length
плешанка *zool.* pied chat *(Oenanthe leucomela)*
плешивость baldness, alopecia, acomia, calvities
~, **краевая** marginal alopecia
~, **лекарственная** drug alopecia
~, **послеродовая** postpartum alopecia
~, **рубцовая** cicatricial alopecia
плеяда Pleiad; pleyade, constellation; galaxy
~ **изотопов** pleyade of isotopes, isotope group
Плеяды *pl. astr., myth.* Pleiades
плимутрок Plymouth Rock chicken
плинианский Plinian
плинт *arch.* plinth; *teleph.* terminal/distributing block; connecting strip
плинтовать 1. smooth, even, level; 2. cleave
плинтус 1. *arch.* plinth, skirting, baseboard; 2. *naut.* skirting board
плиоцен, плиоценовый *geol.* Pliocene, Pleiocene
плис *text.* velveteen, cotton velvet
~, **рубчатый** corduroy
плиска *zool.* wagtail *(Motacilla)*
плиссе accordion pleats; (accordion-)pleated
плита 1. plate, panel, slab; board; 2. cooker, range; stove; 3. *geol.* platform
~, **балластировочная** ballasting plate
~, **бетонная** concrete slab
~, **броневая** armour plate
~, **водобойная** apron plate/slab
~, **газовая** gas range
~, **гибочная** bending slab
~, **дорожная** road/pavement slab
~, **древесная** wallboard
~, **древесноволокнистая** wood fibreboard
~, **древесностружечная** chip wood board, chipboard
~, **жесткая** stiffened slab
~, **камбузная** cooking range/stove
~, **кухонная** hot/stove plate, kitchen range
~, **могильная/гробовая** tombstone, gravestone
~ **настила** floor plate/slab
~, **несущая** support plate; load-bearing slab
~, **облицовочная** facing slab/plate
~, **опорная** base/support/bed/foot plate
~, **отражательная** baffle plate
~, **подъемная** lifting/hoisting plate
~, **правильная/рихтовальная** dressing plate
~, **стеклянноволокнистая** glass-fibre board
~, **транцевая** *naut.* transom plate
~, **упорная** thrust plate
~, **фундаментная** foundation/base plate, bed-

plate
~, **штукатурная** plaster board
плитка 1. tile; plate; thin slab; 2. hot plate, cooker; 3. cake, bar
~, **газовая походная** Primus camp stove
~, **кафельная** glazed tile
~, **подкладная** *tech.* thrust disk/slab
~, **шиферная** slate tile
~ **шоколада** bar of chocolate
~, **шпунтовая** bent/gutter tile
~, **электрическая** hot plate
плиткообразный platelike; tabular; platy
плитняк flagstone
плитообразный tabular; platy, flaggy
плиточный tile *(attr.)*
плица *naut.* blade, paddle, float
плод fruit; *med.* f(o)etus
~, **недоношенный** premature f(o)etus
~, **переношенный** overmature f(o)etus
~, **утробный** f(o)etus
плодник *bot.* pistil
плодниковый *bot.* pistillate
плодный f(o)etal; fertile
плодоведение pomology
плодовитость fertility, fecundity; productivity
~, **естественная** natural reproduction; fertility of noncontracepting couples
~, **пониженная** subfecundity
плодовитый fecund, fertile; *zool.* polytocous
плодовместилище *bot.* apothecium
плодовод fruit grower, pomiculturist
плодоводство fruit growing/gardening
плодоголовка *bot.* varpocephalum
плодожил *zool.* chestnut weevil *(Curculio proboscideus)*
плодожорка *zool.* fir seed moth *(Laspeyresia bracteatana)*
плодолистик *bot.* carpel
плодонос, плодоносец carpophore
плодоношение fructification, fruitage
плодопрорастающий *bot.* blastocarpous
плодосмен crop rotation
плодосъемник fruit picker
плодоядный carpophagous; frugivorous
плозивный *phon.* plosive; explosive
пломба 1. seal; stamp; 2. *med.* filling
~, **временная** provisional/temporary filling
пломбир sealing tongs/tool
пломбировать 1. seal; 2. *med.* fill, stop
плоский flat, plane
~**кая стопа** *med.* flatfoot
плоскобокорезы side-cutting pliers
плоско-вогнутый plano-concave
плоско-выпуклый plano-convex
плосковязальный *text.* flat-knitting
плоскоголов *zool.* flathead *(Platycephalus)*
плоскоголовые *pl. zool.* platycephaloids *(Platycephalidae)*
плоскоголовый platycephalic, platycephalous
плоскогубцы flat(-nosed) pliers
~, **универсальные** combination pliers
плоскодонный flat-bottomed
плоскоклеточный *biol.* planocellular
плоскоклюв *zool.* tody *(Todus)*
плосконосый platyrrhine
плоскопечатная машина flat-bed printing press
плоскостной plane, planar; flat
плоскостность planeness, flatness
плоскостопие *med.* platypodia, flat/splay foot
~, **поперечное** transverse platypodia
~, **продольное** longitudinal platypodia
~, **профессиональное** occupational platypodia
плоскость plane; surface; flatness; sheet; level
~, **азимутальная** azimuth plane
~, **атомная** atomic plane, plane of atoms
~, **базовая** datum/reference/base plane
~, **бесконечно удаленная** *math.* plane at infinity
~, **боковая** profile plane
~ **визирования** sighting plane, plane of vision
~ **вращения** plane of rotation
~ **годографа** locus plane
~, **двойниковая** *cryst.* twinning plane
~, **заданная/нулевая** datum/reference plane
~, **касательная** tangential plane
~ **крена** *av.* rolling plane
~ **мидель-шпангоута** midstation plane
~ **наведения** guidance plane
~, **несущая** carrying/supporting/lifting plane
~ **отражения** reflection/reflecting plane
~ **падения** incidence plane
~ **первого вертикала** *av.* east-west plane
~ **плавания** waterplane, plane of buoyancy
~, **предметная** *opt.* object plane
~, **преломления** refraction plane
~, **продольная** longitudinal [fore-and-aft] plane
~ **продольного движения** pitching plane
~ **раздела фаз** *phys.* interface (plane)
~ **разрыва** plane of fracture; *aero.* discontinuity plane
~ **рыскания** yawing plane
~ **сечения/секущая** sectional plane
~ **скольжения** slip/glide/skidding plane
~ **сноса** drift plane
~, **спрямляющая** rectifying plane
~ **сцепления** bonding plane
~ **тангажа** pitch(ing) plane
~, **узловая** nodal plane
~ **шпангоута** *naut.* transverse plane
плоскоцветковый flat-flowered, planiflorous
плот raft; float
~, **бревенчатый** log raft
~, **буксируемый** tow
~, **надувной** inflatable raft
~, **спасательный** life raft, pontoon lifeboat
плотва *zool.* roach *(Rutilus rutilus)*
плотик 1. small raft/float; 2. *geol.* bedrock, rock floor
плотина dike, dyke, dam, weir
~, **арочная** arch dam

~, водосборная/водосливная spillway/overflow dam
~, высоконапорная high-head dam
~, глухая non-overflow dam; fixed dam
~, деривационная diversion dam
~, каменная rockfill dam
~, контрафорсная buttress dam
~, низконапорная low-head dam
~, противопаводковая flood-control dam, high-water detention dam
плотник carpenter
плотномер densimeter, density gauge/meter, densitometer
~ грунта soil density gauge
~, поплавковый float densimeter
~, статический hydrostatic densimeter
плотность 1. density; compactness, consistency, consistence; massiveness; 2. tightness
~ вероятности probability density
~, весовая specific weight/gravity
~ городской застройки urban saturation
~, гравиметрическая gravity density
~ движения traffic density
~ деления fusion density
~ записи *comp.* writing/recording density
~ застройки жилыми зданиями housing unit density
~, многомерная multivariate density
~ нагрузки load density
~ населения population density
~, насыпная bulk density
~ нейтронного потока neutron flux/current
~, объемная volumetric/volume/bulk density
~ осадков rainfall density
~ относительная relative density, density ratio
~ пассажирских мест seating density
~ по берду *text.* porter of reed
~ по основе *text.* set of the warp, warp texture
~ по утку *text.* set/density of the weft
~ потока flux density
~, равновесная equilibrium density, steady-state density
~ размещения компонентов component density
~ расположения дорожек *comp.* track density
~ распределения frequency distribution; density function
~ растительного покрова density of stand/sward
~ сыпучих тел bulk density
~ тока current density, specific current
~, удельная specific density
~, электронная/электронов electron density
~ ядер nuclear density
~ ядра density of the nucleus
плотный 1. dense, compact, close; 2. tight, leak-proof, impermeable
~ всюду *math.* everywhere dense
~, нигде не *math.* nowhere dense, nondense
плотовод raftsman; log driver
плотофит *bot.* plotophyte, floating plant
плотоход timber chute/slide; raft chute
плотоядные *pl. zool.* carnivores *(Carnivora)*
плотоядный carnivorous, creophagous
плоть flesh
~, крайняя *anat.* prepuce
плохообтекаемый bluff
площадка platform; site, area; ground; plot; pad; floor
~, вертолетная helicopter platform/deck/pad, helipad
~, взлетная take-off pad
~, детская children's playground
~, загрузочная/завалочная charging area/floor
~, лестничная landing
~, моечная washrack
~, монтажная assembly area; erection site
~, перегрузочная transfer platform
~, посадочная landing field/area/ground
~, предангарная apron, tarmac
~, предстартовая run-/warming-up area
~, разгрузочная discharging platform
~, разливочная *met.* casting platform
~, сборочная assembly/assembling area
~, стартовая launching/firing pad
~, стояночная parking area/pad/apron
~, строительная building/construction site
~, шихтовая charge/charging platform
площадь area; space; surface; (city) square
~, водосборная catchment area
~ днища bottom area
~, единичная unit area
~, жилая residential floor area, housing floor space
~, кормовая feed/forage acreage
~ круга area of a circle
~, лобовая frontal area
~ мишени target area
~ опоры support/base area
~, орошаемая irrigated area, area under irrigation
~ парусности sail area
~ пастбищ grazing acreage
~ под зерновыми злаками corn acreage
~ под хлебными злаками cereal acreage
~, посевная sown/planted area, area under crops
~, производственная plant space; operation area
~ разброса *stat.* dispersion area
~ рассеивания dispersion/diffusion/scattering area
~ сеноуборки hay acreage
~ сечения sectional area
~, складская storage space
~ скотного двора yard space
~ соприкосновения contacting area
плуг plough, plow
~, безотвальный mould bordless plough

~, **двухотвальный** double-breasted plough
~, **двухкорпусный** two-bottom plough
~, **двухъярусный** double-depth plough
~, **лемешный** share plough
~, **лущильный** stubble breaking plough
~, **многокорпусный** multibottomed plough; gang plough
~, **навесной** mounted plough
~, **нивелирующий** skimmer
~, **оборотный** reversible plough
~, **однокорпусный** single-bottom plough
~, **отвальный** mould-board plough
~, **плантажный** trenching/deep/subsoil plough
~, **полунавесной** semi-mounted plough
~, **прицепной** pull-type plough
~, **самоходный** self-propelled plough
~, **снеговой** snowplough
плуг-канавокопатель trenching plough
плуг-лущильник stubble plough breaker
плуг-распашник ridging plough
плунжер plunger; ram
~, **выталкивающий** ejector ram
~, **регулирующий** pilot plunger
плутон 1. *geol.* pluton; 2. *myth., astr.* Pluto
~, **плоский** tabular pluton, sill
плутоний *chem.* plutonium, Pu
плутонил *chem.* plutonyl
плутонов Plutonian
плывун quicksand, floating earth, guick ground
плыть swim; float; sail; drift
плювиал pluvial (period)
плювиограмма rain gauge chart
плювиограф pluviograph, hyetograph, recording rain gauge
плювиометр pluviometer, hyetometer, ombrometer, rain gauge
плюмбометрия metallometric survey
плюмерия *bot.* frangipani plant *(Plumeria)*
плюмула *bot.* plumule
плюр pelure *(paper)*
плюрализм pluralism
плюралист, плюралистский pluralist
плюривариантный multivariate
плурижанр *math.* plurigenus
плюрипроизводная pluriderivative
плюска *bot.* cupule, acorn cup
плюсквамперфект *gram.* pluperfect, past perfect
плюсковидный cupular, cupuliform
плюсконосные *pl. bot. Cupuiliferae*
плюсконосный *bot.* cupuliferous, cup-bearing
плюс-модификаторы plus modifiers
плюсна *anat.* metatarsus
плюсневой *anat.* metatarsal
плюсне-фаланговый *anat.* metatarsophalangeal
плюсовать immerse, dip; dye
плюсовка 1. padding dyer; 2. foulard-dyeing machine
плюш plush
~, **двухлицевой** double-sided plush
~, **репсовый** ribbed plush
~, **рубчатый** corded plush
плюшевание *leath.* polishing
Плюшкин *fig.* skinflint, miser, stingy person
плющ *bot.* ivy *(Hedera)*
плющевидный, плющеподобный ivylike, hederaceous
плющение flattening; laminating; crushing; *met.* flatting
плющилка crusher
плющильня *met.* flatting/rolling mill
плющить platten; laminate; crush
пляж beach
пляска dance; dancing
~ **святого Вита** *med.* St. Vitus's dance, chorea
пневма *med.* pneuma
пневмартроз pneumarthrosis
пневматик pneumatic tyre/tire
~, **авиационный** aircraft tyre
~, **баллонного типа** balloon-type tyre
~, **бескамерный** tubeless tyre
~ **с гладким протектором** smooth-tread tyre
~ **с глубоким рисунком протектора** anti-skid tyre
~ **с радиальным расположением слоев корда** radial ply tyre
пневматика 1. pneumatics; 2. pneumatic tools/appliances/apparatuses
пневматический pneumatic, air-operated/powered
пневматоз *med.* pneumatosis
пневматофор *bot.* pneumatophore, respiratory/breathing root
пневматоцеле pneumatocele, pneumo(no)cele
~, **внечерепное/внутричерепное** extracranial/intracranial pneumatocele
пневматурия *med.* pneumaturia
пневмеркатор pneumercator
пневмоавтоматика automated pneumatic appliances, pneumatic controls
пневмоагрегат pneumatics unit
пневмогемоторакс *med.* pneumoh(a)emothorax
пневмограмма pneumogram, pneumatogram
пневмограф pneumograph, pneumatograph
пневмодатчик pressure pickup/gauge
пневмодвигатель air motor
пневмодерматом *med.* pneudermatome
пневмозолоудаление pneumatic ash handling
пневмококк pneumococcus
пневмокониоз *med.* pneumoconiosis
~, **профессиональный** occupational pneumoconiosis
пневмокамера pneumatic cell/chamber
пневмоклапан air-pressure operated valve, pneumatic valve
пневмоколонка air column
пневмолебедка pneumatic hoist
пневмолиз *med.* pneumolysia, pneumolysis
~, **верхний/нижний** superior/inferior pneumolysia
пневмолизин pneumolysin
пневмометр pneometer, pneumometer, pulmo-

meter, pneumatometer
пневмометрия pneumometry, pneometry, pneumatometry
пневмомикоз *med.* pneumomycosis
пневмомолот pneumatic hammer
пневмоника pneumonics
пневмонит *med.* pneumonitis
пневмония pneumonia
~, абсцедирующая abscessing pneumonia
~, апикальная apical/apex pneumonia
~, бронхиальная bronchopneumonia, bronchial pneumonia
~, брюшнотифозная typhoid pneumonia
~, вирусная virus pneumonia
~, гнойная purulent/suppurative pneumonia
~, грибковая mycotic pneumonia
~, двусторонняя double pneumonia
~, застойная congestion pneumonia
~, казеозная caseous/cheesy pneumonia
~, крупноочаговая macrofocal pneumonia
~, липоидная lipoid/lipid/oil pneumonia
~, личиночная larval pneumonia
~, мелкоочаговая microfocal pneumonia
~, монилиазная moniliasis pneumonia
~, неразрешенная unresolved pneumonia
~, обтурационная obturative pneumonia
~, очаговая focal pneumonia
~, плазмоцитоклеточная plasma cell pneumonia
~, пневмоцистозная pneumocystic pneumonia
~, сибиреязвенная anthrax pneumonia, pulmonary anthrax
~, сливная confluent pneumonia
~, центральная central/core pneumonia
~, чумная pneumonic plague, plague pneumonia
~, эфемерная ephemeral pneumonia, pulmonary congestion
~, эфирная ether pneumonia
пневмонэктомия pneumonectomy, pneumectomy, pulmonectomy
пневмопексия pneumopexy, pneumonopexy
пневмоперикард pneumopericardium
пневмоплеврит pneumopleuritis, pneumonopleuritis
пневмоподушка pneumocushion
пневмопочта pneumatic tube message system; pneumatic dispatch system
пневмопривод pneumatic actuator/drive
~, мембранный diaphragm actuator
пневмореле air-pressure operated relay
пневмосепаратор air separator
пневмосклероз pneumosclerosis
~, лучевой radiation pneumosclerosis
~, мышечный myopneumosclerosis
~, очаговый focal pneumosclerosis
~, табачный smoker's pneumosclerosis
~, химиотоксический chemical pneumosclerosis
пневмоскоп pneumoscope, pneoscope, pneumatoscope
пневмосопротивление pneumonic/fluidic resistance
пневмостартер pneumatic starter
пневмотерапия pneumotherapy, pneumatotherapy, pneumonotherapy
пневмоторакс *med.* pneumothorax, aerothorax, auropleura
~, двусторонний double pneumothorax
~, закрытый closed pneumothorax
~, клапанный valvular/tension pneumothorax
~, напряженный tense/strained pneumothorax
~, осумкованный encapsulated pneumothorax
~, хирургический operative pneumothorax
пневмотормоз pneumatic brake
пнемвотрамбовка air/pneumatic rammer
пневмотранспортер pneumatic conveyer
пневмохолецистит pneumocholecystitis
пневмоцилиндр pneumatic (actuating) cylinder
пневмоцефалия pneumocephalus
пневмоэмпиема pneumoempyema
пневмоэнцефалография pneumoencephalography
поаккордная плата piecework rate
побег *bot.* sprout, runner, offshoot, bine; sucker
~, вегетативный innovation/vegetative shoot
~, корневой root sucker, sobole
побежалость iridescence; tarnish
~, медная verdigris
побелка whitening, whitewashing
побои *law* battery
побочный secondary, accessory, subsidiary
поведение behaviour; conduct, demeano(u)r; *tech.* performance
~ во времени time behaviour
~, неправомерное *law* misconduct
~, предписанное specified behaviour
~ системы system behaviour
повелительное наклонение *gram.* imperative mood
поверенный *law* procurator
~ в делах chargé d'affaires
поверка 1. check(ing), check-up, verification; test; 2. roll call
~ калибровки calibration test
поверхностно-активный surface-active
~ное вещество surface-active substance, surfactant
поверхностно-покровный *geol.* suprategular
поверхностно-слоистопокровный *geol.* supratectal
поверхность surface, area; face
~, алгебраическая algebraic surface
~, беговая *(of rail)* running surface
~, боковая lateral surface
~, винтовая *math.* helical/helicoidal surface
~, вихревая vortex surface
~, внешняя/внутренняя outer/inner surface
~ вращения surface of revolution
~ второго порядка quadric (surface)
~, геометрическая geometric surface

~, граничная boundary surface
~ двумерного нормального распределения *stat.* bivariate normal surface
~ забоя *min.* face end
~, ионообменная ion-exchange surface
~ касания contact surface
~ катания roll surface
~ кожи, лицевая grain side
~, ладонная palmar surface
~, линейчатая ruled surface
~, лобовая frontal surface
~ лопасти blade face
~, междолевая interlobar surface
~, межфазовая interface
~ нагрева heating [heat transfer] surface
~ напряжений stress surface
~, несущая *aero.* lifting/supporting surface, aerofoil, airfoil
~, обтекаемая streamlined surface
~, ограничивающая bounding surface
~ одинаковой фазы equiphase surface
~, односвязная simply connected surface
~, опорная supporting/bearing surface
~ охлаждения/охлаждающая cooling surface
~ пятого порядка *math.* quintic (surface)
~ равного давления equipressure surface
~ равного потенциала equipotential surface
~, равнофазная equiphase surface
~, развернутая developed surface
~ раздела фаз interphase
~, ребристая finned surface
~, риманова *math.* Riemann(ian) surface
~, светочувствительная photosensitive [light-sensitive] surface, photosurface
~, седловидная saddle-shaped surface
~, сопряженная/сопрягающаяся mated/mating surface
~, срединная medial surface
~, теплообменная heat-exchange surface
~ теплопередачи heat-transfer surface
~, тормозная/тормозящая braking surface
~ трения friction surface
~, трехсвязная triply-connected surface
~, уровенная datum/reference surface/level/plane
~, фоточувствительная = светочувствительная
~ четвертого порядка *math.* quartic (surface)
~, шаговая pitch surface
~ ядра nuclear surface/wall
повестка 1. notification, notice, call; *law* summons, subpoena; *mil.* call-up papers; buggle/first call
~ дня agenda
~, судебная writ of summons
повив lay *(of cable)*
повилика *bot.* dodder *(Cuscuta)*
повиликовые *pl. bot.* dodder family *(Cuscutaceae)*
повинность, трудовая civilian duty
поводка warping, warpage, deformation
поводник *bot.* habenaria *(Habenaria)*
поводок 1. catch, dog; (lathe) carrier; tappet; 2. *agr.* drag/-tow/-draw-bar
~ сателлитов pinion carrier
повозка vehicle
повой *bot.* glorybind *(Calystegia, Convolvulus)*
поворот 1. turn, turning; rotation; 2. bend
~ кругом turn-about
~, левый/правый left-/right-hand turn
~ плода *med.* version
~ системы координат rotation of a coordinate system
поворотливость manoeuvrability; *naut.* turning ability/qualities; helm handiness
поворотливый manoeuvrable
поворотный rotatable; rotational, rotary, revolving, turning
~ круг *rail.* turntable
повреждение damage; fault; injury, harm, breakage
~, тяжкое телесное *law* grievous bodily harm
~, усталостное *tech.* fatigue damage
поврежденный faulty, defective, broken-down, out of order; damaged, injured
~ в уме mentally deranged
повременный on a time basis; periodic(al)
~ная заработная плата time wage(s)
повтор echo; *telev..* ghost, double/echo image; *mus. lit.* refrain; repetend
повторение repetition; recurrence; replication; reiteration; retry
повторитель *tech.* repeater, follower
~, зарядный charge follower
~, ходовой code repeater
~ кривых curve follower
повторное слушание дела *law* retrial
повторно используемый re-usable
повторяемость recurrence, repetitiveness, repetition rate; recurrence period/interval
повыситель booster; augmenter
повышение rise, rising; elevating, raising; enhancement; increase, boost
~, играть на sell a bull, bull *(at stock exchange)*
~ качества improvement of quality
~ квалификации upgrading/improvement of one's professional qualifications
~ напряжения *el.* stepping up of the voltage
~ по службе promotion
~ по старшинство promotion by seniority
~ тяги thrust uprating
~ цен price increase
повязка dressing; bandage; fillet, band
~, влажная water dressing
~, восьмиобразная figure-of-eight dressing
~, герметичная occlusive dressing
~, гипсовая plaster(-of-Paris) bandage
~, давящая compressing bandage
~, колосовидная spica bandage
~, косыночная triangular/scarf bandage
~, набедренная loincloth
~, первичная first-aid bandage, [emergency]

dressing
~, **пластырная/липкопластырная** adhesive/sanilastic bandage
~, **подвешивающая** suspensory bandage
~, **Т-образная** T-bandage
поганка 1. bot. toadstool, nonedible mushroom; 2. *zool.* grebe, diver *(Podiceps)*
погасание extinction
погашаемый *econ.* redeemable; payable
погашать liquidate, cancel; pay off, redeem
погашение liquidation, cancellation, discharge, redemption; amortization; repayment
погибь camber, crown
~ **бимса** *naut.* camber/crown/round of beam
поглотитель 1. absorbent, absorbing medium; 2. absorber
~ **влаги** water absorber, desiccant
~ **перенапряжений** surge absorber
поглотительный absorbent, absorptive, absorbing
поглощаемость absorptivity, absorption capacity
поглощать absorb, take up; consume
поглощение absorption, absorbing; intake; uptake; imbibition; consumption
~ **альфа-излучения** alpha-absorption
~ **звука** acoustic/sound absorption
~, **избыточное** excess absorption
~ **инфракрасных лучей** infra-red absorption
~ **нейтронов** neutron absorption/capture
~ **нейтронов, вредное** neutron poisoning
~, **поверхностное** adsorption
~, **повторное** re-absorption
~, **преимущественное** preferential absorption
~, **примесное** impurity absorption
~ **света** light/luminous absorption
~, **собственное** self-absorption
~, **сплошное** continuous absorption
~, **удельное** specific absorption
погода weather
~, **летная** flying weather
поголовный 1. universal, general; 2. estimated by the head
~ **налог** poll tax
поголовье number of livestock; *agr.* population
~, **взрослое** adult stock
~ **крупного рогатого скота** cattle population, number of cattle
~, **мелкое** small stock
~ **рабочего скота** work stock
~, **ремонтное** replacement stock
~ **скота** livestock population
погон I *mil.* shoulder-board/-strap
погон II *chem.* distillate, fraction
~, **боковой** side-cut distillate
~, **верхний/переходящий** overhead distillate
погон III 1. *mil.* race (ring); traverse circle; 2. *naut.* horse iron
~, **мачтовый** sail track
погонный running, linear
~ **вес** weight per meter
погоныш *zool.* crake *(Porzana)*
погостемон *bot.* Cablin patchouli *(Pogostemon cablin)*
пограничный boundary, border, frontier *(all attr.)*; marginal
~ **слой** boundary layer
~**ная застава** frontier post
погранохрана *abbr.* frontier guards
погреб vault, cellar, basement
~ **боеприпасов** ammunition room/space/magazine
~, **пороховой** powder room/magazine
~, **снарядный** shell/projectile room
погремок *bot.* rattlepot *(Alectrolophus)*
погрешность error, mistake, fault, inaccuracy
~, **аппаратурная** instrumental/apparatus error
~ **в вычислении** miscalculation; computational error
~, **выборочная** *stat.* sampling error
~, **вычислительная** computational error
~ **градуировки** calibration error
~ **измерения** measurement error, error of measurement
~ **квантования** quantizing/quantization error
~ **курса** heading error
~ **метода** error of procedure/method
~, **неустранимая** irreducible/inherent error
~ **округления** round-off error; rounding error
~, **остаточная** residual error
~ **отсчета** reading error
~, **предельно допустимая** limit of error
~ **прибора** instrument error/inaccuracy
~, **случайная** random error
~, **средняя** mean error
~, **средняя квадратическая** root-mean-square error, standard deviation
~, **тарировочная** calibration error
~, **фазовая** phase/phasing error
~ **юстировки** alignment error
погружаемый submersible
погружать 1. immerse, submerge, dip, sink, plunge; 2. load
погружаться submerge, plunge, sink, dive
погружение sinking, immersion, submergence, submersion, dip(ping), plunge, dive, diving
погружной submersible
~ **насос** deep-water [submersible] pump
погрузка loading; *naut.* embarkation, embarkment, shipping
~ **без упаковки [внавалку, насыпью]** bulk stowage
~ **на суда** embarkation
~ **посредством лихтеров** lighterage
погрузчик (mechanical) loader
~, **боковой** side loader
~, **вилочный** forklift loader
~, **грейферный** clam-type loader
~, **ковшовый** scoop/bucket loader, loading shovel
~, **ленточный** belt loader
~, **самоходный** self-propelled loader; fork-lift truck

погрузчик-штабелер loader-stacker/-piler
под *met.* hearth, bottom, hearthstone, sole;
~ вагранки cupola well/hearth
~, заправленный fritted hearth
~, кислый/основной acid/basic hearth
подаватель feeder, feeding device/carriage
подавать feed, supply, deliver; *(document)* file
подавитель suppressor; depressant
~ колебаний oscillation suppressor
~ помех interference suppressor
~ шумов noise suppressor; *el.* squelch circuit
подавление suppression; elimination; neutralization; quenching; inhibition
~ несущей (частоты) carrier suppression
~ обратной связи feedback/regeneration suppression; antifeedback
~ помех/шумов interference/noise suppression
~ резонансного излучения quenching of resonance
~ радиопомехами jamming
подавлять suppress; quell; neutralize
подагра *med.* gout, podagra
~, атипическая irregular/abarticular gout
~, известковая calcium gout
~, метастатическая transferred gout
~, свинцовая saturnine/lead gout
~, суставная articular/regular gout
подагрический gouty, podagral, podagric, podagrous
подалгебра subalgebra
подалирий *zool.* swallowtail butterfly *(Papilio podalirius)*
подальгия *med.* podalgia, tarsalgia, pododynia
подартрит *med.* podarthritis
податливость pliability; compliance, pliancy; yielding; flexibility
~ крепи pliability/yielding of the supports
податливый compliant, pliable; yielding
податной *hist.* 1. tax *(attr.),* duty *(attr.);* 2. tax collector
~ инспектор assessor of taxes
~ная система taxation
подать *hist.* duty, tax, assessment
~, поголовная/подушная capitation tax
подача 1. feed(ing), delivery, supply(ing); 2. *sport* service, serve
~, баллонная gas pressurization
~ бумаги paper feed
~ вручную manual feed
~, вытеснительная pressure feeding
~, гравитационная gravity feed; drop delivery
~ карт card feeding/movement
~ ленты tape transport
~, магазинная magazine feed
~, механическая power feed
~ насоса pump output/capacity/delivery
~, неправильная misfeed
~, непрерывная continuous feeding/supply
~ охлаждающей жидкости coolant supply
~ под вакуумом vacuum/underpressure feed
~ под давлением pressure feed
~, поперечная cross feed
~, принудительная pressure/force/forced feed
~, ручная manual/hand feed
~ самотеком gravity feed
~ свежего топлива make-up fuel supply
~, турбонасосная turbopump injection
подбел *bot.* andromeda *(Andromeda)*
~, лечебный butterbur *(Petasites officinalis)*
подберезовик *bot.* birch mushroom *(Boletus scaber)*
подбивать line, pad
подбивка lining, padding, quilting
подбой, подбойка 1. lining; 2. *min.* breaking-in hole; 3. set hammer; tamping pick, tamper
подбойник swage
подбор selection, picking up; choice; assortment
~ состава по весу proportioning by weight
~ фракции grading
подборка selection, selecting; *agr.* picking-up
подбородок *anat.* chin
подбородочно-подъязычный geniohyoid
подбородочно-язычный genioglossal
подбородочный *anat.* mental, genial
подборщик pickup attachment, pick-up, picker, crop collector
~ валков windrow pick-up
~, комбайновый pick-up header
~, пневматический suction pick-up
подбрюшинный subperitoneal
подбрюшник 1. boiler cradle; 2. *naut.* cradle/saddle strap; packing timbers; 3. bellyband
подбугорье *anat.* hypothalamus
подвахтенные *mil.* watch off duty
подведомственность jurisdiction; province
подверженность susceptibility
подвес suspension; mounting, mount; hanger
~ гирокомпаса frame of gyrocompass
~ гироскопа gyroscope suspension
~, демпфирующий shockproof suspension
~, кабельный suspender, dropper, sling, hanger
~, карданный gimbal mount/suspension
подвеска 1. hanging up; suspension; 2. suspension bracket/support/arm; suspender; hanger; pendant
~, амортизированная cushioned suspension
~ двигателя *auto.* engine mounting
~ для труб pipe hanger
~, карданная gimbal suspension/mount
~, крыльевая *av.* wing store
~, люлечная bar suspension
~ на амортизаторах damped suspension
~ на винтовых пружинах helical spring suspension
~ на рессорах leaf-spring suspension
~, подкрыльная *av.* underwing store
~, пружинная spring hanger
~, стержневая torsion-bar suspension
~, тормозная brake support/carrier
~, трубная pipe hanger
~, упругая flexible/elastic/dynamic suspension

~, цепная catenary suspension; chain suspender
подвесной outboard *(attr.)*; suspension *(attr.)*
подвесок appendage
подветренный downwind, leeward, lee
~ная сторона lee side
подвешивать suspend; hang up; lift
подвздох pit of the stomach
подвздошно-бедренный *anat.* iliofemoral
подвздошно-крестцовый *anat.* iliosacral, sacroiliac
подвздошно-паховый *anat.* ilioinguinal
подвздошно-подчревный *anat.* iliohypogastric
подвздошно-поясничный *anat.* iliolumbar
подвздошно-реберный *anat.* iliocostal
подвздошно-седалищный *anat.* iliosciatic
подвздошно-слепокишечный *anat.* iliocecal
подвздошный *anat.* iliac
подвигание *min.* advance
подвид 1. subspecies; 2. *gram.* subaspect
подвижка movement; shift; drift
подвижничество *rel.* asceticism; devotion
подвижной moving, mobile, movable; portable
~ состав *rail.* rolling stock
~ное равновесие mobile equilibrium
~ные праздники *eccl.* movable feasts
подвижность mobility; motility; (trans)poratability; manoeuvrability; agility
~, дрейфовая drift mobility
~, миграционная migratory mobility
~ ионов ion(ic) mobility
~ рабочей силы labo(u)r mobility
~ электронов electron(ic) mobility
подвисание hanging
подвисочный *anat.* infratemporal, subtemporal
подвишенник *bot.* plum/clitopilus mushroom *(Clitopilus prunulius)*
подвод supply, feed; intake, inlet; *el.* lead
~ мощности power input
подводить supply, feed; deliver
подводной feeding, supplying
подводный submarine, underwater, undersea; subaqueous, subaquatic; submerged, sunk
~ная лодка submarine
~ное крыло hydrofoil
~ное растение submersed plant
~ное течение undercurrent
подвозбудитель *el.* pilot exciter, subexciter
подвой *bot.* tree stock
подворный household *(attr.)*, homestead *(attr.)*
~ная перепись census of households
~ная подать *hist.* hearth/chimney money
подвывих *med.* incomplete luxation/dislocation, subluxation, semiluxation
подгар burning
подглазничный *anat.* infraorbital, suborbital
подголовник headrest
~ кресла seat headrest
подгонка adjustment, alignment; fit
~ методом наименьших квадратов least-square fit
подгонять adjust; trim; fit
подгорать burn
подготавливать prepare
подготовка 1. preparation; conditioning; (pre)treatment; 2. training
~, авиационная aeronautical training
~, артиллерийская fire preparation; artillery bombardment
~ без отрыва от производства in-service training, on-the-job training
~, боевая combat training
~ воды water treatment
~ данных data preparation/origination
~ к полету flight preparation
~, летная flying training
~ летчика pilot training
~ личного состава crew training
~, наземная ground training
~ на корабле *mil.* shipboard training
~ на тренажере simulator training
~, одиночная individual training
~, профессиональная vocational training
~ самолета, предполетная departure check-out procedure
~, строевая marching drill
~, химическая chemical warfare training
~, штурмансная navigational training
подготовленность degree of training/preparation
подгруздь *bot.* scrobiculate mushroom *(Lactarius scrobiculatus)*
подгруппа subgroup
подгумусовый underhumus *(attr.)*
поддерево subtree
поддержание maintenance, upkeep, keeping up; servicing
поддерживать 1. maintain, keep up; sustain; 2. support, bear, carry, shore up, prop up
поддержка support, backing; prop, stay, rest
~, артиллерийская gun-firing support
~, боевая combat support
~, огневая fire support/cover
поддиапазон sub-range, sub-band
поддиафрагмальный subdiaphragmatic, subphrenic
поддоменник *met.* bottom furnace/house
поддомкрачивать jack up
поддон bottom tray/plate; pallet; drain pan; oil pan/sump
~, грузовой cargo/freight pallet
~ двигателя engine sump
~ картера oil pan
~, сетчатый wiremesh pallet
~ талой воды *(in refrigerator)* drip tray
~, ящичный box pallet
поддон-отстойник sump
поддувало ashpit; wind box
поддьяк *eccl.* subdeacon
поденный by the day, daily
поденщик, поденный day/casual labo(u)rer
поджатие contraction; compression,

precompression
~ воздуха air precompression/ramming
поджелудочный *anat.* pancreatic
поджигание firing, ignition
поджигатель 1. igniter; 2. incendiary, arsonist
~ войны warmonger
подзарядка *el.* additional charging; boost charge
подзаконные акты *law* subordinate legislation; by-laws
подземный underground *(attr.),* subterranean, subsurface; *myth.* chthonic
подзимний early-winter, before winter
подзол 1. podzol, podsol, ashen-gray soil; 2. tannery waste
подзолистый podzolic
подзона subzone
подзор *naut.* counter
~ кормы/кормовой stern counter/overhang
подкаменщик *zool.* miller's thumb *(Cottus gobio)*
подканал subchannel
подкасательная *mat.* subtangent
подкат *met.* semi-finished rolled products/stock
подкачка 1. pumping; 2. swap-in
подкисление acidulation, acidification, acidifying
подкислять acidify; acidulate
подкладка pad, lining, cushion; backing
~, рельсовая rail chair/sole; tie plate
~, стоечная footplate
подкладная шайба packing washer
подклинивание wedging up, ramming up
подклинивать wedge up
подключать link up, connect
подключично-позвоночный *anat.* vertebrosubclavian
подключичный subclavian, subclavicular
подковник *bot.* horseshoe vetch *(Hippocrepis)*
подковообразный, подкововидный U-shaped, horseshoe-shaped, hippocrepiform; *bot.* lecotropal
подковонос *zool.* horseshoe bat *(Rhinolophus)*
подкожица hypoderma; hypodermis
подкожный subcutaneous, hypodermic
~ слой hypodermis
~ укол hypodermic injection
подколенный popliteal
подкоп tunnel, underground passage
подкопытник *zool.* jack snipe *(Lymocryptes minima)*
подкоренной 1. *math.* subradical; 2. under the roots
~ная величина radicand
подкорка *anat.* subcortex
подкорковый subcortical
подкорм extra/additional feed
подкормка *agr.* 1. feeding; fattening; 2. additional fertilization, dressing
~, жидкая liquid/spray dressing
~, листовая foliar dressing/application
~, междурядная side-dressing
~, поверхностная top dressing
~, рядковая row top-dressing
подкормщик (fertilizer) applicator, top dresser
~, аммиачный nitro-shooter
подкоровый subcortical
подкорье sapwood, alburnum
подкос brace, bracing strut, prop, stay
~, диагональный/задний drag strut/brace
~, шарнирный articulated strut
~, поперечный transversal strut; jib stay
подкосный strut-braced
~ мост strut-bridge
подкрепление reinforcement, strengthening; stiffening
подкреплять 1. reinforce, strengthen, fortify; stiffen; 2. support, shore up
подкритический subcritical, below-critical
подкрыльцовый axillary
подкуривать fumigate, smoke
подладанник *bot.* cytinus *(Cytinus)*
подлегарс *naut.* beam stock, riser; rising, wearing
подлежащее *gram.* subject
подлесник *bot.* sanicle *(Sanicula)*
подлет *av.* hop
подлещик *zool.* white/silver bream *(Gustera blicca bjoerkna)*
подливка grout, mortar; grouting
подлинник original
подлинность authenticity, genuineness, originality
подложечный *anat.* epigastric
подложка base, backing, foundation; substrate, substratum, sublayer; *comp.* wafer, substrate
~ интегральной схемы integrated circuit substrate
~ катализатора catalyst carrier
~ печатной схемы printed-circuit substrate
подлокотник arm/elbow rest
подлопаточный *anat.* subscapular
подмагничивание magnetic bias(ing)
подмагничивать magnetize, bias
подмандатный *polit.* mandated, under mandate
подмаренник *bot.* bedstraw *(Galium)*
подмассив *comp.* subarray
подматрица *comp.* submatrix
подмногообразие *math.* submanifold, subvariety
подмножество *math.* subset; *stat.* subpopulation
~, собственное proper subset
~ языка sublanguage
подмодель master pattern
подмодуль *math.* submodule
подмодульный submodular
подмодулятор driver circuit/device, driver
подмолочник *bot.* orange-brown lactarius *(Lactarius volemus)*
подмосковный situated near Moscow
Подмосковье districts/area near Moscow
подмости scaffold(ing), trestle, falsework
~, козловые trestle/framing/horse scaffold
~, передвижные/катучие rolling platform
~, подвесные suspending/cradle scaffold
подмостки scaffold(ing), trestle; *theatre* stage,

boards
подмошник *bot.* subtomentose mushroom *(Boletus subtomentosus)*
подмыв washing away, underwashing; erosion
подмышечный axillary
подмышка axilla, armpit, maschale, axillary space
поднадвиг *geol.* underthrust
поднаем subtenancy, sublease
поднаковальня anvil bed/block
подналадка corrective adjustment
поднаниматель subtenant, underlessee, undertenant
поднесущая *rad.* subcarrier
поднимать lift, raise, hoist; pick up
~ целину open up [reclaim] virgin lands
подниматься rise, ascend; climb; come to the surface
подногтевой hyponychial, subungual
подножка 1. footboard, footstep, step plate; 2. *sport* backheel, trip
поднормаль *math.* subnormal
поднятие raising, lifting; rise, ascent, elevation
~ карты *top.* overprinting
~ перископа periscope exposure
подобабок *bot.* rough-stemmed boletus *(Boletus scaber)*
подобие similarity, similitude, likeness
~, локальное/неполное limited similitude
~, полное complete similitude/similarity
подобласть subregion, subdomain
подобно-изоморфный order-isomorphic
подобно-однолистный *math.* schlichtartig
подоболочка subshell
подобромидроз *med.* podobromidrosis
подограф *med.* podograph
подогрев heating(-up), warming(-up); preheating
подогреватель preheater, heater, heating unit
~ дутья hot blast stove, Cowper stove
~, змеевиковый coil heater
~, огневой combustion heater
~, форсуночный injector-type heater
подоконник window sill
подопечный under wardship/guardianship
~ная территория trust territory
подорешник *bot.* 1. orange-brown lactarius *(Lactarius volemus);* 2. prickly cup, changeable mushroom *(Pholiota mutabilis)*
подорлик *zool.* spotted eagle *(Aquila)*
подорожник *bot.* plantain *(Plantago)*
подосиновик *bot.* orange cup boleuts, aspen mushroom *(Boletus rufus)*
подостемон *bot.* riverweed *(Podostemon)*
подотношение *philos.* subrelation
подотрасль subsector, subbranch; subindustry
подотряд *biol.* suborder
подотчет *fin.* sum paid on account
подотчетность *fin.* accountability
подотчетный accountable
подофилл *bot.* May apple *(Podophyllum)*
подофиллин podophyllin
подоходный *fin.* income-related, according to income
~ налог income tax
подошва foot; sole; base, floor, bottom; *(of wave)* trough, hollow
~ ахтерштевня aftfoot
~ башмака shoe sole
~ дока dock bottom/floor
~ откоса foot of a slope
~ плотины foot of a dam
~ припая ice foot
~ рельса rail base/foot
~ уступа bench bottom
подошвенный *anat.* plantar, pelmatic
подпалубный underdeck
подпаутинный *anat.* subarachnoid
подпергамент imitation parchment
подпестичник *bot.* podogynium
подпестичный *bot.* hypogynous, hypopetalous
подпеченочный *anat.* subhepatic
подписавший signatory
подписка 1. subscription; 2. written undertaking, signed commitment/promise
~, аварийная average bond
подписной subscription *(attr.)*
подпись 1. signature; signing; 2. inscription, caption; legend
~ крестом mark signature
~ по доверенности signature by procuration
~, собственноручная sign manual
подпитка replenishment, replenishing, make-up
подпласток substratum
подплевральный *anat.* subpleurall
подповерхностный subsurface *(attr.)*
подподпрограмма subsubroutine, subsubprogram(me)
подполе *math.* subfield
подпор positive/hydrostatic head, backwater; *av.* overpressure
подпора support, prop, strut, stay, rest, brace; *naut.* poppet, shore
~, днищевая bottom shore
~, наклонная diagonal shore
~, связочная ligament support
~, скуловая bilge shore
подпороговый subliminal
подпорядок suborder
подпоследовательность subsequence
подпочва subsoil; substratum
подпочвенный subsoil *(attr.),* subterranean, underground, subsurface *(attr.)*
подпочечный *anat.* infrarenal
подпрапорщик *hist.* ensign-bearer
подпрессовка prepressing, premoulding
подпрограмма *comp.* subroutine, subprogram(me)
~, аппаратно реализованная hardware subroutine
~, библиотечная library subroutine
~ деления division subroutine
~, закрытая/замкнутая closed subroutine

~, перемещаемая relocatable subroutine
~ печати print subroutine
~, повторно используемая reenterable subroutine
~ пользователя user-supplied subroutine
~, системная system subprogram/subroutine
~, тестовая test subprogram/subroutine
подпрограмма-функция fuction-subprogram
подпроизведение *math.* subproduct
подпространство *math.* subspace
~, векторное/линейное vector subspace
~, дефектное defect subspace
подпротектор *auto.* undertread
подпруга bellyband, saddlegirth
подпруда weir, dam
подпружиненный spring-actuated/loaded/opposed
подпункт subitem, subparagraph
подпупочный *anat.* infraumbilical
подпятник heel socket; thrust/axial bearing
~ качения rolling thrust bearing
~ руля *naut.* rudder carrier
~ скольжения sliding thrust bearing
~ ствола spigot socket
подразбиение *math.* subdivision, subpartition
подраздел subsection, subdivision; subheading
подразделение 1. subdivision, subsection; *mil.* (small) unit, subunit, element, detachment; 2. *math.* refinement
~, авиационное air unit
~, воздушно-десантное airborne unit
~, дежурное alert unit
~, рабочее working party
~ технического обслуживания maintenance unit
~, учебное training unit
~, хозяйственное administrative unit
подразделять subdivide, partition; classify, class
подразумеваемый *law* tacit
подрамник, подрамок underframe, subframe
подрастающее поколение rising/oncoming generation
подреберный *anat.* subcostal
подреберье *anat.* hypochondrium, hypochondriac region
подрегулирование, подрегулировка readjustment, fine adjustment
подрез undercut
подрезка, подрезание facing, trimming, cutting
подрессоривание *tech.* cushioning, springing
подрешетка sublattice
подробность detail; feature
~ти, технические technicalities
подрод subgenus
подродовой infrageneric
подрост undergrowth
подрубрика subheading, subtitle
подруливать *av.* taxi up
подрыв, подрывание blasting, blowing-up; demolition, destruction
подрывать 1. undermine, dig under, dig from beneath; 2. blast, blow up
подрывной 1. undermining; *fig.* subversive; 2. blasting, exploding, demolition *(attr.)*
подрывник *min.* blaster, shotman, shotfirer
подряд I subseries
подряд II contract
подрядчик contractor
~, главный/генеральный prime contractor
~, вторичный associate contractor; subcontractor
подсасывание suction
подсборка subassembly
подсвекольник *bot.* beetroot, redroot, pigweed *(Amaranthus retroflexus)*
подсвет shielded/back/side/bias illumination
подсветка illumimation, lighting; brightening
подсев *agr.* undersowing
подсед 1. undergrowth; 2. fine hair, scud, undercoat; 3. *vet.* malanders
подсека clearing, land cleared of trees/wood
подсекатель curring jaw
подсемейство subfamily
подсеть subnetwork
подсечка *min.* undercutting; *sport* undercut
подсингония *geol.* subsystem
подсистема subsystem
подслаивание undercoating, subcoating
подследственный *law* (person) under investigation
подслеповатость partial blindness, cecutiency
подслизистый submucous
подслой sublayer; precoat
подслойка *bot.* subiculum
подслучай subcase
подслушивание listening, monitoring; interception; *teleph.* wire-tapping
подслушивать monitor, intercept; *teleph.* wiretap
подснежник *bot.* snowdrop *(Galanthus)*
подснежный subniveal
подсовокупность *math.* subset
подсознание, подсознательность subconsciousness
подсознательный subconscious
подсолнечник *bot.* sunflower *(Helianthus)*
подсолнечный subsolar; in the sun
подсос suction, inflow, ingress, inleakage
подсостояние sub-state
подсписок sublist
подсправочник subdirectory
подставимость *math.* substitutability
подставка rest, support, prop, stay, stand, brace, strut, shore; *mus.* bridge *(of stringed instruments)*
~, кормовая skeg shore
~, крестцовая *med.* sacral support
подстановка 1. substitution, replacement *(also math.)*; 2. *math.* permutation
~, нечетная odd permutation
~ с переменой знака exchange with sign reversal
~, четная even permutation
~ Эйлера Euler's substitution

подстановление substitution
подстанция substation
  ~, преобразовательная converter substation
  ~, телефонная sub-exchange; suboffice
  ~, трансформаторная transformer substation
подстилающий underlying, subjacent
  ~ слой underlayer
подстраивать 1. tune, adjust; 2. build on
подстроение substructure
подстройка fine adjustment/tuning
  ~ частоты frequency trim/adjustment/control
подстрока substring
подстрочник interlinear [word-for-word] translation
подструктура substructure
подступенок riser
подсудимый *law* (the) accused, indictee, defendant
подсудность 1. jurisdiction, competence; 2. liability for prosecution
подсудный 1. under the jurisdiction, within the competence; 2. triable, liable for prosecution
подсудок judge's assistant/deputy; court clerk
подсухожильный *anat.* subtendinous
подсушивание, подсушка (pre)drying, dehumidification
подсушивать dehumidify, dry
подсхема *el.* subcircuit
  ~ базы данных subschema
подсчет count(ing), calculation, computation
  ~ карт card coun(ting)
  ~ зон extent count(ing)
подсчитывать compute, count up, calculate
подсыпка 1. pouring in; heaping up; 2. *met.* cushion fill
подтвердительный confirmatory, confirming; corroborative
подтверждать acknowledge; confirm; corroborate
подтверждение acknowledgement, confirmation; corroboration; affirmation
  ~ ввода *comp.* entry approval
  ~ подлинности authentication
  ~ прав доступа *comp.* certification
  ~ правильности validation
  ~ приема сообщения acknowledgement
подтекст implied/covert meaning, implication
подтепловый subthermal
подтоварник *naut.* dunnage
подток seepage; inflow
подтональный subsonic, infrasonic, infra-acoustic, subaudio *(attr.)*
подтопление flooding, underflood
подточка 1. recess; groove; 2. sharpening
подтравливание etching
подтропики subtropics
подтягивать 1. tighten (up); 2. pull up, haul up
подуровень sublevel; sublayer
подуст *zool.* podust *(Chondrostoma nasus)*
подушка cushion, pillow; pad; mat; bed; shoe
  ~, балластная *rail.* ballast bed
  ~, бетонная concrete raft
  ~, виброизолирующая antivibration pad
  ~, воздушная air cushion
  ~ заголовника headrest pad
  ~ матрицы die bed
  ~, паровая steam blanket
  ~, породная rock cushion
  ~, рельсовая rail chair
  ~ сидения seat pad/cushion
  ~, стрелочная *rail.* switch chair
  ~, упорная thrust block/pad
  ~, якорная anchor shoe/bed
подфарник side lamp/light; fender/parking light
подфасциальный subfascial
подформула *math.* corollary formula
подфункция *math.* minorant (function)
подфюзеляжный *aero.* underfuselage, underbelly, ventral
подхват pick-up
подхватывать pick up
подход approach; access
  ~, вероятностный probabilistic approach
  ~, системный systems approach; system concept
подходка *leath.* fleshing/paring/scraping knife
подцарство *biol.* subkingdom; phylum
подцвет complimentary colo(u)r
подцветковый, подцветочный *bot.* subfloral
подцветник *bot.* hypanthium
подцепочка subchain
подцикл subcycle
подчашие *bot.* calycle, epicalyx
подчеканивать caulk
подчелюстной *anat.* submaxillary
подчинение subordination; *gram.* hypotaxis
подчинительно-присоединительный *gram.* subordinating-connective
подчинительно-сочинительный *gram.* subordinating-coordinating
подчистка 1. erasure; cleaning; 2. fraudulent alteration
подчитчик copyholder
подчитывать read a copy *(to a proofreader)*
подчревный eanat. hypogastric
подчревье *anat.* hypogastrium
подшабривать scrape up
подшивка hemming; boarding; lining
подшипник bearing
  ~, агатовый agate bearing
  ~, баббитовый babbit(t)-lined bearing
  ~, бакаутовый pockwood bearing
  ~ без смазки dry bearing
  ~, водонепроницаемый water-sealed bearing
  ~, втулочный sleeve bearing
  ~, головной head bearing
  ~, глухой solid/bush bearing
  ~, гребенчатый collar bearing
  ~, двойной double-thrust bearing
  ~, двухрядный double-row bearing
  ~, дейдвудный *naut.* stern (tube) bearing
  ~, задний rear/aft bearing

~, игольчатый needle bearing
~, качающийся pendulum bearing
~ качения rolling-contact [roller, antifriction] bearing
~ коленчатого вала crankshaft bearing
~, конический роликовый tapered/conical roller bearing
~, консольный bracket bearing
~, коренной main/crankshaft bearing
~, направляющий pilot/guide bearing
~, неразъемный solid/sleeve bearing
~, однорядный single-row bearing
~ одностороннего действия [одинарный] singlethrust bearing
~, опорный carrier/journal/block bearing
~, плавающий floating bearing
~, радиально-упорный radial thrust bearing
~, радиальный radial bearing
~, разборный separable bearing
~, разъемный split bearing
~ распределительного вала camshaft bearing
~, регулируемый adjustable bearing
~, роликовый roller bearing
~, роликовый конический taper/conical roller bearing
~ с баббитовой заливкой babbit(t)-lined bearing
~ с бочкообразными роликами barrel-shaped roller bearing
~ с принудительной смазкой forced-lubrication bearing
~, самосмазывающийся self-lubricating/oiling bearing
~, самоустанавливающийся self-aligning bearing
~, сдвоенный twin bearing
~, сегментный segmenal shoe/bearing
~ скольжения sliding/slipper/plain bearing
~, упорный thrust bearing
~, фланцевый flange bearing
~, шариковый ball-bearing
~, шатунный crankpin bearing
подщелачивать alkalify
подъезд 1. entrance, doorway; 2. approach, access
подъельник *bot.* Indian pipe *(Monotropa)*
подъем 1. rise, rising, ascent; elevation; lifting, hoisting, raising; 2. grade, gradient, slope, acclivity; 3. *fig.* development
~ аэростата balloon ascent
~ клапана valve lift/stroke/travel
~ кровли pitch of the roof
подъемник elevator, hoist, lift; ground jack
~, автомобильный auto lift, lifting platform
~, барабанный elevating drum
~, башенный tower hoist
~, винтовой screw jack; screw hoist
~, вильчатый fork lift
~, грузовой freight hoist
~, консольный gib hoist
~, кресельный chair lift
~, пневматический air hoist
~, рычажный scissor lift
~, шихтовый charging hoist
подъемно-транспортный hoisting-and-transport
подъемный lifting, hoisting, raising;
~ мост lift bridge
подъязычно-глоточный hyopharyngeal
подъязычно-нижнечелюстной hyomandibular
подъязычно-язычный hyoglossal
подъязычный hypoglossal, sublingual
подъяремный subjugular
подыгра *game th.* subgame
подьячий *hist.* scribe, scrivener, clerk
подэлемент subelement
подэпидермальный subepidermal
подэпителиальный subepithelial
поезд train
~, автомобильный road train, truck-trailer combination, truck and trailer(s)
~, воздушный aerial/air train
~, грузовой freight train
~, курьерский express train
~, пассажирский passenger train
~, планерный *av.* glider train
~, пожарный firefighting train
~, почтовый mail train
~, пригородный commuter/local train
~, санитарный ambulance train
~, скорый fast train
~ согласованный с расписанием пароходов boat train
~, строительно-монтажный construction train
~, товаро-пассажирский mixed train
~, транзитный through train
поездограф railway traffic recorder
пожар fire; conflagration
~ лесной forest fire
~, шахтный mine/underground fire
~, локализовать bring the fire under control, contain a fire
пожарно-известительный fire-alarm
пожарный fire *(attr.)*
~ кран fire hydrant
~ насос fire engine/pump
~ная команда fire brigade
пожаробезопасность fire safety
пожароопасность fire hazard/risk, danger of fire
пожароопасный fire-hazardous/dangerous
пожаротушение firefighting, fire suppression
пожелтение yellowing
позадиглоточный postpharyngeal
позадипищеводный retroesophageal
позадичелюстной retromandibular
позвонок *anat.* vertebra
~, второй шейный second cervical vertebra toothed vertebra
~, выступающий seventh cervical vertebra
~, грудной thoracic vertebra
~, копчиковый coccygeal vertebra
~, крестцовый sacral vertebra

~, первый шейный cranial vertebra
~, поясничный lumbar vertebra
~, самый нижний поясничный basilar vertebra
~, шейный cervical vertebra
позвоночник spinal/vertebral columm, spine
позвоночно-грудинный vertebrosternal, sterno-verterbral
позвоночно-крестцовый vertebrosacral
позвоночно-подвздошный vertebro-iliac
пазвоночно-реберный vertebrocostal, costo-vertebral
позвоночно-хрящевой vertebrochondral
позвоночные *pl. zool.* vertebrates *(Vertebrata)*
позвоночный *anat.* vertebral; spinal
~ столб spinal column, spine
позднеледниковый late-glacial
позднетретичный Late Tertiary
позднецветущий late-flowering/blooming, tardiflorous
поздника *bot.* black nightshade *(Solanum nigrum)*
поземельный land *(attr.)*
~ налог land tax
позитив, позитивный positive
позитивизм *philos.* positivism
позитрон *phys.* positron
позитроний positronium
позиционер positioner
позиционирование positioning
позиционный position *(attr.),* positional
~ная война trench warfare
позиция 1. position; 2. item
~ в таблице адресов directory position
~ двоичного разряда bit position
~, исходная waiting position/station
~, ключевая key position/point
~, контрольная check position
~, перфорационная *comp.* punching position
~, огневая fire/firing position
~, стартовая launching position
~, старшая *comp.* high-order position
познаваемость cognizability, cognoscibility, knowability
познаваемый cognizable, cognoscible, knowable
познавательный cognitive, gnostic; epistemic
познавать *philos.* cognize
познание *philos.* cognition; noesis
позолота gilding, gilt; gold-plating
позыв 1. *law, obs.* summons; 2. urge, call
~ на рвоту vomituration, retching
позывной call signal/sign/letter; distinguishing signal
~ для радиосвязи radio call sign
~ корабля ship's number
~ летательного аппарата aircraft call sign
~, микрофонный voice call sign
~ торгового судна mercantile call sign
~ шлюпки, флажный boat-distinguishing flag
позывные *pl.* signal/call letters; pennants; number
~ корабля ship's number/pennant
~ флажным сигналом name-number pennants
поилка drinking/water trough, waterbowl
поименное голосование vote by the roll call
поимущественный property *(attr.)*
поиск search(ing), seeking; scanning, sweep; hunting; *min.* prospecting, exploration
~, ассоциативный associative search
~, быстрый *comp.* fast-access retrieval
~ веером radial search
~, гидролокационный investigation by echo
~ информации data research; information retrieval
~ мин mine hunting
~, надводный surface search
~ неисправностей trouble shooting/tracing,
~ по спирали spiral search
~ по таблице table lookup
~, подводный underwater search
~ с возвратом backtracking
~, случайный random search
~ цифровой информации retrieval of digital information
поисково-спасательный search-and-rescue
поисковый searching; prospecting
~ корабль scout ship
пойкилобласт *physiol.* poikiloblast
пойкилодермия *physiol.* poikiloderma
~ Сиватта Civatte's disease
пойкилотермия *physiol.* poikilothermism
пойкилотромбоцит poikilothrombocyte
пойкилоцитоз poikilocytosis, poikilocythemia
пойма flood plain/land; alluvial flat
~, илистая mud flat
показ display, show, presentation, demonstration
~, предварительный preview
показание 1. indication; reading, readout; 2. testimony, deposition, statement
~ данное под присягой evidence on oath
~, письменное *law* deposition
~ прибора instrument reading
~ с чужих слов *law* hearsay evidence; second-hand evidence
~, свидетельское testimonial evidence
~ свидетеля защиты testimony for the defence
~ свидетеля обвинения testimony for the prosecution
~ шкалы dial reading
показатель index; indicator; pointer, marker; *math.* exponent; index number; coefficient, quotient; measure; rate
~ адаптивности adaptive value/index
~ адиабаты adiabatic exponent, ratio of specific heats
~, белковый protein quotient
~, вероятностный probability rate
~, водородный *chem.* pH value
~ времени ожидания *econ.* waiting measure
~ времени простоя *econ.* down-time index
~, годовый annual rate
~ добротности Q-factor, figure of merit

~ достоверности confidence index
~, дробный fractional exponent
~ жесткости rigidity index
~ за год, наинизший year's low
~ затухания damping coefficient
~ износостойкости wear index
~, ионный ionic index
~ испаряемости evaporative rate
~ качества quality index
~ качества продукции product-quality index
~ качества работ performance index
~, количественный quantitative index
~ корня *math.* index of a root/radical
~ коррозии corrosion index
~ крови, цветной *med.* colo(u)r index, cell-colo(u)r ratio, blood quotient
~, максимизируемый *econ.* maximand
~, минимизируемый *econ.* minimand
~ направленности directivity index
~, натуральный physical indicator
~, невероятностный nonprobabilistic measure
~ недостаточности deficiency indicator
~ нетрудоспособности incapacity rate
~ обученности achievement/educational quotient
~ питательности nutrient index
~ плана plan indicator/target
~ плодовитости fertility measure
~ поглощения absorption coefficient
~ полезности utility indicator
~ политропы polytropic exponent/index
~ преломления refraction/refractive index
~ прибыльности profit performance
~ приоритета measure of priority
~ продуктивности/производительности performance data
~ рассеяния scattering coefficient
~ расширения expansion number
~ рождаемости birth rate
~, росто-весовой *med.* statural-weight value
~ смертности death/mortality/lethality rate
~, совокупный aggregative index
~ степени *math.* exponent; power; exponent/index of a power
~ степени, дробный fractional exponent
~ твердости hardness number
~ текучести flow index
~, технико-экономический technical-and-economic index
~ технологичности processability index
~ умственных способностей intelligence quotient, IQ
~ ухудшения качества demerit index
~, фагоцитарный *med.* phagocytic index
показательный 1. demonstrative, demonstration *(attr.),* model *(attr.);* 2. *math.* exponential
~ суд show trial
показывать indicate, show, read
покатый sloping, slanting
покачивание rocking, swinging
покачивать rock/swing slightly
покашливание tussiculation, hacking cough
покер poker
покаяние *eccl.* penance
покидание abandonment, escape
~, аварийное emergency escape
~ катапультированием ejection escape
~ с парашютом bail-out
~ с помощью катапультного кресла ejection-seat escape
поковка 1. forging; 2. forged piece, forging
покой 1. *phys.* rest; 2. *bot.* dormancy
~, приемный *med.* hospital/casualty/accident ward
~ семян seed dormancy
поколение generation
~, вымершее extinct generation
~, дочернее filial generation
~, подрастающее oncoming/rising generation
~ ядер generation of nuclei
покров cover, blanket, mantle, coat; shroud; canopy; *anat. zool.* integument; tunic; tegumentum; *zool.* vestiture
~, внутренний endothecium
~, кожный integument
~, напочвенный soil cover/mantle
~, облачный cloud cover/canopy
~, плодный f(o)etal integument
~, растительный vegetation cover
~, тектонический overthrust sheet
~ цветка floral envelope, perianth
Покров *eccl.* feast of the Intercession of the Holy Virgin
покровный covering, cover *(attr.); med.* integumentary; tegumental
покрытие cover(ing); coating, facing, lining; *met.* plating; *astr.* occultation
~, анодное anodic/anode coating
~, антикоррозионное corrosion-resistant coating
~, водостойкое/водонепроницаемое watertight [water-resistant] coating
~, гальваническое electroplating
~, горячее hot-dip metal coat(ing)
~, дерновое sod matting
~ дороги/дорожное road covering/surfacing/pavement
~, защитное protective coat(ing)
~, испаряющееся evaporative coating
~, катодное cathodic coat(ing)
~ крыши roofing
~ лаком/лаковое lacquer coating; varnishing
~, маскировочное camouflage coat/cover
~, медное cooper plating/coat(ing)
~, металлокерамическое cermet coating
~, нескользящее non-slip/skid coating
~, огнестойкое fire-resistant coating
~, огнеупорное fireproof/refractory coating
~, отражающее reflecting/reflective coating
~, пенное foam blanket
~ погружением dip coat(ing)

~, противогидролокационное antiasdic/antisonar coating
~, противорадиолокационное antiradar [radarabsorbent] coating
~, противоэрозионное erosion-resistant coating
~, светящееся fluorescenct coating
~, стекловидное viterous coat(ing)
~, теплозащитное heat-/thermal-protective coating
~, теплоотражающее heat-reflective coating
~, теплостойкое heat-resistant coating
~, теплопроводящее conductive coating
~, тугоплавкое refractory coating
покрытоголовый *paleont.* stegocephalous, stegocephalian
покрытоплодный *bot.* angiocarpous
покрытоpотые *pl. zool.* ectoproctans *(Ectoprocta)*
покрытосемянные *pl. bot.* angiosperms *(Angiospermae)*
покрытосемянный *bot.* angiospermous
покрышка cover; mantle; casing; lid; *geol.* cap rock; *anat.* operculum, tegmen; *auto.* tyre cover/casing; outer tyre
~, автомобильная tyre casing, autotyre
~, безбортовая beadless tyre
~, височная *anat.* temporal operculum
~, грузовая truck tyre
~ для зимней эксплуатации winter/snow tyre
~, клинчерная clincher [beaded edge] tyre
~, легковая car/small tyre
~, лобная *anat.* frontal operculum
~, лобно-теменная frontoparietal operculum
~, многослойная multiple cord tyre
~, низкопрофильная oval cover/tyre
~ полуплоской/плоской конфекции flat-/drumbuilt tyre
~, ремонтированная retreaded tyre
~, скоростная high-speed tyre
~, усиленная reinforced [heavy-duty] tyre
покупатель *law* vendee
покупательная сила/способность purchasing power/capacity
покупать buy, purchase
~, на срок [с будущей доставкой] buy ahead
покупка purchase; buying, purchasing
~ в кредит purchase on credit
~ в рассрочку hire purchase
~ за наличные spot buying; cash buying
~ на срок forward purchase
~, окончательная outright purchase
~ ранее других preemption
~ с целью поддерживания курса supporting purchase
~ ценных бумаг security purchase
покупная способность purchasing power
пол I sex
~, гетерогаметный heterogametic/heterozygous/diagametic sex
~, гомогаметный homogametic/homozygous/monogametic sex
пол II floor, flooring
~, мозаичный inlaid floor
~, паркетный parquet floor, parquetry
~, черный subfloor, rough floor
~, шпунтовый tongue floor
полагать *math.* assume, suppose, let
полветра *naut.* wind abeam/abreast
полдень noon, midday
~, звездный sideral noon
поле field, area; space; region, zone; array; margin, edge
~ адреса *comp.* address field/array/panel/range
~, бегущее *el.* travelling field
~, безвихревое irrotational/noncircular field
~, векторное *math.* vector field
~, вихревое vortex/vorticity/eddy field; *math.* rotational field
~, внезародышевое *anat.* extraembrionic area
~, возбуждения *el.* exciting field
~, вращающееся rotating/rotary/revolving field
~ Галуа *math.* Galois field
~ гамма-излучения gamma field
~, гнездовое *teleph.* jack/field/bank
~, гравитационное field of gravity, gravitational field
~ данных *comp.* data array
~ для считывания *comp.* zone to be read
~ допуска *mech.* tolerance zone, margin tolerance
~, звуковое acoustic/sound field
~ Земли, магнитное Earth's magnetic field, geomagnetic field
~ зрения field of view/vision, visual field
~ излучения radiation field
~ индикации viewport
~ команд *coomp.* instruction array, command field
~, конвекционное convective field
~, магнитное, земное geomagnetic [terrestial magnetic] field
~, наборное *comp.* patchboard, plugboard, patch(ing) panel
~, невихревое irrotational field, non-circuital field
~, нейтринное neutrino field
~, непрерывное continuous field
~, нестационарное non-steady field
~ облучения exposure/irradiation/radiation field
~, обонятельное olfactory area
~ операции *comp.* operation area
~, остаточное residual/remanent field
~, очищающее clearing field
~, постоянное permanent/constant field
~, потенциальное векторное conservative/potential vector field
~ преддверия *anat.* vestibular area
~ признака *comp.* flag field
~, рабочее work area
~ разложения многочлена *math.* root field of

a polynomial
~, размагничивающее demagnetizing/degaussing field
~ рассеяния scattering/stray/fringing field
~, самосогласованное self-consistent/-congruent field
~, сенсорное sensorial area
~, силовое field of forces, force field
~ силы тяжести gravity field
~, слуховое sensorial area
~ смещения basic field
~, статическое static/stationary/constant field
~ страницы margin
~, тормозящее retarding/retardation/ brake/deceleration field
~ тяготения field of gravity, gravitational field
~, удерживающее confinig field
~ улитки cochlear area
~, эмбриональное embrionic area
~ ячеек для данных *comp.* data cell array
полевица *bot.* bent grass *(Agrostis)*
полевичка *bot.* love grass *(Eragrostis)*
полевка *zool.* vole *(Microtos; Clethrionomys)*
полевод crop-grower, agriculturist
полеводство crop-growing/cultivation, field husbandry/cropping
полегание *agr.* lodging, beating down
полезащитная лесная полоса forest shelter belt
полезность usefulness, utitlity; practicality
~, выборочная sampling utitlity
~ для потребителя consumer utitlity
~, количественная *econ.* cardinal utility
~, максимаксная maximax utility
~, максиминная maximin utility
~, приростная incremental utility
~, субъективная personalistic utility
полезный useful; helpful; beneficail
~ная нагрузка *tech.* payload
~ное действие *mech.* efficency
~ные ископаемые mineral wealth; useful minerals
полей *bot.* pennyroyal mint *(Mentha pulegium)*
поленика *bot.* acrtic bramble *(Rubus arcticus)*
полет flight; flying
~, активный powered/active flight
~, аэродромный local flight
~, безмоторный unpowered/powerless flight
~, беспилотный unmanned/pilotless flight
~, боевой combat flight
~ боком sideward/sideways flight *(helicopter)*
~, бреющий hedge-hopping [tree-top, low-level] flight
~ в болтанку bumpy flight
~ в восходящих тепловых потоках thermalling
~ в паре twin flight
~ в строю formation flight
~, визуальный visual/contact flight
~, внерейсовый nonscheduled flight
~, внетрассовый off-airway flight
~, внутренний domestic flight
~, высотный high-altitude flight
~, групповой formation flight
~, дневной day(light) flight
~, дозвуковой subsonic flight
~, зафрахтованный charter flight
~, земной terrestial flight
~, испытательный test flight
~, исследовательский exploratory/research flight
~, космический space flight
~, кругосветный round-the-word flight
~, маршрутный en-route flight
~, моделируемый simulated flight
~, моторный power(ed) flight
~ на аэростате baloon flight
~ на автопилоте auto-controlled flight
~ на больших высотах high-altitude flight
~ на большую дальность long distance flight
~ на буксире towed flight
~ на буксировку towing flight
~ на дальность distance [long-range] flight
~ на планере glider flight
~ на привязи tethered flight
~ на продолжительность endurance flight
~ на разведку reconnaissance flight
~ на реактивном самолете jet flight
~ на режиме висения hover(ing) flight
~ на тренажере simulated flight
~ над водным пространством over-water flight
~ над сушей overland flight
~, непилотируемый unmanned flight
~, нерегулярный unscheduled flight
~, околозвуковой transonic flight
~, орнитоптерный flapping flight
~, парящий soaring
~, перевернутый inverted [upside-down] flight
~, пилотируемый manned/piloted flight
~, планирующий gliding flight
~ по инерции coasting (flight)
~ по кругу circling flight
~ по локсодромии rhumb-line flight
~ по маршруту en-route flight
~ по орбите orbital flight
~ по приборам instrument flight, flight on instruments
~ по прямолинейному маршруту point-to-point flight
~ по радиолучу radio-beam flight
~, показной demonstration flight
~, реактивный jet flight
~, рейсовый scheduled flight
~ с выключенным двигателем engine-off [unpowered] flight
~ с дозаправкой топливом в воздухе refuelling flight
~ с инструктором dual flight
~ с креном banked flight
~ с набором высоты climbing/upward flight

~ с неработающим двигателем engine-off flight
~ с первой космической скоростью earth-orbital velocity flight
~ с работающим двигателем powered/propelled flight, engine-on flight
~ с разгоном accelerated/ boosted flight
~ с третьей космической скоростью solar escape velocity flight
~ с экипажем manned flight
~, сбалансированный trimmed flight
~, сверхзвуковой supersonic flight
~, скоростной high-speed flight
~, слепой blind flight; instrument flight
~ со звуковой скоростью sonic flight
~ со скольжением yawed/sliding flight
~ со снижением desscending/downard flight
~ со сносом drift flight
~ строем formation flight
~, трассовый airway flight
~, тренировочный training/practice flight
~ туда и обратно round-trip flight, out-and-return flight
~, фигурный aerobatic/acrobatic flight
~ хвостом вперед reverse/readward flight
полетопригодность airworthiness
полетопригодный airworthy
ползун slide, runner, slide block, crosshead, slider, slipper; sliding bearing
ползучесть creep, creepage
~, установившаяся steady creep
~, сдвиговая creep in shear
ползушка slider; cursor
полиандрический polyandrian
полиандрия polyandry
полиантес *bot.* tuberose *(Polyanthes tuberosa)*
полиантокарп(ий) *bot.* polyanthocarp
полиартрит *med.* polyarthritis
~, острый ревматический acute inflammatory rheumatism
~, туберкулезный tuberculous polyarthritis, pulmonary osteoarthropathy
~ эволютивный evolution polyarthritis
палиархия *polit.* polyarchy
палиархность *bot.* polyarchy
полив irrigation, watering
~ затоплением flooding irrigation
~ по полосам gravity irrigation
полива glaze, glazing, enamel
поливакцина polyvalent/multivalent vaccine
поливалентность polyvalence
поливальная система sprinkler/sprinkling system
поливинил *chem.* polyvinyl
поливинилацетат *chem.* polyvinylacetate
~, эмульсионный suspension polyvinylacetate
поливинилпирролидон *chem.* polyvinyl-pirrolidone
поливинилхлорид polyvinylchloride, PVC
поливитамин polyvitamine
поливитаминный polyvitaminic
поливка I irrigation, watering
поливка II *cul.* gravy
поливной irrigation *(attr.)*
полигамия polygamy
полигамно-двудомный *bot.* polygamodioecious
полигамно-однодомный *bot.* polygamomonoecious
полигамный polygamous
полиганглионеврит *med.* polyganglioneuritis
полигаплоид polyhaploid
полиген polygene
полигенния, полигенность polygeny
полигенный polygenic, polygenetic
полигеосинклиналь polygeosynclinale
полигибрид polyhybrid
полигимния *mith.* Polyhymnia
полигиничный polynous
полигиния polygyny
полигинный *bot.* polygynoecial
полигиперменорея polyhypermenorrhea
полигирия polygyria
полигирный polygyral, polygyric
полиглобулия *med.* polyglobulia, polyglobulism
полиглот polyglot
полиглотный polyglottic, multilingual
полигляндулярный polyglandular, pluriglandular
полигон 1. testing/proving/firing ground; test range; 2. *math.* polygon; *geod.* field traverse, geodetic/triangulation chain
~, ракетный rocket/missile range
~, учебный training/practice range
полигонатум *bot.* Solomon's seal *(Polygonatum)*
полигонометрия *geod.* polygonometry, traverse/polygon measurements, survey/field traverse
полиграфический polygraphic, printing
~кая промышленность printing industry
полиграфия printing/graphic arts, printing industry, polygraphy
полидактилия *med.* polydactylia, polydactylism
полидипсия *med.* polydipsia
полидисплазия *med.* polydysplasia
полиен *chem.* polyene
полиинфекция *med.* polyinfection, multiinfection
поликариотический polykaryotic
поликариоцит polycaryocyte
поликарпий *bot.* polycarp
поликарпик *bot.* polycarpous plant
поликарпический *bot.* polycarpic, polycarpous
поликератоз *med.* polykeratosis
поликистоз *med.* polycystosis
~ почки polycystic kidney, polycystic renal disease
поликистозный *med.* polycystic
поликлиника outpatients' clinic; polyclinic; health centre/center
поликомпонентный multicomponent
поликонденсация polycondensation, condensation polymerization
поликристалл polycrystal
поликристаллический polycrystalline
полилинейный, полилинеарный *math.* multi-

linear
полилогарифм *math.* polylogarithm
полимастия *anat.* polymastia, polymazia
полименорея *med.* polymenorrh(o)ea
полимер *chem.* polymer
~, кремнийорганический organo-silicon polymer
~, ориентированный oriented/directed polymer
~, привитый graft-polymer
~ с разветвленной структурой branched polymer
~ сетчатой структуры cross-linked polymer
~, сшитый cross-linked polymer
~, хелатный chelate polymer
~, элементоорганический hetero-organic polymer
полимеризат polymerization product, polymerizate
полимеризатор polymerization reactor
полимеризовать polymerize
полимерия 1. *chem.* polymerism; 2. *anat.* polymery, polymeria
полимерный polymeric
полимербетон polymer-impregnated concrete
полиметаллизм *fin.* polymetallism
полимикробный polimicrobial
полимиксин *chem.* polymyxin, aerosporin
полимиозит *med.* polymyositis
полиморфемный polymorphemic
полиморфизм polymorphism, pleomorphism
~ сыпи polymorphism of skin eruption
полиморфность polymorphy
полиморфный polymorphic, polymorphous, pleomorphic; multiform
полиморфоцит polymorphocyte, myelocyte
полиморфы *pl.* polymorphs
полиневрит *med.* polyneuritis, multiple neuritis
~, брюшнотифозный typhoid polyneuritis
~, наследственный familial polyneuritis
~, сыпнотифозный typhus polyneuritis
полинезиец, полинезийский Polynesian
полинезийка Polynesian (woman)
полином *math.* polynomial
~ Лежандра Legendre polynomial
~ наименьшего уклонения polynomial of least deviation
полинуклеоз polynucleosis, multinucleosis
полинуклеотид polynucleotide
полиовирус *med.* poliomyelitis virus
полиоз *med.* poliosis, canities
полиомиелит *med.* poliomyelitis
~, непаралитический nonparalytic/aparalytic poliomyelitis
~, передний anterior poliomyelitis
~, понтинный pontile/pontin poliomyelitis
~, смешанный combined polyomyelitis
полиомиелитный *med.* poliomyelitic
полиомиелоэнцефалит poliomyeloencephalitis, polioencephalomyelitis
полионим polyonym
полиопия *med.* polyopia, polyopy, polyopsy, multiple vision
полиоэнцефалит *med.* polioencephalitis
полип 1. *med.* polyp(us); 2. *zool.* polyp, hydranth
~, аденоматозный adenomatous/cellular polyp
~, зубной dental/tooth/pulp polyp
~, кистозный cystic polyp
~, миоматозный myomatous/fleshy polyp
~ носа nasal polyp
~, первичный protopolyp
~, простейший hydrula
~ с костными включениями osseous polyp
~, хоанальный choanal polyp
полипапиллома polypapilloma, multiple papillomas
полипептид *chem.* polypeptide
полипептидаза *bioch.* polypeptidase
полипептидный polypeptide
полипеталия *bot.* polypetaly
полипид *zool.* polypide
полиплазмия *med.* polyplasmia, hydremia
полипласт polyplastic material
полиплоидия polyploidy
полипняк *zool.* polypary
полипноэ polypnea, tachypnea
полиповидный polypiform, polypoid
полипоз *med.* polyposis
полипропилен *chem.* polypropylene
полирибонуклеотид polyribonucleotide
полировать polish; burnish; buff; finish; gloss
полис I *hist.* city-state, polis
полис II *fin.* policy
~, азартный wager policy
~, валютированный valued policy
~, генеральный running/floating policy
~, групповой group/master policy
~ морского страхования marine insurance policy
~ на срок time policy
~, невалютированный/нетаксированный open/unvalued policy
~, разовый named policy
~, срочный time policy
~ страхования жизни life insurance policy
~, страховой insurance policy
~, таксированный [с объявленной стоимостью] valued policy
полисахарид *chem.* polysaccharide
полисемантизм, полисемия polysemy
полисерозит *med.* polyserositis
полисимметрия polysymmetry
полисмен policeman
полисодержатель policy holder
полисомия polysomia, polysomy
полиспаст tackle pulley/block, polyspast
~, цепной chain block
полиспермия polyspermia, polyspermy, polyspermism
полистела *bot.* polystele
полистелия *bot.* polystely
полистиль *arch.* polystyle

полистирол polystyrene
~, ударно-прочный high-impact polystyrene
политбюро *abbr.* Politbureau, Political Bureau
политеизм polytheism
полителия *anat.* polythelia
политен *chem.* polythene, polyethylene
политетрафторэтилен polytetrafluoroethylene, teflon
политехнизация introduction of polytechnical training/education
политехнизм polytechnical education/training
политехникум polytechnical college/school
политзаключенный political prisoner
политизация politicization
политика policy; politics
~ ведущая к инфляции inflationary policy
~, внешняя/иностранная foreign policy
~, внутренняя home policy
~, выжидательная wait-and-see policy
~ невмешательства *econ.* laissez-faire policy; *polit.* policy of non-interference
~ неучастия в блоках *pol.* nonalignment
~ регулирования доходов incomes policy
~ регулирования процентных ставок interest policy
~ "стой-иди" stop-and-go policy
~ цен и доходов prices and incomes policy
~ экономического роста expansionary policy
политинформация political information
политкаторжанин political convict
политкружок political study group
политминимум minimum of political training
политональность *mus.* polytonality
политоничность, политонизм, политония polytonism
политотдел *abbr.* political department
политпросвет, политпросвещение political education
политрихия *med.* polytrichia, polytrichosis
политрихум *bot.* hair-cap moss *(Polytrichum)*
политропа polytropic curve, polytrope
политрук *abbr. mil.* (Soviet) political instructor
политсостав *mil.* (Soviet) political officers
политуправление political administration
политура polishing liquid
политшкола political school
политэкономия political economy/economics
политэмигрант political emigrant
полиуретан *chem.* polyurethane
полиурия *med.* polyuria, hydruria, profuse micturition
Полифем *myth.* Polypheme, Polyphemus
полифалангия *med.* polyphalangism, hyperphalangism
полифилия polyphyly, polyphyletic origin
полифонемный polyphonematic
полифония polyphony
полифосфаткиназа *bioch.* polyphosphate kinase
полифтонг *phon.* polyphthong
полихазий *bot.* polychasium
полихета *zool.* polychaete
полихроичный polychroic, pleochroic
полихромазия, полихроматофилия polychromatophilia, polychromasia, polychromophilia, polychromatia
полихроматический polychromatic, polychromic
полихромный polychromatic, polychrome
полицитемия *med.* polycythemia
поличное *law* material evidence
полиэдр polyhedron, polytope
полиэдрический polyhedral
полиэкран multiscreen
полиэлектрод multiple electrode
полиэнергетический polyenergetic; polyergic
полиэтилен *chem.* pollyethylene, polyethene
полиэфир *chem.* polyether; polyester
полиэция *bot.* poly(o)ecism
полиядерный polynuclear
полк *mil.* regiment
полка rack, shelf; flange; *rail.* berth
~, кабельная cable rack/shelf
поллиний *bot.* pollinium, pollen mass
поллиния *bot.* sugar grass *(Pollinia)*
поллукс, поллуцит *miner.* pollux, pollucite
Поллукс *myth., astr.* Pollux
поллюция *physiol.* pollution, spermatorrh(o)ea
полногранник *cryst.* holohedron
полногранность *cryst.* holohedry, holohedrism
полнокристаллический holocrystalline
полнокровие *med.* plethora
полнометражный full-length, feature-length
полномочие plenary powers, authority, power; *law* proxy; power of attorney; procuration
полномочия *pl.* power(s); authority; terms of reference
~, дискреционные arbitrary power
~, конституционные constitutional/organic powers
~, неограниченные unlimited/plenipotentiary/plenary powers
полномочный, полномощный plenipotentiary; authorized
полнородный *law* full *(brother or sister)*
половец *hist.* Polovtzy, Polovtsian
половодье high water; flood(ing)
половозрелость puberty, sexual maturity
половозрелый puberal, pubertal; sexually mature
половия *mech.* centrode
положение position; location, site, situation; place; state; condition; proposition; postulate
~, вертикальное vertical/upright position
~ включенное on position
~, военное martial law
~, выключенное off position
~, исходное initial/reference/home position; starting point
~, конечное end/final/terminal position
~, крайнее заднее rearmost position
~, крайнее переднее fully-forward position
~ лежа lying position

~ **лежа на животе** prone position
~ **лежа на спине** supine position
~ **на плаву** floating
~, **общественное** social status/standing
~, **правовое** legal provision, provision of law
~, **продовольственное** food situation
~, **пусковое/стартовое** launching position
~, **рабочее** operating position; on-position
~, **равновесное** equilibrium position
~, **семейное** marital status
~, **смежное** adjacent/contiguous position
~, **социально-экономическое** socioeconomic status
~, **стесненное материальное** reduced/strained circumstances
~, **стояночное** static attitude; parking position
~, **существующее** *law* status quo
~, **флюгерное** feathered pitch
~, **чрезвычайное** state of emergency
положительность positivity; positivism
полоз I *zool.* racer *(Coluber)*
полоз II *tech.* slide, skid, runner
~, **спусковой** bilge/sliding way
полоз-удав *zool.* black racer/snake *(Coluber constrictor)*
пололепестник *bot.* habenaria *(Habenaria)*
полок platform; table
поломка breakdown, failure; breaking, breakage
~, **усталостная** fatigue failure
полонизировать Polonize
полоний *chem.* polonium, Po
полоса 1. strip; band; belt; strap; 2. stripe, belt, zone, streak; 3. *met.* bar
~, **береговая** strand, shore
~, **ветрозащитная** windbreak
~, **взлетно-посадочная** runway, take-off strip
~ **гидроаэродрома** alighting channel
~, **дерновая** *aero.* grass strip
~ **движения** traffic lane
~, **жильная** lode, ledge
~ **излучения/испускания** emission band
~, **лесозащитная** forest shelter belt; windbreak
~, **летная** airstrip, landing strip
~ **отчуждения** *law* right-of-way
~ **поглощения** absorption band
~, **полезащитная** shelter belt; windbreak
~, **посадочная** landing strip
~ **пропускания** transmission/pass band; bandwidth
~ **частот** frequency band
~, **фокальная** *math.* focal Monge('s) strip
~, **экваториальная штилевая** doldrums
полосатик *zool.* rorqual *(Balaenoptera)*
полосатый striped, streaky, striate(d), strial; ribboned; zonate; *zool.* fasciate
полоска stria; stripe
~ **молоточка** *anat.* mallear stripe
~, **пограничная** *anat.* terminal stria
~, **сосудистая** vascular stripe
полоскание 1. gargle; 2. rinsing, rinse
~, **вяжущее** astringent gargle
~, **солевое** saline gargle
полосковидный, полоскообразный striate, ribbonlike
полоскун *zool.* North American raccoon *(Procyon lotor)*
полостной cavitary; cavernous; cavity-type
полость I cavity, hollow, void, recess; chamber; space; *biol.* loculus; lumen
~ **брюшины** peritoneal cavity
~, **брюшная** abdominal cavity
~, **гайморова** *anat.* antrum of Highmore
~ **глотки** pharyngeal cavity
~ **зуба** pulp cavity
~, **мозговая** medullary cavity
~, **подкрыльцовая** axillary cavity
~ **рта** oral cavity
~ **среднего уха** tympanic cavity
~, **суставная** joint cavity
~ **таза** pelvic cavity
~, **троичная** trigeminal cavity
~, **щелевая** fissure cavity
полость II 1. *geom.* nappe; 2. travel(l)ing rug
полосчатый striated, streaky, lineate; ribboned
полотнище panel; *av.* gore
~, **боковое** *naut.* goring cloth
~ **купола** canopy gore
~, **опознавательное** identification panel
~, **разрывное** rip panel
~, **сигнальное** signal/marking panel
полотно 1. canvas; cloth; linen; 2. (road)bed 3. *art* painting, canvas; 4. blade; 5. leaf *(of door)*
~ **антенны** sheet of an array
~, **гусеничное** caterpillar band
~ **дороги** roadbed
~, **железнодорожное** railway bed/groundwork
~, **земляное** subgrade, roadbed
~ **конвейера** conveyer belt
~, **кордное** cord fabric
~ **моста** bridge floor/road
полпред *abbr.* ambassador/representative plenipotentiary
полпредство *abbr.* plenipotentiary representation, embassy
полрумба half (point)
полтора one and a half
полуавтомат semiautomatic machine/device/rifle
~, **сварочный** semi-automatic (arc) welding machine
~, **токарный** semi-automatic lathe
~, **фрезерный** semi-automatic milling machine
~, **шлифовальный** semin-automatic grinder
полуаутотрофный semiautotrophic, hemiautotrophic
полубайт *comp.* half-byte, nibble
полубазидия *bot.* hemibasidium
полубак *naut.* forecastle (erection)
полубархат *text.* velveteen
полубезработный semiemployed; half-/part-timer
полубесконечный *math.* semi-infinite

полубессознательный semiconscious; semicomatose
полубилинейный *math.* sesquilinear, semibilinear
полубит *comp.* half-bit
полубимс *naut.* half-beam
полубитуминозный semibituminous, subbituminous
полубог demigod
полубогиня demigoddess
полубочка *aero.* half-roll
полувагон open wagon, gondola car
полувахта *naut.* dogwatch
полувзвод *mil.* half a platoon
полувоенный paramilitary
полуволна half wave
полувыборка *comp.* half-select
полувыпуклый semiconvex
полувычитатель half-subtractor
полугармоника *rad.* harmonic of one-half frequency
полугласный 1. semivocalic; nonsyllabic; 2. semivowel
полугоризонталь *top.* mediate contour
полугранник *geom.* hemihedron
полугруппа semigroup; *math.* monoid
полугруппоид *math.* semigroupoid
полуда *met.* tinning; tin plate
полудополнение semicomplement, semisupplement
полудужье semiarch, semiarc; haunch; semi-intrados
полужабра *zool.* hemibranch
полужесткий semirigid, semihard
полужестокрылые *pl. zool.* hemipterans *(Hemiptera)*
полужестокрылый *zool.* hemipterous
полузаказной *comp.* semicustom
полужидкий semiliquid, semifluid
полузонтик *bot.* cyme
полузонтичный, полузонтиковый *bot.* cymose, cymous
полуинвариант *math.* semiinvariant, cumulant
полуископаемый semifossil, subfossil
полуистина half-truth
полуканал *anat.* semicanal, half canal
полукасательная *math.* semitangent
полуквадрика *geom.* regulus
полуквалифицированный semi-skilled
полуклевер *bot.* black medic *(Medicago lupulina)*
полуклетка *bot.* semicell
полуклубневой semituberous
полуклюз *naut.* open chock; mooring port/pipe, hawser port/pipe
полукокс semicoke
полуколичественный semiquantitative
полуколония semicolonial territory
полуколонна semicolumn
полуколосистый *bot.* semispikate, half-spiked
полукольцо semiring, half-ring
полукома *med.* semicoma
полукоматозный *med.* semicomatose
полукомплект subassembly, subset
полукопытные *pl. zool.* subungulates *(Subungulata)*
полукопытный *zool.* subungulate
полукристаллический semicrystalline
полукровка half-bred/breed, demisang
полукруг semicircle
полукруглый, полукруговый, полукруговой semicircular; semicyclic(al)
полукрылатый half-winged, semialate
полукрыло half-wing
полукупол semidome
полукустарник *bot.* semishrub, subshrub, forb, suffrutex
полукустарниковый suffruticous
полулинейный semilinear
полулинза hemilens
полулогарифмический semilogarithmic, semilog
полулунный semi-lunar, half-moon; crescentiform, crescentic, lunate
полумасса *paper* half-stuff, half-stock
полуматовый dull-/flat-finish *(attr.),* semidull
полумесяц demilune, half-moon; crescent
полумесячный fortnightly
полуметалл semimetal
полумеханизированный semimechanical
полумодуль half-module
полумонокок *aero.* semimonocoque
полумуфта half-coupling
полумягкий 1. semisoft; 2. semisoft consonant
полунельсон *sport* half nelson
полунемой semimute
полунепрерывность semicontinuity
полунепроницаемый semitight; semipermeable
полуница *bot.* hill/green strawberry *(Fragaria viridis/collina)*
полуобезьяны *pl.* prosimians *(Prosimiae)*
полуоборот half-turn; *mil.* half-face
полуобработанный semifinished
полуобращенный, полуобратный half-inverted, hemitropal, hemitropous
полуокружность semicircle
полуось 1. *tech.* axle shaft; half-axle; 2. *geom.* semiaxis
полупалуба *naut.* half deck
полупансионер day boarder
полупаралич *med.* paresis
полупериод semiperiod, half-period; half-cycle
 ~, биологический biological half-life
 ~ распада *phys.* half life
полупетля *aero.* half-loop and a half-roll, Immelmann turn
полуплод *bot.* hemicarp, half fruit
полуплоскость *math.* half-plane
полуповорот *naut.* half-board; *sport* half twist
полуподвал semibasement
полуподвешенный semisuspended
полупозвонок *anat.* semivertebra
полупокрытоплодный *bot.* hemiangiocarpous

полуполба *bot.* emmer *(Triticum dicoccum)*
полупонтон half-pontoon
~, надувной half-float
полупризма hemiprism
полупризнание incomplete admission/confession
полуприцеп semitrailer
~, низкорамный low-bed semitrailer
полуприцеп-контейнеровоз container chassis
полупричастие *gram.* semiparticiple
полупроводник *phys.* semiconductor
~, беспримесный intrinsic/pure semiconductor
~, несобственный extrinsic/impurity semiconductor
~, объемный bulk semiconductor
~, примесный extrinsic/impurity semiconductor
~, чистый pure/intrinsic semiconductor
~, элементарный elemental semiconductor
полупроводниковый semiconductor *(attr.);* semiconducting; transistorized
полупродукт semifinished/intermediate product
полупроницаемый, полупропускающий semipermeable
полупростой *math.* semisimple, semiprime
полупространство half-space
полупрофиль half-profile
полупрямая *math.* half-line, ray
полупустынный semidesertic, semiarid
полупустыня semidesert
полуравнина peneplain
полуразмах *aero.* semispan
полуразность semidifference, half-difference
полуразрез phantom view
полураскос knee brace
полураспад *phys.* half-decay/disintegration
полурифма half-rhyme
полурота *mil.* half-company
полусапрофит hemisaprophyte, semisaprophyte
полусвет twilight; penumbra
полусинусоида half-sinusoid
полусинусоидальный half-sine
полуслово *comp.* half-word
полусогласный *phon.* semiconsonant
полусон somnolence; drowsiness
полусталь *met.* semisteel
полустекловатый subvitreous
полустоячий *bot.* erectopatent
полустроб *rad.* gate
~, второй/задний late gate
~, первый/передний early gate
полустык half-joint
полусумма half-sum; semi-sum
~ крайних значений *stat.* midrange
полусумматор *comp.* half-adder
полусфера hemisphere
полутело half-body
полутяжелый *sport* light-heavyweight
полуударение *phon.* semistress
полуфабрикат semifinished/semimanufactured/intermediate product/article/piece; *pl.* semimanufactures, semiproducts, semis, semifinished/semimanufactured goods; *cul.* ready-to-cook foods
полуфарфор semiporcelain
полуферма *constr.* semi-truss
полуфинал *sport* semifinal
полуформа half-mo(u)ld; mo(u)ld section
полухордовые *pl. zool.* hemichordates *(Hemichordata)*
полуцелый *math.* half-integral
полуциклический semicyclic, hemicyclic
полуцилиндр semicylinder
получатель recipient, receiver; addressee; *law* donatory
~ государственного социального пособия welfare recipient
~ денег payee
~ денежного перевода remittee
~ по трансферту transferee
получать receive; obtain, get
~ деньги по аккредитиву cash a letter of credit
~ деньги по векселю cash a bill
~ наличные *econ.* cash
~ пособие be on welfare
получение receipt, receiving, obtaining; derivation, generation, output, production
~ данных data acquisition
~ информации information acquisition/gaining
полушарие hemisphere
~, доминирующее dominant hemisphere
~, мозговое cerebral hemisphere
~, северное/южное northern/southern hemisphere
полушелк half silk
полушерсть half wool, wool mixture
полуширина half width; half breadth; semiwidth
полуширота *naut.* half-breadth plan
полушник *bot.* quillwort *(Isoetes)*
полуэмпирический semiempirical
полуэскадрон *mil.* half-squadron
полуэтаж low stor(e)y, low/mezzanine floor
полуэталонный semistandard *(attr.)*
полуют *naut.* poop (erection), aftercastle
полуячейка half-cell
полый hollow; *anat.* caval
полынь *bot.* sagebrush *(Artemisia)*
~, горькая common wormwood, absinthium
полынья ice clearing/lane, water opening, polynia
польза advantage; use, benefit, good; utility
~, публичная public convenience
пользование use, utilization
~, потребительское consumptive use
пользователь *comp.* user
~, высокоприоритетный high-priority user
~, зарегистрированный authorized user
~, коммерческий real user
~, конечный end user
~, незарегистрированный unauthorized user
~, низкоприоритетный low-priority user
~ удаленного терминала remote user
~, эпизодический ad hoc user

пользователь-непрограммист *comp.* nonprogrammer user
пользователь-непрофессионал *comp.* lay user
полька 1. Polish woman/girl, Pole; 2. polka
полюдье *hist.* collection of tribute
полюс pole; *el.* terminal
 ~ возбуждения *el.* field pole
 ~ мира celestial pole
 ~, одноименный like pole
 ~, отрицательный negative terminal
 ~, разноименный unlike/opposite pole
 ~, Северный North (Arctic) Pole
 ~, Южный South (Antarctic) Pole
полюсоопределитель polarity indicator
полюсный polar
поляк Pole
поляника *bot.* arctic bramble *(Rubus arcticus)*
поляра polar (line/diagram); polar curve, liftdrag curve, drag polar
поляризатор polarizer
поляризация polarization; polarizing
 ~, гальваническая electrolytic/electrochemical/cell polarization
 ~, доменная domain polarization
 ~, круговая circular polarization
 ~, левая counterclockwise [left-hand circular] polarization
 ~, линейная linear/plane polarization
 ~, миграционная interlayer polarization
 ~, объемная space-charge polarization
 ~, остаточная remanent polarization
 ~, плоская plane polarization
 ~, поперечная cross/transverse polarization
 ~, правая clockwise [right-hand circular] polarization
 ~, сложная aggregate polarization
 ~, электрическая dielectric polarization
поляризованность degree/extent/intensity of polarization
поляризовать polarize
поляризуемость polarizability
полярископ polariscope
 ~, стробоскопический polaristrobometer
полярник polar explorer
полярность polarity
 ~, обратная/обращенная/перевернутая reversed polarity
 ~, прямая normal polarity
полярон *phys.* polaron
помазание *eccl.* anointment, anointing, chrismatory
помазанник *eccl.* the anointed
помазать, помазывать *eccl.* anoint
помароза *bot.* rose apple *(Syzygium jambos)*
помбух *abbr.* assistant bookkeeper
помело *bot.* grapefruit *(Citrus grandis)*
померанец *bot.* sour/bitter/Seville orange *(Citrus aurantium)*
померанский Pomeranian
поместное дворянство *hist.* landed gentry
поместный собор *eccl.* local council
поместье estate
 ~, господское seignorial estate/domain
 ~, крупное latifundium
помесь hybrid; crossbreed, crossbred; mongrel
помеха interference, noise disturbance; encumbrance; jamming; clutting, background
 ~, грозовая lightning/thunderstorm static
 ~, заградительная barrage jamming
 ~, манипуляционная keying noise
 ~ от дождя rain clutter
 ~ от солнечного излучения solar noise
 ~, прицельная selective/spot jamming
 ~хи, атмосферные atmospheric noise/disturbance, statics
 ~хи, импульсные discontinuous jamming
 ~хи, наведенные induced noise
 ~хи, статические statics, atmospheric noise, atmospherics
помехоглушитель noise silencer
помехозащищенность, помехоустойчивость noise immunity
помехозащищенный, помехоустойчивый noiseproof, jamproof, antijamming
помехофильтр noise filter
помешательство insanity, madness; lunacy
 ~, буйное impulsive insanity
 ~, однопредметное monomania
 ~, религиозное theomania
 ~, сутяжническое barratrous/litigious madness
 ~, тихое melancholic insanity; paranoid melancholia
помещение 1. room, premises; space; 2. placing, location, installation; insertion
 ~ бытового обслуживания service space
 ~, дежурное ready room
 ~, жилое living quarters, accommodation(s); *naut.* berthing/living space
 ~ коллективного проживания collective-type dwelling
 ~, производственное production floor/area/space
 ~, складское store room, repository
 ~, служебное service room
 ~, туалетное toilet compartment
 ~ экипажа crew quarters
помещик landlord, landed gentleman, lord of the manor
помещичество, помещики landlordry, landed classes
помещичий landlord's, landlordly
 ~ дом manor house
 ~чья усадьба country estate
помидор *bot.* tomato *(Lycopersicon)*
помилование *law* pardon, grace
поминальник *eccl.* list of dead for whom a prayer is read
поминки *pl.* funeral repast
поминовение *eccl.* prayer for the dead/sick
поммастера *abbr.* foreman's assistant

помол grinding, milling; grind; comminution
Помона *myth.* Pomona
поморник *zool.* skua *(Stercorarius)*
поморье coastal area; littoral region
~, балтийское Pomerania
~, северное White Sea Coast
поморянин inhabitant of coastal region; Pomeranian
помост platform, stage, rostrum
помощник assistant, helper; help; deputy
~ дежурного по кораблю junior/petty officer
~ капитана mate
~ капитана, старший/второй chief/second mate
~ режиссера assistant producer; stage manager
помощь assistance, aid, help; relief; service; *med.* care
~, акушерская obstetric care
~, амбулаторная ambulatory care
~, медицинская medical care/aid; *Am.* medicare
~ на дому domiciliary aid, home visiting service
~, немедленная emergency assistance/aid/treatment/service
~, неотложная хирургическая emergency surgery
~, первая first aid
~, скорая ambulance; first aid
помпа pump
помпаж surging, surge
помпельмус *bot.* pomelo, shaddock *(Citrus maxima)*
помпировать surge
помполит *abbr. mil., obs.* assistant commander for political affairs
помреж *abbr.* stage manager
помутнение clouding, blur, dimming; opacification, opacity; turbidity; dimness; darkening
~ роговой оболочки *med.* keratoleukoma, nebula, albugo
~ хрусталика lenticular opacity, cataract
понедельная зарплата *econ.* weekly wages
понижение decrease, reduction, decline, lowering, fall, drop; diminution; bringing down; *geol.* depression, sibsidence, low
~ горизонта horizon dip/depression
~ давления pressure drop/reduction
~, играть на *fin.* sell a bear, speculate for a fall, sell short
~, играющий на bear
~ по службе demotion
~, эрозионное *geol.* degradation
понизительный reducing; *el.* step-down
понимание understanding, comprehension; perception
~, зрительное visual comprehension
~ речи speech understanding
понос *med.* diarrh(o)ea
~, викариирующий vicarious diarrh(o)ea
~, водянистый watery/serous diarrh(o)ea
~, дизентерийный dysenteric diarrh(o)ea, dysentery
~, жирный fatty diarrh(o)ea
~, летний summer/choleraic diarrh(o)ea
~, медикаментозный drug diarrh(o)ea
~, панкреатический pancreatogenous diarrh(o)ea
понселе *phys.* poncelet
понтедерия *bot.* pickerelweed *(Pontederia)*
понтийский Pontic
понтифик pontifex
понтификат pontificate
понтон pontoon
~, десантный landing/beaching pontoon
~, закрытый caisson
~, моторный propulsion pontoon
~, надувной pneumatic pontoon
~, подъемный float case
~, полуторный tripartite pontoon
~, самоходный propulsion pontoon
~, составной bipartite pontoon
~, судоподъемный salvage/lifting pontoon, camel
понтонер *mil.* pontooner
пончик doughnut
пончо poncho
понятие notion; concept; *philos.* conception
~, числовое numeric concept
понятийный notional, conceptual
попадание 1. hit; 2. entry, penetration, ingress, ingestion
~, прямое direct hit
~ воды entry/ingestion of water
попарно in pairs, pairwise
поп-арт pop art
поперек transverse, crosswise
попеременно-перистый *bot.* alternately pinnate
попеременный alternating, alternate; alternative
поперечина crosspiece, crossbar, crossbeam, traverse, cross-member
~ моста bridge sleeper/joist
поперечник diameter; size across; cross section
поперечно-бороздный transcolpate
поперечно-перегородчатый transseptate
поперечнополосатый cross-striated, transversostriated
поперечноротые *pl. zool.* plagiostomes *(Plagiostomi)*
поперечный transversal, transverse, cross; diametrical; lateral; crosscut, cross-sectional
~ная балка crossbeam
~ная сумма transverse sum/total
~ная пила crosscut saw
попечитель guardian; trustee
попечительский совет board of guardians/trustees
попечительство guardianship, trusteeship; custody; care, charge
поплавковый float *(attr.),* float-operated/controlled, float-type

~ гироскоп (fluid-)floated gyroscope
поплавок float
~ гидросамолета seaplane float
~ карбюратора carburettor float
~, подкрыльный wingtip float; inboard stabilizing float
~, подфюзеляжный fuselage float
~, шаровидный spherical/ball float
поплин *text.* poplin
поповник *bot.* woundwort, oxeye daisy *(Chrysanthemun leucanthemum)*
поползень *zool.* nuthatch, European nutcracker *(Sitta europaea)*
пополнение replenishment, restocking; supplementation; recharge; resupply; replacement
~ горючим refuelling
~ запасов replenishment/resupply operation
~ запасов продовольствия provisioning
~ личным составом personnel replacement
~, непрерывное continuous replacement
~ поголовья *agr.* restocking
~ потерь replacement of casualties
~ расхода, плановое routine replenishment
пополнять replenish; fill up; restock; reinforce
~ запас топлива refuel
поправка 1. *law* amendment; 2. correction, rectification; readjustment; 3. repair(ing), mending; *med.* recovery, recuperation
~, аэродинамическая corection for position error(s), position-error correction
~, барометрическая allowance for elevation
~, боковая lateral/deflection correction
~ Буге Bouguer correction
~, инструментальная instrument(-error) correction
~ к пеленгу bearing correction
~ к показанию прибора instrument reading correction
~, курсовая heading/course correction
~ лага log correction
~ на ветер windage, wind correction
~ на группировку *stat.* correction for grouping, Sheppard's correction
~ на запаздывание lag-error correction
~ на конечность выборки *stat.* finite sampling correction
~ на конечность совокупности *stat.* finite population correction
~ на кривизну меридиана correction for meridian curvature
~ на непрерывность *stat.* continuity correction
~ на параллакс parallax correction
~ на рассеяние scattering correction
~ на скорость speed compensation
~ на снос drift correction
~ на эффект стенки wall correction
~ пеленгов bearing adjustment
~ по тангажу pitch correction
~ по дальности range correction
~, угловая angular correction, correction angle
~, угломестная elevation correction
поправление correction; amelioration; restoration
попугай *zool.* parrot *(Psittacidae)*
популизм Populism
популяционизм populationism
популяционистский populationistic
популяция *biol. math.* population
~, замкнутая intrabreeding population
~, изогенная isogenic population
~, клеточная cell population
~, сбалансированная equilibrium population
~, свободно скрещивающаяся random mating population
попурри *mus.* potpourri
попуск flush, discharge, release; pass; drawdown
пора pore; *bot.* pit; *(defect)* freckle
~, газовая gas pocket
~, ветвистая branching pit
~, млечная lactiferous pore
пораденит *med.* poradenitis
поражаемость vulnerability
поражать destroy, hit
поражение injury, lesion; destruction; *med.* affection; hitting, hit
~ в правах *law* disfranchisement
~, грибковое *med.* mycotic lesion, mycosis
~ желез adenopathy, adenosis
~, лучевое (ir)radiation injury
~, обратимое reversible injury/damage
~, очаговое focal lesion/disease
~, системное systemic lesion
~, скрытое latent injury
~ цели target destruction /damage
поразрядно *comp.* digit-by-digit; bit-by-bit
порайонный by district/region/area
порез cut, slash; cutting
поречник *bot.* mountain parsley *(Peucedanum oreoselinum)*
поречня *zool.* otter *(Lutra)*
поризм(а) *math.* porism
пористость porosity
~ костей osteoporosis
~, сообщающаяся intercommunicating porosity
~, усадочная shrinkage porosity
пористый porous; cellular
порог threshold; sill
~, аварийный emergency sill
~ бедности poverty [low-income] threshold
~ возбудимости excitation threshold
~ восприятия threshold of perceptibility
~, выходной *met.* skimming weir
~ двери door sill
~ деления *nucl.* fission threshold
~, направляющий *hydr.* baffle
~ обнаружения detection limit/threshold
~, обонятельный olfactory threshold
~ разборчивости intelligibility threshold
~ разрешения resolution threshold
~ слышимости threshold/borderline of audibility
~ срабатывания threshold of response

~, **шлаковый** *met.* slag dam
порогамия *bot.* porogamy
пороговый liminal
порода 1. *miner.* rock; 2. breed, race; stock; species; sort
~, **безрудная** barren rock
~, **боковая** wall rock
~ **вскрыши** overburden rock
~, **вспученная** expanded rock
~, **вспучивающаяся** swelling rock
~, **вторичная** secondary/deuterogene rock
~, **вулканическая** igneous/eruptive rock, volcanite
~, **газоносная** gas rock
~, **глинистая** clay/argillaceous rock
~, **древесная** wood species
~, **железистая/железосодержащая** ferruginous rock
~, **изверженная** igneous/volcanic/plutonic rock
~, **кислая** acid/persilicic rock
~, **коренная** bedrock
~ **кровли** roof rock
~, **материнская** parent/mother/source rock
~, **налегающая** overlying rock
~, **наносная** mantle rock; drift
~, **нефтематеринская** mother rock of oil
~, **нефтеносная** oil-bearing (petroliferous) rock
~, **околорудная** wallrock
~, **перекрывающая** overlying/overlapping rock
~, **песчанистая** arenaceous rock
~, **подстилающая** underlying/base/bedding/basement rock
~, **покрывающая** top/cover/cap rock, capping
~, **порошковатая** rock powder
~, **промышленная** pay/commercial rock
~, **пустая** barren/dead rock
~, **растительная** soil
~, **рудная** ore rock
~, **рудоносная** ore-bearing rock
~, **сланцевая** shale rock
~, **соляная** saliferous rock
~, **тальковая** soapstone, steatite
~, **хемогенная** chemogenic/chemical rock
породистость pure breeding; breed, race
породистый thoroughbred, pedigree; pure-bred
породообразующий rock-forming/-building/-making
порожняк empty car, deadhead, empties
пороз *med.* porosis
~, **мозговой** cerebral porosis
порозность porosity
порозный porous
порок 1. flaw; imperfection, defect; *med.* failure; disease; 2. vice
~ **воли** *law* flaw in the will
~, **литейный** *met.* casting defect
~, **клапанный** *med.* valvular disease/defect
~ **развития** developmental defect/anomaly
~ **сердца** heart disease, valvular defect/disease
~ **сердца, врожденный** congenital heart disease
поронот(ус) *zool.* butter/harvest fish, dollarfish *(Poronotus triacanthus)*
порообразный poriform
порообразование pore formation
порообразователь foaming/blowing/expanding agent
поропласт cellular/foamed plastic, rigid foam plastic
порох powder; gunpowder
~, **бездымный** smokeless powder
~, **беспламенный** flashless powder
~, **двухосновный** double-base powder
~, **дымный** black/smoke powder
~, **минный** blasting powder
~, **пироксилиновый** pyroxilyne/nitrocellulose powder
~, **ракетный** propellant powder
~, **трубчатый** tubular powder, cylindrical-and-hollow powder
пороцефалез *med.* porocephaliasis, porocephalosis
порошкование triturating; pulverization
порошковдуватель powder blower/sprayer
порошковидный, порошкообразный powderlike, powdery
порошковый powder *(attr.);* powdered, powdery
порошок powder; dust
~, **алмазный** diamond powder/dust
~, **закалочный** quenching powder
~, **зубной** tooth powder
~, **индикаторный** detecting powder
~, **корундовый** grinding powder
~, **магнезитовый** loose dead-burned magnesite
~, **мыльный** powdered soap
~, **наждачный** emery dust
~, **отбеливающий** bleaching powder
~, **прессованный** powder compact, compacted powder
~, **прессовочный** moulding powder/compound
~, **стиральный** washing powder; powdered detergent
~, **текучий** fluidized powder
~, **урановый** powdered uranium
~, **шлифовальный** grinding powder
порт I port, harbour
~, **баржевый** barging port
~ **ввоза** port of entry
~, **военно-морской** naval port
~ **выгрузки** port of debarkation/discharge
~ **вынужденного захода** port of distress/refuge
~ **захода** port of call
~, **морской** seaport
~ **назначения** port of destination
~ **отгрузки** port of shipment
~ **отправления** departure port
~ **отхода/отшествия** port of departure
~ **погрузки** port of loading/shipment/embarkation, harbour of loading/shipping
~ **пополнения запасов топлива** fuelling port

~ **прибытия** port of arrival
~ **приписки** port of registry/documentation
~ **пришествия** port of destination
~, **речной** river port
порт II *naut.* porthole, port
~, **бортовой** side port
~, **штормовой** freeing/clearing/wash port, storm/ freeing scupper
Порта *hist.* the Sublime Porte
портал 1. *arch.* portal, doorway; 2. *tech.* gantry
портальный portal
~ **кран** gantry crane
портаменто *mus.* portamento
портативность portability
портативный portable
портик 1. *naut.* porthole, port; 2. *arch.* portico, porch, antium
~, **швартовный** mooring port
~, **штормовой** freeing/clearing/wash port, storm/ freeing scupper
порткаустик *med.* portcaustic, porte-caustique
портландцемент portland cement
~, **сульфатостойкий** sulphate/resisting portland cement
портовый harbour *(attr.)*
порто-франко free port/zone
порт-убежище port of distress/refuge
порт-шкентель *naut.* port-lifter
порт-штерт *naut.* port rope
портретист portrait painter, portraitist
португалец, португальский Portuguese
португалка Portuguese (woman)
портулак *bot.* purslane, portulaca *(Portulaca)*
портулакария *bot.* purslane tree *(Portulacaria afra)*
портулаковые *pl. bot.* purslane family *(Portulacaceae)*
портупей-юнкер *mil. hist.* senior cadet
портупея *mil.* sword/shoulder belt
портфель portfolio; brief case bas
~ **векселей** *fin.* bill holdings
~ **заказов** *econ.* order book, stock of orders, work in hand
~, **министерский** ministerial post, portfolio
~ **ценных бумаг** security holdings
портшез sedan chair
поруб *hist.* dungeon
порубка cutting, felling
порубщик woodcutter
порука guarantee, pledge; voucher; bail
~, **круговая** collective responsibility/pledge
~**ки, выпускать на** *law* hold to bail, release on bail
поручатель = поручитель
поручейник 1. *bot.* water parsnip *(Sium);* 2. *zool.* marsh sandpiper *(Tringa stagnatilis)*
поручение commission, assignment, errand; order
~, **инкассовое** collection letter
~ **о переводе (денег)** order for remittance
~, **платежное** payment order
~, **судебное** *law* rogatory letters
поручень handrail, grab handle, rail(ing)
поручик *obs.* lieutenant
поручитель guarantor, guarantee; warrantor; voucher; bail, bailsman; surety; underwriter
поручительство bail; voucher; surety; guarantee
~, **кредитное** credit voucher/guarantee
порфир *petr.* porphyry
порфира 1. purple robe; 2. *bot.* laver *(Porphyra)*
порфирин *biochem.* porphyrin
порфирит *petr.* porphyrite
порфирия *med.* porphyria
порцион ration; allowance
порционирование proportioning
порча spoiling, spoilage, waste, breakage; defect
поршень piston; plunger
~ **гидростатическото пресса** ram
~ **двойного действия** double-acting piston
~, **дозирующий** metering plunger/ram
~, **плоский** flat/disk piston
~ **с двусторонним штоком** double-rod piston
~ **силового цилиндра** ram
~, **скальчатый** plunger
~, **ступенчатый** stepped/differential piston
~ **тормозного цилиндра** brake piston
~, **уравновешивающий** balance/dummy piston
порыв gust; puff
~, **боковой** sidegust, lateral gust
~ **ветра** (wind) gust
~, **восходящий** upgust
~, **встречно-боковой** head-quartering gust
~, **встречный/лобовой** head-on gust
~, **наземный** ground gust
~, **нисходящий** downgust
~, **попутно-боковой** tail-quartering gust
~, **попутный** tail-on gust
~, **сильный** heavy gust
~, **слабый** light gust
порывистый gusty
порядная грамота *hist.* contract, agreement
порядок order, ordering; arrangement; manner; procedure, way; *pol.* regime
~, **алфавитный** alphabetical order
~, **боевой** battle formation, fighting order
~ **величины** order of magnitude
~ **включения** *el.* switching sequence
~, **восходящий/возрастания** ascending order
~, **высший** higher order
~, **двоичный** *comp.* binary exponent
~ **действий** procedure
~, **десятичный** decimal order
~ **зажигания** ignition order/sequence
~ **кривой** order/degree of a curve
~ **малости** *math.* order of smallness
~, **обратный** inverse/reverse order
~ **обращения** *comp.* access
~ **обслуживания** servicing procedure
~, **общественный** public order/peace
~ **очередности** order of priority
~, **произвольный** random order

~ **приближения** order of approximation
~ **пуска** start-up procedure
~ **следования** sequence
~, **старый** *pol.* old regime
~ **точности** order of accuracy
~, **установленный** established order; routine
~ **числа** *math.* exponent (of a number)
~, **шахматный** staggered order
посадка 1. landing; alighting; 2. boarding, embarkation; 3. *mach.* fitting, fit; 4. planting; plantation
~, **аварийная** emergency/distress landing; crashlanding
~ **без зазора и натяга** metal-to-metal fit
~ **без пробега** no-roll landing
~ **в любых метеоусловиях** all-weather landing
~, **вынужденная** forced landing
~, **горячая** shrink(age) fit
~, **грубая** course/rough fit
~ **движения** *mach.* close-running fit, easy-slide fit
~, **мягкая** *space* soft landing
~ **на авианосец** carrier landing
~ **на воду** water landing, alighting on water
~ **на грунт** off-runway landing
~ **на живот** belly (wheels-up) landing
~ **на суда** embarkation, embarkment, boarding
~ **на сушу** *space* dry-land touchdown
~ **на фюзеляж** belly (wheels-up) landing
~, **неподвижная** tight/close/interference/stationary fit
~ **по ветру** downwind landing
~ **по командам с земли** ground-controlled landing
~ **по приборам** instrument landing
~ **при боковом ветре** crosswind landing
~ **при встречном ветре** headwind/upwind landing, landing into a wind
~ **с зазором (свободная)** clearance fit
~ **с козлом** rebound landing
~ **с натягом** interference/stationary fit
~ **с парашютированием** pancake landing
~ **с попутным ветром** tailwind landing
~ **с тормозным парашютом** drag-chute landing
~ **с тросовым подтягом** tethered landing
~, **скользящая** slide/sliding fit, easy-push fit
~ **со сносом** drift landing
~, **ходовая** running/free fit
~, **широкоходовая** loose-running [coarse-clearance] fit, extra-slack running fit
посадские люди *hist.* tradespeople
посев sowing, seeding; *biol.* culture; inoculation
~ **бактериальных культур** inoculation
~, **гнездовой** nest/hole sowing
~, **перекрестный** cross sowing
~, **пунктирный** single-grain sowing
~, **разбросной** broadcast sowing/seeding
~, **рядовой** drill/row sowing, drilling
~ **штрихом** *biol.* streak inoculation
поседение *med.* grayness of the hair, canites, poliosis, achromatosis
~, **кольцевидное** ring-shaped canites
~ **ногтей** leukonynchia
Посейдон *myth.* Poseidon
поселение 1. settling; settlement; 2. deportation
поселок settlement; housing estate/subdivision
~, **дачный** suburban settlement
~ **городского типа** urban village
посеребрение silvering, silver plating
Посидон *myth.* Poseidon
посконник *bot.* thoroughwort, boneset *(Eupatorium)*
Послания *bibl.* the Epistles
посланник envoy, minister
~ **чрезвычайный и полномочный министр** envoy extraordinary and minister plenipotentiary
послабляющий *med.* aperient
послегриппозный postinfluenzal, postgrippal
послед afterbirth, placenta
последействие aftereffect; residual effect
последовательно in succession/turn/sequence; *el. mech.* in series
последовательно-параллельный series-parallel
последовательность sequence, succession; series; successiveness; graduality; order; range
~ **адресов** address chain
~, **алфавитная** alphabetic sequence/order
~, **бесконечно малая** *math.* infinitesimal sequence
~, **возрастающая** ascending sequence
~, **временная** time sequence
~ **выборочных средних** *math.* progressive average
~, **вызывающая** calling sequence
~ **данных** chain data
~, **заданная** specific sequence
~ **импульсов** pulse train, train of pulses
~ **клеточных поколений** cell lineage
~ **кодов** code chain
~ **команд** order/instruction sequence, sequence of instructions, order/instruction/command series
~, **лакунарная** lacunar(y) sequence; gap sequence
~ **параметров** *comp.* flow of parameters
~ **передачи адресов** address transfer sequence
~ **предпочтений** *stat.* preference chain
~, **расходящаяся** *math.* divergent sequence
~, **случайная** random sequence
~ **сортировки** selection/collation sequence
~, **сходящаяся** *math.* convergent sequence
~ **фаз, обратная** negative phase-sequence
~ **фаз, прямая** positive phase-sequence
~, **фундаментальная** fundamental/Cauchy sequence
~, **функциональная** sequence of functions, function sequence
~ **ходов** *game th.* sequence of moves
~ **чередования культур** *agr.* cropping sequence

~, числовая numerical sequence, sequence of numbers
последовательный 1. successive, consecutive, sequential; 2. consistent; 3. *el.* serial, in series
~ по битам *comp.* serial by bit
~ная выборка sequential/serial connection
последовые *pl. zool.* placentals *(Placentalia)*
последовый placental, afterbirth *(attr.)*
последующий *math.* consequent
послежелтушный *med.* postictal, posticteric
послеизображение afterimage, retained image, image retention
послеимпульс afterpulse
послеиндустриальный postindustrial
послеконечный postfinal
послекризисный *physiol.* epicritic
послеледниковый postglacial
послелихорадочный *med.* postfebrile
послелог *gram.* postposition
посленаркозный *med.* postan(a)esthetic
посленачальный postinitial
послеоктябрьский post-October
послеоперационный *med.* postoperative
послеохладитель aftercooler; recooler
послереволюционный post-Revolutionary
послеродовой postnatal, postpartum, puerperal
послесвечение afterglow; *telev.* persistence
~ экрана screen afterglow/persistence
послесловие epilogue; postface
послетифозный posttyphoid
послетретичный *geol.* post-Tertiary
послеусилитель postamplifier
послефокусировка secondary focussing
послойный in/by layers, layer-by-layer
послужной список service record
послушание *eccl.* work of penance
послушник eccl. novice; lay brother
послушничество *eccl.* novitiate
посменный in shifts; by turns
посмертный postmortal, posthumous; portmortem
пособие benefit, allowance; assistance, help; grant; aid; *educ.* manual, textbook
~ бастующим strike benefit
~ в натуральной форме in-kind benefit
~ в связи с потерей кормильца survivors' benefit
~, выходное redundancy payment, severance pay, dismissal/severance/terminal wages
~, единовременное lump-sum grant
~ многодетным семьям family subsidy
~ на жизнь living allowance
~ на рождение ребенка birth/maternity grant
~ на образование education grant
~ навигационное guide to navigation
~, наглядное visual aid
~ по безработице unemployment benefit/relief; dole
~ по болезни sick pay/benefit, sickness benefit
~ по инвалидности disability allowance
~ по многодетности allowance to mothers of large families
~ по социальному обеспечению social welfare
~ по старости old-age benefit
~ роженице maternity benefit
~, страховое insurance benefit
~, суточное daily allowance
~, учебное educational/training aid
пособник *law* accessory; abbetor
посол ambassador
~ для специальных поручений roving ambassador
посольство embassy
посох staff, crook; *eccl.* crosier
пост I 1. post, station; 2. position, office
~, контрольный inspection station/post
~, линейный line-service station
~, министерский ministerial position; portfolio
~, наблюдательный observation post/station
~ наведения guidance/control post
~ обнаружения detection post
~ связи communication post
~ управления control desk/station
пост II fasting; fast
~, великий Lent
поставка delivery, purveyance, shipment; supply
~, будущая *econ.* forward delivery
~ по требованию *econ.* delivery on call
~ хлеба grain delivery (to the government)
поставщик supplier
постамент 1. *naut.* masthouse, mast/boom/derrick table; 2. pedestal, base, postament
постановка setup, arrangement; formulation; setting, placing, putting; *theatre* staging, production
~ в док *naut.* docking
~ в очередь *comp.* enqueueing
~ вопроса the formulation/presentation of the problem
~ голоса *mus.* voice training
~ мин mine-laying/-planting
~ на якорь anchorage, anchoring; mooring
~ опытов experimentation
~ сетей заграждения net laying
~ трала sweep setting/streaming
~ эксперимента experimental setup/conditions
постановление decree, enactment; disposition; resolution, decision; *law* (court) ruling
~ закона provision of law
постановщик *(thetre)* producer; *(cinema)* director
постель bed, bedding; *geol.* bottom, substratum
~ наносов bedrock, bed
~, сборочная *naut.* assembly jig
постельник *hist.* gentleman of the bedchamber, chamberlain
постельница *hist.* lady of the bedchamber
постенница *bot.* pellitory *(Parietaria)*
постинфарктный *med.* postmyocardial infarction
постит *med.* posthitis
постклимактерический postclimacteric, postmenopausal

постлиминарный *law* postliminary
постлиминий *law* postliminy
постлюдия *mus.* postlude
постнистагм *med.* postnystagmus
постоянная *math.* constant
  ~ Больцмана Boltzmann's constant
  ~ времени time constant
  ~, деэлектрическая dielectric constant, permittivity value
  ~ интегрирования constant of integration
  ~, негармоническая nonharmonic component/constant
  ~ Планка Planck's constant
  ~, произвольная arbitrary constant/parameter
  ~ распада decay/disintegration constant
  ~, сглаживающая smoothing constant
  ~ Стефана-Больцмана Stefan-Boltzmann constant
  ~ Холла Hall constant/coefficient
постоянножаберный *zool.* perennibranchiate
постоянный constant, permanent, steady; standing
  ~ ток *el.* direct current
  ~ная армия regular army
  ~ная величина *math.* constant
постплиоцен *geol.* post-Pliocene
постпубертетный postpuberal, postpubertal
постриг *eccl.* taking the monastic vows; *(of woman)* taking the veil
постригать 1. give a haircut; clip, trim; 2. *eccl.* make a monk/nun
построение construction, structure; formation
  ~, аксиоматическое axiomatic structure
  ~ алгоритмов algorithm elaboration
  ~, боевое *mil.* battle formation/order
  ~ кривых curve tracing/plotting
построитель вертикали *av.* horizon sensor
постройком *abbr.* construction committee
постскриптум postscript
поступательно-возвратный *mech.* reciprocating
поступательное движение translational motion
поступления *pl. econ.* receipts, proceeds, earnings, returns; entries
  ~ иностранной валюты exchange earnings
  ~ от невидимых статей экспорта invisible earnings
  ~ по линии туризма tourist receipts
  ~, чистые net earnings
поступь (винта) effective pitch, advance
постфактум post factum, after the fact/event
посудомойка dishwashing machine, dishwasher
посылка I 1. transmission, dispatching, sending; 2. package, parcel; pulse, impulse
посылка II premise, antecedent; hypothesis
посягательство, посягновение encroachment, infringement; *law* offence
поташник *bot.* common Russian thistle *(Salsola* ruthenica)
потемнение darkening, blackening; fogging
потенциал potential
  ~, векторный vector potential
  ~, граничный boundary potential
  ~ действия action potential
  ~, жизненный life potential
  ~ зажигания ignition/firing/starting potential
  ~, обменный exchange potential
  ~, окислительно-восстановительный oxidation-reduction potential, redox potential
  ~, опережающий advanced potential
  ~, опорный reference potential
  ~ отсечки/запирания cut-off potential
  ~, пиковой *physiol.* spike potential
  ~ покоя resting/quescent potential
  ~ поля field potential
  ~ поля сил force-field potential
  ~, пороговый threshold potential
  ~, пробивной/пробоя breakdown potential
  ~, равновесный equilibrium potential
  ~ сил тяготения Newtonian potential
  ~, следовой afterpotential
  ~, собственный self-potential
  ~, суммарный compound (action) potential
  ~ условный нуль datum potential
потенциалис *gram.* potential mood
потенциалоскоп storage tube, electrostatic memory tube
потенциометр potentiometer
  ~, автоматический self-balancing (automatic) potentiometer
  ~, азимутальный azimuth/bearing potentiometer
  ~, емкостный capacitance potentiometer
  ~ задатчика selector potentiometer
  ~, кольцевой ring potentiometer
  ~, ламельный sectionalized potentiometer
  ~, одновитковый/однооборотный single-turn potentiometer
  ~ обратной связи feedback [follow-up] potentiometer
  ~, самопишущий recording potentiometer
  ~ следящей системы follow-up potentiometer
  ~, струнный slidewire potentiometer
  ~, цифровой digital potentiometer
потенцирование 1. potentiation; 2. *math.* taking antilogarithms; raising to higher power
потери *pl.* losses *(see also* **потеря***)*
  ~ в живой силе casualties
  ~ в системе system losses
  ~ из-за разделения separation losses
  ~ на отходы scrap losses
  ~ на трение friction(al) losses
  ~, послеуборочные postharvest losses
  ~ при хранении storage losses
  ~, производственные manufacturing losses
потеря loss
  ~ зрения loss of sight
  ~ летучести devolatilization
  ~ напора head loss, loss of head
  ~ напряжения *el.* voltage loss
  ~ от скольжения slippage loss
  ~ радиосвязи radiocommunications failure

~ **речи** aphasia, alalia, speechlessness
~ **скорости** stall(ing)
~ **трудоспособности** disability
~ **устойчивости** loss of stability; *(of material)* buckling, crippling
~ **утойчивости при растяжении** tension buckling
~ **фазы** phase retardation
~ **чувствительности** *med.* an(a)esthesia, loss of sensation
потница *med.* miliaria, prickly heat, heat rash
потовой sudoral, sudoriferous
потогонный sudorific, diaphoretic
поток flow, current, stream; flux
~, **безвихревой** noneddying/irrotational/noncirculatory flow
~, **безнапорный** gravity flow
~ **ввода** input flow
~ **вектора** vector flux, flux of a vector
~, **вихревой** eddy/vortex/rotational flow
~ **воздуха** air flow/current stream
~, **восходящий** ascending/upward current; upstream
~, **встречный** counterstream
~, **втекающий** inward flow
~, **входящий** incoming flux/current, inflow; *econ.* input (flow)
~ **вызывов** flow of demands/calls/customers
~, **вытекающий** outward flow
~, **выходящий** outgoing flux/stream; effluent; output stream
~ **жидкости** fluid flow
~ **заданий** *comp.* job stream
~, **завихренный/закрученный** swirling flow
~, **заторможенный** stagnant/stagnated/retarded flow
~, **звуковой** sonic flow
~ **излучения** radiation flux, flux of radiation
~, **индуктивный** induced flow
~, **интегральный** integrated flux
~ **информации** data/information flow
~ **информации, обратный** data feedback
~, **кольцевой** annular flow
~, **конвекционный** convective flow, convection current
~ **космического излучения** cosmic-ray flux
~, **краевой** edge flux
~, **лучистый** radiant flux
~, **максимальный** peak flux
~, **марковский входящий** Markov input
~, **метеорный** meteor/meteoroid shower
~, **многопродуктовый** *econ.* multicommodity flow
~, **многоходовой** multipass flow
~, **набегающий** incoming/incident flow, inflow
~, **нейтронный/нейтронов** neutron flux/current
~, **непрерывный** continuous flow
~, **нестационарный** unsteady flow/flux
~, **нисходящий** descending/downward current; downstream
~, **объединенный выходящий** pooled output
~, **обратный** reverse flow
~, **объемный** volumetric flow
~, **одноходовой** one-pass [straight-through] flow
~ **отказов** failure flow
~, **падающий** incident flux
~, **пространственный** three-dimensional flow
~, **прямой** straight-through flow
~, **пуассоновский** Poisson input(s)/arrival(s)
~, **пульсирующий** pulsating/fluctuating/oscillating flow
~, **результирующий** net current/flux/flow
~, **световой** luminous flux
~, **слабый** low flux/flow
~ **смещения** displacement flow
~, **сопутствующий** co-current flow
~, **спиральный** helical flow
~, **стационарный** steady flux/flow
~ **тензорного поля** tensor flux
~, **установившийся** steady-state flow
~ **якоря** *el.* armature flux
потолок ceiling; *met.* crown *(of furnace)*
~, **боевой** *av. mil.* combat/operational ceiling
~ **на банковские кредиты** credit ceiling
~ **нормы прибыли** rate-of-return ceiling
~ **объема производства** production ceiling
~ **процентной ставки** interest rate ceiling
~ **ссуды** lending ceiling
потолочина *min.* ceiling, top; ceiling beam
потоотделение *med.* perspiration, sweating, sudation, hidropoiesis, diaphoresis
~, **мочевое** ur(h)idrosis
~, **одностороннее** hemiphoresis, hemidrosis
~, **повышенное** ephidrosis, hyper(h)idrosis
~, **пониженное** hidroschesis, ischidrosis
потоп flood; deluge
~, **всемирный** the Flood, the Deluge
поточник *bot.* beccabunga speedwell *(Veronica beccabunga)*
поточный flow *(attr.)*
~**ная система** conveyer/conveyor system
~**ное производство** line production
потребитель consumer, user
~, **коммунальный** utility consumer
~, **конечный** final/ultimate consumer
~, **оптовый** large-scale consumer
потребительная стоимость use value
потребительский consumers', consumer *(attr.)*
~**кая кооперация** consumers' cooperative
~**кие товары** consumers' goods
~**кое общество** consumer society
потребление 1. consumption; use; 2. demand; *tech.* load
~, **внутреннее** domestic/home consumption
~ **кислорода** oxygen demand; oxygen consumption
~, **конечное** final/ultimate consumption
~, **личное** personal/private/individual consumption

~ на душу per capita consumption
~, номинальное rated consumption
~ обрабатывающей промышленности manufacturing consumption
~, показное conspicuous consumption
~, производственное productive consumption
~, расточительное wasteful consumption
~ семей household consumption
~, суммарное aggregate consumption
~ тепла heat demand
~ энергии power consumption
потреблять consume; use; expend; demand
потребность demand, need, requirement; necessity, want
~, естественная physical need
~ кислорода oxygen demand
~, разовая one-time demand/requirement
~, энергетическая power demand/requirements
потуги в родах labo(u)rs
потухание extinction
потягивание *med.* pandiculation
поход march; *naut.* cruise; *mil.* campaign; *sport* hike, trip, excursion
~, крестовый *hist.* crussade
походка gait; walk, step
~, атактическая ataxic gait
~, волочащаяся dragging/paraparetic gait
~, качающаяся staggering gait, titurbation
~, лягушечья saltatory gait
~, ножницеобразная scissor [cross-legged] gait
~, петушинная prancing/steppage gait
~, спотыкающаяся stumbling gait
~, утиная waddling gait
походный 1. marching, march *(attr.)*; 2. mobile, field *(attr.)*
~ порядок marching order
~ная кровать field/folding bed
~ная кухня field kitchen
похотник *med.* clitoris
почасовой by the hour
початковидный, початкообразный *bot.* spadiciform
початковый *bot.* spadiceous, earlike
початкодробилка corncob crusher/grinder
початкоотделитель *agr.* corncob snapper
початкоочиститель (corncob) husker, husking box
початкособиратель, початкосрыватель corn(cob) picker
початкоцветковый *bot.* spadix-flowered
початок 1. *bot.* ear; spadix; corn ear, concob; 2. *text.* cop
~, уточный weft/pin cop
почва soil, earth; ground
~, аллювиальная alluvial/silt soil
~, безводная arid soil
~, глинистая argillaceous/clayey soil
~, гумусовая humus soil
~, заболоченная waterlogged soil
~, засоленная salinized/saline soil
~, известняковая limy/chalky soil
~, илистая silty soil
~, каштановая/коричневая chestnut soil; brown earth
~, кислая sour/acid soil
~, лессовая loessal soil, loess
~, луговая meadow soil
~, мерзлотная cryogenic soil
~, наносная alluvial/drift/erratic soil
~, оподзоленная podsolized soil
~, подзолистая podsolic soil
~, рисовая paddy soil
~, серо-коричневая grey-brown soil
~, щелочная alkaline soil
почвенный soil *(attr.)*; edaphic, pedological, pedogenic
~ покров soil cover
почвовед pedologist, soil scientist
почвовед-генетик soil geneticist
почвоведение pedology, edaphology, soil science
~, мерзлотное cryopedology
почвоведческий pedologic(al)
почвозацеп nonskid wheel grip; grouser, lug
почвообрабатывающий soil-cultivating/tilling
почвообразование soil formation/genesis
почвообразовательный, почвообразующий soil-forming
почвоуглубитель subsoil plough/plough/ripper/ cultivator, subsoiler
почвоуглубление subsoil plowing/ploughing
почвоулучшающий soil-improving/conditioning
почвоулучшитель soil conditioner
почвоуплотнитель soil packer, planker
почвоутомление soil fatigue
почвофреза rotary cultivator/plow/plough, rototiller, gyrotiller
почернение blackening, nigrescence, growing black
почесуха *med.* prurigo
~ Гебры Hebra's prurigo
~, детская Hutchinson's prurigo, prurigo of dentition
~, узелковая Hyde's disease
почечка little bud/nodule; gemmule, budlet
~, выводковая *bot.* propagulum, propagule
~ зародыша plumule
почечно-желудочный renogastric
почечно-кишечный renointestinal
почечно-клеточный nephrocellular
почечно-кожный renocutaneous
почечно-легочный renopulmonary
почечно-сердечный nephrocardiac, renicardiac, cardiorenal
почечный I *anat.* renal, nephric
~ камень 1. gallstone; 2. *miner.* nephrite
почечный II bud *(attr.)*; gemmaceous
~ покров perula, leaf-bud scale
~ная чешуйка tegmentum, bud scale
почка I *anat.* kidney
~, амилоидная amyloid/waxy kidney
~, бесформенная lump kidney

~, **блуждающая** wandering/floating kidney
~, **губчатая** sponge kidney
~, **застойная** congestive kidney
~, **подвижная** floating/movable kidney
~, **рубцово-перерожденная** cicatrical kidney
~, **цирозная** cirrhotic kidney
почка II bud; gemma; *min.* nodule
~, **верхушечная** crown/terminal bud
~, **зародышевая** gemma
~, **пазушная** axillary bud
~, **прививочная** *hort.* eye
~, **придаточная** adventitious bud
~, **рудная** nodule
~, **спящая** resting bud
почкование budding, gemmation, plastogamy
~, **верхушечное** terminal budding
~, **внутреннее** endogenous budding
~, **межстенное/промежуточное** intermural increase
почковаться *bot.* pullulate; *biol.* bud, gemmate
почковиднoплодный *bot.* nephrocarpous
почковидный, почковатый, почкообразный 1. reniform, kidney-shaped; 2. gemmiform, gemmaceous, budlike; 3. nodular
почковый bud *(attr.),* gemmaceous
~**вая разновидность** bud mutation/sport
почколистный *bot.* kidney-shaped, gemmiferous
почкосложение, почкосмыкание *bot.* (a)estivation
почленный *math.* term-by-term, termwise
почта 1. mail, post; 2. post-office
~, **воздушная** air mail
~, **дипломатическая** diplomatic pouch
~, **заказная** registered mail
~, **обратная** return mail
~, **пневматическая** airtube [pneumatic tube] message system, pneumatic rabbit
почтамт postmaster office, main post office
почти-кольцо *math.* near-ring
почтовый postal; mail *(attr.);* post-office
~ **вагон** mail car
~ **голубь** carrier/homing pigeon
~ **перевод** postal money order
~ **сбор** postage
~ **ящик** mailbox
~**вая бумага** writing paper
~**вая марка** postage stamp
~**вая оплата** postage
~**вое отделение** branch post office
~**во-посылочная служба** parcel post
~**вые расходы** postage
пошлина duty; toll; customs; levy
~, **ввозная/импортная** import duty
~, **гербовая** stamp duty
~, **дифференциальная** differential tariff; differential/discriminating/preferential duty
~, **запретительная** prohibitive duty
~, **карательная** retaliatory duty
~, **компенсационная** countervailing duty
~, **наследственная** legacy duty; estate/death duty
~, **патентная** patent fee
~, **переменная** variable levy
~, **покровительственная** protective duty
~, **портовая** wharfage
~, **преференциальная** preferential duty
~, **протекционистская** protective duty; protection/protective tariff
~ **с веса** poundage
~, **таможенная** customs
~, **транзитная** transit duty
~, **уравнительная** countervailing duty
~, **фискальная** revenue duty
~, **экспортная** export tax
пошлинник tax collector; collector of customs
поштучный by the piece
~**ная заработная плата** piece wages
поэтапный stage-by-stage
пояс 1. belt, girdle; band, waistband; sash; *(of vessel)* strake; 2. zone, region, belt
~, **бортовой** side strake
~, **брюшной** pelvic girdle
~, **гигиенический** sanitary napkin
~, **килевой** keel strake
~ **конвертера** *met.* trunnion ring
~ **лонжерона** spar boom
~ **пассатов** belt of trade winds
~, **плечевой** *anat.* shoulder girdle
~, **привязной** belly strap
~, **пробковый спасательный** cork belt/buoy
~, **скуловой** *naut.* bilge strake
~, **спасательный** life belt/preserver
~, **тазовый** *anat.* pelvic girdle
~, **тропический** tropical zone
~, **часовой** time zone
~, **швеллерный** channel belt
~, **шпунтовый** gardboard/keel/sand strake
~ **экваториального затишья** calm belt
поясница *anat.* loin
пояснично-грудной thoracolumbar
пояснично-крестцовый lumbosacral
пояснично-подвздошный lumboinguinal, lumboiliac
пояснично-реберный lumbocostal
пояснично-ягодичний lumbogluteal
пояснично-яичниковый lumboovarian
поясничный *anat.* lumbar
поясок little belt/girdle/band; zonule
правдоподобие likelihood
правеж *hist.* forced recovery (of debt) by torture
правила *pl.* rules, regulations; principles; laws
~ **безопасности** safety regulations/rules
~, **валютные** currency regulations
~ **внутреннего распорядка** standing orders
~, **Йорк-Антверпенские** York-Antwerp rules
~, **карантинные** quarantine regulations/laws
~ **обслуживания** service regulations
~ **ограничивающие уровень шума** noise regulations
~ **отбора** selection rules

~, портовые harbour rules, port regulations
~ приемки acceptance rules
~ сделок с ценными бумагами securities regulations
~, таможенные customs rules/regulations
~ техники безопасности safety rules/standards
~ техническото обслуживания maintenance regulations
~ уличного движения traffic rules/regulations
~ установления цен price regulations
~ ухода service regulations
~ эксплуатации operating rules, service regulations

правило rule, regulation; principle; law
~ буравчика *el.* right-hand screw/corkscrew/thumb/Ampere's rule
~ выбора очереди queue-selection rule
~ вывода rule of inference
~ замещения substitution rule
~ знаков *math.* convention/rule of signs
~ левой руки *el.* left-hand rule
~ ложного положения *math.* method/rule of false position, false-position method
~ площадей *aero* area rule
~ правой руки *el.* right-hand rule
~ средней трети middle-third rule
~ суммирования Эйнштейна Einstein's summing convention
~, твердое hard-and-fast rule
~, тройное *math.* rule of three, golden rule

правильный correct, right, regular, proper; true
~ цветок actinomorphic flower
~ная дробь *math.* proper fraction

правка correction; trimming, dressing, straightening; setting; adjustment
~ корректуры proofreading

правление 1. rule, government, governing; management; 2. board of management/administration/directors, management board

право right; title; law
~, абсолютное right in rem
~, авторское copyright
~, административное administrative law
~ ангарии right of angary
~ апелляции (right of) appeal
~ аренды tenant right
~, брачно-семейное matrimonial law
~, вещное real right, right in rem; property law; proprietary interest/right
~ владельца прибрежного участка riparian right
~ владения title
~, внутреннее/внутригосударственное national/domestic law
~, военное military/martial law
~, военно-морское admiralty/naval law
~, воздушное air law, law of the air
~ войны law of war/arms
~, встречное/противостоящее countervailing equity
~ входа ingress
~ выхода egress
~ голоса voting right, suffrage
~ государства (на принудительное отчуждение собственности) right of eminent domain
~, государственное public/constitutional law
~, гражданское civil law
~, давать entitle
~ давности prescription
~, действующее law in force; positive law
~, деликтное law of torts
~, дискреционное option
~, договорное law of contract, contract/conventional law
~, доказательственное law of evidence
~ допуска right of access, access right
~, естественное absolute law
~ заимствования, специалное special drawing right
~, законное/закрепленное преимущественное vested prior right
~ законодательной инициативы power to initiate legislation
~, залоговое lien
~ захвата right of seizure
~, земельное land law
~, избирательное elective franchise
~, каноническое canon law
~, коллизионное (law of) conflict of laws
~, коммерческое commercial law
~, континентально-европейское continental law
~, космическое space law
~ лиц law of persons
~, материальное substantive law, law of substance
~, международное international law, law of nations
~, международное договорное treaty law, international conventional law
~, международное морское international maritime law, international law of the sea
~, международное частное private international law
~, морское law of the sea, maritime/admiralty/sea law, merchant shipping law
~, морское транспортное maritime shipping law
~ на иск right of action, right to claim
~ на отдых right to rest and leisure
~ на спасение right of salvage
~ на узуфрукт usufructory right
~, наследственное succession law, law of succession
~ обратного требования (к индоссату) right of recourse
~ обращения в суд right of access to courts
~, общее (common) law
~ обыска right of search

~, обычное custom/consuetudinary/customary/common/tacit law
~, обязательственное law of obligation
~ осмотра right of visit/inspection
~ первоочередности right of way
~ пересмотра power of review
~ подписи authority to sign
~ подписки на акции stock subscription right
~ посадки landing right
~, преимущественное right of priority
~, прецедентное case/judgement law; judge-made law; judicial legislation
~, призовое prize law
~ присвоения appropriative right
~ проезда/прохода right of passage/way, wayleave
~, процессуальное law of procedure; right to procedure
~, публичное public law
~ регресса right of recourse; right of contribution/relief; remedy over
~ рыболовства right of fishing/fishery; piscary
~, семейное law of domestic relations
~ собраний right of assembly
~ собственности right of property, proprietary right/interest; property law; ownership
~ спасителя на удержание спасенного имущества salvage lien
~ справедливости equity law
~ страны флага судна law of the flag
~, торговое merchant/commercial law, law merchant, law of merchants
~, торговое обычное custom of merchants
~ убежища right of asylum
~ удержания right of retention, lien
~, церковное law spiritual/Christian, ecclesiastical law, Law of the Church
~, частное private law
правобортовой starboard *(attr.)*
правовед lawyer, jurist
правоведение jurisprudence
правоверный 1. orthodox; 2. Moslem
правовинтовой right-handed
правовой legal, law *(attr.);* rightful
правовращающий right-handed; dextrorotory
правоглазие *med.* dextrocularity
правозаступник lawyer
правоинвариантный right-invariant
право-левоглазие *med.* dextrosinistrocularity
правомерность legitimacy
правомерный lawful, legitimate
правомочие competence
правомочность competency, competence
правомочный competent
правонарушение offence, tort, malefaction, infringement/transgression of the law
~ против личности tresspass to the person
правонарушитель offender, lawbreaker, law-breaker, delinquent, wrongdoer, transgressor
правооснование *law* title
правоотношение legal relation
правопорядок law and order
правопреемник *law* assignee, cessionary; successor
правопреемство legal continuity/succession
правопритязание rigít
праворукий dextral, dextromanual, right-handed
правосознание sense of justice
правоспособность legal capacity
правоспособный *law* competent; capable
правофланговый 1. right-flank/wing *(attr.);* 2. right-flank man
правоцентристский *pol.* centre-right
правый right, right-hand *(attr.); naut.* starboard *(attr.)*
празеодим *chem.* praseodymium, Pr
праиндоевропейский proto-Indo-European
праистория primeval history, protohistory
пракрит, пракритский Prakrit
практикум practical work; tutorial
практицизм practicalism; empiricism; pragmatism
прапорщик *mil.* 1. warrant officer; 2. *hist.* ensign
праславянский proto-Slavic
праща sling
преагональный *med.* preagonal, preagonic
преамбула preamble
превалирующий prevalent, prevailing, predominant
превращение conversion, transformation, transmutation; change, metamorphosis
~ в газ gasification
~ в жидкое состояние liquefaction
~, фазовое phase transformation/transition
~, химическое chemical transformation
~ элементов transmutation of elements
превышение excess, exceeding, surpassing
~ давления overpressure
~ кредита overdraft
~ лимита подписки oversubscription
прегипофиз prehypophysis
прегнин pregneninolone, ethisterone
преграда barrier, obstruction, obstacle
~, грудобрюшная diaphragm
предавать суду take to court
предбрюшинный *anat.* preperitoneal
предварительный preliminary; tentative; *law* interlocutory
~ная команда *mil.* preparatory command
~ная продажа advance sale
~но напряженный prestressed
~но программированный preprogrammed
~ное заключение imprisonment before trial, detention under remand
~ное следствие *law* preliminary investigation, inquest
предвыборный election *(attr.),* pre-election *(attr.)*
предвычисление 1. precomputation; 2. *naut.* prediction
предгидролиз prehydrolysis

предгрудинный *anat.* presternal
предгрудник *zool.* omosternum
преддверие vestibule; antrum; entrance
~ ада limbo
~ влагалища entrance to the vagina
~ гортани vestibule of the larynx, false glottis
~ привратника желудка pyloric end of the stomach
преддверно-спинальный *anat.* vestibulospinal
преддетонация predetonation, preknock
предел limit; boundary, bound; extent; capacity; range; end
~ в среднем limit in the mean
~, верхний upper/superior limit
~ видимости visibility range/limit
~ выносливости endurance/fatigue limit
~ дальности range limit
~, допустимый tolerance limit
~, доверительный *math.* confidence limit
~ измерения measuring range
~ насыщения saturation point/limit
~, нижний lower/inferior limit
~ обнаружения limit/threshold of detection
~ погрешности error limit, margin of error
~ ползучести creep strength/limit
~ прочности ultimate strength
~ прочности на разрыв/растяжение ultimate tensile strength
~ скорости velocity ceiling; speed limit
~ слышимости limit of audibility/audition
~ текучести yield strength/limit/point
~ тяги thrust limit
~ упругости limit of elasticity; elastic limit
~ усталости fatigue strength/limit; endurance limit
предельный ultimate, terminal, limiting, limitary, marginal; extreme, utmost, maximum
~ возраст maximum age
~ срок deadline
~ная скорость maximum/top speed
преджелудок forestomach, preventiculus
преджелудочный pregastric
предзавкома *abbr.* chairman of factory (trade-union) committee
предикат *gram., philos.* predicate
предикатный predicative, predicate
предкавказский ciscaucasian
предкамера antechamber, prechamber, precombustion chamber, premixing cavity, burner cap
предклин, предклинье *anat.* precuneous
предконечный penultimate; prefinal
предкрылок *aero.* leading-edge flap/slat
~, убирающийся retractable slat
предлежание presentation
~ плаценты placental presentation, placenta previa
~ плечика плода *med.* shoulder/transverse presentation, transverse lie
~ плода presentation
~ плода, неправильное malpresentation
~ плода, нормальное головное normal cephalic presentation, vertex presentation
~ плода, ягодичное breech presentation
~ пуповины presentation of the cord
предлог *gram.* preposition
предложение I offer, proposal; *econ.* supply; *fin.* bid, bidding
~, встречное counteroffer
~, избыточное *econ.* oversupply
~, ложное sham/straw bid
~, наилучшее highest/best bid
~, наихудшее lowest bid
~ о покупке контрольного пакета акций take-over bid
~, рационализаторское proposal for (technical) innovation; efficiency suggestion
~ цены bidding
предложение II *gram.* clause; sentence; *philos.* proposition
~, главное main clause
~, придаточное dependent clause
~, простое simple sentence
предложно-именной *gram.* propositional-nominal
предложно-местный *gram.* propositional-locative
предложный *gram.* prepositional
предмет 1. object; article, item; thing; *top.* feature; 2. subject; topic, theme; *educ.* course
~ в каталоге catalogue item
~, материальный *law* tangible thing
~, местный landmark, feature of terrain
~, неодушевленный inanimate object
~, одушевленный animate object
~ потребления object of utility
~ продовольственного снабжения food item, item of subsistence
~ снабжения supply item; item of issue
~ спора *law* matter in dispute/issue
~, факультативный elective course
~ты домашнего обихода household goods/articles
~ты первой необходимости basic/prime necessities, staple commodities
~ты ширпотреба consumer goods
предметно-обстоятельственный *gram.* objective-adverbial
предметность concreteness
предметный object *(attr.)*, item *(attr.)*; objective; subject *(attr.)*
~ каталог subject catalog(ue)
~ указатель subject index
~ное стекло microscopic slide
предметодержатель stand, mount; slide
предмостное укрепление bridgehead
преднизолон *pharm.* prednisolone, meticortelone
преднизон *pharm.* prednisone, dehydrocortisone
предоперационный preoperative
предопухолевый *med.* pretumour
предотвращение prevention

предохладитель precooler
предохранение protection; safety
предохранитель safety device/lock/stop/catch safeguard; *el.* fuse, cutout
~, плавкий (safety) fuse, fusible plug/cutout
~, трубчатый cartridge fuse
предохранительный protective, protecting, preservative; preventive; safety *(attr.)*; prophylactic
предохранять protect, safeguard; guard; prevent
предочистка воды water pretreatment
предписание (letter of) instruction, order; ordinance; assignment; *law* breve, precept
~, командировочное travel order
предплечный, предплечевой antebrachial
предплечье *anat.* forearm
предплодие *bot.* procarp
предплужник co(u)lter
предплюсна *anat.* tarsus, sole of the foot
предплюсневой *anat.* tarsal
предплюсне-плюсневой *anat.* tarsometatarsal
предпобег *bot.* protocorm
предположение assumption, supposition, surmise, hypothesis, conjecture
предполетный pre-flight
предположительный presumptive, conjectural
предпосадочный pre-landing *(attr.)*
предпосевной presowing
предпочка *embryol.* pronephros
предприниматель entrepreneur; employer
предпринимателский entrepreneurial
предпринимательство entrepreneurship; business undertakings; (spirit of) enterprise
предприятие enterprise, undertaking; business; agency, works
~, аграрно-промышленное factory-farm enterprise
~, головное parent enterprise
~, государственное state(-owned) enterprise
~, деловое business enterprise
~, коммерческое commercial enterprise
~, коммунальное utility enterprise/plant
~, кустарное/ремесленное artisan enterprise
~, мелкое small-scale enterprise
~, многоотраслевое multi-activity enterprise
~ общественного питания public catering establishment
~,отечественное domestic enterprise
~, прибыльное paying concern
~, семейное family-owned/run enterprise
~, смешанное/совместное 1. joint venture; 2. partnership
~, функционирующее *econ.* going concern
~, хозрасчетное financially autonomous enterprise
предрак precancer, precancerosis
предраковый precancerous, premalignant
предрасположение, предрасположенность predisposition, diathesis
предродовой prenatal, antepartem
председатель chairman, chairperson; president
~ совета директоров chairman of the board
председательствующий chairman, chairperson
предсердечный *anat.* precardiac, precordial
предсердие *anat.* auricle
предсердно-желудочковый *anat.* atrioventricular
предсерийный pre-series, pre-production *(attr.)*
предсказание prediction, forecast, prognostication
~ надежности reliability prediction
~ погоды weather forecasting/forecast
предсказывать predict, forecast, prognosticate
предсмертный premortal, antemortem
предспора forespore
представимый *math.* representable
представитель representative; agent; spokesman
~ власти public official/agent
~, торговый trade representative
представительство representation; agency; representatives
~, генеральное general agency
~, дипломатическое diplomatic representatives/mission
~, торговое trade representative/agency
представление presentation; representation; notation; *theatre* performance
~ в аналоговой форме analog(ue) representation
~ в допольнительном коде complement representation
~ в коде с дополнением до двух two's complement representation
~ в прямом коде true representation
~ в цифровой форме digital representation
~, графическое graphical/graphic/diagrammatic representation; diagram; graphics
~ данных data presentation/display
~, двоично-десятичное bidecimal notation
~ информации data presentation
~, неприводимое *math.* irreducible representation
~ с плавающей запятой/точкой floating-point notation/representation
~ с фиксированной запятой/точкой fixed-point notation/representation
~, символьное character representation
~, цифровое digital notation/presentation
~ числа notation
~ числа, двоичное binary notation
предстарческий presenile
предстательный prostatic
предтеча forerunner, precursor; **Иоанн П.** John the Baptist
предтопок *met.* primary furnace
предумышленность, предумышление premeditation
предумышленный premeditated; deliberate
предупредительный 1. warning, precautionary; 2. preventive
предупреждение 1. warning, caution; 2. prevention

~ беременности/зачатия contraception
~ болезни disease prevention
~ о тумане fog signal
~, официальное formal notice
~ пожара fire prevention
~, раннее early warning
предусиление preamplification
предусилитель preamplifier
предускорение preacceleration
предхрящ *anat.* precartilage
предчашечный *anat.* prepatellar
предшественник predecessor; precursor; *agr.* forecrop, preceding crop
~ запаздывающих нейтронов neutron precursor
предъявитель presenter; *fin.* bearer
~ иска *law* plaintiff
~, чек на check/cheque payable to bearer
предъявление presentation, presenting, producing
~ иска prosecution
~нию, по on presentation, at sight
предъявлять, предъявить present, produce, show; bring forward
~ обвинение bring an accusation, charge
предынтегральный *math.* preintegral
предыонизация preionization
предынфарктный *med.* preceding/before infarction
предыскание preselection
предэклампсия *med.* preeclampsia
предэпилептический *med.* pre-epileptic
преждецветный proteranthous, flowering first
презентант *fin.* bearer
презерватив condom
презумпция presumption
~ вменяемости presumption of sanity
~ доказательства *law* prima facie evidence
~, неопровержимая conclusive/absolute presumption
~, правовая presumption of law
~, фактическая presumption of fact
преимущественный preferential, priority *(attr.)*; pre-emptive
~ное право right of priority; preference
преимущество advantage; preference; *law* privilege; priority
преисподня inferno, hell, the nether world
прейскурант price list
~ сезонных тарифов seasonal rate schedule
~ тарифов tariff book
преканкроз *med.* precancrum
прекапилляр *anat.* precapillary
прекарциноматозный *med.* precarcinomatous
преклимакс preclimax; *med.* preclimacterium
преколлагеновый precollagenous
прекоматозный *med.* precomatose
прекрасноголовчатый *bot.* calocephalous
прекращать, прекратить terminate, discontinue, break off, suspend, sever; stop
прекращение discontinuation, termination, stopping, ending, cessation, stoppage; interruption
~ огня *mil.* cease-fire
~ питания [подачи энергии] power failure
~ работы work stoppage, walkout
~ связи communication failure
прелат prelate
прелатский prelatial, prelatic
прелина *leath.* rotten spot/stain
преломление refraction
~ звука sound/acoustic refraction
преломлять refract
прелюд(ия) prelude
премиальные bonus
премия bonus, premium; prize, reward; award
~ за заслуги merit bonus
~, паушальная страховая lump-sum premium
~, поощрительная incentive bonus/prize
~, страховая insurance premium
~, целевая task bonus
~, экспортная export bonus
премоляр *anat.* premolar tooth
пренебрежимо малый *math.* negligible
пренилтрансфераза prenyltransferase
пренит *miner.* prehnite
прения *pl.* debate(s), discussion(s)
~ сторон *law* oral hearing, hearing of arguments
преображение transformation; *rel.* Transfiguration
преобразование transformation, conversion; transform; reconstruction; transmutation
~, аналого-цифровое analog(ue)-to-digital conversion, AD conversion
~, бесконечно малое infinitesimal transformation
~, булевое Boolean transformation
~ в цифровую форму digitalization
~ данных data conversion, translation of data
~ данных, предварительное data reduction
~ двоичного кода binary code conversion
~, двоичное binary conversion/transformation
~ звезды в треугольник star-to-delta conversion
~ из двоичной системы в десятичную binary-to-decimal conversion
~ из десятичной системы в двоичную decimal-to-binary conversion
~ из параллельной формы в последовательную parallel-serial conversion, serialization
~, каноническое canonical transformation
~ координат coordinate transformation, transformation/conversion of coordinates
~ Лапласа Laplace transform(ation)
~ Лапласа, двустороннее two-sided [bilateral] Laplace transform(ation)
~ Лежандра Legandre('s) transform(ation)
~ многоугольник-звезда *el.* mesh-star conversion
~ непрерывных величин в цифровые analog(ue)-to-digital conversion
~, обратное reverse conversion; inverse transfor-

mation
~, пакетное batched compilation
~, проективное projective transformation
~ проекции *cart.* rectification of projection
~ речевых данных в цифровую форму voice digitalisation
~ свертки *math.* convolution transform
~ случайной величины transformation of variate
~ треугольника в звезду *el.* delta-to-star convertion
~ Фурье, быстрое fast Fourier transformation
~, цифроаналоговое digital-to-analog(ue) conversion
~ частоты frequency conversion
~ энергии energy conversion
преобразователь converter; inverter; transducer; transformer; processor
~, амплитудно-временной pulse-height pulse-width converter
~ аналог-код analog(ue)-to-number converter
~, аналого-цифровой analog(ue)-to-digital converter/encoder, digitizer
~ вида колебаний mode transducer/transformer
~, время-импульсный time-to-pulse converter
~, гидроакустический sonar transducer
~ дальности range converter
~ давления pressure pick-off, pressure transducer
~ данных data converter/transducer/set; modem
~ двоичного кода в десятичный binary-to-decimal converter
~ десятичного кода в двоичный decimal-to-binary converter
~ дискретных данных в непрерывные digital-analog(ue) converter
~ знаков character translator
~ из десятичной системы в двоичную decimal-to-binary converter
~ импульсов pulse converter
~ кода code converter/transducer
~ напряжения в код voltage-to-digital converter
~, независимый off-line converter
~, обратный inverter
~, обратимый reversible transducer
~, одноякорный rotary converter
~, параллельно-последовательный serializer
~ полного сопротивления impedance transformer
~, последовательно-параллельный deserializer
~, ртутный mercury-arc converter
~ системы счисления radix converter
~ строчного стандарта *telev.* line-scan conversion unit
~ угол-код shaft angle encoder
~ углового перемещения в код shaft position-to-digital converter, shaft digitizer
~, функциональный function generator
~, цифроаналоговый digital-to-analog(ue) converter
~, цифровой digitizer
~, электромашинный rotary converter
~, электронно-оптический image converter/tube
~ энергии, термоэлектронный thermionic converter
преобразователь-генератор self-generating transducer
преобразователь-двигатель modulating transducer
преобразовывать convert, invert; transduce, transform, transmute; modify, change; *math.* transpose
~ в цифровую форму digitize
преосвященный 1. Right Reverend; 2. bishop
препарат 1. preparation; compound; 2. specimen, mount, slide; section
~, гистологический histologic specimen
~ железа iron preparation
~, жирующий *leath.* oiling agent
~, лекарственный medicinal preparation, drug
~, лиофилизированный frozen-dried preparation
~, микробиологический microbiologic specimen
~, спазмолитический anticonvulsive drug
~, ферментный enzyme preparation
~ы, галеновые galenicals
~ы, неогаленовые neogalenicals
препаратодержатель mounting screen *(of microscope)*
препаратор laboratory assistant
препинания, знаки *gram.* punctuation marks
преподобие *eccl.* Reverence *(title)*
преподобный holy; Saint; Venerable; *(of priest)* reverend
препуциальный *anat.* preputial
препуций *anat.* prepuce
препятствие obstacle, impediment, obstruction; hindrance; barrier
препятствовать hinder, obstruct; prevent; impede; block
~ правосудию *law* obstruct justice
прерывание interruption, break; *comp.* interrupt
~, аппаратное hardware interrupt
~ беременности abortion
~, векторное vectored interrupt
~, внешнее external interrupt
~, заблокированное disabled interrupt
~, игнорируемое disarmed/ignored interrupt
~ из-за ошибки в программе program(me)-check interrupt
~, информационное data interrupt
~, командное command interrupt
~ луча beam chopping
~, маскированное disabled/masked interrupt
~, маскируемое maskable interrupt
~, обратное reverse interrupt

~ от периферийного устройства peripheral interrupt
~ питания power interruption
~ по вводу-выводу input-output interrupt
~ по времени time/clock interrupt
~ по вызову polling interrupt
~ по вызову супервизора *comp.* supervisor interrupt
~ по запросу query interrupt
~ по защите памяти memory protection interrupt
~ по машинному сбою machine check interrupt
~ по нарушению четности parity interrupt
~ по освобождению not-busy interrupt
~ по сигналу об ошибке error interrupt
~ по таймеру clock interrupt
~, программное programmed/software interrupt
~, программно-управляемое program(me)-controlled interrupt
~, разрешенное enabled interrupt
~ с изменяемым приоритетом priority-shifted interrupt
~ с ожиданием pending interruption
~, системное system-call interrupt; trap
прерыватель interrupter, breaker; *el.* contact/circuit breaker, cutout; chopper
~, импульсный pulse chopper
~, камертонный tuning-fork interrupter
~ пучка beam chopper
~, ртутный mercury interrupter/breaker
~, тиратронный thyratron interrupter/breaker
~, фотоэлектрический photoelectric chopper
прерывать interrupt, break (off), sever, disconnect
прерывность discontinuity
пресбиакузис *med.* presby(a)cusis, presbyacusia
пресбиопия *med.* presbyopia, presbytia, presbytism
пресветлейшество *obs.* Serene Highness
пресветлость *obs.* Highness
пресвитер *eccl.* 1. presbyter; 2. elder
пресвитерианец, пресвитерианский Presbyterian
пресвитерианство Presbyterianism
пресвитерия *rel.* presbytery
пресвятой most holy
~тая Дева *rel.* the Virgin
пресистола *med.* presystole, late diastole
преследование pursuit, chase; persecution, victimization; *law* prosecution
~ по горячим следам hot pursuit
~, судебное (legal) proceedings
~, уголовное criminal proceedings, prosecution
преследовать pursue, chase; persecute, victimize; *law* prosecute, sue, take legal proceedings/steps against
пресмыкающийся *zool.* creeping, crawling, vermigrade
пресноводный freshwater *(attr.),* limnetic
пресс press; punch; stamp
~, вальцовый roller press
~, винтовый screw press
~, вырубной/вырубочный blanking/stamping press
~, высадочный heading press
~, вытяжной drawing press
~, гибочный bending press
~ двойного действия double-action press
~, двухстоечный double-arm press
~ для объемной штамповки die-forging press
~, дыропробивной punching/piercing press
~, клепальный riveting press
~, ковочный/кузнечный forging press
~, коленчатый toggle press
~, кромкогибочный flanging/joggling press
~, обрезной trimming press
~, правильный/рихтовальный straightening press
~ простого действия single-acting press
~, пудлинговый squeezer
~, сенной hay baler/press
~, чеканочный coining press
~, челюстной jaw press
~, червячный screw/worm press
~, шнековый screw extruder
~, штамповочный stamping press
пресс-бетон pressed concrete
пресс-изделие moulded article
пресс-литье die-casting
пресс-масленка 1. grease cup/nipple; 2. forced lubricator, lubricating screw
пресс-масса/материал moulding material
прессовальный pressing
прессование 1. pressing; moulding; 2. extrusion; 3. compaction
прессовать 1. press, compress; compact; 2. mould; 3. extrude
прессовка pressing, compression; compaction; moulding
прессовщик press operator/attendant
пресспапье 1. blotter; 2. paperweight
пресс-подборщик pickup baler
пресс-форма compression mould
~ для литья под давлением injection mould
~, литьевая transfer mould
~, многогнездная multicavity mould
~, секторная collapsible mould
прессшпан pressboard
престиж prestige, image
престижная деятельность public relations
престол 1. throne; 2. *eccl.* see; alter;
~, святейший (апостолический) Holy (Apostolic) See
преступление crime, offence; malefaction; *law* felony
~, должностное malfeasance
~, нераскрытое unexposed crime
~, предумышленное premeditated crime, crime of forethought
~ совершенное в результате упущения/бездействия crime of omission

~ совершенное по небрежности crime of negligence
~, уголовное criminal offence
преступник criminal, offender; malefactor; *law* felon
~, военный war criminal
~, молодой совершеннолетний young adult offender
~, невменяемый criminal lunatic
~, несовершеннолетний juvenile delinquent
преступность criminality; maleficence; crime; crime rate
~, городская urban crime
~ несовершеннолетних juvenile delinquency
~ престарелых senile delinquency
преступный criminal, felonious; maleficent, malfeasant
~ная небрежность criminal neglect
претензия claim; demand; complaint
~ удовлетворяемая в первую очередь senior claim
~ удовлетворяемая после первоочередных junior claim
претерит, претерительный, претеритный *gram.* preterit
претор *hist.* praetor
преторианец, преторианский *hist.* praetorian
преторий, преториум pretorium
префикс *gram.* prefix
префикс платеж *comm.* prior payment
прецедент precedent case
прецизионность precision, accuracy
прецизионный precision *(attr.),* precise
прецирроз *med.* precirrhosis
преюдициальный *law* pre-judicial
приадъективный *gram.* ad-adjectival
приальпийский subalpine
Приап *myth.* Priapus
приапический *myth.* Priapean
прибавочный additional; *econ.* surplus *(attr.)*
~ продукт surplus product
~ная стоимость surplus value
прибайкальский Baikal *(attr.),* adjacent to the Baikal
прибалтийский Baltic, along the Baltic Sea
приближаться approach; *math.* approximate
приближение 1. approach(ing); 2. *math.* approximation
~ более высокого порядка higher approximation
~, борновское Born approximation
~, взвешенное *stat.* weighted approximation
~, грубое rough approximation
~, допустимое valid approximation
~, кусочно-линейное piecewise-linear approximation
~, минимаксное minimax approximation
~, наибольшее nearest approximation
~ наименьших квадратов least-squares approximation
~ первого порядка first-order approximation
~, последовательное successive/progressive approximation, approximation by iteration
~ с весом *stat.* weighted approximation
~, точечное point approximation
приближенный, приблизительный approximate, rough
прибой surf, beating/battering of waves
прибойник rammer
прибор device, appliance, apparatus, instrument, gear; gauge
~, автономный self-contained/reacting device, independent/autonomous unit, off-line equipment
~ автоматического управления automatic control device
~, барометрический pressure-operated instrument
~, бортовой авиационный airborne instrument
~, вспомогательный support/auxiliary instrument
~, выходной end instrument
~, гравиметрический pendulum apparatus
~, дистанционный измерительный telemeter
~, дозиметрический radiation monitoring instrument; radiation protection instrument
~, звукометрический sound analyzer
~, измерительный measuring device, meter
~, индивидуальный дозиметрический personal dosimeter
~, индикаторный reading/indicating instrument
~, ионный gas-filled device/tube
~, калибровочный calibrator
~, контрольный monitoring/check/reference instrument, monitor
~, курсовой heading instrument
~ летчика cockpit instrument
~, манометрический pressure-operated instrument
~, метеорологический weather instrument
~, многопредельный/многошкальный multirange instrument
~, навигационный плановый *aero.* horizontal situation indicator
~, образцовый standard instrument
~ обнаружения detecting instrument
~, оконечный end instrument/device
~, оптикоэлектронный optoelectron device
~ опознавания identification device
~, пилотажный flight instrument
~ регистрации наработки hourmeter instrument
~, регистрирующий/самопишущий recording instrument, recorder; register
~, регулирующий controller
~, резервный stand-by instrument
~ с круглой шкалой dial instrument
~ с непосредственным отсчетом direct-reading meter/instrument

~, сдвоенный dual-type instrument
~, светосигнальный blinking device
~, скважинный down-hole logging instrument
~ слепой посадки blind-landing instrument
~, стрелочный pointer-type instrument
~, сцепной *rail.* draft gear, draw-gear
~, тензометрический strain-gauge instrument
~, термоэлектрический thermocouple instrument
~, точный precision instrument
~, угломерный inclinometer
~, ударный *rail.* buffer gear
~, цифровой измерительный digital instrument
~, шкальный indicating instrument
~, электронно-лучевой cathode-ray tube/device
~, эталонный standard/calibration/reference instrument
приборист instrument operator/technician
приборостроение instrument manufacture/engineering
прибрежник *bot.* shoreweed *(Littorella)*
прибрежноморской coastal, seashore *(attr.)*
прибыль 1. profit, gain, return; 2. *met.* head (metal), deadhead, (lost) head; riser
~, атмосферная *met.* atmospheric riser
~, бухгалтерская *econ.* book profit
~, валовая gross profit
~ до уплаты подоходного налога profit before tax
~, добавочная extra profit
~ за вычетом налога after-tax profit
~, избыточная surplus profit
~, извлекать derive profit
~, литейная casting head, riser
~, монопольная monopoly profit
~ на единицу продукции unit profit
~ на капитале capital gains
~, непредвиденная windfall profit
~, нераспределенная retained/undivided profit
~, нереализованная book profit
~, неустойчивая erratic earnings
~, плановая planned profit
~ подлежащая обложению налогом taxable profit
~, получать make a profit
~ распределяемая на обыкновенные акции equity profit
~ слитка discard head, sinkhead
~, торговая merchant's profit
~, торговая валовая gross profit on sales
~, устойчивая sustained profit
~, чистая net profit/gains
~, чистая текущая net operating profit
прибыльность profitableness
~ капиталовложений earning power of real assets
приваривать weld on/to/together
приват-доцент *hist.* private/unsalaried lecturer
приведение 1. reduction, reducing; 2. bringing
~ в действие actuation; putting into effect
~ в движение setting in motion, starting
~ в створ alignment
~ к абсурду reductio ad absurdum
~ к норме normalization
~ к общему знаменателю reduction to a common denominator
~ к уровню моря reduction to sea level
~ подобных *math.* collecting/combining similar/like terms
~ сил reduction of forces
приведенность *math.* reducibility
привес weight increase, gain in weight
~, суточный daily gain
привесок *biol.* appendage
прививать 1. *bot.* graft, engraft; 2. *med.* inoculate, vaccinate
~ внакрой splice-graft
~ глазком bud-graft
прививка 1. *bot.* graft, grafting, engrafting; 2. *med.* inoculation, vaccination
~ внакрой splice grafting
~ врасчеп split grafting
~ глазком bud grafting, budding
~ лапками whip grafting
~, первичная primary vaccination
~, повторная revaccination
~, противооспенная vaccination
прививок *bot.* 1. graft, scion; 2. grafted sapling
привлекать draw; attract
~ к суду [к судебной ответственности] bring to trial, put on trial, take to court
привод I taking into custody, arrest; bringing to court
привод II drive, gear; transmission; homing
~, асинхронный induction/asynchronous motor drive
~, безредукторный gearless/direct drive
~, бесступенчатый variable-speed drive
~, винтовой screw drive
~, гидравлический силовой hydraulic actuator
~, зубчатый gear drive
~, кулачковый cam drive
~, кулисный link-motion drive
~, механический power drive
~ на четыре колеса four-wheel drive
~ от вала shaft drive
~ от вала отбора мощности power take-off drive
~, реактивный jet-drive
~, реверсивный reversible/reversing drive
~, реечный rack-and-gear (rack-and-pinion) drive
~, ременный belt transmission/drive
~, сельсинный synchro power drive
~, тросовой cable linkage
~, храповой ratchet drive
~, червячный wormgear drive

~, шестеренчатый gear drive
~, эксцентриковый cam drive
приводимость *math.* reducibility
приводимый 1. *math.* reducible; 2. *mech.* driven
приводнение water landing, alighting on water, splash-down
~, аварийное emergency splash-down, ditching
~ космического корабля spacecraft splash-down
приводной 1. driving; actuating; power *(attr.)*; 2. homing
приводящий *anat.* adducting
привой *bot.* graft, scion
привратник желудка *anat.* pylorus
привязка *top.* tie, fixation, conjunction
~аэроснимков photogrammetric bridging; control of aerial photography; plate conjunction
~, геодезическая geodetic fixation/tie
~, топогеодезическая survey control
~, топографическая topographical control/survey
пригар, литейный sticking [burnt-on] sand
пригнутый *bot.* declinate; decumbent
приговор sentence; verdict
~, оправдательный acquittal; verdict of not guilty
пригодность usefulness, fitness, adaptability, serviceability, applicability
~ к обработке workability
пригодный useful, fit, serviceable, applicable, adaptable
пригонка fitting, adjustment, matching; fit
пригонять fit, adjust, match
пригорание scorching, burning; *eng.* sticking, seizing
~ поршневого кольца piston ring sticking
приданое *law* dotal property
придаток addition, supplement; appendage, adjunct; *anat.* appendage, annexum
~, жаберный branchial appendage
~, крыловидный alar projection
~, пузырчатый vesicular appendage
~, семенной *bot.* strophiole, hilium appendage
~, червеобразный vermiform appendage/appendix/process
~ яичка epididymis
~ яичника epoophoron
придаточный additional; accessory; *bot.* adventive, adventitious; *gram.* subordinate, dependent
~ венчик corolla appendage
~ корень secondary/adventitious root
~ное предложение subordinate clause
придонный *geol.* benthic, benthonic
придыхание *phon.* aspiration
придыхательный *phon.* aspirate
прием 1. reception, receipt; acceptance; admission; intake; admittance; 2. method, way, technique, procedure
~ данных data acquisition
~, дуплексный double reception
~ на слух [клопферный] aural reception, sound reading, reception by ear
~ радиолокационных сигналов radar reception
~, фототелеграфный facsimile reception
приемистость acceleration characteristics/capacity;
~ двигателя engine response/acceleration
приемка acceptance; receipt
приемник 1. *rad.* receiver; 2. receptacle, container, receiving vessel, bin, tank; hopper
~, батарейный battery-operated receiver
~, бортовой airborne receiver
~ видеосигналов video-receiver
~ давления pressure pick-up
~, детекторный crystal receiver
~ звука sound/sonic detector
~ излучения radiation detector/sensor
~ ионов ion collector
~, контрольный monitoring receiver
~, ламповый *rad.* valve/tube receiver
~, монофонический monoaural receiver
~, радиолокационный radar receiver
~ с противопомеховой защитой anti-jam receiver
~, сейсмоакустический seismicrophone
~, сетевой mains-/socket-powered receiver
~ сигналов/системы наведения guidance receiver
~ системы самонаведения homing receiver
~ слежения tracking receiver
~, супергетеродинный superhet/superheterodyne receiver
~ теплового излучения infra-red detector/cell
~, фототелеграфный facsimile receiver
приемно-передаточный *rad.* two-way, send-receive
приемно-передающий send-receive
приемный 1. receiving; collecting; reception *(attr.)*; admission, entrance *(attr.)*; 2. adoptive, adopted, foster *(attr.)*
~ день reception day
~ отец foster father
~ сын adopted son
~ экзамен admission examination
~ная комиссия *art* selection committee
~ные испытания acceptance trials
приемоответчик transponder
приемопередатчик *rad.* transceiver
приемо-передающий *rad.* send-receive
приемочный пункт reception centre
приемщик examiner, inspector, receiving clerk/official
приживление *surg.* implantation
приживлять graft; grow together
прижигание cauterization, cautery
~ химическими веществами chemical cauterization, chemocautery
~ электричеством electric cautery, electrocautery
прижигать cauterize; scorch slightly

прижим clamping, pressing; clamp, clip
прижог burn, scorched spot
приз 1. prize; award; 2. *law* capture, prize
приземление landing; touchdown
приземлять(ся) land, touchdown, touch the ground
~ с недолетом touch down short of runway, undershoot
~ с перелетом overshoot
приземной, приземный ground, surface *(attr.)*
призма prism
~, абдукционная *med.* abducent/abducting prism
~, замедляющая moderating column
~, косая/наклонная oblique prism
~, оборотная reversing prism
~, правильная regular prism
~, прямая right prism
~, светоделительная beam splitter
~, ступенчатая echelon prism
~, тепловая thermal column
~, усеченная truncated prism
призмоидальный, призмообразный prismoidal
признавать recognize; acknowledge, admit
~ виновным *law* return guilty
~ иск *law* plead no defence
~ себя виновным plead guilty
~ требование allow a claim
признак sign, characteristic, feature, symptom, trait; mark; attribute; *med.* character; *comp.* flag; *stat. math.* test; criterion
~ Д 'Аламбера d'Alembert's test, generalized ratio test
~ делимости criterion for divisibility
~ занятости *comp.* busy flag
~, интегральный *math.* integral test
~ конца *comp.* terminator
~ Коши Cauchy test/condition
~, наследственный hereditary characteristic/character
~ неисправности fault symptom
~ окончания *comp.* complete/completion flag
~, отличительный salient/distinctive feature
~ ошибки *comp.* error flag
~ просмотра файла file scan feature
~ с отрицательной/положительной связью *stat.* negatively/positively associated attributes
~ сравнения comparison test/feature
~ сходимости test for convergence, convergence test
~ точности *comp.* accuracy control character
признание 1. confession; admission; acknowledgement; recognition; 2. *law* acceptance
призывник *mil.* recruit, draftee, inductee
призывной conscription *(attr.)*
~ возраст *mil.* call-up age
~ пункт enlistment office/centre
прикавказский paracaucasian
приказ order; command; *law* precept; writ
~, боевой operational/combat/field order
~, запретительный судебный writ of prohibition
~ о вызове в суд writ of summons
~ о вводе во владение, судебный writ of possession
~ о задержании warrant for apprehending
~ о заключении под стражу warrant of commitment
~ о платеже *econ.* payment order
~ о прекращении огня *mil.* cease-fire order
~ о принудительном изъятии имущества *law* warrant of distress
~ об открытии конкурсного производства warrant in bankruptcy
~ об отплытии *naut.* sailing/steaming orders
~ об участии в заседании *law* writ of attendance
~ общего характера *law* general warrant
~ по войскам *mil.* order of the day
~, повторный судебный *law* alias writ
~ суда об аресте лица writ of capias
~, судебный *law* writ
прикасаться touch, contact
прикатка roll(ing) on
прикатывать roll down, compact
приклад 1. *mil.* rifle butt, stock, buttstock; 2. trimmings
~, откидной folding stock
приклеивать glue, stick, paste, affix, attach
прикосновение contact, touch
прикрытие screen(ing), cover; protection
прикус *anat.* occlusion, bite; *vet.* crib biting
~, закрытый closed bite
~, неправильный malocclusion, abnormal occlusion
~, открытый open bite
~, передний anterior occlusion
~, перекрестный cross bite
прикуска *vet.* crib biting
прилавок counter; *geol.* rock bench
прилавок-витрина display counter
прилагательное-определение adjectival attribute
прилагательное-числительное adjective numeral
приладка, прилаживание adjustment, fitting, fit
прилаживать adapt, adjust, fit
прилегание fit
прилегать fit; be adjacent
прилегающий adjacent, adjoining; *bot.* accumbent
приледниковый periglacial, extraglacial
прилежащий adjacent, contiguous, adjoining
прилепестник *bot.* parapetalum
прилепестниковый *bot.* parapetalous, parapetaloid
прилет *av.* arrival
прилив 1. rising tide, flood (tide); 2. *met.* lug, boss; 3. *med.* congestion, rush
~ в период солнцестояния solstitial tide

~, геомагнитный geomagnetic/lunar tide
~, двойной полусуточный double half-day tide, agger
~, квадратурный neap/quadrature tide
~ крови *med.* rush of blood
~, лунный lunar tide
~, неправильный суточный mixed diurnal tide
~ новолуния, сизигийный new-moon spring tide
~, основный mother tide
~, полуденный noon tide
~, полусуточный semidiurnal tide
~, сизигийный spring tide
~, смешанный/составной compound tide
~, солнечный solar tide
~, статический equilibrium tide
~, суточный diurnal tide
~, частный partial tide
~, четвертьсуточный quarter-diurnal tide
приливно-отливный tidal, ebb-tide, intertidal; flood-and-ebb *(attr.)*
приливо-отливный tidal, intertidal
прилипало *zool.* pilot sucker *(Echeneis)*
прилипаловые *pl. zool.* suckerfishes, shark suckers *(Echeneidae)*
прилипание attachment, adhesion, adherence, sticking
прилист *bot.* paraphyll
прилистник *bot.* scale leaf, stipule
прилистниковый *bot.* stipular, stipulaceous
прилистничек *bot.* stipel
приложение 1. application, use, usage; 2. annex, addition, supplement, addendum; enclosure; supplement; 3. *gram.* apposition, appositive
~ нагрузки load application
приложение-придаточное *gram.* subordinate apposition
приложимость applicability; application
прилунение lunar landing
прилуняться land on the Moon
прим accent mark, prime; *math.* prime
прима 1. *mus.* prime; first violin; first/top string; 2. *print.* prima
примадонна prima donna, diva
примальный *math.* primal
примарный *math.* prime, primary
примас *eccl.* primate
примат I *zool.* primate
примат II *philos.* primacy
применение application, employment, utilization, use; adaptation
~ в мирных целях peaceful application
применимость applicability, feasibility; adaptability; workability
пример example, instance
примерный suggested; model *(attr.)*
примесь 1. admixture, addition; 2. impurity, foreign matter, contaminant; adulterant
~, вредная deleterious impurity
~, гасящая quenching admixture
примечание note; annotation; comment(s)
примитив, примитивист, примитивистский primitive
примочка wash, lotion, fomentation
~, глазная eye-lotion, collyrium
~, горячая stupe
~, свинцовая lead water, goulard (water)
примула *bot.* primrose *(Primula)*
примулаверин *pharm.* primulaverin
примулин primulin(e)
примус primus stove; camp stove
примыкание contiguity; *gram.* parataxis
примыкать 1. be adjacent, abut, adjoin, border; 2. attach, join, fix
примыкающий adjacent, adjoining; *gram.* paratactic
принадлежности *pl.* accessories, fixtures, attachments; outfit; tackle
~, писчебумажные stationery
~, постельные bedding
~, рыболовные fishing tackle
~, спортивные sports requirements/outfits
~, шлюпочные boat fittings
принадлежность belonging; membership, affiliation
~, государственная/национальная nationality
принайтовливать *naut.* snug
принимать 1. take, receive; accept; 2. assume; 3. adopt
~ требование *law* allow a claim
приноготковник *bot.* nailwort *(Paronychia)*
приносящий *anat.* afferent
приношение gift, offering
~ по обету votive offering
принудительный 1. compulsory, coercive, forced; 2. *tech.* forced; positive
~ труд forced labo(u)r
~ная вентиляция forced/positive ventilation
принуждение compulsion, coercion; *law* duress
~ посредством угроз *law* duress per minas
~, физическое *law* duress
принтер = печатающее устройство
~, строчный line printer, line-at-a-time printer
принц prince
~, наследный prince royal
~, Уэльский Prince of Wales
принцепс princeps
принцип principle; law; tenet; concept
~ безаварийности fail-safe concept
~, блочный modular [building-block] principle
~ взаимности reciprocity principle/theorem; principle of interdependence/reversibility/reciprocity/relationship
~ двойственности/дуальности duality principle
~ действия operating principle, principle of operation
~ доминирования *math.* concept of

dominannce, dominance principle
~ изоморфности моделей principle of analogies
~, мажоритарный majority principle
~ максимального правдоподобия *stat.* principle of maximum likelihood
~ максимума модуля *math.* maximum modulus principle
~ минимизации затрат *econ.* cost-minimizing rule
~, модульный modular approach; mechanical assembly technique, MAT
~ наименьшего действия least action principle, Hamilton principle
~ наименьшей полезности principle of least utility
~ наложения/суперпозиции superposition theorem/principle
~ обслуживания в обратном порядке first-in-last-out principle, FILO principle
~ обслуживания в порядке поступления first-in-first-out principle, FIFO principle
~ обратимости reciprocity principle
~ работы working principle
~, ступенчатый step principle
принц-супруг prince consort
принятие 1. acceptance, adoption; reception; admission, admittance; 2. taking
~ на вооружение introduction into the combat inventory, service introduction
~ присяги taking of the oath; oath taking
~ решения decision making
приобретатель acquisitor; *law* purchaser
~, добросовестный innocent purchaser
~ право собственности по давности владения *law* usucapt
приозерье lakeland
приоритет priority
~ зависящий от состояния системы endogenous priority; state priority
~ не зависящий от состояния системы exogenous priority
~ прерывающий обслуживание preemptive priority
приостанавливать suspend; interrupt, stop, withhold
~ сдачу withhold/ suspend delivery
приостановка suspension, (temporary) halt, stoppage; pause; cessation
~ военных действий suspension of hostilities
приостановление stay, halt, suspension; *law* abeyance
~ исполнения решения *law* stay of execution
~ действия закона suspension of a statute
~ платежей suspension of payments
~ рассмотрение дела *law* stay of action
припадок seizure, fit, paroxysm; attack
~, большой эпилептический major epilepsy, grand mal epilepsy
~, истерический conniption fit
~, малый эпилептический minor epilepsy, petit mal epilepsy
~, сердечный heart attack
~, эпилептический epileptic attack/seizure/fit
припадочный epileptic
припаивать solder, braze
припай 1. coast/shore/land/fast ice, shore ice belt; 2. solder
припайка soldering, brazing
припаривание *med.* poultice, fomentation
припаривать *med.* foment, poultice
припарка *med.* poultice, fomentation, cataplasm
~, горячая hot-pack poultice
~ из льняного семени linseed poultice
припасовка fitting, alignment
припасы supplies, stores
~, боевые ammunition; war supplies
~, съестные food supplies
приписывать 1. assign, attribute; 2. register
приплата extra/additional payment
припой solder
~, легкоплавкий quick solder
~, мягкий soft/common solder
~ с канифолью, трубчатый rosin-core solder
~, твердый hard/brazing/strong solder
~, цинковый brazing spelter
приполярный subpolar, parapolar (circum)polar
припудривать bepowder, powder, dust
припуск *tech.* allowance, surplus, margin, overmeasure
~ на обработку machining allowance
~ на усадку/усушку shrinkage allowance
~ на коррозию corrosion margin/allowance
припухлость *med.* intumescence, swelling
~ конъюнктивы, отечная chemosis
припухлый tumescent, swollen
припыл *met.* parting compound/powder/sand
прирабатывать 1. *mach.* run in; 2. earn extra
приработка *tech.* running-in
приработок extra/additional earnings
приравнивать equal, equate, equalize; adjust
приращение increment, increase, gain; accretion; augment
~, алгебраическое algebraic gain
~, конечное finite increment
~, отрицательное decrement
природа 1. nature; 2. character, nature
~, живая animated nature
~ материи character of matter
природовед natural historian, naturalist
природоведение natural study/history
природопользование nature management
прирост = приращение
приросток *bot.* shoot, sprout, excrescence, growth, paraphysis
прируб butt joint
присадка additive, addition, dope
~, активирующая activator
~, антидетонационная detonation suppressor, anti-knock additive/agent, fuel dope

~, антикоррозионная anti-corrosive additive
~, антиокислительная antioxidant additive
~, антипенная antifoam additive
~, антисварочная antiwelding compound
~ в ковш *met.* ladle addition/additive
~ к маслу oil additive
~, многофункциональная multi-purpose additive
~, моющая detergent additive
~, противоизносная wear-preventive additive
~, противонагарная anticarbon additive
~, противообледенительная anti-icing additive
присадочный additional, additive
присваивание assignment, allocation
~ значений value assignment
~ меток labelling
~ приоритетов prioritization
присваивать appropriate, take possession of; purloin; 2. assign, allocate, award, confer;
~ звание confer the rank of
~ незаконным путем *(money)* embezzle
~ себе звание assume a title
присвоение 1. appropriation; 2. award(ing), conferment
~ адресов *comp.* address assignment
~ незаконное misappropriation; *(of money)* embezzlement
~, общественное social appropriation
~ очередного звания promotion
~ чужих денег defalcation
~ чужой движимости *law* trover
присемянка, присемянник *bot.* aril
прислужник *eccl.* acolyte
присоединение attachment; connection; joining; affiliation; accession
присоединительно-определительный *gram.* connective-determinative
присоединительно-противительный *gram.* connective-adversative
присоединительный connective, adjunctive; *gram.* conjunctive
присос 1. *zool.* sucker; osculum
присоска 1. *zool.* sucker; 2. *bot.* haustorium
приспособление 1. device, contrivance, appliance, equipment, fixture, outfit; 2. adaptation, accommodation, adjustment
~, защитное protective device; safeguard; protector
~, компенсаторное *med.* compensatory adaptation
~, предохранительное safety device; fender
~ сборочное assembly device/jig
приспособленность adaptiveness; fitness, suitability
~ развития developmental adaptation
приспособляемость adaptibility, adaptivity, adjustability, faculty of accommodation
приспущенный half-mast
пристав police officer
~, парламентский Sergeant-at-Arms
~, судебный bailiff; usher
~, частный precinct chief of police
приставка 1. *tech.* attachment; adapter; head; accessory; 2. *gram.* prefix
~, коротковолновая short-wave adapter
~, магнитофонная tape deck; tape-recording accessory
~, противопомеховая anti-jamming unit
~, сопловая nozzle extension
приставочно-суффиксальный *gram.* prefixal-suffixal
приставочный 1. *tech.* attachable; 2. *gram.* prefixal
пристань wharf, landing platform, jetty, pier, bunder; pier, quay
~ для лихтеров lighterage wharf
~, плавучая landing stage, wharfboat, float
пристеночно-висцеральный parietovisceral
пристенный, пристеночный parietal
пристрелка *mil.* adjustment/ranging fire
~, контрольная verification fire
~, фотокинопулемета harmonization of the camera gun/recorder
пристройка annex, wing, lean-to
приступ seizure, paroxysm, attack, fit, spell
присуждение adjudication; award
присуха *bot.* rust
присущий inherent, innate, intrinsic; immanent; implicit; characteristic, proper
присучка *text.* joining/piecing the ends
присяга oath
~, воинская oath of enlistment
~, королевская oath of royal duties
~, ложная perjury false oath
~ при вступлении в должность oath of office
~ге, приводить к swear in, administer the oath to
~гу, принимать take the oath
присягать take oath; swear an oath
~ на верность [в верности] swear allegiance
присяжный 1. sworn; 2. juror, juriman
~ заседатель juriman, juror
~ поверенный counsellor-at-law
присяжные *law* jury
притвор 1. vestibule; (church) narthex; 2. groove, frame/rabbet ledge, fluting
притир lap, lapping tool
притирать lap, grind
притирка lapping, grinding in, abrasion
~ клапанов valve grinding/seating
притирочный lapping, grinding
приток 1. inlux, inflow; 2. tributary, affluent
притолока lintel, platband
приточный inflowing
~ная вентилация forced ventilation, blowing
~ная трава *bot.* Dane's-blood bellflower *(Campanula glomerata)*
притупление blunting, dulling, obtusion
притуплять blunt, dull

притча parable
притчардия *bot.* pritchardia palm *(Pritchardia)*
притяжательно-относительный *gram.* possessive-relative
притяжателный *gram.* possessive
притяжение attraction; gravitation
~, земное gravity; terrestial gravitation
~, колоновское Coulomb attraction
~, лунное lunar/moon gravity
притязание claim
притязать lay claim (to)
прифуговка *wood.* jointing, joint
прихватка 1. tack/temporary weld; 2. clamp, vice
прихватывание seizing
~ тормоза brake dragging
приход I receipts; income; revenues; input
~ и расход receipts and expenditures
приход II *eccl.* parish; curacy
приходный income *(attr.),* revenue *(attr.)*
~ ордер credit order
~ная книга receipts ledger/book/register
приходовать *fin.* enter, credit
приходо-расходный credit and debit *(attr.)*
~ная книга account book
приходящий nonresident, visiting
~ие больные outpatients
прицветник *bot.* bract
прицветничек *bot.* bractlet, bracteole
прицел gunsight, sight
~, зенитный anti-aircraft sight
~, инфракрасный ночной infra-red night sight
~, коллиматорный collimating/reflector sight
~, кольцевой ring sight
~, оптический telescopic/optical sight
~, радиолокационный radar sight, aiming radar
~ со счетно-решающим устройством computing sight
~, стрелковый gun sight
~, телескопический sighting telescope
прицел-визир camera sight
прицеливание aiming, sighting
прицеп trailer
~, низкорамный drop-frame trailer
~, портальный straddle trailer
прицеп-контейнер container trailer
прицеп-лесовоз straddle trailer
прицепник *bot.* false carrot *(Caucalis)*
прицепной hook-on, trailer *(attr.);* tractor-drawn
прицеп-роспуск pole trailer
прицеп-самосвал dump trailer
прицеп-цистерна tank trailer
причал pier, wharf; dockside, moorage; (gantry) berth
~, грузовой loading berth
~, достроечный fitting-out berth
~, плавучий floating landing stage
~, рейдовый mooring berth
~, танкерный tanker terminal/berth
причаливание mooring; *space* tethering
причальный mooring
причастие I *eccl.* communion, the Eucharist
причастие II *gram.* participle
~, действительное active participle
~, страдательное passive participle
причастник *eccl.* communicant
причастно-глагольный *gram.* participial-verbal
причащаться receive the Eucharist, receive communion; communicate
причащение *eccl.* receiving communion
причина reason; cause
~ и следствие cause and effect
причинно-присоединительный *gram.* causal-conjunctive
причинно-следственный causal-resultative, cause-and-effect *(attr.)*
причинность causality; causation
причинный causal, causative; etiologic(al)
~ная связь causal relationship, causality, causation
пришабривать *tech.* scrape, scour; fit together
пришабровка fitting, adjusting; scraping off
пришлифовать grind to size/fit; grind in
прищеп cleft; grafting; grafted plant/tree; graft
прищепка *hort.* grafting
прищеплять 1. *hort.* graft; 2. fasten with clothespin(s)
прищипка, прищипывание *hort.* pinching out
прищипывать nip off, pinch
приют refuge, shelter; lodging; asylum
~, детский orphanage
~, родительный maternity home
проба 1. test, trial; assay; experiment; 2. sample, specimen; 3. *met.* purity, fineness, standard; hallmark
~, аллергическая allergy test
~, биологическая biological assay/test, bioassay
~, бродильная fermentation test
~, гидравлическая hydrostatic test
~, грубая crude test
~, капельная drop test/analysis
~, кожная (intra)cutaneous/skin test, cutireaction test
~ на алкалоиды Bourchardat's test
~ на белок test for protein/albumin
~ на выживание survival test
~ на закалку quench test
~ на знакопеременный изгиб bending-unbending test, reverse-bend test
~ на иприт mustard test
~ на истирание abrasion test
~ на кислотность и щелочность pH test
~ на мышьяк Bettendorff's test
~ на наличие астигматизма Becker's test
~ на наличие беременности Aschheim-Zondek test, Berkovitz test
~ на определение группы крови grouping test
~ на перегиб alternate bend test
~ на садку *leath.* cracking test
~ на сахар test for glucose/sugar

~ **на сплющивание** flattening test
~ **Пирке** Pirquet's test
~, **плавочная** heat sample
~, **расчетная** sample for calculation
~ **с сахарной нагрузкой** glucose/sugar tolerance test
~ **с физической нагрузкой** exercise tolerance test
~, **склерометрическая** scratch test
~, **сухая** dry assay
~, **тигельная** crucible assay
~, **товарная** commercial sample
~, **третейская** umpire analysis
~, **холостая** blank test
пробазидия *bot.* probasidium, hypobasidium
пробег run, range, path, travel; mileage
~ **иона** ion path
~, **испытательный** trial run
~, **конечный** final range
~, **лыжный** ski race
~, **межремонтный** run between repairs
~ **нейтрона** neutron path
~ **по прямой** crowflight range
~, **порожний** empty-car miles
~ **порожняком** deadhead run
~, **послепосадочный** *av.* landing/alighting run
~, **предельный** ultimate range
~, **свободный** free run; *av.* free roll
~, **среднесуточный** average daily run
~, **суточный** day's run
~ **шины** run of a tyre
пробел blank, spacing, interval, blank character, gap, omission; interspace; ignore sign
~ **в праве** *law* gap in law
~, **междусловный** word space
пробивание 1. *mech.* piercing, punching, perforation; penetration; 2. *el.* breakdown, rupture; *comp.* punching
пробивать 1. perforate, pierce, punch; 2. *el.* break through; *comp.* punch
~ **дорогу/путь** open the way
~ **облако** break cloud
пробивка piercing, punching, perforation; punch, hole
~, **многократная** multipunching, overpunching
~ **отверстий** perforation
пробивной piercing, puncturing, punching; penetrative; *el.* disruptive, breakdown *(attr.)*
пробирка test tube
пробирный testing, assaying; analyzing
~ **камень** touchstone
~**ная палата** assay office
~**ное клеймо** mark of assay, hallmark
пробировать test, assay; hallmark
пробирщик assayer
пробит *math.* probit, probability unit
пробка plug, stopper, cork *(also bot.)*; cap
~, **ввертная** screw stopper
~, **винтовая** thread plug
~, **воздушная** air lock
~, **глухая** blind/dummy plug
~, **заправочная** filling plug
~, **калиберная** plug gauge
~, **крошеная** granulated cork
~ **наливного отверстия** filler plug
~, **плавкая** fusible plug
~, **притертая** ground stopper
~, **резьбовая** threaded/screw/thread plug
~, **сливная/спускная** drain plug
~, **смотровая** inspection plug
~, **стравливающая** bleeding plug
~, **транспортная** traffic jam, backup of cars
пробковение *bot.* suberization
пробковый cork *(attr.)*; suberic, suberose
~ **дуб** cork oak *(Quercus suber)*
~ **камбий** cork cambium, phellogen
~**вая кислота** suberic acid
~**ная кора** cork bark suber
пробкосемянный *bot.* cork-seeded, phellospermous
проблема problem
проблематика (range of) problems
пробник 1. tester; probe; 2. sampler
пробный trial *(attr.)*, experimental, tentative; exploratory
~ **камень** touchstone
~ **шар** trial balloon
прободение *med.* perforation, tresis
~ **желудка** gastrobrosis, gastrodialysis
~ **язвы** perforation of the ulcer
прободной perforating; perforated, ruptured
пробоина hole; breach; puncture; *mil.* shot/bullet hole
пробой 1. puncture, breakthrough; rupture; breakdown; 2. *el.* flashover, spark-over
~, **зенеровский** Zener breakdown
~ **оболочки** shell puncture
пробойка 1. puncher, drift; 2. piercing, punching
пробойник punch, puncher, drift, punching pin
пробоотбиратель, пробоотборник sampler
~, **погружной** immersion sampler
пробопечатание *print.* proof printing
проборка основы *text.* looming, looming-in/up
проборщик основ *text.* loomer, warp drawer
пробосцидея *bot.* devil's-claws *(Proboscidea)*
пробочный cork *(attr.)*
~ **пояс** cork jacket, life belt
пробуксовка slip, slipping, skidding, slippage
~ **сцепления** clutch slip(page)
пробуравливать, пробуривать drill, bore
провал 1. failure, flop, reverse, fiasco 2. collapse, falling in, downfall; *geol.* depression, sink, cave-in, caving
~, **карстовый** karst hole
~, **котлообразный** cauldron
~ **на импульсе** pulse valley/dip
~ **памяти** memory failure/blackout
провансалец, провансальский, прованский Provençal
провансаль mayonnaise (dressing)

провар *weld.* penetration; fusion; melting
проваривание, проварка boiling
проверка verification, chek(ing), checkup, inspection; examination; control; test
~ без прерывания работы non-interruptive check
~ в контрольных точках benchmark test
~ в процессе обслуживания service check(ing)
~, выборочная selective check; spot check, random inspection
~ герметичности leak test
~ достоверности validation
~ знаний *educ.* examination
~ инвентаря inventory
~ и исправление данных checking and rectification of data
~, квалификационная proficiency check
~ методом моделирования simulation testing
~ на наличие ошибок error check
~ на занятость *teleph.* busy/engaged test
~ на непротиворечивость consistency check
~ на нечетность parity-odd check
~ на плотность leak check
~ на срыв stall check
~ на четность parity check
~ отчетности *fin.* audit
~, плановая programmed check
~ правильности программы program(me) verification
~, предварительная pre-check
~ признака *comp.* flag test
~, профилактическая preventive check(ing)
~ работоспособности/работы functional check
~ состояние запасов stock check
~ суммированием summation/sum check, total checking
~, текущая runnig/current check
~ технических характеристик performance check
~, эксплуатационная service check
~, эмпирическая empirical verification
~, эпизодическая casual inspection
проверочный checking, verifying; examining, test *(attr.)*
провертывать bore, perforate
проверять check, verify; examine, inspect; test; control
провес sag; dip
проветривать ventilate, air
провешивать plumb, test with a plumb-line
провиант supplies, victuals, provisions
провизор pharmacist, pharmaceutist, druggist
провизорный provisional, temporary
провисание sag(ging), slack
провод wire, conductor, lead, cable
~, адресный *comp.* address wire, A-wire
~, антенный aerial wire
~, батарейный battery lead
~, бронированный armoured wire
~, вводный lead-in conductor/cable
~, воздушный open/overhead/aerial wire
~ высокого напряжения high-voltage wire
~, голый bare wire/conductor
~ зажигания ignition wire
~, заземляющий/заземления earth/ground/earthing/grounding wire
~ записи адреса address-write wire, a-w wire
~, звонковый bell wire
~, контрольный pilot wire
~, ленточный ribbon/strip wire
~ малого сечения light-gauge wire
~, многожильный/многопроволочный stranded wire
~, монтажный hookup wire
~, нулевой neutral wire
~, обмоточный magnet wire
~, обратный return wire/conductor
~, освинцованный lead-covered wire
~, питающий/питания supply/power/feed wire
~, плетеный [с оплеткой] braided wire
~, подводящий lead
~, разрядный *comp.* bit/digit line
~, саперный blasting cable
~, скрученный twisted conductor
~ соединения на массу = заземляющий
~, станционный *teleph.* office wire
~ считывания/считывающий/съемный sense wire
~, удлинительный extension lead
~, установочный installation wire
~, экранированный shielded/screened wire
~, эмалированный enamelled wire
проводимость conductivity, conduction; conductance
~, аберрантная желудочковая ventricular aberration, aberrant ventricular conduction
~, активная conductance
~, антероградная anterograde/forward conduction
~, внутрипредсердная intra-atrial conduction
~, действующая effective conductance
~, динамическая электрическая dynamic conduction
~, дырочная hole-type [*p*-type] conductivity
~ звуковых волн sound conductance
~, желудочковая ventricular conduction
~, магнитная permeance
~, наведенная induced conduction
~ нагрузки, полная load admittance
~, нейтронная neutron conduction
~, нервная nerve conduction
~, объемная *el.* cubic/volume conductance
~, односторонняя unidirectional conduction
~, переходная transient conductance
~, переходная, полная transfer admittance
~, полная admittance
~, предсердная *anat.* atrial conduction
~, примесная impurity/extrinsic conduction

~, прямая forward conductance
~, реактивная susceptance
~, реактивная емкостная capacitive susceptance
~, реактивная индуктивная inductive susceptance
~, собственная intrinsic conduction
~, собственная полная self-admittance
~, скачущая saltatory conduction
~, удельная conductivity, specific conductance
~ утечки leakage conductance
~, фотоэлектрическая photoconductivity
~ четырехполюсника admittance of a two-port
~, электрическая conductance; (electrical) conductivity
~, электронная electron conductivity/conduction, *n*-type conductivity
проводить conduct, carry; pass
проводка 1. *el.* wiring, wires; 2. leading, conducting; *met.* guide; *naut.* pilotage; 3. *fin.* entry
~, вводная entry guide
~, внешняя outside wiring
~, внутренняя indoor/internal wiring
~, встроенная integral wiring
~, дебетовая debit item/entry
~, кредитовая credit entry, entry on the credit side
~, кабельная cable run, cabling
~, обязательная лоцманская compulsory pilotage
~, открытая *el.* surface/open wiring
~, скрытая *el.* conceiled wiring
~, сложная *econ.* combined/compound entry
проводник 1. conductor; 2. guide; 3. *rail.* porter; 4. *naut.* hauling line, hawse rope, messenger
~ второго рода ion/electrolytic conductor
~ первого рода electron/metallic conductor
~, печатный printed conductor
~ тепла heat conductor
провододержатель hanger
провод-рельс conductor/contact rail
провозоспособность carrying capacity
проволока wire
~ вязальная tie/binding/baling wire
~, горячетянутая hot-drawn wire
~, колючая barbed wire
~, контровочная locking/safety wire
~, оттяжная anchor wire
~, струнная piano wire
~, твердотянутая hard-drawn wire
~, шплинтовая pin wire
проволочник *zool.* wireworm
проволочно-волочильный wire-drawing
пловолочно-намотный wire-coiling
проворачивание turning (over), cranking
~ винта *av.* propping-off, propeller swinging
проворачивать jack-over; turn through
прогамета *bot.* progamete
прогаметангий *bot.* progametangium, progametange
прогамный progamic, progamous
прогар burn-through/out, burn, burnt place, blow-hole; scorching
прогарина burnt place, local burn
прогерия *med.* progeria
прогерманский pro-German
прогестерон *pharm.* progesterone, syngesterone
прогиб sag(ging), deflection, flexure; buckling, caving in, depression; *geol.* trough
~ кровли *min.* roof deflection
~ от собственного веса dead-load [self-weight] deflection
~ при нагревании thermal deflection
прогибаться sag, cave in, flex
прогибиционист prohibitionist
прогибомер deflectometer
прогимназия Progymnasium, junior high school
проглоттида *anat.* proglottid, proglottis
прогнатизм *anat.* prognathism
прогнатический prognathic, prognathous
прогноз forecast, prediction; prognosis; projection
~, арифметически взвешенный arithmetic weighted forecast
~ долгосрочный long-range/term/period forecast
~, инерционный persistence forecast
~, краткосрочный short-range/term/period forecast
~, ледовый sea ice forecast
~, медицинский medical prognosis
~, несмещенный unbias(s)ed prediction
~, осторожный conservative prediction/forecast
~ погоды weather forecast
~ полученный методом наименьших квадратов least-squares prediction
~, предварительный tentative forecast
~, среднесрочный medium-term/range/period forecast
~, точечный point prediction
прогнозирование prediction, forecasting, prognostication; prognostics
~ надежности reliability prediction
~ отказов failure/breakdown prediction
~ с адаптацией adaptive prediction
~ смертности mortality prognostics
прогнозировать forecast, predict, prognosticate; prognose
прогнозируемый predictable
прогнозист forecaster
прогностика prognostication
прогон I girder, beam, purlin, baulk, bearer; span; springer
~, тротуарный sidewalk baulk
прогон II run, pass
~ бумаги, холостой *comp.* paper skip
~, имитационный simulation pass
~, контрольный benchmark/check run
~ ленты tape run(ning)
~ модели simulation run
~ перфокарт card passage/run, passage of cards
~, повторный rerun

~, **пробный** trial run
~ **программы** *comp.* program(me) run
~ **теста** test run, running of test routine
~, **холостой** *comp.* skip
прогонка 1. running through; trial run; pass; 2. threading/screw die
~, **повторная** rerun
прогорание burning through/out
прогорклость rancidity, rankness
прогорклый rancid, rank
программа program(me); schedule; plan; *educ.* curriculum; syllabus; *comp.* routine
~, **активная** active routine/program
~ **анализа ошибок** error analysis routine
~, **аппаратно реализованная** hardware program
~ **асинхронного вывода** asynchronous exit routine
~, **библиотечная** *comp.* library program
~ **в абсолютных адресах** absolute program
~ **в реальном масштабе времени** real-time program
~ **ввода** loading routine/program; loader
~ **ввода-вывода** *comp.* input-output routine
~, **ветвящаяся** branching program
~ **визита** visit scedule
~, **воздушно-космическая** aerospace program
~, **вспомогательная** support program
~, **вызываемая** called program
~, **вызывающая** calling program, caller
~, **выписанная** written-out program
~ **вычислительной машины** computer program
~ **генерации растра** grid/raster generation routine
~, **главная** master program
~, **диагностическая** diagnostic [error search] toutine/program, diagnotor
~, **динамическая последовательная** dynamic serial program
~, **долгосрочная** long-term/-range program
~, **жесткая** wired-in program
~ **жилищного строительства, государственная** public housing program
~ **загрузки** load routine/program, loader
~ **записи ошибок канала** channel error recording routine
~ **запроса места** space request routine
~ **затрат** *econ.* spending plan
~, **зашитая** wired program
~ **защищенная авторским правом** copyrighted program
~ **изготовления проектных чертежей** design drawing program
~ **инициализации ядра** nucleus initialization program
~ **испытаний** test/trials program
~, **исполнительная** executive program
~ **исправления ошибок** error-correcting program/routine
~, **исходная** *comp.* source program
~ **капиталовложений** capital investment project; investment plan
~, **кодирующая** code generating routine
~, **компилирующая** compiler, compiling program/routine
~, **компонующая** assembly program/routine
~ **контроля** checker
~ **машинного анализа цепей** network optimization program
~ **метода обращения** access method routine
~, **моделируемая/моделирующая** simulation program/routine
~ **на входном языке** source program
~ **на выходном языке транслятора** target program
~ **на магнитной ленте** magnetic tape program/routine
~ **не включаемая в бюджет** extra-budgetary program
~, **независимая** stand-alone routine
~ **нулевых ошибок** zero defects program
~ **обеспечения качества** quality program
~ **обеспечения надежности** reliability program
~ **обеспечения пенсией по старости** old-age pension scheme
~ **обеспечения сбыта** sales program
~, **обрабатывающая [обработки данных]** processing programme
~ **обработки** *comp.* handler
~ **обслуживающая** service program/routine
~ **обслуживания** support routine
~ **обслуживания массивов данных** file [data-set] utility program
~, **обучающая** teaching program
~ **общественных работ** public works program/plan
~, **объектная** object/target program
~, **оверлейная** overlay program
~ **окончания** end routine
~ **операции над матрицами** matrix operation program
~, **оптимальная** minimum access/latency routine, optimally coded program
~, **оптимизирующая** optimizing program
~ **опытно-конструкторских работ** development programme
~ **организации сбыта** marketing program
~ **освоения космического пространства** space programme
~ **оценки** evaluation program/routine
~, **организующая** administration program, administrator
~ **оценки состояния системы** damage assessment routine
~, **пассивная** inactive program
~ **пенсионного обеспечения** *econ.* pension scheme
~ **переднего плана** foreground program

~, перемещаемая relocatable program
~ печати print routine/program
~ повторения, машинная computer rerun routine
~, повторно используемая reentrant program
~ поиска search/seek routine
~, предварительная preprogram
~, преобразующая change/transformation program
~, прерывающая/прерывания interrupt routine
~, прикладная application program
~, приоритетная foreground program
~ проверки при помощи монитора monitor checking routine
~, проверочная check/test routine/program
~, производственная production program
~, простая single program
~ расходов *econ.* spending plan
~ расширения *econ.* expansion program/scheme
~, редактирующая editor (programme)
~, резидентная resident program
~ с плавающей запятой floating-point routine, program with floating point
~, самозагружающаяся self-loading routine
~ самозагрузки bootsrap (program)
~, сервисная/служебная utility program/routine
~ синтаксического анализа parser
~, скомпонированная assembled program
~ сортировки sorter
~ социальной помощи, государственная public welfare program
~, специальная ad hoc program, special program
~ спуска *space* descent program
~, срочная *econ.* crash program
~, стандартная routine
~, структурированная structured program
~ табличных вычислений spreadsheet program
~, транзитная transit/transient routine
~, трансформирующая change/transformation program
~ трассировки tracing routine
~ уплотнения compressor/condensing program
~ управления control/administrative/steering program
~ управления каталогом catalog(ue) management routine
~, управляемая данными data-flow program
~, управляющая control program
~ учета accounting routine
~, фоновая background program
~, хранимая stored program
~, часто выполняемая high-frequency program
~ экономического развития, национальная national economic plan, macroeconomic plan
~, экспериментальная pilot program
программа-диспетчер *comp.* supervisor/executive program
программа-монитор *comp.* monitor (program)
программа-резидент fixed stored program
программатор programmer
программирование programming
~, автоматическое computer-aided [automatic] programming, autoprogramming, self-programming
~ в абсолютных адресах absolute programming
~ в диалоговом режиме interactive programming
~ в относительных адресах relative programming
~ ввода данных input programming
~, выпуклое convex programming
~ на коммутационной панели switchboard/patchboard programming
~ обнаружения отказов failure detection programming
~, оптимальное optimum coding/programming, minimal access/latency programming, minimum access coding
~, предварительное pre-programming
~, прикладное application programming
~, системное system programming
~ собственными силами in-house programming
~, структурное structured programming
~, целочисленное integer programming
программировать program(me)
программист programmer
программно-управляемый program-controlled
программный program(me) *(attr.);* programmed
~ное обеспечение software
~ные средства software
программоноситель program medium
программотехника software engineering
прогрев, прогревание warming up, warm-up, preheating
прогревать warm up
прогресс progress
~, научно-технический scientific and technological progress
прогрессист progressionist, progressist
прогрессия progression
~, арифметическая arithmetic progression/series
~, геометрическая geometric progression/series
продавление forcing/pressing through; punching, puncturing
продавливать press/force through
продажа sale, selling
~ без упаковки bulk selling
~, быстрая ready sale
~ в кредит credit sale
~ в рассрочку instal(l)ment selling
~ в розницу retail sale
~ вразвес bulk selling
~ вразнос door-to-door [house-to-house] selling
~ груза целиком bulk sale
~ за наличные cash sale

~, **комиссионная** sale by commission
~ **на срок** sale for future delivery, forward sale/selling
~ **оптом/оптовая** wholesale; bulk selling
~, **розничная** retail sale
~ **с немедленной оплатой и поставкой** spot trading
~ **с немедленой поставкой за наличный расчет** spot sale
~ **с рассрочкой платежа** time-payment schemes selling
~ **"так есть"** "as is" sale
~ **через автоматы** automatic machine trading
продвижение advance; progression; movement
продевать thread, feed through
продир *met.* guide scratch
продифференцировать differentiate
продление extension, prolongation; renewal
продмаг *abbr.* food store, grocery (store)
продналог *hist.* tax in kind
Продовольственная и сельскохозяйственная организация ООН Food and Agricultural Organization, FAO
продолжительность duration, length, period, time
~ **активного существования, минимальная** minimum lifetime
~ **безмоторного полета** powerless/unpowered flight endurance
~ **включения** *el.* duty cycle
~ **дня** day length
~ **жизни** life time/span; longevity; *stat.* life expectancy
~ **жизни, средняя** life expectancy
~ **запуска** starting period/time, duration of starting
~ **импульса** pulse duration/length/width
~ **незанятости** *tech.* unoccupied time
~, **номинальная** rated duration
~ **отбоя** *teleph.* releasing time
~ **отстаивания** *chem.* retention time
~ **плавки** heat time
~ **простоя** down/idle time
~ **работы** operation time; period of service
~ **ремонта** repair time
~ **службы** service life
~ **существования** lifetime
~ **существования, предполагаемая** expected/anticipated lifetime
~ **существования, расчетная** design/projected lifetime
~ **технического обслуживания** maintenance time
~ **флуоресценции** fluorescence life time
~ **эксплуатации** in-service time, operating/running time
продольный longitudinal; *naut.* fore-and-aft
продороживание grooving, slotting, recessing
продороживание, продорожка grooving, slotting, channel(l)ing
продотряд *abbr., hist.* food-requisition detachment
продразвертка *abbr. hist.* requisitioning of agricultural produce
продром(а) *med.* prodrome
продромальный prodromal, prodromic, premonitory; precursory
продторг *abbr.* food trade organization
продувание 1. = **продувка;** 2. insufflation, inflation
продувать blow through/down /off/out; scavenge; insufflate; inflate, perflate
продувка blowing through/down/off/out; blow-through/down; scavenging
~ **в аэродиномической трубе** wind-tunnel test
~, **донная** *met.* bottom blowing
~ **кислородом** *met.* oxygen lancing
~ **сверху** *met.* top blowing
продукт product; produce; *see also* **продукты**
~ **активации** *phys.* activation product, activate
~ **аминирования** aminate
~, **бросовый** waste product; reject
~, **валовой** gross product
~, **валовой внутренний** gross domestic product
~, **валовой страны/национальный** gross national product
~, **вторичный** secondary product; by-product; after-product
~ **гидролиза** hydrolysate
~ **гниения** putrilage
~ **горения** combustion product
~, **готовый** finished product
~ **деления** fission product; cleavage product
~ **деления, мгновенный** prompt fission product
~, **диализированный** dialysate
~, **дочерний** daughter product
~ **животноводства** animal product
~ **захвата** capture product
~ **изнашивания** *mach.* grit
~, **исходный** primary product
~ **коксования** coke by-product
~, **конечный** final/end product
~, **летучий** volatile product
~, **надрешеточный** plus material, oversize
~, **немаркированный** unidentified/unlabel(l)ed product
~, **нефтехимический** petrochemical product
~, **нефтяной** petroleum product
~ **обжига** calcinate
~ **обмена веществ** metabolic product, metabolite
~, **оборотный** recycle
~, **общественный** social product
~, **основной** staple
~, **отвальный** waste product
~, **отогнанный** *chem.* overhead product
~, **первичный** prime product
~, **пищевой** foodstuff, food
~, **побочный** by-product, co-product, secondary product
~, **подрешеточный** minus material, undersize

~ **превращения** transmutation product
~, **предельный** marginal product
~, **прибавочный** surplus product
~ **производственного назначения** capitalized product
~, **промежуточный** intermediate product, semiproduct
~, **профилирующий** characteristic product
~ **размола** ground material
~ **распада** decay/disintegration product
~ **расщепления** cleavage/fission/splitting/disintegration product
~ **реакции** reaction product
~ **сгорания** combustion product
~, **сельскохозяйственный** farm produce/product
~, **совокупный национальный/общественный** gross national/social product
~, **товарный** marketable/commodity product
~, **третичный** tertiary product
~ **уноса массы** ablation product
~ **фотолиза** photolyte
~, **целевой** main/prime product
~, **чистый национальный** net national product

продуктивность productivity; efficiency

продукты *pl.* 1. products, produce; 2. foodstuffs, victuals
~ **высшего качества, мясные** fancy-grade meat
~ **деления** fission products
~ **животноводства** livestock products
~, **консервированные** preserved food; canned/tinned food
~, **молочные** dairy products
~, **пищевые** foods, foodstuffs, alimentary products
~, **продовольственные** food products, foodstuffs
~ **реактора, побочные** reactor by-products
~, **сельскохозяйственные** farm products
~, **скоропортящиеся** perishable products/goods

продукция product(s); output, yield; commodity; manufacture; goods
~ **в условиях полного использования производственных мощностей** capacity output
~, **валовая** gross output
~, **вспомогательная** ancillary product(s)
~, **готовая** final/finished/end product/goods
~, **завершенная** finished manufacture(s)
~, **заказанная** stock on order
~, **многономенклатурная** diversified product
~ **не находящая сбыта** unsalable product/output
~, **незавершенная** work/material in process
~, **отечественная** domestic product(s)
~, **печатная** printed matter
~, **побочная** incidental output
~, **сельскохозяйственная** farm products
~, **товарная** commercial/commodity output; salable/marketable products/output

продух, продушина air hole

проезд passage, transit; journey, travel
~ **автотранспортом** travel by road, road travel
~ **воздушным транспортом** travel by air
~ **железнодорожным транспортом** travel by rail
~ **морским транспортом** travel by sea

проект project, design, plan; layout; draft
~ **документа** draft paper
~, **индивидуальный** custom design
~, **концептуальный/предэскизный** conceptual design
~, **рабочий** detailed design/plans
~, **технический** engineering/contract design
~, **типовой** standard design
~, **экспериментальный** pilot project
~, **эскизный** preliminary/rough/sketch design

проектант planner; project engineer

проективный projective
~**ная геометрия** projective geometry

проектирование 1. design(ing), design work; planning; 2. projection, screen display
~, **автоматизированное** computer-aided design
~, **восходящее** bottom-up design
~ **изделия** product design
~, **инженерное** design engineering
~, **комплексное** integrated design
~ **конструкции** structural design
~, **машинное** computer-aided design
~ **на экран** visual projection, screen presentation
~, **нисходящее** top-down design
~, **предварительное** predesign(ing)
~ **промышленного объекта** industrial design
~ **с резервом** overdesign
~ **системы** system design
~, **структурное** structured design
~, **эскизное** initial design

проектировать design, project; plan

проектировщик designer

проектно-конструкторский design-and-engineering

проектно-расчетный/сметный preliminary-estimation

проектный design *(attr.);* designed, rated; designing, planning

проектор projector, projection machine
~, **диапозитивный** slide projector
~, **ионный** field ion projector
~, **картографический** map/chart projector
~ **с автоматической сменой кадров** magazine-fed projector
~, **электронный** field-emission microscope

проекционный projection *(attr.),* projecting
~ **аппарат** projector
~ **фонарь** magic lantern, still projector

проекция projection
~, **аксонометрическая** clinographic/axonometric projection

~ Гаусса, поперечная цилиндрическая Gauss transverse cylindrical projection
~ Гаусса, равноугольная Gauss conformal projection
~, картографическая chart/map/cartographic projection
~, конформная conformal/orthomorphic projection
~, косая/косоугольная oblique/skew projection
~ Меркатора, поперечная transverse/inverse Mercator projection
~, нормальная коническая Lambert projection
~, обратная reciprocal projection
~, орфографическая meridional orthographic projection
~, перспективная geometric projection
~, поперечная меркаторская inverse mercator projection
~, поперечная цилиндрическая равноугольная transverse/inverse Mercator projection
~, произвольная картографическая aphylactic projection
~, равновеликая/равноплощадная equal-area (equivalent) projection
~, равнопромежуточная equidistant projection
~, равноугольная conformal/orthomorphic projection
~ точки *math.* foot
~, центральная gnomonic projection
~, цилиндрическая равноугольная Mercator projection
~, эквивалентная authalic [equal-area] projection
проем opening, aperture
~ двери/дверной doorway, door aperture
~, оконный window opening/aperture
~ световой *arch.* fenestration
проецировать project
прожектор searchlight, floodlight; *electr.* electron gun
~, воспроизводящий reading gun
~, записывающий writing gun
~, зенитный anti-aircraft searchlight
~, ионный ion gun
~, магнетронный magnetron injection gun
~, плазменный plasma gun
~, посадочный landing light; runway floodlight
~, потолочный cloud searchlight, ceilometer
~, радиолокационный searchlight-control radar
~, сигнальный signal light projector
~, электронный electron gun
прожигание burning (through)
прожилок filament; fibre, fiber; *geol.* veinlet, vein, streak
прожиточный минимум living/subsistence wage, subsistence minimum
прожог *met.* burn-through
прозаик, прозаист prose writer, prosaist
прозвучивание sonic test
прозектор *med.* dissector, prosector, anatomist
прозекторская, прозектура dissecting room, prosectorium
прозелит proselyte
прозенхима *bot.* prosenchyma
прозенхиматический, прозенхимный prosenchymatous
прозерин *pharm.* neostigmine methylsulphate
Прозерпина *myth.* Proserpine
прозодежда *abbr.* working clothes, overalls
прозолоть golden hue/tint
прозопальгия *med.* prosopalgia, trigeminal neuralgia, tic douloureux
прозопис *bot.* mesquite *(Prosopis)*
прозоплектенхима *bot.* prosoplectenchyma
прозор gap, spacing; distance
прозрачномер transparency meter
прозрачность 1. transparency, transparence, lucidity, diaphaneity, pellucidity; 2. transmissivity, transmittance
~ для тепловых лучей diathermancy
~ контрастная contrast transmittance *(of the atmosphere)*
~ барьера barrier penetrability
прозрачный transparent; translucent, pellucid, limpid, diaphanous; hyaloid
проигрыватель player, phonograph, sound reproducer
проигрыш penalty; loss; *sport* defeat
произведение 1. work, product; production; 2. *math.* product; composition
~, бесконечное infinite/continued product
~, векторное vector/cross/external/outer product
~, внешнее external/outer/vector/cross product
~, внутреннее internal/inner/dot/scalar product
~, двойное векторное vector product of three vectors
~, декартовое Cartesian/direct product
~ искусства work of art
~, косое *(in algebra)* skew product; *(in topology)* fibre bundle
~, литературное literary work
~, прямое Cartesian/direct product; direct/Cartesian sum
~, скалярное scalar/dot/inner/internal product
~ событий intersection of events
~, тензорное tensor product, composition of tensors
~, частичное partial product
производитель 1. producer, manufacturer, maker; *agr.* grower; 2. procreator; sire; bull; stallion, studhorse
~ с высокими издержками производства high-cost producer
производительность productivity, productive capacity, output, yield, production; efficiency; rating, duty; performance; delivery/output/production rate

~ **в эксплуатации** in-use performance
~ **вентилятора** fan delivery/capacity/duty
~, **годовая** annual production rate
~ **земли** land capacity/productivity
~ **котлоагрегата** rating of a boiler unit
~ **на вспашке** plough capacity
~ **на человеко-час** output per man-hour
~ **нагнетания воздуха** air delivery rate
~ **насоса** pump capacity/delivery/output, pumpage
~, **недостаточная** underproductivity
~ **ниже нормативной** substanard performance
~, **номинальная/расчетная** rated capacity/output
~, **нормальная/нормативная** standard performance
~, **объемная** volume output
~, **паспортная** rated capacity, capacity rating
~, **подъемная** lift/hoisting capacity
~, **почасовая** hourly efficiency/output
~, **предельно достижимая** top performance
~ **рабочего** worker efficiency
~, **расчетная** rated output
~ **системы** system capacity
~ **скважины** well yield
~, **сменная** output per shift
~, **сметная** estimated output/production
~, **текущая** current output
~ **труда** labour productivity/efficiency
~, **удельная** specific capacity/duty/efficiency
~, **целевая** target productivity
~, **часовая** hourly output, output per hour

производительный productive; efficient; yielding
~ **период жизни** reproductive span

производить 1. produce, make, turn out, manufacture; 2. carry out, conduct, perform
~ **в офицеры** promote to the rank of officer
~ **потомство** procreate
~ **следствие** conduct an inquiry/investigation

производная *math.* derivative
~, **верхняя** upper derivative
~ **времени** time derivative
~ **высшего порядка** derivative of higher order
~ **дробного порядка** fractional derivative
~, **левая** left-hand (backward) derivative, derivative on the left
~ **момента** moment derivative
~, **нижняя** lower derivative
~, **обобщенная** generalized derivative
~, **перекрестная** cross-coupling [crosswise, cross] derivative
~ **перемещения, вторая** velocity rate, acceleration
~ **перемещения, первая** velocity
~ **перемещения, третья** acceleration rate
~ **по направлению** directional derivative
~ **по нормали** normal derivative
~ **по скорости** rate derivative
~ **по x** derivative with respect to x
~, **полная** total derivative
~, **правая** right-hand (forward) derivative, derivative on the right
~, **старшая** highest derivative
~ **устойчивости** stability derivative
~, **частная** partial derivative

производное *chem.* derivative

производный derivative, derived; secondary

производственник (production) worker

производственный production *(attr.),* manufacturing, industrial
~ **стаж** production experience; length of service

производство 1. production, manufacture, manufacturing, turning out; 2. carrying out, execution; 3. industry, industrial branch
~ **без выхода на рынок** captive production
~, **бондарное** cooperage
~, **бумажное** papermaking
~ **в опытных масштабах** pilot manufacture/production
~ **газа** gas production/generation
~, **доменное** blast-furnace process
~, **единичное** single-unit (piece-work) production
~, **книжное** book production/publication
~ **кожи** leather manufacture
~, **коксохимическое** coke and by-product process/production
~, **конвейерное** belt-line production/manufacture
~, **конкурсное** *law* proceedings/procedure in bankruptcy
~, **консервное** canning
~, **крупносерийное** large-lot/-scale manufacture
~, **кузнечно-прессовое** press forging
~, **кузнечно-штамповочное** press/die forging
~, **кустарное** artisan/handicraft/domestic/home/craft industry
~, **лесохимическое** wood chemical industry
~, **листопрокатное** plate/sheet rolling
~, **литейное** foundry (operation)
~, **манифактурное** *econ.* manufacturing system
~, **мартеновское** open-hearth process
~, **массовое** mass/quantity/wholesale production
~, **мелкосерийное** small-lot/scale production
~, **многономенклатурное** multiproduct manufacture
~, **модельное** pattern-making
~ **на душу населения** per capita production
~, **надомное** *econ.* cottage industries
~, **незавершенное** goods/work/material in process
~, **непрерывное** continuous process
~ **обуви** shoemaking
~, **опытное/полузаводское** pilot-scale production
~, **отечественное** domestic production
~ **по лицензии** licence manufacture
~ **побочных продуктов** by-product production
~, **поточное** flow-line [in-line] production
~ **предметов потребления** production of

consumer goods, commodity production/manufacture
~, **прокатное** rolling
~, **ремесленное = кустарное**
~, **сварочное** welding work/engineering
~, **сельскохозяйственное** agricultural/rural industry
~, **серийное** lot/serial manufacture/production, series production; production run
~ **стали** steelmaking
~, **собственное** domestic/home manufacture
~, **суммарное** *law* summary proceedings/procedure
~ **тепла** heat generation/production
~, **трубопрокатное** pipe rolling
~, **фабричное** manufacturing, manufacture; industrial output
~, **швейное** clothing manufacture
~, **штучное** piecework production, single-item manufacture
~ **электрической энергии** electricity/power production, generation of electric energy
~ **энергии** power production/generation
произвольный 1. arbitrary; 2. voluntary; optional
произносительный articulatory
произносить pronounce, articulate, spell out, say, utter
~ **неправильно** mispronounce
~ **приговор** pronounce a sentence
~ **речь** deliver a speech
~ **четко** enunciate clearly/distinctly
происхождение origin, genesis; descent; source; birth, parentage
~ **видов** origin of species
~ **Земли** origin of the Earth, geogenesis
~ **растений** phytogeny
происшествие accident; incident; occurrence, event, happening
~, **аварийное** breakdown accident
~, **летное** flight/flying/aircraft accident
~, **морское** shipping casualty
~ **на посадке** landing accident
~ **с человеческими жертвами** fatal accident
~, **чрезвычайное** emergency
прокаженный 1. leprous, leprotic; 2. leper
проказа 1. leprosy, lepra; 2. *bot.* anthracnose, algal spot
~, **бугорчатая** nodular/tubercular leprosy
~, **гладкая** smooth leprosy
~ **крыс** rat/murine/mouse leprosy
~, **нервная** neural/anaesthetic leprosy
~, **обезображивающая** mutilating leprosy
~, **пятнистая** macular leprosy
~, **суставная** articular leprosy
~, **туберкулоидная** tuberculoid/tubercular leprosy
прокаин *pharm.* procaine
прокаин-амид procaine amide
прокаин-пенициллин procaine penicillin
прокаливаемость hardenability
прокаливание calcination
прокаливать calcine, calcinate
прокамбий *bot.* procambium
прокаменноугольный procarboniferous
прокат 1. *met.* rolled metal/products/stock; rolling; 2. release; hire, rental
~, **круглый** round bars, rounds
~ **листовой, толстый** plates
~ **листовой, тонкий** sheets
~, **полосовой** strip
~, **сортовой** bar
~, **стальной** rolled steel
~, **фасонный** structural shapes
прокатка *met.* rolling, milling; lamination
~, **винтовая** screw rolling
~, **петлевая** loop-mill rolling
~ **слябов** slabbing
~ **толстолистовой стали** plate rolling
~, **черновая** roughing
~, **чистовая** finish rolling
прокатный rolled, rolling
~ **стан** rolling mill
прокатолический pro-Catholic
прокатываемость *met.* rollability
прокатывать flatten; laminate; roll
прокачивать run/pump through; draw through
прокачка pumping
~ **масла** oil circulation
прокимен *eccl.* prokeimenon
прокладка gasket, packing, spacer, washer, seal; *naut.* plot(ting); laying
~, **герметическая** pressure packing
~ **головки блока цилиндров** cylinder-head gasket
~, **гребенчатая** serrate gasket
~ **для регулирования зазора** spacer shim
~ **курса** course plotting
~ **пути** plotting
~ **подводного кабеля** cable laying
~, **регулировочная** shim
~, **рельсовая** tie plate
~, **слюдяная** *el.* mica separator
~, **уплотнительная** pressure-seal gasket
прокладчик 1. plotter, tracer; 2. layer
прокладывать plot; lay; trace; drive
проклеивать size, glue, paste
проклеивание, проклейка sizing
проковка hammering, forging
прокол piercing; puncture
прокрустов *myth.* Procrustean
~**о ложе** bed of Procrustes
прокрутка двигателя idling of engine
прокручивать rotate, turn; revolve; run idle
проктальгия *med.* proctalgia, proctagra
проктатрезия *med.* proctatresia, atretoproctia
проктейринтер *med.* procteurynter
проктит *med.* proctitis, rectitis
~, **гнойный** purulent proctitis
~, **застойный** congestive proctitis
~, **подострый** subacute proctitis
~, **язвенный** ulcerous proctitis

проктодиния *med.* proctodynia
проктоклизис *med.* proctoclysis, proctoclysia
проктоколит *med.* proctocolitis, coloproctitis, rectocolitis
проктокольпопластика proctocolpoplasty
проктолог proctologist
проктопаралич proctoparalysis
проктопексия *med.* proctopexy, rectopexy
проктопластика proctoplasty
проктоплегия proctoplegy
проктополип proctopolypus
прокторея proctorrh(o)ea
прокторагия proctorrhagia
проктоскоп proctoscope, rectoscope
проктоскопия proctoscopy, rectoscopy
проктостаз *med.* proctostatis
проктоцистотомия proctocystotomy
проктэктомия proctectomy, rectectomy
прокуратура public prosecutor's office
прокурор public prosecutor, prosecuting attorney
~, главный attorney general
~ округа district attorney
пролактин prolactin
проламин prolamine
пролан prolan
пролапс *med.* prolapse
пролегомен prolegomenon
пролежень *med.* bedsore, decubitus ulcer
пролежневый decubital
пролейкемия leukanemia, proleukemia
пролейкоцит proleucocyte, leucoblast
пролеска *bot.* 1. mercury *(Mercurialis);* 2. bluebell *(Scilla)*
пролесковидный *bot.* scilloid, squill-like
пролескоцветковый *bot.* squill-flowered
пролесник *bot.* mercury *(Mercurialis)*
пролет 1. fly-by; fly-over, overflight; flight; 2. gap; span; aperture; bay
~ арки/арочный chord of an arch, arch span
~, балочный beam span
~, береговой abutment span
~, близкий near miss
~ в свету span, internal width
~ в строю треугольника delta formation pass
~, завалочный *met.* charging bay
~, конвертерный *met.* furnace aisle/bay
~ лестницы flight of stairs
~ моста bridge span
~ на спине *av.* inverted pass
~ по прямой crow flight
~, судоходный navigable pass, fairway arch
~ челнока *text.* shuttle flight
пролетариат proletariat
пролетарий, пролетарка, пролетарский proletarian
пролетать fly past/by/through; fly
пролетка droshky, cab
пролеткульт *abbr. hist.* proletarian culture
пролетный flying; flight *(attr.)*
~ные птицы birds of passage
пролив *geog.* strait, sound, channel
~, Гибралтарский Strait of Gibraltar
~ в территориальных водах territorial strait
пролин proline
проломник *bot.* rock jasmine *(Androsace)*
промазка coating, spreading
промазывать grease, smear, oil, lubricate
промасливать oil, lubricate, grease
промах miss; lapse
промбанк *abbr.* industrial bank
промежностно-влагалищный *anat.* perineovaginal
промежностно-мошоночный perineoscrotal
промежностный *anat.* perineal
промежность *anat.* perineum
промежуток interval; interstice; gap; space; span, distance, clearance, opening
~ времени time interval/span, period; lapse of time
~ дискретности sampling interval
~ интегрирования *math.* interval of integration
~, искровой spark gap
~, межзубный *anat.* interdental space
~, межпальцевой interdigital space
~, межреберный intercostal space
~, пробивной *el.* disruptive distance
~, разрядный discharge gap
~, светлый *psych.* lucid interval
промежуточный intermediate; interim; *law* mesne
~ная сумма sub-total
промер measurement; survey; sounding
~ механическим лотом wire sounding
~ морской offshore survey
~, прибрежный inshore survey
промеран *chem.* chlormerodrine
промерзание freezing; frost penetration
промеристема *bot.* primordial meristem, promeristem
прометазин *chem.* promethazine
прометафаза prometaphase
прометеев, прометеевский Promethean
Прометей *myth.* Prometheus
прометий *chem.* promethium, Pm
промилле per mille; parts per thousand
промкомбинат *abbr.* group of [small/local] industrial enterprises
промкооперация *abbr.* producer's cooperative
промодель mastern pattern
промойный eroded by water; washed away
промпродукт *abbr.* intermediate product
промтовары *abbr.* industrial/manufactured goods
промторг *abbr.* trade organization for manufactured goods
промфинплан *abbr.* industrial and financial plan
промывка, промывание washing, flushing; rinsing, lavage, lavagement; ablution; irrigation; scrubbing *(of gas); med.* bathing; douching
~ бронхов bronchial lavage
~ желудка gastric lavage

~ **воздуха** air scrubbing
~, **противоточная** countercurrent washing
~ **раны** bathing of a wound
~ **уха** ear irrigation
промысел 1. trade, business; 2. catching, hunting
~ **во внутренних водах** inland fishery
~, **горный** mining
~, **китобойный** whaling, whale catching; whale fishery
~, **крабовый** crab fishery
~, **кустарный** domestic/handicraft industry
~, **морской** sea fishery
~, **отхожий** seasonal work
~, **охотничий** hunting
~, **подсобный** *econ.* ancillary craft
~, **прибрежный** in-shore fishing, coast fishery
~, **рыбный/рыболовный** fishing; fishery
промысловик 1. hunter; fisherman; 2. miner
промысловый 1. trade *(attr.);* 2. hunting, game *(attr.);* fishing
~ **зверь** fur-bearing animal
~ **флот** fishing fleet
~**вая кооперация** prdoucers' cooperative
~**вая птица** game bird
~**вая рыба** food fish
промышленник 1. industrialist, manufacturer; 2. hunter; fisherman
промышленность industry; manufacture
~, **авиационная** aircraft/aviation industry; aero-industry
~, **авиационно-космическая** aerospace industry
~, **винодельческая** wine industry
~ **водопользования** water industry
~, **высокотехничная** high-technology industry
~ **вычислительных средств** computer industry
~, **горная/горнодобывающая** mining industry
~, **добывающая** extractive/primary/mining industry
~, **кожевенно-обувная** leather and wootwear industry
~, **коксохимическая** by-product coke industry
~, **консервная** canning industry
~, **космическая** space industry
~, **кустарная** cottage industry
~, **лакокрасочная** paint and varnish industry
~, **легкая** light industry/industries
~, **лесная** forest(ry) industry; wood industry
~, **лесохимическая** wood-chemical industry
~, **ликерно-водочная** alcoholic beverages industry
~, **молочная** dairy industry
~, **моторостроительная** engine industry
~, **мукомольно-крупяная** flour-and-cereals industry
~, **мыловаренная** soap industry
~, **надомная** cottage industry
~, **наукоемкая** high-technology industry
~, **нефтеперерабатывающая** oil refining industry
~, **нефтехимическая** petrochemical industry
~, **нефтяная** oil/petroleum industry
~, **оборонная** defence/defense industry
~, **обрабатывающая** manufacturing/processing industry
~, **отечественная** home industry
~, **пивоваренная** brewing industry
~, **пищевая** food (processing) industry
~, **полиграфическая** printing industry
~ **преимущественно военного назначения** defence-oriented/centered industry
~ **программных средств** software industry
~ **производящая средства производства** capital goods industry
~, **самолетостроительная** aircraft industry
~ **средств связи** communication industry
~, **станкостроительная** machine-tool industry
~, **сталелитейная** steel industry
~, **строительная** construction industry; building trade
~, **судостроительная** shipbuilding industry
~, **тяжелая** heavy industry
~, **угольная** coal mining industry
~, **хлебопекарная** baking industry
~, **целлюлозно-бумажная** wood-pulp and paper industry
~, **швейная** clothing/tailoring industry
~, **электроэнергетическая** electric-power industry
~ **энергетическая** power industry
промышленный 1. industrial; manufacturing; 2. payable
пронатор *anat.* pronator
пронацистский pro-Nazi
пронация *anat.* pronation
проникать penetrate; permeate, pervade
~, **взаимно** interpenetrate
проникновение, проникание penetration, permeation, pervasion; infusion
~, **взаимное** interpenetration
~, **межкристаллитное** intercrystalline/intergranular penetration
проницаемость permeability, perviousness; penetrance; penetration; transmittance; transmittivity
~, **акустическая** acoustic transmittivity
~ **барьера** barrier factor/transparency/penetration
~, **диэлектрическая** dielectric permeability/permittivity; dielectric constant
~ **капилляров** capillary permeability
~, **квантово-механическая** quantum-mechanical penetration
~, **магнитная** magnetic permeability
~ **по индуктивности** inner/intrinsic permeability
~, **повышенная** hyperpermeability
~ **потенциального барьера** potential barrier penetrance/penetration/transparency; barrier factor
~, **электрическая** permittivity

прообраз *math.* inverse image; pre-image; operand
пропавший без вести missing
пропагула *bot.* propagule; propagulum
пропадание loss; disappearance, vanishing
пропазин propazone
пропан propane
пропаривание, пропарка steaming, curing/drying with steam
пропаривать steam
пропарный *paleont.* proparian
пропахивать plough through/between; intertill
пропашка tillage/ploughing between rows, intertillage
пропашник *agr.* furrow plough, row cultivator
пропашной трактор row-crop tractor
пропедевтика propedeutics
пропепсин propepsin, pepsinogen
пропептон propeptone, secondary proteose; deuteralbumose
пропил I *chem.* propyl
пропил II saw cut, slit; notch
пропиламин *chem.* propylamine
пропилгекседрин *pharm.* propylhexedrine, benzedrex
пропилеи *pl. arch.* propylaea
пропилен *chem.* propylene
пропилит *petr.* propylite
пропиливать saw through
пропилитизация *petr.* propylitization; propylitic alteration
пропилиодон propyliodone; dionosil
пропиловый propyl *(attr.)*
пропионат *chem.* propionate
пропионовый propionic, propanoic
пропитка 1. impregnation; steeping; soaking; 2. saturant
~, водоотталкивающая water-repellant impregnation
пропиточный impregnating; saturating
~ное вещество impregnating agent, saturant
проплавка smelting; melting
проплавление penetration; *met.* weld penetration; depth of fusion; joint
~, сквозное through penetration
пропласток *geol.* intercalation, interlayer; parting
~, газовый gas streak
~ глины clay parting
~, надвиговый thrust sheet
~, повторный recurrent parting
~ соли salt stringer
проповедь sermon; homily
~, нагорная *rel.* Sermon on the Mount
пропорционально proportionally; proportionately, in proportion to
~, изменяться обратно vary inversely
~, изменяться прямо vary directly
пропорциональность proportionality, proportion
~, обратная inverse proportionality/ratio
~, прямая direct proportionality/ratio
пропорциональный proportional; proportionate
~, обратно inversely proportional
~, прямо directly proportional
пропорция proportion; ratio
~, непрерывная continued proportion
~, производная proportion derived by addition [by subtraction, by addition and subtraction]
~, разностная arithmetical proportion
~, сложная compound proportion
пропотевание *med.* transudation
проприоцепция *physiol.* proprioception
пропудривать dust/powder thoroughly
пропуск 1. miss; omission, skip; blank; 2. pass; passage, letting through; 3. *mil.* password
~ зажигания misfire
~ занятий absence from school/classes
~ записи (data) record skip
пропускаемость transmittivity, transmittance
пропускание passing through, passage; transmission
~ пучка beam transmission
~ через фильтр filtration
пропускать pass through, let pass/through; pass; let in; admit; *phys.* transmit
~ случай miss an opportunity
~ уроки miss lessons/classes
пропускник disinfection/delousing centre
пропускной passage *(attr.)*; transmission *(attr.)*
~ная бумага blotting paper
~ная способность throughput/carrying capacity
прораб *abbr.* construction superintendent; resident engineer
проран (closure) channel; outlet; passage
прорастаемость germination/sprouting ability, germinability
прорастание 1. *petr.* intergrowth; *geol.* penetration; 2. *bot.* germination, sprouting
прорастать 1. intergrow, grow through, penetrate; 2. germinate; sprout; shoot
проращивание germination, sprouting, shooting
прореживание *agr.* thinning out, gapping
прореживатель *agr.* thinner; gapper; thinner hoe
прореживать thin out
прорезинивание, прорезинение rubberizing, rubberproofing
прорезинивать rubberize, coat with rubber
прорезка cutting; slitting; slotting
прорезной cutting, grooving
прорезывание cutting; *med.* dentition, teething; eruption (of teeth)
~ зуба, неправильное maleruption
~ молочных зубов first/primary/deciduous dentition
~ постоянных зубов secondary dentition
прорезь cut; opening, slit; slot, aperture
проржаветь rust through
проросток *bot.* seedling; sprout, shoot; germ
прорыв breakthrough, break; inrush, burst; outbreak
~ в области наук/техники breakthrough

прорывать 1. break through/out; tear; 2. dig through
просадка slump; settling; settlement
просасываться seep/leak through
просачивание leakage; percolation; infiltration; seepage; transudation; oozing
просачиваться leak, percolate; seep; ooze; filter through, infiltrate; seep out, exude; trickle
просверливание drilling; boring, perforation
просвет clearance; clearance/open space; gap; *anat.* lumen
~, дорожный road clearance
~, железистый glandular lumen
просвечивание translucence; X-raying; radioscopy; exposure to radiation
~ гамма-лучами gamma-raying
~, рентгеноскопическое X-ray examination; X-raying
просвира *eccl.* communion bread
просвирник *bot.* mallow *(Malva)*
просвирчатый *bot.* malvaceous
просеивать sieve, screen
просек *min.* through-cut; cross adit/hole
проскальзывать slip
проскок overshoot; overswing; jump
~ стрелки overshoot of a pointer
проскомидия *eccl.* Offertory
проскрипционный proscriptive
проскрипция proscription
прослаивать interleave; interlay, interstratify
прослеживать trace, track
прослой *geol.* interlayer, intercalation, interbed; parting; band; streak
~, мощный heavy parting
~, секущий cross course
~, сланцевый slate intercalation/band, interstratified slate
прослойка stratum, layer; sheet; interlayer
~, воздушная air space
~, резиновая rubber interlayer
прослушивание 1. audition; monitoring; 2. playback
просмолка tar-coating; tarring
просмотр review, viewing; survey; scanning; examination; run; lookup
~, диэлектрический dielectric scanning
~, повторный rescanning
~ содержимого памяти *comp.* storage scan
~ состояний *comp.* status scan
~ файла *comp.* file scan(ning)
просновка *text.* warping; warp dressing
просо *bot.* millet, panicum *(Panicum)*
просовидный *bot. med.* miliary
просодия prosody
просорушка millet scourer/huller
проспект I 1. folder, booklet; prospectus; 2. summary
проспект II avenue, boulevard
проспорангий *bot.* prosporangium
просроченный overdue; expired
просрочивать delay
~ платежи fall into arrears
просрочка delay, expiration (of a term)
проставка spacer, distance piece
проставление/простановка размеров dimensioning
простата *anat.* prostate
простатальгия *med.* prostatalgia, prostatic neuralgia
простатит *med.* prostatitis
~, застойный congestion
~, трихомонадный mycotic prostatitis
простатодиния prostatodynia
простатолит *med.* prostatic calculus, prostatolith
простатопатия *med.* prostatopathy
простаторея prostatorrh(o)ea
простатотомия prostatomy; prostatotomy
простатэктомия *surg.* prostatectomy
~, залобковая/позадилонная retropubic prostatectomy
~, перинеальная/промежностная perineal prostatectomy
~, чреспузырная transvesical prostatectomy
простейшие *pl. zool.* protozoans *(Protozoa)*
простенок *arch.* pier; wall, partition
~ между окнами interfenestration
простирание *geol.* strike, course
простой I simple, ordinary; common; plain
~тая хрaна plain/simple food
~тое письмо non-registered letter
~тое число *math.* prime number
простой II standstill; idle period/time, downtime; shutting down
~, внеплановый unscheduled downtime
~ оборудования machine idle time
~, плата за demurrage
~ судна/вагона demurrage
простойный demurrage *(attr.)*; idle
простома *zool.* prostomium
простостолбчатый simplibaculate
пространственно-временной space-time *(attr.)*, spaciotemporal, existential
пространственно-гармонический space-harmonic
пространственно-зарядный space-charge *(attr.)*
пространственно-линейный linear-dimensional
пространственно-плоскостный planar-dimensional
пространственность spatiality
пространственный spatial; dimensional
пространство space; area; room
~, адресное *comp.* address space
~ анод-катод anode-cathode path
~, аффинное affine space
~, банахово *math.* Banach space
~, безвоздушное airfree space; vacuum
~, бесконечномерное *math.* infinite-dimensional space
~, бикомпактное *math.* bicompact/compact

space
~ ближнее космическое near space
~, векторное *math.* vector space
~ вероятностей *math.* probability space
~, воздушное airspace
~, вокругсосудистое *anat.* perichorioid space
~, вредное dead space
~ выборок *stat.* sample space
~, гильбертово *math.* Hilbert space
~, глубокое космическое deep space
~ для головы head room
~ для локтей elbow room
~ для ног leg room
~ для хранения stowage room/space
~, евклидово Euclidean/Cartesian space
~, забрюшинное *anat.* retroperitoneal space
~, залобковое *anat.* retropubic space
~, изобарное isobaric space
~ импульсов momentum space
~, ключичное clavicular space
~, корнеобитаемое root medium/zone
~, космическое outer/cosmic space
~, космическое околоатмосферное pre-atmospheric space
~, космическое околоземное circumterrestrial space
~ котла, огневое heating/combustion chamber of a boiler
~, междулонжеронное inter-spar area
~, междупалубное *naut.* tween-deck (space)
~, междустенное intrathecal/endothecal space
~, межклеточное intercellular space
~, межлопаточное *anat.* interscapular space
~, межреберное *anat.* intercostal space
~, метризуемое *math.* metrizable space
~, мировое outer space
~ над головой head room
~, неевклидово non-Euclidean space
~, обобщенное *math.* generalized space
~, околоапертурное aporium
~, околоземное near space
~, окололунное near-lunar space
~, открытое космическое extra-vehicular space
~ оценок *math.* estimation space
~ ошибок *math.* error space
~, паровое steam void
~, перихороидальное perichoroid space
~, подпаутиновое *anat.* subarachnoid space
~, полное *math.* complete space
~, поражаемое beaten space
~, предгильбертово *math.* pre-Hilbert space
~, приполярное apoporium
~ проектных параметров design space
~, промежуточное interspace
~, разреженное rarefied/evacuated space
~ района аэродрома/аэропорта terminal airspace
~, ретциево *anat.* Retzius' space, retropubic space
~ решений *mat.* solution/decision space
~, риманово Riemannian space
~ с мерой measure space
~, сопряженное conjugate/dual/adjoint space
~ стратегии *game th.* policy/strategy space
~, темное астоновское Aston's dark space
~, теноново Tenon's space, interfacial space
~ Траубе, полулунное Traube's space, semilunar space
~, унитарное unitary space, Hermitian vector space
~ условий *math.* requirements space
~, фазовое phase space
~ элементарных событий *math.* sample space
пространство-время space-time
~, искривленное curved space-time
~, конформное conformal space-time
прострация exhaustion, prostration
~, нервная neurasthenia, nervous prostration
прострел 1. shooting-through; 2. *med.* myalgia; myositis; lumbago; 3. *bot.* anemone *(Anemone)*
простуда (common) cold
простудный *med.* catarrhal
проступ tread
просфора *eccl.* communion bread; host; prosphora
просчет 1. miscalculation, blunder; counting loss; 2. computation run
просянка *zool.* corn bunting *(Emberiza calanora)*
протактиний *chem.* protactinium; protoactinium, Pa
проталий, проталлиум *bot.* prothallium
проталкивание pushing through
протамин *biochem.* protamine
протаминаза protaminase
протандрический protandrous
протандрия protandry
протаргол *chem.* silver protein
протаскивание pulling/dragging through
протачивать turn (on a lathe)
протаска towing
протеаза *biochem.* protease, proteolytic enzyme
~, желудочная gastric protease; pepsin
протез prosthesis, prothesis
~ артерии arterial graft
~, биоточный biocurrent prosthesis
~ конечности extremity prosthesis
~, мостовидный зубной bridge prosthesis, dental bridge
~, полный зубной complete denture
~, присосной зубной suction plate
~ сердечного клапана valve prosthesis
~, съемный зубной removable denture
~, слуховой hearing aid
~, частичный зубной partial denture
~, щеленебный cleft palate prosthesis
протезирование prosthetics
протезный prosthetic; orthopedic
протезист (dental) prosthetist
протеид proteid
протеин protein

протеиназа proteinase
протеинемия protein(a)emia
протеинкиназа protein kinase
протеиновый protein *(attr.),* proteic
протеиноз proteinosis
протеинотерапия proteinotherapy
протеинурия proteinuria
 ~ Бенс-Джонса Bence-Jones proteinuria
 ~, внепочечная extrarenal proteinuria
 ~, ложная pseudoproteinuria
протей 1. *bact. myth.* Proteus; 2. *zool.* mud puppy *(Nectarus)*
протектор 1. protective layer/agent/device; protector; 2. tyre tread
 ~, возобновлять re-tread (recap) a tyre
 ~, ошипованный studded trea
 ~, съемный detachable tread
протекционизм *econ.* protectionism
протекционистский protectionist; protecting
протекционный protectionist; protective
протеогормон proteohormone
протеоза *biochem.* proteose
протеоземия proteosemia
протеолиз *biochem.* proteolysis; protein hydrolysis/breakdown
 ~, ожоговой burn proteolysis
протеометаболический proteometabolic
протеотоксикоз proteotoxicosis
протерандрический, протерандричный *bot.* proterandrous; protandrous
протерогенез proterogenesis
протерогинический proterogynous; protogynous
протерогиния proterogyny, protogyny
протерозой *geol.* Proterozoic (era)
протерозойский Proterozoic
протеротип *biol.* photerotype
протест protest; *law* objection
 ~, запоздалый *(of bill)* past due protest
 ~, заявлять lodge/enter/make a protest
 ~, капитанский/морской protest by master, sea/ship's/captain's protest
 ~ о неплатеже по векселю notice of dishonour/protest; protest of a promisory note
протестант 1. *rel.* Protestant; 2. *law* protester
протестантство *rel.* Protestantism
протестовать вексель *law* protest a promisory note, note a bill
протетический prothetic; prefixed
против часовой стрелки counterclockwise
противень tray, pan
противительно-ограничительный *gram.* adversative-restrictive
противительно-разделительный *gram.* adversative-distributive
противительно-уступительный *gram.* adversative-concessive
противник adversary; opponent; *law* counterpart
противоатомный antiatomic, antinuclear
противоболевой analgetic, analgesic
противобродильный antifermentative
противовес counterweight, counterbalance; balance weight/mass; counterpoise
противовирусный antiviral
противовключение connection in opposition
противовоздушный anti-aircraft *(attr.)*
противовспениватель antifoam agent, foam suppressor
противовоспалительный *med.* antiinflammatory, antiphlogistic
противовытяжение *surg.* counterextension, countertraction
противогаз gas-mask
противогазовый, противогазный antigas *(attr.);* gas-defence *(attr.)*
противоглушитель antijamming device
противоглистный antihelmint(h)ic; antivermicular, vermifugous; helmint(h)agogue
противогнилостный antiputrefactive; antirot *(attr.)*
противогрибковый fungicidal; antimycotic
противодавление counterpressure, back pressure
противодействие counteraction, opposite action; reaction; opposition; resistance
противодействовать counteract, react; oppose
противодесантный *naut.* antiamphibious; *av.* antiairborne
противодетонирующий antiknock *(attr.)*
противоемкостный anticapacitance *(attr.)*
противоестественный unnatural, preternatural; perverted
противозагрязнитель antipollutant
противозаконность illegality; unlawfulness
противозаконный unlawful, illegal
противозачаточный contraceptive
противозенитный 1. anti-aircraft; 2. counter-flak
противозмеиный antivenene
противозобный antigoitrogenic
противоизлучение counterradiation
противоион *phys.* counter-ion
противоискровой antispark *(attr.)*
противокатаральный anticatarrhal
противокатерный anti-boat
противокашельный, противокашлевый antitussive
противокислотный antiacid; antacid; acid-proof
противокозелок *anat.* antitragus; antigoat
противокоррозионный anticorrosive
противокражный burglarproof; theftproof
противолежать subtend, be/lie opposite to
противолежащий subtending; lying opposite
противолепестковый *bot.* antipetalous
противолепрозный antileprotic
противолихорадочный antifebrile
противоличиночный larvicide
противолодочный antisubmarine
противолокационный antiradar
противолуна antiselena
противолучевой antiactinic
противомагнитный antimagnetic
противомалярийный antimalarial

противомеховый anti-interference *(attr.),* anti-jamming; antinoise *(attr.)*
противомикробный antimicrobic; antimicrobial
противомольный antimoth *(attr.);* mothproof
противонародный antipopular
противообласть *math.* range
противообледенитель defroster, deicer; anti-icer; deicing [anti-icing] device
~, воздушно-тепловый hot-air anti-icer
~ оперения tail-unit anti-icer
~, химико-жидкостный fluid anti-icer
противообманный cheatproof
противообрастающий antifouling
противообщественный antisocial
противоожоговый antiflash
противоозоностаритель *chem.* antiozonant
противоокислитель *chem.* antioxidant
противоопухолевый antitumo(u)ral, antineoplastic
противоосколочный splintproof
противооспенный antivariolic
противоотечный antioedemic, antioedematous
противооткатный antirecoil, counter-recoil
противопаразитарный antiparasitic
противопехотный *mil.* antipersonnel
противоподлодочный antisubmarine
противопожарный fire-fighting, anti-fire, fire-extinguishing, fireproof
~ная защита fire protection
противопоказание 1. contraindication; 2. *law* contradictory evidence
противопоказанный contraindicated; contraindicative
противополе *phys.* counterfield
противоположение contraposition; antithesis; opposition
противоположительный contrapositive
противоположно contrarily; inversely, conversely
~ прямо antithetically
противоположность contrast; antipode; antithesis; opposite
противоположный opposite; antipodal, antithetic(al); contrary, opposed; convese; unlike
~, взаимно mutually antithetic(al)
~, диаметрально antipodal, diametrically opposite
противопоносный *med.* antidiarrh(o)eal
пропитопоставительный opposing; contrasting
противопоставление 1. contraposition; contrasting; 2. opposition; opposing
противопоставлять 1. oppose; 2. contrast; antithesize; set off
противоправный *law* wrongful; illegal
противопульный bulletproof
противопылный antidust; dustproof
противорадиационный antiradiation *(attr.)*
противорадиация counterradiation; back radiation
противорадиолокационный antiradar; radar-absorbent
противоракета antimissile missile, countermissile
противоракетный antimissile
противораковый anticancerogenic; anticarcinogenic; antineoplastic
противорахитный antirachitic
противорвотный antimimetic
противоревматический antirheumatic
противоречащий contradictory
протеворечивость contradictoriness; inconsistency
противоречивый contradictory; conflicting; inconsistent; antilogous; antinomic(al)
противоречие contradiction; inconsistency; antinomy; antagonism; conflict
противоречить 1. contradict; 2. run counter (to); conflict (with), be contrary (to)
противосамолетный antiaircraft
противосвертывающий anticoagulative; anticoagulating
противосвязь 1. *constr.* counter-tie; 2. negative feedback
противосифилитический antisyphilitic; antiluetic
противосияние *astr.* counterglow; antitwilight
противосмысл antisense
противоснарядный *mil.* shellproof
противосовпадание, противосовпадение anticoincidence
противосолнце *astr.* anthelion
противоспазматический, противоспастический spasmolytic, antispastic, antispasmodic
противоспутник antisatellite missile/killer
противосредство antidote
противоставление contraposition
противостаритель antiager; antioxidant, antidegradant; preservative
противостолбнячный antitetanic, antitetanus
противостояние 1. confrontation; 2. *astr.* opposition
противосудорожный anticonvulsive, antispasmodic
противотанковый anti-tank *(attr.)*
противотело antibody
противосумерки anticrepuscular rays, antitwilight
противотечение countercurrent, counterflow; backwater
противотифозный antityphoid
противоток countercurrent; counterflow
противоточный counterflow *(attr.),* countercurrent; reverse-flow; *chem.* reflux
противотрихомонадный *med.* antimycotic
противотуберкулезный antitubercular; antituberculous
противотяга reverse thrust
противоугон rail anchor, anticreeper
противотуманный anti-fog *(attr.)*
противоудар counterstroke; countercoup
противоударный shockproof
противоусадочный shrink-resistant
противоутомитель antifatigue agent

противофаза antiphase; reverse/opposite phase
противохолерный anticholeraic
противохимический *mil.* antigas *(attr.)*
противоходный countercurrent
противоцинготный antiscorbutic
противофлаттерный anti-flutter
противочесоточный antiscabetic; antiscabious
противошум antiphone
противоюзовый non-skid; anti-skid
противоядие antidote; counterpoison; antitoxin
противоядный antitoxic, antidotal
протий *chem.* protium
протист *biol.* protist, protistan
протистолог protistologist, microbiologist
протистология protistology, microbiology
протоактиний *chem.* protoactinium
протоарийский proto-Arian
протобазидий, протобазидия *bot.* protobasidium
протогалактика protogalaxy
протогем *chem.* protoheme, hematine
протоген protogen(e), lipoic acid
протогиничный, протогинный protogynous
протогиния protogyny
протодиакон, протодьякон *accl.* archdeacon
протодиастолический protodiastolic, early diastolic
протозвезда protostar
протозвездный protostellar
протозевгма *gram.* prozeugma
протозигота protozygote
протозойный protozoal; protozoan
протозоолог protozoologist
протоиерей *eccl.* archpriest
протоиндоевропейский proto-Indo-European
протоистория protohistory, prehistory
проток 1. channel; canal; watercourse; 2. *anat.* duct
~, артериальный/боталов arterial canal; Botallo's duct
~, грудный thoracic duct
~, желточнокишечный umbilical duct; vitelline duct; yolk-stalk
~, кольцевой annular passage
~, общий желчный common bile duct
~, общий печеночный common hepatic duct
~, околоушный parotid duct
~, передний полукружный anterior semicircular duct
~ поджелудочной железы pancreatic duct
~, подчелюстной submandibular duct
~, подъязычный sublingual duct
~ придатка яичка duct of the epididymis
~, пузырный cystic duct
~ семенного пузырька, выводящий excretory duct of the seminal vesicle
~, семявыбрасывающий ejaculatory duct
~, слезно-носовой nasolacrimal duct
~, улитковый cochlear duct
протоканонический *bibl.* protocanonical
протокарион *biol.* protokaryon
протокарпофор *bot.* protocarpophore
протокатехатоксигеназа protocatechuate oxygenase
протоклаз *miner.* protoclase
протококк *bot.* protococcus
протокол 1. minutes, record (of proceedings); statement; 2. *dipl.* protocol; 3. *comp.* protocol
~, вести keep the minutes
~ виртуального канала *comp.* virtual circuit protocol
~, заносить в *law* place on record
~ канального уровня сети *comp.* link protocol
~ квантирования связи *comp.* handshaking protocol
~, маркерный *comp.* buck-passing protocol
~ обмена сигналами signalling protocol
~ пакетной связи *comp.* packet protocol
~, полицейский police charge-sheet
~ с фиксированным маршрутом fixed-path protocol
~ сеансового уровня сети session protocol
~, сетевой network protocol
~ эстафетного доступа token-access protocol
протоколировать record
протоколист record-keeper; *law* court clerk
протокон *zool.* protocone
протоконюль *zool.* protoconule
протокристаллический protocrystalline
протоксид *chem* protoxide
протоксилема *bot.* protoxylem
протолиз protolysis
протолитический protolytic
протолоф protoloph
протомеристема *bot.* protomeristem
протон proton
~ космического излучения cosmic-ray proton
~ отдачи recoil proton
~, падающий incident proton
~, налетающий impacting/colliding proton
протонический *ling.* protonic; pretonic
протонный proton *(attr.)*; protonic
протонотарий prot(h)onotary
протообраз protoplast
протоперитеций *bot.* protoperithecium
протоплазма protoplasm
протоплазматический, протоплазмический, протоплазмовый protoplasmatic; protoplasmic
протоплазмодий *biol.* protoplasmodium
протоподит *zool.* protopodite
протоплазмолиз protoplasmolysis, achromatolysis
протопорфирин *biochem.* protoporphyrin
протопорфирия protoporphyria
протопресвитер *eccl.* protopresbyter
протопризма protoprism
проторакс *zool.* prothorax
проторозавр *paleont.* protorosaur *(Protorosaurus)*
протосолнце protosun
протосома protosome

протоспутник protosatellite
протостела *bot.* protostele
прототаллом *bot.* protothallium; protothallus
прототека *zool.* prototheca
прототетика protothetics
прототип prototype; archtype; primary/original standard
~ метра, международный international prototype meter
прототроф prototroph
прототрофный prototrophic
протофит protophyte
протофибриллы *pl.* protofibrils, protofibrillae
протофлоэма *bot.* protophloem
протохлорофилл protochlorophyll
протоцеркальный protocercal
протоцефалон protocephalon
проточек *anat.* ductule
~, блуждающий aberrant ductule
проточка 1. turning (on a lathe); 2. drilling/boring through
проточный 1. flowing, running; circulating; 2. *phon.* fricative
протоэпифит *bot.* protoepiphyte
протоэритроцит protoerythrocyte
протрава 1. *met.* pickle; mordant; pickling; 2. *leath.* mordant
~, квасцовая alum mordant
~, сумаховая sumac mordant
~, таннидная tannin mordant
~, хромовая chrome mordant
протравитель 1. mordanting/pickling machine; 2. mordant; *agr.* dresser
~ семян seed disinfectant/dresser
протравка pickling; etching
протравливание 1. *leath.* treatment with mordant; mordanting; 2. *met.* pickling
протравливать 1. *leath.* treat with a mordant; mordant; 2. *met.* pickle
протрактор 1. *nav.* plotter, station pointer; protractor; 2. *surg.* protractor; 3. *anat.* protractor muscle
~ с тремя линейками three-arm protractor
протромбин prothrombin; prothrombase
протромбиназа prothrombinase
протромбиновый prothrombin *(attr.)*
протромбиноген prothrombinogen
протромбокиназа prothrombokinase
протуберанец *astr.* prominence
протягивать stretch, extend; draw; haul; pull
протяженность extent; length; expanse, span; range
протяжка 1. broaching; drawing; 2. broach; 3. transit; advance; transport
проушина lug, ear, eye
профаг prophage; symbiotic/latent phage
профаза *biol.* prophase
~ с остаточной спиралью relic spiral prophase
профактив *abbr.* the most active members of a trade union
профактивист *abbr.* trade-union activist
профбилет *abbr.* trade-union card
профбюро *abbr.* trade-union bureau
профгрупорг *abbr.* trade-union group organizer
профгруппа *abbr.* trade-union group
профдвижение *abbr.* trade-union movement
профермент *bioch.* proenzyme
профессионалист professional; specialist
профессиональный vocational; occupational; professional
~ союз trade union; labour/professional union
~ная подготовка vocational training
~ное заболевание occupational disease
~ное обучение occupational training
профессура professorship
профзаболевание *abbr.* occupational disease
профилактик maintenance technician
профилактика 1. *med.* prophylaxis; 2. *tech.* preventive checkup/maintenance
профилактический prophylactic; preventive; precautionary
профилакторий *med.* preventorium
профилирование contouring, profiling; *constr.* grading; blading; *met.* forming; shaping
профилировать profile; shape, form; *(road)* grade
профилограф profile recorder; profilograph
~ самолетный airborne profile recorder
профиль 1. contour; profile; outline; side view; 2. section; shape
~, аэродинамический aerofoil section/profile
~, вертикальный elevation profile
~, гнутый roll-formed section
~, гидродинамический hydrofoil profile/section
~, двутавровый I-section; H-section
~ дна bathymetric chart, bottom profile
~, желобчатый fluted section
~, зетовый Z-section
~ зуба tooth flank
~, кованый forged section
~, круглый *met.* rounds
~, крупносортный heavy section
~ кулака/кулачка cam profile/contour
~, ламинарный laminar-flow profile/section
~, линзообразный lenticular (double-arc) profile/section
~, магнитный magnetic profile/traverse
~, мелкосортный *met.* light section
~, обтекаемый streamline section
~, омегообразный hat section
~, периодический die-rolled section
~, поперечный transverse/bench section
~ пути, поперечный cross-section of a road
~ пути, продольный vertical alignment of a road
~, ромбовидный double-wedge profile/section
~, S-образный reflexed profile/section
~, сводный compound section
~, стальной steel section

~, тавробульбовый T-bulb section
~, тавровый T-section/shape, tee section
~, тянутый drawn section
~, углобульбовый *met.* angle-bulb section; bulb angle
~, угловой/уголковый angle
~, фасонный *met.* shape; shaped metal
~, холоднокатаный cold-rolled section
~, чечевицеобразный = линзообразный
~, швеллерный channel
~, штампованный formed shape
профильно-шлифовальный profile-grinding
профильный profile; section(al)
~ная сталь sectional steel
профком *abbr.* trade-union committee
профорг *abbr.* trade-union organizer
профорганизация *abbr.* trade-union organization
профориентация *abbr.* vocational counselling/guidance
профос *mil. hist.* provost
профработа *abbr.* trade-union work
профработник *abbr.* trade-union worker/activist/official
профсобрание *abbr.* trade-union meeting
профсоюз *abbr.* trade union
профсоюзный trade-union, union *(attr.)*
~ное движение trade-union movement, (trade) unionism
профтехнический *abbr.* professional and technical
профтехучилище *abbr.* vocational training school
профтехшкола *abbr.* trade/technical school
профуполномоченный *abbr.* trade-union representative
профшкола *abbr.* trade-union school
прохлорит *miner.* prochlorite
проход 1. passage, pass; passageway, gangway, alley; alleyway, walk(way); aisle; 2. passing, run, pass; 3. breach; gap; opening, aperture, inlet; channel; 4. *anat.* duct; passage, meatus
~ ассемблера assembly pass
~, горный mountain pass
~, мирный *naut. law* innocent passage
~, наружный слуховой external acoustic/auditory meatus
~, повторный *comp.* rescanning
~ сортировки sort pass
проходимость passability; trafficability; crosscountry ability; *med.* permeability, patency
проходимый passable; permeable; patent
проходить pass, go through/past/along/by
проходка *min.* heading; driving; penetration; cutting; working; sinking
~ колодца well sinking
~ по породе stone drifting
~ туннеля tunnel(l)ing
проходная entrance checkpoint/gate
проходной passing *(attr.),* going/passing through
проходчик *min.* tunnel(l)er; shaft sinker
прохождение passage, passing; going through; transit; transmission
~ на ветре *naut.* weathering
~ перигея perigee passage
~ сигнала signal propagation/transmission
~ частот transmission of frequencies
~ шлюзов *naut.* negotiation of the locks
процедура procedure; *med.* treatment
~, водная hydrotherapeutic procedure
~, встроенная built-in procedure
~ входа в систему log-on procedure
~, вызванная invoked procedure
~ выхода из системы log-off procedure; exiting procedure/handling
~, грязевая *med.* peloid treatment
~, лечебная treatment (procedure)
~, многомерная решающая *math.* multiple decision procedure
~ периода компиляции *comp.* compile-time procedure
~ с постоянной памятью fixed store procedure
~, судебная legal proceedings
~, таможенная procedure of customs
~ торгов bidding procedure
процедура-функция function procedure
процедурная medical treatment room
процедурный procedural; procedure-oriented
процеживать strain; filter; percolate
процент per cent; percent; percentage, rate; *fin.* interest (rate)
~ аварий wreck rate
~ больных sickness rate
~ брака rejection rate, percentage of defective goods
~, весовой weight percent
~, выравненный *fin.* equated interest
~ грамотного населения literacy rate
~ занятости employment rate
~ извлечения recovery/extraction ratio
~ к уплате outstanding interest
~, капитализированный capitalized interest
~, молекулярный molecular percent
~, молярный mole percent
~, невзысканный back interest
~ неграмотного населения illiteracy rate
~, объемный volume percent
~ отсева dropout rate
~, отсроченный deferred interest
~ подлежащий погашению due interest
~ потерь percentage loss
~ при авансировании *econ.* lombard rate
~ пустот void factor
~, рентный royalty interest
~ смертности mortality rate/percentage
~, средний *fin.* equated interest
~, ссудный cost of borrowing
~, суммарный cumulative percentage
~ убыточности dagree of damage
~ установленный законом statutory interest
~, учетный discount rate
~ы, накопленные/наросшие начисленные

accrued interest/charges
~ы по закладной ипотеке mortgage burden
~ы, сложные compound interest
процентиль *stat.* percentile; centile
процентный per cent, percental
~ные ценные бумаги interest-bearing securities
процесс 1. process; operation; 2. *law* (legal) proceedings; lawsuit; case; trial
~, активный *comp.* awake process
~, авторегрессионный *stat.* autoregressive process
~, аммиачно-содовый Solvay process
~, бездоменный direct ore-reduction process
~, бессемеровский Bessemer process
~ в твердой фазе solid-phase process
~, ваграночный cupola process
~, вероятностный probabilistic process
~, восстановительный reduction process
~, вырожденный *stat.* degenerate process
~, гангренозный *med.* mortification
~, групповой batch process
~ деления fission process
~, доменный blast-furnace process
~, ждущий *comp.* asleep/sleeping process
~ зависящий от времени time-dependent process
~, итерационный *comp.* iterative process
~, кинетический rate process
~, кислородноконвертерный oxygen-converter [basic oxygen] process
~, колебательный oscillating/oscillatory process
~, контролируемый controlled process
~, космогенный cosmogenetic process
~, круговой cycle/cycling process
~, магматический magmatic/orthotectic process
~, марковский *math.* Markov(ian) process
~, мартеновский *met.* open-hearth process
~ массового обслуживания *comp.* queueing process
~, многоактный multievent process
~, многократный plural process
~, многоступенчатый multistage process
~, множественный multiple process
~, необратимый irreversible process
~, непрерывный continuous process
~, неравновесный non-equilibrium process
~, нестационарный non-steady process
~, неустановившийся unsteady-state process
~, обратимый reversible process
~, обратный inverse process
~, одноступенчатый single-stage process
~, окислительно-восстановительный redox process
~ остановки shut-down process/procedure
~, параллельный concurrent process
~ переноса transfer/transport process
~, переходный transient process
~ пуска start-up procedure
~, подчиненный *comp.* slave process
~, пользовательский *comp.* user process
~ принятия решений decision-making process
~, равновесный equilibrium process
~, регулируемый controlled process
~, случайный random/stochastic process
~, сольвеевский Solvay process
~, стационарный stable/steady/stationary process
~, судебный judicial/legal proceedings, lawsuit
~, технологический manufacturing/technology process
~, томасовский basic Bessemer process
~, уголовный criminal procedure
~, умственный operation of the mind, mentation
~, установившийся steady-state process
~, цепной chain process
процесс-адресат receiving process
процесс-источник producer process
процесс-отправитель sending process
процессно-объектный *gram.* object-process
процессор processor
~, автономный stand-alone processor
~, ассоциативный content-addressable processor
~ базы данных backend processor
~ ввода-вывода input/output processor
~, ведущий master processor
~, вспомогательный support processor
~, главный host processor
~ для обработки элементов element processor
~ изображений image processor
~, конвейерный pipelined processor
~, лингвистический language processor
~, матричный array processor
~ межсетевого сопряжения gateway processor
~, однокристальный single-chip processor
~ очередей queue processor
~, подчиненный slave processor
~, присоединенный attached processor
~, разрядно-модульный bit-slice processor
~ речевых сигналов speech processor
~ с распараллеливанием операций concurrent processor
~, связной communications processor
~ узла сети node processor
~, центральный central processing unit, CPU; main frame/unit
процессуальный *law* procedural, procedure *(attr.)*
Процион *astr.* Procyon
прочес, прочесывание carding, combing out; *mil.* mopping up
проческа screening
прочесывать comb; *(flax)* hackle
прочность strength; resistance; stability; endurance; fastness; toughness
~ в мокром состоянии wet strength
~ в сухом состоянии dry strength
~ в холодном состоянии cold strength

~ грунта bearing power of soil
~ конструкции structural strength/integrity
~ на изгиб bending/flexural strength
~ на излом fracture/breaking strength
~ на износ resistance to wear; wearability
~ на истирание abrasion resistance
~ на кручение torsional strength
~ на пробой *el.* breakdown/disruptive strength
~ на раздавление crushing strength
~ на разрыв breaking/rupture strength
~ на раскалывание splitting/cleavage strength
~ на растяжение tensile strength
~ на сдвиг shear(ing) strength
~ на сжатие compressive/compression strength
~ на скручивание torsional strength
~ на срез shear(ing) strength
~ на удар impact strength/resistance
~, объемная bulk strength
~ под давлением pressure resistance
~ сварного шва weld strength
~, ударная impact strength
~, удельная specific strength
~, усталостная fatigue strength/resistance
прошедшее время *gram.* past tense
прошивать 1. *met.* broach; 2. sew; stitch
прошивень *met.* broach; die; punch
прошивка 1. broaching; piercing; 2. sewing; stitching; 3. (tool) broach, broaching bit
прощупывать palpate, touch
проэкт = проект
проэструс *physiol.* pro-oestrus
проявитель *phot.* developer
~, дубящий tanning developer
~ контрастный high-contrast developer
~, мягкоработающий low-contrast developer
~, фиксирующий developer-fixer
проявление 1. manifestation; show, appearance; 2. *phot.* development
проявлять 1. manifest, show, evince; 2. *phot.* develop
пруд pond; pool
~, выростной nursery pond
~, маточный stock pond
~, нерестовый spawning/hatchery pond
прудонизм *polit.* Proudhonism
пруд-садок nurse pond
пружина spring
~, амортизационная/амортизирующая coushioning/buffer/counter spring
~, боевая *mil.* firing spring
~, арретировочная retaining/stop spring
~, буферная buffer/cushioning/damping spring
~, винтовая helicoiodal spring
~, витая coil spring
~, возвратная return/opposing spring
~, главная mainspring
~, двойная dual spring
~ кручения torsion spring
~, листовая/пластинчатая leaf spring
~, нажимная compression spring
~, натяжная tension spring
~, противодействующая antagonistic/opposing spring
~ створок door spring
~, тарельчатая disk spring
~ ударника striker spring
~ установочная adjusting spring
~, ходовая mainspring
пружинконосец *bot.* elaterophore
пруригиозный *med.* pruringious, itching
прус *zool.* Italian locust *(Calliptamus italicus)*
прусак *zool.* German cockroach *(Blattella germanica)*
прусс *hist.* Prussian
пруссак Prussian
пруссачество Prussianism
прусский Prussian
~кая синяя Prussian blue
прут, пруток bar; rod
~, присадочный welding/filler rod
прутовидный, прутьевидный rod-shaped
прыгун 1. *zool.* springbuck, springbok *(Antidorcas marsupialis)* 2. *bot.* touch-me-not *(Impatiens noli-me-tangere)*
прыжок jump; leap, spring, saltation
~ в воду dive
~ в высоту high jump
~ в длину broad jump
~, вынужденный *av.* bail-out
~, затяжной парашютный delayed parachute jump
~ на точность приземления jump for touch-down accuracy
~, опорный vault
~ с места standing jump
~ с парашютом parachute jump
~ с разбега running jump
~ с шестом pole vault
~ углом rear vault
прыщинец *bot.* lesser spearwort; spearwort buttercup *(Ranunculus flammula)*
прядение spinning
~, вигоневое vigogne spinning
~, гребенное worsted spinning
~, грубогребенное Bradford spinning
~ из расплава melt spinning
~ из раствора solvent spinning
~, камвольное worsted spinning
~ основной пряжи warp spinning
~, угарное waste spinning
~ уточной пряжи weft spinning
~, фильерное funnel spinning
~ шерсти, полугребенное semi-worsted spinning
прядильный spinning
~ное растение fibre/textile plant
прядильня spinning mill; spinnery
прядь strand
пряжа yarn
~, вышивальная embroidery yarn

~, основная warp yarn
~, уточная weft yarn
~, штапельная spun yarn
прямая straight line; *aero.* leg
~, бесконечно удаленная = несобственная
~, базисная straight-line basis, base line
~, несобственная ideal/improper/infinite line, line at infinity
~, предпосадочная *aero.* base leg
~, проективная projective line
~, расчетная *aero.* planned straight leg
~ сходимости axis of convergence
~, числовая numerical line, number line/scale
прямобороздный orthocolpate
прямоветвистый *bot.* orthocladous
прямовозбуждаемый *el.* directly excited
прямоволосый orthotrichous
прямодействующий direct-action; direct-acting
пряможилковатый rectinerved; rectivenous
прямозубчатый orthodentate
прямозубый straight-toothed
прямой straight; direct
~ билет through ticket
~мая кишка *anat.* rectum
~мая речь *gram.* direct speech
прямокишечный *anat.* rectal
прямоклювый *zool.* rectirostral
прямокрылые *pl. zool.* orthopterans *(Orthoptera)*
прямокрылый *zool.* orthopteral, orthopterous
прямолинейновозвратный reciprocating
прямолинейность rectilinearity
прямолинейный rectilinear
прямолопастный ortholobate
прямопараллельный *bot.* fastigate
прямоперегородочный rectimurate
прямопереходный direct-transitive
прямопропорциональность direct proportionality
пряморастущий autoorthotropous
прямоток parallel flow/current; concurrent flow
прямоточный uniflow, direct-flow; single-flow/-pass, concurrent; *(of engine)* ram-jet
прямоугольник *math.* rectangle
прямоугольный rectangular; orthogonal; rightangled
прямоходящий *zool.* orthograde
прямоцветковый *bot.* rectiflorous
прясть spin
псалмопевец 1. psalm singer; 2. the Psalmist
псалом psalm
псаломщик *eccl.* psalm reader; sexton
псалтерион 1. *eccl.* Psalter; 2. *mus.* psaltery
псаммит *petr.* psammite
псаммома *med.* psammoma; brain-sand tomo(u)r
псаммофил *zool.* psammophile
псаммофит *bot.* psammophyte, sandplant
псевдоадрес pseudoaddress; dummy/floating address
псевдоакация *bot.* robinia *(Robinia pseudoacacia)*
псевдоанкилоз *med.* pseuodankylosis, false/fibrous ankylosis
псевдоатаксия *med.* pseudoataxia; pseudotabes
псевдоблепсия *med.* pseudoblepsia, pseudoblepsis, parablepsia, perverted/false vision
псевдобугорчатый pseudopunctate
псевдовектор pseudovector, axial vector
псевдовкрапленник *petr.* metacryst
псевдогаллюцинация pseudohallucination
псевдогермафродитизм pseudohermaphroditism; anorogynism
псевдогонорея *med.* pseudogonorrh(o)ea, catarrhal urethritis
псевдоглей pseudogley; similigley
псевдоготика, псевдоготический pseudo-Gothic
псевдодвойник pseudotwin
псевдодвоичный *comp.* pseudo-binary
псевдодуга pseudoarc
псевдоизомерия pseudoisomerism
псевдокарп(ий) *bot.* pseudocarp, false fruit
псевдокарпогон(ий) *bot.* pseudocarpogone, pseudocarpogonium
псевдокоманда *comp.* pseudoinstruction
псевдоквадратный pseudotetragonal
псевдокислота *chem.* pseudo acid
псевдоклассицизм pseudoclassicism
псевдокод pseudo-code, abstract code
псевдокривая pseudocurve
псевдокристалл pseudo crystal
псевдокристаллический pseudocrystalline
псевдокубический pseudoisometric
псевдокруп *med.* pseudocroup; catarrhal croup
псевдолейкемия pseudoleukemia; aleukemia; pseudoleukocythemia; Hodgkin's disease
~, миелогенная myelomatosis
псевдолюэс *med.* pseudolues
псевдомаквис, псевдомаккия pseudomaquis
псевдомарксистский pseudo-Marxist
псевдомера *math.* pseudomeasure
псевдометрика *math.* pseudometric
псевдомиктический, псевдомиктичный, псевдомиктный pseudomictic
псевдомицелий *bot.* pseudomycelium
псевдомногообразие *math.* pseudomanifold
псевдомногочлен *math.* pseudopolynomial
псевдоморфный pseudomorphic; pseudomorph(ous)
псевдоморфоз pseudomorphism
псевдоморфоза *miner.* pseudomorph
~, частичная epimorphosis
псевдонарцисс *bot.* common daffodil
псевдонародный pseudonational; pseudo-folkloristic
псевдонаука pseudo science
псевдоним pseudonym, alias, pen/stage name
псевдоожижаемость fluidizability
псевдоожижать fluidize
псевдоожижение fluidization; quasi-liquefaction
псевдоокаменелость pseudofossil

псевдооснование pseudo base
псевдоотладка *comp.* bebugging
псевдооценка *math.* pseudovaluation
псевдопеременная *math.* pseudovariable
псевдопаралич *med.* pseudoparalysis
~, врожденный атонически congenital atonic pseudoparalysis
псевдопамять *comp.* pseudo-memory/store
псевдопериант(ий) *bot.* pseudoperianth
псевдоперигоний *bot.* pseudoperigone
псевдопериод quasi-period
псевдопрограмма pseudo-program(me); pseudo-routine
псевдореализм pseudo realism
псевдореалист pseudo realist
псевдореалистический pseudorealistic
псевдорозеточность *bot.* pseudo-rosette disease; pupation disease
псевдоромантизм pseudo romanticism
псевдосимвол *comp.* idle character
псевдоскаляр *math.* pseudoscalar (quantity)
псевдослоистость pseudoschistosity; pseudo-lamination
псевдосфера pseudosphere
псевдотабес *med.* pseudotabes, pseudotaxia
псевдотрихиноз pseudotrichiniasis, pseudo-trichinosis
псевдотуберкулезный pseudotuberculous, pseudotubercular
псевдофибрин pseudofibrin, parafibrinogen
псевдофилософ pseudo philosopher; philoso-phaster
псевдофилософия pseudo philosophy; philoso-phastry
псевдохолера meliodosis; pseudoglanders
псевдоцирроз *med.* pseudocirrhosis
псевдочернозем pseudochernozem
псилотический psilotic; depilatory
псилосин psilocin
псилотовые *pl. bot.* psylotaceans *(Psilotacae)*
псилофит *bot.* psilophyte
псилофитовые *pl.; bot.* psilophytaleans *(Psilophytales)*
псилоцитин psilocytin
пситтакоз *med.* psittacosis, parrot disease
пси-функция psi-function
психагог psychagogos, psychagogue
психальгия psychalgia; psychalgalia; mind/soul pain, algopsychalia
психанопсия *med.* psychanopsia; mind blindness
психастеник, психастенический *med.* psychasthenic
психастения *med.* psychasthenia
психатаксия *med.* psychataxia; mental confusion
психея 1. psyche; 2. **(П).** Psyche
психиатр psychiatrist, psychopathist
психиатрия psychiatry, psychiatrics; psychiatria
~ подростков adolescent psychiatry
~, судебная forensic psychiatry
психика psychics; psyche; psychology
психический psychical, psychic; mental
~ недостаток mental infirmity
психоанализ psych(o)analysis
психоаналитик psychoanalyst
психоаудиторный psychoauditory
психогенез(ис), психогения psychogenesis, psychogenia, psychogeny
психогенный psychogenetic; psychogenic
психогностика psychognosis
психогигиена mental hygiene
психоделическиий psychodelic; psychedelic
психодиагностика psychodiagnosis
психоз psychosis
~, военный war psychosis
~, старческий senile psychosis
психозойский *geol.* Psychozoic
психокинез psychokinesis
психолингвистика psycholinguistics
психолог psychologist
психологизм psychologism
психологический, психологичный psychologi-cal
психология psychology
~ инженерная human engineering
психометрика psychometrics
психометрия psychometry
психомонизм *philos.* psychomonism, solipsism
психомоторика psychomotor system
психомоторный psychomotor *(attr.)*
психоневроз psychoneurosis; neuropsychosis, neuropsychopathy
психоневролог psychoneurologist
психоневрология psychoneurology; neuropsy-chiatry
психоневропатология psychoneuropathology
психономика psychonomics; psychonomy
психопарез psychoparesis
психопат psychopath
~, безвольный weak-willed psychopath
~, гипертимический hyperthymic psychopath
~, паранойяльный paranoic psychopath
психопатический, психопатичный psycho-pathic
психопатия psychopathy; psychinosis; psychono-sis
психопатолог psychopathologist
психоплегия psychoplegy
психополовой psychosexual
психоприемник lunatic asylum
психопрофилактика psychoprophylaxis
психорексия psychorrhexis
психоритмия psychorrhythmia
психосенсорный psychosensorial, psychosensory
психосердечный psychocardiac
психосоматика psychosomatic medicine
психостимулятор psychostimulant
психотерапевт psychotherapeutist
психотерапия psychotherapy; psychotherapeutics
~, отвлекающая counter-attracting psycho-therapy

психотехника psychotechnics
психотехнический psychotechnical
  ~ тест ability/intelligence test
психотомиметический psychotomimetic, psychosomimetic
психотроника psychotronics
психотрия *bot.* wild coffee *(Psychotria)*
психофармакология psychopharmacology, neuropharmacology
психофизиология psychophysiology
психофильный *bot.* butterfly-pollinated
психохирургия psychosurgery
психоэпилепсия psychoepilepsy, psychic epilepsy
психрометр psychrometer; wet-and-dry bulb hygrometer
  ~, аспирационный ventilation/aspiration/aspirated psychrometer
  ~, дистанционный telepsychrometer, distant-reading psychrometer
псоас-симптом psoas symptom
псоит *med.* psoitis
  ~, серозно-гнойный seropurulent proitis
психрофил psychrophile
психрофит psychrophyte
псовые *pl. zool.* canids *(Canidae)*
псоралея *bot.* scurfy pea *(Psoralea)*
псориаз *med.* psoriasis
  ~, бородавчатый verrucous psoriasis
  ~, буллезный bullous psoriasis
  ~, воронкообразный infundibuliform psoriasis
  ~, пустулезный pustular psoriasis
псориазный psoriatic; psoriasic
псофометр psophometer; circuit noise meter
птармика *bot.* sneezewort yarrow *(Achillea ptarmica)*
птелея *bot.* hop tree *(Ptelea)*
птенохория *bot.* ptenochory
птеригий *med. anat. zool.* pterygium
  ~, неразрастающийся inactive perygium
птеригиумоподобный pterygoid
птеригомаксиллярный pterygomaxillary
птеригомандибулярный pterygomandibular
птерин pterin; rhizopterin
птериндезаминаза pterin deaminase
птерис *bot.* brake *(Pteris)*
птеродактиль *paleont.* pterodactyl, ornithosaurian
птерозавр *paleont.* pterosaur *(Pterosaurus)*
птероилглютаминовый pteroylglutamic
птерокактус *bot.* wing cactus *(Pterocactus)*
птерокарпус *bot.* padauk *(Pterocarpus)*
птеропод *zool.* pteropod; pteropodan
птероподовые *pl. zool.* pteropods, pteropodans *(Pteropoda)*
птероподовый *zool.* pteropod, pteropodial
птероцелтис *bot.* wing celtis *(Pteroceltis)*
птиализм *med.* ptyalism; salivation
птиалин *bioch.* ptyalin
птиалолит ptyalolith; sialolith
птиалолитиаз ptyalolithiasis; sialolithiasis
птиалолитотомия ptyalolithotomy; sialolithotomy
птиалоцеле ptyalocele; salivary cyst
птигматит *miner.* ptygmatite
птифур *cul.* petit four
птихопаридный *paleont.* ptychoparioid
птица bird; fowl *pl. coll.* poultry
  ~, бегающая ratite
  ~, болотная/ходячая wading bird, wader
  ~, водоплавающая waterfowl
  ~, воловья/коровья cowbird *(Molothrus)*
  ~, домашняя *coll.* poultry
птица-носорог *zool.* hornbill *(Buceros)*
птицебойня poultry-dressing farm
птицеводство poultry farming/husbandry/breeding/raising; aviculture
птицеед *zool.* 1. bird spider *(Avicularia avicularia);* 2. bird of prey
птицезвери *pl. zool.* monotremes *(Monotremata)*
птицеклювый ornithorynchous
птицекомбинат poultry processing plant; poultry breeding plant/farm
птицемлечник *bot.* star-of-Bethlehem *(Ornithogalum)*
птиценогие *pl.* ornithopods *(Ornithopoda)*
птиценожка *bot.* serradella; bird's foot *(Ornithopus sativa)*
птицеобразный, птицевидный, птицеподобный birdlike; aviculoid
птицеопыление bird pollination; ornithogamy, ornithophily
птицеопыляемый ornithophilous; bird-pollinated
птицетазовые *pl.* ornithischians *(Ornithischia)*
птицефабрика = птицекомбинат
птицеферма poultry farm
птоз *med.* ptosis
  ~, утренний morning/waking ptosis
птозный *med.* ptotic
птолемеев Ptolemaic
птомаин *bioch.* ptomaine
пуаз *phys.* poise
пуанзеция *bot.* poinsettia *(Poinsettia)*
пуансон punch; puncheon; punch pin/stamp; upper die
пуантилизм *art* pointillism; divisionism
пубитомия *surg.* pubiotomy
пугач 1. *zool.* eagle owl *(Bubo bubo);* 2. popgun; toy pistol
пуговичник 1. *bot.* common tansy *(Tanasetum vulgare);* 2. button maker
пуд pood (16.38 kg)
пудлингование puddling
пудлинговать puddle
пудра powder; dust
  ~, пироксилиновая guncotton dust
  ~, сахарная powdered sugar
пудрилка dusting/powdering device, duster
пудровка dusting, powdering
пуерария *bot.* kuozu bean/vine *(Pueraria)*
пузанок *zool.* clupeid herring *(Caspialosa)*
пузыревидный, пузыреобразный bubblelike;

vesicular; cystiform
пузырек vesicle, follicle; bubble; blister
~, глазной lens/lenticular vesicle
~ граафов Graafian vesicle/follicle
~, зародышевый embryo/germinal vesicle
~, зрительный ocular/optic vesicle
~, мальпигиев Malpighian vesicle
~, обонятельный olfactory vesicle
~, семенной seminal vesicle; *bot.* seed sac
~, слуховой acoustic/auditory/otic vesicle
~ уровня level bubble
пузыреногие *pl. zool.* thrips *(Thysanoptera)*
пузыреносный vesiculiferous; physophorous
пузыреплодник *bot.* ninebark *(Physocarpus)*
позыреплодный *bot.* bladder-fruited, cystocarpous
пузыресемянник *bot.* Alexandrian laurel *(Danae racemosa)*
пузырник *bot.* 1. senna *(Colutea);* 2. bladder fern *(Cystopteris)*
пузырно-брюшной *anat.* vesicoabdominal
пузырно-мочеточниковый vesicoureteral
пузырно-пупочный vesicoumbilical
пузырно-шеечный vesicocervical
пузырный vesical, vesicular, cystic
пузырчатка 1. *med.* pemphigus; 2. *bot.* bladderwort *(Utricularia)*
пузырчатковые *pl. bot.* bladderwort family *(Utriculariaceae)*
пузырчатость vesiculation
пузырчатый vesicular; bubbly; *geol.* porous
пузырь bubble; blister; *anat.* vesicle, cyst; bladder; bulla
~, водяной bleb, blister
~, газовой gas hole/pocket/cavity
~, желчный gallbladder; bile cyst
~, зародышевый blastocyst; blastula
~, мозговой cerebral/encephalic vesicle
~, мочевой urinary bladder
~, решетчатый ethmoidal bulla
~, плавательный *zool.* air/swimming bladder; pneumatophore
~, слуховой auditory bulla
пузырьковидный, пузырькообразный vesicular; utricular; utriculiform; bladder-shaped
пузырьковый *med.* phlyctenular
пузырьконосный vesiculiferous, utriculiferous
пул pool
пулазан *bot.* pulasan; capulasan *(Ephelium mutabile)*
пулемет *mil.* machine gun
~, зенитный anti-aircraft machine gun
~, турельный turret machine'gun
~, фюзеляжный fuselage-mounted machine gun
пуленепробиваемый, пуленепроницаемый, пулестойкий bullet-proof
пульверизатор atomizer; sprayer
пульверизировать atomize, spray
пульмональный pulmonary, pulmonic
пульмонит *med.* pulmonitis; pneumonitis
пульпа pulp
~ зуба dental/dentinal pulp
~ зуба, живая vital/living pulp
~ зуба, корневая root pulp
~ зуба, коронковая coronal pulp
~ зуба, мертвая dead/nonvital pulp
~ зуба, разложившаяся putriscent pulp
~ зуба, умерщвленная devitalized/devital pulp
~ селезенки splenic pulp
пульпальгия *med.* pulpalgia
пульпарный *anat.* pulpal
пульпит *med.* pulpitis
~, гнойный suppurative pulpitis
пульповидный pulpiform, pulpy
пульповой pulpal, pulpar
пульпозный pulpy
пульпообразование pulpifaction
пульпопровод sludge line/pipe
пульс pulse; pulse rate
~, аритмичный arhythmic pulse
~ недостаточного наполнения deficient pulse
~, неравномерный unequal pulse
~, нерегулярный/неправильный irregular pulse
~, низкий small pulse
~, нитевидный thready/filiform pulse, small fine pulse
~, обратнопарадоксальный reversed paradoxical pulse
~, печеночный hepatic pulse
~, подскакивающий water-hammer [trip-hammer] pulse, cannonball/collapsing/piston pulse
~ пуповины funic pulse
~, равномерный equal pulse
~, учащенный/частый rapid pulse
~ хорошего наполнения full pulse
~, центральный артериальный systemic arterial pulse
пульсар *astr.* pulsar
пульсатор pulser, pulsator
пульсация pulsation; beating, throbbing; fluctuation; pulse; oscillation
~, надчревная epigastric pulsation
~ несущей *el.* carrier flutter/fluctuation
~, сердечная heart beating, cardiac pulsation
~, систолическая systolic pulsation
~ сонной артерии carotid pulsation
~, хаотичная irregular spikes
~, яремная jugular pulse
пульсировать pulsate, beat, throb; fluctuate
пульсометр pulsometer; pulsimeter
пульсотахометр pulsotachometer; cardiotachometer
пульс-реле pulsing relay
пульт desk; console; control/panel board; music stand
~, дирижерский conductor's stand
~, измерительный measurement panel
~, испытательный test set; tester

~, клавишный keyboard
~, контрольный monitoring panel/console
~, микшерский mixer console
~ настройки tuning panel
~, оператора operator's console
~ перезаписи re-recording console
~ управления control desk/panel/console/board
~, экранный *comp.* visual display unit
пуля bullet; projectile
~, трассирующая tracer bullet
пума *zool.* puma; cougar *(Felis* concolor*)*
пумело *bot.* pomelo, shaddock *(Citrus grandis)*
пумит *geol.* pumice
пуниец *hist.* Carthaginian
пунический *hist.* Punic; Carthaginian
пункт 1. point; spot; place; 2. station, post, centre; 2. item; point
~, базисный reference point
~ дезактивации decontamination room/station
~, диспетчерский master station
~, донорский blood donor centre; blood bank
~, заправочный fuelling point/station
~, избирательный polling station
~, исходный starting/datum point
~, командно-диспетчерский *av.* traffic control tower
~, контрольный monitoring point/station
~, кульминационный climax
~ маршрута (en-)route point
~, медицинский medical aid station/centre
~ наведения direction/vectoring post
~ назначения point of destination
~, населенный town or village; population cluster; inhabited/populated locality
~ обработки данных data-processing terminal
~ опознавания identification point
~, опорный 1. point of reference; 2. *mil.* fortified stronghold
~ остановки stop-off point
~, отличительный landmark
~ первой помощи first-aid station/post; emergency station
~, перевязочный dressing station
~ приема receiving centre/station
~, радиолокационный radar point/station
~, радиопеленгаторный radio direction-finding station
~ сбора rendevous/assembly point/area
~ связи communication centre
~, сторожевой watch point
~ технического обслуживания maintenance point
~, узловой *rail.* junction point
~, усилительный *com.* repeater station/office
~, учебный training station/centre
пунктат *med.* punctate
пунктация *med.* stippling
пунктир dotted line; *art* stipple, stippling
пунктирный dotted; spot *(attr.)*
~ная сварка spot welding
пунктирник dotting instrument/device
пунктирование 1. *med.* puncturing, puncturatio; 2. dotting
пунктировка dotting; stippling
пунктуация *gram.* punctuation
пункционный *surg.* paracentetic
пункция *med.* puncture, paracentesis, punctura; tapping
~ барабанной перепонки paracentesis of the middle ear [of the drum membrane]; myringotomy
~ головного мозга paracentesis of the head/scull; cephalocentesis
~ живота paracentesis of the abdomen; abdominocentesis
~, надлобковая suprapubic puncture
~ перикарда [околосердечной сумки] paracentesis of the pericardium; pericardiocentesis
~, плевральная pleurocentesis, pleuracentesis
~, поясничная lumbar/spinal puncture
~ сердца paracentesis of the heart; cardiocentesis
~ яичников ovariocentesis, oothecocentesis
пуночка *zool.* snow bunting *(Plectrophenax nivalis)*
пунсон punch; die
пунширование punching; perforating
пупавка, пупавник *bot.* camomile *(Anthemis)*
пупинизация *el.* pupinization; coil loading
пуповина *anat.* umbilical cord
пуповинный funic; umbilical
пупок umbilicus; omphalos; omphalus; navel; umbo
~ фораминифер umbo
пупочник *bot.* navel seed *(Omphalodes)*
пупочно-кишечный omphalomesenteric; omphalomesaric
пупочный *anat.* umbilical, omphalic
~ канатик umbilical cord
пуританин, пуританский Puritan
пуританство Puritanism
пурка *agr.* grain-unit scale; grain tester
пурпур purple
~, зрительный visual/eye purple, rhodopsin
пурпура *med.* purpura
~, зудящая itching puprura
~, мозговая cerebrospinal fever, malignant puprura
~, оспенная malignant smallpox
~, сосудистая vascular purpura
пуск start, starting, startup; run; launching
~ в эксплуатацию bringing/placing into service, putting into operation; commissioning
~, контрольный check run
~ по адресу address start
~, повторный restart
~, пробный trial start
~ программ первой очереди foreground initiation; foreground program(me) start
~ ракеты missile launch; rocket firing
~ реактора reactor startup
пускание starting, start
~ крови bloodletting; phlebotomy

пускатель starter
пускать 1. start (up); switch/turn on; run; 2. let; allow; permit; let go; release
~ в ход start; set in motion/operation
~ в эксплуатацию put in operation; commission
~ кровь bleed; phlebotomize
пусковой starting; launching; actuating
пуста puszta *(in Hungary)*
пустота emptiness; void, blankness; vacuum; hollow, cavity
~, торричеллиева Torricellian vacuum
пустотелый hollow
пустотность 1. hollowness; void; 2. vacuum
пустоцвет *bot.* barren flower
пустоцветный barren-/sterlie-flowered
пустошка *zool.* hoopoe *(Upupa epops)*
пустоягодник *bot.* strawberry clover *(Trifolium fragiferum)*
пустула *med.* pustule; pustula
~, злокачественная malignant pustule, cutaneous anthrax
~, посмертная postmortem pustule
пустулезный *med.* pustular; pustulant
пустуловидный, пустулообразный pustuliform
пустыневед desertologist
пустынник, пустынножитель anchorite, hermit; eremite
пустынноколосник *bot.* desert rod *(Eremostachys)*
пустырник *bot.* motherwort *(Leonurus)*
путемер pedometer; odometer
путеочиститель track cleaner
потеподъемник track lifter, track-raising machine
путепровод viaduct
~ над дорогой overpass; overbridge
путепрокладчик tracklayer, tracklaying machine
путерихтовщик track liner/maintainer
путеукладчик track-laying machine; tracklayer
пути *pl.* routes; lanes, tracks; tracts
~, дыхательные respiratory tracts
~, лимфатические lymphatic viae
~, мочевыводящие urinary tracts
~ подвоза supply routes
~, родовые maternal passages
~ сообщения communication routes; communications
~ сообщения, морские sea lanes
путч *polit.* Putsch
путь way, path; route; lane; tract; course
~ в графе graph path
~, внутренний водный inland waterway
~, водный waterway
~ доступа access path
путь, железнодорожный railway track
~, бесстыковой continuous welded track
~, главный main track; trunk line
~, двойной double track
~, тупиковый dead-end track; stub track
~, узкоколейный narrow-gauge track
~, ширококолейный wide-gauge track
путь *(continued)*
~, запасный alternate route; *rail.* sidetrack
~, зрительный optic tract
~ луча ray path
~, Млечный *astr.* Milky Way
~, морской shipping lane
~, обходный/окружный circuitous course
~, объездной detour
~, одиночный *anat.* solitary tract
~, пирамидный *anat.* pyramidal/corticospinal tract
~ подвода supply route
~, подкрановый crane track
~ подхода approach lane/path
~, подъездной access road
~, проводящий conduction tract
~, сочетательный *anat.* association tract
~, торговый trade route
~, тормозной braking/stopping distance
~ урагана storm track
пуффин *zool.* puffin; shearwater *(Puffinus)*
пухляк *zool.* willow tit *(Parus)*
пуховник *bot.* prairie dogbane *(Apocynum sibiricum)*
пухоед *zool.* 1. hair moth; 2. bird/biting louse
пухоотделитель linter
пуццолан(а) *constr.* pozz(o)ulana; pozz(o)ulan
пучковатый fascicular; fasciculate; bunched; clustered; tufted
пучковидный, пучкообразный fascicular, fasciculate; clustered; tufted; clusterlike
пучковый fascicular, fasciculate; bunched, bundled
пучконосный *bot.* fascicle-/fascicule-bearing
пучность antinode; loop; crest
~ тока current loop
пучок 1. bunch, bundle; cluster; tuft; fascicle; 2. beam; pencil
~, восходящий emerging/emergent beam
~ гамма-излучения gamma beam
~, импульсный pulsed beam
~, кабельный multiwire/bunch cable, loom
~, крючковатый uncinate fascicle
~ лучей pencil of rays
~, мезонный meson beam
~, мигающий pulsed beam
~, мышечный muscle bundle
~, мю-мезонный muon beam
~, непрерывный steady/continuous beam
~, остронаправленный pencil (beam)
~, отогнутый retroflex fascicle
~, падающий incident beam
~ пи-мезонов pion beam
~, постоянный steady/continuous beam
~ проводящей ткани strand of conducting tissue
~ прямых *math.* pencil of lines
~, пульсирующий pulsed beam
~, световой pencil of light
~, сосудистый vascular bundle

пушбол *sport* pushball
пушистоветвистый *bot.* downy-branched
пушистоколосый *bot.* downy-eared
пушистолистный *bot.* dasyphyllous
пушистоплодный *bot.* dasycarpous; hebecarpous
пушистость fluffiness; downiness
пушистоцветковый *bot.* dasyanthous
пушица *bot.* cotton sedge/grass *(Eriophorum)*
пушка cannon; gun
~, зенитная antiaircraft gun
~, кобальтовая cobalt bomb, cobalt irradiation unit
~, многоствольная multiple-barrel gun
~, неподвижная fixed-mount gun
~, плазменная plasma gun
~ с вращающимися створами rotary cannon/gun
~, скорострельная rapid-firing gun/cannon
пушнина fur skins, peltry
пушница *bot.* canna *(Canna)*
пушной fur-bearing; furred
~ товар furs
пушонка *agr.* hydrated/powder lime
пуэрилизм puerilism; second childhood
пуэрперальный puerperal, puerperous; puerperant
пуэрперий puerperium
пуэрториканец, пуэрториканский Puerto Rican
пфальцграф *hist.* palsgrave; count palatine
пфальцграфский *hist.* palatine
пфальцграфство *hist.* palatinate, county palatine
пчелиный apian; bee *(attr.)*
~ воск beeswax
~ная матка queen bee
пчелка *bot.* bee orchid *(Ophrys afifera)*
пчеловод beekeeper, apiarist
пчеловодство beekeeping; apiculture, bee culture
пчелоед *zool.* honey buzzard *(Pernis apivorus)*
пчелолечение apiotherapy
пчельник, пчельная apiary; bee garden
пшат *bot.* Russian olive *(Elaeagnus angustifolia)*
пшеница *bot.* wheat *(Triticum)*
~, озимая winter wheat
~, твердая durum (wheat)
~, яровая spring wheat
пшеничный wheaten, wheat *(attr.);* frumentaceous
пыжик young reindeer
пыжьян *zool.* Siberian whitefish
пылеватый dustlike; powdery; pulverous, dusty
пылезащитный, пылезащищенный dustproof
пыленепроницаемый dust-tight, dustproof
пылеосадитель dust precipitator/catcher/arrester
пылеотвод dust flue
пылеотделитель dust separator/remover
пылепитатель pulverized-coal feeder
пылесос vacuum cleaner
пылеуловитель dust collector/separator/trap
пыль dust; powder
~, алмазная diamond dust; diamantine
~, буровая drilling dust
~, взвешенная airborne dust
~, зольная fly ash
~, колошниковая blast-furnace dust
~, наждачная emery powder/dust; powder(ed) emery
~, снежная powder snow
~, тонкая/тонкодисперсная fine dust
пыльник *bot.* anther
~, взрывающийся dehiscent anther
пыльниковый *bot.* antheral, anther *(attr.)*
~ мешок pollen sac
~ная ямка anther bed
пыльца 1. *bot.* pollen; 2. *zool.* farina
пыльцевой pollen *(attr.),* pollinic
~вая трубка/трубочка pollen tube
пыльцевход *bot.* micropyle
пыльцеголовник *bot.* phantom orchid *(Cephalanthera)*
пыльцелистник *bot.* microsporophyll
пыльценосный polliniferous
пыльцеуловитель *bot.* pollen trap
пырей *bot.* wheat grass, quack grass, couch-grass *(Agropyrum)*
пьедестал pedestal; stand
пьедра *med.* piedra
пьезогенный *petr.* authigenous
пьезовибратор piezoelectric vibrator
пьезодатчик piezosensor, piezoelectric transducer
пьезокардиограмма piezocardiogram
~, пищеводная esophageal piezocardiogram
пьезокварц piezoelectric quartz crystal, piezoquartz
пьезоманометр piezoelectric pressure gauge
пьезометр piezometer
пьезомикрофон piezoelectric microphone
пьезопереход piezojunction
пьезотерапия piezotherapy, artificial pneumothorax
пьезотензометр piezoelectric strain gauge
пьезоэлектрик piezoelectric material/crystal
пьезоэффект piezoelectric effect
пьезоячейка piezocell
пьемонтец, пьемонтский Piedmontese
пьянство, запойное dipsomania
пяденица *zool.* inchworm
пястно-запястный *anat.* carpometacarpal
пястно-фаланговый *anat.* metacarpophalangeal
пястный *anat.* metacarpal
пясть *anat.* metacarpus
пята *tech.* thrust/vertical journal, foot, heel; abutment, pivot
~, кольцевая ring pivot
~ плотины heel of a dam
~, шаровая ball journal
пятерично-двоичный quibinary
пятеричный quinary; quintuple, fivefold
пятиатомный pentatomic, pentavalent
пятибалльный five-point *(attr.)*
пятиборец *sport* pentathlonist
пятиборье *sport* pentathlon
пятивалентный *chem.* pentavalent

пятигранник *geom.* pentahedron
пятигранный pentahedral
пятидесятник 1. *hist.* *mil.* commander of fifty men; 2. *rel.* Pentecostalist
пятидесятница *rel.* Pentacost
Пятикнижие Penteteuch
пятиконечный five-pointed
пятилетка five-year plan
пятилистник *bot.* marsh cinquefoil *(Potentilla palustris)*
пятилучевой, пятилучистый pentaclinal
пятиместный five-seater *(attr.),* with five seats
пятиокись *chem.* pentoxide
пятиосновный pentabasic
пятипалый pentadactyl, five-finger(ed)
пятиполье *agr.* five-field crop rotation
пятирядный quinqueserial, pentastichous
пятислойный five-ply
пятистишие pentastich
пятистопник *lit.* pentameter line
пятистопный *lit.* pentameter *(attr.)*
 ~ ямб iambic pentameter
пятисторонний pentahedral; five-sided
пятисторонник pentahedron
пятистрочие five-line stanza
пятитомник five-volume edition/work/publication
пятиугольник *geom.* pentagon
пятиугольный pentagonal, five-cornered
пятихлорный *chem.* pentachloride
пятицветный pentachromic
пятка heel; sole
 ~ иглы *text.* needle butt
 ~ якоря *naut.* anchor crown
пятнадцатиугольник *geom.* pentadecagon; quinquedecagon
пятнистость spottiness; spotting, mottling; maculation
 ~ листьев leaf spot, chlorosis
пятнистый spotted, spotty, speckled; mottled, maculate; macular; maculous
пятно spot, stain; blot, blotch; patch; macule
 ~, белое blank
 ~, воздушное *leath.* air stain, blister
 ~, голубое *med.* blue spot
 ~, гонорейное *med.* Saenger's macula
 ~, желтое *med.* macula lutea; yellow spot
 ~, жирное/масляное grease stain
 ~, зародышевое germinal spot
 ~, родимое birthmark; mother's mark
 ~, синее *med.* livor
 ~, слепое *anat.* blind spot, optic papilla
 ~, фокусное focal point/spot
пятновыводитель spot/stain remover
пяточно-большеберцовый calcaneotibial
пяточно-ладьевидный calcaneonavicular
пяточно-таранный calcaneoastragaloid
пяточный *anat.* calcaneal; calcanean

Р, р

рабат *econ.* rebate
рабда *zool.* rhabdus
рабдолит *geol.* rhabdolith
рабдомиосаркома *pathol.* rhabdomyosarcoma
рабкооп *abbr.* workers' cooperative store
рабкор *abbr.* worker correspondent
рабкрин *abbr.* *obs.* workers' and peasants' inspectorate
раблезианский Rabelaisian
рабовладелец slaveowner, slaveholder
рабовладельческий slaveowning, slaveholding
рабовладельчество, рабовладение slaveholding; slavery
работа work, working, operation; performance; activity, activities; functioning; job; labour; *see also* **работы**
 ~, автономная 1. independent [self-sufficient] operation; 2. *comp.* off-line operation
 ~, административно-хозяйственная administration
 ~, аккордная piece work, work by the piece, task work, piecework job
 ~ без надзора оператора unattended operation
 ~ без обратной связи open-loop operation
 ~, безаварийная trouble-free operation/running
 ~, безотказная/бесперебойная failure-/trouble-free operation, no-failure operation
 ~, безошибочная error-free operation
 ~, бригадная team work
 ~ в аварийном режиме emergency operation
 ~ в две смены double-shift [two-shift] operation
 ~ в диалоговом режиме *comp.* conversational operation
 ~ в импульсном режиме pulsed operation
 ~ в интерактивном режиме interactive operation
 ~ в критическом режиме critical operation
 ~ в одну смену one-shift operation, one shift work
 ~ в открытом космосе extravehicular activity
 ~ в пакетном режиме batch operation
 ~ в помещении inside /indoor work
 ~ в пошаговом режиме one-step operation
 ~ в пусковом режиме starting-mode operation
 ~ в режиме реального времени real-time operation
 ~ в связке clustered operation
 ~ в течение всей жизни life work
 ~ внутри космического аппарата intravehicular activity
 ~ вне цикла out-cycle work/operation
 ~ вразнос racing, runaway
 ~, высокоприоритетная *comp.* foreground

~ группы team work
~, двухсменная double shift [two-shift] work
~, дипломная graduation thesis
~ для выполнения по субподряду given-out work
~, договорная contract/contractual work
~, индикаторная rated work
~, каторжная hard labour
~, конторская/канцелярская office/clerical work
~ ключом *rad.* keying
~, круглосуточная round-the-clock operation; 24-hour service
~, многозадачная *comp.* multitasking
~ на борту космического корабля spaceborne operation
~ на дому outwork
~ на клавиатуре keying
~ на комиссионной основе commission work
~ на малых оборотах idling, idle operation
~ на мощности power operation
~ на небольшой мощности low-power operation
~ на полной мощности full-throttle operation
~ на пониженной мощности partial throttle operation
~ на расчетном режиме design operation
~ на холостом ходу idling, idle operation
~, научная scientific paper
~, научно-исследовательская research (work)
~, научно-исследовательская и опытно-конструкторская research-and-development work /activity, R&D work
~, незавершенная in-process job
~ неполный рабочий день part-time job
~, непостоянная casual work
~, нерегулярная intermittent work
~, опытно-конструкторская research and development (work); development engineering effort/work
~, отказоустойчивая fault-tolerant operation
~, периодическая intermittent/batch operation
~ по контракту contractual employment
~ по найму wage work; working for hire
~ по наряду task work, work by task
~, повседневная routine work
~, поденная work by the day
~, подсобная subsidiary work
~, полезная useful work
~ полную смену full-time operation
~ полный рабочий день full-time work/job
~, почасовая timework
~, прерывистая intermittent operation
~ при полной нагрузке full-capacity operation
~, ровная *eng.* smooth running
~ с машинным темпом machine-paced work
~ с несколькими заданиями multijob/multitask operation, multitask(ing)
~ с обратной связью closed-loop operation
~ с полной нагрузкой full-load operation
~ с полной тягой full-thrust operation
~ с разделением времени time-sharing operation
~ с сокращенным рабочим днем short-time work
~, сверхсменная/сверхурочная overtime work
~, сдельная piecework, task work; piecework job
~ сердца cardiac output
~, синфазная in-phase operation
~, сменная shift work
~ со сбоями malfunction
~ со словами переменной длины *comp.* variable length operation
~, спешная/срочная rush job
~ строго по правилам work-to-rule
~, умственная mental work, brainwork
~, управленческая managerial/administrative work
~, учебная 1. study; 2. teaching
~, фиктивная dummy activity
~, фоновая background activity
~, черная common labo(u)r
~, штукатурная plastering

работать work; operate; function; run; perform; be in operation; be open
~ неполное количество рабочих часов [неполную рабочую неделю] work on part time

работник worker, workman; operator; employee
~ занятый неполный рабочий день part-time employee/worker
~ занятый полный рабочий день full-time employee/worker
~ имеющий низкий доход low-income worker
~, инженерно-технический technologist, technical and engineering employee; white-collar worker
~, квалифицированный skilled worker/operator
~, научно-технический technologist
~, научный research worker; scientist
~, непроизводственный nonproduction worker
~ умственного труда brain/nonmanual worker
~ физического труда manual worker

работница woman worker, workwoman
~, домашняя maid, help, servant

работодатель employer

работорговец *hist.* slave trader, slaver

работорговля slave trade

работоспособность efficiency, working capacity, capacity for work; *(of machine)* serviceability

работы *pl.* operations, work, activities; *see also* **работа**
~, аварийно-спасательные salvaging (operations)
~, арматурные installation of reinforcement
~, бетонные concreting, placement of concrete
~, буровзрывные drilling and blasting (opera-

tions)
~, взрывные blasting (operations)
~, водолазные diving work
~, вскрышные overburden mining/operations
~, вспомогательные bywork
~, горноспасательные rescue work/operations
~, дноуглубительные/землечерпательные dredging
~, дорожно-строительные road-building
~, достроечные outfitting work
~, исследовательские, конструкторские, доводочные и технологические research, development, test and evaluation work, RDTE work
~, кровельные roofing
~, монтажные installation/assembly work
~, научно-исследовательские research
~, научно-исследовательские и опытно-конструкторские research and development, R&D
~, общественные public works
~, опытно-конструкторские development work
~, погрузочно-разгрузочные material handling (operations); freight/cargo handling
~, поисково-спасательные search and rescue operations
~, полевые field work
~, промысловые fishing operations
~, профилактические *tech.* preventive maintenance work
~, поисковые *geol.* prospecting operations
~, разведочные exploration work; prospecting
~, регламентные routine/scheduled operations
~, ремонтные repairs
~, сельскохозяйственные farm work
~, спасательные salvage operations, salvaging, rescue work
~, спасательно-восстановительные rescue-and-recovery operations
~, столярные woodworking
~, строительные construction work, civil engineering work
~, такелажные rigging
~, уборочные harvesting
~, штукатурные plastering; stuccoing
рабочие *pl.* workers, workpeople; labo(u)r, labour force, manpower; *see also* **рабочий I**
~, занятые employed workers
~, излишние redundant workers
~, незанятые idle/unemployed labour
~, постоянные regular workers
~, уволенные redundant/discharged workers
рабочий I worker, workman, labo(u)rer, operator, hand; working man; *see also* **рабочие**
~, верховой derrickman
~, высококвалифицированный highly skilled worker
~, забойный face worker, faceman
~ занятый неполный рабочий день part-time worker
~, квалифицированный skilled worker
~, кочующий migratory worker
~, малоквалифицированный low-skill worker
~, надомный outworker
~, наемный wage worker, hired worker
~, неквалифицированный unskilled/general worker
~, обученный trained worker
~ по добыче *min.* pit/quarry worker, pit miner
~, поверхностный *min.* surface worker
~, подземный *min.* underground worker, miner
~, поденный daysman
~, подменный relief worker
~, подсобный helper, shop hand; maintenance worker
~, полуквалифицированный semi-skilled worker
~ получающий заработную плату wage employee/earner
~, производственный factory/manufacturing/production worker; *colloq.* blue-collar worker
~, сдельный pieceworker
~, сельскохозяйственный farm hand/worker, agricultural/farm/rural worker
~ сталелитейной промышленности steel worker
~, торговый наемный commercial wage worker
~ член профсоюза union workman
рабочий II workers'; working; working-class
~ день working hours/day
~ день, неполный short time
~ диапазон температуры working/operating temperature range
~ скот draught/draft animals
~ ход *eng.* power/working stroke
~ чертеж working drawing
~чая сила manpower; labo(u)r
~чее место workplace; job
~чие пчелы worker bees
рабочком *abbr.* workers' committee
рабселькор *abbr.* worker-peasant correspondent
рабсила *abbr.* manpower; labo(u)r force, labo(u)r
рабство slavery, servitude, bondage
рабфак *abbr.* workers' school, ''rabfak''
рабфаковец workers' school student
рабыня bondmaid, slave
равви, раввин rabbi
равенала *bot.* travel(l)er's tree *(Ravenala)*
равендук, равентух raven's-duck, ravens
равенство equality, parity; *math.* equation
~, приближенное approximate equation
~ва, знак equal-sign
равнение alignment; levelling; smoothing
равноапостольный coequal with the apostles
равнобедренный *geom.* isosceles
равнобокий, равнобочный equilateral, equal-sided
равновеликий equidimensional; equivalent

равновероятность equiprobability
равновероятный equiprobable, equally probable
равновершинный *bot.* fastigiate
равновесие equilibrium, equipoise, balance
~, **безразличное** indifferent/neutral equilibriun
~ **в режиме штопора** *aero.* spinning equilibrium
~, **вековое** secular equilibrium
~ **двух фаз** bi-phase equilibrium
~, **демографическое** population equilibrium
~ **доходов и расходов бюджета** budgetary balance
~, **застывшее** frozen equilibrium
~, **изостатическое** isostacy, isostatic equilbrium
~, **кислотно-щелочное** acid-base equilibrium
~ **между деньгами и товарами** money-commodity equilibrium
~ **между запасами и снабжением** stock-and-provision balance
~ **моментов** moment balance
~, **нарушенное** disturbed equilibrium
~, **неустойчивое** unstable/labile equilibrium
~ **обмена веществ** metabolic equilibrium
~, **окислительно-восстановительное** redox equilibrium
~, **переходное** transient equilibrium
~, **подвижное** mobile equilibrium
~, **популяционное** population/genetic equilibrium
~ **при минимизации издержек** least-costs equilibrium
~ **при покое** equilibrium at rest
~ **сил** equilibrium of forces
~, **случайное** stochastic balance
~ **спроса и предложения** supply-and-demand equilibrium
~, **тепловое/термическое** thermal/heat equilibrium
~, **упругое** elastic equilibrium
~, **установившееся** steady-state equilibrium; terminal equilibrium
~, **устойчивое** stable/true equilibrium
равновесность equilibrium, balance
равновесный equilibrium *(attr.)*, equilibrial; equiponderant, equiponderous
равновесомый of equal weight, equal-weight
равновластие isocracy
равновременный isochronous, isochronal; tautochronous; simultaneous, synchronous
равновычетный *math.* equiresidual
равнодействующая *phys.* resultant (force)
~ **сил** force resultant
равноденственный equinoctial, equidiurnal
равноденствие equinox
~, **весеннее** vernal equinox
~, **осеннее** autumnal equinox
равнозернистый equigranular; isometric
равнозначность equivalence
равнозначный equivalent
равнокруговой homocyclic
равнокрылый *zool.* homopterous
равнолепестный *bot.* homopetalous, isopetalous
равнолистный *bot.* isophylous
равнолопастный equilobate; *zool.* isocercal
равномерно evenly, uniformly; equally
~ **распределенный** uniformly distributed, equidistributed
~ **созревающий** even-maturing
равномерность evenness; uniformity
равномерный even, uniform; steady
равномощность *math.* equivalence; equipollence
равнонаправленный equidirectional
равноногие *pl. zool.* isopods, isopodans (*Isopoda*)
равноногий *zool.* isopodous
равнообразие homeomorphism
равнообъемный equal-volume
равноосный *phys.* equiaxial
равноотстоящий equidistant, equally spaced
равноплодник *bot.* isopyrum *(Isopyrum)*
равнополярный isopolar
равнопотенциальный equipotential
равноправие equality of rights
равноправный of/enjoying/possessing equal rights; *comp.* peer-to-peer
равнопромежуточный equidistant
равноразветвленный isotomous
равноресничный *zool.* holotrichous
равнораспределение equidistribution
равносильность, равносилие equivalence, equipotence
равноотстоящий equidistant
равносигнальный equisignal, equal-signal
равноразмерный equidimensional, isometric
равносильный equivalent, equipotent
равнослоговой, равносложный parisyllabic, isosyllabic
равностворчатый *zool.* equivalve
равностепенный *math.* equipotential; equivalent
равносторонний equilateral, isolateral, equal-sided
равносходимость equiconvergence
равносходящийся equiconvergent
равнотемпературный isothermal
равноугольник *geom.* isogon
равноугольный equiangular, isoangular
равноудаленный equidistant; equispaced
равноударенный *phon.* even-stressed
равноускоренный uniformly accelerated
равнофазный equiphase
равноцветный homochromatic
равноценность equivalence, equal value, equality in number
равноценный equivalent, of equal value/worth
равночастотный equifrequential
равночисленный of equal number, equal
равноэнергетический equal-energy, equienergy *(attr.)*
рагит *geol.* rhagite
рагу *cul.* ragout
рад 1. *phys.* rad (unit); 2. *psych.* rad

рада *hist.* Rada; council, soviet
Радамант *myth.* Rhadamanthus
радар, радарный radar *see also* **РЛС**
раджа rajah
раджпут Rajpoot, Rajput
радиак *phys.* radiac
радиалия *zool.* radiale
радиаль *zool.* radial plate
радиально-лучевой radial-beam
радиально-лучистый radiate-fibrous
радиально-сверлильный radial-drilling
радиальный radial, radiate
радиан *math.* radian
радиант *astr.* radiant
~ **космического ливня** shower radiant
радиатор radiator; heat exchanger; cooler; heat-sink; emitter
~, **водо-масляный** water-oil cooler; water-oil heat exchanger
~ **водяного отопления** hot-water heating radiator
~ **для последующего охлаждения** after-cooler
~, **змеевиковый** coiled radiator
~, **кольцевой** annular radiator
~, **масляный** oil cooler/radiator
~ **парового отопления** steam-heating radiator
~, **пластинчатый/ребристый** finned/ribbed radiator
~, **плинтусный** baseboard radiator
~, **промежуточный** intercooler
~, **ребристо-трубчатый** gilled-tube [finned, tubular] radiator
~, **сотовый** honeycomb/cellular radiator
~, **трубчато-пластинчатый/ребристый** fin-and-tube radiator; secondary-surface radiator
~, **трубчатый** tube/tubular/multitubular radiator
радиационный radiative
радиация radiation; emission
~, **ионизирующая** ionizing radiation
~, **космическая** space/cosmic radiation
~, **остаточная** residual radiation
~, **отраженная** back-scattered radiation
~, **первоначальная** initial/nuclear radiation
~, **проникающая** penetration/ionizing radiation
~, **рассеянная** diffuse/scattered radiation
~, **солнечная** solar radiation
радий *chem.* radium, Ra
~, **бромистый** radium bromide
~, **сернокислый** radium sulphate
~, **углекислый** radium carbonate
~, **хлористый** radium chloride
радийсодержащий radium-bearing
радикал I *chem.* radical; *math.* radical/root sign
~, **кислотный** acid radical
~ **кольца** *math.* radical ideal
~, **основной** *chem.* base radical
~, **цветовой** colo(u)r radical, chromophore
радикал II *polit.* radical
радикальность radicalism
радикотомия *surg.* radicotomy, rhizotomy
радикула *bot.* radicle, radicule, rootlet
радикулальгия *med.* radiculalgia
радикулит *med.* radiculitis, nerve root syndrome
~, **пояснично-крестцовый** lumbosacral radiculitis
~, **шейно-грудной** thoracocervical radiculitis
радикуломенингомиелит radiculomeningo-myelitis, rhizomeningomyelitis
радикуломиелопатия radiculomyelopathy
радио radio; wireless
радиоавтограф autoradiograph, radioautograph
радиоавтография autoradiography, radioautography
радиоактивность radioactivity
~ **в воде** waterborne radioactivity
~ **в воздухе/атмосфере** airborne radioactivity
~, **искусственная/наведенная** artificial/induced radioactivity
~ **окружающей среды** environmental radioactivity
~, **сильная** high-level radioactivity
~, **слабая** low-level radioactivity
~, **удельная** specific radioactivity
радиоактиний *chem.* radioactinium, thorium-227
радиоальтиметр *aero.* radio altimeter, terrain clearance indicator
радиоантенна radio/wireless aerial
радиоаппарат radio set
радиоаппаратура radio equipment
радиоастрокоррекция radio stellar monitoring
радиоаутограмма radioautogram, autoradiogram
радиоаутография radioautography, autoradiography
радиобашня radio tower
радиобиолог radiobiologist
радиобиология radiobiology
~ **клетки** cellular radiobiology
радиоблок radio component/unit
радиобуй radio buoy
радиовахта radio watch
радиобуря radio/noise storm
радиовещание (radio) broadcasting
радиовзрыватель radio/electronic fuse/detonator
радиовидение radiovision, panoramic radar; wave visualization/imaging
радиовидимость radio visibility/contact
радиовилка radio bifurcation
радиовизор thermo-microwave visualization system
радиовойна radio/electronic warfare
радиоволна radio wave
~, **внеземная** extraterrestial radio wave
~, **земная** ground radio wave
~, **пространственная** sky radio wave; space radio wave
радиовсплеск radio pulse/outburst
радиовызов radio call
радиовысотомер radio altimeter
~ **с частотной модуляцией** frequency-modu-

lated [FM] radio altimeter
радиовышка radio tower
радиогенный, радиогеничный radiogenic
радиогеоразведка radioprospecting
радиогидроакустический radiosonic
радиоглаз *astr.* radio telescope
радиоглушение radio jamming
радиогоризонт radio horizon
радиограмма radiogram(me), radiotelegram(me)
радиограф radiograph
радиографировать radiograph
радиография radiography
~, бетатронная betatron radiography
~, перекрестная cross radiography
~, поточная mass radiography
~, рентгеновская X-ray radiography
~, следовая track radiography
радиодальномер radio range-finder, electronic/radio telemeter
~, геодезический tellurometer
~ с частотной модуляцией frequency-modulated [FM] radio range-finder
радиоданные radio data
радиодевиация deviation of radio azimuth/bearing
радиодермит *med.* radiodermitis
радиодеталь radio part/component
радиодезинформация *mil.* radio/electronic deception
радиодефектоскопия X-ray [microwave] flaw detection, radiographic inspection, X-ray defectoscopy
радиодиагностика radiodiagnosis
радиодонесение radio report/message
радиозасечка radio fix
радиозатмнение radioeclipse
радиозонд radioprobe, radiosonde, radar sonde
~, аэростатный balloon-borne radiosonde
радиоизлучение radio-emission, radio waves
радиоизотоп radioisotope, radioactive isotope
~ дочерний daughter radioisotope
~, исходный/материнский parent isotope
~, исчезающий extinct radioisotope
радиоимпульс radio-frequency pulse
радиоиндуцированный radiation-induced
радиоиод *chem.* radioiodine
радиоиодиндикация radioiodine indication
радиоиодированный radio-iodinated
радиоисточник radiation source
радиоканал radio channel/link, radio-communication channel
радиокомпас radio compass; direction finder, homing device/instrument
радиоконтроль radio monitoring
радиоконтур radio circuit
радиокурабильный radiocurable, curable by X-irradiation
радиокурсоуказатель radio direction finder
радиола radiogramophone
радиолампа radio/vacuum/electron tube/valve
радиолечение radiotherapy
радиолиз radiolysis, radiolytic decomposition
радиолиния radio line/link
радиолог radiologist
радиология radiology
~, медицинская radiation/nuclear medicine, medical radiology
радиолокатор, радиолокационная станция radar (set/station)
~, автоматическая automatic/unattended radar
~ автоматического сопровождения automatic tracking radar
~ аэропорта, обзорная airport surveillance radar, ASR
~, бортовая *aero.* airborne radar
~, всепогодная all-weather radar
~ дальнего обнаружения early-warning radar
~, диспетчерская aerodrome control radar
~, двухпозиционная bistatic radar
~ для обнаружения спутников satellite search radar
~, доплеровская Doppler radar
~, импульсная pulse-modulated radar
~, когерентная coherent/coho radar
~ кругового обзора all-round looking radar, surveillance radar
~, метеорологическая weather/meteorological radar
~, многолучевая stacked-beam radar
~, многорежимная multimode radar
~ наблюдения surveillance radar
~ наведения guidance/vectoring radar
~ наведения ракет по лучу beam-transmitter/rider radar
~, наземная ground/surface radar
~ непрерывного излучения continuous-wave radar
~, обзорная surveillance/acquisition radar
~ облучения цели (target) illuminating radar
~ обнаружения acquisition/search radar
~ опознавания цели target identification radar
~ определения траектории цели flight-path radar
~, панорамная panoramic/pictorial radar
~ переднего обзора forward-looking radar
~ перехвата interception radar
~ подсвета цели illuminating radar
~, подфюзеляжная ventral radar
~, поисковая search radar
~, посадочная landing radar, precision approach radar, PAR
~ предупреждения столкновений anti-collision [colision-warning] radar
~ с высокой разрешающей способностью high-resolution/-grain/-definition radar
~ с различением истинных целей decoy discrimination/sorting radar
~ с секторным обзором sector-scanning radar
~ с частотной модуляцией frequency -modulated [Md] radar
~, самолетная airborne/aircraft radar

~ сопровождения tracking radar
~ управления воздушным движением air-traffic control radar
радиолокационный radar *(attr.)*
радиолокация radio location and ranging, radiolocation; radar
~, импульсная pulse(d) radar/radiolocation
~ непрерывного излучения continuous wave radar
радиолуч radio/electronic beam
~, ведущий guiding beam
~, глиссадный glide-path/-slope beam
~, курсовой localized beam
~, посадочный landing radio beam
~, приводной homing beam
радиолярия *zool.* radiolarian *(Radiolaria)*
радиомаркер (radio) marker; marker (beacon)
~, ближний middle (radio) marker, MM
~, дальний outer (radio) marker, OM
радиомачта radio mast/tower
радиомаяк radio beacon; radio range
~, аварийный survival/recovery radio beacon
~, аварийный приводной crash locator radio beacon
~, береговой shore-based radio beacon
~, ближнего действия short-range radio beacon
~, ближний маркерный middle radio marker
~, всенаправленный omnidirectional radio range, omnirange
~, глиссадный glide-path transmitter/beacon
~, дальний маркерный outer (radio) marker
~, длинноволновый low-frequency radio beacon
~ зоны ожидания holding beacon
~, индивидуальный приводной personal locator beacon
~, курсовой посадочный localizer [runway localizing] beacon
~, маршрутный route radio beacon
~, направленный radio range (beacon)
~, ненаправленный non-directional radio beacon
~, пеленговый radio direction-finding beacon
~ с визуальной индикацией visual radio range
~ системы слепой посадки blind-landing beacon
~ со звуковой индикацией aural radio range
~, ультракоротковолновый very-high-frequency [VHF] radio range
~, фазовый phase-shift radio range
радиомегафон loud hailer
радиометр radiometer, radiation meter; *med.* ratemeter
~, гамма-каротажный gamma-logging radiometer
~, клинический clinical tracer dose ratemeter
~, поисковый survey radiometer
~ с высокой разрешающей способностью high-resolution radiometer
~, цифровой digital radiometer
радионаведение electronic/radio guidance
радионепроницаемый radioopaque
радиообмен radio traffic
радионуклид radionuclide, radioactive nucleus
радиообнаружение radio detection/acquisition
радиоориентировка radio direction finding; radio fixing
радиоотражение radio echo/reflection
радиопеленг radio bearing
~, наблюденный observed radio bearing
~, обратный reciprocal radio bearing
радиопеленгатор radio direction finder
радиопеленгация, радиопеленгирование radio direction finding; radio homing
радиопередатчик radio transmitter, sending set
~, аварийный radio distress transmitter
~ команд наведения guidance transmitter
~ направленного действия radio beam transmitter
радиопередача radio transmission/program(me), broadcast(ing)
радиопилюля *med.* radiopill
радиоперехват radio interception/monitoring
радиоподслушивание radio monitoring
радиополукомпас hand-operated direction finder; fixed-/rotatble-loop radio compass
радиопомехи radio interference/noise
~, атмосферные atmospherics
~, заградительные barrage jamming
радиоприемник radio receiver/set, receiving set
~, автомобильный car radio
~ батарейного питания battery-operated radio
~, бортовой airborne receiver; on-board receiver
~, курсовой localized receiver
~, ламповый valve/tube radio set
~, монофонический monoaural radio receiver
~ с универсальным питанием ac/dc radio receiver
~ сетевого питания mains-operated radio
~, супергетеродинный superheterodyne receiver, superhet
радиоприцел radio sight
радиопрожектор searchlight radar
радиопрослеживание radio tracking
радиопротиводействие electronic/radio countermeasures
радиоразведка electronic/radio reconnaissance
радиорезистентность radioresistivity, radioresistance
радиореле radio relay
радиорепортаж running commentary
радиорефрактометр microwave refractometer
радиорупор loudspeaker
радиосвечение *phys.* radioluminescence
радиосвязь radio communication/contact
~ дальняя long-range radio communication
~, двусторонняя two-way radio communication
~, односторонняя one-way radio communica-

tion
~, магистральная point-to-point radio communication
~, мобильная vehicular/mobile radio communication
~, тропосферная tropospheric-scatter radio communication
~, фототелеграфная facsimile radio link
радиосенсибилизация radiosensitization, radiosensibilization
радиосеть radio network
радиосигнал radio signal
~ времени radio time signal
~, туманный radio fog signal
радиостанция radio station
~, аварийная survival/crash radio set
~, авиационная aeronautical radio station
~, бортовая *av.* aircraft radio station; *naut.* ship radio station
~, ведомая slave radio set
~, ведущая master radio set
~, всеволновая comprehensive radio set
~, коротковолновая high-frequency radio set
~, плавучая radio ship
~, приемо-передающая two-way radio set
~, ранцевая man-pack radio set
~, судовая ship radio station
радиостойкость radioresistance
радиостудия broadcasting studio
радиосхема radio/electronic circuit
радиотаблетка *med.* radiopill
радиотелеграф radiotelegraph
радиотелеграфист radiotelegraph operator, radiotelegrapher, radiotelegraphist
радиотелеграфия radiotelegraphy, wireless telegraphy
радиотелетайп radio teletype
радиотелеизмерение, радиотелеметрия radiotelemetry
радиотелескоп radio telescope
~, подвижный steerable radio telescope
радиотелеуправление radio telecontrol, radio remote control
радиотелефон radiotelephone
радиотелефонист radiotalker
радиотелефония radio telephony, voice radio
радиотень radio shadow
радиотеплолокатор thermo-microwave imager
радиотеплолокация thermo-microwave imaging radar
радиотерапевт radiotherapist, radiation therapist
радиотерапевтический radiotherapeutic
радиотерапия radiotherapy, radiotherapeutics
~, внутритканевая interstitial radiotherapy
~, вращательная/ротационная rotation radiotherapy
радиотермия radiothermy
радиотехник radio mechanic/technician, radioman
радиотехника radio engineering
радиотоксичность radiotoxicity
радиоторий *chem.* radiothorium, RaTh, thorium-228
радиоточка local radiospeaker, wired-radio outlet/terminal
радиотрансляция radio relay system, (re)broadcasting
радиотуманность radio nebula
радиоуглерод radiocarbon
радиоузел radio relay centre/center, broadcasting centre/center
радиоуправление radio control/guidance
радиоуправляемый radio-controlled/guided
радиоусилитель radio amplifier
радиоустановка radio set/installation/plant/unit
радиоустойчивость radiation resistance, radioreristance
радиоустройство electronic/radio device/set/installation
радиофарфор radio-engineering porcelain
радиофикация installation of loudspeaker system(s), installation of public-address system
радиофицировать equip/provide with radio(s); install a loudspeaker system in
радиофон radiotelephone, radiophone
радиофония radiophony
радиофотоснимок radiophotogram
радиофототелеграмма radiophotogram(me)
радиофототелеграфия facsimile transmission, wireless picture transmission
радиофургон radio van
радиохимия radiochemistry; nuclear chemistry
радиохлор *chem.* radiochlorine
радиохром radiochrome
радиоцентр broadcasting/radio centre
радиоцепь radio circuit
радиохроматография 1. radiochromatography; 2. radiochromatograph
радиочастоты radio frequencies
~, высокие high frequencies, HF
~, гипервысокие hyperhigh frequencies, HHF
~, крайне высокие extremely high frequencies, EHF
~, низкие low frequencies, LF
~ очень высокие very high frequencies, VHF
~, очень низкие very low frequencies, VLF
~, сверхвысокие superhigh frequencies, SHF
~, сверхнизкие extremely low frequencies, ELF
~, ультравысокие ultrahigh frequencies, UHF
радиочистый radio-pure
радиочувствительность *med.* radiosensitiveness, radiosensibility
радиошум radio noise/interference
радиоэкранирование radio screening/shielding
радиоэлектроника communications/radio electronics
~, авиационная aviation electronics, avioelectronics
радиоэлемент radioactive chemical element, radioelement
радист radio operator, radioman

~, вахтенный *naut.* watch-standing radio operator
~, воздушный *aero.* flight radio operator
~, наземный ground radio operator
радиумтерапия *med.* radium therapy, radiotherapy
~, наружная external radium theraypy, radiumization
радиус 1. radius, semidiameter; 2. range; scope
~, боровский Bohr radius
~ виража *aero.* radius of banked turn
~ витка штопора radius of spin
~ воздушного винта propeller radius
~ действия radius of action, range of operation, effective range; *naut.* cruising radius
~ действия, небольшой short radius of action
~ действия, боевой combat radius
~ закругления скулы *naut.* bilge radius
~ иона ionic radius
~ качения шины rolling radius of a tyre
~ кривизны radius of curvature
~ кривошипа crank radius/throw
~ критической части сопла nozzle-throat radius
~ кручения radius of torsion
~, кулоновский Coulomb radius
~, мгновенный instantaneous radius
~ несущего винта *aero.* rotor radius
~ поворота *auto* turning radius
~ сходимости radius of convergence
~, тактический tactical radius
~ экранирования shielding distance
~ ядра nuclear radius
радиус-вектор radius/position vector
радон radon, radium emanation, niton, Rn
радон-219 radon 219, actinon
радон-220 radon 220, thoron
радоновый radon *(attr.)*
радуга rainbow
~, лунная moonbow
~ над морем/океаном marine rainbow
~ туманная fogbow, Ulloa's ring
радужка, радужина *anat.* iris of the eye
радула *zool.* radula
разбавитель diluent, thinner; extender
разбавление dilution, thinning
~ изотопное isotopic/isotope dilution
~, предельное limit dilution
~ рядов *math.* dilution of series
~ ядерного топлива dilution/denaturing of nuclear fuel
разбавлять dilute
разбаланс, разбалансировка disbalance, unbalance, out-of-balance/trim
разбалансировать bring out of balance, trim out
разбалансироваться get out of balance/trim
разбег *aero.* take-off running/rolling; take-off run/roll; taking-off
разбивать break (down, up); fracture; split, partition
разбивка 1. breaking, shattering; 2. dividing, splitting, breaking up/down; *tech.* laying/marking out, laying off, lofting
~ базиса *geod.* base adjustment
~, весовая weight breakdown
~ на плазе/плазовая lofting
~ по странам breakdown by countries
~ программы *comp.* program(me) layout
~ пути pegging of a track
~ сети опорных точек control work, stationing
~ цены и издержек breakdown of price and costs
разбиение 1. breaking; fragmentation, division; splitting; 2. *math.* decomposition, subdivision, partitioning
~, информационное *game th.* information partition
~ на логические блоки *comp.* logical partitioning
~ на модули modular decomposition
~ на области regionalization
~ на равные части equipartition
~ на слои layering
~ памяти на страницы memory paging
~ по игрокам *game th.* player partition
~ со страничной организацией pageable partitioning
~ текста на страницы pagination
разбирательство *law* examination, inquiry, investigation
~, арбитражное arbitration
~, судебное court examination, assize
разбирать disassemble, dismantle, disjoint, take apart/down, strip, pull/knock down
разбитие = разбиение
разблокировать release, unlatch, unlock
разболтовывать, разболчиват unbolt
разбор anslysis, critique, examination, discussion; investigation; *gram.* parsing
~, водить/проводить hold a critique; discuss
~ дела *law* investigation of a case
~, клинический clinical discussion
~ учения post-exercise critique
разборка disassembly, dismantling, taking apart, stripping, dismounting
~, неполная partial disassembly
разборный dismountable, collapsible; knock-down, sectional; prefabricated
разборочный disassembling, dismantling, stripping
разборчивость distinctness, intelligibility; legibility
разбрасывание spreading, scattering
разбрасыватель spreader, disperser, scatterer
разброс dispersal, spread(ing), scattering; *fig.* diversity, variation, variance
~, аппаратурный instrumental spread
~ величин spread in values
~ данных *stat.* data scattering, variability of data
~ ионизации ionization straggling
~ по вертикали vertical dispersion

~ по времени time spread/straggling
~ по импульсам momentum spread
~ по длинам волн wavelength spread
~ по скоростям velocity spread
~ по энергии energy spread/straggling
~ погрешностей error spread
~, приборный instrumental spread
~ рабочих характеристик spread in performance
~, случайный straggling
~ точек *math.* spread of points
~, угловой angle spread
разбрызгивать splash, sputter, sprinkle, spray, atomize; spill
разбуривать drill out/through, bore; enlarge
разбухание swelling; bloating; inflation
развал 1. break-up/down, collapse; split; 2. *naut.* flam, flare, falling-out; 3. camber *(of wheel)*
~, дифракционный diffraction break-up
~, кулоновский Coulomb break-up
~ оболочки shell break-up
~ тока current decay
~ ядра muclear break-up
развальцовка, развальцовывание beading; expanding, expansion; flaring; lamination
развальцовывать bead; expand, open out; flare; laminate
разведение 1. breeding, rearing, raising; growing, cultivating, cultivation; 2. opening, moving apart, swinging open; 3. diluting, thinning
~ животных animal husbandry
~, изотопное isotopic dilution
~, линейное line breeding
~ черенками *hort.* cutting propagation
разведенный divorced; separated
разведенная divorcee
разведка 1. exploration, survey, prospecting; *mil.* reconnaissance, intelligence
~, аэрогеофизическая airborne geophysical prospecting
~, аэрорадиометрическая airborne radiometric survey
~ боем *mil.* reconnaissance in force
~, воздушная aerial/air reconnaissance
~, воздушно-космическая aerospace reconnaissance
~, гравитационная/гравиметрическая *geol.* gravitational prospecting/exploration
~, космическая space reconnaissance
~, магнитная magnetometric prospecting
~, метеорологическая weather reconnaissance
~ мин mine detection/location
~, морская undersea exploration/prospecting
~, оперативная operational intelligence
~, предварительная *geol.* prospecting
~, радиометрическая *geol.* radiometric/radioactive prospecting
~, радиорная radiore survey
~, радиотехническая electronic reconnaissance
~, рыбопромысловая fish reconnaissance
~ с фотографированием, воздушная aerial photographic reconnaissance
~ сейсмографом seismographic prospecting
~ скважинами borehole surveying
~, тактическая combat/tactical intelligence
разведочный intelligence *(attr.),* reconnaissance *(attr.);* exploratory, prospecting
разведсводка *abbr.* intelligence report
разведслужба *abbr.* intelligence service
разведчик 1. intelligence officer, secret service man/agent; 2. *mil.* scout
разведывать reconnoiter, scout
разведывательный = разведочный
развертка 1. scan(ning), sweep; 2. *math.* development, evolvement; involute, evolvent; 3. *mech.* reamer; reaming
~, винтовая helical scanning
~ дальности range sweep
~, двухсторонняя bilateral scanning
~, ждущая slave/driven/triggered sweep; triggered timebase
~, кадровая vertical/frame scanning
~, кольцевая circular sweep
~, коническая tapered reamer
~ по дальности range sweep
~, последовательная line-sequential scanning
~ пучком beam scanning
~, разжимная expansion reamer
~ с винтовыми зубьями helically-fluted reamer
~ с компенсацией compensated scan
~ с кривой метацентров *math.* metacentric involute
~, секторная sector scan; sector display
~, спиральная spiral sweep
~, чистовая finishing reamer
развертывание 1. reaming; 2. unfolding, unrolling, unwrapping; uncoiling; 3. development, evolution; 4. *mil.* deployment
развертывать 1. *tech.* ream; 2. unfold, unroll, unwind, uncoil; unfurl; 3. develop; expand; 4. *rad.* scan, sweep; 5. *mil.* deploy; launch
развес weighing
разветвитель splitter, bifurcating device
разветвление branching, ramification; bifurcation, fork; *fig.* divergence
~ дороги road fork
~ распада branching of decay
~ цепи chain branching
разветвленность ramification, ramified structure
разветвлять ramify, divaricate, branch, fork, bifurcate
развешивать weigh out
развивать 1. develop; generate; evolve; 2. unwind, untwist, untwine
~ чрезмерно overdevelop
развилина, развилка bifurcation, fork
развинчивать unscrew
развитие development; evolution, progress; growth

~, **видовое** phylogenesis
~, **запоздалое** backwardness
~, **индивидуальное** ontogenesis, ontogeny
~, **недостаточное** underdevelopment
~, **ненормальное/неправильное** abnormal development, malformation; dysplasia
~, **обратное** regression; involution
~ **плода** *med.* f(o)etation; *bot.* fructification
~, **порочное** dysontogenesis, monstrosity
~, **преждевременное** precocity
~, **экзогенное** exogeny
~, **эмбриональное** embryonic development
~, **эндогенное** endogeny
развод I saw tooth setting; set of sawteeth
развод II posting, mounting, stationing
~ **караулов/часовых** posting of sentries, guard mounting
развод III breeding
разводить breed, raise; rear; cultivate, grow
разводка 1. tooth setter, saw set; 2. saw setting
разводящий *mil.* corporal of the guerd
развоз(ка) delivery, transport
разворачивание 1. unwrapping; unwinding, unrolling; 2. *math.* development; 3. turning, swinging
разворачивать(ся) turn (around), swing
разворот 1. turn; 2. *print.* centre/center spread, facing pages
~, **крутой** sharp/close/tight turn
развязка 1. decoupling; undoing; 2. overhead crossing, overpass
~ **частотная** frequency isolation
разгар erosion, wear, scoring
разгерметизация depressurization, decompression; seal failure, loss of sealing/pressure
~ **кабины** cabin depressurization/decompression
разгерметизировать depressurize
разгибать unbend, straighten; *anat.* extend
разговенье *rel.* breaking one's fast
разгон acceleration, speed-up, speeding-up, booost
~ **ветра/волны** fetch
~ **до сверхзвуковой скорости** supersonic acceleration
~ **на стартовом участке** launching boost
~ **реактора** 1. reactor riding/going up; 2. reactor runaway
~ **реактора, аварийный/неуправляемый** reactor runaway
разгонка 1. *chem.* distillation; 2. acceleration, speeding up; 3. *print.* spacing
разгонять 1. accelerate, speed up, boost; 2. drive away, disperse, scatter; 3. *print.* space out
разгоняться pick up speed, run/speed up
разграничительный differentiating; delimitative
разграфлять rule, mark with lines
разгружать unload, discharge; off-load, relieve
разгрузка unloading, discharging; load relief
~ **в порту** port discharge
~ **на плаву** afloat discharge
~ **памяти** *comp.* dump
~ **посредством лихтеров** lighterage
~, **рейдовая** off-shore unloading
~ **самотеком** gravity discharge
разгрузочное судно *naut.* lighter
разгруппирование debunching
раздавливание crushing, smashing, squashing
раздатчик dispenser, distributor
раздача 1. distribution, dispensation, dispensing; 2. widening, enlarging, expansion, flaring
раздваивать divide in two, halve, bisect, bifurcate
раздваиваться bifurcate, split, fork, divaricate
раздвоение bifurcation, branching, forking, divaricating; splitting
~ **аорты** aortic bifurcation
~ **личности** *psych.* dual personality
~ **сонной артерии** carotid bifurcation
раздвоенный bifurcate, forked, split; bicornate, bicornuate, bicornate, bicornous; bifid; *bot.* dichotomous; bicrural
раздел 1. division; separation, partition; 2. part, section
~ **вычислительного процесса** *comp.* partition
~ **документации** document section
~ **программы** program(me) division
разделение division; separation, partition, disjunction; splitting
~ **блоков на записи** deblocking
~, **вилообразное** forking
~ **во времени** timing separation
~ **водном методом** aqueous separation
~ **времени** *comp.* time-sharing
~, **гравитационное** gravity separation
~, **диффузионное** diffusion/diffusive separation
~ **изотопов** isotope separation
~ **каналов** *rad.* channel spacing/separation
~ **каналов, временное** *comp.* time demultiplexing
~ **методом отдачи** recoil separation
~ **методом противотока** counter-current separation
~ **на две части** *biol.* dichotomy
~ **на каналы** channeling
~ **на отсеки** compartmentation
~ **по амплитуде** amplitude separation
~ **по весу** gravity separation
~ **по импульсу** momentum separation
~ **по крупности** sizing, classification
~ **по частоте** frequency division
~ **под действием силы тяжести** gravity separation
~ **продуктов деления** fission-product separation
~ **слов** *comp.* word separation
~ **труда, международное** international division/differentiation of labo(u)r
~ **труда, общее** division of labour in general
~ **труда, общественное** social division of labour
~ **труда, частное** division of labour in particular
~ **центрифугированием** centrifugal separation

~ **экстрагированием** extraction separation
~, **электролитическое** electrolytic fractionation
~ **ядерных изомеров** nuclear isomer separation
разделенный divided; separated; *comp.* shared
~ **цилиндр** split cylinder
разделитель separator; separating agent; divider
~ **изотопов, электромагнитный** calutron
~, **циклонный** cyclone separator
разделительный separating, dividing; partitioning; spacing; disconnecting; *gram.* disjunctive
~ **союз** disjunctive conjunction
разделка 1. dressing, finishing; preparation; 2. *tech.* splicing; termination *(of cable)*
~ **лома** cropping
~, **оконная** glazing bead
раздельнокарпельный *bot.* dialycarpellary
раздельнолепестковый, раздельнолепестный *bot.* choripetalous, dialypetalous
раздельнолистный *bot.* choriphyllous, dialyphyllous, segmented
раздельноногий *zool.* fissiped
раздельнополость dicliny
раздельночашелистиковый *bot.* dialysepalous, chorisepalous
раздельный separate
~**ная уборка** *agr.* two-stage harvesting
раздражать 1. irritate, annoy; 2. *med.* stimulate
раздражение 1. irritation; 2. *med.* stimulation
~ **болевое** painful stimulation
~, **нервное** nervous irritation
~, **световое** photostimulation, photic stimulation; photoirritation
~ **шумом** stimulation by noise
раздражитель stimulus, stimulant; irritant
~, **безусловнорефлекторный** unconditioned stimulus
~, **гетерологический** heterologous stimulus
~, **звуковой** sound stimulus
~, **зрительный** visual stimulus
~, **кожный** skin irritant
~, **обстановочный** situation stimulus
~, **подкрепляющий** reinforcing stimulus
~, **подпороговый** subthreshold/subliminal stimulus
~, **речевой/словесный** verbal stimulus
~, **световой** photic stimumus
~, **слуховой** auditory stimulus
~, **тормозящий** inhibiting stimulus
~, **условнорефлекторный** conditioned stimulus
раздражительность irritability
~, **болезненная** erethism
раздробление breaking, crushing; fragmentation, splintering; *math.* reduction
~ **мочевых камней** *med.* lithotrity, lithotripsy
раздроблять crush, break to pieces, disintegrate; *math.* reduce
раздувать inflate, distend
раздутость bulge, swelling; distension
разжелобок valley
разжижать dilute, thin
разжижение dilution, thinning, rarefaction
разжижитель diluent; thinning agent
разлагать(ся) decompose, dissociate, disintegrate, break down; *math.* expand
разлив 1. flood, inundation; 2. bottling
разливать pour; cast
разливка pouring, casting
~ **сверху** top pouring/casting, downhill teeming
~ **сифоном** bottom pouring/casting, uphill teeming
разливной pouring *(attr.)*; bottling *(attr.)*
~**ное вино** unbottled wine, wine from the tap
различие difference, dissimilarity, contrast, discrepancy; diversity; distinction
~**ия в ценах** price differentials
~**ия, видовые** specific differences
разложение decomposition, disintegration; dissociation; *phys.* resolution; *math.* development; expansion
~ **бинома** binomial expansion
~ **в гармонический ряд** harmonic expansion
~ **в ряд Тейлора** Taylor expansion
~ **в ряд Фурье** expansion/development in a Fourier series
~ **временного ряда** time-series decomposition
~, **гнилостное** putrefaction, decay
~, **каталитическое** catalytic decomposition
~ **личности** dissociation of personality
~ **логических функций** logical function decomposition, decomposition of logical functions
~ **Лорана** Laurent's expansion
~ **на множители** factorization, factoring
~ **на простейшие дроби** resolution/decomposition into partial fractions
~ **на простые множители** prime factorization
~ **натурального числа, каноническое** standard form of a natural number
~, **пиролитическое** pyrolytic/thermal decomposition
~ **по степеням** power expansion
~ **пород** rock decay
~, **посмертное/трупное** postmortem/cadaveric decomposition
~ **света** dispersion of light
~ **сил** resolution of forces
~, **спектральное** spectrum analysis
~ **темпорального ряда** time-series decomposition
~, **ферментное** enzymic decomposition
разложимый на множители *math.* factorable
разлом fracture; *geol.* fault
размагничивание demagnetization; *naut.* deperming, wiping; degaussing
размагничиватель demagnetizer
~ **судна** degausser
размагничивать demagnetize; *naut.* degauss, deperm
размазывать smear (out)
разматывать unwind, unreel, uncoil, reel out,

wind/reel off; pay out
размах scale, swing, sweep; *stat.* range
~ выборки *stat.* sampling range
~ колебаний double amplitude
~ крыла wing span
~, межквартильный *stat.* interquartile range
~ напряжения *el.* peak-to-peak amplitude/value
~ регулятора *mech.* throw of a governor
~ стабилизатора *aero.* tailplane span
размельчать comminute; break, crush, triturate
размер dimension, size; gauge
~ в свету clear/inner dimension
~, внешний overall dimension
~ выборки *stat.* sample size
~ выгула *agr.* floor space
~, монтажный fixing dimension
~, предельный limit(ing) size
~, присоединительный mounting dimension
~, сопряженный mating dimension
~, установочный mounting dimension
~ частиц particle size
~ ячейки сита mesh dimension, screen size
~ы в дюймах imperial dimensions
~ы, общие/внешние overall dimensions
размерность dimensionality; dimension
~, большая high dimensionality
~ векторного пространства *math.* dimension of a vector space
~ массива *comp.* array dimension
~, переменная adjustable dimensions
размерный dimensional
разметка marking (out, off); setting; formatting
разметчик marker, layer-off, marker-off
размещать place, locate, put, position; accommodate, billet, quarter; *econ.* invest
~ облигационный заем float a bond issue
размещение placement, disposition, accommodation; arrangement, layout; *math.* permutation, arrangement; 3. investment; 4. billeting, quartering
~ багажа luggage stowage
~ груза freight positioning/arrangement; *naut.* trim
~ заказов placing of orders
~ информации information configuration
~ контрактов placement of contracts
~ с повторениями permutation/arrangement with repetitions
~ сидений *auto.* seat layout
~ сил allocation of forces
разминирование mine clearance/removal/cleaning
размножаться propagate, multiply, breed; proliferate
размножение 1. reproduction, propagation; breeding; multiplication; 2. duplication, reproduction, copying; mimeographing
~ без оплодотворения parthenogenesis
~, бесполое asexual reproduction
~, девственное parthenogenesis
~ делением fissiparity
~ документов copy-making, duplication
~ ленты duplication of tapes
~ нейтронов neutron multiplication
~ нейтронов источника source-neutron multiplication
~, половое sexual reproduction, gamogenesis, amphigony
~ почкованием blastogenesis
~ спорами sporiparity
~ топлива fuel breeding
~ топлива на тепловых нейтронах thermal breeding of fuel
~ черенками propagation by cuttings
~ чертежей reproduction of drawings
~ ядерного горючего fuel breeding
размножитель breeder, breeding reactor
размозжение crushing
размол grinding, milling; pulverization
~ бумажной массы pulp beating stage
~, ударный impact grinding
размораживание defrosting
размотка unwinding, unreeling
размыв, размывание wash-out; erosion by water
~ берега shore cutting
~ горизонта fouling of the horizon
~ изображения blurring
~ фронта *meteor.* frontolysis
размыкание *el.* disconnection, breaking, interruption
~ цепи interruption/opening of a circuit
размыкатель *el.* circuit breaker, release
размыкать cut off, disconnect, break, turn off, interrupt
рьзмытие diffusion; blurring
размягчать soften; mollify; malaxate
размягчение softening; malacia, mollities
~ артерии arteromalacia
~, белое white/an(a)emic softening
~ желудка gastromalacia
~ костей osteomalacia
~, красное red/h(a)emorrhagic softening
~ легких pneumomalacia
~ мозга encephalomalacia, softening of the brain
~ мышц myomalacia
~ ногтевой пластинки onchylomalacia
~ пищевода esophagomalacia
~ черепа craniomalacia
разнесение 1. delivery, conveyance; 2. separation; spacing; scattering, dispersing; 3. recording, entering
разновероятный unequally probable
разнодомность *bot.* dioecism
разнодомный *bot.* dioecious
разнозернистый inequigranular
разноименность heteronymy
разноименный heteronymic; *el.* of opposite sign
разнолепестник, разнолистник *bot.* candytuft *(Iberis)*
разнолепестный *bot.* heteropetalous
разнолистный, разнолистый *bot.* heterophyl-

lous, diversifolious
разноногие *pl., zool.* amphipods *(Amphipoda)*
разноорешник *bot.* popcorn flower *(Allocarya)*
разноплодие, разноплодность *bot.* heterocarpy
разнополовой *bot.* heterosexual
разнополость *bot.* dicliny
разнополый 1. *biol.* heterosexual; 2. *bot.* diclinous
разнополярный heteropolar
разнорабочий general/unskilled worker, odd-job worker, odd-jobman
разнородность heterogeneity, heterogeneousness
разнородный heterogeneous, heterogenic
разнос 1. spacing, separation, offset; 2. overspeeding, racing, runaway; 3. delivery, carrying; 4. entering, recording
~ каналов *rad.* channel spacing/separation
~ частот frequency spacing/separation
разносить 1. space apart, separate; 2. race, run away; 3. deliver, carry round, distribute; 4. enter, record
разносклоняемость *gram.* heteroclisis
разносклоняемый *gram.* heteroclitic, of irregular declension
разнослоговой heterosyllabic
разносный delivery *(attr.)*
~ная торговля street peddling/hawking
разноспоровость *bot.* heterospory
разноспоровый heterosporous
разноспрягаемый *gram.* irregularly conjugated
разностеномер, радиоактивный radiation pipe-thickness gauge
разностолбиковый, разностолбчатый *bot.* heterocyclic, heterostylous, heterostyled
разностолбчатость *bot.* heterostylism, heterostyly
разносторонний inequilateral; *fig.* versatile; *math.* scalene
~нее развитие экономики diversification of the economy
разность difference; diversity, variety; divergence
~ арифметической прогрессии (common) difference of an arithmetic(al) progression
~ вперед (для интерполирования) *math.* forward/descending difference
~ высот difference of height/altitude
~ долгот difference in/of longitude
~, конечная *math.* finite difference
~, левая *math.* backward difference
~ меридиональных частей meridional difference
~ множеств *math.* difference of sets
~ назад (для интерполирования) backward/ascending difference
~ напряжений voltage difference
~, обратная *math.* reciprocal difference
~, первая *math.* first difference
~, правая *math.* forward difference
~, средняя mean difference
~, суммарная gross difference
~ углов angular difference
~, часовая hourly difference
~, частная partial difference
~ широт difference of/in latitude
~ широт, северная northing
~ широт, южная southing
разнотипный polytypical
разнотипность diversity/variety of types
разнотравые herbage, mixed herbs
разнотычинковый *bot.* heteran drous
разнотычиночный *bot.* heterostemonous
разноформенный heteromorphous
разнохозяйность, разнохозяйственность *biol.* heteroecism; metoecism
разноцветковый *bot.* diversiflorous
разночерешковый *bot.* heteropodous
разночинец *hist.* self-employed plebeian
разнощитковые *pl., paleont.* heterostracans *(Heterostraci)*
разноячеистый heterobrochate
разобщать disconnect, disengade, uncouple, release
разобщитель disconnector, cutout
разогрев warming up, heat-up; preheating
разрабатывать work out; develop; *min.* exploit, mine, work; *agr.* cultivate
~, тщательно elaborate
разрабатываться *mech.* run/work out of true, run/work untrue
разработка working out, elaboration, development; *min.* mining, working, exploitation; *agr.* cultivation
~ алгоритмов *math.* algorithm elaboration
~, дражная dredging
~ изделия product development
~, изыскательская exploratory development
~, конструкторская engineering development
~ недр mining
~ опытного образца prototype development/design
~, открытая *min.* openpit/opencast/strip/surface mining
~ проблемы problem developmen/elaboration
~ рудных местонахождений ore mining
~ки, исследования и research and development
~ки, подземные underground operations
~ки, технологические development work
разработчик development engineer/centre
разращения *med.* vegetations; proliferations
~ аденоидной ткани adenoid vegetations
разрегулированность, разрегулировка maladjustment, misalignment
разрегулировать put out of alignment
разрежать 1. rarefy; exhaust, evacuate; 2. *agr.* thin out
разрежение rarefaction; vacuum
~ костной ткани osteoporosis
разреженность rarefaction, suction pressure; vacuity
разрез 1. cutting, cut, incision, discission, section; slit, slot, rip; 2. sectional view, profile, (cross-)section, plan; elevation

~, боковой lateral section; side elevation
~ брюшной стенки abdominal section, celiotomy
~, вертикальный vertical section, elevation, profile
~, внебрюшинный *surg.* extraperitoneal incision
~, воротниковый collar incision
~, горизонтальный horizontal section, plan
~ грудной клетки *surg.* thoracotomy
~, задний rear/back elevation
~, кожный *surg.* skin incision
~, колонковый columnar section
~, крестообразный *surg.* crucial incision
~ лонного сочленения symphysiotomy, symphyseotomy
~ мочеточника ureterotomy
~, нейтронный *min.* neutron log
~, опорный key section
~, передний front elevation
~ печени *surg.* hepatomy
~, полулунный *surg.* semilunar incision
~, поперечный lateral/cross/transverse section
~ почки nephrotomy
~, продольный/профильный longitudinal/lengthwise section
~, промежностный *surg.* perineal section
~ рентгеновский *min.* gamma-ray log
~, сводный/суммарный composite section
~, схематический diagrammatic section
~, частичный cut-away drawing/view
~ шейки матки *surg.* discission of the cervix
~ электрохирургической иглой acusection
разрезать cut; incise; slot, slit; dissect
разрезной 1. slotted, split; 2. cutting *(attr.);* 3. *bot.* laciniate(d)
~ная гайка split nut
разрешать 1. allow, permit, let; authorize; 2. resolve, separate; distinguish, discriminate
разрешающая способность resolution; discrimination
~ в метрах на линию resolution in metres/line
~ на местности ground resolution
~ по азимуту azimuth resolution
~ по времени time resolution
~ по дальности range resolution/discrimination
~ по направлению/пеленгу bearing resolution
~ по углу места elevation resolution
разрешение 1. permission, permit; authorization, licence, clearance; 2. resolution, resolving power; 3. *comp.* enable *(signal)*
~, аппаратурное instrumental resolution
~, временное time resolution
~, высокое high resolution
~, диспетчерское *aero.* air traffic control [ATC] clearance
~, идеальное infinite resolution
~ на ввоз import licence
~ на взлет take-off clearance
~ на вступление в брак marriage licence
~ на выход leave clearance
~ на доступ access authorization
~ на жилищное строительство housing permit
~ на использование программы *comp.* program(me) authorization
~ на посадку landing permit/authorization
~ на право работы labo(u)r permit
~ на строительство building permit/licence
~ на передачу *comp.* credit
~, низкое low resolution
~ прибора instrumental resolution
~ таможни на вывоз товара из склада bond note
~ таможни на погрузку stamp note
~, угловое angular resolution
разрешимость *math.* solvability
разрешимый solvable
разрушать destroy, demolish; wreck, ruin
разрушаться break down; fail; collapse
разрушение destruction, demolition, breakdown; ruin; failure
~, избирательное preferred destruction
~ конструкции structural failure
~, коррозионное corrosion damage
~ костной ткани *med.* caries
~, межкристаллитное intercrystalline failure
~, мгновенное instantaneous failure
~, пластическое plastic collapse/fracture; ductile failure
~, постепенное progressive failure
~ при изгибе bending failure
~ при кручении torsion failure
~ при растяжении tensile failure
~ при сжатии compression failure/fracture
~ при срезе shearing failure
~, усталостное fatigue/endurance failure
~, хрупкое brittle failure/fracture
разрыв 1. break, breaking, breakup, rupture; breaking off, severance; gap; *med.* laceration, rhexis, tearing; *math.* (point of) discontinuity; 2. burst, explosion; 3. *geol.* fault, rent
~, бесконечный *math.* infinite discontinuity
~ вены *med.* phleborrhexis
~ дипломатических отношений severance of diplomatic relations
~, инфляционный inflationary gap
~ каната tearing of a rope
~, конечный *math.* finite discontinuity
~ матки *med.* hysterorrhexis, metrorrhexis
~ мозга brain laceration
~ мышцы myorrhexis
~ непрерывности discontinuity
~ первого/второго рода *math.* discontinuity of the first/second kind
~ пласта seam fracture
~ плодных оболочек *med.* breaking of the waters
~ пневматика tyre blow-out
~, поперечный *geol.* cross fault

~ **потока** flow discontinuity
~ **сердца** cardiorrhexis
~ **скальпа** scalp laceration
~ **следа** track gap
~, **сложный** composite rupture
~, **устранимый** *math.* removable discontinuity
~ **функции** discontinuity of a function
~, **хрупкий** brittle fracture
разрывать break off, break, rupture, disrupt; interrupt; lacerate; blow up, burst; *el.* disconnect, interrupt, open
разрывность discontinuity
разрывный 1. discontinuous; 2. explosive, bursting
~**ная переменная** *math.* discontinuous variable
разрыхление loosening (up)
~ **ногтей** *med.* onycholysis
разрыхлитель fluffer; *agr.* scarifier, ripper
~ **теста** leavening agent, leaven
разрыхлять loosen; fluff up; scarify
разряд 1. *comp.* digit; 2. *el.* discharge; 3. order, position, rank, place; category, bracket
~ **адреса** *comp.* address bit
~ **в газе [газовый]** gaseous/gas discharge
~ **в дейтерии** deuterium discharge
~ **в магнитном поле, дуговой** penning discharge
~, **грозовой** lightning discharge
~, **двоичный** bit, binary digit
~, **дежурный** keep-alive discharge
~, **десятичный** decimal position
~ **десятков** tens order
~, **дуговой** arc discharge
~ **единиц** units order
~ **знака, двоичный** sign bit
~, **знаковый** sign position
~, **значащий** significant digit
~, **избыточный** redundant bit
~, **изгибающийся** wriggling discharge
~, **импульсный** impulsive discharge
~, **информационный** data bit
~, **искровой** spark discharge
~, **кистевой** brush discharge
~, **колебательный** oscillatory discharge
~, **колонный** columnar discharge
~, **кольцевой** ring discharge
~, **коммутационный** switching discharge
~, **контрольный** check bit
~, **крайний левый** leftmost position
~, **крайний правый** rightmost position
~, **коронирующий/коронный** corona (discharge)
~ **между облаками** cloud-to-clod discharge
~, **младший** low order/position
~ **молнии** lightning discharge
~, **незначащий** nonsignificant digit
~, **несамостоятельный** non-self-maintained discharge
~ **от облака к земле** cloud-to-ground discharge
~, **отшнурованный** pinched discharge
~ **переноса** carry bit
~, **поверхностный** surface discharge
~, **ползучий** creeping discharge
~, **пробивной** disruptive discharge
~, **самосжатый** self-constricted discharge
~, **самостоятельный** self-maintained discharge
~, **самый младший** lowest order
~, **самый старший** highest order
~, **сжатый газовый** pinched gas discharge
~, **сильноточный** high-current discharge
~, **спиральный** helical discharge
~, **старший** high order/position
~ **сумматора** adder order
~ **сумматора, двоичный** accumulator bit
~, **тарифный** *econ.* wage category
~, **тихий** silent/glow discharge
~, **тлеющий** glow discharge
~, **тороидальный** toroidal discharge
~, **флаговый** flag bit
~ **числа** digit position/place; bit place/position; symbol rank
~ **числа, более старший** higher order [more significant] digit/position
~ **числа, младший** less significant [lower-order] digit/position
~ **числа, низший** the least significant digit/position
~ **числа, следующий по старшинству** the next higher order digit/position
~**ы, атмосферные** atmospherics
~**ы, тарифные** *econ.* worker's rates
разрядка 1. *print.* spacing; 2. discharging; 3. *fig.* detente, easing of tension
~ **международной напряженности** easing of international tension
разрядник *el.* (discharge) arrester, discharger, spark gap; lightning arrester
~, **антенный** aerial lightning protector
~, **газовый** gas-discharge (lightning) arrester
~, **грозовой** lightning arrester
~, **искровой** spark discharger/gap
~, **коаксиальный** coaxial gas-filled switching tube
~, **линейный** line-type lightning arrester
~, **резонансный** gas-filled switching tube
~, **роговой** horn arrester
~ **статического электричества** precipitation static discharger
~, **стержневой электростатический** rod-type static discharger
~, **стреляющий** expulsion-type (lightning) arrester
~, **фитильный электростатический** wick-type static discharger
~, **шаровой** sphere gap
разрядность *comp.* word length/size; capacity
разряжение unloading; discharging
разряжать unload; discharge
разубоживать *min.* impoverish, downgrade; dilute
разуплотнение seal failure
разупорядочение disordering, randomizing; dis-

order
~ **решетки** lattice disorder
разъедать corrode, erode; attack; pit
разъединение disconnection, disengagement, uncoupling; release; breaking
~ **линии** clearing
разъединитель disconnector, circuit breaker, (disconnecting) switch, cutout; isolator
~, **заземленный** earthing switch
~, **линейный** line isolator
~ **мощности** power isolator/cutout
~, **секционный** sectionalizer
~, **трехполюсный** three-pole isolator/switch
~, **шинный** bus isolator
разъединять 1. disjoin, disunite, separate, detach, disengage; release, uncouple; 2. *el.* disconnect, cut/switch off, break
разъезд 1. *rail.* side/switching track; 2. departure; trip, travel; 3. *mil.* mounted patrol
разъездной 1. travel(l)ing; 2. sidetrack *(attr.)*
разъем connector, joint, connecting device; *el.* plug and socket unit
~, **быстродействующий** quick-acting connector
~, **герметичный** sealed connector
~, **многоштырьковый штепсельный** multipin connector
~, **отрывной** break-away coupling
~, **технологический** production break
~, **штепсельный** plug-and-socket joint, plug-in connector
разъемный detachable, separable; split *(attr.)*
райграс *bot.* rye grass *(Lolium)*
рай-дерево *bot.* Aaron's beard, common smoke tree *(Rhus cotinus)*
райжилотдел *abbr.* district housing department
райисполком *abbr.* district executive committee
райком *abbr.* district committee
район area, zone, region, district; field
~ **вечной мерзлоты** permafrost
~ **военных действий** war zone
~ **возникновения** breeding ground
~ **встречи** rendezvous area
~ **выгрузки** discharging area
~ **закрытый для плавания** closed waters
~, **запретный** prohibited/restricted area; off-limits area
~ **города, деловой/коммерческий** downtown business district
~, **жилой** residential area
~ **монокультуры** *agr.* one-crop country
~, **населенный** populated area
~, **ненаселенный** uninhabited area
~, **нефтеносный** oil field
~ **оживленного судоходства** crowded waters
~, **отдаленный сельский** backcountry district
~ **падения** impact area
~ **плавания** navigation area, operational waters
~ **приводнения** water alighting area; *space* splashdown area
~, **приморский** littoral, maritime area
~ **рыболовства/рыбопромысловый** fishing area/zone/grounds
~ **свалки грунта** spoil ground
~, **столичный** metropolitan area
~, **укрытый** shelter area
районирование zoning, zonation, geographical demarcation; regionalization
районо *abbr.* regional department of public education
райские птицы birds of paradise *(Paradiseidae)*
райсовет *abbr.* district soviet
райя rayah
рак 1. crawfish, crayfish; *med.* carcinoma, cancer; 3. *astr.* Cancer, Crab
~, **базальноклеточный** basal cell carcinoma/epithelioma
~, **базоспиноцеллюлярный** basosquamous cell carcinoma
~, **брюшной** intra-abdominal cancer
~, **вторичный** secondary/metastatic carcinoma
~, **гипернефроидный** hypernephroma, hypernephroid carcinoma, carcinoma of the kidney
~, **железистый** glandular carcinoma/cancer, adenocarcinoma
~, **лучевой** radiation carcinoma
~, **медуллярный** medullary/encephaloid/spongy carcinoma, soft cancer
~ **молочной железы, угревидный** comedocarcinoma
~ **мошонки** scrotal carcinoma
~, **ороговевающий** keratinizing carcinoma
~, **панцирный** cancer en cuirasse
~, **паренхиматозный** liver cell carcinoma, hepatoma
~, **плоскоклеточный** squamous cell carcinoma
~, **почечно-клеточный** renal cell carcinoma, hypernephroma
~, **профессиональный** occupational cancer
~, **радиевый/рентгеновский/рентгенологов** radiation carcinoma
~, **рассеянный** disseminated cancer
~, **речной** *zool.* crawfish *(Astacus)*
~, **семейный** cancer à deux
~, **твердый/фиброзный** scirrhous carcinoma
~, **ултрафиолетовый** radiation carcinoma
~, **усоногий** *zool.* barnacle *(Cirripedia)*
~, **целомический** mesothelioma
~, **цилиндроклеточный** columnar cell carcinoma
~ **шейки матки** cervical carcinoma, carcinoma of uterine cervix
~, **эпидермальный** epidermoid carcinona/cancer
~, **язвенно-инфильтративный** ulceroinfiltrative carcinoma
ракель *print.* doctor blade, scraper
ракета rocket; flare; missile; *naut.* hydrofoil vessel
~, **авиационная управляемая** aircraft guided missile
~, **аэродинамическая** aerodynamic [cruise-

type] missile
~, баллистическая ballistic missile
~, беспилотная unmanned/pilotless rocket
~ ближнего действия short-range missile
~, боевая combat/military missile
~, высотная high-altitude rocket
~ дальнего действия long-range missile/rocket
~, двухступенчатая two-stage missile/rocket
~, диверсионная diversionary missile
~, дозвуковая subsonic rocket/missile
~, жидкостная liquid-propellant missile/rocket
~, заградительная barrage rocket
~ запускаемая с корабля sea-launched missile
~ запускаемая с подводной лодки submarine-launched missile
~, зенитная управляемая guided antiaircraft missile
~, исследовательская research rocket
~ класса "воздух-воздух" air-to-air missile
~ класса "воздух-земля/поверхность" air-to-surface missile
~ класса "воздух-космос" air-to-space missile
~ класса "воздух-подводная цель" air-to-underwater missile
~ класса "земля-воздух" surface-to-air missile
~ класса "земля-земля" surface-to-surface missile
~ класса "корабль-подводная цель" ship-to-underwater missile
~ класса "космос-земля" space-to-surface missile
~ класса "подводная лодка-воздух" underwater-to-air missile
~ класса "подводная лодка-земля" underwater-to-surface missile
~ класса "подводная лодка-корабль" underwater-to-ship missile
~ класса "подводная лодка-подводная цель" underwater-to-underwater missile
~, корабельная ship-borne/-launched missile
~, корабельная противолодочная ship-to-underwater missile, surface-to-underwater missile
~, космическая space rocket
~, крылатая winged/cruising missile, cruise missile
~, лодочная submarine-based missile
~, межконтинентальная intercontinental missile
~ многократного использования non-expendable rocket
~, многоступенчатая multistage missile/rocket
~ на жидком топливе liquid-propellant rocket
~ на твердом топливе solid-propellant rocket
~ на химическом топливе chemical rocket/missile
~ небольшой дальности действия short-range rocket
~, неспасаемая nonrecoverable rocket
~, неуправляемая unguided rocket/missile
~, низковысотная low-altitude missile
~, низколетящая low-trajectory missile
~ одноразового использования expendable/nonrecoverable rocket/missile
~, одноступенчатая single-stage missile/rocket
~, оперенная finned missile/rocket
~, осветительная flare
~ пакетной схемы, составная cluster missile/rocket
~, парашютная parachute flare
~, планирующая glide rocket
~, пороховая powder rocket
~, посадочная landing flare
~, почтовая mail(-carrying) rocket
~, противоградовая anti-hail rocket
~, противокорабельная anti-shipping missile
~, противолодочная anti-submarine missile
~, противорадиолокационная antiradar missile
~, противоракетная antimissile missile
~, противотанковая anti-tank missile/rocket
~, разведывательная reconnaissance rocket
~ с автономным управлением self-dependent [independent-controlled] missile
~ с атомным двигателем atomic-powered rocket
~ с атомным зарядом atomic(-armed) missile
~ с боевым зарядом warhead-equipped missile
~ с головкой самонаведения target-seeking missile
~ с ионным двигателем ion rocket
~ с крестообразным крылом cruciform-winged rocket
~ с поперечным делением, составная [с последовательным расположением ступеней] tandem rocket
~ с пороховым двигателем powder rocket
~ с продольным делением, составная clustered rocket
~ с самоликвидатором self-destruct-equipped rocket
~ с треугольным крылом delta-wing rocket
~ с ядерным двигателем nuclear-powered rocket
~ с ядерным зарядом muclear(-armed) missile
~, самонаводящаяся homing missile
~, сверхзвуковая supersonic missile
~, связная liaison rocket
~, сигнальная signal flare/rocket
~ со стартовыми двигателями boosted rocket
~, составная multistage rocket
~ средней дальности medium-range missile
~, твердотопливная solid-propellant rocket/missile
~, тормозная brake rocket, retrorocket
~, экономичная low-cost rocket
ракета-зонд rocket-probe, probe rocket
ракета-ловушка decoy missile
ракета-мишень target missile
ракета-носитель booster rocket, launch vehicle,

rocket booster vehicle
ракета-перехватчик intercepter missile
ракета-торпеда rocket(-assisted) torpedo
~, лодочная submarine rocket torpedo
~, противолодочная antisubmarine rocket, antisubmarine torpedo missile
~, самонаводящаяся rocket-boosted homing torpedo
ракетка *sport* racket
ракетница flare/signal pistol
~, бортовая signal rocket discharger
ракетный rocket *(attr.)*
ракетодержатель flare carrier
ракетодинамика rocket/missile dynamics
ракетодром rocket airfield/range, launching site
ракетоносец missile-carrier
~, дальний long-range missile-carrier
~, фронтовой tactical missile-carrier
ракетоносный missile-carrying
ракетоплан rocket plane/glider, boost-glide aircraft, rocket-propelled aircraft
ракетостроение rocket manufacture/production
ракитник *bot.* broom *(Cytisus);* broom plantation
ракля doctor blade, cleaning doctor; stripper
раковидный cancerous; cancriform
раковина 1. *met.* cavity, blister, pit, bubble; 2. *anat.* concha; 3. shell; *zool.* test; 4. sink, washbowl; 5. *naut.* quarter, quarterpoint
~, газовая *met.* blow hole
~, двустворчатая *zool.* bivalve test
~, кормовая *naut.* after quarterpoint
~, коррозионная corrosion cavity
~, левая *naut.* port quarter
~, морская sea shell
~, носовая *anat.* turbinated bone
~, песочная *met.* sand hole
~, правая *naut.* starboard quarter
~ стетоскопа/телефона earpiece
~, точечная *met.* pin-hole
~, усадочная *met.* contraction/shrinkage cavity, shrink hole, blowhole
~, ушная *anat.* floor of the auricle
~, шлаковая slag blowhole
раковиновидный, раковинообразный conchiform, shell-like; cochleate
раковинолистный *bot.* shell-leaved
раковистый shelly; shell *(attr.),* shell-like; conchiform; conchoidal
раковый *med.* carcinomatous, cancerous
ракообразное *zool.* crustacean, shellfish
ракообразные *pl., zool.* crustaceans *(Crustacea)*
ракообразный 1. *med., bot.* cancroid; 2. *zool.* crustacean, crustaceous
ракорд ленты, ведущий leader
ракорд ленты, задний trailer
ракоскорпионы *pl., zool.* merostomes, gigantostracans *(Merostomata, Gigantostraca)*
рак-отшельник *zool.* hermit crab *(Eupagurus);* Diogenes crab *(Cenobita diogenes)*
ракоустойчивый cancer-resistant
ракс *naut.* hank
ракс-бугель *naut.* mast travel(l)er, parrel (truss)
ракурс aspect angle; camera angle; *art* foreshortening
ракушечник shell rock, coquina
ракушка (small) shell
ракушковые *pl., zool.* ostracods *(Ostracoda)*
ракша *zool.* roller *(Coracias garrulus)*
рама 1. frame; casing; 2. chassis, carriage
~ автомобиля chassis frame
~ гондолы nacelle frame
~, дверная door casing/frame
~ двигателя/моторная/подмоторная engine mount/bed/frame
~, двухпролетная double frame
~ затвора block carrier
~, карданная gimbal frame
~, лесопильная saw frame, frame saw; gang mill
~ окна/оконная window frame, sash
~, опорная supporting/bed frame
~, растровая screen mount/holder
~, рулевая rudder frame
~, сварная welded frame
~ светлого люка skylight sash
~ тележки *rail.* bogie/truck frame
~, фундаментная foundation/bed frame
~ шпангоутная frame ring, all-round frame
рамазан, рамадан Ramadan, Ramazan
рамбурсировать reimburse
рамбутан *bot.* rambutan *(Nephelium lappaceum)*
рами *bot.* false nettle *(Boehmeria)*
рамикотомия *surg.* ramicotomy, ramisection
рамка 1. (small) frame; 2. *el.* coil, loop, loop antenna/aerial; 3. *pl.* framework, limits
~, визирная *geod., phot.* finder frame
~ гироскопа gyroscope gimbal frame/ring
~ карданного подвеса gimbal frame/ring
~, кадровая *phot.* gate mask
~ карты map margin/border
~, катушечная *text.* creel
~, копировальная *phot.* contact printing frame
~, поворотная rotatable loop
~ радиокомпаса radio loop
~ радиопеленгатора radio loop
~ с гнездами *teleph.* jack strip
~ с ключами *teleph.* key shelf
~, соединительная *teleph.* terminal/connecting strip
рамноза *chem.* rhamnose
рампа 1. manifold; 2. *theatre* footlights
~, кислородная oxygen manifold/header
рамтила *bot.* Ethiopian niger seed *(Guizotia abyssinica)*
рамулиспороз *bot.* ramulispora leaf spot
рамуляриоз *bot.* ramularia leaf spot
рамфоринх *paleont.* rhamphorynch *(Rhamphorhynchus)*
рана wound, injury, trauma, vulnus, traumatism
~, гнойная purulent/suppurating wound

~, **инфицированная** infected/septic wound
~, **огнестрельная** guhshot wound
~, **поверхностная** flesh wound
~, **рваная** lacerated wound
~, **резаная** incised wound, cut
~, **сквозная** perforating wound
~, **ушибленная** contused wound, bruise
ранверсман *aero.* renversement, reversement
~ **на вертикали** vertical reversement, cartwheel
ранг rank; class; category
~ **матрицы** matrix rank, rank of a matrix
~, **столбцовый** *math.* column rank
~, **строчный** row rank
~ **тензора** *math.* tensor rank/order
рангоут *naut.* masts and spars, masting, spars, sparring
~ **и такелаж** rig, harness
~ **и такелаж, верхний** top-hamper
рангоутное дерево *naut.* bum, spar
рандбалка marginal/wall/edge beam
рандомизировать randomize
раневой wound *(attr.),* vulnerary
ранение wounding; wound, injury
~, **касательное** tangential/gutter wound
~, **осколочное** fragmentation effect/wound
ранет *bot.* rennet *(apple)*
ранетка *bot.* variety of Siberian apple
ранец pack, knapsack, haversack; satchel
~, **нагрудный** chest pack
~, **наспинный** back pack
~ **парашюта** parachute bag
ранжировать range, rank
раннезрелый, раннеспелый early-ripening/maturing
раннетретичный *geol.* early-tertiary
ранник *bot.* common polypody *(Polypodium vulgare)*
ранорасширитель *surg.* (wound) retractor, dilator
рант border, rim, edge; *(of shoe)* welt
рантье rentier
ранцевый backpack *(attr.)*
рапа salt water, brine
рапидофиллум *bot.* needle palm *(Rhapidophyllum)*
рапира foil, rapier
рапирист *sport* foilsman
рапис *bot.* lady palm *(Rhapis)*
рапонтика *bot.* garden rhubarb *(Rheum rhaponticum)*
рапорт, рапортовать report
раппорт *text.* repeat
рапс *bot.* rape *(Brassica napus)*
рапунцель *bot.* rampion bellflower *(Campanula rapunculus)*
расамала *bot.* rasamala *(Altingia excelsa)*
расизм racism, racialism
расист, расистский racist, racialist
раскаленность incandescence; glow
раскаленный incandescent, glowing
~ **добела** white-hot, incandescent
~ **докрасна** red hot
раскалывание cleavage, cleaving; splitting; cracking; breaking
раскалять make burning hot, incandesce
раскат 1. *met.* feed, workpiece; 2. roll, peal
раскаталогизировать *comp.* uncatalogue
раскатка 1. unrolling, unreeling, uncoiling; 2. flattening out, rolling out
раскачивание, раскачка swinging, swaying, rocking, reeling; overswing, hunting; oscillation
расквартировать billet, quarter
раскидывание scattering; spreading out
раскисление *chem.* deoxidation, reduction
раскислитель deoxidizer, deoxidizing agent
раскислять *chem.* deoxidize, reduce
раскладка allotment; distribution; apportioning; breakdown
расклепывать unshackle
расковка hammering out, spreading, flattening
раскодировать decode, decipher
раскол cleft, crevice, crack, split; division; break; *rel.* schism, dissidence
раскалачивание smashing, breaking; rapping
расколотолепестный *bot.* schizopetalous
расколотолистный *bot.* fissifolious, schizophyllous
расколотоплодный *bot.* schizocarpic, schizocarpous
раскомпенсация decompensation
расконсервация 1. degreasing, deslushing, demothballing; 2. reactivation, putting back into operation
расконсервировать 1. degrease, depreserve, deslush; 2. reactivate, put back into operation
расконтрить unlock; release
раскорчевка uprooting, stubbing, stumping
раскос *constr.* cross/angle brace, strut, truss
~, **вынесенный** outrigger
~, **обратный** counter brace
раскраска, раскрашивание painting
раскрой, раскройка cutting out
раскрутка *aero.* spin-up; overspeeding
~, **ускоренная** boosted run-up
раскручивать untwine, untwist, uncoil
раскрыв opening, aperture
раскрывание opening; *bot.* dehiscence, bursting
раскрытие opening, disclosure; exposure, uncovering; *bot.* dehiscence
~ **неопределенностей** *math.* evaluation of indeterminate expressions/forms
~ **скобок** *math.* removal of brackets/parentheses; removal of grouping signs
раскулачивание *hist.* dispossession of the kulaks
раскулачивать dispossess the kulaks
распад decay, decomposition, disintegration, breakdown, breakup; dissociation, degeneration, lysis
~, **атомный** atomic decay/disintegration
~ **белка** albuminolysis
~, **взрывной** explosive disintegration

~, двухфотонный two-photon/quanta decay
~, искусственный artificial/induced disintegration
~ лейкоцитов leucolysis
~, многоступенчатый/множественный multiple decay
~, последовательный series disintegration
~ почечной ткани nephrolysis
~, радиационный radiative decay
~, разветвленный branching decay
~ с захватом электрона electron-capture decay
~, самопроизвольный spontaneous decay, autodecomposition
~ сопровождающийся гамма-излучением radiative decay
~, цепной chain decay
~ эритроцитов erythrocytolysis
~, язвенный exulceration, ulceration
распадаться disintegrate, decompose, fall apart, break down; decay, dissociate
распаивать unsolder
распаковка unpacking; *comp.* unpack
~ блока данных unblocking, deblocking
распатор, распаторий *med.* raspatory
~, глазной ophthalmic raspatory
~, позвоночный raspatory for spinal surgery
распатор-скребок scraping rasparatory
распахивать plough, plow (up)
~ землю break the ground
распашка ploughing, plowing
распечатка *comp.* listing; dump; print, printout
~, выборочная selective printout
~ оперативной памяти short-memory dump
~ содержимого основной памяти main memory dump, main store printout
~ содержимого памяти memory printout
~ содержимого регистров register printout
~, сплошная nonselective printout
~ файла file print
распечатывать 1. *comp.* print out; 2. unseal, break the seal of
распиливать saw
распиловка sawing
~, продольная length cutting
~, радиальная rift sawing
~, тангенциальная slash sawing
расписание timetable; schedule; bill; list
~, аварийное emergency schedule; damage-control bill
~, вахтенное *naut.* watch bill
~ дежурств duty schedule/list
~ на случай столкновения *naut.* collision bill
~ остановок реактора reactor shut-down schedule
~ полетов flight schedule/board
~, пожарное fire roll/bill
~, пожарно-спасательное fire and rescue bill
~, штатное manning table; table of organization
расписка receipt; voucher
~ в получении денег cash voucher
~, временная interim/temporary receipt
~, двойная [в двух экземплярах] receipt in duplicate
~, долговая debenture
~, ломбардная pawn ticket
~ о получении займа loan receipt
~, обратная return receipt
~, складская warehouse receipt
~, сохранная deposit receipt
расплав melt, fusion
~, загазованный gassy melt
~, питающий feeding melt
~, шлаковый melted slag
расплавлять melt down, fuse, liquefy
распластовать, распластывать separate/divide into layers; laminate
расплата payment
~ по долговым обязательствам discharge of liabilities
расплачиваться pay off; settle accounts
расплескивать spill; splash, swash, spatter
расплетать unplait, unbraid, untwist
расплетаться come unplaited/undone/untwined
расплод breeding, reproduction
расплываться spread; run, flow
расплющивать flatten/hammer out; crush flat
распознаваемость recognizability
распознаваемый recognizable, identifiable, discernable
~ машиной machine-readable
распознавание discernment, discerning; recognition, identification
~ адресных меток address mark/marker recognition
~ знаков character identification
~ зрительных образов visual recognition
~ изображения/образов pattern recognition
~ речи speech recognition
~ символов character recognition
~ сигнала на фоне помех target-noise discrimination
~ слов word recognizing/recognition
располагать place, dispose, position; arrange
расположение positioning, location, disposition; arrangement, layout, position; situation; configuration; site
~, веерообразное fan structure
~, географическое geographical setting
~ информации information configuration
~ комнат layout of rooms
~ кресел, последовательное tandem seating
~ кресел по четыре в ряд four-abreast seating
~, кольцевое cyclic arrangement
~, кулисообразное echelon arrangement
~, крестообразное cruciform configuration
~ плода *med.* attitude of f(o)etus
~, полосатое banded arrangement
~ с интервалами spacing
~ сердца, обратное dextrocardia, cardiac

heterotaxia
~ сидений seating arrangement
~, смежное juxtaposition
~, совмещенное co-location
~ стержней rod pattern
~, ступенчатое stepped arrangement
~ технологических каналов piping geometry
~ урана, блочное lumping of uranium
~ ускорителя boost configuration
~ цилиндров cylinder arrangement
~, шахматное staggered arrangement
распор thrust; pressure
распорка spacer, distance piece/bar, strut, brace; crossbar, spreader; stud, link bar
~, пружинная spring spacer
распорядитель manager; administrator
~ ресурсов *comp.* resource administrator/manager
распорядок standing order, routine; regulation(s), instruction, order
~ дня daily routine
распоряжение instruction, order, precept; decree; directive; ordinance
~ суда *law* bench warrant
распределение distribution; division; allocation, assignment, allotment
~ адресов address assignment
~, апостериорное posterior distribution
~, априорное prior distribution
~, беспорядочное random distribution
~, биномиальное *math.* binomial/frequency distribution
~ близкое к нормальному near-normal distribution
~ Больцмана Boltzmann distribution
~ в Земле terrestrial distribution
~ в пространстве spatial distribution
~ вероятностей probability distribution
~ вихрей vorticity distribution
~ волн allocation of frequencies
~, возрастное age distribution
~, выборочное sampling distribution
~, высотное altitudinal distribution
~ выхода по массам yield-mass distribution
~ Гаусса Gaussian distribution
~ генеральной совокупности population distribution; parent distribution
~ генов, независимое independent gene assortment
~, геометрическое geometric distribution
~ Гиббса Gibbs canonical distribution
~, гипергеометрическое hypergeometric distribution
~, граничное marginal distribution
~ груза *naut.* trimming
~, двумерное *math.* bivariate distribution, two-dimension distribution
~, двухвершинное bimodal distribution
~ деформации strain distribution
~ диполей dipole/doublet distribution
~, дискретное discrete distribution
~, доверительное fiducial distribution
~ жесткостей elastic distribution
~ зажигания ignition timing
~ импульсов momentum distribution
~ каналов *rad.* channel allocation
~ корней *math.* location of the roots *(of a polynomial)*
~ Коши *math.* Cauchy distribution
~ кривизны camber distribution
~ Максвелла Maxwell(ian) distribution
~ машинного времени *comp.* machine-time distribution
~, многомерное *math.* multivariate/multidimensional distribution
~ наборов данных *comp.* data set allocation
~ нагрузки load distribution/sharing
~ напряжений stress distribution
~ населения по возрастным группам age distribution
~ невязки *geod.* error distribution
~, неосевое off-axis distrubution
~, неправильное maldistribution
~, непрерывное continuous distribution
~, несферическое aspherical [non-spherical] distribution
~, нормальное normal/Gaussian distribution
~, нормальное логарифмическое lognormal distribution
~, одномерное univariate distribution
~, однородное uniform distribution
~, оптовое wholesale distribution
~ ответственности allocation of responsibilities
~ основной памяти *comp.* main store allocation
~ памяти memory/storage/store allocation
~ памяти на сердечниках core store allocation
~, первичное *econ.* primary distribution
~ плотности density distribution
~ плотности энергии, спектральное energy density spectrum
~ по зонам zonation
~ по импульсам momentum distribution
~ по крупности size distribution
~ по направлениям directional distribution
~ по поверхности surface distribution
~ по профессиям occupational distribution
~ по размаху *aero.* spanwise distribution
~ по районам regional distribution
~ по странам breakdown by countries
~ по хорде chordwise distribution
~ подъемной силы lift distribution
~ поля field distribution/pattern
~, показательное exponential distribution
~ потока flux distribution/pattern
~, предельное limiting distribution
~ прибылей allocation of profits
~, пропорциональное apportionment, proportional distribution/allocation
~ Пуассона Poisson distribution/law; law of

small numbers
~, равновесное equilibrium distribution
~, равномерное uniform distribution
~, разрывное discontinuous distribution
~ ресурсов resource distribution/allocation
~, решетчатое lattice distribution
~ рождений, временное birth timing, timing of child bearing
~ Рэлея Rayleigh distribution
~ сведений message switching
~ сил force distribution
~ силовых линий line-of-force distribution
~ скоростей velocity distribution
~, случайное random distribution
~, совместное joint/simultaneous distribution
~ сопротивления drag distribution
~ Стьюдента Student's distribution, (Student's) T-distribution
~ температуры по глубине vertical temperature distribution
~, угловое angular/directional distribution
~ Ферми Fermi distribution
~ Фишера F-distribution, Fisher's distribution
~ хи-квадрат chi-squared distribution
~ частиц по размерам particle-size distribution
~ частот frequency allocation; *stat.* frequency array
распределитель distributor; *el.* allotter, allotting switch
~, вращающийся rotary distributor
~ зажигания ignition timer/distributor
~ импульсов pulse distributor
~ шихты *met.* stock distributor/spreader
распределять distribute; divide; allot, allocate, assign; apportion; share
распрессовка pressing-out
распродажа bargain/clearance sales; sell-off
распроданный out-of-stock
распространение propagation, spreading, diffusion; emission; extension; range
~ в природе occurrence in nature, natural abundance
~, волноводное waveguide propagation
~, географическое geographic range
~ звука sound/acoustic propagation
~ звука в водной среде propagation of underwater sound
~ звукового удара propagation of sonic boom
~, многолучевое multiple path transmission
~ поверхностной волны ground-wave propagation
~ радиоволн radio-wave propagation
~ радиоволн в пределах прямой видимости line-of-sight wave propagation
~ радиоволн, ионосферное ionospheric wave propagation
~ радиоволн, многолучевое/многопутевое multi-path wave propagation
~, радиоволн, поверхностное ground wave propagation
~ тепла thermal diffusivity, flow of heat
распространенность 1. occurrence, currency, prevalence; diffusion, dispersion; spread; 2. range; extent; 3. abundance
распространенный widespread, dispersed, distributed; prevalent; extended
~, повсеместно cosmopolitan
~ное предложение *gram.* extended sentence
распространительный extended; broad
~ное толкование broad interpretation
распространяться spread, propagate; diffuse
распускаться 1. *bot.* open, come/bloom out, blossom; 2. spread; come loose/undone
распыление, распыливание atomization, pulverization; spraying; dispersal, scattering; *met.* sputtering
~, катодное cathode sputtering
~ топлива atomization of fuel
распыливать atomize, pulverize; spray; sputter
распылитель atomizer, diffuser, sprayer, pulverizer; spray gun; nozzle
~ жидкости atomizer
~ порошков powder nozzle
~, пароструйный steam-jet atomizer/sprayer
расса *zool.* small Indian civet *(Viverricula indica)*
рассада *agr.* seedlings
рассадник *bot.* nursery, seed-plot, seedbed
рассадопосадочная машина seedling planter
рассасывание *med.* resorption, resolution
рассасываться *med.* resolve
рассверливать bore/drill out, ream
рассев 1. screening, sieve-sizing, sieving; 2. sifter, bolter; 3. sowing, dissemination
~, кривошипный gyratory sifter
~, тонкий fine screening
рассеивание dissemination, dispersion, dispersing, dispersal, diffusion, dissipation
~ тепла heat dissipation
рассеиватель scatterer; diffuser
~, избирательный selective diffuser
~, комптоновский Compton scatterer
~, неточечный extended scatterer
~, объемный volume/bulk scatterer
~, реальный actual scatterer
~, рэлеевский Rayleigh scatterer
~, точечный point scatterer
~ фары head lamp lens/glass
рассеивать scatter, disperse; dissipate; diffuse
рассеиваться scatter, disperse; dissipate
рассекатель dissector
рассекать dissect
рассекречивать declassify
расселение resettlement; displacement (of population)
рассечение dissection, sectio
~ влагалища colpotomy, vaginotomy, coleotomy
~ грыжевого мешка herniotomy, kelotomy, celotomy
~ мочевого пузыря cystotomy, vesicotomy
~ сердечной стенки cardiomyotomy

~ спаек adhesiotomy
~ теноновой капсулы capsotomy, capsulotomy
~ шейки матки trachelotomy
рассеяние dispersion, dispersal, scattering; diffusion; dissipation; spread, dissemination; *stat.* variance
~, беспорядочное random scattering
~, брэгговское Bragg scattering
~ внутрь inscattering, scattering-in
~, дифракционное diffraction/shadow/ordered scattering
~ звука acoustical scattering, scattering of sound
~ значений value dispersion
~, ионосферное ionospheric scattering, ionoscatter
~, квазиупругое quasi-elastic scattering
~, Клейн-Нишиновское Klein-Nishina scattering
~, комбинационное Raman scattering
~, комптоновское Compton scattering
~, ложное spurious scattering
~, магнитное magnetic leakage/scattering
~, многократное multiple/plural scattering
~ мощности power dissipation
~, некогерентное incoherent scattering
~, неупругое inelastic scattering
~, обратное backward/back scattering
~ пи-мезонов pion scattering
~, рамановское Raman scattering
~, резерфордовское Rutherford scattering
~, рэлеевское Rayleigh scattering
~, теневое shadow scattering
~ тепла heat dissipation
~ тумана fog dissipation
~, турбулентное turbulent dissipation
~, чистое pure scattering
~ энергии energy dissipation
рассеянный diffuse; scattered, dispersed, diffused
рассеянноцветковый *bot.* sparsiflorous
рассинхронизация mistiming
расслабление slackening, lassitude, enfeeblement, weakening; relaxation, relaxing; debilitation
~ сосудов vasorelaxation
~ сфинктера sphincteric incontinence
расслабленный slack, atonic, limp; relaxed
расслаблять slacken, weaken, enfeeble; debilitate
расслаиваться delaminate, exfoliate; segregate
рассланцевание schist formation
расследование investigation, inquiry; examination
~, предварительное preliminary inquiry
расследовать investigate, inquire into, examine
расслоение stratification, foliation, lamination, exfoliation
~, гравитационное gravity segregation
~ по кливажу cleavage foliation
~ по плотности density stratification/foliation
~ слоистого материала delamination
~ эмульсии emulsification
расслой *met.* sliver, spill; seam
рассмотрение consideration; treatment
рассогласование mismatch, misalignment; mistuning
рассогласованный unmatched; untuned
рассоединение uncoupling, disengaging, disconnection
рассол brine; pickle, salt water/liquor
~, аммиачный ammoniacal brine
~, маточный mother brine
рассоление desalinization, desalting
рассредоточение dispersal, dispersion
рассрочка instal(l)ment plan
~ку, платеж в payment by instal(l)ments
~ку, покупка в hire purchase
расстановка placing, positioning; disposition, arrangement; order
расстеклование devitrification
расстояние distance; range; interval, space, spacing; run
~, апогейное apogee distance
~, афелийное aphelion distance
~, безопасное safe distance/clearance/gap
~, близкое close quarters/range
~ в милях mileage
~ вдоль axial distance
~ видимости sight distance
~, зенитное zenith distance, coaltitude
~, кодовое code/signal distance
~, межатомное interatomic distance, atomic spacing
~ между зрачками *anat.* pupillary distance
~ между отпечатанными знаками printing character spacing
~ между ядрами internuclear separation
~, межионное interion [ion-ion] distance
~, межосевое centre-to-centre distance
~, меридиональное meridional/meridian distance
~, ортодромическое ground distance/range
~, перигелийное perihelion distance
~ по локсодромии rhumb-line distance
~ по ортодромии great-circle distance
~ по оси axial distance
~ по прямой crow-flight distance
~ по радиусу radial distance
~ по сетке координат grid distance
~ по траверзу althwart distance
~ по трассе down-range distance
~ по хорде chordwise distance
~, полярное codeclination, polar distance
~ прямого видения line-of-sight distance
~, пройденное по лагу run by log
~, разрядное *el.* arcing distance
~, среднеквадратичное root-mean-square distance
~, тормозное braking distance
~, фокусное focal distance/length
расстройка detuning; misalignment, maladjustment
расстригать *eccl.* unfrock

расстройство disturbance, derangement, disorder; disruption; confusion; impairment, distress
~ **артикуляции** *med.* dysarthria, logophasia
~ **всасывания** malabsorption
~ **глотания** dysphagia
~, **двигательное** dyskinesia
~, **диспептическое** dyspeptic disorder, dyspepsia
~, **душевное** mental derangement, insanity, alienation
~ **желудка** stomach upset, gastric disturbance
~ **жирового обмена** lipodystrophy
~ **зрения** visual disturbance, dysopsia
~, **кишечное** intestinal disorder
~ **координации движений** incoordination of muscular action, ataxia
~ **обмена веществ** metabolic disturbance
~ **письма** dysgraphia
~, **психическое = душевное**
~ **речи** speech disturbance, alalia, dyslalia, dysphasia
расстыковать disengage, detach; *space* undock
растачивание boring
~ **на конус** taper boring
~, **черновое** rough boring
~, **чистовое** finish boring
раствор 1. solution; 2. opening, aperture, span, spread
~, **бедный** weak solution
~, **буровой** drilling fluid
~, **водный** aqueous solution
~, **глинистый** *constr.* clay mortar
~, **дезактивационный** *med.* decontamination solution
~, **дезинфицирующий** disinfectant solution
~, **дубящий** tanning solution
~, **доэвтектический** hypoeutectic solution
~, **жидкий** *constr.* grout, grouting
~ **иода, спиртовой** iodine solution
~, **кладочный** masonry mortar
~, **маточный** mother liquor
~, **молярный** molar [gram-molecular] solution
~, **мыльный** suds
~, **низкозамерзающий** antifreezing solution
~, **обращающий** *phot.* reversing bath
~, **останавливающий** *phot.* stop bath
~, **отработанный** spent solution
~, **отстоявшийся** supernatant (liquor)
~ **перекиси водорода** hydrogen peroxide solution
~ **Рингера** Ringer's solution, compound solution of sodium chloride
~, **солевой** saline/salt solution, saline
~, **спиртовой** alcoholic solution
~, **стандартный** standard(ized) solution
~, **строительный** mortar
~, **титрирующий** titrating solution, titrant
~, **тонирующий** *phot.* toner
~, **травильный** *met.* pickling solution
~, **травящий** *print.* etching solution
~, **физиологический** physiologic salt solution, saline
~ **фовлеров** Fowler's solution, potassium arsenate solution
~, **штукатурный** plaster; stucco
растворимый soluble, dissoluble, dissolvable
~ **в жирах** liposoluble
~ **в уксусной кислоте** acetosoluble
растворитель solvent; diluent, thinner; menstruum
~, **безводный** anhydrous solvent
~ **для краски** paint thinner
~, **экстрагирующий** extractive solvent
растворонасос mortar pump
растекание spreading
растение plant
~, **бобовое** legume
~, **болотное** holophyte
~, **ветроопыляемое** wind-pollinated plant
~, **вечнозеленое** evergeen (plant)
~, **водовоздушное** amphiphyte
~, **водоплавающее** aquatic plant
~, **водяное** water/hydrophilic plant
~, **высшее** higher plant
~, **вьющееся** climbing plant, climber
~, **двудомное** didecious/didecian plant
~, **двухлетнее** biennial (plant)
~, **древесное** wood/lignaceous plant
~, **кормовое** fodder plant
~, **красильное** dye plant
~, **лекарственное** medicinal plant
~, **многолетнее** perennial (plant)
~, **насекомоопыляемое** insect-pollinated plant
~, **низшее** lower plant
~, **однолетнее** annual (plant)
~ **озер** limnophyte
~, **пищевое** food plant
~, **ползучее** creeper; vinaceous plant
~, **равноспоровое** homosporous plant
~, **разноспоровое** heterosporous plant
~, **слоевищное** thallophyte
~, **стелющееся** trailer
~, **сухолюбивое** xerophilous plant
~, **теплолюбивое** thermophilic plant
~, **травянистое** herbaceous plant
~, **холодолюбивое** cryophyte
~, **широколиственное** broad-leaved plant
растениеводство plant growing
~, **тепличное** glasshouse cropping
растение-хозяин host plant
растигор *bot.* chamaedrys germander *(Teucrium chamaedrys)*
растирать grind, pulverize, rub to powder
растительность vegetation; flora
~, **бентоническая** benthos
~, **болотная** bog vegetation
~, **водная** aquatic vegetation
~, **дикая** natural growth
~, **степная** steppe-land [xerophilous] vegetation
~, **чахлая** stunted vegetation

растительный vegetable *(attr.),* plant *(attr.),* vegetative, vegetal
~ альбумин vegetable albumin
~ное сообщество plant community, coenosis
растление seduction
растлитель seducer
расторгат abrogate, dissolve; rescind; cancel, annul
расторжение abrogation, cancellation; rescission, annulment; dissolution
растормаживание 1. *tech.* brake release; 2. *psych.* disinhibition
расторопша *bot.* holy thistle *(Silybum)*
расточка boring/drilling out, bore, recess
растр raster; grid; *print.* grid
~, точечный dot pattern
растрата 1. embezzlement, defalcation; 2. squandering; spending
растрескивание cracking; splitting, fracturing; *bot.* dehiscence
раструб mouthpiece, bellmouth, funnel, bell;
~ трубы bell [female end] of a pipe
растягивание stretching, extending, extension, expansion; tension;
~ дозы *phys.* dose protraction
~ импульса pulse stretching
растяжение tension; strain, sprain; distention
~ влагалища colpectasia, colpectasis
~ мочевого пузыря cystectasia, bladder distention, dilation/distention cyst, megalobladder
~ сустава strain
~ сухожилий strained muscles
~, чрезмерное hyperdistention, overextension
растяжка 1. tension member, brace, guy; tie rod; 2. stretching, extending, expansion
~, анкерная anchor line
~, канатная guy rope/cable
~, проволочная anchor/tension wire
~ развертки sweep expansion
расфасовка packing, prepacking
расфокусировка defocus(s)ing
расфазировка misphasing
расход 1. expense, expenditure; outlay; cost(s); 2. consumption; 3. flow rate; discharge
~, весовой weight/mass flow rate
~ воды water consumption; water discharge
~, дополнительный surcharge
~, километровый consumption per kilometre
~ кислорода oxygen consumption
~ кислорода, секундный oxygen flow rate
~ корма на грамм привеса feed/gram gain
~, массовый mass flow rate
~,мгновенный instantaneous flow rate
~, номинальный rated consumption
~, объемный volume-flow, volumetric flow
~, расчетный theoretical/rated consumption
~ реки river discharge
~ тепла, удельный heat rate
~, секундный flow rate
~, удельный specific consumption
~, фактическмй actual flow
~, часовой hourly consumption, consumption per hour
~ы, авансом оплаченные prepaid expenses
~ы, административные general expenses/burden
~ы, амортизационные depreciation costs/charges
~ы, военные military/defence spending
~ы, государственные public expenses/outlay; government/public expenditures
~ы, дополнительные extra charges
~ы, командировочные travelling expenses
~ы, мелкие petty expenses
~ы на жизнь living expenses
~ы на жилищное строительство housing expenditures
~ы на здравоохранение health expenditures
~ы на оборону defence expenditures/spending
~ы на обслуживание service expenses
~ы на предметы роскоши luxury expenditures
~ы на уход maintenance costs
~ы, накладные overhead costs/expenses/charges, overheads
~ы, неэксплуатационные nonoperating expenses
~ы, общефирменные general burden/expenses
~ы, организационные establishment charges; organizational expenses
~ы, отложенные на будущие годы deferred charges
~ы по выгрузке landing charges
~ы по дисконтированию discount expenses
~ы по инкассированию collection charges
~ы по отправке груза forwarding charges
~ы по охране окружающей среды environmental expenditures
~ы по перевозке carriage charges
~ы по погрузке shipping charges
~ы по сбыту продукции distribution expenses
~ы по стимулированию сбыта promotional expenses
~ы по страхованию insurance cost(s)
~ы по швартовке *naut.* mooring charge/bill
~ы, постоянные fixed charges
~ы, потребительские consumer spending
~ы, почтовые postage
~ы предусмотренные сметой budgeted amounts
~ы, проездные travelling expenses
~ы, производительные productive expenditures
~ы, ремонтные repair cost(s)
~ы, текущие operational expenditure(s), operation cost(s)
~ы, сверхнакладные super burden
~ы, социально-культурные welfare expenditure(s)
~ы, транспортные transportation charges
~ы, цеховые shop costs

~ы, цеховые накладные departmental charges
~ы, эксплуатационные operating/operational/working/upkeep/running/maintenance cost(s)
расходимость divergence
~ ряда *math.* divergence of a series
~, угловая angular divergence/spread
расходомер flowmeter, flow-rate indicator; volumetric meter
~ воздуха air-flow indicator
~, вертушечный turbine flowmeter
~, весовой gravimetric flowmeter
~, массовый mass-rate flowmeter
~, напорный head flowmeter
~, объемный volumetric flow indicator
~, открытый free flowmeter
~, поплавковый float-type flowmeter
~, регистрирующий recording flowmeter, flow recorder
~ с вертушкой/крыльчаткой propeller/turbine flowmeter, rotameter
~, сильфонный bellows flowmeter
расходящийся divergent, diverging
расхождение divergence; discrepancy; disagreement; disjunction; *space* separation; *stat.* inequality
~, альтернативное disjunctional separation, alternative disjunction
~ на встречных курсах *naut.* passing on opposite courses
~, неслучайное nonrandom disjunction
~ струи jet divergence
~ стыка separation of a joint
~, угловое angular divergence/spread
расцветка colouring, colour, tinting
расценка estimate
расцепитель release/tripping mechanism, trip, release
расцепка uncoupling; release
расцепление disengagement, tripping, release
расцеплять uncouple, disengage, release, trip
расчаливание wire-bracing
расчалка (bracing) wire
~, лобовая drag wire
~, несущая lift wire
~, противофлаттерная anti-flutter wire
расчеканивать caulk
расчет 1. calculation, estimation, computation; estimate; reckoning; analysis; 2. design; 3. *econ.* settlement, payment; 4. crew, team
~, безналичный payment by check/cheque
~, бухгалтерский accounting calculation
~ вариационным методом variational calculation
~ времени timing
~ деформации strain/deformation analysis
~ за наличный cash down
~ затрат и результатов cost-benefit calculation
~ издержек statement/reckoning of cost(s)
~ комиссионных commission statement
~ конструкции structural analysis
~ матричным методом matrix calculation
~ методом динамического программирования dynamic programming analysis
~ методом дискретных элементов finite element analysis
~ методом Монте-Карло Monte Carlo calculation
~ методом плоских сечений strip analysis
~ на прочность strength analysis/calculation; structural design
~ на середину месяца mid-month account
~, наличный cash payment
~ напряжений stress analysis
~, неверный miscalculation
~, незавершенный outstanding account
~ обтекания flow-field analysis
~, орудийный *mil.* gun's crew
~ по окончании срока службы separation payment
~, прожекторный searchlight party
~ процентов interest statement
~ прочности strength calculation/analysis
~, прямой straight-forward calculation
~ с большим запасом conservative design
~ с запасом overdesign
~ систем systems analysis
~, строгий rigorous calculation
~, точный exact/precision design
~ устойчивости stability analysis
~ характеристик performance prediction/design
~, хозяйственный profit-and-loss basis, self-supporting basis
~ цепей network analysis
расчетный design *(attr.)*, planned, estimated; computational, calculation *(attr.)*
~ баланс balance of payments
~ банк clearance bank
~ная ведомость payroll
~ная книжка paybook
~ная палата clearing house
~ная сила rated/design force
расширение expansion, extension, broadening, widening; *med.* dilatation, dilation, distention, ectasia, ectasis
~, алгебраическое algebraic extension
~ аорты aortectasia, aortectasis
~ артерий arteriectasia, dilatation of the arteries
~ вен phlebectasia, varicosity, varicosis
~, доплеровское Doppler broadening
~ желудка dilatation of the stomach, gastrectasia, gastrectasis
~ записи (data) record extension
~ запроса *comp.* query enhancement
~ импульса pulse stretching/widening
~ капилляров trichangiectasia
~ легких lung expansion
~ набора данных data set extension
~, объемное cubical/volume/volumetric expansion

~ **оператора** *math.* extension of an operator
~, **остаточное** permanent expansion
~ **памяти** memory expansion
~ **при постоянном давлении** constant-pressure expansion
~ **при постоянном объеме** constant-volume expansion
~ **программы** program(me) extension
~ **прямой кишки** proctectasia, proctectasis
~ **сердца** cardiomegaly, megalocardia, dilation of the heart
~ **системы** system expansion
~ **сосудов** vasodilatation
~ **спектра** spectrum broadening
~ **струи** jet spreading
~, **тепловое** thermal expansion
~ **толстой кишки** megacolon
~ **языка** language extension
расширенный expanded, extended, broadened, widened; dilated; enlarged
~**ное воспроизводство** expanded/extended reproduction
расширитель expander; widener; stretcher; *med.* dilator, bougie; dilating forceps
~, **буровой** reaming drill
~, **гортанный** laryngeal bougie
~ **для слезного канала** lacrimal dilator
~ **импульсов** pulse stretcher
~ **скважин** hole reamer
~ **шейки матки** uterine dilator
расширительное толкование broad interpretation
расширяться expand, extend, widen, broaden; dilate
расшифровывать decode, decipher; interpret
расшифровка deciphering, decoding; interpretation
расштыбовщик *min.* gummer
расщелина *med.* hiatus, cleft; fissure, aperture
~ **верхней губы** cleft lip, cheiloschisis
~, **верхнечелюстная** maxillary hiatus
~, **крестцовая** sacral hiatus
расшнуровывать unlace
расщеп split; *met.* sliver
расщепитель splitter
расщепление splitting, cleaving, cleavage; splintering; decomposition, disintegration, breaking up; *phys.* fission, splitting
~ **белков** proteolysis
~ **жира** lipolysis, adipolysis
~ **кисти** cleft/split hand, lobster-claw hand
~ **культи** kineplasty
~ **мошонки** scrotal fissure
~ **ногтей** schyzonychia
~ **пептидов** peptidolysis
~, **психическое** mental splitting, sejunction
~ **резонанса** resonance splitting
~ **сахара** glycolysis
~ **стопы** cleft foot
~ **хроматиды** chromatid break
~ **ядра** nuclear splitting/disintegration
расщепленнозубчатый *bot.* cleft-toothed
расщепленнозубые *pl. zool.* fissidentaceans, schizodonts *(Schizondonta)*
расщепленноклювый fissirostral
расщепленнолистный *bot.* cleft-leaved
расщепленноногие *pl. zool.* fissipeds, schizopods *(Schizopoda)*
расщепленный cleaved, cleft; split; disintegrated
~ **надвое** bifid
расщеплять cleave, split; splinter; break up
расщепляющий cleaving, splitting; decomposing
~ **белки** proteolytic
~ **глюкозиды** glucosidolytic
~ **жиры** lipolytic, lipoclastic, adipolytic
расщепляющийся *phys.* fissionable, fissile
раувольфия *bot.* devil pepper *(Rauwolfia)*
рафания *med.* r(h)aphania
рафаэлевский Raphaelesque
рафаэлизм Raphaelism
рафиды *pl. bot.* raphides
рафинад lump sugar
рафинат refined petroleum/oil
рафинация refining
рафинирование refining
~ **отгонкой** distillation refining
~, **пирометаллургическое** fire refining
~ **сульфидированием** sulfidizing refining
~ **хлорированием** chlorination refining
раф(ф)ия *bot.* raffia palm *(Raphia)*
раффиноза *chem.* raffinose, melitose, meliotriose
раффлезия *bot.* rafflesia, monster flower *(Rafflesia)*
рахат-локум Turkish delight
рахиальгия *med.* rachialgia
рахиоплегия *med.* rachioplegia, spinal paralysis
рахисхиз *med.* rachischisis
рахит *med.* rachitis, rickets
~ **взрослых/поздний** late/adult rickets
~, **детский** infantile rickets
~ **острый** acute rickets, infantile scurvy
рахитический rachitic, rickety
рахотомия *surg.* rachitomy, rachiotomy
рационализатор innovator; efficiency expert
рационализаторский innovator's, innovation *(attr.)*
~**кое предложение** innovation proposal
рационализаторство efficiency/innovation drive
рационализация innovation; streamlining
рационализировать innovate, streamline, raise the efficiency of
рационализм *philos.* rationalism, rationality
рация (portable) radio set/station
рацпредложение *abbr.* proposal for innovation, efficiency/streamlining proposal
рацион *agr.* diet, menu
~, **безбелковый** nonprotein diet
~, **полноценный** adequate/complete diet
~, **суточный** daily allowance
рашкуль *art* charcoal pencil

рашпиль rasp (file)
~ по дереву wood rasp
рванина *met.* fissure
рваный lacerated
рваться fray
рвота vomit, emesis, vomiting, vomition, vomitus
~, кровавая black vomit, h(a)ematemesis
~, сухая dry vomiting, retching
рвотный vomitive, emetic
~ корень ipecacuanha
~ное средство emetic
рдест *bot.* pondweed *(Potamogeton)*
реагент *chem.* reagent, reactant
~, дефлокулирующий deflocculant
~, жироудаляющий fat extractant
~, коагулирующий coagulant
~, комплексообразующий complexing agent
реагировать react; respond
реактанс *el.* reactance
реактив *chem.* reactant, reagent
~, меченый labelled reagent
~, осаждающий precipitant, precipitating agent
реактивация reactivation
~, множественная multiple reactivation
~ скрепливанием cross reactivation
~, тепловая heat reactivation
реактивировать reactivate; revive
реактивность reactivity; *el.* reactance
~, вносимая insertion reactance
~, запаздывающая delayed reactivity
~, мгновенная prompt reactivity
~, остаточная shut-down [residual] reactivity
~ реактора reactor/pile reactivity
реактивный reactive; jet *(attr.)*
реактопласт thermosetting plastic
реактор reactor; pile; *chem.* reaction vessel
~, авиационный aircraft reactor
~, автономный autonomous [self-contained] reactor
~, агрегатированный integral reactor
~, атомный atomic reactor/pile
~ бакового типа tank-type reactor
~, башенный column/tower reactor
~ без воспроизводства топлива non-regenerative reactor, burner
~ без замедлителя unmoderated reactor
~ без отражателя bare/naked/unreflected/untamped reactor
~, бесконечный плоский infinite slab reactor
~, биологический fermenter
~, блочный lumped reactor
~ большой мощности high-power reactor
~, быстрый fast reactor
~, виртуальный virtual/image reactor
~, водный aqueous reactor
~, водо-водяной water-moderated water-cooled reactor
~, воспроизводящий regenerative reactor
~, высокопроизводительный/высокоэффективный high-performance reactor
~, гелиевый helium-cooled reactor
~ двойного назначения [двухцелевой] dual-/double-purpose reactor
~, двухзонный two-region reactor
~, дейтерий-натриевый deuterium-sodium reactor
~ для производства плутония plutonium-production [plutonium-only] reactor
~ для производства плутония и энергии plutonium-and-power reactor
~, кипящий boiling-water [water-boiler] reactor
~ колодезного типа well-type reactor
~, корабельный ship/marine/naval/propulsion reactor
~ малой мощности low-power reactor
~, медленный slow-neutron reactor
~, мнимый image/virtual reactor
~, многозонный multiple-region reactor
~, многосекционный multiple-section reactor
~ на быстрых нейтронах fast-neutron reactor
~ на мгновенных нейтронах prompt-neutron reactor
~ на медленных нейтронах slow-neutron reactor
~ на надтепловых нейтронах epithermal-neutron reactor
~ на промежуточных нейтронах intermediate-neutron reactor, epithermal reactor
~ на резонансных нейтронах resonance reactor
~ на суспензионном топливе suspension/slurry reactor; fluidized reactor
~ на твердом топливе solid-fuel(led) reactor
~ на торий-урановом топливе thorium-uranium fuelled reactor
~, надкритический supercritical [above-critical] reactor
~, надтепловой epithermal reactor
~, наземный land-based reactor
~, неотравленный unpoisoned/clean reactor
~, односкоростной one-velocity reactor
~, опытный pilot/developmental reactor
~ охлаждаемый газом под давлением pressurrized gas-cooled reactor
~, плоский slab reactor
~, погруженный/погружной swimming-pool reactor
~ под давлением pressurized reactor
~, промежуточный intermediate/epithermal reactor
~, промышленный industrial/commercial reactor
~, прямоточный direct-flow [once-through] reactor
~, размножающий breeder [reactor]
~, регулируемый controlled reactor
~ с бериллиевым замедлителем beryllium-moderated reactor
~ с бериллиевым отражателем beryllium-reflected reactor

~ с большой плотностью потока high-flux reactor
~ с водой под давлением pressurized-water reactor
~ с воспроизводством топлива converter; breeder
~ с графитовым замедлителем graphite-moderated reactor
~ с двойным циклом dual-cycle reactor
~ с двумя активными зонами two-core reactor
~ с жидким замедлителем liquid-moderated reactor
~ с замедлителем moderated reactor
~ с кипящей водой boiling-water reactor, water boiler reactor
~ с коэффициентом воспроизводства больше единицы breeder [reactor]
~ с коэффициентом воспроизводства меньше единицы converter, non-breeder reactor
~ с малой плотностью потока low-flux reactor
~ с однократным циклом топлива once-through reactor
~ с отражателем reflected/tamped reactor
~ с перегревом, кипящий superheating boiling reactor
~ с переменным уровнем замедлителя variable-moderator reactor
~ с повторным циклом recycling reactor
~ с расширенным воспроизводством топлива breeder (reactor)
~ с твердой активной зоной solid-core reactor
~ с тесной решеткой closely packed reactor
~, саморазгоняющийся autocatalytic reactor
~, сборный package(d) reactor
~, суспензионный suspension/slurry reactor; fluidized reactor
~, термоядерный thermonuclear/fusion reactor
~, тяжеловодный heavy-water reactor, deuterium-moderated reactor
~, универсальный multi-purpose [general-purpose] reactor
~, уран-графитовый uranium-graphite [graphite-uranium, carbon-uranium] reactor
~, учебный training reactor
~, форсированный spiked reactor
~, цепный chain reactor
~ циклического действия batch reactor
~, чистый unpoisoned/clean reactor
~, шламовый slurry reactor
~, энергетический power reactor
~, ядерный nuclear reactor, atomic pile
реактор-двигатель propulsion reactor
реактор-конвертер размножитель breeder-converter
реактор-облучатель reactor-irradiator
реактор-размножитель breeder [reactor]
~ для производства плутония plutonium breeder
~ на быстрых нейтронах fast breeder
~ на тепловых нейтронах thermal breeder
~ с внешней зоной воспроизводства external breeder
~ с внутренней зоной воспроизводства internal breeder
~ с торий-урановым циклом thorium-uranium breeder
~, энергетический power breeder
реакторостроение reactor construction/engineering
реакционер reactionary
реакционноинертный *chem.* reactionless; inert
реакционноспособный reactive
реакционность reactionary spirit, reactionism
реакционный 1. *chem.* reaction *(attr.),* reacting, reactional; 2. reactionary, retrograde, reactionist *(attr.)*
реакция reaction, response; test
~ Абдергальдена Abderhalden's reaction
~ агглютинации agglutination test
~ аннигиляции annihilation reaction
~, анодная anodic reaction
~, биуретовая biuret reaction/test
~, болевая pain reaction
~ Борде-Жангу Bordet-Gengou test, B.D. test
~ в реальном времени real-time response
~, вазомоторная vasomotor/vasomotorial/vasomotoric/vasomotory reaction
~ Вассермана Wassermann reaction/test
~, внутрикожная intracutaneous/intradermal reaction
~, восстановительная reducing reaction
~, вторичная second-order [secondary] reaction
~ вызванная дейтронами deuteron-induced reaction
~ вызванная столкновением collision-induced reaction
~, вялая sluggish response
~ Герксгеймера Herxheimer's reaction
~ горения combustion reaction
~, двигательная motor reaction, mobile response
~ деления fission reaction
~ деления, цепная fission chain reaction
~ дивергенции divergence/divergency test
~ замещения substitution/replacement reaction
~, заторможенная deferred reaction
~ захвата capture reaction
~, защитная defence reaction
~, зеркальная mirror reaction
~, зрачковая pupillary reaction
~ излучения radiation/radiative reaction
~ исправной схемы *comp.* fault-free response
~, капельная drop reaction, spot test
~, кожная cutaneous/skin reaction, cutireaction
~, кожно-гальваническая cutaneogalvanic reaction
~, кожно-туберкулиновая cutituberculin/

dermotuberculin reaction
~, конечная end reaction
~, конкурирующая competitive/competing/concurrent reaction
~, конъюнктивальная ophthalmoreaction, oculoreaction, conjunctival reaction
~, лейкоцитарная leucocytic reaction
~, локальная local/focal reaction
~, лучевая radiation reaction
~ Манту Mantoux reaction/test
~, миастеническая myasthenic reaction
~ на белок albumin test/reaction
~ на быстрые нейтроны fast-neutron response
~ на быстрых нейтронах fast-neutron reaction
~, незатухающая sustained reaction
~ нейтрализации neutralization test
~, непроизвольная *med.* involuntary/spontaneous reaction
~, неравновесная *chem.* non-equilibrium reaction
~, неуправляемая/неконтролируемая/нерегулируемая цепная uncontrolled chain reaction
~, нингидриновая ninhydrin reaction
~ нулевого порядка zero-order reaction
~, оборонительная defence reaction
~, обратимая reversible reaction
~, обратная inverse/reverse/back reaction
~, окислительно-восстановительная oxidation-reduction reaction, redox reaction
~ опоры bearing/support reaction
~ оседания эритроцитов erythrocyte sedimentation reaction/test, ESR
~, ответная response
~ отдачи recoil reaction
~ отклонения комплемента complement deviation test
~ Пирке *med.* Pirquet's reaction/test
~ по крену *naut.* roll response
~ по рысканию *naut.* yaw response
~ по тангажу *naut.* pitch response
~ побочная side reaction/effect
~ покраснения *med.* erythematous reaction
~, равновесная balanced reaction
~, разветвленная цепная branching chain reaction
~, ранняя лучевая early radiation response
~ самовоспламенения hypergolic reaction
~, самоподдерживающаяся self-sustaining reaction
~, световая light reaction
~ связывания комплемента complement fixation test, CFT
~ синтеза fusion reaction
~ скалывания-деления spallation-fission reaction
~ скалывания, фотоядерная photospallation
~, сложная complex/composite reaction
~, содружественная consensual reaction
~, сопряженная coupled reaction
~, ступенчатая staircase reaction
~ схемы с неисправностью *comp.* failing response
~, твердотельная solid-state reaction
~, термоядерная thermonuclear/fusion reaction
~, установившаяся steady-state response
~ фиксации алексина alexin fixation test
~ флоккуляции flocculation reaction, cephalin-cholesterol flocculation test
~, цепная chain reaction
~, частная partial reaction
~ человеческого организма human reaction/response
~ Шика Schick's test
~, экзотермическая exothermic/exoergic reaction
~, экзоэнергетическая exoergic reaction
~, эндотермическая endothermic/endoergic reaction
~, эндоэнергетическая endoergic reaction
~, ядерная nuclear reaction
~ ядерного синтеза thermonuclear reaction
~ якоря *el.* armature reaction
реал I *print.* composing frame, rack
реал II *hist.* rial, real *(Spanish coin)*
реализатор *physiol.* realizer, effector
~ пола sex realizer, sex genes differential
реализация 1. realization, implementation; substantialization; actualization; 2. sale, disposal
реализовать, реализировать 1. realize, implement, bring about, carry into effect; 2. sell; cash
реализуемость 1. feasibility; practicability; 2. marketability
реализуемый 1. feasible, realizable; practicable; 2. marketable, cashable
реалия realty, something real; *pl.* realia
реальгар *miner.* realgar, red arsenic glass
реально-уступительный *gram.* real-concessive
реанимация *med.* resuscitation, revivification, revivescence
реанкилоз reankylosis
ребек *mus.* rebec(k)
ребенок child, infant, baby
~, внебрачный natural/illegitimate child
~, грудной infant, suckling, nursing baby
~, дефективный disturbed child
~ дошкольного возраста preschool child
~, законнорожденный legitimate child
~, недоношенный premature infant/baby
~ отнятый от груди weanling
~, родившийся в срок full-term baby
~, родной own child
~, узаконенный legitimized child
реберно-грудинный *anat.* costosternal
реберно-ключичный *anat.* costoclavicular
реберно-лопаточный costoscapular
реберно-мечевидный costoxiphoid
реберно-позвоночный costovertebral, costocentral
реберно-хрящевой costochondral

реберный *anat.* costal
реборда rim, flange; bead
ребристый ribbed, finned
ребро rib; *tech.* fin; *math.* edge
~ атаки *aero.* leading edge
~, брюшное sternal rib
~ графа *math.* edge/link of a graph
~ жесткости reinforcing/stiffening/strengthening rib, ribbed stiffener
~, колеблющееся *anat.* floating rib
~, крестовидное transverse rib
~, крестцовое sacral rib
~, ложное *anat.* false rib
~, охлаждающее cooling fin
~, режущее cutting end
~, шейное cervical rib
~, усиливающее reinforcing/strengthening rib
реброобразный costiform
ревальвация revaluation; valorization
ревальвировать revalue
реваншизм *polit.* revanchism
реввоенсовет *abbr., hist.* Revolutionary Military Council
ревень *bot., pharm.* rhubarb *(Rheum)*
реверс 1. reverse, reversal; 2. reversing gear, reverser
~ воздушного винта propeller reversal
~ рулей *aero.* control reversal
~ тяги в полете in-flight thrust reverse
~ управления reversal of control
~ элеронов aileron reversal
реверсирование reversal, reverse, reversing
~ подачи feed reversal
~ тяги thrust reversing
~ шага воздушного винта propeller pitch reversing
реверсировать reverse
реверсор reverser
~, двухстворчатый clamshell-door reverser
~ тяги thrust reversal unit, thrust reverser
~ тяги внутреннего двигателя inboard thrust reverser
~ тяги, створчатый clamshell thrust reverser
реверс-редуктор reverse-reduction gear
ревизия auditing, audit; inspection
~ баланса balance-sheet audit
~, внешняя independent/external audit
~ кассовых остатков cash audit
~ счетов examination of accounts
ревизор inspector; examiner
~, программный *comp.* code auditor
ревком *abbr.* revolutionary committee
ревмартрит *med.* rheumarthritis, articular rheumatism
ревмартроз rheumarthrosis
ревматальгия *med.* rheumatalgia
ревматизм rheumatism, rheumatic fever
~, гонорейный/гонококковый gonorrheal/gonococcal rheumatism
~ пустыни desert rheumatism, coccidioidomycosis
~, суставной articular rheumatism
~, туберкулезный tuberculous rheumatism
~, узловатый nodose rheumatism
ревматический rheumatic, rheumatismal
ревматолог rheumatologist
ревмокардит *med.* rheumatic carditis
революция revolution
~, научнотехническая technological revolution
ревун hooter, howler, klaxon; *zool.* howler, howling monkey *(Alouatta)*
~, сигнальный warning/alarm hooter
регата regatta
регби Rugby football
регбист Rugby football player
регенерат reclaimed product/rubber; reclaim, regenerate
регенератор reclaimer; regenerator
регенерация regeneration, recovery, reclaiming; refreshing; reprocessing
~, водная aqueous reprocessing
~ изображения display refreshing
~ масла oil reclaiming/reclamation
~ отбросного тепла waste-heat recovery
~ отходов waste recovery
~, пирохимическая pyrochemical reprocessing
~ растворителя solvent recovery
~ урана uranium regeneration/recovery/reprocessing
~ формовочной смеси reclamation of moulding sand
регенерировать regenerate, recover, restore, reprocess; reclaim
регионарный regional
регент regent
регентство regency
регистр 1. register; list; journal; *econ.* account book; 2. *tech.* damper; 3. *mus.* stop, register; 4. *comp.* case *(of printer)*
~ адреса запоминающего устройства memory address register, store register
~ адреса канала channel address register
~ адреса памяти store/memory search register
~ адреса ячейки памяти memory address register
~, адресуемый addressable register
~ аналогового вывода, буферный analogue output buffer register
~, аппаратный machine register
~ арифметического устройства arithmetic register
~, базовый base register
~, буквенный *comp.* letters case
~, буферный buffer register
~, бухгалтерский ledger
~ ввода-вывода input-output register
~, верхний upper case
~ возврата return register
~ временного хранения holding register
~ второго слагаемого addend register
~, всасывающий *tech.* suction damper

~, входной receiving register
~ вызовов/запросов call[ing] register
~, выпускной *tech.* discharge damper
~ данных data register/cartridge
~ данных файла file data register
~ делимого numerator register
~, доступный пользователю user-visible register
~, дымовой *tech.* funnel/smoke damper
~ занятости *comp.* vacant register
~, запоминающий *comp.* storage register
~ знака sign register
~ защиты памяти memory protection register
~ кода операции opcode/op register
~, кодовый code register
~ команд instruction register
~, конвейерный *comp.* pipeline register
~, контрольный check register
~ Ллойда Lloyd's Register *(of shipping)*
~ множимого multiplicand register
~ множителя multiplier register
~ на линиях задержки delay-line register
~ на триггерах flip-flop register
~, накапливающий accumulator register
~ направления передачи данных *comp.* data direction register
~, нижний *print.* lower case
~ обмена exchange/transfer register
~ общего назначения general-purpose register
~, органный *mus.* organ stop
~ остатка remainder register
~ памяти memory register
~ патентов patent register/record
~ первого слагаемого augend register
~ передача данных *comp.* transfer register
~ пересылки данных, выходной destination register
~, последовательно-параллельный serial/parallel register
~ прерываний interrupt register
~ признака *comp.* search register
~ произведения product register
~ пуска по адресу address start register
~ расширения extension register
~ сдвига/сдвигающий/сдвиговый shift register
~ сдвига, кольцевой circulating shift register
~ себестоимости *econ.* cost record
~ синхронизации доступа timed-access register
~ сложения адресов address addend register
~ состояния status register
~ сравнения coincidence register
~ сумматора accumulator register
~, суммирующий adder
~ считывания данных data read register
~ считывания и записи read-write register
~ точки входа entry point register
~, триггерный flip-flop register
~, тумблерный switch register
~ указателя pointer register
~ умножения и деления multiplicand divisor register
~, управляющий control register
~ учета запасов *econ.* stock ledger
~ хранения адресов address holder
~, цифровой *print.* figures case
~ частичной суммы partial sum register
~ частного quotient register
~ числа number register
регистр-защелка latch register
регистратор 1. recorder, recording system/device; *comp.* logger; register; monitor; 2. record keeper, registrar, registering clerk
~ акций *econ.* registrar
~ аналоговых данных analogue data recorder
~ времени задержки/запаздывания delay-time recorder
~ интенсивности пучка beam-monitor
~ использованного времени elapsed-time recorder
~ кода code recorder
~ облигаций registrar
~ потока flux monitor
~, фольговый foil monitor
~ уровня излучения radiation monitor
~, цифровой digital recorder
~, электроннолучевой cathode-ray recorder/display
регистратура registration office; registry
регистрация recording, registration, registering; writing
~ в журнале logging
~, выборочная sample registration
~ данных data recording/logging
~ изображений на экране radar recording
~ машинного времени machine time logging
~, непрерывная continuous recording
~ ошибок *comp.* error logging
~ показания приборов instrumentation recording
~ смерти death record
~ фона background monitoring
~ хода вычислений *comp.* execution logging
регистрировать record, register
регистрироваться register
~ в полиции report to the police; register with the police
регламент 1. regulations, order, rule, standing orders; 2. time limit *(at meetings)*
~, военный articles of war
~, соблюдать observe the time limit, keep within the time limit
~, судебный rules of court, court rules
регламентация, регламентирование regulation
регламентировать regulate; set, establish, fix
регма *bot.* regma
регмаглипты *astr.* rhegmaglypts
реголит *geol.* regolith

регресс regression, retroversion, regress; *law* recourse
регрессия regression
  ~, взвешенная *stat.* weighted regression
  ~, квадратическая *stat.* quadratic regression
  ~, множественная *stat.* multiple regression
  ~ потомков filiar regression
  ~, среднеквадратическая *stat.* mean-square regression
  ~, частная *stat.* net regression
регулирование regulation; control; adjustment, setting, tuning; modulation
  ~, автономное independent [off-line] control
  ~, астатическое floating [non-corresponding] control
  ~ без обратной связи open-loop control
  ~ балансировки trim control
  ~ валютного курса exchange-rate adjustment/control
  ~ величины тяги thrust control/modulation
  ~ влажности humidity/moisture control
  ~ вручную hand control
  ~, грубое coarse/shim control
  ~ движения traffic control
  ~, двухпозиционное on-off control
  ~ денежных выплат по заработкам money-wage adjustment
  ~, дистанционное remote control
  ~ дифферента trim control
  ~ качества quality control
  ~, минимаксное minimax control
  ~, многосвязное multivariate control
  ~ напора воды headwater control
  ~ напряжения voltage control
  ~, непрерывное continuous control
  ~ параметров окружающей среды environmental control
  ~ перепада давления differential pressure control
  ~, плавное smooth/modulating control
  ~ по возмущению disturbance-compensating control
  ~ по замкнутому циклу closed-cycle/loop control
  ~ по знаку sign control
  ~ по интегралу integral control
  ~ по нагрузке load control
  ~ по нескольким параметрам multivariable control
  ~ по (первой) производной derivation/derivative control
  ~ по производной derivative control
  ~ по разомкнутому циклу open-loop control
  ~ по угловой скорости angular rate control
  ~ по ускорению acceleration control, second derivation/derivative control
  ~, позиционное point-to-point control
  ~, правовое legal regulation
  ~, прерывистое discontinuous/intermittent control
  ~, программное program[med] control
  ~, прямое direct control; self-acting control
  ~ разомкнутым контуром open-loop control
  ~ расхода flow-rate control
  ~, релейное discontinuous-type control
  ~ рождаемости birth/family/population/fertility control
  ~ рождаемости, внутрисемейное family planning
  ~ русла реки river correction
  ~ с запаздыванием retarded control
  ~ с обратной связью feedback control
  ~, связанное interacting control
  ~ скорости speed control
  ~ скорости вращения rotational speed control
  ~, следящее servo control
  ~ состава воздуха atmosphere control
  ~ ставок заработной платы wage control
  ~, ступенчатое step/incremental control
  ~, точное/тонкое fine control
  ~ тяги thrust control
  ~ уличного движения street traffic control
  ~, упреждающее anticipatory control
  ~ усиления gain control
  ~ центровки centre-of-gravity control
  ~, цифровое digital control
  ~ шага blade/pitch control
  ~, шаговое step-by-step control
  ~, шиберное damper control
регулировать regulate, control, govern; adjust, set
регулировка regulation, control; adjustment, tuning, setting, correction
  ~ впрыска injection timing
  ~ громкости volume/audio control
  ~, грубая rough/coarse adjustment
  ~, заводская shop adjustment, factory-made adjustment
  ~ зажигания ignition timing/adjustment
  ~ зазора clearance adjustment
  ~ кадровой синхронизации *telev.* vertical hold control
  ~ линейности *telev.* linearity control
  ~ оборотов малого газа idle-speed adjustment
  ~ обратной связи feedback adjustment
  ~ освещения lighting control
  ~ от руки [ручная] hand/manual adjustment
  ~, плавная stepless [infinitely variable] regulation
  ~ по дальности range adjustment
  ~, продольная fore-and-aft adjustment
  ~, стендовая bench alignment
  ~ тембра *rad.* tone control
  ~ тона tone correction
  ~, точная fine adjustment
  ~ усиления gain control
  ~ установки колес wheel alignment, wheel setup
  ~ холостого хода idle speed adjustment
  ~ чувствительности по времени sensitivity time control
  ~ яркости brightness control

регулируемый adjustable, controlled
~ термостатом thermostatically controlled
регулярность regularity
регулировщик traffic controller
регулятор regulator, control(ler); governor, adjuster
~, астатический floating controller; integral action controller
~, барометрический altitude/barometric pressure control
~, баростатический barostat
~ воздушного винта propeller governor
~, временной time-schedule controller
~, всережимный all-speed governor
~ выдержки времени timing controller
~ громкости volume control
~ генератора dynamo/generator governor
~ горения combustion controller
~ давления pressure controller/regulator/governor
~, двухпозиционный two-position [on-off, bang-bang] controller
~, двухрежимный two-speed governor
~ дискретного действия sampled-data [sampling, periodic] controller
~, дроссельный throttle governor
~, задающий master controller
~ замедления deceleration control
~ запуска двигателя engine start controller
~, изодромный proportional-plus-integral controller
~ импульсного действия = дискретного действия
~ качества смеси mixture control
~ косвенного действия indirect/relayed controller
~ максимальных оборотов maximum-speed governor
~, маятниковый pendulum governor
~, мембранный diaphragm governor
~ нагрузки load governor/regulator; *el.* load/charge controller
~ наддува boost control
~ наддува скафандра suit controller
~ напряжения voltage control/regulator
~ непосредственного/прямого действия direct-acting [self-actuated] controller
~ непрерывного действия continuous-action controller
~ непрямого действия indirect-control regulator
~ оборотов r.p.m. governor
~ оборотов двигателя engine speed governor
~ оборотов малого газа idle-speed governor
~, пневматический air-operated [pneumatic] controller
~ подачи воздуха air supply regulator
~ подачи топлива fuel-feed controller
~, подчиненный slave controller
~, поплавковый float-type governor/controller
~ постоянства расхода constant flow regulator
~ прерывистого действия sampled-data controller
~, программный program[med] controller
~ раскрутки overspeed control
~ расхода flow controller/governor/regulator
~, релейный = двухпозиционный
~, реостатный resistance controller/regulator
~ с гибкой обратной связью variable feedback controller
~ с двумя устойчивыми состояниями bistable controller
~ скорости speed regulator/governor/controller
~, статический static [type-O] controller
~, ступенчатый step regulator
~ тембра *rad.* tone control
~ температуры, автоматический thermostat
~ турбокомпрессора turbocompressor/turbosupercharger regulator
~ удара *comp.* typebar pressure regulator, touch regulator
~ усиления gain/amplification control; attenuator
~, форсажный reheat controller
~ холостого хода idle-speed governor
~, центробежный centrifugal control; gravity regulator/governor
~ частоты строк *telev.* horizontal hold control
~ числа оборотов r. p. m. control, speed controller/regulator
~ числа оборотов, предельный overspeed protection governor
~ шага воздушного винта propeller-pitch control unit
~, шаговый step-by-step controller
~, шаровой ball governor
~ яркости *telev.* brightness control; dimmer
регуляция regulation; adjustment; control
~, гормональная hormonal control
~, нервно-мышечная neuromuscular/nervimuscular control
~, нервно-сосудистая neurovascular regulation, neurocirculatory control
регургитация *med.* regurgitation
редактор editor
~, главный editor-in-chief
~, графический/изображений graphics editor
~, диалоговый *comp.* interactive editor
~ связей linker
~, строковый line editor
~, экранный full-screen editor
редан *naut.* planing step
редингот redingote
редис *bot.* garden radish *(Raphanus sativus)*
редкоземельный rare-earth
редколлегия *abbr.* editorial board
редкометаллический, редкометальный rare-metal
редкоцветковый *bot.* rariflorous, sparsely flowered
редокс *chem.* redox, reduction-oxidation
редокс-потенциал redox potential

редокс-система redox system
редрессация *med.* redressment, correction; redressement *(Fr.)*
~, силовая redressement forcé
редуктаза *bioch.* reductase
редуктор 1. *mech.* reduction gear, reducer; reduction gearbox; reducing valve; 2. *chem.* reducing agent, reducer; 3. pressure regulator/reducer
~, ацетиленовый *weld.* acetylene pressure regulator
~, выносный remotely-mounted reduction gear
~, газовый *weld.* gass pressure regulator
~ давления pressure-reducing valve, pressure reducer
~, двухступенчатый double-reduction gear
~, зубчатый reduction gear
~, кислородный oxygen pressure regulator/reducer
~, одноступенчатый single-reduction gear
~, планетарный planetary/epicyclic reduction gearbox
~, рамповый manifold/central pressure regulator
~, реверсивный reverse-reduction gear
~ с цилиндрическими шестернями spur reduction gearbox/gearing
~, угловой angled gearbox
редукция reduction
редуцировать reduce
рéдька *bot.* radish *(Raphanus)*
редЮит *mil.* redoubt, reduit *(Fr.)*
реек *naut.* lug yard
реестр register, list, roll, roster; registry
~, судовой register book
режим condition[s], mode, duty; routine; procedure; performance, working/operating conditions; rate; *med.* regimen; discipline; *polit.* regime
~, аварийный emergency conditions/operation
~, автономный off-line mode/conditions/operation
~ автоматической стабилизации autostabilizer mode
~ авторотации *aero.* windmilling/autorotation regime
~ адресации *comp.* addressing mode
~, базисный (power) base load operation
~ безопасности safety procedures
~, буферный floating service
~ быстрых электронов *telev.* high-velocity-beam operation, high-velocity scanning
~ ввода с клавиатуры keyboard mode
~ ветрянки *aero.* windmilling regime
~ висения *aero.* hover mode, hovering
~, вихревой eddy flow
~, водный water regime, hydrolicity
~ воспроизведения playback mode
~, временной time behaviour
~ вставок insert/insertion mode
~, вынужденный forced conditions
~ горения, детонационный knocking combustion
~, графический graphics mode
~ движения жидкости, напорный forced flow
~, двухсменный two-shift work schedule
~ демпферования damping regime
~, диалоговый *comp.* conversational mode
~, диетический/пищевой diet, dietary regimen
~, динамический dynamic conditions
~, длительный long-term regime
~ дозаписи *comp.* append mode
~, докритический subcritical regime
~ доступа access mode
~ доступа, страничный *comp.* page mode
~ дублирования duplicate mode
~ заглавных букв capitals mode
~ загрузки load mode
~ задания начальных условий *comp.* reset mode
~, закритический supercritical regime
~ записи recording mode
~ заполнения форм *comp.* forms mode
~, запросно-ответный *comp.* query mode
~ запуска start-up conditions, starting regime, wakeup mode
~, заторможенный decelerated mode
~ захвата цикла cycle-steal mode
~ захода на посадку *av.* approach regime
~ защищенного символа *comp.* protected character mode
~, знаковый *comp.* character mode
~, импульсный pulse mode, pulser operation conditions
~ использования, монопольный exclusive usage mode
~, испытательный test run/regime
~, колебательный oscillatory regime
~, командный *comp.* command mode
~ конвейерной обработки pipeline mode
~, консультативный *comp.* help mode
~ контроля verification mode
~, кормовой feeding schedule
~, крейсерский cruise conditions/mode/regime
~, критический critical conditions/regime, criticality
~, ламинарный laninar-flow regime, laminar/streamline conditions
~ локализаuии locate mode
~, льготный налоговый preferential tax treatment
~ максимальной нагрузки peak-load conditions
~ малого газа idling conditions
~, международно-правовой international legal regime/status
~, многозадачный multitask mode
~, многопоточный *comp.* multithreading
~ молчания no-signal condition/state
~, мультиплексный multiplex mode
~ наведения guidance mode
~ нагрузки under-load conditions

~, надкритический supercriticality
~ наибольшего благоприятствования most favoured nation treatment
~, наивыгоднейший optimum conditions
~, напряженный heavy duty
~, нарушенный perturbed regime
~, неравномерный non -equilibrium conditions
~, нерасчетный off-design conditions
~, неустановившийся unsteady regime
~, неустойчивый unstable regime
~, номинальный rating; design condition(s), nominal rating conditions
~ нулевой тяги zero-thrust conditions
~ обучения *comp.* training mode
~ обработки данных data mode
~ огня *mil.* rate of fire
~ ожидания *comp.* waiting mode; *av.* holding pattern
~, оперативный *comp.* on-line mode/operation
~, оптимальный optimum regime/conditions
~ остановки shutdown conditions
~ отладки *comp.* debugging mode
~, пакетный batch mode
~ парашютирования *av.* pancaking regime
~, переходный transitional regime, transient conditions/state
~, периодический periodic duty/pattern
~, питьевой *med.* water discipline
~, пищевой *med.* diet, dietary regimen
~ планирования *aero.* glide conditions
~, подкритический subcritical conditions, subcriticality
~ подстановки substitute mode
~, подчиненный *comp.* slave mode
~ поиска search mode
~ покоя quiescent conditions
~, полиэкранный split-screen mode
~, пользовательский user(-operating) mode
~ попеременного доступа *comp.* alternate mode
~, постельный *med.* bed rest
~ потока flow conditions/pattern/regime
~, пошаговый single-step mode
~, правовой legal regime
~, прерывистый intermittent regime
~, преференциальный *law* preferential treatment
~, привилегированный privileged mode
~ приема *rad.* receive condition/pattern/mode
~, приоритетный foreground mode
~ прогона run mode
~ прогрева warm-up (mode)
~ продувки blow-down (mode)
~, прозрачный transparent mode
~ прокатки *med.* rolling schedule
~ простоя unoperable mode
~, пультовый console mode
~ работы operating/working conditions; mode/type of operation
~ работы магнитной ленты magnetic tape mode
~ работы, номинальный rated duty
~ работы, параллельный concurrent mode
~ работы, переменный varying duty
~ работы, периодический periodic duty
~ работы при постоянной нагрузке steady-load conditions
~ работы, устойчивый steady operation
~ работы, холостой no-load operation
~, рабочий operating mode/conditions, operational regime, performance, duty
~, равновесный equilibrium conditions
~ разделения времени time-sharing
~, расчетный design conditions
~ реального времени real-time mode
~, релейный on-off mode
~ решения задач problem mode
~, ручной manual mode
~ с резервированием standby mode
~ самовращения *aero.* windmilling/autorotation regime
~ самонаведения homing mode
~, санаторный sanatorium regimen
~, сверхкритический supercritical conditions, supercriticality
~ свободного доступа free running mode
~ слежения/сопровождения tracking mode
~ смазки relubrication intervals
~, собственный native mode
~ стабилизации hold/stabilization mode
~, стартстопный start-stop mode
~ стока regime of run-off
~ струи jet behaviour
~, температурный temperature/heat conditions; temperature rate
~, тепловой thermal conditions
~ технического обслуживания maintenance conditions
~ течения flow regime, flow
~, типовой standard conditions
~ торможения braking regime
~, трудовой work routine
~, турбулентный turbulent flow regime; turbulent operating conditions
~, тяжелый heavy duty
~ управления control mode
~, установившийся steady-state operating conditions, steady state/regime
~, фоновый background mode
~ холостого хода idle mode, no-load conditions
~, циклическмй repetitive mode
~, широковещательный broadcast mode
~, эксплуатационный operational/operating/working conditions

режиссер *theatre* producer; stage director; *cin.* director

рез cut; kerf

резак 1. cutting jaws; cutter; 2. *weld.* cutting torch; 3. *agr.* coulter
~, ацетилено-кислородный oxy-acetylene

cutter
~, **газовый** gas cutter, gas-cutting torch
~, **двухпламенный/двухфакельный** two-nozzle cutter
~ **для пилюль** pill cutter
~ **для фотобумаги** print trimmer
~, **зачистной** deseaming/scarfing torch
~, **подводный** underwater gas cutter
резание cutting; slicing
~, **продольное** slitting
~, **ударное** impact cutting
резать cut, slit; incise
резеда *bot.* mignonette *(Reseda)*
резекция *med.* resection; exsection
~ **десны** gum resection, gingivectomy
~ **легкого** pneumonectomy
~ **мочевого пузыря** cystectomy
~ **мочеиспускательного канала** urethrectomy
~ **носовых раковин** conchotomy
~, **окончатая** window resection
~ **почки** nephrectomy
~ **предстательной железы** prostatectomy
резен-киль *naut.* hog piece, upper false keel
резен-линия *naut.* rising line
резерв reserve; stockpile; standby, backup; redundancy
~, **валютный** exchange/currency reserve
~ **времени** time slack
~ **для покрытия эксплуатационных издержек** provision for maintenance
~ **для уплаты налога** provision for tax
~, **золотой** gold holding
~, **ликвидный [ликвидных активов]** liquidity reserve
~, **наличноденежный (банка)** counter/till cash
~ **на амортизацию** depreciation reserve
~, **неамортизированный** qualifying reserve
~ **по переоценке активов** *acc.* valuation reserve
~ **предусмотрительности** contingency reserve
~ **с логической схемой включения** logic-gate redundancy
~ **страховых взносов** policy reserve fund
~ **установленный законом** statutory reserve
~**ы, трудовые** manpower reserves
резерват 1. game reserve; 2. reservation
резервация 1. reservation; 2. game reserve
резервирование 1. redundancy; backup; 2. changeover to stand-by facility; 3. reservation
~, **аварийное** standby redundancy
~, **аппаратное** *comp.* hardware redundancy
~, **встроенное** built-in redundancy
~, **групповое** subassemblies redundancy
~ **замещением** replacement/standby/active redundancy
~ **мест** seat/space reservation
~, **многократное** reiterative/multiple redundancy
~, **модульное** modular redundancy
~ **на уровне устройств** device redundancy
~, **однократное** single redundancy
~, **постоянное** passive redundancy
~ **путем дублирования** standby redundancy
~, **тройное** triple redundancy
резервировать 1. reserve; 2. provide duplicate/dual/standby facilities
резервный reserve, redundant, spare, standby, backup *(attr.)*
~**ные технические средства** standby facilities
резервуар reservoir, tank, vessel, hopper, cistern, container, receptacle
~, **вакуумный** evacuated vessel
~, **водонапорный** pressure/head/elevated water reservoir/tank
~ **воды** water reservoir; *geol.* water body
~ **высокого давления** pressure vessel, high-pressure tank
~ **для масла** oil tank
~, **нижний** sump
~, **отстойный** settling tank
~, **питательный** feed tank
~, **складской** storage reservoir
~ **со сжатым воздухом** air-pressurized tank
~, **уравнительный** surge tank
резерпин *pharm.* reserpine, rauloydin, raurine, sandril, serpasil, reserpoid, serpate
резерфорд rutherford, Rd, rd
резец cutter, cutting tool; incisor
~, **алмазный** diamond tool
~, **буровой** boring bit
~, **гравировальный** engraving stylus/tool
~, **долбежный** slotting tool
~, **изогнутый** gooseneck tool
~, **канавочный** grooving/recessing tool
~, **левый** left-cut tool
~, **плужный** coulter
~, **правый** right-cut tool
~, **прорезной/пазовый** grooving/slotting tool
~, **радиусный** round-nose tool
~, **расточный** boring tool/cutter
~, **резьбонарезной** threading tool/cutter
~, **строгальный** planer/planing tool
~, **твердосплавный** carbide-tipped tool
~, **токарный** turning/lathe tool; turner's chisel
~, **фасонный** shaping/forming tool
~, **фрезерный** milling cutter
~, **чашечный** cup-tip tool
~, **черновой** roughing tool
~, **чистовой** finishing tool
резидент resident
резидентный resident
~**ная часть памяти** *comp.* memory-resident part
~**ное устройство системы** *comp.* system residence unit/device
резина rubber
~, **амортизационная** shock-absorbing rubber
~, **армированная тканью** fabric-reinforced rubber
~, **беговая** tread rubber

~, **бензостойкая** petrol-/gasoline-resistant rubber
~, **бессажевая** pure-gum rubber
~, **брекерная** cushion rubber
~, **вулканизированная** cured/vulcanized rubber
~, **губчатая** sponge/porous/expanded rubber
~, **керосиностойкая** kerosene-resistant rubber
~, **листовая** sheet rubber
~, **маслостойкая** oil-resistant rubber
~, **микропористая** microporous rubber
~, **пенистая** air-foam rubber
~, **пористая** porous/cellular rubber
~, **роговая** hard rubber, ebonite
~, **саженаполненная** carbon-black rubber
~, **силиконовая** silicone rubber
~, **электропроводящая** conductive rubber
резистентность resistance
резистор resistor
~, **вилитовый** vylite resistor
~, **гасящнй** damping/quenching resistor
~, **герметизированный** encapsulated resistor
~, **керметовый/металлокерамический** cermet resistor
~, **кремниевый** silicon resistor
~ **микросхемы/микроминиатюрный** resistor microelement
~, **объемный** bulk resistor
~, **печатный** printed resistor
~, **пленочный** film resistor
~, **проволочный** wire-wound resistor
~, **тиритовый** thyrite resistor
~, **углеродистый** carbon resistor
резка cutting; shearing
~, **ацетилено-кислородная** (oxy-) acetylene cutting
~ **в защитном газе** gas-shielded cutting
~, **воздушно-дуговая** air-arc cutting
~, **газовая** gas/autogenous cutting
~, **газодуговая** gas-arc cutting
~, **газопламенная** torch/flame/oxygen cutting
~, **дуговая** (electric) arc cutting
~, **кислородная** oxygen/flame cutting
~, **кислородно-дуговая** oxygen-arc[oxy-arc, oxy-electric] cutting
~ **кислородным копьем** oxygen-lance cutting
~, **контактная электрическая** resistance cutting
~ **ножницами** shearing
~, **плазменно-дуговая** plasma arc cutting
~, **подводная** underwater cutting
~, **продольная** slitting
~, **профильная** profiling, profile cutting
~ **угольной дугой** carbon-arc cutting
~, **электродуговая** arc cutting
~, **электроискровая** spark cutting
резкость изображения picture sharpness
резнатрон *el.* resnatron
резол resol, a-stage resin
резольвента resolvent; resolvent kernel
~ **интегрального уравнения** resolvent (kernel) of an integral equation
резольвометр resolution meter
резонанс resonance
~ **в атомном пучке, магнитный** atomic-beam magnetic resonance
~, **двугорбый** double-humped resonance
~, **земной** ground resonance
~, **индиевый** indium resonance
~, **квадрупольный** quadrupole resonance
~, **ложный** spurious resonance
~ **напряжений** voltage resonance
~, **неразрешенный** unresolved resonance
~, **нижний** low-lying resonance
~, **основной** main/major resonance
~, **паразитный** spurious resonance
~, **параллельный** parallel/current resonance, antiresonance
~, **последовательный** series resonance
~ **реакции деления** fission resonance
~, **собственный** natural resonance
~, **спиновый** spin resonance
~, **ядерный магнитный** nuclear magnetic resonance, NMR
резонансный resonant, resonance *(attr.)*
резонирующий resonant
резонатор resonator; cavity
~, **атомнолучевой** atomic-beam resonator
~ **бегущей волны** travelling-wave resonator
~, **волноводный** waveguide resonator
~, **входной** input resonator
~, **высокочастотный** radio-frequency cavity
~, **выходной** output resonator, catcher
~ **Гельмгольца** Helmholtz resonator
~, **кварцевый** quartz crystal vibrator, quartz resonator, crystal vibrator
~, **кольцевой** ring vibrator
~ **ленточного типа** strip-type resonator
~, **нагруженный ферритом** ferrite-loaded resonant cavity
~, **объемный** cavity resonator, resonant cavity
~, **проходной** reentrant/transmission resonator
~, **сложный** compound resonator
~ **ускорителя** accelerator cavity
~, **эталонный** reference cavity
резонировать resonate
резонирующий resonating, resonant
резорбция resorption; re-absorption
~ **десны** gingival resorption/recession
~ **корня зуба** root resorption
резорцин resorcinol
результант *math.* resultant
результат result; outcome
~, **конечный** end result
~ **моделирования** simulated result
~, **численный** numerical result
результативность efficiency
результирующая *math.* resultant
рззультирующий resulting, resultant
резус 1. *bioch.* Rhesus factor, Rh, rh, Rh-Hr; 2. *zool.* rhesus monkey *(Acaca mulatta)*
резус-агглютиноген Rh-Hr agglutinogen

резус-антиген Rh-Hr antigen
резус-антисыворотка Rh-Hr antiserum
резус-антитело Rh antibody, Hr antibody
резус-гаптен Rh hapten
резус-ген, Rh gen, rh gen, Hr gen, hr gen
резус-несовместимый Rh-incompatible
резус-отрицательный Rh-negative
резус-положительный Rh-positive
резус-реакция Rh testing/typing, Rh-Hr typing
резус-система Rh-Hr system
резус-фактор Rh factor, Rhesus factor, Rh-Hr factor, Hr-factor, hr-factor
резуха *bot.* rock cress *(Arabis)*
резцедержатель toolholder
резьба 1. thread; threading; 2. carving, engraving
  ~, винтовая (screw) thread
  ~ Витворта/дюймовая Whitworth/inch thread
  ~, внешняя external/male thread
  ~, внутренняя internal/female thread
  ~, газовая gas thread
  ~, двухзаходная double-start thread
  ~, коническая taper thread
  ~, крупная coarse thread
  ~, левая left-hand(ed) thread
  ~, мелкая fine thread
  ~, механическая machine-cut thread
  ~, многозаходная multiple-start thread
  ~, наружная external/male thread
  ~, неполная flattened/imperfect/truncated thread
  ~, однозаходная single-start thread
  ~ по дереву wood carving
  ~, правая right-hand(ed) thread
  ~, прямоугольная square thread
  ~, трапецеидальная trapezoidal thread
  ~, трехзаходная triple-start thread
  ~, упорная buttress thread
  ~, цилиндрическая parallel/cylindrical/straight thread
  ~, червячная worm thread
резьбонарезной thread-cutting
резьбофрезерный thread-milling
резьбошлифовальный thread-grinding
реимпорт reimportation
реимплантация reimplantation, replantation
реин *chem.* rheine
рей *naut.* yard
  ~ для паруса sailyard
рейд *naut.* roads, roadstead
  ~, внешний outer roads
  ~, внутренний inner roads
  ~, выходить на lay out
  ~, закрытый closed roads
  ~, защищенный sheltered/safe roads
  ~, незащищенный unsafe roads
  ~е, стоять на lie out
рейзистор raysistor
рейка batten, ribband; lath; edging; rod, stick
  ~, бордюрная cant strip
  ~, визирная sighting batten
  ~, водомерная staff/depth gauge, water-level staff; tide staff
  ~, дальномерная/геодезическая stadia/telemeter/range rod
  ~, зубчатая gear/toothed rack
  ~, мерная measuring batten/staff, surveyor's rod; *naut.* sounding stick
  ~, нивелирная levelling/target rod
  ~, пазовая seam batten
  ~, промерная gauging/graduated rod
  ~, снегомерная snow stake
  ~, стыковая joint strip
  ~, уплотняющая packing strip
  ~, уровенная *naut.* water gauge
реймский Rhemish
рейнвейн Rhine wine
рейнский Rhenish, Rhine *(attr.)*
рейс 1. *naut.* voyage, trip; passage; 2. *av.* flight; 3. *auto.* run, trip; 4. *rail.* run
  ~ в один конец single trip
  ~ вне расписания non-scheduled flight
  ~, грузовой *av.* cargo/freighting flight
  ~, грузо-пассажирский freight-passenger flight
  ~, заграничный foreign/outward voyage
  ~, каботажный *naut.* coastal voyage
  ~, космический space trip
  ~, обратный homeward/return voyage, home passage; *av.* return flight; *rail.* homeward/return run; *auto.* return/homeward trip
  ~, первый *naut.* mayden trip/voyage
  ~ по расписанию scheduled flight
  ~, почтовый mail flight
  ~, пробный trial trip/ride/flight
  ~ реактивного самолета jet flight
  ~, регулярный scheduled flight/run/service
  ~, транзитный through flight/run/passage
  ~, чартерный charter flight
  ~ туда и обратно round trip
  ~ы грузовых судов cargo boat service
  ~ы, ежедневные daily service
  ~ы, обратные homeward trades/voyages/ runs
  ~ы судов груженных углем coal trades
рейсмас, рейсмус surface/marking/shifting gauge
рейсовый scheduled
рейстрек race track
рейсфедер drawing/ruling pen; pencil holder
рейсшина T-square, tee-square
рейтер rider, jockey weight
рейх Reich
  ~, третий Third Reich
рейхсканцлер Chancellor of the Reich
рейхсрат *hist.* Reichsrat
рейхстаг Reichstag
река river; stream
  ~, вытекающая effluent river
  ~ забвения Lethe
  ~, затопленная drowned river
  ~, зрелая mature river
  ~, исчезающая intermittent river/stream
  ~, каменная stone/rock river

~, консеквентная consequent/accordant river
~, меридиональная longitudinal river
~, молодая juvenile river
~, несудоходная unnavigable river
~, первичная/первоначальная original river
~, перехваченная captured river
~, подземная subterranean river
~, подледная/подледниковая subglacial river
~, подчиненная tributary
~, приливо-отливная tidal river
~, судоходная navigable river
~, умирающая drying-up river
~, утопленная drowned river
реквием *mus., rel.* requiem
реквизировать *mil.* commandeer, requisition
реквизит *theatre* properties, props; *law* essential element; data
реклама advertisement; advertising; publicity; promotion
~, газетно-журнальная media advertising
~ действующая на подсознание subliminal advertising
~, повторная repeated/reminder advertising
~, потребительская consumer advertising
~, световая electrical/illuminated advertising
рекламация claim, complaint; protest; unsatisfactory equipment claim/report, UER
рекламировать 1. send an UER, make a claim; 2. advertise; publicise
рекламный advertising, publicity *(attr.),* promotional
рековедение potamology
рекогносцировать *mil.* reconnoitre, reconnoiter
рекогносцировка reconnaissance
реконвалесцент convalescent
реконвалесценция convalescence, recuperation, recovery
реконверсия civilian conversion
реконсервация reprocessing for storage
реконструировать 1. redesign; update, modernize; 2. reconstruct, rebuild
реконструкция 1. redesign, remodelling; reorganization; rearrangement; 2. reconstruction
рекредитив *dipl.* letter of recall
рексия *bot.* meadow beauty *(Rhexia)*
ректально-брюшной rectoabdominal
ректально-влагалищный rectovaginal
ректит *med.* rectitis, proctitis
ректификат rectified spirit/alcohol, rectificate
ректификатор rectifier
ректификация rectification; (fractional) distillation, fractionation; purification
ректифицировать rectify; fractionate, distil(l) by fractionation
ректоколит *med.* rectocolitis, proctocolitis
ректопексия *med.* rectopexy, proctopexy
ректор (university) president, rector
ректорат university administration; (university) president's office; rectorate
ректорство rectorship, rectorate
ректосигмоид, ректосигмоидальный rectosigmoid
ректосигмоидэктомия *surg.* proctosigmoidectomy
ректоскоп rectoscope, proctoscope
ректоскопия rectoscopy, proctoscopy
ректоцеле rectocele, proctocele
ректэктомия *surg.* rectectomy, proctectomy
рекуперативный recuperative, regenerative; recovery *(attr.)*
рекуперация recuperation, recovery
рекуперировать recuperate, recover; restore, regenerate
рекурсивно-перечислимый recursively-enumerable
рекурсия *phon.* final glide, off-glide
релаксант *med.* relaxant
~, периферический peripheral relaxant
релаксация relaxation
~ диафрагмы *med.* (o)esophaglal relaxation
~ напряжений stress relief/relaxation
~, перекрестная cross relaxation
~, спин-решеточная spin-lattice relaxation
~, ядерно-магнитная nuclear-magnetic relaxation
реле relay
~ активного сопротивления resistance relay
~ активной мощности active-power relay
~, барометрическое barometric switch, baroswitch
~ бдительности *rail.* acknowledging relay
~ без замедления instantaneous relay
~, бесконтактное noncontact/static relay;
~, бета-электронное beta-electronic relay
~, блокировочное interlock/hold/safety relay
~, быстродействующее high-speed [quick-acting, fast-operate] relay
~, быстроотпускающее quick-release relay
~, биметаллическое bimetallic/strip relay
~, включающее cut-in relay
~ времени time relay/switch, timer
~ выдержки времени time/timing/delay relay
~, выключающее shut-down relay; tripping relay
~ гашения поля field-decelerating relay
~, главное master relay
~ давления pressure-sensitive relay
~, двухъякорное double armature relay
~ задержки delay relay
~, замедленное delayed/slug relay
~, замедляющее delaying [time-lag] relay
~ защиты/защитное protective relay
~, импульсное impulsing relay
~, ионное ionic relay
~, исполнительное control/actuator relay
~, клопферное *tel.* sounder relay
~ контроля supervisory relay
~, коромысловое beam relay
~ короткого замыкания short-circuit relay

~ максимального напряжения overvoltage relay
~ максимального тока over-current relay
~, максимальное maximum relay; overload/overcurrent/overvoltage relay
~, максимально-минимальное over-and-under relay
~ мгновенного действия instantaneous relay
~, минимальное undervoltage/undercurrent relay
~ напряжения voltage relay
~ номерника annunciator relay
~ обрыва line/wire break relay
~ обрыва поля field-failure relay
~ освещенности lighting control relay
~ перегрузки overload relay
~ переменного тока a. c. relay
~ перемещения distance relay
~, плавкое fuse relay
~ повреждения failure/fault relay
~ пожарной сигнализации fire-detector switch
~ полного сопротивления impedance relay
~, полупроводниковое semiconductor relay; solid-state relay
~, поплавковое float switch
~ постоянного тока d.c. relay
~ производной derived relay
~, промежуточное slave relay; pilot relay
~, пусковое starting relay
~, путевое track relay
~, размыкающее cut-off relay
~ разряжения vacuum relay
~ расхода flow-rate relay
~, расширительное thermal expansion relay
~ реактивного сопротивления reactance relay
~ резонанса resonance relay
~, резонансное tuned relay
~ с выдержкой времени time-lag/delay relay
~ с защелкой latching [lock-up] relay
~ с ручным возвратом hand-reset relay
~ с самоблокировкой latching [lock-up] relay
~ с самовозвратом self-reset[ting] relay
~ с торможением biased relay
~ сигнализации annunciator relay
~, сигнальное alarm/signal relay
~, силовое power relay
~ силы force relay
~ скорости speed-sensitive relay
~ скорости течения flow velocity relay
~, тепловое thermal relay/switch
~ течения flow relay
~ управления control relay
~ ускорения accelerating relay
~, ускоренное accelerated relay
~, шаговое stepping relay
~, электротепловое electrothermal relay
~, язычковое reed relay
реле-счетчик meter/counting relay
рельеф relief; topography; configuration; *geogr.* terrain
~ берега coastal features
~, волнистый undulating relief; rolling topography
~, высокогорный alpine relief
~ дна bottom contour/configuration
~ местности lay of land, configuration of terrain; natural features
~, моренный morainic/morainal topography
~ морского дна submarine relief/topography
~, невыраженный faint relief
~, неровный rugged topography
~, низкий low relief; bas-relief
~, обращенный obsequent relief
~, палимпсестовый inherited relief
~, пересеченный broken ground
~, сглаженный subdued relief
~, спокойный smooth relief
~, техногенный man-made [technogenic] relief
~, топографический topography
~, увалистый ridgy [spur-like] topography
рельефность relief
рельс rail
~, бесстыковой сваркой welded rail
~, контактный third rail
~, направляющий guide rail
~, пошерстный trailing rail
~, противошерстный facing rail
~, Т-образный T-rail, tee rail
~, ходовый running/track rail
рельсопрокатный rail-rolling
~ завод rail mill
рельсоукладчик track-laying machine
реляксатор relaxation oscillator
релятивизм *philos.* relativity, relativism
релятивистический, релятивистичный, релятивистский relativistic
релятивно-атрибутивный *gram.* relative-attributive
релятивно-предикативный *gram.* relative-predicative
релятивность relativity
рема *ling.* rheme
рембрандтовский Rembrandtesque
реманентность remanence
ремедиум tolerance of mint, remedy; remedy allowance
ремез *zool.* penduline titmouse *(Remis pendulinis)*
ременный belt *(attr.)*
~ привод belt drive
ремень belt; strap; *see also* **ремни**
~ безопасности *auto.* safety belt; seat belt
~ для правки бритвы razor strop
~, зарядный charging belt
~, клиновидный V-belt, vee belt
~, кордтканевый multicord belt
~, кордшнуровый single-cord belt
~, натяжной retractable belt
~, плечевой shoulder strap
~, плоский flat drive belt

~, поясной waist belt/strap
~, приводной transmission/drive belt
~, привязной seat/safety belt
~, страховочный safety strap
ремень-рыба *zool.* oarfish *(Regalecus glesne)*
ремерия *bot.* Asia poppy *(Roemeria)*
ремесленник craftsman; artisan, handicraft worker
ремесло handicraft; trade; craft; occupation
ремешок (small) strap
ремиз, ремиза, ремизка *text.* harness; heald, heddle; shaft, leaf
~, кромочная skeleton heald
ремиссия *med.* remission, remittence; *econ.* remittance
ремитент *econ.* remitter, remittee, payee
ремнелепестник *bot.* lizard orchis *(Himanthoglossum)*
ремнелистный *bot.* strap-leaved, lorifolious
ремнеткачество belt/strap weaving
ремнецветник *bot.* loranth *(Loranthus)*
ремни belts, straps, harness; *see also* **ремень**
~, затягивать привязные fasten the seat belts
~, привязные safety harness; seat belts
ремонт I repair[s], overhaul(ing); maintenance
~, аварийный emergency repair(s)
~, агрегатный unit repair, repair by replacement of units
~ в полевых условиях field repair[s]
~, внеплановый off-schedule [unscheduled] repair/maintenance
~, восстановительный restoring repair, reconditioning
~, гарантийный warranty repair
~ двигателя, капитальный engine overhaul
~, капитальный overhaul, major repair/overhaul
~, косметический face lifting
~, мелкий minor repair(s)
~, планово-предупредительный scheduled preventive maintenance
~, плановый scheduled repair[s]
~, профилактический preventive repair[s]
~, средний medium repair
~, текущий current/routine/running/regular repair/maintenance
~, ямочный patching *(of road surface)*
ремонт II *agr.* restocking, replenishment, replacement; *mil.* remount
ремонтантный *bot.* flowering again, remontant, everbearing
ремонтировать I repair, overhaul; refit; maintain
ремонтировать II *mil., agr.* remount; restock, replenish
ремонтируемость repairability
ремонтно-механический mechanical-repair
ремонтный I repair *(attr.),* maintenance *(attr.)*
~ная летучка emergency repair truck/vehicle
~ная мастерская repair shcp
~ные работы repairs
ремонтный II *mil.* remount; *agr.* for breeding/replacement
~ное поголовье replacement stock
ремонтопригодность maintainability, serviceability
ремулад *cul.* remo[u]lade
ренальный *anat.* renal
Ренессанс the Renaissance
ренет queen apple, renet
рений *chem.* rhenium, Rh
ренин *bioch.* renin
ренклод *bot.* Reine Claude, greengage *(Prunus italica)*
реннин *bioch.* rennin, rennet, pexin, cymosin
ренниноген *bioch.* renninogen, rennogen, prorennin, prochymosin, pexinogen
ренопатия *med.* renopathy, nephropathy
ренопривный *med.* renoprival
рено-ренальный *med.* renorenal
рента rent
~, выплачиваемая current annuity
~, государственная пожизненная government annuity
~, ежегодная annuity
~, земельная land/ground rent
~, натуральная rent in kind
~, отсроченная deferred annuity
~, пожизненная life annuity
~ продуктами natural rent
~, условноначисленная imputed rent
рентабельность cost effectiveness; profitability, profitableness
рентабельный cost-effective; profitable, profit-making, paying
рентген 1. *phys.* roentgen, R; 2. X-ray apparatus/unit
рентгенизация X-raying
рентгенизировать X-ray
рентгенметр roentgenometer, radiation meter, R-meter
рентгеноанализ X-ray examination
рентгеноанатомия radioanatomy
рентгеноастрономия X-ray astronomy
рентгенов Roentgen *(attr.)*
~ы лучи X-rays
рентгеновский Roentgen *(attr.),* X-ray *(attr.)*
~ кабинет X-ray room
рентгенограмма roentgenogram[me], X-ray photograph/pattern
рентгенограф roentgenograph, X-ray photograph
рентгенография roentgenography, radiography, X-ray photography
~, импульсная impulse roentgenography
~, контактная close-focus [contact] roentgenography
~ плода f[o]etography, embryography
~, послойная sectional roentgenography, tomography
рентгенодефектоскопия X-ray flaw detection, radiomateriology, X-ray testing/inspection
рентгенодиагностика X-ray diagnosis, radiodiagnosis, diagnostic roentgenology

рентгенокимография roentgenkymography
~, однощелевая monofissure roentgenkymography
рентгеноконтрастный radiopaque
рентгенолог radiologist, roentgenologist
рентгенология radiology, roentgenology
рентгенометр roentgenometer
рентгенометрия roentgenometry, X-ray dosimetry
рентгенопроницаемый radiotransparent
рентгенопросвечивание roentgenoscopy, fluoroscopy
рентгеносвечение radioluminescence
рентгеноскоп roentgenoscope, fluoroscope
рентгеноскопия roentgenoscopy, fluoroscopy; X-ray inspection
рентгеноснимок roentgenogram[me], X-ray photograph
рентгеноспектроскопия X-ray spectroscopy
рентгеностереопельвиметрия stereoradiopelvimetry
рентгеносъемка X-ray photography
рентгенотерапевт radiotherapist
рентгенотерапия roentgenotherapy, X-ray therapy, roentgenization, radiation therapy
~, близкофокусная close-focus roentgenotherapy
~, дистанционная long-focus roentgenotherapy
рентгенотехник X-ray technician, radiographer
рентгенотехника X-ray technology; radiology
рентгенофлюорография fluororontgenography, photofluorography
рентгенофотография roentgenophotography
рентгеночувствительный radiosensitive
рентген-эквивалент roentgen-equivalent
~, биологический man/medical roentgen-equivalent
~, физический physical roentgen-equivalent
реобаза rheobase
реовирус reovirus
реограмма rheogram
реограф rheograph
реография rheography
~ под давлением pressure rheography
реокардиография rheocardiography
реокисление reoxidation
реология rheology
реометр rheometer, flow-rate meter, flowmeter
реоморфизм rheomorphism
Реомюр Reaumur
реопирометр resistance pyrometer
реопланктон rheoplankton, potamoplankton
реоплетизмография impedance plethysmography
реоскоп rheoscope, current detector
реостат rheostat
~ возбуждения field rheostat
~ дальности range/distance rheostat
~, пусковой starter rheostat
~, угольный carbon resistor
реостатный rheostatic
реострикция pinch effect, rheostriction (effect)
реотаксис rheotaxis
реотан rheotan; resistor wire
реотахиграфия rheotachygraphy
реотропизм rheotropism
реофит *bot.* rheophyte
реохорд[а] *el.* slide-wire, rheostat with slide
реоэнцефалография rheoencephalography
репа *bot.* turnip *(Brassica rapa)*
репак *bot.* bird rape *(Brassica campestris)*
репарации *pl.* reparations, compensation, damages
~, военные war reparations
репатриант repatriate
репатрировать repatriate
репей *bot.* bur *(Xantium)*
репейник *bot.* 1. great burdock *(Arctiumn lappa);* 2. bur, flower head of burdock
репейница *zool.* painted lady *(Pyrameis cardui)*
репейничек *bot.* agrimony *(Agrimonia)*
репеллент repellent, insectifuge
репер reference point, bench/reference/datum mark
~, вековой fundamental bench mark
~, вспомогательный/промежуточный intermediate bench mark
~, потайной secret bench mark
~, сборочный *mach.* location mark
~ чертежа datum line
реперфоратор reproducer, reproducing punch
репертуар repertoire, repertory
репетир repeater
репетирование *theatre* rehearsal
репетировать *theatre* rehearse
репетитор 1. coach, tutor; reader; 2. rehearser
репетиция 1. rehearsal; tryout; 2. repeater mechanism
~, генеральная dress rehearsal
репешок *bot.* agrimony *(Agrimonia)*
репитер repeater; repeating/repeater unit
~ гидролокатора asdic/sonar repeater
~ гирокомпаса repeating (gyro) compas, gyro repeater
~ компаса repeater compass
~ устройства наблюдения search sensor repeater
реплантация *med.* reimplantation, replantation; *bot.* replantation, replanting
реплика 1. *opt.* replica; 2. remark, retort; *theatre* cue
реплицировать *theatre* give the cue
реплум *bot.* replum
репник *bot.* rapistrum *(Rapistrum)*
репница *zool.* cabbage butterfly *(Pieris rapae)*
реповидный, репообразный turnip-shaped. napiform
реполов *zool.* linnet *(Carduelis cannabina)*
репортер correspondent; reporter
репрезентативность *stat.* representation, representativeness
репрессалия reprisal; retaliation
репрессировать submit to respression
репродуктор 1. *comp.* reproducing puncher, repro-

ducer; 2. *el.* loudspeaker
~, **перфорационный** reproducing puncher
репродуцировать reproduce; copy
репс *text.* rep, repp, reps
репшнур *sport* reepschnur, rope
репяшок *bot.* argimony *(Agrimonia)*
ресивер receiver; plenum/pressure chamber
ресинтез resynthesis
реслинг *sport* wrestling
ресница *bot.* cilium; *anat.* eyelash
ресничатый ciliate, ciliary
ресничка *bot., physiol.* cilium (*pl.* cilia), flagellum (*pl.* flagella)
реснички *pl.* cilia
~, **двигательные** kinetocilia
~, **клеточные** cellular cilia
~, **соединительные** desmocilia
ресничный ciliate, ciliary
респиратор respirator, breathing apparatus/mask
~ **для опрыскивания** spray mask
~ **с подачей воздуха** air-line respirator
респираторный respiratory
республика republic
~, **братская** sister republic
~, **союзная** union republic
рессора spring
~, **пружинная** coil spring
~, **торсионная** torsion[al] spring
~, **шарнирно-закрепленная** articulated spring
рестарт restart
~ **после восстановления питания** power-fail restart
~ **с контрольной точки** checkpoint restart
ресурс 1. resource; 2. safe/operational/service life; overhaul period
~, **блокированный** locked resource
~, **вычислительный** computational resource
~ **долговечности** (operational) life
~, **многократно используемый** reusable resource
~, **назначенный** specified life
~, **общий** public resource
~, **расходуемый** consumable resource
~ **системы** system resource
~, **средний** mean life
~, **эксплуатационный** operating life
~**ы, внутренние** domestic/home resources
~**ы, водные** water supplies
~**ы, людские** manpower
~**ы, незанятые** idle resources
~**ы, неиспользованные природные** untapped natural resources
~**ы, свободные** uncommitted resources
~**ы, трудовые** labour force[s], manpower
ретикула reticulum
ретикулез *med.* reticulosis
~ **кожи** cutaneous reticulosis
~, **липомеланотический** lipomelanic reticulosis
~, **миелоидный** myeloid reticulosis
~, **опухолевый** tumoral reticulosis
ретикулез-лейкоз reticular cell leukemia
ретикулин *bioch.* reticulin[e]
ретикулогистиоцитарный reticulohistiocytic
ретикулоклеточный reticulocellular
ретикулоксантоматоз *med.* reticuloxanthomatosis, lipoid granulomatosis of bone
ретикулопения reticulopenia
ретикулосаркома reticulum cell sarcoma, reticulosarcoma
ретикулосаркоматоз *med.* reticulum cell sarcomatosis
ретикуло-таламический reticulothalamic, thalamoreticular
ретикулоэндотелий *anat.* reticuloendothelium
ретикулоэндотелиоз *med* . reticuloendotheliosis
~, **системный** systemic reticuloendotheliosis
ретикулум reticulum
ретина *anat.* retina, optomeninx
ретинен, ретинин *bioch.* retinene
ретиненредуктаза *bioch.* retinene reductase
ретинит *med.* retinitis
~ **беременных** gravidic retinitis
~, **гипертонический** angiospastic retinitis
~, **гнойный** purulent retinitis
~, **лейкемический** leak(a)emic/splenic retinitis
~, **полосатый** striated retinitis
~, **разлитой** diffuse retinitis
~, **серозный** serous/simple retinitis, edema of the retina
~, **точечный** punctate retinitis
ретинобластома retinoblastoma, glioma of the retina
ретинол *chem.* retinol, resinol, rosin oil
ретинопапиллит retinopapillitis, papilloretinitis
ретинопатия *med.* retinopathy
~, **гипертоническая** hypertensive retinopathy
~, **почечная** renal retinopathy
ретиноскопия retinoscopy, skiascopy, shadow test
ретинохороидит retinochor(i)oiditis, chorioretinitis
ретопорителий retopelithelium
ретороманский Rhaeto-Romanic, Rhaetian
реторсия *law* retorsion, retortion; retaliation
реторта *chem.* retort
ретотелий retothelium
ретотельсаркома reticulum cell sarcoma, reticulosarcoma
ретракция retraction
~ **нижней челюсти** mandibular retraction
ретранслировать retransmit, relay, rebroadcast
ретранслятор retransmitter, relay station, repeater
ретрансляция retransmission, relaying
ретратта *fin.* redraft
ретроактивность retroactivity; retroaction
ретроград retrograde person; reactionary
ретроградация retrogradation; reversion
ретроградный retrograde; backward
ретроградство retrogradism
ретрография retrography, mirror writing

ретролентальный retrolental, retrolenticular
ретроназальный retronasal, posterior nasal
ретроокулярный retroocular, retrobulbar
ретроорбитальный postorbital, retroorbital
ретрополяция *econ.* backcast
ретроспектива retrospective review
ретроспективно in reprospect, retrospectively
ретрофлексивный, ретрофлексный retroflex
ретрофлексия *med.* retroflexion, retroflection
ретроцедировать *law* retrocede
ретроцессионер *law* retrocessionaire
ретроцессия *law* retrocession
ретур *chem.* recycle
ретушевка *phot.* retouching
ретушер *phot.* retoucher
ретушировать *phot.* retouch
рефакция *econ.* recompense, fret
реферат 1. abstract; synopsis; 2. paper
реферативный журнал abstract journal
реферировать abstract *(a book, etc.),* review
референт 1. consultant; assistant; adviser; expert 2. reviewer, reader
референц-сфероид spheroid of reference
референц-эллипсоид ellipsoid/spheroid of reference, reference ellipsoid/spheroid
рефинансировать refund
рефлекс reflex
~, **ахиллов** Achilles tendon reflex
~, **бедренно-брюшной** ferroabdominal reflex
~, **безусловный** unconditioned reflex, inborn/instinctive reflex
~, **брюшинно-сердечный** abdominocardiac reflex
~, **вазоконстрикторный** vasopressor/vasoconstrictor reflex
~, **вдыхательный** inspiratory reflex
~, **выдыхательный** expiratory reflex
~, **выпрямительный** righting reflex, erectorspinal reflex
~, **гипогастральный** hypogastric reflex, femuroabdominal reflex
~, **глотательный** swallowing/deglutition reflex
~, **запястно-пальцевой** carpophalangeal reflex
~, **защитный** defense/defence reflex
~, **зрачковый** pupillary reflex
~, **кардиальный** cardiac reflex
~, **кашлевой** cough/laryngeal reflex
~, **кожный** cutaneous reflex
~, **коленный** knee-jerk reflex, patellar-tendon reflex
~, **ладонный** palmar reflex
~, **лобковый** pubic reflex
~, **локтевой** elbow/triceps reflex
~, **мандибулярный** mandibular/jaw/masseter reflex
~, **мигательно-челюстной** jaw-working reflex, jaw winking syndrome
~, **мигательный** wink/opticofacial reflex
~, **миостатический** myostatic/stretch reflex
~ **мочеиспускания** micturition/urinary/bladder reflex
~, **небный** palatal/palatine reflex
~, **оборонительный** defence/defense reflex
~, **пателлярный = коленный**
~ **растяжения** stretch/myostatic reflex
~, **слюнный** salivary reflex
~, **сосательный** sucking reflex
~, **условный** conditioned reflex, trained/behavio(u)r reflex
~, **условный запаздывающий** delayed conditioned reflex
~, **условный следовой** trace conditioned reflex
~, **установочный** righting reflex
~, **шаговый** step reflex
~, **ягодичный** gluteal reflex
рефлексивность, рефлективность reflectivity
рефлектор reflector; mirror; speculum
~, **лобный** head/frontal mirror
рефлекторный reflex *(attr.);* reflecting
рефлюкс reflux, backward flow; regurgitation
~, **маточно-венозный** uterovenous reflux
рефляция *econ.* reflation
рефонологизация *phon.* rephonemicization
реформа reform
~, **валютная/денежная** currency/monetary reform
~, **земельная** land reform
реформат *eccl.* reformist
реформаторий reformatory, training school for juvenile delinquents
реформаторский reformative, reformatory
реформатский reformist; Protestant
~**кая церковь** Reformed Church
реформационный reformational
рефрактерный refractory, nonsusceptible; obstinate
рефрактометр refractometer
~, **погружной** dipping refractometer
~ **с фазовым контрастом** phase contrast refractometer
рефрактор refracting telescope
рефракция refraction
~, **акустическая** acoustical/sound refraction
~ **глаза** ocular refraction
~, **ионосферная** ionospheric refraction
рефулер discharge pipe; dredging pump, flushing dredge
рефулерный насос dredging pump
рецензент reviewer, reader; critic
рецензия review; critique
рецепт 1. formulation, mix formula, recipe; 2. *med.* prescription
рецептакул[а], рецептакулум *bot.* receptacle
рецептор *anat.* receptor, ceptor
~, **внешний** exteroceptor
~, **внутренний** interoceptor
~, **зрительный** visual ceptor
~, **кожный** cutaneous ceptor
~, **нервный** nerve ceptor
~, **обонятельный** olfactory ceptor

~ **равновесия** equilibrium ceptor
~, **слуховой** auricular ceptor
~, **температурный** thermal ceptor
~, **тепловой** thermoreceptor
рецепторный receptor *(attr.)*
рецептура 1. formulation; 2. execution of a prescription
рецептурный prescription *(attr.);* formulation *(attr.)*
рецессия recession
рецидив 1. *law* recidivism, second/repeated offence; 2. *med.* relapse, recurrence
рецидивировать *med.* relapse, recur
рецидивирующий *med.* recurrent, relapsing
рецидивист *law* recidivist, repeated offender
рециркулировать recirculate, recycle
рециркуляция recirculation, recycling
~, **непрерывная** continuous recycling
~ **отходов** waste recycling
речевой vocal, speech *(attr.);* conversational
речник 1. river gravel; 2. river transport worker/official
речной river *(attr.),* riverine, fluvial
~ **планктон** potamoplankton
речь speech; speaking, talking, talk; language
~ **внутренняя** internal/inner speech
~, **гнусавая** nazalized speech, rhinolalia
~, **замедленная** bradyarthria, bradyglossia
~, **мозжечковая** cerebellar speech
~, **отчетливая** distinct/articulate/well-enunciated speech
~, **пищеводная** esophageal speech
~, **синтезированная** synthetic speech
~, **скандированная** staccato/syllabic speech
~, **слитная** continuous speech
~, **тронная** speech from the throne, King's/Queen's speech
решать 1. decide; resolve; determine; 2. *math.* solve; 3. *tech.* design
решатель *tech.* solver
решение 1. decision; *law* ruling; judgement; verdict; 2. *math.* solution, solving
~, **автомодельное** similar solution
~ **администрации** managerial decision
~, **арбитражное** arbitration award
~, **бесповоротное** decision past recall
~ **в аналитическом виде** closed-form solution
~ **в целых числах** integer solution
~ **в чистых стратегиях** *game th.* pure-strategy solution
~ **в явном виде** explicit solution
~, **вероятностное** probabilistic solution
~, **взрывное** *math.* explosive solution
~ **зависящее от времени** time-dependent solution
~, **затухающее** *math.* damped solution
~, **исходное** exit decision
~, **квазиавтомодельное** quasi-similar solution
~, **компромиссное** trade-off [compromise] decision/solution
~, **контрольное** check solution
~, **кусочное** piecewise solution
~ **методом конечных разностей** finite-difference solution
~ **методом малых возмущений** small-perturbation solution
~ **методом Монте-Карло** Monte-Carlo solution
~ **методом наименьших квадратов** least-squares solution
~ **методом проб и ошибок** trial-and-error solution
~, **невырожденное** non-degenerate solution
~, **неединодушное** split decision
~, **неоднозначное** ambiguous solution
~, **неявное** implicit solution
~, **нулевое** zero/null solution
~ **обратной геодезической задачи** inverse position computation
~, **однозначное** unique solution
~, **окончательное** *stat.* terminal solution
~, **особое** singular solution
~, **проектное** design
~ **руководства** managerial decision
~, **стационарное** steady-state solution
~, **стратегическое** *econ.* policy decision
~, **строгое** rigorous solution
~, **техническое** engineering solution
~, **частное** particular solution
~, **численное** numerical solution
~, **явное** explicit solution
решетка grid; lattice; grille, grating, grate
~ **аккумуляторной пластины** plate grid
~ **активной зоны** *nucl.* core lattice
~, **антенная** aerial/antennal array
~, **атомная** atomic lattice
~ **без воды** dry lattice
~ **без отражателя** bare/unreflected lattice
~, **вихревая** vortex lattice
~, **гранецентрованная** face-centered lattice
~ **громкоговорителя** louvre of a loudspeaker
~, **групповая** cluster lattice
~, **двойная** dual lattice
~, **двойниковая** twinning lattice
~, **деионная** *el.* deion/deionization lattice
~, **дифракционная** diffraction grating
~ **из урана и замедлителя** *nucl.* uranium-moderator lattice
~ **из урановых стержней** uranium-rod lattice
~, **интерференционная** interference grating
~, **кольцевая** annular cascade
~, **конечная** finite lattice
~, **клапанная** flap valve grid
~, **кристаллическая** crystal lattice
~, **люковая** hatch grating
~, **набухающая** expanding lattice
~, **нарушенная** perturbed lattice
~, **обратная** reciprocal lattice
~, **объемноцентрированная** body-centered lattice
~, **опорная** support grid

~, ослабляющая attenuator grating
~, пластинчатая slab lattice
~, пространственная spatial/space lattice
~, прямая direct/straight lattice
~ с большим шагом widely-spaced lattice
~ с водным замедлителем *nucl.* water-moderated lattice
~ с водой wet lattice
~ с замедлителем *nucl.* moderated lattice
~ с избыточным замедлением overmoderated lattice
~ с малым шагом close-spaced lattice
~ с недостаточным замедлением *nucl.* undermoderated lattice
~ с урановыми блоками lumped-uranium lattice
~, стержневая rod lattice
~, ступенчатая echelon grating
~, точечная point lattice
~, убирающаяся retractable grill
~, уран-тяжеловодная uranium-heavy-water lattice
~, успокоительная quiescence grating
решетнник *bot.* shallow-pored boletus *(Boletus bovinus)*
решетовидный, решетчатовидный cribriform
решетчатоплодный *bot.* lattice-/grilled-fruited
решетчатый grilled; cribrate; lattice *(attr.)*
рештак *min.* chute, pan, trough
реэкспорт re-exportation, re-export
ржаветь rust, become/grow rusty
ржавление rusting, corrosion
ржавчина rust *(also bot.)*
~, медная verdigris
ржавчиноустойчивый rustproof, rust-resistant
ржанка *zool.* plover *(Charadrius)*
ржаной rye *(attr.)*
рибекит *geol.* riebeckite
рибитил ribityl
рибит[ол] ribitol
рибоза *chem.* ribose
рибозид riboside, nucleoside
рибокиназа ribokinase
рибонуклеаза ribonuclease, ribonucleinase
рибонуклеат, рибонуклеинат ribonucleate
рибонуклеиновый ribonucleic
рибонуклеопротеид ribonucleoprotein
рибонуклеотид ribonucleotide
рибосома *biol.* ribosome
рибофлавин riboflavin
риванол *pharm.* rivanol, ethoxydiaminoacridine lactate
ригель crossbar, span piece; collar beam
Ригель *astr.* Rigel
ригидность rigidity, stiffness, rigo(u)r
~, посмертная postmortem rigidity
~, трупная cadaveric rigidity
ригсдаг *polit.* Rigsdag
ридберг *phys.* rydberg
риза *eccl.* vestment, chasuble, cope
ризница *eccl.* vestry, sacristy, diaconicon
ризоидный rhizoid, rhizomorphoid
ризодерма rhisodermis, epiblem[a]
ризокарповые *pl. bot.* rhizocarps *(Rhizocarpeae)*
ризолит rhizolith
ризом(а) *bot.* rhizoma, rhizome, rootstock
ризопод *zool.* rhizopod
ризоподия rhizopodium
ризотаксис *bot.* rhizotaxis, root arrangement
ризофит *bot.* rhyzophyte
ризофор *bot.* rhizophore
ризофра *bot.* mangrove, rhizophore *(Rhizophora mangle)*
рик *abbr.* district executive committee
рикардианство Ricardianism
рилинг-стан reeling mill
риккетсиоз *med.* rickettsiosis
~, везикулезный/осповидный rickettsial pox
~, крысиный murine/endemic typhus fever
риккетсиозный, риккетсиальный ricketsial
рикошет ricochet, bounce, skip
рикошетировать ricochet
риксдаг *polit.* Riksdag
рикша ricksha(w)
римлянин Roman
римлянка Roman (woman)
римский Roman
~ папа the Pope
~кая церковь Roman Catholic Church
римскокатолический Roman Catholic
ринальгия *med.* rhinalgia, rhinodynia
ринальный rhinal, nasal
ринит *med.* rhinitis
~, гнойный purulent rhinitis
~, гриппозный influenzal rhinitis
~, дифтерийный diphtheric rhinitis
~, нервнорефлекторный neuroreflex rhinitis
~, острый acute rhinitis, coryza
~, туберкулезный scrofulous rhinitis
~, фибринозный fibrinous/membranous/croupous rhinitis
ринолалия rhinolalia, rhinophonia, nazalized speech
риноларингит rhinolaryngitis
риноларингология rhinolaryngology
ринолит rhinolith, rhinolite, nasal calculus
ринопластика rhinoplasty
ринопсия rhinopsia, convergent squint
ринорафия rhinorrhaphy
ренорея rhinorrh(o)ea, nasal hydrorrh(o)ea
риноррагия rhinorrhagia, epitaxis, nose bleed
риносинусопатия rhinosinusopathy
риноскоп rhinoscope, nasoscope, nasal speculum
риноскопия rhinoscopy
риностеноз rhinostenosis, rhinocleisis, nasal obstruction
ринофарингит rhinopharyngitis, nasopharyngitis, epipharyngitis
ринофима rhinophyma, toper's/whisky nose
ринхозавр *paleont.* rhynchosaur *(Rhynchosaurus)*

ринхоспора *bot.* beak rush *(Rhynchospora)*
ринхоцефал *zool.* rhynchocephalian
риолит *petr.* rhyolite, rhyolith
  ~, кальциевый calcirhyolite
  ~, щелочный alkali-rhyolite
риотрон *el.* ryotron
риотаксис rhyotaxis
рипария *bot.* riverbank grape *(Vitis riparia)*
рипсалис *bot.* mistletoe cactus *(Rhipsalis)*
рирпроектор background projector
рирпроекция rear projection, rear-projected film
рирэкран back-projection screen
рис *bot.* rice *(Oryza)*
рисберма *constr.* rear apron, downstream apron/floor
риск risk, hazard; venture; peril, danger
  ~, апостериорный a posteriori risk
  ~, байесовский *econ.* Bayes(ian) risk
  ~ владельца owner's risk
  ~, допустимый tolerated risk
  ~, исключенный excepted peril
  ~, минимаксный *game th.* minimax risk
  ~, морской maritime/marine risk, risk at sea
  ~ неплатежа по ссуде credit risk
  ~, нестрахуемый uninsurable/noninsurable risk
  ~, расчетный calculated risk
  ~, совокупный integrated/overall risk
  ~, средний average/mean risk
  ~, страховой insurance risk
  ~, страхуемый insurable risk
  ~, суммарный total risk
  ~, условный conventional risk
  ~, чрезвычайный страховой abnormal insurance risk
  ~и, смешанные страховые miscellaneous insurance risks
риска scratch index, notch, tracing/graduation line
рисовидка *bot.* rice grass *(Oryzopsis)*
рисоводство rice growing
рисорджименто Risorgimento
рисошелушитель rice sheller
ристоцетин ristocetin, spontin
рисунок 1. drawing; sketch; figure; 2. pattern, design
  ~ покрышки tread pattern/design
  ~ шрифта type design
ритардандо *mus.* ritardando
ритенуто *mus.* ritenuto
ритидом *bot.* rhytidome, bark
ритм rhythm, beat; pace; cycle
  ~, биологический biologic cycle
  ~, возвратно-отраженный reciprocating rhythm
  ~, коронарный синусовый *anat.* coronary sinus rhythm
  ~, нижнеузловой lower nodal rhythm
  ~, отраженный reciprocal rhythm
  ~, сердечный cardiac rhythm
  ~, синусовый sinus rhythm
  ~, среднеузловой midnodal rhythm
  ~, суточный circadian cycle
  ~, удвоенный bigeminal/coupled rhythm, bigeminy
  ~, утроенный triple rhythm
ритмика rhythmicity; rhythmics
ритмично rhythmically; at a steady pace
ритмичность rhythm; rhythmicity; intermittency
риторика rhetoric
риф reef; ledge
  ~, биогенный organic reef
  ~, водорослевый algal reef
  ~, задний backreef
  ~, коралловый coral reef
  ~, мшанковый bryozoan reef
  ~, надводный коралловый uplifted reef
  ~, окаймляющий/прибрежный fringing reef
  ~, поднятый uplifted/elevated reef
рифамицин rifamycin
рифление knurling, corrugating, channelling, grooving; knurl, groove, flute
рифленый fluted, corrugated, channeled
рифма rhyme
риформинг *chem.* reforming
риформинг-бензин reformed gasoline/petrol
риф-штерт *naut.* knittle
рихардия *bot.* calla lily *(Richardia)*
рихтовка straightening, trueing, alignment
рицин 1. *chem.* ricin; *med.* castor oil; 2. *bot.* castor-oil plant *(Ricinus communis)*
рицинин ricinine
ришта *zool.* Guinea worm *(Dracunculus mediensis)*
РЛС [радиолокационная станция] radar
  ~ автоматического сопровождения target-track radar
  ~, бортовая shipborne/on-board/vehicle-borne/airborne radar
  ~, диспетчерская airfield control radar
  ~ кругового обзора all-round looking radar
  ~, метеорологическая weather radar
  ~ обнаружения detection radar
  ~ поиска и сопровождения search and track radar
  ~, посадочная landing assist radar
  ~, судовая marine radar
робало *zool.* snook *(Centropomus)*
роббер *cards* rubber
робиния *bot.* locust *(Robinia)*
робот robot
  ~, следящий slave-robot
робот-инспектор inspection robot
роботехниа robotics
роботизация robotization; introduction/use of robots
ров ditch, pit
  ~, крепостной moat
ровит rauvite
рог 1. horn; antler; 2. *tech.* hornlike protrusion, beak, arm; 3. bugle
  ~, альпийский alp(en)horn
  ~ изобилия horn of plenty

~, **кожный** *anat.* cutaneous horn
~ **компенсатора** balance horn
~, **копчиковый** coccygeal horn
~, **крестцовый** sacral horn
~, **разрядный** arcing horn
~ **якоря** anchor arm
рогатик *bot.* goatsbeard *(Clavaria)*
рогатиковые *pl. bot.* coral fungi *(Clavariaceae)*
рогачка *bot.* rocket weed *(Erucastrum)*
роговидный hornlike; ceratoid; *bot.* cornute, corniculate
роговик *miner.* hornfels
~, **известково-кремнистый** calc-flinta
~, **известково-силикатный** calc-silicate hornfels
роговица *anat.* cornea
роговичный *anat.* corneal
роговой horny, horn *(attr.),* corneous, corneal, cornified; keratic, keratoid
~**вая обманка** *miner.* hornblende
~**вая оболочка** *anat.* cornea
рогоз *bot.* reed mace, cattail *(Typha)*
рогозовые *pl. bot.* cattail family *(Typhaceae)*
рогозуб *zool.* Mary-River salmon *(Neoceratodus forsteri)*
роголистник *bot.* hornwort *(Ceratophyllum)*
роголистниковые *pl., bot.* hornwort family *(Ceratophyllaceae)*
рогообразный corniform, hornlike
рогообразование cornification, hornification, keratogenesis
рогоподобный hornlike; corniculate
рогохвост *zool.* horntail *(Sirex)*
рогоцветковый *bot.* horn-flowered
рогулька *text.* flyer
рогульник *bot.* water chestnut, Jesuit nut *(Trapa natans)*
род 1. genus; race; 2. tribe; clan; family; kin; 3. kind, class, category, genre, type, sort; 4. *gram.* gender
~ **авиации** aviation category
~ **войск/оружия** *mil.* arm of the service
~, **средний** *gram.* neuter gender
родамин *chem.* rhodamine
родан *chem.* rhodane
роддом *abbr.* maternity home/hospital
родентицид rodenticide
родиевый *chem.* rhodic
родий *chem.* rhodium, Rh
родильница *med.* puerpera
родильный puerperal; maternity *(attr.)*
~ **дом** maternity home/hospital
~**ное отделение** *med.* delivery room/ward
родинка birth mark
родиола *bot.* stonecrop *(Rhodiola)*
родирование *chem., met.* rhodium plating
родительный *gram.* genitive
~ **разделительный/части** partitive genitive
~ **определительный** genitive of definition
родительский parental; paternal
родник spring
родничок *anat.* fontanel[le]
~, **клиновидный** sphenoidal fontanelle
~, **передний** anterior/frontal fontanelle
~, **сосцевидный** mastoid/posterolateral fontanelle
~ **черепа** cranial fontanelle
родовой 1. family *(attr.);* tribal; 2. patrimonial, ancestral; 3. *med.* labo(u)r *(attr.),* birth *(attr.),* puerperal, parturient; *biol.* generic; 5. *gram.* gender *(attr.)*
родовспомогательный obstetric(al)
родовспоможение obstetric aid
рододендрон *bot.* rhododendron *(Rhododendron)*
родолит *miner.* rhodolite
родомирт *bot.* rose myrtle *(Rhodomyrtus)*
родонат *chem.* thiocyanate
родонит *miner.* rhodonite
родоразрешение *med.* delivery
~ **посредством кесарева сечения** delivery by cesarian section
~ **посредством наложения щипцов** forceps delivery
родотипус *bot.* jetbead *(Rhodotypus)*
родоускоряющий *med.* parturifacient, oxytocic
родофилл *bot.* rhodophyll
родохрозит *miner.* rhodochrosite
родственник relative, relation; collateral; kinsman
~, **ближайший** next-of-kin
~, **дальний** remote/distant relative
~ **по мужской линии** agnate
родство kinship, relationship
~ **кровное** consanguinity, blood relationship
~ **по боковой линии** collateral relationship
~ **по отцу [по мужской линии]** agnation
~ **по мужу или жене** affinity
~, **побочное** collateral relationship
роды delivery, childbirth, birth, accouchement, parturition, labo(u)r
~, **быстрые** rapid parturition, oxytocia
~, **искусственные** induced labo[u]r
~, **множественные** multiple birth
~, **несостоявшиеся** missed labo(u)r
~, **преждевременные** premature birth/delivery
~ **при головном предлежании** head delivery
~ **при затылочном предлежании** occipital delivery
~ **при ножном предлежании** foot delivery
~ **при поперечном положении плода** cross birth
~ **при ягодичном предлежании** breech delivery
~, **своевременные/срочные** labo(u)r at term, partus maturus
~, **стремительные** precipitated labo(u)r
~, **сухие** dry labo(u)r, xerotocia
~, **трудные** difficult childbirth/parturition/labo(u)r, dystocia, mogitocia
рожа *med.* erysipelas, rose, St. Anthony's fire
~, **блуждающая/мигрирующая/ползучая**

ambulant/wandering erysipelas
~, **свиная** swine erysipelas
рожать give birth to
рожающая parturient
рождаемость birth rate; natality; reproduction
~**ти, коэффициент** birth/childbearing rate
~**ти, регулирование** birth/family/fertility/population control
рождение birth, delivery; *nucl.* production, generation
~ **вне брака** out-of-wedlock [extramarital] birth
~ **нежизнеспособного плода** miscarriage
~, **нерадиационное** radiationless [non-radiative] generation
~ **пар** pair generation
~, **парное** pair production
~ **пи-мезонов** pion production
~, **радиационное** radiative generation
~, **связанное/совместное** associated production
~ **уродов** monstriparity
рождественский Christmas *(attr.)*
~ **сочельник** Christmas Eve
Рождество Christmas
роженица woman in childbirth, lying-in woman
рожистый *med.* erysipelatous
рожковый pronged, horn-type/-like
~**вое дерево** *bot.* carob tree *(Ceratonia siliqua)*
рожок 1. small horn, hornet; cornet; *mus.* horn, clarion, bugle; 2. *anat.* corniculum, cornicle
~, **английский** *mus.* English horn
~, **пастуший** alpen/alp horn
~, **слуховой** ear trumpet
роза 1. *bot.* rose *(Rosa)*; 2. *math.* rosette, rose
~ **ветров** wind rose
~ **повторяемости ветров** frequency wind rose
розанные *pl. bot.* rose family *(Rosaceae)*
розелла *bot.* roselle, Jamaica sorrel *(Hibiscus sabdariffa)*
розенкрейцер *hist.* Rosicrucian, Brother of the Rosy Cross
розеола *med.* roseola
розетка 1. rosette; *anat.* rosula, rosella; 2. *arch.* rose window, rosace; 3. *el.* socket, wall plug, female connector, receptacle
~ **листьев** leaf rosette
~, **штепсельная** plug socket/receptacle, wall socket
~, **трехполюсная** three-pin socket
розеточница *bot.* stonecrop *(Rosularia)*
розетта *arch.* rosette, rosace
розмарин *bot.* rosemary *(Rosmarinus)*
розница, розничный *econ.* retail
розовые, розоцветные *pl. bot.* rose family *(Rosaceae)*
розыск search; investigation, inquiry
~, **уголовный** criminal investigation
~**и вагона** *rail.* car locating activities
рой swarm; nest
~, **ионный** *nucl.* ion cluster
рокамболь *bot.* rocambole, giant garlic *(Allium scorodoprasum)*
рокер rocker arm
рокировать[ся] *chess* castle
рокировка *chess* castling
~, **длинная** queen's side castling
РОКК *abbr.* Russian Red Cross Society
рокфор Roquefort cheese
рол roll; cylinder; reel
ролик roller, roll; pulley, sheave; *(of furniture)* castor, caster
~ **ведомый** driven roller
~, **ведущий** drive/transport/leader roller; guide pulley
~ **каретки** track roller
~, **натяжной** tensioning/tension roller
~, **прижимной** clamping/pinch/pressure roller; capstan idler
~ **с датой** date roller
~, **сварочный** seam-welding electrode wheel, welding roll
~, **транспортирующий** conveyer roller
~, **тросовый** cable roller
роликоподпятник roller thrust bearing
роликоподшипник roller bearing
~, **конический** tapered/conical roller bearing
~, **радиально-упорный** radial thrust roller bearing
~ **с витыми роликами** spiral roller bearing
~, **упорный** thrust roller bearing
ролл hollander, beater
~, **массный** paper-bulb beater, hollander
~, **отбельный** bleacher
рольганг roller conveyer/bed/table
~, **входной** inlet table
~, **выпускной** outlet/runout/delivery table
~ **загрузки** charging table
~, **качающийся** shaking/oscillating table
~, **отводящий** runout/delivery table
~, **сборочный** assembly table
роман ram
романизировать, романизовать Romanize
романистика Romance philology
романский Romance, Romanic, Romanesque
~ **стиль** *arch.* Romanesque
~**кие языки** Romance languages
романтизация romanticization
романтизировать romanticize
романтизм romanticism
романтик 1. romantic; 2. romanticist
романтика romanticism
романцемент Roman cement
романцовия *bot.* mistmaiden *(Romanzoffia)*
ромашка *bot.* matricary, mayweed *(Matricaria)*
ромашник *bot.* pyrethrum, chrysanthemum *(Pyrethrum)*
ромб *geom.* rhombus, rhomb; lozenge, diamond
~, **гребенчатый** pectinate rhomb
ромбический rhombic, rhomb-shaped, rhomboid[al]

ромбовидный *tech.* double-wedge, diamond-shaped
ромбододеказдр rhombic dodecahedron
ромбоид *geom.* rhomboid
ромбоидальный, ромбоидный rhomboid[al]
ромбоэдр rhombohedron
ромбоэдрический rhombohedral, rhombohedric
ромстек, ромштекс *cul.* rump steak
ромэн *bot.* romaine lettuce *(Lactuca scariola longifolia)*
рондо 1. *mus.* rondo; 2. *lit.* rondeau, rondel
ронжа *zool.* Siberian jay *(Cractes infaustus)*
ронка *zool.* snorer *(Haemulon bonariense)*
ропалия *zool.* rhopalium
рорквал *zool.* rorqual *(Balaenoptera)*
роса 1. dew; 2. *bot.* downy mildew
росение *arg.* dewretting
росичка *bot.* finger grass *(Digitaria)*
росограф drosograph
росомаха *zool.* wolverine, glutton *(Gulo gulo)*
роспись 1. *art* . mural; 2. signal document
российский Russian
росомер drosometer
россыпь placer deposit, placer
~, береговая beach placer
~, золотоносная auriferous gravel, gold placer
~, ископаемая fossil placer
~, ледниковая glacial drift
~, морская submarine placer
~, подводная subaqueous placer
~, промышленная commercial placer
рост 1. height; length; stature; size; 2. growth, increase, rise; development, advance
~ давления pressure build-up
~, исполинский gigantism
~, карликовый nanism, dwarfishness
~ курсов *fin.* rally
~, непропорциональный *med.* disproportional stature
~, низкий short stature
~ по экспоненциальному закону exponential increase
~ стоимости капитального имущества capital gains
~, умеренный *econ.* moderate/healthy increase
~ экономики economic growth/advance
ростбиф *cul.* roast beef
ростильня *agr.* germination dish
ростомер height meter; *bot.* auxanometer
ростр 1. *zool.* rostrum; 2. *hist., naut.* rostrum, beak
~, имеющий rostrate
ростр-блок *naut.* boat chock
ростры *naut.* skids
росянка *bot.* sundew *(Drosera)*
рот mouth; *anat.* os, stoma
~, вторичный deuterostoma
~, первичный archaeostoma
рота *mil.* company
~, медицинская surgical company
~, стрелковая rifle company
роталоидный *zool.* rotaliform
ротанг *bot.* rotang, rattan palm *(Calamus rotang)*
ротапленка duplicating stencil paper
ротапринт offset duplicator
ротативный rotative, rotating, rotary
ротатор duplicating machine
ротационный rotary, rotatory, rotational
ротация 1. *print.* rotary press; 2. *agr.* crop rotation
ротмистр *mil., obs.* cavalry captain
ротный *mil.* company *(attr.)*
ротовой oral, buccal, stomatic, stomatal
ротоглотка *anat.* stomatopharynx
ротожелудочный *anat.* stomatogastric
ротоногий *zool.* stomatopodous
ротоносовой *anat.* orinasal
ротопленка rotofilm
ротор 1. rotor; 2. *math.* curl, rotation
~, барабанно-дисковый disk-drum rotor
~, барабанный drum(-type) rotor
~ вектора *math.* curl of a vector
~, герметизированный canned rotor
~ гироскопа gyrowheel
~, гладкий unslotted/smooth rotor
~, глубокопазный deep bar rotor
~ градиента *math.* curl of a gradient
~, дисковый disk(-type) rotor
~, короткозамкнутый *el.* squirrel-cage rotor
~, массивный solid rotor
~, многофазный *el.* polyphase rotor
~, однокаскадный *el.* single-spool rotor
~ осевых ступеней axial rotor
~ с беличьей клеткой *el.* squirrel-cage rotor
~ с распределенной обмоткой distributed pole rotor
~, сварной welded rotor
~, фазный *el.* wound rotor
~, явнополюсный *el.* salient pole rotor
роторасширитель mouth dilator/gag
~, винтовой ratchet-action mouth gag
роульс roller
роуландит rowlandite
роющий fossorial
роцелла, рочелла *bot.* archill *(Rocella)*
роштейн raw matte
РОЭ *med.* erythrocyte sedimentation rate, ESR
ртутистый *chem.* mercurous
ртутить coat with quicksilver
ртутник mercury-arc rectifier
ртутно-индиевый *chem.* indium-mercuric
ртутно-таллиевый mercury-thallium
ртутный mercuric, mercurial, mercurous, mercury *(attr.)*
ртуть mercury, quicksilver, hydragyrum, Hg
~, азотнокислая mercuric nitrate
~, амидохлорная mercuric ammonium chloride
~, гремучая mercuric fulminate, fulminating mercury
~, двуиодистая mercuric iodide
~, двухлористая mercuric chloride, mercury

bichloride/perchloride
~, **иодистая** mercurous iodide, mercury protoiodide, yellow mercury iodide
~, **металлическая** mercury, quicksilver
~, **односернистая** protosulphide of mercury
~, **оксицианистая** mercuric oxycyanide
~, **салициловокислая** mercuric salicylate, mercury subsalicylate
~, **самородная** native mercury
~, **селеновая** mercuric selenide, tiemannite
~, **сернистая** mercuric sulphide
~, **хлористая** mercurous chloride
~, **хлорная** mercuric chloride
~, **цианистая** mercuric cyanide, cyanide of mercury
рубанок plane
~, **двойной** double-iron plane
~, **фасонный** moulding plane
рубашка jacket; shirt; *tech.* casing
~ **водяного охлаждения** water cooling jacket
~ **отражателя** reflector liner
~, **паровая** steam jacket
~ **паруса** bunt
~, **смирительная** strait-jacket
рубеж line, boundary, border; barrier
~, **исходный** line of departure, starting line
~ **обнаружения** detection/acquisition line
~ **перехвата** line of interception, interception line
~, **промежуточный** intermediate objective
~ **радиолокационного обнаружения** radar barrier
рубеоз *med.* rubeosis
рубероид roofing paper/material
рубец cicatrix, cicatrice; scar; gush
~, **глиозный** glial cicatrix
~, **деформирующий** vicious cicatrix
~, **келоидный** keloid/cheloid cicatrix
~, **листовой** *bot.* leaf scar, cicatrix
~, **мозговой** brain cicatrix
~ **разрыва** tension gush
рубидиевый *chem.* rubidic
рубидий *chem.* rubidium, Rb
Рубикон Rubicon
рубильник blade/knife/lever switch
~ **на два направления** double-throw blade switch
~ **силовой сети** power switch
рубин ruby; *print.* agate
~, **восточный** oriental ruby
~, **капский** Cape ruby
рубицелл *miner.* rubicelle, ruby spinel
рубка I *naut.* (deck) house, room
~, **гидроакустическая** asdic/sonar room
~, **кормовая** aft house
~, **оперативная** operational/operations room
~, **палубная** deck house
~, **радиолокационная** radar room
~, **рулевая** rudder/wheel house
~, **флагманская** flag plot
~, **ходовая** pilot house/room
~, **штурманская** chartroom, charthouse, navigation room
рубка II felling, cutting; lacking, chopping
~, **сплошная** clear felling
рубро-кортикальный rubrocortical
руброфития *med.* tinea, ringworm
рубцевание *med.* cicatrization, scarring
рубцеваться cicatrize, scar
рубцовый cicatrical, scarry
рувиллит *miner.* rouvillite
ругоза *paleont.* tetracoral
руда ore
~, **бедная** lean [low-grade, poor] ore
~, **белая кварцовая** alunite
~, **белая свинцовая** white lead ore; cerussite
~, **блеклая медная** grey copper ore
~, **блеклая мышьяковая** tennantite
~, **блеклая сурьмяная** grey antimony, stibnite
~, **брекчиевидная** breccia ore
~, **богатая** high-grade [rich] ore
~, **желтая медная** yellow copper ore, chalcopyrite
~, **жильная** vein/lode ore
~, **зеленая железная** green iron ore, dufrenite
~, **игольчатая** needle ore, alkinite
~, **кобальто-марганцевая** asbolite, asbolane
~, **кокардовая** cocarde/sphere ore
~, **колокольная** bell-metal ore, stannite
~, **колчеданная** sulphide ore, kies
~, **красная железная** hematite
~, **красная медная** cuprite
~, **красная свинцовая** red lead ore, crocoite
~, **красная сурьмяная** red antimony, kermesite
~, **кремнистая цинковая** calamine, hemimorphite
~, **крупнокусковая** lump ore
~, **кусковатая** lumpy ore
~, **мелкая** fine ore
~, **несортированная** run-of-mine ore
~, **нетронутая** solid ore
~, **необожженная** raw ore
~, **обожженная** roasted ore
~, **оспенная** variolotic ore
~, **полиметаллическая** complex ore
~, **промышленная** pay ore
~, **пятнистая** spotted ore
~, **рыхлая** friable ore
~, **рядовая** run-of-mine ore
~, **свинцовая** lead/galenical ore
~, **серая медная** grey copper ore, tetrahedrite
~, **серая серебряная** polybasite
~, **серая сурьмяная** grey antimony, stibnite
~, **серная** sulphur ore
~, **сернистая/сульфидная** sulphide ore, kies
~, **сложная/смешанная** complex ore
~, **смоляная** pitchblende
~, **стекловатая медная** vitreous copper, copper glance, chalcocyte
~, **стеклянная/стекловатая серебряная** vitreous silver, silver glance, argentite

~, **сырая** raw/crude ore
~, **углекислая** carbonate ore
~, **урановая смоляная** pitchblende, uraninite
~, **хромовая** chrome iron ore, chromite
рудбекия *bot.* coneflower *(Rudbeckia)*
рудерпис *naut.* rudder piece
рудерпост *naut.* rudder/outer post; rudderstock
~, **съемный** rudder shaft
рудимент rudiment, vestige
рудиментарный rudimentary, vestigial; inchoate
рудник mine, pit
~, **газовый** gassy/fiery mine
~, **угольный** coal mine/pit, colliery
рудный ore *(attr.)*
рудовоз *naut.* ore carrier
~, **крупнотоннажный** super ore-carrier
~, **подводный** submarine ore-carrier
рудодробилка ore crusher
рудоносность ore content
рудоносный ore-bearing, metaliferous
рудоотложение ore deposition
рудоподготовка ore dressing/benefication/concentration
рудоразборка ore sorting
рудосодержащий ore-bearing
ружейник gunsmith
ружье gun, rifle, handgun
~, **гладкоствольное** plain-barrel gun
~, **двуствольное** double-barreled gun
~, **дробовое** shotgun
~, **духовое** air gun
~, **нарезное** rifled gun, rifle
~, **промысловое охотничье** hunting gun
~, **спортивное охотничье** sporting gun
рука 1. *anat.* arm, upper extremity, brachium, upper arm; hand; 2. *tech.* arm; lever
~, **копирующая** *tech.* slave arm
~, **управляющая** *tech.* master arm
рукав sleeve; arm; chute; hose, pipe, tube
~ **аварийного слива** dump chute
~, **армированный** armoured/reinforced hose
~, **бездорновый** long-length hose
~, **всасывающий** suction hose
~, **газонаполнительный** inflation sleeve
~ **дельты** delta arm
~ **моря** arm of the sea, estuary
~, **мусорный** garbage chute/slop
~, **напорный** pressure/delivery hose
~, **отводящий** ejection chute
~, **пластмассовый** plastic sock/bag
~, **пожарный** fire hose
~ **реки** channel, branch; creek
~ **с оплеткой** braided hose
~, **слепой** blind/dead channel; oxbow, bayou
~, **сливной** drain chute/sleeve/hose
~, **тормозной** brake coupling pipe
~, **узкий** firth
~, **фурменный** *met.* tuyere pipe
рукоблудие masturbation, onanism
руководитель manager; director, administrator, head; leader; instructor; supervisor
~ **высокого ранга** high/senior executive
~ **крупной промышленной фирмы** captain of industry
~, **классный** class manager, supervising teacher
~ **научно-исследовательских работ** research manager/director, director of studies
~, **научный** supervisor of studies
~ **низшего звена на производстве** first-line manager
~ **ответственный за конструирование и производство изделия** product/brand manager
~ **ответственный за совершенствование изделия** product development manager
~ **отдела** head of section/department
~ **отдела рекламы** publicity manager
~ **отдела сбыта** sales manager
~ **полетов** flying control officer
~ **посадки** landing signal officer
~ **программы** program[me] director/manager
~ **проекта** project leader/manager
~ **службы материально-технического снабжения** inventory manager
~, **технический** engineering manager
руководство 1. management, administration, supervision; direction; leadership; guidance 2. manual, handbook, guide; 3. governing body, management
~ **для плавания** sailing directions
~ **кадрами** personnel/manpower management
~, **оперативное** operational management/control
~ **по летней эксплуатации самолетов** flight manual
~ **по морской практике** manual of seamanship
~ **по ремонту** repair manual
~ **по уходу и обращению** care and maintenance handbook
~ **по уходу и техническому обслуживанию** maintenance manual
~ **по эксплуатации** operating/service manual
~ **полетами** flight operations control
~ **работами** work supervision
~, **техническое** technical management/control
рукокрылые *pl.* chiropterans *(Chiroptera)*
руконогий *zool.* pedimanous
рукополагать *eccl.* ordain
рукоположение *eccl.* ordainment
рукоятка handle, grip; hand lever, crank handle
~ **блокировки** locking handle
~ **крана** cock handle
~ **крена** *aero.* roll control knob
~, **основная** primary grip/handle
~, **поворотная** twist grip/knob
~ **тангажа** *aero.* pitch knob
~, **тормозная** brake arm
~ **центрирования** centering knob
рукоятник *min.* cager; banksman
рукоять arm; stick; lever *see also* **рукоятка**
рулада *mus.* roulade, run
рулевой 1. *adj.* rudder *(attr.);* steering; 2. *n.* steersman, helmsman, coxswain; pilot

~ привод steering gear
рулежный *av.* for taxiing, taxi *(attr.)*
рулена *zool.* wrasse *(Crenilabrus tincta)*
руление, рулежка *av.* taxiing, taxying
~ по ветру taxiing downwind
~ при боковом ветре cross-wind taxying
~ против ветра taxying into the wind
рулет *cul.* roll
рулетка 1. roulette; 2. tape measure, measuring reel
рули *aero., naut.* control surfaces, controls
~, брошенные free controls
~, носовые canard controls
~ струйные control jets
~, успокоительные stabilizing fins
рулить *aero.* taxi, steer
рулон roll; reel, coil; *text.* bolt
~ полосы *met.* strip coil, coil of strip
~ холоднокатаной полосы cold-rolled coil
рулоноразматыватель *met.* decoiler
руль 1. *naut.* rudder; helm; *auto.* steering wheel; 2. *aero.* control surface; wane
~, аварийный jury rudder
~, аэродинамический aerial rudder, air vane, aerodynamic control surface
~, воздушный aerodynamic rudder
~, вертикальный vertical/steering rudder
~ ветродвигателя wind arm
~ высоты *av.* elevator
~, газовый gas/jet/exhaust vane
~, графитовый carbon graphite vane
~, компенсированный compensated/balanced control surface
~ крена roll control surface/tab
~, навесной hinged rudder
~ направления rudder
~, носовой bow rudder; *aero.* canard control
~, обтекаемый streamlined rudder
~, обыкновенный unbalanced rodder
~, подвесной underhung/suspended/hanger rudder
~ системы стабилизации вращения spin control jet
~, струйный control jet
~, трехсекционный triplicated rudder
руль-тали rudder tackle
румб *naut.* compass point, bearing
~, главный cardinal point
~, четвертной interquadrinal [quadrantal, half-cardinal, semi-cardinal] point
румбатрон *phys.* rhumbatron
румпель *naut.* rudder/steering tiller
~, аварийный energency tiller
~, простой single-arm tiller
~, прямой straight tiller
~, секторный rudder/steering quadrant, quadrant tiller
румпельтали *naut.* steering tackle
румын, румынский Rumanian, Romanian
румынка Rumanian/Romanian (woman)
румянец *med.* blush, flush; *bot.* erubescence
румянка *bot.* viper's bugloss *(Echium)*
руна *ling.* rune
рунец *zool.* ked, sheep tick *(Melophagus ovinus)*
рунический *ling.* runic
руно fleece
~, золотое *myth.* the Golden Fleece
рупия I *med.* rupia
рупия II rupee
рупор megaphone horn, speaking trumpet
~, отражательный reflex horn
рупорно-зеркальный horn-reflector *(attr.)*
рупорнообразный funnel-/horn-shaped, flared
русак *zool.* European hare *(Lepus europaeus)*
русалии spring holiday in commemoration of the dead
русальная неделя Sunday before Whitsuntide
русизм *ling.* Russicism
русин, русинский Ruthenian
русинка Ruthenian (woman)
русист specialist in Russian philology
русистика Russian philology/studies
русифицировать Russify, Russianize
руслени *naut.* chainwales
русло bed, channel, waterway
~, лимфатичное *med.* lymphatic channel
руслоуглубительный channel-deepening/dredging
русофил Russophile
русофильство Russophilism
русофоб Russophobe
русофобия, русофобство Russophobia
русофобский Russophobe *(attr.)*
русский Russian
руссоизм Rousseauism
руссоист Rousseauite, Rousseauist
руссофил, руссофоб = русофил, русофоб
рустика, рустовка *arch.* rustic work/masonry, rustication
рустов *naut.* shank painter
Русь *hist.* Rus, Kievan Russia
рута *bot.* rue *(Ruta)*
рутениевый *chem.* ruthenic
рутений *chem.* ruthenium, Ru
~, двухлористый ruthenous chloride
рутенистый *chem.* ruthenous, ruthenious
рутзерфордин rutherfordine
рутидоз *med.* rutidosis, rhytidosis
рутил *miner.* rutile
~, игольчатый acicular rutile
~, сетевидный reticulated rutile, sagenite
рутовые *pl. bot.* rue faimily *(Rutaceae)*
рухляк *miner.* rottenstone, marl
~, гипсоносный gypseous marl
рухляковый marlaceous, marly
ручательство guarantee, guaranty, voucher, warrant[y]
ручей 1. *tech.* groove; channel; impression; 2. brook, creek, stream
ручейник *zool.* caddis fly *(Trichoptera)*
ручка handle, grip, holder; knob; arm; brachium

~, автоматическая fountain pen
~ дистанционного управления remote control stick
~ замка latch handle
~ коррекции adjustment/correction knob
~ настройки tuning knob
~, ременная подъемная lift strap
~ согласования synchronizing knob
~ управления 1. control handle; 2. joystick
~, шариковая ball-point pen
рыба fish; *see also* **рыбы**
~, икряная gravid fish
~, икромечущая egg-laying fish
~, красная cartilaginous fish, sturgeons
рыба-бабочка *zool.* butterfly-fish *(Chaetodon)*
рыба-гадюка *zool.* viper fish *(Chauliodus)*
рыбалот fish lead
рыба-лоцман *zool.* pilot fish *(Aucrates ductor)*
рыба-луна *zool.* sunfish *(Mola)*
рыба-самец milter
рыба-самка spawner
рыбец *zool.* vimba *(Vimba vimba)*
рыбий fish *(attr.)*; piscine
~ жир cod-liver oil
~ клей fish glue
рыбина *naut.* diagonal/rising/ribband line, diagonal; ribband; batten
~, закрепленная footing, footling
~, скуловая bilge diagonal, rising line
~, спусковая launching [ground-way] ribband
~, строительная shoring/framing ribband
рыбинси *pl. naut.* hold battens, sparring
рыбоводство fish farming/breeding/culture, pisciculture
рыбозмея *zool.* caecilian, coecilian *(Ichthyopsis)*
рыбокомбинат fish factory, fishworks
рыбокоптильный fish-smoking, fish-curing
рыболов 1. fisherman; angler; 2. *zool.* fisher, pekan *(Martes pennanti)*
рыболовство fishery; fishing
~, промышленное commercial fishing
рыболокатор fish finder
рыбообрабатывающий fish-processing
рыбоохрана fish conservancy/protection/preservation
рыбопитомник fish farm
рыбоплавниковые *pl. zool.* ichthyopterygians *(Ichthyopterygia)*
рыбопромышленность fishing industry
рыбоход fish pass/chute/ladder; fishway
рыбочистка fish scaler, fish scaling/descaling machine
рыбоядный ichthyophagous, piscivorous
рыбы *pl.* fishes; *astr.* Pisces
~, акулоподобные cladoselachians *(Cladoselacheie)*
~, бесчелюстные cephalaspids *(Aspidocephalii)*
~, двоякодышащие dipnoan/dipnoous fishes *(Dipnoi)*
~, ильные mudfishes
~, костистые teleosts *(Teleostei)*
~, легочные lung-fishes
~, панцирные placoderms *(Placodermi)*
~, хоановые choanate fishes *(Choanichthyes)*
~, хрящевые cartilaginous fishes
~, хрящекостные chondrosteans *(Chondrostei)*
рыжик *bot.* 1. saffron milk cap, delicious lactarius *(Lactarius deliciosus)*; 2. false flax *(Camelina)*
рыльце *bot.* stigma
~, перистое plumose stigma
рым *naut.* eye, lug, ring, eyebolt, rim
~, буксировочнй/буксирный towing lug
~, подъемный hoisting/lifting eye
~, швартовный mooring ring
рым-болт ring/eye bolt
рынок market; market place
~, валютный exchange/currency market
~, внешний foreign market
~, внутренний home/domestic/internal market
~, денежный money market
~, дисконтный discount market
~, единый single market
~ кредита credit market
~ кредита, обильный easy/cheap money
~, международный евровалютный Eurocurrency market
~, международный евродолларовый Eurodollar market
~ наличного товара spot market
~, насыщенный glutted market
~ недвижимости real estate market
~ облигаций bond morket
~, Общий the Common Market
~, оживленный active/broad market
~, организованный formal market
~, перенапреженный overbought market
~ по сделкам на срок forward/settlement market
~ подержанных товаров secondhand market
~ ссуд loan market
~ труда labour/job market
~, устойчивый steady/stiff/strong market
~, учетный discount market
~ ценных бумаг securities/placement market
~, черный black market
рысиха *zool.* female lynx
рыск *nav.* yawing
рыскание yaw, yawing; hunting; griping, restlessness, yaw instability
~ на курсе *naut.* rambling, wabbling on the course
рыскать *naut.* yaw; careen off; slew
рыскливый *naut.* restless, ardent
рысь *zool.* lynx *(Felis lynx, Lynx lynx)*
рыхление loosening; hoeing
~ подпочвы subsoiling
рыхлитель *agr.* ripper, scarifier; cultivator
~, почвенный soil aerator
рыхлить *agr.* rip, loosen, make friable
рыхлозернистый loosely grained

рыхлота *met.* microporosity
рыцарь knight
~ печального образа *lit.* Knight of the Rueful Countenance
~, странствующий knight errant
рычаг lever, arm, stick, joystick
~, вильчатый fork lever
~, дроссельный/газа throttle lever
~, задающий/командный master arm
~, исполнительный/копирующий slave arm
~ первого рода lever of the first order
~, поворотный pivoted lever
~, рулевой steering arm
~ стеклоочистителя windscreen wiper arm
~, стопорный locking lever
~ стояночного тормоза parking brake lever
~ съема *comp.* access arm
~, тормозной braking lever
~ управления control lever
рэнд rand *(monetary unit)*
рюкзак knapsack, rucksack
рябизна *met.* pitted/spotted surface
рябина *bot.* mountain ash *(Sorbus)*
рябинник I *bot.* 1. false spirea *(Sorbaria);* 2. grove of mountain ash
рябинник II *zool.* fieldfare *(Turdus pilaris)*
рябишник *bot.* comnon tansy *(Tanacetum vulgare)*
рябок *zool.* sand grouse *(Pterocles)*
рябуха *plant pathol.* wildfire
рябчик I *zool.* hazel grouse/hen *(Tetrastes bonasia)*
рябчик II *bot.* fritillaria, fritillary *(Fritillaria)*
рябь ripples, rippled surface/marks, rippling
~, ветровая wind ripples
~, волновая undulation ripples
~, звуковая acoustic ripples
~, песчаная sand ripples
ряд 1. series; set; sequence, order, succession, progression, continuum; 2. row, line; rank; file; 3. *tech.* family, series
~, ароматический aromatic series
~, безусловно сходящийся unconditionally convergent series
~, бесконечный infinite series
~, биномиальный binomial series
~, вариационный variational series
~, временной time/temporal series
~, всюду сходящийся степенной *math.* everywhere/permanently convergent power series
~, выровненный smoothed series
~, гармонический/гармоничный harmonic series
~, геометрический geometric series/progression
~, гомологический homologous series
~, гранулоцитарный granulocytic/myeloid series
~ данных sequence of data
~, дискретный discrete sequence
~, жирный fatty series
~, знакопеременный/знакочередующийся alternating series
~, изодиаферный isodiapheric sequence
~ импульсов pulse train
~ кирпичной кладки course of brickwork
~, конечный finite series
~, лимфоцитарный lymphocytic/lymphoid series
~, линейный linear series, lineage
~ Лорана Laurent series
~, мажорирующий dominant series
~ Маклорена Maclaurin series
~, масштабный *cart.* succeeding scales
~, одномерный временной univariate time series
~, олефиновый olefin family
~ ошибок error range
~, парафиновый paraffin family
~, планетарный *mech.* planetary gear set
~ по степеням x series in powers of *x*
~, показательный *math.* power series
~ превращений transmutation series
~, прерывистый discontinuous series
~, равномерно сходящийся uniformly convergent series
~, радиоактивный radioactive/decay chain; radioactive series
~ распадов decay series
~, расходящийся divergent series
~ решетки lattice row
~ свай piling, row of piles
~, сглаженный smoothed series
~, случайный random series
~, собственно расходящийся properly divergent series
~ состояний, дискретный sequence of discrete states
~ состояний, непрерывный continuum of states
~, статистический statistical series/array
~, стационарный временной stationary time series
~, степенной power series
~, сходящийся convergent series
~ счетчиков counter tray
~ Тейлора Taylor series
~ триангуляции triangulation chain/strip
~, тригонометрический trigonometric series
~, транскюриевый transcurium series
~, трансурановый transuranium series
~, тромбоцитарный thrombocytic series
~, упорядоченный ordered series
~ урана uranium series
~, усеченный truncated series
~, условно сходящийся conditionally convergent series
~, филогенетический phylogenetic sequence, lineage
~, функциональный *math.* series of functions
~ Фурье Fourier/harmonic series
~ цилиндров *tech.* cylinder bank/row
~, числовой numerical/number series

~, экономический временной economic time series
~, эритроцитарный erythrocytic series
рядовка *bot.* blewits *(Tricholoma personatum)*
рядовой 1. *adj.* ordinary, common; rank-and-file; 2. inrows; 3. *n.* *mil.* private [soldier]
ряж crib, cribwork, bankhead
ряпушка *zool.* European whitefish, vendace *(Coregonus albula)*
ряска *bot.* duckweed *(Lemna)*
рясковые *pl., bot.* duckweed family *(Lemnaceae)*
рясt *bot.* squill, wild hyacinth *(Scilla)*

С, с

саам Lapp, Laplander
саамка Lapp woman
саамский Lappish
сабадилла *bot.* sabadilla *(Sabadilla)*
сабадиллин *chem.* sabadilline, cevadilline
сабаль *bot.* palmetto *(Sabal)*
сабеизм Sabaism, star worship
сабеист Sabaist, star worshipper
сабей, сабейский Sabaean
сабейка Sabaean (woman)
сабельник *bot.* comarum *(Comarum)*
сабинский *hist.* Sabine
сабир Sabir
саблевиднолистный *bot.* acinacifolious
саблевидный, саблеобразный *bot.* aciniform, acinaciform
саблезубые *pl.* *paleont.* machairodonts *(Machairodontidae)*
сабля saber
сабо sabot
сабур *pharm.* aloe *(Aloe)*
сабуровый *pharm.* aloetic
саван shroud, cerement
саванна savanna(h)
савка *zool.* Oxyura
савой *bot.* savoy cabbage *(Brassica oleracea sabauda)*
сагайдак *hist.* (bow and) quiver
сагиб sahib
сагитта *zool.* sagitta *(Sagitta)*
саго sago flour
саговник *bot.* cycas, sago palm *(Cycas)*
саговниковые *pl.* *bot.* cycas family *(Cycadaceae)*
саговниковый *bot.* cycadaceous, cycadeous
сагуин *zool.* sagoin, marmoset *(Hapale jachus)*
саддукей *rel.* Sadducee
садиться 1. alight, land; set; 2. board, embark; 3. *text.* shrink
~ по ветру land downwind
~ против ветра land into the wind
садка *met.* charge; load
садовик *bot.* plum mushroom/agaric *(Agaricus prunulis)*
садоводство gardening; horticulture
~, декоративное landscape gardening
~, промышленное/товарное market gardening
сажа soot; *tech.* carbon black
~, белая silica white/filler
~, газовая carbon/gas black
~, канальная channel/impingement black
~, печная furnace black
~, термическая thermal black
~, форсуночная oil-based furnace black
сажалка planting machine, planter, plant settler
сажать plant
саженки *sport* overarm stroke, crawl
сажень sazhen, sagene (2.13 metres)
~, морская fathom (1.83 metres)
сажесдуватель soot blower
сажка *bot.* sooty blotch disease, smut
саза *bot.* sasa *(Sasa)*
сазан *zool.* wild carp, sazan *(Cyprinus carpio)*
саиб sahib
саидский Sahidic
сайга *zool.* saiga, sheeplike antelope *(Saiga tatarica)*
сайда *zool.* sea salmon, pollack *(Pollachius* virens*)*
сайзель *bot.* sisal *(Agave sisalina)*
сайка *zool.* Arctic cod *(Boreogadus saida)*
сайодин *pharm.* sajodin
сайра *zool.* Pacific saury *(Cololabis saire)*
сакродиния *med.* sacrodynia
сакроилеит *med.* sacroileitis
~, гнойный purulent sacroileitis
сакрококсит sacrocoxitis, sacrocoxalgia
сакрококцигеальный sacrococcygeal
сакролюмбальный sacrolumbar
сакротомия *surg.* sacrotomy, sacretomy
сакс *hist.* Saxon
саксонец Saxon
саксонка Saxon (woman)
саксонский Saxon; Saxonic
~ фарфар Meissen china
саксофон saxophone
саксуал *bot.* saxual, haloxylon *(Haloxylon)*
сактосальпинкс *med.* sactosalpinx
сакура *bot.* Oriental cherry *(Prunus serrulata)*
сал *bot.* sal shorea *(Shorea robusta)*
салазки carriage, sliding block, slide rails, skids, sled(s); *sport* toboggan
~, крестовые cross slide
~ револьверного суппорта turret carriage
~ сиденья seat slide/rail
~, спусковые *naut.* launching cradle
~ суппорта carriage saddle
~, фрезерные cutter slide
салака *zool.* 1. sprat *(Clupea sprattus);* 2. Baltic herring *(Clupea narengus membrans)*
саламандра *zool.* salamander
салангана *zool.* salangane *(Collocalia)*

салат *bot.* lettuce *(Lactuca)*
салеп salep
саливация salivation, ptyalism
саливогландулез *med.* salivoglandulosis
Салим *bibl.* Salem
салинг *naut.* (topmast) crosstrees
салипирин *pharm.* salipyrine, antipyrine salicylate
салицил *chem.* salicylate
салициламид *chem.* salicylamide
салицилат *chem.* salicylate
  ~ натрия solidum salycilate
салицилизм *med.* salicylism
салициловый *chem.* salicylic
салический *hist.* Salic
сало tallow, fat; *anat.* sebum
  ~ кожное sebum cutaneum, cutaneous fatty secretion
салон *naut.* mess; *av.* cabin, compartment
салол *pharm.* salol, phenyl salicylate
салолин, саломас hydrogenated fat
сальварсан *pharm.* arsphenamine, salvarsan
сальвия *bot.* sage *(Salvia)*
сальдирование *econ.* balancing, balance
  ~, прямое direct/net balance
сальдировать balance; sum up
сальдо *econ.* balance
  ~, активное active/positive balance
  ~ банковского счета bank balance
  ~, выводить strike the balance
  ~, дебетовое debtor balance, balance due
  ~ к переносу balance carried forward
  ~ клиринга clearing balance
  ~, конечное closing balance
  ~, кредитовое credit balance
  ~ миграции net migration
  ~ на базе ликвидности balance on liquidity basis
  ~, начальное opening balance
  ~, общее overall balance
  ~, перенесенное balance brought from last account
  ~ по контокорренту current balance
  ~ по текущим расчетам, активное current surplus
  ~, пассивное adverse/passive balance
  ~ с переноса balance brought forward
  ~ счетчика counter balance
  ~ торгового баланса, активное export balance of trade, export surplus
сальмонеллез *med.* salmonellosis
сальмонеллы *pl.* Salmonella
сальник 1. (packing) gland, stuffing box, gasket, oil-seal, O-ring seal; 2. *anat.* omentum
  ~, большой *anat.* greater omentum, gastrocolic omentum
  ~, водоплотный/жидкоплотный watertight gland, stuffing box
  ~, дейдвудный stern tube stuffing box
  ~, малый *anat.* lesser omentum, gastrohepatic omentum
  ~, опрокинутый inverted stuffing box
  ~, переборочный bulkhead stuffing box
  ~, расширительный expansion stuffing box
  ~ с угольной набивкой carbon-packed gland
сальниковый 1. packing *(attr.)*; 2. *anat.* omental, epiploic
сальный sebaceous, sebiferous; greasy
сальпингит *med.* salpingitis
  ~, гнойный purulent salpingitis
  ~, катаральный catarrhal salpingitis
  ~, узловатый nodular salpingitis
  ~, язвенный ulcerous/ulcerative salpingitis
сальпингостомия *med.* salpingostomy, salpingostomatomy
самарий *chem.* samarium, Sm
самаритянин, самаритянский Samaritan
самаритянка Samaritan (woman)
самбо *abbr.* unarmed self-defence
самбук *bot.* elder *(Sambucus)*
самбука *mus.* sambuke
самка *zool.* female
самнит, самнитский Samnite
самоампутация self-amputation, autotomy
самоанализ self-examination, introspection
самоанец, самоанский Samoan
самоблокировка self-locking, automatic interlocking
самовводящийся self-loading
самовентиляция self-ventilation/aeration
самовключатель circuit recloser
самовключение automatic switching
самовластие autocracy, absolute power
самовнушение autosuggestion, self-suggestion
самовозбуждение self-excitation, auto-excitation
самовозврат self-reset
самовозгорание spontaneous combustion/ignition
самовоспламенение spontaneous/hypergolic ignition
самовоспламеняющийся self-igniting, hypergolic
самовосстановление self-recovery
самовращение autorotation; *av.* windmilling
самовсасывающий self-priming *(of pump)*
самовулканизация self-vulcatization/curing
самогасящий(ся) self-quenching
самогипноз autohypnosis, idiohypnosis
самодвижущийся self-propelled/powered
самодельный self-made, homemade
самодержец autocrat
самодовлеющий self-sufficing/sufficient
самозажигание spontaneous ignition, selfignition
самозапускающийся self-starting/triggering
самозаражение *med.* autoinfection
самозарождающийся autogenic, autogenous
самозарождение spontaneous generation; autogenesis, autogeny
самозатягивающийся self-sealing
самоизлучение spontaneous radiation
самоиндуктивность self-inductance
самоиндукция self-induction

самоконтролирующийся self-testing/checking
самоисправляющийся self-correcting
самоконтроль *comp.* self-test/check/testing
самолет aircraft, *Eng.* aeroplane, *Am.* airplane, plane
~, аэрофотосъемочный aerial photographic aeroplane, photographic survey aeroplane
~ базирующийся на авианосец carrier-based aeroplane
~, беспилотный pilotless aeroplane, drone
~ беспилотный радиоуправляемый remotely controlled drone aeroplane
~, боевой combat/operational aeroplane, warplane
~ вертикального взлета vertical take-off [VTO] aeroplane
~ вертикального взлета и посадки vertical take-off and landing [VTOL] aeroplane
~, винтовой propeller/driven aeroplane
~, всепогодный all-weather aeroplane
~, гигантский giant/jumbo aeroplane
~, гигантский реактивный jumbo-jet
~, грузовой freight/cargo aeroplane
~ дальнего действия long-range aeroplane
~, двухдвигательный twin-engine aeroplane
~, двухместный two-seat aeroplane
~, двухпалубный double-deck aeroplane
~, десантно-транспортный *mil.* assault transport aeroplane; troop carrier
~ для дальних перевозок long-haul aircraft
~ для научных исследований research aeroplane
~, дозвуковой subsonic aeroplane
~ изменяемой геометрии variable-geometry aeroplane
~, корабельный ship-based/borne aeroplane
~ короткого взлета и посадки short take-off and landing [STOL] aeroplane/aircraft
~, короткорейсовый short-haul aeroplane
~, лодочный flying boat
~, магистральный trunk-route aeroplane
~, маневренный manoeuvrable/aerobatic aeroplane
~, однодвигательный single-engined aeroplane
~, одноместный single-seat aeroplane
~, однопалубный single-deck aeroplane
~ охранения escort aeroplane
~, палубный carrier-based/borne aeroplane
~, поплавковый float aeroplane
~, поршневой piston-engined aeroplane
~, почтовой mail-carrying aeroplane
~, радиоуправляемый radio-controlled aeroplane
~, реактивный jet aeroplane
~ с верхним расположением крыла shoulderwing aeroplane
~ с высокой скороподъемностью fast-climbing aeroplane
~ с высокыми летными качествами high-performance aeroplane
~ с двумя двигателями twin-engine [twin-jet, twin-turboprop] aeroplane
~ с изменяемой геометрией variable-geometry [VG] aeroplane
~ с колесным шасси wheeled aeroplane
~ с кольцевым крылом annular-wing aeroplane
~ с коротким разбегом и пробегом short takeoff and landing aeroplane
~ с крылом изменяемой стреловидности swing-wing [variable-sweep] aeroplane
~ с лыжным шасси skiplane
~ с поршневым двигателем piston-engined aeroplane
~ с ракетным двигателем rocket-powered aeroplane
~ с толкающим винтом pusher aeroplane
~ с треугольным крылом delta-wing aeroplane
~ с турбовинтовыми двигателями turboprop aeroplane
~ с убирающимся шасси retractable-gear aeroplane
~ с шасси на воздушной подушке ground-effect take-off and landing [GETOL] aeroplane
~, санитарный air ambulance
~, сверхзвуковой supersonic aeroplane
~ сопровождения escort aeroplane
~ со стреловидным крылом swept-wing aeroplane
~, спортивный sporting aeroplane
~, среднерейсовый medium-haul aeroplane
~ типа "утка" canard aeroplane
~, турбовентиляторный turbofan aeroplane
~, турбовинтовой turboprop aeroplane
~, турбореактивный turbojet aeroplane
~, учебно-тренировочный trainer aeroplane
~, учебный basic/primary trainer
самолет-амфибия amphibious aeroplane, amphibian
самолет-буксировщик tow(ing) aeroplane
самолет-вертолет compound helicopter
самолет-заправщик tanker/refuelling aeroplane, flying tanker
самолет-истребитель fighter (aeroplane)
самолет-носитель carrier/mother/launch aeroplane
самолетовождение aeronavigation, avigation
самолето-вылет *av.* sortie
самолетостроение aircraft manufacture/industry
самолет-перехватчик interception aeroplane
самолет-разведчик reconnaissance aeroplane
самолет-ракетоносец missile-carrier aeroplane
самолет-снаряд flying bomb
самолет-топливозаправщик tanker/refuelling aeroplane, flying tanker, aerial refueller
самолет-штурмовик attact/strike/strafing aeroplane
самолечение self-treatment, autotherapy
самоликвидатор self-destructor
самолюс *bot.* brookweed *(Samolus)*
самомодуляция self-modulation

самонаведение self-guidance, homing guidance
~, оптическое optical homing
~ по инфракрасному излучению infra-red homing
~ по световому излучению light homing
~ по электромагнитному излучению electromagnetic homing
~, радиолокационное radar homing
~, тепловое heat(-seaking) homing
самонаводящийся homing, self-guided/guiding
самонагрев(ание) spontaneous heating
самообеспеченность *econ.* self-sufficiency
самообслуживание self-service
самообучение *comp.* self-learning
самоокисление spontaneous oxidation
самоокупаемость *econ.* self-repayment
самооплодотворение self-fertilization, autogamy
самоопределение *pol.* self-determination
самоотжиг self-annealing
самоотпуск *met.* self-tempering
самопереваривание self-digestion, autolysis
самописец recording instrument, recorder
~, барабанный drum recorder
~, бортовой airborne/flight recorder
~, быстродействующий fast-response recorder
~ давления pressuregraph, pressure recorder
~, двухкоординатный XY-recorder
~, двухперьевой dual pen recorder
~, дискретный digital recorder
~, дистанционный remote recorder
~, ленточный 1. tape recorder; 2. stripchart recorder
~, многоперьевой multipen recorder
~, многоточечный multiple-point recorder
~, осциллографический trace recorder
~, перьевой/чернильный pen/ink recorder
~, проволочный wire recorder
~ с магнитной лентой magnetic-tape recorder
~, цифровой digital recorder
~ эхолота depth recorder
самоподача *tech.* self-feeding
самоподготовка independent/unsupervised study
самопознание self-knowledge/-actualization
самопрививка *bot.* self-grafting
самораздражение *biol.* self-stimulation
саморазложение *chem.* autolysis
саморазряд self-discharge, standing loss
самораспад autodecomposition
саморастворение autolysis
самородок nugget; native metal
самосвал dump truck, dumper
~ с разгрузкой в бок side dump truck
~ с разгрузкой на две стороны two-way dump truck
~ с разгрузкой назад rear-dump truck
самосветящийся self-luminous
самосжатие self-constriction
самосопряженный *math.* self-adjoint/conjugate
самосский Samian
самострел I self-inflicted wound
самострел II *hist.* arbalest, crossbow
самостробирование self-gaiting
самосуд lynch/gibbet law
самотаска drag conveyer
самотек gravity flow
самотеком by gravity
самоуплотняющийся self-sealing/packing
самоуправление 1. *tech.* automatic control; 2. autonomy, home rule
самофазировка autophasing
самоход *mech.* power feed; self-feed
~, поперечный cross power feed
~, продольный longitudinal power feed
самоходный self-propelled, automotive
самоцвет *bot.* chaff-flower *(Achyranthes)*
самум simoom, sand/dust storm
самурай Samurai
самшит *bot.* box *(Buxus)*
санбат *abbr. mil.* medical battalion
сангвинария *bot.* bloodroot *(Sanguinaria)*
сангвиник sanguine person
сангвинический sanguine
сандал I *naut.* sandal
сандал II *bot.* sandalwood tree *(Santalum album)*
сандаловый sandalwood *(attr.),* santalaceous
сандарак 1. *bot.* cypress pine *(Callitris);* 2. sandarac resin
санджак *hist.* sanjak
сандружина *abbr.* medical detachment/unit
санидин *miner.* sanidine
~, стекловидный rhyacolite
санинспекция *abbr.* sanitary inspection
санитария sanitation, practical hygiene
санкюлот *hist.* sabsculotte
санпросвет *abbr.* health education
сансевиерия *bot.* bowstring hemp (Sansevieria)
санскрит Sanscrit
санскритолог Sanscrit scholar, Sanscritist
санскритология Sanscritology
санскритский Sanscrit *(attr.)*
сантехник *abbr.* sanitary engineer
сантехника *abbr.* sanitary engineering/equipment
сантибар centibar
сантибел centibel
сантиграмм centigram
сантим centime
сантиметр 1. centimeter; 2. tape measure
сантипуаз centipoise
сантистокс centistokes
сантолина *bot.* lavender cotton *(Santolina)*
санузел *abbr.* rest room; toilet facilities
санчасть *abbr. mil.* medical unit
сап *med.* glanders, malleus, equinia
~, кожный farcy
~, ложный pseudoglanders, malleoidosis, melioidosis
сапажу *zool.* capuchin monkey, sapajou *(Cebus)*
сапан *bot.* sapanwood *(Caesalpinia sappan)*
сапер *mil.* combat engineer
сапиндус *bot.* soapberry *(Sapindus)*

сапной *med.* glanderous
саподилла *bot.* sapodilla *(Achras zapota)*
сапонария *bot.* soapwort *(Saponaria)*
сапонат soapstone
сапонит *miner.* saponite, pionite
сапота *bot.* sapote, sapodilla *(Calocarpum sapota)*
сапрогенный saprogenous, saprogenic
сапроколь saprocoll
~, ископаемый saprocollite
сапролит soapstone, saprolite
~, известковый terra rossa, red ocher
сапропель *geol.* sapropel; organic slime
~, затвердевший sapropelite, sapropelith
~, известковый talc-sapropel
~, илистый slimy sapropel
сапрофаг saprophage, saprophagan
сапрофит saprophyte
сапрофитный saprophytic, saprophilous
сапсан *zool.* peregrine falcon *(Falco peregrinus)*
сапун breather, air vent
сапфир sapphire
~, звездчатый star/asteriated sapphire
сапфический *lit.* Sapphic
сарана *bot.* tawny day lily *(Hemerocallis fulva)*
саранча *zool.* locust
~, перелетная migratory/Asiatic locust
сарацин *hist.* Saragen
сарацинка *hist.* Saragen (woman)
сарацинский Saragen, Saragenic(al)
сараган *zool.* garfish *(Belone belone)*
~, морской needle garfish
саргановые *zool.* needlefishes *(Beloniadae)*
саргассы *pl.* gulfweeds, saragasso weeds
сарда *zool.* bonito *(Sarda)*
сардина *zool.* sardine, pilchard (Sardina)
сардинец, сардинский Sardinian
сардинка *zool.* scaled sardine
сардоникс sardonix
саркоаденома *med.* sarcoadenoma, adenosarcoma
саркодес *bot.* snow plant *(Sarcodes)*
саркозиноксидаза sarcosine oxidase
саркоидоз sarcoidosis
саркокарпый *bot.* sarcocarp
сарколейкоз sarcoleucosis, sarcoleukemia
сарколеммный sarcolemmic, sarcolemmous
сарколизин sarcolysine, sarkolysine, melfalan
сарколог sarcologist
саркома *path.* sarcoma
~, ангиолитическая angiolithic sarcoma, psammona
~, веретеноклеточная spindle cell sarcoma, fascicular sarcoma
~, круглоклеточноя round cell sarcoma, encephaloid sarcoma
~, мукозная mucoid sarcoma
саркомер *anat.* sarcomere
саркомный *med.* sarcomatous
саркомоподобный sarcoid
саркоплазматический sarcoplasmic
саркопсиллез sarcopsyllosis
саркоспоридиоз sarcosporidiosis, sarcocystosis
сармат, сарматский *hist.* Sarmatian
саррацения *bot.* pitcher plant *(Sarracenia)*
сарценхима *bot.* sarcenchyma
сарыч *zool.* buzzard *(Buteo)*
сассапарель *bot.* greenbrier *(Smilax)*
сателлит 1. satellite; 2. *tech.* satellite gear/wheel, planet pinion; 3. complementary rock
~ дифференциала differential pinion
~, нуклеарный nuclear satellite, paranucleolus
сатинирование hot-pressing, glazing, satinizing
сатир *myth., zool.* satyr
сатириаз *med.* satyriasis
Сатурн *astr., myth.* Saturn
сатурнизм *med.* saturnism, lead poisoning, plumbism
сауриаз *med.* sauriasis, alligator skin
сауроматум *bot.* lizard arum *(Sauromatum)*
сафический *lit.* Sapphic
сафлор 1. *bot.* safflower *(Carthamus tinctorius);* 2. safflower dye/oil
сафранин *chem.* safranine, aniline pink
сафьян morocco leather, saffian
сахар sugar
~, виноградный grape sugar
~, головной sugar loaf
~, желтый brown sugar
~ крови *med.* blood sugar
~, кусковый lump sugar
~, молочный *chem.* milk sugar, lactose
~, пиленый cube sugar
~, плодовый *chem.* fructose, levulose
~, свинцовый *chem.* sugar of lead
сахараза *bioch.* saccharase, invertase
сахариметр saccharimeter, saccharometer
сахарин *chem.* saccharin, benzosulfimide
сахарный sugar *(attr.);* saccharic, glucaric
~ завод sugar refinery
~ песок granulated sugar
~ная кукуруза sweet corn
~ная пудра powdered/confectioner's sugar
сахароза saccharose, cane sugar, sucrose
сахаромицеты *pl. Saccharomyces*
сахарообразование saccharification
сахар-песок granulated sugar
сахар-рафинад refined sugar
сахар-сырец raw sugar
сбалансированность balanced state
~, бухгалтерская accounting equation
сбережение economy; conservation; *pl.* savings
~ия договорного характера contractual savings
~ия, фактические ex post savings
сберкасса *abbr.* savings bank
сберкнижка *abbr.* savings book
сбой malfunction, failure, error, check
~ в интерфейсе *comp.* interface check/error
~ в канале channel failure
сбойка *min.* crosscut, connection, holing
сбор 1. collection, collecting, gathering; *comp.* ac-

quisition, logging, origination; 2. duty, toll, charges, dues; receipts, returns; yield; 3. *pharm.* species
~, акцизный excise tax
~ винограда *agr.* vintage
~, гербовый stamp duty
~, грудной pectoral species/tea
~ данных data acquisition/collection
~ данных в графической форме graphic data/information acqisition
~ данных на магнитную ленту magnetic tape data recording
~ данных с помощью внешних устройств peripheral data acquisition
~, дополнительный surcharge
~ за стоянку *naut.* moorage
~, лихтерный *naut.* lighterage
~, мочегонный diuretic species/tea
~, мягчительный emollient species
~ первичных данных primary data acquisition/collection
~, полный *theatre* full house
~, почтовый postage
~, причальный berthage, wharfage, quayage, berthing dues, berth charge
~, слабительный laxative species/tea
~, страховой insurance fee
~, таможенный custom charges
~ урожая harvest, harvesting
~, шлюзовый *naut.* lockage toll
~, якорный anchor dues
~ы, портовые harbour dues, port charges
сборка 1. assembly, assembling, mounting, fitting, joining, setting up, erection; assemblage; 2. unit, assembly
~, агрегатная sectional assembly
~ мусора garbage collection
~ на месте field assemply
~, предварительная pre-assembly; prefarbication, preliminary assembly
~, секционная unit assembly
~, узловая subassembly
~, шинная *el.* busbar assembly
сборник 1. collector, receiver; storage tank, header; accumulator; 2. collection, miscellany
~, выпускной exaust collector
~, тарифный tariff book
сборочный assembly *(attr.),* assembling
сборщик 1. assembly worker, filter; 2. collector
~ налогов tax collector/receiver
сбраживание fermentation
сбраживающий fermentative
сбрасывание throwing off/down, dropping, disposal, discarding, dumping, discharging, discharge
~, принудительное/аварийное *av.* jettisoning
сбрасыватель tripper device; release
сбрасывать discard, dump; drop
сброс 1. *see* **сбрасывание**; drop, release; 2. *geol.* fault; 3. *comp.* reset
~, аварийное *av.* jettisoning
~ бытовых сточных вод household effluent
~, обратный *geol.* reverse fault
~, осевой pivotal fall
~ системы *comp.* system reset
~, складчатый folded fault
сбрособразование *geol.* faulting
сброс-сдвиг strike-slip fault
сбыт *econ.* sale, marketing; market
~ плохой slow/poor market
~, хороший ready sale/market
сваб *oil* swab
сваебоец pile driver
сваевыдергиватель pile extractor
свайка *naut.* splicing fid/tool
сваливание dop(ping), stall(ing)
~ самолета stalling
сваливаться fall, drop; *av.* stall
свалка dumping ground; dump
свариваемость weldability
сваривать weld; weld together
сварка welding
~, автогенная autogenous welding
~, автоматическая machine welding
~, ацетиленовая oxy-acetylene welding
~ в защитных газах gas-shielded welding
~ в нижнем положении down-hand welding
~ в среде углекислого газа carbon-dioxide shielded welding
~, верхняя overhead welding
~, водородно-кислородная oxyhydrogen welding
~ встык butt welding
~, высокочастотная dielectric welding
~, газовая gas/autogeneous welding
~, газопрессовая pressure-gas welding
~, дуговая arc welding
~, дуговая импульсная pulsed arc welding
~, контактная resistance welding
~, кузнечная forge welding
~, линейная seam/line welding
~, металлодуговая metallic arc welding
~ методом сопротивления resistance welding
~, механизированная machine welding
~, многопроходная multipass welding
~ на подъем uphill welding
~ на спуск downhill welding
~ неплавящимся электродом non-consumable electrode welding
~, нижняя downhill/flat welding
~, печная furnace pressure welding
~ плавлением fusion/fuse welding
~ плавящимся электродом consumable electrode welding
~, плазменная plasma-arc/jet welding
~ под флюсом submerged-arc welding
~, порошковая powder welding
~, потолочная overhead/inverted welding
~ прихваточным швом tack welding
~, рельефная projection/relief welding
~, роликовая seam welding

~ **с накладкой** tie welding
~ **с накоплением энергии** energy-storage welding
~ **сопротивлением** resistance welding
~, **термитная** thermit(e) welding
~ **токами высокой частоты** high frequency welding
~, **торцовая** edge welding
~, **точечная** spot welding
~ **угловым швом** fillet welding
~, **угольно-дуговая** carbon arc welding
~, **ударная** percussion/percussive welding
~ **узким швом** bead welding
~, **шовная** seam welding
~ **электродуговая** (electric) arc welding
~, **электроннолучевая** electron-beam welding
сварной welded
сварочный welding *(attr.)*
~ **пистолет** welding gun
~**ная головка** welding head/cap
~**ная машина** welding machine, welder
сварщик welder
свастика swastika
свая pile, post
~, **винтовая** boring pile
~, **деревянная** timber pile
~, **маячная** gauge/guide/leading pile
~, **отбойная** *naut.* fender pile, pile fender
~, **сборная** precast pile
~, **фундаментная** foundation pile
~, **шпунтовая** sheet/grooved pile
свая-оболочка tubular pile
свая-стойка end-bearing pile
сведа *bot.* seepweed *(Sueda)*
свежезамороженный chilled, fresh-/quick-frozen
свекла *bot.* beet *(Beta);* beetroot
~, **кормовая** fodder beet
~, **сахарная** sugar beet
~, **столовая (красная)** red beet
свекловодство sugar-beet growing
свеклокомбайн beet combine
сверление drilling, boring
сверлилка hand drill; drilling machine
сверлило *zool.* borer
сверлильно-долбежный drilling-and-mortizing
сверлильно-нарезной boring-and-threading
сверлильно-фрезерный boring-and-milling
сверлить drill, bore, perforate
~ **начерно** rough-drill
~ **начисто** rebore
сверло drill, borer, perforator; auger; bit
~ **алмазное** diamond drill
~, **бурильное** drill bit
~, **винтовое** twist drill, worm bit
~, **горное** rock drill
~, **ожекторное** chip ejection drill
~, **оснащенное твердым сплавом** carbide-tipped drill
~, **ружейное** gun drill
~, **центровочное** centre drill
~, **черновое** first bit
~, **чистовое** finishing bit/drill
сверлянка *zool.* borer *(Pholas tullus)*
свернувшийся coagulated, clotted
свернутость convolution; twist
свернутый convolute, coiled; intort
свертка *math.* convolution; fold(ing)
~ **двух функций** resultant of two functions
~ **тензора** contraction of a tensor; strangling of an index
свертка-развертка rollout-rollin
сверток 1. *med.* clot, coagulum; 2. bundle
свертываемость *med.* coagulability
свертывание 1. coagulation; 2. rolling up, folding; 3. *math.* convolution; transvection; 4. winding up, closure
свертывающий coagulant, coagulating, coagulative
сверхатмосферный superatmospheric
сверхбаллон doughnut [super balloon] tyre
свербыстрый superfast, superhigh-speed
сверхвысоковольтный supervoltage *(attr.)*
сверхвысокочастотный superhigh-frequency, microwave *(attr.)*
сверхдавление overpressure, superpressure
сверхдержава superstate, superpower
сверхестественный supernatural
сверхжесткий ultrahard
сверхзатухание overdamping
сверхзанятость *econ.* overemployment
сверхвид *biol.* superspecies
сверхзвезда superstar; quazar
сверхзвук ultrasound, supersound
сверхзвуковой supersonic, ultrasonic
сверхизлучение superradiant radiation
сверхигра *game th.* supergame
сверхиндустриализация overindustrialization
сверхкомплектный supernumerary
сверхкомпрессия supercompression
сверхконсерватор ultraconservative
сверхкорона *astr.* supercorona
сверхкраткый extrashort
сверхкритический supercritical, above-critical
сверхкритичность supercriticality
сверхлегирование *met.* overdoping
сверхлегкий extralight, superlight
сверхлимитный *fin.* extrabudgetary
сверхлистовый *bot.* epiphyllous
сверхмалый midget *(attr.)*
сверхматериальность supersubstantiality
сверхматериальный supersubstantial
сверхмерный oversize
сверхминиатюрный microminiature, subminiature
сверхмодуляция overmodulation
сверхмощный superpower *(attr.)*
сверхнагрузка surplus/extra load
сверхнациональный supranational
сверхнизкий ultralow
сверхплавкий hyperfusible

сверхплановый over and above the plan
сверхплоскость *math.* hyperplane
сверхплотный superdense
сверхприбыль superprofit, extra profit
свехпроводимость superconductivity
сверхпроводник superconductor
сверхпроводящий superconducting
сверхразмерный overdimensioned; oversize
сверхранний *agr.* very early
сверхрасширение overexpansion
сверхрегенерация superregeneration
сверхрезонанс overresonance
сверхрефракция superrefraction
сверхсветовой exceeding the speed of light
сверхсекретный top-secret
сверхсжатие overcompression
сверхскопление *astr.* supercluster
сверхскоростной superhigh-speed, superfast
сверхсменный extrashift; overtime
сверхсметный not budgeted for, extrabudget
сверхсовременный ultramodern
сверхсопряжение hyperconjugation
сверхсрочнослужащий *mil.* re-enlistee
сверхсрочный *mil.* re-enlisted
~ная служба re-enlistment
сверхстандартный superstandard
сверхстоимость excess cost
сверхструктура superstructure
сверхсходимость overconvergence
сверхсчетный odd
сверхтвердый superhard, extrahard
сверхтекучий superfluid
сверхтермометр hyperthermometer
сверхток *el.* overcurrent, excess current
сверхтонкий hyperfine, ultrafine, ultrathin
сверхупругий hyperelastic, superelastic
сверхурбанизация overurbanization
сверхурочный overtime; *pl.* overtime pay
~ная работа overtime (work)
~ное время overtime
сверхусиление overamplification
сверхустойчивость superstability
сверхфильтр ultrafilter
сверхчеловек superman
сверхчеловеческий superhuman; transhuman
сверхчистый super-pure, ultrapure
сверхчувствительность supersensitivity
сверхчувствительный ultrasensitive, supersensitive; *phot.* ultraspeed *(attr.)*
сверхштатный supernumerary
сверхъестественный supernatural, preternatural
сверх-я *psych.* superego
сверция *bot.* swertia *(Swertia)*
сверчковые *pl. zool. Grylloidea*
сверчок *zool.* field cricket *(Gryllus)*
~, домовый house cricket
~, обыкновенный grasshopper warbler *(Locustella naevia)*
свес overhang; projection
~ кормы *naut.* after/stern overhang; stern counter
~, носовой *naut.* forward overhang
свет light
~, ближний *auto.* lower beam, passing light
~, верхний *constr.* skylight
~, дальний *auto.* high/upper beam, driving light
~ заднего хода *auto.* backup light
~, задний *auto.* tail light
~, заливающий floodlight
~, контровый *phot.* back light
~ накачки лазера laser pumping light
~, незамаскированный naked light
~, падающий incident light
~, посторонний stray light
~, солнечный sunshine
~, стояночный *auto.* parking light
светило luminary; heavenly body
светильник luminaire, lamp, lighting fixture
~, бестеневой shadowless illuminator
~, головной *min.* caplamp
~, потолочный ceiling luminaire
~, рудничный miner's lamp
светимость luminance, radiance, luminosity, luminous density/emittence
светлейший *obs.* his/her Highness
светлость 1. Grace; Serene Highness; Serenity; Ladyship; 2. brightness
световод, светопровод
светляк *zool.* glowworm, firefly *(Lampyrus)*
светобоязливый photophobic, heliophobic
светобоязнь photophobia
световод, светопровод light guide/pipe
световой 1. luminous; 2. photic, light *(attr.)*
~ год light year
светограмма *mil.* flash/blinker/light message
светоделитель beam splitter, dichroic
светодиод *el.* light-emitting diode
светокоагулятор photocoagulator
светокопирование blueprinting, photostating, light-sensitive duplication
светокультура *agr.* photoculture
светолечебница heliotherapy institute
светолечение heliotherapy, phototherapy
светолюбивый *bot.* photophilous, heliophilous
светомаскировка blackout, darkening; dim-out
светомасса = светосостав
светомаяк light beakon
~, ориентирный landmark beakon
светометр photometer
светонепроницаемость opacity
светонепроницаемый lightproof, opaque
светоотдача light/lumen efficiency, light output
светоощущение photoperception
светопоглотитель light absorber/trap
светопоглощаемость luminous absorptivity
светопоглощение luminous/light absorption
светопреломление refringence, refraction of light
светопреломляющий light-refracting
светоприемник light detector
светопровод light guide/pipe/conductor/conduit
~, волоконный fibre-optic light conductor

светопроводящий photoconductive
светопрозрачность translucency
светопрозрачный translucent
светопроницаемость transparency; translucence
светопроницаемый transparent, translucent, light-permeable; diaphanous
светопрочный photostable; fast *(of colours)*
светорассеяние diffusion/scattering of light
светосигнал light/visible signal
светосигнализация light signalling
светосигнальный light-signal (attr.)
светосила 1. candlepower, illuminative power; 2. *phot.* speed *(of lens),* aperture ratio; transmission *(of spectrometer)*
~ объектива lens aperture/speed
~ телескопа light-gathering power of a telescope
светосильный *phot.* high-power, rapid, fast
светосостав, светомасса phosphor, luminous compound
светостойкий light-resistant/proof/fast, photostable
светостойкость light fastness/resistance
светосхема illuminated chart
светотень *art* chiaroscuro
светотехника illumination/lighting engineering
светотравление light etching
светофильтр optical/light/colour filter, shade, heliofilter
~, съемочный camera filter
~ цветоделительный colour separation filter
светофор traffic light(s)
~, аэродромный ground traffic signal lights
светочувствительность photosensitivity, sensitivity to light; actinism; *phot.* speed
светочувствительный light-sensitive/reactive, photosensitive
~ная бумага sensitized/photographic paper
светоэлектрический photoelectric
свеча 1. candle; 2. spark plug; 3. candle-power; watt; 4. *aero.* vertical climb; 5. *med.* suppository
~ зажигания/запальная ignition/spark plug
~, зажигательная fuse lighter
~, курительная incense candle
~ накаливания glow candle
~, новая *opt.* new candle, candela
свечение glow, luminescence, fluorescence; radiance, glimmer
~, белое white heat, incandescence
~ моря sea fire
~ на концах мачт *naut.* corposant
~ ночного неба night glow
~, полярное auroral luminescence
~ разряда discharge glow
свидетель witness; attestor; *geol.* monadnock
~ защиты witness for the defence
~ обвинения witness for the prosecution
свидетель-очевидец eyewitness
свидетельский testimonial
свидетельство 1. certificate; licence; bill; 2. evidence, testimony
~, авторское certificate of authorship
~, брачное marriage licence
~, врачебное health certificate
~, временное interim certificate
~, досмотровое inspection certificate
~, залоговое/ипотечное mortgage instrument, hypothecation letter, mortgage deed
~, карантинное *naut.* bill of health
~ качества certificate of quality
~, квалификационное certificate of competency
~ летчика pilot's licence
~, морское navicert
~, неподтвержденное *law* uncorroborated testimony
~ о браке marriage lines/certificate
~ о приемке acceptance certificate
~ о происхождении certificate of origin
~ об испытании test certificate
~ общественной необходимости certificate of necessity
~, охранное морское navicert
~, проверочное *(of instrument)* certificate of calibration
~, пробирное assay certificate
~, страховое insurance certificate
свидетельствовать testify *(to, against);* witness, attest, certify
свидина *bot.* bloody/red dogwood, blood-twig dogwood *(Cornus sanguinea)*
свинец *met.* lead, Pb
~, азотнокислый lead nitrate
~, актиниевый actinium lead, actinium D
~, белый white lead
~, бромистый lead bromide
~, ванадиевокислый vanadinite
~, губчатый spongy lead
~, зейгерованный liquated lead
~, красный red lead, minium
~, металлический lead metal, metallic lead
~, обычный common lead
~, первичный primeval lead
~, подовый bottom lead
~, радиоактивный radiolead
~, радиевый radium lead, radium G
~, сернистый lead sulphide, galena, halenite
~, сернокислый lead sulphate
~, сурьмянистый antimonial/hard lead
~, теллуристый lead telluride
~, ториевый thorium lead, thorium D
~, углекислый lead carbonate
~, уксуснокислый lead acetate, sugar of lead
~, фосфорнокислый lead phosphate
~, хлористый lead chloride
~, хромовокислый lead chromate
~, черновой black lead, lead bullion
~, чушковый pig lead, lead pig
свинецалкил *chem.* lead alkyl
свинецсодержащий lead-bearing, plumbiferous
свинка I *med.* mumps, epidemic parotitis
свинка II piglet

~, морская *zool.* guinea pig *(Cavia porcellus)*
свиноотkорм pig-raising/fattening
свиновод pig farmer
свиноводство pig husbandry/farming
свиноферма pig farm
свинорой *bot.* dogtooth grass *(Cynodon)*
свинушка *bot.* paxil *(Paxillus)*
свинцевание lead plating/coating
свинцовистый plumbeous, leaden
свинчатка *bot.* leadwort, plumbago *(Plumbago)*
свинья pig, swine, hog; boar; sow
~, морская porpoise *(Phocaena)*
свип-генератор sweep-frequency generator
свиристель *zool.* waxwing *(Bobmycilla)*
свисток blast
свистулька *zool.* cornetfish *(Zistularia corneta)*
свистун *zool.* white-lipped frog *(Leptodactylus labialis)*
свистящий *phon.* sibilant
свита 1. retinue, entourage, suite, attendants; 2. *geol.* formation, suite, series
~, нефтеносная oil-bearing formation
свищ 1. flaw; *leath.* grub, warble; hole; 2. *med.* fistula
~, бранхиогенный branchial fistula
~, брюшно-маточный abdominouteral fistula
~, гранулирующий granulation fistula
~, желудочный gastric/gastrocutaneous fistula, gastrostoma
~, искусственный stoma
~ мочевого пузыря vesical fistula
~, одонтогенный alveolar fistula
~, печеночный hepatic/biliary fistula
~, слюнный salivary fistula, sialosyrinx
~, толстокишечный colonic fistula
~, хилезный chylous fistula
~, шейный cervical fistula
свищевой *med.* fistulous
свиязь *zool.* wigeon *(Anas penelope)*
свобода freedom, liberty
~ вероисповедания freedom of worship
~ вращения в вертикальной плоскости tilting freedom
~ вращения в горизонтальной плоскости veering freedom
~ вращения вокруг собственной оси spinning freedom
~ личности personal freedom
~ морей/мореплавания/судоходства freedom of the seas
~ объединений freedom of association
~ открытого моря freedom of high seas
~ печати press freedom, freedom of the press
~ рыболовства freedom of fishery
~ слова freedom of speech
~ собраний freedom of assembly
~ совести freedom of conscience; freedom of worship, religious liberty
~ стачек freedom of strike
свободнолепестный *bot.* apopetalous
свободнолистный *bot.* aposepalous
свободномыслящий free-thinking
свободнопестичный *bot.* apocarpous
свободноплавающий free-floating, planktonic
свободноопертый simply-supported
свободностоящий self-supporting, stand-alone
свободный 1. free; 2. vacant, blank, empty; 3. available; spare; 4. disengaged, idle; free
~ землевладелец *hist.* freeholder
свод I vault; *anat.* fornix
~ глотки vault, of the pharynx
~, куполовидный convex crown
~ неба *anat.* vault of the palate
~, небесный heavenly dome, firmament, canopy of heaven
~, обратный inverted arch
~, сегментный segmental arch
~ слезного мешка fornix of the lacrimal sac
~ пещеры cave roof
свод II code; collection; summary
~ законов войны articles of war
сводимость reducibility
сводка report, bulletin; summary; *print.* final proof; revise
~ данных collected/consolidated data
~ из вахтенного журнала abstract logbook
~, метеорологическая meteorological/weather report
сводчатый arched, vaulted
свойлачивание *text.* felting, fulling
свойство 1. relationship by marriage, legal/affinal relationship; affinity; 2. characteristic, property
~, моющее detergency
~ва, эксплуатационные service/performance characteristics
СВЧ (сверхвысокие частоты) microwave frequencies, microwaves
связанный bonded; linked; connected; associated, related; affinitive; *bot.* fasciated; *chem.* fixed, bound; combined
связист communication man/officer, signalman
связка 1. bundle, sheaf, pack; cluster; 2. *math.* connective; net; 3. *anat.* ligament; fold; retinaculum; 4. *gram., log.* copula
~, безличная *gram.* impersonal link verb
~, венозная venous ligament
~, возвратная reflex ligament
~, выйная nuchal ligament
~, голосовая vocal ligament/cord
~, грудино-ключичная sternoclavicular ligament
~ двигателей cluster of engines
~, легочная pulmonary ligament
~, межостистая interspinous ligament
~, межпоперечная intertransverse ligament
~ надколенника patellar ligament
~, окольная collateral ligament
~, паховая inguinal ligament
~, половопаховая genitoinguinal ligament
~, преддверная vestibular ligament

~, пропозициональная *mat.* propositional connective, connective symbol
~, раздвоенная bifurcated ligament
~ ракетных ускорителей boost(er) cluster
связник 1. contact man, messenger; 2. *bot.* connective
связной 1. liaison; 2. messenger, orderly
связность *math.* connectivity, connection
связочно-именной *gram.* copulative-nominal
связочно-сказуемый copulative-predicative
связочный 1. *anat.* ligamentous; *gram.* copulative
связующее binder; adhesive
связывание binding; linking; fixation; bonding
связывать 1. connect, couple, link, tie, bind, bond; 2. *chem.* fix, combine
связь 1. communication(s), connection(s); liaison; contact, touch; 2. binding, bond, coupling, tie, union, link, linkage; 3. *mech.* constraint; 4. *tech.* brace, link, tie; *el.* coupling; 5. *chem.* bond
~, автотрансформаторная *el.* tapped-coil coupling
~, авиационная aeronautical communication
~, акустическая обратная acoustic feedback
~, анкерная tie bolt/tie
~, анодная anode coupling
~, атомная atomic bond
~, балочная tie beam
~, ветровая wind tie
~, взаимная interconnection
~, гальваническая direct coupling
~, геометрическая geometric constraint
~, гликозидная *chem.* glycosidic bond
~, голономная *mech.* holonomic constraint
~, дальняя long-range/distance communication, telecommunication
~, двусторонняя 1. two-way communication; 2. bilateral constraint
~, дистанционная telecommunication;
~, дифференциальная *math.* differential constraint
~, емкостная capacitance coupling
~, звуковая audio communication
~, звукоподводная underwater sonar communication
~, зрительная visual communication
~, инерционная inertia coupling
~, ионная ionic/electrovalent bond, ion-dipole bond
~, катодная cathode coupling
~, ковалентная covalent bond
~, корреляционная correlation
~, кратная multiple bond
~, кристаллическая crystal binding
~, лазерная laser-beam communication
~, макроэргическая high-energy bond
~, межатомная interatomic binding
~, международная international traffic
~, межмашинная intercomputer communication
~, межнуклонная internucleonic coupling
~, местная local traffic
~, металлическая metallic bond
~, мнимая *stat.* illusory association
~, многократная multiple bond/linkage; multiplex communication
~, молекулярная molecular bond/link
~, морская naval communication
~ на маршруте en-route communication
~, наземная ground communication
~, направленная beam communication
~, нестационарная *mech.* rheonomous constraint
~, обратная feedback; regenerative coupling
~, обменная exchange coupling
~, односторонняя 1. one-way communication; 2. unilateral constraint
~, ограничивающая *stat.* constraint
~, основная *auto.* main coupling
~, отрицательная обратная negative/degenerative feedback
~ по положению, обратная position feedback
~ по производной, обратная rate feedback
~, паразитная stray/spurious coupling
~, перекрестная cross coupling
~, пептидная peptide bond
~, позиционная обратная position feedback
~, положительная обратная positive feedback
~, полная *stat.* complete association
~, поперечная 1. *constr.* cross brace; 2. cross bond, cross-linkage
~, почтовая postal service
~, постоянная continuous communication
~, причинная causality
~, проводная wire communication
~, продольная longitudinal member
~, прямая feedforward
~ Рассела-Сондерса Russel-Saunders coupling, RS-coupling
~, реостатная *el.* resistance coupling
~, речевая voice communication
~ СВЧ microwave communication(s)
~, сильная close coupling
~, слабая loose coupling
~, сопряженная conjugated link
~, спин-орбитальная spin-orbit coupling
~, телеграфная telegraphy
~, телефонная telephony, telephone service
~, удерживающая *mech.* bilateral constraint
~, фототелеграфная telephoto/facsimile service/communication
~, химическая chemical bond
~, циркулярная/циркуляторная conference connection
~, цифровая digital communication
~, частная *stat.* partial association
~ человека с машиной man-machine communication
~ через искусственный спутник Земли satellite-assisted communication
~, широкополосная wide-band communication

~, **щелевая** slot coupling
~, **электрическая** electrical bonding
святейшество, ваше *eccl.* Your Holiness
святейший отец *eccl.* Most Holy Father
святки *pl.* Christmastide
святой holy; sacred
~ **престол** Holy See
~**тая неделя** Holy Week
~**тая святых** holy of holies
святокупец simoniac, simonist
священноначалие hierarchy
священный holy, sacred, sanctified; hieratic
~ **и неприкосновенный** sacrosanct
~ **союз** *hist.* the Holy Alliance
~**ное писание** Scripture, Holy Writ
сгиб, сгибание bend, flexure; fold; bending
сгибать bend, flex
сглаживание smoothing; planing; burnishing
~ **методом наименьших квадратов** least-squares smoothing
~ **суммированием** smoothing by addition
сговор agreement; collusion; prevarication
~ **в ущерб третьей стороне** *law* covin
~ **против правосудия** conspiracy to defeat justice
сгораемость combustibility
сгорание combustion; burning; *nucl.* burn-up
~, **детонационное** *eng.* knocking combustion
~, **предкамерное** precombustion
~ **с разложением** destructive combustion
сгорать burn up/down/out
сграффито *art*. sgraffito
сгуститель thickener; densifier, condenser
~, **гравитационный** gravity thickener
~, **многокамерный** multicell thickener
~ **периодического действия** batch thickener
сгусток cluster, bunch; clot; coagulate
~, **ионный** ion cluster
~ **крови, предсмертный** antemortem clot
сгущение thickening; condensation
~ **желчи** pachycholia
~ **крови** pachyemia, pachemia, pachychemia
~ **химуса** pachychymia
сдавать hand over/in, turn in/over; give up, relinquish
~ **в аренду** lease
~ **в лом** scrap
~ **внаем** rent out
~ **дежурство** be relieved from one's watch
сдаточный delivery *(attr.)*
сдача 1. delivery; handing in/over; 2. surrender
~, **безусловная** unconditional surrender
~ **в аренду** tenancy, leasing
сдваивание doubling, pairing
~ **частот** frequency pairing
сдвиг shift, displacement; *geol.* fault; shear; *math.* translation; *econ.* bias
~ **в структуре потребления** bias in consumption
~ **в структуре продукции** shift in product mix
~ **влево** *comp.* left shift, shift left
~ **вправо** *comp.* right shift, shift right
~ **времен** *gram.* shifting of tenses
~, **доплеровский** Doppler shift
~, **изотопический** isotope shift
~ **нуля** zero shift
~, **радиальный** radial shear
~ **фазы** phase lag
~, **циклический** *comp.* circular/cyclic shift
сдвигать shift; move; displace
сдвоеннобороздный *bot.* geminicolpate
сделка transaction, deal; bargain; agreement
~ **за наличные [за наличный товар]** spot transaction
~ **за наличный расчет** cash transaction
~, **компенсационная** offsetting transaction/deal; barter deal
~ **на наличный товар** spot transaction
~ **на срок** forward/future transaction, futures deal
~, **товарообменная** swap(ping)
~**ки на срок [срочные]** futures
сдельность *econ.* piecework
сдельный *econ.* piecework *(attr.),* by the piece/job
~ **рабочий** pieceworker
~**ная оплата труда** piece/productivity pay, piecework wages
~**ная работа** piecework
сдельщик pieceworker; rate worker
сдельщина piece-rate basis/system; piecework
сеанс 1. session; 2. sitting; 3. seance, performance; 4. treatment; 5. run
~ **передачи** transmission session
~ **работы на терминале** *comp.* terminal session
~ **связи** communication session
себестоимость *econ.* 1. cost; 2. prime cost, cost of production/manufacture
~ **единицы продукции** unit cost
~, **заводская** manufacturing cost
~ **на единицу товара** single cost
~, **нормативная** standard value of production
~ **продукции** product cost
~, **расчетная/сметная** estimated cost
~, **фабрично-заводская** manufacturing cost
~, **цеховая** shop cost
~, **штучная** unit cost
себолит *med.* sebolith
себорея *med.* seborrhea, seborrhagia, steatorrhea
~ **головы** seborrhea of the scalp
~, **густая** concrete seborrhea
~ **новорожденных, чешуйчатая** seborrheic dermatitis in infants
сев = посев
север north
~, **истинный** true north
~ **координатной/картографической сетки** grid north
~, **крайний** the Arctic
северница *bot.* brookweed *(Samolus)*
северный northern, north *(attr.),* boreal; *cart.*

north; *(of wind)* northerly
североамериканский North American
североатлантический North Atlantic
северо-восток north-east
северо-восточный north-east(ern), north-easterly
северо-запад north-west
северо-западный north-west(ern), north-westerly
северокавказский North Caucasian
североморской North Sea (attr.)
северо-северо-восток north-northeast
северо-северо-запад north-northwest
севозапашник drill grubber
севооборот crop rotation/succession; shift of crops; rotation cropping
~, паровой fallow rotation
~, пятипольный five-course rotation
~, травопольный grassland-crop rotation
севосмен crop rotation
севр, севрский Sevres
севрюга *zool.* stellate/starred sturgeon *(Acipenser stellatus)*
сегарс *naut.* mast loop
сегидилья seguidilla
сегмент segment; section;
~, зубчатый segment gear
~, кодовый (en)coding section
~, параболический segment of a parabola
~, плоский segment of a circle
~ с литерными рычагами segment with type levers
~, сферический segment of a sphere
~, тактовый cadence segment
сегментарный segmental, segmentary
сегментарно-рефлекторный reflexosegmental
сегментно-цепной chain-and-segment
сегментный 1. segmental, segmentary; 2. *zool.* metameric
сегнетокерамика ferroelectric ceramics
сегнетомагнетик magnetoelectric
сегнетоэластик ferroelastic
сегнетоэлектрик ferroelectric material
сегнетоэлектричество ferroelectricity
седалище *anat.* ischium, clunes, hip bone
седалищно-анальный *anat.* ischioanal
седалищно-влагалищный ischiovaginal
седалищно-прямокишечный ischiorectal
седалищный *anat.* ischiatic, ischiadic
седаш *bot.* hemp agrimony/eupatorium *(Eupatorium cannabinum)*
седил *print.* cedilla
седло saddle; *tech.* seat; *meteor.* col; *math.* saddle point
~, барическое col
~, вьючное pack saddle
~, дамское side-saddle
~ клапана valve seat
~ рессоры spring seat
~, спортивное flat saddle
~, флексурное *geol.* saddle bend
седловатость *naut.* sheer *(of deck)*
седловидный saddle-shaped, selliform
седловина saddle; *meteor.* col
сезам *bot.* Oriental sesame *(Sesamum indicum)*
сезамовый sesame *(attr.); pl. Pedaliaceae*
седогептулоза *med.* sedoheptulose
сезень *naut.* nipper
сезина *law* seizin
сезон season
сезонник seasonal worker
сезонность seasonal character/fluctuation
сейвал *zool.* sei whale, fish whale *(Balaenoptera borealis)*
сейм the Sejm; diet
~, местный dietine
сеймик *hist.* regional diet/assembly
сейна seine
сейнер *naut.* seiner
сейроспора *bot.* seirospore
сейсмический, сейсмичный seismic(al)
сейсмичность seismicity
~, слабая minor seismicity
сейсмогеология seismic geology
сейсмограф seismograph
~ давления pressure-sensitive seismograph
~, крутильный torsion seismograph
~, скважинный well seismograph
сейзмозондирование seismic-refraction sounding
сейсмолог seismologist
сейсмокаротаж seismic well logging
сейсмология seismology
~, инженерная structural seismology
сейсмонастия *bot.* seismonasty
сейсмоприемник seismic detector
сейсмопрофилирование seismic profile shooting
сейсморазведка seismic prospecting/survey
сейсмостойкий antiseismic, earthquakeproof
сейсмотерапия *med.* seismotherapy, sismotherapy
сейсмостойкость seismic stability
сейф safe
сейш(а) seiche, standing wave
~, двуузловая binodal seiche
сейшельский Seychelles
СЕКАМ *telev.* SECAM
секанс *math.* secant
секатор pruning/lopping shears, pruner
секвенция 1. sequence; 2. *math.* sequent
~, средняя miosequent
секвестр 1. *law* sequestration; 2. *med.* sequestrum
секвестральный *med.* sequestral
секвестрация sequestration
секвестровать *law* sequestrate, sequester
секвестротомия *surg.* sequestrotomy
секвестроэктомия *surg.* sequestrectomy
секвойя *bot.* sequoia *(Sequoia)*
~ вечнозеленая redwood *(Sequoia empervirens)*
секрет *physiol.* secretion, secretum
секретариат secretariate; office
секретарь I secretary

~, **генеральный** secretary-general; general secretary
~, **Государственный** Secretary of State
секретарь II *zool.* secretary bird *(Sagittarius serpentarius)*
секретер writing desk/cabinet, secretary
секретность secrecy; privacy
секретный secret, confidential, classified
~ **замок** combination lock
~**но, совершенно** top secret
секреторный *physiol.* secretory
секреция *physiol.* secretion
~ **желчи** biliary secretion
~ **инсулина, повышенная** hyperinsulinism
~ **молока, усиленная** superlactation
~, **повышенная** hypersecretion
~, **пониженная** huposecretion
~, **экзокринная/внешняя** exocrine/external secretion, excretion
~, **эндокринная/внутренняя** endocrine/internal secretion, incretion
сексолог sexologist
сексопатология sexual pathology
сексот *abbr.* secret/undercover agent
секста *mus.* sixth
секстан(т) sextant
~, **авиационный** air/aircraft sextant
~, **гидрографический** surveying sextant
~, **гироскопический** gyro sextant
~, **космический** spaceborne sextant
~, **портативный** box sextant
~, **промерный** hydrographic/sounding/surveying sextant
~, **пузырьковый** bubble sextant
секстет *mus.* sextette
секстиллион, секстильон *Am.* sextillion, *Eng.* thousand trillions
сексуальность sexuality
сектант sectarian; member of a sect
сектор sector; quadrant; area, region
~ **видимости** arc of visibility
~, **государственный** public/state sector
~, **зубчатый** gear/toothed quadrant
~, **круговой** circular sector, sector of a circle
~, **нетоварный** noncommodity sector
~, **обслуживающий** *econ.* tertiary sector
~ **пеленга** bearing sector
~, **рулевой** *naut.* rudder quadrant
~, **сферический/шаровой** spherical sector, sector of a sphere
~, **частный** private sector
секторальный, секторный sectoral, sectorial
секуляризировать secularize, render secular
секунда second *(also mus., geom.)*
~ **дуги** second of arc
~, **чрезмерная** *mus.* sesquitone
секундант second
секундомер stopwatch, seconds counter, timer
~, **самопишущий** chronograph
секундометрист timekeeper, timer
секуригера *bot.* hatchet vetch *(Securigera)*
секуринега *bot.* securinega *(Securinega)*
секущая *math.* secant, transversal
секционирование sectionalization, sectioning
секционированный segmented
секционировать sectionalize
секционная *med.* dissecting room, prosectorium
секционный sectional
секция section; segmeut; panel; unit, cell
~ **кипения** *chem.* boiling section
~ **культиватора** cultivator gang/rig
~, **моторвагонная** *rail.* motor car unit
~, **обедняющая** *chem.* stripping/exhausting/scrubbing section
~, **обогатительная** enriching section
~, **отпарная** *chem.* stripping section
~, **палубная** *naut.* deck panel
~ **перегрева** superheat section
~ **подогрева** preheat section
~ **предварительной сборки** prefabricated section
~,**промывная** washing/scrubbing section
~, **укрепляющая** *chem.* rectifying section
~, **успокоительная** *chem.* calming section
~, **центральная** core section
селахии *pl. zool.* selachians *(Selachii)*
селезенка *anat.* spleen
~, **блуждающая** floating/mobile spleen
~, **ветчинная** waxy/lardaceous spleen
~, **дольчатая** lobular spleen
~, **порфирная** diffuse waxy spleen
~, **малярийная** malarial/ague spleen
~, **пятнистая** flecked spleen
селезеночник *bot.* golden saxifrage *(Chrysosplenium)*
селезеночно-печеночный *anat.* hepatolienal
селезеночный splenic, lienal
селективность selectivity; discrimination
селективный selective; discriminative
селектирование *electr.* strobing, gating
селектор selector; separator
~, **временной** time gate
~ **импульсов** pulse discriminator
~ **импульсов, амплитудный** pulse/height discriminator/selector
~ **импульсов совпадении** coincidence sorter
~, **нейтронный/нейтронов** neutron-velocity selector, neutron chopper
~ **нейтронов, быстродействующий/высокоскоростной** fast neutron selector
селекционер selectionist; breeder
селекционировать *agr.* select
селекционный *agr.* selection *(attr.);* selective
селекция 1. *agr.* selection; breeding; 2. *tech.* gating, selection; 3. discrimination
~, **внутриорганизменная** interselection
~ **растений** plant breeding
~, **угловая** *laser* angular mode selection; beam angle narrowing
селен *chem.* selenium, Se

~, радиоактивный radioselenium
~, серный selenium sulphide/sulfide
Селена *myth.* Cynthia
селенил *chem.* selenyl
селенистый *chem.* selenious, selenous
селеноводород *chem.* hydrogen selenide
селеновый selenium *(attr.)*, selenic
селенодезический selenodetic
селенодонтый *zool.* selenodont
селенолит *miner.* selenolite
селенофомоз *bot.* Selenophoma leaf spot
селеопасность mudflow danger/hazard
селепровод mudflow conduit
селесброс mudflow flume
селин *bot.* selinum *(Selinum)*
селитебная территория residential district
селитра saltpetre, saltpeter, nitre
~, аммиачная ammonium nitrate
~, известковая calcareous nitrate
~, калиевая potassium nitrate
~, самородная чилийская caliche
~, стенная calcium nitrate, wall saltpetre
~, чилийская/натронная Chile/Chilean saltpetre, sodium niter
селитряный nitric; nitrogenous; saltpetrous
селитрянка *bot.* nitrebrush, niterbush, lotus tree *(Nitraria)*
сель mud/earth flow, mud stream
сельвасы *pl. geog.* selva
сельдевый *pl. zool. Clupeidae*
сельдеобразный *zool.* clupeoid, herringlike
сельдерей *bot.* celery *(Apium)*
сельдерейнолистный *bot.* apiifolious
сельджук, сельджукский Seljuk
сельдь *zool.* herring
селькор *abbr.* rural/village correspondent
селькуп, селькупский Selkup
сельмаг *abbr.* village store
сельпо *abbr.* village cooperative stores
сельсин selsyn (motor), synchro (motor), autosyn
~, безредукторный direct-driven selsyn/synchro
~, бесконтактный non-contact selsyn, magslip
~, ведущий exciter selsyn
~, грубый [грубого отсчета] coarse [low-speed] synchro/selsyn
~ дальности range selsyn
~, двухотсчетный two-speed synchro/selsyn
~, контактный slip ring synchro/selsyn
~ принимающий дистанции range reciving selsyn
~, точный [точного отчета] fine selsyn
сельсин-анализатор selsyn/synchro analyzer
сельсин-датчик transmitting selsyn, synchro transmitter, synchro pick-off
сельсин-приемник receiving selsyn, selsyn receiver, synchro motor/receiver
~, дифференциальный synchro-differential motor
сельский rural, country *(attr.)*, village *(attr.)*
~кое хозяйство, нетоварное subsistence farming/agriculture
сельскохозяйственный agricultural, farming
сельсовет *abbr.* village soviet/council
сельтерская вода Selzer water
сельхозартель *abbr.* agricultural cooperative
сельхозбанк *abbr.* agricultural bank
сельхозинвентарь *abbr.* agricultural implements
сельхозмашина *abbr.* agricultural machine
сельхозугодья *abbr.* farming lands/fields
семантика semantics; meaning
семантико-стилистический semantic-stylistic
семасиолог semasiologist
семасиология semasiology, semantics
сематология sematology
семафор semaphore, arm signal
семга *zool.* salmon *(Salmo salar)*
семейка, дружная *bot.* aspidistra *(Aspidistra)*
семейно-наследственный *med.* heredofamilial
семейный family *(attr.)*, familial; domestic
~ патронат foster-family home
~ное положение family/marital status
семейственность nepotism
семейство 1. family; 2. collection, set, aggregate; family
~ бобовых pea family
~ вычислительных машин computer family
~ кривых *math.* family/set/assemblage of curves
~ логических элементов logic family
~, нептуния neptunium family
~ оценок *stat.* family of estimations
семема *ling.* sememe
семенник 1. *biol.* testis; 2. *bot.* pericarp; seed plant
семенной 1. seed *(attr.)*; 2. *biol.* spermatic, seminal
семенные *pl. bot.* spermatophytes
семеноводство seed growing/farming/production
семенодольный *bot.* cotyledonous
семенодоля *got.* cotyledon, seed lobe/leaf
семенозачаток *bot.* ovule
семеноложе *bot.* pericarp, sporangium, seedbed
семеносец *bot.* mother tree
семеносный *bot.* seminiferous, seed-bearing
семеричный septenary; septuple
семечко *bot.* pip, stone, pyrene
семивалентный heptavalent, septivalent
семиборье *sport* heptathlon
семигранник heptahedron
семигранный heptahedral
семизвездие The Pleiades; the seven stars
семизначный *nath.* seven-place/element
семиинтерквартильный *math.* semiinterquartile
семик *eccl.* the seventh Thursday after Easter
семилепестный *bot.* septempetalous, heptapetalous
семилетка *econ.* seven-year plan
семинар seminar; workshop
семинарист seminarian, seminarist
семинария seminary
~, духовная theological seminary

семиномный seminomatous
семинульский *geol.* seminulan
семиология *ling.* sem(e)iology
семиотика *med.* sem(e)iotics, symptomatology
семисложный septenary
семистопный *lit.* heptametrical
~ **ямб** iambic heptameter
семисторонний septilateral
семит Semite
семита *zool.* semita, fasiole
семитизм Semitism
семитолог Semitologist, Semitist
семитский, семитический Semitic
семиугольник heptagon
семиугольный heptagonal, septangular
семолина *cul.* semolina
семсовхоз *abbr.* seed-growing state farm
семссуда *abbr.* seed loan for planting
семфонд *abbr.* seed fund/stock
семья family; *ent.* colony; *fig.* generation, kin
~, **незавершенная** growing household
~, **полная** stationary household
семя 1. *biol.* sperm, semen; 2. *bot.* seed; 3. *fig.* progeny, descendants
семявместилище *bot.* seed coat; testa
семявыбрасывающий ejaculatory
семядольный *bot.* cotyledonous
семядоля *bot.* cotyledon, seed lobe
семяед *zool.* seed-eater
семязачаток *bot.* ovule
семяизвержение, семяизлияние ejaculation
~, **преждевременное** premature ejaculation
семяложе *bot.* seedbed
семянка *bot.* achene, akene, achenium
семяножка *bot.* ovule/seed stalk, funicle
семяножковый *bot.* funiculose
семянос(ец) *bot.* placenta
семяносный *bot.* seminiferous, seed-bearing
семяношение *bot.* placentation
семяпочка seedbud, ovule
семяпровод *biol.* seminal duct; spermaduct
семясушилка seed drier
семячко *bot.* pip, pyrene
семячковый плод pip fruit
сенаж haylage
сенбернар St. Bernard *(dog)*
сенега *bot.* senega *(Polygala senega)*
сенегалец, сенегальский Senegalese
сенегалка Senegalese (woman)
сенильность *med.* senility
сенильный *med.* senile
сенна *bot.* senna *(Casia)*
сено hay
сеновал hayloft
сенобрикетировщик hay wafering machine
сеноворошилка, сеноворошитель hay/swath tedder
сеновязалка hay baler
сенодробилка hay grinder/crusher
сеноизмельчитель hay chopper
сенокопнитель hay stacker
сенокос hay-making; hay-making time
сенокосилка hay/field mower, mowing machine
сеноман *geol.* Cenomanian (stage)
сенон *geol.* Senonian
сеноподборщик hay pick-up
сеноподъемник hay elevator
сенопресс hay press/baler
сеносоломорезка roughage hay cutter
сеноуборка hay-making/harvesting
сенохранилище hay storage
сенсибилизатор 1. sensitizer; 2. *med.* sensitinogen, allergen
сенсибилизация 1. *med.* sensitization, sensibilization; allergization; 2. *phot.* sensitization
~, **аутоэритроцитарная** autoerythrocyte sensitization
~, **перекрестная** crossed sensitization
~, **повторная** resensitization
сенсибилизин, сенсибилин sensibilisin, sensibilin, toxogenin
сенсибилизировать sensitize
сенсилла *zool.* sensilla
сенсимонизм Saint-Simonism
сенситивность sensitivity
сенсорный sensory; sensorial
сенсуалист *philos.* sensationalist, sensualist
сенсуалистский *philos.* sensualistic; sensual
сенсуальный *philos.* sensual, sensational
сентимо centimo
сеньор *hist.* liege lord, seigneur
сеньория *hist.* seigneury, seigniory
сепалодия *bot.* sepalody
сепаративный separative; separatist *(attr.)*
сепаратист separatist; secessionist
сепаратистский separatist; separatistic
сепаратор separator
~, **аккумуляторный** *el.* cell/battery separator
~, **винтовой** spiral separator
~ **изотопов** isotope/mass separator
~, **молочный** cream separator, skimming machine/centrifuge
~, **отстойный** separating tank
~, **проточный** flow-type separator
~, **циклонный** cyclone separator
сепарация, сепарирование separation; segregation; sorting
сепарировать separate, segregate; isolate
сепиолит *miner.* sepiolite
~, **аморфный** beta-sepiolite
~, **волокнистый** alpha-/para-sepiolite
сепион *zool.* sepion, cuttlebone
сепия 1. *zool.* cuttlefish *(Sepia)*; 2. sepia
сепсис *med.* sepsis
~, **акушерский** obstetric sepsis
~, **грибковый** mycotic sepsis
~, **криптогенный** cryptogenetic/cryptogenic sepsis
~, **пневмококковый** pneumococcal/pneumococcic sepsis

~, **послеродовый** puerperal sepsis
~, **риногенный** rhinogenous sepsis
~, **ротовой** oral sepsis
~, **синегнойный** pyocyanic/pyocyanogenic sepsis
~, **тонзиллогенный** tonsillar(y) sepsis
~, **урологический** urosepsis, urologic sepsis
~, **фузоспириллезный** fusospirochetal sepsis
септ *hist.* sept
септа *biol.* septum
~, **главная** major septum
~, **малая** minor septum
~, **первичная** protoseptum
~, **плечевая** alar septum
~, **складчатая** fluted septum
~, **срединная** median septum, metaseptum
~ **третьего порядка** tertiary septum
~, **шиповатая** acanthine septum
септаккорд *mus.* seventh
септалий *zool.* septal neck
септария *geol.* septarium, septarian nodule/boulder, turtle stone
септет *mus.* septet(te)
септик septic tank
септикопиемия *med.* septicopy(a)emia
септикофлебит septicophlebitis
септильон septillion
септима *mus.* seventh
септиметрит *med.* septimetritis
септиневрит *med.* septineuritis
септированный septate
септицемия *med.* septic(a)emia
~, **акушерская** obstetric septicemia
~, **мокротная** sputum septicemia
~, **послеродовая** puerperal septicemia
~, **сапная** glanderous septicemia
~, **чумная** septicemic plague
септицидный *bot.* septicidal
септичность septicity
септорий *bot.* septoria *(Septoria)*
септориоз *bot.* septoria spot
септотом *surg.* septotome
септофлавин septoflavine
септула *biol.* septulum
септуор *mus.* septuor, septet(te)
сера sulphur, sulfur, S; brimstone
~, **бромистая** sulphur bromide
~, **комовая** sulphur in lumps
~, **общая** total sulphur
~, **осажденная** precipitated sulphur
~, **очищенная** purified/bright sulphur
~, **ромбическая** orthorhombic sulphur
~, **самородная** native sulphur, brimstone
~, **селенистая** selenosulphur, volcanite
~, **сублимированная** sublimed sulphur
~, **ушная** cerumen, ear wax
~, **черенковая** cane brimstone, rod sulphur
~, **чешуйчатая** flake sulphur
сераделла *bot.* serradella, bird's foot *(Ornithopus)*
серак *geol.* serac
сераль seraglio
серао, серау *zool.* serow *(Capricornis)*
серапис Serapis
сераскир seraskier
серафим seraph
серафимский, серафический seraphic(al)
серб Serb, Serbian
сербка Serbian woman
серболужицкий Lusatian, Sorbian
сербословенский Serbo-Slovenian
сербохорватский Serbo-Croatian
сербский Serbian
серв serf
серваж serfdom, serfage
сервал *zool.* serval *(Felis serval)*
сервелат *cul.* cervelat, saveloy
сервис service; catering; *sport* service
сервитут *law* easement, servitude
~ **в целях удобства** easement of convenience
~, **земельный** land/real/predial servitude
~, **необходимый** easement of necessity
~, **частный** private easement
серводвигатель, сервомотор servomotor
~ **вертикальных рулей** azimuth servomotor
~, **гидравлический** fluid servomotor
~, **пусковой** starting servomotor
сервоклапан servo-actuated valve, pilot valve
сервокомпенсатор balance/balancing tab
~, **кинематический** geared balance tab
~, **пружинный** spring tab
сервокомпенсация servo balance
сервомеханизм servo, servo unit, servomechanism, servo actuator, servosystem
сервомуфта servo clutch
сервопривод servo drive/motor/actuator; servo, servodyne
~, **азимутальный** azimuth servo
~ **антенны** aerial servo
~, **основной** primary servo
~, **угломестный** elevation servo
серворуль servo/assisted steering
сервосистема servo system
сервотормоз servo/power brake
сервоуправление servo-control
сервоусилитель servo-amplifier
сердар sirdar
сердечник I core
~, **воздушный** air core
~ **каната** rope core
~, **многодырочный** multi-aperture core
~, **С-образный** C-core
~, **ферритовый** magnetic core
~, **чашеобразный** cup core
~, **Ш-образный** E-core
~, **шихтованный** laminated core; imbricated/interleaved core
сердечник II *bot.* bitter cress *(Cardamine)*
сердечниколистный *bot.* cardamine-leaved
сердечно-аортальный cardioaortic
сердечнодыхательный cardiopneumatic
сердечно-легочный cardiopulmonary

сердечномышечный cardiomuscular
сердечно-печеночный cardiohepatic
сердечно-почечный cardiorenal, cardionephric
сердечно-сосудистый cardiovascular
сердечный *med.* cardiac, heart *(attr.)*
сердолик *miner.* carnelian, cambay stone
 ~, желтый canary stone
сердце heart
 ~, блуждающее movable heart
 ~, висячее suspended/hanging heart
 ~, волосатое hairy heart, shaggy pericardium
 ~, жирное/ожиревшее fatty/adipose heart
 ~, зобное goiter heart
 ~, капельное pendulous/drop heart, cardioptosia
 ~, кифосколиотическое kyphoscoliotic heart
 ~, ладьевидное sabot heart, wooden-shoe heart
 ~, легочное pulmonary heart
 ~, микседематозное myxedema heart
 ~, митральное luxus heart
 ~, однокамерное unilocular heart
 ~, панцирное armoured heart
 ~, периферическое skin heart
 ~, пивное beer-heart
 ~, промежуточное intermediate heart
 ~, симметрично удлиненное tear drop heart
 ~, системное systemic heart
 ~, солдатское soldier's/irritable heart
 ~ спортсмена athletic heart
 ~, тиреотоксическое thyroid heart
 ~, фиброзное fibroid heart
сердцебиение palpitation; heartbeat rate
 ~, замедленное bradycardia
 ~, нервное nervous palpitation
 ~, учащенное tachycardia
сердцевидка *zool.* cockle *(Cardium)*
сердцевиднолистный *bot.* cordifolious
сердцевидный heart-shaped, cordiform, cordate
сердцевина pith, core, heart
 ~, губчатая porous/spongy pith
 ~ древесины pith
 ~, отталкивающая *nucl.* repulsive core
 ~, твердая *nucl.* hard core
 ~ ядра nuclear core
сердцевинный pithy; medullary
сердцелистный *bot.* cardiophyllous
серебреный silver-plated
серебристолистный silver-leaved, argyrophyllous
серебристость silveriness
 ~ метелки *bot.* silvertop
серебро silver, argentum, Ag
 ~, азотнокислое silver nitrate
 ~, белковое (strong) silver protein, protargol
 ~, бромистое silver bromide
 ~ в слитках silver bullion
 ~ высокой пробы fine silver
 ~, гремучее fulminating silver
 ~, йодистое silver iodide
 ~, медистое copper silver
 ~, металлическое silver metal, metallic silver
 ~, молочнокислое silver lactate
 ~, монетное sterling/standard silver
 ~ низкой пробы low-assay silver
 ~, самородное native silver
 ~ сернистое сурьмянистое silver-antimony sulphide, aerosite, pyragyrite
 ~, сернистокислое silver sulphite
 ~, сернокислое sulver sulphate
 ~, сурьмянистое dyscrasite
 ~, столовое silverware, silver
 ~, сусальное silver leaf
 ~, теллуристое telluric silver, hessite
 ~, уксуснокислое silver acetate
 ~, фосфорнокислое silver phosphate
 ~, фтористое silver fluoride
 ~, хлористое silver chloride
 ~, хлоробромистое silver chlorobromide
 ~, цианистое silver cyanide
сереброносный, серебросодержащий argentiferous
сереброплавильный silver-melting/smelting
серебряник 1. silver coin/piece; 2. silversmith
серебряный argental, argentic, argentine, silver *(attr.)*
середина middle, centre, center; midst
середняк middle-class peasant
середокрестная неделя the fourth week of Lent
сережка *bot.* catkin, amentum, willow pussy
сережковидный *bot.* amentiform
сережкоцветный *bot.* amentaceous
серж *text.* serge
сержант sergeant
серийный serial; normal, standard
 ~ное изделие production-line item
 ~ное производство serial/mass manufacture
серин *chem.* serine
сериндегидраза serine dehydrase
сериндезаминаза serine deaminase
серир serir, candy-pebble desert
серистый *chem.* sulphureous
серицит *miner.* sericite, potash mica
серия series, set, train, run, range, row; part
 ~ вихрей vortex train
 ~ волн wave train
 ~, головная pilot series
 ~ измерений run/set of measurements
 ~ импульсов pulse train
 ~ испытаний test series
 ~, колебательная vibrational series
 ~ спектра spectral series
сермесоак icecap, continental/inland ice
серна *zool.* chamois *(Rupicapra rupicapra)*
сернистокислый sulphite (of)
 ~ натрий sodium sulphite
сернистость sulphur content
сернистый sulphide (of); sulphurous
 ~ азот nitrogen sulphide
 ~ натрий sodium sulphide
сернобык *zool.* oryx *(Oryx gazella)*
серноватистокислый thiosulphate (of)
серноватистый thiosulphuric, hydrosulphurous

серноватокислый dithionate, hyposulphate (of)
серноватый dithionic, hyposulphuric
сернокислый sulphate, sulfate (of)
~ барий barium sulphate
~ натрий sodium sulphate
серный sulphuric, sulphur *(attr.)*
~ колчедан sulphur pyrite
~ цвет *chem.* flowers of sulphur
сероводород hydrogen sulphide, sulphureted hydrogen
сероводородный hydrosulphuric
серодиагностнка *med.* serodiagnosis
серозит *med.* serositis, orrhomeningitis
~, множественный multiple serositis
серозногнойный *med.* seropurulent
серознолимфатический serolymphatic
серозный *med.* serous
серолог serologist
серология serology, orrhology
серомицин seromycin
серомукозный seromucous
серопневмоперикард *med.* pneumoseropericardium, hydropneumopericardium
серопневмоторакс pneumoserothorax, hydropneumothorax
серопрофилактика seroimmunity, seroprophylaxis, seroprevention
серореакция seroreaction, orrhoreaction
сероспинка *zool.* walleyed herring *(Pomolobus pseudoharengus)*
серотерапия serum therapy, serotherapy, orrhotherapy
серотип serotype
серотонин serotonin, s-hydroxytryptamine
сероторакс *med.* serothorax, hydrothorax
сероуглерод carbon disulphide/bisulphide
серп 1. sickle; 2. *anat.* falx
~ большого мозга *falx cerebri*
~ мозчечка falcula
~, паховый inguinal aponeurotic fold, conjoined/conjoint tendon
серпазил, серпазол *pharm.* reserpine, serpasil, sandril
серпантин hairpin curve
серпентин *miner.* serpentine
~, благородный precious/noble serpentine
~, коротkостолбчатый sunguilite
~, листоватый marmolite
~, смолистый resinous serpentine, retinalite
серпник *bot.* samphire *(Crithmum)*
серповидноветвистый, серповетвистый *bot.* sickle-branched, drepanocladous
серповидноклеточный *med.* sickle-/crescent-cell
серповиднолопастный *bot.* sickle-lobed, drepanolobed
серповидный crescent-/sickle-shaped, crescentic, falciform, falcate
серпоклюв *zool.* sicklebill
серпокрылка *zool.* hook-tip moth *(Drepanidae)*
серполистный *bot.* sickle-leaved,drepanophyllous
серпула *zool.* serpula *(Serpula)*
серпуха *bot.* sawwort *(Serratula)*
сертификат certificate
~, аудиторский certificate of accounts
~, закладной mortgage certificate
~, именной registered certificate
~, морской navicert
~, морской судовой ship navicert
~ на предъявителя bearer/nominee certificate
~ о мореходности certificate of seaworthiness
~, предаваемый вкладной negotiable certificate of deposit
~ сортности grading certificate
~, транзитный shipper's certificate
~ы, государственные оберегательные National Savings Certificates
серфинг surfing, surfboarding
серьга shackle, earring, eartab, eyelet, lug, strap, stirrup, link; eyebar
~, буксирная tow ring
серьяния *bot.* supplejack *(Serjania)*
сесамовидный sesamoid
сисамоидит *vet.* sesamoiditis
сесбания *bot.* sesbania *(Sesbania)*
сеслерия *bot.* moor grass *(Sesleria)*
сессия session; sitting; term
сестерций, сестерция *hist.* sesterce
сестра sister; *med.* nurse
~, двоюродная first cousin
~, медицинская hospital nurse
~, молочная foster sister
~, операционная scrub nurse
~, палатная charge nurse
~, патронажная visiting nurse
~, родная full sister
~, сводная stepsister
~, троюродная second cousin
сестра-обследовательница visiting nurse
сета *bot.* seta
сетевидный, сетеобразный netlike, retiform
сетка net, network, mesh, grid; nettng; gauze lattice; reticle; graticule; (luggage) rack; grate, grating; *zool.* reticulum; *anat.* plexus
~, антидинатронная suppressor grid
~, арматурная reiforcing fabric
~, багажная luggage net/rack
~, географическая geographic grid/reticle/net/graph
~, градусная *cart.* grade grid
~, картографическая chart grid/graticule
~, катодная *el.* cathode (space-charge) grid
~, координатная reference/coordinate grid, grid system, map graticule
~ лампы *el.* grid
~, масштабная graticule
~, оптическая optical reticle
~, прицельная aiming reticle
~, противодинатронная suppressor grid
~, прямоперегородочная rectimurate reticulum

~, **растровая** *print.* cross-line screen
~, **тарифная** rate schedule; scale of rates/charges
~, **тканая** wire gauze net
~, **управляющая** control grid
~ **фильтра** filter screen
~ **частот** frequency spectrum
~, **экранирующая** *rad.* screen grid
сетковидный, сеткообразный reticular, retiform, netlike, netted
сеточка reticulum
~, **водяная** *bot.* water net *(Hydrodictyon)*
сетчатка *anat.* retina; 2. mesh, netting
сетчато-жилковатый retinerved, net-veined
сетчатокрылые *pl., zool.* neuropterans *(Neuroptera)*
сетчатонервный retinerved, net-veined
сетчатоплодный *bot.* net-fruited, dictyocarpous
сетчатость reticulation, nettin
сетчатый reticular, reticulate, retiform, latticed, meshed, netlike, netted
сеть 1. net; 2. network, system; 3. mains; supply
~, **артериальная** arterial network
~, **базисная** *geod.* base net
~, **водопроводная** water-sypply system
~, **вычислительная** computer network
~, **геодезическая** geodetic network
~ **для грузовых поддонов** pallet net
~ **дождемерных пунктов** precipitation/rainfall network
~, **дорожная** highway system
~, **железнодорожная** railway network/system, railways
~ **заграждения/защитная** protecting net
~ **заземления** *el.* earthing/grounding system
~, **звездообразная** star network
~, **канализационная** sewerage network; drainage system
~, **кольцевая** ring network
~, **маскировочная** camouflage net
~, **многопунктовая** multipoint network
~, **осветительная** lighting system/network
~, **отопительная** heating system
~ **передачи данных** data communication network
~, **питающая** power line, mains
~, **подошвенная венозная** plantar venous network
~, **полигонометрическая** polygon network
~, **радиальная** star-wired network
~, **рыболовная дрифтерная** drift net
~, **рыболовная жаберная/объячеивающая** gill net
~, **рыболовная отцеживающая** impounding net
~ **с ведущим узлом** *comp.* host-based network
~ **с коммутацией пакетов** packet-switched network
~ **с маркерным доступом** token-passing network
~ **с равноправными узлами** peer-to-peer network
~ **связи** communication network
~, **силовая** mains
~, **соматическая** skeleton network
~, **сосудистая** vascular network
~, **спасательная** life-saving het
~ **специализированных магазинов** specialty chain
~, **съемочная** *geod.* survey network
~, **телефонная** telephone system/network
~, **тепловая** heating system
~, **траловая** trawl, trail net
~ **универмагов** department store chain
~, **электрическая** electric/power network
сеть-волокуша dragnet
сехиум *bot.* chayote *(Sechium edule)*
сечение section; cross-section; sectional view; profile; *med.* sectio
~, **атомное** atomic cross-section
~ **в направлении потока** streamwise section
~ **в направлении хорды** chord/chordwise section
~, **входное** inlet/entry section
~, **выходное** outlet/exit section
~ **горизонталей** *top.* contour interval
~ **деления** fission cross-section
~ **для быстрых/медленных/тепловых/холодных нейтронов** fast-/slow-/thermal-/cold-neutron cross-section
~ **захвата** capture cross-section
~, **золотое** *math.* golden section, extreme and mean ratio
~, **миделевое** *naut.* midship section; *av.* frontal/maximum cross-section
~, **кесарево** *surg.* cesarean section/operation, abdominohysterotomy
~, **комптоновское** Compton cross-section
~, **круговое** circular cross-section
~ **лонжерона** spar section
~ **мочевого пузыря** cystotomy
~ **мочевого пузыря, высокое/надлобковое** suprapubic cystotomy
~ **на быстрых нейтронах** fast-neutron crosssection
~ **ослабления** attenuation cross-section
~ **по потоку** streamwise section
~ **по хорде** chordwise/chord section
~, **парное** pair cross-section
~ **переноса** transport/transmission crosssection
~ **перехода** transition cross-section
~, **полное** total/overall cross-section
~, **поперечное** cross-section, transversal section
~, **продольное** longitudinal section
~, **рабочее** net section
~ **радиационного захвата** radiative-capture cross-section
~, **расчетное** reference section
~, **рельефа** *cart.* contour/vertical interval
~, **сложное** compound cross-section
~ **сопла, выходное** nozzle exit section

~ **сопла, критическое** nozzle throat
~ **сопла, начальное** nozzle inlet section
~ **соударения/столкновения** collision crosssection
Сечь *hist.* Cossack host
сеялка sowing machine, sower, seeder
~, **гнездовая** hill-drop planter
~, **дисковая** disk drill
~, **зерновая** grain drill
~, **катковая** roller planter/seeder
~, **пунктирная** single-seed drill
~, **разбросная** seed broadcaster, broadcast sower
~, **рядовая** seed drill, drill seeder
~, **секционная** gang drill
~, **туковая** fertilizer distributor/spreader
~, **широкорядная** wide-row drill
сеялка-культиватор cultivating drill, drill and thinner
сеяние 1. *agr.* sowing; 2. screening, sifting; 2. *biol.* inoculation
сеять *agr.* sow; *(in furrows)* drill
сжатие compression, contraction, constriction; oblateness; pressing, pressure; squeezing
~, **боковое** lateral compression
~, **внецентренное** eccentric compression
~, **всестороннее** uniform compression
~ **данных** data compression/reduction
~ **жилы** pinching
~ **Земли** 1. contraction of the Earth; 2. oblateness of the Earth
~ **земной коры** crustal contraction
~ **импульса** pulse compression
~, **одноосное** uniaxial compression
~, **предварительное** precompression
~, **температурное** thermal shrinkage
~, **убывающее** declining compression
~, **ударное** shock compression
сжатогортанный *phon.* constricted
сжатый 1. compressed; constricted; condensed; pinched; 2. compact; concise, brief, succinct
сжигание burning (up), combustion; incineration
~ **в кислоте** acid digestion
сжигать burn (up), incinerate; cremate; scorch
~ **мусор** incinerate rubbish
сжижаемость liquefaction; condensability
сжижать liquefy, liquate
сжижение liquefaction, liquation
сжимаемость compressibility, contractibility, contractility; condensability
сжимаемый compressible, contractile
сжимание compression, contraction; shrinkage
сжиматель compressor; *anat.* sphincter
сжимать compress; press; contract; constrict
сжиматься contract, shrink; compress
сжимающий compressive
си *mus.* B, si
сиализм *med.* sialism, sialismus
сиаллит *geol.* siallite
сиалоаденит *med.* sialoadenitis, sialodenitis
сиаловый sialic, sialine, salivary
сиалогенный sialogenous, sialogogic, sialogogue
сиалодохит, сиалодуктит sialodochitis
сиалоз sialosis
сиалолит sialolith
сиалолитиаз sialolithiasis
сиалолитомия sialolithomy
сиалопротеид sialoproteid, sialoprotein
сиалорея sialorrh(o)ea
сиалостаз sialoschesis, sialoporia
сиаль *geol.* sial
сиаманг *zool.* siamang *(Hylobates syndactylus)*
сиамец, сиамский Siamese
~**мские близнецы** Siamese twins
сиамка Siamese (woman)
сибарит sybarite
сибаритствовать lead the life of a sybarite
Сибилла *myth.* Sibyl
сибилянт, сибилянтный *phon.* sibilant
сибиреязвенный *med.* anthracic
сибирка 1. *bot.* sibirea *(Sibiraea);* 2. sibirka *(waist-length caftan)*
сибирский Siberian
~**кая кошка** Persian cat
~**кая язва** *med.* anthrax
сибиряк Siberian
сибирячка Siberian (woman)
сибсы *pl.* siblings
сивец *bot.* succisa *(Succisa)*
Сивилла *myth.* Sibyl
сивуч *zool.* eared seal *(Eumetopias jubatus)*
сиг *zool.* cisco, whitefish *(Coregonus)*
сигара cigar
сигарета 1. ciragette; 2. small cigar, cigarillo
~ **без фильтра** plain cigarette
~ **с фильтром** filter-tipped cigarette
сигесбекия *bot.* St. Paul's wort *(Siegesbeckia)*
сигиллярии *pl. paleont.* sigillarias *(Sigillaria)*
сигмовидный sigmoid
сигмоида sigmoidal fold
сигмоидальный sigmoid(al)
сигмоидит *med.* sigmoiditis
сигмоидопексия *med.* sigmoidopexy
сигмоидоректостомия, сигмоидопроктостомия sigmoidorectostomy, sigmoidoproctostomy
сигмоидоскоп, сигмоскоп sigmoidoscope, sigmoscope
сигмоидотомия sigmoidotomy
сигнал signal; call; sign
~, **аварийный** alarm/emergency signal
~, **акустический** acoustic/audible/audio signal
~ **бедствия** distress signal/call/message
~ **бдительности** warning signal
~ **в цифровом виде** digital signal
~ **ведомой станции** slave signal
~, **воздействующий** actuating signal
~ **времени** time tick/signal
~, **входной** input/incoming signal
~ **вызова/вызывной** call signal
~, **выходной** output/outgoing signal
~, **гасящий** blanking signal

~ горном buggle call
~ готовности ready signal
~, групповой composite signal
~ гудком hoot call
~ датчика signal from the sensor
~ дальности range signal
~ дудкой *naut.* boatswain's signal
~ единого времени common timing signal
~ забоя ошибки *tel.* erasure signal
~ задатчика/задающий demand signal
~ закрытия связи closing-down signal
~ занятости busy signal
~ записи *comp.* write signal
~, запрещающий inhibit(ing) signal
~ запрещения посадки closed-field signal
~ запроса/опроса request/interrogation signal
~, запускающий trigger signal
~, звуковой sound/sonic/audio/audible signal; *auto.* motor car horn
~ звуковой частоты audio signal
~, зрительный visual/optical signal
~ изображения *telev.* video/picture signal
~, импульсный pulse signal; sampled signal
~, информационный data signal
~, исполнительный signal of execution
~ кадровой синхронизации frame/vertical sync signal
~ клотиковьм фонарем *naut.* flashing signal
~, кодовый code sign
~, командный command/demand signal, command
~, контрольный monitoring signal
~ коррекции correction/corrective signal; restoring signal
~, курсовой on-course signal
~, ложный spurious/false/ghost signal
~, максимальный peak signal
~, мигающий световой flashing attention signal
~, минимално обнаруживаемый minimum detectable/discernable signal
~ минутной готовности one-minute signal
~, модулированный по амплитуде/фазе/частоте amplitude-/phase-/frequency-modulated signal
~, модулирующий 1. modulating signal/wave; 2. baseband signal
~, морской nautical signal
~ на входе датчика sensor input signal
~ на выходе датчика sensor output signal
~ наведения guidance signal
~ начала блока *comp.* start-of-block signal
~ начала передачи commencing signal
~ недоступности номера *teleph.* trouble [out-of-order] tone
~ неисправности fault signal
~, ненаправленный omnidirectional signal
~ нормального режима reference signal
~ о повторении repeat signal
~ о санаторном состоянии quarantine signal
~ о течи leak indicator/alarm
~ об уходе с заданного курса off-course alarm/signal
~, обнаруживаемый detectable signal
~ обратной связи feedback signal
~ ограниченный по частоте band-limited signal
~, однополосный single-sideband signal
~ окончания передачи сообщения end-of-message signal
~ операции ввода inbound operation signal
~ опознавания цвета *telev.* colour identification signal
~, опознавательный identification/recognition signal
~, опорный reference/comparison signal
~ опроса polling signal
~, оптический optical/visual signal
~ остановки stop (signal)
~ отбоя sign-off [ringing-off] signal
~, ответный response/answer signal
~, ответный радиолокационный radar reply
~, отменительный annulling/obliterator signal
~ отметки времени timing/time signal
~, отраженный reflected/return/echo signal; radar echo
~ ошибки error signal
~, паразитный spurious signal
~ паузы interval signal
~ пеленга bearing signal
~ переноса *comp.* carry signal
~ переполнения overflow signal/alarm
~, побочный *rad.* spurious signal
~ повреждения/пробоя breakdown signal
~, подводный звуковой submarine sound signal
~ подводным колоколом submarine-bell signal
~ подтверждения acknowledgement signal
~, пожарный fire alarm (signal)
~, позывной *rad.* call sign/signal
~, полезный valid/desired signal
~, полный *telev.* composite colour signal
~, полуденный *naut.* noon bell
~ помехи jamming signal; *(video)* ghost signal; electronic countermeasure signal
~, предупредительный warning/cautionary signal
~ прерывания interrupt signal
~, прерывистый intermittent signal
~, проблесковый blink/flashing signal
~ радиологической опасности radiation signal
~, разделительный spacing signal
~, разрешающий enabling signal
~ разрешения посадки landing signal
~ распознавания recognition/recognizing signal/message
~ рассогласования error signal
~, речевой voice signal

~ с активным высоким уровнем active-high signal
~ с активным низким уровнем active-low signal
~ с доплеровским сдвигом Doppler-shifted signal
~ с пассивной паузой return-to-zero signal
~ самонаведения homing signal
~, световой light/luminous/flashing/blinker/optical signal
~, сельсинный synchro signal
~, синфазный in-phase [common-mode] signal
~, синхронизирующий synchronizing/locking/timing signal
~ сиреной, туманный fog signal siren
~, случайный random signal
~ сопровождения tracking signal
~ старта starting signal
~ стирания ошибки erasure signal
~, стробирующий gate/gating signal
~ считывания sense/read/reading signal
~, тактовый cadence signal
~, тональный audio signal
~ точного времени standard time signal
~ требующий уступить дорогу right-of-way signal
~ тревоги/тревожный alarm signal
~, туманный fog signal/alarm
~ указывающий курс/пеленг compass signal
~ уклонения от курса off-course warning
~, управляющий control/pilot/steering/drive/actuating signal
~, упреждающий anticipatory signal
~, условный code signal/word
~, хаотический random signal
~ химической тревоги gas alarm
~, хронирующий timing signal
~ цветности *telev.* chrominance signal; chroma [modulated subscriber] signal
~ цветных полос, испытательный *telev.* colour-bar-chart signal
~, цифровой numeral/number signal
~, частотно-модулированный frequency-modulated signal
~ шума noise signal
~, эталонный reference signal
сигнализатор signal(l)ing/warning device, signal/warning indicator; annunciator
~, визуальный annunciator, visual indicator
~, высотный altitude warning unit
~ давления pressure warning indicator
~, дозиметрический gamma alarm unit
~, звуковой audio/audible annunciator
~ метана fire-damp (methane) alarm unit
~ обледенения ice detector/indicator
~ остатка топлива low-level fuel indicator
~ падения давления low-pressure indicator
~ перегрузки, световой overload light
~ перегрева overheat warning indicator
~ повреждений fault alarm
~ пожара fire detector
~ появления стружки chip detector
~ прекращения питания [пропадания сети] power failure indicator
~ радиации radiation warning monitor
~ раскрутки overspeed warning device/unit
~, световой annunciator
~ течи/утечки leakage indicator/detector
сигнализация 1. signal(l)ing; warning; indication; 2. signal/signalling system
~, аварийная alarm signalling/system
~ высоты *aero.* altitude warning
~ замыкания на землю earth/ground fault signalling
~, звуковая sound signalling
~, зрительная visual signalling
~ об отказах failure warning
~ обледенения ice warning
~ перегрева temperature-rise alarm
~, подводная subaqueous signalling
~, пожарная fire alarm system/equipment
~ положения шасси *aero.* undercarriage warning
~ полотнищами *av.* signal panel warning
~, предупредительная световая warning lights
~, тревожная alarm system
сигнализировать signal; indicate; warn
сигнальный signal *(attr.)*, signal(l)ing
~ экземпляр advance copy
~ пистолет signal pistol; flare gun
сигнатура *print., med.* signature, libel; *math.* signature
сидальцея *bot.* checker/prairie mallow *(Sida sidalceae)*
сиденье seat
~ водителя driver's seat
~, катапультируемое pilot's ejection seat
~, общее bench seat
~, поворачивающееся swivelling seat
~, подвесное hammock seat
~, регулируемое adjustable seat
сидерит *miner.* siderite, chalybite
~, глинистый argillaceous siderite
~, углистый blackband ironstone
сидеритис *bot.* ironwort, iron woundwort *(Siderites)*
сидеритовый *miner.* sideritic
сидерический *astr.* sideral
сидероз *med.* siderosis
~ легких pulmonary siderosis, scissors grinder's disease, grinder's asthma
сидероксилон *bot.* jungle plum *(Sideroxylon)*
сидеролит siderolite
сидеропения *med.* sideropenia, hyposiderosis
сидеропизолит *miner.* sideropisolite, pseudopisolite
сидерофиброз siderofibrosis
сидерофил siderophile
сидерофилин siderophilin

сидерофиллит *miner.* siderophyllite
сидерофилия siderophilia
сидерофильный siderophilous, siderophile
сидерофир *miner.* siderophyre
сидерофон siderophone
сидерофор siderophore
сидероцит *biol.* siderocyte
сидр cider
сидячеглазый *zool.* sessile-eyed
сидячелистный *bot.* sessifolious, sessile-leaved
сидячецветковый *bot.* sessile-flowered, sessiflorous
сидячий sitting; sit-sown; *bot.* sessile, assidenous
  ~чая забастовка sit-down strike
сиена, сиенна *miner.* sienna
  ~, жженая burnt sienna
сиенит *petr.* syenite, Thebaic marble
  ~, авгитовый augite-syenite
  ~, кальциевый calcisyenite
  ~, кварцевый quartz-syenite
  ~, плагиоклазовый calcisyenite
  ~, плауенский plauenite
  ~, слюдяной mica-syenite
сиенский Sienese
  ~кая земля sienna
сиерра sierra
сизаль sisal; *bot.* sisal agave *(Agave sisalana)*
сизигийный syzygial; spring *(attr.)*
сизигиум *bot.* rose apple (Syzygium *jambos)*
сизигия *astr.* syzygy; spring
Сизиф *myth.* Sisyphus
сизифов *myth.* Sisyphean
  ~ труд Sisyphean toil/labour
сизоворонка *zool.* roller *(Coracias garrulus)*
сизон *bot.* honewort, stone parsley *(Sison)*
сиккатив paint drier; drying agent, siccative
сикл sheckel
сикоз, сикозис *med.* sycosis
  ~ ноздрей nostril sycosis, tinea naris
сикомор *bot.* sycamore fig; *(Ficus sycomoris);* American plane tree *(Planatus occidentalis)*
сиконий *bot.* syconium
сикофант sycophant, informer
сикстинский Sistine
сикх Sikh
сила force, strength; power; *phys.* energy; intensity; *pl. mil.* forces; *law* authority, validity; *see also* **силы**
  ~, архимедовая buoyancy force
  ~, астазирующая labilizing force
  ~, боковая lateral/side/transverse force
  ~, валентная valence force
  ~ веса force of weight, weight force
  ~ включения switch-on force
  ~, внецентренная eccentric force
  ~, внешняя external/superposed/applied/impressed force
  ~ внутреннего трения viscous force
  ~, внутренняя internal force
  ~, возмущающая disturbing/exciting force
  ~ волнения magnitude of waves
  ~, вращающая rotary/rotational/turning force; angular force
  ~, вязкая viscous force
  ~ вязкого сопротивления viscosity force
  ~, гидродинамическая подъемная hydrodynamic lift
  ~, движущая motive/driving/propulsive force
  ~, демпфирирующая damping force
  ~, доказательная *law* probative force
  ~, живая kinetic force
  ~ звука sound intensity
  ~ земного притяжения terrestrial gravity force, Earth's gravitational force
  ~, знакопеременная alternating force
  ~, изгибающая bending force
  ~ излучения intensity of radiation
  ~, импульсная/импульсивная impulsive force
  ~ инерции force of inertia, inertia(l) force
  ~, исковая *law* actionability
  ~, касательная tangential force
  ~, контурная boundary force
  ~, кориолисова Coriolis force
  ~, коэрцитивная coercive force
  ~ крыла, подъемная wing lift
  ~, кулоновская Coulomb force
  ~ лобового сопротивления drag (force)
  ~ лопасти, подъемная *aero.* blade lift
  ~, лоренцова Lorentz force
  ~, лошадиная horsepower, hp
  ~ магнита magnet power
  ~ магнитного поля magnetic force/intensity; magnetic field strength
  ~, магнитодвижущая magnetomotive force
  ~, мгновенная instantaneous force
  ~ нажатия compressing force
  ~ направленная вверх/вниз/вперед/назад upward/downward/forward/rearward force
  ~ натяжения tension force
  ~, непреодолимая *law* act of God, force majeure, superior/irresistible force
  ~, обобщенная generalized force
  ~ обусловленная вихрями, подъемная vortex lift
  ~, обратная *law* retroactive effect
  ~, оптическая focal/length power
  ~, осевая axial thrust
  ~ отдачи recoil force
  ~, отрывающая pull
  ~ отталкивания repulsive/repelling force
  ~ плавучести buoyancy/buoyant force
  ~, подъемная lift (force); lifting capacity
  ~, полезная подъемная useful lift
  ~, покупательная purchasing power
  ~, поперечная lateral/side/transverse force; *naut.* athwartship force; *mech.* shearing force
  ~, приложенная superposed/superimposed/applied force
  ~ притяжения force of attraction, attractive/

attracting force
~, пробивная penetration force
~, продольная longitidinal [fore-and-aft] force
~ противодействия opposing force
~, противоэлектродвижущая counterelectromotive force
~, рабочая labour/work force, labour, manpower
~, равнодействующая resultant force
~ разгруппирования debunching force
~ разрешающая resolving power, resolution
~, разрушающая breaking force
~, растягивающая tensile force
~, расчетная rated force
~, расчетная подъемная design lift
~ расширения expansive force
~ резонанса strength of resonance
~, результирующая resultant force
~ света 1. luminous intensity; 2. candle power
~ связи binding force; *chem.* bonding force
~ сжатия compression/compressive force
~, скручивающая torsional force
~ создающая крутящий момент torsional force
~ сопротивления resistance
~, спин-орбитальная spin-orbit force
~ сцепления adhesive/cohesive/bonding force
~, термоэлектродвижущая thermoelectromotive force
~ тока strength of current; current intensity
~ торможения braking/stopping force
~ трения force of friction, friction(al) force
~ тяги thrust (force), pulling force
~ тяги воздушного винта propeller thrust
~ тяжести (force of) gravity
~, тянущая tractive force, pull
~, удерживающая confining/holding force
~ упругости elastic force
~, уравновешенная balanced/equilibrium force
~, уравновешивающая balancing/equilibrant force
~, фотоэлектродвижущая photoelectromotive force
~, центробежная centrifugal force
~, центростремительная centripetal force
~, эйлерова critical load
~, эйлерова критическая buckling force
~, электродвижущая electromotive force
~ электростатического поля electrostatic force
~, юридическая validity; effect
~ якоря, держащая anchor hold
силан *chem.* silane, silicon hydride
Силен *myth.* Silenus
силена *bot.* silene *(Silene)*
силикагель *chem.* silica gel
силикагель-осушитель silicagel-dehumidifier
силикальцит silicalcite
силикат *chem* silicate
~ алюминия aluminium silicate
~, безводный anhydrous silicate
~, водный hydrous silicate
~, вязанный inosilicate
~, кольцевой cyclosilicate
~, листовой/покровный phyllosilicate
~ натрия sodium silicate
~, предельный saturated silicate
~, цепной inosilicate
силикатирование silication, silicatization
силикатобетон lime concrete
силикатоз *med.* silicatosis
силикоз *med.* silicosis, stone mason's disease
силиколит *miner.* silicolite, flinty rock
силикомарганец silicomanganese
силикон *chem.* silicone
силикотермия silicothermic process/reduction
силикотический *med.* silicotic
силикотуберкулез silicotuberculosis
силикохром ferrochrome silicon
силицид *chem.* silicide
силицизация *chem.* silicification
силиций *chem.* silicium, silicon
силицирование silicon impregnation, siliconizing
силл *geol.* sill
~, сложный composite/multiple sill
силлабема syllabeme
силлабика syllabics
силлабический syllabic
силлабо-тонический syllabic-accentual
силлепсис *gram.* syllepsis
силлогизм syllogism
~, приведенный reduced syllogism
силлогистика syllogistic
силлогистический, силлогический syllogistic
силовой power *(attr.);* dynamic
~ кабель power cable
~вая линия *phys.* line of force
~вая станция power station; powerhouse
~вое поле *phys.* field of force
силоксан *chem.* siloxane
силон *text.* silon
силос *agr.* 1. silo; 2. silage; ensiled fodder
силосовать ensile, silage, silo
силосораздатчик *agr.* silage feeder
силосорезка silage cutter/chopper/shredder
силосоуборочный silage-harvesting
силосохранилище silo
силумин *met.* silumin
силур *geol.* Silurian
силуэт silhouette; outline, profile
силы forces; *see also* **сила**
~ авианосной авиации sea-based air forces
~, авианосные carrier forces
~, близкодействующие short-range forces
~, валентные valence forces
~, взаимно-пересекающиеся concurrent forces
~, военно-воздушные air force
~, военно-морские naval forces, navy
~, воздушно-космические aerospace forces
~, вооруженные armed forces

~, **десантные** landing force(s)
~, **ко(м)планарные** coplanar forces
~, **межнуклонные** *phys.* internucleonic forces
~ **морской пехоты** marine troops
~, **надводные** *naut.* surface force
~, **небоевые** noncombatant forces
~, **неко(м)планарные** non-coplanar forces
~, **непересекающиеся** non-concurrent forces
~, **обменные** exchange forces
~, **объемные** volumetric forces
~ **охранения** escort/patrol force(s)
~, **ударные** striking/strike force
сильвинит I *geol.* sylvinite
сильвинит II *med.* inflammation of the sylviduct
сильнодействующий potent, powerful
сильнолегированный heavily alloyed/doped
сильномочковатый *bot.* multifasciculate
сильнопахучий highly odorous, graveolent
сильносвязанный tightly bound
сильноскладчатый acutely/sharply folded
сильноточный heavy-current; *el.* high-current
сильф *myth.* sylph
сильфида 1. *pl. zool.* carrion/burying beetles *(Silphidae)*; 2. *myth.* sylph
сильфия, сильфиум *bot.* rosin weed, compass plant *(Silphium)*
сильфон bellows, sylphon
~ **обратной связи** feedback bellows
~, **управляющий** control bellows
сима *geol.* sima
симаруба *bot.* simaruba *(Simaruba)*
симбатно symbasically
симбатность *biol.* symbasis
симбатный *biol.* symbasic
симбиоз *biol.* symbiosis
~ **бактерий** bacterial symbiosis
~ **водорослей** algal symbiosis
симбионт symbion(t), symbiote; aulophite
симбиотрофный *bot.* symbiotrophic
симблефарон *med.* symblepharon
~, **неполный** partial/incomplete symblepharon
~, **полный** total/complete symblepharon
симблефароптеригиум symblepharopterygium
символ symbol; character; emblem; image; digit
~, **буквенно-цифровой** alphanumeric character
~ **веры** *rel.* creed
~ **выхода** *comp.* excape character
~, **графический** graphic character
~, **двоичный** *comp.* binary digit
~ **игнорирования** ignore character
~, **информационный** data symbol
~ **исключения** cancel character
~ **квитирования** acknowledge character
~, **кодовый** code character
~ **конца передачи** end-of-transmission character
~ **Кристофеля** *math.* Christoffel's symbol
~ **Кронекера** *math.* Kronecker's symbol
~ **Лежандра** *math.* Legendre's symbol
~, **магнитный** magnetic character
~ **новой строки** new line character
~ **нуля** null character
~ **отношения** *comp.* relation character, relator
~, **плавающий** drifting character
~ **присваивания** assignment symbol
~ **пробела** space character
~ **проверки условия** decision symbol
~ **пропуска** blank character, ignore sign
~, **разделительный** separator symbol
~ **стирания ошибки** delete/erase character
~, **управляющий** control character
~, **фиктивный** idle character
~ **Якоби** Jacobi's symbol
символика *math.* symbolism, symbolics, notations
~, **матричная** matrix notations
~ **микроблоков** microblock symbolism
сименс *phys.* siemens, S
симментальский Simment(h)al
симметрирование symmetrization; balancing
~, **емкостное** *el.* capacitance balancing
симметричность symmetry
симметричный symmetrical; balanced
симметрия symmetry; balance
~, **зеркальная** bilateral/mirror symmetry; dissymetry; plane symmetry
~ **относительно оси** axial symmetry
~ **относительно плоскости** plane symmetry
~ **относительно точки** point symmetry
~, **перистая** pinnate symmetry
~, **ромбоэдрическая** rhombohedral symmetry
~, **циклическая** cyclosymmetry
симоген *chem.* cymogene
симония *hist.* simony
симпатико-адреналовый *med.* sympathetico-adrenal
симпатиконеврит *med.* sympathiconeuritis
симпатикотомия *med.* sympathicotomy
симпатикотония *med.* sympathicotonia
симпатикотрипсия *med.* sympathicotrypsy
симпатикотропный *med.* sympathicotropic
симпатин *biochem.* sympathin
симпатический sympathic, sympathetic
~**кая нервная система** sympathetic nervous system
~**кие чернила** invisible ink
симпато-адреналовый sympathoadrenal
симпатобласт sympathoblast, sympathicoblast
симпатобластома sympathoblastoma
симпатогония sympathogonia
симпатолитический sympatholytic, sympathocolytic, sympathoparalytic
симпатомиметический sympathomimetic
симпатопатия sympathicopathy
симпатрия *biol.* sympatry
симпатэктомия sympathectomy, sympathetectomy
~, **поясничная** lumbar sympathectomy
~, **шейно-грудная** thoracocervical sympathectomy
симпетальный *bot.* gamopetalous, sympetalous
симплазма *biol.* symplasm
симпласт symplast

симплекс *math., el.* simplex
~, правильный regular symplex
симплектит symplektite
симплектитовый symplektitic
симплектический symplectic
симплока *lit.* symploce, simploce
симплокарпус *bot.* skunk cabbage *(Symplocarpus)*
симплоковые *pl. bot.* sweetleaf family *(Symplocaceae)*
симподиальный *bot.* sympodial
симподий *bot.* sympode, sympodium
симпозиум symposium
симптом symptom, sign
~ абстиненции abstinence/withdrawal symptom
~, глазной ocular sign
~, дополнительный accessory/concomitant/consecutive symptom
~, очаговый focal symptom
~, пирамидный pyramid sign
симптоматика, симптоматология symptomatology
симптоматокомплекс symptom-complex/group, complex of symptoms, syndrome
симпьезометр sympiezometer
симулировать simulate; copy, imitate
симультанность simultaneity, simultaneousness
симфалангия symphalangism, syndactylism
симург *myth.* simurgh
симфиз symphysis
симфизарный syphysial, symphysis *(attr.)*
симфизиопатия symphyseopathy, symphysiopathy
симфизиорафия symphyseorrhaphy, symphysiorrhaphy
симфизиотомия symphyseotomy, symphisiotomy
симфизит symphisitis
симфизодактилия symphysodactilia, syndactylia
симфиогенез *bot.* symphyogenesis
симфониетта *mus.* symphonette
симфонический symphonic, symphony *(attr.)*
симфония 1. symphony; 2. *eccl.* concordance
симфориокарпос *bot.* snowberry *(Symphoriocarpos)*
симфратизм *geol.* symphratism, regional metamorphism, dynamometamorphism
синагога synagogue
синайский Sinaitic; Sinai (attr.)
синаксарий synaxarion, synaxarium
синалар fluocinolone acetonide, synalar
синалефа *gram.* synal(o)epha
синальбин *chem.* sinalbin
синальгия *med.* synalgia
синангий *bot.* synangium
синандрий synandrium
синандрия synandry
синантезис *bot.* synanthesis
синантетический, синантический *geol.* synantetic
синантия *bot.* synanthy
синантроп *anthr.* Peking man
синапс *med.* synapse
~, возбуждающий stimulating synapse
~, тормозящий inhibitory synapse
синапсидный *zool.* synapsidean
синапсис synapsis
синаптена synaptene, zygotene
синаптикула *zool.* synapticulum
синаптих *paleont.* synaptichus
синаптический synaptic
синаптоспермия synaptospermy
синартроз synarthrosis
синафия synaphe(i)a
сингал, сингалезский, сингальский Sin(g)halese
сингалка, сингалезка Sin(g)halese woman
сингамия *biol.* syngamy
сингамный syngamic, syngamous
сингенез *biol.* syngenesis
синглет *phys.* singlet
сингомоклиналь synhomocline
сингония syngony, crystal system
~, квадратная tetragonal system
~, кубическая/правильная cubic system
~, моноклинная monoclinic system
~, ромбическая orthorhombic system
сингулярность *math.* singularity
синдактилический *med., zool.* syndactylous, syndactil(e)
синдактилия syndactyly, syndactylia, syndactylism
синдесмит *med.* syndesmitis
синдесмоз *anat.* syndesmosis
синдесмология *anat.* syndesmology
синдесмома *pathol.* syundesmoma
синдесмопластика syndesmoplasty
синдесмоплексия syndesmoplexy
синдесморафия syndesmorrhaphy
синдесмотомия syndesmotomy
синдесмэктомия *surg.* syndesmectomy
синдесмэктопия syndesmectopia
синдетикон seccotine *(glue)*
синдик syndic
синдиплоидия syndiploidy
синдицированный *econ.* syndicated
синдром *med.* syndrome
~ абстиненции abstinence syndrome
~, апоплексический apoplectic/stroke syndrome
~ атаксии ataxia syndrome
~, врожденный congenital syndrome
~, гепаторенальный hepatorenal/hepatonephric syndrome
~, гиперспленический hypersplenic syndrome, hypersplenism
~, гипофизарный hypophysis syndrome
~ длительного раздавливания crush syndrome; compression syndrome
~ дуги аорты aortic arch syndrome
~, желудочно-сердечный gastrocardiac syndrome
~, ишемический ischemic syndrome
~, лучевой radiation syndrome

~ навязчивости obsessional syndrome
~, ногте-пателлярный nail-patella syndrome, arthroonychodysplasia
~ отсутствия пульса aortic arch syndrome
~ перенапрежения overtension syndrome
~ позвоночного диска disk syndrome
~ потеря соли salt depletion syndrome
~ приобретенного иммунного дефицита (СПИД) acquired immune deficiency syndrome (AIDS)
~, рентгенологический X-ray sindrome
~, сердечно-легочный cardiopulmonary syndrome
~ усилия effort syndrome; anxiety neurogis
~, шейно-плечевой cervicobrachial syndrome
~, эпифизарный epiphysial syndrome
синдромный syndromic
синди Sindhi
синева *bot.* blue rot, bluing
синеглазка *bot.* common/Asiatic dayflower *(Commelina communis)*
синегнойный *med.* pyocyanic
синеголовка *zool.* figure-of-eight moth *(Diloba coeruleocephala)*
синеголовник *bot.* eryngo, eryngium *(Eryngium)*
синедрион *hist.* Sanhedrim, Sanhedrin
синезис *gram.* synesis
синекдоха *lit.* synecdoche
синекдохический synecdochial
синеклиза *geol.* syneclise, tectonic depression
синеломкость blue brittleness/shortness
синель *text.* chenille
синема *bot.* synema
синематека film library
синергида *bot.* synergid
синергизм, синергия synergism; synnergy
синергический synergetic, synergic, synergistic
синереза *phon.* synaeresis
синерезис *phys.* syneresis
синерод *chem.* cyanogen
синеспинка *zool.* blueback herring *(Pomolobus aestivalis)*
синестезия *psych.* synesthesia
синехия *med.* synechia, iritic adhesion
~, кольцевидная annular synechia
~, полная total synechia
синец *zool.* zope *(Abramis ballerus)*
синешейка *zool.* bluethroat *(Luscinia svecica)*
синзоохория synzoochory
синигрин *chem.* sinigrin, potassium myronate
синильный *chem.* hydrocyanic
~ная кислота hydrocyanic/prussic acid
синистрин *chem.* sinistrin
синистроторсия sinistrotorsion, sinistrorotation, sinistrogyration
синица *zool.* titmouse, tomtit, tit *(Parus)*
синицеза *lit.* synizesis
синкарион *biol.* synkaryon, syncaryon
синкарп, синкарпий *bot.* syncarp(ium)
синкарпия *bot.* syncarpy
синкарпный *bot.* syncarpous
синкинезический *med.* synkinetic
синкинезия *med.* synkinesis, syncinesis
синклаза *geol.* synclase
синклиналь *geol.* syncline
~, килевая carinate syncline
~, котловинообразная basin
~, сложная composite syncline
синклинорий *geol.* synclinorium, synclinore
синклит council, synod
синкопа *mus., gram.* syncope
синкопировать *mus.* syncopate
синкотилия *bot.* syncotyly
синкретизировать syncretize
синкретичность syncretism; syncretic nature
синкумар acenocoumarol, sintrom
синнема *bot.* synnema
синоатриальный *anat.* sino-atrial/auricular
синовиальный *anat.* synovial, synoviparous
синовит *med.* synovitis, arthromeningitis
~, бурситный bursal synovitis, bursitis
~, грибковый fungous synovitis/arthritis
~, подагрический gouty synovitis
синовия *med.* synovia, joint oil
синод synod
~, Святейший правительственный Holy Governing Synod
синодик *eccl.* death bill
синодический *astr.* synodic(al)
синодный, синодский synadal
синолог Sinologist
синология Sinology
синонимика, синонимия synonymy
синонимический synonymous, synonymic(al)
синопсис synopsis
синоптик weather forecaster
синоптический synoptic(al)
~кая карта weather map
~кие условия meteorological conditions
синорхидия *med.* synorchism, synorchidism
синостоз *med.* synostosis
~, костнопластический osteoplastic synostosis
~ позвонков vertebral synostosis
синостозный *med.* synost(e)otic
синосхиз *anat.* synoscheos
синофтальмия synophthalmia, cyclopia
синрабдосома *geol.* synrhabdosome
синсепальный *bot.* synsepalous
синсоматический synsomatic, protosomatic
синспермия synspermy
синтагма *ling.* syntagma
синтагмема *ling.* syntagmeme
синтаксис syntax
синтаксический, синтаксичный, синтактический syntactic(al)
синтактика syntactics
синтез synthesis
~ ядер nuclear fusion
синтезатор synthesizer
синтезирование synthesization, synthetisation

синтезировать synthesize, synthetize
синтетаза *bioch.* synthetase
синтетик synthesizer, synthesist
синтетика synthetics, synthetic materials
синтетический 1. synthetic(al); 2. *log.* ampliative
синтетичность synthetical character/nature
синтип *biol.* syntype
синто *rel.* Shinto
синтоизм *rel.* Shintoism
синтоист, синтоистка Shintoist
синтомицин, синтомицетин synthomycin
синтонизировать syntonize, tune
синтонин *biochem.* syntonin, acid albumin
синтопия syntopy
синтрофизм syntrophysm, syntrophy
синузия *bot.* synusia, synusium
синуит *med.* sinuitis, sinusitis
синупаллиатный *zool.* sinupalliate, sinupallial
синус I *anat.* sinus
  ~, мантийный/паллиальный pallial sinus
  ~ перикарда oblique sinus
  ~ плевры pleural sinus
  ~ селезенки splenic sinus
синус II *math.* sine
  ~ амплитуды amplitude sine
  ~, гиперболический hyperbolic sine, sinh, sh
  ~, интегральный sine integral function, integral sine
  ~, обращенный versed sine
синус-верзус versed/reversed sine
синус-гальванометр sine galvanometer
синусит *med.* sinuitis, sinusitis
синусно-косинусный sine-cosine *(attr.)*
синусно-предсердный *anat.* sinuatrial, sinoatrial
синусовый *anat.* sinusal
синусоид *anat.* sinusoid
синусоида *math.* sinusoid, sine curve
  ~, затухающая damped sinusoid
синусоскоп *med.* antroscope
синусоскопия antroscopy
синусотомия *surg.* sinusotomy, sinuotomy
синусофарингоскоп *med.* antropharyngoscope
синус-преобразование sine transform
синус-счетчик sine meter
синустромбоз *med.* sinus thrombosis
синфазирование *el.* phase synchronization, synphasing
синфазность *el.* phase synchronism/coincidence
синфазный *el.* cophased, cophasal, inphase
синхейлия *med.* synchilia
синхизис synchysis
синхондроз *anat.* synchondrosis
  ~, грудинный sternal synchondrosis
  ~, передний внутризатылочный anterior intraoccipital synchondrosis
синхондрозотомия *surg.* synchondroseotomy
синхондротомия *surg.* synchondrotomy
синхорология synchorology
синхрогенератор synchronizing signal generator, sync generator
синхроимпульс sync pulse
синхрона *math.* synchrone
синхронизатор synchronizer
синхронизация synchronization; timing
  ~ вспомогательных сигналов pilot synchronization
  ~, импульсная impulse/pulse synchronization
  ~ импульсов pulse timing
  ~, грубая rough synchronization
  ~, кадровая *telev.* frame/vertical synchronization
  ~, строчная *telev.* line/horizontal synchronization
  ~, фазовая phase locking
синхронизировать synchronize, time
синхронизм, синхронность synchronism
синхроничность synchroneity, synchronism
синхрония synchrony, synchronism
синхронно synchronously, synchronically
синхронность synchronism, synchrony
синхронный synchronous, synchronal, synchronic
синхротрон *phys.* synchrotron
  ~, протонный proton synchrotron, synchrophasotron
синхросигнал synchronizing signal
синхросмесь *telev.* composite sine pulse
синхрофазирование phase synchronization
синхроциклотрон *phys.* synchrocyclotron
синцианин syncyanin
синцитиальный *biol.* syncytial
синцитий, синцит *biol.* syncytium
синцитиолизин syncytiolysin, placentolysin
синцитиома syncytioma, deciduoma, chorioma
синцитиотоксин syncytiotoxin
синцитиотрофобласт syntrophoblast, syncytiotrophoblast
синь blue
  ~, кобальтовая/Тенара cobalt blue
  ~, парижская French blue
  ~, прусская Prussian/Berlin blue
синьга *zool.* scoter *(Melanitta)*
синьория *hist.* signoria
синька 1. blueprint (paper); 2. blu(e)ing, blue
синэкология *biol.* synecology
синэнклитика synenclitic
синэргетика synergetics
синэстрол hexestrol
синэция synoecy, synoeky, synoecism
синюха 1. *bot.* Greek valerian, polemonium *(Polemonium);* Jacob's ladder *(Polemonium caeruleum);* 2. *med.* cyanosis, blue disease
синюховые *pl. bot.* phlox family *(Polimoniaceae)*
синюшность *med.* cyanosis
  ~ конечностей acrocyanosis
синюшный *med.* cyanotic, cyanosed
синявка *bot.* moor grass *(Molonia caerulea);* amethyst agaric *(Lepista nuda)*
синяк 1. *med.* ecchymosis, bruise, black-and-blue spot; 2. *bot.* viper's bugloss *(Echium);* 3. *bot.* blewit *(Oyroporus cyanescens)*
сионизм Zionism
сионист, сионистка, сионистский Zionist

сионский Zion
сип *zool.* vulture *(Gyps)*
~, белоголовый griffon vulture *(Gyps fulvus)*
сипуха *zool.* barn owl *(Tyto alba)*
сирен *zool.* Congo eel *(Siren)*
сирена 1. siren, syren, hooter, howler; 2. *myth., paleont.* siren; 3. *med.* sirenomelus
сиреновые *pl. zool.* mud eels *(Sirenidae)*
сирень *bot.* lilac *(Syringa)*
сирены *pl. zool.* 1. sirenians, sea cows *(Sirenia);* 2. sirens *(Sirenidae)*
сириец, сирийский Syrian
сирийка Syrian (woman)
сирийско-романский Syro-Roman
сирин *myth.* siren
~, домовый *zool.* little awl *(Athene noctua)*
сирингит *med.* syringitis
сирингобульбия *med.* syringobulbia
сирингомиелит *med.* syringomyelitis
сирингоменингоцеле *med.* syringomeningocele
сирингомиелит *med.* syringomyelitis
сирингомиелия syringomyelia, cavitary myelitis
сирингомиелоцеле syringomyelocele, hydrorrhachis interna
сиринготомия syringotomy, fistulotomy
сирингоцистаденома *med.* syringocystadenoma
сирингоэнцефаломиелия syringoencephalomyelia
сиринкс Pandean pipe, panpipe, syrinx
Сириус *astr.* Sirius, Dog Star
сирокко *meteor.* sirocco
сироп syrup, sirup
~, инвертный invert syrup
~, маточный mother syrup
система system; method; arrangement, organization; scheme; formation
~, аварийная emergency system
~ аварийного выключения реактора emergency shut-down system
~ аварийной защиты safety system
~ автоматизированного проектирования computer-aided design system, CAD system
~ автоматизированной разработки computer-aided engineering system, CAE system
~, автономная autonomous [stand-alone] system
~, аккордно-премиальная task-and-bonus plan
~, банковская banking system
~ без резервирования non-redundancy system
~ блокировки interlocking system
~, блочная modular [inter-block] system
~ Боме Baumé system
~ в виде банка данных data-bank system
~, вегетативная нервная vegetative/involuntary nervous system
~, вегетативно-эндокринная vegetoendocrinic/vegetoendocrinous system
~ векторов левая/правая left-/right-handed system of vectors
~ весов weighting system
~ водопитания feedwater system
~ водоснабжения, оборотная in-plant recirculation
~ воспроизведения reproduction system
~, восьмиричная octal system
~, вторая сигнальная second signal(l)ing system
~ выборки *comp.* selection system
~, выделительная excretory system
~, вытяжная exhaust system
~ вычетов *math.* residue system
~, вычислительная computer/computing system
~, герметичная leak-tight system
~ громкоговорящей связи public-address system
~, двойная binary system
~ двойного учета dual standard cost system
~, девонская *geol.* Devonian system
~, денежная monetary system
~, диалоговая call-reply system
~ дистанционного управления remote control system
~, дифферентная *naut.* trim system
~ для руководства management operating system, MOS
~ дождевания sprinkling system; overhead irrigation system
~, дополнителная запоминающая backing/auxiliary store system/unit
~, дублированная dual system
~, дублирующая backup system
~, дыхательная respiratory system
~ единиц, международная (СИ) international system of units, SI
~ единиц МКС MKS system of units
~ единиц СГС CGS system of units
~ жизнеобеспечения life support system
~ зажигания, полупроводниковая transistorized ignition system
~ заземления earth/ground system
~, закрытая/замкнутая closed system
~, замедляющая *el.* slow-wave structure; slow-wave guide
~ записи writing system
~ заработной платы, поощрительная wage incentive plan, task-and-bonus plan
~, зубная dentition
~ измерений, имперская imperial measures
~, иммуная immune system
~ индексации, десятичная decimal system of coding
~, инерциальная inertial system
~, информационная data system
~, информационно-поисковая data/information retrieval system
~ искусственного интеллекта artificial intelligence system, AI system
~, испольная sharecropping
~ калькуляции издержек cost system
~, канализационная sewer/sewerage system

~, карточная rationing system
~, квантовая quantized system
~, кембрийская Cambrian system
~ клиринговых расчетов clearing system
~ кодирования coding/encoding system
~, колебательная vibrating/vibratory system; oscillatory/resonant system
~ коллективного пользования multiuser system
~, коллоидная colloid system
~ команд instruction set
~ контроля monitoring/control system
~ координат coordinate system, system of coordinates
~ координат, декартовая Cartesian coordinate system
~ координат, косоугольная oblique coordinate system
~ координат, левая/правая left-handed/right-handed coordinate systen
~ координат, прямоугольная rectangular coordinate system
~ кормления feeding policy
~ криволинейных координат curvilinear coordinate system
~ критериев качества quality system
~, кровеносная circulatory system; vascular system
~, кроветворная h(a)ematopoietic system
~, марковская Markov(ian) system
~ массового обслуживания service system
~ массового обслуживания, марковская Markov(ian) queue
~ массового обслуживания, разомкнутая open-end-queue
~ материального стимулирования incentive scheme
~, меловая *geol.* Cretaceous system
~ Менделеева, периодическая periodic chart/table
~ мер/метрическая metric system
~, микропроцессорная microprocessor-based system
~, многоканальная many-server [multichannel, multiple queue] system
~, многомерная multivariable system
~, многопродуктовая multiitem system
~, многоуровневая multilevel system
~, многоэлементная *math.* multiple system
~, модульная modular system
~, мочеполовая *med.* urinogenital system
~, мышечная muscular system
~ наемного труда wage-labour system
~ народного образования educational system
~, неспецифическая *med.* nonspecific system, reticular activating system
~, нормированная normalized system
~ нумерации numbering scheme
~, обводная by-pass system
~ обозначения notation
~ обработки данных data processing system
~ обратной связи feedback system
~, окислительно-восстановительная redox system, oxidation-reduction system, OR system
~, оперативно доступная on-line system
~, оперативно недоступная off-line system
~, ордовикская *geol.* Ordovician system
~, оросительная irrigation system/project
~ орошения sprinkling system
~, ортогональная orthogonal set
~ освещения lighting (system), illumination
~ отсчета reference frame/system, frame of reference; reading system
~ оценки appraisal plan
~, палеогеновая *geol.* Paleogene system
~ памяти *comp.* memory/storage system
~, первая сигнальная first signal(l)ing system
~ передачи данных data transmission system
~, периферическая нервная peripheral nervous system
~, пермская *geol.* Permian system
~ питания feed system; fuel system
~, пищеварительная digestive/alimentary system
~, подчиненная *comp.* slave system
~ поощрительных вознаграждений bonus plan, financial incentive plan
~ премиальных ставок premiun system
~ прерывания программы program(me) interrupt system
~ программирования, основная basic programming system
~ программного обеспечения software system
~, противопожарная fire-extinguishing system
~ прямоугольных координат Cartesian/rectangular coordinate system
~ пуска/пусковая start-up system
~ работы по сменам shift system
~ равновесия equilibrium system
~ раздачи кормов *farm.* feeding system
~ разработки method/system of mining
~, рассредоточенная multisite system
~, реакторная reactor system
~, реакционная reacting system
~ реального времени real-time system
~, резервированная redundant system
~, респираторная = дыхательная
~, ретикулоэндотельная reticuloendothelial system
~, речная river system
~, рифейская *geol.* Riphean system
~ родственных связей kinship pattern
~ рычагов leverage
~ с дистанционным доступом remote-access system
~ с обратной связью feedback system

~ с ожиданием *math.* queueing system
~ с отказами loss system
~ с разделением времени time-sharing system
~ с человеком manned system
~, самонастраивающаяся adaptive system
~, самообучающаяся self-learning system
~, самоуравновешивающаяся self-balancing system
~, самоусовершенствующаяся evolutionary system
~ связанных уравнений *math.* coupled equations
~ связи communication system
~ связи, уплотненная multiplex communication system
~, сельсинная synchro system
~, сердечно-сосудистая cardiovascular system
~ сил force system
~, силурийская *geol.* Silurian system
~, симпатическая нервная sympathetic nervous system
~, синийская *geol.* Sinian system
~ синхронизации timing/synchronizing system
~, следящая follow-up system
~ слежения tracking system
~ смазки/смазочная lubrication (system)
~ сметных оценок estimating-cost system
~, солнечная solar system
~ сопровождения tracking system
~, сопряженная conjugated/coupled system
~, сосудистая = кровеносная
~ счисления number system
~ счисления, восьмеричная octal system
~ счисления, двоичная binary system
~ счисления, десятичная decimal system
~ счисления, позиционная radix/positional number system
~ счисления, шестнадцатиричная hexadecimal system
~, телеметрическая telemetering system
~, топливная fuel system
~, тройная trinary system; ternary system
~, трюмная *naut.* bilge system
~ тяг linkage
~, тяго-дутьевая draught system
~ убоя *farm.* slaughter policy
~ управления control
~, управляемая controlled system, system under control
~ уравнений system/set of equations
~ участия в прибылях bonus scheme
~ учета издержек cost system
~ учета и отчетности, дистанционная teleaccounting system
~ учета, калькуляционная cost accounting plan, interlocking cost plan
~, ферментная enzyme system
~ функций set of functions
~ ценностей value system
~, центральная нервная central nervous system
~ цифрового управления digital control system
~, человекомашинная man-machine system
~, шарнирно-стержневая hinged-rod-system
~ элементов, периодическая = Менделеева
~, эндокринная endocrine system
~, энергетическая единая power grid
~, энергетическая объединенная interconnected power grid
систематизация systematization; classification
систематизировать systematize; classify
систематика 1. systematics; classification, systematization; 2. taxonomy
систематический 1. systematic, methodic(al); systemic; 2. taxonomic
систематичный systematic, methodical
системно-транзитивный *math.* set-transitive
системно-функциональный systemic-functional
системный systemic
системотехника systems engineering
систола *med., lit.* systole
~ желудочков ventricular systole
~ предсердий atrial/auricular systole
систолический *med.* systolic
систр sistrum
ситалл glass ceramic; polycrystalline glass
ситар *mus.* sitar
ситатунга *zool.* sitatunga *(Tragelaphus spekii)*
ситник *bot.* rush *(Juncus)*
ситняг *bot.* spike sedge *(Eleocharis)*
Сити *econ.* the City
сито sieve, strainer, sifter; cullender; bolter
~, барабанное trommel screen
~, вращающееся rotary/revolving screen
~, крупноячеистое coarse screen
~, мелкоячеистое coarse screen
ситовидный sievelike, cribriform, cribrose, sieve *(attr.)*
ситовник *bot.* pycreus *(Pycreus)*
ситология sitology, dietetics
ситомания *med.* sitomania
ситостан sitostane
ситостерин *chem.* sitosterol
ситотаксис, ситотропизм sitotaxis, sitotropism
ситотерапия sitotherapy, dietotherapy
ситуация situation; condition(s)
~, конфликтная collision; contention
~, терминальная *comp.* terminal condition
~, тупиковая clinch; deadlock
ситцепечатание calico/cotton printing
сиу Sioux
сиф cif, cost, insurance and freight
сифема sifema
сифилид *med.* syphilid, syphiloderm
~, буллезный bullous/pemphigoid syphilid
~, гуммозный/бугорковый gummatous/nodular/tubercular syphilid, cutaneous gummas

~, **ладонный** palmar syphilid
~ **кожи** syphiloderma
~, **подошвенный** plantar syphilid
~, **ползущий** serpiginous syphilid
~, **эктиматозный** ecthymatous syphilid
сифилидолог *med.* syphilologist
сифилидология *med.* syphilology
сифилис *med.* syphilis, lues
~, **бессимптомный** latent syphilis
~, **врожденный** hereditary/congenital syphilis
~ **нервной системы** lues nervosa
~, **свежий** early syphilis
сифилитический syphilitic
сифилома syphiloma, siphylophyma
сифиломный syphilomatous
сифон 1. siphon, syphon; 2. *zool.* siphon
~, **вводной** inhalent/incurrent syphon
~, **внешний** ectosolenian tube
~, **внутренний** entosolenian tube
~, **выводной/клоачный** exhalent/excurrent siphon
~ **головоногих** siphuncle
сифонировать siphon
сифоногамный *bot.* siphonogamic, siphonogamous
сифоноглиф *zool.* siphonoglyph
сифонопора *zool.* siphonopore
сифоностела, сифоностель *bot.* siphonostele
сифоностомный *zool.* siphonostomatous, syphonostomous
сифонофор(а) *zool.* siphonophoran, siphonophore
сифункл *zool.* suphuncle
сицилианский, сицилийский Sicilian
сициос *bot.* bur cucumber *(Sicyos)*
сияние glow, radiance; aurora, aureola
~, **полярное** polar aurora, aurora polaris
~, **северное полярное** aurora borealis
~, **Южное полярное** aurora australis
сиятельство *(title)* Excellency
скабиоза *bot.* scabious, mourning bride, gypsy rose *(Scabiosa)*
сказуемный *gram.* predicational, predicative
сказуемо-бесподлежащий *gram.* predicational-subjectless
сказуемое predicate
сказуемостный predicative, predicational
сказуемость predicativity
скайтерьер Skye terrier
скакун *zool.* tiger beetle *(Cicindela)*
скакунчик *zool.* leaping spider *(Saltigradae)*
скала rock; stone; cliff
~, **подводная** reef
~, **отвесная** cliff
скаленоэдр *cryst.* scalenohedron
скаленоэдрический scalenohedral
скалывание spallation; splitting/chopping off; shear/sliding fracture
скальд *lit., hist.* scald, skald, bard
скальный rocky, rock *(attr.)*; rupicolous
~**ное растение** rupicolous/saxicolous plant, petrophyte
скальп scalp
скальпель scalpel
~, **брюшистый** general operating knife
~, **глазной** ophthalmic scalpel
скальпирование scalping
скаляр(а), скалярный *math.* scalar
скамоний *bot.* scammony *(Convolvulus scammonia); pharm.* scammony
скаммонин *pharm.* scammonin, jalapin, orizabin
скамья bench
~ **подсудимых** (prisoner's) dock
~ **присяжных** jury-box
скандий *chem.* scandium, Sc
скандинав(ец), скандинавский Scandinavian
скандинавизм Scandinavianism
скандинавка Scandinavian (woman)
сканирование scanning, scan
~, **винтовое** screw scanning
~, **зигзагообразное** line-by-line scanning
~, **ленточное** stripwise scanning
~, **привязанное** captive scan
~ **с обращением** inversion scanning
сканировать scan
скапливаться accumulate, agglomerate, aggregate
скарабей *zool.* scarabaeus, scarab *(Scarabaeus)*
скарификатор 1. *med.* scarificator; 2. *agr.* scarifier
скарифицировать scarify
скарлатина *med.* scarlatina, scarlet fever
~, **атипичная/скрытая** latent scarlatina
~, **токсическая [токсико-септическая]** scarlatina maligna
скарлатинозный scarlatinal
скарлатиноподобный scarlatiniform
скарпель pitching (stonecutter's) chisel
скат I slope, gradient, incline, slant, descent; *eng.* chute, slide, skid; *anat.* clivus
~ **крыши, готический** gothic pitch
скат II *zool.* skate *(Raja)*
~, **слепой электрический** electric skate
~, **электрический** torpedo
скат III skat *(card game)*
скатоль *chem.* scatole
скатофагия scatophagy, rhypophagy
скатывать rollup; roll down
скаут Boy scout, scout
скафандр full-pressure suit; *naut.* diving suit
~, **автономный космический** full extravehicular suit
~, **водолазный** diving-suit/dress
~, **высотный** (high-altitude) full-pressure suit
~ **для работы в открытом космосе** extravehicular activity suit, outdoors space suit
~, **космический** spacesuit, spaceflight suit
~, **космический, для работы на борту** intravehicular space suit
~, **спасательный** survival suit
скафиринх *zool.* shovelnose sturgeon *(Scaphirhynchus)*

скафопод *zool.* scaphopod *(Scaphopoda)*
скафоцефалический scapocephalic, scapocephalous
скафоцефалия *med.* scaphocephalism, scaphocephaly
скачивание rinsing, draining, flushing; slagging
скачкообразный saltatory; intermittent
скачок jump, saltation, leap, spring, bound; sudden/abrupt change; discontinuity
 ~ в развитии saltation, mutation
 ~ давления pressure jump
 ~, квантовый quantum jump
 ~ плотности *phys.* density discontinuity
 ~ поглощения absorption discontinuity
 ~, температурный temperature jump
 ~ уплотнения shock (wave)
 ~ уплотнения, прямой normal shock (wave)
скашивать 1. chamfer, bevel; 2. cut, mow
скважина well, hole; gap, aperture; chink, slit
 ~, артезианская artesian well
 ~, безрезультатная barren well
 ~, буровая borehole
 ~, взрывная blast hole
 ~, водоразборная water-supply well
 ~, дренажная release/relief hole
 ~, замочная keyhole
 ~, морская off-shore well
 ~, наблюдательная observation/inspection hole
 ~, нагнетательная injection well
 ~, нефтяная oil well
 ~, поисковая/разведочная prospecting hole
 ~, разгрузочная relief well
 ~ сбросовых вод effluent-disposal well
 ~, эксплуатационная operating hole
скважистость, скважность porosity; void content, pore space
скважистый, скважный porous, cavernous
скважность dyty rate/factor; on-off time ratio
сквален *chem.* squalene
скваленгидроксилаза squalene hydroxylase
сквамозигоматит squamozygomatitis
сквамозный squamous, squamate, scaly
скваттер squatter
скворец *zool.* starling *(Sturnus vulgaris)*
скег *naut.* skeg
 ~, бортовой side skeg, sideboard, sidewall
скелет skeleton; frame; carcass
 ~, висцеральный visceral skeleton, splanchonoskeleton
 ~, внутренний endoskeleton
 ~, наружный exoskeleton, dermoskeleton
скелетный skeletal, skeleton *(attr.)*
скена skene, scena
скенеит sken(e)itis
скеннер scanner
скеннирование scanning
скерда *bot.* hawk's beard *(Crepis)*
скерцо *mus.* scherzo
скиаграмма skiagram, skiagraph
скиаграфия skiagraphy
скиаскопия skiascopy, retinoscopy, shadow test
скиатический sciatic
скидка deduction, discount, rebate, reduction
 ~ на амортизацию depreciation allowance
 ~ на большой заказ large-order discounting
 ~ на провоз багажа free baggage allowance
 ~ на утечку tret
 ~ при сделке за наличные cash discounting
 ~ с налога tax rebate
 ~ со страхового взноса premium discounting
 ~, тайная ценовая covert price reduction
 ~, ценовая price discounting/rebate
скиатрон skiatron
скидывание 1. reduction, deduction; 2. throwing off; unloading
скиния *hist.* Tabernacle
скин-слой skin layer
скин-эффект skin effect
скиофильный *bot.* sciophilous, shade-loving
скипидар turpentine
скип (charging) skip
скипоклеть skip-cage, cage and skip
скирд(а) stack, rick
 ~ сена haystack
скирдование stacking, ricking
скирр *med.* scirrhus, fibrous/hard cancer, scirrhous carcinoma
скит small monastery
скитник hermit, anchorite; monk
скитский monasterial
скиф I *naut.* skiff
скиф II *hist.* Scythian
скифомедуза *zool.* discophoran, scyphomedusan *(Scyphomedusae)*
скифский *hist.* Scythian
склад storehouse, store, storeroom, depot; storage; depository; warehouse; stock
 ~, базовый плавучий base depot ship
 ~, вещевой *naut.* clothing depot
 ~ вооружения ordnance depot
 ~ горючесмазочных материалов fuel-oil depot
 ~ для транзитных грузов entrepot *(Fr.)*
 ~, дровяной lumber yard
 ~ заготовок stockyard
 ~, лесной timber-yard
 ~, открытый ground/outdoor storage
 ~, полевой dump
 ~ снабжения supply depot
 ~, технический maintenance depot
 ~, товарный warehouse, depository for goods
 ~, топливный fuel/fuelling depot
складирование storage, storekeeping, storing, warehousing
складировать store, keep in storage/warehouse
складка fold, crumple, plication; corrugation
 ~, голосовая *anat.* vocal fold
 ~ молоточка *anat.* malleoral fold
 ~ наковальни *anat.* incudal fold
 ~, носо-губная nasolabial fold

~, **подкрыльцовая** axillary fold
~, **слезная** lacrimal fold
~ **стремени** stapedial fold
~**ки, ресничные** ciliary folds
складковыпрямление *geol.* unfolding
складкообразный plaitlike, pliciform
складкообразование *geol.* folding
складочный storage *(attr.),* warehouse *(attr.)*
склад-холодильник refrigerated warehouse; cold storage
складчатодольный *bot.* diplecolobous
складчатолистный *bot.* ptychophyllous
складчатоплодный *bot.* ptychocarpous
складчатосеменной *bot.* ptychospermous
складчатость plication; *geol.* folding
~, **архейская** Archean folding
~, **киммерийская** Cimmerian folding
~, **ларамийская** Laramide folding
~, **постумная** posterior folding
~, **протерозойская** Proterozoic folding
~, **рифейская** Riphean folding
~, **таконийская** Taconian folding
складчатый folded, plicate(d); plaited
складываемое *math.* addend
складывание folding
складывать add/lay/put together; sum up
складывающийся foldable
склеивать glue/paste/bond together; splice
склеивающий adhesive, bonding
склейка 1. adhesive; 2. gluing/pasting together
склепывать rivet (together)
склера *anat.* sclera, sclerotica, sclerotic coat, white of the eye
склераденит *med.* scleradenitis
склеральный scleral, sclerotic
склередема *med.* scleredema
склереида *bot.* grit/stone cell, sclereid
склерема *med.* sclerema
склеренхима *bot.* sclerenchyma
склеренхимный, склеренхиматозный *bot.* sclerenchymatous
склериаз *med.* scleriasis
склерит 1. *med.* scleritis, sclerotitis; 2. *zool.* sclerite
~, **кольцевидный** annular scleritis
склеробазидия *bot.* sclerobasidium
склерогонидия *bot.* sclerogonidium
склеродактилия *med.* sclerodactyly, sclerodactylia, acroscleroderma
склеродермит sclerodermitis, sclerodermatitis
склеродермия, склеродерма sclerodermia, scleroderma; morphea
~, **ограниченная** circumscribed scleroderma, morphea
склероз *med.* sclerosis
~ **артерии** arterial sclerosis, arterosclerosis
~ **вен** phlebosclerosis
~ **венечных артерий** coronary sclerosis
~, **возрастной** presenile sclerosis
~ **желез** adenosclerosis
~ **кожи** scleroderma
~, **костный** bone sclerosis, eburnation
~ **легких** pneumosclerosis
~, **мышечный** myosclerosis
~, **очаговый** focal/insular sclerosis
~ **почек** nephrosclerosis
~, **рассеянный** multiple/disseminated sclerosis
~ **сердечной мышцы** cardiosclerosis
~ **сосудов** vascular sclerosis, angiosclerosis
~ **спинного мозга** myelosclerosis, spinal sclerosis
~ **уха** otosclerosis, sclerotitis
~ **хрусталика** phacosclerosis
склерозироваться sclerose, undergo sclerosis
склерозный sclerous, sclerosal, scleroid
склерозпродуцирующий *path.* sclerogenic
склероирит *med.* scleroiritis
склерокарп(ий) *bot.* sclerocarp
склерокератит sclerokeratitis
склерокератоз sclerokeratosis
склерокератоирит sclerokeratoiritis
склерокистозный sclerocystic
склерома *med.* scleroma
~ **гортани** laryngoscleroma
~ **губы** labioscleroma
~ **носа** rhinoscleroma
склеромаляция scleromalacia
склеромер *anat.* scleromere
склерометр sclerometer
склерон *met.* Scleron *(alloy)*
склеронихия *med.* scleronychia
склерономный scleronomous
склероотит sclerotitis, otosclerosis
склероофорит *med.* sclero-oophoritis, sclerothecitis
склеропаренхима *bot.* scleroparenchyma
склеропелит *geol.* scleropelite
склероперикератит scleroperikeratitis
склеропластика scleroplasty
склеропротеин scleroprotein, fibrous protein
склеропункция scleronyxis, scleroticonyxis, scleroticopuncture
склеророговичный sclerocorneal
склеросепта *zool.* scleroseptum
склеророговичный sclerocorneal
склероспороз *bot.* downy mildew
склеростеноз sclerostenosis
склеротомия sclerotomy
склеростомы *pl. Sclerostomata*
склеротиниоз, склеротиноз *bot.* sclerotiniose
склеротиния *bot.* sclerotinia *(Sclerotinia)*
склеротиоз *bot.* charcoal rot
склеротический sclerotic, sclerous, sclerosal
склеротом *surg.* sclerotome
склеротомия *surg.* sclerotomy, scleroticotomy
склеротрихия sclerotrichia, sclerothrix
склерофилл *bot.* sclerophyll, sclerophyllous plant
склерофилльный *bot.* sclerophyllous, hard-leaved
склерофит *bot.* sclerophyte
склерохориоидит sclerochorioiditis

склероциальный *bot.* sclerotial
склероций, склероция *bot.* sclerotium
склероктомия sclerectomy
склеренцефалия sclerencephaly, sclerencephalia
склиз slide; slipway
склон slope; decline; hillside
~ волны, наветренный back face of wave
~ волны, подветренный lee face of wave
склонение declination; variation; *gram.* declension
~, магнитное variation (of the compass), magnetic declination
~ звезды declination of a star
~, солнечное solar declination
~ компаса compass declination
склянка bottle, flask; phial; *naut.* hourglass
~, реактивная reagent flask/bottle
~ с притертой пробкой stoppered bottle
~, флорентийская separating flask
скоба clamp, cramp; stirrup; bracket, clip; brace, fastening, clincher; shackle; staple
~, буксирная towing shackle
~, вертлюжная chain swivel
~ водосточного желоба gutter hanger
~ дискового тормоза brake caliper
~ для крепления труб pipe clip
~, железная cramp iron
~, зажимная binding clip
~, замыкающая locking clamp
~, калиберная/измерительная caliper/snap gauge
~, крепежная bracket
~ минрепа *naut.* mooring shackle
~, плотничная dog anchor
~, подъемная lifting yoke
~, поручневая grab handle
~, предохранительная safety clip
~, соединительная/такелажная connecting/joining/joiner shackle
~, сцепная coupling link
~, треугольная heart shackle
~, фертоинговая/швартовная mooring shackle
~, шарнирная hinged clip
~, якорная anchor chain shackle
скобель *wood.* barking/drawing/scraper knife
скобка 1. *dim. of* **скоба**; 2. bracket, parenthesis; 3. *med.* agraffe; 4. *zool.* rotula
скобки *pl.* parentheses, brackets
~, квадратные square brackets
~, круглые parentheses
~, угловые French quotes
~, фигурные braces
скоб-трап *naut.* spar ladder
скобяной hardware *(attr.)*
сколекоспора *bot.* scolecospore
сколекотрихоз *bot.* brown stripe
сколекс *zool.* scolex
сколимус *bot.* oyster plant, golden thistle
сколиоз *med.* scoliosis
~, грудной thoracic/thoracal scoliosis
~, офтальмический ocular/ophthalmic scoliosis
~, пояснично-грудной thoracolumbar scoliosis
~, поясничный lumbar/coxitic scoliosis
~, рубцовый cicatrical scoliosis
~, шейно-грудной thoracocervical scoliosis
сколиозный scoliotic
сколиорахитический scoliorachitic
сколит 1. *miner.* skolite; 2. *paleont.* worm tube, scolite
сколофор *zool.* scolophore
скольжение slide, slip, sliding, slipping, slippage; gliding; creep
~, боковое *aero* side-slip(ping)
~ винта propeller slip
~, внутреннее slip(ping)
~ ледника glacial slide
~, наружное skid(ding)
~ пневматика tyre skidding
~ ремня belt slip
скользить slide; slip
сконденсировать condense
скопа *zool.* osprey, fish hawk *(Pandion haliaetus)*
скопец eunuch
скопление accumulation, concentration, aggregation, congestion; cluster
~ атомов aggregate of atoms
~ гибридов hybrid swarm/flock
~ дислокаций dislocation pile-up
~ капель blob of droplets
~ льда ice accretion
~ радиоактивных атомов accumulation of radioactive atoms, ''hot spot''
~ транспортных средств traffic conjestion
скополамин *chem.* scopolamine
~, бромистоводородный scopolamine hydrobromide, hydrobromide of scopolamine
скопулит *petr.* scopulite
скопчество eunuchry, spadonism
скорбут *med.* scurvy; scorbutus
~, детский infantile scurvy, acute rickets
~, морской sea scurvy
~, сухопутный land scurvy
скорбутный scorbutic
скорбь, мировая *lit.* Weltschmerz
скорлупа shell; husk, hull
скорлуповидный crustaceous; conchiform
скороварка pressure cooker
скорода *bot.* chive *(Allium schoenoprasum)*
скородействующий quick-acting/operating, highspeed *(attr.)*
скоропашка coulter plow/plough
скороплодный early maturing/ripening
скороподъемность 1. climbing rate, rate of climb; 2. time-to-climb
скоропортящийся perishable
скороспелость early ripening; precocity
скороспелый *agr.* early, precocious, early/fast ripening
скоростемер velocity meter

скоростник high-speed worker/performer; highly efficient operative
скоростной high-speed *(attr.)*, rapid
~ бег на коньках speed skating
скорострельность firing rate, rate of fire
скорострельный quick-firing, rapid-firing
скорость velocity; speed, pace; rate; *tech.* gear
~, абсолютная absolute speed
~, большая high speed
~ в подводном положении submerged speed
~ ввода данных data input rate
~, вертикальная vertical velocity
~, весовая mass flow rate
~ восстановления recovery rate
~ вращения rotational speed/velocity
~ вычислений computation speed
~ двигателя (transmission) gear
~ деформации rate of deformation; strain rate
~, дозвуковая subsonic speed
~, докритическая subcritical speed
~, доплеровская Doppler velocity
~, дрейфовая *naut.* drift speed
~, закритическая supercritical speed
~ записи writing/recording speed
~ затухания rate of decay
~ звука speed of sound, acoustic velocity
~ изменения азимута azimuth rate
~ изнашивания wear rate
~, исходная reference speed
~, коммерческая *econ.* schedule speed
~ коррозии corrosion rate
~, космическая space velocity; cosmic speed
~, космическая вторая escape velocity
~, космическая, первая orbital/circular velocity
~, космическая, третья solar escape velocity
~, крейсерская cruising speed
~ кровотока circulation rate
~, круговая circular velocity
~ круговорота turnover rate
~, линейная linear velocity
~, мгновенная угловая instantaneous angular velocity
~ на входе inlet/entrance velocity
~ на выпуске/выходе outlet/exit/exhaust velocity
~ набора высоты, вертикальная rate of climb
~, надводная surface speed
~ нарастания rate of rise
~ обмена rate of exchange
~ обнаружения detection rate
~ оборота rate of turnover
~ обработки processing speed
~, окружная peripheral/circumferential velocity
~ опускания lowering speed
~ осадкообразования sedimentation rate
~ оседания rate of sibsidence
~ оседания эритроцитов erythrocyte sedimentation rate, ESR
~, относительная speed of relative movement, relative speed
~ отрыва *aero.* lift-off speed
~ охлаждения cooling rate
~ передачи transmission rate
~ передачи в бодах baud rate
~ передачи в цифровых системах symbol rate
~ передачи данных data-transmission/flow rate
~ по лагу *naut.* speed by log, log speed
~ поглощения absorption/uptake rate
~ погружения rate of dive; sinking speed
~ подачи feed rate
~ подвигания advance rate
~ подводного хода submarine/underwater/submerged speed
~ подъема hoisting speed
~ поиска search rate
~ полета flight/flying speed
~ ползучести creep rate
~ поступательного движения translational velocity
~ потока flow rate, stream velocity/rate
~ потока, массовая mass flow rate
~ потока, объемная volumetric flow rate
~ потока, суммарная total flow rate
~, предельная top speed; speed limit
~ прерывания chopping speed
~ прокатки *met.* rolling rate
~ развертки/развертывания scanning speed/velocity; sweep speed
~ распада decay/disintegration rate
~ распространения propagation velocity, speed of spreading
~ расхода 1. rate of flow, flow rate; 2. rate of consumption
~, расчетная design speed
~ реакции 1. reaction rate; 2. speed of response
~, результирующая resultant velocity/speed
~, релятивистская relativistic velocity
~ роста growth rate
~, сверхзвуковая supersonic speed
~ света velocity/speed of light
~ снижения *aero* speed of descent
~ срабатывания speed of response
~, среднеквадратичная root-mean-square [rms] velocity
~ течения velocity of flow, flow rate
~, ударная shock velocity
~, фазовая phase velocity
~ хода *naut.* speed
~ хода, полная full speed
~ хода, эксплуатационная service speed
~ частицы particle velocity
~, экономическая *naut.* cruising/endurance speed
~, эксплуатационная *aero.* operating speed
~, эскадренная collective speed
скоротечность transiency, transience; short duration

скоротечный transient; brief, fast; *med.* fulminating, fulminant; galloping
скорпена *zool.* scorpene *(Scorpaena porcus)*
скорпеновые *pl. zool.* scorpion fishes
скорпион 1. *zool.* scorpion *(Androctonus)*; 2. *astr.* Scorpio
скорпионница *bot.* scorpion's tail *(Scorpiurus)*
скорца *miner.* scorza, epidote
скос chamfer(ing), bevel; scarf; taper(ing); skew(ing)
скосарь *zool.* scarred snout beetle *(Otiorhynchus)*
скот livestock; cattle
~, **крупный рогатый** cattle
~ **мелкий рогатый** sheep and goats
~, **молочный** dairy cattle
~, **мясо-молочный** dual-purpose cattle
~, **нагульный** fattening cattle
~, **откормочный** feeder
~, **племенной** breeding stock
~, **рабочий** draught/draft animals, workstock
~, **рогатый** horned livestock
скотобойня slaughterhouse
скотовод cattle breeder, stock-farmer
скотобаза livestock depot
скотоводство cattle breeding/raising/husbandry, stock-raising
~, **молочное** dairy farming
~, **мясное** beef raising
скотовоз cattle truck
скотография scotography, skiagraphy
скотоложство bestiality, sodomy
скотома *med.* scotoma
~, **мерцательная** scintillating scotoma
~, **островковая** insular scotoma
~, **физиологическая** physiologic scotoma, blind spot
скотоместо *farm.* room for one animal
скотомный *med.* scotomatous
скотоскопия *med.* scotoscopy, skiascopy
скотофилия scotophilia, nyctophilia
скотофобия scotophobia
скоция *arch.* scotia
скошенный tapered, chamfered, bevelled; skew
скрап scrap (metal), salvage
~, **заводской** mill scrap
~, **ковшовый** ladle scrap
~, **оборотный** plant/domestic/home scrap
скребень *zool.* proboscis worms *(Acanthocephala)*
скребница *bot.* scale fern *(Ceterach)*
скребок scraper; scratcher
скрепа clamp, clip, brace, tie
скрепер scraper
~, **гребковый** rake scraper
~, **канатный** cable drag scraper
~, **прицепной** pull-type scraper
~, **ящичный** box-type scraper
скрепер-волокуша dragging scraper
скреперование scraping
скрепка (paper) clip; clamp, brace
скреплят fasten; tighten, strengthen
скрещение intersection; interlacing, junction; *ling.* blending, contamination
скрещивание intersecting, crossing; junction; *biol.* crossing, hybridization; cross
~, **аналитическое** test cross
~, **внутривидовое** intraspecific crossing
~, **возвратное** backcross
~, **диаллельное** diallele crossing
~, **линейное** linecrossing
~, **межвидовое** interspecific crossing/hybridization
~, **межлинейное** interstrain crossing
~, **обратное** reciprocal crossing, recrossing
~, **родственное** inbreeding
~ **сибсов** sibbing, sib mating
~, **сложное** composite crossing
~, **сортирующее** assortative mating
~, **тройное** three-way cross
скрипач, скрипачка violinist
скрипичный violin *(attr.)*
~ **ключ** G clef, treble clef
скрипка violin
скрипун *zool.* linden borer *(Saperda vestita)*
скротальный *anat.* scrotal, oscheal
скротит *med.* scrotitis
скротопластика scrotoplasty
скрофулез *med.* scrofula, scrofulosis
скрофулезный scrofulous
скрофулодерма scrofuloderma, scrofulide
скруббер scrubber, gas cleaner/washer
~, **безнасадочный** spray/jet scrubber
~, **насадочный** packed-bed scrubber
~ **центробежный** rotary/centrifugal scrubber
скругление rounding off; radius
скрутка twist, twisting
скрученник *bot.* lady's-tresses *(Spiranthes autumnalis)*
скрученноволосистый *bot.* streptopilose
скрученнолепестный *bot.* streptopetalous
скрученнолистный *bot.* streptophyllous
скрученноплодный *bot.* streptocarpous
скрученностебельный *bot.* streptocaulis
скрученностебельчатый *bot.* streptocaulis
скручивание twisting; torsion, twist
скручивать twist
скрытница *bot.* prickle grass *(Crypsis)*
скрытнохоботник *zool.* snout beetle *(Cryptorhynchus)*
скрытовулканический cryptovolcanic
скрытогенетический cryptogenetic
скрытографический cryptographic
скрытожаберник *zool.* hellbender *(Cryptobranchus)*
скрытокристаллический cryptocrystalline
скрытолистный *bot.* endophyllous, cryptophyllous
скрытоплодный *bot.* cryptocarpous
скрытоpотный cryptostomatous
скрытосемянные *pl. zool.* angiosperms *(Angiospermae)*
скрытосемянный *bot.* angiospermous

скрытошейные *pl. zool.* cryptodirans *(Cryptodira)*
скрытый latent; occult; cryptic; hidden
скула 1. *naut.* bilge; shoulder; chine; 2. *anat.* zygoma, cheekbone
 ~, кормовая after shoulder
 ~, носовая forward shoulder
 ~, острая sharp bilge; sharp/hard chine
скуло-верхнечелюстной zygomaticomaxillary
скуло-височный zygomaticotemporal
скуловой 1. *naut.* bilge *(attr.)*; 2. *anat.* zygomatic
скуло-глазничный zygomatico-orbital
скуло-лобный zugomaticofrontal
скумбрещука *zool.* saury *(Scombresox saurus)*
скумбрия *zool.* mackerel, scomber *(Scomber)*
скумпия *bot.* smoke tree *(Cotinus)*
скунс *zool.* skunk *(Mephitis mephitis)*; skunk fur
скупщина *polit.* Skupshtina
скутер scooter
слабина *naut.* slackness, slack
слабительное cathartic, purgative; purge
 ~, легкое laxative, mild cathartic, aperient
 ~, солевое saline purgative, Epsom salt
слабительный purgative
слаблинь *naut.* earing, lace piece, lacing line
слабоветвистый *bot.* subramose
слабовидящий asthenopic
слабоволие ab(o)ulia; weak will
слабодифтонгический *phon.* diphthongoidal
слабозасоленный subsaline, weekly salinized
слабоквалифицированный poorly skilled
слабокислотный, слабокислый subacid
слаборазвитый underdeveloped; embryonic
слабосолонцеватый weekly/slightly alkalinized
слабость weakness, asthenia, debility, infirmity, feebleness; adynamy
 ~ зрения weak/poor sight, asthenopia
 ~, мышечная myasthenia, muscular debility
 ~, нервная neurasthenia
 ~ родовой деятельности uterine inertia
 ~, сердечная cardiac weakness, cardiasthenia
слаботочный *el.* weak-current *(attr.)*
слабоумие feeble-mindedness, mental deficiency, moronity, dementia, hypophrenia, anoesia, anoia
 ~, врожденное congenital mental deficiency
 ~, ложное pseudodementia
 ~, предстарческое presenile dementia, Alzheimer's disease
 ~, старческое senile dementia
славист Slavist, Slavicist; Slavonic scholar
славистика Slavic/Slavonic studies
славистический Slavic, Slavonic, Slavistic
славка *zool.* warbler *(Sylvia)*
славянизация Slavization, Slavification
славянин Slav
славянка Slav (woman)
славяновед, славянист Slavic/Slavonic scholar
славяноведение Slavic/Slavonic studies
славяно-греческий Slavo-Greek
славяно-русский (Old Church) Slavic/Slavonic and Russian
славянофил, славянолюб Slavophile
славянофильство Slavophilism
славянофоб Slavophobe, Slavophobist
славянофобство Slavophobia
славянство Slavdom, the Slavic peoples
слагаемое component, summand, term, item
 ~, второе *math.* augend
 ~, первое *math.* addend
слагающая component, constituent
сладкокорень *bot.* common polypody *(Polypodium vulgare)*
сладконожник *bot.* Japanese raisin tree *(Hovena dulcis)*
сладкоплодный hedycarpous, sweet-fruited
слайд 1. slide; 2. *sport* sliding/rower's seat
слалом *sport* slalom
 ~, гигантский giant slalom
сланец shale, schist; slate
 ~, батистовый cambric schist
 ~, горючий oil/combustible shale
 ~, кианитовый cyanite schist
 ~, кремневый flinty slate, chert
 ~, кремнистый siliceous schist
 ~, кровельный roofing slate
 ~, мергелистый marl slate
 ~, нефтеносный oil shale
 ~, слюдяной mica schist/slate
сланоягодник *bot.* seaberry *(Haloragis)*
сланцеватость schistosity; cleavage, foliation
 ~, плоская platy schistosity
сланцеватый, сланцевый shaly, slaty, schistose
слева on/from/to the left; *naut.* on the port (side)
след trace, track; wake; mark; footprint, footstep; *anat.* vestige, rudimentary structure
 ~, инверсионный/конденсационный condensation trail, contrail
 ~ за крылом *aero.* wing wake
 ~ матрицы spur
 ~, метеорный meteor trail
 ~ перескопа *naut.* periscope wake
следить follow, trace, track; watch
следователь investigator; investigating officer/magistrate; coroner
следственный investigating, investigatory
следствие 1. *law* investigation, examination, inquest; 2. consequence, sequel; result
 ~, естественное corollary
 ~, предварительное preliminary (judicial) examination
 ~, причина и cause and effect
 ~, производить hold an investigation/inquiry
следующий *naut.* bound
 ~ на восток eastbound
 ~ на запад westbound
 ~ на север northbound
 ~ на юг southbound
слежение tracking, following; tracing
 ~ за спутником satellite tracking
 ~ по дальности elevation tracking

~ по углу места range tracking
~ с помощью радиосредств radio tracking
слеживание caking; slumping
слеживаться become caked/compressed, cake, slump
слезник 1. *anat.* lacrimal sac; 2. *arch.* drip, dripstone; 3. *bot.* Job's tears *(Coix lacrimajobi)*
слезно-носовой *anat.* nasolacrimal
слезный *anat.* lacrimal, tear *(attr.)*
слезовыделение lacrimation, tear secretion
слезоточивый lacrimatory; running (of eyes)
слеминг *naut.* slamming
сленг slang
слепен *zool.* horsefly; gadfly *(Tabanus)*
слепец 1. blind man; 2. *zool.* mole rat *(Spalax)*
слепнуть become/get blind, lose one's sight
слепоглухонемой blind deafmute
слепоглухонемота deafmutism with blindness
слепозмейка *zool.* blind snake/worm *(Typhlops)*
слепой blind; cecal; sightless
слепок mould, cast; copy
слепорожденный congenitally blind
слепота blindness, loss of sight, ablepsia, typhlosis
~, дневная day blindness, hemeralopia
~, корковая cortical blindness
~, куриная night blindness, nyctalopia
~, музыкальная note/music blindness
~, полная complete blindness, amaurosis
~, профессиональная occupational blindness
~, психическая psychic blindness
~, словесная word/text blindness, alexia
~, снежная snow blindness, niphablepsia
~, тактильная astereognosis, stereoagnosis
~, цветовая colour blindness, daltonism
~, частичная цветовая partial colour blindness
слепушонка *zool.* northern mole vole *(Ellobius talpinus)*
слепыш *zool.* mole rat *(Spalex)*
слесарь locksmith; metalworker
слесарь-инструментальщик toolmaker
слесарь-монтажник/сборщик fitter
слив discharge, discharging; drain; runoff
~ воды water drain
~ топлива, аварийный fuel jettisoning
слива *bot.* plum tree *(Prunus)*
сливать 1. drain, empty, pour out/off; 2. pour together, blend, mix
сливаться flow/run together; fuse; merge; conflow
сливающийся confluent
сливки cream
~, сбитые whipped cream
сливкоподготовитель pre-churner
сливной draining, drain *(attr.)*
слизевик *bot.* slime mould, myxomycete *(Myxomyces)*
слизевой mucous, mucic, mucosal, mucilaginous
слизень *zool.* slug *(Limax)*
слизеобразующий mucigenous, muciparous
слизеподобный mucoid
слизетечение mucilaginous discharge; *med.* blennorrh(o)ea
слизисто-гнойный *med.* mucopurulent
слизесто-кровянистый *med.* mucosanguineous
слизесто-серозный *med.* mucoserous
слизистый mucous, mucic, mucilaginous, mucosal
слизневидка *zool.* hag moth *(Probetron pithecum)*
слизь mucus; mucilage; phlegm
~, вязкая tenaceous mucus
~, лекарственная mucilage, mucilago
~, носовая nasal mucus
слип *naut.* slip, slipway, marine railway
слипание adherence, adhesion, cohesion
слипаться adhere, stick together
слипинг-док railway slip
слитие fusion, merging, merger, merge
слитковоз ingot buggy/bogie/car
слиток ingot, pig, bar
~ кипящей стали rimmed ingot
~, непрерывный continuously cast ingot
~, фасонный shape ingot
слияние 1. confluence; junction; coalescence; 2. merging, blendimg, amalgamation, fusion, union, merger
~ банков consolidation of banks
~, вертикальное *econ.* vertical merger
~ гамет gametic syngamy
~, горизонтальное *econ.* horizontal merger
~ клеток cell fusion
~ компаний/предприятий amalgamation of business
~ ледников glacier coalescence
~, нисходящее downstairs merger
~ промышленных предприятий industrial amalgamation/merger
~, прямое outright merger
~, центрическое centric fusion
~ ядер nuclear fusion
словак Slovak
словарь dictionary; vocabulary, glossary
словацкий Slovak(ian)
словачка Slovak (woman)
словенец Slovene
словенка Slovene (woman)
словенский Slovene, Slovenian
словесно-образный pictorial-verbal
словесно-семантический word-semantic
словинец Slovinetz, Kashube, Kashubian
словинский Kashubian
слово word
~, адресное *comp.* address word
~ адреса канала channel address word
~, алфавитное alphabetic item
~, арифметическое информационное arithmetic item, arithmetic data word
~, двоичное binary code word, binary item, bit word
~ для диска, адресное *comp.* disk address word, disk control field, DCF
~, зависимое *comp.* contiguous item
~ из таблицы адресов directory word

~, **информационное** *comp.* item, data word
~, **ключевое** index word
~, **кодовое** *comp.* code word
~, **командное** order/command/instruction word
~, **машинное** machine/information/computer word
~, **последнее** *law* last statement/plea
~, **равнозвучащее** homonym
~, **равнозначащее/равнозначное** synonym
~ **состояния канала** channel status word
~ **состояния программы** program(me) status word
~, **управляющее** control word
~, **честное** *law* parole
~ **функциональное** *comp.* function word
~, **цифровое** numeric word
словоизменение *gram.* inflection
словолитня type foundry
словообразование word formation
словообразовательный, словообразующий, словопроизводный word-formative/forming, derivational
словопроизводство word formation, derivation
словосложение *gram.* composition
словосочетание phrase, combination of words
~, **устойчивое** set phrase
словотворчество creation of new words; word formation
слог syllable
~, **заударный** post-tonic syllable
~, **побочный** weakly stressed syllable
~, **ударный** stressed/accented syllable
слогоделение syllabication
слоевище *bot.* thallome, thallus, frond
слоение stratification
сложение 1. *math.* adding, addition; 2. composition, structure, constitution
~ **векторов** composition of vectors
~ **волн** wave interference
~, **геометрическое** vectorial addition
~ **движений** composition of motions
~, **двоичное/троичное** binary/ternary addition
~, **листовое** lamelliform structure
~ **наказаний** *law* cummulative sentence
сложноподчиненное предложение complex sentence
сложнопроизводный *gram.* complex-derivative, parasynthetic
сложнорефлекторный complex-reflex
сложносокращенное слово acronym
сложносочиненное предложение compound sentence
сложноцветный *bot.* composite
сложноцикличный multicyclic
слоисто-покровный *geol.* tectate
слоистость lamination, stratification
слоистый laminated, stratified, layered, sandwich *(attr.)*
слой layer, stratum, bed; sheet, lining; coating, coat; lamina, lamination; *constr.* course
~, **вихревой** vortex layer
~, **граничный** boundary layer
~, **двухфазный** binary layer
~, **дренажный** drainage course
~, **кипящий** fluidized bed
~, **легированный** doped layer
~, **лучистый** radiate layer
~, **наплавленный** built-up layer
~ **нестационарный** unsteady layer
~, **нижележащий** sublayer, underlayer
~ **общества, средний** middle class
~ **огнегасящей пены** foam blanket
~, **осадочный/отложений** bedding plane
~, **пигментный** pigmented layer
~, **плазменный** plasma sheath
~, **пограничный** boundary layer
~, **подстилающий** underlayer
~, **подпахотный** subsoil
~ **Пуркинье** Purkinje's layer
~, **роговой** *anat.* horny layer
~, **связующий** binder course
~, **слизистый** *anat.* mucosal layer
~, **токовый** current sheath
~, **ударный** shock layer
~, **уносимый** ablation layer
~, **фильтрирующий** filter bed/course
~ **Хевисайда** Heavyside layer
~ **штукатурки, выравнивающий** floating coat
~, **энтропийный** entropy layer
~ **эпидермиса, ростковый** germinative layer
~, **эпидермиса, шиповидный** pricle-cell layer, spinous layer
слойчатость lamination
слом breaking up, scrapping; demolition
слон elephant *(Elephas)*
~, **морской** elephant seal *(Mirounga)*
слоник *zool.* snout beetle, weevil *(Curculionidae)*
слониха elephant-cow
слоновость *med.* elephantiasis, pachydermia
~ **десны** gingival elephantiasis
~ **половых органов** genital elephantiasis
слон-отшельник rogue elephant
служащий office worker, clerk, white-collar worker; employee; official, officer
~, **государственный** public/government employee; civil/public servant
~, **канцелярский/конторский** office employee, clerical worker; file clerk
~ **получающий жалованье** salaried employee
служба service; office; job, work; employment; department
~, **аэродромная диспетчерская** airport air traffic control (service)
~, **брандвахтенная** *naut.* examination service
~, **внутренняя** *mil.* routine service
~ **времени** accurate hour's service
~ **движения** traffic/movement department
~, **действительная** *mil.* active service
~, **грузовая** freight service

~ дезактивации Contamination Service
~, диспетчерская *av.* air traffic control
~, дозиметрическая radiation monitoring service
~, дозорная patrol service/duty
~ живучести damage-control service
~, зубоврачебная dental department
~, интендантская supply branch/department
~, караульная guard duty
~, корабельная duty aboard [on board]
~, маркшейдерская surveying
~, метеорологическая weather/meteorological service
~, морская maritime service
~, морская спасательная sea rescue service
~ морских перевозок sea transport service, ferry service
~ на море service at sea
~, нестроевая *mil.* special duty
~ обеспечения полетов flight service
~, противопожарная fire service
~, радиолокационная radar service
~ радиосвязи radio service
~ снабжения supply/issuing service
~, спасательная rescue/salvage servi ce
~ пути *rail.* track maintenance
~ тыла *mil.* logistics
~, почтовопосылочная parcel post
~ управления воздушным движением air traffic control (service)
~ эфемерид ephemeris survey
служебник *eccl.* diurnal, missal
служебный official; office *(attr.)*
~ное время working hours
~ное слово *ling.* connective/syntactic word
слух hearing, audition; ear
~, абсолютный absolute/perfect ear
~, вкусовой gustatory audition
~, внутренний conductive hearing
~, музыкальный musical ear
~, цветовой colo(u)r hearing, chromatic audition, pseudochromasthesia
слуховой acoustic, auditory; hearing *(attr.)*
~ аппарат hearing aid
~ рожок ear trumpet
слухозрительный aural-visual
слухопроизносительный audiolingual
случай case; event; occasion; opportunity
~ аварийный emergency (case)
~, несчастный accident, casualty
~, предельный limiting case
~, частный particular case
случайность chance; randomness
~ти, критерий *stat.* test of randomness
случайный accidental; random, fortuitous; casual, chance *(attr.)*; stochastic
~ная величина variate; random quantity
случной mating, pairing
слушание *law* hearing
слышимость audibility
слюда *miner.* mica
~, калиевая potash mica, muscovite
~, коллекторная *el.* commutator mica
~, листовая sheet mica
~, серебристая silvery mica, cat silver
~ урановая uran-mica, uranite
~, чешуйчатая scale mica
слюдизация micasization, micatization
слюдит *miner.* glimmerite
слюдяной micaceous, mica *(attr.)*
слюна saliva, spittle
слюннокаменный sialolithic, ptyalolithic
слюнный salivary, sialic, sialine
слюногон *med.* sialagogue
слюногонный sialagogic, salivant, salivator
слюноотделение salivation
~ избыточное hypersalivation, ptyalism, sialorrh(o)ea, ptyalorrh(o)ea
~, недостаточное oligosialia
слюноотсос salivary ejector
слюнотечение hypersalivation, sialorrh(o)ea, ptyalorrh(o)ea
сляб *met.* slab
сляб-болванка dummy slab
слябинг *met.* slabbing mill
смазка 1. lubricant; grease, oil; 2. lubrication, greasing, oiling
~, барботажная splash lubrication
~, графитовая graphite grease
~, капельная drop-feed lubrication
~, картерная oil-bath lubrication
~, кольцевая ring lubrication
~, консистентная lubricating grease, consistent lubricant
~, первородная/сыровидная/творожистая *anat.* vernix caseosa
~ погружением immersion lubrication
~ под давлением [принудительная] forced-feed lubrication
~ разбрызгиванием splash lubrication
~, сажевая carbon-black lubricant
~ самотеком gravity lubrication
~, суставная *anat.* synovia
~, сухая dry film lubrication
~, трансмиссионная gear grease
~, тугоплавкая thermally inert lubricant
~, универсальная multi-purpose lubricant
~, фитильная wick-fed lubrication
~ шприцем squirt-gun lubrication
смазочный lubricating, lubrication *(attr.)*
смазывание lubrication; oiling; greasing
смазывать lubricate; oil; grease
смальта smalt
смарагд emerald, smaragd
смарида *zool.* pickarel *(Smaris melanura)*
смачивание wetting, moistening
смачиваемость wettability; water affinity
смачиватель wetting agent
смачивать wet, moisten
смегма *physiol.* smegma, sebum

смежность contiguity, adjacency; juxtaposition
смежный adjacent, contiguous; adjoining
смена 1. change, changing; replacement; 2. shift; 3. relief
 ~ караула/часовых guard mounting; changing of the guard
 ~, ночная (mid)night shift
 ~, рабочая shift
сменность 1. interchangeability; 2. shift system
сменный replaceable, interchangeable, renewable
смерзание congelation
смерзаться congeal
смертность death/mortality/fatality rate; mortality, fatality
 ~ детей/детская infant mortality
 ~ новорожденных neonatal mortality
 ~, профессиональная occupational mortality
 ~ ранняя детская neonatal mortality
смерть death, exitus; decease
 ~, внутриутробная/антенатальная/плода f(o)etal death
 ~, дыхательная respiratory death
 ~, мозговая cerebral death
 ~, насильственная violent death
 ~ от лучевого поражения radiation death
смерч spout, whirlwind
 ~, водяной waterspout
 ~, песчаный sand spout
смесеобразование mixing
 ~, внешнее carburation
 ~, предварительное premixing
смесимость miscibility
смеситель mixing unit/device, mixer
 ~, барабанный drum/barrel mixer
 ~ периодического действия batch-type [intermittent] mixer
 ~, поточный flow nixer
 ~, предварительный premixer
смесительный mixing, blending
смеска blend, mixture
смесь mixture, blend, mix; composition; stock; compound
 ~, азеотропная *chem.* azeotropic mixture, azeotrope
 ~, бедная lean/poor/weak/rare mixture
 ~, бетонная concrete mix
 ~, двойная binary mixture
 ~, литическая lytic mixture
 ~, незамерзающаяся antifreeze mixture
 ~, рабочая *eng.* combustible mixture
 ~ распределений *stat.* mixture of distributions
 ~, резиновая rubber stock/mix
 ~, резиновая шинная tyre stock
 ~, самовоспламеняющаяся hyperbolic mixture, self-igniting/inflammable mixture
 ~, тройная ternary mixture
 ~, формовочная *met.* moulding sand
смета (cost) estimate, estimated cost; cost sheet; budget; calculation
 ~ в натуральных единицах physical budget
 ~ капиталовложений capital budget
 ~, общая финансовая complete/master budget
 ~, оперативная operating budget
 ~, перспективная forecast budget/estimate
 ~ по капитальному имуществу fixed/capital assets budget/estimate
 ~ расходов и прибылей cost-benefit calculation
 ~ финансирования программы program(me) budget
сметчик estimator, calculator
смешение mixing; blending; merging
смешиваемость miscibility
 ~, взаимная intermiscibility
смещать displace, shift; misalign; bias
смещение displacement, shift; removal; dislocation, dislodgement; bias; offset; *med.* ectopia, ectopy, heterotopia
 ~, автоматическое self-bias
 ~, базовое base-line offset
 ~ влево sinistroposition
 ~ вперед anteposition
 ~ вправо dextroposition
 ~ изображения image shift
 ~ кзади retro displacement, retrodeviation
 ~ книзу infradisplacement
 ~, комбинационное Raman displacement
 ~ кости *med.* ostectopy, osteectopia
 ~ кпереди antedisplacement
 ~ матки *med.* metrectopia, metrectopy
 ~ несущей *telev.* carrier offset
 ~ нуля [нуль-пункта] zero creep/drift
 ~ оценки *stat.* bias of estimation
 ~ пучка beam translation
 ~, фазовое phase shift
смещенный shifted, displaced, dislocated, displaced; offset; heterotopic, ectopic, atopic
смилакс *bot.* greenbrier, smilax *(Smilax)*
смилацина *bot.* Solomon's-plume *(Smilacina)*
смог smog, city fog
смокинг dinner jacket
смоковница, священная sacred fig tree, pipal tree *(Ficus religiosa)*
смола resin, gum; *miner.* tar, pitch
 ~ анионообменная anion-exchange resin, anionite
 ~, бакелитовая phenol-formaldehyde resin
 ~, древесная rosin, wood resin/tar
 ~, жидкая pitch, tar
 ~, заливочная *el.* potting resin
 ~, ионообменная ion-exchange resin, ionite
 ~, каменноугольная coal tar
 ~, карбамидная urea resin
 ~, катионобменная cation-exchange resin, cationite
 ~, кремнийорганическая silicone resin
 ~, литьевая *el.* cast resin
 ~, пентафталевая pentaerythritol-modified phthalic resin
 ~, поливинилхлоридная polyvinylchloride

resin, PVC resin
~, полиэтиленовая polyethylene resin
~, полиэфирная polyester resin
~, твердая rosin
~, термореактивная thermosetting resin, heat-convertible resin; resinoid
~, фенольная phenolic resin
~ холодного отверждения cold-curing resin
~, черная pitch; tar
~, эпоксидная epoxy resin
смолевка 1. *bot.* silene, campion *(Silene)*; 2. *zool.* balsam bark weevil *(Pissodes dubius)*
смолевколистный *bot.* catchfly-/campion-leaved
смоление 1. *chem.* resinification; 2. tarring
смолка frankincense, incense
смолница smolnitza, Mediterranean black soil
смолокурение tar/turpentine distillation
смолоносница *bot.* giant fennel *(Ferula)*
смолообразование resinification
смолянка *bot.* bitter cress *(Cardamine)*
смонтировать = монтировать
смородина *bot.* currant *(Ribes)*
сморчок *bot.* morel *(Morchella)*
сморщивание puckering; corrugation; *med.* kraurosis, cirrhosis
сморщиватель *anat.* corrugator
смутное время *hist.* Time of Troubles
смушка lamb/astrakhan fur
смыв washout, washoff
смывать wash away/out/off; flush
смывка 1. washing off; erosion; flushing; 2. paint remover, cleaning fluid
смыкать link, couple, connect, joint; interlock
смыслоразличающий *ling.* distinctive
смычка 1. *naut.* chain length, shot; 2. linking, joint; union; *phon.* occlusion, closure, stop
~, якорная outboard/bending shot
смычно-взрывной *phon.* occlusive-plosive
смычно-дрожащий *phon.* occlusive-trilled
смычно-проходный *phon.* occlusive-fricative
смычный *phon.* occlusive; obstruent
смягчать soften; mollify; *mech.* cushion; modify; temper
смягчающий softening; alleviating; *med.* emollient
~ие вину обстоятельства *law* extenuating circumstances
смягчение softening; mollification, mitigation; abatement; moderation; *phon.* palatalization
~ напряженности relaxation of tension
смягчитель softener; *chem.* plasticizer
снабжение supply, provision; delivery
~, избыточное oversupply
~, материально-техническое logistics; supply of material and equipment
~ продовольствием victualling, provisioning
~, шлюпочное *naut.* boat's gear
~ электроэнергии (electrical) power supply
снаряд 1. instrument, tool, apparatus, contrivance; gear; 2. shell, projectile, missile; 3. dredge, dredger
~, грейферный clamshell dredge
~, дноуглубительный mud/floating dredge
~, землесосный suction(-tube) dredge
~, землечерпательный (bucket) dredge
~, камнечерпательный rock dredge
~, многоковшовый/многочерпаковый elevator dredge
~, осколочный *mil.* fragmentation shell
~, практический *mil.* drill shell
~, рефулерный землесосный pipeline dredge
~, рыболовный fishing tackle/gear
~, скважный borehole tool/equipment
~, учебный *mil.* dummy shell/missile
~, шаландовый (землечерпательный) barge-loading dredge
~, штанговый dipper dredge
снаряжение equipment, outfit; gear; tackle; *naut.* harness rigging; *mil.* loading; munition
~, водолазное diving outfit/equipment/gear
~, спасательное survival kit/equipment
~, шлюпочное boat furnishing
снасть tackle; implements, tools; *naut.* rope, rigging
снашивание wearing out; abrasion
~ основного капитала capital consumption
снег snow
~, вечный perpetual/perennial snow
~, метелевый/перемещающийся drifting snow
~, талый slush, melting snow
снегирь *zool.* bullfinch *(Pyrrhula pyrrhula)*
снегозадержание *agr.* snow retention
снегогенератор snow-making machine/plant
снегозащита snow protection/fence; protection against snowdrifts
снегомер snow gauge/sampler
снегоочиститель snow remover, snowplough, snowplow, snow-fighting vehicle
~, плужный snow crab, snowplough, snowplow
~, пневматический snow blower
~, роторный rotary snow blower/plough, snow thrower
~, щеточный snow brush
снегоочистка, снегоуборка snow clearing/removal
снегопах snowplough, snowplow
снегоход snowmobile, snow tractor, oversnow vehicle
снедок *bot.* salad chervil *(Anthriscus cerefolium)*
снежник firn basin/bank
снежноягодник *bot.* snowberry *(Symphoricarpos)*
снежные альпийцы *bot.* dealpine species
снеток *zool.* European smelt, sparling *(Osmerus eperlanus)*
снижать lower, bring down; reduce; lessen
~ расходы cut down (trim, reduce) expenses
снижение 1. descent, lowering; 2. reduction, decrease, drop(ping), lowering; deterioration
~ давления depressurizing, loss of pressure
~ заработной платы reduction of wages

~ издержек *econ.* cost saving
~ по приборам *aero.* instrument letdown
~ рождаемости birth inclination
~ скорости slowing down
~ шума noise abatement/reduction/suppression
снимок photograph, (still) picture, camera shot
~ крупным планом close-up
~, моментальный snapshot
~, радиографический radiograph, radiogram
~, рентгеновский roentgenograph, roentgenogram, radiogram, X-ray photograph/picture
снобизм snobbery
снование, сновка *text.* warping
сноп 1. sheaf; bundle; 2. beam, shaft; cone
сноповязалка sheaf-binding machine
снос *nav.* drift; shift
~, боковой leeway, sidle; sway
~ ветром wind drift
~ влево counterclockwise drift
~ вправо clockwise drift
сноситься drift
~ влево drift to port
~ вправо drift to starboard
снотворное hypnotic, soporific
снотворный soporific, somnifacient; hypnotic
сныть *bot.* goutweed *(Aegopodium)*
снятие removal; dismantling, taking down; lifting; relieving
~ блокировки unblocking
~ возбуждения *el.* de-excitation
~ заусеницев deburring
~ оболочки desheating, dejacketing
~ полуды detinning
~ с креста *art* Deposition, Descent from the Cross
~ с мели *naut.* refloating
~ с работы dismissal; relieving
~ с эксплуатации putting out of service
соавтор coauthor, joint author
соаренда cotenancy, joint tenancy
соарендатор cotenant, co-lessee, joint tenant
собака dog, hound
~, гиеновая African hunting dog
~, ездовая Siberian husky
~, морская dogfish *(Scylliorhinus caniculus)*
собака-рыба *zool.* puffer *(Tetraodon)*
собакозубые *pl., zool.* cynodonts *(Cynodonta)*
собачка *tech.* dog, cam, detent, catch; pawl
~, откидная catch hook
~, предохранительная safety catch/dog
собашник *bot.* great burdock *(Arctium lapa)*
собес *abbr.* social security (department)
собирать 1. *tech.* assemble; erect; 2. collect
собкор *abbr.* our own correspondent
соблюдение observance; keeping, maintenance; holding
~ дистанции distance holding
~ интервалов spacing
~ порядка maintenance of order
соболь *zool.* sable *(Martens zibellina)*
собор 1. cathedral; 2. *hist.* council, synod, assembly
~, вселенский (o)ecumenical council
~, земский Assembly of the Land
~ кардиналов *rel.* the sacred college
соборный synodic(al)
~ное уложение *hist.* Council Code
соборовать *eccl.* administer extreme unction
собрание 1. gathering, congregation, meeting, assembly; 2. collection
~ законов code of laws
~, закрытое closed meeting/session
~, общее general meeting/assembly
~ сочинений complete works
~, учредительное constituent assembly
собственник owner, proprietor, possessor
~, доверительный *law* trustee
~, земельный landed proprietor; landowner
~, мелкий земельный small holder
~, полноправный rightful owner
собственно 1. properly; strictly; 2. proper
~ кожа *anat.* dermis, true skin
собственность 1. possession, ownership; 2. property
~, акционерная corporate ownership
~, арендованная leasehold
~, верховная eminent domain
~, государственная state property; public domain/property/ownership
~, движимая *law* movable estate
~, доверительная trust
~, земельная landed estate/property
~, личная personal/individual property
~, материальная corporeal property
~, недвижимая real property
~, общая common property; joint/collective ownership
~, общественная public ownership/property
~, общинная communal property
~, реальная tangible property
~, частная private ownership/property
собственный own; proper; eigen-; *tech.* natural
~ вес own/dead weight
~кое имя *gram.* proper name
~ная величина eigen value
~ные колебания *phys.* natural oscillation
событие event *(also math.)*; occurance
~, достоверное certain event
~, исходное/первичное *stat.* ancestor
~, несовместимое incompatible/antithetical/disjoint event; *pl.* mutualy exclusive events
~, противоположное *math.* complementary/contrary event
~, сложное compound/complex event
~, случайное random/stochastic event
сова *zool.* owl
совдеп *abbr., hist.* Soviet of Deputies
совершение fulfilment, performance; feasance
совершеннолетие legal/full/adult/lawfull age, majority, legal adulthood
~ия, достичь come of age

совершеннолетный of age, major, adult
совершенноротые *pl. zool.* teleostomes (*Teleostomi)*
совершенный *gram* perfective
совершенствование further improvement; perfection
совесть conscience
  ~ти, свобода freedom of worship
совет Soviet, soviet; council; board; meeting
  ~, административный board of administration
  ~ Безопасности Security Council
  ~, Верховный Supreme Soviet
  ~, городской town soviet; town council
  ~ директоров board of directors
  ~ Национальностей *obs.* Soviet of Nationalities
  ~, научный research council
  ~, опекунский board of trustees
  ~ управляющих board of governors
  ~ Экономической Взаимопомощи, СЭВ *hist.* Council for Mutual Economic Assistance, CMEA, COMECON
советизация sovietization
советник adviser; counsel(l)or, council(l)or
  ~ посольства embassy counsel(l)or
советовед, советолог sovietologist
советоведение, советология sovietology
совещание conference; consultation; meeting
  ~ на верхах summit (conference)
  ~, производственное conference on production matters, production conference
совещательный consultative, deliberative; advisory
  ~ голос deliberative vote
совзнак *abbr., hist.* Soviet bank note
Совинформбюро *abbr., obs.* Soviet Information Bureau
совка *zool.* 1. screetch owl *(Megascops);* 2. earworm *(Heliothis);* 3. black witch moth *(Erebus odora);* 4. cutworm *(Noctuidae)*
совкаин *pharm.* dibucaine hydrochloride, cinchocaine hydrochloride
совладелец co-owner/tenant, joint owner/tenant
совладение joint ownership/property; condominium; tenancy in common
совместимость compatibility; adaptibility
  ~ аппаратная *comp.* hardware compatibility
  ~ на системном уровне system compatibility
  ~, перекрестная cross compatibility
  ~ по штепсельным разъемам plug-to-plug compatibility
  ~, программная software compatibility
  ~ сверху вниз downward compatibility
  ~ снизу вверх upward compatibility
  ~ соединения connecting compatibility
совместимый compatible; consistent; adaptable
совместитель holder of more than one office; pluralist
совместительство holding of more than one office, pluralism
совместно jointly, in common; simultaneously
  ~ и порознь jointly and severally
  ~ подписавший joint signatory
совместно-орудийный *gram.* sociative-instrumental
совместность уравнений consistency of equations
совместный joint, combined; cooperative; *math.* simultaneous; compatible
  ~ное обитание cohabitation
  ~ное обучение coeducation
  ~ное предприятие joint venture
совмещать align; match; register
совмещение *math.* superposition, bringing into coincidence; *tech.* alignment match
  ~, боковое side alignment
  ~ цветов colo(u)r registration
совмин *abbr.* Coincil of Ministers
совнарком *abbr., hist.* Council of People's Commissars
совнарком *abbr., hist.* Council of National Economy
совокупительный coital
совокупление coitus, coition; copulation
  ~, задержанное coitus reservatus
  ~, прерывистое/прерванное coitus interruptus
совокупность aggregate, totality; assemblage, collection, set; combination; *stat.* population
  ~, бесконечная *stat.* infinite/unlimited population, infinite set
  ~ выборов totality of choices
  ~, выборочная *stat.* sampled population, sample
  ~, генеральная *stat.* total population, universe of population, universal set
  ~ данных data aggregate
  ~, двумерная bivariate population
  ~, дискретная discrete population
  ~ доказательств cumulative evidence; body of evidence
  ~, истинная true population
  ~, исходная parent population
  ~ качеств set of characteristics
  ~, конечная finite population
  ~ конкретной реальности *philos.* abstract universal
  ~, контрольная test population
  ~, непрерывная continuous population
  ~ организмов *biol.* biota
  ~ преступлений *law* cumulative offences
  ~ приговоров *law* cumulative sentences
  ~ работающих employed population
  ~ рабочей силы manpower population
  ~, реальная existent/real population
  ~ точек points totality
  ~ тычинок androecium
  ~, упорядоченная ordered set
  ~ условий set of conditions
совпадение coincidence, concurrence; matching, match; congruence

~, быстрое fast coincidence
~ во времени time coincidence
~, двойное double coincidence
~, двухкратное twofold coincidence
~, запаздывающее/сдвинутое delayed coincidence
~ импульсов pulse coincidence
~, медленное slow coincidence
~, случайное accidental/chance/random coincidence
совсод *abbr.* assistance council
совхоз State farm, sovkhoz
согдиец, согдийский Sogdian
согласие agreement, accord; consent, assent
~, молчаливое *law* tacit consent, sufferance, acquiescence
~ при выравнивании *stat.* goodness of fit
согласный *gram.* consonant(al); *(noun)* consonant
~, глухой tenuis
~, звонкий sonant
~, фрикативный/щелевой spirant
согласование coordination. agreement; matching; adjustment; congruence, correlation; *gram.* agreement, concord, congruence
~, временное time correlation
~ нагрузок matching of load, impedance matching
согласовывать coordinate, correlate; match, fit
~ во времени synchronize
согласующее устройство adapter
соглашатель *polit.* compromiser, appeaser
соглашательство *polit.* appeasement, compromise
соглашение agreement, covenant; contract; settlement, arrangement; *polit.* convention
~, валютное monetary agreement
~, джентльменское/устное gentlemen's agreement
~, дополнительное collateral agreement
~, лицензионное licence/licensing agreement
~ о надбавках к заработной плате pay package
~ о ставках заработной платы wage settlement
~ о рыболовстве fishery agreement
~ о товарообороте/торговле и платежах trade and payment(s) agreement
~, платежное payments agreement
~ по вопросам заработной платы wage package
~, преференциальное торговое preferential trade agreement
~, прямое express agreement
~, резервное standby agreement
~, устное verbal agreement
согнутозубчатый *bot.* campylodentate
согнутоиглый *bot.* campylocanthous
согнутоплодный *bot.* camptocarpous
согнутый curved, bent; angled; *bot.* campylotropous
сода soda; sodium carbonate; baking soda
~, двууглекислая sodium bicarbonate
~, едкая sodium hydrate
~, кальцинированная soda ash
~, каустическая caustic soda, sodium hydroxide
~, кристаллическая crystal carbonate
~, питьевая/пищевая drinking/edible soda, sodium bicarbonate
~, технологическая soda ash
содалит *miner.* sodalite
содалитит *miner.* sodalithite
соддит, соддиит *miner.* soddyite
содержание 1. content; contents; 2. maintenance, upkeep, care; 3. allowance, pay, wages
~ жира fat content
~ загрязняющих веществ, остаточное residual waste load
~, зимнее *farm.* wintering
~ золы ash content
~, информационное data content
~, краткое summary
~ минеральных примесей/солей saline load
~ органического вещества organic content
~ памяти *comp.* store/memory contents
~ по массе content by mass
~ по объему content by volume
~, процентное percentage
содержать contain, hold; include; carry; keep
содержимое content(s)
содоклад supplementary paper/report
содоку *med.* rat bite fever/disease, sodoku
содолжник joint debtor, co-debtor
содомия sodomy
соединение 1. connection, joint, bond, coupling, join, link, union; joining, fastening, jointing; 2. *chem.* compound; combination; 3. *mil.* formation, force, large unit; 4. *stat.* confounding; 5. *astr.* conjunction; 6. *math.* complexion
~, безрезьбовое slip joint
~, болтовое bolted connection/joint; bolting
~ вгладь flush joint
~, вертлюжное swivel coupling/joint
~, вильчатое fork link
~, винтовое screw(ed) joint
~ внахлестку lap joint
~, водное hydrate
~, водородное hydride
~, волокнистое *anat.* fibrous joint
~ враструб bell-and-spigot joint, faucet joint
~ встык butt/edge/abutment joint
~, высокомолекулярное high-molecular (weight) compound
~, высокополимерное high-polymeric compound
~, галоидное halide
~, герметичное tight joint
~, гетероцепное heterochain compound
~, гибридное hybrid junction
~, глухое dead joint
~ гомологического ряда homologous com-

pound
~, двойное binary compound
~, жесткое rigid coupling
~ звезда-звезда *el.* star-to-star connection
~ звездой *el.* star connection
~ звеньев chain joint
~, ионное ionic compound
~ исков *law* joinder of actions/claims
~, карданное universal/cardan/Hooke's joint
~, клеммовое terminal connection
~, клешневидное chelate compound
~ костей union of bones
~, межсухожильное intertendinous connection
~, металлоорганическое organometallic compound
~, меченое labelled/tagged compound
~, мышьяковистое arsenic compound
~ на корпус *el.* frame connection
~ на массу earth/ground connection
~, неорганическое inorganic compound
~, неподвижное fixed joint/coupling
~, неправильное *surg.* malunion
~, непредельное unsaturated compound
~, неразъемное permanent joint/coupling
~, нервно-мышечное neuromuscular junction
~, низкомолекулярное low-molecular(-weight) compound
~, ниппельное nipple connection
~, паяное soldered/brazed joint
~, перекисное peroxide
~, подвижное movable/sliding joint; movable coupling
~, последовательное *el.* series connection
~, предельное saturated conpound
~ проводами wire connection; wiring
~, пространственное three-dimensional connection
~, разъемное 1. detachable joint; 2. *el.* plug connection
~, резьбовое threaded connection
~, сварное welded joint
~, синовиальное *anat.* synovial/movable joint; diarthrosis
~, скользящее slide joint
~, смешанное *el.* series-parallel connection
~, стыковое butt joint
~, тавровое tee joint
~, торцовое edge joint, end-to-end connection
~, точечное spot weld, spot-welded joint
~, транзитное *el.* through connection
~ треугольник-треугольник *el.* delta-delta connection
~ треугольником *el.* delta connection
~, угловое corner/angle joint
~, фланцевое flange(d) joint
~, хрящевое *anat.* cartilaginous joint, amphiarthrosis
~ цепей chain coupling
~, циклическое cycle/cyclic/ring compound
~, шарнирное hinge/articulated joint
~, шаровое шарнирное ball-and-socket joint
~, шлицевое splined joint
~, шомпольное piano-hinge connection
~, шпоночное keyed joint
~, шпунтовое groove-and-tongue joint, match/joggle joint
~, штепсельное plug-and-socket connection
~, штифтовое pin-hole connection/joint
~, штуцерное nipple joint
~, штыковое bayonet joint
соединитель connector, coupler; bond; fastener
соединительнотканный *anat.* connective
соединительный connective, connecting; coupling; conjunctive, joining
~ союз *gram.* copulative conjunction
~ная ткань *anat.* connective tissue
соединять connect, join, bond, unite, couple; *chem.* combine
~ в параллель *el.* connect in parallel
~ встречно *el.* connect in opposition
~ на корпус/землю *el.* connect to chassis/ground
~ параллельно *el.* connect in parallel
~ последовательно *el.* connect in series
созвездие *astr.* constellation, asterism
сознательный conscious; deliberate; *law* voluntary; *psych.* supraliminal
~, классово class-conscious
созревание maturation, ripening; *food* ageing
~ бетона curing of concrete
~, неполное semimaturation
~, половое puberty, pubescence
~, преждевременное половое pubertas precox
созревать mature, ripen
соизменимость *math.* covariance
соизмеримость commensurability, commensuration
соизмерять commensurate
соисполнитель *law* accessory; co-author
сойка *zool.* jay *(Garrulus)*
сок juice; liquor
~, желудочный gastric juice
~, кишечный intestinal juice
~, млечный *bot.* latex
~, отработанный exhausted liquor
~ предстательной железы prostatic fluid
сокатализатор cocatalyst
сокирки *pl. bot.* forking lakrspur *(Delphinium consolida)*
соклетие *biol.* syncytium, syncite
сокогонный *physiol.* succagogue, secretagogue
сокол *zool.* falcon *(Falco)*
соколиный falcoline
~ная охота falconry
соколовушка juice trap
Сократ Socrates
сократимый 1. *math.* reducible; cancel(l)able; 2. *anat.* retractile; contractile
сократимость contractility

~, **мышечная** myotility
сократительный contractile
сократовский, сократический Socratic(al)
сокращаемость contractility, contractibility
сокращаемый = сократимый
сокращать 1. reduce, cut short/down; minimize; 2. shorten; abbreviate; 3. *math.* cancel
сокращение 1. reduction, cutting down; curtailment; 2. shortening; abridgement; abbreviation; 3. contraction; 4. *math.* cancellation
~, **голодное** *physiol.* hunger contraction
~ **денежной массы** monetary contraction
~ **дроби** reduction/simplification of a fraction
~ **желудков** ventricular contraction
~ **земной коры** crustal shortening, contraction of the earth's crust
~, **лоренцово** *phys.* Lorentz contraction
~ **матки, тетаническое** uterine tetanus
~ **подобных членов** cancellation of like terms
~ **хромосом** contraction/condensation of chromosomes
сокредитор joint creditor
сокрытие concealment
~ **краденого** receiving stolen goods
сокристаллизация cocrystallization
соланидин solanidine, solatubine
соланин solanine, solatunine
солеварня saltern
солевой saline
солегатарий *law* collegatary
солемер salinometer, salimeter, salt gauge, brine meter
соленоид solenoid
соленостела *bot.* solenostele
соленость salinity, saltiness, saltness
солерос *bot.* glasswort *(Salicornia)*
солеустойчивость salt resistance/endurance; salinity tolerance
солея *eccl.* solium
солидарность 1. *law* collective responsibility 2. solidarity
солидарный *law* solidary, joint *(attr.)*
солидарно *law* jointly and severally
солидол cup grease, solid oil
солидонагнетатель grease gun/pump/squirt, lubricating gun
солитер *zool.* taenia, tapeworm *(Taenia)*
~, **бычий** beef tapeworm
~, **свиной** pork tapeworm
солль *geol.* kettle hole
солнечник *zool.* dory (Zeus)
солнечно-звездный solar-sidereal
солнце sun, Sun
~, **возмущенное** disturbed Sun
~, **истинное** real/true/apparent Sun
~, **ложное** parhelion, fictitious/mock Sun
~ **спокойное** quiet/undisturbed Sun
~, **среднее** mean sun
солнцелюбивый heliophilous, sun-loving
~**вое растение** heliophyte
солнцелечение heliotherapy
солнцепоклонник Sun worshipper, heliolater
солнцепоклонничество heliolatry, Sun worship
солнцестояние solstice
~, **зимнее** winter solstice
~, **летнее** summer solstice
солнцецвет *bot.* sun rose, rockrose *(Helianthemum)*
солод malt
~, **винокуренный** distiller's nalt
~, **пивоваренный** brewer's malt
солодка licorice, liquorice *(Glycyrrhiza)*
солодковый liquorice, licorice
солодовый malt *(attr.)*
солодоращение, соложение malting
соломовяз(алка) straw binder/trusser
соломопрес straw baler/press
соломоцвет *bot.* chaff-flower *(Achyranthes)*
солонец solonetz, alkali/sodic soil
солонцевато-солончаковый solonetz-solonchak, saline-alkali
солончак 1. solonchak, saline soil, while alkali soil; 2. salt bottom, saltierra; salt lake/spring
соль I salt
~, **азотистокислая** nitrite
~, **азотнокальциевая** nitrite of lime
~, **английская/горькая** Epsom salt, bitter salt, magnesium sulphate
~, **бертолетова** Berthollet's salt, potassium chlorate
~, **бромистоводородная** bromide
~, **водная** hydrous salt
~ **виннокислая** tartrate
~, **глауберова** Glauber's salt, sodium sulphate
~ **гремучей кислоты** fulminate
~, **двойная** double salt
~, **двууглекислая** bicarbonate
~, **двуугленатриевая** sodium bicarbonate
~, **двухромовокислая** bichromate, dichromate
~, **донная/иловая** bottom salt
~ **железа** iron salt
~, **каменная** rock salt
~, **карлсбадская/карловарская** Carlsbad salt
~, **кислая** acid salt
~, **красная кровяная** potassium ferricyanide
~, **кремнекислая** siliceous salt
~ **лимонной кислоты** citrate
~ **мочевой кислоты** urate
~ **мышьяковистой кислоты** arsenite
~ **мышьяковой кислоты** arsenate, arseniate
~, **основная** basic salt
~, **поваренная/столовая** table/common salt
~ **радия** radium salt
~, **сегнетовая** Seignette's salt, potassium sodium tartrate
~, **сернистокислая** sulphide
~, **серноватистокислая** hyposulphite
~, **сернокислая** sulphate
~, **серноникелевая** nickel sulphate
~, **средняя** neutral salt

~ **сурьмяной кислоты** antimonate, antimoniate
~, **углекалиевая** potash
~, **углекислая** carbonate
~, **уксуснокислая** acetate
~, **уксусномедная** copper acetate
~, **хромовокислая** chromate
~, **циановокислая** cyanate
соль II *mus.* G, sol
~ **бемоль** G flat
~ **диез** G sharp
соль III *zool.* sole *(Solea solea)*
соль IV sol *(silver Coin of Peru)*
сольват *chem.* solvate
сольватация *chem.* solvation
сольданелла *bot.* alpenclock *(Soldanella)*
сольфеджио *mus.* solfeggio, sol-fa
солюсульфон solasulphone
солюсурьмин sodium antimonyl gluconate
солянка *bot.* Russian thistle, saltwort *(Salsola)*
солянокислый *chem.* hydrochloric, hydrochlorate (of)
солярий solarium
солярит *med.* solar/celiac plexitis
сом *zool.* freshwater catfish *(Silurus)*
сома *anat.* soma
сомалиец Somali
сомалийка Somali (woman)
сомалийский Somalian
соматом somatone, somite
соматомегалия *med.* somatomegaly, gigantism
соматотропный somatotrop(h)ic
соматофит *bot.* somatophyte
сомик *zool.* black bullhead *(Ameirus melas)*
сомит somite
~, **грудной** thoracic somite
сомнамбула somnambulist, somnambule, sleep-walker
сомнамбулизм somnambulism, somnambulance, sleep-walking
сомножество *math.* co-set, corresponding set
сомножитель *math.* factor; multiplier
сон I *phys.* sone
сон II sleep
~ **гипнотический** hypnotic sleep, hypnosis
~, **медикаментозный** medication sleep
~, **неестественно глубокий** sopor
~, **пароксизмальный/эпилептический** paroxysmal sleep, sleep epilepsy
сонаниматель joint rentee, cotenant
сонаследник coheir, parcener, joint heir
сонаследница coheiress
сонаследование *law* coparcenary
сонет *lit.* sonnet
сонливость somnolence, somnolency, drowsiness
сонный *med.* carotic; sleepy
~**ная артерия** carotid (artery)
сонометр sonometer, phonometer
сонорность sonorousness, sonority
сонохимия sonochemistry
соня *zool.* dormouse *(Gliridae)*
сообвиняемый codefendant
сообитание cohabitation
сообщать report; inform; communicate; announce
сообщение 1. message, communication, notice, report; announcement; information; release; 2. traffic, service
~ **в цифровом виде** digital message
~, **воздушное** air service/communication
~, **входное** input message
~, **железнодорожное** rail service
~, **журнальное** log message
~ **морем** maritime communication(s)
~ **о бедствии** distress message
~ **о состоянии работ** progress report
~ **об ошибке** *comp.* error message
~, **паромное** ferry service
~ **по расписанию** scheduled service
~, **речевое** voice message
~, **товарно-пассажирское** mixed traffic
~, **шифрованное** cipher message
сообщество community
~, **Европейское** European Community
~, **ископаемое** fossil assemblage
~ **пород** *geol.* rock assemblage/association
~, **растительное** plant community; coenosis
сообщник accomplice; accessory
сообщничество complicity
соозначение *log.* connotation
соопека *law* joint guardianship
сооружать erect; install
сооружение 1. structure; facility; installation, plant; construction, edifice; 2. erecting, building, constructing, construction
~, **земляное** earthwork
~, **каменное** masonry (structure)
~, **мостовое** bridgework
сооружения installations, facilities, structures, plant
~, **ангарные** hangarage
~, **аэродромные** airfield facilities
~, **внелинейные** off-line installations
~, **коммунальные** public utilities
~, **линейные** on-line installations
~, **путевые** permanent way equipment
~, **складские** storage facilities
~, **стационарные** fixed structures
соосаждение co-precipitation
соосность coaxiality, (coaxial) alignment
соосный coaxial
соответствие correspondence; conformity, congruence, adequacy; correlation; compliance
~ **техническим условиям** conformance to specification
соответствовать correspond; match; suit
соответчик *law* co-respondent, co-defendant
соотнесение correlation, correlating
соотносительность correlation; correlativity
соотносительный correlative
соотношение correlation, relationship, relation; rate; proportion, ratio
~ **весов** weight ratio

~, **масштабное** scaling relation
~ **между затратами и эффективностью** cost-effectiveness balance
~ **неопределенностей** uncertainty relation
~, **объемное** volumetric ratio
~ **полов** sex ratio
~ **сил** balance of forces
~ **тяг** thrust ratio
~ **цен** price ratio
сопло nozzle
~, **ввертное** threaded nozzle
~, **воздушное** air-blast nozzle
~, **входное** inlet nozzle
~, **выхлопное** exhaust nozzle
~, **выходное** exit nozzle
~, **впрыскивающее** injector nozzle
~, **дозирующее** metering nozzle
~, **звуковое** sonic nozzle
~ **Лаваля** Laval nozzle
~, **кольцевое** annular/ring nozzle
~, **концевое** tip nozzle
~, **кормовое** aft nozzle
~ **малой тяги** low-thrust nozzle
~, **мерное** Venturi tube
~, **нагнетательное** delivery nozzle
~, **нерегулируемое** fixed-area nozzle
~, **носовое/переднее** fore nozzle
~, **побудительное** booster nozzle
~, **продольное** longitudinal (fore-and-aft) nozzle
~, **профилированное** contoured/shaped nozzle
~, **расходомерное** flow/Venturi nozzle
~, **реактивное** jet/propulsive/propulsion/propelling nozzle
~ **с внутренним телом** spike nozzle
~ **с жалюзи** louvred nozzle
~ **с затвором** shut-off nozzle
~ **сверхзвуковое** supersonic/Laval nozzle
~, **смывное** sluice/hydrojet nozzle
~ **управления по крену** roll-control nozzle
~ **управления по курсу** yaw-control nozzle
~ **управления по тангажу** pitch-control nozzle
~ **ускорителя** booster nozzle
~, **форсажной камеры** afterburner nozzle
~ **форсунки** injector/blast nozzle
соплодие *bot.* infructescence, fruit system
соподчинение *gram.* collateral subordination
сополимер *chem.* copolymer
~, **привитой** graft copolymer
сополимеризация copolymerization
сопроложение juxtaposition
сопоручитель *law* co-surety, cosignor
сопостановщик *(film)* codirector; *theatre* coproducer
сопоставлять juxtapose; compare; correlate
сопредельный adjacent, contiguous, adjoining
сопредседатель co-chairman
соприкасающийся contiguous
соприкосновение contact; contiguity; touch
сопровождать accompany, escort; track, follow
сопровождение accompanying, accompaniment; convoy, escort; following, tracking
~ **по дальности** range tracking
~ **по углу места** tracking in elevation
~, **радиолокационное** radar tracking
сопромат *abbr., educ.* strength of materials
сопротивление 1. resistance; drag; opposition; *phys.* strength; 2. *el.* resistor
~, **активное** resistance; *el.* effective resistance
~, **аэродинамическое** aerodynamic/air drag, air resistance
~, **балансировочное** trim drag
~ **в цепи сетки** grid resistance/resistor
~, **ветровое** wind drag/resistance
~, **вихревое** vortex drag/resistance
~ **воздуха/воздушное** air drag/resistance
~, **волновое** characteristic/wave impedance
~, **вредное** parasitic drag
~, **временное** ultimate resistance/strength
~, **входное** input resistance
~, **выходное** output resistance
~, **гасящее** quenching resistor; voltage dropping resistor
~, **граничное** boundary resistance
~ **давления** pressure drag/resistance
~, **дифференциальное** incremental resistance
~, **добавочное** *el.* series resistor/resistance
~, **емкостное** capacitive/capacity reactance, capacitance
~ **затвора** gate resistance
~ **изгибу** bending/flexural strength
~ **излому** folding strength
~ **излучения** radiation resistance/impedance
~, **измерительное** instrument resistance
~ **износу** wear resistance
~, **импульсное** *naut.* momentum drag/resistance
~, **индуктивное** 1. induced/inductive drag; 2. *el.* inductive reactance/impedance
~, **керамическое** *el.* ceramic resistor
~, **комплексное** *el.* complex impedance
~ **кручению** torsional strength/rigidity
~, **лобовое** drag, head/frontal resistance
~, **магнитное** magnetic/magneto resistance, reluctance
~ **магнитное удельное** specific reluctance, reluctivity
~, **малое** low resistance
~ **материалов** strength of materials
~ **на износ** wear resistance
~ **нагрузки** load impedance
~, **нагрузочное** load resistor
~ **обшивки** skin resistance
~, **объемное** volume/cubic resistance
~, **объемное удельное** volume resistivity
~, **отсасывающее** bleeder resistor
~ **ползучести** creep resistance
~, **полное** 1. total/overall drag; 2. *el.* impedance
~, **полное входное** input impedance
~, **полное выходное** output impedance
~, **полное электрическое** electrical impedance

~ при изгибе, временное ultimate bending strength
~ раздавливанию crushing strength
~ разрыву rupture/breaking strength
~, разрядное 1. discharging resistor; 2. discharge resistance
~, растяжению tensile strength
~, реактивное reactance, reactive impedance
~ сдвигу shear strength/resistance
~, сеточное grid resistor
~ сжатию compressive/compression strength
~ скалыванию cleavage strength
~ скольжению slip resistance
~ скручиванию torsional resistance/strength
~ следа wake drag
~ срезу shear stress/strength/resistance
~, собственное internal resistance
~ трения friction(al) drag/resistance
~, тяговое draught resistance
~, угольное *el.* carbon resistor
~ удару shock/impact resistance/strength
~, удельное specific resistance; resistivity
~ усталости fatigue strength/resistance/endurance
~ формы form drag/resistance
~ шероховатости roughness drag/resistance
сопротивляемость resistivity, resistibility, capacity to resist; *el.* resistivity
сопрягаемость compatibility
сопрягать mate; integrate; *comp.* interface; *math.* conjugate; *tech.* join, combine, gang
сопряжение ganging, integration, mating; linking, union, junction; *comp.* interface, interfacing
сопряженность association; conjugacy; contingency; interlinking; *biol.* conjugation
сопряженный conjugate; adjoint; associated; linked, paired, ganged
сорбит sorbitol, sorbite
сорбитизировать sorbitize
сорбция sorption
~, химическая chemisorption
сорлинь *naut.* boggin line, rudder chain
сорога *zool.* roach *(Rutilus rutilus)*
сорока *zool.* magpie *(Pica pica)*
сорокоуст forty days prayers (for the dead)
сорочай *zool.* oyster catcher
сорт 1. grade, quality; brand; 2. variety; 3. kind, type, sort
~, высокоурожайный high-yielding variety
~, высший top grade
~, культурный *agr.* variety, cultivar, cultivated variety
~, скороспелый early ripening variety
~, тенический industrial/technical grade
~ товарный commercial/market grade
сортамент, сортимент range of products/sizes, assortment; *tech.* wire gauge
сортировать grade, classify, sort, assort
~ по крупности/размерам size
сортировка 1. grading, classifying; sorting, assortment; 2. grader, sorter, classifier; sorting/grading machine
сортность grade, quality
сортовой high-quality/grade, special-quality
~вая сталь section/structural steel
сортопрокатный *met.* section-rolling
~ стан shape/section rolling mill
сортоучасток *agr.* variety/cultivar test plot
сорфинг *sport* surfing
сосальщик fluke
~, печеночный liver fluke
соскабливание scraping, curettage
соскабливать scrape, shave off
сосковидный, соскообразный papilliform, nippleshaped; mammiliform
сосковый papillary, mammilary
сослагательный *gram.* subjunctive
сословие 1. estate, (social) class; 2. corporation
~, дворянское nobility, noble estate
~, духовное the clergy
~, купеческое the mercnants
~, помещичье squiralty
сословный class *(attr.),* estate *(attr.)*
~ная монархия limited monarchy
сослужение *eccl.* joint performance (of church service)
сосна *bot.* pine tree, pine *(Pinus)*
сосновик *zool.* parrot crossbill *(Loxia pytyopsittacus)*
сосновый pinetree *(attr.),* pine *(attr.),* pinewood *(attr.)*
~вая хвоя pine needle
~вая шишка pine cone
сособственник joint owner/proprietor
сособственность joint ownership
сосок papilla, nipple; teat
~, грудной mammilla
~, запавший retracted/crater nipple
~ зрительного нерва optic papilla/disk
сосочек papilla; mammilla
~, валикообразный vallate/circumvallate papilla
~, волосяной hair papilla
~, зубной dental papilla/bulb
~, околоушный parotid papilla
~, резцовый incisive/palatine papilla
~, слезный lacrimal papilla
~ языка lingual papilla
сосочковый papillary, papillate; papilliform
соссюрит *miner.* saussurite
состав 1. composition, content, make-up; 2. *tech.* compound; mixture; 3. formula, formulation; 4. analysis; 5. staff, membership, personnel, body, strength
~, антикоррозионный corrosion prevention compound
~ бетона concrete mix
~, герметизирующий sealing compound, sealant
~, гранулометрический grain-size analysis/composition, fineness ratio, granulometric composi-

tion
~, **закрепляющий** fixative, fixing agent
~, **инженернотехнический** technical staff
~, **инструкторский** training personnel/staff
~ **корма** *agr.* feed composition
~, **личный** personnel; staff
~ **народонаселения** population structure
~, **объемный** volumetric composition
~, **офицерский** officers
~, **пенообразующий** foam composition
~, **подвижной** *rail.* rolling stock
~ **почвы** soil composition
~, **преподавательский** teaching staff, lecturers
~ **преступления** *law* corpus delicti
~, **профессиональный** occupational pattern
~ **рабочих, списочный** workers on payroll
~, **рядовой** rank and file; *mil.* ranks
~, **ситовый** sieve analysis
~ **совокупности** structure
~, **списочный** active roll
~, **технический** technical/maintenance personnel, engineers and technicians
~, **товарный** *rail.* freight train
~, **химический** 1. chemical composition; 2. chemical compound
~ **экипажа** crew component
~, **элементарный** elemental composition
~ **ядра** constitution of the nucleus

составление drawing up; formulation; compilation; composition; formation; working out
~ **проектов** designing, project-making
~ **сметы** cost estimation
~ **счетов** billing
~ **таблиц** tabulation

составлять 1. draw up; formulate; work out; write/make out; compose, design; compile, form; 2. constitute, be, make, represent, add up to

составляющая constituent, component
~, **активная** active/power/energy component
~, **вращательная** rotational component
~, **гармоническая** harmonic/sinusoidal component
~, **единичная** unique constituent
~ **затрат** costs component
~, **касательная** tangential component
~, **кососимметричная** skew-symmetric component
~, **периодическая** cyclical component
~, **сингулярная** *math.* function of singularities
~, **синфазная** in-phase component
~ **скорости** velocity component
~, **случайная** random component
~ **тока, активная** resistive current component
~ **тяги** thrust component

составной 1. composite, compound; sectional; 2. constitutive, component

состояние state, condition; position; stage
~, **агональное** agonal/moribund state
~ **агрегатное** state of aggregation
~ **бездействия** *geol.* dormant state
~, **бессознательное** unconsciousness
~, **болезненное** diseased state
~, **бредовое** delirious state, delirium
~ **в момент выделения** nascent state
~, **взвешенное** suspension
~, **включенное** on position
~, **возмущенное** perturbed/disturbed state
~ **войны** state of war
~, **выключенное** off position
~, **газообразное** gaseous state
~, **галлюцинозное** hallucinosis
~ **готовности** standby condition
~, **депрессивное** depression; *econ.* doldrums
~, **детерминованное** *cyb.* determinate state
~, **деформированное** strained state, state of strain
~, **долгоживущее** long-lived state
~, **жидкое** liquid state, fluidity, liquidity
~ **занятости** occupied state
~, **захватное** capture state
~ **излучения** radiant state
~, **исправное** serviceable condition, good working order
~, **исходное** initial state
~, **квазисвязанное** quasi-bound state
~, **квазистационарное** quasi-steady/stable/stationary state
~, **квантовое** quantum state
~, **колебательное** vibrational state
~, **коллоидное** colloidal state
~, **короткоживущее** short-lived state
~, **кристаллическое** crystalline state
~, **критическое** critical state; criticality
~, **марковское** Markov behaviour
~, **мгновенно-критическое** prompt-critical state
~, **надкритическое** above-critical state, supercriticality
~ **невесомости** weightlessness
~, **неисправное** disrepair
~, **нерабочее** idle state/condition
~, **неравновесное** nonequilibrium/disequilibrium state
~, **несвязанное** unbound state
~, **нестационарное** non-steady state
~, **одночастичное** single-particle state
~ **ожидания** *comp.* wait state
~ **отказа** *comp.* failed state
~, **подкритическое** subcriticality
~ **покоя** rest; quescent/dormant state
~, **предсмертное** terminal state
~ **при поставке** as-received condition
~ **пригодности к работе** good working order
~, **псевдоожиженное** quasi-liquid state, fluidized state
~, **психическое** mental state
~, **рабочее** operative/working/service condition, working order
~ **равновесия** equilibrium; balance; equiponderant state; stability

~, **реакционное** reactive state
~ **резерва** standby condition
~ **с большей энергией** higher-energy state
~ **с большой энергией** high-energy state
~ **с малой энергией** low-energy state
~, **сверхкритическое** supercriticality
~, **связанное** bound state
~, **скрытое** latent state
~, **собственное** eigenstate
~, **стандартное** reference state
~, **стационарное** steady/stationary state
~, **тупиковое** *comp.* failed state
~, **тяжелое** *med.* grave/poor condition
~, **умственное** mental condition
~, **установившееся** steady state
~ **эксплуатационной готовности** serviceable condition
~, **энергетическое** energy state
состоятельность 1. *math., stat.* consistency; 2. *fin.* solvency
состоятельный 1. consistent; 2. solvent
сострахование co-insurance
состраховщик co-insurer
состязание *law* controversy
состязательность *law* adversary nature
состыковать mate, join; *space* dock
сосуд vessel, container, receptacle
~, **выносящий** *anat.* efferent vessel
~ **Дьюара** Dewar flask, vacuum jacketed flask
~, **измерительный/мерный** measuring vessel, graduate
~, **коронарный** coronary vessel
~, **кровеносный** blood vessel
~, **млечный** lactiferous vessel
~, **приемный** receiver; receptacle
~, **приносящий** *anat.* afferent vessel
~, **разделительный** separation vessel
~, **реакционный** *chem.* reaction vessel
сосудисто-нервный *anat.* neurovascular
сосудистый *anat.* vascular
сосудодвигательный *anat.* vasomotory, vasomotor, vasomotorial, vasomotoric
сосудорасширитель vasodilator
сосудорасширяющий vasodilative
сосудосуживатель vasoconstrictor
сосудосуживающий vasoconstrictive
сосун suction/dredging/drag pipe
сосуществование coexistence
сосцевидно-теменной *anat.* parietomastoid
сосцевидный mastoid(al), mammilliform, mammiform; nipple-shaped/-like
сотенник, сотенный *hist.* sotnia commander
сотня 1. *hist.* sotnia, cavalry squadron; 2. one hundred
сотовидный honeycomb(ed); celluar
сотовый cellular, honeycomb *(attr.)*
сотопласт honeycombed plastic
сотрясение tremor; shock; *med.* concussion, commotio
соты honeycomb cells
соударение collision, impingement, encounter; percussion; impact
соударяться collide; encounter
соумышленник accomplice, accessory
соустье *anat.* anastomosis
соучастие participation; complicity
соучастник accomplice, accessory
~ **по/после выполнению преступления** *law* accessory before/after the fact
софизм sophistry, sophism
софистика sophistry; casuistry
соф(ф)ит 1. studio lights/lamps; 2. *arch.* soffit
софора *bot.* pagoda tree *(Sophora)*
софокусный confocal
сохранение 1. conservation, preservation; retention, reservation; 2. upkeep, maintenance, care; 3. saving
~ **импульса** momentum conservation
~ **массы** mass conservation
~ **материи** matter conservation
~ **углов** preservation of angles
~ **энергии** conservation of energy
сохранять 1. conserve, preserve; retain; 2. save; 3. keep, take care of
соцветие *bot.* inflorescence
соцдоговор *obs.* socialist emulation agreement
социалдемократия social democracy
социализм socialism
~, **гильдейский** guild socialism
~, **развитой** advanced socialism
~, **фабианский** Fabian socialism
~, **утопический** utopian socialism
социал-империализм socialistic imperialism
социалистический socialist *(attr.)*
социал-соглашатель *polit.* compromiser
социально-бытовой social
социально-классовый social-class *(attr.)*
социально-культурный sociocultural
социально-политический sociopolitical
социально-хозяйственный/экономический socioeconomic
социатив *gram.* sociative(-instrumental) case
социнианство *hist.* Socinianism
социолингвистика sociolinguistics
социолог sociologist
соцобязательство *abbr.* social commitment
соцсоревнование *abbr., obs.* socialist emulation
соцстрах *abbr.* social insurance
соцстроительство *abbr.* socialist construction; the building of socialism
сочевичник *bot.* bitter vetch *(Orobus)*
сочетание combination; *gram.* word combination, phrase
~ **хромосом** apposition of chromosomes
сочетательность associativity
сочетательный associative; combinative
сочетать(ся) combine, blend; unite
сочленение 1. joint, coupling; 2. articulation
~, **шарнирное** articulated/hinged joint
~, **управляющее** master joint

сочленять join, connect, link; articulate
сочнолистный *bot.* chylophyllous
сочноплодный *bot.* chylocarpous
сочносемянный *bot.* chylospermous
сочностебельный *bot.* chylocaulous
сошник 1. *anat.* vomer; 2. *agr.* ploughshare, plowshare; coulter
сошниково-носовой *anat.* vomeronasal
сошниково-основный *anat.* vomerobasilar
сошниковый *anat.* vomerine
союз I union, alliance; coalition; league
~, **брачный** conjugal/family/marital union
~, **денежный** monetary union
~, **Советский** *hist.* the Soviet Union
~, **таможенный** customs/tariff union
союз II *gram.* conjunction
~, **противительный** adversative conjunction
~, **разделительный** disjunctive conjunction
~, **сочинительный** coordinative conjunction
союзно-наречный *gram.* conjunctive-adverbial
союзно-сочиненный *gram.* conjunctive-coordinated
союзный *gram.* conjunctive, conjunctional
соя *bot.* soybean *(Glycine soja)*
спагетти spaghetti
спаги spahi, spakee
спад decrease, drop, decline, fall; decay; *econ.* recession, slump
~ **активности** activity decay
~ **в темпах роста** crowth recession
~ **деловой активности** business recession/decline/slow down
~, **линейный** linear fall-off [decrease]
~ **тяги** thrust decay
спадать decrease, fall, drop; abate, diminish
спаечный adhesive; commissural
спазм spasm
~ **артерии** arteriospasm
~ **век** *med.* plepharospasm
~, **глоточный** pharyngismus, pharyngospasm
~, **губо-язычный** glossolabial spasm
~ **кишечника** colic
~ **мимической мускулатуры** mimic spasm/tick
~, **рефлекторный** reflex spasm
~ **сосудов** vasospasm, angiospasm
~ **языка** glossospasm
спазматический spastic, spasmodic
спазмолитический spasmolytic
спай joint, juncture; seam, seal; soldering
~ **термопары** thermojunction
спайка commissure; adhesion; *tech.* soldering, brazing; seam, splice, joint
спайнолепестность *bot.* gamopetaly, sympetaly
спайнолепестный *bot.* gamopetalous, sympetalous
спайнолистный *bot.* gamophyllous, symphyllous
спайностолбиковый *bot.* gamostylous
спайность *cryst.* cleavage
спайнотычиночный *bot.* gamostaminate
спайный = спаечный; *cryst.* cleavable
спальник *hist.* chamberlain of the bedchamber
спараксис *bot.* wandflower *(Sparaxis)*
спардек *naut.* spardeck
спаржа *bot.* asparagus *(Asparagus)*
спаривать pairing, mating; coupling; ganging
Спартак Spartacus
спартакиада field day, sports meeting
спартанец, спартанский Spartan
спартина *bot.* cord grass *(Spartina)*
спарто *bot.* esparto needle grass, alfa grass *(Stipa tenacissima)*
Спас 1. *rel.* Saviour; 2. *eccl.* Transfiguration
спасаемый *rock.* recoverable
спасать rescue, save, salve, salvage; recover
спасание rescue; *tech.* recovery
спаситель rescuer; lifesaver; *rel.* Saviour
спастический spastic, spasmodic
спек cake, sinter, sintered mass
спекаемость caking capacity
спекание caking, sintering; agglomeration
~ **контактов** *el.* sticking of contacts
~ **микрофона** *el.* packing of a microphone
~ **угля** clinkering
спекать(ся) cake, sinter; bake; clinker; *med.* coagulate, curdle
спектр spectrum
~, **видимый** visible/light spectrum
~ **высокого разрешения** fine spectrum
~, **дебаевский** Debye spectrum
~ **деления** fission spectrum
~ **доплеровских частот** Doppler spectrum
~, **дуговой** arc spectrum
~ **звука** acoustic/sound spectrum
~ **излучений** radiation/emission spectrum
~ **комбинационного рассеяния** Raman spectrum
~, **линейчатый** line spectrum
~, **нейтронный/нейтронов** neutron spectrum
~, **непрерывный** continuous spectrum, continuum
~ **обтекания** flow pattern/field
~ **первого запрещения** first-forbidden spectrum
~, **полосатый** band spectrum
~ **помех** interference spectrum
~, **прерывистый** discontinuous spectrum
~ **пробегов** range spectrum
~, **равноэнергетический** equienergy spectrum
~, **рентгеновский** X-ray (roentgen) spectrum
~ **сверхтонкого расщепления** hyperfine spectrum
~ **скоростей** velocity spectrum/distribution
~, **смягченный** softened/degraded spectrum
~, **сплошной** continuous spectrum, continuum
~, **ударный** shock spectrum
~ **ультрафиолетового излучения** ultraviolet spectrum
спектроанализатор spectrum analyzer
спектрогелиограмма spectroheliogram(me)

спектрограф spectrograph
~, призменный prism spectrograph
спектрометр spectrometer
~, быстросканирующий rapid-scanning (high-speed) spectrometer
~, высокочастотный radio-frequency (r.f.) spectrometer
~, двухлучевой double-beam spectrometer
~, кристаллический crystal spectrometer
~ на совпадениях coincidence spectrometer
~ нейтронов реактора *nucl.* pile-neutron spectrometer
~ с разверткой sweep-type spectrometer
~ с самописцем recording spectrometer
~, электронный electron spectrometer
~, ядерно-резонансный nuclear-resonance spectrometer
спектроскоп spectroscope
~, пламенный flame spectroscope
~ прямого зрения direct-vision spectroscope
спектроскопия spectroscopy
~ излучений radiation spectroscopy
~, мезонная meson spectroscopy
спектрофотометр spectrophotometer
спектрохимия spectrochemistry
спекулянт profiteer, speculator; jobber
~, вексельный bill jobber
~ играющий на повышение bull/long jobber
~ играющий на понижение bear/short seller
~ на фондовой бирже (stock) jobber
~ по покупке и продаже земельных участков land-jobber
спекулярия *bot.* Venus's looking glass *(Specularia)*
спелеолог speleologist
спелеологический speleological; spelaean
спелость ripeness; maturity
~, восковая waxy ripeness, yellow-ripe stage, middle dough ripeness
~, молочная milky ripeness
~, полная complete ripeness, full-ripe stage
~, съемная/уборочная picking maturity/ripeness
спель *met.* refining foam
~, графитная kish
спельта *bot.* spelt *(Triticum spelta)*
спенсерианский Spencerian
сперма sperm, semen
сперматангий *bot.* spermatangium
сперматека *zool.* spermatheca
сперматида spermatid
сперматобласт spermatoblast, spermoblast, sperm cell
сперматогенез spermatogenesis, spermatogeny
сперматогенный spermatogenic, spermatogenetic
сперматогония, сперматогоний *bot.* spermatogonium, spermatogone, spermatospore, spermatophore
сперматозоид spermatozoon; *pl.* spermatozoa
сперматозоидный spermatozoan, spermatozoal, spermatozoidal
сперматоидный spermatoid
сперматолиз spermatolysis, spermolysis
сперматопоэтический *biol* spermatopoietic
сперматорея *med.* spermatorrh(o)ea
сперматомер spermatomere
сперматофит *bot.* spermatophyte, seed plant
сперматофлебэктазия spermatophlebectasia, varicocele
сперматоцид spermatocide, spermicide
сперматоцист spermatocyst
сперматоцистит spermatocystitis
сперматоцит spermatocyte
сперматоцитный spermatocytal
сперматоцитома spermatocytoma, seminoma
сперматурия seminuria, spermaturia
спермацет spermaceti, cetaceum
спермациофор *bot.* spermatiophore, sporophore
спермацит *bot.* pollinoid
спермий spermium, spermatozoon; *pl.* spermia, spermatozoa
спермин, спермокрин, спермолиенол spermine
сперминоксидаза spermine oxidase
спермобласт spermoblast, spermatoblast
спермогоний *bot.* spermogonium
спермодерма *bot.* spermoderm, episperm
спермокарпий *bot.* spermocarp
спермолиз spermolysis, spermatolysis
спермолизин spermatolysin
спермолит spermolith
спермолитический spermatolytic
спермофит *bot.* spermatophyte, seed plant
спермофор spermophore, spermatophore, placenta
спецификация 1. specification; 2. list of parts; bill; schedule
~ груза freight note
~ действий action specification
~ материалов list/bill of materials
~ на запасные части list of spares, list of service/spare parts
~, отгрузочная shipping specification
спецификум specific/characteristic features
спецкор *abbr.* special correspondent
спецкурс *abbr.* special course of studies
спецлиния *abbr.* unlisted/special line
спецодежда work(ing) clothes, overalls
спецотдел *abbr.* special department
спецподготовка *abbr.* special training
спецслужба *abbr.* special service
спецставка *abbr.* special wage/salary/rate
спецшкола *abbr.* special school
СПИД (синдром приобретенного иммунного дефицита) AIDS (acquired immune deficiency syndrome)
спидобарограф VH recorder
спидограф airspeed recorder
спидола portable radio
спидометр speedometer
спизелла-крошка *zool.* field sparrow *(Spizella*

pusilla)
Спика *astr.* Spica
спиккато *mus.* spiccato
спикула *bot., zool.* spicula, spicule
спилантес *bot.* spot flower *(Spilanthes)*
спилок *leath.* split (leather)
~, лицевой grain split
спин *phys.* spin
~, полуцелый half-integral spin
~, целый integral spin
~ четно-нечетного ядра even-odd spin
~ четно-четного ядра even-even spin
спина back; dorsum
спинакер-гик *naut.* spinnaker boom
спинацен spinacene, squalene
спинно-аддукторный spino-adductor
спинно-брюшной dorsoventral
спинной dorsal; spinal
~ мозг *anat.* spinal cord
~ хребет *anat.* spinal column
спинно-лобный dorsofrontal
спинно-мозговой cerebrospinal, spinal
спинозизм Spinozism
спино-мозжечковый cerebellospinal
спинорбитальный spin-orbital
спинорог *zool.* filefish, triggerfish *(Balistes)*
спино-спинальный spinospinal
спино-таламический spinothalamic
спин-решеточный spin-lattice *(attr.)*
спинтарископ *phys.* spinthariscope, scintilloscope
спира loop, coil
спиралевидный, спиралеобразный helicoid, spiroid
спирализация spiralling, coiling
~, левая left-handed spiralling
~, правая right-handed spiralling
спираль spiral, helix; coil; *arch.* scroll
~ Архимеда *math.* spiral of Archimedes
~, гиперболическая reciprocal/hyperbolic spiral
~, двухнитевая double strand spiral
~ Корню Cornu spiral
~, ленточная tape helix
~, остаточная/реликтовая relic spiral
~, сжатая close spiral
спирально-конический spiral-screwed, trochoid
спирально-плоскостной planispiral
спиральноплодный *bot.* spirocarpous
спирально-свернутый coiled
спиральность helicity, spirality
спиральный spiral, helical; coiled; helicoid
спирема spireme
спирея *bot.* spiraea *(Spiraea)*
спирилла spirillum
спириллез *med.* spirillosis
спиритический spiritualistic; spiritistic
спиричуал *mus.* spiritual (song)
спирогира *bot.* spirogyra, pond scum *(Spirogyra)*
спирометр spirometer, pulmometer
спиротека spirotheca
спирохета spirochete
спирохетоз spirochetosis
~, желудочный gastric spirochetosis
~, клещовый tick spirochetosis
спирохетозный spirochetotic, spirochetal
спирохеты *pl.* Spirochaeta
спирт alcohol, spirit, spirits
~, безводный anhydrous spirit/alcohol
~, винный etnyl alcohol, ethanol
~, денатурированный denatured/methylated alcohol
~, древесный methyl alcohol
~, камфорный camphor spirit
~, метиловый methyl alcohol, wood spirit, methanol
~, нашатырный spirit of hartshorn, ammonia spirit, ammonia solution
~, стандартный proof spirit
~, твердый solid alcohol
~, этиловый ethyl alcohol, ethanol
спиртаза *biochem.* alcoholase
спиртовка spirit lamp, alcohol burner
спиртовой alcoholic, spirituous
спиртомер alcoholometer
спиртоустойчивый alcohol-fast
список 1. list; roll; bill, sheet; 2. manuscript copy
~ адресов mailing list
~ грузов/груза cargo sheet
~ запрещенных товаров banned list
~ избирателей/избирательный poll; ballot
~ команд instruction/order/command list
~ корабельных запасов allowance list
~ личного состава muster roll; personnel roster
~ магазинный, обратный [прямой] *comp.* push-down [push-up] list
~, налоговый tax roll
~ облагаемых объектов tax roll
~ опечаток errata, corrigenda
~ очередности заказов/заявок waiting list
~ пассажиров waybill
~, послужной employment sheet; statement of service
~, трудовой service record, statement of service
списывать write-off
спица 1. spoke; 2. knitting needle
спич speech, public address
сплав I floating, rafting
сплав II alloy; composition, compound; blend
~, белый white metal
~, высоколегированный high alloy
~, двойной/двухкомпонентный binary alloy
~, доэвтектический hypoeutectic alloy
~, жаропрочный high-temperature alloy
~, жаростойкий heat-resistant alloy
~, железный ferroalloy
~, заэвтектический hypereutectical alloy
~, износостойкий wear-/abrasion-resistant alloy
~, конструкционный structural alloy
~, коррозионностойкий corrosion-resistant alloy
~, легкоплавкий fusible (low-melting) alloy

~, **литейный** cast/casting alloy
~, **магнитомягкий** magnetically soft alloy
~, **металлокерамический** metal-ceramic alloy, cermet
~, **низколегированный** low alloy
~, **низкоплавящийся** low-melting/temperature alloy
~, **подшипниковый** bearing alloy/metal
~, **сверхтвердый** extra-hard alloy
~, **типографский** type-metal alloy
~, **трекомпонентный/тройной** ternary alloy
~ **цветных металлов** non-ferrous alloy
~, **четвертной/четырехкомпонентный** quaternary alloy
сплавина floating mat/bog/island
сплавление fusion; alloying; (re)melting
сплавлять 1. alloy; melt, fuse; 2. float, raft
~ **лес** drift/raft logs
сплавщик raftman, rafter, wood-floater
спланхноплевра splachnopleura
спланхноптоз *med.* splachnoptosis
спланхнотом splanchnotome
спланхноцеле splachnocele
спленит *med.* splenitis, lienitis
спленомаляция splenomalacia, lienomalacia
спленомегалия *med.* splenomegaly, megalosplenia
спленопатия *med.* splenopathy, lienopathy
спленорафия splenorrhaphy
спленоцеле splenocele, lienocele
сплесень *naut.* splice
~, **кабельный** mariner's splice
сплеснивать *naut.* splice, raddle
сплетать braid, plait; splice; interweave
сплетение 1. interweaving, interlacing; plaiting; meshing; tangle, entanglement; *anat.* plexus; *gram.* suppletion
сплошной continuous, unbroken; *bot.* farctate
сплошность continuity
сплюшка *zool.* scops owl *(Otus scops)*
сплющенный flattened, oblate
сплющивать flatten
сподий *agr.* spodium
спожинки *eccl.* Lent before Assumption
спойлер *av.* spoiler, interceptor
спонгиобластома *med.* spongioblastoma, gliosarcoma
спонгиоз *med.* spongiosis
спонгиоза *med.* spongiosa
спонгиозит *med.* spongiositis
спонгиозный spongoid, spongiform, spongy, spongiose
спонгиоцит *anat.* spongiocyte
спондеический *lit.* spondaic
спондиас *bot.* mombin *(Spondias)*
спондилез, спондилоз *med.* spondylosis, vertebral ankylosis
спондилит *med.* spondylitis
спондилоартрит spondylarthritis
спондилоартроз spondylarthrosis
спондилопатия spondylopathy, rachipathy
спондилосинтез *med.* spondylosyndesis, spinal fusion
спондилосхиз *med.* spondyloschisis, rachischisis
спондильный spondylous
спонсон *naut.* sponson
спора spore
~, **бесполая** agamospore
~, **ячеистая** alveolar spore
спорангиевый *bot.* sporangial
спорангиеносный sporangiferous, sporangiophorous
спорангий *bot.* sporangium, spore case/sac
спорангиола *bot.* sporangiole, sporangiolum
спорангиофор *bot.* sporangiophore
споридиеносный *bot.* sporidiferous
споридий, споридия *bot.* sporidium
спороболус, спороболюс *bot.* dropseed *(Sporobolus)*
споровик *zool.* sporozoan
споровместилище sporangium, spore case/sac/capsule
споровый sporous; cryptogamic, cryptogamous
спорогенез *biol.* sporogenesis, sporogeny
спорогон, спорогоний *bot.* sporogonium, moss capsule
спорогония sporogony
спорокучка *bot.* sorus
спороносец *bot.* sporophore
спороносный sporiferous, sporophorous
спороношение sporulation
спорообразный sporelike, sporoid
спорообразование sporulation, sporogenesis
спороплазма sporoplasm
спороп очка *bot.* sporophydium
споротека sporotheca
споротрихоз sporotrichosis
спорофит *bot.* sporophyte
спорофор *bot.* sporophore
спороцист(а) sporocyst
спороцит *bot.* sporocyte
спортзал *abbr.* gymnasium, gym hall
спортплощадка *abbr.* playground; athletic field
спорула *biol.* sporule
спорынья *bot.* ergot *(Claviceps puprurea)*
спорыш *bot.* knotgrass *(Polygonum aviculare)*
спорящие стороны *law* contending parties
способ manner, way, mode, method, technique, procedure, process
~ **адресации** *comp.* addressing mode
~ **выборки** access technique
~, **обратноступенчатый** *weld.* step-back method
~, **открытый** *min.* strip-mining
~ **последовательных приближений** step-by-step method
~ **приближения** approximation method
~ **производства** production process/technique, technology
~ **разработки** mining method

способность capacity, capability, ability, power
~, абсорбционная absorptive/absorption/absorbing power/capacity
~, аккумулирующая accumulating/storage capacity
~, восстанавливающая 1. reducing power; 2. recovery characteristics
~, восстановительная reducing power
~, временная разрешающая time resolution
~, высокая разрешающая high/fine resolution
~, грузопропускная freight transit capacity
~ делиться fissionability, fissibility
~, диспергирующая dispersive capacity/power
~ дороги, пропускная road capacity
~, задерживающая retentivity, retentiveness
~, избирательная selectivity
~, излучательная emittance
~, ионизационная ionizing power
~, испускательная emissivity, emissive/radiating/radiation capacity
~ к сжатию compressibility
~ к усадке shrinkability
~, нагрузочная load-carrying capacity
~, несущая bearing [load-carrying] capacity; *aero.* lift/lifting ability
~, окислительная oxidation power
~ осаждаться precipitability
~, отражательная 1. reflection/reflecting power; 2. reflectance
~, покупательная purchasing/buying power
~, полная излучительная total emittance
~, производственная producing capacity
~ проникания penetrative quality
~ пропитываться impregnability
~, пропускная thrughput (capacity)
~, реакционная reactivity
~, разрешающая resolution, resolving power
~, теплотворная calorific power/value, heat capacity/value
~, экстрагирующая extractive power, extractivity
~, эмиссионная emissive power, emissivity
справа at/on/from the right; *naut.* on the starboard side
справедливый *math.* valid, correct, true
справочник reference book, manual, directory; handbook, guide
~ телефонный telephone directory
спрекелия *bot.* Aztec/Jacobean lily *(Sprekelia)*
спринцевание syringing, sprinkling; irrigation
спринцевать syringe, irrigate
спринцовка syringe
спрос *econ.* demand; quest
~, будущий prospective/future demand
~, вероятностный probabilistic demand
~, взвешенный скользящий weighted moving demand
~, внутренний domestic demand
~ и предложение supply and demand
~ на рабочую силу labour demand
~, неудовлетворенный pent-up demand
~, оживленный brisk demand
~, повышенный keen demand
~, постоянный/устойчивый steady/stable/stationary demand
~, потребительский consumer demand
~, предельный limit of demand
~, рыночный market requirement
~, экстренный emergency demand
спру *med.* sprue, tropical diarrh(o)ea
спрут *zool.* octopus *(Octopus)*
спряжение *gram.* conjugation
спрямление rectification, straightening; righting; alignment
спумулит foam/spumous rock, spumulite
спуск 1. lowering; descent, descending; 2. slope, incline, downgrade; 3. release; discharge; *naut.* launch; 4. chute, slide; 5. detent, trigger; 6. escapement
~, винтовой spiral chute
~, донный bottom discharge
~, ливневый gutter, gully
~, пожарный tube-fire escape
~ троса *naut.* lay of a rope
~ флага *mil.* evening colours
спутанность confusion; disarrangement
~ сознания mental confusion, obfuscation
спутник 1. satellite; 2. accessory
~, беспилотный unmanned satellite
~, маневрирующий manoeuvrable/manoeuvring satellite
~, метеорологический weather satellite
~, обитаемый manned satellite
~, связной communication satellite
~ планеты planetary satellite
~, разведывательный reconnaissance satellite
~, управляемый monoeuvrable satellite
срабатывание 1. actuation, operation, functioning; 2. response; 3. wear(ing)
~ вследствие трения abrasive wear
срамной *anat.* pudendal
срастание accretion; intergrowth; concretion, coalescence, concrescence
~, двойниковое *miner.* twinning
~ кровеносных сосудов inosculation
~ переломов *med.* knitting consolidation
сращение adhesion, symphysis; concrescence
сращивание jointing, splicing; joining; splice
сращивать join; splice
среда medium; environment, milieu, surroundings; *chem.* agent; atmosphere
~, агрессивная corrosive medium
~, безграничная unbounded medium
~, водная aqueous medium
~, дисперсная dispersion medium
~ Дорсе Dorset's culture/egg medium
~, Дьедонне Diedonné's medium
~, естественная *biol.* habitat; natural environment
~, жидкая liquid medium

~, закалочная quenching medium; hardening compound
~, защитная shielding atmosphere/medium
~, культуральная *biol.* culture medium
~ Леффлера Loeffler's medium
~, науглероживающая carburizing atmosphere
~, неоднородная inhomogeneous (non-homogeneous, non-uniform) medium
~, несущая carrying agent
~, окружающая environment; surroundings
~, охлаждающая refrigerating/cooling medium, coolant
~, питательная nutrient/culture medium
~, проводящая conducting medium/agent
~, рабочая working environment/milieu
~, регулируемая controlled atmosphere
~, сплошная continuous medium, continuum
средиземноморский Mediterranean
срединный median; middle, mesal; medial
~ная Империя *hist.* the Middle Kingdom
среднеазиатский Central Asian/Asiatic
среднеанглийский Middle English
среднеарифметический arithmetic-mean
средневековый medieval, Middle-Age
средневзвешенное weighted mean
средневольтный *el.* medium-voltage
средневосточный Middle Eastern
сроднегодовой 1. annual average, average annual; 2. mid-annual
среднегрудь mesothorax
сроднедолевой midlobar, mediolobar
среднее average, mean
~, апостериорное aposteriori mean
~ арифметическое arithmetic mean/average, simple mean/average
~ взвешенное weighted mean/average
~ выборочное sample mean
~ гармоническое harmonic average/mean
~ геометрическое geometric mean/average
~ квадратичное quadratic mean, root mean square
~ по времени time average
~ по множеству ensemble average
~ по одной выборке single sample mean
~ приведенное adjusted mean
~ пропорциональное mean proportional
среднеевропейский Central European
среднезадний medioposterior
среднезернистый medium-grained
среднеквадратичный, среднеквадратический *math.* root-mean-square
среднеледниковье Middle Glacial Period
среднерусский Central Russian
среднесибирский mid-Siberian
среднеспелый *agr.* midseason-ripening
среднеспинка *zool.* mesonotum
среднестолбиковый *bot.* mesostylous
среднестрадательный *gram.* mediopassive
среднесуточный average daily
среднетвердый medium-hard
среднечастотный *el.* mid-frequency
среднеширотный *geog.* mid-latitude; temperate
среднеязычный *phon.* mediolingual, mediopalatal
средний 1. average, mean; 2. middle, medial, median, medium; central, centre
~ род *gram.* neuter (gender)
~ная арифметическая (величина) arithmetic mean
~яя история history of the Middle Ages
средостение *anat.* mediastinum
~ яичка septum of the testis
средостенный mediastinal
средства means, facilities, aids; equipment, apparatus; *fin.* funds; *see also* **средство**
~, аппаратные *comp.* hardware
~, бюджетные budgetary means/funds
~, государственные public funds
~, графические *comp.* graphics
~, денежные cash resources; funds
~ для отдыха и развлечения recreational facilities
~ для хранения storehouse facilities
~ доказывания *law* evidence
~, заемные loan proceeds
~, замороженные материальные dead/inactive stock
~ к существованию livelihood, sustenance
~, лазерные laser aids
~, ликвидные means of liquidity
~ математического обеспечения *comp.* software
~, навигационные navigation aids
~ обеспечения посадки landing aids
~, оборотные working/current capital
~, основные fixed assets; fixed capital
~ отладки *comp.* debug(ging) aids/tools
~, программно-аппаратные *comp.* firmware
~, программные *comp.* software
~ производства means of production
~, резервные standby funds
~, резервные технические standby facilities
~ связи communication facilities
~, спасительные rescue facilities
~, ссужаемые loan funds
~ существования means of subsistence; livelihood
~, транспортные transport vehicles; transportation facilities
~, финансовые funds; capital; resources
средство 1. means, facility; tool; aid; appliance; 2. *chem.* agent; medium; 3. *med.* remedy
~, абсорбирующее absorbent
~, адреноблокирующее *med.* adrenoreceptor blocking agent
~, активирующее activator, activating agent
~, акустиковизуальное audio-visual aid
~, анестезирующее anesthetic
~, антисептическое antiseptic
~, антитоксическое antitoxic
~, бактерицидное bactericide

~, **болеутоляющее** analgesic, analgetic
~, **визуальное** visual aid
~, **водоотнимающее** dehydrant, dehydrating agent
~, **водоотталкивающее** water repellent
~, **восстанавливающее** reducing agent
~, **вспенивающее** frothing/foaming agent
~, **газообразующее** blowing agent
~, **ганглиоблокирующее** ganglionic blockader
~, **гемостатическое/кровоостанавливающее** h(a)emostatic, antih(a)emorrhagic, styptic
~, **дезактивационное** decontaminant, decontamination agent
~, **дезинсекционное** desinsectant
~, **дезинфицирующее** disinfectant
~, **дезодорирующее** deodorant, deodorizer
~, **денатурирующее** denaturant
~, **жаропонижающее** antipyretic, febrifuge
~, **желчегонное** cholagogue
~, **законное платежное** legal tender
~, **консервирующее** preservative; curing agent
~, **мочегонное** diuretic, uragogue
~, **моющее** detergent
~, **обезвоживающее** dehydrating agent
~, **обезволашивающее** depilator
~, **обезжиривающее** degreaser, degreasing agent
~, **обесцвечивающее** decolorant
~ **обмена/обращения** medium of exchange
~, **окислительное** oxidizer
~, **омыляющее** saponifier, saponifying agent
~, **осаждающее** precipitant
~, **осветительное** illuminant
~ **отладки** *comp.* debugging facility
~, **отпугивающее** repellent
~, **отхаркивающее** *med.* expectorant
~, **пенообразующее** frothing/foaming agent
~, **перевозочное** *naut.* ferry craft
~, **плавучее** craft
~ **платежа** medium of payment
~, **платежное законное** legal tender
~, **покупательное** means of purchase
~, **потогонное** *med.* diaphoretic, sudorific
~, **прижигающее** *med.* cautery, caustic
~, **противовоспалительное** antiphlogistic
~, **противодетонирующее** antiknock dope, antidetonant
~, **противозачаточное** contraceptive
~, **противомольное** moth-proofing agent
~, **противообледенительное** anti-icing [de-icing] compound
~ **связи** means of communication
~, **связывающее** binder
~, **слабительное** purgative, laxative
~, **слюногонное** *med.* salivant, ptyalogogue
~, **смазочное** lubricant
~, **сосудорасширяющее** vasodilator
~, **сосудосуживающее** vasoconstrictor
~, **спазмолитическое** antispasmodic
~, **спасательное** life-saving appliance/apparatus
~, **транспортное** vehicle
~, **уплотняющее** sealing medium/compound
~, **успокаивающее** tranquilizing agent

срежиссировать *cin.* direct; *theatre* produce

срез 1. section, cut; 2. shear(ing), truncation
~, **косой** scarf

срезать cut; truncate; shear

Сретение *eccl.* the Purification of St. Mary, Candlemass Day

сродство affinity; cognation, relationship

срок 1. period, term, time interval, time; 2. date, term
~ **амортизации** depreciation period
~ **векселя** maturity/term of a bill
~ **выполнения** date due
~ **гарантии/гарантийный** guarantee/warranty period
~ **годности** working/effective/useful life
~ **годности при хранении** shelf/storage life
~ **давности/давностный** *law* limitation period
~ **действия** period of validity; duration; period of action/operation
~ **доставки** time of delivery
~ **завершения** finishing/completion date, deadline
~ **исковой давности** *law* limitation period
~, **испытательный** probationary/test period
~, **конечный** deadline
~, **льготный** grace period, period of grace
~, **межремонтный** period between repairs
~, **на неопределенный** *law* without day, sine die *(Lat.)*
~ **начисления полного износа** write-off period
~ **ожидания** *econ.* demurrage
~ **окупаемости** payback time, time of recovery/outlay
~, **оплатить в** meet the due date
~ **оплаты** due date
~ **опротестования** time for protesting
~ **отгрузки** date of shipment
~, **плановый** schedule date
~ **платежа** date due for payment, payment date
~ **платежа по векселю** time of maturity
~ **погашения ссуды** maturity of loan
~ **подписки** subscription period
~ **полезной службы** useful life
~ **полномочий** term of office
~, **последний** closing date
~ **поставки** time/date of delivery
~, **предельный** 1. limitation; 2. deadline
~ **приобретательной давности** time of prescription
~ **сдачи** time of delivery
~ **службы** operational/service life, operating/running/working/lifetime, lifespan, lifetime, durability; retirement time
~ **службы, ожидаемый/расчетный** life expectancy
~ **службы, эксплуатационный** operating

service life
~ **хранения** storage/shelf life
~ **хранения файлов** *comp.* file expiration date
~ **ценной бумаги** time of maturity
срост adhesion; *med.* symphysis
сростноголовчатый *bot.* syncephalic
сростнолепестный *bot.* sympetalous, gamopetalous
сростнолистный gamophyllous, symphyllous
сростноплодный *bot.* syncarpous, gamocarpous, symphycarpous
сростночелюстной *zool.* plectognathic, plectognathous
сросток 1. concretion, aggregation, coalescence, growth; 2. splice; joint
сруб felling
сруливать turn off
срыв 1. breakdown, collapse, frustration; stall; disruption, derangement; 2. stripping, tearing off; 3. *aero.* stall
~, **волновой** shock stall
~ **забастовки** strikebreaking
~, **концевой** tip stall
~ **нейтрона/протона** neutron/proton stripping
~ **пламени** flame-out
~ **самолета** aeroplane stall(ing)
срывать 1. disturb, upset, frustrate; 2. strip, tear down/off
ссадина excoriation, abrasion; raw sore, scratch
ссуда loan; lending; allowance; subsidy
~, **банковская** bank loan/accommodation
~, **безвозвратная** nonrepayable subsidy
~, **беспроцентная** loan without interest
~ **до востребования** money/loan at call, call/demand loan
~ **долгосрочная** long-term loan
~, **краткосрочная** short-term loan, money at short call
~, **натуральная** cash loan
~, **однодневная** overnight/morning/daily loan
~, **онкольная** loan at call, call/demand loan
~ **погашаемая по требованию** loan at notice
~ **под закладную** mortgage loan
~ **под недвижимость** real estate loan
~ **под обеспечение** secured loan/advance
~, **потребительская** consumer/consumption loan
~, **просроченная** bad loan
~, **среднесрочная** medium-term loan
~, **срочная** term loan
~, **суточная** day-to-day money, overnight money
ссудодатель lender, loaner, bailor
ссудополучатель loan recipient
ссудо-сберегательный loan-and-savings *(attr.)*
ссужать lend, loan
ссылаться allege, plead; allude; refer
~ **на незнание** *law* plead ignorance
ссылка reference; quotation, citation
~ **на закон** *law* pleading a statute, count upon a statute
~, **перекрестная** cross reference
ссыльнокаторжный hard labour camp inmate
ссыльнополитический political deportee
ссыльный exile, deportee; convict
ссыпать discharge, pour; spill
ссыпной пункт grain-collecting centre/station
ссыхаться dry up, shrink, shrivel
стабилизатор 1. stabilizer; regulator, balancer; 2. *chem.* inhibitor; 3. *aero.* tail, fin, (balancing) plane
~, **гиросиловой** flywheel
~, **крыльевой** fin stabilizer
~, **курсовой** directional/heading stabilizer
~ **напряжения** voltage regulator/stabilizer
~, **неуправляемый** *aero.* fixed-incidence tailplane
~ **ракеты** rocket fin
~ **самолета** aeroplane tailplane
~ **тока** *el.* current stabilizer/regulator
~, **управляемый** variable-incidence tailplane
~, **хвостовой** tailplane, stabilizing tail
~ **частоты** frequency regulator/stabilizer, frequency control circuit
~, **щелевой** slotted tailplane
стабилизация stabilization; regulation
~ **вращением** spin stabilization
~ **курса/курсовая** heading hold/lock
~ **по тангажу [при килевой качке]** pitch stabilization
~, **поперечная [при бортовой качке]** roll stabilization
~, **пространственная** *naut.* attitude stabilization
~ **рыскания** yaw stabilization
~ **скорости** speed hold
~ **тангажа** pitch stabilization
~ **угла тангажа** pitch attitude hold
~, **угловая** attitude stabilization
~ **усиления** gain stabilization
стабилитрон *el.* Stabilitron, stabilizer diode, voltage stabilizing tube
ставка rate; *game th.* stake
~, **базисная** basic/prime rate
~ **дисконта** discount rate
~, **единообразная** flat rate
~ **местного налога** local rate
~ **накладных расходов** cost rate
~, **отступная** fallback rate
~, **переучетная** rediscount rate
~, **повременная** time rate
~, **поденная** per diem rate
~ **процента** interest rate
~, **сдельная** piece rate
~ **ссудного процента** lending rate
~ **страховой премии** rate of insurance
~ **таможенной пошлины** rate of duty
~, **учетная** discount rate
~, **фрахтовая** shipping rate
ставленная грамота *eccl.* certificate of ordination

ставрида *zool.* 1. scad *(Trachurus);* 2. jack mackerel *(Trachurus symmetricus)*
ставридовые *n. pl. zool.* carangids *(Carangidae)*
ставрогамия staurogamy, cross-fertilization/pollination
стагачка *bot.* adder's-mouth orchard *(Malaxis)*
стагнация stagnation
~ при одновременной инфляции stagflation
стагноспороз *bot.* purple brown blotch, Stagnospora leaf spot
стадий *hist.* stadium *(unit of length)*
стадийность occurrence in stages; stage structure
стадийный stage *(attr.);* phasic; in stages
стадия stage, step, phase
~, зачаточная embryonic stage
~ испытаний testing stage
~ проектирования blueprint/design stage
~ разработки development phase
стадо herd, drove
~, молочное dairy herd
~, товарное commercial herd
стаж length of service
~, испытательный probationary period, probation
стажер probationer; trainee
стажирование 1. probation service; 2. on-the-job training
стажировать work on probation
стаз *med.* stasis
~, желчный cholestatis, cholestasia
~, каловый coprostasia, constipation
~, кровяной h(a)emostasis, h(a)emostasia
стакан cup, glass, beaker; tumbler; socket, sleeve, bushing; housing, box
~ подшипника bearing body
~, химический beaker
стаксель *naut.* staysail
сталагмит stalagmite
сталактит stalactite
сталеалюминиевый steel-aluminium *(attr.)*
сталеалюминий steel aluminium
сталебетонный steel-concrete *(attr.)*
сталевар steelmaker, steel founder
сталелитейный steel-casting *(attr.)*
сталеплавильный steel-smelting *(attr.)*
~ная печь steel furnace
сталепрокатный steel-rolling *(attr.)*
сталийное время lay days
сталинец follower of Stalin
сталинский Stalin's, of Stalin
сталия lay days
сталкиваться collide, come into collision; impinge
сталь steel
~, азотированная nitrided steel
~, алмазная extra-hard steel
~, арматурная reinforcing bars
~, аустенитовая austenitic steel
~, бессемеровская Bessemer steel
~, бористая/борная boron steel
~, булатная Damascus steel, damascene
~, бульбовая bulb steel
~, быстрорежущая high-speed steel
~, ванадиева vanadium steel
~, вольфрамовая tungsten steel
~, вороненая blue steel
~, высоколегированная high-alloy steel
~ высокопрочная high-strength/tensile steel
~, высокоуглеродистая high-carbon steel
~, дамасская Damascus steel, damascene
~, динамная dynamo steel
~, доэвтектоидная hypoeutectoid steel
~, жаропрочная/жаростойкая/жароупорная high-temperature steel
~, закаленная hardened steel
~, заклепочная rivet steel
~, заэвтектоидная hypereutectoid steel
~, инструментальная tool steel
~, кислотостойкая acid-resistant steel
~, кислотоупорная acid-proof steel
~, кобальтовая cobalt steel
~, ковкая forging-grade steel
~, конструкционная structural steel
~, корпусная *naut.* hull plate
~, коррозионностойкая corrosion-resistant steel
~, кремнистая silicon steel
~, круглая rounds
~, легированная alloy(ed) steel
~, листовая sheet steel
~, литая cast steel
~, малоуглеродистая low-carbon steel
~, марганцевая manganese steel
~, мартеновская open-hearth steel
~, мартенситная martensitic steel
~, молибденовая molybdenum steel
~, мягкая mild/soft steel
~, нелегированная plain steel
~, немарочная off-grade steel
~, нержавеющая stainless steel, corrosion-resisting steel
~, низколегированная low-alloy steel
~, низкоуглеродистая low-carbon steel
~, обручная hoop iron
~, отпущенная tempered steel
~, оцинкованная galvanized steel
~, перлитная pearlitic steel
~, плакированная clad steel
~, полособульбовая flat-bulb steel
~, полосовая strip/flat steel
~, полуспокойная semi-killed steel
~, профильная steel shapes; section steel
~, прутковая rod/bar steel; rods
~, самозакаливающаяся air-hardening steel
~, сварочная weld steel
~, сверхвысокопрочная ultra-high-strength steel
~, спокойная killed steel
~, среднелегированная medium-alloy steel
~, тавробульбовая tee-bulb steel
~, текстурованная grain-oriented steel

~, **толстолистовая** plate steel; steel plates
~, **тонколистовая** sheet steel; steel sheet(s)
~, **топочная** fire-box steel
~, **углеродистая** carbon steel
~, **углобульбовая** bulb-angle steel
~, **угловая** angle steel
~, **фасонная** structural shapes; shaped/sectional steel
~, **ферритная** ferritic steel
~, **хромистая** chromium steel
~, **хромованадиевая** chrome-vanadium steel
~, **хромомарганцевокремнистая** chrome-manganese-silicium steel
~, **хромоникелевая** chrome-nickel steel
~, **цементированная** cemented (carburized, case-hardened) steel
~, **швеллерная** channel steel, shannels
~, **штамповая** die steel
сталь-армко Armco steel, ingot/Armco iron
стальбетон steel concrete
стальник *bot.* restharrow *(Ononis)*
стальной steel *(attr.);* steely
~ **канат** steel-wire rope
~**ная камера** safe
стамед, стамет stamin
стамеска chisel
~, **долбежная** carving chisel
~, **желобчатая** fluted chisel
~, **ригельная** mortise chisel
стаминодий *bot.* staminode, staminodium
стан mill
~, **балочный** beam mill
~, **бандажный** tyre/tire mill
~ **горячей прокатки** hot-rolling mill
~, **дрессировочный** skin-rolling mill
~, **заготовочный** billet mill
~, **калибровочный** sizing mill
~, **листопрокатный** sheet mill
~, **мелкосортный** small-section mill
~, **непрерывный** continuous rolling mill
~, **обжимный** blooming/cogging mill
~, **петлевой** looping mill
~, **полосовой** strip mill
~, **прокатный** rolling mill
~, **профилировочный** shaping mill
~, **прутковый** wire-rod mill
~, **сортовой** section mill
~, **ткацкий** loom
~, **толстолистовой** plate mill
~, **тонколистовой** sheet mill
~, **трубный** tubular mill
~, **трубопрокатный** tube/pipe mill
стангерия *bot.* fern cycad *(Stangeria)*
стандарт standard
~ **безопасности** safety standard
~, **бумажный** *fin.* paper basis
~, **временный** tentative/temporary standard
~, **государственный общесоюзный** All-Union State Standard
~, **золотовалютный** gold-exchange standard
~, **золотомонетный** gold-coin standard
~, **золотослитковый** gold-bullion standard
~, **необязательный** permissive standard
~, **общесоюзный** All-Union standard
~, **обязательный** mandatory standard
~, **отраслевой** branch standard
~ **предприятия** factory standard
~, **производственный** shop practice standard
~, **технический** engineering standard
~, **узаконенный** statutory standard
стандартизация standardization; calibration
стандерс *naut.* stand, standard
станина bed plate/frame, mounting, stand, bench
станиоль tin foil
станкостроение machine-tool industry/manufacture
станкостроительный machine-tool *(attr.)*
станнат *chem.* stannate
станнит *chem.* stannite
станнометрия stannometric survey
станок 1. machine, machine tool; bench; rig; lathe; 2. mounting
~, **болтонарезный** bolt-threading machine
~, **буровой** drilling rig
~, **бурозаправочный** jackmill
~, **винтонарезный/винторезный** thread-/screw-cutting machine
~, **волочильный** drawing bench/machine
~, **гайконарезный** nut-cutting machine
~, **гибочный** bending machine
~, **давилный** spinning lathe
~, **долбежный** *met.* slotting machine; *wood.* mortising machine
~, **дыропробивной** punching machine
~, **закаточный** beading nachine
~, **заточный** sharpening [tool-grinding] machine
~, **затыловочный** relieving machine
~, **зенковальный** reaming machine
~, **зубодолбежный** gear shaper
~, **зуборезный** gear-cutting machine
~, **зубострогальный** gear planer
~, **зубофрезерный** gear-hobbing machine
~, **зубошлифовальный** gear-grinding machine
~, **карусельный** vertical lathe
~, **клепальный** rivetting machine/press
~, **консольный сверлильный** bracket drilling machine
~, **копировально-фрезерный** copying/profile milling machine
~, **копировальный** copying/profiling/duplicating machine
~, **корректурный** galley press
~, **кромкозагибочный** flanging/edging machine
~, **кромкообрубочный** trimmer
~, **кромкострогальный** edge-planing machine
~, **круглопильный** circular saw
~, **ленточно-шлифовальный** band-grinding machine
~, **листоправильный** plate/straightening/levelling machine, plate straightener

~, металлообрабатывающий (metal-removal) machine tool
~, металлорежущий (metal-cutting) machine tool
~, металлоштамповочный metal forming machine
~, многошпиндельный сверлильный multiple drilling machine
~, мультипликационный animation stand
~, обрезный edging/trimmimg machine
~, одноколонный продольно-строгальный open-side planer
~, одночелночный ткацкий one-shuttle loom
~, пазовальный groove-cutting machine
~, патронный токарный chuck lathe
~, поперечно-строгальный shaping machine, shaper
~, портальный сверлильный gantry-drilling machine
~, правильный straightening/levelling machine, straightener
~, притирочный lapping machine
~, продольно-режущий slitting machine, slitter
~, продольно-строгальный planing machine, planer
~, протяжной/прошивочный broaching machine
~, прядильный spinning loom
~, расточный boring machine, borer
~, рашкетный *print.* proof press
~, револьверный turret lathe
~, резьбонарезный thread-/screw-cutting machine, threading machine
~, резьбофрезерный thread-milling machine
~, рифтовочный riffling machine
~, рихтовальный straightener; levelling machine
~ с программным управлением program-/computer-controlled machine
~ с ручным управлением manual machine tool
~ с цифровым управлением numerically controlled machine
~, сверлильный drill, drilling machine
~, строгальный planer, planing machine
~, ткацкий *text.* loom
~, токарно-винторезный screw-cutting lathe
~, токарно-лобовой faceplate/surfacing lathe
~, токарно-револьверный turret lathe
~, токарный lathe
~, торцовочный cross-cut [cut-off, butting] saw
~, точильный sharpening/grinding machine
~, трубоволочильный tube-drawing bench
~, трубонарезной pipe-threading machine
~, фасонный токарный shaping/forming lathe
~, фланцезагибочный flanging machine
~, фрезерный milling machine
~, фуговальный jointing machine
~, центровальный centering machine
~, центровой токарный centre lathe
~, цилиндрорасточный cylinder-boring machine
~, цилиндрошлифовальный cylinder-grinding machine
~, шипорезный dovetailing machine; tenonmaking machine
~, шлифовально-притирочный [хонинговальный] honing machine, honer
~, шлифовальный grinding machine, grinder
~, шпоночно-долбежный keyway-slotting machine
~, шпоночно-фрезерный keyway-milling machine
~, шпунтовочный groove-cutting machine

станок-автомат automatic machine/lathe
станок-качалка pumping unit
станок-тренога tripod mount
станочник (machine) operator
станс *lit.* stanza
станция station; post; *tech.* plant
~, атомная atomic/nuclear power station/plant
~, атомная паросиловая atomic steam power station/plant
~, атомная энергетическая/энергосиловая atomic power station/plant
~, бензозаправочная petrol station; *Am.* filling station
~, бортовая vehicle-borne station; *naut.* ship-borne station; *aero.* airborne station
~, буйковая *naut.* buoy station
~, бункерная fuelling station
~, ведомая slave station
~, ведущая master station
~, вертолетная helicopter station, heliport
~, водоочистительная waste water reclamation plant; waterworks
~, гелиоэлектрическая solar power station
~, гидроакустическая asdic (set); *(Am.)* sonar (system)
~, гидролокационная echo-ranging system, active sonar
~, гидроэлектрическая hydroelectric power station/plant
~ железнодорожная railway/railroad station
~, железнодорожная товарная goods station, freight terminal
~, железнодорожная тупиковая stub/head station
~, железнодорожная узловая junction/multiple station
~, каротажная (well-)logging set/unit
~, контрольная monitoring station
~, космическая space station/probe
~, обитаемая space manned station
~ наведения vectoring/guidance station
~, паросиловая steam power plant
~ питания power source unit
~ помех/радиопомех jamming station, jammer
~, радиовещательная radio broadcast station

~, радиолокационная radar installation/unit
~, радиометрическая radiometry station
~, ралиопеленгаторная radio direction-finding station
~, радиотелеметрическая radio-telemetering station
~, рейдовая *naut.* inshore station
~, селекционная plant-breeding station
~, силовая power station/plant
~ скорой и неотложной помощи first-aid station
~ слежения tracking station
~, сортировочная *rail.* shunting yard, switchyard
~, спасательная *naut.* lifeboat [life-saving] station
~, телеграфная telegraph office/station
~, телеметрическая telemetry station
~, телефонная telephone exchange
~, телефонная автоматическая automatic/dial telephone exchange
~, телефонная ручного обслуживания manually operated exchange
~, тепловая thermal power plant
~, товарная freight terminal
~, широковещательная broadcasting station
~, шумопеленгаторная passive sonar
~, электрическая electric power station
стапелия *bot.* carrion flower *(Stapelia)*
стапель *naut.* shipbuilding berth/ways, slipway; stocks, building slips
~, балансирный rocking ways
стапель-блок building-block, bilge log
стапель-палуба pontoon deck, dock floor
стапедопластика *med.* stapedioplasty
стапедотенотомия *med.* stapediotenotomy
стапедэктомия *ned.* stapedectomy
старение ageing; senescence
~, демографическое ageing of population
~, деформационное strain ageing/hardening
~, механическое [от наклепа] strain ageing
~, преждевременное presenilation
~ радиоактивных отходов storage for decay of radioactive waste
~, световое light ageing
старн-кница *naut.* stern(post) knee
старн-пост *naut.* sternpost, propeller/screw/stern/inner post
старн-тимберс *naut.* stern timber
старовер, староверец *rel.* Old Believer
стародубка *bot.* Siberian adonis *(Adonis sibiricus)*
старообрядец *rel.* Old Ritualist/Believer
старопашка *agr.* old-arable land
старопечатный early printed, old-printed
старославянский *ling.* Old Church Slavonic
староста elder; head; leader
~, церковный churchwarden
~, цеховой shop steward/deputy
старпом *abbr., naut.* chief mate
старт start; *aero.* launch, take-off, blast-off
стартер starter (unit)
~, пороховой cartridge starter
~, ручной hand crank starter
стартер-генератор dynamo starter, starter-generator
~, аэродромный ground starter
~, воздушный compressed-air starter
~, пиротехнический/пиропатронный/пороховой cartridge starter
стартовый starting; *aero.* launching, take-off
старческий senile; gerontal, geriatric; decrepit
старшина *mil.* master/first sergeant; *naut.* petty officer, mate
~, дежурный duty petty officer
~ шлюпки *naut.* coxswain
старшина-гидроакустик *naut.* asdic rating
старшина-рулевой *naut.* quartermaster
старшина-сигнальщик signal rating
старшинство seniority; precedence
статив frame, rack, stand
статика statics
~ жидкостей hydrostatics
~ жидкостей и газов statics of fluids
статист *theatre* mute, extra, super
статистик statistician
статистика statistics
~ банковских операций banking statistics
~, биологическая biostatiscics
~ больших выборок macrostatistics
~, вариационная analysis of variance
~, внутризаводская plant statistics
~, вспомогательная ancillary statistics
~ движения населения vital/registration statistics
~, демографическая vital statistics
~ занятости employment figures
~ здравоохранения health statistics
~, коммерческая business/commercial statistics
~ крайних *math.* statistics of extremes
~ малых выборок microstatistics
~, математическая statistical analysis
~, многомерная multivariate statistics
~, непараметрическая distribution-free statistics
~, порядковая order statistics
~ профессиональной занятости occupation statistics
~ рождаемости birth statistics
~ хи-квадрат chi-square statistics
~, экономическая statistical economics
~, экстремальная порядковая extreme order statistics
статокония statoconium, statolith
статор *el.* stator
~, регулируемый variable stator
~ сельсина synchro stator
статоскоп altitude control, statoscope
статоспора *bot.* resting spore, statospore
статус *law* status
~ государства statehood

статус-кво *law, Lat.* status quo
статут statute, articles; ordinance
статутный statute *(attr.);* statutable
~ная миля statute mile
~ное право statute law
статья 1. item, entry; clause, paragraph; 2. article, paper
~ баланса balance sheet account
~ бюджета/бюджетная budget item
~, дополнительная rider
~, заказная invited paper
~ описания файла file description entry
ставрогамия *bot.* cross-fertilization, staurogamy
стафилит *med.* staphylitis
стафилоангина staphyloangina, staphylococcic pharyngitis
стафилококк staphylococcus (*pl.* staphylococci)
~, белый Staphylococcus albus
~, золотистый Staphylococcus aureus
стафилококковый staphylococcic, staphylococcal
стафилококкция *ned.* staphylococcia
стафилолизин staphylolysin, staphylococcolysin
стафилома staphyloma
~, кольцевидная annular staphyloma
~ роговой оболочки corneal staphyloma, keratoconus, keratoglobus
стафиломный staphylomatous
стафилоптоз staphyloptosis, staphyloptosia
стафилорафия staphylorrhaphy
стафилотомия staphylotomy, uvulatomy
стафилоэктомия staphylectomy
стахановец, стахановка, стахановский *hist.* stakhanovite
стахиоспермный, стахиоспермовый stachyospermous
стахиоспория stachiospory
стахиоспоровый stachyosporous
стационар 1. hospital; 2. permanent establishment
стационарность stationarity, stationariness
стационарный 1. stationary, fixed, static, steady, steady-state; permament; 2. hospital
~ больной hospital patient, inpatient
~ное лечение hospital treatment, hospitalization
стачка *econ.* strike
~, дикая wildcat strike
~, итальянская/польская stay-in [stay-down] strike
ствол 1. trunk; stem; 2. barrel *(of weapon);* nozzle; monitor; 3. shaft; 4. *rad.* basic broadband channel
~, блуждающий *anat.* vagal trunk
~ нервный nerve trunk
~, пенный foam monitor
~, плечеголовной brachicephalic trunk, innominate artery
~, реберно-шейный costocervical trunk
~, яремный *anat.* jugular trunk
створ alignment, range; line
створаживать coagulate
створка 1. *anat.* cusp; 2. door, flap, gate, shutter; fold, leaf; 3. *bot.* glume
~, брюшная *zool.* ventral valve
~ радиатора radiator flap
~ сердечного клапана septal cusp
~ шлюзных ворот sluice gate leaf
створчатый 1. folding; leaved; 2. cuspidate(d); 3. valved, valvate
~тые двери folding door
стеапсин steapsin
стеапсиноген steapsinogen
стеарин stearin(e)
стеариновый stearic; stearine *(attr.)*
стеародермия *med.* stearodermia
стеародин stearodine
стеароптен *chem.* stearoptene
стеатогранулема *med .* steatogranuloma
стеатитовый steatitic
стеатоз *med.* steatosis
стеатолиз steatolysis
стеатолитический steatolytic
стеатоматоз steatomatosis
стеатонекроз steatonecrosis
стеатопатия *med.* steatopathy
стеатопигия steatopyga, steatopygia
стеаторея steatorrh(o)ea
стеатоцеле steatocele
стебелек stalklet; petiole; pedicle, peduncle
стебель stalk, stem; shank; handle; *zool.* notochord
стебельный stalk *(attr.)*; *bot.* caulescent
стебельчатоглазые *pl. zool.* stylommatophorans *(Stylommatophora)*
стебельчатоглазый *zool.* stalk-eyed, podophthalmous
стебельчатый petiolate(d), pedunculate, pedicillate; cauline; stalky, stalked; stemlike
стеблевание *bot.* bolting, shooting
стеблевидный stem-shaped, stemlike, stalklike, cauliform
стеблевый stem *(attr.)*; pedicellate; cauline
стеблекорневой *bot.* caulor(r)hizous
стеблелист *bot.* cohosh *(Caulophyllum)*
стеблеобъемлющий *bot.* amplexicauline, stem--clasping
~ лист clasping leaf
стеблеплод *bot.* caulocarp, stem crop
стеблеплодный *bot.* caulocarpous, caulocarpic
стеблеруб *agr.* stalk cutter
стеблестой *agr.* plant stand
стеблеизмельчитель stalk cutter
стеблеотвод stalk ejector
стегать stitch; quilt
стегозавр *paleont.* stegosaurus
стегокарпный *bot.* stegocarpous
стегоцефалы *paleont.* stegocephalians *(Stegocephalia)*
стежок stitch
стек stack
~ магазинного типа *comp.* push-down stack
стеклить 1. glaze; 2. *leath.* polish, burnish
стекло glass; window panel; *fig.* glassware
~, армированное 1. wire glass; 2. armoured

glass
~, **безосколочное** shatterproof [safety, nonsplintering] glass
~, **борное** borax glass
~, **броневое** bulletproof glass
~, **ветровое** windshield, windscreen
~, **витражное** stained glass
~, **витринное** plate glass
~, **водомерное/водоуказательное** water-level gauge glass
~, **жидкое** water glass, sodium silicate solution
~, **защитное** shielding/protective glass
~, **зеркальное** plate glass
~, **известковое** lime glass
~, **калийное** potassium glass
~, **кварцевое** quartz glass
~, **кремнийорганическое** silicone glass
~, **ламповое [керосиновой лампы]** lamp chimney
~, **листовое** sheet glass
~, **лобовое** windscreen, windshield
~, **мерное** gauge glass
~, **многослойное** laminated glass
~, **молочное** opal glass
~, **натриевое** soda [sodium silicate] glass
~, **небьющееся** shaterproof/safety glass
~, **неодимовое** neodymium(-doped) glass
~, **оконное** window glass/pane
~, **окрашенное** stained glass
~, **органическое** acrylic plastic
~, **очковое** ocular glass/lens, spectacle lens
~, **перлитовое** perlitic glass
~, **полупрозрачное** translucent glass
~, **покровное** cover glass
~, **превращается в** vitrify
~, **предметное** slide
~, **пуленепробиваемое/пулестойкое** bulletproof glass
~, **растворимое** soluble glass, silicate of potassium
~, **рифленое** rippled/ribbed/corrugated glass
~, **свинцовое** lead glass
~, **силикатное** soda-lime glass
~, **смотровое** sight/port glass
~, **строительное** structural glass
~, **сульфатное** sulphate of soda glass
~, **тарное** container glass
~, **техническое** industrial glass
~, **тугоплавкое** refractory glass
~, **увеличительное** magnifying glass
~, **увиоловое** uviol glass
~, **фасонное** moulded glass
~ **фонаря** canopy glass
~, **часовое** watch glass/crystal
~, **электровакуумное** electron-tube glass
~, **ячеистое** cellular glass
стеклобетон glass concrete
стеклоблок glass sandwich/block
стеклобой broken glass, cullet
стеклование vitrification
стекловар glassmaker
стекловарение glass manufacture/production
стекловата glass wool
стекловать(ся) vitrify
стекловатый glassy, hyaline, vitreous
стекловидный glasslike, vitreous, vitriform, glassy; *anat.* hyaline, hyaloid
стокловолокнистый glass-fibre/fiber
стекловолокнит glass-fibre/fiber laminate
стекловолокно glass bibre/fiber
стекловыдувальщик, стеклодув glass blower
стеклограф 1. glass-marking pencil; 2. collotype press
стеклографировать collotype, print on collotype press
стеклография collotype, collotypy
стеклокерамика glass ceramics
стеклолакоткань varnished glass cloth
стеклолинолеум glass-fabric-based linoleum
стекломасса molten mass, glass mass
стекломат glass mat
стеклонаполнитель glass filler
стеклообогреватель defroster, deicer, windshield heater
стеклонить fibreglass thread
стеклоомыватель windscreen washer
стеклоочиститель windscreen/windshield wiper
~, **вращающийся** clear-vision/view screen wiper
стеклоплав glassmaker
стеклоплавильный glassmaking *(attr.)*
~ **завод** glassworks, glass foundry
стеклопласт fibreglass (glass-fibre) material
стеклопластик glass-fibre-reinforced plastic
стеклорез 1. glazier's diamond; 2. glass cutter
стеклорогожка glass-fibre mat
стеклорубероид fibreglass roofing material
стеклотара glass containers
стеклотекстолит glass-fabric-base laminate
стеклоткань glass fabric/cloth
стеклофанера glass-fibre veneer
стеклофирный vitrophyric
стеклохолст glass-fibre mat
стеклочерепица glass-fibre tile
стеклошифер fibreglass roofing material
стеклошпон laminated glass-fibre sheet
стеклуемость vitrification
стеклуемый vitrifiable
стекляница *zool.* clearwing moth *(Aegeriidae)*
стеклянный glass *(attr.),* vitreous
стекольный glass *(attr.),* vitreous
~ **бой** cullet
~ **завод** glassworks, glass factory
стекольщик glazier
стеллаж (storage) rack, stack, shelving, stand, shelves, stilling, stillage
стеллаж-холодильник *met.* cooling bed
стелларит stellarite, stellar coal
стеллария *bot.* starwort *(Stellaria)*
стеллатэктомия *med.* stellectomy
стеллит *met.* Stellite

стель *bot.* stele
стелька insole, inner sole; welt
~, ортопедическая corrective insole
стелющийся *bot.* decumbent, prostrate, creeping, trailing
стем *naut.* stem piece
стемсон *naut.* stemson
стемфилиоз *bot.* Stemphylium leaf spot
стен *phys.* sthene
стена wall
~, глухая dead/blank wall
~, заполняющая filler/panel wall
~, защитная shielding wall
~, капитальная main wall
~, монолитная cast-in-place/situ wall
~, несущая load-bearing wall
~, подпорная retaining wall
~, разгрузочная relieving wall
~ ручной кладки masonry wall
~, самонесущая self-bearing wall
~, сборная prefabricated wall
~, фундаментная foot/foundation wall
~, шпунтовая sheet-pile [piling] wall
стенгазета *abbr.* wall newspaper, bulletin-board newspaper
стенд stand; bench; (exhibition) stall
~, вибрационный shaker device/unit
~, вращающийся whirling stand
~, выставочный display stand/stall
~, динамометрический dynamometer test cell, thrust-measuring stand
~, испытательный test bench/stand/bed/jig
~, летающий пилотажный in-flight simulator
~, нивелировочный alignment jig/fixture
~, разливочный *met.* casting stand
~, сборочный assembly stand/jig
стенка wall; web; *naut.* dockside, quay
~ балки web of a beam, beam web
~, водобойная baffle plate
~, волноотбойная seawall
~, гимнастическая wall bars
~, достроечная *naut.* outfitting quay
~, клеточная cell wall
~, ложная *zool.* pseudotheca
~ лонжерона spar web
~, набережная sea/wharf wall
~, причальная moorage wall, quay
~, сосудистая vascular wall
~, шпунтовая sheet-pile wall
стенкоотрывной *bot.* septifragal
стенкоположный *bot.* parietal
стенкораздельный *bot.* septicidal
стенник *bot.* candytuft *(Iberis)*
стенница *bot.* wall pellitory *(Parietaria officinalis)*
стенной wall *(attr.)*; mural; *biol.* parietal
стенобатный *zool.* stenobathic
стенобитный, стенобойный *hist.* battering
~ таран battering-ram
стенограмма shorthand report
стенограф stenographer
стенографировать write in shorthand
стенография shorthand, stenography
стеноз *med.* stenosis
~ гортани laryngostenosis
~ желудка gastrostenosis
~ привратника pyloric stenosis, pylorostenosis
~ слезных протоков stenochoria
~, субаортальный subaortic stenosis
стенозированный *med.* stenosed
стенозный stenosis *(attr.)*, stenosal
стенокардия *med.* stenocardia, angina pectoris, constriction of the heart
стенокарпус *bot.* fire-wheel tree *(Stenocarpus)*
стенолаз *zool.* wall creeper *(Trichodroma muraria)*
стеноп stenopeic camera
стенописец muralist
стенопись mural (painting)
стеносинклиналь stenosyncline
стеноспермный stenospermous
стенотермность stenothermy
стенотермный, стенотермический stenothermic, stenothermous, stenothermal
стенотипия stenotypy
стенофагия stenophagy
стенофилльный *bot.* stenophyllous, angustifoliate
стенохория stenochory
стеньга *naut.* topmast
стень-шкентель *naut.* burton pendant
степенный *math.* exponential; power *(attr.)*
степень 1. degree, grade, extent, level; rate, ratio; modulus; 2. *math.* power; exponent; index
~ влияния фактора *stat.* level of factor
~ достоверности confidence figure
~ дробления fineness of dispersion
~, дробная *math.* fractional power
~ живучести degree of survivability, survival probability
~ зависимости *stat.* degree of dependence
~ запрещенности/запрета degree/strength of forbiddenness
~ износа rate of wear
~ комплексирования degree of integrity
~ корня index of a radical
~ кровного родства degree of consanguinity
~ многочлена *math.* degree of a polynomial
~ надежности degree of reliability
~ наполнения degree of admission
~ обеспеченности margin of safety
~ обесценивания rate of depreciation
~ предпочтения *game th.* strength of preference
~ приближения degree of approximation
~ распада rate of decay
~ расширения expansion ratio, degree of expansion
~ редукции *mech.* reduction gear ratio
~ резервирования redundancy level
~ свободы range/extent of discretion; *mech.* degree of freedom
~ свободы, однонуклонная single-nucleon degree of freedom

~ **связанности** *stat.* degree of association
~ **сжатия** compression/pressure ratio
~ **скашивания** rate of wear and tear
~ **случайности** degee of randomness
~ **уравнения** degree/order of an equation
~ **устаревания** obsolescence rate
~ **целостности** *math.* degree of wholeness
~ **числа** power of a number
степс *naut.* mast step/stool; poppet
стер stere, cubic meter
стерад, стерадиан *math.* steradian, sterad
стервятник *zool.* carrion vulture/buzzard *(Neophron)*
стереида *bot.* stereid, grit/stone cell
стереоадаптер stereocontrol unit
стереобластула stereoblastula, morula
стереогноз stereognosis
стереограмма axonometric chart
стереограмметрия stereogrammetry
стереодальномер stereoscopic range finder
стереозвук stereophonic/binaural sound
стереоизомер *chem.* stereoisomer, stereomer
стереоизомерия stereoisomerism
стереокартирование stereomapping
стереокартограф stereoplanigraph, stereocartograph
стереокинез stereokinesis
стереокино stereoscopic motion-picture theatre
стереом *bot.* stereome, mechanical tissue
стереометр paralax box, stereometer
стереометрия solid geometry, stereometry
стереомный stereomatic
стереомодель stereoscopic model, three-dimensional presentation, stereomodel
стереопара stereopair, stereoscopic/double image, stereogram(me)
стереоплазма *biol.* stereoplasm
стереопланиграф stereoplanigraph, stereoscopic plotting instrument
стереопроектор stereoprojector
стереорентгенограмма stereoscopic X-ray picture, stereogram(me)
стереорентгенография stereoradiography, stereoroentgenography, stereoskiagraphy
стереорентгеноскопия stereoscopic X-ray examination
стереорисовка stereo compilation
стереоскоп stereoscope
~, **зеркальный** mirror stereoscope
стереоскопичность stereoscopism; stereoscopy
стереоснимок, стереофотография stereoscopic photograph, stereogram
стереоспондиловый *zool.* stereospondylous
стереотаксис stereotaxis, stereotaxy
стереотаксический stereotaxic, stereotactic
стереотип stereotype; *print.* master form
стереотипирование stereotyping
стереотипировать stereotype
стереотипия stereotyping; stereotypy; stereotype
стереотипный stereotype *(attr.)*
стереотруба stereoscopic telescope
стереоупорядоченный stereospecific
стереофильм stereoscopic [three-dimensional] film
стереофонический stereophonic, binaural
стереофоничность stereophonism
стереофония stereophonic effect, stereophonism, stereophonics
стереофотограмметрия stereophotogrammetry
стереохимия stereochemistry
стереохромия stereochromy
стереоэффект stereoscopic effect
стерженщик *met.* core-maker
стержень 1. bar, rod, stem, shank, bolt, spindle, pin, shaft; 2. *constr.* column; 3. *met.* core; 4. *med.* modiolus
~, **аварийный [аварийной защиты]** safety/security rod, emergency shut-down rod
~ **автоматического регулирования** *nucl.* automatic control rod
~ **активной зоны** *nucl.* core rod
~, **арматурный** reinforcing rod
~ **болта** shank/body of bolt
~, **графитовый** electrographic pen, conductive pencil
~, **дистанционирующий** spacer rod
~ **заклепки** rivet shank
~ **иглы** needle stem/shank
~ **клапана** valve spindle/stem
~, **компенсирующнй [компенсационно-аварийный]** *nucl.* shim-safety rod
~, **компенсирующий/корректировочный** *nucl.* shim rod
~, **контактный** contact pin
~, **литейный** mould core
~, **литерный** type/print bar
~, **магнитный** *el.* magnetic core
~ **мушки** front-sight post
~, **направляющий** guide pin/rod
~, **нивелировочный** sight/sighting rod
~ **поплавка** float spindle
~ **початка кукурузы** corncob
~, **приводной** drive rod
~, **распорный** stay rod
~ **сильно поглощающий нейтроны** poison rod
~, **сплошной** solid bar
~, **тонкий** 1. pin-type rod; 2. slender column
~, **топливный** fuel rod
~, **тяговый** drawbar, pull bar
~ **улитки** *med.* modiolus
~ **фурункула** core
стерилизатор sterilizer
~, **аптечный** pharmacy sterilizer
~, **сухожарный** dry-heat sterilizer
стержневидный rodlike, rod-shaped; *bot.* rachiform; scapiform
стерилизатор-укладка sterilizing case
стерилизация 1. sterilization; 2. castration
~, **дробная** fractional sterilization
~, **лучевая/облучением** (ir)radiation steriliza-

tion
~ токами высокой частоты high-frequency sterilization
~ электронами electron-beam sterilization
стерилизировать sterilize
стерильность sterility; barrenness; infertility
~, генная genic sterility
~, межвидовая interspecific sterility
~, мужская male sterility; pollen sterility
~, перекрестная intersterility, cross sterility
стерильный sterile
стерин *chem.* sterol
~, животный animal sterol, zoosterol
~, растительный phytosterol, sitosterol, plant sterol
стеркобилин stercobilin, urobilin
стеркобилиноген stercobilinogen, urobilinogen
стерколит stercolith, fecalith
стеркулиевые *pl. bot.* chocolate family *(Sterculiaceae)*
стеркулия *bot.* bottle tree *(Sterculia)*
стерлинг sterling
~ов, фунт pound sterling
стерлядь *zool.* sterlet *(Acipenser ruthenus)*
стернальгия sternalgia, sternodynia
стернит *zool.* sternite
стероид *chem.* steroid, sterid
стерх *zool.* Asiatic white crane *(Grus leucogeranus)*
стеснение, стесненность *econ.* stringency
стетоскоп *med.* stethoscope, stethophone
~, акушерский fetoscope
~, детский pediatric stethoscope
стетофонендоскоп phonendoscope
стефанандра *bot.* stephanandra *(Stephanandra)*
стехиология stoichiology
стехиометрический stoichiometric(al)
стехиометрия stoichiometry
стибин *chem.* stibine
стибнит stibnite, antimonite, grey antimony, antimony glance
стивидор stevedore
~, главный head stevedore
~, старший foreman stevedore
~, трюмный hold man
стигийский Stygian
стигма *med., zool.* stigma
стигмовидный *bot.* stigmatoid, stigma-shaped
стигмастерин stigmasterol
Стикс Styx
стилбестрол *pharm.* stilbestrol, diethylstilbestrol
стилет style, stilet, stilette, stiletto
~, ядовитый stinger
стилизатор stylist
стилизация stylization
стилизировать stylize
стилист stylist, master of literary style
стилистика stylistics
стилистический stylistic
стилобат *arch.* stylobate
стилолит *geol.* stylolite
стилометр Steelometer, metallurgical photometric spectroscope
стилоидит *med.* styloiditis
стиломицин *med.* stylomycin
стилоскоп Steeloscope, metallurgical arc spectroscope
стиль style; fashion, manner; genre
~, новый New Style, Gregorian calendar
~, старый Old Style, Julian calendar
стильб *opt.* stilb
стильбен *chem.* stilbene
стильбит *miner.* stilbite
стильный stylish; style *(attr.)*
~ная мебель period furniture
стимул stimulus; *econ.* incentive, inducement; drive
~, безусловный unconditioned stimulus
~, звуковой audio/auditory stimulus
~, зрительный/визуальный visual stimulus
~, материальный financial incentive
~, нейтральный/иидифферентный neutral stimulus
~, нематериальный nonfinacial incentive
~, подпороговый subthreshold/subliminal stimulus
~, пороговый threshold/liminal stimulus
~, реактогенный eliciting stimulus
~, световой light stimulus
~, сложный/многокомпонентный complex stimulus
~, условный conditioned stimulus
~, финансово-материальный financial incentive
стимулирование 1. stimulation; 2. encouragement, promotion; motivation; *econ.* incentive
~ спроса demand promotion
~ экономики stimulation/priming of the economy
стимулировать stimulate; encourage; promote
стимулятор stimulant; stimulator, stimulating agent
~ лейкопоэза leucopoietic stimulant
~ роста growth factor/stimulant
~, сердечный cardiostimulant
~ сердца pacemaker
~ сердца, вживляемый implantable pacemaker
~, тканевый tissue factor
стимуляция stimulation; drive
стипелла *bot.* stipel
стипендиант scholarship holder/student
стипендия scholarship, fellowship; grant; stipend
стипль-чез steeplechase
стиптицин cotarnine chloride
стираемость erasability
стиракс *bot.* snowbell, storax, styrax *(Styrax)*
стирание 1. erasing, erasure, deletion, obliteration, defacement; rubbing/wiping off; 2. cancellation, cancel(l)ing
~ зубов *med.* dental erasion
~ массива данных file deletion
~ памяти *comp.* store/memory erasing

~ ультрафиолетовым излучением ultraviolet erasing
стирать erase, delete, obliterate; wipe/rub off; clear, clean; wash
стирол *chem.* styrene
стиропор expanded polystyrene
стирофлекс styroflex, polystyrene film
стирофом styrofoam
стих verse; line
~, белый blank verse
стихарь *eccl.* sticharion, cotta, surplice
стихидий *bot.* stichidium
стихира *eccl.* canticle, hymn
стихирарь *eccl.* hymnbook, hymnal
стихобазидия *bot.* stychobasidium
стиховна *eccl.* psalms sung at vespers
стихосложение versification; *lit.* prosody
стихотворный versicular
~ размер *lit.* meter
~ная речь poetic diction/speech
стог *agr.* stack, rick
стогование, стогометание *agr.* stacking
стожить *agr.* to stack, to rick
стоимость 1. cost; 2. value
~, амортизируемая depreciable cost
~ в деньгах money's worth
~ в наличных деньгах cash value
~ в стране происхождения home value
~, валовая gross (book) value
~ валюты value of currency
~, внутренняя intrinsic value
~, восстановительная replacement/physical/sound value; replacement cost
~, выкупная redemption/surrender value
~, действительная intrinsic/real value; real/net cost
~, денежная cash value
~ добавленная value added
~ доставки cost of delivery
~ жизни cost of living
~ забракованной продукции scrap cost
~, завышать overrate
~, избыточная прибавочная extra/excess surplus value
~, индивидуальная subjective value
~, конечная final/ultimate cost
~ кредита cost of borrowing
~, ликвидационная liquidation/residual/disposal/salvage value; salvage cost
~, меновая exchange/market value
~, неамортизированная undepreciated value
~, номинальная/нарицательная denominational/nominal/face value; nominal cost
~, облагаемая rat(e)able value
~, общественная потребительная social use value
~, объявленная stated/declared value
~, остаточная depreciated cost
~, оценочная assessed/appraised value
~ по оценке assessed/appraised value
~, первоначальная 1. gross (book) value; 2. original/acquisition cost
~, полная (капитала) gross (book) value
~ пополнения запасов replenishment cost
~, потребительная use value
~, предельная marginal cost
~, прибавочная surplus value
~ провоза freight charge
~ работы working/operational costs
~ рабочей силы manpower cost; payroll cost
~ разработки development cost
~, рыночная market/commercial value
~, сметная estimated cost
~, совокупная aggregate value
~, страхование, фрахт cost, insurance and freight
~, страхование, фрахт и комиссия (посредника) cost, insurance, freight and commission
~, страхование, фрахт и курсовая разница cost, insurance, freight and exchange
~, страховая insurance value
~, суммарная integrated cost
~ технического обслуживания maintenance/service cost
~ товарно-материальных запасов inventory cost
~, условно-начисленная imputed value
~ утиля scrap cost
~ хранения на складе holding/storage cost(s)
~, чистая net/real value
~, эксплуатационная operating cost
стоицизм stoicism
стоический stoic; *fig.* stoical
стойка 1. stay, strut, prop, support, stand column, post, pillar; brace; rack; pedestal; *naut.* stanchion. poppet; *mech.* fixed/stationary member; 2. stance; posture; stand; 3. *(hunting)* set
~, амортизационная shock-absorber strut
~ антенны aerial mast/strut
~, дверная door pillar/post
~, леерная *naut.* rail/handrail stanchion
~ капота cowling strut
~ на руках *sport* handstand
~ перил baluster, rail post
~ подкрыльная outrigger unit
~, рудничная pit prop/jack
~, тентовая awning stancheon
~ Ферми vertical beam
~ шасси undercarriage leg
стойкий stable, durable; resistant; fast
~ на свету lightproof
~ против атмосферных условий weatherproof
~ против сотрясений shockproof
~кая краска fast colo(u)r
стойкость durability, stability; endurance; resistance; strength
~ к истиранию abrasion resistance
~, радиационная radiation resistance/stability, radioresistance, radiotolerance

~, термическая heat resistance/stability
~, химическая chemical stability
сток 1. flow; runoff, drainage, discharge; effluent; 2. drain; cutter; sewer; sink; *med.* confluence
~, безнапорный gravity flow
~, бытовой domestic waters, household effluent
~, возвратный regeneration flow
~, дождевой rainfall run-off
~, дренажный drainage effluent
~, канализационный sewage effluent
~, ливневый shower/storm runoff
~ нейтронов neutron sink
~, паводковый flood runoff/flow
~, плоский *aero.* two-dimensional sink
~ энергии energy sink
стокомер flowmeter; runoff meter
стокс *phys.* stokes
стол 1. table; desk; 2. meals, board
~, вибрационный vibrating/vibration/shake table
~, встряхивающий jolting table
~, красочный *print.* ink table
~, операционный operating table
~, письменный desk
~ поворотного стенда tilt stand table
~, поворотный rotatable table; slewing/swivel table
~, подъемный lift table
~, приставной side table
~, раздвижной table with leaves, lengthening table
~, секционный *med.* dissecting table
~, сотрясательный shaking table
~, справочный information desk
~, чертежный drawing table/desk
~, шведский smorgasbord
~, штурманский chart table
стол-планшет plotting table
столб column, pillar; pole, post, mast; pylon; pier; *el.* pile; *bot.* stele; *see also* **столбы**
~, водяной/воды water column
~, подорожный milestone
~, позвоночный vertebral/spinal column, backbone, spine
~, придорожный signpost
~, ртутный mercury column
~ свода *anat.* column of the fornix
~ спинного мозга, боковой/задний/передний lateral/posterior/anterior column
~, термоэлектрический thermopile
~, фундаментный foundation/bearing post
столбец 1. column; 2. *bot.* carpophore
столбик (small) column; pole, post;
~, разрешающий *math.* pivot(al) column
~, указательный signpost
~ определителя *math.* column of a determinant
столбняк *med.* tetanus; *fig.* stupor
~, идиопатический idiopathic/medical tetanus
~, истерический imitative tetanus
~, местный local/modified tetanus
~, мозговой cerebral/cephalic/head/hydrophobic tetanus
~, послеродовой puerperal tetanus
столбняковый stuporous, stuporose
столбнячный tetanic
столбовидный, столбообразный columnar
столбочек *bot.* column, gynostemium
столбчатый columnar; *zool.* paxillate
столбы *pl.* columns; *see also* **столб**
~, заднепроходные anal columns
~, почечные renal columns
~ спинного мозга, серые gray columns
столетник century plant, agave, American aloe *(Agave americana)*
столик small table/desk
~ микроскопа, предметный microscope stage, objective table
столкновение collision; impingement; impact; encounter; clash; conflict
~ в воздухе mid-air collision
~, касательное/скользящее grazing collision
~, лобовое head-on collision
~, молекулярное molecular encounter
~, надтепловое epithermal collision
~ с судном, стоящим неподвижно allision
столон *bot.* stolon, offshoot; tiller, runner
столотека *bot.* stolotheca
столпник *eccl., hist.* stylite
столпничество *eccl., hist.* stylitism
столпотворение devil's delight
~, вавилонское babel
столыпинский *hist.* Stolypin's
стольник *hist.* stolnik
стома *anat., zool.* stoma
стомагальгия *med.* stomatalgia, stomatodynia
стоматит *med.* stomatitis, sore mouth
~, афтозный aphthous stomatitis
~, грибковый mycotic stomatitis
~, дрожжевой thrush
~, свинцовый lead stomatitis
~, эпидемический epidemic stomatitis, foot-and-mouth disease
~, язвенно-гангренозный ulcerogangrenous stomatitis
~, язвенный ulcerative stomatitis
стоматический stomatal, oral
стоматолог stomatologist, dentist
стоматология stomatology, dentistry
~, ортопедическая prosthodontics, prosthetic dentistry
~, терапевтическая preventive dentistry
~, хирургическая operative dentistry
стоматомикоз stomatomycosis
стоматопластика stomatoplasty
стоматоррагия stomatorrhagia
стомиевые *pl. zool.* scaly fishes *(Stomiatidae)*
стоножка *zool.* centipede *(Scolopendra cingulata)*
стопа I heap, stack, pile; ream
стопа II foot
~, вогнутая talipes varus

~, гигантская macropodia, pes gigas
~, конская tip foot, talipes equinus
~, косолапая club/reel foot, pes contortus
~, ногтеобразная claw foot
~, обращенная внутрь talipes varus
~, обращенная кнаружи talipes valgus
~, отвислая drop/dangle foot
~, плоская flat/splay foot, pes/talipes planus
~, поперечноплоская broad/spread foot, talipes transversoplanus, metatarsus latus
~, расщепленная cleft/split foot
стоп-анкер kedge/stream/stern anchor
стопка stack, pile; heap; rouleau *(of coins)*
стопкадр *cin.* still
стоп-кран shutdown cock/valve, stopping yalve
стоповидно-жилковатый *bot.* pedatinerved
стоповидно-лопастный *bot.* pedatilobate
стоповидно-надрезный *bot.* pedatifid
стоповидно-раздельный *bot.* pedatipartite
стоповидный *bot.* pediform, pedatiform, footlike
стопор stopper/stop lock, catch, arrester, detent, retainer, stopper, holding/locking device
~, винтовой screw stopper
~ для стальных тросов wire stopper/gripper
~, предохранительный safety stop/lock
~, пружинный spring lock
~ резака cutter stop
~, рулевой *naut.* rudder brake
~ рулей *aero.* control lock
~, храповый ratchet stop
стопорезка paper cutter, guillotine
стопорить 1. lock; catch; stop; check; 2. plug, stopper
стоп-сигнал stoplight; *auto.* brake light
стоп-стержень *nucl.* shut-down [shut-off] rod, scram rod
сторнировать *econ.* reverse an entry
сторожевой корабль escort/patrol ship, patrol/escort vessel/craft; destroyer escort
сторона 1. side; 2. party; 3. direction; 4. flank; 5. country, land
~, боковая lateral side
~, виновная party at fault
~, воюющая belligerant (party), combatant
~ всасывания suction side
~ входа entry side
~ выигравшая дело prevailing party
~ выпуска outlet/tapping side
~, договаривающаяся contracting party
~, заинтересованная party concerned
~ кожи, бахтарменная/мездровая flesh side of leather
~ левая left-hand side; *naut.* port side
~, лицевая face/upper side, face; front
~ Луны, обратная far side of the Moon
~, наветренная windward/upwind side
~ нагнетания pressure/delivery side
~ не выполняющая своих обязанностей defaulter
~ плотины, верховая/напорная upstream side of a dam
~ плотины, низовая downstream side of a dam
~ по делу *law* party in the case
~, подветренная leeward/downwind side, leeside
~, подписавшаяся *law* signatory
~ потерпевшая ущерб damaged party
~, правая right-hand side; *naut.* starboard side
~ равенства member of an equality
~ разрежения suction side
~ реактора, лицевая pile face
~ сбегания *(of belt conveyer)* slack side
~, третья *law* third party/person
~ треугольника side/leg of a triangle
~ угла arm/side of an angle
стохастический stochastic; random
стохастичность stochasticity
сточный drainage *(attr.)*, sewage *(attr.)*
~ная канава ditch
~ная труба sewer pipe
~ные воды discharge/waste waters, sewage
стояк 1. rising pipe, riser, standpipe; 2. post, stancheon, strut; stay; 3. downcomer
~ дождевателя sprinkler riser
~, дымовой chimney
~, литниковый down gate, sprue
~, обратный return riser
стояние stand, standing; halt, stop
~, густое *agr.* close stand
~ полной воды high-water level/stand
~, редкое *agr.* thin stand
~ уровня прилива stand of tide
стоянка stop, stand; parking; *naut.* anchorage, berth, moorage; parking/stopping place/lot
~ в доке docking
~ гидросамолетов, якорная seaplane anchorage
~ на одном якоре/бочке swinging mooring
~ на якоре riding at anchor
~, самолетная aircraft park
~ такси taxi stand
~, укрытая *naut.* anchorage
~, якорная anchorage
стоячесть stationariness; *fig.* stagnation, passivity, inactivity
стоячий 1. stagnant, stationary, still; inactive, passive; 2. erect, standing, upright, orthostatic, vertical
~чая волна *phys.* standing wave
страбизм *med.* strabismus, heterotropia, squint
~, аккомодационный accomodative strabismus
~, вертикальный vertical squint
~, внешний external/divergent squint, exotropia
~, внутренний internal/convergent squint
~, монокулярный uniocular strabismus/squint
~, перемежающийся alternating strabismus
~, скрытый/подавленный latent/suppressed squint
~, содружественный concomitant squint
страбизмометр strabismometer, strabometer
стработомия *surg.* strabotomy

стравливание release, releasing, letting out
страдательно-возвратный *gram.* passive-reflexive
страдательность *gram.* passivity
страдательный *gram.* passive
 ~ залог passive voice
страдивариус Stradivarius (violin)
страна country, land; region
 ~, неприсоединившаяся nonaligned country
 ~ пребывания country of sojourn
 ~, слаборазвитая undeveloped country
странгурия *med.* strangury
страннохвост *zool.* palm civet *(Paradoxurus)*
страноведение regional/area studies
страсти *pl. rel.* Passion Music
 ~ Господни/Христовы the Passion
страстной :
 ~ четверг Maundy Thursday
 ~ная (неделя) Holy Week
 ~ная пятница Good Friday
страстоцвет *bot.* passionflower *(Passiflora)*
страстоцветные *n. pl. bot.* passionflower family *(Passifloraceae)*
стратагема, стратегема stratagem
стратег strategist
стратегия strategy
 ~, байесовская Bayes(ian) strategy
 ~, беспроигрышная safe strategy
 ~, генеральная master strategy
 ~, марковская Markov(ian) strategy
 ~, морская naval/maritime strategy
 ~, непересекающаяся nonoverlapping strategy
 ~ поиска search strategy
 ~ равновесия equilibrium strategy
 ~ с обратной связью feedback strategy
 ~, смешанная mixed/randomized strategy
 ~, устойчивая *game th.* equalizer
 ~, частично оптимальная suboptimal strategy
 ~, экономичная thrifty strategy
стратиг strategos, strategus
стратиграф stratigrapher
стратисфера lithosphere
страто-вулкан strato-volcano, stratified cone
стратоизогипса *geol.* structure contour
стратонавт stratonaut, stratosphere flier
стратоплан stratospheric aircraft/plane
стратостат stratospheric (high-altitude) balloon
стратосферический, стратосферный stratospheric(al)
страус *zool.* ostrich *(Struthio camelus)*
страусник *bot.* ostrich fern *(Struthiopteris germanica)*
страх fear, fright, dread, scare; anxiety
 ~, болезненный morbid fear, phobia
страхделегат *abbr.* insurance delegate
страхкасса *abbr.* insurance office/fund
страхование insurance; *naut.* underwriting
 ~ авиапассажиров aircraft passenger insurance
 ~ авиаперевозок air transport insurance
 ~, государственное national/state insurance
 ~ гражданской ответственности liability insurance
 ~ жизни life insurance
 ~ жизни двух/более лиц joint life insurance
 ~ жизни на определенный срок fixed term [term life] insurance
 ~, имущественное property insurance
 ~, морское marine insurance/underwriting
 ~ на случай нетрудоспособности disability insurance
 ~ на случай потери кормильца survirors' benefits insurance
 ~ на срок term insurance
 ~ на случай старости [по старости] old-age insurance
 ~ на случай болезни sickness insurance
 ~ от болезней и несчастных случаев health and accident insurance
 ~ от военных рисков war (risk) insurance
 ~ от несчастного случая accident/casualty insurance
 ~ от пожара fire insurance
 ~ от финансовых потерь suretyship/fidelity insurance
 ~ пассажиров в автомашинах motor vehicle passenger insurance
 ~ по восстановительной стоимости replacement insurance
 ~ по нетрудоспособности disability insurance
 ~ по старости old age (and survivors) insurance
 ~ профессиональной ответственности professional liability insurance
 ~ расходов на образование educational insurance
 ~ сельскохозяйственных культур crop insurance
 ~ фермерского хозяйства farm insurance
 ~, чрезмерное overinsurance
страхователь insurer, insurant; underwriter
страховка insurance; insuring
страховой insurance *(attr.)*
 ~ полис insurance policy
 ~вая премия insurance premium
страховать insure (against); ensure
страховщик insurer; *naut.* underwriter
страхфонд *abbr.* insurance fund
Страшный суд Day of Judgement, Doomsday
стрейнер strainer
стрекающие *n. pl. zool.* cnidarians *(Cnidaria)*
стрекоза *zool.* dragon fly *(Anisoptera)*
стрела 1. arrow; shaft; pointer, indicator; 2. *tech.* boom, jib; 3. dip, sag
 ~, боковая side boom
 ~, выносная outrigger
 ~, грузовая derrick, cargo boom
 ~ копра ram guide
 ~ крана jib/derrick arm, crane beam
 ~, поворотная swinging boom
 ~ прогиба sag; bending deflection; sagitta
 ~ прогиба рессоры spring camber/deflection

стрелец *astr.* Sagittarius, Archer
стрелиция *bot.* bird-of-paradise flower *(Strelizia)*
стрелка 1. pointer, indicator, needle; *(of watch)* hand; 2. arrow; 3. *bot.* spear, scape; 4. *rail.* switch
~, **глиссадная** glide-slope needle
~, **глухая** blind switch
~ **компаса** compass/magnetic needle
~, **курсовая** course pointer
~, **минутная** minute hand
~, **секундная** second hand
~, **часовая** hour hand
~**ке, по часовой** clockwise
~**ки, против часовой** counterclockwise
стрелкование *bot.* shooting, bolting
стрелковидный scapiform
стрелковый 1. shooting, rifle *(attr.)*; 2. *mil.* infantry, rifle *(attr.)*
~ **спорт** marksmanship, shooting
~**вая рота** infantry company
стрелкоцветковый *bot.* scapiflorous
стреловидность *aero.* sweep
~ **задней кромки** sweepback of trailing edge
~ **крыла** wing sweepback
~, **обратная/отрицательная** sweep forward
~ **передней кромки** sweepback of leading edge
~, **переменная** varying sweep
~, **прямая/положительная** sweepback
стреловидный arrow-shaped; sagittal, sagittate; *bot.* hastate
стрелок gunner
стрелолист *bot.* arrowhead *(Sagittaria)*
стрелочник *rail.* switchman
стрельба firing, shooting; fire
~, **зенитная** anti-aircraft fire
~ **по площади** *mil.* area fire
стрелочный pointer, arrow, switch *(all attr.)*
стрельбище shooting/rifle/target range
стрельчатка *zool.* dagger moth *(Acronycta)*
стрельчатый sagittate, sagittary, arrow-shaped, lanced, lancet *(attr.)*
стремиться *math.* tend *(to)*; approach
~ **к пределу** tend to a finite limit
стремя stirrup *(also anat.)*
стремянка step/rung ladder, steps
~ **рессоры** spring clip, U-bolt
стремянный 1. *anat.* stapedial; 2. *hist., n.* groom
стренатка *zool.* yellow-breasted bunting *(Embreriza aureola)*
стренга, стрендь *naut.* strand
стрепет *zool.* little bustard *(Otis tetrax)*
стрептобактерии streprobacteria
стрептобактерия streptobacterium
стрептобацилла streptobacillus
стрептобациллез streptobacillosis
стрептодерматит *med.* streptodermatitis
стрептодермия *med.* streptoderma
стрептокиназа-стрептодорназа streptokinase-streptodornase, varidase
стрептокарпус *bot.* Cape primrose *(Streptocarpus)*
стрептококк streptococcus
стрептококки *pl.* streptococci, *Streptococcus*
стрептококковый streptococcal, streptococcic
стрептолизин streptolysin, streptococcolysin, streptocolysin
стрептомицеты *pl.* streptomycetes, *Streptomyces*
стрептомицин *pharm.* streptomycin
стрептосептицемия streptococcemia, streptosepticemia, strepticemia
стрептотрихоз streptotrichosis, streptotrichiasis
стрептотрицин *pharm.* streptothricin
стрептоцид (белый) sulfanilamide
стрессор stress factor
стретто *mus.* stretto; stretta
стриарный striatal
стриатум *anat.* (corpus) striatum
стригаль, стригальщик shearer
стригун *zool.* pine-shoot beetle *(Myelophilus)*
стридор *med.* stridor
~, **гортанный** laryngeal stridor
стридорозный *med.* stridulous
стриж *zool.* swift *(Apus)*
стрижка *farm.* shearing
стриктейнер articulated container trailer
стриктура stricture
~ **прямой кишки** proctencleisis
~, **спазмотическая** spasmodic/functional stricture
~, **стойкая** permanent/organic stricture
стрикция striction
стример streamer
стрингер *naut., aero.* stringer, longitudinal
~, **бортовой** side stringer/longitudinal
~, **монолитный** integral stringer
~, **неразрезной** continuous stringer
~ **Н-образного сечения** H-girder
~, **прессованный** extruded stringer
~, **скуловой** bilge stringer/longitudinal
стриопаллидарный striopallidal
стриппер *met.* (ingot) stripper
стрихнин *chem.* strychnine
~, **азотнокислый** strychnine nitrate
стрихнинизм *med.* strychninism, strychnism
стричься have one's hair cut, cut one's hair
~ **в монахи** take the monastic vows
строб gate, strobe
~, **задержанный** delayed gate
~, **ультраузкий** narrow-narrow gate
стробил *bot.* strobile, strobilus; cone
стробила 1. *zool.* strobila, strobile; 2. *bot.* strobile
стробильный *bot.* strobilous, strobilate
строб-импульс gate/gating/strobe pulse
стробирование gating, strobing
стробировать gate, strobe
стробоскоп stroboscope
стробоскопия stroboscopy
стробирующий gating
строб-реле relay gate
строгальный planing, shaving
~ **станок** planing/shaving machine, planer

строгать plane; shave
строганый planed, shaven
строгий *math.* rigorous
строгость *math.* rigour, rigor
строение 1. construction, erection, building; 2. structure; composition; configuration, constitution; texture, fabric; pattern
~ **белка** protein structure
~, **верхнее** 1. *rail.* track structure, permanent way; 2. *(of bridge)* superstructure
~, **дендритное** dendritic structure
~, **дольчатое** lobular structure
~, **клеточное** cellular structure
~, **нижнее** 1. *rail.* ground work; 2. *(of bridge)* substructure
~ **пролетное (моста)** framework/superstructure (of bridge)
~, **сетчатое** reticulate structure
~, **химическое** chemical constitution
строжка 1. *weld.* gouging; 2. *leath.* slicking
~, **ацетилено-кислородная** oxyacetylene-flame gouging
~ **угольной дугой** carbon-arc gouging
строенный triplicated, triplex, triple
строительно-монтажный building and assembly
строительный building, construction *(attr.),* structural
~ **материал** building material
~**ная механика** structural mechanics
строительство building, construction, erection; *fig.* organization, setting up, development
~, **дорожное** highway construction/engineering
~, **жилищное** housing construction
~, **капитальное** capital construction/development
~, **муниципальное** public construction
~, **нежилищное** nonresidential construction work
~, **незавершенное** construction in progress
~ **"под ключ"** turnkey construction
~, **полносборное** prefabrication construction
строить 1. *build.* construct; 2. *geom.* plot; 2. *fig.* set up, form
строй 1. system, order, regime; 2. *mil.* formation; disposition; 3. structure; 4. *mus.* key, pitch; 5. *tech.* commission, operation
~, **боевой** *mil.* fighting formation
~, **вступить в** go into operation, become operational, be commissioned, go on line
~, **высокий** high pitch
~ **кильватера** *naut.* single line ahead, column
~, **общественный/социальный** social structure/system
~ **поиска** scouting formation
~, **сомкнутый** close order/formation
~ **"треугольник"** delta formation
~ **"утка"** *aero.* canard/duck formation
~ **уступа** *naut.* bow-and-quarter line
~, **феодальный** feudal system
~ **фронта** *naut.* line abeam/abreast
~, **экономический** economic system
стройбанк *abbr.* construction/building bank
стройбат *abbr., mil.* construction battalion
стройка construction, building; development; *fig.* construction site
стройконтора *abbr.* construction office
строка *comp., math.* line; row; string
~ **битов** *comp.* bit string
~ **знаков** *comp.* character string
~ **команды** command line
~ **матрицы** matrix row
~ **определителя** row of a determinant
~ **(перфо)карты** card row/line
~ **примечаний** comment line
~ **развертки** *telev.* scan line
~, **разрешающая** pivot(al) row
~ **символов** *comp.* character string
~ **таблицы** table row
строма *anat., bot.* stroma, framework
строматолиз, стромолиз stromatolysis
стромболианский *geol.* Strombolian
стромовидный stromatoid
стронгилоидоз strongyloidiasis, strongyloidosis
стронцианит *miner.* strontianite
стронциевый *chem.* strontic; strontian
стронций strontium, Sr
стронциоапатит *miner.* strontian apatite
строп (rope) sling; strop
~, **бочечный** *naut.* barrel/cask/butt sling, hogshead hooks
~, **вертолетный спасательный** helicopter's rescue sling
~ **для погрузки тюков** bale string
~, **каболочной** rope strop
~, **проволочный стальной** wire-rope sling
~ **резака** cutter span
~, **сеточный** net sling
стропа line, (top) cord; *see also* **строп**
~ **парашюта** parachute shroud line; cord
стропальщик slinger
стропилина, стропило rafter, truss; beam
стропить sling, attach slings to
стропка *naut.* becket, lanyard
~, **крепящая** retaining wire
строповка *naut.* slinging
строфа *lit.* stanza, strophe
~, **алкеева** Alcaic
строфант *bot.* strophanthus *(Strophanthus)*
строфантидин strophanthidin
строфантин *pharm.* strophanthin
~ **Г** g-strophanthin, oubain
~ **К** strophanthin K
строфиола *bot.* strophiole
строфический *lit.* strophic, stanzaic
строфоида *math.* strophoid
строфотрон strophotron
строфулус, строфулюс *med.* strophulus, gum rash
строчка stitch; stitching
строчок *bot.* saddle fungus *(Helvella)*
струбцина (screw) clamp, holdfast
струг 1. plow, plough, grader, level(l)er; 2. shaving

knife, drawknife; plane; adze
~ **для желобков** fluting plane
~, **фанерный** veneer planer
струговидный *bot.* runcinate
стружка chip; shaving
стружкодробитель, стружколом chip breaker
стружкоудаление chip disposal/removal
струйка jet; filament; *geol.* Stria, costella
~, **обратная** reentrant jet
~ **тока** stream filament
струйник *aero.* control jet
~ **крена** roll-control jet
~ **рыскания** yaw-control jet
~ **тангажа** pitch-control jet
структура structure, pattern; composition; organization; texture; configuration; framework
~ **альфа-фазы** alpha structure
~, **внутризеренная** intragranular structure
~, **возрастно-половая** age-sex pattern
~ **данных** data structure
~, **двухбуквенная** *math.* digram
~ **доходов** revenue composition
~ **занятости** occupational pattern
~, **институциональная** institutional framework
~, **конвейерная** *comp.* pipeline organization
~ **команды** *comp.* instruction format
~, **крупнозернистая** coarse-grained structure
~, **мелкозернистая** fine grain structure
~ **металл-диэлектрик-полупроводник** metall-insulator semiconductor [MIS] structure
~ **металл-окисел-полупроводник** metal-oxide-semiconductor [MOS] structure
~, **нуклонной оболочки** nucleon shell structure
~, **оболочечная** shell structure
~ **поверхности** surface texture/pattern
~ **потока** flow pattern
~ **потребления** pattern of consumption
~, **профессиональная** occupational composition
~ **резонанса** structure of a resonance
~, **резонансная** resonance structure
~, **сверхтонкая** hyperfine structure
~, **тонкая** fine structure
~, **упорядоченная** ordered structure
~ **файла** file organization
~, **ячеистая** honeycomb/cellular structure
структурирование structurization, structure formation
структурировать structurize
структурно-чувствительный structure-sensitive
структурный structural; structured; *geol.* geotectonic
структурообразователь structure-forming agent, cross-linking agent; soil conditioner
струма *med.* struma, goiter; *bot.* struma
~, **базедовая** exophthalmic/Basedow's goiter
~, **загрудинная** substernal goiter/struma
~, **твердожелезная/деревянистая** cast-iron struma, ligneous struma, Riedel's disease
~, **узловатая** adenoma of the thyroid
~, **фолликулярная** colloid parenchymatous goiter
струмит *med.* strumitis, thyroiditis
~, **железоподобный** ligneous thyroiditis
струмоген *med.* goitrogen
струмогенный *med.* goitrogenic
струна string; *anat.* cord
~, **барабанная** cord of the tympanum
~, **косая** oblique cord, oblique [round ligament] of the elbow joint
~, **спинная** notochord
струнец *zool.* ascaris *(Ascaris lumbricoides)*
струнный stringed, string *(attr.)*
~ **инструмент** *mus.* stringed instrument
струнобетон reinforced concrete
струнодержатель *mus.* tailpiece
струнокорневой *bot.* chordorhizal, string-rooted
струп crust, crusta; scab
струпьевидный squarrous, squarrose
стручковидный, стручковатый *bot.* siliquiform
стручковый *bot.* siliculose; leguminous; podded
~ **перец** capsicum; cayenne (pepper)
~**вая фасоль** string beans
стручок *bot.* pod; follicle; silique
~, **короткий** silicle
~, **цареградский** carob *(Ceratonia siliqua);* carob bean
стручочковидный, стручочкообразный *bot.* silicular, siliculous, siliculose, silicle-shaped
струя jet, spurt, spout; flow, current, stream
~, **выхлопная** exhaust jet
~ **горючего** fuel spray
~, **дозвуковая** subsonic jet/stream
~ **за винтом** backwash
~, **закрученная** swirling jet
~ **звуковая** sonic jet/stream
~, **кильватерная** (ship's) wake
~, **кольцевая** annular jet
~ **космического излучения** cosmic-ray jet
~, **обратная** backward/reentrant jet
~, **плазменная** plasma jet
~, **плоская** planar [two-dimensional] jet
~, **попутная** wake, back eddy
~, **спутная** wake
~, **суживающаяся** convergent jet
~, **эжектирующая** inducing jet
стряхивать shake off/down; dislodge
студент undergraduate university/college student
студент-второгодник repeater
студент-медик medical student
студент-юрист law student
студень gelatine; gel, jelly
~, **гремучий** blasting gelatine
~, **превращать в** jellify, gelatinate, gel
~, **коллоидный** colloidal gel
студия studio
~ **звукозаписи** sound-recording studio
~, **радиовещательная** radio broadcast studio
~, **театральная** drama school

студнеобразный, студнеподобный jellylike, gelatinous
стук knock(ing), rap(ping), tap(ping), ping(ing)
стул 1. stool; chair; 2. *med.* stool(s), feces, defecation
~ наковальни anvil block
ступа mortar
ступенчато stepwise, in steps/stages
ступенчатость gradation, graduation
ступень 1. step; 2. stage, level, grade, phase, step
~ давления pressure stage
~ заднего хода *naut.* astern stage
~ лестницы stair step
~, маршевая *av.* sustainer stage
~, откачивающая scavenge stage/unit
~ переднего хода ahead stage
~ равновесия equilibrium stage
~, реактивная reaction stage
~, стартовая *rock.* zero stage
~, фазоразделяющая phase-splitting stage
ступенька step
ступка mortar
стык joint; butt; junction; splice; interface
~, косой chamfered joint
~, прямой butt joint
~, сварной welded joint
~, ступенчатый step/zigzag joint
~, технологический construction joint
стыкование butt-jointing; splicing
стыковать join, splice, butt
стыковаться *space* dock, link; rendezvous
стыковка *space* docking, link-up; rendezvous
стыковой junctural; butt *(attr.)*
~вая сварка butt welding
стэн *phys.* sthene
стюард major-domo; *naut., aero.* steward; *aero.* cabin attendant
стюардесса air hostess; stewardess, cabin attendant
стягивание contraction, shrinkage; constriction; shrinking; concentration
стягивать 1. contract; constrict; 2. subtend
стяжка brace, tie (rod), stay; coupling
~, вагонная car coupler
~, винтовая turnbuckle
~, проволочная stay/brace wire
~, рессорная spring clamp/hook
су *Fr.* sou
суахили Swahili
субаренда subtenancy, sublease; subletting
субарендатор subtenant, sublessee, undertenant
субарктика subarctic region
субатлантический sub-Atlantic
субацидность subacidity, hypoacidity
суббота Saturday
~, Великая Holy Saturday
субботник 1. Sabbatarian; 2. *obs.* subbotnik
субгармоника *phys.* subharmonic; suboctave
субгедральный *miner.* subhedral, hypidomorphic
субгеоклиналь subgeocline
субгимениальный *bot.* subhymenial
субгумидный subhumid
субделянка subplot of land
субдиафрагмальный subdiaphragmatic
субдоговор subcontract
субериниэация suberization, suberification, suberinization
субклеточный subcellular
субклимакс subclimax
субконтрагент *law* subcontractor
сублетальный sublethal
сублимировать sublimate, sublime
сублитораль sublittoral area/zone
субмаксиллит submaxillaritis, submaxillitis
субмаксиллярный *anat.* submaxillary
субмедиальный submedial, submedian
субнивальный subniveal
субнуклонный sub-nucleonic
субподряд subcontract
субподрядчик subcontractor
~ низшей ступени lower-/second-tier subcontractor
субпоставщик *law* subcontractor
субпродукт *food* by-product; offal
субрегиональный subregional, subareal
субретка *theatre* soubrette
суброгировать *law* subrogate
субсидировать subsidize
субсидия grant, subsidy, grant-in-aid, appropriation-in-aid; subvention
субстантив substantive, noun
субстрат substrate; substratum, basement; *philos.* substance
~, исходный original substrate
~, почвенный bedrock
субстрат-пептид peptide substrate
субсчет control account
субталамический *anat.* subthalamic
субтилопептидаза *biochem.* subtilopeptidase
субтитр subtitle
субтропики subtropics, subtropical zone
субфебрилитет *med.* subfebrile condition
субфебрильный subfebrile
субфилиал subbranch
субфонемный subphonemic, infraphonemic
субхондриальный subchondrial, subcartilaginous
субширотный sublatitudinal
субштам substrain
субщелочный subalkaline
субъединица subunit
субъект subject
~ права 1. person of law; 2. holder of a right
~ преступления perpetrator
~, юридический legal entity
субъективизм 1. subjectivity, subjective attitude; 2. *philos.* subjectivism
субъективист, субективистический subjectivist
субъективистский subjectivistic
субъективно-идеалистический subjective-

idealistic
субъективно-речевой subjective-speech *(attr.)*
субъективность subjectivity, subjectiveness
субъективный subjective; personal
~ная полезность personalistic utility
субъектно-объектный subject-object *(attr.)*
субэндотелий subendothelium
субэпителиальный subepithelial
субъядерный subnuclear
суверен sovereign
суверенитет sovereignty
~, божественный thearchy
суворовец student at the Suvorov Military College
суглинистый loamy; argillaceous
суглинон loam, loamy/clayey soil
~, жирный clay loam
~, известковый marl
~, илистый silty loam
~, карбонатный carbonaceous loam
~, лессовый loess/loessial loam
суд (law) court, court of justice; *fig.* trial
~, апелляционный court of appeal(s), court of appelate jurisdicticn
~, арбитражный arbitration tribunal/court
~ ассизов assizes
~ Божий *hist.* trial by ordeal
~, военно-полевой court martial
~ второй инстанции court of appelate jurisdiction
~ выбранный по соглашению сторон *Lat.* forum prorogatum
~, вызов в *law* subpoena
~ домициля *Lat.* forum domicilii
~, епархиальный/епископский diocesan court
~, канцлерский chancery
~, кассационный court of cassation
~, консисторский consistory/diocesan court
~, Международный (ООН) International Court of Justice
~ места заключения договора *Lat.* forum contractus
~ места правонарушения *Lat.* forum delicti
~ места совершения акта *Lat.* forum rei gestae
~ места совершения действия *Lat.* forum actus
~ местонахождения вещи *Lat.* forum rei
~ местонахождения наследства *Lat.* forum patrimonii
~ местонахождения предмета спора *Lat.* forum rei sitae
~, морской maritime court; Court of Admiralty
~, отдавать под prosecute
~ первой инстанции court of first instance
~ присяжных jury
~, приходский parish court
~, рыцарский Court of Chivalry
~, светский civil court
~, Страшный Day of Judgment, Doomsday
~ суммарной юрисдикции court of summary jurisdiction
~, третейский court of arbitration; reference tribunal
~, уголовный апелляционный Court of Criminal Appeal
~, церковный ecclesiastical/spiritual court
~ шерифа Sheriff's court
судак *zool.* pike perch; zander *(Lucioperca)*
судан sudan *(pigment)*
суданец, суданский Sudanese
суданка Sudanese (woman)
судебник code of laws
судебно-медицинский medicolegal
судебный judicial; forensic; legal
~ приказ injunction
~ная медицина forensic medicine
~ная химия forensic chemistry
судейский judicial, judicatory; *sport* referee's
судимость criminal record, conviction(s)
судить judge; *law* try; adjudicate
~ заочно try in absentia
судно vessel, ship, craft
~, атомное atom-powered ship, atom ship, A-ship
~, аварийное injured vessel
~, аварийно-ремонтное emergency repair ship
~, аварийно-спасательное salvage-and-rescue vessel
~, беспалубное undecked vessel
~, брандвахтенное examination vessel
~, буксирное tug, tugboat, towboat
~, винтовое propeller/screw vessel
~, водолазное diving ship
~, водометное jet-propelled vessel
~, водоналивное water-tank boat/ship, water ship/carrier
~, воздушное aircraft
~, вспомогательное auxiliary/service ship
~, грузовое freighter, freight/cargo carrier, cargo vessel, freight/cargo ship
~, грузопассажирское cargo and passenger ship, passenger-cargo vessel/carrier, combination vessel
~ для перевозки грузов в контейнерах containership
~, дизельное motor ship/vessel
~ для массовых грузов bulk carrier
~ для насыпных грузов bulk cargo ship
~ для погрузки и выгрузки на побережье без причала beach-loading vessel
~ дневного плавания day-service ship
~, дноуглубительное dredging craft
~, досмотровое boarding/examination vessel
~, железобетонное reinforced-concrete vessel
~ заграничного плавания foreign-going vessel
~, загрузочное feeder vessel
~, заднеколесное stern-wheeler
~, зверобойное sealer
~, кабельное cable-laying ship/vessel

~, каботажное coaster, shoaler, coasting ship/vessel, home-trade vessel
~, карантийное quarantine vessel/ship
~, китобойное whaler, whaling vessel/boat
~, колесное paddle-wheel vessel
~, контейнерное container ship/vessel
~, крановое crane/lift vessel
~, круизное cruise ship
~, крупнотонажное supership
~ ледового плавания ice ship
~, ледокольное icebreaker (ship)
~ лежащее в дрейфе stopped ship
~, лоцманское pilot boat/vessel
~, лоцмейстерское lights servicing vessel, buoyage/mark vessel, danlayer, danner
~, морское sea boat, sea-craft, deep-sea vessel
~ на воздушной подушке hovercraft, hovership, air-cushion vessel
~ на подводных крыльях hydrofoil ship/vessel/craft
~, надводное surface vessel
~, наливное tanker
~, научно-промысловое fishery research vessel
~, немореходное unseaworthy ship
~, несамоходное non-self-propelled vessel/ship, non-powered vessel/ship
~, нефтебуровое drilling vessel
~, нефтеналивное oil tanker
~ ночного плавания overnight vessel/ship
~ обслуживания service ship
~ обязанное изменить курс при расхождении burdened [giving-way, obligated] vessel, altering craft
~, одновинтовое single-screw vessel/ship
~, однопалубное single-deck vessel, one-decker
~, паровое steamship
~, паромное ferrycraft
~, парусное sailing vessel/ship
~, парусно-моторное motor-sailing ship
~, пиратское pirate ship
~ плавающее под удобным флагом convenience flag ship
~, подводное submarine (vessel)
~, пожарное firefighting craft, fire vessel
~, портовое harbour craft
~ потерпевшее аварию shipwreck
~ потерявшее управление disabled vessel
~, прогулочное pleasure soat
~, промерное sounding vessel
~, промысловое catching vessel
~, промыслово обрабатывающее factory ship
~, противопожарное fire-fighting ship/craft
~, рейдовое roader
~, рейсовое liner
~ речного и морского плавания river-marine vessel
~, рыболовное fishing vessel/craft
~, санитарное ambulance ship
~ с горизонтальным способом грузовых операций roll-on/roll-off vessel, Ro-Ro vessel
~, спасательное rescue ship/vessel, salvage craft/boat
~, сторожевое picket boat
~ стоящее на рейде roader
~ стоящее на якоре vessel at anchor
~, сухогрузное dry-cargo ship/vessel
~, торговое merchant ship/vessel, merchantman
~, тронковое trunk-deck vessel
~, туерное chain vessel/tug
~, универсальное грузовое all-cargo vessel
~, фланговое wing boat
~, шельтердечное shelterdecker, shelter-deck vessel
судно-база base/mother/parent ship
судно-доставщик feeder ship
судоверфь shipyard
судовладелец shipowner
судоводитель mariner, operator of a vessel, navigator
судовождение navigation, pilotage
судовой ship *(attr.);* naval, maritime, marine
~ патент certificate of registry
судооборот shipping traffic
судоподъемник ship elevator, ships lift
судопроизводство *law* court/legal procedure, legal/judicial proceedings
~, военное military justice
~, гражданское civil justice
судопромышленник shipbuilder
судорабочий stevedore
судоремонтный ship-repairing
судорога cramp, spasm, convulsion
~, гипналогическая hypnalogic cramp
~ жевательных мышц masseteric cramp
~, зевательная yawning cramp
~, писчая writer's cramp, chirospasm
~-, профессиональная occupation spasm/neurosis
судороги *pl.* convulsions, spasm, cramp
~, кивательные nodding spasm
~, клонические clonic convulsions/spasm, clonospasm
~, корковые central convulsions
~, перемежающиеся intermittent cramp, tetany
~, тепловые heat cramp
~, тонико-клонические tonoclonic spasm
~, тонические tonic/tetanic convulsions, tonic spasm
судорожность convulsiveness, spasmodicalness
судорожный spastic, spasmodic, convulsive; cramping, crampy
судостроене shipbuilding
~, малотонажное boatbuilding
судостроитель shipbuilder, shipwright
судостроительный shipbuilding
судоустройство judicial organization/system
судоходность navigability
судоходный 1. navigable; 2. shipping

судоходство shipping, navigation, seaborne traffic
~, каботажное coasting/coastwise trade
~, морское high-sea shipping
~ на внутренных морях inland shipping
~, озерное lake shipping
~, речное river navigation
~, торговое merchant shipping
~, трамповое tramp navigation
судья 1. judge; 2. *sport* referee, umpire
~, мировой Justice of the Peace
~, третейский arbitrator
суждение judjement; assessment; opinion
~, дизъЮнктивное *logic* disjunction
~, негативное *logic* apophasis
сужение constriction, narrowing, contraction, reduction; narrowing, necking; taper (ratio); *med.* stenosis
~ аорти aortostenosis, aortarctia, narrowing of the aorta
~ артерии arteriostenosis
~ бронха bronchostenosis, bronchoconstriction
~ влагалища colpostenosis
~ глотки pharyngostenosis
~ крыла wing taper (ratio)
~ мочеиспускательного канала urethrostenosis, stricture of the urethra
~ пищепровода esophagostenosis, stricture of the esophagus
~ полости рта stenostomia
~ струи jet contraction
~ трахеи tracheostenosis
~ устья аорти subaortic stenosis
суженнобороздной, суженнобороздчатый constricticoplate
суженный constricted, contracted, narrowed; tapered; *med.* stenosed, stenotic, strictured
суживатель *anat.* constrictor, sphincter
суживать constrict, narrow; taper; take in
суживаЮщийся tapered, tapering, converging
сукновальня fullery, fulling mill
сукровица *med.* sanies
суровичный *med.* sanious
сукцинат *chem.* succinate
сукциндегидрогеназа succinic dehydrogenase
сулема bichloride of mercury, corrosive sublimate
сулемовый sublimate(d)
сулица *hist.* short spear/lance
сулки two-wheeled carriage, sulky
сулой overfall, lop, rip, tide race
~, приливный tide rip
сулок *eccl.* vexillum
султан I sultan
султан II 1. *bot.* tassel, anthurus; 2. plume, panache
султанка *zool.* 1. sultana bird *(Porphyrio);* 2. goatfish *(Mullus)*
сульгин sulphaguanidine, sulphanilylguanidine
сульфазин *pharm.* sulphadiazine, sulphamidopyrimidine
сульфаметазин sulphamethazine
сульфаметин sulphamethin
сульфамид *chem.* sulphamide
сульфаниламид *pharm.* sulphanilamide, sulpha drug
сульфанилилгуанидин = сульгин
сульфат *chem.* sulphate, sulfate
~, водный hydrous sulphate
~ железа ferrous sulphate
~ магния magnesium sulphate
~ радия radium sulphate
~ тория thorium sulphate
~ урана uranium sulphate
сульфатаза *bioch.* sulphatase
сульфатация, сульфатирование sulphation, sulphating
сульфатировать sulphate
сульфатканкринит *miner.* wishnewite, sulphatic cancrinite
сульфатный sulphatic
сульфатолит *miner.* sulphatolite
сульфацил sulphacetamide
~, растворимый sulphacetamide sodium
сульфгидрильный sulphydryl
сульфид *chem.* sulphide, sulfide
~ натрия sodium sulphide
~ уранила uranyl sulphide
сульфидин *pharm.* sulphapyridine, sulphidine, dagenan
сульфидирование sulphidizing
сульфидно-стронциевый strontium-sulphide
сульфидный *chem.* sulphidic
сульфимид *chem.* sulphimide
сульфирование *chem.* sulphurization; sulphonation
сульфировать sulphurate, sulphurize, sulphonate
сульфит *chem.* sulphite, sulfite
сульфитатор sulphiter, sulphitation tank
сульфитация *chem.* sulphitation
сульфитировать sulphite
сульфитный sulphitic
сульфобромид *chem.* sulphonic-acid bromide
сульфогалит *miner.* sulphohalite
сульфозол sulphosol
сульфокислота sulphonic acid
сульфомиксин suphomyxin
сульфон *chem.* sulphone
сульфонал *pharm.* sulphonal
сульфонамид *chem.* sulphonamide, sulphonic acid amide
сульфонметан *chem.* sulphonmethane, sulphona
сульфосалициловый sulphosalicylic, salicylsulphonic
сульфосоль sulpho salt
сульфохлорид sulphonic-acid chloride
сульфурил *chem.* sulphuryl, sulphonyl
сумасшествие insanity, madness, lunacy
суматрийский Sumatran
сумах *bot.* sumac(h) *(Rhus)*
сумеречный *zool.* crepuscular
сумерки twilight, dusk

~, **гражданские** civil twilight
~, **навигационные** navigation/nautical twilight
сумка 1. bag; 2. *anat.* bursa; 3. *bot.* sac, ascus
~, **локтевая** ulnar bursa
~, **околосердечная** pericardium, pericardial sac
~, **парашютная** parachute bag/container
~, **пастушья** *bot.* shepherd's purse *(Capsella bursa-pastoris)*
сумковидный, сумкообразный baglike; bursiform
сумконосный pouch-bearing; ascophorous
сумкоспора *bot.* ascospore
сумма sum; amount; total
~ **в распоряжении** available amount
~ **векторов/геометрическая** vector sum
~, **валовая** gross amount, gross/total sum
~ **векселей** stock of bills of exchange
~, **денежная** sum of money
~, **депонированная** deposited sum
~ **иска/исковая** amount in dispute, claim amount
~ **иска, случайная** random claim amount
~, **итоговая** sum total
~, **капитальная** principal amount
~, **контрольная** checksum, control total
~, **номинальная** face amount
~, **общая** gross/total sum, total amount
~, **округленная** round sum
~ **отклонений** variance total
~, **поперечная** transverse sum/total
~, **причитающаяся** amount due
~, **промежуточная** sub-total
~, **прямая** *math.* direct/Cartesian sum
~ **ряда** *math.* sum of series
~, **степенная** *math.* sum of powers
суммарность additivity; cumulativeness
суммарный summary; cumulative, additive; total, overall; combined
~**ное действие** cumulative effect
сумматор summer; adder, totalizer; adding machine; accumulator
~ **адресов** address accumulator
~,**арифметический** nonalgebraic adder
~, **двоичный** binary(-digit) adder
~, **десятичный** decimal adder
~ **зон** zone adder
~, **импульсно-компенсационный** pulse-bucking adder
~ **индексов** index accumulator
~ **на три входа** three-input adder
~, **накапливающий** accumulator
~, **одноразрядный** one-/single-digit adder
~ **порядков чисел** exponent adder
~ **последовательного действия** serial adder
~ **с выдачей на печать** adding-listing machine
~ **с двумя входами** two-input adder
~ **с округлением, накапливающий** roundoff accumulator
~ **с плавающей запятой/точкой,** ~, **накапливающий** floating-point accumulator
~ **с ускоренным переносом** anticipated carry adder
~ **со сквозным переносом** ripple-carry adder
~ **чисел** number adder
~, **электромеханический** electric adder module, electric addition mechanism
~, **электроннолучевой** cathode-ray tube adder
сумматор-вычитатель adder-subtracter
сумматор-регистратор cumulative recorder
суммирование sumuation, summing up, addition, adding; accumulation; recapitulation
~, **промежуточное** subtotalling
суммировать 1. sum up; add; accumulate; 2. summarize, recapitulate
суммируемость summability
суммируемый summable
сумочник *bot.* capsella *(Capsella)*
~, **пастуший** shepherd's purse *(Capsella bursa-pastoris)*
сумчатый 1. *bot.* ascogenous, ascomycetous, utricular; 2. *zool.* marsupial
сунна Sunna(h)
суннит, суннитка, суннитский Sunnite
суньятсенизм Sun Yat-senism
суоми Suomi, Finn
суперавиация high-altitude aviation
суперарбитр chief arbitrator; umpire
супервидикон supervidicon
супервизор supervisor; monitor
~ **страничного обмена** *comp.* paging supervisor
~, **оверлейный** overlay supervisor
суперген supergene
супергенез supergenesis
супергенный *geol.* supergene
супергетеродин *el.* superheterodyne, superheterodyne receiver, superhet
суперессив *gram.* superessive (case)
суперинтендент *eccl.* superintendent
суперкапиллярный supercapillary
суперлайнер superliner
супермен superman
супермицелла supermicelle
суперобложка book cover/jacket, dust jacket
суперортикон *telev.* superorthicon, image orthicon
суперпаразит superparasite, hyperparasite
суперпозиция superposition, overlapping, overlaying
суперрегенератор superregenerative receiver
суперсегментный supersegmental
суперсексы supersexes
суперсинтагма supersyntagma
суперстрат *ling.* superstratum
супертанкер supertanker
суперфикс *ling.* suprafix
суперфициарий *law* superficiary
суперфонемный superphonemic
суперфосфат, суперфосфатный *chem.* superphosphate
~, **аммиачный** ammoniac superphosphate
~, **аммонизированный** ammoniated superphos-

phate
~, гранулированный granulated/granular superphosphate
~, двойной triple/double superphosphate
~, осадочный sludge superphosphate
суперэлита *agr.* highest quality seed(s)
супесчаная почва sandy-loam soil
супесь sandy loam/soil
супин *gram.* supine
супинатор 1. *anat.* supinator; 2. arch/instep support, instep insole
суппозиторий *med.* suppository, suppositorium
суп(п)орт 1. *mech.* carriage; 2. rest, support
~, боковой side-hung carriage
~, верхний top rest
~, поворотный swivel carriage, pivoting slide rest
~, поперечный reach-over carriage
~, револьверный turret
~ шлифовального станка wheel-head
суп-пюре *cul.* cream soup
супрavергенция supravergence
супрадентия *med.* supradentia
супраренин *pharm.* suprarenin(e), epinephrine
супрасегментный suprasegmental
супрастин *pharm.* chloropyramine, allergan S, halopyramine, suprastin, sinopen
супрахороидальный *anat.* suprachoroidal
супремум *math.* least upper bound, supremum
супрессор suppressor
~ кроссинговера crossing-over suppressor
супротивнолистный *bot.* opposite-leaved, oppositifolious
супротивноперистый *bot.* oppositipinnate
супротивноцветковый oppositiflorous, opposite-flowered
супротивный opposite
~ное листорасположение decussate leaf arrangement
супружество conjugability wedlock, matrimony
сура *rel.* sura(h), chapter of the Koran
сургуч sealing wax
сурдина, сурдинка *mus.* sordine, mute
сурдобарокамера isolation vacuum chamber
сурдокамера isolation chamber
сурдомутизм surdomutism, deaf-mutism
сурдопедагогика *educ.* surdopedagogy
сурепица *bot.* bird/summer rape, coleseed *(Brassica campestris)*
сурепка *bot.* winter cress *(Barbarea)*
сурепное масло coleseed oil
суржа *agr.* mixture of rye and wheat
сурик minium, red lead
~, железный iron minium, red ocher
~, свинцовый red lead
сурико-красный cinnabar-red
суринамский Surinam
суровый *text.* unbleached, grey, gray, brown; raw
сурок *zool.* marmot *(Marmota)*
суррогат substitute; ersatz
сурьма *chem.* antimony, stibium, Sb
~, белая valentinite
~, игольчатая needle antimony
~, мышьяковая arsenical antimony, allemontite
~, пятисернистая antimony pentasulphide
~, сернистая antimony sulphide
~, трехсернистая antimony trisulphide
сурьмянистосвинцовый antimony-lead *(attr.)*
сурьмянистый antimonious, antimonial, antimonide, stibious
сурьмяно-цезиевый cesium-antimony, antimony-cesium *(attr)*.
сурьмяный antimonic, antimonial, stibic, stibial
сусак *bot.* flowering rush *(Butomus)*
сусаль (artificial) gold leaf
суслик *zool.* spermophile, suslic *(Citellus)*
суспензировать, суспендировать suspend
суспензия, суспензий suspension; slurry
~, водная aqueous suspension
~, жидкая liquid suspension
суспензорий suspensory (bandage); supporter
суссекс *agr.* Sussex cattle
сустав joint, articulation
~, блоковидный hinge joint
~, винтовой screw/cochlear joint
~, голеностопный talocrural articulation, ankle joint; mortise joint
~, грудно-ключичный sternoclavicular joint
~, двухкамерный binocular joint
~, двухосевой biaxial joint
~, коленный knee joint
~, комбинированный complex joint
~, ложный false joint
~, лучезапястный wrist joint, radiocarpal articulation
~, мыщелковый condylar articulation
~ наковальни и молоточка incudomallear joint
~ наковальни и стремени incudostapedial articulation
~, плечевой humeral articulation, shoulder joint
~, плече-лучевой humeroradial joint
~, плоский plane/gliding/arthrodial joint
~, подтаранный subtalar joint
~, пяточно-кубовидный calcaneocuboid joint
~, седловидный saddle joint
~, сложный compound joint
~, тазобедренный hip joint
~, цилиндрический trochoid/rotary joint
~, шаровидный spheroid/cutyloid joint, ball-and-socket joint, enarthrosis
суставной articular; articulate(d); arthral, arthrous, arthroidal
суставчатый articulate(d)
сутки day, day and night, twenty-four hours
~, гражданские legal/civil day
~, звездные sidereal day
~, солнечные solar day
~, солнечные истинные apparent solar day
~, солнечные средние mean solar day

~, круглые tweny-four hours straight
~, лунные lunar day
~, световые light day
~, приливные tidal day
суточные *pl.* daily allowance, per diem
суточный daily, twenty-four-hours *(attr.),* diurnal, round-the clock; *biol.* circadian
сутулиться bend, stoop, become bent
сутунка *met.* slab/sheet billet, sheet bar
сутура *bot., zool.* suture
суфле *cul.* soufflé
суфлер 1. *tech.* breather; 2. *theatre* prompter
суфлирование venting, breathing
суфляр puffing/spouting hole; *min.* blower, feeder, piper
суфражизм suffragette movement
суфражистка suffragette
суффигирование suffixation
суффикс *gram.* suffix
~, расширенный extended suffix
суффлейный *phon.* breathed sound
суффозия suffosion, undermining; subsoil/internal erosion
суффузия *med.* suffusion
сухарь 1. rusk, cracker; bisquit; *pl.* crumbs; 2. *tech.* slide/sliding block; 3. *bot.* whitish russula *(Russula delica);* 5. refractory clay
~ кардана universal-joint trunnion
~ щековый дробилки toggle seat
суховей hot dry wind
сухогруз 1. dry-cargo ship/carrier; 2. dry cargo
сухожилие *anat.* tendon; sinew
~, ахиллово/пяточное Achilles tendon
~, перстне-пищеводное cricoesophageal tendon
сухожильный tendinous
сухой dry, dried-up, dried; arid, droughty
~ закон prohibition, dry law
~хие роды *med.* dry labour
~хим путем by land, overland
сухолистный *bot.* xerophyllous
сухолюб *bot.* xerophyte, xerophylous plant
сухонос *zool.* swan goose *(Anser cygnoides)*
сухопарник steam dome/heater/box
сухопутно-морской terraqueous *(journey)*
сухопутный 1. land *(attr.),* overland, by land; 2. *bot., zool.* terricole, terricolous
сухость dryness; aridity
~ волос xerasia
~ глаза xerothalmus
~ кожи xeroderma
~, патологическая xerosis
~ слизистой оболочки носа xeromycteria
~ слизистой оболочки рта xerostomia
сухотка *med.* tases
~ спинного мозга [спинная] locomotor ataxia, posterior spinal sclerosis, tabes dorsalis
сухоустойчивый siccostable, xeric
сухоцветковый *bot.* xeroflorous
сухоцвет(ник) *bot.* immortelle *(Xeranthemum)*
~ однолетный common immortelle
сучильный twisting
сучильщик *text.* twister
сучение twisting, roving; (silk) throwing
сучить spin, twist; (silk) throw
сучок knot
суша (dry) land
сушеница *bot.* cudweed *(Gnaphalium)*
сушилка drier, dryer; desiccator
~, башенная drying tower
~, дымогазовая waste-gas drier
~, индукционная high-frequency drier
~, конвейерная continuous through-drier
~, конвективная convection drier
~, коридорная tunnel drier
~, ленточная conveyer-band (continuous-belt) drier
~, петлевая loop drier
~, противоточная counter-current drier
~ прямого действия direct-heat drier
~, прямоточная co-current drier
~, радиационная infrared drier
~, распылительная spray drier
~ с замкнутой циркуляцией recirculation [closed-circuit] drier
~ с кипящем слоем fluidized bed drier
~, сублимационная freeze drier
~, тарельчатая plate drier
~, теоретическая plate drier
~, туннельная tunnel drier
~, шахтная shaft drier
сушило drying stove/oven
~, ямное drying pit
сушитель drier, desiccator, drying agent
сушка drying, desiccation, dehydration; curing, seasoning
~, аэрофонтанная flash drying
~, индукционная high-frequency drying
~, конвективная convection drying
~, радиационная radiant(-heat) drying
~, распылительная spray drying
существительное *gram.* substantive, noun
~, нарицательное common noun
~, отглагольное verbal noun
~, причастное participial noun
сущность essence; substance, nature; core; *philos.* entity, ens
~, внутренняя inmost self
суягная овца ewe in yean
сфагнум *bot.* sphagnum, bog/peat moss
сфазирование phasing-in
сфалерит *miner.* sphalerite, zinc blende
сфен *miner.* sphene, titanite
~, бурый lederite
сфенодон *zool.* sphenodont *(Sphenodon)*
сфеноидальный sphenoid(al)
сфеноидит *med.* sphenoiditis
сфенолит *geol.* sphenolith
сфеноптеридные *n. pl. paleont.* sphenopterides
сфеностела sphenostele

сфенотрипсия sphenotripsy
сфенофилловые *n. pl. paleont. Sphenophyllales*
сфеноцефал sphenocephalus
сфеноцефалия sphenocephaly
сфера 1. sphere; 2. field, realm, domain; range, scope; line
~, единичная *math.* unit sphere
~ компетенции authority
~ распространения incidence
сференхима *bot.* sphaerenchyma
сфериальные *n. pl. bot.* sphere fungi *(Sphaeriales)*
сфериолит spheriolite
сферит *miner.* sphaerite
сферичность sphericity
сфероид *math.* spheroid
~,вытянутый prolate spheroid
сферокарповый *bot.* sphaerocarpous
сферокобальтит *miner.* sphaerocobaltite
сферолит 1. *miner.* spherolite; 2. spherolith
сферолитовый spherulitic
сферосидерит *miner.* sphaerosiderite
сферосома spherosome
сферотека *bot. Sphaerotheca*
сферула *zool.* spherule
сферулиноз *bot.* Sphaerulina leaf spot
сферофакия *med.* spherophakia, microphakia
сфероцит spherocyte
сфецида *zool.* digger wasp, dauber *(Sphecidae)*
сфигмограф sphygmograph, pilsimeter
сфигмокардиограф spygmocardiograph, sphygmocardioscope
сфигмоманометр sphygmomanometer, sphygmometer
~, мембранный aneroid sphygmomanometer
сфигмосистола sphygmosystole
сфингозин sphingosine
сфинголипидоз sphingolipidosis
сфингомиелиназа sphingomyelinase
сфингофильный sphingophilous
сфинкс 1. sphinx; 2. *zool.* sphinx baboon *(Papio sphinx);* 3. *zool.* hawk moth *(Sphinx)*
сфинктер *anat.* sphincter
~ заднего прохода sphicter ani
~ зрачка sphincter pupillae
~ мочеиспукательного канала sphincter urethrae
сфинктерит shphincteritis
сфинктеропластика sphincteroplasty
сфорцандо *mus.* sforzando
сфорцато *mus.* sforzato, sforzando
сфрагистика sphragistics
схватки spasm, contractions, fit
~, родовые labour/birth pains, birth pangs
~, родовые судорожные spasmodic labour pains, clonus uteri
схватывание 1. gripping; clutching; 2. setting, hardening
схватываться set, harden
схема diagram, chart; design; plan, layout, scheme; schematic model; pattern; outline, sketch; circuit, network
~, автодинная autodyne circuit
~ аксиом *math.* axiom scheme
~, алгоритмическая algorithmic diagram
~, аналоговая analog(ue) circuit
~ анодного повторителя see-saw circuit
~ антисовпадений anticoincidence circuit
~, балансная balanced circuit
~ Бернулли *math.* Bernoulli trials
~ блокирозки interlocking circuit
~, блочная block diagram
~, быстродействующая fast (high-speed) circuit
~, вентильная gate (circuit), gating circuit
~, вероятностная probability sheme
~ включений connection diagram
~ выборки данных data selection circuit
~, выключающая trip(ping) circuit; shut-down circuit
~ вычисления pattern of calculation; computational sheme, computing plan
~ вычитания subtract/subtraction circuit
~, гасящая quench(ing) circuit
~, герметизированная potted circuit
~ движения населения population movement schedule
~, двухтактная push-pull circuit
~ деления dividing circuit
~, древовидная tree chart/diagram
~ задержки delay circuit
~, заказная custom circuit
~ замещения equivalent circuit
~, заостряющая peaking circuit
~, запоминающая memory/storage circuit
~ запрета inhibitor circuit, inhibit gate
~ запуска trigger circuit
~, зарядная charging circuit
~ И AND circuit/gate, coincidence gate
~ И - И AND - to - AND circuit
~ И - ИЛИ AND - to - OR circuit
~ ИЛИ OR circuit/gate
~ изменчивости variability chart
~ ИЛИ - ИЛИ OR - ELSE circuit
~, импульсная pulse circuit
~ И - НЕТ inhibitory circuit/gate
~, интегральная integrated circuit
~, интегрирующая integrating circuit/network
~, информационная *game th.* pattern of information
~ калибровки calibrating circuit
~, кинематическая mechanical diagram
~, ключевая gating circuit, gate
~, кольцевая ring circuit
~, кольцевая десятичная ring-of-ten circuit
~, кольцевая пересчетная ring scaler
~, ламповая tube circuit
~, логическая logic (circuit)
~, мажоритарная majority circuit
~, маршрутная (plant) flow diagram

~, мнемоническая mnemonics, mimic diagram/flowsheet
~ моделирования simulation scheme
~, моделирующая analogous/analog(ue) circuit
~, монтажная assembly/erection diagram; *el.* wiring diagram
~, мостиковая/мостовая *el.* bridge circuit
~ на дискретных компонентах discrete component circuit
~ на основе твердого тела [на твердых телах] solid circuit
~ НЕ NOT circuit/gate
~ несовпадения non-coincidence [anticoincidence] circuit
~, обостряющая peaking circuit
~ обратной связи feedback circuit
~ объединения OR circuit/gate
~, ограничивающая clipping circuit
~, однотактная single-ended circuit
~ однократного цикла (технологическая) once-through flowsheet
~ опробавания sampling design
~ опрокидывания flip-flop (circuit)
~, организационная organization chart
~, пересчетная scaler, scaling circuit/unit
~, печатная printed/plated circuit
~ повторителя OR circuit/gate
~, последовательностная sequential diagram
~, потенциальная level circuit
~ потока flow pattern
~, принципиальная principal scheme, basic/key/block/circuit diagram
~, пояснительная key diagram
~ программы, иерархическая hierarchy chart
~ производственного процеса process layout; flow diagram/chart, flowsheet
~ прокатки *met.* rolling schedule
~ пропускания gating circuit, gate
~, пусковая trigger circuit; start-up system
~ разноименности exclusive OR circuit/element
~, разностная difference scheme
~ распада decay/disintegration scheme
~ расположения (общая) layout
~, расчетная design diagram
~ с одним устойчивым состоянием monostable circuit
~ самонастройки (self-)adaptive circuit
~ сброса blanking circuit
~ связи communication diagram
~ сглаживания/сглаживающая smoothing circuit
~, селекторная gating circuit, gate
~ сил force diagram
~, скелетная block/skeleton diagram
~ сложения adder/adding circuit
~ смазки lubrication chart/diagram
~, смесительная mixer circuit
~ совпадения match/mixer circuit; AND circuit/gate, coincidence gate
~ стирания blanking circuit
~, сторожевая *comp.* watchdog
~, стробирующая gating circuit, gate
~, структурная block diagram
~, суммирующая adder/adding/summing circuit
~, счетная counter/counting circuit
~, технологическая [технологического процесса] flow diagram/chart, flowsheet
~, Т-образная T-circuit/network
~, транзисторная transistor(ized) circuit
~, тригерная *comp.* flip-flop circuit
~ трубопроводов piping layout/plan
~ управления control circuit
~, формирующая shaping circuit, shaper
~, функциональная functional/block diagram
~, хронирующая timing circuit
~, электрическая electric circuit, circuitry, circuit diagram
~, электронная electronic circuit
~ электропроводки wiring diagram
схематизировать outline, present as an outline
схематизм sketchiness; schematism
схематика, схемотехника circuitry
схематичность sketchiness
схематичный, схематический schematic, diagrammatic; sketchy; oversimplified
схемный diagram, chart, circuit *(all attr.)*; diagrammatic
схемотехника circuit engineering
схенус *bot.* bog rush *(Schoenus)*
схизма *eccl.* schism
схизматик, схизматический schismatic
схизогенез schizogenesis
схизогенный schizogenetic, schizogenic, schizogenous
схизогнатический *zool.* schizognathous
схизокарпий *bot.* schizocarp
схизокарпный schizocarpic, schizocarpous
схизокодон *bot.* fringe bell *(Schizocodon)*
схизокотилия *bot.* schizocotyly
схизолизигенный *bot.* schizolysigenous
схизолит *miner.* schizolite
схизомицеты *bot.* schyzomycetes
схизоморфный schizomorphic
схизоневра *bot.* schizoneuric plant
схизоспора *bot.* schizospore
схизостела *bot.* schizostele
схизостелия *bot.* schizostely
схизостилис *bot.* crimson flag *(Schizostylis)*
схизофит *bot.* schizophite
схизофрагма *bot.* hydrangea vine *(Schizophragma)*
схизофрения = шизофрения
схима *eccl.* schema
схистоцерка *zool.* *Schistocerca*
схистоцит *med.* schistocyte
схлопывание collapse
сход departure; take off
сходимость convergence, converging
~ в среднем *math.* convergence in the mean

~ по вероятности convergence in probability, probability convergence
~ по мере convergence in measure
~ пучка beam convergence
~, равномерная uniform convergence
сходиться converge
сходящийся convergent
схождение convergence
схоласт(ик) scholastic; doctrinaire
схоластика scholasticism
сцеживание decanting, decantation
сцеживать decant, pour off; elutriate
сценарий scenario; film script
сценарист script/scenario writer
сценография scenography
сцеп 1. train, tow; 2. coupling, hitching
сцепка coupling; connection; tow, hitch
~, многозвенная articulated coupling
сцепление 1. adhesion; engagement, link, coupling; 2. *auto.* clutch; 3. *math.* concatenation
~ алгоритма algorithm concatenation
~, ленточное band coupling
~ с полом sex linkage
~, совместное linkage in conjunction
~ шины с дорогой road grip of a tyre
сцепляемость adhesiveness; cohesion
сцеплять link, couple, clutch, connect; concatenate; engage, mesh
сцепной linking, coupling; link *(attr.)*
сцепщик *rail.* coupler
сциадопитис *bot.* umbrella pine *(Sciadopitis)*
сциена *zool.* drumfish *(Sciaena)*
сцилла *bot.* squill *(Scilla)*
Сцилла *myth.* Scylla
сциллaремин scillaremin
сцилларен scillaren
сцилларенин scillarenin
сциллин scillin
сциллипикрин scillipicrin
сциллитин scillitin
сциндапсус *bot.* ivy arum *(Scindapsus)*
сцинк *zool.* skink *(Scincus)*
сцинтиграфия scintigraphy
сцинтиллирование scintillation
сцинтиллировать scintillate
сцинтиллограф scintiscanner, scintillograph
сцинтиллография scintiscan, scintillography
сцинтиллятор scintillator, scintillation counter; scintillometer; scintilloscope; scintillating phosphor
сциофилл skiophyll, sciophyll
сциофильный skiophilous, sciophilous
сциофит *bot.* skiophyte, sciophyte, skiaphyte, shade-loving plant
сцифистома *zool.* scyphistoma
сцифовидный, сцифообразный *bot.* scyphiform
сцифоидные *n. pl. zool.* scyphozoans *(Scyphozoa)*
сцифомедуза *zool.* scyphomedusan *(Scyphomedusa)*
сцифостома *zool.* scyphostoma, scyphistoma
сцифула *bot.* scyphulus
счеканивать engrave, emboss
счет 1. counting, count, reckoning; account; 2. bill; check, cheque; 3. expense, account; 4. *mus.* time
~ альфа-частиц alpha-counting
~, ассигнованный appropriation account
~, брокерский discretionary account
~ в банке bank account
~ в гроссбухе general account; ledger
~ в обратную сторону countdown
~, вести operate an account
~, взаимный контрольный reciprocal control account
~ времени, предстартовый *rock.* countdown
~ гамма-квантов gamma-counting
~, двойной *stat.* double counting; duplication
~ делений *phys.* fission counting
~, детализированный itemized account
~, доверительный trust account
~ за телефон telephone bill
~ за топливо fuel bill
~ за фрахт freight note
~ записей record count(ing)
~ издержек cost account
~, итоговый summary account
~ капитала assets account
~ кассы cash account
~ лоро loro account
~, неоплаченный outstanding account
~, неурегулированный unsettled account
~ ностро nostro account
~ облигаций bond account
~, обратный count-down
~ общий/объединенный joint account
~ основных фондов fixed assets account
~, особый specified/special account
~ по которому допущен овердрафт overdrawn acccunt
~ по которому начисляются проценты compound interest account
~, переводный transfer account
~, повторный *stat.* double counting
~ подотчетных сумм expense account
~, поставить на charge to account
~ производственных издержек manufacturing account
~, промежуточный interim/intermediate account
~, просроченный overextended account
~, прямой count-up
~, распределительный clearing account
~ расходов bill of costs/charge(es)
~, расчетный settlement account
~ с дебетовым сальдо debit account
~, сводный closing account
~ сомнительных дебиторов suspense account
~ ссуд loan account
~, текучий current account; checking account
~, урегулированный/оплаченный settled

account
~, фиктивный simulated account
~ ценных бумаг securities/stock account
счетверенный quadruple(d)
счетно-вычислительный computer *(attr.)*
счетно-решающий computing *(attr.)*
счетность countability
счетный counting; calculating; accounting, accounts *(attr.)*
~ная книга accounts book
~ная линейка slide rule
~ная машина calculating/adding machine
~ное существительное noun of quantity
счетовод accountant; bookkeeping/accounts clerk
~ по заранее оплаченным счетам advance-payment clerk
счетовод-контролер audit clerk
счетоводный bookkeeping, accounting *(attr.)*
счетоводство bookkeeping, accounting
счет-фактура invoice
~, сличенный audited voucher
счетчик 1. counter; register, meter; 2. teller; census taker; 3. *electr.* counter/counting tube
~ адресов address/location counter
~ альфа-частиц alpha-counter
~ активной энергии *el.* watt-hour meter
~ ампер-часов ampere-hour meter
~ бета-частиц beta-counter
~, борный boron counter
~, бромноаргоновый argon-bromine counter
~, бромный bromine-quenched counter
~, быстродействующий high-speed (fast) counter
~ ватт-часов watt-hour meter
~ времени timer-counter; time totalizer
~, вычитающий down counter
~, газовый 1. gas meter; 2. gas counter
~, газоразрядный gas-discharge counter
~ гамма-квантов gamma-counter
~ Гейгера-Мюллера Geiger-Mueller counter, G-M tube
~, двоичный binary counter
~, декадный decade counter tube, decatron
~ делений fission counter
~ делений, сдвоенный back-to-back fission counter
~, защищенный shielded counter
~, десятичный decade counter
~ записей record counter
~ излучений radiation counter
~, импульсный pulsed counter
~ импульсов pulse counter/meter
~ инфракрасного излучения infrared quantum counter
~ карт card counter
~ команд instruction/command counter
~ количества электричества ampere-hour meter
~, кольцевой ring/annular counter
~, контрольный monitor counter
~ координат coordinate counter, dead-reckoning counter
~, кристаллический crystal counter
~, многонитный multiwire counter
~ на триггерах flip-flop counter
~, накапливающий accumulating counter; adder, accumulator
~, направленный directional counter
~ нейтронов, борный boron neutron counter
~, ненаправленный non-directional counter
~ оборотов revolution/cycle counter, tachometer
~ ожидания *comp.* wait counter
~, острийный point counter
~, погружной dip/dipping/immersion counter
~, поисковый prospecting counter
~ полной энергии *el.* volt-ampere-hour counter
~ при переписи населения census interviewer
~, проточный flow(-type) counter
~, пусковой start-up counter
~ с высокой разрешающей способностью high-resolution counter
~, самогасящийся self-quenching/quenched counter
~, реверсивный bidirectional counter
~ со схемой опрокидывания flip-flop counter
~, стильбеновый stilbene counter
~, суммарный total counter
~, суммирующий totalizing meter; up-counter
~, сцинтилляционный scintillation counter
~, тканеэквивалентный tissue-equivalent counter
~, торцевой end-window counter
~ фотонов photon/quantum counter
~, циркуляционный проточный loop-line flow counter
~, циферблатный dial counter
~, цифровой digit counter
~, электронный декадный electronic decade counter; decade counting tube
~ электроэнергии electricity meter
~, эталонный standard meter
счетчик-накопитель storage counter
счетчик-расходомер flowmeter counter
счетчик-сумматор adding/storage counter
счеты abacus, counting frame
счисление 1. counting, enumeration, numbering; 2. reckoning; 3. notation, number system
~ пути *nav.* dead reckoning
счислитель computer, counter; reckoner
считать 1. count; compute, calculate; reckon; 2. consider; regard
~ в обратном направлении count down
~ в прямом направлении count up
считчик proofreader
считывание read-out, reading; sensing
~ аналоговых данных analog(ue) read-out
~ знаков character sensing
~ информации information read-out
~, контрольное test reading
~ маркеров labelling/mark/tagging reading

~ массива read-out of file
~, повторное rereading
~, проверочное proofreading
~, радиочастотное r.f. read-out [sensing]
~ с диэлектрических карт dielectric card reading
~ с диэлектрической перфоленты dielectric punched tape reading
~ с (перфо)карт card reading
~ с разрушением информации destructive reading
~ с синхронизацией clock-actuated read-out
~, стирающее destructive read-out
~ файла file scanning
~, фотоэлектрическое photoelectric reading/sensing
считыватель *comp.* reader
считывать read out/off, read; sense; scan
считывающее устройство reader
сшивание sewing together; *med.* suture, suturing; *comp.* joining
сшивать sew/stitch together; *med.* suture; *comp.* join
съезд 1. conference, convention, congress, rally ; 2. *rail.* access/connecting/crossing track
съем 1. removal, withdrawal; extraction; dismantling; 2. *text.* doff(ing); 3. *met.* dross
съемка survey(ing), mapping, plotting; filming, photography
~, авиарадиометрическая/аэрорадиометрическая aerial radiometric survey
~, аэрофототопографическая aerosurveying, mapping photography
~, буссольная compass survey
~, газовая emanation survey
~, геодезическая geodetic survey
~, глазомерная eye sketch, rough/approximate survey, field sketching
~, инструментальная instrment survey
~, кадастровая/местности land survey
~, маркшейдерская mine/underground survey
~, маршрутная route survey
~, мензульная/планшетная plane-table survey
~, площадная aerial survey
~, поверочная revision survey
~ под углом canted shot
~, снегомерная snow survey
~, фототеодолитная photomapping
съемник stripping device, stripper, extractor, removal, puller
~ шин колес tyre removal rig
съемный detachable, removable, withdrawable
съемочный filming, picture-taking, shooting
съемщик 1. leaseholder, leaser, tenant, lessee; 2. remover, stripper; *text.* doffer
сыворотка 1. *biol.* serum; 2. whey
~, антимикробная antibacterial serum
~, иммунная immune serum; antiserum, specific serum
~, кроличья rabbit serum
~, лечебная curative/immune serum
~, молочная serum lactis, whey
~, противовирусная antivirus serum
~, противогриппозная antigrippal serum
~, розовая purpura
~, сифилитическая syphiloderma
~, скарлатинозная scarlet rash
~, сывороточная serum rash/eruption
~, угревидная acneiform eruption
~, уртикарная urticaria
~, экземоподобная eczematous eruption
сыреть grow damp/moist
сырец raw product/material
сырный cheese *(attr.),* cheesy; caseous
~ая неделя *eccl.* Shrovetide
сыроежка *bot.* russula *(Russula)*
сырой 1. raw, crude; 2. damp, moist; soggy, wet
сыропуст *eccl.* Shrove Sunday
сыропустная неделя *eccl.* shrove week
сыропустное воскресенье Shrove Sunday
сырье raw material(s); raw produce
~, алмазное rough diamonds
~, исходное feedstock
~, минеральное mineral products
~, пищевое edible raw material
сыть *bot.* cyperus *(Cyperus)*
сыч *zool.* owl
сыссертскит *miner.* syssertskite
сымуг *zool.* abomasum, rennet bag
сычужина rennin, rennet
СЭВ *abbr., hist.* CMEA, Council for Mutual Economic Assistance, Comecon
сэр sir
сэбин *phys.* sabin
сюзерен *hist.* suzerain, overlord
сюзеренитет suzerainty
сяг, сяжок *zool.* antenna, feeler

Т, т

Т *av.* tee, T
~, посадочное landing T/tee
табак 1. tobacco; 2. tobacco plant, tabacum *(Nicotiana tabacum)*
~, жевательный chewing tobacco
~, курительный/культурный smoking tobacco
~, нюхательный snuff tobacco
~, трубочный pipe tobacco
табаковод tobacco grower/planter/farmer
табаководство tobacco cultivation/growing
табакоз *med.* tabacosis, tabacism, chronic tobacco poisoning
табель 1. table; list; chart; 2. time/attendance board, time clock card
табельная time board; clock-house

табес *med.* tabes
~, детский juvenile tabes
~, спинномозговый tabes dorsalis/spinalis, locomotor ataxia
табетичный tabetic, tabic, tabid
таблетирование, таблетировочный tableting
таблетка tablet, lozenge; pill, troche, pastil
таблица table, chart; list; schedule; diagram
~ адресов address table/file/notebook; directory
~ входов entry table
~ выборочных данных sample table
~ вычитания subtraction table
~, гипсометрическая altitude-pressure table
~, градуировочная calibration chart
~ дальностей distance/range table
~ доз dosologic table
~ затрат и выпуска input-output table
~ издержек производства cost schedule
~, испытательная *telev.* test pattern/chart
~, истинностная/истинности *math.* truth table
~ кодов code chart
~, комбинационная cross-tabulation
~, контрольная checklist
~ курсов *fin.* list of quotations
~ логарифмов table of logarithms
~ межотраслевого баланса input-output table
~ Менделеева periodic chart/table
~ начислений износа depreciation schedule
~ отбора рисков *stat.* select table
~ паспортных данных header table
~, переводная/пересчетная/преобразования conversion table
~ прерываний interrupt table
~, просмотровая lookup table
~, психрометрическая humidity table
~, расчетная computational table
~ с двумя входами double-entry table
~, сводная summary table
~ сложения addition/add table
~ смертности *stat.* life table
~ смертности для страхования жизни actuarial life table
~ содержания тома *comp.* volume table of contents, data carrier contents
~, справочная reference table
~, тарировочная calibration table
~ умножения multiplication table
~ экспозиции *phot.* exposure table
~ элементов Менделеева, периодическая periodic chart/table
табличка *anat., zool.* plate; *av.* placard
~ чашечки calyx plate
~, паспортная nameplate
табличный tabular; tabulated
табло panel; board; display
~, индикаторное advisory annunciator panel
~, информационное information board
~, ламповое lamp display
~, путевое (световое) *rail.* illuminated indicator board
~, световое light annunciator, illuminated indicator board
~ сигнализации отказов failure indicator
табопаралич *med.* taboparesis
таборит *hist.* Taborite
табулатура, табулятура *mus.* tabulature
табулировать tabulate
табу taboo
табулограмма *comp.* totals printout, tabulation
табулярный *hist.* tabulary
табулятор tabulator, tabulating machine
~, установочный setting tabulator
табун *chem.* tabun
таволга *bot.* spirea *(Spiraea)*
тавот (consistent) grease, cup grease
тавотник, тавотница grease cup
тавотонагнетатель, тавот-пресс grease gun
таврить brand
таврический Taurian
тавробульб tee bulb bar
тавровый T-shaped
тавтология tautology
тавтоним *bot., zool.* tautonym
таг *hist.* tagus
тагальский Tagalog
таган trivet
тагетес *bot.* marigold *(Tagetes)*
тагмема *ling.* tagmeme
тагмемика tagmemics
тагос *hist.* tagus
тагуан *zool.* taguan *(Petaurista petaurista)*
таджик, таджикский Tadzhik, Tadjik
таджичка Tadzhik/Tadjik woman
таз 1. pan; (wash) basin; 2. *med.* pelvis
~, женский gynecoid pelvis
~, инфантильный juvenile pelvis
~ мужского типа android/funnel/masculine pelvis
~, расщепленный split pelvis
~, рахитический rachitic/malocostean pelvis
~, узкий contracted pelvis
тазик *paleont.* coxa
тазобедренный *anat.* coxofemoral
тазовертельный *anat.* pelviotrochanteric
тазовый *anat.* pelvic
тазомер pelvimeter
таинство *rel.* sacrament
тай Thai, Siamese
тайванский, тайваньский Taiwanese
Тайвань Taiwan
Тайланд Thailand
тайкун tycoon
тайм *sport* half *(game)*
таймен *zool.* huchen *(Hucho hucho);* salmon trout *(Salmo trutta)*
таймер timer, timing unit, clock
тайм-чартер time-charter
таймшит time-sheet
тайна secret; secrecy; mystery

~ **переписки** privacy of correspondence
тайник *bot.* listera *(Listera)*
тайнобрачие 1. secret marriage; 2. *bot.* cryptogamy
тайнобрачный *bot.* cryptogamic, cryptogamous
тайпин *hist.* Taiping
тайский Thai
тайфун typhoon
такелаж *naut.* rigging, gear
~, **бегучий** running gear/rigging
~ **грот-мачты** main rigging
~, **запасный** rechange
~ **фок-мачты** fore/head rigging
такелажник rigger
такелажный *naut.* rigging
~, **подвесной** suspension rigging
~, **привязной** captive rigging
такин *zool.* takin *(Budorcas taxicolor)*
такса toll, fee, rate (of charge), fixed rate; tariff
таксатор valuer, estimator, appraiser
~, **официальный** qualified appraiser
таксация valuation, estimation
таксема *ling.* taxeme
таксировать estimate, evaluate, assess; fix the price of; *for.* cruise
таксин taxine
таксис *biol.* taxis
таксит *petr.* taxite
таксовый fee *(attr.),* rate-fixing
таксодиум, таксодий *bot.* common bald cypress *(Taxodium distichum)*
таксометр taximeter
таксон taxon
таксономия taxonomy; categorization
таксопарк *abbr.* taxi depot/pool
таксофон coin-box telephone, coin/pay telephone
такусса *bot.* ragi millet *(Eleusine tacussa)*
такт 1. *tech.* stroke, cycle; 2. *mus.* time, beat, step, bar, rhythm; measure
~ **впуска/всасывания** admission/induction/inlet/intake/suction stroke
~ **выпуска/выхлопа** exhaust stroke
~ **расширения** expansion stroke
~ **сжатия** compression stroke
тактик tactician
тактика tactics
тактильный tactile
тактирование clocking; time beating
тактовый time, bar *(attr.);* timing
~**вая дорожка** clock/timing track, clock marker track
~**вая черта** *mus.* bar
талальгия *med.* talalgia, pternalgia
таламический *anat.* thalamic
таламо-круральный thalamocrural, thalamopeduncular
таламо-маммилярный thalamomammillary
таламо-рубральный thalamorubral, rubrothalamic
таламо-руброоливарный thalamorubroolivary
таламус thalamus; optic thalamus; *bot.* receptacle
талассемия *med.* thalassemia, thalassanemia
талассический thalassic, thalassal
талассотерапия thalassotherapy
талассохтонный *geol.* oceanic
талаттоген thalattogen
талаттозавр *paleont.* thalattosaur
талаума *bot.* talauma *(Talauma)*
талер thaler *(coin)*
тали pulley blocks/tackle, hoisting tackle, purchase, block and tackle
~, **вспомогательные/горизонтальные** fleeting tackle
~ **с цепью** hoist chain
~ **над люком [грузовые]** hatch tackle
талидомид *chem.* thalidomide
Талия *myth.* Thalia
таллий *chem.* thallium, T
таллиевый *chem.* thallic
таллиман tallyman
таллоидный *bot.* thalloid, thallose
таллом *bot.* thallome, thallus
талломный *bot.* thalline, thallodal, thallodic
таллоспора *bot.* thallospore
таллотерофит *bot.* thallotherophite
таллотоксикоз *med.* thallotoxicosis
таллофит *bot.* thallophyte
таллофитный *bot.* thallophytic
таллус = таллом
талмуд Talmud
талмудист Talmudist; *fig.* doctrinaire, pedant
талмудистский talmudistic; *fig.* doctrinaire
талмудический Talmudic(al); *fig.* doctrinaire
талон coupon, card, ticket, token; stub; voucher; *econ.* stock certificate
~, **посадочный** boarding card
талреп *naut.* turnbuckle; lanyard
~, **винтовой** rigging/adjuster screw, pipe/sleeve turnbuckle, screw/wire stretcher
таль hoist, lifting tackle
тальвег thalweg; valley line
тальк talc; talcum powder
~, **листоватый** foliated talc
~, **отвердевший** indurated talc
~, **чешуйчатый** micaceous talc
тальковый talcous; talcose; talcum *(attr.)*
талькоз *med.* talcosis
талькозный *med.* talcosal
тамада master of ceremonies; toastmaster
тамаринд *bot.* tamarind *(Tamarindus)*
тамбур 1. vestibule, lobby; 2. *arch.* tambour; 3. *naut.* companion; *rail.* covered platform
~, **выходной** exit lock
тамил Tamil
тамилка Tamil (woman)
тамильский Tamil, Tamilian
таможенный customs *(attr.)*
таможня custom house/office, customs
тампала *bot.* tampala *(Amaranthus tricolor)*
тамплиер *hist.* Knight Templar

тампон plug, wad; *med.* tampon
тампонада *med.* tamponade, tamponage
тампонаж *tech.* plugging, tamping
тампонирование 1. *tech.* plugging, tamping; 2. *med.* tamponing, tamponment, tamponade
тампонировать 1. *tech.* plug up, backfill; 2. *med.* tampon, insert/place a tampon
тамтам *mus.* tom-tom
тамус *bot.* black bryony, tamus *(Tamus)*
тан *hist.* thane
тамурия *med.* thamuria, pollakiuria
танагра *zool.* tanager *(Thraupidae)*
Танагра Tanagra figurine
танальбин albutannin
тананаривский Tananarive
танатобиологический thanatobiologic
танатогенез thanatogenesis
танатоз thanatos
танатология thanatology
Танатос Thanatos
танатоценоз thanatocoenosis
танацетон *chem.* tanacetol, tanacetone
танбур *mus.* tamboura, tambura
тангаж *aero., naut.* pitch, pitching
танганьиканский, танганьикский Tanganyikan
тангенс tangent, tan, tg
тангенс-буссоль tangent compass
тангенсоида *math.* tangent curve, tangensoid
тангента *rad.* speaking key/switch, press-/push-to-talk key
тангентальный tangent
тангенциальный tangential
тангут, тангутский Tangut
тандем tandem
тандер *aero.* turnbuckle
танец dance; dancing
  ~ на пуантах toe dance
Танжер Tangier
танжерин *bot.* tangerine *(Citrus deliciosa)*
танжерский Tangerine
танзаниец, танзанийский Tanzanian
танзанийка Tanzanian (woman)
танин tannin
танк 1.cistern, container, hopper, tank; 2. *mil.* tank
  ~, закалочный freezing tank
  ~, сливной collecting tank
танкаж *agr.* meat-and-bone meal, tankage
танкер tanker, tank ship/vessel, oil tanker/carrier, oiler
  ~, подводный submarine tanker/oiler
  ~, портовый yard oiler
  ~ снабжения tanker supply ship
  ~, универсальный versatile tanker
танкер-заправщик refuelling/replenishment tanker
танкер-рудовоз oil and ore-carrying ship
танкетка *mil.* tankette, small tank
танкист tankman
танковый tank *(attr.);* armour, armoured
танк-тральщик *mil.* mine-clearing tank
танназа *bioch.* tannase
таннат *chem.* tannate
таннин, таннид tannin
танниноносный *bot.* tanniferous
тантал *chem.* tantalum, Ta
Тантал *myth.* Tantalus
танталат, танталатный *chem.* tantalate
танталовый tantalic
тантиль *stat.* tantile
тантьема bonus
тапетум *bot.* tapetum, tapetal layer
тапиока tapioca
тапир *zool.* tapir *(Tapirus)*
тапсия *bot.* deadly carrot *(Thapsia)*
тар *zool.* tahr, thar *(Hemitragus)*
тара packing, package, wrapping/packing material; container; *comm.* tare
  ~, возвратная/инвертная/многооборотная returnable/reusable container
  ~, групповая multipack
  ~ не подлежащая возврату throwaway tare
  ~, порожняя empties
  ~, разовая disposable/throwaway container
  ~, решетчатая crate
  ~, складная collapsible container
таракан *zool.* cockroach *(Blatta)*
тараксацерин *chem.* taraxacerin
тараксацин *chem.* taraxacin
таран I ram; *hist.* battering ram; *mil.* ramming, ram(ming) attack
таран II *bot.* alpine fleeceflower *(Polygonum alpinum)*
таранит ram (through)
таранно-большеберцовый *anat.* astragalotibial, talotibial
таранно-ладьевидный *anat.* astragaloscaphoid, talonavicular, taloscaphoid
таранно-малоберцовый talofibular, astragalofibular
таранно-пяточный talocalcaneal, talocalcanean, astragalocalcanean
таранный 1. ram *(attr.);* 2. *anat.* astragular
тарантизм tarantism, dancing mania
тарантул(а) *zool.* tarantula
тарань *zool.* roach *(Rutilus rutilus)*
тарбаган *zool.* Mongolian bobak *(Marmota bobak sibirica)*
тарденуазский Tardenoisian
тарелка plate; tray; disk
  ~, буферная *rail.* buffer disk/head
  ~ клапана/клапанная valve plate/disk
  ~, испаряющая boiler tray
  ~, колпачковая bubble-cap tray, bubble plate
  ~. летающая flying saucer
  ~, обогащающая enriching/rectifying plate
  ~, отбойная baffle/deflecting plate
  ~, теоретическая theoretical plate/stage
тарелочник *mus.* cymbalist
тарельчатый dish-shaped, platelike

~ **клапан** disk valve
тарзальгия *med.* tarsalgia, podalgia
тарзальный *anat.* tarsal
тарзиоид *zool.* tarsioid *(Tarsius)*
тарзит *med.* tarsitis
тарзо-глазничный tarso-orbital
тарзо-метатарзальный tarsometatarsal
тарзо-тарзальный tarsotarsal, mediotarsal
тарзо-тибиальный tarsotibial, tibiotarsal
тарзо-фалангеальный tarsophalangeal
тари *mus.* tari
тарирование, тарировка calibration, gauging; taring; calibration test/check
тарировать calibrate, tare, gauge
тариф tariff, rate; charge, duty
~ **большой скорости** express rate
~, **грузовой** cargo tariff
~, **дифференциальный** differential rate/tariff
~ **для импортных грузов** import rate/tariff
~ **для мелких отправок** less-than-carload rate
~, **единый** single/consolidated tariff
~, **зональный** area/zone rate
~, **карательный** retaliatory/retaliation/penalty tariff
~, **льготный** reduced tariff
~, **многоколонный** multilinear tariff
~, **начальный** *fin.* basing rate
~ **обратного пробега** backhaul rate
~, **ограничительный** discriminatory tariff
~, **одноколонный** unilinear [single-schedule] tariff
~, **основный** standard rate
~, **пассажирский** passenger tariff, fare
~, **переходный** bridge rate
~, **повременный** time rate
~, **покилометровый/помильный** mileage rate
~, **предпочтительный/преференциальный** preferential tariff
~, **простой** unilinear [single-schedule] tariff
~, **протекционистский** protection/protective/ protectionist tariff
~ **Райта** *econ.* Wright demand rate
~, **сквозной** through rate
~, **смешанный** compound tariff
~, **стоимостный** ad valorem tariff
~, **ступенчато-пропорциональный** block meter rate
~, **таможенный** custom tariff/duties/schedule
~, **транспортный** transportation rate
~, **трехколонный** three-tier tariff/rates
~ **установленный законом** statutory tariff
~, **фискальный** revenue tariff
тарификация rate fixing; job rating
тарифицировать tariff, rate (a job)
тарифный tariff, rate *(attr.)*
~**ная сетка** scale of rates/charges/dues
тарный container, packing *(attr.)*
тароматериал container/packing material
тарпан *zool.* tarpan, wild horse
тарпон *zool.* silverfish, tarpon *(Megalops atlanticus)*
тарра *bot.* snake melon *(Cucumis melo flexuosus)*
тарро *bot.* elephant's-ear, taro *(Colocasia antiquorum)*
тартар *myth.* Tartarus
тартронилмочевина *chem.* tartronyl urea
тархун *bot.* tarragon *(Artemisia dracunculus)*
тасманец, тасманийский, тасманский Tasmanian
ТАСС *abbr., obs.* Tass
тастатура *tel.* keyboard, key assembly/set
татарин Tatar, Tartar
таста key
татарка I Tatar (woman)
татарка II *bot.* Welsh onion *(Alium fistulosum)*
татарник *bot.* cotton thistle *(Onopordon)*
татарский Tatar, Tatarian
~**кое мыло** *bot.* cross campion, scarlet lychnis, Jerusalem cross *(Lychnis chalcedonica)*
татарщина *hist.* Tatar rule; tribute due to Tatar rulers
татуировка tattooing; tattoo
таурин *chem.* taurine
тау-мезон tau-meson
тауметр tau-meter
таурин *chem.* taurine
таурохолевый *chem.* taurocholic
тауроциамин *chem.* taurocyamine
тауроциаминкиназа taurocyamine kinase
тау-сагыз *bot.* Russian dandelion *(Scorzonera tau-saghyz)*
таутога *zool.* tautog *(Tautoga onitis)*
таутомерический tautomeric, tautomeral
таутомерия *chem.* tautomerism
таутохрона *math.* tautochrone
тафонимия taphonymy
тафофилия taphophilia
тафоценоз taphocoenosis
тафридиоз *bot.* taphridium leaf spot
тафрогенез taphrogenesis
тафрогения taphrogeny
тафта *text.* taffeta
тахеометр tacheometer, tachymeter
тахеометрия tacheometry, tachymetry
тахигенез *biol.* tachygenesis
тахикардия *med.* tachycardia, tachyrhythmia
~, **атрио-вентрикулярная [нодальная]** nodal tachycardia
~, **желудочковая** ventricular tachycardia
~, **предсердная** atrial/auricular tachycardia
~, **синусовая** sinus tachycardia
тахилалия tachylalia, rapid speech
тахилит *miner.* tachylite, basaltic glass
тахиметр tachymeter
тахиметрия tachymetry
тахина *zool.* tachina fly *(Tachinidae)*
тахипноэ *med.* tachypnea, polypnea
тахистерин *bioch.* tachysterol
тахистоскоп tachistoscope

тахистоскопия tachistoscopy
тахителический tachytelic
тахифагия tachyphagia, rapid eating
тахифемия tachyphemia
тахифилаксия tachyphylaxis
тахифразия tachyphrasia
тахогенератор tachometer generator, tachogenerator
тахограмма tachogram
тахограф tachograph, recording tachometer
таходинамо tachogenerator
тахометр tachometer; revolution counter
  ~ двигателя engine tachometer
  ~, дистанционный teletachometer, distant-reading tachometer
  ~, емкостный capacitor tachometer
  ~, индукционный drag-cup tachometer
  ~, инерционный inertia tachometer
  ~, магнитоиндукционный magnetic drag-type tachometer
  ~, самопишущий recording tachometer
  ~, стробоскопический stroboscopic tachometer, strobotac
  ~, часовой clockwork tachometer
тахометрический tachometric
тахометрия tachometry
тацет *bot.* polyanthus narcissus *(Narcissus tazetta)*
тачать stitch
тачка I wheelbarrow
тачка II stitching
ташизм *art* tachisme
таяние melting, thaw, thawing
таять thaw; melt
тванг nasal accent, twang
твердение 1. hardening; 2. solidification, setting; consolidation
твердеть 1. harden, grow hard; 2. solidify, set; 3. cure, mature
твердокожие *med.* scleroderma
твердомер hardness gauge/tester, durometer
твердонебный hard-palatal
твердоплавкий infusible
твердоплодный *bot.* sclerocarpous
твердосемянный *bot.* hard-seeded
твердостекловатый durovitreous
твердость hardness; rigidity, stiffness
  ~, маятниковая pendulum hardness
  ~ на вдавливание indentation hardness
  ~ на истирание abrasion hardness
  ~ по Бринеллю Brinell hardness
  ~ по Виккерсу Vickers hardness
  ~ по Кнупу Knoop hardness
  ~ по конусу conical indentation hardness, Ludwik hardness
  ~ по Мартенсу scratch hardness
  ~ по пирамиде pyramid/diamond/Vickers hardness
  ~ по Роквеллу Rockwell hardness
  ~ по склероскопу scleroscope/Shore/rebound hardness
  ~ по царапанию scratch hardness
  ~ по шкале Моса Mohs' scratch hardness
  ~ по Шору Shore/scleroscope hardness
твердотопливный solid-fuel
твердотельный solid-state
твердотянутый hard-drawn
твердоэрный *phon.* hard **r**
твердый hard; solid; firm; fixed; rigid
  ~ срок fixed time limit
  ~дое правило hard-and-fast rule
  ~дые цены fixed prices
твид tweed
твиндек *naut.* tweendeck
  ~, верхний/нижний upper/lower tweendeck
творец *rel.* Creator
творительный instrumental
  ~ падеж *gram.* instrumental case
творило mortar trough; lime pit/slaker
теантропизм theanthropism, theantropy
театр theatre, theater
  ~, анатомический *med.* operating theatre; dissecting room
  ~, зеленый outdoor theatre
  ~ военных действий theatre of operations
  ~ военных действий, морской theatre of naval operations
театровед theatre/drama specialist
театроведение theatre/drama study
тебаин *chem.* thebaine
тевтон(ец) *hist.* Teuton
тевтонический, тевтонский *hist.* Teutonic
тевяк *zool.* gray seal *(Halichoerus grypus)*
тег *comp.* tag
тегмен *bot., zool.* tegmen
тегментальный tegmental, tegmentary
тегменто-таламический thalamotegmental
тегула tegule
тезавратор *econ.* hoarder
тезаврация, тезаврирование hoarding, accumulation; treasure
тезауроз *med.* thesaurosis
тезаурус thesaurus
Тезей *myth.* Theseus
тезис *philos.* thesis
тезисный thesial
теин *chem.* theine, caffeine
теинизм *med.* theinism, theism
теист theist
теистический theistic; theist
тейлоризм Taylorism, Taylor system
тека *bot.* theca, moss capsule
текаспора *bot.* thecaspore, ascospore
текинец Turkmen, Turkoman, Tekke
текинка Turkmen/Tekke (woman)
текинский Turkmen *(attr.)*
текодонт, текодонтный thecodont
текодонтизм *med.* thecodontia
текома *bot.* trumpetbush *(Tecoma)*
текоматоз *med.* thecomatosis
текориза thecorhiza

текостегноз *med.* thecostegnosis, thecostegnosia
тексроп texrope
текст text; *mus.* words, libretto
~, **дикторский** script
текстиль textiles
текстильный textile
~ **комбинат** textile mill(s)
текстолит textolite, cloth-/fabric-based laminate
текстообработка *comp.* word processing
текстура texture, structure, fabric; *met.* grain orientation
~, **алевролитовая** *geol.* aleurolitic texture
~, **беспорядочная** directionless/irregular texture/structure
~, **кокардовая/кольчатая** cocarde structure
~, **комковатая** lump structure
~, **корковая** bread-crust structure
~, **кристаллическая** crystal structure
~, **ленточная** banded structure
~, **лучистая** beam structure
текстурированный textured; *met.* grain-oriented
тектит *geol.* tektite
~, **стекловидный** tektite glass
тектовулканический vulcanotectonic
тектоген tectogene
тектогенез tectogenesis
тектоморфный tectomorphic
тектона *bot.* teak *(Tectona)*
тектоника tectonics, structural geology
~ **выжимания** ejective tectonics
~, **надвиговая** thrust tectonics
~, **новейшая** neotectonics, recent tectonics
тектоносфера tectosphere, asthenosphere
текторий tectorium
текто-спинальный tectospinal
тектостела tectostele
текто-таламический tectothalamic
тектум *anat.* tectum
текучесть fluidity, liquidity, flowability; flow; *econ.* turnover, fluctuation; *met.* yield
тела *pl. med.* bodies; *see also* **тело**
~, **антиимунные** anti-immune bodies
~, **защитные/иммунные** immune bodies, antibodies
~, **хроматиновые** chromatinic bodies
~, **центральные** mid bodies
телальгия *med.* telalgia
телангиит *med.* telangiitis
телангиоз *med.* telangiosis
телангиэктазия *med.* telangiectasia, telangiectasis
~, **бородовчатая** angiokeratoma
~, **лимфатическая** lymphangiectasia
~, **пауковидная** spider telangiectasia
телангиэктатический telangiectatic
телеавтоматика automatic remote control, teleautomatics
телеамперметр teleammeter
телеателье television service shop
телебашня television tower
телевещание television broadcasting, telecasting
телевидение television
~, **замкнутое [без выхода в эфир]** closed-circuit television
~, **космическое** space television
~, **монохромное [черное-белое]** monochrome [black-and-white] television
~, **объемное** three-dimensional television
~, **учебное** educational television
~, **цветное** colour/color television
~, **школьное** classroom television
телевизор television receiver/set, TV set
телегамма-аппарат *med.* teletherapy gamma-apparatus
телегамматерапия telecurie therapy
телегамия *bot.* telegamy
телегония *biol.* telegony
телеграмма telegram(me), cable, wire
~, **входящая** incoming telegram
~, **доплатная** "collect" telegram
~, **исходящая** outgoing telegram/message
~ **с оплаченным ответом** telegram with prepaid reply
~, **срочная** urgent telegram/message
телеграмма-молния express/urgent telegram
телеграф 1. telegraph; 2. telegraph office
~, **самопишущий** teleprinter
~, **тональный** voice-frequency telegraph
~, **швартовый** *naut.* docking/mooring telegraph
телеграфирование telegraphy
~, **двухполюсное** double-current transmission/signalling; *Am.* polar working
~, **дуплексное** duplex telegraphy
~, **многократное** multiplex telegraphy
~, **надтональное** superaudio telegraphy
~, **однополюсное** single-current signalling/transmission
~, **тональное** voice-frequency carrier telegraphy
~, **фазовое** phase-shift keying
~, **частотное** frequency-shift keying
телеграфировать telegraph, cable, wire, send a telegram/cable/wire
телеграфист telegraph operator
телеграфия telegraphy
~, **беспроводная** radio/wireless telegraphy
~, **проводная** wire telegraphy
телеграфный telegraph *(attr.)*
теледатчик remote pickup
теледвигатель telemotor
теледиапроектор slide scanner
тележка trolley, carriage, truck, cart, hardcart
~, **аккумуляторная** storage-battery truck
~, **буксирная/буксировочная** towing carriage/trolly
~, **буксируемая** towed cart
~ **в магазине самообслуживания** shopping cart
~ **для внутризаводских перевозок** floor truck
~, **загрузочная** charging car
~, **крановая** crane trolley

~, **насосная** hydrant cart
~ **оператора** *cin., telev.* dolly
~, **опрокидывающаяся** tipping car
~, **подъемно-транспортная** lift truck
~, **пусковая/стартовая** launching trolley
~, **разливочная** *met.* casting car, teeming truck
~, **ручная** hand truck
~, **скиповая** skip car
~ **тендера** tender truck
~, **технологическая** handling trolley/dolly
~, **транспортная** freight/industrial truck
~ **шасси** *av.* bogie
~, **шлепперная** *met.* drag carriage
тележурнал serialized TV program(me)
телезапись television/TV recording
телезритель television viewer, televiewer; *pl.* TV audience(s)
телеизмерение remote measurement, telemetering, telemetry
телеинтервью television interview
телеиспытание teletest
телейтогонидия teleutogonidium
телейтокучка, телейтосорус *bot.* teleutosorus, telium
телейтопустула *bot.* telium
телейтоспора *bot.* teleutospore, winter spore
телейтостадия teleutostage
телекамера television/TV camera
телекардиостимулятор *med.* telecardiostimulator
телекинез telekinesis
телекино television cinema/film
телекиноканал telecine channel
телекинопроектор telecine/television projector; film scanner
телекиноустановка telecine unit
телекия *bot.* oxeeye *(Telekia)*
телекомментатор television/TV commentator
телекомпания television company
телеконтроль remote control
телеконференцсвязь teleconferencing
телекс, телексный telex
телекюритерапия telecurie therapy
телелюбитель television fan
телеметрический telemetric, telemetering
телеметрия telemetry, telemetering, remote metering
~, **биологическая** biotelemetry, biological telemetry
~ **сердца** cardiac telemetry
телемеханизация introduction of remote control
телемеханика remote control, teleautomatics
телемост long-distance TV communication (via satellite)
теленабор telegraphic typesetting
теленовости television/TV news
телеобработка данных data teleprocessing
телеобъектив telephoto/telescopic lens
телеозавр *paleont.* teleosaur *(Teleosaurus)*
телеолог teleologist
телеология teleology
телеоператор television cameraman
телепат telepathist
телепередатчик television transmitter
телепередача television/TV broadcast, telecast
телеперфоратор telepunching unit
телепостановка teleplay
телеприемник television receiver/set, TV set
телерадиотерапия telecurie/teleradium therapy, radium beam therapy
телерегистрация remote registration
телерегулирование remote control
телерегулируемый remote-controlled
телерентгенография teleroentgenography, teleradiography
телерентгенотерапия teleroentgenotherapy, X-ray teletherapy
телерепортаж live television broadcast
телерепортер television reporter
телесамописец telerecorder
телесвязь telecommunication
телескоп telescope
~ **Галилея** Galilean telescope
~, **зеркально-линзовый** mirror-lens [reflecting-refracting] telescope
~, **линзовый** lens/refractive/refracting telescope
~, **менисковый** meniscus/Maksutov telescope
~, **пробежно-ионизационный** range-ionization telescope
телескопия telescopy
телесный corporal; somatic; physical; solid
~ **угол** solid angle
телеспектакль teleplay
телесперма *bot.* green thread *(Thelesperma)*
телестанция television broadcasting station
телестих *lit.* telestich
телестудия television/telecasting studio
телесъемка telephotography
телетайп teleprinter, teletype
телетайпист teleprinter operator, teletypist
телетеатр television/TV theatre/theater
телеуправление telecontrol, remote/distance/distant/supervisory control
телеуправляемый remote-controlled, remotely operated, telecontrolled
телеучет telemetering
телефикация installation of television equipment/network
телефильм television/TV film, telefilm
телефиум *bot.* orpine *(Telephium)*
телефон telephone; telephone set/instrument
~, **автоматический** dial telephone
~, **безбатарейный** voice-powered/operated telephone
~, **гидроакустический** underwater telephone
~, **головной** headphone, headset, earphone
~, **междугородный** long-distance telephone service/system
~, **международный** international telephone service

~ **системы МБ** local-battery telephone set
~ **системы ЦБ** common-battery telephone set
~, **чашечный** earpiece, ear-cap
телефон-автомат public [coin-box] telephone, telephone booth
телефонизация installation of telephones
телефонирование telephoning; telephony
~, **многоканальное** multiplex telephony
~, **однополюсное** side-band telephone
телефонировать telephone, phone
телефонист, телефонистка switchboard/telephone operator
телефония telephony
~, **автоматическая** dial telephony
~, **тональная** voice-frequency telephony
телефонный telephone *(attr.)*
~**ная книжка** telephone directory
~**ная станция** telephone exchange
телефоновидение phonovision
телефонограмма telephone message; telegram sent by telephone
телефоровые *n. pl. bot.* leather fungi; *(Thelephoraceae)*
телефото telephoto(graph)
телефотоаппарат long-range camera
телефотометр telephotometer
телехория telechory
телефотосъемка telephotography
телец 1. calf; 2. *astr.* Taurus
~, **золотой** golden calf
телецентр television/TV centre/center
телеэлектрокардиограмма teleelectrocardiogram
телеэлектростимулятор teleelectrostimulator, telestimulator
телибласт *biol.* thelyblast, feminonucleus
телигения thelygeny
телигонум *bot.* dog cabbage *(Thelygonum cynocrambe)*
телий *bot.* telium
теликарион thelykaryon
телиоспора *bot.* teliospore, teleutospore, winter spore
телиоспорангий *bot.* teliosporangium
телиплазма thelyplasm
телиптерис *bot.* wood fern *(Thelypteris)*
телит *med.* thelitis, mammary papilitis, mammillitis, inflammation of the nipple
телитокия *biol.* thely(o)toky
теллур *chem.* tellurium, Te
теллуризм *med.* tellurism
теллурий 1. *astr.* tellurion, tellurian; 2. *chem.* tellurium
теллуристый *chem.* tellurous, telluric
теллуритовый telluric
теллурический telluric, tellurian, terrestrial
теллуровый *chem.* telluric
Теллус *myth.* Tellus
тело body; *phys.* solid; *tech.* shank; *see also* **тела**
~, **абсолютно-черное** absolute/ideal blackbody
~, **вольфово** *anat.* Wolfian body, mesonephros
~ **без протока** closed body
~ **грудины** body of the sternum, mesosternum, gladiolus
~, **желтое** *anat.* yellow body, corpus luteum
~ **желудка** body of the stomach
~, **жировое** fat body, corpus adiposum
~, **инородное** foreign body
~, **изверженное** *geol.* eruptive/igneous body
~, **интрузивное** intrusion, intrusive body
~ **литеры** *print.* type body
~ **локтевой кости** shaft of the ulna
~, **материнское** *geol.* parent body
~, **миндалевидное** *anat.* amygdaloid nucleus
~ **множеств** *math.* field of sets
~, **мозолистое** *anat.* corpus callosum
~, **насадочное** *chem.* packing
~, **натурное** full-scale body
~, **небесное** celestial/heavenly body
~, **обтекаемое** streamlined body
~, **опорное** reference body
~ **плечевой кости** shaft of the humerus
~, **постороннее** foreign body
~, **проводящее** conducting body
~, **рабочее** working medium/substance
~, **ресничное** *anat.* ciliary body
~, **рудное** ore body
~, **спасаемое** recovery body
~, **стекловидное** vitreous body
~, **твердое** solid (body); *mech.* rigid body
~, **трубчатое** *geol.* chimney deposit
~, **черное** black body, perfect/complete radiator
~, **шишковидное** *anat.* pineal body/gland, conarium
телобласт *embryol.* teloblast
телодонты *paleont.* thelodonts *(Thelodontidae)*
телогены telogenes
телокинез *biol.* telokinesis
телолемма *anat.* telolemma
телолецитальный *zool.* telolecithal
телом *bot.* telome
теломеры telomeres
теломитический *bot.* telomitic
телоподобный *math.* fieldlike
телорез *bot.* water soldier *(Stratiotes)*
телосинапс telosynapsis
телосложение constitution, habitus, build, bodily structure; stature
телотит *geol.* thelotite
телотроха *zool.* telotroch(a)
телофаза telophase
телофюзен telofusain
телохранитель bodyguard
телофрагма *anat.* telophragma, Krause's membrane
телохромомер telochromomere
телохромосома telochromosome
телоцентрический *biol.* telocentric
тельсон *zool.* telson
тельфайрия *bot.* oil vine *(Telfairia)*
тельфер telpher; hoist

тельца corpuscles, bodies
~, мальпигиевы/почечные renal/Malpighian corpuscles
~ Негри Negri bodies/corpuscles
~, околоаортальные para-aortic bodies
~, пластинчатые lamellated corpuscles
тельце corpuscle
~, белое кровяное leucocyte, white blood cell
~, копчиковое coccygeal body
~, краевое marginal sensory body
~, красное кровяное erythrocyte, red blood cell
~, луковицеобразное bulboid/bulblike corpuscle, end-bulb of Krause
~, осязательное tactile/touch/oval corpuscle, tactor
тема theme; subject, topic; *ling.* base, theme
тематизация *ling.* base formation
тематика subject matter, subjects, topics
тематически thematically
тематический subject *(attr.),* topical; *gram., mus.* thematic
тембр timbre
темворс Tamworth (swine)
теменно-височный parietotemporal
теменно-затылочный parietooccipital
теменной *anat.* parietal
теменно-клиновидный parietosphenoid
теменно-копчиковый parietococcygeal
теменно-лобный parietofrontal
теменно-сосцевидный parietomastoid
теменно-чешуйчатый parietosquamous
темляк *mil.* sword knot
темник *hist.* temnik *(Tatar general)*
темнитель *phot.* dimmer
темный dark, dim, dusky; turbid
~ная вода *med.* amaurosis
темп rate, pace; *mus.* tempo
~ изменения rate of change
~ прироста/роста rate of increase/growth, growth rate
~ производства production rate, rate of manufacture
~ развития инфляции rate of inflation
~ разгона acceleration rate
~ стрельбы rate of fire, firing rate
темпера tempera
темперамент temperament; temper
~, меланхолический melancholic/atrabilious temperament
~, сангвинический sanguine/sanguineous temperament
~, флегматический phlegmatic/lymphatic temperament
~, холерический choleric/bilious temperament
температура temperature
~ белого каления white heat
~ взрыва explosive temperature
~ воспламенения ignition temperature
~ вспышки flash/ignition temperature
~ горения combustion temperature
~, дневная day-time temperature
~ замерзания freezing temperature/point
~ застывания congelation/solidification temperature
~ земной коры crustal temperature
~ инеобразования hoar-frost temperature
~ испарения vaporization temperature
~, исходная reference temperature
~ каплепадения dropping point, drop temperature
~ Кельвина Kelvin temperature
~ кипения boiling point
~, комнатная room temperature
~ конденсации condensation point
~ красного каления red heat
~ крашения dyeing temperature
~ кристаллизации solidification temperature
~ Кюри Curie temperature
~, минусовая subzero temperature
~ на входе inlet/intake temperature
~ на выходе outlet temperature
~ начала кипения bubble point, initial boiling point
~ ниже нуля subzero temperature
~, ночная night-time temperature
~ ожижения liquefaction temperature
~ окружающей среды ambient temperature; environmental temperature
~ отжига annealing temperature
~ отпуска tempering temperature
~ парообразования steaming/vaporization temperature
~ плавления melting point, fusing temperature
~, плюсовая above-zero temperature
~ по влажному/мокрому/смоченному термометру wet-bulb [moist-bulb] temperature
~ по Кельвину Kelvin temperature
~ по Реомюру Reaumur temperature
~ по сухому термометру dry-bulb temperature
~ по Фаренгейту Fahrenheit temperature
~ по Цельсию centigrade temperature
~, повышенная higher/elevated temperature
~, пониженная lowered temperature
~ помутнения cloud point/temperature
~, предельная limiting/maximum/ceiling temperature; minimum temperature
~, рабочая operating temperature
~ равновесия equilibrium temperature
~, расчетная design/estimated temperature
~ самовоспламенения temperature of spontaneous ignition, autoignition [self-ignition] temperature
~ синтеза ядер nuclear-fusion temperature
~ спекания sintering temperature
~, средняя годовая average yearly temperature
~, средняя суточная mean daily temperature
~ схватывания setting temperature
~ таяния melting point
~ текучести flow temperature/point
~, точечная spot/point temperature

~ точки росы dew-point temperature
~, установившаяся steady-state temperature
~ хранения storage temperature
~ Цельсия Celsius/centigrade temperature
~ ядерного синтеза fusion/thermonuclear temperature
~, яркостная luminance/brightness temperature
температурно-зависимый temperature-dependent
температурно-регулированный temperature-controlled
температурный temperature *(attr.)*
~ шов expansion joint
температуропроводность thermal diffusivity/conductivity
температуростойкость temperature stability
температурочувствительный temperature-/heat-sensitive
темперация temperament, temperation
темперирование tempering
темперировать temper; *tech.* bring to the appropriate temperature
темплет template, templet
темя *anat.* vertex; sincipit, crown
тенакль *print.* copyholder, tenaculum
теналгия *med.* tenalgia, tenodynia
тенденция trend; tendency; inclination, bent; bias; leaning, drive
~ к повышению upward trend/tendency, rising/stronger tendency; *(shares)* bullish tendency
~ к понижению downward trend, downtrend; *(shares)* bearish tendency
~ роста *econ., stat.* growth trend
~ сбыта sales trend
тендер *rail.* tender; *naut.* cutter, tank, tender
тендовагинит *med.* ten(d)osynovitis, ten(d)ovaginitis, tentothecitis, tentolemmitis
~, гнойный purulent tenosynovitis
теневой umbral; shade *(attr.),* shady
~вое растение sciophyte
теневыносливый shade-enduring/resistant/tolerant
тенезмы *med.* tenesmus
тенелюб *bot.* sciophilous [shade-loving] plant
тенелюбивый *bot.* sciophilous, shade loving
~вое растение sciophyte
тензиметр tensimeter
тензиометр tensi(o)meter
~, мышечный miotensiometer
тензограф recording strain gauge
тензодатчик strain gauge, tensometer
~, компенсационный dummy strain gauge
~ крутящего момента torsion strain gauge
~, наклеиваемый bonded strain gauge
~, пленочный film strain gauge
~, привариваемый weldable strain gauge
~, проволочный wire strain gauge
~, рабочий active strain gauge
~, самокомпенсирующийся self-balancing strain gauge
~ сопротивления resistance strain gauge
~, фольговый foil strain gauge
тензометр = тензодатчик
тензометрирование strain gauging
тензометрировать measure strain
тензометрический strain-gauge
тензометрия strain measurement
тензор tensor
~, антисимметрический antisymmetric/alternating tensor
~, аффинный affine tensor
~, ковариантный covariant tensor
~, контравариантный contravariant tensor
~, кососимметрический skew-symmetric [antisymmetric] tensor
~, смешанный mixed tensor
~, транспонированный conjugate tensor
тенидий *zool.* taenidium
тениидоз *med.* infestation with taenia
тинилидон *med.* tenylidone
тениодонтый *pl., paleont.* taeniodonts *(Taeniodonta, Taeniodontia)*
тениоз *med.* teniasis
тения taenia
теннецетин tennecetin
теннис tennis
~, настольный table tennis, ping-pong
теннисист tennis player
тенобурсит *med.* tenobursitis
тенодез *med.* tenodesis
тенолиз tenolysis
теномиопластика tenomyoplasty, tenotomyoplasty
теномиотомия *surg.* myotenotomy, tenomyotomy
тенонит *med.* tenonitis
тенонэктомия tenonectomy
тенопластика tenontoplasty, tenoplasty
тенорафия tenorrhaphy, tendinosuture, tenosuture
тенорецептор tenoceptor, tenoreceptor
теностоз *med.* tenostosis
тенотом *surg.* tenotome, tendotome
тенотомировать *surg.* tenotomize
тенотомия *surg.* tenotomy, tendotomy
~ при косоглазии curb tenotomy
тенрек *zool.* tenrec *(Tenrec)*
тенсиометр tensiometer
тентакулиты *pl., paleont.* tentaculites *(Tentaculitidae)*
тент canvas; awning; canopy
тень shade; shadow
~, аэродинамическая air/wind shadow, dead air
теобромин *chem.* theobromine
теогония theogony
теодицея theodicy
теодолит theodolite, transit (instrument)
~, маршрутный/съемочный transit instrument
~, оптический micrometer/microscopic/microoptic theodolite
~, повторительный repeating theodolite

~, **простой** direction theodolite
~, **универсальный** double-centre theodolite
теодолит-нивелир theodolite level
теодолитный theodolitic
теокразия theocrasy
теократ theocrat
теократический theocratic(al), thearchic
теократия theocracy, thearchy
теолог theologian
теология theology
теорема theorem
~ **взаимности** reciprocity theorem
~ **двойственности/дуальности** duality theorem
~ **единственности** uniqueness theorem
~ **запаздывания** lag theorem
~ **касания** tangency theorem
~ **количеств движения** momentum theorem
~ **косинусов** cosine theorem
~, **обратная** reciprocal/converse theorem
~ **о доминировании** *game th.* dominance theorem
~ **о магистрали** turnpike theorem
~ **о минимаксе** minimax theorem
~ **о седловой точке** saddle-value theorem
~ **о среднем** mean-value theorem
~ **отсчетов** sampling theorem
~ **Пифагора** Pythagoras theorem
~ **подобия** similarity/similitude theorem
~, **предельная** central limit theorem
~, **противоположная** contrary theorem
~ **равновероятности** equal probability theorem
~ **разложения** expansion theorem
~ **синусов** sine theorem, law of sines
~ **сложения** addition theorem
~ **Стокса** Stokes theorem
~ **существования и единственности** existence and uniqueness theorem
~ **умножения** product rule
~ **эффективности** efficiency theorem
теоремный theoremic, theorematic
теоретизировать theoterize
теоретик theoretician, theorist; theorizer
теоретико-вероятностный probability-theoretic, probabilistic
теоретико-игровой game-theoretic
теоретико-множественный set-theoretic(al), theoretical-set
теоретико-функциональный function-theoretical
теоретико-числовой number-theoretic
теоретически in theory, theoretically
теоретический theoretical; speculative
теоретичность theoretical/speculative nature
теоретичный abstract, theoretical
теория theory; theorics, theoretics
~ **автоматов** automata theory
~ **алгорифмов** theory of algorithms
~, **атомистическая** atomism
~ **атомного ядра** nuclear theory
~ **благосостояния, экономическая** welfare economics
~ **больших выборов** *stat.* theory of large sample
~ **булевых алгебр** Boolean calculus
~ **вероятностей** theory of probability, probability theory, theory of chances
~ **вихревая** vortex theory; circulation theory
~ **внешних эффектов** theory of externalities
~ **возмущений** *math.* perturbation theory
~, **волновая** wave theory
~ **выборочного метода** sampling theory
~ **вычетов** residue theory
~, **генная** gene theory
~ **границы** frontier theory
~ **графов** graph theory
~ **групп** theory of groups, group theory
~ **группирования** bunching theory
~ **денег** monetary economics; theory of money
~ **денег, количественная** quantity theory of money
~ **денег, товарная** commodity theory of money
~ **денег, хартальная** *econ.* chartalism
~ **занятости** employment theory
~ **заработной платы, договорная** bargaining theory of wages, bargaining-power theory
~ **доказательства** proof theory
~ **зародышевой плазмы** germ-plasm theory
~ **игр** *math.* game theory, theory of games
~ **интегрирования** integration theory, theory of integration
~ **информации** information theory
~, **квантовая** quantum theory
~, **квантовомеханическая** wave-mechanical theory
~, **кейнсианская** Keynesian theory
~, **кейнсианская экономическая** Keynesian economics
~, **клеточная** cell theory
~ **количественной полезности** theory of cardinal utility
~ **количественных признаков** theory of attributes; *stat.* theory of variables
~ **конических сечений** conics
~ **конфликтных ситуаций** *game th.* theory of conflict
~ **крыла** airfoil theory
~, **лучевая** ray theory
~ **массового обслуживания** queueing theory; theory of waiting lines, theory of queues
~ **матриц** matrix theory/calculus, theory of matrices
~, **матричная** template theory
~, **мезонная** meson theory
~ **меры** *math.* measure theory
~ **минимума заработной платы** subsistence theory of wages
~ **мишени** target theory
~ **множеств** theory of sets, set theory
~ **множественного рождения** multiple pro-

duction theory
~, монетарная экономическая monetary economics
~ надежности reliability theory
~ наибольших деформаций ultimate strain theory
~ народонаселения, мальтузианская Maltusian population theory
~ о свободе воли libertanianism
~ относительности theory of relativity
~ относительности, общая general/unifield theory of relativity, general relativity
~ относительности, специальная/частная special theory of relativity
~ оценок estimation theory
~ очередей waiting-line [queueing] theory
~ ошибок error analysis
~ поверхностей theory of surfaces
~ подобия theory of similitude, similarity theory
~ познания theory of knowledge
~ полезности utility theory
~ полезности, кардиналистская theory of cardinal utility
~ полезности, ординалистская theory of ordinal utility
~ полета theory of flight
~ поля field theory, theory of fields
~ предельной полезности marginal utility theory
~ предельной производительности marginal productivity theory
~ представлений theory of representations, representation theory
~ принятия решений decision theory
~ прогнозирования prediction theory
~ промежуточной связи intermediate-coupling theory
~ процента, неденежная nonmonetary theory of interest
~ размерностей dimensional theory
~ размещения occupancy/allocation theory
~ регулирования control theory
~ света, волновая wave theory of light
~ света, квантовая quantum theory of light
~ связи communications theory
~ систем system theory; *math.* systems theory
~ случайных процессов theory of stochastic processes
~ соединений combinatorics, combinatorial mathematics
~ спаривания mating theory
~ сравнений congruence theory
~, строгая rigorous theory
~, струйная *naut.* momentum theory
~ течений flow theory
~ управления control theory; management science/theory
~ упругости theory of elasticity, theory of elastic strength
~, феноменологическая phenomenological/formal theory
~ хиазматипии chiasmatype theory
~ цветности theory of colour value
~ цен price theory
~ цепей network/circuit theory
~ чисел number theory, theory of numbers
~ эволюции transmutation/evolution theory
теосинт(е) *bot.* teosinte *(Euchlaena)*
теософ theosoph, theosophist
теософический, теософский theosophic(al)
теософия theosophy
теофедрин *pharm.* theophedrine
теофиллин, теоцин theophylline, theocine
тепидарий *hist.* tepidarium
теплица hothouse, forcing house; greenhouse
тепло heat; warmth
~, джоулево Joule heat
~, конвективное convection heat
~, лучистое radiant heat
~, остаточное residual heat; afterheat; shutdown heat
~, отбросное/отходящее waste heat
~, побочное by-product heat
~, полезное net heat
~, регенерируемое recoverable heat
теплоаккумулирующий heat-retaining/storing
тепловоз diesel locomotive
~ с электрической передачей diesel electric locomotive
тепловой thermal, thermic, heat; calorific, caloric; heating *(attr.)*
~, почти near-thermal
~ удар *med.* heatstroke
тепловыделение heat release/generation/liberation
тепловыделяющий heat-generating/releasing
теплогашение thermoquenching
теплоемкость thermal/heat/calorific capacity, heat storing capacity
~, изобарная heat capacity at constant pressure
~, изохорная heat capacity at constant volume
~, мольная molar heat capacity
~, объемная volumetric heat capacity; heat capacity per unit volume
~, удельная specific heat
теплозащита heat/thermal insulation; heat shield
теплоизлучение heat/thermal radiation, thermal emission
теплоизолятор heat-insulating material, heat insulator
теплоизоляция thermal/heat insulation
теплокровный *zool.* warm-blooded, hematothermal
теплолабильный *chem.* thermolabile
теплокамера heat chamber
теплолечение thermotherapy
теплолюб *biol.* thermophile
теплолюбивый thermophilic, thermophilous, caloritropic, heat-loving
тепломер calorimeter; thermometer; heat flow meter

теплонапряженность thermal stress, heat density, calorific intensity; thermal factor
теплонепроницаемый heatproof
теплоноситель heat-transfer medium/agent, heat carrier; coolant; thermophore
~, водяной water coolant
~, жидкой heat-transfer fluid
~, жидкометаллический liquid-metal coolant, liquid-metal heat-transfer fluid
~, металлический heat-transfer metal
~ реактора reactor coolant
~, холодильный coolant
теплоноситель-замедлитель coolant-moderator
теплообмен heat exchange; heat transfer
~, лучистый radiant/radiative heat exchange
~, возвратный countercurrent heat exchange
теплообменник heat exchanger; heat converter
~, абонентский domestic heat converter
~, водо-водяной water-to-water heat exchanger
~, кожухотрубный shell-and-tube heat exchanger
~, обводный by-pass heat exchanger
~, оросительный spray-type heat exchanger
~, парогазовый stream-to-gas heat exchanger
~, противоточный counter-current heat exchanger
~ с перекрестным током cross-flow heat exchanger
~ с противотоком counterflow heat exchanger
~, трубчато-пластинчатый fin-and-tube heat exchanger
теплообразование thermogenesis, heat generation
теплообразователь heat generator/producer
теплообразующий thermogenic, thermogenetic, calorigenic, calorigenetic, heat-producing/generating
теплоотбор heat take-off
теплоотвод 1. heat abstraction/removal; 2. heat sink; heat-conducting path
теплоотдача heat emission/irradiation/transfer/output/exchange/removal; thermal emissivity
теплоотдающий exothermic, heat-releasing
теплоотражающий heat-reflecting
теплопеленгатор infrared radar, heat direction finder, heat seaker
теплопередача heat transfer/transmission/exchange
~ излучением radiative heat transfer
~, конвективная convection/convective heat transfer
~, нестационарная unsteady-state heat transfer
теплоперенос, теплопереход heat transfer
теплопоглотитель heat absorber
теплопоглотительный, теплопоглощающий heat-absorbing
теплопоглощение heat/thermal absorption
теплоподвод heat input/supply
теплопоступление heat input
теплопотеря heat loss
теплопоток heat/thermal flux
теплоприемник heat receiver/absorber; heat radiation sensor
теплоприход heat input
теплопровод heat pipeline; heating main; hot-water system
теплопроводник heat conductor
теплопроводность, теплопроводимость heat/thermal conductivity/conduction, transcalency
теплопроводный heat-condicting,transcalent
теплопроводящий heat-condicting/carrying; diathermal, diathermic
теплопрозрачность diathermancy, diathermy
теплопрозрачный diathermal, diathermic
теплопроизводительность calorific/heat value, thermal/heat output; heating capacity/efficiency
теплопроизводящий heat-producing/generating
теплопроницаемость diathermancy, diathermy
теплорегулирующий heat-regulating/regulative, thermotaxic
теплорегуляция thermoregulation, thermotaxis
теплосеть *abbr.* heating system
теплосмена thermal cycling
теплоснабжение heat supply
~, централизованное district heating
теплосодержание enthalpy, heat content
теплостабилизировать thermostabilize
теплостарение heat aging
теплостойкий heatproof, heat-resistant/resisting/stable, heatfast
теплостойкость heat/thermal endurance, heat resistance/stability
теплосъем heat removal/extraction
теплота heat
~ внутреннего трения viscous heat
~ деления fission heat
~ испарения heat of evaporation
~, лучистая radiating/radiation heat
~ отвердевания heat of solidification
~ парообразования heat of evaporation/vaporization
~ плавления melting heat, heat of fusion
~ радиоактивного распада radioactive-decay heat
~ разбавления heat of diffusion
~ распада продуктов деления fission-product decay heat
~ сжижения heat of liquefaction
~, скрытая latent heat
~ сублимации heat of sublimation
~, удельная specific heat
~, физическая sensible heat
теплотворность calorific/heat/heating/thermal value/power
теплотворный calorific, heat-producing
теплотехник heating engineer/technician; heat/combustion engineer
теплотехника calorifics, thermotechnics, heat engineering/technology
теплотехнический heat-engineering
теплотрасса heating main
теплоудерживающий heat/retaining/conserving
теплоупорный heatproof

теплоустойчивость = теплостойкость
теплоустойчивый = теплостойкий
теплофизика thermophysics, thermal physics
теплофикация heat supply; introduction/laying of a heating system
теплофицировать provide/introduce a heating system
теплоход motor vessel/ship, motorship
теплоход-контейнеровоз container motorship
теплоцентраль heating plant; steam plant
теплочувствительность heat/thermal sensitivity
теплочувствительный heat-sensitive
теплоэлектрический thermoelectric
теплоэлектростанция, теплоэлектроцентраль thermoelectric power station/plant
теплоэнергетика heat-and-power engineering
терагерц terahertz
терапевт therapeutist, therapist
терапевтический therapeutic(al)
терапевтика therapeutics
терапия therapy, therapeutics, therapia
~, гормональная hormonotherapy
~, заместительная substitution/replacement therapy
~, лучевая radiotherapy, radiation therapy
~, метамерная metameric therapy
~, нейтронзахватная neutron capture therapy
~, неспецифическая aspecific therapy
~, трудовая occupational therapy
~, шоковая shock treatment/therapy
~, электроимпульсная electroimpulsive therapy
~, электросудорожная electroconvulsive therapy, electric shock therapy
терапсид *paleont.* therapsid
тератобластома *med.* teratoblastoma
тератогенез teratogenesis, teratogeny
тератозооспермия teratozoospermia
тератология teratology
~ растений phytoteratology
тератома *med.* teratoma
~, гистоидная histoid teratoma
~, забрюшинная retroperitoneal teratoma
~, зрелая differentiated teratoma
~ крестцовой области sacral teratoma
~, организмоидная embryoma
тератоспермия teratospermia
тербий *chem.* terbium, Tb
тергит *zool.* tergite
тергум *zool.* tergum
тердесиен terra di Sienna, Sienna
теребилка (flax) puller
теребление plucking, pulling
терескен *bot.* winter fat *(Eurotia)*
тередо shipworm
терилен *text.* terylene
териодонт *paleont.* theriodont
терм *math.* term
~ Буля/логический Boolean/logical term
терма I hot/thermal spring
терма II *phys.* therma
термакадам tar macadam, tarmac
термалгезия *med.* therm(o)algesia
термалгия *med.* thermalgia
термантида *miner.* thermantide, porcelanite
термаллой Fe-Al-Ni alloy
термализация thermalization
термидор *hist.* Thermidor
термидорианец, термидорианский *hist.* Thermidorian
термин I term; special/technical term
термин II *myth.* Terminus
терминал *comp.* terminal; end device, data communication/transmission terminal
~ ввода/вывода input-output terminal
~, графический graphics terminal
~, дистанционный remote terminal
~ для ввода данных data-entry terminal
~, защищенный security terminal
~, интеллектуальный intelligent terminal
~, неавтономный on-line terminal
~, неинтеллектуальный dumb terminal
~ пакетной обработки batch terminal
~ передачи данных data-communication terminal
~ с разделением времени time-sharing terminal
~, сенсорный touch-sensitive terminal
~, сетевой network terminal
~, экранный CRT terminal
терминалия *bot.* tropical almond *(Terminalia catalpa)*
терминативный *gram.* terminative, terminate, conclusive
терминирование *ling.* naming
терминировать name, term
терминология terminology, nomenclature
терминообразование formation of terms, termformation
термион thermion
термистор *el.* thermistor, thermal resistor
термистор-болометр thermistor-bolometer
термит I *zool.* thermite *(Isoptera)*
термит II *chem.* thermit(e)
термитник nest of termites, termitary
термический thermal, thermic
термоакустический thermoacoustic
термоамперметр *el.* thermoammeter
термоанализ thermal anlysis; pyrolysis
термоанемометр hot-wire/-film anemometer, thermoanemometer
термоанестезия thermoanesthesia, thermoanalgesia
термобарокамера thermal vacuum/pressure chamber
термобатарея thermopile, thermal battery
~, атомная atomic heat battery
~, дисковая stacked-disk thermopile
термобрикетирование hot briquetting
термоваттметр *el.* thermocouple-type wattmeter

термовесы thermobalance
термовольтметр thermocouple-type voltmeter
термовыключатель temperature/thermal/overheat thermostatic switch, thermoswitch
термовысвечивание thermal glow, thermoluminescence
термогенез thermogenesis
термогенный thermogenous, thermogenic
термогигрограф thermohygrograph, recording thermohygrometer
термогигростат thermohygrostatic chamber
термограф thermograph, recording thermometer, temperature recorder
термодатчик temperature sensor
термодвигатель thermomotor, heat engine
термодвижущий thermomotive
термоделение thermofission
термодетектор temperature detector
термодиагностика thermodiagnosis
термодинамика thermodynamics
~, **техническая** engineering thermodynamics
~, **прикладная** applied thermodynamics
термозонд thermal probe, bathythermograph
термоизоляция heat/thermal insulation
термоионный thermionic
термокаротаж thermal logging
термокатод thermionic/hot cathode
термокаустика *med.* thermocautery, thermoelectric cautery
термокаутер cautery (iron)
термоклина thermocline
термокомпрессия thermal compression
термометр thermometer
~, **бесшкальный** non-indicating thermometer
~, **биметаллический** bimatal(lic) thermometer
~, **влажный** wet-bulb thermometer
~, **дистанционный** telethermometer, distant-reading thermometer
~ **для низких температур** low-reading thermometer
~, **максимальный/минимальный** maximum/minimum thermometer
~, **медицинский** clinical thermometer
~, **пращевой** sling thermometer
~, **ротовой** oral thermometer
~, **самопишущий** recording thermometer, thermograph
~, **сигнальный** alarm thermometer
~ **сопротивления** resistance thermometer
~, **стержневой** stick thermometer
~, **сухой** dry-bulb thermometer
~, **технический** industrial thermometer
термометрия thermometry, temperature measurement
термометр-щуп thermometer-probe
термонастия *bot.* thermonasty
термоневроз *med.* thermoneurosis
термообработка heat treatment
~, **циклическая** thermal cicling
термоокислительный thermal-oxidative
термоосмос thermoosmosis
термопара thermocouple, thermopile
~, **бесконтактная** insulated thermocouple
~, **борная** boron thermocouple
~, **защищенная** sheathed thermocouple
~, **малоинерционная** quick-response thermocouple
~, **погружная** immersion-type thermocouple
термопатрон temperature/thermal bulb
термопласт(ик) thermosoftening/thermoplastic material, thermoplast(ic)
термопластикат thermally softened material
термопластикация thermal plasticization, heat softening
термопластичность thermal plasticity
термополимеризация heat polymerization
термопреобразователь thermal converter
термоприемник temperature detector
термопсис *bot.* thermopsis *(Thermopsis)*
термореактивный thermoreactive, thermosetting
терморегулирующий thermotaxic, thermotactic
терморегулятор temperature controller/regulator, thermoregulator; thermostat
терморегуляция temperature control/regulation, thermoregulation; thermotaxis
терморезистор temperature-sensitive resistor
термореле thermal/temperature relay; thermal switch
термос thermos bottle
термосифон thermal siphon, thermosiphon
термоспай thermojunction
термостабильность thermal/heat stability
термостабильный heat-proof, thermostable
термостат 1. constant-temperature cabinet; 2. thermostat, thermostatic regulator/switch
термостатика thermostatics
термостатирование thermostatting; thermostatic control
термостойкий thermostable, heat-resistant
термостойкость heat/thermal resistance, thermal stability
термостолбик thermopile
термоток thermoelectric current
термоупругость thermoelasticity
термоустойчивость thermal/heat stability
термохромирование thermal chrome plating
термочувствительный heat-/temperature-sensitive
термошкаф heat chamber/cabinet
термощуп thermosound, thermal probe
термоэлектродвижущий thermoelectromotive
термоэлектронный thermoelectronic, thermionic
термоэлемент thermoelement, thermopile, thermocouple, thermoelectric cell
термоядерный thermonuclear
терн *bot.* sloe, blackthorn *(Prunus spinosa)*
тернарный ternary
тернев Newfoundland dog
тернослив(а) *bot.* bullage (plum), damson (plum) *(Prunus insititia)*

тероморфа *paleont.* theromorph
терофит *bot.* therophyte
терпен *chem.* terpene
терпентин *chem.* turpentine
терпентинный, терпентиновый terebinthine, terebinthinate
терпин *chem.* terpine
терпингидрат terpin hydrate
терпинеол *chem.* terpineol
терпинол *chem.* terpinol
Терпсихора *myth.* Terpsichore
терпуг *zool.* greenling *(Hexagrammos)*
терракота terra cotta
террамицин *pharm.* Terramycin, oxytetracycline
террарий terrarium
терраса terrace; platform; balcony; *geol.* bench
~ сброса *geol.* fault escarpment/scarp
террасовидный, террасообразный terraciform
терренкур *med.* terrain cure
терригенный *geol.* terrigenous, terrigene
терриконик *min.* refuse/spoil/waste heap, refuse/waste dump
территориальность territoriality
территория territory; area, region, district
~, подопечная trust territory
~, приречная streamside
~, разведанная proven territory
терц tierce
терцерон, терцеронка terceron, mulatto
терцина *lit.* terza rima
терциомицин tertiomycin
терция *mus., math.* third; *print.* 16-point type
~, большая/малая major/minor third
тесать cut, trim, dress, hew, square, shape
Тесей *myth.* Theseus
тесла *phys.* tesla, T
тесло adze
тессера *arch.* tessera
тесситура *mus.* tessitura, texture
тест test
~ на выносливость endurance test
~ на оперативную память immediate recall test
~ умственных способностей intelligence test
теста *bot.* testa, seed coat
тестер tester, checker, test probe
~ для интегральных схем integrated circuit tester, IC tester
~ для проверки микропроцессоров microprocessor exerciser
~, отбраковочный go/no-go tester
тестикулярный testicular, testiculate
тесто dough; pastry; paste
~, глиняное puddled clay
~, жидкое цементное grout
~, известковое lime paste/putty
тестирование testing
~, нисходящее top-down testing
тестомесилка dough mixer, kneader
тестостерон testosterone
тест-программа check/test routine, test program(me)
тест-таблица test chart/pattern
тестудинария *bot.* tortoise plant *(Testudinaria)*
тета theta
тетанизирующий *med.* tetanizing, tetanigenous
тетания *med.* tetany, tetania
~, детская infantile tetany
~, скрытая latent tetany
~, тиреоидная thyroprival tetany
~, транспортная transport tetany, rail sickness, railroad disease
тетанолизин tetanolyzin
тетанус *med.* tetanus; tetanic contraction
~ матки uterine/puerperal tetanus
тета-ритм theta rhythm
тетартоэдр tetartohedron
тетартоэдрический tetartohedral
тета-функция theta function
тет-де-пон *mil.* bridgehead
тетерев *zool.* black grouse, blackcock *(Lyrurus tetrix)*
тетива 1. bowstring; 2. stringer, side rail, stringboard
~ лестницы notchboard
тетрагексаэдр tetrahexahedron
тетрагидрофуран *chem.* tetrahydrofurane
тетрагир tetragyre
тетрагон tetragon
тетрагональ quadrangle
тетрагонально-скаленоэдрический tetragonal-scalenohedral
тетрагония *bot.* New Zealand spinach *(Tetragonia expansa)*
тетрагонолобус *bot.* deer vetch *(Tetragonolobus)*
тетрада tetrad
тетрадимит *miner.* tetradymite
тетрадиплоид quadruple diploid
тетрамерный *bot.* tetramerous
тетраметр *lit.* tetrameter
тетраном tetranomial
тетрануклеотид tetranucleotide
тетраплоид, тетраплоидный *biol.* tetraploid
тетраплоидия tetraploidy
тетрапод tetrapod
тетраподия tetrapody
тетрарх *hist.* tetrarch
тетрархия *hist.* tetrarchy
тетрасома *biol.* tetrasome
тетрасоматия *biol.* tetrasomaty
тетраспорангий *bot.* tetrasporangium
тетрастиль *arch.* tetrastyle
тетратионовый *chem.* tetrathionic
тетратоэдр tetratohedron
тетратоэдрический tetratohedral
тетрафтонг *phon.* tetraphthong
тетрафторметан *chem.* tetrafluoromethane
тетрахлорметан *chem.* tetrachloromethane, carbon tetrachloride

тетрахорд *mus.* tetrachord
тетрахотомический *bot., zool.* tetrachotomous
тетрациклин *pharm.* tetracycline
тетраэдр *geom.* tetrahedron
тетраэдрит *miner.* tetrahedrite
тетраэдрический tetrahedral
тетраэтилсвинец *chem.* tetraethyl lead
тетрил *chem.* tetryl, tetranitromethylaniline
тетробол tetrobolon
тетрод tetrode
 ~, двойной twin/double tetrode
 ~, лучевой beam tetrode
 ~, полевой field-effect tetrode
 ~, полупроводниковый semiconductor [transistor, solid-state] tetrode
тетроза *chem.* tetrose
теург theurgist
теургия theurgy
Тефия *astr.* Thetis
тефлон Teflon
тефрозия *bot.* tephrosia, hoary pea *(Tephrosia)*
тефф *bot.* love grass *(Eragrostis)*
техасец, техасский Texan
техкадры *abbr.* technical/engineering manpower
техминимум *abbr.* minimum of technical knowledge
техниций *chem.* technetium, Te
технизация technicalization
техник technician
 ~, авиационный aircraft technician
 ~, зубной dental technician, prosthetist
 ~ по приборам instrument technician
 ~ связи communication technician
техника 1. technology; engineering; 2. technique, method; procedure; 3. equipment, hardware , machines and installations, technical equipment; *mil.* materiel
 ~ авиационная 1. aeronautical engineering; 2. aircraft equipment; aviation materiel
 ~, авиационная электронная aviation electronics, avionics
 ~, авиационная-космическая aerospace engineering
 ~, атомная [атомного ядра] nucleonics
 ~ безопасности 1. safety precautions/regulations; accident prevention (measures); 2. safety engineering
 ~, боевая *mil.* combat materiel
 ~, высокочастотная radio-frequency engineering
 ~, вычислительная 1. computer equipment/facilities; 2. computer science/technology/engineering
 ~, дезактивационная decontamination equipment; decontamination technique
 ~, дорожная 1. road engineering; 2. roadbuilding machinery/machines/equipment
 ~ запоминания store/memory technique
 ~, криогенная cryogenics
 ~, организационная office equipment/aids; management/managerial/clerical aids
 ~, радиолокационная radar engineering; radar equipment/installations
 ~, ракетная rocket/missile technology
 ~, реактивная jet technology
 ~ реакторной безопасности (nuclear) reactor safety engineering
 ~ связи communications engineering/equipment
 ~, струйная fluidics
 ~, термоядерная thermonuclear engineering, thermonucleonics
 ~, ультразвуковая ultrasonics
 ~ управления management technology/technique
 ~, холодильная refrigerating equipment
 ~, электронная electronics
 ~, ядерная 1. nuclear technology/engineering; nucleonics; 2. nuclear equipment/plant
техникум technical secondary school
технический technical; engineering
 ~кие культуры *agr.* industrial crops
 ~кие средства *comp.* hardware
технократ technocrat
технократия technocracy
технологический technological
 ~ маршрут flow route
 ~кая карта process sheet
технологичность technological effectiveness; adaptability to efficient manufacture
 ~, ремонтная repairability
 ~, эксплуатационная serviceability
технологичный adaptable to efficient manufacture
технология technology; process/production engineering; production/manufacturing process; technique, know-how
 ~, автоматизированная automated process/technology; computer technology
 ~ изготовления manufacturing method/process/practice
 ~, информационная *comp.* information processing technology
 ~, поведенческая behavioural engineering
 ~ производства production process/practice
 ~, химическая chemical engineering, industrial chemistry
 ~, ядерная nuclear technology/engineering; nucleonics
технорук *abbr.* technical manager/director
техобслуживание *abbr.* (technical) maintenance
техперсонал *abbr.* technical personnel/staff
техпромфинплан technical, industrial and financial plan
техред *abbr.* technical editor
техуход *abbr.* technical maintenance
тецидеидный, тецидоидный thecidean
теций *bot.* thecium
течебезопасный leak-tight/-proof
течеискатель leak detector/hunter/tester/localizer
течение flow; current, stream; flux; course; flowing, running
 ~, безвихревое irrotational [non-eddying, vor-

tex-free] flow
~, **безнапорное** gravity flow
~, **безотрывное** attached flow
~, **верхнее** *geog.* upper reaches
~, **ветровое** wind current
~, **вихревое** vortex/eddying/rotational flow
~, **волновое** wave stream/flow
~, **вращательное** rotatory stream
~, **встречное** counterflow, back(ward) flow
~, **вынужденное** forced flow
~, **выходящее** outgoing stream
~, **вязкое** viscous/frictional flow
~, **дрейфовое** drift current
~ **идеальной жидкости** ideal flow
~, **кавитационное** cavity/cavitating flow
~, **квазиустановившееся** quasi-steady flow
~, **молекулярное** molecular/Knudsen flow
~, **непрерывное/неразрывное** continuous flow
~, **несвободное** restricted flow
~, **нестационарное** non-steady-state flow
~, **неустановившееся** nonstationary/unsteady/transient flow
~, **нижнее** *geog.* lower reaches; *meteor.* undercurrent; *naut.* lower current
~, **обратное** counter/backward flow
~, **отливное** ebb tide/stream/current
~, **пассатное** *naut.* trade current
~, **плоское** (two-dimensional) flow
~ **по направлению ветра** *naut.* lee current
~, **побочное** subsidiary flow
~ **под действием силы тяжести** gravity flow
~, **подповерхностное** undercurrent
~, **придонное** bottom current
~, **приливное** tidal stream/current
~, **разрывное** discontinuous/noncontinuous flow
~ **реальной жидкости** real flow
~ **реки** watercourse
~, **сейшевое** seiche current
~ **сжатия** compressive flow
~, **скользящее** slip flow
~ **сплошной среды** continuum flow
~, **срывное** separated flow
~, **стационарное** steady flow
~ **Стокса** Stokes flow
~, **струйное** jet stream/flow
~, **установившееся** steady (state) flow
течка I chute, spout
течка II *zool.* (o)estrus, heat
течь 1. *n.* leakage, leak; 2. *v.* leak, escape; run, flow, stream
тиазин *chem.* thiazin(e)
тиазол *chem.* thiazol(e)
тиамин *chem.* thiamin(e), vitamin B_1, aneurine
тиамин-хлорид thiamine/aneurine hydrochloride
тиара tiara
тиарелла *bot.* foamflower *(Tiarella)*
тибетец, тибетский Tibetan
тибетка Tibetan (woman)
тибио-навикулярный *anat.* tibionavicular, tibioscaphoid
тибиотарзальный *anat.* tibiotarsal, tarsotibial
тибио-феморальный *anat.* tibiofemoral
тибио-фибулярный tibiofibular, tibioperoneal
тигель 1. *met.* crucible, melting pot; 2. *print.* platen
тиглиновый *chem.* tiglic
тигмоморфоз *biol.* thigmomorphosis
тигмонастия thigmonasty
тигмотаксис thigmotaxis
тигмотропизм thigmotropism
тигр *zool.* tiger *(Panthera tigris)*
~, **малайский** Javan tiger
~, **саблезубый** saber-toothed tiger
тигридия *bot.* tiger flower *(Tigridia)*
тигроид, тигроидный *biol.* tigroid
тизанолена *bot.* tiger grass *(Thysanolaena)*
тик I teak(-wood)
тик II *med.* tic; spasm
~, **болевой** trigeminal neuralgia
~, **лицевой** mimic/facial/convulsive tic, facial spasm
~, **речевой** speech tic
тиксотропия *chem.* thixotropy
тила *bot.* tylosis, tylose
тилакоидный thylakoid
тиллея *bot.* pygmyweed *(Tillaea)*
тиллит *miner.* tillite; teallite; (glacial) till
тилль *geol.* till, boulder-clay
тилозин tylosin
тильда tilde
тималия *zool.* babbler *(Timalia)*
тимберс *naut.* timber; beam
тимелеевые *n. pl. bot.* mezereon family *(Thimelaeaceae)*
тимелея *bot.* sparrowwort *(Thymelaea)*
тимидин thymidine
тимидинкиназа thymidine kinase
тимин *bioch.* thymine
тиминоза thyminose
тимит *med.* thymitis
тимократия timocracy
тимол *chem.* thymol
~, **голубой** thymol blue
~, **йодистый** thymol iodite
~, **карбонатный** thymol carbonate, thymotal
тимоловый thymol
тимолфталеин *chem.* thymolphthalein
тимома *med.* thymoma
тимонуклеиновый thymonucleic
тимотрофный thymotrophic
тимофеевка *bot.* timothy, herd's-grass *(Phleum)*
тимоцит *med.* thymocyte
тимпан *mus.* timbrel; *arch.* tympan(um)
тимпанальный, тимпанический tympanal, tympanic
тимпанит *med.* tympanitis, myringitis
тимпания tympany
тимпаномастоидит *med.* tympanomastoiditis
тимпанопластика *med.* tympanoplasty
тимпанопункция tympanopuncture

тимпанотомия tympanotomy, myringotomy
тимьян *bot.* thyme *(Thymus)*
тимьяновый 1. *bot.* thymy, thyme; 2. *chem.* thymic
тимэктомия *surg.* thymectomy
тина mud, ooze, slime, slit; blanket algae
тинаму *zool.* tinamou *(Rhynchotus)*
тинктура tincture
тинол soldering paste, tinol
тиоацетазон thioacetazone
тиобарбитал thiobarbital
тиогуанин thioguanin(e)
тиодифениламин thiodiphenylamine, phenothiazine
тиокислота thioacid
тиоксолон thioxolone
тиокол 1. *chem.* thiocol; 2. *plast.* thiokol
тиол *chem.* thiol
тиолоксидаза thiol oxidase
тиолютин thiolutine
тиомочевина thiourea, thiocabramide
тионил *chem.* thionyl
тионин *chem.* thionine, amidophenthiazone
тиопневмокониоз sulphur pneumoconiosis, sulphopneumoconiosis
тиосемикарбазид thiosemicarbazide
тиосерный thiosulphuric, hyposulphurous
тиосоединение thiocompound
тиосульфат thiosulphate
тиоурацил thiouracil
тиофен *chem.* thiophene
тиофильный thiophile
тиофос *chem.* thiophos
тиохром thiochrome
тиоцианат thiocyanate
тиоцинат sulphocyanide
тиоэфир *chem.* thioether
тип type; mode; class; form; *tech.* pattern, model, design
 ~ вида *biol.* holotype
 ~ деления *nucl.* fission mode
 ~ распада decay mode, mode of disintegration
 ~ рода *biol.* genotype
 ~ шрифта font style
типаж range of models; *art* type(s)
типизация, типизирование typification
типизировать typify
типик *eccl.* tipicon
типовый standard, model; routine, ordinary
типогенез *biol.* typogenesis
типограф I printer
типограф II *zool.* bark beetle *(Ips typographus)*
типолог typologist
типология typology
типоморфный typomorphic
типоним *bot., zool.* typonym
типоразмер standard/typical size
типун *zool., pathol.* pip
типчак *bot.* sheep's fescue *(Festuca sulcata)*
тир I shooting range/gallery
тир II pitch, tar
тирамин *bioch.* tyramine
тираннозавр tyrannosaur *(Tyrannosaurus)*
тиратрон thyratron
 ~, импульсный pulse thyratron
 ~, логический thyratron gate
 ~, ртутный mercury-vapour thyratron
 ~ тлеющего разряда glow thyratron
тиреогенный thyrogenic, thyrogenous
тиреоглобулин thyroglobulin
тиреоидизм thyroidism
тиреоидин thyroidin
тиреоидный thyroid
тиреоидит thyroiditis
тиреоидектомия thyroidectomy
тиреостатический thyrostatic
тиреотоксикоз *med.* thyrotoxicosis
тиреотоксин thyrotoxin
тиреотоксический thyrotexic
тиреотропин, тиреотрофин thyrotrophin, thyroid stimulating hormone
тиреотропный thyrotropic
тиреохондротомия thyrochondrotomy
тириотеций thyriotecium
тиристор thyristor
 ~, мощный/силовый power thyristor
тирит Thyrite, tyrite, silicon carbide varistor
тиркушка *zool.* pratincole *(Glareola)*
тирозин *bioch.* tyrosine
тирозиназа tyrosinase
тирозинурия tyrosinuria
тироидин thyroiodine, iodothyrine
тиролец, тирольский Tyrolean, Tyrolese
тиролька 1. Tyrolean (woman); 2. Tirolese/Tyrolese hat
тироксин thyroxin
тиромедан thyromedan
тиротоксин *bioch.* thyrotoxin
тиротрицин tyrothricin
тироцидин tyrocidin
тирс 1. *bot.* thyrse, thyrsus; 2. *myth.* thyrsus
тис *bot.* yew *(Taxus)*
тискание *print.* proof-making
тиски vise, vice
 ~, верстачные/настольные/слесарные bench vice
 ~, винтовые screw vice
 ~, зажимные grip vice
тиснение *print.* stamping
тисовидный *bot.* taxoid
тисолистный *bot.* taxifolious, yew-leaved
титан I *chem.* titanium, Ti
титан II *myth., astr.* Titan
титанат *chem.* titanate
титанида *myth.* Titaness
титанистый titanous, titaniferous
титанит *miner.* sphene, titanite
титанотерий *paleont.* titanothere
титло titlo, tilde *(for omitted syllable)*
титр I *chem.* titre, titer

титр II *cin.* subtitle, caption
титр III *text.* metric number
титриметрия, титрометрия titrimetry
титрирование, титрование titration
~, **индикаторное** colour-indication titration
~, **кислотно-основное** acid-base titration
~, **кулонометрическое** coulometric titration
~, **обратное** back titration
~, **окислительно-восстановительное** redox titration
~ **по замещению** substitution titration
~ **по остатку** precipitation titration
~, **предварительное** forward titration
титровать *chem.* titrate
титрометрический titrimetric
титул I title; title page
титул II *law* title
~, **вещноправовой** estate
~ **основанный на приобретательной давности** title by prescription
~, **правовой** title
титульный лист title page
тиурам thiuram
тиф *med.* typhus
~, **брюшной** typhoid/enteric fever, abdominal typhoid, typhia
~, **возвратный** relapsing/recurrent fever
~, **крысиный** murine/endemic typhus
~, **рецидивный сыпной** recrudescent typhus
~, **сыпной** epidemic [exanthematous, petechial, louse-borne, classical] typhus, jail/prison fever
тифдрук *print.* mezzotint
тифлит *med.* typhlitis, typhloenteritis, cecitis
тифлолитиаз *med.* typhlolithiasis
тифлопедагогика methods of teaching the blind
тифлоптоз typhloptosis, cecoptosis
тифлорафия typhlorrhaphy, cecorrhaphy
тифлостомия typhlostomy, cecostomy
тифлоцеле typhlocele, celocele, cecal hernia
тифлоэктазия typhloectasia
тифлоэктомия typhlectomy, cecectomy
тифобактерин typhobacterin, thyphoid vaccine
тифозный typhous; typhoid, typhoidal, typhodes
тифоид *med.* typhoid
~, **желтушный** bilious typhoid
тифон 1. *naut.* typhon; 2. *myth.* Typhon
тифоподобный paratyphoid, typhus-like
тифулез *bot.* Typhyla blight, snow/winter scald, snow mould
тихоокеанский Pacific
тихоходный *naut.* slow-going
ткальный weaving; weaver's
тканеэквивалентный tissue-equivalent
ткань fabric, cloth, material; *anat.* tissue
~, **брекерная** breaker fabric
~, **бурая жировая** brown fatty tissue
~, **воздухоносная** aerenchyma, respiratory tissue
~, **выстилающая** epithelial tissue, epithelium
~, **грубая джитовая** gunny
~, **губчатая** cavernous/erectile tissue
~, **декоративная** tapestry material
~, **джерсевая** jersey cloth
~ **для сит** bolting cloth
~, **жгутовая** roving fabric
~, **жировая** adipose/fatty tissue
~, **камвольная** worsted cloth
~, **камчатная** damask
~, **кроветворная** hematopoietic tissue
~, **лимфатическая** lymphatic/lymphoid tissue
~, **махровая** terry cloth
~, **нервная** nervous tissue
~, **обвивочная** upholstery fabric
~, **одежная** apparel fabric
~, **переплетная** bookbinder's/binding cloth
~, **пигментная** pigmental tissue
~, **пропитанная** impregnated/proofed fabric
~, **прорезиненная** rubberrized fabric
~, **слизистая** mucoid tissue
~, **соединительная** connective tissue
~, **сосудистая** vascular tissue; protoxylem, trachenchyma
~, **суровая** grey cloth, crude fabric
~, **техническая** industrial fabric
~, **узорчатая** fancy cloth
~, **фильтровальная** filter cloth
~, **хлопчатобумажная** cotton cloth
~, **шелковая** silk cloth
~, **шерстяная** woolen cloth, wool fabric
~, **эмбриональная** embryonic tissue
~, **эпителиальная** epithelial tissue, epithelium
ткачик *zool.* weaver *(Ploceus)*
тладианта *bot.* tuber gourd *(Thladiantha)*
тление 1. smouldering; 2. decay, decomposition
тля *zool.* aphid, plant louse *(Aphididae)*
тмема *bot.* tmema
тмин *bot.* caraway *(Carum)*
тобогган toboggan
т-образный T-shaped
товар commodity, merchandise, article; goods, wares; *see also* **товары**
~, **громоздкий** bulky product
~, **дефицитный** commodity in short supply, scarce commodity/item
~, **индивидуализованный** *law* ascertained goods
~, **наличный** spot goods
~, **некондиционный** merchandise off condition
~, **нереализованный** unsold goods
~, **основной** staple
~ **отпускаемый в кредит** credit stock
~ **по кассовым сделкам** spot goods
~ **с немедленной сдачей** spot goods
товарищество association, society; partnership; company
~, **бессрочное** partnership at will
~, **коммандитное** limited partnership
~, **кредитное** credit/loan society
~, **полное** general partnership
товарно-денежный commodity-money *(attr.)*

товарность *econ.* marketability
товарный goods, commodity *(attr.);* freight
~ знак trade mark
~ поезд/состав freight/goods train
~ склад warehouse
~ная наличность stock in trade
~ная станция freight station
~ная теория денег commodity theory of money
товаровед 1. goods manager; 2. commodity researcher/expert
товароведение commodity research
товаровладелец goods/commodity owner
товарообмен 1. commodity/goods exchange; 2. trade
товарооборот (goods) turnover, commodity circulation; sales
~ на единицу площади магазина sales density
~, розничный retail turnover
товароотправитель consigner
товаро-пассажирский goods-and-passenger
~ поезд mixed train
товарополучатель consignee
товаропроизводитель commodity producer
товары *pl.* goods, commodities, wares
~, бакалейно-гастрономические grocery
~, второстепенные nonessentials
~, галантерийные fancy goods
~, дефицитные scarce commodities, goods/items in short supply
~ длительного пользования durable/hard goods
~ залежалые unsalable stock
~, инвестиционные capital/investment goods
~, канцелярские office supplies; stationery
~ кратковременного пользования nondurables, soft goods
~, массовые staples
~, модные style-and-fashion goods
~, наличные goods on hand
~, неметаллические soft goods
~, непродовольственные nonfoods
~, неходкие slow-moving goods, unmarketable goods
~, объемистые bulky items
~, парфюмерные perfumery, perfumes
~ первой необходимости essential goods/commodities
~, полотняные white goods
~, потребительские consumer commodities
~, продовольственные food commodities
~, промышленные industrial/manufactured goods
~, реимпортированные reimports
~ с ограниченным сроком годности/пользования perishable goods/commodities, semidurables
~, транзитные goods in transit
~, ходкие quick-selling lines/goods, fast-moving goods
~ широкого потребления consumer goods
~, штучные piece goods
тогда и только тогда *math.* if and only if, iff; when and only when
тоголезец, тоголезский Togolese
тоголезка Togolese (woman)
тодди-пальма *bot.* toddy/jaggery/fishtail palm *(Caryota urens)*
тождество identity
ТОЗ *abbr.* agricultural association
ток current; stream, flow
~, анодный anode/plate current
~ белого поля *phot.* white current
~, блуждающий stray current
~, вихревой eddy current
~ возбуждения exciting/field/excitation current
~, встречный back/reverse current
~, входной input current
~, выпрямленный rectified current
~, выходной output current
~, действующий root-mean-square current
~ дрейфа drift current
~, дырочный hole current
~, емкостный capacitive current
~ зажигания/запальный ignition current
~ запрета *comp.* inhibit current
~, зарядный charging current
~, затухающий decaying current
~, земной earth/terrestrial/telluric current
~, катодный cathode/cathodic current
~, колебательный oscillating current
~, контурный loop current
~ короткого замыкания short-circuit current
~ крови blood flow/stream, circulation
~, круговой ring current
~, максимальный peak/maximum/threshold current
~ накала filament heating current
~, номинальный rated current
~, обратный reverse/back current
~, объемный steady-volume current
~, однофазный single-phase current
~, остаточный residual current, aftercurrent
~, паразитный spurious/stray/parasitic current
~, переменный alternating current
~, пилообразный saw-tooth current
~ питания supply/feed current
~ плазмы plasma current
~ подмагничивания bias current
~ покоя 1. *telegr.* spacing current; 2. *rad.* quiescent current
~, постоянный direct current; constant/continuous current
~, прерывистый intermittent current
~ пучка beam current
~, реактивный reactive/wattless current
~, синфазный in-phase current
~ смещения 1. bias current; 2. displacement current
~, сторонний extraneous current

~, **темновой** dark current
~, **термоэлектрический** thermocurrent
~, **термоэлектронный** termionic/ thermoelectronic current
~, **трехфазный** three-phase current
~, **установившийся** steady(-state) current
~ **утечки** leakage/insulation current
~ **Фуко** Foucault/eddy current
~ **электронного пучка** cathode-ray current
~ **якоря** armature current
токай, токайский Tokay
токарный *mech.* turning
~ **станок** lathe
токарь lathe operator, turner
токката *mus.* toccata
токовращатель pole changer
токология tocology, obstetrics
токонесущий current-carrying, live
токоограничитель current limiter, current-limiting circuit-breaker
токоотвод current tap
токоподвод current lead
токопрерыватель current interrupter
токоприемник 1. current collector; 2. load; consumer
~, **дуговой** collector bow
~, **пантографский** pantograph collector
~, **роликовый** trolley
токопроводящий current-carrying/conducting
токосниматель = токоприемник
токосъемник current collector; sliding contact/ shoe, trolley, slip ring
токоферол *bioch.* tocopherol, vitamin E
токофобия *med.* tocophobia, fear of childbirth
токсальбумоза toxalbumose, poisonous albumose
токсаскаридоз toxascaridosis
токсафен *chem.* toxaphene
токсемия tox(a)emia, toxicemia, blood poisoning
токсидермия, токсикодермия toxiderma, toxidermatosis
токсикоз *med.* toxicosis
~, **ожоговый** burn toxicosis
токсикоинфекция tox(o)infection, toxiinfection
токсиколог toxicologist
токсил toxyl
токсимуцин toxi(co)mucin
токсин toxin
~, **ботулинический** botulinum/botulinus/botulinal toxin
~ **животного происхождения** animal toxin, zootoxin
~, **инактивированный** inactivated/detoxicated toxin
~ **коринебактерий** tuberculous toxin
~, **ожоговый** burn toxin
~ **растительного происхождения** plant toxin, phytotoxin
~, **столбнячный** tetanotoxin
~ **утомления** fatigue toxin, ponogen, kintoxin
токсинемия = токсемия
токсинообразующий toxigenic
токсический, токсичный toxic, toxicant, poisonous
токсичность toxicity
~ **излучения** radiotoxicity
~, **повышенная** hypertoxicity
~, **пониженная** hypotoxicity
~ **радиоактивных изотопов** radiotoxicity
токсодонт *paleont.* toxodont
токсоплазмоз toxoplasmosis
токсэмия *med.* tox(a)emia
тол *chem.* trinitrotoluene, trotyl, tolite
толбутамид *chem.* tolbutamide, orinase
толикаин *chem.* tolycaine
толкатель pusher, push rod, plunger, follower; tappet
~, **винтовой** screw-and-nut pusher
~ **клапана** valve tappet/lifter/follower/plunger; cam follower
~ **ядра** *sport* shot-putter
толкач 1. ram; jack; 2. *rail., naut.* pusher, pushboat
толкование interpretation, interpreting, construing, construction; elucidation
~, **ограничительное** restrictive/strict/narrow interpretation/construction
~ **расширительное** broad/loose/latitudinarian/ liberal/extensive interpretation/construction
~ **слов** oneiromancy
толковать interpret; construe
~ **ограничительно** interpret narrowly/strictly/ restrictively
~ **расширительно** interpret broadly/liberally
толокнянка *bot.* manzanita *(Arctostaphylos)*
толстоветвистый *bot.* thick-branched/twigged, pachycladous
толстовец, толстовка Tolstoyan
толстовство, толстовщина Tolstoyism
толстостенный thick-walled; pachypleurous
толстянка *bot.* crassula *(Crassula)*
толстянковые *n. pl. bot.* orpine family *(Crassulaceae)*
толуидин toluidine, aminotoluene
толуилен toluylene, stilbene
толуол toluene, methylbenzene, toluol
толчея 1. choppy/chopping sea, chop; 2. pounding/ crushing/pestle mill; pounder
толчок shock, impact, bump, push, jolt; kick
~, **верхушечный** apex beat
~, **ионизационный** ionization burst
~, **окраинный** marginal shock
~, **последующий** aftershock
~, **сердечный** cardiac beat
~, **тектонический** tectonic tremour
толщина thickness; width; depth; gauge
~ **свода** depth of a vault
~ **скин-слоя** skin depth
~ **чертежной линии** width/weight of a line
толщиномер thickness gauge, calipers
~, **бета-лучевой** beta thickness gauge

~, изотопный radioisotope thickness gauge
~, рентгеновский X-ray thickness gauge
~, радиоактивный radiation thickness gauge
толь roofing felt; (tarred) roofing paper
том volume
~ прямого доступа *comp.* direct-access volume
~, рабочий (с файлами) scratch volume
~а, опознавание (data) volume recognition
томасирование Thomas process; basic Bessemer process
томасшлак Thomas slag/phosphate
томбуй *naut.* anchor/cable buoy
томизм Thomism
томильный soaking; malleablizing
томист Thomist; Thomean
томить malleablize; soak; graphitize
томкод *zool.* tomcod *(Microgadus)*
томление табака tobacco withering
томление чугуна graphitizing, malleablizing; soaking
томленая цементная сталь blister steel
томограмма tomogram
~, прицельная localized tomogram
томограф tomograph, laminograph
томография tomography, laminography, bodysection roentgenography, planigraphy
томофлюорограф fluorographic planigraph
томофлюорография fluorographic planigraphy
томсенолит *miner.* thomsenolite
тон 1. tone; note; 2. hue, tone, tint
~, высокий/низкий high-pitch/low-pitch tone
~ сердца cardiac/heart sound
~ сердца, предсердный atrial sound
~, чистый *med.* pure tone
тональность tonality
тонарм pickup/tone arm
тонафазия tonaphasia, vocal amusia
тонвал capstan
тонзиллит *med.* tonsillitis
тонзиллолит tonsillolith, tonsillith
тонзиллотом tonsillotome, amygdalotome
тонзиллотомия *surg.* tonsillotomy, amygdalotomy
тонзиллэктомия tonsillectomy
тонзиллярный tonsillar, tonsillary, amygdaline
тонзура tonsure
тонизировать tone up; stimulate
тонизирующий *med.* tonic, roborant
~ сердце cardiotonic
тоника *mus.* keynote, tonic
тонический 1. *mus., lit.* tonic; 2. *med.* tonic, restorative, reconstituent
тонковетвистый slender-branched, leptocladous
тонковласник *bot.* filmy fern *(Trichomanes)*
тонкодисперсный finely dispersed
тонкожилковатый fine-/closely veined
тонкозернистый fine-grained, finely granular
тонкоизмельченный finely pulverized/ground
тонкокишечный small-intestine
тонкокожий thin-skinned, leptodermatous
тонколистник *bot.* filmy fern *(Hymenophyllum)*
тонколистный thin-leaved, leptophyllous
тонколистоватый thinly laminated; tenuifolious
тонколучник *bot.* fleabane *(Stenactis)*
тонконог *bot.* koeleria *(Koeleria)*
тонкопанцирные, тонкоскорлупные *n. pl. zool.* leptostracans *(Leptostraca)*
тонкосетчатый subreticulate
тонкослоистый fine-stratified, thin-laminated
тонкослойный thin-layer
тонкостебельный fine-/thin-stalked
тонкостенный thin-walled
тонкотекстурный fine-textured
тонкотел *zool.* leaf monkey *(Presbytis)*
тонкотелый slim-bodied, lean
тонкотянутый thin-drawn
тонмейстер sound monitor/mixer/engineer
тонна ton
~, весовая weight ton
~ водоизмещения displacement ton
~, длинная (английская) gross/long/English ton (1016 kg)
~, короткая (американская) short ton (907 kg)
~, метрическая metric ton (1000 kg)
~, налоговая rateable ton
~, обмерная measurement/freight/cargo ton (1.13 cubic metres)
~, регистровая register/gross/vessel ton (2.83 cubic metres)
~, фрахтовая freight/cargo ton
тоннаж tonnage
~, валовой регистровый gross register tonnage
~, грузовой cargo tonnage
~, каботажный coastal tonnage
~, морской sea-going tonnage
~, общий aggregate tonnage
~, сухогрузный dry-cargo tonnage
~, регистровый registered tonnage
тонна-сила ton-force
тонно-километр kilometer-ton
тонно-миля ton-mile, mile-ton
тоноклонический *med.* tonoclonic, tonioclonic
тоносигнал audio/tone signal
тонось capstan
тонофибрилла tonofibril
тонстудия sound studio
тонус *physiol.* tonus, tonicity; tension
~, жизненный vitality
~, мышечный muscular tension
~, нервно-мышечный neuromuscular tension
~, сосудистый vascular tension
тонфильм sound film
тончастота audio/tone frequency
тонштихель *print.* engraver's hand tool
топ *naut.* top; head
~ фок-мачты foremast-head
топаз *miner.* topaz
~, дымчатый smoky topaz
топазолит *miner.* topazolite

топенант *naut.* span rope, topping lift/line; lift, jigger
~ грузовой стрелы derrick span
~, основной working topping lift
топенант-блок topping lift block
топенант-тали span/lift tackle
топенант-шкентель topping line
топика *philos.* topic
топинамбур *bot.* Jerusalem artichoke *(Helianthus tuberosus)*
топить I heat, fire, stoke
топить II sink, send to the bottom; drown
топка I 1. furnace; firebox; 2. firing, heating; stoking; 3. stoker; 4. melting down
~ котла boiler furnace
~, механичная mechanical stoker
~, нефтяная oil-fired furnace
~, факельная torch furnace
~, циклонная cyclone furnace
~, шахтная shaft-type furnace
топка II cap; cup
~ картушки компаса compass cup
топкий 1. combustible; 2. easily melting
топливный fuel *(attr.)*
топливо fuel; combustible; *rock.* propellant
~, беззольное ash-free fuel
~, высококалорийное high-energy fuel
~, высокооктановое high-octane fuel
~, газообразное ракетное propellant gas, gas propellant
~, двухкомпонентное ракетное bipropellant
~, дизельное diesel fuel/oil
~ для реактивных двигателей jet fuel/propellant
~, желатинизированное/загущенное thickened fuel
~, жидкометаллическое liquid-metal fuel
~, ископаемое fossil fuel
~, коллоидное colloidal fuel
~, малосернистое low-sulphur fuel
~, моторное engine fuel
~, недетонирующее antiknock fuel
~, нефтяное котельное boiler oil
~, однокомпонентное ракетное monopropellant
~, отработанное spent/reject/depleted fuel
~, ракетное (rocket) propellant
~, реактивное jet fuel
~ с присадками doped fuel
~, сернистое sulphur-bearing fuel
~, твердое solid fuel
~, эталонное reference fuel
~, этилированное ethylized fuel
~, ядерное nuclear/atomic/reactor fuel
топливозаправка (re)fuelling
топливозаправочный fuelling
топливозаправщик fuel truck, (re)fueller
топливомер fuel-quantity/-contents/-level indicator
топливоподача fuel supply/feed
топливоподогреватель fuel oil heater
топливопровод fuel line/pipe/pipeline
топливоснабжение fuel supply
топливохранилище fuel storage depot
топляк sunken log, deadhead, snag, sinker
топогнозия topognosis, topognosia
топограф topographer; surveyor
топография topography; surveying
~, барическая baric/pressure topography
~, подводная submarine topography
топодим topodeme
тополог topologist
топология topology; layout
~, алгебраическая algebraic topology
~, комбинаторная combinatorial topology
~ системы system topology
тополь *bot.* poplar, cottonwood *(Populus)*
топонимика toponymy, toponymics, place names
топонимия toponymy
топономастика place-name study
топор axe
~, бондарный hollow adze
~, лесорубочный felling axe, wood-chopper
~, пожарный fire/fireman's axe
~ с гвоздодером claw hatchet
~, плотничный broad axe
топорики *pl. bot.* forking larkspur *(Delphinium consolida)*
топороногие *n. pl. zool.* pelecypods *(Pelecypoda)*
топ-палуба *naut.* top deck
топрик *naut.* davit span
топсель *naut.* topsail
~, гафельный gaff topsail
топ-тимберс *naut.* top timber
топь mire, swamp, marsh, bog, morass
~, приливная tidal marsh
тор I *phys.* torr
тор II *math.* torus
тор III *bot.* torus
Тор IV *myth.* Thor(r)
тора *rel.* Toran, The Pentateuch
торазин chlorpromazine hydrochloride, thorazine hydrochloride
торакальный thoracic, thoracal
торако-абдоминальный thoracoabdominal
торако-акромиальный acromiothoracic, thoracoacromal
торакоалгиа thoracalgia, thoracodynia
тораколюмбальный thoracolumbar
торакомиодиния thoracomyodynia
торакопластика thoracoplasty
торакоскоп *med.* thoracoscope, stethoscope
торакотомия *surg.* thoracotomy
~, задне-боковая posterolateral thoracotomy
~, нижняя inferior thoracotomy
~, передне-боковая anterolateral thoracotomy
тораконентез *med.* thoracocentesis, thoracentesis
торакс thorax
торг *abbr.* trading organization
торги *pl.* tender; bid; auction

~, закрытые sealed bid; auction by tender
торговец trader, merchant, tradesman, dealer
~ акциями share broker
~ бакалейными товарами grocer
~, биржевой dealer
~ живым товаром white slaver
~ канцелярскими товарами stationer
~, мелкий petty dealer
~, оптовой wholesale dealer, wholesaler
~, розничный retail dealer/trader, retailer
~, странствующий itinerant trade
торговля trade, commerce, trading, marketing; traffic
~, бартерная barter trade
~, внешняя foreign trade
~, внутренняя home/domestic trade
~ за наличные cash trade
~, запрещенная illicit trade
~, каботажная coasting/coastwise trade
~, меновая barter (trade)
~, оживленная brisk trade
~, оптовая wholesale trade, wholesaling
~ оружием traffic in arms, arms traffic
~, пограничная frontier trade
~, посылочная mail-order business/trade
~, развозная retail delivery trade
~, транзитная entrepot trade
~ ценными бумагами security trading
торгово-промышленный commercial and industrial
торговый trading, trade, commercial; mercantile
~ знак trademark
~ флот merchant marine
~вая палата chamber of commerce
~вая сеть shops
~вое право commercial law
торгоут Torgot, Torgut
торгпред *abbr.* trade representative
торгпредство *abbr.* trade mission; commercial organization *(abroad);* trade representation
тордилиум *bot.* hartwort *(Tordylium)*
тореадор toreador
торевтика toreutics
торец end face, (butt) end
торзиометр torsion meter, torquemeter
торзионный torsional
тори *polit.* Tory
торианит *miner.* thorianite
торид *chem.* thoride
ториевый *chem.* thoric
торизм *polit.* Torism
торий *chem.* thorium, Th
~, азотистый thorium nitride
~, азотнокислый thorium nintrate
~, сернокислый thorium sulphate
~, хлористый thorium chloride
торилис *bot.* hedge parsley *(Torilis)*
торировать thoriate
торит *miner.* thorite
торица *bot.* spurry *(Spergula)*
торичеллиев Torricellian
торический toric; torus *(attr.)*
торичник *bot.* sand spurry *(Spergularia)*
торкрет cement gun
торкрет-бетон gunite, gunned concrete
торкретирование guniting, concrete spraying
торкретировать gunite, gun-concrete
торкрет-пушка pneumatic concrete gun
торможение braking; deceleration, retardation; drag, stagnation; *physiol.* inhibition
~, аварийное emergency braking
~, аэрофинишером *aero.* arrestment
~, безусловное unconditioned inhibition
~, вязкое viscious drag
~ двигателем engine braking
~, конкурентное competitive inhibition
~, охранительное protective inhibition
~, реактивное jet braking
~ с аккумулированием энергии energy storage braking
~ тягой thrust braking
~, угасительное extinctive inhibition
~, условное conditioned inhibition
тормоз brake
~, аварийный emergency brake
~, бензельный rope stop
~ воздушного винта propeller brake
~, воздушный pneumatic/air brake
~, гидравлический hydraulic/fluid/liquid/oil brake
~, двухколодочный double-block (two-shoe) brake
~, канатный cable/rope brake
~, клиновый wedge brake
~, колодочный block/shoe brake
~, ленточный band/belt/strap brake
~, ножной/педальный foot/pedal/toe brake
~, ручной lever/hand brake
~ с гидравлическим приводом hydraulically operated brake
~, стояночный parking brake
~, фрикционный friction brake
~, храповичный ratchet brake
тормозить brake, apply the brakes on; decelerate, slow down, retard; arrest
тормозной braking; retarding, decelerating; arresting; *(physiol.)* inhibitory
тормозящий braking; retarding; inhibiting
торогумит thorogummite
тороид thoroid
тороидальный toroidal
тороидный toroid(al)
торолит *miner.* thoreaulite
торон thoron, thorium emanation
торотунгистит thorotungistite
тороуранинит thorouraninite
торос ice hummock/reef
торпеда torpedo
~, авиационная aerial torpedo
торпедировать 1. torpedo; 2. *min.* shoot *(oil*

wells)
торпедный torpedo *(attr.)*
~ аппарат torpedo tube
~ катер motor torpedo boat
торпедо 1. *zool.* dark electric ray *(Torpedo nobiliana);* 2. *naut. see* **торпеда**
торпедометание torpedo bombing
торпедоносец torpedo bomber/plane/carrier
торпекс torpex
торпотрон torpotron
торр *phys.* torr
торрея *bot.* torreya *(Torreya)*
торричеллиев Torricellian
~а пустота Torricellian vacuum
торс 1. torso; trunk; 2. *math.* torse
торсиометр torsiometer, torquemeter
торсион torsion bar/shaft
торсионный torsional; torsion *(attr.)*
торсия torsion
торулез *med.* torulosis, cryptococcosis
торф peat; turf
~, болотный bog peat
~, гуминовый humic peat
~, жирный bituminous peat
~, луговой meadow peat
~, плотный stone peat
торфокомпост peat compost
торфоразработка peat mining
торфорез peat/turf cutter
торфсыпец crumbling/milled peat
торфяник, торфянка peat bog; peatland
торфяной peaty; turfy, turfen; peat *(attr.)*
~ное болото peat bog
торцевание, торцовка facing, squaring up
торшер floor lamp
тосканец, тосканский Tuscan
Тот *myth.* Thoth
тоталитаризм, тоталитарность totalitarianism
тотара *bot.* totara *(Podocarpus totara)*
тотем totem
тофильдия *bot.* false asphodel *(Tofieldia)*
точение 1. turning *(on a lathe);* 2. sharpening, grinding
точеный 1. turned, machined; 2. sharpened
точечно pointwise, point-, spot-
точечно-опорный point-supported
точечно-сварный spot-welded
точечный punctuate, dotted; point, spot *(attr.)*
~ная сварка spot welding
~ное оценивание point estimation
точило I grindstone, whetstone, oilstone
точило II wine press
точильный sharpening, grinding
~ брусок/камень whetstone
~ станок tool-grinding machine
точильщик 1. sharpener, grinder; 2. *zool.* furniture beetle *(Anobium)*
точить 1. turn *(on a lathe);* 2. sharpen, grind, whet; strop; hone
точка I point, dot; spot; period
~, асимптотическая asymptotic point
~, бесконечно удаленная point at infinity
~ безубыточности *econ.* break-even point
~, верхняя мертвая top/upper dead centre
~ ветвления branch/ramification point
~ ввода entry/entrance point
~ весеннего равноденствия vernal equinox point, first point of Aries
~ визирования sighting/aiming point
~ воспламенения ignition/fire point
~ вспышки flash point
~ встречи impact point
~ входа entry/entrance point
~ возврата return point; cuspidal point
~ вызова invocation point
~ выхода exit point
~, граничная boundary point
~, десятичная decimal point
~ зажигания firing point
~ замерзания freezing point
~, зеркальная mirror point
~ инея frost point
~ испарения evaporation/vaporization point
~, исходная starting/initial/reference point, origin
~ касания point of tangency
~ кипения boiling point
~, конечная terminal point, end-point
~ кода Морзе Morse dot
~, контрольная checkpoint; reference point, benchpoint
~, крайняя extremity
~, кратная multiple point
~ Мак-Бернея *med.* McBurney's point
~, материальная mass/material point, particle
~, мертвая dead point/centre
~, надглазничная *anat.* supraorbital point, ophrion
~, наружная мертвая outer dead centre/point
~ насыщения saturation point
~, начальная initial/starting point
~ неопределенности *math.* ambiguous point
~, неособоя *math.* nonsingular/regular point
~, несобственная *math.* improper point
~, нижняя мертвая bottom/lower dead centre
~, нулевая datum point, point of reference
~, огневая *mil.* weapon/fire emplacement
~ округления *math.* umbilical point, umbilic
~, опорная check/control point
~ опоры point of support/bearing, supporting point; fulcrum
~ осаждения precipitation point
~ осеннего равноденствия autumnal equinox point, first point of Libra
~, особая *math.* singular point
~ падения в воду splash point
~ перегиба inflection point
~ перезапуска return point
~ пересечения point of intersection, crossing/intersection point
~ перехода к жидкую фазу liquefaction point

~ плавления melting/fusion/fusing point
~ плотности density point
~ поворота turning point
~, пограничная boundary point
~ покоя stationary/rest point, point of rest
~ ползучести creep point
~ помутнения cloud point
~, предельная *math.* limit/cluster/accumulation point
~ прерывания breakpoint
~ пуска launching point
~ прикосновения point of tangency/contact/adherence; *geom.* tacpoint
~ размягчения softening point
~ разрыва point of discontinuity
~, регулярная особая *math.* regular singular point, inessential singularity
~, реперная reference/datum point
~ роста *bot.* apical/growing point
~ росы dew point
~ с запятой semicolon
~ самоокупаемости *econ.* break-even point
~ самоприкосновения point of osculation, tacnode, double cusp
~, сварная weld spot
~, седловая *game th.* equilibrium/saddle point
~ солнцестояния solstitial point
~, соответственная *top.* conjugate/matching point
~ срабатывания trip point
~ стеклования glass-transition point
~ стояния *top., geod.* point of observation/sight/view
~, существенно особая essential/irregular/singular point, irregular/essential sungularity
~ таяния melting point
~ текучести flow point
~, узловая *math.* nodal point, node
~, устранимая особая *math.* removable singular point, removable singularity
~, фазовая phase point/vector
~ фокуса focal point
~ экстремума bending/turning/bend point
точка II 1. sharpening, grinding; whetting; stropping; honing; 2. turning *(on a lathe)*
точковидный, точкообразный punctiform
точно exactly, precisely; accurately; punctually
точность precision, exactness, exactitude; accuracy; punctuality; fidelity
~ воспроизведения fidelity/accuracy of reproduction
~, высокая high accuracy, great precision
~, допустимая feasible accuracy
~ метода, собственная intrinsic accuracy of a method
~ по азимуту azimuth accuracy
~ по углу места elevation accuracy
~, предельная extreme accuracy
~, расчетная design accuracy
~, фактическая actual accuracy

точный accurate, precise, exact; correct; punctual
тошнота nausea, sickness
~ беременных nausea of pregnancy
~, двигательная motion/car sickness
~, обменная metabolic nausea
~, рефлекторная reflex nausea
тошнотворный, тошнотный nauseous, nauseant, nauseating
тощекишечный *anat.* jejunal
трабекула *anat., bot.* trabecula, crossbar
трава grass; herbage, herb
~, богородская mother-of-thyme *(Thymus serpyllum)*
~, морская common eelgrass *(Zostera marina)*
~, кормовая посевная grass crop
~, почечуйная lady's thumb *(Polygonum persicaria)*
~, суданская Sudan grass *(Sorghum vulgare sudanese)*
траверз *naut.* beam
~, левый port beam
~, наветренный weather beam
~, подветренный lee beam
~, правый starboard beam
~а, впереди before the beam
~а, позади abaft the beam
~е, на abeam, abreast, athwart, on the beam
~е, прямо на right abeam
траверс *geod.* traverse
траверса crosspiece, crossarm, crossbeam, the beam
травильность etchability
травильный etching
травить 1. etch; 2. *met.* pickle; 3. *naut.* veer out, pay/slack away, slacken; 4. poison, kill, exterminate
травка 1. etching; 2. *naut.* paying/letting out
травление etching; pickling
~ точек dot-etching
~, эмульсионное powderless etching
травма trauma, injury, wound
~, акустическая/шумовая acoustic trauma, noise injury
~, огнестрельная gunshot wound
~, производственная employment/occupational/industrial trauma
травматизм traumatism, traumatosis; accident frequency
~, профессиональный/производственный occupational traumatism
травматолог traumatologist
травматотропический, травматотропный traumatotropic
травматропизм, травмотропизм traumotropism
травмировать *med.* traumatize; injure
травмоид traumoid
травник *zool.* redshank *(Tringa totanus)*
травовед herbalist
травовидный graminiform, grasslike
травокосилка grass/lawn mower
травооборот *agr.* grass rotation

траиополье grassland agriculture
травопольный grassland *(attr.)*
~ **севооборот** grassland crop rotation
травостой herbage, grass stand; grass crop
травоядность herbivority
травоядный 1. herbivorous, graminivorous, phytophagous; 2. *n.* herbivore, graminivore, phytophagan
травянистый herbaceous, grassy, grass *(attr.)*
травянка 1. dye obtained from plants; 2. *bot.* maiden pink *(Dianthus deltoides)*
травяной grassy, grass *(attr.)*
трагакант tragacanth(a)
трагедийный *theatre* tragic; Thespian
трагус *bot.* bur grass *(Tragus)*
традесканция *bot.* spiderwort *(Tradescantia)*
траектограф trajectory tracer
траектория trajectory; path; track, trace
~, **восходящая** ascending path/trajectory
~ **наведения** guidance trajectory
~, **настильная** flat trajectory
~, **нисходящая** descending path/trajectory
~, **орбитная** orbital path
~ **планирования** glide path
~, **плоская** planar trajectory
~, **пологая** flat trajectory
~ **прямого выстрела** grazing trajectory
~, **расчетная** computed/estimated trajectory
~ **свободного падения** free-fall path/trajectory
тракаулон *bot.* tearthumb *(Tracaulon)*
тракт route, path; channel; line; circuit; track; *el.* section; *anat.* tract, tractus
~, **дыхательный** respiratory tract
~, **желудочно-кишечный** gastroenteric/gastrointestinal tract
~, **замкнутый** closed loop
~, **звуковой** sound section/channel
~, **мочеполовой** urinary tract
~ **перфокарт** *comp.* card track/channel
~, **пищеварительный** digestive/alimentary tract, alimentary canal
~ **сверхвысокой частоты** microwave section
~, **сосудистый** uveal tract
~, **циркуляционный** circulation circuit
~, **фидерный** feeder
трактат treatise, tractate
трактовка treatment; interpretation
трактор tractor
~, **высококлиренсный** high-clearance tractor
~, **гусеничный** caterpillar/crawler tractor
~, **колесный** wheel tractor
~, **крутосклонный** hillside tractor
~, **лесохозяйственный** forestry tractor
~, **низкоклиренсный** low-clearance tractor
~, **пропашный** row-crop [cultivator] tractor
тракторист tractor driver
тракторостроение tractor industry/manufacture
трактор-снегоход snow tractor
трактор-тягач tow/haulage tractor
трактриса *math.* tractrix
трал sweep, drag; trawl
~, **авиационный** airplane sweep
~, **близнецовый** twin/pair trawl
~, **вертолетный** helicopter sweep
~, **змейковый** kite sweep
~, **квартропный** quarter rope trawl
~, **минный** mine-sweeping gear
~, **мягкий** wire sweep
~, **придонный** bottom sweep
траление 1. trawling; 2. mine sweeping
~, **бортовое** side trawling
~, **кормовое** stern trawling
трал-искатель searching sweep
тралить sweep; trawl
траловый 1. trawling; 2. mine-sweeping
тральщик 1. trawler; 2. minesweeper
трамбование tamping, ramming, compacting
трамбовать tamp, ram, beat down, compact
трамбовка 1. = **трамбование**; 2. rammer, tamper, beetle, tamping bar/plate, compactor
трамвай tram; *Am.* street car
трамп *naut.* tramp vessel/boat, tramp
трамплин springboard
транец *naut.* transom
~, **наклонный** raking transom
транзистор transistor; transistor radio
~, **бусинковый** bead transistor
~, **вакуумно-осажденный** vacuum-deposited transistor
~ **для логической схемы** logic transistor
~, **интегральный** integrated transistor
~, **кремниевый** silicon transistor
~, **легированный** doped transistor
~, **МДП-** insulated-gate field-effect transistor, IGFET
~, **МОП-** metal-oxide-semiconductor transistor, MOS-transistor
~, **мощный** power transistor
~ **на кристаллике** chip transistor
~, **однопереходный** unijunction transistor
~, **оптикоэлектронный** electrooptical transistor, optotransistor
~, **плоскостный** junction transistor
~, **полевой** field-effect transistor, FET
~, **пролетный** transit-time transistor
~ **с гетеропереходом** heterojunction transistor
~ **с плавным переходом** graded-junction transistor
~, **СВЧ** microwave transistor
~, **точечный** point-contact transistor
транзисторный transistorized; transistor *(attr.)*
~, **полностью** all-transistor, fully transistorized
транзит transit (traffic)
транзитивность transitivity
транзитный transit
~**ные товары** goods in transit
транзиторный transitory
транквилизатор tranquilizer, ataractic, ataraxic
транс trance
~, **гипнотический** hypnotic trance, hypnotism
~, **сомнабулический** somnambulistic trance

трансальдолаза *bioch.* transaldolase
трансаминаза *bioch.* transaminase
трансаминирование transamination
трансатлантик transatlantic liner
трансбордер *rail.* transfer table, traverser
трансвестизм transvesti(ti)sm
трансгрессивный transgressive; overlapping
трансгидрогеназа *bioch.* transhydrogenase
трансдуктор transducer
трансевропейский trans-European
трансекта transect
трансепт *arch.* transept
трансзвуковый transonic
трансиллюминация transillumination, diaphanoscopy
транскрибирование transcription
транскрибировать transcribe
транскрипция transcription
транслатив *gram.* translative case
транслировать relay, transmit, retransmit, broadcast
транслитерировать transliterate
транслятор *comp., tel.* translator; *rad.* repeater
трансляционный relaying, retransmitting; translational
трансляция 1. transmission, relaying, broadcasting; 2. repeating station, repeater
 ~, оконечная terminal repeater
 ~, простая non-regenerative repeater
 ~, регенеративная regenerator
транскюриевый transcurium, transcurial
трансмембранный transmembrane
трансметилаза transmethylase
трансметилирование transmethylation
трансмиссионный transmission *(attr.),* transmissive
трансмиссия transmission; *mech.* power drive
 ~, ременная belt transmission/drive
 ~, ступенчатая fixed-ratio transmission
 ~, фрикционная friction drive/transmission
трансмутантный transmutative
транспарант transparency; transparent poster
трансплантат transplant, graft
трансплутониевый *phys.* transplutonium *(attr.)*
трансположение transarrangement, transconfiguration
транспонирование *mus.* transposition, transposing
транспонировать *mus.* transpose
транспорт 1. transport; transfer; 2. transport vehicles/craft/vessel(s); 3. carrying forward
 ~, авиационный/воздушный air transport
 ~, автомобильный road/motor/automobile transport
 ~, безрельсовый off-track vehicles/transport
 ~, внутренний inland transport
 ~, внутризаводский on-site transport; marelials-handling equipment
 ~, внутрицеховой shop transport and hoisting equipment
 ~, городский urban/town transport
 ~, железнодорожный rail transport
 ~, морской seaborne/maritime/marine transport
 ~, общественный public transport
 ~, речной river transport
 ~, трубопроводный pipeline transport
 ~, холодильный refrigerated transport
транспортабельный carriageable, transportable
транспортер 1. *mech.* conveyer; 2. *auto.* carrier, transporter
 ~, дисковый disc conveyer
 ~, ленточный belt conveyer
 ~, плавающий amphibious carrier
 ~, рольганговый roller conveyer
транспортер-вездеход offroad transporter
транспортир protractor
 ~, маркшейдерский surveyor's level
транспортирование, транспортировка transportation, conveying, carriage, haulage
транспортировать 1. transport, convey, carry, haul; 2. carry forward
транспортник 1. transport worker; 2. transport plane
транспортно-десантный troop-carrying
транссибирский trans-Siberian
транссудация transudation
транстихоокеанский trans-Pacific
трансуран *chem.* transuranium
трансфераза *bioch.* transferase
трансферрин transferrin, siderophilin
трансфер(т) *comm.* transfer
трансфершина *el.* transfer bus
трансфокатор *opt.* variable focus lens
трансформа, трансформанта *math.* transform
 ~ Фурье Fourier transform
трансформатор *el.* transformer
 ~, бустерный/вольтодобавочный booster transformer
 ~, возбудительный excitation transformer
 ~, втулочный bushing transformer
 ~, входной input transformer
 ~, выходной output transformer
 ~, двухобмоточный double-wound transformer
 ~, емкостный capacitor transformer
 ~, звонковый bell transformer
 ~, измерительный instrument transformer
 ~, импульсный pulse transformer
 ~, линейный *com.* line transformer, repeating coil
 ~, маслонаполненный oil-filled transformer
 ~, масляный oil-immersed transformer
 ~ напряжения voltage transformer
 ~, отсасывающий sucking transformer
 ~, переходный interstage transformer
 ~, повышающий step-up transformer
 ~ полного сопротивления impedance transformer
 ~, понижающий step-down transformer
 ~ постоянного тока constant-current transformer

~, преобразовательный converter transformer
~, промежуточный intervening/intermediate transformer
~, проходной bushing-type transformer
~, путевой *rail.* track transformer
~, регулируемый variable-ratio transformer
~, резонансный tuned transformer
~ с отводами tapped transformer
~, сварочный welding transformer
~ связи coupling transformer
~, силовой power transformer
~, стержневой core-type transformer
~ строчной развертки *telev.* line transformer
трансформация transformation, conversion; modification
трансформировать transform; convert; change
трансфосфатаза transphosphatase, phosphokinase
трансфосфорилирование transphosphorylatyon
трансфузия transfusion
трансцендента *math.* transcendental function
трансцедентальность *philos.* transcendance, transcendancy, transcedentality
трансцедентальный, трансцендентный transcendental
трансэкваториальный transequatorial
транцевый *naut.* transom *(attr.)*
траншеекопатель trenching machine, trencher
траншея trench, ditch; *min.* pit
~, капитальная *min.* permanent trench
~, разрезная *min.* working trench
трап I stairway, gangway, ladder, stairs
~, аварийный emergency escape ladder
~, забортный accommodation/gangway/side ladder
~, лоцманский pilot ladder
~, надувной *av.* inflatable escape slide/chute
~, пассажирский passenger ramp/step
~, передвижной mobile steps/stairway
~, посадочный *av.* boarding ramp
~, самодвижущийся пассажирский selfpropelled passenger stairs
~, сходной companion ladder/way
трап II trap; retainer; separator
трап-балка *naut.* ladder davit
трапецевидный, трапецеидальный trapezoid(al), trapeziform
трапеция 1. trapezium; *Am.* trapezoid; 2. *astr.* Trapezium; 3. *sport* trapeze
~, равнобедренная/равнобочная isoscecles trapezium
трапецоид *geom.* trapezoid
трапецоэдр trapezohedron
трапецоэдрический trapezohedral
трапп *geol.* trap
~, стекловатый glassy diabase, sordawalite
трапповый *geol.* trappean
трас(с) *miner.* trass
трасса route, course, line, run, path, direction
~, воздушная airway, air route
~ воздушного транспорта air lane
~ записи recording track
~ трубопровода pipeline route
трассант *fin.* drawer
трассат *fin.* drawee
трассер flare; tracer
~, тепловой infra-red flare
трассировать I mark/lay out, trace
трассировать II *fin.* draw a bill
трассировка tracing, layout; routing; location
~, приближенная *comp.* loose routing
~ соединений wire routing
трассировщик router
траст *econ.* trust
~, голосующий voting trust
~, дискреционный discretionary trust
~, завещательный testamentary trust, trust under will
~, непреднамеренный resulting trust
~, пенсионный pension/retirement trust
~ по соглашению trust under agreement
~ установленный по решению суда trust under decree, court trust
тратта bill of exchange, draft, bill
~, акцептованная acceptance bill
~, встречная redraft
~ выставленная банком на другой банк bank exchange
~, документарная documentary/secured bill
~, домицилированная domiciled bill
~ оплачиваемая по прибытии товара arrival draft
~ подлежащая оплате по предъявлении demand draft
~, рамбурсная reimbursement draft
~, срочная time draft
~, срочная по предъявлении sight/presentation draft/bill/exchange
траулер trawler
~, бортовой [с бортовым тралением] side-set trauler
~, кормовой stern trauler
~, морозильный freezer trawler, full-freezer
~, рыболовный fish trawler
траурница *zool.* mourning cloak butterfly *(Nymphalis antiopa)*
трафарет stencil; template, templet; mask
траф(ф)ик traffic
трахеалгия *med.* trachealgia
трахеальный, трахейный *anat.* tracheal
трахеида *bot.* tracheid
~, тяжевая strand tracheid
трахеит *med.* tracheitis, trachitis, tracheal catarrh
трахелит *med.* trachelitis, cervicitis
трахелиум *bot.* throatwort *(Trachelium)*
трахелопексия trachelopexia, trachelopexy
трахелопластика tracheloplasty
трахелорафия trachelorrhaphy
трахелотомия trachelotomy, cervicectomy
трахеобронхит tracheobronchitis
~, спастический spasmodic tracheobronchitis

трахеобронхоскопия (tracheo)bronchoscopy
трахеола *zool.* tracheole
трахеоларинготомия tracheolaryngotomy
трахеомаляция tracheomalacia
трахеопатия tracheopathia
трахеопластика *surg.* tracheoplasty
трахеорафия tracheorrhaphy
трахеоррагия tracheorrhagia
трахеостомия tracheostomy
трахеотомировать *surg.* tracheotomize
трахеотомия *surg.* tracheotomy
~, верхняя superior tracheotomy
~, нижняя inferior tracheotomy
~, срединная median tracheotomy
трахеоцеле trache(i)ocele, bronchocele
трахея trachea, windpipe
трахикарпус *bot.* windmill palm *(Trachycarpus)*
трахилобий *bot.* rasp pod *(Trachylobium)*
трахинотус *zool.* pompano *(Trachinotus)*
трахиптерус *zool.* ribbonfish *(Trachipteridae)*
трахит *miner.* trachyte
~, кварцевый quartz trachyte, rhyolite
траходонт *paleont.* trachodon *(Trachodon)*
трахома *med.* trachoma, contagious granular conjunctivitis
треба religious rite/ceremony
требник *rel.* prayer book; formulary
требование requirement; request, claim; demand; *pl.* specifications; *law* action
~, акцессорное ancillary suit
~, встречное *law* counter-claim
~ по ипотеке hypothecary claim
~ повышения заработной платы wage demand
~, преимущественное lien; prior claim
~ признавать allow a claim
~ рынка market demand
~ия, технические specifications; system requirements
~ия, технологические process/technological requirements
тревога alarm, alert
~, воздушная air alert
~, химическая gas alarm/alert
трегалоза *chem.* trehalose
трегерит *miner.* troegerite
треджен *sport* trudgen stroke
тред-юнион trade union
тред-юнионистский trade-unionist, trade-union *(attr.)*
трезвучие *mus.* triad
трезубец trident
трезубчатый tridentate, three-toothed/pronged
трейлер trailer
трейлеровоз *naut.* trailership
трек track; trace, path
трель trill; warble
трематода *zool.* trematode, trematoid
трематодоз *med.* infestation with trematodes
тремоло *mus.* tremolo
тремор tremor, trembling, shaking
~, волевой volitionl tremor
~, динамический kinetic tremor
~, ртутный mercurial tremor
~, свинцовый saturnine tremor
~, старческий senile tremor
~, тонкий fine tremor
тренаж drill; training
тренажер trainer; simulator
~, авиационный aircraft simulator
~, космический space (mission) simulator
~, пилотажный flight simulator/trainer
тренд I *naut.* trend, anchor crown
тренд II *stat., math.* trend
~, возрастающий upward trend
~ дисперсии *stat.* variance trend
~, квадратический/кубический/линейный *stat.* quadratic/cubic/rectilinear trend
~, нелинейный *stat.* curvilinear trend
~, периодический cyclic trend
~ роста upward trend
~, убывающий *stat.* downward trend
трензел 1. tumbling/tumbler gear; 2. *mus.* triangle
трение friction, rubbing; attrition
~ в подшипниках bearing friction
~ верчения pivoting friction
~, внутреннее internal friction, viscosity
~ второго рода rolling friction
~, вязкое viscous friction
~, гидродинамическое fluid/flow friction
~, граничное boundary friction
~ движения/кинетическое kinetic/dynamic friction, friction of motion
~ качения rolling friction
~ первого рода sliding friction
~ покоя/статическое static friction, friction of rest
~ скольжения sliding friction
~, тормозящее drag friction
тренога, треножник tripod
треонин threonine
трепан *surg.* trepan, trephine
трепанатор *surg.* trepanner, trephiner
трепанация *surg.* trepanation, trephination
~ склеры sclerotomy
~ черепа cranial trepanation, craniotomy
трепанг *bot.* trepang, sea cucumber *(Holothuria edulis)*
трепанировать *surg.* trepan, trephine, trepanize
трепанобиопсия *med.* trepanobiopsy
трепел diatomite, fossil flour, tripoli (earth)
трепетание palpitation, flutter
~ предсердий atrial/auricular flutter
треск crackling, crackle, crack; crepitus
~, оловянный *met.* tin cry
~, суставной articular crepitus, grating of a joint
треска *zool.* cod *(Gadus morrhua);* pout *(Gadus luscus)*
тресковый cod; pout *(attr.);* gadoid
~ жир cod-liver oil

трескучка *bot.* regma
трескучник *bot.* yellow mignonette *(Reseda luteola)*
трест *econ.* trust
~, мозговой brain trust
трестирование *econ.* trustification, formation of trusts
трестировать trustify, combine/organize into a trust
третейский arbitration *(attr.)*
~ суд arbitration court/tribunal
~ судья arbitrator, arbiter
~кое решение arbitration; award
третичный tertiary, ternary; *geol.* Tertiary
треугольник triangle; trigone
~, бедренный femoral/subinguinal triangle
~, верхний сонный *anat.* superior carotid triangle
~, вписанный inscribed triangle
~, затылочный *anat.* occipital triangle
~, косоугольный oblique triangle
~, крестцовый *anat.* sacral triangle
~, лопаточно-ключичный subclavian triangle
~ невязки *geod.* error triangle, triangle of errors
~, обонятельный *anat.* olfactory trigone
~, описанный circumscribed triangle
~, остроугольный acute-angled triangle
~, прямоугольный right-angled triangle
~, равнобедренный isosceles triangle
~, равносторонний equilateral triangle
~, разносторонний scalene triangle
~, тупоугольный obtuse-angled triangle
~, сонный *anat.* carotid triangle
~ скоростей triangle of velocities, speed triangle
треугольный triangular, trigonal
треф *met.* wobbler
трефин *med.* trephine
трехатомный triatomic
трехбрюшковый trigastric
трехбугорчатый trituberculate
трехбуквенный three-letter, trigram
трехвалентность *chem.* trivalence
трехвалентный *chem.* trivalent
трехверстный three-versts
трехветвистый triramose; trifurcate
трехвидный trimorphic
трехвинтовой triple-screw
трехгибридный trihybrid
трехглавый tricephalous, tricipital
~вая мышца *anat.* triceps
трехгранник *geom.* trihedron, trihedral
трехгранный *geom.* trihedral; *bot.* triquetrous
трехгубый trilabiate
трехдневный three-day
~вная лихорадка tertian fever/ague
трехдолевой trilobate, trilobed
трехдольный *bot.* tricotyledonous
трехдольчатый trilobiate
трехдомность *bot.* trioecism
трехдомный *bot.* trioecious
трехжильный 1. *el.* three-core; 2. *bot.* trinervate
трехзамещенный trisubstituted
трехзначный three-digit/-figure/-place
~ная функция three-value function
трехзубка *bot.* heath grass *(Sieglingia decumbens)*
трехзубчатый three-pronged/-toothed, tridentate
трехзубый three-toothed, tridentate
трехкаскадный three-stage
трехкилевой three-keeled; tri-finned
трехклапанный trivalve
трехколесный three-wheeled
~ велосипед three-wheeled bicycle
трехколосый *bot.* tristachyous
трехкомпонентный three-component
трехконтактный three-pin/-point/-prong
трехкоробчатый *bot.* three-capsular
трехкосточник *bot.* horse gentian *(Triosteum)*
трехкрайник trihedron
трехкратный triple, threefold
трехкруговой tricyclic
трехкрылый *bot.* trilate, tripterous
трехкулачковый three-jawed; three-lobe
трехламповый *el.* three-tube
трехлемешный *agr.* three-share
трехлетний *bot.* triennial
трехлинейный trilinear; three-way
трехлистный trifoliate
трехлисточковый *bot.* trifoliolate
трехлонжероновый three-spar
трехлопастный three-bladed; *bot.* trilobate, three-lobed
трехлучевой triradial, triradiate, triactinal
трехмачтовый three-masted
трехмерность tridimensionality
трехмерный three-dimensional; trivariate; *mus.* triple
~ такт triple time
трехместный 1. three-seater; 2. *math.* triadic
трехмесячный trimestrial, trimensual
трехмешковый trisaccate
трехморфный trimorphous; triformed
трехмоторный three-engined
трехногий three-legged; tripedal
трехокись trioxide
трехосевой triaxial
трехосновный *chem.* tribasic
трехосный three-axle; triaxial
трехпалый tridactylous; three-fingered
трехпальчатый tridigitate, tridactyl
трехпарный tergeminate; trijugate
трехперегородчатый, трехперегородочный triseptate
трехперестка *zool.* button quail *(Turnix)*
трехпестичный *bot.* trigynous
трехпластинчатый trilaminar
трехпозиционный three-position/-step
трехполье, трехпольщина three-field system
трехпроводный *el.* three-wire/-conductor
трехпутный three-lane/-track/-way
трехпыльниковый *bot.* triantherous

трехразветвленный trifurcate(d)
трехраздельный tripartite; three-cleft, trifid
трехразмерный three-dimensional; steric
трехразрядный *math.* three-digit/-position
трехреберный three-ribbed/finned, tricostate
трехребристый tricostate
трехрогий three-horned, tricornute
трехручный trimanual
трехсемядольный *bot.* tricotyledonous
трехсекционный three-segment
трехслоговой, трехсложный trisyllabic
трехслойный three-ply/-layer
трехстабильный tristable
трехстворчатый 1. three-sash, three-leaved; 2. *anat.* tricuspidate, tricuspid(al); 3. *biol.* three-valved
трехстишие *lit.* tercet
трехстопный *lit.* three-feet, of three feet
  ~ ямб iambic trimeter
трехступенчатый three-stage/-step(ped)
трехтактный three-phase/-cycle/-stroke
трехтрубный *naut.* three-stack(ed)
трехфазный three-phase
трехходовой three-way/-pass/-throw
трехцветка *bot.* wild pansy *(Viola tricolor)*
трехцветность trichromatism
трехцветный trichromatic, tricolo(u)red
трехчлен *math.* trinomial
трехчленный 1. three-term, trimembral; ternary; 2. *math.* trinomial; 3. *bot.* trimerous
трехщелевой tripple-slotted
трехъязычный trilingual
трехэтажный three-story/-storied
треченто trecento
трещина crack, crevasse, fissure, chink, crevice, cleft; fracture; split; *(in wood)* shake
  ~ в леднике crevasse
  ~, волосная hairline crack
  ~ высыхания desiccation fissure, sun crack
  ~, кольцевая *(in timber)* ring/cup shake
  ~, радиальная/разветвленная star crack
  ~ разрыва rupture, rent
  ~, сердцевинная heart shake *(in wood)*
  ~ сжатия compression joint
  ~ усушки *wood.* shrinkage crack
трещиностойкий crackproof, fracture-proof
трещиностойкость crack/fracture resistance
трещотка 1. *tech.* ratchet; 2. rattle
триада triad; trine, ternion; ternary
триадический triadic
триадный triad; ternary
триакантин *chem.* triacanthine hydrochloride
триакистетраэдр *geom.* triakistetrahedron
триаллиламин triallylamine
триампизин triampyzine
триамсинолон, триамцинолон triamcinolone, aristocort, kenacort
триангуляция triangulation, trigonometrical survey, horizontal triangulation
  ~ первого класса first-order [main] triangulation
триархия triarchy
триас, триасовый *geol.* Triassic
триб, трибка *tech.* pinion
триба *biol.* tribe, clan
трибадия *med.* tribadism, tribady
триболюминесценция *phys.* triboluminescence
трибометр tribometer, friction gauge
трибрахий *lit.* tribrach
трибунал tribunal; court
  ~, военно-морской naval court-martial
  ~, военный court-martial, military court
трибунат *hist.* tribunate
трибунский *hist.* tribunitian
трибутирин *chem.* tributyrin, glyceryl tributyrate
трибутириназа *bioch.* tributyrinase
тривакцина trivaccine, triple vaccine
тривиальность triviality
тригамия trigamy
тригатрон trigatron
триггер trigger; flip-flop
  ~, дежурный set-reset flip-flop
  ~, запрещающий inhibit flip-flop
  ~ на электронных лампах vacuum tube toggle
  ~, насыщенный saturating flip-flop
  ~, ненасыщенный nonsaturating flip-flop
  ~, переключающий relay flip-flop
  ~ переноса carry flip-flop
  ~ прерывания interrupt flip-flop
  ~, противодребезговый debounce flip-flop
  ~ с защелкой latched flip-flop
  ~ с раздельными входами R-S flip-flop, set-reset flip-flop
  ~ со счетным входом complementing flip-flop
  ~, счетный [счетного типа] trigger/complement/toggle/T flip-flop
  ~, тактируемый clocked flip-flop
тригеминия *med.* trigeminy
тригидрат trihydrate
тригла *zool.* gurnard *(Trigla)*
триглиф *arch.* triglyph
тригонит I *miner.* trigonite
тригонит II *med.* trigonitis
тригонометрия trigonometry
  ~ на плоскости plane trigonometry
  ~, сферическая spherical trigonometry
тригпункт triangulation/trigonometrical point
триер *agr.* grain cleaner/grader/separator
триеровать *agr.* grade, screen, sift
тризм *med.* trismus, lockjaw
  ~, челюстной lockjaw
тризна *hist.* funeral feast
трикирий tricerion, trikerion
триклиний *hist.* triclinium
трико 1. tricot; 2. tights
  ~ телесного цвета flesh tights
трикотаж knitted fabrics/goods; knitted wear
трилистник *bot.* trefoil *(Trifolium)*
трилистный trifoliate
триллион *Brit.* billion; *Am.* trillion

триллиум *bot.* wake-robin, trillium *(Trillium)*
трилогия trilogy
трима *bot.* tryma
тримаран *naut.* trimaran
триместр trimester, term
триметилксантин trimethylxanthine, caffeine
триммер 1. *aero.* trimmer, trimming tab; 2. *mech.* trimming machine; 3. *el.* trimmer capacitor; trimmer conveyer
тримонэция trimonoecy
триморфный trimorphous, trimorphic
тринакс *bot.* thatch palm *(Thrinax)*
тринарный ternary
тринитрин glyceryl trinitrate
тринитросоединение trinitro compound
тринитротолуол trinitrotoluene, trotyl, TNT
~, плавленый cast TNT
тринитрофенол trinitrophenol, picric acid
триния *bot.* trinia *(Trinia)*
трином, триномиальный *math.* trinomial
триод *el.* triode
~, двойной twin/double triode
~, ионный gas-filled triode
~, одноцокольный single-end triode
~, полевой/канальный field-effect transistor
~, полупроводниковый transistor
~, пусковой trigger triode
триокись trioxide
триоктаэдр trioctahedron
триоль *mus.* triplet
триостеум *bot.* horse gentian *(Triosteum)*
триостренник *bot.* pod grass *(Triglochin)*
трип velveteen
трипаносома *zool.* trypanosome
трипаносомоз *med.* trypanosomiasis
~, африканский African trypanosomiasis, sleeping sickness
триплан triplane
триплегия *med.* triplegia
триплекс 1. (laminated) triplex glass; 2. triplex (process); 3. triplexer
триплет, триплетный triplet
триплоид, триплоидный *biol.* triploid
триплоидия *biol.* triploidy
триплокаулический, триплокаульный *bot.* triplocaulous
триподия *lit.* tripody
триппер *med.* gonorrh(o)ea
трипперный *med.* gonorrh(o)eal
трипс *zool.* thrips *(Thysanoptera)*
трипсакум *bot.* gama grass *(Tripsacum)*
трипсин *chem.* trypsin
трипсиноген trypsinogen, trypsogen
триптаза *bioch.* tryptase
триптих *art.* triptych
триптический *physiol.* tryptic
триптот *gram.* triptote
триптофан *bioch.* tryptophan
триптофансинтетаза tryptophan synthetase
трипутник *bot.* swertia *(Swertia)*
трирадиальный triradial, triradiate
трираздельный trifid
трирема trireme
трисахарид trisaccharide
трисель *naut.* trysail
трисеменный *bot.* trispermous
трисемянка *bot.* triachenium
трисиллаб trisyllable
трисомики *pl.* trisomics
трисомия trisomy
трисоставный three-part, tripartite
тританопия tritanopia, blue blindness
тритий *chem.* tritium, T
тритицин *chem.* triticin
тритома *bot.* torch lily *(Tritoma)*
тритон I *phys.* triton, T; *mus.* tritone
~ отдачи recoil triton
тритон II *zool.* triton, newt *(Triturus)*; *myth.* Triton
тритония *bot.* tritonia, montbretia *(Tritonia)*
тритринакс *bot.* trithrinax palm *(Trithrinax)*
триумвир *hist.* triumvir
триумвират *hist.* triumvirate
трифан *miner.* triphane, spodumene
трифолиата *bot.* trifoliate orange *(Poncirus trifoliata)*
трифоль *bot.* marsh trefoil *(Menyanthes trifoliata)*
трифорий *arch.* triforium
трифтонг *phon.* triphthong
трифторид trifluoride
трихазий *bot.* trichasium
трихиаз *med.* trichiasis
трихидий *bot.* trichidium
трихина,трихинелла *zool.* trichina *(Trichinella spiralis)*
трихинеллез,трихиноз *med.* trichinelliasis, trichinellosis, trichinosis, trichiniasis
трихинеллезный trichinellous, trichinous
трихинозный trichinal
трихобласт *bot.* trichoblast
трихоген *zool.* trichogen
трихогиалин trichohyaline
трихогина *bot.* trichogyne
трихоглосия trichoglossia, hairy tongue
трихозант(а) *bot.* edible snake gourd *(Trichosanthes anguina)*
трихокардия trichocardia, hairy heart
трихокарпий *bot.* trichocarp
трихом(а) *bot.* trichome, plant hair
трихоматоз *med.* trichomatosis, trichoma
трихоматозный trichomatose, trichomatous
трихомикоз trichomycosis, tricomycetosis
трихомицин trichomycin
трихомонада *zool.* trichomonad *(Trichomonas)*
трихомоноз *med.* trichomoniasis, trichomonosis
трихорексис *med.* trichorrhexis, trichoschisis
трихоспория *med.* trichosporosis
трихоталлический *bot.* trichothallic
трихотецин trichothecin
трихотомический trichotomous, trichotomic

трихотомия trichotomy
трихофития *med.* trichophytosis, (crusted) ringworm, favus, tinea
трихофитоз *med.* trichophytosis
трихоцефалез *med.* trichocephaliasis
трихроичность trichroism
трихроичный trichroic
трихроматический trichromatic, trichromic
трихуриаз,трихуроз *med.* trichuriasis
трицепс *anat.* triceps
трициртис *bot.* toad lily *(Tricyrtis)*
трищетинник *bot.* trisetum *(Trisetum)*
триэдр trihedron, trihedral
триэдрический trihedral
триэтиламин triethylamine
триэцичный *bot.* tri(o)ecious
триэция tri(o)ecism
троакар *surg.* troc(h)ar
трог trough valley; trough
троглодит troglodyte, cave man
троеборье *sport* triathlon
троебрачие, троеженство trigamy
троеперстие *eccl.* making the sign of the cross with three of the fingers
Троица *eccl.* Trinity; Whitsun(day)
троичность triplicity, triality; triunity
троичный 1. triune; 2. *eccl.* Trinity *(attr.)*; 3. *math.* radix-three, base-three, ternary
тройник tee-joint, T-joint, tee (pipe)
тройничный 1. triple; ternary; 2. *anat.* trigeminal; trifacial
тройной threefold, triple, triplicate, ternary, treble; triplex; trine
 ~ное правило *math.* the rule of three
тройственный tripartite, triple; trinal; ternary
 ~ союз *hist.* Triple Alliance
 ~ное согласие *hist.* Triple Entente
тролль *myth.* troll
тромб I tornado; whirlwind
 ~, пыльный dust devil
тромб II *med.* thrombus, blood clot
 ~, агональный agony thrombus
 ~, гиалиновый hyaline/agglutinative thrombus
 ~, предсмертный antemortem thrombus
 ~, пристеночный parietal/mural thrombus
тромбоангиит *med.* thromboangiitis
 ~, облитерирующий Buerger's disease
тромбоген thrombogen, prothrombin, plasmozyme
тромбоз *med.* thrombosis
 ~, внутрисердечный intracardiac thrombosis
 ~ глубоких вен deep vein thrombosis
 ~, дилятационный dilation thrombosis
 ~, коронарный coronary thrombosis
 ~, мигрирующий/скачущий jumping thrombosis
 ~, послеродовой puerperal thrombosis
 ~, тромбоцитарный plate/platelet thrombosis
 ~, церебральный cerebral thrombosis
тромбозный *med.* thrombotic
тромбокиназа *bioch.* thrombokinase
тромболизин thrombolysine
тромболитический thrombolytic, thromboclastic
тромбообразование thrombogenesis
тромбообразующий thrombogenic
тромбопластин thromboplastin, thrombozyme
тромбопластинообразование thromboplastinopoiesis
тромбопоэз thrombopoiesis
тромботонин thrombotonin, thrombocytin
тромбофлебит *med.* thrombophlebitis
 ~, послеродовой (iliofemoral) postpartum thrombophlebitis, phlegmasia, white leg
тромбоцит thrombocyte, blood platelet
тромбоцитоз thrombocytosis
тромбоцитопения *med.* thrombocytopenia, low platelet count
тромбоцитопоэз thrombocytopoiesis
тромбоэмболия thromboembolism
тромп *arch.* trumpet vault
тронный throne *(attr.)*
 ~ная речь king's speech
тропакокаин thropacocaine, benzoylpseudotropeine
тропарь *eccl.* hymn, anthem
тропафен *chem.* thropaphen
тропеин *chem.* tropeine
тропик tropic
 ~ Козерога tropic of Capricorn
 ~ Рака tropic of Cancer
тропин tropine
тропофилл tropophyll
тропофильный tropophilous
тропофит tropophyte
тропофитный tropophytic
трос rope, cable, line, hawser
 ~ аэрофинишера *av.* arresting wire; *(on aircraft carrier)* cross-deck pendant
 ~, буксирный towing hawser/rope/line
 ~, вытяжной rip cord, withdrawal line
 ~ гидролокатора sonar cable
 ~, джутовый jute rope
 ~, кокосовый grass line, coir cable
 ~, льняной flax/linen rope
 ~, манильский Manila/abaca rope
 ~, натяжной guy rope, stretching wire
 ~, несмоленый white rope
 ~, несущий carrier/bearer cable; suspension strand/cable
 ~ оттяжки guy rope
 ~, пеньковый hemp rope
 ~, плетеный plaited rope
 ~, привязной captive cable
 ~, причальный mooring rope/line/cable
 ~, разрывной rip wire
 ~, смоленый tarred/black rope
 ~, трала/тралящий sweep rope/wire
 ~, тяговый pull rope
 ~, швартовый tie-down cable
тростник *bot.* reed, cane *(Phragmites)*
тростниковидный *bot.* arundinaceous
тростниковый cane *(attr.)*; cany, reedy;

juncareous, calamiform
~ **сахар** cane sugar
~**вая заросль** reed thicket
тростянка 1. *bot.* river/hard grass *(Sclerocloa);* 2. reed pipe
тротил trotyl, trinitrototluene, TNT
трофика trophicity, trophism
трофобластома trophoblastoma, chorioepithelioma
трофогенный trophogenous
трофоневроз trophoneurosis
~, **мышечный** muscular trophoneurosis, progressive muscular atrophy
трофотаксис trophotaxis, trophotropism
трофофит *bot.* trophophyte
трохантерит trochanteritis
трохантеритный trochanterian, trochanteric
трохантин trochantin(e)
трохеический *lit.* trochaic
трохей *lit.* trochee
трохоида *math.* trochoid, trochoidal curve
трохотрон trochotron, cycloidal-focussing mass-spectrometer
трохоцефалия trochocephalia, trochocephaly, roundheadededness
троцкизм Trotskyism
троцкист, троцкистка Trotskyite, Trotskyist
троцкистский Trotskyist
трощение *text.* doubling; twisting; throwing
троюродный брат second cousin
троянец, троянский Trojan
труба pipe, tube; duct; piping, tubing; stack, chimney; tunnel
~, **аэродинамическая** wind/air tunnel
~, **безнапорная** nonpressure pipe
~, **бесшовная** seamless pipe
~, **брекчиева** *geol.* breccia chimney
~, **вводная** lead-in tube/pipe
~, **вентиляционная/вестовая** vent tube, air conduit, ventilation/air duct
~, **вертикальная** riser/rising pipe, riser
~, **визирная** sighting telescope
~, **водопроводная** water supply pipe
~, **водопускная** downcomer
~, **водосточная/водоотводная** drain-pipe
~, **воздухозаборная** air induction pipe
~, **воздухоотводная** air-escape pipe
~, **волоченая** drawn pipe
~, **впускная** inlet/intake/feed/supply/admission pipepipe
~, **выпускная/выхлопная** outlet/exgaust/exit/discharge pipe; exhaust duct/stack
~, **вытяжная** draft tube
~, **газоотводная** exhaust-gas duct
~, **гельмпортовая** *naut.* rudder trunk/well
~, **горячекатаная** hot-rolled pipe
~, **горячетянутая** hot-drawn pipe
~, **дейдвудная** *naut.* stern/shaft tube
~, **дымовая** chimney/smoke stack, funnel
~, **дымогарная** fire/flue/smoke tube
~, **евстахиева** *anat.* auditory/Eustachian tube
~, **жаровая** flame/fire/flue/furnace tube
~, **заборная** intake tube
~, **замкнутая** return-flow tunnel
~, **зрительная** telescope; field glass
~, **кавитационная** cavitation tunnel
~, **канализационная** sewer (pipe)
~, **катаная** rolled pipe
~, **кипятильная** steaming [steam-generating] pipe
~, **коллекторная** manifold pipe
~, **колонковая** core barrel; coring tube
~, **компенсационная** expansion pipe
~, **маточная** *anat.* uterine/fallopian tube, oviduct, salpinx
~, **морская зрительная** naval telescope
~, **нагнетательная** delivery tube
~, **напорная** pressure pipe
~, **наставная** adapter
~, **обводная** bypass pipe
~, **обратная** return pipe
~, **обсадная** casing
~, **опускная** down-flow pipe
~, **отводящая** outlet/discharge pipe
~, **отсасывающая** draught/draft tube
~, **пароотводная** steam-outlet/-escape tube
~, **переговорная** speaking tube
~, **переполнительная** overflow pipe
~, **перепускная** bypass tube
~ **перископа** *naut.* periscope mast
~, **питающая** supply tube/duct
~, **подающая** delivery/feed/supply pipe
~, **подводящая** inlet pipe
~, **подъемная** rising/lifting pipe, riser
~, **приточная** influent pipe
~, **пылеотводная** dust trunk
~, **пыльцевая** pollen tube
~, **расходомерная** flow-rate measuring tube
~, **ребристая** finned pipe
~, **сливная** drain pipe
~, **слуховая** *anat.* auditory/Eustachian tube, otopharyngeal/pharyngotympanic tube
~, **спускная** discharge/outlet pipe
~, **сточная** scupper pipe; sewer
~ **суфлирования** breather tube
~, **торпедная** *mil.* torpedo tube
~ **управления** control tube
~, **фаллопиева = маточная**
~, **фановая** sewage/sewer pipe
~, **холоднокатаная** cold-rolled pipe
~, **холоднотянутая** cold-drawn pipe
~, **якорная** hawse pipe
трубач 1. trumpeter; 2. *zool.* stentor *(Stentor);* trumpeter swan; trumpeter pigeon
трубачик *zool.* tree cricket *(Oecanthus)*
труб-бакштаг *naut.* funnel backstay
трубиться *bot.* boot, shoot
трубка tube, pipe; *el.* valve, tube; *see also* **труба**
~ **Вентури** Venturi tube
~, **вихревая** vortex tube

~, **волосная** capillary tube
~, **газовая счетная** gass-filled counter tube
~, **газозаборная** gas probe
~, **дыхательная** breather; snorkel
~, **жиклерная** jet tube
~, **зажигательная** igniter/blasting fuse
~, **запальная** primer
~, **индикаторная** *el.* display tube
~, **ионо-ускорительная** ion-accelerating tube
~, **капиллярная** capillary tube
~, **катодно-лучевая** cathode-ray tube
~, **микротелефонная** handset
~, **мышечная** myotube
~, **напорная** pressure tube
~, **нервная** neural tube
~, **отводящая** discharging tube
~ **Пито** Pitot tube
~ **подпора** back-pressure tube
~ **поля** field tube; tube of force
~ **Прандтля** Prandtl tube
~ **Престона** Preston tube
~, **расходомерная** flow tube
~, **рентгеновская** X-ray tube
~, **сердечная** cardiac tube
~, **силовая** tube of force; field tube
~ **силовых линий** field tube
~, **стеклодувная** blow-pipe
~ **суфлера** breather pipe
~, **счетная** counter tube
~, **телевизионная передающая** camera tube
~, **телевизионная приемная** kinescope, picture tube
~, **телефонная** handset
~, **трехлучевая/трехпрожекторная** three-gun (picture) tube
~, **ударная** percussion primer
~, **электроннолучевая** cathode-ray tube, CRT, electron-beam/-ray tube
~, **электроннолучевая знаковая** character display cathode-ray tube

трубкование *bot.* booting, shooting, stem extension
трубковерты *zool.* leaf-rolling weevils *(Attelabidae)*
трубковидный,трубкообразный tubiform, tubular, pipelike, tubelike
трубкозуб *zool.* aardvark (*Orycteropus afer)*
трубкокрылы *pl. zool.* solenostomids *(Solenostomidae)*
трубконос *zool.* tube-nosed bat
трубконосный tubiferous
трубкоротый *zool.* siphonostomatous
трубкоцветный *bot.* tubiliflorous
трубно-брюшинный tuboperitoneal
трубно-брюшной tuboabdominal
трубно-влагалищный tubovaginal
трубно-маточный tubo-uterine
трубно-яичниковый tuboovarian
трубный 1. pipe, tube *(attr.); anat.* tubal; 2. trumpet *(attr.)*
трубогиб pipe bender
трубонарезной pipe-threading
трубообразный pipe-shaped, pipelike
трубопровод pipeline, main, piping, tubing, duct, manifold
~, **водяной** water-supply pipe(line)
~, **нагнетательный/напорный** delivery piping
~, **перепускной** crossover pipeline
~, **питательный/подводящий** supply/feed pipeline
~, **противопожарный** fire main/pipeline
~, **спускной** drainage pipeline

трубопроводчик pipe fitter
трубопрокатка tube/pipe rolling
труборез pipe/tube cutter
трубоукладчик pipe-layer; pipe laying machine
трубоцвет *bot.* common trumpet creeper *(Tecoma radicans)*
трубочка tubule, tube
~, **железистая** glandular tubule
~, **костная** osseous tubule

трубчатый tubular, tubulate, tubulous, pipelike
труб-штаг *naut.* funnel shroud/stay/guy
труд labo(u)r, work; *pl.* proceedings, transactions, publications, works
~, **барщинный** corvée
~, **детский** child labour
~, **живой** direct/living labour
~, **квалифицированный** skilled labour
~, **крепостной** peonage
~, **наемный** hired/wage labour
~, **неквалифицированный** unskilled labour
~, **непроизводительный** nonproductive labour
~, **ночной** night work
~, **обобществленный** associated labour
~, **общественно-необходимый** socially necessary labour/work
~, **общественно-полезный** socially useful work
~, **овеществленный** embodied/materalized labour
~, **подневольный** peonage
~ **подростков** juvenile labour
~, **прибавочный** surplus labour
~, **принудительный** forced labour
~, **производительный** productive labour
~, **ручной** manual labour
~, **совокупный общественный** aggregate/collective labour of society
~, **умственный** mental work
~, **физический** manual/hand/physical labour

труддисциплина *abbr.* labour discipline
трудновоспламеняемый flash-resistant
труднолетучий difficult of volatilization, involative
трудноплавкий hard to melt, hardly fusible
труднорастворимый hardly soluble, almost insoluble
трудноуправляемый difficult to handle/control,

unhandy
трудовой work, labo(u)r; working
 ~ договор labour contract
 ~ доход earned income
 ~ стаж record of work
 ~вая книжка work-record book
трудодень workday *(unit of work)*
трудоемкий labour-intensive/-consuming, time-taking
трудоемкость labour input; labour-consuming nature; labour-output ratio; labour intensity/coefficient
трудоинтенсивность labour intensity
трудоспособность capacity/ability to work, employability; working capacity/ability
трудоспособный able-bodied, employable, capable of working
трудотерапия *med.* occupational/work therapy, ergotherapy
трудоустраивать find a job for, place in a job
трудоустройство provision of employment
трунцит *med.* truncitis
труп cadaver, corpse, carcass; dead body
трупиал *zool.* oriole *(Icterus)*
трупный cadaver, cadaverous, cadaveric
трутовик, трутовица *bot.* polypore *(Lactiporus);* sponk *(Polyporus)*
трюизм truism; platitude
трюк stunt; trick
трюм *naut.* bilge; hold
 ~, грунтовой hopper
 ~, кормовой aft/after hold
 ~, носовой fore hold
 ~, тросовый hawser store
трюфель *bot.* truffle *(Tuber)*
трясинник *bot.* earth moss *(Phascum)*
тряска jolting, shaking; buffeting; vibration
трясконстойкий vibration-proof, shockproof
трясогузка *zool.* wagtail *(Motacilla)*
трясунка *bot.* quaking grass *(Briza)*
трясуны *pl.* Holy Rollers *(religious sect)*
трясучка *tech.* shaker, vibration unit/table
тсуга *bot.* hemlock *(Tsuga)*
туаз *hist.* toise
туаллерген tularemic allergen
туба 1. tube; 2. *mus.* tuba
туберидий *bot.* tuberidium
туберкул(а) tubercula
туберкулез tuberculosis, phthisis, tb
 ~ глаз ocular tuberculosis
 ~ гортани tuberculosis of the larynx, laryngeal phthisis, laryngophthisis
 ~, диссеминированный disseminated/miliary tuberculosis
 ~ кожи cutaneous tuberculosis, tuberculoderma, scrofuloderma
 ~ легких pulmonary tuberculosis, phthisis, consumption, tuberculosis of the lungs
 ~ лимфатических узлов scrofula, lymphatic tuberculosis
 ~, милиарный miliary tuberculosis; general tuberculosis
 ~ мозговых оболочек meningeal tuberculosis, tuberculous meningitis
 ~ надпочечников adrenal tuberculosis, Addison's meningitis
 ~, очаговый focal tuberculosis
 ~ позвоночника spinal tuberculosis
 ~, птичий avian/fowl tuberculosis
 ~, циротический cirrhotic tuberculosis
 ~, юношеский juvenile tuberculosis
туберкулезный tuberculous
туберкулема tuberculoma, tuberculome
туберкулид tuberculid, tuberculoderma
туберкулин *med.* tuberculin
туберкулома = туберкулема
тубероза *bot.* tuberose *(Polianthes)*
туберозный tuberous
тубокурарин (d-)tubocurarine
тубоварит tubo-ovaritis, salpingoophoritis
тубоотит *med.* tubo-otitis
тубулярный tubular, tubiliform
тубус tube, barrel; *naut.* twill trunk
 ~ микроскопа drawtube
 ~ объектива lens barrel
тувинец, тувинский Tuvinian
тувинка Tuvinian (woman)
тугоплавкий high-melting, refractory, infusible
тугоплавкость refractoriness, infusibility
тугоподвижность rigidity
тугоподвижный rigid
тугоухость dullness of hearing, hypoacusis, ambliacousia, bradyacousia
тугрик tugrik *(Mongolian unit of currency)*
тугун 1. *zool.* tugun *(Corregonus tugun);* 2. *hist.* tuchun, war lord
туер chain tug/vessel
тузик (two-oar) dinghy, scull, sculler
тук (solid) mineral fertilizer
тукан *zool.* toucan *(Rhamphastos)*
туканчик *zool.* toucanet *(Aulacorhynchus)*
тукодробилка fertilizer crusher/grinder/disintegrator
тукоразбрасыватель fertilizer spreader
тулес *zool.* gray plover *(Squatarola squatarola)*
тулий *chem.* thulium, Tm
туловище torso, trunk
туловищный truncal
тульский from [made in] Tula
туляремийный, туляремический *med.* tular(a)emic
туляремия *med.* tular(a)emia
туман I fog; mist, naze
 ~, адвективный advection mist
 ~, легкий haze, mist
 ~, ледяной ice/frost fog
 ~, масляный oil mist
 ~, моросящий drizzling/wet fog
 ~, сильный thick fog
 ~, слабый thin fog

~ **склона** upslope fog
~, **сухой** dry fog, haze
~, **сырой** wet fog
туман II toman *(Persian gold coin)*
туманность 1. nebula, nebulosity; 2. haziness
~, **досолнечная** presolar nebula
туманный nebulous; hazy, misty, foggy
тумба *naut.* stand; post
~, **причальная** dolphin, bollard, deadhead, landfast, makefast
~, **швартовая** mooring/checking bollard
тумблер toggle/tumbler switch
~ **включения** engage switch
~ **останова** stop switch
~ **отключения** disengage switch
~ **отцепки** release switch
туна *bot.* tuna *(Opuntia tuna)*
тунбергия *bot.* clock vine *(Thunbergia)*
тунг *bot.* tung tree *(Aleurites)*
тунгус, тунгусский Tungus
тунгуска Tungus (woman)
тундра tundra
~, **венковая** wreath tundra
~, **кустарниковая** underbush tundra
~, **медальонная/пятнистая** mottled tundra
тундрянка *zool.* rock ptarmigan *(Lagopus mutus)*
тунец *zool.* tuna *(Thunnus)*
туника tunic, tunica; *bot.* tunic flower *(Tunica)*
тунисец, тунисский Tunisian
туниска Tunisian (woman)
туннелеобразный tunnellike, tunnel-shaped
туннелирование tunnelling
туннель tunnel; duct; underpass
~, **напорный** power conduit/tunnel
~, **пешеходный** underpass, subway
тупик I *leath.* blunt knife
тупик II *zool.* puffin *(Fratercula)*
тупоконечный blunt-ended, obtuse
туполистный *bot.* amblyophyllous
тупоугольный obtuse-angled
тур I 1. turn; 2. *sport* round
тур II *zool.* 1. mountain goat *(Capra);* 2. aurochs, urus *(Bos primigenius)*
тура *chess* castle, rook
турако *zool.* touraco *(Turacus)*
туранга *bot.* Asiatic poplar *(Populus diversifolia)*
турач *zool.* francolin *(Francolinus francolinus)*
турачка *naut.* warping drum
турбаза *abbr.* tourist centre/hostel
турберия *bot.* Thurber cotton *(Gossypium thurberi)*
турбидиметр turbidimeter, opacimeter
турбидиметрия turbidimetry
турбина turbine
~, **активно-реактивная** impulse-reaction turbine
~, **двухпоточная** double-flow/-pass turbine
~, **двухроторная** two-spool turbine; birotary [counter-rotating] turbine
~ **заднего хода** *naut.* astern/reverse turbine
~, **ковшовая** Pelton turbine
~, **нереверсивная** unidirectional turbine
~, **однопоточная** single-flow/-pass turbine
~, **осевая** axial-flow turbine
~, **парогазовая** steam-gas turbine, hydrogen peroxide turbine
~ **переднего хода** *naut.* ahead turbine
~, **поворотно-лопастная** Kaplan turbine
~, **предвключенная** superposed/topping turbine; high-exhaust-pressure turbine
~, **радиальная** radial-flow turbine
~, **реактивная** reaction turbine
~, **реверсивная** reversible turbine
~ **с противодавлением** back-pressure turbine
~, **силовая** power/output turbine
~, **судовая** marine turbine
~, **ускорительная** booster turbine
~, **центробежная** cenrifugal turbine, outward-flow turbine
турбина-компаунд compound turbine
турбоагрегат turbine unit, turbo-generator set
турбобур turbo-drill; rotary drill
турбовентилятор turbo-blower, turbo-fan
турбовинтовой turbo-propeller *(attr.)*
турбогенератор turbine(-driven) generator, turbo-generator
~ **переменного тока** turboalternator
~ **постоянного тока** turbodynamo
турбодетандер turbo-expander, expansion turbine
турбокомпрессор turbo-compressor; turbo-supercharger; gas generator/producer
турбонагнетатель turbo-supercharger
турбонаддув turbo-supercharging
турбонасос turbo-pump
турбопреобразователь turboconverter
турбореактивный turbojet
турбоход turbo-vessel
турбоэлектровоз turbo-electric locomotive
турбоэлектроход turbo-electric ship
турбулентность turbulence
~, **грозовая** storm turbulence
турбулентный turbulent
турбулизатор turbulence stimulator, vortex generator, turbulator
турбулизация agitation
турбулиметр turbulence meter
турель turret
~, **верхняя** *av.* top turret
~, **нижняя носовая** chin turret
~, **хвостовая** tail/rear turret
турецкий Turkish
~ **барабан** kettledrum, bass drum
турецко-русский Turco-Russian, Turkish-Russian
туризм 1. tourism; hiking; 2. tourist trade
турион *bot.* turio(n)
турист tourist; hiker
туркестанец, туркестанский Turkestani
туркестанка Turkestani (woman)
туркмен, туркменский Turkmen
туркменка Turkmen (woman)
турмалин *miner.* tourmaline

~, бесцветный achroite
~, бурый/дравитовый dravite
~, зеленый Brazilian emerald
~, индиго-синий indicolite
~, фиолетово-красный siberite
турник *sport* horizontal bar
турникет 1. turnstile; 2. *med.* tourniquet
турнипс *bot.* turnip *(Brassica rapa)*
турнир tournament
турпан *zool.* scoter *(Melanitta)*
турпет *bot.* turpeth root *(Operculina turpethum)*
турпоход *abbr.* hike
турухтан *zool.* ruff *(Philomachus pugnax)*
турча *bot.* featherfoil *(Hottonia)*
тускарора *bot.* wild rice *(Zizania aquatica)*
тутовник *bot.* mulberry *(Morus)*
тутоводство mulberry growing
тутолистный *bot.* mulberry-leaved, morifolious
тутовый mulberry *(attr.)*
~ шелкопряд silkworm
~вая ягода mulberry
тутор I *med.* splint
тутор II *hist.* tutor
туф *geol.* tuff, tufa
~, изверженный eruptive tuff
~, известковый tufa
~, кремневый siliceous/quartz sinter
~, осколочный fragmentary/clastic tuff
~, пепловый ash tuff/rock
~, стекловатый vitric tuff
туфелька *zool.* paramecium *(Paramecium)*
туфли *pl.* shoes; slippers
~ с шипами *sport* sprint shoes
туфобетон tuffcrete
туфобрекчия *petr.* tuff breccia
туфовидный, туфовый tuffaceous, tuff *(attr.)*
тухолит thucholite
тучность obesity
тучный obese
туш *mus.* flourish
туша carcass
туше 1. *mus.* touch; 2. *sport* touché
тушение extinguishing, putting out; quenching, suppressing, smothering
тушить extinguish, put out; quench; suppress
тушканчик *zool.* jerboa *(Dipus)*
туя *bot.* arborvitae *(Thuja)*
тыква *bot.* gourd, pumpkin *(Cucurbita)*
тыл rear; back; *mil.* rear services/structure
~ и снабжение *mil.* logistics
~ кисти back of the hand
~ стопы upper surface of the foot
тыловой rear; logistic
тыльный rear, back, dorsal
тырса *bot.* common cattail *(Typha latifolia)*
тысяцкий *hist.* chiliarch; chief, leader; village elder
тысячелистник *bot.* yarrow *(Achillea)*
тысячелистниковый *bot.* achileafolious
тысячелистный millefoliate
тысяченожка *zool.* millipede, millepede *(Blanjulus guttalatus)*
тысячник *hist.* captain, leader
тычинка *bot.* stamen
~, бесплодная sterile stamen, staminodium
~, двубратственная diadelphous stamen
~, двусильная didynamous stamen
~, однобратственная monadelphous stamen
тычинконосный staminiferous
тэйлоризм Taylor system
тэк *zool.* Asiatic/Siberian ibex *(Capra sibirica)*
тэмми *text.* tammy
тэта theta
ТЭЦ thermoelectric power plant
тюк bale
тюковать bale
тюленевый seal, sealskin *(attr.);* phocid
тюлень *zool.* seal *(Phoca)*
тюльпан *bot.* tulip *(Tulipa)*
тюрк Turki
тюрколог Turkologist, specialist in Turkish philology
тюямунит *miner.* tyuyamunite, calciocarnotite
тяга 1. traction; draught; haulage; pulling, towing; *aero.* thrust; 2. connecting/linking/tie/pull rod
~, аккумуляторная battery traction
~, буксирная tow-rod/-bar
~, взлетная takeoff thrust
~ воздуха air draught/draft
~ воздушного винта propeller thrust
~, горизонтальная horizontal thrust
~ двигателя engine thrust
~, конная horse traction
~ несущего винта rotor thrust
~, паровая steam traction
~, продольная fore-and-aft-rod
~, пружинная spring rod
~, реактивная jet thrust/propulsion
~, реверсивная retrothrust, reverse(d) thrust
~, соединительная connecting rod
~, удельная specific thrust/impulse
~, ускорительная booster thrust
~, форсажная afterburning thrust
~, форсированная blast draft
тягач (truck) tractor, tow car; prime mover
тягло *hist.* 1. impost, tax; dues; 2. household taxed
тяговооруженность thrust-to-weight ratio
тяговый tractive, traction *(attr.);* hauling
тягомер draught gauge/meter/indicator
тяготение *phys.* gravity, gravitation
~, всемирное universal gravitation
тяготеть gravitate
тягучесть 1. tenacity, toughness; 2. ductility, malleability; 3. viscosity
тягучий 1. ductile, tensile; malleable; 2. tough; 3. viscous
тяж 1. drawing/tension bar, rod; 2. *anat.* band, cord, streak, strand, funis
~, мышечный muscle fibre/fiber, muscular band
~, сосудистый anginoid streak

~, сцепной drag bar
тяжба *law* litigation, lawsuit, suit
~бу, вести be at law *(with)*
~бу, возбуждать bring a suit *(against)*
тяжебный *law* litigatory
тяжелоатлет *sport* weightlifter
тяжелоатлетический *sport* weightlifting *(attr.)*
тяжеловоз 1. heavy truck/lorry; 2. heavy draft horse
тяжелораненый critically wounded/injured
тяжелоходы *paleont.* gravigrades *(Gravigrada)*
тяжесть 1. weight; gravity; 2. load, burden
тяжущийся *law* litigant
тянульный stretching, drawing
тянуть stretch, draw; pull
тяпка chopper, cleaver; how; mattock

У, у

уайомингит *miner.* wyomingite
уайт-спирит white/mineral spirit
уакари *zool.* uakari *(Cacajao)*
убавлять diminish, reduce; lessen; shorten
убежище shelter; *fig.* refuge; asylum; *mil.* dugout
~, газозащитное gas(-proof) shelter
~, герметизированное sealed shelter
~, противоатомное atomic shelter
~, противохимическое anti-gas shelter
убиквист ubiquist; *rel.* Ubiquitarian, Ubiquist
убирать 1. remove; retract; 2. *agr.* harvest
убирающийся retractable
ублюдок hybrid; halfbreed, mongrel
убой slaughter(ing); butchering
убойный slaughter *(attr.)*
~ вес dressed carcass weight; carcass/dead/slaughter weight
~ скот cattle for slaughter
уборка 1. removal; taking away; disposal; *tech.* retraction; 3. *tech.* recovery; stowing; 4. *agr.* gathering in; reaping, harvesting; 5. cleaning up
~, раздельная *agr.* two-phase/-stage harvesting, swath/windrow harvesting
~ шасси landing-gear retraction
уборщик 1. harvester; 2. caretaker, janitor; cleaner; sweeper
~, самоходный self-propelled harvester
убывание decrease; diminution; reduction
убыль loss; diminution; decline
убыток 1. loss; deficit; 2. *pl.* damages
убыточный unprofitable; unremunerative
уваливаться под ветер *naut.* drop off
увальность, увальчивость *naut.* leewardiness, lee-helm; slackness
увар cooking losses
УВД *abbr.* air traffic control, ATC
увеальный *anat.* uveal
уведомление advice, notice; notification; letter of advice
уведомлять advise, give notice to, notify, inform
увеит *med.* uveitis
~, передний anterior uveitis
~, содружественный sympathetic uveitis
увеличение 1. magnification; enlargement; *opt.* magnifying power; 2. increase, augmentation, enhancement, growth; rise; amplification, expansion, extension; *phys., math.* increment
~ жесткости hardening
~ изображения zooming
~ мошонки *med.* oscheoncus
~ ногтей onychauxis
увеличивать 1. magnify, enlarge; 2. increase; augment; raise
~ продажу push up sales
увеличитель enlarger; magnifier; augmenter
увеопаротит *med.* uveoparotitis
увеосклерит *med.* uveoscleritis
увиолевый uviol
увиостойкий uviofast, uvioresistant
увлажнение moistening, damp(en)ing; wetting; *(air)* humidification
увлажнитель humidifier; dampener, moistener
увлажнительный dampening; moistening; wetting
увлекать entrain; carry off/along
увлечение *phys.* entrainment; carrying away
увод drift; displacement
увулярия *bot.* bellwort *(Uvularia)*
УВЧ-терапия *med.* U.N.F. therapy
увязывать correlate, coordinate; link; match
угар burning/melting loss; *text.* waste; *met.* carbon-monoxide fumes
угасание extinction; decline; vanishing
угашение extinguishing; suppression; stifling
углевод *chem.* carbohydrate
углеводород *chem.* hydrocarbon
~, ароматический aromatic hydrocarbon
~, насыщенный/предельный saturated hydrocarbon
~, ненасыщенный/непредельный unsaturated hydrocarbon, olefine (hydrocarbon)
углеводородистый, углеводородный hydrocarbonic
углевоз coal carrier; *naut.* collier
углевыжигательный charring
~ная печь charcoal kiln
угледобывающий coal-mining
угледобыча coal output
углежжение charcoal burning
углекислота carbonic acid, carbon dioxide
~, твердая dry ice
углекислый carbonate *(of)*
угленосный coal-bearing, carboniferous
углепетрография coal petrography; anthracography
углеподобный anthracoid
углеразрез strip-mining of coal
углерод *chem.* carbon, C

~ **графита/графитовый** graphitic carbon
~, **двусернистый** carbon disulphide
~ **закала** hardening carbon
~ **карбида** carbide carbon
~, **меченый** carbon tracer
~, **несвязанный** free carbon
~, **радиоактивный** radiocarbon
~, **связанный/фиксированный** fixed carbon
~, **сернистый** carbon bisulphide/disulphide
~, **химически связанный** combined carbon
~, **четырехбромистый** tetrabromomethane
~, **четырехйодистый** tetraiodomethane
~, **четырехфтористый** tetrafloromethane
~, **четыреххлористый** carbontetrachloride
углеродистый carbonaceous; carboniferous; carbon-bearing; carbide *(of)*
углеродный carbonic
углесодержащий coal-bearing; carbonaceous
углесос coal pump
углефикация carbonification
углистый coaly; carbonaceous
угловатоветвистый *bot.* goniocladous
угловатолистный *bot.* goniophyllous
угловатоплодный *bot.* goniocarpous
угловатосемянный *bot.* goniospermous
угловатость angularity; angulation
угловатый angular; angled, angulate
угловой angular, angled; angulated; angle; corner
~ **градус** degree of arc
~ **удар** *sport* corner kick
~**вая минута** minute of arc
~**вое железо** angle steel
углокрыльница *zool.* anglewing *(Polygonia)*
угломер protractor; goniometer; graphometer
~, **зеркальный** mirror goniometer
~, **нониусный** vernier protractor
угломер-квадрант goniometric sight
угломерный goniometric; azimuthal; goniometer *(attr.)*
угломер-транспортир level protractor
углубитель *naut.* depressor
углубление 1. deepening; 2. excavation, depression; cavity; recess; hollow; pouch; naut. draught; draft; 4. *anat.* fossa
~, **грушевидное** *anat.* piriform recess/fossa
~, **зрительное** *anat.* optic recess
~, **карстовое** pothole, sink
~ **киля** *naut.* keel draught
~, **пузырно-маточное** vesicouterine/uterovesical pouch
~, **селезеночное** splenic recess
~, **улитковое** cochlear recess
угодье land; ground; area
~, **лесное** forest
~, **лугопастбищное** grassland
~, **сельскохозяйственное** farm/farming land
~, **сенокосное** hay land
угол angle; corner
~ **атаки** angle of attack/incidence
~, **балансировочный** trim(ming) angle
~ **бокового отклонения** azimuth deviation
~, **большой** wide angle
~ **бросания** angle of departure, epoch angle
~, **брэгговский** Bragg angle
~, **валентный** valence angle
~ **веерности** divergence angle
~, **векторный** vectorial angle
~ **взмаха лопасти** *aero.* flapping angle
~ **визирования** angle of sight/view, sighting angle
~ **виража** banked-turn angle
~, **внешний/внутренний** exterior/interior angle
~, **вогнутый** concave corner
~ **возвышения** elevation angle; angular altitude
~ **возмущений** Mach angle
~ **волнового конуса** Mach angle
~ **встречи** angle of impact/contact; striking angle
~ **входа** entrance/inlet/entry/reentry angle
~, **выпуклый** convex corner
~ **выхода** outlet/exit/takeoff angle
~ **глаза** *anat.* angle of the eye; canthus
~ **глиссады** glide slope
~ **головки зуба** addendum angle
~ **граничный** limiting angle
~ **грудины** *anat.* sternal angle
~, **двугранный** dihedral angle; dihedron
~ **дифферента** *naut.* trim angle
~, **докритический** subcritical angle
~ **дополнительный** 1. *(to 90)* complementary angle; 2. *(to 180)* supplementary angle; 3. *(to 360)* conjugate angle
~ **дрейфа** *naut.* drift/crab/sideslip angle
~, **закритический** supercritical angle
~ **закручивания** twist/torsion angle
~ **запаздывания** lag/retardation angle
~ **засечки** *geod.* intersection angle, angle of crossing
~, **затылочный** *anat.* occipital angle
~ **захвата** capture angle
~ **зрения** visual angle; vision/viewing angle; angle of view
~, **зрительный** optic angle
~ **испускания** emission angle
~ **касания** contact angle
~ **килеватости** angle of floor, rise-of-floor angle, deadrise angle
~, **координатный** quadrantal angle; quadrant
~ **крена** 1. *naut.* heeling angle; rolling angle; 2. *av.* angle of bank/roll; 3. *auto.* roll/lean angle
~ **крутки** twist/warp angle
~, **крутой** acute angle; steep angle
~ **курса** heading/course bearing
~, **курсовой** relative bearing/direction/azimuth
~ **магнитного отклонения** angle of dip
~ **Маха** Mach angle
~ **места** (angle of) elevation
~ **многогранника** solid angle
~, **многогранный** polyhedral angle
~ **набора высоты** angle of climb, climb angle
~ **наклона** slope angle; dip angle, angle of incli-

nation; slope
~ **наклонения** angle of inclination
~, **накрест лежащий** alternate angle
~, **небольшой** shallow angle
~ **опережения** angle of lead/advance; advance angle
~, **острый** acute angle
~ **отдачи** recoil angle
~ **отклонения** deflection angle
~ **откоса** angle of slope/repose
~ **отражения** reflection angle; angle of reflection
~ **отставания** angle of lag; lag/delay angle
~ **падения** angle of incidence/impingement/arrival; incidence/incident/impingement angle
~, **плоский** plane angle
~ **поворота** turning angle
~, **полетный** flying angle
~, **полный** perigon; round angle
~, **пополнительный** supplementary angle
~, **посадочный** landing angle
~, **предельный** critical/limit angle
~ **преломления** refraction angle
~ **при вершине треугольника** vertex angle
~ **при основе треугольника** base angle of a triangle
~, **прилежащий** adjacent angle
~, **пространственный** solid angle
~, **противолежащий** opposite angle
~, **прямой** right angle
~ **разворота** angle of turn
~ **рассеяния** angle of deflection
~ **расхождения** divergence angle
~ **рыскания** angle of yaw
~ **сбрасывания** dropping angle
~ **сваливания** *aero.* angle of stall
~ **скольжения** *(of propeller)* slip angle
~ **слежения** tracking angle
~, **смежный** adjacent angle
~ **сноса** angle of sideslip/drift; crab/sideslip/drift angle
~ **столкновения** angle of impingement, impact angle
~, **срезанный** truncated angle
~ **сужения** convergent angle
~ **сходимости/схождения** convergence angle
~, **телесный** solid angle
~, **трехгранный** trihedral angle
~, **тупой** obtuse angle
~ **упреждения** lead angle
~ **установки лопасти** blade angle
~, **фазовый** phase angle; angle of lag
~, **часовой** *astr.* hour angle
~, **черепно-лицевой** *anat.* craniofacial angle
~, **шаговый** *(of propeller)* pitch angle
~, **экваториальный** equator angle
уголовный criminal
~ **преступник** criminal
уголок 1. (metal) angle; angle piece; corner
~, **красный** *obs.* decorated "red corner"
~, **крепящий** fastening angle
~, **загнутый** *met.* turned-in angle
~, **неравнобокий** unequal angle
уголь coal; carbon
~, **активный/активированный** active carbon, activated charcoal
~, **беспламенный** blind coal
~, **бурый** brown coal; lignite
~, **деревянистый** lignite
~, **длиннопламенный** long-flame coal; candle coal
~, **древесный** charcoal; xylanthrax
~, **дуговой** arc carbon
~, **жирный** rich/fat coal
~, **каменный** (hard) coal
~, **коксующийся** coking coal
~, **костяной** bone coal/char
~, **кузнечный** forge coal
~, **обнаженный** cropping coal
~, **полосчатый** banded coal
~, **ретортный** retort carbon
~, **рядовой** run-of-mine coal
~ **с высоким содержанием летучих** high-volatile coal
~, **смолистый** resinous/pitch coal
~, **спекающийся** baking/caking/sintering coal
~, **сырой** run-of-mine coal
~, **торфяной** peat coal
~, **тощий** lean coal
~, **флотский** steamboat coal
~, **электродный** electrode carbon
угольник 1. triangle; 2. angle piece; 3. elbow
угольный coal; carbon *(attr.);* coaly; *chem.* carbonic
~ **бассейн** coal field/basin
уголь-орешек nut/chestnut coal
уголь-сланец coaly shale
угольсодержащий coal-bearing; carbonaceous
угорь I *zool.* eel *(Anguilla)*
угорь II *med.* acne, comedone, blackhead
угребычковые *n. pl. zool.* eelgobies *(Taenioididae)*
угреватый pimpled; pimply
угревидный *med.* acneiform
угревые *zool.* freshwater eels *(Anguillidae)*
угреобразные *n. pl. zool. Anguilliformes*
угрица *zool.* follicle mite *(Denodex folliculorum)*
угроза threat; menace
~ **здоровью** health hazard
угрозыск *abbr.* Criminal Investigation Department
угро-финский Ugro-Finnic, Finno-Ugrian
удав *zool.* boa (constrictor) *(Constrictor constrictor)*
удавовые *n. pl. zool.* boa family *(Boidae)*
удавчик *zool.* sand snake *(Eryx)*
удаление 1. removal; disposal; abscission; withdrawal; ablation; deletion; 2. distance; range
~ **бытовых отбросов** sewage disposal
~ **заусениц** *met.* deburring
~ **льда** ice removal, de-icing
~ **отбросов/отходов** waste disposal

~ препятствий clearing/removal of obstacles
~ символа character deletion
~ смазки degreasing
~ страницы из памяти *comp.* page-out
удалять remove; eliminate; withdraw; dispose of
удар blow; stroke; impact; shock; knock; attack; *sport* kick; shot; *med.* beat; stroke
~, апоплектический apoplectic stroke
~, гидравлический/гидродинамический hydraulic shock
~, звуковой sonic bang/boom
~, лобовой head-on impact
~ молнии lightning stroke; thunderbolt
~, обратный *mech.* kickback
~, ответный retaliation; retaliatory attack
~, посадочный *av.* landing impact
~, прямой knock-on impact
~, пульсовый *med.* pulse beat
~, свободный *sport* free kick
~, солнечный sunstroke; insulation; siriasis
~, таранный attack collision
~, тепловой heat stroke/apoplexy; heat/thermal shock
~, угловой *sport* corner kick
~, штрафной *sport* penalty kick
~, штыковой *mil.* bayonet charge
ударение accent; stress
ударник I *obs.* shock worker
ударник II firing pin; striker
ударный 1. striking; percussive; percussion; 2. *mus.* percussion *(attr.)*; 3. shock-working, shock; 4. *mil.* shock; striking; 5. very urgent, pressing
~ механизм percussive/firing mechanism
~ная бригада *obs.* shock brigade
~ные части shock troops
удароглушитель shock absorber
ударопрочный, ударостойкий shock-resistant; impact-proof, high-impact *(attr.)*
удваивание, удвоение doubling; duplication
удвоенноперестый *bot.* bipinnate
удвоенносемянный *bot.* dissospermous
удвоенноцветковый *bot.* twin-flowered
удвоитель doubler; duplicator
удельный I *phys.* specific
~ вес specific gravity
удельный II *hist.* apanage
~ период period of apanage/independent principalities
удержание, удерживание containment, retention, confinement; keeping; *fin.* deduction
~ вещи *law* retention
~, незаконное detainer
~ магнитным полем magnetic confinement
~ продуктов деления fission-product retention
удерживатель *med.* retinaculum
удерживать contain; confine; retain; hold; restrain, keep back; deter
удильщик angler *(also zool)*
удильщиковые *n. pl. zool.* angler-fishes *(Lopiidae)*
удлинение elongation; lengthening, prolongation, extension, stretching
~, критическое ultimate elongation
~, остаточное remanent elongation
~, относительное unit elongation
~, тепловое thermal extension
~, удельное elongation per unit length
удлиненный oblong; elongated; lengthened
удлинитель lengthener; extension piece/cord; stretcher
~ стержня extension rod
удобоваримый digestible; digestive
удобоисполнимый feasible; easy to carry out
удобообрабатываемый workable; easily worked/processed/machined
удобопонятный comprehensible; intelligible
удобоуправляемый manoeuvrable; easily controlled
удобочитаемый legible; readable
удобрение 1. manure; fertilizer; 2. application/use of fertiliser(s), manuring; fertilization
~, азотное nitrogen fertilizer
~, зеленое green manure
~, комплексное compound/combined fertilizer
~, промышленное commercial fertilizer
~, фосфорное phosphate fertilizer
удобство convenience, comfort
~ обслуживания serviceability
~ осмотра inspectability
удой yield of milk; milking
удойливый yielding much milk
~вая корова good milker
удойность yield of milk
удостоверение certification; attestation; certificate; licence
~ личности identity/identification card
~ о пригодности (к эксплуатации) certificate of fitness
удостоверять certify; attest
удушение *med.* strangulation; suffocation
удушье asphyxia
ужовка *zool.* money cowrie *(Cyprea moneta)*
ужовник *bot.* adder's-tongue/fern *(Ophioglossum)*
узаконивать legalize, legitimize
узанс *law* usage
узарин *chem.* uzarin
узбек, узбекский Uzbek
узбечка Uzbek (woman)
уздечка *anat.* frenulum
узел 1. knot; node; *anat.* ganglion, knob; 2. *tech.* unit, assembly; group; block; component, subassembly, subunit, node, nodule; joint; 3. junction; hub, centre; 4. *geol.; bot.* node; 5. *naut.* knot; *see also* **узлы**
~, барабанный tympanic ganglion
~, варикозный *anat.* varix
~, выбленочный clove hitch
~, геморроидальный hemorrhoid
~, жесткий rigid joint
~, железодорожный railway junction

~, **интеркаротидный** *anat.* carotic gland
~, **карданный** gimbal joint
~, **коммутационный** switching centre/point; exchange
~, **конструктивный** structural assembly
~, **лимфатический** lymph node
~, **морской** knot
~ **нагрузки** load node/centre
~, **нервный** nerve ganglion
~, **неразвязывающийся** snarl knot
~, **позвоночный** vertebral ganglion
~ **привода** drive unit
~, **ретрансляционный** relay centre
~ **решетки** *phys.* lattice point
~, **санитарный** lavatory; toilet; bathroom
~, **сборочный** assembly/subassembly unit
~, **сварной** welded assembly/unit
~ **связи** communications centre
~ **сетки** mesh point
~, **шарнирный** hinged joint
узел-восьмерка figure-eight knot
узелки *pl. med.* nodules
узелковый, узловидный nodular ; nodulate(d) , nodous
узелок *bot., anat.* small knot; nodule; tubercle, tuber
~, **корневой** radicle apex; radiculodium
узик *bot.* tormentilla; cinquefoil *(Potentilla erecta/ tormentilla)*
узковетвистый *bot.* stenocladous
узкоголовчатый *bot.* stenocephalous
узкоголовый *anat.* stenocephalous; stenocephalic
узкогрудость stenothorax
узкогрудый narrow-chested
узкое место *fig.* bottleneck, gooseneck
узкоколейка narrow-gauge railway
узкокрылка *zool.* wharf borer *(Nacerdes melanura)*
узколистный, узколистый angustifolious; angustifoliate; stenophyllous
узколопастный narrow-lobed; stenolobate
узколучевой narrow-beam
узконадкрылые *n. pl. zool.* false blister beetles *(Oedemeridae)*
узкополосица *agr.* strip cultivation
узкополосный narrow-band
узкорядный *agr.* narrow-row(ed); close-sown
узкоспециализированный highly specialized
узкость narrow waters/passage; narrows; strait(s), bottleneck; throat
узкотелка *zool.* cylindrical bark beetle *(Colydiidae)*
узловатость nodosity; knottiness
узловатый knotty; knotted; nodosal; nodulous; nodose; nodulous; *anat.* gangliform
узловязальный *tech.* knotting
узловязатель knotter
узловой 1. *tech.* junction *(attr.);* 2. nodal, nodose; nodular
узлообразование nodulation; knotting
узлы *anat.* nodes; ganglions, ganglia; varices; *see also* **узел**
узор pattern; design
~, **сетчатый** reticulation
~, **фасонный** fancy pattern
узукапия *law* usucapion
узурпатор usurper
узус *law* usage, custom
узуфрукт *law* usufruct
узуфруктуарий *law* usufructuary
уинтатерии *pl. paleont.* uintatheres *(Uintatheria)*
уйгур Uig(h)ur
уйгурский Uig(h)ur, Uig(h)uric, *Uig(h)urian*
указ decree; edict
указание instruction; direction; indication
указатель 1. indicator; pointer; designator; cursor; 2. index; 3. directory, guide
~ **авиагоризонта** attitude indicator
~ **времени** timing indicator
~ **высоты** altitude indicator; altimeter
~ **давления** pressure indicator/gauge
~ **дальности** range indicator
~, **дистанционный/вынесенный** remote indicator
~, **дорожный** road sign
~, **железнодорожный** railway guide
~ **курса** course/heading/direction indicator
~ **маршрута** routing indicator
~ **места/местонахождения** position indicator
~ **пеленга** bearing indicator
~ **расхода** flow indicator
~, **сельсинный** selsyn indicator
~, **систематический** classified index
~ **сноса** drift indicator
~ **торможения** decelerometer
~ **уровня** level indicator/gauge
~ **ускорения** acceleration indicator, accelerometer
~ **фирм** trade directory
~ **числа оборотов** tachometer, r.p.m. indicator
указательный indicating, indicatory; *gram.* demonstrative
укатывать roll; compact by rolling
укладка 1. packing; stacking; 2. placing; laying, placement
укладчик 1. spreader; placer; 2. stacker; piler; 3. packer; 4. compacting machine
уклон gradient; inclination; grade; slope; pitch
~ **вверх** upward gradient
~, **под** downgrade; downhill
~, **пологий** gentle gradient
уклонение deviation, digression, deflection; departure; declination; evasion, avoidance; evasive action/manoeuvre
~ **от уплаты налогов** tax evasion
уклонизм *polit.* deviationism
уклонист *polit.* deviationist
уклономер inclinometer
уклоноуказатель *rail.* gradient board/indicator
уключина *naut.* oarlock; rowlock

~, **поворотная** swivel rowlock
уковка *met.* forging reduction
укол puncture; prick, jab; piercing; injection
~, **ответный** *sport* riposte
укомплектовать 1. equip; furnish *(with);* 2. bring up to full strength, staff; man
укорачивать shorten; abbreviate; lessen; diminish; curtail
укоренение taking/striking root; radication
укорочение shortening, reduction; curtailment
~ **меандра** short-cut
~, **относительное** unit shortening
укосина jib; cantilever; boom; strut, brace
~, **перекидная** reversible jib
Украина Ukraine
украинец Ukrainian
украинка Ukranian (woman)
укрепление 1. reinforcement, strengthening; fastening, fixing, propping up; consolidation; 2. *mil.* fortification
~ **берегов** bank protection /reinforcement
~, **береговое** coastal fortification
~ **грунтов** soil stabilization
~ **ферм** farm consolidation
укреплять 1. strengthen, reinforce, fasten, prop/shore up; fix ; make fast; consolidate; *mil.* fortify
укроп *bot.* dill *(Anethum)*
~, **аптечный** common fennel *(Foeniculum vulgare)*
~, **душистый/огородный/пахучий** common dill *(Anethum graveolens)*
укрупнение 1. enlargement; extension; growth; 2. amalgamation; merger; 3. *chem.* crosslinking; coalescence
укрупнять enlarge; combine into larger units; amalgamate; consolidate
укрыватель краденого receiver of stolen goods
укрывать cover; shelter; conceal; habour
~ **преступника** receive offender
укрытие shelter, cover; concealment
~ **для личного состава** personnel shelter
уксус vinegar
~, **плодный** cider vinegar
~, **солодовый/пивной** malt vigenar
~, **свинцовый** vinegar of lead
уксусник *bot.* staghorn sumac *(Rhus typhina)*
уксуснокислый acetous; acetate (of)
уксусный acetic; vinegary
укупориватель capping machine, capper
укупорка 1. packing; crating; 2. corking
улавливание catching; trapping; capture; entrapment; collecting; recovery; reclamation
улавливать catch; trap; capture; entrap; recover, reclaim
улан *mil.* uhlan
улар *zool.* snow cock *(Tetroagallus)*
улекс *bot.* gorse; furze *(Ulex)*
улетучивание volatilization; evaporation
улетучиваться volatilize
улика (inferential) evidence
~, **косвенная** circumstantial evidence
Улисс *myth* Ulysses
улитка 1. *zool.* snail *(Helix);* 2. *anat.* cochlea
~, **перепончатая** membranous cochlea
улитковидный, улиткообразный conchiform; cochleiform; trochiform; helical; trochispiral
улитково-преддверный *anat.* cochleovestibular
улитковый 1. helical; 2. *anat.* cochlear
улит-отшельник *zool.* solitary sandpiper (*Tringa solitaria)*
уличение conviction; exposure; accusation
улов catch; haul; take
уловитель catcher, trap, collector; precipitator
улус *hist.* ulus; domain; settlement
улучшение improvement; amelioration; refinement
ультимо *econ.* end month account
ультраакустика ultrasonics; supersonics
ультравысокочастотный ultrahigh-frequency
ультразвук ultrasound; subsonic [ultra-audible] sound, supersound
ультразвуковой ultrasonic; supersonic
ультразвукоскопия ultrasonoscopy
ультракоротковолновой ultrashort-wave
ультракороткий ultrashort
ультрамикровесы ultramicrobalance
ультрамикроопределение ultramicrodetermination
ультрамикроскоп ultramicroscope
ультрамикроскопия ultramicroscopy
ультрамикротом ultramicrotome
ультрамикрохимия ultramicrochemistry
ультраосновной ultrabasic
ультраустойчивый ultrastable
ультрафиолетовый ultraviolet
ультрацентрифуга ultracentrifuge; high-speed centrifuge
~, **угловая** angle-head ultracentrifuge
ультрацентрифугирование ultracentrifugation
ультращелочной peralkaline
умалишенный insane; mad
умбиликус *bot.* navelwort *(Umbilicus)*
умбра 1. umber; terra umbra; 2. *bot.* mud minnow *(Umbra limi)*
~, **обоженная** burnt umber
~, **природная** raw umber
умение skill; know-how; ability; capacity
уменьшаемое *math.* minuend
уменьшать(ся) diminish; decrease, reduce
уменьшение decrease; reduction, diminution, lessening; lowering; drop; abatement
~ **амплитуды** amplitude contraction
~, **дробное** proportional reduction
~ **потоотделения** anaphoresis
~ **раздражения** abirritation
~ **размеров** miniaturization
уменьшительно-ласкательный *gram.* diminutive-hypocoristic
уменьшительно-пренебрежительный *gram.* diminutive-pejorative
уменьшительный diminishing; *gram.* diminutive

~ное имя familiar first name
умеренный moderate; temperate; medium
умерщвление mortification, killing, destruction
умерщвлять kill; destroy
умирающий moribund; dying
умлаут, умляут *ling.* umlaut; mutation
умножать 1. *math.* multiply; 2. increase; augment
умножение 1. multiplication; 2. increase; augmentation; rise
~, аппаратное *comp.* hardware multiplication
~, время-импульсное *comp.* time-division multiplication
~ на обратную величину reciprocal multiplication
~, параметрическое reactance multiplication
~ с накапливанием accumulative multiplication
~, сокращенное abridged multiplication; short (-cut) multiplication
~ частоты frequency multiplication
~ ядерного топлива breeding of nuclear fuel
умножитель multiplier
~, время-импульсный time-division multiplier
~, итерационный iterative multiplier
~, матричный *comp.* array multiplier
~, моделирующий analog(ue) multiplier
~, сдвоенный paired amplifier
~, фотоэлектронный photomultiplier tube
~ частоты frequency multiplier
~, цифровой digital multiplier
~, электронный electron multiplier (tube)
умножитель-аккумулятор multiply-accumulator
умножитель-сумматор multiply-adder
умозрение *philos.* contemplation; speculation
умозрительно speculatively; theoretically
умозрительный speculative; contemplative
умолчание *comp.* default
умопомрачение (temporary) insanity
уморение *leath.* killing
умственный mental; intellectual
~ труд brainwork
умформер *el.* rotary converter; dynamotor
умысел design; intention; *law* dolus
~, злой malice (prepensed); premeditated malice; ill-will; wicked design
умышленный deliberate; intentional, wilful; voluntary; designed
~ное убийство premeditated murder
умягчать soften
умягчитель softener
униат, униатка member of the Uniate Church
униатский Uniate
универмаг *abbr.* department store
универсально-применимый all-purpose *(attr.)*
универсальность universality; universalism; versatility; multipurposeness; generality
универсальный universal; all-purpose; general-purpose/-service/-utility, multipurpose; versatile
~ магазин department store; general shop
универсам supermarket
универсиада (World) Student Games
универсум universe
уникальный unique
уникогерентный unicoherent
уникум unique object/person
унионида *zool.* freshwater clam/mussel *(Unio)*
унитаз water-closet pan
унитарный unitary; unitarian
унитарность unitarity
униформизация *math.* uniform convergence; uniformization
уничтожение annihilation; destruction; extermination; *math.* cancellation
~ иррациональности *math.* removal of an irrationality; rationalization
уния *hist. eccl.* union
унос 1. priming; 2. carrying/taking away; entrainment
~, капельный drop priming
~ массы mass loss; ablation
унтер-офицер *mil.* noncommissioned officer
унтертон subharmonic tone
унция ounce
у-образный U-shaped
Уолл-стрит Wall Street
упадочный decadent
упаковка 1. pack; package; packing material; 2. packaging, packing
упаковочный packing
упаковывание packaging
упаковывать pack
упаривание concentration by boiling/evaporation
упаривать boil down; evaporate
~ досуха evaporate to dryness
упас *bot.* upas tree *(Antiaris toxicaria)*
упек baking loss
упланд *bot.* upland cotton *(Gossypium hirsutum)*
уплата payment; settlement; refundment
~ долга acquittance
~ наличными деньгами cash payment
~ налогов при получении заработной платы pay-as-you-earn
~ процентов interest payment
уплотнение 1. gasket; seal; packing; gland; 2. packing; gasketing; sealing, compaction; 3. *med.* induration, consolidation; 4. *el.* multiplexing
~, бессальниковое packless seal
~, воздухонепроницаемое airtight seal
~, газонепроницаемое gas-tight seal
~, герметическое hermetic/pressure/tight/vacuum seal
~, гидравлическое hydraulic/water seal, liquid packing
~, гравитационное gravitational compaction
~ грунта soil stabilization
~, двустороннее double seal
~, застойное *med.* cyanotic induration
~, застывшее frozen seal
~ каналов multiplexing
~, качающееся wobble seal

~, **кольцевое** O-ring seal
~, **конусное** wedge-type seal
~ **легкого** *med.* pulmomary compaction
~ **линии связи** multiplexing; multiplex operation/working; multi-channel operation
~, **мембранное** diaphragm seal
~, **ножевое** featheredge seal
~ **рабочего дня** filling up of the working day; efficient work during the day
~, **сальниковое** gland sealing; packing gland
~, **сильфонное** bellows seal
~, **угольное** carbon packing
~, **щелевое** slit gasket
уплотненный 1. packed; sealed; compact(ed); compressed; condensed; 2. *med.* indurate; 3. multiplexed
~ **график** tight schedule
уплотняемость compactibility
уплотнять compact; compress, pack; seal; *el.* apply multiplex equipment *(to a line)*
~ **рабочий день** fill up the working day; work efficiently during the day
уплощение flattening
уполномоченный 1. *adj.* authorized; empowered; plenipotentiary; commissioner; authorized agent, assignee; proxy
~ **профсоюзной организации** steward
уполномочивать authorize; empower; entrust, *Am.* intrust
уполномочие authorization; power; authority
упор stop; rest; support; stay; shore; thrust
~ **для ног** foot rest
~, **переставной** adjustable support/stop
~, **стрелять в** fire pointblank
упорка 1. support; stay; 2. tool rest; 2. *naut.* footboard; foot spar
упорядочение ordering; adjustment; regulation; ranking; putting in proper order
упорядоченный *stat.* ordered, ordrerly
~ **по времени** time-ordered
управдел *abbr.* office manager
управдом *abbr.* manager of a block of flats, house manager
управление 1. administration, management, control; guidance; government; 2. steering; operation, control; driving; 3. board, office, agency, department, administration, authority; 4. *gram.* government
~, **автоматизированное** 1. automated/automatic control; 2. computer control
~, **автономное** 1. off-line control; 2. internal control, self-contained control
~, **административное** managerial control
~ **базой данных** database management
~ **без обратной связи** open-loop control
~, **безынерционное** infinitely fast control
~, **бустерное** power-assisted control
~ **в автономном режиме** off-line control
~ **в интерактивном режиме** interactive control
~ **в оперативном режиме** on-line control
~ **в реальном времени** real-time control
~ **вводом-выводом** input-output control
~, **водохозяйственное** basin authority
~ **воздушным движением** air traffic control
~ **выборкой** access control
~, **главное** central board/administration
~ **гражданской авиации** *(U.S.A.)* Civil Aeronautics Administration
~, **голосовое** voice-activated control
~ **данными** data management
~ **данными, монопольное** exclusive data control
~ **движением** traffic control
~, **диспетчерское** supervision; supervisory control; *rail.* dispatcher control
~, **дистанционное** remote control
~ **дифферентом** trim control
~ **заданиями** job management
~ **запасами** stock/inventory control
~, **импульсное** 1. sampled-data control; 2. pulse control
~ **качеством** quality control
~, **кнопочное** pushbutton /pressbutton control ; finger-tip control
~, **клавишное** keyboard control
~, **комплексное** integrated control
~ **международной торговли** *(U.S.A.)* Office of International Trade
~, **микропрограммное** microprogramming control
~, **монопольное** *comp.* exclusive control
~ **на расстоянии** remote control
~, **оперативное** 1. on-line control; real-time control; 2. day-to-day management
~ **параллелизмом** concurrency control
~, **плановое** planned control
~, **плохое** maladministration; poor control
~ **по азимуту** azimuth guidance/control
~ **по аэронавтике и исследованию космического пространства, Национальное** *(U.S.A.)* National Aeronautics and Space Administration
~ **по дальности** range control
~ **по курсу** directional/yaw control
~, **позиционное** point-to-point control
~, **полудистанционное** semiremote control
~ **посредством (перфо)карт** card control
~, **программное** programmed control
~, **прямое цифровое** direct digital control
~ **пуском** start-up control
~, **рулевое** steering (control)
~ **рулем** steering; steerage; rudder/helm control
~, **ручное** manual control/operation
~ **с обратной связью** feedback control
~ **с прогнозированием** feedforward control
~, **сельсинное** selsyn control
~, **сенсорное** feather-touch control
~, **совместное/смешанное** joint management
~, **совмещенное** integrated control

~ стержнями rod-type control
~ стрельбой *mil.* fire control
~ счетчиками counter control
~, таможенное board of customs
~ технологическим процессом process control
~, тиристорное thyristor control
~, центральное статистическое central statistical office; central board of statistics
~, цифровое программное digital control
~, числовое numerical control
управляемость 1. manoeuvrability, steerability; directional control; 2. controllability, manageability
~, боковая/поперечная lateral controllability
управляемый controllable; operable; directed; controlled; guided, operated; driven
~ вручную hand-operated/-controlled, manually controlled; manual-control *(attr.)*
~ данными data-controlled/-directed
~, дистанционно remote-controlled; telecontrolled
~ лентой tape-controlled/-operated
~ по радио radio-controlled
управлять 1. administer, manage; rule; govern; 2. control, guide; operate, drive; steer
управляющий 1. managing; governing; ruling; 2. controlling; guiding; driving; steering
~ делами executive officer; office manager
~ имением land steward; estate agent
~ фермой/хозяйством (farm) bailiff
управхоз *abbr.* service and supply manager
упразднение abolition, cancellation; annulment
упреждение lead, advance; prediction
упрочнение strengthening; reinforcement; hardening; stabilization
~, деформационное/механическое strain hardening
~ закалкой quench hardening
~, термохимическое surface impregnation
~ цианированием cyanide hardening
~ цементацией case-hardening
упрощение simplification
упрощенство oversimplification; simplism
упругий elastic; resilient
~, идеально completely elastic
упругопластический elastic-plastic, elastoplastic
упругость elasticity, resilience; resiliency; *(of steam)* pressure; *mech.* flexibility
~ второго рода elasticity of shear(ing)
~, вязкая/замедленная viscous elasticity
~ на изгиб transverse elasticity; elasticity of flexure
~ на растяжение elasticity of elongation; longitudinal elasticity
~ на сдвиг elasticity of shear
~ на сжатие elasticity of compression
~, объемная bulk/volume/cubical elasticity; compressibility; bulk modulus
~ пара steam/vapour pressure
~ первого рода longitudinal elasticity, elasticity of elongation
~ при изгибе elasticity of flexure
~ при кручении elasticity of torsion; torsional elasticity
~ при растяжении elasticity of elongation/extension
~ при сдвиге elasticity of shearing
~ при сжатии elasticity of compression
упругоустойчивый elastically stable
упряжь harness; gear; *rail.* draft/coupling gear
упущение omission; oversight; dereliction
~, виновное/преступное *law* culpable omission
уравнение 1. *math.* equation; 2. equalization
~ Абеля Abel's equation
~ адиабаты adiabatic equation
~ авторегрессии *stat.* autoregression/autoregressive equation
~, базовое fundamental/base/basic equation
~ Бернулли Bernoulli('s) equation
~, биквадратное biquadratic equation
~ в неявной форме implicit equation
~ в явной форме explicit equation
~, вековое secular equation
~ вихревой системы vorticity equation
~ возмущения perturbation equation
~, волновое wave equation
~ времени equation of time
~ Гамильтона Hamilton's equation, canonical equation of motion
~ Гиббса-Гельмгольца Gibbs-Helmholtz equation, equation of maximum work
~, двучленное binomial equation; two-term equation
~, диофантово diophantine equation
~, дифференциальное с частными производными partial differential equation
~ для оценок *stat.* estimation equation
~ издержек costs equation
~, квадратное quadratic equation
~ Клапейрона Clapeyron equation
~ колебаний oscillation/pendulum equation
~ количества движения momentum equation
~, конформно-инвариантное conform-invariant equation
~ Коши-Римана Cauchy-Riemann equation
~ критерия criterion equation
~, кубическое cubic equation; third-degree equation; equation of the third degree
~ Лагранжа Lagrange('s) equation, d'Alembert('s) equation
~ Лапласа Laplace('s) equation
~ Лежандра Legendre('s) equation
~, линейное linear equation
~, масштабное *comp.* transformation equation
~ максимального правдоподобия maximum likelihood equation
~, машинное computer/machine equation
~ Навье-Стокса Navier-Stokes equation
~, нелинейное nonlinear equation

~, неоднородное nonhomogeneous/inhomogeneous/heterogeneous ecuation
~ неопределенное indeterminate equation
~ непрерывности continuity equation
~, неприводимое irreducible equation
~ неразрывности continuity equation
~ обмена equation of exchange
~, однородное homogeneous equation
~, основное fundamental equation
~ первого порядка first-order equation
~ первого рода equation of the first kind
~ первой степени linear/simple equation
~ плоскости equation of a plane
~ правдоподобия likelihood equation, equation of maximum likelihood
~, приведенное квадратное quadratic equation in normal/reduced form
~ прямой equation of a straight line
~ Пуассона Poisson's equation; adiabatic equation
~ пятой степени quintic equation
~ равновесия equilibrium equation
~ размерностей dimensional equation
~, разностное difference equation
~ регрессии regression equation
~, резольвентное *stat.* resolvent equation
~ Рейнольдса Reynolds equation
~ с частными производными partial differential equation
~, самосопряженное self-adjoint equation
~, скоростное rate/kinetic equation
~, сопряженное adjoint equation
~ состояния equation of state
~ сохранения conservation equation
~, стационарное steady-state equation
~, степенное exponential equation
~ теплопроводности heat equation
~ типа свертки convolution type equation
~ траектории trajectory/path equation
~, транспонированное transposed equation
~ третьей степени cubic equation
~ четвертой степени quartic equation
~ Шредингера Schroedinger equation
~ Эйлера Euler's equation
уравнивание equalization; compensation; adjustment
~, повторное readjustment
~ погрешностей compensation of errors
уравнивать adjust; equalize
уравниватель equalizer; balancer; regulator
уравниловка egalitarianism; wage-levelling
уравнитель 1. equalizer; balancer; compensator; adjuster; 2. *naut.* guide ship
уравнительный 1. equalizing; compensating; balancing; 2. egalitarian
уравновешенность balance, equilibrium
уровновешивание balancing; equilibration
уравновешивать balance; counterbalance
ураган hurricane; tornado
ураза *rel.* Ramadan, Ramazan
урало-алтайский *ling.* Turanian, Uralo-Altaic
уральский Uralian, Ural, Urals *(attr.)*
уран *chem.* uranium, U; *astr., myth.* Uranus
~, азотнокислый uranium nitrate
~, металлический uranium metal; metallic uranium
~, пятихлористый uranium pentachloride
~, сернокислый uranium sulphate
~, треххлористый uranium trichloride
~, фтористый uranium fluoride
~, четыреххлористый uranium tetrachloride
~, шестифтористый uranium hexafluoride
~, щавелевокислый uranium oxalate
уранил *chem.* uranyl
~, азотнокислый uranyl nitrate
~, уксуснокислый uranyl acetate
~, хлористый uranyl chloride
уранинит *miner.* uraninite; pitch ore/blende; nivenite; nasturan
уранит *miner.* uranite; uran-mica
~, известковый/кальциевый calcium uranite; autunite
~, медный copper uranite; torbernite
Урания *myth.* Urania
урановый uranic
уранопластика uraniscoplasty, uranoplasty, palatoplasty
ураноплегия uranoplegia; palatoplegia
урансодержащий uranium-containing/-bearing; uraniferous
урартский Urartrean, Urartic
урарты *n. pl.* Urartreans
урат *chem.* urate
уратоксидаза urate oxidase
уратолиз uratolysis
урацил uracil
урацилдегидрогеназа *biochem.* uracil dehydrogenase
урбанизм urbanism; urban theory
урбанистический, урбанический urbanistic
ургидроз urhidrosis; uridrosis
урд *bot.* mungo bean; black gram; urd (*Phaseolus mungo)*
урду Urdu
уреаза *biochem.* urease
уредий, урединий *bot.* uredinium; uredosorus
урединология uredinology; study of rusts
уредиоспора, урединиоспора uredospore; urediniospore
уредоконидия *bot.* uredoconidium
уредокучка, уредоложе *bot.* uredinium
уредостадия *bot.* uredostage; summer stage
урез cutting off; cut
~ воды water edge/line/boundary
~ малой воды low-water line
~ полной воды high-water line
уреид *chem.* ureide
уреидоизомасляный ureidoisobutyric
уреидосукциназа ureidosuccinase
уреидоянтарный ureidosuccinic

уремия *med.* ur(a)emia
уреометр ureometer, ureameter
уреометрия ureametry
уретан urethan(e)
уретерит *med.* ureteritis
~, виллозный villous ureteritis
~, кистозный cystic ureteritis
уретеролиз ureterolysis, ureterodialysis
уретеролит ureterolith
уретеролитиаз ureterolithiasis
уретеропатия ureteropathy
уретеропиелит ureteropyelitis
уретеропиоз ureteropyosis
уретеропластика ureteroplasty
уретерорафия ureterorrhaphy
уретероррагия ureterorrhagia
уретероцеле *med.* ureterocele
уретра *anat.* urethra
~, спонгиозная spongy urethra
уретральгия urethralgia
уретрит urethritis
~, венерический/гонорейный gonorrh(o)ea
~, трихомонадный mycotic urethritis
уретропластика *surg.* urethroplasty
уретропоэз urethropoiesis
уретрорафия urethrorrhaphy
уретрорея urethrorrh(o)ea
уретроррагия urethrorrhagia, urethremorrhagia
уретроскоп urethroscope
уретростомия *surg.* urethrostomy
уретротомия surg. urethrotomy
уриал *zool.* urial; red sheep *(Ovis ammon vignei)*
уридиловый uridylic
уридин uridine
уридинтрифосфат *chem.* uridine triphosphate
уриказа uricase; urico-oxidase
урикемия, урикацидемия uricemia, uricacidemia; lithemia
уриколиз *bioch.* uricolysis
урина urine
урна 1. urn; 2. ballot box; 3. *bot.* moss capsule, urn
урночка *bot.* urn
уробилин *bioch.* urobilin
уровень level *fig.* standard; *tech.* gauge; level; layer
~ воды water level/plane/stage
~ воды, средний mean-water level
~ возбужденный exicited/exicitation level
~, высший higher level
~, глубокий low-lying level
~ грунтовых вод ground water table/level
~ деления/делительный *nucl.* fission level
~ дефицита shortage rate
~, доверительный *math.* confidence level
~ жизни/жизненный standard of living; living standard(s)
~ занятости (населения) occupational level
~ запасов *econ.* inventory stratum
~, заполненный/занятый occupied/filled level
~, захватный capture level
~ земли ground level
~ значимости significance level, level of significance
~ излучения radiation level
~, исходный reference level/plane
~ квалификации proficiency level
~ квантования quantization level
~, квантовый quantized level
~ кислотности acidity level
~, колебательный vibrational level
~, контрольный reference level
~ логического нуля O-level
~ логической единицы 1-level
~ моря sea level
~ моря, средний mean sea level
~ мощности power level
~, начальный datum level
~, номинальный доверительный *stat.* nominal confidence level
~, низкий low level
~, низший lowest level
~, нулевой datum/reference/zero level
~, одиночный/отдельный single level
~ образования educational status
~ отсчета reference level; base
~ ощущения sensation level
~ помех noise/interference level
~ пола floor height/level
~, пороговый threshold level
~ прерывания interrupt level
~ развития state-of-the-art
~ с отвесом carpenter's level
~, сетевой network layer
~ смертности morbidity level/rate
~, средний mean/average level
~, сферический universal level
~, тепловой thermal level
~ управления control level; *comp.* level of the hierarchy
~ фактора *stat.* level of factor
~ фона hum level; background noise level
~ шума, стационарный steady-state noise
~, энергетический energy level
~ энергии power level
~, эталонный index level; reference level
~ эффективности (труда) performance standard
~, ядерный энергетический nuclear energy level
уровнеграф level recorder
уровнемер level gauge/indicator/meter
~, мембранный diaphragm level indicator
~, поплавковый float-type level indicator
~, радиоизотопный radioactive level gauge
~, электрический electric-contact level indicator
урогастрон urogastrone
урогематин *bioch.* uroh(a)ematin
урогематома *med.* uroh(a)ematoma
урогематопорфирин uroh(a)ematoporphyrin
урогенитальный urogenital; urinogenital; genito-

urinary
уро-генито-ано-ректальный urogenitoanorectal
урогенний *physiol.* urinogenous; urogenous
урогепарин uroheparin
уроглюкозиметр urinoglucosimeter
урография urography
~, выделительная/экскреторная excretory urography
урод monster
уродливость *med.* monstrosity; malformation; 2. *geol.* teratologic form; monstrosity
уродливый monstrous; teratic
уродство monstrosity; malformation; abnormality; teratism
~, врожденное congenital malformation
~, частичное demimonstrosity
урожай 1. harvest; 2. crop; 3. yield
~, богатый rich harvest; heavyweight crop
~, валовой croppage
урожайность yield, output; productivity
урожайный yielding, productive; fertile
~ год bumper-year crop
урокимограмма urokymogram
урокимография urokymography
урокиназа *bioch.* urokinase
урокортизол dihydrocostisol, urocortisol
урокортизон dihydrocortisone; urocortisone
урокриоскопия urinocryoscopy
уроксантин *bioch.* uroxanthin
уролагния urolagnia
уролит urolith, urinary calculus
уролитиаз urolithiasis
уролитиазный urolithic
уролог urologist; urinologist
урология urology; urinology
урометр urinometer, urometer, urogravimeter
уронефроз uronephrosis; hydronephrosis
уросома *zool.* urosome
уроспермум *bot.* sheep's beard *(Urospermum)*
уростиль *zool.* urostyle
уросульфан sulphacarbamide
уротеобромин urotheobromine
уроторакс urothorax
уротропин urotropine; methenamine; hexanethylenetetramine
урохорд *zool.* urochord
уроцианин urocyanin
уроцианоз urocyanosis
уроцистит urocystitis
уроэритрин uroerythrin
урсулинка Ursuline nun
уртикарный urticarial; urticarious
уругваец, уругвайский Uruguayan
уругвайка Uruguayan (woman)
уруть *bot.* parrot feather *(Myriophyllum)*
ус 1. tongue, tab; nib; feather; 2. *zool.* feeler ; antenna; 3. *bot.* tendril, awn
~, китовый whalebone; baleen
усадка shrinkage
~, литейная casting shrinkage
~ по длине longitudinal shrinkage
усадочность shrinking (property)
усатый 1. *bot.* awned; barbate; 2. *zool.* whiskered
усач *zool.* 1. barbel *(Barbus)*; 2. capricorn beetle *(Cerambycidae)*
усвоение 1. assimilation; 2. mastering; adoption; learning
усвояемость 1. assimilability; 2. conprehensibility
усекать truncate
усечение 1. truncation; 2. *gram.* apocope
усик 1. *bot.* tendril; awn; runner; 2. *zool.* feeler; antenna; cirrus; barbel
усиковидный; усикообразный cirriform; tendril-shaped; antenniforn
усиление 1. srengthening; reinforcement; reinforcing; potentiation; 2. amplification; gain; magnification; intensificaton
~ антенны aerial gain/amplification
~ высокой частоты high-frequency amplification
~ звука sound amplification; audio gain
~ напряжения voltage gain/amplification
~ низкой частоты low-frequency amplification; audio gain
~, предварительное preamplification
~, резонансное resonant amplification
усиливать 1. strengthen, reinforce; potentiate; 2. amplify; intensify; boost
усиливаться become/grow stronger; intensify; increase
усилие 1. effort; exertion; 2. force; stress
~, ветровое wind force
~, давящее push force
~, единичное unit force
~ зажима clamping force
~, изгибающее bending force
~, крутящее/скручивающее torsional/twisting force
~, ломающее ultimate stress
~ отвода return force
~ подачи forward force
~, рабочее operating force
~, разрывное breaking stress/load
~, растягивающее tensile force
~, сдвигающее shearing force
~, срезывающее shearing force
~, толкающее push force
~, тяговое tractive/traction force
~, тянущее pull force, tractive effort
~, умственное mental effort
~, физическое physical effort/strain
усиливающий *med.* intensifying; tonic
~ тонус сосудов vasotonic
усилитель amplifier; booster; intensifier
~, антенный aerial amplifier
~, апериодический untuned/nonselective amplifier
~ бегущей волны travelling-wave amplifier
~ видеоимпульсов video amplifier
~ воспроизведения playback amplifier

~ вращающего момента torque amplifier
~ высокой частоты high-frequency amplifier; *Am.* radio-frequency amplifier
~гармоник harmonic amplifier; selective amplifier
~, гидравлический hydraulic booster
~, двухкаскадный two-stage amplifier
~, двухконтурный double-tuned amplifier
~, двухтактный push-pull amplifier
~ звука sound/audio amplifier
~ звуковой частоты audio-frequency amplifier
~, золотниковый spool-valve amplifier
~ импульсов/импульсный pulse amplifier
~, кадровый frame/vertical amplifier
~, канаповый window amplifier; channel amplifier
~, ламповый valve/tube amplifier
~, линейный linear amplifier; line amplifier
~ линии связи, промежуточный line repeater
~, масштабный scaler
~, минеральный *plast.* mineral filler
~, многокаскадный multistage amplifier
~ мощности power amplifier
~ на дросселях choke-coupled amplifier; *Am.* impedance-coupled amplifier
~ на интегральных схемах integrated circuit amplifier
~ на криотронах cryotron amplifier
~ на полупроводниках semi-conductor amplifier
~ на транзисторах transistor(ized) amplifier
~ напряжения voltage amplifier
~ низкой частоты low-frequency amplifier; *Am.* audio-frequency amplifier
~ обратной связи feedback amplifier
~, одноканальный single-channel amplifier
~, однокаскадный one-stage amplifier
~, одноконтурный single-tuned amplifier
~, однотактный single-ended amplifier
~, оконечный final amplifier
~, полупроводниковый semiconductor amplifier; solid-state amplifier
~ постоянного тока direct current amplifier, d.c. amplifier
~, предварительный pre-amplifier
~ промежуточной частоты intermediate-frequency amplifier
~ развертки sweep amplifier; timebase amplifier
~, резистивный/реостатный RC-coupled amplifier
~, резонансный tuned amplifier
~ с обратной связью feedback amplifier
~ с отрицательной обратной связью degenerative amplifier; negative-feedback amplifier
~ с положительной обратной связью regenerative amplifier; positive-feedback amplifier
~ с разделенной нагрузкой split-load amplifier
~ сверхвысокой частоты microwave amplifier
~ сигналов коррекции slaving amplifier
~ слабых токов low-current amplifier
~, стробированный gated amplifier
~, срочный *telev.* line/horizontal amplifier
~, струйный fluidic amplifier
~, твердотельный solid-state amplifier
~, телефонный telephone repeater
~, трансформаторный transformer-coupled amplifier
~, транзисторный transistor(ized) amplifier
~ узких селекторных импульсов narrow-gate amplifier
~, узкополосный narrow-band amplifier
~, широкополосный broad-/wide-band amplifier; fast amplifier
~ яркости изображения image-brightness tube

усилитель-ограничитель amplifier-limiter

усилитель-прерыватель chopping amplifier

ускорение acceleration; speeding-up; quickening
~ в переходном режиме transient acceleration
~, вековое *astr.* secular acceleration
~, возрастающее progressive acceleration
~ вращательного движения rotary acceleration
~, долговременное sustained acceleration
~ дрейфа drift/side acceleration
~, земное acceleration of the Eearth
~, кажущееся phantom acceleration
~, касательное tangential acceleration
~, кориолисово complementary/Coriolis acceleration
~, нарастающее progressive acceleration
~, неравномерное non-uniform acceleration
~, отрицательное deceleration; negative acceleration
~, поворотное = кориолисово
~ поступательного движения translatory acceleration; acceleration of translation
~, продольное axial acceleration
~ работ speeding-up of operations/work
~, равнодействующее resultant acceleration
~, равномерное uniform acceleration
~ равнопеременного движения uniform acceleration
~ свободного падения free fall acceleration
~ силы тяжести acceleration of gravity; gravitational/gravity acceleration
~ сноса drift/side acceleration
~ тангажа pitching acceleration
~, ударное impact acceleration
~, установившееся steady acceleration; steady state acceleration
~ частиц particle acceleration

ускоритель 1. *mech.* accelerator; booster; 2. *chem.* accelerant; promoter; accelerator
~ ван-де-Граафа Van de Graaf accelerator/generator
~ взлета assisted take-off unit
~ взлета, ракетный rocket-assisted take-off unit/bottle

~ **взлета, реактивный** jet-assisted take-off unit/bottle
~ **заряженных частиц** charged-particle accelerator
~ **ионов** ion accelerator
~ **катализа** catalysis promoter/accelerator
~, **импульсный** pulse(d) accelerator
~, **индукционный** induction accelerator; betatron
~ **Кокрофта-Уолтона** Cockroft-Walton accelerator
~ **коррозии** corrosion accelerant
~, **малоактивный** *plast.* slow-acting accelerator
~, **многокамерный** multiple-cavity accelerator
~ **на высокие энергии** high-energy accelerator
~, **плазменный** plasma accelerator
~, **ракетный** rocket booster; boost/booster motor; accelerating rocket engine
~, **реактивный** jet booster
~ **с бегущей волной** travelling-wave accelerator
~ **с волноводом** wave-guide accelerator
~ **с переменным полем** alternating-field accelerator
~ **с постоянным полем** constant-field accelerator; fixed-field accelerator
~ **с резонаторами** cavity-type accelerator
~ **самовоспламенения** self-ignition promoter
~, **сбрасываемый** jettisonable booster
~, **сильноточный** high-current accelerator
~, **стартовый** launching booster
~ **схватывания** setting accelerator
~, **циклический** *nucl.* circular-orbit accelerator
~, **циклический резонансный** circular resonance accelerator
~, **электронный** electron accelerator; electronic accelerator
~, **электронный циклический** electronic circular accelerator

ускорительный accelerating; boost (*attr.*)

ускорять accelerate, speed up; boost

условие condition; term; provision; stipulation; clause; *see also* **условия**
~ **Вульфа-Брэгга** *phys.* Bragg equation/law
~ **гармонического баланса** Barkhauen criterion for oscillation
~, **конечное** terninal condition
~ **критичности** criticality condition
~ **максимализации прибыли** *econ.* best-profit condition; profit maximizing condition
~ **максимальности** *math.* maximum condition; ascending claim condition
~ **минимакса** minimax condition
~ **минимальности** *math.* minimum condition; descending chain condition
~ **непрерывности/неразрывности** continuity condition
~, **ограничивающее** constraint; limiting condition; boundary condition
~ **оптимальности** optimality condition
~, **отлагательное** *law* suspensive condition
~, **предельное** limiting condition
~ **трансверсальности** *math.* transversality condition

условия *pl.* conditions; terms; provisions; stipulation; clauses; articles; *see also* **условие**
~, **аварийные** emergency
~ **безветрия** no-wind conditions; still-air conditions
~ **выдачи кредита** lending terms
~, **граничные** boundary conditions, constraints
~, **договорные** terms of contract; contractual terms; treaty provisions/clauses
~ **достигнутые в результате переговоров** negotiated terms
~ **жилищнобытовые** living conditions
~, **жилищные** housing conditions
~ **задачи** stetement of a problem
~, **запретительные** *econ.* exclusionary terms
~, **имитированные** simulated conditions
~, **климатические** *tech.* environmental conditions
~ **космического пространства** space environnent
~, **краевые** boundary conditions
~, **льготные** easy terms
~, **метеорологические** weather conditions
~, **наихудшие** worst-case conditions
~ **невесомости** weightlessness conditions
~ **непрерывности/неразрывности** continuity conditions
~, **нерасчетные** off-design conditions
~ **ограничительные** *econ.* exclusionary terms
~, **окружающие** the environment; ambient conditions
~, **оптимальные** optimum conditions
~ **опыта** experimental conditions/setup
~ **осуществимости** feasibility conditions
~ **платежа** terms of payment
~, **предельные** extreme/boundary/marginal conditions
~ **приема** *rad.* receiving conditions
~, **природные** natural environment
~, **производственные** working conditions
~, **рабочие** operational environment
~ **равновесия** equilibrium conditions; conditions of equilibrium
~, **расчетные** design conditions
~ **средних широт** temperate conditions
~ **страхования** insurance terms and conditions
~ **страхования, типовые** standard policy conditions
~, **технические** specifications; technical requirements
~ **труда** working conditions
~, **установившиеся** steady-state conditions
~ **хранения** storage conditions
~ **эксперимента** experimental conditions
~, **эксплуатационные** operating/operative/operation/working/service conditions
~, **экстремальные** extreme conditions

условно 1. conventionally; by convention; 2. conditionally; 3. tentatively
~, осудить *law* put on probation
условно-желательный *gram.* optative; conditional-subjunctive
условно-рефлекторный conditioned-reflex
условно-следственный conditional-consequential
условность 1. conventionality; convention; conventionalism; 2. conditionality
условно-уступительный conditional-concessive
условный 1. conventional; agreed upon; 2. conditional; 3. tentative; provisional; *law* qualified; 4. *law* suspended
~ знак conventional sign
~ приговор *law* suspended sentence
~ рефлекс conditioned reflex
услуги service(s)
~, вычислительные computer service
~, дополнительные 1. fringe services; 2. value-added service
~, коммунальные (public) utilities; public services
усниновый *chem.* usninic; usnic
усовершенствование improvement; updating; *educ.* advanced/postgraduate training
усоногие *n. pl. zool.* cirripeds; barnacles (*Cirripedia)*
Успение *rel.* Assumption
успокаивать 1. damp(en); 2. *met.* kill
успокаивающее *med.* sedative
успокоитель damper; stabilizer
~ качки *naut.* roll stabilizer; ship stabilizer
~ колебаний deoscillator; dash-pot
усреднение 1. averaging; smoothing; 2. blending
~ по времени time averaging
~ по множеству ensemble averaging
~ руд ore blending
усреднитель *hydr.* equalizing reservoir
устав rules; regulations; statutes; articles; standing orders
~, боевой field manual
~ внутренней службы interior service regulations
~, военный articles of war
~, дисциплинарный disciplinary regulations
~, корабельный navy regulations
~, строевой drill regulations
уставка *comp.* setpoint; control point
уставный statutory; prescribed
~ капитал authorized capital stock
усталость fatigue
~ при изгибе bending fatigue
~ при кручении torsional fatigue
~ при растягивающем напряжении tensile fatigue
~ при ударе impact fatigue
устанавливать 1. mount; install; set up; place; position; assemble; erect; 2. establish, determine, ascertain; 3. adjust; regulate
установка 1. equipment; installation; plant; unit; set; apparatus; 2. mounting, placing, installing; erecting; setting up, positioning; 3. adjustment; alignment; 4. precept; directive
~, автономная self-contained unit
~, атомная силовая atomic power plant
~, буровая drilling rig
~ валков roll adjustment/setting
~, вакуумная vacuum-pumping assembly
~, ветроэнергетическая wind-driven electric plant
~, водоочистительная water purifier
~, воздуходувная blower installation
~ высотомера altimeter setting
~, гамма-терапевтическая gamma therapy unit
~, гидросиловая water-power unit
~, гидроэнергетическая hydroelectric power plant
~, головная pilot/prototype plant
~, гребная *naut.* propulsion unit
~, грубая coarse adjustment
~, двигательная power plant/unit; propulsion system/unit
~, дегазационная degassing unit
~, дезактивационная decontamination plant/installation
~ для облучения irradiation facility/unit
~, дождевальная overhead irrigation system; sprinkler installation/system
~, дозиметрическая health-monitoring installation
~, дробеструйная shot-blast unit
~ интервалов печати line adjustment
~, испытательная test plant
~ коллективного использования multiuser site
~ кондицирования воздуха air conditioning installation; air conditioner
~, лабораторная laboratory-scale plant
~ ловушек в программе *comp.* trap setting
~, моделирующая simulator
~, мусоросжигательная incinerator
~ на место positioning
~ на нуль zero adjustment/setting, zeroing
~ на фокус focussing
~, насосная pump assembly
~, неправильная misadjustment; misalignment
~ нивелира level setting, level set-up
~ нуля zero adjustment/setting
~, обеспыливающая dust-collecting installation/plant; dust-catcher
~, обжигательная *nucl.* calciner
~, опреснительная water-distillating plant
~, опытно-промышленная pilot plant
~, откачная pumping assembly
~, паросиловая steam power plant
~, перегонная distilling plant
~, пересчетная scaler
~, поворотная swivelling mount

~, **предварительная** presetting
~, **пулеметная** machine-gun mount
~, **пусковая** launcher
~, **радиолокационная** radar installation
~ **режима** mode setting
~, **резервная технологическая** standby plant
~, **силовая** power plant/unit; propulsion system
~ **системы в начальное состояние** system reset
~ **столбов** pole setting/erecting; poling
~, **стоматологическая** dental unit
~, **телевизионная** TV camera unit
~ **тома** *comp.* volume mounting; data carrier mounting
установщик fitter; mounter; installer; setter; setting device
устой 1. abutment; buttress; 2. pillar; support
~, **выносной** projecting abutment
~, **обсыпной** buried abutment
~, **ряжевый** log-crib abutment
устойчивость 1. stability; 2. resistance; immunity; tolerance
~ **в полете** *aero.* inflight stability
~, **вращательная** rotational stability
~ **к истиранию** abrasion resistance
~ **к коррозии** corrosion resistance
~ **к окружающим условиям** environmental tolerance
~ **к отказам** fault-tolerance
~ **к утомлению** fatigue resistance
~ **кадра** picture steadiness
~ **на курсе следования** course-keeping stability
~ **по курсу** heading stability
~ **по крену** roll/rolling stability
~ **по скорости** speed stability
~, **поперечная** stability in roll; roll/rolling stability
~ **при продольном изгибе** buckling stability
~ **при хранении** storage stability shelf life
~, **радиационная** radiation resistance
~, **собственная** inherent stability
~, **ферментативная** resistance to enzymes
~, **химическая** chemical stability/resistance
устойчивый 1. stable, steady, sustained; firm; 2. resistant
устранение elimination; removal
~ **девиации компаса** compass compensation
~ **неисправностей/повреждений** clearing/removal of faults; trouble-shooting
устранять eliminate; remove; suppress; clear
устрица *zool.* oyster *(Ostrea)*
устрицевод oyster farmer/culturist
устричник oyster bed
устричный oyster *(attr.)*; ostreiform
устройство 1. device, apparatus; mechanism; instrument; (item of) equipment; gear; unit; facility; means; 2. arrangement; structure; system; setup; layout; design
~, **аварийно-защитное** emergency/safety device
~ **ввода** input device
~, **внешнее** *comp.* peripheral equipment
~, **входное** input device/unit
~, **вычислительное** computer; computing device
~, **вычитающее** subtractor
~, **государственное** state system
~ **графического ввода** graphic input device
~, **графическое регистрирующее** plotting device; plotter
~, **декодирующее** decoder
~, **дисковое запоминающее** disk unit
~, **дистанционное контрольное** remote monitor
~ **для обработки данных** data processor
~ **для считывания данных** data reader
~, **заборное** intake
~, **задающее** setting device
~, **записывающее** recorder; writer; graphic instrument
устройство, запоминающее, *abbr.* **УЗ** storage; memory, store
УЗ, автономное off-line storage/nemory
УЗ, без разрушения информации nondestructive storage/memory
УЗ, быстродействующее quick-/rapid-access storage/memory
УЗ, внешнее external storage/memory
УЗ, внутреннее internal storage/memory
УЗ, долговременное permanent storage/memory
УЗ, магнитострикционное magnetostrictive storage/memory
УЗ на барабанах drum storage
УЗ на дисках disk storage/memory, disk store unit
УЗ на лентах tape storage/memory
УЗ на магнитных картах magnetic card memory/storage
УЗ на магнитных сердечниках magnetic-core storage/memory; core memory/store
УЗ на перфолентах punch tape storage
УЗ на триггерах flip-flop storage/memory
УЗ на ферритовых сердечниках ferrite core storage/nemory
УЗ на электроннолучевых трубках cathode-ray tube memory/store
УЗ, одностороннее read-only memory
УЗ, оперативное on-line storage/memory
УЗ, последовательное serial-access storage/memory
УЗ с быстрой выборкой quick-access memory; rapid-access memory
УЗ с запирающей сеткой barrier-grid electrostatic storage/memory/store
УЗ с нестираемой записью nonerasable storage/memory
УЗ с поразрядной выборкой bit-organized memory
УЗ с произвольной выборкой random access

storage/memory
УЗ с прямой выборкой word-organized storage/memory; word-selection storage/memory
УЗ с разрушением информации volatile storage/memory
УЗ с совпадением токов coincident-current storage/memory
УЗ с центральным процессором on-line storage/memory
УЗ со стираемой записью erasable storage/memory
УЗ, триггерное flip-flop storage/memory
УЗ, циклическое circulating/cyclic storage; circulating/cyclic memory
УЗ, электроннолучевое cathode-ray tube memory

устройство:
~, зарядное (аккумуляторное) battery charger
~, звукозаписывающее sound recorder
~, измерительное measuring device
~, индикаторное display/indicator unit; indicator
~ исправляющее ошибки error-correcting device/feature
~, клавишное keyboard unit
~, клавишное перфорационное key punch unit
~, кодирующее encoding/coding device; coder;
~, контрольное monitor; monitoring device
~, корректирующее 1. error correcting device 2. *auto.* compensating device/network; compensator
~, леерное *naut.* life lines/rails
~, логическое logical unit/device
~, моделирующее simulator
~, модульное modular unit
~, обслуживающее server
~, оконечное terminal (unit)
~, оперативное запоминающее read-write memory
~ отображения display device
~, передающее и приемное данных transceiver
~, периферное peripheral device
~, печатающее printer
~, печатающее буквенно-цифровое alphanumerical printer
~, печатающее колесного типа wheel printer
~, печатающее построчно line printer
~ печати и перфорации printing punch
~, питающее feeder, feeding device
~ подачи feeder
~, постоянное запоминающее read-only memory
~, предупреждающее warning device; alarm
~, программирующее programmator
~, противоугонное anti-theft device
~, пусковое starter, starting device
~, развертывающее scanner
~ распознавания слов word recognizer, word recognition machine
~, распределительное *el.* switch-gear
~, растровое сканирующее raster-scan device
~, регистрирующее recorder; logger
~, резервное backup device
~, решающее resolver
~ с прямым доступом direct-access unit
~ сбора данных data acquisition device
~, сдвигающее shifter
~, сенсорное touch-sensitive device
~, сканирующее scanner
~, следящее tracer
~ сопряжения interface unit
~, сортировальное sorter
~, сравнивающее comparator
~, суммирующее summator; summer; adder
~ считывания/считывающее reader
~ считывания печатного текста print type reader
~ считывания буквенно-цифровой информации alphanumeric/alphameric reader
~ считывания, постраничное page reader
~ считывания с ленты tape reader
~ считывания с перфокарт card reader
~, сцепное coupler, coupling device
~, счетно-решающее computer, computing device
~ умножения multiplier unit
~ управления control unit/device
~, файловое запоминающее file unit
~, фотопечатающее photoprinter
~, хронирующее timer; synchronizer
~, цифровое digital device
~, читающее reader, reading machine/device

уступ ledge, shelf; bench, scarp, escarpment; step; *met.* shoulder
уступительно-противительный *gram.* concessive-adversative; concessive-contrastive
уступительный *gram.* concessive
уступка 1. concession; cession; 2. *econ.* (price) reduction; rebate
устье mouth; orifice; opening; aperture; *anat.* ostium
~ аорты aortic ostium
~, сложное compound month
~ фурмы *met.* nose of a tuyere
устьице small mouth/orifice ; *bot.* stoma
устьичный *bot.* stomatal; stomatic
усушка loss of moisture/weight
усы *pl.* whiskers; moustache
усыновитель *law* adopter
усыновленный *adj.* adopted; *n.* adoptee
усыновлять *law* adopt
усыхать dry up; shrink, shrivel
утаивать conceal; hide; appropriate
утварь utensils; tinware
~, церковная church plate
утвердительно-противительный *gram.* affirmative-adversative
утверждать confirm; assert, affirm; ratify; vali-

date; strengthen; consolidate
утеропластика *med.* uteroplasty
утесник *bot.* gorse; furze (*Ulex*)
утечка leakage; escape; leak; spillage; loss
~ внутрь inward leakage
~ давления pressure loss
~ излучения radiation leakage
~ капитала capital outflow
~ мозгов brain drain
~ на землю ground leak/leakage
~ на корпус leakage to frame/chassis
~ нейтронов neutrone escape/leakage/streaming
~ рабочей силы *econ.* manpower drain
~ радиоактивности escape of radioactivity
утилизация recovery; recycling; salvaging; utilization
утилизировать recover; recycle; salvage; utilize
утилитаризм utilitarianism
утилитарист utilitarian
утилитарность utalitarian attitude; utility
утилитарный utilitarian
утильзавод recycling plant; salvage factory
утильное железо scrap iron/metal
утильрезина scrap/waste rubber
утильсырье utility refuse/waste; scrap (material); recoverable refuse
утильцех salvage/recovery department
утка 1. *zool.* duck; 2. *aero.* canard configuration 3. *naut.* cleat
~, крепительная belaying/snatch cleat
~, швартовая mooring cleat
утка-пеганка *zool.* shelldrake; shellduck (*Tadora*)
утка-широконоска *zool.* spoonbill; shoveller (*Anas clypeata*)
утконос *zool.* duckbill, duck-billed platypus (*Ornithorhynchus anatinus*)
утлегарь *naut.* jib boom; outrigger
уток *text.* weft; woof
утолщать thicken; make thicker
утолщение thickening; thickened part; enlargement; *med.* intumescence; swelling
~, варикозное varix
~, годичное *bot.* annual ring/band
~, мозолистое callosity; callus
~ раковины *zool.* varix
утомление fatigue; tiredness; exhaustion
~, слуховое aural fatigue
~, цветовое colour fatigue
утомляемость fatiguability; fatigue
утончаться grow/become thin(ner); thin out
утоп sunken/lost logs
утопический, утопичный utopian
утопичность utopianism
утопия utopia
утопленный countersunk; recessed; buried; embedded, flush (with), sunk; submerged
уточина *text.* weft (thread), wefting
уточнение specification, clarification; closer definition; refinement; elaboration
уточнять specify; make more exact/precise; elaborate
утраивать treble; triple; triplicate
утраквист *eccl.* Utraquist; Calixtine
утрамбовывать rad; pack; tamp; beat down
утрата loss; forfeiture
утрикул *bot.* utricle
утроение tripling; trebling
утроенндвойчатый *bot.* thrice-twin; tergeminate
утроенноперистый *bot.* thrice-pinnate; tripinnate
утроеннотройчатый *bot.* thrice-ternate; triternate
утроенный triple; trebled, threefold; triplicate
утроитель trebler, tripler
утруска spillage
утюг (flat) iron; pressing iron
~, дорожный road/drag planer; road drag
утюжение ironing; pressing
утяжеление weighting; loading; charging
утяжелитель weighting material/compound; filler
утяжелять make heavy; weight; charge, load
ухват holder, clip; grip; oven prongs/fork
ухо 1. lug, ear; 2. *anat.* ear
~, внутреннее internal ear
~, наружное external ear
~, среднее middle ear
уховертка *zool.* earwig (*Forficula auricularia*)
уход I care, tending; upkeep; maintenance; servicing; handling; attending; attendance
~ за больными nursing
~ за почвой soil maintenance
~, постоянный running maintenance
уход II departure; leaving; resignation; quitting; *tech.* drift
~, массовый *fig.* exodus
~ от цели target escape
~ по азимуту azimuth wander
~ с орбиты exit/abort from orbit
~ частоты frequency drift
ухудшение deterioration, worsening; aggravation; degradation; decline, decadence; impairment
участие participation; part; share; *econ.* stake, interest
участок section; area, zone; district; site; *(land)* plot; lot; parcel; tract; *(length)* leg; *mil.* sector
~, водосборный catch land
~ заряжения contaminated area
~, золотоносный gold field
~, избирательный polling station
~ маршрута *av.* route leg
~, мелководный *naut.* high spot
~, нерестовый spawning/breeding ground
~, производственный production division
~ фронта sector of the front
~ цепи *el.* subcircuit
учащение acceleration; quickening
учебно-воспитательный educational
учебно-производственный work-study
учебный teaching; training, instructional, school;

academic
~ **год** school/academic year
~ **план** curriculum
~ **предмет** (school) subject
~ **самолет** training aircraft
~**ное время** school hours
учение 1. learning; studying; study; studies; training; teaching; instruction; 2. doctrine; teaching; theory; 3. *mil.* exercise; drill
~, **боевое** emergency drill
~ **о землетрясениях** study of earthquakes; seismology
ученичество apprentice-training
ученый-атомник/-ядерник atomic scientist
учесть *see* **учитывать**
учет 1. account; record accounting; record-keeping; 2. stock-taking; inventory; 3. calculation; 4. *fin.* discount
~ **банка** bank discount
~, **бухгалтерский** bookkeeping
~, **бухгалтерский при помощи перфокарт** punched card accounting
~ **в натуральном выражении** accounting on the basis of quantities
~, **глазомерный** occular estimate
~ **запасов** store accounting
~ **издержек в торговле** distribution cost accounting
~ **коммерческих операций** business accounting
~, **машинный/механизированный** machine accounting
~ **населения** civil registration
~, **неполный** undercount
~ **по видам хозяйственной деятельности** functions accounting
~, **первичный** primary accounting
~, **производственный** cost accounting
~ **товарно-материальных запасов** inventory account
~, **функциональный бухгалтерский** activity accounting
учетверение quadrupling; quadruplication
учетно-медицинская карточка medical record/card
учетчик record keeper; accountant
учетчик-статистик statistical clerk
училище school
~, **военное** military school/academy; officers training school
~, **военно-морское** naval college/academy
~, **производственно-техническое, ~, ремесленное** vocational school
~, **Суворовское** Suvorov Military College
учитывать *fin.* discount
~ **вексель** discount a bill
~ **вексель в банке** negotiate a bill to a bank
учком *abbr.* student committee
учлет *abbr.* student pilot
учпедгиз *abbr.* State Publishing House of Educational and Pedagogical Literature
учредитель founder
учредительный constituent; constitutive
~**ное собрание** constituent assembly
учреждение 1. establishment, institution; institute, office; agency; 2. instituting; setting up; founding; establishment
~, **государственное** government office/institution/agency
~, **научное** scientific institution
~, **клиринговое** clearing agency
ушан *zool.* long-eared bat (*Plecotus auritus*)
ушиб contusion; bruise; injury
ушибать contuse, bruise; injure; hurt
ушибленно-рваный contused-and-lacerated
ушивание sewing; taking in; *med.* suturing
~ **аорты** aortorrhaphy
~ **мышцы сердца** *surg.* cardiorrhaphy
ушиватель suture appliance; stapling machine
уширение broadening, widening
~, **аппаратурное** instrumental broadening
~ **импульсов** pulse stretching
~ **пучка** beam spread/divergence
уширитель expander; broadener; widener; reamer
ушко lug; ear, eye; *anat.* auricle
~, **монтажное** mounting lug
~ **на конце троса** cable eye
~ **рессоры** spring eye
ушковидный, ушкообразный auriculate; earlike
ушной ear *(attr.);* auricular; otic
~**ная раковина** *anat.* auricle, helix; pinna
ущелье gorge; ravine, canyon
ущемление *med.* strangulation; incarceration
ущерб damage; harm; injury; loss; detriment; prejudice
~, **возмещать** indemnify
~, **наносить** damage, harm, prejudice
~ **от несчастного случая** casualty
ущербный, ущербленный waning (*of the moon*)
уэльсец Welshman
уэльский Welsh, Cymric
уязвимость vulnerability (to)
уязвимый vulnerable; open to attack/injury

Ф, ф

фа *mus.* fa
фабелла *anat.* fabella
фабзавком *abbr.* factory committee
фабзавуч *abbr.* factory workshop school
фабиана *bot.* Peru false heath *(Fabiana imbricata)*
фабианец, фабианский Fabian
фабианизм Fabianism
фабком *abbr.* factory committee
фаблио, фабльо *lit.* fabliau
фабрика factory; plant; mill

~, кондитерская confectionery
~, консервная cannery
~, обогатительная dressing/concentrating mill/plant
~, обувная shoe/footware factory
~, прядильная spinning mill
~, ткацкая weaving mill
~, хлопкопрядильная cotton spinning mill
фабрикант factory/mill owner, manufacturer, industrialist
фабрикат manufactured product, ware
фабрично-заводской factory *(attr.)*; industrial
фабричный 1. factory, plant *(attr.)*; industrial, manufacturing; 2. factory-made
фабула *lit.* plot, story
фавеола faveolus
фавиди *pl.* favids
фавизм *med.* favism, fabism
фавн *myth.* faun; wood spirit
фавус *med.* favus
фаг *bact.* phage, bacteriophage
фаговый phage *(attr.)*
фагодиагностика phage diagnosis
фагокариоз phagokariosis
фагот *mus.* bassoon
фаготист *mus.* bassoonist, bassoon player
фаготроф phagotroph
фагоустойчивость phage resistance
фагоцит *biol.* phagocyte
~, иммунный educated phagocyte
фагоцитарный, фагоцитный phagocytic
фагоцитин phagocytin
фагоцитировать phagocytize, phagocytose
фагоцитобласт phagocytoblast
фагоцитоз phagocytosis
~, незавершенный incomplete phagocytosis
фагоцитолиз phagocytolysis, phagolysis
фагоцитолический phagocytolytic, phagolytic
фагочувствительность phagosensitivity
фадом fathom
фаевый *text.* made of faille
фаза phase; stage
~, беспорядочная random phase
~ вдоха inspiratory phase
~, водная aqueous phase
~ восковой спелости *agr.* dough phase, phase of wax ripeness
~, временная time phase
~ выборки команды *comp.* fetch phase
~ выдоха expiratory phase
~, газовая/газообразная gaseous/gas phase
~, граничная boundary phase
~, делящаяся fissile phase
~, диспергированная/дисперсная disperse(d) phase
~, жидкая liquid state
~ записи *comp.* write phase
~ затмения phase of eclipse
~ захвата locking phase
~ исполнения *comp.* execute phase
~ компиляции *comp.* compile phase
~ компоновки программы assembly phase
~ Луны lunar/moon phase
~ несущей carrier phase
~, обратная reversed phase
~, однородная continuous/homogeneous phase
~, опорная reference phase
~, основная master phase
~, отрицательная minus phase
~, паровая/парообразная vapour phase
~ первого прогона программы object phase
~, переходная transition phase
~ плодоношения fruiting/fruition stage
~ покоя dwell phase
~, положительная plus phase
~, промежуточная intermediate phase; interstitial phase
~, противоположная opposite/counter/reversed phase, antiphase
~ рассеяния scattering phase
~, сверхпроводящая superconducting phase
~, стекловидная vitreous phase
~, твердая solid phase
фазан *zool.* pheasant (*Phasianus*)
фазеолин *bioch.* phaseolin
фазирование phasing-in; phase locking; paralleling
фазировать (bring in) phase
фазис phase
фазный phase *(attr.)*; phasic
фазово-импульсный pulse-phase
фазово-контрастный phase contrast
фазово-модулированный phase-modulated
фазовращатель phase shifter/inverter, phase-changing unit, phase-shifting network
фазовый phase *(attr.)*; phasic
фазовыравнитель phase equalizer/compensator
фазоинвертор phase inverter
фазоиндикатор phase monitor/indicator
фазокомпенсатор phase advancer/equalizer/compensator
фазометр phase-meter
фазоразделитель phase separator
фазорасщепитель phase splitter/separator
фазорегулятор phase regulator/shifter
фазосдвигающий, фазосмещающий phase-shifting
фазотрон synchrocyclotron, frequency-modulated cyclotron, phasotron
~, кольцевой annular synchrocyclotron
фазотропия *chem.* phasotropy
фай *text.* faille
файдешин *text.* faille de Chine
файл *comp.* file
~, архивированный archived file
~, байт-ориентированный character-type file
~, блок-ориентированный block-type file
~ большой емкости high-capacity file
~, вспомогательный optional file; scratch file
~, дистанционный remote file

~, **дублирующий** backup file
~, **загрузочный** EXE file
~ **задания** job file
~ **замкнутых процедур** closed-transaction file
~ **изменений** amendments file
~, **исходный** father file
~, **командный** command file
~ **на диске** disk file
~ **на (магнитной) ленте** tape file
~ **независимый от устройства** device-independent file
~, **неиспользуемый** dead file
~, **обновляемый** update file
~, **общедоступный** public file
~, **оперативно доступный** on-line file
~, **основной** master file
~, **плоский** flat file
~, **последовательный** sequential file
~, **рабочий** work/scratch file; activity file
~ **с произвольной выборкой** random-access file
~ **с прямым доступом** direct file
~ **с разделами** partitioned file
~ **связей** link file
~, **скрытый** hidden file
~ **совместного доступа** shareable file
~, **справочный** directory file
~ **трассировки** trace file
~, **учетный** accounting file
~, **цепной** chained file
~, **эталонный** memo file (for replication)

файрфильдит *miner.* fairfieldite
факел 1. torch, flare; 2. flame, jet/tongue of flame; glow
~, **дальнобойный** long flame
~, **пусковой** starting torch/flame, starter/pilot/ignition flame
~ **ракетного двигателя** rocket blast
~, **солнечный** solar flare; facula
факельное шествие torchlight procession
факит *med.* phacitis, phakitis, phacoiditis
факоанафилаксия *med.* phacoanaphylaxis
факолиз phacolysis
факолит 1. *min.* phacolite; 2. *geol.* phacolith
факолитический phacolytic
факома phacoma
факомаляция *med.* phacomalacia
факосклероз phacosclerosis, hard cataract
факоцеле phacocele
факоцистит phacocystitis
факсимиле facsimile ; replica
факт fact
~, **доказанный** *law* fact in evidence
~, **ноторный/общеизвестный** *law* fact of common knowledge/notoriety
~, **совершившийся** fait accompli
фактив *gram.* factive
фактивный *gram.* factive, factitive
фактис factice
фактически in fact, actually, virtually
~ **невиновный** *law* innocent in fact
фактический actual; virtual, real
~ **объем производства** effective output
фактография factual account
фактор factor; agent, agency; element; *math.* coefficient, factor
~, **аппаратурный** instrumental factor
~ **внешней среды** environmental factor
~, **масштабный** scale factor
~, **обуславливающий/определяющий** conditioning/determining/governing factor
~ **ослабления** reduction/attenuation factor
~ **повторяемости** multiplicity factor
~, **сдерживающий** deterrent
~, **структурный** structure factor
~ **формы** form factor
~, **человеческий/субъективности** human factor
фактор-алгебра quotient algebra
фактор-группа *math.* factor/quotient group
фактор-замкнутый *math.* quotient-/factor-closed
факториал *math.* factorial
~, **двойной** double factorial
факторизуемый *math.* factorable
фактория trading station
фактор-кольцо *math.* quotient ring
фактор-мера *math.* quotient measure
фактор-многообразие *math.* quotient manifold
фактор-множество *math.* quotient/factor set
фактор-модуль *math.* difference/factor module
факторный factorial
фактор-пространство *math.* quotient/factor/coset space
фактор-символ *math.* quotient symbol
фактура 1. *comm.* invoice, bill (of parcels); 2. texture, facture; finish; pattern
~, **ориентировочная/предварительная/примерная** proforma invoice
фактурирование billing
фактурный автомат invoicing automaton/machine
факультативность optionality
факультативный optional; elective
факультет faculty, department; school
факультетский faculty *(attr.)*
фал *naut.* halyard
~, **тентовый** crowfoot halyard
фаламицин *pharm.* phalamycin
фаланга 1. *anat.* phalanx; 2. *hist.* phalanx; *polit.* Falange
~, **дистальная** terminal/distal phalanx
~, **колосовидная** tufted phalanx
~, **ногтевая** ungual phalanx, phalangette
фалангеальный phalangeal
фалангит *med.* phalangitis
фалангэктомия phalangectomia
фалангист 1. *hist.* phalangite, soldier/member of a phalanx; 2. phalansterian; 3. *polit.* Falangist
фаланстер(ия) phalanstery
фалернский Falernian

фалинь *naut.* boat/gift rope, boat's painter
~, кормовой stern fast
~, носовой bow painter, grapnel
фаллопиева труба *anat.* Fallopian tube, oviduct
фаллопластика *med.* phalloplasty
фаллос phallus
фаллосома *zool.* phallosome
фаллус 1. *bot.* stinkhorn *(Phallus impudicus);* 2. phallus
фалреп *naut.* side rope, manrope
фалрепный *n.* side boy; *pl.* gangway party/staff
фалстарнпост inner sternpost
фалунит *miner.* fahlunite
фальстем stem apron, cutwater, false stem
фальстарт *sport* false start
фальц rabbet, groove; fold; edge
фальцаппарат folding machine, folder
фальцгобель rabbeting plane, fillister
фальцевальный 1. folding; 2. rabbeting
фальцевать 1. fold; 2. *wood.* rabbet, chamfer, groove
фальцет *mus.* falsetto (voice)
фальцмашина (sheet-)folding machine; book folder
~, двухтетрадная/двойная double-sixteen folder
фальцовка 1. folding; seaming; 2. grooving; 3. rabbet-plane, fillister
~ перпендикулярная right-angle folding
фальшборт *naut.* bulwark
фальшвал false shaft
фальшкиль *naut.* false keel, keel shoe
фальшфейер *naut.* blue light/fire, false fire
фанариот Phanariot(e)
фанера 1. plywood; 2. veneer
~, клееная plywood
~, лущеная peeled/cut veneer
~, металлизированная metal-faced veneer
~, многослойная multiply plywood
~, ножевая/резаная/строганая sliced veneer
~, пиленая sawn/sawed veneer
фанерит *miner.* phanerite
фанеритовый phaneric, phanerocrystalline
фанерование veneering
фанерогам *bot.* phanerogam, seed plant
фанерогамный *bot.* phanerogamous
фанерогенный *med.* phanerogenic
фанерогиалиновый phanerohyaline
фанеромерный *miner.* phaneromerous
фанеростемонный phanerostemonous
фанерострогальный veneer-cutting
фанерофит *bot.* phanerophyte
фанза foulard
фантазия fantasy; imagination; *mus.* fantasia
фантастика *lit.* fantastic tales; fiction
~ научная science fiction
фантастрон phantastron
фантом phantom
фаопланктон *bot.* phaoplankton, surface plankton
фаосома *zool.* phaosome
фара headlight, headlamp
~ ближнего света auxiliary headlight
~ дальнего света headlight
~, задняя *auto.* reversing light
~, посадочная landing light
~, рулежная *av.* taxying lamp/light
~, убирающаяся retractable headlamp
фарада *phys.* farad, F
фарадей *phys.* faraday, Faraday unit/constant
фарадизация faradization, faradism
фарадметр faradmeter
фарадотерапия faradotherapy, faradization
фарандола *mus.* farandole
фараон Pharaoh
фарбитис *bot.* morning-glory *(Ipomea)*
фарватер *naut.* fairway, waterway, channel, shipway; deep stream
~, огражденный buoyed channel
~, подходный approach channel
~, судоходный ship channel
Фаренгейт Fahrenheit
фарингит pharyngitis
~, сухой atrophic pharyngitis
фарингоамигдалит pharyngotonsillitis
фарингокератоз pharyngokeratosis
фаринголарингит pharyngolaryngitis
фаринголит pharyngolyth
фарингомикоз pharyngomycosis
фарингооральный pharyngo-oral
фарингопатия pharyngopathia, pharyngopathy
фарингопластика pharyngoplasty
фарингоплегия pharyngoplegia
фарингоринит pharyngorhinitis
фарингоскопия pharyngoscopy
фарингоспазм pharyngospasm, pharyngismus
фарингостеноз pharyngostenosis, stricture of the pharynx
фаринготомия *surg.* pharyngotomy
фарисей *hist.* Pharisee
фарисейский Pharisaic; pharisaic(al)
фарисейство Pharisaism; hypocrisy
фарлонг furlong
фармакодинамика pharmacodynamics, pharmacokinetics
фармаколог pkarmacologist
фармакопедика pharmacopedia, pharmacopedics
фармакопейный pharmacopoeial, official
фармакопея Pharmacop(o)eia
фармакопсихоз pharnacopsychosis, drug addiction
фармакотерапия pharmacotherapy
фармакохимия pharmacochemistry
фармацевт pharmacist, pharmaceutist, druggist, apothecary
фармацевтика pharmaceutics
фармацевтический pharmaceutic(al)
фармация pharmaceutics, pharmacy
фарнохинон *chem.* farnoquinone
фартук apron; cover
~ молотилки bluff board; deflector curtain
фарфор porcelain, china; chinaware

~, китайский China porcelain
~, костяной bone china
~, литой/формованный wet-process porcelain
~, прессованный dry-process porcelain
~, фриттовый fritted porcelain
~, химический acid-proof porcelain
~, яшмовый jasperated china
фарфоровый porcelain, china *(attr.)*; ceramic
~ завод porcelain/ceramic works
~ сервиз china set
~вая глина china clay, kaolin
фарш forcemeat, sausage meat; stuffing
фаршированный *cul.* stuffed
фаршировка *cul.* stuffing
фас front; face
фасад facade, front, face; elevation, view
~, боковой side view/elevation
~, выступающий avant-corps
~, передний front elevation
фасетка facet
фаска facet, face, level/chamfered edge, chamfer
фасовать pack, package; fill
фасовка packing
фасолевидный, фасолеобразный phaseoliform, fabiform, bean-shaped
фасоль *bot.* bean (plant) *(Phaseolus)*
~, ароматная caracol bean
~, обыкновенная French/kidney bean
фасонирование shaping, forming
фасонный shaped; moulded
фасциальный *anat.* fascial
фасциит *med.* fasciitis, fascitis
фасциола *zool.* fasciole; fasciola
~, печеночная liver fluke *(Fasciola hepatica)*
фасциолез *med., vet.* fascioliasis
фасцит = фасциит
фасция *anat.* fascia
~ бедра, широкая broad fascia
~, височная temporal fascia
~, жевательная masseteric fascia
~, орбитальная orbital fascia, periorbita
~, подколенная popliteal fascia
~, пояснично-грудная thoracolumbar fascia
фата-моргана fata morgana; mirage
фауна fauna
~, птичья avifauna
фаунистика faunology; zoogeography
фаут *wood.* flaw
фахак *zool.* toadfish, swellfish *(Tetrodon)*
фахверк framed structure/building, framework
фация facies
фашенпис *naut.* fashion piece/timber
фашизация Fascistization
фашизировать Fascistize, Fascisticize
фашизм Fascism
фашина fascine, faggot; brush
фашист, фашистский Fascist
фаэтон 1. phaeton; 2. *auto.* convertible, tourer, drop-head coupé; 3. *myth.* Phaeton; 4. *zool.* tropic bird *(Phaeton)*
фаянс faience, earthenware
~, делфтский delftware
~, известковый lime faience
~, санитарный sanitary ware/faience
~, художественный ornamental faience
феак Phaeacian
Феб *myth.* Phoebus
Феба *myth.* Phoebe
фебрилитет febris, fever
фебрильный febrile, feverish; pyretic
фебрифугальный febrifugal, antipyretic
фебрифугин *med.* febrifuge
февронианский *hist., eccl.* Febronian
фединг *rad.* fading
федия *bot.* African valerian *(Fedia cornucopiae)*
фейербахианец adherent of Ludwig Feuerbach
фейерверк firework(s); firecracker
фейхоа *bot.* feijoa *(Feijoa)*
фекалии feces, excrements, stools
фелиция *bot.* felicia *(Felicia)*
феллах fellah
феллема *bot.* cork tissue, phellem
феллодерма *bot.* phelloderm, secondary cortex
фелон *law* felon
фелония *law* felony, felonious act
фелука *naut.* feluc(c)a
фелонь chausible
фельдмаршал field marshal
фельдсвязь communication by courier
фельдфебель *obs.* sergeant major
фельдшер medical/doctor's assistant
фельдшпатоид, фельдшпатид *petr.* feldspathoid
фельдъегерь messenger, courier
фельетон feuilleton
фельзит *petr.* felsite
~, шаровой spherolithic felsite
фельзофир *petr.* felsophyre
фелюга, фелюка *naut.* feluc(c)a
Фемида *myth.* Themis
фен I hair drier, drying fan
фен II *chem.* phene
фен III *meteor.* foehn
фенадон *pharm.* methadone
фенакит *miner.* phenacite, phenakite
фенамин *pharm.* phenocoll, phenamine, aminoacetophenetidin, glycollphenetidin
фенантрахинон phenanthraquinone
фенантрен *chem.* phenanthrene
фенантролин *chem.* phenanthroline
фенацемид *chem.* phenacemide, phenylacetylurea
фенацетин *pharm.* acetophenetidin, phenacetin
фендлера *bot.* fendler bush *(Fendlera)*
фенек *zool.* fennec *(Fennecus zerda)*
фенианский *hist.* Fenian
фенианство *hist.* Fenianism
феникс 1. *myth.* Phoenix; 2. *astr.* Phoenix; 3. *bot.* date palm *(Phoenix dactylifera)*
фенил *chem.* phenyl
фенилаланин *bioch.* phenylalanine,

phenylaminopropionic acid
фениламин *chem.* phenylamine, phenylamide, aniline
фенилацетил *chem.* phanylacetyl
фенилгидразин *chem.* phenylhydrazine
фенилендиамин *chem.* phenylenediamine
фенилмасляный phenylbutyric
фенилмолочный phenyllactic
фенилсалицилат phenyl salicylate, satol
фенилсерный phenylsulphuric
фенилтиомочевина *chem.* phenylthiourea
фенилуксусный phenylacetic
фенилуретан phenylurethan(e), phenylethyl carbamate, ethylcarbamilate
фенит *miner.* fenite
фенитизация fenitization
феницит *miner.* phoenicite
фенобарбитал phenobarbital, phenobarbitone, luminal, phenylethylbarbituric acid
фенобарбитал-натрий phenobarbital/lumimal sodium, sodium phenylethyl-barbiturate
феногенетика *biol.* phenogenetics, developmental genetics
фенодем, фенодим phenodeme
фенодин phenodin, hematin
фенокопия phenocopy
фенокрист(алл) *geol.* phenocryst
феноксазин *chem.* phenoxazine
фенол phenol, phenyl alcohol, phenic/carbolic acid
  ~, жидкий liquefied phenol, liquefied carbolic acid
  ~, камфорный camphorated phenol, camphorated carbolic acid
фенолаза = фенолоксидаза
фенолит phenolic material, phenolite
фенолкислота phenolic acid
фенолмонобромат *chem.* monobromated phenol
феноловый *chem.* phenolic, phenol *(attr.)*
фенолог phenologist
фенология *biol.* phenology
фенолоксидаза *bioch.* phenol oxydase, tyronidase
фенолспирт *chem.* phenolic alcohol
фенолсульфонфталеин, фенолрот phenolsulphonphthalein, phenol red
фенолфталеин *chem.* phenolphthalein
фенольный phenolic
фенолят *chem.* phenolate, phenate
феном phenome
феномен phenomenon
феномный phenomic
фенопласт phenolic plastic, phenoplast
фенотиазин *chem.* phenothiazine
фенотип phenotype
фенофаза phenophase, phenological stage
фентоламин phentolamine hydrochloride, regitine hydrochloride
фенхель *bot.* fennel *(Foeniculum)*
феод *hist.* feud, fief
феодал feudal lord
феодальность feudality
феодальный feudal; feudalistic
феомеланин pheomelanin
феопласт *bot.* phaeoplast
феоспоровый *bot.* phaeosporous
феофилл *bot.* phaeophyll
феофитин *biochem.* phaeophytin
ферма I truss, girder; framework
  ~, арочная arched girder, arch truss
  ~, висячая suspension girder
  ~, консольная cantilever truss
  ~ лонжерона spar truss
  ~, одностоечная king-post truss
  ~, плоская truss web
  ~, раскосная girder frame
  ~, решетчатая trellice/lattice girder
  ~, строительная erection truss
  ~, стропильная principal
  ~, хвостовая *aero.* tail boom/truss
ферма II farm
  ~, арендованная tenant farm
  ~, животноводческая stock-raising/-rearing farm; livestock/animal farm; cattle ranch
  ~, звероводческая beast breeding farm, fur farm
  ~, коневодческая horse breeding farm
  ~, молочная dairy farm
  ~, мясо-молочная meat and milk production farm
  ~, нетоварная subsistence farm
  ~, овощеводческая vegetable farm
  ~, овцеводческая sheep-rearing farm
  ~, плодоводческая fruit-growing farm
  ~, птицеводческая poultry farm
  ~, растениеводческая crop-growing farm
  ~, свиневодческая hog/pig farm
  ~, товарная cash crop farm
  ~, товарно-молочная market-milk dairy
фермата, фермато *mus.* fermata
фермент *biol.* enzyme; ferment
  ~, дезаминирующий deamidizing enzyme
  ~, дыхательный respiratory enzyme
  ~, инвертирующий invertase, invertin, inverting enzyme
  ~, ингибирующий inhibitory enzyme, antienzyme
  ~, липолитический lipase, lipolytic enzyme
  ~, мочевой urease, urea enzyme
  ~, окислительный oxidase, oxidation enzyne
  ~, осахаривающий saccharifying enzyme
  ~, химический chemical ferment
ферментативный enzymatic, enzymic; fermentative
ферментатор fermenter
ферментационный fermentative; fermentation *(attr.)*
ферментация fermentation
  ~, полухроматидная halfchromatid fermentation
ферментировать ferment
ферментный enzymatic; fermental

ферментолиз enzymolysis
ферментология enzymology; fermentology
ферментопатический enzymopathic
ферментоподобный enzymoid
фермент-субстратный enzyme-substrate *(attr.)*
фермер farmer
~ **ведущий нетоварное хозяйство** subsistence farmer
~, **крупный** big farmer; rancher, ranchman
~, **мелкий** small farmer; yeoman
фермер-арендатор tenant farmer
фермер-животновод stock-farmer
фермий *chem.* fermium, Fm
фермион *phys.* fermion
фермовоз girder-/frame-carrier truck
фернамбук 1. *bot.* brazilwood *(Caesalpinia brasiliensis)*; 2. *chem.* brazilein
фернико *met.* fernico
ферония *bot.* wood apple *(Feronia)*
феррат *chem.* ferrate
ферратин *bioch.* ferratin, hematinic
ферредоксин ferredoxin
ферригемоглобин *bioch.* ferrih(a)emoglobin, methemoglobin
феррисол ferrisol
феррит *chem.* ferrite
феррит-гранат iron garnet
ферритный, ферритовый ferrite *(attr.)*
феррицитохром *bioch.* ferricytochrome
ферроалюминий ferroaluminium
ферробор ferroboron, iron boron
феррованадий ferrovanadium
ферровольфрам ferrotungsten
феррогранат ferrogarnet
феррокарт ferrocart
феррокремниевый ferrosilicon *(attr.)*
ферромагнетик ferromagnetic, ferromagnet, ferromagnetic material
ферромагнитный ferromagnetic
ферроманган, ферромарганец ferromanganese
ферросиликохром ferrochrome silicon
ферросиликоцирконий zirconium ferrosilicone
ферросилиций ferrosilicon
ферросплав ferroalloy
ферротипия *phot.* ferrotype
ферроторит ferrothorite
феррохелатаза ferrochelatase
феррохром ferrochromium, ferrochrome
ферроцианид ferrocyanide
ферроэлектрик, ферроэлектрический ferroelectric
фертилизин *bioch.* fertilizin
фертильный период childbearing span
фертоинг *naut.* mooring swivel
ферула *bot.* giant fennel *(Ferula)*
феруловый *chem.* ferulic
фестон 1. *arch.* festoon; 2. *met.* slag screen
фетальный fetal, foetal
Фетида *myth., astr.* Thetis
фетишизация fetishization
фетишизировать fetishize, make a fetish of
фетишизм fetishism
~, **денежный** fetishism of money
~, **товарный** fetishism of commodities
фетография *med.* f(o)etography; embryography
фетоскоп *med.* f(o)etoscope
фетр felt
фехраль ferro-aluminium high-resistance alloy
фехтование *sport* fencing
фея fairy
~ **Моргана** Fata Morgana
ФЗО *abbr.* industrial/factory training
ФЗУ *abbr.* factory/workshop education
фиакр fiacre, small coach/carriage
фиалка *bot.* violet *(Viola)*
фианкетто fianchetto *(in chess)*
фибра fibre, fiber
~, **листовая** fibreboard
фибремия fibremia, inosemia
фибрилла fibril, fibrilla
~, **мышечная** muscular fibril, myofibril
~, **эластиновая** elastinic fibril
фибриллярный fibrillar, fibrillary, fibrilate(d)
фибрин *bioch.* fibrin
фибринация *med.* fibrination, fibrinosis
фибриновый, фибринный fibrinous
фибриноген *bioch.* fibrinogen
фибриногенный fibrinogenic, fibrogenous
фибринозно-клеточный fibrinocellular
фибринолиз *bioch.* fibrinolysis
фибринообразующий fibrinogenic, fibrinogenous
фибринопластин fibrinoplastin
фибринопластиновый fibrinoplastic
фибрин-пептид fibrinopeptide
фибрин-фермент *bioch.* thrombin, plasmase
фиброангиома fibroangioma
фибробетон fibrous concrete
фиброгемосидерический fibrohemosideric
фиброглиома fibroglioma, inoglioma
фиброглия *anat.* fibroglia
фиброз *med.* fibrosis
~, **заместительный** replacement fibrosis
~ **легких** pneumososclerosis
фиброзит fibrositis, inositis, interstitial myositis
~, **узелковый** nodular fibrositis
~, **шейный** cervical fibrositis
фиброзно-жировый fibroadipose, fibrofatty
фиброзно-известковый fibrocalcerous
фиброзно-очаговый fibrofocal
фиброзный fibrous, fibrotic
фиброкистома fibrocystoma, inocystoma
фибролит 1. fibreboard, fibre mat; 2. *miner.* fibrolite
фиброма *med.* fibroma, inoma
~ **кожи** dermatofibroma
~ **матки** uterine fibroid tumour
~ **нерва** neurofibroma
фиброматоз fibromatosis
фибромиома fibromyoma, inomioma

фибромный fibromatous
фибромукозный fibromucoid
фиброневрома fibroneuroma, inoneuroma
фиброостеома fibro-osteoma
фиброполип *med.* fibropolypus
фибросаркома *med.* fibrosarcoma
фиброторакс fibrothorax
фиброхондрит *med.* fibrochondritis, inochondritis
фиброцит *anat.* fibrocyte, inocyte
фиброэластоз *med.* fibroelastosis
фиброэпителиома *med.* fibroepithelioma
фибула fibula; broach, clasp
фиванский, фивейский, фивский Thebaic, Thebaid; Theban
фига *bot.* fig *(Ficus carica)*
фигура figure, configuration; form, shape
  ~, вписанная inscribed figure
  ~ высшего пилотажа aerobatic manoeuvre
  ~, крупная *fig.* eminent/outstanding figure
  ~, описанная circumscribed figure
  ~, порошковая powder pattern/figure
фигуральный figurative; metaphorical
фигурант *theatre* super, extra; *(in ballet)* figurant
фигурист figure skater
фидель *mus.* fiddle
фидер feeder, transmission/feeder line; main; *agr.* feeder
фиджиец Fiji
фиджийский Fijian
фидуциальный fiducial; fiduciary
фидуциар *law* fiduciary
фидуциарный fiduciary; fiducial
физа *zool.* *Physa*
физалин physalin
физалис *bot.* ground cherry *(Physalis)*
физзарядка *abbr.* physical exercises
физиатр = физиотерапевт
физиатрия = физиотерапия
физиды *pl. zool.* pond snails *(Physidae)*
физик physicist
физика physics
  ~ атмосферы aerophysics, atmospheric physics
  ~, атомная atomic physics
  ~ атомного ядра nuclear physics
  ~, биологическая biophysics
  ~ высоких давлений high-pressure physics
  ~ высоких энергий high-energy physics
  ~ деления fission physics
  ~ Земли geophysics
  ~, квантовая quantum physics
  ~, космическая space/cosmic physics
  ~ космических лучей cosmic-ray physics
  ~ моря marine physics
  ~, нейтронная neutron physics, neutronics
  ~ твердого тела solid state physics
  ~, химическая chemical physics
  ~, электронная electron physics
  ~ элементарных частиц elementary-particle physics
  ~, ядерная nuclear physics
физик-атомник nuclear physicist
физико-математический physicomathematical
физико-морфический physicomorphic
физико-теология physicotheology
физико-химический physicochemical
физикохимия physicochemistry
физик-радиолог radiological physicist
физиогения *biol.* physiogeny
физиогномика = физиономика
физиограф physiographer
физиография physiography, physiographic geology
физиократ physiocrat
физиократия physiocracy
физиолог physiologist
физиологический physiologic(al)
  ~ раствор saline, physiological salt solution, sodium chloride solution
физиология physiology
  ~ животных animal physiology
  ~ органов чувств esthesiophysiology
  ~, патологическая physiopathology, pathologic physiology
  ~ развития developmental physiology
  ~ растений plant physiology, phytophysiology
  ~ человека human physiology
физиолого-анатомический physiologico-anatomical
физиономика, физиономистика physiognomy
физиономист physiognomist
физиономический physiognomical
физиономия physiognomy; facial expression, face
физиопатический physiopathic, physiopathologic
физиопатия physiopathy
физиопрофилактика physioprophylaxis
физиотерапевт physiatrist, physical therapist, physiotherapist
физиотерапевтический physiotherapeutic
физиотерапия physiotherapy, physiatrics
физиохирургия physiosurgery
физический 1. physical; corporeal; 2. physics *(attr.),* physical
  ~ труд manual labour
  ~ кабинет physics laboratory
  ~кая культура physical education; gymnastics
  ~кая химия physical chemistry
  ~кое лицо *law* physical person
физкультура physical education; gymnastics
  ~, лечебная remedial gymnastics, exercise therapy
физмат *abbr.* physicomathematical department
физогастрия *zool.* physogastry
физодермоз *bot.* physoderma disease, brown leaf spot
физокарпный physocarpous
физорг *abbr.* physical training organizer
физостегия *bot.* lion's-heart *(Physostegia)*
физостигма *bot.* Calabar bean *(Physostigma)*
физостигмин *chem.* physostigmine, eserine
физподготовка *abbr.* physical training
физрук *abbr.* physical training instructor

фикобилипротеин phycobiliprotein
фикоксантин phycoxanthin
фикология phycology, algology
фикомицет *bot.* phycomycete
фикопиррин phycopyrrin
фикотека phycotheca
фикофеин phycophaeine
фикохризин phycochrysin
фикоцецидий phycocecidium
фикоциан(ин) *bioch.* phycocyanin
фикоэритрин *bioch.* phycoerythrin
фиксаж fixative; *phot.* fixing agent/bath/solution
фиксатор fixing device/arm, lock, catch, stop, detent; index pin; latch
~ **переключения передач** gear lock
фиксация fixation, fixing; clamping; holding; immobilization
~, **внутрикостная** *med.* internal fixation
~ **лопатки** scapulopexy
~ **печени** hepatopexy
~ **селезенки** splenopexy, splenopexis
~, **толстой кишки** colonopexy, colopexy, colofixation
~ **уровня вооружений** weapons freeze
фиксировать fix, immobilize; clamp; hold; locate; lock, restrain, secure, make fast
фиктивность fictitiousness
фиктивный fictitious; unreal; dummy, fake
фикус *bot.* 1. rubber plant *(ornamental);* 2. India-rubber fig *(Ficus elastica)*
фикусоподобный *bot.* ficoid
фила 1. *hist.* phyle; 2. *biol.* phylum
филакобиоз *zool.* phylacobiosis
филактерия *rel.* phylactery
филамент filament
~, **ретикулиновый** reticulinic filament
~, **хвостовый** caudal filament; spermatic filament
филандер *zool.* four-eyed opossum *(Philander opossum)*
филант *zool.* bee-killer wasp *(Philanthus)*
филантроп philanthropist
филантропия philanthropy
филармония 1. philharmonic society; 2. philharmonic orchestra
филарх phylarch
филателия philately
филема *bot.* fileme
филенка panel; slat
филиал branch office; subsidiary; division; affiliated organization, affiliate
филиальный branch *(attr.);* subsidiary, affiliated, divisional
филигрань filigree
филин *zool.* eagle owl *(Bubo bubo)*
филиппика philippic
филиппинец Filipino
флиппинка Filipina
филиппинский 1. Phillipine; 2. Filipino
филирея *bot.* phillyrea *(Phillyrea)*
филистер, филистерский Philistine
филистерство Philistinism
филистимлянин, филистимлянский *hist.* Philistine
филицин *chem.* filicin, filicin anhydride, filicinic acid
филициновый filicinic, filicic
филия *eccl.* branch
филлад *bot.* phyllade
филлантус *bot.* leaf flower *(Phyllanthus)*
филлахороз *bot.* black blotch
филлидий *bot.* phyllidium
филлит *miner.* phyllite
филлитис *bot.* hart's-tongue *(Phyllitis scolopendrium)*
филлицин *chem.* phyllicin, theophylline-calcium salycillate
филлогенетический *bot.* phyllogenetic
филлодий *bot.* phyllode, phyllodium, phyllody
филлодоце *bot.* mountain heath *(Phyllodoce)*
филлокактус *bot.* leaf cactus, phyllocactus, epiphyllum *(Phyllocactus, Epiphyllum)*
филлокладий *bot.* phylloclade, leaflike branch
филлокладус *bot.* celery pine *(Phyllocladus)*
филлоксантин *biochem.* phylloxanthin, xanthophyll
филлоксера *zool.* phylloxera; root louse
филлом *bot.* phyllome
филлопирол phyllopyrrole
филлоподий *bot.* phyllopodium, phyllopode
филлопорфирин *bioch.* phylloporphyrin
филлориза phyllorhiza
филлостахис *bot.* phyllostachys *(Phyllostachys)*
филлостиктоз *bot.* phyllosticta leaf spot
филлотаксис *sot.* phyllotaxis, phyllotaxy, leaf arrangement
филлохинон *bioch.* phylloquinone
филлоэритрин *bioch.* phylloerythrin
филлярия *bot.* phyllary
филоварикоз filovaricosis
филогенез(ис) phylogenesis, phylogeny
филогенетика phylogenetics
филогенетический phylogenetic, phylogenic, phyletic
филогения phylogeny
~ **органов** organophyly
филодендрон *bot.* philodendron *(Philodendron)*
Филомела *myth.* Philomela
филоподия *zool.* filopodium
философский камень philosopher's stone
филотения *bot.* xanthosoma *(Xanthosoma lindensi)*
филоценогенез phylocoenogenesis
филум *biol.* phylum
фильдеперс Persian thread
фильера 1. *met.* die, draw plate; 2. *text.* spinneret(te)
~, **твердосплавная** carbide die
фильм film, motion picture; movies
~, **мультипликационный** animated film

~, **немой** silent film
~, **объемный** three-dimensional film
~, **озвученный** sound-track film
~, **полнометражный** full-length film
~, **хроникальный** newsreel
~, **художественный** feature film
~, **широкоэкранный** wide-screen film
фильмистор metal-film resistor
фильмоскоп film-strip projector
фильмотека film library
фильтр filter; strainer
~, **анионитовый** anion exchanger
~, **безнапорный** gravity-type filter
~ **верхних частот** *el.* high-pass filter
~, **взвешивающий** weighting filter
~, **воздушный** air filter/strainer/cleaner
~, **входной** inlet/entry filter
~, **выходной** output/exit filter
~, **гребенчатый** comb filter
~, **грубый** rough/coarse filter, strainer
~, **губчатый** sponge filter
~, **дискретный** discrete/quantized filter
~, **жидкостный** fluid filter
~, **заграждающий/запирающий** rejection/suppression filter
~ **индустриальных помех** radio-interference suppression filter
~, **ионообменный** ion-exchange filter
~, **капельный** drop/trickling filter
~, **катионовый/катионообменный** cation exchanger
~, **кварцевый** crystal/quartz filter
~, **конусный** wedge filter
~, **корректирующий** compensating/correcting filter; equalizer
~, **линейный** line filter
~, **мелкопористый** fine filter
~, **многозвенный** ladder-type filter; iterative filter
~ **нижних честот** low-pass filter
~, **обратный** inverse/reciprocal filter
~, **параллельный** bypass filter
~ **периодического действия** batch-type filter
~, **подстраиваемый** adjustable filter
~, **полосовой** band-pass filter
~, **порошкообразный** powdered filter
~, **противопомеховый** noise/interference filter
~, **прямоточный** straight-type filter
~, **пылеулавливающий** dust filter
~, **пьезокерамический** piezoelectric ceramic filter
~, **разделительный** isolation/separation filter
~, **резонансный** resonant/tuned filter
~, **ротационный** rotary filter
~, **рукавный** bag filter
~, **самонастраивающийся** adaptive filter
~, **СВЧ/сверхвысокочастотный** microwave filter
~, **сетчатый** gauze filter
~, **складчатый** plaitet/folded filter
~ **сточных вод** sewage filter
~, **твердотельный** solid-state filter
~, **толстослойный** deep-bed filter
~ **тонкой очистки** fine filter
~, **угольный** charcoal filter
~, **улавливающий** entrainment/trap filter
~, **фетровый** felt filter
~, **широкополосный** wide-band filter
~, **щелевой** (knife-)edge filter
~, **электростатический** electrostatic precipitator
фильтр-влагоотделитель filter-drier
фильтр-ловушка entrainment filter
фильтр-отстойник gravity filter
фильтр-поглотитель filter-absorber
фильтрат filtrate; percolate
фильтрация filtration, filtering; percolation
~, **клубочковая** glomerular filtration
~, **пространственная** multidimensional filtering
фильтровальный filter *(attr.),* filtering
фильтровать filtrate, filter, percolate, strain
фильтруемость filterability
фильц felt
филяриатоз, филяриоз *med.* filariasis
филярный filar, threadlike
фимиам incense
фимоз *med.* phimosis
фимозный phimotic
финагент *abbr.* revenue agent/inspector
финансирование financing
~ **капиталовложений** capital financing
~, **промежуточное** interim financing
финансист 1. financier; 2. financial expert
финансовый financial; finance *(attr.)*
~ **год** fiscal year
~ **инспектор** revenue inspector
~ **капитал** financial/banking capital
финвал *zool.* fin whale, finback *(Balaenoptera physalus)*
финигляциал *geol.* Finiglacial subepoch
финик 1. date (palm); date *(fruit)*
~, **морской** *zool.* peapod shell *(Botula)*
финикиец, финикийский *hist.* Phoenician
финикийка *hist.* Phoenician (woman)
финиковая пальма date palm *(Phoenix)*
фининспектор *abbr.* revenue/finance inspector
финишер finishing machine, finisher
финка Finn (woman), Finnish (woman)
финляндец Finlander
финляндский Finish
финн Finn
финна 1. pork measles *(Cysticercus cellulosae);* 2. pork/armed tapeworm *(Taenia solium)*
финно-венгерский Finno-Hungarian
финноз *med.* measles
финнозный measly
финно-угорский Finno-Ugric, Finno-Ugrian
финно-угроведение Finno-Ugrian studies
финно-угры *pl.* Finno-Ugrians

финотдел *abbr.* finance department
финплан *abbr.* financial plan
финсектор financial sector
финский Finnish, Finnic
финт *sport* feint; trick
финта *zool.* shad *(Alosa fallax)*
фиоль *bot.* wallflower *(Cheiranthus cheiri)*
фион phyone, growth hormone
фиорд fiord, fjord
фиоритура *mus.* fioritura, grace
фиппсия *bot.* icegrass *(Phippsia)*
фирма firm, business house; company; concern
~, акционерная incorporated concern
~, брокерская brokerage house/office; commission company
~, дочерняя subsidiary
~, консультативная professional consultants
~, маклерская commission company; brokerage house
~, неакционерная unincorporated concern
~, оптовая distributor
фирман *hist.* firman
фирменный знак trademark
фирн firn, névé
фирнизация firnification
фисгармония, фисгармоника *mus.* harmonium, physharmonica
фиск *fin.* fisc, state treasury
фисташка *bot.* pistache, pistachio tree *(Pistacio)*
фистула I *med.* fistula
~ печени hepatic/biliary fistula
~ почки renal fistula
фистула II *mus.* falsetto
фистулезный fistulose
фистульный fistulous
фистулярный fistular, fistulous
фитаза *bioch.* phytase
фителефас *bot.* ivory palm *(Phytelephas)*
фитеума *bot.* mixed flower *(Phyteuma)*
фитин *biochem.* phytin, phytate
фитинг pipe fitting, adapting/junction piece
~, переходный adapter; bushing
фитобезоар *med.* phytobezoar
фитобентос *bot.* phytobenthos
фитогенез phytogenetis, phytogeny
фитогенетика phytogenetics, plant genetics
фитогенный phytogenous, phytogenic
фитогеография phytogeography, plant geography
фитогормон phytohormone, plant hormone
фитография phytography, descriptive botany
фитоз *med.* phytosis
фитозавры *paleont.* phytosaurs, phytosaurians (*Phytosauria)*
фитоксилин phytoxylin
фитол *bioch.* phytol
фитолакка *bot.* pokeberry *(Phytolacca)*
фитология phytology; botany
фитономия phytonomy
фитономус *zool.* alfalfa weevil *(Phytonomus)*
фитонцид phytoncide
фитопалеонтология paleobotany, paleophytology
фитопатолог phytopathologist
фитопатология phytopathology, plant pathology
фитоплазма phytoplasm, plant protoplasm
фитопланктон phytoplankton
~, сетный net phytoplankton
фитостерин *chem.* phytosterol, phytosterin, phytostearin, phytocholesterol
фитосфингозин phytosphingosine
фитотерапия phytotherapy
фитотоксин phytotoxin, plant/vegetable toxin
фитотомия phytotomy, plant anatomy
фитотрихобезоар phytotrichobezoar, trichophytobezoar
фитотрон phytotron, climatizer
фитофаг *zool.* phytophagan
фитофармакология phytopharmacology
фитофизиология phytophysiology, plant physiology
фитофильный phytophilous
фитофтора *bot.* phytophthora *(Phytophthora)*
фитофтороз *bot.* phytophthora rot, late blight, gummosis
фитохимия phytochemistry
фитохория phytochory
фитохром *bot.* phytochrome
фитоценогенез phytocoenogenesis
фитоценоз phytocoenosis, plant community/association
фитоценология phytocoenology, phytosociology
фитоценотип phytocoenotype
фитоцид phytocide
фифи *zool.* wood sandpiper *(Tringa glareola)*
фихтеанец, фихтеанский Fichtean
фихтеанство Fichteanism
фиш *naut.* fish
фиш-балка *naut.* fish-davit
фиш-блок *naut.* fish-block
фиш-гак *naut.* fishhook
фишка peg, plug; *el.* connector
фиш-тали fish tackle
флаван flavan
флаванолон flavanolon(e)
флаванон flavanon(e)
флавенсомицин flavensomycin
флавин *chem.* flavin(e)
флавин-фермент flavoenzyme
флавокиназа *biochem.* flavokinase
флавопротеид, флавопротеин flavoprotein
флаг flag; banner; *mil.* colours, colors, *naut.* ensign
~ бедствия distress flag
~, брандвахтенный guard flag
~ Великобритании, военноморской White Ensign
~ Великобритании, государственный Union Flag
~, военно-морской man-of-war flag
~ заднего хода astern flag
~, кормовой ensign

~, лоцманский pilot flag
~ на гафеле *naut.* flag at the peak
~, парламентерский flag of truce
~, приспущенный flag at half-mast
~ США, военноморской United States ensign
~, "удобный" flag of convenience
флагдук, флагдух bunting, flag cloth
флагелла flagellum
флагеллиспора *biol.* flagellula
флагман *naut.* flagship
флаг-фал *naut.* halyard
флагшток flagstaff, flagpole
флажок small flag, hand flag
флажолет *mus.* flageolet
флакон flask, vial, phial, bottle
фламандец, фламандка Fleming
фламандский Flemish
фламбирование flaming, flame sterilization/treatment
фламинго *zool.* flamingo
фланг flank; wing
фланговый flank *(attr.)*; flanking
фланец flange; collar
~, глухой blind/blank flange
~, закрытый dead flange
~ крепления mounting flange
~, наварной/приварной welded flange
~, опорный bearing flange
~, открытый live flange
~, перестановочный/установочный adjusting flange
~, переходный adapter flange, adapter
~, соединительный connecting/coupling flage
фланжировать, фланцевать flange
фланцевый flange *(attr.)*; flanged, collared
флат flat/sheet paper
флаторезка *print.* sheeting machine
флаттер flutter; vibration
~ на гиперзвуковых скоростях hypersonic flutter
~, срывной stalling flutter
флебит *med.* phlebitis
~, гнойный suppurative phlebitis
~, подагрический gouty phlebitis
~, послеродовой puerperal phlebitis, white/milk leg
флеболиз *med.* phlebolysis
флеболит *med.* phlebolith
флеболитиаз *med.* phlebolithiasis
флеборафия phleborrhaphy
флеборексис phleborrhexis
флеборрагия phleborrhagia, venous h(a)emorrhage
флеботромбоз phlebothrombosis
флебэктазия *med.* phlebectasia
флебэктомия *med.* phlebectomy
флебэктопия *med.* phlebectopia, phlebectopy
флегма *chem.* reflux; *med.* phlegm
флегматизатор *chem.* retarder, deterrent; phlegmatizer; stabilizer
флегмона *med.* phlegmon
~, гнилостная putrefactive phlegmon
флейта flute
флейтист flutist
флейторыл *zool.* trumpet fish *(Aulostomus maculatus)*
флейц wide paintbrush, flat
флексионный *gram.* inflexional
флексия 1. flexion, flection; 2. *gram.* inflection
~, дорсальная *anat.* dorsiflexion
~, подошвенная plantar flexion
флективно-аналитический *gram.* inflectional-analytical
флективность *gram.* inflectedness
флективный *gram.* inflected, inflectional
флектировать *gram.* inflect
флеоспороз *bot.* phleospora leaf spot
флеп flap
флерница *zool.* chrysop *(Chrysopa)*
флерон *arch.* fleuron, poppyhead
флигель *arch.* wing; outbuilding
флигель-адъютант *mil.* aide-de-camp
фликтена *med.* phlyctena
фликтеноз *med.* phlyctenosis, phlyctenulosis
фликтенозный *med.* phlyctenar, phlyctenous
фликтенула *med.* phlyctenule, phlyctenula
фликтенулезный phlyctenular
флинт *miner.* flint
флинтглас flintglass, lead glass
флиш *geol.* Flysch
флобафен *chem.* phlobaphene
флогопит *miner.* phlogopite, amber mica
флогоцитоз *med.* phlogocytosis
флокен *met.* flake
флоккулировать flocculate
флоккулянт flocculant
флокс *bot.* phlox *(Phlox)*
фломастер felt-/soft-tip pen
флор *naut.* floor
~, переборочный bulkhead floor
флора flora; vegetation
~ и фауна biota
~ и фауна морского дна benthos
~, кишечная intestinal flora
~, культурная cultivated flora
флорентийка Florentine (woman)
флорентинец, флорентинский Florentine
флоресценция florescence
флороглюцин *chem.* phloroglucin, phloroglucinol, phloroglucol
флортимберс *naut.* floor timber, ground/first/lower futtock
флот fleet; marine; navy
~, атомный nuclear-powered fleet
~, военно-морской navy, war fleet
~, морской marine, sea fleet
~, морской торговый maritime fleet, sea-going merchant navy
~, океанский ocean-going fleet
~, оперативный task/operational fleet
~ резервный idle fleet, moth-ball fleet

~, **рыболовный** fishing fleet
~, **торговый** merchant marine/navy/fleet/shipping, mercantile navy/marine
~, **транспортный** cargo fleet
флотация flotation
~, **беспенная** nonfrothing flotation
~, **выборочная/избирательная/селективная** selective/differential flotation
~, **грубая/основная** rough flotation
~, **коллективная** bulk flotation
~, **окисленная** oxide flotation
~, **пенная** foam/froth flotation
~, **пневматическая** air flotation
флотилия flotilla
флотировать subject to flotation, float
флотский fleet *(attr.)*
флоэма *bot.* phloem, bast tissue
флуктуация fluctuation
~, **беспорядочная** random fluctuation
~, **броуновская** Brownian fluctuation
~ **потерь энергии** energy-loss fluctuation
~, **случайная/хаотическая** random fluctuation
флуктуировать fluctuate
флуоресцентный, флуоресцирующий fluorescent
флуоресценция fluorescence
~, **сверхизлучительная** superradiant fluorescence
~, **ударная** impact fluorescence
флуоресцировать fluoresce
флуорограф photofluorograph, photoroentgenograph
флуороскоп fluoroscope
~, **просвечивающий** transmission fluoroscope
~, **рентгеновский** X-ray fluoroscope
флювиометр fluviometer
флюгарка 1. *naut.* badge, (ship's) emblem; 2. wind vane
~ **датчика** transmitting wind vane
флюгельгорн *mus.* flugelhorn
флюгер weather/wind vane, weathercock
флюгирование feathering
флюидизированный fluidized
флюкс flux; flow
флюксметр fluxmeter, gaussmeter, magnetic-flow meter
флюктуация = флуктуация
флюктуировать = флуктуировать
флюмазин flumethiazide, ademol
флюоресцентный fluorescent
флюоресценция = флуюресценция
флюоресцировать fluoresce
флюорид *chem.* fluoride
флюоридизация fluoridization
флюоризация fluoridation, fluoridization
флюорограф photoroentgenograph, photofluorograph, fluorograph
флюорография photoroentgenography, radiophotography
флюороз *med.* fluorosis
флюс 1. flux; 2. *med.* alveolar abscess, gumboil
~, **высококремнистый** high-silica flux
~, **глиноземистый** alumina flux
~ **для цинкования** galvanizing flux
~, **кремнеземистый** silica flux
~, **сварочный** welding flux
флюсование, флюсовка fluxing
флютбет flood bed, spillway, apron
фляга flask, can; container; *mil.* water bottle
фляжолет = флажолет
флянец = фланец
фоб *econ.* free on board, f.o.b.
фовеальный foveate(d), pitted
фойникохроит *miner.* phoenicochroite, phoenicite
фок *naut.* foresail
~, **гоночный** balloon foresail
~, **косой** jib foresail
фока- брас forebrace
фока-ванты fore-lower rigging
фока-галс fore tack
фока-гик fore-boom
фока-рей fore yard
фока-стаксель fore-staysail
фока-шкот foresail sheet, foresheet
фока-штаг fore stay, headstay
фок-мачта foremast
фокстерьер fox terrier
фок-топсель *naut.* fore-topsail
фокус focus; focal point
~, **главный** principal focus, focal point
~, **задний** rear/back focus
~, **мнимый** apparent/virtual focus
~, **острый** fine focus
~, **точечный** point focus
фокусировать focus; focalize; bring in focus
фокусировка focus(s)ing; focalization
~, **жесткая** strong focussing
~, **знакопеременная** alternating-gradient focussing
~ **на бесконечность** infinity focussing
~ **по направлению** directional focussing
~, **по скорости** velocity focussing
~, **резкая** sharp focussing
~, **спиральная** helical focussing
~, **точечная** point focussing
~, **фазовая** phase focussing
фол *sport* foul
фолада *zool.* European piddock *(Pholas dactylus)*
фолклендский Falkland *(attr.)*
фолликул *anat.* follicle
~ **волоса** hair follicle
~, **лимфатический** lymph follicle
фолликулез folliculosis
фолликулит *med.* folliculitis
фольварк farm; grange
фольга foil
~, **алюминиевая** aluminum foil
~, **золотая** gold foil/leaf
~, **контрольная** monitoring foil
~, **ленточная** foil tape

~, **чистая** blank foil
фольклор folklore
фольклористика folklore studies, study of folklore
Фома Thomas
~ **неверный** *colloq.* doubting Thomas
фон 1. background; 2. hum, background noise; 3. *phys.* phon
~ **гамма-излучения** gamma background
~, **звездный** stellag/star background
~, **инструментальный/прибора** instrument background
~, **радиоактивный** radiation background
~, **рассеянный** diffuse background
~, **стационарный** steady background
~ **шумов** noise background
фонареглазые *n. pl. zool.* lantern eye fishes (*Anomalopidae)*
фонарик small lamp/lantern
~, **электрический** flashlight
фонарница *zool.* lantern fly (*Fulgora)*
фонарь 1. lantern, lamp; light; 2. *av.* canopy, hood
~, **батарейный** flashlight
~, **боковой/габаритный** side-lamp/-light marker light
~, **взрывобезопасный** safety light
~ **заднего хода** *auto* back-up light
~, **задний** tail/rear lamp; *av.* rear canopy
~, **зенитный** garret window, clerestory
~ **кабины** *av.* cockpit canopy
~, **карманный** flashlight
~, **проблесковый** flasher
~, **проекционный** projector
~, **рудничный** miner's lamp
~, **сбрасываемый** jettisonable hood/canopy
~, **сигнальный** call/flashing lamp, signal lantern, blinker tube/device/gun
~, **стояночный** parking light/lamp
~, **стрелочный** *rail.* switch lamp
фонастения *med.* phonasthenia, functional vocal fatigue
фонация phonation
~, **ослабленная** hypophonia, subenergetic phonation
~, **усиленная** hyperphonia, superenergetic phonation
фонд fund; stock; resources; *see also* **фонды**
~, **амортизационный** depreciation fund
~, **благотворительный** endowment/welfare fund
~ **валютного регулирования** exchange stabilization fund
~, **возобновляемый** revolving fund
~ **выдачи пособий** relief fund
~, **выкупной** sinking-fund reserve
~ **выравнивания** equalization fund; exchange stabilization fund
~, **жилищный** (amount of) housing, housing stock/resources, residential housing
~ **заработной платы** wage fund; payroll
~ **зданий** stock of buildings; buildings available
~, **земельный** land(s) available
~, **золотой** gold reserves
~ **капиталовложений** investment fund
~ **личного потребления** personal consumption fund
~, **Международный валютный** International Monetary Fund, IMF
~ **накопления** accumulation fund
~, **оборотный** working capital
~, **основной** capital fund
~, **пенсионный** superannuation fund
~ **погашения** sinking/redemption fund
~ **пожертвования** endowment/welfare fund
~ **потребления** consumption fund
~, **премиальный** bonus fund
~ **рабочего времени** working time fund
~, **рабочий** labour fund
~, **резервный** *fin.* reserve; emergency/reserve fund
~, **семенной** seed stock
~ **социального обеспечения** social security fund
~, **уравнительный** equalization fund, exchange stabilization fund
фондовая биржа stock exchange; stock market
фондовооруженность (одного рабочего) fixed capital per worker
фондоемкий capital-intensive
фондоемкость capital intensity/coefficient; capital-output ratio
фондоотдача output-capital ratio, capital productivity (ratio)
фонды funds; stock; assets; resources; capital; *see also* **фонд**
~, **замороженные** frozen capital/funds
~, **кредитные** credit resources
~ **материальных активов** tangible capital stock
~, **неделимые** nondistributable assets
~, **оборотные** working/current capital
~, **основные** fixed assets/capital
~, **основные производственные** basic production assets, fixed capital stock, fixed assets
~, **основные промышленные** fixed industrial assets
~ **открытого типа, инвестиционные** open-end funds
~, **переходящие** carry-over assets
~ **потребления, общественные** public/social consumption funds
~, **производственные** productive/production assets
фонема *ling.* phoneme
фонематика phonematics
фонематический phonemic, phonematic
фонемика, фонемология phonemics
фонетист phonetician
фониатрия *pathol.* phoniatrics, phoniatry
фонограмма phonogram(me), sound recording
фонолит *miner.* phonolite

фонолог phonologist, phonemicist
фонологизация phonologization, phonemicization
фонология phonology; phonemics
фономашинистка audiotypist
фонометр phonometer, acoustimeter
фонон *phys.* phonon, lattice-vibration quantum
фонопсия *med.* phonopsia
фонотека *mus.* record library; tape library
фонтан spout, spurt, fountain
~, газовый gas blowout
~, нефтяной oil gusher
фонтиналис *bot.* water moss *(Fontinalis)*
форамен *geol.* foramen
фораминифера *zool.* foraminifer
~, двухрядно-однорядная biserial-uniserial foraminifer
~, прободенная perforate foraminifer
форвакуум initial vacuum, forevacuum
форвард *sport* forward
фордевинд *naut.* wind astern/aft, stern wind
фордун *naut.* backstay
форез phoresis
форезия *zool.* phoresy
форель *zool.* trout *(Salmo)*
~, радужная rainbow trout
~, ручьевая brook trout
форзац *print.* end sheet/leaf/paper; flyleaf
форзиция *bot.* forsythia *(Forsythia)*
форинт forint
форкамера precombustion chamber, injection/burner cup, prechamber, premixing cavity
форкиль *av.* dorsal fin
фор-люк *naut.* forehatch
форма 1. form, shape; configuration; pattern; 2. *met.* mould, cast; 3. uniform; 4. blank
~, билинейная bilinear form
~ волны wave form
~, высказывательная propositional/sentence form
~ заказа ordering form
~ записи *math.* notation
~, литейная casting/foundry mould
~, литейная глиняная/земляная loam mould
~, машиночитаемая machine-readable form
~, металлическая (литейная) metal/chill mould, chill
~, многократная (литейная) permanent mould
~, нормальная *math.* normal/canonical form
~, обтекаемая aerodynamic/streamline form
~, парадная *mil.* full-dress uniform
~ печати *comp.* printer layout
~, печатная printing plate
~, печатная текстовая/шрифтовая type form
~, полилинейная *math.* multilinear form
~, поршневая (литьевая) plunger mould
~, продолговатая/сигарообразная prolate form
~, разовая (литейная) temporary mould
~ распада mode of disintegration/decay
~, стержневая (литейная) core mould
~, сырая green-sand mould
~ типа меню *comp.* menu format
~, тупиковая *log.* irredundant form
~, цифровая digital form
~, эрмитова *math.* Hermitian form
формализация formalization
формализировать formalize
формализм *math.* formalism
~, волновой wave formalism
~, гамильтоновский Hamiltonian formalism
~, корпускулярный particle formalism
~, лагранжевский Lagrangian formalism
~, матричный matrix formalism
формалин formol, formalin
формалистичность formalism
формальдегид *chem.* formaldehide, formic aldehyde, methyl aldehyde
формальность formality; procedure; ceremony
~,таможенная costom-house formality
форманта *phon.* formant
фор-марсель *naut.* fore-topsail
формат size; form; *comp.* format
~, выходной output format
~ данных data format
~ записи record format
~ издания format
~ информационного слова data word format
~, неупакованный nonpacked format
~, полиэкранный split-screen format
~, упакованный packed format
форматирование formatting
~ изображения на экране screen formatting
формирование forming, formation, shaping
формация formation; structure
~, гипокампусная hippocampal formation
~, растительная plant formation; phytium
формикарий ant hill, formicary
формил *chem.* formyl
формилирование *chem.* formylation
формирование forming, formation, shaping
~ адреса *comp.* address generation
~ задания *comp.* job setup
~ изображения picture generation
~ импульсов pulse shaping/forming
~ кадра framing
~ поездов making-up of trains
~, предварительное preshaping
~ файла file composition
~ цепочки chaining
формирователь shaper, former; reshaper; shaping/forming unit; *comp.* driver
~, адресный address driver
~ записи write driver
~ импульсов pulse former
~, разрядный digit driver
~ сигналов signal conditioner
~ сигналов записи write driver
~ сигналов считывания read driver

~ **тактовых импульсов** clock driver, timing-pulse distributor
~, **шинный** bus driver
формировать form, shape; set up
~ **поезд** marshal a train, make up a train
формицин phormycin, formaldehyde acetamide
формование forming, formation, shaping; moulding
~, **выдувное/раздувом** blow moulding
~ **заливкой** slush moulding
~ **маканием/окунанием** dip moulding
~ **спеканием** sinter moulding
~ **шприцгуссом** extrusion moulding
~, **электролитическое** electroforming
формовать mould; shape
формовка *met.* moulding; shaping
~ **в глине** loam/clay moulding
~ **в опоках** flask/box moulding
~ **в песчано-глинистой смеси** sand moulding
~ **в стержнях** core moulding
~, **опочная** flask/box moulding
~ **по сухому** dry sand moulding
~ **по сырому** green sand moulding
~ **по шаблону** template moulding
~, **почвенная/ямная** pit moulding
формовочный moulding, casting; shaping
формоловый, формольный formol *(attr.)*
формообразование 1. forming, shaping; 2. *biol.* morphogenesis; 3. *gram.* inflection
формообразовательный formative; *gram.* formbuilding, inflectional
формуемость mouldability
формула formula; formulation; *math.* equation
~ **Бейеса** Bayes formula
~ **Бернулли** Bernoulli distribution
~ **в явном виде** explicit formula
~, **валовая** *chem.* molecular formula
~ **Гаусса** Gauss formula
~ **де Бройля** de Broglie relation/equation
~, **интерполяционная** interpolation equation
~, **исходная** assumption formula
~, **итерационная** iterative formula
~, **колесная** *eng.* wheel arrangement
~, **квадратурная** *math.* quadrature formula
~, **кубатурная** *math.* cubature formula
~ **крови** *med.* blood count
~, **лейкоцитарная** *med.* differential blood count
~, **опытная** empirical/experimental formula
~, **приближенная** approximation/approximate formula
~, **пропозициональная** sequential/propositional formula
~ **прямолинейной зависимости** straight-line formula
~ **размерности** dimensional formula
~, **расчетная** design formula
~ **строения/структурная** structural formula
~ **трапеций** trapezoid(al) rule/formula
~ **усреднения** averaging formula
~ **Френеля** Fresnel's reflection formula
~, **химическая** chemical composition formula
формулировка formulation, formulating; wording
формуляр log sheet/book, record book/sheet; data card; form; blank; library card
формфактор *phys.* formfactor, shape factor
форонида *zool.* phoronid
фороцит phorocyte
фороцитоз phorocytosis
форпик *naut.* forepeak
форполимер *chem.* prepolymer
форпост *mil.* advanced post; outpost
форсаж 1. intensification, power augmentation, boosting; acceleration; 2. reheat(ing), afterburning
форсирование forcing, boosting, augmentation; overload operation; *mil.* forced crossing
форсировать force, boost; augment; (reactor) spike; (engine) reheat, boost; *mil.* force
форсировка forcing; boosting
форсиция = форзиция
форс-мажор force majeure, act of God, (force of the) elements
фор-стеньга fore-topmast
форсунка fuel injector/atomizer/sprayer; injecting/injection/jet nozzle; burner
~, **вихревая** vortex/swirl injector, atomizing/spray/swirl nozzle
~, **газовая** gas burner
~ **горючего** fuel nozzle
~, **двухкомпонентная** bipropellant/doublet injector
~, **двухсопловая** dual injector
~, **мазутная** fuel atomizer/nozzle
~, **нефтяная** oil atomizer/burner
~, **однокомпонентная** monopropellant injector
~, **пароструйная** steam jet injector/atomizer
~, **смесительная** mixing nozzle
~, **топливная** fuel spray nozzle, fuel burner
~, **штифтовая** pintle nozzle
~, **щелевая** slot atomizer, split jet injector
форте *mus.* forte
фортепьяно *mus.* piano
фортификация *mil.* fortification
~, **долговременная** permanent fortification
форточка ventilation/openable window, vent
ФОРТРАН FORTRAN
фортуна *myth.* Fortuna
форштевень *naut.* stem, stempost
~, **брусковый** bar stem
~, **листовой** plate/soft stem
~, **наклонный** raking/overhung stem
форцепс *med.* forceps
форшлаг *mus.* grace note
фосген *chem.* phosgene, carbonyl chloride, carbon oxychloride
фосса *zool.* fossa, foussa *(Cryptoprocta ferox)*
фоссильный fossil *(attr.)*
фосфаген *chem.* phosphagen, creatine phosphate
фосфамид *chem.* phosphoamide
фосфат *chem.* phosphate
~ **аммония** ammonium phosphate

~ **калия** potassium phosphate
~, **коллоидный** colloidal/bone phosphate
~, **минеральный** inorganic phosphate
~, **плавленый** fused phosphate
~ **свинца** lead phosphate
фосфатаза *biochem.* phosphatase
фосфатдиабет *med.* phosphatic diabetes
фосфатид *bioch.* phosphatide, phospholipid
фосфатидил phosphatidyl
фасфатидилглицерин phosphatidylglycerol
фосфатидилсерин phosphatidylserine
фосфатидный phosphatidic
фосфатирование phosphating, phosphatization, phosphatizing; phosphate treatment/coating
фосфатировать phosphatize
фосфатный phosphate *(attr.),* phosphatic
фосфатурия *med.* phosphaturia, phosphuria
фосфен *physiol.* phosphene
фосфид *chem.* phosphide
фосфин phosphine
фосфит phosphite
фосфоамидаза phosphoamidase
фосфоаргинин phosphoarginine
фосфогексоизомераза phosphohexose isomerase
фосфогексокиназа phosphohexokinase, phosphofructokinase
фосфогидролаза phosphohydrolase
фосфогипс phosphogypsum
фосфоглицерид phosphoglyceride
фосфоглицерил phosphoglyceryl
фосфоглицериновый phosphoglyceric
фосфоглицеромутаза phosphoglyceromutase, glycerate phosphomutase
фосфоглюкомутаза phosphoglucomutase, glucose phosphomutase
фосфоглюконат phosphogluconate
фосфоглюконовый phosphogluconic
фосфоинозитид phosphoinositide
фосфокетолаза phosphoketolase
фосфокетопентоэпимераза phosphoketopentoepimerase
фосфокиназа phosphokinase
фосфокреатин phosphocreatine, creatine phosphate
фосфолипид phospholipid, phosphatide
фосфоний phosphonium
фосфопантотеновый phosphopantothenic
фосфопентоза phosphopentose isomerase
фосфопептон phosphopeptone
фосфопиридоксаль pyridoxal phosphate
фосфопротеид, фосфопротеин phosphoprotein
фосфор 1. *chem.* phosphorus, P; 2. phosphor, luminophor
~, **пурпуровый** purple/red phosphorus
~, **пятихлористый** phosphorus pentachloride
~, **треххлористый** phosphorus trichloride
фосфоргидроз phosphorhidrosis, phosphorescent sweating
фосфоресценция phosphorescence
~ **моря** sea-fire
фосфоресцировать phosphoresce
фосфоресцирующий, фосфоресцентный phosphorescent
фосфорибозил phosphoribosyl
фосфорилаза phosphorylase
фосфорилировать phosphorylate
фосфорилирование phosphorylation
~, **окислительное** oxidative phosphorylation
фосфорилхолин phosphorylcholine
фосфористый phopshorous
фосфорноватистокислый hypophosphite (of); hypophosphorous
фосфорноватистый hypophosphorous
фосфорнокислый phosphate (of); phosphatic, phosphated
~ **кальций** calcium phosphate
фосфорный phosphoric, phosphorous
фосфорография phosphor printing, phosphorophotography
фосфорорганический organophosphorous
фосфосфингозид phosphosphingoside
фосфотрансфераза phosphotransferase
фосфофераза phosphopherase
фосфофермент phosphoenzyme
фосфоэнолпировиноградный phosphoenolpyruvic
фот *opt.* phot, centimeter-candle, ph
фотальгия *med.* photalgia, photodynia
фотарий *radiol.* photarium, radiation therapy room
фотикон *telev.* photicon
фотоавтомат coin-operated photograph booth
фотоанализатор photoanalyzer
фотоаналитик photoanalyst
фотоаппарат (photographic) camera
~, **зеркальный** reflex camera
~, **малоформатный** compact/miniature camera
~, **микроформатный/миниатюрный** subminiature camera
~, **регистрирующий** camera recorder, photorecorder
~, **складной** folding camera
~, **широкопленочный** roll-film camera
фотоаппаратура photographic equipment
фотоателье photographer's studio
фотобатерия photoelectric battery
~, **солнечная** solar-cell battery
фотобачок *phot.* developing tank
фотобомба photoflash bomb, flash/photo bomb
фотобумага photographic paper
~, **высокочувствительная** fast paper
~, **глянцевая** glossy paper
~, **контрастная** contrast paper
~, **малочувствительная** slow paper
~, **матовая** matt paper
~, **сверхконтрастная** high-contrast paper
~, **цветная** colour photographic paper
фотовозбуждение photoexcitation
фотовосстановление photoreduction
фотовспышка flashlight, flashtube; flash bulb

фотогальванический photovoltaic
фотогелиограф photoheliograph
фотогидролиз photohydrolysis
фотоглифия photoglyphy
фотоголовка photoelectric head
фотогравировальный photoengraving
фотогравюра photogravure, photoengraving, photoglyph
фотограмметрия *surv.* photogrammetry, photographic survey
  ~, наземная terrestrial photogrammetry
фотограф photographer
фотографизм photographic accuracy
фотографирование photography, photographing
  ~, воздушное aerial/air photography
  ~, маршрутное воздушное strip aerial photography
  ~, подводное underwater photography
фотографировать photograph, take a picture of
фотография photography
  ~, дистанционная telephotography
  ~, моментальная snapshot photography
  ~, теневая shadowgraph
  ~, художественная artistic photography
  ~, цветная colour photography
  ~, черно-белая black-and-white photography
фотодатчик photosensor
фотодейтрон photodeutron
фотоделение photofission
фотодерматит, фотодерматоз photodermatitis, photodermatosis, photodermia
фотодиагноскоп *med.* photodiagnoscope
фотодинамика *bot.* photodynamics
фотодиод photodiode
фотодиссоциация photodissociation, photolysis
фотодисфория *med.* photodysphoria
фотодокумент documentary photograph; photograph of a document
фотодонесение *mil.* photo-intelligence report, reconnaissance photograph
фотодромия photodromy
фотодыхание photorespiration
фотожурналист newspaper photographer
фотозапись photographic record; protorecording
фотозатвор camera shutter
фотозвезда *astr.* photostar
фотозвукопродуктор photographic sound recorder
фотоизображение fascimile
фотоимпульс photoimpact
фотоисточник photosource
фотокалька paper negative
фотокамера (photo) camera, still camera
  ~, скоростная high-speed camera
  ~, съемочная surveying camera
фотокарта photographic map, photomap
фотокассета camera cassette/magazine
фотокатализ photocatalysis
фотокатализатор photocatalyst
фотокатод photocathode, photoelectric/photosensitive cathode
фотокератоскоп photokeratoscope
фотокимограф photokymograph
фотокинез photokynesis
фотокинопулемет aerial camera-gun
фотокинотеодолит cinetheodolite
фотоклеистогамия *bot.* photocleistogamy
фотокомната *phot.* dark room
фотокопирование photocopying
фотокорреспондент camera reporter; press photographer
фотоксилин photoxylin
фотолиз photolysis, photodissociation
фотолитография photolithography
фотолюбитель amateur photographer
фотолюк camera window/hatch
фотолюминесценция photoluminescence
фотолюминофор photoluminophor
фотомезон *phys.* photomeson
фотометр photometer
  ~, зрительный/визуальный visual photometer
  ~, мерцающий/мигающий flicker photometer
  ~, рентгеновский X-ray photometer
  ~, теневой shadow photometer
  ~, трубчатый tube photometer
  ~, шаровой sphere/globe photometer
фотометрирование photometric measurement, photometry
фотометрировать scan photometrically
фотометрия photometry
  ~, звездная stellar photometry
  ~ пламени/пламенная flame photometry
  ~, рентгеновская X-ray photometry
  ~, субъективная/зрительная visual photometry
фотомиоклонус *med.* photomyoclonus
фотомонтаж composite photograph, photomontage, mosaic assembly
фотон *phis.* photon
  ~, падающий incident photon
  ~ распада decay photon
  ~, рассеянный scattered photon
фотонабор phototypesetting, photosetting, photocomposition
  ~ная машина photocomposer
фотонастия *bot.* photonasty
фотонегатив photographic negative
фотонейтрон *phys.* photoneutron
фотообъектив camera lens, (photographic) objective
фотоокисление, фотооксидирование photooxidation
фотоотщепление photodetachment
фотоофсет photo-offset
фотоохота hunting with a camera
фотопатология *med.* photopathy, photonosus
фотопередатчик picture/facsimile transmitter
фотопечать photographic printing, photo-offset printing; photocopying
фотопик photopeak, photoline

фотоплан photographic map/plan, photomap; aerophotographic plan
фотопластинка photographic plate, photoplate
~, высокочувствительная fast plate
~, диапозитивная slide/lantern plate, slide
~, малочувствительная slow plate
фотопленка photographic film
~, катушечная/роликовая roll film
~, плоская/форматная cut/flat film
фотополимер photopolymer
фотопочта photomail
фотоприемник photodetector
фотопроводимость photoconductivity; protoconduction
фотопроводник photoconductor
~, объемный bulk photoconductor
фотопсия photopsia, photopsy
фотопулемет camera gun
фоторазведка photographic reconnaissance/intelligence, photoreconnaissance
фоторазложение photodissociation, photodecomposition
фотораспад, фоторасщепление photodisintegration
фоторегистрограмма streak picture
фоторегуляция photocontrol
фоторедуктор reduction printer
фоторезист photoresist (material)
фотореле photorelay, light/photoelectric relay
~ времени photoelectric timer, phototimer
~, резистивное photoresistance-cell relay
~, селеновое selenium-cell relay
фоторепортаж documentary film, picture story/report
фоторепортер press photographer
фоторепродукция photographic reproduction/duplication
фоторетинит *med.* photoretinitis
фоторецептор photoreceptor, photoceptor
фоторецепторный photoreceptive, photoperceptive
фоторождение *phys.* photoproduction
~ звезд photostar production
~ мезонов photoproduction of mesons
~, парное double photoproduction
~, упругое elastic photoproduction
фоторужье camera gun
фотосенсибилизация photosensitization
фотосинтез photosynthesis
фотосинтезировать photosynthetize
фотосинтетический photosynthetic
фотосинхронизация phototiming
фотослой photosensitive layer
фотоснимок photograph, picture
фотосопротивление 1. photoresistance; 2. photoresistor, photoconductive/photoresistor cell, dry photocell
фотостарение light ageing
фотостудия photographic studio
фотосхема photographic sketch/plan
фотосчетчик photoelectric counter
фотосъемка photography, photographing; photosurveying
фототаксис *bot.* phototaxis, photokinesis
фототека photograph/stills library, photograph collection
фототелеграмма radiophotogram, phototelegram, wirephoto, facsimile
фототелеграф phototelegraph; facsimile
фототелеграфия phototelegraphy, telephoto, wirephoto system, facsimile telegraphy
фототеодолит phototheodolite, field camera, ground photogrammetric camera
фототипировать photoengrave
фототипическое издание phototype edition
фототипия 1. phototype, collotype; 2. phototypy
~, цветная chromocollotype
фототок photoelectric current, photocurrent
фототиристор light-activated thyristor
фототранзистор electrooptical light transistor, phototransistor
фототрансформатор plotting/rectifying camera, rectifier, transformation/rectifying apparatus
фототриангуляция phototriangulation, aerotriangulation
~, плоская/плоскостная plane aerotriangulation
~, пространственная spatial aerotriangulation
фототропизм phototropism, heliotropism
фототропия phototropy
фототрофия phototrophy
фотоувеличитель photographic enlarger
фотоудар photoimpact
фотоумножитель photomultiplier (cell), multiplier phototube
~, герметизированный potted photomultiplier
~, электронный electron photomultiplier
фотоупругость photoelasticity
фотоусилитель photoamplifier
фотоустановка camera mounting
фотофильный *bot.* photophilous, luciphile
фотофильтр camera filter
фотофиниш *sport* photo finish
фотофорез photophoresis
фотофосфорилирование photophosphorylation, photosynthetic phosphorylation
фотохимия photochemistry
фотоцинкография photozincography, photoetching
фоточувствительность photosensitivity, photoelectric sensitivity
фотоэкспонометр *phot.* exposure meter
фотоэлектрет *phys.* photoelectret
фотоэлектродвижущий photoelectromotive
фотоэлектроника photoelectronics
фотоэлемент photocell, phototube, photoelectric cell/tube, photo-element
~, вентильный rectifier photocell; barrage photocell, barrier-type cell
~, газонаполненный gas photocell, gas-filled/-

discharge photocell
~, контрольный monitor photocell
~, плоскостной junction photocell
~, расщепленный split photocell
~, серно-кадмиевый cadmium-sulphide photocell
~, точечный point-contact photocell
~, эмиссионный enission photocell, outereffect photocell, photoemissive cell
фотоэмиссия 1. photoemission; 2. photoemissivity
фотоэмульсия photographic emulsion, photoemulsion
~, загруженная loaded photoemulsion
фотоэритема *med.* photoerythema
фотоэффект photoeffect, photoemissive/photoelectric effect
~, вентильный barrier-layer effect
~, внешний extrinsic photoeffect
~, внутренний intrinsic photoeffect
фотоядерный photonuclear
фрагмент fragment
~ данных data sample
фрагментальный fragmental
фрагментарность fragmentariness
фрагментарный fragmentary, fragmental
фрагментация fragmentation
~, ядерная nuclear fragmentation
~ ядрышка nucleolar fragmentation
фрагмобазидия phragmobasidium
фрагмокон *paleont.* phragmocone
фрагмоспора phragmospore, septate spore
фразеограмма phraseological unit
фразеологизм phraseological unit, phrase
фразировка *mus.* phrasing
фразовый phrase *(attr.)*, phrasal
фрак tails, tail coat
фракиец, фракийский Thracian
фракийка Tracian (woman)
фрактура I *med.* fracture
фрактура II *print.* black letter, Gothic/fracture type
фракционер *polit.* factionalist, factionist, factionary
фракционерство factionalism, fractionalism
фракционирование fractionation, fractionating; grading
~, вытеснительное displacement fractionation
~, мелкое superfractionation
фракционировать fractionate; grade
фракционность factionalism, factiousness
фракционный 1. *chem.* fractional, fractionating, fractionary; 2. *polit.* factional, factious; fractional
~ная перегонка fractional distillation
фракция 1. *chem.* fraction; portion, part; 2. *polit.* faction, group; fraction
~, высокодисперсная fine fraction
~, выходящая/отбираемая effluent fraction
~, исходная/материнская parent fraction
~, концевая/хвостовая residual/tail fraction
~, крупная coarse fraction; oversize
~, мелкая small/fine fraction, fines
~, молярная mole fraction
~, обедненная stripped/depleted fraction
фрамбезиома *med.* frambesioma, yaw
фрамбезия *med.* frambesia, yaws
фрамицетин framycetin
фрамуга transom, fanlight; casement
франгулин *chem.* frangulin
франк I *hist.* Frank
франк II franc
франкения *bot.* sea heath *(Frankenia)*
франкирование prepayment; franking
франкировать prepay
франкмасон freemason
франкмасонский freemasonic(al)
франкмасонство freemasonry, freemasonism
франко- *comm.* free, free of charge
франко-американский Franco-American
франко-борт free on board, f.o.b.
франко-бункер free bunker
франко-вагон free on rail/car, f.o.r.
франко-грузовик free on truck
франко-завод ex mill/works
франкоканадец French-Canadian
франколин *zool.* francolin *(Francolinus)*
франко-набережная free on quay, ex quay
франко-платформа free on truck
франко-порт отправления free port of departure
франко-пристань ex quay
франко-русский Franco-Russian
франко-склад ex warehouse
франко-строп судно ex-ship
франко-строп судна в порту разгрузки free overside
франко-судно free on board, f.o.b.
франкофил Francophile
франкоязычный French-speaking
франкский *hist.* Frankish
франций *chem.* francium, Fr
францисканец, францисканский Franciscan
францисцея *bot.* rain tree *(Brunfelsia)*
француженка French woman
француз Frenchman
французский French
~ ключ adjustable spanner, monkey wrench
~кая булавка safety pin
фрахт freight
~ в оба конца out-and-home freight
~, обратный home(ward) freight
~ по чартеру charter freight
фрахтование freightage; affreightment; chartering
фрахтователь freighter; affreighter; charterer
фрахтовать freight; charter
фрахтовка freightage; chartering
фреатофит *bot.* phreatophyte
фребелевский Froebelian
фрегат 1. *naut.* frigate; 2. *zool.* frigate bird *(Fregata)*
~, ракетный guided-missile frigate
фреза, фрезер (milling) cutter; mill

~, **двухугловая** double-angle milling cutter
~, **дисковая** side mill, side milling cutter
~, **дорожная** road harrow
~, **зуборезная** gear (milling) cutter
~, **зуботехническая** *med.* dental cutter
~, **пазовая** grooving cutter
~, **почвенная** rotary cultivator/tiller
~, **почвенная дисковая** rotary disk hoe/tiller
~, **торфяная** peat cutter
~, **торцовая** face milling cutter
~, **угловая** angle milling cutter
~, **фуговальная** jointing cutter
~, **шпунтовая** tonguing cutter
фрезерный milling; cutting
~ **станок** milling machine
фрезерование milling, cutting; *agr.* rotary cultivation/tillage
~ **по подаче** down milling
~ **против подачи** conventional/upmilling
~, **торцовое** face milling
~, **фасонное** form milling
фрезеровать mill; cut; *agr.* till, scarify
фрезеровка, фрезование = **фрезерование**
фрезеровщик milling-machine operator
фрейдизм Freudism, Freudianism
фрейдист, фрейдистский, фрейдовский Freudian
фрейлина *hist.* maid of honour, lady-in-waiting
фреквентатив *gram.* frequentative/iterative aspect
фреквентативный *gram.* frequentative, iterative
френальгия *med.* phrenalgia
френикотрипсия *med.* phrenicotripsy, phreniclasia
френикоэкзерез phrenicoexeresis, phrenicectomy
френозин *bioch.* phrenosin(e)
френолог phrenologist
френология phrenology
френоспазм phrenospasm, cardiospasm
фреон *chem.* freon, chlorofluoromethane
фреска, фреско fresco painting; fresco
фригидность frigidity
фригидный frigid
фригиец, фригийский Phrygian
фригольд free tenure, frank-tenament, freehold
фригольдер *hist.* freeholder, frank-tenant, charterer
фригория refrigerating unit
фриз I *arch.* frieze; rib, border
фриз II Frisian
фризер (ice-cream) freezer
фризский Frisian
фрикадель, фрикаделька *cul.* meat ball, quenelle
фрикасе, фрикассе *cul.* fricassee
фрикативный *phon.* fricative
фрикцион friction coupling/clutch
~, **многодисковый** multiple-disk clutch
~, **однодисковый** single-disk clutch
фрикционный frictional, friction *(attr.)*
фрима *bot.* lopseed *(Phryma)*
фримациевые *n. pl. bot.* lopseed family *(Phrymaceae)*
фринин phrynin
фритредер free trader, free-tradist
фритредерство free trade system
фритта (glass) frit
фриттер acoustic shock absorber
фриттование fritting
фритюр *cul.* deep fat *(for frying)*
фритюрница *cul.* deep-frier
фришевать refine, purify; freshen
фронда *hist.* Fronde
фрондер *hist.* Frondeur
фронт front
~, **волновой/волны** wave-front
~ **импульса, задний** trailing edge
~ **импульса, передний** leading edge
~ **синхроимпульса** clock edge
~ **разгрузки** unloading area
~ **шквалов** squall front
фронтиспис *print.* frontispiece
фронтит *med.* frontal sinusitis
фронтон *arch.* fronton, pediment
фронто-окципитальный frontooccipital
фруктан fructan
фруктоза fructose, levulose, fruit sugar
фруктозамин fructosamine
фруктозидаза fructosidase
фруктозилтрансфераза fructosyltransferase, transfructosylase
фруктозофосфат fructose phosphate
фруктопираноза fructopyranose
фруктофураноза fructofuranose
фруктуроновый fructuronic
фталазол phthalsulphathiazole, sulphothalidine
фталевый *chem.* phthalic
фталеин *chem.* phthalein
фталимид *chem.* phthalimid(e)
фтивазид phthivazid, vanicide, vanilaberon
фтизиатр *med.* phthisiologist, phthisiotherapist
фтизиатрия phthisiology
фтириаз *med.* phthiriasis
фтор *chem.* fluorine, F
фторапатит fluor apatite
фторацетамид fluoroaceamide
фторацетат fluoroacetate
фторгидрокортизон fluorohydrocortisone
фторгидролаза fluorohydrolase
фтординитробензол fluorodinitrobenzene
фторид *chem.* fluoride
~ **бора** boron fluoride
~ **урана** uranium fluoride
фторировать 1. *chem.* fluoridize; 2. *(water)* fluorinate, fluoridate
фтористоводородный *chem.* fluorohydric, hydrofluoric
фтористый *chem.* fluoric; fluoride (of)
~ **водород** hydrogen fluoride
~ **натрий** sodium fluoride
фторкаучук fluoroelastomer, fluorine rubber
фторный *chem.* fluoric

~ная медь cupric fluoride
фтороводород hydrogen fluoride
фторокись oxyfluoride
~ урана uranium oxyfluoride
фторопласт fluoroplastic; teflon
фторполимер fluoropolymer; teflon
фторотан halothane, fluothane
фтороформ trifluoromethane
фторэластомер fluoroelastomer
фуга I *mus.* fuge
~, трехголосая triple fuge
фуга II groove, slot, rabbet
фуганок jointer/jointing/bench plane
фугас *mil.* land mine, field charge, fougasse
фугат centrifuge effluent
фугато *mus.* fugato
фугетта *mus.* fughetta
фуговальный/фуговочный jointing, mortising; planing
~ станок jointing machine
фуговать 1. joint, mortise; 2. smooth, plane, surface; even
фузариоз bot. fusariosis, fusarial rot
фузея *hist.* fusil, flintlock musket
фузилер, фузелер *hist.* fusilier, fusileer
фузит *geol.* fusite, fusain
фузия *ling.* fusion
фузоспирохетный *med.* fusospirochetal
фузоспирохетоз *med.* fusospirochetosis
фукоза *chem.* fucose
фукозан *bioch.* fucosan
фукоксантин fucoxanthin
фукосиллактоза fucosyllactose
фукостерин fucosterol
фуксин fuchsin, rosaniline monohydrochloride
фуксия *bot.* fuchsia *(Fuchsia)*
фукус *bot.* fucus, rockweed *(Fucus)*
фульват fulvate
фульвовый fulvic
фульгурит *geol.* fulgurite
фулькра fulcrum
фульминат fulminate
фуляр foulard
фумарамовый fumaramic
фумарилацетоацетаза fumarylacetoacetase
фумаровый fumaric
фумарола *geol.* fumarole
~, временная intermittent fumarole
~, нашатырная/щелочная alkaline fumarole
~, первичная hypogene fumarole
~, сернистая solfatara
фумарольный fumarolic
фумигатор fumigator, gassing machine
фумигацин *chem.* fumigacin, helvonic acid
фунгицид fungicide
фунгицидин fungicidin
фунгицидность fungicidity
фунгицидный fungicidal
фунгозный fungous, fungal, fungoid
фундальный *anat.* fundal
фундамент foundation, base, groundwork; bed, substructure; footing, seat(ing)
~, консольный cantilever footing
~ машин engine bearer/seating
~, мелкозаложенный shallow foundation
~, сейсмостойкий earthquake-proof foundation
~, ступенчатый benched foundation
фундаментализм fundamentalism
фундаментальность fundamentality
фундаментальный fundamental; basic
фундировать 1. ground, substantiate, base; 2. *fin.* fund, consolidate
фундук *bot.* giant filbert *(Corylus maxima)*
фундулюс *zool.* killifish; mayfish *(Fundulus)*
фундус *bot.* collum, fundus
фуникулер funicular, cable/cogwheel railway
фуникулит *med.* funiculitis
фуникулопексия *med.* funiculopexy
функия *bot.* plantain lily *(Funkia)*
функционал *math.* functional
функционально-полный functionally complete
функциональный functional; operating, working
функционер functionary; official
функционирование functioning; operation; performance
функционировать function, operate, work
функция function
~, абелева Abelian function
~, автоморфная automorphic function
~, алгебраическая algebraic function
~, аналитическая analytic function
~, аннулирующая nullifier
~, аппроксимирующая approximation function
~, бесконечнозначная infinite-valued function
~, бесселева/Бесселя Bessel('s) function
~, библиотечная library function
~, булева Boolean function
~, векторная vector function
~, вероятностная probability/distribution function
~, весовая weight/weighting function
~ взаимной корреляции cross-correlation function
~ влияния weighting function, kernel
~ влияния, гауссова Gaussian kernel
~ влияния, энергетическая energy kernel
~, возбудительная arousal function
~ возбуждения excitation/drive function
~, возмущающая perturbation/disturbing function
~, возрастающая increasing function
~, волновая wave function
~ времени/временная time function, function of time
~, выборочная *stat.* sample function
~ выигрыша *stat.* payoff/gain function
~ выигрыша и потерь gain-and-loss functiom
~, выпуклая convex function
~, вырожденная confluent function
~, вычислимая computable function

~ Гамильтона Hamiltonian function
~ Гиббса Gibbs function
~, гиперболическая hyperbolic function
~, гипергеометрическая hypergeometric function
~, гладкая smooth function
~, голоморфная holomorphic function
~, двойственная dual function
~, двоякопериодическая doubly periodic function
~, действительная/действительнозначная real-valued function
~ действия *mech.* action function
~ Дирака Dirac delta function
~ Дирихле Dirichlet('s) function
~, дискретная discrete function
~, дифференцируемая differentiable function
~ дожития survivorship function
~, дробно-линейная linear fractional function, homographic function
~, дробно-рациональная fractional rational function
~ занятости *econ.* employment function
~ затрат, вогнутая concave cost
~ затрат, линейная linear cost
~ издержек cost function
~, интегрируемая integrable function
~, интегрирующая integrator
~ источника source function
~, исходная *log.* assumed function
~, классифицирующая discriminator, discriminant function
~ ковариации *stat.* covariance function
~, комплекснозначная complex valued function
~, конечно аддитивная finitely additive function
~, корреляционная correlation function
~ коэффициента полезного действия efficiency function
~, критическая test/critical function
~, кусочно-линейная piece-linear function
~, кусочно-непрерывная piecewise/sectionally/locally continuous function
~ Лагранжа Lagrangian function
~ Лежандра Legendre function
~, линейная linear/ramp function
~, логарифмическая logarithmic function
~, логическая logical function
~, логическая И AND function
~, логическая НЕ NOT function
~, мажорирующая majorant
~, мажоритарная majority function
~ Матье Mathieu('s) function
~, мероморфная meromorphic function
~, многозначная multiple-value(d) function, many-value(d) function, multivalued function
~, многолистная multivalent function
~ множеств function of sets
~, монотонная monotone/monotonic function
~ напряжения stress function
~, невозрастающая non-increasing function
~, невычислимая noncomputable function
~, неопределенная indeterminate funtion
~ неопределенности ambiguity/Woodward function
~, непрерывная continuous function
~, непрерывно дифференцируемая continuously differentiable function
~, неубывающая non-decreasing function
~, нечетная odd function
~, неявная implicit function
~ нулевого порядка zero-order function
~, обобщенная generalized function, distribution
~, обратная inverse function
~, обратная гиперболическая inverse hyperbolic function, arc-hyperbolic function
~, общая показательная general exponential function
~, общерекурсивная general recursive function
~ ограничений constraint function
~, ограниченная bounded function
~, однозначная single-value(d) function
~, однолистная simple/univalent/schlicht/uniforn function
~, опорная support/supporting function
~, ортонормированная orthonormal function
~ отдачи *econ.* response function
~, первообразная antiderivative, primitive
~, передаточная transfer function
~, переходная transition function/probability
~, периодическая periodic function
~, пилообразная saw-tooth function/curve
~ Пирса dagger/Pierce function
~, подынтегральная integrand (function), function to be integrated
~, показательная exponential (function)
~ полезности *econ.* utility function
~, полилинейная multilinear function
~, полная аналитическая monogenic/complete analytic function
~, полунепрерывная semicontinuous function
~, порождающая generator
~ потерь *stat.* loss function
~ правдоподобия likelihood function
~, предельная limit function
~ преобразования transfer function
~, преобразованная transform
~ прибыли *econ.* return/profit function
~, присоединительная associated function
~, пробная trial function
~, производная derived function
~, производственная production function
~, производящая generating function
~, произвольная arbitrary function
~, равномерно непрерывная equicontinuous function, uniformly continuous function
~ разбиения partition function
~ разрешения resolution function
~ разрешения треугольной формы triangu-

lar resolution function
~, разрывная/прерывистая discontinuous function
~ распределения distribution/frequency function
~ распределения вероятности probability distribution function
~ распределения, интегральная distribution function
~ распространения propagator, propagation function
~ распространенности abundance function
~ рассеяния scattering function
~ регрессии regression function
~, решающая/решения decision function
~ с ограничением constrained function
~, сеточная lattice function
~, силовая force function
~, синусоидальная sine function
~ скачков saltus-function
~, скачкообразная step function
~ скорости velocity function
~, словная composite function
~, случайная random/stochastic function
~, собственная eigenfunction; fundamental/characteristic function
~, сопряженная adjoint function
~ состояния state function, function of state
~, степенная power function
~, ступенчатая step/staircase/jump function
~, суммируемая summable function
~, счетно аддитивная countably additive function
~, табличная table function
~, точечная/точки point function
~, трансцендентная transcendental function
~, тригонометрическая trigonometric function
~, убывающая decreasing function
~ управления/управляющая control function; control vector
~ условного риска conditional risk function
~, характеристическая characteristic/fundamental function, eigenfunction
~ хеширования hashing function
~, целая entire/integral function
~, целевая objective/preference function; efficiency/effectiveness function; criterion/goal/target function
~, цилиндрическая cylindrical/cylinder function
~, четная even function
~, шаровая solid/spatial spherical function
~ Эйлера Euler function
~, эмпирическая fitted function
~ энергии Гиббса Gibbs function
~, явная explicit function
функция-аргумент inner function
функция-оригинал original (function)
фунт pound (453.6 g)
~ стерлингов pound (sterling)
фураж fodder
фуражир *obs., mil.* forager
фуражка *mil.* service cap
фуражный fodder *(attr.)*
фуран *chem.* furan
фураноза furanose
фуранокумарин furanocoumarin
фурацилин nitrofurazone, furacin
фургон van, caravan; truck
~, багажный luggage van, *Am.* baggage van
~, герметичный sealed van
~, изотермический refrigerator van
фурил *chem.* furyl
фурия *myth.* Fury
фурма *met.* tuyere
~, верхняя overhead tuyere
~, нижняя bottom tuyere
~, шлаковая tapping/slag tuyere
фурмарьерит fourmarierite
фурнитура accessories, fittings
фуроил *chem.* furoyl
фуроин *chem.* furoin
фурохинолин *chem.* furoquinoline
фурункул *med.* furuncle, boil
фурункулез *med.* furunculosis
фурункулезный furuncular, furunculous
фурфурамид *chem.* furfuramid(e)
фурфуран *chem.* furfurane
фурфурол *chem.* furfural, furfurol(e)
фурьеризм Fourierism
фурьерист Fourierist
фусцин *bioch.* fuscin
фут foot (30,48 cm)
футарк futhorc, futhork, futharc, the runic alphabet
футбол football, soccer
футеровать line; *met.* fettle
футеровка lining; *met.* firebrick/refractory lining; brickwork
~, динасовая timber lining
~, набивная rammed/tamped lining
~ печи kiln lining
~, спеченная sintered lining
~ топки furnace lining
~, углеродистая carbon lining
~, шамотная fireclay lining
футеровочный lining *(attr.)*
футляр housing, casing, case, jacket, sheath
футокс *naut.* futtock (timber), foothook
футор 1. patch, lap; 2. bushing, sleeve, liner; lining
футроп *naut.* footrope
футуристический, футуристский futurist(ic)
футуролог futurologist, futurist
футурология futurology, future research/study
футурум *gram.* future
футшток *naut.* graduated/sounding pole, sounding/gauging rod, tide/depth gauge, foot rule/gauge
фуцин *chem.* fucin
фьельд *geog.* fjeld
фьорд = фиорд
фьюмингование fuming
фюзеляж *aero.* fuselage; body

~, **балочный** beam-type fuselage
~, **герметизированный** pressurized fuselage
~ **круглого сечения** circular fuselage
~ **лодочного типа** hull-type fuselage
~, **ферменный** truss fuselage
фюзеляжный fuselage *(attr.),* fuselage-mounted
фюзен *geol.* fusain, fusite
фюзенизация fusainization
фюзитоподобный fusitoid
фюрер *hist.* Fuehrer

Х, х

хабазит *miner.* chabasite, chabazite
хабу *zool.* habu *(Trimesurus flavoviridus)*
хавбек *sport* halfback
хавтайм *sport* half time
хавтонометр *print* half-tone meter
хагиография hagiography
хадакристалл *geol.* chadacryst
хадж pilgrimage to Mecca, hadj
хаджи hadji, hajji
хадрозавр *paleont.* hadrosaur *(Hadrosaurus)*
хаз *leath.* (horse) butt
хазматоплазма *biol.* chasmatoplasm
хазмогамия *bot.* chasmogamy
хазмогамный chasmogamic, chasmogamous
хазмоклейстогамия chasmocleistogamy
хазмопеталия chasmopetaly
хазмофил *bot.* chasmophile
хазмофит chasmophyte
хайбол highball
хакан *hist.* chagan, khan
хакассит *geol.* khakassite
халазия chalazia
халазальный, халазовый chalazal, chalazian
халат 1. overall; smock; 2. dressing gown, housecoat; robe
~, **докторский** doctor's smock
~, **маскировочный** *mil.* camouflage cloak
~, **хирургический** surgical coat
халва *cul.* halva(h), halavah, Turkish sweetmeat
халдей Chaldean
халдейка Chaldean (woman)
халдейский Chaldaic(al), Chaldean
халезия *bot.* silver bell *(Halesia)*
халибит *miner.* chalybite, siderite, ferrous carbonate
халибы *pl. hist.* Chalybes
халикоз *med.* flint disease, chalicosis
халикотерии *pl. paleont.* chalicotheres
халистатический halistatic
халиф *hist.* calif, caliph, khalif
халифат *hist.* califate, caliphate
халифский *hist.* caliphal
халифство *hist.* caliphate
халоболит *geol.* halobolite, pelagite
халотропный halotropic
халцедон *miner.* chalcedony
~ **в кубах** cubosilicite
~, **зеленый** chrysoprase
~, **красный** sard (onyx)
халцедоновый chalcedonic, chalcedonous
халькография chalcography, mineragraphy
халькоз *med.* chalcosis
~ **глаза** chalcosis lentis
халькозин *miner.* chalcocite, chalcosine
~, **кубический** digenite
халькозинизация chalcositization
хальколит *miner.* torbernite, chalcolite
халькон *chem.* chalcone
халькопирит *miner.* chalcopyrite
~, **почковидный** blistered copper ore
халькофильный chalcophile, chalcophilous
хальмугра *bot.* chaulmoogra, chaulm(a)ugra *(Taractogenus kurzii)*
хальцида *zool.* chalcid fly
хамамелис *bot.* witch hazel *(Hamamelis)*
хамедафна *bot.* leatherleaf *(Chamaedaphne)*
хамедорея *bot.* pacaya *(Chamaedorea)*
хамелеон 1. *zool.* chameleon *(Chamaeleon); fig.* turncoat; 2. potassium permanganate; chameleon mineral; 3. *astr.* Chamaeleon
хамелеонский chameleonic; *fig.* changeable, inconsistent, fickle
хаменерий, хаменериум *bot.* willowweed *(Epilobium)*
хамерлок *sport* hammer lock
хамеропс *bot.* Mediterranean/hair palm *(Chamaerops)*
хамефит *bot.* surface plant, chamaephyte
хамит Hamite
хамитка Hamite (woman)
хамитский Hamitic
хамовые *n. pl. zool.* chamids *(Chamida)*
хамса *zool.* Black Sea khamsa, anchovy *(Engraulis encrasicholus ponticus)*
хамсин *met.* khamsin
хан khan
Ханаан Canaan
хангул *zool.* hangul *(Cervus elaphus affinis)*
хандбол *sport* handball
ханжа Chinese vodka
ханжеский, ханжеской hypocritical, bigoted; sanctimonious
ханкорния *bot.* mangabeira *(Hancornia speciosa)*
ханство khanate
ханты, хантэ Khanty
хантыйский, хантэйский Khanty
хаос *myth.* Chaos
хаотический, хаотичный chaotic(al)
хаотично-неоднородный randomly inhomogeneous
хаотичность chaotic character/state; state of chaos; randomness

хара *bot.* stonewort *(Chara)*
харакири hara-kiri
характер character, nature; pattern; temper; type; feature, mode, state
~, гидродинамический flow pattern(s)
~ грунта *nav.* quality/type of the bottom
~ группирования составных частей *geol.* manner of aggragation
~ дна nature of the bottom
~ долины valley pattern
~, дурной bad temper
~ залегания mode of occurrence
~ замедления, возрастной *nucl.* age character of slowing-down *(of neutrons)*
~ излома appearance of fracture
~ износа wear pattern
~, классический classicality
~ кривой nature of curve
~, мягкий gentle disposition
~ местности relief
~, неглавный *math.* nonprincipal character
~ неисправности type of malfunction
~ обтекания flow pattern
~ огня маяка *nav.* light characteristics
~ отдельности *geol.* jointing
~ переноса металла metal transfer characteristics
~ пламени flame condition
~ поведения behaviour pattern
~ поверхности relief
~ погоды weather character
~ прилива type of tide
~ работы mode of operation/functioning
~ результатов pattern of results
~, случайный *math.* irregular pattern, random character/nature
~, тяжелый difficult nature
~ укладки частиц arrangement of particles
~, эпилептический epileptic temperament
характеризировать characterize, define, describe
характериограф curve tracer
характеристика 1. characteristic; description; performance *(of machine);* 2. reference; personal information file; 3. *math.* index of logarithm, characteristic; 4. *tech.* (characteristic) curve; rating; properties, qualities; response; *see also* **характеристики**
~, акустическая acoustic response
~ акустического поля acoustic signature
~, амплитудная amplitude response (curve), input-output characteristic
~, аналитическая analytical characteristic
~, анодная anode/*Am.* plate characteristic
~, анодно-сеточная mutual characteristic, *Am.* transfer characteristic
~ антенны field pattern
~, антидетонационная antiknock rating/value
~, апертурная *telev.* resolution characteristic
~ без нагрузки unloaded characteristic, no-load characteristic
~, безразмерная dimensionless characteristic
~, веберамперная flux-current characteristic
~ вентилятора fan characteristic; fan performance curve
~, весовая weight curve/characteristic
~, винтовая (двигателя) *naut.* (engine) power(-propeller) revolutions curve
~, влагоразрядная moisture discharge characteristic
~, внешняя external characteristic
~, возрастающая upward/rising characteristic (curve)
~, вольтамперная volt-ampere characteristic(s), voltage-current characteristic
~, временная time response/behaviour
~, времятоковая current-time curve
~, входная input characteristic
~ высвечивания scintillation response
~, выходная output characteristic
~ гидродинамического поля (корабля) *naut.* pressure signature
~, графическая characteristic curve
~ группирования *com.* bunching characteristic
~, детонационная (топлива) knock rating/value
~, динамическая dynamic characteristic/response; *auto.* performance curve
~, диодная diode characteristic
~ добротности Q characteristic
~, дроссельная thrust versus r. p. m. curve
~, жесткая *el.* flat characteristic
~ зажигания firing characteristic
~ запирания *electr.* cut-off characteristic
~, зарядная charge characteristic
~ затухания attenuation characteristic
~ затухания звука при распространении propagation loss characteristic
~ звукопоглощения sound absorption characteristic
~ избирательности *rad., telev.* selectivity characteristic
~ изделия *econ.* product performance
~, калибровочная calibration curve/equation
~ камеры *nucl.* chamber response/performance
~ качества measure of quality, quality characteristic
~, качественная qualitative/attribute characteristic
~, квадратичная square-law characteristic
~ квазиконформного отображения *math.* dilatation ratio
~, кинетическая motion characteristic
~, колесная system/arrangement of wheels, wheel arrangement
~ компрессора по расходу воздуха compressor airflow versus r. p. m. curve
~, коррозионная corrosion performance
~, криволинейная/нелинейная *math.* curved characteristic
~, кулон-вольтная charge-voltage characteristic

~, кусочно-линейная piecewise linear characteristic
~, линейная linear characteristic/response
~, логарифмическая log-log characteristic
~, магнитная magnetic characteristic, B-H curve
~ магнитного поля (корабля) *nav.* magnetic signature
~, механическая speed-torque characteristic
~, модуляционная drive/modulation characteristic
~, мягкая *el.* drooping characteristic
~, нагрузочная load characteristic/curve
~ направленности directional pattern/response/characteristic, directivity/beam pattern
~ насоса head-capacity characteristic, pump characteristic
~, насьщающая/насыщения *phys.* saturation characteristic
~, нелинейная non-linear characteristic/response
~, номинальная rating, nominal performance
~ области применения applicational characteristic
~, обратная диода reverse characteristic
~, общая overall performance curve
~, относительная relative response
~ отражения (звука) reflection/echoing characteristic
~, падающая *el.* dropping/falling characteristic
~, перегрузочная overload characteristic; *av.* g-load curve
~ передачи *telev.* transfer characteristic
~ передачи полутонов *telev.* gray-/half-tone response
~ перекрывающего разряда *el.* flashover characteristic
~, переходная, ~ переходного процесса transient response; unit-step (function) response
~, плоская flat response/characteristic
~ поглощения absorption characteristic
~, полная total characteristic
~, пологая quiet (characteristic) curve
~ полосы пропускания band-width characteristic, band-pass response/shape
~, полутоновая half-tone characteristic
~ поля давления pressure signature
~ помех noise characteristic
~ помехоустойчивости noise-cancelling characteristic
~ послесвечения decay/persistence characteristic
~ потока flow conditions
~ прибора instrument response curve
~, приведенная reduced characteristic
~ приемника receiver response
~ производительности output/performance rating
~, профиля, аэродинамическая air-foil characteristic
~, прямолинейная straight characteristic
~, пусковая starting characteristic
~, рабочая performance (curve), operating/working/performance characteristic
~ разгона *chem.* transient response
~, размольная (угля) grind-ability index
~ ракетного топлива propellant performance
~, расходная flow/discharge characteristic
~, расчетная estimated performance/characteristic
~ реактора reactor performance
~ реакции response
~ реакций системы response (characteristic/curve) of a system
~ регулирования control characteristic/performance, control law curve
~ режима воздушного винта *av.* advance(-diameter) ratio
~ режима несущего винта *av.* tip-speed ratio
~, резонансная resonance response/curve
~, реологическая *hydr.* flow characteristic
~, световая *opt.* light characteristic; *telev.* light transfer characteristic
~, сериесная *el.* series/rising characteristic
~ сети *therm.* system head curve
~, сеточная *el.* grid characteristic
~, сеточно-анодная control characteristic; inverse grid-plate/-anode characteristic
~ системы system characteristic
~ системы, переходная *auto.* unit-step response (function)
~ системы, частотная мнимая *auto.* imaginary frequency response of a system
~, сквозная *cyb.* through characteristic
~, скоростная thrust versus speed curve; velocity characteristic
~, скоростная двигателя power-speed curve
~, скоростная шины speed performance
~, собственная inherent characteristic
~ сохраняемости storage characteristic
~ социално-экономического положения socioeconomic characteristic
~, спадающая downward sloping characteristic
~, спектральная spectral (response) characteristic, spectral response (function), spectrum pattern
~ спектральной чувствительности spectral-sensitivity characteristic
~, срывная *aero.* stalling characteristic
~ стационарного состояния steady-state response
~, стендовая test-bench characteristic
~, ступенчатая staircase characteristic
~, счетная counter characteristic curve; counting response; counting-rate-voltage characteristic
~ счетчика counter response/performance
~ телекамеры, спектральная spectral/taking characteristic of a TV camera
~, температурная temperature characteristic
~, тепловая thermal characteristic/response
~, техническая technical data, specifications
~, токовая current characteristic
~ турбины steam consumption diagram

~, **тяговая** tractive characteristic
~ **управления** control characteristic
~, **усредненная** averaged characteristic
~, **усталостная** fatigue characteristic
~, **фазовая** phase characteristic/response
~ **фазового сдвига** phase-shift characteristic
~ **фазово-частотная**, ~ **фазочастотная** phase(-frequency) characteristic
~, **фоновая** background characteristic
~ **холостого хода** *el.* no-load characteristic
~, **частотная** *electr.* frequency response/characteristic, dynamic-response characteristic
~, **числовая** numerical characteristic
~ **чувствительности** response curve, sensivity characteristic
~, **шумовая** noise performance
~, **шунтовая** shunt characteristic
~, **эквивалентная** total/lumped characteristic
~, **эксплуатационная** operating characteristic; performance specification; in-use performance, operational performance
~, **энергетическая** energy response

характеристики *pl.* performance, qualities, characteristics; *see also* **характеристика**
~, **аэроупругие** aeroelastic characteristics
~, **балансировочные** trim characteristics
~ **взаимозаменяемости** interchangeability characteristics
~, **взлетно-посадочные** airfield performance
~, **виражные** circling performance
~, **всепогодные** all-weather capability
~ **вывода из штопора** spin-recovery characteristics
~ **выносливости** fatigue performance, fatigue-resistant/-resistance qualities
~, **геометрические** geometrical/dimensional characteristics
~, **гидродинамические** hydrodynamic performance/parameters
~ **горизонтального полета** level-flight performance
~, **двигателя** engine performance/characteristics
~, **дезактивизационные** decontamination characteristics
~ **излучения** radiation characteristics
~, **конструктивные** design characteristics
~ **конструкции, динамические** structural dynamics/response
~, **летные** flight performance
~ **на висении** hovering performance
~ **на дозвуковых скоростях** subsonic performance
~ **на докритических режимах** prestall characteristics
~ **на малых высотах, летные** low-level performance
~ **набора высоты** climb performance
~ **несущего винта** *av.* rotor performance
~, **номинальные** rating data
~, **оптимальные** optimum performance
~, **пилотажные** handling qualities, flight handling characteristics
~ **планирования** *av.* gliding performance
~ **повышенной живучести (конструкции)** failsafe characteristics
~ **поглощения нейтронов** neutron-absorbing characteristics
~, **помпажные** surge characteristics
~ **послесвечения** afterglow characteristics
~ **потока** flow conditions
~ **при выполнении разворота** turning performance
~, **пропульсивные** propulsive performance
~, **прочностные** strength properties
~, **рабочие** performance specifications
~ **ракеты-носителя** launcher/booster characteristics
~ **распада** decay characteristics
~, **расчетные** estimated/calculated performance
~ **самолета в колебательном движении** aeroplane oscillatory qualities
~ **самолета, летно-технические** aircraft performance
~ **свечения** fluorescent characteristics
~ **системы управления** control characteristics
~ **скороподъемности** climb performance
~, **срывные** stall(ing) characteristics
~, **статические** static performance
~, **стендовые** bench performance
~, **технические** specifications
~, **технологические** manufacturing/processing characteristics
~ **течения** flow properties
~, **тяговые** thrust performance
~ **удельной прочности** strength-weight properties
~ **управляемости** handling qualities, control characteristics
~, **физические** physical characteristics
~, **ходовые** *rail.* rolling characteristics
~ **человека как звена управления** human performance
~, **эксплуатационные** service/operating/operational/performance characteristics

характеристический characteristic(al)

характерный 1. typical, representative; characteristic; 2. distinctive; specific; 3. *art* character *(attr.)*

характрон *telev.* shaped-beam character display tube

харациновые *n. pl. zool.* characinids, characines *(Characinidae)*

харза *zool.* yellow-throated marten *(Martes flavigula)*

Харибда *myth.* Charybdis

хариджан Harijan, of the caste of untouchables

харизма charism(a), spiritual power

харита *myth.* Charis

хариус *zool.* grayling *(Thymalus)*

харкать spit, expectorate

Харон *myth.* Charon
харподон *zool.* Bombay duck *(Harpodon nehereus)*
харталиэм *econ.* chartalism
хартия 1. charter; 2. manuscript, codex; scrolls, parchment, papyrus
~ вольностей, Великая *hist.* Magna C(h)arta, (the) Great Charter
хартли *math.* hartley
хартшифер *geol.* hartschiefer
харчо *cul.* kharcho, Caucasian mutton soup
хары *pl. bot.* Chara algae, stoneworts *(Chara)*
хасид 1. Hasid(a)ean; 2. devout Jew
хасидский Hasidic
хастеллой *met.* nickel(-chrome)-molybdenum alloy, Hastelloy
хатьма *bot.* tree mallow *(Lavatera)*
хаус *zool.* jungle cat, chaus *(Felix chaus)*
хаусхолд *econ.* household
хваление *eccl.* laudation, glorification, praising, eulogy
хватательный prehensile
хватка 1. grasp, grip; 2. skill
~, мертвая death/mortal/irop grip
хват-тали *naut.* tackle burton, handy tackle
хвоевидный *bot.* needlelike, aciculiform
хвоелистный *bot.* pine-leaved, pinifolious
хвоинка *bot.* needle *(of a conifer),* acicula
хвойник *bot.* ephedra *(Ephedra)*
хвойниковые *n. pl. bot.* joint-fir family *(Ephedraceae)*
хвойный *bot.* coniferous, cone-bearing
хворост brushwood, dry twigs/branches
хвост 1. tail (piece); *tech.* shank, stem, shaft; 2. *math.* remainder; 3. *pl. min.* tailings, rejects; 4. *med.* cauda
~, вильчатый stradle-type root
~, имеющий caudal, caudate, caudiferous
~ импульса pulse tail
~, кнутообразный whiplash tail
~ кометы comet's tail
~, конский *anat.* bundle of spinal nerve roots, horse's tail
~ кривой tail of a curve
~, ласточкин dovetail
~ ленты *comp.* tape trailer
~ лопатки турбины blade root
~ напильника shank of a file
~ придатка яичка *anat.* tail of the epididymis, globus minor of the epididymis
~ сплава *for.* tail end
~, Т-образный T-shape root
~ угольного пласта *min.* spoon end
~ хвостатого ядра *anat.* tail of the caudate nucleus
~, хлыстовидный whip tail
~ы, крупные coarse tailings
~ы, отвальные final/discardable tailings/rejects, residual discard
~ы отсадки jig(ging) tailings
~ы перечистки cleaner tailings
~ы, суммарные overall tailings
~ы флотации flo(a)tation tailings/rejects
хвостатый (large)tailed, urodelous, caudate, having a (large) tail; *n. pl. zool.* urodelans, urodeles, caudates *(Urodela, Caudata)*
хвостец *anat.* coccyx
~ амфибий urostyle
хвостизм *polit.* khvostism, political minimalism
хвостик 1. *bot.* caudicle; 2. tag, tab; *naut.* trailing-edge piece
~ крыла *av.* wing trailing-edge assembly
хвостикодробилка *food* beet tail chopper
хвостиколовушка beet tailing catcher/separator
хвостист 1. *polit.* opportunist, timeserver; khvostist, person following in the tail of events; 2. *educ.* student who is behind in his assignments/ exams
хвостник *bot.* mare's tail *(Hippuris)*
хвостовидный caudiform, taillike, tail-shaped
хвостовик *tech.* shank, stem, shaft, butt
~ автосцепки *rail.* coupler socket/bar
~ бура bit shank
~ вала, ведущий *auto.* drive/driving end
~ контакта wiring tail
~ лопатки (турбины) blade root
~ нервюры *av.* trailing-edge rib
~ тарелки клапана stem
~ форштевня *naut.* stem foot
~, цилиндрический pintle
~ штампа die shank
~ электрического соединителя terminator point
хвостовка *zool.* long-tailed titmouse *(Aegithalos caudatus)*
хвостовой, хвостовый 1. caudal, tail *(attr.),* rear, posterior; 2. residual
~ плавник *zool.* caudal/tail fin
хвостозубый urodentale
хвостокол *zool.* stingray *(Dasyatis)*
хвостолистный *bot.* urophyllous
хвостоплодный *bot.* urocarpous
хвостохранилище tailing(s) dump/pond, slurry pond
хвостцовый *anat.* coccyx *(attr.)*
~вая кость *anat.* coccyx
хвощ *bot.* horsetail, scouring rush, horsepipe *(Equisetum)*
хвощевидный *bot.* equisetum-like; *n. pl.* Equisetales; *also* **хвощеобразные**
хвощовый equisetic, equisetaceous; *n. pl.* horsetail family *(Equisetaceae),* equisetaceous plants
хвоя needles
~, шиловидная awl-like needles
хевея *bot.* rubber tree, hevea *(Hevea)*
хедер *agr.* header, cutting table
~, шнековый auger cutting table
хедив *hist.* khedive, viceroy of Egypt
хедихиум *bot.* ginger lily *(Hedychium)*
хейлез *med.* chilosis
хейлит *med.* ch(e)ilitis

~, гнойный impetiginous chilitis, pyoderma of lips
~, микотический commissural chilitis
хейлогнатосхизис chilognathoschisis
хейлопластика *surg.* chiloplasty
хейлорафия *surg.* chilorrhaphy
хейлоцистида cheilocystidium
хейрантус *bot.* wallflower *(Cheiranthus)*
хейростробовые *n. pl. zool.* cheirostrobils *(Cheirostrobila)*
хейротип *zool. bot.* chirotype
хейхера *bot.* alumroot *(Heuchera)*
хек *zool.* hake *(Merluccius)*
хела *zool.* chela
хелат chelating/complexing agent, chelate
хелатометрия chelatometry
хелатообразование chelation
хелатор chelating/complexing agent
хелидонин *chem.* chelidonin(e)
хелипеда *zool.* cheliped
хелицера *zool.* chelicera, biting appendage
хелицероносные *n. pl. zool.* chelicerates *(Chelicerata)*
хелюхраун *geol.* helluhraun lava
хелобиальный helobious
хелоне *bot.* turtlehead *(Chelone)*
хелониас *bot.* swamp pink *(Helonias)*
хельсиния *bot.* baby's tears *(Helxine)*
хельсинкский Helsinki *(attr.)*
хемантус *bot.* blood lily *(Haemanthus)*
хемерокаллис *bot.* day lily *(Hemerocallis)*
хемилюминесценция chemi(co)luminescence
хемисорбция chemisorption
хемоаутотрофия chemoautotrophy
хемобактерия chemobacter
хемогенный chemogenic, chemogenous
хемодектома *med.* chemodectoma, carotid body tumo(u)r, receptoma
хемодифференциация chemodifferentiation
хемоиммунитет chemoimmunity
хемокаустика chemocautery
хемокинез(ис) chemokinesis
хемолазер chemical laser
хемопауза chemopause
хемореология chemorheology
хеморецептор *physiol.* chemo(re)ceptor
хемосеротерапия chemoserotherapy
хемосинтез chemosynthesis
хемосмос chemosmosis
хемосорбент chemical adsorbent, chemisorbent
хемосорбция chemical absorption/adsorption, chemisorption, chemosorption
хемосорбция-хемодесорбция two-site chemisorption
хемосфера chemosphere
хемотаксис *biol.* chem(i)otaxis, chemotropism
хемоталамэктомия *surg.* chemothalamectomy
хемотерапия *med.* chemotherapy
хемофармакодинамический chemopharmacodynamic, biochemorphic
хемохирургия chemosurgery
хемоядерный chemonuclear
хемулен *zool.* muffle *(Haemulon bonariense)*
хендс *sport* handball
хенекен *bot.* henequen agave *(Agave fourcroydes)*
хенна *bot.* henna *(Lawsonia inermis);* henna dye
хенодезоксихолевый *bioch.* chenodesoxycholic
хеноподиевый chenopodiaceous
хепатус *zool.* lancet fish *(Alepisaurus ferox)*
хер Slavonic name of the letter x
херес sherry
хермес *zool.* chermes *(Adelges)*
~желтый spriuce gall aphid *(Chermes abietis)*
херсофит *bot.* chersophyte
хертель hurdle (in horse races)
херувизм *med.* cherubism
херувим *rel., fig.* cherub
херувимская *eccl.* cherubic hymn
херувимский cherubic
хескер *agr.* (corn) husker
хескер-пиккер *agr.* husker-picker
хескер-шеллер *agr.* husker-sheller
хескер-шредер *agr.* husker-shredder
хета chaeta, seta
хетогнаты *pl. zool.* chaetognaths *(Chaetognatha)*
хетосепториоз *plant paleont.* chaetoseptoria leaf spot
хетт, хеттей *hist.* Hittite
хеттейский, хеттский Hittite
хеттеянка Hittite (woman)
хеширование *comp.* hashing
хеш-таблица *comp.* hash table
хеш-функция *comp.* hashing function
хиазм(а) 1. *anat., biol.* chiasm(a); 2. *ling.* chiasmus
~, перевернутая reverse chiasma
~, тройная/четверная triple/quadruple chiasma
~мы, дистанционные distance chiasms
хиазматипия chiasmatypy
~, полная total chiasmatypy
хиазмод *zool.* swallower *(Chiasmodon)*
хиастобазидия chiastobasidium
хиастолит *miner.* chiastolite, macle
хиастоневрия *zool.* chiastoneury
хиатус hiatus; gap, opening
хибогнат(ус) *zool.* plains minnow *(Hybognathus)*
хибопсис *zool.* silver chub *(Hybopsis)*
хиднокарпус *bot.* chaulmoogra tree *(Hydnocarpus)*
хиджра hegira, Mohammedan calendar, hej(i)ra, hijra
хикама *bot.* yam bean, wayaka *(Pachyrhizus angulatus)*
хи-квадрат *math.* chi-square
~ распределение chi-square(d) distribution
хилангиома *med.* chylangioma
хилария *bot.* hilaria *(Hilaria)*
хилезный *physiol.* chylous, chylaceous
хилемия *med.* chyl(a)emia
хилиазм *eccl.* chiliasm

хилиаст *eccl.* chiliast, believer in a coming happier time on Earth
хилидроз chylidrosis
хилификация *physiol.* chylification
хилодактилус *zool.* morwong *(Cheilodactylus)*
хиломикрон *bioch.* chylomicron
хиломиктерус *zool.* spiny boxfish *(Chilomycterus schoepfi)*
хилоперикард *med.* chylopericardium, chylopericarditis
хилоперитонеум *med.* chyloperitonem, chylous ascites
хилопневмоторакс chylopneumothorax
хилопоэз *physiol.* chylopoiesis, chylification, chylifaction
хилопоэтический *physiol.* chylopo(i)etic, chyliferous
хилоторакс *med.* chylothorax, chylopleura
~, врожденный congenital chylothorax
хилоцеле *med.* chylocele
хилурия *med.* chyluria, albiduria
~, филяриозная filarial chyluria
хилус *physiol.* chyle
химаза *bioch.* chymase, rennin
химахима *zool.* yellow-headed caracara, chimachima *(Milvago chimachima)*
химводоочистка chemical water treatment
хименокаллис *bot.* spider lily *(Hymenocallis)*
химера 1. *myth., biol.* chim(a)era; 2. *zool.* ratfish *(Chimaera)*
~, клеточная cytochimera
~, хромосомная chromosomal chimera
химерический, химеричный, химерный chimeric(al), fanciful
химеровые *n. pl. zool.* chim(a)eras *(Holocephali, Chimaeridae)*
химзавод *abbr.* **химический завод** chemical plant/factory
химизация introduction/use of chemicals
химизм chemism, chemical property/activity, chemistry, (chemical) behaviour
химик 1. chemist; 2. chemical industry worker
химикалии *pl.* chemicals
химикат chemical
химико-технологический chemical-engineering
химико-фармацевтический chemopharmaceutic, chemical and pharmaceutical
химик-технолог chemical-process engineer
химик-фармацевт pharmaceutical chemist
химик-энергетик chemist specialist in power engineering
химиобиотический chemobiotic
химиопрофилактика *med.* chemoprophylaxis
химиорезистентность chemoresistance
химиосинтез chemosynthesis
химиотаксис chemi(o)taxis
химиотаксический chemiotaxic, chemotactic
химиотерапевтический chemotherapeutic
химиотерапия *med.* chemotherapy, chemotherapeutics
химиотропизм *biol.* chemi(o)tropism
химиофизика physicochemistry
химический chemical
~ карандаш undelible/ink pencil
~кая чистка dry cleaning/cleaner
химия chemistry
~, агрономическая agriculture chemistry
~, биологическая biochemistry, bilogic(al) chemistry
~ воды chemistry of water
~, геологическая geochemistry
~, гистологическая microchemistry, histologic chemistry
~ горячих атомов hot-atom chemistry
~(изотопных) индикаторов tracer chemistry
~, коллоидная colloid(al) chemistry, chemistry of colloids
~, космическая cosmochemistry, cosmic chemistry
~, лабораторная experimental chemistry
~, неорганическая inorganic chemistry
~ низких температур cryogenic chemistry, low-temperature chemistry
~, общая general chemistry
~, органическая organic chemistry
~ отходов waste chemistry
~, прикладная applied chemistry
~, радиационная radiochemistry
~ ракетных топлив propellant chemistry
~ сельского хозяйства agricultural chemistry
~, синтетическая synthetic/constructive chemistry
~, судебная forensic chemistry
~ твердого тела chemistry of solids
~, техническая industrial chemistry
~ тяжелых элементов heavy-element chemistry
~ удобрений chemistry of fertilizers
~, фармацевтическая pharmaceutical chemistry
~, физиологическая metabolic/physiologic chemistry
~, физическая physical chemistry
~, ядерная nuclear chemistry
химозин *bioch.* chymosine, rennet, rennin
химозиноген chymosinogen
химопапаин chymopapain
химопсин *bioch.* chymotrypsin with trypsin, chymolase oral, chymoral, haugase, orenzyme
химостойкость chemical endurance/stability
химотрипсин *bioch.* chymotrypsin, avazyme, chymar, chymolaze, enzeon, quimotrase, zolyse
химотрипсиноген *bioch.* chymotrypsinogen
химотроника chemotronics
химраствор *abbr.* **химический раствор** chemical solution
химснаряд *abbr.* **химический снаряд** gas shell, chemical projectile
химус *physiol.* chyme
химусный chymous

химфак *abbr.* **химический факультет** department of chemistry
химчистка *abbr.* **химическая чистка** 1. dry cleaner; 2. dry/chemical cleaning
хина cinchona bark, quina(quina)
хиназолин *chem.* quinazoline
хинакрин *chem.* quinacrine/mepacrine hydrochloride
хинамин quinamin(e)
хинат chinate
хинафтол quinaphthol
хингамин chloroquine (di)phosphate/sulphate
хингидрон *chem.* quinhydrone
хинди Hindi
хинду Hindu, Hindoo
хиндустанец Hindu, Indian
хиндустани 1. Hindu, Indian; 2. Hindustani (language), Hindoostanee
хинидин *chem.* quinidine, chinidine, chinidina, conquinine, beta-quinine
~, **сернокислый** quinidine sulphate
хинин(а) quinine, quinina
~, **двухлористоводородный** quinine dihydrochloride
~, **сернокислый** quinine sulphate
хиницин chinicin(e)
хиннодубильный, хинодубильный cinchotannic, quinotannic
хинный quinic, cinchonic
~**ная корка** cinchona bark
хинозол quinosol, chinosol, 8-hydroxy-quinolone sulphate
хинолин *chem.* quinolin(e), chinoline, leucoline
хинолиновый quinolinic
хинон *chem.* quinone
хинопирин *pharm.* quinopyrine
хиноформ *pharm.* quinoform
хинтерланд *geogr.* hinterland
хиодон *zool.* river whitefish *(Hiodon);* mooneye, white shad *(Hiodon tergisus)*
хиолит *geol.* chiolite
хионантус *bot.* fringe tree *(Chionanthus)*
хионодокса *bot.* glory-of-the-snow, chionodoxa *(Chionodoxa)*
хионофил snow-favo(u)red plant
хионофильный snow-loving, chinophylous
хионофит chionophyte
хионофоб chionophobe
хиосец, хиосский Chian
хипеаструм *bot.* amaryllis, knight's star *(Amaryllis, Hippeastrum)*
хиперолофус *zool.* sandy sprat *(Hyperolophus)*
хиппи hippie
хипсипопс *zool.* garibaldi *(Hypsypops rubicundus)*
хирагра *med.* chiragra
хиральность chirality, handedness
хирограф ch(e)irograph; ch(e)irographist
хиромант ch(e)iromantist, chiromancer, palmist
хиромантия ch(e)iromancy, palmistry
Хирон *myth.* Chiron
хироним *bot.* chironym
хиропластика h(e)iroplasty
хиропот *zool.* saki *(Chiropotes)*
хироптерофил ch(e)iropterophilous plant
хироптерохория ch(e)iropterochory
хироскоп chiroscope
хироскопия chiroscopy
хиротип chirotype
хиротонисать *eccl.* ordain, impose hands on
хиротония *eccl.* ordination, holy orders, consecration
хирург surgeon; chirurgeon
~, **лечащий** surgeon in charge
~, **синий** *zool.* blue tang *(Acanthurus coeruleus)*
хирургический surgical, chirurgical, operative
~ **нож** surgery knife
хирургия surgery, chirurgery
~, **большая** major surgery
~ **военно-полевая** field/military surgery
~, **восстановительная** restorative surgery
~, **глазная** eye/ophthalmic surgery
~, **грудная** thoracic surgery
~, **зубная** dental surgery
~, **малая** minor surgery
~, **неотложная** urgent/emergency/accident surgery
~, **общая** general surgery
~, **пластическая** plastic surgery
~, **сердечная** cardiosurgery, cardiac surgery
~, **торакальная** thoracic surgery
~, **ультразвуковая** ultrasonic surgery
~, **челюстно-лицевая** maxillofacial surgery
хирург-консультант surgical consultant
хирурговые *n. pl. zool.* unicorn fishes *(Acanthuridae)*
хирург-стоматолог dental/oral surgeon
хитин *bioch.* chitin
хитиназа *bioch.* chitinase
хитинизация chitinization
хитиново-фосфатный chitinous-phosphate
хитиновый chitinous
хитино-песчанистый chitino-arenaceous
хитобиаза chitobiase
хитозан *bioch.* chitosan
хитон 1. *hist.* chiton, (Greak) tunic; 2. *zool.* marine mollusc, chiton *(Chiton)*
хитрость slyness, craftiness; ruse, strtegem, trick
~, **военная** *law* ruse of war, war strategem
хищник beast/bird of prey; carnivore; predator; *pl., zool.* carnivores *(Carnivora);* staphylinids, rove-beetle family *(Staphylinidae)*
~**и, древние** *paleont.* creodonts
~**и, наземные** *zool.* fissiped(e)s
хищники-крошки *pl. zool.* flower bugs *(Anthocoridae)*
хищнический 1. rapacious, predatory; 2. harmful, injurious
~ **инстинкт** predatory instinct
~ **лов рыбы** robbing the sea of fish
хищничество 1. predatoriness, rapaciousness; 2.

embezzlement, misappropriation
хищный predatory, raptorial; vulturine
~ные птицы birds of prey
~ные растения carnivorous plants
хладагент, хладоагент cooling medium/agent, coolant, refrigerant, refrigerating medium
~, байпассированный by-passed refrigerant
~, низкокипящий low-boiling refrigerant
хладноломкий *tech.* cold-short/-brittle
хладноломкость *tech.* cold shortness/brittleness
хладнопрессованный cold-pressed
хлад(н)остойкий cold-resistant, antifreezing
хладобойня slaughterhouse with refrigerators
хладопроизводительность refrigerating capacity
хладостойкость cold resistance, resistance to cold
хладотекучесть *plast.* cold flow
хладотехника refrigerating technique/engineering/equipment
хладотранспорт cold/refrigerated transport
хламида *hist.* chlamys
хламидогонидия chlamydogonidium
хламидомонада *bot.* *Chlamydomonas*
~, снежная red-snow alga *(Chlamydomonas nivalis)*
хламидоспора *bot.* thick-walled spore, chlamydospore
хламидоциста chlamydocyst
хлеб 1. bread; 2. *pl.* corn, cereals; (grain) crops
~, белый/пшеничный wheaten bread
~, выпекать bake/make bread
~, деревенский cottage bread
~, докторский bran bread
~, зарабатывать себе на earn one's living
~, кислый sour/leavened bread
~ на корню *agr.* standing corn/grain/crop
~, насущный *rel.* daily bread
~, пеклеванный fine rye bread
~а, озимые *agr.* winter grain crops, winter cereals
~а, полеглые *agr.* laid/lodged corn
~а, яровые *agr.* spring grain crops
хлебный 1. bread *(attr.)*; panary; 2. cereal, frumentaceous, grain *(attr.)*; 3. *agr.* grain-producing
~ амбар granary
~ квас bread kvas
~ край grain-producing country
~ магазин baker's shop
~ паек bread ration
~ спирт grain/ethyl alcohol
~ экспорт grain export
~ная торговля grain trade
~ное дерево *bot.* breadfruit tree *(Artocarpus)*
~ные дрожжи baker's yeast
~ные зерновые растения grain crops
~ные злаки cereals
хлебозавод bakery, bread-baking plant
хлебозавод-автомат automatic bakery
хлебозаготовитель grain buyer (for State grain stock pile)
хлебозаготовка collection of grain (for State grain stock pile)
хлебокопнитель *agr.* stacker
хлебопахотный farming, agricultural
хлебопашенный grain-growing; farming, agricultural
хлебопашество grain husbandry/growing
~м, заниматься be a farmer, till the soil
хлебопашец *agr.* grain grower, ploughman, *Am.* plowman
хлебопекарня bakehouse, bakery, baking room
хлебопоставка grain delivery (to the State)
хлебопродукт grain product
хлебопроизводящий grain-producing
хлебопромышленник grain dealer/broker/merchant
хлеборез(ка) bread slicer/cutter, bread slicing machine
хлеборезный bread-slicing/-cutting
хлебороб farmer, grain producer
хлеборобный grain-producing; farming, agricultural
хлеборобство grain production; farming
хлебородие abundant grain harvest; fertility
хлебородный rich, producing much grain
~ год bumper-crop year
хлебосдача grain delivery (to the State)
хлебостой *agr.* standing crop, stand of grain
~ перед уборкой standing crop
хлеботорговец corn chandler, grain merchant
хлеботорговля grain trade
хлебоуборка grain harvest(ing)
хлебоуборочный harvesting
~ комбайн combine harvester
хлебофураж fodder/feed grain
хлебоэкспорт grain export
хлев cowshed, cattle shed
~ для овец, ~, овечий sheephouse
~, свиной hogpen, pigsty
хлоазма *med.* liver spots, mask of pregnancy, chloasma
~ беременных chloasma uterinum, chloasma of the face
~, бронзовая tropical mask, chloasma bronzium
хлопальщик *hunting* beater
хлопинит *geol.* chlopinite, khlopinite
хлопки *pl.* cotton waste, combings
хлопковидный cottonlike
хлопковод cotton planter/grower
хлопководство cotton farming/growing/culture
хлопководческий cotton-growing
~ район cotton-growing area
хлопковый cotton *(attr.)*
~вая промышленность cotton industry
~вое масло cottonseed oil
хлопкозавод ginnery
хлопкозаготовительный cotton-collecting/-storing
хлопкозаготовка cotton collecting/storing

хлопкокомбайн cotton harvester/combine
хлопкоочески *pl.* cotton waste
хлопкоочиститель cotton cleaner, gin
хлопкоочистительный cotton-cleaning
~ная машина cotton gin
хлопкоочистка ginning
хлопкопрядение cotton spinning
хлопкопрядильщик cotton spinner
хлопкороб cotton grower
хлопкосажалка cotton planter
хлопкосеющий cotton-growing
хлопкосеяние cotton growing, cultivation of cotton
хлопкоткачество cotton weaving
хлопкоуборка cotton picking/harvesting
хлопкоуборочный cotton-picking
хлопок I 1. *bot.* cotton plant *(Gossypium);* cotton; 2. *pl.* cotton waste, combings
~, длинноволокнистый long-staple cotton
хлопок II crack; thump; patter; bang; puff
~, сверхзвуковой supersonic bang/boom
~пки в карбюраторе superfiring/popping in the carburettor
хлопок-сырец cotton wool, raw cotton
хлопунец *bot.* mountain chick-pea *(Astragalus cicer)*
хлопушка 1. *cin.* sound slate, clapper board; 2. *chem., therm.* explosion door; gate valve; 3. *bot.* bladder catchfly *(Silene inflata)*
хлопчатка *text.* cotton (fabric)
хлопчатник 1. *bot.* cotton *(Gossypium);* 2. cotton field
~, бразильский kidney cotton *(G. brasiliense)*
~, индийский common cotton *(G. neglectum)*
хлопчатобумажный cotton *(attr.)*
~ порох guncotton
хлопчатый flaky, flocculent
хлопьевидно-опушенный *bot.* floccose
хлопьевидный flaky, flocculent, floccular, flocculate(d); *bot.* floccose, woolly
хлопьеобразование flocculation
хлопья *pl.* flocks, flakes; flocci
~, кукурузные corn flakes
~, превращать в flocculate
хлор *chem.* chlorine, Cl
~, активный available/active chlorine
~, атомарный elemental chlorine
~ в цепи chlorine in chain
~, газообразный chlorine gas, gaseous chlorine
~, общий total chlorine
~, остаточный residual chlorine
~, связанный fixed chlorine
~, трехфтористый chlorine trifluoride
хлора *bot.* yellowwort *(Chlora)*
хлорагогенный *zool.* chloragogen
хлоразид *chem.* chlorazide
хлорал *chem.* chloral
хлоралгидрат, хлоральгидрат chloral hydrate
хлоралоза *med.* chloralose, glucochloral
хлораль chloral, trichloracetic aldehyde, trichloraldehyde
хлоралюминит chloral(l)uminite
хлорамбуцил *pharm.* chlorambucil, leukeran
хлорамин *chem.* chloramine
~ Б chloramine B, sodium benzene sulphonchloramine
~ Т chloramine T
хлорамфеникол chlor(o)amphenicol, chloromycetin, levomycetin
хлорамфеникол-сукцинат chloramphenicol sodium succinate, chloromycetin succinate
хлорамфибол chloramphibole, dashkesanite
хлоранемия *med.* chloran(a)emia
хлорантия *bot.* chloranthy
хлорантовые *n. pl. bot.* chloranth family *(Chloranthaceae)*
хлораргирит *geol.* chlorargyrite, cerargyrite, horn silver
хлорат *chem.* chlorate
~ калия potassium chlorate
хлоратор chlorinator
хлорацетат *chem.* chlor(o)acetate
хлорацетофенон chloracetophenone
хлорбензол chlor(o)benzene, phenyl chloride
хлорбутин *pharm.* chlorambucil, leukeran
хлорвинил vinyl chloride
хлоргидрия *med.* chlorhydria
хлоргидрохинон chlorhydroquinone
хлорелла *bot.* green freshwater algae, chlorella *(Chlorella)*
хлорефидроз *med.* chlorephidrosis
хлор-замещенный chlorine-substituted
хлорид *chem.* chloride
~ аммония ammonium chloride
~ гидроксиламина hydroxylamine chloride
~ железа ferric chloride
~ натрия sodium chloride, rock salt
~ серебра argentic/silver chloride
хлоридин pyrimethamine, daraprim
хлоридовозгонка chloridation
хлорин chlorinated polyvinyl, chloride fibre/fiber
хлорипривный chloroprivic
хлорирование chlorination
~ воды water chlorination
~ до точки перелома breakpoint chlorination
~ на свету photochlorination
хлорис *bot.* windmill grass *(Chloris)*
хлористоводородный hydrochloric, muriatic, hydrochloride (of)
хлористокислый *chem.* chlorous; chlorite (of)
хлористый *chem.* chlorous; chloride (of)
~ натрий sodium chloride
~тая кислота chlorous acid
хлоритизация chloritization
хлоритоид *miner.* chloritoid, chlorite spar
хлоркаучук chlorinated rubber
хлорлейкемия, хлорлейкоз *med.* chloroma, chloroleukemia
хлормицетин chlor(o)amphenicol, levomycetin,

chloromycetin
хлорноватистокислый *chem.* hypochlorous; hypochlorite (of)
хлорноватистый hypochlorous
хлорноватокислый *chem.* chlorate (of), chloric
хлорноватый chloric
хлорнокислый chlorate (of)
хлорность chlorinity, chloride content
хлорный, хлоровый chloric, chlorous; chloride (of)
~ная известь chloride of lime
~ная кислота chloric acid
хлоробромистый *chem.* chlorobromide
хлорогонидия chlorogonidium
хлороз *bot., med.* chlorosis
~, египетский Egyptian chlorosis
~ жилок veinal chlorosis
хлорозный chlorotic
хлорокись *chem.* oxychloride
~ висмута bismuth oxychloride/white
хлорокруорин *bioch.* chlorocruorin
хлоролейкемия *med.* chloroleukemia
хлоролейцит *bot.* chloroleucit, chloroplast, chlorophyll body
хлорома *med.* chloroleukemia, chloroma
хлоромицетин chloramphenicol, chloromycetin
хлоронитрат *chem.* chloronitrate
~ кобальта, водный hydrated cobalt chloronitrate
хлоропал *miner.* chloropal
~, желтый nontronite
хлоропенический chloropenic
хлоропения *med.* chloropenia, hypochloremia
хлоропласт *bot.* chloroplast, chlorophyll body, chloroleucite
хлороплатинат *chem.* chloroplatinate
хлоропоглощаемость chlorine absorbing capacity
хлоропривный chloroprivic
хлоропротиксен chloroprothixin, taractan
хлоротический, хлоротичный chlorotic
хлоруксусный *chem.* chlor(o)acetic
хлорофан mineral chlorophane
хлорофилл *bot.* chlorophyl(l)
хлорофиллаза *bioch.* chlorophyllase, chlorophyll esterase
хлорофиллан *bot.* chlorophyllan
хлорофиллид *bioch.* chlorophyllide
хлорофиллит *miner.* chlorophyllite
хлорофилловый, хлорофил(л)ьный chlorophyllous, chlorophyllaceous
хлорофиллоносный chlorophylliferous,
~ная паренхима chlorenchyma
хлорофит chlorophyte
хлорофора *bot.* fustic tree *(Chlorophora)*
хлороформ *chem.* chloroform(um), trichloromethane, methylene trichloride
хлороформирование *med.* chloroformization
хлороформировать chloroform
хлороформный chloroformic
хлорохин chloroquine (di)phosphate/sulphate
хлорошпинель *miner.* chlorospinel
хлорпикрин *chem.* chloropicrine, trichloronitromethane, nitrochloroform
хлорпрокаин *chem.* chlorprocaine/nescaine hydrochloride
хлорпромазин chlorpromazine, thorasine
хлорпропамид chlorpropamide, diabinese
хлорсодержащий chlorine-containing
хлортен chlorothen citrate, chloromethapyrilene citrate, tagathen
хлортетрациклин *pharm.* chlor(o)tetracycline, aureomycin
~, кальциевый aureomycin calcium
хлортетрациклин-гидрохлорид aureomycin hydrochloride
хлортиазид *pharm.* chlorothiazide, diuril
хлортиомочевина *chem.* chlorothiourea
хлортрианизен chlorotrianisene
хлоруксусный chloroacetic
хлорфенол chlor(o)phenol
хлорциан cyanogen chloride
хлоршпинель *miner.* chlorospinel
хлорэтил *chem.* ethyl chloride, chlorethyl
хлорэтиловый chlorethylic
хлыст I 1. (horse) whip; 2. *for.* trunk, full-length log, tree length; felled tree; 3. *zool.* virga
хлыст II Khlyst
хлыстовидный whip-like, rod-shaped; *zool.* virgate; *bot.* vimineous
хлыстовик *zool.* whipworm *(Trichocephalus trichiurus)*
хлыстовство, хлыстовщина Khlysts
хмелевка *agr.* kind of shag tobacco
хмелевод hop grower
хмелеводство hop growing
хмелевой hop *(attr.)*
~ побег/стебель hopbine, hopvine
~вая шишка hop cone
хмелеграб *bot.* hop hornbeam *(Ostrya)*
хмелелистный *bot.* hop-leaved, humulifolious
хмелеотделитель hop separator
хмелесеровня hop sulphitation chamber
хмелесушилка hop kiln/drier
хмелецедильник hop strainer
хмель 1. *bot.* hop *(Humulus);* hop cone, hops; 2. intoxicating beverage
хмельник hop field
хмельное *n.* intoxicating beverage, liquor, alcohol
хна *bot.* henna *(Lawsonia inermis);* henna dye
хоамолит *miner.* choamolith, hoamolite
хоана *anat.* choana
хоаноцит flagellate cell, choanocyte
хобби hobby
хобот 1. *zool.* trunk; proboscis; 2. *tech.* yoke; arm, snout, toolholder; bar, ram
~ лафета *mil.* trail of a gun carriage
~ передвижного упора *met.* gauge head
хоботковидный snout-shaped, proboscidiform; rostelliform

хоботный rostral; *n. pl. zool.* proboscideans, proboscidians *(Proboscidea)*
хоботок 1. *zool.* proboscis, promuscis; 2. *bot.* rostrum
~ гастроподы proboscis
~, сосательный sucking beak, suctorial proboscis
хоботообразный proboscidiform
хоботье *agr.* glumes, chaff
хогтонит *geol.* haughtonite
хогфиш *zool.* hogfish *(Lachnolaimus maximus)*
ход 1. motion, move(ment), run(ning), going, travel; speed, pace; 2. passage, way; duct, channel; traverse; *med.* meatus; 3. *tech.* throw, stroke; lead; 4. wheelbase, runners *(of sled);* 5. development, progress, course; functioning, operation; 6. cycle; variation, rate; *game th.* round; *eccl.* procession; 7. move; manoeuvre, maneuver
~, азимутальный azimouth(al) motion
~ амортизатора travel
~ батана *text.* path of lay
~ без толчков smooth motion
~, бесшумный silent/noiseless running
~ вверх upstroke, upward/ascending stroke
~, висячий полигонометрический hanging/open(ed) traverse
~ вниз downstroke, downward/inward/descending stroke
~ вперед, полный *naut.* full speed ahead
~ впуска/всасывания *auto.* suction/admission/intake/charging/injection stroke
~, временной time dependence/variation
~ выпуска *auto.* outstroke, exhaust stroke
~, высотный *phys.* altitude curve
~, давать set in motion, get underway
~, давать задний put into reverse; back up
~, двойной double stroke
~ доменной печи run/operation of a blast furnace
~ зависимости variation; dependence
~, задний reverse movement, reverse/backward running; *rail.* return motion, moving back; *auto.* back stroke; *nav.* sternway, going astern, backward going
~, замкнутый *geod.* closed circuit
~, зольный *leath.* lime round
~ иглы *auto.* needle lift
~ каретки *comp.* carriage movement; *text.* pitch of the coil
~ каретки, обратный carriage return
~ клапана valve stroke
~ контактов contact travel
~ корабля в штилевой полосе ship leeway
~, крестный *rel.* religious procession
~ кривой shape of a curve, trend/shape/run of a curve
~, лесосплавный floating route
~, литниковый sprue
~ луча *opt.* ray path/length
~ луча, строить trace a ray, set up a ray
~, магистральный *geod.* main/primary/principal traverse
~, малый *nav.* low/slow speed; *Am.* one-third speed
~, маточный *anat.* egg/mother gallery
~, мензульный plane-table traverse
~, мертвый backlash, play, lost motion, free travel, slack
~ нагнетания *auto.* pressure stroke
~ нейтронной плотности distribution of neutron density
~, неравномерный irregular/discontinuous/uneven running
~, нерестовый spawning run
~, нивелирный level(ling) line
~, носовой nasal passage
~, обратный reverse/return/rearward motion, reverse, backward running, back stroke; return movement/trace
~, одинарный single stroke
~ педали *auto.* pedal stroke/travel
~, передний forward motion/running; *nav.* aheading, advancing
~, перекидной *met.* cross-over flue
~ печи run/operation/working of a furnace
~ пилообразного напряжения *el.* stroke of a sawtooth voltage
~, плавный smooth running
~, польный *nav.* full speed
~ поршня piston stroke
~, потайной secret passage
~, прямой forward line, direct route
~, рабочий *auto.* working/power stroke
~ развертки sweep motion; trace of scan
~ расширения expansion/working/combustion/firing stroke
~ расширения амортизатора extension
~ регулирующего стержня control-rod travel
~, рекогносцировочный reconnaissance traverse
~, самый малый *nav.* dead slow/creep speed
~, самый полный *nav.* flank/emergency speed
~, свободный free/easy running, free travel; free wheeling
~ сжатия compression/pressure/bump stroke
~ сжатия (амортизатора) contraction
~, смоляной resin/gum duct, gum canal
~ событий course of events
~, спокойный smooth/quiet running
~, суточный diurnal/daily variation
~, тактический tactical move
~, теодолитный transit/field/theodolite traverse
~, топочный (furnace) flue
~ уровня моря variation of sea level
~, хитрый syllogism
~, холостой idling, idle/free/light/loose running, idling/blank cycle, idle stroke, no-load running/stroke
~, цепной chain track
~ цикла cycle path
~ часов, отрицательный/положительный rate of losing/gaining

~ часов, суточный watch/clock rate
~ червя worm track
~, черный rear entrance
~ электронного луча, обратный flyback, return trace, retrace
~ якоря armature travel
ходатай solicitor; intercessor, mediator
~ по делам solicitor
ходатайство 1. petition, application, solicitation, pleading; entreaty; 2. intercession
ходатайствовать solicit, petition, apply; intercede
ходкий best-selling, marketable
ходкость 1. popularity, marketableness; 2. *tech.* performance coefficient, speed ratio
~ судна propulsive quality of a ship
ходовой 1. *tech.* running, operational, working; *naut.* free, movable; 2. pedestrian, for walking; 3. *zool.* migrating, migratory
~ механизм running gear
~ парк park/pool of machinery in working order
~вая дорога pedestrian walk
~вые испытания performance/running tests
~вые качества *auto.* road performance
ходок 1. *min.* passage(way), manway, travelling way, footway; 2. walker; goer
~, вентиляционный/воздушный air passage/conduit/course
~, людской passageway, manway
~, плохой poor walker; *nav.* bad sailor *(of ship)*
ходоуменьшитель reducing/reduction gear
ходулочник *zool.* stilt *(Himantopus himantopus)*
ходьба walking, pacing
ходячий 1. walking; 2. current, popular; in demand
~ больной ambulant case
~чая монета legal tender, currency
хождение 1. walking; going; wandering; pilgrimage; 2. *econ.* circulation
~ за ручкой *av.* responsiveness
~ картушки компаса swing of the card
~ по мукам *rel.* ordeal, purgatory
хозар, хозарский Khazar, Chazar
хозарка Khazar (woman)
хозрасчет *abbr.* **хозяйственный расчет** *econ.* self-financing, self-supporting basis, financial autonomy
~, перевести на put on a self-supporting basis, make financially autonomous
хозрасчетный self-supporting/-financing
хозяева-переносчики *pl.* carrier hosts
хозяин 1. landlord, owner, proprietor; 2. boss, master; manager; 3. *biol.* host; *min.* host mineral
~, вставочный insert host
~, дополнительный supplementary host
~, облигатный/факультативный obligate/optional host
~, окончательный primary/definitive/final host
~, промежуточный intermediate/intermediary/secondary host
хозяйственник business executive/manager
хозяйственный 1. economic(al); 2. household *(attr.)*
~ год economic year
~ инвентарь household furnishings
~ магазин household/hardware store
~ расчет *see* **хозрасчет**
~ная тукосмесь *agr.* farm-mixed fertilizer
хозяйство 1. economy; 2. equipment, facilities; services; supply; 3. *agr.* farm, holding; farming; 4. house(hold); housekeeping
~, весовое weighing equipment/facilities
~, водное water management
~, водопроводное water supply (equipment)
~, водяное water handling facilities
~, газовое gas equipment/supply; gas-fuel handling facilities
~, городское municipal services
~, денежное money economy/supply; money
~, домашнее household, housekeeping
~, зерновое grain husbandry/farming
~, издольное *agr.* share farming
~, инструментальное tool stock
~, коммунальное public services/utilities
~, лесное forestry; forest management
~, лугопастбищное *agr.* allgrass farming, grassland agriculture, grazing/pasture/grass(land) farming
~, многоотраслевое diversified agriculture/farming, multiple product farming
~, многоукладное mixed economy
~, молочное dairy farming/unit
~, монокультурное *agr.* monocrop economy, monocultural farming, one-crop farming
~, народное macroeconomy, national economy
~, натуральное natural/subsistence economy
~, натуральное крестьянское *econ.* self subsistince peasant economy
~, неспециализированное *agr.* multiple product farming, diversified farming
~, нетоварное сельское subsistence farming
~, одноотраслевое monocultural farming, one/single-crop farming, single-product farming
~, односемейное домашнее one-family household
~, откормочное feeding farm
~, пастбищное *agr.* pasture husbandry
~, патриархальное household system
~, плановое planned economy
~, плодоводческое fruit-growing farm
~, полунатуральное *econ.* seminatural economy
~, прудовое pond pisciculture
~, путевое *rail.* track facilities
~, ремонтное maintenance services/facilities
~, рыбное fish industry; fisheries
~, севооборотное rotation farming
~, сельское agriculture, rural economy, farming
~, скотоводческое cattle farm
~, специализированное monocultural farming, one-crop farming

~, **топливное** fuel handling equipment/facilities
~, **травопольно-зерновое** lea farming
~, **транспортное** transport equipment, transportation facilities
хозяйствование managing, management
хоккеист hockey player
хоккей *sport* hockey
~ **с мячом** field hockey, bandy
~ **с шайбой** ice hockey
хоккейный hockey *(attr.)*
~**ная клюшка** hockey stick
~**ная шайба** puck
холамин *chem.* ethanolamine, cholamine, betahydroxyethylamine
холангиоаденома *med.* cholangioadenoma
холангиокарцинома *med.* malignant cholangioma
холангиолит *med.* cholangiolitis
холангиома *med.* cholangioma
~, **доброкачественная** benign cholangioma
~, **злокачественная** malignant cholangioma, cholangiocarcinoma
холангит *med.* cholangitis, angiocholitis
холангитический cholang(e)itic
холановый cholanic
холдинг-компания *econ.* holding/controlling company
холевый *bioch.* cholic
холегематин cholehematin
холедохит choledochitis
холекальциферол *pharm.* cholecalciferol, vitamin D_3
холелитиазис *med.* biliary calculus
холемия *med.* chol(a)emia
холера *med., vet.* cholera
~, **детская** cholera infantum
~, **куриная** chicken/fowl cholera
~, **молниеносная** fulminating/malignant cholera
~ **свиней** hog cholera, swine fever
холерез choleresis
холерик choleric/bilious temperament; choleric person
холерина cholerine, bilious cholera
холеричность iracundity
холерический choleric; iracund
холерный choleraic; *n. med.* cholera patient
~ **вибрион** cholera bacillus
холероподобный choleriform, choleroid, choleralike
холестаз cholestasis, cholestasia
холестеатома *med.* pearl tumo(u)r, cholesteatoma
холестерин, холестерол *bioch.* cholesterol, cholesterin
холестеринемия *med.* cholester(a)emia, cholesterin(a)emia, cholesterolemia
холестероз cholestero(lo)sis
холестеролоксидаза *bioch.* cholesterol oxidase
холетелин *bioch.* choletelin, bilixanthin
холецистит *med.* cholecystitis
~, **гнойный** purulent cholecystitis
~, **калькулезный** calculous cholecystitis
~, **хронический калькулезный** Courvoisier's gallbladder
холецистоеюностомия *surg.* cholecystojejunostomy
холецистокинин cholecystokinin
холизм *philos.* (w)holism
холил cholyl
холинацетилаза choline acetylase
холинэргический cholinergic
холиямб holiamb, variety of iambic trimeter
холл foyer, vestibule, (entrance) hall
холм hill, knoll, hillock; mound
~, **вулканический** volcanic knoll/cone
~, **изрезанный** broken hill
~, **лонный** *anat.* pubic mound
~, **могильный** tumulus, grave mound
~, **океанический** oceanic hill
~, **остаточный** monadrock, residual hill
~, **плоский** table rock, mesa
~, **подводный** seaknoll
~, **рифовый** reef knoll
~, **щебневый** kame
холмик hillock, hummock, mound; 2. *anat.* monticulus, monticle, colliculus, hillock
~, **верхний/нижный** superior/inferior colliculus
~, **лицевой** facial hillock
~, **семенной** seminal hillock/colliculus, caput gallinaginis, verumontanum
холмисто-моренный monticulate-morainic
холмистость hilliness
холмистый hilly, hillocky, hummocky; collinal; rolling, undulating
холмовой hill-growing; collinal
холмогорье monticulate area, hill country
холобазидия *bot.* holobasidium
хологамия *bot.* hologamy
холод cold
~, **искусственный** refrigeration
холодец *cul.* 1. jellied meat/fish; 2. cold borshch
холодильник 1. refrigerator, cooler; cold-air unit; cold storage (warehouse), refrigeration warehouse/store; 2. *met.* cooling bed/bank/table; *foundry* chill; *tech.* condenser
~, **абсорбционный** absorption refrigerator
~ **автономного расхолаживания (реактора)** let/shut-down cooler
~, **башенный** cooling tower
~, **встроенный** built-in refrigerator
~, **вторичный** after cooler
~, **главный** main condenser
~ **горна** *met.* hearth cooling stave
~, **двусторонний** *met.* double cooling bed
~, **двухкамерный** two-compartment refrigerator
~, **домашний** domestic/household refrigerator
~ **доменной печи, желобный** channel cooler
~ **заплечиков доменной печи** bosh cooler, bosh cooling plate/box
~, **масляный** lubricating-oil cooler
~, **многоэтажный** multistorey cold storage

~, низкотемпературный frozen food refrigerator, deep-freeze refrigerator
~, оросительный drip cooler
~ предварительного охлаждения precooler
~, промежуточный intercooler
~ реечного типа *met.* racking type cooling bed, cooling conveyer rack
~, трубчатый *met.* shell-and-tube cooler
~, фурменный tuyere cooler
~, цепной *met.* chain skid (cooling) bank
~ шлаковой летки cinder cooler
холодильный cooling, refrigerating, refrigeratory, frigorific
~ная техника refrigeration engineering, refrigeration equipment
~ная установка refrigerating unit
~ное дело refrigeration
холодискус *bot.* rock spirea *(Holodiscus)*
холодник *cul.* cold beetroot soup, cold borshch
холодное *n. cul.* jellied meat/fish
холоднокатаный *tech.* cold-rolled
холоднокровный *zool.* poikilothermic, h(a)ematocryal, cold-blooded; *n. pl.* coldblooded animals, poikolotherma, poikolotherms
холодноломкий *tech.* cold-short
холоднолюбивый cold-loving, psychrophilic
холоднообработанный cold-worked/-wrought
холодностойкий cold-resistant/-hardy
холоднотянутый cold-drawn/-finished, hard-drawn
холодноформированный *tech.* cold-worked/-finished
холодный cold, chilly; frigid; algid; cool
холодолюбивый psychrophilic, psychrophilous, cryophilous, cold-loving
холодоноситель coolant, (secondary) refrigerant, refrigerating/cooling medium
холодопроизводительность refrigerating capacity/power/effect, cooling power/effect, refrigeration output
холодоснабжение refrigeration supply
холодостойкий frigostable, cold-resistant/-hardy/-enduring
холодостойкость, холодоустойчивость cold/winter hardness, cold endurance/constancy
холокауст holocaust, *hist.* Holocaust
хололит *med.* chololith
холоп *hist.* bond slave, villein; serf
холопка *hist.* bondmaid, bondwoman
холопство *hist.* bondage, bond slavery, servitude, villeinage
холостить castrate, emasculate, gelt; caponize
холостой 1. unmarried, single; alone 2. *bot., zool.* barren, sterile; *3. mil.* blank, dummy; 4. *tech.* idle, free-running; 5. castrated, emasculated, gelded
~ волк rogue wolf
~ ход *tech.* idling, idle motion
холоторакс *med.* cholothorax
холощение castration, emasculation; gelding
холощеный castrated, emasculated, gelded
~ конь gelding
~ бык castrated bull, bullock
холст 1. linen; (sack)cloth; burlap; 2. *art* canvas
~, грубый sacking
~, клееный buckram, bocasine
~, мешочный sackcloth
~, наждачный emery cloth
хомофит *bot.* chomophyte
хомут 1. *tech.* yoke; clamp, clip, ring; stirrup; 2. *agr.* (horse-)collar
~ автосцепки, тяговой coupler yoke
~ выбрасывателя extractor collar
~ для брони armour grip
~ для крепления трубопровода pipe strap
~ для шланга hose clip
~, направляющий guide yoke
~, подвесной suspension clip
~, ремонтный repair clamp
~, рессорный spring stirrup
~, трубный casing/pipe clamp
~, шарнирный hinged yoke
хомутик clamp, clip
~ для крепления tension band
~, обычный токарный ordinary lathe dog
~ прицела elevating slide, rear-sight slide
~ спускового крючка trigger yoke
хомье *pl.* buccal rings, chomata
хомяк *zool.* hamster *(Cricetus)*
хомячок *zool.* small hamster *(Cricetulus)*
~, скорпионовый grasshopper mouse *(Onychomys)*
хон hone, honing head
хонван *pharm.* phosphestrol
хондра *miner.* chondre, chondrule
хондриевидный chondroid
хондрилла, хондрилля *bot.* skeletonweed *(Chondrilla)*
хондриокинез chondriokinesis
хондриоконт *biol.* chondriocont
хондриомит *biol.* mitochondrion, chondriomite
хондриосома *biol.* mitochondrion, chondriosome
хондрит 1. *geol.* chondrite; 2. *med.* chondritis
~, реберный costal chondritis, Tietze's disease
~, сетчатый net chondrite, lithosiderobolite
хондрогенез *med.* chondrogenesis
хондродиния *med.* chondrodynia
хондродисплазия, хондродистрофия *med.* chondrodystrophia, chondrodystrophy, chondrodysplasia
~, атипическая/эпифизарная Morquio's disease
хондроз *physiol.* chondrosis
хондроид *med.* cartilaginoid, chondroid
хондроитин *bioch.* chondroitin
хондроитиновый *bioch.* chondroitic
хондроитинсерный chondroitin-sulphuric
хондроитинсульфат chondroitin sulphate, betaheparin
хондрокарцинома chondrocarcinoma
хондрома *med.* cartilaginous growth, chondroma
~, саркоматозная chondrosarcoma

хондромаляция *med.* chondromalacia
~, внутриутробная chondromalacia fetalis
хондромер chondromere
хондромиксома *med.* chondromyxoma
хондромиома *med.* chondromyoma
хондромукоид, хондромукопротеид *bioch.* chondromucin, chondromucoid
хондромукопротеин *bioch.* chondromucoprotein
хондропластика chondroplasty
хондропротеид *bioch.* chondroprotein
хондротом *surg.* chondrotome, cartilage knife
хондрофит *med.* chondrophyte
хондрофор *zool.* chondrophore
хондроцит chondrocyte
хондрус *bot.* Irish moss *(Chondrus)*
хонинговальный honing
хонингование honing (process)
хонолит *geol.* chonolith
хоппер *rail.* hopper car
хоппер-дозатор hopper-batcher
хор choir; chorus
хорал *mus.* choral(e)
хорват, хорватский Croatian
хорда 1. *math.* chord; 2. *anat., zool.* chorda, c(h)ord, notochord
~ в направлении потока streamwise chord
~, дорсальная/аксиллярная dorsal/axillary chord
~ дуги span of an arc
~, концевая/корневая *av.* tip/root chord
~, средняя аэродинамическая mean aerodynamic chord, second mean chord
~, средняя геометрическая standard/first mean chord, mean geometric chord
~ стягивающая угол *math.* subtense
~, фокальная focal chord
~ центроплана centre-wing chord
хордальный (noto)chordal
хордовые *n. pl. zool.* chordates *(Chordata)*
хордометр *mus.* string gauge
хордотомия *surg.* tractomy, chordotomy
~, бульбарная bulbar tractomy
хореатический choreatic, choreal, choreic
хореeподобный *med.* choreiform, choreoid
хореиформный *med.* choreiform
хореический 1. *med.* choreal, choreatic, choreic; 2. *lit.* trochaic
хорей I *lit.* trochee, choree, chorreus
хорей II long staff/pole/stick
хорек 1. *zool.* polecat *(Putorius);* 2. fitchew, polecat pelt
~, африканский ferret *(Mustela putorius furo)*
~, кошачий cacomistle cat *(Bassariscus)*
хореоатетоз *med.* choreoathetosis
хореоатетозный choreoathetoid
хореограф choreographer
хореография choreography
хореоподобный *med.* choreiform
хорея *med.* (Sydenham's) chorea, St. Vitus'(s) dance
~ беременных chorea gravidarum
~ взрослых, наследственная hereditary/Huntington's chorea
~, врожденная congenital chorea
~, инфекционная Sydenham's/juvenile chorea
~, постгемиплегическая athetosis, posthemiplegic chorea
~, сенильная senile chorea, paralysis agitans
~, эпидемическая choreomania
хориальный *embryol.* chorionic
хориза, хоризис *bot.* chorisis
хоризмовый *med.* chorismic
хориоидальный *anat.* choroid(al), chorioid(al)
хориоидея *anat.* chor(i)oid
хориоидит *med.* choroiditis
~, ареолярный areolar/Forster's choroiditis
~, диссеминированный diffuse/disseminated choroiditis
хориоидкарцинома *med.* choriocarcinoma, chorioma, chorionic epithelioma, chori(n)epithelioma
хориоидный *anat.* chor(i)oid
хориокарцинома *med.* chorionic carcinoma
хориома *med.* chorioma
~, злокачественная choriocarcinoma
хориоменингит *med.* choriomeningitis
хорион *med.* chorion
~, ветвистый shaggy chorion
~, истинный smooth chorion, chorion leve
~, первичный primitive/previllous chorion
хорионэпителиома *med.* choriocarcinoma, chorionic carcinoma, chorionepithelioma
хориоретинит *med.* chorioretinitis, retinochoroiditis, choroidoretinitis
хориоциклит chorioidocyclitis
хорипетальный *bot.* dialypetalous, choripetalous
хорист chorist(er); member of a choir/chorus
хористка chorus girl, chorist(er)
хориямб *lit.* choriamb, a foot of four syllables
хормейстер chorus/choir master
хорнпайп *mus.* hornpipe, musical instrument; English dance
хоровод round dance
хоровой choral, choristic(al)
~ кружок singing circle, chorus
~вая песня choral piece
хорога *bot.* Chinese artichoke, chorogi *(Stachys affinis/tuberifera)*
хороидный chor(i)oid
хорология chorology, geographical distribution
~ растительных сообществ vegetation horology
хороплета choropleth
хорт(ый) greyhound
хоругвеносец *eccl.* gonfalonier, gonfalon bearer
хоругвь 1. *eccl.* gonfalon, banner; 2. *hist., mil.* standard, ensign
хорунжий *n. hist., mil.* standard-bearer; cornet
хоры *pl. arch.* gallery, balcony; choir/organ loft
хота jota, Spanish folk dance
хохлатка 1. crested bird/hen; 2. *zool.* redhumped

caterpillar *(Schizura cincinna);* 3. *bot.* corydalis *(Corydalis)*
хохлатки *pl. zool.* notodontians *(Notodontidae)*
хохлатник *bot.* pappus grass *(Pappophorum)*
хохлатый cristate; crested, tufted; comate; pappose
хохлач *zool.* 1. crested seal *(Cystophora cristata);* 2. crested bird
хохлообразный tuft/bunch-shaped
хохол crest; topknot; *bot.* coma, bunch, tuft
хохолок *bot.* coma; aigrette; floccus
~, семенной seed tuft/coma
хохотун joker, laugher
~, черноголовый *zool.* great black-headed gull *(Larus ichthyaetus)*
хохотунья *zool.* herring gull *(Larus argentatus)*
хохуля *zool.* Russian desman *(Desmana moschata)*
храм temple, church; *fig.* sanctuary
храмовник *hist.* Knight Templar
храмовой church *(attr.)*
~ праздник *eccl.* patronal festival
~ рыцарь *hist.* Knight Templar
хранение storage, storing, stowage, holding, (safe)keeping
~, архивное archiving
~, бессрочное dead storage
~, бестарное bulk storage
~, бункерное bin storage
~ в законсервированном состоянии dead storage
~ в массиве/файле filing
~ в холодильнике cold storage
~ в цифровом виде digital storage
~ времени time-keeping
~ грузов на пристани *naut.* wharfage
~ данных data storage
~ данных, промежуточное holding
~ документов record keeping
~, долгосрочное prolonged storage, long-term storage
~, закрытое indoor storage
~ запасов на плаву *naut.* floating storage
~, краткосрочное short-time storage
~ на корабле shipboard stowage
~ на складе warenhouse storage, warehousing
~ на холоду cold storage
~, навальное/насыпью bulk storage
~, низкотемпературное freezer storage
~ под открытым небом outside-air storage, open storage
~, сдать на check, store, leave in a checkroom
~, складское warehousing, warehouse storage
хранилище storehouse, depot, depository; reservoir, storage reservoir/vessel; *paleont.* repository
~, банковское (bank) vault
~ данных repository
~ из синтетической пленки, надувное bubble barn
~, крытое shed storage
~ радиоактивных отходов radioactive waste storage
хранимость shelf life
хранитель keeper, custodian; curator; *law* depositary
хранить store, hold; maintain, preserve; keep
~ в массиве file
~ при влажном/сухом режиме store moist/dry
храп 1. catch, cramp; can hook; 2. *naut.* clip hook; crampo(o)n
храпение *med.* snore, stertor, rhonchus
храпеть snore; snort *(of animal)*
храповик *tech.* 1. ratchet (wheel/gear); ratchet-and-pawl gear 2. suction basket
~ и собачка ratchet and pawl
~ номеронабирателя *com.* dial ratchet
~ пусковой рукоятки *auto.* starting crank jaw/claw
~, тормозной brake ratchet
храпок *tech.* strainer, strum; suction basket
храпцы *pl. naut.* clasp hook, match hooks, crampo(o)n, sail-hanks
хребет 1. range, crest, (mountain) ridge; 2. *paleont.* spine, spinal column
~ в открытом океане mid-ocean ridge
~, горный mountain range/ridge
~, изрезанный rugged ridge
~, передовой forerange
хребетный spinal, dorsal, tergal
хребтовый 1. central, middle; 2. spinal, ridge *(attr.)*
хребтообразный ridged
хрен *bot.* horseradish *(Armoracia)*
хрестоматия *lit.* reading book, reader
хризалида chrysalis, pupa, aurelia
хризалидокарпус *bot.* butterfly palm *(Chrysalidocarpus)*
хризантема *bot.* chrysanthemum *(Chrysanthemum)*
хризантемоцветковый chrysanthemum-flowered
хризаробин *pharm., chem.* chrysarobin
хризиаз chrysiasis
хризобаланус *bot.* coco plum *(Chrysobalanus)*
хризоберилл *miner.* chrysoberyl
хризогонидий, хризогонидия chrysogonidium
хризодерма *med.* chrysoderma, aurochromoderma
хризоколла *miner.* chrysocolla
~, глиноземистая pilarite
~, коллоидальная cornuite
хризолит *miner.* chrysolite, olivine
~, драгоценный peridot
~, желто-зеленый Brazilian chrysolite
хризомонады *pl. zool.* chrysomonads *(Chrysomonadina)*
хризопраз *miner.* chrysoprase, green chalcedony
хризотамнус *bot.* rabbit brush *(Chrysothamnus)*
хризотил *miner.* chrysotile
~, железистый xylotile
хризофилл *bot.* chrysophyll
хризофиллум *bot.* star apple *(Chrysophyllum)*

хризохром chrysochrome
хризоцианоз *med.* chrysocyanosis
хрип wheeze, wheezing sound
~, предсмертный death rattle
хрипение wheezing; rhonchus
хрипы *pl. med.* rale; crepitation, rhonchus
~, ателектатические atelectatic/border/marginal rale
~, булькающие gurgling rale
~, горловые guttural rale
~, крепитирующие crepitant/vesicular rale
~, мелкопузырчатые fine bubbling rale
~, свистящие sibilant/whistling rale
Хриссеида *myth.* Chryseis
христианизация Christianization, conversion/adaptation to Christianity
христианин, христианский Christian
христианка Christian (woman)
христианство Christianity; *coll.* Christiandom
~, обращать в convert to Christianity
христов Christ's, (of) Christ
~ день *eccl.* Easter day
христолюбивый *eccl.* pious; religious; godly; devout; Christian
Христос Christ
христосование *rel.* exchange of a triple kiss (on Easter)
хрия *ling.* chria
хром I *chem.* chromium, Cr; chrome pigment
~, хлористый chlorine chromium, chromous chloride
хром II box calf; chrome leather
хромаль chrome-aluminium alloy
хромат *chem.* chromate
хроматиды *pl. biol.* chromatids
~, несестринские nonsister chromatids
хроматизм 1. *phys.* chromatism, chromatic aberration; 2. *mus.* chromatic scale
хроматин *biol.* chromatin
~, половой sex chromatin
хроматиновый *biol.* chromatin(ic)
хроматирование chromate treatment, chrom(at)izing, chromating
хроматический, хроматичный chromatic
~ кая нить *biol.* spirem(e)
хроматичность chromaticity, chrominance
хроматограмма chromatogram
~, двумерная two-dimensional chromatogram
~, круговая circular chromatogram
хроматограф chromatograph
хроматография chromatography, stratographic analysis/adsorption, adsorption analysis
~, бумажная paper chromatography
~ в потоке stream chromatography
~, восходящая ascending chromatography
~, вытеснительная displacement chromatography
~, газоабсорбционная gas-absorption chromatography
~, газовая/газораспределительная gas chromatography
~, двухмерная dimeric chromatography
~, жидкостная liquid chromatography
~, ионообменная ion exchange chromatography
~, капиллярная capillar/column chromatography
~, кольцевая annular/circular chromatography
~, комплексообразовательная chelation chromatography
~, осадочная precipitation/sedimentation chromatography
~, проявительная elution chromatography
~, радиальная radial/circular chromatography
~, распределительная partition chromatography
~, совместная cochromatography
~, тонкослойная thin-layer chromatography
~, хемосорбционная chemisorption chromatography
~, элюентная elution chromatography
хроматопсия colo(u)red vision, chromatopsia
хроматофилия chrom(at)ophilia
хроматофильный chrom(at)ophil(e), chrom(at)ophilic, chrom(at)ophilous
хроматофор *bot., zool.* chromatophore, pigment cell, pigment-bearing body, chromoplast
хроматофороматоз *med.* chromatophoromatosis, melanomatosis
хромаффин *anat., zool.* chromaffin, chromaphil
хромаффиноцитома *med.* pheochromocytoma
хромель chromium-nickel alloy, Chromel
хромидроз *med.* chromidrosis
хромиевый *chem.* chromic
хромирование chromium/chrome plating; chroming; *leath.* chrome tanning
~, диффузионное chromizing
~, диффузионное жидкостное chromizing in a molten bath
~, диффузионное твердое chromizing by the powder process
хромированный chromium-plated; chromized; *leath.* chrome-tanned
хромировать chromium-plate, cromeplate
~, диффузионным способом cromize, chromatize
хромировка chromium-work
хромировщик chrome plater
хромис *zool.* blacksmith, blackfish, damselfish *(Chromis punctipinnis)*
хромистый chromic, chromous
~тая сталь chromium steel
хромит *miner.* chromite, chromic iron; chrome-iron ore
хромоалитирование chrome-aluminizing
хромобактерия chromobacterium
хромовокислый chromate (of)
~ калий potassium chromate
хромовщик chrome tanner
хромовый I chromium *(attr.); chem.* chromic
хромовый II box calf
~ вая кожа box calf

хромогенный chromogenous, chromogenic
хромогравер chromolithographer
хроможелатин bichromated gelatine
хромозома *see* **хромосома**
хромоизомерия *chem.* chromoisomerism
хромолейцит *bot.* chromoleucite
хромолиз chrom(at)olysis
хромолипоид *bioch.* lipochrome, chromolipoid
хромолитографический, хромолитографский chromolithographic
хромолитография chromo(lithograph); chromolithography
хромомеры *pl. biol.* chromomeres
 ~, сложные compound chromomeres
 ~, центромерные centromeric chromomeres
хромомагнезит chrome magnesite
хромометр colorimeter, chromometer
хромон *chem.* chromon(e)
хромопексия chromopexy
хромопласт *biol.* chromoplast, chromatophore, pigment cell, pigment-bearing body
хромопротеид *bioch.* chromoprotein
хроморинорея *med.* chromorrhinorrhea
хромосома *biol.* chromosome; *see also* **хромосомы**
 ~, гаплоидная haploid chromosome
 ~, гигантская giant chromosome
 ~, добавочная accessory chromosome
 ~, кольцевая/кольцевидная ring chromosome
 ~, малая small chromosome, m-chromosome
 ~, непарная unpaired chromosome
 ~, слабо конъюгирующая loosely paired chromosome
 ~, филадельфийская Philadelphia/ Ph^1 chromosome
хромосомный chromosomal, chromosome *(attr.)*
 ~ набор chromosomal complement
хромосомы *pl. biol.* chromosomes; *see also* **хромосома**
 ~, дочерние daughter chromosomes
 ~, маркированные gene tagged chromosomes
 ~, ограниченные полом sex limited chromosomes
 ~, остаточные residual chromosomes
 ~, половые множественные multiple/compound sex chromosomes
 ~, разветвленные branched chromosomes
 ~, сложные compound chromosomes
 ~ слюнных желез salivary gland chromosomes
 ~, телоцентрические telocentric chromosomes
 ~ типа ламповых щеток lamp brush chromosomes
 ~, точечные dot chromosomes
хромосфера *astr.* chromosphere
хромосферно-фотосферный chromospheric-photospheric
хромота *med.* lameness, limp(ing), dysbasia, claudication
 ~, перемежающая intermittent claudication
хромотрихия chromotrichia, colo(u)red/pigmented hair
хромотропия chromoisomerism, chromotropy
хромофильный *biol.* chromophilic, chromophilous
хромофор chromophore
хромофорный chromophoric, chromophorous
хромофотография chromophotography, colo(u)r photography; chromophotograph
хромофотометр colorimeter, chromophotometer
хромофототипия chromocollotype, chromotypography
хромоцентр *biol.* karyosome, chromocenter
хромоцит *anat.* chromocyte
хромоцитометр h(a)emoglobinometer, chromocytometer
хромпик *chem.* potassium dichromate/bichromate
хромсодержащий chromiferous
хромшпинель *miner.* chrome spinel, picotite
хронаксия *physiol.* chronaxia, chronaxie, chronaxy, chronaxis
хронема *ling.* chroneme
хронизатор *electr.* timer, synchronizer, synchronization equipment
 ~ с внешней синхронизацией, ~ с внешним генератором master-oscillator timer
 ~ с внутренней синхронизацией self-synchronous timer
хронизация synchronization, timing
хроника 1. *hist., lit.* chronicle; 2. news items; *cin.* newsreel; historical film
 ~, светская gossip/society column
хроникер reporter; chronicler
хрониосепсис *med.* chronic septicemia
 ~, эритематозный erythematous chronic septicemia
хронирование timing, synchronization
 ~, избирательное selective timing
 ~ импульсов pulse timing
хронировать time, synchronize
хронист *hist.* chronicler, chronographer
хронический chronic(al)
хронограф I *hist.* chronicle, chronogram
хронограф II chronograph
 ~, печатающий printing chronograph
 ~ с двойной секундной стрелкой split-second chronograph
 ~ с минутным счетчиком flyback minute recorder
хронография 1. chronography; 2. chronicle, chronogram
хронокарта timecard, clock card
хронолог chronologist
хронологизация dating
хронологизирующий dating, timing
хронологический chronologic(al)
хронология chronology, order of time
 ~ экономического цикла business cycle chronology
хронометр chronometer, timer, time keeper

~, **авиационный** aerochronometer
~, **звездный** sidereal chronometer
~, **кварцевый** crystal clock
~, **контрольный** standard chronometer
хронометраж time metering, timing, chronometry, timekeeping
~, **выборочный** pick-out timing, timing at random
~, **сплошной** continuous timing
хронометражист timer, timekeeper, time study engineer/man/specialist, time checker
хронометрирование timing, chronometration, timekeeping, (motion and) time study
~ **отдельных операций** flyback timing
~ **трудовых движений** motion time analysis
хронометрист timer, timekeeper
хронометрия chronometry, timekeeping
хронореле timing relay
хроносчетчик time/hour meter
хронотропизм *physiol.* chronotropism
~, **отрицательный** retardation of the movement, negative chronotropism
~, **положительный** acceleration of the movement, positive chronotropism
хроококковые, хроококкообразные *n. pl. bot. Chroococcales*
хрупкий fragile, brittle
хрупколомкий short-brittle
хрупкость fragility, brittleness, frangibility, *med.* fragilitas; embrittlement; friability; *met.* shortness
~, **водородная** hydrogen embrittlement, pickle brittleness
~, **коррозионная** corrosion/corroding brittleness, corrosion embrittlement
~ **костей** *med.* brittleness of the bones, fragilitas ossium
~ **ногтей, ненормальная** *med.* onychorrhexis
~ **отпуска/отпускная** temper brittleness
~ **пленки** *leath.* finish break
~, **радиационная** radiation embrittleness
~, **тепловая** thermal/hot embrittlement/brittleness
~ **угля** coal brittleness
~, **ударная** impact brittleness
~, **щелочная** caustic (em)brittlement
хруплявик *bot.* polycnemum *(Polycnemum)*
хруст *med.* bony crepitus, crepitation
хрусталик *anat., zool.* (crystalline) lens
хрусталиковидный phacoid(al)
хрусталиковый *anat.* lenticular
хрусталь crystal, flint/cut glass
~, **английский** crystal/flint glass
~, **горный** *geol.* mountain/rock crystal
хрустальный crystalline, cut-glass, crystal *(attr.)*
~**ная посуда** cut/flint-glass ware
хрустан *zool.* dotterel *(Charadrius morinellus)*
хрущ *zool.* chafer *(Melolontha)*
~, **майский** May beetle
~, **шерстистый** goldsmith beetle
хрущак *zool.* flour beetle
~, **рогатый** broad-horned flour beetle *(Gnathocerus cornutus)*
хрущеобразный scarabaeiform
хрущик *zool.* chafer, beetle
~, **сосновый** pine chafer
~, **японский** Japanese beetle
хрящ I *anat.* cartilage, gristle
~ **века** palpebral cartilage
~, **внутрисуставной** interarticular cartilage
~, **гиалиновый** hyaline/true cartilage
~, **гиоидный** hyoid cartilage
~ **гиомандибулярный** hyomandibular cartilage
~, **клиновидный** cuneiform cartilage, cartilage of Wrisberg
~, **меккелев** Meckel's/mandibular cartilage
~, **надгортанный** epiglottic cartilage
~, **небноквадратный** palatoquadrate bar
~, **перстневидный** cricoid/annular cartilage
~, **подвижный** floating/loose cartilage
~, **превращать в** chondrify
~, **рожковидный** corniculate/Santorini's cartilage
~, **сошниково-носовой** vomeronasal/vomerine/Jacobson's cartilage
~, **суставной** articular/arthroidal cartilage
~, **фиброзный** fibrocartilage
~, **черпаловидный** arytenoid cartilage
~, **щитовидный** thyroid cartilage
~, **эпифизарный** epiphyseal cartilage
хрящ II *geol.* gravel, grit
хрящеватолистный *bot.* chondrophyllous
хрящеватость I cartilaginousness, gristliness
хрящеватость II gravelliness
хрящеватый I *anat.* cartilaginous, gristly, chondroid
хрящеватый II 1. *geol.* gravelly, gravellish, calculous, arenose; 2. *bot.* glareous
~ **грунт** gravelly ground
хрящевидный *anat.* cartilaginoid, chondroid
хрящевина cartilage, gristle
хрящевинный gristly, cartilaginous
хрящевой I cartilaginous, gristly
хрящевой II *geol.* gravelly
хрящекостный chondro-osseous
хрящеперый *zool.* selachian
хрящеподобный chondroid, cartilaginoid
хрящецветник *bot.* illecebrum *(Illecebrum)*
хтонический *myth.* chthonic, chthonian
худоба leanness, thinness
художественность high artistic value, artistry
художественный 1. of art, of the arts; 2. artistic; tasteful
~ **фильм** feature film
~ **вкус** artistic taste
~**ная литература** fiction, belles-lettres
~**ная самодеятельность** amateur art activities
художество art; artistry
художник artist; painer
~ **по интерьеру** interior decorator
~ **по тканям** pattern designer
художник-виньетист vignettist

художник-декоратор decoreographer
художник-колорист colo(u)rist
художник-конструктор industrial/engineering designer
художник-концептуалист conceptual artist, conceptualist
художник-копиист copyist painter
художник-минималист minimal artist, minimalist
художник-монументалист muralist
художник-мультипликатор movie cartoonist
художник-профессионал professional painter
художник-юморист cartoonist
худосочие *med.* cachexy, malnutrition
худосочный *med.* cachectic
худрук *abbr.* **художественный руководитель** art adviser/instructor
хук *boxing* hook
хукер *naut.* hooker
хула abuse, insult, affront, vituperation
хуление abuse, insult; censure
хулиган hooligan, hoodlum; rowdy, ruffian
хулиганство hooliganism
хулок *zool.* hoolock gibbon *(Hylobates hoolock)*
хульман *zool.* entellus langur *(Presbytis entellus)*
хумулин lupulin, humulin
хуннемания *bot.* golden cup *(Hunnemannia)*
хунта junta
хунтилит *geol.* huntilite
хунхуз *hist.* robber, cutthroat, bandit
хурал *polit.* hural
хурма 1. *bot.* persimmon *(Diospyros);* 2. date plum
хустония *bot.* bluets *(Houstonia)*
хутия *zool.* hutia conga *(Capromys pilorides);* con(e)y *(Geocapromys)*
хутор farm(stead)
хуторской farm *(attr.)*
  ~кое хозяйство individual farm
хуторянин farmer
хьюмея *bot.* humea *(Humea)*
хьюэттит *miner.* hewettite
хэк *zool.* hake *(Merluccius)*
хэкер *comp.* hacker
хэтодиптерус *zool.* spadefish *(Chaetodipterus faber)*

Ц, ц

цадик saddik
цанга *mach.* collet (chuck), fingered bushing
  ~, конусная conical grip
цанги *pl.* tongs, pliers
цанговый collet *(attr.),* tongs *(attr.)*
  ~ патрон collet/split chuck, draw-in chuck
ЦАП *abbr.* **цифроаналоговый преобразователь** *comp.* digital-to-analog(ue) converter, DAC
цапка *agr.* (square) hoe
цапкование *agr.* hoeining
цапля *zool.* heron *(Adrea)*
цапон-лак nitrocellulose lacquer/varnish, zapon
цапфа journal, neck; pin, trunnion; gudgeon, pivot
  ~ башмака slipper pin
  ~, ведущая *rail.* driving pin
  ~ гироскопа pivot of a gyro
  ~, гребенчатая collar journal
  ~ ковша trunnion of a ladle
  ~ конвертера trunnion of a converter
  ~, коническая conical/taper(ing)/tapered journal
  ~, концевая end journal
  ~ кривошипа crankpin
  ~, орудийная gun trunnion
  ~ подшипника bearing trunnion
  ~, пустотелая hollow trunnion
  ~, упорная thrust journal
  ~, шаровая spherical journal, ball pin/gudgeon
царапание scratching; abrasion
царапина scratch (mark); score; abrasion
  ~, ледниковая glacial groove
царапка scratcher *(tool)*
царга side-bar/-beam
царевич *hist.* tsarevich, czarevitch, czar's son
царевна *hist.* tsarevna, czarevna, czar's daughter
царедворец courtier
царение reign, rule
цареубийственный regicidal
цареубийство regicide *(action)*
цареубийца regicide *(agent)*
царизм czarism, tsarism
царица 1. *hist.* czarina, tsarina, czar's wife, empress of Russia; 2. *ent.* queen
  ~ ночи *bot.* queen of the night *(Cereus grandiflorus)*
царский czar's, tsar's; imperial, royal; czarist, tsarist
  ~кая вена *anat.* basilic vein
  ~кая водка *chem.* aqua regia
  ~кие врата/двери *eccl.* holy gates
царство kingdom, domain, realm; reign
  ~ ему небесное *rel.* God rest his soul
  ~, животное animal kingdom/realm
  ~, растительное vegetable/plant kingdom
  ~, сонное land of Nod
царствование reign
царствующий ruling, reigning
царь czar, tsar; king, ruler
  ~ и Бог *rel.* God Almightly, the Almighty
  ~ небесный *rel.* God, Heavenly Father
  ~ царей *rel.* King of Kings, God
цаушнерия *bot.* fire chalice *(Zauschneria)*
цвет I colo(u)r; tint; hue
  ~, ахроматический achromatic colour
  ~, голубой *telev.* cyan
  ~ лица complexion
  ~, нелиняющий fast colour
  ~, предельный *geol.* extreme/axial colour
  ~, пурпурный purple (colour); magenta
  ~, реальный *telev.* real/physical colour

~, серый *opt.* achromatic/grey colour
~, сложный compound colour
~, составляющий component colour
~, теоретический nonphysical colour
~, фиктивный fictitious/nonphysical colour
~, фоновый background colour
~ черты streak (colour)
~, эталонный reference/standard colour
~а побежалости temper colours
~а сурика miniate, miniaceous
цвет II 1. *geol.* flower; *coll.* blossom; 2. blossom-time
~, очный *bot.* pimpernel *(Anagallis)*
~, серный sulphur flowers
~, цинковый zinc flower/bloom
цветение florescence; blooming, blossoming; florification; anthesis; *geol.* efflorescence
~ ногтей *med.* leukonychia
цветень *bot.* pollen
цветковидный flowerlike, floriform, anthoid
цветковый *bot.* floral, flowering; *n. pl.* seed/flowering plants *(Spermatophyta, Anthophyta)*
цветколистный *bot.* anthophyllous
цветневой pollen *(attr.)*
цветник *hort.* flower bed/garden
цветной 1. chromatic, colo(u)r *(attr.),* (multi)colo(u)red, tinted; 2. *tech.* nonferrrous
~ная капуста cauliflower
~ное стекло stained glass
цветнокожая *n.* colo(u)red woman
цветнокожий *n.* colo(u)red man
цветнослепой colo(u)r-blind
цветность chromacity, colo(u)r, chromaticness; chrominance
цветный (multi)colo(u)red, variegated
цветоведение colo(u)r science, chromatics
цветовод flower grower, floriculturist
цветоводство flower growing, floriculture
цветовой chromatic
~вая гамма colo(u)r spectrum
~вая слепота colo(u)r blindness
цветовосприятие colo(u)r perception/sensation
цветовоспроизведение colo(u)r reproduction
цветовоспроизводящий chromatogenic, colo(u)r-reproducing
цветовыделение component colo(u)r selection, chromatic selection
цветоделение colo(u)r separation
цветоед *zool.* curculio *(Anthonomus)*
цветозамещаемый allochromatic
цветоизбирательный colo(u)r-selective
цветоизменение, цветоизменяемость allochroism, chang of colo(u)r
цветоизменяющий versicolo(u)r
цветок flower; *pl.* blosson
~, бесплодный barren flower
~, верхушечный apical/terminal flower
~, околопестичный perigynous flower
~, полевой field/wild flower
~ с верхней завязью hypogynous flower
~, сложный composite flower
~ со средней завязью perigynous flower
~, стебельковый scapose flower
~, язычковый ligulate flower
цветокорректор colo(u)r scanner/corrector
цветоложе *bot.* torus, receptacle
~, общее anthoclinium
цветомер colorimeter, tintometer
~ для муки flour colour grader
цветометрия colorimetry
цветонасыщенность colo(u)r saturation
цветоножка *bot.* pedicle, pedicel, peduncle, stalk
цветонос *bot.* flower stalk, peduncle
цветоносец *bot.* anthophore
цветоносный *bot.* flower-bearing, floriferous, anthophorous
цветообозначение colo(u)r designation
цветоощущение colo(u)r perception/sensation
цветопередача colo(u)r rendition/rendering
цветоразличение colo(u)r discrimination
цветорасположение *bot.* inflorescence, flower disposition
~, зонтичное umbel
~, колосовое spike, ear
~, кольчатое whorl, verticil
цветорассеяние chromatic dispersion
цветосмеситель colo(u)r mixer
цветостойкий, цветоустойчивый colo(u)rfast
цветостойкость, цветоустойчивость colo(u)rfastness
цветочный floral, flower *(attr.)*
~ покров *bot.* flower envelope, perianth
~ стебель flower stalk
~ узор flowery pattern
~ чай flower tea
~ щиток *bot.* corymb, flattish-topped raceme
цветочувствительность colo(u)r/chromatic sensitivity
цветуха *bot.* bolting
цветущий flowering, blossoming, florescent, blooming; *fig.* prosperous
~ днем *bot.* hemeranthic, hemeranthous
~ ночью *bot.* noctiflorous, nyctanthous
цветы flowers; *see also* **цветок**
~, железные flowers of iron
~, кобальтовые erythrite, cobalt bloom
~, серные flowers of sulphur
~, сурьмяные flowers of antimony, antimony bloom, valentinite
~, урановые zyppeite, uranobloom
~, цинковые flowers of zinc, zinc bloom, hydrocincite
цвинглианец *rel.* Zwinglian(ist)
цвинглианство *rel.* Zwinglianism
цвиттер *geol.* zwitter
цеаксантин zeaxanthin
цеанотус *bot.* ceanothus, redroot *(Ceanothus)*
цеболлит *geol.* cebollite
цевка 1. *tech.* cog, cogwheel/pin tooth, reel, bobbin, spool; spindle, (driving) pin; 2. *zool.* tarsometatar-

sus
~, деревянная wooden bobbin
~, роликовая roller tooth
цевочный 1. *tech.* cog (wheel) *(attr.),* bobbin *(attr.); zool.* tarsometatarsal
цевье 1. fore-end *(of rifle stock);* 2. pivot
цедент *law* cessionary, assignor, grantor, transferor
цедилька filter, strainer, colander
цедра *cul.* (dried) lemon/orange peel
цедрат *bot.* citron *(Citrus medica)*
цедрела *bot.* cedrela, Spanish cedar *(Cedrela odorata)*
цеж, цежь strained solution, filtrate
цежение filtering, straining; percolation
цезальпиниевые *n. pl. bot.* senna family *(Caesalpiniaceae)*
цезальпиния *bot.* brasiletto *(Caesalpinia)*
цезаризм Caesarism
цезарист Caesarian, Caesarean
Цезарь Caesar
цезиево-сурьмяный c(a)esium-antimony
цезий *chem.* c(a)esium, Cs
цезированный c(a)esium-coated
цезура *lit., mus.* caesura
цезурный caesural, caesuric
цейба *bot.* Malabar simal tree, simal *(Salmalia malabarica)*
цейлонец Ceylonese
цейлонка Ceylonese (woman)
цейлонский (of) Ceylon, Singalese
цейс Zeiss optical instrument
цейтлупа high-speed slow-motion camera
цейтнот *chess* time trouble
цейхауз, цейхгауз *mil.* armory, arsenal, stores
цекование *mach.* spotfacing
цековка spotfacing *(operation);* spotfacer *(tool)*
цекопексия *med.* c(a)ecopexy, c(a)ecofixation, typhlopexia
цекорафия *med.* c(a)ecorraphy, typhlorrhaphy
цекэктомия *surg.* c(a)ecectomy, typhlectomy
цекропия *bot.* snakewood tree, trumpet tree; pump wood *(Cecropia)*
цекум *anat., zool.* c(a)ecum, blind sac
целакант *zool.* coelacanth *(Latimeria chalumnae)*
целанид *pharm.* lantoside C
целаструс *bot.* staff tree, bittersweet *(Celastrus)*
целебность curative/healing properties
целебный curative, medicinal, healing; salubrious, sanative, healthy
~ные источники mineral springs
~ные травы medicinal herbs
целевой target *(attr.);* having a special purpose
~ продукт prime product
целенаправленность purposefulness, singlemindedness, purpose
целенаправленный purposive, purposeful, singleminded; goal-directed, telic, teleological
це́ление aiming *(a weapon)*
целе́ние curing, healing, treatment
целесообразность expediency, expedience advisability
целесообразный appropriate, advisable, expedient, suitable
целестин *miner.* celestite, celestine
целеуказание target designation/indication
целеустановка aim, object(ive)
целиакия *med.* celiac/Gee's disease, Gee-Herter syndrome, intestinal infantilism
целибат celibacy
це́лик *mil.* (back)sight; *opt.* sight(ing point)
цели́к 1. *min.* pillar; 2. *see* **целина**
~, междукамерный interchamber/intervening pillar
~, междуэтажный level pillar
~, опорный support pillar
~, предохранительный protective/protecting/safety pillar
~, рудный ore pillar
~ у штрека stub pillar
целиком *econ.* in bulk, all, wholly; as a unit
целина *agr.* virgin land/soil, old idle land
целинный *agr.* virgin-land
~ная земля old idle land
целиоскопия *med.* cel(i)oscopy
целительный medicinal, curative, healing, sanative
целла *arch.* cella
целлофан cellulosic packing material/film; cellophane
целлофанирование laminating
целлулоза *see* **целлюлоза**
целлулоид celluloid
целлулоидный, целлулоидовый celluloid
целлюлит *med.* cellulitis
~, тазовый parametritis, pelvic cellulitis
целлюлоза cellulose (pulp)
~, вискозная rayon/viscose cellulose
~, длинноволокнистая long-fibre cellulose
~, древесная wood cellulose
~ для производства бумаги woodpulp
~, кормовая fodder cellulose
~, коротковолокнистая short-fibered cellulose
~, слизистая mucocellulose
~, товарная market/commercial cellulose
~, хлопковая cotton/linters pulp, cotton linters
~, щелочная alkali(ne)/sodium cellulose
целлюлозно-бумажный pulp-and-paper
целлюлозоразлагающий cellulose-decomposing
целлюлозоразрушающий cellulose-digesting
целлюлярный cellular
целобласт *bot.* coeloblast
целобластула coeloblastula
целовальник *hist.* tax collector; innkeeper
целование kissing; *hist.* oath (by kissing the cross)
~, крестное oath taking
целогаструла coelogastrula
целое *n.* the whole; *math.* integer, whole number; *econ.* entier
~, замкнутое *math.* closed entity
~, кратное integral multiple

целозамкнутость complete closure
целозамкнутый integrally/completely closed
целозамыкание complete closure
целозия *bot.* cockscomb *(Celosia)*
целококкус *bot.* sago palm *(Coelococcus)*
целократный integral multiple
целом *zool.* coelom(e), celom
~, внеэмбриональный extraembryonic celom
целомные *n. pl. zool.* coelomates *(Coelomata)*
целомодукт *zool.* coelomoduct
целомоцит coelomocyte
целомудрие chastity
целостат *astr.* coelostat
целость 1. wholeness, integrity, entireness; 2. safety
целотелиома mesothelioma
целочисленный integer-valued, integral, whole-number(ed)
~ многогранник integer polyhedron
цель 1. target, object; 2. objective, purpose; service; aim, goal, intent
~, береговая *nav.* shoreside target
~, бить/попадать в hit the mark
~, быстроисчезающая transient target
~, внеплановая target of opportunity
~, воздушная air/aerial/airborne/aircraft/atmospheric/flyable/flying target
~, групповая multiple/complex/complicated/formation target
~, живая personnel target
~, кажущаяся virtual target
~, ложная decoy (target), false/dummy target
~, маневрирующая manoevring/turning/evasive target
~, морская waterborne target
~, надводная/наземная surface/ground target
~, назначенная designated/programmed/predetermined target
~, невидимая out-off-sight target
~, неопознанная unidentified target
~, неподвижная fixed/stationary/steady target
~, новая fresh target
~, одиночная individual/single target
~, орбитальная ложная orbital decoy
~, отдельная discrete target
~, радиолокационная reflecting/radar target
~, реальная actual/live target
~, точечная (pin-)point target
~, уклоняющаяся dodging/evasive/evading target
~, учебная training/trial target
~, флюктуирующая релеевская Rayleigh fluctuating target
цельноголовые *n. pl. zool.* holocephal(i)ans *(Holocephali)*
цельнодеревянный all-wood
цельнокатаный seamless-/solid-rolled, one-piece rolled
цельнокованый seamless-/solid-forged, one-piece forged
цельнокостный holostean, holosteous; *n. pl. zool., paleont.* holosteans *(Holostei)*
цельнокрайний smooth-edged/-margin; entire
цельнолепестный *bot.* integripetalous, holopetalous
цельнолистный *bot.* integrifolious
цельнолитой, цельнолитый cast in block, unit-/all-cast
цельнометаллический all-metal
цельномолочный whole-/unskimmed-milk
цельнооформленный inseparable, continuous, uninterrupted; integral; whole
цельноплодный *bot.* holocarpous, holocarpic
цельноповоротный all-moving
цельнопозвонковый holospondylous
цельнорезиновый all-rubber
цельносварной, цельносварный all-welded
цельностальной all-steel
цельнотянутый seamless, weldless, solid-drawn
цельный 1. one-/single piece, solid, integral; entire; unbroken, unitary; 2. whole, undiluted, unskimmed *(of milk)*; balanced *(of a person)*
Цельсий Celsius
~сия, термометр centigrade thermometer
цемент *tech., anat.* cement
~ без добавки plain cement
~, безусадочный noncontracting cement
~, быстросхватывающийся rapid-/quick-/fast-setting cement
~, быстротвердеющий early-strength cement
~, высокопрочный high-strength cement
~, гипсовый gypsum cement
~, гипсошлаковый gypsum-slag cement
~, глинистый argillaceous cement
~, глиноземистый alumina cement
~, дорожный road/pavement cement
~, железистый ferruginous cement
~ зуба/зубной (tooth) cement
~, зубоврачебный dental/copper/temporary/eugenol cement
~, известково-шлаковый lime-slag cement
~, известковый calcareous cement
~, кислотоупорный acid-proof cement
~, кладочный masonry cement
~, клинкерный neat/straight cement
~, медленнотвердеющий low-early-strength cement
~, напрягающий self-stressing cement
~, огнеупорный refractory/fire cement
~, пленочный filmy cement
~ пор interstitial cement
~, пуццолановый puzzolan cement
~, расширяющийся expanding cement
~ с добавками blended cement
~ с замедлителем retarded cement
~ с малым содержанием извести low-limed cement
~ Сореля Sorel cement
~, стекловатый vitreous cement
~, схватывающийся set cement
~, тампонажный oil-well cement

~, **товарный** finished cement
~, **тяжелый** weighted cement
~, **шлакоизвестковый** slag-lime cement
цементация 1. *constr.* (pressure) grouting, grout injection; cementation, cementing; 2. *met.* carburizing, casehardening
~, **вторичная** re-cementation
~, **газовая** gas carburizing
~, **жидкостная** liquid carburizing
~ **накипи** scale hardening
~ **цветных металлов** electrolytic precipitation
цементирование *constr.* (pressure) grouting; cementation; 2. casehardening
цементированный, цементованный 1. cemented; 2. *met.* carburized, casehardened
~**ная сталь** casehardened steel
цементировать 1. *constr.* grout; cement; 2. *met.* caseharden, carburize; precipitate by electrolysis
цементирующий cementing; cementitous, binding
цементит iron carbide, cementite
~, **глобулярный/зернистый** globular/divorced cementite
цементный cement(al)
~ **бетон/раствор** cement concrete/mortar
цементование cementing, cementation
цементовка cementing
цементовоз cement truck/ *naut.* carrier
цемент-пушка cement/concrete gun
цена 1. price, cost; 2. value, worth; *see also* **цены**
~, **аукционная** tender price; forced sale price
~ **базисного года** base-year price
~, **верхняя (игры)** *game th.* upper (pure) value
~ **включающая пошлину** duty-paid value
~, **выгодная для покупателей** buyer's/bargain price
~ **деления верньера/нониуса** vernier division
~ **деления гальванометра** galvanometer constant, sensitivity of a galvanometer
~ **деления шкалы (прибора)** multiplying/scale factor, value of a division, scale factor/multiplier, scale spacing/division, least division/graduation
~, **договорная** negotiated/contract price
~ **дуэли** *game th.* value of duel
~, **единая** single price
~ **единицы товара** unit price
~, **завышенная** overcharge
~, **заготовительная** procurement price
~, **закупочная** purchasing/purchase price
~, **за наличные (при продаже)** cash price
~, **запретительная** prohibitive price
~ **игры** *math., comp.* game value, value/worth of (a) game
~, **искусственно поддерживаемая** pegged price
~, **конкурентная** competitive price
~, **крайняя** bedrock price
~ **краткосрочного равновесия** short-run equilibrium price
~ **локо** loco price
~, **максимальная** price ceiling, ceiling/maximum/peak price
~, **минимальная** bottom/minimum/floor/lowest price
~, **нарицательная** par value
~, **начальная** *fin.* starting price
~, **недоступная** prohibitive price
~ **нетто** net(t) price
~, **нормированная** normalized price
~ **обеспечивающая сбыт** salable price
~, **одинаковая** flat price
~, **оптовая** wholesale price
~, **отправная** *fin.* starting price
~, **отпускная** disbursing price
~, **официально объявленная** posted price
~, **очень низкая** rock-bottom price, bedrock price
~, **паритетная** parity price
~, **паушальная** lump-sum price, all-round price
~, **плановая** target price
~ **покупателя** bid price
~, **покупная** buying/purchase price
~ **по прейскуранту** list/sticker price
~ **по ценнику** tag price
~, **прейскурантная** list/sticker price
~ **препятствующая появлению конкурентов** entry preventing price
~ **при поступлении** receiving price
~, **продажная** selling price
~ **производства** cost price
~, **равновесная** equilibrium price
~, **расчетная** accounting/settlement price, making-up price
~, **розничная** retail price
~, **рыночная** market price
~ **с выгрузкой на берег** landed price
~ **с надбавкой** premium price
~ **с пошлиной** duty-paid price
~ **сиф** price c.i.f., price cost, insurance and freight
~, **скрытая** inplicit/shadow price
~, **согласованная** negotiated price
~ **со скидкой** discount price
~, **твердая** firm/set price
~ **устанавливаемая с целью вытеснения конкурента** elimination price
~, **фабричная** prime cost
~, **фактическая** actual price
~ **фактической продажи** realization value
~ **фас** price f.a.s., price free alongside ship
~ **фирмы-лидера** leader's price
~ **фоб, ~ франко-борд** price/quotation f.o.b., price free on board
~ **франко-вагон** f.o.r. price
~ **франко вдоль борта судна** price f.a.s., price free alongside ship
~ **франко-завод** mill-net price, quatation ex mill, f.o.b. mill price
~ **франко-местонахождения товара** loco price
~ **франко шахта** f.o.b. mine price
~ **черного рынка** curb price, black market price

~, штучная unit price
ценантий *bot.* coenanthium, clinanthium
цененхима *zool.* coenenchyma, gelatinous uniting material, sclerobase
ценз 1. qualifications, requirements; 2. enumeration, inventory; *hist.* census; 3. list, register; roll
~, возрастной age qualification
~, имущественный property qualification
~ оседлости residence/residential qualification
цензитарий *hist.* censitaire
цензор censor
цензорский censor(ial), censorian, censorious
цензорство censorship
цензура censorship
цензурировать *math., stat.* censor
цензурность censorship
цензурный 1. censorial, censorian, censorious; 2. decent; printable
ценитель 1. connoisseur, expert; judge; 2. appraiser, valuer
ценник price ticket/list
ценностный valuable; value *(attr.)*
ценность 1. value, worth; 2. *pl.* valuables; values
~, белковая *agr.* protein value
~, кормовая *agr.* food value
~, племенная *agr.* breeding value
~, предполагаемая prospective value
~, промышленная commercial value
~ системы system value
~сти, духовные intellectual/cultural values
~сти, учитываемые bankable securities
ценный 1. valuable; precious; costly, expensive; 2. with a stated value
~ная посылка insured parcel
~ные бумаги *fin.* securities, stock
~ные бумаги, беспроцентные noninterest - bearing securities
~ные бумаги, быстрореализуемые securities realizable at short notice
~ные бумаги во внебиржевом обороте unlisted securities, over-the-counter securities
~ные бумаги, государственные public stock(s)
~ные бумаги, заложенные pledged securities
~ные бумаги, именные registered securities
~ные бумаги, незарегистрированные на бирже off-board securities
~ные бумаги, необлагаемые налогом nontaxable securities, tax-exempt securities
~ные бумаги, нерыночные nonmarketable securities
~ные бумаги, обесцененные wildcat securities
~ные бумаги, особо надеждные, ~ные бумаги, первоклассные gilt-edged securities, highgrade securities
~ные бумаги, передаваемые negotiable securities
~ные бумаги, поддельные wildcat securities
~ные бумаги, процентные interest-bearing securities
~ные бумаги, рыночные marketable securities, stock-exchange securities
ценобиальный coenobiar
ценобий coenobium
ценобласт *biol.* coenoblast
ценовид coenospecies
ценогамета *bot.* coenogamete
ценогенез(ис) *biol.* c(o)enogenesis
ценоз *biol.* c(o)enosis, biocoenosis
~, животный zooc(o)enosis
~, растительный phytoc(o)enosis
ценозигота coenozygote
ценокарпий *bot.* coenocarp(ium)
ценомоноэция *bot.* coenomonoecia, coenomonoecium
ценон coenon
ценообразование establishing of prices, pricing
~ по принципу "средние издержки плюс прибыль" rule-of-thumb pricing, cost plus pricing, full cost pricing
ценосарк *zool.* coenosarc
ценостеум *zool.* coenosteum, calcareous skeleton
ценотип *biol.* coenotype
ценоций *zool.* coenocium
ценоцит *biol.* coenocyte
цент cent
центал cental, hundredweight
центибар centibar
центибел centibel
центипуаз centipoise
центифолия *bot.* cabbage rose *(Rosa centifolia)*
центнер (metric) centner, hundredweight
центр 1. centre, *Am.* center; central point, focus; 2. capital (city); 3. central office
~, активный active centre/nucleus
~ антициклона high-pressure centre, centre of anticyclone/high
~ Брока *anat.* speech/Broca's centre
~, ведущий (токарный) driving centre
~ Вернике *anat.* sensory speech centre, Wernicke's centre
~ вихря centre of the vortex
~ внедрения новшеств innovation centre
~, внешний триангуляционный surface mark
~, внутренний триангуляционный subsurface mark
~ возмущения centre of perturbation
~ вращения centre of rotation
~ выборки *stat., math.* centre of a sample
~, вычислительный computing/computation/computer centre, computer bureau
~, главный метеорологический main meteorological office
~, главный питательный main feeding point
~ глотания *anat.* deglutition centre
~ гомологии *math.* centre of homology/perspectivity, ray centre
~ гомотетии homothetic centre, centre of similitude

~ грани face centre
~ дальней космической связи distant-/deepspace radio centre
~, двигательный *anat.* motor centre
~ движения *mech.* centre of motion
~, деловой business centre
~ деформации strain centre
~, дыхательный respiratory centre
~ жесткости *av.* elastic/flexural centre
~ жиросчетов *econ.* giro centre
~, задерживающий *anat.* inhibitory centre
~, задний токарный tailstock centre
~, закладной *geod.* lower/underground centre
~ захвата trapping site
~ захвата электрона electron trapping site
~ землетрясения seismic/earthquake focus/origin
~ земного шара geocentre
~ зоны генерации клистрона centre of a klystron
~, зрительный *anat.* visual centre
~ издержек *econ.* cost centre
~ изгиба flexural centre
~ изображения centre of image
~ инерции centre of mass/inertia
~, информационно-вычислительный dataprocessing (computer) centre, dp/information centre
~, информационный documentation centre
~ качаний/колебаний centre of oscillation
~, кашлевой *anat.* coughing centre
~, клеточный cell centre
~, коммутационный *com.* switching centre/office
~ конденсации condensation centre/nucleus
~ конденсации, остаточный *nucl.* reevaporation nucleus
~ концентрации focus of concentration
~, корковый *anat.* cortical centre
~ крена roll centre
~ кривизны centre of curvature
~ кристаллизации crystallization nucleus, *cryst.* grain of crystal
~ кручения *mech.* centre of twist, shear centre
~ люминесценции luminescence centre
~ масс centre of mass/inertia
~, мгновенный instant(aneous) centre
~ метаболизма *anat.* metabolic centre
~ насыщения centre of saturation
~, научно-исследовательский research centre, R&D centre
~, неподвижный *tech.* dead/fixed centre
~ обрабатывающей промышленности manufacturing centre
~ обнаружения и наведения *nav.* detection-and-control centre
~, обонятельный *anat.* olfactory centre
~ обработки данных data-processing/-reduction centre
~, обратный (токарный) reverse/female centre
~ окостенения *anat.* centre of ossification
~ оледенения, лаврентьевский *geol.* Laurentian centre of glaciation
~ оседания *geol.* focus of foundering
~ описанной окружности *math.* circumcentre
~ парусности *naut.* centre of effort/sail
~ питания *el.* main substation
~ плавучести centre of buoyancy
~ повреждения damage centre
~ подобия centre of similitude, homothetic centre
~, полуовальный *anat.* semioval/medullary centre
~ потока core of flow
~ предварительной обработки информации information filter centre
~ прилипания trap(ping) centre
~ приложения тектонических сил *geol.* thrust centre
~, примесный impurity site
~ проектирования/проецирования *math.* centre of projection, station point
~ промежности, сухожильный *anat.* perineal body
~, промежуточный intermediate centre
~ проявления development nucleus/centre
~, психомоторный *anat.* psychomotor/psychocortical centre
~ пучка *math.* vertex of a pencil
~ равновесия centre of equilibrium
~ распространения centre of distribution/spread
~ рассеивания centre of dispersion
~, рассеивающий scattering centre
~, рвотный *anat.* vomiting centre
~, рефлекторный reflex centre
~ речи *anat.* speech/Broca's centre
~ ряда *math.* centre of a series
~ сбора информации information collector
~ сбора и обработки метеорологической информации weather record-processing centre
~ связи communications centre
~, сенсорный *anat.* sensory/sense centre, sensorium
~ сил/силовой centre of force
~ симплекса midpoint of a simplex
~, слуховой *anat.* auditory centre
~, слюноотделительный *anat.* salivary centre
~ сосредоточения населения *econ.* population centre
~, сосудодвигательный *anat.* vasomotor centre
~, сосудорасширяющий *anat.* vasodilator centre
~, сосудосуживающий *anat.* vasoconstrictor centre
~ стяжения contraction centre
~, теплорегулирующий *anat.* heat/calorifacient centre
~ технического обслуживания maintenance

centre
~, **торговый** shopping centre; mart
~ **тяжести** centre of gravity; c.g., c-of-g
~ **тяжести фигуры** centroid
~ **удара** centre of percussion/impact/shock
~ **урагана** storm centre
~, **учебный** training centre
~ **циклона** centre of cyclone/low
~, **чихательный** *anat.* sneezing centre
~ **шарнира** fulcrum
~ **эякуляции** *anat.* ejaculation centre
централ *hist.* central prison
централизатор *math.* centralizer; *rail.* interlocker
централизация centralization; interlocking; integration
~, **диспетчерская** centralized traffic control
~, **маршрутная** interlocking system, route control interlocking
~, **релейная** all-relay interlocking
централизм *polit.* centralism
централизованный centralized; one-/singlepoint; interlocked
централь *tech.* central unit, main/power line; *math.* central
~, **гироскопическая** central gyro(scopic) unit
~ **скорости и высоты** airspeed and height central unit
центральновершинный circumflex
центрально-разностный *math.* central-difference
центрально-симметричный centrally symmetric
центральный central, centre *(attr.)*
~ **нападающий** *sport* centre-forward
~**ная печать** leading newspapers
~**ное отопление** central heating
~**ные газеты** newspapers published in the capital
центрантус *bot.* centranth; Jupiter's-beard centranthus *(Centranthus ruber)*
центратор *opt.* centring mount
центризм *polit.* centrism
центриола, центриоль *biol.* centriole
центрипетальный centripetal
центрирование cent(e)ring; alignment
~ **изображения** image cent(e)ring
~ **луча** *telev.* beam alignment
центрировка centering
центрифуга 1. centrifuge, centrifugal machine; 2. spin drier
~, **бесситовая** solid-bowl centrifuge
~, **быстроходная** ultra-centrifuge
~, **ионная** magnetron, ionic centrifuge
~, **осадительная/отстойная** centrifugal sedimentor
~, **осветлительная** polisher *(in sugar industry)*
~, **противоточная** counter-current centrifuge
~, **проточная/прямоточная** concurrent centrifuge, co-current centrifuge, flow-through centrifuge
~, **рефрижераторная/охлаждаемая** cold/refrigerate centrifuge
~ **с вибрационной выгрузкой** vibrating/oscillating centrifuge
~, **саморазгружающаяся** self-discharging centrifuge
~ **со шнековой выгрузкой** scroll-discharge centrifuge
~, **фильтрующая** centrifugal filter
~, **шламовая** slurry centrifugal separator
центрифугал centrifugal reel
центрифугат centrifugalized deposit, centrifugate
центрифугирование centrifuging, centrifugation, centrifugalization, centrifugal action/separation
~, **отстойное** sedimentation centrifuging
~, **противоточное** counter-streaming centrifuging
центрифугировать centrifuge, centrifugalize
центрический centric(al)
центроаффинный centered-affine, centroffine
центробежный centrifugal, axifugal; *physiol.* efferent, eisodic
центровальный center(ing)
центрование 1. dotting, marking (with centre punch); centering; 2. *mach.* centre-drilling
центровать 1. centre, align; 2. *mach.* drill centres
центровка 1. *mach.* centre-drilling; centre drill; 2. *av.* centre-of-gravity position/location, c.g. position; 3. alignment; centring; 4. dotting, marking
~, **задняя** aft c.g. position/location
~, **нейтральная** neutral point
~, **передняя** forward c.g. position/location
~ **самолета** aeroplane c.g. position
~, **фактическая** actual c.g. position
центровой centric, central, centre *(attr.)*
~**вое сверло** centering drill
центровочный centering
центрогенный centrogenic, centrogene
центроид(а) centroid, centrode
~, **неподвижная** fixed centrode
~, **подвижная** body centrode, centrode of body
центроидный centroidal
центроклиналь *geol.* centrocline
центролецитальный *zool.* centrolecithal
центромер(а) *biol.* centromere
~, **множественный** multiple centromere
~, **полулокализованный** semilocalized centromere
центроплан *av.* centre(-wing) section, centre(section) wing
центросома *biol.* centrosome
~ **яйцеклетки** ovocentre
центросимметричный centrally symmetrical, centrosymmetrical
центростремительный centripetal; *physiol.* afferent, abterminal
центросфера *biol.* centrosphere, attraction sphere; *geol.* barysphere, pyrosphere
центротехника hierarchical control
центрфорвард *sport* centre-forward
центумвир *hist.* centumvir, Roman judge for civil

suits
центумвират *hist.* centumvirate
центурий *chem.* centurium, Cn
центурион *hist.* centurion, commander of a century
центурия *hist.* century
ценур *zool.* coenurus
ценуроз *vet.* coenurosis, turn sickness, blind staggers
ценхрус *bot.* bur grass, sandbur *(Cenchrus)*
цены *pl.* prices; *see also* **цена**
~ в магазине store prices
~ внутреннего рынка domestic/internal prices
~, действующие established prices
~, единые uniform prices
~, множественные multiple prices
~, неокругленные odd prices
~ при продаже за наличные spot prices
~, приведенные adjusted prices
~, сметные estimate prices
~, установленные established prices
~, устойчивые высокие hard prices
цеолит *miner.* zeolite
цеолитный, цеолитовый zeolitic
цеофиллит zeophyllite
цеп *agr.* flail (knife)
~, измельчающий chopper flail
~, молотильный threshing flail
~, шарнирный free-swinging flail
цепенелый benumbed, numb; torpid
цепень *zool.* tapeworm *(Taenia)*
~, бычий beef/hookless/unarmed tapeworm
~, карликовый dwarf tapeworm
~, свиной armed/pork/salitary tapeworm
~, эхинококковый hydatid tapeworm
цепеобразный cateniform, catenate, catenulate; chainlike
цепкий 1. catching, adhesive, clinging, clingy; 2. tenacious, prehensile; scansorial; 3. keen; perceptive, retentive
~кое растение *bot.* clinging plant, climber
цепной catenary, chain(omatic)
~ мост chain bridge
~ная линия ladder network; *math.* catenary (line)
~ная реакция *chem.* chain reaction
~ная собака watchdog
цепнореагирующий chain-reacting
цепочечный chainlike, catenate, catenulate
цепочка (small) chain; *el.* network; series; *comp.* file
~ адресов address chain
~ векторов vector train
~ вызовов/запросов call(er) chain
~ двоичных разрядов bit chain/string
~, дифференцирующая *el.* differentiating network
~, кодовая code chain
~ командных слов command word chaining
~ логических схем logic chain
~ логического вывода inference chain
~ подчинения *comp.* owner-member chain (in a data-base)
~, радиоактивная radioactive chain/series
~ радиоактивных распадов, ~ распада decay/disintegration chain
~ рассуждения reasoning chain
~ хеширования hash chain
цепочковидный, цепочкообразный catenulate, catenate, cateniform, chainlike
цепочконосный cateniferous
цеппелин *hist.* Zeppelin (airship)
цепь 1. *el.* (electric) circuit; cable network; 2. *mech., chem.* chain; *math.* edge train, trail, chain progression; 3. *geol.* catena; chain, series; range; row; 4. *mil.* file, line
~ аварийной защиты *el.* safety/protective circuit
~, автоколебательная astable circuit
~, аукционная auctioneer circuit (responding to the maximum signal)
~, барханная chain of sand dunes
~ без контрафорсов/распорок studless chain
~, бесконечная continuous/endless chain
~ блокировки *el.* blocking/holding/lockout circuit
~, боевая firing circuit
~, боковая *chem.* side chain
~, брюшная нервная ventral cord
~, буксирная tow(ing) chain
~, вертлюжная buckle chain
~ взрывателя fuse chain
~, взрывная explosive train
~ возбуждения *el.* excitation/exciting/drive circuit
~ воздействий control/actuating path
~ воздействий, основная main control chain/loop
~, воздушная *el.* overhead circuit, open-wire circuit
~, выбирать якорную *naut.* heave on the chain
~ выключения по периоду period trip circuit
~, высевающая *agr.* feed chain
~ высокого напряжения *el.* high-tension/-voltage circuit
~, гамильтонова *math.* Hamilton's arc
~ главной передачи *auto.* final drive chain
~, Г-образная *el.* L-(section) network
~, горная mountain range/chain
~, горная дугообразная mountain arc
~, грузовая/грузоподъемная lifting/loading chain
~, гусеничная track/crawler chain
~, демпфирующая damping/antihunt/buffer circuit
~ держащего якоря riding chain/cable
~, задерживающая/задержки delay/lag line/circuit
~ задержки тактовой дорожки clock-track delay circuit
~ зажигания ignition circuit

~, зажимная gripping chain
~, закороченная short(ed) circuit
~, замкнутая ring chain; *el.* closed/completed circuit, loop
~ защиты protection/protective/guard circuit
~, звеньевая link chain
~, землемерная *geod.* surveyor's chain
~, зубчатая toothed chain
~ инстанций chain of command
~, искусственная несимметричная simplexed circuit, half-/earthed-phantom circuit
~, искусственная симметричная *com.* phantom circuit
~, ковшовая (экскаватора) bucket chain
~, компенсирующая compensating network
~, конвейерная conveyer chain
~, контрольная/контроля *el.* monitoring/test/control circuit
~ контроля уровня нейтронного потока neutron-flux control circuit
~, корректирующая *el.* compensating circuit
~ коррекции гироскопа erection circuit
~, крючковая hook-link chain
~, логическая logic(al) circuit
~, ложная parasitic circuit
~ Маркова/марковская *math.* Markov's/Markovian chain
~ межкаскадной связи interstage circuit
~, мерная *geod.* surveyor's/poll chain
~ мертвого якоря mooring chain
~, многозвенная *el.* iterated/ladder network
~, многофазная *el.* polyphase circuit
~, мостовая full-bridge network
~ нагрузки load circuit
~ накала *el.* filament/heater circuit
~, направляющая guide chain
~, незамкнутая *el.* open/incomplete circuit
~, неразветвленная *el.* series circuit; *chem.* unbranched chain
~, неуплотненная *com.* single-channel circuit
~, нечистая якорная foul cable/hawse
~ обвязочная (для подъема грузов) lashing/sling chain
~, обесточенная *el.* dead circuit
~ обратной связи *el.* feedback, reaction circuit; *comp.* feedback path
~, огневая explosive/ignition train
~ озер series of lakes
~ опережения lead circuit
~ оповещения alert circuit
~, основная *el.* main circuit; *chem.* main chain
~ ответвления *com.* tap circuit
~ отключения *el., electr.* disabling circuit
~ отпирания *el., electr.* enabling circuit
~ переменного тока *el.* a.c. circuit, alternating-current circuit
~ переноса *comp.* carry circuit
~, печатающая print chain
~ Пикара *com.* simplexed circuit, half/earthed-phantom circuit
~ питания supply/power/feed circuit, power-supply circuit
~, пластинчатая leaf/laminated chain
~, П-образная *el.* pi-(section) network
~ подводных гор *geol.* seamount chain
~, подводящая gathering chain
~, подъемная hoisting chain
~, последовательная *el.* series circuit
~ постоянного тока d.c. circuit, direct-current circuit
~, приводная driving/sprocket chain
~, простая *math.* path, simple chain progression
~ противоскольжения antiskid/tyre chain, nonskid chain
~, пусковая starting/trigger circuit
~ равенств *math.* continued equality
~, разборная dismountable/detachable chain
~, разветвленная *el.* forked/bifurcated/tree/parallel circuit; *chem.* branched/branching chain
~, развязывающая *el.* isolation/isolating network
~ размыкания маршрутов *rail.* route release circuit
~, разомкнутая *comp.* open circuit
~ реагирующих веществ reaction chain
~ регулирования *auto., comp.* control circuit; *nucl.* control loop
~, режущая *min.* cutting chain
~, резервирующая *mech.* redundant circuit
~, резервная stand-by circuit
~, релейная relay circuit
~, рельсовая track circuit, ground return
~, решающая computing network
~, роликовая roller chain
~ с контрафорсами/распорками studded chain
~ самоблокировки *el.* self-holding circuit
~, сварочная welding circuit
~ связи communication/coupling circuit
~, сглаживающая levelling circuit
~ сетки grid circuit
~ сдвига *comp.* shift(ing) circuit
~ синхронизации *electr.* sync circuit
~, сквозная *com.* built-up circuit, through circuit
~, скребковая flight chain
~ смещения *electr.* bias circuit
~ сопряжения *chem.* conjugated chain
~, составная *el.* composite/compound circuit
~ стоп-анкера stream chain
~, стробирующая *electr.* gate circuit
~, суперфантомная *com.* double phantom circuit
~ считывания *comp.* read(ing) circuit
~ толкателя *min.* haul chain
~, тормозная drag/locking chain; *hydr.* braking circuit
~, транспортерная conveyer chain
~, тяговая hauling/haulage/pull chain
~ управления control circuit
~, фазовращающая/фазосдвигающая

phase-shift filter, phase-shifting circuit
~, хронирующая *electr.* clock/timing circuit
~ цепного канавокопателя, режущая trenching chain
~, шарнирная articulated-link chain, pintle chain
~, штыревая pintle chain
~, эйлерова *math.* Eulerian chain
~, якорная *naut.* anchor chain/cable
~ якоря *el.* armature circuit
цепь-волокуша *timb.* brake chain
церазин *chem.* cerasin; cherry-tree gum
церамогалит *geol.* ceramohalite, keramohalite
цератозавр *paleont.* ceratosaur *(Ceratosaurus)*
цератозамия *bot.* horncone *(Ceratozamia)*
цератоптерис *bot.* horn/water fern *(Ceratopteris)*
цератофиллин *bot.* ceratophylline
Цербер *myth.* Cerberus
церва *bot.* dyer's rocket/weed *(Reseda luteola)*
цервикальный *anat.* cervical
цервико-везикальный *anat.* cervicovesical
цервикопластика *surg.* cervicoplasty
цервицит *med.* cervicitis, trachelitis
~, трихомонадный mycotic cervicitis
церебеллярно-рубральный *med.* cerebellorubral
церебеллярный cerebellar
церебральный cerebral
цереброваскулярный cerebrovascular
церебрастения *med.* psychasthenia, cerebrasthenia
цереброн cerebron, phrenosin
церебропатия *med.* encephalopathy, cerebropathy
церебросклероз *med.* encephalosclerosis, cerebrosclerosis
цереброскопия *med.* encephaloscopy, cerebroscopy
цереброспинальный *anat.* cerebrospinal, cerebrorrhachidian, encephalorrhachidian
церебростерин cerebrosterol
церезин ceresin(e) (wax)
церемониал ceremonial, ritual
~ом, пройти *mil.* march past
церемониальный ceremonial, ritual, solemn
~ марш march past
церемониймейстер master of ceremonies
церемония ceremony; *pl.* formality, formalities
Церера *astr., myth.* Ceres
цереус *bot.* cereus *(Cereus)*
церий *chem.* cerium, Ce
~, йодноватокислый ceric iodate
церин cerin
церка *zool.* cercus
церкарий, церкария *zool.* cercaria, larval stage
церковник 1. clergyman; 2. churchgoer, churchman
церковнический ecclesiastic(al); clerical, church *(attr.)*
церковнокнижный church-literary
церковноприходский *eccl.* parish
церковнорелигиозный church-religious
церковнославянский Church Slavonic
церковнослужитель priest, minister, clergyman
церковный ecclesiastic(al); church(ly); clerical
~ приход parish
~ староста churchwarden
~ сторож sexton
~ное право canon/ecclesiastical law
~ные суды *hist.* spiritual courts
церкопода *zool.* cercopod
церкосепториоз *bot., pathol.* cercoseptorial leaf spot
церкоспореллез *bot., pathol.* cercosporella leaf spot
церкоспороз cercosporosis
церонограф ceraunograph
церопластика ceroplastics
церцис *bot.* Judas tree *(Cercis siliquastrum)*
цесаревич *hist.* cesarevitch, heir to the Russian throne
цесарка *zool.* guinea fowl *(Numida meleagris)*
цесарский *hist.* Caesarean, Caesarian
цессатив *ling.* terminative
цессионарий *law* cessionary, assign(ee), transferee
цессия *law* cession, transfer, assignment
цестода *zool.* cestoid, cestode
~ды, настоящие polyzootic cestodes *(Cestoda)*
цестодообразные *n. pl. zool.* monozootic cestodes *(Cestodaria)*
цеструм *bot.* jessamine, cestrum *(Cestrum)*
цестус *hist.* cestus, ancient boxing-glove
цетрарин *chem.* cetrarinic acid, cetrarin
цетрария *bot.* Iceland moss *(Cetraria islandica)*
цефалантус *bot.* button bush *(Cephalanthus)*
цефаласпида *paleont.* cephalaspis *(Cephalaspis)*
цефализация cephalization
цефалин *bioch.* cephalin, kephalin
цефалогематома *med.* cephal(o)hematoma, cephal(o)hematocele
цефалон *zool.* head shield, cephalon
цефалоподы *pl. zool.* cephalopod(an)s *(Cephalopoda)*
цефалоспориоз 1. *med.* infestation with Cephalosporium; 2. *bot., pathol.* Cephalosporium stem streak
цефалотаксус *bot.* plum yew *(Cephalotaxus)*
цефалотом *surg.* cephalotome
цефалоторакс *zool.* cephalothorax
цефальгия *med.* headache, cephalalgia, cerebralgia
Цефей *astr.* Cephei; *myth.* Cepheus
цех 1. shop, department, section; shed; factory workshop; plant 2. *hist.* guild, corporation
~, бессемеровский Bessemer plant
~ блюминга blooming mill department
~, бондарный cooperage shop
~, брошюровочный book-stitching shop
~, вспомогательный auxiliary/service shop, service department
~ глубокой печати gravure department
~, достроечный fitting-out shop
~, дубильный fan room, fanyard
~, заготовительный blanking shop

~, зольный *leath.* lime yard
~, инструментально-штамповый tool-and-die shop
~, инструментальный tool(maker's) shop, toolroom
~, консервный canning/preserving shop
~, корпусный hull/structural shop
~, кузнечный forge/blacksmith shop
~, литейный foundry
~, лудильный tinning plant
~, мартеновский open-hearth plant
~, механический machine shop
~, модельный model/pattern shop
~, монтажный assembly/erecting/fitting/ *el.* wiring shop
~, наборный composing room
~ обработки и удаления отходов *nucl.* waste processing and disposal plant
~, опытный development/experimental/pilot shop
~, основной producing/production department
~ отгрузки готовой продукции shipping department
~, отделочный finishing shop/department; *leath.* currying shop
~, передельный *met.* rerolling department
~, переплетный book bindery, bookbinding department
~, печатный printing department, pressroom
~, подготовительный stock-preparation shop
~ предварительной сборки prefabrication shop/shed
~, прокатный rolling-mill shop
~, разливочный *met.* casting plant
~, ремонтно-механический mechanical repair shop
~, ремонтный repair/maintenance shop
~, сборочный assembling/assembly/erecting shop, assembling department
~, сталеплавильный steel-making plant
~, такелажный rigging shop
~, формный *print.* plateroom, plate department
~, формовочный moulding shop
~, чугунолитейный iron foundry
цехин zecchino, zecchine, sequin
цехком(итет) *abbr.* **цеховой комитет** shop committee
цеховой (pertaining to the) shop; *fig.* limited
~вая нормаль shop practice standard
цеце *zool.* tsetse (fly) *(Glossina morsitans)*
цецидий cecidium, gall
циан *chem.* cyanogen *(gas);* cyan
~, хлористый cyanogen chloride
цианазид *chem.* cyanogen azide
цианамид *chem.* cyanamid(e)
~ кальция calcium cyanamide, lime nitrogen
циангидрин *chem.* cyanhydrin, cyanalcohol
цианидин *bioch.* cyanidin
цианизация cyanidation, cyanization
цианин cyanine
цианирование cyanidation, cyanide process; *met.* cyaniding; casehardening with cyanides
~ полным иловым процессом all-slime cyanidation/leaching
цианистоводородный hydrocyanic, hydrocyanide (of)
цианистый *chem.* cyanic, cyanide (of)
~ калий potassium cyanide
циановодород hydrogen cyanide
циановокислый cyanate (of)
цианоз *med.* cyanosis, cyanoderma, blue disease
~ конечностей acrocyanosis
~, поздний cyanose tardive
цианозит cyanosite, chalcanthite
цианокобаламин *bioch.* cyanocobalamin, vitamin B_{12}
цианотипия cyanotype
цианотический, цианотичный cyanotic
цианотрихит *miner.* cyanotrichite, lettsomite
цианофильный cyanophil(e)
цианофицин *bioch.* phycocyanin, cyanophycin
циануровый cyanuric
циатея *bot.* tree fern *(Cyathea)*
циатий *bot.* cyathium
цибетин civet, zibetum
циботиум *bot.* Scythian lamb *(Cibotium barometz)*
циветта *zool.* civet *(Viverra)*
цивилизованный civilized
цивилист *law* civilian, specialist in civil law
цивилистика *law* civil law
цивильный civil(ian)
~ лист civil list
цивический *hist.* lay, secular; civil(ian)
цигаденус *bot.* grassy death camas *(Zygadenus intermedius)*
цигер cheese albumin
цигетропин cyhetropine
цикада *zool.* cicada *(Cicadidae)*
цикадка *zool.* leafhopper *(Cicadellidae)*
цикадовые *n. pl. zool.* cycads *(Cycadaceae)*
цикадофиты *pl. bot.* cycadophytes *(Cycadophyta)*
цикатрикс, цикатриса cicatrix, cicatrice
цикл cycle; period; *math.* cyclic/circular permutation; 2. iteration, run; 3. loop
~, базисный reference cycle
~, бесконечный endless loop
~ Брайтона Brayton cycle
~ в программе program cycle/loop, loop of command/instruction/order
~ в цикле loop-within-loop
~, вековой secular cycle
~, вложенный nested loop
~, водородный *nucl.* hydrogen-helium cycle
~ воспроизводства breeding cycle
~, вулканический volcanic cycle/sequence
~ выборка-исполнение *comp.* fetch-execute cycle
~ выборки access cycle
~ выборки команды *comp.* instruction cycle
~ выгорания burn-out cycle
~ горообразования *geol.* orogenic cycle

~, двойной dual/binary cycle
~, двухтактный two-stroke cycle
~ дезактивации decontamination cycle
~, действительный actual/real cycle
~ долговечности life cycle
~, долговременный secular cycle
~, жевательный chewing/masticating cycle
~, жизненный (системы) activity/life cycle; life history
~ заданной длительности timed loop
~, закрытый/замкнутый closed cycle/loop
~ записи *comp.* write cycle, cycle of write
~, захваченный *comp.* stolen cycle
~, знакопеременный alternating cycle
~, идеальный ideal/perfect cycle
~, итерационный iteration loop
~ Карно/Отто Carnot/Otto cycle
~ колебаний cycle of fluctuation
~, командный *comp.* instruction cycle
~, кондиционный conditioning cycle
~ коррекции (гироскопа) erection cycle
~ лавы *min.* wall cycle
~ летных испытаний flight-test cycle
~, листовой foliar cycle
~, лунный node/lunar cycle
~, магнито-гидродинамический MHD cycle
~ максимального срока службы maximum-life cycle
~, машинный *comp.* machine/computer cycle
~ на быстрых нейтронах fast-neutron cycle
~ на тепловых нейтронах slow-neutron cycle
~ напластования *geol.* stratification cycle
~ напряжений *mech.* stress cycle
~, незамкнутый/открытый open cycle
~, необратимый irreversible cycle
~, непрерывный uninterrupted cycle
~ обеззараживания decontamination cycle
~, обратимый reversible cycle
~ обращения к памяти *comp.* memory/storage cycle
~ обслуживания *econ.* service cycle
~, овариальный *med.* ovarian cycle
~, однократный once-through operation
~, одноступенчатый single-stage cycle
~ ожидания waiting cycle/loop
~ ожидания освобождения (устройства) *comp.* busy-wait loop
~, окислительно-восстановительный oxidation-reduction cycle, redox cycle
~, орнитиновый *med.* Kreb's urea/ornithine cycle
~, основной *therm.* basic/ *comp.* main/ *geol.* major cycle
~, охватывающий outer loop
~ очистки decontamination cycle
~ памяти memory cycle
~, пароводяной water-flow cycle; water-steam circuit
~ переработки processing cycle
~ печати print cycle
~, повторный recycle
~ поиска *comp.* search cycle
~, посадочный *min.* cycle of caving
~ последовательности cycle of sequence/succession
~, предельный *el.* limit cycle
~, предстартовый countdown cycle
~, приливо-отливный tidal cycle
~, программный program-routine/loop
~ программы *comp.* loop of instructions
~, производственный production cycle
~, протонный *nucl.* proton-proton chain
~, проходческий sinking cycle
~, прямоточный once-through cycle
~ пуска launch cycle
~, пусковой *auto.* starting cycle
~, рабочий working/operational/operating/running cycle; *comp.* operation loop
~ размножения топлива *nucl.* fuel-breeding cycle
~, разомкнутый open cycle/loop
~ разрыв-слияние-мост breakage-fusion-bridge cycle
~ Ранкина/Ренкина *therm.* Rankine cycle
~, регенеративный/регенерации regenerative/recovery/regeneration/reprocessing cycle; heater-extraction cycle
~ роста *econ.* rate-of-growth cycle
~ с воспламенением от сжатия Diesel cycle
~, сборочный assembly cycle
~, симметричный symmetrical cycle
~, сложный loop-within-loop; *therm.* compound cycle
~ со вторичным перегревом пара reheat cycle
~ со сгоранием при постоянном давлении Diesel cycle
~ со сгоранием при постоянном объеме Otto cycle
~, стационарный stedy-state cycle
~ стирания *comp.* erase cycle
~, суточный daily/diurnal cycle
~ считывания (и записи) read(-write) cycle
~, тепловой thermal/heat cycle
~, топливный *nucl.* fuel cycle, burn-out cycle
~ трикарбоновых кислот *chem.* tricarboxylic acid cycle, citric acid cycle; *med.* Krebs cycle
~, тройной ternary cycle
~ турбореактивного двигателя turbojet cycle
~, углеродный *nucl.* carbon(-nitrogen) cycle, Bethe cycle
~ хронирования *electr., comp.* timing cycle
~, четырехтактный four-stroke cycle
~, частичный minor/partial cycle
~, эстральный *anat.* (o)estrous cycle
цикламен *bot.* cyclamen *(Cyclamen)*
цикламид, цикламин *chem.* cyclamin(e)
циклахена *bot.* sump weed *(Iva)*
циклевание, циклевка scraping

циклида *math.* cyclide
циклизация *chem.* ring formation, cyclization
~, внутримолекулярная intrachain cyclization
~, межмолекулярная interchain cyclization
циклизин cyclizine, marezine
циклит *med.* cyclitis
~, гнойный purulent cyclitis
~, неосложненный pure/uncomplicated cyclitis
~, серозный serous/simple cyclitis, punctate keratitis, descemetitis
циклический cyclic(al), reccuring, repetitive; *math.* ignorable
~ код recurrent/cyclic code
~кая команда *comp.* repetitive command/order/ instruction
цикличность cyclicity, cyclic recurrence, cycle of opeartions
цикличный, цикловой cyclic(al)
циклобарбитал *pharm.* cyclobarbital, cyclobarbitone, phanodorm
циклогексан *chem.* cyclohexane
циклогексанол *chem.* inosital, cyclohexanol, cyclohexitol
циклогенез *meteor.* cyclogenesis
циклограмма sequence/timing diagram, cyclogram
~, круглая cyclogram
~ пуска countdown graph
циклодол benzhexol (hydrochloride)
циклодром *sport* cycle race track
циклоз(ис) *biol.* cyclosis, circulation
циклоида *math.* cycloid
~, обратная inverted cycloid
циклоидальный cycloidal
циклоидный cycloid
циклокаучук cyclorubber
цикломер cycle timer/counter, cyclometer
циклометиазид cyclopenthiazide
циклометр cyclometer, cycle timer/counter
цикломицин *pharm.* tetracycline, aureomycin, terramycin
циклон 1. *meteor.* cyclone, depression, area of low pressure, low(-pressure area); 2. *mach.* cyclone separator/collector
~, батарейный multiclone
~, внутрибарабанный internal cyclone steam separator
~ в тылу за антициклоном wake cyclone
~, высокий upper cyclone
~, зараждающийся incipient/nascent cyclone
~, неподвижный permanent low
~, развивающийся developing cyclone/low
~ с замкнутыми изобарами closed low
~, сепарационный cyclone steam separator
~, углубляющийся deepening cyclone
циклонический, циклонный cyclonic(al)
циклообразование *meteor.* cyclogenesis
циклоп 1. *myth.* Cyclops; 2. *zool.* cyclops, freshwater copepod *(Cyclops)*
циклопический 1. *myth.* Cyclopean, Cyclopian; 2. *zool.* cyclopean
циклопия *med.* cyclopia, synophthalmia
циклоплегия *med.* paralysis of accommodation, cycloplegia
циклополимеризация cyclic polymerization, cyclopolimerization
циклопропан *chem.* cyclopropane, trimethylene
циклорама cyclorama, circular panorama
циклорасщепитель cycle splitter
циклостомовый *zool.* cyclostomate
циклотема *geol.* cyclothem
циклотимический cyclothymi(a)c
циклотимия *med.* cyclothymia
циклотрон *phys.* cyclotron
~, импульсный/мигающий pulsed cyclotron
~, классический conventional cyclotron
~, малый omegatron
~, однодуантный one-dee cyclotron
~, сверхмощный bevatron
циклофон *electr.* beam-switching tube, electron-beam commutator
циклофория *med.* cyclophoria
~, аккомодационная accommodative cyclophoria
циклофосфамид, циклофосфан *pharm.* cytoxan, cyclophosphamide
циклофрения *med.* cyclothymia, cyclophrenia
циклоэнцефалия *med.* cyclocephalia, cyclencephalia
цикля *tech.* scraper; *leath.* slecker, slicker, spokeshave
цикнохес *bot.* swan orchid *(Cycnoches)*
цикoвка *tech.* spot-facing
цикорий, цикория, цикорник *bot.* chic(c)ory, succory *(Cichorium);* chicory (root)
цикута *bot.* water/poison hemlock, conium, spotted cowbane poison, spotted parsley *(Cicuta)*
цикутотоксин *chem.* cicutoxin
цилиарный *biol.* ciliary
цилиндр 1. *math.* cylinder; *tech.* cylinder, barrel, drum; roll(er); *med.* cast; 2. top hat
~ Бенс Джонса *med.* Bence Jones cylinder
~, вертикальный vertical/upright cylinder
~, восковидний waxy cast
~, вращающийся rotating cylinder
~, выносной external/independent cylinder
~, выпускной *text.* discharging/delivery roller
~, высокого давления high-pressure cylinder
~, вытяжной *text.* drawing roller
~, гиалиновый hyaline cast
~, гидравлический hydraulic/water cylinder
~, гидравлический силовой hydraulic jack/ ram
~ гидроподъемника hydraulic lift(ing)/ram cylinder
~, градуированный graduated/measuring cylinder
~, жировой fatty cast
~ золотника valve barrel
~, колонковый core barrel
~, коматозный coma cast, Kulz's cylinder

~, **контрольный** check cylinder
~, **кровяной** blood cast
~, **круговой** *math.* cylinder of revolution, circular cylinder
~, **ложный** spurious/false/mucous cast, pseudocast
~, **мерный** graduated/measuring cylinder
~, **мочевой** urinary cast
~, **нажимной** pressure roller
~, **наклонный** oblique/inclined cylinder
~ **насоса** pump barrel
~, **неподкрепленный** unstiffened cylinder
~, **обтекаемый со скольжением** yawed cylinder
~, **оребренный** finned/ribbed cylinder
~, **осевой** axis cast, axon
~, **отлитый отдельно** separately/single-cast cylinder
~, **офсетный** offset/blanket/transfer cylinder
~, **передаточный** *print.* offset/blanket/transfer cylinder
~, **печатающий** character cylinder, print roll
~, **печатный** impression cylinder
~, **питающий** *text.* feed roller
~, **подкрепленный** stiffened cylinder
~, **подкрепленный кольцами** ring-stabilized cylinder
~ **подъема жатки** table lift ram, platform adjustment cylinder
~, **почечный** renal cast
~, **прессующий** *met.* injection/shot cylinder
~ **привода питателя** feed drive cylinder
~ **привода пресса, гидравлический** *agr.* ram drive cylinder
~, **прижимный (штампа)** hold-down cylinder
~, **разгрузочный** unloading/discharging cylinder
~, **раскатный** *print.* drum roller
~, **ребристый** ribbed/finned cylinder
~ **с водяной рубашкой** *aut.* (water-)jacketed cylinder
~, **силовой** (actuating) power cylinder, actuating cylinder, ram, jack
~, **силовой подъемный** elevating ram
~, **тормозной** brake cylinder
~, **тормозной главный** brake master cylinder, main brake cylinder
~ **уборки (шасси)** *av.* retraction cylinder
~ **управления** *mech.* control cylinder
~ **управления жаткой** table control ram
~, **усеченный** *math.* truncated cylinder
~ **Фарадея** *el.* Faraday cup/cylinder
~, **фибринозный** fibrinous cast
~, **формный** *print.* form/plate cylinder
~ **экструдера** barrel cylinder
~, **эпителиальный** epithelial cast
цилиндрический cylindric(al), cylinder-shaped; terete
цилиндровый cylindric(al)
цилиндроид 1. *math.* cylindroid; 2. *med.* mucous/spurious/false cast, pseudocast
цилиндроклеточный columnar-celled
цилиндроконический cylindro-conical
цилиндроспориоз *bot., pathol.* cylindrosporium leaf spot
цилиндрурия *med.* cylindruria
цимбалы *pl. mus.* cymbals
цимбалярия *bot.* basket ivy *(Cymbalaria)*
цимбиморфоз *vet.* cymbimorphiasis
цимозный *bot.* cymose, cymous
цимоидный cymoid
цимофан *miner.* cat's eye, chrysoberyl, cymophane
цинанхум *bot.* swallowwort *(Cynanchum)*
цинародий cynarrhodion
цинга *med.* scurvy, scorbutus
~ **у детей** infantile scurvy, Barlow's/Moeller's/Cheadle's disease, acute rickets
цингибер ginger
цинготный scorbutic, scurvied
~ **больной** scorbutic
цинерариелистный *bot.* cinerariaefolious, cineraria-leaved
цинерария *bot.* cineraria *(Cineraria)*
цинерит *geol.* volcanic tuff, cinerite
цинизм, циничность cynism
цинический, циничный cynical
цинк *chem.* zinc, Zn
~, **азотнокислый** zinc nitrate
~, **гранулированный** mossy/granulated zinc
~, **листовой** sheet zinc
~, **рафинированный** high-grade zinc
~, **рубиновый** ruby zinc/blende, zincite
~, **сернистокислый** zinc sulfate
~, **технический** commercial zinc
~, **углекислый** carbonate of zinc, smithsonite
~, **хлористый** zinc chloride
цинкит *miner.* zincite, red zinc ore, ruby zinc/blende, spartalite
цинкование galvanizing; sherardizing; zinc plating, zincification
~ **горячим способом** hot(-dip) galvanizing
~, **электролитическое** galvanizing by electrodeposition
цинковать (coat with) zinc, galvanize
цинковый zincic, zincous; zinciferous; zincoid
цинкогравюра zincogravure
цинкограф zincographer
цинкография zincography; zincographer's shop
цинкографский zincographic(al)
цинкосодержащий zinciferous
цинна *bot.* wood reed *(Cinna)*
цин(н)ия *bot.* zinnia *(Zinnia)*
циновка mat
цинубель tooth/scraper plane
цинхонин *chem.* cinchonine
циперус *bot.* flat sedge *(Cyperus)*
цирик Mongolian soldier
цирк 1. *geol.* cirque, corrie; 2. circus
~, **ледниковый** corrie glacier, glacial cirque
~, **лунный** lunar cirque
цирконий *chem.* zirconium, Zr
~, **губчатый** spongy zirconium

циркулировать circulate, revolve, rotate; circle
циркуль (pair of) compasses; dividers
~, **делительный** (a pair of) divider(s)
~, **измерительный/круговой** bow/caliber/calliper compasses, divider(s)
~, **падающий чертежный** drop-pen compasses
~, **пропорциональный** bisecting compasses, proportional dividers
~, **рычажный** beam compasses
~, **школьный** board compasses
циркуль-измеритель divider(s), (pair of) compasses
циркульный (pertaining to) compasses; circular
~**ная пила** circular saw
циркулянт *math.* circulant
циркуляр 1. *tel.* conference call/circuit; 2. instruction
циркулярка circular saw
циркулярный I circular, circulated
~ **аккредитив** circular letter of credit
циркулярный II circulating, circulatory, cyclic
~ **психоз** circular insanity
циркулятор *electr.* circulator
~ **СВЧ** microwave circulator
~, **трехплечий** three-port circulator
циркуляционный circulating, circulatory, circulation *(attr.)*
циркуляция circulation; *chem.* recycling; *comp.* looping; *nav.* gyration, circulation path, turning circle/motion
~, **безвихревая** irrotational circulation
~ **гена** gene flow
~, **естественная/самотеком** natural/gravity circulation
~ **крови** (blood) circulation
~, **магнитогидродинамическая** MHD circulation
~, **наименьшая** helm circle
~, **неустановившаяся** variable motion
~, **принудительная** forced/induced circulation
~ **(судна)** circulation path, turning circle (of a vessel)
~, **установившаяся** steady turning motion
циркумвалационный circumvallate
циркумдукция *physiol.* circumduction
циркумнутация *bot.* revolving nutation, circumnutation
циркумфлекс, циркумфлексный, циркумфлексовый circumflex
циркумцизия circumcision, circumcising
цирripедия *pl. zool.* barnacles, cirripeds *(Cirripedia)*
цирриия *meteor.* cloud of curling fibres, cirrus
цирроз *med.* cirrhosis
~ **желудка** sclerotic gastritis
~ **легких** fibrous/indurative pneumonia, Hamman-Rich syndrome
~ **печени** hepatocirrhosis, cirrhosis (of the liver)
~ **печени, билиарный** biliary/Hanot's cirrhosis
~ **печени, гипертрофический** unilobar/Todd's cirrhosis
~ **печени, жировой** fatty cirrhosis
~ **печени, пигментный** h(a)emochromatosis
~ **печени, портальный** portal/Laennec's cirrhosis, hobnail liver
~ **почки** granular kidney
цирроз-рак *med.* primary cancer of the liver
цирротический cirrhotic
циррус *see* **цирриия**
циртохоанитический, циртохоанитовый cyrtochoanitic
цирцея *bot.* enchanter's nightshade, circaea *(Circaea)*
цирюльник barber
~, **севильский** *mus.* The Barber of Seville
цис-положение cis-position/arrangement
цис-соединение cis-compound
циссоида *geom.* cissoid
циссус *bot.* treebine *(Cissus)*
циста *med.* cyst
цистаденома *med.* cyst(o)adenoma
цистамин *chem.* viscous oil, cystamine
цистатионаза cystathionase
цистеин cystein(e)
цистерна 1. tank, cistern; 2. *rail.* tank car; 3. *anat.* cistern(a); *see also* **цистерны**
~, **автомобильная** tank car
~, **балластная** *naut.* balast/flooding tank
~, **бескингстонная** *naut.* free-flooding tank
~ **водяной защиты реактора** shield tank
~, **грузовая** cargo(-fuel/-oil) tank
~, **дифферентная** *naut.* trim(ming) tank
~ **для навозной жижи** *agr.* liquid-manure tank
~ **для удобрений** *agr.* fertilizer tanker
~, **масляная** lub(ricating) oil tank
~, **междупалубная** *naut.* tweendeck tank
~ **млечного сока** *anat.* chyle/Pequet's cistern, chylocyst
~, **напорная** pressurizing/header/charging/gravity tank
~, **нефтяная** *rail.* oil tank wagon
~ **перекреста** *anat.* cistern of the chiasm
~, **подпаутинная** *anat.* subarachnoid cistern
~, **расходная** *naut.* daily supply/service tank
~, **сточная** drain (collecting) tank, bilge tank
~, **фекальная** sewage (retention) tank
цистерна-глушитель muffler tank
цистерна-жижерасбрасыватель *agr.* liquid manure spreading tank
цистерна-цементовоз cement tank wagon
цистерны *pl.* tanks; *see also* **цистерна**
~, **активизированные успокоительные** activated tanks
~, **пассивные успокоительные** passive anti-rolling tanks
~ **Фрама, успокоительные** Frahm anti-rolling tanks
цистернэпифит cisternepiphyte
цистерцианец *eccl.* Cistercian (monk)
цистида *bot.* cystid(ium)

цистинурия *med.* cystinuria
~, наследственная familial cystinuria
цистит *med.* inflammation of the bladder, cystitis
~, лучевой radiation cystitis
~, шеечный cystitis colli
~, язвенный Hunner's ulcer
цистицерк *zool.* cysticercus *(Cysticercus)*
цистография cysto(radio)graphy
~ во время мочеиспускания micturating cystography
цистодуоденостомия *med.* (chole)cystoduodenostomy
цистоидеи *pl. zool.* cystoideans, cystids *(Cystoidea)*
цистоилеопластика *med.* cystoileoplasty
цистокарп(ий) *bot.* cystocarp
цистоколостомия *med.* (chole)cystocolostomy
цистокучка cystosorus
цистолит *bot.* cystolith
цистома *med.* cystic tumo(u)r, cystoma
~, миксоидная myxoid cystoma, myxocystoma
цистообразный cystous, cystic
цистоптерис *bot.* bladder fern *(Cystopteris)*
цистосорус cystosorus
цистотил(а) cystotyle
цистотомия *surg.* cystotomy
~, надлобковая suprapubic cystotomy, epicystotomy
~, промежностная perineal cystotomy, hypocystotomy
~, шеечная cystauchenotomy, cystotrachelotomy
цистоэпителиома *med.* cystocarcinoma, cystoepithelioma
цистра *mus.* citole, mediaval cithern
цис-транс-изомерия *chem.* cis-trans-isomerism
цитадель citadel, fortress; *fig.* stronghold
цитарексилум *bot.* fiddlewood *(Citharexylum)*
цитата citation, quotation
цитварный *bot.* santonic, wormseed
~ная полынь santonica *(Artemisia cina)*
~ное семя wormseed, santonin
цитидин *pharm.* cytidine, cytisine
цитидиндифосфат *chem.* cytidine diphosphate, CDP
цитидинмонофосфат *chem.* cytidine monophosphate, CMP
цитирование citing, quoting
цитобиология cell(ular) biology, cytobiology, cytology
цитобласт *biol.* cell nucleus, blastocyte, cytoblast
цитобластема *med.* (cyto)blastema
цитобластома *med.* blasto(cyto)ma
цитогенез cytogenesis, cytogeny
цитогенетика cytogenetics
цитогиалоплазма *med.* cytohyaloplasm, cytolymph
цитодерма cell wall, cytoderm
цитодиагностика *med.* cytodiagnosis
цитозим *bioch.* cytozyme, thromboplastin
цитокинез cell division, cytokinesis, cytocinesis, cytodieresis
цитокинин cytokinin
цитолизин *bioch.* cytolysin
цитолог cytologist
цитология cytology, cell(ular) biology, cytobiology
~ пункционная puncture cytology
~ растений plant cytology
цитомегалия *med.* cytomegalia (infantum)
цитомембрана cell membrane, cytomembrane
цитопатогенный cytopathogenic
цитопатология cellular pathology, cytopathology
цитоплазма *biol.* cell protoplasm, cytoplasm
цитоплазматический cytoplasm(at)ic
цитопоэз cytopoiesis
цитоспороз *bot., pathol.* cytospora die back
цитостом *zool.* cell mouth, cytostome
цитотоксический cytotoxic(al)
цитофильный cytophil(e), cytophilous
цитофотометрия cytophotometry
цитохимера *biol.* cytochimera
цитохимический cytochemical
цитохимия cytochemistry
цитоядерный cytonuclear
цитра *mus.* zither
цитраль *chem.* citral
цитрат *chem.* citrate
~ меди cuprous citrate
~ натрия sodium citrate
цитратнорастворимый citric-soluble
цитратный citrated
цитрин I *bioch.* citrin, vitamin P
цитрин II *geol.* citrine, false topaz
цитринин citrinin(e)
цитроворум-фактор leucovorin calcium, citrovorum factor
цитрол citrol
цитрон *bot.* citron *(Citrus medica)*
цитронелла *bot.* citronella grass *(Cymbopogon ardus)*
цитрус *bot.* citrus *(Citrus)*
~ бергамот bergamot orange *(C. bergamia)*
цитрусовод citrus grower, citriculturist
цитрусоводство citrus growing
цитрусовый citrous; *n. pl. bot.* citrus plants *(Aurantioideae)*
~вые культуры citrus fruit crops
цитула *zool.* cytula, impregnated ovum
цифелла *bot.* cyphella
циферблат dial (plate); clock dial, hour plate
~, светящийся (self-)luminous dial
цифирница *print.* numbers case
цифра figure, numeral, number, cypher, cipher; digit
~, арабская Arabic figure/numeral
~, вводимая input digit
~, верная valid figure/digit
~, восьмеричная octal digit
~ второго разряда second-order/-position/place digit
~ второго слагаемого addend/augend digit
~, двоичная binary digit/character/bit

~ двоичного кода bit, binary digit
~, двоично-кодированная binary-coded digit
~, десятичная decimal digit
~, десятичная двоично-кодированная binary coded decimal digit, BCD digit
~, достоверная valid digit
~, знаковая sign digit
~, значащая significant/valid/nonzero digit, significant figure
~, значащая, самая старшая highest-order digit
~, контрольная check digit, control/planned/cheduled figure; *econ.* target figure
~ младшего разряда low-order digit
~ набора, десятичная *com.* ten digit
~ набора, единичная *com.* unit digit
~, ненулевая nonzero digit
~, правая high-order digit
~ разряда десятков tens digit
~ разряда единиц unit digit
~, римская Roman figure/numeral
~ старшего разряда high-order digit
~, точечная dot-matrix digit
~ числа, первая leading digit
~ры, контрольные *econ.* planned targets
цифратор digitizer, converter
~ времени time-to-number converter
~ с периодическим счетом chronometric digitizer
~ угла angular-position quantizer
цифрация numbering
цифроаналоговый digital-analog(ue)
цифрователь digitizer
цифровка numbering
цифровой numerical, digital
~ преобразователь digitizer
~вые данные figures
цицания *bot.* wild rice *(Zizania)*
цицеро *print.* cicero, pica
Цицерон *hist.* Cicero
цицероновский Ciceronian
ЦК *abbr.* **Центральный Комитет** the Central Committee
цмин *bot.* everlasting *(Helichrysum)*
цоизия *bot.* lawn grass *(Zoysia)*
цоколевка *electr.* basing, base-/pin-connection diagram
цоколь 1. *arch.* socle, plinth, foundation; pedestal; 2. *el.* cap, *Am.* base
~ антенны antenna mounting
~ без колонн stereobate
~, винтовой/резьбовой screw base
~, восьмиштырьковый octal base
~ острова island base/socle
~ Свана, штыковый bayonet base
~, семиштырьковый seven-pin base
~ трубы chimney base
цокольный (pertaining to) socle/plinth
~ этаж ground floor
цокор *zool.* zokor *(Myotalpa aspalax)*
Ц-система centre-of-mass system
цуг 1. line/train of vehicle; 2. flight of woodcocks
~ волн wave train/packet
~ импульсов pulse train/group
цунами tsunami, seismic seawave
цуцугамуши *med.* Japanese river fever, sutsugamushi disease
цынга *med.* scurvy, scorbutus; *see* **цинга**

Ч, ч

чабер *bot.* savory *(Satureia)*
чабрец *bot.* thyme *(Thymus)*
чавыча *zool.* king/Chinook salmon *(Oncorhynchus tschawytscha)*
чага *bot.* birch/bracket fungus *(Polyporus betulinus)*
чагерак *bot.* camel's thorn *(Alhagi)*
чадра veil, yashmak
чаеводство tea culture/farming/growing
чаедробилка tea cutter, tea rolling machine
чаеуборочный tea-picking/-harvesting
чаеупаковочный tea-packing
чай 1. *bot.* tea(-plant) *(Thea);* thea, tea (leaves); *med.* species; 2. tea party
~, ароматизованный scented tea
~, грудной pectoral species/tea
~, желчегонный cholagogue species/tea
~, зеленый green tea, thea viridis
~, кирпичный brick tea
~, липовый limeleaf tea
~, мочегонный diuretic species/tea
~, плиточный brick tea
~, потогонный sudorific species/tea
~, почечный Java tea *(Orthosiphon)*
~, рвотный Appalachian tea *(Ilex vomitoria)*
~, слабительный laxative species/tea, St. Germain tea
~, цветочный flower/rose tea
~, черный black tea, thea nigra
чайка I *zool.* gull *(Larus)*
чайка II *hist.* boat
чайная *n.* tearoom; teahouse; teashop
чайный tea *(attr.); n. pl. bot.* tea family *(Theaceae)*
~ная колбаса *cul.* bologna sausage
чайот *bot.* chayote *(Sechium edule)*
чайрикер *hist.* farm hand; hired man, metayer *(in Central Asia)*
чайхана, чайханэ chaikhana, tearoom
чакан *bot.* cattail, reed mace *(Typha)*
чакона *mus.* chaconne
чал *naut.* mooring rope
чалиться *naut.* moor, fasten by cable/anchor
чалка *naut.* 1. tie/mooring rope; 2. mooring, tying up
чалма turban
чан vat, tank, tub

~, бродильный fermentation vat, fermenting tub, fermenter
~, бучильный bucking tub, steeping vat
~, винодельческий wine vat
~ для выдержки *food.* holding vat/tank
~ для вымачивания steeping vat/tank, soaking vat
~, дрожжерастильный yeast-growing vat
~, дубильный tanning vat
~, замочный soaking vat, steeping tank
~, запарный steamer, steaming vat
~, заторный mash vat/tub
~, зольный *leath.* liming pit
~, красильный *text.* dye vat
~, купажный *food.* blend tank
~, моечный washing vat/tank
~, осадочный *min.* precipitator, precipitation/sedimentation tank
~, отбельный *text.* bleaching tank/vat
~, отмочный *leath.* soaking pit
~, отстойный settler
~ сокового хода *leath.* colo(u)ring vat, colo(u)r pit
~, фильтрационный filter vat, straining tank
~, шпарильный scalding vat/tank
чаполоть *bot.* sweet grass *(Hierochloe odorata)*
чарльстон Charleston (dance)
чартер *econ.* charter (party)
~, банковский bank charter
~ на срок time charter
~, рейсовый trip/voyage charter
чартерный chartered
чартер-партия *econ., naut., av.* charter (party)
чартизм *hist.* Chartism
чартист, чартистский *hist.* Chartist
час 1. hour; o'clock; 2. time, period; duty; *see also* **часы II**
~, академический *educ.* class period
~, комендантский *mil.* curfew
~, летный flight/flying hour
~, обратный inverse hour, inhour
~, учебный class period
часовник *eccl.* breviary, Book of Hours
часовой I 1. hour-long, lasting one hour; 2. per hour, by the hour, hourly
~ пояс hour/time zone
~ угол horary/hour angle
часовой II clock *(attr.),* watch *(attr.)*
~ механизм clockwork
часовой III *n.* guard, sentry, sentinel
~ у знамени colo(u)r sentinel
часозанятие *tel.* traffic unit, call-hour
часомер hourmeter
часослов *eccl.* Book of Hours, breviary
части *pl.* parts; *see also* **часть**
~, выступающие *naut.* appendages, projecting parts, projections
~, запасные spare/repair parts, spares; tools, parts and accessories
~, комплектующие supporting spares; list of standard equipment
~, тыловые *mil.* service support troops
частица 1. particle, bit, fraction, fragment; corpuscle, grain, speck; 2. small part, portion; scintilla; 3. *chem.* species; 4. *gram.* particle
~, активная reactive species
~, алевритовая silt particle
~, бесконечно малая infinitesimal particle
~, более короткоживущая shorter living particle
~, бомбардируемая target/struck particle
~, бомбардирующая projectile, bombarding particle
~, быстрая high-speed particle, fast/swift particle
~, взвешенная particle of suspension
~, взвешенная в воздухе aerated particle
~, влетающая incoming/entering particle
~, вылетающая outgoing/emerging/exit particle
~, горячая hot particle
~, дираковская Dirac particle, fermion
~, длиннопробежная long-range particle
~, дочерняя product particle
~, заряженная charged particle
~, инородная foreign particle
~, исходная progenitor
~, короткоживущая short-living/-lived particle
~ космического излучения cosmic projectile
~ кристалла, первоначальная ultimate crystal particle
~, лавинная cascade particle
~, левовинтовая left(-handed) particle
~ малой энергии low-energy particle
~, мельчайшая ultimate particle
~, многозарядная multicharged particle
~, наиболее короткоживущая shortest-lived particle
~, налетающая incident/impinging/incoming particle
~, невылетающая confined particle
~, незаряженная uncharged/neutral particle
~, несвязанная unbound/uncoupled particle
~, несмачиваемая unwettable particle
~, неустойчивая decaying/unstable particle
~, обугленная coarbonized fragment
~ отдачи recoil particle
~, плитовидная slab-shaped particle
~, правовинтовая right(-handed) particle
~ примеси foreign particle
~, проникающая penetrating particle
~ пыли dust particle
~, раскаленная incandescent particle
~ распада decay/disintegration particle
~, реакционноспособная reactive species
~, свободная unbound/free/uncoupled particle
~, связанная bound particle
~, смачиваемая wettable particle
~, стабилизированная trapped/stabilized particle
~, тонкая fine/small particle
~, третичная tertiary(particle)

~, **ударяемая** struck particle
~, **ударяющая** bombarding/incident/striking/ colliding particle
~, **устойчивая** stable particle
~ **Хиггса** Higgs particle
~, **элементарная** fundamental/elementary particle
~ **ядерно-активная** nuclearly active particle
частичка particulate
частично-коаксиальный partial-coaxial
частично-рекурсивный partially-recursive
частично-стелющийся *bot.* erectopatent
частичность incompleteness
частично-фокусированный partially focus(s)ed
частичный partial; fractional; particulate
частник *econ.* private trader/craftsman
частнодифференциальный partial-differential
частное *n.* 1. *math.* quotient; 2. the particular
~, **левое** left-hand quotient
~, **неполное** partial denominator
частноотрицательный specifically-negative
частнопрактикующий privately practicing
частнопредпринимательский private-enterprise
частнособственнический private-ownership
частность detail; particular feature; relative frequency
частный 1. particular; partial; peculiar, special, individual; 2. private, privately-owned; personal; 3. *hist.* district; 4. *n.* *hist.* chief of police
~ **вывод** partial conclusion
~ **дом** *hist.* distict police station
~ **пристав** *hist.* chief of police
~ **случай** special case
~**ная собственность** private property/ownership
частокол picket fence
частость frequency, rate
~ **событий** *tech.* relative frequency
частота 1. frequency, rate; 2. density, closeness
~ **аберрации** aberration rate
~ **биений, собственная** natural beat frequency
~, **весьма высокая** very high frequency, VHF
~ **взятия отсчетов** sampling rate
~, **возбуждающая** exciting/master frequency
~ **возбуждения** *rad.* exciting/driving frequency
~ **волновода, критическая** wave-guide cut-off frequency
~ **волны** wave frequency
~ **вращения** *mech.* rotational speed; frequency of revolution; revolutions per minute, r.p.m.
~ **выборки** sampling rate
~ **вызова/вызовов** *com.* calling frequency/rate
~ **выключений двигателей** shut-down rate of the engines
~ **вынужденных колебаний** forced frequency
~, **высокая** high/radio frequency, HF
~ **гармоник** harmonic frequency
~, **гасящая** *rad.* quenching frequency
~ **гена** *med.* gene frequency
~ **генерации** *electr.* oscillation/oscillating frequency
~, **генерации лазера/мазера** lazer/mazer frequency
~, **гироскопическая** gyrofrequency, cyclotron frequency
~, **граничная** frequency limit, limiting/boundary frequency
~, **групповая** *stat.* cell/class frequency
~ **двусторонней связи** duplex frequency
~, **дежурная** guard frequency
~ **дециметрового диапазона волн** ultra-high frequency, UHF
~, **дискретная фиксированная** *rad.* spot/ *Am.* fixed frequency
~ **для передачи сигналов бедствия** distress/ emergency frequency
~ **для связи на маршруте** enroute frequency
~, **доплеровская** Doppler frequency/shift
~, **дополнительная** secondary frequency
~, **дробная** fractional frequency
~ **дыханая** *med.* respiration/breathing rate
~, **задающая** driving/master frequency
~ **замирания (радиосигнала)** rate of fading
~, **запасная** alternate/guard frequency
~ **заполнения импульса** basic/carrier frequency
~, **звуковая** *rad.* audio/sonic/acoustical/sound frequency, af, AF
~, **звуковая низкая** bass (frequency)
~ **звуковых колебаний** audio/acoustical frequency
~ **излучения** emission/radiation frequency
~, **импульсная** impulse frequency, pulse rate frequency
~ **импульсов** (im)pulse rate/frequency
~, **инфразвуковая** infrasonic/subsonic frequency
~ **кадров/кадровая** *telev.* frame/vertical frequency, frame rate
~ **кадросмен** *cin.* picture frequency, film speed
~ **квантования (сигнала по времени)** sampling rate
~ **кварца** crystal frequency
~ **килевой качки** pitch frequency
~ **киносъемки** camera frequency/speed
~ **колебаний** oscillation/ *mech.* vibration frequency
~, **комбинационная** combination/heterodyne frequency
~, **контрольная** pilot frequency
~, **крайне высокая** extremely-high frequency, EHF
~, **крайне низкая** extremely-low frequency, ELF
~, **критическая волновода** cut-off frequency of a waveguide
~, **круговая** circular/angular/cyclic/radian frequency; angular velocity
~ **летных происшествий** *av.* accident rate
~ **линии связи "Земля-космос", рабочая** uplink frequency

~ линии связи "космос-Земля", рабочая downlink frequency
~, максимально применимая, МПЧ *rad.* maximum usable frequency, MUF
~, маршрутов density of traverses
~, международная вызывная *rad.* International calling frequency
~ мерцаний *telev.* flicker frequency
~ метрового диапазона волн very-high frequency, VHF
~, микроволновая microwave frequency
~, минимально применимая *rad.* lowest-useful (high) frequency, LUF
~ накачки *electr.* pump frequency
~ настройки *rad.* resonant/tuning frequency
~ немодулированного сигнала centre frequency
~, несущая (разрешенная) (authorized) carrier frequency
~, низкая low/audio frequency, LF
~ обращений revolution/rotation frequency
~ обращения к памяти operation rate of storage
~ односторонней связи simplex frequency
~, опорная reference frequency
~ опроса (датчиков) sampling rate
~, основная base/basic/fundamental frequency/harmonic
~ отказов *mech.* failure rate
~ отсечки cut-off frequency
~, очень высокая very-high frequency, VHF
~, очень низкая very-low frequency, VLF
~ ошибок error rate
~, паразитная spurious frequency
~ переключений switching rate/frequency
~, переходная *electr.* crossover/transition frequency
~ повторения repetition frequency/rate, recurrence rate
~ повторения запускающих импульсов trigger-repetition rate
~, поднесущая subcarrier (frequency)
~, подтональная subsonic/subaudio/infrasonic frequency
~ полетов flying rate
~ полукадров/полей *telev.* field frequency
~ поперечных колебаний *aero.* roll frequency, frequency in roll
~, пороговая treshold frequency
~ порывов ветра gust frequency
~ поступления *comp.* arrival rate (of messages/queries)
~ появления грубых ошибок blunder rate, coarse error rate
~ появления неисправностей rate of malfunctions
~, предельная limit(ing)/cutoff frequency
~ преобразования conversion frequency
~, присвоенная allocated/authorized frequency
~ продольных колебаний frequency in pitch
~, промежуточная intermediate frequency
~, промышленная commercial/mains frequency
~, прослушиваемая incoming frequency
~, пространственная spatial frequency
~ пульса *med.* pulse rate
~ пусков launching rate
~, рабочая working/operating/operational frequency
~, рабочая оптимальная optimum traffic frequency, OTF; optimum working frequency, OWF
~ развертки изображения *telev.* scanning frequency
~ развертки луча *telev., electr.* sweep(-generator) frequency
~, разговорная voice frequency
~, разностная difference/differential frequency
~ рассеяния scattering frequency
~ регенерации refresh rate
~, резонансная resonance/resonant frequency
~ рекомбинаций recombination rate
~ рисунка *com.* picture frequency
~ сантиметрового диапазона волн super-high frequency, SHF
~, сверхвысокая super-high frequency, SHF
~ света light frequency
~ свободных колебаний natural resonant frequency, free/natural frequency
~ сердечных сокращений rate of heart's beat, heart rate
~ сети network frequency
~ силовой сети mains frequency
~, синхронизирующая sync(hroinizing)/ *comp.* clock frequency
~, синхронная synchronous frequency
~ скольжения *el.* slip frequency
~ следования spacing frequency
~ следования импульсов pulse recurrence/repetition rate, pulse repetition/recurrence frequency
~ слияния мельканий *med.* flicker/critical fusion frequency
~ случайных импульсов chance-pulse rate
~, слышимая audio/acoustical frequency
~, собственная allowed/free/natural/fundamental frequency
~ сокращения желудочков *med.* ventricular rate
~, сопряженная break/corner frequency
~ соударений collision frequency
~, средняя medium frequency, MF
~ среза (фильтра) *electr.* cut-off frequency
~ срыва колебаний *rad.* quench frequency
~, стабилизированная кварцем crystal-controlled frequency
~ строк/строчная *telev.* line/horizontal frequency
~ суперизации *rad.* quench frequency
~, тактовая *comp.* clock rate/frequency
~, тональная voice/speech/audio frequency, VF
~ точек *telev.* dot frequency
~ трещин *geol.* density of fractures

~, угловая angular/radian/circular frequency, angular velocity
~, ультравысокая ultra-high frequency, UHF
~, ультразвуковая ultrasonic/supersonic/superaudio frequency
~ уходов на второй круг rate of overshoot
~ флаттерных колебаний flutter oscillation frequency
~ формы волны *nav.* wave number
~ хиазм *med.* chiasma frequency
~, холостая idler frequency
~, хронирующая timing frequency
~, центральная central/centre frequency
~, циклическая angular velocity, angular/radian/circular/cyclic frequency
~, эталонная standard/reference frequency
~ явлений frequency of occurrence
частота-подставка *rad.* injection/conversion frequency
частотно-временной time-and-frequency
частотно-зависимый frequency-sensitive/dependent
частотно-избирательный frequency-selective
частотно-импульсно-модулированный pulsed frequency-modulated
частотно-модулированный frequency-modulated, FM
частотно-селективный frequency-selective
частотность frequency (of occurrences)
частотно-фазовый phase-frequency
частотный frequency *(attr.)*
частотограмма periodogram
частотомер frequency meter/teller, period meter
~, вибрационный *el.* (vibrating) reed frequency meter
~, гетеродинный heterodyne frequency meter; heterodyne wavemeter
~, камертонный tuning-fork frequency meter
~, резонансный tuned-circuit-type frequency meter
~ СВЧ microwave wavemeter
~, тональный audio-frequency meter
~, цифровой digital frequency meter, frequency counter
~, язычковый (vibrating-)reed frequency meter
частотомерный frequency-measuring
частотопреобразовательный frequency-converting
частуха *bot.* water plantain *(Alisma)*
частухолистный *bot.* alisma-leaved
часть 1. part, piece; portion, fraction; share; quota; fragment; stake; component; 2. field, sphere; branch, section, department; 3. member, side; 4. *mil.* unit; 5. administrative district of a city; police station
~, авиаремонтная aircraft repair unit
~, авиационная air/aviation unit
~, аликвотная aliquot (part)
~, аллогенная составная allogenic/allothigenous/derived constituent
~, беззольная ash-free part
~, весовая part by weight
~, выдвижная extension piece
~ высокого давления high-pressure section
~, выступающая prominent/projecting part; protuberance, protrusion; *pl. naut.* appendages
~, главная составная major/essential constituent
~ головы насекомого, главная *zool.* epicranium
~ горы, передняя mountain front
~ грудины, средняя midsternum
~, губчатая *anat.* cavernous/spongious part
~ диафрагмы, поясничная *anat.* lumbar/vertebral part of the diaphragm
~ дороги, проезжая roadway
~, дробная *math.* fractional part
~ жилы, краевая *geol.* vein selvage
~ жилы, нерудная *geol.* gangue
~, задняя rear part; afterbody
~, замещающая составная substitute constituent
~, запасная spare/replacement part
~, зарамочная edge; border; margin
~, земной коры, верхняя outer crust of the Earth
~, изношенная материальная worn-out materiel
~ изображения *comp.* subimage
~, кодовая (en)coding section
~ (команды) адресная address part
~ комплексного числа, действительная *math.* real part of a complex number
~, концевая tip (section)
~, кормовая (судна) *naut.* stern
~, корневая root (section)
~ корпуса, кормовая/носовая *naut.* afterbody/forebody
~ массива subarray
~, материальная hardware; equipment, materiel, physical facilities, supplies
~, машинная *comp.* hardware
~ моря, глубоководная abyssal sea
~ мульды, внутренняя *geol.* trough core
~, нарастающая incremental portion
~ наследства, незавещанная *law* residuary estate
~, неисправная материальная malfunctioning materiel
~, немашинная/неаппаратная software
~, неотъемлемая *law* part and parcel; integral part
~, неподвижная stationary/static part
~, нераздельная integral part
~ низкого давления (турбины) low-pressure section
~, носовая nose (section)
~, обедняющая depleting/stripping section
~ обмотки, лобовая *el.* coil end
~, обогатительная enriching section

~, опорная (конструкции) bearing part/member
~, основная main portion/part
~, откидная hingeing/swing section
~, передняя front, forepart
~, поврежденная материальная damaged materiel
~, подфюзеляжная *av.* underfuselage/belly section
~, пожарная fire brigade
~, предлежащая presenting part
~ программы, вводная *comp.* prologue
~ программы, заключительная *comp.* postlude
~ программы, резидентная program resident
~, проточная (парового котла) flow passage
~, проточная (турбины) blading section
~, рабочая working part/section; test section
~, ракетная missile unit
~, режущая cutting end/unit
~, резидентная *comp.* resident
~ системы subsystem
~ складки, сводовая crest
~, сменная replacement part
~, соединительная connector; connecting piece; *pl.* fittings
~, составная constituent (element), ingredient, component (part), integral part
~, сотая centesimal
~, спектра, видимая visible(-light) spectrum
~ схемы *comp.* subcircuit
~ тела, передняя/хвостовая forebody/afterbody
~ угля, органическая organic portion of coal, potential ash
~ уравнения side of an equation
~, устойчивая составная fixed constituent
~, учебная teaching department
~, хвостовая tail piece/cone, afterbody, tail
~, ходовая *auto.* running gear, undercarriage
~ цикла, исполнительная/командная *comp.* execute/instruction part of cycle
~ числа, дробная/целая fractional/integer part of a number
~ числа, цифровая mantissa
~ шины, беговая/бортовая tread/bead section of a tyre
~, эндогенная составная *geol.* endogenous/cognate inclusion

часы I *pl.* clock, watch; timepiece; meter
~, астрономические astronomical clock
~, атомные atomic clock
~, бортовые aircraft/space clock
~, газовые wet-gas meter
~, звездные sidereal clock
~, индикаторные dial indicator
~, камертонные tuning fork clock
~, кварцевые crystal/quartz clock, crystal-controlled clock
~, маятниковые (free) pendulum clock
~, наручные wristwatch
~, песочные hourglass, sandglass
~, радиомаячные time-signal radio, time-signal broadcasting station
~ реального времени real-time clock
~ с боем striking clock, clock with chimes
~ с самозаводом self-winding electric clock
~, солнечные sun-clock, sundial
~, сухие газовые bellows-type gas flow-meter
~, цифровые digital time unit, digital clock
~ штурманские navigation watch
~, электронные electronic timer

часы II *pl.* hours; *mil.* guard duty; *eccl.* canonical hours
~ пик rush hours; peak traffic
~, приемные office hours
~, присутственные official hours
~ работы банка banking hours
~ работы учреждения office hours
~ слабой нагрузки slack hours
~, служебные hours of attendance

часы-хронометр time-keeper-piece, chronometer watch

чахотка *med.* phthisis, consumption
~, горловая laryngeal phthisis, tuberculosis of the larynx
~ костного мозга panmyelophthisis
~, легочная phthisis, pulmonary consumption
~, скоротечная acute tuberculosis, galopping consumption, phthisis florida
~, черная black/miner's/collier's phthisis, anthracosis

чахоточный consumptive, phthisic(al)
~ румянец hectic flush

чачбан face veil

чаша 1. bowl, cup; pan, dish; *eccl.* communion cup, chalice; 2. *geol.* basin
~ весов scale
~, литниковая pouring basin
~ поилки *agr.* drinking trough

чашевидый, чашеобразный bowl-/cup-shaped; crateriform; caliciform; scyphiform; patelliform; cyathiform, calicular

чашелистник *bot.* sepal, calyx lobe
~и, имеющий sepaled, sepalous

чашецветник *bot.* sweet shrub *(Calycanthus)*

чашечка 1. *med.* calycle, calyculus; 2. *bot.* scyphulus, calyx; 3. small cup
~, глазная optic cup
~ иглокожих theca
~ капсюла primer cup
~, коленная patella, kneecap, kneepan
~ коралла prototheca, calyx
~, наружная epicalyx
~, опадающая deciduous calyx
~, сростнолистная calyx with united segments
~, цельная gamosepalous calyx

чашечковидный, чашечкообразный patellate, patelliform, calyciform, calycine, calyx-shaped

чашечковый calycine, chaliced

чашечкоплодный *bot.* calycicarpous
чашечный cup-shaped, calycine
~ телефон earphone, earpiece
чашка 1. cup; bowl; pan, scale; 2. dish
~ барометра barometer cistern
~ весов balance/scale pan
~, выпарительная evaporating dish
~ Даппена Dappen dish
~ звонка bell gong
~, (над)коленная kneepan, kneecap, patella
~ Петри/Стендера/Эсмарха Petri/Stender/Esmarch dish
~ сидения *auto., av.* seat pan/bucket
чашник *hist.* cellarer
чаща *for.* thicket
чеглок *zool.* hobby falcon *(Falco subbuteo)*
чедер *cul.* Cheddar (cheese)
чек I cheque, *Am.* check; bill; receipt, sales cheque
~, банковский banker's cheque
~ без права передачи nonnegotiable cheque
~, возвратный redemption cheque
~, выписывать draw a cheque
~ датированный будущим числом post-dated cheque
~ для оплаты определенных счетов ear-marked cheque
~ для получения наличных денег open cheque
~, дорожный *Am.* travel(l)er's check
~ имеющий секретную пометку marked cheque
~, кроссированный cross(ed) cheque
~ на получение зарплаты pay cheque
~ на получение процентов и дивидендов interest-and-dividend cheque
~ на предъявителя *law* cheque to bearer, *Am.* bearer check
~ не имеющий покрытия kite/rubber cheque
~, незаполненный blanc cheque
~, некроссированный uncrossed cheque
~ непринятый к оплате банком dishono(u)red cheque
~, ордерный order cheque
~, открытый/некроссированный open cheque
~ передаваемый по индоссаменту negotiable cheque
~ перенесенный на перфокарту punched-card cheque
~, погашенный cancelled cheque
~ подписанный казначеем treasurer's cheque
~ подписанный служащим банка officer's cheque
~, подтвержденный marked cheque
~, представительский bearer cheque
~, просроченный stale cheque
~, расчетный clearing house cheque
~ со специальным кроссированием specifically cross(ed) cheque
~ только для безналичного расчета cheque only for account
~, туристический travel(l)er's cheque
~, удостоверенный certified cheque
~, устаревший stale cheque
~, фиктивный kite/rubber cheque
чек II *agr.* check basin
чек-расписка *econ.* voucher cheque
чека I cotter (pin), forelock, (linch)pin, splint pin, catch/lock pin
~ безопасности *min.* safety/cotter pin
~, блокировочная interlocking pin
~, боевая *mil.* arming/firing pin
~, вытяжная *av.* rip/pull pin
~, коническая tapered cotter
~, натяжная tightening key, adjusting cotter/wedge, gib
~, предохранительная packing stop; safety pin
~, разводная split cotter
~, срезная shear pin
~ сцепления coupling pin
Чека II *abbr.* **Чрезвычайная Комиссия по борьбе с контрреволюцией, саботажем и спекуляцией** *hist.* Cheka, Soviet secret police
чекан I (embossing/coining) die, stamp, punch(eon); 2. caulking iron/chisel/tool
чекан II, чеканка *zool.* chat *(Saxicola)*
чеканить 1. *met.* emboss, coin, mint; engrave, stamp; 2. *tech.* caulk; peen; 3. *agr.* prune; pinch out
чеканка 1. *met.* embossing, engraving, stamping, toreutics; coining, coinage, minting, mintage; 2. *met.* caulking, peening
~, калибровочная sizing
~ монет mintage, coinage
~, ручная hand caulking
~ шва seam caulking
чеканный embossed, engraved, stamped; stamping, engraving
чеканочный embossing, engraving, stamping
~ пресс coining press
чеканщик caulker; stamper, engraver; coiner, minter
чекер *agr.* checker
чекист secret service agent, security officer; *hist.* official of Cheka
чекмень *hist.* uniform of a Kazakh officer
чековый cheque *(attr.), Am.* check
~вая книжка cheque-book
чекодатель *fin.* drawer (of cheque)
челеста *mus.* celesta, glockenspiel
челновидный, челнообразный boat-/canoe-shaped, scaphoid
челнок 1. shuttle; 2. canoe, dugout
~, ворсоткацкий pile-weaving shuttle
~, дуговой/изогнутый bow/curved shuttle
~ пневмопочты rabbit; shuttle
~, роликовый fly/flying shuttle
~, скользящий sliding shuttle
~, шпрынковый spindle shuttle

челночный shuttle *(attr.)*
~ маршрут shuttle service
чело *tech.* stoking hole (in furnace/stove)
челобитная *hist.* petition
челобитчик *hist.* petitioner
челобитье *hist.* (presenting/delivery of a) petition
человек man, human being, person, Homo
~, вадиакский Wadjak man
~, гейдельбергский Heidelberg man, Palaeanthropous
~, древний early man
~, ископаемый hominid, fossil man
~, конфинский Confins man
~, кроманьонский Cro-Magnon man
~, неандертальский Neanderthal man
~, обезьяноподобный ape-man
~, пекинский Peking man *(Sinanthropus pekinensis)*
~, первобытный primitive/prim(a)eval man
~, пещерный cave man/dweller, troglodyte
~, пильтдаунский Piltdown man
~, снежный yeti, Abominable Snowman
~, тингтсунский Tingtsun man
~, яванский Java man
человек-бог theanthropus
человек-волк/-медведь werewolf/werebear
человековедение, человекознание study of human nature
человеко-день *econ.* man-day
человеко-годы man years
человеколюбие philanthropy
человеко-минута man-minute
человеконенавистничество misanthropy
человекообразный anthropoid, anthropomorphous, humaniform; *n. pl. zool.* anthropoids *(Anthropoidea)*
человекоподобный resembling a human being
человеко-смена man-shift
человекоубийство *law* homicide, manslaughter
человекоубийца murderer, homicide
человеко-час man-hour
человек-реклама *econ.* sandwich-man
человеческий human; humane
человечество humanity, mankind, the human race
человечный humane
челюстевидный maxilliform, mandibuliform
челюстно-зубной *anat.* maxillodental
челюстной *anat.* maxillary, gnathic; jaw *(attr.)*
челюстно-лицевой *anat.* maxillofacial
челюстно-небный *anat.* maxillopalatine
челюстно-подъязычный *anat.* mylohyoid
челюстно-раковинный *anat.* maxilloturbinal
челюстноротый *zool.* guathostomatous; *n. pl.* guathostomes *(Guathostomata)*
челюстно-скуловой *anat.* maxillojugal, zygomaticomaxillary
челюстно-слезный *anat.* maxillolacrimal
челюсть 1. *anat.* jaw(bone), maxilla; 2. dental plate, dentures, set of artificial/false teeth; 3. *mach.* jaw(plate)
~, буксовая horn block/plate, pedestal horn
~, вставная denture set of false teeth
~ грейфера grab bucket shell/scoop
~, нижняя *anat.* mandible, lower jaw(bone)
челядин *hist.* servant, domestic, menial
челядь *hist., coll.* men, servants, menials
чемерица, чемерка *bot.* (false) hellebore *(Veratrum)*
чемодан suitcase
~, матросский sea bag
чемперти *law* champerty, illegal bargain
чемпион champion
чемпионат *sport* championship
~ по высшему пилотажу aerobatic championship
чемпионство champion's title
чемыш *bot.* salt tree *(Halimodendron)*
чепе *abbr.* **чрезвычайное происшествие** unusual incident/event
чепрак *leath.* saddle-cloth, butt, double bend; shabrack
черва *coll., zool.* larvae (of bees)
червевидный, червеобразный vermiform, vermicular, wormlike; *n. pl. zool.* wormlike forms *(Vermoidea)*
черветочина *wood.* wormhole, vermiculation
червец *zool.* scale insect, coccid, mealybug
~, лаковый Indian lac insect *(Coccus lacca)*
червеядный vermivorous, worm-eating
черви *zool., pl.* worms; maggots *(Vermes); see also* **червь**
~, кольчатые annelids, annelid(i)ans *(Annelida)*
~, колючеголовые spiny-/thorny-headed worms *(Acanthocephala)*
~, круглые round worms, nemathelminths *(Nemathelminthes)*
~, ленточные tapeworms, cestoids *(Cestoda)*
~, морские nereids, marine worms
~, нитевидные thread worms, thread-like nematodes, filariids *(Filaria)*
~, плоские flatworms, platyhelminths *(Platyhelminthes)*
~, щетинконогие chaetopods *(Chaetopoda)*
~, щетинкочелюстные chaetognaths *(Chaetognatha)*
червивый wormy, maggoty, grubby, worm-eaten, vermiculous, infested with grubs
червобой *agr.* worm-damage
червовод *agr.* specialist in rearing silkworms
червоводня silkworm nursery
червоводство rearing of silkworms
червонец *hist.* chervonets, gold coin; *colloq.* ten-ruble banknote
червоточина *wood.* vermiculation, wormhole, maggot hole, pith fleck
червоточный worm-eaten
червоядный vermivorous, worm-eating
червь I 1. worm; maggot; grub; 2. *anat.* vermis; *see also* **черви**
~, дождевой earthworm

~ **мозжечка** vermis (cerebelli)
~, **шелковичный** silkworm
червь II old name of the letter **ч**
червяги *pl.* apods *(Apoda)*
червяк *tech.* endless screw, worm
~, **архимедов** Archimedean worm
~ **винтового насоса, рабочий** drive/actuating helix, drive/actuate impeller
~, **глобоидный** hour-glass worm, Hindley screw
~, **двухзаходный/двухходовой** two-/double-thread worm
~, **левооборотный/левый** left-hand worm
~, **многозаходный/многоходовой** multi-threaded worm
~, **однозаходный/одноходовой** one-/single-thread worm
~, **подающий** feed screw
~, **самотормозящий** irreversible worm
~, **транспортирующий** conveyer worm
~, **эвольвентный** involute worm
червячный worm *(attr.)*
~**ная передача** wormgear
~**ная фреза** hob cutter
червячок *anat.* vermis
чердак *arch.* attic, garret; loft
череда I chain, sequence, file (of people)
череда II *bot.* beggar-ticks, bur marigold *(Bidens)*
чередование 1. alternation, alternate arrangement, rotation; 2. *comp.* interleaving, interlacing, sequence; interchange
~ **адресов памяти** memory interleaving
~ **гласных и согласных** *gram.* vocalic and consonantal ablaut
~ **градиента** alternation of gradient
~ **кодов** code interlace/interleave
~ **культур** *agr.* succession cropping, diversification
~ **нагрузок** load sequence
~ **оледенения** *geol.* glacial sequence
~ **поколений** metagenesis, heterogenesis, alternation of generations
~ **секторов (диска)** disc interleaving
~ **слоев** alternation of beds/strata
~ **службы личного состава** *mil.* rotation
~ **строк, последовательное** line sequence
~ **точек** dot interlace
~ **точек, последовательное** dot sequence
чередовать 1. alternate, rotate; 2. *comp.* interleave, interlace
чередующийся alternate, alternating
черемуха *bot.* 1. cherry *(Padus);* 2. bird-cherry
черемша *bot.* ramson *(Allium ursinum)*
черенкование *hort.* propagation by cutting, grafting
~, **зеленое** propagation by softwood cutting
~ **листом** propagation by leaf
черенковать *hort.* graft; take a cutting (of)
черенок 1. shank; haft, grip, handle; shaft; 2. *hort.* cutting, slip, graft
~, **вызревший/невызревший** hardwood/softwood cutting
~, **корневой/стеблевой** root/stem cutting
~, **морский** *zool.* razor clam *(Ensis)*
~, **прививочный** scion
череп skull, cranium; *anat.* brain pan
~, **башенный** tower/steeple skull, oxycephaly
~, **бесхвостый** anuran skull
~ **долихоцефалический** dolichocephalism, dolichocecephaly
~ **и кости** skull and crossbones
~, **конический** acrocephaly
~, **лицевой** bones of the face, visceral cranium
~, **лодкообразный** cymbocephaly
~, **макроцефалический** macrocephalic/megocephalic skull, cephalonia
~, **микроцефалический** microcephalic/nanocephalic skull, microseme, nanocephaly
~, **мозговой** cerebral cranium
~, **пропорциональный** metrocephalic/orthocephalic skull
~, **размягченный** craniotabes, craniomalacia
~ **среднего размера** mesocephalic skull
~, **суженный** stenobregmatic/stenocephalous/leptocephalous skull
~, **удлиненный** hypsicephaly
~, **удлиненный и расширенный** hypsibrachycephalic skull
~, **укороченный** brachycephalism, brachycephaly
~, **уплощенный** platycephaly, platycrania
~, **утолщенный** pachycephalic skull, pachycephalia
~, **хрящевой** chondrocranium, cartilaginous/embryonic skull, f(o)etal cranium
~**а, лишенный** acranius, acraniate
черепаха *zool.* turtle, tortoise, chelonian
~, **гигантская** giant tortoise *(Testudo)*
~, **зеленая/суповая** edible/green turtle *(Chelonia mydas)*
~, **мягкокожая** soft-shelled turtle, trionychid
~, **наземная** land turtle, tortoise
~, **пресноводная** fresh-water turtle
~ **с кожистым панцирем** leatherback turtle
черепаховый turtle *(attr.),* testudinate, tortoiseshell
~ **суп** *cul.* turtle soup
черепашка *zool.* tortoise beetle *(Eurygaster integriceps)*
черепитчаторасположенный imbricate(d)
черепитчатый 1. imbricate, tegular; 2. made of tiles
черепица 1. tile, tiling; 2. *zool.* tegula
~, **бортовая** margin tile
~, **вальмовая** hip tile
~, **желобчатая** channel tile
~, **коньковая** ridge/gable tile
~, **кровельная** roof(ing) tile
~, **плоская** plane tile, shingle
~, **шпунтовая** gutter tile
черепицеобразноналегающий incubous
черепицеобразный imbricate(d); tegular

~ное покрытие imbrication
черепичный tile(d); tegular
~ завод tile factory
~ная крыша tiled roof
черепно-глоточный *anat.* craniopharyngeal
черепной *anat.* cranial; *n. pl. zool.* vertebrates *(Vertebrata)*
~ная коробка cranium
черепно-лицевой *anat.* craniofacial
черепно-мозговой *anat.* craniocerebral
черепные *n. pl. zool.* vertebrates *(Vertebrata)*
череподержатель skull clamp
черепокожный *zool.* testaceous
черепообразный skull-like
чересло *agr.* co(u)lter, cutter
~, ножевидное knife colter
~ плуга plough coulter
~, подвесное hanging cutter
чересполосица *agr.* strip farming
череш *bot.* desert candle *(Eremurus)*
черешковидный petiolate(d)
черешковый petiolar, petiolate, pediculate, stalked
черешня *bot.* cherry (tree) *(Cerasus); coll.* sweet cherries
черешок *bot.* petiole, pedicel, (foot)stalk
~ листа leafstalk
~, листовиднорасширенный phyllode
черешочек *bot.* leaflet/small petiole, petiolule
черкас *hist.* Cossack, Ukrainian
черкес, черкесский Circassian
черкеска Circassian coat
черкешенка Circassian (woman/girl)
чермакит *geol.* oligoclase
черна *zool.* cherna *(Alphester afer)*
~, американская red grouper *(Epinephelus morio)*
черная *n.* colo(u)red woman, black, Negress
чернение blackening
чернеть *zool.* scaup duck *(Aythya)*
черника *bot.* 1. blueberry *(Vaccinium);* 2. bilberry, whortleberry *(V. myrtillus);* 3. *coll.* blueberries, bilberries, huckleberries, whortleberries
черниколистный *bot.* vaccinifolious
чернила *pl.* ink
~, дефектоскопические flaw-detection ink
~ для нанесения данных coding ink
~, несмываемые/химические indelible ink
~, симпатические invisible/secret/sympathetic ink
~, токопроводящие electrographic/conductive ink
чернильница inkwell, inkholder; inkstand
~, телеграфная Morse inker
чернильный ink *(attr.)*
~ карандаш indelible pencil
~ мешок, ~ная железа *biol.* ink sac
~ орешек *bot.* gallnut, oak gall, nutgall
~ная война war of words
чернобыльник *bot.* mugwort, wormwood *(Artemisia vulgaris)*
черновик first/rough draft
черноглазка *zool.* shrimplike crustacean, euphausiid *(Euphausia)*
черноголовка 1. *bot.* selfheal *(Prunella);* 2. *zool.* blackcap *(Sylvia atricapilla)*
черноголовник *bot.* 1. selfheal *(Prunella);* 2. burnet *(Poterium)*
черноголовый black-headed/-haired
~вая цапля *zool.* black-necked heron *(Ardea melanocephalia)*
черногорец, черногорский Montenegrin
черногорка Montenegrin (woman)
чернозелье *bot.* black pea (vine) *(Lathyrus niger)*
чернозем *agr.* chernozem, black earth
~, мощный deep chernozem
~, оподзоленный degraded black earth
черноземовидный chernozemic, chernozem-like
чернозобик *zool.* black-bellied sandpiper, dunlin *(Calidris alpina)*
чернозобый *zool.* having a black crop *(of birds)*
~ дрозд black-throated thrush *(Turdus astrogularis)*
черноклен *bot.* Tatarian maple *(Acer tataricum)*
черноклюв *zool.* boat-billed heron *(Cochlearis cochlearis)*
чернокнижество, чернокнижие *hist.* black magic, sorcery, witchcraft
чернокожая *n.* black woman
чернокожий colo(u)red, black(-skinned); *n.* black, colo(u)red man
чернокорень *bot.* hound's tongue *(Cynoglossum)*
чернокрылый melanopterous, black-winged
чернолистный melanophyllous, black-leaved
чернолицый dark-complexioned
чернолоз *bot.* laurel willow *(Salix pentandra)*
черноокрашенный melanochroic, melanochrous
чернопленчатость *bot., pathol.* bacterial stripe, bacterial black chaff
черноплодный *bot.* melanocarpous
чернорабочий unskilled; *n.* unskilled worker
черноризец monk
чернорубашечник *hist.* Black Shirt, Italian fascist
черносемянный *bot.* melanospermous
чернослив *coll.* prunes
черносотенец, черносотенник *hist.* member of the Black Hundred, extreme reactionary
черносотенство *polit.* ideology of the Black Hundreds
черносошник *hist.* peasant (farmer)
черносошный *hist.* residing on State land; liable to tax, taxed
черноспинка *zool.* black-backed Caspian herring *(Caspialosa kessleri)*
чернотал 1. *bot.* laurel willow, bay-leaf willow *(Salix pentandra);* 2. *miner.* turgite, hydroh(a)ematite
чернотелка *zool.* darkling beetle *(Tenebrionidae)*
~, рогатая forked fungus beetle *(Boletotherus cornutus)*
черноточечный nigropunctate, black-spotted

черноцветковый *bot.* melananthous, black-flowered
черночашечковый *bot.* melanocalyx *(attr.)*
чернушка *bot.* nutmeg flower *(Nigella sativa)*
черный 1. black; dark; 2. unskilled, heavy *(of work);* 3. *tech.* semifinished, half-finished, rough; 4. *hist.* State-owned, taxed, liable to tax; 5. *n.* black; *fig.* the Devil
~ **двор** backyard
~ **металл** ferrous metal
~ **народ** *hist.* the common people
~**ная икра** black caviar
~**ная металлургия** iron and steel industry; ferrous metallurgy
~**ная пелена** *av., med.* blackout, temporary loss of consciousness
~**ная работа** menial work
~**ное дерево** ebony, blackwood
~**ное духовенство** monks; celibates
черныш *zool.* sandpiper *(Tringa ochropus);* blackcock, heathcock
чернь I *hist.* the common people
чернь II 1. black (pigment); 2. niello, metal engraving; 3. *for.* deciduous forest
~ **кобальтовая/костяная** cobalt/bone black
~, **медная** black copper, melaconite
черпак scoop; bucket, pail; *med.* arytena
~, **шлюпочный** *naut.* (boat) bailer, draw
черпаловидный *anat.* arytenoid
черпало-надгортанный *anat.* aryepiglottic, aryteno-epiglottidean, aryteno-epiglottic
черпальный scooping, ladling, dredging, bailing
~**ная машина** bucket conveyor
черпальщик ladler, scooper
черпание scooping, bailing (out), ladling
черпать scoop (up/out), bail (out); ladle
черт 1. devil; 2. *miner.* chert, compact flinty chacedony
~, **морской** *zool.* angler *(Lophius piscatorius)*
черта 1. line; dash, bar, stroke; *miner.* streak; *math.* (over)line, (over)bar, vinculum; 2. feature, trait, characteristic; 3. boundary; *see also* **черты**
~, **береговая** coastline
~ **города** city line
~, **делительная** *print.* virgule
~, **дробная** *math.* fraction bar/stroke, fraction(al) (horizontal) line, line in a fraction
~, **косая** slash
~, **курсовая (на компасе)** lubber mark/line
~ **над символом** *math.* overscribed bar
~, **наследственная** hereditary feature
~, **обратная косая** backslash
~ **оседлости** *hist.* the (Jewish) Pale
~, **характерная** idetifying/salient feature
чертеж drawing; draft, sketch; blueprint, design, plan
~ **в натуральную величину** full-size/-scale drawing
~ **в разрезе** sectional/cutaway drawing
~, **габаритный** dimensional/outline drawing, dimensional outline
~, **главный монтажный/сборочный** key sheet/drawing, assembly blueprint
~ **детали** part drawing
~, **деталированный/детальный/деталировочный** detail(ed) drawing
~, **договорный** contract plan
~, **компоновочный** (general-)arrangement drawing, design/layout drawing
~, **конструктивный** structural/construction plan/drawing
~ **мидель-шпангоута** *naut.* midship section plan
~, **монтажный** assembly/erection/mounting drawing
~ **общего вида** general view/arrangement drawing
~ **общего расположения** (general) arrangement plan
~ **оснастки** tooling drawing
~, **подписывать** letter a drawing
~ **поперечного сечения** cross sectional drawing
~, **производственный** production drawing
~, **рабочий** design/working/(work)shop/detail drawing, detail/working plan
~, **разметочный** layout drawing
~ **расположения жилых помещений** accommodation plan
~, **сборочный** assembly drawing
~, **строительный** construction plan
~, **схематический** sketch, schematic diagram(me)
~, **технический** mechanical/technical drawing
~ **технологических разъемов** exploded drawing
~, **узловой** sub-assembly drawing
~, **установочный** installation drawing
~, **центровочный** c.g. positions drawing, centre-of-gravity positions drawing
~, **черновой** rough drawing
~, **эскизный** sketch, draft
~, **эталонный** master drawing
чертежная *n.* drawing/drafting room
чертежник draftsman, draughtsman
чертежный drawing, drafting, graphic(al)
~**ная линейка** ruler
чертилка scriber, marking tool/awl, scratch awl
чертить draw, design, sketch, make a blueprint
чертогон, чертополох *bot.* thistle *(Carduus)*
чертополоховка *bot.* painted lady *(Pyrameis cardui)*
черты *pl.* features; *see also* **черта**
~ **земли, основные** larger features of the Earth
~ **лица** features
~**тах, в общих** in generalities, roughly
~**тах, в основных** in the main, basically
черчение drawing
~, **машиностроительное** machine/mechanical drawing

~, **полевое** field sketching
~ **по синьке, ~ по синей копии** blueprint drawing
~, **техническое** engineering drawing
~, **чистовое** fair drawing
чесалка combing machine; hackle; ripple
чесальня combing mill, hackling room
чесание carding; combing, rippling, hackling
чеснок *bot.* garlic *(Allium sativum)*
чесночник *bot.* 1. hedge garlic *(Alliaria);* 2. garlic-scented agaric *(Marasmius scorodonius)*
~, **обыкновенный** garlic mustard *(Alliaria vulgaris)*
чесночница *zool.* spade-footed toad, pelobatid *(Pelobates)*
чесночный *bot.* alliaceous, garlic(ky)
чесотка *med.* scabies; mange; itch, rash
~, **зерновая** grain itch
~, **норвежская** Norwegian scabies
чесотный, чесоточный scabious, scabby; mangy
~ **клещ** *zool.* itch mite *(Sarcoptes scabei)*
честер *cul.* Cheshire cheese
честный 1, hono(u)red, worthy; honest; 2. *eccl.* sainted, saintly, sanctified
~**ная игра** *econ.* fair play
~**ное слово** *law* parole
честь hono(u)r; credit
~, **воинская** salute
~, **отдавать** salute
чет *math.* even (number)
~ **и нечет** odd and even
четверик *hist.* chetverik *(old Russian unit of measure)*
четверка 1. *educ.* (school grade) four; 2. *el.* quad; 3. *naut.* four-oar(ed boat), jolly (boat); 4. *av.* (flight of) four aircraft; 5. quarter, fourth part
~ **точек, гармоническая** *math.* harmonic quadruplet
четверник *cryst.* fourling
четверной fourfold, quadruple; quaternary, quaternate; tetradic
четвероевангелие *eccl.* the four Gospels
четверокурсник *educ.* senior student, fourth year student
четвероногий *zool.* quadruped(al), tetrapod(ous), four-legged/-footed; *n. pl.* quadrupeds, tetrapods *(Tetrapoda)*
четверорукий *zool.* quadrumanous; *n. pl.* fourhanded, quadrumanes *(Quadrumana)*
четверостишие *lit.* quatrain, stanza of four lines
четвероугольник quadrangle
четверохолмие *anat.* quadrigeminal bodies
четвертичный *geol.* Quaternary
четвертная *n. hist.* 25 ruble bank note
четвертной quarter
~ **билет** *hist.* 25 ruble bank note
~**ная нота** *mus.* quarter/fourth note
четвертование *hist.* quartering
четверть 1. quarter; 2. *mus.* fourth/quarter note; 3. term; 4. *constr.* closer (brick); 5. *hist.* old Russian dry measure (210 litres); old Russian liquid measure (3.1 litres)
~ **круга** quadrant
четвертьволновый quarter-wave (length)
четвертьквадратный quarter-square
четвертьоборотный quarter-turn
четвертьфинал *sport* quarterfinal
четки *pl.* 1. *med.* beading; 2. *eccl.* rosary
~, **рахитические** rachitic rosary, beading of the ribs
четкий sharply/well defined; clear(-cut); precise, efficient; distinct, legible; *el.* positive (of relay)
четковидный, четкообразный *med.* beaded, moniliform; *bot.* granose, lenticular
~ **стручок** *bot.* loment(um)
четкость sharpness; definition; clearness, distinctness; fine detail resolution; contrast, acutance
~, **вертикальная/хоризонтальная** vertical/horizontal definition
~ **в углах изображения** *telev.* corner detail
~ **изображения** image sharpness, photographic acutance
~ **срабатывания** *el.* positive action
чет-нечет *math.* odd-even
четно-нечетный even-odd
четность 1. evenness; 2. parity
~, **внутренняя** intrinsic parity
~, **отрицательная/положительная** odd/even parity
~ **состояния** parity of a wave function
четно-четный even-even
четный even(-numbered)
четочник *bot.* rosary pea *(Abrus)*
четырехатомный tetratomic, tetrabasic; tetraacid
четырехвалентность tetravalence
четырехвалентный tetravalent, quadrivalent
четырехвалковый four-roller
~ **стан** four-high mill
четырехвинтовый *naut.* quadruple-screw
четырехглавый 1. *anat.* quadricept; 2. *arch.* having four domes
четырехглазка *zool.* four-eyed fish *(Anableps tetrophthalmus)*
четырехгнездный quadrilocular, tetrathecal
четырехголосный *mus.* four-part
четырехгранник *geom.* tetrahedron
четырехгранный four-sided, tetrahedral
~ **колос** *bot.* four-rowed spike, quadrangular spike
четырехдвигательный four-engine(d)
четырехдневный quartan; recurring every fourth day; four-day
четырехдольчатый quadrifid
четырехжаберные *n. pl. zool.* tetrabranchs, tetrabranchiates *(Tetrabranchia(ta))*
четырехжилковый four-veined/-nerved
четырехжильный four-wire
четырехзамещенный tetrasubstituted
четырехзвенник four-link chain
четырехзначный four-unit/-aspect/-digit/-figure/-valued/-place

четырехзубчатый quadridentate, tetradentate
четырехкамерный four-chambered, quadrilocular
четырехканальный four-channel/-track/-terminal
четырехколесный four-whell(ed)
четырехконечный four-rayed/-point, tetracrepid, quadriradiate
четырехкосточковый *bot.* tetrapyrenous
четырехкратный fourfold, quadruple
четырехкруговой tetracyclic
четырехкрылый quadrialate, quadripinnate, quadripennate, tetrapterous
четырехкрыльник *bot.* winged pea *(Tetragonolobus)*
четырехлепестковый, четырехлепестный *bot.* quadripetalous, tetrapetalous
четырехлистный four-leaved, quadrifolious, tetraphyllous; *math.* four-sheeted
четырехлопастный *bot.* quadrilobate, quadrilobed; *tech.* four-bladed
четырехлучевой four-beam; quadriradiate, tetraradiate
четырехмерный four-dimensional
четырехмутовчатый tetracyclic
четырехногий four-legged/-footed, quadruped
четырехокись *chem.* tetroxide
четырехосновный tetrabasic, quadribasic, tetratomic
четырехосный quadriaxial, tetraaxial
четырехостный *bot.* quadriaristate
четырехпалый four-fingered/-toed, quadridigitate, tetradactyl(ous)
четырехпарный quadrijugal, quadrijugate, quadrijugous
четырехперегородчатый quadriseptate
четырехперистый quadripinnate
четырехпестичный tetragynous
четырехплодолистиковый tetracarpellary
четырехполье *agr.* four-plot rotation
четырехполюсник four-terminal network, twoport, two-terminal pair network, quadripole
~, активный active two-port, active four-terminal network
~, выравнивающий compensating network
~, Г-образный L-section four-terminal network
~, компенсирующий compensating network
~, невзаимный nonreciprocal two-port terminal
~, обратимый reciprocal/reversible four-terminal network
~, пассивный passive two-port, passive four terminal network
~, П-образный pi-section two-port
~, симметричный bilateral network
~, Т-образный T-section four-terminal network
~, уравновешенный balanced two-terminal pair network
четырехполюсный quadripolar, quadripole, tetrapolar, four-pole
четырехпроводный four-wire
четырехпыльниковый *bot.* tetrantherous
четырехраздельный quadrifid, quadripartite, tetrachotomous
четырехразрядный *math.* four-digit/-place
~ное число four-figure number
четырехрамочный four-gimbal
четырехрассеченный quadrisected
четырехрезцовые *n. pl. zool.* duplicidentates, duplicidents *(Duplicidentata)*
четырехрогий tetracerous; *bot.* quadricorn
четырехручный *zool.* quadrumanous; *mus.* fourhand(ed)
четырехрядный four-row(ed), tetraserial, quadriserial; *bot.* tetrastichous, quadrifarious
четырехсемянка *bot.* tetrachaenium
четырехсемянный tetraspermous
четырехсильный *bot.* tetradynamous
четырехскатный *arch.* with four sloping surfaces *(of roof)*
четырехсложный four-syllable, tetrasyllabic; *bot.* quaternate
четырехслойный four-layer/-ply
четырехспоровый tetrasporous, quadrisporous
четырехстолбиковый tetrastylic, tetrastylous
четырехстопный *lit.* tetrameter, verse of four measures/dipodies/feet
четырехсторонний four-sided, quadrilateral; *polit.* quadripartite
четырехсторонник *geom.* quadrilateral
четырехструнный four-stringed
четырехступенчатый four-stage/-step(ped)
четырехтактный *tech.* four-cyle/-stroke; *mus.* four-beat
четырехточечный four-point
четырехтычинковый, четырехтычиночный *bot.* tetrastemonous, tetrandous
четырехугольник quadrangle, quadrilateral, tetragon; square; box *(on diagram)*
~, вписанный/описанный quadrilateral inscribed/circumscribed (in circle)
~, геодезический geodetic quadrangle, braced quadrilateral
~, неплоский warped/skew quadrilateral
четырехугольный tetragonal, quadrangular, four-angled/-cornered
четырехударный *phon.* four-stress
четырехусый *zool.* four-bearded
четырехфазный four-phase, quadriphase
четыреххлористый *chem.* tetrachloride (of); tetrachlorated
~ углерод carbon tetrachloride
четырехходовой *tech.* four-way/-direction
четырехцветковый quadriflorous
четырехцветный four-colo(u)r, quadricolo(u)r
четырехчастный four-part
~ такт *mus.* quadruple time/rhythm
четырехчлен quadrinominal
четырехчленный four-membered/-termed; tetramerous
~ цветок tetramerous flower
четырехшпорцевый quadricalcarate
четырехштырковый four-pin/-prong

четырехщетинковый quadrisetose, quadrisetaceous
четырехъядерный quadrinucleate
четырехэлементный four-element
четь *hist.* 1. quarter, one fourth; 2. old Russian dry measure (210 litres)
четья минея *eccl.* menology; martyrology
чех Czech
чехарда leapfrog; *fig. polit.* reshuffling
~, министерская cabinet reshuffling
чехлик covering, (moss)cap; calyptra
~, корневой *bot.* rootcap
чехликовидный, чехликообразный *bot.* hood-shaped, calyptriform
чехловидный sheathlike
чехлоноска *zool.* casebearer *(Coleophora)*
чехол cover, case, casing, jacket; bag; sheath; housing, hood; mantle
~, брезентовый canvas bag, tarpaulin cover
~ защитный protective/storage cover
~ коралла *zool.* epitheca
~ купола парашюта parachute sleeve/sock, parachute deployment bag, parachute inner pack
~ на матрац mattress cover
~, огнеупорный fire-resistant/-proof cover
~, осадочный sedimentary cover/mantle
~, парусный sail bag/cover
~, платформенный plarform mantle
~, рангоутный mast cover/cloth
~ тепловыделяющего элемента can, jacket, sheath
~, термоизоляционный thermal(-insulation) cover
~, шлюпочный boat cover, tilt
чехольчик *biol.* indusium, protective membrane/scale
чехонь, чехоня *zool.* sicklefish *(Polecus cultratus)*
чехословак Czechoslovak
чехословацкий Czechoslovak(ian)
чехословачка Czechoslovak (woman)
чечевица 1. *obs.* lens; 2. *bot.* lentil *(Lens)*
~, большая *zool.* great rose finch *(Carpodactus rubicilla)*
~ руля *naut.* foot step
чечевицевидный, чечевицеобразный lens-shaped; lentiform, lenticular
чечевичка *bot.* lenticel, cortical pore
чечевичный lenticular
чечен(ец), чеченский Chechen
чеченка Chechen (woman)
чечетка *zool.* redpoll *(Carduelis)*
чечна *bot.* Ural p(a)eony *(Paeonia anomala)*
чешский Czech(ish)
чешско-лужицкий Checho-Lusatian
чешуевидный, чешуеобразный scale-shaped, scalelike, squamiform
чешуеколосый *bot.* lepidostachyous
чешуекрылый *zool.* lepidopteral, lepidopterous; *n. pl.* lepidopterans, lepidopterids *(Lepidoptera)*
чешуелистный *bot.* scaly-leaved, squamifolious, lepidophyllous
чешуеног *zool.* salt-march snake, legless lizard, slowworm *(Pygopus)*
чешуеотделитель *food ind.* descaler
чешуеплодный *bot.* scaly-fruited, lepidocarpous
чешуехвостник *bot.* sickle-grass *(Pholiurus)*
чешуечка squamula, squamule
чешуйка scale; flake; perule, ramentum; *med.* squamule
~ коралла *zool.* exotheca
~, кроющая bract/cover scale
~, ороговевшая/роговая horny scale
~, полая arched scale, fornix
~ слюды mica flake
~, цветковая lemma, flowering glume
чешуйница *zool.* fish moth, silverfish *(Lepisma saccharina)*
чешуйчатка *bot.* pholiota *(Pholiota)*
чешуйчатник *paleont.* lepidodendrid *(Lepidodendron)*
чешуйчатолистный *bot.* leptophyllous
чешуйчатость scaliness, squamosity, squamation
~ чашелистников imbrication of sepals
чешуйчатый scaly, scaled; scale-like; flaky; lamelar; *med.* squamous, squamate, squamosal; *n. pl. zool.* squamates *(Squamata)*
~ лист scale/scaly leaf
~тая кора scale/scaly bark
чешуя scale, flake, scute; plate, lamella; *med.* squama; *bot.* palea
~, верхняя upper/superior palea
~, ганоидная ganoid scale
~, заостренная колосовая awn pointed glume
~ злаков grass palea
~, колосковая glume
~, костная/циклоидная cycloid scale
~, кроющая bract/cover scale, chaff
~ лобной кости frontal squama
~, межсеменная interseminal scale
~, плодоносящая ovuliferous scale
~, семенная fruit scale
~, тектоническая *geol.* wedge scales; schuppen
~, третья колосковая sterile lemma
~, цветковая lemma, flowering glume
~шую, имеющий squamate, squamiferous
~шую, сбрасывать exuviate
чибис *zool.* lapwing, pewit *(Vanellus vanellus)*
чиггер *zool.* chigoe *(Sarcopsylla penetrans)*
чиж(ик) *zool.* siskin *(Spinus);* goldfinch, thistle finch *(Spinus tristis)*
чизелевание *agr.* chisel(l)ing, subsoiling
чизель(-культиватор) *agr.* chisel (cultivator), chisel plough, subsoiler
чий *bot.* chee grass *(Stipa splendens)*
чикл the gum of the sapodilla tree, chickle
чикса *naut.* (mast) cheek
чикчиры *pl. hist.* tight-fitting trousers
чилибуха *bot.* strychnine, poison nut *(Strychnos nux vomica)*
чилига *bot.* pea shrub/tree *(Caragana)*

чилиец, чилийский Chilean
чилийка Chilean (woman)
чилим *bot.* water chestnut *(Trapa natans)*
чин 1. rank; 2. official; 3. *hist.* rite, ritual, ceremony
чина *bot.* pea vine, vetchling *(Lathyrus)*
~, душистая sweet pea *(Lathyrus odoratus)*
чинар(а) *bot.* Oriental plane tree *(Platanus orientalis)*
чингиль *bot.* salt tree *(Halimodendron)*
чинить I repair, mend
чинить II point, sharpen
чиновник official, officer, functionary
~, акцизный excise officer
~, таможенный custom(s)/revenue officer
чиноположение hierarchy; statutes, regulations, rules
чинопроизводство promotion in rank
чинш *hist.* quitrent, rent in money/kind
чиншевик *hist.* copyholder
чип *electr.* chip
~, безкорпусный unpackaged chip
~, кремниевый silicon chip
~, перевернутый, ~ с защелком flip chip
~ с интегральными схемами integrated circuit chip
чиполино *miner.* cipol(l)ino, cipolin
чипсы *pl. cul.* potato chips
чирей *med.* boil, furuncle; abscess
чирок *zool.* teal *(Anas)*
чирус *zool.* small mackerel *(Scomber)*
числа *pl. math.* numbers; *see also* **число**
~ Бернулли *math.* Bernoulli('s) numbers
~, взаимно простые relatively prime numbers, coprime numbers
~, относительные directed/signed/algebraic numbers
~, пифагоровы Pythagorean numbers/triples
~, попарно простые mutually/pairwise relatively prime numbers, coprime numbers
~, предпочтительные preferred numbers
~, равноостаточные congruent numbers
~, случайные random numbers
~, сопряженные комплексные conjugate complex numbers
~, фигурные figurate numbers
числа-близнецы *pl. math.* number-/prime twins
численность (total) number; quantity, size, count; numerosity, numerousness; *mil.* strength; *naut.* complement
~ выборки *stat.* sample size
~ команды/экипажа в мирное время peace complement
~ населения, валовая *econ.* gross population
~, общая total number
~ рабочей группы crew size
~ рабочих и служащих labo(u)r/work force
~ю in number, numbering
численный 1. numerical; 2. quantitative
числитель *math.* numerator
~ дроби term of fraction
~ отношения antecedent
числительное *n. gram.* numeral; number
~, количественное cardinal number
~, неопределенно-количественное indefinite numeral
~, порядковое ordinal (numeral/number)
числительное-местоимение *n. gram.* indefinite numeral
числительное-прилагательное *n. gram.* ordinal numeral
числительное-существительное *n. gram.* noun numeral
числительный quantitative
~ное, имя *gram.* numeral
числить count, reckon, calculate; regard as
числиться be reckoned/recorded/listed/considered
~ больным be on the sick list
~ за/под номером be itemized under No.
число 1. number; figure, numeral; 2. quantity; 3. date, day; *see also* **числа**
~ Аббе Abbe number
~ Авогадро Avogadro('s) number/constant
~, арифметическое arithmetic/absolute number
~, атомное atomic/charge number
~ без знака *comp.* unsigned number
~, безразмерное dimensionless/nondimensional/pure number
~ Вебера Weber number
~, вещественное real number
~ в обратном двоичном коде *comp.* one's complement number
~, водородное hygrogen number
~, водяное (калориметра) water equivalent
~, волновое wave number
~, восьмеричное octal number
~ выводов *comp.* pin count
~, вычитаемое subtrahend
~ Грасгофа Grashoff number
~ Греца Graetz number
~, двоично-десятичное BCD number, binary coded decimal (number)
~, двоичное binary number/digit, bit
~, двоично-кодированное dual-/binary coded number
~, двоично-пятеричное biquinary number
~ двойной длины double-length number
~ двойных ходов в минуту strokes per minute, s.p.m.
~, действительное real number
~ делений, общее *nucl.* total fissions
~, десятичное decimal (number)
~, десятичное упакованное packed/packaged decimal number
~, докритическое *mech.* subcritical number
~, дробное fraction, fractional/broken number
~, единственное *gram.* singular
~, закритическое *mech.* beyond-critical number, supercritical number
~ заполнения оболочки *nucl.* shell occupation number

~, золотое the golden number
~ избыточных нейтронов neutron excess number
~, именованное concrete/denominate(d) number
~ импульсов pulse count
~, иррациональное irrational/surd (number)
~, йодное iodine number/value
~, кардинальное cardinal (number), cardinality
~ Кармана Karman number
~, кислотное acid/neutralization number
~, кодированное code number
~, колебательное квантовое oscillation quantum number
~, координационное coordination number
~, кратное multiple, aliquot
~, круглое round number
~, логарифмируемое number under the logarithm
~ Локка Lock number
~ Лошмидта/Льюиса Loschmidt/Lewis number
~, массовое mass/nucleon number, A
~ М/Маха Mach (number), M number
~ Маха, критическое по росту сопротивления drag-rise Mach number
~ мест seating capacity
~, мнимое pure(ly) imaginary (number)
~, многозначное multidigit/multiple/multiplace number
~, множественное *gram.* plural
~ Мозли Moseley number
~, недостоверное массовое uncertain mass number
~, неименованное absolute number
~ нейтронов на акт деления/поглощения *nucl.* neutrons per fission/absorption
~ несущей способности, калифорнийское *av.* California bearing ratio, CBR value
~, n-значное n-digit number
~, нормализованное standard/normalized number, number in normal form
~ Нуссельта Nusselt number
~ оборотов rotational speed, number of revolutions
~ оборотов вала shaft speed, shaft r.p.m.
~ оборотов в минуту revolutions per minute, r.p.m. speed
~ оборотов на режиме малого газа idle/idling speed
~, обратное inverse number
~, однозначное/одноразрядное digit, simple number, one-figure number
~, октановое octane number/rating/value
~ омыления saponification number/value
~, отвлеченное unnamed/abstract/absolute/non-dimensional/dimensionless/pure number
~ отсчетов в единицу времени count(ing) rate
~ мест seating capacity
~ Пекле Peclet number
~, переводное transference number
~, передаточное *mech.* gear/transmission ratio; gearing; *el.* gain
~ переменной длины *comp.* multi-length number
~ переноса transfer/transport number; *math.* carry quantity
~ периодов в секунду cycles per second
~ повторений repeat count
~, подкоренное radicant
~, полуцелое half-integer
~, порядковое ordinal/serial number, ordinal
~, почти магическое near-magic number
~ Прандтля Prandtl number
~ превышения out-of-range number, overflow number, capacity exceeding number
~ продуба *leath.* tanning number
~, произвольное (по Дини) Dini's derivate, derived number (of a function)
~, простое prime (number)
~ разрядов в регистре register length
~, рациональное rational (number)
~ Рейнольдса, (критическое) (transition/critical) Reynolds number
~ Рейхерта-Мейссля Reichert-Meissl number
~, родановое thiocyanogen number/value
~ с плавающей/фиксированной запятой/точкой *comp.* floating-/fixed-point number
~ следов track population
~ со знаком signed number
~, собственное eigenvalue
~, совершенное perfect number
~, составное composite number
~, спиртовое alcohol number
~ степени, алгебраическое algebraic number of degree
~ столкновений collision frequency
~ Строухала/Стэнтона Strouhal/Stanton number
~ твердости hardness number
~ твердости по Бринеллю/Виккерсу Brinell/Vickers (hardness) number
~ твердости по Роквеллу/Моосу Rockwell/Mohs (hardness) number
~, уменьшаемое minuend
~, упакованное десятичное *comp.* packed decimal
~, фагоцитарное *biol.* phagocytic index
~ Фарадея Faraday constant, faraday
~, флегмовое *chem.* reflux ratio
~ Фруда по глубине погружения *naut.* Froude depth number, submergence-Froude number
~ Фруда по водоизмещению *naut.* V-volume Froude number
~ Фруда по ширине b-/beam-Froude number
~, целое integer, integral/whole number
~, целое комплексное complex/Gaussian integer
~ Циолковского mass ratio, initial-to-final-weight ratio, Ziolkovsky number

~ **человеко-лет, общее** *stat.* total lifetime
~ **Шервуда/Шмидта** Sherwood/Schmidt number
~, **шестнадцатиричное** hexadecimal number
~ **Эйлера** Euler number
~ **экранирования** screening number
~ **элементов растра** raster count
~, **эфирное** ester number/value
~**м, помечать** date
~**м, помечать задним** antedate
чистец *bot.* marsh/clown's woundwort, betony *(Stachys)*
чистик I *tech.* scraper
чистик II *zool.* guillemot *(Cepphus)*
чистилище *rel.* purgatory
чиститель cleaner, machine for cleaning
чистить 1. clean(se); scour, scrub; *met.* bran; 2. *polit.* purge
~ **наждаком** grind with emery
~ **щеткой** brush
чистка clean(s)ing, clearing, cleanup; dredging; *polit.* purge
~ **голья** *leath.* scudding
~ **подводной части** *naut.* graving
чистовик clean/fair copy
чистоводородный pure-hydrogen
чистовой clean, fair; *tech.* finishing, smoothing
~ **резец** finishing tool
~**вая рукопись** fair/final manuscript
чистодел *tech.* burnisher, burnishing tool
чистокровка *agr.* thoroughbred animal, purebred, pureblood
чистокровный, чистоплеменный pure-blooded, thoroughbred
чистописание *educ.* calligraphy, chirography, penmanship
чистопородный pure breed, thoroughbred
чистопробный pure *(of gold/silver),* of the highest quality
чистота clean(li)ness, purity; clarity; fineness
~, **видимая** apparent purity
~ **зоны Френеля** *rad.* Fresnel zone clearance
~ **поверхности** surface finish
~, **сортовая** *agr.* purity of a variety/cultivar
чистотел *bot.* celandine *(Chelidonium)*
чистоуст *bot.* royal fern, osmund *(Osmunda)*
чистый 1. clean, blank; neat; fair; 2. pure, homogenous, undiluted; sterling *(of silver);* 3. *rel.* innocent; chaste; 4. *math.* net; 5. honest, disinterested; unsullied
~ **вес** net weight
~ **понедельник** *eccl.* first Monday in Lent
~ **спектроскопически** specpure
~ **экземпляр** fair copy
~**тая отставка** final retirement
~**тая правда** naked truth
~**тая прибыль** net profit/gain
чистяк *bot.* pilewort *(Ficaria)*
читабельность *comp.* readability
читальня reading room/hall
читальщик *eccl.* reader of psalms for the dead
читать 1. read; 2. *educ.* teach; deliver, give *(a lecture, etc.)*
~ **курс истории** teach a course in history
~ **лекцию** give a lecture
~ **проповедь** preach a sermon
чихание sneezing, sternutation
чихательный sneezing, sternutative, sternutatory
чихирь insufficiently fermented (Caucasian) red wine
чихотник *bot.* yarrow *(Ptarmica)*
чичерон(е) guide, cicerone
чищение clean(s)ing; scrubbing; polishing
член 1. member, part of a whole; fellow; *math.* term; component; *gram.* article; 2. *anat.* limb; *see also* **члены**
~ **более высокого порядка** higher(-order) term
~, **возмущающий** disturbing term
~, **выражающий тренд** trend term
~, **высший** leading/highest (degree) term
~, **главный** dominant term
~, **дизъюктивный/конъюктивный** component/element of a disjunction/conjunction
~, **дополнительный** additive term
~, **затухающий** transient term
~ **коллегии защитников** *law* member of the bar
~, **конечный** end/extreme member
~ **множества** element/member of a set
~, **мужской половой** *anat.* penis
~, **неисчезающий** non-vanishing/-zero term
~, **общий** general term
~, **определенный** *gram.* definite article
~ **определителя** *math.* element/entry of a determinant
~, **основной** dominant term
~, **остаточный** remainder (term)
~ **отношения** term of ratio/relation
~ **отношения, первый/предыдущий** antecedent (of a ratio); first component (of an implication); previous term (of a series)
~ **отношения, последующий** succedent, succesor, consequent
~ **парламента** member of parliament, M.P.
~, **перенести** transpose a term
~, **поправочный** correction/corrective term
~ **последовательности** member of a sequency
~, **почетный** honorary member
~ **пропорции** proportional, term of proportion
~ **пропорции, крайний** extreme (term of a proportion)
~ **пропорции, средний** mean (term of a proportion), mean proportional
~ **ряда (общий)** (general) term of series
~, **свободный** absolute/constant/free term
~, **связывающий** coupling term
~, **старший** leading term, highest degree(d) term
~, **старший многочлена** leading coefficient
~ **уравнения, (свободный)** (absolute) term

of an equation
~ **экипажа, (дублирующий)** (back-up) crew member
членение articulation, segmentation, division into parts
члеНИК *anat., bot., zool* segment; joint
~ **криноидеи/лилии** columnal
~ **лапки насекомого, нижний** tarsus
~ **руки** brachium
~ **стебля** ossicle, joint of a stalk
членикообразный articulus-shaped, segmentary
членимость divisibility, divisibleness
членимый articulated, segmented, divided into parts
членистоногие *n. pl. zool.* arthropods *(Arthopoda)*
~, **бескрылые** wingless arthropods, apterans *(Aptera)*
членистоногий *zool.* arthropodous, arthropodal, arthropodan
членистопальчатый arthrodactylose
членистоперистый *bot.* articulate-pinnate
членистоплодный *bot.* condylocarpous
членистосложный articulate-compound
членистостебельные *n. pl. bot.* arthrophytes *(Arthrophyta)*
~, **первичные** calamophytes
членистый *anat., bot., zool.* articulate(d), jointed, segmented; lomentaceous
~ **плод** jointed fruit, loment(um)
членить articulate, segment, divide into parts
членкор, член-корр *abbr.* **член-корреспондент** Corresponding/Associate Member (of an Academy)
членовредительство *law* mutilation, maiming; deliberate self-mutilation
членораздельность articulation, articulateness
~ **речи** word intelligibility
членораздельный articulate; clear, distinct; *n. pl. zool* articulates *(Articulata)*
член-основатель charter member; founder
членский membership *(attr.)*
~ **билет** membership card
~ **взнос** membership fee/dues
членство membership
члены *pl. math.* terms; *see also* **член**
~ **королевской семьи** royalty
~, **непарные** no-pair terms
~, **неподобные** dissimilar terms
~, **перекрестные** cross terms
~, **подобные** similar/like terms
~ **предложения** *gram.* parts of a sentence
чоглок *zool.* hobby, small species of falcon *(Falco subbuteo)*
чокер *for.* choker
чолья *bot.* cholla *(Cylindropuntia)*
чомга *zool.* great crested grebe *(Podiceps cristatus)*
чоппер *mech.* chopper
чосеровский *lit.* Chaucerian
чоха Circassian coat
чревный *ant.* c(o)eliac
чрево *rel.* womb
чревовещание ventriloquy
чревовещатель ventriloquist
чревовещательный ventriloquous; ventriloquistic, ventriloquial
чревосечение *surg.* celiotomy, laparotomy, ventrotomy
чревоугодник glutton, gourmand, belly-god
чревоугодничество gluttony
чрезбрюшинный *anat.* transperitoneal, transabdominal
чрезвлагалищный *anat.* transvaginal
чрезвычайность uncommon occurrence; extraordinariness; exceptionality
чрезвычайный 1. extraordinary, exceptional; extreme; 2. emergency, special
~ **декрет** decree extraordinary
~ **и полномочный посол** Ambassador Extraordinary and Plenipotentiary
~ **представитель** ambassador extraordinary
~ **съезд** emergency convention
~ **фонд помощи детям при ООН, ЮНИСЕФ** United Nations International Children's Emergency Fund, UNICEF
~**ное положение** state of emergency
~**ное собрание/заседание** special/extraordinary meeting
~**ные меры** extraordinary/emergency measures
~**ные полномочия** extraordinary/emergency powers
чрезкожный *anat.* percutaneous, transdermic, transcutaneous
чрезкостный perosseous
чрезлобковый transpubic
чрезмерный excessive, extreme; redundant
чрен *geol.* salt pan
чресплевральный *anat.* transpleural, transthoracic
чреспузырный transvesical
чтение reading; reading matter
~ **карты** map/chart reading
~, **контрольное/проверочное** proofreading
~ **лекций** lecturing
~ **стихов** recitation
чтец reader; reciter, elocutionist
чубук I chibouk, chibouque, Turkish pipe
чубук II grape stalk
чубушник *bot.* mock orange *(Philadelphus)*
чуваш, чувашский Chuvash
чувашка Chuvash (woman)
чувственно-двигательный sensory-motor
чувственность sensuality; carnality, animalism; sensibility
чувственный 1. sensual, sensuous; sensitive; carnal; sexy; libidinous; 2. *philos.* perceptible, sensible; aesthetic
~ **объект** the sensible
~**ное восприятие** perception
~**ное ощущение** sense impression
чувствительность 1. sensitivity, sensitivenes;

response; sensibility; sentience; *phot.* speed; 2. apprehensibility, perceptibility; 3. sentimentality
~, болевая *med.* algesthesia, algesthesis, appreciation of pain
~, вегетативно-висцеральная *med.* splanchnesthesia, splanchnic sensation, visceral sense/sensation; seveth sense
~, весов sensitivity/sensitivenes of balance
~, вибрационная *med.* vibratory sensibility, pallesthesia
~, выключенная *med.* analgesia, an(a)esthesia, hemianalgesis, hemian(a)esthesia
~, глубокая deep/mesoblastic sensibility
~, двухмерно-пространственная *med.* posture sense
~, дискриминационная discriminating/pressure sense
~, измененная *med.* paresthesia, dysesthesia, cacesthesia
~, изотопическая abundance sensitivity
~ к закадмиевым нейтронам epicadmium efficiency
~ к ИК излучению infrared sensitivity, IR-sensitivity
~ к кислороду aerotropism
~ к нейтронам *nucl.* neutron sensitivity/response
~ к облучению radiosensitivity, radiosusceptibility, dosage sensitivity, irradiance responsivity
~ к освещенности *phot.* illumination responsivity
~ к свету light sensitivity, photosensitivity
~, кожная *med.* cutaneous sensitivity
~, костная *med.* bone/pallesthetic sensibility, pallesthesia
~, локализованная кожная *med.* topesthesia, topognosis
~, мышечная *med.* kinesthesia, muscular sense, deep sensibility
~, нелинейная nonlinear response
~, относительная relative/reference sensitivity
~ по скорости speed sensitivity
~, повышенная *med.* hyperesthesia, hypersusceptibility, hypersensitivity, acroesthesia, oxyesthesia
~, повышенная болевая *med.* hyperalg(es)ia
~, повышенная холодовая *med.* hypercryesthesia, hypercryalgesia
~, пониженная *med.* hypesthesia, bradyesthesia, obtusion
~, пониженная болевая *med.* hypalg(es)ia
~, поясная (зональная) *med.* zonesthesia, strangalesthesia, girdle sensation
~, пороговая threshold sensitivity/pickup
~, предельная threshold/ultimate/limiting sensitivity
~, приведенная (измерительного прибора) factor of merit
~ приемника, реальная minimum discernible signal of a receiver
~ пусковой схемы triggering sensitivity
~, световая light/luminous sensitivity
~, спектральная spectral response/sensitivity
~, статическая static sensitivity/responsivity
~ счетчика по мощности starting power of a meter
~, тактильная tactile sense, sense of touch
~, температурная *med.* therm(o)esthesia, warmth/thermal/temperature/thermic sense
~, тепловая thermal/thermic sense, thermal response
~ уха ear sensitivity
~ фотоэлемента phototube sensitivity
~, холодовая cryesthesia, cryalgesia
~, хроматическая *med.* colo(u)r sense, chromatopsia, colo(u)red vision
~, чрезмерно высокая oversensitivity, excessive sensitivity
~, электроакустическая *com.* electroacoustic response
~, электрокожная electrocutaneous sensibility
~, электромышечная electromuscular sensitivity
чувствительный 1. sensitive, sensible; sensorium, sentient; 2. appreciable, perceptible, noticeable; 3. sentimental
~ к излучению radiosensitive
~ к изменению энергии energy-sensitive
чувство 1. sense, sensation, feel(ing); 2. affection; love
~ взаимного доверия и симпатии rapport
~ времени time sense
~ давления baresthesia, pressure sense
~ движения kinesthetic sense
~, материнское motherly love
~ мучительного беспокойства anxiety
~, мышечное kinesthesia, muscular/deep sense
~ осязания tactile sense
~ покалывания pin sensation
~ ползания мурашек formication
~, пространственное space sense
~ равновесия sense of balance/ equilibrium, static sense
~, седьмое seventh/visceral sense
~ страха phobia
~ управления control feel, feel of controls
~, шестое sixth sense, cenesthesia
чувствование sensation, feeling; esthesia
чувяки *pl.* soft leather shoes, slippers
чугаль aluminium cast iron
чугун 1. cast/pig iron; 2. cast-iron pot
~ в болванках/чушках pig iron
~, высоколегированный high-alloy cast iron
~ горячего хода hot-blast/-blown pig iron
~, закаленный chilled cast iron
~, зеркальный mirror/specular iron, spiegeleisen
~, ковкий malleable cast iron
~, крупнозернистый (very) open-grained pig iron
~, легированный alloy(ed) cast iron

~, линейный foundry pig iron; cast iron
~, мелкозернистый fine-grained pig iron
~, модифицированный inoculated/modified cast iron
~, некондиционный off(-analysis/-grade) pig iron
~, немарочный ungraded pig iron
~, низкокремнистый dry pig iron
~, передельный (conversion) steel-making pig iron
~, серый grey (cast) iron
~, твердый cold pig iron
чугунный cast-iron
~ная дорога *hist.* railway
чугуновоз hot-metal ladle/car
чугунолитейный, чугуноплавильный iron *(attr.)*
~ завод iron foundry
чудо miracle; wonder, marvel; wonderwork; *(of person),* paragon, prodigy
чудо-богатыр hero
чудовище monster; abomination, monstrosity
чудовищный monstrous; villainous; *biol.* teratoid
чудодейственный wonder-working, miraculous
чудо-мальчик wonder boy
чудотворец miracle man, wonder-monger/-worker; *rel.* thaumaturge, thaumaturgus
чудотворный miracle-/wonder-working, miraculous; *rel.* thaumaturgic
чудотворство *rel.* thaumaturgics, thaumaturgy
чудь *coll.* old Russian name of Estonians and other Finnish tribes
чужеземец alien, outlander, foreigner, stranger
чужеземный foreign, alien, outlandish; peregrine; tramontane
чужеопыление *bot.* allogamy, cross-pollination/-fertilization
чужепланетный from another planet
чужеродец 1. stranger, alien; 2. person belonging to a different tribe/family
чужеродный alien, foreign(-born); allogenic; heterologous
чужестранец alien, outlander, foreigner, stranger
чужеядность parasitism
чужеядный *bot.* parasitic
~ное растение parasitic plant, phytoparasite
чужой foreign, alien; extraneous; ectogenic, ecdemic; epigene; someone else's, another's; *n.* foreigner, stranger
чукотский, чукча Chukchi
чукучан *zool.* common sucker *(Catostomus commersonii)*
чукчанка Chukchi (woman)
чулок sleeve
~, кабельный cable grip
чум tent, tepee
чума plague, pest(is), black death/plague, pestilence
~, амбулаторная *med.* ambulant/ambulatory plague, pestis minor, larval pestis
~, бубонная bubonic/glandular plague, pestis major, pestis fulminans
~, водяная *bot.* elodea *(Elodea canadensis)*
~, геморрагическая h(a)emorrhagic plague, black death, h(a)emorrhgic form of bubonic plague
~ диких животных, бубонная sylvatic plague
~, кишечная intestinal plague
~, кожная cutaneous plague
~ крупного рогатого скота cattle plague, rinderpest
~ кур fowl pest/plague
~ свиней hog cholera, swine pest
~, септицемическая/септическая septicemic plague, pesticemia, lememia, pestis siderans
~, собачья canine distemper
чумной pestilential, pestiferous, lemic, loimic, plague-stricken
~ная палочка plague bacillus
чурак cut log
чурек *cul.* unleavened bread (baked in the Caucasus)
чуткий sensitive, responsive; keen
~ сон leight sleep
чутье flair, feeling; intuition; instinct; *(of animals)* scent
чуфа *bot.* chufa earth almond, chufa flat sedge *(Cyperus esculentus)*
чушка pig, ingot, bar
~ для балласта ballast pig
~ки, отливать pig, cast into pigs

Ш, ш

шабазит *miner.* chabasite, chabazite
шабаш *rel.* Sabbath
~ ведьм wiches' sabbath
шабдар *bot.* Persian clover *(Trifolium resupinatium/suaveolens)*
шабер scraper, scraping/shaving knife
~, желобчатый fluted scraper
~, изогнутый hook scraper
~, механический power scraper
~, пневматический air scraper
~, полукруглый hollow/concave/curved scraper
~, слесарный machinist's scraper
~, трехгранный three-cornered scraper, cant scraper
шабли Chablis (wine)
шаблон 1. *mach.* (profile) gauge, template, templet, (master)form, pattern, model; mo(u)ld; 2. *comp.* picture; mask; 3. *polit.* cliché, stereotype
~, вращающийся *foundry* turning strickle
~, галтельный fillet gauge
~ для заточки и установки резьбовых резцов centre gauge
~ для обрезки доски *wood.* face mould
~ для пил saw gauge
~, кабельный *teleph.* lacing board

~ **калибра** *met.* pass templet
~, **клиновой** angular/taper gauge
~, **контурный** outline template
~, **крестообразный** *text.* croos gauge
~, **плазовый** *naut.* mocking-up mould
~, **поверочный** check templet
~, **протяжной** *foundry* sweep (template), drawing strickle/template
~, **профильный** template, form/shape/profile gauge
~, **путевой** *rail.* gauge template, rail/track gauge
~, **разметочный** face template
~, **резьбовой** screw pitch/thread gauge, thread template
~, **сварочный** welding gauge
~, **скребковый** *foundry* sweep, drawing strickle/template
~, **установочный** adjuster gauge
~, **формовочный** strickle board, sweep template
~, **центровой** *mach.* centre gauge
~, **шпангоутный** *naut.* frame mould/set
шаблон-высотомер *mach.* centre-height gauge
шаблонирование *foundry* strickling
шаблонодержатель template holder
шаблонщик pattern/stencil maker, templater
шабот anvil block, stock anvil
шабрение, шабровка scraping, scouring
шабреный scraped, scoured, shaved
шаг 1. step, pace; 2. spacing; pitch; 3. *comp.* iteration
~, **большой** *mach.* coarse/high pitch
~ **винта** lead of a screw
~ **винтовой линии** lead of a helix
~ **воздушного/гребного винта** propeller pitch
~ **воздушного винта, динамический** zero-thrust pitch
~ **воздушного винта, тормозящий** braking pitch
~ **вычисления** algebraic step
~ **голландский** dutch roll
~ **гребного винта, конструктивный** design pitch
~ **дисков** spacing of disks
~ **задания** *comp.* job step
~ **задания, возобновлять** restart a job step
~ **заклепочного соединения/шва** rivet pitch/spacing, pitch of rivets
~ **зацепления** pitch of a gear, tooth pitch
~ **зубьев** tooth pitch
~, **изменяемый** variable/adjustable pitch
~ **индукции/интегрирования** *math.* induction/integration step
~, **итеративный** *comp.* iteration
~ **кадра** *cin.* pull-down
~ **катушки** *el.* coil pitch
~ **квантования по амплитуде** quantization step
~ **квантования по времени** sampling interval
~ **коллектора** *el.* commutator pitch
~ **контактор** contact spacing
~ **кристаллической решетки** crystal lattice spacing
~, **крупный** *mach.* coarse pitch
~ **лопаток** blade spacing, blading pitch
~, **малый/мелкий** *mach.* fine pitch
~ **между дорожками** *comp.* track pitch
~ **между осями** centre spacing
~, **нулевой** zero pitch
~ **обмотки** *el.* winding pitch
~ **оплетения** pitch of braiding
~ **перфораций** *comp.* feed pitch
~ **пилы** tooth spacing
~ **подачи** *tel.* line advance
~ **подвигания (подрывки)** *min.* (ripping) web
~ **поиска (гидролокатора)** step angle
~, **полюсный/полюсов** *el.* pole pitch
~, **поперечный** *mach.* transverse pitch
~, **пробный** exploratory step/move
~ **программы** program step
~, **продольный** *mach.* long pitch
~ **разбиения** *math.* subinterval
~ **развертки** *tel., phot.* line advance
~ **разности** *math.* difference interval
~ **распределения** *math.* span of a distribution, distribution step
~ **резьбы (винта)** thread/screw pitch, pitch of a (screw) thread
~ **решетки** lattice spacing/pitch
~ **сетки** *math.* interval/pitch of a lattice; *comp.* array pitch
~ **скрутки (кабеля)** lay/pitch of a strand
~ **спирали** lead of a helix
~ **таблицы** interval/step/spacing of a table
~, **торцовый** circumferential pitch
~ **транспозиции** *el.* transposition interval
~, **угловой** angle/angular pitch
~ **ходового винта** *mach.* lead-screw pitch
~, **хордальный зацепления** chord(al) pitch
~ **цепи** *mach.* pitch of a chain, chain pitch
шагающий walking; self-propelled
шаговый *tech.* spacing
~**вая работа** step-by-step operation
шагомер 1. pitch-measuring device; pitch gauge/counter; 2. pedometer; distance recorder
шагреневый shagreen; alutaceous
~**вая кожа** shagreen leather
шагренирование *leath.* pebbling, shagreening, pebble-graining
шагренировать pebble
шагрень 1. shagreen (leather); 2. *text.* sharkskin
шадрик potash
шайба 1. *tech.* washer; spacer; plate; 2. *sport* puck
~ **вертикального оперения** *av.* endplate fin
~ **Гровера** spring/bent washer
~, **дистанционная** distance plate/washer; spacing washer/plate
~, **дроссельная** throttling orifice, orifice plate

~, замковая safety washer
~, измерительная orifice plate
~, изоляционная/изолирующая *el.* insulating bush/washer
~, контровочная/контрящая lock washer
~, концевая end/boundary plate
~, кулачковая/кулачная cam (plate)
~, литерная type disk
~, нажимная, конусная *el.* vee-ring, V-ring
~, нажимная сальника packing washer
~, подкладная (plain) washer
~, предохранительная safety washer
~, пружинная/пружинящая spring washer
~, пылезащитная *agr.* sand collar
~, разрезная horseshoe/slot(ted) washer, C-type washer
~, распорная spacing washer/plate
~, распыливающая *agr.* nozzle orifice/spray disk; sprayer plate
~, регулировочная adjusting washer
~, стопорная lock washer
~, тяговая *mach.* sheave
~, уплотнительная sealing washer
~, установочная spacer
~, флюсовая ferrule
~, центрующая centring/centering ring
~, эксцентриковая eccentric disk
шайка I gang, band
шайка II small washbasin
шайтан 1. *myth.* evil spirit, Satan; 2. ancestral spirit; totem
шакал *zool.* jackal *(Canis aureus)*
шаланда *naut.* scow, barge, lighter
~, грунтоотвозная dump scow, hopper barge
~для вывоза нечистот sewage vessel
~, малая punt
~, саморазгружающаяся dump scow
шалашник *zool.* bowerbird *(Ptilonorhynchus)*
шале *arch.* chalet
шалот *bot.* shallot *(Allium ascalonicum)*
шалфей *bot.* sage *(Salvia)*
шаль shawl
~, кашемировая India/Cashmere shawl
шамаль *met.* shamal
шаман *rel.* shaman
шаманизм, шаманство shamanism
шамбелян *hist.* chambellan, chamberlain
шамбертен Chambertin (wine)
шамозит *miner.* chamo(i)site
шамот chamotte, grog, fire clay; firebrick
~, обжигать burn/calcine/fire grog
шамотобетон castable refractory
шампанизация champagnization
шампанское, шампанское вино champagne
шампиньон *bot.* true/common/meadow mushroom, agaric *(Agaricus/Psalliota campestris)*
шампиньоновые *n. pl. bot.* gill fungi *(Agariceae)*
шамплен *geol.* Champlain
шампунь shampoo
шандра *bot.* hoarhound, horehound *(Marrubium)*
шанец *hist.* fieldwork; entrenchment
шанжан *text.* chatoyant silk
шанкерный *med.* chancrous
шанкр *med.* chancre
~, мягкий chancroid, soft/simple/eating/ Celsus'/noninfecting chancre
~, смешанный mixed chancre
~, твердый/гунтеровский Hunter's/hard/true/indurated/sulcus chancre
шанкроид *med.* chancroid, soft/simple chancre
шанкроподобный *med.* chancriform
шанс chance; *pl. math.* odds
~ы, неравные long odds
~ы, почти равные short odds
~ы, равные equal odds
шантаж blackmail
шантажист(ка) blackmailer
шапирограф *print.* hectograph, gelatone pad for copies
шапка hat; cupola; *math.* cap; *print.* headline; *geol.* oil cap rock
~ брожения *chem.* cap of fermentation pomace
~, музыкальная *rad.* signature
~, облачная scarf cloud
шар I 1. sphere; ball; globe; 2. ballot; vote
~, белый/черный yes/no vote
~, воздушный (air) balloon
~, запальный/калильный hot bulb
~, земной the Globe, terrestrial/Earth's globe
~, избирательный ballot
~ мягкого железа (у компаса) quandrantal sphere
~, огненный (ядерного взрыва) fire ball
~, светомерный integrating/Ulbricht sphere
~, сигнальный *nav.* signal/sweeping ball
~, угольный *geol.* coal ball/apple
~, уравновешенный *meteor.* constant-level balloon
шар II *geol.* strait, sound
шарабан 1. *hist.* charabanc; 2. cabriolet; chariot
шарж caricature, cartoon; travesty; grotesque
~, дружеский harmless caricature, friendly jest
шаржир-машина charging machine
шаржирование, шаржировка 1. *theatre* caricaturing, lampooning, overacting; 2. *met.* charging
шар-зонд sounding balloon
~, автоматический unmanned/automatic balloon
~, уравновешенный constant-level/pressure balloon
шарик ball; *med.* globule, pellet; spherule; bead; bulb
~, жировой *med.* lipoplast
~, красный кровяной *med.* erythrocyte
~, молочный milk corpuscle/globule
шариковидный, шарикообразный globe-/globule-shaped, globular
шариковый globular
~ подшипник ball bearing

~вая (авто)ручка ball-point pen, biro
шарикоподшипник *tech.* ball bearing
шарка *bot.* virus pot disease, streak and ring spot variegation
шарлатан charlatan, fraud, quack
шарлот *bot.* shallot *(Allium ascalonicum)*
шарлотка *cul.* charlotte
шарманка *mus.* street/barrel organ
шарнир *tech.* hinge, (pivoting) joint; *geol.* bend
~ антиклинали *geol.* upper bend, apex, crest
~, болтовый pin hinge/joint
~, вертикальный *av.* lag/drag/alpha hinge
~, врезной blind hinge
~, горизонтальный *av.* flapping/delta hinge
~, Гука/карданный Hooke's/cardan/universal joint
~, двухзвенный torque link(s)/arms, scissors, nutcracker
~, качающийся swivel joint
~, осевой *av.* feathering hinge
~, расширительный expansion hinge
~ седла *geol.* upper/saddle/arch bend
~ синклинали *geol.* lower/trough bend
~ складки *geol.* flexure, fold bend
~, узловой tie-bar joint, multiple joint
~, универсальный universal joint/coupling
~, шаровой ball-and-socket joint
~ шасси *av.* landing-gear joint
~, шкворневой swivel joint
~, шомпольный piano hinge
шарнирно-укрепленный hinged
шарнирный articulated, hinged, jointed
~ механизм linkwork
~ная цепь link chain
шаровидноголовчатый round-headed, sphaerocephalous
шаровидноколосный *bot.* round-spiked
шаровидноскученный conglobate
шаровидность globosity, spheroidity, sphericity, rotundity
шаровидноцветковый *bot.* spheric-/globe-flowered
шаровидный ball-/globe-shaped, globular, globate(d), spheric(al), spheroid(al)
шаровка I *agr.* blind/preemergence cultivation
шаровка II *zool.* fingernail clam *(Sphaerium transversum)*
шаровница *bot.* globe daisy *(Globularia)*
шаровой globular, spherical, ball *(attr.)*
~ клапан ball valve
~вая молния ball lightning
шарообразный *see* **шаровидный**
шаропилот, шар-пилот pilot balloon
~ для определения высоты облачности ceiling balloon
шарошка 1. *mach.* burr mill, milling/rolling cutter, toothed wheel; 2. roller/rotary bit; friction roll
шартрез Chartreuse (liqueur)
шарф scarf
шарьяж *geol.* overthrust sheet, mass overthrust
шасла *agr.* Chasselas, sweetwater (grapes)
шасси *aut.* chassis; *av.* undercarriage, *Am.* landing gear; *mil.* alighting/running gear
~, вездеходное, ~ высокой проходимости cross-country chassis
~, велосипедное *av.* zero-track landing gear, tandem-type landing gear
~, гусеничное track-/caterpillar-type landing gear
~, длиннобазное long-wheelbase chassis
~, лыжное ski landing gear
~, многоколесное bogie landing carriage
~ необычной схемы unorthodox landing gear
~, одноколесное mono-wheel landing gear
~, ориентирующееся castoring undercarriage
~ платформенного типа platform chassis/frame
~, прицепное trailer chassis
~, сбрасываемое detachable landing gear
~, тележечное bogie undercarriage
~, трехколесное tricycle landing gear
~, убирающееся retractable landing gear
~ экскаватора power-shovel mounting
шасталка *agr.* awner
шастанье *agr.* deawning
шастать *agr.* awn, hull; polish *(grain)*
шатер 1. *constr.* hip(ped) roof; 2. *agr.* shock, conical stack; 3. tent; marquee
шатировать *art* shade
шато-икем Chateau d'Yquem (wine)
шатрообразный, шатровидный tentlike
шатун *tech.* connecting rod, con-rod
~, вилкообразный/вильчатый forked connecting rod
~, главный master (connecting) rod
~, прицепной knuckle/articulated (connecting) rod
~, смещенный offset connecting rod
шафран 1. *bot.* crocus, saffron *(Crocus);* 2. *agr.* variety of apples
шах 1. Shah; 2. *chess* check
~ и мат checkmate
шахиншах the Shah of Iran
шахматист(ка) chess player
шахматный 1. chess *(attr.);* 2. chessboard-shaped; checkered, checkerboard; staggered; tessellated
~ные фигуры chessmen
шахматообразный checkered; staggered
шахта 1. *min.* mine, pit; colliery; 2. *tech.* shaft; stack; 3. *naut.* trunk; well; *rock.* silo
~, антенная antenna trunk
~, бездействующая idle/abandoned mine
~, вентиляционная *naut.* ventilating/ventilator/blower/air trunk; air pit/shaft
~, воздухозаборная induction trunk
~, воздухоприемная air-supply duct
~, газовая gassy/gaseous/foul mine
~, дозиметрическая radiation monitoring pit
~, естественная *geol.* deep karst channel
~ зерносушилки *agr.* grain column
~, каменноугольная bituminous coal mine

~ люка *naut.* hatch(way) trunk
~ подъемника elevator well/shaft
~, пометная manure pit
~, постановочная laying well
~, пусковая launch(ing) silo
~, разведочная exploring mine, test pit, prospecting shaft
~, расширительная (танкера) expansion trunk
~, стартовая silo; launching/starting tube
~ сушилки *agr.* drying column
~ трапа *naut.* staircase, stair-well
~, угольная coal mine, colliery
~, шлаковая/шлакосмывная *met.* ash pit
шахта-хранилище storage well
шахтер miner
шахтоуправление mine management/administration
шашель, шашень *zool.* pileworm, shipworm *(Teredo navalis)*
шашечница chessboard, checkerboard
шашечный checker *(attr.),* checkered
~ная доска checkerboard, chessboard
шашка I 1. checker piece; *pl.* checkers (game); 2. blasting cartridge; block/slab/charge of explosive; 3. *rock.* grain
~, бронированная restricted(-burning) grain
~, дымовая smoke pot/candle
~, запальная primer
~, пироксилиновая guncotton slab
~, подрывная blasting cartridge, demolition block/slab
~, пороховая powder grain
~ промежуточного заряда detonation pellet
~, сигнальная *av.* flare pot
~, стержневая rod grain
~, тетриловая tetryl pellet
~, тринитротолуоловотетриловая TNT-tetryl block
~, трубчатая tubular grain
шашка II saber, (cavalry) sword
шашлык *cul.* shish kebab, shashlik
шашлычная shashlik house/restaurant
шваб, швабский Swabian
швабка Swabian (woman)
швабра *naut.* mop, swap
шваннома *mad.* neurolemma, junction nervus, schwannoma
шварт *med.* adhesion, induration
швартов 1. *naut.* (shore-)fast, mooring fast/hawser/line/rope; *pl.* mooring lines; 2. *av.* picket, tie-down
~, кормовой after rope/spring, stern fast/line
~, носовой bow line/rope/spring, head mooring
~, стальной mooring wire
швартование mooring
швартовка 1. making fast, mooring; 2. *av.* picket(t)ing, tie-down
~ бортом, ~ к борту alongside landing
швартов(н)ый mooring, fastening
~ бакан mooring buoy
~ шпиль warping capstan
швед, шведка Swede
шведский Swedish
швейник garment industry worker, sewer, worker of the clothing industry
швейная *n.* sewing workshop
швейный sewing
~ная фабрика garment/clothes factory
~ная промышленность clothing industry
швейцар doorkeeper, doorman, hall porter
швейцарец, швейцарский Swiss
швейцарка Swiss (woman)
швейцарская *n.* doorman's lodge
швелевание low-temperature carbonization/distillation, semicoking
швел(л)ер channel (section), U-beam; longitudinal sleeper
~, автомобильный car-building channel
~, гнутый roll-formed channel
швеллерный channel(-type)
швельгаз low-temperature distillation gas
швельшахта low-temperature carbonization shaft
шверт *naut.* centreboard, drop/sliding keel
швертбот centreboard boat, centreboarder
шверц *naut.* leebord, side board
швея-мотористка electric sewing-machine operator
швея-ручница hand finisher
швицевание *leath.* tainting (process)
швырялка 1. *agr.* throwing-fork potato raiser, potato spinner; 2. grain cleaning and grading machine
швыряние tossing, throwing, hurling
шевалье *hist.* chevalier
шевер *mach.* shaver
шевингование *mach.* shaving
шевиот cheviot (cloth); Cheviot (sheep)
~, чистошерстяной wool cheviot
шеврет chevrette, lamb/sheepskin leather
шевро glazed kid leather, kidskin, goat leather
шеврон *mil.* chevron, stripe
шевронный herringbone *(attr.)*
шед *zool.* (American) shad *(Alosa sapidissima)*
шеддок *bot.* shaddock *(Citrus grandis/maxima)*
щедевр masterpiece, chef d'oeuvre
шеечно-влагалищный *anat.* cervicovaginal
шеечно-губной *anat.* cervicolabial
шеечно-пузырный *anat.* cervicovesical
щеечно-щечный *anat.* cervicobuccal
шеечно-язычно-осевой *anat.* cervicolinguoaxial, axiolinguocervical
шеечный *anat.* cervical
шезлонг chaise longue
шейка 1. neck; collum; 2. *tech.* pin, (pivot) journal, gudgeon; web (of rail); 3. *zool.* crawfish tail
~ вала shaft neck, (neck) journal
~ зуба *anat.* dental neck
~ коленчатого вала (bearing) journal, crank journal
~ коленчатого вала, шатунная crankpin

~, **корневая** root collar/neck
~ **матки** *anat.* cervix, neck of the uterus/womb
~, **осевая** axle journal
~ **прокатного вала** roll neck
~ **разделения** scission neck (of a nucleus)
~ **рельса** rail web
~ **ствола** *min.* shaft mouth
шейкер cocktail shaker
шейкер-культура shake-culture
шейма anchor chain
шейно-грудной *anat.* cervicothoracic
шейно-затылочный *anat.* cervico-occipital
шейно-лицевой *anat.* cervicofacial
шейно-плечевой *anat.* cervicobrachial
шейный cervical, jugular
~ **позвонок** cervical vertebra
шейх sheik(h), Arab chief
шейх-уль-ислам sheikh ul Islam
шекспировед Shakespearean
шекспироведение Shakespeare scholarship/studies
шекспировский Shakespearian
шекспироман, шекспиропоклонник Bardolater, idolizer of Shakespeare
шекспиропоклонничество Bardolatry
шелк silk fiber/thread/fabric; rayon
~, **ацетатный** (cellulose-)acetate rayon
~, **вискозный** viscose rayon
~ **в мотках** skein silk
~, **вышивальный** embroidery silk
~, **искусственный** artificial silk, rayon
~, **крученый** reeled-and-thrown silk, twisted silk
~, **крученый натуральный** thread silk
~, **натуральный** natural/pure silk
~, **отваренный** boiled-off silk
~, **переливчатый** chatoyant silk
~, **травчатый** figured silk
~, **триацетатный** triacetate rayon
шелковидный silky, silklike
шелковина silk thread/strand/filament/fibre
шелковинка silk thread
шелковистолепестный *bot.* sericopetalous
шелковистолистный *bot.* silky-leaved, sericophyllous
шелковистоцветковый *bot.* silky-flowered, sericanthous
шелковица mulberry(-tree) *(Morus)*
шелковичный mulberry *(attr.)*
~ **червь** silkworm *(Bombyx mori)*
~ **кокон** *zool.* silk cocoon
шелковка *print.* silk paper
шелковник *bot.* water crowfoot *(Ranunculus aquatilis)*
шелковод silkworm breeder
шелководственный seri(ci)cultural
шелководство silkworm breeding, seri(ci)culture, silk farming
шелковолистный *bot.* silky-leaved, sericophyllous
шелковый silk(y), silken; sericeous
шелкография silkscreen printing/process
шелкокрасильный silk-dyeing
шелкокрутильный silk-spinning
шелкокрутильщик silk spinner, throwster
шелкокручение silk spinning/throwing, spinning of raw silk, silk reeling and throwing
шелкомоталка silk reel/filature, silk reeling machine
шелкопряд *zool.* silkworm *(Bombyx mori)*
шелкопрядение silk spinning
шелкопрядильный silk-spinning
~**ная фабрика** silk mill
шелкопрядильня silk mill
шелкоткацкий silk-weaving
~ **станок** silk-weaving loom
~**кая фабрика** silk-weaving mill
шелкоткачество silk weaving
шелк-сырец raw silk; floss
шеллак shellac
шеллер *agr.* sheller, husker, huller
шеллингианец, шеллингист *philos.* follower of Schelling
шеллингианство *philos.* Schellingi(ani)sm
шелль *geol.* Chelles
шелльский *geol.* Chellean, Chellian
шелушение 1. *agr.* hulling, husking, shelling; 2. *med.* desquamation; peeling (off), scaling; furfuration
~, **отрубевидное** defurfuration, scaly/branny desquamation; dandruff, scurf
шелушильный husking, hulling, podding, shelling; descaling
шелушинка chaffy scale; ramentum
шелушистый ramentaceous, furfuraceous
шелушитель *agr.* huller, sheller, hulling machine
шелушиться *med.* desquamate, scale off, shred, peel
шельга *arch.* vault head, extrados, crown
шельмование *hist.* public dishono(u)r; *fig.* defamation
шельмовать 1. *hist.* subject to public dishono(u)r; 2. *fig.* defame, abuse; run down
шельтердек *naut.* shelter deck
шельф *geogr.* shelf, submarine platform
~, **броневой** armour shelf
~, **бугристый** hillocky shelf
~, **выровненный** levelled shelf
~, **затопленный** submerged shelf
~, **материковый** continental shelf
~ **переборки** bulkhead shelf/girder
~, **транцевый** transom shelf
шельфовый shelf(y)
шелюга *bot.* silver willow *(Salix acutifolia)*
шеннон *com.* shannon
шепелявение, шепелявость *med.* parasigmatism, lisping, hissing
шепелявить lisp
шепинг *tech.* shaper, shaping machine
шепит *geol.* schoepite
шептало sear-notch lug, hammer catch

~ **спуска** trigger sear
шептание *med.* whispering
~, **беззвучное** mussitation
шерардизация *met.* sherardization, sherardizing
шерардия *bot.* field madder *(Sherardia)*
шербет *cul.* sherbet, fruit-juice drink
шергень *nav.* declivity board
шериф sherif(f); shereef, Moslem chief magistrate
шерл *miner.* schorl, black tourmaline
~, **малиновый** red schorl, rutile
шерохование buffing
шероховатка *bot.* bottle-brush grass *(Asperella)*
шероховатозубчатый *bot.* rough-toothed, trachydentate
шероховатолистный *bot.* rough-leaved, trachyphyllous
шероховатоплодный *bot.* rough-fruited, trachycarpous
шероховатосемянный *bot.* rough-seeded, trachyspermous
шероховатость *tech.* (degree of) roughness, unevenness; *bot., zool.* scabrousness
~ **покрытия** coating roughness
шероховатый rough, coarse; asper(ate); rugged, uneven; *bot.* scabrous
шероховка buffing
шерп(а) Sherpa
шерстемойка 1. wool washing/scouring (process); 2. wool washer, wool-scouring plant, woolwashing mill
шерстемойня wool-scouring plant, wool-washing will
шерстеносный laniferous, lanigerous
шерстеобразный woollike, laniform
шерстеподобный wool(l)y, woollike
шерстепрядение wool spinning
шерстепрядильня wool-spinning mill
шерстепрядильщик wool spinner
шерстестебельник *bot.* pipewort *(Eriocaulon)*
шерстечесальный wool-carding
шерстистоветвистый *bot.* eriocladous
шерстистоколосый *bot.* eriostachyous
шерстистолистный *bot.* woolly/downy-leaved, eriophyllous
шерстистоплодный *bot.* eriocarpous
шерсистосемянный *bot.* eriospermous
шерсистостебельчатый *bot.* eriocaulis
шерстистость wool(l)iness
шерстистоцветковый wooly-flowered, laniflorous, erianthous
щерстисточешуйчатый *bot.* eriolepis
шерстистый wool(l)y, fleecy; lanate, lanose, lanuginose; urotrichous
шерстобит wool beater
шерстобитня wool-beating mill/instrument/device
шерстобойня wool-beating mill
шерстокрыл *zool.* flying lemur *(Cynocephalus); pl.* lemurs *(Dermoptera)*
шерстомоечный, шерстомойный wool-washing
шерстомойка wool-washing machine/plant
шерстоносный wool-bearing, lanigerous, laniferous
шерстопрядение wool spinning
шерстопрядильня wool-spinning mill
шерстостебельник *bot.* pipewort *(Eriocaulon)*
щерстоткацкий wool-weaving
шерстоткачество wool weaving
шерсточесалка wool card; wool-carding establishment/unit
шерсть 1. hair (of animals); 2. wool; 3. woolen material
~, **верблюжья** camel hair
~ **высокого качества** quality wool
~, **заводская** dead/pelt wool
~, **извитая** curly wool
~, **козья** goat hair
~, **мериносовая** merino wool
~, **минеральная** mineral/cinder wool
~, **натуральная** virgin wool
~, **немытая** *agr.* greasy wool
~, **нестандартная** bulk grade wool
~, **овечья** sheep wool
~, **овечья шубная** pelt wool
~, **отборная** first-class wool
~, **рунная** fleece wool
~, **чумная** murrain wool
шерстяк *bot.* cup grass *(Eriochloa)*
шерт *geol.* chert, flinty chalcedony
шертинг shirting (cloth)
шерхебель rough/jack plane, round-nose plane
шершаволистный *bot.* trachyphyllous
шершавость roughness; hirsuteness; shagginess
шершень *zool.* (giant) hornet *(Vespa scabro)*
шест post; quanting/punting/barge pole; quant; rod, stick
~, **мерный** measuring stick
~, **отпорный** barge pole
~ **с подрывным зарядом** charge-placing pole
шествие procession, train
~, **погребальное** funeral procession
шестерка 1. (the) six, group of six; hexad; 2. *naut.* six-oar(ed) boat
~, **академическая парная** *sport* six scull
шестерной sixfold, sextuple, senary
шестерня *tech.* pinion, gear (wheel), cogwheel
~, **ведомая** driven gear
~, **ведущая** drive/driving gear, pinion
~, **вытяжная** *text.* draft pinion/wheel
~ **заднего хода** reverse gear/pinion
~, **коническая** bevel gear/pinion; mitre gear
~, **косозубая** helical gear
~, **мотальная** *text.* winding/star/ratchet wheel
~, **насадная** pinion ring
~, **паразитная** idle(r)/intermediate gear
~, **передвижная** sliding gear
~, **планетарная** planet(ary) gear, planet wheel
~, **полуосевая** differential (side) gear, axle shaft gear
~ **постоянного зацепления** constant-mesh

gear
~, промежуточная *auto* intermediate/idle gear, idler
~, распределительная timing gear
~, солнечная/центральная sun gear/wheel
~, цилиндрическая *mach.* spur gear/pinion
шестиатомный hexatomic, hexahydric
шестиборец *sport* hexathlon participant
шестиборье *sport* hexathlon
шестивалентный hexavalent, hexabasic
шестивесельный *naut.* six-oared
шестигнездный six-celled, sexlocular
шестигранник *geom.* hexahedron
шестигранный hexahedral
~ колос *bot.* six-rowed spike
шестидвигательный six-engine(d)
шестидесятиричный, шестидесятеричный *math.* hexagesimal, base-60
шестидесятник *hist.* man of the sixties
щестизарядный *mil.* with six cartridges
~ револьвер six-shooter
шестизвенный six-member/-unit
шестизначный six-figure/-unit
шестизубчатый six-toothed, sexidentate
шестикратный sixfold, sextuple
шестикруговой hexacyclic
шестикрючный hexacanth(ous)
шестилопастный six-lobed/-blade
шестилучевой six-rayed, hexactinal, hexactine, sexradiate
шестимачтовый six-masted
шестимерный six-dimensional
шестиместный six-seater, with berths/seats for six
шестиногий *zool.* hexapodous, hexapodal; *n. pl.* hexapod(an)s *(Hexapoda)*
шестиосновный hexabasic, hexatonic
шестипалый sixdigitate, hexadactylic, hexadactyle
шестипарный sejugous, sejugate
шестипестиковый, шестипестичный hexagynous
шестиполье *agr.* six-field crop rotation
шестипольный six-field
~ севооборот *agr.* six-field crop rotation
шестиполюсник six-pole/-terminal network
шестипроводной six-wire
шестираздельный sexpartite
шестиразмерный six-dimensional
шестиребристый sexcostate
шестиричный *math.* senary
шестирядный six-row, sexfarious, hexastichous
шестисемянный *bot.* hexaspermous
шестислойный six-layer/-ply
шестиспоровый sexsporous, hexasporous
шестиствольный six-barrel(led)
шестистворчатый six-valved, sexivalvate
шестистишие *lit.* sextain
шестистопный *lit.* six-foot
~ ямб iambic hexameter
шестисторонний six-sided, hexagonal
шеститактный six-step
шеститычинковый, шеститычиночный *bot.* hexandrous, hexastemonous
шестиугольник *geom.* hexagon, sexangle
шестиугольный six-sided, hexagonal, sexangular
шестицветковый *bot.* six-flowered, hexanthous
шестнадцатиричный, шестнадцатеричный hexadecimal, sedecimal
шестнадцатая *n. mus.* semiquaver, sixteenth note
шестоватый columnar, rodlike, polelike
шестовик pole jumper/vaulter
шестовина pole
шестовой *n. naut.* punter
шестопсалмие *eccl.* six psalms
шеф 1. patron, sponsor; 2. *cul.* head cook, chef
шефердия *bot.* buffalo berry *(Shepherdia)*
шеф-монтаж contract supervision
шеф-пилот chief pilot
шеф-повар head cook, chef
шефство patronage, sponsorship
шея neck; *med.* trachelos; cervix
~ бизонья/бычья buffalo/bull neck
~, крыловидная webbed neck
ши *bot.* shea butter tree/seed *(Butyrospermum parki)*
шибболет shibboleth, peculiarity of speech
шибер I 1. *tech.* gate, slide (valve), damper; 2. *print.* looper
~, воздушный air damper
~, выдвижной sliding gate
~, дозирующий proportioning gate
~, донный discharge gate
~ дымовой трубы stack/chimney damper, chimney (slide) valve
~, запорный shut-off gate
~ люка *min.* stopper board
~, отсекающий/отсечной gate-type shut-off valve, cut-off gate
~, перекидной flap gate
~, регулирующий regulating door
шибер II *econ.* large-scale speculator
шизандра *bot.* magnolia vine *(Schisandra)*
шизантус *bot.* butterfly flower *(Schizanthus)*
шизогенез *biol.* schizogenesis, reproduction by fission
шизогенетический schizogenous, schizogenetic
шизогенный schizogenous
шизогония *biol.* schizogony
~, тканевая tissue schizogony
шизокарпий *bot.* schizocarp
шизолит *miner.* schizolite
шизоморфный schizomorphic, deuteromorphic
шизотемия *med.* schizothemia, schizothymia
шизофит *bot.* fission plant, schizophyte
шизофрения *med.* schizophrenia
~, злокачественная process schizophrenia
~, ранняя dementia precox
шизоцит *med.* schistocyte, schizocyte
шизоцитоз *med.* schistocytosis, schizocytosis
шиизм *rel.* Shiism
шиит Shiite, Shiah
шиитский Shiitic, Shiite

шикиматдегидрогеназа *bioch.* shikimate dehydrogenase
шикимовый shikimic
шикша *bot.* crowberry *(Empetrum)*
шикшевые *n. pl. bot.* crowberry family *(Empetraceae)*
шиллер-шпат *geol.* schiller spar, bastite
шиллинг shilling
шило awl, prick(er)
~, плоское bradawl
~, сапожное cobbler's awl
~, типографское bodkin
шиловатый, шиловидный awl-shaped; subuliform, subulate(d); aculeiform; *med.* styloid, styliform, peg-shaped
шилоглоточный *anat.* stylopharyngeal
шилоклювка, шилоноска *zool.* avocet *(Recurvirostra avosetta)*
шилолистный *bot.* awl-leaved
шилонижнечелюстной *anat.* stylomandibular, stylomaxillary
шилоподъязычный *anat.* stylohyoid, stylohyal
шилососцевидный *anat.* stylomastoid
шилохвость *zool.* pintail, water pheasant *(Anas acuta)*
шильдик designation strip; *el., rad.* circuit/control label
шильник, шильница *bot.* awlwort *(Subularia)*
шимми *av.* shimmy, vibration
шиммирование shimming
шимпанзе *zool.* chimpanzee *(Anthropopithecus troglodites)*
шина 1. tyre, *Am.* tire; 2. *el.* bus(bar); 3. *comp.* wire, line; 4. *med.* splint, frame; 5. *print.* ink rail
~ аварийного питания essential busbar
~ адреса, кодовая address wire/bus
~, адресная *comp.* a-wire, address wire/line
~ адресных меток address mark line
~, арочная ribbed tyre
~, балканская *med.* Balkan frame/splint
~, бескамерная tubeless tyre
~, бескордная cordless tyre
~, большегрузная heavy-duty tyre
~ в сборе busbar assembly
~ ввода-вывода input/output bus
~ видеосигналов video bus
~ возбуждения drive wire/line
~, входная input-/in-line, bus-in
~ выборки *comp.* drive wire, select(ion) line
~, высокочастотная HF bus
~, гипсовая plaster splint
~, главная main bus
~ главного процессора host bus
~ графических данных graphics bus
~, грузовая lorry tyre, *Am.* truck tyre
~ данных data bus
~ Дидрихса, транспортная *med.* Diedrichs' transportation splint
~, дублированная dual-redundant bus
~, заземляющая earthing busbar
~, запасная *auto.* spare tyre
~ записи адреса *comp.* address write wire, a-w wire
~, импульсная pulse bus
~, интерфейсная interface bus
~, информационная data path/bus
~, камерная tubed tyre
~ Крамера, проволочная лестничная Cramer's splint
~, клинчерная clincher tyre, beaded edge tyre
~, магистральная *comp.* highway
~, местная branch bus
~ младших разрядов данных lower data bus
~, монтажная *comp.* backplane bus (on a motherboard)
~, мотоциклетная motorcycle tyre
~, мультиплексная multiplexed bus
~, обходная transfer busbar
~, общая common bus; *comp.* unified bus
~, основная backbone bus
~, ответвительная tee-off bus, branch bus
~, отходящая outgoing bus
~ передачи чисел number-transfer bus
~ переноса carry line
~ питания power line
~, пневматическая air/pneumatic tyre
~ повышенной прочности reniforced tyre
~, полноразрядная *comp.* full bus
~, проколостойкая puncture-proof tyre
~ радиального типа radial tyre
~, распределительная distribution busbar
~ с временным разделением timeshared bus
~ с грунтозацепами adhesive(-type) tyre, traction-type tyre, ground-grip tyre
~ с последовательным опросом daisy-chain bus
~, самозаклеивающаяся self-sealing tyre
~, сборная collecting bus
~, сигнальная signal wire/line
~ синхронизации *comp.* clock bar
~, собирательная collecting bar
~ старших разрядов данных upper data bus
~ считывания sense wire/line
~, фидерная feeder (bus)bar
~, числовая number(-transfer) line/bus, word line
~ чтения read bus
~ широковещательной рассылки broadcast bus
~, электрическая busbar
~ электростанции station bus
шина-аэроплан *med.* airplane splint
шина-баллон balloon tyre
шина-лира *med.* banjo splint
~, отводящая abduction splint
~ Т-образная T-splint
~ Томаса Thomas' splint
~, шейная cervical splint
шина-эластик *auto.* cushion tyre
шинель greatcoat; overcoat
шинельная *n.* cloakroom

шинирование *med.* splinting, splintage
шинка *tech.* tie
~, контактная в реле contact arm
~ металлизации bonding conductor/braid
шинковать *cul.* shred, chop, mince
шинковка *cul.* 1. chopping, shredding; 2. shredder, chopper; slaw cutter
шиногиб husbar bender
шинодержатель 1. *el.* bus support, busbar clamp; 2. *auto.* tyre holder
шинопровод *el.* busduct, busway
~, закрытый metalclad busduct
~, штепсельный plug-in busduct
шинус *bot.* pepper tree *(Schinus)*
шиншилла 1. *zool.* chinchilla *(Chinchilla laniger);* 2. chinchilla fur/pelt
шип I 1. (anti-skid) spike/stud nail, crampon; 2. *tech.* tenon, tongue, (end) journal, pin, dowel, pivot, lug; 3. *bot.* barb, prickle, spine, thorn; 4. *zool.* acantha, spine
~, вставной inserted tenon
~ выдающийся advanced spine
~, головной cephalic spine
~ для соединения модели *foundry* pattern dowel
~, забитый driven-in pin
~, конический conical/tapered journal
~, косой tapered tenon
~ ласточкин хвост dovetail
~, направляющий guide pin/tenon
~, острый acicular spine
~ поворотного стола pin/roller of an index(ing) table
~, посадочный locating pin
~, потайной stub tenon
~, сквозной through tenon
~, хвостовой caudal spine
~ шаровой ball/spherical end journal
~, ядовитый poisonous spine
шип II *zool.* sturgeon *(Acipenser nudiventris)*
шипение sizzle; fizzing; sibilation; effervescence; sputtering; *acous.* hissing/rushing sound
шипик minute spine, spinule; thornlet, small prickle
шиповатость spinosity; hispidity
шиповато-щетинистый prickly-bristled
шиповатый spiny, spinous, prickly, thorny, acanthaceous, aculeate
шиповидный, шипообразный spiniform, spine-like, acanthoid, aculeiform
шиповки *pl. sport* spikes *(shoes)*
шиповник *bot.* 1. brier, wild/hedge rose *(Rosa);* 2. hip
шиповой tenon *(attr.),* spike *(attr.),* spine *(attr.);* mortising
~вое долото mortise chisel
~вое соединение mortise/tenon joint
шипорезный tenon-cutting, dovetailing
~ станок dovetailing/tenoning machine
шипохвост *zool.* spiny-tailed lizard *(Uromastix)*
шипун *zool.* 1. mute swan '*(Cygnus olor)*

шипучий fizzling, fizzy, sizzling, sparkling, effervescent
шипящий fricative, hissing; *n. phon.* sibilant, hissing sound
ширение stretching
~ ткани *text.* cloth stretching/tentering
ширилка stretcher, stretching machine
~, конвейерная belt stretcher
ширильный stretching
ширина width, breadth; broadness; wideness; *naut.* beam
~ авиатрассы width of airway
~ борозды *agr.* furrow width
~ в свету inside width, width in the clear
~, габаритная *mach.* overall width; *naut.* overall beam/breadth
~, гауссовая Gaussian width
~ двигателя engine width
~, делительная fission width
~, диффузионная diffusion breadth
~, доплеровская Doppler width
~ зазора gap depth/length, slit/gap width
~ затухания damping width
~ захвата, рабочая effective/working width
~ захвата сельскохозяйсвенного орудия width, coverage
~ зуба *mach.* tooth width, face width of a tooth
~ импульса pulse width/length
~ калибра *met.* width of a groove
~ колеи tread; *auto.* wheel spacing; *rail.* track gauge
~ колеи шасси *av.* wheel track
~ конвергенции amplitude of convergence
~ ленты *comp.* tape width, bandwidth
~ линии записи trace width
~ линии излучения лазера/мазера laser/maser linewidth
~ линии спектра, естественная intrinsic spectrum-line width
~ литеры *print.* width of a type
~ лопасти, относительная blade-width ratio
~ луча beam width
~ обода *auto* rim width
~ очка *print.* width of type
~ полосы (пропускания/частот) *rad., electr.* (frequency) bandwidth, BW
~ пропила saw kerf
~ протектора tread width
~ пучка beam width
~, радиационная radiation/radiative width
~ разрешения *opt., electr.* resolution width
~ распределения *stat.* distribution range
~ резонанса resonant width
~ резонансного уровня resonance level width
~ русла bed/channel width
~ следа, диффузная diffusion breadth of a track
~ спектральной линии, естественная intrinsic line width
~ стробимпульса/стробирования gate width

~ строки *telev.* line width
~ судна beam athwartship
~ судна, теоретическая moulded beam/breadth of a vessel
~, угловая angular width
~ уровня, парциальная partial level width
~ фарватера width of the channel
~ фронта погрузки *min.* clean-up range
~ хода tread
~ энергетического уровня energy level width
ширма 1. *opt.* screen, blind, shield; 2. *therm.* platen
широкий 1. wide, broad; big, extensive; 2. lavish, generous
~кая колея *rail.* wide/broad gauge
широковетвистый branchy, curycladous
широковещание broadcast(ing)
широковыемчатый *bot.* repandous
широкоголовчатый platycephalous
широкоголовый brachycephalic, chamecephalous, platycephalous
широкогорлый wide-mounthed/-necked
широкодиапазонный wide-range, broad-band
широкозахватный wide-span
широкоизлучатель 1. wide-angle luminaire; *cin.* fload lamp; 2. *auto.* wide-/broad-beam (head)lamp, wide-spread light
ширококолейный *rail.* broad-gauge
ширококолокольчик *bot.* balloonflower *(Platycodon)*
ширококолосый *bot.* broad-eared/-spiked, platystachyous
широколист(н)ый *bot.* broad-leaved, platyphyllous, latifoliate
~ лес broad-leaved forest
широколопастный *tech.* broad-/wide-bladed; *bot.* broad-lobed, latilobate
широконоска *zool.* shoveler *(Anas clypeata)*
широконосый broad-/flat-nosed; *zool.* platyrrhine, platyrrhiniar
широкоперегородчатый latimurate, latiseptate
широкополосный broadband, wide-band/-strip
широкорот *zool.* roller *(Eurystomus)*
широкосифонный eurysiphonate
широкоскулый *anthr.* with prominent cheekbones
широкотравье broad-leaved grasses
широкоугольный broad-/wide-angled
широкоцветковый *bot.* broad-flowered, latiflorous, platyanthous
широкоязычковый broad-tongued, platyglossal, platyglossate
широта 1. *geogr.* latitude; 2. breadth; *math.* range, width
~, астрономическая astronomical/celestial latitude
~ в косой картографической проекции oblique latitude
~ выборки sample range
~, географическая geographic(al)/terrestrial latitude
~, геодезическая geodetic/spheroidal latitude
~, изометрическая isometric/conformal latitude
~, параметрическая/приведенная parametric/reduced/geometric(al) latitude
~, промежуточная middle latitude
~ пункта отшествия latitude left
~ пункта пришествия latitude in
~, северная/южная northern/southern latitude
~, счислимая assumed latitude
~, эквивалентная equal-area latitude
~, эклиптическая ecliptic/celestial latitude
~ты, конские horse latitudes
~ты, штилевого пояса horse latitudes
широтно-импульсный pulse-width/-length
широтно-модулированный width-modulated
ширпотреб *coll.* consumer(s)/consumption goods, commodities; mass consumption
ширстрек *naut.* sheer/gunwale strake
шистозоматоз *med.* schistosomiasis
~, кишечный intestinal schistosomiasis
~, мочеполовой urinary schistosomiasis, endemic/Egyptian hematuria, bilharziasis
~, японский Asiatic schistosomiasis, kabure, urticarial fever, katayama disease
шистометр *med.* schistometer
шистосома *zool.* blood fluke, schistosome *(Schistosoma)*
шистоцерка *zool.* desert locust *(Schistocerca gregaria)*
шитье 1. needlework; sewing, stitching; 2. embroidery; embroidering
~, брошюрное *print.* centre-fold stitching
~ внакидку проволокой centre/saddle stitching
~ втачку block/flat/side stitching
~, переплетное centre-fold stitching of cased-in books
~, спиральное wire binding
шифер slate; corrugated asbestos board
~, кровельный roofing slate
шиферный slatelike, slaty
шифон *text.* chiffon
шифр 1. code, cipher, cypher; key; 2. call number, pressmark *(in libraries);* 3. monogram, initials
шифратор (data) (en)coder
~, дифференциальный incremental coder
~, матричный matrix encoder
~ положения position coder
шифрация coding
шифровальщик, шифровщик cipher clerk/officer, code clerk, encoder; cryptographer
шифрование encoding, encryption
шифрованный in code, coded, (en)ciphered
шифровать (en)code, (en)cipher; (en)crypt
шифровка 1. cipher communication, code message; 2. (en)ciphering
шифрограмма cipher telegram/message
шифрпост coding room/board/office
шихта *met.* charge, blend, mixture; burden, (furnace) load; batch, stock
~, агломерационная sintering mixture, sinter

burden
~, ваграночная cupola mixture
~ доменной печи blast-furnace burden
~ с высоким содержанием лома/агломерата high-scrap/-sinter charge
~, самоплавкая/самофлюсующая self-fluxing charge
~, углеводородная hydrocarbon charge
шихтарник *met.* stockyard, bedding plant, mixing place/shed
шихтование blending, mixing; burdening; charge/mixture calculation
шихтовать blend, mix; restore bent/distorted parts to shape
шихтовка 1. blending, mixing, burdening; 2. interleaving; layering; laminating
шихтовый charge *(attr.)*, mixture *(attr.)*, burden *(attr.)*
~ двор mixing place
шихтосмеситель reclaiming unit, reclaimer
шиш *hist.* 1. brigand, ruffian; 2. informer, spy
шишак *hist.* spiked helmet
шишельник *foundry* core maker
шишельный coremaking *(attr.)*
~ ящик core box
шишечка microstrobile, little cone
шишечник I *bot.* hawknut *(Bunium)*
шишечник II *zool.* knight fish *(Monocentrus)*
шишечниковые *n. pl. zool.* pinecone fishes *(Monocentridae)*
шишка 1. *bot.* cone, strobile; 2. boss, tubercle, knob; 3. *foundry* (mould) core, kernel
~, висячая pendulous cone
~, выпуклая raised boss
~, кипарисовая galbulus
~, сосновая pine cone
~, сосновообразная pinaceous cone
~, спороносная strobile, strobilus
шишковидный conelike, cone-shaped; strobilaceous; tubercular; *med.* pineal
~ная железа pineal body, pituitary gland
шишковник *bot.* medlar *(Mespilus germanica)*
шишконосный *bot.* cone-bearing, coniferous, strobiliferous; *n. pl.* gymnosperms, conifers *(Coniferae)*
шишкообразующий *bot.* cone-forming
шишкоплодный *bot.* conocarpous
шишкосушилка *foundry* core drier
шкала scale(-plate); dial
~ авиагоризонта altitude indicator scale
~ азимута azimuth dial
~ бальности землетрясений magnitude scale, scale of earthquake intensities
~, безнониусная direct-drive dial
~ Блит-Сернандера Blytt-Sernander scale
~ Боме Baume (hydrometer) scale
~ Бофорта/ветровая Beaufort wind scale
~ в тысячных mil scale
~ Верньера Vernier scale
~, верньерная vernier scale/dial
~ видимости visibility scale
~ времени time scale/base
~ времени, подвижная triggered time base
~ высот *cart.* altitude scale, scale of height
~ высот на карте, цветная elevation tints
~ геологического времени stratigraphic/geological time-table
~ грохочения mesh scale
~, градуированная calibrated scale
~, грузовая/дедвейта deadweight scale
~ давления высотомера altimeter/millibar subscale
~ дальности range scale
~, двусторонняя centre-zero scale
~ долгот longitude scale
~ заработной платы scale of wages, wage scale
~ звездных величин magnitude scale
~, индикаторная display scale
~, квадратичная square-low scale
~ Кельвина/Реомюра Kelvin/Reaumur scale
~ кислотности acidity scale
~, колориметрическая colo(u)r scale
~ коррозионной стойкости corrosion-resistance scale
~ кренов bank scale
~, круглая/круговая circular scale, (round) dial
~, курсовая *av., nav.* heading dial
~, логарифмическая log-type scale, logarithmic scale
~ Мооса Mohs' scale
~ налоговых ставок tax schedule
~ настройки *rad.* tuning scale
~ нивелира limb
~, нониусная vernier dial
~ обобщения понятия ladder of abstraction
~ осадки draft/inmersion scale
~ отношений *stat.* ratio scale
~ отсчета reference scale
~ оценок rating scale
~, оценочная estimation scale
~, переводная conversion table
~ погашения *econ.* repayment schedule
~ подоходного налога *econ.* income tax schedule
~ полезности utility scale
~, полукруглая semicircular/fan dial
~ полутонов *comp.* grey scale
~ прибора instrument dial
~ процентов table of interest
~, равномерная uniform scale, evenly divided scale
~ расстояний *cin.* scale of distance
~, растянутая expanded scale
~ расценок rate/rating scale
~ с нониусом vernier scale
~, серая, ~ серых тонов *opt.* grey scale
~ сит mesh gauge
~, скользящая sliding scale
~, сменная snap-in scale

~ ставок налогового обложения scale of taxes
~, стоградусная centigrade/Celsius scale
~ тарифных ставок rate/rating scale
~, точная accurate/fine scale
~ твердости hardness scale
~, температурная международная практическая 1968 года, МПТШ-68 international practical temperature scale of 1968, IPTS-68
~, термодинамическая Кельвина Kelvin's thermodynamic temperature scale
~, угломестная elevation angle dial
~ Фаренгейта Fahrenheit scale
~ цветности воды Forel scale
~ цветов colograde, colourgrade
~, цветовая *print.* colour/chromatic scale
~ цветов, гипсометрическая subdued colours, graduation of tints
~ Цельсия Celsium/centigrade/centesimal scale
~ широт latitude scale
~, энергетическая energy scale
шкалик *hist.* shkalik *(liquid measure)*
шканцы *pl. naut.* quarterdeck
шкаторина *naut.* edge of a sail trimmed with soft rope
~, боковая leech
~, задняя after leech
~, нижняя foot
~, передняя luft, fore/foreward leech
~, стоячая mast leech
шкатула small trunk
шкатулка box, case; cassette
шкаф locker, cabinet; closet, case
~, батарейный *com.* battery box/cabinet
~, вводный *el.* entrance bay
~, витринный display case
~, воздушно-сушильный air-drying cabinet
~, вытяжной *tech.* exhaust/vent/fume hood, fume cupboard/cabinet
~, картотечный file cabinet
~, клемный terminal-block cabinet
~, книжный bookcase
~, несгораемый safe, strong-box
~, питающий *com.* power supply cabinet
~, платяной wardrobe
~, распределительный (кабельный) distribution cabinet
~, расстойный *food.* proofer
~, релейный *com.* relay box
~ с фильмотекой film-storage cabinet
~, сушильный desiccator, drier, drying oven; *cin., phot.* drying cabinet/box
~, стенной wall closet, built-in closet
~, тепловой *food.* thermal cabinet
~, холодильный refrigerated cabinet, reach-in cabinet
~, холодильный низкотемпературный food freezer/froster, deep freezer
~, холодильный с автоматическим оттаиванием self-defrost cabinet
шквал squall; tornado; *mil.* barrage
~ без осадков white squall
~, градовой squall of hail
~, грозовой thunder-squall
~ с градом и мокрым снегом squall of hail and sleet
шквальный squally
~ огонь *mil.* mass barrage, heavy fire
шквара 1. scoria, dross; 2. *cul.* cracklings
шкварный *food* low-grade *(of fat)*
шкворень pivot, (centre/king) pin, bolt, pintle; dragbolt, coupling pin
~ поворотного кулака steering knuckle pivot
~, поворотный king/knuckle/axle pin, kingbolt
шкентель *naut.* pendant
~ выдвижного киля centre-board pendant
~ выстрела boom pendant
~, грузовой cargo hoist, runner, whipping line
~лем, поднимать whip
шкентель-тали *naut.* hoist purchase
шкерт *naut.* lanyard
шкив sheave; pulley
~, вариаторный variable-speed pulley
~, ведомый driven pulley
~, ведущий drive/driving pulley
~, выпуклый crown pulley
~, желобчатый sheave (wheel)
~, канатный rope pulley/sheave
~, консольный end pulley
~, кормовой/носовой *naut.* stern/bow sheave
~, направляющий guide pulley
~, натяжной tension/tightening pulley
~, приводной/рабочий drive/driving pulley
~, разъемный split pulley
~ ременной передачи belt pulley
~, теребильный *agr.* pulling drum
~, тормозной brake pulley, friction plate
~, холостой idle/loose/free/dead pulley
шкив-маховик flywheel pulley
шкимушгар *naut.* spun yarn/line
шкимушка *naut.* marline, nettle, selvagee, spun line/yarn, coarse yarn
шкипер *naut.* skipper, master; captain; boatswain's storekeeper
шкиперская *naut.* boatswain's store
школа 1. school; 2. training, schooling; 3. *hort.* nursery
~, вечерняя evening school/classes
~, высшая higher school; university
~, заочная correspondence school
~, коммерческая B-school, bussiness/commercial school
~, начальная primary school
~, планерная glider school
~, рикардианская Ricardian school
~, сестринская nursing school
~ совместного обучения coeducational/mixed school
~, средняя high/secondary school
~, старшая средняя senior high school

~ фабрично-заводского обучения factory school
~, фритрейдерская free trade school
~, чикагская Chicago school
школа-интернат boarding school
шкот *naut.* sheet
~ пластыря lowering line
~, потравленный flowing sheet
шкотовый sheet *(attr.)*
~ узел sheet bend, netting knot
~ угол clew
шкура *leath.* hide; skin; pelt
~, борушистая ribbed hide
~, овечья/оленья sheepskin/buckskin
~, парная fresh hide
~, пресно-сухая sun-dryed hide
~ру, тузликовать soak hides in brine
шкурить *mach.* work with emery cloth, sandpaper
шкурка 1. pelt; skin, fell; 2. sand/abrasive/emery paper; abrasive/emery cloth
~, шлифовальная emery paper/cloth
~, шлифовальная, песчаная sand paper/cloth
шлаг *naut.* bight, loop
~ с простым узлом magnus hitch
шлагбаум roadway/lifting gate, bascule barrier, traffic-control barrier, turnpike
шлагтов *naut.* fid
шлак residue, remainder, rest; slag, scoria; cinder; dross; *nucl.* poison
~, ваграночный cupola slag
~, верхний flushing/early slag
~, волокнистый mineral wool
~, восстановительный reducing slag
~, вулканический *geol.* scoria
~, вязкий heavy/viscous slag
~, глиноземистый aluminous slag
~, горшечный first-run slag
~, горячий (спелый) rich slag
~, железистый ferriferrous slag
~, защитный protecting slag
~, зеленый fresh slag
~, известковистый limey slag
~, известковый lime slag
~, конечный tap cinder, final/finishing/remaining/tap slag
~, кричный refinery slag
~, кружевной lacy slag
~, легкоплавкий fusible slag
~, мартеновский open-hearth slag
~, медный copper scum
~, нижний roughing slag
~, оборотный return slag
~, окислительный oxidizing slag
~, оловянный tin dross/skimmings
~, остаточный remaining slag
~, первичный first/early slag
~, подовый hearth cinder
~, рафинированный refinery slag
~, сварочный hearth cinder
~ со спелью kish slag
~, спелый matured slag
~, стекловидный glassy/scoury slag
~, томасовский/основной basic slag
~, топочный cinder
~, тугоплавкий refractory slag
~, химически активный reactive slag
~и, выгорающие *nucl.* burnable poison
~и, реакторные reactor poison, fission(-product) poison
шлакобетон slag concrete
шлакоблок breeze/cinder block
шлакование 1. slagging, slag formation, scorification; 2. clinkering
шлаковата slag/cinder wool
шлаковать slag, scorify, clinker
шлаковидный slaggy, drossy, scoriaceous
шлаковик slag chamber/pocket; dirt trap, dust catcher
шлаковина *met.* slag patch/inclusion/pocket
шлаковня skimmer, slag pot
шлаковоз slag car
шлаковойлок slag felt
шлаковый 1. slag(gy); scoriaceous; *med.* residual; 2. *n.* cinder man, slagger
шлаковыпор relief spruce
шлакодробилка, шлаколоматель slag/clinker breaker; cinder mill
шлакообразование slagging, slag/clinker formation
шлакообразователь slag-former
шлакоотвал ash-disposal area
шлакоотделение slagging, slag removal
шлакоотделитель slag separator
шлакоотстойник slag basin
шлакопортландцемент portland blastfurnace slag cement
шлакоприемник slag catcher
шлакопровод slag sluice/discharge/pipeline
шлакосниматель skim gate, slag separator
шлакоудаление slag disposal/removal
шлакоуловитель slag catcher, skim(ming) gate
шлакоустойчивость slag-resistance
шлакоцемент slag cement
шлам slime, sludge, mud, slurry, pulp; silt; residue, sediment
~, буровой drilling/bore mud, drilling sludge, muck
~, губчатый slime sponge
~, каустизационный *paper* lime sludge
~, котловой boiler sludge/mud
~, первичный *min.* primary slime
~, рудный ore slime/slurry
~, угольный coal sludge
шламообразный slimy, slimelike, sludgelike
шламообразование sliming, sludging, slurrying
шламоотборник sludge/cuttings sampler
шламоотделитель sludge remover, slime separator
шламоотстойник slime/slurry tank, slime thickener, mud-setting pit

шламосборник sludge setting tank, slurry sump
шламосодержание sludge content
шламоуловитель sludge trap
шланг hose (pipe), flexible pipe/tube
~, армированный reinforced hose
~, бронированный wire-armoured hose
~, всасывающий suction hose
~, гибкий flexible hose/pipe
~, гофрированный corrugated/concertina hose
~, грузовой (на танкере) cargo hose
~, дюритовый rubberized-fabric hose, durite hose
~, заправочный *av.* filling/trailed hose
~, матерчатый fabric hose
~, нагнетающий pressure delivery hose
~, подающий delivery hose
~, пожарный fire hose
~, поливной spray hose
~, резиновый rubber hose
~, топливозаправочный fuelling hose, fuel-delivery hose
шлангбалка *naut.* hose davit
шлангокабель flexible drill string
шлейка breeching, breech band, breastband
шлейф 1. plume, column; train, trail; 2. *agr.* float, slicker, drag; 3. *electr.* stub; loop, circuit; 4. (mirror) galvanometer; 5. *math.* tail
~, валунный *geol.* boulder-train
~ выходящих газов exhaust plume
~, газовый bubble train
~, глубоководный deep-sea fan
~ дисперсии *math.* tail of variance
~, емкостный capacitive stub
~, закорачивающий shorting stub
~, замкнутый closed loop
~, измерительный measurung loop
~, коаксиальный coaxial stub
~, лавовый lava train
~, ледниковый trail
~, настроечный resonant/tuning stub
~, согласующий matching stub
~, тыловой backing stub
~, цветной coloured plume
~, широкополосный broad-band stub
шлейф-борона *agr.* clod smasher, drag, sweeper/smoothing harrow
шлейф-вибратор/-диполь *electr.* folded dipole
шлейф-волокуша *agr.* band sweep
шлейфование *agr.* smoothing, harrowing (with a float)
шлейфовать smooth, slick, level, plane
шлем I helmet, head-piece; 2. *bot.* hood, galea
~, апоневротический/сухожильный *anat.* epicranial aponeurosis
~, водолазный diver's/diving helmet/hood
~, высотно-компенсирующий partial-pressure helmet
~, герметический *av.* pressurized helmet
~, защитный crash helmet
~ конвертера *met.* nose
~, летный flying/flight helmet
~, мягкий soft hat
~ парашютиста-десантника paratroop helmet
~ сварщика, защитный welder's/welding helmet
~, тропический tropical helmet, topee
~, усиленный strengthened helmet
шлем II *cards* slam
~, большой/малый grand/little slam
шлем-каска hardhat
шлемник *bot.* scullcap *(Scutellaria)*
шлемовидный, шлемообразный helmetlike; galeate(d), galeiform; cassideous
шлемофон communication/earphone helmet, intercom/interphone headset; combined headset and laryngophone
шленка *agr.* merino; Silesian sheep
шлеппер transfer skid/table/bed; dragging device; transfer (handling) gear
~, вводящий pull-on transfer
~, выводящий/отводящий pull-off transfer
~, канатный rope transfer
~ с захватками dog transfer, sweep arm transfer
~, цепной chain transfer
шлея *agr.* breeching, breech band, breastband
шликер *met.* slip; dross
шлирен-фотография schlieren photography
шлиры *pl. geol.* schlieren
шлиф (thin) slide, (micro)section
~ косой angle lap(ping)
~, плоскостный planar section
~, стеклянный glass grind
шлифовальник grinding tool/instrument, grinder
шлифовальный grinding; polishing, burnishing, buffing
~ круг abrasive disk
~ станок grinding machine, grinder
шлифовальщик grinder; polisher
шлифование 1. grinding; sanding, polishing, abrading, abrasion; filing; 2. *leath.* smoothing, buffing
~, высокопроизводительное high-velocity grinding
~, грубое/обдырочное coarse grinding
~ на конус taper grinding
~, плоское surface grinding
~ с маслом oil grinding
~, чистое finish grinding
шлифовать grind; sand; abrade, file, smooth; polish, burnish
~ начерно foregrind
~ начисто grind for finishing
шлифовка 1. polishing, grinding, burnishing; abrading, abrasion; 2. polishing/grinding instrument; 3. polish, smooth surface
~ кристаллов lapping of crystals
шлих heavy cocentrate, alluvial sand
~, черный black sand
шлиховать *met.* planish
шлиховой pertaining to alluvial sand
~ анализ panning

~вое золото alluvial/placer gold
шлихт *leath.* moon/perching knife
шлихта *text.* size, dressing
шлихтик smoothing plane
шлихтовальный sizing
~ная машина (warp-)sizing machine, sizer
шлихтовальщик *text.* sizer
шлихтование *text.* sizing, dressing
~ основы/пряжи warp/yarn sizing
шлихтовать 1. *text.* (treat with) size; 2. smooth, plane
шлихтовка 1. sizing, dressing; 2. smoothing
шлиц *mach.* 1. slit, slot; 2. spline, feather, groove
~, крестообразный Philips slot
~ под отвертку, ~, регулировочный screw-driver slot (of a control)
шлицевать slot, slit, make slits, groove
шлицевый slotted, slit
~вая фреза slot/slitting cutter
шлык 1. *hist.* pointed cap/hat; hood; 2. headgear of Orthodox monk
шлюз 1. sluice, floodgate; *naut.* lock; *av.* air lock (chamber); 2. *min.* sluice box
~, амальгамационный apron plate, plate sluice/amalgamator
~ в кессоне pneumatic chamber
~ в почве *min.* ground sluice
~, верхний head lock
~, водоспускной discharge sluice
~, впускной inlet sluice
~, двухкамерный/двухступенчатый tandem lock, double-lift lock
~, двухниточный double-lane lock
~, камерный canal lock
~ космического корабля spacecraft air lock
~, многокамерный/многоступенчатый flight lock
~ на канале canal lock
~, нижний tail lock
~, однокамерный/одноступенчатый single-lift lock
~, однониточный single-lane lock
~, осушительный emptying lock
~, паводковый flood/tide gate
~, парный twin locks
~, приливный full-tide lock
~, рыбоходный fish(-passing) lock
~ с опускными щитами drop gate sluice
~, спускной floodgate
~, судоходный shipping/navigation lock
~ сухого дока drydock lock
шлюзование sluicing, locking through, lockage
шлюзовать sluice, lock through
шлюзовик sluicer, sluice keeper, lockman
шлюп *naut.* sloop; shallop
шлюпбакштаг *naut.* davit guy
шлюпбалка *naut.* (boat) davit
~, заваливающая/откидная luffing davit
~, поворотная/радиальная slewing/swinging/radial/rotary davit
~, S-образная swan-necked davit
шлюпка boat, launch
~ береговой спасательной службы inshore rescue boat
~, быстроходная гребная fast-pulling boat
~, гребная rowboat, rowing boat
~, дежурная guard/ready/emergency boat
~, дежурная спасательная accident/quarter boat
~, дозорная watch boat
~, моторная powerboat
~, надувная inflatable boat
~, парусная sail(ing) boat
~, прогулочная pleasure boat
~ с подвесным мотором outboard motor boat
~, спасательная lifeboat, rescue boat
шлямбур jumper, drift
шляпа 1. hat, bonnet; 2. *min.* vein cap, hat
~, железная iron hat, gossan
~, каменная cap rock
~, соляная salt cap
шляпка 1. head, cap; 2. *bot.* pileus
~ гвоздя nail head
~, маленькая pileolus
~, плоско-выпуклая flattened-round head
~, полупотайная rounded countersunk head
шляпковидный, шляпкообразный hat-shaped; *bot.* pileiform
шляхетство 1. *hist.* szlachta, Polish gentry; 2. *hist.* (Russian) nobility, gentry
шляхта *hist.* Polish gentry, szlachta
шляхтич *hist.* member of Polish gentry
шмальта smalt
шмальтит *geol.* smaltite, smaltine, tin-white cobalt
шмелевидно-цветковый *bot.* bumblebee-flowered
шмель *zool.* bumblebee *(Bombus)*
шмуцроль *print.* travelling tympan
шмуцтитул *print.* half/false/bastard title
шнек 1. auger; screw; 2. screw/worm/auger/spiral conveyer
~, аспирационный aspiration worm
~, вертикальный лопастный vertical blade auger
~, выгрузной delivery/discharge/emptying/unloading screw/spiral conveyer
~, двухходовой two-start screw conveyer
~, дозирующий metering screw/spiral conveyer
~, загрузочный loading auger, feed(ing)/loading screw/spiral conveyer
~, колосовой *agr.* tailings screw/spiral conveyer
~, разгрузочный emptying/unloading screw/spiral conveyer
~, раздаточный/распределительный *agr.* distributing screw/spiral conveyer
~, режущий *agr.* cutting auger conveyer
шнека *hist.* galley
шнековый screw *(attr.)*, worm *(attr.)*
~ конвейер screw conveyer
~вая машина screw extruder

шнекодержатель augerholder
шнитт-лук *bot.* chive *(Allium schoenoprasum)*
шницель *cul.* schnitzel, fillet of pork/veal
шноркель, шнорхель snorkel, snort-tube
шнур 1. cord, string; lace; 2. cable, electric cord; 3. fuse
~, амортизационный shock/bungee cord
~ аппарата *teleph.* instrument cord
~, бикфордов safety/Bickford/blaster fuse
~, воспламенительный igniter cord
~, вытяжной extraction/rip cord
~, вязальный packing cord
~, гибкий *tel.* tinsel cord
~, детонирующий demolition/detonating cord, (instantaneous) detonating fuse
~, коммутационный *teleph.* patching cord, patchcord
~, линейный плазменный linear pinch
~, медленногорящий огнепроводный slow-burning fuse, slow match
~, многожильный multiple cord
~, набивочный *mach.* packing cord
~, обрывной weak tie/link
~, огнепроводный Bickford/blaster/safety/ blasting fuse
~, одножильный single cord
~, переносный *teleph.* jack cord
~ питания *el.* power cord
~, плазменный plasma filament/column/cord, pinch
~, предохранительный safety tie
~, розеточный *teleph.* instrument cord
~, сетевой *el.* power/line cord
~, тороидальный плазменный toroidal pinch
~, трассировочный marking/tracing tape
~, удлинительный *el.* extension cord
~, уплотняющий *mach.* packing cord
~, штепсельный patchcord
шнурковый shoestring *(attr.)*
шнуровальный lacing, tying
шнуровидный cordlike, stringlike, ropelike
шов 1. *tech.* seam, weld; joint, junction; 2. *surg.* suture, raphe; stich
~ Альберта, кишечный Albert's suture
~, барьерный *naut.* crack arrester/stopper
~ Бейкера Baker abdominal closure suture
~, бороздной *geol.* channeled suture
~ века, боковой lateral palpebral raphe
~, внутрикожный subcuticular suture
~, Вольфлера, кишечный Wolfler's suture
~, восьмиобразный harelip/pin/twisted suture, figure-of-8 suture
~ впритык flat-batted seam
~ вровень butted seam
~ Гели, кишечный Gely's suture
~, герметичный pressure-tight weld
~, гладкий/затертый flush joint
~, глубокий сближающий approximation suture
~, декоративный fancy stich
~, заклепочный rivet(ed) joint/seam/row
~, заклепочный шахматный staggered rivet(ed) joint
~ заподлицо flush joint
~, зубчатый serrate/dentate suture
~, истинный true suture
~, кетгутовый хирургический catgut suture
~, кисетный purse-string suture
~, кишечный intestinal suture
~ кладки masonry joint
~, клеевой glue line
~, кожный cutaneous suture
~, кожный сближающий apposition/coaptation suture
~, костный хирургический osteosuture
~, краевой marginal suture
~ Кушинга, непрерывный Cushing's suture, continous intestinal suture
~, лобный frontal suture
~, ложный pseudoraphe
~, матрасный mattress/guilted suture
~, межлобный metopic suture
~ мошонки raphe scroti
~, накладной lapping/overhead seam
~, непрерывный continuous/glover's/spiral/un-interrupted suture
~, непроницаемый/плотный tight seam
~, обвивной непрерывный blanket suture
~ Панкоаста, пластический Pancoast's suture
~ Паре, некровавый Pare's suture
~, первично-вторичный отсроченный primosecondary suture, delayed suture
~, плоский plane/harmonic suture
~, погружной buried suture
~, подрубочный hem stitch
~, поперечный girth seam
~, постельный bed joint
~, потайной blind stitch
~, провизорный relaxation/tension suture
~, продольный main seam
~, простой узловой plain interrupted suture
~, пуговчатый button suture
~ Рихтера, узловой Richter's suture
~, сапожный *surg.* cobbler's suture
~, сварной weld; welded seam
~, сварной без усиления flush weld
~, сварной многопроходный multipass/ multirun weld
~, сварной нижний downhand/flat weld
~, сварной поперечный circumferential/girth weld
~, сварной стыковой butt/groove weld
~ сварной таврового соединения tee/T-weld
~, семенной *bot.* raphe
~, скользящий/температурный expansion joint
~, слоевой joint
~, соединительный commissure
~, стреловидный sagittal/interparietal/rhabdoid

suture
~ **стыковой сварной** butt weld
~, **сухожильный** tendon suture
~ **твердого неба** palatine raphe
~, **температурный (на расширение)** expansion joint
~, **температурный (на сжатие)** contraction joint
~, **технологический** construction joint
~, **точечный сварной** spot weld
~, **угловой сварной** fillet weld
~, **углубленный** excavated suture
~, **узловой** interrupted suture
~, **усиленный** convex/reinforced weld
~, **усадочный** shrinkage strip
~, **хирургический** surgical/operative suture
~ **Хольстеда** Halsted's suture
~ **черепа** cranial suture
~ **Черни, кишечный** Czerny's suture
~, **чешуйчатый** squamous suture
шовинизм *polit.* chauvinism; jingoism
шовинистический, шовинистский chauvinist(ic), jingoistic
шовный sutural, commissural, seam *(attr.)*
~**ная сварка** seam welding
шодди *text.* shoddy (fabric)
шок *med.* shock
~, **анафилактоидный** anaphylactoid shock/crisis
~, **болевой** pain shock
~, **военный** shell shock
~, **геморрагический** h(a)emorrhagic shock
~, **гуморальный** humoral shock
~, **лекарственный** medicinal/drug shock
~, **лучевой** radiation shock
~, **нефрогенный** nephrogenetic shock
~, **нитритный** nitroid shock
~, **ожоговый** burn shock
~, **олигемический** oligemic shock
~, **послеоперационный** surgical/deferred/delayed shock
~, **предоперационный** preoperative shock
~, **протеиновый** protein shock
~, **родовой** labo(u)r shock
~, **электрический** electroplexy, electroshock (therapy), E.S.T., electric convulsive therapy
шоковый (pertaining to) shock
шоколадник *bot.* cacao tree *(Theobroma cacao)*
шокотерапия *med.* shock therapy
шомпол *mil.* cleaning rod; *tech.* dip/ram rod, ramrod
шоопирование shooping, Shoop/metallization process, metal spraying
шор *geol.* salina, salt lake, shor
шорня saddlery, saddler's shop
шортгорн *agr.* Shorthorn, Durham, Teeswater
шоссе highway; macadam road
~, **гудронированное** tarred/tarmac road
шоссеец highway engineer
шоссировать surface, macadamize, metal
шотландец Scotsman, Scot
шотландка I Scotswoman
шотландка II *text.* tartan, plaid, checkered fabric
шотландский Scottish, Scotch
~ **горец** Highlander
~ **плед** tartan plaid
шот-эффект *electr.* shot/Schottky noise, shot effect
шофер (motor-car) driver, chauffeur
шоферское свидетельство driver's licence
шоферство drivership
шпага sword, rapier
шпагат 1. (binder) twine, string, cord; 2. *sport* split
шпажист swordsman, fencer
шпажник *bot.* gladiolus *(Gladiolus)*
шпаклевание puttying; caulking
шпаклевка 1. underpaint/hard putty, filler; 2. putty knife, spattle; 3. *naut.* spaddle; 4. filling with putty
шпаклевщик puttier
шпала 1. *rail.* sleeper, crosstie, *Am.* tie; 2. *mil.* stripe bar
~, **железобетонная** concrete sleeper
~, **стыковая** joint sleeper
~, **цельнобрусковая** rectangular sleeper
шпалера 1. trellis, latticework, espalier; 2. line of tress; hedge; 3. *pl. mil.* lines, rows
шпалерный trellis *(attr.)*
~ **кустарничек** trailing dwarf shrub, ground shrub
шпалоподбойка tie tamper, sleeper tamping machine
шпалоподтяжка tie nipper, tie nipping machine
шпалоразгрузчик tie handler
шпалоукладчик tie layer, tie-laying machine
шпангоут *naut.* ring, frame, rib
~, **бортовой** shell/side frame
~ **в плоскости лонжерона** spar frame
~, **вспомогательный** secondary frame
~, **кольцевой** (attachment) ring, circular/ring frame
~ **мотогондолы** *av.* engine nacelle frame
~, **твиндечный** tweendeck frame
~, **теоретический** station/transverse section, frame line
~, **трюмный** hold frame
~, **усиленный** deep frame
~ **швеллерного сечения** U-section frame
шпанка 1. *bot.* black/sweet cherry *(Prunus avium);* 2. *agr.* merino sheep; 3. *zool.* Spanish fly *(Lytta vesicatoria);* 4. *pharm.* cantharides
шпаритель *food.* scalder, scalding machine
шпарутка *text.* temple
шпат I *miner.* spar
~, **алмазный** adamantine spar
~, **атласный** satin spar
~, **болонский** Bologna spar/stone
~, **горький** bitter spar, magnesian lime, dolomite
~, **железный** spatic iron, siderite
~, **жемчужный** pearl spar
~, **исландский** Iceland spar/crystal
~ **керамический** ceramic fel(d)spar
~, **малиновый/марганцевый** rhodochrosite,

dialogite
~, **пенистый** foam spar, earth foam, aphrite
~, **плавиковый** fluor spar, fluorite
~, **полевой** fel(d)spar
~, **синий** blue spar, lazulite
~, **тяжелый** heavy spar, barite
~, **цинковый** zinc spar, smithsonite, calamine
~, **щелочной полевой** alkali fel(d)spar
шпат II *vet.* (bone-)spavin
шпателевидный spatulate
шпатель *art* palette/pallet knife; *constr.* putty knife; *print.* slice; *med.* spatula
~, **аптечный** chemist's spatula
~, **двусторонний** double-ended spatula
~ **для языка** tongue spatula
~, **стоматологический/зуботехнический** dental spatula
шпатель-предохранитель spatula-protector
шпахтель=шпатель
шпация 1. *print.* space; 2. *naut.* (frame) spacing; timber-and-space
~, **полукруглая** end/nut quad
~, **практическая** frame spacing
~, **теоретическая** station spacing
шпейза *met.* speiss
шпенек pin, peg
шпергель *bot.* corn spurry *(Spergula arvensis)*
шпигат *naut.* drain/scupper hole, scupper; flood(ing) hole/port
~, **палубный** deck scupper
~, **штормовой** freeing/clearing/wash port, storm scupper
шпигель 1. *miner.* specular/spiegel iron ore, hematite; 2. *leath.* shell
шпиговать 1. *cul.* (inter)lard; stuff; 2. *naut.* thrum
шпиговка 1. *cul.* (inter)larding; larding pin; 2. *naut.* thrum
шпик I *cul.* lard
шпик II plain-clothes detective, sleuth
шпиль 1. *naut.* capstan; 2. needle, steeple, spire
~, **буксирный** towing capstan
~, **швартовный** warping/gipsy/gypsy capstan
~, **якорный** anchor capstan
шпиль-гат *naut.* bar hole
шпилька *mach.* stud (bolt), dowel, (screw) pin; tack
~ **вытяжного троса** *av.* rip pin
~ **для катушек** *text.* bobbin skewer
~, **контактная** *el.* contact stud
~, **контрольная** set pin
~ **курка** hammer pin
~, **направляющая** locating pin
~, **предохранительная** *mach.* safety-lock pin
~ **с резьбой** threaded pin
~ **спускового рычага** trigger(-lever) pin
~, **стекольная** *constr.* glaziers' point
~, **стопорная** stop pin
~, **формовочная** moulder's brad
шпильман *hist.* wandering minstrel
шпинат *bot.* spinach *(Spinacia)*
шпингалет (door/window) fastener, bolt, latch, catch
шпиндель spindle; (rotor)shaft, arbo(u)r; axle
~ **веретена** *text.* spindle blade
~ **клапана** valve stem
~, **кулачковый** cam spindle
~ **прокатного стана, зубчатый** gear-type spindle
~ **прокатного стана, трефовый** wobbler spindle
~, **расточный** boring spindle
~, **установочный** setting spindle
~, **хлопкоуборочный** cotton-picking spindle
шпиндельный spindle *(attr.)*
~**ная бабка** spindle box, headstock
шпинель *miner.* spinel
~, **железистая/черная** iron spinel, hercynite
~, **оранжево-красная** rubicelle
~, **розово-красная** balas(-ruby)
~, **фиолетовая** almandine spinel, almandite
шпион spy
шпиц I Pomeranian, spitz(-dog)
шпиц II spire, steeple
шпицрутен *hist.* rod; instrument of punishment
шплинт cotter/split/clinch/forelock pin, lockpin
шплинтовать cotter(-pin), secure/fasten/lock with a cotter pin
шпон 1. (sheet of) veneer; 2. *print.* lead, composing/setting rule, slug, space-line, riglet
~, **лущеный** rotary-cut veneer
шпонка 1. (bushing) key, cotter key, spline, dowel; gib; 2. *print.* lead, space-line, riglet
~, **врезная** sunk key
~, **врезная, клиновая** draw key, gib-head key
~, **закладная** inserted/loose key
~, **затяжная** interference-fit key
~, **клиновая** taper(ed) key
~, **касательная** tangent key
~, **направляющая** guide/feather/spline key
~, **натяжная** gib key
~ **с контрклином** fox key, gib and cotter
~, **сегментная** semicircular/Woodruf key
~, **скользящая** sliding key
~, **фрикционная** friction/saddle key
~, **цилиндрическая/круглая** round pin
шпоночный leed *(attr.)*, key *(attr.)*
~ **паз** keyslot, keyway
~**ное соединение** key joint
шпор *naut.* foot, heel
~ **грузовой стрелы** derrick foot/heel
шпора 1. lug; *agr.* grouser, grouter, wheel lug; 2. *bot., anat., med.* spur, calcar
~, **зимняя** ice grouser
~, **костная** bony spur
~, **птичья** calcar avis
~, **пяточная** calcaneal spur
шпорец, шпорца *bot.* spur, calcar
шпорник *bot.* larkspur *(Delphinium)*
шпорниколистный *bot.* larkspur-leaved

шпороцветник *bot.* spurflower *(Plectranthus)*
шпорце *bot.* spur, calcar
~, нектароносное nactariferous spur
шпорцевидный calcariform, spurlike
шпорцевый *zool.* spurred, with a clawlike process
шпренгель *constr.* subdiagonal, strut frame
шпринг *naut.* spring line/rope
~, ставить судно на clubhaul
~, становиться на anchor with a spring
шпрингбок *zool.* springbok, springbuck *(Antidorcus marsupialis)*
шпринтов *naut.* sprit
шпринтовый парус spritsail
шприц injector; (grease) gun; *med.* syringe; *zool.* antlia
~, ампульный ampule injector, ampin
~, гортанный laryngeal syringe
~ для внутриматочных вливаний intrauterine irrigator
~ для воды water syringe
~ для густой смазки grease/squirt-gun
~ для промывания полостей irrigation syringe
~, заливочный primer
~, ирригационный fountain syringe, rubber bulb syringe
~, колбасный *food.* sausage stuffer
~, медицинский автоматический auto-injector
~, неразборный one-piece syringe
~, посолочный *food* pickle injector
~, смазочный lubricating gun
~, смазочный ручной hand lubricator
шприцгусс *met.* die casting
шприцевание 1. injection, injecting; *auto.* gungreasing; 2. *plast.* extrusion, extruding, forcing
~ колбасным фаршем sausage stuffing
шприц-машина *plast.* extruder, extruding/forcing machine, (screw) extrusion machine
~, двухшнековая twin-screw extruder
шприц-пистолет spray gun
шприц-тюбик *med.* syrette
шприцуемость *plast.* extrudability
шпрот *zool.* sprat *(Sprattus)*
шпрюйт *naut.* bridle; sling
~ отводителя трала otter bridle
~ углубителя depressor bridle
~, уравнительный equalizing bridle
~, цепной chain sling
шпрынка челнока shuttle peg/cock
шпуледержатель, шпулярник bobbin holder
шпулька bobbin, spool
~, вышивная swivel bobbin
~, челночная shuttle bobbin
шпульник spooler
шпуля *text.* spool, bobbin
~, основная warp bobbin
~, уточная weft/filling bobbin
шпунт I 1. *naut.* rabbet; groove, tongue; 2. plug, tap; 3. *constr.* sheet piling/pile, cutoff wall; 4. *met.* piling bar
~, забивать drive a sheet pile
~, килевой keel rabbet
~ рудерписа rudder bearing
~ с замками lock sheet piling bar
~, соединение в tongue and groove joint
шпунт II 1. sculptor's/stonecutter's chisel; 2. tonguing/grooving/rabbet plane
шпунтгубель = шпунт II, 2.
шпунтина sheet pile
шпунтовальный grooving, rabbeting
~ станок tonguing-and-grooving machine
шпунтование, шпунтовка rabbeting, grooving (and tonguing)
шпунтовать rabbet, tongue
шпунтовик match plane; hammer-head chisel
шпунтовой tongue-and-groove
шпунтубель tonguing/grooving/rabbet plane
шпур 1. *min.* blasthole, borehole; 2. *math.* spur, trace *(of matrix);* 3. *met.* taphole
~, врубовый cut/snubber hole
~, забитый, ~ с забивкой stemmed hole
~, забуривать collar a blasthole
~, камерный chambered hole
~, котловой gopher hole
~, наклонный angled/inclined hole
~, отбойный main/outer hole
шрам scar; slash; cicatrix
~, ледниковый glacial scar/scratch/score
~, листовой cicatrix
шрапнель *mil.* shrapnel
шратт *geol.* creg, crete
шредер shredding machine, shredder
шрекингерит *geol.* schroekingerite
шрифт 1. (printing) type, print, type face; lettering (on drawings); 2. handwriting, script
~, готический Gothic type, Old English type, black-letter type
~, гражданский Russian/civil type
~, декоративный fancy/ornamental/decorative type
~, дидотовский Didot (type)
~ для оптического считывания, стандартный *comp.* OCRA font
~, египетский square serif
~, дубовый extra bold grotesque
~ жирного начертания heavy faced type, extra bold type
~, жирный bold face
~, заголовочный headline/title type
~, контурный contour/skeleton/outline type
~, косой slanting/oblique type
~ курсивного начертания Italic (type)
~, латинский, ~ прямого начертания Roman type
~, матричный *comp.* matrix font
~, машинописный/машинный typewriter type
~, наклонный sloping type
~, нормальный/стандартный standard type

~ **полужирного начертания** medium/bold-faced type
~, **прямой** Roman type
~, **светлого начертания** light/lean-faced type
~, **титульный** headline/title type
~, **узкий** condensed type
~ **чужой** wrong fount type
~, **штрихованный** shaded type
шрифтолитейный type-casting
шрифтонаборный type-setting
шрот *agr.* (ground) oil-cake
штаб staff, headquarters
~ **ВВС, главный** main airforce headquarters
~ **инженерных войск** engineer headquarters
~, **морской** naval headquarters
штабдив *abbr.* **штаб дивизии** division headquarters
штабелевание, штабелевка, штабелирование piling, stacking, stockpiling
штабелер stacker, piler; palletizer
~ **для досок** lumber stacker
~, **разгрузочный** unstacker, unpiler
штабелеразборщик unstacker, unpiler
штабелеукладчик piler, stacker
штабелировать pile/stack up, stockpile
штабель stock, pile
~ **из микромодулей** micromodule stack
~ **пиломатериалов** lumber pile
~, **смесительный** mixing heap
~ **труб** stockpile of pipes
штаб-квартира *mil.* headquarters
штабкор *abbr.* **штаб корпуса** corps headquarters
штаб-лекарь *hist.* regimental surgeon
штаб-офицер *hist.* field officer
штабс-капитан *hist.* staff/junior captain
штаб-трубач *mil.* trumpet-major
штаг *naut.* stay, shroud
~ **котла** boiler stay
штаг-карнак *naut.* cap stay
штакетник picket boards
шталмейстер *hist.* Master of the Horse, equerry
штамб *bot.* tree trunk, stem, bole
штамбовый *hort.* standard *(attr.)*
штамм *biol.* strain, (stock) culture
~, **адаптированный к РКЭ** avianized strain, egg-adapted strain
~, **вирусный** viral strain
~, **дикий/исходящий/исходный** wild strain
~, **измененный под действием фага** phage-changed strain
~, **индикаторный** indicator/detector strain
~, **лизогенный** lysogenic strain
~, **псевдолизогенный** pseudolysogenic/carrier strain
~, **родительский** parent strain
штамм-носитель бактериофага virus-carrying bacterial phage strain
штамп 1. (blanking/punching) die; press tool; 2. title block; stamp, impress; 3. *fig.* stock phrase, cliché
~ **выдавливания** extruding die
~, **вырезной/вырубной** blanking die
~, **высадочный** heading/upsetting/upsetter die
~, **высечной** notching die
~ **для перфорации** punch pin/stamp
~, **завивочный** coiling die
~, **загибочный** seaming die
~, **ковочный** (drop-)forging die
~, **кромко(за)гибочный** beading/flanging die
~, **многоручьевой** multiple impression die
~, **обжимной** swaging die
~, **обрезной** trimming die
~, **открытый** open die, open press tool
~, **пробивной/прошивочный** piercing die
~, **разделительный** shearing die
~, **сборочный** assembling die
~, **склепочный** riveting die
~, **скручивающий** curling die
~, **фальцовочный** seaming die
~, **чеканочный** embossing/coining die, coining press tool
штампмашина *food.* stamp(ing) machine, stamper
штампованьный punching, pressing, stamping,
штамповать *tech.* (die-)stamp; blank, press, punch
~ **в горячем виде** swage
штамповка *tech.* pressing, punching, stamping, blanking
~ **взрывом** explosive forming
~ **выдавливанием** extruding
~, **горячая** swaging, hot stamping
~, **листовая** sheet/plate forming; sheet-metal stamping
~, **объемная** die forging
~, **фасонная** profile stamping
~, **холодная** cold stamping/forging
штамповочный pressing, stamping, punching
~ **молот** swage hammer
~ **пресс** stamping press
штамповый die *(attr.),* punch *(attr.)*
штамподержатель die holder
штанга 1. *tech.* (push/pull) rod, bar, beam; boom; 2. *sport* crossbar; barbell, weight
~, **бурильная/боровая** bore/boring/drill rod
~, **ведущая** guide rod
~, **измерительных приборов** *av.* instrumentation boom
~, **изолирующая** *el.* insulating/hook stick, insulating rod
~, **нажимная** push rod
~, **направляющая** guide rod
~, **направляющая антенны** boom of an array
~, **носовая** nose probe/boom
~ **опрыскивателя** *agr.* spray boom/bar
~, **печатающая** *comp.* print bar
~ **печного выталкивателя** *met.* peel bar
~, **распорная** *mach.* spreader (bar)
~, **соединительная** coupling bar/rod
~ **сталкивателя (обрезков проката)** push-

off arm
~, стыковочная docking probe
~ толкателя *met.* push rod, pusher bar/ram, push-out bar
~, торсионная torsion bar
~, тяговая dragbar, pull rod
~, ударная sinker bar, drill stem
~, управляющая actuation/control rod
~ штанговой крепи bolt roof
штангенглубиномер depth gauge
штангензубомер gear-tooth gauge/vernier
~ с нониусом vernier gear-tooth caliper
штангенинструмент beam-type measuring tools
штангенрейсмус height gauge
~ с нониусом vernier height gauge
штангенциркуль caliper square, slide gauge, slide/vernier/sliding calipers; beam compass; trammel(s)
~ антропометрический anthropometric sliding calipers
~, разметочный trammel(s)
~ с нониусом vernier calipers
штангист *sport* weight lifter
штангодержатель rod holder
штангоизвлекатель rod puller
штандарт *stat.* standard deviation
штанец cutting/specimen die, stamping knife, cutting tool, knife blade
штанцевание stamping/punching-out
штапелеизмеритель stapling apparatus
штапелирование *text.* stapling
штапель *text.* staple
~ шерсти, ~ шерстяного волокна wool staple
штапельный staple *(attr.)*
~ная ткань spun fabric
~ная ткань, вискозная spun-rayon fabric
штапик strip, mould(ing fillet); (glazing) bead; glazier's mould
штат I state
~ы, Генеральные *hist.* States General
штат II staff, establishmen, personnel; *pl.* list of staff members
~ы военного времени war-time establishment
штатгальтер *hist.* stad(t)holder
штатив stand; support, tripod, holder; *chem.* upright, rack
~ для пипеток pipet(te) stand/support
~ для пробирок test-tube rack
~ для тиглей crucible holder
~ для фильтрования filter stand
~, нагрудный *cin.* shoulder-pod
~, универсальный universal support
штатный 1. issued, standard *(attr.);* 2. on the staff; permanent; regular
~ сотрудник газеты staff writer
~ная должность permanent appointment
~ное расписание manning table, list of staff members
штатский *n.* civilian
штатское (платье) civilian clothes
штауфер (screw-feed) grease cup
штевень *naut.* (stem and) sternpost
ште(й)гер *min.* foreman miner
штейн matte
~, бедный/богатый low/high-grade matte
~, обогащенный concentrated matte
~, сырой green matte
~, тощий low-grade matte
штек(к)ер *el.* (jack) plug
~ коммутационного шнура patchtip, patchplug
штемпелевальный stamping, cancelling
штемпелевать stamp
штемпель (rubber) stamp
~ времени time stamp
~, почтовый postmark
штемпельный stamp *(attr.)*
~ная подушка ink pad
штенгель exhaust tube
штепселевание *el.* plugging-in
штепсель *el.* plug, adapter, socket
~, вызывной *teleph.* calling/signal(l)ing/ringing plug
~, двухполюсный double-contact plug, two-pin plug
~, зарядный charging plug
~, испытательный test plug
~, однополюсный one-pole plug, single-pin plug
~, опросный *tel.* answering plug
~, переключающий change-over plug
~, рельсовый *rail.* bond plug
~ с гнездом plug and jack
~, трехконтактный three-pin plug
~, штифтовой pin plug
штепсельный plug *(attr.),* socket *(attr.)*
~ шнур patchcord
~ная розетка socket
~ное соединение plug connection
штернбергия *bot.* fall daffodil *(Sternbergia lutea)*
штерт *naut.* lanyard; becket
~ ведра bucket lanyard, bail rope
~ уключины rowlock lanyard
штивать *naut.* stow, even out
штивка *naut.* stowage, stowing
штилевать *naut.* becalm, be becalmed
штилевые полосы *nav.* the doldrums
штиль I *naut.* calm
~, мертвый dead calm
~, попадать в fall calm, be becalmed
штиль II *mus.* stem
штирборт *naut.* starboard
штифт pin, dowel, stud; peg, sprig
~, выталкивающий ejecting plug
~ из крови homoplastic rod
~, конический taper(ed) pin
~, копирный follower, tracer
~, направляющий guide pin
~, предохранительный safety/retention/preventer/shear pin
~, приводной *print.* register pin

~, призонный set/fitting/steady pin
~, резьбовой threaded pin
~, реперный locating dowel
~ с потайной головкой counter sunk-head pin
~ с пропилом slotted/notched pin
~, соединительный dowel/slotted pin; glut (pin)
~, стопорный lock/catch/retention/safety pin
~, считывающий *comp.* pecker
~, упорный stop pin
~, установочный/центрирующий register/set/locating pin; spigot
~ цоколя *el.* base pin
штифтовать fix by pins/dowels
штихель burin, graver, engraving tool; stylus; *med.* chisel
~, долотообразный bit-shaped chisel
~, желобоватый grooved chisel
штихмас(с) 1. inside micrometer; 2. measuring rod
шток 1. *tech.* rod; stem; 2. *naut.* stock, pole; 3. *geol.* stock, boss
~ бойка striker rod
~, выдвижной *mach.* retractable rod
~ золотника valve spindle/rod/shaft, slide valve/shaft/rod
~, интрузивный intrusive/plutonic stock
~ клапана valve stem/rod
~, лежачий lying/flat stock
~, направляющий guiding rod
~ поршня piston rod
~, простой simple stock/boss
~ пружины spring stem
~, рудный ore body/stockwork
~, сложный composite/multiple stock
~, соляной salt stock/core/plug
~, сопутствующий satellitic/satellite stock
~ толкателя клапана *auto.* tappet rod
~, тормозной brake push rod
~ якоря anchor stock
штокверк *min.* stockwork
штоклак stick lac
штокроза *bot.* hollyhock *(Althaea rosea)*
шток-тали *pl. naut.* stock tackle
штокфиш stockfish, dried cod
штольня *min.* adit, side drift; tunnel; gallery
~, водоотливная drainage adit
~, ответвленная side adit
штопальный darning
штопка 1. darning; mending; 2. darning thread
~, незаметная/художественная invisible mending
~, шерстяная mending/darning wool
штопор 1. corkscrew; 2. *av.* (corkscrew) spin
~, восходящий upward spin
~, входить в go/get into a spin
~, колебательный/неустойчивый oscillatory spin
~, крутой steep spin
~, левый/правый left/right(-handed) spin
~, нормальный normal/tail/erect(ed) spin
~, перевернутый inverted/outside spin
~ с отрицательной/положительной перегрузкой outside/inside spin
~, трехвитковый 3-turn spin
~, управляемый controlled/precision spin
~, устойчивый stable spin
~, учебный practice spin
штопорение *av.* spinning
штопорить *av.* spin, descend in a spin
штопоровидный, штопорообразный cork-screw-shaped; spiraliform, spiral, helical
штопфер *med.* plugger
штора blind, shutter; shade
~ противопожарная fire screen
~ радиатора radiator blind
~, свертывающаяся roll-up shutter
шторка blind, curtain
~, затемняющая/светомаскировочная black out curtain
~ кабины (летчика) *av.* cockpit hood
~, лицевая face curtain/blind
~, маскирующая camouflage blind
шторм *nav.* (strong) gale
~, сильный whole/heavy gale
~ом, задержанный weather/storm-bound
штормовка *naut.* weather proof jacket
штормовой gale *(attr.),* storm *(attr.),* stormy
~ ветер gale
~ костюм weatherproof clothing
~ парус storm sail, smiter
~ сигнал storm-warning signal, gale signal
штормтрап *naut.* Jacob's/storm/rope/pilot/sea ladder, abandon-ship ladder
~, кормовой stern ladder
~ с балясинами round-rung ladder
штос(с) fencing pass, thrust
штоссель *tech.* ram
штоф I *hist.* shtof *(liquid measure)*
штоф II *text.* damask, brocade
штраба *mach., arch.* toothing, batter
штраф fine, forfeiture, penalty; surcharge
~, брать fine
~ вследствие дефицита shortage penalty, stock-out penalty
~у, подвергаться incur a penalty
~у, подлежащий subject to penalty
штрафбат *abbr.* **штрафной батальон** penal battalion
штрафник *mil.* convicted soldier
штрафной penal, penalty *(attr.)*
~ бросок *sport* free throw
~ удар *sport* penalty kick
~ная площадка *sport* penalty area
штрафование fining, imposing a penalty
штрафовать fine, impose a penalty, penalize
штрейкбрехер strikebreaker, scab
штрек *min.* drift, heading, entry
~ висячего бока hanging-wall drift
~, откаточный haulage drift
~, скреперный scram drift

штрековый drift *(attr.)*
штрипка footstrap
штрипс *met.* strip; skelp
~ со скошенной кромкой beveled-edge skelp
~, стальной strip steel
штрипсовый strip *(attr.)*
~вая сталь skelp steel
штрих 1. dash (line); mark(ing); dash; stria; *opt.* line; *cart.* hachure; *math.* prime, the mark (‘); 2. slit, ruling; *print.* stroke; 3. *fig.* feature, trait
~ буквы, основной thick/heavy stroke
~, двойной double prime/accent
~ диффракционной решетки line ruling, ruled groove
~, ледниковый *geol.* glacial score/scratch/scar
~, начальный (на логарифмической линейке) initial index
~, нитяной spider line
~ Шеффера *math.* Sheffer(‘s) stroke (operation), symbol of alternative negation
штрихование (cross) hatching; lining, shading
штриховатость striation; *bot., pathol.* single-virus streak, streak/stripe disease
штриховатый striate(d), hatched
штриховать crosshatch, (section-)line; *cart.* shade
штриховка 1. hatching, shading; 2. *cart.* hachure; *geol.* striation, striature
~, бороздчатая grooving
~ диффракционной решетки ruling, slit
~, ледниковая glacial striation/scoring
штрихозапись facsimile recording
штрихопередача facsimile transmission
штрихпунктир dot(-and)-dash line
штрихточечный dash-and-dot, dot-and-dash
штудирование studying
штука piece; item
штукатур plasterer, stuccoworker
штукатурение plastering, stuccowork
штукатурить (apply) plaster, finish in/with plaster
~ снаружи здания stucco, finish in/with stucco
~ под шубу rough-cast
штукатурка plaster(ing)
~, гипсовая gypsum plaster/cement
~, звукопоглощающая acoustic(al) plaster
~, наружная stucco
~, орнаментная pargeting
~, сухая wall/plaster board
штукатурный (covered with) plaster
~ные работы plastering; plaster work, stuccowork
штуковка fine drawing; invisible mending
штунда *eccl., hist.* shtunda *(Russian religious sect)*
штундистский, штундист Stundist
штурвал *tech.* handwheel; steering/control wheel
~, валоповоротный turning wheel
~ высот height wheel
~ горизонтальных рулей diving wheel, plane handwheel
~ управления *av.* control wheel
штурвальный *n.* helmsman, steersman, pilot
штурм assault, storm, attack
штурман navigating mate, sailing master; *mil., av.* navigator, navigating/navigation officer
~, авиационный air/flight navigator; second officer
~ наведения aircraft controller
~, старший senior navigator
штурман-радист radio operator/navigator
штурманская *n. nav.* charthouse, chartroom
штурманский navigator’s
~кая рубка *nav.* charthouse, chartroom
штурмовик 1. *av.* strafing (aero)plane, lowflying attack aircraft; attack/strike aeroplane/aircraft; 2. member of an assault party, stormer; 3. *hist.* Nazi storm trooper
~, палубный carrier-based strike aircraft
штурмовка *mil.* stafing, low-level attack, lowflying air attack
штурмовой attack *(attr.),* assault *(attr.)*
~вая авиация attack aviation; assault aircraft, low-flying attack aircraft
~вая группа assault group/party
~вая лестница *hist.* scaling ladder
штурмовщина rush(ed) work
штуртрос *naut.* steering/tiller line, wheel/steering/tiller rope
~, цепной steering/tiller chain
штуф *geol.* lump of coal/ore
штуцер 1. (connection) point, (pipe) connection/union; nipple; coupling; sleeve; 2. sporting rifle; carbine
~, впускной inlet connection
~, дренажный venting connection
~, заливной filler
~ заправки водой water-servicing point
~ заправки кислородом oxygen-charging point
~, заправочный fill point
~, зарядный charging point/connector
~, конденсатный condenser tube
штучный single, individual, by the piece; custom-made
~ пол parquet/inlaid floor
~ товар piece goods, goods sold by the piece
~ная оплата/продажба payment/sale by the piece
~ная работа piecework
штыб *min.* culm, slack, coal dust, breeze, cinders; *geol.* chippings, rubble
штык 1. bayonet; 2. *naut.* bend, knot; 3. *met.* bar, pig, ingot; 4. spade’s depth; 5. *tech.* ram rod
~, задвижной mooring knot
~, игольчатый rod bayonet
~, клинковый knife/saber bayonet
~, простой bend-clinch
~, рыбацкий buoy-rope bend
штыка *met.* bar, ingot, pig
штык-болт *naut.* reef earing
штыкование, штыковка digging up at spade’s depth

штыковой bayonet *(attr.)*
штык-тесак sword bayonet
штылек handle, grip, hilt
штырек dowel *(attr.)*, pin *(attr.)*
~ банановый banana pin
~ цоколя base pin
штырь pin, dowel, pintle
~ вертлюга cradle pintle
~, волноводный waveguide post
~, контактный contact pin
~, настроечный *rad.* tuning post
~ разъемного контакта male contact
~, рулевой *nav.* rudder pintle
~ связи coupling probe
~, стыковой joining tenon
~, сцепной *agr.* drawbar/hitch pin
~ цоколя base pin
~, штепсельный pin
шуба I fur(-lined) coat; fur, pelt
шуба II *constr.* rock face; rough-faced rocks
шубный fur-bearing
~ клей carpenter's glue
шуга ice gruel/scum, slush/sludge/cream ice
~, смерзшая packed slush
шугосброс ice chute/slide, sludge ice chute
шудра Sudra, the lowest Hindu caste
шум noise; sound; bruit; souffle; *rad.* hum; interference; *med.* murmur; *see also* **шумы**
~, анемический anemic/hemic murmur
~, белый, ~ дельта-коррелированный white/pure noise
~ в боковой плоскости lateral noise
~ в голове head noises, tinnitus cerebri
~ в ушах tinnitus/susurrus aurium, murmur/noise(s)/ringing in the ears
~, венозный venous hum
~, вибрационный structure-generated noise
~, вихревой turbulence/vortex noise
~, внешний external/ambient noise
~, внутрисердечный cardiac murmur
~ волчка nun's murmur, humming-top murmur, venous hum in the neck, bruit de diable
~ всасывания intake noise
~, высокочастотный r.f. noise, microwave noise
~ выхлопа *auto.* exhaust noise/roar
~, гауссов(ский)/нормальный Gaussian noise
~, голосистолический holosystolic/pansystolic murmur
~ гребного винта propeller/screw/propulsion noise
~ двигателя engine noise
~, дельта-коррелированный delta-correlated noise, white/pure noise
~, дующий bellows/blowing murmur
~, дыхательный respiratory murmur
~ Дюрозье Durozier's murmur
~ изгнания крови из серца ejection murmur
~, кардиальный cardiac murmur
~, каротидный carotid bruit
~ квантования quantizing/quantization noise
~, квантовый quantum noise
~, клапанный valvular murmur
~ лампы valve/tube noise
~, линейный circuit noise
~, маточный uterine souffle
~ механизмов engine/machinery noise
~, модуляционный modulation noise, noise behind the signal
~ молочной железы mammary souffle
~ моря ambient sea/water noise
~ на входе intake noise
~, неклапанный nonvalvular murmur
~, нормальный normal/Gaussian noise
~ окружающей среды ambient/extraneous noise
~ от сети (питания) power-supply noise, mains hum
~, побочный additional noise
~ пограничного слоя boundary-layer noise
~, посторонний ambient/extraneous noise
~, пресистолический presystolic/atriosystolic murmur, late diastolic murmur
~, псевдохаотический pseudo-random noise
~ пуповины fetal/funic(ular)/umbilical sound
~ реактивной струи jet noise
~ сердца murmur, susurrus
~ сердца, воркующий dove-coo murmur
~, систоло-диастолический seesaw murmur, to-and-fro murmur
~, случайный random noise
~, собственный internal/inherent noise, self-generated noise
~, статистический random noise/waves
~, структурный structure-borne/-generated noise
~, стенотический stenosal murmur
~, суперный *rad.* syperregenerative hiss
~, тепловой thermal/Johnson noise
~ трения friction sound
~ трения плевры pleural murmur
~ Фишера, церебральный Fischer's cerebral murmur
~ Флинта, стенотический Flint's murmur
~, фоновый background/hum noise, hum
~, функциональный functional/inorganic murmur
~, хаотический random noise
~, хлопающий flag/flapping sound
~ цепи *rad.* circuit noise; *auto.* chain rattle
шумер, шумерский Sumerian
шумовик 1. *theatre* sound-effects man; 2. player of a percussion instrument
шумовка I skimmer, strainer; straining/perforated spoon
шумовка II crackle, crackling
шумовой noise *(attr.)*, sound *(attr.)*
~ оркестр percussion band
~ фон *rad.* background noise, hum
~вые эффекты sound effects

шумо(за)глушитель *el.* noise silencer/limiter/suppressor/attennator, sound damper, muffler, noisekiller
~, аэродромный ground muffler
шумозащитный noise-protective
шумоизлучатель noise maker, sound device
шумоизмеритель noise analyser
шумомер sound (level) meter, phonometer, audionoise meter
шумопеленгатор *nav.* passive sonar/system, sound locator, hydrophone, direction/underwater listening device, direct-listening (sonar) system
~, автоматический robot submarine listening post
шумопеленгование (direct) listening
~, направленное directional listening
шумоподавление noise cancelling/reduction/suppression
шумоподавляющий noise-cancelling/-suppressing
шумостойкость noise stability
шумостойкий noiseproof
шум-фактор noise factor/figure, signal-to-noise merit
шумы *pl.* noise; *see also* **шум**
~ Галактики galactic noise
~ квантования quantization noise
~ несущей carrier noise
~ окружающей среды ambient noise
~ СВЧ microwave noise
~, собственные internal/intrinsic noise
~ солнечного радиоизлучения solar (radio) noise
шумящий noisy
шунгит *miner.* schungite
шунт *el., surg.* shunt, by-pass, bypass
~, встроенный built-in shunt
~, желудочково-сосцевидный ventriculomastoid shunt
~, измерительный instrument shunt
~ межжелудочковый право-левосторонний right-to-left shunt, reversed shunt
~, портакавальный portacaval shunt, Eck fistula
~, сердечно-легочный cardiopulmonary bypass
шунтирование shunting, bypassing, bridging
~ кровеносных сосудов vascular shunting
шунтировать shunt, bypass, bridge
шуровать poke, stir, stoke; rabble
шуровка 1. poking, rabbling; 2. stirrer, rabble, poker
шурпа *cul.* shurpa, soup of mutton, vegetables and rice
шуруп *naut.* (wood) screw
~, шлюпочный boat screw
шуруповерт screwdriver
шурф *geol., min.* exploring/prospecting shaft; bore pit, boring
~, взрывной blast hole
~ для сжигания газа flare pit
~, пробить sink a boring
~, разведочный prospecting/test pit
~, эксплуатационный working pit
шурфование, шурфовка *min.* sinking/digging a hole, pitting
шурфовать *min.* excavate/sink a prospect hole
шурфовщик prospector, digger
шут 1. *hist.* fool, jester; 2. clown, buffoon
шутка 1. joke, jest; 2. *theatre* farce
шутовской колпак fool's cap
шхеры *pl. geogr.* skerries, skerry, rocky islets, stacks
шхуна schooner
~, марсельная topsail schooner
шхуна-баржа schooner barge

Щ, щ

щавелевокислый *chem.* oxalate (of)
~ калий/натрий potassium/calcium oxalate
щавелевоуксусный *chem.* oxaloacetic, oxaloacetate (of)
щавелевый 1. *chem.* oxalic; 2. *bot.* sorrel *(attr.)*
щавелек *bot.* sheep sorrel *(Rumex acetosella)*
щавелелистный *bot.* dock-/sorrel-leaved
щавелинка sorrel leaf
щавель *bot.* dock, sorrel *(Rumex)*
щавельник *cul.* sorrel soup
щебень 1. crushed/broken rock/stone; (brokenstone) ballast; road metal; 2. *geol.* detritus, rock debris/waste
~, коралловый coral sand
~, мелкий chippings
~, шлаковый broken slag
~бнем, мостить macadamize, metal
щебнебоец stone breaker
щебнечиститель *rail.* ballast cleaner
щебнистый covered with ballast; *geol.* detrital schistose
щебрец *bot.* thyme *(Thymus)*
щебрушка *bot.* basil thyme *(Acinos)*
щеврица *zool.* pipit, titlark *(Anhus)*
щегол *zool.* goldfinch *(Carduelis carduelis)*
щелоль *zool.* spotted redshank *(Tringa erythropus)*
щека 1. *tech.* cheek(-piece); web; jaw; side (piece), side wall; 2. *geol.* high and steep slope; cliffy riverbank
~ автосцепки coupler guard arm
~ блока cheek of pulley
~ дробилки jaw plate, crusher jaw
~ дробилки, подвижная movable/swing(ing) jaw
~, зажимная gripping jaw
~ коленчатого вала crank web/cheek
~ кривошипа crank web, (crank) cheek
~, накаточная roller jaw

~, откидная snatch cheek
~ початкоотрывателя stripper bar
~ тисков vice jaw
щековой *tech.* cheek *(attr.),* jaw *(attr.)*
~вая дробилка jaw breaker
~вая стена side wall
щеколда latch, catch, bar, cam, dog, lock(ing bar); (window) fastener
~, пружинная spring latch
~ якоря мины mine anchor tripping latch/finger
щелеватый slitty, fissured, fissury, cracked, chinked
щелевидный slit-/chink-shaped, slitlike, rimiform, fissural
щелевой 1. slotted; fissured, fissural; cracked; rimose, rimate; 2. *phon.* fricative
щелезуб *zool.* solenodon(t) *(Solenodon)*
щелеобразный slot-shaped, slitlike
щелеобразование fissuration
щелина large crack/fissure
щелинный *phon.* fricative
щелистый fissured, cracked, rimose, chinky
щелкун *zool.* click/snapping beetle, elater *(Elater)*
щелок lye, liquor, alkaline solution
~, варочный cooking liquor
~, едкий alkali lye
~, калийный caustic potash
~, крепкий strong liquor
~, натровый/натронный soda lye
~, отработанный spent lye
~, промывной wash liquor
щелочение alkaline washing, alkalization; lixiviation; bucking; steeping/washing in lye
щелочерастворимый alkali-soluble
щелочепрочность alkali resistance
щелочепрочный, щелочеупорный, щелочеустойчивый alkali-proof/-resistant, alkali-resisting; lyeproof
щелочестойкость alkali resistance
щелочноземельный alkali(ne)-earth
щелочной alkali(ne), alkalinous; lixivial, lixivious
~ металл/раствор alkali metal/solution
щелочность alkalinity, causticity, alkalescence, basicity
~, общая total alkali(nity)
~, повышенная hyperalkali(nity)
~, резервная alkaline reserve
~, третья ternary/third alkalinity
щелочь *chem.* alkali; *pl. med.* alkali(e)s
~, едкая caustic alkali; alkali lye
~, химически связанная fixed alkali
щель *anat., geol.* slit, slot, cleft, fissure; chink, scissure; crack, crevice; rima; aperture, gap, rift;
~, анальная anal/gluteal cleft
~, аэродинамическая aerodynamic slot
~, барабанно-сосцевидная tympanomastoid fissure
~, визирная *geod.* aperture sight
~, возбуждаемая волноводом waveguide-fed slot
~, волноводная waveguide slot
~, вскрытая *geol.* dehiscence, dehiscent slit
~, входная entrance slit
~, высевная *agr.* seeding slot
~, выходная output/exit slit
~, глазная lid slit
~ глазной мишени peep-sight
~, голосовая (true) glottis, rima vocalis
~, двойная double slit, slit pair
~ детектора detector slit
~ для монет *teleph.* coin slot
~ для пролета ионов ion slit
~ для прохождения нейтронов neutron channel
~ для сдува blowing slit
~ для фольги foil slot
~, жаберная branchial cleft, gill slit/cleft
~, зарубная *min.* kerf, slot
~ источника source slit
~, коллимирующая/ограничивающая collimating slit
~, кольцевая ring slot; *rad.* annular slit/slot
~ купола *astr.* dome slit, shutter opening
~, листовая leaf gap
~, мантийная pallial slit
~ мозжечка postcentral fissure
~ мозжечка, горизонтальная great horizontal fissure, horizontal sulcus of the cerebellum
~, ограничивающая collimating slit
~, окаймленная pitted slit
~, отсасывающая suction slot
~, полукруглая crescent slot
~, поперечная *rad.* transverse slot
~ преддверия гортани false glottis
~, продольная *rad.* longitudinal slot
~, пупковая umbilical slit
~ разверзания (однолучевая) (single) slit/line of dehiscence, dehiscence cleft
~, развертывающая scanning slot
~, регулируемая controllable/adjustable slot
~, ротовая mouth slit
~ связи coupling slot
~, сифональная siphonal notch
~, сканирующая scanning slit
~, смотровая sighting/eye/vision/viewing/observation slit, peephole
~, срамная pudental/vulvar slit, urogenital cleft
~, считывающая scanning slit/gate
~, третичная tertiary slit
~, фреатическая phreatic slot
~, эквационная equational split
~, энергетическая *phys.* energy gap
щенок pup(py); cub; whelp
щепа 1. *wood.* chips, splinter, sliver; *coll.* kindling; 2. hog fuel; 3. shingle
~, кровельная roof shingle
щепальный cleaving, splintering, splitting, chipping
щепаный cleaved, splintered, chipped
щепить 1. split, splinter, cleave, chop, chip; 2. *hort.* (en)graft
щепотник *rel.* follower of official Russian Ortho-

dox Denomination
щербина 1. pockmark; 2. chipped spot; indentation, pit, breach, notch
щетина bristle
щетинистоволос(ист)ый hispid
щетинистозубчатый *bot.* chaetodont, setosedentate
щетинистолистный *bot.* chaetophyllous
щетинистолопастный *bot.* chaetolobate
щетинистопильчатый *bot.* setaceoserrate
щетинистость hispidity, hirsuteness, bristliness
щетинистый hispid, hirsute, bristly, bristling, setaceous; strigose
щетинка seta, bristle; striga
~, короткая setula, setule
щетинковидный, щетинкообразный setiform, setaceous, bristlelike
щетинковый chaetopodous, chaetopodan; setal, setaceous; *n. pl. zool.* chaetopods *(Chaetopoda)*
щетинконосный setiparous, setiferous; aristate; setulous
щетинкочелюстной *zool.* chaetognathous; *n. pl.* chaetognaths *(Chaetognatha)*
щетинник *bot.* bristle grass *(Setaria)*
щетиновидный, щетинообразный setiform, setaceous; bristle-shaped, bristlelike
щетинозуб *zool.* chaetodont(id) *(Chaetodon)*
щетинозубые *n. pl. zool.* chaetodontids *(Chaetodontidae)*
щетинолистный *bot.* chaetophyllous
щетинохвостка *zool.* bristletail *(Thysanura)*
щетиночка setule, fine bristle
щетиночконосный setulous, setuliferous, setule-bearing
щетка 1. *el.* brush; 2. *min.* druse; 3. fetlock
~, бутылочная *bot.* lemon bottle brush *(Callistemon lanceolatus)*
~ для механической чистки труб *met.* go-devil
~ для очистки решет *agr.* screen brush
~ для прочистки каналов duct cleaner
~ искателя, параллельная *teleph.* bridging wiper
~, кардная *text.* card file
~, коллекторная slip-ring brush
~, контактная brush, contact wiper/brush
~, контрольная *el.* checking/pilot/control brush
~ кристалла druse
~, многослойная *el.* sandwich brush
~, неразрезная *el.* solid brush
~, отрицательная/положительная *el.* cathodic/anodic brush
~, ощупывающая *comp.* sensing brush
~, половая broom
~, проволочная wire brush
~, разрезная *el.* split brush
~, слоистая *el.* dual grade brush
~, собирательная *el.* collecting brush
~ стеклочистителя windshield wiper blade
~, считывающая *comp.* read(ing) brush
~, трубчатая duct brush/cleaner
щетковидный, щеткообразный brushlike, scopiform, strigillate
щеткодержатель *el.* brush holder/carrier
~, коробчатый box-type brush holder
~, патронный cartridge brush holder
щеткоподъемный brush-lifting
щеткохвост *zool.* brush-tailed phalanger *(Trichosurus vulpecula)*
щеточка *zool.* strigil(lis), (bee's) pollen brush, scopa
щечно-глоточный *anat.* buccopharyngeal
щечно-губной *anat.* buccolabial
щечно-десенный *anat.* buccogingival
щечно-дистальный *anat.* buccodistal
щечно-осевой *anat.* buccoaxial
щечно-шейный *anat.* buccocervical
щечно-язычный *anat.* buccolingual
щечный *anat.* buccal, genal
щи *cul.* shchi, cabbage soup
щипец 1. *arch.* gable (end); 2. *zool.* muzzle
щипковый *mus.* pizzicato, plucked
щиповка *zool.* spined/mud loach *(Cobitis taenia)*
щипцы *pl.* pair of tongs/pincers, (cutting) pliers, nippers; *med.* forceps; extractor
~, абортные abortzang, abortion forceps
~, акушерские axis-traction forceps, obstetrical/ Barton's/Chamberlen/Hodge's/Kjelland's/Levret's/ Tarnier's/Tucker-McLean forceps
~, двузубчатые double-prong forceps
~ для завивки волос curling irons
~ для льда ice tongs
~ для маркировки ушей (у животных) ear marker
~ для орехов nutcracker
~ для разделения тканей tissue forceps
~ для разрезания зубных коронок crown cutters
~ для сгибания бюгелей clasp bending pliers
~ для тиглей crucible tongs
~, зуботехнические dental forceps
~, каминные fire irons/tongs
~, кишечные intestinal forceps
~, кожно-головные craniodermal forceps
~, костные bone(-cutting) forceps, osteophore, Liston's forceps
~, костные акушерские bone-cutting obstetrical forceps
~, крампонные nippers
~, лапчатые gripping forceps
~, ложечные spoon-shaped forceps
~ Мишеля Michel's forceps
~, носовые тампонные nasal packing forceps
~, обжимные *met.* crimper
~, окончатые tenestrated forceps
~, ортодонтические orthodontic forceps
~, пулевые bullet forceps
~, тигельные crucible tongs
щипцы-иглодержатели *pl.* needle forceps/holder

щипцы-кусачки Liston's forceps, bone-cutting forceps
щипчики *pl.* 1. forceps, pincers, tweezers; 2. *zool.* pedicellariae
щирица *bot.* amaranth, redroot *(Amaranthus)*
щирицевидный *bot.* amaranthoid
щирицевые *n. pl. bot.* amaranth family *(Amaranthaceae)*
шит I *bot.* shieldwort *(Peltaria)*
щит II 1. screen, guard, shield; buckler; baffle; shade; 2. panel (display), board, plate; 3. (sluice) gate; slide valve; 4. *zool.* carapace (tortoise) shell; scutum; clypeus; 5. *astr.* The Shield, Scutum
~, аварийный распределительный emergency switchboard
~, брюшной *zool.* ventral shield
~, буксируемый *mil.* towed practice target
~, ветровой windscreen, wind board/guard
~ водоспуска emptying sluice
~, воздухораздельный air baffle
~, выставочный stand/display rack
~, главный управления master/main control board
~, глазной *zool.* ocular scale/shield
~, головной *zool.* cephalon, head shield
~, грязевой mudguard, splash guard/board
~, демонстрационный scoreboard
~, диспетчерский instrument board/panel
~, дозиметрический dosimetric board
~, зарядный charging/accumulator switchboard
~, измерительный meter board
~, коммутационный switchboard; *comp.* patchboard
~, краевой marginal shield
~, ледяной ice sheet
~, облачный cloud canopy
~, отводящий *naut.* otterboard, paravane
~ отличительного огня light board, lamp screen
~ переключений *teleph.* distributing frame, cross-connecting frame
~ подборщика *agr.* pickup cover
~, подшипниковый *el.* end bracket/shield
~, предохранительный safety screen
~, проходческий tunnel shield
~, распределительный *el.* distributing switchboard, distributing/distribution board/frame, feeder switchboard, distribution/switch panel
~, расстилочный *agr.* spreading board
~, снегозащитный snow shed/fence
~, струеотбойный *av.* blast screen/fence, jet deflector screen
~, туловищный *zool.* thorax
~ управления control/operating panel
~, шандровый dam walling
~, швартовый rat guard
~ом, покрытый *zool.* scutiferous
щитик *bot.* stylostegium
щитки *av., pl.* flaps; *see also* **щиток** 3.
~, выпускать extend the flaps
~, выпущенные deflected flaps
~, убранные retracted flaps
щитковатый *bot.* corymbose
щитковидный 1. *bot.* shield-shaped, corymbous, corymbiform; 2. *zool.* scutellate, scutelliform
щитковый scutal, scutellar; *n. pl. zool.* ostracoderms *(Ostracodermi)*
щитконосный *bot.* corymbiferous
щитни *pl. zool.* tadpole shrimps *(Triopsidae; Notostraca)*
щитник *zool.* stinkbug *(Pentatoma)*
щитница *bot.* clypeola *(Clypeola)*
щитовидно-глоточный *anat.* thyropharyngeal
щитовидно-гортанный *anat.* thyrolaringeal
щитовидно-надгортанный *anat.* thyroepiglottic
щитовидно-пальчатый *bot.* peltatogitate
щитовидно-подъязычный *anat.* thyrohyoid, hyothyroid
щитовидно-язычный/языковый *anat.* thyroglossal, thyrolingual
щитовидный 1. *bot.* shield-shaped, peltate(d), scutate, scutellate, scutiform; clypeate; 2. *anat.* thyroid
~ лист peltate leaf
~ хрящ thyroid cartilage
щитовка *zool.* parlatoria *(Parlatoria); pl.* scale insects, coccids *(Coccoidea)* armo(u)red scale insects *(Diaspididae)*
щитовник *bot.* shield/male fern, filix mas, aspidium *(Dryopteris)*
щитовой *zool.* scutal, scutellar
щитогрудые *n. pl. zool.* thoracostracans *(Thoracostraca)*
щитодержатель *hist.* charger
щиток 1. blind, shield; *auto.* dashboard; 2. panel; nameplate, escutcheon; 3. *av.* flap; sub-pane; 4. *zool.* corselet, thorax; scute, scutum; clypeus; *paleont.* carapace; 5. *bot.* corymb, cyme; 6. *sport* shin guard; *see also* **щитки**
~, автоматически опускающийся, смотровой *av.* automatically descending visor
~ автоматической системы управления *av.* automatic flight control system control panel, AFCS control panel
~, аппаратный *com.* patchboard
~, аэродинамического тормоза *av.* air-brake panel
~ бедствия distress panel
~, вершинный apical panel
~, взлетно-посадочный split flap
~ воздушного тормоза *av.* air-brake flap
~, воздушный air flap/screen
~ вспомогательных систем utility panel
~, выдвижной *av.* extension/extending split flap
~ гермошлем, смотровой helmet visor
~ двустворчатых моллюсков *zool.* escutcheon
~ для направления недомолота tail-board
~ для отвода ботвы vine tender

~ для отвода корневищ trash tender
~ зародыша *bot.* scuttelum
~ интерцептора spoiler panel
~ коленной чашечки *zool.* scutum
~ колеса шасси *av.* wheel flap
~, контрольный, ~ контрольных приборов *el.* instrument/ *av.* monitoring panel
~ Крюгера, носовой Krueger flap
~, маленький *zool.* scutella
~, носовой *av.* leading-edge flap
~, отражательный target door
~ переднеспинки *zool.* pronotal disk
~, пикирующий *av.* dive/diving brake
~, пламезащитный glare shield
~, плечевой shoulder plate
~, подопочный foundry bottom plate
~, посадочный lading (split) flap
~, предохранительный face shield
~ приборной доски instrument sub-panel
~, приборный instrument dashboard
~, противослепляющий auto-dazzling flap
~, противосолнечный sunblind
~ радиста radio operator's panel
~ регулирования вырабатывания топлива fuel management panel
~ регулировки ширины валка *agr.* wind-rowing curtain
~ руля, ~ рулевой поверхности tab
~, рылцевой/ростральный rostrum
~ с надписью escutcheon plate
~ с номинальными данными, фирменный rating plate
~, сварочный face shield, welding hand screen
~ сигнализации warning/caution/indicator panel
~, сложный *bot.* compound corymb
~, спинной *zool.* dorsal shield
~, тормозной *av.* brake/deceleration flap; air-/speed-brake flap
~ управления control panel, controller
~ управления системой пеленгации selective identification feature control panel, CIF control panel
~ управления топливной системой fuel-management panel
~ управления триммерами trim-tab control panel
~ управления электросистемой electrical management panel
~, фирменный nameplate
~, формовочный moulder screen
~, цветочный *bot.* corymb
~, центральный centre panel
~ черепахи *zool.* mail
~ черепахи, нижний plastron
~ черепахи, роговой horny scute
~ шлема *av.* helmet visor
щитолистник *bot.* pennywort *(Hydrocotyle)*
щитолистный *bot.* scutifolious, peltate-leaved
щитолопастный *bot.* peltatilobate
щитомордник *zool.* moccassin *(Ancistrodon)*
щитонадгортанный *anat.* thyroepiglottic
щитоносец 1. *hist.* armo(u)r-/shield-bearer; 2. *zool.* pichiciago *(Chlamyphorus)*
щитоноска *zool.* tortoise beetle *(Cassida)*
щитоносный *zool.* shield-bearing, scutiferous; *bot.* corymbiferous; peltigerous
щитообразный shield-shaped; *zool.* scutiform, scutate, clypeate; *bot.* peltate, peltiform
щитораздельный *bot.* peltatipartite
щиторассеченный *bot.* peltatisected
щиточерпаловидный *anat.* thyroarytenoid
щитошейный *anat.* thyrocervical
щит-углубитель *naut.* otterboard
щука *zool.* pike *(Esox)*
~, морская ling *(Molva vulgaris)*
щуковые *n. pl. zool.* pikes, piskerels, muskellunges *(Esocidae)*
щукообразные *n. pl. zool.* *(Esociformes)*
щуп 1. *tech.* feeler (gauge), clearance/feel/thickness gauge, probe; sounding borer; dip rod/stick; 2. *tech.* (test) prod; tracer; 3. *mil.* mine probe
~, донный bottom sampler
~, масляный oil gauge rod, oil dipstick
~, минный *mil.* mine-probing rod
~, опробовательный sampling probe, sounding(s) borer
~, ручной handspike
щупальце *zool.* feeler, tentacle; antenna
~, краевое marginal tentacle
~ мшанки lophophore
~, нижнегубное labium tentacle
щупальцевидный, щупальцеобразный feeler-shaped, tentaculiform, tentaculate, tentaculoid
щупальцевый tentacled, tentacular, having feelers/antennae; palpate
щупальцежвалы *pl. zool.* chelicerae
щупальценосец *zool.* lophophore
щупальценосный tentacled, tentacular
щупать feel, touch, palpate; probe, sound
~ глазами scan
~ пульс *med.* feel the pulse
щупик feeler, palpus, tentacle
~, максиллярный maxillary palpus/tentacle
щупло *text.* feeler
~ вязальной машины knitting feeler
~ ткацкого навоя beam feeler
~, фотоэлементное photoelectric feeler
щуплость зерна blind-seed disease
щуплый 1. *agr.* underdeveloped *(of seeds);* 2. undersized, puny, sickly, weak; frail
щур *zool.* grosbeak *(Pinicola)*
щуренок *zool.* young pike, pikelet *(Esox)*
щурка *zool.* bee eater *(Merops)*
щучка *bot.* (tufted) hair grass *(Deschampsia)*

Э, э

эбельманит *geol.* ebelmanite
эбен ebony, blackwood
эбеновый ebony *(attr.);* ebeneous; *n. pl., bot.* ebony family *(Ebenaceae)*
~вое дерево Bombay ebony *(Diospyros ebenum)*
эбонит ebonite, hard rubber, vulcanite
~, микропористый cellular/expanded/microporous ebonite
эбуллиоскоп ebullioscope
эбуллиоскопия ebullioscopy
эбурнеация кости *med.* eburnation
эв *abbr.* **электрон-вольт** electron volt
эвабисит *geol.* euabyssite
эвакогоспиталь *abbr.* **эвакуационный госпиталь** evacuation hospital
эвакопункт *abbr.* **эвакуационный пункт** casualty-clearing station; evacuation centre
эвакуатор *med.* evacuator
~ большой кривизны large curvature evacuator
эвакуация evacuation; withdrawal
~, аварийная emergency evacuation
~ больных и раненых по воздуху aeromedical evacuation
эвакуированный evacuated; *n.* evacuee
~ по воздуху aeromedical evacuee
эвакуировать evacuate; exhaust, empty
эвакуируемый *n.* evacuee
эвальвация evaluation; calculation
эванемохория eunanemochory
эвансит *miner.* evansite
эвапоратор vaporizer, evaporator
эвапорация evaporation; desiccation
эваporograф recording evaporimeter, evaporograph, atmograph
эвапорометр evaporimeter, evaporometer
эвапоспория *bot.* euapospory
эввитрен, эввитрин *geol.* amorphous/structureless vitrain
эвгедральност automorphism, idiomorphism
эвгедральный euhedral, amorphous, idiomorphic
эвгемеризм, эвгемеристика euhemerism
эвгенол *chem.* eugenol
эвгеосинклиналь eugeosyncline
эвглена *biol.* Euglena
эвгранитовый *geol.* eugranitic
эвдемонизм eud(a)emonia, eudemonism
эвдиалит *miner.* eudialyte
эвдиометр *chem.* eudiometer, absorption tube
эвдиометрический *chem.* eudiometric(al)
эвдиофит eudiophite
эвекция *astr.* evection, lunar inequality
эвенк, эвенкийский Evenk
эвенкийка Evenk (woman)
эвентрация *med.* eventration, protrusion from the belly/abdomen
эверглейд, эверглэйд *geogr.* everglade
эверния *bot.* staghorn evernia *(Evernia prunastri)*
эверс *naut.* ever, ewer
эвикция *law* eviction, dispossession
эвисцерация *surg.* evisceration, exenteration
эвкалипт *bot.* eucalyptus, gum-tree *(Eucalyptus)*
эвкалиптовый eucalyptic, eucalyptus *(attr.)*
эвклаз *miner.* euclase
Эвклид Euclid
эвклидов *math.* Euclidean, Euclidian
эвклимакс true/climatic climax, euclimax
эвколлоид *chem.* true colloid, eucolloid
эвкриптит *geol.* eucryptite
~, низкотемпературный pseudoeucryptite
эвкристаллический *geol.* eucrystalline
эвксерофильный *bot.* euxerophilous
эвлизит *geol.* eulysite
эвмезофильный *bot.* eumesophilous
ЭВМ *abbr.* **электронная вычислительная машина** (electronic) computer
~, большая mainframe
~, бортовая onboard computer
~, бытовая home computer
~ второго поколения second-generation computer
~, высокопроизводительная high-performance computer
~, главная host computer
~, дорожная laptop computer
~ коллективного использования multiuser computer
~, комбинированная mixed computer (system), combined A/D computer system
~, макетная breadboard computer
~ младшей модели low-end computer
~, монопольно используемая single-user computer
~ на ИС IC [integrated-circuit] computer
~, настольная desktop computer
~, обучающая training computer
~, параллельная concurrent computer
~ переднего края front-end computer
~, персональная PC, personal computer
~ поочередного действия serial computer
~ последовательного действия sequential computer
~, потоковая data-flow computer
~ производительностью несколько миллионов команд в секунду multi-MIPS-computer
~ промышленного использования industrial computer
~ пятого поколения FGC, fifth generation computer
~ работающая в реальном масштабе времени real-time computer
~, резервная standby computer
~ с байтовой адресацией byte-addressable computer
~ с жесткой программой fixed-program(me) computer
~ с кассетными ЗУ cassette-based computer

~ с мандатной адресацией capability machine
~ с полным набором команд CISC computer, complete instruction set computer
~ с разделением времени time-sharing computer
~ с сокращенным набором команд RISC, reduced instruction set computer
~ с хранимой программой stored-program computer
~, сверхбыстродействующая everfaster computer
~, связная communications computer
~ сети networked computer
~ со страничной организацией памяти paged machine
~, совместимая compatible computer
~, специализированная dedicated computer
~ средней производительности mid-range computer
~ старшей модели high-end computer
~, универсальная mainframe
~, управляемая потоком команд control-flow computer
~, управляющая control computer
~, учрежденческая office computer
~, целевая target computer
~, цифровая digital computer
~, шлюзовая gateway computer *(in a network)*
эвмагматический segregated
Эвмениды *pl., myth.* Eumenides, Erinyes, Furies
эвольвента *math.* evolvent, involute (curve)
~ кривой метацентров metacentric involute
эвольвентный involute
эвольвентомер involute measuring machine
эволюта *math.* evolute
~, метацентрическая curve of (transverse) metacentres, metacentric evolute
эволюционирование evolvement, development, evolution
эволюционистский evolutionist(ic)
эволюционный evolutionary, evolutional
~ процесс process of evolution
эволюция 1. evolution; development; 2. *mil.* manoeuvre
~, морфологическая morphogenesis, morphogeny
~, необратимая irreversible evolution
~, неопределенная indeterminate evolution
~, скачкообразная saltatory evolution
~, случайная accidental evolution
эвонимус *bot.* euonymus, evonymus *(Euonymus)*
эвпатрид *hist.* eupatrid, member of the Greek aristocracy
эврибатный eurybathic, eurybenthic
эвривалент *bot.* eurytopic plant, euryvalent
эвригалинный *biol.* euryhaline
Эвридика *myth.* Eurydice
эврика (h)eureka
эвримезофильный eurymesophilous
эвриптер(ид), эвриптерус *paleont.* eurypterid *(Eurypterus)*
эвристика *math.* heuristics
эвристический *math.* heuristic(al)
эврит *geol.* eurite, felsite
~, порфироидный felsite-porphyry
эвритерм eurythermic plant, eurytherm
эвритермический, эвритермный eurythermic, eurythermal
эвритмия eur(h)ythmy
эвритовый *geol.* euritic, felsitic
эвритопный *bot.* eurytopic
~ное растение eurytopic plant
эврициста euricyst
эвстазия *geol.* eustasy
~, ледниковая glacial eustasy
эвстела eustele
эвтектика 1. *foundry* eutectic; eutectic solution/alloy; 2. eutectic (brine) ice
~, тройная ternary eutectic; three-component eutectic
эвтеника euthenics
Эвтерпа *myth.* Euterpe, Muse of music
эвтрофия *ecol.* eutrophy
эвфемизм euphemism
эвфиллоидный *bot.* euphylloid
эвфонизм euphonism
эвфонический euphonic(al)
эвфония, эвфоника euphony
эвфорбин *pharm.* euphorbium
эвфория euphoria, euphory
Эвфросина *myth.* Euphrosyne
эвфуистический euphuistic(al)
эвхаристия *rel.* Euharist
эвхроит *miner.* euchroite
эвшель zaffer, zaffre
эгалитаризм egalitarianism
эгалитарист egalitarian
эгейит *geol.* egueiite
эгида aegis, patronage; protection
эгилозавр *paleont.* aigialosaur *(Aigialosaurus)*
эгилопс *bot.* goat grass *(Aegilops)*
эгирин *geol.* aegirine, aegirite, acmite
эгитогнатный *zool.* aegithognathous
эгоизм egoism, egotism, selfishuess
эгоистический egoistic(al), egotistic(al), selfish
эготизм egotism; egomania
эгофония *med.* egophony, tragophony
эгоцентризм egocentrcity, egocentrism
эгоцентрический, эгоцентричный egocentric(al)
эгрессивно-щелкающий *phon.* egressive-click
эгрет(ка) aigrette, egret plume
эгутер dandy roll
эдафический edaphic
эдафон soil organism, edaphon
эдафофит edaphophyte
Эдда *lit.* Edda
~, Младшая Younger/Prose Edda
~, Старшая Older/Poetic Edda

эдельвейс *bot.* edelweiss *(Leontopodium)*
Эдем *bibl.* Eden; *fig.* paradise
эдема *med.* (o)edema; dropsy
эдемский Edenic; paradisiac(al)
эджевортия, эджеворция *bot.* paper bush *(Edgeworthia)*
эджер *met.* edger (mill)
~ встроенного типа integral-type edger
эдикт *hist.* edict
эдикула *arch.* aedicula
эдил *hist.* (a)edile; important official
Эдип *myth.* Oedipus
эдипов Oedipean
эдификатор *bot.* edificator plant, environment-forming plant
эдогоний *bot.* oedogonium *(Oedogonium)*
эдомитянин *hist., bibl.* Edomite
эдс *abbr.* **электродвижущая сила** electromotive force, emf
~, обратная back/counter emf
эжектор ejector, jet pump
~, (водо)струйный (water-)jet ejector
~, двухступенчатый two-stage ejector
~, мусорный garbage ejector
~, нисконапорный low-head ejector
~, отсасывающий exhaust ejector
~, пусковой starting ejector
~, трюмный bilge ejector
эжектор-смеситель mixing ejector
эзельгофт head hoop, cap
эзерин *chem.* physostigmine, eserine
~, салициловокислый physostigmine/eserine salicylate
~, сернокислый physostigmine/eserine sulphate
эзерфоль *print.* roll-leaf
эзкуррит ezcurrite
Эзоп *myth.* Aesop
эзопов, эзоповский Aesopian
эзотерический esoteric(al)
эзофагит *med.* esophagitis
~, пептический/язвенный peptic/ulcerous esophagitis
эзофагодуоденостомия *med.* esophagoduodenostomy, Schlatter's operation
эзофагопластика *med.* esphagoplasty
эзофагостомиаз *med.* esophagostomiasis, nodular disease
эзофагэктазия *med.* esophagectasia, esophagectasis
эзофория *med.* esophoria, convergent squint
эйгенвектор *math.* eigenvector
эйгенвеличина *math.* eigenvalue
эйгенфункция *math.* eigenfunction
эйдология eidolo(lo)gy
эйдофор *tel.* light-modulator large-screen TV system
эйективный ejective
эйекция *phon.* ejection
эйзотропный eisotropic, centripetal
эйконал *opt.* eikonal
~, точечный point eikonal
эйконометр eiconometer, eikonometer
эйлеровый *math.* Eulerian
эйнштейний *chem.* einsteinium, Es
эйнштейновский Einsteinian
эйфория *med.* euphoria, euphory
эйххорния *bot.* water hyacinth *(Eichhornia)*
эквадорец, эквадорский Ecuadorian
эквадорка Ecuadorian (woman)
экватор *geogr.* equator, the Line
~ глазного яблока *med.* equator of the eyeball
~, земной earth's/terrestrial equator
~, истинный true equator
~, магнитный magnetic/dip equator, aclinic line
~, небесный celestial/equinoctial equator, equinoctial line
~, средний mean equator
~ хрусталика *med.* equator of the crystalline lens
экваториал *astr.* equatorial (telescope)
экваториально-апертурный plan-aperturate
эквиангармонический *math.* equianharmonic
эквиангармоничность *math.* equianharmonicity
эквиаффинный *math.* equiaffine
~ное преобразование equiaffinity
эквивалент equivalent; arificial device, dummy
~, азотный nitrogen equivalent
~ антенны artificial/dummy aerial
~, белковый *agr.* protein equivalent
~ в долларах *econ.* dollar equivalent
~ в киловатт-часах kilowatt-hour equivalent
~, денежный money equivalent
~ диода/лампы dummy diode/tube
~ затухания attenuation equivalent
~, зерновой grain equivalent
~, кормовой *agr.* fodder/food equivalent
~ нагрузки *el., rad.* dummy load
~ передачи *com.* transmission equivalent
~ рентгена, физический, фэр roentgenequivalentphysical, rep
~, световой lumen equivalent
~ тепла/теплоты, механический mechanical equivalent of heat, Joule's equivalent, conversion constant
~, тепловой thermal/heat equivalent
~, тормозной stopping equivalent
~, тротиловый trinitrotoluene/TNT equivalent
~, угольный coal equivalent
~, целочисленный *comp.* integral equivalent
эквивалентность 1. equivalence, equivalency; equivalence relation; 2. match
~ массы и энергии relativistic equivalence of mass and energy, mass-energy equivalence
эквивалентный equivalent
~, морфологически homoeomorphous
эквидистанта equidistant curve
эквидистантность equidistance
эквилибрирование tightrope walking
эквилибрировать (walk on a) tightrope
эквилибрист tightrope walker, equilibrist
эквилибристика tightrope walking

эквимолекулярный, эквимолярный equimolecular, equimolar
эквиполлентность equipollence, equipollency
эквипотенциал *geod.* equipotential (surface)
эквипотенциаль equipotential line
эквипроективный *math.* equiprojective
эквиустойчивость equistability
эквифациальный equifacial
эквихордальный equichordal
экдемит ecdemite, heliophyllite
экер *surv.* right-angle mirror, cross-staff, optical square; *see also* **еккер**
экзальтационный exaltative
экзальтация exaltation; elation
~ молекулярной рефракции optical exaltation
экзамен exam(ination); test
~, выпускной finals, schoolleaving examination
~, держать/сдавать take an examination
~ на аттестат зрелости final school examination, scoolleaving examination
~, повторный repeat/resit examination
~, приемный entrance/admission examination
~ы, государственные finals, graduation examinations/exams
~ы, переводные promotion exam(ination)s, end-of-year exam(ination)s
экзаменатор 1. *educ.* examiner; tester; 2. scoring device, grader
~ уровней *geod.* level checker, bubble-level tester
экзаменационный examination *(attr.)*, test *(attr.)*
~ билет examination (question)paper
~ная комиссия examining board
~ная сессия examination period
экзаминуемый, экзаминующийся examinee
экзаминатор examiner; tester
экзантема exanthem(a), exanthesis
экзарх exarch
экзархат exarchate
эксегеза *bibl.* exegesis, Biblical interpretation
эксегетика exegetics
эксекватура exequatur
экзекутор 1. administrator; 2. one who administers corporal punishment
экзекуция 1. *law* execution; 2. punitive expedition
экзема *med.* eczema, tetter, skin eruption
~, гипертрофическая lichenified eczema
~, детская infantile eczema
~, застойная stasis eczema
~, корочковая eczema crustosum
~, краевая eczema marginatum, tinea cruris
~, мокнущая moist/exudative/weeping eczema
~ пекарей baker's eczema
~ сгибательных поверхностей суставов flexural eczema
~, себорейная seborrheic eczema/dermatitis
~, солнечная solar eczema, eczema solare
~, сухая dry eczema
~, чешуйчатая scaly eczema, eczema squamosum
экзематический afflicted with eczema
экзематозный eczematous, tetterous
экземовидный eczematoid
экземпляр 1. copy; 2. example, specimen
~, опытный text example, prototype
~, основной master copy
~, пробный specimen
~, серийный production example
~, сигнальный press proof
~ах, печатать в трех/четырех type in triplicate/quadruplicate
экзентерация *surg.* evisceration, exenteration
экзенцефалический exencephalic, exencephalous
экзерсис exercise, practice
экзерциргауз *obs.* drill room/hall
экзерциция 1. *mil.* drill; 2. exercise, practice
экзина *bot.* exine, extine, exosporium
~, бородавчатая verrucate exine
~, бугорчатая tubercular exine
~, многослойная stratified exine
~, перегородчатая murate exine
~, полосатая striated exine
~, шиповатая echinate/spine exine
~, ямчатая pitted/scrobiculate exine
~, ячеистая alveolar exine
экзистенциальный existentialist(ic)
экзоантиген exoantigen, ectoantigen
экзогамета *biol.* exogamete
экзогамный exogamic, exogamous
экзогенный exogenous, exogenic, exogenetic
~ная спора ectospore
экзогидатогенный exohydatogenous
экзогормон *bioch.* exohormone
экзодерма *bot.* exoderm(is)
экзокардиальный exocardiac, exocardial
экзокардия *med.* exocardia, ectocardia
экзокарп(ий) *bot.* exocarp, epicarp
экзоплазма *biol.* exoplasm, ectoplasm
экзопневмопексия *med.* exopneumo(no)pexy
экзополигенный exopolygenous, exopolygenic, exopolygene
экзосвязь intermolecular bond
экзосепта exoseptum
экзоскелет *zool.* exoskeleton
экзосмос exosmosis, exosmose
экзоспора *bot.* exospore
экзоспорий *bot.* exosporium
экзоспоровый exosporous, exosporal
экзостоз *med.* exostosis
~, зубной dental exostosis
~, костно-суставной exostosis bursata
~, костно-хрящевой osteocartilaginous exostosis
~, множественный multiple exostosis
~, подногтевой subungual exostosis
~, хрящевой exostosis cartilaginea
экзосфера exosphere, region of escape
экзот exotic (organism/plant)

экзотека exotheca
экзотермический, экзотермичный *chem.* exothermic, exothermal
экзотермичность exothermicity, exothermic nature
экзотизм, экзотика exotic character/object
экзотичность exoticism
экзотоксин *bioch.* exotoxin, ectotoxin
экзотропия *med.* external squint, exotropia
экзофермент exoenzyme
экзофтальм *med.* exophthalmus, exophthalmos, ocular proptosis, exorbitism, ophthalmocele
~, злокачественный malignant exophthalmos
~, прогрессирующий progressive/advancing exophthalmos
~, пулсирующий pulsating exophthalmos
экзофтальмометр exophthalmometer
~, зеркальный mirror exophthalmometer
экзохорда *bot.* pearlbush *(Exochorda)*
экзоэнергетический, экзоэнергический exoe(ne)rgic, exoenergetic
экзоэнзим exoenzime
экзэнцефалический exencephalic, exencephalous
экивок 1. equivoque, quibble, ambiguity, evasion; 2. subtlety, intricacy
экипаж I carriage
экипаж II 1. *naut.* company; *av.* crew; 2. *rail.* locomotive underframe
~, боевой combat crew
~ в полном составе full crew
~, гражданский civil aircrew
~ из одного человека one-man crew
~ корабля ship's company
~ космического корабля spacecrew, spacecraft crew
~ космонавтов astronaut crew
~, летный flying crew, aircrew
~, минимальный skeleton crew
~, основной aircrew on duty, flight/operating crew
~, резервный stand-by crew
~, сменяющий relief crew
~, флотский naval barracks/depot, receiving station/unit, training centre
экипаж-дублер stand-by crew
экипажная *n.* carriage shed/works
экипировка 1. equipping/fitting out; 2. outfit, equipment, equipage
эккер *surv.* optical square, right-angle mirror, cross-staff
~, двухзеркальный double optical square
~, призменный prism square
экклезиаст *bibl.* Ecclesiastes; Ecclesiasticus
экклезия *hist.* ecclesia, popular assembly
экклиметр clinometer, elevation meter
эклампсия *med.* eclampsia, eclampsy
~ без судорог crampless eclampsia
~ беременных eclampsia of pregnancy
~, детская infantile eclampsia
~, послеродовая puerperal eclampsia
эклектика, эклектицизм, эклектичость eclecticism
эклер *cul.* éclair; cake with cream filling
эклиметр elevation meter, clinometer
эклиптика *astr.* ecliptic
~, истинная/средняя true/mean ecliptic
эклога *lit.* eclogue, short pastoral poem
эковид *bot., zool.* ecospecies
экозит ecosite, ecoparasite
экологический ecological
~ баланс ecological balance/equilibrium
~ вид ecospecies
~ ряд ecological sere
экология ecology, bionomics, ethology
~ животных/растений animal/plant ecology
~ (растительных) сообществ synecology
эконом 1. housekeeper, steward; 2. economist
экономайзер *tech.* economizer
~, водяной (feed-water) economizer
~, воздушный air heater
~, змеевиковидный continuous economizer
~, кипящий steaming economizer
~, кислородный oxygen-demand regulator, oxygen economizer
~, низкотемпературный low-level economizer
~, ребристый extended-surface economizer
экономгеография *abbr.* **экономическая география** economic geography
эконометрика, эконометрия econometrics
экономизм *hist.* economism
экономика 1. economics; 2. economy, economic structure
~, атомистическая atomistic economy
~, бартерная nonmonetary/barter economy
~ в условиях дефицита ресурсов shortage economy
~, внутренняя domestic economy
~, военная war/defence/military economy
~ городов urban economics
~, денежная monetary economy
~, дефицитная economy of scarcity
~, застойная stagnant economy
~ землепользования land economics
~, изолированная closed economy
~ имущественного неравенства economics of inequality
~, инженерная engineering economics
~ интенсивных культур intensive crop economics
~, контролируемая controlled/directed economy, command (directed) economy
~, контролируемая государством statism
~, мировая international/global economics, world economy
~ мирового хозяйства world economy
~, монокультурная *agr.* single-crop economy
~, нездоровая ailing economy
~ неравновесия disequilibrium economics
~, нерыночная controlled/directed economy
~, несельскохозяйственная nonfarm

economy
~, **плановая** planned economics/economy
~, **плюралистическая** pluralistic economy
~ **потребления** consumer economics
~ **предприятий** business economics
~, **прикладная** applied economics
~ **промышленности** industrial economics
~, **развивающаяся** expanding economy
~ **развивающихся стран** economics of underdevelopment; economics of developing countries
~, **рыночная** market economy
~, **рыночно-управляемая** market-directed economy
~ **с избытком рабочей силы** labo(u)r-surplus economy
~ **свободного предпринимательства** free enterprise economy
~ **свободной конкуренции** laissez-faire economy
~ **сельского хозяйства** agraian/agricultural/farm/rural economics, rural economy
~, **смешанная** mixed economy/society
~, **стабильная/устойчивая** stationary economy, steady-state economy
~ **управления** managerial economics
~ **централизованно планируемая** centrally planned economy
~, **частнопредпринемательская** private enterprise economy
~ **ядерной энергетики** nuclear power economics
~**ки, демилитаризация** civilian conversion
~**ки, дополняющие** *pl.* complementary economies

экономист-калькулятор economic analyst
экономист-плановик economic planner
экономист-статистик statistician
экономить economize; spare; save; cut down expenses
~ **на кассовых остатках** economize on cash balances

экономически самостоятельный selfsufficient
экономический economic(al)
~ **застой** economic stagnation
~ **и социальный совет (ООН)** Economic and Social Council, ECOSOC
~**кая комиссия ООН для стран Азии и Дальнего Востока, ЭКАДВ** United Nations Economic Commission for Asia and the Far East, ECAFE
~**кая Комиссия ООН для стран Европы, ЭКЕ** United Nations Economic Commission for Europe, ECE
~**кая комиссия ООН для стран Латинской Америки, ЭКЛА** United Nations Economic Commission for Latin America, ECLA
~**кая комиссия ООН для стран Среднего Востока, ЭКСВ** United Nations Economic Commission for the Middle East, ECME
~**кие науки** economics
~**кое отставание** economic lag

экономичность economy; *math.* economic operation; efficiency
экономичный economical, efficient; economy-type
экономия 1. economy; 2. thrift; saving
~ **в стоимости производства** saving in cost
~ **мощности** saving of power/energy
~ **на издержках** cost saving
~, **политическая** political economy/economics
~ **топлива** fuel economy/saving
~**мии, режим** economy drive/effort, antiwaste program(me)

экономка housekeeper
экономполитика *abbr.* **экономическая политика** economic policy
экосез *mus.* écossaise, Scottish dance (tune)
экосистема ecosystem, ecological system
экосфера physiological atmosphrere, ecosphere
экотип *ecol.* ecotype, habitat type
экразер *surg.* écraseur
экран 1. *phys., tech.* screen; shield; blind; barrier; baffle; visor; 2. *met.* water wall; 3. *print.* copyholder; 4. *arch.* reredorse, reredos(se)
~, **акустический** acoustic(al)/sound baffle, acoustic shield/screen, sound panel
~, **безбликовый** nonglare screen
~ **без послесвечения** nonpersistent screen
~, **биологический (защитный)** biological shield
~, **борный** *nucl.* boron shield
~, **водо-железный** water-iron shield
~ **вулканической пыли** veil of volcanic dust
~, **гладкотрубный** bare-tube water wall
~, **дифрагирующий** diffracting screen
~ **для защиты от (быстрых) нейтронов** (fast-)neutron shield
~ **для защиты от гамма-излучения** gamma-shield
~ **для защиты от прямого облучения** direct exposure shield
~ **для защиты от рентгеновского излучения** X-ray screen
~, **заземленный** earth screen; *Am.* ground shield
~, **звуковой (громкоговорителя)** baffle
~, **защитный** protective screen/shield
~ **из уносимых материалов, теплозащитный** ablative heat shield
~ **индикатора** display screen
~ **кинескопа** picture-tube face, kinescope face; phosphor (screen)
~, **кольцевой** ring(-shaped) shield
~, **люминесцентный** luminescent/fluorescent/phosphor screen
~, **металлизированный** metal-backed screen
~ **монитора** monitor screen
~, **настенный** water wall
~, **огнестойкий** fire-resistant shield
~, **отражательный** baffle plate

~, отражающий reflecting screen; reflection barrier
~, ошипованный refractory-faced water wall, studded-tube water wall
~ первичной (биологической) защиты *nucl.* primary shield
~, пластмассовый plastic shield
~ плотины screen
~, плотный membrane(-type) water wall
~, поглощающий absorbing barrier
~, проекционный projection screen
~, прозрачный transparent screen
~, просвечивающий fluorescent screen
~, противовзрывной blast shield
~ радиационной защиты radiation shield
~, радиолокационный radar screen/scope
~, разделенный на области split screen
~, рассеивающий diffusing/diffuser screen
~, рентгеновский X-ray screen
~ с большим временем послесвечения long-persistence phosphor (screen)
~ с послесвечением afterglow/persistent screen
~ с темновой записью dark-trace screen
~ с темным изображением dark-trace screen
~ с тубусом beaded screen
~ с тяжелым наполнителем heavy-aggregate shield
~, свинцовый lead shield/screen
~, сенсорный touch(-sensitive) screen
~, слоистый laminated shield
~, сплошный bulk shield
~, тепловой thermal/heat screen; *nucl.* thermal shield
~, теплозащитный heat shield
~, теплоизоляционный/теплоизолирующий heat-shield blanket
~, тонкий защитный curtain
~, топочный (котла) water wall
~, усиливающий intensifying screen
~, утепленный refractory-faced water wall
~, фазосдвигающий phase-changing screen
~ фотоувеличителя enlarger easel
~ ЭЛТ, алюминизированный aluminium-backed (phosphor) screen

экранизация 1. filming, screening; 2. screen version
экранирование shielding, screening, screenage
~ зажигания ignition shielding
~, электромагнитное electromagnetic screening
~ электропроводки electrical-wiring shielding
~, электростатическое electrostatic screening
экранировка = экранирование
~ кабеля cable braid(ing)
экранизирующий shielding, screening
~щая сетка screen grid
экраноплан *naut.* ram-wing craft, winged vehicle, ground effect craft/machine
эксгаустер, эксгаустор exhauster, suction/exhaust fan
~, мокрый wet suction fan
эксгибиционизм exhibitionism
эксгумация exhumation, disinterment
эксгумировать exhume, disinter
эксикатор exsiccator, desiccator, drier
эксикаты *pl., bot.* exsiccatae
экситон *phys.* exciton, bound hole-electron pair
экситрон *tech.* excitron
экскаватор excavator, excavating machine, earth-moving machine, power shovel
~, вскрышной stripping shovel
~, грайферный clam-shell excavator
~, гусеничный crawler-mounted power shovel; crawler-mounted wheel-type trenching machine
~, колесный rubber-tyre-mounted power shovel
~, лопато-колесный bucketwheel excavator
~, многоковшовый (цепной) ladder/wheel-type trenching machine
~, многочерпаковый multibucket excavator, land dredger
~ на автомобильном шасси truckmounted power shovel
~ на гусеничном ходу caterpillar excavator, crawler(-mounted) power shovel
~, одноковшовый single-bucket excavator; (back)hoe, back/pull shovel
~, одноковшовый типа прямая лопата push/forward shovel
~, отвальный stripping shovel
~, роторный отвальный wheel-excavator
~, траншейный trencher, ditcher, trenching machine, trench excavator
~, универсальный convertible shovel
~, шагающий walking excavator
экскаватор-драглайн dragline excavator
экскаватор-канавокопатель ditch-and-trench excavator
экскаваторщик excavator operator
экскавация excavation
~, гидравлическая hydraulicking, hydraulic excavation
экскoнъюгант exconjugant
экскременты *pl.* excrement, f(a)eces
~, ископаемые fossilized excrements, coprolites
экскреторный excretory
экскреты *pl., physiol.* excreted material, excreta
экскреция *physiol.* excretion; egestion
экскурс excursus, digression
экскурсант excursionist, sightseer
экскурсия 1. excursion, trip, sightseeing tour; outing; 2. group of sightseers, excursion party
экскурсовод(ка) tour guide
экслибрис bookplate, ex-libris
экспандер (automatic volume) expander
экспансивность expansiveness, expansivity, effusiveness
экспансионистский expansionist
~кая политика policy of expansion
экспатриант expatriate
экспатриация expatriation; exile

экспедирование dispatching, expediting
экспедитор 1. forwarder, forwarding agent, dispatcher, shipping clerk; 2. *obs.* head clerk
экспедиция 1. expedition; party; 2. dispatching, forwarding; shipping; 3. dispatch room/office; forwarding service
~, китобойная whaling expedition
~, океанская overseas expedition
~, поисково-спасательная search-and-rescue expedition
~, спасательная rescue party; salvage expedition
эксперимент experiment, test, trial
~ в подкритическом режиме subcritical experiment
~, выборочный *stat.* sampling experiment
~, дополнительный/параллельный replicate experiment
~, единичный *stat.* single experiment
~, контрольный control (experiment), check experiment
~, модельный model experiment/test
~, натурный full-scale experiment
~, факторный *stat.* factorial experiment
~, факторный с двумя переменными two-factor factorial experiment
экспериментальный experimental; tentative, developmental; pilot *(attr.)*
экспериментатор experimentor, experimenter, experimentalist
экспериментирование experimentation, experimenting
эксперт expert; specialist; examiner
~, судебно-медицинский forensic medical examiner
~, судовой *naut.* surveyor
эксперт-заседатель *law* assessor
~, морской nautical assessor
экспертиза 1. expert examination/opinion/statement; expertise, examination; 2. commission of experts
~, врачебно-летная flight-surgeon's appraisal/judgement
~, врачебно-трудовая medical labo(u)r examination
~ временной нетрудоспособности temporary invalidity examination
~, судебно-психиатрическая forensic psychiatric examination
экспертный expert *(attr.)*
~ная комиссия commission of experts
экспираторный expiratory, expirational
~ное ударение *phon.* expiratory stress
экспирация exhalation, exhaling, expiration; breathing out
эксплантант explant
эксплантация *med.* explantation
экспликация *cart.* legend
эксплозивный *phon.* plosive
эксплуатант *av.* operator
эксплуататор exploiter
эксплуатационный *n. av.* operator; *pl.* the operating staff
эксплуатационный *adj.* operational, operating; working
~ная надежность/пригодность operational reliability
~ное время operating/running time
~ные расходы operating expenses
~ные испытания field/service trial(s)
эксплуатация 1. operation, (operational) use, usage; utilization; 2. operating, running, working, exploitation; maintenance
~, безаварийная accident-free operation
~, длительная/продолжительная long-term usage/service/operation
~, круглогодичная year-round operation
~, опытная tentative/trial operation
~, промышленная commercial operation
~, техническая operation
~ции, нормальные условия normal operational/working conditions
эксплуатирование exploiting; using; working
эксплуатировать 1. use, operate, run; work; 2. attend to, maintain, service
~ жестоко sweat
эксплуатируемый operated; worked; used
~ частной фирмой privately operated
экспо world fair, exhibition
экспозиметр exposure meter
экспозитура *polit.* foreign agent
экспозиция 1. *cin., opt.* exposure; *lit.* exposition; 2. arrangement, display; layout; exhibition
~, повторная double exposure
~, смертельная *med.* lethal exposure
~ фотопленки photographic film exposure
экспонат exhibit
~, павильонный stand exhibit
экспонент 1. exhibitor; 2. *math.* exponent, index
экспонента *math.* exponential (curve), exponent
~, затухающая damped exponential
экспоненциал *math.* exponential
экспоненциальный exponential
~, логарифмически log-exponential
~ ряд *math.* power-law series, exponential series
экспонирование 1. *phot.* exposure; 2. exhibiting, showing, displaying
~, двойное/многократное double/multiple exposure
экспонировать 1. *phot.* expose; 2. exhibit
экспонометр *phot.* exposure meter
~, визуальный/оптический visual exposure meter
~, рентгеновский X-ray intensitometer
экспонометрия exposure metering
экспорт *econ.* export(s)
~, бросовой dumping
~, невидимый invisible exports
~ продовольствия food export
экспортирование exportation

экспортный export *(attr.)*
экспресс express (train/bus/ship)
экспресс-анализ proximate analysis
экспрессивность expressivity, expressiveness
экспрессионизм *art* expressionism
экспресс-метод proximate method, quick test
экспрессность fastness, rapidity
экспрессом (by) special delivery
экспромпт 1. extemporisation; improvisation, extempore; 2. *lit.* extempore poem; *mus.* impromtu
экспромптом off-hand, impromptu; suddenly
экспроприация dispossessing, expropriation
экссолюция exsolution
экссудат *med.* exudate, exudation
~, геморрагический h(a)emorrhagic exudate
~, гнойный purulent exudate
~, раневой wound exudate
экссудативный, экссудационный exudative
экссудация exudation
экстаз ecstasy, enthusiasm, exaltation, enravishment
~, впадать/приходить в fall/go into ecstasies
экстатический, экстатичный ecstatic
экстензометр extensometer
~, зеркальный mirror extensometer
экстенсивность extensiveness, extensity
экстерн 1. *educ.* external student; 2. *obs.* externe, unpaid hospital doctor
экстернат *educ.* external studies
экстеро(ре)цептор *physiol.* exteroceptor
экстерриториальность exterritoriality, extraterritoriality
экстерриториальный exterritorial, extraterritorial
экстерьер *zool.* appearance, form, build *(of an animal)*
экстина *bot.* extine, exine, outer spore membrane
экстинкция extinction
~, межзвездная interstellar extinction
экстирпатор 1. *surg.* extirpator; 2. *agr.* weeder
экстирпация extirpation; *agr.* weed extermination
экстраагглютинин extra-agglutinin
экстраверзия *med.* extroversion, extraversion, eversion, extrophy, turning inside out
экстрагент extractant, extracting agent
экстрагирование, экстракция extracting, extraction
~ с помощью летучих растворителей volatile solvent extraction
экстрагировать extract
~, совместно co-extract
экстрагируемость extractability
экстрагляциальный extraglacial
экстражирный *print.* extrabold
экстракласс special class/category
экстракорпоральный extracorpor(e)al
экстракратонный *geol.* extracratonal
экстраксилярный extraxylar, extraxylemic
экстракт 1. extract, essence; *med.* extract(um); 2. *obs.* abstract, resumé, summary, précis
~, аллергический allerg(en)ic extract
~, бесклеточный cell-free extract
~, водный aqueous extract
~ для наполнения *leath.* filling extract
~, дубильный tanning extract
~, жидкий fluid/liquid extract
~ задней доли гипофиза pituitary extract
~ коры надпочечника adrenal cortex extract
~ паращитовидной железы parathyroid extract, parathormone
~, печеночный liver extract
~, растительный vegetable extract
~, спирто-водный hydroalcoholic extract
~, хвойный spruce extract
~, эфирный ethereous extract
экстрактивный extractable; extractive
экстрактный, экстрактовый for production of extracts
экстрактор *tech.* extractor, extraction apparatus/unit
~, вакуумный vacuum extractor
~ типа смеситель-осадитель mix-and-settle extractor
экстракционный extractive; extraction *(attr.)*
экстракция extraction
~ в индикаторных количествах tracer-scale extraction
~, вторичная re-extraction, strip(ping)
~ жидкостная/двухжидкостная liquid-liquid extraction
~, избирательная/селективная selective extraction
~ катаракты *med.* flap extraction
~, периодическая batch extraction
~, расщепительная split extraction
~, совместная co-extraction
экстралингвистика metalinguistics, external linguistics
экстраординарный extraordinary, unusual, uncommon
экстрапар vapour
экстрапирамид(аль)ный extrapyramidal
экстраплевральный extrapleural
экстраполяционный extrapolating; extrapolation *(attr.)*
экстраполяция extrapolation
~, линейная arithmetic/linear extrapolation
экстрапочта express mail, special delivery
экстрасенс psychic
экстрасистола *med.* extrasystole
~, атрио-вентрикулярная, ~, узловая A-V extrasystole, functional/atrioventricular extrasystole
~, возвратная return extrasystole
~, желудочковая infranodal/ventricular extrasystole
~ напряжения/покоя extrasystole of tension/rest
~, предсердная atrial/auricular extrasystole
экстрасистолия *med.* extrasystole
экстратемпоральный timeless, nontemporal
экстраток extra current

экстратонкий *print.* extra-condensed
экстремаль *math.* extremal, extremum
~, ломаная broken(-line) extremal
~, особая singular extremal
~ с углубленными точками extremal with corners
экстремальный extremal, extremum *(attr.)*
экстремум 1. *math.* extremum, bend/extreme point; 2. extreme
~, локальный local/relative extremum
~ упорядоченного ряда turning point of an ordered series
~ы погоды weather extremes
экстренность special character; urgency
экстренный 1. urgent, emergency *(attr.);* 2. extra, extraordinary; special
~ное издание special edition
экструдат extruded article, extrudate
экструдер *tech.* extruding machine, extrusion press, extruder
~, двухчервячный twin-screw extruder
экструдирование extrusion
экструзивный extrusive
экструзионный extruding, extrusion *(attr.)*
экструзия extrusion (moulding)
~, перемежающаяся intermittent extrusion
~ с закалкой quench-tank extrusion
экссудат exudate, exudation
эксфолиативный exfoliative
эксцедент *econ.* surplus *(of insurance)*
эксцементоз *med.* excementosis
эксцентрик *tech.* eccentric, (disk) cam
~ заднего/обратного хода reverse/rear/backward eccentric
~, регулируемый variable/adjustable eccentric
эксцентриковый pertaining to a cam/eccentric
~ пресс cam/eccentric press
эксцентриситет, эксцентрицитет eccentricity
~ вектора тяги *rock.* thrust misalignment
~ вилки шасси *av.* axle offset
~ лимба eccentricity of circle
~ нагрузки eccentricity of load
эксцентрический 1. odd, peculiar; queer; 2. *tech.* off-centre, eccentric
эксцентричность 1. aberration, queerness, eccentricity; 2. *tech.* centering error, misalignment, eccentricity
эксцеребрация *med.* excerebration, eccephalosis
эксцесс *math.* kurtosis, excess
~ плотности распределения kurtosis of a frequency curve
~ом меньше нормального, с platykurtic
эксцизия *surg.* excision, cutting out/off
эксципул *bot.* exciple, excipulum
эктазия *med.* ectasia, ectasis
эктима *med.* ecthyma
эктимовидный *med.* ecthymatiform, ecthymiform
эктипография braille (writing), ectypography
эктлипсис *phon.* ecthlipsis, suppression of a sound
эктоаск *bot.* ectoascus
эктобласт *biol.* ectoblast
эктогенез *biol.* ectogenesis
эктогония ectogony, ectogeny
эктодерма ectoderm
эктодермальный ectodermal, ectodermic
эктомезенхима ectomesenchyma
эктомия *surg.* ectomy
эктопаразит *zool.* ectoparasite, external parasite
~, животный epizoon
~, растительный epiphyte
эктопия *med.* ectopia, ectopy
~ зрачка ectopia pupillae
~ хрусталика ectopia lentis
~ яичка ectopia testis
эктоплазматический ectoplasmic, ectoplasmatic
эктотрофный ectotrophic
~ная микориза ectotrophic mycor(r)hiza
эктофит *bot.* ectophyte
эктофлоедный *bot.* ectophloedic
эктофлойный *bot.* ectophloic
эктоэнзим *bioch.* ectoenzyme
эктропион *med.* ectropion, ectropium
эктэкзина ectexine, exoexine
экуменизм ecumenicity, ecumenicalism, ecumenicism
экуменический ecumenical
экхимоз *med.* ecchymosis
~ Н-образный H-shaped ecchymosis
экхондрома *med.* ecchondroma
экю ecu (European currency unit)
элайосом(а) *bot.* elaiosome
эласмобранхий *zool.* elasmobranch fish
эласмотерий *paleont.* elasmothere *(Elasmothereium)*
эластаза *bioch.* elast(in)ase
эластик stretch thread/yarn/fabric
эластичность elasticity, resilience, flexibility
~ на отскок rebound resilience
~, перекрестная cross-elasticity
~ спроса demand elasticity
~, точечная point elasticity
эластичный elastic, flexible, resilient, pliant
эластовискозиметр elastovisco(si)meter
эластоз *med.* elastosis
~, солнечный actinic elastosis
эластокинетика elastokinetics
эластомер *chem.* elastomer
~, фторсодержащий fluorine-containing elastomer
эластометр elastometer
эластотонометр elastotonometer
~, глазной ophtalmic elastotonometer
элатерофор elaterophore
элатив *gram.* elative
элеат, элеатский Eleatic
элевайф *zool.* walleyed herring, alewife *(Alosa pseudoharengus)*
элеватор 1. elevator, hoist, lift; *agr.* grain elevator; 2. *med.* molar forceps
~ ГБ depth-charge hoist

~, **ковшовой** bucket elevator
~, **хирургический** periosteum elevator
~, **шнековый** auger-type elevator, screw elevator
элевон *av.* elevon, ailavator
~, **зависающий** drooped elevon
элевсинский *hist.* Eleusinian
Элевтерий *myth.* Eleutherian
элевтерозой *zool.* eleutherozoan
элегаз SF_6 gas
элегиачеcкий, элегический *lit., mus.* elegiac(al); epicedian; epicedial
элегия *lit., mus.* elegy
элеидин *bioch.* eleidin, keratohyalin
элейский Eleatic
элейтероморф *geol.* eleutheromorph, idioblast
электор elector, voter
электоральный electoral, voting
электризация 1. electrification, electrization; 2. *med.* electrotherapy
~ **бумаги** static-charge accumulation by paper
~ **влиянием** electrization by induction
~ **трением** triboelectrification
электризовать electrize, electrify
электрик electrician
электрификатор electrical engineer
электрификация electrification
~ **сельского хозяйства** use/introduction of electric power in farming
электрифицированный provided with electric power; joined to the grid
~ **полностью** all-electric
электрифицировать adapt to (the use of) electric power, provide with electric power
электрический electric(al)
~, **отрицательно** electronegative
~, **положительно** electropositive
электричество 1. electricity; 2. electric light/ power
~, **грозовое** thunderstorm electricity
~, **связанное** latent electricity
~, **статическое отрицательное** resinous/ negative electricity
~, **статистическое положительное** vitreous/ positive electricity
~ **трения** frictional electricity
электричка *colloq.* electric train/railway
электроагрегат generating set
~, **бортовой** *av.* auxiliary power unit
электроакустика electroacoustics
электроанализ electroanalysis
электроаналогия electrical analogy
электроанемометр hot-wire anemometer
электроаппарат electrical device/apparatus
электроаппаратура electric(al) apparatus/equipment
электроаэрозоль electrically charged/ionized aerosol
электробензомер electric fuel gauge
электроблокировка electrical locking/interlock
электробормашина electric drill(ing machine)
электробритва electric(-powered) shaver
электробур electric drill
электробытовой household electrical
~**вые приборы** electrical appliances
электровагонетка electric truck
~, **грузовая** electric industrial truck
электровалентность electrovalency, electrovalence
электровзвешивание electric weighing
электровзрыватель electric fuse
электровибратор electric vibrator, electrovibrator
электровлагомер electric hygrometer
электроводокачка electric pump
электровоз electric locomotive
~, **аккумуляторный** storage-battery locomotive
~, **контактный** trolley-wire locomotive
~, **маневровый** switching/shunting electric locomotive
электровозбудимость electroexcitability
электроволнограф electric wave recorder
электровоспламенитель *min., nav.* electric fuse, primer, (electrical) igniter; flash/electric fuse
электровосстановление *chem.* electroreduction, electrolytic reduction
электровыделение *met.* electrowinning
электровыключатель electrical cut-off-switch, electrical interrupter
электровязкостный electroviscous
электрогазодинамика electrogasdynamics
электрогемостаз *med.* electroh(a)emostasis
электрогенератор electric generator
электрогидравлика electrohydraulics
электрогидроагрегат electrohydraulic unit
электрогидродинамика electrohydrodynamics
электрогидроклапан solenoid-operated hydraulic valve
электрогидротерапия electrohydrotherapy
электрогипноз electrohypnosis
электрогитара electric guitar
электроглянцеватель hot glazer
электрографический electrographic
~ **способ печати** xerographic printing, xerographic print process
электрод electrode
~, **активный** active/discharge electrode; *med.* therapeutic/exciting/localizing electrode
~, **вживленный** *med.* implanted electrode
~, **возбуждающий** launching coil
~ **вспомогательного разряда** keep-active electrode, ignitor
~, **вторично-эмиттирующий** dynode, secondary-emission electrode
~, **газовый** gas cell
~, **голый**, ~ **без покрытия** bare/uncoated electrode
~, **графитированный/графитовый** graphitized/graphite electrode
~ **для наплавки** *weld.* building-up electrode
~ **зажигания** starting electrode
~, **замедляющий** decelerating electrode

~, игольчатый *med.* point electrode
~, изогнутый *weld.* offset electrode
~, индифферентный *med.* indifferent/dispersing electrode
~, каломельный calomel electrode
~, коленообразный *weld.* cranked electrode
~, ленточный *weld.* strip electrode
~, неплавящийся nonconsumable electrode
~, неполяризующийся non-polarizable electrode
~, непрерывный continuous electrode
~, обратимый reversible electrode
~, отводящий exploring electrode
~, отклоняющий deflecting electrode
~, отрицательный negative electrode, cathode
~, очищающий sweeping electrode
~, плавящийся consumable electrode
~, пластинчатый plate electrode
~, покрытый (бором) (boron-)coated/-covered electrode
~, положительный positive electrode, anode
~, послеускоряющий post-deflection/-accelerating electrode
~ присоса suction electrode
~, промежуточный intermediate electrode
~, ртутный (капельный) (dropping) mercury electrode
~ с толстым покрытием, ~, толстообмазанный heavy-coated electrode
~ с тонким покрытием, ~, тонкообмазанный thin-coated electrode
~, самоспекающийся self-baking/-sintering electrode
~, сварочный welding electrode/rod
~, сварочный армированный sheathed electrode
~, сварочный для дуговой сварки arc-welding electrode
~, собирающий collecting electrode, collector
~ стока drain (electrode)
~, топливный fuel/hydrogen electrode
~, тормозящий retarding/decelerating electrode
~, трубчатый *weld.* tubular/core(d) electrode
~, угольный carbon electrode
~ умножителя, промежуточный multiplied electrode, dynode
~, управляющий control electrode, intensity-controlling electrode
~, ускоряющий accelerating/acceleration electrode
~, фокусирующий focus(sing) electrode
~, формирующий пучок beam-forming electrode
~, экранирующий shield(ing) electrode
~ы, плоскопараллельные parallel-plate electrodes

электродвигатель (electric) motor, electromotor
~, асинхронный induction/asynchronous motor
~, асинхронный с фазным ротором slip-ring induction motor, *Am.* phase-wound(-rotor) motor
~, безколлекторный/бесщеточный commutatorless/brushless motor
~, брызгозащищенный splashproof motor
~, вентильный thyratron motor
~, взрывозащищенный explosion-/flame-proof motor
~, включать start a motor, switch/turn on a motor
~, водозащищенный deluge-proof/-tight motor
~, встраиваемый built-in motor
~, газонепроницаемый gas-proof/-tight motor
~, герметический canned motor
~, гистерезисный hysteresis motor
~, гребной propulsion/propelling/main motor
~, закрытый sealed-in motor, enclosed(-type) motor
~, искрозащищенный sparkproof motor
~, колекторный commutator motor
~, компаундный compound(-wound) motor
~, конденсаторный capacitor(-start) motor
~, короткозамкнутый (squirrel-)cage motor
~, многоскоростной multispeed motor
~, многофазный polyphase motor
~, неявнополюсный nonsalient-pole motor, round-rotor motor
~ открытого исполнения, ~ открытый open motor
~ переменного/постоянного тока a.c./d.c. motor
~, погружной submersible motor
~, приводной drive/actuating motor
~, реактивный reaction motor
~, редукторный gearmotor
~, репульсионный repulsion induction motor
~ с беличьей клеткой squirrel-cage motor
~ с жесткой характеристикой flat-response motor, constant-voltage motor
~ с мягкой характеристикой drooping-response motor
~ с печатной обмоткой printed circuit motor
~ с последовательным возбуждением series(-wound) motor
~ с регулируемой скоростью вращения adjustable/variable speed motor
~ с самовентиляцией self-ventilated motor
~, самосинхронизирующийся/сельсинный selsyn/autosynchronous motor
~, сериесный series(-wound) motor
~ сигнального привода electric signal motor
~, следящий servo motor
~ триммера trim-tab motor
~, тяговый traction motor
~, фланцевый flange(-mounted) motor
~, шаговой/шаговый step(ping) motor, step-by-step motor
~, шунтовый shunt(-wound) motor
~, явнополюсный salient-pole motor

электродвигательный, электродвижущий electromotive

электродепарафинизация electrical dewaxing

электродвижение electric propulsion
электродетонатор electic(al) detonator/fuse, electric (blasting) cap
электродиагностика electrodiagnostics
электрод-заземитель earth/ground electrode
электродиализ electrodialysis
эелктродиспергирование, электродисперсия electrodispersion
электро(до)держатель electrode holder/carrier
~, дуговой welding rod holder
~ клещевого типа tong-type electrode holder
электродоение electric milking
электродойка, электродоилка electric milker
электрод-оправка *weld.* electrode bar
электродрель electric drill
электродренаж electroosmotic dewatering
электродуга electric arc
электроемкость *phys.* electrocapacity
электрозакалка electrohardering, electric hardening, electrotempering
электрозаклепка electric rivet, plug lap joint
~, несквозная plug weld
~, сквозная rivet weld
электрозамок, электрозамычка electric lock
электрозапал electric primer/fuse/igniter, electric blasting cap
электрозапись electrical recording
электрозащелка electric lock
электрозондирование *geol.* electrical sounding
электроизгородь *agr.* electric fence
электроизоляционный dielectric, electrical-insulation
электроизоляция electrical insulation
электроинструмент electro-driven tools, power/electric tool(s)
электроинтегратор differential analyser
электрокамин electrical fireplace
электрокар(а) battery-powered/-operated/-driven truck, storage-battery truck; on-site factory truck
электрокар-тягач battery tractor
электрокардиограмма *med.* electrocardiogram, ECG
электрокардиограф electrocardiograph
~, многоканальный multichannel electrocardiograph
~ с фотозаписью photographic recording electrocardiograph
~ с чернильной записью pen electrocardiograph
электрокардиография electrocardiography
~ плода f(o)etal electrocardiography
~, подводная submerged electrocardiography
электрокардиостимулятор pacemaker
электрокаротаж electric(al) logging/coring
электрокартон electric-grade cardboard
электрокаустика, электрокаутер electrocautery
электрокимография electrokymography
электрокинетика electrokinetics
электроклапан electric-actuated valve, solenoid-operated valve
электроклепка electric riveting
электроклиширование electrotyping
электрокоагуляция (surgial) diathermy, electrothermic coagulation, electrocoagulation, diathermocoagulation
электрокожный electrodermal
электроконвектор electric convection heater
электрокорунд manufactured/synthetic corundum, alundum, electrolytically produced corundum
~, нормальный aluminium-oxide abrasive
электрокотел electric boiler
электрокофемолка electric coffee grinder
электрола electric record player
электролампа electric lamp/bulb
электролебедка electric hoist/winch
электролечение electrotherapy, electrotherapeutics, electrical treatment
электролиз electrolysis
электролизер electrolytic bath/cell, electrolyzer
электролизный electrolytic(al)
электролизуемый electrolyzable
электролиния electric power line
электролит electrolyte
~, амфотерный amphoteric electrolyte
~, обработанный foul/spent electrolyte
~, тройной ternary electrolyte
~, щелочной alcaline electrolyte
электроловушка для насекомых electrocutor trap
электролужение electrotinning
электромагнетизм electromagnetics, electromagnetism
электромагнит electromagnet
~, броневой pot electromagnet, shell-type electromagnet
~, включать energize an electromagnet
~, включающий closing electromagnet
~ вращения (шагового искателя) horizontal magnet
~, грузоподъемный cargo-lifting electromagnet
~, исполнительный driving/control electromagnet
~, крановый/подъемный lifting electromagnet
~ подъема vertical magnet
~, расцепляющий release magnet
~ с откидным якорем clapper-type electromagnet
~, сцепляющий clutch (electro)magnet
~, тяговый tractive/traction electromagnet
электромашина electric machine
электромашинка electric typewriter
электромашиностроение electric machine industry/engineering
электрометр electrometer
~, капиллярный capillary/Lippmann electrometer
~, ламповый (vacuum) tube electrometer
~, многокамерный multiple electrometer
~, самопишущий electrograph

~, струнный string/filament electrometer
~, язычковый (вибрационный) vibrating-reed electrometer
электромеханик electrician
электромеханика electromechanics
электромиография electromyography
~, стимуляционная stimulant electromyography
электромобиль battery-driven vehicle, electric automobile, electric motor car
электромолотилка electrically powered thresher
электромонтаж electrical installation, assembly of electrical equipment
электромонтажник construction/erection electrician
электромонтер linesman, wireman, electrician
электромотор electric motor; *see also* **электродвигатель**
элекромоторный electromotive
электромотыга electrical tiller
электромясорубка electric meat grinder/mincer
электрон I electron, *Am.* Dowmetal, aluminium-magnesium alloy; gold-silver alloy
электрон II *phys.* electron, negatron
~, антисвязывающий antibonding electron
~ бета-распада beta-decay electron, disintegration electron
~, блуждающий vagabonding/roaming electron
~ большой/малой скорости high/low-velocity electron
~, быстродвижущийся fast-moving electron
~, быстрый fast(-moving) electron, high-speed electro
~ валентной зоны valence band electron
~, влетающий incoming/oncoming/ingoing electron
~, внешний outer(-shell)/peripheral electron
~, внеядерный extranuclear electron
~, внутрений inner(-shell) electron
~, вторичный secondary/multiplier electron
~, вращающийся spinning electron
~, выбитый knock-on electron, knocked-out electron
~, вылетающий outgoing/outcoming electron
~, вырожденный degenerate electron
~ высокой энергии high-energy electron
~, жесткий hard electron
~, захваченный trapped electron
~, избыточный excess electron
~, испущенный emitted electron
~, комптоновский Compton/recoil electron
~ космического излучения cosmic-ray electron
~, медленный slow(-moving) electron, low-speed electron
~, мягкий soft electron
~, наиболее глубокой innermost electron
~, налетающий incident electron
~, недостающий missing electron
~, несвязанный/неспаренный unbound/unpaired electron
~, оболочечный shell electron
~, одиночный lone electron
~ Оже Auger electron
~, орбитальный orbital/planetary electron
~ отдачи recoil/Compton electron
~, падающий incident electron
~, почти свободный quasi-free electron
~ проводимости conduction/conductivity electron
~ распада decay/disintegration electron
~, рассеивающий scattering electron
~, рассеянный scattered/stray electron
~, свободный free/unbound/mobile electron
~, связанный bound electron
~ связи bonding electron
~, туннелирующий tunneling electron
~, убегающий runaway electron
~, фоновый background electron
~ы куперовской пары Cooper pair electrons
~ы, моноэнергетические monoenergetic/monokinetic electrons
~ы, связанные bound/fixed electrons
электронагрев electric heating
электронагреватель electric heater
электронасос electric pump, motor-driven pump
электрон-вольт electron-volt, eV
электроник specialist in electronics
электроника electronics
~, авиационная avionics, aviation electronics
~, аэрокосмическая aerospace electronics
~, геофизическая geoscience electronics
~, интегральная integrated electronics
~, квантовая quantum electronics
~, космическая astrionics, space electronics
~, криогенная cryogenic electronics, low-temperature electronics
~, молекулярная moletronics, molecular electronics
~, полупроводниковая semiconductor electronics
~, промышленная industrial electronics
~ СВЧ microwave electronics
~, силовая power electronics
~, техническая engineering electronics
~, тонкопленочная (thin-)film electronics
~, ядерная nuclear electronics
электронно-возбужденный/наведенный electron-induced
электронно-вычислительный/-счетный electronic-computer
электронно-лучевой electron-ray
электронный electronic
~, полностью all-electronic
~ная вычислительная машина electronic computer
~ная лампа electron/vacuum tube
электронограмма electron diffraction pattern/picture
электронограф electron diffractometer, electron-diffraction apparatus/camera/instrument, cathodo-

graph
электронож electrotome, electric scalpel/knife
электроноотталкивающий electron-repelling
электронщик specialist in electronics
электрообеспыливание electric dust suppression
электрообеспыливатель electric dust suppressor/catcher
электрообогрев electric heating
электрооборудование electrical equipment
~, генерирующее electric generating equipment
электроожог electric burn
электроокраска electrocoating, electrostatic painting
электроосаждение electrodeposition, electrolytic precipitation
электроосветитель illuminator, lightman
электроосмос electro(endo)osmosis
электроотопление electric heating
электроотрицательность electronegativity
электроочистка electrical cleaning/precipitation
электропайка electric soldering/brazing
электропастух *agr.* electric fence
электропахота electric ploughing/plowing
электропаяльник electric soldering iron
электропередача electricity/power transmission
электроперенос *phys.* electrotransport
электропечь electric furnace
~, индукционная тигельная crucible/coreless induction furnace
~, плазменно-дуговая plasma furnace
электропила electric/power saw
электропирозапал, электропиропатрон electric squib
электропильщик electric/power saw operator
электропитание (electric) power supply
электропитающий power-supply
электроплавка electric melting
электроплетизмография electroplethysmography
электроплит(к)а electric stove/cooker, hot plate
электроплуг electric plough/plow
электропневмоклапан solenoid-operated pneumatic valve
электроподогреватель electric heater
электроподушка hot pad
электроподъемник electric elevator
электропоезд electric train
~, пригородный commuter/suburban electric train
электропокрытие electroplating, electrodeposition; electroplate, electrodeposit
электрополировка electrolytic polishing, electropolishing
электроположительность electropositivity
электрополотер electric floor polisher
электропредохранитель *el.* safety fuse
электроприбор electrical appliance
~, бытовой household electrical appliance
электропривод electric/motor drive
~, вентильный rectifier drive
~, ионный gas-tube drive
~, регулируемый controlled-velocity electric drive
~, следящий servo drive
~, шаговый step motor drive
электропровод electrical wire; electricity/power cable
электропроводимость 1. electric(al) conduction; 2. conductance; conductivity
электропроводка (electrical) wiring
~, внутренняя indoor wiring
~, вторичная control/secondary wiring
~ на клицах cleat(ed) wiring
~ на роликах reel-insulator wiring
~, открытая open/surface wiring
~, скрытая flush/buried wiring
~, щитовая panel wiring
электропроводность conductance; conductivity
~ горных пород rock electric conductivity
~, дырочная hole conduction
~, примесная impurity conduction
~, собственная intrinsic conduction
электропрогрев electric heating
электропроигрыватель 1. turntable; 2. electric record player, electric gramophone
электропроизводительность electrical efficiency
электропромышленность electrical industry
электропунктура, электропункция *med.* galvanopuncture, electropuncture
электропылесос (electric) vacuum cleaner
электроразведка geophysical prospecting by electric equipment
электрораздражение *med.* electrostimulation; electrical irritation
электроразъем electrical connector
электрорасщепление electrodisintegration
электрорезка *met.* electrical cutting, electric arc cutting
электрореограф electrorheograph
электрорентгенография X-ray xerography, xeroradiography
электросварка electric(-arc) welding
~, контактная resistance/contact welding
электросварщик electric welder
электросверло electric drill
электросвечение electroluminescence
электросвязь electric communication, electrocommunication, telecommunication
электросепарация electrostatic separation
электросеть electricity supply network
~, городская urban network
~, многофазная polyphase network
~, петлевая loop network
~ с изолированной нейтралью isolated neutral system
электросила electric power
электросковорода electric frying pan
электроскоп electroscope
~, дозиметрический, ~ дозиметр radioscope

~, интегрирующий condensing electroscope, capacitor-type electroscope
~ с бузинными шариками pith-ball electroscope
~ с зеркальным отсчетом reflection reading electroscope
~ с золотыми листками Lauritzen electroscope
~ с кварцевой нитью quartz-fibre electroscope
~, экранированный shielded electroscope
электрослесарь electrical fitter
электроснабжение (electric-)power supply
~ городов municipal power supply
~ от сети общего пользования commercial power supply
электросократимость *med.* electrocontractility
электросон *med.* electric sleep
электроспуск electric(al) trigger
электросталь electric (furnace) steel
электростанция electric power plant/station, generating station
~, атомная atomic/nuclear power plant, atomic energy plant
~, базисная base-load power station/plant
~ большой мощности high-power station
~, газомазутная gas-and-oil-burning power station/plant
~, гидроаккумулирующая hydroelectric pumped-storage power plant
~ на жидком топливе oil-burning power plant
~ на обычном топливе conventional-fuel power plant
~ на твердом топливе coal-burning power plant
~ на ядерном топливе nuclear-fuel power plant
~, пиковая peak(-load) power station
~, приливная tidal power plant
~, промышленная industrial/service power plant
~, резервная standby power plant
~ с магнитогидродинамическим генератором magnetohydrodynamic power plant
~, тепловая thermal power station, steam power plant, fuel-fired steam plant, steam (-electric) plant
~, теплофикационная heat-electric generating plant
~, термоядерная thermonuclear power plant
~, ядерная nuclear power/energy plant
электростатика electrostatics
электросторож electric fence
электрострижка *agr.* electrical shearing
электроструя electrojet
электросудорожный *med.* electroconvulsant
электросушилка *food* electric drier
электротабло electronic announcement board; electric scoreboard
электроталь electric monorail hoist, electric wire rope hoist
электротележка electric cart, battery-operated vehicle
электротензомесдоза strain-gauge dynamometer
электротерапия electrotherapeutics, electrotherapy
электротермический electrothermic, electrothermal
электротермия electrothermics, electrothermancy
электротехник electrician; electrical engineer
электротехника electrical engineering, electric technology, electrotechnics
электротипия *print.* electrotyping, electrotypy; electrotype
электротон(ус) *physiol.* electrotonus
электротравма *med.* electric trauma
электротрепан *surg.* electric trephine
электротропизм *biol.* electrotropism, galvanotropism, electrotaxis
электротяга electric traction
электроуправление electrical control
электроустановка electrical installation/plant/set
~, закрытая/открытая indoor/outdoor electrical installation
~, силовая electric(al) power plant/unit
электрофарфор electric-grade porcelain
электрофизика electrophysics
электрофильность electrophilicity
электрофильтр electrostatic/electric(al) precipitator, electric filter
~, многопольный multistage (electrostatic) precipitator
~, мокрый irrigated (electrostatic) precipitator
~, трубчатый pipe(-type) [tubular] electrostatic precipitator
электрофон electric gramophone; electrophone; audiosystem
электрофорез *med.* electrophoresis
~ в крахмальном геле starch gel electrophoresis
~, иммунный immunoelectrophoresis
~, лекарственный medicinal electrophoresis
электроформовка electroforming
электрофотография electro-photography, xerography
электрофотометр photoelectric/electronic photometer
электрохимия electrochemistry
электрохирургия electrosurgery
электроход diesel/-turbo-electric ship
электрохозяйство electrical equipment
электроцентраль electric power station
электрочайник electric kettle
электрочасы *pl.* electric clock
~, вторичные slave (electric) clock
электрочувствительный electro-sensible/-sensitive
электрошлифмашина electric grinder
электрошлифование electrolytic grinding
электрошнур electric cable/cord
электрошок electric shock, electroshock
электроштабелер electric stacker
электрощиток power panel, electrical distribution box
электроэндоосмос electro(endo)osmosis

электроэнергетика 1. electrical power engineering; 2. electric power; power industry
электроэнергия electric power/energy
~, произведенная в ватт-часах watt-hours generated power
электроэнцефалограмма *med.* electroencephalogram, EEG, electrocorticogram
электроядерный electronuclear
электрум electrum *(virgin metal)*
электуарий *pharm.* electuary
элемент 1. element, unit, component; member; 2. *el.* cell; 3. *math.* quantity; *geom.* part; 4. *comp.* entry, item; *see also* **элементы**
~, аккумуляторный accumulator cell, storage (-battery) cell
~ активной зоны *nucl.* core element
~, активный active element/component
~, анатомический anatomical/morphologic element
~ антенны, активный radiating element, directly excited element, active aerial
~ антенны, пассивный parasitically excited element, passive aerial
~, арифметический arithmetic element
~, армирующий *constr.* reinforcing element
~ баланса *econ.* balance sheet/item
~, бета-неустойчивый/устойчивый beta-labile/stable element
~, блокирующий blocking element
~, боковой side member
~, весовой balance unit
~ Вестона Weston standard cell
~ внезапности surprise element
~, воспринимающий detecting/sensing/sensitive element, sensing unit, sensor
~, входной/выходной input/output element
~ выборки sample unit
~, вычислительный computer/computing element
~, газовый gas cell
~, гальванический galvanic cell/element
~ главной диагонали определителя *math.* leading element in a determinant
~ гидрографии water feature
~, главный *math.* pivot element
~, готовый fabricated part
~ данных data item/element
~, двоичный *comp.* binary cell
~, делящийся fissible/fissionable element
~, детекторный/детектирующий detecting element
~, дискретный discrete element/component
~, дочерний *phys.* daughter element
~ жесткости stabilizing member
~ жидкости fluid element
~, жидкостный wet cell
~ задержки delay element/component
~ залегания пласта *geol.* bed position
~, замедляющий moderating element
~, запальный тепловыделяющий seed fuel element
~, записывающий recording element
~ запоминающего устройства storage/memory/store element
~, запоминающий storage cell
~, звукоприемный sound pick-up element
~ И AND element
~ игры *econ.* game element
~, избыточный redundant element
~, измерительный sensor, measuring element
~ изображения *telev.* picture/image element, image detail, elemental area
~ ИЛИ OR element
~, индуктивный чувствительный inductive sensing element
~, искусственный artificial element, man-made element
~, исполнительный actuator; final element
~, исходный *phys.* parent/original element
~, канонический *math.* regular function element
~ команды order/instruction/command element
~, консольный cantilever member
~, конструктивный/конструкции structural element/member
~, контактный *el.* contact element/electrode
~, коррозионный corrosion cell
~, легирующий alloying/doping element
~, линейный *math.* line/linear element
~, логический logic(al) component, logic element, gate
~, логический комбинационный combinational/decision/memoryless element, gate
~, ложный тепловыделяющий dummy fuel element
~ мажоритарной логики majority element
~, материнский *phys.* parent element
~, матричный *math.* element of a matrix, matrix element
~ матрицы array cell
~, местный local (galvanic) cell
~, меченый tagged element
~, микрологический/микроминиатюрный micrologic element
~ микросхемы integrated-circuit element, IC element
~, многоэлектродный multielectrode cell
~ множества set member
~ набора 1. *comp.* member; 2. *naut.* framing member
~, навесной *electr.* interconnection element
~, нагревательный heating/heater element
~, направляющий pivot element
~ НЕ NOT component
~, недиагональный off-diagonal element, nondiagonal element
~, неметаллический non-metal
~, необязательный optional member
~, нерегламентированный *econ.* unrestricted element
~, несущий carrier; *constr.* bearing member

~, нетипичный foreign/random element
~, нормальный standard cell
~, нулевой zero/null element, zero
~, обедненный depleted element
~, обогащенный enriched element
~, обратный *math.* inverse (element)
~ объема *math.* element/differential of volume
~, одноизотопный monoisotopic element
~, однородный homologue
~, опорный support member
~, опорный (отсчета) reference element
~, оптический релейный photoelectric/light relay, photorelay, photoswitch
~ орбиты orbital element
~, осколочный fission(-product) element
~, особый линейный *math.* singular line element
~ отдельности parting unit
~, отработанный тепловыделяющий spent fuel element
~, петрогенный/породообразующий *geol.* rockforming element
~, печатающий print member
~, печатный *comp.* printed component
~, пленочный *electr.* (thin-)film component
~ поверхности *math.* surface element
~, поглощающий *electr.* dissipative element
~ подвески spring unit; suspension component
~, подкрепляющий stiffener
~, полосковый *electr.* strip element
~, поперечный transverse/cross-member
~ последовательности *comp.* sequent
~ примесный impurity element
~, пробельный *print.* spacing material
~ программного прерывания *comp.* program interruption element
~, продольный *constr.* longitudinal member
~ производственной операции operation element
~, противоположный *math.* additive inverse (element)
~, процессорный processing element
~, пусковой *auto.* starting element
~, работающий в предельном режиме marginal component
~, работающий на изгиб/кручение *constr.* member in bending/torsion
~, работающий на растяжение/срез *constr.* member in tension/shear
~, работающий на сжатие compressional member, (compression) strut
~, радиоактивный radioelement
~, развертывающий *telev.* picture element, elemental area
~, разделительный separating/separation element
~, разрешающий *math.* pivot, pivot(al) element
~ разрешения resolution cell/element
~, рассеивающий scattering element, scatterer
~ растра/растровый *telev.* picture element, elemental area
~, регистрирующий recording element
~, регламентированный restricted element
~, редкоземельный rare earth element
~, резервный redundant element
~ риска risk element
~, родоначальный parent element
~ с жидким электролитом wet cell
~, самородный native element
~ связи *rad.* coupling element
~, связующий *chem.* binder
~, сегнетоэлектрический ferroelectric element
~, силовой *mach.* load-bearing element; *constr.* primary (structural) member, main/primary load-carrying/-bearing member
~ системы управления control element
~, следовой trace element
~ следящей системы servo element, servosystem component; *comp.* follower
~, случайный stochastic element
~, сменный фильтрующий filtering cartridge, filter insert
~, соединительный *comp.* connective
~, солнечный sollar cell
~ сопряжения с абонентом custom adapting element
~, сопряженный *math.* transform
~, сопутствующий accompanying element
~, стабилизирующий stabilizing element/member, stabilizer
~, стандартный *comp.* library cell
~, строковый *comp.* string item
~, струйный *auto.* fluidic element
~, суммирующий *comp.* adding element
~, сухой dry cell
~, схемный circuit element
~, тепловыделяющий/топливный *nucl.* fuel cell/element
~, тензометрический чувствительный straingauge sensing element
~, термочувствительный temperature-sensitive element
~, термоэлектрический thermocouple, thermojunction
~, управляемый *auto.* controlled element
~, управляющий *auto.* control element
~, усиливающий *constr.* reinforcing member; stiffener
~, фильтрующий strainer/filter(ing) element
~, фильтрующий войлочный felt filter element
~ формата format item
~, форменный *geol.* tissue element
~, фотовольтаический/фотогальванический photovoltaic cell
~, фотоэлектрический photocell, photoelectric cell
~, функциональный functional element; *comp.* functor

~, **химический с большим/малым атомным номером** high-/low-Z element
~, **холостой тепловыделяющий** *nucl.* inert/dummy fuel element
~ **цепи** circuit element
~, **чувствительный** sensing/sensitive element, sensing unit; sensor; pickup
~, **шариковый** *comp.* ball-point pen
~, **электронагревательный/электрообогревательный** electric-heating element
элементарно-геометрический elementary-geometric
элементарность elementary character/quality/nature; elementariness; simplicity
элементарный elemental, fundamental; *math.* elementary; partial *(of fractions)*; 2. simple; rudimentary
элемент-индикатор indicator element
элементный *el.* cell *(attr.)*
элемент-примесь admixture element
элементы *pl.* elements; details, features; *see also* **элемент**
~, **бесконечно удаленные** *math.* points at infinity
~ **букв, выступающие/свисающие** *print.* ascenders/descenders
~ **гидрографии открытого моря** off shore features
~ **книги, титульные** front matter
~ **корабля, маневренные** nautical characteristics/properties, manoeuvring properties
~, **несобственные** *math.* points at infinity, ideal points
~, **печатающие** *print.* printing areas
~, **редкоземельные** rare earths
~ **рельефа** terrain details, ground features
~, **титульные (книги)** *print.* front matter
эленхост *logic* elench(us), refutation; sophism
элеодендрон *bot.* false olive *(Eleodendron)*
элеокарпус *bot.* eleocarpus *(Eleocarpus)*
элеолит *miner.* el(a)eolite, nepheline
элерон *av.* aileron
~, **зависающий** drooped/drooping aileron
~, **концевой** wingtip aileron
~, **отклонять** apply/move/deflect the aileron
~, **поднятый** up aileron
~ **правой/левой консоли** starboard/port aileron
~, **щелевой** slot-lip aileron
~**ы, секционные** divided ailerons
элерон-закрылок flap-aileron
элерон-щиток split-edge aileron
элефантиаз(ис) *med.* elephantiasis, barbados leg, pachydermia
Элизей, Элизий, Элизиум *myth.* Elysium, the abode of the blessed dead; *obs.* paradise
элизия *ling.* elision; omission/suppression of a vowel/syllable
эликсир *med.* elixir
~, **ароматный** aromatic/simple elixir
~, **грудной** antitussive, terpin hydrate (and codein) elixir
~, **зубной** dentifrice water
элиминация elimination
~, **клеточная** cellular elimination
элинвар *met.* elinvar (alloy)
элионный electron-ionic
элита 1. *agr.* best specimens; seed stock; 2. elite, chosen/select part
элитный highest-quality, select, choice *(attr.)*
элитра *zool.* elytron, elytrum
Эллада *hist.* Hellas, Greece
элладотерий *paleont.* helladotherium
эллин Helene, ancient Greek
эллинг 1. *naut.* covered slip/ways, ship-repair shed, slipway; 2. *av.* airship/dirigible hangar; dock, shed
~, **заводить в** dock
~, **судоподъемный** slipway, railway slip, ship railway
~, **судостроительный** covered-in berth
эллинизм 1. *hist.* Hellenistic period; 2. Hellenism
эллинистический *hist.* Hellenic, Hellenistic
эллинский Hellenic, ancient Greek
эллинство Hellenism
эллипс 1. *geom.* ellipse; 2. *gram.* ellipsis
~ **Гоманна** Hohmann ellipse
~, **напряженный** ellipse of stresses
~ **ошибок** ellipse of errors, probability ellipse
~ **перехода** transfer ellipse
~ **рассеивания/рассеяния** dispersion/concentration ellipse, ellipse of dispersion/concentration; shot group
эллипсный elliptic(al), ellipse *(attr.)*
эллипсовидный, эллипсообразный ellipsoid(al), ellipsoid-shaped
эллипсоид ellipsoid, spheroid
~ **вращения** spheroid, ellipsoid/spheroid of revolution
~, **вытянутий/сплющенный** prolate/oblate ellipsoid
~ **деформации/напряжения** strain ellipsoid
~, **земной** terrestrial ellipsoid
~, **земной принятый** accepted earth ellipsoid
~ **инерции** inertia/Poinsot ellipsoid
~ **отнесения** ellipsoid of reference
~ **скоростей** velocity ellipsoid
эллиптичность ellipticity
~ **волновода** waveguide ellipticity, axial ratio
элодея *bot.* waterweed, elodea *(Elodea)*
элонгация elongation, stretch
элопс *zool.* giant herring *(Elops)*; chiro *(E. saurus)*
ЭЛТ *abbr.* **электроннолучевая трубка** cathode-ray tube, CRT
~, **знакопечатающая, ~ со знаковой индикацией** character-display CRT, shaped-beam tube
~, **индикаторная** radar display/indicator CRT
~, **осциллографическая** oscilloscope CRT
~ **прямого видения** electrical-to-visual storage tube, direct-view storage tube, display storage tube

~ с барьерной сеткой barrier-grid tube
~ с записью темной трассой dark-trace CRT
~ с полуускорением луча postacceleration CRT
~, телевизионная приемная (TV) picture tube, kinescope
~ типа графекон scan-conversion tube
эльзасец, эльзасский Alsatian
эльзаска Alsatian (woman)
эльсвортит *miner.* ellsworthite
эльф *myth.* elf
~, добрый kewpie
элюация, элюирование, элюция elution, elutriation
эмалевидный enamellike
эмалевый enamel(ed)
эмалепроволока enamel(ed) wire
эмалирование enamel(l)ing
эмалированный enamel(l)ed
эмалировка 1. enamel(l)ing; 2. enamel
эмаль 1. (vitreous) enamel, porcelain enamel; 2. (organic) enamel
~, глухая opaque (vitreous) enamel
~ горячей сушки, ~ печной сушки baking (organic) enamel
~, грунтовая ground/under-coat (vitreous) enamel
~, зубная/зубов *anat.* enamel
~ зубная, волнистая *anat.* curled enamel
~, масляная oleoresin (organic) enamel
~, покровная cover-coat (vitreous) enamel
~, прозрачная transparent (vitreous) enamel
~ холодной сушки air-drying (organic) enamel
эмальпровод enamel (insulated) wire
эманация 1. emanation; 2. *chem.* radon
~, вредная deleterious/noxious emanation
~ газа gas exhalation
~, металлоносная metalliferous emanation
~ радия radium emanation, radon-222, Rn
~ тория thorium emanation, thoran, radon-220, Tn
эманометр emanometer, emanation meter
эмаскуляция *med.* emasculation, castration
эмбарго embargo; prohibition, ban
~, накладывать (lay) embargo, place an embargo
~, снять take off an embargo
эмблема emblem; symbolic device/badge; *mil.* insignia
~, фирменная identifying plaque
эмблика *bot.* emblic, East Indian tree *(Phyllantus emblica)*
эмбол *med.* embolus, embolum
~, воздушный air embolus
~, жировой fat/oil embolus
~, клеточный cellular embolus
~, мягкий bland embolus
~, парадоксальный crossed/paradoxical embolus
~, седловидный saddle/pantaloon/riding/straddling embolus
эмболия *med.* embolism
~ брыжечных артерий mesenteric embolism
~, воздушная air embolism, aeroembolism
~ кровеносных сосудов h(a)ematogenous embolism
~ легких pulmonary embolism
~, лимфатическая lymph/lymphogenous embolism
~, множественная multiple/miliary embolism
~, несептическая bland/simple/nonseptic embolism
~, парадоксальная crossed/paradoxical embolism
~, пиемическая pyemic/infective embolism
~, послеродовая puerperal embolism
эмболалия *med.* embololalia, emboloph(r)asia
эмбриогенез *biol.* embryogenesis, embryogeny
эмбриогенетический, эмбриогенный embryogen(et)ic
эмбрион *biol.* embryo; *obs.* embryon
эмбриотега, эмбриотегий *bot.* embryotega
эмеритальный old-age-pension *(attr.)*
~ная касса old-age insurance saving bank
эмеритура *obs.* 1. old-age benefit/pension; 2. old-age insurance saving bank
эмигрант emigrant, emigré
эмиграция emigration; *coll.* emigrants, emigrés
~, массовая *econ.* exodus
эмилия *bot.* tassel flower *(Emilia)*
эминенция *eccl.* eminence
эмир emir, ameer, emeer
эмират emirate, office/jurisdiction/state of an emir
эмиссар emissary
эмиссионный emitting, emissive
~ банк bank of issue
эмиссия *phys.* emission, emittance; *econ.* emission of money
~, автоэлектронная autoelectronic/field/cold emission
~, анодная anode/positive(-ion) emission
~, вторичная/первичная secondary/primary emission
~ долговых обязательств *econ.* debenture issue
~, термоионная/термоэлектронная thermoionic emission, thermal electron emission
~, удельная specific emission
~, фотоэлектронная photoemission, external photoelectric effect
~, холодная autoelectronic/field/cold emission
эмитент emitter
эмитировать *phys.* emit, give off; *econ.* issue
эмиттер *electr.* emitter
~ неосновных/основных носителей minority/majority emitter
~, термоэлектронный thermoionic emitter
~, торированный вольфрамовый thoriated-tungsten emitter
эммер *bot.* amelcorn, emmer *(Triticum dicoccum)*
эмоциональность emotionality, emotionalism

эмоциональный emotional; emotive; affective
~ подход к явлениям emotionalism
эмоция emotion
~, радостная elation
эмпиема *med.* empyema
~ брюшной полости pyoperitoneum
~, доброкачественная latent/benign empyema
~ желчного пузыря cholecystempyesis
~ мошонки empyocele
~ перикарда purulent pericarditis, pyopericardium
~ плевры pyothorax, purulent pleurisy
~ сустава arth(o)empyesis
эмпиризм, эмпирика empiricism
эмпириокритик *philos.* empiriocritic(ist)
эмпирист empiricist
эмпирический empiric(al); *philos.* empiricist; descendental
эмпирия empiricism; (human) experience
эму *zool.* emu, emeu *(Dromiceius)*
эмульгатор emulsifying agent, emulsifier, emulgator
~, анионоактивный anionic emulsifying agent
~, жирнокислый fatty-acid emulsifying agent
эмульгация, эмульгирование emulsification
эмульгируемость emulsifiability
эмульсатор emulsor, emulsifier
эмульсин *bioch.* emulsin, synaptase
эмульсификатор emulsifier
эмульсия emulsion
~, битумная emulsified asphalt
~, водяная aqueous emulsion
~, высокочувствительная *phot.* fast emulsion
~, жировая *leath.* fat liquor
~, крупнозернистая/мелкозернистая *phot.* coarse-/fine-grain(ed) emulsion
~, малочувствительная *phot.* slow emulsion
~, масляная для опрыскивания *agr.* oil (-based) spray
~ на подложке backed emulsion
~, насыщенная дейтерием deuterium-loaded emulsion
~, обратная W/O emulsion, water-in-oil emulsion
~, проложенная sandwiched emulsion
~, прямая O/W emulsion, oil-in-water emulsion
~, расслоившаяся broken emulsion
~, смазочно-охлаждающая *mach.* cutting fluid
~, снимающаяся strip(pable)/stripping emulsion
~, электронно-чувствительная electron-sensitive emulsion
эмульсол cutting emulsion
эмфаза *ling.* emphasis; stress
эмфатический *ling.* emphatic(al)
эмфизема *med.* emphysema
~, кишечная intestinal emphysema
~ легких pulmonary emphysema
~ легких, старческая senile/Jenner's emphysema
~, подкожная (sub)cutaneous emphysema
~ средостения mediastinal emphysema, pneumomediastinum
~ стеклодувов glass blower's emphysema
~, трупная cadaveric emphysema
~, экстракраниальная subgaleal/scalp emphysema, extracranial pneumatocele
эмфизематозный emphysematous
эналлогенный enallogene, exogenetic
энантема *med.* enenthem(a)
энантиотропия enantiotropy, enantiotropism
энантовый *chem.* oenanthic
энвироника, энвирология environmental engineering, envirology
энгидрит *miner.* enhydrite
энгранаж depth
эндамеба *zool.* endam(o)eba *(Endamoeba)*
эндаортит *med.* endaortitis, endo-aortitis
эндартериит *med.* end(o)arteritis
~, бактериальный/септический bacterial endarteritis
эндекаэдр *cryst.* hendecahedron
эндем endemic plant/form/animal
эндемик endemic form/species
эндемический, эндемичный *biol., med.* endemic(al), endemial
эндемия endemic disease, endemia
эндиадиойн *lit.* hendiadys, hendiadyoin
эндивий *bot.* endive *(Cichorium endivia)*
эндобатолический endobatholithic
эндобласт *biol.* endoblast, hypoblast
эндоваскулит *med.* endangeitis, endangiitis, endovasculitis
эндовибратор endovibrator, cavity resonator
эндогамический endogamous, endogamic
эндогамия endogamy, inbreeding
эндогенез *biol.* endogenesis, endogeny
эндогенетический endogenetic
эндогенный endogenous, endogenic
эндогидатогенный endohydatogenous
эндодерма *zool.* endoderm; *bot.* endodermis
эндодермис *bot.* endodermis
эндокард(ий) *anat.* endocardium
эндокардит *med.* endocarditis
~, бородавчатый verrucous/vegetative endocarditis
~, возвратный recurrent endocarditis
~, врожденный/фетальный congenital endocarditis
~, септический bacterial/septic/malignant/infections/infective endocarditis
~, фибропластический endomyocardial fibroelastosis, Loeffler fibroplastic endocarditis
~, хирургический surgical/postoperative endocarditis
~, язвенный ulcerous endocarditis
эндокарп(ий) *bot.* endocarp
эндоконтактный endomorphic
эндокринный endocrine, endocrinous, endocrinic
эндокринолог endocrinologist
эндокриптия endocryptic process

эндолизин *med.* leukin, endolysin
эндолимфа *anat.* endolymph
эндлимфатический endolymph(at)ic
эндолитический, эндолитный endolithic
эндометрий *anat.* endometrium
эндометриоз *med.* endometriosis
~, позадишеечный retrocervical endometriosis
эндометрит *med.* endometritis
~, железистый glandular endometritis
~, послеродовой puerperal endometritis
эндоморфоз endomorphism
эндоневрий *med.* endoneurium, Henle's sheath
эндопаразит endoparasite, internal parasite
эндорадиозондирование endoradioprobing
эндоризный *bot.* endorhizal, monocotyledonous
эндосепта *biol.* endoseptum
эндосифон endosiphon; anus siphon
эндосмос ensdosmose, endosmosis
эндосперм *bot.* endosperm
~, клеточный cellular endosperm
~, руминированный ruminate endosperm
эндост *anat.* internal periosteum, endosteum
эндостеит *med.* endosteitis, endostitis
эндосэкзина endosexine
эндотека *bot.* endothecium, endotheca
эндотелий *anat.* endothelium
эндотелиома *med.* endothelioma
~ мужского полового члена penial endothelioma
~ яичника ovarian endothelioma
эндотелиотоксин *med.* endotheliolysin, endotheliotoxin
эндотермический endothermic, endothermal
эндотермичность endothermicity
эндотермия endothermy, diathermy
эндотрофный *bot.* endotrophic
эндофермент *bioch.* endoenzyme
эндофилльный endophyllous
эндофильный endophilic
эндоэнергетический, эндоэргический endoergic, endoenergic, endothermic
эндсы *pl., timb.* ends of boards
эндшпиль *chess* end game
эндэкзина endexine
Энеида *lit.* The Aeneid
энеолит Aeneolithic/Eneolithic age
энергейя *philos.* energeia
энергетик power engineer
энергетика 1. power engineering; energetics; 2. power (generation/industry)
~, атомная/ядерная nuclear/atomic power (engineering), atomic/nuclear power industry
~, тепловая thermal power
энергетический energy-/power-production
~кое хозяйство power economy
энергичный energetic, vigorous, strong
энергия energy, power
~, атомная atomic power/energy, nuclear power
~ атомной связи atomic-buiding energy
~ бета-распада beta-disintegration energy
~ бета-спектра, граничная beta end-point energy
~, бомбардирующей частицы bombarding-particle energy
~ в импульсе pulse energy
~ в системе центра масс centre-of-mass energy
~, взаимная mutual energy
~ взаимодействия interaction energy
~, внутренняя internal/intrinsic/inner energy
~, внутриатомная nuclear power/energy
~, водная water/hydroelectric energy
~ возбуждения (атомов) excitation energy
~ вращения rotation(al) energy
~ вращения, кинетическая rotational/angular kinetic energy
~ входа/выхода *mech.* intake/exit energy
~ гамма-излучения gamma-energy
~ гамма-перехода gamma-transition energy
~, геотермическая geothermal energy
~, гидравлическая hydraulic energy
~, граничная limiting energy, end-point energy
~, движущая propulsive energy
~ деления fission energy/power
~ деления, критическая fission barrier
~ деформации (потенциальная) strain energy
~, джоулева Joule energy
~, жизненная biotic/vital energy
~, запасенная stored energy
~ запуска triggering energy
~, звуковая sound/sonic energy
~, избыточная excess/extra energy
~, излучаемая radiant/radiated energy
~ излучения radiation energy
~ ионизации ionizing/ionization energy
~, исходная reference energy
~, кинетическая (вращения) (angular) kinetic energy
~ колебаний vibrational/vibratory/oscillation energy
~ кулоновского взаимодействия Coulomb energy
~, ложная spurious energy
~, лучистая radiant/radiative energy
~, магнитная magnetic field energy
~, магнитоупругая magnetoelastic energy
~, максимальная maximum/peak energy
~ массы покоя rest-mass energy
~, метаболическая metabolizable energy
~ на перекачку pumping power
~, надтепловая epithermal energy
~, накопленная accumulated/stored energy
~ насыщения saturation energy
~, нулевая zero(-point) energy
~, обменная exchange/metabolizable energy
~, объемная volume energy
~, остаточная residual energy
~ отдачи recoil energy
~ отделения/отрыва separation energy

~ отталкивания repulsive/repulsion energy
~ пары дырка-электрон hole-electron energy
~ перехода transition energy
~, пиковая peak/maximum energy
~, пластовая energy of a seam
~ поверхностного натяжения capillary energy
~, поглощенная absorbed energy
~, подтепловая sub-thermal energy
~ покоя (частицы) rest energy
~, полезная useful energy/power, usable energy
~ пороговая threshold energy; cut-off energy
~ поступательного движения translational/translatory energy
~, потенциальная energy of a rest, energy of position; potential energy
~, потерянная lost energy
~, потребляемая 1. energy input; 2. power consumption
~ притяжения attraction energy
~, произвольная arbitrary energy
~ пучка beam energy
~ развала break-up energy
~ разрыва, усталостная fatigue tensile energy
~ распада disintegration/decay energy
~, располагаемая avaiable energy
~ расщепления disintegration energy, break-up energy
~, резонансная resonance/resonant energy
~ рельефа *geol.* scope of relief
~ рентгеновского излучения X-ray energy
~ решетки lattice energy
~ роста growing capacity
~, световая luminous/light energy
~ светопреломления refractive energy
~, свободная free/available energy
~, связанная bound/unavailable energy
~ связи/связывания binding energy
~ связи на нуклон binding energy per nucleon
~, скрытая latent energy
~, собственная self-energy, intrinsic/proper energy
~, солнечная solar energy/power
~ сопряжения conjugation energy
~, средняя 1. mean/average energy; 2. medium/moderate energy
~ столкновений collision(al) energy, energy of collision
~ сцепления cohesive/cohesion energy
~, тепловая 1. heat/thermal energy; 2. thermal power
~, термоядерная thermonuclear power; fusion energy
~ тормозного излучения bremsstrahlung energy
~ удара impact energy, energy of impact
~, удельная specific energy
~ упругости elastic (strain) energy
~, фермиевская Fermi energy
~, химическая chemical energy/power
~, электрическая electric energy/power
~, ядерная nuclear power
~ ядерной реакции nuclear-reaction energy
энергобаза (electric) power base, source of power supply
энергобаланс power balance
энергоблок power(-generating) unit
энерговооруженность 1. *av.* power to weight ratio; 2. *econ.* power supply per production unit; installed power per employee
~ труда power availability per worker/man, (electric) power supplied per job
энерговыделение engery release/liberation
энергоемкий power-intensive/-consuming, energy-consuming
энергоемкость energy content; energy/power capacity; power input; energy-output ratio
энергозависимость *comp.* volatility
энергозатрата energy expenditure
энерголиния power line
энергомашиностроение 1. power-plant engineering; 2. power-plant industry
энергообмен power interchange
энергооборудвание power equipment
энергообъединение power pool
энергоотдача energy liberation/release
энергопитание power supply
энергопоезд railway mobile power plant, mobile power-generation unit
энергопотеря loss of energy, power loss
энергопотребление power consumption
энергоресурсы *pl.* power resources
энергосеть mains, power supply system
энергосистема power (supply) system
~, единая power grid
~, объединенная electric power pool
~, районная power supply region, single-area interconnection, single-area interconnected power system
энергоснабжение power/energy supply
энергосодержание energy content
энергостанция power plant/station
энергоустановка power plant
энергочувствительный energy-sensitive
энергоэквивалентный power-equivalent
энзе *abbr.* **неприкосновенный запас** *mil.* emergency ration
энзим *bioch.* enzyme, ferment
энзиматический enzymatic, enzymic
энзоотия enzootic (disease)
энигма enigma; riddle
энигматический, энигматичный enigmatic(al)
энкарпа *arch.* encarpus, frieze ornament
энкаустика encaustic (painting), wax painting
энклав enclave; enclosure
энклиза, энклитичность enclisis
энкомиенда, энкомьенда *hist.* encomienda
энозис enosis, union
энокарпус *bot.* bacaba palm *(Oenocarpus)*
эномотия *hist.* enomoty, Spartan subdivision

энотера *bot.* evening primrose *(Oenothera)*
энофтальм *med.* enophthalmos, enophthalmia
энсин *naut.* ensign
энтада *bot.* sea bean, entada *(Entada scandens)*
энтальпия *therm.* enthalpy, heat content, sensible/total heat
~ заторможенного потока stagnation enthalpy
~ плавления enthalpy of fusion
энтамеба *zool.* entam(o)eba
энтелехия *philos.* entelechy, distinctness
энтерит *med.* enteritis
~, изолированный regional/terminal enteritis, chronic cicatrizing enteritis
~, мембранозный (muco)membranous enteritis
~, псевдомембранозный pseudomembranous/mucous enteritis
энтероанастомоз *med.* enteroanastomosis, intestinal anastomosis
энтерогепатит *med.* enterohepatitis
~, инфекциозный histomoniasis
энтеродиния *med.* enteralgia, enterodynia
энтерокиназа *bioch.* enterokinase, enteropeptidase
энтероколит *med.* enterocolitis
~, лямблиозный giardiasis, lambliasis
энтеролит *med.* enterolith, intestinal concretion
энтеролобиум *bot.* ear-pod tree *(Enterolobium)*
энтерорафия *med.* enterorrhaphy
энтимема *logic* enthymeme
энтимематический enthymematic(al)
энтодерма *zool.* entoderm, endoderm, hypoblast
энтозоон *zool.* entozoon
энтомогамия insect pollination, entomogamy, entomophily
энтомолог(ист) entomologist
энтомофаг *zool.* insect eater *(Entomophaga)*
энтомофил *bot.* entomophilous plant
энтомофильный entomophilous
~ное опыление entomophilous pollination
энтомофит *bot.* entomophyte
энтропия entropy
~, обобщенная generalized entropy
~, остаточная residual entropy
~, парциальная partial entropy
~ сообщения message entropy
~, удельная specific entropy
~, условная conditional/conditioned entropy
энтузиазм enthusiasm
энуклеация *med.* enucleation
~ глаза ophthalmectomy
энурез *med.* bed-wetting, enuresis
~, ночной nocturnal enuresis
энцефалартос *bot.* Kaffir bread *(Encephalartos)*
энцефалит *med.* encephalitis
~, аллергический hype(re)rgic encephalitis
~, герпетический zoster encephalitis
~, гнойный suppurative/purulent encephalitis
~, диффузный периаксиальный Schilder's disease, periaxial encephalitis
~, клещевой, ~ весенне-летний Russian tickborne encephalitis, vernal/woodcutter's encephalitis
~, комариный mosquito-borne encephalitis
~, концетрический периаксиальный Balo's disease
~, летаргический lethargic encephalitis, nona
~ лошадей, энзоотический Borna disease
~ новорожденных encephalitis neonatorum
~, таежный = ~, клещевой
~, эпидемический influenzal encephalitis
энцефаломалация *med.* softening of the brain, encephalomalacia
энцефалопатия *med.* encephalopathy
энциклика *eccl.* encyclic(al)
энциклопедизм encyclop(a)edic learning, encyclop(a)edism, erudition
энциклопедист 1. *hist.* Encyclop(a)edist; 2. person of encyclop(a)edic learning, erudite
энциклопедический encyclop(a)edic
эогея Eogaea
эодевон *geol.* Eodevonian
эозин *chem.* eosin(e), tetrabromfluorescein
~, растворомый soluble eosin
эозин-метиленблау eosin-methylene blue, EMB
эокембрий *geol.* Eocambrian
эоклимакс primeval climax, eoclimax
Эол *myth.* Aeolus
эолов Aeolian
~а арфа Aeolian harp
эолово-обломочный anemoclastic
эоловый (a)eolian, atmogenic, wind-borne/laid
эон *geol.* (a)eon
эопалеозой *geol.* Eopal(a)eozoic
эоплейстоцен *geol.* Eopleistocene
Эос *myth.* Eos
эоцен *geol.* Eocene (system/period/strata)
эпаналепсис *lit.* epanalepsis; repetition
эпарх *hist.* eparch, governor of province; metropolitan
эпейрогенезис, эпейрогения *geol.* continent-building, epeirogenesis, epeirogeny
эпейрофорез continental drift/displacement
эпексегезис *lit.* epexegesis, addition of explanatory words
эпендимома *med.* ependymoma
~, злокачественная/клеточная malignant/cellular ependymoma
эпибласт *biol.* epiblast
эпиболия *embryol.* epiboly, epibole
эпигеальный, эпигейный epig(a)eal, epig(a)ean, epig(a)ous, growing/living close to the ground
эпигенез(ис) *biol.* epigenesis
эпигенный *bot.* epigenous; *cryst., geol.* epigene
эпигея *bot.* trailing arbatus *(Epigaea repens)*
эпиглоттис *anat.* epiglottis
эпиглоттит *med.* epiglottiditis, epiglottitis
эпидемиологический epidemiologic(al)
эпидемический epidemic(al)
эпидемия epidemic, widespread outbreak
эпидерма, эпидермис *biol.* epiderm(a), cuticle

эпидермальный, эпидермический epidermal, epidermic, epidermatic
эпидермоид *med.* epidermoid (cyst)
эпидермолиз *med.* epidermolysis
~, буллезный epidermolysis/acantholysis bullosa, Goldscheider's disease
эпидермофития *med.* epidermophytosis, derm(at)ophytosis, derm(at)omycosis
~, паховая tinea cruris, ringworm of the thighs and perineal region
~ стоп tinea/dermatomycosis pedis, ringworm of the feet, athlete's foot
эпидиаскоп overhead projector, epidiascope
эпидидимит 1. *geol.* epididymite; 2. *med.* epididymitis
~, гонорейный gonorrheal epididymitis
эпидиктический *lit.* epideictic
эпидот *geol.* epidote, pistacite
~, марганцевый manganese epidote, piedmontite
~, норвежский arendalite
эпидурит *med.* external meningitis, peripachymeningitis
эпизиопластика episioplasty
эпизодический, эпизодичный episodic(al), occasional, sporadic, incidental
эпизона epizone
эпизоотик specialist in epizooty
эпизоотия epizootic, epizooty
эпизоотология epizoot(i)ology
эпик epic poet, epicist
эпика *lit.* epics, epic poetry
эпикард(ий) *anat.* epicardium
эпикондилальгия *med.* tennis elbow, epicondylalgia
эпикондилит *med.* epicondylitis
~, профессиональный occupational epicondylitis
эпикотиль *bot.* epicotyl
эпикотильный epicotylar
эпикриз *med.* 1. catamnesis; 2. epicrisis
эпикристаллический epicrystalline
эипкуреец, эпикуреист Epicurean, epicure
эпикурейство Epicureanism, epicurism
эпилепсия *med.* epilepsy, epilepsia
~, вызваная activated epilepsy
~, генуинная idiopathic epilepsy
~, детская epilepsia nutans
~, джексоновская/корковая/кортикальная cortical/Jacksonian epilepsy
~, малая abortive/latent/masked epilepsy, petit mal
~ Оппенгейма Oppenheim's/acoustic epilepsy
~, подкорковая subcortical epilepsy
~, поздняя delayed/tardy epilepsy
~, прокурсивная procursive/accelerative epilepsy
~, психомоторная temporal lobe epilepsy, psychomotor epilepsy
~, рефлекторная reflex/peripheral epilepsy
~, травматическая (post)traumatic epilepsy
эпилептиформный epileptoid, epileptiform
эпилептический epileptic(al)
эпилептогенный epileptogenic, epileptogenous
эпилирование (d)epilation
эпилировка *leath.* topping
эпилитораль epilittoral
эпилятор epilatory, depilatory
эпиляция (d)epilation
~, лучевая radiation epilation
эпиморфизм *math.* epimorphic mapping, epimorphism
эпинефрин *bioch.* adrenaline, epinephrine
эпинефрома *med.* epinephroma, hypernephroma
эпиникий epinikion, epinicion, song of victory
эпипалеолит *geol.* Epipal(a)eolithic, Mesolithic
эпипараклаз *geol.* overthrust (fault)
эпипорода epirock
эпископ episcope
~ с наружным отражателем overhead projector
эписланец *geol.* epischist
эписпадия *med.* epispadia(s)
эписперм(ий) *bot.* seed coat, episperm
эписпорий *bot.* epispore; *zool.* episporium
эпистилий, эпистиль *arch.* architrave, epistyle
эпистофея *anat., zool.* odontoid (process)
эпистрофей *anat.* epistropheus, axis
эпитаксия epitaxy
эпиталама, эпиталамия epithalamium, epithalamion
эпиталамус *anat.* epithalamus
эпиталлом *bot.* epithallus
эпитафия *archeol.* epitaph, tombstone inscription
эпитеза *gram.* epithesis
эпитека *bot., zool.* epitheca
эпителий *anat.* epithelium
~, десенный gingival epithelium
~, железистый glandular epithelium
~, кубический cubical/cuboidal epithelium
~, мерцательный ciliated epithelium
~, многослойный stratified/laminated epithelium
~, многослойный плоский stratified squamous epithelium
~, мостовидный pavement epithelium
~, мышечный myoepithelium, muscle epithelium
~, однослойный single-layer epithelium
~, плоский squamous epithelium
~ простой плоский simple pavement/squamous epithelium
~ хрусталика epithelium of the lens
~, цилиндрический columnar/cylindrical epithelium
эпителиоидножелезистый epithelioglandular
эпителиоклеточный epitheliocellular
эпителиолитический epitheliolytic
эпителиома *med.* epithelioma
~, базальноклеточная basal cell carcinoma/epithelioma

~, **обызвествляющаяся** Malherbe's calcifying epithelioma
~, **саморубцующаяся, ~ язвенно-рубцовая** Darier's disease, psorospermosis follicularis
эпитема *bot., zool.* epithem(a), epitheme
эпитет epithet, descriptive term
эпитеций *bot.* epithecium
эпифеномен epiphenomenon
эпифиз(а), эпифизис *anat.* epiphysis
эпифизарный epiphysial, epiphyseal
эпифилл *bot.* epiphyll
эпифиллум, эпифиллюм *bot.* leaf cactus *(Epiphyllum)*
эпифит *bot.* epiphyte
~ **на деревьях** tree epiphyte
~, **ложный** pseudoepiphyte
эпифитный, эпифитовый epiphytic(al)
эпифокус epifocus, epicentre
эпифора *med.* watery eye, epiphora
эпихлоргидрин *chem.* epichlorhydrin
эпихолестерин *chem.* epicholesterol
эпицентр epicentre, zero point
~ **взрыва** ground/surface zero
эпицикл *astr., math.* epicycle
эпициклический epicyclic
~**кая передача** epicyclic gear
эпициклоида *geom.* epicycloid
~, **удлиненная** inflected epicycloid
~, **укороченная** quirked epicycloid
эпициклоидальный, эпициклоидный epicycloidal
эпицистотомия *surg.* suprapubic cystotomy, epicystotomy
эпонихий *med.* eponychium, perionychium
эпический epic
эпичность epicism
эпиэвгеосинклиналь *geol.* epieugeosyncline
эпоксид *chem.* epoxy (resin)
эпоксидно-эфирный epoxy-ether
эпоксисмола epoxy resin
эпоксисоединение epoxy compound, epoxide
эполемент *mil.* epaulement
эполет(а) *mil.* epaulet(te)
эпопея *lit.* epopee, epic, epopoeia
эпос *lit.* epic literature, epos
эпоха (geological/historical) age, era, epoch
~, **вюрмская** Wurmian epoch
~, **геохимическая** geochemical epoch
~, **ледниковая** Glacial/Pleistocene epoch
~, **манлийская** Manlius epoch
~, **междождливая** interpluvial epoch
~, **новейшая/современная** Recent epoch
~, **плиоценовая/сангамонская** Pliocene/Sangamon
~ **свайных построек** lacustrine age, lake-dwelling period
Эпрон *abbr.* Expedition for Special Underwater Operations
эпсилон epsilon
~ **энтропия** epsilon entropy
эпулис *med.* epulis, gum tumo(u)r
~ **беременных** epulis gravidarum
эпюр(а) 1. orthographic epure/representation, projection; 2. curve, diagram
~ **весовой нагрузки** weight curve
~ **крутящих моментов** torque diagram, twisting-moment diagram
~ **моментов** moment curve/diagram
~ **нагрузки** load (distribution) curve/diagram
~ **напряжений** stress diagram
~ **напряжения (тока)** voltage (current) waveform
~ **распределения давления** pressure profile
~ **распределения нагрузки** loading diagram
~ **распределения скоростей** velocity profile/diagram/curve
эра era
~, **архейская** Archean (era)
~, **археозийская** Archeozoic (era)
~, **кайнозойская** Cenozoic/Cainozoic (era)
~, **мезозойская** Mezozoic (era)
~, **палеозойская** Pal(a)eozoic (era)
~, **протерозойская** Proterozoic (era)
~, **психозойская** Psychozoic (era)
~**ры, до нашей** B.C., before Christ
~**ры, нашей** A.D., Anno Domini
Эрата, Эрато *myth.* Erato
эрбий *chem.* erbium, Er
эрг *phys.* erg
эргастерий ergasterion
эргастул ergastulum, cell, dungeon
эргодичность *math.* ergodicity
эргометр ergometer
~, **велосипедный** bicycle ergometer
эргоновин ergonovine, ergometrine, ergobasine, ergostetrine
эргономика, эргономия ergonomics; biotechnology; *Am.* human engineering
эргостерин *bioch.* ergosterol, ergosterin, provitamin D_2
~, **активированный** activated ergosterol in oil
эрготамин ergotamine
~ **тартрат** ergotamine tartrate, gynegren
эрдельтерьер Airedale terrier
Эреб *myth.* Erebus
эрекция *physiol.* erection
~, **болезненная** paniful erection, chordee
эремурус *bot.* desert candle *(Eremurus)*
эрепсин *bioch.* ereptase, erepsin
эретизм *med.* erethism
эрзац ersatz, substitute
эриантус *bot.* plume grass *(Erianthus)*
эрдокс *min.* airdox
Эрида *myth.* Eris
эрика *bot.* heath *(Erica)*
эриколистный heath-leaved
Эриния *myth.* Erinys
эритема *med.* erythema
~, **геоморрагическая экссудативная** Henoch's purpura, hemorrhagic exudative erythema

~, индуративная Bazin's disease, Darrier-Roussy sarcoid
~, кольцевидная erythema annulare/iris
~, лучевая radiation erythema
~, полиморфная Hebra's disease, erythema multiforme/polymorphe
~, рубцующаяся cicatrizant erythema
~, солнечная sunburn, erythema solare
~, тепловая erythema caloricum/abigne
эритредема *med.* erythredema, (infantile) acrodynia, Swift's/pink disease
эритремия erythr(a)emia, Vaquez' disease, polycyth(a)emia rubra
эритрин *chem.* erythrin; *miner.* erythrine
эритробласт *anat.* erythroblast
~, мегалобластоидный megaloblast
~, поздний/ранний late/early megaloblast
эритрогемометр *med.* erythrohemometer
эритродермия *med.* erythroderma, erythrodermia, erythrodermatitis
~, десквамативная Leiner's disease
эритродонтоз *med.* erythrodontia
эритроизомераза *med.* erythrose isomerase
эритроксилон *bot.* cocain tree *(Erythroxylum)*
эритролейкемия *med.* erythroleukemia
эритромиелоз, эритролейкоз *med.* erythroleukosis, leukan(a)emia, panmyelosis
эритромицин *pharm.* erythromycin, ilotycin
эритроплазия *med.* erythroplasia of Queyrat
эритропоэз *med.* erythro(cyto)poiesis
эритропсин erythropsin, visual purple, rhodopsin
эритропсия erythrop(s)ia, red vision
эритрофил erythrophil(e)
эритрофилл *bot.* erythrophyll
эритрофор *zool.* erythrophore, allophore
эритроцит *anat.* erythrocyte, red corpuscle
~, крупный macrocyte, macroblast, megalocyte, megaloblast
~, мелкий microcyte, microblast
~, рудиментарный erythroblast, h(a)ematoblast, microblast
эритроцитобласт erythro(cyto)blast
эритроцитоз *med.* erythrocytosis, polycyth(a)emia, erythr(a)emia
эритроцитолиз *med.* erythrocytolysis, h(a)emolysis
эритрулоза *chem.* erythrulose
эркер *arch.* bay window, oriel, o'riel-window
эрланг *com.* erlang
эрлифт airlift
Эрмий *myth.* Hermes
эрмитаж 1. hermitage, secluded residence; 2. (Э.) the Hermitage Museum
эрмитов, эрмитовский, эрмитовый Hermitian, hermitian
эрмитово-кососимметричный skew-Hermitian
эрмитовость hermicity, Hermitian character/property/nature
эродируемость, эродированность, эродийность erodibility
эрозивный erosive; wearing
эрозиеустойчивый, эрозионноустойчивый erosion-resistant
эрозионный erosional, erosive
эрозия erosion
~, ветровая wind erosion, deflation
~, геморрагическая *med.* h(a)emorrhagic erosion
~, глубинная river-bed erosion
~, интенсивная rapid/intense erosion
~, овражная gully erosion
~, отступающая/пятящаяся/ретрогрессивная retrogressive/backward erosion
~, плоскостная/поверхностная sheet/surface erosion
~ поверхности уносимого покрытия ablative surface erosion
~ потока, наступательная headwater erosion
~ почвы soil erosion
~, русловая channel erosion
~ слизистой оболочки желудка *med.* Daenlafoy's erosion
~, субаэральная sub-aerial erosion
~ хромосом chromosome erosion/mottling
~ шейки матки cervical erosion
Эрос, Эрот *myth.* Eros
эротема *lit.* rhetorical question, erotema, eroteme, erotesis
эротизм, эротичность eroticism, erotism
эротика 1. sensuality; 2. erotic literature, erotica
эротический, эротичный erotic(al), ero(to)genic, ero(to)genous; sensual
эротогенный ero(to)genous, ero(to)genic
эротоман *med.* sexual maniac, erotomaniac
эротомания *med.* erotomania, morbid sexual passion
эротопатия *med.* erotopathy
эрратический *geol.* erratic
эрстед *el.* oersted
эрстедметр oersted meter
эрудиция learning, erudition
эруптивный eruptive
эрупция *astr.* (solar) eruption
эрцгерцог archduke
эрцгерцогиня archduchess
эрцгерцогство archduchy, archdukedom
эрес *abbr.* **реактивный снаряд** *mil.* rocket shell, self-propelled missile
эсдек *hist.* Social Democrat
эсер *hist.* Social Revolutionary
эсеровский *hist.* socialist-revolutionary
эсеровщина, эсерство *polit., hist.* ideology of the Social Revolutionary Party
эсесовец = эсэсовец
эскадра *naut.* squadron
~ кораблей охранения escort squadron
~ крейсеров cruiser squadron
~ эсминцев destroyer squadron
эскадренный fleet *(attr.)*

~ миноносец destroyer
эскадрилья *av.* squadron
~, авиационная air(craft) squadron
~ аэрофоторазведка photographic-reconnaissance squadron
~, боевая combat squadron
~, вертолетная helicopter squadron
~, истребительная fighter squadron
~, разведывательная reconnaissance squadron
~ самолетов-топливозаправщиков tanker squadron
~ связи liaison squadron
~, учебная training squadron
эскадрон *mil.* (cavalry) squadron, troop
эскалада *hist.* escalade, escalado
эскалатор moving staircase, escalator
эскалация escalation
эскалоп *cul.* cutlet
эскапистский *lit.* escapist
эскарп scarp, escarp(ment)
~ сбросовый *geol.* fault scarp, cliff of displacement
~, флексурный *geol.* monoclinal fold scarp
эсквайр esquire; landed proprietor; *hist.* shielbearer
эскер *geol.* esker, ridge of gravel
эскеровый, эскероподобный eskerine
эскиз diagrammatic sketch, rough drawing, study; outline, draft
~ от руки freehand sketch
эскизный чертеж outline sketch
эскимо *cul.* chocolate-covered ice-cream
эскимос Eskimo
эскимоска Eskimo (woman)
эскимосский Eskimo(an)
эскорт *mil.* escort; armed guard
~, почетный escort of hono(u)r
эскортирование *mil.* escorting; attendance
эскудо *fin.* escudo *(currency unit)*
Эскулап *myth.* Aesculapius
эскулин *chem.* esculin
эсминец *abbr.* **эскадренный миноносец** *naut.* destroyer
эспадрон cutting sword, backsword, spadroon
эспарто *bot.* esparto needle grass *(Stipa tenacissima)*
эспарцет *bot.* sainfoin, holy clover *(Onobrychis)*
эспарцетовидный *bot.* onobrychioid
эсперантист(ка) Esperantist
эсперанто Esperanto
эсперантский Esperantic
эспиналь, эспинар evergreen thorn bush, espinal, espinar
эссе *lit.* essay
эссеист essayist
эссенция essence
эссив *gram.* essive (case)
эст 1. *obs.* Estonian; 2. Chud(e)
эстакада 1. *constr.* trestle; scaffold bridge; viaduct, overpass, flyover; 2. *naut.* pier
~, балочная beam/girder trestle
~, бункерная bin trestle, high-line bunkers
~, железнодорожная railway overpass
~, железобетонная reinforced cocrete pier
~, конвейрная conveyer gallery
~, крановая crane trestle
~, морская offshore pier
~, пешеходная pedestrian overcrossing/overpass
~, разборная передвижная portable sectional trestle
~, рамная frame trestle
~, тоннельная tunnel trestle
эстамп *art* print, engraving, plate
эстафет(а) 1. *sport* relay race; 2. baton *(used in a relay race)*
эстафетный relay-race
~ бег relay race
~ная палочка *sport* baton
эстезиология (a)esthesiology
эстераза *bioch.* esterase
эстеридный *zool.* estherian
эстерификация *chem.* esterification
эстет (a)esthete
эстетизм, эстетичность (a)estheticism
эстетик (a)esthetician
эстетика 1. (a)esthetics; 2. design
~ производства industrial design
эстиоменовый *med.* esthiomenous, corroding, ulcerating, phagedenic
эстонец, эстонский Estonian
эстонка Estonian (woman)
эстрагон *bot.* tarragon *(Artemisia dracunculus)*
эстрада 1. platform, stage; 2. *theatre* variety (art), vaudeville
эстрадник variety performer, artiste
эстрадный vaudeville *(attr.)*, variety *(attr.)*
~ артист variety performer
эстральный *physiol.* (o)estrous, (o)estrual
эстран *bioch.* estrane
эстрин *bioch.* estrin
эстриол *bioch.* estriol, theelol, folliculin/estrone hydrate, trihydroxyestrin
эстроген *bioch.* estrogen
эстрогенный *med.* estrogenic, estromimetic
эстрон *pharm.* estrone, theelin, folliculin, follicular hormone, ketohydronyestrin
эструс *zool.* (o)estrus, (o)estrum
эстуарий, эстуария estuary, firth, frith
~ с баром closed estuary
Эсфирь *bibl.* Esther
эсхатология *theol.* eschatology, doctrine of the last/final things
эсэсовец *hist.* SS man
эсэсовский SS
этаж floor; storey, story; *geol.* group, stage, level
~, верхний upper storey/floor; *(of bus)* upper/top deck
~ водоносных пород aquifer layer, waterbearing level
~, второй first/ *Am.* second floor

~, **нижний** lower storey; ground/lower floor
~, **первый** first storey; *Eng.* ground/ *Am.* first floor
~, **подвальный** lower storey, basement
~, **структурный** structural stage, layer of geology
этажерка bookcase; rack, set of shelves; *electr.* stack
этажность number of stories
этазол *pharm.* sulphaethidole
~, **растворимый** suphaethidole sodium
этакридин rivanol, ethoxydiaminoacridine lactate
эталий *bot.* aethalium
эталон 1. (reference/primary) standard; 2. calibrating device, calibration instrument; 3. *opt.* etalon
~ **времени** time standard
~, **вторичный** substandard, secondary standard
~, **государственный** state/national standard
~, **групповой** collective standard
~, **жидкий** liquid standard
~, **насыпной** powder standard
~, **одиночный** individual standard
~, **первичный** primary/basic standard
~, **переносный** transfer standard
~, **рабочий** working standard
~ **сравнения** comparison standard
~ **частоты** frequency standard
эталонирование calibration (against a basic standard), standardization, standardizing
~ **источника** source calibration
эталонировать calibrate, standardize
эталон-копия transfer standard
эталонный standard *(attr.),* calibration *(attr.)*
~ **прибор** calibration/reference instrument
~**ная картотека** reference file
эталон-свидетель witness standard
этаминал-натрий *pharm.* pentobarbital (sodium), nembutal
этан *chem.* ethane
этанол ethanol, ethyl/grain alcohol
этап 1. stage, step, phase; 2. *mil.* halting place, étape; 3. stockade *(for deported prisoners);* 4. *sport* lap; 5. *fig.* milestone
~ **возвращения в атмосферу** re-entry phase
~ **высадки** landing phase
~, **заключительный** final phase/stage
~ **захода на посадку** *av.* approach
~ **изготовления** manufacturing stage
~ **использования** operational phase
~, **конечный** final stage
~ **конструирования** engineering stage
~ **конструирования и разработки** design-and-development phase
~ **маршрута** *av.* route leg/stage
~ **наведения** guidance phase
~, **начальный** incipient/initial stage
~ **планирования** planning stage/phase
~ **погрузка** embarkation phase
~ **поиска** search phase
~ **приемки** acceptance phase
~, **промежуточный** interim stage
~ **самонаведения** homing phase
~ **сближения** approach/closing phase; *rock.* rendezvous stage
~ **сборки** assembly stage
~ **снижения/спуска** descent phase
~ **создания опытного образца/прототипа** pilot/prototype stage
~ **эксплуатации** operational phase
этаперазин *pharm.* perphenazine, trilafon
этапирование transportation under guard
этапник *hist.* convict in transit
этапные *n. pl.* convicts in transit
этапный phase *(attr.),* stage *(attr.)*
~**ным порядком, отправить/послать** *hist.* deport/transport under guard
этвеш *phys.* eötvös
этен *chem.* ethene, ethylene, olifiant gas
этеогенез etheogenesis, ethiogenesis
Этеокл *myth.* Eteocles
этерификация etherification
этернит asbestos shingle(s); eternit, asbestos cement
этернитовый asbestos-cement
этика ethics
~, **врачебная** medical ethics
этикет I etiquette, manners; propriety, decorum
~, **морской** sea manners
этикет II, этикетка label, nameplate, tag, tally card
этикетаж (preparation of) labels; label(l)ing
этикетирование, этикетировка label(l)ing, tagging
этикетировать label, tag, provide with labels/tags
этикетка tally card, label, nameplate; tag
~ **с ценой** price ticket
этикетный pertaining to etiquette, etiquettical
этикето-захватчик label picker
этико-социальный ethicosocial
этил *chem.* ethyl
~, **бромистый** ethyl bromide, hydrobromic ether
~, **хлористый** ethyl chloride, hydrochloric ether, chloral an(a)esthetic, kelene
этилацетат acetic ether, ethyl acetate
этилдибромарсин ethylarsine dibromide
этилен *chem.* ethylene, ethene, olifiant gas
этилморфин-гидрохлорид ethylmorphine (hydro)chloride, dionin
этиловый ethyl *(attr.)*
~ **спирт** ethyl alcohol
этимологический etymologic(al)
этимология etymology
этимон *ling.* etymon, root word
этинил *chem.* ethynyl, ethinyl
этинил-тестостерон *pharm.* ethisterone, anhydrohydroxy progesterone, pregneninolone
этиологический (a)etiological
этиология *med.* (a)etiology
этионамид *pharm.* ethionamide, trecator
этичность ethicality, ethicalness
этмоидальный ethmoid(al)
этнарх ethnarch, ruler, governor

этнический ethnic
этноботаника ethnobotany
этногенез ethnogeny
этнографический ethnographic(al)
этнодемография ethnodemography
этнолингвистика ethnolinguistics
этнолог ethologist
этнологический ethnologic(al)
этноцентризм ethnocentrism
этология ethology
этос ethos
этотоин *pharm.* ethotoine, peganone
этриоскоп *meteor.* aethrioscope
этрурия *hist.* Etruria
этруск, этрусский Etruscan
этруска Etruscan (woman)
этюд *lit., art* sketch, study; *mus.* exercise, etude
этюдник *art* painter's case
эубитум eubitumen
эукаин *pharm.* eucaine (hydrochloride)
эулевомицетин *pharm.* levomycetine sulphate
эулитораль eulittoral (zone)
эуплоидия *biol.* euploidy
эутерофит *bot.* eutherophyte
эутимия *med.* euthymia, joyfulness
эуфиллин *pharm.* euphilline, aminophylline, metaphylline
эуфорбия 1. *bot.* euphorbia *(Euphorbia);* 2. ephorbium, gum resin
эухромосома euchromosome
эфа *zool.* carpet viper *(Echis carinatus)*
эфармоз *ecol.* epharmosis
эфеб *hist.* ephebe, ephebus, young Greek citizen
эфедра *bot.* ephedra, ma-huang, sea-grape *(Ephedra)*
эфедрин *pharm.* ephedrine
~, хлористоводородный ephedrine hydrochloride
эфедровые *n. pl. bot.* joint-fir family *(Ephedraceae)*
эфелидоз *med.* lentigo, lenticula, freckle, ephelis
~, злокачественный malignant lentigo/freckle
эфемер(а) *bot.* ephemeral, ephemeron
эфемерида 1. *zool.* ephemerid; 2. *astr.* ephemeris, ephemeride, *pl.* ephemerides
эфемиридный ephemerical
эфемерность ephemerality
эфемерный short-lived, ephemerous, ephemeral
эфемерофит *bot.* ephemerophyte
эфенди effendi, sir
эфесец, эфесский *hist.* Ephesian
эфиоп, эфиопский Ethiopian
эфиопка Ethiopian (woman)
эфир 1. *chem.* ether; 2. *rad.* the air
~ азотной кислоты nitrate
~ алюминиевой кислоты aluminate
~, виниловый vinyl ether/ester
~, выходить в go on the air
~ для наркоза an(a)esthesic ether
~, петролейный petroleum ether
~, простой ether
~, сложный ester
~, этиловый (diethyl) ether, ethyl/sulphuric ether
эфира *zool.* ephyra
эфирный 1. *chem.* ether(ic); 2. ethereal, light, airy
~ное масло essential/volatile oil
эфиромасличный essential-/volatile-oil bearing
~ное растение, эфироносс essential-oil-bearing plant
эфироносный ether-bearing
эфициллин *pharm.* ephycillin, penemat, bronchocillin
эфор *hist.* class of magistrates, ephor
эффект 1. effect; *math.* term; 2. *pl. theatre* effects
~ аддитивности/аддитивный additive effect
~, антенный aerial/antenna effect
~, армирующий reinforcing effect
~, арочный *met.* bridging
~ Барнета Barnett effect
~, береговой shore/coast effect
~, бинауральный binaural/stereophonic effect
~ близости proximity effect
~ взаимной компенсации cancellation, cancelling effect
~ взаимодействия 1. interaction term/effect; 2. cross-coupling effect
~ взрыва blast effect
~, внутрихромосомный intrachromosomal effect
~, возмущающий disturbing effect
~, вторичный secondary effect
~ второго/высокого порядка second/high-order effect
~ выгорания дырок *electr.* hole burning effect
~, высотный altitude effect
~, гальваномагнитный galvanomagnetic/magnetogalvanic effect
~, гироскопический gyroscopic action/effect
~ горячих электронов hot-electron effect
~, гравитационный gravity/gravitation effect
~, граничный *phys.* boundary effect
~, двумерный two-dimensional effect
~ деления на быстрых нейтронах fast-fission effect
~ Джоуля-Томсона Joule-Thomson effect/coefficient
~, динатронный *electr.* secondary electron effect, dinatron effect
~, дисперсионный *geol.* dispersion effect
~, дистанционный remote effect
~, длительный long-term effect
~, долготный longitude effect
~ дозы dosage effect
~, донный afterbody effect
~ Доплера/доплеровский Doppler effect
~ Дорна Dorn effect
~, дробовой *electr.* shot effect/noise, Schottky effect
~, дробящий shattering effect
~, дроссельный throttling effect

~ Жамена Jamin effect
~ замедления slowing-down effect
~ затенения shadow/blanketing effect
~ затухания damping effect
~ захвата capture effect
~ Зеемана Zeeman effect
~ земной подушки *av.* ground effect
~, индукционный induction/inductive effect
~, ионизационный ionization/ionizing effect
~ искажения цены price dirtorting effect
~, каландровый *plast.* calender effect
~, каналовый channe(l)ling effect
~, канальный *rad.* ducting
~, квадратичный quadratic effect
~, квантовомеханический quantum-mechanical effect
~ Коанда Coanda effect
~, коллективный cooperative effect
~ Комптона Compton effect/recoil
~ кондиционирования conditioning effect
~, концевой/краевой end effect
~ Кориолиса Coriolis effect
~ Коттон-Мутона *opt.* Cotton-Mouton effect/birefringence
~, краевой fringe/edge effect
~, кратковременный short-term effect
~, кумулятивный cumulative effect
~ Лодочникова Lodochnikov/dispersion effect
~, ложный spurious effect
~, магнитомеханический magnetomechanical effect
~, магнитострикционный magnetostriction, magnetostriction/magnetostrictive effect
~, масштабный/масштаба scale/size effect
~, маховой *mech.* flywheel effect
~ Мейснера Meissner effect
~, межхромосомный interchromosomal effect
~, механический mechanic effect/action
~, микрофонный microphonic effect, microphonism, microphonics, microphonicity, microphony
~ накопления build-up effect, storage effect
~ направленного проявления, ~ направленности *cin.* directional effect
~ насыщения *astr.* effect of confusion
~ непарных спинов unpaired-spin effect
~, непосредственный immediate/direct effect
~, неядерный non-nuclear effect
~ облучения, соматический/генетический somatic/genetic effect of radiation
~, обменный *phys., chem.* exchange effect
~, оболочечный shell effect
~ обратной связи feedback effect
~, обратный inverse/converse effect; boomerang effect
~, общий cumulative effect
~ объемного заряда space-charge effect
~, объемный volume effect
~ ограничения спроса demand damping effect
~ Оже Auger effect
~ Оппенгеймера-Филлипса Oppenheimer-Phillips effect
~ Орбели Orbeli effect
~, островной *electr.* shadow forming effect *(in a tube)*
~ отдачи recoil effect
~ охлаждающий cooling/chilling effect
~ памяти memory effect
~, парниковый greenhouse effect
~, парный pair(ing) effect
~, пастеровский Pasteur's effect
~, переходный transition/transient effect
~ Пигу *econ.* real balance effect, hard-money effect, wealth-income effect, Pigou effect
~, пироэлектрический pyroelectric effect
~ плотности density effect
~, побочный side effect
~, поверхностный *mech.* surface effect; *el.* skin effect; surface action
~ Пойнтинга-Робертсона *astr.* Poynting-Robertson effect
~, поляризационный/поляризации polarization effect
~, пороговый *electr.* threshold effect
~ последействия aftereffect
~ последующего возмущения after-disturbance effect
~ прерывистого освещения *cin.* intermittency effect
~ пространственного заряда space-charge effect
~, прямой direct effect
~, пустотный void effect
~, пьезомагнитный piezomagnetic effect
~, пьезоэлектрический piezoelectric effect
~, радиационный radiative effect
~ разрушения destructive effect
~, распространяющийся spreading effect
~ реакции, тепловой heat of reaction
~ реальных кассовых остатков real balance effect, hard-money effect, wealth-income effect, Pigou effect
~, результирующий net effect
~ Ричардсона Richardson effect
~ самопоглощения self-absorption effect
~ свертывания coagulating effect
~ световой luminous effect
~ сжимаемости compressibility effect
~ симметрии symmetry effect
~ слипания sticky effect
~, смачивающий wetting effect
~ снижения конкуренции anticompetitive effect
~, солевой salt(ing) effect
~, спиновый spin effect
~ стадности *econ.* bandwag(g)on effect
~, стеночный/стенки wall effect
~, стереоскопический relief/stereoscopic effect
~, стереофонический binaural/stereophonic effect
~, суточный day-night effect, diurnal effect

~, теневой shadow effect
~, тензорезистивный tensoresistance, tensoresistive effect
~, тепловой/термический thermal/heating effect
~, термоупругий thermoelastic effect
~, тормозной braking effect
~, третичный tertiary effect
~, туннельный tunnel(ing)/barrier effect
~ увеличения/уменьшения жесткости hardening/softening effect
~ уменьшения плотности reduced-density effect
~ усталости fatigue effect
~ уширения broadening effect
~, флюгерный *av.* weathercocking, weathercock effect
~, фотоэлектрический photoelectric effect, photoeffect
~, фотоэлектрический вентильный photovoltaic effect
~, фотоэлектрический внешний outer photoelectric/photoemissive effect
~, фотоэлектрический внутренний inner photoelectric/photoconductive effect
~, фотоядерный photonuclear effect
~ химической связи chemical-binding effect
~ Холла Hall effect
~, центробежный centrifugal effect
~, широтный latitude effect
~ Шотки *electr.* Schottky/shot effect, shot noise
~ Штарка Stark effect, electric field effect
~ экранирования screening effect
~ экранирования стержней *nucl.* rod-shadow effect
~, экранирующий damping effect
эффективность efficiency, effectiveness, efficacy; performance
~, асимптотическая *stat.* asymptotic(al) efficiency
~, биологическая biological efficiency/effectiveness
~, боевая combat effectiveness
~, геометрическая (лазерного осветителя) light-trapping efficiency
~ закрылков flap efficiency
~ закупок purchasing efficiency
~ захвата capture/catching efficiency
~ использования кормов feed-conversion effectiveness
~ наплавки deposition rate
~ облучения irradiation efficiency
~, общая overal efficiency
~ органов управления control power/effectiveness
~ орошения irrigation efficiency
~ очистки decontamination efficiency
~ применения operational effectiveness
~ регулятора governor effectiveness
~ рулей control power/effectiveness
~ системы system effectiveness
~, совместная *stat.* joint efficiency
~ счета counting efficiency
~ торможения braking action
~ усилителя *rad.* gain-bandwidth product
~, эксплуатационная operational effectiveness
эффективный effective, efficient, efficacious; instrumental, powerful
эффектный spectacular, impressing, striking
эффектор *physiol.* effector
эфферентный efferent, conveying outward/away
эффузив *geol.* effusive rock, effusion
эффузивный effusive
эффузный *bot.* effuse
эхин *arch.* echinus
эхинодорус *bot.* burhead *(Echinodorus)*
эхиноидея *zool.* echinoid, sea urchin
эхинококк *zool.* echinococcus *(Echinococcus)*
эхинококкоз *med.* echinococcosis
~, легкого pulmonary hydatid disease, pulmonary echinococcosis
эхинопанакс *bot.* devil's-club *(Echinopanax)*
эхинопсин *chem.* echinopsin
эхо 1. echo; reflected sound; 2. echoing
~, ближнее close/near(-end) echo
~, вторичное second-trace/-trip echo, second time around echo
~, гармоническое harmonic echo
~, ложное false echo
~, многократное multiple trip/reflection echo, flutter echo
~, постоянное permanent echo
~, радиолокационное radar echo
~, спиновое (nuclear) spin echo
эхо-айсбергомер iceberg detector
эхо-глушитель echo-suppressor
эхограмма echogram, (echo-sound) record
~ маршрутного промера track/line of sounding
эхография echography, echographia
эхо-заградитель echo suppressor
эхозонд echometer
эхо-индикатор (Doppler) echo indicator
эхо-камера echo chamber
эхолалия *med.* echolalia, echophrasia
эхоледомер ice/inverted/topside fathometer, upward-beamed/-scanning fathometer, underice sonar, ice-detecting equipment
~ для определения границ полыньи polynya delineator
эхо-локатор active sonar
эхо-локация echolocation, echo ranging
эхолот *nav.* echo sounder, sonic/echo/acoustic depth finder, fathometer, echo sounding instrument; *av.* acoustic altimeter
~, глубоководный deep-sounding instrument
~ для съемки рельефа дна bottom-charting fathometer
~, навигационный navigational echo sounder
~ с самописцем self-recording echo sounder,

depth recorder, echograph
эхолотирование echo sounding, echolocation
эхомимия *med.* echopathy, echomimia
эхо-отображение *comp.* echoing
эхопеленгование *nav.* echo ranging
эхопечать *comp.* echo-printing
эхопраксия *med.* echopraxia, echomatism, echomotism
эхопроверка *comp.* echo-checking
эхо-резонатор echo box
эхо-сигнал target/echo signal, return
~ (от) цели target echo
эхосчитывание *comp.* read-back
эхофония *med.* echophonia, echophony
эхоэнцефалография *med.* echoencephalography
~, ультразвуковая ultrasound/ultrasonic echoencephalography
эци(ди)альный *bot.* aecial, aecidial
эцидий, эцидиоложе, эцидиосорус *bot.* aecidium, aecium
эцидиоспора *bot.* aecidiospore, aeciospore
эций *bot.* aecium
эциоспора *bot.* aeciospore
эшара *med.* eschar, dead/disorganized tissue
эшафот scaffold
эшелетт *opt.* echelette
~, нарезать rule an echelette
эшелле echelle grating
эшелон 1. *opt.* *echelon;* 2. *av.* flight level, cardinal altitude; 3. *mil.* echelon; special/troop train; *auto.* convoy, column
~ автомобилей train of vehicles
~ вертолетного десанта helicopter wave
~, верхний upper flight level
~, воинский military train
~, второй rear echelon
~, отражательный reflection echelon
~, переходный transition level
~, пропускающий transmission echelon
~ с углем coal train
~, тыловой rear echelon
эшелонирование *av.* separation; *mil.* depth of formation, echelonment, distribution in depth
~, боковое lateral separation
~, вертикальное, ~ по высоте altitude/level/vertical separation
~ во времени separation in time
эшелонировать *mil.* distribute in depth, echelon
эшинит *miner.* (a)eschynite
эш(ш)ольция *bot.* gold poppy *(Eschscholtzia)*
эякулят *physiol.* ejaculate
эякуляция *med.* ejaculation, ejaculatio
~, преждевременная premature ejaculation, ejaculatio pr(a)ecox

Ю, ю

юань yuan, Chinese monetary unit
юбея *bot.* (syrup) jubaea *(Jubaea spectabilis)*
юбилей anniversary (celebration), jubilee
~, двадцатипятилетний twenty-fifth anniversary, silver anniversary
~ столетний centenary
юбилейный anniversary *(attr.)*
юбиляр person/institution whose anniversary/jubilee is celebrated
юбка *tech.* skirt; *rock.* flare
~ газогенератора fuel bell
~ изолятора petticoat, shell, cup, shed
~ капота cowl(ing) shutter
~ купола парашюта canopy skirt
~ поршня (piston) skirt/cup
~ поршня с прорезом split/slotted piston skirt
~, стабилизирующая *rock.* flare, stabilizing skirt
ювелир jewel(l)er
ювенильный *med.* juvenile
юг south
~, географический/истинный true south
~у, обращенный к south-faced
юган *bot.* hay plant *(Prangos pabularia)*
юго-восток south-east
юго-восточный southeast(ern)
юго-западнославянский South-West Slavic
юго-западный southwest(ern)
югослав Yugoslav
югославянка Yugoslav (woman)
югославский Yugoslav(ian)
юго-юго-восток south-southeast
юго-юго-запад south-southwest
югум *bot., zool.* jugum, yoke
югурт yoghourt, yoghurt, yaourt
юдофил Judophile
юдофильство Judophilism
юдофоб anti-Semite
юдофобия, юдофобство Judophobia, anti-Semitism
южноаравийский South Arabic
южноафриканский South African
южнокореец South Korean
южнорусский South Russian
южнославянский South Slavic
юз 1. skid(ding); 2. *com.* Hughes' printing telegraph
~, парировать correct a skid
~ пневматика tyre skidding
юзинь *naut.* houseline
юзист operator of Hughes' printing telegraph
юинтаит *miner.* uinta(h)ite, gilsonite
юит, юитский Eskimo
юитка Eskimo (woman)
юкагир, юкагирский Yukaghir
юкка 1. *bot.* yucca, Spanish dagger *(Yucca)*; 2. *text.* yucca fibre
юкколистный *bot.* yucca-leaved

юкола dry-cured fish, sun-dried fish
юкстагломерулярный juxtaglomerular
юла *zool.* wood lark *(Lullula arborea)*
юлианский Julian
~ календарь Julian calender
юмор humo(u)r
~, мрачный gallows humo(u)r
юмореска *mus., lit.* humoresque
юморист comic writer, writer of comic storis, humo(u)rist
юмористический, юмористичный funny, comic
юморка *bot.* sickle alfalfa *(Medicago falcata)*
юнга sea/ship('s)/sailor/cabin boy; sea cadet
юнга-горнист boy bugler
юнга-музыкант boy musician
юнга-радист boy telegraphist
Юнеско *abbr.* **Организация Объединенных Наций по вопросам образования, науки и культуры** UNESCO, United Nations Educational, Scentific and Cultural Organization
юниор *sport* junior
~ов, команда junior team
юница *agr.* heifer
юнкер *hist.* 1. cadet; 2. Junker, Prussian landowner
юнкерский cadet *(attr.)*
~кое училище military school
юнкерство *hist.* 1. *coll.* Junkerdom, Prussian landowners; 2. cadetship
юнко *zool.* snow-bird, junco *(Junco)*
юнкор *abbr.* **юный корреспондент** young correspondent
юннат *abbr.* **юный натуралист** young naturalist
Юнона *myth., astr.* Juno
юность 1. adolescence, juvenility, juvenescence; 2. youth, young people
юноша youth, young fellow
юношеский juvenile, adolescent; youthful
юношество 1. *coll.* teenagers; 2. youth, juvenility
юный young, juvenile, adolescent
~, топографически topographically young
Юпитер I *astr., myth.* Jupiter
юпитер II floodlight
юра *geol.* Jurassic (period)
юридический legal, judicial, juridical, juristic(al)
~ факультет department of law, law school
~кая консультация legal advice office
~ких наук, доктор Doctor of Laws
~кое лицо juridical person
юрисдикция *law* jurisdiction, extent of power, legal authority
~, апелляционная appellate jurisdiction
~ в области регулирования regulatory jurisdiction
~, исключительная sole jurisdiction
~, консульская consular jurisdiction
~, местная local/territorial jurisdiction
~, надзорная supervisory jurisdiction
~, уголовная criminal jurisdiction
юрисконсульт legal adviser, jurisconsult
юриспрудент *obs.* lawyer
юриспруденция jurisprudence, (science of) law
юрист lawyer; *colloq.* student of law
юрод(ивец) *rel.* God's fool
юродивость mental alienation, folly, craziness
юродивый 1. *n. rel.* God's fool; 2. feeble-minded, foolish, cracked, crazy
юродский touched, feeble-minded, crazy
юродство 1. aberration, foolishness; 2. idiocy, madness, craziness
юрок *zool.* bramble finch, brambling *(Fringilla montifringilla)*
юрский *geol.* Jurassic
юрт 1. allotment of land of a Cossack village/farm; 2. *hist.* territory of a Tartar principality
юрта yurt(a), jurt, nomad's tent
Юрьев день *rel.* St. George's day
юссив *gram.* construction expressing command; jussive
юстировальный aligning, tunning, adjusting
юстирование alignment, adjustment; positioning; tuning, regulating
юстировать align, adjust; tune; regulate
юстировка adjustment, alignment; tuning; positioning
~, боковая side alignment
~ номинала adjustment/trimming to value
~ камеры camera adjustment/ *telev.* alignment, lining up of camra
~ магнита adjustment of a magnet
~ объектива lens/objective adjustment
~ пучка *electr.* positioning of a beam
~, точная fine adjustment
юстировочный adjusting, aligning, regulating
юстировщик tuner; adjuster, aligner
юстиция justice; jurisdiction
ют *naut.* poop, quarter-deck; (erection) aftercastle
ютландский Jutlandish
юферс *naut.* deadeye, dead man's eye
юфть Russian leather, yuft
~, яловая Russian calf leather
ююба *bot.* jujube, buckthorn *(Zizyphus)*
~ унаби common jujube *(Zizyphus jujuba)*

Я, я

я ego, the I
~, второе alter ego
яблоко 1. apple; 2. apple tree
~, адамово *anat.* Adam's apple, pomum Adami, laryngeal prominence
~, глазное *anat.* globe of the eye, eyeball
~, звездное *bot.* star apple *(Chrysophyllum caimito)*
~ раздора *fig.* bone of contention, apple of discord
~ старнпоста *naut.* sternpost bossing, stern boss

~ **шарнира Гука** casing of a joint
яблоковидный, яблокообразный maliform, pomiform, apple-shaped
яблоневый apple tree *(attr.); n. pl. bot.* apple family *(Malaceae, Pomaceae)*
яблоня apple tree *(Malus)*
яблочко bull's eye
яблочный pomaceous, apple *(attr.)*
~ **пирог** *cul.* apple pie
~**ная кислота** malic acid
яванец Javanese
яванка Javanese (woman)
яванский Javanese
яванский Javanese
~ **человек** Java man, Pithecanthropus
явантроп *anthr.* Java man
явка 1. appearance, presence, attendance; 2. secret meeting place, secret address/meeting, safe house
~ **в суд** appearance in court
~ **обязательна** attendance is compulsory
явление 1. phenomenon, appearance; occurrence; event; effect; 2. *theatre* scene; *see also* **явления**
~, **атмосферное** atmospheric phenomenon
~, **вредное** ill effect
~ **вулканизма** manifestation of volcanism
~ **гало** halo phenomenon
~ **гистерезиса** creeping magnetism, hysteresis effect
~, **дейтерическое** deuteric effect
~ **Жамена** Jamin effect
~, **закономерное** natural phenomenon
~, **запаздывания** lag effect
~ **кавитации** cavitation effect
~, **карстовое** *geol.* karst/solution feature
~, **контактное** contact phenomenon
~, **массовое** mass phenomenon
~ **мертвой фазы** *met.* dead effect
~ **напряжений** strain phenomenon
~, **неядерное** non-nuclear phenomenon
~, **опасное атмосферное** atmospheric hazard
~ **остаточного магнетизма** remanence effect
~ **отталкивания** *naut.* cushioning effect
~ **переноса** transport/transfer phenomenon
~, **позднемагматическое** deuteric effect, late-magmatic effect
~, **предварительное** forerunning phenomenon
~, **приливно-отливное** tidal effect, tide
~ **природы** natural phenomenon
~ **резонанса** resonance condition/effect
~, **сейсмическое** seism
~, **селевое** mudflow phenomenon
~, **сопутствующее** attendant phenomenon
~, **тектоническое** tectonic phenomenon/event
явления effects, phenomena; *see also* **явление**
~, **болезненные** symptoms of disease
~, **вторичные** second-order effects
~, **коллективные** cooperative phenomena
~, **метеорные** meteoritic phenomena
~ **на поверхности (образца)** surface phenomena
~ **общественной жизни** social phenomena
~, **остаточные** residual effects
~, **побочные** side-effects
~, **происходящие в пограничном слое** boundary-layer phenomena
~, **трехмерные** three-dimensional effects
явленный *rel.* appearing miraculously
являться 1. appear, present oneself; *mil.* report; 2. arrive, show/turn up; 3. occur, arise
~ **в суд** *law* appear in court
явнобрачный *bot.* phanerogamous, phanerogamic; *n. pl.* flowering plants, phanerogams *(Phanerogamae)*
явновыраженный clearly defined, noticeable, salient; distinct
явнокристаллический phanerocrystalline, eucrystalline
явнолепестный *bot.* phaenopetalous
явнопыльниковый *bot.* phanerantherous
явность clearness, obviousness, evidence; visibleness; patency
явный 1. distinct, evident; patent, manifest, open, clear; plain; 2. obvious; *math.* explicit
~**ная небрежность** sheer carelessness
~**ное недоразумение** evident misunderstanding
явор *bot.* sycamore (maple) *(Acer pseudoplatanus)*
явочный 1. secret; 2. *mil.* recruiting; reporting
~ **пункт** recruiting office
~**ная квартира** secret address
явственный distinct(ive), manifest
явствовать be clear/obvious/apparent
явь reality
ягдташ game/fowling bag
ягель *bot.* cup/reindeer moss *(Cladonia rangiferina),* lichen
ягнение lambing
ягниться yean, lamb
ягнятник *zool.* lammergeyer, lammergeier *(Gypaetus barbatus)*
ягода berry
~, **бузинная** elderberry
~, **буйволова** *bot.* buffalo berry
~, **винная** dried fig
ягодицы *pl. anat.* buttocks
ягодичный *anat.* gluteal; *zool.* pygal
ягодка *bot.* drupelet
ягодковые *n. pl. bot.* Thymelaeaceae
ягодник *agr.* berry field/plantation/patch; berry bush
ягодовидный, ягодообразный berrylike, bacciform, berry-shaped
ягодоносный bacciferous, berry-bearing
ягодомойка berry washer, berry washing machine
ягуар *zool.* jaguar *(Felis/Panthera onca)*
ягурт *cul.* yaghourt, yog(h)urt
яд poison; venom; toxin
~, **быстродействующий** quick-acting poison
~, **бытовой** household poison
~ **для грызунов** rodenticide
~, **змеиный** snake venom

~, каталитический anticatalyst
~ кобры ophiotoxin
~ комплексного действия complex action poison
~, кровяной blood poison/toxin
~, крысиный ratsbane
~ насекомого insect poison
~, некумулятивный noncumulative poison
~, производственный/промышленный industrial poison
~, пчелиный bee venom
~, радиоактивный radiation/radioactive poison
~, растительный plant/vegetable poison
~, трупный ptomaine
ядар *geog.* yadar, subpolar meadow
ядерно-резонансный nuclear-resonance
ядерно-химическй chemonuclear
ядерно-цитоплазматический nucleocytoplasmatic
ядерно-чистый nuclear-pure
ядерный 1. nuclear; 2. nucleate(d); core *(attr.)*
~ сок caryolymph
~ реактор nuclear reactor
~ная оболочка nuclear membrane, karyotheca
ядовитый poisonous, toxic, venomous; vitriolic
ядозуб *zool.* beaded lizard *(Heloderma)*
ядоносность toxicity, venomousness, poisonousness
ядоносный poison-bearing; toxiferous; venomous
ядообразующий toxigenic
ядопровод *agr.* toxic-matter tube
ядохимикат *agr.* pesticide
ядра nuclei; *see also* **ядро**
~, бугорковые/бугорные *med.* tuberal nuclei
~ Вигнера/зеркальные mirror(-image) nuclei, Wigner nuclei
~, внутрипластинчатые intralaminar nuclei
~, выстроенные oriented nuclei
~, добавочные accessory nuclei
~, дугообразные arcuate nuclei
~, комплементарные complementary nuclei, co-nuclei
~, концевые terminal nuclei
~ моста *med.* nuclei of the pons
~, начальные nuclei of origin
~, ориентированные oriented nuclei
~, полузеркальные semimirror nuclei
~, преддверные *anat.* vestibular nuclei, nuclei of the vestibular nerve
~, противоположно выстроенные oppositely-aligned nuclei
~, (само)сопряженные self-conjugate nuclei
~, слуховые/улитковые *anat.* cochlear nuclei
ядреный 1. having a well-developed kernel; juicy, succulent; vigorous, healthy; 2. strong *(of drinks);* 3. fresh, envigorating
ядрица *agr.* unground buckwheat
ядро 1. kernel; 2. core, nucleus; 3. *sport, hist., mil.* ball, shot
~, автономное accessory nucleus
~, альфа-частичное alpha-particle nucleus
~, аппаратное *comp.* hardcore
~, ароматическое aromatic ring/nucleus
~, атомное (atomic) nucleus
~ безопасности *comp.* security kernel
~, бесспиновое *phys.* spin-zero nucleus
~ блокового нерва nucleus of the trochlear nerve
~, боковое lateral nucleus
~, бомбардируемое struck/bombarded nucleus
~, верхнее superior nucleus
~ вихря vortex core
~, возбужденное excited nucleus
~, вращающееся spinning nucleus
~, вставочное *med.* Staderini's nucleus
~, вторичное secondary nucleus
~, вытянутое prolate nucleus
~, галактическое galactic nucleus
~ гелия helion; alpha-particle
~, голое bare/naked nucleus
~ гомоморфизма *math.* kernel of a homomorphism
~, грудное thoracic nucleus, Clarke's column
~, дважды магическое double-magic nucleus
~, двигательное motor nucleus
~ дейтерия deut(e)ron, deuton
~, делящееся fissioning/fissile/fissionable nucleus
~, деформированное deformed nucleus
~ дислокации *cryst.* dislocation core
~, диффузионное diffusion kernel
~ добавочного нерва nucleus of the accessorial nerve
~, добавочное accessory nucleus
~, дочернее daughter/product nucleus, daughter
~ древесины core of wood, duramen, heartwood
~, заднее posterior nucleus
~ замедления slowing-down kernel
~ зародыша *cryst.* germ nucleus
~, захватывающее capturing nucleus
~ Земли (внутреннее) (inner) core of the Earth
~ зубчатое dentate nucleus
~, изобарное/изомерное isobaric/isomeric nucleus
~ интегрального уравнения *math.* kernel of an integral equation
~ инфузорий, генеративное micronucleus
~, исходное original/parent/mother nucleus
~ кавитации cavitation core/nucleus
~, клеточное nucleus
~, клиновидное cuneate nucleus
~, кокардовое crusted kernel
~ кометы cometary nucleus, nucleus of a comet
~, конечное final nucleus
~ континента archicontinent
~, континентальное continental nucleus
~, короткоживущее short-lived nucleus
~, красное red/tegmental nucleus
~, легкое light nucleus

~ линейного преобразования *math.* null space of linear transformation
~, лишенное внешних электронов stripped nucleus
~, магическое *math.* magic(-number) nucleus
~, материнское *math.* parent/mother/original nucleus
~, межножковое interpeduncular ganglion
~, металлическое metal core
~, мыльное kettlewax, neat soap
~, мякотное nucleus pulposus/gelatinosus
~, надзрительное supraoptic nucleus
~, наружное окаменелости external cast
~, немагическое non-magic nucleus
~, нестабильное unstable nucleus
~, нечетное odd-mass nucleus
~, нечетно-нечетное odd-odd nucleus
~, нечетно-четное odd-even nucleus
~, образовавшееся product nucleus
~ окостения ossification number
~ оператора *math.* kernel of an operator
~ ослабления attenuation kernel
~, остаточное residual nucleus
~ основания *constr.* core of a foundation
~ отдачи recoil nucleus
~, отделившееся/отжатое/отслоившееся/пережатое detached core
~, первичное primary nucleus
~, переднее anterior nucleus
~, перекрывающее overlap kernel
~ плазмы plasma core
~ пламени flame cone; *rock.* kernel
~, плоское plane kernel
~ плотины core of a dam
~ поводка habenular nucleus, nucleus of the habenula
~, подбугорное subthalamic/hypothalamic nucleus, nucleus/body of Luys
~, покоящееся resting nucleus
~ половой клетки germ/reproductive nucleus
~ последовательности *math.* core of a sequence
~ потока *hydr.* flow core
~ почки nucellus
~, почти магическое near-magic nucleus
~, предкрышечное pretectal nucleus
~, пробковидное emboliform nucleus, embolus
~ программы *comp.* program kernel
~ продукт деления fission-product nucleus
~, промежуточное compound nucleus; *med.* interstitial/intermediate nucleus
~, пушечное *mil.* cannon ball
~, радиоактивное radioactive nucleus, radionuclide
~, разрешающее *math.* resolvent/reciprocal kernel, resolvent
~, распадающееся disintegrating/decaying nucleus
~, рассеивающее scattering nucleus
~ рассеяния scattering kernel
~, расщепляющееся disintegrating nucleus
~, резонансное resonant nucleus
~, результирующее resultant/product nucleus
~ с избытком нейтронов, ~ с избыточным числом нейтронов neutron-rich nucleus
~ с нечетным зарядом odd-charge nucleus
~ сварной точки spot weld nugget
~, сверхзвуковое supersonic core
~ свода arch core
~, сегментарное/сегментационное segmentation/clevage nucleus
~ семяпочки ovule nucleus/nucellus
~ сечения *mech.* core of a (cross-)section
~ синклинали core of syncline
~ складки core of fold
~, скульптурное external cast
~, сложное compound/complex nucleus
~ смещения displacement kernel
~, соляное salt core
~, составное compound nucleus
~, сплюснутое oblate nucleus
~, способное к делению fissionable/fissile nucleus
~, среднетяжелое medium(-weight) nucleus, intermediate nucleus
~ сублимации nucleus of sublimation
~ тепловыделяющего элемента fuel element core
~, тонкое nucleus of the slender funiculus, nucleus of Goll's column
~, транспортное transport kernel
~ третьего поколения, дочернее granddaughter
~ трития triton
~ тяжелое/сверхтяжелое heavy/superheavy nucleus
~ управляющей программы control program nucleus
~ упругого рассеяния elastic-scattering kernel
~, хвостатое caudate nucleus
~ хрусталика nucleus of the lens
~, четное even-mass nucleus
~, четно-четное even-even nucleus
~, чувствительное sensory nucleus
~, шаровидное spherical nucleus
~ шатра roof nucleus, nucleus of the fastigium
~ экранированное shielded nucleus
~ эндосперма endosperm nucleus
~ Юкавы Yukawa/diffusion kernel
~ ядрышка nucleolus, nucleolinus, nucleolonucleus, nucleus of a nucleolus
ядровый core *(attr.),* nucleus *(attr.),* kernel *(attr.)*
~вая древесина duramen, heartwood
~вое мыло high-grade soap
ядро-мишень target nucleus
ядроносный nucleiferous
ядроподобный nucleiform, nucleoid, kernel-shaped
ядро-продукт daughter/product nucleus

ядротехника nucleonics, nuclear engineering/technology
ядрышко *biol.* nucleolus
~, двойное amphinucleolus, amphinucleus
~, ложное false nucleolus
~, хроматичное/хроматиновое chromatin nucleolus
язва ulcer, sore; *med.* ulcus
~, аденская Aden/tropical/Malabar ulcer, Naga/desert sore
~ анастомоза anastomotic ulcer
~, атероматозная atheromatous ulcer
~, быстро распадающаяся cancrum, canker
~, варикозная varicose/stasis/gravitational ulcer
~, венерическая venereal ulcer
~, воспалительная inflamed ulcer
~, восточная = пендинская
~ вульвы, острая Lipschutz' disease, acute ulceration of the vulva
~, глубокая hollow ulcer
~ двенадцатиперстной кишки duodenal (peptic) ulcer
~, декубитальная decubitus ulcer, bedsore
~ желудка gastric ulcer, ulcer of the stomach
~, ишемическая ischemic ulcer
~ кишечника intestinal ulcer, enterelcosis
~ кожи, (лучевая) (radiation) cutaneous ulcer
~ кожи, раковая rodent/Jacob's ulcer, noli me tangere
~, коррозионная *met.* (corrosion) pit
~, кровоточащая h(a)emorrhagic ulcer
~, медленно заживающая atonic ulcer
~, моровая plague
~, омозолелая callous/indolent ulcer
~, пендинская oriental/date/Aleppo/Delhi/godovnic/Bagdad/Natal/tropical boil; endemic/Turkestan/Annam ulcer; Aleppo veil, Biskra button, cutaneous leishmaniasis
~ пищевода esophageal ulcer
~, ползучая serpiginous/creeping ulcer
~, прободающая perforating ulcer
~ пупка omphalelcosis
~, разъедающая phagodenic/perambulating ulcer
~, сибирская anthrax
~, сифилитическая syphilitic/hard/autochthonous ulcer, chancre
~ тощей кишки, пептическая jejunal/stomal ulcer
~, хромовая chrome/tanner's ulcer
~, щелевидная fissured ulcer
язвенник *bot.* anthyllis, kidney-vetch *(Anthyllis)*
язвенниколистный *bot.* anthyllidifolious
язвенно-вегетирующий ulcerovegetative
язвенно-железистый ulceroglandular
язвенно-некротический ulceronecrotic
язвенно-пленчатый ulceromembranous
язвенный ulcerous; *obs.* caustic, biting
~ная болезнь stomach ulcer
язвина pit, indentation; corrosion spot; dimple
язвительный caustic, sarcastic; acrimonious
язык 1. *anat.* tongue; 2. language; 3. clapper *(of bell)*
~, алгоритмический algorithmic/programming language
~, американский морской *zool.* American sole *(Achirus fasciatus)*
~ ассемблера assembly program language, assembler language
~, базовый core language
~ без присваиваний assignment-free language
~, близкий к естественному near-natural language
~, болезненный *med.* glossalgia, glossodynia, glossagra
~, бороздчатый fissured tongue
~ булевых операторов Boolean-based language
~, волосатый hairy tongue
~, ворсинчатый villous tongue
~, входной input/source language
~ высокого уровня higher-level language
~, выходной object/target language
~, географический geographical/mappy tongue
~ глетчерного льда glacier tongue
~ графических символов graphics language
~, двойной diglossia
~, дескрипторный descriptor language
~ диалога conversational language
~, естественный natural language
~ запросов query language
~ интерактивного взаимодействия interactive language
~ интерпретируемый interpreted language
~ исходный source language
~ ключевых слов keyword-oriented language
~ команд(ный) instruction/command language
~ конечного пользователя end-user language
~, копченый *cul.* smoked tongue
~ ледника *geol.* ice/glacier tongue
~ макроассемблера macro language
~ макрокоманд macroinstruction language
~ манипулирования данными data manipulation language
~ математики mathematical language/terms
~, машинно-независимый computer-independent language
~, машинно-ориентированный computer-oriented language
~, машинный machine language/code
~ машины, собственный native language
~, младописменный language recently put into writing
~ мнемосхем mnemonic language
~ моделирования simulation language
~ моделирования задач SIMSCRIPT, simulation language for operations research problem
~, морской *zool.* 1. (common) sole *(Solea vulgaris);* 2. tongue sole *(Symphurus)*
~, недоразвитый *med.* ateloglossia

~, непроцедурный nonprocedural language
~ низкого уровня lower-level language
~, обложенный *med.* coated/furred tongue
~ обработки списков list-processing language
~, общий вычислительный common computer language
~, объектно-ориентированный object-oriented language
~, объектный object/target language
~ определения данных data definition language
~ оригинала source language
~, основанный на логике предикатов predicate logic language
~ перевода target language
~ пламени body of flame, tail
~ постановки задач problem-statement language
~ предикатов predicate language
~ представления знаний knowledge representation language
~, проблемно-ориентированный problem-oriented language
~ программирования program(ming) language
~ программирования, удобный для восприятия человеком human-readable language
~ продукций production language
~ проектирования design language
~, процедурный procedural language
~, раздвоенный *anat.* bifid/cleft tongue
~, реляционный relational language
~, родной native language
~ с гнездовой структурой nested language
~ с укороченной уздечкой *anat.* tongue-tie, ankyloglossia, lingua fraenata
~ сверхвысокого уровня super language
~, символический symbolic language
~ системы искусственного интеллекта AI [artificial intellect] language
~, скарлатинозный малиновый *med.* (dark red) strawberry/raspberry tongue
~, складчатый furrowed/fluted/grooved tongue
~, средневерхненемецкий Middle High German
~, старописьменный language with an old system of writing
~, сухой dry tongue
~, табличный tabular language
~ тепла *meteor.* warmth tongue
~, транслируемый compiled language
~ транслятора, входной source language
~ транслятора, выходной object language
~, увеличенный *med.* macroglossia, megaloglossia
~, уменьшенный *med.* microglossia
~ управления заданиями job control language
~, условный code language
~, формальный formal/symbolic language
~, фреймовый frame-based language
~ функционального программирования functional language
~ холода *meteor.* cold tongue
~ целевой машины target language
~, черный *med.* black (hairy) tongue, glossophytia, lingua nigra
~, эталонный reference language
языковед linguist
языковедение linguistics
языковедческий linguistic(al)
языковидный, языкообразный linguiform, lingulate, tongue-shaped
языковой linguistic, language *(attr.)*
языковый *anat.* lingual, glossal, tongue *(attr.)*
языко-глоточный *anat.* glossopharyngeal
якыкодержатель tongue-holding forceps, tongue depressor
~, пружинный spring tongue depressor
языкознание linguistics, science of language
язык-основа primitive language
языкотворец person coining new words/expressions
языкотворчество formation of new words/expressions
языкотруб *bot.* salpiglossis *(Salpiglossis)*
язык-посредник mediator/intermediary/intermediate language
языческий pagan, heathen, heathenish
язычество paganism, heathenism; heathendom
язычковидный lingulate, linguliform
язычковолистный, язычколистный *bot.* ligulifolious
язычковый ligulate(d), *anat.* uvular; *n. pl. zool. Pentastomida*
~ цветок *bot.* semifloret, ray flower
язычник pagan, heathen
язычно-небный *anat.* glossopalatine
язычно-язычковый *anat.* glossostaphyline
язычок *anat.* lingula, tongue, uvula; *bot.* ligule, ligula; *tech.* lug, catch, tab; *mus.* reed
~ вязальной иглы latch
~ замка lock catch
~ мозжечка tongue of the cerebellum
~ нижней челюсти mandubular tongue
язь *zool.* ide *(Leuciscus idus)*
яичко *anat.* testicle; *bot.* ovule
~, не опустившееся в мошонку monorchidism, monorchidia, retained/undescended testicle
~, перевернутое inverted testis
яичковидный, яичковый testicular
яичник *anat.* ovary, oarium, ootheca, oophoron
~, большой серый polycystic ovary, enlarged cystic ovary
~, добавочный third/accessory ovary
~, раздвоенный ovarium bipartitum/disjunctum
~, размягченный oothecomalacia, oophoromalacia
~, складчатый ovarium gyratum

~, увеличенный oothecauxe, oophorauxe
~, узловатый mulberry ovary
яичниковый ovarian, oaric
яичница *cul.* fried eggs
яичница болтунья/глазунья *cul.* scrambled eggs
яичный 1. egg *(attr.)* 2. yellow; 3. egg-shaped
~ белок egg white
~ желток yolk of egg
~ порошок powdered/dried eggs
яйла mountain pasture
яйцевидно-копьеобразный/ланцетовидный ovate-lanceolate
яйцевидный egg-shaped, oviform, ovoid, oval
яйцевод *anat.* oviduct
яйцеводный oviducal
яйцевой *biol.* oval, ovular
яйцеед *zool.* egg parasite
яйцеживородящий ovoviviparous
яйцеживорождение *zool.* ovoviviparity
яйцеклад *zool.* ovipositor
яйцекладка oviposition
яйцекладущий oviparous, ootocous, egg-laying
яйцеклетка *biol.* ovum, female sexual cell, oocyte, ovocyte, egg cell
яйценоский egg-laying
~кие куры *agr.* good laying hens
яйценоскость *agr.* egg-laying capacity/qualities, laying
яйценосный oviferous, oviparous
яйцеобразный oviform, ovoid, oval, egg-shaped
яйцеобразование ovogenesis
яйцепровод *anat.* oviduct
яйцеродность *zool.* oviparity
яйцеродный oviparous
яйцеродящий oviparous, ootocous
яйцерождение *zool.* oviparity
яйцо egg; *biol.* ovum
~, алецитальное alecithal ovum
~ всмятку soft-boiled egg
~, голобластическое holoblastic ovum
~, крутое hard-boiled egg
~, куриное hen's egg
~, лецитальное lecithal/meroblastic ovum
~, оплодотворенное fertilized egg
~, первичное ooblast, primitive ovum
~, центролецитальное centrolecithal ovum
як *zool.* yak *(Poephagus grunniens)*
якаранда *bot.* jacaranda *(Jacaranda)*
якать pronounce unaccented **е** as **я** and **о** as **а**
якобиан *math.* Jacobian, functional/Jacobian/determinant
якобинец *hist., polit.* Jacobin, extreme radical, extremist/radical in politics
якобинизм, якобинство Jacobinism
якобинский Jacobinic(al)
якобит *hist.* Jacobite
якобсит *miner.* jacobsite
якоревидный anchorlike
якорек anchor-shaped fishhook
якорница *naut.* anchor-bearing boat
якорный anchor(ing)
~ корень anchoring root
~ная стоянка anchorage
якорцы *pl. bot.* caltrop *(Tribulus)*
якорь 1. *naut.* anchor; 2. *el.* armature
~, адмиралтейский admiralty/common anchor
~, барабанный drum/cylindrical armature
~ без сердечника coreless armature
~, бесштоковый stockless anchor
~, винтовой мертвый screw anchor
~, гладкий smooth-core armature
~, грибовидный mushroom anchor
~, дву(х)лапый double-fluked anchor
~, держащий riding anchor
~, держащий судно при отливе ebb anchor
~, дисковый disk(-type) armature
~, дискообразный *el.* discoidal armature
~, заведенный по траверсу breast anchor
~, импровизированный jury anchor
~, кольцевой ring armature
~, ледовый ice anchor/drag
~, мертвый dump mooring, mooring anchor/block, permanent mooring anchor, mooring clump
~ мины mine anchor
~, неполяризованный nonpolarized/neutral armature
~, нечистый foul anchor
~, неявнополюсный non-salient-pole armature
~, однолапый one-fluke anchor
~, однорогий one-arm(ed) anchor
~, осевой axial armature
~, плавучий drogue, dift/floating/sea/water anchor, kite/sea drag
~, поворотный hinged armature
~, подветренный lee anchor
~, полый hollow armature
~ реле armature of relay
~ с вентиляционными каналами hole/perforated armature
~ с вращающимися лапами swinging-fluke anchor
~ с пазами slotted armature
~ с простой обмоткой single-wound armature
~ со складными лапами folding/portable anchor
~ со стержневой обмоткой bar-wound armature
~, становой bower (anchor)
~ Хола Hall's anchor
~, шаровой ball armature
~, шлюпочный boat anchor
~, штоковый stock(ed) anchor
~, явнополюсный pole/radial armature, (salient-)pole armature
якорь-зонт screen anchor
якорь-опора anchor leg
якорь-трап anchor trap
якулятор jaculator, dart-thrower
якут, якутский Yakut

ял *naut.* rowing boat, yawl
ялапа *bot.* jalap, marvel of Peru *(Mirabilis jalapa)*
ялик *naut.* scull(er), skiff, wherry, dinghy
~, плоскодонный punt
яличник ferryman, wherryman
яловка = яловица 1. *agr.* barren/dry cow; barren sheep; 2. *leath.* buff(s), heifer, cowhide
~, краснодубная bark-tanned side leather
~, хромовая chrome kips
яловость barenness, dryness
яловый barren, dry
ям *hist.* mail staging post, post coach station
яма 1. pit, pot, hole; bunker; well; depression, hollow; *anat.* fossa; 2. *obs.* prison
~, аккумуляторная *nav.* battery tank/well
~, височная temporal fossa
~, воздушная air pocket
~, волчья *mil.* traphole, trou-de-loup
~, выгребная cesspool, cesspit
~, гауссовская потенциальная Gaussian potential well
~, глиняная clay pit/hole
~, глубоко потенциальная deep potential well
~ для выжигания угля charring pit
~ для окалины scale pit
~, долговая debtors prison
~, защитная shield pit
~, йодная iodine pit
~, компостная *agr.* compost pit
~, кочегарная cinder pit
~, литейная casting/foundry pit
~, отстойная settling pit
~, петлевая *met.* looping pit
~, подводная submarine pit/well
~, потенциальная *el.* potential well
~, прямоугольная square well
~, седалищно-прямокишечная Velpean's/ischiorectal fossa
~, силосная silo pit
~, скиповая *met.* skip pit
~, смотровая inspection pit
~, соляная salt pan
~, угольная coal bunker
~, шлаковая *met.* slag/cinder pit
ямайский Jamaican, Jamaica *(attr.)*
ямайское *n.* Jamaica rum
яма-отстойник *met.* settling pit
ямаскит *geol.* yamaskite
ямб *lit.* iambus, iambic verse
~, четырехстопный iambic tetrameter
ямбический iambic
ямбоза *bot.* rose apple, jambosa *(Syzygium jambos)*
ямболан *bot.* duhat, jambolan (plum), Java plum *(Syzygium jambolanum)*
ямка *anat.* fossa, fovea, pit, cap-shaped depression; dimple; socket, cavity
~ атланта atlas cavity
~ атланта, верхняя суставная superior articular fovea of the atlas
~, блоковая trochlear pit
~, большая надключичная greater supraclavicular fossa
~ большого мозга, боковая lateral fossa of the brain, fossa of Sylvius
~, венечная coronoid fossa
~, вертельная trochanteric/digital fossa
~ вертлюжной впадины acetabular fossa
~, вертлюжная acetabulum
~, височная temporal fossa
~, гипофизарная hypophysial fossa, pituitary fossa of the sphenoid bone
~, двубрюшная digastric fossa
~ желчного пузыря gall-bladder fossa
~ затылочной кости, мыщелковая condylar fossa
~, зубная tooth socket
~, клиновидной кости, ладьевидная scaphoid fossa of the sphenoid bone
~, крыловидная pterygoid fossa/pit
~, локтевая cubital fossa, chelidon
~ локтевого отростка olecranon fossa
~, лучевая radial fossa
~, малая надключичная lesser supraclavicular fossa, Zang's space
~, медиальная паховая medial inguinal fossa
~, межножковая interpeduncular fossa
~ миндаликовая tonsillar fossa
~ мочеиспускательного канала, ладьевидная navicular fossa of the urethra
~, надминдаликовая supratonsilar fossa
~, надостная supraspinous fossa
~, надпузырная supravesical fossa
~, надчревная epigastric fossa, pit of stomach
~ наковальни fossa for the incus
~, нижнечелюстная mandibular fossa, glenoid fossa/cavity
~, нижняя fovea inferior
~, овальная oval fossa
~, подвздошная iliac fossa
~, подвисочная infratemporal/zygomatic fossa
~, поддуговая subarcuate/floccular fossa
~, подколенная politeal fossa
~, подкрыльцовая axillary fossa, armpit
~, подложечная epigastric fossa
~, подлопаточная subscapular fossa
~, подостная infraspinous fossa
~, подчелюстная submandibular fossa
~, посадочная *agr.* hole
~, резцовая incisive fossa
~, ротовая stomodeum
~, связочная ligament pit
~, септальная septal groove
~ сетчатки, центральная central pit
~, собачья canine fossa
~, стекловидная patellar fossa
~ черепа cranial fossa
~, яремная jugular fossa
ямкоделатель *agr.* dibble, dibber
ямкообразный foveate(d)
ямокопатель *agr.* hole digger, hole/earth borer,

earth drill, pitting machine
~, горный hillside earth borer
~, садовый orchard earth borer
ямочка *anat.* pit, foveola, minute fovea
~, желудочная gastric pit
~, зернистая granular pit, Pacchionian depression
ямочно-гребенчатый pitted-combed
ямс *bot.* yam *(Dioscorea)*
ямчато-точечный foveolate-punctate
ямчатый pitted, foveate, lacunose; scrobiculate
янки Yankee
янозит *miner.* janosite, copiarite
янсенизм *eccl.* Yansenism
янсенист *eccl.* Yansenist
янтарный succinic, succinous; amber(-colo(u)red)
янтарь amber, succinite
~, голубоватый blue amber
~, канадский Canadian/Canada amber
~, костяной osseous/bone amber
~, черный jet, black amber
янтинит *miner.* ianthinite, janthinite
Янус *myth.* Janus; *fig.* Janus-face
яничар *hist.* janissary, janizary
японец Japanese
японист, японовед specialist on Japan
японистика, японоведение Japanese studies
японка I Japanese (woman)
японка II cap sleeve
японский Japanese
~кая борьба jiu-jitsu, ju-jitsu
яр I *geol.* 1. gill, ravine; 2. steep bank
яр II *physiol.* heat
яранга skin tent, yaranga; tepee
ярд yard *(measure)*
ярданг *geol.* yardang, ridge formation by wind erosion
яремный *anat.* jugular
яриться *physiol.* be in heat
ярка *agr.* young ewe
яркомер brihtness meter
~, объективный/фотоэлектрический photoelectric photometer
~, субъективный visual photometer
яркостно-модулированный intensity-modulated
яркость 1. brightness, *telev.* luminance, luminosity; radiance, brilliance; 2. *fig.* vividity, flamboyance
~ изображения image bightness
~ освещения intensity of the lighting
~ света, полная full-light intensity
~ свечения, эффективная effective intensity of luminosity
~ фона background brightness
ярл *hist.* (Scandinavian) noble/schief(tain); jarl, earl
ярлык 1. label, tag; 2. *hist.* edict, yarlyk
~, адресный address tag
~, багажный baggage label
~ с названием/номером tally card
~ с указанием необходимого ремонта repair tag
ярмарка (trade) fair
ярмо 1. *el.* field frame, yoke; 2. oxbow
~ рабства yoke of slavery
~ трансформатора transformer yoke
ярмозубые *n. pl.* zeuglodonts *(Zeuglodontia)*
ярмозубый *paleont.* zeuglodont
яровизация *agr.* vernalization, jarovization
яровизировать *agr.* vernalize, jarovize
яровина, яровое *agr.* spring crop
яровой *agr.* (a)estivial, spring *(attr)*
~вые культуры/хлеба spring crops
ярость fury, rage, frenzy
ярочка *agr.* ewe lamb
ярунок miter/bevel square
ярус 1. tier, layer; story, storey; *obs.* floor; 2. *theatre* circle; 3. *geol.* stage, stratum, layer; 4. long line *(for fishing)*
~ верхний overstor(e)y upper layer; upper deck
~, второй second balkony
~ кустарников shrub layer
~ листового полога plant canopy, canopy level
~ облаков cloud level/sheet
~, структурный *geol.* structural stage
ярусность layering, stratification
ярусный tiered, floored, layered; graduated, steplike, stepped
ярутка *bot.* pennycress *(Thlaspi)*
ярыга *hist.* constable
ярь(-медянка) verdigris, basic cupric acetate
яс jar, jarring device
~, бурильный drilling jar
~, длинноходовой lond-stroke jar
~, ловильный fishing jar
ясак 1. *hist.* tribute. tax, duty; assessment; 2. *obs.* signal
ясачник *hist.* tax/tribute collector
ясельничий *n. hist.* superintendent of the stables
ясеневый, ясенный ash(en)
~ лес ash forest
ясенец *bot.* dittany, burning-bush *(Dictamnus)*
ясенка *bot.* chinaberry *(Melia)*
ясень *bot.* ash(tree) *(Fraxinus)*
ясколка *bot.* chickweed, cerastium *(Cerastium)*
ясли *pl.* 1. *agr.* feeding rack; hayrack; crib, manger, trough; 2. day nursery
ясменник *bot.* woodruff *(Asperula)*
ясменниколистный *bot.* woodruff-leaved
ясмин *bot.* jasmin(e), jessamin(e) *(Jasminum)*
ясновидение clairvoyance
ясновидец, ясновидящий clairvoyant
яснотка *bot.* dead nettle *(Lamium)*
яснотколистный *bot.* dead-nettle-leaved
ясный bright, clair; fine; lucid, distinct, coherent; obvious, apparent
яспис *obs.* jasper
ястреб *zool.* hawk *(Accipiter)*
ястребинка *bot.* hawkweed *(Hieracium)*
ястребинный hawkish, accipitral; vulturine
~ная охота falconry
~ная сова hawk owl *(Surnia ulula)*

ястреб-перепелятник *zool.* sparrow hawk *(Accipiter nisus)*
ястребятник falconer, hawker
ястык 1. fish eggs in the ovarian membrane; 2. roe film, ovarian membrane
ятаган yatag(h)an, long Turkish dagger
яталит *geol.* yatalite
ятрен *pharm.* chiniofon, (sodium) anayodin, yatren
ятрогения iatrogeny
ятрофа *bot.* nettle spurge *(Jatropha)*
ятрофизика iatrophysics
ятрохимия iatrochemistry
ятрышник *bot.* orchis *(Orchis)*
ятрышниковые *n. pl. bot.* orchird family *(Orchidaceae)*
ятрышниковый orchidaceous
ятрышникоцветный *bot.* orchidiflorous
яфетид Japhetide
яфетический yaphetic, of European race
яхонт *obs.* ruby; sapphire
яхта yacht
~, гоночная racing yacht
~, крейсерская cruising yacht
~, парусная sailing yacht
~, прогулочная pleasure yacht
яхта-кеч *naut.* ketch-yacht
яхт-клуб yacht club
яхтсмен yachtsman
ячеечный alveolate
ячеисто-пленочный cellulomembranous
ячеисто-сетчатый foveo-reticulate
ячеистый cellular, alveolar, alveolate; porous, favose, honeycomb(ed)
ячейка 1. cell, alveolus, alveole; 2. *comp.* location *(of memory)*; 3. *math.* mesh; 4. *mil.* foxhole
~, анодная anode kettle
~, базовая master cell
~, биполярная bipolar cell
~ волновода waveguide cell
~, выбранная selected cell
~ высевающего диска seed cell
~, газоразрядная gas-discharge cell
~ граптолита theca
~ двоичная binary cell
~, декадная decade stage
~, единичная unit/single cell
~, запальная seed unit
~ запоминающего устройства memory/storage cell
~ зоны воспроизводства blanket unit
~ ЗУ storage cell
~ И AND-circuit/-gate, coincidence gate
~ Керра Kerr cell
~ кристаллической решетки crystal-lattice cell
~ мшанки zooecium
~, начальная stating location
~ НЕ-И NAND-network
~ памяти 1. storage location/register; 2. memory/store/storage cell
~ памяти, двоичная binary store cell
~, памяти, рабочая *comp.* working register
~ памяти, стандартная standard location/register
~ прерываний interrupt location
~, размножающая multiplying unit
~, растровая *telev.* screen cell
~, релейная *el.* relay unit
~ решетки lattice cell
~ сита sieve mesh
~, седержащая столон stolotheca
~, стандартная *comp.* library cell
~, стрелковая *mil.* rifle pit, foxhole
~ триггерная trigger circuit
~ триера *agr.* pocket
~, ултразвуковая ultrasonic cell
~ устройства накопления информации information memory/store cell
~ фазового пространства phase-space cell
~ фильтра *mach.* filter cell
~ шрифтовой кассы *print.* compartment
~, электролитическая electrolytic cell
~, элементарная *phys.* lattice/unit/elementary cell
ячейковый cellular
ячейкообразный cellular, cellulated, honeycombed; compartmented
ячея 1. *biol.* cell; 2. mesh
ячменевидный barleylike
ячменеотбойник *agr.* barley cylinder/separator
ячменка fine ground barley; barley porridge
ячмень 1. *bot.* barley *(Hordeum)*; 2. *med.* hordeolum, sty(e)
ячневый made from barley
~вая крупа fine-ground barley
яшма jasper
~, египетская Egyptian jasper/pebble
~, красная jasper
~, ленточная ribbon jasper
~, опаловая opal jasper, chloropal
~, полосатая striped jasper
~, фарфоровая porcelain jasper, porcellanite
~, шаровая globular jasper
яшмовидный jasperlike
яшмовый jaspidean, jasper *(attr.)*
ящер *zool.* pangolin, scaly ant-eater; *(Manis)*
ящерица **1.** *zool.* lizard *(Lacerta)*; 2. *astr.* Lizard, Lacerta
~, жабовидная horned lizard *(Phrynosoma)*
ящерица-ядозуб *zool.* Gila monster *(Heloderma Suspectum)*
ящерицеобразный lizardlike, lizard-shaped
ящеричный saurian, lizard *(attr.)*
ящерогады *pl. zool.* rhynchocephalians *(Phynchocephalia)*
ящероголов *zool.* snakefish *(Trachinocephalus)*
ящероголовые *n. pl. zool.* lizard fishes *(Synodontidae)*
ящероногие *n. pl. paleont.* sauropods *(Sauropoda)*

ящероногий *zool.* sauropod(ous)
ящеротазовые *n. pl. zool.* saurischians *(Saurischia)*
ящерохвостые *n. pl. zool.* saururans *(Saururae)*
ящеры *pl. zool.* saurians *(Sauria)*
~, летающие pterosaurians, pterosaurs *(Pterosaria)*
ящик 1. box, case, crate, container; 2. drawer, chest, cabinet, locker
~, агитационный *met.* agitation section
~, аккумуляторный accumulator box/case
~, балластный *agr.* ballast box
~, вещевой *ant.* glove compartment
~, воздушный *naut.* air box/case, air-buoyancy tank
~, выдвижной drawer
~ для аккумуляторной батареи battery container
~ для ветоши cotton-waste box
~ для зарядных гильз cartridge-case box
~ для инструментов instrument/tool box
~ для ключей key locker
~ для отжига *met.* annealing pot
~ для сигнальных флагов *naut.* flag/signal chest/box/locker
~ для уборочных материалов broom box
~ для цементации *met.* carburizing box, casehardening box
~, зерновой *agr.* grain box
~, инструментальный tool/instrument box, tool kit/chest/tray
~, калильный *met.* annealing box
~, канатный *naut.* cable/chain locker
~ комода, верхний top drawer of the chest
~, красочный *print.* ink fountain
~, мусорный dustbin, garbage can
~, обогатительный *met.* separation box
~, палубный deck trunk
~ первой помощи, санитарный first-aid box
~, почвенный soil box
~, почтовый mailbox; letter/pillar-box
~, прозрачный *comp.* glass box
~, распределительный *el.* distribution box
~, рассадный seedling box
~ с подрывными средствами demolition box
~ секстана sextant case
~ сеялки drill box
~, сигнальный *rail.* signal case
~, сортировочный *met.* separation box
~, стержневой *met.* core box
~, стержневой разъемный split core box
~, теплый *therm.* hot well/tank
~, укладочный towage box
~, упаковочный packing/carrying case
~, цементационный *met.* carburizing box, casehardening box
~, черный *comp.* black box
~, шпальный *rail.* ballast box, tie space
ящикообразный boxlike, U-box shaped
ящикоукладчик case loader
ящур *vet.* foot-and-mouth disease, aphthous fever, aphthosa, epidemic stomatitis
ящурный (infected with) foot-and-mouth disease